Detection of Random Signals in Dependent Gaussian Noise

Antonio F. Gualtierotti

Detection of Random Signals in Dependent Gaussian Noise

Springer

Antonio F. Gualtierotti
HEC and IDHEAP
University of Lausanne
Lausanne, Switzerland

ISBN 978-3-319-22314-8 ISBN 978-3-319-22315-5 (eBook)
DOI 10.1007/978-3-319-22315-5

Library of Congress Control Number: 2015951830

Mathematics Subject Classification (2010): 60-02, 60B11, 60G15, 60G25, 60G30, 60G35, 60H05, 60H10, 62M07, 94A13, 46E22

Springer Cham Heidelberg New York Dordrecht London

Printed on acid-free paper

Springer International Publishing AG Switzerland is part of Springer Science+Business Media (www.springer.com)

To those dead or alive whose life has or has room for mine

Preface

The statistical signal processing developments described in this [book] [112] *are important to engineers working in many areas of communications because non-Gaussian signals are encountered in digital communications, as well as in areas such as seismology, radio astronomy, and sonar. For those in the field of statistical signal processing itself, the problem of detecting non-Gaussian signals is paradigmatic in that it provides a useful framework for discussing other current research areas, such as wavelet decompositions, neural networks, and higher order spectral analysis* [94]. And ... *the most fundamental problem of signal detection* [is] *the determination of the likelihood ratio for detecting signals against a noise background* [130, 148]. It is the first purpose of this book to explain, for the first time perhaps, how, for a basic, fairly general, fairly realistic signal and noise model, the likelihood may be computed, that is, given an analytic expression. The noise shall be a Gaussian process, continuous in quadratic mean, and the signal, a random process which depends on the noise, but whose law is unknown. The second purpose is to gather some of the mathematics that enter the likelihood's computation and possibly enlighten it. The end result is a nice mathematical story in the sense of T. Tao [256], in which rather diverse streams of mathematics merge to produce the result one hopes for.

The prototype application of such mathematics is SONAR [181] (the acronym for SOund NAvigation and Ranging), that is, the use of acoustic pressure waves to ascertain the presence of objects, military or other, in an ocean environment. The model requiring the book's mathematics is that *of a non-Gaussian signal in additive and dependent Gaussian noise* [which] *can be viewed as the canonical detection problem for active sonar in a reverberation-limited environment (when reverberation is the predominant source of noise)* [17].

In principle, for acoustic signal detection, one should resort to acoustic wave propagation equations, that is, partial differential equations describing the behavior of pressure when there is, and when there is not, an object to detect. The laws of those pressures would be derived, and the likelihood computed. Unfortunately, the parameters entering those equations are random fields whose law is unknown. Here is the verdict of a specialist [246, p. 152]: *Except for very simple cases, determination of the probability distributions of the wave field in a stochastic medium (also in weakly inhomogeneous stochastic medium) is impossible.* Middleton [194, 195] has developed a series of statistical-physical models, trying to overcome the hurdle, while keeping the physics. Unfortunately, again there are difficulties with such models. Middleton obtained only the one-dimensional laws, and attempts to get the two-dimensional ones [183] yield, given many assumptions and simplifications, exceedingly complicated expressions. Furthermore, estimating the parameters in the models is no mean task, and again unwanted assumptions, such as independence of observations, must be made. The road chosen for this book is thus that of finding a generic, purely statistical, likelihood, and to use it as a guide to algorithms.

This book is organized around a small number of facts which determine the existence of the likelihood. For it to be, the signal must be smoother than the noise, and the right amount of smoothness is obtained when it is required that the signal belong to the reproducing kernel Hilbert space of the noise. The first part of this book is thus devoted to such spaces. One of the few available tools for the production of likelihoods, when the law of the signal is unknown, is Girsanov's theorem, valid for diverse types of martingales. But processes with the properties of martingales are seldom found in applications, definitely not in SONAR, and one must thus devise a way to bridge the gap that exists between general Gaussian noises and Gaussian martingales, those of Girsanov's theorem. One tool that has proved effective for so doing is the Cramér-Hida decomposition (or representation) of second order processes. The second part of this book is thus devoted to such decompositions. The third and last part of this book deals with the likelihood, first for Gaussian martingales, and then for Gaussian processes other than martingales. Though the book is limited to the Gaussian noise case, its methods have somewhat wider scope and may be extended to cover, in particular, spherically invariant noises [19], which cover some of Middleton's models mentioned above, and causally filtered, independent, weighted Wiener and Poisson processes [59]. In an intermediate position, in the same vein, one finds skew-normal processes as "signal-plus-Gaussian noise" models [123].

The resulting book is large, and one may wonder whether the game is worth the candle, as the mathematics represent but a rather small step towards obtaining workable algorithms. One answer is as follows [20]: *Many discrete-time, finite-sample detection algorithms are obtained from consideration of only the discrete-time (and finite-dimensional) problem. If this is done, and the data represent discretized, continuous-time data, then the problem of developing an optimally effective algorithm is akin to that which the blind man faces in describing the elephant. It is obviously preferable, if possible, to develop a discrete-time algorithm based on approximations of the likelihood ratio for the continuous-time problem.* The likelihood that is mathematically determined by the model serves thus as a benchmark. One finds that same point of view in other areas of current research: *The guiding principle underpinning the specific development of the subject of Bayesian inverse problems in this article is to avoid discretization until the last possible moment. This principle is enormously empowering throughout numerical analysis* [253].

Statistical communication theory is a blend of disciplines, especially analysis, probability, statistics, engineering, and numerics. That means many skills to master, and few are those willing, or simply having the time, to work out the details of a streamlined presentation. The choice has thus been made to provide at least a part of those details. That decision has some advantages. The first relates to applications: the pertinence of a mathematical theory, for applications, is often hidden in the details, and presenting those may help focus on where the mathematics hinges, and thus on its relevance. The second is *that most errors in mathematics are concealed under the surreptitious cover of terseness, whereas a fuller exposure leaves less to pitfalls* [56, p. vi]. So one hopes that details will facilitate reading, and especially correction, when needed! There is also more material than strictly necessary for full understanding of the likelihood, but that may prove useful for adequate application, as neighboring results often help having things in perspective.

Three persons have to be thanked for their generous help. M. Émery (Strasbourg) and L.D. Pitt (Virginia) were kind enough to explain to the author part of their work, even when the questions where about "elementary topics." K.D. Schmidt (Dresden), a friend, has always welcomed questions about measure theory and provided essential help with producing the manuscript.

It is C.R. Baker (UNC, Chapel Hill) who anticipated the potential of combining Girsanov's theorem with the Cramér-Hida decomposition. He and I (the author) worked out many of the details during a more than one-year long sunny visit at the University of Texas at Austin. C.R. Baker has also been my thesis supervisor: I wish here to thank him for letting me in on the pleasures of likelihood calculation and, last but not least, for his friendship.

Yens, Switzerland Antonio F. Gualtierotti
2015

Prolog

Possibly the simplest description of the detection problem is that found in [130, p. 76]: *The voltage at the input terminals of a radar or communications receiver is always fluctuating in a random manner because of the chaotic thermal motions of the surroundings. Any signals that are present—echoes from a distant target or information-bearing communication signals—are added to this noisy background; and if the strength of the signals is small, they are difficult to distinguish from the noise. The task of an observer is to decide whether signals of some specified type are present in the total input voltage received during a certain interval of time. No matter what procedure he uses to make these decisions, there is always a chance that he may be wrong, declaring a signal is present when there is none, or vice versa. He seeks a way of handling the receiver input so that the decisions are made with the greatest possible success in a series of observations.* In the case of sonar, things are even murkier [140, p. 2]: *One may say that "signal" is what one wants to observe and noise is anything that obscures the observation. Thus, a tuna fisherman who is searching the ocean with the aid of sonar equipment will be overjoyed with sounds that are impairing the performance of a nearby sonar system engaged in tracking a submarine. Quite literally, one man's signal is another man's noise.*

Let thus the waveform $s = \{s(t), t \in [0, T]\}$ be observed. If it is a noise waveform, it is an observed path of the stochastic process N whose probability law is P_N. If it is a waveform due to the presence of a signal S, it is an observed path of a functional of S and N, with probability law $P_{S,N}$. Since physical signals have finite energy, the square of the waveforms must be integrable and the laws P_N and $P_{S,N}$ thus "sit" on selected subspaces of $L_2[0, T]$, the Hilbert space of square integrable functions, or on one of its manifolds such as that of continuous ones. The "noise only" option is called the H_0 hypothesis, and the "signal-and-noise" one is labeled H_1. The observer's task is to ascribe s either to H_0 or to H_1.

A (mathematical) solution to the detection problem requires at least that one answers the following questions [10]:

1. Is the mathematical model reasonable?
2. What is the optimum operation on the observed waveform?

3. Given a specific procedure for deciding between H_0 and H_1, what is the performance of that procedure?

The first requirement, for the answer to the first question to be yes, is that the model not be singular, which means in practice that detection cannot be achieved without error. One detailed justification for such a requirement may be found in [225]. Its mathematical translation is that $P_{S,N}$ should have an absolutely continuous component with respect to P_N. As to the second question, the optimum operation on the data according to several criteria (e.g., Bayes, Neyman-Pearson) is to compute the value of the likelihood ratio at the observation and compare this value to a threshold. If the threshold is exceeded, one decides that a signal is indeed present. The likelihood ratio is in fact the Radon-Nikodým derivative of $P_{S,N}$ with respect to P_N, evaluated at s, the observation. The two problems of singularity and computation of the likelihood ratio go hand in hand. The latter is thought by many to be a more practically important problem than the former. However, one uses essentially the same techniques to attack both problems, and it would seem difficult to obtain the likelihood while ignoring the various results on (non-)singular detection.

Practical signal detection problems frequently involve non-Gaussian signals in additive, dependent noise. A good example is the canonical problem of detection by active sonar in a reverberation-limited environment (especially volume reverberation) [194]. The noise in such situations can frequently be regarded as arising from reflections by many small scatterers, which can be reasonably assumed to have statistically independent behavior. The central limit theorem then gives a Gaussian process. The signal process, however, will frequently be dominated by reflections from a few large scatterers, such as the sonar dome. These scatterers each give rise to a non-Gaussian random process, which gets summed at the receiver to give a non-Gaussian process. The problem of detecting very quiet submarines, emanating primarily non-Gaussian broadband signals, is analogous in nature.

The objective here is to obtain a general solution to such problems as they arise in a radar–sonar context. A general solution is defined to be one of a general form having parameters that are functions of the data, and involving a minimal set of assumptions on data properties, furthermore being susceptible of being reasonably approximated. "Reasonable" refers to a discrete-time approximation that can be implemented once the data parameters are known, together with procedures for estimating those parameters, and which converges to the continuous-time solution as the number of samples increases. Such a general solution is contained in this book. The only meaningful assumptions are that the Gaussian noise process be mean-square continuous, which is no restriction in practice, and that the signal paths belong to the reproducing kernel of the noise, a necessary requirement for the detection problem to be non-singular.

As the law $P_{S,N}$ is usually unknown, answers to the third question can only be empirical.

References [22, 23] provide evidence that the program sketched above can be successfully completed.

Credits and Comments

A serious effort is made below to point at the principal sources for the material in the book (more references are to be found directly in the text). Since the book is a large one, potential for omission, and mistaken attribution, is abundant, and an apology is offered here to those who may feel their work has not been given the proper attention. The references were chosen with no effort to reduce their number, or provide the earliest ones: the "method of convenience" has been used, and, when a result was needed, the first adequate reference found was the one actually employed.

Chapter 1

Obtaining the functions of an RKHS as maps of the following form [(Proposition) 1.1.15]: $t \mapsto \langle h, F(t)\rangle_H$ is a very convenient procedure, when applicable (as it requires "guessing" H and F), and is used throughout the "*exposé*." It was found in Saitoh's book [232]. Example 1.3.15 illustrates well its advantages. The material on supports of reproducing kernel Hilbert spaces is from [82]. Membership properties are from the lecture notes of Neveu [202] and from the book of Fortet [106]. The material on covariances has its sources in the lecture notes of V.S. Mandrekar, delivered at the Polytechnic of Lausanne, in the middle 70s. That eventually materialized into book expression [52]. The two Propositions 1.3.20 and 1.3.21, relating the RKHS of a second order process, and the range of the square root of the covariance operator that the covariance of that process determines, are from the lecture notes of C.R. Baker, produced over the years, at the Statistics Department of the University of North Carolina at Chapel Hill. An analogous result may be found in J. Neveu's notes already cited [202]. Section 2.8 of Chap. 2 provides a general treatment of the question. The material on triangular covariances is from [82]. It provides an interesting illustration of what may hide behind some of the seemingly simple concepts of the RKHS theory. The results on separable RKHS's have diverse sources, in particular [35, 106, 196, 198]. The representation of the covariance in (Remark) 1.5.13 supports the modeling of Chap. 17, though the latter requires a canonical representation. Example 9.2.1 shows that (Remark) 1.5.13 is, to that end, not sufficient. Projections are considered in [106, 202]. Example 1.6.9 is from [118]. Result (Proposition) 1.6.19 stems from [188], and is analogous, for RKHS's, to Kolmogorov's result on existence of stochastic processes. The material on operators is from [146], and results on covariance operators in RKHS's are from [35].

Chapter 2

Chapter 2 is based, to a large extent, on [50]. It is in line with result (Proposition) 1.1.15: the functions of the RKHS should reflect the properties of F. Having such a "duality" may help "understand" one of its terms using the other. The same authors have obtained the "same" results for functions with values in vector spaces [51]. The representation of RKHS's as L_2 spaces, and related embeddings, are from [106]. The example using RKHS's to define inner products for measures is from [35, 254]. It illustrates the versatility of RKHS's, given some ingenuity. Result (Proposition) 2.4.30 is from [232], and is the type of result one needs for applications as in Sect. 17.6. Result (Proposition) 2.4.38 is from [134]. Example 2.6.12 is from [82].

Chapter 3

To a large extent, Sect. 3.1 is based on [106]. The systematic use of the map $J_{2,1}$, rather than $J_{2,1}^{\star}J_{2,1}$, as is often the case, has some advantages, as seen in the proof of (Proposition) 3.1.34 (where $J_{2,1}$ is L_F, for a particular case). Compactness of $J_{2,1}$ is relevant for Chap. 4, and thus for the existence of the likelihood. Restriction to subsets [(Proposition) 3.1.18] is important for computations. Section 3.2 is from [233]. The end result [(Corollary) 3.2.25] has some interest, as shall be reiterated, for the theory of mismatched channels [13], that is, those channels for which the power constraint on the transmitted signal is more restrictive than that imposed by the channel noise, in that it provides "a measure" of the mismatch. Section 3.3 is again from [106] and bears again on the existence of the likelihood, as the latter depends on the intersection of the RKHS's involved being "large" [Chap. 5]. The last section is from [197, 198]. It is of relevance for mismatched channels also, as well as for Gaussian detection based on simultaneous reduction of covariance operators [9, 147].

Chapter 4

The material in that chapter is mostly from [176]. It explains, in terms of the RKHS's involved, the meaning of the requirement that, for the likelihood to exist, the signals must be in the RKHS of the noise. Some ideas and results from [176] may already be found in the references from earlier times [82, 106, 188]: the procedure amounts to, as is often the case, going through compatible finite procedures over sets of points which are dense in some sense. No effort has been made to sort out the contributions of each author as the aim was to show that much is articulated around the compatibility result of Meschkowski [188].

Chapter 5

The material is from [106], and provides what is perhaps the best, though not the shortest, explanation of the (RKHS) conditions which secure the existence of a likelihood, and in particular, the Gaussian likelihood. Fortet furthermore sets the problem within the standard context of the Lebesgue decomposition of measures, and relates equivalence to domination of probabilities on certain sub-manifolds. That the intersection of RKHS's is important for Gaussian discrimination emerges also in [99]. Another explanation for the appearance of a compact perturbation of the identity in such matters may be found in [104]. The shortest, and most general, derivation of the Gaussian likelihood is perhaps that of Vakhania and Tarieladze [261]. Reference [52] is a very nice survey, with ampler scope, along lines similar to those of Vakhania and Tarieladze [261]. The extension to mixtures is from [243]. Most derivations of the Gaussian likelihood result in an infinite product of exponentials, not the most useful of representations. There is at least one exception, that of Rao and Varadarajan [219], which provides, in terms of properties of the covariance operators involved, conditions for the likelihood to be the exponential of a quadratic form. The "Gaussian RKHS way" to the likelihood should be compared to the "Girsanov's way" of Part III, the difference being that in the first case, one knows the probability laws, but not in the second.

Chapters 6, 7, and 9

Chapters 6, 7, and 9, yield, what are perhaps, successively, the most elementary and instructive, the most elegant, and the deepest studies of the same topic, that of the representation of second order processes as superpositions of causal transformations of "white noises." The first presentation results from a "hands on" approach based on the fact that the elements of the linear space of a process with orthogonal increments have an integral representation, and proceeds by exhaustion. It is in such a development that one sees most immediately, and clearly perhaps, why the result is ... what it is. The second presentation constructs the linear space of a second order process as a direct integral, and then proceeds to the Cramér-Hida representation by choosing a basis in it. Knight's presentation uses the projection process to reveal the martingale structure of the linear space of the process. All representations use similar "tricks," especially to obtain the absolute continuity of the measures involved.

Chapter 6 is essentially based on [15]. The material on orthogonally scattered measures is explained in [182]. The representation of functions as integrals with respect to an orthogonally scattered measure is from [122]. The material on separability, and existence of limits, is due to [48]. At the end of Baker [15], one finds the Hellinger-Hahn theorem as a corollary. The reverse procedure (proof of the Hellinger-Hahn theorem, followed by the Cramér-Hida representation) is presented in [143]. The freedom provided by (Proposition) 6.4.34 is essential to the derivation of the likelihood. The other two approaches are silent on the topic. Reference [133] uses that fact.

Chapter 7 is from [200]. Though the references in the latter book list all the papers of Cramér and Hida on multiplicity, the term does not appear in there, and the pendant to the Cramér-Hida representation is in terms of a "Hilbertian stochastic integral." In [68], Neveu's approach is used to produce a recursive expression for a Gaussian likelihood. It should perhaps be noted that Neveu's approach yields the Cramér representation, and not the Hida one.

Chapter 9 has two main sources: [157, 158]. To have a complete picture, one must insert [161, 193]. The results of Knight require some extra hypotheses, smoothness ones in particular. Knight has given the prediction process a much wider scope in [160], but practical uses for such power remain, it seems, yet to be found. There is however a close connection between the Cramér-Hida and Knight's approaches: the first is, in the Gaussian case, the wide-sense version of the second [159], and that fact is essential in obtaining the likelihood.

The Cramér-Hida representation is a way to prediction, and the latter is, in turn, conditional expectation, which may be seen as an operator. Multiplicity is at the core of operator theory, in Hilbert spaces at least [45, 126]. So it is not surprising that the Cramér-Hida representation may be seen as a problem of Hilbert subspaces (resolution of the identity). Examples are [72, 132, 143].

Chapter 8

Sections 8.1 and 8.2 are from [227]. The remaining part of the chapter is devoted to multiplicity one, the only case which one can reasonably expect to be used for practical purposes, with the exception perhaps of Goursat processes [Sect. 8.4], as analytical Cramér-Hida representations stemming from a covariance are, at best, hard to obtain, and for the time being nonexistent. Section 8.3 is mostly from Cramér [63, 64]. The material on approximation by processes of multiplicity one is from [137]. Section 8.4 is based on [212], which provides what are, possibly, the most explicit results on Cramér-Hida representations. Related results may be found in [179, 180]. Goursat processes [212] may serve, especially for data adjustment questions, as surrogate Cramér-Hida representations for processes whose representation one may not pinpoint.

Chapter 10

The chapter is a jumble of results which underlie the sequel and are often passed over quickly. Two topics, sets of measure zero, and inverses of monotone functions are spelled out in detail to sooth the author's conscience. It should however be noticed that the need for more detail in such matters comes from the masters [70]. The material on inverses of monotone functions has two sources [88, 125]. The material on exponentials of martingales has its origin in [189], its extension to martingales with values in a Hilbert space, in [206]. A rather elegant treatment of Gaussian processes with independent increments, their exponentials, and the attendant Girsanov's theory, may be found in [172].

Chapters 11, 12, 13, 14, and 15

Those chapters cover what may be termed the Girsanov's theory of the likelihood for Cramér-Hida processes, that is, processes with values in l_2, continuous paths, and independent components and increments. What may be the most lucid, and complete, exposition of the topic, in the case of a vector noise made of a finite number of independent, standard Wiener processes, is that of Memin [187]. Coverage of parts of the same material (dimension one), with some bonuses, may be found in [172]. Chapters 11 to 15 fill the gap which exists between the finite and the infinite frameworks. Part of that gap was reduced in [58]. Chapter 12 is an adaptation, almost verbatim, to the infinite dimensional context, of results found in [38]. The "latest" on the "moment condition" may be found in [156].

Chapter 16

The original Girsanov's theorem combines what may be seen as two distinct results: invariance of quadratic variation, and invariance in law, under random drift translation, and absolutely continuous change of measure. That result seems to have at first attracted little attention, except from electrical engineers, and in particular T. Kailath, of Stanford, who saw in it a way to obtain a non-Gaussian likelihood. Then the "martingale community" saw that the invariance of quadratic variation part of Girsanov's theorem is central to martingale theory, and law invariance was sort of forgotten, though potentially important for modeling non-Gaussian noise. The question has been settled in [31], and the answer confirms, in part, an assertion found in [40, p. 116]. It basically says that, with the exception of "white noise," continuous martingales do not provide interesting noise models. To get to that conclusion much research had to be done. The results are documented in the following papers: [90], for Vershik's theorem on lacunary isomorphism, completed with [89, 205], for Ocone martingales; [265], for the relations between Ocone's martingales and exponential ones. Reference [31] is based on [89, 90, 205, 265]. There is a shorter way to Vershik's theorem on lacunary isomorphism than that used here, constructive and technical, which is from [90], and its consequence, from [89]: it is to be found in [169], and uses a martingale convergence theorem. But again, the shorter way is opaquer than the longer, at least for the amateurs, to which the present author belongs.

Chapter 17

The theoretical part of Chap. 17 is essentially based on two papers: [18, 20]. Those made the assumption that multiplicity is finite, a serious restriction, as multiplicity is hard to obtain from the covariance of the process, and the importance of the Cramér-Hida maps was somewhat in the background. One fact worth noticing about the form of the likelihood that has been obtained is that it is "computable" at the signal as received, without further ado. That is not the case with the "white noise likelihood" as was acknowledged by J.M.C. Clark [57], who saw that the likelihood had to be "tweaked" to accommodate actual signals. He then introduced the notion of "robust likelihood." An interesting, wide ranging complement to Sect. 17.5 (scope of the SPGN) may be found in [42, 101, 102, 104]. The comments on applications are from [14], the remark on inverse problems, from [253]. Finally, the essential reason why the method used here to obtain the likelihood works may be found in [159, p. 113].

Notation and Terminology

General Notation

$\sim$	Almost sure equality, identity in law
$\ll$	Absolute continuity of measures
$\equiv$	Mutual absolute continuity of measures, identity of the elements, respectively to the left, and to the right, of the symbol
$\oplus$	Direct sum
$\otimes$	Tensor product, product of σ-algebras
$\overline{\otimes}$	Completion of product of σ-algebras
$\uplus$	Disjoint union
$\subseteq_b$	Contained boundedly
$\subseteq_c$	Contained contractively
$a \longleftarrow b$	a is replaced with b
Δ	Symmetric difference
$\vee$	Maximum, operation of generating an object (σ-algebra, vector space)
ι	$\sqrt{-1}$
$\tilde{S} = \lvert S \rvert$	Cardinal of set S
$\Delta(S^2)$, $\Delta(S)$	Diagonal of set $S \times S = \{(s, s), s \in S\}$
$S[x_1]$	When S is a subset of a product space, section of S at the first component x_1
d_X	Distance on space X, related to object X
$[1 : n]$	Integers $1, \ldots, n$
I, J	Set of indices, intervals
t_l, t_r, t_u	Left, lower, right, upper limits of an interval
0_S, 1_S	Function that is 0 on space S, respectively, 1

$s(f)$	Function which delivers the sign of f (1, when $f \geq 0$, -1, when $f < 0$)
χ_S	Indicator function of set S
I_S	Equivalence class of χ_S
$f^{\mid S}$	Restriction of f to S
$[f]$	Equivalence class of the function f
$\dot{f}$	Member of the equivalence class f
$\mathcal{D}[f], \mathcal{N}[f], \mathcal{R}[f]$	Domain, null space (kernel), and range of map f
$f^{(i)}$	i-th derivative of f
$f[x_1]$	When f has two arguments, x_1, and x_2, map $x_2 \mapsto f[x_1](x_2) = f(x_1, x_2)$
$\mathcal{L}(f)$	Laplace transform of f
$D_1 f(D_t f), D_2 f(D_x f)$	When f has two arguments, derivative with respect to first, respectively, second argument
D_τ	A matrix of Radon-Nikodým derivatives
δ	Point mass, eigenvalue function
$\delta_{i,j}$	One, when $i = j$, and zero otherwise function which is one when two elements are identical, zero otherwise
id_S	Identity over set S
$\mathcal{F}$	Class of functions
$C_0[0, 1]$	Continuous functions on $[0, 1]$, 0 at the origin
$C(T)$	Continuous functions on T
$C_c(T)$	Continuous functions on T with compact support
$C_K(T)$	Continuous functions with support in K (K generally compact)
$C_0(T)$	Continuous functions on T that vanish at infinity
$\mathcal{P}(S)$	Subsets of set S
$\mathcal{B}(X)$	Borel sets of topological X
$\mathcal{C}(X)$	Cylinder sets generated on X by functionals
$\mathcal{D}(X)$	σ-algebra on X generated by functionals
$\mathcal{E}(X)$	σ-algebra on X generated by evaluation maps
$\mathcal{K}(X)$	Compact sets of X
$\mathcal{O}(X)$	Open sets of X
$\mathcal{E}$	A family of maps defined on set S, with real values
$\mathcal{S}$	σ-algebra generated on set S using the family $\mathcal{E}$ above
$\mathcal{L}(\mathcal{E})$	Linear manifold generated by $\mathcal{E} = (V[\mathcal{E}])$
$\mathcal{M}(\mathcal{S})$	Linear space of functions adapted to $\mathcal{S}$ and $\mathcal{B}(\mathbb{R})$
$V(\mathcal{M})$	Linear manifold of $\mathcal{M}(\mathcal{S})$
$\mathcal{Q}(\mathcal{E})$	Quadratic forms made of elements in $\mathcal{E}$
$\overline{V(\mathcal{M})}^P$	Closure in $L_2(S, \mathcal{S}, P)$ of manifold $V(\mathcal{M})$
$V_1(\mathcal{M})$	Subset of $V(\mathcal{M}) \cap L_2(S, \mathcal{S}, P)$ of elements of norm 1
$\lambda_{P,V(\mathcal{M})}(\alpha)$	An upper bound for $V_1(\mathcal{M})$

C, Γ, Σ	Covariances
D	Diagonal matrix
L	Triangular matrix of ones (summation operator)
M	Matrix
I_n	Identity matrix of $\mathbb{R}^n$
$E(\lambda)$	(Spectral) projection
$E(d\lambda)$	Measure whose values are projections
$\hat{E}(f)$	$= \int f(t)E(dt)$
H_B	Range of B as Hilbert space
P, Q	Probability, projection
$P^{\mid S}$	Probability P restricted to family of sets $\mathcal{S}$
Q_{aP}, Q_{sP}	In the Lebesgue decomposition, absolutely continuous, respectively, singular part of Q with respect to P
$Q \preceq P[V(\mathcal{M})]$	On $V(\mathcal{M})$, P dominates Q
O_H, I_H	Null, respectively, identity operator of Hilbert H
T_m	In linear space, translation by m
B, C, D, J, M	Operators (typically: J inclusion, M multiplication)
R, S, U, V, W	Idem (R covariance, U unitary)
$\mathcal{L}(H, K)$	Bounded, linear operators from H to K
$\mathcal{B}_1(H)$	Operators of H with finite trace
$\mathcal{B}_2(H)$	Hilbert-Schmidt operators of H
$[B]$	Abbreviation for $(B^*B)^{1/2}$
$t(B)$	Abbreviation for $\sum_{i\in I}\langle B[e_i], e_i\rangle_H$, e_i's complete, orthonormal
$\tau(B)$	Trace norm of operator B, equal to $t([B])$
$\sigma(B)$	Spectrum of operator B
R_C	Covariance operator associated with the covariance C
X, Y, Z	Random variables ($E[X]$, $E_P[X]$, $V[X]$, $V_P[X]$ expectation and variance)
$X \sim Y$	Random variables with the same law
$X \sim \mathcal{N}(0, 1)$	Standard normal random variable
$IID(X)$	Independent, identically distributed, with law that of X
$\mathcal{L}(X)$	Law of random variable X
U	Uniform random variable on $[0, 1]$
$M, \langle M\rangle$	Martingale, respectively, associated quadratic variation
$\underline{X}, \underline{Y}, \underline{Z}$	Vectors of random variables
$\sigma(X)$	σ-algebra generated by element X
${}^\circ\sigma(X)$	$\sigma(X)$ enlarged to contain measurable sets of measure zero
$\sigma^\circ(X)$	$\sigma(X)$ enlarged to contain measurable sets of measure zero, and their subsets
F_X	Distribution function of random variable X
$X[\omega]$	When X is a process, the path of X at ω
$\boldsymbol{X}$	When X is a process, the map $\omega \mapsto X[\omega]$
X_t	When X is a process, the equivalence class of $\omega \mapsto X(\cdot, t)$

P_X	When P is a probability, X a process, the probability generated by P and X, $P \circ X^{-1}$
Π	A property, a projection, a product
$(\Omega, \mathcal{A}, P), (\Theta, \mathcal{B}, Q)$	Probability spaces
$P_{A\|B}$	(Regular) conditional law for P of object A given object B
$\hat{\mathcal{A}}$	Universally measurable sets for $\mathcal{A}$
$\mathcal{N}(\mathcal{A}, P)$	The sets of $\mathcal{A}$ which have, for P, measure zero
$\underline{\mathcal{A}}$	Filtration of $\mathcal{A}$
$\mathcal{A}_\infty$	σ-algebra generated by the filtration $\underline{\mathcal{A}}$
$\underline{\mathcal{A}}_N$	Trace of $\underline{\mathcal{A}}$ over the complement of set N
$\mathcal{A}_t^{\circ N}$	Completion of $\mathcal{A}_t$ with respect to the subsets of N
$(\Omega, \underline{\mathcal{A}}, P)$	$(\Omega, \mathcal{A}, P)$ with a filtration of $\mathcal{A}$
$\mathcal{L}_p(\Omega, \mathcal{A}, P)$	Adapted functions whose p-th power is integrable (when $p = 0$, adapted maps, almost surely finite)
$L_p(\Omega, \mathcal{A}, P)$	Equivalence classes, for P, of adapted functions whose p-th power is integrable
$L_p^S(\Omega, \mathcal{A}, P)$	Equivalence classes, for P, of adapted functions with values in space S, for which the p-th power of some norm is integrable
$\int_A^{(w)}$ $(\int_A^{(s)})$	Symbol for weak (strong) integral on A, result being $h^w_{\xi,A}$ $(h^s_{\xi,A})$

Part I: Reproducing Kernel Hilbert Spaces

$B(c, r)$	Ball centered at c, of radius r, open or closed
$c_\wedge(\cdot \wedge \cdot)c_\vee(\cdot \vee \cdot)$	Triangular covariance
$\mathcal{D}[C_0]$	Family of covariances dominated by covariance C_0
r_C	$= c_\vee / c_\wedge$
r^C	$= 1/r_C$
$\mathcal{E}_t$	Evaluation map at t
$h[t]$	Element of H determined by t
$h^{\|S}$	Function h restricted to set S
L, H, K, K_F	Hilbert spaces
H_F	Linear closure of a map with range in H, subspace orthogonal to the kernel of L_F, denoted $\mathcal{N}[L_F]$
$\mathcal{H}, \mathcal{K}, \mathcal{K}_F$	Reproducing kernels
$\Delta_{\mathcal{H}}(t_1, t_2)$	$\mathcal{H}(t_1, t_1) - 2\mathcal{H}(t_1, t_2) + \mathcal{H}(t_2, t_2)$
$d^2_{\mathcal{H}}(t_1, t_2)$	$\\|\mathcal{H}(\cdot, t_1) - \mathcal{H}(\cdot, t_2)\\|^2_{H(\mathcal{H},T)}$
$\mathcal{H} \ll \mathcal{K}$	Domination of kernels (covariances)
$\mathcal{H} \ll_\kappa \mathcal{K}$	Domination of kernels (covariances) with a finite trace property
$J_{\mathcal{K},\mathcal{H}}$	$\mathcal{K}(\cdot, t) \mapsto \mathcal{H}(\cdot, t)$ when $\mathcal{H} \ll \mathcal{K}$

$H(f; T_n)$	The elements h of $H(\mathcal{H}, T)$ for which $h(t_i) = f(t_i), t_i \in T_n$
$h_{f;T_n}$	$= \sum_{i=1}^{n} \phi_i \mathcal{H}(\cdot, t_i)$
$\mathcal{H}(\cdot, t), \mathcal{K}_F(\cdot, t)$	Maps obtained from reproducing kernels fixing one argument
$H(\mathcal{H}, T), H(\mathcal{K}_F, T)$	Reproducing kernel Hilbert spaces
$\mathcal{C}(\mathcal{H}, T), \mathcal{B}(\mathcal{H}, T)$	Cylinder, respectively Borel, sets of $H(\mathcal{H}, T)$
$H(\mathcal{H}, T) \sqsubseteq H(\mathcal{K}, T)$	$H(\mathcal{H}, T) \subseteq H(\mathcal{K}, T)$ and, on $H(\mathcal{H}, T)$, $\Vert h \Vert_{H(\mathcal{K},T)} \leq \Vert h \Vert_{H(\mathcal{H},T)}$
L_F	Linear map whose range, denoted $\mathcal{R}[L_F]$, is a reproducing kernel Hilbert space
$[n, \alpha, (t, T)]$	Shorthand for $n \in \mathbb{N}, \{\alpha_1, \ldots, \alpha_n\} \subseteq \mathbb{R}, \{t_1, \ldots, t_n\} \subseteq T$
T	Domain of a function, set of indices
T_c	Subset of T determined by constraint c, countable subset of T
T_d	Determining set for a covariance
T_n	Finite subset of T with n elements
$T_{\mathcal{H}}$	Support of a reproducing kernel
$T^<$	$= T \cap]-\infty, t[$
T^C	Subset of T over which a triangular covariance is not zero
$\mathcal{T}$	σ-algebra on T
	τ Measure on $\mathcal{T}$
$V[\mathcal{F}]$	Linear space generated by a family $\mathcal{F}$ of elements
$\rho_{T_{\mathcal{H}}}$	Restriction map to $T_{\mathcal{H}}$
$\rho_{T_{\mathcal{H}}}^{\leftarrow}$	"Inverse" of $\rho_{T_{\mathcal{H}}}$
$\Sigma_{\mathcal{H},T_n}$	Covariance matrix obtained when restricting $\mathcal{H}$ to $T_n \times T_n$

Part II: Cramér-Hida Representations

T	Interval of $\mathbb{R}$
t_l, t_r	$\inf T$, respectively, $\sup T$
f	Unless otherwise stated, a map with T as domain, and range in H, a real Hilbert space
F	$F(t_2) - F(t_1) = \Vert f(t_2) - f(t_1) \Vert_H^2$
C_f	The function with domain $T \times T$ and values $\langle f(t_1), f(t_2) \rangle_H$
f^+, f^-	When f has limits to the right, respectively, to the left, the function with those limits as values
$L_t[f]$	(Closed) subspace generated linearly in H by the values of $f^{\vert\{T \cap]-\infty, t]\}}$
$L_t^{(+)}[f]$	(Closed) subspace generated by $L_t[f]$ and $f^+(t)$
$L_t^{\delta}[f]$	$L_{t+\delta}[f] \cap L_t[f]^{\perp}$
P_t	Projection onto $L_t[f]$
H_t	All h such that $P_t[h] = 0_H$, and, for $\theta > t$, $P_\theta[h] = h$
$f_h(t)$	$P_t[h]$, $h \in H$

$L_{\cap T}[f]$	Intersection of all subspaces $L_t[f]$
$P_\cap$	Projection onto $L_{\cap T}[f]$
$L_{\cup T}[f]$	Subspace generated by the union of all subspaces $L_t[f]$
$P_\cup$	Projection onto $L_{\cup T}[f]$
$L_t^-[f]$	Subspace generated by the subspaces $L_\theta[f]$ contained in $L_t[f]$
P_t^-	Projection onto $L_t^-[f]$
$L_t^+[f]$	Subspace generated by the subspaces $L_\theta[f]$ containing $L_t[f]$
P_t^+	Projection onto $L_t^+[f]$
$H_t^\star$	All h such that $P_t[h] = 0_H$, and $P_t^+[h] = h$
$\Delta_t[P]$	$P_t^+ - P_t$
$M[t]$	$\dim\{\mathcal{R}[\Delta_t[P]]\}$
$M[+\|t], M[t\|+]$	Dimension of $L_t^{(+)}[f] \cap L_t[f]^\perp$, respectively, $L_t^+[f] \cap L_t^{(+)}[f]^\perp$
$\Delta[P]$	$\sum_{t\in T_d} \Delta_t[P]$
$L_d[f]$	Projection onto $\Delta[P]$
$f_d(t)$	Projection of $f(t)$ onto $L_d[f]$
$L_c[f]$	(Closed) subspace orthogonal of $L_d[f]$
$f_c(t)$	Projection of $f(t)$ onto $L_c[f]$
T_d	Subset of T of elements for which $\Delta_t[P] \neq O_H$, respectively, some (left, right) limit of f does not exist
$\mathcal{P}[S]$	Pre-ring of subsets of S
$\mathcal{R}[S]$	Ring of subsets of S
$\sigma(\mathcal{P}[S])$	σ-ring generated by $\mathcal{P}[S]$
m	Measure on some family of subsets of S, with values in H
$L[m]$	(Closed) subspace generated linearly in H by the values of m
$\int [f]\,dm$	Integral of (equivalence class) of numerical f with respect to vector measure m
M	$M(S_0) = \|m(S_0)\|_H^2$
$m_\mathcal{S}$	m restricted to the family $\mathcal{S}$
$M_\mathcal{S}$	M restricted to the family $\mathcal{S}$
$\mathcal{S}_f$	The sets in $\mathcal{S}$ that have a finite measure, for some specified measure
μ_h	$\mu_h(S_0) = \langle h, m(S_0)\rangle_H$
$\|\mu\|$	Total variation of the measure μ
m_f	Vector measure with values in H, $m_f(S_0) = \int_{S_0} [f]\,dm$, or vector measure such that $m_f(]t_1, t_2]) = f(t_2+) - f(t_1+), f : T \longrightarrow H$
m_B	Vector measure in K, $m_B(S_0) = B(m(S_0))$, $B : H \longrightarrow K$, linear and bounded
$\boldsymbol{m}$	Vector whose components are of type m
H_S	Product of Hilbert spaces indexed by $(S, \mathcal{S})$
$H_\mathcal{S}$	Measurable field of Hilbert spaces in H_S
H_S^Λ	A subset of H_S
$H_\mathcal{S}^\Lambda$	Measurable field generated by H_S^Λ
$H_\mathcal{S}^I$	Given a measure σ, subset of $H_\mathcal{S}$ of square integrable functions
$H_\mathcal{S}^\sigma$	Given a measure σ, equivalence classes in $H_\mathcal{S}^I$

$\int_S^{\oplus} H_s \sigma(ds)$	The space H_S^{σ} with an Hilbertian structure, a direct integral of the family $\{H_s, s \in S\}$
$(B, V[H], K)$	Generalized process on H
$F^{\underline{h}}_{i,j}(t)$	$\langle h_i(t), h_j(t)\rangle_H$
$M^{\underline{h}}_{i,j}$	Measure determined by $F^{\underline{h}}_{i,j}$
$\boldsymbol{M}_{\underline{h}}$	Matrix measure with entries $M^{\underline{h}}_{i,j}$
$\mathcal{L}_2(T, \mathcal{T}, \boldsymbol{\tau})$	Vector valued functions whose "square" is integrable with respect to a matrix measure
$L_2(T, \mathcal{T}, \boldsymbol{\tau})$	Hilbert space of equivalence classes of functions in $\mathcal{L}_2(T, \mathcal{T}, \boldsymbol{\tau})$
$[\underline{a}(t), \underline{h}(t)]$	The sum $\sum_{i=1}^n a_i(t) h_i(t)$
$\int [\underline{a}, \underline{m}]$	The sum $\sum_{i=1}^n \int a_i dm_i$
$A[t \mid t_1, \dots, t_n]$	A matrix made of the evaluations of $\underline{a}$ at n points larger than t, non-singular when associated with a Goursat non-singular map
$\mathcal{L}[t_1, t_2]$	$\{P^f_{t_1}[f(t)], t \geq t_2\}$
$L[t_1, t_2]$	$\overline{V[\mathcal{L}[t_1, t_2]]}$
$\Gamma_f(\lambda)$	$\int_T e^{-\lambda\theta} C_f(\theta, \theta) d\theta$
Π^a_α	Exponential law with parameter α, and $[a, \infty[$ as support
$j(t, \lambda)$	$\int_T e^{-\lambda\theta} f(t + \theta) d\theta$
$q(t, \lambda)$	$\lambda \int_T e^{-\lambda\theta} P_t [f(t + \theta)] d\theta$
$h_\lambda(t)$	Knight's martingale
$\mathcal{L}^\Phi_t(\lambda)$	$\int_0^\infty e^{-\lambda\theta} \Phi(t + \theta, t) d\theta = \int_0^\infty e^{-\lambda\theta} \Phi_t(\theta) d\theta$
Δ_t	$\{\theta \in]0, t[: \text{WRT} Leb, \Phi_\theta(\tau) = 0, a.e. \tau\}$
Λ_t	$\{(\lambda, \theta) \in]0, \infty[\times]0, t[: \mathcal{L}^\Phi_\theta(\lambda) = 0\}$
$m_X[t]$	Index of multiplicity of X at t
$s_X[t]$	Index of stationarity of X at t

Part III: Likelihoods

$f \diamond g$	Given $f(x, y)$ and $g(x, y)$, $f \diamond g(x, y) = f(x, g(x, y))$
$f \square g$	Given $f(x, y)$ and $g(x, y)$, $f \square g(x, y) = f(g(x, \cdot), y)$
$\underline{B}$	Cramér-Hida process (with values in l_2, and components B_n)
$B(\cdot, t)$	$\Vert \underline{B}(\omega, t) \Vert^2_{l_2}$
b_n	Variance of B_n
b	Sum of the b_n's
M_n	Measure determined by b_n
$C(t)$	Covariance operator of $\underline{B}(\cdot, t)$
$\langle\langle \underline{B} \rangle\rangle$	Quadratic variation of $\underline{B}$
$E_\alpha(\omega, t)$	Exponential of a martingale, depending on parameter α
V	Process with paths of bounded variation
W_M, S_M	Wiener process, respectively, change of time, associated with continuous martingale
S, Σ	Stopping times

$\underline{\alpha}(t)$	$\inf\{\alpha(\theta) \geq t\}$
$\overline{\alpha}(t)$	$\inf\{\alpha(\theta) > t\}$
$I_{\underline{B}}\{\underline{a}\}$	Integral of $\underline{a}$ with respect to $\underline{B}$
$I^P_{\underline{B}}\{\underline{a}\}$	Integral of $\underline{a}$ with respect to $\underline{B}$, when there are several probabilities of which one must take account
$\mathcal{L}_2[b_n], L_2[b_n]$	Space of (equivalence classes of) square integrable functions for the measure M_n
$L_2[\underline{b}]$	Direct sum of the spaces $L_2[b_n]$
$\underline{a}_{\vert t}$	$\underline{a}$ multiplied by $\chi_{[0,t]}$
$\underline{a}_{\vert S}$	$\underline{a}$ multiplied by $\chi_{[\![0,S]\!]}$
$\mathcal{I}_0[\underline{b}]$	Classes of stochastic processes with values in $\mathbb{R}^\infty$ and paths almost surely in $L_2[\underline{b}]$
$\mathcal{I}^P_0[\underline{b}]$	Classes of stochastic processes with values in $\mathbb{R}^\infty$ and, with respect to P, paths almost surely in $L_2[\underline{b}]$
$\mathcal{I}_2[\underline{b}]$	Elements in $\mathcal{I}_0[\underline{b}]$ whose norm squared has finite expectation
$\mathcal{I}^{loc}_2[\underline{b}]$	Elements in $\mathcal{I}_0[\underline{b}]$ for which there exists a localizing sequence such that the elements, restricted to the corresponding stochastic intervals, yield elements of $\mathcal{I}_2[\underline{b}]$
$H_n(t,x)$	Hermite polynomial
s	Either l_2 or $\mathbb{R}^\infty$
$\Vert\cdot\Vert_s$	$\Vert\cdot\Vert_{l_2}$, when $s = l_2$, Fréchet quasinorm, when $s = \mathbb{R}^\infty$
C	$\prod_{n=1}^{\infty} C_n[0,1]$, $C_n[0,1] = C[0,1]$
$\underline{c}$	An element in C
$\Vert c\Vert$	sup-norm of c in $C[0,1]$
$\Vert\underline{c}\Vert_p$	$\left\{\sum_{n=1}^{\infty} \Vert c_n\Vert^p\right\}^{1/p}$
$\Vert\underline{c}\Vert_{C_{l_2}}$	$\sup_{t\in[0,1]} \Vert\underline{c}(t)\Vert_{l_2}$
$\Vert\underline{c}\Vert_{C_F}$	$\sum_{n=1}^{\infty} \frac{2^{-n}\Vert c_n\Vert}{1+\Vert c_n\Vert}$
C^p_S, C_{l_2}, C_F	C with respective norm $\Vert\underline{c}\Vert_p$, $\Vert\underline{c}\Vert_{C_{l_2}}$, $\Vert\underline{c}\Vert_{C_F}$
K	C when it is either C_{l_2} or C_F
$\mathcal{K}$	The Borel sets of K
$\underline{\mathcal{K}}$	Filtration of K, that is generated by the evaluation maps
P^K	Probability on K
$\mathcal{E}_\mu(\underline{k},t)$	$\underline{k}(t)$, evaluation map at t, for K, when measure μ prevails
$\underline{a}^K$	Process of type $\underline{a}$ defined on K
$\underline{X} = \underline{S}[\underline{a}] + \underline{B}$	Received signal in the form of a signal, determined by $\underline{a}$, plus additive noise
$\Phi_X(\omega,t)$	$(\underline{X}[\omega], t)$
$\underline{\mathcal{A}} \approx \underline{\mathcal{B}}$	Isomorphic filtrations
$\underline{\mathcal{A}} \sqsubseteq \underline{\mathcal{B}}$	Filtration $\underline{\mathcal{A}}$ included in filtration $\underline{\mathcal{B}}$
$\underline{\mathcal{A}} \succ \underline{\mathcal{B}}$	Filtration $\underline{\mathcal{A}}$ includable in filtration $\underline{\mathcal{B}}$
$\mathcal{A}[A]$	σ-algebra of subsets of $\mathcal{A}$ included in $A \in \mathcal{A}$
$\mathcal{P}[\mathcal{A}]$	Finite partitions of Ω with sets in $\mathcal{A}$
$\wp$	An element of $\mathcal{P}[\mathcal{A}]$
$S(\wp \mid \mathcal{A}_0)$	Support of $\wp$ with respect to $\mathcal{A}_0$

$mult_P[\mathcal{A} \mid \mathcal{A}_0]$	Conditional multiplicity of $\mathcal{A}$ given $\mathcal{A}_0$, for probability P
$Q_S(A)$	For measurable A and S, $P(A \cap S)/P(S)$
$\underline{\mathcal{A}} \vee \underline{\mathcal{B}}$	Filtration of elements $\mathcal{A}_t \vee \mathcal{B}_t$
$\mathcal{F}_{\mathcal{I}}(\underline{\mathcal{A}})$	Family of filtrations immersed in $\underline{\mathcal{A}}$
$\mathcal{S}(\underline{\mathcal{A}})$	Family of σ-algebras of $\mathcal{A}$ that are saturated for $\underline{\mathcal{A}}$
K	Compact metric space with distance d_K
$\mathcal{P}_K$	Space of probability measures defined on the Borel sets of K
$\mathcal{V}(\mathcal{A}, K)$	Family of random elements with values in K, adapted to $\mathcal{A}$
$d_{\mathcal{V}(\mathcal{A},K)}(X, Y)$	$E_P\left[d_K(X, Y)\right]$, distance between two elements X and Y of $\mathcal{V}(\mathcal{A}, K)$
d_{KR}	Kantorovich-Rubinshtein distance on $\mathcal{P}_K$
M^-	Generalized inverse of matrix M
$\mathbb{Z}_0^-$	$\{\ldots, -2, -1, 0\}$
$I_M\{f\}, J_M\{\phi\}$	Integral of f, respectively ϕ, with respect to martingale M
$P^{\omega}_{M\mid\langle M\rangle}$	For $\omega \in \Omega$, fixed, but arbitrary, probability $C \mapsto P_{M\mid\langle M\rangle}(\omega, C)$
$\mathcal{M}$	A class of martingales
$\mathcal{P}$	A class of laws for a subset of $\mathcal{M}$
$F_\epsilon[f](t)$	$\chi_{[\epsilon,\infty[}(t)f(T_{-\epsilon}(t))$
$G_\epsilon[f](t)$	$f(T_\epsilon(t))$
N	Gaussian noise
$\underline{B}_N$	Cramér-Hida process obtained from N
P_N	Probability induced on the cylinder sets of $\mathbb{R}^{[0,1]}$ by N
P_N^2	Probability induced on the Borel sets of $L_2[0, 1]$ by N
P_N^c	Probability induced on the Borel sets of $C[0, 1]$ by N
Φ, Φ_2, Φ_c	Cramér-Hida maps for, respectively, real, square integrable, and continuous paths

Contents

Part II Cramér-Hida Representations

Part I
Reproducing Kernel Hilbert Spaces

The sequence of words *"reproducing kernel Hilbert space"* shall often be abbreviated in the sequel using the acronym *"RKHS."* As shall be seen [Part III], when the noise is Gaussian, the signals which may be detected using the likelihood belong to the RKHS associated with the specific noise with which one has to deal. The study of RKHS's is thus the study of the spaces of signals available for detection.

One has chosen to develop RKHS's systematically as ranges of continuous, linear operators with domain a Hilbert space [Chap. 1]. Such an approach has many practical advantages as can be ascertained all through the book, as, for example, when one establishes which kinds of functions belong to a specific RKHS [Chap. 2]. The ability of detecting signals is based on the relations of inclusion that may or may not exist between the RKHS's of the signal and the noise, as well as on their "size" [Chap. 3], and it depends furthermore on the conditions (compactness of linear operators) that insure that the paths of a signal are in the RKHS of the noise [Chap. 4]. Finally, to appreciate what is achieved in Chap. 17, one must know how detection is studied in a Gaussian context [Chap. 5].

Chapter 1
Reproducing Kernel Hilbert Spaces: The Rudiments

This chapter provides and illustrates, following the table of contents, the basic tools of the theory of reproducing kernel Hilbert spaces.

1.1 Definition and First Properties

Let T be a set, and H be a Hilbert space of functions, defined on T, with values in $\mathbb{R}$ (one shall presently see that such spaces exist), whose inner product, for $(h_1, h_2) \in H \times H$, fixed, but arbitrary, shall be denoted

$$\langle h_1, h_2 \rangle_H.$$

Given a function of two arguments, $\mathcal{H} : T \times T \longrightarrow \mathbb{R}$, $\mathcal{H}(\cdot, t)$ shall be, for fixed, but arbitrary $t \in T$, the function $x \mapsto \mathcal{H}(x, t)$. It shall always be assumed that $\mathcal{H}$ is not identically zero.

Definition 1.1.1 A reproducing kernel for H, if it exists, is a function of two arguments, $\mathcal{H} : T \times T \longrightarrow \mathbb{R}$, such that, for $(h, t) \in H \times T$, fixed, but arbitrary,

1. $\mathcal{H}(\cdot, t) \in H$;
2. $h(t) = \langle h, \mathcal{H}(\cdot, t) \rangle_H$.

Definition 1.1.2 A reproducing kernel Hilbert space is a Hilbert space of real valued functions for which there exists a reproducing kernel (one shall presently see that such spaces exist).

If, in (Definition) 1.1.2, H is the Hilbert space of functions, T the domain of these, and $\mathcal{H}$ the reproducing kernel, the notation shall usually be $H(\mathcal{H}, T)$. The acronym for *reproducing kernel Hilbert space* shall be RKHS. When the reproducing kernel has domain $T \times T$, one shall say that it is a reproducing kernel over T.

A.F. Gualtierotti, *Detection of Random Signals in Dependent Gaussian Noise*,
DOI 10.1007/978-3-319-22315-5_1

Remark 1.1.3 Suppose $H(\mathcal{H}, T)$ is an RKHS. Given $(t, x) \in T \times T$, fixed, but arbitrary, (Definition) 1.1.1 yields that

$$\langle \mathcal{H}(\cdot, x), \mathcal{H}(\cdot, t)\rangle_{H(\mathcal{H},T)} = \mathcal{H}(t, x),$$
$$\langle \mathcal{H}(\cdot, t), \mathcal{H}(\cdot, x)\rangle_{H(\mathcal{H},T)} = \mathcal{H}(x, t),$$

so that $\mathcal{H}(t, x) = \mathcal{H}(x, t)$ (reproducing kernels are symmetric).

In the sequel, the triad $[n, \alpha, (t, T)]$ shall represent the following elements:

$$n \in \mathbb{N},$$
$$\{\alpha_1, \ldots, \alpha_n\} \subseteq \mathbb{R},$$
$$\{t_1, \ldots, t_n\} \subseteq T.$$

$\{t_1, \ldots, t_n\}$ shall also be written as T_n.

Remark 1.1.4 Suppose $H(\mathcal{H}, T)$ is an RKHS. Definition 1.1.1 and the previous remark also yield that, for $[n, \alpha, (t, T)]$, fixed, but arbitrary,

$$\sum_{i=1}^{n}\sum_{j=1}^{n} \alpha_i \alpha_j \mathcal{H}\left(t_i, t_j\right) = \left\| \sum_{i=1}^{n} \alpha_i \mathcal{H}(\cdot, t_i)\right\|^2_{H(\mathcal{H},T)}.$$

Consequently the matrix $\Sigma_{\mathcal{H},T_n}$, with entries $\mathcal{H}\left(t_i, t_j\right)$ is positive definite, and strictly positive definite when the functions $\{\mathcal{H}(\cdot, t_1), \ldots, \mathcal{H}(\cdot, t_n)\}$ are linearly independent. In particular, for $t \in T$, fixed, but arbitrary,

$$\mathcal{H}(t, t) = \|\mathcal{H}(\cdot, t)\|^2_{H(\mathcal{H},T)}$$

Proposition 1.1.5 *One has, for the RKHS $H(\mathcal{H}, T)$, that*

1. *$\mathcal{H}$ is symmetric and positive definite* (it is thus a covariance: (Definition) 1.3.1)*;*
2. *for $(t_1, t_2) \in T \times T$, fixed, but arbitrary,*

$$|\mathcal{H}(t_1, t_2)| \leq \mathcal{H}^{1/2}(t_1, t_1)\mathcal{H}^{1/2}(t_2, t_2);$$

3. *the family $\mathcal{H}[T] = \{\mathcal{H}(\cdot, t),\ t \in T\}$ is total* [266, p. 32] *in $H(\mathcal{H}, T)$, so that the linear manifold $V[\mathcal{H}]$ generated by that family is dense in $H(\mathcal{H}, T)$;*
4. *for $(h, t) \in H(\mathcal{H}, T) \times T$, fixed, but arbitrary, and $\mathcal{E}_t : \mathbb{R}^T \longrightarrow \mathbb{R}$ defined using the following relation: $\mathcal{E}_t[h] = h(t)$,*

$$|\mathcal{E}_t[h]| = |h(t)| \leq \mathcal{H}^{1/2}(t, t)\,\|h\|_{H(\mathcal{H},T)},$$

so that each evaluation map is a continuous linear functional of $H(\mathcal{H}, T)$.

Proof It suffices to invoke (Remarks) 1.1.3, 1.1.4, and the properties of inner products [266]. □

Proposition 1.1.6 *Let $\mathcal{K}$ be a reproducing kernel for the RKHS $H(\mathcal{H}, T)$. Then $\mathcal{K} = \mathcal{H}$, that is, RKHS's have a unique reproducing kernel.*

Proof As $\mathcal{K}(\cdot, t) \in H(\mathcal{H}, T)$,

$$\mathcal{K}(x, t) = \langle \mathcal{K}(\cdot, t), \mathcal{H}(\cdot, x) \rangle_{H(\mathcal{H},T)}.$$

But $\mathcal{K}$ being a kernel for $H(\mathcal{H}, T)$ and $\mathcal{H}(\cdot, x)$ being an element of $H(\mathcal{H}, T)$,

$$\mathcal{H}(t, x) = \langle \mathcal{H}(\cdot, x), \mathcal{K}(\cdot, t) \rangle_{H(\mathcal{H},T)}.$$

Thus, using (Remark) 1.1.3 for the equality below, for fixed, but arbitrary elements $(x, t) \in T \times T$,

$$\mathcal{K}(x, t) = \mathcal{H}(t, x) = \mathcal{H}(x, t).$$

□

Proposition 1.1.7 *Suppose that one has two RKHS's with the same domain and the same kernel, say $H(\mathcal{H}, T)$ and $K(\mathcal{H}, T)$. Then $H(\mathcal{H}, T)$ and $K(\mathcal{H}, T)$ are identical as Hilbert spaces. Consequently, one domain and one kernel determine a unique RKHS.*

Proof Since $H(\mathcal{H}, T)$ and $K(\mathcal{H}, T)$ have same kernel and domain, for $[n, \alpha, (t, T)]$ and $[p, \beta, (\theta, T)]$ fixed, but arbitrary,

$$\langle \sum_{i=1}^{n} \alpha_i \mathcal{H}(\cdot, t_i), \sum_{j=1}^{p} \beta_j \mathcal{H}(\cdot, \theta_j) \rangle_{H(\mathcal{H},T)} =$$

$$= \sum_{i=1}^{n} \sum_{j=1}^{p} \alpha_i \beta_j \mathcal{H}(t_i, \theta_j)$$

$$= \langle \sum_{i=1}^{n} \alpha_i \mathcal{H}(\cdot, t_i), \sum_{j=1}^{p} \beta_j \mathcal{H}(\cdot, \theta_j) \rangle_{K(\mathcal{H},T)}.$$

Consequently, for $\{h_1, h_2\} \subseteq V[\mathcal{H}]$, fixed, but arbitrary,

$$\langle h_1, h_2 \rangle_{H(\mathcal{H},T)} = \langle h_1, h_2 \rangle_{K(\mathcal{H},T)}.$$

$V[\mathcal{H}]$ is dense in $H(\mathcal{H}, T)$ because of fact (Proposition) 1.1.5. So, given a fixed, but arbitrary $h \in H(\mathcal{H}, T)$, there exists $\{h_n, n \in \mathbb{N}\} \subseteq V[\mathcal{H}]$ such that, in $H(\mathcal{H}, T)$, $\lim_n h_n = h$. $\{h_n, n \in \mathbb{N}\}$ is thus a Cauchy sequence of $H(\mathcal{H}, T)$ in $V[\mathcal{H}]$, and, consequently, also of $K(\mathcal{H}, T)$. Let k be the corresponding limit. Then

$$\begin{aligned} h(t) &= \langle h, \mathcal{H}(\cdot, t)\rangle_{H(\mathcal{H},T)} \\ &= \lim_n \langle h_n, \mathcal{H}(\cdot, t)\rangle_{H(\mathcal{H},T)} \\ &= \lim_n \langle h_n, \mathcal{H}(\cdot, t)\rangle_{K(\mathcal{H},T)} \\ &= \langle k, \mathcal{H}(\cdot, t)\rangle_{K(\mathcal{H},T)} \\ &= k(t), \end{aligned}$$

so that, as functions, $h = k$, and, as sets, $H(\mathcal{H}, T) \subseteq K(\mathcal{H}, T)$. The respective roles of $H(\mathcal{H}, T)$ and $K(\mathcal{H}, T)$ are, in the preceding argument, interchangeable. Thus, as sets, they are equal. The norms in both spaces are furthermore equal:

$$\|h\|_{K(\mathcal{H},T)} = \|k\|_{K(\mathcal{H},T)} = \lim_n \|h_n\|_{K(\mathcal{H},T)} = \lim_n \|h_n\|_{H(\mathcal{H},T)} = \|h\|_{H(\mathcal{H},T)}.$$

□

Proposition 1.1.8 *A Hilbert space H of real valued functions over a set T is an RKHS if, and only if, all evaluation maps are continuous linear functionals of H.*

Proof Suppose that H is the RKHS $H(\mathcal{H}, T)$.

That the evaluation maps are continuous linear functionals of $H(\mathcal{H}, T)$ has already been asserted in property (Proposition) 1.1.5.

Proof Suppose that all evaluation maps are continuous linear functionals of H.

Fix $t \in T$ arbitrarily. By the Riesz representation theorem [266, p. 64], there is a unique $h[t] \in H$ such that $\mathcal{E}_t[h] = \langle h, h[t]\rangle_H$, $h \in H$. It then suffices to set, for fixed, but arbitrary $(t, x) \in T \times T$,

$$\mathcal{H}(x, t) = h[t](x).$$

Indeed $\mathcal{H}(\cdot, t) = h[t] \in H$, and, for $h \in H$, fixed, but arbitrary,

$$\langle h, \mathcal{H}(\cdot, t)\rangle_H = \langle h, h[t]\rangle_H = \mathcal{E}_t[h] = h(t).$$

□

Proposition 1.1.9 *Let $H(\mathcal{H}, T)$ be an RKHS, and $c \in]0, \infty[$ be a fixed, but arbitrary constant. Let $T_c = \{t \in T : \mathcal{H}(t, t) \leq c\}$. Then:*

1. [Point-wise convergence] *when $\{h_n, n \in \mathbb{N}\} \subseteq H(\mathcal{H}, T)$ is a sequence which converges weakly to $h \in H(\mathcal{H}, T)$, for fixed, but arbitrary $t \in T$, $\lim_n h_n(t) = h(t)$;*

2. [Uniform convergence] *when* $\{h_n,\ n \in \mathbb{N}\} \subseteq H(\mathcal{H}, T)$ *is a sequence which converges in norm to* $h \in H(\mathcal{H}, T)$, *given* $\epsilon > 0$, *fixed, but arbitrary, there exists* $n(\epsilon) > 0$ *such that, for any* $n > n(\epsilon)$ *and any* $t \in T_c$, $|h_n(t) - h(t)| < \epsilon$.

Proof From facts (Definition) 1.1.1 and (Proposition) 1.1.5, one has that

$$|h(t) - h_n(t)| = \left|\langle h - h_n, \mathcal{H}(\cdot, t)\rangle_{H(\mathcal{H},T)}\right| \leq \mathcal{H}^{1/2}(t, t)\, \|h - h_n\|_{H(\mathcal{H},T)}\,.$$

The equality proves item 1, and the inequality, item 2. □

Since $V[\mathcal{H}]$ is dense in $H(\mathcal{H}, T)$ [(Proposition) 1.1.5], one has the following corollary:

Corollary 1.1.10 *Every* $h \in H(\mathcal{H}, T)$ *is the pointwise limit of a sequence* $\{h_n,\ n \in \mathbb{N}\} \subseteq V[\mathcal{H}]$.

Corollary 1.1.11 *When* $H(\mathcal{H}, T)$ *is separable, for every complete orthonormal set* $\{e_n,\ n \in \mathbb{N}\} \subseteq H(\mathcal{H}, T)$, *and fixed, but arbitrary* $(h, t) \in H(\mathcal{H}, T) \times T$,

1. $h(t) = \sum_{n=1}^{\infty} \langle h, e_n\rangle_{H(\mathcal{H},T)}\, e_n(t)$,
2. $\left|\sum_{n=1}^{\infty} \langle h, e_n\rangle_{H(\mathcal{H},T)} e_n(t)\right| \leq \mathcal{H}^{1/2}(t, t)\, \|h\|_{H(\mathcal{H},T)}$.

The series converges at every $t \in T$, *and uniformly on the sets* T_c *of result* (Proposition) 1.1.9.

Proof Let $h_n = \sum_{i=1}^{n} \langle h, e_n\rangle_{H(\mathcal{H},T)} e_n$. $\{h_n,\ n \in \mathbb{N}\}$ converges in norm to h, and

$$h_n(t) = \langle h_n, \mathcal{H}(\cdot, t)\rangle_{H(\mathcal{H},T)} = \sum_{i=1}^{n} \langle h, e_n\rangle_{H(\mathcal{H},T)}\, e_n(t)\,.$$

Thus item 1 follows from (Proposition) 1.1.9, and item 2, from item 1 and (Proposition) 1.1.5. □

Remark 1.1.12 ([35, p. 34]) Suppose T has a topology, and that the map $F : T \longrightarrow H(\mathcal{H}, T)$, computed as $F(t) = \mathcal{H}(\cdot, t)$, is continuous. Let $\{h_n, n \in \mathbb{N}\} \subseteq H(\mathcal{H}, T)$ be weakly convergent to $h \in H(\mathcal{H}, T)$. Then, given a fixed, but arbitrary compact subset $T_c \subseteq T$, $\left\{h_n^{|T_c}, n \in \mathbb{N}\right\}$ converges uniformly to $h^{|T_c}$.

Indeed, H_c, the image of T_c by F, is compact [84, p. 224]. For $t \in T_c$, fixed, but arbitrary, let

$$B(t, \delta) = \left\{h \in H(\mathcal{H}, T) : \|h - \mathcal{H}(\cdot, t)\|_{H(\mathcal{H},T)} < \delta\right\}.$$

One thus obtains an open covering of H_c from which a finite one, corresponding say to

$$\left\{t_1^{(\delta)}, \ldots, t_{n[\delta]}^{(\delta)}\right\} \subseteq T_c,$$

may be extracted. Thus, for $t \in T_c$, fixed, but arbitrary, there is an element

$$t_i^{(\delta)}[t] \in \left\{t_1^{(\delta)}, \ldots, t_{n[\delta]}^{(\delta)}\right\}$$

such that

$$\left\|\mathcal{H}(\cdot, t) - \mathcal{H}\left(\cdot, t_i^{(\delta)}[t]\right)\right\|_{H(\mathcal{H},T)} < \delta.$$

As a weakly convergent sequence is bounded [266, p. 79], let

$$\kappa = \sup_n \|h_n\|_{H(\mathcal{H},T)} < \infty.$$

Given $\epsilon > 0$, fixed, but arbitrary, let $\delta = \epsilon(4\kappa)^{-1}$. Using weak convergence, choose [(Proposition) 1.1.9] n_ϵ such that, for $n \geq n_\epsilon$ and $i \in [1 : n[\delta]]$, fixed, but arbitrary,

$$\left|h\left(t_i^{(\delta)}[t]\right) - h_n\left(t_i^{(\delta)}[t]\right)\right| < \frac{\epsilon}{2}.$$

Then, for $t \in T_c$, fixed, but arbitrary, since $h(t) - h_n(t)$ may be written as

$$\left(h(t) - h\left(t_i^{(\delta)}[t]\right)\right) + \left(h\left(t_i^{(\delta)}[t]\right) - h_n\left(t_i^{(\delta)}[t]\right)\right) + \left(h_n\left(t_i^{(\delta)}[t]\right) - h_n(t)\right),$$

its absolute value is dominated by

$$\left|h\left(t_i^{(\delta)}[t]\right) - h_n\left(t_i^{(\delta)}[t]\right)\right| + \left|\{h(t) - h_n(t)\} - \left\{h\left(t_i^{(\delta)}[t]\right) - h_n\left(t_i^{(\delta)}[t]\right)\right\}\right|$$

which is strictly less than

$$\frac{\epsilon}{2} + \left|\langle h - h_n, \mathcal{H}(\cdot, t)\rangle_{H(\mathcal{H},T)} - \left\langle h - h_n, \mathcal{H}\left(\cdot, t_i^{(\delta)}[t]\right)\right\rangle_{H(\mathcal{H},T)}\right|.$$

Thus, using the fact [266, p. 78] that, in the presence of weak convergence,

$$\|h\|_{H(\mathcal{H},T)} \leq \liminf_n \|h_n\|_{H(\mathcal{H},T)} \; (\leq \sup_n \|h_n\|_{H(\mathcal{H},T)} = \kappa),$$

one has that

$$\begin{aligned}
|h(t) - h_n(t)| &< \frac{\epsilon}{2} + \left|\left\langle h - h_n, \mathcal{H}(\cdot, t) - \mathcal{H}\left(\cdot, t_i^{(\delta)}[t]\right)\right\rangle_{H(\mathcal{H},T)}\right| \\
&\leq \frac{\epsilon}{2} + \|h - h_n\|_{H(\mathcal{H},T)} \left\|\mathcal{H}(\cdot, t) - \mathcal{H}\left(\cdot, t_i^{(\delta)}[t]\right)\right\|_{H(\mathcal{H},T)} \\
&< \frac{\epsilon}{2} + (2\kappa)\left(\frac{\epsilon}{4\kappa}\right) \\
&= \epsilon.
\end{aligned}$$

Remark 1.1.13 Suppose T has a topology, and $F : T \longrightarrow H(\mathcal{H}, T)$, computed as $F(t) = \mathcal{H}(\cdot, t)$, is continuous. As, for $\{t_1, t_2\} \subseteq T$, fixed, but arbitrary,

$$\begin{aligned} |h(t_1) - h(t_2)| &= \left|\langle h, \mathcal{H}(\cdot, t_1) - \mathcal{H}(\cdot, t_2)\rangle_{H(\mathcal{H},T)}\right| \\ &\leq \|h\|_{H(\mathcal{H},T)} \|F(t_1) - F(t_2)\|_{H(\mathcal{H},T)}, \end{aligned}$$

bounded sets in $H(\mathcal{H}, T)$ are made of equicontinuous functions.

Remark 1.1.14 Another version of result (Corollary) 1.1.11 is as follows. Suppose that

$$\{e_i, i \in I \subseteq \mathbb{N}\} \subseteq H(\mathcal{H}, T)$$

is a fixed, but arbitrary orthonormal set, and $\underline{\alpha} \in l_2$, a fixed, but arbitrary element. Then, for $t \in T$, fixed, but arbitrary,

$$\left|\sum_{i \in I} \alpha_i e_i(t)\right| = \left|\left\langle \sum_{i \in I} \alpha_i e_i, \mathcal{H}(\cdot, t)\right\rangle_{H(\mathcal{H},T)}\right| \leq \mathcal{H}^{1/2}(t, t) \|\underline{\alpha}\|_{l_2}.$$

Many RKHS's can be obtained rather quickly with the help of the following result which will be of use repeatedly. $\mathcal{R}$ denotes range, and $\mathcal{N}$, kernel or null space. The content of an adjacent straight parenthesis indicates of which map one considers the range or the kernel. A contraction is a bounded, linear operator whose norm is at most one.

Proposition 1.1.15 *Let T be a set, and H, a real Hilbert space. Suppose one is given a function $F : T \longrightarrow H$. Let*

$$H_F = \overline{V[\mathcal{R}[F]]} \subseteq H,$$

and P_{H_F} denote the corresponding projection of H onto H_F. Let the map $L_F : H \longrightarrow \mathbb{R}^T$ be defined using the following relation: for $(h, t) \in H \times T$, fixed, but arbitrary,

$$L_F[h](t) = \langle h, F(t)\rangle_H.$$

L_F shall be taken as a map from H onto its range. On $\mathcal{R}[L_F]$, for elements $(h_1, h_2) \in H \times H$, fixed, but arbitrary, let the formula

$$\langle L_F[h_1], L_F[h_2]\rangle_{\mathcal{R}[L_F]} = \langle P_{H_F}[h_1], P_{H_F}[h_2]\rangle_H \tag{$\star$}$$

designate a tentative inner product. Then:

1. *L_F is linear, so that $\mathcal{R}[F]$ is a linear manifold in $\mathbb{R}^T$;*
2. *the tentative inner product formula ($\star$) makes of $\mathcal{R}[F]$ a Hilbert space, so that L_F is a contraction;*

3. *$\mathcal{N}[L_F]$ is closed and equal to $H_F^{\perp}$;*
4. *$\mathcal{R}[L_F]$ is an RKHS whose reproducing kernel $\mathcal{K}_F$ is given by the following relation:*

$$\mathcal{K}_F(t_1, t_2) = \langle F(t_1), F(t_2)\rangle_H;$$

5. *L_F has the following further properties:*

   (i) *it is a partial isometry, with H_F as initial set, and $\mathcal{R}[L_F]$ as final set;*
   (ii) *the restriction $U_F = L_F^{|H_F}$ of L_F to H_F is unitary, and*

$$L_F = U_F P_{H_F};$$

   (iii) *when it is an injection, L_F is unitary;*
   (iv) *$L_F[F(t)] = \mathcal{K}_F(\cdot, t)$, and $L_F^{\star}[\mathcal{K}_F(\cdot, t)] = F(t)$.*

Proof *L_F is linear:*

$$\begin{aligned} L_F[\alpha_1 h_1 + \alpha_2 h_2](t) &= \langle \alpha_1 h_1 + \alpha_2 h_2, F(t)\rangle_H \\ &= \alpha_1 \langle h_1, F(t)\rangle_H + \alpha_2 \langle h_2, F(t)\rangle_H \\ &= \alpha_1 L_F[h_1](t) + \alpha_2 L_F[h_2](t) \\ &= \{\alpha_1 L_F[h_1] + \alpha_2 L_F[h_2]\}(t). \end{aligned}$$

Proof *The expression ($\star$) for $\langle L_F[h_1], L_F[h_2]\rangle_{\mathcal{R}[L_F]}$ produces an inner product on $\mathcal{R}[L_F]$:*

- it is bilinear as, using successively the linearity of L_F and the definition,

$$\begin{aligned} &\langle \alpha_1 L_F[h_1] + \alpha_2 L_F[h_2], L_F[h]\rangle_{\mathcal{R}[L_F]} = \\ &\quad = \langle L_F[\alpha_1 h_1 + \alpha_2 h_2], L_F[h]\rangle_{\mathcal{R}[L_F]} \\ &\quad = \langle P_{H_F}[\alpha_1 h_1 + \alpha_2 h_2], P_{H_F}[h]\rangle_H \\ &\quad = \alpha_1 \langle P_{H_F}[h_1], P_{H_F}[h]\rangle_H + \alpha_2 \langle P_{H_F}[h_2], P_{H_F}[h]\rangle_H \\ &\quad = \alpha_1 \langle L_F[h_1], L_F[h]\rangle_{\mathcal{R}[L_F]} + \alpha_2 \langle L_F[h_2], L_F[h]\rangle_{\mathcal{R}[L_F]}; \end{aligned}$$

- it is strictly positive definite: indeed, by definition,

$$\|L_F[h]\|^2_{\mathcal{R}[L_F]} = \|P_{H_F}[h]\|^2_H \geq 0;$$

  but $\|P_{H_F}[h]\|_H$ is zero when $h \in H_F^{\perp}$; then, by the definition of L_F, $L_F[h] = 0_T$, and the bilinear form is thus strictly positive definite.

Proof The manifold $\mathcal{R}[L_F]$ with the given tentative inner product ($\star$) is complete:
let indeed $\{L_F[h_n], n \in \mathbb{N}\} \subseteq \mathcal{R}[L_F]$ be Cauchy; then the sequence

$$\{P_{H_F}[h_n], \in \mathbb{N}\} \subseteq H_F$$

is Cauchy, and there exists a unique $h \in H_F$ such that, in H,

$$\lim_n P_{H_F}[h_n] = h;$$

but then, by the definition of the (RKHS) inner product,

$$\lim_n \|L_F[h_n] - L_F[h]\|_{\mathcal{R}[L_F]} = \lim_n \|P_{H_F}[h_n] - h\|_H = 0,$$

and every Cauchy sequence in $\mathcal{R}[L_F]$ has a unique limit in $\mathcal{R}[L_F]$.

Proof The manifold $\mathcal{N}[L_F]$ is closed: because of [266, p. 35],

$$\mathcal{N}[L_F] = \mathcal{R}[F]^{\perp} = V[\mathcal{R}[F]]^{\perp} = \overline{V[\mathcal{R}[F]]}^{\perp} = H_F^{\perp}.$$

Proof The Hilbert space of functions $\mathcal{R}[L_F]$ is an RKHS:
let $\mathcal{K}_F(\cdot, t) = L_F[F(t)] \in \mathcal{R}[L_F]$. Then

$$\mathcal{K}_F(x, t) = L_F[F(t)](x) = \langle F(t), F(x)\rangle_H$$

defines a candidate kernel; but, for $k = L_F[h]$,

$$\begin{aligned}
\langle k, \mathcal{K}_F(\cdot, t)\rangle_{\mathcal{R}[L_F]} &= \langle L_F[h], L_F[F(t)]\rangle_{\mathcal{R}[L_F]} \\
&= \langle P_{H_F}[h], F(t)\rangle_H \\
&= \langle h, F(t)\rangle_H \\
&= L_F[h](t) \\
&= k(t),
\end{aligned}$$

and $\mathcal{K}_F$ has the reproducing property.

Proof (5.iv) By definition

$$L_F[h](t) = \langle h, F(t)\rangle_H;$$

but, since $L_F[h] \in H(\mathcal{K}_F, T)$, also

$$L_F[h](t) = \langle L_F[h], \mathcal{K}_F(\cdot, t)\rangle_{H(\mathcal{K}_F, T)} = \langle h, L_F^{\star}[\mathcal{K}_F(\cdot, t)]\rangle_H.$$

□

Remark 1.1.16 Result (Proposition) 1.1.15 shall often be used in the sequel so that one shall adopt a perhaps more convenient, certainly more flexible notation than that used so far. One shall keep $\mathcal{K}_F$, L_F, and H_F, but replace $\mathcal{R}[L_F]$ with K_F. When convenient one shall also write $\mathcal{K}, L$ and K, and sometimes also $H(\mathcal{K}_F, T)$ for K_F, and $H(\mathcal{K}, T)$ for K.

Remark 1.1.17 ([209, p. 1]) Every RKHS may be obtained by the procedure of (Proposition) 1.1.15.

Let indeed $H(\mathcal{H}, T)$ be an RKHS. As seen [(Proposition) 1.1.5], $\mathcal{H}$ is a covariance. But each covariance gives rise as follows [273, p. 238] to a Gaussian process whose mean is the zero function, and whose covariance is $\mathcal{H}$. Let $\Omega = \mathbb{R}^T$, and $\mathcal{A}$ be the σ-algebra generated by the evaluation maps $\{\mathcal{E}_t, t \in T\}$. Then the requirement that, for fixed, but arbitrary $n \in \mathbb{N}$ and $\{t_1, \ldots, t_n\} \subseteq T$,

$$\left(\mathcal{E}_{t_1}, \ldots, \mathcal{E}_{t_n}\right) \sim \mathcal{N}\left(\underline{0}_{\mathbb{R}^n}, \Sigma_{\mathcal{H},T_n}\right)$$

determines a probability measure $P_{\mathcal{H}}$ on $\mathcal{A}$ (the entries of $\Sigma_{\mathcal{H},T_n}$ are those of the following set: $\{\mathcal{H}\left(t_i, t_j\right), \{i, j\} \subseteq [1:n]\}$). The stochastic process is produced setting

$$X(f, t) = \mathcal{E}_t[f].$$

Let X_t be the equivalence class of $X(\cdot, t)$ with respect to $P_{\mathcal{H}}$, and set

$$F : T \longrightarrow L_2(\Omega, \mathcal{A}, P_{\mathcal{H}}), \quad F(t) = X_t.$$

An application of (Proposition) 1.1.15 yields $H(\mathcal{H}, T)$.

Remark 1.1.18 To use (Proposition) 1.1.15, it suffices, allowing for (Proposition) 1.1.7, to identify the Hilbert space H and the function F.

Remark 1.1.19 When T has "structure," and F of (Proposition) 1.1.15 is "adapted" to it, the functions of $H(\mathcal{H}, T)$ will "reflect" the properties of F and T. This statement has already been evidenced in (Remark) 1.1.12 and in (Remark) 1.1.13. The chapter which follows shall make that same statement systematically explicit.

The following examples of RKHS's illustrate the use, and efficiency, of (Proposition) 1.1.15.

Example 1.1.20 Let $\underline{u}$ and $\underline{v}$ be fixed, but arbitrary elements of $\mathbb{R}^n$ with respective components $u_1, \ldots, u_n$ and $v_1, \ldots, v_n$. The usual inner product of $\mathbb{R}^n$ shall be denoted, for fixed, but arbitrary $(\underline{u}, \underline{v}) \in \mathbb{R}^n \times \mathbb{R}^n$,

$$\langle \underline{u}, \underline{v} \rangle_{\mathbb{R}^n} = \sum_{i=1}^{n} u_i v_i.$$

Let $\underline{e}_i$ be the i-th standard basis vector of $\mathbb{R}^n$, and M be a symmetric, strictly positive definite matrix of dimension n. Define an inner product on $\mathbb{R}^n$ using the following relation, valid for fixed, but arbitrary $\{\underline{h}_1, \underline{h}_2\} \subset \mathbb{R}^n$:

$$\langle \underline{h}_1, \underline{h}_2 \rangle_M = \langle M\left[\underline{h}_1\right], \underline{h}_2 \rangle_{\mathbb{R}^n}.$$

Let H_M denote the corresponding Hilbert space: thus when $M = I_n$, the identity matrix, $H_{I_n} = \mathbb{R}^n$. B shall be any $n \times n$ matrix.

Let $T = [1 : n]$ be the set of integers running from 1 to n, and define

$$F_B : T \longrightarrow H_M \text{ using } F_B(i) = B\underline{e}_i.$$

Then (writing for example L_B for L_{F_B})

$$L_B\left[\underline{h}\right](i) = \langle \underline{h}, F_B(i) \rangle_M = \langle B^\star M\left[\underline{h}\right], \underline{e}_i \rangle_{\mathbb{R}^n},$$

so that $L_B = B^\star M$. Thus

$$\mathcal{N}[L_B] = \mathcal{N}[B^\star M] \text{ and } K_B = \mathcal{R}[B^\star M].$$

Furthermore, since, for any bounded, linear operator L [266, pp. 35,71],

$$\mathcal{N}[L]^\perp = \overline{\mathcal{R}[L^\star]},$$

and that finite dimensional subspaces are closed [266, p. 48],

$$H_B = \mathcal{N}[L_B]^\perp = \overline{\mathcal{R}[(B^\star M)^\star]} = \mathcal{R}[M^\star B] = \mathcal{R}[MB],$$

and thus

$$\begin{aligned}\langle L_B\left[\underline{h}_1\right], L_B\left[\underline{h}_2\right] \rangle_{K_B} &= \langle P_{\mathcal{R}[MB]}\left[\underline{h}_1\right], P_{\mathcal{R}[MB]}\left[\underline{h}_2\right] \rangle_M \\ &= \langle M P_{\mathcal{R}[MB]}\left[\underline{h}_1\right], P_{\mathcal{R}[MB]}\left[\underline{h}_2\right] \rangle_{\mathbb{R}^n},\end{aligned}$$

and also

$$\mathcal{K}_B(i,j) = \langle F_B(i), F_B(j) \rangle_M = \langle B^\star M B\left[\underline{e}_i\right], \underline{e}_j \rangle_{\mathbb{R}^n}.$$

As a particular case one may choose $B = I_n$ and $M = \Sigma$, a strictly positive definite covariance matrix. Then:

- $L_B = M = \Sigma$,
- $\underline{e}_i = L_B\left[\underline{h}_i\right]$ has the solution $\underline{h}_i = \Sigma^{-1}\left[\underline{e}_i\right]$,
- $P_{\mathcal{N}[L_B]^\perp} = I_n$,

so that

$$\begin{aligned}\langle \underline{e}_i, \underline{e}_j\rangle_{K_B} &= \langle L_B\left[\underline{h}_i\right], L_B\left[\underline{h}_j\right]\rangle_{K_B}\\ &= \langle \underline{h}_i, \underline{h}_j\rangle_{\Sigma}\\ &= \langle \Sigma\left[\underline{h}_i\right], \underline{h}_j\rangle_{\mathbb{R}^n}\\ &= \langle \Sigma^{-1}\left[\underline{e}_i\right], \underline{e}_j\rangle_{\mathbb{R}^n}.\end{aligned}$$

Consequently $\langle \underline{a}, \underline{b}\rangle_{K_B} = \langle \Sigma^{-1}\left[\underline{a}\right], \underline{b}\rangle_{\mathbb{R}^n}$.

Example 1.1.21 Let T be a set, and $f : T \longrightarrow \mathbb{R}$ be a function different from the null function. Write f_α for the function $f_\alpha(t) = \alpha f(t)$. Then $H = \{f_\alpha, \alpha \in \mathbb{R}\}$ is a Hilbert space with inner product $\langle f_\alpha, f_\beta\rangle_H = \alpha\beta$, isomorphic to $\mathbb{R}$. In particular $\|f\|_H = \|f_1\|_H = 1$.

Let $F(t) = f(t) f = f_{f(t)} \in H$. Then

$$L_F[f_\alpha](t) = \langle f_\alpha, f_{f(t)}\rangle_H = \alpha f(t) = f_\alpha(t),$$

so that L_F is the identity. Furthermore $\mathcal{N}[L_F] = \{f_0\}$ and L_F is unitary, so that $K_F = H$. H is thus an RKHS with kernel

$$\mathcal{K}_F(t_1, t_2) = \langle F(t_1), F(t_2)\rangle_H = \langle f_{f(t_1)}, f_{f(t_2)}\rangle_H = f(t_1) f(t_2).$$

Example 1.1.22 Let V be a manifold of $\mathbb{R}^T$ of dimension $n \in \mathbb{N}$, generated by the following linearly independent set: $\{v_1, \ldots, v_n\}$. Let M be a symmetric, strictly positive definite matrix of size n. For $\{t_1, t_2\} \subseteq T$, fixed, but arbitrary, let

$$\mathcal{H}(t_1, t_2) = \langle M\underline{v}(t_1), \underline{v}(t_2)\rangle_{\mathbb{R}^n},$$

where $\underline{v}(t) \in \mathbb{R}^n$ has components $\{v_1(t), \ldots, v_n(t)\}$. $\mathcal{H}$ is positive definite. Since

$$\mathcal{H}(x, t) = \langle M\underline{v}(x), \underline{v}(t)\rangle_{\mathbb{R}^n} = \langle \underline{v}(x), M\underline{v}(t)\rangle_{\mathbb{R}^n} = \sum_{i=1}^{n} \langle M\underline{v}(t), \underline{e}_i\rangle_{\mathbb{R}^n} v_i(x),$$

it follows that

$$\mathcal{H}(\cdot, t) = \sum_{i=1}^{n} \langle M\underline{v}(t), \underline{e}_i\rangle_{\mathbb{R}^n} v_i \in V.$$

For $i \in [1:n]$, fixed, but arbitrary, let $c_i(t) = \langle M\underline{v}(t), \underline{e}_i\rangle_{\mathbb{R}^n}$ and $\underline{c}$ have components $c_i,\ i \in [1:n]$. Then

$$\underline{c}(t) = M\underline{v}(t) \qquad \text{and} \qquad \mathcal{H}(\cdot, t) = \sum_{i=1}^{n} c_i(t)\, v_i.$$

For $\{h_1, h_2\} \subseteq V$ (and $t \in T$), fixed, but arbitrary, with

$$h_1(t) = \sum_{i=1}^{n} \alpha_i^{(1)} v_i(t) \text{ and } h_2(t) = \sum_{i=1}^{n} \alpha_i^{(2)} v_i(t),$$

set

$$\langle h_1, h_2\rangle_V = \langle M^{-1}\underline{\alpha}_1, \underline{\alpha}_2\rangle_{\mathbb{R}^n},$$

where, for example, $\underline{\alpha}_1$ has components

$$\left\{\alpha_1^{(1)}, \ldots, \alpha_n^{(1)}\right\}.$$

One thus defines an inner product. Since

$$\|h_1 - h_2\|_V^2 = \langle M^{-1}\left(\underline{\alpha}_1 - \underline{\alpha}_2\right), \left(\underline{\alpha}_1 - \underline{\alpha}_2\right)\rangle_{\mathbb{R}^n},$$

with that inner product, V is complete, and thus a Hilbert space of functions.

Let $F : T \longrightarrow V$ be defined using the following relation: $F(t) = \mathcal{H}(\cdot, t)$. Then, for $h \in V$, fixed, but arbitrary, $h = \sum_{i=1}^{n} \alpha_i v_i$,

$$\begin{aligned} L_F[h](t) &= \langle h, \mathcal{H}(\cdot, t)\rangle_V \\ &= \langle M^{-1}\underline{\alpha}, \underline{c}(t)\rangle_{\mathbb{R}^n} \\ &= \langle M^{-1}\underline{\alpha}, M\underline{v}(t)\rangle_{\mathbb{R}^n} \\ &= \langle \underline{\alpha}, \underline{v}(t)\rangle_{\mathbb{R}^n}, \end{aligned}$$

so that

$$L_F[h](t) = \sum_{i=1}^{n} \alpha_i v_i(t) = h(t).$$

L_F is thus the identity. Furthermore,

$$\mathcal{K}_F(t_1, t_2) = \langle \mathcal{H}(\cdot, t_1), \mathcal{H}(\cdot, t_2)\rangle_V = \langle M^{-1}M\underline{v}(t_1), M\underline{v}(t_2)\rangle_{\mathbb{R}^n} = \mathcal{H}(t_1, t_2).$$

Hence $V = H(\mathcal{H}, T)$.

One may notice that, for $\{i,j\} \subseteq [1:n]$, fixed, but arbitrary,

$$\langle v_i, v_j\rangle_{H(\mathcal{H},T)} = \langle M^{-1}\underline{e}_i, \underline{e}_j\rangle_{\mathbb{R}^n}.$$

Consequently, when one chooses $M = I_n$, the identity matrix of dimension n, $\{v_1,\ldots,v_n\}$ is an orthonormal set in $H(\mathcal{H},T)$.

Suppose that M is only positive definite, and that the vectors are not linearly independent. Let a largest linearly independent subset of $\{v_1,\ldots,v_n\}$ be denoted $\{u_1,\ldots,u_p\}$, $p < n$, and the remaining vectors be denoted $\{w_1,\ldots,w_{n-p}\}$. Let $M_{w,v}$ be the matrix that expresses the w's as linear combinations of the u's. There is then a permutation matrix [127, p. 86] P such that, for $t \in T$, fixed, but arbitrary,

$$\underline{v}(t) = P^\star \begin{bmatrix} I_p \\ M_{w,v} \end{bmatrix} \underline{u}(t) = M_{v,u}\left[\underline{u}(t)\right].$$

Then

$$\langle M\left[\underline{v}(t_1)\right], \underline{v}(t_1)\rangle_{\mathbb{R}^n} = \langle\left\{M_{v,u}^\star M M_{v,u}\right\}\left[\underline{u}(t_1)\right], \underline{u}(t_2)\rangle_{\mathbb{R}^p}.$$

Let m be the rank of $M_{v,u}^\star M M_{v,u}$. When $m < p$, let,

- for orthonormal $\{\underline{m}_1,\ldots,\underline{m}_m\}$, $M_{v,u}^\star M M_{v,u} = \sum_{i=1}^m \mu_i \underline{m}_i \otimes \underline{m}_i$,
- M_d be the diagonal matrix with successive entries $\mu_1,\ldots,\mu_m$,
- for $i \in [1:m]$, fixed, but arbitrary, $\tilde{v}_i(t) = \langle \underline{m}_i, \underline{u}(t)\rangle_{\mathbb{R}^p}$.

Then

$$\langle\left\{M_{v,u}^\star M M_{v,u}\right\}\left[\underline{u}(t_1)\right], \underline{u}(t_2)\rangle_{\mathbb{R}^p} = \langle M_d\left[\underline{\tilde{v}}(t_1)\right], \underline{\tilde{v}}(t_2)\rangle_{\mathbb{R}^m}.$$

The functions $\tilde{v}_1,\ldots,\tilde{v}_m$ are linearly independent. Suppose indeed that,

$$\sum_{i=1}^m \alpha_i \tilde{v}_i = 0.$$

Then, for $t \in T$, fixed, but arbitrary,

$$\left\langle \sum_{i=1}^m \alpha_i \underline{m}_i, \underline{u}(t)\right\rangle_{\mathbb{R}^p} = 0.$$

One has that $\{\underline{u}(t), t \in T\}$ generates $\mathbb{R}^p$ since

$$0 = \langle \underline{x}, \underline{u}(t)\rangle_{\mathbb{R}^p} = \sum_{i=1}^p x_i u_i(t)$$

implies $\underline{x} = \underline{0}_{\mathbb{R}^p}$ as the u_i's are linearly independent. Consequently

$$\sum_{i=1}^{m} \alpha_i \underline{m}_i = \underline{0}_{\mathbb{R}^p},$$

that is $\alpha_1 = \cdots = \alpha_m = 0$. It follows that one may always assume that M is strictly positive definite and that $v_1, \ldots, v_n$ are linearly independent.

Example 1.1.23 ([7, p. 43]) Let $\{a, b\} \subseteq \mathbb{R}$, $a < b$, be fixed, but arbitrary, T be $[a, b]$, and P_T^n be the set of real polynomials of degree n over T. P_T^n is a vector space. Let $c \in T$ be fixed, but arbitrary, and let every $p \in P_T^n$ have the (Taylor) representation

$$p(t) = \sum_{i=0}^{n} p^{(i)}(c) \frac{(t-c)^i}{i!},$$

where $p^{(i)}(c)$ is the i-th derivative of p evaluated at c. Let $H = \mathbb{R}^{n+1}$ with its usual inner product, and let $F : T \longrightarrow H$ be defined using the following relation: for $i \in [0 : n]$, fixed, but arbitrary,

$$\langle F(t), \underline{e}_i \rangle_H = \frac{(t-c)^i}{i!},$$

so that

$$L_F[\underline{x}](t) = \langle \underline{x}, F(t) \rangle_H = \sum_{i=0}^{n} x_i \frac{(t-c)^i}{i!}, \quad \text{and thus } L_F[\underline{x}] \in P_T^n.$$

L_F is a bijection, so that

$$\langle L_F[\underline{x}], L_F[\underline{y}] \rangle_{\mathcal{R}[L_F]} = \langle \underline{x}, \underline{y} \rangle_H.$$

Furthermore, for $\{t_1, t_2\} \subseteq T$, fixed, but arbitrary,

$$\mathcal{K}_F(t_1, t_2) = \langle F(t_1), F(t_2) \rangle_H = \sum_{i=0}^{n} \frac{(t_1-c)^i}{i!} \frac{(t_2-c)^i}{i!}.$$

P_T^n is thus an RKHS with kernel $\mathcal{K}_F$.

Example 1.1.24 Let $T = \mathbb{N}$, and $F : T \longrightarrow l_2$ be defined using the following relation: $F(n) = \underline{e}_n$, the n-th element of the standard basis of l_2. Then, for $\underline{h} \in l_2$, fixed, but arbitrary, $L_F[\underline{h}](n) = h_n$, the component of $\underline{h}$ in position n. Obviously the null space of L_F is the zero sequence so that L_F is the identity. Finally, for fixed, but arbitrary $\{n_1, n_2\} \subseteq \mathbb{N}$,

$$\mathcal{H}_F(n_1, n_2) = \langle F(n_1), F(n_2)\rangle_H = \delta_{n_1, n_2},$$

and $\mathcal{H}_F(\cdot, n) = \underline{e}_n$. Thus $l_2 = H(\mathcal{H}_F, \mathbb{N})$.

Example 1.1.25 Let *Leb* denote Lebesgue measure, and $L_2[0,1]$, the following space: $L_2([0,1], \mathcal{B}[0,1], Leb)$. Let the equivalence class of the indicator function of $[0,t]$, $\chi_{[0,t]}$, be denoted I_t, and that of the function h, $[h]$. Choose for H, $L_2[0,1]$, and, for F, the map defined using the following equality: $F(t) = I_t$. Then

$$L[[h]](t) = \langle [h], I_t\rangle_{L_2[0,1]} = \int_0^t h(x)\,dx.$$

L is thus the standard Volterra operator [236, p. 384]. Also $\mathcal{N}[L] = \{[0_{[0,1]}]\}$ and

$$\langle L[[h_1]], L[[h_2]]\rangle_K = \langle [h_1], [h_2]\rangle_{L_2[0,1]} = \int_0^1 h_1(x)\,h_2(x)\,dx.$$

Finally

$$\mathcal{K}(t_1, t_2) = \langle F(t_1), F(t_2)\rangle_{L_2[0,1]} = \langle I_{t_1}, I_{t_2}\rangle_{L_2[0,1]},$$

so that $\mathcal{K}(t_1, t_2) = t_1 \wedge t_2$, the covariance of the standard Wiener process.

Let $f_1, f_2 : [0,1] \longrightarrow \mathbb{R}$ be defined using the following relations: for $t \in [0,1]$, fixed, but arbitrary,

$$f_1(t) = 1,\ f_2(t) = t.$$

Then f_1 does not belong to $H(\mathcal{K}, [0,1])$, but f_2 does, and its norm is equal to one.

Example 1.1.26 Let $(\Omega, \mathcal{A}, P)$ be a probability space, and $L_2(\Omega, \mathcal{A}, P)$ be the Hilbert space of equivalence classes of random variables whose square is integrable. Let $X : \Omega \times T \longrightarrow \mathbb{R}$ be a second order stochastic process, that is, for $t \in T$, fixed, but arbitrary, $X(\cdot, t)$ is adapted to $\mathcal{A}$ and $E_P[X^2(\cdot, t)] < \infty$. The equivalence class of $X(\cdot, t)$ shall be denoted X_t, and $L_P[X]$ shall be the subspace of $L_2(\Omega, \mathcal{A}, P)$ generated linearly by $\{X_t,\ t \in T\}$, that is, $\overline{V[\{X_t,\ t \in T\}]}$.

Let $F : T \longrightarrow L_2(\Omega, \mathcal{A}, P)$ be defined using the following assignment: $F(t) = X_t$. Then $H_F = L_P[X]$. F allows one to define an RKHS K using the ensuing assignment: for $Y \in L_2(\Omega, \mathcal{A}, P)$, fixed, but arbitrary, $\dot{Y}$, one of its members,

$$L_F[Y](t) = \langle Y, X_t\rangle_{L_2(\Omega, \mathcal{A}, P)} = E_P\left[\dot{Y} X(\cdot, t)\right].$$

The reproducing kernel $\mathcal{K}$ is obtained using the following expression:

$$\mathcal{K}(t_1, t_2) = \langle X_{t_1}, X_{t_2} \rangle_{L_2(\Omega,\mathcal{A},P)} = E_P[X(\cdot, t_1) X(\cdot, t_2)],$$

and the inner product, the following equality:

$$\langle L_F[Y_1], L_F[Y_2] \rangle_K = \langle P_{L_P[X]}[Y_1], P_{L_P[X]}[Y_2] \rangle_{L_2(\Omega,\mathcal{A},P)}.$$

Let

$$\begin{aligned} I &= [1_\Omega]_{L_2(\Omega,\mathcal{A},P)}, \\ I_X &= P_{L_P[X]}[I], \\ h_I &= L_F[I], \\ h_Y &= L_F[Y]. \end{aligned}$$

Then:

- *The element h_I has norm less than one.* Indeed,

$$\begin{aligned} \|h_I\|_K^2 &= \|L_F[I]\|_K^2 \\ &= \left\|P_{L_P[X]}[I]\right\|_{L_2(\Omega,\mathcal{A},P)}^2 \\ &\le \|I\|_{L_2(\Omega,\mathcal{A},P)}^2 \\ &= E_P\left[1_\Omega^2\right] \\ &= 1; \end{aligned}$$

- *The expectation of certain random variables can be computed as an RKHS inner product.*

  Indeed,

$$\langle h_Y, h_I \rangle_K = \langle P_{L_P[X]}[Y], I_X \rangle_{L_2(\Omega,\mathcal{A},P)},$$

  so that, when $Y \in L_P[X]$,

$$\langle h_Y, h_I \rangle_K = \langle Y, I \rangle_{L_2(\Omega,\mathcal{A},P)} = E_P\left[\dot{Y}\right].$$

  In particular, when $Y = X_t$, one gets $L_F[X_t](x) = \mathcal{K}(x, t)$, and, furthermore,

$$h_I(t) = \langle h_I, \mathcal{K}(\cdot, t) \rangle_K = \langle h_I, h_{X_t} \rangle_K = E_P[X(\cdot, t)].$$

  Thus the mean function of X belongs to the RKHS it generates, and, in it, has a norm at most equal to one. It is the image of the class of random variables that are almost surely equal to one.

- *The variance of certain random variables can be expressed in terms of an RKHS inner product.*

  Indeed, when $Y \in L_P[X]$,

$$V_P\left[\dot{Y}\right] = E_P\left[\dot{Y}^2\right] - E_P^2\left[\dot{Y}\right] = \|h_Y\|_K^2 - \langle h_Y, h_I\rangle_K.$$

Suppose that one, instead of F, uses $F_0 : T \longrightarrow L_2(\Omega, \mathcal{A}, P)$, defined using the following relation:

$$F_0(t) = X_t - [E_P[X(\cdot, t)]]_P.$$

The kernel $\mathcal{K}_0$ is then the covariance of X. H_{F_0} is generated by the family $\{X_t - [E_P[X(\cdot, t)]]_P,\ t \in T\}$. Furthermore

$$L_{F_0}[Y](t) = \langle Y, X_t - [E_P[X(\cdot, t)]]_P\rangle_{L_2(\Omega, \mathcal{A}, P)},$$

so that

$$h_I(t) = L_{F_0}[I](t) = \langle I, X_t - [E_P[X(\cdot, t)]]_P\rangle_{L_2(\Omega, \mathcal{A}, P)} = 0.$$

The elements of H_{F_0} have then a mean that is equal to zero as h_I is the zero element, and, for $Y \in L_P[X]$,

$$\begin{aligned} 0 &= \langle L_{F_0}[Y], L_{F_0}[I]\rangle_K \\ &= \langle P_{H_{F_0}}[Y], P_{H_{F_0}}[I]\rangle_{L_2(\Omega, \mathcal{A}, P)} \\ &= \langle Y, I\rangle_{L_2(\Omega, \mathcal{A}, P)} \\ &= E_P\left[\dot{Y}\right]. \end{aligned}$$

Consequently $L_{F_0}[Y](t) = \operatorname{cov}\left(\dot{Y}, X(\cdot, t)\right)$.

This example may be extended to include "white noise" [124, p. 251].

Let O be an open subset of $\mathbb{R}$, and $\mathcal{D}[O]$ be the vector space of functions with values in $\mathbb{R}$, and domain O, whose support is contained in a compact subset of O, and which are infinitely differentiable. A sequence $\{f_n, n \in \mathbb{N}\} \subseteq \mathcal{D}[O]$ converges towards zero when the supports of the elements in the sequence are subsets of the same compact set of O, and that, for $i \in \mathbb{N}$, fixed, but arbitrary, $f_n^{(i)}$ being the i-th derivative of f_n,

$$\lim_n \max_t \left|f_n^{(i)}(t)\right| = 0.$$

A generalized, second order process is a linear map

$$X : \mathcal{D}[O] \longrightarrow L_2(\Omega, \mathcal{A}, P).$$

The process is continuous when the map X is continuous. When the map X has the following form:

$$X[f] = \text{(class of)} \int_T X(\omega, t) f(t)\, dt,$$

for locally integrable X (the integral of the paths of X exists on compact sets), one says that the generalized process is represented by an "ordinary" one. A generalized process is Gaussian when its range is contained in a subspace of Gaussian random variables.

In what follows no distinction between a random variable and its equivalence class shall be made. Let W be a standard Wiener process, and $W[f]$ be the value at f of the generalized process it represents. When $T = [0, 1]$, one shall restrict attention to functions f that have support in T, so that their values, as well as those of their derivatives, at 0 and 1, are zero. One has that $E_P[W[f]] = 0$, and that

$$E_P[W[f]\, W[g]] = \int_0^1 \int_0^1 \{t \wedge \theta\} f(t)\, g(\theta)\, dt d\theta.$$

One shall need below another expression for that covariance which is obtained as follows. Let F be a primitive of f, and G, of g :

- [step 1] using $2\{s \wedge t\} = s + t - |s - t|$, splitting the unit square into two parts, $s \leq t$ and $s > t$, and integrating separately the $s + t$ and the $|s - t|$ parts, one obtains that

$$\begin{aligned} 2 \int_0^1 \int_0^1 \{s \wedge t\} f(s)\, g(t)\, ds dt &= \int_0^1 sf(s)\, ds \int_0^1 g(s)\, ds \\ &+ \int_0^1 sg(s)\, ds \int_0^1 f(s)\, ds \\ &- \int_0^1 f(s)\, ds \int_0^s (s - t)\, g(t)\, dt \\ &- \int_0^1 g(s)\, ds \int_0^s (s - t) f(t)\, dt; \end{aligned}$$

- [step 2] let $\phi(t) = t$ and $\psi(t) = \int_t^1 g(s)\, ds$; then $[\phi(t)\, \psi(t)]_0^1 = 0$, and, $\dot{\phi}$ and $\dot{\psi}$ denoting derivatives,

$$\begin{aligned} \int_0^1 sg(s)\, ds &= -\int_0^1 \phi(s) \dot{\psi}(s) ds \\ &= [\phi(s)\, \psi(s)]_0^1 - \int_0^1 \phi(s)\, \dot{\psi}(s)\, ds \end{aligned}$$

$$= \int_0^1 \dot{\phi}(s)\,\psi(t)\,ds$$

$$= \int_0^1 dt \int_t^1 g(s)\,ds$$

$$= \int_0^1 \{G(1) - G(t)\}\,dt;$$

- [step 3] one has that

$$\int_0^s (s-t)\,g(t)\,dt = [(s-t)\,G(t)]_0^s - \int_0^s (-1)\,G(t)\,dt = \int_0^s G(t)\,dt,$$

so that

$$\int_0^1 f(s)\,ds \int_0^s G(t)\,dt = \left[F(t)\int_0^t G(s)\,ds\right]_0^1 - \int_0^1 F(s)\,G(s)\,ds$$

$$= F(1)\int_0^1 G(s)\,ds - \int_0^1 F(s)\,G(s)\,ds$$

$$= \int_0^1 \{F(1) - F(s)\}\,G(s)\,ds.$$

Assembling the parts one gets that

$$2\int_0^1\int_0^1 \{s \wedge t\} f(s)\,g(t)\,ds dt = F(1)\int_0^1 \{G(1) - G(s)\}\,ds$$

$$+ G(1)\int_0^1 \{F(1) - F(s)\}\,ds$$

$$- \int_0^1 \{F(1) - F(s)\}\,G(s)\,ds$$

$$- \int_0^1 \{G(1) - G(s)\}\,F(s)\,ds$$

$$= 2\int_0^1 \{F(1) - F(s)\}\{G(1) - G(s)\}\,ds.$$

Thus

$$E_P[W[f]\,W[g]] = \int_0^1 \{F(1) - F(s)\}\{G(1) - G(s)\}\,ds. \qquad (\star)$$

Let now $\dot{W}$ be the derivative of W in the sense of distributions, and ϕ be a test function (here $\dot{\phi}$ shall denote the derivative of ϕ rather than an element in an equivalence class):

$$\dot{W}[\phi] = -W\left[\dot{\phi}\right],$$

so that (($\star$) with $F(1) = G(1) = 0$)

$$C_{\dot{W}}(\phi, \psi) = E_P\left[W\left[\dot{\phi}\right]W\left[\dot{\psi}\right]\right] = \int_0^1 \phi(t)\,\psi(t)\,dt.$$

Let $F_{\dot{W}} : \mathcal{D}[T] \longrightarrow L_2(\Omega, \mathcal{A}, P)$ be defined using the following relation:

$$F_{\dot{W}}[\phi] = \dot{W}[\phi],$$

and $L_{F_{\dot{W}}} : L_2(\Omega, \mathcal{A}, P) \longrightarrow \mathbb{R}^{\mathcal{D}[T]}$, using the following one:

$$L_{F_{\dot{W}}}[X](\phi) = E_P\left[X\dot{W}[\phi]\right].$$

Now, as $W[\phi] = \int_T W(\cdot, t)\phi(t)\,dt$,

$$E_P[XW[\phi]] = \int_0^1 E_P[XW(\cdot, t)]\,\phi(t)\,dt.$$

But $E_P[XW(\cdot, t)]$ belongs to $H(C_W, T)$, the RKHS of (Example) 1.1.26, so that [(Example) 1.1.25], for some $f_X \in L_2[0, 1]$, with $V[f_X](t) = \int_0^t f_X(\theta)\,d\theta$,

$$E_P[XW[\phi]] = \int_0^1 V[f_X](t)\,\phi(t)\,dt.$$

Thus

$$\begin{aligned} E_P\left[X\dot{W}[\phi]\right] &= E_P\left[X\left(-W\left[\dot{\phi}\right]\right)\right] \\ &= -\int_0^1 V[f_X](t)\,\dot{\phi}(t)\,dt \\ &= \int_0^1 f_X(t)\,\phi(t)\,dt, \end{aligned}$$

and

$$H\left(C_{\dot{W}}, \mathcal{D}[T]\right) = \left\{\Phi_f : \phi \mapsto \int_0^1 \phi(t)f(t)\,dt,\ \phi \in \mathcal{D}[T],\ f \in L_2[0, 1]\right\}.$$

Since, as a consequence of the covariance computation,

$$E_P\left[\left(\dot{W}[\phi]-\dot{W}[\psi]\right)^2\right]=E_P\left[\left(W[\dot{\psi}]-W[\dot{\phi}]\right)^2\right]=\int_0^1(\phi(t)-\psi(t))^2\,dt,$$

the linear space spanned by $\dot{W}$ is isomorphic to $L_2[0,1]$ so that

$$\langle\Phi_f,\Phi_g\rangle_{H(C_{\dot{W}},\mathcal{D}[T])}=\langle f,g\rangle_{L_2[0,1]}.$$

The identification of Φ_f with f allows one to look at the space $L_2[0,1]$ as if it were an RKHS.

The following example illustrates the role of the reproducing kernel in determining the geometry of an RKHS.

Example 1.1.27 Let $H(\mathcal{H},T)$ be an RKHS and, for fixed, but arbitrary $t\in T$ such that $\mathcal{H}(t,t)>0$, consider the following subsets and functions:

$$\begin{aligned}H[t:1]&=\{h\in H(\mathcal{H},T):\ \mathcal{E}_t[h]=1\}\quad&,\text{ and } f_t(x)&=\tfrac{\mathcal{H}(x,t)}{\mathcal{H}(t,t)},\\ S_{H(\mathcal{H},T)}&=\left\{h\in H(\mathcal{H},T):\ \|h\|_{H(\mathcal{H},T)}=1\right\}&,\text{ and } g_t(x)&=\frac{\mathcal{H}(x,t)}{(\mathcal{H}(t,t))^{1/2}}.\end{aligned}$$

Then:

1. $f_t\in H[t:1]$, and $\min_{h\in H[t:1]}\|h\|_{H(\mathcal{H},T)}=\|f_t\|_{H(\mathcal{H},T)}=\{\mathcal{H}(t,t)\}^{-1/2}$;
2. $g_t\in S_{H(\mathcal{H},T)}$, and, for $h\in S_{H(\mathcal{H},T)}$, fixed, but arbitrary,

$$|h(t)|\le|g_t(t)|=(\mathcal{H}(t,t))^{1/2}.$$

By definition indeed, $f_t(t)=1$, and thus $f_t\in H[t:1]$. Furthermore [(Remark) 1.1.4],

$$\|f_t\|_{H(\mathcal{H},T)}=\frac{\|\mathcal{H}(\cdot,t)\|_{H(\mathcal{H},T)}}{\mathcal{H}(t,t)}=\frac{1}{(\mathcal{H}(t,t))^{1/2}}.$$

But the generally valid relation [(Proposition) 1.1.5] $|h(t)|\le(\mathcal{H}(t,t))^{1/2}\|h\|_{H(\mathcal{H},T)}$ yields, when $h(t)=1$,

$$\frac{1}{(\mathcal{H}(t,t))^{1/2}}\le\|h\|_{H(\mathcal{H},T)}.$$

As

$$\|g_t\|_{H(\mathcal{H},T)}=\frac{\|\mathcal{H}(\cdot,t)\|_{H(\mathcal{H},T)}}{(\mathcal{H}(t,t))^{1/2}}=\frac{(\mathcal{H}(t,t))^{1/2}}{(\mathcal{H}(t,t))^{1/2}}=1,$$

$g_t \in S_{H(\mathcal{H},T)}$. Again, as $|h(t)| \leq (\mathcal{H}(t,t))^{1/2} \|h\|_{H(\mathcal{H},T)}$, when $\|h\|_{H(\mathcal{H},T)} = 1$,

$$|h(t)| \leq (\mathcal{H}(t,t))^{1/2} = \frac{\mathcal{H}(t,t)}{(\mathcal{H}(t,t))^{1/2}} = |g_t(t)|.$$

The following concepts, definitions, and results are useful for the computational side of RKHS's. In particular they allow one to often restrict attention to that part of T over which $\mathcal{H}$ is strictly positive definite.

Let $\mathcal{H}$ be a reproducing kernel on T that is not the zero kernel, and $\mathcal{F}[\mathcal{H}]$ be the family of subsets of $\{\mathcal{H}(\cdot,t),\ t \in T\}$ with the following property: $\mathcal{F}_0 \in \mathcal{F}[\mathcal{H}]$ if, and only if, every nonempty finite subset of $\mathcal{F}_0$ contains only linearly independent elements. $\mathcal{F}[\mathcal{H}]$ is thus a family of finite character [131, p.13], and, as such, contains a maximal element [131, p. 14], say $\mathcal{F}_m$. Let

$$T_{\mathcal{F}_m} = \{t \in T : \mathcal{H}(\cdot,t) \in \mathcal{F}_m\}.$$

The definition which follows then makes sense.

Definition 1.1.28 A support for $\mathcal{H}$ is a set of the form $T_{\mathcal{F}_m}$. It shall be denoted $T_{\mathcal{H}}$.

The supports of $\mathcal{H}$ have the following properties:

Property 1.1.29 *The set* $\{\mathcal{H}(\cdot,t),\ t \in T_{\mathcal{H}}\}$ *is a Hamel basis for* $V[\mathcal{H}]$.

Property 1.1.30 $T_{\mathcal{H}} \subseteq \{t \in T : \mathcal{H}(\cdot,t) \neq 0_{\mathbb{R}^T}\}$.

Property 1.1.31 *Suppose that* $t \in T \setminus T_{\mathcal{H}}$, *and* $\mathcal{H}(\cdot,t) \neq 0_{\mathbb{R}^T}$. *Since a support is the index set of a Hamel basis, there exists then a finite subset of distinct elements* $\{t_1,\ldots,t_n\} \subseteq T_{\mathcal{H}}$ *such that the family*

$$\{\mathcal{H}(\cdot,t), \mathcal{H}(\cdot,t_1), \ldots, \mathcal{H}(\cdot,t_n)\}$$

is linearly dependent (otherwise $T_{\mathcal{H}}$ *would not be maximal).*

Property 1.1.32 *Suppose that* $t \in T \setminus T_{\mathcal{H}}$, *and* $\mathcal{H}(\cdot,t) \neq 0_{\mathbb{R}^T}$. *There exists* $t_0 \in T_{\mathcal{H}}$ *such that*

$$\{T_{\mathcal{H}} \setminus \{t_0\}\} \cup \{t\}$$

is a support of $\mathcal{H}$.

Indeed, since a support yields a Hamel basis, there is a unique, linearly independent set (depending on t)

$$\{\mathcal{H}(\cdot,t_1), \ldots, \mathcal{H}(\cdot,t_n),\ \{t_1,\ldots,t_n\} \subseteq T_{\mathcal{H}},\ \text{distinct}\}$$

such that

$$\mathcal{H}(\cdot, t) = \sum_{i=1}^{n} \alpha_i \mathcal{H}(\cdot, t_i),$$

none of the α's being zero. As

$$\beta_1 \mathcal{H}(\cdot, t) + \beta_2 \mathcal{H}(\cdot, t_2) + \beta_3 \mathcal{H}(\cdot, t_3) + \cdots + \beta_n \mathcal{H}(\cdot, t_n) =$$
$$= \beta_1 \alpha_1 \mathcal{H}(\cdot, t_1) + (\beta_1 \alpha_2 + \beta_2) \mathcal{H}(\cdot, t_2) + \cdots + (\beta_1 \alpha_n + \beta_n) \mathcal{H}(\cdot, t_n),$$

the set $\{\mathcal{H}(\cdot, t), \mathcal{H}(\cdot, t_2,) \mathcal{H}(\cdot, t_3), \ldots, \mathcal{H}(\cdot, t_n)\}$ is linearly independent and may replace $\{\mathcal{H}(\cdot, t_1), \ldots, \mathcal{H}(\cdot, t_n)\}$.

Property 1.1.33 *Consider the following property* $\Pi_{\mathcal{H}}$:
⌈*the requirements*

- $n \in \mathbb{N}$,
- $\{t_1, \ldots, t_n\} \subseteq T$ *distinct,*
- *for* $i \in [1 : n]$, $\mathcal{H}(\cdot, t_i) \neq 0_{\mathbb{R}^T}$,

have, as consequence, that $\{\mathcal{H}(\cdot, t_1), \ldots, \mathcal{H}(\cdot, t_n)\}$ *is a linearly independent family.*⌋

A unique support for $\mathcal{H}$ *exists if, and only if, property* $\Pi_{\mathcal{H}}$ *obtains, and, in that case,* $T_{\mathcal{H}} = \{t \in T : \mathcal{H}(\cdot, t) \neq 0_{\mathbb{R}^T}\}$.

Indeed, from (Fact) 1.1.32, when there is a unique support, say $T_{\mathcal{H}}$, the relation $\mathcal{H}(\cdot, t) \neq 0_{\mathbb{R}^T}$ implies $t \in T_{\mathcal{H}}$. Property $\Pi_{\mathcal{H}}$ thus obtains. Conversely, when property $\Pi_{\mathcal{H}}$ obtains, there is a unique support which is the set of indices $\{t \in T : \mathcal{H}(\cdot, t) \neq 0_{\mathbb{R}^T}\}$.

Property 1.1.34 *Whenever*

- $n \in \mathbb{N}$,
- $\{t_1, \ldots, t_n\} \subseteq T$ *distinct,*
- $\{\mathcal{H}(\cdot, t_1), \ldots, \mathcal{H}(\cdot, t_n)\}$ *is a linearly independent family,*

there is a support that contains $\{t_1, \ldots, t_n\}$.

Definition 1.1.35 The pseudo-distance(-metric) on T, associated with $\mathcal{H}$, is defined, for fixed, but arbitrary $(t_1, t_2) \in T \times T$, using the following relation:

$$d_{\mathcal{H}}(t_1, t_2) = \|\mathcal{H}(\cdot, t_1) - \mathcal{H}(\cdot, t_2)\|_{H(\mathcal{H}, T)}.$$

$d_{\mathcal{H}}$ is indeed symmetric, positive, and equal to zero when $t_1 = t_2$. When the family $\{\mathcal{H}(\cdot, t), t \in T\}$ is linearly independent, $d_{\mathcal{H}}$ is a distance.

Definition 1.1.36 Any subset $T_0 \subseteq T$ such that $\{\mathcal{H}(\cdot, t), t \in T_0\}$ is a Hamel basis for $V[\mathcal{H}]$ is called a Hamel subset of T for $\mathcal{H}$.

$\{\mathcal{H}(\cdot, t), t \in T_0\}$ is then total in $H(\mathcal{H}, T)$, and $(T_0, d_{\mathcal{H}})$ is a metric space. A Hamel subset is a support.

Remark 1.1.37 "Hamel subset" and "support" are thus two names for essentially the same concept.

Lemma 1.1.38 *Let H and K be real Hilbert spaces, H_0 and K_0, total subsets. Let $U : H_0 \longrightarrow K_0$ be a surjection such that, in H_0,*

$$\langle U[h_1], U[h_2]\rangle_K = \langle h_1, h_2\rangle_H.$$

U may then be extended (uniquely) to a unitary map from H to K.

Proof U is an injection. Suppose indeed that, on H_0, $U[h_1] = U[h_2]$. Then, on H_0, from $\langle U[h_1], U[h]\rangle_K = \langle U[h_2], U[h]\rangle_K$ and the assumption,

$$\langle h_1, h\rangle_H = \langle h_2, h\rangle_H.$$

But, since H_0 is total, $h_1 = h_2$.

Let $h \in V[H_0]$, that is, $h = \sum_{i=1}^{n} \alpha_i h_i, h_i \in H_0, i \in [1 : n]$. Set:

$$U[h] = \sum_{i=1}^{n} \alpha_i U[h_i].$$

Such an extension of U, from H_0 to $V[H_0]$, is well defined. Let indeed

$$\sum_{i=1}^{n} \alpha_i U[h_i] = \sum_{j=1}^{p} \alpha_j^\star U[h_j^\star].$$

Then, for $h \in H_0$, fixed, but arbitrary, successively,

$$\left\langle \sum_{i=1}^{n} \alpha_i U[h_i], U[h] \right\rangle_K = \left\langle \sum_{j=1}^{p} \alpha_j^\star U[h_j^\star], U[h] \right\rangle_K,$$

$$\sum_{i=1}^{n} \alpha_i \langle U[h_i], U[h]\rangle_K = \sum_{j=1}^{p} \alpha_j^\star \left\langle U[h_j^\star], U[h]\right\rangle_K,$$

$$\sum_{i=1}^{n} \alpha_i \langle h_i, h\rangle_H = \sum_{j=1}^{p} \alpha_j^\star \langle h_j^\star, h\rangle_H,$$

$$\left\langle \sum_{i=1}^{n} \alpha_i h_i, h \right\rangle_H = \left\langle \sum_{j=1}^{p} \alpha_j^\star h_i^\star, h \right\rangle_H.$$

Again, as H_0 is total,

$$\sum_{i=1}^{n} \alpha_i h_i = \sum_{j=1}^{p} \alpha_j^\star h_i^\star .$$

U has thus a linear extension to $V[H_0]$ which maintains the inner product relation. It can thus be extended by continuity to H. □

Corollary 1.1.39 *1. When K_0 is not total, the isometry shall be between H and $\overline{V[K_0]}$.*

2. Let H and K be real Hilbert spaces, H_0 and K_0 dense manifolds in these, and $U : H_0 \longrightarrow K_0$, an isometric bijection of H_0 onto K_0. U has then a unitary extension to H and K.

Proposition 1.1.40 *Let $T_{\mathcal{H}}$ be a support for the reproducing kernel $\mathcal{H}$ defined over T. The restriction of $\mathcal{H}$ to $T_{\mathcal{H}} \times T_{\mathcal{H}}$ shall be denoted $\mathcal{H}^{|T_{\mathcal{H}}}$, and the restriction to $T_{\mathcal{H}}$ of a function f in $\mathbb{R}^T$, $f^{|T_{\mathcal{H}}}$. Let $\rho_{T_{\mathcal{H}}} : \mathbb{R}^T \longrightarrow \mathbb{R}^{T_{\mathcal{H}}}$ be the restriction map:*

$$\rho_{T_{\mathcal{H}}}(f) = f^{|T_{\mathcal{H}}}.$$

The restriction of $\rho_{T_{\mathcal{H}}}$ to $H(\mathcal{H}, T)$ is unitary onto $H\left(\mathcal{H}^{|T_{\mathcal{H}}}, T_{\mathcal{H}}\right)$.

Proof Since $V[\mathcal{H}]$ and $V\left[\mathcal{H}^{|T_{\mathcal{H}}}\right]$ are dense in the RKHS's they generate, one may restrict attention to those two manifolds [(Corollary) 1.1.39].

Let

$$f = \sum_{i=1}^{n_f} \alpha_i^f \mathcal{H}\left(\cdot, t_i^f\right)$$

be fixed, but arbitrary in $V[\mathcal{H}]$. Now, for fixed, but arbitrary $i \in \left[1 : n_f\right]$,

$$\mathcal{H}\left(\cdot, t_i^f\right) = \sum_{j=1}^{n(t_i^f)} \beta_j(t_i^f) \mathcal{H}\left(\cdot, t_j(t_i^f)\right)$$

where, for $j \in \left[1 : n\left(t_i^f\right)\right]$, fixed, but arbitrary, $t_j\left(t_i^f\right) \in T_{\mathcal{H}}$. Thus f has the representation

$$f = \sum_{i=1}^{n_f} \sum_{j=1}^{n(t_i^f)} \alpha_i^f \beta_j(t_i^f) \mathcal{H}\left(\cdot, t_j(t_i^f)\right).$$

Its restriction to $T_{\mathcal{H}}$ yields an element of $V\left[\mathcal{H}^{|T_{\mathcal{H}}}\right]$. Furthermore, as, for $(i,j) \in \left[1 : n_f\right] \times \left[1 : n_f\right]$, fixed, but arbitrary,

$$
\begin{aligned}
&\mathcal{H}\left(t_i^f, t_j^f\right) = \\
&\quad = \left\langle \mathcal{H}\left(\cdot, t_i^f\right), \mathcal{H}\left(\cdot, t_j^f\right)\right\rangle_{H(\mathcal{H},T)} \\
&\quad = \sum_{k=1}^{n\left(t_i^f\right)} \sum_{l=1}^{n\left(t_j^f\right)} \beta_k\left(t_i^f\right) \beta_l\left(t_j^f\right) \langle \mathcal{H}\left(\cdot, t_k\left(t_i^f\right)\right), \mathcal{H}\left(\cdot, t_l\left(t_j^f\right)\right)\rangle_{H(\mathcal{H},T)} \\
&\quad = \sum_{k=1}^{n\left(t_i^f\right)} \sum_{l=1}^{n\left(t_j^f\right)} \beta_k\left(t_i^f\right) \beta_l\left(t_j^f\right) \mathcal{H}\left(t_k\left(t_i^f\right), t_l\left(t_j^f\right)\right) \\
&\quad = \sum_{k=1}^{n\left(t_i^f\right)} \sum_{l=1}^{n\left(t_j^f\right)} \beta_k\left(t_i^f\right) \beta_l\left(t_j^f\right) \mathcal{H}^{|T_{\mathcal{H}}}\left(t_k\left(t_i^f\right), t_l\left(t_j^f\right)\right) \\
&\quad = \left\langle \sum_{k=1}^{n\left(t_i^f\right)} \beta_k\left(t_i^f\right) \mathcal{H}^{|T_{\mathcal{H}}}\left(\cdot, t_k\left(t_i^f\right)\right), \sum_{l=1}^{n\left(t_j^f\right)} \beta_l\left(t_j^f\right) \mathcal{H}^{|T_{\mathcal{H}}}\left(\cdot, t_l\left(t_j^f\right)\right)\right\rangle_{H\left(\mathcal{H}^{|T_{\mathcal{H}}}, T_{\mathcal{H}}\right)},
\end{aligned}
$$

it follows that

$$
\begin{aligned}
\|f\|_{H(\mathcal{H},T)}^2 &= \sum_{i=1}^{n_f} \sum_{j=1}^{n_f} \alpha_i^f \alpha_j^f \mathcal{H}\left(t_i^f, t_j^f\right) \\
&= \left\| \sum_{i=1}^{n_f} \sum_{k=1}^{n\left(t_i^f\right)} \alpha_i^f \beta_k\left(t_i^f\right) \mathcal{H}^{|T_{\mathcal{H}}}\left(\cdot, t_k\left(t_i^f\right)\right) \right\|_{H\left(\mathcal{H}^{|T_{\mathcal{H}}}, T_{\mathcal{H}}\right)}^2 \\
&= \left\|\rho_{T_{\mathcal{H}}}(f)\right\|_{H\left(\mathcal{H}^{|T_{\mathcal{H}}}, T_{\mathcal{H}}\right)}^2 .
\end{aligned}
$$

Suppose now that

$$
f = \sum_{i=1}^{n} \alpha_i^f \mathcal{H}^{|T_{\mathcal{H}}}\left(\cdot, t_i^f\right) \in V\left[\mathcal{H}^{|T_{\mathcal{H}}}\right]
$$

is fixed, but arbitrary. One has that

$$f = \rho_{T_{\mathcal{H}}}\left(\tilde{f}\right), \tilde{f} = \sum_{i=1}^{n} \alpha_i^f \mathcal{H}\left(\cdot, t_i^f\right) \in V[\mathcal{H}].$$

Furthermore $\tilde{f}$ is the unique element of $V[\mathcal{H}]$ whose restriction to $T_{\mathcal{H}}$ yields f. Suppose indeed that $\hat{f} \in V[\mathcal{H}]$ is such that

$$\rho_{T_{\mathcal{H}}}\left(\hat{f}\right) = f.$$

Since $\tilde{f}$ and $\hat{f}$ agree on $T_{\mathcal{H}}$, one must only consider points $t \in T \setminus T_{\mathcal{H}}$. For such a point,

$$\mathcal{H}(\cdot, t) = \sum_{i=1}^{n(t)} \alpha_i(t) \mathcal{H}(\cdot, t_i(t)), \text{ some } \{t_1(t), \ldots, t_n(t)\} \subseteq T_{\mathcal{H}}.$$

But then

$$\begin{aligned}
\tilde{f}(t) &= \langle \tilde{f}, \mathcal{H}(\cdot, t)\rangle_{H(\mathcal{H},T)} \\
&= \sum_{i=1}^{n(t)} \alpha_i(t) \tilde{f}(t_i(t)) \\
&= \sum_{i=1}^{n(t)} \alpha_i(t) \hat{f}(t_i(t)) \\
&= \langle \hat{f}, \mathcal{H}(\cdot, t)\rangle_{H(\mathcal{H},T)} \\
&= \hat{f}(t).
\end{aligned}$$

□

Remark 1.1.41 Result (Proposition) 1.1.40 is related to (Proposition) 1.6.3, which is a result about projections using subsets of T.

Remark 1.1.42 Let $f \in \mathbb{R}^{T_{\mathcal{H}}}$ be fixed, but arbitrary. It may be extended to a map $\tilde{f} \in \mathbb{R}^T$ as follows. For $t \in T_{\mathcal{H}}$, fixed, but arbitrary, set

$$\tilde{f}(t) = f(t).$$

When $t \in T \setminus T_{\mathcal{H}}$, one has uniquely, for some

$$n(t) \in \mathbb{N}, \left\{t_1^{(t)}, \ldots, t_{n(t)}^{(t)}\right\} \subseteq T_{\mathcal{H}}, \left\{\alpha_1^{(t)}, \ldots, \alpha_{n(t)}^{(t)}\right\} \subseteq \mathbb{R} \setminus \{0\},$$

$$\mathcal{H}(\cdot,t)=\sum_{i=1}^{n(t)}\alpha_i^{(t)}\mathcal{H}\left(\cdot,t_i^{(t)}\right).$$

Set then

$$\tilde{f}(t)=\sum_{i=1}^{n(t)}\alpha_i^{(t)}f\left(t_i^{(t)}\right).$$

Suppose now that $f\in\mathbb{R}^T$, that $f^{|T_{\mathcal{H}}}$ is its restriction to $T_{\mathcal{H}}$, and that $\tilde{f}_{T_{\mathcal{H}}}$ is the extension just defined. One has then that

$$f(t)-\tilde{f}_{T_{\mathcal{H}}}(t)=\begin{cases}0 & \text{when } t\in T_{\mathcal{H}}\\ f(t)-\sum_{i=1}^{n(t)}\alpha_i^{(t)}f\left(t_i^{(t)}\right) & \text{when } t\in T\setminus T_{\mathcal{H}}\end{cases}.$$

One shall write

$$\tilde{f}_{T_{\mathcal{H}}}=\rho_{T_{\mathcal{H}}}^{\leftarrow}\left[f^{|T_{\mathcal{H}}}\right],$$

and the following result obtains.

Proposition 1.1.43 *When $T_{\mathcal{H}}\subseteq T$ is a support of $\mathcal{H}$, a reproducing kernel on T, and $f\in\mathbb{R}^T$ is fixed, but arbitrary, then $f\in H(\mathcal{H},T)$ if, and only if,*

1. $\rho_{T_{\mathcal{H}}}[f]\in H\left(\mathcal{H}^{|T_{\mathcal{H}}},T_{\mathcal{H}}\right)$,
2. $\rho_{T_{\mathcal{H}}}^{\leftarrow}\rho_{T_{\mathcal{H}}}[f]=f$.

Proof When $f\in H(\mathcal{H},T)$,

$$f^{|T_{\mathcal{H}}}\in H\left(\mathcal{H}^{|T_{\mathcal{H}}},T_{\mathcal{H}}\right),$$

since [(Proposition) 1.1.40] the restriction of $\rho_{T_{\mathcal{H}}}$ to $H(\mathcal{H},T)$ is onto $H\left(\mathcal{H}^{|T_{\mathcal{H}}},T_{\mathcal{H}}\right)$. Also $\rho_{T_{\mathcal{H}}}^{\leftarrow}\rho_{T_{\mathcal{H}}}[f]=f$ since then

$$f(t)-\sum_{i=1}^{n(t)}\alpha_i^{(t)}f\left(t_i^{(t)}\right)=\left\langle f,\mathcal{H}(\cdot,t)-\sum_{i=1}^{n}\alpha_i^{(t)}\mathcal{H}\left(\cdot,t_i^{(t)}\right)\right\rangle_{H(\mathcal{H},T)}=0.$$

Suppose now that $f\in\mathbb{R}^T$ is such that

$$f^{|T_{\mathcal{H}}}\in H\left(\mathcal{H}^{|T_{\mathcal{H}}},T_{\mathcal{H}}\right),\text{ and that }\rho_{T_{\mathcal{H}}}^{\leftarrow}\rho_{T_{\mathcal{H}}}[f]=f.$$

Then, since for $t \in T$, fixed, but arbitrary, $\rho_{T_{\mathcal{H}}}^{\leftarrow} \rho_{T_{\mathcal{H}}} [f] (t)$ has [(Remark) 1.1.42] the generic form

$$\sum_{i=1}^{n[t]} \alpha_i [t] f^{|T_{\mathcal{H}}} (t_i [t]) ,$$

letting $\rho_{T_{\mathcal{H}}}^{\star}$ be the unitary adjoint of the restriction of $\rho_{T_{\mathcal{H}}}$ to $H(\mathcal{H}, T)$,

$$\begin{aligned}
f(t) &= \sum_{i=1}^{n[t]} \alpha_i [t] f^{|T_{\mathcal{H}}} (t_i [t]) \\
&= \sum_{i=1}^{n[t]} \alpha_i [t] \left\langle f^{|T_{\mathcal{H}}}, \mathcal{H}^{|T_{\mathcal{H}}} (\cdot, t_i [t]) \right\rangle_{H\left(\mathcal{H}^{|T_{\mathcal{H}}}, T_{\mathcal{H}}\right)} \\
&= \left\langle \rho_{T_{\mathcal{H}}} [f] , \sum_{i=1}^{n[t]} \alpha_i [t] \, \mathcal{H}^{|T_{\mathcal{H}}} (\cdot, t_i [t]) \right\rangle_{H\left(\mathcal{H}^{|T_{\mathcal{H}}}, T_{\mathcal{H}}\right)} \\
&= \left\langle \rho_{T_{\mathcal{H}}} [f] , \sum_{i=1}^{n[t]} \alpha_i [t] \, \rho_{T_{\mathcal{H}}} [\mathcal{H} (\cdot, t_i [t])] \right\rangle_{H\left(\mathcal{H}^{|T_{\mathcal{H}}}, T_{\mathcal{H}}\right)} \\
&= \left\langle \rho_{T_{\mathcal{H}}} [f] , \rho_{T_{\mathcal{H}}} \left[\sum_{i=1}^{n[t]} \alpha_i [t] \, \mathcal{H} (\cdot, t_i [t]) \right] \right\rangle_{H\left(\mathcal{H}^{|T_{\mathcal{H}}}, T_{\mathcal{H}}\right)} \\
&= \langle \rho_{T_{\mathcal{H}}} [f] , \rho_{T_{\mathcal{H}}} [\mathcal{H} (\cdot, t)] \rangle_{H\left(\mathcal{H}^{|T_{\mathcal{H}}}, T_{\mathcal{H}}\right)} \\
&= \langle \rho_{T_{\mathcal{H}}}^{\star} \rho_{T_{\mathcal{H}}} [f] , \mathcal{H} (\cdot, t) \rangle_{H(\mathcal{H}, T)} \\
&= \rho_{T_{\mathcal{H}}}^{\star} \rho_{T_{\mathcal{H}}} [f] (t) ,
\end{aligned}$$

which is the value at t of an element in $H(\mathcal{H}, T)$. □

Remark 1.1.44 One has the following situation, writing U for $\rho_{T_{\mathcal{H}}}^{|H(\mathcal{H},T)}$:

$$\begin{array}{ccc}
\mathbb{R}^T & \overset{\rho_{T_{\mathcal{H}}}}{\longrightarrow} & \mathbb{R}^{T_{\mathcal{H}}} \\
H(\mathcal{H}, T) & \overset{U}{\longrightarrow} & H\left(\mathcal{H}^{|T_{\mathcal{H}}}, T_{\mathcal{H}}\right) \\
H(\mathcal{H}, T) & \overset{U^{\star}}{\longleftarrow} & H\left(\mathcal{H}^{|T_{\mathcal{H}}}, T_{\mathcal{H}}\right) \\
\mathbb{R}^T & \overset{\rho_{T_{\mathcal{H}}}^{\leftarrow}}{\longleftarrow} & \mathbb{R}^{T_{\mathcal{H}}}
\end{array}$$

Thus, in that sense, $\rho_{T_{\mathcal{H}}}^{\leftarrow}$ is an extension of $\rho_{T_{\mathcal{H}}}^{\star}$.

Proposition 1.1.45 *Let the support of* $\mathcal{H}$ *be* $T, f \in \mathbb{R}^T$, *and*

$$T_n = \{t_1, \ldots, t_n\} \subseteq T$$

be fixed, but arbitrary. Let

$$H(f; T_n) = \{h \in H(\mathcal{H}, T) : h(t_i) = f(t_i), \ i \in [1 : n]\}.$$

Then:

1. $H(f; T_n) \neq \emptyset$;
2. *there exists a unique* $h_{f;T_n} \in H(\mathcal{H}, T)$ *such that*

$$\left\| h_{f;T_n} \right\|_{H(\mathcal{H},T)} = \min_{h \in H(f;T_n)} \|h\|_{H(\mathcal{H},T)};$$

3. $h_{f;T_n} \in V[\{\mathcal{H}(\cdot, t_1), \ldots, \mathcal{H}(\cdot, t_n)\}]$;
4. *when* $f \in H(\mathcal{H}, T)$, $h_{f;T_n}$ *is the orthogonal projection of* f *onto the (closed) subspace* $V[\{\mathcal{H}(\cdot, t_1), \ldots, \mathcal{H}(\cdot, t_n)\}]$.

Proof One may assume that $\{t_1, \ldots, t_n\}$ are distinct. Then the functions of the set $\{\mathcal{H}(\cdot, t_1), \ldots, \mathcal{H}(\cdot, t_n)\}$ are linearly independent, so that the matrix $\Sigma_{\mathcal{H},T_n}$, with entries $\{\mathcal{H}(t_i, t_j), \{i, j\} \subseteq [1 : n]\}$, is nonsingular. Let $\underline{f}$ be the vector with entries $f(t_i)$. The system

$$\Sigma_{\mathcal{H},T_n} \boxed{x} = \underline{f}$$

has thus a unique solution $\underline{\phi}$, and $\sum_{i=1}^{n} \phi_i \mathcal{H}(\cdot, t_i)$ belongs to $H(f; T_n)$, so that item 1 obtains.

As, for $t \in T$ and $h \in H(\mathcal{H}, T)$, fixed, but arbitrary, $h(t) = \langle h, \mathcal{H}(\cdot, t) \rangle_{H(\mathcal{H},T)}$, and that $\{\mathcal{H}(\cdot, t_1), \ldots, \mathcal{H}(\cdot, t_n)\}$ are linearly independent, item 2 obtains because of a general optimization result in Hilbert space [175, p. 65], from which items 3 and 4 follow also. In fact

$$h_{f;T_n} = \sum_{i=1}^{n} \phi_i \mathcal{H}(\cdot, t_i).$$

□

Remark 1.1.46 When the support of $\mathcal{H}$ is strictly contained in T, for (Proposition) 1.1.45 to be valid, one must add the requirement that $\rho_{T_{\mathcal{H}}}^{\leftarrow}(\rho_{T_{\mathcal{H}}}(f)) = f$.

Remark 1.1.47 One has that

$$\begin{aligned}\left\|h_{f;T_n}\right\|^2_{H(\mathcal{H},T)} &= \sum_{i=1}^{n}\sum_{j=1}^{n}\phi_i\phi_j\mathcal{H}\left(t_i,t_j\right)\\ &= \left\langle\Sigma_{\mathcal{H},T_n}\left[\underline{\phi}\right],\underline{\phi}\right\rangle_{\mathbb{R}^n}\\ &= \left\langle\Sigma^{-1}_{\mathcal{H},T_n}\left[\underline{f}\right],\underline{f}\right\rangle_{\mathbb{R}^n}.\end{aligned}$$

1.2 Membership in a Reproducing Kernel Hilbert Space

This section contains some tools which, sometimes, allow one to establish that a given function does, or does not, belong to a specific RKHS.

Proposition 1.2.1 *Let $H(\mathcal{H},T)$ be a fixed, but arbitrary RKHS, and suppose that the support of $\mathcal{H}$ is T. Let $f\in\mathbb{R}^T$ be fixed, but arbitrary. The following statements are equivalent:*

1. $f\in H(\mathcal{H},T)$;
2. *there exists a finite $\kappa(f)\geq 0$, depending on f only, such that, for fixed, but arbitrary $n\in\mathbb{N}$ and $\{t_1,\ldots,t_n\}\subseteq T$, the notation being that of* (Proposition) 1.1.45,
$$\left\|h_{f;T_n}\right\|_{H(\mathcal{H},T)}\leq\kappa(f);$$
3. *there exists a finite $\kappa(f)\geq 0$, depending on f only, such that, for fixed, but arbitrary $[n,\alpha,(t,T)]$,*
$$\left|\sum_{i=1}^{n}\alpha_i f(t_i)\right|\leq\kappa(f)\left\|\sum_{i=1}^{n}\alpha_i\mathcal{H}(\cdot,t_i)\right\|_{H(\mathcal{H},T)}.$$

When item 1, or one of its equivalents, obtains,

$$\|f\|_{H(\mathcal{H},T)}=\sup_{T_n\subseteq T}\left\|h_{f;T_n}\right\|_{H(\mathcal{H},T)}.$$

Proof (1 $\Rightarrow$ 2) From (Proposition) 1.1.45, one has that $h_{f;T_n}$ is the orthogonal projection of f onto $V[\{\mathcal{H}(\cdot,t_1),\ldots,\mathcal{H}(\cdot,t_n)\}]$. Consequently,

$$\left\|h_{f;T_n}\right\|_{H(\mathcal{H},T)}\leq\|f\|_{H(\mathcal{H},T)},$$

and one may choose $\kappa(f)=\|f\|_{H(\mathcal{H},T)}$.

Proof (2 ⇒ 3) By definition, and because of (Proposition) 1.1.45,

$$h_{f;T_n}(t_i) = f(t_i),\ i \in [1:n],$$

and

$$h_{f;T_n} \in V[\{\mathcal{H}(\cdot,t_1),\ldots,\mathcal{H}(\cdot,t_n)\}].$$

Thus

$$\sum_{i=1}^{n} \alpha_i f(t_i) = \sum_{i=1}^{n} \alpha_i h_{f;T_n}(t_i) = \left\langle h_{f;T_n}, \sum_{i=1}^{n} \alpha_i \mathcal{H}(\cdot,t_i) \right\rangle_{H(\mathcal{H},T)}. \quad (\star)$$

Consequently

$$\left| \sum_{i=1}^{n} \alpha_i f(t_i) \right| \le \left\| h_{f;T_n} \right\|_{H(\mathcal{H},T)} \left\| \sum_{i=1}^{n} \alpha_i \mathcal{H}(\cdot,t_i) \right\|_{H(\mathcal{H},T)}$$
$$\le \kappa(f) \left\| \sum_{i=1}^{n} \alpha_i \mathcal{H}(\cdot,t_i) \right\|_{H(\mathcal{H},T)}.$$

Proof (3 ⇒ 1) Let $L_f : V[\mathcal{H}] \longrightarrow \mathbb{R}$ be defined using the following relation:

$$L_f\left[\sum_{i=1}^{n} \alpha_i \mathcal{H}(\cdot,t_i) \right] = \sum_{i=1}^{n} \alpha_i f(t_i).$$

Since, whenever $\{t_1,\ldots,t_n\} \subseteq T$ are distinct, $\{\mathcal{H}(\cdot,t_1),\ldots,\mathcal{H}(\cdot,t_n)\}$ are linearly independent, L_f is well defined, linear, and unique [46, p. 26]. Furthermore

$$\left| L_f\left[\sum \alpha_i \mathcal{H}(\cdot,t_i) \right] \right| = \left| \sum_{i=1}^{n} \alpha_i f(t_i) \right| \le \kappa(f) \left\| \sum_{i=1}^{n} \alpha_i \mathcal{H}(\cdot,t_i) \right\|_{H(\mathcal{H},T)}.$$

L_f is thus bounded, and has a bounded linear extension [266, p. 62], say $\tilde{L}_f$, which is thus a continuous linear functional, and, as such, has the following representation [266, p. 64]:

$$\tilde{L}_f[h] = \langle h, h_f \rangle_{H(\mathcal{H},T)},\ h_f \in H(\mathcal{H},T).$$

When $t \in T$ and $h = \mathcal{H}(\cdot,t)$ are fixed, but arbitrary, by definition,

$$\tilde{L}_f[h] = L_f[\mathcal{H}(\cdot,t)] = f(t).$$

But, using the linear functional representation, one has also that

$$\tilde{L}_f[h] = \langle \mathcal{H}(\cdot,t), h_f \rangle_{H(\mathcal{H},T)} = h_f(t).$$

Consequently $f = h_f \in H(\mathcal{H},T)$.

Proof (Norm of f) One has that [266, p. 62]

$$\|f\|_{H(\mathcal{H},T)} = \|h_f\|_{H(\mathcal{H},T)} = \sup_{h\in V[\mathcal{H}],h\neq 0_{H(\mathcal{H},T)}} \frac{|L_f[h]|}{\|h\|_{H(\mathcal{H},T)}}.$$

Let $\Sigma_{\mathcal{H},T_n}$ be the matrix with entries $\{\mathcal{H}(t_i,t_j), \{i,j\} \subseteq [1:n]\}$, and $\underline{f}$ have entries $f(t_i)$, $i \in [1:n]$. Let S be the set of elements of the form $[n,\alpha,(t,T)]$ with $\underline{\alpha} \neq \underline{0}_{\mathbb{R}^n}$ and elements $t_1, \ldots, t_n$ distinct. Then

$$\begin{aligned}
\|f\|^2_{H(\mathcal{H},T)} &= \sup_{h\in V[\mathcal{H}],h\neq 0_{H(\mathcal{H},T)}} \frac{|L_f[h]|^2}{\|h\|^2_{H(\mathcal{H},T)}} \\
&= \sup_S \frac{\left|\sum_{i=1}^n \alpha_i f(t_i)\right|^2}{\left\|\sum_{i=1}^n \alpha_i \mathcal{H}(\cdot,t_i)\right\|^2_{H(\mathcal{H},T)}} \\
&= \sup_S \frac{\left\langle \underline{\alpha},\underline{f}\right\rangle^2_{\mathbb{R}^n}}{\langle \Sigma_{\mathcal{H},T_n}[\underline{\alpha}],\underline{\alpha}\rangle_{\mathbb{R}^n}}.
\end{aligned}$$

But, since it is assumed that the support of $\mathcal{H}$ is T,

$$\left\langle \underline{\alpha},\underline{f}\right\rangle_{\mathbb{R}^n} = \left\langle \Sigma^{1/2}_{\mathcal{H},T_n}[\underline{\alpha}], \Sigma^{-1/2}_{\mathcal{H},T_n}\left[\underline{f}\right]\right\rangle_{\mathbb{R}^n}.$$

Thus, letting S_0 denote the family

$$\{n \in \mathbb{N}, \{t_1,\ldots,t_n\} \subseteq T, \text{ distinct}\},$$

one has that

$$\begin{aligned}
\sup_S \frac{\left\langle \underline{\alpha},\underline{f}\right\rangle^2_{\mathbb{R}^n}}{\langle \Sigma_{\mathcal{H},T_n}[\underline{\alpha}],\underline{\alpha}\rangle_{\mathbb{R}^n}} &= \sup_S \left\langle \frac{\Sigma^{1/2}_{\mathcal{H},T_n}[\underline{\alpha}]}{\left\|\Sigma^{1/2}_{\mathcal{H},T_n}[\underline{\alpha}]\right\|_{\mathbb{R}^n}}, \Sigma^{-1/2}_{\mathcal{H},T_n}\left[\underline{f}\right]\right\rangle^2_{\mathbb{R}^n} \\
&= \sup_{S_0} \left\langle \Sigma^{-1}_{\mathcal{H},T_n}\left[\underline{f}\right],\underline{f}\right\rangle_{\mathbb{R}^n}.
\end{aligned}$$

But, because of (Remark) 1.1.47, the latter inner product is

$$\left\|h_{f;T_n}\right\|^2_{H(\mathcal{H},T)}.$$

□

Remark 1.2.2 When T is not the support of $\mathcal{H}$, one must, in addition, require that

$$\rho_{T_{\mathcal{H}}}^{\leftarrow}\left(\rho_{T_{\mathcal{H}}}[f]\right) = f.$$

Remark 1.2.3 It is sometimes useful to remember that, in (Proposition) 1.2.1,

$$\|f\|_{H(\mathcal{H},T)}^2 = \sup_{S_0} \left\langle \Sigma_{\mathcal{H},T_n}^{-1}\left[\underline{f}\right], \underline{f}\right\rangle_{\mathbb{R}^n}.$$

Proposition 1.2.4

1. *Let $H(\mathcal{H},T)$ be an RKHS, and $h \in H(\mathcal{H},T)$ be a fixed, but arbitrary function that is not identically zero. Then the kernel $\mathcal{H}_h$ defined, for $(t_1,t_2) \in T \times T$, fixed, but arbitrary, using the following relation:*

$$\mathcal{H}_h(t_1,t_2) = \mathcal{H}(t_1,t_2) - \frac{h(t_1)\,h(t_2)}{\|h\|_{H(\mathcal{H},T)}^2},$$

is positive definite.

2. *Let $f \in \mathbb{R}^T$ be a function that is not identically zero, and for which there exists $\kappa > 0$ such that the kernel $\mathcal{H}_{\kappa,f}$ defined, for $(t_1,t_2) \in T \times T$, fixed, but arbitrary, using the following relation:*

$$\mathcal{H}_{\kappa,f}(t_1,t_2) = \mathcal{H}(t_1,t_2) - \kappa f(t_1) f(t_2),$$

is positive definite. Then

(i) $f \in H(\mathcal{H},T)$,
(ii) $\kappa \leq \|f\|_{H(\mathcal{H},T)}^{-2}$.

Proof The following remark shall be of use in the ensuing proof. As seen in the proof of (Proposition) 1.2.1 (relation ⋆), the following relation, flagged as (⋆⋆), obtains:

$$\frac{\left|\sum_{i=1}^n \alpha_i f(t_i)\right|}{\left\|\sum_{i=1}^n \alpha_i \mathcal{H}(\cdot,t_i)\right\|_{H(\mathcal{H},T)}} =$$

$$= \left|\left\langle h_{f;t_1,\ldots,t_n}, \frac{\sum_{i=1}^n \alpha_i \mathcal{H}(\cdot,t_i)}{\left\|\sum_{i=1}^n \alpha_i \mathcal{H}(\cdot,t_i)\right\|_{H(\mathcal{H},T)}}\right\rangle_{H(\mathcal{H},T)}\right|.$$

Furthermore, when $f \in H(\mathcal{H},T)$, in that latter equality, $h_{f;T_n}$ may be replaced with f since it is then [(Proposition) 1.1.45] the projection of f onto the subspace generated by the family $\{\mathcal{H}(\cdot,t_1),\ldots,\mathcal{H}(\cdot,t_n)\}$.

Suppose that $h \in H(\mathcal{H}, T)$. Then, given $[n, \alpha, (t, T)]$, fixed, but arbitrary, from the relation $(\star\star)$ above, with f being h,

$$\frac{\left\{\sum_{i=1} \alpha_i h(t_i)\right\}^2}{\left\|\sum_{i=1}^n \alpha_i \mathcal{H}(\cdot, t_i)\right\|_{H(\mathcal{H},T)}^2} \leq \|h\|_{H(\mathcal{H},T)}^2,$$

which may be rewritten as

$$\|h\|_{H(\mathcal{H},T)}^2 \sum_{i=1}^n \sum_{j=1}^n \alpha_i \alpha_j \mathcal{H}(t_i, t_j) - \sum_{i=1}^n \sum_{j=1}^n \alpha_i \alpha_j h(t_i) h(t_j) \geq 0,$$

or

$$\sum_{i=1}^n \sum_{j=1}^n \alpha_i \alpha_j \mathcal{H}_h(t_i, t_j) \geq 0,$$

that is, one has confirmed statement 1.

Suppose now that $\mathcal{H}_{\kappa,f}$ *is positive definite.* Then

$$\begin{aligned} 0 &\leq \sum_{i=1}^n \sum_{j=1}^n \alpha_i \alpha_j \mathcal{H}_{\kappa,f}(t_i, t_j) \\ &= \sum_{i=1}^n \sum_{j=1}^n \alpha_i \alpha_j \mathcal{H}(t_i, t_j) - \kappa \sum_{i=1}^n \sum_{j=1}^n \alpha_i \alpha_j f(t_i) f(t_j). \end{aligned}$$

Consequently

$$\frac{\left\{\sum_{i=1}^n \alpha_i f(t_i)\right\}^2}{\left\|\sum_{i=1}^n \alpha_i \mathcal{H}(\cdot, t_i)\right\|_{H(\mathcal{H},T)}^2} \leq \frac{1}{\kappa}.$$

But then again (Proposition) 1.2.1 says that $f \in H(\mathcal{H}, T)$. The remark at the beginning of the proof then yields that

$$\left\langle f, \frac{\sum_{i=1}^n \alpha_i \mathcal{H}(\cdot, t_i)}{\left\|\sum_{i=1}^n \alpha_i \mathcal{H}(\cdot, t_i)\right\|_{H(\mathcal{H},T)}} \right\rangle_{H(\mathcal{H},T)}^2 \leq \frac{1}{\kappa}.$$

□

Example 1.2.5 ([92]) Let X and Y be second order processes over the same probability space. Let S be the index set of X and T that of Y. Suppose that the means of X and Y are zero. Set, for $(s, t) \in S \times T$, fixed, but arbitrary,

$$C_{X,Y}(s, t) = E_P[X(\cdot, s) Y(\cdot, t)].$$

C_X and C_Y shall denote the covariances of X and Y respectively.

Let $s \in S$ be fixed, but arbitrary, and $f_s : T \longrightarrow \mathbb{R}$ be the map whose defining relation is: $f_s(t) = C_{X,Y}(s,t)$. Then

$$\frac{\left|\sum_{i=1}^{n} \alpha_i f_s(t_i)\right|^2}{\sum_{i=1}^{n}\sum_{j=1}^{n} \alpha_i \alpha_j C_Y(t_i, t_j)} = \frac{E_P^2\left[X(\cdot, s)\left(\sum_{i=1}^{n} \alpha_i Y(\cdot, t_i)\right)\right]}{E_P\left[\left(\sum_{i=1}^{n} \alpha_i Y(\cdot, t_i)\right)^2\right]}.$$

Using the Cauchy-Schwarz inequality on the right-hand side numerator of the latter equality, one sees, from (Proposition) 1.2.1, that $f_s \in H(C_Y, T)$.

Using the Cauchy-Schwarz inequality on the expression

$$E_P^2\left[X(\cdot, s)\left(\sum_{i=1}^{n} \alpha_i Y(\cdot, t_i)\right)\right],$$

one has, for $Y_0 \in L_P[Y]$, fixed, but arbitrary,

$$\left|E_P\left[X(\cdot, s)\dot{Y}_0(\cdot)\right]\right|^2 \leq C_X(s,s)\left\|Y_0\right\|_{L_P[Y]}^2.$$

The map

$$Y_0 \mapsto E_P\left[X(\cdot, s)\dot{Y}_0(\cdot)\right]$$

is thus a continuous linear functional of $L_P[Y]$, and there exists consequently a unique $Y[s] \in L_P[Y]$ such that

$$E_P\left[X(\cdot, s)\dot{Y}_0(\cdot)\right] = \langle Y[s], Y_0\rangle_{L_2[Y]}.$$

That allows one to define $B : L_P[X] \longrightarrow L_P[Y]$ using the assignment:

$$B[X_s] = Y[s].$$

As, for $[n, \alpha, (s, S)]$, fixed, but arbitrary,

$$\begin{aligned} E_P\left[\left\{\sum_{i=1}^{n} \alpha_i X(\cdot, s_i)\right\}\dot{Y}_0(\cdot)\right] &= \sum_{i=1}^{n} \alpha_i E_P\left[X(\cdot, s_i)\dot{Y}_0(\cdot)\right] \\ &= \sum_{i=1}^{n} \alpha_i \langle Y[s_i], Y_0\rangle_{L_P[Y]} \\ &= \left\langle \sum_{i=1}^{n} \alpha_i Y[s_i], Y_0\right\rangle_{L_P[Y]}, \end{aligned}$$

one has that $B\left[\sum_{i=1}^{n}\alpha_i X_{s_i}\right]=\sum_{i=1}^{n}\alpha_i B\left[X_{s_i}\right]$, and also that

$$\begin{aligned}\left|\left\langle B\left[\sum_{i=1}^{n}\alpha_i X_{s_i}\right], Y_0\right\rangle_{L_P[Y]}\right| &= \left|E_P\left[\left\{\sum_{i=1}^{n}\alpha_i X\left(\cdot, s_i\right)\right\}\dot{Y}_0\right]\right| \\ &\leq \left\|\sum_{i=1}^{n}\alpha_i X_{s_i}\right\|_{L_P[X]}\left\|Y_0\right\|_{L_P[Y]}.\end{aligned}$$

Consequently,

$$\left\|B\left[\sum_{i=1}^{n}\alpha_i X_{s_i}\right]\right\|_{L_P[Y]} \leq \left\|\sum_{i=1}^{n}\alpha_i X_{s_i}\right\|_{L_P[X]},$$

and B has a continuous linear extension to a contraction. For $X_0 \in L_P[X]$ and $Y_0 \in L_P[Y]$, fixed, but arbitrary, the following equalities obtain:

$$E_P\left[\dot{X}_0(\cdot)\,\dot{Y}_0(\cdot)\right] = \langle B\left[X_0\right], Y_0\rangle_{L_P[Y]},$$

and, in particular,

$$C_{X,Y}(s,t) = \langle B\left[X_s\right], Y_t\rangle_{L_P[Y]}.$$

An analogous development yields also a contraction $C: L_P[Y] \longrightarrow L_P[X]$ such that

$$E_P\left[\dot{X}_0(\cdot)\,\dot{Y}_0(\cdot)\right] = \langle X_0, C\left[Y_0\right]\rangle_{L_P[X]},$$

and, in particular,

$$C_{X,Y}(s,t) = \langle X_s, C\left[Y_t\right]\rangle_{L_P[Y]}.$$

As a consequence $C = B^\star$.

Using the map $L\left[X_t\right] = C_X(\cdot, t)$, one obtains contractions between the associated RKHS's. One thus finds, in a particular context, the cross-covariance operators of [12].

The following example shows that checking for membership may be difficult, and in particular that the function $t \mapsto \mathcal{H}(t,t)$ need not belong to $H(\mathcal{H}, T)$.

Example 1.2.6 ([188]) Let $H(\mathcal{H}, T)$ be an RKHS. The function $f(t) = \mathcal{H}(t,t)$, $t \in T$, need not belong to $H(\mathcal{H}, T)$.

Let $\{a_n,\ n \in \mathbb{N}\} \subseteq \mathbb{R}$ be such that

$$0 < a_n < 1/2,\ a_{n+1} < a_n,\ n \in \mathbb{N},\ \text{and}\ \lim_n a_n = 0.$$

Let then $\{b_n,\ n \in \mathbb{N}\} \subseteq \mathbb{R}$ be such that

$$1/2 < b_n < 1,\ b_n < b_{n+1},\ n \in \mathbb{N},\ \text{and}\ \lim_n b_n = 1.$$

Finally let $f_n : \,]-1,+1[\, \longrightarrow \mathbb{R}$ be defined by the following rules:

$$\alpha_n = \frac{a_n + a_{n+1}}{2},\quad \beta_n = \frac{b_n + b_{n+1}}{2},$$

$$f_n(x) = \begin{cases} 0 & \text{on }]-1, a_{n+1}], \\ 2\left((x - a_{n+1}) / (a_n - a_{n+1})\right) & \text{on }]a_{n+1}, \alpha_n], \\ -2\left((x - a_n) / (a_n - a_{n+1})\right) & \text{on }]\alpha_n, a_n], \\ 0 & \text{on }]a_n, b_n], \\ 2\kappa_n\left((x - b_n) / (b_{n+1} - b_n)\right) & \text{on }]b_n, \beta_n], \\ -2\kappa_n\left((x - b_{n+1}) / (b_{n+1} - b_n)\right) & \text{on }]\beta_n, b_{n+1}], \\ 0 & \text{on }]b_{n+1}, 1[. \end{cases}$$

By construction f_n is a continuous function whose support is $[a_{n+1}, a_n] \cup [b_n, b_{n+1}]$. Each part of the support of f_n is the base of a triangle whose other sides are determined by the graph of f_n on the support. The first triangle has height one, the second, height κ_n. Thus, almost surely with respect to Lebesgue measure, $f_m f_n = 0$ for $n \neq m$. Consequently, in that case,

$$\int_{]-1,+1[} f_m(x) f_n(x)\, dx = 0.$$

The constants κ_n shall be chosen so that

$$\int_{]-1,+1[} f_n^2(x)\, dx = 1.$$

But a calculation yields that

$$\int_{]-1,+1[} f_n^2(x)\, dx = \frac{(a_n - a_{n+1}) + \kappa_n^2 (b_{n+1} - b_n)}{3},$$

so that

$$\kappa_n = \left\{ \frac{3 - (a_n - a_{n+1})}{b_{n+1} - b_n} \right\}^{1/2} \uparrow \infty.$$

The equivalence classes in $L_2(\,]-1,+1[\,)$, denoted $[f_n]$, of the terms that form the sequence $\{f_n,\ n \in \mathbb{N}\}$, are thus orthonormal.

Let H be the Hilbert subspace generated by those equivalence classes, that is,

$$H = \left\{ \sum_{n=1}^{\infty} \alpha_n \, [f_n] \, , \; \{\alpha_n, \; n \in \mathbb{N}\} \in l_2 \right\} .$$

Let $T = \,]-1, +1[\,$, and define $F : \; T \longrightarrow H$ using $F(t) = \sum_{n=1}^{\infty} f_n(t) \, [f_n]$. This definition makes sense since, in the expression

$$\sum_{n=1}^{\infty} f_n^2(t) \, ,$$

at most one of the terms is different from zero. Then

$$L \left[\sum_{n=1}^{\infty} \alpha_n \, [f_n] \right] = \sum_{n=1}^{\infty} \alpha_n f_n \in K,$$

$\mathcal{N}[L] = \left\{ 0_{L_2(\,]-1,+1[\,)} \right\}$, and

$$\mathcal{K}(t_1, t_2) = \sum_{n=1}^{\infty} f_n(t_1) f_n(t_2) \, .$$

Let $\epsilon > 0$ be small and I_ϵ denote the interval $]-1, 1-\epsilon[$. All intervals of the form $[a_{n+1}, a_n]$ are in I_ϵ, but only a finite number of those of the form $[b_n, b_{n+1}]$ are. By construction,

$$0 \le f_n(t) \le 1 \text{ on } [a_{n+1}, a_n] \, ,$$
$$0 \le f_n(t) \le \kappa_n \text{ on } [b_n, b_{n+1}] \, ,$$

and f_n is zero outside $[a_{n+1}, a_n] \cup [b_n, b_{n+1}]$. Consequently, the function $t \mapsto f_n^2(t)$ is uniformly bounded on I_ϵ, and so must be the function $t \mapsto \mathcal{K}(t, t)$, since, for $t \in T$, fixed, but arbitrary, $\mathcal{K}(t, t) = f_n^2(t)$ for some $n \in \mathbb{N}$. It follows that the series $\sum_{n=1}^{\infty} \alpha_n f_n$, which is norm convergent in K, the RKHS of $\mathcal{K}$, is uniformly convergent on I_ϵ, for $\epsilon > 0$, and, since it is a series of continuous functions, the limit is a continuous function. K thus contains only continuous functions, since ϵ is arbitrary. But the function $t \mapsto \mathcal{K}(t, t)$ is not continuous since

$$\mathcal{K}(0, 0) = 0, \; \mathcal{K}(\alpha_n, \alpha_n) = 1, \; \lim_n \alpha_n = 0.$$

1.3 Covariances and Reproducing Kernel Hilbert Spaces

The reason RKHS's are useful for the study of second order processes is that the second order properties of those processes are determined by their covariance, which is a reproducing kernel.

Definition 1.3.1 Let T be a set. A covariance C on T is a function of two arguments, $C : T \times T \longrightarrow \mathbb{R}$, which is symmetric and positive definite, that is, such that

1. for fixed, but arbitrary $(t_1, t_2) \in T \times T$,

$$C(t_1, t_2) = C(t_2, t_1);$$

2. for fixed, but arbitrary $[n, \alpha, (t, T)]$

$$\sum_{i=1}^{n} \sum_{j=1}^{n} \alpha_i \alpha_j C\left(t_i, t_j\right) \geq 0.$$

Remark 1.3.2 As seen [(Proposition) 1.1.5], the reproducing kernel of an RKHS is a covariance, and one shall see [(Proposition) 1.3.4] that covariances are reproducing kernels of specific RKHS's.

Remark 1.3.3 The fact that C is positive definite does not imply that it is symmetric. For example, letting $T = \{1, 2\}$ and

$$\begin{bmatrix} C(1,1) & C(1,2) \\ C(2,1) & C(2,2) \end{bmatrix} = \begin{bmatrix} 1 & 0 \\ 1 & 1 \end{bmatrix},$$

one has that

$$\sum_{i=1}^{2} \sum_{j=1}^{2} \alpha_i \alpha_j C(i,j) = \left(\alpha_1 + \frac{\alpha_2}{2}\right)^2 + \frac{3}{4}\alpha_2^2 \geq 0.$$

The proposition which follows has already been proved in (Remark) 1.1.17. The proof there relies on the properties of Gaussian processes. The proof below is "from first principles."

Proposition 1.3.4 *Let C be a covariance on T. C is then the reproducing kernel of a unique RKHS denoted $H(C, T)$.*

Proof One uses the following general result of linear algebra [46, p. 30] to define an inner product on $V[C]$.

Suppose

- $U,\ V,\ W$ are real linear spaces,
- $U_0 \subseteq U,\ V_0 \subseteq V$ are subsets,

- $\Phi : U_0 \times V_0 \longrightarrow W$ is a map.

There exists then a bilinear $\Psi : U \oplus V \longrightarrow W$ such that Ψ restricted to $U_0 \times V_0$ is Φ if, and only if, both conditions which follow obtain:

1. given $[n, \alpha, (u, U_0)]$, fixed, but arbitrary,

$$\sum_{i=1}^{n} \alpha_i u_i = 0$$

implies that

$$\sum_{i=1}^{n} \alpha_i \Phi\,(u_i, v) = 0,\ v \in V_0;$$

2. given $[p, \beta, (v, V_0)]$, fixed, but arbitrary,

$$\sum_{j=1}^{p} \beta_j v_j = 0$$

implies that

$$\sum_{j=1}^{p} \beta_j \Phi\left(u, v_j\right) = 0,\ u \in U_0.$$

Let now

- $U = V = V\,[C]$,
- $W = \mathbb{R}$,
- $U_0 = V_0 = \mathcal{C} = \{C\,(\cdot, t)\,,\ t \in T\}$,
- $\Phi\,[C\,(\cdot, t_1)\,, C\,(\cdot, t_2)] = C\,(t_1, t_2)\,.$

Then, for example,

$$\sum_{i=1}^{n} \alpha_i \Phi\,(C\,(\cdot, t_i)\,, C\,(\cdot, t)) = \sum_{i=1}^{n} \alpha_i C\,(t_i, t)\,,$$

so that, if $\sum_{i=1}^{n} \alpha_i C\,(t_i, \cdot) = 0$, certainly

$$\sum_{i=1}^{n} \alpha_i C\,(t_i, t) = 0,\ t \in T.$$

There exists thus Ψ which shall be written as an inner product:

$$\begin{aligned}\langle h_1, h_2 \rangle &= \left\langle \sum_{i=1}^{n} \alpha_i^{(1)} C\left(\cdot, t_i^{(1)}\right), \sum_{j=1}^{p} \alpha_j^{(2)} C\left(\cdot, t_j^{(2)}\right) \right\rangle \\ &= \sum_{i=1}^{n} \sum_{j=1}^{p} \alpha_i^{(1)} \alpha_j^{(2)} C\left(t_i^{(1)}, t_j^{(2)}\right).\end{aligned}$$

The definition shows that $\langle h, h \rangle \geq 0$, so that, to have an inner product, one must show that $\langle h, h \rangle = 0$ implies that $h = 0$. But, for

$$h = \sum_{i=1}^{n} \alpha_i C(\cdot, t_i),$$

$$\langle h, C(\cdot, t) \rangle = \sum_{i=1}^{n} \alpha_i C(t, t_i) = h(t),$$

and, since one works with positive bilinear forms [266, p. 11],

$$h^2(t) = \langle h, C(\cdot, t) \rangle^2 \leq \langle C(\cdot, t), C(\cdot, t) \rangle \langle h, h \rangle = C(t, t) \langle h, h \rangle.$$

Let H denote the Hilbert space completion of $V[C]$ with respect to the inner product just defined.

Define $F : T \longrightarrow H$ using $F(t) = [C(\cdot, t)]$, the equivalence class of $C(\cdot, t)$ in H. K of (Remark) 1.1.16 is then $H(C, T)$. □

Covariances are obtained from maps into Hilbert spaces, that is according to pattern (Proposition) 1.1.15, and that is the content of the proposition to follow.

Proposition 1.3.5 *Let T be a set. The following statements are equivalent.*

1. *C is a covariance on T.*
2. *There exist a Hilbert space H and a map $F : T \longrightarrow H$ such that, for $(t_1, t_2) \in T \times T$, fixed, but arbitrary,*

$$C(t_1, t_2) = \langle F(t_1), F(t_2) \rangle_H.$$

3. *There exists a family of functions $\{f_\lambda : T \longrightarrow \mathbb{R},\ \lambda \in \Lambda\}$ such that, for $\{t, t_1, t_2\} \subseteq T$, fixed, but arbitrary,*
   (i) $\sum_{\lambda \in \Lambda} f_\lambda^2(t) < \infty$,
   (ii) $C(t_1, t_2) = \sum_{\lambda \in \Lambda} f_\lambda(t_1) f_\lambda(t_2)$.

Proof (1 ⇔ 2) Suppose first that C is a covariance on T. Let $H = H(C, T)$ and $F : T \longrightarrow H$ be defined using $F(t) = C(\cdot, t)$. Then C has the form given in item 2.

Conversely, if statement 2 is true, statement 1 is also true because of the properties of the inner product of a real Hilbert space.

Proof (2 ⇔ 3) Suppose now C has the form given in statement 2, and let $\{e_\lambda,\ \lambda \in \Lambda\}$ be an orthonormal basis for H. Define, for $(\lambda, t) \in \Lambda \times T$, fixed, but arbitrary,

$$f_\lambda(t) = \langle F(t), e_\lambda \rangle_H.$$

Parseval's formulae [266, p. 44] then yield that

$$\sum_{\lambda \in \Lambda} f_\lambda^2(t) = \|F(t)\|_H^2 < \infty;$$

$$C(t_1, t_2) = \langle F(t_1), F(t_2) \rangle_H$$

$$= \sum_{\lambda \in \Lambda} \langle F(t_1), e_\lambda \rangle_H \langle F(t_2), e_\lambda \rangle_H$$

$$= \sum_{\lambda \in \Lambda} f_\lambda(t_1) f_\lambda(t_2).$$

Thus statement 2 implies statement 3. Suppose conversely that statement 3 is true. Choose a real Hilbert space H of cardinality Λ (such spaces are known to exist: [266, p. 49]), and, in H, a complete orthonormal basis $\{e_\lambda,\ \lambda \in \Lambda\}$. Set then

$$F(t) = \sum_{\lambda \in \Lambda} f_\lambda(t)\, e_\lambda.$$

The assumptions of statement 3 have as consequences that

$$F(t) \in H$$

and that

$$\langle F(t_1), F(t_2) \rangle_H = \sum_{\lambda \in \Lambda} f_\lambda(t_1) f_\lambda(t_2) = C(t_1, t_2).$$

Consequently statement 3 implies statement 2. □

The following examples illustrate (Proposition) 1.3.5.

Example 1.3.6 Let $\gamma_n > 0,\ n \in \mathbb{N}$. Define $T = \mathbb{N}$, and

$$C(m, n) = \begin{cases} \gamma_m \text{ when } n = m \\ 0 \text{ when } n \neq m \end{cases}.$$

Let $H = l_2$ be the Hilbert space of square summable sequences, and e_n be the n-th element of the standard basis of l_2. Let $F : T \longrightarrow l_2$ be defined using

$$F(n) = \gamma_n^{1/2} e_n.$$

Then C is a covariance as, letting $\delta_{m,n} = \begin{cases} 1 \text{ when } m = n \\ 0 \text{ when } m \neq n \end{cases}$,

$$\langle F(m), F(n) \rangle_{l_2} = \gamma_m^{1/2} \gamma_n^{1/2} \delta_{m,n} = C(m,n).$$

Furthermore, with $\underline{h}$ having components $h_1, h_2, h_3, \ldots$,

$$L[\underline{h}](n) = \langle \underline{h}, F(n) \rangle_{l_2} = \gamma_n^{1/2} h_n,$$

and $\mathcal{N}[L] = \{\underline{0}_{l_2}\}$. Consequently, L is unitary and

$$\langle L[\underline{h}_1], L[\underline{h}_2] \rangle_K = \langle \underline{h}_1, \underline{h}_2 \rangle_{l_2}.$$

Letting some of the γ's be zero, one gets an RKHS that is isomorphic to a subspace of l_2 generated by a subset of the standard basis.

Example 1.3.7 Let $(\Omega, \mathcal{A}, P)$ be a probability space, and $T = \mathcal{A}$. Denote $L_2[P]$ the space $L_2(\Omega, \mathcal{A}, P)$. Let $I_A = [\chi_A]_{L_2[P]}$ be the equivalence class of the indicator of A, and define

$$F : \mathcal{A} \longrightarrow L_2[P] \text{ using } F(A) = I_A.$$

Then

$$\langle F(A_1), F(A_2) \rangle_{L_2[P]} = P(A_1 \cap A_2)$$

is a covariance and, $\dot{f}$ denoting one function in the equivalence class f,

$$L[f](A) = \langle f, F(A) \rangle_{L_2[P]} = \int_A \dot{f}(\omega) P(d\omega).$$

Again $\mathcal{N}[L] = \{[0]_{L_2[P]}\}$, and

$$\langle L[f], L[g] \rangle_K = \int_\Omega \dot{f}(\omega) \dot{g}(\omega) P(d\omega).$$

Example 1.1.26 is a particular case of the latter example.

One could also define

$$F : \mathcal{A} \longrightarrow L_2[P] \text{ using } F(A) = I_A - [P(A)]_{L_2[P]}.$$

Then

$$\langle F(A_1), F(A_2)\rangle_{L_2[P]} = P(A_1 \cap A_2) - P(A_1)P(A_2)$$

is also a covariance, and

$$L[f](A) = \langle f, F(A)\rangle_{L_2[P]} = \int_A \dot{f}(\omega)P(d\omega) - P(A)E_P[\dot{f}].$$

The next proposition helps recognize and build covariances.

Proposition 1.3.8 *Let* $\{C_n,\ n \in \mathbb{N}_0\}$ *be a family of covariances on the same set* T, $f : T \longrightarrow \mathbb{R}$ *be a fixed, but arbitrary function, and* $\gamma \in \mathbb{R}_+$, *be a fixed, but arbitrary constant. The following equalities, valid for* $(t_1, t_2) \in T \times T$, *fixed, but arbitrary, define covariances:*

1. $C(t_1, t_2) = \gamma$;
2. $C(t_1, t_2) = f(t_1)f(t_2)$; [(Example) 1.1.21]
3. $C(t_1, t_2) = \gamma C_0(t_1, t_2)$;
4. $C(t_1, t_2) = C_1(t_1, t_2) + C_2(t_1, t_2)$;
5. $C(t_1, t_2) = C_1(t_1, t_2) \times C_2(t_1, t_2)$;
6. $C(t_1, t_2) = \lim_n C_n(t_1, t_2)$, *provided* $\lim_n C_n(t_1, t_2)$ *exists.*

Proof All statements, except the fifth, follow directly from the definition of a covariance [(Definition) 1.3.1]. Statement 5 follows from (Proposition) 1.3.5. Let indeed, for fixed, but arbitrary $\{t, t_1, t_2\} \subseteq T$,

$$C_1(t_1, t_2) = \sum_{i\in I} f_i^{(1)}(t_1)f_i^{(1)}(t_2) \text{ where } \sum_{i\in I} \left(f_i^{(1)}\right)^2(t) < \infty,$$

$$C_2(t_1, t_2) = \sum_{j\in J} f_j^{(2)}(t_1)f_j^{(2)}(t_2) \text{ where } \sum_{j\in J} \left(f_j^{(2)}\right)^2(t) < \infty.$$

One then sets, for fixed, but arbitrary $(i, j) \in I \times J$ and $t \in T$,

$$f_{i,j}(t) = f_i^{(1)}(t) \times f_j^{(2)}(t).$$

Since

$$\sum_{i\in I}\sum_{j\in J} f_{i,j}^2(t) < \infty,$$

and that

$$C_1(t_1, t_2) \times C_2(t_1, t_2) = \sum_{i\in I}\sum_{j\in J} f_{i,j}(t_1)f_{i,j}(t_2),$$

$C = C_1 \times C_2$ is a covariance. □

Corollary 1.3.9 *For fixed, but arbitrary $n \in \mathbb{N}$, let $\alpha_n \in \mathbb{R}_+$ be fixed, but arbitrary. Let $f : \mathbb{R} \supseteq D_f \longrightarrow \mathbb{R}$ be defined using*

$$f(x) = \sum_{n=1}^{\infty} \alpha_n x^n.$$

Let C be a covariance on T such that, for $(t_1, t_2) \in T \times T$, fixed, but arbitrary, $C(t_1, t_2) \in \mathcal{D}_f$. Then $f \circ C : T \times T \longrightarrow \mathbb{R}$ is a covariance.

Example 1.3.10 The following are functions to which (Corollary) 1.3.9 applies.

1. $f(x) = e^{\alpha x}$, $\alpha > 0$.
2. $f(x) = \frac{1}{(1-x)^{\alpha}}$, $\alpha > 0$,
   as, for $|x| < 1$, $\frac{1}{(1-x)^{\alpha}} = 1 + \sum_{n=1}^{\infty} \frac{\alpha(\alpha+1)\cdots(\alpha+n-1)}{n!} x^n$.
3. $f(x) = \arcsin(x)$,
   as, for $|x| < 1$, $\arcsin(x) = \sum_{n=1}^{\infty} \frac{1\times 3\times 5\times\cdots\times(2n-1)}{2\times 4\times 6\times\cdots\times 2n\times(2n+1)} x^{2n+1}$.

Example 1.3.11 The following is an example of use of (Corollary) 1.3.9. Let $T = H$, a real Hilbert space, and $F : T \longrightarrow H$ be defined using $F = I_H$, the identity. Then $C(h_1, h_2) = \langle h_1, h_2 \rangle_H$ is, by definition, a covariance, and $K_F = H^{\star}$. One obtains a covariance C_α on $T = H$ when setting

$$C_\alpha(h_1, h_2) = e^{\alpha \langle h_1, h_2 \rangle_H}, \ \alpha > 0.$$

Example 1.3.12 (RKHS Generated by a Scalar Multiple of a Covariance) Let C be a covariance on T, and γ be a strictly positive real number. Denote C_γ the covariance

$$C_\gamma(t_1, t_2) = \gamma C(t_1, t_2), \ (t_1, t_2) \in T \times T,$$

and define

$$F : T \longrightarrow H(C, T) \text{ using } F(t) = \gamma^{1/2} C(\cdot, t).$$

Then

$$L[h](t) = \langle h, F(t) \rangle_{H(C,T)} = \gamma^{1/2} h(t),$$

so that $L[h] = \gamma^{1/2} h$. Thus L is a bijection and

$$\langle L[h_1], L[h_2] \rangle_K = \langle h_1, h_2 \rangle_{H(C,T)}.$$

Consequently

$$\langle h_1, h_2 \rangle_K = \langle L\left[\frac{h_1}{\gamma^{1/2}}\right], L\left[\frac{h_2}{\gamma^{1/2}}\right] \rangle_K = \gamma^{-1} \langle h_1, h_2 \rangle_{H(C,T)},$$

and

$$\mathcal{K}(t_1, t_2) = \langle F(t_1), F(t_2) \rangle_{H(C,T)} = \gamma C(t_1, t_2) = C_\gamma(t_1, t_2).$$

Thus, as sets, $H\left(C_\gamma, T\right) = H(C, T)$, but

$$\langle h_1, h_2 \rangle_{H(C,T)} = \gamma \langle h_1, h_2 \rangle_{H\left(C_\gamma, T\right)}.$$

Example 1.3.13 (RKHS Generated by a Sum of Covariances) Such a covariance could arise as the covariance of the sum of two independent stochastic processes.

Let $H = H(C_1, T) \oplus H(C_2, T)$ be the direct sum [266, p. 39] of the RKHS's $H(C_1, T)$ and $H(C_2, T)$. Define

$$F : T \longrightarrow H \text{ using } F(t) = (C_1(\cdot, t), C_2(\cdot, t)).$$

Then

$$L[(h_1, h_2)](t) = \langle (h_1, h_2), (C_1(\cdot, t), C_2(\cdot, t)) \rangle_H = (h_1 + h_2)(t).$$

The functions of the RKHS $H(C_1 + C_2, T)$ are thus those obtained from summing the elements of the manifolds $H(C_1, T)$ and $H(C_2, T)$. Furthermore

$$\mathcal{N}[L] = \{(h, -h),\ h \in H(C_1, T) \cap H(C_2, T)\}.$$

A calculation confirms that the associated RKHS has, as kernel, $C_1 + C_2$. It is here perhaps that one sees most immediately the value of defining RKHS's using (Proposition) 1.1.15 (compare, for example, with [35, p. 24]; see also (Example) 1.3.15). One may notice that:

- The square norm of the projection of (h_1, h_2) onto H_F is the minimum of expressions of the form

$$\left\|\tilde{h}_1\right\|^2_{H(C_1,T)} + \left\|\tilde{h}_2\right\|^2_{H(C_2,T)}$$

  for which $\tilde{h}_1 + \tilde{h}_2 = h_1 + h_2$.

  Let indeed $P_{H_F}[(h_1, h_2)] = (h_1^\star, h_2^\star)$. Then

$$\|L[(h_1, h_2)]\|^2_{K_F} = \|P_{H_F}[(h_1, h_2)]\|^2_H = \|h_1^\star\|^2_{H(C_1,T)} + \|h_2^\star\|^2_{H(C_2,T)},$$

  and, $(h_1^\star, h_2^\star)$ being the projection,

$$\langle (h_1, h_2) - (h_1^\star, h_2^\star), (C_1(\cdot, t), C_2(\cdot, t)) \rangle_{H(C_1,T) \oplus H(C_2,T)} = 0,$$

so that $h_1(t) + h_2(t) = h_1^\star(t) + h_2^\star(t)$. Furthermore

$$\|P_{H_F}[(h_1, h_2)]\|^2_{H(C_1,T)\oplus H(C_2,T)} \leq \|(h_1, h_2)\|^2_{H(C_1,T)\oplus H(C_2,T)}$$
$$= \|h_1\|^2_{H(C_1,T)} + \|h_2\|^2_{H(C_2,T)} .$$

- When $H(C_1, T) \cap H(C_2, T) = \{0_{\mathbb{R}^T}\}$, the map L is an isomorphism of the direct sum onto $H(C_1 + C_2, T)$. One shall see in Chap. 3 that

$$H(C_1, T) \subseteq H(C_1 + C_2, T),$$
$$H(C_2, T) \subseteq H(C_1 + C_2, T).$$

  Consequently, one can understand

$$H(C_1 + C_2, T) = H(C_1, T) \oplus H(C_2, T)$$

  as an orthogonal decomposition.

Example 1.3.14 (Illustration of (Example) 1.3.13*)* Let C be a covariance on T, and $f : T \longrightarrow \mathbb{R}$ be a fixed, but arbitrary function. Let $f \otimes f$ be the covariance function defined, for fixed, but arbitrary $(t_1, t_2) \in T \times T$, using [(Example) 1.1.21]:

$$[f \otimes f](t_1, t_2) = f(t_1) f(t_2),$$

and let $C_f = f \otimes f + C$. Then:

1. *When $f \in H(C, T)$,*
   (i) $H(f \otimes f, T) \cap H(C, T) = H(f \otimes f, T)$,
   (ii) $\mathcal{N}[L] = \{(\alpha f, -\alpha f), \alpha \in \mathbb{R}\}$.

   The projection onto $\mathcal{N}[L]$ is obtained as

$$P_{\mathcal{N}[L]}[(\alpha f, h)] = \frac{\langle(\alpha f, h), (f, -f)\rangle_{H(f\otimes f,T)\oplus H(C,T)}}{\|(f, -f)\|^2_{H(f\otimes f,T)\oplus H(C,T)}}(f, -f).$$

   Let $\kappa(\alpha, h) = \frac{\alpha - \langle h, f\rangle_{H(C,T)}}{1+\|f\|^2_{H(C,T)}}$, so that

$$P_{\mathcal{N}[L]}[(\alpha f, h)] = \kappa(\alpha, h)(f, -f).$$

   The elements of $H(C_f, T)$ are of the form $\alpha f + h$, whose norm is

$$\|L[(\alpha f, h)]\|^2_{H(C_f,T)} =$$
$$= \left\|\left\{I_{H(f\otimes f,T)\oplus H(C,T)} - P_{\mathcal{N}[L]}\right\}(\alpha f, h)\right\|^2_{H(f\otimes f,T)\oplus H(C,T)}$$

$$= \|(\alpha f, h) - \kappa(\alpha, h)(f, -f)\|^2_{H(f \otimes f, T) \oplus H(C,T)}$$
$$= \|([\alpha - \kappa(\alpha, h)]f, h + \kappa(\alpha, h)f)\|^2_{H(f \otimes f, T) \oplus H(C,T)}$$
$$= [\alpha - \kappa(\alpha, h)]^2 + \|h + \kappa(\alpha, h)f\|^2_{H(C,T)}.$$

A calculation then shows that

$$\|f\|^2_{H(C_f,T)} = \frac{\|f\|^2_{H(C,T)}}{1 + \|f\|^2_{H(C,T)}} < 1.$$

2. *When* $f \in H(C,T)^c$, $H(C_f, T)$ is isomorphic to $H(f \otimes f, T) \oplus H(C, T)$.

Second order stochastic processes provide an illustration of the present example. Let thus X be a second order process with index set T, on some probability space $(\Omega, \mathcal{A}, P)$. Let also, for $\{t, t_1, t_2\} \subseteq T$,

$$\mu_X(t) = E_P[X(\cdot, t)],$$
$$\Gamma_X(t_1, t_2) = E_P[X(\cdot, t_1) X(\cdot, t_2)],$$
$$C_X(t_1, t_2) = E_P[\{X(\cdot, t_1) - \mu_X(t_1)\}\{X(\cdot, t_2) - \mu_X(t_2)\}]$$
$$= \Gamma_X(t_1, t_2) - \mu_X(t_1)\mu_X(t_2).$$

Then $\Gamma_X = C_X + \mu_X \otimes \mu_X$, and what precedes applies *mutatis mutandis*.

Example 1.3.15 (RKHS Generated by a Product of Covariances) Such a covariance could arise as the covariance of the product of two independent stochastic processes.

The most general case is that of two covariances C_S and C_T defined on different sets S and T (which could be the same set in two "versions"). Let $\otimes$ denote the Hilbert spaces tensor product [266, p. 51], and

$$F : S \times T \longrightarrow H(C_S, S) \otimes H(C_T, T)$$

be defined using

$$F(s,t) = C_S(\cdot, s) \otimes C_T(\cdot, t).$$

Then

$$L[h_S \otimes h_T](s,t) = \langle h_S \otimes h_T, C_S(\cdot, s) \otimes C(\cdot, t)\rangle_{H(C_S,S) \otimes H(C_T,T)}.$$

But, by definition of the Hilbert spaces tensor product,

$$\langle h_S \otimes h_T, C_S(\cdot, s) \otimes C(\cdot, t)\rangle_{H(C_S,S) \otimes H(C_T,T)}$$

equals

$$\langle h_S, C_S(\cdot, s)\rangle_{H(C_S,S)} \langle h_T, C_T(\cdot, t)\rangle_{H(C_T,T)},$$

so that

$$L[h_S \otimes h_T](s,t) = h_S(s)\, h_T(t).$$

Since $\{C_S(\cdot, s),\ s \in S\}$ generates the RKHS $H(C_S, S)$, and $\{C_T(\cdot, t),\ t \in T\}$, the RKHS $H(C_T, T)$, the range of F generates $H(C_S, S) \otimes H(C_T, T)$ [266, p. 52], and L is an isomorphism. Furthermore

$$\begin{aligned}\mathcal{K}((s_1, t_1), (s_2, t_2)) &= \langle F(s_1, t_1), F(s_2, t_2)\rangle_{H(C_S,S)\otimes H(C_T,T)} \\ &= C_S(s_1, s_2)\, C_T(t_1, t_2).\end{aligned}$$

Suppose now one has two covariances C_1 and C_2 on the same set T. One can then also define $F : T \longrightarrow H(C_1, T) \otimes H(C_2, T)$ using

$$F(t) = C_1(\cdot, t) \otimes C_2(\cdot, t).$$

Then

$$L[h_1 \otimes h_2](t) = h_1(t)\, h_2(t).$$

L is no longer an isomorphism, but

$$\mathcal{K}(t_1, t_2) = C_1(t_1, t_2)\, C_2(t_1, t_2).$$

Definition 1.3.16 Let H be a real Hilbert space. A (weak) covariance operator on H is a linear and bounded operator of H which is positive and self-adjoint. When the operator has finite trace, it is a (strong) covariance operator.

Covariance operators often result from covariances as described farther. One uses repeatedly the following fact.

Fact 1.3.17 ([Square Root Theorem][162, p. 27]) *Let H be a real Hilbert space, and R, a linear, bounded, and positive operator. R has a unique square root, denoted, $R^{1/2}$, that is, the latter is the unique linear and bounded S such that $R = S^2$. Furthermore, the square root commutes with every linear and bounded operator which commutes with R. One has that $\|R^{1/2}\| = \|R\|^{1/2}$. R and its square root have the same null space, and the closures of their ranges are equal.*

When R has the following form: $R(h) = \sum_i \lambda_i \langle h, e_i\rangle_H\, e_i$, where the λ's are strictly positive, and the e_i's, orthonormal, letting H_e be the (closed) subspace generated by the e_i's, one has that $\mathcal{N}[R] = H_e^{\perp}$, and that $\overline{\mathcal{R}[R]} = H_e$.

Example 1.3.18 (The RKHS of a Covariance Operator) Let R be a covariance operator of the real Hilbert space H. As seen below the following facts obtain.

The RKHS K_R of R is obtained as the range of $UR^{1/2}$, where $U : H \longrightarrow H^\star$ is defined using $U[h](x) = \langle x, h\rangle_H$. Furthermore [266, pp. 35 and 71], if $H_0 = \mathcal{N}[R^{1/2}]$, $H_0^\perp = \overline{\mathcal{R}[R^{1/2}]}$, and

$$\langle UR^{1/2}[h_1], UR^{1/2}[h_2]\rangle_{K_R} = \langle P_{H_0^\perp}[h_1], P_{H_0^\perp}[h_2]\rangle_H.$$

The reproducing kernel is, for fixed, but arbitrary $(h_1, h_2) \in H \times H$,

$$\mathcal{K}_R(h_1, h_2) = \langle R[h_1], h_2\rangle_H.$$

Let indeed $T = H$, and $F : T \longrightarrow H$ be defined using $F(h) = R^{1/2}[h]$. Then

$$L[h](x) = \langle h, F(x)\rangle_H = \langle h, R^{1/2}[x]\rangle_H = \langle R^{1/2}[h], x\rangle_H,$$

so that $L[h] = UR^{1/2}[h]$. Furthermore $\mathcal{N}[L] = \mathcal{N}[R^{1/2}]$. The reproducing kernel $\mathcal{K}_R$ is

$$\mathcal{K}_R(h_1, h_2) = \langle F(h_1), F(h_2)\rangle_H = \langle R[h_1], h_2\rangle_H.$$

As $\mathcal{K}_R(x, h) = \langle x, R[h]\rangle_H = \langle x, R^{1/2}[R^{1/2}[h]]\rangle_H$, one has that

$$\mathcal{K}_R(\cdot, h) = L[F(h)] = L\left[R^{1/2}[h]\right] = UR[h].$$

Remark 1.3.19 With a given covariance, there may thus be associated two RKHS's, that defined by the covariance acting as a reproducing kernel, and that defined by the covariance operator determined by the covariance. The relation between these two RKHS's is illustrated below.

The context shall be as follows. Let $\{a, b\} \subseteq \mathbb{R}$, $a < b$, $T = [a, b]$, $\mathcal{T}$ be the σ-algebra of Borel sets, and *Leb* be Lebesgue measure. Let C be a covariance that belongs to the manifold $\mathcal{L}_2(T \times T, \mathcal{T} \otimes \mathcal{T}, Leb \otimes Leb)$. Defining $R_C[f]$ to be the equivalence class in $L_2(T, \mathcal{T}, \tau)$ of

$$t \mapsto \int_T C(t, x)\dot{f}(x)\,dx,$$

one gets [266, pp. 135 and 163] a Hilbert-Schmidt operator R_C on the Hilbert space $L_2(T, \mathcal{T}, Leb)$, which has a representation of the following form:

$$R_C[f] = \sum_{i \in I} \lambda_i \langle f, e_i\rangle_{L_2(T,\mathcal{T},Leb)} e_i,$$

with I at most countable, $\lambda_i > 0$, $i \in I$, $\sum_{i \in I} \lambda_i^2 < \infty$, and the set $\{e_i, \ i \in I\} \subseteq L_2(T, \mathcal{T}, Leb)$ orthonormal.

Proposition 1.3.20 *Let $T = [a, b]$, $\mathcal{T}$ be the σ-algebra of Borel sets, and Leb be Lebesgue measure. Let C be a covariance in $\mathcal{L}_2(T \times T, \mathcal{T} \otimes \mathcal{T}, Leb \otimes Leb)$, and $R_C = \sum_{i \in I} \lambda_i [e_i \otimes e_i]$ be the associated Hilbert-Schmidt operator. Assume that there exists a choice of $\dot{e}_i$ in e_i, $i \in I$, for which $\sum_{i \in I} \lambda_i \dot{e}_i(t_1) \dot{e}_i(t_2)$ converges for all $(t_1, t_2) \in T \times T$. Let, for fixed, but arbitrary $(t_1, t_2) \in T \times T$,*

$$\Gamma(t_1, t_2) = \sum_{i \in I} \lambda_i \dot{e}_i(t_1) \dot{e}_i(t_2) .$$

Then:

1. *Γ is a covariance kernel* (according to [245, p. 115], almost surely equal to C in the space $\mathcal{L}_2(T \times T, \mathcal{T} \otimes \mathcal{T}, Leb \otimes Leb)$), *and*

$$H(\Gamma, T) \subseteq \mathcal{L}_2(T, \mathcal{T}, Leb) .$$

2. *Let*

$$[H(\Gamma, T)] = \left\{[h]_{L_2(T, \mathcal{T}, Leb)} \in L_2(T, \mathcal{T}, Leb), \ h \in H(\Gamma, T)\right\},$$

that is, the family of equivalence classes obtained by taking the equivalence class in $L_2(T, \mathcal{T}, Leb)$ of each function of $H(\Gamma, T)$. Then also:

 (i) $\mathcal{R}[R_C^{1/2}] = [H(\Gamma, T)]$, *and, given h_1 and h_2 in $H(\Gamma, T)$,*

$$[h_1]_{L_2(T, \mathcal{T}, Leb)} \neq [h_2]_{L_2(T, \mathcal{T}, Leb)} \text{ if, and only if, } h_1 \neq h_2.$$

 (ii) *When $C = \Gamma$, $H(C, T)$ can be represented as the square root of R_C, and conversely. This will happen in particular when C is continuous, because of Mercer's proposition* [245, p. 128].

Proof Define $F : T \longrightarrow L_2(T, \mathcal{T}, Leb)$ using the following relation:

$$F(t) = \sum_{i \in I} \lambda_i^{1/2} \dot{e}_i(t) \, e_i.$$

The requirement that Γ be a convergent series makes such a definition legitimate. Define $L_F : L_2(T, \mathcal{T}, Leb) \longrightarrow \mathbb{R}^T$ using the following relation:

$$L_F[f](t) = \langle f, F(t) \rangle_{L_2(T, \mathcal{T}, Leb)}.$$

Then

$$L_F[f](t) = \sum_{i\in I} \lambda_i^{1/2} \langle f, e_i \rangle_{L_2(T,\mathcal{T},Leb)} \dot{e}_i(t),$$

$$[L_F[f]]_{L_2(T,\mathcal{T},Leb)} = \sum_{i\in I} \lambda_i^{1/2} \langle f, e_i \rangle_{L_2(T,\mathcal{T},Leb)} e_i = R_C^{1/2}[f] \in \mathcal{R}[R_C^{1/2}],$$

and

$$\langle F(t_1), F(t_2) \rangle_{L_2(T,\mathcal{T},Leb)} = \Gamma(t_1, t_2).$$

As the range of L_F is $H(\Gamma, T)$, and that the equivalence classes of elements in $H(\Gamma, T)$ are contained in the range of the square root of R_C, a manifold of $L_2(T, \mathcal{T}, Leb)$, item 1 obtains. But every element

$$R_C^{1/2}[g] \in \mathcal{R}[R_C^{1/2}]$$

is of the form

$$\sum_{i\in I} \lambda_i^{1/2} \langle g, e_i \rangle_{L_2(T,\mathcal{T},Leb)} e_i = \left[\sum_{i\in I} \lambda_i^{1/2} \langle g, e_i \rangle_{L_2(T,\mathcal{T},Leb)} \dot{e}_i \right]_{L_2(T,\mathcal{T},Leb)}$$

$$= [L_F[g]]_{L_2(T,\mathcal{T},Leb)},$$

so that, as sets,

$$\left\{ [L_F[f]]_{L_2(T,\mathcal{T},Leb)}, f \in L_2(T, \mathcal{T}, Leb) \right\} = \mathcal{R}[R_C^{1/2}].$$

Suppose finally that, for fixed, but arbitrary $t \in T$, $L_F[f](t) = 0$. Its equivalence class in $L_2(T, \mathcal{T}, Leb)$ is thus

$$R_C^{1/2}[f] = 0_{L_2(T,\mathcal{T},Leb)}.$$

But, since $R_C^{1/2}[f] = \sum_{i\in I} \lambda_i^{1/2} \langle f, e_i \rangle_{L_2(T,\mathcal{T},\tau)} e_i$, using [266, pp. 35 and 71] and (Fact) 1.3.17,

$$f \in \overline{V[\{e_i, i \in I\}]}^{\perp} = \overline{\mathcal{R}[R_C^{1/2}]}^{\perp} = \mathcal{N}[R_C^{1/2}],$$

so that $\mathcal{N}[L_F] \subseteq \mathcal{N}[R_C^{1/2}]$. But, when $f \in \mathcal{N}[R_C^{1/2}]$, as seen above, independently of i, $\langle f, e_i \rangle_{L_2(T,\mathcal{T},Leb)} = 0$. Thus, because of the representation of L_F, the latter inclusion is an equality, and

$$\mathcal{N}[L_F]^{\perp} = \overline{\mathcal{R}[R_C^{1/2}]}.$$

Consequently, whenever f and g are in the range of $R_C^{1/2}$,

$$\begin{aligned}\|L_F[f] - L_F[g]\|_{H(\Gamma,T)} &= \|L_F[f-g]\|_{H(\Gamma,T)} \\ &= \left\| P_{\overline{\mathcal{R}[R_C^{1/2}]}}[f-g] \right\|_{L_2(T,\mathcal{T},Leb)} \\ &= \|f-g\|_{L_2(T,\mathcal{T},Leb)}\,.\end{aligned}$$

□

Proposition 1.3.21 *$T, \mathcal{T}$, and Leb are as in* (Proposition) 1.3.20. *Let C be a covariance on T that is adapted to $\mathcal{T} \otimes \mathcal{T}$ and $\mathcal{B}(\mathbb{R})$. Suppose that the function*

$$t \mapsto C(t,t)$$

(which is adapted to $\mathcal{T}$ and $\mathcal{B}(\mathbb{R})$ [138, p. 92]) *is integrable with respect to Leb. C is then a square integrable kernel, and there is an associated Hilbert-Schmidt operator $R_C = \sum_{i \in I} \lambda_i [e_i \otimes e_i]$.*

One has that

$$\mathcal{R}[R_C^{1/2}] = [H(C,T)]\,.$$

Furthermore, given h_1 and h_2 in $H(C,T)$, simultaneously,

$$h_1 \neq h_2 \text{ and } [h_1]_{L_2(T,\mathcal{T},Leb)} \neq [h_2]_{L_2(T,\mathcal{T},Leb)}$$

if, and only if, for $t \in T$, fixed, but arbitrary, setting

$$\tilde{e}_i(t) = \lambda_i^{-1} \left\langle [C(\cdot,t)]_{L_2(T,\mathcal{T},Leb)}, e_i \right\rangle_{L_2(T,\mathcal{T},Leb)},$$

one has that

$$C(t,t) = \sum_{i \in I} \lambda_i \tilde{e}_i^2(t)\,.$$

Proof As [(Proposition) 1.1.5] $C^2(t_1,t_2) \le C(t_1,t_1)\,C(t_2,t_2)$, one has that C is square integrable.

Let $h \in H(C,T)$: it then follows, as presently seen, that $h \in \mathcal{L}_2(T,\mathcal{T},Leb)$, and that

$$\|h\|^2_{L_2(T,\mathcal{T},Leb)} \le \left\{ \int_T C(t,t)\,dt \right\} \|h\|^2_{H(C,T)}\,.$$

Indeed, for $t \in T$, fixed, but arbitrary, $x \mapsto C(x,t)$ is adapted, and

$$\begin{aligned}
\int_T \left\{ \sum_{i=1}^n \alpha_i C(\cdot, t_i) \right\}^2 (t)\, dt = \\
&= \int_T \left\langle \sum_{i=1}^n \alpha_i C(\cdot, t_i), C(\cdot, t) \right\rangle_{H(C,T)}^2 dt \\
&\leq \left\| \sum_{i=1}^n \alpha_i C(\cdot, t_i) \right\|_{H(C,T)}^2 \int_T \| C(\cdot, t) \|_{H(C,T)}^2 dt \\
&= \left\| \sum_{i=1}^n \alpha_i C(\cdot, t_i) \right\|_{H(C,T)}^2 \int_T C(t,t)\, dt \\
&< \infty.
\end{aligned}$$

Consequently the operator that sends $h \in H(C,T)$ to its equivalence class in $L_2(T, \mathcal{T}, Leb)$ is well defined and continuous. Let J denote that operator. For $f \in L_2(T, \mathcal{T}, Leb)$, fixed, but arbitrary,

$$\begin{aligned}
J^\star[f](t) &= \langle J^\star[f], C(\cdot, t) \rangle_{H(C,T)} \\
&= \langle f, J[C(\cdot, t)] \rangle_{L_2(T, \mathcal{T}, Leb)} \\
&= \int_T C(x,t) \dot{f}(x)\, dx.
\end{aligned}$$

Consequently, $JJ^\star = R_C$, the Hilbert-Schmidt operator determined by C. But then, by Douglas's proposition [80],

$$\mathcal{R}[J] = \mathcal{R}[R_C^{1/2}].$$

Since [266, pp. 35 and 71] $\mathcal{N}[J] = \overline{\mathcal{R}[J^\star]}^\perp$, J is an injection if, and only if, $\mathcal{R}[J^\star]$ is dense in $H(C,T)$, or, if, and only if,

$$C(\cdot, t) \in \overline{\mathcal{R}[J^\star]}.$$

The polar decomposition [266, p. 186] yields that $J^\star = UR_C^{1/2}$, with U a partial isometry with

$$\overline{\mathcal{R}[R_C^{1/2}]} \text{ as initial set, and } \overline{\mathcal{R}[J^\star]} \text{ as final set.}$$

Since $\overline{\mathcal{R}[R_C^{1/2}]}$ is generated by $\{e_i,\ i \in I\}$, $\{J^\star[e_i],\ i \in I\}$ generates $\overline{\mathcal{R}[J^\star]}$.

Now, since

$$\begin{aligned}\left\langle J^{\star}\left[\frac{e_i}{\sqrt{\lambda_i}}\right], J^{\star}\left[\frac{e_j}{\sqrt{\lambda_j}}\right]\right\rangle_{H(C,T)} &= \left\langle JJ^{\star}\left[\frac{e_i}{\sqrt{\lambda_i}}\right], \frac{e_j}{\sqrt{\lambda_j}}\right\rangle_{L_2(T,\mathcal{T},Leb)} \\ &= \left\langle R_C\left[\frac{e_i}{\sqrt{\lambda_i}}\right], \frac{e_j}{\sqrt{\lambda_j}}\right\rangle_{L_2(T,\mathcal{T},Leb)} \\ &= \sqrt{\frac{\lambda_i}{\lambda_j}}\langle e_i, e_j\rangle_{L_2(T,\mathcal{T},Leb)},\end{aligned}$$

$C(\cdot, t) \in \overline{\mathcal{R}[J^{\star}]}$ if, and only if,

$$C(\cdot, t) = \sum_{i \in I}\left\langle C(\cdot, t), J^{\star}\left[\frac{e_i}{\sqrt{\lambda_i}}\right]\right\rangle_{H(C,T)} J^{\star}\left[\frac{e_j}{\sqrt{\lambda_j}}\right]$$

or

$$\begin{aligned}C(t, t) &= \sum_{i \in I}\left\langle C(\cdot, t), J^{\star}\left[\frac{e_i}{\sqrt{\lambda_i}}\right]\right\rangle^2_{H(C,T)} \\ &= \sum_{i \in I}\left\{\frac{1}{\sqrt{\lambda_i}}\langle J\left[C(\cdot, t)\right], e_i\rangle_{L_2(T,\mathcal{T},Leb)}\right\}^2 \\ &= \sum_{i \in I}\left\{\frac{1}{\sqrt{\lambda_i}}\int_T C(x, t)\, \dot{e}_i(x)\, dx\right\}^2 \\ &= \sum_{i \in I}\lambda_i \tilde{e}_i^2(t).\end{aligned}$$

□

Remark 1.3.22 When, in (Proposition) 1.3.21, J is an injection, $H(C, T)$ is separable as $J^{\star}$ is compact [8, p. 291].

1.4 Triangular Covariances

Covariances (reproducing kernels), which are the product of functions evaluated at, respectively, the minimum and maximum of two indices, are the source of many not so obvious illustrative examples. They are intimately related to the Markovian properties of processes [200, p. 53] and to Goursat processes (Sect. 8.4).

In the immediate sequel T shall be a subset of $\mathbb{R}$ and, given $t \in T$, fixed, but arbitrary, the following sets shall be needed:

$$T_t^{<} = T \cap]-\infty, t[\,,\ T_t^{\leq} = T \cap]-\infty, t]\,,\ T_t^{>} = T \cap]t, \infty[\,,\ T_t^{\geq} = T \cap [t, \infty[\,.$$

Definition 1.4.1 Let $C : T \times T \longrightarrow \mathbb{R}$ be a symmetric map. It has a factorization when there exist maps $c_\wedge : T \longrightarrow \mathbb{R}$ and $c_\vee : T \longrightarrow \mathbb{R}$ such that, for $\{t_1, t_2\} \subseteq T$, fixed, but arbitrary,

$$C(t_1, t_2) = c_\wedge(t_1 \wedge t_2)\, c_\vee(t_1 \vee t_2)\,.$$

When C is also a covariance, it is called a triangular covariance.

When C has a factorization, T^C shall denote the following set:

$$\{t \in T : c_\wedge(t)\, c_\vee(t) \neq 0\} = \{t \in T : C(t, t) \neq 0\}\,,$$

and $r_C : T^C \longrightarrow \mathbb{R}$ shall denote the following map:

$$r_C(t) = \frac{c_\vee(t)}{c_\wedge(t)}\,.$$

Since, when C is a covariance, $C(t, t) \geq 0$, then, when C is a covariance,

$$T^C = \{t \in T : C(t, t) \neq 0\} = \{t \in T : C(t, t) > 0\}\,.$$

Remark 1.4.2 Let C have a factorization with components $c_\wedge$ and $c_\vee$, continuous, and of bounded variation on $[0, 1]$. Suppose that $c_\wedge$ is strictly positive on $]0, 1]$, $c_\vee$, strictly positive, and $c_\wedge / c_\vee$, strictly increasing, with associated measure M. Let $F : [0, 1] \longrightarrow L_2([0, 1], \mathcal{B}([0, 1]), M)$ be the map computed using $F(t) = c_\vee(t) I_t$. Then $L_F[h](t) = c_\vee(t) \int_0^t h(\theta) M(d\theta)$, and

$$\langle L_F[h_1], L_F[h_2] \rangle_{H(C,[0,1])} = \langle h_1, h_2 \rangle_{L_2([0,1]\mathcal{B}([0,1]),M)}\,.$$

In what follows, fewer restrictions are imposed on $c_\wedge$ and $c_\vee$, and more information about M and its L_2 space shall be provided. Supports are in that matter crucial.

Proposition 1.4.3 *Let $C : T \times T \longrightarrow \mathbb{R}$ be a symmetric map with the following factorization: for $\{t_1, t_2\} \subseteq T$, fixed, but arbitrary,*

$$C(t_1, t_2) = c_\wedge(t_1 \wedge t_2)\, c_\vee(t_1 \vee t_2)\,.$$

C is a covariance if, and only if,

1. *for $t \in T^C$, fixed, but arbitrary, $r_C(t) > 0$;*
2. *for $\{t_1, t_2\} \subseteq T^C$, $t_1 < t_2$, fixed, but arbitrary, $r_C(t_1) \geq r_C(t_2)$;*

3. *when* $t \in T$ *and* $c_\wedge(t) = 0$,

$$\text{either } c_\vee(t) = 0,$$

$$\text{or} \quad c_\wedge^{|T_t^{\leq}} \equiv 0;$$

4. *when* $t \in T$ *and* $c_\vee(t) = 0$,

$$\text{either } c_\wedge(t) = 0,$$

$$\text{or} \quad c_\vee^{|T_t^{\geq}} \equiv 0.$$

Proof *Suppose that* C *is a covariance.*

Let $t \in T^C$ be fixed, but arbitrary. Then, as $c_\wedge(t)\, c_\vee(t) = C(t,t) > 0$,

$$r_C(t) = \frac{c_\vee(t)}{c_\wedge(t)} = \frac{c_\wedge(t)\, c_\vee(t)}{c_\wedge^2(t)} > 0.$$

Let $\{t_1, t_2\} \subseteq T$ be fixed, but arbitrary. One has that ((Proposition) 1.1.5)

$$C^2(t_1, t_2) \leq C(t_1, t_1)\, C(t_2, t_2).$$

Thus

- when $\{t_1, t_2\} \subseteq T^C$, $t_1 \leq t_2$,

$$c_\wedge^2(t_1)\, c_\vee^2(t_2) \leq c_\wedge(t_1)\, c_\vee(t_1)\, c_\wedge(t_2)\, c_\vee(t_2),$$

  so that, since none of $c_\wedge(t_1), c_\wedge(t_2), c_\vee(t_1)$ and $c_\vee(t_2)$ is zero, and thus $c_\wedge(t_1)\, c_\vee(t_2) \neq 0$,

$$|c_\wedge(t_1)|\, |c_\vee(t_2)| \leq |c_\wedge(t_2)|\, |c_\vee(t_1)|$$

  which translates into $r_C(t_2) \leq r_C(t_1)$;
- when $\{\theta, t\} \subseteq T$, $\theta \leq t$, $c_\wedge(t) = 0$, $c_\vee(t) \neq 0$, as above,

$$|c_\wedge(\theta)|\, |c_\vee(t)| \leq |c_\wedge(\theta)|^{1/2}\, |c_\vee(\theta)|^{1/2}\, |c_\wedge(t)|^{1/2}\, |c_\vee(t)|^{1/2},$$

  and, since the right-hand side of the latter inequality is zero, and that $c_\vee(t)$ is assumed different from zero, $c_\wedge(\theta) = 0$;
- when $\{\theta, t\} \subseteq T$, $t \leq \theta$, $c_\vee(t) = 0$, $c_\wedge(t) \neq 0$, as above,

$$|c_\wedge(t)|\, |c_\vee(\theta)| \leq |c_\wedge(t)|^{1/2}\, |c_\vee(t)|^{1/2}\, |c_\wedge(\theta)|^{1/2}\, |c_\vee(\theta)|^{1/2} = 0,$$

  and, since $c_\wedge(t) \neq 0$, $c_\vee(\theta) = 0$.

Proof Suppose that items 1 to 4 obtain.

To see what happens, it is simplest to use explicit expressions. Let thus

$$\{t_1, t_2, t_3, t_4, t_5\} \subseteq T$$

be fixed, but arbitrary. One may suppose, without restriction, that

$$t_1 < t_2 < t_3 < t_4 < t_5.$$

Let C_5 be the matrix with entries

$$c_\wedge (t_i \wedge t_j)\, c_\vee (t_i \vee t_j)\,,\ \{i,j\} \subseteq [1:5]\,.$$

Then

$$C_5 = \left[\begin{array}{c|c|c|c|c} c_\wedge (t_1) c_\vee (t_1) & c_\wedge (t_1) c_\vee (t_2) & c_\wedge (t_1) c_\vee (t_3) & c_\wedge (t_1) c_\vee (t_4) & c_\wedge (t_1) c_\vee (t_5) \\ \hline c_\wedge (t_1) c_\vee (t_2) & c_\wedge (t_2) c_\vee (t_2) & c_\wedge (t_2) c_\vee (t_3) & c_\wedge (t_2) c_\vee (t_4) & c_\wedge (t_2) c_\vee (t_5) \\ \hline c_\wedge (t_1) c_\vee (t_3) & c_\wedge (t_2) c_\vee (t_3) & c_\wedge (t_3) c_\vee (t_3) & c_\wedge (t_3) c_\vee (t_4) & c_\wedge (t_3) c_\vee (t_5) \\ \hline c_\wedge (t_1) c_\vee (t_4) & c_\wedge (t_2) c_\vee (t_4) & c_\wedge (t_3) c_\vee (t_4) & c_\wedge (t_4) c_\vee (t_4) & c_\wedge (t_4) c_\vee (t_5) \\ \hline c_\wedge (t_1) c_\vee (t_5) & c_\wedge (t_2) c_\vee (t_5) & c_\wedge (t_3) c_\vee (t_5) & c_\wedge (t_4) c_\vee (t_5) & c_\wedge (t_5) c_\vee (t_5) \end{array}\right].$$

Suppose $c_\wedge (t_4) = 0$. As either $c_\vee (t_4) = 0$ or $c_\wedge (t_1) = c_\wedge (t_2) = c_\wedge (t_3) = 0$, the fourth row and column of C_5 are made of zeros. Suppose $c_\vee (t_1) = 0$. As either $c_\wedge (t_1) = 0$ or $c_\vee (t_2) = c_\vee (t_3) = c_\vee (t_4) = c_\vee (t_5) = 0$, the first row and column of C_5 are made of zeroes. One may thus assume that none of the entries of C_5 is zero, and thus that $\{t_1, t_2, t_3, t_4, t_5\} \subseteq T^c$.

Let $D_\wedge^{(5)}$ be the diagonal matrix whose diagonal entries are $c_\wedge (t_1), \ldots, c_\wedge (t_5)$, and L_5 be the following matrix:

$$L_5 = \begin{bmatrix} 1\,0\,0\,0\,0 \\ 1\,1\,0\,0\,0 \\ 1\,1\,1\,0\,0 \\ 1\,1\,1\,1\,0 \\ 1\,1\,1\,1\,1 \end{bmatrix}.$$

It may be checked that

$$L_5^{-1} = \begin{bmatrix} 1 & 0 & 0 & 0 & 0 \\ -1 & 1 & 0 & 0 & 0 \\ 0 & -1 & 1 & 0 & 0 \\ 0 & 0 & -1 & 1 & 0 \\ 0 & 0 & 0 & -1 & 1 \end{bmatrix}.$$

Let $D_{r_C}^{(5)}$ be the matrix

$$\{L_5^{-1}\}^\star \{D_\wedge^{(5)}\}^{-1} C_5 \{D_\wedge^{(5)}\}^{-1} L_5^{-1}.$$

It can be checked that $D_{r_C}^{(5)}$ is a diagonal matrix with diagonal entries equal to, successively,

$$r_C(t_1) - r_C(t_2), r_C(t_2) - r_C(t_3), r_C(t_3) - r_C(t_4), r_C(t_4) - r_C(t_5), r_C(t_5).$$

$D_{r_C}^{(5)}$ is thus positive definite and thus so is C_5. C is thus positive definite and consequently a covariance. □

Remark 1.4.4 Assume that (Proposition) 1.4.3 obtains. As, for $\{t_1, t_2\} \subseteq T$, fixed, but arbitrary,

$$c_\wedge(t_1 \wedge t_2) c_\vee(t_1 \vee t_2) = C(t_1, t_2) = \langle C(\cdot, t_1), C(\cdot, t_2)\rangle_{H(C,T)},$$

the matrix C_n (as C_5 in (Proposition) 1.4.3) is the Gram matrix of $C(\cdot, t_1), \ldots, C(\cdot, t_n)$. Thus these functions and elements of $H(C, T)$ are linearly independent if, and only if, the matrix C_n has an inverse. But, as seen, that is secured by the following conditions:

- no rows and columns of zeros: $t_i \in T^C$ or $c_\wedge(t_i) c_\vee(t_i) > 0$, so that

$$r_C(t_i) > 0;$$

- $D_{r_C}^{(n)}$ (as $D_{r_C}^{(5)}$ in (Proposition) 1.4.3) has an inverse: for $t_i \neq t_j$,

$$r_C(t_i) \neq r_C(t_j).$$

Remark 1.4.5 Assume that (Proposition) 1.4.3 obtains. For $\{\theta, t\} \subseteq T$, fixed, but arbitrary, by definition,

$$C(\theta, t) = \chi_{T_t^\leq}(\theta) c_\wedge(\theta) c_\vee(t) + \chi_{T_t^>}(\theta) c_\wedge(t) c_\vee(\theta).$$

Suppose that $c_\wedge(t) = 0$. Then, since either $c_\vee(t) = 0$ or $c_\wedge(\theta) = 0$ for $\theta \leq t$, $C(\cdot, t) = 0_{H(C,T)}$. That remains true when $c_\vee(t) = 0$. Consequently, when $C(\cdot, t) \neq 0_{H(C,T)}$, $t \in T^C$.

Remark 1.4.6 Assume that (Proposition) 1.4.3 obtains. Let T_C be a support for C, and $t \in T_C$ be fixed, but arbitrary. Then [(Fact) 1.1.30] $C(\cdot, t) \neq 0_{H(C,T)}$, so that [(Remark) 1.1.4]

$$C(t, t) = \|C(\cdot, t)\|_{H(C,T)}^2 > 0,$$

and thus $t \in T^C$. Consequently $T_C \subseteq T^C$.

Remark 1.4.7 Assume that (Proposition) 1.4.3 obtains. When $\{t_1, t_2\} \subseteq T^C$, $t_1 < t_2$, and $t \in [t_1, t_2] \cap T$ are fixed, but arbitrary, either both $c_\wedge(t) = c_\vee(t) = 0$, or $t \in T^C$. Indeed, when $c_\wedge(t) = 0$ but $c_\vee(t) \neq 0$, $c_\wedge(\theta) = 0$ for $\theta \in T_t^{\leq}$, fixed, but arbitrary, giving $c_\wedge(t_1) = 0$, which is impossible, as $t_1 \in T^C$.

Remark 1.4.8 Assume that (Proposition) 1.4.3 obtains. Let $T_{r_C} \subseteq T^C$ be a set of points at which r_C is strictly decreasing. Then, restricting r_C to that set,

- one obtains, for the diagonal elements of $D_{r_C}^{(n)}$ [(Proposition) 1.4.3], strictly positive numbers, that is, that latter matrix has an inverse, and

$$C_n = D_\wedge^{(n)} L_n^\star D_{r_C}^{(n)} L_n D_\wedge^{(n)}$$

$$C_n^{-1} = \left\{D_\wedge^{(n)}\right\}^{-1} L_n^{-1} \left\{D_{r_C}^{(n)}\right\}^{-1} \left\{L_n^{-1}\right\}^\star \left\{D_\wedge^{(n)}\right\}^{-1};$$

- letting
  - $r^C = \frac{1}{r_C}$,
  - $D_\vee^{(n)}$ be the diagonal matrix with successive diagonal entries

$$c_\vee(t_i),\ i \in [1:n],$$

  - $D_{r^C}^{(n)}$ be the diagonal matrix with, in the diagonal, successive entries

$$r^C(t_1), r^C(t_2) - r^C(t_1), \ldots, r^C(t_n) - r^C(t_{n-1}),$$

  one has similarly that

$$C_n = D_\vee^{(n)} L_n^\star D_{r^C}^{(n)} L_n D_\vee^{(n)},$$

$$C_n^{-1} = \left\{D_\vee^{(n)}\right\}^{-1} L_n^{-1} \left\{D_{r^C}^{(n)}\right\}^{-1} \left\{L_n^{-1}\right\}^\star \left\{D_\vee^{(n)}\right\}^{-1}.$$

Remark 1.4.9 Assume that (Proposition) 1.4.3 obtains. C_n^{-1} is a tridiagonal matrix with the following entries, made explicit presently:

- $\alpha_1, \ldots, \alpha_n$ in the diagonal,
- $\beta_1, \ldots, \beta_{n-1}$ above and below the diagonal.

The α's and the β's are obtained as follows. Let, for $\{i,j\} \subseteq [1:n]$,

$$\gamma_{i,j} = c_\wedge(t_i)\, c_\vee(t_j) - c_\wedge(t_j)\, c_\vee(t_i).$$

Suppose that $\gamma_{i,j} = 0$: then $c_\wedge(t_i)\, c_\vee(t_j) = c_\wedge(t_j)\, c_\vee(t_i)$, which means that $r_C(t_i) = r_C(t_j)$, a case that has been excluded by assumption [(Remark) 1.4.4].

So the following assignments make sense and may be checked to be correct:

$$\alpha_1 = -\frac{c_\wedge(t_2)}{c_\wedge(t_1)}\frac{1}{\gamma_{1,2}},$$

$$\alpha_i = \frac{\gamma_{i+1,i-1}}{\gamma_{i-1,i}\,\gamma_{i,i+1}} \quad (\text{for } i \in [2:n-1]),$$

$$\alpha_n = -\frac{c_\vee(t_{n-1})}{c_\vee(t_n)}\frac{1}{\gamma_{n-1,n}},$$

$$\beta_i = \frac{1}{\gamma_{i,i+1}} \quad (\text{for } i \in [1,n-1]).$$

Proposition 1.4.10 *Let C be a covariance on $T \subseteq \mathbb{R}$, with a factorization. A subset $T_s \subseteq T$ is a support of C if, and only if, it is a maximal subset such that $r_C^{|T_s}$ is strictly positive and strictly decreasing.*

Proof Let $T_s \subseteq T$ be a support, that is, a maximal subset such that, when in T_s, $t_1 < \cdots < t_n$, then $C(\cdot,t_1),\ldots,C(\cdot,t_n)$ are linearly independent. Then, from (Remark) 1.4.4, $r_C^{|T_s}$ is strictly positive, and strictly decreasing. If T_s is a maximal subset such that $r_C^{|T_s}$ is strictly positive, and strictly decreasing, it is a support, again because of (Remark) 1.4.4. □

Remark 1.4.11 The supports of C are thus the maximal sets over which r_C is strictly positive and decreasing.

Proposition 1.4.12 *Let C be a covariance on $T \subseteq \mathbb{R}$, with a factorization. Let $\{t_1,t_2\} \subseteq T^C$, $t_1 < t_2$, be two points such that $r_C(t_1) = r_C(t_2)$. Let $t \in [t_1,t_2] \cap T^C$ be fixed, but arbitrary, and let*

$$\kappa_1 = c_\wedge(t)/c_\wedge(t_1),$$

$$\kappa_2 = c_\vee(t)/c_\vee(t_1).$$

Then $\kappa_1 = \kappa_2 = \kappa$, and

$$C(\cdot,t) = \kappa C(\cdot,t_1).$$

Proof Since r_C is monotone decreasing, $r_C(t_1) \geq r_C(t) \geq r_C(t_2)$, so that $r_C(t) = r_C(t_1)$. But then, as seen in (Remark) 1.4.4, $C(\cdot,t)$ and $C(\cdot,t_1)$ are linearly dependent, so that, for some κ, $C(\cdot,t) = \kappa C(\cdot,t_1)$. But, when $\theta \leq t_1$, $\theta \in T$,

$$C(\theta,t) = c_\wedge(\theta)\,c_\vee(t) = \frac{c_\vee(t)}{c_\vee(t_1)}c_\wedge(\theta)\,c_\vee(t_1) = \kappa_2 C(\theta,t_1).$$

As $C(t_1,t) = \kappa\, C(t_1,t_1) = \kappa_2\, C(t_1,t_1)$, and, since $t_1 \in T^C$, then $C(t_1,t_1) > 0$, so that $\kappa = \kappa_2$. Furthermore, for $\theta \geq t_1$,

$$C(\theta,t_1) = c_\wedge(t_1)\, c_\vee(\theta) = \frac{c_\wedge(t_1)}{c_\wedge(t)} c_\wedge(t)\, c_\vee(\theta) = \frac{1}{\kappa_1} C(\theta,t),$$

and thus $\kappa_1\, C(t_1,t_1) = C(t_1,t) = \kappa_2 C(t_1,t_1)$, or $\kappa_1 = \kappa_2$.

Corollary 1.4.13 *Let C be a covariance on $T \subseteq \mathbb{R}$, with a factorization, $h \in H(C,T)$ be fixed, but arbitrary, and $\{t_1,t_2\} \subseteq T^C$, $t_1 < t_2$, be two points such that $r_C(t_1) = r_C(t_2)$. Then the functions $\frac{h}{c_\wedge}$ and $\frac{h}{c_\vee}$ are constant on the set $[t_1,t_2] \cap T^C$.*

Proof One has, for $t \in [t_1,t_2] \cap T^C$, fixed, but arbitrary,

$$h(t) = \langle h, C(\cdot,t)\rangle_{H(C,T)} = \kappa \langle h, C(\cdot,t_1)\rangle_{H(C,T)} = \kappa\, h(t_1).$$

Example 1.4.14 Let $T = [0,3]$,

$$c_\wedge(t) = \begin{cases} t & \text{when } 0 \leq t \leq 1 \\ 1 & \text{when } 1 < t \leq 2 \\ t-1 & \text{when } 2 < t \leq 3 \end{cases} \quad\text{and}\quad c_\vee(t) = \begin{cases} t & \text{when } 0 \leq t \leq 1 \\ \frac{3-t}{2} & \text{when } 1 < t \leq 2 \\ \frac{t-1}{2} & \text{when } 2 < t \leq 3 \end{cases}.$$

One has that $T^C =]0,3]$ and

$$r_C(t) = \begin{cases} 1 & \text{when } 0 \leq t \leq 1 \\ \frac{3-t}{2} & \text{when } 1 < t \leq 2 \\ \frac{1}{2} & \text{when } 2 < t \leq 3 \end{cases}.$$

Since supports of C are determined by r_C and that the latter is constant on $[0,1]$ and $[1,2]$, strictly decreasing on $[1,2]$, any support T_s is of the form

$$T_s = \{a\} \cup]1,2[\cup \{b\}, \; a \in]0,1] \text{ and } b \in [2,3].$$

The simplest support is $[1,2]$.

Remark 1.4.15 Let C be a covariance on T, with factorization made of $c_\wedge$ and $c_\vee$. Let $\{t,t_1,\cdots,t_n\} \subseteq T$, $t_1 < \cdots < t_n$, be fixed, but arbitrary. Suppose that $C(\cdot,t_1),\ldots,C(\cdot,t_n)$ are linearly independent, which means, as seen [(Remark) 1.4.11], that $r_C(t_i) > 0$, $r_C(t_i) > r_C(t_{i+1})$, and that

$$C(\cdot,t) = \sum_{i=1}^{n} \lambda_i C(\cdot,t_i).$$

Let

$$\Gamma(\theta) = \sum_{i=1}^{n} \lambda_i C(\theta, t_i) = \sum_{i=1}^{n} \lambda_i c_\wedge(\theta \wedge t_i) c_\vee(\theta \vee t_i),$$

and, for $i \in [1:n]$, fixed, but arbitrary,

$$c_i = \sum_{j=1}^{i} \lambda_j c_\wedge(t_j), \quad \gamma_i = \sum_{j=1}^{i} \lambda_j c_\vee(t_j).$$

Thus, for example, when $t_j \leq \theta \leq t_{j+1}$,

$$\begin{aligned}
&\sum_{i=1}^{j} \lambda_i c_\wedge(t_i) c_\vee(\theta) + \sum_{i=j+1}^{n} \lambda_i c_\wedge(\theta) c_\vee(t_i) = \\
&\quad = c_\vee(\theta) c_j + \left\{ \sum_{i=1}^{n} \lambda_i c_\wedge(\theta) c_\vee(t_i) - \sum_{i=1}^{j} \lambda_i c_\wedge(\theta) c_\vee(t_i) \right\} \\
&\quad = c_\vee(\theta) c_j + \left\{ c_\wedge(\theta)\gamma_n - c_\wedge(\theta)\gamma_j \right\}.
\end{aligned}$$

Consequently,

$$\Gamma(\theta) = \begin{cases}
\gamma_n c_\wedge(\theta) & \text{when } \theta \leq t_1 \\
(\gamma_n - \gamma_1) c_\wedge(\theta) + c_1 c_\vee(\theta) & \text{when } t_1 \leq \theta \leq t_2 \\
\vdots & \vdots \\
(\gamma_n - \gamma_i) c_\wedge(\theta) + c_i c_\vee(\theta) & \text{when } t_i \leq \theta \leq t_{i+1} \\
\vdots & \vdots \\
(\gamma_n - \gamma_{n-1}) c_\wedge(\theta) + c_{n-1} c_\vee(\theta) & \text{when } t_{n-1} \leq \theta \leq t_n \\
c_n c_\vee(\theta) & \text{when } \theta \geq t_n
\end{cases}.$$

Furthermore,

- choosing $\theta \leq t \wedge t_1$, the following equation: $C(\theta, t) = \Gamma(\theta)$ yields that

$$c_\wedge(\theta) c_\vee(t) = \sum_{i=1}^{n} \lambda_i c_\wedge(\theta) c_\vee(t_i) = c_\wedge(\theta)\gamma_n,$$

and, when one is interested in the value of $r_C(\theta)$, θ belongs to T^c, so that $c_\wedge(\theta) > 0$ and one gets that $c_\vee(t) = \gamma_n$;

- similarly, choosing $\theta \geq t \vee t_n$, $C(\theta, t) = \Gamma(\theta)$ yields

$$c_\wedge(t)\, c_\vee(\theta) = \sum_{i=1}^{n} \lambda_i c_\wedge(t_i)\, c_\vee(\theta) = c_\vee(\theta) c_n,$$

so that $c_\wedge(t) = c_n$.

Suppose that

- $t < t_1$: for $t \leq \theta \leq t_1$,

$$C(\theta, t) = c_\wedge(t)\, c_\vee(\theta) = \gamma_n c_\wedge(\theta),$$

so that, using $c_\wedge(t) = c_n$,

$$r_C(\theta) = \frac{c_\vee(\theta)}{c_\wedge(\theta)} = \frac{\gamma_n}{c_n};$$

- $t = t_1$: for $t \leq \theta \leq t_2$,

$$C(\theta, t) = c_\wedge(t)\, c_\vee(\theta) = (\gamma_n - \gamma_1)\, c_\wedge(\theta) + c_1 c_\vee(\theta),$$

so that, using $c_\wedge(t) = c_n$,

$$r_C(\theta) = \frac{\gamma_n - \gamma_1}{c_n - c_1};$$

- $t_1 < t < t_2$: for $t \leq \theta \leq t_2$,

$$C(\theta, t) = c_\wedge(t)\, c_\vee(\theta) = (\gamma_n - \gamma_1)\, c_\wedge(\theta) + c_1 c_\vee(\theta)$$

so that, using $c_\wedge(t) = c_n$,

$$r_C(\theta) = \frac{\gamma_n - \gamma_1}{c_n - c_1};$$

- $t = t_2$: for $t_2 \leq \theta \leq t_3$,

$$C(\theta, t) = c_\wedge(t)\, c_\vee(\theta) = (\gamma_n - \gamma_2)\, c_\wedge(\theta) + c_2 c_\vee(\theta)$$

so that, using $c_\wedge(t) = c_n$,

$$r_C(\theta) = \frac{\gamma_n - \gamma_2}{c_n - c_2};$$

- ...

One sees that, whatever the positioning of t with respect to $t_1, \ldots, t_n$, there are intervals of constancy of r_C, containing t, so that, in the representation of $C(\cdot, t)$ with respect to $C(\cdot, t_1), \ldots, C(\cdot, t_n)$, n must be one [(Proposition) 1.4.12].

Remark 1.4.16 When dealing with covariances with a factorization, the representation of the covariance that one uses may be of relevance as seen in (Remark) 1.4.2. Thus, as seen there, the representation that helps identify the RKHS of the covariance has the following form. Assume that $T = T^C = [t_l, t_r]$, $t_l \geq 0$. Let

$$r^C = \frac{c_\wedge}{c_\vee}.$$

Then, for $\{t_1, t_2\} \subseteq T$, fixed, but arbitrary,

$$C(t_1, t_2) = c_\vee(t_1)\, c_\vee(t_2)\, r^C(t_1 \wedge t_2).$$

The function r^C is monotone increasing since it is the reciprocal of r_C.

The map r^C may be extended to a monotone increasing function on $\mathbb{R}$ by letting it be equal to $r^C(t_l)$ when $t \leq t_l$, and equal to $r^C(t_r)$ when $t \geq t_r$. One thus obtains a measure τ_C on the Borel sets of $\mathbb{R}$ using, for fixed, but arbitrary reals a and b, with $a < b$, the assignment

$$\tau_C(]a, b[) = r^C(b-) - r^C(a+).$$

Then $\tau_C(]-\infty, t_l[) = 0$, and $\tau_C(]t_r, \infty[) = 0$.

For the definition of τ_C, it is no restriction to suppose that r^C is continuous to the right on $]t_l, t_r[$ [46, p. 139]. The consequence is that one may write, for fixed, but arbitrary $t \in]t_l, t_r]$,

$$\tau_C([t_l, t]) = r^C(t) - r^C(t_l).$$

Let δ_κ be the measure defined, for fixed, but arbitrary Borel B, using the following relation:

$$\delta_\kappa(B) = \kappa\, \chi_B(t_l),$$

and let then $\tilde{\tau}_C = \tau_C + \delta_\kappa$, with $\kappa = r^C(t_l)$. It then follows that

$$r^C(t) = \tilde{\tau}_C([t_l, t]),$$

so that

$$\begin{aligned} r^C(t_1 \wedge t_2) &= \langle I_{[t_l, t_1]}, I_{[t_l, t_2]} \rangle_{L_2(T, \mathcal{T}, \tilde{\tau}_C)} \\ &= r^C(t_l) + \langle I_{[t_l, t_1]}, I_{[t_l, t_2]} \rangle_{L_2(T, \mathcal{T}, \tau_C)}. \end{aligned}$$

Let $F : T \longrightarrow L_2(T, \mathcal{T}, \tilde{\tau}_C)$ be defined using the following relation:

$$F(t) = I_{[t_l, t]}.$$

Then, as just seen,

$$\langle F(t_1), F(t_2)\rangle_{L_2(T,\mathcal{T},\tilde{\tau}_C)} = r^C(t_1 \wedge t_2),$$

and

$$L_F[f](t) = \langle f, I_{[t_l,t]}\rangle_{L_2(T,\mathcal{T},\tilde{\tau}_C)} = r^C(t_l)\dot{f}(t_l) + \langle f, I_{[t_l,t]}\rangle_{L_2(T,\mathcal{T},\tau_C)}.$$

That one need not distinguish $\dot{f}$ from f follows from the fact that there is a point mass lurking in the background. Finally

$$\langle L_F[f], L_F[g]\rangle_{H(r^C,T)} = r^C(t_l) f(t_l) g(t_l) + \langle f, g\rangle_{L_2(T,\mathcal{T},\tau_C)}.$$

Now, when $c_\vee$ is strictly different from zero, one has, as shall be seen [(Example) 1.6.8], that

$$H(C,T) = \left\{c_\vee h,\ h \in H\left(r^C, T\right)\right\},$$

and that

$$\langle c_\vee h_1, c_\vee h_2\rangle_{H(C,T)} = \langle h_1, h_2\rangle_{H(r^C,T)}.$$

Remark 1.4.17 Let C and T be as in the previous remark [(Remark) 1.4.16], and $f \in \mathbb{R}^T$ be fixed, but arbitrary. Set

$$\Phi = \frac{f}{c_\wedge} \text{ and } \Psi = \frac{f}{c_\vee}.$$

The following relation [(Remark) 1.1.47]:

$$\left\|h_{f;T_n}\right\|^2_{H(C,T)} = \left\langle \Sigma^{-1}_{C,T_n}\left[\underline{f}\right], \underline{f}\right\rangle_{\mathbb{R}^n}$$

may be expressed, using (Remark) 1.4.9, as presently seen, as follows:

$$\begin{aligned}\left\|h_{f;T_n}\right\|^2_{H(C,T)} &= \sum_{i=1}^{n-1} \frac{\{\Phi(t_i) - \Phi(t_{i+1})\}^2}{r_C(t_i) - r_C(t_{i+1})} + \frac{\Phi^2(t_n)}{r_C(t_n)} \\ &= \frac{\Psi^2(t_1)}{r^C(t_1)} + \sum_{i=1}^{n-1} \frac{\{\Psi(t_{i+1}) - \Psi(t_i)\}^2}{r^C(t_{i+1}) - r^C(t_i)}.\end{aligned}$$

Let indeed $\Delta_i = r_C(t_i) - r_C(t_{i+1})$, and $c_\wedge(t_i) = c_i$. Then, for example, the product

$$\begin{bmatrix} \frac{1}{c_1} & 0 & 0 & 0 \\ 0 & \frac{1}{c_2} & 0 & 0 \\ 0 & 0 & \frac{1}{c_3} & 0 \\ 0 & 0 & 0 & \frac{1}{c_4} \end{bmatrix} \begin{bmatrix} 1 & 0 & 0 & 0 \\ -1 & 1 & 0 & 0 \\ 0 & -1 & 1 & 0 \\ 0 & 0 & -1 & 1 \end{bmatrix} \begin{bmatrix} \frac{1}{\Delta_1} & 0 & 0 & 0 \\ 0 & \frac{1}{\Delta_2} & 0 & 0 \\ 0 & 0 & \frac{1}{\Delta_3} & 0 \\ 0 & 0 & 0 & \frac{1}{\Delta_4} \end{bmatrix} \times$$

$$\times \begin{bmatrix} 1 & -1 & 0 & 0 \\ 0 & 1 & -1 & 0 \\ 0 & 0 & 1 & -1 \\ 0 & 0 & 0 & 1 \end{bmatrix} \begin{bmatrix} \frac{1}{c_1} & 0 & 0 & 0 \\ 0 & \frac{1}{c_2} & 0 & 0 \\ 0 & 0 & \frac{1}{c_3} & 0 \\ 0 & 0 & 0 & \frac{1}{c_4} \end{bmatrix}$$

yields the matrix

$$\begin{bmatrix} \frac{1}{c_1^2\Delta_1} & -\frac{1}{c_1c_2\Delta_1} & 0 & 0 \\ -\frac{1}{c_1c_2\Delta_1} & \frac{1}{c_2^2}\left(\frac{1}{\Delta_1}+\frac{1}{\Delta_2}\right) & -\frac{1}{c_2c_3\Delta_2} & 0 \\ 0 & -\frac{1}{c_2c_3\Delta_2} & \frac{1}{c_3^2}\left(\frac{1}{\Delta_2}+\frac{1}{\Delta_3}\right) & -\frac{1}{c_3c_4\Delta_3} \\ 0 & 0 & -\frac{1}{c_3c_4\Delta_3} & \frac{1}{c_4^2}\left(\frac{1}{\Delta_3}+\frac{1}{\Delta_4}\right) \end{bmatrix}.$$

When that matrix gets pre- and post-multiplied by a vector whose components are $f(t_1), f(t_2), f(t_3), f(t_4)$, combining these values with the c's, one gets the expression

$$\frac{\Phi^2(t_1)}{\Delta_1} - 2\frac{\Phi(t_1)\Phi(t_2)}{\Delta_1} + \left(\frac{1}{\Delta_1}+\frac{1}{\Delta_2}\right)\Phi^2(t_2) - 2\frac{\Phi(t_2)\Phi(t_3)}{\Delta_2} + \cdots,$$

that is, one of the two displayed above.

Those expressions determine membership of f in $H(C,T)$, and, considering the case of that with Ψ, may be understood in the following way. Let T have finite length $|T|$, and $\phi : T \longrightarrow \mathbb{R}$ be a map. Allow for the following notation:

$$\begin{aligned} I_{n,0} &= \chi_{\{t_l\}}, \\ I_{n,i} &= \chi_{\left]|T|\frac{i-1}{2^n}+t_l,|T|\frac{i}{2^n}+t_l\right]}, \\ \phi_{n,0} &= \phi(t_l), \\ \phi_{n,i} &= \phi\left(\tfrac{i}{2^n}\right), \\ r^C_{n,i} &= r^C(t_{n,i}), \\ \phi_n(\theta) &= \left\{\tfrac{\phi_{n,0}}{r^C(t_l)}\right\} I_{n,0}(\theta) + \textstyle\sum_{i\in[1:n]} \left\{\tfrac{\phi_{n,i}-\phi_{n,i-1}}{r^C_{n,i}-r^C_{n,i-1}}\right\} I_{n,i}(\theta). \end{aligned}$$

Then

$$\frac{\phi_{n,0}^2}{r^C(t_l)} + \sum_{i\in[1:n]} \frac{(\phi_{n,i} - \phi_{n,i-1})^2}{r_{n,i}^C - r_{n,i-1}^C} = \int_T \phi_n^2(\theta)\,\tilde{\tau}_C(d\theta).$$

Consequently f belongs to $H(C,T)$ when Ψ has a square that is integrable with respect to $\tilde{\tau}_C$.

The nature of the range of r_C is determining for the identification of $H(C,T)$ as shall be seen. Some of its properties are listed below.

Definition 1.4.18 Let C be a covariance with factorization terms $c_\wedge$ and $c_\vee$, $T_C \subseteq T$, a support of C, and $r_C = \frac{c_\vee}{c_\wedge}$, whose domain is T^C. The range of r_C when restricted to T_C shall be denoted

$$\mathcal{R}[r_C^{|T_C}].$$

Remark 1.4.19 Since [(Remark) 1.4.6] $T_C \subseteq T^C$, $\mathcal{R}[r_C^{|T_C}] \subseteq \mathcal{R}[r_C]$.

Proposition 1.4.20 *For any support T_C, $\mathcal{R}[r_C^{|T_C}] = \mathcal{R}[r_C]$.*

Proof Let $\mu > 0$ be fixed, but arbitrary. Suppose $\mu \in \mathcal{R}[r_C]$, that is, μ is a value taken by r_C. Since T_C is a maximal set over which r_C is strictly decreasing [(Remark) 1.4.11], were $T_C \cap r_C^{-1}(\mu) = \emptyset$ to obtain, T_C would not be maximal. Thus, for any T_C and $0 < \mu \in \mathcal{R}[r_C]$,

$$T_C \cap r_C^{-1}(\mu) \neq \emptyset.$$

Let $T_C^{(1)}$ and $T_C^{(2)}$ be two supports, not equal, and suppose that

$$\mu \in \mathcal{R}\left[r_C^{|T_C^{(1)}}\right] \subseteq \mathcal{R}[r_C].$$

By definition $\mu > 0$ and thus $T_C^{(2)} \cap r_C^{-1}(\mu) \neq \emptyset$, so that there exists $t_\mu \in T_C^{(2)}$ such that $r_C(t_\mu) = \mu$. Consequently

$$\mathcal{R}\left[r_C^{|T_C^{(1)}}\right] \subseteq \mathcal{R}\left[r_C^{|T_C^{(2)}}\right].$$

Since the argument is independent of the indices,

$$\mathcal{R}\left[r_C^{|T_C^{(2)}}\right] = \mathcal{R}\left[r_C^{|T_C^{(1)}}\right],$$

and the range of the restriction of r_C to a support is independent of that support. Let $\mathcal{R}$ denote that unique range.

If now $t \in T^C$, $C(\cdot, t) \neq 0_{H(C,T)}$ as $C(t,t) = c_\wedge(t)\, c_\vee(t) \neq 0$. There exists thus a support T_C containing t. Consequently $\mathcal{R}[r_C] \subseteq \mathcal{R}$. □

Remark 1.4.21 The importance of the range of r_C is illustrated by the following considerations.

Suppose that $\mathcal{R}[r_C] = \mathcal{R}\left[r_C^{|T_C}\right] = \left[\rho_l^C, \rho_r^C\right]$, $\rho_l^C \geq 0$. Then one has that $r_C^{-1}\left(\left[\rho_l^C, \rho_r^C\right]\right) = T_C \subseteq T^C$ (since $r_C^{|T_C}$ is strictly decreasing), and one may define

$$\Upsilon_C : \mathbb{R}^{T_C} \longrightarrow \mathbb{R}^{\left[\rho_l^C, \rho_r^C\right]}$$

using the following relation: for $f \in \mathbb{R}^{T_C}$ and $t \in \left[\rho_l^C, \rho_r^C\right]$, fixed, but arbitrary,

$$\mathcal{E}_t\left(\Upsilon_C[f]\right) = \frac{f}{c_\wedge}\left(r_C^{-1}(t)\right) = \Phi\left(r_C^{-1}(t)\right),$$

that is

$$\Upsilon_C[f] = \frac{f}{c_\wedge} \circ r_C^{-1} = \Phi \circ r_C^{-1}.$$

As seen [(Proposition) 1.2.1 and (Remark) 1.4.17], the condition for f to belong to $H\left(C^{|T_C}, T_C\right)$ is that the following sums, for fixed, but arbitrary $\{t_1, \ldots, t_n\} \subseteq T_C$, $t_1 < \cdots < t_n$, be bounded by a (finite) bound depending only on f :

$$\sum_{i=1}^{n-1} \frac{\{\Phi(t_i) - \Phi(t_{i+1})\}^2}{r_C(t_i) - r_C(t_{i+1})} + \frac{\Phi^2(t_n)}{r_C(t_n)}.$$

For $i \in [1:n]$, fixed, but arbitrary, let $r_C(t_i) = \theta_i$, so that $\theta_n < \cdots < \theta_1$. Then

$$\Phi(t_i) = \Phi\left(r_C^{-1} \circ r_C(t_i)\right) = \Phi\left(r_C^{-1}(\theta_i)\right) = \Upsilon_C[f](\theta_i),$$

so that the latter sum rewrites as

$$\sum_{i=1}^{n-1} \frac{\{\Upsilon_C[f](\theta_i) - \Upsilon_C[f](\theta_{i+1})\}^2}{\theta_i - \theta_{i+1}} + \frac{\Upsilon_C[f]^2(\theta_n)}{\theta_n}.$$

Letting $\tilde{\theta}_1 = \theta_n, \ldots, \tilde{\theta}_i = \theta_{n-i+1}, \ldots, \tilde{\theta}_n = \theta_1$, the sum becomes

$$\frac{\Upsilon_C[f]^2\left(\tilde{\theta}_1\right)}{\tilde{\theta}_1} + \sum_{i=1}^{n-1} \frac{\left\{\Upsilon_C[f]\left(\tilde{\theta}_{n-i+1}\right) - \Upsilon_C[f]\left(\tilde{\theta}_{n-i}\right)\right\}^2}{\tilde{\theta}_{n-i+1} - \tilde{\theta}_{n-i}} =$$

$$= \frac{\Upsilon_C[f]^2\left(\tilde{\theta}_1\right)}{\tilde{\theta}_1} + \sum_{i=1}^{n-1} \frac{\left\{\Upsilon_C[f]\left(\tilde{\theta}_{i+1}\right) - \Upsilon_C[f]\left(\tilde{\theta}_i\right)\right\}^2}{\tilde{\theta}_{i+1} - \tilde{\theta}_i}.$$

Consequently, letting C_W be the covariance of the standard Wiener process (for which $r_{C_W}(t) = t^{-1}$), $f \in H\left(C^{|T_C}, T_C\right)$ if, and only if,

$$\Upsilon_C[f] \in H\left(C_W^{|[\rho_l^C, \rho_r^C]}, [\rho_l^C, \rho_r^C]\right).$$

Furthermore one has that the corresponding norms are equal.

One has the identity

$$f = c_\wedge \times \{\Upsilon_C[f] \circ r_C\}.$$

Consequently, when $h \in H\left(C_W^{|[\rho_l^C, \rho_r^C]}, [\rho_l^C, \rho_r^C]\right)$,

$$c_\wedge \times \{h \circ r_C\} \in H\left(C^{|T_C}, T_C\right), \quad \text{and} \quad f = \rho_{T_C}^{\leftarrow}[c_\wedge \times \{h \circ r_C\}].$$

The norms are equal.

Remark 1.4.22 Given appropriate measurability and integrability conditions, the covariance operator associated, on $[0, 1]$, with the triangular covariance

$$c_\wedge(s \wedge t) c_\vee(s \vee t),$$

has the following form:

$$M_{c_\vee} V M_{c_\wedge} + \{M_{c_\vee} V M_{c_\wedge}\}^\star,$$

where V is the Volterra operator, and M_f is the operator of multiplication by f.

1.5 Separable Reproducing Kernel Hilbert Spaces

In the context of RKHS's, separable means, basically, finitely computable, an important feature.

Proposition 1.5.1 *Let $H(\mathcal{H}, T)$ be a separable RKHS. There exists then a countable $V_c \subseteq V[\mathcal{H}]$ which is total in $H(\mathcal{H}, T)$.*

Proof It is enough to consider the countably infinite case. Let thus the set $H_c = \{h_n,\ n \in \mathbb{N}\} \subseteq H(\mathcal{H}, T)$ be countable and total. Since [(Proposition) 1.1.5] $V[\mathcal{H}]$ is dense in $H(\mathcal{H}, T)$, given $n \in \mathbb{N}$, fixed, but arbitrary, there exists $k_{n,p} \in V[\mathcal{H}]$ such that

$$\left\|h_n - k_{n,p}\right\|_{H(\mathcal{H},T)} < p^{-1}.$$

Let $V_c = \{k_{n,p}, (n,p) \in \mathbb{N} \times \mathbb{N}\}$, and $h_0 \in H(\mathcal{H}, T)$ and $\epsilon > 0$ be fixed, but arbitrary. Choose both

- $p \in \mathbb{N}$ such that $p > \frac{2}{\epsilon}$,
- $h_n \in H_c$ such that $\|h_0 - h_n\|_{H(\mathcal{H},T)} < \frac{\epsilon}{2}$.

Then

$$\left\|h_0 - k_{n,p}\right\|_{H(\mathcal{H},T)} \leq \|h_0 - h_n\|_{H(\mathcal{H},T)} + \left\|h_n - k_{n,p}\right\|_{H(\mathcal{H},T)} \leq \frac{\epsilon}{2} + \frac{1}{p} < \frac{\epsilon}{2} + \frac{\epsilon}{2}.$$

□

Definition 1.5.2 Let $H(\mathcal{H}, T)$ be an RKHS. A set $T_d \subseteq T$ is determining when, given that $h \in H(\mathcal{H}, T)$ and $h^{|T_d} = 0_{T_d}$, it follows that $h = 0_T$.

Remark 1.5.3 Let $T_d \subseteq T$ be determining for $H(\mathcal{H}, T)$. Then

$$\mathcal{H}[T_d] = \{\mathcal{H}(\cdot, t), t \in T_d\}$$

is total in $H(\mathcal{H}, T)$ as $h(t) = \langle h, \mathcal{H}(\cdot, t)\rangle_{H(\mathcal{H},T)}$. Provided there is at least one $t \in T$ such that $\mathcal{H}(\cdot, t) \neq 0_T$, which is thus linearly independent, there is a Hamel basis, say $\mathcal{H}[T_S], T_S \subseteq T_d$, for $V[\mathcal{H}]$, which belongs to $\mathcal{H}[T_d]$ [46, p. 26]. T_S is also determining, and then T_S is a Hamel subset of T, and consequently a support for $\mathcal{H}$. One may thus take that determining set, Hamel subset, and support are one and the same concept. According to case, one or the other of these avatars will prove more convenient to use than the others.

Remark 1.5.4 Suppose $H(\mathcal{H}, T)$ is separable, and $\{V_i, i \in I\} \subseteq V[\mathcal{H}]$ is a countable set which is total in $H(\mathcal{H}, T)$ [(Proposition) 1.5.1]. Let T_I be the union of the indices corresponding to the elements $\mathcal{H}(\cdot, t_j^{(i)})$ which subtend the V_i's. They form a countable, determining set.

Proposition 1.5.5 *Let $H(\mathcal{H}, T)$ be an RKHS. When there is a countable, determining $T_{dc} \subseteq T$, $H(\mathcal{H}, T)$ is separable.*

Proof Let $h \in \{\mathcal{H}(\cdot, t), t \in T_{dc}\}^{\perp}$. Then (orthogonality), for $t_{dc} \in T_{dc}$,

$$h(t_{dc}) = \langle h, \mathcal{H}(\cdot, t_{dc})\rangle_{H(\mathcal{H},T)} = 0.$$

But, T_{dc} being determining, $h = 0$. Consequently

$$\{\mathcal{H}(\cdot, t), t \in T_{dc}\}^{\perp} = \{0_{H(\mathcal{H},T)}\},$$

so that

$$\overline{V\left[\{\mathcal{H}(\cdot,t)\,,\ t\in T_{dc}\}\right]} = H(\mathcal{H},T)\,.$$

□

Proposition 1.5.6 *Let $H(\mathcal{H},T)$ be an RKHS.*

1. *When $H(\mathcal{H},T)$ is separable, for any complete orthonormal system*

$$\{h_i,\ i\in I\}\,,$$

for $t\in T$, fixed, but arbitrary, in $H(\mathcal{H},T)$,

$$\mathcal{H}(\cdot,t) = \sum_{i\in I} h_i(t)\,h_i.$$

2. *Suppose there exists an orthonormal system $\{h_i,\ i\in I\}$ such that, for t in T, fixed, but arbitrary, in $H(\mathcal{H},T)$,*

$$\mathcal{H}(\cdot,t) = \sum_{i\in I} h_i(t)\,h_i.$$

Then $\{h_i,\ i\in I\}$ is complete, and $H(\mathcal{H},T)$, separable.

3. *When either item 1 or 2 obtains, for $(t_1,t_2)\in T\times T$, fixed, but arbitrary, in $\mathbb{R}$,*

$$\mathcal{H}(t_1,t_2) = \sum_{i\in I} h_i(t_1)h_i(t_2).$$

Proof When the assumptions of item 1 are true, one has, in $H(\mathcal{H},T)$, that

$$\mathcal{H}(\cdot,t) = \sum_{i\in I}\langle\mathcal{H}(\cdot,t),h_i\rangle_{H(\mathcal{H},T)}h_i.$$

But $\langle\mathcal{H}(\cdot,t),h_i\rangle_{H(\mathcal{H},T)} = h_i(t)$. So item 1 is true.

Let now the assumption of item 2 obtain, and $h\in H(\mathcal{H},T)$ be fixed, but arbitrary. Then

$$h(t) = \langle h,\mathcal{H}(\cdot,t)\rangle_{H(\mathcal{H},T)} = \sum_{i\in I} h_i(t)\langle h,h_i\rangle_{H(\mathcal{H},T)}.$$

That latter expression rewrites as ($\mathcal{E}_t$ is continuous)

$$\mathcal{E}_t(h) = \mathcal{E}_t\left(\sum_{i\in I} h_i\langle h,h_i\rangle_{H(\mathcal{H},T)}\right).$$

Since $t \in T$ is arbitrary, $h = \sum_{i \in I} h_i \langle h, h_i \rangle_{H(\mathcal{H},T)}$, and, since h is arbitrary, $\{h_i,\ i \in I\}$ is complete, and thus $H(\mathcal{H}, T)$ is separable. Result 3 follows using series expansion. □

Remark 1.5.7 ([77]) Let $\mathcal{F}[T]$ be a pre-Hilbert space of functions [8, p. 7] for which the evaluation maps are continuous, and let H be the completion of $\mathcal{F}[T]$ [8, p. 54]. The functional $\mathcal{E}_t$ has then an extension to H, say L_t, which, because of the Riesz representation theorem [8, p. 209], has the following representation: $L_t[h](\theta) = \langle h, h[t] \rangle_H$, some $h[t] \in H$. Define $F : T \longrightarrow H$ using the following relation: $F(t) = h[t]$. Let H_F be the (closed) subspace generated linearly by the set $\{h[t], t \in T\}$, P_F be the associated projection, and the map $L_F : H \longrightarrow \mathbb{R}^T$ be defined using the following relation: $L_F[h](t) = \langle h, h[t] \rangle_H$. $\mathcal{R}[L_F]$ is an RKHS $H(\mathcal{H}, T)$, with inner product given by the following relation:

$$\langle L_F[h_1], L_F[h_2] \rangle_{H(\mathcal{H},T)} = \langle P_F[h_1], P_F[h_2] \rangle_H,$$

and kernel, by the following one:

$$\mathcal{H}(t_1, t_2) = \langle F(t_1), F(t_2) \rangle_H = \langle h[t_1], h[t_2] \rangle_H.$$

Using the way the completion of $\mathcal{F}[T]$ is obtained, one has that

$$f(t) = \mathcal{E}_t(f) = L_t\left[[f]\right] = \langle [f], h[t] \rangle_H = L_F\left[[f]\right](t).$$

$\mathcal{F}[T]$ is thus a subset of $H(\mathcal{H}, T)$, which is its functional completion [117, p. 185] if, and only if, $H_F = H$, or L_F is unitary.

Let now the reproducing kernel $\mathcal{H}$ have the following representation:

$$\mathcal{H}(t_1, t_2) = \sum_{i \in I} h_i(t_1) h_i(t_2),$$

for $\{h_i, i \in I \subseteq \mathbb{N}\}$ linearly independent, and such that, for $t \in T$, fixed, but arbitrary, $\sum_{i \in I} h_i^2(t) < \infty$.

Let $\mathcal{F}[T] = V\left[\{h_i, i \in I\}\right]$. As the h_i's are linearly independent, each $f \in \mathcal{F}[T]$ has a unique representation in the following generic form: $f = \sum_{i=1}^{n} \alpha_i h_i$. Let

$$\|f\| = \sum_{i=1}^{n} \alpha_i^2.$$

With that norm, $\mathcal{F}[T]$ is a pre-Hilbert space, and, because of the Cauchy-Schwarz inequality, and the "summability" assumption on the h_i's, the evaluation maps are continuous. Let H be the completion of $\mathcal{F}[T]$ as described above. The specific form of the inner product of $\mathcal{F}[T]$ has, as consequence, that $\{h_i, i \in I\}$ is a complete orthonormal set in H. One has thus that:

The functions $\{h_i, i \in I\}$ *form a complete orthonormal set in* $H(\mathcal{H}, T)$ *if, and only if, given that* $\sum_{i \in I} \alpha_i h_i(t) \equiv 0$ *and* $\underline{\alpha} \in l_2$, *then* $\underline{\alpha} = \underline{0}_{l_2}$.

One may use the preceding result to check that, in [123], where one to require the skew-normal processes defined there to be stationary, they then would be Gaussian.

Proposition 1.5.8 *If* T *is a separable topological space, and* $H \subseteq \mathbb{R}^T$ *is a Hilbert space of continuous functions that is not separable, then* H *cannot be an RKHS.*

Proof Suppose *a contrario* that there is a kernel $\mathcal{H}$ which is reproducing for H. Fix then a countable, dense subset T_{dc} of T. It is a determining set, and, according to (Proposition) 1.5.5, H should be separable. But that contradicts the assumption. □

Corollary 1.5.9 *An RKHS of continuous functions defined on a separable topological space is itself separable, and an RKHS of functions defined on a separable topological space that is not separable must contain functions that are not continuous.*

The following notation is required for what follows. Given $H(\mathcal{H}, T)$ and $(t_1, t_2) \in T \times T$, set

$$\Delta_{\mathcal{H}}(t_1, t_2) = \mathcal{H}(t_1, t_1) - 2\mathcal{H}(t_2, t_1) + \mathcal{H}(t_2, t_2).$$

One has thus that

$$\Delta_{\mathcal{H}}(t_1, t_2) = \|\mathcal{H}(\cdot, t_1) - \mathcal{H}(\cdot, t_2)\|^2_{H(\mathcal{H},T)} = d^2_{\mathcal{H}}(t_1, t_2).$$

Proposition 1.5.10 ([106, p. 142]) $H(\mathcal{H}, T)$ *is separable if, and only if, given a fixed, but arbitrary* $\epsilon > 0$, *there exists a countable partition of* T, *say* $\{T_i[\epsilon], i \in I\}$ *such that, whatever* $i \in I$, $(t_1, t_2) \in T_i[\epsilon] \times T_i[\epsilon]$,

$$\Delta_{\mathcal{H}}(t_1, t_2) < 4\epsilon^2.$$

Proof Suppose first that $H(\mathcal{H}, T)$ *is separable.*

Let $\{h_i, i \in I\} \subseteq H(\mathcal{H}, T)$ be a countable, dense subset of $H(\mathcal{H}, T)$. Define then

$$S_i = \left\{t \in T : \|\mathcal{H}(\cdot, t) - h_i\|_{H(\mathcal{H},T)} < \epsilon\right\}.$$

Since $\{h_i, i \in I\}$ is dense, $\cup_{i \in I} S_i = T$. One then defines recursively

$$\begin{aligned}
T_1[\epsilon] &= S_1, \\
T_2[\epsilon] &= S_2 \setminus [S_2 \cap T_1[\epsilon]], \\
&\vdots \\
T_{n+1}[\epsilon] &= S_{n+1} \setminus \left[S_{n+1} \cap \left[\cup_{i=1}^n T_i[\epsilon]\right]\right], \\
&\vdots
\end{aligned}$$

The sets $T_i[\epsilon]$ are, by construction, disjoint, and

$$\cup_{i=1}^{n} T_i[\epsilon] = \cup_{i=1}^{n} S_i.$$

Thus $\{T_i[\epsilon], \; i \in I\}$ is a partition of T. By construction, whenever t_1 and t_2 belong to $T_i[\epsilon]$, one has that $\|\mathcal{H}(\cdot, t_1) - \mathcal{H}(\cdot, t_2)\|_{H(\mathcal{H},T)} < 2\epsilon$, so that

$$\Delta_{\mathcal{H}}(t_1, t_2) = \|\mathcal{H}(\cdot, t_1) - \mathcal{H}(\cdot, t_2)\|^2_{H(\mathcal{H},T)} < 4\epsilon^2.$$

Proof *Suppose conversely that the latter inequality obtains for any* $\epsilon > 0$ *and associated countable partition* $\{T_i[\epsilon], \; i \in I\}$ *of* T.

Choose then, for every $i \in I$ and every $n \in \mathbb{N}$,

$$\epsilon = n^{-1}, \text{ and, in the set } T_i[\epsilon], \text{ a point } t_i^{(n)}.$$

Fix $t \in T$, and suppose that this t belongs to $T_i\left[\frac{1}{n}\right]$. Set then $i_n[t] = i$. One then has that

$$\left\|\mathcal{H}(\cdot, t) - \mathcal{H}\left(\cdot, t_{i_n[t]}^{(n)}\right)\right\|_{H(\mathcal{H},T)} < 2n^{-1}.$$

That shows that the family

$$\mathcal{F} = \left\{\mathcal{H}\left(\cdot, t_i^{(n)}\right), \; i \in I, \; n \in \mathbb{N}\right\}$$

is a countable, dense set in $\mathcal{G} = \{\mathcal{H}(\cdot, t), t \in T\}$, that is, $\mathcal{G} \subseteq \overline{\mathcal{F}}$. Thus

$$V[\mathcal{G}] \subseteq V[\overline{\mathcal{F}}] \subseteq \overline{V[\mathcal{F}]},$$

so that

$$H(\mathcal{H}, T) \subseteq \overline{V[\mathcal{F}]} = \overline{V_{\mathbb{Q}}[\mathcal{F}]},$$

where the index $\mathbb{Q}$ denotes rational linear combinations. As $H(\mathcal{H}, T) = \overline{V_{\mathbb{Q}}[\mathcal{F}]}$, that the latter is separable [266, p. 32], $H(\mathcal{H}, T)$ is separable. □

Corollary 1.5.11 *Let* $H(\mathcal{H}, T)$ *be an RKHS such that* $d_{\mathcal{H}}$ *is a metric on* T. *Then* $H(\mathcal{H}, T)$ *is separable if, and only if,* $(T, d_{\mathcal{H}})$ *is separable.*

Proof If $H(\mathcal{H}, T)$ is separable, from (Proposition) 1.5.10, one knows that, whatever $\epsilon > 0$, there is a countable partition of T, say $\{T_i[\epsilon], \; i \in I\}$, such that, whatever $i \in I$ and $(t_1, t_2) \in T_i[\epsilon] \times T_i[\epsilon]$,

$$d^2_{\mathcal{H}}(t_1, t_2) = \Delta_{\mathcal{H}}(t_1, t_2) \le 4\epsilon^2.$$

For $\epsilon_i = i^{-1}$, $i \in \mathbb{N}$, choose arbitrarily $t_{i,j} \in T_j[\epsilon_i]$: the family of these $t_{i,j}$'s constitutes a countable, dense subset.

Suppose now that $(T, d_{\mathcal{H}})$ is separable, and that $\{t_i, i \in I\}$ is a dense set. Then, since

$$\|\mathcal{H}(\cdot, t) - \mathcal{H}(\cdot, t_i)\|^2_{H(\mathcal{H},T)} = d^2_{\mathcal{H}}(t, t_i),$$

$\{\mathcal{H}(\cdot, t_i), i \in I\}$ is dense in $\{\mathcal{H}(\cdot, t), t \in T\}$. That $H(\mathcal{H}, T)$ is separable follows as in the proof of (Proposition) 1.5.10. □

Proposition 1.5.12 ([198]) *Let C be a covariance on T. $H(C, T)$ is separable if, and only if, there exists a Gaussian process $G : [0,1] \times T \longrightarrow \mathbb{R}$, with base $([0,1], \mathcal{B}[0,1], Leb)$, whose mean is the zero function, and whose covariance is C.*

Proof *Suppose first that the process G exists.*

Let then G_t be the equivalence class of $\omega \mapsto G(\omega, t)$, and H_G be the subspace of $L_2([0,1], \mathcal{B}[0,1], Leb)$ generated by the family $\{G_t, t \in [0,1]\}$. As the space $L_2([0,1], \mathcal{B}[0,1], Leb)$ is separable [46, p. 174,376], and $H(C,T)$ is isomorphic to $H_G \subseteq L_2([0,1], \mathcal{B}[0,1], Leb)$ [(Example) 1.1.26], $H(C,T)$ is separable [266, p. 32].

Proof *Suppose conversely that $H(C,T)$ is separable (one has assumed $C \neq 0$).*

Let $\{h_i, i \in I\} \subseteq H(C,T)$ be a countable orthonormal basis. For $i \in I$, let $\Pi_i = (\mathbb{R}, \mathcal{B}(\mathbb{R}), P)$ be the probability space of the standard normal random variable, and $\Pi = \otimes_{i \in I} \Pi_i$ be the product of these spaces. Thus

$$\Pi = \left(\mathbb{R}^I, \mathcal{B}\left(\mathbb{R}^I\right), \otimes_{i \in I} P\right).$$

The evaluation maps $\mathcal{E}_i : \mathbb{R}^I \longrightarrow \mathbb{R}$, $i \in I$, form a family of independent standard normal random variables which generate, in $L_2(\Pi)$, a Gaussian subspace, say $H_{\mathcal{E}}$. The map

$$U : h_i \mapsto [\mathcal{E}_i]_{L_2(\Pi)}, \; i \in I,$$

is unitary. For each $t \in T$, choose a representative $G(\cdot, t) \in U[C(\cdot, t)]$. G is a Gaussian process on Π whose mean is zero and whose covariance is equal to

$$E_{\otimes_{i \in I} P}[G(\cdot, t_1), G(\cdot, t_2)] = \langle C(\cdot, t_1), C(\cdot, t_2)\rangle_{H(C,T)} = C(t_1, t_2).$$

Π is not atomic, and it is separable, since $\otimes_{i \in I} \mathcal{B}(\mathbb{R})$ is generated by a countable family of subsets. Thus it is isomorphic to $L_2([0,1], \mathcal{B}[0,1], Leb)$ [226, p. 323].

□

Remark 1.5.13 When $H(C,T)$ is separable, C has thus a representation of the form

$$C(t_1, t_2) = \int_0^1 G(x, t_1) G(x, t_2)\, dx, \; G(\cdot, t) \in \mathcal{L}_2([0,1], \mathcal{B}[0,1], Leb), \; t \in T.$$

In other words, as shall be seen farther [(Definition) 2.3.1], C has an L_2 representation.

1.6 Subspaces of Reproducing Kernel Hilbert Spaces and Associated Projections

The simplest estimations and predictions are obtained as projections. In RKHS's certain useful projections, those that correspond to restrictions of the observation's domain, have particularly simple expressions. A few facts about domination of covariances, denoted $\ll$, are required. Those are to be found in Sect. 3.1.

Let $H(\mathcal{H}, T)$ be an RKHS, and Ξ be a closed subspace with projection P_Ξ. Let $F_\Xi : T \longrightarrow H(\mathcal{H}, T)$ be defined using the following relation:

$$F_\Xi(t) = P_\Xi\left[\mathcal{H}(\cdot, t)\right],$$

and H_Ξ be the closed, linear subspace generated by $\mathcal{R}[F_\Xi]$. It is Ξ : indeed, by definition $\mathcal{R}[F_\Xi] \subseteq \Xi$, and $\xi \in \Xi$, $\xi \perp \mathcal{R}[F_\Xi]$, implies that $\xi = 0_{H(\mathcal{H},T)}$.

Let now $L_\Xi : H(\mathcal{H}, T) \longrightarrow \mathbb{R}^T$ be defined using the following relation:

$$L_\Xi[h](t) = \langle h, F_\Xi(t)\rangle_{H(\mathcal{H},T)} = \langle P_\Xi[h], \mathcal{H}(\cdot, t)\rangle_{H(\mathcal{H},T)} = P_\Xi[h](t),$$

so that $\mathcal{R}[L_\Xi] = \Xi$. $\Xi = H(\mathcal{H}_\Xi, T)$ is an RKHS with inner product

$$\langle \xi_1, \xi_2\rangle_{H(\mathcal{H}_\Xi,T)} = \langle P_\Xi[\xi_1], P_\Xi[\xi_2]\rangle_{H(\mathcal{H},T)} = \langle \xi_1, \xi_2\rangle_{H(\mathcal{H},T)},$$

and kernel

$$\mathcal{H}_\Xi(t_1, t_2) = \langle P_\Xi[\mathcal{H}(\cdot, t_1)], \mathcal{H}(\cdot, t_2)\rangle_{H(\mathcal{H},T)}.$$

In the relation $L_\Xi = U_\Xi \circ P_\Xi$ [(Proposition) 1.1.15], restricted to Ξ, one obtains that U_Ξ is the identity since

$$\xi(t) = \langle \xi, \mathcal{H}(\cdot, t)\rangle_{H(\mathcal{H},T)} = \langle \xi, P_\Xi[\mathcal{H}(\cdot, t)]\rangle_{H(\mathcal{H},T)} = L_\Xi[\xi](t).$$

Finally

$$\begin{aligned}\sum_{i=1}^{n}\sum_{j=1}^{n} \alpha_i\alpha_j\mathcal{H}_\Xi\left(t_i, t_j\right) &= \left\| P_\Xi\left[\sum_{i=1}^{n} \alpha_i \mathcal{H}\left(\cdot, t_i\right)\right]\right\|^2_{H(\mathcal{H},T)} \\ &\le \left\|\sum_{i=1}^{n} \alpha_i \mathcal{H}\left(\cdot, t_i\right)\right\|^2_{H(\mathcal{H},T)} \\ &= \sum_{i=1}^{n}\sum_{j=1}^{n} \alpha_i\alpha_j\mathcal{H}\left(t_i, t_j\right).\end{aligned}$$

One may thus state:

Proposition 1.6.1 *Let $H(\mathcal{H}, T)$ be an RKHS, and Ξ be a fixed, but arbitrary (closed) subspace of $H(\mathcal{H}, T)$. Let the orthogonal projection onto Ξ be denoted P_{Ξ}. Ξ is then an RKHS with kernel $\mathcal{H}_{\Xi}$ given by the following expression: for $(t_1, t_2) \in T \times T$, fixed, but arbitrary,*

$$\mathcal{H}_{\Xi}(t_1, t_2) = \langle P_{\Xi}[\mathcal{H}(\cdot, t_1)], P_{\Xi}[\mathcal{H}(\cdot, t_2)]\rangle_{H(\mathcal{H},T)};$$

the inner product of $(\xi_1, \xi_2) \in \Xi \times \Xi$, fixed, but arbitrary, is given by the following expression:

$$\langle \xi_1, \xi_2\rangle_{H(\mathcal{H}_{\Xi},T)} = \langle \xi_1, \xi_2\rangle_{H(\mathcal{H},T)}.$$

In particular,

1. for $(t_1, t_2) \in T \times T$, fixed, but arbitrary,

$$\mathcal{H}_{\Xi}(t_1, t_2) = P_{\Xi}[\mathcal{H}(\cdot, t_2)](t_1);$$

2. for $(h, t) \in H(\mathcal{H}, T) \times T$, fixed, but arbitrary,

$$P_{\Xi}[h](t) = \langle h, P_{\Xi}[\mathcal{H}(\cdot, t)]\rangle_{H(\mathcal{H},T)};$$

3. $\mathcal{H}_{\Xi} \ll \mathcal{H}$.

Remark 1.6.2 Suppose that K is a real Hilbert space, and that the RKHS $H(\mathcal{H}, T)$ is a (closed) subspace of K. Then the formula

$$k_{\mathcal{H}}(t) = \langle k, \mathcal{H}(\cdot, t)\rangle_K$$

provides the projection in K onto $H(\mathcal{H}, T)$. Indeed, the orthogonal decomposition of k with respect to $H(\mathcal{H}, T)$ has the form $k = h_k \oplus h_k^{\perp}$, $h_k \in H(\mathcal{H}, T)$, so that

$$\begin{aligned} k_{\mathcal{H}}(t) &= \langle k, \mathcal{H}(\cdot, t)\rangle_K \\ &= \langle h_k \oplus h_k^{\perp}, \mathcal{H}(\cdot, t)\rangle_K \\ &= \langle h_k, \mathcal{H}(\cdot, t)\rangle_K \\ &= \langle h_k, \mathcal{H}(\cdot, t)\rangle_{H(\mathcal{H},T)} \\ &= h_k(t). \end{aligned}$$

The most useful projections in an RKHS are those obtained by restriction of the index set. Here are their properties.

Proposition 1.6.3 *Let $H(\mathcal{H}, T)$ be an RKHS, and $S \subseteq T$ be a fixed, but arbitrary subset. Let $\mathcal{H}_S$ denote the restriction of $\mathcal{H}$ to $S \times S$, and H_S, the closed subspace of $H(\mathcal{H}, T)$ generated by $\{\mathcal{H}(\cdot, s), s \in S\}$. P_S shall denote the projection onto H_S. The map $U_S : H(\mathcal{H}, T) \longrightarrow \mathbb{R}^S$ which "sends" the element $h \in H(\mathcal{H}, T)$ to its* *restriction to S, that is* $U_S[h] = h^{|S}$, *is a partial isometry with* H_S *as initial set, and* $H(\mathcal{H}_S, S)$ *as final set. In particular, for* $h \in H(\mathcal{H}, T)$, *fixed, but arbitrary:*

1. $\|U_S[h]\|_{H(\mathcal{H}_S,S)} \leq \|h\|_{H(\mathcal{H},T)}$;
2. $U_S[P_S[h]] = U_S[h]$;
3. $H_S^{\perp} = \{h \in H(\mathcal{H}, T) : h(s) = 0, \; s \in S\}$;
4. *for* $s \in S$, *fixed, but arbitrary,* $P_S[h](s) = h(s)$.

Proof Let $F_S : S \longrightarrow H(\mathcal{H}, T)$ be defined using $F_S(s) = \mathcal{H}(\cdot, s)$. Then

$$L_S[h](s) = \langle h, F_S(s)\rangle_{H(\mathcal{H},T)} = \langle h, \mathcal{H}(\cdot, s)\rangle_{H(\mathcal{H},T)} = h(s),$$

so that $L_S = U_S$.

Furthermore $\mathcal{N}[L_S] = \{h \in H(\mathcal{H}, T) : h(s) = 0, \; s \in S\}$, and, since, by the definition of a reproducing kernel, $h(t) = \langle h, \mathcal{H}(\cdot, t)\rangle_{H(\mathcal{H},T)}$, $\mathcal{N}[L_S] = H_S^{\perp}$. Thus

$$\begin{aligned}\langle L_S[h_1], L_S[h_2]\rangle_{H(\mathcal{H}_S,S)} &= \langle P_{\mathcal{N}[L_S]^{\perp}}[h_1], P_{\mathcal{N}[L_S]^{\perp}}[h_2]\rangle_{H(\mathcal{H},T)} \\ &= \langle P_S[h_1], P_S[h_2]\rangle_{H(\mathcal{H},T)}.\end{aligned}$$

Also, for fixed, but arbitrary $(s_1, s_2) \in S \times S$,

$$\mathcal{H}_S(s_1, s_2) = \langle F_S(s_1), F_S(s_2)\rangle_{H(\mathcal{H},T)}.$$

Consequently the range of U_S is $H(\mathcal{H}_S, S)$, and U_S, restricted to H_S, is unitary.

Finally, for fixed, but arbitrary $h \in H(\mathcal{H}, T)$, since $h - P_S[h] \perp H_S$, one has that $h(s) = P_S[h](s)$, $s \in S$. □

Remark 1.6.4 Suppose that $h \in H_S$, and that $h^{|S} = 0$. Then, as $h \in H_S^{\perp}$ [(Proposition) 1.6.3,3.], $h = 0$.

Remark 1.6.5 If one chooses, for Ξ in (Proposition) 1.6.1, H_S in (Proposition) 1.6.3, one obtains that H_S is a reproducing kernel Hilbert space isomorphic to $H(\mathcal{H}_S, S)$. The functions of the RKHS H_S have domain T, while those of $H(\mathcal{H}_S, S)$ have domain S.

Remark 1.6.6 The fact that $h(s) = P_S[h](s), s \in S$, does not mean that $h(t) = 0, t \in T \setminus S$. Let, for example, S be the interval $[0, 1/2]$, and Π_S, the projection of $L_2[0, 1]$ defined using $\Pi_S([f]) = [\chi_S f]$ [129, p. 103]. Let, as in (Example) 1.1.25, $F(t) = I_t \in L_2[0, 1]$, $t \in [0, 1]$. $L_F \Pi_S L_F^{\star}$ is then a projection of $H(C_W, [0, 1])$ (self-adjoint and idempotent). Given $t \in [0, 1]$, fixed, but arbitrary, as $L_F^{\star}(C_W(\cdot, t)) = I_t$ [(Example) 1.1.25 and (Proposition) 1.1.15], and $\Pi_S(I_t) = [\chi_{[0,1/2]}\chi_{[0,t]}]$, $L_F \Pi_S L_F^{\star}(C_W(\cdot, t))$ is, when $t \leq 1/2$, $C_W(\cdot, t)$. Thus $P_S = L_F \Pi_S L_F^{\star}$, and, when $s \in S$, fixed, but arbitrary,

$$P_S(C_W(\cdot, s))(x) = C_W(x, s),$$

but, for $x > (1/2)$ and $s \neq 0$, $C_W(x, s) = s > 0$.

Example 1.6.7 In (Example) 1.3.15, two RKHS are defined from the same tensor product of RKHS's $H(C_S, S) \otimes H(C_T, T)$. The first has kernel $C_S \otimes C_T$, the second $C_1 \times C_2$. Let $S = T$, $C_S = C_1$, $C_T = C_2$ and write $C_1 \otimes C_2$ for the tensor product covariance function on $T \times T$. Consider the restriction of functions in $H(C_1 \otimes C_2, T \times T)$ to the diagonal of $T \times T$. One then gets the RKHS with kernel $C_1 \times C_2$, the product of the covariances. The definition by projection yields more information on this RKHS than the direct definition.

Example 1.6.8 Let $H(\mathcal{H}, T)$ be an RKHS, and $f \in \mathbb{R}^T$ be a fixed, but arbitrary function. One may obtain, as follows, an RKHS $H(\mathcal{H}_f, T)$.

Define $F_f : T \longrightarrow H(\mathcal{H}, T)$ using $F_f(t) = f(t)\mathcal{H}(\cdot, t)$. Then

$$L_f[h](t) = \langle h, F(t)\rangle_{H(\mathcal{H},T)} = \langle h, f(t)\mathcal{H}(\cdot, t)\rangle_{H(\mathcal{H},T)} = f(t)h(t) = [fh](t).$$

Thus $L_f[h]$ is multiplication of h by f in $\mathbb{R}^T$. Furthermore

$$\mathcal{N}[L_f] = \{h \in H(\mathcal{H}, T) : h(t) = 0,\ t \in \mathcal{N}[f]^c = T \setminus \mathcal{N}[f]\},$$

and

$$\mathcal{H}_f(t_1, t_2) = \langle F_f(t_1), F_f(t_2)\rangle_{H(\mathcal{H},T)} = f(t_1)f(t_2)\mathcal{H}(t_1, t_2).$$

One has that

$$V\left[\mathcal{R}[F_f]\right] = V\left[\{\mathcal{H}(\cdot, t), t \in \mathcal{N}[f]^c\}\right].$$

Let, in $H(\mathcal{H}, T)$,

$$H_f = \overline{V\left[\{\mathcal{H}(\cdot, t), t \in \mathcal{N}[f]^c\}\right]}.$$

Then

$$\langle L_f[h_1], L_f[h_2]\rangle_{H(\mathcal{H}_f,T)} = \langle [fh_1], [fh_2]\rangle_{H(\mathcal{H}_f,T)} = \langle P_{H_f}[h_1], P_{H_f}[h_2]\rangle_{H(\mathcal{H},T)}.$$

In (Proposition) 1.6.3, let $S = \mathcal{N}[f]^c$. The inner product of $H(\mathcal{H}_S, S)$ is obtained using the following formula:

$$\langle L_S[h_1], L_S[h_2]\rangle_{H(\mathcal{H}_S,S)} = \langle P_S[h_1], P_S[h_2]\rangle_{H(\mathcal{H},T)}.$$

But $P_S = P_{H_f}$. Consequently $H(\mathcal{H}_f, T)$ and $H(\mathcal{H}_S, S)$ are isomorphic. The isomorphism is

$$[fh] \mapsto h^{|\mathcal{N}[f]^c}.$$

Suppose now that $S \subseteq T$, and that $f(t) = \chi_S(t)$. The construction just achieved provides an isomorphism between $\chi_S h$, which is defined on T, and $h^{|S}$, which is defined on S.

Example 1.6.8 may be of use in the following situation. Let $X = AG$ with A and G independent stochastic processes. Then

$$P(X \in S) = \int P(X \in S \mid A = a) P_A(da) = \int P(aG \in S) P_A(da).$$

The process aG, a an ordinary function, has an RKHS driven by the covariance $a(t_1) a(t_2) C_G(t_1, t_2)$, C_G the covariance of G.

The next example is of interest in the context of the Cramér-Hida representation covered in Part II of this book.

Example 1.6.9 ([118, p. 498]) Let $T = [a, b]$, and $H(\mathcal{H}, T)$ be an RKHS. Let H_t be the subspace of $H(\mathcal{H}, T)$ generated by $\{\mathcal{H}(\cdot, \theta), a \leq \theta \leq t\}$, and let P_t denote the associated projection. Then:

1. $P_b = I_{H(\mathcal{H},T)}$;
2. when $h \in H(\mathcal{H}, T)$ and $\theta \in [a, t]$, one has that $P_t[h](\theta) = h(\theta)$;
3. when $(h_1, h_2) \in H(\mathcal{H}, T) \times H(\mathcal{H}, T)$ and $h_1(\theta) = h_2(\theta)$ for all $\theta \in [a, t]$, one has that $P_t[h_1] = P_t[h_2]$;
4. when $(h, t) \in H(\mathcal{H}, T) \times T$ and

$$S(h, t) = \{h' \in H(\mathcal{H}, T) : h'(\theta) = h(\theta), \theta \in [a, t]\},$$

one has that

$$\|P_t[h]\|_{H(\mathcal{H},T)} = \min_{h' \in S(h,t)} \|h'\|_{H(\mathcal{H},T)};$$

5. when $(h, h') \in H(\mathcal{H}, T) \times H(\mathcal{H}, T)$, $h' \in S(h, t)$, and $h' \neq P_t[h]$,

$$\|P_t[h]\|_{H(\mathcal{H},T)} < \|h'\|_{H(\mathcal{H},T)}.$$

Suppose indeed that $(h, t) \in H(\mathcal{H}, T) \times T$, is fixed, but arbitrary, and that $h \perp H_t$. Then, simultaneously, for $\theta \in [a, t]$,

$$\langle h, \mathcal{H}(\cdot, \theta) \rangle_{H(\mathcal{H},T)} = h(\theta),$$
$$\langle h, \mathcal{H}(\cdot, \theta) \rangle_{H(\mathcal{H},T)} = 0.$$

Thus $h(\theta) = 0$, $\theta \in [a, t]$. In particular, when $t = b$, $h(t) = 0$, $t \in T$, and then $P_b = I_{H(\mathcal{H},T)}$. This is item 1.

For item 2, given $\theta \in [a, t]$, one has that,

$$\begin{aligned} P_t[h](\theta) &= \langle P_t[h], \mathcal{H}(\cdot,\theta)\rangle_{H(\mathcal{H},T)} \\ &= \langle h, P_t[\mathcal{H}(\cdot,\theta)]\rangle_{H(\mathcal{H},T)} \\ &= \langle h, \mathcal{H}(\cdot,\theta)\rangle_{H(\mathcal{H},T)} \\ &= h(\theta). \end{aligned}$$

For item 3, it suffices to set $h = h_1 - h_2$. Then, by item 2,

$$P_t[h_1](\theta) - P_t[h_2](\theta) = P_t[h](\theta) = h(\theta) = 0,\ \theta \in [a,t],$$

which is enough because of (Remark) 1.6.4.

A consequence of item 3 is that $P_t[h'] = P_t[h]$ whenever $h' \in S(h,t)$ (defined in the statement of item 4).

For item 4, one may write

$$\|h'\|^2_{H(\mathcal{H},T)} = \|P_t[h']\|^2_{H(\mathcal{H},T)} + \|(I_{H(\mathcal{H},T)} - P_t)[h']\|^2_{H(\mathcal{H},T)} \geq \|P_t[h']\|^2_{H(\mathcal{H},T)}.$$

But, when $h' \in S(h,t)$, as stated, $P_t[h'] = P_t[h]$, and, replacing $P_t[h']$ by $P_t[h]$ in the right-hand side of the latter inequality, one has that

$$\|h'\|_{H(\mathcal{H},T)} \geq \|P_t[h]\|_{H(\mathcal{H},T)},\ h' \in S(h,t).$$

But item 2 yields that $P_t[h'](\theta) = h'(\theta)$, $\theta \in [a,t]$. As $h' \in S(h,t)$, it follows that $h'(\theta) = h(\theta)$, $\theta \in [a,t]$. But then $P_t[h'] \in S(h,t)$, and thus $P_t[P_t[h']] = P_t[h]$, that is $P_t[h'] = P_t[h]$. So there is an element in $S(h,t)$ whose norm is equal to that of $P_t[h]$.

For item 5, when $h' \in S(h,t)$, $P_t[h'] = P_t[h]$, so that, using the assumption,

$$0 < \|h' - P_t[h]\|^2_{H(\mathcal{H},T)} = \|h' - P_t[h']\|^2_{H(\mathcal{H},T)}.$$

Consequently

$$\|h'\|^2_{H(\mathcal{H},T)} > \|P_t[h']\|^2_{H(\mathcal{H},T)} = \|P_t[h]\|^2_{H(\mathcal{H},T)}.$$

Computations require finite sets, and the results which follow provide the framework within which a reduction to finite dimensional subspaces is possible. The first result restates and completes facts about the role of supports already explained (starting at (Definition) 1.1.28).

Proposition 1.6.10 ([176]) *Let $H(\mathcal{H},T)$ be an RKHS, and $S \subseteq T$ be fixed, but arbitrary. The following statements are equivalent:*

1. S is a determining set [(Definition) 1.5.2] *for $H(\mathcal{H},T)$;*

2. *$U_S : H(\mathcal{H}, T) \longrightarrow H(\mathcal{H}_S, S)$, defined using $U_S[h] = h^{|S}$, is unitary;*
3. *every $h_S \in H(\mathcal{H}_S, S)$ has a unique extension to an $h \in H(\mathcal{H}, T)$ for which*

$$\|h\|_{H(\mathcal{H},T)} = \|h_S\|_{H(\mathcal{H}_S,S)};$$

4. *the family $\{\mathcal{H}(\cdot, s),\ s \in S\}$ is total in $H(\mathcal{H}, T)$;*
5. *the family of continuous linear functionals*

$$\mathcal{E}_s = \left\{\langle \cdot, \mathcal{H}(\cdot, s)\rangle_{H(\mathcal{H},T)},\ s \in S\right\}$$

separates points in $H(\mathcal{H}, T)$.

Proof Let $H_S = \overline{V[\mathcal{H}_S]}$, the closure being in $H(\mathcal{H}, T)$.

[1 ⇒ 2] As seen in (Proposition) 1.6.3, $U_S = L_S$, and $\mathcal{N}[L_S] = H_S^{\perp}$. But, when $h \in H_S^{\perp}$,

$$h(s) = \langle h, \mathcal{H}(\cdot, s)\rangle_{H(\mathcal{H},T)} = 0,\ s \in S.$$

Since S is determining, $h(t) = 0,\ t \in T$, and $\mathcal{N}[L_S] = \{0_T\}$. L_S is thus unitary, and so is U_S.

[2 ⇒ 3] Let $h_S \in H(\mathcal{H}_S, S)$ be fixed but arbitrary. Since U_S is onto, the possible extensions of h_S form the set $U_S^{-1}[h_S]$. Suppose that

$$U_S[h_1] = U_S[h_2] = h_S,\ \{h_1, h_2\} \subseteq H(\mathcal{H}, T).$$

Then $0_S = U_S[h_1 - h_2]$. But U_S is unitary and thus $\|h_1 - h_2\|_{H(\mathcal{H},T)} = 0$.

[3 ⇒ 4] Let $h \in H_S^{\perp}$ be fixed, but arbitrary. Then $h(s) = 0,\ s \in S$. Since h has a unique extension preserving norms, its extension must be the zero function, and item 4 is proved.

[4 ⇒ 5] Let $\{h_1, h_2\} \subseteq H(\mathcal{H}, T)$ be such that $h_1 \neq h_2$. Then the following equality: $h_1(t) = h_2(t),\ s \in S$, implies that $\langle h_1 - h_2, \mathcal{H}(\cdot, s)\rangle_{H(\mathcal{H},T)} = 0, s \in S$, so that $h_1 - h_2 = 0$, which is impossible by assumption. There is thus $s \in S$ such that $h_1(s) \neq h_2(s)$, or $\mathcal{E}_s[h_1] \neq \mathcal{E}_s[h_2]$.

[5 ⇒ 1] Suppose $h \in H(\mathcal{H}, T)$ and $h(s) = 0,\ s \in s$. Then $\mathcal{E}_s[h] = 0,\ s \in S$, which means, by assumption, that $h = 0$. □

The following lemmas and proposition, which result in a sort of Kolmogorov extension theorem [5, p. 191] for RKHS's, all require the same set of assumptions that are stated separately, as a fact, for convenience.

Fact 1.6.11 *1. $\{T_n,\ n \in \mathbb{N}\}$ is an increasing family of subsets of some common set S ($T_n \subseteq T_{n+1} \subseteq S$), and $T = \cup_n T_n$.*

2. $\{\mathcal{H}_n : T_n \times T_n \longrightarrow \mathbb{R},\ n \in \mathbb{N}\}$ *is a family of reproducing kernels such that, for* $p > n$ *in* $\mathbb{N}$, *and* h *in* $H\left(\mathcal{H}_p, T_p\right)$, *fixed, but arbitrary,*

$$h^{|T_n} \in H\left(\mathcal{H}_n, T_n\right), \quad and \quad \left\|h^{|T_n}\right\|_{H(\mathcal{H}_n,T_n)} \leq \|h\|_{H(\mathcal{H}_p,T_p)}.$$

$\mathcal{H}_{p|n}$ shall denote the restriction of $\mathcal{H}_p$ to T_n.

Remark 1.6.12 Assume (Fact) 1.6.11, and fix $p > n$. Let

$$V_{p,n} : H(\mathcal{H}_p, T_p) \longrightarrow H(\mathcal{H}_n, T_n)$$

be the restriction map of (Fact) 1.6.11. Let $H_{p,n}$ be the subspace of $H(\mathcal{H}_p, T_p)$ generated by $\left\{\mathcal{H}_p(\cdot, t), t \in T_n\right\}$. From (Proposition) 1.6.3, one has a map

$$L_{p,n} : H(\mathcal{H}_p, T_p) \longrightarrow H(\mathcal{H}_{p|n}, T_n)$$

with $L_{p,n} = U_{p,n}P_{H_{p,n}}$, and $U_{n,p} : H_{n,p} \longrightarrow H(\mathcal{H}_{p|n}, T_N)$, unitary. Let

$$J_{p,n} : H(\mathcal{H}_{p|n}, T_n) \longrightarrow H(\mathcal{H}_n, T_n)$$

be the inclusion, so that

$$V_{p,n} = J_{p,n}L_{p,n}.$$

Fact 1.6.11 says that $V_{n,p}$ is a contraction. One furthermore knows that $L_{p,n}$ is continuous.

Temporarily, for convenience, one shall sometimes use the following notation $(p > n)$:

- H_n for $H\left(\mathcal{H}_n, T_n\right)$;
- $\|\cdot\|_n$ for $\|\cdot\|_{H(\mathcal{H}_n,T_n)}$;
- $h^{(p)}$ for an element of H_p, and $h^{(p|n)}$ for the restriction of $h^{(p)}$ to T_n, also written $h^{(p)|T_n}$.

Lemma 1.6.13 *The assumptions are those of (Fact)* 1.6.11. *Suppose* $f : T \longrightarrow \mathbb{R}$ *is a function with the property that, for* $n \in \mathbb{N}$, $f^{|T_n} \in H_n$. *The sequence* $\left\{\left\|f^{|T_n}\right\|_n, n \in \mathbb{N}\right\}$ *has then a limit (which may be infinite).*

Proof For $p > n$ in $\mathbb{N}$,

$$\left\|f^{|T_n}\right\|_n = \left\|\left\{f^{|T_p}\right\}^{|T_n}\right\|_n.$$

By assumption $f^{|T_p} \in H_p$ so that, by assumption again,

$$\left\|\left\{f^{|T_p}\right\}^{|T_n}\right\|_n \leq \left\|f^{|T_p}\right\|_p.$$

Consequently $p > n$ implies

$$\left\|f^{|T_n}\right\|_n \leq \left\|f^{|T_p}\right\|_p.$$

□

Lemma 1.6.14 *The assumptions are those of (Fact)* 1.6.11. *Let*

$$H = \left\{f \in \mathbb{R}^T : \textit{for all } n \in \mathbb{N}, f^{|T_n} \in H_n, \textit{ and } \lim_n \left\|f^{|T_n}\right\|_n < \infty\right\}.$$

For $\{f_1, f_2\} \subseteq H$, *fixed, but arbitrary, the following relation defines an inner product:*

$$\langle f_1, f_2 \rangle_H = \lim_n \langle f_1^{|T_n}, f_2^{|T_n} \rangle_n.$$

With that inner product, H is a Hilbert space.

Proof One must notice that the formula for the inner product yields, for $f \in H$,

$$\|f\|_H = \lim_n \left\|f^{|T_n}\right\|_n,$$

which exists, and is finite, by assumption.

Proof H is a vector space.

Let indeed $\{\alpha_1, \alpha_2\} \subseteq \mathbb{R}$ and $\{f_1, f_2\} \subseteq H$ be fixed, but arbitrary, and consider $f = \alpha_1 f_1 + \alpha_2 f_2$. Then

$$f^{|T_n} = \alpha_1 f_1^{|T_n} + \alpha_2 f_2^{|T_n} \in H_n.$$

Furthermore, using (Lemma) 1.6.13 on f_1 and f_2,

$$\begin{aligned}
\left\|f^{|T_n}\right\|_n &= \left\|\alpha_1 f_1^{|T_n} + \alpha_2 f_2^{|T_n}\right\|_n \\
&\leq |\alpha_1| \left\|f_1^{|T_n}\right\|_n + |\alpha_2| \left\|f_2^{|T_n}\right\|_n \\
&\leq |\alpha_1| \|f_1\|_H + |\alpha_2| \|f_2\|_H \\
&< \infty.
\end{aligned}$$

Since $\left\{\left\|f^{|T_n}\right\|_n, \ n \in \mathbb{N}\right\}$ is an increasing sequence (again (Lemma) 1.6.13), f belongs to H.

Proof The statement's formula determines an inner product.

The limit that defines the tentative inner product exists: indeed, as H is a vector space, the following expression is legitimate:

$$\langle f_1^{|T_n}, f_2^{|T_n} \rangle_n = \frac{1}{4}\left\{\left\|(f_1 + f_2)^{|T_n}\right\|_n^2 - \left\|(f_1 - f_2)^{|T_n}\right\|_n^2\right\},$$

and the right-hand side has a limit.

- *The function* $(f_1, f_2) \mapsto \langle f_1, f_2 \rangle_H$ *is bilinear.*
  Indeed, for $\{\alpha_1, \alpha_2\} \subseteq \mathbb{R}$, $\{f, f_1, f_2\} \subseteq H$, fixed, but arbitrary,

$$\begin{aligned}\langle \alpha_1 f_1 + \alpha_2 f_2, f \rangle_H &= \lim_n \langle \alpha_1 f_1^{|T_n} + \alpha_2 f_2^{|T_n}, f^{|T_n} \rangle_n \\ &= \lim_n \left\{ \alpha_1 \langle f_1^{|T_n}, f^{|T_n} \rangle_n + \alpha_2 \langle f_2^{|T_n}, f^{|T_n} \rangle_n \right\} \\ &= \alpha_1 \langle f_1, f \rangle_H + \alpha_2 \langle f_2, f \rangle_H.\end{aligned}$$

- *The condition* $\langle f, f \rangle_H = 0$ *means that* $f = 0_H$.
  Since, for $n \in \mathbb{N}$, fixed, but arbitrary, $\|f\|_H \geq \left\|f^{|T_n}\right\|_n$, $f^{|T_n} = 0$, so that, for $t \in T$, $f(t) = 0$.

Proof H is complete.

Suppose $\{f_n, \ n \in \mathbb{N}\}$ is a fixed, but arbitrary Cauchy sequence in H. Having fixed $\epsilon > 0$, suppose that, for $n, p \geq n(\epsilon)$,

$$\left\|f_n - f_p\right\|_H < \epsilon.$$

Since, for fixed, but arbitrary $q \in \mathbb{N}$, $f_n^{|T_q} \in H_q$ and

$$\left\|f_n^{|T_q} - f_p^{|T_q}\right\|_{H_q} \leq \left\|f_n - f_p\right\|_H,$$

the sequence

$$\left\{f_n^{|T_q}, \ n \in \mathbb{N}\right\}$$

is a Cauchy sequence in H_q. As such it has a limit $f^{(q)} \in H_q$. Now, for $m \in \mathbb{N}$, fixed, but arbitrary, since $f^{(q+m)} \in H_{q+m}$, by assumption $f^{(q+m)|T_q} \in H_q$. Consequently

$$\begin{aligned}\left\|f^{(q+m)|T_q} - f^{(q)}\right\|_q &= \lim_n \left\|f^{(q+m)|T_q} - f_n^{|T_q}\right\|_q \\ &= \lim_n \left\|f^{(q+m)|T_q} - \left\{f_n^{|T_{q+m}}\right\}^{|T_q}\right\|_q.\end{aligned}$$

But, by assumption,

$$\left\|f^{(q+m)|T_q} - \left\{f_n^{|T_{q+m}}\right\}^{|T_q}\right\|_q \leq \left\|f^{(q+m)} - f_n^{|T_{q+m}}\right\|_{q+m}.$$

It follows that $\left\|f^{(q+m)|T_q} - f^{(q)}\right\|_q = 0$.

The requirement that $f = f^{(n)}$ on T_n defines thus unambiguously a function on T.

It remains to check that f belongs to H. But, since, by definition, $f^{|T_n} = f^{(n)}, f^{|T_n} \in H_n$, $n \in \mathbb{N}$. One must thus only prove that $\lim_n \|f^{(n)}\|_n < \infty$, since the corresponding sequence of norms is increasing. But

$$\begin{aligned}\|f^{(n)}\|_n &\le \|f^{(n)} - f_p^{|T_n}\|_n + \|f_p^{|T_n}\|_n \\ &\le \|f^{(n)} - f_p^{|T_n}\|_n + \|f_p\|_H .\end{aligned}$$

Since a Cauchy sequence is bounded, there exists $\kappa \in \mathbb{R}_+$ such that, independently of p, $\|f_p\|_H \le \kappa$. Furthermore, given $\epsilon > 0$, there is a $p(\epsilon, n)$ such that

$$\left\|f^{(n)} - f_{p(\epsilon,n)}^{|T_n}\right\|_n < \epsilon.$$

Thus, independently of n,

$$\|f^{(n)}\|_n \le \epsilon + \kappa < \infty.$$

□

Lemma 1.6.15 *The assumptions are those of (Fact) 1.6.11. Let $p > n$ in $\mathbb{N}$ be fixed, but arbitrary. Let $\mathcal{H}_{p|n}$ be the restriction of $\mathcal{H}_p$ to $T_n \times T_n$, and $H_{p|n}$ be the associated RKHS with norm $\|\cdot\|_{p|n}$. Then*

1. *as sets, $H_{p|n} \subseteq H_n$;*
2. *for $h \in H_p$, fixed, but arbitrary,*

$$\left\|h^{|T_n}\right\|_n \le \left\|h^{|T_n}\right\|_{p|n}.$$

Proof $H_{p|n}$ is obtained as in (Proposition) 1.6.3. In particular, $H_{p|n}$ results from the restrictions to T_n of the maps in H_p. Letting $H_{p,n}$ denote the (closed) subspace generated linearly in H_p by the family $\{\mathcal{H}_p(\cdot, t),\ t \in T_n\}$, and $P_{H_{p,n}}$, the associated projection, one has, for $\{h, h_1, h_2\} \subseteq H_p$, fixed, but arbitrary, that

- $\langle h_1^{|T_n}, h_2^{|T_n}\rangle_{p|n} = \langle P_{H_{p,n}}[h_1], P_{H_{p,n}}[h_2]\rangle_p$;
- $P_{H_{p,n}}[h]^{|T_n} = h^{|T_n}$.

Then, since, by assumption, for $h \in H_p$, fixed, but arbitrary, $h^{|T_n} \in H_n$, and that the maps $h^{|T_n}$ form $H_{p|n}$, $H_{p|n} \subseteq H_n$. Furthermore, for $h \in H_p$, fixed, but arbitrary,

$$\left\|h^{|T_n}\right\|_n = \left\|P_{H_{p,n}}[h]^{|T_n}\right\|_n.$$

But, by assumption, since $P_{H_{p,n}}[h] \in H_p$,

$$\left\|P_{H_{p,n}}[h]^{|T_n}\right\|_n \le \left\|P_{H_{p,n}}[h]\right\|_p = \left\|h^{|T_n}\right\|_{p|n}.$$

Consequently, for $h \in H_p$, fixed, but arbitrary, $\left\|h^{|T_n}\right\|_n \le \left\|h^{|T_n}\right\|_{p|n}$. □

Remark 1.6.16 In the formalism of (Remarks) 1.6.12, 1.6.15 has the following meaning:

1. $\left\|V_{p,n}[h]\right\|_n = \left\|J_{p,n}L_{p,n}[h]\right\|_n \leq \left\|L_{p,n}[h]\right\|_{p|n}$;
2. $J_{p,n}$ is not only an inclusion, but also a contraction.

Lemma 1.6.17 *The assumptions are those of (Fact)* 1.6.11. *For* $p > n$ *in* $\mathbb{N}$ *and* $t \in T_n$, *fixed, but arbitrary,*

$$\mathcal{H}_p(t,t) \leq \mathcal{H}_n(t,t).$$

Proof Let $\mathcal{H}_{p|n}$ denote the restriction of $\mathcal{H}_p$ to $T_n \times T_n$. Lemma 1.6.15 yields [(Proposition) 3.1.23] that $\mathcal{H}_{p|n}$ is dominated by $\mathcal{H}_n$, that is, for fixed, but arbitrary $[n, \alpha, (t, T_n)]$,

$$\begin{aligned}\sum_{i=1}^{m}\sum_{j=1}^{m}\alpha_i\alpha_j\mathcal{H}_{p|n}\left(t_i,t_j\right) &= \left\|\sum_{i=1}^{m}\alpha_i\mathcal{H}_{p|n}(\cdot,t_i)\right\|_{p|n}^2 \\ &\leq \left\|\sum_{i=1}^{m}\alpha_i\mathcal{H}_{n}(\cdot,t_i)\right\|_{n}^2 \\ &= \sum_{i=1}^{m}\sum_{j=1}^{m}\alpha_i\alpha_j\mathcal{H}_{n}\left(t_i,t_j\right).\end{aligned}$$

The lemma's claim is a particular case of the latter inequality due to the fact that $\mathcal{H}_{p|n} = \mathcal{H}_p^{|T_n \times T_n}$. □

Lemma 1.6.18 *The assumptions are those of (Fact)* 1.6.11. *Set* $\mathcal{H}_{p|n} = \mathcal{H}_p^{|T_n \times T_n}$. *Then, for* $p > n > m$ *in* $\mathbb{N}$ *and* $t \in T_m$, *fixed, but arbitrary,*

$$\left\|\mathcal{H}_{n|m}(\cdot,t) - \mathcal{H}_{p|m}(\cdot,t)\right\|_m^2 \leq \mathcal{H}_n(t,t) - \mathcal{H}_p(t,t).$$

Proof Let

$$h = \mathcal{H}_n(\cdot,t) - \mathcal{H}_{p|n}(\cdot,t).$$

As $\mathcal{H}_n(\cdot,t) \in H_n$, and, by assumption, $\mathcal{H}_{p|n}(\cdot,t) \in H_n$, $h \in H_n$. Consequently, by assumption, as $t \in T_m$,

$$h^{|T_m} = \mathcal{H}_{n|m}(\cdot,t) - \mathcal{H}_{p|m}(\cdot,t) \in H_m,$$

and $\left\|h^{|T_m}\right\|_m \leq \|h\|_n$. Thus

$$\left\|\mathcal{H}_{n|m}(\cdot,t) - \mathcal{H}_{p|m}(\cdot,t)\right\|_m \leq \left\|\mathcal{H}_n(\cdot,t) - \mathcal{H}_{p|n}(\cdot,t)\right\|_n.$$

But, expanding the square of the right-hand side of the latter inequality, one has that

$$\left\|\mathcal{H}_n(\cdot,t)-\mathcal{H}_{p|n}(\cdot,t)\right\|_n^2=\mathcal{H}_n(t,t)-2\mathcal{H}_{p|n}(t,t)+\left\|\mathcal{H}_{p|n}(\cdot,t)\right\|_n^2.$$

and then, because of (Lemma) 1.6.15,

$$\left\|\mathcal{H}_{p|n}(\cdot,t)\right\|_n^2\leq\left\|\mathcal{H}_{p|n}(\cdot,t)\right\|_{p|n}^2=\mathcal{H}_{p|n}(t,t).$$

The claim of (Lemma) 1.6.18 is thus warranted as, when $t\in T_m$,

$$\mathcal{H}_{p|n}(t,t)=\mathcal{H}_p(t,t).$$

□

Proposition 1.6.19 ([188, p. 111]) *The assumptions are those of (Fact)* 1.6.11. *Then:*

1. *Given* $(t_1,t_2)\in T\times T$, *fixed, but arbitrary,* $\lim_n\mathcal{H}_n(t_1,t_2)$ *exists and is the value at* (t_1,t_2) *of a reproducing kernel* $\mathcal{H}$.
2. H *of (Lemma)* 1.6.14 *is the RKHS* $H(\mathcal{H},T)$.

Proof For $m\in\mathbb{N}$ and $t\in T_m$, fixed, but arbitrary, the positive sequence $\{\mathcal{H}_n(t,t),n\in\mathbb{N}\}$ is decreasing [(Lemma) 1.6.17], and thus convergent. Lemma 1.6.18 then says that

$$\left\{\mathcal{H}_{n|m}(\cdot,t),n\in\mathbb{N},n>m\right\}$$

is a Cauchy sequence in H_m. Let $h_{m,t}\in H_m$ be its limit. Then, since norm convergence implies pointwise convergence, for $(t,x)\in T_m\times T_m$, fixed, but arbitrary,

$$h_{m,t}(x)=\lim_n\mathcal{H}_{n|m}(x,t)=\lim_n\mathcal{H}_n(x,t).$$

Let $h_{m+p,t}\in H_{m+p}$ be defined analogously. Then, for $(t,x)\in T_{m+p}\times T_{m+p}$, fixed, but arbitrary,

$$h_{m+p,t}(x)=\lim_n\mathcal{H}_{n|m+p}(x,t)=\lim_n\mathcal{H}_n(x,t).$$

Consequently, for (t,x) in $T_m\times T_m$, fixed, but arbitrary, $h_{m+p,t}(x)=h_{m,t}(x)$, so that the assignment

$$\mathcal{H}(x,t)^{|T_m\times T_m}=h_{m,t}(x)$$

is unambiguous.

One must keep in mind for the sequel the following facts: for $m\in\mathbb{N}$, $t\in T_m$, fixed, but arbitrary,

- $\lim_n \mathcal{H}_{n|m}(\cdot, t) = h_{m,t} = \mathcal{H}(\cdot, t)^{|T_m} \in H_m$;
- $\lim_n \mathcal{H}_n(x, t) = \lim_n \mathcal{H}_{n|m}(x, t) = h_{m,t}(x) = \mathcal{H}(x, t)$.

$\mathcal{H}$ is symmetric as, for $(t, x) \in T_m \times T_m$, fixed, but arbitrary,

$$\mathcal{H}(x, t) = \lim_n \mathcal{H}_{n|m}(x, t) = \lim_n \mathcal{H}_{n|m}(t, x) = \mathcal{H}(t, x).$$

$\mathcal{H}$ is also positive definite for, given $\{t_1, \ldots, t_p\} \subseteq T$, there exists $m \in \mathbb{N}$ such that $\{t_1, \ldots, t_p\} \subseteq T_m$. And then

$$\begin{aligned}\sum_{i=1}^{p} \sum_{j=1}^{p} \alpha_i \alpha_j \mathcal{H}\left(t_i, t_j\right) &= \sum_{i=1}^{p} \sum_{j=1}^{p} \alpha_i \alpha_j \lim_n \mathcal{H}_{n|m}\left(t_i, t_j\right) \\ &= \lim_n \sum_{i=1}^{p} \sum_{j=1}^{p} \alpha_i \alpha_j \mathcal{H}_{n|m}\left(t_i, t_j\right) \\ &\geq 0.\end{aligned}$$

$\mathcal{H}$ is thus a reproducing kernel.

It remains to prove that $\mathcal{H}$ reproduces H. Two requirements, a) and b) below, must be fulfilled:

a) $\mathcal{H}(\cdot, t)$ must belong to H which means that

- $\mathcal{H}(\cdot, t)^{|T_n} \in H_n$ for arbitrary $n \in \mathbb{N}$,
- and that $\lim_n \left\| \mathcal{H}(\cdot, t)^{|T_n} \right\|_n < \infty$.

Let $t \in T_m$ and $n \in \mathbb{N}$ be fixed, but arbitrary. There is $p \in \mathbb{N}$ such that $T_m \subseteq T_p$ and $T_n \subseteq T_p$. By definition, for $(t, x) \in T_p \times T_p$, fixed, but arbitrary,

$$\mathcal{H}(x, t) = h_{p,t}(x), \ h_{p,t} \in H_p.$$

Thus

$$\mathcal{H}(\cdot, t)^{|T_n} = h_{p,t}^{|T_n}.$$

But, by assumption, $h_{p,t}^{|T_n} \in H_n$ and thus $\mathcal{H}(\cdot, t)^{|T_n} \in H_n$.

From (Lemma) 1.6.18 one gets, for $m \in \mathbb{N}$, $t \in T_m$, fixed, but arbitrary, $p > n > m$,

$$\left\| \mathcal{H}_{n|m}(\cdot, t) - \mathcal{H}_{p|m}(\cdot, t) \right\|_m^2 \leq \mathcal{H}_n(t, t) - \mathcal{H}_p(t, t).$$

Taking the limit on p, it follows that

$$\left\| \mathcal{H}_{n|m}(\cdot, t) - \mathcal{H}(\cdot, t)^{|T_m} \right\|_m^2 \leq \mathcal{H}_n(t, t) - \mathcal{H}(t, t). \qquad (\star)$$

Now

$$\left\|\mathcal{H}(\cdot,t)^{|T_m}\right\|_m \leq \left\|\left\{\mathcal{H}(\cdot,t)^{|T_m}\right\} - \mathcal{H}_{n|m}(\cdot,t)\right\|_m + \left\|\mathcal{H}_{n|m}(\cdot,t)\right\|_m. \qquad (\star\star)$$

Since also [(Lemma) 1.6.15]

$$\left\|\mathcal{H}_{n|m}(\cdot,t)\right\|_m \leq \left\|\mathcal{H}_{n|m}(\cdot,t)\right\|_{n|m} = \mathcal{H}_n^{1/2}(t,t), \qquad (\star\star\star)$$

combining both inequalities ($(\star)$ and $(\star\star\star)$ in $(\star\star)$),

$$\left\|\mathcal{H}(\cdot,t)^{|T_m}\right\|_m \leq \{\mathcal{H}_n(t,t) - \mathcal{H}(t,t)\}^{1/2} + \mathcal{H}_n^{1/2}(t,t),$$

so that, taking the limit on n,

$$\left\|\mathcal{H}(\cdot,t)^{|T_m}\right\|_m \leq \mathcal{H}^{1/2}(t,t) < \infty.$$

b) One must have, for $f \in H$ and $t \in T$, fixed, but arbitrary,

$$\langle f, \mathcal{H}(\cdot,t)\rangle_H = f(t).$$

But, for n large enough, as $f^{|T_n} \in H_n$,

$$\begin{aligned} f(t) &= f^{|T_n}(t) \\ &= \langle f^{|T_n}, \mathcal{H}_n(\cdot,t)\rangle_n \\ &= \langle f^{|T_n}, \mathcal{H}(\cdot,t)^{|T_n}\rangle_n \\ &+ \langle f^{|T_n}, \mathcal{H}_n(\cdot,t) - \mathcal{H}(\cdot,t)^{|T_n}\rangle_n, \end{aligned}$$

so that, taking the limit on n, one has the reproducing property characteristic of RKHS's.

□

Example 1.6.20 Let $H(\mathcal{H}, T)$ be an RKHS, and $\{T_n, n \in I \subseteq \mathbb{N}\}$ be an increasing sequence of subsets of T whose union is T_0. Let also $\mathcal{H}_n = \mathcal{H}^{|T_n \times T_n}$.

Given $p > n$ in $\mathbb{N}$ and h in $H(\mathcal{H}_p, T_p)$, fixed, but arbitrary, because of (Proposition) 1.6.3, one has that $h^{|T_n} \in H(\mathcal{H}_n, T_n)$, and that

$$\left\|h^{|T_n}\right\|_{H(\mathcal{H}_n,T_n)} \leq \|h\|_{H(\mathcal{H}_p,T_p)},$$

so that (Fact) 1.6.11 obtains. Furthermore, for $p > n$ in $\mathbb{N}$ and h in $H(\mathcal{H}, T)$, fixed, but arbitrary, H_n being the (closed) subspace of $H(\mathcal{H}, T)$ generated by the family

$\{\mathcal{H}(\cdot,t), t \in T_n\}$,

$$\begin{aligned}\left\|h^{|T_n}\right\|_{H(\mathcal{H}_n,T_n)} &= \|P_{H_n}[h]\|_{H(\mathcal{H},T)} \\ &\le \left\|P_{H_p}[h]\right\|_{H(\mathcal{H},T)} \\ &= \left\|h^{|T_p}\right\|_{H(\mathcal{H}_p,T_p)} \\ &\le \|h\|_{H(\mathcal{H},T)},\end{aligned}$$

so that (Proposition) 1.6.19 obtains, and the resulting RKHS is $H(\mathcal{H}_0, T_0)$, since one obviously has that $\mathcal{H}_0 = \lim_n \mathcal{H}_n$. This should not be surprising as an increasing sequence of projections has a limit which is a projection [266, p. 85]. Result (Proposition) 1.6.19 is thus an extension of the latter example.

Corollary 1.6.21 ([176]) *Let $H(\mathcal{H},T)$ be a separable RKHS, and T_0 be a Hamel subset of T* [(Definition) 1.1.36]. *There exists then* [(Proposition) 1.5.1] *$T_c \subseteq T_0$, a countable dense set. Its elements shall be denoted $\{t_{c,1}, t_{c,2}, t_{c,3}, \ldots\}$. $T_{c,n}$ shall be the set*

$$\{t_{c,1}, t_{c,2}, t_{c,3}, \ \ldots \ , t_{c,n}\},$$

$\mathcal{H}_{c,n}$ shall be the restriction of $\mathcal{H}$ to $T_{c,n} \times T_{c,n}$, and

$$U_{c,n} : H(\mathcal{H},T) \longrightarrow H(\mathcal{H}_{c,n}, T_{c,n}),$$

the partial isometry of (Proposition) 1.6.3. *$\mathcal{H}_0$ and U_0 are defined similarly. Then:*

1. *$(T_0, d_{\mathcal{H}_0})$ is a separable metric space.*
2. *T_c is a determining set* [(Definition) 1.5.2] *for $H(\mathcal{H},T)$.*
3. *For any $h \in H(\mathcal{H},T)$, the sequence*

$$\left\{\|U_{c,n}[h]\|_{H(\mathcal{H}_{c,n},T_{c,n})}, \ n \in \mathbb{N}\right\}$$

is monotone increasing, and

$$\lim_n \|U_{c,n}[h]\|_{H(\mathcal{H}_{c,n},T_{c,n})} = \|h\|_{H(\mathcal{H},T)}.$$

4. *Let $\mathcal{H}_c$ be the restriction of $\mathcal{H}$ to $T_c \times T_c$, and suppose that $f : T \longrightarrow \mathbb{R}$ is a function such that*
   (a) $\left\|f^{|T_{c,n}}\right\|_{H(\mathcal{H}_{c,n},T_{c,n})} \le \left\|f^{|T_{c,n+p}}\right\|_{H(\mathcal{H}_{c,n+p},T_{c,n+p})}$,
   (b) $\lim_n \left\{\left\|f^{|T_{c,n}}\right\|_{H(\mathcal{H}_{c,n},T_{c,n})}\right\} < \infty$.

Then $f^{|T_c} \in H(\mathcal{H}_c, T_c)$, *and there exists a unique* $h \in H(\mathcal{H}, T)$ *such that*

$$h^{|T_c} = f^{|T_c}, \text{ and } \|h\|_{H(\mathcal{H},T)} = \left\|f^{|T_c}\right\|_{H(\mathcal{H}_c,T_c)} = \lim_n \left\|f^{|T_{c,n}}\right\|_{H\left(\mathcal{H}_{c,n},T_{c,n}\right)}.$$

Proof (1) $d_{\mathcal{H}_0}$ is a metric as T_0 is a Hamel set. $H(\mathcal{H}_0, T_0)$ is separable as the unitary image of a subspace of a separable space [(Proposition) 1.6.3]. But then item 1 is true because of (Corollary) 1.5.11.

Proof (2) On the basis of item 1, let T_c be a (countable) dense subset of the separable $(T_0, d_{\mathcal{H}_0})$, and let $h \in H(\mathcal{H}, T)$ be fixed, but arbitrary. Suppose that $h(t) = 0, \ t \in T_c$.

$h^{|T_0}$ belongs to $H(\mathcal{H}_0, T_0)$ [(Proposition) 1.6.3] and is continuous for $d_{\mathcal{H}_0}$ [(Proposition) 2.6.9]. Since it is zero on a dense subset, it is zero on T_0. Since T_0 is a determining set for $H(\mathcal{H}, T)$ [(Remark) 1.5.3], h must be the zero function. So item 2 obtains.

Proof (3) Let $h \in H(\mathcal{H}, T)$ be fixed, but arbitrary. Because of (Proposition) 1.6.10, it is sufficient to obtain as limit the norm in $H(\mathcal{H}_c, T_c)$. But that is true as seen in (Example) 1.6.20.

Proof (4) Because of (Proposition) 1.6.19, the set of functions $f : T_c \longrightarrow \mathbb{R}$ for which

- $f^{|T_{c,n}} \in H(\mathcal{H}_{c,n}, T_{c,n})$,
- $\left\|f^{|T_{c,n}}\right\|_{H\left(\mathcal{H}_{c,n},T_{c,n}\right)} \leq \left\|f^{|T_{c,n+p}}\right\|_{H\left(\mathcal{H}_{c,n+p},T_{c,n+p}\right)}$,
- $\lim_n \left\{\left\|f^{|T_{c,n}}\right\|_{H\left(\mathcal{H}_{c,n},T_{c,n}\right)}\right\} < \infty$,

is an RKHS K whose kernel $\mathcal{K}$ is such that $\mathcal{K}^{|T_{c,n}\times T_{c,n}} = \mathcal{H}_{c,n}$. K is thus $H(\mathcal{H}_c, T_c)$. The function $f^{|T_c}$ belongs then to $H(\mathcal{H}_c, T_c)$, and, as such, has a unique extension $g \in H(\mathcal{H}_0, T_0)$ to T_0, with the same norm [(Proposition) 1.6.10]. But then it has a unique extension $h \in H(\mathcal{H}, T)$ to T with the same norm [(Proposition) 1.6.10]. □

The results to follow show that projections in RKHS's have particularly simple expressions when they arise from the restriction of functions to finite sets. Suppose thus that T_0 is finite, say $T_0 = \{t_1, \ldots, t_n\}$. The elements of $H(\mathcal{H}_0, T_0)$ have the form

$$t \mapsto \sum_{i=1}^{n} \alpha_i \mathcal{H}(t, t_i), \ \{\alpha_1, \ldots, \alpha_n\}, \ \{t, t_1, \ldots, t_n\} \subseteq T_0.$$

Let L be the map whose range is $H(\mathcal{H}_0, T_0)$ [(Proposition) 1.6.3]. Since then $L[h] = h^{|T_0}$, and

$$L[h](t_i) = \sum_{k=1}^{n} \alpha_k^h \mathcal{H}(t_i, t_k), \quad h^{|T_0}(t_i) = h(t_i),$$

one has the following relation (which may be construed as a linear system with the α's as unknowns):

$$\begin{bmatrix} h(t_1) \\ \vdots \\ h(t_n) \end{bmatrix} = \begin{bmatrix} \mathcal{H}(t_1, t_1) \cdots \mathcal{H}(t_1, t_n) \\ \vdots \qquad\qquad \vdots \\ \mathcal{H}(t_n, t_1) \cdots \mathcal{H}(t_n, t_n) \end{bmatrix} \begin{bmatrix} \alpha_1^h \\ \vdots \\ \alpha_n^h \end{bmatrix}.$$

That relation may be written in the following form: $\underline{h} = \Sigma_{\mathcal{H},T_0}\left[\underline{\alpha}_h\right]$. Consequently, in case $\Sigma_{\mathcal{H},T_0}$ is invertible,

$$\underline{\alpha}_h = \Sigma_{\mathcal{H},T_0}^{-1}\left[\underline{h}\right], \; t \in T_0.$$

Furthermore

$$\begin{aligned} &\langle L[h_1], L[h_2]\rangle_{H(\mathcal{H}_0,T_0)} = \\ &= \sum_{i=1}^{n}\sum_{j=1}^{n} \alpha_i^{h_1}\alpha_j^{h_2}\mathcal{H}\left(t_i, t_j\right) = \langle \Sigma_{\mathcal{H},T_0}\left[\underline{\alpha}_{h_1}\right], \underline{\alpha}_{h_2}\rangle_{\mathbb{R}^n} \\ &= \langle \Sigma_{\mathcal{H},T_0}^{-1}\left[\underline{h}_1\right], \underline{h}_2\rangle_{\mathbb{R}^n}, \end{aligned}$$

so that, using the definition of the inner product in (Proposition) 1.1.15,

$$\langle P_0[h_1], P_0[h_2]\rangle_{H(\mathcal{H},T)} = \langle \Sigma_{\mathcal{H},T_0}^{-1}\left[\underline{h}_1\right], \underline{h}_2\rangle_{\mathbb{R}^n}.$$

Let finally

$$P_0[h](t) = \sum_{i=1}^{n} a_i^h \mathcal{H}(t, t_i), \; t \in T, \; \{t_1, \ldots t_n\} = T_0.$$

Since the restriction of $P_0[h]$ to T_0 equals the restriction of h to T_0 [(Proposition) 1.6.3], $\underline{a}_h = \Sigma_{\mathcal{H},T_0}^{-1}\underline{h}$. Consequently, with $\underline{e}_i$, the i-th vector of the standard basis of $\mathbb{R}^n$, and $\underline{\sigma}(t)$, the vector with components $\mathcal{H}(t, t_i)$, $1 \le i \le n$,

$$P_0[h](t) = \sum_{i=1}^{n}\langle \Sigma_{\mathcal{H},T_0}^{-1}\underline{h}, \underline{e}_i\rangle_{\mathbb{R}^n}\mathcal{H}(t, t_i) = \langle \Sigma_{\mathcal{H},T_0}^{-1}\underline{h}, \underline{\sigma}(t)\rangle_{\mathbb{R}^n}.$$

Those latter remarks lead to the following statement.

Proposition 1.6.22 ([202]) *Let $H(\mathcal{H}, T)$ be an RKHS, and $\{t_1, \ldots, t_n\} \subseteq T$ be a fixed, but arbitrary finite subset of T, denoted T_n. Let $\Sigma_{\mathcal{H},T_n}$ be the matrix with entries $\mathcal{H}\left(t_i, t_j\right)$, $\left\{t_i, t_j\right\} \subseteq T_n$, $1 \le i,j \le n$, and H_n be the subspace of $H(\mathcal{H}, T)$ generated by $\{\mathcal{H}(\cdot, t_i), \; t_i \in T_n, \; 1 \le i \le n\}$. Assume that $\Sigma_{\mathcal{H},T_n}$ is invertible.*

Given $\{h_1, h_2\} \subseteq H(\mathcal{H}, T)$ fixed, but arbitrary, let, for $j \in \{1, 2\}$,

$$h_i^{(j)} = h_j(t_i), \quad \text{and } \underline{h}_j \text{ have components } h_i^{(j)}, \ 1 \leq i \leq n.$$

Let also $\sigma_i(x) = \mathcal{H}(x, t_i)$, and $\underline{\sigma}(x)$ be the vector whose components, for $1 \leq i \leq n$, have value $\sigma_i(x)$. Then:

1. $P_{H_n}[h_1](t) = \langle \Sigma_{\mathcal{H},T_n}^{-1}[\underline{h}_1], \underline{\sigma}(t)\rangle_{\mathbb{R}^n}, \ t \in T$;
2. $\langle P_{H_n}[h_1], P_{H_n}[h_2]\rangle_{H(\mathcal{H},T)} = \langle \Sigma_{\mathcal{H},T_n}^{-1}[\underline{h}_1], \underline{h}_2\rangle_{\mathbb{R}^n}$.

Corollary 1.6.23 *Let X be a second order process indexed by T, defined on a probability space $(\Omega, \mathcal{A}, P)$, with a mean that is equal to zero, and a covariance denoted C_X. Let $L_P[X]$ denote the subspace of $L_2(\Omega, \mathcal{A}, P)$ generated linearly by the family $\{X_t, \ t \in T\}$, where X_t is the equivalence class of $X(\cdot, t)$ in $L_2(\Omega, \mathcal{A}, P)$. Since* [(Example) 1.1.26] *the equivalence class X_t of $X(\cdot, t)$ is in unitary correspondence with $C_X(\cdot, t)$, the projection of $Y \in L_P[X]$ onto the subspace generated linearly by $\{X_{t_1}, \ldots, X_{t_n}\}$ is expressed as follows:*

$$\langle \Sigma_{C_X,T_n}^{-1}\underline{h}, \underline{X}\rangle_{\mathbb{R}^n},$$

where

(i) Σ_{C_X,T_n} *is the matrix with entries* $C_X(t_i, t_j)$, $\{i, j\} \subseteq [1:n]$,
(ii) $h(t) = E_P\left[\dot{Y}X(\cdot, t)\right]$,
(iii) $\underline{X}$ *has components* $X_{t_1}, \ldots, X_{t_n}$.

Proof Let Σ_{C_X,T_n}^{-1} have entries $\gamma_{i,j}$, $1 \leq i, j \leq n$, and H_n be generated by $\{C_X(\cdot, t_1), \ldots, C_X(\cdot, t_n)\}$. As

$$P_{H_n}[h] = \sum_{i=1}^{n} \left\{ \sum_{j=1}^{n} \gamma_{i,j} h(t_j) \right\} C_X(\cdot, t_i),$$

the unitary correspondence between $C_X(\cdot, t)$ and X_t [(Example) 1.1.26] proves the result. □

1.7 Operators in Reproducing Kernel Hilbert Spaces

Operators in RKHS's are determined by kernels, and some of the details are spelled out in this section. A few facts about domination of covariances, denoted $\ll$, are required. Those are to be found in Sect. 3.1.

1.7.1 Bounded Linear Operators

One finds below the main properties of operators between reproducing kernel Hilbert spaces that are linear and bounded.

Proposition 1.7.1 *The following properties obtain:*

1. *Let $H(\mathcal{H}_1, T_1)$ and $H(\mathcal{H}_2, T_2)$ be RKHS's, and*

$$B: H(\mathcal{H}_1, T_1) \longrightarrow H(\mathcal{H}_2, T_2)$$

be a bounded linear operator. For fixed, but arbitrary $t^{(2)} \in T_2$, let

$$H(\mathcal{H}_1, T_1) \ni h^{(1)}[t^{(2)}] = B^\star\left[\mathcal{H}_2\left(\cdot, t^{(2)}\right)\right].$$

Then, for fixed, but arbitrary $(h^{(1)}, t^{(2)}) \in H(\mathcal{H}_1, T_1) \times T_2$,

$$\mathcal{E}_{t^{(2)}}\left[B\left[h^{(1)}\right]\right] = \langle h^{(1)}, h^{(1)}[t^{(2)}]\rangle_{H(\mathcal{H}_1, T_1)},$$

and, for $\left(t_1^{(2)}, t_2^{(2)}\right) \in T_2 \times T_2$, fixed, but arbitrary,

$$\mathcal{H}_B\left(t_1^{(2)}, t_2^{(2)}\right) \stackrel{def}{=} \langle h^{(1)}[t_1^{(2)}], h^{(1)}[t_2^{(2)}]\rangle_{H(\mathcal{H}_1, T_1)} \ll \|B\|^2 \mathcal{H}_2\left(t_1^{(2)}, t_2^{(2)}\right).$$

2. *Let, for fixed, but arbitrary $\left\{t^{(2)}, t_1^{(2)}, t_2^{(2)}\right\} \subseteq T_2$,*

$$h^{(1)}[t^{(2)}] \in H(\mathcal{H}_1, T_1),$$

and

$$\mathcal{H}_B\left(t_1^{(2)}, t_2^{(2)}\right) \stackrel{def}{=} \langle h^{(1)}[t_1^{(2)}], h^{(1)}[t_2^{(2)}]\rangle_{H(\mathcal{H}_1, T_1)}.$$

Suppose there exists $\kappa > 0$ such that

$$\mathcal{H}_B\left(t_1^{(2)}, t_2^{(2)}\right) \ll \kappa \mathcal{H}_2\left(t_1^{(2)}, t_2^{(2)}\right).$$

There exists then a bounded, linear B as in item 1 such that

$$B^\star\left[\mathcal{H}_2\left(\cdot, t^{(2)}\right)\right] = h^{(1)}[t^{(2)}].$$

Proof (1) The first assertion follows from the following equalities:

$$\begin{aligned} \langle h^{(1)}, h^{(1)}[t^{(2)}]\rangle_{H(\mathcal{H}_1, T_1)} &= \langle h^{(1)}, B^\star\left[\mathcal{H}\left(\cdot, t^{(2)}\right)\right]\rangle_{H(\mathcal{H}_1, T_1)} \\ &= \langle B\left[h^{(1)}\right], \mathcal{H}\left(\cdot, t^{(2)}\right)\rangle_{H(\mathcal{H}_2, T_2)} \\ &= B\left[h^{(1)}\right]\left(t^{(2)}\right), \end{aligned}$$

and the second from these:

$$\begin{aligned}
\sum_{i=1}^{n}\sum_{j=1}^{n}\alpha_i\alpha_j\mathcal{H}_B\left(t_i^{(2)},t_j^{(2)}\right) &= \left\|\sum_{i=1}^{n}\alpha_i h^{(1)}[t_i^{(2)}]\right\|_{H(\mathcal{H}_1,T_1)}^2 \\
&= \left\|B^\star\left[\sum_{i=1}^{n}\alpha_i\mathcal{H}_2\left(\cdot,t_i^{(2)}\right)\right]\right\|_{H(\mathcal{H}_1,T_1)}^2 \\
&\le \|B^\star\|^2\left\|\sum_{i=1}^{n}\alpha_i\mathcal{H}_2\left(\cdot,t_i^{(2)}\right)\right\|_{H(\mathcal{H}_2,T_2)}^2 \\
&= \|B\|^2\sum_{i=1}^{n}\sum_{j=1}^{n}\alpha_i\alpha_j\mathcal{H}_2\left(t_i^{(2)},t_j^{(2)}\right).
\end{aligned}$$

Proof (2) Let $\mathcal{H}_2^\kappa = \kappa\mathcal{H}_2$. Consider the following maps.

- $J : H\left(\mathcal{H}_2^\kappa, T_2\right) \longrightarrow H\left(\mathcal{H}_B, T_2\right)$, defined using the following relation: for $t^{(2)} \in T_2$, fixed, but arbitrary,

  $$J\left[\mathcal{H}_2^\kappa\left(\cdot,t^{(2)}\right)\right] = \mathcal{H}_B\left(\cdot,t^{(2)}\right).$$

  It is a contraction, and its transpose is an inclusion [(Proposition) 3.1.5].
- $L_F : H\left(\mathcal{H}_1, T_1\right) \longrightarrow H\left(\mathcal{H}_B, T_2\right)$ is the map corresponding to

  $$F : T_2 \longrightarrow H\left(\mathcal{H}_1, T_1\right),$$

  defined, for $t^{(2)} \in T_2$, fixed, but arbitrary, using relation (Proposition) 1.1.15,

  $$F\left(t^{(2)}\right) = h^{(1)}[t^{(2)}].$$

  In particular $L_F^\star\left[\mathcal{H}_B\left(\cdot,t^{(2)}\right)\right] = F\left(t^{(2)}\right)$.
- $\tilde{J} : H\left(\mathcal{H}_2^\kappa, T_2\right) \longrightarrow H\left(\mathcal{H}_2, T_2\right)$ is the identity [(Example) 1.3.12].
- $B = \tilde{J}J^\star L_F$.

Then, for $t^{(2)} \in T_2$, fixed, but arbitrary,

$$B^\star\left[\mathcal{H}_2\left(\cdot,t^{(2)}\right)\right] = L_F^\star J\tilde{J}^\star\left[\mathcal{H}_2\left(\cdot,t^{(2)}\right)\right].$$

But, since [(Example) 1.3.12]

$$\begin{aligned}
\langle h_1^{(2)},\tilde{J}^\star\left[h_2^{(2)}\right]\rangle_{H\left(\mathcal{H}_2^\kappa,T_2\right)} &= \langle\tilde{J}\left[h_1^{(2)}\right],h_2^{(2)}\rangle_{H(\mathcal{H}_2,T_2)} \\
&= \langle h_1^{(2)},h_2^{(2)}\rangle_{H(\mathcal{H}_2,T_2)} \\
&= \kappa\langle h_1^{(2)},h_2^{(2)}\rangle_{H\left(\mathcal{H}_2^\kappa,T_2\right)} \\
&= \langle h_1^{(2)},\kappa h_2^{(2)}\rangle_{H\left(\mathcal{H}_2^\kappa,T_2\right)},
\end{aligned}$$

$\tilde{J}^\star$ is the identity times κ. Thus

$$\begin{aligned} L_F^\star J \tilde{J}^\star \left[\mathcal{H}_2\left(\cdot, t^{(2)}\right)\right] &= L_F^\star J \left[\kappa \mathcal{H}_2\left(\cdot, t^{(2)}\right)\right] \\ &= L_F^\star J \left[\mathcal{H}_2^\kappa\left(\cdot, t^{(2)}\right)\right] \\ &= L_F^\star \left[\mathcal{H}_B\left(\cdot, t^{(2)}\right)\right] \\ &= F\left(t^{(2)}\right) \\ &= h^{(1)}[t^{(2)}]. \end{aligned}$$

□

Remark 1.7.2 (Kernel Associated with an Operator in an RKHS) Since, for fixed, but arbitrary $t^{(2)} \in T_2$, the element $h^{(1)}[t^{(2)}]$ is a function defined on T_1, one customarily regards the set $\{h^{(1)}[t^{(2)}],\ t^{(2)} \in T_2\}$ as a kernel

$$\mathcal{B} : T_1 \times T_2 \longrightarrow \mathbb{R}$$

with, for $(t^{(1)}, t^{(2)}) \in T_1 \times T_2$, fixed, but arbitrary,

$$\mathcal{B}\left(t^{(1)}, t^{(2)}\right) = h^{(1)}[t^{(2)}]\left(t^{(1)}\right).$$

Then

$$\begin{aligned} \mathcal{B}\left(t^{(1)}, t^{(2)}\right) &= \langle B^\star\left[\mathcal{H}_2\left(\cdot, t^{(2)}\right)\right], \mathcal{H}_1\left(\cdot, t^{(1)}\right)\rangle_{H(\mathcal{H}_1, T_1)} \\ &= \langle B\left[\mathcal{H}_1\left(\cdot, t^{(1)}\right)\right], \mathcal{H}_2\left(\cdot, t^{(2)}\right)\rangle_{H(\mathcal{H}_2, T_2)} \\ &= B[\mathcal{H}(\cdot, t^{(1)})](t^{(2)}) \\ &= B^\star[\mathcal{H}(\cdot, t^{(2)})](t^{(1)}). \end{aligned}$$

Also

$$\begin{aligned} \left\langle \mathcal{B}\left(\cdot, t^{(2)}\right), h^{(1)}\right\rangle_{H(\mathcal{H}_1, T)} &= \left\langle B^\star\left[\mathcal{H}_2\left(\cdot, t^{(2)}\right)\right], h^{(1)}\right\rangle_{H(\mathcal{H}_1, T)} \\ &= \left\langle \mathcal{H}_2\left(\cdot, t^{(2)}\right), B\left[h^{(1)}\right]\right\rangle_{H(\mathcal{H}_2, T)} \\ &= B\left[h^{(1)}\right]\left(t^{(2)}\right). \end{aligned}$$

Remark 1.7.3 The kernel of the identity operator of $H(\mathcal{H}, T)$ is $\mathcal{H}$.

Example 1.7.4 In (Example) 1.2.5, let $L_X : L_P[X] \longrightarrow H(C_X, S)$ be the unitary map of (Example) 1.1.26, and L_Y be defined analogously. Then, for $(s, t) \in S \times T$, fixed, but arbitrary,

$$\begin{aligned} C_{X,Y}(s,t) &= \langle X_s, B^\star[Y_t]\rangle_{L_P[X]} \\ &= \langle L_X^\star[C_X(\cdot, s)], B^\star\left[L_Y^\star[C_Y(\cdot, t)]\right]\rangle_{L_P[X]} \end{aligned}$$

$$= \langle C_X(\cdot, s), \{L_X B^\star L_Y^\star\}[C_Y(\cdot, t)]\rangle_{H(C_X,S)}$$
$$= \{L_Y B L_X^\star\}^\star [C_Y(\cdot, t)](s),$$

and thus $C_{X,Y}$ is the kernel of the operator $L_Y B L_X^\star$.

Proposition 1.7.5 *Let $H(\mathcal{H}_1, T_1)$ and $H(\mathcal{H}_2, T_2)$ be RKHS's, and*

$$B : H(\mathcal{H}_1, T_1) \longrightarrow H(\mathcal{H}_2, T_2)$$

be a bounded linear operator with kernel $\mathcal{B}$. Let $\mathcal{B}^\star$ denote the kernel of $B^\star$. Then $\mathcal{B}^\star : T_2 \times T_1 \longrightarrow \mathbb{R}$, and, for $(x_2, t_1) \in T_2 \times T_1$, fixed, but arbitrary,

$$\mathcal{B}^\star(x_2, t_1) = \mathcal{B}(t_1, x_2).$$

Proof One has that $B^{\star\star} = B$. Referring to (Remark) 1.7.2, the kernel of $B^\star$ is given using the following relation:

$$\mathcal{B}^\star(x_2, t_1) = (B^\star)^\star [\mathcal{H}_1(\cdot, t_1)](x_2) = B[\mathcal{H}_1(\cdot, t_1)](x_2).$$

But

$$B[\mathcal{H}_1(\cdot, t_1)](x_2) = \langle B[\mathcal{H}_1(\cdot, t_1)], \mathcal{H}_2(\cdot, x_2)\rangle_{H(\mathcal{H}_2,T_2)}$$
$$= \langle \mathcal{H}_1(\cdot, t_1), B^\star[\mathcal{H}_2(\cdot, x_2)]\rangle_{H(\mathcal{H}_1,T_1)}$$
$$= \mathcal{B}(t_1, x_2).$$

□

Corollary 1.7.6 *Let $H(\mathcal{H}, T)$ be an RKHS, and $B : H(\mathcal{H}, T) \longrightarrow H(\mathcal{H}, T)$ be a bounded linear operator with associated kernel $\mathcal{B}$. B is self-adjoint if, and only if, $\mathcal{B}$ is symmetric.*

Proof Let $(x, t) \in T \times T$ be fixed, but arbitrary. If B is self-adjoint, from (Propositions) 1.7.1 and 1.7.5,

$$\mathcal{B}^\star(x, t) = B[\mathcal{H}(\cdot, t)](x) = B^\star[\mathcal{H}(\cdot, t)](x) = \mathcal{B}(x, t).$$

But, according to (Proposition) 1.7.5, $\mathcal{B}^\star(x, t) = \mathcal{B}(t, x)$. Consequently $\mathcal{B}(t, x) = \mathcal{B}(x, t)$.

Conversely, let $(x, t) \in T \times T$ be fixed, but arbitrary, and suppose that $\mathcal{B}(t, x) = \mathcal{B}(x, t)$. Then $\mathcal{B}^\star(x, t) = \mathcal{B}(x, t)$, or $B[\mathcal{H}(\cdot, t)](x) = B^\star[\mathcal{H}(\cdot, t)](x)$. It follows that, for $t \in T$, fixed, but arbitrary, $B^\star[\mathcal{H}(\cdot, t)] = B[\mathcal{H}(\cdot, t)]$, and thus, since $\{\mathcal{H}(\cdot, t), t \in T\} \subseteq H(\mathcal{H}, T)$ is total, that $B^\star = B$. □

Proposition 1.7.7 *Let $H(\mathcal{H}_1, T_1)$ and $H(\mathcal{H}_2, T_2)$ be RKHS's, and*

$$B_1, B_2 : H(\mathcal{H}_1, T_1) \longrightarrow H(\mathcal{H}_2, T_2)$$

be bounded linear operators with respective kernels $\mathcal{B}_1$ and $\mathcal{B}_2$. Form

$$B = \alpha_1 B_1 + \alpha_2 B_2,$$

and let $\mathcal{B}$ be the kernel of B. Then

$$\mathcal{B} = \alpha_1 \mathcal{B}_1 + \alpha_2 \mathcal{B}_2.$$

Proof $B^\star = \alpha_1 B_1^\star + \alpha_2 B_2^\star$ so that

$$\begin{aligned}\mathcal{B}(x_1, t_2) &= B^\star [\mathcal{H}_2(\cdot, t_2)](x_1) \\ &= \alpha_1 B_1^\star [\mathcal{H}_2(\cdot, t_2)](x_1) + \alpha_2 B_2^\star [\mathcal{H}_2(\cdot, t_2)](x_1) \\ &= \alpha_1 \mathcal{B}_1(x_1, t_2) + \alpha_2 \mathcal{B}_2(x_1, t_2).\end{aligned}$$

□

Proposition 1.7.8 *Let $H(\mathcal{H}_1, T_1)$, $H(\mathcal{H}_2, T_2)$, and $H(\mathcal{H}_3, T_3)$ be RKHS's. Let*

$$B_1 : H(\mathcal{H}_1, T_1) \longrightarrow H(\mathcal{H}_2, T_2),$$

$$B_2 : H(\mathcal{H}_2, T_2) \longrightarrow H(\mathcal{H}_3, T_3)$$

be bounded linear operators with respective kernels $\mathcal{B}_1$ and $\mathcal{B}_2$. Let $B = B_2 B_1$ have kernel $\mathcal{B}$. Then

$$\mathcal{B}(x_1, t_3) = \langle \mathcal{B}_1(x_1, \cdot), \mathcal{B}_2(\cdot, t_3)\rangle_{H(\mathcal{H}_2, T_2)}, \quad (x, t) \in T_1 \times T_3.$$

Proof $B^\star = B_1^\star B_2^\star$ so that

$$\begin{aligned}\mathcal{B}(x_1, t_3) &= B^\star [\mathcal{H}_3(\cdot, t_3)](x_1) \\ &= B_1^\star B_2^\star [\mathcal{H}_3(\cdot, t_3)](x_1) \\ &= \langle B_1^\star B_2^\star [\mathcal{H}_3(\cdot, t_3)], \mathcal{H}_1(\cdot, x_1)\rangle_{H(\mathcal{H}_1, T_1)} \\ &= \langle B_2^\star [\mathcal{H}_3(\cdot, t_3)], B_1 [\mathcal{H}_1(\cdot, x_1)]\rangle_{H(\mathcal{H}_2, T_2)}.\end{aligned}$$

But $B_2^\star [\mathcal{H}_3(\cdot, t_3)] = \mathcal{B}_2(\cdot, t_3)$, and

$$B_1 [\mathcal{H}_1(\cdot, x_1)] = \left(B_1^\star\right)^\star [\mathcal{H}_1(\cdot, x_1)] = \mathcal{B}_1^\star(\cdot, x_1) = \mathcal{B}_1(x_1, \cdot).$$

□

Proposition 1.7.9 *Let $H(\mathcal{H}, T)$ be an RKHS, and*

$$B : H(\mathcal{H}, T) \longrightarrow H(\mathcal{H}, T)$$

be a bounded linear operator. Then B is positive, that is, for $h \in H(\mathcal{H}, T)$, fixed, but arbitrary,

$$\langle B[h], h\rangle_{H(\mathcal{H},T)} \geq 0,$$

if, and only, if the associated kernel $\mathcal{B}$ is positive definite.

Proof One has for $(t_1, t_2) \in T \times T$, fixed, but arbitrary,

$$\begin{aligned}\langle B[\mathcal{H}(\cdot, t_1)], \mathcal{H}(\cdot, t_2)\rangle_{H(\mathcal{H},T)} &= \langle \mathcal{H}(\cdot, t_1), B^\star[\mathcal{H}(\cdot, t_2)]\rangle_{H(\mathcal{H},T)}\\ &= \langle \mathcal{H}(\cdot, t_1), \mathcal{B}(\cdot, t_2)\rangle_{H(\mathcal{H},T)}\\ &= \mathcal{B}(t_1, t_2).\end{aligned}$$

Consequently, for fixed, but arbitrary $[n, \alpha, (t, T)]$ and $h = \sum_{i=1}^n \alpha_i \mathcal{H}(\cdot, t_i)$, one has that

$$\langle B[h], h\rangle_{H(\mathcal{H},T)} = \sum_{i=1}^n \sum_{j=1}^n \alpha_i \alpha_j \mathcal{B}\left(t_i, t_j\right).$$

Thus B is positive on $V[\mathcal{H}]$ if, and only if, $\mathcal{B}$ is positive definite, and B is positive if, and only if, it is positive on $V[\mathcal{H}]$. □

Proposition 1.7.10 *Let B and $\{B_n, n \in \mathbb{N}\}$ be bounded, linear operators from $H(\mathcal{H}_1, T_1)$ to $H(\mathcal{H}_2, T_2)$, and $\mathcal{B}$ and $\{\mathcal{B}_n, n \in \mathbb{N}\}$ be their respective kernels. Let, for $c_1 \in \mathbb{R}_+$, fixed, but arbitrary,*

$$T_{c_1}^{(1)} = \left\{t^{(1)} \in T_1 : \mathcal{H}_1\left(t^{(1)}, t^{(1)}\right) \leq c_1\right\}.$$

$T_{c_2}^{(2)}$ is defined similarly. Then:

1. when, for $h^{(1)} \in H(\mathcal{H}_1, T_1)$, fixed, but arbitrary,

$$\left\{B_n\left[h^{(1)}\right], n \in \mathbb{N}\right\} \text{ converges weakly to } B\left[h^{(1)}\right],$$

one has that, for $\{t^{(1)}, t^{(2)}\} \subseteq T_1 \times T_2$, fixed, but arbitrary,

$$\lim_n \mathcal{B}_n\left(t^{(1)}, t^{(2)}\right) = \mathcal{B}\left(t^{(1)}, t^{(2)}\right);$$

2. *when* $\lim_n B_n = B$ *(in operator norm), one has, for fixed, but arbitrary* $\{c_1, c_2\} \subseteq \mathbb{R}_+$, *that*

$$\{\mathcal{B}_n(\cdot,\cdot), n \in \mathbb{N}\} \text{ converges uniformly, on } T^{(1)}_{c_1} \times T^{(2)}_{c_2}, \text{ to } \mathcal{B}(\cdot,\cdot).$$

Proof The kernel $\mathcal{B}$ of B is obtained, for $(t^{(1)}, t^{(2)}) \in T_1 \times T_2$, fixed, but arbitrary, as

$$\begin{aligned}\mathcal{B}\left(t^{(1)}, t^{(2)}\right) &= B^\star\left[\mathcal{H}_2\left(\cdot, t^{(2)}\right)\right]\left(t^{(1)}\right)\\ &= \langle B^\star\left[\mathcal{H}_2\left(\cdot, t^{(2)}\right)\right], \mathcal{H}_1\left(\cdot, t^{(1)}\right)\rangle_{H(\mathcal{H}_1, T_1)}\\ &= \langle \mathcal{H}_2\left(\cdot, t^{(2)}\right), B\left[\mathcal{H}_1\left(\cdot, t^{(1)}\right)\right]\rangle_{H(\mathcal{H}_2, T_2)}.\end{aligned}$$

Similarly,

$$\mathcal{B}_n\left(t^{(1)}, t^{(2)}\right) = \langle \mathcal{H}_2\left(\cdot, t^{(2)}\right), B_n\left[\mathcal{H}_1\left(\cdot, t^{(1)}\right)\right]\rangle_{H(\mathcal{H}_2, T_2)}.$$

Item 1 is now obvious. Furthermore

$$\begin{aligned}&\left|\mathcal{B}\left(t^{(1)}, t^{(2)}\right) - \mathcal{B}_n\left(t^{(1)}, t^{(2)}\right)\right| =\\ &= \left|\langle \mathcal{H}_2\left(\cdot, t^{(2)}\right), [B - B_n]\left[\mathcal{H}_1\left(\cdot, t^{(1)}\right)\right]\rangle_{H(\mathcal{H}_2, T_2)}\right|\\ &\leq \left\|\mathcal{H}_2\left(\cdot, t^{(2)}\right)\right\|_{H(\mathcal{H}_2, T_2)} \|B - B_n\| \left\|\mathcal{H}_1\left(\cdot, t^{(1)}\right)\right\|_{H(\mathcal{H}_1, T_1)}\\ &= \left\{\mathcal{H}_1\left(t^{(1)}, t^{(1)}\right)\mathcal{H}_2\left(t^{(2)}, t^{(2)}\right)\right\}^{1/2} \|B - B_n\|.\end{aligned}$$

That is item 2. □

Remark 1.7.11 Let

$$S = \left\{h^{(1)} \in V[\mathcal{H}_1], h^{(2)} \in V[\mathcal{H}_2] : \left\|h^{(1)}\right\|_{H(\mathcal{H}_1, T_1)} = \left\|h^{(2)}\right\|_{H(\mathcal{H}_2, T_2)} = 1\right\}.$$

As the range of B is contained in the closure of $V[\mathcal{H}_2]$, restricting B to $V[\mathcal{H}_1]$, one has that [266, p. 60]

$$\sup_S \left|\langle B\left[h^{(1)}\right], h^{(2)}\rangle_{H(\mathcal{H}_2, T_2)}\right| = \|B\|.$$

Thus, with now

$$S_1 = \left\{\{\alpha_1, \ldots, \alpha_n\} \subseteq \mathbb{R}, \left\{t^{(1)}_1, \ldots, t^{(1)}_n\right\} \subseteq T_1, n \in \mathbb{N}\right\},$$

and

$$S_2 = \left\{ \{\beta_1, \ldots, \beta_p\} \subseteq \mathbb{R}, \left\{t_1^{(2)}, \ldots, t_n^{(2)}\right\} \subseteq T_2, p \in \mathbb{N} \right\},$$

one has that

$$\sup_{S_1 \times S_2} \sum_{i=1}^{n} \sum_{j=1}^{p} \alpha_i \beta_j \mathcal{B}\left(t_i^{(1)}, t_j^{(2)}\right) = \|B\|.$$

Remark 1.7.12 When

$$\sup_n \sup_{S_1 \times S_2} \sum_{i=1}^{n} \sum_{j=1}^{p} \alpha_i \beta_j \mathcal{B}_n\left(t_i^{(1)}, t_j^{(2)}\right) < \infty,$$

and, for $t^{(1)} \in T_1$ and $t^{(2)} \in T_2$, fixed, but arbitrary, $\{\mathcal{B}_n(t^{(1)}, t^{(2)}), n \in \mathbb{N}\}$ is Cauchy, $\{B_n, n \in \mathbb{N}\}$ is weakly Cauchy. That follows from [266, p. 80].

1.7.2 Unitary Operators

As shall be seen, unitary operators in separable spaces have kernels corresponding to series formed from orthonormal elements.

Proposition 1.7.13 *Let $H(\mathcal{H}, T)$ be a separable RKHS, and*

$$U: H(\mathcal{H}, T) \longrightarrow H(\mathcal{H}, T)$$

be a unitary operator with kernel $\mathcal{U}$. There exist then two complete orthonormal sets, $\{e_i, i \in I\}$ and $\{f_i, i \in I\}$, both in $H(\mathcal{H}, T)$, such that, for $(t_1, t_2) \in T \times T$, fixed, but arbitrary,

$$\mathcal{U}(t_1, t_2) = \sum_{i \in I} e_i(t_1) f_i(t_2).$$

Proof For fixed, but arbitrary $t \in T$, one has that $\mathcal{U}(\cdot, t) \in H(\mathcal{H}, T)$, so that, for any complete orthonormal set $\{e_i, i \in I\} \subseteq H(\mathcal{H}, T)$, $\mathcal{U}(\cdot, t)$ is represented as

$$\mathcal{U}(\cdot, t) = \sum_{i \in I} \langle \mathcal{U}(\cdot, t), e_i \rangle_{H(\mathcal{H},T)} e_i.$$

Consequently, for fixed, but arbitrary $x \in T$,

$$\mathcal{U}(x, t) = \langle \mathcal{U}(\cdot, t), \mathcal{H}(\cdot, x) \rangle_{H(\mathcal{H},T)} = \sum_{i \in I} \langle \mathcal{U}(\cdot, t), e_i \rangle_{H(\mathcal{H},T)} e_i(x).$$

Now [(Remark) 1.7.2]

$$\begin{aligned}\langle \mathcal{U}(\cdot,t), e_i\rangle_{H(\mathcal{H},T)} &= \langle U^\star[\mathcal{H}(\cdot,t)], e_i\rangle_{H(\mathcal{H},T)}\\ &= \langle \mathcal{H}(\cdot,t), U[e_i]\rangle_{H(\mathcal{H},T)}\\ &= U[e_i](t).\end{aligned}$$

Let $f_i = U[e_i]$. Since U is unitary, $\{f_i,\ i \in I\} \subseteq H(\mathcal{H},T)$ is a complete orthonormal set, and the result follows. □

Proposition 1.7.14 *Let $H(\mathcal{H},T)$ be an RKHS, and $\{e_i,\ i \in I\}$ and $\{f_i,\ i \in I\}$ be two complete orthonormal sets, both in $H(\mathcal{H},T)$. Set, for $(t_1,t_2) \in T \times T$, fixed, but arbitrary,*

$$\mathcal{U}(t_1,t_2) = \sum_{i \in I} e_i(t_1) f_i(t_2).$$

$\mathcal{U}$ is then the kernel of a unitary operator of $H(\mathcal{H},T)$.

Proof One has that

$$\begin{aligned}e_i(t_1) &= \langle e_i, \mathcal{H}(\cdot,t_1)\rangle_{H(\mathcal{H},T)},\\ f_i(t_2) &= \langle f_i, \mathcal{H}(\cdot,t_2)\rangle_{H(\mathcal{H},T)},\end{aligned}$$

so that

$$\sum_{i \in I} e_i^2(t_1) = \sum_{i \in I} \langle e_i, \mathcal{H}(\cdot,t_1)\rangle^2_{H(\mathcal{H},T)} = \|\mathcal{H}(\cdot,t_1)\|^2_{H(\mathcal{H},T)} = \mathcal{H}(t_1,t_1),$$

and, similarly, $\sum_{i \in I} f_i^2(t_2) = \mathcal{H}(t_2,t_2)$. Consequently

$$\left\{\sum_{i \in I} |e_i(t_1)|\,|f_i(t_2)|\right\}^2 \le \sum_{i \in I} e_i^2(t_1) \sum_{i \in I} f_i^2(t_2) = \mathcal{H}(t_1,t_1)\,\mathcal{H}(t_2,t_2) < \infty.$$

The kernel $\mathcal{U}$ is thus well defined. Now $\mathcal{U}(\cdot,t) = \sum_{i \in I} f_i(t)\, e_i$, so that one may define

$$U[h](t) = \langle h, \mathcal{U}(\cdot,t)\rangle_{H(\mathcal{H},T)} = \sum_{i \in I} f_i(t)\,\langle h, e_i\rangle_{H(\mathcal{H},T)}.$$

Consequently

$$U[h] = \sum_{i \in I} \langle h, e_i\rangle_{H(\mathcal{H},T)} f_i,$$

$$\|U[h]\|^2_{H(\mathcal{H},T)} = \sum_{i\in I} \langle h, e_i\rangle^2_{H(\mathcal{H},T)} = \|h\|^2_{H(\mathcal{H},T)}.$$

Since $\mathcal{R}[U]$ contains a complete orthonormal set, U is unitary. □

1.7.3 Hilbert-Schmidt Operators

As shall be seen, Hilbert-Schmidt operators are at the core of the L_2 theory of stochastic processes, and the following result is often used. Properties of Hilbert-Schmidt operators are listed in [266, pp. 133–135,163].

Proposition 1.7.15 *Let $\mathcal{H}_1$ and $\mathcal{H}_2$ be reproducing kernels on T_1 and T_2 respectively, $T = T_1 \times T_2$, and $\mathcal{H}$ be the reproducing kernel on T obtained as follows* [(Example) 1.3.15]*: for $t_1 = \left(t_1^{(1)}, t_1^{(2)}\right)$ and $t_2 = \left(t_2^{(1)}, t_2^{(2)}\right)$ in T, fixed, but arbitrary,*

$$\mathcal{H}(t_1, t_2) = \mathcal{H}_1\left(t_1^{(1)}, t_2^{(1)}\right)\mathcal{H}_2\left(t_1^{(2)}, t_2^{(2)}\right).$$

Let $B : H(\mathcal{H}_1, T_1) \longrightarrow H(\mathcal{H}_2, T_2)$ be a Hilbert-Schmidt operator, and $\mathcal{B}$ be its kernel. Then

$$\mathcal{B} \in H(\mathcal{H}, T),$$

and every $\mathcal{B} \in H(\mathcal{H}, T)$ is the kernel of a Hilbert-Schmidt operator from $H(\mathcal{H}_1, T_1)$ to $H(\mathcal{H}_2, T_2)$. Furthermore, $\|\cdot\|_{HS}$ denoting the Hilbert-Schmidt norm,

$$\|B\|_{HS} = \|\mathcal{B}\|_{H(\mathcal{H},T)}.$$

Proof Suppose that B is Hilbert-Schmidt.

Let $\{e_{\lambda^{(1)}},\ \lambda^{(1)} \in \Lambda_1\}$ and $\{e_{\lambda^{(2)}},\ \lambda^{(2)} \in \Lambda_2\}$ be complete orthonormal sets in, respectively, $H(\mathcal{H}_1, T_1)$ and $H(\mathcal{H}_2, T_2)$. Since $\mathcal{B}$ is a kernel for B, for fixed, but arbitrary $t_2 \in T_2$, $\mathcal{B}(\cdot, t_2) \in H(\mathcal{H}_1, T_1)$, and then

$$\mathcal{B}(\cdot, t_2) = \sum_{\lambda^{(1)}\in\Lambda_1} \langle \mathcal{B}(\cdot, t_2), e_{\lambda^{(1)}}\rangle_{H(\mathcal{H}_1,T_1)}\, e_{\lambda^{(1)}},$$

so that

$$\begin{aligned}\mathcal{B}(t_1, t_2) &= \langle \mathcal{B}(\cdot, t_2), \mathcal{H}_1(\cdot, t_1)\rangle_{H(\mathcal{H}_1,T_1)} \\ &= \sum_{\lambda^{(1)}\in\Lambda_1} \langle \mathcal{B}(\cdot, t_2), e_{\lambda^{(1)}}\rangle_{H(\mathcal{H}_1,T_1)}\, e_{\lambda^{(1)}}(t_1).\end{aligned}$$

But, using the definition of $\mathcal{B}$ as a kernel for B,

$$\begin{aligned}\langle \mathcal{B}(\cdot, t_2), e_{\lambda^{(1)}} \rangle_{H(\mathcal{H}_1,T_1)} &= \\ &= B[e_{\lambda^{(1)}}](t_2) \\ &= \langle B[e_{\lambda^{(1)}}], \mathcal{H}_2(\cdot, t_2)\rangle_{H(\mathcal{H}_2,T_2)} \\ &= \sum_{\lambda^{(2)}\in\Lambda_2} \langle B[e_{\lambda^{(1)}}], e_{\lambda^{(2)}}\rangle_{H(\mathcal{H}_2,T_2)} \langle e_{\lambda^{(2)}}, \mathcal{H}_2(\cdot, t_2)\rangle_{H(\mathcal{H}_2,T_2)} \\ &= \sum_{\lambda^{(2)}\in\Lambda_2} \langle B[e_{\lambda^{(1)}}], e_{\lambda^{(2)}}\rangle_{H(\mathcal{H}_2,T_2)} e_{\lambda^{(2)}}(t_2).\end{aligned}$$

Now, as B is Hilbert-Schmidt,

$$\sum_{\lambda^{(1)}\in\Lambda_1}\sum_{\lambda^{(2)}\in\Lambda_2} \langle B[e_{\lambda^{(1)}}], e_{\lambda^{(2)}}\rangle^2_{H(\mathcal{H}_2,T_2)} = \sum_{\lambda^{(1)}\in\Lambda_1} \|B[e_{\lambda^{(1)}}]\|^2_{H(\mathcal{H}_2,T_2)} < \infty.$$

Furthermore, using the representation of $H(\mathcal{H}, T)$ as the range of the map L of (Example) 1.3.15, since $\{e_{\lambda^{(1)}} \otimes e_{\lambda^{(2)}}, (\lambda^{(1)}, \lambda^{(2)}) \in \Lambda_1 \times \Lambda_2\}$ is a complete orthonormal set in $H(\mathcal{H}_1, T_1) \otimes H(\mathcal{H}_2, T_2)$,

$$\left\{L[e_{\lambda^{(1)}} \otimes e_{\lambda^{(2)}}], \left(\lambda^{(1)}, \lambda^{(2)}\right) \in \Lambda_1 \times \Lambda_2\right\}$$

is a complete orthonormal set in $H(\mathcal{H}, T)$. But

$$L[e_{\lambda^{(1)}} \otimes e_{\lambda^{(2)}}](t_1, t_2) = e_{\lambda^{(1)}}(t_1)\, e_{\lambda^{(2)}}(t_2).$$

Thus, as

$$\begin{aligned}\mathcal{B}(t_1, t_2) &= \sum_{\lambda^{(1)}\in\Lambda_1} \langle \mathcal{B}(\cdot, t_2), e_{\lambda^{(1)}}\rangle_{H(\mathcal{H}_1,T_1)} e_{\lambda^{(1)}}(t_1) \\ &= \sum_{\lambda^{(1)}\in\Lambda_1}\sum_{\lambda^{(2)}\in\Lambda_2} \langle B[e_{\lambda^{(1)}}], e_{\lambda^{(2)}}\rangle_{H(\mathcal{H}_2,T_2)} e_{\lambda^{(2)}}(t_2)\, e_{\lambda^{(1)}}(t_1),\end{aligned}$$

that latter series represents an element of $H(\mathcal{H}, T)$, and consequently $\mathcal{B}$ belongs to $H(\mathcal{H}, T)$. Since

$$\mathcal{B} = L\left[\sum_{\lambda^{(1)}\in\Lambda_1}\sum_{\lambda^{(2)}\in\Lambda_2}\langle B\left[e_{\lambda^{(1)}}\right], e_{\lambda^{(2)}}\rangle_{H(\mathcal{H}_2,T_2)}\, e_{\lambda^{(1)}}\otimes e_{\lambda^{(2)}}\right],$$

one has furthermore that

$$\|\mathcal{B}\|^2_{H(\mathcal{H},T)} =$$

$$= \left\|\sum_{\lambda^{(1)}\in\Lambda_1}\sum_{\lambda^{(2)}\in\Lambda_2}\langle B\left[e_{\lambda^{(1)}}\right], e_{\lambda^{(2)}}\rangle_{H(\mathcal{H}_2,T_2)} e_{\lambda^{(1)}}\otimes e_{\lambda^{(2)}}\right\|^2_{H(\mathcal{H}_1,T_1)\otimes H(\mathcal{H}_2,T_2)}$$

$$= \sum_{\lambda^{(1)}\in\Lambda_1}\sum_{\lambda^{(2)}\in\Lambda_2}\langle B\left[e_{\lambda^{(1)}}\right], e_{\lambda^{(2)}}\rangle^2_{H(\mathcal{H}_2,T_2)}.$$

The latter, as seen, is the Hilbert-Schmidt norm of B.

Proof Suppose now that $\mathcal{B}$ *is a fixed, but arbitrary element of* $H(\mathcal{H}, T)$.
As already seen, $\mathcal{B}$ has a representation of the form

$$\mathcal{B} = \sum_{\lambda^{(1)}\in\Lambda_1}\sum_{\lambda^{(2)}\in\Lambda_2}\alpha_{(\lambda^{(1)},\lambda^{(2)})} L\left[e_{\lambda^{(1)}}\otimes e_{\lambda^{(2)}}\right],$$

with

$$\sum_{\lambda^{(1)}\in\Lambda_1}\sum_{\lambda^{(2)}\in\Lambda_2}\alpha^2_{(\lambda^{(1)},\lambda^{(2)})} < \infty,$$

so that, for fixed, but arbitrary $(t_1, t_2) \in T_1 \times T_2$,

$$\mathcal{B}(t_1, t_2) = \sum_{\lambda^{(1)}\in\Lambda_1}\sum_{\lambda^{(2)}\in\Lambda_2}\alpha_{(\lambda^{(1)},\lambda^{(2)})} e_{\lambda^{(1)}}(t_1)\, e_{\lambda^{(2)}}(t_2).$$

Now, using the inequality of Cauchy-Schwarz, for a fixed, but arbitrary element $h_1 \in H(\mathcal{H}_1, T_1)$,

$$\left\{\sum_{\lambda^{(1)}\in\Lambda_1}\sum_{\lambda^{(2)}\in\Lambda_2}\left|\alpha_{(\lambda^{(1)},\lambda^{(2)})}\right|\left|e_{\lambda^{(2)}}(t_2)\right|\left|\langle h_1, e_{\lambda^{(1)}}\rangle_{H(\mathcal{H}_1,T_1)}\right|\right\}^2 =$$

$$= \left\{\sum_{\lambda^{(1)}\in\Lambda_1}\left\{\sum_{\lambda^{(2)}\in\Lambda_2}\left|\alpha_{(\lambda^{(1)},\lambda^{(2)})}\right|\left|e_{\lambda^{(2)}}(t_2)\right|\right\}\left|\langle h_1, e_{\lambda^{(1)}}\rangle_{H(\mathcal{H}_1,T_1)}\right|\right\}^2$$

$$\leq \sum_{\lambda^{(1)}\in\Lambda_1}\left\{\sum_{\lambda^{(2)}\in\Lambda_2}\left|\alpha_{(\lambda^{(1)},\lambda^{(2)})}\right|\left|e_{\lambda^{(2)}}(t_2)\right|\right\}^2\sum_{\lambda^{(1)}\in\Lambda_1}\langle h_1,e_{\lambda^{(1)}}\rangle^2_{H(\mathcal{H}_1,T_1)}$$

$$\leq \sum_{\lambda^{(1)}\in\Lambda_1}\left\{\sum_{\lambda^{(2)}\in\Lambda_2}\alpha^2_{(\lambda^{(1)},\lambda^{(2)})}\sum_{\lambda^{(2)}\in\Lambda_2}e^2_{\lambda^{(2)}}(t_2)\right\}\|h_1\|^2_{H(\mathcal{H}_1,T_1)}$$

$$= \sum_{\lambda^{(1)}\in\Lambda_1}\sum_{\lambda^{(2)}\in\Lambda_2}\alpha^2_{(\lambda^{(1)},\lambda^{(2)})}\mathcal{H}_2(t_2,t_2)\,\|h_1\|^2_{H(\mathcal{H}_1,T_1)}.$$

Consequently, and analogously,

$$\mathcal{B}(\cdot,t_2) = \sum_{\lambda^{(1)}\in\Lambda_1}\left\{\sum_{\lambda^{(2)}\in\Lambda_2}\alpha_{(\lambda^{(1)},\lambda^{(2)})}e_{\lambda^{(2)}}(t_2)\right\}e_{\lambda^{(1)}}\in H(\mathcal{H}_1,T_1),$$

and one may then compute

$$\langle h_1,\mathcal{B}(\cdot,t_2)\rangle_{H(\mathcal{H}_1,T_1)} = \sum_{\lambda^{(1)}\in\Lambda_1}\sum_{\lambda^{(2)}\in\Lambda_2}\alpha_{(\lambda^{(1)},\lambda^{(2)})}e_{\lambda^{(2)}}(t_2)\,\langle h_1,e_{\lambda^{(1)}}\rangle_{H(\mathcal{H}_1,T_1)}.$$

But h_2, defined using

$$h_2 = \sum_{\lambda^{(2)}\in\Lambda_2}\left\{\sum_{\lambda^{(1)}\in\Lambda_1}\alpha_{(\lambda^{(1)},\lambda^{(2)})}\langle h_1,e_{\lambda^{(1)}}\rangle_{H(\mathcal{H}_1,T_1)}\right\}e_{\lambda^{(2)}},$$

belongs to $H(\mathcal{H}_2,T_2)$, so that

$$t_2\mapsto\langle h_1,\mathcal{B}(\cdot,t_2)\rangle_{H(\mathcal{H}_1,T_1)} = \langle h_2,\mathcal{H}_2(\cdot,t_2)\rangle_{H(\mathcal{H}_2,T)}$$

is a function of $H(\mathcal{H}_2,T)$. One may then define, for $(h_1,t_2)\in H(\mathcal{H}_1,T_1)\times T_2$, fixed, but arbitrary,

$$B[h_1](t_2) = \langle h_1,\mathcal{B}(\cdot,t_2)\rangle_{H(\mathcal{H}_1,T_1)},\ \text{ and }\ B[h_1] = h_2.$$

One thus obtains a linear and bounded operator from the RKHS $H(\mathcal{H}_1,T_1)$ to the $H(\mathcal{H}_2,T_2)$ one, and the set of inequalities that have been exhibited above establishes that B is also Hilbert-Schmidt. □

Example 1.7.16 In (Example) 1.7.4, the operator B shall be Hilbert-Schmidt whenever

$$C_{X,Y}\text{ belongs to the RKHS whose kernel is }C_X\times C_Y.$$

That last remark has been used to define canonical correlations for stochastic processes [92].

1.7.4 Covariance Operators

Since useful signals often have paths in RKHS's, they induce, on these, laws for which covariance operators are an essential parameter.

In this section $H(\mathcal{H}, T)$ shall be a *separable* RKHS, $(\Omega, \mathcal{A}, P)$, a probability space, and $\xi : \Omega \longrightarrow H(\mathcal{H}, T)$ a map. On $H(\mathcal{H}, T)$, one shall consider the following σ-algebras:

$\mathcal{B}(\mathcal{H}, T)$: the Borel sets (σ-algebra generated by the open sets);

$\mathcal{C}(\mathcal{H}, T)$: the cylinder sets (σ-algebra generated by the continuous linear functionals);

$\mathcal{D}(\mathcal{H}, T)$: the σ-algebra generated by the following family of continuous linear functionals:

$$\left\{\langle \cdot, h_i\rangle_{H(\mathcal{H},T)},\ h_i \in H(\mathcal{H}, T),\ i \in I\right\},$$

where $H(I) = \{h_i,\ i \in I\} \subseteq H(\mathcal{H}, T)$ is a total set;

$\mathcal{E}(\mathcal{H}, T)$: the σ-algebra generated by the evaluation maps.

From the definitions one has that

$$\mathcal{D}(\mathcal{H}, T) \subseteq \mathcal{C}(\mathcal{H}, T) \subseteq \mathcal{B}(\mathcal{H}, T),$$

and that

$$\mathcal{E}(\mathcal{H}, T) \subseteq \mathcal{C}(\mathcal{H}, T) \subseteq \mathcal{B}(\mathcal{H}, T).$$

Lemma 1.7.17 $\mathcal{E}(\mathcal{H}, T) = \mathcal{B}(\mathcal{H}, T)$

Proof Since the evaluation maps are continuous linear functionals [(Proposition) 1.1.5], $\mathcal{E}(\mathcal{H}, T) \subseteq \mathcal{B}(\mathcal{H}, T)$. So one must prove the converse inclusion.

Let $H(I) = \{h_i,\ i \in I\} \subseteq V[\mathcal{H}]$ be a countable set that is dense in $H(\mathcal{H}, T)$ [(Proposition) 1.5.1], and $B(h, \alpha) \subseteq H(\mathcal{H}, T)$ be the closed ball centered at h, with radius α.

When $h \in V[\mathcal{H}]$, $h = \sum_{i=1}^{n} \alpha_i \mathcal{H}(\cdot, t_i)$, so that

$$\langle x, h \rangle_{H(\mathcal{H},T)} = \sum_{i=1}^{n} \alpha_i \mathcal{E}_{t_i}(x),$$

and thus $x \mapsto \langle x, h \rangle_{H(\mathcal{H},T)}$ is adapted to $\mathcal{E}(\mathcal{H}, T)$. Now [266, p. 61], for fixed, but arbitrary $h_0 \in H(\mathcal{H}, T)$,

$$\|h_0\|_{H(\mathcal{H},T)} = \sup_{h \in B(0,1)} \left| \langle h_0, h \rangle_{H(\mathcal{H},T)} \right| = \sup_{h \in B(0,1) \cap H(I)} \left| \langle h_0, h \rangle_{H(\mathcal{H},T)} \right|.$$

Thus, for $h \in B(h_0, \alpha)$, and $\tilde{h} \in B(0, 1) \cap H(I)$, fixed, but arbitrary,

$$\left| \langle h - h_0, \tilde{h} \rangle_{H(\mathcal{H},T)} \right| \leq \|h - h_0\|_{H(\mathcal{H},T)} \leq \alpha.$$

But the supremum of a countable set of expressions, adapted to $\mathcal{E}(\mathcal{H}, T)$, dominated by α, is also adapted to $\mathcal{E}(\mathcal{H}, T)$, and dominated by α. Consequently:

$$B(h_0, \alpha) = \bigcap_{\tilde{h} \in B(0,1) \cap H(I)} \left\{ h \in H(\mathcal{H}, T) : \left| \langle h - h_0, \tilde{h} \rangle_{H(\mathcal{H},T)} \right| \leq \alpha \right\}.$$

So, $B(h_0, \alpha)$ belongs to $\mathcal{E}(\mathcal{H}, T)$. But the closed balls generate $\mathcal{B}(\mathcal{H}, T)$. □

Lemma 1.7.18 $\mathcal{D}(\mathcal{H}, T) = \mathcal{B}(\mathcal{H}, T)$

Proof By assumption, $V[H(I)]$ is dense in $H(\mathcal{H}, T)$. One may show, as in (Proposition) 1.5.1, that $V[H(I)]$ contains a countable dense set, and then adjust the proof of (Lemma) 1.7.17 to this latter case. □

Proposition 1.7.19 *ξ is adapted to $\mathcal{A}$ and $\mathcal{B}(\mathcal{H}, T)$ if, and only if, for $t \in T$, fixed, but arbitrary, $\mathcal{E}_t \circ \xi$ is adapted to $\mathcal{A}$*

Proof This result is a direct consequence of (Lemma) 1.7.17. □

Definition 1.7.20 Let $\xi : \Omega \longrightarrow H(\mathcal{H}, T)$ be adapted, and $A \in \mathcal{A}$ be fixed, but arbitrary. Then ξ is said to be weakly integrable on A when

1. for $h \in H(\mathcal{H}, T)$, fixed, but arbitrary,

$$E_P\left[\chi_A(\cdot) \left| \langle \xi(\cdot), h \rangle_{H(\mathcal{H},T)} \right| \right] = \int_A \left| \langle \xi(\omega), h \rangle_{H(\mathcal{H},T)} \right| P(d\omega) < \infty,$$

2. there exists $h_{\xi,A}^{w} \in H(\mathcal{H}, T)$ such that, for $h \in H(\mathcal{H}, T)$, fixed, but arbitrary,

$$E_P\left[\chi_A(\cdot) \langle \xi(\cdot), h \rangle_{H(\mathcal{H},T)}\right] = \langle h_{\xi,A}^{w}, h \rangle_{H(\mathcal{H},T)}.$$

$h^w_{\xi,A}$ is called the weak integral of ξ on A, and one writes, according to convenience,

$$h^w_{\xi,A} = E_P[\chi_A \xi] = \int_A^{(w)} \xi \, dP.$$

Definition 1.7.21 When ξ is weakly integrable on A for all $A \in \mathcal{A}$, it is said to be weakly, or Pettis integrable.

Definition 1.7.22 Let $\xi : \Omega \longrightarrow H(\mathcal{H}, T)$ be adapted, and $A \in \mathcal{A}$ be fixed, but arbitrary. Then ξ is said to be strongly integrable on A when

$$\int_A \|\xi(\omega)\|_{H(\mathcal{H},T)} P(d\omega) < \infty.$$

One writes, for that integral, either $\int_A \xi \, dP$ or, when useful, $\int_A^{(s)} \xi \, dP$.

Definition 1.7.23 When ξ is strongly integrable on Ω, it is said to be Bochner integrable, so that, when ξ is Bochner integrable, it is strongly integrable on every $A \in \mathcal{A}$.

Remark 1.7.24 As $\left|\langle \xi(\omega), h\rangle_{H(\mathcal{H},T)}\right| \leq \|\xi(\omega)\|_{H(\mathcal{H},T)} \|h\|_{H(\mathcal{H},T)}$, when ξ is strongly integrable on A,

$$\int_A \left|\langle \xi(\omega), h\rangle_{H(\mathcal{H},T)}\right| P(d\omega) \leq \left\{\int_A \|\xi(\omega)\|_{H(\mathcal{H},T)} P(d\omega)\right\} \|h\|_{H(\mathcal{H},T)},$$

so that $h \mapsto \int_A \langle \xi(\omega), h\rangle_{H(\mathcal{H},T)} P(d\omega)$ is a continuous linear functional on $H(\mathcal{H}, T)$, and then, by the Riesz representation theorem [266, p. 64], there exists $h^s_{\xi,A} \in H(\mathcal{H}, T)$ such that

$$\int_A \langle \xi(\omega), h\rangle_{H(\mathcal{H},T)} P(d\omega) = \langle h, h^s_{\xi,A}\rangle_{H(\mathcal{H},T)}.$$

One writes

$$h^s_{\xi,A} = \int_A^{(s)} \xi(\omega) P(d\omega).$$

Remark 1.7.25 Obviously, when ξ is strongly integrable on A, it is weakly integrable also, and the two integrals result in the same element of $H(\mathcal{H}, T)$, that is $h^w_{\xi,A} = h^s_{\xi,A}$.

Example 1.7.26 Let $\Omega = \mathbb{N}$ and $\mathcal{A}$ be the subsets of $\mathbb{N}$. Let $P(\{n\}) = \frac{1}{2^n}$.

The space l_2 is an RKHS [(Example) 1.1.24]. Let $\underline{\xi} : \mathbb{N} \longrightarrow H(\mathcal{H}, T) = l_2$ be defined using the following relation:

$$\underline{\xi}(n) = \left\{\frac{2^n}{n}\right\} \underline{e}_n.$$

$\underline{\xi}$ is adapted as, for $p \in \mathbb{N}$, fixed, but arbitrary,

$$\mathcal{E}_p\left[\underline{\xi}(n)\right] = \begin{Bmatrix} \frac{2^n}{n} \text{ when } p = n \\ 0 \text{ when } p \neq n \end{Bmatrix} = \frac{2^n}{n} \langle \underline{e}_n, \underline{e}_p \rangle_{l_2}.$$

$\underline{\xi}$ is not Bochner integrable since

$$\int_{\mathbb{N}} \left\|\underline{\xi}(n)\right\|_{l_2} P(dn) = \sum_{n \in \mathbb{N}} \left\|\underline{\xi}(n)\right\|_{l_2} P(\{n\}) = \sum_{n \in \mathbb{N}} \frac{2^n}{n} \frac{1}{2^n} = \infty.$$

Let $A \subseteq \mathbb{N}$ be fixed, but arbitrary, and let $l_A \subseteq l_2$ be the subspace spanned by $\{\underline{e}_n, n \in A\}$. Let $\underline{h}_0$ be defined using the following relation:

$$h_n^{(0)} = \langle \underline{h}_0, \underline{e}_n \rangle_{l_2} = \frac{1}{n}.$$

Then

$$P_{l_A}\left[\underline{h}_0\right] = \sum_{n \in A} h_n^{(0)} \underline{e}_n,$$

and, for fixed, but arbitrary $\underline{h} \in l_2$,

$$\langle \underline{h}, P_{l_A}\left[\underline{h}_0\right] \rangle_{l_2} = \sum_{n \in A} \frac{h_n}{n}.$$

Consequently

$$\int_A \langle \underline{h}, \underline{\xi}(n) \rangle_{l_2} P(dn) = \sum_{n \in A} \frac{h_n}{n} = \langle \underline{h}, P_{l_A}\left[\underline{h}_0\right] \rangle_{l_2}.$$

$\underline{\xi}$ is thus weakly integrable on A.

Remark 1.7.27 When ξ is weakly integrable on A, $X(\cdot, t) = \mathcal{E}_t \circ \xi(\cdot)$ is integrable on A, and the map

$$t \mapsto E_P\left[\chi_A(\cdot)\, \mathcal{E}_t \circ \xi(\cdot)\right]$$

belongs to $H(\mathcal{H},T)$. By definition indeed there exists $h^w_{\xi,A} \in H(\mathcal{H},T)$ such that, for fixed, but arbitrary $h \in H(\mathcal{H},T)$,

$$\int_A \langle h, \xi(\omega)\rangle_{H(\mathcal{H},T)} P(d\omega) = \langle h, h^w_{\xi,A}\rangle_{H(\mathcal{H},T)}.$$

Letting $h = \mathcal{H}(\cdot,t)$ one gets that, for fixed, but arbitrary $t \in T$,

$$\int_A X(\omega,t)\, P(d\omega) = \int_A \mathcal{E}_t[\xi(\omega)]\, P(d\omega) = h^w_{\xi,A}(t).$$

Remark 1.7.28 The fact that X in (Remark) 1.7.27 is integrable does not imply that ξ is weakly integrable, as shown by the following example. In (Example) 1.7.26 let $\underline{\xi}(n) = 2^n \underline{e}_n$. Then

$$\langle \mathcal{H}(\cdot,p), \underline{\xi}(n)\rangle_{l_2} = 2^n \delta_{n,p}.$$

Consequently $\int_{\mathbb{N}} \langle \mathcal{H}(\cdot,p), \underline{\xi}(n)\rangle_{l_2} P(dn) = 1$. But $1_{\mathbb{N}}$ does not belong to l_2 which is, in this case, $H(\mathcal{H},T)$. Thus, by its very definition, ξ cannot be weakly integrable.

Proposition 1.7.29 *Let $A \in \mathcal{A}$ be fixed, but arbitrary, and $X(\cdot,t) = \mathcal{E}_t \circ \xi(\cdot)$. The following two conditions are then equivalent:*

1. *the map $\xi_A : T \longrightarrow \mathbb{R}$ defined using*

$$t \mapsto E_P[\chi_A(\cdot) X(\cdot,t)]$$

belongs to $H(\mathcal{H},T)$;

2. *the map $\Xi_A : V[\mathcal{H}] \longrightarrow \mathbb{R}$ defined using the following assignment:*

$$h \mapsto E_P\left[\chi_A(\cdot)\langle h, \xi(\cdot)\rangle_{H(\mathcal{H},T)}\right]$$

is a continuous linear functional on $H(\mathcal{H},T)$.

When both statements obtain, one has that

$$\Xi_A[h] = \langle h, \xi_A\rangle_{H(\mathcal{H},T)}.$$

Proof (1 ⇒ 2) Let $h = \sum_{i=1}^n \alpha_i \mathcal{H}(\cdot,t_i)$. Then

$$E_P\left[\chi_A(\cdot)\langle h, \xi(\cdot)\rangle_{H(\mathcal{H},T)}\right] = \sum_{i=1}^n \alpha_i E_P[\chi_A(\cdot) X(\cdot,t_i)]$$
$$= \sum_{i=1}^n \alpha_i \xi_A(t_i)$$

$$= \sum_{i=1}^{n} \alpha_i \langle \xi_A, \mathcal{H}(\cdot, t_i) \rangle_{H(\mathcal{H},T)}$$
$$= \langle h, \xi_A \rangle_{H(\mathcal{H},T)}.$$

Proof (2 ⇒ 1) By the Hahn-Banach theorem [266, p. 62], there exists a continuous linear functional on $H(\mathcal{H}, T)$, say $\tilde{\Xi}_A$, such that

$$\tilde{\Xi}_A^{|V[\mathcal{H}]} = \Xi_A, \quad \text{and} \quad \left\| \tilde{\Xi}_A \right\| = \left\| \Xi_A \right\|.$$

By the Riesz representation theorem [266, p. 64], there exists $h_A \in H(\mathcal{H}, T)$ such that, for $h \in H(\mathcal{H}, T)$, fixed, but arbitrary,

$$\tilde{\Xi}_A[h] = \langle h, h_A \rangle_{H(\mathcal{H},T)}.$$

Then

$$\begin{aligned} h_A(t) &= \langle h_A, \mathcal{H}(\cdot, t) \rangle_{H(\mathcal{H},T)} \\ &= \tilde{\Xi}_A[\mathcal{H}(\cdot, t)] \\ &= \Xi_A[\mathcal{H}(\cdot, t)] \\ &= E_P[\chi_A(\cdot) X(\cdot, t)] \\ &= \xi_A(t). \end{aligned}$$

Consequently $\xi_A = h_A \in H(\mathcal{H}, T)$. □

Proposition 1.7.30 *Let $\xi : \Omega \longrightarrow H(\mathcal{H}, T)$ be an adapted map, and, for $t \in T$, fixed, but arbitrary, $X(\cdot, t) = \mathcal{E}_t \circ \xi$. Let $A \in \mathcal{A}$ be such that, for $t \in T$, fixed, but arbitrary, $E_P[\chi_A(\cdot) X(\cdot, t)]$ exists. For fixed, but arbitrary $h \in V[\mathcal{H}]$, let*

$$\Phi_A[h] = \int_A \left| \langle h, \xi(\omega) \rangle_{H(\mathcal{H},T)} \right| P(d\omega).$$

Φ_A is well defined by assumption. The following statements are then equivalent:

1. *Φ_A is continuous at the origin;*
2. *Φ_A is continuous;*
3. *Φ_A is Lipschitz-continuous.*

Proof Since, by definition, item 3 implies item 2, which implies item 1, it is enough to prove that item 1 implies item 3.

Because of the continuity at the origin assumption, there exists $\epsilon > 0$ such that, whenever

$$h \in V[\mathcal{H}] \text{ and } \|h\|_{H(\mathcal{H},T)} \leq \epsilon,$$

then

$$\Phi_A(h) \leq 1.$$

Thus, when $h \in V[\mathcal{H}]$ and $\|h\|_{H(\mathcal{H},T)} \neq 0$,

$$\left\| \frac{h}{\|h\|_{H(\mathcal{H},T)}} \epsilon \right\|_{H(\mathcal{H},T)} \leq \epsilon, \quad \text{and} \quad 0 \leq \Phi_A\left[\frac{h}{\|h\|_{H(\mathcal{H},T)}} \epsilon\right] \leq 1,$$

so that

$$0 \leq \Phi_A(h) \leq \frac{\|h\|_{H(\mathcal{H},T)}}{\epsilon}.$$

Let now $\{h_1, h_2\} \subseteq V[\mathcal{H}]$ be fixed, but arbitrary. Then, using firstly the inequality $||a| - |b|| \leq |a - b|$, and, secondly, the inequality just obtained,

$$\begin{aligned}
|\Phi_A[h_1] - \Phi_A[h_2]| &= \left|E_P\left[\chi_A(\cdot)\left\{\left|\langle h_1, \xi(\cdot)\rangle_{H(\mathcal{H},T)}\right| - \left|\langle h_2, \xi(\cdot)\rangle_{H(\mathcal{H},T)}\right|\right\}\right]\right| \\
&\leq E_P\left[\chi_A(\cdot)\left|\left\{\left|\langle h_1, \xi(\cdot)\rangle_{H(\mathcal{H},T)}\right| - \left|\langle h_2, \xi(\cdot)\rangle_{H(\mathcal{H},T)}\right|\right\}\right|\right] \\
&\leq E_P\left[\chi_A(\cdot)\left|\langle h_1 - h_2, \xi(\cdot)\rangle_{H(\mathcal{H},T)}\right|\right] \\
&= \Phi_A[h_1 - h_2] \\
&\leq \frac{1}{\epsilon}\|h_1 - h_2\|_{H(\mathcal{H},T)}.
\end{aligned}$$

□

Proposition 1.7.31 *Let* $A \in \mathcal{A}$ *be fixed, but arbitrary. When* ξ_A *of* (Proposition) 1.7.29 *belongs to* $H(\mathcal{H}, T)$, *and* Φ_A *of* (Proposition) 1.7.30 *is continuous at the origin,* ξ *is weakly integrable on A, and*

$$\int_A^{(w)} \xi \, dP = \xi_A.$$

Proof Let $h \in H(\mathcal{H}, T)$ be fixed, but arbitrary. One must prove that

- $\langle h, \xi\rangle_{H(\mathcal{H},T)}$ is integrable on A;
- $E_P\left[\chi_A(\cdot)\langle h, \xi(\cdot)\rangle_{H(\mathcal{H},T)}\right] = \langle h, \xi_A\rangle_{H(\mathcal{H},T)}$.

Let $\{h_n, n \in \mathbb{N}\} \subseteq V[\mathcal{H}]$ converge in $H(\mathcal{H}, T)$ to h. One has that

$$\int_A \left|\langle h_n, \xi(\omega)\rangle_{H(\mathcal{H},T)} - \langle h_p, \xi(\omega)\rangle_{H(\mathcal{H},T)}\right| P(d\omega) = \Phi_A\left(h_n - h_p\right).$$

But there exists [(Proposition) 1.7.30] $\kappa \geq 0$ such that

$$\Phi_A\left(h_n - h_p\right) \leq \kappa \left\|h_n - h_p\right\|_{H(\mathcal{H},T)}.$$

The equivalence classes of the sequence $\left\{\langle h_n, \xi\rangle_{H(\mathcal{H},T)},\ n \in \mathbb{N}\right\}$ form thus a Cauchy sequence in $L_1\left(A, \mathcal{A} \cap A, P^{|\mathcal{A}\cap A}\right)$. But, since, for fixed, but arbitrary $\omega \in \Omega$,

$$\lim_n \langle h_n, \xi(\omega)\rangle_{H(\mathcal{H},T)} = \langle h, \xi(\omega)\rangle_{H(\mathcal{H},T)},$$

that Cauchy sequence converges in $L_1\left(A, \mathcal{A} \cap A, P^{|\mathcal{A}\cap A}\right)$ to $\langle h, \xi\rangle_{H(\mathcal{H},T)}$. Consequently

$$\lim_n E_P\left[\chi_A(\cdot)\langle h_n, \xi(\cdot)\rangle_{H(\mathcal{H},T)}\right] = E_P\left[\chi_A(\cdot)\langle h, \xi(\cdot)\rangle_{H(\mathcal{H},T)}\right].$$

Now, by (Proposition) 1.7.29, $E_P\left[\chi_A(\cdot)\langle h_n, \xi(\cdot)\rangle_{H(\mathcal{H},T)}\right] = \langle h_n, \xi_A\rangle_{H(\mathcal{H},T)}$, so that

$$E_P\left[\chi_A(\cdot)\langle h, \xi(\cdot)\rangle_{H(\mathcal{H},T)}\right] = \langle h, \xi_A\rangle_{H(\mathcal{H},T)}.$$

□

Definition 1.7.32 Let $\mu : \mathcal{A} \longrightarrow H(\mathcal{H}, T)$ be a map. It is absolutely continuous with respect to P when, given $\epsilon > 0$, fixed, but arbitrary, there exists $\eta(\epsilon) > 0$ such that

$$A \in \mathcal{A} \text{ and } P(A) < \eta(\epsilon) \Longrightarrow \|\mu(A)\|_{H(\mathcal{H},T)} < \epsilon.$$

Proposition 1.7.33 *Assume that ξ is weakly integrable, and define a map $\mu : \mathcal{A} \longrightarrow H(\mathcal{H}, T)$ using the following assignment:*

$$\mu(A) = \int_A^{(w)} \xi(\omega) P(dx).$$

The map μ is then absolutely continuous with respect to P.

Proof Since, for fixed, but arbitrary $h \in H(\mathcal{H}, T)$

$$\langle h, \mu(A)\rangle_{H(\mathcal{H},T)} = \int_A \langle h, \xi(\omega)\rangle_{H(\mathcal{H},T)} P(d\omega),$$

μ is weakly countably additive. Furthermore, when $P(A) = 0$,

$$\langle h, \mu(A)\rangle_{H(\mathcal{H},T)} = 0, \text{ all } h \in H(\mathcal{H}, T),$$

that is, $\mu(A) = 0_{H(\mathcal{H},T)}$. But then [135, p. 76] μ is absolutely continuous with respect to P. □

Proposition 1.7.34 *$\xi : \Omega \longrightarrow H(\mathcal{H}, T)$ is weakly integrable if, and only if, both statements that follow obtain:*

1. *for $A \in \mathcal{A}$, fixed, but arbitrary, the map*

$$t \mapsto E_P[\chi_A(\cdot) X(\cdot, t)] = E_P[\chi_A(\cdot) \mathcal{E}_t \circ \xi(\cdot)]$$

belongs to $H(\mathcal{H}, T)$;

2. *the map μ of* (Proposition) 1.7.33 *is absolutely continuous with respect to P.*

Proof Suppose ξ is weakly integrable.

According to (Remark) 1.7.27, item 1 obtains, and item 2 is (Proposition) 1.7.33.

Proof Suppose that assertions 1 and 2 obtain.

ξ is weakly measurable as, for fixed, but arbitrary $t \in T$, $X(\cdot, t)$ is (implicitly) adapted because of item 1: one then uses (Lemma) 1.7.17.

If one proves that, for fixed, but arbitrary $A \in \mathcal{A}$, Φ_A of (Proposition) 1.7.30 is continuous at the origin, weak integrability shall follow because of (Proposition) 1.7.31.

Let thus $A \in \mathcal{A}$ be fixed, but arbitrary, $\{h_n, n \in \mathbb{N}\} \subseteq V[\mathcal{H}]$ converge to $0_{H(\mathcal{H},T)}$ in $H(\mathcal{H}, T)$, and $\epsilon > 0$ be fixed, but arbitrary. Because of item 2, given $\sqrt{\epsilon}$, there exists $\eta(\epsilon)$ such that, for $A_0 \in \mathcal{A}$,

$$P(A_0) < \eta(\epsilon) \text{ implies that } \|\mu_A(A_0)\|_{H(\mathcal{H},T)} < \sqrt{\epsilon}.$$

As, for fixed, but arbitrary $\omega \in \Omega$,

$$\lim_n \langle \xi(\omega), h_n \rangle_{H(\mathcal{H},T)} = 0,$$

$\left\{\langle \xi(\omega), h_n \rangle_{H(\mathcal{H},T)}, n \in \mathbb{N}\right\}$ converges in probability to zero, and thus there exists $n(\epsilon) \in \mathbb{N}$ such that, for $n \geq n(\epsilon)$,

$$\|h_n\|_{H(\mathcal{H},T)} < \frac{\sqrt{\epsilon}}{4}, \text{ and } P\left(\omega \in \Omega : \left|\langle \xi(\omega), h_n \rangle_{H(\mathcal{H},T)}\right| > \frac{\epsilon}{2}\right) < \eta(\epsilon).$$

Let, for $n \in \mathbb{N}$,

$$A_n^+[\epsilon] = A \cap \left\{\omega \in \Omega : \left|\langle \xi(\omega), h_n \rangle_{H(\mathcal{H},T)}\right| > \frac{\epsilon}{2}\right\}$$
$$\cap \left\{\omega \in \Omega : \langle \xi(\omega), h_n \rangle_{H(\mathcal{H},T)} \geq 0\right\},$$

and

$$A_n^-[\epsilon] = A \cap \left\{\omega \in \Omega : \left|\langle \xi(\omega), h_n \rangle_{H(\mathcal{H},T)}\right| > \frac{\epsilon}{2}\right\}$$
$$\cap \left\{\omega \in \Omega : \langle \xi(\omega), h_n \rangle_{H(\mathcal{H},T)} < 0\right\}.$$

Then

$$\begin{aligned}
\Phi_A(h_n) &= \\
&= \int_A \left| \langle \xi(\omega), h_n \rangle_{H(\mathcal{H},T)} \right| P(d\omega) \\
&\leq \int_{A \cap \left\{ \left| \langle \xi, h_n \rangle_{H(\mathcal{H},T)} \right| > \frac{\epsilon}{2} \right\}} \left| \langle \xi(\omega), h_n \rangle_{H(\mathcal{H},T)} \right| P(d\omega) + \frac{\epsilon}{2} \\
&= \int_{A_n^+[\epsilon]} \langle \xi(\omega), h_n \rangle_{H(\mathcal{H},T)} P(d\omega) - \int_{A_n^-[\epsilon]} \langle \xi(\omega), h_n \rangle_{H(\mathcal{H},T)} P(d\omega) + \frac{\epsilon}{2} \\
&= \left\langle h_n, \mu\left(A_n^+[\epsilon]\right) \right\rangle_{H(\mathcal{H},T)} - \left\langle h_n, \mu\left(A_n^-[\epsilon]\right) \right\rangle_{H(\mathcal{H},T)} + \frac{\epsilon}{2} \\
&\leq \|h_n\|_{H(\mathcal{H},T)} \left\{ \left\| \mu\left(A_n^+[\epsilon]\right) \right\|_{H(\mathcal{H},T)} + \left\| \mu\left(A_n^-[\epsilon]\right) \right\|_{H(\mathcal{H},T)} \right\} + \frac{\epsilon}{2}.
\end{aligned}$$

As

$$A_n^+[\epsilon] \subseteq \left\{ \omega \in \Omega : \left| \langle \xi(\omega), h_n \rangle_{H(\mathcal{H},T)} \right| > \frac{\epsilon}{2} \right\},$$

and

$$A_n^+[\epsilon] \subseteq \left\{ \omega \in \Omega : \left| \langle \xi(\omega), h_n \rangle_{H(\mathcal{H},T)} \right| > \frac{\epsilon}{2} \right\},$$

one has that, for $n \geq n(\epsilon)$,

$$\Phi_A(h_n) \leq \frac{\sqrt{\epsilon}}{4} \left\{ 2\sqrt{\epsilon} \right\} + \frac{\epsilon}{2} = \epsilon.$$

□

Example 1.7.35 (Covariance Operator in an RKHS) Assume that $\omega \mapsto \|\xi(\omega)\|_{H(\mathcal{H},T)}$ is adapted, and that

$$E_P\left[\|\xi(\cdot)\|^2_{H(\mathcal{H},T)} \right] < \infty.$$

Then the map $\omega \mapsto \langle h, \xi(\omega) \rangle_{H(\mathcal{H},T)} \xi(\omega)$ is adapted, and Bochner integrable. The operator $R_\xi : H(\mathcal{H},T) \longrightarrow H(\mathcal{H},T)$ defined using the following assignment:

$$h \mapsto E_P\left[\langle h, \xi(\omega) \rangle_{H(\mathcal{H},T)} \xi(\omega) \right]$$

is well defined, and continuous, since the second moment of the norm of ξ exists, and that, for Bochner integrals, the norm of the integral is less than the integral of the norm [135, p. 82]. R_ξ is, by definition, the covariance operator of ξ. Its kernel

$\mathcal{K}_\xi$ is obtained from the following calculation:

$$\begin{aligned}\mathcal{K}_\xi(x,t) &= R_\xi^\star[\mathcal{H}(\cdot,t)](x)\\ &= \langle \mathcal{H}(\cdot,x), R_\xi^\star[\mathcal{H}(\cdot,t)]\rangle_{H(\mathcal{H},T)}\\ &= \langle R_\xi[\mathcal{H}(\cdot,x)], \mathcal{H}(\cdot,t)\rangle_{H(\mathcal{H},T)}\\ &= \langle E_P\left[\langle \mathcal{H}(\cdot,x), \xi(\omega)\rangle_{H(\mathcal{H},T)}\xi(\omega)\right], \mathcal{H}(\cdot,t)\rangle_{H(\mathcal{H},T)}\\ &= E_P\left[\langle \mathcal{H}(\cdot,x), \xi(\omega)\rangle_{H(\mathcal{H},T)}\langle \mathcal{H}(\cdot,t), \xi(\omega)\rangle_{H(\mathcal{H},T)}\right]\\ &= E_P[X(\cdot,x)X(\cdot,t)].\end{aligned}$$

Thus the kernel of the operator R_ξ is, when the mean of ξ is zero, the covariance of the stochastic process with paths in $H(\mathcal{H},T)$ that ξ determines.

The operator R_ξ is also compact. Indeed, when $\{h_n,\ n\in\mathbb{N}\}\subseteq H(\mathcal{H},T)$ converges weakly to zero, that sequence is bounded in norm by some constant $\kappa\geq 0$ [266, p. 79], $\lim_n\langle h_n,\xi(\omega)\rangle_{H(\mathcal{H},T)}\xi(\omega)=0_{H(\mathcal{H},T)}$, and

$$\left\|\langle h_n,\xi(\omega)\rangle_{H(\mathcal{H},T)}\xi(\omega)\right\|_{H(\mathcal{H},T)}\leq\kappa\left\|\xi(\omega)\right\|_{H(\mathcal{H},T)}^2,$$

which is integrable. Consequently [135, p. 83]

$$\begin{aligned}\lim_n R_\xi[h_n] &= \lim_n\int_\Omega\langle h_n,\xi(\omega)\rangle_{H(\mathcal{H},T)}\xi(\omega)\,P(d\omega)\\ &= \int_\Omega\lim_n\left\{\langle h_n,\xi(\omega)\rangle_{H(\mathcal{H},T)}\xi(\omega)\right\}P(d\omega)\\ &= 0.\end{aligned}$$

Chapter 2
The Functions of a Reproducing Kernel Hilbert Space

The functions of an RKHS, that is, the signals one works with, may not be easy to handle *per se*. In those cases, it may sometimes be useful to know that they belong to spaces in which computation may be easier, such as L_2 spaces. This chapter deals with such questions. The correspondence between RKHS's and other spaces of functions follows naturally from the construction of (Proposition) 1.1.15.

The behavior of functions in an RKHS reflects that of the map F which determines its reproducing kernel [(Proposition) 1.1.15]. Since $F(t)$ can be $\mathcal{H}(\cdot, t)$ (so that L_F is the identity), the behavior of the kernel also determines that of the functions in its RKHS. Conditions on F are often simpler to state, and understand, than those bearing, equivalently, on the corresponding reproducing kernel. One concentrates here on measurability, integrability, continuity, and range inclusion which are of most interest in signal detection. But differentiability properties are also ubiquitous [7, 35].

It should be noted that a separability assumption is often necessary. It usually serves to have that F is measurable.

The concepts which follow often find easy illustrations through the use of covariances that have a factorization [Sect. 1.4].

2.1 Kernels and the Operators They Determine

The map F being that mentioned above, the definitions which follow specify those of its weak properties that shall be considered in the sequel, as well as their meaning, in terms of the associated RKHS's.

A.F. Gualtierotti, *Detection of Random Signals in Dependent Gaussian Noise*,
DOI 10.1007/978-3-319-22315-5_2

Definition 2.1.1 q is a quasi-norm, on the vector space E, when the following obtain, for $\{\lambda, \lambda_n, n \in \mathbb{N}\} \subseteq \mathbb{R}$ and $\{x, y\} \subseteq E$, fixed, but arbitrary:

1. $q(x) = 0$ if, and only if, $x = 0$;
2. $q(\lambda x) \leq q(x)$ when $-1 \leq \lambda \leq 1$;
3. $q(x + y) \leq q(x) + q(y)$;
4. $\lim_n q(\lambda_n x) = 0$ whenever $\lim_n \lambda_n = 0$.

Example 2.1.2 The function $q : \mathbb{R} \longrightarrow \mathbb{R}$ defined as $q(x) = \frac{|x|}{1+|x|}$ is a quasi-norm as the map $t \geq 0$, $t \mapsto \frac{t}{1+t}$ is increasing, continuous, and sub-additive.

Definition 2.1.3 ([46, p. 364]) Let T be a set, $\mathcal{T}$ be a σ-algebra of subsets of T, and τ be a σ-finite measure on $\mathcal{T}$. Then:

1. $\mathcal{L}_0(T, \mathcal{T}, \tau)$ is the real vector space of functions adapted to $\mathcal{T}$ and $\mathcal{B}(\mathbb{R})$. When convergence in $\mathcal{L}_0(T, \mathcal{T}, \tau)$ is convergence in measure on subsets of finite measure, one has there a topological vector space.
2. Given $p \in [1, \infty[$, fixed, but arbitrary, $\mathcal{L}_p(T, \mathcal{T}, \tau)$ is the pseudo(semi)normed space of those functions f in $\mathcal{L}_0(T, \mathcal{T}, \tau)$ for which the function $t \mapsto |f(t)|^p$ is integrable.
3. $\mathcal{L}_\infty(T, \mathcal{T}, \tau)$ is the pseudo(semi)-normed space of those functions f in $\mathcal{L}_0(T, \mathcal{T}, \tau)$ for which there exists $N_f \in \mathcal{T}$ such that $\tau(N_f) = 0$, and f, restricted to N_f^c, is bounded.
4. When f and g belong to $\mathcal{L}_0(T, \mathcal{T}, \tau)$, and $\{t \in T : f(t) \neq g(t)\}$ has measure zero for τ, one writes $f \sim g$, and obtains, so doing, an equivalence relation. The vector space of equivalence classes of functions in $\mathcal{L}_0(T, \mathcal{T}, \tau)$ is denoted $L_0(T, \mathcal{T}, \tau)$. $L_p(T, \mathcal{T}, \tau)$ and $L_\infty(T, \mathcal{T}, \tau)$ have a similar definition, and are Banach spaces.
5. The following relation:

$$\left\| [f]_{L_0(T,\mathcal{T},\tau)} \right\|_A = \int_A \frac{|f(t)|}{1 + |f(t)|} \tau(dt), \; A \in \mathcal{T}, \; \tau(A) < \infty,$$

   defines a family of quasinorms on $L_0(T, \mathcal{T}, \tau)$ such that

$$\lim_n \left\| [f]_{L_0(T,\mathcal{T},\tau)} - [f_n]_{L_0(T,\mathcal{T},\tau)} \right\|_A = 0$$

   if, and only if, $\{f_n, \; n \in \mathbb{N}\}$ converges in measure to f on A.

Definition 2.1.4 Let T be a set, H be a real Hilbert space, and $F : T \longrightarrow H$ be a map. Let $\mathcal{T}$ be a σ-algebra of subsets of T. F is weakly measurable with respect to $\mathcal{T}$ and $\mathcal{B}(\mathbb{R})$ when, for $h \in H$, fixed, but arbitrary, the real valued map $t \mapsto L_F[h](t) = \langle h, F(t) \rangle_H$ is adapted to $\mathcal{T}$ and $\mathcal{B}(\mathbb{R})$.

Definition 2.1.5 Let T be a set, H be a real Hilbert space, and $F : T \longrightarrow H$ be a map. Let $\mathcal{T}$ be a σ-algebra of subsets of T, and τ be a σ-finite measure on $\mathcal{T}$.

1. Let $p \in [1, \infty[$ be fixed, but arbitrary. F is weakly p-integrable with respect to $\mathcal{T}$, $\mathcal{B}(\mathbb{R})$, and τ, when, for $h \in H$, fixed, but arbitrary, the real valued map $t \mapsto L_F[h](t) = \langle h, F(t)\rangle_H$ belongs to $\mathcal{L}_p(T, \mathcal{T}, \tau)$.
2. F is weakly in $\mathcal{L}_\infty$ with respect to $\mathcal{T}$, $\mathcal{B}(\mathbb{R})$, and τ, when, for $h \in H$, fixed, but arbitrary, the real valued map $t \mapsto L_F[h](t) = \langle h, F(t)\rangle_H$ belongs to $\mathcal{L}_\infty(T, \mathcal{T}, \tau)$.

Definition 2.1.6 Let T be a set, H be a real Hilbert space, and $F : T \longrightarrow H$ be a map. Let T be a topological space. F is weakly continuous with respect to the topology of T, when, for $h \in H$, fixed, but arbitrary, the real valued map $t \mapsto L_F[h](t) = \langle h, F(t)\rangle_H$ is continuous.

Definition 2.1.7 Let T be a set, H be a real Hilbert space, and $F : T \longrightarrow H$ be a map. Let $\mathcal{T}$ be a σ-algebra of subsets of T, and τ be a σ-finite measure on $\mathcal{T}$. Let $B(h, \epsilon)$ be the open ball, centered at h, whose radius is $\epsilon > 0$. When F is adapted to $\mathcal{T}$ and $\mathcal{B}(H)$, let

$$\mathcal{R}[F, \tau] = \left\{h \in H : \tau\left(F^{-1}\{B(h, \epsilon)\}\right) > 0, \text{ all } \epsilon > 0\right\}.$$

The essential range of F (for $\mathcal{T}$ and τ) is the closure, in H, of $V[\mathcal{R}[F, \tau]]$, the manifold linearly generated from $\mathcal{R}[F, \tau]$. It shall be denoted $H_{F,\tau}$.

Example 2.1.8 In the context of (Remark) 1.4.16, H is the space $L_2(T, \mathcal{T}, \tilde{\tau}_C)$, and $F(t)$, $I_{[t_l, t]}$. Thus

$$\begin{aligned}\|F(t_0) - F(t)\|_H^2 &= r^C(t_0) - 2r^C(t_0 \wedge t) + r^C(t)\\ &= (\text{when } t_0 \le t)\ r^C(t) - r^C(t_0)\\ &= (\text{when } t_0 > t)\ r^C(t_0) - r^C(t).\end{aligned}$$

The map $t \mapsto F(t)$ is thus continuous when r^C is, and

$$\{t \in T : \|F(t_0) - F(t)\|_H < \epsilon\}$$

is open and has positive Lebesgue measure. Consequently $F(t_0) \in \mathcal{R}[F, Leb]$, and the essential range of F is H.

Proposition 2.1.9 *Let H_F be the (closed) subspace of H linearly generated by F, that is, $H_F = \overline{V[\mathcal{R}[F]]}$. $\mathcal{R}[F, \tau]$ is a closed subset of H, contained in H_F, so that $H_{F,\tau} \subseteq H_F$.*

Proof Let $\{h_n, n \in \mathbb{N}\} \subseteq \mathcal{R}[F, \tau]$ converge to $h \in H$. For fixed, but arbitrary $\epsilon > 0$, eventually, $h_n \in B(h, \epsilon)$. But then there exists $\delta > 0$ such that $B(h_n, \delta) \subseteq B(h, \epsilon)$. Consequently,

$$\tau\left(F^{-1}(B(h, \epsilon))\right) \ge \tau\left(F^{-1}(B(h_n, \delta))\right) > 0,$$

and $h \in \mathcal{R}[F, \tau]$.

Suppose that $h \in \mathcal{R}[F, \tau]$. Then, for $n \in \mathbb{N}$, fixed, but arbitrary, there is $t_n \in T$ such that $F(t_n) \in B\left(h, \frac{1}{n}\right)$. Thus $H_F \ni \lim_n F(t_n) = h$. □

The next definition introduces the kernels which allow one to define integral operators between L_p-spaces.

Definition 2.1.10 Let T and X be sets. Suppose $\mathcal{T}$ and $\mathcal{X}$ are σ-algebras of subsets respectively of T and X, and τ and μ are σ-finite measures on $\mathcal{T}$ and $\mathcal{X}$ respectively. Let $1 \leq p, q \leq \infty$ be fixed, but arbitrary numbers, and $\mathcal{K} : T \times X \longrightarrow \mathbb{R}$ be a map adapted to $\mathcal{T} \otimes \mathcal{X}$ and $\mathcal{B}(\mathbb{R})$.

When $\mathcal{K}$ has the following properties for all fixed, but arbitrary

$$f \in \mathcal{L}_q(X, \mathcal{X}, \mu),$$

one says that $\mathcal{K}$ is a (p, q)-bounded kernel for $L_p(T, \mathcal{T}, \tau)$ and $L_q(X, \mathcal{X}, \mu)$:

1. there exists $N_f \in \mathcal{T}$ such that
   a) $\tau(N_f) = 0$,
   b) for $t \in N_f^c$, $x \mapsto \mathcal{K}(t, x) f(x)$ belongs to $\mathcal{L}_1(X, \mathcal{X}, \mu)$;
2. the map $\phi : T \longrightarrow \mathbb{R}$ defined using the following relation:

$$\phi(t) = \begin{cases} \int_X \mathcal{K}(t, x) f(x) \, \mu(dx) & \text{when } t \in N_f^c \\ 0 & \text{when } t \in N_f \end{cases}$$

   belongs to $\mathcal{L}_p(T, \mathcal{T}, \tau)$.

The assignment

$$[f]_{L_q(X, \mathcal{X}, \mu)} \mapsto [\phi]_{L_p(T, \mathcal{T}, \tau)}$$

produces a linear operator

$$L_{\mathcal{K}} : L_q(X, \mathcal{X}, \mu) \longrightarrow L_p(T, \mathcal{T}, \tau).$$

The following lemma prepares (Proposition) 2.1.12.

Lemma 2.1.11 *Let $p \in [1, \infty]$ be fixed, but arbitrary, and $\{g, g_n, n \in \mathbb{N}\}$, a subset of $\mathcal{L}_p(X, \mathcal{X}, \mu)$, be such that*

$$\lim_n \left\| [g]_{L_p(X, \mathcal{X}, \mu)} - [g_n]_{L_p(X, \mathcal{X}, \mu)} \right\|_{L_p(X, \mathcal{X}, \mu)} = 0.$$

One may then assume that there exists a positive $\gamma \in \mathcal{L}_p(X, \mathcal{X}, \mu)$, and a set $G \in \mathcal{X}$, such that

1. $\mu(G) = 0$,

2. *for* $x \in G^c$, *fixed, but arbitrary,*
   a) $|g_n(x)| \leq \gamma(x)$,
   b) $\lim_n g_n(x) = g(x)$.

Proof It relies on standard proofs of the Fischer-Riesz theorem on the completeness of L_p-spaces [113, p. 159].

The case of $p = \infty$: The definition of the essential supremum norm [155, p. 168] says that there exists $G \in \mathcal{X}$ such that $\mu(G) = 0$, and, for $x \in G^c$, fixed, but arbitrary,

$$|g(x) - g_n(x)| \leq \left\| [g]_{L_\infty(X,\mathcal{X},\mu)} - [g_n]_{L_\infty(X,\mathcal{X},\mu)} \right\|_{L_\infty(X,\mathcal{X},\mu)}.$$

The latter inequality thus suffices to yield the lemma.

The case of $p \in [1, \infty[$: One chooses first a subsequence $\{g_{n_p}, p \in \mathbb{N}\}$ such that

$$\left\| \left[g_{n_{p+1}}\right]_{L_p(X,\mathcal{X},\mu)} - \left[g_{n_p}\right]_{L_p(X,\mathcal{X},\mu)} \right\|_{L_p(X,\mathcal{X},\mu)} \leq \frac{1}{2^p}.$$

One then sets

$$\gamma_p(x) = \sum_{i=1}^{p-1} \left|g_{n_{i+1}}(x) - g_{n_i}(x)\right|, \quad \gamma(x) = \sum_{i=1}^{\infty} \left|g_{n_{i+1}}(x) - g_{n_i}(x)\right|.$$

One has that [113, p. 160]

$$\left\| [\gamma]_{L_p(X,\mathcal{X},\mu)} \right\|_{L_p(X,\mathcal{X},\mu)} \leq \sum_{i=1}^{\infty} \left\| \left| [g_{n_{i+1}}]_{L_p(X,\mathcal{X},\mu)} - [g_{n_i}]_{L_p(X,\mathcal{X},\mu)} \right| \right\|_{L_p(X,\mathcal{X},\mu)}.$$

The choice made for the subsequence yields that the norm of γ is finite, and, in particular, that γ is almost surely finite. Consequently the series with general term $g_{n_{i+1}}(x) - g_{n_i}(x)$ is absolutely convergent, almost surely with respect to μ. But then

$$g_{n_p}(x) = \sum_{i=1}^{p-1} \left\{g_{n_{i+1}}(x) - g_{n_i}(x)\right\} + g_{n_1}(x)$$

is almost surely convergent, necessarily to g, and

$$\left|g_{n_p}(x)\right| \leq \sum_{i=1}^{\infty} \left|g_{n_{i+1}}(x) - g_{n_i}(x)\right| + |g_{n_1}(x)| = \gamma(x) + |g_{n_1}(x)|.$$

But the latter is an $\mathcal{L}_p$-function. □

Proposition 2.1.12 *Let $\mathcal{K}$ be a (p,q)-bounded kernel for $L_p(T,\mathcal{T},\tau)$ and $L_q(X,\mathcal{X},\mu)$, and $L_{\mathcal{K}} : L_q(X,\mathcal{X},\mu) \longrightarrow L_p(T,\mathcal{T},\tau)$ be the associated linear operator* [(Definition) 2.1.10]. *$L_{\mathcal{K}}$ is bounded.*

Proof Since a closed, linear operator, defined everywhere, is bounded [266, p. 94], it will suffice to prove that $L_{\mathcal{K}}$ is closed. Let thus the sequence $\{g_n, n \in \mathbb{N}\} \subseteq L_q(X,\mathcal{X},\mu)$ converge, in $L_q(X,\mathcal{X},\mu)$, to g, and the sequence $\{f_n = L_{\mathcal{K}}[g_n], n \in \mathbb{N}\} \subseteq L_p(T,\mathcal{T},\tau)$ converge, in $L_p(T,\mathcal{T},\tau)$, to f. One must prove that $L_{\mathcal{K}}[g] = f$.

Lemma 2.1.11 allows one to assume that there exists

- a positive $\gamma \in \mathcal{L}_q(X,\mathcal{X},\mu)$, and a $G \in \mathcal{X}$ such that
  - $\mu(G) = 0$,
  - for $x \in G^c$, $|\dot{g}_n(x)| \le \gamma(x)$, and $\lim_n \dot{g}_n(x) = \dot{g}(x)$,
- a positive $\phi \in \mathcal{L}_p(T,\mathcal{T},\tau)$, and an $F \in \mathcal{T}$ such that
  - $\tau(F) = 0$,
  - for $t \in F^c$, $\left|\dot{f}_n(t)\right| \le \phi(t)$, and $\lim_n \dot{f}_n(t) = \dot{f}(t)$.

The assumptions on the kernel $\mathcal{K}$ yield that there exists

- $T_0 \in \mathcal{T}$ such that
  - $\tau(T_0) = 0$,
  - for $t \in T_0^c$, the equivalence class of the map $x \mapsto \mathcal{K}(t,x)\gamma(x)$ belongs to $L_1(X,\mathcal{X},\mu)$,
- for $n \in \mathbb{N}$, $T_n \in \mathcal{T}$ such that
  - $\tau(T_n) = 0$,
  - for $t \in T_n^c$, the equivalence class of the map $x \mapsto \mathcal{K}(t,x)\dot{g}_n(x)$ belongs to $L_1(X,\mathcal{X},\mu)$.

Let $N_0 = \cup_{n \ge 0} T_n$. Then, outside of G,

$$\lim_n \mathcal{K}(t,x)\dot{g}_n(x) = \mathcal{K}(t,x)\dot{g}(x), \text{ and } |\mathcal{K}(t,x)\dot{g}_n(x)| \le |\mathcal{K}(t,x)|\gamma(x),$$

and, outside of N_0, by dominated convergence [113, p. 163],

$$\lim_n \dot{f}_n(t) = \lim_n \int_X \mathcal{K}(t,x)\dot{g}_n(x)\mu(dx) = \int_X \mathcal{K}(t,x)\dot{g}(x)\mu(dx).$$

But, outside of F, $\lim_n \dot{f}_n(t) = \dot{f}(t)$, and, furthermore, $\lim_n f_n = f$ in $L_p(T,\mathcal{T},\tau)$, by dominated convergence. Consequently,

$$f = L_{\mathcal{K}}[g].$$

□

Among the kernels of integral operators, Carleman kernels are those that have particularly useful properties. Their definition follows.

Definition 2.1.13 Let T and X be sets. Suppose $\mathcal{T}$ and $\mathcal{X}$ are σ-algebras of subsets respectively of T and X, and τ and μ are σ-finite measures on $\mathcal{T}$ and $\mathcal{X}$ respectively. Let $1 \leq p, q \leq \infty$ be fixed, but arbitrary numbers, and $\mathcal{K} : T \times X \longrightarrow \mathbb{R}$ be a (p, q)-bounded kernel. Let

$$\alpha(q) = \begin{cases} \infty \text{ when } q = 1 \\ \frac{q}{q-1} \text{ when } 1 < q < \infty \\ 1 \text{ when } q = \infty \end{cases}.$$

When there exists $T_0 \in \mathcal{T}$ such that

1. $\tau(T_0) = 0$,
2. for $t \in T_0^c$, $x \mapsto \mathcal{K}(t, x)$ belongs to $\mathcal{L}_{\alpha(q)}(X, \mathcal{X}, \mu)$,

$\mathcal{K}$ is called a Carleman, (p, q)-bounded kernel.

Remark 2.1.14 When $p = \frac{q}{q-1}$, $\frac{1}{p} + \frac{1}{q} = 1$.

Remark 2.1.15 Square integrable kernels are examples of $(2, 2)$-bounded Carleman kernels. For properties of Carleman kernels, one may consult [266, p. 132].

Remark 2.1.16 Let $C = c_\wedge c_\vee$ be a covariance with a factorization [(Definition) 1.4.1]. When $c_\wedge$ and $c_\vee$ are both adapted, C is adapted. As

$$\begin{aligned} \int_T |C(x,t)|\,|f(x)|\,\mu(dx) &= |c_\vee(t)| \int_{[t_l,t]} |c_\wedge(x)|\,|f(x)|\,\mu(dx) \\ &\quad + |c_\wedge(t)| \int_{]t,t_r]} |c_\vee(x)|\,|f(x)|\,\mu(dx) \\ &= \phi(t) + \psi(t), \end{aligned}$$

that

$$\{\phi(t) + \psi(t)\}^p \leq 2^{p-1}\{\phi^p(t) + \psi^p(t)\},$$

because $p \geq 1$, $\phi(t) \geq 0$, $\psi(t) \geq 0$, and that

$$\phi^p(t) = |c_\vee(t)|^p \left\{\int_{[0,t]} |c_\wedge(x)|\,|f(x)|\,\mu(dx)\right\}^p,$$

one sees that, when choosing appropriately integrable $c_\wedge$ and $c_\vee$, (Proposition) 2.1.12 will hold.

2.2 Reproducing Kernel Hilbert Spaces of Measurable Functions

One is really interested here in RKHS's whose functions are integrable. As a preliminary step, one must be concerned with their measurability.

Proposition 2.2.1 *Let $H(\mathcal{H}, T)$ be an RKHS, and suppose that $\mathcal{T}$ is a σ–algebra of subsets of T. Whenever the family $\{\mathcal{H}(\cdot, t), t \in T\}$ is made of functions adapted to $\mathcal{T}$, $H(\mathcal{H}, T)$ too consists of functions that are adapted to $\mathcal{T}$.*

Proof By assumption, $V[\mathcal{H}]$ is made of functions that are adapted to $\mathcal{T}$. But every function of $H(\mathcal{H}, T)$ is the pointwise limit of functions in $V[\mathcal{H}]$ [(Corollary) 1.1.10], and so it must be adapted to $\mathcal{T}$. □

Proposition 2.2.2 *Let T be a set, and $\mathcal{T}$ be a σ-algebra of subsets of T. Let H be a real Hilbert space, and $F : T \longrightarrow H$ be a map. Suppose that H_F is separable. The following statements are then equivalent:*

1. *F is weakly measurable;*
2. *F is adapted to $\mathcal{T}$ and $\mathcal{B}(H)$;*
3. *$\mathcal{H}_F$ is adapted to $\mathcal{T} \otimes \mathcal{T}$ and $\mathcal{B}(\mathbb{R})$;*
4. *for $t \in T$, fixed, but arbitrary, $\theta \mapsto \mathcal{H}_F(\theta, t)$ is adapted to $\mathcal{T}$ and $\mathcal{B}(\mathbb{R})$.*

When τ is a σ-finite measure on $\mathcal{T}$, and the latter statements obtain, then $H(\mathcal{H}_F, T) = \mathcal{R}[L_F] \subseteq \mathcal{L}_0(T, \mathcal{T}, \tau)$, so that one may define

$$J_{F,0} : H(\mathcal{H}_F, T) \longrightarrow L_0(T, \mathcal{T}, \tau)$$

using the following assignment:

$$J_{F,0}[h] = [h]_{L_0(T,\mathcal{T},\tau)}.$$

Then:

(i) *$J_{F,0}$ is an operator which is linear and bounded;*
(ii) *$\Lambda_{F,0} : H \longrightarrow L_0(T, \mathcal{T}, \tau)$ defined using $\Lambda_{F,0} = J_{F,0} L_F$ is linear and bounded;*
(iii) *$\mathcal{N}[\Lambda_{F,0}] = H_{F,\tau}^{\perp}$* [(Definition) 2.1.7].

Proof (1 ⇒ 2) Let $\{e_i, i \in I\} \subseteq H_F$ be a countable, complete orthonormal set. Then, for fixed, but arbitrary $t \in T$,

$$F(t) = \sum_{i \in I} \langle F(t), e_i \rangle_H e_i = \sum_{i \in I} L_F[e_i](t)\, e_i.$$

F has thus separable range. Since it is furthermore weakly measurable, it is (strongly) measurable [260, p. 89].

Proof (2 ⇒ 3) Let Φ be the map on $H \oplus H$ defined using the following relation: for fixed, but arbitrary $(h_1, h_2) \in H \times H$,

$$(h_1, h_2) \mapsto \langle h_1, h_2 \rangle_H,$$

and Ψ be the map on $T \times T$ defined using the following relation: for fixed, but arbitrary $(t_1, t_2) \in T \times T$,

$$(t_1, t_2) \mapsto (F(t_1), F(t_2)).$$

Φ is continuous, and thus adapted, and Ψ is adapted by assumption. Since $\mathcal{H}_F = \Phi \circ \Psi$, it is adapted too.

Proof (3 ⇒ 4) The sections of a measurable map are measurable.

Proof (4 ⇒ 1) Let $h = \sum_{i=1}^{n} \alpha_i F(t_i)$. Then

$$L_F[h](t) = \langle h, F(t) \rangle_H = \sum_{i=1}^{n} \alpha_i \mathcal{H}_F(t, t_i),$$

and thus $t \mapsto L_F[h](t)$ is adapted to $\mathcal{T}$ and $\mathcal{B}(\mathbb{R})$. But H_F is the closure in H of the set of expressions of the form h above, and thus, since convergence in H implies point-wise convergence in $\mathcal{R}[L_F]$, $t \mapsto L_F[h](t)$ is adapted to $\mathcal{T}$ and $\mathcal{B}(\mathbb{R})$, for all $h \in H_F$. For $h \in H_F^{\perp}$, $t \mapsto L_F[h](t) = 0$ is adapted to $\mathcal{T}$ and $\mathcal{B}(\mathbb{R})$. Consequently $t \mapsto L_F[h](t)$ is adapted to $\mathcal{T}$ and $\mathcal{B}(\mathbb{R})$ for all $h \in H$.

To see that $J_{F,0}$ is continuous, let $\{h_n,\ n \in \mathbb{N}\} \subseteq H(\mathcal{H}_F, T)$ converge to h. Then, for every $t \in T$, $\lim_n h_n(t) = h(t)$ [(Proposition) 1.1.9]. The function q, met in [(Example) 2.1.2], that is, $q : t \mapsto \frac{t}{1+t}$, t in $[0, \infty[$, has the following properties:

1. $q(0) = 0$;
2. $q(t) \leq 1$;
3. q is continuous and strictly increasing;
4. $q(t+u) \leq q(t) + q(u)$.

Consequently, for $A \in \mathcal{T}$ such that $\tau(A) < \infty$,

$$\lim_n \int_A q(h(t) - h_n(t))\, \tau(dx) = 0$$

(dominated convergence), so that, for convergence in measure [(Definition) 2.1.3, 5],

$$\lim_n J_{F,0}(h_n) = J_{F,0}(h).$$

$\Lambda_{F,0}$ is then continuous as the composition of continuous functions.

It remains to characterize the kernel of $\Lambda_{F,0}$. By definition

$$H_F^{\perp} = \mathcal{N}[L_F] \subseteq \mathcal{N}[\Lambda_{F,0}],$$

and, by (Proposition) 2.1.9, $H_{F,\tau} \subseteq H_F$ so that

$$H_F^{\perp} \subseteq H_{F,\tau}^{\perp}.$$

It is thus no restriction, given the assertion to be proved, to assume that $H_F^{\perp}$ reduces to the zero element, that is, that $H_F = H$, so that the latter is separable. Furthermore, since [266, p. 35]

$$\mathcal{R}[F,\tau]^{\perp} = V\left[\mathcal{R}[F,\tau]\right]^{\perp} = \overline{V\left[\mathcal{R}[F,\tau]\right]}^{\perp} = H_{F,\tau}^{\perp},$$

it is sufficient to prove that $\mathcal{N}[\Lambda_{F,0}] = \mathcal{R}[F,\tau]^{\perp}$.

Let $T_0 = \{t \in T : F(t) \in \mathcal{R}[F,\tau]^c\}$. Since $\mathcal{R}[F,\tau]^c$ is open [(Proposition) 2.1.9], it belongs to $\mathcal{B}(H)$, the Borel sets of H, and, since F is adapted to $\mathcal{T}$ and $\mathcal{B}(H)$, $T_0 \in \mathcal{T}$.

One shall first prove that $\tau(T_0) = 0$, which is needed below. To that end, let $t_0 \in T_0$ be fixed, but arbitrary. Let $B(h,\epsilon)$ denote the open ball centered at h, of radius $\epsilon > 0$. By the definition of T_0, there must be $\epsilon_0 > 0$ such that

$$\tau\left(F^{-1}\{B(F(t_0),\epsilon_0)\}\right) = 0.$$

Since H is separable, it has a countable base of open sets $\{O_i,\ i \in I\}$ [270, p. 112], and

$$B(F(t_0),\epsilon_0) = \cup_{O_i \subseteq B(F(t_0),\epsilon_0)} O_i.$$

Let $J \subseteq I$ be the set of indices for which there exists $t_0 \in T_0$ and $\epsilon_0 > 0$ with

$$\tau\left(F^{-1}\{B(F(t_0),\epsilon_0)\}\right) = 0, \text{ and } F(t_0) \in O_j \subseteq B(F(t_0),\epsilon_0).$$

Then $T_0 \subseteq \cup_{j \in J} F^{-1}\{O_j\}$, and, consequently,

$$\tau(T_0) \leq \sum_{j \in J} \tau\left(F^{-1}\{O_j\}\right) = 0.$$

Let now $h \in \mathcal{R}[F,\tau]^{\perp}$. Suppose $t \in T_0^c$. Then $F(t) \in \mathcal{R}[F,\tau]$, and, since h and $F(t)$ are then orthogonal, $L_F[h](t) = \langle h, F(t)\rangle_H = 0$. Since $\tau(T_0) = 0$, $\Lambda_{F,0}[h] = J_{F,0}(L_F[h]) = [0]_{L_0(T,\mathcal{T},\tau)}$. Thus

$$\mathcal{R}[F,\tau]^{\perp} \subseteq \mathcal{N}[\Lambda_{F,0}].$$

Suppose next that $h \in \mathcal{N}[\Lambda_{F,0}]$, and that there exists $k_0 \in \mathcal{R}[F, \tau]$ such that h and k_0 are not orthogonal, that is, $|\langle h, k_0 \rangle_H| > 0$. Since the map $k \mapsto \langle h, k \rangle_H$ is continuous, there exists $\epsilon > 0$ such that

$$|\langle h, k \rangle_H| > 0, \ k \in B(k_0, \epsilon).$$

Let $t \in F^{-1}\{B(k_0, \epsilon)\}$ be fixed, but arbitrary. Then $F(t) \in B(k_0, \epsilon)$, so that $|\langle h, F(t) \rangle_H| > 0$. Since $k_0 \in \mathcal{R}[F, \tau]$, $\tau\left(F^{-1}\{B(k_0, \epsilon)\}\right) > 0$, and it follows thus that $|L_F[h](t)| > 0$ on a set of strictly positive measure: h cannot belong to $\mathcal{N}[\Lambda_{F,0}]$, a contradiction. One must thus have that h belongs to $\mathcal{R}[F, \tau]^{\perp}$, and consequently

$$\mathcal{N}[\Lambda_{F,0}] \subseteq \mathcal{R}[F, \tau]^{\perp}.$$

□

Remark 2.2.3 When H_F is not separable, the proposition may not hold. Here is an example.

Let $T = [0, 1]$, $\mathcal{T}$ be the Lebesgue sets, and τ be Lebesgue measure. Let H be the Hilbert space of functions $f : T \longrightarrow \mathbb{R}$ for which $\sum_{t \in T} |f(t)|^2 < \infty$, with the inner product $\langle f, g \rangle_H = \sum_{t \in T} f(t) g(t)$ [8, 266, pp. 21, 27, 34, respectively 144]. Let A be a set that is not measurable [155, p. 93], and

$$\delta_t(x) = \begin{cases} 1 \text{ when } x = t \\ 0 \text{ when } x \neq t \end{cases}.$$

Then

$$\langle \delta_{t_1}, \delta_{t_2} \rangle_H = \begin{cases} 1 \text{ when } t_1 = t_2 \\ 0 \text{ when } t_1 \neq t_2 \end{cases}.$$

Let

$$F(t) = \begin{cases} \delta_t & \text{when } t \in A \\ 0_{\mathbb{R}^T} & \text{when } t \in A^c \end{cases}.$$

If $\mathcal{H}_F : (t_1, t_2) \mapsto \langle F(t_1), F(t_2) \rangle_H$ were measurable, $t \mapsto (t, t)$ composed with $\mathcal{H}_F$ would be measurable. But the latter is χ_A. Furthermore $\|F(t)\|_H^2 = \chi_A(t)$, so that, for $0 < \epsilon < 1$,

$$F^{-1}(B(0, \epsilon)) = \left\{t \in T : \|F(t)\|_H^2 < \epsilon^2\right\} = A^c.$$

Consequently, neither $\mathcal{H}_F$ nor F are measurable. However,

$$\langle h, F(t) \rangle_H = \left\{ \begin{array}{lll} \langle h, \delta_t \rangle_H & = h(t) & \text{when } t \in A \\ \langle h, 0_T \rangle_H & = 0 & \text{when } t \in A^c \end{array} \right\} = \chi_A(t) h(t).$$

Since $h(t) \neq 0$ for at most a countable number t's, F is thus weakly measurable.

Remark 2.2.4 As seen in (Proposition) 2.2.1, item 4 of (Proposition) 2.2.2 is enough to have the functions in the RKHS measurable.

Remark 2.2.5 Suppose that $f : T \longrightarrow \mathbb{R}$ is adapted to $\mathcal{T}$ and $\mathcal{B}(\mathbb{R})$, and let $\tilde{F} = f \times F$. When F is adapted to $\mathcal{T}$ and $\mathcal{B}(H)$, so is $\tilde{F}$, as it is obviously weakly adapted.

2.3 Representations of Reproducing Kernels

All reproducing kernels have a Hilbert space representation [(Proposition) 1.3.5]. Such a representation is particularly useful when the Hilbert space is an L_2 space. Reproducing kernels that have an L_2 representation as defined below are isomorphic to subspaces of L_2 spaces.

Definition 2.3.1 Let $\mathcal{K}$ be a reproducing kernel on T. It has an L_2 representation when there exists $\{k_t,\ t \in T\} \subseteq L_2(X, \mathcal{X}, \mu)$, with μ a σ-finite measure, such that, for $(t_1, t_2) \in T \times T$, fixed. but arbitrary,

$$\mathcal{K}(t_1, t_2) = \langle k_{t_1}, k_{t_2} \rangle_{L_2(X, \mathcal{X}, \mu)}.$$

Example 2.3.2 Result (Proposition) 1.5.12 (see (Remark) 1.5.13) shows that every kernel whose RKHS is separable has an L_2 representation.

Example 2.3.3 Every RKHS obtained with the help of a map

$$F : T \longrightarrow L_2(X, \mathcal{X}, \mu)$$

has an L_2 representation.

Example 2.3.4 Remark 1.4.16 contains cases of L_2 representations for covariances with a factorization.

Example 2.3.5 ([103]) A strictly positive measure for a topological space X is a Borel measure for which every open, not void subset has strictly positive measure, and every point is contained in an open set of finite measure.

Let X be a metric space, and μ be a strictly positive measure on X. Let C be a continuous covariance on X that defines a positive operator on $L_2(X, \mathcal{X}, \mu)$, with the property that $x \mapsto C(x, x)$ is integrable with respect to μ. C has then a representation of the following form:

$$C(x_1, x_2) = \sum_{i=1}^{\infty} \lambda_n e_n(x_1) e_n(x_2),$$

where the convergence is in $L_2(X \times X, \mathcal{X} \otimes \mathcal{X}, \mu \otimes \mu)$, but absolute and uniform on compact subsets of $X \times X$; the functions in the series are orthonormal in

$L_2(X, \mathcal{X}, \mu)$, and the pairs (λ_n, e_n) are the eigenvalue-eigenvector pairs of the covariance operator R determined by C.

Set

$$\Gamma(x_1, x_2) = \sum_{i=1}^{\infty} \lambda_n^{1/2} e_n(x_1) e_n(x_2).$$

One thus gets the kernel of $R^{1/2}$, whose range contains continuous, square integrable functions only (with respect to μ). Furthermore

$$C(x_1, x_2) = \int_X \Gamma(\xi, x_1) \Gamma(\xi, x_2) \mu(d\xi).$$

As a particular case of (Proposition) 1.1.15, one has the following result:

Proposition 2.3.6 *Let $\mathcal{K}$ be a reproducing kernel on T, with the L_2 representation $\{k_t, t \in T\} \subseteq L_2(X, \mathcal{X}, \mu)$. The resulting RKHS $H(\mathcal{K}, T)$ is then the set of functions*

$$\left\{h(t) = \langle k, k_t \rangle_{L_2(X, \mathcal{X}, \mu)}, \ t \in T, \ k \in L_2(X, \mathcal{X}, \mu)\right\}.$$

Let K_0 be the (closed) linear subspace generated by $\{k_t, t \in T\}$ in $L_2(X, \mathcal{X}, \mu)$, and P_{K_0} be the associated projection. The following functions

$$h_1(t) = \langle k^{(1)}, k_t \rangle_{L_2(X, \mathcal{X}, \mu)}, \quad \text{and} \quad h_2(t) = \langle k^{(2)}, k_t \rangle_{L_2(X, \mathcal{X}, \mu)}$$

have then the following RKHS inner product

$$\langle h_1, h_2 \rangle_{H(\mathcal{K}, T)} = \langle P_{K_0}\left[k^{(1)}\right], P_{K_0}\left[k^{(2)}\right] \rangle_{L_2(X, \mathcal{X}, \mu)}.$$

The map $L : L_2(X, \mathcal{X}, \mu) \longrightarrow H(\mathcal{K}, T)$ defined using the following assignment:

$$L[k_t] = \mathcal{K}(\cdot, t)$$

is a partial isometry with K_0 as initial set, and $H(\mathcal{K}, T)$ as final set.

Proof Define $F : T \longrightarrow L_2(X, \mathcal{X}, \mu)$ using the following assignment:

$$F(t) = k_t, \ t \in T.$$

Then

$$L[k](t) = \langle k, k_t \rangle_{L_2(X, \mathcal{X}, \mu)},$$

and

$$\langle L\left[k^{(1)}\right], L\left[k^{(2)}\right]\rangle_K = \langle P_{K_0}\left[k^{(1)}\right], P_{K_0}\left[k^{(2)}\right]\rangle_{L_2(X,\mathcal{X},\mu)}.$$

Also

$$\langle F(t_1), F(t_2)\rangle_{L_2(X,\mathcal{X},\mu)} = \langle k_{t_1}, k_{t_2}\rangle_{L_2(X,\mathcal{X},\mu)},$$

so that $\mathcal{K}$ is indeed the reproducing kernel of $H(\mathcal{K}, T)$. □

The proposition to follow says that for separable RKHS's which have an L_2 representation, that representation happens through measurable kernels.

Proposition 2.3.7 *Let* $(T, \mathcal{T})$ *be a measurable space, and suppose that the reproducing kernel* $\mathcal{K}$ *on* T *is adapted* $\mathcal{T} \otimes \mathcal{T}$, *and has the* L_2 *representation* $\{k_t,\ t \in T\} \subseteq L_2(X, \mathcal{X}, \mu)$. *Suppose furthermore that* K_0 [(Proposition) 2.3.6] *is separable. Then* $\mathcal{K}$ *has also an* L_2 *representation* $\{\kappa_t,\ t \in T\} \subseteq L_2(X, \mathcal{X}, \mu)$ *with*

$$\kappa_t = [\kappa(t, \cdot)]_{L_2(X,\mathcal{X}\mu)},$$

κ, *a function which is adapted to* $\mathcal{T} \otimes \mathcal{X}$. *Furthermore, for fixed, but arbitrary* $t \in T$, $\kappa_t = k_t$.

Proof One must notice that, since $\mathcal{K}$ is adapted to $\mathcal{T} \otimes \mathcal{T}$, $H(\mathcal{K}, T)$ is made of measurable functions [(Propositions) 2.2.1, 2.2.2]. Also, since K_0 is separable by assumption, and unitarily isomorphic to the RKHS $H(\mathcal{K}, T)$ [(Proposition) 2.3.6], the latter is separable.

Let $n \in \mathbb{N}$ be fixed, but arbitrary. Since the RKHS $H(\mathcal{K}, T)$ is separable, there exists [(Proposition) 1.5.10]

$$\left\{T_i^{(n)} \in \mathcal{T},\ i \in \mathbb{N}\right\}$$

such that,

- for fixed, but arbitrary $\{i, j\} \subseteq \mathbb{N}$, $i \neq j$, $T_i^{(n)} \cap T_j^{(n)} = \emptyset$,
- $\cup_i T_i^{(n)} = T$,
- for $\left(t_1^{(n,i)}, t_2^{(n,i)}\right) \in T_i^{(n)} \times T_i^{(n)}$, fixed, but arbitrary,

$$\left\|\mathcal{K}\left(\cdot, t_1^{(n,i)}\right) - \mathcal{K}\left(\cdot, t_2^{(n,i)}\right)\right\|_{H(\mathcal{K},T)}^2 < \frac{1}{2^n}.$$

Let, in each set $T_i^{(n)}$, a fixed, but arbitrary point $t_i^{(n)}$ be chosen, and

$$\{e_j,\ j \in J \subseteq \mathbb{N}\}$$

be a complete orthonormal set in $H(\mathcal{K}, T)$. Since

$$\sum_{j\in J} e_j^2(t) = \sum_{j\in J} \langle e_j, \mathcal{K}(\cdot, t)\rangle_{H(\mathcal{K},T)}^2 = \|\mathcal{K}(\cdot, t)\|_{H(\mathcal{K},T)}^2 = \mathcal{K}(t,t) < \infty,$$

for $t_i^{(n)}$, fixed, but arbitrary, there exists $p_{n,i} \in \mathbb{N}$ such that

$$\sum_{j>p_{n,i}} e_j^2\left(t_i^{(n)}\right) < \frac{1}{2^n}.$$

Thus, for $t \in T_i^{(n)}$, fixed, but arbitrary,

$$\begin{aligned}
\sum_{j>p_{n,i}} e_j^2(t) &= \sum_{j>p_{n,i}} \left\{e_j(t) - e_j\left(t_i^{(n)}\right) + e_j\left(t_i^{(n)}\right)\right\}^2 \\
&\leq 2\left[\sum_{j>p_{n,i}} \left\{e_j(t) - e_j\left(t_i^{(n)}\right)\right\}^2 + \sum_{j>p_{n,i}} e_j^2\left(t_i^{(n)}\right)\right] \\
&\leq 2\left[\sum_{j>p_{n,i}} \left\{e_j(t) - e_j\left(t_i^{(n)}\right)\right\}^2 + \frac{1}{2^n}\right].
\end{aligned}$$

But, still for $t \in T_i^{(n)}$,

$$\begin{aligned}
\frac{1}{2^n} &> \left\|\mathcal{K}(\cdot, t) - \mathcal{K}\left(\cdot, t_i^{(n)}\right)\right\|_{H(\mathcal{K},T)}^2 \\
&= \sum_{j\in J} \langle \mathcal{K}(\cdot, t) - \mathcal{K}\left(\cdot, t_i^{(n)}\right), e_j\rangle_{H(\mathcal{K},T)}^2 \\
&= \sum_{j\in J} \left\{e_j(t) - e_j\left(t_i^{(n)}\right)\right\}^2 \\
&= \sum_{j\leq p_{n,i}} \left\{e_j(t) - e_j\left(t_i^{(n)}\right)\right\}^2 + \sum_{j>p_{n,i}} \left\{e_j(t) - e_j\left(t_i^{(n)}\right)\right\}^2 \\
&\geq \sum_{j>p_{n,i}} \left\{e_j(t) - e_j\left(t_i^{(n)}\right)\right\}^2,
\end{aligned}$$

so that

$$\sum_{j>p_{n,i}} e_j^2(t) < 2\left\{\frac{1}{2^n} + \frac{1}{2^n}\right\} = \frac{4}{2^n}.$$

Thus, for $t \in T_i^{(n)}$, there exists $p_{n,i} \in \mathbb{N}$ such that

$$\sum_{j>p_{n,i}} e_j^2(t) < \frac{4}{2^n}.$$

Let K_0 and L be as in (Proposition) 2.3.6. Since $L^\star$ is a partial isometry with $H(\mathcal{K}, T)$ as initial set, and K_0 as final set,

- $L^\star L$ is the projection onto the initial set of L [266, p. 86], that is, is the identity of K_0,
- $\{f_j = L^\star[e_j], j \in \mathbb{N}\}$ is a complete orthonormal set in K_0.

Thus, since $L[k_t] = \mathcal{K}(\cdot, t)$,

$$k_t = L^\star L[k_t] = L^\star[\mathcal{K}(\cdot,t)] = L^\star\left[\sum_{j\in J}\langle e_j, \mathcal{K}(\cdot,t)\rangle_{H(\mathcal{K},T)} e_j\right] = \sum_{j\in J} e_j(t) f_j.$$

Set, for $t \in T_i^{(n)}$, fixed, but arbitrary,

- $\kappa_t^{(n)} = \sum_{j\le p_{n,i}} e_j(t) f_j \in K_0$,
- $\dot{\kappa}_t^{(n)}(x) = \sum_{j\le p_{n,i}} e_j(t) \dot{f}_j(x)$.

$\dot{\kappa}_t^{(n)}(x)$ is thus a function which is adapted to $\mathcal{T} \otimes \mathcal{X}$,

$$\left[\dot{\kappa}_t^{(n)}\right]_{L_2[X,\mathcal{X},\mu]} = \kappa_t^{(n)},$$

and, in $L_2(X, \mathcal{X}, \mu)$, $\lim_n \kappa_t^{(n)} = k_t$.

Let now $t \in T$, be fixed, but arbitrary, and n_1 and n_2 be distinct, fixed, but arbitrary integers in $\mathbb{N}$. Let i_1 be the integer for which

$$t \in T_{i_1}^{(n_1)},$$

and i_2 be that for which

$$t \in T_{i_2}^{(n_2)}.$$

Let $m_1 = p_{n_1,i_{n_1}} \wedge p_{n_2,i_{n_2}}$, and $m_2 = p_{n_1,i_{n_1}} \vee p_{n_2,i_{n_2}}$. Then

$$\kappa_t^{(n_1)} - \kappa_t^{(n_2)} = \sum_{j=m_1+1}^{m_2} e_j(t)\dot{f}_j,$$

so that

$$\left\|\kappa_t^{(n_1)} - \kappa_t^{(n_2)}\right\|^2_{L_2[X,\mathcal{X},\mu]} = \sum_{j=m_1+1}^{m_2} e_j^2(t)$$
$$< \begin{cases} \frac{4}{2^{n_1}} \text{ when } m_1 = p_{n_1,i_{n_1}} \\ \frac{4}{2^{n_2}} \text{ when } m_1 = p_{n_2,i_{n_2}} \end{cases}$$
$$\le \frac{4}{2^{n_1 \wedge n_2}}.$$

One has thus a Cauchy sequence in $L_2[X, \mathcal{X}, \mu]$, and, because of the Fischer-Riesz theorem [113, p. 160], considering, when necessary a subsequence, an element $\kappa_t \in L_2[X, \mathcal{X}, \mu]$ such that, respectively in $L_2[X, \mathcal{X}, \mu]$, and almost surely with respect to μ (in terms of equivalence classes),

$$\lim_n \kappa_t^{(n)} = \kappa_t.$$

Set, for $(t, x) \in T \times X$,

$$\kappa(t, x) = \limsup_n \dot{\kappa}_t^{(n)}(x).$$

κ is thus adapted to $\mathcal{T} \otimes \mathcal{X}$. Furthermore, for $t \in T$, fixed, but arbitrary, μ-almost surely, $\kappa(t, \cdot) = \dot{\kappa}_t$, so that, for $(t_1, t_2) \in T \times T$, fixed, but arbitrary,

$$\left\langle [\kappa(t_1, \cdot)]_{L_2[X,\mathcal{X},\mu]}, [\kappa(t_2, \cdot)]_{L_2[X,\mathcal{X},\mu]}\right\rangle_{L_2[X,\mathcal{X},\mu]} = \langle \kappa_{t_1}, \kappa_{t_2}\rangle_{L_2[X,\mathcal{X},\mu]}$$
$$= \lim_n \left\langle \kappa_{t_1}^{(n)}, \kappa_{t_2}^{(n)}\right\rangle_{L_2[X,\mathcal{X},\mu]}.$$

Now, given t_1, t_2, and n, fixed, but arbitrary, there are i_1 and i_2 such that $t_1 \in T_{i_1}^{(n)}$ and $t_2 \in T_{i_2}^{(n)}$. Set $m_n = p_{n,i_1} \wedge p_{n,i_2}$. Then

$$\langle \kappa_{t_1}^{(n)}, \kappa_{t_2}^{(n)}\rangle_{L_2[X,\mathcal{X},\mu]} = \sum_{j \le m_n} e_j(t_1)\, e_j(t_2).$$

Consequently [(Proposition) 1.5.6],

$$\langle [\kappa(t_1, \cdot)]_{L_2[X,\mathcal{X},\mu]}, [\kappa(t_2, \cdot)]_{L_2[X,\mathcal{X},\mu]}\rangle_{L_2[X,\mathcal{X},\mu]} = \mathcal{K}(t_1, t_2)$$
$$= \langle k_{t_1}, k_{t_2}\rangle_{L_2[X,\mathcal{X},\mu]}.$$

□

Remark 2.3.8 When $L_2(X, \mathcal{X}, \mu)$ is separable, since $H(\mathcal{K}, T)$ is the unitary image of a subspace of $L_2(X, \mathcal{X}, \mu)$, it is separable, and (Proposition) 2.3.7 obtains.

When H is an L_2 space, $\Lambda_{F,0}$ of (Proposition) 2.2.2 is an integral operator as seen in the next corollary.

Corollary 2.3.9 *Let $(T, \mathcal{T}, \tau)$ and $(X, \mathcal{X}, \mu)$ be σ-finite measure spaces. Suppose that:*

(a) $F : T \longrightarrow L_2(X, \mathcal{X}, \mu)$ *is weakly measurable;*
(b) $\overline{V[\mathcal{R}[F]]}$ *is separable.*

Then:

1. *there exists $\mathcal{K}_F : T \times X \longrightarrow \mathbb{R}$ such that*

   (i) *$\mathcal{K}_F$ is adapted to $\mathcal{T} \otimes \mathcal{X}$ and $\mathcal{B}(\mathbb{R})$;*
   (ii) *for $t \in T$, the map $x \mapsto \mathcal{K}_F(t, x)$ belongs to (the class) $F(t)$.*

2. *$\Lambda_{F,0} : L_2(X, \mathcal{X}, \mu) \longrightarrow L_0(T, \mathcal{T}, \tau)$ is an integral operator whose kernel is $\mathcal{K}_F$;*
3. *for $(t_1, t_2) \in T \times T$, $\mathcal{H}_F(t_1, t_2) = \int_X \mathcal{K}_F(t_1, x) \mathcal{K}_F(t_2, x) \mu(dx)$.*

Proof $\mathcal{H}_F$ is adapted to $\mathcal{T} \otimes \mathcal{T}$ and $\mathcal{B}(\mathbb{R})$ [(Proposition) 2.2.2]. Thus (Proposition) 2.3.7 applies. □

Example 2.3.10 (Inner Products for Measures Using RKHS's [254]*)* This example shows the versatility of RKHS's (provided one has the proper insight!). It has applications to central limit theory. The measures considered below are assumed to have finite total variation.

Let T be a metric space, $\mathcal{T} = \mathcal{B}(T)$, the Borel sets of T, $\mathcal{M}$, the family of signed measures on $\mathcal{T}$. A signed measure μ is the difference of two positive and bounded measures. The Hahn-Jordan decomposition of μ is denoted (μ^+, μ^-), and its total variation measure, $|\mu| = \mu^+ + \mu^-$. Suppose that, for $\{t_1, t_2\} \subseteq T$, fixed, but arbitrary,

$$\mathcal{K}(t_1, t_2) = \int_S \kappa(t_1, s) \kappa(t_2, s) \sigma(ds),$$

where

- σ is a positive measure on $\mathcal{S}$, a σ-algebra of subsets of S,
- $\kappa : T \times S \longrightarrow \mathbb{R}$ is adapted to $\mathcal{T} \otimes \mathcal{S}$, and has the property that

$$\kappa_T = \sup_{t \in T} \|\kappa(t, \cdot)\|_{L_2(S, \mathcal{S}, \sigma)} < \infty.$$

$\mathcal{K}$ is thus a bounded reproducing kernel.

`Particular cases:`

(a) Let $\{f_n : T \longrightarrow \mathbb{R},\ n \in \mathbb{N}\}$ be a family of functions such that, for $t \in T$, fixed, but arbitrary,

$$\sum_{n \in \mathbb{N}} f_n^2(t) < \infty.$$

Let $\mathcal{K}(t_1, t_2) = \sum_{n \in \mathbb{N}} f_n(t_1) f_n(t_2)$. It is a reproducing kernel. Let σ be the counting measure on $\mathcal{P}(\mathbb{N})$, and $\kappa(t, n) = f_n(t)$. Then

$$\mathcal{K}(t_1, t_2) = \int_{\mathbb{N}} \kappa(t_1, n)\, \kappa(t_2, n)\, \sigma(dn).$$

(b) Let $S = T = [0, 1]$, $\kappa(t, s) = \chi_{[0,t]}(s)$, $\sigma(ds) = ds$. Then

$$t_1 \wedge t_2 = \int_T \kappa(t_1, s)\, \kappa(t_2, s)\, \sigma(ds).$$

(c) Let $S = T = \mathbb{R}_+$, $\kappa(t, s) = e^{-ts}$, σ a positive, bounded measure on the Borel sets of $\mathbb{R}_+$. Then

$$\mathcal{K}(t_1, t_2) = \int_T e^{-(t_1+t_2)s} \sigma(ds).$$

`Fact` *When μ is a signed measure on $\mathcal{T}$, then, for almost every $s \in S$, with respect to σ, $t \mapsto \kappa(t, s)$ is integrable with respect to μ.*

It suffices to consider μ positive and bounded. Then, for $\{t_1, t_2\} \subseteq T$, fixed, but arbitrary,

$$\begin{aligned}
|\mathcal{K}(t_1, t_2)| &\leq \int_S |\kappa(t_1, s)\, \kappa(t_2\, s)|\, \sigma(ds) \\
&\leq \|\kappa(t_1, \cdot)\|_{L_2(S, \mathcal{S}, \sigma)} \|\kappa(t_2, \cdot)\|_{L_2(S, \mathcal{S}, \sigma)} \\
&\leq \kappa_T^2 \\
&< \infty.
\end{aligned}$$

Consequently

$$\int_T \int_T \left\{ \int_S |\kappa(t_1, s)\, \kappa(t_2\, s)|\, \sigma(ds) \right\} \mu \otimes \mu(dt_1, dt_2) < \infty.$$

Using Fubini's theorem, one gets that, for almost every $s \in S$, with respect to σ,

$$\int_T \int_T |\kappa(t_1, s)\, \kappa(t_2, s)|\, \mu \otimes \mu\,(dt_1, dt_2) < \infty.$$

But the left-hand side of the latter inequality is $\left\{\int_T |\kappa(t, s)|\, \mu(dt)\right\}^2$.

ASSUMPTION CM[1] *When, for almost every $s \in S$, with respect to σ,*

$$\int_T \kappa(t, s)\, \mu(dt) = 0$$

then $\mu = 0$.

Some definitions and facts:

1. *With the assumptions made so far, the following relation establishes an inner product on $\mathcal{M}$:*

$$\langle \mu_1, \mu_2 \rangle_{\mathcal{K}} = \int_T \int_T \mathcal{K}(t_1, t_2)\, \mu_1 \otimes \mu_2\,(dt_1, dt_2).$$

   That relation makes indeed sense as $\mathcal{K}$ is bounded, and the measures have finite total variation. It is "structurally" symmetric and bilinear. Since, as above,

$$\int_T \int_T \mathcal{K}(t_1, t_2)\, \mu \otimes \mu\,(dt_1, dt_2) = \int_S \left\{ \int_T \kappa(t, s)\, \mu(dt) \right\}^2 \sigma(ds),$$

   $\langle \mu, \mu \rangle_{\mathcal{K}} \geq 0$. The assumption CM insures that $\langle \mu, \mu \rangle_{\mathcal{K}} = 0$ implies that μ is identically zero.
2. Let $F : T \longrightarrow L_2(S, \mathcal{S}, \sigma)$ be defined using the following relation:

$$F(t) = [\kappa(t, \cdot)]_{L_2(S, \mathcal{S}, \sigma)},$$

   and $L_F : L_2(S, \mathcal{S}, \sigma) \longrightarrow \mathbb{R}^T$ using the following one:

$$L_F[k](t) = \int_S \kappa(t, s)\, \dot{k}(s)\, \sigma(ds).$$

   The range of L_F is the RKHS with domain T and kernel $\mathcal{K}$. Let H_F denote the linear subspace generated by $\{F(t), t \in T\}$.

[1] C:= *caractérisation*, M:= *mesures*.

3. *Define, for* $\mu \in \mathcal{M}$, *fixed, but arbitrary,* $B[\mu]$ *as the equivalence class in* $L_2(S, \mathcal{S}, \sigma)$ *of the map* $s \mapsto \int_T \kappa(\theta, s)\, \mu(d\theta)$. *It is obviously linear and injective, because of assumption* CM. *It is thus well defined.*
4. One has that

$$\begin{aligned} \langle B[\mu_1], B[\mu_2] \rangle_{L_2(S,\mathcal{S},\sigma)} &= \\ &= \int_S \left\{ \int_T \kappa(t,s)\, \mu_1(dt) \right\} \left\{ \int_T \kappa(\theta,s)\, \mu_2(d\theta) \right\} \sigma ds \\ &= \int_T \int_T \mathcal{K}(t,\theta)\, \mu_1 \otimes \mu_2(dt, d\theta) \\ &= \langle \mu_1, \mu_2 \rangle_{\mathcal{K}}. \end{aligned}$$

5. Let $f \in H_F^{\perp}$. Then

$$\begin{aligned} \langle B[\mu], f \rangle_{L_2(S,\mathcal{S},\sigma)} &= \int_S \dot{f}(s) \left\{ \int_T \kappa(\theta,s)\, \mu(d\theta) \right\} \sigma(ds) \\ &= \int_T \left\{ \int_S \dot{f}(s)\, \kappa(\theta,s)\, \sigma(ds) \right\} \mu(d\theta) \\ &= 0. \end{aligned}$$

   Thus $\mathcal{R}[B] \subseteq H_F$.
6. Let $B_F = L_F B$. Then

$$\begin{aligned} B_F[\mu](t) &= L_F[B[\mu]] \\ &= \int_S k(t,s)\, \overbrace{B[\mu]}^{\cdot}(s)\, \sigma(ds) \\ &= \int_S \kappa(t,s) \int_T \kappa(\theta,s)\, \mu(d\theta)\, \sigma(ds) \\ &= \int_T \mathcal{K}(\theta,t)\, \mu(d\theta). \end{aligned}$$

   When taking $\mu = \delta_{t_0}$, one obtains that $B_F[\delta_{t_0}] = \mathcal{K}(\cdot, t_0)$. Furthermore

   - since in an RKHS, $|h(t)| \leq \mathcal{H}^{1/2}(t,t)\, \|h\|_{H(\mathcal{H},T)}$,

$$\sup_{t \in T} |B_F[\mu](t)| \leq \sup_{t \in T} \mathcal{K}^{1/2}(t,t)\, \|B_F[\mu]\|_{H(\mathcal{H},T)};$$

   - since $\mathcal{K}$ has a representation in L_2,

$$\sup_{t \in T} \mathcal{K}^{1/2}(t,t) = \sup_{t \in T} \|\kappa(t,\cdot)\|_{L_2(S,\mathcal{S},\sigma)}.$$

7. One has that

$$\begin{aligned}\langle B_F[\mu_1], B_F[\mu_2]\rangle_{H(\mathcal{K},T)} &= \\ &= \langle \int_T \mathcal{K}(\theta,\cdot)\,\mu_1(d\theta), \int_T \mathcal{K}(\theta,\cdot)\,\mu_2(d\theta)\rangle_{H(\mathcal{K},T)} \\ &= \langle P_{H_F} B[\mu_1], P_{H_F} B[\mu_2]\rangle_{L_2(S,\mathcal{S},\sigma)} \\ &= \langle B[\mu_1], B[\mu_2]\rangle_{L_2(S,\mathcal{S},\sigma)}.\end{aligned}$$

Since L_F is unitary on H_F, B_F is a unitary map.

8. For $\{\mu, \mu_0\} \subseteq \mathcal{M}$, fixed, but arbitrary,

$$\begin{aligned}\int_T B_F[\mu](t)\,\mu_0(dt) &= \int_T \left\{\int_T \mathcal{K}(\theta,t)\,\mu(d\theta)\right\}\mu_0(dt) \\ &= \langle B[\mu], B[\mu_0]\rangle_{L_2(S,\mathcal{S},\sigma)} \\ &= \langle B_F[\mu], B_F[\mu_0]\rangle_{H(\mathcal{K},T)}.\end{aligned}$$

But, since B_F is unitary, one has, for $h \in H(\mathcal{K},T)$, $h = B_F[\mu]$, some μ, and thus

$$\int_T h(t)\,\mu_0(dt) = \langle h, B_F[\mu_0]\rangle_{H(\mathcal{K},T)}.$$

`Particular cases:`

1. Consider particular case (a) above. Then $L_2(S,\mathcal{S},\sigma) = l_2$, and $\kappa(t,n) = f_n(t)$. The required conditions become:

   - for $\sup_{t\in T}\|\kappa(t,\cdot)\|_{L_2(S,\mathcal{S},\sigma)} < \infty$: $\sup_{t\in T}\sum_{n=1}^{\infty} f_n^2(t) < \infty$;
   - CM: for $n \in \mathbb{N}$, fixed , but arbitrary, $\int_T f_n(t)\,\mu(dt) = 0$ implies $\mu = 0$.

   The map $F(t) = [\kappa(t,\cdot)]_{L_2(S,\mathcal{S},\sigma)}$ becomes $F(t) = \{f_n(t), n \in \mathbb{N}\}$, so that, for $\underline{\alpha} \in l_2$ and $t \in T$, fixed, but arbitrary,

$$L_F[\underline{\alpha}](t) = \sum_{n=1}^{\infty} \alpha_n f_n(t).$$

   Consequently $H(\mathcal{K},T) = \left\{\sum_{n=1}^{\infty}\alpha_n f_n, \underline{\alpha} \in l_2\right\}$. H_F is the closed subspace generated by the family $\{\underline{f}(t), t \in T\}$, and

$$\langle L_F[\underline{\alpha}_1], L_F[\underline{\alpha}_2]\rangle_{H(\mathcal{K},T)} = \langle P_{H_F}[\underline{\alpha}_1], \underline{\alpha}_2\rangle_{l_2}.$$

$B[\mu]$ is the map $n \mapsto \int_T f_n(t)\,\mu(dt)$, and

$$B_F[\mu] = \sum_{n=1}^{\infty} \left\{ \int_T f_n(t)\,\mu(dt) \right\} f_n.$$

Finally

$$\begin{aligned} \langle \mu_1, \mu_2 \rangle_{\mathcal{K}} &= \langle B[\mu_1], B[\mu_2] \rangle_{l_2} \\ &= \langle B_F[\mu_1], B_F[\mu_2] \rangle_{H(\mathcal{K},T)} \\ &= \sum_{n=1}^{\infty} \int_T f_n(t)\,\mu_1(dt) \int_T f_n(t)\,\mu_2(dt). \end{aligned}$$

2. Consider particular case (c) above. Then $\kappa(t,s) = e^{-ts}$. The required conditions become:

   - for $\sup_{t\in T} \|\kappa(t,\cdot)\|_{L_2(S,\mathcal{S},\sigma)} < \infty$: $\sup_{t\in T} \int_S \kappa^2(t,s)\,\sigma(ds) < \infty$;
   - CM: for $s \in S$, fixed , but arbitrary, $\int_T e^{-st}\mu(dt) = 0$ implies $\mu = 0$.

   The map $F(t) = [\kappa(t,\cdot)]_{L_2(S,\mathcal{S},\sigma)}$ becomes

$$F(t) = \left[e^{-t\times\cdot}\right]_{L_2(S,\mathcal{S},\sigma)},$$

so that, for $f \in L_2(S,\mathcal{S},\sigma)$ and $t \in T$, fixed, but arbitrary,

$$L_F[f](t) = \int_S e^{-ts} \dot{f}(s)\,\sigma(ds).$$

Consequently

$$H(\mathcal{K},T) = \left\{ \phi(t) = \int_S e^{-ts} \dot{f}(s)\,\sigma(ds),\, [f]_{L_2(S,\mathcal{S},\sigma)} \in L_2(S,\mathcal{S},\sigma) \right\}.$$

H_F is the closed subspace generated by the family

$$\left\{ \left[e^{-t\times\cdot}\right]_{L_2(S,\mathcal{S},\sigma)}, t \in T \right\},$$

and

$$\langle L_F[f_1], L_F[f_2] \rangle_{H(\mathcal{K},T)} = \langle P_{H_F}[f_1], f_2 \rangle_{L_2(S,\mathcal{S},\sigma)}.$$

$B[\mu]$ is the equivalence class in $L_2(S, \mathcal{S}, \sigma)$ of the map $\{s \mapsto \int_T e^{-st}\mu(dt)\}$ and

$$B_F[\mu] = \int_S e^{-sx\cdot} \left\{ \int_T e^{-s\theta} \mu(d\theta) \right\} \sigma(ds).$$

Finally

$$\langle \mu_1, \mu_2 \rangle_{\mathcal{K}} = \langle B[\mu_1], B[\mu_2] \rangle_{l_2}$$

$$= \langle B_F[\mu_1], B_F[\mu_2] \rangle_{H(\mathcal{K},T)}$$

$$= \int_S \left\{ \int_T e^{-st}\mu_1(dt) \right\} \left\{ \int_T e^{-st}\mu_2(dt) \right\} \sigma(ds).$$

Remark When one allows complex valued functions, letting T and S be euclidian spaces of the same dimension n, and $\kappa(\underline{t}, \underline{s})$, the exponential of $i\langle \underline{t}, \underline{s} \rangle_{\mathbb{R}^n}$, one obtains analogous results for Fourier transforms.

The Hilbert structure for signed measures that has been described is useful for studying convergence of measures as shown by the following proposition whose proof needs a preliminary fact. $C_0(T)$ denotes the space of continuous functions that vanish at infinity.

Fact *Suppose* CM *obtains, and, for $t \in T$, fixed, but arbitrary, $\mathcal{K}(\cdot, t) \in C_0(T)$, then $H(\mathcal{K}, T)$ is dense in $C_0(T)$.*

Proof $\mathcal{M}$ is the dual of $C_0(T)$ [178, p. 97]. Thus, a fixed, but arbitrary $\mu \in \mathcal{M}$ is a continuous linear functional on $C_0(T)$, and, since, by assumption, for $t \in T$, fixed, but arbitrary, $\mathcal{K}(\cdot, t) \in C_0(T)$, it makes sense to consider the following set of relations: for $t \in T$, fixed, but arbitrary,

$$\int_T \mathcal{K}(\theta, t)\mu(d\theta) = 0.$$

Now

$$\int_T \mathcal{K}(\theta, t)\mu(d\theta) = \int_T \left\{ \int_S \kappa(\theta, s)\kappa(t, s)\sigma(ds) \right\} \mu(d\theta)$$

$$= \int_S \kappa(t, s) \left\{ \int_T \kappa(\theta, s)\mu(d\theta) \right\} \sigma(ds).$$

Since $\int_T \kappa(\theta, s)\mu(d\theta)$ belongs to H_F, the former zero requirement means that the latter integral is zero, and then the CM assumption that $\mu = 0$. Consequently $\{\mathcal{K}(\cdot, t), t \in T\}$ is a total family in $C_0(T)$, and the vector space it generates is dense. But limits in $H(\mathcal{K}, T)$ of sequences in $V[\{\mathcal{K}(\cdot, t), t \in T\}]$ are also limits in $C_0(T)$, since convergence in $H(\mathcal{K}, T)$ implies uniform convergence, given the assumptions ($\mathcal{K}$ is then uniformly bounded). Consequently

$$V[\{\mathcal{K}(\cdot, t), t \in T\}] \subseteq H(\mathcal{K}, T) \subseteq C_0(T),$$

with the left-hand side dense in the right-hand side.

`Fact` *Suppose the following obtain:*

(a) *T is a separable, locally compact metric space;*
(b) *$\mathcal{K}$ is continuous;*
(c) *for $t \in T$, fixed, but arbitrary, the map $\mathcal{K}(\cdot, t)$ has the following property:*
for fixed, but arbitrary $\epsilon > 0$, there is a compact $C_{t,\epsilon} \subseteq T$ such that, for θ in $T \setminus C_{t,\epsilon}$, $|\mathcal{K}(\theta, t)| < \epsilon$, that is $\mathcal{K}(\cdot, t) \in C_0(T)$;
(d) *P and, for $n \in \mathbb{N}$, fixed, but arbitrary, P_n are probabilities.*

Then the following assertions are equivalent:

(1) $\{P_n, n \in \mathbb{N}\}$ converges weakly to P, that is, for $f \in C_0(T)$, fixed, but arbitrary,

$$\lim_n \int_T f(t) P_n(dt) = \int_T f(t) P(dt),$$

(2) $\lim_n \|P - P_n\|_{\mathcal{K}} = 0$,
(3) weakly in $H(\mathcal{K}, T)$, $\lim_n B_F[P_n] = B_F[P]$.

Proof (3 ⇒ 1) One has seen that, for $h \in H(\mathcal{K}, T)$, fixed, but arbitrary,

$$\langle h, B_F[\mu]\rangle_{H(\mathcal{K},T)} = \int_T h(t) \mu(dt).$$

Consequently, when $\lim_n B_F[P_n] = B_F[P]$, weakly in $H(\mathcal{K}, T)$,

$$\lim_n \int_T h(t) P_n(dt) = \int_T h(t) P(dt).$$

But, since $H(\mathcal{K}, T)$ is dense in $C_0(T)$, and that the P_n's are probabilities, weak convergence follows [178, p. 98].

Proof (1 ⇒ 2) Because of

- the assumption of weak convergence,
- the inclusion $H(\mathcal{K}, T) \subseteq C_0(T)$,
- the relation $\langle h, B_F[\mu]\rangle_{H(\mathcal{K},T)} = \int_T h(t) \mu(dt)$,

one has that, weakly in $H(\mathcal{K}, T)$, $\lim_n B_F[P_n] = B_F[P]$. Furthermore

$$\|B_F[\mu]\|^2_{H(\mathcal{K},T)} = \int_T \int_T K(t_1, t_2) \mu \otimes \mu(dt_1, dt_2).$$

Now from the weak convergence of P_n to P follows that of $P_n \otimes P_n$ to $P \otimes P$ [208, p. 57]. But $\mathcal{K}$ is continuous and uniformly bounded, so that

$$\lim_n \int_T \int_T K(t_1, t_2) P_n \otimes P_n (dt_1, dt_2) = \int_T \int_T K(t_1, t_2) P \otimes P (dt_1, dt_2),$$

or $\lim_n \|B_F [P_n]\|^2_{H(\mathcal{K},T)} = \|B_F [P]\|^2_{H(\mathcal{K},T)}$. Since weak convergence plus convergence of norms imply, in Hilbert space, strong convergence, the claim is valid as B_F is unitary.

Since B_F is unitary, obviously item 2 implies item 3, and the proof is finished.

2.4 Embeddings of Reproducing Kernel Hilbert Spaces

The embeddings considered shall be into L_2 spaces, so that Hilbert space operators may be used.

Definition 2.4.1 Suppose $(T, \mathcal{T}, \tau)$ is a σ-finite measure space, and $\mathcal{H}$, a reproducing kernel on T such that each $h \in H(\mathcal{H}, T)$ is adapted to $\mathcal{T}$, and its square is integrable with respect to τ. One then says that $H(\mathcal{H}, T)$ is imbedded in $L_2(T, \mathcal{T}, \tau)$. $J_{\mathcal{H},\tau} : H(\mathcal{H}, T) \longrightarrow L_2(T, \mathcal{T}, \tau)$ shall be the "inclusion map" defined using the following relation: for $h \in H(\mathcal{H}, T)$, fixed, but arbitrary,

$$J_{\mathcal{H},\tau}[h] = [h]_{L_2(T,\mathcal{T},\tau)}.$$

Definition 2.4.2 The reproducing kernel $\mathcal{H}$ has property Π_J for $(T, \mathcal{T}, \tau)$ whenever

1. for $t \in T$, fixed, but arbitrary, $\mathcal{H}(\cdot, t)$ is adapted to $\mathcal{T}$,
2. $t \mapsto \mathcal{H}(t, t)$ is adapted to $\mathcal{T}$,
3. $\kappa_{\mathcal{H},\tau} = \int_T \mathcal{H}(t, t)\, \tau(dt) < \infty$.

Remark 2.4.3 Let $T = [0, 1]$ and $\mathcal{H}(t_1, t_2) = t_1 \wedge t_2$. The elements of $H(\mathcal{H}, T)$ have the following generic form [(Example) 1.1.25]:

$$\phi(t) = \int_0^t f(\theta)\, d\theta,\ f \in \mathcal{L}_2([0, 1]),$$

and

$$\|\phi\|^2_{L_2[0,1]} = \int_0^1 \langle I_{[0,t]}, [f]_{L_2[0,1]} \rangle^2_{L_2[0,1]} dt \le \frac{1}{2} \|\phi\|^2_{H(\mathcal{H},T)}.$$

$H(\mathcal{H}, T)$ is thus embedded in $L_2[0, 1]$. What follows extends the properties of this example to a larger class of kernels.

Remark 2.4.4 Item 1 of (Definition) 2.4.2 does not imply item 2 as $t \mapsto \mathcal{H}(t,t)$ may not belong to $H(\mathcal{H},T)$ [(Example) 1.2.6].

Remark 2.4.5 When C is a covariance with a factorization $c_\wedge c_\vee$, for C to have property Π_J it suffices that $c_\wedge$ and $c_\vee$ both belong to $\mathcal{L}_2(T,\mathcal{T},\tau)$.

Remark 2.4.6 Property Π_J was used in (Proposition) 1.3.21.

Remark 2.4.7 Suppose that, for some $F : T \longrightarrow H$, for fixed, but arbitrary $(t,x) \in T \times T$,

$$\mathcal{H}(x,t) = L_F[F(t)](x) = \langle F(x), F(t)\rangle_H.$$

Definition 2.4.2 means then that F is weakly measurable, and that

$$t \mapsto \|F(t)\|_H^2$$

is measurable and integrable. When $\overline{V[\mathcal{R}[F]]}$ is separable, F is strongly (Bochner) measurable [(Proposition) 2.2.2], and thus the definition means that F is strongly (Bochner) integrable [207, p. 114].

Remark 2.4.8 When $H(\mathcal{H},T)$ is separable, and Π_J obtains, $\mathcal{H}$ is adapted. It is indeed enough to apply (Proposition) 2.2.2 to $F : t \mapsto \mathcal{H}(\cdot,t)$.

Remark 2.4.9 When $\mathcal{H}$ has an L_2 representation, and $H(\mathcal{H},T)$ is separable, condition $\int_T \mathcal{H}(t,t)\,\tau(dt) < \infty$ means that [(Proposition) 2.3.6]

$$\int_T \tau(dt) \int_X \mathcal{K}^2(x,t)\,\mu(dx) < \infty,$$

and thus (Definition) 2.4.2 is a stronger requirement than (Corollary) 2.3.9.

Remark 2.4.10 Item 2 of the following lemma says that $H(\mathcal{H},T)$ is a Hilbertian subspace of $L_2(T,\mathcal{T},\tau)$ [35, p. 224].

Lemma 2.4.11 *Suppose the reproducing kernel $\mathcal{H}$, defined on T, has property Π_J for $(T,\mathcal{T},\tau)$. Then* [$\kappa_{\mathcal{H},\tau}$ is defined in (Definition) 2.4.2]:

1. *$H(\mathcal{H},T)$ is embedded in $L_2(T,\mathcal{T},\tau)$;*
2. *given $h \in H(\mathcal{H},T)$, fixed, but arbitrary,*

$$\left\|[h]_{L_2(T,\mathcal{T},\tau)}\right\|^2_{L_2(T,\mathcal{T},\tau)} \leq \kappa_{\mathcal{H},\tau}\,\|h\|^2_{H(\mathcal{H},T)};$$

in particular, given $t \in T$, fixed, but arbitrary,

$$\left\|[\mathcal{H}(\cdot,t)]_{L_2(T,\mathcal{T},\tau)}\right\|^2_{L_2(T,\mathcal{T},\tau)} \leq \kappa_{\mathcal{H},\tau}\,\mathcal{H}(t,t);$$

3. *the inclusion map $J_{\mathcal{H},\tau}$ is linear and bounded.*

Proof Since for $t \in T$, fixed, but arbitrary, $\mathcal{H}(\cdot, t)$ is adapted to $\mathcal{T}$, all elements of $H(\mathcal{H}, T)$ are adapted to $\mathcal{T}$ [(Proposition) 2.2.1]. Furthermore

$$\begin{aligned}\int_T h^2(t)\,\tau(dt) &= \int_T \langle h, \mathcal{H}(\cdot, t)\rangle^2_{H(\mathcal{H},T)}\tau(dt)\\ &\leq \|h\|^2_{H(\mathcal{H},T)} \int_T \|\mathcal{H}(\cdot, t)\|^2_{H(\mathcal{H},T)}\,\tau(dt)\\ &= \kappa_{\mathcal{H},\tau}\,\|h\|^2_{H(\mathcal{H},T)}\,.\end{aligned}$$

□

Proposition 2.4.12 *When the reproducing kernel $\mathcal{H}$, defined on T, has property Π_J for $(T, \mathcal{T}, \tau)$, the following assignment: for $f \in L_2(T, \mathcal{T}, \tau)$, fixed, but arbitrary,*

$$h_f(t) = \int_T \mathcal{H}(x, t)\,\dot{f}(x)\,\tau(dx),$$

defines the value at f of an operator $B_{\mathcal{H},\tau} : L_2(T, \mathcal{T}, \tau) \longrightarrow H(\mathcal{H}, T)$ which is linear and bounded [$B_{\mathcal{H},\tau}[f](t) = h_f(t)$: the operator values may thus be computed as ordinary integrals]. *Furthermore $B^{\star}_{\mathcal{H},\tau} = J_{\mathcal{H},\tau}$.*

Proof Due to (Lemma) 2.4.11, $H(\mathcal{H}, T)$ is a subset of $\mathcal{L}_2(T, \mathcal{T}, \tau)$. Thus, for fixed, but arbitrary $f \in L_2(T, \mathcal{T}, \tau)$, still because of (Lemma) 2.4.11, for $h \in H(\mathcal{H}, T)$, fixed, but arbitrary,

$$\Lambda_f(h) = \int_T h(t)\,\dot{f}(t)\,\tau(dt),$$

is a well defined linear functional on $H(\mathcal{H}, T)$. Furthermore, since

$$|h(t)| = \left|\langle h, \mathcal{H}(\cdot, t)\rangle_{H(\mathcal{H},T)}\right| \leq \|h\|_{H(\mathcal{H},T)}\,\|\mathcal{H}(\cdot, t)\|_{H(\mathcal{H},T)},$$

by Cauchy-Schwarz's inequality,

$$\begin{aligned}&\left|\int_T h(t)\,\dot{f}(t)\,\tau(dt)\right| \leq\\ &\leq \|h\|_{H(\mathcal{H},T)} \int_T \|\mathcal{H}(\cdot, t)\|_{H(\mathcal{H},T)}\,\left|\dot{f}\right|(t)\,\tau(dt)\\ &\leq \|h\|_{H(\mathcal{H},T)} \left\{\int_T \|\mathcal{H}(\cdot, t)\|^2_{H(\mathcal{H},T)}\,\tau(dt)\right\}^{1/2} \left\{\int_T \dot{f}^2(t)\,\tau(dt)\right\}^{1/2}\\ &= \kappa^{1/2}_{\mathcal{H},\tau}\,\|f\|_{L_2(T,\mathcal{T},\tau)}\,\|h\|_{H(\mathcal{H},T)}\,.\end{aligned}$$

The linear functional Λ_f is thus continuous, and so there exists, by Riesz's representation theorem [266, p. 64], a unique element $h[f] \in H(\mathcal{H}, T)$ such that, for $h \in H(\mathcal{H}, T)$, fixed, but arbitrary,

$$\langle h, h[f]\rangle_{H(\mathcal{H},T)} = \Lambda_f(h) = \int_T h(t)\, \dot{f}(t)\, \tau(dt). \tag{$\star$}$$

Let $U_{\mathcal{H},T}$ be the map $h \mapsto \langle \cdot, h\rangle_{H(\mathcal{H},T)}$ which identifies $H(\mathcal{H}, T)$ with its dual. One sets

$$B_{\mathcal{H},\tau}[f] = U^{\star}_{\mathcal{H},T} \circ \Lambda_f.$$

Then [266, p. 60], using the latter inequality, and taking into account that h has norm one,

$$\begin{aligned}
\|B_{\mathcal{H},\tau}[f]\|_{H(\mathcal{H},T)} &= \left\|U^{\star}_{\mathcal{H},T} \circ \Lambda_f\right\|_{H(\mathcal{H},T)} \\
&= \|h[f]\|_{H(\mathcal{H},T)} \\
&= \sup_{\|h\|_{H(\mathcal{H},T)}=1} \left|\langle h[f], h\rangle_{H(\mathcal{H},T)}\right| \\
&= \sup_{\|h\|_{H(\mathcal{H},T)}=1} \left|\int_T h(t)\, \dot{f}(t)\, \tau(dt)\right| \\
&\le \kappa^{1/2}_{\mathcal{H},\tau} \|f\|_{L_2(T,\mathcal{T},\tau)}.
\end{aligned}$$

$B_{\mathcal{H},\tau}$ is thus an operator which is linear and bounded. Furthermore, using $(\star)$,

$$\begin{aligned}
B_{\mathcal{H},\tau}[f](t) &= \langle U^{\star}_{\mathcal{H},T} \circ \Lambda_f, \mathcal{H}(\cdot, t)\rangle_{H(\mathcal{H},T)} \\
&= \langle h[f], \mathcal{H}(\cdot, t)\rangle_{H(\mathcal{H},T)} \\
&= \int_T \mathcal{H}(x, t)\, \dot{f}(x)\, \tau(dx) \\
&\overset{def}{=} h_f(t),
\end{aligned}$$

and, using the definition of the adjoint of an operator [266, p. 70],

$$\begin{aligned}
\langle f, B^{\star}_{\mathcal{H},\tau}[h]\rangle_{L_2(T,\mathcal{T},\tau)} &= \langle B_{\mathcal{H},\tau}[f], h\rangle_{H(\mathcal{H},T)} \\
&= \langle U^{\star}_{\mathcal{H},T} \circ \Lambda_f, h\rangle_{H(\mathcal{H},T)} \\
&= \langle h[f], h\rangle_{H(\mathcal{H},T)} \\
&= \int_T h(t)\, \dot{f}(t)\, \tau(dt) \\
&= \langle f, [h]_{L_2(T,\mathcal{T},\tau)}\rangle_{L_2(T,\mathcal{T},\tau)}.
\end{aligned}$$

Consequently $B^{\star}_{\mathcal{H},\tau} = J_{\mathcal{H},\tau}$. □

Remark 2.4.13 Since, formally and intuitively,

$$\begin{aligned}\langle h, h[f]\rangle_{H(\mathcal{H},T)} &= \int_T h(t)\,\dot{f}(t)\,\tau(dt)\\ &= \int_T \langle h, \mathcal{H}(\cdot,t)\rangle_{H(\mathcal{H},T)}\,\dot{f}(t)\,\tau(dt)\\ &= \langle h, \int_T \mathcal{H}(\cdot,t)\,\dot{f}(t)\,\tau(dt)\rangle_{H(\mathcal{H},T)},\end{aligned}$$

$h[f]$ can be "read" as a weak (Pettis) integral [207, p. 114] of the function

$$t \mapsto \left[\dot{f}(t)\,\mathcal{H}(\cdot,t)\right]_{L_2(T,\mathcal{T},\tau)}:$$

$$h[f] = \int_T^{(w)} \mathcal{H}(\cdot,t)\,\dot{f}(t)\,\tau(dt) \in H(\mathcal{H},T).$$

Remark 2.4.14 $B^{\star}_{\mathcal{H},T}B_{\mathcal{H},\tau}$ is an integral operator of $L_2(T,\mathcal{T},\tau)$ with kernel $\mathcal{H}$. When $\mathcal{H}$ is taken in its covariance guise, $B^{\star}_{\mathcal{H},T}B_{\mathcal{H},\tau}$ is the associated covariance operator, say $R_{\mathcal{H}}$. Then, as in (Proposition) 1.3.21, setting $J^{\star} = B_{\mathcal{H},\tau}$,

$$\mathcal{R}[B^{\star}_{\mathcal{H},\tau}] = \mathcal{R}[R^{1/2}_{\mathcal{H}}].$$

The results which follow yield thus, in the case of covariances, information on $R_{\mathcal{H}}$.

Proposition 2.4.15 *When the reproducing kernel $\mathcal{H}$, defined on T, has property Π_J for $(T,\mathcal{T},\tau)$,*

$$\tilde{B}_{\mathcal{H},\tau} = B^{\star}_{\mathcal{H},\tau}B_{\mathcal{H},\tau} = J_{\mathcal{H},\tau}B_{\mathcal{H},T}$$

is Hilbert-Schmidt, and one has, for its Hilbert-Schmidt norm, denoted using an index HS,

$$\left\|\tilde{B}_{\mathcal{H},\tau}\right\|_{HS} \le \kappa_{\mathcal{H},\tau}.$$

Proof Let L_0 be a (closed) subspace of $L_2(T,\mathcal{T},\tau)$ with an at most countable orthonormal basis $\{e_i,\ i \in I\}$. Let $\mathcal{H}^{L_0}_t$ be the projection of $[\mathcal{H}(\cdot,t)]_{L_2(T,\mathcal{T},\tau)}$ onto L_0, so that

$$\mathcal{H}^{L_0}_t = \sum_{i\in I} \left\langle [\mathcal{H}(\cdot,t)]_{L_2(T,\mathcal{T},\tau)}, e_i\right\rangle_{L_2(T,\mathcal{T},\tau)} e_i.$$

Let $f_i = B_{\mathcal{H},\tau}[e_i]$. Then, using the definition of $B_{\mathcal{H},\tau}$,

$$\begin{aligned} f_i(t) &= \langle B_{\mathcal{H},\tau}[e_i], \mathcal{H}(\cdot, t)\rangle_{H(\mathcal{H},T)} \\ &= \left\langle e_i, B^\star_{\mathcal{H},\tau}[\mathcal{H}(\cdot, t)]\right\rangle_{L_2(T,\mathcal{T},\tau)} \\ &= \langle e_i, J_{\mathcal{H},\tau}[\mathcal{H}(\cdot, t)]\rangle_{L_2(T,\mathcal{T},\tau)} \\ &= \left\langle e_i, [\mathcal{H}(\cdot, t)]_{L_2(T,\mathcal{T},\tau)}\right\rangle_{L_2(T,\mathcal{T},\tau)}. \end{aligned}$$

Thus, using (Lemma) 2.4.11, item 2, for the last inequality,

$$\sum_{i\in I} f_i^2(t) = \left\|\mathcal{H}_t^{L_0}\right\|^2_{L_2(T,\mathcal{T},\tau)} \leq \left\|[\mathcal{H}(\cdot, t)]_{L_2(T,\mathcal{T},\tau)}\right\|^2_{L_2(T,\mathcal{T},\tau)} \leq \kappa_{\mathcal{H},\tau}\mathcal{H}(t,t).$$

But then, since $f_i = B_{\mathcal{H},\tau}[e_i]$,

$$\begin{aligned} \left\|\tilde{B}_{\mathcal{H},\tau}[e_i]\right\|^2_{L_2(T,\mathcal{T},\tau)} &= \left\|B^\star_{\mathcal{H},\tau} B_{\mathcal{H},\tau}[e_i]\right\|^2_{L_2(T,\mathcal{T},\tau)} \\ &= \left\|B^\star_{\mathcal{H},\tau}[f_i]\right\|^2_{L_2(T,\mathcal{T},\tau)} \\ &= \|J_{\mathcal{H},\tau}[f_i]\|^2_{L_2(T,\mathcal{T},\tau)} \\ &= \int_T f_i^2(t)\,\tau(dt), \end{aligned}$$

so that, using the inequality $\sum_{i\in I} f_i^2(t) \leq \kappa_{\mathcal{H},\tau}\mathcal{H}(t,t)$ obtained above, and the integrability of $t \mapsto \mathcal{H}(t,t)$,

$$\sum_{i\in I} \left\|\tilde{B}_{\mathcal{H},\tau}[e_i]\right\|^2_{L_2(T,\mathcal{T},\tau)} = \sum_{i\in I}\int_T f_i^2(t)\,\tau(dt) = \int_T \sum_{i\in I} f_i^2(t)\,\tau(dt) \leq \kappa^2_{\mathcal{H},\tau}.$$

The latter inequality suffices to establish that $\tilde{B}_{\mathcal{H},\tau}$ is a Hilbert-Schmidt operator [266, p. 133]. □

Remark 2.4.16 As (Proposition) 2.4.12 proves that $\tilde{B}_{\mathcal{H},\tau}$ is an integral operator with $\mathcal{H}$ as kernel, if $\mathcal{H}$ were adapted to $\mathcal{T} \otimes \mathcal{T}$ (see (Remark) 2.4.8), it would suffice to check that its square is integrable [119, p. 70].

Corollary 2.4.17 *Let the reproducing kernel $\mathcal{H}$, defined on T, have property Π_J for $(T,\mathcal{T},\tau)$, and $\tilde{B}_{\mathcal{H},\tau}$ be the operator of* (Proposition) 2.4.15. *$\tilde{B}_{\mathcal{H},\tau}$ has thus* [266, p. 163] *a standard decomposition*

$$\tilde{B}_{\mathcal{H},\tau}[f] = \sum_{i\in I} \beta_i \langle f, b_i\rangle_{L_2(T,\mathcal{T},\tau)} b_i,$$

with $\beta_i > 0,\ i \in I$, *and* $\{b_i,\ i \in I\}$, *an at most countable orthonormal set. Then*

$$\left\{\tilde{b}_i = B_{\mathcal{H},\tau}\left[\beta_i^{-(1/2)} b_i\right],\ i \in I\right\}$$

is an orthonormal set in $H(\mathcal{H}, T)$.

Proof With

$$\delta_{i,j} = \begin{cases} 1, & i = j \\ 0, & i \neq j \end{cases},$$

one has that

$$\begin{aligned} &\langle B_{\mathcal{H},\tau}\left[\beta_i^{-(1/2)} b_i\right], B_{\mathcal{H},\tau}\left[\beta_j^{-(1/2)} b_j\right]\rangle_{H(\mathcal{H},T)} = \\ &= \langle B_{\mathcal{H},\tau}^{\star} B_{\mathcal{H},\tau}\left[\beta_i^{-(1/2)} b_i\right], \beta_j^{-(1/2)} b_j\rangle_{L_2(T,\mathcal{T},\tau)} \\ &= \beta_j^{-(1/2)} \beta_i^{1/2} \delta_{i,j}. \end{aligned}$$

□

Corollary 2.4.18 *Let the reproducing kernel* $\mathcal{H}$, *defined on* T, *have property* Π_J *for* $(T, \mathcal{T}, \tau)$, *and* $\tilde{B}_{\mathcal{H},\tau}$ *be the operator of* (Proposition) 2.4.15. *Then*

$$\mathcal{N}\left[\tilde{B}_{\mathcal{H},\tau}\right]^{\perp} = \overline{\mathcal{R}[B_{\mathcal{H},\tau}^{\star}]}.$$

Proof Fix arbitrarily $f \in \mathcal{N}\left[\tilde{B}_{\mathcal{H},\tau}\right]$. Then, because

$$0 = \langle \tilde{B}_{\mathcal{H},\tau}[f], f\rangle_{L_2(T,\mathcal{T},\tau)} = \|B_{\mathcal{H},\tau}[f]\|_{H(\mathcal{H},T)}^2,$$

$f \in \mathcal{N}[B_{\mathcal{H},\tau}]$, and thus, for every $h \in H(\mathcal{H}, T)$,

$$\langle f, B_{\mathcal{H},\tau}^{\star}[h]\rangle_{L_2(T,\mathcal{T},\tau)} = \langle B_{\mathcal{H},\tau}[f], h\rangle_{H(\mathcal{H},T)} = 0.$$

Consequently $\mathcal{N}\left[\tilde{B}_{\mathcal{H},\tau}\right]$ is orthogonal to the range of $B_{\mathcal{H},\tau}^{\star}$, so that

$$\mathcal{N}\left[\tilde{B}_{\mathcal{H},\tau}\right] \subseteq \left\{B_{\mathcal{H},\tau}^{\star}[H(\mathcal{H}, T)]\right\}^{\perp},$$

and thus [266, pp. 35 and 37]

$$\mathcal{N}\left[\tilde{B}_{\mathcal{H},\tau}\right]^{\perp} \supseteq \left\{B_{\mathcal{H},\tau}^{\star}[H(\mathcal{H}, T)]\right\}^{\perp\perp} = \overline{B_{\mathcal{H},\tau}^{\star}[H(\mathcal{H}, T)]}.$$

Now fix f arbitrarily in (for the equality: [266, p. 71])

$$\mathcal{N}\left[\tilde{B}_{\mathcal{H},\tau}\right]^{\perp} = \overline{\mathcal{R}\left[\tilde{B}_{\mathcal{H},\tau}\right]}.$$

Since the b_i's of (Corollary) 2.4.17 are in the range of $\tilde{B}_{\mathcal{H},\tau}$, one has then that $f = \lim_n f_n$, for some sequence

$$\left\{f_n = \sum_{i=1}^{p(f_n)} \alpha_{i(f_n)} b_{i(f_n)},\ n \in \mathbb{N}\right\} \subseteq L_2\left(T, \mathcal{T}, \tau\right).$$

But, as $\tilde{B}_{\mathcal{H},\tau}[b_i] = \beta_i b_i$,

$$f_n = \sum_{i=1}^{p(f_n)} \alpha_{i(f_n)} B^{\star}_{\mathcal{H},\tau}\left[\beta^{-1}_{i(f_n)} B_{\mathcal{H},\tau}\left[b_{i(f_n)}\right]\right] = B^{\star}_{\mathcal{H},\tau}\left[\sum_{i=1}^{p(f_n)} \frac{\alpha_{i(f_n)}}{\beta_{i(f_n)}} B_{\mathcal{H},\tau}\left[b_{i(f_n)}\right]\right],$$

so that

$$\mathcal{N}\left[\tilde{B}_{\mathcal{H},\tau}\right]^{\perp} \subseteq \overline{B^{\star}_{\mathcal{H},\tau}\left[H\left(\mathcal{H}, T\right)\right]}.$$

□

Corollary 2.4.19 *Let the reproducing kernel $\mathcal{H}$, defined on T, have property Π_J for $(T, \mathcal{T}, \tau)$, and $\tilde{B}_{\mathcal{H},\tau}$ be the operator of* (Proposition) 2.4.15. *It has a (finite) trace.*

Proof From [(Corollary) 2.4.17]

$$\tilde{B}_{\mathcal{H},\tau}\left[b_i\right] = B^{\star}_{\mathcal{H},\tau} B_{\mathcal{H},\tau}\left[b_i\right] = \beta_i b_i, \text{ and } \tilde{b}_i = B_{\mathcal{H},\tau}\left[\beta_i^{-(1/2)} b_i\right],$$

one gets that

$$B^{\star}_{\mathcal{H},\tau}\left[\tilde{b}_i\right] = \beta_i^{1/2} b_i.$$

Let P_0 be the projection of $H\left(\mathcal{H}, T\right)$ whose range is generated by

$$\left\{\tilde{b}_i,\ i \in I\right\}.$$

Then, since $B^{\star}_{\mathcal{H},\tau}$ is the inclusion map [(Proposition) 2.4.12], almost surely with respect to τ, referred to when using the symbol $\sim$,

$$\sum_{i \in I} \beta_i \dot{b}_i^2\left(t\right) \sim \sum_{i \in I} \tilde{b}_i^2\left(t\right) = \sum_{i \in I} \langle \tilde{b}_i, \mathcal{H}\left(\cdot, t\right)\rangle^2_{H(\mathcal{H},T)} = \left\|P_0\left[\mathcal{H}\left(\cdot, t\right)\right]\right\|^2_{H(\mathcal{H},T)}.$$

Consequently, almost surely with respect to τ,

$$\sum_{i \in I} \beta_i \dot{b}_i^2\left(t\right) \leq \mathcal{H}\left(t, t\right).$$

Integrating over T yields $\sum_{i \in I} \beta_i \leq \kappa_{\mathcal{H},\tau}$. □

Remark 2.4.20 Property Π_J thus allows one to embed an RKHS into an L_2 space and relate the two using operators with finite trace.

Definition 2.4.21 Let $H(\mathcal{H}, T)$ be embedded in $L_2(T, \mathcal{T}, \tau)$ [*(Definition)* 2.4.1]. When

$$\|J_{\mathcal{H},\tau}[h]\|_{L_2(T,\mathcal{T},\tau)} = 0 \text{ means that } h(t) = 0, \ t \in T,$$

that is, when $J_{\mathcal{H},\tau}$ is an injection, the embedding is said to be regular.

Proposition 2.4.22 *Let $H(\mathcal{H}, T)$ be embedded in $L_2(T, \mathcal{T}, \tau)$. Suppose T is a topological space, $\mathcal{T}$ is the family of Borel sets of T, and τ is such that $\tau(O) = 0$, O open, implies $O = \emptyset$. Then, when $\mathcal{H}$ is continuous for the product topology, the embedding is regular.*

Proof The elements in $H(\mathcal{H}, T)$ are continuous [(Proposition) 2.6.1]. Suppose that

$$[h]_{L_2(T,\mathcal{T},\tau)} = 0.$$

Then $T_p = \{t \in T : |h(t)| > 0\}$ is an open, measurable set, and, for that set, $\tau(T_p) = 0$. T_p is thus empty, and $h(t) = 0, \ t \in T$. □

Proposition 2.4.23 *Let the reproducing kernel $\mathcal{H}$, defined on T, have property Π_J for $(T, \mathcal{T}, \tau)$, and $\tilde{B}_{\mathcal{H},\tau}$ be the operator of* (Proposition) 2.4.15. *Suppose that the resulting embedding is regular. Then:*

1. *$H(\mathcal{H}, T)$ is separable;*
2. *$\mathcal{H}$ has the following representation* [(Corollary) 2.4.17]*: for $(t_1, t_2) \in T \times T$, fixed, but arbitrary,*

$$\mathcal{H}(t_1, t_2) = \sum_{i \in I} \tilde{b}_i(t_1) \tilde{b}_i(t_2);$$

3. $\sum_{i \in I} \beta_i = \kappa_{\mathcal{H},\tau}$ [(Definition) 2.4.2, Corollary 2.4.17].

Proof Let $h \in H(\mathcal{H}, T)$ be fixed, but arbitrary, and suppose that, for $i \in I$, fixed, but arbitrary,

$$\langle h, \tilde{b}_i \rangle_{H(\mathcal{H},T)} = 0.$$

Since [(Corollary) 2.4.17]

$$\langle h, \tilde{b}_i \rangle_{H(\mathcal{H},T)} = \langle h, B_{\mathcal{H},\tau}\left[\beta_i^{-(1/2)} b_i\right] \rangle_{H(\mathcal{H},T)} = \langle B^{\star}_{\mathcal{H},\tau}[h], \beta_i^{-(1/2)} b_i \rangle_{L_2(T,\mathcal{T},\tau)},$$

$B^{\star}_{\mathcal{H},\tau}[h]$ is orthogonal to the subspace generated by the b_i's, that is, using successively [266, p. 71] and (Corollary) 2.4.18,

$$B^{\star}_{\mathcal{H},\tau}[h] \in \overline{\mathcal{R}[\tilde{B}_{\mathcal{H},\tau}]}^{\perp} = \mathcal{N}[\tilde{B}_{\mathcal{H},\tau}] = \overline{\mathcal{R}[B^{\star}_{\mathcal{H},\tau}]}^{\perp},$$

so that $B^{\star}_{\mathcal{H},\tau}[h] = 0$ in $L_2(T, \mathcal{T}, \tau)$. Since the embedding is regular, $h = 0_{H(\mathcal{H},T)}$, that is, $J_{\mathcal{H},\tau}$ is an injection. But [266, p. 71]

$$\{0_{H(\mathcal{H},T)}\} = \mathcal{N}\left[B^{\star}_{\mathcal{H},\tau}\right] = \overline{\mathcal{R}\left[B_{\mathcal{H},\tau}\right]}^{\perp},$$

so that

$$\overline{\mathcal{R}[B_{\mathcal{H},\tau}]} = H(\mathcal{H}, T).$$

Since $\mathcal{R}[B_{\mathcal{H},\tau}]$ contains the orthonormal set $\{\tilde{b}_i,\ i \in I\}$, item 1 is true.

Item 2 follows from (Proposition) 1.5.6.

Finally

$$\begin{aligned}
\int_T \mathcal{H}(t,t)\, \tau(dt) &= \sum_{i \in I} \int_T \langle \mathcal{H}(\cdot, t), \tilde{b}_i \rangle^2_{H(\mathcal{H},T)}\, \tau(dt) \\
&= \sum_{i \in I} \int_T \tilde{b}_i^2(t)\, \tau(dt) \\
&= \sum_{i \in I} \left\| B^{\star}_{\mathcal{H},\tau} \tilde{b}_i \right\|^2_{L_2(T,\mathcal{T},\tau)} \\
&= \sum_{i \in I} \left\| B^{\star}_{\mathcal{H},\tau} B_{\mathcal{H},\tau}\left[\beta_i^{-(1/2)} b_i\right] \right\|^2_{L_2(T,\mathcal{T},\tau)} \\
&= \sum_{i \in I} \beta_i.
\end{aligned}$$

□

Definition 2.4.24 Let $H(\mathcal{H}, T)$ be embedded in $L_2(T, \mathcal{T}, \tau)$. This embedding is complete whenever, for fixed, but arbitrary $(h_1, h_2) \in H(\mathcal{H}, T) \times H(\mathcal{H}, T)$,

$$\langle h_1, h_2 \rangle_{H(\mathcal{H},T)} = \langle J_{\mathcal{H},\tau}[h_1], J_{\mathcal{H},\tau}[h_2] \rangle_{L_2(T,\mathcal{T},\tau)}.$$

Remark 2.4.25 The inclusion map is, for a complete embedding, an isometry [266, p. 66], and $J_{\mathcal{H},\tau}[H(\mathcal{H}, T)]$ is thus closed in $L_2(T, \mathcal{T}, \tau)$ [266, p. 86].

Proposition 2.4.26 *$H(\mathcal{H}, T)$ is completely embedded in $L_2(T, \mathcal{T}, \tau)$ if, and only if, for $\{t, t_1, t_2\} \subseteq T$, fixed, but arbitrary,*

1. $[\mathcal{H}(\cdot, t)]_{L_2(T,\mathcal{T},\tau)} \in L_2(T, \mathcal{T}, \tau)$,
2. $\mathcal{H}(t_1, t_2) = \int_T \mathcal{H}(x, t_1)\, \mathcal{H}(x, t_2)\, \tau(dx)$.

Furthermore a complete embedding is regular.

Proof *Suppose $H(\mathcal{H}, T)$ is completely embedded in $L_2(T, \mathcal{T}, \tau)$.*

It is then, by definition, embedded, and item 1 is trivially true. Item 2 expresses, for $h_1 = \mathcal{H}(\cdot, t_1)$ and $h_2 = \mathcal{H}(\cdot, t_2)$ the fact that the embedding is complete.

Proof *Suppose conversely that items 1 and 2 obtain.*

Item 2 yields that, for $(h_1, h_2) \in V[\mathcal{H}] \times V[\mathcal{H}]$, fixed, but arbitrary,

$$\langle h_1, h_2 \rangle_{H(\mathcal{H},T)} = \langle J_{\mathcal{H},\tau}[h_1], J_{\mathcal{H},\tau}[h_2] \rangle_{L_2(T,\mathcal{T},\tau)}.$$

$J_{\mathcal{H},\tau}$ is thus an isometry which "sends" a function of $H(\mathcal{H}, T)$ to its equivalence class in $L_2(T, \mathcal{T}, \tau)$. □

Proposition 2.4.27 *Suppose $H(\mathcal{H}, T)$ is completely embedded in $L_2(T, \mathcal{T}, \tau)$. The operator $J^\star_{\mathcal{H},\tau}$, which is a partial isometry, with initial set $J_{\mathcal{H},\tau}[H(\mathcal{H}, T)]$ and final set $H(\mathcal{H}, T)$ (and is thus onto), is the integral operator obtained, for $[f]_{L_2(T,\mathcal{T},\tau)} \in L_2(T, \mathcal{T}, \tau)$, fixed, but arbitrary, using the following relation:*

$$h(t) = \int_T \mathcal{H}(x, t) f(x)\, \tau(dx).$$

Proof One has that

$$\begin{aligned} J^\star_{\mathcal{H},\tau}[f](t) &= \langle J^\star_{\mathcal{H},\tau}[f], \mathcal{H}(\cdot, t) \rangle_{H(\mathcal{H},T)} \\ &= \langle f, J_{\mathcal{H},\tau}[\mathcal{H}(\cdot, t)] \rangle_{L_2(T,\mathcal{T},\tau)} \\ &= \int_T \mathcal{H}(x, t)\, \dot{f}(x)\, \tau(dx). \end{aligned}$$

□

Remark 2.4.28 Since, for a complete embedding, $J_{\mathcal{H},\tau}$ is an isometry, $J^\star_{\mathcal{H},\tau} J_{\mathcal{H},\tau}$ is the identity of $H(\mathcal{H}, T)$, and $J_{\mathcal{H},\tau} J^\star_{\mathcal{H},\tau}$, the projection onto $J_{\mathcal{H},\tau}[H(\mathcal{H}, T)]$ [266, p. 86].

One shall see that signals arising in detection have representations as outputs of filters, that is, are elements in the range of integral operators from L_2 spaces to RKHS's. It is thus important to be able to invert these operators. The following result states some conditions under which such an inversion is possible. But a preliminary lemma is needed.

Lemma 2.4.29 *Let $H(\mathcal{H}, T)$ be an RKHS which is completely embedded into $L_2(T, \mathcal{T}, \tau)$, and $S \subseteq T$, be a fixed, but arbitrary subset. Then $H(\mathcal{H}_S, S)$ is completely embedded into $L_2(S, \mathcal{T} \cap S, \tau^{|\mathcal{T} \cap S})$.*

Proof The elements of $H(\mathcal{H}_S, S)$ are obtained as the restrictions to S of elements in $H(\mathcal{H}, T)$, and $\mathcal{H}_S$ is the restriction to S of $\mathcal{H}$ [(Proposition) 1.6.3]. One then applies (Proposition) 2.4.26 which only requires a transcription. □

Proposition 2.4.30 ([232, p. 85]) *Let $\mathcal{H}$, defined over T, have, for (t_1, t_2) in $T \times T$, fixed, but arbitrary, the following L_2 representation:*

$$\mathcal{H}(t_1, t_2) = \int_X \kappa(x, t_1)\, \kappa(x, t_2)\, \mu(dx),$$

where κ is a function defined on $X \times T$, adapted to $\mathcal{X} \otimes \mathcal{T}$, and such that

$$\int_X \kappa^2(x, t)\, \mu(dx) < \infty, \ t \in T.$$

κ_t shall denote the equivalence class of $\kappa(\cdot, t)$ and H_κ the (closed) subspace of $L_2(X, \mathcal{X}, \mu)$ generated linearly by $\{\kappa_t,\ t \in T\}$.

Suppose that $H(\mathcal{H}, T)$ is completely embedded in $L_2(T, \mathcal{T}, \tau)$, and let h in $H(\mathcal{H}, T)$ be fixed, but arbitrary. It has, for $t \in T$, fixed, but arbitrary, a representation of the following form:

$$h(t) = \int_X \kappa(x, t) f_h(x)\, \mu(dx),$$

where $\int_X f_h^2(x)\, \mu(dx) < \infty$.

Suppose there exists $\{T_n,\ n \in \mathbb{N}\} \subseteq \mathcal{T}$ such that

(A) *$T_n \subseteq T_{n+1}$, $n \in \mathbb{N}$ and $\cup_n T_n = T$,*
(B) *for $n \in \mathbb{N}$, fixed, but arbitrary,*
   (a) $\int_{T_n} \mathcal{H}(t, t)\, \tau(dt) < \infty$,
   (b) $\int_X \left\{ \int_{T_n} \kappa(x, t)\, h(t)\, \tau(dt) \right\}^2 \mu(dx) < \infty$.

One has then that, in $L_2(X, \mathcal{X}, \mu)$,

$$P_{H_\kappa}[f_h] = \lim_n \left[\int_{T_n} \kappa(\cdot, t)\, h(t)\, \tau(dt) \right]_{L_2(X, \mathcal{X}, \mu)}.$$

Proof The following calculation is needed for a subsequent interchange of integrals. Let thus $g \in L_2(X, \mathcal{X}, \mu)$ be fixed, but arbitrary. Then

$$\int_{T_n} \int_X \tau(dt)\, \mu(dx)\, |\dot{g}(x)|\, |h(t)|\, |\kappa(x, t)| \leq$$

$$\leq \int_{T_n} \tau(dt)\, |h(t)|\, \|g\|_{L_2(X, \mathcal{X}, \mu)}\, \|\kappa_t\|_{L_2(X, \mathcal{X}, \mu)}$$

$$= \|g\|_{L_2(X,\mathcal{X},\mu)} \int_{T_n} \tau\,(dt)\,|h\,(t)|\,\mathcal{H}\,(t,t)^{1/2}$$

$$\leq \|g\|_{L_2(X,\mathcal{X},\mu)} \left\|[h]_{L_2(T,\mathcal{T},\tau)}\right\|_{L_2(T,\mathcal{T},\tau)} \left\{\int_{T_n} \mathcal{H}\,(t,t)\,\tau\,(dt)\right\}^{1/2}$$

$$< \infty.$$

Let L_κ be the defining map which sends $L_2(X,\mathcal{X},\mu)$ onto $H(\mathcal{H},T)$, that is, the map defined using $L_\kappa[k](t) = \langle k,\kappa_t\rangle_{L_2(X,\mathcal{X},\mu)}$, and H_κ be the subspace generated in $L_2(X,\mathcal{X},\mu)$ by the family $\{\kappa(\cdot,t), t \in T\}$. Let L_n be the defining map which sends $H(\mathcal{H},T)$ onto $H(\mathcal{H}_{T_n},T_n)$, and H_n be the subspace generated in $H(\mathcal{H},T)$ by the family $\{\mathcal{H}(\cdot,t), t \in T_n\}$. Let finally $J_{\mathcal{H}_{T_n},\tau^{|T_n}}$ be the embedding of (Lemma) 2.4.29.

Assumption (B,b) says that the equivalence class k_n of

$$x \mapsto \int_{T_n} \kappa(x,t)h(t)\tau(dt)$$

belongs to $L_2(X,\mathcal{X},\mu)$. One has that $L_\kappa^\star L_n^\star[L_n[h]] = k_n$. Indeed,

$$\begin{aligned}
&\langle L_\kappa^\star L_n^\star[L_n[h]], g\rangle_{L_2(X,\mathcal{X},\mu)} = \\
&= \langle L_n[h], L_n\,[L_\kappa[g]]\rangle_{H(\mathcal{H}_{T_n},T_n)} \\
&= \langle J_{\mathcal{H}_{T_n},\tau^{|T_n}}\,[L_n[h]]\,, J_{\mathcal{H}_{T_n},\tau^{|T_n}}\,[L_n\,[L_\kappa\,[g]]]\rangle_{L_2(T_n,\mathcal{T}\cap T_n,\tau^{|T_n})} \\
&= \int_{T_n} \tau\,(dt)\,h(t)\langle g,\kappa_t\rangle_{L_2(X,\mathcal{X},\mu)} \\
&= \int_{T_n} \tau\,(dt)\,h\,(t) \int_X \mu\,(dx)\,\dot{g}\,(x)\,\kappa\,(x,t) \\
&= \int_X \mu\,(dx)\,\dot{g}\,(x) \int_{T_n} \tau\,(dt)\,\kappa\,(x,t)\,h\,(t) \\
&= \langle k_n, g\rangle_{L_2(X,\mathcal{X},\mu)}.
\end{aligned}$$

Let P_n be the projection in $H(\mathcal{H},T)$, onto H_n. Then $L_\kappa^\star\left[L_n^\star\,[L_n\,[h]]\right] = k_n$ rewrites as $L_\kappa\,[P_n\,[h]] = k_n$, so that, taking the limit in $L_2(X,\mathcal{X},\mu)$, and taking into account that P_n tends to the identity, $L_\kappa^\star[h] = \lim_n k_n$. But

$$L_\kappa^\star[h] = L_\kappa^\star\,[L_\kappa\,[f_h]] = P_{H_\kappa}\,[f_h]\,.$$

□

Example 2.4.31 ([232, p. 89]) Let $F : [0, \infty[\longrightarrow L_2[0,1]$ be defined using the following relation

$$F(t) = \left[x \mapsto \sqrt{2}\sin(\pi t x)\right]_{L_2[0,1]}.$$

Since $\{F(n),\ n \in \mathbb{N}\}$ is a complete orthonormal set in $L_2[0,1]$ [134, p. 37], the map

$$L[f](t) = \langle f, F(t)\rangle_{L_2[0,1]}$$

is unitary between $L_2[0,1]$ and the RKHS $H(\mathcal{H}, [0,\infty[)$ with the following kernel [120, p. 139]:

$$\mathcal{H}(t_1,t_2) = 2\int_0^1 \sin(\pi t_1 x)\sin(\pi t_2 x)\,dx = \frac{\sin(\pi[t_1 - t_2])}{\pi(t_1 - t_2)} - \frac{\sin(\pi[t_1 + t_2])}{\pi(t_1 + t_2)}.$$

Now $\int_0^\infty \mathcal{H}(x,t_1)\,\mathcal{H}(x,t_2)\,dx =$

$$2\int_0^1\int_0^1 d\theta d\eta \sin(\pi t_1 \theta)\sin(\pi t_2 \eta)\left\{2\int_0^\infty dx \sin(\pi\theta x)\sin(\pi\eta x)\right\},$$

and (as above)

$$2\int dx \sin(\pi\theta x)\sin(\pi\eta x) = \frac{\sin(\pi[\theta - \eta]x)}{\pi(\theta - \eta)} - \frac{\sin(\pi[\theta + \eta]x)}{\pi(\theta + \eta)}.$$

Since [152, p. 68] $\lim_{\alpha\to\infty} \frac{\sin(\alpha x)}{\pi x} = \delta(x)$, the last but one expression (with, for example, x taking the part of $\theta - \eta$, and πx, that of α), evaluated at 0 and at ∞, yields a delta function whose value is different from zero only when $\theta = \eta$. The conditions (Proposition) 2.4.23 for a complete embedding thus obtain, and the solution of

$$h(t) = \sqrt{2}\int_0^1 \sin(\pi t x) f(x)\,dx$$

with unknown f is the equivalence class of $\lim_n \sqrt{2}\int_{1/n}^n \sin(\pi x t)\,h(t)\,dt$.

One shall gather here a number of definitions and facts about Borel and Radon measures on Hausdorff, locally compact spaces that shall be needed repeatedly. Most definitions and facts are taken from [275, p 450]. The same Hausdorff space X obtains throughout.

Let X denote a Hausdorff space. The following notation shall be used:

- $\mathcal{B}(X)$: Borel sets of X;
- $\mathcal{O}(X)$: open sets of X;
- $\mathcal{K}(X)$: compact sets of X.

Definition 2.4.32 A Borel measure μ on $\mathcal{B}(X)$ is a map whose domain is $\mathcal{B}(X)$, whose values belong to $\overline{\mathbb{R}_+}$, which is zero at the empty set and is σ-additive.

Let μ be a Borel measure on $\mathcal{B}(X)$, and $B \in \mathcal{B}(X)$ be fixed, but arbitrary. μ is

1. outer regular for B when $\mu(B) = \inf\{\mu(O),\ O \supseteq B,\ O \in \mathcal{O}(X)\}$;
2. inner regular for B when $\mu(B) = \sup\{\mu(K),\ K \subseteq B,\ K \in \mathcal{K}(X)\}$;
3. regular for B when it is outer and inner regular for B.

When $\mathcal{S}(\mathcal{B}(X))$ is an arbitrary collection of subsets of $\mathcal{B}(X)$, μ is outer, inner, or plain regular for $\mathcal{S}(\mathcal{B}(X))$ when it is, respectively, outer, inner, or plain regular for each of the sets in $\mathcal{S}(\mathcal{B}(X))$. When the class of sets is $\mathcal{B}(X)$, then μ is respectively, outer, inner or plain regular.

A Radon measure μ on $\mathcal{B}(X)$ is an outer regular Borel measure such that

1. $\mu(K) < \infty,\ K \in \mathcal{K}(X)$;
2. $\mu(O) = \sup\{\mu(K),\ K \subseteq O,\ K \in \mathcal{K}(X)\},\ O \in \mathcal{O}(X)$.

A signed Radon measure is a signed measure whose Jordan decomposition is the difference of two positive Radon measures.

Fact 2.4.33 ([275, p. 452]) *Let μ be a Radon measure on $\mathcal{B}(X)$. Then:*

1. *when $B \in \mathcal{B}(X)$ and $\mu(B) < \infty$, B is inner regular for μ;*
2. *when μ is σ-finite, every $B \in \mathcal{B}(X)$ is inner regular for μ, and thus regular.*

Example 2.4.34 ([263, p. 136]) Every (positive) Radon measure on the real line is a Lebesgue-Stieltjes measure obtained from a right (left) continuous, monotone increasing function.

Fact 2.4.35 ([167, p. 334]) *Let $C_c(X)$ denote the linear space of functions with domain X, range in $\mathbb{R}$, and compact support. Let $\Lambda : C_c(X) \longrightarrow \mathbb{R}$ be linear and positive* [$\Lambda(f) \geq 0$ when $f \in C_c(X)$ and $f \geq 0$]. *There exists then a unique Radon measure on $\mathcal{B}(X)$ such that*

$$\Lambda(f) = \int_X f(x)\,\mu(dx).$$

Fact 2.4.36 *Let $K \in \mathcal{K}(X)$ be fixed, but arbitrary. Let $C_K(X)$ denote the linear space of functions with domain X, range in $\mathbb{R}$, and support in K, and let Λ : $C_c(X) \longrightarrow \mathbb{R}$ be linear and positive. Then*

1. [275, p. 471] *$C_K(X)$ is a Banach space;*
2. [167, p. 323] *Λ is bounded on $C_K(X)$.*

Fact 2.4.37 ([263, p. 160]) *Let $\{\mu_n,\ n \in \mathbb{N}\}$ be a family of Radon measures on $\mathcal{B}(X)$ such that, for fixed, but arbitrary $x_0 \in X$,*

1. *for every compact neighborhood N_{x_0} of x_0, $\lim_n \mu_n(N_{x_0}) = 1$,*
2. *for every compact set K in the complement of $\{x_0\}$, $\lim_n \mu_n(K) = 0$.*

Then, for every continuous function f on X which has compact support,

$$\lim_n \int_X f(x)\,\mu_n(dx) = f(x_0).$$

One way to recognize RKHS's in L_2 spaces is as follows.

Proposition 2.4.38 ([134, p. 54]) *Let T be a Hausdorff, locally compact, first countable space, $\mathcal{T}$ be its Borel sets, and τ be a Radon measure on $\mathcal{T}$ that gives strictly positive measure to open sets. Suppose H is a Hilbert subspace of $L_2(T, \mathcal{T}, \tau)$ such that each of its equivalence classes contains a continuous function. One shall write, for the equivalence classes of H,*

$$[h]_{L_2(T,\mathcal{T},\tau)},$$

with h being its continuous member [if h and h' are both continuous,

$$\left\{t \in T : \left|h(t) - h'(t)\right| > 0\right\}$$

is open, and thus has strictly positive τ measure, so that h and h' cannot belong to the same equivalence class]. *Let*

$$H_c = \left\{h,\ [h]_{L_2(T,\mathcal{T},\tau)} \in H\right\}$$

and define, for $(h_1, h_2) \in H_c \times H_c$, fixed, but arbitrary,

$$\langle h_1, h_2\rangle_{H_c} = \langle [h_1]_{L_2(T,\mathcal{T},\tau)}, [h_2]_{L_2(T,\mathcal{T},\tau)}\rangle_{L_2(T,\mathcal{T},\tau)}.$$

H_c is then an RKHS completely embedded in $L_2(T, \mathcal{T}, \tau)$.

Proof Let $t \in T$ be fixed, but arbitrary. Since T is first countable, let $\{O_n,\ n \in \mathbb{N}\}$ be a countable base of open neighborhoods at t. One may assume that $O_{n+1} \subseteq O_n,\ n \in \mathbb{N}$. Since T is Hausdorff, $\{t\} = \cap_n O_n$ [84, p. 156]. Since T is Hausdorff and locally compact, there are open sets $C_n,\ n \in \mathbb{N}$, such that $t \in C_n \subseteq \overline{C}_n \subseteq O_n$, with $\overline{C}_n$ compact [84, p. 238]. But then there is, for each $n \in \mathbb{N}$, a continuous $f_n : T \longrightarrow [0, 1]$ such that its support $S(f_n) \subseteq O_n$ is compact, and $f_n(t) = 1,\ t \in \overline{C}_n$ [263, p. 14]. In particular, $\overline{C}_n \subseteq S(f_n)$. One may assume that

$$O_1 \supset S(f_1) \supset \overline{C}_1 \supset C_1 \supset O_2 \supset S(f_2) \supset \overline{C_2} \supset C_2 \supset O_3 \ldots$$

One last important fact is that compact sets have, for Radon measures, finite measure.

Since open sets are measurable, and $\tau(C_n) > 0$, $\delta_n = \tau(C_n)^{-1}\chi_{C_n}$ is an integrable function, with integral equal to one, so that $d\tau_n = \delta_n d\tau$ is a probability measure. If N_t is a compact neighborhood of t, there is an n_t such that, for $n \geq n_t$, $O_n \subseteq N_t$. Consequently

$$1 \geq \tau_n (N_t) \geq \tau_n (O_n) \geq \tau_n (C_n) = 1,$$

so that $\lim_n \tau_n (N_t) = 1$.

If K is a compact set that does not contain t, since a Hausdorff, locally compact space is completely, regular [84, p. 238], there is, by definition, a continuous $\phi : T \longrightarrow [0,1]$ such that $\phi(t) = 1$, and $\phi(k) = 0,\ k \in K$. There is thus an open set containing t that is disjoint from K. But since the τ_n's have no support outside the O_n's, one must have $\lim_n \tau_n (K) = 0$. Consequently (Proposition) 2.4.37 applies.

Now, since $\int_T \delta_n^2 (t)\, \tau (dt) = \{\tau (C_n)\}^{-1}$, the map L_n, defined using the following relation:

$$L_n \left([h]_{L_2(T,\mathcal{T},\tau)} \right) = \langle [h]_{L_2(T,\mathcal{T},\tau)}, [\delta_n]_{L_2(T,\mathcal{T},\tau)} \rangle_{L_2(T,\mathcal{T},\tau)},$$

is a continuous linear functional on H, and, since $f_1 \delta_n = \delta_n$ and hf_1 is continuous with compact support,

$$\begin{aligned} \lim_n L_n \left([h]_{L_2(T,\mathcal{T},\tau)} \right) &= \lim_n \int_T h(x)\, \tau_n (dx) \\ &= \lim_n \int_T h(x) f_1 (x)\, \tau_n (dx) \\ &= h(t). \end{aligned}$$

Consequently, for fixed, but arbitrary $[h]_{L_2(T,\mathcal{T},\tau)}$, there is a finite constant, depending on the class of h and on t, such that

$$\sup_n L_n \left([h]_{L_2(T,\mathcal{T},\tau)} \right) \leq \kappa \left([h]_{L_2(T,\mathcal{T},\tau)}, t \right) < \infty.$$

Because of the Banach-Steinhaus theorem [266, p. 76], there exists $\tilde{\kappa}$ such that $\|L_n\| \leq \tilde{\kappa} < \infty,\ n \in I\!N$. But then, for $n \in \mathbb{N}$, fixed, but arbitrary,

$$\left| L_n \left([h]_{L_2(T,\mathcal{T},\tau)} \right) \right| \leq \tilde{\kappa} \left\| [h]_{L_2(T,\mathcal{T},\tau)} \right\|_{L_2(T,\mathcal{T},\tau)},$$

so that, taking the limit,

$$|h(t)| \leq \tilde{\kappa} \left\| [h]_{L_2(T,\mathcal{T},\tau)} \right\|_{L_2(T,\mathcal{T},\tau)} = \tilde{\kappa} \|h\|_{H_c}.$$

So the evaluation maps are continuous linear functionals on the Hilbert space H_c, which is so an RKHS. □

2.5 Reproducing Kernel Hilbert Spaces of Functions with Integrable Power

The main differences between this section and the preceding one are:

- one considers L_p spaces and not L_2 ones exclusively;
- all results involve the F function;
- a separability condition obtains, allowing one to have that F is (strongly) adapted.

The proof of next proposition requires a lemma that is stated and proved first.

Lemma 2.5.1 *Let $(T, \mathcal{T}, \tau)$ be a σ-finite measure space, H be a real Hilbert space, and $F : T \longrightarrow H$ be a weakly measurable map for which $H_F = \overline{V[\mathcal{R}_F]}$ is separable.*
There exists then a measurable partition of T, say $\{T_n,\ n \in \mathbb{N}_0\} \subseteq \mathcal{T}$, such that

1. $T_0 = \emptyset$,
2. $\tau(T_n) < \infty$, $n \in \mathbb{N}$,
3. $F_n = \chi_{T_n} F$ *is (strongly) measurable, and, for fixed, but arbitrary* $n \in \mathbb{N}$,

$$\|F_n(t)\|_H \leq n,\ t \in T_n,$$

so that, for fixed, but arbitrary $n \in \mathbb{N}$, F_n is strongly (Bochner) integrable.

Proof By assumption and (Proposition) 2.2.2, F is (strongly) measurable. Since the norm of a strongly measurable function is measurable as the composition of a measurable function and a continuous one, for $n \in \mathbb{N}$, fixed, but arbitrary, one may set

$$A_n = \{t \in T : \|F(t)\|_H \leq n\} \in \mathcal{T},$$

so that $A_n \subseteq A_{n+1}$ with $T = \cup_{n \in \mathbb{N}} A_n$. Since τ is σ-finite, there exists a partition of T, say $\{B_n,\ n \in \mathbb{N}\} \subseteq \mathcal{T}$, such that

- $B_n \subseteq B_{n+1}$, $n \in \mathbb{N}$,
- $\cup_n B_n = T$,
- $\tau(B_n) < \infty$, $n \in \mathbb{N}$.

Let $C_n = A_n \cap B_n \in \mathcal{T}$. Since $A_n \subseteq A_{n+1}$ and $B_n \subseteq B_{n+1}$, $C_n \subseteq C_{n+1}$. Furthermore, for fixed, but arbitrary $t \in T$, eventually $t \in C_n$, and, consequently, $\cup_n C_n = T$. Let $C_0 = T_0 = \emptyset$, and

$$T_n = C_n \setminus C_{n-1},\ n \in \mathbb{N}.$$

Those T_n's are disjoint, belong to $\mathcal{T}$, "sum" to T, and $F^{|T_n}$ is bounded in norm by n.
For $n \in \mathbb{N}$, fixed, but arbitrary, let

$$F_n = \chi_{T_n} F.$$

Since F_n is weakly measurable [(Remark) 2.2.5], and has range in H_F, a separable (closed) subspace, it is (strongly) measurable. A function with values in a Banach space which is (strongly) measurable is strongly (Bochner) integrable when its norm is integrable [207, p. 114], which is the case of F_n, since its norm is uniformly bounded on a set of finite measure, and zero outside that set. □

Proposition 2.5.2 *Let $(T, \mathcal{T}, \tau)$ be a σ-finite measure space, H be a real Hilbert space, and $F : T \longrightarrow H$ be a map. Suppose H_F is separable. Let $p \in [0, \infty]$ be fixed, but arbitrary. The following statements are then equivalent:*

1. *F is weakly p-integrable;*
2. *$\mathcal{H}_F = \langle F(\cdot), F(\cdot)\rangle_H$ is (p, q)-bounded with $q = \frac{p}{p-1} = \alpha(p)$* [(Definition) 2.1.13].

When these conditions obtain,

(i) *the map $J_{F,p} : H(\mathcal{H}_F, T) \longrightarrow L_p(T, \mathcal{T}, \tau)$ defined using the following relation:*

$$J_{F,p}[h] = [h]_{L_p(T,\mathcal{T},\tau)}$$

is linear and bounded, and the operator $\Lambda_{F,p} = J_{F,p}L_F$ is linear and bounded;

(ii) *for $p \in [1, \infty]$, $\Lambda^{\star}_{F,p} : L^{\star}_p(T, \mathcal{T}, \tau) \longrightarrow H^{\star}$ has the following representation in terms of a weak (Pettis) integral* [207, p. 114]: *when, for fixed, but arbitrary $f \in L_q(T, \mathcal{T}, \tau)$ and $g \in L_p(T, \mathcal{T}, \tau)$,*

$$\Phi_f(g) = \int_T \dot{f}(t)\, \dot{g}(t)\, \tau(dt),$$

then

$$\Lambda^{\star}_{F,p}[\Phi_f] = \left\langle \cdot, \int_T^{(w)} (fF) d\tau \right\rangle_H$$

(when $p = \infty$, $q = 1$, the weak integral yields the restriction of $\Lambda^{\star}_{F,p}$ to $L_1(T, \mathcal{T}, \tau)$ considered as a subset of the dual of $L_\infty(T, \mathcal{T}, \tau)$);

(iii) *let $U_H : H \longrightarrow H^{\star}$ be defined, for $h \in H$, fixed, but arbitrary using the following relation: $U_H[h] = \langle \cdot, h\rangle_H$. $\mathcal{H}_F$ is a Carleman kernel, and $\Lambda_{F,p}U^{\star}_H\Lambda^{\star}_{F,p} = L_{\mathcal{H}_F}$, the latter being as defined in* [(Definition) 2.1.10], *that is, as an integral operator with kernel $\mathcal{H}_F$.*

Proof (1 ⇒ 2) Suppose thus that F is weakly p-integrable, so that every function in $H(\mathcal{H}_F, T)$ is p-integrable, and the map $J_{\mathcal{H},p}$ is well defined.

As the proof uses items (i) to (iii) of the statement's second part, one starts by proving those.

Proof (i) To have that $J_{\mathcal{H},p}$ is continuous, it suffices to prove that $J_{F,p}$ is a closed map [266, p. 94]. Let thus

$$\{h_n,\ n \in \mathbb{N}\} \text{ converge to } h \text{ in } H(\mathcal{H}_F, T),$$

and

$$\left\{[h_n]_{L_p(T,\mathcal{T},\tau)},\ n \in \mathbb{N}\right\} \text{ converge to } f \text{ in } L_p(T, \mathcal{T}, \tau).$$

Then, for $t \in T$, fixed, but arbitrary, $\lim_n h_n(t) = h(t)$ [(Proposition) 1.1.9], and

$$\left\{[h_n]_{L_p(T,\mathcal{T},\tau)},\ n \in \mathbb{N}\right\}$$

converges weakly to f [275, p. 390], so that it is bounded [85, p. 68]. Consequently,

$$\left\{[h_n]_{L_p(T,\mathcal{T},\tau)},\ n \in \mathbb{N}\right\}$$

being bounded, and $\{h_n,\ n \in \mathbb{N}\}$ converging everywhere to h,

$$\left\{[h_n]_{L_p(T,\mathcal{T},\tau)},\ n \in \mathbb{N}\right\}$$

converges weakly to $[h]_{L_p(T,\mathcal{T},\tau)}$ [275, p. 391]. Since weakly convergent sequences can have only one limit [85, p. 68],

$$f = [h]_{L_p(T,\mathcal{T},\tau)},$$

and $J_{F,p}$ is closed. $\Lambda_{F,p}$ is then linear and bounded, as the composition of two such maps.

Proof (ii) Since F is weakly measurable, and H_F is separable, F is (strongly) measurable [(Proposition) 2.2.2], and so is fF, for scalar and measurable f [(Remark) 2.2.5].

Suppose first that $p \in [1, \infty[$. Since, by assumption, $q = p/(p-1)$, p and q are conjugate, and Hölder's inequality yields that [229, p. 67], for fixed, but arbitrary $f \in L_q(T, \mathcal{T}, \tau)$, $h \in H$, and $T_0 \in \mathcal{T}$,

$$\int_{T_0} \left|\langle h, F(t)\rangle_H \dot{f}(t)\right| \tau(dt) \leq \|f\|_{L_q(T,\mathcal{T},\tau)} \left\|\Lambda_{F,p}[h]\right\|_{L_p(T,\mathcal{T},\tau)}.$$

Since $\Lambda_{F,p}$ is continuous,

$$\left\|\Lambda_{F,p}[h]\right\|_{L_p(T,\mathcal{T},\tau)} \leq \left\|\Lambda_{F,p}\right\| \|h\|_H,$$

so that

$$h \mapsto \int_{T_0} \langle h, F(t)\rangle_H \dot{f}(t)\, \tau(dt)$$

is a continuous linear functional of H. There exists thus [266, p. 64] an element of H, represented as

$$\int_{T_0}^{(w)} (fF)\, d\tau \in H,$$

such that, for fixed, but arbitrary $h \in H$,

$$\langle h, \int_{T_0}^{(w)} (fF)\, d\tau \rangle_H = \int_{T_0} \tau(dt) \langle h, \dot{f}(t) F(t) \rangle_H = \int_{T_0} \langle h, F(t) \rangle_H \dot{f}(t)\, \tau(dt).$$

Let now T_0 be T. When E and F are Banach spaces, and $R : E \longrightarrow F$ is an operator which is linear and bounded, the map conjugate to R, $R^\star : F^\star \longrightarrow E^\star$ is obtained letting [8, p. 281]

$$R^\star [f^\star][e] = f^\star [R[e]], \ e \in E, f^\star \in F^\star.$$

If one now makes the following assignments:

- for $R : E \longrightarrow F, \quad \Lambda_{F,p} : H \longrightarrow L_p(T, \mathcal{T}, \tau)$,
- for $R^\star : F^\star \longrightarrow E^\star$, $\Lambda_{F,p}^\star : L_p^\star(T, \mathcal{T}, \tau) \longrightarrow H^\star$,

one has, for some $f \in L_q(T, \mathcal{T}, \tau)$ (since F stands for an L_p space, an element of $F^\star$ is obtained as

$$\Phi_g[f] = \int \dot{f}\dot{g}\, d\tau$$

for an element g in the corresponding L_q space),

- for $R[e]$, $\Lambda_{F,p}[h] = J_{\mathcal{H},p}[\langle h, F(\cdot) \rangle_H]$,
- for $f^\star[R[e]]$,

$$\begin{aligned}
\Phi_f\left[\Lambda_{F,p}[h]\right] &= \int_T \dot{f}(t) \langle h, F(t) \rangle_H \tau(dt) \\
&= \int_T \langle h, \dot{f}(t) F(t) \rangle_H \tau(dt) \\
&= \langle h, \int_T^{(w)} (fF)\, d\tau \rangle_H.
\end{aligned}$$

Consequently, for $R^\star[f^\star](h)$, one has that

$$\Lambda_{F,p}^\star\left[\Phi_f\right](h) = \langle h, \int_T^{(w)} (fF)\, d\tau \rangle_H.$$

Since the dual of L_∞ contains L_1 strictly [85, 113, p. 296, respectively, p. 198], when $p = \infty$, $\Lambda^\star_{F,p}$ must be restricted to L_1.

Proof (iii and item 2) One has that

$$\begin{aligned}
\Lambda_{F,p} U^\star_H \Lambda^\star_{F,p} \left[\Phi_f\right] &= \Lambda_{F,p} \left[\int_T^{(w)} (fF)\, d\tau \right] \\
&= J_{F,p} L_F \left[\int_T^{(w)} (fF)\, d\tau \right] \\
&= J_{F,p} \left[\langle \int_T^{(w)} (fF)\, d\tau, F(\cdot) \rangle_H \right] \\
&= J_{F,p} \left[\int_T \dot{f}(t) \langle F(t), F(\cdot) \rangle_H \tau(dt) \right] \\
&= J_{F,p} \left[\int_T \mathcal{H}_F(\cdot, t) \dot{f}(t)\, \tau(dt) \right] \\
&= L_{\mathcal{H}_F}[f].
\end{aligned}$$

$\Lambda_{F,p} U^\star_H \Lambda^\star_{F,p}$ is thus an integral operator whose kernel is $\mathcal{H}_F$. For it to be "Carleman-(p, q)," the following properties must obtain:

1. the equivalence class of $\mathcal{H}_F(t, \cdot)$ must belong to $L_{\alpha(q)}(T, \mathcal{T}, \tau)$, and that is true because p has been chosen to be $\alpha(q)$, and because the equivalence class of $\mathcal{H}_F(t, \cdot)$ is $\Lambda_{F,p}[F(t)]$, which belongs to $L_p(T, \mathcal{T}, \tau)$ ($\mathcal{H}_F$ is thus a Carleman kernel);
2. $\mathcal{H}_F(t, \cdot)\dot{f}$ must be integrable for almost all t, with respect to τ, but that follows from the following inequality:

$$\int_{T_0} \left| \langle h, F(t) \rangle_H \dot{f}(t) \right| \tau(dt) \leq \|f\|_{L_q(T,\mathcal{T},\tau)} \left\| \Lambda_{F,p}[h] \right\|_{L_p(T,\mathcal{T},\tau)},$$

   whose validity has already been checked (proof of (ii));
3. the map $t \mapsto \int_T \mathcal{H}_F(t, \theta) \dot{f}(\theta)\, \tau(d\theta)$ must belong to $\mathcal{L}_p(T, \mathcal{T}, \tau)$, but that is true because its equivalence class is in the range of $\Lambda_{F,p} U^\star_H \Lambda^\star_{F,p}$.

Proof (2 ⇒ 1) Suppose thus that $\mathcal{H}_F$ is (p, q)-bounded, with $q = \frac{p}{p-1}$. One must prove that F is weakly p-integrable.

By assumption H_F is separable, and furthermore $\mathcal{H}_F$ is adapted [(Definition) 2.1.10]: that is sufficient [(Proposition) 2.2.2] to have that F is weakly adapted. It thus remains to prove that the map $t \mapsto \langle h, F(t) \rangle_H$ belongs to $\mathcal{L}_p(T, \mathcal{T}, \tau)$.

Let $\{T_n,\ n \in \mathbb{N}_0\} \subseteq \mathcal{T}$ be the decomposition of T in (Lemma) 2.5.1. Let $f \in L_q(T, \mathcal{T}, \tau)$ be fixed, but arbitrary, and

$$\tilde{F}_n : T \longrightarrow H$$

be defined using the following relation:

$$\tilde{F}_n(t) = \chi_{T_n}(t)\, \dot{f}(t)\, F(t) = \chi_{T_n}(t) \dot{f}(t)\, F_n(t).$$

$\tilde{F}_n$ is (strongly) adapted [(Proposition) 2.2.2], and

$$\left\|\tilde{F}_n(t)\right\|_H \le n\, \chi_{T_n}(t)\, \left|\dot{f}(t)\right|.$$

For $n \in \mathbb{N}$, fixed, but arbitrary, let the measure τ_n be defined using the following relation: $d\tau_n = \chi_{T_n} d\tau$. Then

$$\tau_n(T) < \infty, \quad \text{and} \quad \dot{f} \in \mathcal{L}_q(T, \mathcal{T}, \tau_n),$$

so that, since for finite measures, when $p < q$, $L_q \subseteq L_p$ [113, p. 190],

$$n\left|\dot{f}\right| \in \mathcal{L}_1(T, \mathcal{T}, \tau_n), \quad \text{or} \quad n\, \chi_{T_n} \left|\dot{f}\right| \in \mathcal{L}_1(T, \mathcal{T}, \tau).$$

Since

$$\int_T \left\|\tilde{F}_n(t)\right\|_H \tau(dt) < \infty,$$

the integral $\int_T \tilde{F}_n(t)\, \tau(dt)$ is well defined as a strong (Bochner) integral [207, p. 114]. Thus $R_n : L_q(T, \mathcal{T}, \tau) \longrightarrow H$ defined using the following relation:

$$R_n[f] = \int_T \tilde{F}_n(t)\, \tau(dt) = \int_{T_n} \dot{f}(t)\, F(t)\, \tau(dt)$$

makes sense, and, consequently, using the norm property of the Bochner integral [207, p. 114], and then Hölder's inequality [229, p. 67],

$$\|R_n[f]\|_H \le \int_T \left\|\tilde{F}_n(t)\right\|_H \tau(dt) \le n\, \{\tau(T_n)\}^{1/p}\, \|f\|_{L_q(T, \mathcal{T}, \tau)}.$$

R_n is thus linear and bounded. Furthermore, since a strongly (Bochner) integrable function is weakly (Pettis) integrable [207, p. 115],

$$\begin{aligned}
\|R_n[f]\|_H^2 &= \langle \int_T F_n(t)\, \tau(dt), \int_T F_n(u)\, \tau(du) \rangle_H \\
&= \int_T \int_T \tau(dt)\, \tau(du)\, \langle F_n(t), F_n(u) \rangle_H
\end{aligned}$$

$$= \int_T \int_T \tau(dt)\, \tau(du)\, \chi_{T_n}(t)\, \chi_{T_n}(u)\, \dot{f}(t)\, \dot{f}(u)\, \langle F(t), F(u) \rangle_H$$

$$= \int_T \tau(dt)\, \chi_{T_n}(t)\, \dot{f}(t) \int_T \mathcal{H}_F(t,u)\, \chi_{T_n}(u)\, \dot{f}(u)\, \tau(du).$$

Let f_n be the equivalence class of

$$\chi_{T_n} \dot{f}.$$

Then, since, by assumption, $\mathcal{H}_F$ is (p,q)-bounded, the equivalence class of

$$\int_T \mathcal{H}_F(t,u)\, \chi_{T_n}(u)\, \dot{f}(u)\, \tau(du)$$

is $L_{\mathcal{H}_F}[f_n]$. It is an element of $L_p(T, \mathcal{T}, \tau)$, and thus, with the help of Hölder's inequality [229, p. 67],

$$\|R_n[f]\|_H^2 \le \|L_{\mathcal{H}_F}[f_n]\|_{L_p(T,\mathcal{T},\tau)} \|f_n\|_{L_q(T,\mathcal{T},\tau)}.$$

Since $L_{\mathcal{H}_F}$ is continuous [(Proposition) 2.1.12],

$$\|R_n[f]\|_H^2 \le \|L_{\mathcal{H}_F}\| \, \|f_n\|_{L_q(T,\mathcal{T},\tau)}^2.$$

Consequently, since $\|f_n\|_{L_q(T,\mathcal{T},\tau)} \le \|f\|_{L_q(T,\mathcal{T},\tau)}$,

$$\sup_n \|R_n\| \le \|L_{\mathcal{H}_F}\|^{1/2} < \infty,$$

and thus also [8, p. 282]

$$\sup_n \|R_n^\star\| \le \|L_{\mathcal{H}_F}\|^{1/2} < \infty.$$

For $h \in H$ and $f \in L_q(T, \mathcal{T}, \tau)$, fixed, but arbitrary, the duality relation for operators [8, p. 281] yields that $R_n^\star[U_H[h]](f) = \langle h, R_n[f] \rangle_H$. But strongly (Bochner) integrable functions are weakly (Pettis) integrable with the same value [207, p. 115], and thus

$$\langle h, R_n[f] \rangle_H = \int_T \chi_{T_n}(t)\, \dot{f}(t)\, \langle h, F(t) \rangle_H\, \tau(dt) = \Phi_{F_n^h}(f),$$

where F_n^h denotes the equivalence class, in $L_p(T, \mathcal{T}, \tau)$, of the following map:

$$t \mapsto \chi_{T_n}(t)\, \langle h, F(t) \rangle_H.$$

It follows that

$$R_n^{\star}\left[U_H\left[h\right]\right](f) = \Phi_{F_n^h}(f), \ \Phi_{F_n^h} \in L_q^{\star}\left(T, \mathcal{T}, \tau\right).$$

Thus, as [229, p. 136] $\left\|\Phi_{F_n^h}\right\| = \left\|F_n^h\right\|_{L_p(T,\mathcal{T},\tau)}$,

$$\begin{aligned}\left\|F_n^h\right\|_{L_p(T,\mathcal{T},\tau)} &= \left\|\Phi_{F_n^h}\right\| \\ &= \left\|R_n^{\star}\left[U_H\left[h\right]\right]\right\| \\ &\leq \left\|R_n^{\star}\right\| \left\|U_H\left[h\right]\right\|_{H^{\star}} \\ &= \left\|R_n^{\star}\right\| \left\|h\right\|_H \\ &\leq \left\|L_{\mathcal{H}_F}\right\|^{1/2} \left\|h\right\|_H.\end{aligned}$$

The sequence of functions

$$\left\{\dot{F}_n^h, \ n \in \mathbb{N}\right\}$$

converges, almost surely, with respect to τ, to the function

$$\dot{F}^h : t \mapsto \langle h, F(t)\rangle_H,$$

and the sequence of functions

$$\left\{\left|\dot{F}_n^h\right|, n \in \mathbb{N}\right\}$$

increases, almost surely, with respect to τ, to the absolute value of the function $\dot{F}^h$. Then, for fixed, but arbitrary $h \in H$, because of the last set of inequalities,

- when $p = \infty$,

$$\left\|F_n^h\right\|_{L_\infty(T,\mathcal{T},\tau)} \leq \left\|F_{n+1}^h\right\|_{L_\infty(T,\mathcal{T},\tau)} \leq \left\|L_{\mathcal{H}_F}\right\|^{1/2} \left\|h\right\|_H,$$

- when $1 \leq p < \infty$, fixed, but arbitrary,

$$\left\|F_n^h\right\|_{L_p(T,\mathcal{T},\tau)} \leq \left\|F_{n+1}^h\right\|_{L_p(T,\mathcal{T},\tau)} \leq \left\|L_{\mathcal{H}_F}\right\|^{1/2} \left\|h\right\|_H.$$

$\dot{F}^h$ is thus essentially bounded when $p = \infty$, and, because of the monotone convergence theorem, in $\mathcal{L}_p\left(T, \mathcal{T}, \tau\right)$, when $1 \leq p < \infty$. □

Remark 2.5.3 When $p = 2$, one may replace Φ_f of (Proposition) 2.5.2 with f.

Corollary 2.5.4 *Let $(T, \mathcal{T}, \tau)$ be a σ-finite measure space, H be a real Hilbert space, and $F : (T, \mathcal{T}, \tau) \longrightarrow H$ be a weakly adapted map. Let, for a fixed, but arbitrary $p \in [1, \infty]$, $\mathcal{H}_F(t, \theta) = \langle F(t), F(\theta) \rangle_H$, and suppose that*

$$\int_{T \times T} |\mathcal{H}_F(t, \theta)|^p \, \tau \otimes \tau\,(dt, d\theta) < \infty.$$

F is then weakly (Pettis) p-integrable.

Proof Fubini's theorem [229, p. 150], and Hölder's inequality [229, p. 67], imply that $\mathcal{H}_F$ is a (p, q)-bounded kernel, and then (Proposition) 2.5.2 applies. □

Proposition 2.5.5 *Let $(T, \mathcal{T}, \tau)$ be a σ-finite measure space, H be a real Hilbert space, and $F : T \longrightarrow H$ be a map. Then:*

1. *when F is weakly (Pettis) integrable, and H_F is separable, F has an additive decomposition, $F = F_1 + F_2$, such that F_1 has countable range, and F_2 is strongly (Bochner) integrable;*
2. *when $p \in [1, \infty[$, and F is strongly (Bochner) p-integrable, $\Lambda_{F,p}$* [(Proposition) 2.5.2] *is compact;*
3. *when F is weakly (Pettis) integrable, and H_F is separable, $\Lambda_{F,1}$ is compact*

Proof (1) The map F_n of (Proposition) 2.5.2 is strongly (Bochner) integrable, so that there is [276, p. 132] a map $\tilde{F}_n : T \longrightarrow H$ with finite range such that

$$\int_T \left\| F_n(t) - \tilde{F}_n(t) \right\|_H \tau\,(dt) \leq \frac{1}{2^n}.$$

Taking, if necessary, $\chi_{T_n} \tilde{F}_n$ [(Lemma) 2.5.1] instead of $\tilde{F}_n$, one may assume that

$$\left\{ t \in T : \tilde{F}_n(t) \neq 0 \right\} \subseteq T_n.$$

Let

$$F_1(t) = \sum_n \chi_{T_n}(t)\, \tilde{F}_n(t).$$

F_1 is well defined, (strongly) adapted, and has, at most, a countable range. One may set $F_2 = F - F_1$. Now since F and F_1 are (strongly) adapted, F_2 is (strongly) adapted, and

$$\begin{aligned}\int_T \|F_2(t)\|_H \, \tau\,(dt) &= \int_T \|F(t) - F_1(t)\|_H \, \tau\,(dt) \\ &= \sum_n \int_{T_n} \|F(t) - F_1(t)\|_H \, \tau\,(dt)\end{aligned}$$

$$= \sum_n \int_{T_n} \left\| F_n(t) - \tilde{F}_n(t) \right\|_H \tau(dt)$$

$$\leq \sum_n \frac{1}{2^n}.$$

F_2 is thus strongly (Bochner) integrable [207, p. 114].

Proof (2) One must show that $\Lambda_{F,p}$ transforms weakly convergent sequences into strongly convergent ones (it suffices to consider sequences converging to zero). To that end, let $\{h_n, n \in \mathbb{N}\} \subseteq H$ be a sequence which converges weakly to zero. Because of weak convergence, one has that [(Proposition) 1.1.9], for fixed, but arbitrary $t \in T$, $\lim_n \langle F(t), h_n \rangle_H = 0$. But

$$|\langle F(t), h_n \rangle_H| \leq \|F(t)\|_H \|h_n\|_H \leq \|F(t)\|_H \sup_n \|h_n\|_H.$$

Since a weakly convergent sequence is bounded [266, p. 79],

$$|\langle F(t), h_n \rangle_H| \leq \kappa \|F(t)\|_H,$$

and the right-hand side is p-integrable. So, by dominated convergence,

$$\lim_n \left\| \Lambda_{F,p}[h_n] \right\|_{L_p(T,\mathcal{T},\tau)} = 0.$$

Proof (3) By item 1, $F = F_1 + F_2$, where F_1 has countable range. One may thus assume [167, p. 233] that there exists a countable index set I, a family of disjoint sets $T_i \in \mathcal{T}$ such that $\tau(T_i) < \infty$, $i \in I$, and elements $h_i \in H$, $i \in I$, all such that

$$F_1(t) = \sum_{i \in I} h_i \chi_{T_i}(t).$$

Let $\mathcal{E}_i$ be the evaluation map on $\mathbb{R}^I$ that returns the i-th element of a sequence, and let $R : H \longrightarrow l_1$ be defined using the following relation:

$$\mathcal{E}_i \{R[h]\} = \tau(T_i) \langle h, h_i \rangle_H.$$

R is well defined as F and F_2, and thus $F_1 = F - F_2$, are all weakly integrable, and that

$$\int_T |\langle F_1(t), h \rangle_H| \tau(dt) = \sum_{i \in I} \tau(T_i) |\langle h_i(t), h \rangle_H|.$$

But the left-hand side of that latter equality is

$$\left\| \Lambda_{F_1,1}[h] \right\|_{L_1(T,\mathcal{T},\tau)},$$

and the right-hand side is $\|R[h]\|_{l_1}$, so that

$$\|R[h]\|_{l_1} = \left\| \Lambda_{F_1,1}[h] \right\|_{L_1(T,\mathcal{T},\tau)}.$$

Since $\Lambda_{F_1,1}$ is bounded [(Proposition) 2.5.2], R is bounded. If now $\{h_n, n \in \mathbb{N}\} \subseteq H$ converges weakly to zero, $\{R[h_n], n \in \mathbb{N}\} \subseteq l_1$ converges weakly to zero [266, p. 81]. But, in l_1, weakly convergent sequences converge strongly [236, p. 194], and thus $\Lambda_{F_1,1}$ must be compact. $\Lambda_{F_2,1}$ is compact since, as seen, F_2 is strongly integrable (item 2). Thus $\Lambda_{F,1}$ is compact as the sum of two compact operators. That completes the proof. □

Remark 2.5.6 When $p > 1$, H_F is separable, and F is only weakly p-integrable, $\Lambda_{F,p}$ may fail to be a compact operator, as shown by the example which follows. However, when $\tau(T) < \infty$,

$$L_p(T,\mathcal{T},\tau) \subseteq L_1(T,\mathcal{T},\tau),$$

and thus $\Lambda_{F,p}$ will be compact as an operator with values in $L_1(T,\mathcal{T},\tau)$.

Example 2.5.7 Let $(T,\mathcal{T},\tau)$ be a σ-finite measure space, and H be a separable, infinite dimensional, real Hilbert space. There exists thus a partition of T by subsets $\{T_n, n \in \mathbb{N}\} \subseteq \mathcal{T}$, such that, for $n \in \mathbb{N}$, $\tau_n = \tau(T_n) \in]0,\infty[$. Let $\{e_n, n \in \mathbb{N}\} \subseteq H$ be an orthonormal basis for H, and $\{\alpha_n, n \in \mathbb{N}\} \subseteq \mathbb{R}$ be a sequence of positive numbers. Let $F : T \longrightarrow H$ be defined using the following relation:

$$F(t) = \sum_n \chi_{T_n}(t)\, \tau_n^{-(1/2)} \alpha_n\, e_n.$$

Then

$$\langle h, F(t)\rangle_H = \sum_n \chi_{T_n}(t)\, \tau_n^{-(1/2)} \alpha_n \langle h, e_n\rangle_H.$$

F is thus weakly adapted. Since it has countable range, it is (strongly) adapted [(Proposition) 2.2.2]. Furthermore

$$\int_T \langle h, F(t)\rangle_H^2\, \tau(dt) = \sum_n \alpha_n^2 \langle h, e_n\rangle_H^2.$$

Let $\{x_n, n \in \mathbb{N}\}$ be a real sequence in l_1. Then $h = \sum_n |x_n|^{1/2} e_n$ belongs to H, and $\langle h, e_n \rangle_H^2 = |x_n|$. Consequently, $s(x)$ denoting the sign of x,

$$\sum_n \alpha_n^2 x_n = \sum_n \alpha_n^2 s(x_n) \langle h, e_n \rangle_H^2,$$

so that, if F is weakly square integrable, as the α's are taken to be positive, $\sup_n \alpha_n < \infty$ [257, p. 185]. When the latter is true, F is obviously weakly square integrable. It shall thus be assumed that $\sup_n \alpha_n < \infty$.

Let

$$f_n(t) = \begin{Bmatrix} \tau_n^{-(1/2)} & \text{when } t \in T_n \\ 0 & \text{when } t \in T_n^c \end{Bmatrix} = \tau_n^{-(1/2)} \chi_{T_n}(t).$$

One gets thus an orthonormal sequence in $L_2(T, \mathcal{T}, \tau)$, and

$$F(t) = \sum_n \alpha_n f_n(t) e_n.$$

Then $\Lambda_{F,2}$ is well defined as $\Lambda_{F,2}[h]$ is the equivalence class of the series

$$\sum_n \alpha_n \langle h, e_n \rangle_H f_n.$$

Letting $a \otimes b[x] = \langle x, b \rangle a$, one has that

$$\Lambda_{F,2} = \sum_n \alpha_n \left\{ [f_n]_{L_2(T,\mathcal{T},\tau)} \otimes e_n \right\}.$$

Thus [235] $\Lambda_{F,2}$ is

- a compact operator if, and only if, $\lim_n \alpha_n = 0$,
- a Hilbert-Schmidt operator if, and only if, $\sum_n \alpha_n^2 < \infty$,
- an operator with finite trace if, and only if, $\sum_n \alpha_n < \infty$.

Finally [235]

$$\Lambda_{F,2}^{\star} = \sum_n \alpha_n \left\{ e_n \otimes [f_n]_{L_2(T,\mathcal{T},\tau)} \right\};$$

$$\Lambda_{F,2}^{\star} \Lambda_{F,2} = \sum_n \alpha_n^2 \{ e_n \otimes e_n \}.$$

It should perhaps be noticed that in that example U_H may be omitted in the left-hand side of the latter equality.

Remark 2.5.8 When $p = \infty$ and F is essentially bounded, $\Lambda_{F,\infty}$ may fail to be compact as shown by the example which follows.

Example 2.5.9 Let $T = \mathbb{N}$, $\mathcal{T}$ be the subsets of T, $\tau(\{n\}) = p_n > 0$, and $H = l_2$. The p-th element of the standard basis of l_2 is $\underline{e}_p$. Let

$$\alpha_{p,n} = \begin{cases} 0 & \text{when } n > p \\ (p+1-n)^{-1} & \text{when } n \leq p \end{cases},$$

and

$$F(n) = \sum_{p=1}^{\infty} \alpha_{p,n} \underline{e}_p = \sum_{j=1}^{\infty} j^{-1} \underline{e}_{n-1+j}.$$

As

$$\|F(n)\|_H^2 = \sum_{j=1}^{\infty} j^{-2} < \infty,$$

$F : \mathbb{N} \longrightarrow l_2$ is well defined. Furthermore, for $p \in \mathbb{N}$, fixed, but arbitrary,

$$\Lambda_{F,\infty}\left[\underline{e}_p\right](n) = \langle \underline{e}_p, F(n)\rangle_H = \alpha_{p,n}.$$

The essential supremum of this latter function is one, so that

$$\left\|\Lambda_{F,\infty}\left[\underline{e}_p\right]\right\|_{L_\infty(T,\mathcal{T},\tau)} = 1.$$

But $\left\{\underline{e}_p, p \in \mathbb{N}\right\}$ converges weakly to zero.

One should perhaps notice that latter example does not require that one distinguishes a function from its equivalence class.

Result (Proposition) 2.5.2 is about $\Lambda_{F,p} U_H \Lambda_{F,p}^{\star}$. The following corollary is about $\Lambda_{F,2}^{\star} \Lambda_{F,2}$: it is closely related to (Proposition) 1.3.20 and the previous section, and it puts those results into perspective. Again there is no need to take into account duals [(Remark) 2.5.3].

Corollary 2.5.10 *Let $(T, \mathcal{T}, \tau)$ be a σ-finite measure space, and H be a real Hilbert space. Let $F : T \longrightarrow H$ be weakly square-integrable, H_F be separable, and $\mathcal{H}_F = \langle F(\cdot), F(\cdot)\rangle_H$. Then:*

1. *one has that*

$$\Lambda_{F,2}^{\star} \Lambda_{F,2}[h] = \int_T \langle h, F(t)\rangle_H F(t)\, \tau(dt) = \int_T \{F(t) \otimes F(t)\}[h]\, \tau(dt),$$

and the integral is a weak (Pettis) integral [207, p. 114]*;*

2. *$\Lambda_{F,2}$ is a Hilbert-Schmidt operator if, and only if, F is strongly (Bochner) square-integrable* [207, p. 114], *that is, if, and only if, it obtains that*

$$\int_T \mathcal{H}_F(t,t)\,\tau(dt) < \infty;$$

3. *when $\int_T \mathcal{H}_F(t,t)\,\tau(dt) < \infty$, the representation*

$$\Lambda_{F,2}^{\star}[f] = \int_T \dot{f}(t)\,F(t)\,\tau(dt)$$

is through a strong (Bochner) integral [207, p. 114], *and the representation*

$$\Lambda_{F,2}^{\star}\Lambda_{F,2}[h] = \int_T \langle h, F(t)\rangle_H F(t)\,\tau(dt) = \int_T \{F(t)\otimes F(t)\}\,[h]\tau(dt)$$

is now through a strong (Bochner) integral in the space of trace-class operators [235, p. 36]*;*

4. *F is strongly (Bochner) square-integrable if, and only if, the operator $L_{\mathcal{H}_F}$ is a trace-class operator.*

Proof One knows [(Proposition) 2.5.2, Remark 2.5.3] that $\Lambda_{F,2}^{\star}[f] = \int_T \dot{f}(t)\,F(t)\,\tau(dt)$, where the integral is a weak (Pettis) integral. Choosing $\Lambda_{F,2}[h]$ for f, one has that

$$\overbrace{\Lambda_{F,2}[h]}^{\cdot} = \langle h, F(t)\rangle_H,$$

hence item 1 obtains.

A proof of item 2 proceeds as follows. Since $H_{F,\tau}^{\perp}$ is the null space (kernel) of $\Lambda_{F,2}$ [(Proposition) 2.2.2], it suffices to restrict attention to $H_{F,\tau}$, and, since one has that $H_{F,\tau} \subseteq H_F$ [(Proposition) 2.1.9], it suffices to restrict attention to H_F. Let thus $\{e_i, i \in I\}$ be a complete orthonormal set for H_F (by assumption, I is at most countable). Then, using successively item 1, then the properties of the Pettis integral [207, p. 114], and finally Fubini's theorem [229, p. 150],

$$\begin{aligned}
\sum_{i\in I} \|\Lambda_{F,2}[e_i]\|^2_{L_2(T,\mathcal{T},\tau)} &= \sum_{i\in I} \langle \Lambda_{F,2}^{\star}\Lambda_{F,2}[e_i], e_i\rangle_H \\
&= \sum_{i\in I} \left\langle \left\{ \int_T \langle e_i, F(t)\rangle_H F(t)\,\tau(dt) \right\}, e_i \right\rangle_H \\
&= \sum_{i\in I} \int_T \langle e_i, F(t)\rangle_H^2\,\tau(dt) \\
&= \int_T \sum_{i\in I} \langle e_i, F(t)\rangle_H^2\,\tau(dt)
\end{aligned}$$

$$= \int_T \|F(t)\|_H^2 \tau(dt)$$

$$= \int_T \mathcal{H}_F(t,t) \tau(dt).$$

Item 2 is thus proved.

One now proves item 3. The assumptions of weak measurability and separability say that F is (strongly) adapted [(Proposition) 2.2.2]. Let $\mathcal{TC}[H]$ denote the Banach space of the trace-class operators of H [235, p. 36]. The maps $F_1 : T \longrightarrow H$ and $F_2 : T \longrightarrow \mathcal{TC}[H_F]$, defined using respectively the following assignments:

$$t \mapsto \dot{f}(t) F(t) \text{ and } t \mapsto F \otimes F(t) \stackrel{def}{=} F(t) \otimes F(t),$$

produce then (strongly) adapted maps (weakly adapted with values in a separable subspace [(Proposition) 2.2.2]): when H_F is generated by the orthonormal basis $\{e_i, i \in I \subseteq \mathbb{N}\}$, $F(t) = \sum_{i \in I} \alpha_i(t) e_i$, so that

$$F(t) \otimes F(t) = \sum_{i \in I} \sum_{j \in I} \alpha_i(t) \alpha_j(t) e_i \otimes e_j,$$

and thus $H_{F \otimes F}$ is generated by $\left\{e_i \otimes e_j, \{i,j\} \subseteq I \subseteq \mathbb{N}\right\}$). Furthermore

$$\left\|\dot{f}(t) F(t)\right\|_H = \left|\dot{f}(t)\right| \{\mathcal{H}_F(t,t)\}^{1/2},$$

so that the map

$$t \mapsto \left\|\dot{f}(t) F(t)\right\|_H$$

belongs to $\mathcal{L}_1(T, \mathcal{T}, \tau)$. Since operators of the form $a \otimes a$ are self-adjoint,

$$\left\{[a \otimes a]^\star [a \otimes a]\right\}^{1/2} = a \otimes a,$$

so that [235]

$$\begin{aligned}
\|F(t) \otimes F(t)\|_{\mathcal{TC}[H_F]} &= \sum_{i \in I} \langle \left\{[F(t) \otimes F(t)]^\star [F(t) \otimes F(t)]\right\}^{1/2} [e_i], e_i \rangle_H \\
&= \sum_{i \in I} \langle F(t) \otimes F(t) [e_i], e_i \rangle_H \\
&= \sum_{i \in I} \langle F(t), e_i \rangle_H^2 \\
&= \|F(t)\|_H^2 = \mathcal{H}_F(t,t).
\end{aligned}$$

The map $t \mapsto \|F(t) \otimes F(t)\|_{\mathcal{TC}[H_F]}$ belongs thus to $\mathcal{L}_1(T, \mathcal{T}, \tau)$, and item 3 is then true.

For item 4, one starts with (Proposition) 2.5.2 and the polar decomposition [266, p. 186]

$$\Lambda_{F,2} = W\left(\Lambda_{F,2}^{\star}\Lambda_{F,2}\right)^{1/2},$$

to obtain that , for some partial isometry W,

$$L_{\mathcal{H}_F} = \Lambda_{F,2}\Lambda_{F,2}^{\star} = W\Lambda_{F,2}^{\star}\Lambda_{F,2}W^{\star}.$$

Thus, for item 4 to be true, one must have that $\Lambda_{F,2}^{\star}\Lambda_{F,2}$ be trace-class [235, p. 38]. However, since F is strongly square integrable if, and only if, $\Lambda_{F,2}$ is Hilbert-Schmidt (that is item 2), and that $\Lambda_{F,2}$ is Hilbert-Schmidt if, and only if, $\Lambda_{F,2}^{\star}\Lambda_{F,2}$ is trace-class [61, p. 88], the claim is validated. □

Remark 2.5.11 Example 2.5.7 shows that the assumption that F is weakly square integrable does not entail that $\Lambda_{F,2}$ is compact, and that the latter can be compact without F being strongly square-integrable as

$$\int_T \|F(t)\|_H^2 \, \tau(dt) = \sum_n \alpha_n^2.$$

Example 2.5.12 Let $\mathcal{H}$ be a reproducing kernel, and $F : T \longrightarrow H(\mathcal{H}, T)$ be defined using $F(t) = \mathcal{H}(\cdot, t)$. The assumption "$F$ weakly square-integrable" means that $H(\mathcal{H}, T) \subseteq \mathcal{L}_2(T, \mathcal{T}, \tau)$, and that of "$H_F$ separable," that $H(\mathcal{H}, T)$ is separable. $\Lambda_{F,2}$ is the map $h \mapsto \left[h_{L_2(T,\mathcal{T},\tau)}\right]$. The condition $\int_T \mathcal{H}(t,t)\, \tau(dt) < \infty$ will be true for instance when $T = [0, 1]$, and $\mathcal{H}$ is continuous.

Proposition 2.5.13 *Let $(T, \mathcal{T}, \tau)$ and $(X, \mathcal{X}, \mu)$ be σ-finite measure spaces, the latter separable (so that the associated L_p-spaces, $1 \leq p < \infty$, are separable* [46, p. 376]. *Let $\mathcal{K} : T \times X \longrightarrow \mathbb{R}$ be a Carleman $(p, 2)$-bounded kernel with $p \in [1, \infty]$, fixed, but arbitrary. Let $F : T \longrightarrow L_2(X, \mathcal{X}, \mu)$ be defined using $F(t) = [\mathcal{K}(t, \cdot)]_{L_2(X,\mathcal{X},\mu)}$. F is then weakly p-integrable.*

Proof Since $\mathcal{K}$ is Carleman, F is weakly measurable [(Proposition) 2.2.2], and, since $L_2(X, \mathcal{X}, \mu)$ is separable, F is (strongly) adapted [(Proposition) 2.2.2]. Furthermore

$$\langle f, F(t)\rangle_{L_2(X,\mathcal{X},\mu)} = \int_X \mathcal{K}(t,x)\, \dot{f}(x)\, \mu(dx) = L_{\mathcal{K}}[f](t).$$

But the range of $L_{\mathcal{K}}$ is contained in $L_p(T, \mathcal{T}, \tau)$ since the kernel is $(p, 2)$. □

2.6 Reproducing Kernel Hilbert Spaces of Continuous Functions

One examines here the pendant, for continuous functions, of Sect. 2.5.

Proposition 2.6.1 *Let $H(\mathcal{H}, T)$ be an RKHS. Suppose that T is a topological space, and that $\mathcal{H}$ is continuous for the product topology. Then $H(\mathcal{H}, T)$ consists of continuous functions.*

Proof Since

$$\|\mathcal{H}(\cdot, t_1) - \mathcal{H}(\cdot, t_0)\|^2_{H(\mathcal{H},T)} = \{\mathcal{H}(t_1, t_1) - 2\mathcal{H}(t_1, t_0) + \mathcal{H}(t_0, t_0)\},$$

the map $F : T \longrightarrow H(\mathcal{H}, T)$ defined using $F(t) = \mathcal{H}(\cdot, t)$ is continuous. Consequently, for fixed, but arbitrary $h \in H(\mathcal{H}, T)$, the map $L_F[h] : T \longrightarrow \mathbb{R}$ defined using $L_F[h](t) = \langle h, F(t)\rangle_H = \langle h, \mathcal{H}(\cdot, t)\rangle_H = h(t)$ is continuous. □

Proposition 2.6.2 *Let $H(\mathcal{H}, T)$ be an RKHS. Suppose that T is a topological space, that $\{\mathcal{H}(\cdot, t), t \in T\}$ is a family of continuous functions, and that $\sup_{t\in T} \mathcal{H}(t, t) < \infty$. Then $H(\mathcal{H}, T)$ consists of continuous functions.*

Proof By assumption $V[\mathcal{H}]$ consists of continuous functions. The reproducing kernel being bounded on the diagonal of $T \times T$, every function in $H(\mathcal{H}, T)$ is the uniform limit of continuous functions and is thus continuous [(Proposition) 1.1.9]. □

Next proposition weakens the conditions of (Proposition) 2.6.2 for metric spaces of indices.

Proposition 2.6.3 *Let $H(\mathcal{H}, T)$ be an RKHS for which T is a metric space with distance d_T. $H(\mathcal{H}, T)$ is made of continuous functions if, and only if, for $t \in T$, fixed, but arbitrary,*

1. *$\mathcal{H}(\cdot, t)$ is continuous,*
2. *there exists $\rho(t) > 0$ such that the restriction of $t \mapsto \mathcal{H}(t, t)$ to the open ball $B(t, \rho(t))$ is bounded.*

Proof Suppose that $H(\mathcal{H}, T)$ is made of continuous functions.

Item 1 certainly obtains. Suppose it is not the case for item 2. There exists then $t_0 \in T$ such that, for $n \in \mathbb{N}$, fixed, but arbitrary, one can find t_n in $B\left(t_0, \frac{1}{n}\right)$ such that $\mathcal{H}(t_n, t_n) \geq n$. Now, for $h \in H(\mathcal{H}, T)$, fixed, but arbitrary,

$$\langle h, \mathcal{H}(\cdot, t)\rangle_{H(\mathcal{H},T)} = h(t) = \lim_n h(t_n) = \lim_n \langle h, \mathcal{H}(\cdot, t_n)\rangle_{H(\mathcal{H},T)},$$

so that, weakly,

$$\lim_n \mathcal{H}(\cdot, t_n) = \mathcal{H}(\cdot, t).$$

However

$$\|\mathcal{H}(\cdot, t_n)\|_{H(\mathcal{H},T)}^2 = \mathcal{H}(t_n, t_n) \geq n,$$

which is impossible in the presence of weak convergence.

Proof Suppose that items 1 and 2 obtain.

Let $h \in H(\mathcal{H}, T)$ and $t \in T$ be fixed, but arbitrary, and suppose that the sequence $\{t_n,\ n \in \mathbb{N}\} \subseteq T$ converges to t. For $h_p \in H(\mathcal{H}, T)$ fixed, but arbitrary, one has that

$$|h(t) - h(t_n)| \leq \left|h(t) - h_p(t)\right| + \left|h_p(t) - h_p(t_n)\right| + \left|h_p(t_n) - h(t_n)\right|.$$

Let now

$$\sup_{t_0 \in B(t,\rho(t))} \mathcal{H}(t_0, t_0) \leq \kappa < \infty,$$

and $\{h_p,\ p \in \mathbb{N}\} \subseteq V[\mathcal{H}]$ converge in $H(\mathcal{H}, T)$ to h. One has then that

$$\left|h(t) - h_p(t)\right| = \left|\langle h - h_p, \mathcal{H}(\cdot, t)\rangle_{H(\mathcal{H},T)}\right| \leq \{\mathcal{H}(t,t)\}^{1/2} \left\|h - h_p\right\|_{H(\mathcal{H},T)},$$

and, similarly, that

$$\left|h_p(t_n) - h(t_n)\right| \leq \{\mathcal{H}(t_n, t_n)\}^{1/2} \left\|h - h_p\right\|_{H(\mathcal{H},T)}.$$

Given $\epsilon > 0$, choose $p \in \mathbb{N}$ such that

$$\left\|h - h_p\right\|_{H(\mathcal{H},T)} < \frac{\epsilon}{2\kappa^{1/2}}.$$

Then

$$|h(t) - h(t_n)| \leq \frac{\epsilon}{4\kappa^{1/2}} \left\{\{\mathcal{H}(t,t)\}^{1/2} + \{\mathcal{H}(t_n, t_n)\}^{1/2}\right\} + \left|h_p(t) - h_p(t_n)\right|.$$

Since p is fixed, and h_p is continuous, letting $n \in \mathbb{N}$ increase indefinitely, one gets that

$$|h(t) - h(t_n)| < \epsilon.$$

□

Example 2.6.4 Let $T \subseteq \mathbb{R}$ be fixed, but arbitrary, and C be a covariance with the factorization $c_\wedge c_\vee$.

When $c_\wedge$ and $c_\vee$ are continuous, since the maps $x \mapsto x \wedge t$ and $x \mapsto x \vee t$ are continuous, the map $\theta \mapsto C(\theta, t)$ is continuous. Furthermore, as the map which sends t to $C(t,t) = c_\wedge(t)\, c_\vee(t)$ is continuous, it is locally bounded. Consequently $H(C,T)$ is made of continuous functions.

Suppose that $H(C,T)$ is made of continuous functions. Since it is assumed that C is not the zero function, T^C is not void. For $\{t, \theta_1, \theta_2\} \subseteq T$, fixed, but arbitrary,

$$C(\theta_1,t) - C(\theta_2,t) = \begin{cases} c_\vee(t)\,[c_\wedge(\theta_1) - c_\wedge(\theta_2)] & \text{when } \theta_1 \vee \theta_2 < t \\ c_\vee(t)\, c_\wedge(\theta_1) - c_\wedge(t)\, c_\vee(\theta_2) & \text{when } \theta_1 \leq t, \theta_2 \geq t \\ c_\wedge(t)\,[c_\vee(\theta_1) - c_\vee(\theta_2)] & \text{when } \theta_1 \wedge \theta_2 \geq t \end{cases}. \qquad (\star)$$

Let $t_l^C = \inf T^C$ and $t_r^C = \sup T^C$. Suppose $t \in T \cap]t_l^C, t_r^C[$. Then, $c_\wedge(t)$ and $c_\vee(t)$ are different from zero, and, given $(\star)$, $c_\wedge$ is continuous on $T\cap]-\infty, t[$, and $c_\vee$, on $T\cap]t, \infty[$. Since t is arbitrary, $c_\wedge$ is continuous on $T\cap]-\infty, t_r^C[$, and $c_\vee$, on $T\cap]t_l^C, \infty[$. Suppose that $t \in T, t > t_r^C$. Since there are values of $\theta \in T^C, \theta < t_r^C$, such that $c_\wedge(\theta) \neq 0$, then $c_\vee(t) = 0$ [(Proposition) 1.4.3]. Consequently, since $c_\vee$ is continuous at t_r^C, $c_\vee(t_r^C) = 0$, and t_r^C does not belong to T^C. The same is true for t_l^C. Thus

- $c_\wedge$ is continuous on $T \cap \left]-\infty, t_r^C\right[$;
- $c_\vee$ is continuous on $T \cap \left]t_l^C, \infty\right[$;
- t_l^C and t_r^C do not belong to T_C.

That means that $c_\wedge$ and $c_\vee$ are continuous on T_C.

Corollary 2.6.5 *Let $H(\mathcal{H}, T)$ be an RKHS for which T is a compact metric space. Suppose that $\mathcal{H}$ is continuous for the product topology. Then, for fixed, but arbitrary $\{t_1, t_2\} \subseteq T$,*

$$\mathcal{H}(t_1, t_2) = \sum_{i \in I} h_i(t_1)\, h_i(t_2),$$

where $\{h_i,\ i \in I\}$ is a countable, orthonormal system, of uniformly continuous functions, bounded by the square root of $\sup_{t\in T} \mathcal{H}(t,t)$, and the convergence of the series is uniform.

Proof $H(\mathcal{H}, T)$ is made of continuous functions [(Proposition) 2.6.1]. T is separable as it is metric and compact [85, p. 22]. $H(\mathcal{H}, T)$ is thus a separable space of continuous functions [(Corollary) 1.5.9]. The representation of the kernel in the form of the given series then follows [(Proposition) 1.5.6], and [(Proposition) 1.1.5]

$$|h_i(t)| \leq \|h_i\|_{H(\mathcal{H},T)} \{\mathcal{H}(t,t)\}^{1/2}.$$

One may thus suppose that I is an "interval" of $\mathbb{N}$ starting at 1. Since $\mathcal{H}$ is continuous on a compact set [84, p. 224], it is bounded [84, p. 224]. The only fact to prove is thus the uniform convergence of the series. To that end, let

$$s_n(t) = \sum_{i=1}^{n} h_i^2(t).$$

On thus obtains an increasing sequence of continuous functions whose limit is the continuous function $\mathcal{H}(t,t)$. By Dini's theorem [111, p. 336], the convergence is uniform as T is compact. Since

$$\left| \sum_{i\in I, i\geq n} h_i(t_1) h_i(t_2) \right|^2 \leq \sum_{i\in I, i\geq n} |h_i(t_1)|^2 \sum_{i\in I, i\geq n} |h_i(t_2)|^2,$$

it follows that the series $\sum_{i=1}^{n} h_i(t_1) h_i(t_2)$ converges uniformly to $\mathcal{H}(t_1,t_2)$. The proof is complete. □

Proposition 2.6.6 *Let $H(\mathcal{H},T)$ be an RKHS for which T is an interval of $\mathbb{R}$. Suppose that $\{\mathcal{H}(\cdot,t),\ t\in T\}$ is a family of functions which are continuous to the right, and that $\sup_{t\in T}\mathcal{H}(t,t)<\infty$. Then $H(\mathcal{H},T)$ consists of functions which are continuous to the right.*

Proof By assumption $V[\mathcal{H}]$ consists of functions which are continuous to the right. Let $h\in H(\mathcal{H},T)$ be fixed, but arbitrary. There exists [(Proposition) 1.1.5] a sequence $\{h_n,\ n\in\mathbb{N}\}\subseteq V[\mathcal{H}]$ such that $h=\lim_n h_n$ in $H(\mathcal{H},T)$. But then

$$\lim_{u\downarrow t}|h(t)-h(u)| \leq \lim_{u\downarrow t}\{|h(t)-h_n(t)|+|h_n(t)-h_n(u)|+|h_n(u)-h(u)|\}.$$

Now, using (Proposition) 1.1.5, as in (Proposition) 2.6.3, and the continuity to the right of h_n,

$$|h(t)-h_n(t)|^2 \leq \sup_{t\in T}\mathcal{H}(t,t)\,\|h-h_n\|^2_{H(\mathcal{H},T)},$$

and, for $n\in I\!N$, fixed, but arbitrary,

$$\lim_{u\downarrow t}|h_n(t)-h_n(u)|=0,$$

and that suffices to establish the result. □

Corollary 2.6.7 *Let $H(\mathcal{H},T)$ be an RKHS for which T is an interval of $\mathbb{R}$. Suppose that $\{\mathcal{H}(\cdot,t),\ t\in T\}$ is a family of functions which are continuous to the right, and that $\sup_{t\in T}\mathcal{H}(t,t)<\infty$. Then $H(\mathcal{H},T)$ is separable.*

Proof $H(\mathcal{H},T)$ is made of functions which are continuous to the right [(Proposition) 2.6.6]. Let T_d be a countable dense subset of T which contains the right end-point of T when that right end-point belongs to T. Suppose that $h\in H(\mathcal{H},T)$

is fixed, but arbitrary, and that $\langle h, \mathcal{H}(\cdot, t_d)\rangle_{H(\mathcal{H},T)} = 0,\ t_d \in T_d$. One then has that $h(t_d) = 0,\ t_d \in T_d$. Let $t \in T \setminus T_d$. There exits then

$$\left\{u_d^{(n)} > t,\ n \in \mathbb{N}\right\} \subseteq T_d$$

such that $\lim_{n\to\infty} u_d^{(n)} = t$. But h being continuous to the right,

$$0 = \lim_{n\to\infty} h\left(u_d^{(n)}\right) = h(t).$$

Consequently $h(t) = 0,\ t \in T$, and $\{\mathcal{H}(\cdot, t_d),\ t_d \in T_d\} \subseteq H(\mathcal{H}, T)$ is a countable dense set. □

Remark 2.6.8 Results (Proposition) 2.6.6 and (Corollary) 2.6.7 are true, *mutatis mutandis,* when the functions considered are continuous to the left.

Proposition 2.6.9 *Let $H(\mathcal{H}, T)$ be an RKHS. Its functions are continuous on $(T, d_{\mathcal{H}})$.*

Proof For fixed, but arbitrary $h \in H(\mathcal{H}, T)$ and $(t_1, t_2) \in T \times T$, one has that [(Proposition) 1.1.5]

$$\begin{aligned}|h(t_1) - h(t_2)| &= \langle h, \mathcal{H}(\cdot, t_1) - \mathcal{H}(\cdot, t_2)\rangle_{H(\mathcal{H},T)} \\ &\leq \|h\|_{H(\mathcal{H},T)} \|\mathcal{H}(\cdot, t_1) - \mathcal{H}(\cdot, t_2)\|_{H(\mathcal{H},T)} \\ &= \|h\|_{H(\mathcal{H},T)}\, d_{\mathcal{H}}(t_1, t_2).\end{aligned}$$

□

The result which follows yields another version of (Proposition) 2.6.3.

Proposition 2.6.10 *Let T be Hausdorff and locally compact, and let $C(T)$ be the vector space of continuous functions $f : T \longrightarrow \mathbb{R}$. $C(T)$ is given the topology of uniform convergence on compact sets* [154, p. 229]. *Let H be a real Hilbert space, and $F : T \longrightarrow H$ be a map. The following two statements are then equivalent:*

1. *F is weakly continuous;*
2. *for $t \in T$, fixed, but arbitrary, $\mathcal{H}_F(\cdot, t)$ is continuous, and $\mathcal{H}_F$ is locally bounded in the sense that, for each compact $K \subseteq T$, there is a finite and positive constant κ_K such that, for $(t_1, t_2) \in K \times K$, fixed, but arbitrary*

$$|\mathcal{H}_F(t_1, t_2)| \leq \kappa_K,\ (t_1, t_2) \in K \times K.$$

When those conditions obtain,

(i) $\mathcal{R}[L_F] \subseteq C(T)$;
(ii) when $\Lambda_{F,c} : H \longrightarrow C(T)$ is defined, for $h \in H$, fixed, but arbitrary, using the following relation: $\Lambda_{F,c}[h] = L_F[h]$, $\Lambda_{F,c}$ is then continuous.

Finally, when T is separable, so is H_F.

Proof It may be noticed that item (i) means that, for $h \in H$, fixed, but arbitrary, $L_F[h]$ is a continuous function. As $L_F[h](t) = \langle h, F(t)\rangle_H$, F is thus weakly continuous. And, when F is weakly continuous, the range of L_F is made of continuous functions. Thus items 1 and (i) state the same fact.

Proof Suppose F weakly continuous.

Let $t \in T$ be fixed, but arbitrary. For $x \in T$, fixed, but arbitrary, the function $x \mapsto \mathcal{H}_F(x,t) = \langle F(t), F(x)\rangle_H$ is continuous as F is assumed to be weakly continuous. Now let $K \subseteq T$ be a compact set, fixed, but arbitrary, and consider the following family of continuous linear functionals:

$$\mathcal{L}_K = \{\Lambda_t(h) = L_F[h](t) = \langle h, F(t)\rangle_H,\ t \in K\} \subseteq H^\star.$$

For $h \in H$, fixed, but arbitrary, $t \mapsto L_F[h](t) = \langle h, F(t)\rangle_H$ is a continuous function which is thus bounded when restricted to K. There is thus a finite, positive constant $\kappa(h)$ such that

$$|\Lambda_t[h]| \leq \kappa(h),\ t \in K.$$

The Banach-Steinhaus theorem [266, p. 76] allows one to then assert that there is a finite, positive constant κ_K such that

$$\|F(t)\|_H = \|\langle \cdot, F(t)\rangle_H\| = \|\Lambda_t\| \leq \kappa_K,\ t \in K.$$

Consequently, for fixed, but arbitrary $(t_1, t_2) \in K \times K$,

$$|\mathcal{H}_F(t_1,t_2)| = |\langle F(t_1), F(t_2)\rangle_H| \leq \|F(t_1)\|_H \|F(t_2)\|_H \leq \kappa_K^2,$$

and $\mathcal{H}_F$ is locally bounded.

Proof Suppose now that $x \mapsto \mathcal{H}_F(x,t)$ is continuous for all fixed t's, and that $\mathcal{H}_F$ is locally bounded.

Let $h = \sum_{i=1}^n \alpha_i F(t_i)$. As

$$L_F[h] = \sum_{i=1}^n \alpha_i L_F[F(t_i)] = \sum_{i=1}^n \alpha_i \mathcal{H}_F(\cdot, t_i),$$

$L_F[h]$ is a continuous function.

Since, for $h \in H_F^\perp$, $L_F[h]$ is the zero function, $t \mapsto \langle h, F(t)\rangle_H$ is continuous. Let now $h_0 \in H_F$ and $t_0 \in T$ be fixed, but arbitrary. One shall show that $L_F[h_0]$ is continuous at t_0 : since $L_F[h_0](t_0) = \langle h_0, F(t_0)\rangle_H$, that will establish the weak continuity of F at t_0.

Let thus, to that end, K be a compact neighborhood of t_0 : it exists as T is locally compact by assumption. Since $\mathcal{H}_F$ is locally bounded, there is a finite, positive κ_K

such that

$$\sup_{u \in K} \|F(u)\|_H = \sup_{u \in K} \{\mathcal{H}_F(u,u)\}^{1/2} \leq \kappa_K.$$

Let $\epsilon > 0$ be fixed, but arbitrary. There exists [(Proposition) 1.1.5] $h = \sum_{i=1}^{n} \alpha_i F(t_i)$ such that

$$\|h_0 - h\|_H \leq \epsilon.$$

Since $L_F[h]$ is continuous, and thus continuous at t_0, and since T is locally compact, there is a compact neighborhood of t_0, say K_0, such that, for $t \in K_0$, fixed, but arbitrary,

$$|L_F[h](t) - L_F[h](t_0)| \leq \epsilon.$$

Then

$$\begin{aligned} |L_F[h_0](t) - L_F[h_0](t_0)| &\leq |L_F[h_0](t) - L_F[h](t)| \\ &\quad + |L_F[h](t) - L_F[h](t_0)| \\ &\quad + |L_F[h](t_0) - L_F[h_0](t_0)|. \end{aligned}$$

But

$$|L_F[h_0](t) - L_F[h](t)| = |\langle h_0 - h, F(t)\rangle_H| \leq \|F(t)\|_H \|h_0 - h\|_H.$$

A similar inequality applies to the third term of the last but one expression. Consequently, for $t \in K \cap K_0$, fixed, but arbitrary,

$$\begin{aligned} |L_F[h_0](t) - L_F[h_0](t_0)| &\leq |L_F[h](t) - L_F[h](t_0)| \\ &\quad + (\|F(t)\|_H + \|F(t_0)\|_H) \|h_0 - h\|_H \\ &\leq \epsilon (1 + 2\kappa_K). \end{aligned}$$

Proof (ii) The topology of uniform convergence on compact sets is given by the family of seminorms of the form $p_K : f \mapsto \sup_{t \in K} |f(t)|$, for compact K [111, p. 344], and a linear map $L : H \longrightarrow C(T)$ is continuous if, and only if, $p_K \circ L$ is continuous [269, p. 217]. Thus, to see whether $\Lambda_{F,c}$ is a bounded operator in case F is weakly continuous, one fixes an arbitrary compact subset K of T. Since $\mathcal{H}_F$ is locally bounded, there is a finite, positive κ_K such that

$$\|F(t)\|_H^2 = \mathcal{H}_F(t,t) \leq \kappa_K^2, \ t \in K.$$

Consequently

$$\sup_{t\in K} |\Lambda_{F,c}[h](t)| = \sup_{t\in K} |\langle h, F(t)\rangle_H| \le \|h\|_H \sup_{t\in K} \|F(t)\|_H \le \kappa_K \|h\|_H .$$

Proof *H_F is separable when T is.*

One chooses arbitrarily a countable, dense subset T_c of T. Let H_c be the subspace of H_F spanned by the set $\{F(t),\ t \in T_c\}$. H_c is separable. Let $h \in H_F$ be orthogonal to H_c. Then, for fixed, but arbitrary $t \in T_c$,

$$L_F[h](t) = \langle h, F(t)\rangle_H = 0.$$

Since $L_F[h]$ is continuous, and T_c is dense in T,

$$L_F[h](t) = \langle h, F(t)\rangle_H = 0,\ t \in T,$$

so that $h \in H_F^{\perp}$. But then $h = 0$. Thus $H_c = H_F$ which is thus separable. □

The following proposition explains the consequences of kernel continuity.

Proposition 2.6.11 *Let T be a Hausdorff, locally compact space, and $C(T)$ be the set of continuous functions $f : T \longrightarrow \mathbb{R}$. $C(T)$ is given the topology of uniform convergence on compact sets* [154, p. 229]. *Let H be a real Hilbert space, and $F : T \longrightarrow H$ be a map. The following statements are then equivalent:*

1. *$\mathcal{H}_F$ is continuous;*
2. *the diagonal of $T \times T$ belongs to the set of points at which the map $(t_1, t_2) \mapsto \mathcal{H}_F(t_1, t_2)$ is continuous;*
3. *F is continuous;*
4. *F is weakly continuous, and $t \mapsto \mathcal{H}_F(t, t)$ is continuous;*
5. *$\Lambda_{F,c} = L_F$ is compact.*

Proof *(1 ⇒ 2)* Since $\mathcal{H}_F$ is continuous on $T \times T$, it is continuous on the diagonal.

Proof *(2 ⇒ 3)* One has, for fixed, but arbitrary $(t_1, t_2) \in T \times T$,

$$\|F(t_1) - F(t_2)\|_H^2 = \mathcal{H}_F(t_1, t_1) - 2\mathcal{H}(t_1, t_2) + \mathcal{H}_F(t_2, t_2),$$

but, when $|t_1 - t_2|$ is small, the right-hand side of the latter expression is small too, because of item 2.

Proof *(3 ⇒ 1)* For fixed, but arbitrary $(t_1, t_2) \in T \times T$ and $(h_1, h_2) \in H \oplus H$, one may define the following continuous maps:

$$\Phi : T \times T \longrightarrow H \oplus H \text{ using } \Phi(t_1, t_2) = (F(t_1), F(t_2)),$$

and

$$\Psi : H \oplus H \longrightarrow \mathbb{R} \text{ using } \Psi(h_1, h_2) = \langle h_1, h_2 \rangle_H.$$

Since one has that

$$\mathcal{H}_F(t_1, t_2) = \Psi \circ \Phi(t_1, t_2),$$

$\mathcal{H}_F$ is continuous.

Proof (3 ⇔ 4) When F is continuous, it is weakly continuous, and the map $t \mapsto \|F(t)\|_H$ is also continuous. Thus item 4 obtains. Suppose now that item 4 is true, and that one has a convergent net in T : $\lim_\nu t_\nu = t$. Since, for fixed, but arbitrary $h \in H$, $t \longrightarrow \langle h, F(t)\rangle_H$ is continuous as well as $t \longrightarrow \|F(t)\|_H$, as limits of nets, $\lim_\nu \langle h, F(t_\nu)\rangle_H = \langle h, F(t)\rangle_H$ and $\lim_\nu \|F(t_\nu)\|_H = \|F(t)\|_H$ [270, p. 75]. Since convergent nets are Cauchy nets [270, p. 260], one can build, as in [270, p. 261], a subsequence of the net, $\{t_{\nu_n}, n \in \mathbb{N}\}$, converging to t. But then [69, p. 98], since it obtains that $\lim_n \langle h, F(t_{\nu_n})\rangle_H = \langle h, F(t)\rangle_H$, and also that $\lim_n \|F(t_{\nu_n})\|_H = \|F(t)\|_H$, one has that $\lim_n F(t_{\nu_n}) = F(t)$. Consequently [270, p. 261], $\lim F(t_\nu) = F(t)$.

Proof (3 ⇔ 5) Let $B_H(0,1)$ denote the closed unit ball in H : to have that $\Lambda_{F,c}$ is compact, it suffices to prove that $\mathcal{F} = \Lambda_{F,c}[B_H(0,1)]$ is a compact subset of $C(T)$ [222, p. 152]. But compactness in $C(T)$ of the set of functions $\mathcal{F}$ is regulated by Ascoli's theorem [154, p. 233] which states the following necessary and sufficient conditions:

- $\mathcal{F}$ is closed in $C(T)$;
- the closure of $\mathcal{F}[t] = \{f(t), f \in \mathcal{F}\}$ is compact for each $t \in T$;
- $\mathcal{F}$ is equicontinuous.

Let $\{f_\nu = \Lambda_{F,c}[h_\nu], h_\nu \in B_H(0,1)\}$ be a net converging to f in $C(T)$. Since $B_H(0,1)$ is weakly compact [129, p. 185], there is a sub-net $\{h_{\nu_\eta}\}$ converging weakly to some $h \in B_H(0,1)$ [154, p. 136]. Since F is continuous, it is weakly continuous (item 4), and thus $\Lambda_{F,c}$ is continuous [(Proposition) 2.6.10]. But then it is continuous for the weak topology [222, p. 48], so that the net $\{\Lambda_{F,c}[h_{\nu_\eta}]\}$ converges to $\Lambda_{F,c}[h]$. Since T is Hausdorff, $C(T)$ is Hausdorff also [84, p. 258]. But a net in a Hausdorff space can only have one limit [154, p. 67], that is $f = \Lambda_{F,c}[h]$. In other words, the image of the closed unit ball is closed.

Since $\Lambda_{F,c}$ is continuous, the image of a bounded set by a continuous function being bounded [269, p. 180], $\mathcal{F}$ is bounded. The evaluation map at t is continuous, since the topology is that of uniform convergence on compact sets, and thus $\mathcal{F}[t]$ is bounded. Since $\mathcal{F}$ is closed, $\mathcal{F}[t]$ is closed. $\mathcal{F}[t]$ is thus compact.

To see that $\mathcal{F}$ is equicontinuous, one must notice that

$$\begin{aligned}
\sup_{f\in\mathcal{F}} |f(t_1) - f(t_2)| &= \sup_{h\in B_H(0,1)} |\Lambda_{F,c}[h](t_1) - \Lambda_{F,c}[h](t_2)| \\
&= \sup_{h\in B_H(0,1)} |L_F[h](t_1) - L_F[h](t_2)| \\
&= \sup_{h\in B_H(0,1)} |\langle h, F(t_1)\rangle_H - \langle h, F(t_2)\rangle_H| \\
&= \sup_{h\in B_H(0,1)} |\langle h, F(t_1) - F(t_2)\rangle_H| \\
&= \|F(t_1) - F(t_2)\|_H .
\end{aligned}$$

Thus $\mathcal{F}$ is equicontinuous if, and only if, F is continuous.

Now, when $\Lambda_{F,c}$ is compact, $\mathcal{F}$ is equicontinuous, which as just seen, means that F is continuous. □

Example 2.6.12 (A Family of RKHS's of Continuous Functions) The derivative of a function f shall be denoted f'. Let $0 \le t_l < t_r < \infty$, and T, the interval $[t_l, t_r]$. For $\{t_1, t_2\} \subseteq T$, fixed but arbitrary, let $C_W(t_1, t_2) = t_1 \wedge t_2$. It is a covariance whose factorization corresponds to $c_\wedge(t) = t$ and $c_\vee(t) = 1$. Let $H(C_W, T)$ be the associated RKHS.

Let C be a covariance with a factorization $C = c_\wedge c_\vee$ such that

- $T_C = T$,
- $c_\wedge$ and $c_\vee$ are twice continuously differentiable,
- r_C has a strictly negative derivative.

In particular r_C is continuous, and thus its range is a finite, closed interval which shall be denoted $I = [a, b]$. One has then, since r_C is decreasing, that $r_C(t_l) = b$ and $r_C(t_r) = a$. One then has also a unitary map $U : H(C, T) \longrightarrow H(C_W, I)$, as obtained in (Remark) 1.4.21:

$$U[h] = \left[\frac{h}{c_\wedge}\right] \circ r_C^{-1}.$$

Furthermore $U^\star[h] = c_\wedge [h \circ r_C]$.

The norm of elements in $H(C, T)$ may be obtained as follows. As, with $\Phi = h/c_\wedge$,

$$\|h\|^2_{H(C,T)} = \|U[h]\|^2_{H(C_W,I)} = \left\|\left[\frac{h}{c_\wedge}\right] \circ r_C^{-1}\right\|^2_{H(C_W,I)} = \left\|\Phi \circ r_C^{-1}\right\|^2_{H(C_W,I)}, \qquad (\star)$$

and that, for $h_W \in H(C_W, I)$, fixed, but arbitrary, both

$$h_W(t) = h_W(a) + \int_I \chi_{[a,t]}(x)\,\eta(x)\,dx, \quad \int_I \eta^2(x)\,dx < \infty,$$

and

$$\|h_W\|_{H(C_W,I)} = \frac{h_W^2(a)}{a} + \int_I \eta^2(x)\,dx$$

obtain, it follows, from the inverse function formula:

$$(f^{-1})'(x) = (f^{-1})'(f[f^{-1}(x)]) = \frac{1}{f'(f^{-1}(x))},$$

that

$$\eta = \left\{\Phi \circ r_C^{-1}\right\}' = \left\{\Phi' \circ r_C^{-1}\right\}\left(r_C^{-1}\right)' = \frac{\Phi' \circ r_C^{-1}}{r_C' \circ r_C^{-1}}.$$

Consequently, using the following formula [262, p. 249]:

$$\int_I f(x)\,dx = \int_T f(r_C(t))\left|r_C'(t)\right| dt,$$

with $f = \left\{\frac{\Phi' \circ r_C^{-1}}{r_C' \circ r_C^{-1}}\right\}^2$, one gets that

$$\int_I \eta^2(x)\,dx = -\int_T \frac{\{\Phi'(t)\}^2}{r_C'(t)}\,dt.$$

Suppose that h of $(\star)$ can be differentiated (more on that below): then

$$\Phi' = \frac{h'c_\wedge - hc_\wedge'}{c_\wedge^2}, \quad \text{and} \quad r_C' = \frac{c_\vee'c_\wedge - c_\vee c_\wedge'}{c_\wedge^2},$$

so that

$$\infty > \int_I \eta^2(x)\,dx = \int_T \frac{\left\{h'(t) - \frac{c_\wedge'(t)}{c_\wedge(t)}h(t)\right\}^2}{\{c_\wedge'(t)\,c_\vee(t) - c_\wedge(t)\,c_\vee'(t)\}}\,dt.$$

Let

$$D = \frac{1}{\{c_\wedge'(t)\,c_\vee(t) - c_\wedge(t)\,c_\vee'(t)\}}.$$

The denominator of D is a continuous function, and thus, because of the last inequality, D has a strictly positive lower bound. Expanding the right-hand side in that same inequality, one gets that

$$\int_I \eta^2(x)\,dx = \int_T \{h'(t)\}^2 D(t)\,dt$$
$$+ \int_T \left\{\frac{c'_\wedge(t)}{c_\wedge(t)}\right\}^2 h^2(t)\,D(t)\,dt$$
$$- 2\int_T \left\{\frac{c'_\wedge(t)}{c_\wedge(t)}\right\} h(t)\,h'(t)\,D(t)\,dt\,.$$

Now

$$2\int_T \left\{\frac{c'_\wedge(t)}{c_\wedge(t)}\right\} h(t)\,h'(t)\,D(t)\,dt =$$
$$= \int_T D(t)\left\{\frac{c'_\wedge(t)}{c_\wedge(t)}\right\} d\{h^2(t)\}$$
$$= D(t)\left\{\frac{c'_\wedge(t)}{c_\wedge(t)}\right\} h^2(t)\bigg|_{t_l}^{t_r}$$
$$- \int_T h^2(t)\,d\left\{D(t)\left\{\frac{c'_\wedge(t)}{c_\wedge(t)}\right\}\right\}$$
$$= D(t)\left\{\frac{c'_\wedge(t)}{c_\wedge(t)}\right\} h^2(t)\bigg|_{t_l}^{t_r}$$
$$- \int_T h^2(t)\left\{D(t)\left\{\frac{c'_\wedge(t)}{c_\wedge(t)}\right\}\right\}'\,dt,$$

so that

$$\int_I \eta^2(x)\,dx = \int_T \{h'(t)\}^2 D(t)\,dt$$
$$+ \int_T h^2(t)\,G(t)\,dt - \left\{D(t)\left\{\frac{c'_\wedge(t)}{c_\wedge(t)}\right\} h^2(t)\bigg|_{t_l}^{t_r}\right\},$$

where

$$G = D\left\{\frac{c'_\wedge}{c_\wedge}\right\}^2 + \left\{D\left\{\frac{c'_\wedge}{c_\wedge}\right\}\right\}'$$
$$= D\left\{\frac{c'_\wedge}{c_\wedge}\right\}^2 + D'\frac{c'_\wedge}{c_\wedge} + D\left\{\frac{c''_\wedge c_\wedge - \{c'_\wedge\}^2}{c^2_\wedge}\right\}$$
$$= D'\frac{c'_\wedge}{c_\wedge} + D\frac{c''_\wedge}{c_\wedge}.$$

As $D' = \{c_\wedge c''_\vee - c''_\wedge c_\vee\} D^2$,

$$G = \{c_\wedge c''_\vee - c''_\wedge c_\vee\} D^2 \frac{c'_\wedge}{c_\wedge} + D \frac{c''_\wedge}{c_\wedge}$$
$$= \frac{D^2}{c_\wedge} \{\{c_\wedge c''_\vee - c''_\wedge c_\vee\} c'_\wedge + \{c'_\wedge c_\vee - c_\wedge c'_\vee\} c''_\wedge\}$$
$$= \frac{D^2}{c_\wedge} \{c_\wedge c'_\wedge c''_\vee - c_\wedge c''_\wedge c'_\vee\}$$
$$= D^2 \{c'_\wedge c''_\vee - c''_\wedge c'_\vee\}.$$

In particular, G is continuous. Furthermore, given that

$$\int_I \eta^2 (x)\, dx =$$
$$= \int_T D(t) \{h'(t)\}^2 dt + \int_T G(t) h^2(t)\, dt - \left\{ D(t) \left\{ \frac{c'_\wedge (t)}{c_\wedge (t)} \right\} h^2(t) \Big|_{t_l}^{t_r} \right\},$$

h' has a square that is integrable.

Now $a = r_C(t_r) = \frac{c_\vee(t_r)}{c_\wedge(t_r)}$, and

$$h_W(a) = \frac{h}{c_\wedge} \circ r_C^{-1}(a) = \frac{h}{c_\wedge} \circ r_C^{-1}(r_C(t_r)) = \frac{h(t_r)}{c_\wedge(t_r)},$$

so that

$$\frac{h_W^2(a)}{a} = \frac{h^2(t_r)}{c_\wedge^2(t_r)} \frac{c_\wedge(t_r)}{c_\vee(t_r)} = \frac{h^2(t_r)}{c_\wedge(t_r)\, c_\vee(t_r)}.$$

Finally

$$\frac{h_W^2(a)}{a} - D(t) \left\{ \frac{c'_\wedge(t)}{c_\wedge(t)} \right\} h^2(t) \Big|_{t_l}^{t_r} =$$
$$= D(t_l) \frac{c'_\wedge(t_l)}{c_\wedge(t_l)} h^2(t_l) + \frac{h^2(t_r)}{c_\wedge(t_r)\, c_\vee(t_r)} - D(t_r) \frac{c'_\wedge(t_r)}{c_\wedge(t_r)} h^2(t_r)$$
$$= D(t_l) \frac{c'_\wedge(t_l)}{c_\wedge(t_l)} h^2(t_l) + \left\{ \frac{1}{c_\wedge(t_r)\, c_\vee(t_r)} - D(t_r) \frac{c'_\wedge(t_r)}{c_\wedge(t_r)} \right\} h^2(t_r).$$

But

$$\frac{1}{c_\wedge(t_r)\,c_\vee(t_r)} - D(t_r)\,\frac{c'_\wedge(t_r)}{c_\wedge(t_r)} =$$

$$= \frac{1 - D(t_r)\,c'_\wedge(t_r)\,c_\vee(t_r)}{c_\wedge(t_r)\,c_\vee(t_r)}$$

$$= \frac{D(t_r)\left\{c'_\wedge(t_r)\,c_\vee(t_r) - c_\wedge(t_r)\,c'_\vee(t_r)\right\} - D(t_r)\,c'_\wedge(t_r)\,c_\vee(t_r)}{c_\wedge(t_r)\,c_\vee(t_r)}$$

$$= -D(t_r)\,\frac{c'_\vee(t_r)}{c_\vee(t_r)}.$$

The norm of h is thus given by the following expression:

$$\|h\|^2_{H(C,T)} = \int_T \left\{D\{h'\}^2 + Gh^2\right\}(t)\,dt + \left\{D\frac{c'_\wedge}{c_\wedge}h^2\right\}(t_l) - \left\{D\frac{c'_\vee}{c_\vee}h^2\right\}(t_r),$$

where h' is the generalized derivative of h, and the usual derivative when it is differentiable. The procedure that has just been described hinges on the fact that all continuous $h \in H(C,T)$ may be differentiated in the appropriate sense. The remarks which follow explain why one may consider that such is the case, and that, furthermore, the derivative has an integrable square. Another way to justify the computation is to do it for usually differentiable functions, and then use the fact that these are dense in the appropriate space.

Beppo-Levi space [a particular case]

Let $I = [a,b] \subseteq \mathbb{R}$ be fixed, but arbitrary. A continuous function $f : I \longrightarrow \mathbb{R}$ admits a generalized derivative g when it can be expressed as

$$f(t) = f(a) + \int_a^t g(\theta)\,d\theta, \text{ with } \int_a^b g^2(\theta)\,d\theta < \infty.$$

The class of g is the generalized derivative of f. $\mathcal{H}^1[a,b]$ denotes the vector space of functions with a generalized derivative. On $\mathcal{H}^1[a,b]$ let

$$[f_1, f_2] = f_1(a) f_2(a) + \int_a^b g_1(\theta)\,g_2(\theta)\,d\theta.$$

One thus defines a Hilbert space, in fact, the RKHS obtained using the following function: $F : I \longrightarrow \mathbb{R} \oplus L_2[a,b]$, defined through the following relation:

$$F(t) = \left(\alpha, I_{[a,t]}\right).$$

Sobolev space [a particular case]

Let $I = [a,b] \subseteq \mathbb{R}$ be fixed, but arbitrary. Suppose that f and g are locally integrable. g is the weak derivative of f when, for all functions ϕ that are indefinitely and continuously differentiable,

$$\int_I f(\theta)\,\phi'(\theta)\,d\theta = -\int_I g(\theta)\,\phi(\theta)\,d\theta.$$

Write $D[f]$ for g. Then

$$H^1[a,b] = \left\{ f \in L_2[a,b] : \int_a^b D[f]^2(\theta)\,d\theta < \infty \right\}.$$

One has that [7, p. 28] to every element $f \in H^1[a,b]$ there corresponds a continuous $\tilde{f}$ such that, almost surely, $\tilde{f} = \dot{f}$ and

$$\tilde{f}(t) = \tilde{f}(c) + \int_a^t D[f](\theta)\,d\theta.$$

When f is continuous, $f = \tilde{f}$, and thus every element of $H(C,T)$ has a generalized derivative, and the above calculation makes sense.

2.7 Spectral Theory: A Vademecum

In the sequel, repeated use of the spectral decomposition of an operator shall occur. All results upon which one shall rely may be found in [266]. Here is a list of the relevant facts.

One needs first an appropriate theory of integration. Mentioned will only be the facts that are needed to understand correctly the meaning of that theory as used in [266, pp. 175, 185–186, 335–350].

Let $\mathcal{I}$ be the family of bounded intervals (open, half-open, closed, made of one point, empty) of $\mathbb{R}$ (the theory works identically for Euclidean space), and $\tilde{\mathcal{I}}$ be the family of finite unions of such intervals. A function defined for intervals, called an intervals function, is a map $\mu : \mathcal{I} \longrightarrow \mathbb{R}$ which is monotone increasing with respect to inclusion, and additive over disjoint intervals. μ has a natural extension to $\tilde{\mathcal{I}}$, it is positive, and zero at the empty set. Intervals functions are regular when their values on arbitrary intervals may be approximated arbitrarily closely from the inside by closed intervals, and from the outside, by open ones. A regular intervals function is called a measure.

Let N be a set of reals. It has measure zero for μ when, for arbitrary $\epsilon > 0$, there exists $\{I_n(N,\epsilon),\ n \in \mathbb{N}\} \subseteq \mathcal{I}$ such that

$$N \subseteq \cup_n I_n(N,\epsilon), \ \text{ and } \sum_n \mu\left(I_n(N,\epsilon)\right) < \epsilon.$$

When μ accompanies a relation, such as $f =_{\mu} g$, it means that the relation holds for all arguments, of f and g in the example, outside a set of μ-measure zero.

A step function is a map of the following form:

$$f(x) = \sum_{i=1}^{n} \alpha_i \chi_{I_i}(x), \; n \in \mathbb{N}, \{\alpha_1, \ldots, \alpha_n\} \subseteq \mathbb{R}, \; \{I_1, \ldots, I_n\} \subseteq \mathcal{I}.$$

Let $\mathcal{S}$ denote the family of all possible step functions. The integral of $f \in \mathcal{S}$ is the number

$$\int f d\mu = \sum_{i=1}^{n} \alpha_i \mu(I_i).$$

Such assignments make sense and have the properties an integral should have. One then extends that notion of integral to functions which are obtained as follows, and whose family is denoted $\mathcal{S}_1$. Let $f : \mathbb{R} \longrightarrow \mathbb{R}$ be a function with the following property: there exists $\{f_n, \; n \in \mathbb{N}\} \subseteq \mathcal{S}$ such that

- $\{f_n, \; n \in \mathbb{N}\}$ is increasing and $\lim_n f_n =_{\mu} f$;
- the sequence $\left\{\int f_n d\mu, \; n \in \mathbb{N}\right\}$ is bounded.

One then defines

$$\int f d\mu = \lim_n \int f_n d\mu.$$

Again such a definition makes sense and has the properties of an integral. The final extension of the integral is obtained by taking differences of functions in $\mathcal{S}_1$ to obtain $\mathcal{S}_2$: for $\{f_1, f_2\} \subseteq \mathcal{S}_1$, let

$$f = f_1 - f_2, \text{ and } \int f d\mu = \int f_1 d\mu - \int f_2 d\mu.$$

Again such assignments make sense and have the properties an integral should have. In particular the usual limit theorems obtain, and the following result yields the gist of the theory: given $f \in \mathcal{S}_2$, there is $\{f_n, \; n \in \mathbb{N}\} \subseteq \mathcal{S}$ such that

$$\lim_n f_n =_{\mu} f, \; \lim_n \int |f - f_n| \, d\mu = 0, \text{ and } \lim_n \int f_n d\mu = \int f d\mu.$$

That done, one must produce measurable functions. The function f is measurable for μ when there exists $\{f_n, \; n \in \mathbb{N}\} \subseteq \mathcal{S}$ with $\lim_n f_n =_{\mu} f$. The usual properties of measurable functions then follow, and, in particular, continuous functions, and functions in $\mathcal{S}_2$ are measurable for μ. Borel measurable functions are also measurable for μ. A set is measurable for μ when its indicator function is, so that the sets in $\tilde{\mathcal{I}}$ are measurable for μ, and have finite measure for μ. Sets that are measurable for μ and have measure zero for μ are said to be sets of measure zero for μ. When a set is measurable for μ, but its indicator function is not integrable, then that set is given infinite measure for μ. When the indicator is integrable, the measure, for μ, of the corresponding set, is its integral with respect to μ.

Let H be a real Hilbert space, and $\mathcal{L}(H)$ be its bounded, linear operators (whose domain is H).

Definition 2.7.1 A spectral family on H is a map $E : \mathbb{R} \longrightarrow \mathcal{L}(H)$ such that *(the prefix s-indicates strong convergence* [266, p. 77]*)*

1. $E(t)$ is a projection for every $t \in \mathbb{R}$;
2. when $t_1 \leq t_2$, $E(t_1) \leq E(t_2)$;
3. for $t \in \mathbb{R}$, fixed, but arbitrary, $s - \lim_{\theta \downarrow 0} E(t+\theta) = E(t)$,
   which means that, for $h \in H$, fixed, but arbitrary,
$$\lim_{\theta \downarrow 0} \| E(t+\theta)[h] - E(t)[h] \|_H = 0;$$
4. $s - \lim_{t \downarrow -\infty} E(t) = 0_H$, and $s - \lim_{t \uparrow \infty} E(t) = I_H$.

Let $h \in H$ be fixed, but arbitrary. The function
$$F_h^E(t) = \| E(t)[h] \|_H^2$$
is monotone increasing, continuous to the right, bounded, and
$$0 = \lim_{t \downarrow -\infty} F_h^E(t) \leq F_h^E(t) \leq \lim_{t \uparrow \infty} F_h^E(t) = \|h\|_H^2 .$$

The measure defined using F_h^E shall be denoted μ_h^E. A map $\phi : \mathbb{R} \longrightarrow \mathbb{R}$ is measurable for E when, for $h \in H$, fixed, but arbitrary, it is measurable for μ_h^E. Step, point-wise limits of such, continuous, and Borel measurable functions are examples of functions measurable for E. Bounded functions, measurable for E, belong to $L_2\left[\mu_h^E\right]$, for every $h \in H$. Since
$$\begin{aligned}
\langle E(t)[h_1], h_2 \rangle_H &= \\
&= \langle E(t)[h_1], E(t)[h_2] \rangle_H \\
&= \frac{\| E(t)[h_1+h_2] \|_H^2 - \| E(t)[h_1-h_2] \|_H^2}{4} \\
&= \frac{F_{h_1+h_2}(t) - F_{h_1-h_2}(t)}{4},
\end{aligned}$$
one has similarly a measure μ_{h_1,h_2}^E.

Let $E(t-) = s - \lim_{\theta \downarrow 0} E(t-\theta)$. One sets:

$$E_{]t_1,t_2]} = E(t_2) - E(t_1),$$
$$E_{]t_1,t_2[} = E(t_2-) - E(t_1),$$
$$E_{[t_1,t_2]} = E(t_2) - E(t_1-),$$
$$E_{[t_1,t_2[} = E(t_2-) - E(t_1-).$$

Those assignments allow one to define an integral of step functions, with respect to E, that produces an operator in $\mathcal{L}(H)$: for $f \in \mathcal{S}$, fixed, but arbitrary,

$$\int f(t)\, E(dt) = \sum_{i=1}^{n} \alpha_i E(I_i),$$

and then the following obtains, for fixed, but arbitrary $h \in H$:

$$\left\| \left\{ \int f(t)\, E(dt) \right\} [h] \right\|_H^2 = \int f^2(t)\, \mu_h^E(dt).$$

A limiting argument allows one to define an operator which integrates, with respect to the spectral measure E, functions f which are measurable for E, and whose square is integrable for μ_h^E. That operator is denoted

$$\hat{E}(f) = \int f(t)\, E(dt).$$

Functions with such properties shall be denoted ϕ, ψ etc. in the sequel.

Operators thus defined have the following properties:

Fact 2.7.2 ([266, pp. 175]) *Let $\mathcal{D}[B]$ denote the domain of the operator B.*

1. *When the equivalence class of ϕ, with respect to μ_h^E, belongs to the space $L_2(\mu_h^E)$, then h belongs to the domain of $\hat{E}(\phi)$.*
2. *$\hat{E}(\phi)$ is self-adjoint* (that is, densely defined, and equal to its adjoint).
3. *Let χ_n be the indicator function of the interval $[-n, n]$, and*

$$\phi_n(t) = \chi_n(\phi(t))\, \phi(t):$$

then, for

$$h_\phi \in \mathcal{D}\left[\hat{E}(\phi)\right], \quad \text{and } h_\psi \in \mathcal{D}\left[\hat{E}(\psi)\right],$$

$$\langle \hat{E}(\phi)[h_\phi], \hat{E}(\psi)[h_\psi] \rangle_H = \lim_n \int \phi_n(t)\, \psi_n(t)\, \mu_{h_\phi, h_\psi}^E(dt)$$

(the notation shall be either

$$\int \phi(t)\,\psi(t)\,\mu^{E}_{h_\phi,h_\psi}(dt) \text{ or } \int \phi(t)\,\psi(t)\,\langle E(dt)\left[h_\phi\right], h_\psi\rangle_H).$$

4. *For* $h \in \mathcal{D}\left[\hat{E}(\phi)\right]$, $\left\|\hat{E}(\phi)[h]\right\|_H^2 = \int \phi^2(t)\,\mu_h^E(dt)$.
5. *When* ϕ *is bounded,* $\hat{E}(\phi) \in \mathcal{L}(H)$ *and* $\left\|\hat{E}(\phi)\right\| \leq \sup_{t\in\mathbb{R}} |\phi(t)|$.
6. *When* $\phi(t) = 1, t \in \mathbb{R}$, $\hat{E}(\phi) = I_H$.
7. *For* $h \in H$, *and* $h_\phi \in \mathcal{D}\left[\hat{E}(\phi)\right]$,

$$\langle \hat{E}(\phi)\left[h_\phi\right], h\rangle_H = \int \phi(t)\,\langle E(dt)\left[h_\phi\right], h\rangle_H.$$

8. *When* $\phi(t) \geq \kappa, t \in \mathbb{R}$, *for* $h \in \mathcal{D}\left[\hat{E}(\phi)\right]$, *fixed, but arbitrary,*

$$\langle \hat{E}(\phi)[h], h\rangle_H \geq \kappa\,\|h\|_H^2.$$

9. $\mathcal{D}\left[\hat{E}(\phi)\right] \cap \mathcal{D}\left[\hat{E}(\psi)\right] = \mathcal{D}\left[\hat{E}(\phi) + \hat{E}(\psi)\right] = \mathcal{D}\left[\hat{E}(|\phi| + |\psi|)\right]$.
10. $\mathcal{D}\left(\hat{E}(\phi) + \hat{E}(\psi)\right) \subseteq \mathcal{D}\left(\hat{E}(\alpha\phi + \beta\psi)\right)$.
11. $\hat{E}(\alpha\phi + \beta\psi)$ *extends* $\alpha\hat{E}(\phi) + \beta\hat{E}(\psi)$, *that is*
    (i) $\mathcal{D}\left[\alpha\hat{E}(\phi) + \beta\hat{E}(\psi)\right] \subseteq \mathcal{D}\left[\hat{E}(\alpha\phi + \beta\psi)\right]$,
    (ii) $\left[\hat{E}(\alpha\phi + \beta\psi)\right]^{|\mathcal{D}[\alpha\hat{E}(\phi)+\beta\hat{E}(\psi)]} = \alpha\hat{E}(\phi) + \beta\hat{E}(\psi)$.
12. $\mathcal{D}\left[\hat{E}(\phi)\,\hat{E}(\psi)\right] = \mathcal{D}\left[\hat{E}(\psi)\right] \cap \mathcal{D}\left[\hat{E}(\phi\psi)\right]$.
13. $\hat{E}(\phi\psi)$ *extends* $\hat{E}(\phi)\,\hat{E}(\psi)$, *that is*
    (i) $\mathcal{D}\left[\hat{E}(\phi)\,\hat{E}(\psi)\right] \subseteq \mathcal{D}\left[\hat{E}(\phi\psi)\right]$,
    (ii) $\left[\hat{E}(\phi\psi)\right]^{|\mathcal{D}[\hat{E}(\phi)\hat{E}(\psi)]} = \hat{E}(\phi)\,\hat{E}(\psi)$.
14. *A set* S *is measurable for* E *when* χ_S *is. Then one often writes* E_S *for* $\hat{E}[\chi_S]$. *One has furthermore the following set of equalities:*

$$E_{\mathbb{R}\setminus S} = \hat{E}\left(1_{\mathbb{R}} - \chi_S\right) = I_H - \hat{E}\left(\chi_S\right) = I_H - E_S.$$

The operators E_S *are projections.*

Fact 2.7.3 ([266, pp. 185–186]) *Suppose now B is a self-adjoint operator on H, that is $\overline{\mathcal{D}[B]} = H$, and, on $\mathcal{D}[B]$, $B = B^\star$. There is then a unique spectral family E^B such that*

$$B = \hat{E}^B(\mathrm{id}_{\mathbb{R}}), \ \mathrm{id}_{\mathbb{R}}(t) = t, \ t \in \mathbb{R}.$$

The following obtain:

1. *Let ϕ be a measurable function for E^B. Then*

$$\phi(B) = \hat{E}^B(\phi).$$

   In particular, given S measurable for E^B, $E^B_S = \chi_S(B)$.
2. *The spectral family associated with $\phi(B)$, say $E^{\phi(B)}$, is obtained as follows (S* below is a Borel set)*:*

$$E^{\phi(B)}(t) = E^B_{\phi^{-1}(]-\infty,t])};$$

$$E^{\phi(B)}_S = E^B_{\phi^{-1}(S)}.$$

3. *Let S be measurable for E^B. Then BE^B_S extends $E^B_S B$.*
4. *Let $D(S,B) = E^B_S(\mathcal{D}[B])$. Then*

   (i) *for $h \in D(]-\infty,\kappa[,B) \setminus \{0_H\}$,*

$$\langle B[h], h\rangle_H < \kappa \|h\|_H^2;$$

   (ii) *for $h \in D(]-\infty,\kappa],B) \setminus \{0_H\}$,*

$$\langle B[h], h\rangle_H \le \kappa \|h\|_H^2.$$

5. *$\langle B[h], h\rangle_H \le \kappa \|h\|_H^2$ for all $h \in \mathcal{D}[B]$ if, and only if, for all $t \ge \kappa$, $E^B(t) = I_H$* (identity operator).
6. *$\langle B[h], h\rangle_H \ge \kappa \|h\|_H^2$ for all $h \in \mathcal{D}[B]$ if, and only if, for all $t < \kappa$, $E^B(t) = O_H$* (zero operator).
7. *Let I be a bounded interval. Then $\mathcal{R}[E^B_I] \subseteq \mathcal{D}[B]$, and*

$$BE^B_I = \hat{E}(\chi_I \mathrm{id}_{\mathbb{R}}) \in \mathcal{L}(H).$$

8. *Fix arbitrarily $t \in \mathbb{R}$ and $\theta > 0$. Then, for $h \in \mathcal{R}[E^B(t+\theta) - E^B(t-\theta)]$,*

$$h \in \mathcal{D}[B], \ \ \text{and} \ \ \|(B - tI_H)[h]\|_H \le \theta \|h\|_H.$$

9. *B is bounded if, and only if, there are constants κ_1 and κ_2 such that*

$$E^B(t) = \begin{cases} O_H \text{ when } t < \kappa_1 \\ I_H \text{ when } t \geq \kappa_2. \end{cases}.$$

One may choose

(i) $\kappa_1 = m = \inf_{\{h \in \mathcal{D}[B]:\|h\|_H = 1\}} \{\langle B[h], h\rangle_H\}$;
(ii) $\kappa_2 = M = \sup_{\{h \in \mathcal{D}[B]:\|h\|_H = 1\}} \{\langle B[h], h\rangle_H\}$.

When $t \in]m, M[$, $E^B(t) \neq O_H$, and $E^B(t) \neq I_H$.

Remark 2.7.4 One sometimes meets the following notation:

$$f(B) = \int_{m-}^{M} f(t)\, E^B(dt),$$

where the minus in $m-$ means that $\dot{f}(m)\, E(m)$ may be different from the zero operator.

Fact 2.7.5 ([266, p. 194]) *Let H_p be the subspace of H generated by the eigenvectors of B, and $H_c = H_p^{\perp}$. H_{cs} is the subspace of elements $h \in H_c$ for which there exists a Borel $N \subset \mathbb{R}$, of Lebesgue measure zero, with the property that*

$$E_N^B[h] = h.$$

Finally $H_{ca} = H_c \ominus H_{cs}$, and $H_s = H_p \oplus H_{cs}$. Then the following facts obtain:

1. $H_p = \{h \in H : \exists S \subseteq \mathbb{R}$ *with* $|S| \leq \aleph_0$ *and* $\mu_h(\mathbb{R} \setminus S) = 0\}$;
2. $H_c = \{h \in H : \mu_h(t) = 0,\ t \in \mathbb{R}\}$;
3. $H_s = \{h \in H : \exists N$, *Borel, of Lebesgue measure zero, with* $\mu_h(\mathbb{R} \setminus N) = 0\}$;
4. $H_{ca} = \{h \in H : \mu_h(N) = 0, N$ *Borel, of Lebesgue measure zero.*$\}$

2.8 Reproducing Kernel Hilbert Spaces as Images of Ranges of Square Roots of Linear Operators

When the functions of an RKHS are square integrable, one may sometimes be able to show that this RKHS is basically the range of the square root of the integral operator determined by its reproducing kernel (Propositions 1.3.20 and 1.3.21). One shall find below a further context in which this assertion proves true.

Definition 2.8.1 Let X be a topological space, and $\mathcal{X}$ be a σ-algebra of subsets of X containing $\mathcal{B}(X)$, the Borel sets. Let μ be a measure on $\mathcal{X}$. Its support is denoted $S(\mu)$, and $x \in S(\mu)$ when $\mu(V_x) > 0$ for open neighborhoods V_x of x. $S(\mu)$ is closed since, whenever the net x_λ in $S(\mu)$ converges to x, and V_x is an open neighborhood of x, V_x is eventually an open neighborhood of x_λ. As the complement of $S(\mu)$ is open, any of its measurable subsets has measure zero for μ.

Remark 2.8.2 The following facts are used in the sequel. Second countable means separable [84, p. 176]. In metric spaces, second countable and separable are equivalent [84, p. 187]. A Hausdorff space is locally compact if, and only if, each of its points has a compact neighborhood [270, p. 130].

Remark 2.8.3 Suppose that T is Hausdorff, locally compact, and second countable. It is thus separable because second countable. Let τ be a positive, Radon measure on $\mathcal{T} = \mathcal{B}(T)$. Suppose that $F : T \longrightarrow H$ is weakly continuous. Then H_F is separable [(Proposition) 2.6.9], so that [(Proposition) 2.2.2]

$$\mathcal{N}[\Lambda_{F,0}] = H_{F,\tau}^{\perp}.$$

However, since $[f]_{L_0(T,\mathcal{T},\tau)} = 0$ if, and only if, $f(t) = 0,\ t \in S(\tau)$,

$$\mathcal{N}[\Lambda_{F,0}] = \{h \in H : \langle h, F(t)\rangle_H = 0,\ t \in S(\tau)\} = \overline{V[F(t),\ t \in S(\tau)]}^{\perp}.$$

Consequently, $H_{F,\tau} = \overline{V[F(t),\ t \in S(\tau)]}$.

The continuity assumption may be replaced by an assumption on H_F : that it be separable. When the range of F is in an RKHS, that amounts to the assumption that the RKHS itself be separable. One may then drop the assumption that T be second countable.

Remark 2.8.4 Suppose that, in (Remark) 2.8.3, T is compact. Then τ is finite. Since F is weakly continuous, for $h \in H$, fixed, but arbitrary, $t \longrightarrow \langle h, F(t)\rangle_H$ is continuous, and there are thus t_m and t_M in T such that, for $t \in T$, fixed, but arbitrary,

$$\langle h, F(t_m)\rangle_H \le \langle h, F(t)\rangle_H \le \langle h, F(t_M)\rangle_H.$$

Thus

$$\begin{aligned}|\langle h, F(t)\rangle_H| &\le |\langle h, F(t_m)\rangle_H| \vee |\langle h, F(t_M)\rangle_H| \\ &\le \|h\|_H \{\|F(t_m)\|_H \vee \|F(t_M)\|_H\}.\end{aligned}$$

Thus (Banach-Steinhaus theorem, [266, p. 76]) F is bounded. It is thus strongly p-integrable, and $\Lambda_{F,p}$ is compact [(Proposition) 2.5.5].

However, for $\Lambda_{F,c}$ to be compact, it is necessary, and sufficient, that F be strongly continuous [(Proposition) 2.6.10].

Proposition 2.8.5 *Let $(T, \mathcal{T}, \tau)$ be a σ-finite measure space, and $H(\mathcal{H}, T)$ be a separable RKHS. The following statements are then equivalent:*

1. $H(\mathcal{H}, T)$ is a manifold in $\mathcal{L}_2(T, \mathcal{T}, \tau)$;

2. *$\mathcal{H}$ is a $(2,2)$-bounded kernel.*

When these equivalent conditions obtain, the following operator:

$$L_{\mathcal{H}_F} = \Lambda_{F,2}\Lambda^{\star}_{F,2} = J_{F,2}J^{\star}_{F,2}$$

is the integral operator with kernel $\mathcal{H}_F = \mathcal{H}$.

Proof This result is a particular case of (Proposition) 2.5.2.

Suppose indeed that $H(\mathcal{H}, T)$ is a manifold in $\mathcal{L}_2(T, \mathcal{T}, \tau)$. Let the map $F : T \longrightarrow H(\mathcal{H}, T)$ be defined using $F(t) = \mathcal{H}(\cdot, t)$, $t \in T$. Then L_F is the identity, $H_F = H(\mathcal{H}, T)$ is separable, and $\mathcal{H}_F = \mathcal{H}$. The assumption (item 1) then says that F is weakly 2-integrable, so that, by (Proposition) 2.5.2, $\mathcal{H}_F$ is $(2,2)$-bounded (in fact a Carleman kernel). The validity of the last part of the statement follows also from (Proposition) 2.5.2. That item 2 implies item 1 follows from (Proposition) 2.5.2 as well. □

Remark 2.8.6 Statement (Proposition) 2.8.5 plus Douglas's theorem [80] yield that $J_{F,2}$ and $L^{1/2}_{\mathcal{H}}$ have the same range. The result which follows gives more information on this relation. But two preliminary lemmas are needed for its proof.

Lemma 2.8.7 *Let T be Hausdorff, locally compact, second countable; $\mathcal{T}$ be $\mathcal{B}(T)$; and τ be a (regular) σ-finite Radon measure on $\mathcal{T}$. Assume that τ has full support, and that $H(\mathcal{H}, T)$ is separable, and a manifold in $\mathcal{L}_2(T, \mathcal{T}, \tau)$. Let $F : T \longrightarrow H(\mathcal{H}, T)$ be defined using $F(t) = \mathcal{H}(\cdot, t)$, $t \in T$, and $L_{\mathcal{H}}$ be the integral operator with kernel $\mathcal{H}$. Then*

1. *$\Lambda_{F,2} = J_{F,2}$ is injective;*
2. *$J_{F,2}[h] = [h]_{L_2(T,\mathcal{T},\tau)} = L^{1/2}_{\mathcal{H}} W^{\star}[h]$, $h \in H(\mathcal{H}, T)$, where $W^{\star}$ is a partial isometry whose initial and final sets are, respectively,*

$$H(\mathcal{H}, T) \text{ and } \overline{\mathcal{R}[L^{1/2}_{\mathcal{H}}]};$$

3. *$L^{1/2}_{\mathcal{H}}$ is injective on the range of $W^{\star}$, and thus*

$$L^{-1/2}_{\mathcal{H}}\left[[h]_{L_2(T,\mathcal{T},\tau)}\right] = W^{\star}[h], \; h \in H(\mathcal{H}, T),$$

so that

$$\left\|L^{-1/2}_{\mathcal{H}}\left[[h]_{L_2(T,\mathcal{T},\tau)}\right]\right\|_{L_2(T,\mathcal{T},\tau)} = \|h\|_{H(\mathcal{H},T)};$$

4. *furthermore*

$$\mathcal{N}[L^{1/2}_{\mathcal{H}}] = \overline{\mathcal{R}[J_{F,2}]}^{\perp},$$
$$\mathcal{R}[L^{1/2}_{\mathcal{H}}] = \mathcal{R}[J_{F,2}].$$

Proof It was noticed, in (Remark) 2.8.3, that, when one assumes that $S(\tau) = T$,

$$H_{F,\tau} = \overline{V[F(t), t \in S(\tau)]} = H_F = H(\mathcal{H}, T).$$

Consequently, since [(Proposition) 2.2.2] $\mathcal{N}[\Lambda_{F,2}] = H_{F,\tau}^{\perp}$, $\mathcal{N}[\Lambda_{F,2}] = \mathcal{N}[J_{F,2}] = \{0_{H(\mathcal{H},T)}\}$.

Because of (Proposition) 2.8.5, $\Lambda_{F,2}\Lambda_{F,2}^{\star} = L_{\mathcal{H}}$. The polar decomposition [266, p. 186] of $\Lambda_{F,2}^{\star}$ thus yields that

$$\Lambda_{F,2}^{\star} = W\left\{\left[\Lambda_{F,2}^{\star}\right]^{\star}\Lambda_{F,2}^{\star}\right\}^{1/2} = W\left\{\Lambda_{F,2}\Lambda_{F,2}^{\star}\right\}^{1/2} = WL_{\mathcal{H}}^{1/2},$$

where W is a partial isometry whose initial and final sets are, respectively,

$$\overline{\mathcal{R}[\Lambda_{F,2}\Lambda_{F,2}^{\star}]^{1/2}]} = \overline{\mathcal{R}[L_{\mathcal{H}}^{1/2}]}, \text{ and } \overline{\mathcal{R}[\Lambda_{F,2}^{\star}]}.$$

Since generally [266, p. 71], for any operator A, $\overline{\mathcal{R}[A^{\star}]}^{\perp} = \mathcal{N}[A]$,

$$\overline{\mathcal{R}[\Lambda_{F,2}^{\star}]}^{\perp} = \mathcal{N}[\Lambda_{F,2}],$$

and, since it has just been acknowledged that $\Lambda_{F,2}$ is an injection, one has that

$$\overline{\mathcal{R}[\Lambda_{F,2}^{\star}]}^{\perp} = \{0_{H(\mathcal{H},T)}\},$$

so that

$$\overline{\mathcal{R}[\Lambda_{F,2}^{\star}]} = H(\mathcal{H}, T).$$

But then [266, p. 86], $W^{\star}$ is a partial isometry whose initial and final sets are, respectively,

$$H(\mathcal{H}, T) \text{ and } \overline{\mathcal{R}[L_{\mathcal{H}}^{1/2}]}.$$

Consequently, as asserted in item 2,

$$J_{F,2} = \Lambda_{F,2} = L_{\mathcal{H}}^{1/2}W^{\star}.$$

Suppose now that $L_{\mathcal{H}}^{1/2}[f] = 0$, $f \in \mathcal{R}[W^{\star}]$. There is $h \in H(\mathcal{H}, T)$ such that $W^{\star}[h] = f$. But then, because of items 1 and 2, $\Lambda_{F,2}[h] = 0$, so that, since $\Lambda_{F,2}$ is injective, $h = 0$ and thus $f = 0$. Item 3 is proved.

Finally, from $\Lambda_{F,2}^{\star} = WL_{\mathcal{H}}^{1/2}$, since the initial set of W is the closure of the range of the square root of $L_{\mathcal{H}}$, the kernel of that square root is that of $\Lambda_{F,2}^{\star}$. But

$$\mathcal{N}[\Lambda_{F,2}^{\star}] = \overline{\mathcal{R}[\Lambda_{F,2}]}^{\perp} = \overline{\mathcal{R}[J_{F,2}]}^{\perp}.$$

Suppose that $f = L_{\mathcal{H}}^{1/2}[g]$. Because of [266, p. 71], and what precedes (item 2),

$$\mathcal{N}[L_{\mathcal{H}}^{1/2}] = \overline{\mathcal{R}[L_{\mathcal{H}}^{1/2}]}^{\perp}, \text{ and } \overline{\mathcal{R}[L_{\mathcal{H}}^{1/2}]} = \mathcal{R}[W^{\star}].$$

Thus $g = g_1 + g_2$, where

$$g_1 \in \mathcal{N}[L_{\mathcal{H}}^{1/2}],\ g_2 \in \overline{\mathcal{R}[L_{\mathcal{H}}^{1/2}]},\ g_2 = W^{\star}[h(g_2)],\ \text{some } h(g_2) \in H(\mathcal{H}, T).$$

Consequently (item 2)

$$f = L_{\mathcal{H}}^{1/2} W^{\star}[h(g_2)] = J_{F,2}[h(g_2)].$$

Thus $\mathcal{R}[L_{\mathcal{H}}^{1/2}] \subseteq \mathcal{R}[J_{F,2}]$. But, from item 2, one has the reverse inclusion. □

Remark 2.8.8 Let $E^{\mathcal{H}}$ denote the spectral family of $L_{\mathcal{H}}$. One has that:

$$\mathcal{N}[L_{\mathcal{H}}] = \mathcal{R}[E^{\mathcal{H}}(0)].$$

Indeed, for $0 < t_1 < t_2$, fixed, but arbitrary, because of (Fact) 2.7.3, items 3 and 7,

$$\begin{aligned}\left\|E^{\mathcal{H}}_{]t_1,t_2]} L_{\mathcal{H}}[h]\right\|^2_{L_2(T,\mathcal{T},\tau)} &= \int \chi_{]t_1,t_2]}(t)\, t^2 \mu_h^{\mathcal{H}}(dt) \\ &\geq t_1^2 \left\{\mu_h^{\mathcal{H}}(t_2) - \mu_h^{\mathcal{H}}(t_1)\right\} \\ &= t_1^2 \left\{\|E^{\mathcal{H}}(t_2)[h]\|^2_{L_2(T,\mathcal{T},\tau)} - \|E^{\mathcal{H}}(t_1)[h]\|^2_{L_2(T,\mathcal{T},\tau)}\right\} \\ &= t_1^2 \|\{E^{\mathcal{H}}(t_2) - E^{\mathcal{H}}(t_1)\}[h]\|^2_{L_2(T,\mathcal{T},\tau)}.\end{aligned}$$

Thus, when $h \in \mathcal{N}[L_{\mathcal{H}}]$, and $0 < t_1 < t_2$,

$$\|\{E^{\mathcal{H}}(t_2) - E^{\mathcal{H}}(t_1)\}[h]\|_{L_2(T,\mathcal{T},\tau)} = 0.$$

Choosing $t_2 > \|L_{\mathcal{H}}\|$ and letting $t_1 \downarrow 0$, one has that

$$\|\{I_H - E^{\mathcal{H}}(0)\}[h]\|_{L_2(T,\mathcal{T},\tau)} = 0.$$

Consequently $h \in \mathcal{R}[E^{\mathcal{H}}(0)]$. Now, when $h \in \mathcal{R}[E^{\mathcal{H}}(0)]$, because of (Fact) 2.7.3, item 4, $h \in \mathcal{N}[L_{\mathcal{H}}^{1/2}] = \mathcal{N}[L_{\mathcal{H}}]$.

Lemma 2.8.9 *Let T be Hausdorff, locally compact, and second countable; $\mathcal{T} = \mathcal{B}(T)$; and τ be a (regular) σ-finite Radon measure on $\mathcal{T}$. Assume that τ has full support, and that $H(\mathcal{H}, T)$ is separable, and a manifold in $\mathcal{L}_2(T, \mathcal{T}, \tau)$. Let $F : T \longrightarrow H(\mathcal{H}, T)$ be defined using $F(t) = \mathcal{H}(\cdot, t)$, $t \in T$, and $L_{\mathcal{H}}$ be the integral operator with kernel $\mathcal{H}$. Then, as $L_{\mathcal{H}}$ is linear, positive, bounded, and self-adjoint, with domain $L_2(T, \mathcal{T}, \tau)$* [(Proposition) 2.8.5], *it has a spectral decomposition with*

$m = 0$ and $M = \|L_{\mathcal{H}}\|$. Let $E^{\mathcal{H}}$ denote the associated spectral family. One has then that, for W defined in (Lemma) 2.8.7,

$$\{E^{\mathcal{H}}(0) - E^{\mathcal{H}}(0-)\} W^{\star} = E^{\mathcal{H}}(0) W^{\star} = 0.$$

Proof $E^{\mathcal{H}}(0-) = O_H$ because of (Fact) 2.7.3, item 9, and, since

- on one hand, because of [266, p. 71] and (Remark) 2.8.8,

$$\overline{\mathcal{R}[L_{\mathcal{H}}^{1/2}]} = \mathcal{N}[L_{\mathcal{H}}^{1/2}]^{\perp} = \mathcal{R}[E^{\mathcal{H}}(0)]^{\perp},$$

- on the other hand, because of (Lemma) 2.8.7, item 2,

$$\mathcal{R}[W^{\star}] = \overline{\mathcal{R}[L_{\mathcal{H}}^{1/2}]},$$

$E^{\mathcal{H}}(0) W^{\star} = 0.$ □

Remark 2.8.10 Let $I = [0, \|L_{\mathcal{H}}\|]$, and $I_0 = I \setminus \{0\}$. One then has, because of (Lemma) 2.8.9, that, for appropriate f, in the range of $W^{\star}$,

$$\int f dE^{\mathcal{H}} = \int_{0-}^{\|L_{\mathcal{H}}\|} f dE^{\mathcal{H}} = \int_I f dE^{\mathcal{H}} = \int_{I_0} f dE^{\mathcal{H}}.$$

Proposition 2.8.11 *Let T be Hausdorff, locally compact, and second countable; $\mathcal{T} = \mathcal{B}(T)$; and τ be a (regular) σ-finite Radon measure on $\mathcal{T}$. Assume that τ has full support, and that $H(\mathcal{H}, T)$ is separable, and a manifold in $\mathcal{L}_2(T, \mathcal{T}, \tau)$. Let $F : T \longrightarrow H(\mathcal{H}, T)$ be defined using $F(t) = \mathcal{H}(\cdot, t)$, $t \in T$, and $L_{\mathcal{H}}$ be the integral operator with kernel $\mathcal{H}$. Then, as $L_{\mathcal{H}}$ is linear, positive, bounded, and self-adjoint* [(Proposition) 2.8.5], *with full domain, it has* [(Fact) 2.7.3] *a spectral decomposition denoted*

$$L_{\mathcal{H}} = \int_I \lambda E^{\mathcal{H}}(d\lambda).$$

Then:

1. *The set*

$$\left\{ f \in L_2(T, \mathcal{T}, \tau) : \int_I \lambda^{-1} \mu_f^{\mathcal{H}}(d\lambda) < \infty \right\}$$

which is the range of $L_{\mathcal{H}}^{1/2}$, and the domain of $L_{\mathcal{H}}^{-1/2}$, is obtained as the equivalence classes in $L_2(T, \mathcal{T}, \tau)$ of the elements in $H(\mathcal{H}, T)$, and, for $h \in H(\mathcal{H}, T)$, fixed, *but arbitrary,*

$$\|h\|^2_{H(\mathcal{H},T)} = \int_I \lambda^{-1} \mu_h^{\mathcal{H}}(d\lambda) = \left\| L_{\mathcal{H}}^{-1/2} \left[[h]_{L_2(T,\mathcal{T},\tau)} \right] \right\|^2_{L_2(T,\mathcal{T},\tau)}.$$

2. *For fixed, but arbitrary* $(t_1, t_2) \in T \times T$, *there exists a Radon measure* μ_{t_1,t_2} *on* $\mathcal{B}(I_0)$ *such that*

   (i) $\mathcal{H}(t_1, t_2) = \int_{I_0} \lambda\, \mu_{t_1,t_2}(d\lambda)$,

   (ii) *for* $B \in \mathcal{B}(I_0)$ *such that* $\overline{B} \subseteq I_0$, *there exists a subset* $\left\{h_i^B,\ i \in I^B\right\}$, *in* $H(\mathcal{H}, T)$, *such that*

   - $\left\{\left[h_i^B\right]_{L_2(T,\mathcal{T},\tau)},\ i \in I^B\right\}$ *is, for the range of* $E_B^{\mathcal{H}}$, *an orthonormal basis, and*
   - $\mu_{t_1,t_2}(B) = \sum_{i \in I^B} h_i^B(t_1)\, h_i^B(t_2)$.

3. *For fixed, but arbitrary* $t \in T$, *there exists a positive Borel measure* μ_t *on* $\mathcal{B}(I_0)$, *and a family* $\{h_i,\ i \in I\}$, *in* $H(\mathcal{H}, T)$, *such that*

   (i) $\left\{[h_i]_{L_2(T,\mathcal{T},\tau)},\ i \in I\right\}$ *is an orthonormal basis for* $\mathcal{N}[L_{\mathcal{H}}]^{\perp}$,

   (ii) μ_t *is finite if, and only if,* $[\mathcal{H}(\cdot, t)]_{L_2(T,\mathcal{T},\tau)} \in \mathcal{R}[L_{\mathcal{H}}]$,

   (iii) *when* μ_t *is finite,* $\mu_t(I_0) = \sum_{i \in I} h_i^2(t)$.

Proof (1) The assertion's validity follows from (Lemma) 2.8.7, and the fact that

$$\infty > \left\| L_{\mathcal{H}}^{-1/2} \left[[h]_{L_2(T,\mathcal{T},\tau)} \right] \right\|^2_{L_2(T,\mathcal{T},\tau)} = \int_I \lambda^{-1} \mu_h^{\mathcal{H}}(d\lambda).$$

Proof (2) Let $\{f, g\} \subseteq L_2(T, \mathcal{T}, \tau)$ be fixed, but arbitrary, and ϕ be a continuous function whose support is compact, and contained in I_0. The map

$$\phi \mapsto \int_{I_0} \frac{\phi(\lambda)}{\lambda} \mu_{f+g}^{\mathcal{H}}(d\lambda)$$

is linear and positive. There exists thus [263, p. 25] a unique (regular) Radon measure ν_{f+g}, on the Borel sets of I_0, such that

$$\int_{I_0} \frac{\phi(\lambda)}{\lambda} \mu_{f+g}^{\mathcal{H}}(d\lambda) = \int_{I_0} \phi(\lambda)\, \nu_{f+g}(d\lambda).$$

There is consequently [263, pp. 26,28] a (regular) signed Radon measure $\nu_{f,g}$, on the Borel sets of I_0, such that

$$\int_{I_0} \frac{\phi(\lambda)}{\lambda} \mu_{f,g}^{\mathcal{H}}(d\lambda) = \int_{I_0} \phi(\lambda)\, \nu_{f,g}(d\lambda).$$

But continuous functions with compact support are dense in L_1 [263, p. 123], so that the latter equality remains true for generally integrable functions ϕ. Applying the formula with

- $\phi(\lambda) = \lambda$,
- $f = W^\star[F(t_1)] = W^\star[\mathcal{H}(\cdot, t_1)]$,
- $g = W^\star[F(t_2)] = W^\star[\mathcal{H}(\cdot, t_2)]$,

one gets, since $L_\mathcal{H}$ is injective on the range of $W^\star$ [(Lemma) 2.8.7], and thus [(Lemma) 2.8.9],

$$\{E^\mathcal{H}(0) - E^\mathcal{H}(0-)\} W^\star = 0,$$

that, since [(Fact) 2.7.2] $\hat{E}(\phi) = I_H$ when $\phi \equiv 1$,

$$\begin{aligned}
\int_{I_0} \nu_{f,g}(d\lambda)\,\lambda &= \int_{I_0} \mu_{f,g}^\mathcal{H}(d\lambda) \\
&= \int_{I_0} \langle E^\mathcal{H}(d\lambda)[f], g\rangle_{L_2(T,\mathcal{T},\tau)} \\
&= \int_{I_0} \langle E^\mathcal{H}(d\lambda)(W^\star[F(t_1)]), W^\star[F(t_2)]\rangle_{L_2(T,\mathcal{T},\tau)} \\
&= \int_{I} \langle E^\mathcal{H}(d\lambda)(W^\star[F(t_1)]), W^\star[F(t_2)]\rangle_{L_2(T,\mathcal{T},\tau)} \\
&= \langle W^\star[F(t_1)], W^\star[F(t_2)]\rangle_{L_2(T,\mathcal{T},\tau)} \\
&= \langle F(t_1), F(t_2)\rangle_{H(\mathcal{H},T)} \\
&= \mathcal{H}(t_1, t_2),
\end{aligned}$$

which is claim (i) of item 2.

Let now $B \subseteq I_0$ be a Borel set such that $\overline{B} \subseteq I_0$. Since [(Fact) 2.7.3]

$$E_B^\mathcal{H} = \chi_B(L_\mathcal{H}) = \int \chi_B\, dE^\mathcal{H},$$

and since B contains only elements larger than a strictly positive number, there is a finite $\kappa \geq 0$ such that $\chi_B(\lambda) \leq \kappa\lambda$, and, for fixed, but arbitrary $f \in L_2(T, \mathcal{T}, \tau)$,

$$\begin{aligned}
\langle E_B^\mathcal{H}[f], f\rangle_{L_2(T,\mathcal{T},\tau)} &= \int_B \langle E^\mathcal{H}(d\lambda)[f], f\rangle_{L_2(T,\mathcal{T},\tau)} \\
&\leq \kappa \int_I \lambda \langle E^\mathcal{H}(d\lambda)[f], f\rangle_{L_2(T,\mathcal{T},\tau)} \\
&= \kappa \langle L_\mathcal{H}[f], f\rangle_{L_2(T,\mathcal{T},\tau)}.
\end{aligned}$$

Consequently [80]

$$\mathcal{R}[E_B^{\mathcal{H}}] \subseteq \mathcal{R}[L_{\mathcal{H}}^{1/2}].$$

Because of (Lemma) 2.8.7,

$$\mathcal{R}[L_{\mathcal{H}}^{1/2}] = \mathcal{R}[J_{F,2}], \text{ and } \overline{\mathcal{R}[L_{\mathcal{H}}^{1/2}]} = \mathcal{R}[W^{\star}].$$

Thus

$$\mathcal{R}[E_B^{\mathcal{H}}] \subseteq \mathcal{R}[L_{\mathcal{H}}^{1/2}] = \mathcal{R}[J_{F,2}] \subseteq \mathcal{R}[W^{\star}]. \qquad (\star)$$

Since $\overline{B} \subseteq I_0$, χ_B is integrable with respect to $\nu_{f,g}$. From (Fact) 2.7.2, item 13, (Fact) 2.7.3, item 3, and appropriate ϕ (for instance, measurable and bounded, on a support strictly included into $]0, \|L_{\mathcal{H}}\|]$), it follows that

$$\phi(L_{\mathcal{H}})\,\chi_B(L_{\mathcal{H}}) = \int \phi(\lambda)\,\chi_B(\lambda)\,E^{\mathcal{H}}(d\lambda) = \hat{E}^{\mathcal{H}}\left[\chi_B\phi\right],$$

and thus that

$$E_B^{\mathcal{H}}\phi(L_{\mathcal{H}}) = E_B^{\mathcal{H}}\phi(L_{\mathcal{H}})\,E_B^{\mathcal{H}} = \phi(L_{\mathcal{H}})\,E_B^{\mathcal{H}}. \qquad (\star\star)$$

Consequently

$$\langle\{\phi(L_{\mathcal{H}})\,E_B^{\mathcal{H}}\}\,[f]\,, g\rangle_{L_2(T,\mathcal{T},\tau)} = \langle\{\phi(L_{\mathcal{H}})\}\left[E_B^{\mathcal{H}}[f]\right], \left[E_B^{\mathcal{H}}[g]\right]\rangle_{L_2(T,\mathcal{T},\tau)},$$

or

$$\langle\hat{E}^{\mathcal{H}}\left(\chi_B\phi\right)[f]\,, g\rangle_{L_2(T,\mathcal{T},\tau)} = \langle\hat{E}^{\mathcal{H}}[\phi]\left[E_B^{\mathcal{H}}[f]\right], \left[E_B^{\mathcal{H}}[g]\right]\rangle_{L_2(T,\mathcal{T},\tau)},$$

which, letting $\phi(\lambda) = \lambda^{-1}$, yields that

$$\int_I \frac{\chi_B(\lambda)}{\lambda}\,\mu_{f,g}^{\mathcal{H}}(d\lambda) = \int_I \frac{1}{\lambda}\langle E^{\mathcal{H}}(d\lambda)\left[E_B^{\mathcal{H}}[f]\right], \left[E_B^{\mathcal{H}}[g]\right]\rangle_{L_2(T,\mathcal{T},\tau)}.$$

As a consequence

$$\int_{I_0} \chi_B(\lambda)\,\nu_{f,g}(d\lambda) = \int_I \frac{1}{\lambda}\langle E^{\mathcal{H}}(d\lambda)\left[E_B^{\mathcal{H}}[f]\right], E_B^{\mathcal{H}}[g]\rangle_{L_2(T,\mathcal{T},\tau)}.$$

Choose $f = W^{\star}[F(t_1)]$, and $g = W^{\star}[F(t_2)]$. Using $(\star)$, one may write, for example,

$$E_B^{\mathcal{H}}[f] = L_{\mathcal{H}}^{1/2}W^{\star}[k].$$

Then, since, on the range of $W^\star$, the operator $L_{\mathcal{H}}^{1/2}$ is injective [(Lemma) 2.8.7],

$$\int_{I_0} \chi_B(\lambda)\, \nu_{f,g}(d\lambda) = \langle L_{\mathcal{H}}^{-1/2}\{E_B^{\mathcal{H}}[f]\}, L_{\mathcal{H}}^{-1/2}\{E_B^{\mathcal{H}}[g]\}\rangle_{L_2(T,\mathcal{T},\tau)}.$$

Since $E_B^{\mathcal{H}}$ is a projection, let (because of $(\star)$)

$$\left\{\left[h_i^B\right]_{L_2(T,\mathcal{T},\tau)} = L_{\mathcal{H}}^{1/2} W^\star\left[h_i^B\right],\ h_i^B \in H(\mathcal{H},T),\ i \in I^B\right\}$$

be a complete orthonormal set in its range. Then, since, as seen above $(\star\star)$, $L_{\mathcal{H}}^{1/2}E_B^{\mathcal{H}} = E_B^{\mathcal{H}}L_{\mathcal{H}}^{1/2}E_B^{\mathcal{H}}$, the following inner product

$$\langle L_{\mathcal{H}}^{-1/2}\{E_B^{\mathcal{H}}[W^\star(F(t_1))]\}, L_{\mathcal{H}}^{-1/2}\{E_B^{\mathcal{H}}[W^\star(F(t_2))]\}\rangle_{L_2(T,\mathcal{T},\tau)}$$

equals

$$\sum_{i\in I^B}\langle L_{\mathcal{H}}^{-1/2}\{E_B^{\mathcal{H}}[W^\star(F(t_1))]\}, \left[h_i^B\right]_{L_2(T,\mathcal{T},\tau)}\rangle_{L_2(T,\mathcal{T},\tau)} \times$$
$$\times \langle L_{\mathcal{H}}^{-1/2}\{E_B^{\mathcal{H}}[W^\star(F(t_2))]\}, \left[h_i^B\right]_{L_2(T,\mathcal{T},\tau)}\rangle_{L_2(T,\mathcal{T},\tau)}.$$

Furthermore, the following one,

$$\langle L_{\mathcal{H}}^{-1/2}\{E_B^{\mathcal{H}}[W^\star(F(t_1))]\}, \left[h_i^B\right]_{L_2(T,\mathcal{T},\tau)}\rangle_{L_2(T,\mathcal{T},\tau)}$$

is equal to

$$\begin{aligned}
\langle E_B^{\mathcal{H}}L_{\mathcal{H}}^{-1/2}[W^\star(F(t_1))], \left[h_i^B\right]_{L_2(T,\mathcal{T},\tau)}\rangle_{L_2(T,\mathcal{T},\tau)} &= \\
&= \langle L_{\mathcal{H}}^{-1/2}[W^\star(F(t_1))], \left[h_i^B\right]_{L_2(T,\mathcal{T},\tau)}\rangle_{L_2(T,\mathcal{T},\tau)} \\
&= \langle L_{\mathcal{H}}^{-1/2}[W^\star(F(t_1))], L_{\mathcal{H}}^{1/2}W^\star\left[h_i^B\right]\rangle_{L_2(T,\mathcal{T},\tau)} \\
&= \langle W^\star(F(t_1)), W^\star\left[h_i^B\right]\rangle_{L_2(T,\mathcal{T},\tau)} \\
&= \langle F(t_1), h_i^B\rangle_{H(\mathcal{H},T)} \\
&= h_i^B(t_1).
\end{aligned}$$

Consequently

$$\nu_{f,g}(B) = \int_{I_0} \chi_B(\lambda)\, \nu_{f,g}(d\lambda) = \sum_{i\in I^B} h_i^B(t_1)\, h_i^B(t_2),$$

and part (ii) of item 2 obtains.

Proof (3) Let $B_n = \left[\frac{1}{n}, \|L_{\mathcal{H}}\|\right]$. Then, as already computed for the proof of (ii), item 2,

$$\nu_{f,f}(B_n) = \left\| L_{\mathcal{H}}^{-1/2} E_{B_n}^{\mathcal{H}} \left[W^{\star} \left[F(t_1) \right] \right] \right\|_{L_2(T,\mathcal{T},\tau)}^2 .$$

Consequently, since the sequence $\left\{ \chi_{B_n},\ n \in \mathbb{N} \right\}$ increases to $I_{]0,\|L_{\mathcal{H}}\|]}$, and the sequence

$$\left\{ E_{B_n}^{\mathcal{H}} = I_{L_2(T,\mathcal{T},\tau)} - E_{[0,\frac{1}{n}[}^{\mathcal{H}},\ n \in \mathbb{N} \right\}$$

increases, on the range of $W^{\star}$, to the identity operator,

$$\lim_n \nu_{f,f}(B_n) = \lim_n \left\| L_{\mathcal{H}}^{-1/2} E_{B_n}^{\mathcal{H}} \left[W^{\star} \left[F(t_1) \right] \right] \right\|_{L_2(T,\mathcal{T},\tau)}^2 ,$$

which is finite if, and only if, $W^{\star}[F(t_1)] \in \mathcal{R}[L_{\mathcal{H}}^{1/2}]$. But

$$\begin{aligned} W^{\star}[F(t_1)] \in \mathcal{R}[L_{\mathcal{H}}^{1/2}] \Leftrightarrow W^{\star}[F(t_1)] &= L_{\mathcal{H}}^{1/2}[f] \\ \Leftrightarrow L_{\mathcal{H}}^{1/2} W^{\star}[F(t_1)] &= L_{\mathcal{H}}[f] \\ \Leftrightarrow J_{F,2}[F(t_1)] &= L_{\mathcal{H}}[f] \\ \Leftrightarrow [F(t_1)]_{L_2(T,\mathcal{T},\tau)} &= L_{\mathcal{H}}[f] . \end{aligned}$$

Let

$$\left\{ [h_i]_{L_2(T,\mathcal{T},\tau)},\ h_i \in H(\mathcal{H},T),\ i \in I \right\} \subseteq L_2(T,\mathcal{T},\tau)$$

be a complete orthonormal set in the range of the square root of $L_{\mathcal{H}}$ (as a separable, inner product space, this range contains an orthonormal basis [266, p. 47].) Then, when $W^{\star}[F(t_1)] \in \mathcal{R}[L_{\mathcal{H}}^{1/2}]$,

$$\begin{aligned} \left\| L_{\mathcal{H}}^{-1/2} W^{\star}[F(t_1)] \right\|_{L_2(T,\mathcal{T},\tau)}^2 &= \sum_{i \in I} \left\langle L_{\mathcal{H}}^{-1/2} W^{\star}[F(t_1)], [h_i]_{L_2(T,\mathcal{T},\tau)} \right\rangle_{L_2(T,\mathcal{T},\tau)}^2 \\ &= \sum_{i \in I} \left\langle L_{\mathcal{H}}^{-1/2} W^{\star}[F(t_1)], L_{\mathcal{H}}^{1/2} W^{\star}[h_i] \right\rangle_{L_2(T,\mathcal{T},\tau)}^2 \\ &= \sum_{i \in I} \left\langle F(t_1), h_i \right\rangle_{H(\mathcal{H},T)}^2 \\ &= \sum_{i \in I} h_i^2(t_1) . \end{aligned}$$

□

Remark 2.8.12 When $S(\tau)$ is strictly contained in T, (Proposition) 2.8.11 remains true as long as one replaces T with $S(\tau)$, and $H(\mathcal{H}, T)$ with $H\left(\mathcal{H}_{S(\tau)}, S(\tau)\right)$.

Corollary 2.8.13 *The assumptions are those of* (Proposition) 2.8.11. *When $L_{\mathcal{H}}$ has pure point spectrum, there exists an orthonormal family*

$$\left\{[h_i]_{L_2(T,\mathcal{T},\tau)},\ h_i \in H(\mathcal{H}, T),\ i \in I\right\} \subseteq L_2(T, \mathcal{T}, \tau),$$

and strictly positive constants $\{\eta_i,\ i \in I\}$, such that

1. $L_{\mathcal{H}} = \sum_{i \in I} \eta_i \left\{[h_i]_{L_2(T,\mathcal{T},\tau)} \otimes [h_i]_{L_2(T,\mathcal{T},\tau)}\right\}$;
2. $H(\mathcal{H}, T) = \left\{h \in \mathcal{L}_2(T, \mathcal{T}, \tau) : \sum_{i \in I} \frac{1}{\eta_i} \left\| [h]_{L_2(T,\mathcal{T},\tau)} \right\|^2_{L_2(T,\mathcal{T},\tau)} < \infty\right\}$;
3. $\|h\|^2_{H(\mathcal{H},T)} = \sum_{i \in I} \frac{1}{\eta_i} \langle [h]_{L_2(T,\mathcal{T},\tau)}, [h_i]_{L_2(T,\mathcal{T},\tau)} \rangle^2_{L_2(T,\mathcal{T},\tau)}$;
4. $\mathcal{H}(t_1, t_2) = \sum_{i \in I} \eta_i h_i(t_1) h_i(t_2)$ *is an absolutely convergent series.*

Proof Items 1 to 3 follow directly from (Proposition) 2.8.11.

The series representation of item 4 has the following justification. When $L_{\mathcal{H}}$ has pure point spectrum, μ_{t_1,t_2} is a discrete measure located at the eigenvalues. Furthermore [(Proposition) 2.8.11],

$$\mu_{t_1,t_2}\left(\left[\eta_j\right]\right) = \sum_{i \in I_j} h_i^{(j)}(t_1)\, h_i^{(j)}(t_2),$$

where

$$\left\{\left[h_i^{(j)}\right]_{L_2(T,\mathcal{T},\tau)},\ i \in I_j\right\}$$

is orthonormal and spans the eigenspace associated with η_j. Since the series is an integral, it converges absolutely. □

Remark 2.8.14 When one assumes that $H(\mathcal{H}, T)$ is also a manifold of continuous functions, or, equivalently [(Proposition) 2.6.10], that $\mathcal{H}$ is locally bounded, and that $x \mapsto \mathcal{H}(x, t)$ is continuous for $t \in T$, that is, that F is weakly continuous, the series in item 4 of (Corollary) 2.8.13 is uniformly convergent on compact sets if, and only if, F is continuous.

Indeed, since the elements in $H(\mathcal{H}, T)$ are continuous, uniform convergence on compact sets makes $\mathcal{H}$ continuous [111, p. 336], and that, in turn, is equivalent to the continuity of F [(Proposition) 2.6.10]. Conversely, when F is continuous, $\mathcal{H}$ is continuous, and it is Dini's theorem [154, p. 239] that insures uniform convergence on compact sets:

$$\left|\mathcal{H}(t_1, t_2) - \sum_{i \leq i_n} \eta_i h_i(t_1)\, h_i(t_2)\right| \leq \sum_{i > i_n} \eta_i \left|h_i(t_1)\, h_i(t_2)\right|.$$

Mercer's theorem is (Corollary) 2.8.13 with the continuity assumptions.

Remark 2.8.15 Proposition 1.3.21 follows directly from (Corollary) 2.8.13. Indeed the assumption on C, that is, that C is adapted to $\mathcal{T} \otimes \mathcal{T}$, and such that $\int_T C(t,t)\,\tau(dt) < \infty$, makes it $(2,2)$-bounded, so that F is weakly square integrable [(Proposition) 2.5.2]. Consequently $H(C,T) \subseteq \mathcal{L}_2(T,\mathcal{T},\tau)$. The definition of F yields that $H(C,T)$ is separable (it is isomorphic to a subspace of a separable space). Thus (Lemma) 2.8.9 and (Proposition) 2.8.11 apply. One should also notice that C has property Π_J [(Definition) 2.4.2].

Chapter 3
Relations Between Reproducing Kernel Hilbert Spaces

When one claims that the signal must be in the RKHS of the noise, for detection to be nonsingular, one in fact means that the family of its paths should be contained, as a set, in the RKHS of that noise. One shall see in the next chapter that such a requirement entails a specific inclusion of related RKHS's, and the entire topic may be seen as a partial answer to the following question: given an RKHS of signals, which noises does it accommodate? What follows covers thus relations between RKHS's and, in particular inclusions and intersections.

3.1 Order for Covariances

As shall be seen, order for covariances is equivalent to inclusion of the associated RKHS's.

Definition 3.1.1 Let C_1 and C_2 be covariances on the same set T. C_2 dominates C_1 when $C_2 - C_1$ is a covariance on T. One writes $C_2 \gg C_1$. C_1 and C_2 are disjoint when, given a covariance C such that $C \ll C_1$ and $C \ll C_2$, C is then the zero covariance.

Remark 3.1.2 One shall see that disjoint covariances are covariances whose RKHS's have an intersection whose only element is the zero function.

Example 3.1.3 Let $C = c_\wedge c_\vee$ and $\Gamma = \gamma_\wedge \gamma_\vee$ be two covariances on the same set T of indices. Let [(Definition) 1.1.28] T_C and T_Γ be supports of C and Γ respectively, that is [(Proposition) 1.4.10], maximal sets of indices at which respectively r_C and r_Γ are strictly positive and decreasing. Let $\{t_1, t_2, t_3\} \subseteq T_C \cap T_\Gamma$, $t_1 < t_2 < t_3$ be fixed, but arbitrary (one thus supposes that three such points exist). Let C_3 and Γ_3 be the matrices with, for $\{i, j\} \subseteq [1 : 3]$, fixed, but arbitrary, respective entries $C(t_i, t_j)$ and $\Gamma(t_i, t_j)$. C_3 and Γ_3 have the following respective representations

A.F. Gualtierotti, *Detection of Random Signals in Dependent Gaussian Noise*,
DOI 10.1007/978-3-319-22315-5_3

[(Remark) 1.4.8]:

$$C_3 = D_{\wedge}^{(3)} L_3^{\star} D_{rC}^{(3)} L_3 D_{\wedge}^{(3)}, \quad \text{and} \quad \Gamma_3 = \Delta_{\wedge}^{(3)} L_3^{\star} \Delta_{r\Gamma}^{(3)} L_3 \Delta_{\wedge}^{(3)},$$

which may be written in the following respective forms:

$$C_3 = \left\{ D_{\wedge}^{(3)} L_3^{\star} \left\{ D_{rC}^{(3)} \right\}^{1/2} \right\} \left\{ D_{\wedge}^{(3)} L_3^{\star} \left\{ D_{rC}^{(3)} \right\}^{1/2} \right\}^{\star} = \tilde{C}_3 \tilde{C}_3^{\star},$$

$$\Gamma_3 = \left\{ \Delta_{\wedge}^{(3)} L_3^{\star} \left\{ \Delta_{r\Gamma}^{(3)} \right\}^{1/2} \right\} \left\{ \Delta_{\wedge}^{(3)} L_3^{\star} \left\{ \Delta_{r\Gamma}^{(3)} \right\}^{1/2} \right\}^{\star} = \tilde{\Gamma}_3 \tilde{\Gamma}_3^{\star}.$$

Consider the matrix equation which, when it has a solution, insures that Γ dominates C [37, p. 277], that is, $\tilde{C}_3 = \tilde{\Gamma}_3 X$: using for example c_1 for $c_\wedge(t_1)$, one has that

$$\begin{bmatrix} c_1 (r_1 - r_2)^{1/2} & c_1 (r_2 - r_3)^{1/2} & c_1 r_3^{1/2} \\ 0 & c_2 (r_2 - r_3)^{1/2} & c_2 r_3^{1/2} \\ 0 & 0 & c_3 r_3^{1/2} \end{bmatrix} =$$

$$\begin{bmatrix} \gamma_1 (\rho_1 - \rho_2)^{1/2} & \gamma_1 (\rho_2 - \rho_3)^{1/2} & \gamma_1 \rho_3^{1/2} \\ 0 & \gamma_2 (\rho_2 - \rho_3)^{1/2} & \gamma_2 \rho_3^{1/2} \\ 0 & 0 & \gamma_3 \rho_3^{1/2} \end{bmatrix} \begin{bmatrix} x_{1,1} & x_{1,2} & x_{1,3} \\ x_{2,1} & x_{2,2} & x_{2,3} \\ x_{3,1} & x_{3,2} & x_{3,3} \end{bmatrix}$$

X is thus an upper triangular matrix whose diagonal elements are respectively:

$$\frac{c_1}{\gamma_1} \left\{ \frac{r_1 - r_2}{\rho_1 - \rho_2} \right\}^{1/2}, \frac{c_2}{\gamma_2} \left\{ \frac{r_2 - r_3}{\rho_2 - \rho_3} \right\}^{1/2}, \frac{c_3}{\gamma_3} \left\{ \frac{r_3}{\rho_3} \right\}^{1/2}.$$

Since the eigenvalues of X are its diagonal elements [121, p. 193], and that the eigenvalues of $XX^{\star}$ are the squares of those diagonal elements [121, p. 194], using the following inequality [37, p. 277]: $XX^{\star} \leq \lambda_{max} (XX)^{\star} I$, one has that

$$C_3 \ll \max \left\{ \left\{ \frac{c_1}{\gamma_1} \right\}^2 \frac{r_1 - r_2}{\rho_1 - \rho_2}, \left\{ \frac{c_2}{\gamma_2} \right\}^2 \frac{r_2 - r_3}{\rho_2 - \rho_3}, \left\{ \frac{c_3}{\gamma_3} \right\}^2 \frac{r_3}{\rho_3} \right\} \Gamma_3.$$

One thus sees that domination of C by Γ is determined by the values of the following expression ($\theta > 0$):

$$\left\{ \frac{c_\wedge(t)}{\gamma_\wedge(t)} \right\}^2 \frac{r_C(t) - r_C(t+\theta)}{r_\Gamma(t) - r_\Gamma(t+\theta)}.$$

Analogously, one obtains expressions of the following form:

$$\left\{\frac{c_\vee(t)}{\gamma_\vee(t)}\right\}^2 \frac{r^C(t+\theta)-r^C(t)}{r^\Gamma(t+\theta)-r^\Gamma(t)},$$

which explain, when investigating the equivalence of Gaussian measures whose covariances have a factorization, the presence of conditions of the following type [146, p. 50]:

$$c_\vee^2\, dr^C = \gamma_\vee^2\, dr^\Gamma .$$

Proposition 3.1.4 *Let C_1, C_2 and C_3 be covariances on T. Then:*

1. *if C_1 dominates C_2 and C_2 dominates C_1, C_1 and C_2 are two notations for the same covariance;*
2. *if C_3 dominates C_2 which dominates C_1, C_3 dominates C_1.*

Proof To prove item 1, one may proceed as follows. When C_1 dominates C_2, for $t \in T$, fixed, but arbitrary, $C_1(t,t) \geq C_2(t,t)$. Thus, because of the assumption, $C_1(t,t) = C_2(t,t)$, $t \in T$. Now, whenever C_1 dominates C_2, $C = C_1 - C_2$ is a covariance and (by Cauchy-Schwarz's inequality: see (Proposition) 1.1.5)

$$C^2(t_1,t_2) \leq C(t_1,t_1)\, C(t_2,t_2) .$$

The right-hand side of the latter inequality is zero, and thus C is the zero covariance. For item 2, one has that $C_3 - C_1 = (C_3 - C_2) + (C_2 - C_1)$, which is, as the sum of two covariances, a covariance [(Proposition) 1.3.8]. □

Proposition 3.1.5 *Let C_1 and C_2 be covariances on T, with associated RKHS's $H(C_1,T)$ and $H(C_2,T)$ respectively. Suppose C_2 dominates C_1, and define $J_{2,1}$: $H(C_2,T) \longrightarrow H(C_1,T)$ using, for $t \in T$, fixed, but arbitrary, the following assignment:*

$$J_{2,1}\left[C_2(\cdot,t)\right] = C_1(\cdot,t) .$$

Then:

1. *$J_{2,1}$ can be extended to be a contraction.*
2. *$J_{2,1}^\star$ is the inclusion map of $H(C_1,T)$ into $H(C_2,T)$, so that, as sets, $H(C_1,T) \subseteq H(C_2,T)$. Furthermore, for $h \in H(C_1,T)$, fixed, but arbitrary,*

$$\|h\|_{H(C_2,T)} \leq \|h\|_{H(C_1,T)} .$$

   $J_{2,1}^\star$ is thus a contraction.
3. *The kernel of $J_{2,1}^\star J_{2,1}$ is C_1.*

Proof The fact that $J_{2,1}$ has a unique linear extension to $V[C_2]$ follows from a general result in linear algebra [46, p. 26]. Indeed, the required condition is that, for any $[n, \alpha, (t, T)]$, as soon as $\sum_{i=1}^{n} \alpha_i C_2(\cdot, t_i) = 0$,

$$\sum_{i=1}^{n} \alpha_i J_{2,1}[C_2(\cdot, t_i)] = 0.$$

But that follows from the assumption of domination as

$$\sum_{i=1}^{n}\sum_{j=1}^{n} \alpha_i\alpha_j C_1(t_i, t_j) \le \sum_{i=1}^{n}\sum_{j=1}^{n} \alpha_i\alpha_j C_2(t_i, t_j)$$

can be written in the following form:

$$\left\| \sum_{i=1}^{n} \alpha_i J_{2,1}[C_2(\cdot, t_i)] \right\|_{H(C_1,T)}^2 \le \left\| \sum_{i=1}^{n} \alpha_i C_2(\cdot, t_i) \right\|_{H(C_2,T)}^2. \qquad (\star)$$

Then

$$\sum_{i=1}^{n} \alpha_i J_{2,1}[C_2(\cdot, t_i)] = J_{2,1}\left[\sum_{i=1}^{n} \alpha_i C_2(\cdot, t_i) \right],$$

and $(\star)$ shows that $J_{2,1}$ can be extended to be a linear and bounded operator whose norm is less or equal to one. So item 1 obtains.

Now, for $(t, u) \in T \times T$, fixed, but arbitrary, on one hand

$$\langle J_{2,1}[C_2(\cdot, u)], C_1(\cdot, t)\rangle_{H(C_1,T)} = C_1(t, u),$$

and, on the other hand,

$$\langle J_{2,1}[C_2(\cdot, u)], C_1(\cdot, t)\rangle_{H(C_1,T)} = \left\langle C_2(\cdot, u), J_{2,1}^{\star}[C_1(\cdot, t)]\right\rangle_{H(C_2,T)}.$$

Consequently

$$C_1(u, t) = C_1(t, u) = J_{2,1}^{\star}[C_1(\cdot, t)](u), \text{ that is, } C_1(\cdot, t) = J_{2,1}^{\star}[C_1(\cdot, t)],$$

and $J_{2,1}^{\star}$ is the inclusion map. Since its norm is less than or equal to one [266, p. 71]), it is also a contraction. Thus, for $h \in H(C_1, T)$, fixed, but arbitrary, one has that

$$\|h\|_{H(C_2,T)} = \left\|J_{2,1}^{\star}[h]\right\|_{H(C_2,T)} \le \|h\|_{H(C_1,T)}.$$

The kernel $\mathcal{B}$ of an operator $B : H(\mathcal{K}, T_{\mathcal{K}}) \longrightarrow H(\mathcal{H}, T_{\mathcal{H}})$ is given by the following equality [(Remark) 1.7.2]:

$$\mathcal{B}\left(t^{\mathcal{K}}, t^{\mathcal{H}}\right) = \left\langle B\left[\mathcal{K}\left(\cdot, t^{\mathcal{K}}\right)\right], \mathcal{H}\left(\cdot, t^{\mathcal{H}}\right)\right\rangle_{H(\mathcal{H}, T_{\mathcal{H}})}.$$

Letting $\mathcal{K} = \mathcal{H} = C_2$, $T_{\mathcal{K}} = T_{\mathcal{H}} = T$, $B = J_{2,1}^{\star} J_{2,1}$, one gets:

$$\mathcal{B}(t_1, t_2) = \left\langle J_{2,1}^{\star} J_{2,1}\left[C_2(\cdot, t_1)\right], C_2(\cdot, t_2)\right\rangle_{H(C_2, T)} = C_1(t_1, t_2).$$

□

Remark 3.1.6 The range of $J_{2,1}$ is dense in $H(C_1, T)$. It is $H(C_1, T)$ if, and only if, $J_{2,1}^{\star}$ has bounded inverse [228, p. 97]. In that case, $H(C_1, T)$ is norm closed in $H(C_2, T)$, and then, on $H(C_1, T)$, both norms are equivalent.

Remark 3.1.7 The map $J_{2,2} = J_{2,1}^{\star} J_{2,1}$ may be looked at as the map of $H(C_2, T)$ resulting from the following assignment: $C_2(\cdot, t) \mapsto C_1(\cdot, t)$. It is then a positive contraction.

Remark 3.1.8 Since $\mathcal{N}[J_{2,1}] = \overline{\mathcal{R}[J_{2,1}^{\star}]}^{\perp}$ [266, p. 71], the null space of $J_{2,1}$ is the complement of the closure, in $H(C_2, T)$, of $H(C_1, T)$.

Remark 3.1.9 Let $H(C, T)$ be an RKHS, and H_0 be a subspace of $H(C, T)$. H_0 is an RKHS with kernel [(Proposition) 1.6.1]

$$\mathcal{H}_0(t_1, t_2) = P_{H_0}\left[C(\cdot, t_2)\right](t_1), \ (t_1, t_2) \in T \times T.$$

Furthermore, $\mathcal{H}_0$ is dominated by C. The map $J_0 : H(C, T) \longrightarrow H(\mathcal{H}_0, T)$ is defined using the following relation:

$$J_0\left[C(\cdot, t)\right] = \mathcal{H}_0(\cdot, t) = P_{H_0}\left[C(\cdot, t)\right].$$

Consequently $J_0 = P_{H_0}$.

Remark 3.1.10 Suppose that $C_0 \ll C_1$, and also that $C_0 \ll C_2$. Then, $H(C_1, T) \supseteq J_{1,0}^{\star}[H(C_0, T)]$, and $H(C_2, T) \supseteq J_{2,0}^{\star}[H(C_0, T)]$. Let $sH(C_0, T)$ be the family of functions which make up $H(C_0, T)$. Then:

$$H(C_1, T) \cap H(C_2, T) \supseteq sH(C_0, T).$$

Example 3.1.11 Let K be a real Hilbert space, C_1 and C_2 be two injective covariance operators on K, and U be the isometry between K and its dual. One has seen [(Example) 1.3.18] that, for example, the RKHS $H(\mathcal{K}_1, K)$ related to C_1, is the range of $UC_1^{1/2}$, with kernel $\mathcal{K}_1(k_1, k_2) = \langle C_1[k_1], k_2\rangle_K$. The requirement that $\mathcal{K}_1 \ll \mathcal{K}_2$ is equivalent to the requirement that $\langle C_1[k_1], k_2\rangle_K \leq \langle C_2[k_1], k_2\rangle_K$. But the latter is equivalent to [11] $\mathcal{R}[C_1^{1/2}] \subseteq \mathcal{R}[C_2^{1/2}]$, and to $C_1^{1/2} = C_2^{1/2}B$, where B is some bounded, linear operator of K. Furthermore:

$$B = BP_{\overline{\mathcal{R}[C_1^{1/2}]}} = P_{\overline{\mathcal{R}[C_2^{1/2}]}} BP_{\overline{\mathcal{R}[C_1^{1/2}]}} . \qquad (\star)$$

Let $J : H(\mathcal{K}_2, K) \longrightarrow H(\mathcal{K}_1, K)$ be defined using the following assignment:

$$J[UC_2^{1/2}[k]] = UC_2^{1/2}BB^\star[k] = UC_1^{1/2}B^\star[k].$$

Then [(Example) 1.3.18]:

$$\begin{aligned} J[\mathcal{K}_2(\cdot,k)] &= J[UC_2[k]] \\ &= J[UC_2^{1/2}[C_2^{1/2}[k]] \\ &= UC_2^{1/2}BB^\star C_2^{1/2}[k] \\ &= UC_1[k] \\ &= \mathcal{K}_1(\cdot,k). \end{aligned}$$

Thus J is $J_{\mathcal{K}_2,\mathcal{K}_1}$, and the properties of J and B are closely related. Suppose thus, for example, that K is separable, and that $\{e_n, n \in \mathbb{N}\}$ is a complete orthonormal set. Then [(Example) 1.3.18]:

$$\begin{aligned} \sum_n \left\|J_{\mathcal{K}_2,\mathcal{K}_1}[UC_2^{1/2}[e_n]]\right\|^2_{H(\mathcal{K}_1,K)} &= \sum_n \left\|UC_1^{1/2}[B^\star[e_n]]\right\|^2_{H(\mathcal{K}_1,K)} \\ &= \sum_n \left\|P_{\overline{\mathcal{R}[C_1^{1/2}]}} B^\star[e_n]\right\|^2_K \\ &\overset{(\star)}{=} \sum_n \|B^\star[e_n]\|^2_K , \end{aligned}$$

and $J_{\mathcal{K}_2,\mathcal{K}_1}$ is Hilbert-Schmidt if, and only if, B is.

Suppose that N is a Gaussian element with values in K, mean zero, and covariance C_N. Let S be a signal, with values in K, independent of N, with covariance C_S. Then, among sufficient conditions which insure that P_{S+N} and P_N, the probabilities related to $S + N$ and N, respectively, are mutually absolutely continuous, and thus that the related detection problem is not singular, one finds [11] $C_S = C_N^{1/2} D C_N^{1/2}$, D trace-class. That means, for B or $J_{\mathcal{K}_2,\mathcal{K}_1}$ above, to be Hilbert-Schmidt.

Example 3.1.12 Let T be an interval, say $T = [t_l, t_r]$, $t_l \geq 0$, and suppose that $C = c_\wedge c_\vee$ and $\Gamma = \gamma_\wedge \gamma_\vee$ are two covariances on T with a factorization. Assume that T is the support of both covariances. Write these as [(Remark) 1.4.16]

$$\begin{aligned} C(t_1,t_2) &= c_\vee(t_1)\, c_\vee(t_2)\, r^C(t_1 \wedge t_2), \\ \Gamma(t_1,t_2) &= \gamma_\vee(t_1)\, \gamma_\vee(t_2)\, r^\Gamma(t_1 \wedge t_2). \end{aligned}$$

For fixed, but arbitrary $\{t_1,t_2\} \subseteq T$, let

$$\mathcal{H}_C(t_1,t_2) = r^C(t_1 \wedge t_2), \quad \text{and} \quad \mathcal{H}_\Gamma(t_1,t_2) = r^\Gamma(t_1 \wedge t_2)$$

(they are reproducing kernels as the covariances of processes whose form is that of $W_{r^C(t)}$). Define $F_C : T \longrightarrow H(\mathcal{H}_C, T)$ using the assignment

$$F_C(t) = c_\vee(t)\, \mathcal{H}_C(\cdot,t).$$

F_Γ is defined analogously: $F_\Gamma(t) = \gamma_\vee(t)\, \mathcal{H}_\Gamma(\cdot,t)$. Then, for instance,

$$L_{F_C}[c_\vee(t)\, \mathcal{H}_C(\cdot,t)](\theta) = \langle c_\vee(t)\, \mathcal{H}_C(\cdot,t), F_C(\theta)\rangle_{H(\mathcal{H}_C,T)} = C(\theta,t),$$

so that the resulting RKHS is $H(C,T)$. Furthermore [(Proposition) 1.1.15]

$$L^\star_{F_C}[C(\cdot,t)] = F_C(t) = c_\vee(t)\, \mathcal{H}_C(\cdot,t).$$

Suppose that $\mathcal{H}_C \ll \mathcal{H}_\Gamma$, that is, for fixed, but arbitrary $[n,\alpha,(t,T)]$,

$$\left\| \sum_{i=1}^n \alpha_i I_{[t_l,t_i]} \right\|^2_{L_2(T,\mathcal{T},\tilde{\tau}_C)} \leq \left\| \sum_{i=1}^n \alpha_i I_{[t_l,t_i]} \right\|^2_{L_2(T,\mathcal{T},\tilde{\tau}_\Gamma)}.$$

Let

$$J : H(\mathcal{H}_\Gamma, T) \longrightarrow H(\mathcal{H}_C, T)$$

be the map $J_{2,1}$ of (Proposition) 3.1.5. Then

$$\begin{aligned} C(\cdot,t) &= L_{F_C}[c_\vee(t)\, \mathcal{H}_C(\cdot,t)] \\ &= L_{F_C}[c_\vee(t)\, J[\mathcal{H}_\Gamma(\cdot,t)]] \\ &= L_{F_C}\left[c_\vee(t)\, J\left[\gamma_\vee(t)^{-1}\gamma_\vee(t)\, \mathcal{H}_\Gamma(\cdot,t)\right]\right] \\ &= L_{F_C}\left[c_\vee(t)\, J\left[\gamma_\vee(t)^{-1} L^\star_{F_\Gamma}[\Gamma(\cdot,t)]\right]\right] \\ &= L_{F_C} J L^\star_{F_\Gamma}\left[c_\vee(t)\, \gamma_\vee(t)^{-1}\, \Gamma(\cdot,t)\right]. \end{aligned}$$

Let M be the following assignment:

$$M\left[\Gamma\left(\cdot,t\right)\right]=\frac{c_{\vee}\left(t\right)}{\gamma_{\vee}\left(t\right)}\Gamma\left(\cdot,t\right).$$

Since the $\Gamma\left(\cdot,t\right)$'s are linearly independent, M extends linearly uniquely to $V\left[\Gamma\right]$. Furthermore, for fixed, but arbitrary $\left[n,\alpha,\left(t,T\right)\right]$, $t_1<\cdots<t_n$, and matrices $\boldsymbol{r}^{\Gamma}$ and $D_{c_{\vee}}$ to be specified,

$$\begin{aligned}\left\|M\left[\sum_{i=1}^{n}\alpha_i\Gamma\left(\cdot,t_i\right)\right]\right\|^2_{H(\mathcal{H}_{\Gamma},T)}&=\sum_{i=1}^{n}\sum_{j=1}^{n}\alpha_i\alpha_j\frac{c_{\vee}\left(t_i\right)}{\gamma_{\vee}\left(t_i\right)}\frac{c_{\vee}\left(t_j\right)}{\gamma_{\vee}\left(t_j\right)}\Gamma\left(t_i,t_j\right)\\&=\sum_{i=1}^{n}\sum_{j=1}^{n}\alpha_i\alpha_jc_{\vee}\left(t_i\right)c_{\vee}\left(t_j\right)r^{\Gamma}\left(t_i\wedge t_j\right)\\&=\left\langle\boldsymbol{r}^{\Gamma}D_{c_{\vee}}\underline{\alpha},D_{c_{\vee}}\underline{\alpha}\right\rangle_{\mathbb{R}^n}.\end{aligned}$$

To see what happens, let $n=3$. Then

$$\boldsymbol{r}^{\Gamma}=\begin{bmatrix}r^{\Gamma}\left(t_1\right)&r^{\Gamma}\left(t_1\right)&r^{\Gamma}\left(t_1\right)\\r^{\Gamma}\left(t_1\right)&r^{\Gamma}\left(t_2\right)&r^{\Gamma}\left(t_2\right)\\r^{\Gamma}\left(t_1\right)&r^{\Gamma}\left(t_2\right)&r^{\Gamma}\left(t_3\right)\end{bmatrix},\quad D_{c_{\vee}}=\begin{bmatrix}c_{\vee}\left(t_1\right)&0&0\\0&c_{\vee}\left(t_2\right)&0\\0&0&c_{\vee}\left(t_3\right)\end{bmatrix}.$$

Writing r_i for $r^{\Gamma}\left(t_i\right)$, and c_i for $c_{\vee}\left(t_i\right)$, one has that $\boldsymbol{r}^{\Gamma}=\boldsymbol{r}_{\Gamma}\boldsymbol{r}_{\Gamma}^{\star}$ with

$$\boldsymbol{r}_{\Gamma}=\begin{bmatrix}r_1^{1/2}&0&0\\r_1^{1/2}&\left(r_2-r_1\right)^{1/2}&0\\r_1^{1/2}&\left(r_2-r_1\right)^{1/2}&\left(r_3-r_2\right)^{1/2}\end{bmatrix}.$$

Thus $D_{c_{\vee}}\boldsymbol{r}^{\Gamma}D_{c_{\vee}}=D_{c_{\vee}}\boldsymbol{r}_{\Gamma}\left\{D_{c_{\vee}}\boldsymbol{r}_{\Gamma}\right\}^{\star}$, and to check that it is dominated in terms of covariances by a multiple of $\boldsymbol{r}^{\Gamma}$, one must solve for X the following matrix equation: $D_{c_{\vee}}\boldsymbol{r}_{\Gamma}=\boldsymbol{r}_{\Gamma}X$ [37, p. 277]. The solution is a lower triangular matrix whose diagonal elements are those of $D_{c_{\vee}}$. Consequently, as in (Example) 3.1.3,

$$\begin{aligned}\left\langle\boldsymbol{r}^{\Gamma}D_{c_{\vee}}\underline{\alpha},D_{c_{\vee}}\underline{\alpha}\right\rangle_{\mathbb{R}^n}&\le\max\left\{c_1^2,c_2^2,c_3^2\right\}\left\langle\boldsymbol{r}^{\Gamma}\underline{\alpha},\underline{\alpha}\right\rangle_{\mathbb{R}^n}\\&=\max\left\{c_1^2,c_2^2,c_3^2\right\}\left\|\sum_{i=1}^{n}\alpha_i\Gamma\left(\cdot,t_i\right)\right\|^2_{H(\Gamma,T)}.\end{aligned}$$

Finally, as J, L_{F_C} and $L^{\star}_{F_\Gamma}$ are contractions [(Propositions) 1.1.15 and 3.1.5] when $c_\vee$ is bounded on T,

$$\begin{aligned}\sum_{i=1}^{n}\sum_{j=1}^{n}\alpha_i\alpha_j C\left(t_i,t_j\right) &= \left\|\sum_{i=1}^{n}\alpha_i C\left(\cdot,t_i\right)\right\|^2_{H(C,T)}\\
&= \left\|L_{F_C}JL^{\star}_{F_\Gamma}M\left[\sum_{i=1}^{n}\alpha_i\Gamma\left(\cdot,t_i\right)\right]\right\|^2_{H(C,T)}\\
&\leq \left\|L_{F_C}JL^{\star}_{F_\Gamma}\right\|^2\left\|M\left[\sum_{i=1}^{n}\alpha_i\Gamma\left(\cdot,t_i\right)\right]\right\|^2_{H(\mathcal{H}_\Gamma,T)}\\
&\leq \left\{\sup_T c^2_\vee\right\}\left\|\sum_{i=1}^{n}\alpha_i\Gamma\left(\cdot,t_i\right)\right\|^2_{H(\Gamma,T)}\\
&= \left\{\sup_T c^2_\vee\right\}\sum_{i=1}^{n}\sum_{j=1}^{n}\alpha_i\alpha_j\Gamma\left(t_i,t_j\right)\end{aligned}$$

Thus, using (Example) 1.3.12, $H(C,T)\subseteq H(\Gamma,T)$.

Corollary 3.1.13 *Let C_1 and C_2 be covariances on T, with associated RKHS's $H(C_1,T)$ and $H(C_2,T)$ respectively. Suppose C_2 dominates C_1, and let E be a weakly closed subset of $H(C_2,T)$. Then*

1. *$E\cap H(C_1,T)$ is a weakly closed subset of $H(C_1,T)$;*
2. *when E is furthermore convex (for example, a subspace), it follows that $E\cap H(C_1,T)$ is a strongly closed subset of $H(C_1,T)$.*

Proof Let $\{h_n,\ n\in\mathbb{N}\}\subseteq E\cap H(C_1,T)$ be a weak Cauchy sequence in $H(C_1,T)$, with weak limit $h_1\in H(C_1,T)$. Then, for fixed, but arbitrary t in T, $\{h_n(t),\ n\in\mathbb{N}\}\subseteq\mathbb{R}$ is a Cauchy sequence, so that, since [(Proposition) 3.1.5] $H(C_1,T)\subseteq H(C_2,T)$, for fixed, but arbitrary $t\in T$,

$$\left\{\langle h_n,C_2(\cdot,t)\rangle_{H(C_2,T)},\ n\in\mathbb{N}\right\}\subseteq\mathbb{R}$$

is a Cauchy sequence. For the same reason [(Proposition) 3.1.5],

$$\|h_n\|_{H(C_2,T)}\leq\|h_n\|_{H(C_1,T)},$$

and, weakly convergent sequences being bounded [266, p. 79], there exists κ in $[0,\infty[$ such that

$$\|h_n\|_{H(C_1,T)}\leq\kappa.$$

Consequently [8, p. 235] $\{h_n,\ n \in \mathbb{N}\} \subseteq E \subseteq H(C_2, T)$ is a weak Cauchy sequence with a limit $h_2 \in E$. But, for fixed, but arbitrary $t \in T$,

$$\begin{aligned}
h_2(t) &= \langle h_2, C_2(\cdot, t)\rangle_{H(C_2,T)} \\
&= \lim_n \langle h_n, C_2(\cdot, t)\rangle_{H(C_2,T)} \\
&= \lim_n h_n(t) \\
&= \lim_n \langle h_n, C_1(\cdot, t)\rangle_{H(C_1,T)} \\
&= \langle h_1, C_1(\cdot, t)\rangle_{H(C_1,T)} \\
&= h_1(t),
\end{aligned}$$

so that $h_1 \in E$, and thus $h_1 \in E \cap H(C_1, T)$. It is a general result that weakly closed convex sets are strongly closed [60, p. 126]. As $E \cap H(C_1, T)$ is convex as soon as E is, the proof is complete. □

Corollary 3.1.14 *Let C_1 and C_2 be covariances on T, with associated RKHS's $H(C_1, T)$ and $H(C_2, T)$ respectively. Suppose C_2 dominates C_1. Then, when $H(C_2, T)$ is separable, so is $H(C_1, T)$.*

Proof Let $\{e_n^{(2)},\ n \in \mathbb{N}\} \subseteq H(C_2, T)$ be a complete orthonormal set. Set, for $n \in \mathbb{N}$, fixed, but arbitrary,

$$e_n^{(1)} = J_{2,1}\left[e_n^{(2)}\right] \in H(C_1, T).$$

Suppose $h_1 \in H(C_1, T)$ and $\langle h_1, e_n^{(1)}\rangle_{H(C_1,T)} = 0,\ n \in \mathbb{N}$. Then

$$\langle J_{2,1}^{\star}[h_1], e_n^{(2)}\rangle_{H(C_2,T)} = 0,\ n \in \mathbb{N},$$

so that $J_{2,1}^{\star}[h_1] = 0$, that is, h_1 is the zero function. That means that the family

$$\{e_n^{(1)},\ n \in \mathbb{N}\}$$

is total in $H(C_1, T)$ [266, pp. 40,45], and thus that the latter is separable. □

Corollary 3.1.15 *Let $H(\mathcal{H}, T)$ be an RKHS, and Λ be a symmetric kernel on T. Λ corresponds to a linear, bounded, self-adjoint operator L on $H(\mathcal{H}, T)$ with*

- *lower bound $\alpha_L > -\infty$,*
- *upper bound $\beta_L < \infty$,*

if, and only if, $\alpha_L \mathcal{H} \ll \Lambda \ll \beta_L \mathcal{H}$.

Proof Suppose that L is a linear, bounded, and self-adjoint operator of $H(\mathcal{H}, T)$, with bounds α_L and β_L, and kernel Λ. Then [266, p. 187]

$$\alpha_L \|h\|^2_{H(\mathcal{H},T)} \leq \langle L[h], h\rangle_{H(\mathcal{H},T)} \leq \beta_L \|h\|^2_{H(\mathcal{H},T)},$$

that is, $L - \alpha_L I_{H(\mathcal{H},T)}$ and $\beta_L I_{H(\mathcal{H},T)} - L$ are positive operators. Since, for example,

$$\left\langle \left\{L - \alpha_L I_{H(\mathcal{H},T)}\right\}[\mathcal{H}(\cdot,t)], \mathcal{H}(\cdot,\theta)\right\rangle_{H(\mathcal{H},T)}$$

equals

$$\langle L[\mathcal{H}(\cdot,t)], \mathcal{H}(\cdot,\theta)\rangle_{H(\mathcal{H},T)} - \alpha_L \mathcal{H}(\theta,t) = \Lambda(\theta,t) - \alpha_L \mathcal{H}(\theta,t),$$

$\Lambda - \alpha_L \mathcal{H}$ is the kernel of $L - \alpha_L I_{H(\mathcal{H},T)}$, and, since the latter is positive, the former is positive definite because of (Proposition) 1.7.9. The condition is thus necessary.

The condition is also sufficient. One may assume that $\alpha_L < \beta_L$, for, otherwise, $\alpha_L \mathcal{H} = \Lambda$, and Λ corresponds to $\alpha_L I_{H(\mathcal{H},T)}$ [(Remark) 1.7.3]. Let thus

$$\mathcal{K} = (\beta_L - \alpha_L)^{-1}(\Lambda - \alpha_L \mathcal{H}).$$

Then, by assumption,

$$0 \ll \mathcal{K} \ll (\beta_L - \alpha_L)^{-1}(\beta_L \mathcal{H} - \alpha_L \mathcal{H}) = \mathcal{H}.$$

Because of (Proposition) 3.1.5, item 2, the kernel $\mathcal{K}$ determines an operator $L_\mathcal{K}$ of $H(\mathcal{H}, T)$ that is linear, bounded, self-adjoint, and positive, since $L_\mathcal{K} = J^\star_{\mathcal{H},\mathcal{K}} J_{\mathcal{H},\mathcal{K}}$, where $J_{\mathcal{H},\mathcal{K}} : \mathcal{H}(\cdot,t) \mapsto \mathcal{K}(\cdot,t)$. $L_\mathcal{K}$ is then a contraction as a product of contractions [(Proposition) 3.1.5], so that $\|L_\mathcal{K}\| \leq 1$. Now $\Lambda = (\beta_L - \alpha_L)\mathcal{K} + \alpha_L \mathcal{H}$, which is the kernel of the operator $L = (\beta_L - \alpha_L) L_\mathcal{K} + \alpha_L I_{H(\mathcal{H},T)}$, has the required bounds since $\|L_\mathcal{K}\| \leq 1$ and

$$\langle L[h], h\rangle_{H(\mathcal{H},T)} = (\beta_L - \alpha_L)\langle L_\mathcal{K}[h], h\rangle_{H(\mathcal{H},T)} + \alpha_L \|h\|^2_{H(\mathcal{H},T)}.$$

□

Proposition 3.1.16 *Let C_1 and C_2 be covariances on T, with associated RKHS's $H(C_1, T)$ and $H(C_2, T)$ respectively. Suppose C_2 dominates C_1. When $J_{2,1}$ is compact,*

1. *$H(C_1, T)$ is separable;*
2. *there exists a covariance C on T with $H(C, T)$ separable, and*

$$H(C,T) \subseteq H(C_2,T);$$

3. *$J_{2,1}[C(\cdot,t)] = C_1(\cdot,t),\ t \in T;$*

4. *the map* $J : H(C,T) \longrightarrow H(C_1,T)$ *defined using the following assignment:*

$$J[C(\cdot,t)] = C_1(\cdot,t)$$

is compact.

Proof When $J_{2,1}$ is compact, $\mathcal{R}[J_{2,1}]$ is separable [266, p. 129]. But then $\overline{\mathcal{R}[J_{2,1}]}$ is separable [266, p. 32], and, since [(Remark) 3.1.6]

$$\overline{\mathcal{R}[J_{2,1}]} = H(C_1,T),$$

the latter is separable.

Let now $H_0 = \overline{\mathcal{R}[J_{2,1}^\star]}$. As $J_{2,1}^\star$ is compact [266, p. 128], $\mathcal{R}[J_{2,1}^\star]$ is separable [266, p. 129], and thus so is H_0 [266, p. 32]. Let

$$C(x,t) = P_{H_0}[C_2(\cdot,t)](x).$$

The latter relation defines a reproducing kernel dominated by C_2 [(Proposition) 1.6.1], and, consequently [(Proposition) 3.1.5], $H(C,T) \subseteq H(C_2,T)$. But, since $H(C,T) = H_0$ [(Proposition) 1.6.1], $H(C,T)$ is separable. Finally

$$\begin{aligned}
\langle J_{2,1}[C(\cdot,t)], C_1(\cdot,x)\rangle_{H(C_1,T)} &= \\
&= \left\langle C(\cdot,t), J_{2,1}^\star[C_1(\cdot,x)]\right\rangle_{H(C_2,T)} \\
&= \left\langle P_{H_0}[C_2(\cdot,t)], J_{2,1}^\star[C_1(\cdot,x)]\right\rangle_{H(C_2,T)} \\
&= \left\langle C_2(\cdot,t), P_{H_0}\left[J_{2,1}^\star[C_1(\cdot,x)]\right]\right\rangle_{H(C_2,T)} \\
&= \left\langle C_2(\cdot,t), J_{2,1}^\star[C_1(\cdot,x)]\right\rangle_{H(C_2,T)} \\
&= \langle J_{2,1}[C_2(\cdot,t)], C_1(\cdot,x)\rangle_{H(C_1,T)}.
\end{aligned}$$

Thus $J_{2,1}[C(\cdot,t)] = J_{2,1}[C_2(\cdot,t)] = C_1(\cdot,t)$. Consequently $J = J_{2,1}P_{H_0}$ is compact as the composition of a compact operator with one that is bounded [266, p. 128]. □

The properties which follow cover the cases for which finite dimensional approximations are possible. Furthermore, the finite trace properties cover the question of which noises are admissible for which signals.

Consider again the situation for which C_1 and C_2 are covariances on T, with C_2 dominating C_1, and fix an arbitrary subset $T_0 \subseteq T$. If C_1^0 and C_2^0 denote the restrictions of, respectively, C_1 and C_2 to $T_0 \times T_0$, then C_2^0 dominates C_1^0, and one has thus [(Proposition) 3.1.5] the following maps:

$$J_{2,1} : H(C_2, T) \longrightarrow H(C_1, T),$$

$$J_{2,1}^0 : H\left(C_2^0, T_0\right) \longrightarrow H\left(C_1^0, T_0\right).$$

Let $H_1^0 \subseteq H(C_1, T)$ be the subspace generated by $\{C_1(\cdot, t_0),\ t_0 \in T_0\}$, and $H_2^0 \subseteq H(C_2, T)$ that which is generated by $\{C_2(\cdot, t_0),\ t_0 \in T_0\}$.

Let $J_1^0 : H(C_1, T) \longrightarrow H\left(C_1^0, T_0\right)$ be the map that restricts a function of $H(C_1, T)$ to the subset T_0 [(Proposition) 1.6.3]. It is a partial isometry, with initial set H_1^0, and final set, $H\left(C_1^0, T_0\right)$, with the property that

$$J_1^0 [C_1(\cdot, t_0)] = C_1^0(\cdot, t_0),\ t_0 \in T_0.$$

J_2^0 is defined analogously, and has, *mutatis mutandis,* the same properties as J_1^0.

Lemma 3.1.17 *Let H be a real Hilbert space, and A be an operator of H with finite trace.*

1. *For any unitary operator U of H, UA and AU have a finite trace norm equal to that of A.*
2. *For any partial isometry W of H, WA and AW have a finite trace norm smaller than, or equal to that of A.*

Proof Write $[A]$ for $(A^\star A)^{1/2}$. Let $\{e_i,\ i \in I\} \subseteq H$ be any complete orthonormal set. One has [235, pp. 37–39], for the trace norm $\tau(A)$, that

$$\tau(A) = t([A]) \text{ where } t(A) = \sum_{i\in I} \langle A[e_i], e_i\rangle_H.$$

If now $B = UA$, as $U^\star U = UU^\star = I_H$ [266, p. 87],

$$\tau(B) = t\left(\left[(UA)^\star (UA)\right]^{1/2}\right) = t\left([A^\star U^\star UA]^{1/2}\right) = t\left([A^\star A]^{1/2}\right) = \tau(A).$$

A has a representation in the following form [235, p. 42]:

$$\sum_{i\in I} \alpha_i \{e_i \otimes f_i\},$$

where $\{a \otimes b\}[h] = \langle h, b\rangle_H\, a$, $\alpha_i > 0$, $i \in I, |I| \leq \aleph_0$, and $\{e_i,\ i \in I\} \subseteq H$ as well as $\{f_i,\ i \in I\} \subseteq H$ are sets of orthonormal elements. Furthermore [235, p. 42], $\tau(A) = \sum_{i\in I} \alpha_i$. Consequently, as

$$AU[h] = \sum_{i\in I} \alpha_i \langle U[h], f_i\rangle_H\, e_i$$

$$= U\left\{\sum_{i\in I}\alpha_i \langle h, U^\star [f_i]\rangle_H U^\star [e_i]\right\}$$

$$= \left\{U\sum_i \alpha_i \{U^\star [e_i]\otimes U^\star [f_i]\}\right\}[h],$$

AU has a representation as UC, where C has a finite trace norm equal to that of A. It thus suffices to apply the result for UA to AU in the form UC.

The second statement follows from [235, p. 39], and the fact that partial isometries have norm one. □

Proposition 3.1.18 *Let C_1 and C_2 be covariances on T, with associated RKHS's $H(C_1, T)$ and $H(C_2, T)$ respectively. Assume that C_2 dominates C_1. Let $T_0 \subseteq T$ be a fixed, but arbitrary subset, C_1^0 and C_2^0 be the restrictions of respectively C_1 and C_2 to $T_0\times T_0$, and $J_{2,1}^0 : H\left(C_2^0, T_0\right) \longrightarrow H\left(C_1^0, T_0\right)$ be defined using the following relation:*

$$J_{2,1}^0\left[C_2^0(\cdot, t_0)\right] = C_1^0(\cdot, t_0),\ t_0 \in T_0.$$

Let $J_1^0[h] = h^{|T_0}$, $h \in H(C_1, T)$. J_2^0 is defined analogously. Then

$$J_1^0 J_{2,1}\left(J_2^0\right)^\star = J_{2,1}^0.$$

Furthermore, when

1. *$J_{2,1}$ is Hilbert-Schmidt, then $J_{2,1}^0$ is Hilbert-Schmidt;*
2. *$J_{2,1}$ has finite trace, then $J_{2,1}^0$ has finite trace, and, with $\tau\left(J_{2,1}^0\right)$ and $\tau(J_{2,1})$ denoting the respective trace norms of the corresponding operators, one has that*[235, p. 38]

$$\tau\left(J_{2,1}^0\right) \leq \tau(J_{2,1}).$$

When T_0 is a determining set [(Definition) 1.5.2] *for $H(C_2, T)$, $\tau\left(J_{2,1}^0\right) = \tau(J_{2,1})$.*

Proof By definition, for $(t_0, t) \in T_0 \times T$, fixed, but arbitrary,

$$\left\langle J_1^0[C_1(\cdot, t)], C_1^0(\cdot, t_0)\right\rangle_{H\left(C_1^0, T_0\right)} = C_1(t_0, t).$$

But, by definition of the adjoint,

$$\left\langle J_1^0[C_1(\cdot, t)], C_1^0(\cdot, t_0)\right\rangle_{H\left(C_1^0, T_0\right)} = \left\langle C_1(\cdot, t), \left(J_1^0\right)^\star\left[C_1^0(\cdot, t_0)\right]\right\rangle_{H(C_1, T)}.$$

Thus, for $t \in T$, fixed, but arbitrary,

$$\left(J_1^0\right)^{\star}\left[C_1^0(\cdot, t_0)\right](t) = C_1(t_0, t) = C_1(t, t_0).$$

In other words,

$$\left(J_1^0\right)^{\star}\left[C_1^0(\cdot, t_0)\right] = C_1(\cdot, t_0), \quad \left(J_2^0\right)^{\star}\left[C_2^0(\cdot, t_0)\right] = C_2(\cdot, t_0).$$

Consequently, for $(t_0, x_0) \in T_0 \times T_0$, fixed, but arbitrary, one has that

$$\begin{aligned} J_1^0 J_{2,1}\left(J_2^0\right)^{\star}\left[C_2^0(\cdot, t_0)\right](x_0) &= J_1^0 J_{2,1}\left[C_2(\cdot, t_0)\right](x_0) \\ &= J_1^0\left[C_1(\cdot, t_0)\right](x_0) \\ &= C_1^0(x_0, t_0). \end{aligned}$$

Thus $J_1^0 J_{2,1}\left(J_2^0\right)^{\star} = J_{2,1}^0$.

Then, if $J_{2,1}$ is Hilbert-Schmidt, $J_{2,1}^0$ is Hilbert-Schmidt as the product of a Hilbert-Schmidt operator by two bounded ones [235, p. 30]. The same is true if "is Hilbert-Schmidt" is replaced by "has finite trace" [235, p. 38]. The inequality on the trace norms follows from the fact that [235, p. 39]

$$\tau\left(J_{2,1}^0\right) = \tau\left(J_1^0 J_{2,1}\left(J_2^0\right)^{\star}\right) \leq \left\|J_1^0\right\| \left\|\left(J_2^0\right)^{\star}\right\| \tau\left(J_{2,1}\right) \leq \tau\left(J_{2,1}\right)$$

as the operators J_1^0 and J_2^0 are contractions [(Proposition) 1.6.3].

Finally, when T_0 is a determining set for $H(C_2, T)$, it is also a determining set for $H(C_1, T)$, since, as sets, $H(C_1, T) \subseteq H(C_2, T)$. But then J_1^0 and J_2^0 are unitary [(Proposition) 1.6.10], and, because of (Lemma) 3.1.17,

$$\tau\left(J_{2,1}^0\right) = \tau\left(J_{2,1}\right).$$

□

Corollary 3.1.19 *Let C_1 and C_2 be covariances on T, with associated RKHS's $H(C_1, T)$ and $H(C_2, T)$ respectively. Let $\mathcal{F}_T$ denote the family of finite subsets of T. When C_2 dominates C_1, $J_{2,1}$ has finite trace if, and only if,*

$$\sup_{T_0 \in \mathcal{F}_T}\left\{\tau\left(J_{2,1}^0\right)\right) < \infty,$$

and, in that case,

$$\tau\left(J_{2,1}\right) = \sup_{T_0 \in \mathcal{F}_T}\left\{\tau\left(J_{2,1}^0\right)\right\}.$$

Proof Suppose that $J_{2,1}$ has finite trace, that is, $\tau(J_{2,1}) < \infty$. Then, since $\tau\left(J_{2,1}^0\right) \leq \tau(J_{2,1})$ [(Proposition) 3.1.18],

$$\sup_{T_0 \in \mathcal{F}_T} \left\{\tau\left(J_{2,1}^0\right)\right\} < \infty.$$

Suppose now that $\sup_{T_0 \in \mathcal{F}_T} \left\{\tau\left(J_{2,1}^0\right)\right\} < \infty$. Since, for example, J_1^0 is a partial isometry with H_1^0 as initial set, and $H\left(C_1^0, T_0\right)$ as final set [(Proposition) 1.6.3], result [(Proposition) 3.1.18] yields that

$$\left(J_1^0\right)^\star J_1^0 J_{2,1}\left(J_2^0\right)^\star J_2^0 = \left(J_1^0\right)^\star J_{2,1}^0 J_2^0,$$

and thus [266, p. 86] that

$$P_{H_1^0} J_{2,1} P_{H_2^0} = \left(J_1^0\right)^\star J_{2,1}^0 J_2^0,$$

so that [235, p. 39]

$$\tau\left(P_{H_1^0} J_{2,1} P_{H_2^0}\right) \leq \tau\left(J_{2,1}^0\right).$$

In their respective Hilbert spaces the families

$$\left\{P_{H_i^\alpha},\ T_\alpha \in \mathcal{F}_T\right\},\ i = 1, 2,$$

are nets of projections such that, for $T_\alpha \subset T_\beta$, fixed, but arbitrary,

$$O_{H(C_i,T)} \leq P_{H_i^\alpha} \leq P_{H_i^\beta} \leq I_{H(C_i,T)}.$$

There is then [81, p. 108], in the respective spaces, an operator P_i, linear and bounded, which is the strong limit of the respective nets, and which has the property that

$$O_{H(C_i,T)} \leq P_{H_i^\alpha} \leq P_i \leq I_{H(C_i,T)}.$$

P_1 and P_2 are projections [266, p. 85], and, since $I_{H(C_i,T)} - P_i$, for $i = 1, 2$, is orthogonal to all $P_{H_i^\alpha}$'s, it is the zero operator. As

$$\begin{aligned}
&\left\|J_{2,1}\left[h^{(2)}\right] - P_{H_1^\alpha} J_{2,1} P_{H_2^\alpha}\left[h^{(2)}\right]\right\| = \\
&\quad = \left\|\left\{J_{2,1} - P_{H_1^\alpha} J_{2,1} + P_{H_1^\alpha} J_{2,1} - P_{H_1^\alpha} J_{2,1} P_{H_2^\alpha}\right\}\left[h^{(2)}\right]\right\| \\
&\quad \leq \left\|\left\{I_{H(C_1,T)} - P_{H_1^\alpha}\right\}\left[J_{2,1}\left[h^{(2)}\right]\right]\right\| + \left\|P_{H_1^\alpha} J_{2,1}\left\{I_{H(C_2,T)} - P_{H_2^\alpha}\right\}\left[h^{(2)}\right]\right\|,
\end{aligned}$$

it follows that

$$\lim_{T_0 \in \mathcal{F}_T} P_{H_1^0} J_{2,1} P_{H_2^0} = J_{2,1}$$

in the sense of strong convergence.

Let now

$$\left\{e_i^{(2)},\ i \in I\right\}$$

be an orthonormal basis of $H(C_2, T)$, and $I_0 \subseteq I$ be a finite subset. Write $[J_{2,1}]$ for

$$\left(J_{2,1}^{\star} J_{2,1}\right)^{1/2}.$$

Then, using $[J_{2,1}] = W^{\star} J_{2,1}$ (polar decomposition [266, p. 188])

$$\begin{aligned}
\sum_{i \in I_0} \left\| [J_{2,1}]^{1/2} [e_i] \right\|_{H(C_1,T)}^2 &= \sum_{i \in I_0} \left| \langle W^{\star} J_{2,1} [e_i], [e_i] \rangle_{H(C_2,T)} \right| \\
&= \lim_{T_0 \in \mathcal{F}_T} \sum_{i \in I_0} \left| \left\langle W^{\star} P_{H_1^0} J_{2,1} P_{H_2^0} [e_i], [e_i] \right\rangle_{H(C_2,T)} \right| \\
&\leq \sup_{T_0 \in \mathcal{F}_T} \tau\left(W^{\star} P_{H_1^0} J_{2,1} P_{H_2^0}\right) \\
&\leq \sup_{T_0 \in \mathcal{F}_T} \tau\left(P_{H_1^0} J_{2,1} P_{H_2^0}\right) \\
&\leq \sup_{T_0 \in \mathcal{F}_T} \tau\left(J_{2,1}^0\right).
\end{aligned}$$

The first of the latter inequalities is based on [61, p. 91], the second on (Lemma) 3.1.17, and the last by what has been acknowledged at the beginning of this proof. The family

$$\left\{\langle [J_{2,1}] [e_i], [e_i] \rangle_{H(C_2,T)},\ i \in I\right\}$$

is thus summable [210, p. 19], and, consequently, $J_{2,1}$ has finite trace. The equality of trace norm and supremum follows from the proof. □

Corollary 3.1.20 *Let C_1 and C_2 be covariances on T, with C_2 dominating C_1. Suppose that $H(C_2, T)$ is separable, and that $J_{2,1}$ has finite trace.*

Let T_0 be a Hamel subset of T [(Definition) 1.1.36], *and*

$$T_c = \{t_{c,1}, t_{c,2}, t_{c,3}, \ldots\}$$

be a subset of T_0 *which is dense in* (T_0, d_{C_2}) [(Definition) 1.1.35, (Corollaries) 1.5.11, 1.6.21]. *The symbol* $T_{c|n}$ *denotes the subset*

$$\{t_{c,1}, t_{c,2}, t_{c,3}, \ \ldots \ , t_{c,n}\},$$

and $\Sigma_n^{(1)}$ *and* $\Sigma_n^{(2)}$ *represent the matrices obtained when restricting respectively* C_1 *and* C_2 *to* $T_{c|n}$. *Then*

$$\tau(J_{2,1}) = \lim_n \tau\left(\Sigma_n^{(1)}\left\{\Sigma_n^{(2)}\right\}^{-1}\right).$$

Proof T_c is a determining set for $H(C_2, T)$ [(Corollary) 1.6.21]. One may then assume, because of (Proposition) 3.1.18, item 2, that $T = T_c$. The supremum in (Corollary) 3.1.19 may then be computed over the sets $T_{c|n}$. Now, by definition, the 0-exponent of (Corollary) 3.1.19 being replaced by $|T_{c|n}$,

$$J_{2,1}^{|T_{c|n}}\left[C_2^{|T_{c|n}}(\cdot, t_{c,i})\right](t_{c,j}) = C_1^{|T_{c|n}}(t_{c,j}, t_{c,i}). \qquad (\star)$$

For fixed, but arbitrary $i \in T_{c|n}$, the map $C_1^{|T_{c|n}}(\cdot, t_{c,i})$ is in fact a vector in $\mathbb{R}^n$ with components

$$C_1^{|T_{c|n}}(t_{c,1}, t_{c,i}), \ldots, C_1^{|T_{c|n}}(t_{c,n}, t_{c,i}).$$

One has thus a map that sends the columns of

$$\begin{bmatrix} C_2^{|T_{c|n}}(t_{c,1}, t_{c,1}) & \cdots & C_2^{|T_{c|n}}(t_{c,1}, t_{c,n}) \\ \vdots & & \vdots \\ C_2^{|T_{c|n}}(t_{c,n}, t_{c,1}) & \cdots & C_2^{|T_{c|n}}(t_{c,n}, t_{c,n}) \end{bmatrix}$$

to those of

$$\begin{bmatrix} C_1^{|T_{c|n}}(t_{c,1}, t_{c,1}) & \cdots & C_1^{|T_{c|n}}(t_{c,1}, t_{c,n}) \\ \vdots & & \vdots \\ C_1^{|T_{c|n}}(t_{c,n}, t_{c,1}) & \cdots & C_1^{|T_{c|n}}(t_{c,n}, t_{c,n}) \end{bmatrix}.$$

That map may be represented as a matrix, say M, and the initial equality $(\star)$ may be read as

$$M\Sigma_n^{(2)} = \Sigma_n^{(1)}.$$

Since $T_c \subseteq T_0$, a Hamel subset, $\Sigma_n^{(2)}$ is invertible [(Remark) 1.1.4], and thus

$$M = \Sigma_n^{(1)} \left\{ \Sigma_n^{(2)} \right\}^{-1}.$$

The trace of $J_{2,1}$ restricted to $T_{c|n}$ is that of M. □

The following proposition yields a purely analytic condition for (Definition) 4.2.8 to hold.

Proposition 3.1.21 *Let C_1 be a covariance over T whose RKHS is separable. There exists a covariance C_2 over T that dominates C_1, and for which the operator $J_{2,1}$: $H(C_2, T) \longrightarrow H(C_1, T)$, defined using the following relation:*

$$J_{2,1}\left[C_2(\cdot, t)\right] = C_1(\cdot, t),$$

is Hilbert-Schmidt if, and only if, there exists an injective Hilbert-Schmidt operator $J_1 : H(C_1, T) \longrightarrow H(C_1, T)$ such that the family $\{h_t,\ t \in T\}$ of solutions to the following family of equations:

$$\{C_1(\cdot, t) = J_1[h_t],\ t \in T\}$$

generates $H(C_1, T)$.

Proof *The condition that has been stated is necessary.*

Let indeed C_1 and C_2 be covariances on T, with associated RKHS's $H(C_1, T)$ and $H(C_2, T)$ respectively. Let C_2 dominate C_1, and the operator $J_{2,1}$ be Hilbert-Schmidt. One may then assume that $H(C_1, T)$ and $H(C_2, T)$ are separable [(Proposition) 3.1.16]. Furthermore $J_{2,1}^\star$ is Hilbert-Schmidt [266, p. 133], and, since the composition of Hilbert-Schmidt operators yields operators with finite trace [266, pp. 165–167], $J_{2,1}^\star J_{2,1}$ is a self-adjoint, positive operator of $H(C_2, T)$, with finite trace. As such it can be written in the following form [266, p. 163]:

$$J_{2,1}^\star J_{2,1}[h_2] = \sum_{i \in I} \lambda_i \langle h_2, e_i \rangle_{H(C_2,T)}\, e_i,$$

where

- $\{h_2, e_i,\ i \in I \subseteq \mathbb{N}\} \subseteq H(C_2, T)$,
- $\lambda_i > 0,\ \sum_{i \in I} \lambda_i < \infty$,
- $\langle e_i, e_j \rangle_{H(C_2,T)} = \delta_{i,j},\ (i,j) \in I \times I$.

One may assume, by restricting attention, if necessary, to $\mathcal{N}[J_{2,1}]^\perp$ [266, pp. 35,71], that $\{e_i,\ i \in I\}$ forms a complete orthonormal set in $H(C_2, T)$. Then

$$\begin{aligned} C_1(t_1, t_2) &= \langle C_1(\cdot, t_1), C_1(\cdot, t_2) \rangle_{H(C_1,T)} \\ &= \langle J_{2,1}[C_2(\cdot, t_1)], J_{2,1}[C_2(\cdot, t_2)] \rangle_{H(C_1,T)} \end{aligned}$$

$$= \left\langle J_{2,1}^{\star} J_{2,1} \left[C_2 \left(\cdot, t_1 \right) \right], C_2 \left(\cdot, t_2 \right) \right\rangle_{H(C_2,T)}$$
$$= \sum_{i \in I} \lambda_i \left\langle C_2 \left(\cdot, t_1 \right), e_i \right\rangle_{H(C_2,T)} \left\langle e_i, C_2 \left(\cdot, t_2 \right) \right\rangle_{H(C_2,T)}$$
$$= \sum_{i \in I} \lambda_i e_i \left(t_1 \right) e_i \left(t_2 \right).$$

One may notice that

$$C_1 \left(t_1, t_2 \right) - \lambda_n e_n \left(t_1 \right) e_n \left(t_2 \right) = \sum_{i \in I, i \neq n} \lambda_i e_i \left(t_1 \right) e_i \left(t_2 \right)$$

is positive definite, so that $e_n \in H \left(C_1, T \right)$ [(Proposition) 1.2.4, item 2]. Thus, since $J_{2,1}^{\star}$ is the inclusion map [(Proposition) 3.1.5],

$$J_{2,1}^{\star} J_{2,1} \left[e_n \right] = \lambda_n e_n = \lambda_n J_{2,1}^{\star} \left[e_n \right] = J_{2,1}^{\star} \left[\lambda_n e_n \right],$$

so that

$$J_{2,1}^{\star} \left[J_{2,1} \left[e_n \right] - \lambda_n e_n \right] = 0, \text{ or } J_{2,1} \left[e_n \right] = \lambda_n e_n.$$

Set $f_i = \lambda_i^{1/2} e_i$. Then

$$\left\langle f_i, f_j \right\rangle_{H(C_1,T)} = \lambda_i^{-1/2} \lambda_j^{-1/2} \left\langle \lambda_i e_i, \lambda_j e_j \right\rangle_{H(C_1,T)}$$
$$= \lambda_i^{-1/2} \lambda_j^{-1/2} \left\langle J_{2,1} \left[e_i \right], J_{2,1} \left[e_j \right] \right\rangle_{H(C_1,T)}$$
$$= \lambda_i^{-1/2} \lambda_j^{-1/2} \left\langle J_{2,1}^{\star} J_{2,1} \left[e_i \right], e_j \right\rangle_{H(C_2,T)}$$
$$= \lambda_i^{1/2} \lambda_j^{-1/2} \left\langle e_i, e_j \right\rangle_{H(C_2,T)}$$
$$= \delta_{i,j}.$$

Thus the f_i's are orthonormal in $H \left(C_1, T \right)$. They are also complete as, for $h \in H \left(C_1, T \right)$, fixed, but arbitrary,

$$0 = \left\langle h, f_i \right\rangle_{H(C_1,T)}$$
$$= \lambda_i^{-1/2} \left\langle h, \lambda_i e_i \right\rangle_{H(C_1,T)}$$
$$= \lambda_i^{-1/2} \left\langle h, J_{2,1} \left[e_i \right] \right\rangle_{H(C_1,T)}$$
$$= \lambda_i^{-1/2} \left\langle J_{2,1}^{\star} \left[h \right], e_i \right\rangle_{H(C_2,T)}.$$

The relation $J_1 \left[f_i \right] = \lambda_i^{1/2} f_i$ defines an injective, positive, self-adjoint, Hilbert-Schmidt operator of $H \left(C_1, T \right)$ [235, pp. 16,41], and $C_1 \left(\cdot, t \right) \in \mathcal{R}[J_1]$. Indeed, as, as seen above,

$$C_1(x,t) = \sum_{i \in I} \lambda_i e_i(x) e_i(t),$$

and furthermore that

$$\sum_{i \in I} e_i^2(t) = \sum_{i \in I} \langle e_i, C_2(\cdot,t)\rangle_{H(C_2,T)}^2 = C_2(t,t) < \infty,$$

one has that

$$C_1(\cdot,t) = \sum_{i \in I} \lambda_i e_i(t) e_i = \sum_{i \in I} e_i(t) J_1[f_i] = J_1\left[\sum_{i \in I} e_i(t) f_i\right].$$

The latter calculation shows also that the functions

$$\left\{ h_t = \sum_{i \in I} e_i(t) f_i,\ t \in T \right\}$$

are solutions of the equations $\{C_1(\cdot,t) = J_1[h],\ h \in H(C_1,T),\ t \in T\}$. It remains to prove that these solutions generate $H(C_1,T)$. To that end, let $h \in H(C_1,T)$ be fixed, but arbitrary, and suppose that, for $t \in T$, fixed, but arbitrary,

$$\langle h, h_t\rangle_{H(C_1,T)} = 0.$$

Using the definition of h_t, since

$$\sum_{i \in I} \langle h, f_i\rangle_{H(C_1,T)}^2 = \|h\|_{H(C_1,T)}^2 < \infty,$$

it follows that, for $t \in T$, as, as sets, $H(C_1,T) \subseteq H(C_2,T)$,

$$\begin{aligned} 0 &= \langle h, h_t\rangle_{H(C_1,T)} \\ &= \sum_{i \in I} e_i(t) \langle h, f_i\rangle_{H(C_1,T)} \\ &= \sum_{i \in I} \langle e_i, C_2(\cdot,t)\rangle_{H(C_2,T)} \langle h, f_i\rangle_{H(C_1,T)} \\ &= \left\langle \sum_{i \in I} \langle h, f_i\rangle_{H(C_1,T)}\, e_i, C_2(\cdot,t) \right\rangle_{H(C_2,T)}. \end{aligned}$$

But then $\sum_{i \in I} \langle h, f_i \rangle_{H(C_1,T)} e_i = h = 0_{H(C_2,T)}$, which means that, for $i \in I$, one obtains that $\langle h, f_i \rangle_{H(C_1,T)} = 0$, or $h = 0_{H(C_1,T)}$.

Proof The proposition's condition is also sufficient.

Suppose indeed that C_1 is a covariance over T for which $H(C_1, T)$ is separable, and that J_1 is an injective Hilbert-Schmidt operator of $H(C_1, T)$ such that the family of solutions $\{h_t,\ t \in T\}$ to the set of equations

$$\{C_1(\cdot, t) = J_1[h_t],\ t \in T\}$$

generates $H(C_1, T)$.

Define

$$F : T \longrightarrow H(C_1, T) \text{ using } F(t) = \|J_1\| h_t.$$

Because of (Proposition) 1.1.15, the range K_F of the map $L_F : H(C_1, T) \longrightarrow \mathbb{R}^T$, defined using the following assignment:

$$L_F[h](t) = \langle h, F(t) \rangle_{H(C_1,T)} = \|J_1\| \langle h, h_t \rangle_{H(C_1,T)},$$

is an RKHS, with reproducing kernel given, for fixed, but arbitrary (t_1, t_2) in $T \times T$, by the following formula:

$$\mathcal{K}_F(t_1, t_2) = \langle F(t_1), F(t_2) \rangle_{H(C_1,T)} = \|J_1\|^2 \langle h_{t_1}, h_{t_2} \rangle_{H(C_1,T)}.$$

Furthermore, since the family $\{h_t,\ t \in T\}$ generates $H(C_1, T)$, the condition $L_F[h](t) = 0,\ t \in T$, means that $h = 0_{H(C_1,T)}$, and, consequently, still because of (Proposition) 1.1.15, that

$$\langle L_F[h_1], L_F[h_2] \rangle_{K_F} = \langle h_1, h_2 \rangle_{H(C_1,T)}.$$

As

$$\begin{aligned}
\sum_{i=1}^{n} \sum_{j=1}^{n} \alpha_i \alpha_j C_1(t_i, t_j) &= \sum_{i=1}^{n} \sum_{j=1}^{n} \alpha_i \alpha_j \left\langle C_1(\cdot, t_i), C_1(\cdot, t_j) \right\rangle_{H(C_1,T)} \\
&= \sum_{i=1}^{n} \sum_{j=1}^{n} \alpha_i \alpha_j \left\langle J_1[h_{t_i}], J_1\left[h_{t_j}\right] \right\rangle_{H(C_1,T)} \\
&= \left\| J_1 \left[\sum_{i=1}^{n} \alpha_i h_{t_i} \right] \right\|_{H(C_1,T)}^2
\end{aligned}$$

$$\leq \|J_1\|^2 \sum_{i=1}^{n} \sum_{j=1}^{n} \alpha_i \alpha_j \left\langle h_{t_i}, h_{t_j} \right\rangle_{H(C_1,T)}$$

$$= \sum_{i=1}^{n} \sum_{j=1}^{n} \alpha_i \alpha_j \mathcal{K}_F \left(t_i, t_j\right),$$

C_1 is dominated by $\mathcal{K}_F$. Choose thus $C_2 = \mathcal{K}_F$, and $H(C_2, T) = K_F$. $J_{2,1}$ is then well defined. One has also that

$$L_F [\|J_1\| h_t] (x) = \langle \|J_1\| h_t, \|J_1\| h_x \rangle_{H(C_1,T)} = \mathcal{K}_F (x, t).$$

The relation $J_{2,1} [\mathcal{K}_F (\cdot, t)] = C_1 (\cdot, t)$ yields then $J_{2,1} [L_F [\|J_1\| h_t]] = J_1 [h_t]$, so that

$$\|J_1\| J_{2,1} L_F = J_1.$$

So, if $\{e_i, i \in I \subseteq \mathbb{N}\}$ is a complete orthonormal set in $H(C_1, T)$, the family $\{L_F [e_i], i \in I\}$ is a complete orthonormal set in K_F, and

$$\sum_{i \in I} \|J_{2,1} [L_F [e_i]]\|_{K_F}^2 = \frac{1}{\|J_1\|} \sum_{i \in I} \|J_1 [e_i]\|_{H(C_1,T)}^2 < \infty.$$

$J_{2,1}$ is thus Hilbert-Schmidt. □

Remark 3.1.22 Let $\{a \otimes b\} [x] = \langle x, b \rangle a$, and $J_1 = \sum_n \lambda_n \{f_n \otimes e_n\}$, with $\{e_n, n \in \mathbb{N}\}$ and $\{f_n, n \in \mathbb{N}\}$ complete orthonormal sets. The following couple of relations:

$$C_1 (t_1, t_2) = \langle J_1 [h_{t_1}], J_1 [h_{t_2}] \rangle_{H(C_1,T)}$$
$$= \sum_n \lambda_n^2 \langle h_{t_1}, e_n \rangle_{H(C_1,T)} \langle h_{t_2}, e_n \rangle_{H(C_1,T)},$$

$$C_2 (t_1, t_2) = \|J_1\|^2 \sum_n \langle h_{t_1}, e_n \rangle_{H(C_1,T)} \langle h_{t_2}, e_n \rangle_{H(C_1,T)},$$

yield the most general representations of covariances for which (Proposition) 3.1.21 is true.

Proposition 3.1.23 *Let C be a covariance on T whose associated RKHS is $H(C, T)$. Suppose $H \subseteq H(C, T)$ is a real Hilbert space of functions such that, for fixed, but arbitrary $h \in H$,*

$$\|h\|_{H(C,T)} \leq \|h\|_H. \qquad (\star)$$

H is then an RKHS whose kernel $\mathcal{H}$ is dominated by C.

Proof Since $H(C,T)$ is an RKHS, the evaluation maps are continuous linear functionals [(Proposition) 1.1.5, item 4], and thus, for fixed, but arbitrary $t \in T$,

$$|\mathcal{E}_t[h]| \leq \kappa(t)\, \|h\|_{H(C,T)}, \ h \in H(C,T).$$

The restriction of the latter inequality to H yields, using the assumption ($\star$) on the norms, that

$$|\mathcal{E}_t[h]| \leq \kappa(t)\, \|h\|_H, \ h \in H.$$

The evaluation maps are thus continuous linear functionals of H, which is thus an RKHS [(Proposition) 1.1.8]. Let $\mathcal{H}$ denote its reproducing kernel. Since, for $t \in T$, fixed, but arbitrary, $\mathcal{H}(\cdot,t) \in H \subseteq H(C,T)$, one has, for fixed, but arbitrary $(t_1,t_2) \in T \times T$, that

$$\mathcal{H}(t_1,t_2) = \langle \mathcal{H}(\cdot,t_2), C(\cdot,t_1)\rangle_{H(C,T)}.$$

Consequently, given $[n,\alpha,(t,T)]$, fixed, but arbitrary,

$$\left\{\sum_{i=1}^{n}\sum_{j=1}^{n}\alpha_i\alpha_j\mathcal{H}(t_i,t_j)\right\}^2 = \left\langle \sum_{i=1}^{n}\alpha_i\mathcal{H}(\cdot,t_i), \sum_{i=1}^{n}\alpha_i C(\cdot,t_i)\right\rangle_{H(C,T)}^2.$$

The right-hand side of the latter equality is dominated by

$$\left\|\sum_{i=1}^{n}\alpha_i\mathcal{H}(\cdot,t_i)\right\|_{H(C,T)}^2 \left\|\sum_{i=1}^{n}\alpha_i C(\cdot,t_i)\right\|_{H(C,T)}^2.$$

Since

$$\left\|\sum_{i=1}^{n}\alpha_i\mathcal{H}(\cdot,t_i)\right\|_{H(C,T)}^2 \leq \left\|\sum_{i=1}^{n}\alpha_i\mathcal{H}(\cdot,t_i)\right\|_{H}^2 = \sum_{i=1}^{n}\sum_{j=1}^{n}\alpha_i\alpha_j\mathcal{H}(t_i,t_j),$$

and that

$$\left\|\sum_{i=1}^{n}\alpha_i C(\cdot,t_i)\right\|_{H(C,T)}^2 = \sum_{i=1}^{n}\sum_{j=1}^{n}\alpha_i\alpha_j C(t_i,t_j),$$

one may cancel the term $\sum_{i=1}^{n}\sum_{j=1}^{n}\alpha_i\alpha_j\mathcal{H}(t_i,t_j)$, when it is different from zero. But, when it is zero, there is nothing to prove. Thus C dominates $\mathcal{H}$. □

Corollary 3.1.24 *For covariances over T, one has that $C_1 \ll C_2$ if, and only if, as manifolds, $H(C_1, T) \subseteq H(C_2, T)$, and, for $h \in H(C_1, T)$, fixed, but arbitrary,*

$$\|h\|_{H(C_2,T)} \leq \|h\|_{H(C_1,T)}\,.$$

When useful, that situation shall be denoted $H(C_1, T) \sqsubseteq H(C_2, T)$.

Corollary 3.1.25 *Let C be a covariance on T whose associated RKHS is $H(C, T)$. Suppose $H \subseteq H(C, T)$ is a real Hilbert space of functions such that, for fixed, but arbitrary $h \in H$,*

$$\|h\|_{H(C,T)} \leq \|h\|_H\,.$$

Let [(Proposition) 3.1.23] *$H = H(\mathcal{H}, T)$, and $\mathcal{H} \ll C$. Let, mutatis, mutandis, $J_{C,\mathcal{H}}$ be the $J_{2,1}$ map defined in* (Proposition) 3.1.5. *The map*

$$J = I_{H(C,T)} - J^{\star}_{C,\mathcal{H}} J_{C,\mathcal{H}}$$

is positive as $J_{C,\mathcal{H}}$ and its adjoint are contractions, so that it makes sense to compute its square root. One has that

$$K = \mathcal{R}[J^{1/2}]$$

is an RKHS, with reproducing kernel $\mathcal{K} = C - \mathcal{H}$, and norm

$$\left\|J^{1/2}[h]\right\|_K = \left\|P_{\mathcal{N}[J^{1/2}]^{\perp}}[h]\right\|_{H(C,T)}\,.$$

Proof Since [266, pp. 35,72]

$$\overline{\mathcal{R}[J^{1/2}]} = \mathcal{N}[J^{1/2}]^{\perp},$$

the following relation:

$$\left\|J^{1/2}[h]\right\|_K = \left\|P_{\mathcal{N}[J^{1/2}]^{\perp}}[h]\right\|_{H(C,T)}$$

says that $J^{1/2}$ restricted to K is unitary, and thus that K is a Hilbert space. Using the fact that $J^{\star}_{C,\mathcal{H}}$ is an inclusion,

$$J[C(\cdot,\theta)](t) = C(t,\theta) - \mathcal{H}(t,\theta)\,,$$

and, for $t \in T$, fixed, but arbitrary,

$$C(\cdot,t) - \mathcal{H}(\cdot,t) = J^{1/2}\left[J^{1/2}[C(\cdot,t)]\right] \in K.$$

Furthermore, for $k \in K$, $k = J^{1/2}[h]$, fixed, but arbitrary,

$$\begin{aligned} k(t) &= \langle C(\cdot,t), k\rangle_{H(C,T)} \\ &= \left\langle C(\cdot,t), J^{1/2}[h]\right\rangle_{H(C,T)} \\ &= \left\langle J^{1/2}[C(\cdot,t)], h\right\rangle_{H(C,T)} \\ &= \left\langle P_{\mathcal{N}[J^{1/2}]^{\perp}}\left[J^{1/2}[C(\cdot,t)]\right], P_{\mathcal{N}[J^{1/2}]^{\perp}}[h]\right\rangle_{H(C,T)} \\ &= \left\langle J^{1/2}\left[J^{1/2}[C(\cdot,t)]\right], J^{1/2}[h]\right\rangle_{K} \\ &= \langle C(\cdot,t) - \mathcal{H}(\cdot,t), k\rangle_K . \end{aligned}$$

$C - \mathcal{H}$ is thus a reproducing kernel for K. □

Remark 3.1.26 There is, in (Corollary) 3.1.25, a direct way to obtain $J_{C,\mathcal{H}}$ which does not transit through (Proposition) 3.1.5.

Let indeed $h \in H(C,T)$ be fixed, but arbitrary, and let $\Lambda_h : H(\mathcal{H},T) \longrightarrow \mathbb{R}$ be the continuous linear functional on $H(C,T)$ determined by h:

$$\Lambda_h[k] = \langle k, h\rangle_{H(C,T)} .$$

As, for $k \in H$,

$$|\Lambda_h[k]| \leq \|k\|_{H(C,T)} \|h\|_{H(C,T)} \leq \|k\|_H \|h\|_{H(C,T)} ,$$

Λ_h is a continuous linear functional of H, and there exists a unique $k[h] \in H$ for which, for $k \in H$, fixed, but arbitrary,

$$\Lambda_h[k] = \langle k, k[h]\rangle_H .$$

Thus the assignment $B[h] = k[h]$ yields $B : H(C,T) \longrightarrow H$, a well-defined linear map. By definition

$$\langle B[h], k\rangle_H = \langle h, k\rangle_{H(C,T)} .$$

Furthermore, with

$$S = \left\{h \in H(C,T), k \in H, \|h\|_{H(C,T)} = \|k\|_H = 1\right\},$$

one has that [266, p. 60]

$$\begin{aligned} \|B\| &= \sup_S |\langle B[h], k\rangle_H| \\ &= \sup_S \left|\langle h, k\rangle_{H(C,T)}\right| \end{aligned}$$

$$\leq \sup_S \|h\|_{H(C,T)} \|k\|_{H(C,T)}$$
$$\leq \sup_S \|k\|_H$$
$$= 1.$$

B is thus a contraction. Since

$$\langle h, B^\star [k] \rangle_{H(C,T)} = \langle B [h], k \rangle_H = \langle h, k \rangle_{H(C,T)},$$

$B^\star$ is the inclusion map. As, for $t \in T$, fixed, but arbitrary,

$$\begin{aligned} B [h] (t) &= \langle B [h], \mathcal{H} (\cdot, t) \rangle_{H(\mathcal{H},T)} \\ &= \langle h, B^\star [\mathcal{H} (\cdot, t)] \rangle_{H(C,T)} \\ &= \langle h, \mathcal{H} (\cdot, t) \rangle_{H(C,T)}, \end{aligned}$$

choosing $C (\cdot, \theta)$ for h,

$$B [C (\cdot, \theta)] (t) = \langle C (\cdot, \theta), \mathcal{H} (\cdot, t) \rangle_{H(C,T)} = \mathcal{H} (\theta, t) = \mathcal{H} (t, \theta).$$

B is thus the map $J_{C,\mathcal{H}}$ of (Proposition) 3.1.5.

Remark 3.1.27 Let $h \in H (C, T)$, be fixed, but arbitrary. Then, by definition,

$$\begin{aligned} \|J [h]\|_K^2 &= \left\|J^{1/2} \left[J^{1/2} [h]\right]\right\|_K^2 \\ &= \left\|J^{1/2} [h]\right\|_{H(C,T)}^2 \\ &= \|h\|_{H(C,T)}^2 - \|J_{C,\mathcal{H}} [h]\|_{H(\mathcal{H},T)}^2. \end{aligned}$$

In particular,

$$\begin{aligned} \|h\|_{H(C,T)}^2 &= \|J_{C,\mathcal{H}} [h]\|_{H(\mathcal{H},T)}^2 + \|J [h]\|_K^2 \\ &= \|J_{C,\mathcal{H}} [h]\|_{H(\mathcal{H},T)}^2 + \left\|\left\{I_{H(C,T)} - J_{C,\mathcal{H}}^\star J_{C,\mathcal{H}}\right\} [h]\right\|_{H(C-\mathcal{H},T)}^2, \end{aligned}$$

which looks like the result of an orthogonal decomposition. The drawback of that decomposition is that a third RKHS must be considered. In the next chapter that complication will vanish.

Remark 3.1.28 The range of J is dense in K. Indeed, K is the range of $J^{1/2}$, J and $J^{1/2}$ have the same range closure [162, p. 27], and $\mathcal{R}[J] \subseteq \mathcal{R}[J^{1/2}]$.

Remark 3.1.29 ([105, p. 259]) $\mathcal{R}[J]$ and $\mathcal{R}[J^{1/2}]$ are simultaneously closed, and, when they are, are equal. It should be noted that closed range for compact operators means range of finite dimension [228, p. 98].

Suppose that the range of J is closed. Since

$$\mathcal{R}[J] \subseteq \mathcal{R}[J^{1/2}] \subseteq \overline{\mathcal{R}[J^{1/2}]} = \overline{\mathcal{R}[J]},$$

the range of $J^{1/2}$ is closed.

Suppose that $\mathcal{R}[J^{1/2}]$ is closed, and that $J[h_n] \to h$. Then

$$\left\langle J^{1/2}[h_n], J^{1/2}[k]\right\rangle_H \to \langle h, k\rangle_H,$$

that is,

$$\left\{J^{1/2}[h_n], n \in \mathbb{N}\right\}$$

is weakly convergent on $\mathcal{R}[J^{1/2}]$. Let $J^{1/2}[k_0]$ be its unique weak limit. Then, weakly, $J[h_n] \to J[k_0]$. Thus $h = J[k_0]$, and the range of J is closed.

When $\mathcal{R}[J]$ and $\mathcal{R}[J^{1/2}]$ are closed, because they have the same closure, they are equal. Suppose now that $\mathcal{R}[J] = \mathcal{R}[J^{1/2}]$. By Douglas's theorem [80], for some invertible operator B,

$$J^{1/2} = J^{1/2} B J^{1/2}.$$

Consequently, when $\lim_n J^{1/2}[h_n] = h$, $h = J^{1/2}B[h]$, and the range of $J^{1/2}$ is closed, that is,

$$\overline{\mathcal{R}[J^{1/2}]} = \mathcal{R}[J^{1/2}].$$

Then

$$\mathcal{R}[J] = \mathcal{R}[J^{1/2}] = \overline{\mathcal{R}[J^{1/2}]} = \overline{\mathcal{R}[J]},$$

and $\mathcal{R}[J]$ is closed.

To establish closure of range, one may sometimes use the Banach closed range theorem [277, p. 210], which says that a bounded linear operator B, from H_1 to H_2, has closed range if, and only if, there is a $\kappa > 0$ such that, for $h \in H_1$, fixed, but arbitrary,

$$\|B[h]\|_{H_2} \geq \kappa\, \delta_{H_1}(h, \mathcal{N}[B]),$$

where δ_{H_1} denotes distance in H_1.

Remark 3.1.30 Because of [228, pp. 96–97], in (Corollary) 3.1.25, $\mathcal{R}[J] = H(C,T)$ if, and only if,

$$\|J[h]\|_{H(C,T)} \geq \alpha\, \|h\|_{H(C,T)},$$

some $\alpha > 0$ (and then J has closed range). Consequently [(Fact) 2.7.3], for $t < \alpha^2$,

$$E_t^J = O_{H(C,T)},$$

the lower end of the spectrum is at least α^2, and 0 is not in the spectrum of J. Since, letting

$$S = \left\{h \in H(C,T) : \|h\|_{H(C,T)} = 1\right\},$$

one has that

$$0 < m = \inf_S \langle J[h], h\rangle_{H(C,T)} \leq \|h\|^2_{H(C,T)} - \|J_{C,\mathcal{H}}[h]\|^2_{H(C,T)},$$

it follows that

$$\left\langle J^\star_{C,\mathcal{H}} J_{C,\mathcal{H}}[h], h\right\rangle_{H(C,T)} + m \leq \|h\|^2_{H(C,T)},$$

and 1 is not in the spectrum of $J^\star_{C,\mathcal{H}} J_{C,\mathcal{H}}$.

Usually the inclusion of $\mathcal{R}[J]$ in $\mathcal{R}[J^{1/2}]$ is strict. Here is an example for which

$$\|h\|^2_{H(C,T)} - \|J_{C,\mathcal{H}}[h]\|^2_{H(C,T)} = 0.$$

Example 3.1.31 Let f and g be two linearly independent, bounded, measurable functions, with domain $T = [0, 1]$. Let

$$F(t) = \int_0^t f(\theta)\, d\theta,$$

G be defined analogously, and

$$h = \alpha F + \beta G.$$

Let $H = H(\mathcal{H}, T)$ be built as in (Example) 1.1.22, with

$$M^{-1} = \begin{pmatrix} m_1 & m_0 \\ m_0 & m_2 \end{pmatrix},$$

and $m_1 m_2 > m_0^2$. Then

$$\|h\|^2_{H(\mathcal{H},T)} = \left\langle M^{-1}\begin{bmatrix}\alpha \\ \beta\end{bmatrix}, \begin{bmatrix}\alpha \\ \beta\end{bmatrix}\right\rangle_{\mathbb{R}^2} = m_1\alpha^2 + 2m_0\alpha\beta + m_2\beta^2.$$

Also (omitting the usual distinction between a function and its equivalence class, and letting C_W be the covariance of the standard Wiener process),

$$\begin{aligned}\|h\|^2_{H(C_W,T)} &= \int_0^1 \{\alpha f(\theta) + \beta g(\theta)\}^2 \, d\theta \\ &= \alpha^2 \|f\|^2_{L_2[0,1]} + 2\alpha\beta \langle f, g\rangle_{L_2[0,1]} + \beta^2 \|g\|^2_{L_2[0,1]} \\ &= \left\langle \begin{pmatrix} \|f\|^2_{L_2[0,1]} & \langle f, g\rangle_{L_2[0,1]} \\ \langle f, g\rangle_{L_2[0,1]} & \|g\|^2_{L_2[0,1]} \end{pmatrix} \begin{bmatrix} \alpha \\ \beta \end{bmatrix}, \begin{bmatrix} \alpha \\ \beta \end{bmatrix} \right\rangle_{\mathbb{R}^2}.\end{aligned}$$

The requirement that

$$\|h\|_{H(C_W,T)} \leq \|h\|_{H(\mathcal{H},T)}$$

becomes

$$\begin{pmatrix} m_1 & m_0 \\ m_0 & m_2 \end{pmatrix} - \begin{pmatrix} \|f\|^2_{L_2[0,1]} & \langle f, g\rangle_{L_2[0,1]} \\ \langle f, g\rangle_{L_2[0,1]} & \|g\|^2_{L_2[0,1]} \end{pmatrix}$$

is positive definite, and that means two further requirements:

- $m_1 \geq \|f\|^2_{L_2[0,1]}$,
- $\left\{m_1 - \|f\|^2_{L_2[0,1]}\right\}\left\{m_2 - \|g\|^2_{L_2[0,1]}\right\} \geq \left\{m_0 - \langle f, g\rangle_{L_2[0,1]}\right\}^2$.

It follows that one must also have $m_2 \geq \|g\|^2_{L_2[0,1]}$.

$\mathcal{H}$ is computed as follows, letting $\mu = \left\{m_1 m_2 - m_0^2\right\}^{-1}$:

$$\begin{aligned}\mathcal{H}(x,t) &= \mu \left\langle \begin{pmatrix} m_2 & -m_0 \\ -m_0 & m_1 \end{pmatrix} \begin{bmatrix} F(x) \\ G(x) \end{bmatrix}, \begin{bmatrix} F(t) \\ G(t) \end{bmatrix} \right\rangle_{\mathbb{R}^2} \\ &= \mu \{[m_2 F(x) - m_0 G(x)] F(t) + [m_1 G(x) - m_0 F(x)] G(t)\} \\ &= \mu \{[m_2 F(t) - m_0 G(t)] F(x) + [m_1 G(t) - m_0 F(t)] G(x)\}.\end{aligned}$$

Since $F(t) = \left\langle \chi_{[0,t]}, f\right\rangle_{L_2[0,1]}$, the map

$$J_{C_W,\mathcal{H}}\left[C_W(\cdot,t)\right] = \mathcal{H}(\cdot,t)$$

has the following representation:

$$\begin{aligned}\mu^{-1} J_{C_W,\mathcal{H}}\left[\int_0^{\cdot} \chi_{[0,t]}(\theta)\, d\theta\right] &= \\ &= \left\langle \chi_{[0,t]}, m_2 f - m_0 g\right\rangle_{L_2[0,1]} F + \left\langle \chi_{[0,t]}, m_1 g - m_0 f\right\rangle_{L_2[0,1]} G\end{aligned}$$

which extends to

$$\mu^{-1} J_{C_W,\mathcal{H}} \left[\int_0^{\cdot} \phi(\theta)\, d\theta \right] =$$
$$= \langle \phi, m_2 f - m_0 g \rangle_{L_2[0,1]} F + \langle \phi, m_1 g - m_0 f \rangle_{L_2[0,1]} G.$$

Let $\tilde{\phi} = \phi \, \|\phi\|^{-1}_{L_2[0,1]}$. Then

$$\mu^{-2} \left\| J^{\star}_{C_W,\mathcal{H}} J_{C_W,\mathcal{H}} \left[\int_0^{\cdot} \tilde{\phi}(\theta)\, d\theta \right] \right\|^2_{H(C_W,T)} =$$
$$= \|f\|^2_{L_2[0,1]} \left\langle \tilde{\phi}, m_2 f - m_0 g \right\rangle^2_{L_2[0,1]}$$
$$+ 2 \langle f, g \rangle^2_{L_2[0,1]} \left\langle \tilde{\phi}, m_2 f - m_0 g \right\rangle_{L_2[0,1]} \left\langle \tilde{\phi}, m_1 g - m_0 f \right\rangle_{L_2[0,1]}$$
$$+ \|g\|^2_{L_2[0,1]} \left\langle \tilde{\phi}, m_1 g - m_0 f \right\rangle^2_{L_2[0,1]}.$$

Choose

- $\tilde{\phi} = 1, f = \lambda$,
- over $\left[0, \frac{1}{2}\right]$, $g = 1$,
- over $\left]\frac{1}{2}, 1\right]$, $g = -1$.

The expression for the norm yields then

$$\left\| J^{\star}_{C_W,\mathcal{H}} J_{C_W,\mathcal{H}} \left[\int_0^{\cdot} \tilde{\phi}(\theta)\, d\theta \right] \right\|^2_{H(C_W,T)} = \mu^2 \left\{ m_2^2 \lambda^4 + m_0^2 \lambda^2 \right\}.$$

The relation $\mu^2 \left\{ m_2^2 \lambda^4 + m_0^2 \lambda^2 \right\} = 1$ requires that

$$\lambda^2 = \frac{2}{\mu^2 m_0^2 + \left\{ \mu^4 m_0^4 + 2\mu^2 m_2^2 \right\}^{1/2}}.$$

Choosing, for example, integers n for m_0^{-1} and p for m_1 and m_2, one obtains a λ that decreases and the relations

$$m_1 m_2 - m_0^2 > 0, (m_1 - \lambda^2)(m_2 - 1) \geq m_0^2$$

obtain.

Corollary 3.1.32 *Let $H(C, T)$ be an RKHS. To each decomposition of the covariance C into the sum of two covariances, say $C = C_1 + C_2$, there corresponds a decomposition of $I_{H(C,T)}$ into the sum $I_{H(C,T)} = J_1 + J_2$ of linear, bounded, positive, and self-adjoint operators such that:*

1. $J_1 = J^{\star}_{C,C_1} J_{C,C_1}$, and $J_1[h](t) = \langle h, C_1(\cdot, t) \rangle_{H(C,T)}$;

2. $J_2 = J^{\star}_{C,C_2} J_{C,C_2}$, *and* $J_2 [h] (t) = \langle h, C_2 (\cdot, t)\rangle_{H(C,T)}$;
3. $\mathcal{R}[J_1^{1/2}] = H (C_1, T)$, *and* $\mathcal{R}[J_2^{1/2}] = H (C_2, T)$;
4. *the restriction of* $J_1^{1/2}$ *to* $\mathcal{N}[J_1]^{\perp}$ *is a unitary map from* $\mathcal{N}[J_1]^{\perp}$ *onto* $H (C_1, T)$ *so that, when* $h \in \mathcal{N}[J_1]^{\perp}$,

$$\left\|J_1^{1/2} [h]\right\|_{H(C_1,T)} = \|h\|_{H(C,T)} ;$$

5. *the restriction of* $J_2^{1/2}$ *to* $\mathcal{N}[J_2]^{\perp}$ *is a unitary map from* $\mathcal{N}[J_2]^{\perp}$ *onto* $H (C_2, T)$, *so that, when* $h \in \mathcal{N}[J_2]^{\perp}$,

$$\left\|J_2^{1/2} [h]\right\|_{H(C_2,T)} = \|h\|_{H(C,T)} .$$

Conversely, to each decomposition of $I_{H(C,T)}$ *into the sum*

$$I_{H(C,T)} = J_1 + J_2$$

of linear, bounded, positive, and self-adjoint operators, there exist RKHS's $H (C_1, T)$ *and* $H (C_2, T)$ *whose norms are defined above. The corresponding kernels are obtained as* $C_1 (\cdot, t) = J_1 [C (\cdot, t)]$, *and* $C_2 (\cdot, t) = J_2 [C (\cdot, t)]$, *where* $C = C_1 + C_2$.

Proof Suppose that $C = C_1 + C_2$. Then $C_1 \ll C$ and $C_2 \ll C$, so that the operators J_{C,C_1} and J_{C,C_2} of (Proposition) 3.1.5 are well defined. Set $J_1 = J^{\star}_{C,C_1} J_{C,C_1}$ and $J_2 = J^{\star}_{C,C_1} J_{C,C_1}$. One has that, in $H (C, T)$, for $\theta \in T$, fixed, but arbitrary,

$$J^{\star}_{C,C_1} J_{C,C_1} [C (\cdot, \theta)] = C_1 (\cdot, \theta) ,$$

so that, for $t \in T$, fixed, but arbitrary,

$$\mathcal{E}_t [J_1 [C (\cdot, \theta)]] = J_1 [C (\cdot, \theta)] (t) = C_1 (t, \theta) = \langle C_1 (\cdot, t) , C (\cdot, \theta)\rangle_{H(C,T)} ,$$

and thus, for $h \in H (C, T)$ and $t \in T$, fixed, but arbitrary,

$$\mathcal{E}_t [J_1 [h]] = J_1 [h] (t) = \langle h, C_1 (\cdot, t)\rangle_{H(C,T)} .$$

Consequently, $(J_1 + J_2) [h] (t) = \langle h, C (\cdot, t)\rangle_{H(C,T)} = h (t)$. The other items reexpress (Corollary) 3.1.25. The converse follows from the assumptions and the first part, as $C = C_1 + C_2$. □

Remark 3.1.33 The ensuing sections provide finer detail on those decompositions.

Set inclusion of RKHS's coincides with domination of norms as seen below.

Proposition 3.1.34 *Let* $H (C_1, T)$ *and* $H (C_2, T)$ *be RKHS's such that, as sets of functions,*

$$H (C_1, T) \subseteq H (C_2, T) .$$

There exists then $\kappa \geq 0$ such that κC_2 dominates C_1, and, for $h \in H(C_1, T)$, fixed, but arbitrary,

$$\|h\|_{H(C_2,T)} \leq \kappa \|h\|_{H(C_1,T)} .$$

Proof Let $F : T \longrightarrow H(C_2, T)$ be the following assignment: $F(t) = C_1(\cdot, t)$, H_F, the subspace of $H(C_2, T)$ generated linearly by $\mathcal{R}[F]$, and P_{H_F}, the associated projection. Then [(Proposition) 1.1.15] $L_F : H(C_2, T) \longrightarrow \mathbb{R}^T$, representing the following assignment: $L_F[h](t) = \langle h, F(t)\rangle_{H(C_2,T)}$ is such that $K_F = \mathcal{R}[L_F]$ is an RKHS with kernel

$$\mathcal{K}_F(t_1, t_2) = \langle C_1(\cdot, t_1), C_1(\cdot, t_2)\rangle_{H(C_2,T)},$$

and inner product

$$\langle L_F[h_1], L_F[h_2]\rangle_{K_F} = \langle P_{H_F}[h_1], P_{F_H}[h_2]\rangle_{H(C_2,T)}.$$

One has, in particular, that

$$L_F[C_2(\cdot, t)](\theta) = \langle C_2(\cdot, t), C_1(\cdot, \theta)\rangle_{H(C_2,T)} = C_1(\theta, t),$$

so that $L_F[C_2(\cdot, t)] = C_1(\cdot, t) \in H(C_2, T) \cap K_F$. Furthermore, the family $\{C_1(\cdot, t), t \in T\}$ is total in K_F. Suppose indeed that, for $t \in T$, and $k \in K_F$, fixed, but arbitrary, $\langle k, C_1(\cdot, t)\rangle_{K_F} = 0$. Let $k = L_F[h(k)]$. Then, as

$$\begin{aligned}
\langle k, C_1(\cdot, t)\rangle_{K_F} &= \langle L_F[h(k)], L_F[C_2(\cdot, t)]\rangle_{K_F} \\
&= \langle P_{H_F}[h(k)], P_{H_F}[C_2(\cdot, t)]\rangle_{H(C_2,T)} \\
&= \langle P_{H_F}[h(k)], C_2(\cdot, t)\rangle_{H(C_2,T)},
\end{aligned}$$

$P_{H_F}[h(k)] = 0_T$, and thus $h(k) \in H_F^{\perp}$, or $k = L_F[h(k)] = 0_T$.

Let $V : K_F \longrightarrow K_F$ denote the following assignment: $V[C_1(\cdot, t)] = \mathcal{K}_F(\cdot, t)$. Then the following equalities, tagged $(\star)$, obtain:

$$\begin{aligned}
\left\| \sum_{i=1}^{n} \alpha_i V[C_1(\cdot, t_i)] \right\|_{K_F}^2 &= \left\| \sum_{i=1}^{n} \alpha_i \mathcal{K}_F(\cdot, t_i) \right\|_{K_F}^2 \\
&= \sum_{i=1}^{n} \sum_{j=1}^{n} \alpha_i \alpha_j \mathcal{K}_F(i, j) \\
&= \sum_{i=1}^{n} \sum_{j=1}^{n} \alpha_i \alpha_j \langle C_1(\cdot, t_i), C_1(\cdot, t_j)\rangle_{H(C_2,T)}
\end{aligned}$$

$$= \left\| \sum_{i=1}^{n} \alpha_i C_1(\cdot, t_i) \right\|_{H(C_2,T)}^2 .$$

Suppose now that $\left\| \sum_{i=1}^{n} \alpha_i C_1(\cdot, t_i) \right\|_{K_F} = 0$. Since $C_1(\cdot, t) = L_F[C_2(\cdot, t)]$, the latter, equal to zero, expression means that

$$\sum_{i=1}^{n} \alpha_i C_2(\cdot, t_i) \in H_F^{\perp},$$

or that, for $\theta \in T$, fixed, but arbitrary,

$$\left\langle \sum_{i=1}^{n} \alpha_i C_2(\cdot, t_i), C_1(\cdot, \theta) \right\rangle_{H(C_2,T)} = \sum_{i=1}^{n} \alpha_i C_1(\theta, t_i) = 0.$$

It follows from ($\star$) that $\left\| \sum_{i=1}^{n} \alpha_i V[C_1(\cdot, t_i)] \right\|_{K_F}^2 = 0$, which means that V is well defined, and linear on $V[\{C_1(\cdot, t), t \in T\}]$. One should note that $V[C_1(\cdot, t)] = L_F[C_1(\cdot, t)]$, but that V is a map of K_F.

Now, as $C_1(\cdot, t) \in H(C_2, T)$,

$$\sum_{i=1}^{m} \alpha_i C_1(\theta, t_i) = \sum_{i=1}^{m} \alpha_i \langle C_2(\cdot, \theta), C_1(\cdot, t_i) \rangle_{H(C_2,T)}$$
$$= \left\langle C_2(\cdot, \theta), \sum_{i=1}^{m} \alpha_i C_1(\cdot, t_i) \right\rangle_{H(C_2,T)} .$$

Consequently,

$$\sum_{j=1}^{n} \sum_{i=1}^{m} \alpha_i \beta_j C_1(\theta_j, t_i) =$$
$$= \left\langle \sum_{j=1}^{n} \beta_j C_2(\cdot, \theta_j), \sum_{i=1}^{m} \alpha_i C_1(\cdot, t_i) \right\rangle_{H(C_2,T)}$$
$$= \left\langle P_{H_F} \left[\sum_{j=1}^{n} \beta_j C_2(\cdot, \theta_j) \right], P_{H_F} \left[\sum_{i=1}^{m} \alpha_i C_1(\cdot, t_i) \right] \right\rangle_{H(C_2,T)}$$
$$= \left\langle L_F \left[\sum_{j=1}^{n} \beta_j C_2(\cdot, \theta_j) \right], L_F \left[\sum_{i=1}^{m} \alpha_i C_1(\cdot, t_i) \right] \right\rangle_{K_F}$$

$$= \left\langle \sum_{j=1}^{n} \beta_j C_1(\cdot, \theta_j), \sum_{i=1}^{m} \alpha_i \mathcal{K}_F(\cdot, t_i) \right\rangle_{K_F}$$

$$= \left\langle \sum_{j=1}^{n} \beta_j C_1(\cdot, \theta_j), V\left[\sum_{i=1}^{m} \alpha_i C_1(\cdot, t_i)\right] \right\rangle_{K_F}.$$

But, as $C_1(\cdot, t) \in K_F$ also,

$$\sum_{i=1}^{m} \alpha_i C_1(\theta, t_i) = \left\langle \mathcal{K}_F(\cdot, \theta), \sum_{i=1}^{m} \alpha_i C_1(\cdot, t_1) \right\rangle_{K_F},$$

so that

$$\sum_{j=1}^{n} \sum_{i=1}^{m} \alpha_i \beta_j C_1(\theta_j, t_i) = \left\langle V\left[\sum_{j=1}^{n} \beta_j C_1(\cdot, \theta_j)\right], \sum_{i=1}^{m} \alpha_i C_1(\cdot, t_i) \right\rangle_{K_F}. \qquad (\star\star)$$

It follows that, for h_1 and h_2 in $V[\{C_1(\cdot, t), t \in T\}]$, fixed, but arbitrary,

$$\langle V[h_1], h_2 \rangle_{K_F} = \langle h_1, V[h_2] \rangle_{K_F}.$$

But then [228, p. 334] V is continuous, and, from $(\star\star)$ above,

$$\begin{aligned}
\sum_{i=1}^{n} \sum_{i=1}^{n} \alpha_i \alpha_j C_1(t_i, t_j) &= \left\langle V\left[\sum_{i=1}^{n} \alpha_i C_1(\cdot, t_i)\right], \sum_{i=1}^{n} \alpha_i C_1(\cdot, t_i) \right\rangle_{K_F} \\
&\le \|V\| \left\| \sum_{i=1}^{n} \alpha_i C_1(\cdot, t_i) \right\|_{K_F}^2 \\
&= \|V\| \left\| \sum_{i=1}^{n} \alpha_i L_F[C_2(\cdot, t_i)] \right\|_{K_F}^2 \\
&\le \|V\| \left\| P_{H_F}\left[\sum_{i=1}^{n} \alpha_i C_2(\cdot, t_i)\right] \right\|_{H(C_2,T)}^2 \\
&\le \|V\| \left\| \sum_{i=1}^{n} \alpha_i C_2(\cdot, t_i) \right\|_{H(C_2,T)}^2 \\
&= \|V\| \sum_{i=1}^{n} \sum_{i=1}^{n} \alpha_i \alpha_j C_2(t_i, t_j).
\end{aligned}$$

So $\|V\|\, C_2$ dominates C_1. Consequently, on $H(C_1, T)$,

$$\|h\|_{H(\|V\| C_2, T)} \leq \|h\|_{H(C_1, T)} \,.$$

But $\|h\|_{H(\|V\| C_2, T)} = \|V\|^{-1} \|h\|_{H(C_2, T)}$. □

Remark 3.1.35 On its own, H_F is an RKHS [(Proposition) 1.6.1], denoted here $H(\mathcal{H}_{FF}, T)$, obtained using the following map:

$$L_{FF} = P_{H_F} : H(C_2, T) \longrightarrow \mathbb{R}^T \text{ with } L_{FF}[C_2(\cdot, t)] = P_{H_F}[C_2(\cdot, t)],$$

and resulting in the reproducing kernel:

$$\mathcal{H}_{FF}(t_1, t_2) = \langle P_{H_F}[C_2(\cdot, t_1)], P_{H_F}[C_2(\cdot, t_2)]\rangle_{H(C_2, T)},$$

and the inner product:

$$\langle P_{H_F}[h_1], P_{H_F}[h_2]\rangle_{H(\mathcal{H}_{FF}, T)} = \langle P_{H_F}[h_1], P_{H_F}[h_2]\rangle_{H(C_2, T)}.$$

Since [(Proposition) 1.1.15] $L_F^\star L_F = P_{H_F}$, one has that

$$L_{FF}[C_2(\cdot, t)] = P_{H_F}[C_2(\cdot, t)] = L_F^\star L_F[C_2(\cdot, t)],$$

so that $L_{FF} = L_F^\star L_F$.

The proof of (Proposition) 3.1.34 just given, with the exception of reference to [228, p. 334], is elementary, but allows one to identify κ as $\|V\|$, with V, an explicitly defined operator, the restriction of L_F to K_F. It should be noted that $\|L_F\| \leq 1$ (it is a contraction), but that one has no estimation of the value of $\|V\|$. There are more sophisticated proofs of that same result, which are presented in the following remarks, and which may prove more instructive as to the structure of the situation. The one using the closed graph theorem is also quite shorter, but makes a *détour* through (Proposition) 3.1.23!

Remark 3.1.36 ([4]) Here is a preliminary observation.

Let $H(C_1, T)$ and $H(C_2, T)$ be RKHS's, and

$$H_0 = H(C_1, T) \cap H(C_2, T)\,.$$

Let $H_0^{C_1}$ be H_0 considered as a subset of $H(C_1, T)$, and $H_0^{C_2}$ be defined similarly. Let

$$I : H_0^{C_1} \longrightarrow H_0^{C_2}$$

be the identity. It is a closed map.

Proof Let $\{h_n, n \in \mathbb{N}\} \subseteq H_0^{C_1}$ converge to h_{C_1} in $H(C_1, T)$, and to h_{C_2} in $H(C_2, T)$ (that is, $\lim_n h_n = h_{C_1}$ and $\lim_n I[h_n] = h_{C_2}$). Since convergence in an RKHS implies pointwise convergence,

$$\begin{aligned} h_{C_1}(t) &= \langle h_{C_1}, C_1(\cdot, t)\rangle_{H(C_1,T)} \\ &= \lim_n \langle h_n, C_1(\cdot, t)\rangle_{H(C_1,T)} \\ &= \lim_n h_n(t) \\ &= \lim_n \langle h_n, C_2(\cdot, t)\rangle_{H(C_2,T)} \\ &= \langle h_{C_2}, C_2(\cdot, t)\rangle_{H(C_2,T)} \\ &= h_{C_2}(t). \end{aligned}$$

Thus $h_{C_1} \in \mathcal{D}_I$, and $I[h_{C_1}] = h_{C_2}$, as $h_{C_1} = h_{C_2} \in H_0$. □

The proof of (Proposition) 3.1.34 then proceeds as follows:

Proof As H_0 of the observation just made is $H(C_1, T)$, and that the identity map is closed, it is bounded by the closed graph theorem [266, p. 94]. Thus, for $h \in H(C_1, T)$, fixed, but arbitrary,

$$\|h\|_{H(C_2,T)} = \|I[h]\|_{H(C_2,T)} \leq \kappa \|h\|_{H(C_1,T)}.$$

Consequently [(Example) 1.3.12]

$$\|h\|_{H(\kappa C_2,T)} = \kappa^{-1} \|h\|_{H(C_2,T)} \leq \|h\|_{H(C_1,T)}.$$

One then uses result (Proposition) 3.1.23 to see that the claimed domination of covariances obtains. □

Remark 3.1.37 ([106, p. 156]) The second alternate proof of (Proposition) 3.1.34 requires the following preliminaries [111, pp. 275,315]:

Preliminaries:

1. Let V be a real vector space. $C \subseteq V$ is convex when $\lambda v_1 + (1-\lambda) v_2 \in C$, whatever $\lambda \in [0, 1]$, $(v_1, v_2) \subseteq V \times V$.
2. Let V be a real vector space. $B \subseteq V$ is balanced when $\lambda v \in B$ whatever $\lambda \in [-1, +1]$ and $v \in B$.
3. Let V be a real vector space. $A \subseteq V$ absorbs $B \subseteq V$ when there exists $\kappa > 0$ such that $\lambda B \subseteq A$ whatever $\lambda \in [-\kappa, +\kappa]$.
4. Let V be a real vector space. $A \subseteq V$ is absorbing in V when A absorbs $\{v\}$, all $v \in V$.
5. Let V be a real vector space. $S \subseteq V$ is symmetric when $-S = S$. Balanced sets are symmetric. The empty set is convex and balanced.

6. Let V be a real vector space. $B \subseteq V$ is a barrel when it is closed, convex, absorbing in V, and balanced.
7. A locally convex topological vector space V is barrelled when everyone of its barrels is a neighborhood of the zero vector. Fréchet and Banach spaces are barrelled.

Proof Let

- B_2 denote the closed unit ball of $H(C_2, T)$, and
- $B_{2|1}$ denote $B_2 \cap H(C_1, T)$.

The first step of the proof consists in checking that $B_{2|1}$ is closed in $H(C_1, T)$ (Corollary 3.1.13 is a similar result, but requires that C_2 dominates C_1, a fact that is not available in the present context).

To that end, one considers a Cauchy sequence in $H(C_1, T)$, say $\{h_n,\ n \in \mathbb{N}\}$, contained in $B_{2|1}$. As such, it has a limit $h^{(1)} \in H(C_1, T)$. That sequence is however bounded in $H(C_2, T)$, since it is contained in B_2. As such [266, p. 79], it contains a subsequence $\{h_{n_p},\ p \in \mathbb{N}\} \subseteq \{h_n,\ n \in \mathbb{N}\}$ which has a weak limit in $H(C_2, T)$, say $h^{(2)}$. Furthermore, as for all weak limits [266, p. 78],

$$\left\|h^{(2)}\right\|_{H(C_2,T)} \leq \liminf_p \left\|h_{n_p}\right\|_{H(C_2,T)},$$

$\|h^{(2)}\|_{H(C_2,T)} \leq 1$. Thus $h^{(2)}$ belongs to B_2. Finally, for fixed, but arbitrary $t \in T$,

$$h^{(2)}(t) = \left\langle h^{(2)}, C_2(\cdot, t)\right\rangle_{H(C_2,T)} = \lim_p \left\langle h_{n_p}, C_2(\cdot, t)\right\rangle_{H(C_2,T)} = \lim_p h_{n_p}(t).$$

Now, since $h_{n_p} \in H(C_1, T)$, $h_{n_p}(t) = \left\langle h_{n_p}, C_1(\cdot, t)\right\rangle_{H(C_1,T)}$, and

$$\lim_p h_{n_p}(t) = \left\langle h^{(1)}, C_1(\cdot, t)\right\rangle_{H(C_1,T)} = h^{(1)}(t).$$

Consequently $h^{(1)}$ belongs to $B_{2|1}$ which is thus closed in $H(C_1, T)$.

One finishes the proof using the geometric properties of $B_{2|1}$ as follows. One first notices that $B_{2|1}$ is a barrel: it is closed as proved; it is convex as the intersection of two convex sets; it is absorbing as the intersection of two absorbing sets; it is balanced as the intersection of two balanced sets. Since a Hilbert space is barrelled [234, p. 60], $B_{2|1}$ is a neighborhood of the origin. Consequently $B_{2|1}$ contains a closed ball $B_0 \subseteq H(C_1, T)$ containing the origin. That means [269, p. 73] that the topology of $H(C_1, T)$ is stronger than the relative topology of $H(C_2, T)$ on $H(C_1, T)$. But this in turn means [269, p. 57] that there exits $\kappa \geq 0$ such that, for $h \in H(C_1, T)$, fixed, but arbitrary,

$$\|h\|_{H(C_2,T)} \leq \kappa \|h\|_{H(C_1,T)}.$$

One finishes as in (Example) 3.1.36. □

Remark 3.1.38 $H(C_1,T) \sqsubseteq H(C_2,T)$ [(Corollary) 3.1.24] if, and only if, there exists a reproducing kernel $\mathcal{H}$ such that $H(C_2,T) = H(C_1+\mathcal{H},T)$. When $\mathcal{H}$ exists, it is unique, and one then writes $H(\mathcal{H},T) = H(C_2,T) \ominus H(C_1,T)$.

Suppose indeed that $H(C_1,T) \sqsubseteq H(C_2,T)$. Then $C_1 \ll C_2$: set

$$\mathcal{H} = C_2 - C_1 \ll C_2.$$

Let $F : T \longrightarrow H(C_1,T) \oplus H(\mathcal{H},T)$ be defined using the following relation: $F_{\mathcal{H}}(t) = (C_1(\cdot,t), \mathcal{H}(\cdot,t))$. Then $L_{\mathcal{H}}$ has range equal to

$$H(C_1+\mathcal{H},T) = H(C_2,T).$$

The kernel $\mathcal{H}$ is also unique, for $H(C_1+\mathcal{H}_1,T) = H(C_1+\mathcal{H}_2,T)$ implies that $C_1+\mathcal{H}_1 = C_1+\mathcal{H}_2$ [(Proposition) 1.1.6]. Conversely, and for the same reason, when $H(C_2,T) = H(C_1+\mathcal{H},T)$, for a reproducing kernel $\mathcal{H}$, $C_2 \gg C_1$, and then $H(C_1,T) \sqsubseteq H(C_2,T)$.

Since, when $C_2 = C_1 + \mathcal{H}$, L_F is unitary, and thus $H(C_2,T)$ is isomorphic to $H(C_1,T) \oplus H(\mathcal{H},T)$, the statement's notation makes sense.

Lemma 3.1.39 $H(C,T) = \{0_{\mathbb{R}^T}\}$ *if, and only if,* $C = 0_{\mathbb{R}^{T\times T}}$.

Proof Suppose that $H(C,T)$ contains only the zero function. Then, since $C(\cdot,t) \in H(C,T)$, for $\{t,\theta\} \subseteq T$, fixed, but arbitrary, $C(\theta,t) = 0$. Conversely, when C is the zero function, $V[C]$ is reduced to the zero function, and so is its closure. □

Lemma 3.1.40 *Given two covariances* C_1 *and* C_2 *over* T,

$$H(C_1,T) \cap H(C_2,T) = \{0_{\mathbb{R}^T}\}$$

if, and only if, C_1 *and* C_2 *are disjoint* [(Definition) 3.1.1].

Proof Suppose indeed that the intersection of the two RKHS's is the zero function, and that the covariance C on T is dominated by both C_1 and C_2. Then $H(C,T) \subseteq H(C_1,T) \cap H(C_2,T) = \{0_{\mathbb{R}^T}\}$, so that $H(C,T)$ is reduced to the zero function, and its covariance must be zero. C_1 and C_2 are thus disjoint.

Suppose conversely that C_1 and C_2 are disjoint and let

$$H = H(C_1,T) \cap H(C_2,T).$$

The following relation, valid for $\{h_1,h_2\} \subseteq H$, fixed, but arbitrary, determines an inner product on H:

$$\langle h_1,h_2\rangle_H = \langle h_1,h_2\rangle_{H(C_1,T)} + \langle h_1,h_2\rangle_{H(C_2,T)},$$

and makes of H a Hilbert space. Suppose indeed that $\{h_n, n \in \mathbb{N}\} \subseteq H$ is a Cauchy sequence. It is then also Cauchy in $H(C_1,T)$ and $H(C_2,T)$, with respective limits

h_1 and h_2. But

$$\begin{aligned} h_1(t) &= \langle h_1, C_1(\cdot,t)\rangle_{H(C_1,T)} \\ &= \lim_n \langle h_n, C_1(\cdot,t)\rangle_{H(C_1,T)} \\ &= \lim_n h_n(t) \\ &= \lim_n \langle h_n, C_2(\cdot,t)\rangle_{H(C_2,T)} \\ &= \langle h, C_2(\cdot,t)\rangle_{H(C_2,T)} \\ &= h_2(t). \end{aligned}$$

Since, for $h \in H \subseteq H(C_1,T) \cap H(C_2,T)$, and $i \in \{1,2\}$, fixed, but arbitrary, $|\mathcal{E}_t(h)| \leq \kappa_i \|h\|_{H(C_i,T)}$, it follows that

$$\begin{aligned} |\mathcal{E}_t(h)| &\leq (\kappa_1+\kappa_2)\left\{\frac{\kappa_1}{\kappa_1+\kappa_2}\|h\|_{H(C_1,T)} + \frac{\kappa_2}{\kappa_1+\kappa_2}\|h\|_{H(C_2,T)}\right\} \\ &\leq (\kappa_1+\kappa_2)\|h\|_H \end{aligned}$$

(one may assume that $\kappa_1+\kappa_2 > 0$, otherwise there is nothing to prove). Thus H is an RKHS. Let $\mathcal{H}$ denote its kernel. By construction, for $h \in H$, fixed, but arbitrary,

$$\|h\|_{H(C_1,T)} \leq \|h\|_{H(\mathcal{H},T)}, \quad \text{and} \quad \|h\|_{H(C_2,T)} \leq \|h\|_{H(\mathcal{H},T)}.$$

Thus [(Corollary) 3.1.24] $\mathcal{H} \ll C_1$, and $\mathcal{H} \ll C_2$. But C_1 and C_2 have been assumed to be disjoint, so that H is reduced to the zero function. □

Proposition 3.1.41 *$H(C_1,T)$ is a (closed) subspace of $H(C_2,T)$ if, and only if, C_1 and $C_2 - C_1$ are disjoint.*

Proof When $H(C_1,P)$ is a subspace of $H(C_2,T)$, because of (Proposition) 1.6.1,

$$C_1(\cdot,t) = P_{H(C_1,T)}[C_2(\cdot,t)],$$

and thus

$$\{C_2 - C_1\}(\cdot,t) = \left\{I_{H(C_2,T)} - P_{H(C_1,T)}\right\}[C_2(\cdot,t)],$$

so that $H(C_1,T)$ and $H(C_2 - C_1,T)$ are orthogonal in $H(C_2,T)$, and thus their intersection is the zero function. C_1 and $C_2 - C_1$ are thus disjoint [(Lemma) 3.1.40]. When C_1 and $C_2 - C_1$ are disjoint, the intersection $H(C_1,T) \cap H(C_2 - C_1,T)$ is made of the zero function, and thus $H(C_2,T) = H(C_1 + [C_2 - C_1],T)$ is isomorphic to $H(C_1,T) \oplus H(C_2 - C_1,T)$. □

3.2 Contractive Inclusions of Hilbert Spaces

When, on T, the covariance C_2 dominates the covariance C_1, one knows that, as sets, $H(C_1, T) \subseteq H(C_2, T)$, and that, for $h \in H(C_1, T)$,

$$\|h\|_{H(C_2,T)} \leq \|h\|_{H(C_1,T)} .$$

It is of interest to measure the gap separating the two norms, for example when they represent "energy constraints" [13]. It is the aim of this chapter to obtain the value of

$$\|h\|^2_{H(C_1,T)} - \|h\|^2_{H(C_2,T)} .$$

One has already met an analogous description in (Remark) 3.1.27, but there, one has an expression for

$$\|h\|^2_{H(C_2,T)} - \|J_{C_2,C_1}[h]\|^2_{H(C_1,T)} .$$

Since $J^{\star}_{2,1}$ is an inclusion and a contraction, the material in this section informs also on $J^{\star}_{2,1}$, and, consequently, on $J_{2,1}$.

3.2.1 *Definition and Properties of Contractive Inclusions*

The material which follows expands on (Proposition) 1.1.15.

Definition 3.2.1 Let H_1 and H_2 be real Hilbert spaces, and suppose that H_1 is a *(linear)* submanifold of H_2. Denote $J_{1,2}$ the inclusion map of H_1 into H_2. Then:

1. when $J_{1,2}$ is continuous, H_1 is said to be contained boundedly in H_2, and one sometimes writes $H_1 \subseteq_b H_2$;
2. when $J_{1,2}$ is a contraction, H_1 is said to be contained contractively in H_2, and one sometimes writes $H_1 \subseteq_c H_2$.

Thus, in both cases, H_1 is the range of an operator [105].

The following result is very much like that of (Proposition) 1.1.15, and the proof is similar.

Proposition 3.2.2 *Let H_1 and H_2 be real Hilbert spaces, and B, a bounded linear operator from H_1 into H_2. Let $H_B = \mathcal{R}[B] \subseteq H_2$. Then*

1. *H_B is a real Hilbert space for the inner product*

$$\langle B\left[h_1^{(1)}\right], B\left[h_2^{(1)}\right]\rangle_{H_B} = \langle P_{\mathcal{N}[B]^{\perp}}\left[h_1^{(1)}\right], P_{\mathcal{N}[B]^{\perp}}\left[h_1^{(1)}\right]\rangle_{H_1};$$

2. *H_B is contained boundedly in H_2;*

3. *when B is a contraction, H_B is contained contractively in H_2.*

Proof Since H_B is the range of a linear operator, it is a linear manifold. Thus one must first check that the following map:

$$\left(h_1^{(1)}, h_2^{(1)}\right) \mapsto \langle B\left[h_1^{(1)}\right], B\left[h_2^{(1)}\right]\rangle_{H_B}$$

is an inner product. It is bilinear as, for example, given $\{\alpha_1, \alpha_2\} \subseteq \mathbb{R}$, and $\left\{h^{(1)}, h_1^{(1)}, h_2^{(1)}\right\} \subseteq H_1$, fixed, but arbitrary,

$$\begin{aligned}
&\langle \alpha_1 B\left[h_1^{(1)}\right] + \alpha_2 B\left[h_2^{(1)}\right], B\left[h^{(1)}\right]\rangle_{H_B} = \\
&= \langle B\left[\alpha_1 h_1^{(1)} + \alpha_2 h_2^{(1)}\right], B\left[h^{(1)}\right]\rangle_{H_B} \\
&= \langle P_{\mathcal{N}[B]^\perp}\left[\alpha_1 h_1^{(1)} + \alpha_2 h_2^{(1)}\right], P_{\mathcal{N}[B]^\perp}\left[h^{(1)}\right]\rangle_{H_1} \\
&= \alpha_1 \langle P_{\mathcal{N}[B]^\perp}\left[h_1^{(1)}\right], P_{\mathcal{N}[B]^\perp}\left[h^{(1)}\right]\rangle_{H_1} + \alpha_2 \langle P_{\mathcal{N}[B]^\perp}\left[h_2^{(1)}\right], P_{\mathcal{N}[B]^\perp}\left[h^{(1)}\right]\rangle_{H_1} \\
&= \alpha_1 \langle B\left[h_1^{(1)}\right], B\left[h^{(1)}\right]\rangle_{H_B} + \alpha_2 \langle B\left[h_2^{(1)}\right], B\left[h^{(1)}\right]\rangle_{H_B}.
\end{aligned}$$

Furthermore $\|B\left[h^{(1)}\right]\|_{H_B} = 0$ implies that

$$\left\|P_{\mathcal{N}[B]^\perp}\left[h^{(1)}\right]\right\|_{H_1} = 0,$$

and thus that

$$B\left[h^{(1)}\right] = B\left[P_{\mathcal{N}[B]^\perp}\left[h^{(1)}\right]\right] = 0_{H_2}.$$

One has thus indeed defined an inner product. It remains to check that, for that inner product, H_B is complete. Suppose thus that

$$\lim_{m,n} \left\|B\left[h_m^{(1)}\right] - B\left[h_n^{(1)}\right]\right\|_{H_B} = 0.$$

Then, by definition,

$$\lim_{m,n} \left\|P_{\mathcal{N}[B]^\perp}\left[h_m^{(1)}\right] - P_{\mathcal{N}[B]^\perp}\left[h_n^{(1)}\right]\right\|_{H_1} = 0,$$

so that there exists $h^{(1)} \in \mathcal{N}[B]^\perp$ such that, in H_1,

$$\lim_n P_{\mathcal{N}[B]^\perp}\left[h_n^{(1)}\right] = h^{(1)}.$$

But then

$$\lim_n \left\|B\left[h_n^{(1)}\right] - B\left[h^{(1)}\right]\right\|_{H_B} = \lim_n \left\|P_{\mathcal{N}[B]^\perp}\left[h_n^{(1)}\right] - h^{(1)}\right\|_{H_1} = 0.$$

Let $J_{B,2}$ denote the inclusion of H_B into H_2. Since

$$\left\|J_{B,2}\left[B\left[h^{(1)}\right]\right]\right\|_{H_2} = \left\|B\left[P_{\mathcal{N}_B^\perp}\left[h^{(1)}\right]\right]\right\|_{H_2} \leq \|B\| \left\|P_{\mathcal{N}_B^\perp}\left[h^{(1)}\right]\right\|_{H_1},$$

and that

$$\left\|P_{\mathcal{N}_B^\perp}\left[h^{(1)}\right]\right\|_{H_1} = \left\|B\left[h^{(1)}\right]\right\|_{H_B},$$

one has finally that

$$\left\|J_{B,2}\left[B\left[h^{(1)}\right]\right]\right\|_{H_2} \leq \|B\| \left\|B\left[h^{(1)}\right]\right\|_{H_B}.$$

Thus items 2 and 3 follow. □

Example 3.2.3 Suppose that (Proposition) 3.1.5 obtains, and choose

$$H_1 = H(C_2, T), \quad H_2 = H(C_1, T), \quad \text{and } B = J_{2,1}.$$

Then [266, p. 71]

$$\mathcal{N}[B]^\perp = \mathcal{N}[J_{2,1}]^\perp = \overline{\mathcal{R}[J_{2,1}^\star]} = \overline{H(C_1, T)},$$

the closure being in $H(C_2, T)$. H_B is $\mathcal{R}[J_{2,1}] = H(C_1, T)$, and

$$\langle J_{2,1}\left[h_1^{(2)}\right], J_{2,1}\left[h_2^{(2)}\right]\rangle_{H_B} = \langle P_{\overline{H(C_1,T)}}\left[h_1^{(2)}\right], P_{\overline{H(C_1,T)}}\left[h_2^{(2)}\right]\rangle_{H(C_2,T)}.$$

Since $J_{2,1}$ is a contraction, H_B is contained contractively in $H(C_1, T)$, that is, the inclusion of H_B into $H(C_1, T)$ is a contraction, so that

$$\left\|J_{2,1}\left[h^{(2)}\right]\right\|_{H(C_1,T)} \leq \left\|J_{2,1}\left[h^{(2)}\right]\right\|_{H_B} = \left\|P_{\overline{H(C_1,T)}}\left[h^{(2)}\right]\right\|_{H(C_2,T)},$$

which, when (with closure in $H(C_2, T)$) $\overline{H(C_1, T)} \subset H(C_2, T)$, is an inequality tighter than that given by the fact that $J_{2,1}$ is a contraction.

Choose now $H_1 = H(C_1, T)$, $H_2 = H(C_2, T)$, $B = J_{2,1}^\star$. Then [266, p. 71]

$$\mathcal{N}[B]^\perp = \mathcal{N}[J_{2,1}^\star]^\perp = \overline{\mathcal{R}[J_{2,1}]} = H(C_1, T).$$

Consequently $H_B = H(C_1, T)$, and

$$\langle J_{2,1}^\star\left[h_1^{(1)}\right], J_{2,1}^\star\left[h_2^{(1)}\right]\rangle_{H_B} = \langle h_1^{(1)}, h_2^{(2)}\rangle_{H(C_1,T)},$$

so that, as Hilbert spaces, $H_B = H(C_1, T)$, and the inclusion result yields the norm inequality of (Proposition) 3.1.5.

Proposition 3.2.4 *Let H_1 and H_2 be real Hilbert spaces, and $B : H_1 \longrightarrow H_2$ be a bounded linear operator. Let H_B denote the Hilbert space built in* (Proposition) 3.2.2. *For fixed, but arbitrary $h^{(2)} \in H_2$, let $\langle h^{(2)}, \cdot\rangle_{H_2}^{|H_B}$ denote the restriction of $\langle h^{(2)}, \cdot\rangle_{H_2}$ to H_B. It is a continuous linear functional on H_B, and*

$$\langle h^{(2)}, \cdot\rangle_{H_2}^{|H_B} = \langle BB^\star \left[h^{(2)}\right], \cdot\rangle_{H_B}.$$

Proof For fixed, but arbitrary $h^{(1)} \in H_1$, the definitions yield that

$$\langle h^{(2)}, B\left[h^{(1)}\right]\rangle_{H_2} = \langle B^\star \left[h^{(2)}\right], h^{(1)}\rangle_{H_1}.$$

As [266, p. 71] $\mathcal{N}[B]^\perp = \overline{\mathcal{R}[B^\star]}$, using (Proposition) 3.2.2, item 1,

$$\begin{aligned}\langle B^\star \left[h^{(2)}\right], h^{(1)}\rangle_{H_1} &= \langle P_{\mathcal{N}[B]^\perp} \left[B^\star \left[h^{(2)}\right]\right], P_{\mathcal{N}[B]^\perp} \left[h^{(1)}\right]\rangle_{H_1} \\ &= \langle B\left[B^\star \left[h^{(2)}\right]\right], B\left[h^{(1)}\right]\rangle_{H_B}.\end{aligned}$$

Thus

$$\langle h^{(2)}, B\left[h^{(1)}\right]\rangle_{H_2} = \langle BB^\star \left[h^{(2)}\right], B\left[h^{(1)}\right]\rangle_{H_B}.$$

□

The next result is a version of Douglas's theorem [80]. The notation $B_1 \leq B_2$, for operators B_1 and B_2 on H, means that, for $h \in H$, fixed, but arbitrary,

$$\langle B_1 [h], h\rangle_H \leq \langle B_2 [h], h\rangle_H.$$

Proposition 3.2.5 *Let H, H_1, H_2 be real Hilbert spaces, and $B_1 : H_1 \longrightarrow H$ and $B_2 : H_2 \longrightarrow H$ be bounded linear operators. Then,*

$$B_1 B_1^\star \leq B_2 B_2^\star$$

if, and only if, there exists a contraction $C : H_1 \longrightarrow H_2$ such that $B_1 = B_2 C$, (which means in particular that $\mathcal{R}[B_1] \subseteq \mathcal{R}[B_2]$).

Proof Suppose that there exists a contraction C such that $B_1 = B_2 C$.
Then, for fixed, but arbitrary $h \in H$,

$$\begin{aligned}\langle B_1 B_1^\star [h], h\rangle_H &= \langle B_1^\star [h], B_1^\star [h]\rangle_{H_1} \\ &= \|C^\star B_2^\star [h]\|_{H_1}^2 \\ &\leq \|C^\star\|^2 \langle B_2 B_2^\star [h], h\rangle_H.\end{aligned}$$

Since $\|C^\star\| = \|C\|$ [266, p. 71], C being a contraction, $C^\star$ is a contraction, and thus $B_1 B_1^\star \leq B_2 B_2^\star$.

Proof Suppose now that $B_1B_1^\star \leq B_2B_2^\star$.

Define $D : \mathcal{R}[B_2^\star] \longrightarrow \mathcal{R}[B_1^\star]$ using the following assignment:

$$D\left[B_2^\star[h]\right] = B_1^\star[h],\ h \in H.$$

One has, for $\{\alpha_1, \alpha_2\} \subseteq \mathbb{R}$, and $\{h, h_1, h_2\} \subseteq H$, fixed, but arbitrary, that

$$\begin{aligned} D\left[\alpha_1 B_2^\star[h_1] + \alpha_2 B_2^\star[h_2]\right] &= D\left[B_2^\star[\alpha_1 h_1 + \alpha_2 h_2]\right] \\ &= B_1^\star[\alpha_1 h_1 + \alpha_2 h_2] \\ &= \alpha_1 B_1^\star[h_1] + \alpha_2 B_1^\star[h_2] \\ &= \alpha_1 D\left[B_2^\star[h_1]\right] + \alpha_2 D\left[B_2^\star[h_2]\right], \end{aligned}$$

and that

$$\begin{aligned} \left\|D\left[B_2^\star[h]\right]\right\|_{H_1}^2 &= \|B_1^\star[h]\|_{H_1}^2 \\ &= \langle B_1B_1^\star[h], h\rangle_H \\ &\leq \langle B_2B_2^\star[h], h\rangle_H \\ &= \|B_2^\star[h]\|_{H_2}^2. \end{aligned}$$

This shows, firstly, that D is well defined since, whenever $B_2^\star[h_1] = B_2^\star[h_2]$, $(h_1, h_2) \in H \times H$,

$$\left\|D\left[B_2^\star[h_1]\right] - D\left[B_2^\star[h_2]\right]\right\|_{H_1}^2 \leq \|B_2^\star[h_1] - B_2^\star[h_2]\|_{H_2}^2 = 0,$$

and, secondly, that D has an extension to a contraction defined on $\overline{\mathcal{R}[B_2^\star]}$. To have H_2 as domain for D, one sets

$$D\left[h^{(2)}\right] = 0 \text{ for } h^{(2)} \in \overline{\mathcal{R}[B_2^\star]}^\perp.$$

Then, given $(h, h^{(1)}) \in H \times H_1$, fixed, but arbitrary,

$$\begin{aligned} \langle B_2\left[D^\star\left[h^{(1)}\right]\right], h\rangle_H &= \langle h^{(1)}, DB_2^\star[h]\rangle_{H_1} \\ &= \langle h^{(1)}, B_1^\star[h]\rangle_{H_1} \\ &= \langle B_1\left[h^{(1)}\right], h\rangle_H. \end{aligned}$$

To have the required result, it suffices to choose $C = D^\star$, which is a contraction since D is one [266, p. 71]. □

Proposition 3.2.6 *Let H, H_1, H_2 be real Hilbert spaces, and $B_1 : H_1 \longrightarrow H$ and $B_2 : H_2 \longrightarrow H$ be bounded linear operators. H_{B_1} is contained contractively in H_{B_2} if, and only if, $B_1B_1^\star \leq B_2B_2^\star$.*

Proof Suppose that H_{B_1} *is contained contractively in* H_{B_2}.

Given (Proposition) 3.2.5, it will suffice to prove that $B_1 = B_2 C$, $C : H_1 \longrightarrow H_2$, a contraction. Let $h^{(1)} \in H_1$ be fixed, but arbitrary. By assumption, one has that $\mathcal{R}[B_1] \subseteq \mathcal{R}[B_2]$, so that there exists $h^{(2)} \in H_2$ such that $B_2 [h^{(2)}] = B_1 [h^{(1)}]$. Furthermore $h^{(2)}$ can be chosen to belong to $\mathcal{N}[B_2]^{\perp}$. Let then $C : H_1 \longrightarrow H_2$ be the following assignment:

$$C\left[h^{(1)}\right] = h^{(2)}.$$

Let $\{\alpha_1, \alpha_2\} \subseteq \mathbb{R}$, and $\left\{h^{(1)}, h_1^{(1)}, h_2^{(1)}\right\} \subseteq H_1$, be fixed, but arbitrary, and

$$\left\{h_1^{(2)}, h_2^{(2)}\right\} \subseteq \mathcal{N}[B_2]^{\perp}$$

be such that

$$B_2\left[h_1^{(2)}\right] = B_1\left[h_1^{(1)}\right] \text{ and } B_2\left[h_2^{(2)}\right] = B_1\left[h_2^{(1)}\right].$$

Then

$$\begin{aligned} B_2\left[\alpha_1 h_1^{(2)} + \alpha_2 h_2^{(2)}\right] &= \alpha_1 B_2\left[h_1^{(2)}\right] + \alpha_2 B_2\left[h_2^{(2)}\right] \\ &= \alpha_1 B_1\left[h_1^{(1)}\right] + \alpha_2 B_1\left[h_2^{(1)}\right] \\ &= B_1\left[\alpha_1 h_1^{(1)} + \alpha_2 h_2^{(1)}\right], \end{aligned}$$

so that

$$C\left[\alpha_1 h_1^{(1)} + \alpha_2 h_2^{(1)}\right] = \alpha_1 h_1^{(2)} + \alpha_2 h_2^{(2)} = \alpha_1 C\left[h_1^{(1)}\right] + \alpha_2 C\left[h_2^{(1)}\right].$$

Then indeed $B_2 C [h^{(1)}] = B_2 [h^{(2)}] = B_1 [h^{(1)}]$, the required equality. Also one has that

$$\begin{aligned} \left\|C\left[h^{(1)}\right]\right\|_{H_2} &= \left\|P_{\mathcal{N}[B_2]^{\perp}}\left[h^{(2)}\right]\right\|_{H_2} \\ &= \left\|B_2\left[h^{(2)}\right]\right\|_{H_{B_2}} \\ &= \left\|B_1\left[h^{(1)}\right]\right\|_{H_{B_2}} \\ &\leq \left\|B_1\left[h^{(1)}\right]\right\|_{H_{B_1}} \\ &= \left\|P_{\mathcal{N}[B_1]^{\perp}}\left[h^{(1)}\right]\right\|_{H_1} \\ &\leq \left\|h^{(1)}\right\|_{H_1}. \end{aligned}$$

That proves that C is well defined, and a contraction.

Proof *Suppose that* $B_1 B_1^{\star} \leq B_2 B_2^{\star}$.

Because of (Proposition) 3.2.5, $B_1 = B_2 C,\ C : H_1 \longrightarrow H_2$, a contraction. A first consequence is that $\mathcal{R}[B_1] \subseteq \mathcal{R}[B_2]$. Also, for $h^{(1)} \in H_1$, fixed, but arbitrary,

$$\begin{aligned}
\left\|B_1\left[h^{(1)}\right]\right\|_{H_{B_2}} &= \left\|B_2 C\left[h^{(1)}\right]\right\|_{H_{B_2}} \\
&= \left\|P_{\mathcal{N}[B_2]^{\perp}} C\left[h^{(1)}\right]\right\|_{H_2} \\
&= \left\|P_{\mathcal{N}[B_2]^{\perp}} C P_{\mathcal{N}[B_1]}\left[h^{(1)}\right] + P_{\mathcal{N}[B_2]^{\perp}} C P_{\mathcal{N}[B_1]^{\perp}}\left[h^{(1)}\right]\right\|_{H_2}.
\end{aligned}$$

Since $B_1 P_{\mathcal{N}[B_1]}[h^{(1)}] = 0$, $B_2 C P_{\mathcal{N}[B_1]}[h^{(1)}] = 0$, so that

$$C P_{\mathcal{N}[B_1]}\left[h^{(1)}\right] \in \mathcal{N}[B_2], \text{ and thus } P_{\mathcal{N}[B_2]^{\perp}} C P_{\mathcal{N}[B_1]}\left[h^{(1)}\right] = 0.$$

Consequently

$$\begin{aligned}
\left\|B_1\left[h^{(1)}\right]\right\|_{H_{B_2}} &= \left\|P_{\mathcal{N}[B_2]^{\perp}} C P_{\mathcal{N}[B_1]^{\perp}}\left[h^{(1)}\right]\right\|_{H_2} \\
&\leq \left\|P_{\mathcal{N}[B_1]^{\perp}}\left[h^{(1)}\right]\right\|_{H_1} \\
&= \left\|B_1\left[h^{(1)}\right]\right\|_{H_{B_1}}.
\end{aligned}$$

H_{B_1} is thus contained contractively in H_{B_2}. □

Corollary 3.2.7 *Let* H, H_1, H_2 *be real Hilbert spaces, and* $B_1 : H_1 \longrightarrow H$ *and* $B_2 : H_2 \longrightarrow H$ *be bounded linear operators.* H_{B_1} *and* H_{B_2} *designate the same Hilbert space if, and only if,* $B_1 B_1^{\star} = B_2 B_2^{\star}$. *In particular the Hilbert spaces* H_{B_1} *and* $H_{[B_1 B_1^{\star}]^{1/2}}$ *are identical.*

Proposition 3.2.8 *Let* H_1 *and* H_2 *be real Hilbert spaces, and* $B : H_1 \longrightarrow H_2$ *be a bounded linear operator.* H_B *is a* (closed) *subspace of* H_2 *if, and only if,* B *is a partial isometry.*

Proof *Suppose* B *is a partial isometry.*

$\mathcal{R}[B]$ is then closed in H_2. One must check that $\mathcal{R}[B] = H_B$. But [266, p. 86]

$$BB^{\star} = P_{\mathcal{R}[B]},$$

and

$$P_{\mathcal{R}[B]} = P_{\mathcal{R}[B]} P_{\mathcal{R}[B]}^{\star},$$

so that, using (Corollary) 3.2.7, H_B and $H_{P_{\mathcal{R}[B]}}$ designate the same Hilbert space. But $H_{P_{\mathcal{R}[B]}} = \mathcal{R}[B]$.

Proof *Suppose H_B is an ordinary subspace of H_2.*

Then $\mathcal{R}[B]$ is closed, which is the first of a set of two conditions for B to be a partial isometry [266, p. 86]. Also, by the definition of H_B,

$$\langle B[h_1], B[h_2]\rangle_{H_B} = \langle P_{\mathcal{N}[B]^\perp}[h_1], P_{\mathcal{N}[B]^\perp}[h_2]\rangle_{H_1},$$

and, because H_B, which is a Hilbert space, is a closed subspace of H_2, it must have the same inner product, that is

$$\langle B[h_1], B[h_2]\rangle_{H_B} = \langle B[h_1], B[h_2]\rangle_{H_2}.$$

Thus $B^\star B = P_{\mathcal{N}[B]^\perp}$, which is the second of the same set of two conditions for B to be a partial isometry [266, p. 86]. □

Example 3.2.9 In (Example) 3.2.3, choose $B = J^\star_{2,1}$. Then $H_B = H(C_1, T)$, so that

$$\begin{gathered} H(C_1, T) \textit{ is a closed subspace of } H(C_2, T) \\ \textit{if, and only if}, J^\star_{2,1} \textit{ is a partial isometry.} \end{gathered}$$

Then $H(C_1, T)$ is the initial set of $J^\star_{2,1}$ as well as its final set. $J_{2,1}$ is also a partial isometry with the same initial and final set. $J^\star_{2,1}J_{2,1}$ is the identity of $H(C_1, T)$ and $J_{2,1}J^\star_{2,1}$ the projection onto it in $H(C_2, T)$.

3.2.2 Complementary and Overlapping Spaces

As seen in (Remark) 3.1.27, the ingredients of a norm decomposition along RKHS's involve an operator B and an operator of type $I - B^\star B$. Here one shall preferentially use the operator $I - BB^\star$ which realizes the "switch" mentioned at the beginning of Sect. 3.2. Complementary and overlapping spaces reflect the relations that may exist between H_B and $H_{I-B^\star B}$. Complementary reminds one of orthogonal complement, and complementary spaces are substitutes for orthogonal complements: when B is a partial isometry, $BB^\star$ and $B^\star B$ are projections, and so is their difference from the identity.

Definition 3.2.10 Let $B : H_1 \longrightarrow H_2$ be a real Hilbert spaces contraction. Define the following operators:

$$\begin{aligned} \text{on } H_1 : \; B_1 &= (I_{H_1} - B^\star B)^{1/2}, \\ \text{on } H_2 : \; B_2 &= (I_{H_2} - BB^\star)^{1/2}. \end{aligned}$$

H_{B_2} is called the space complementary to H_B. The overlapping space is the intersection $H_B \cap H_{B_2}$.

Remark 3.2.11 Suppose that $H_1 = H_2 = H$. Write temporarily D_B for B_1, and $D_{B^\star}$ for B_2. It is a consequence of the definition of D_B and $D_{B^\star}$ that, for $h \in H$, fixed, but arbitrary,

$$\|h\|_H^2 = \|B[h]\|_H^2 + \|D_B[h]\|_H^2 = \|B^\star[h]\|_H^2 + \|D_{B^\star}[h]\|_H^2.$$

D_B and $D_{B^\star}$ are called the defect operators of B [107, p. 130], and they give a measure of how far B and $B^\star$ are from being isometric. Thus B is isometric if, and only if, $D_B = 0$, and similarly for $B^\star$. The sequel describes an extension of that idea to products of operators between different Hilbert spaces.

Remark 3.2.12 One has, for positive R, that [162, p. 27] $\mathcal{N}[R^{1/2}] = \mathcal{N}[R]$. Thus an element $h^{(2)}$ in $\mathcal{N}[B_2]$ is such that $h^{(2)} = BB^\star[h^{(2)}]$.

Remark 3.2.13 For fixed, but arbitrary $h^{(1)} \in H_1$, one has, for example, that

$$\frac{\langle (I_{H_1} - B^\star B)[h^{(1)}], h^{(1)}\rangle_{H_1}}{\|h^{(1)}\|_{H_1}^2} = 1 - \frac{\|B[h^{(1)}]\|_{H_2}^2}{\|h^{(1)}\|_{H_1}^2},$$

and thus $I_{H_1} - B^\star B$ and $I_{H_2} - BB^\star$ are nonnegative, self-adjoint contractions. Consequently so are B_1 and B_2. Furthermore the spectrum of these operators, since they are nonnegative contractions, is contained in the interval $[0, 1]$.

Remark 3.2.14 When H_B is an ordinary subspace, that is when B is a partial isometry, $BB^\star$ and $I_{H_2} - BB^\star$ are complementary projections, and

$$H_{B_2} = H_B^\perp.$$

Proposition 3.2.15 *When $B : H_1 \longrightarrow H_2$ is a Hilbert spaces contraction,*

$$BB_1 = B_2 B.$$

Proof First of all, $B(I_{H_1} - B^\star B) = (I_{H_2} - BB^\star)B$.

Suppose then that, for some integer $n \in \mathbb{N}$,

$$B(I_{H_1} - B^\star B)^n = (I_{H_2} - BB^\star)^n B.$$

One has then that

$$\begin{aligned} B(I_{H_1} - B^\star B)^{n+1} &= B(I_{H_1} - B^\star B)(I_{H_1} - B^\star B)^n \\ &= (I_{H_2} - BB^\star)B(I_{H_1} - B^\star B)^n \\ &= (I_{H_2} - BB^\star)(I_{H_2} - BB^\star)^n B \\ &= (I_{H_2} - BB^\star)^{n+1} B. \end{aligned}$$

Consequently, for any polynomial p, and operators $A_1 = p(I_{H_1} - B^\star B)$ and $A_2 = p(I_{H_2} - BB^\star)$, on H_1 and H_2 respectively, $BA_1 = A_2 B$.

Now a standard result [129, p. 232] says that, if A is a bounded, self-adjoint, linear operator, with spectrum $\sigma(A)$, f is a continuous function, and $\{p_n,\ n \in \mathbb{N}\}$ a sequence of polynomials such that

$$\lim_n \left\{ \sup_{x \in \sigma(A)} |f(x) - p_n(x)| \right\} = 0,$$

then, in operator norm, $f(A) = \lim_n p_n(A)$. Furthermore, for any bounded linear operator B such that $AB = BA$, one has that $f(A)B = Bf(A)$. Consequently, for any sequence $\{p_n,\ n \in \mathbb{N}\}$ of polynomials that converges uniformly, on the interval $[0, 1]$, to the square root function [230, p. 159], taking into account that the spectra of the operators one considers are subsets of $[0, 1]$, one has that

$$\lim_n p_n(I_{H_1} - B^\star B) = (I_{H_1} - B^\star B)^{1/2} = B_1,$$

and

$$\lim_n p_n(I_{H_2} - BB^\star) = (I_{H_2} - BB^\star)^{1/2} = B_2,$$

which yields the required equality. □

Corollary 3.2.16 $B^\star B_2 = B_1 B^\star$

Proof It suffices to write the formula of (Proposition) 3.2.15 for $B^\star : H_2 \longrightarrow H_1$. □

Proposition 3.2.17 *For $B : H_1 \longrightarrow H_2$, a Hilbert spaces contraction,*

1. *$h \in H_{B_2}$ if, and only if, $B^\star[h] \in H_{B_1}$;*
2. *for fixed, but arbitrary $\{h_1, h_2\} \subseteq H_{B_2}$,*

$$\langle h_1, h_2 \rangle_{H_{B_2}} = \langle h_1, h_2 \rangle_{H_2} + \langle B^\star[h_1], B^\star[h_2] \rangle_{H_{B_1}}.$$

Proof [1] Suppose first that $h \in H_{B_2}$. Then, for some $h^{(2)} \in H_2$, $h = B_2[h^{(2)}]$. Consequently, because of (Corollary) 3.2.16,

$$B^\star[h] = B^\star B_2\left[h^{(2)}\right] = B_1\left[B^\star\left[h^{(2)}\right]\right] \in H_{B_1}.$$

Conversely, suppose that $h^{(2)} \in H_2$ is such that $B^\star[h^{(2)}] = B_1[h^{(1)}]$, some $h^{(1)} \in H_1$. Then

$$h^{(2)} = (I_{H_2} - BB^\star)\left[h^{(2)}\right] + BB^\star\left[h^{(2)}\right] = B_2^2\left[h^{(2)}\right] + BB_1\left[h^{(1)}\right].$$

But (Proposition) 3.2.15 yields that $BB_1 = B_2B$, and thus that

$$h^{(2)} = B_2^2\left[h^{(2)}\right] + B_2B\left[h^{(1)}\right] = B_2\left\{B_2\left[h^{(2)}\right] + B\left[h^{(1)}\right]\right\} \in H_{B_2}.$$

Proof [2] One must first notice that, because of item 1, the expression in item 2 makes sense.

Let h belong to H_{B_2}, that is $h = B_2\left[h^{(2)}\right],\ h^{(2)} \in H_2$. Then

$$B^\star[h] = B^\star B_2\left[h^{(2)}\right],$$

so that, using (Corollary) 3.2.16,

$$B^\star[h] = B_1B^\star\left[h^{(2)}\right] = B_1\left[P_{\mathcal{N}[B_1]^\perp}\left[B^\star\left[h^{(2)}\right]\right]\right].$$

Thus, given $h \in H_{B_2}$, there exists $h^{(1)} \in H_1$ such that

$$B^\star[h] = B_1\left[h^{(1)}\right],\ h^{(1)} \in \mathcal{N}[B_1]^\perp. \qquad (\star)$$

One may then check, as in the proof of item 1, that, given $h \in H_{B_2}$, one may write

$$h = B_2\left\{B_2[h] + B\left[h^{(1)}\right]\right\},\ h^{(1)} \in \mathcal{N}[B_1]^\perp \subseteq H_1.$$

One has furthermore that $B_2[h] + B[h^{(1)}] \perp \mathcal{N}[B_2]$. Indeed [266, p. 71],

- firstly,

$$B_2[h] \in \mathcal{R}[B_2] \subseteq \overline{\mathcal{R}[B_2]} \perp \mathcal{N}[B_2^\star] = \mathcal{N}[B_2],$$

- and, secondly, as $h^{(1)} \in H_1$, and $h^{(1)} \perp \mathcal{N}[B_1]$,

$$h^{(1)} \in \overline{\mathcal{R}[B_1^\star]} = \overline{\mathcal{R}[B_1]},$$

  so that, for some sequence $\left\{h_n^{(1)},\ n \in \mathbb{N}\right\} \subseteq H_1$, using (Proposition) 3.2.15 in the second step,

$$\begin{aligned} h^{(1)} &= \lim_n (H_1)\, B_1\left[h_n^{(1)}\right], \\ B\left[h^{(1)}\right] &= \lim_n (H_2)\, BB_1\left[h_n^{(1)}\right] \\ &= \lim_n (H_2)\, B_2B\left[h_n^{(1)}\right] \\ &\in \overline{\mathcal{R}[B_2]} \perp \mathcal{N}[B_2^\star] = \mathcal{N}[B_2]. \end{aligned}$$

The facts just proved allow one to write, for $(h_1, h_2) \in H_{B_2} \times H_{B_2}$, and corresponding

$$\left(h_1^{(1)}, h_2^{(1)}\right) \in H_1 \times H_1, \text{ with } h_1^{(1)} \perp \mathcal{N}[B_1] \text{ and } h_2^{(1)} \perp \mathcal{N}[B_1],$$

$$\begin{aligned}
\langle h_1, h_2\rangle_{H_{B_2}} &= \\
&= \langle B_2 \{B_2 [h_1] + B[h_1^{(1)}]\}, B_2 \{B_2 [h_2] + B[h_2^{(1)}]\}\rangle_{H_{B_2}} \\
&= \langle P_{\mathcal{N}[B_2]^\perp} \{B_2 [h_1] + B[h_1^{(1)}]\}, P_{\mathcal{N}[B_2]^\perp} \{B_2 [h_2] + B[h_2^{(1)}]\}\rangle_{H_2} \\
&= \langle \{B_2 [h_1] + B[h_1^{(1)}]\}, \{B_2 [h_2] + B[h_2^{(1)}]\}\rangle_{H_2} \\
&= \langle B_2 [h_1], B_2 [h_2]\rangle_{H_2} + \langle B_2 [h_1], B[h_2^{(1)}]\rangle_{H_2} \\
&\quad + \langle B[h_1^{(1)}], B_2 [h_2]\rangle_{H_2} + \langle B[h_1^{(1)}], B[h_2^{(1)}]\rangle_{H_2}.
\end{aligned}$$

Now, using the definition of the different ingredients, one has, for the first term to the right of the latter equality, that

$$\begin{aligned}
\langle B_2 [h_1], B_2 [h_2]\rangle_{H_2} &= \langle B_2^2 [h_1], h_2\rangle_{H_2} \\
&= \langle (I_{H_2} - BB^\star) [h_1], h_2\rangle_{H_2} \\
&= \langle h_1, h_2\rangle_{H_2} - \langle B^\star [h_1], B^\star [h_2]\rangle_{H_1}.
\end{aligned}$$

Also, using (Proposition) 3.2.15, and the already exhibited (see $\star$) following relation:

$$B^\star [h] = B_1 [h^{(1)}], \; h \in H_{B_2}, \; h^{(1)} \in \mathcal{N}[B_1]^\perp \subseteq H_1,$$

one obtains, for the second term, that

$$\begin{aligned}
\langle B_2 [h_1], B[h_2^{(1)}]\rangle_{H_2} &= \langle h_1, B_2 B[h_2^{(1)}]\rangle_{H_2} \\
&= \langle h_1, BB_1 [h_2^{(1)}]\rangle_{H_2} \\
&= \langle B^\star [h_1], B_1 [h_2^{(1)}]\rangle_{H_1} \\
&= \langle B^\star [h_1], B^\star [h_2]\rangle_{H_1}.
\end{aligned}$$

Similarly, one has, for the third term, that

$$\langle B[h_1^{(1)}], B_2 [h_2]\rangle_{H_2} = \langle B^\star [h_1], B^\star [h_2]\rangle_{H_1}.$$

Thus

$$\langle h_1, h_2\rangle_{H_{B_2}} = \langle h_1, h_2\rangle_{H_2} + \langle B^\star [h_1], B^\star [h_2]\rangle_{H_1} + \langle B[h_1^{(1)}], B[h_2^{(1)}]\rangle_{H_2}.$$

Now, still using the same relation (that is, $(\star)$): $B^\star[h] = B_1[h^{(1)}]$, one has that

$$\langle B^\star[h_1], B^\star[h_2]\rangle_{H_1} = \langle B_1[h_1^{(1)}], B_1[h_2^{(1)}]\rangle_{H_1} = \langle B_1^2[h_1^{(1)}], h_2^{(1)}\rangle_{H_1}.$$

Recalling the definition of B_1, one has that

$$\begin{aligned}\langle B_1^2[h_1^{(1)}], h_2^{(1)}\rangle_{H_2} &= \langle h_1^{(1)}, h_2^{(1)}\rangle_{H_2} - \langle B^\star B[h_1^{(1)}], h_2^{(1)}\rangle_{H_2}\\ &= \langle h_1^{(1)}, h_2^{(1)}\rangle_{H_2} - \langle B[h_1^{(1)}], B[h_2^{(1)}]\rangle_{H_2}.\end{aligned}$$

Thus finally

$$\langle h_1, h_2\rangle_{H_{B_2}} = \langle h_1, h_1\rangle_{H_2} + \langle h_1^{(1)}, h_2^{(1)}\rangle_{H_1}.$$

As $h_1^{(1)} \perp \mathcal{N}[B_1]$ and $h_2^{(1)} \perp \mathcal{N}[B_1]$,

$$\begin{aligned}\langle h_1^{(1)}, h_2^{(1)}\rangle_{H_1} &= \langle P_{\mathcal{N}[B_1]^\perp}[h_1^{(1)}], P_{\mathcal{N}[B_1]^\perp}[h_2^{(1)}]\rangle_{H_1}\\ &= \langle B_1[h_1^{(1)}], B_1[h_2^{(1)}]\rangle_{H_{B_1}}\\ &= \langle B^\star[h_1], B^\star[h_2]\rangle_{H_{B_1}},\end{aligned}$$

□

which establishes item 2.

Corollary 3.2.18 *When $B : H_1 \longrightarrow H_2$ is a Hilbert spaces contraction, the overlapping space has the following representation:*

$$H_B \cap H_{B_2} = BH_{B_1}.$$

Proof ($BH_{B_1} \subseteq H_B \cap H_{B_2}$)
Certainly $BH_{B_1} \subseteq H_B$. Let thus $h_1 \in H_{B_1}$, and choose $h_1^{(1)} \in H_1$, with $h_1 = B_1[h_1^{(1)}]$. Then (Proposition) 3.2.15 yields that

$$B[h_1] = BB_1[h_1^{(1)}] = B_2B[h_1^{(1)}],$$

so that $B[h_1] \in H_B \cap H_{B_2}$.

Proof ($H_B \cap H_{B_2} \subseteq BH_{B_1}$)
Definition 3.2.10 applied to $B^\star : H_2 \longrightarrow H_1$ says that

- $[B^\star]_2$ is the map on H_1 defined using $[B^\star]_2 = (I_{H_1} - B^\star B)^{1/2} = B_1$,
- $[B^\star]_1$ is the map on H_2 defined using $[B^\star]_1 = (I_{H_2} - BB^\star)^{1/2} = B_2$.

One then remarks that (Proposition) 3.2.17, used with $B^\star$ in place of B, yields, with, for example, $[B^\star]_1$ being the pendant of B_1 for $B^\star$, that (iff $\equiv$ if, and only if,)

$$h_1 \in H_{[B^\star]_2} \text{ iff } (B^\star)^\star[h_1] \in H_{[B^\star]_1}, \text{ that is, iff } B[h_1] \in H_{[B^\star]_1},$$

which translates to

$$h_1 \in H_{B_1} \text{ iff } B\left[h_1\right] \in H_{B_2}.$$

Consequently, when $h_2 \in H_B \cap H_{B_2}$, there are $h_2^{(1)} \in H_1$ and $h_2^{(2)} \in H_2$ such that

$$h_2 = B\left[h_2^{(1)}\right] = B_2\left[h_2^{(2)}\right].$$

That means, in particular, that $h_2^{(1)} \in H_{B_1}$, and thus that $h_2 \in BH_{B_1}$. □

Corollary 3.2.19 *When $B : H_1 \longrightarrow H_2$ is a real Hilbert spaces contraction, then $H_B \cap H_{B_2} = \{0_{H_2}\}$ if, and only if, H_B and H_{B_2} are ordinary subspaces of H_2 orthogonal to each other.*

Proof The null operator from the Hilbert space H_1 to the Hilbert space H_2 shall be denoted O_{H_1,H_2}, and O_H, when $H_1 = H_2 = H$.

Proof Suppose that H_B and H_{B_2} are ordinary subspaces of H_2 orthogonal to each other.

Then [266, p. 84] $P_{H_B}P_{H_{B_2}} = P_{H_{B_2}}P_{H_B} = O_{H_2}$, so that $P_{H_B \cap H_{B_2}} = O_{H_2}$, or

$$H_B \cap H_{B_2} = \{0_{H_2}\}.$$

Proof Suppose that $H_B \cap H_{B_2} = \{0_{H_2}\}$.

Given that [(Corollary) 3.2.18] $H_B \cap H_{B_2} = BH_{B_1}$, one has that

$$H_{B_1} \subseteq \mathcal{N}[B], \text{ so that } BB_1 = O_{H_1,H_2}, \text{ and thus } BB_1^2 = O_{H_1,H_2}.$$

Consequently, $B\left(I_{H_1} - B^\star B\right) = O_{H_1,H_2}$, or $B = BB^\star B$, so that

$$\mathcal{R}[B] = \mathcal{R}[BB^\star],$$

and, furthermore, that

$$BB^\star = \left(BB^\star\right)^2, \text{ and } B^\star B = \left(B^\star B\right)^2.$$

$BB^\star$ and $B^\star B$ are thus idempotent. Since they are moreover selfadjoint, they are projections [266, p. 83]. The consequence is [266, p. 86] that B is a partial isometry whose initial set is the range of $B^\star B$, and whose final set is the range of $BB^\star$. It follows that [266, p. 86] $BB^\star = P_{H_B}$, and that H_B is closed. Since, by definition,

$$B_2^2 = I_{H_2} - BB^\star = I_{H_2} - P_{H_B},$$

it is a projection orthogonal to P_{H_B} [266, p. 83]. But then [266, p. 86] B_2 is a partial isometry, and H_{B_2} is thus a subspace. Since $P_{H_B} = BB^\star$, and

$$P_{H_{B_2}} = B_2 B_2^\star = B_2^2 = I_{H_2} - BB^\star,$$

one has that [266, p. 84]

$$P_{H_B} P_{H_{B_2}} = P_{H_{B_2}} P_{H_B} = P_{H_B \cap H_{B_2}} = P_{\{0_{H_2}\}} = O_{H_2},$$

so that $H_B \perp H_{B_2}$. □

Fact 3.2.20 *Let H and K be real Hilbert spaces, and $H \oplus K$, their direct sum. Let*

$$\Pi_H : H \oplus K \longrightarrow H \text{ be defined using } \Pi_H[(h,k)] = h,$$

$$P_H : H \oplus K \longrightarrow H \oplus K \text{ be the projection onto } H \times \{0_K\}.$$

Then $\Pi_H^\star[h] = (h, 0_K)$, and $\Pi_H^\star \Pi_H = P_H$.

Proof One has that $\langle \Pi_H[(h,k)], h_0 \rangle_H = \langle h, h_0 \rangle_H = \langle (h,k), (h_0, 0_K) \rangle_{H \oplus K}$.

Lemma 3.2.21 *Let H, H_1 and H_2 be real Hilbert spaces, and*

$$C : H_1 \longrightarrow H,\ D : H \longrightarrow H_2, \text{ be contractions, and } B = DC.$$

Set

$$B_2 = (I_{H_2} - BB^\star)^{1/2},\ C_2 = (I_H - CC^\star)^{1/2},\ D_2 = (I_{H_2} - DD^\star)^{1/2}.$$

Define

$$S_1 : \quad H_2 \longrightarrow H_2 \text{ using } S_1[h_2] = D_2[h_2],$$

$$S_2 : \quad H \longrightarrow H_2 \text{ using } S_2[h] = DC_2[h],$$

$$S : H_2 \oplus H \longrightarrow H_2 \text{ using } S = S_1 \Pi_{H_2} + S_2 \Pi_H.$$

1. One has the following equalities:

(i) $S^\star = (D_2, C_2 D^\star) = (S_1, S_2^\star)$,
(ii) $SS^\star = I_{H_2} - BB^\star = B_2^2$,
(iii) $H_S = H_{B_2}$.

2. Let H^0 and H_2^0 be two closed subspaces of H and H_2 respectively, and

$$H_0 = H_2^0 \times H^0.$$

Then

$$P_{H_0} = (P_{H_2^0}, P_{H^0}).$$

Proof One has, using (Fact) 3.2.20, that

$$S^\star = \Pi_{H_2}^\star S_1^\star + \Pi_H^\star S_2^\star = \Pi_{H_2}^\star D_2 + \Pi_H^\star C_2 D^\star,$$

which yields (i) of item 1. Now

$$\begin{aligned} SS^\star &= S(D_2, C_2D^\star) \\ &= S_1D_2 + S_2C_2D^\star \\ &= D_2^2 + DC_2^2D^\star \\ &= (I_{H_2} - DD^\star) + D(I_H - CC^\star)D^\star \\ &= I_{H_2} - DCC^\star D^\star \\ &= I_{H_2} - BB^\star \\ &= B_2^2. \end{aligned}$$

That, because of (Corollary) 3.2.7, yields the last equality of item 1.

One has that

$$H_0 = H_2^0 \times \{0_H\} + \{0_{H_2}\} \times H^0 = \Pi_{H_2}^\star[H_2^0] + \Pi_H^\star[H^0]$$

which yields

$$P_{H_0} = \Pi_{H_2}^\star P_{H_2^0} + \Pi_H^\star P_{H^0}.$$

□

Proposition 3.2.22 *Let the assumptions and the notation be those of (Lemma)* 3.2.21. *Then:*

1. $H_{B_2} = H_{D_2} + DH_{C_2}$ (in particular $H_{D_2} \subseteq H_{B_2}$).
2. *If* $h_2 \in H_{B_2}$ *has the representation* $h_2 = d + D[c]$, *with* $d \in H_{D_2}$, *and* $c \in H_{C_2}$, *then*

$$\|h_2\|_{H_{B_2}}^2 \leq \|d\|_{H_{D_2}}^2 + \|c\|_{H_{C_2}}^2 .$$

3. *For each* $h_2 \in H_{B_2}$, *there is a unique* $(c,d) \in H_{C_2} \times H_{D_2}$ *such that*

$$h_2 = d + D[c], \quad \text{and} \quad \|h_2\|_{H_{B_2}}^2 = \|d\|_{H_{D_2}}^2 + \|c\|_{H_{C_2}}^2 .$$

4. H_{D_2} *is contained contractively in* H_{B_2} (see item 1).
5. *Let* $D^{|H_{C_2}}$ *be* D *restricted to* H_{C_2}. *It is a contraction into* H_{B_2}.

Proof [1] Because of (Lemma) 3.2.21, item 1, result (iii), one has that $H_{B_2} = H_S$. The definition of S, that is $= D_2\Pi_{H_2} + DC_2\Pi_H$, yields the claim as

$$S[(h,k)] = D_2[h] + DC_2[k]. \qquad (\star)$$

Proof [2 + 3] Let $h_2 = B_2\left[h_2^{(1)}\right]$, $d = D_2\left[h_2^{(2)}\right]$ and $c = C_2[h]$. The following relation: $h_2 = d + D[c]$ may then be expressed in the ensuing manner:

$$\begin{aligned} B_2\left[h_2^{(1)}\right] &= D_2\left[P_{\mathcal{N}[D_2]^\perp}\left[h_2^{(2)}\right]\right] + DC_2\left[P_{\mathcal{N}[C_2]^\perp}[h]\right] \\ &= S\left(P_{\mathcal{N}[D_2]^\perp}\left[h_2^{(2)}\right], P_{\mathcal{N}[C_2]^\perp}[h]\right). \end{aligned}$$

Thus, since $H_{B_2} = H_S$ [(Lemma) 3.2.21], by definition of H_S,

$$\begin{aligned} (a)\quad \left\|B_2\left[h_2^{(1)}\right]\right\|_{H_{B_2}}^2 &= \left\|S\left(P_{\mathcal{N}[D_2]^\perp}\left[h_2^{(2)}\right], P_{\mathcal{N}[C_2]^\perp}[h]\right)\right\|_{H_S}^2 \\ &= \left\|P_{\mathcal{N}[S]^\perp}\left(P_{\mathcal{N}[D_2]^\perp}\left[h_2^{(2)}\right], P_{\mathcal{N}[C_2]^\perp}[h]\right)\right\|_{H_2\oplus H}^2. \end{aligned}$$

Let H_0 be as in (Lemma) 3.2.21, with $H^0 = \mathcal{N}[C_2]$, and $H_2^0 = \mathcal{N}[D_2]$. One then has that

$$P_{H_0^\perp} = I_{H_2\oplus H} - P_{H_0} = (I_{H_2}, I_H) - \left(P_{\mathcal{N}[D_2]}, P_{\mathcal{N}[C_2]}\right) = \left(P_{\mathcal{N}[D_2]^\perp}, P_{\mathcal{N}[C_2]^\perp}\right),$$

and, since $H_0 = \mathcal{N}[D_2] \times \mathcal{N}[C_2]$, that (because of $(\star)$)

$$H_0 \subseteq \mathcal{N}[S], \text{ and consequently that } H_0^\perp \supseteq \mathcal{N}[S]^\perp.$$

Thus [266, p. 84]

$$\begin{aligned} (b)\quad P_{\mathcal{N}[S]^\perp}\left(P_{\mathcal{N}[D_2]^\perp}\left[h_2^{(2)}\right], P_{\mathcal{N}[C_2]^\perp}[h]\right) &= P_{\mathcal{N}[S]^\perp}P_{H_0^\perp}\left[\left(h_2^{(2)}, h\right)\right] \\ &= P_{H_0^\perp}P_{\mathcal{N}[S]^\perp}\left[\left(h_2^{(2)}, h\right)\right], \end{aligned}$$

and

$$(c)\quad \left\|P_{\mathcal{N}[S]^\perp}P_{H_0^\perp}\left[\left(h_2^{(2)}, h\right)\right]\right\|_{H_2\oplus H}^2 \le \left\|P_{H_0^\perp}\left[\left(h_2^{(2)}, h\right)\right]\right\|_{H_2\oplus H}^2.$$

Let $P_{\mathcal{N}[S]^\perp}\left(h^{(2)}, h\right) = \left(\tilde{h}_2^{(2)}, \tilde{h}\right)$. Then, from (a) and (b), one gets that

$$\begin{aligned} \left\|B_2\left[h_2^{(1)}\right]\right\|_{H_{B_2}}^2 &= \left\|P_{H_0^\perp}\left[\left(\tilde{h}_2^{(2)}, \tilde{h}\right)\right]\right\|_{H_2\oplus H}^2 \\ &= \left\|P_{\mathcal{N}[D_2]^\perp}\left[\tilde{h}_2^{(2)}\right]\right\|_{H_2}^2 + \left\|P_{\mathcal{N}[C_2]^\perp}\left[\tilde{h}\right]\right\|_{H}^2, \end{aligned}$$

and, from (a), (b), and (c), that

$$\begin{aligned}\left\|B_2\left[h_2^{(1)}\right]\right\|_{H_{B_2}}^2 &\leq \left\|P_{H_0^{\perp}}\left[\left(h_2^{(2)}, h\right)\right]\right\|_{H_2 \oplus H}^2 \\ &= \left\|P_{\mathcal{N}[D_2]^{\perp}}\left[h_2^{(2)}\right]\right\|_{H_2}^2 + \left\|P_{\mathcal{N}[C_2]^{\perp}}[h]\right\|_H^2 .\end{aligned}$$

Consequently

$$\begin{aligned}\left\|B_2\left[h_2^{(1)}\right]\right\|_{H_{B_2}}^2 &= \left\|P_{\mathcal{N}[D_2]^{\perp}}\left[\tilde{h}_2^{(2)}\right]\right\|_{H_2}^2 + \left\|P_{\mathcal{N}[C_2]^{\perp}}\left[\tilde{h}\right]\right\|_H^2 \\ &\leq \left\|P_{\mathcal{N}[D_2]^{\perp}}\left[h_2^{(2)}\right]\right\|_{H_2}^2 + \left\|P_{\mathcal{N}[C_2]^{\perp}}[h]\right\|_H^2 .\end{aligned}$$

It is then sufficient to notice that, by definition [(Proposition) 3.2.2],

$$\begin{aligned}\left\|P_{\mathcal{N}[D_2]^{\perp}}\left[\tilde{h}_2^{(2)}\right]\right\|_{H_2}^2 &= \left\|D_2\left[\tilde{h}_2^{(2)}\right]\right\|_{H_{D_2}}^2 , \\ \left\|P_{\mathcal{N}[D_2]^{\perp}}\left[h_2^{(2)}\right]\right\|_{H_2}^2 &= \left\|D_2\left[h_2^{(2)}\right]\right\|_{H_{D_2}}^2 , \\ \left\|P_{\mathcal{N}[C_2]^{\perp}}\left[\tilde{h}\right]\right\|_H^2 &= \left\|C_2\left[\tilde{h}\right]\right\|_{H_{C_2}}^2 , \\ \left\|P_{\mathcal{N}[C_2]^{\perp}}[h]\right\|_H^2 &= \left\|C_2[h]\right\|_{H_{C_2}}^2 .\end{aligned}$$

Proof [4 + 5] Item 1 says that $H_{D_2} \subseteq H_{B_2}$, and item 2, with $c = 0$, yields that

$$\|d\|_{H_{B_2}} \leq \|d\|_{H_{D_2}} .$$

Item 5 is item 2 with $d = 0$. □

Corollary 3.2.23 *Let the assumptions and notation be those of* (Proposition) 3.2.22, *and suppose that* $H_{D_2} \cap D\left[H_{C_2}\right] = \{0_{H_2}\}$. *Then:*

1. $H_{B_2} = H_{D_2} + DH_{C_2}$ *is an orthogonal direct sum.*
2. H_{D_2} *is contained isometrically in* H_{B_2} (see item 1).
3. $D^{|H_{C_2}}$, *the restriction of* D *to* H_{C_2}, *is a partial isometry from* H_{C_2} *to* H_{B_2} (from item 1, $DH_{C_2} \subseteq H_{B_2}$) *with initial set* $C_2\left[\mathcal{N}[DC_2]^{\perp}\right]$.

Proof [1] By definition, for fixed, but arbitrary $(h_2, h) \in H_2 \times H$,

$$S(h_2, h) = S_1[h_2] + S_2[h], \quad \text{with } S_1[h_2] = D_2[h_2],\ S_2[h] = DC_2[h] .$$

Since $H_{D_2} \cap D\left[H_{C_2}\right] = \{0_{H_2}\}$,

$$\mathcal{N}[S] = \mathcal{N}[S_1] \times \mathcal{N}[S_2] = \left(\mathcal{N}[S_1] \times \{0_H\}\right) + \left(\{0_{H_2}\} \times \mathcal{N}[S_2]\right),$$

an orthogonal decomposition. Consequently [266, p. 84], using (Lemma) 3.2.21, item 2, with $H^0 = \mathcal{N}[S_2]$ and $H_2^0 = \mathcal{N}[S_1]$,

$$\begin{aligned} P_{\mathcal{N}[S]^\perp} &= I_{H_2 \oplus H} - \left(P_{H_2^0}, P_{H^0}\right) \\ &= (I_{H_2}, I_H) - \left(P_{\mathcal{N}[S_1]}, P_{\mathcal{N}[S_2]}\right) \\ &= \left(P_{\mathcal{N}[S_1]^\perp}, P_{\mathcal{N}[S_2]^\perp}\right), \end{aligned}$$

and, for fixed, but arbitrary $(h_2, h) \in H_2 \times H$,

$$\begin{aligned} \langle D_2 [h_2], DC_2 [h] \rangle_{H_{B_2}} &= \langle S (h_2, 0_H), S (0_{H_2}, h) \rangle_{H_S} \\ &= \langle P_{\mathcal{N}[S]^\perp} (h_2, 0_H), P_{\mathcal{N}[S]^\perp} (0_{H_2}, h) \rangle_{H_2 \oplus H} \\ &= \langle \left(P_{\mathcal{N}[S_1]^\perp} [h_2], 0_H\right), \left(0_{H_2}, P_{\mathcal{N}[S_2]^\perp} [h]\right) \rangle_{H_2 \oplus H} \\ &= 0. \end{aligned}$$

Proof [2] One has, for fixed, but arbitrary $h_2 \in H_2$, since $H_S = H_{B_2}$ [(Lemma) 3.2.21],

$$\begin{aligned} \|D_2 [h_2]\|^2_{H_{D_2}} &= \left\|P_{\mathcal{N}[D_2]^\perp} [h_2]\right\|^2_{H_2} \\ &= \left\|P_{\mathcal{N}[S_1]^\perp} [h_2]\right\|^2_{H_2} \\ &= \left\|P_{\mathcal{N}[S]^\perp} (h_2, 0_H)\right\|^2_{H_2 \oplus H} \\ &= \|S (h_2, 0_H)\|^2_{H_S} \\ &= \|D_2 [h_2]\|^2_{H_{B_2}}. \end{aligned}$$

Proof [3] Let $D^{|H_{C_2}}$ denote the restriction of D to H_{C_2}. As

$$\mathcal{N}[D] \cap H_{C_2} \subseteq \mathcal{N}[D^{|H_{C_2}}],$$

one has that, in H_{C_2},

$$(\mathcal{N}[D] \cap H_{C_2})^\perp \supseteq \mathcal{N}[D^{|H_{C_2}}]^\perp.$$

Thus, in proving item 3, one may restrict attention to $(\mathcal{N}[D] \cap H_{C_2})^\perp$, the orthogonal complement being in H_{C_2}. Suppose thus that

$$h \in (\mathcal{N}[D] \cap H_{C_2})^\perp \subseteq H_{C_2}, \text{ with } h = C_2 \left[\tilde{h}^\perp\right], \text{ and } \tilde{h}^\perp \in \mathcal{N}[C_2]^\perp \subseteq H.$$

Since $\mathcal{N}[C_2] \subseteq \mathcal{N}[DC_2]$, in H, $\mathcal{N}[C_2]^{\perp} \supseteq \mathcal{N}[DC_2]^{\perp}$, so that one may consider, successively,

$$\tilde{h}^{\perp} \in \mathcal{N}[DC_2]^{\perp} \text{ and then } \tilde{h}^{\perp} \in \mathcal{N}[DC_2].$$

Suppose thus at first that $\tilde{h}^{\perp} \in \mathcal{N}[DC_2]^{\perp} = \mathcal{N}[S_2]^{\perp}$. Then, since $H_S = H_{B_2}$, using the various definitions, and in particular (Proposition) 3.2.2 and (Lemma) 3.2.21,

$$\begin{aligned}
\|D[h]\|_{H_{B_2}}^2 &= \left\|DC_2\left[\tilde{h}^{\perp}\right]\right\|_{H_{B_2}}^2 \\
&= \left\|S\left(0_{H_2}, \tilde{h}^{\perp}\right)\right\|_{H_S}^2 \\
&= \left\|P_{\mathcal{N}[S]^{\perp}}\left(0_{H_2}, \tilde{h}^{\perp}\right)\right\|_{H_2 \oplus H}^2 \\
&= \left\|\left(0_{H_2}, P_{\mathcal{N}[S_2]^{\perp}}\left[\tilde{h}^{\perp}\right]\right)\right\|_{H_2 \oplus H}^2 \\
&= \left\|P_{\mathcal{N}[S_2]^{\perp}}\left[\tilde{h}^{\perp}\right]\right\|_{H}^2 \\
&= \left\|\tilde{h}^{\perp}\right\|_{H}^2 \\
&= \left\|P_{\mathcal{N}[C_2]^{\perp}}\left[\tilde{h}^{\perp}\right]\right\|_{H}^2 \\
&= \left\|C_2\left[\tilde{h}^{\perp}\right]\right\|_{H_{C_2}}^2 \\
&= \|h\|_{H_{C_2}}^2,
\end{aligned}$$

so that there is preservation of norms when

$$h \in \left(\mathcal{N}[D] \cap H_{C_2}\right)^{\perp} \subseteq H_{C_2} \text{ has the form } C_2\left[\tilde{h}^{\perp}\right],\ \tilde{h}^{\perp} \in \mathcal{N}[DC_2]^{\perp}.$$

Suppose then that $\tilde{h}^{\perp} \in \mathcal{N}[DC_2]$. Since

$$DC_2\left[\tilde{h}^{\perp}\right] = 0,\ C_2\left[\tilde{h}^{\perp}\right] \in \mathcal{N}[D],\ \text{so that } C_2\left[\tilde{h}^{\perp}\right] \in \mathcal{N}[D] \cap H_{C_2},$$

and thus the elements $\tilde{h}^{\perp} \in \mathcal{N}[DC_2]$ are sent to elements of H_{C_2} in the kernel of D. Thus the cases for which $\tilde{h}^{\perp} \in \mathcal{N}[DC_2]$ may be ignored.

For item 3 to obtain, it remains to prove that $C_2\left[\mathcal{N}[DC_2]^{\perp}\right]$ is a closed subset of H_{C_2}. But, for fixed, but arbitrary $\{h_m, h_n\} \subseteq \mathcal{N}[DC_2]^{\perp}$,

$$\|C_2[h_m] - C_2[h_n]\|_{H_{C_2}} = \left\|P_{\mathcal{N}[C_2]^{\perp}}[h_m] - P_{\mathcal{N}[C_2]^{\perp}}[h_n]\right\|_H.$$

Let $\lim_n (H)\, P_{\mathcal{N}[C_2]^{\perp}}[h_n] = h \in \mathcal{N}[C_2]^{\perp}$. Then

$$\lim_n \|C_2[h] - C_2[h_n]\|_{H_{C_2}} = 0.$$

□

From the point of view that prevails here, the main result in this section is the following proposition.

Proposition 3.2.24 *Suppose that* $B : H_1 \longrightarrow H_2$ *is a Hilbert spaces contraction, and that* $B_2 = (I_{H_2} - BB^\star)^{1/2}$. *Then:*

1. $H_2 = H_B + H_{B_2}$;
2. *for* $h_2 \in H_2$ *such that* $h_2 = B[h_1] + h_{B_2}$, *with* $h_1 \in H_1, h_{B_2} \in H_{B_2}$, *one has that*

$$\|h_2\|_{H_2}^2 \leq \|h_1\|_{H_1}^2 + \|h_{B_2}\|_{H_{B_2}}^2 ;$$

3. *equality in item 2 is achieved when using* $h_1 = B^\star[h_2]$ *and* $h_{B_2} = B_2^2[h_2]$.

Proof Those statements are a consequence of (Proposition) 3.2.22 when the following choices are made:

- H is chosen to be H_1,
- C is chosen to be $O_{H_1,H_1} = O_{H_1}$, the null operator,
- D is chosen to be B (of statement (Proposition) 3.2.24, not of (Proposition) 3.2.22).

The operator B of (Proposition) 3.2.22 is then O_{H_1,H_2}.

As a consequence of those choices, H_{B_2} becomes H_2, H_{C_2} becomes H_1, and H_{D_2} becomes H_{B_2}, and thus

- $H_{B_2} = H_{D_2} + DH_{C_2}$ translates into

$$H_2 = BH_1 + H_{B_2} = H_B + H_{B_2},$$

- $h_2 = D[c] + d,\ h_2 \in H_{B_2},\ c \in H_{C_2},\ d \in H_{D_2}$, translates into

$$h_2 = B[h_1] + h_{B_2},\ h_1 \in H_1,\ h_2 \in H_2,\ h_{B_2} \in H_{B_2},$$

- $\|h_2\|_{H_{B_2}}^2 \leq \|c\|_{H_{C_2}}^2 + \|d\|_{H_{D_2}}^2$ translates into

$$\|h_2\|_{H_2}^2 \leq \|h_1\|_{H_1}^2 + \|h_{B_2}\|_{H_{B_2}}^2 .$$

These are items 1 and 2. Also

- $S_1 = D_2$ translates into

$$S_1 = B_2,$$

- $S_2 = DC_2$ translates into

$$S_2 = B,$$

- $S^\star = (D_2, C_2 D^\star)$ translates into

$$S^\star = (B_2, B^\star) .$$

Thus

$$\begin{aligned}\|S^\star[h_2]\|^2_{H_2 \oplus H_1} &= \|(B_2[h_2], B^\star[h_2])\|^2_{H_2 \oplus H_1} \\ &= \|B_2[h_2]\|^2_{H_2} + \|B^\star[h_2]\|^2_{H_1} \\ &= \langle B_2^2[h_2], h_2\rangle_{H_2} + \langle BB^\star[h_2], h_2\rangle_{H_2} \\ &= \|h_2\|^2_{H_2} .\end{aligned}$$

$S^\star$is thus an isometry, and [266, p. 71] $\mathcal{N}[S]^\perp = \overline{\mathcal{R}[S^\star]} = \mathcal{R}[S^\star]$. Since, given the definition of B_2,

$$h_2 = B_2^2[h_2] + BB^\star[h_2] ,$$

and that the latter may be written as

$$h_2 = B_2[B_2[h_2]] + B[B^\star[h_2]] = S(B_2[h_2], B^\star[h_2]) ,$$

which is the translation of

$$B_2\left[h_2^{(1)}\right] = D_2\left[h_2^{(2)}\right] + DC_2[h] = S\left(h_2^{(2)}, h\right) ,$$

one has equality in item 2 when replacing (see the proof of (Proposition) 3.2.22) $\left(h_2^{(2)}, h\right)$ with $P_{\mathcal{N}[S]^\perp}\left(h_2^{(2)}, h\right)$. But here

$$P_{\mathcal{N}[S]^\perp}(B_2[h_2], B^\star[h_2]) = P_{\mathcal{R}[S^\star]} S^\star[h_2] = S^\star[h_2] = (B_2[h_2], B^\star[h_2]) .$$

□

Corollary 3.2.25 *Let C_1 and C_2 be covariances on T, with C_2 dominating C_1. Then, for fixed, but arbitrary $h^{(1)} \in H(C_1, T)$,*

$$\left\|h^{(1)}\right\|^2_{H(C_1,T)} = \left\|J_{2,1}^\star\left[h^{(1)}\right]\right\|^2_{H(C_2,T)} + \left\|\left\{I_{H(C_1,T)} - J_{2,1}J_{2,1}^\star\right\}^{1/2}\left[h^{(1)}\right]\right\|^2_{H(C_1,T)} .$$

Analogously, one has also that, for fixed, but arbitrary $h^{(2)} \in H(C_2, T)$,

$$\left\|h^{(2)}\right\|^2_{H(C_2,T)} = \left\|J_{2,1}\left[h^{(2)}\right]\right\|^2_{H(C_1,T)} + \left\|\left\{I_{H(C_2,T)} - J_{2,1}^\star J_{2,1}\right\}^{1/2}\left[h^{(2)}\right]\right\|^2_{H(C_2,T)} .$$

Proof In (Proposition) 3.2.24 choose

- $H_1 = H(C_2, T)$,
- $H_2 = H(C_1, T)$,
- $B = J_{2,1}$.

Then for fixed, but arbitrary $h^{(1)} \in H(C_1, T)$, trivially,

$$h^{(1)} = J_{2,1}J_{2,1}^{\star}\left[h^{(1)}\right] + \left\{I_{H(C_1,T)} - J_{2,1}J_{2,1}^{\star}\right\}\left[h^{(1)}\right] = BB^{\star}\left[h^{(1)}\right] + B_2^2\left[h^{(1)}\right],$$

so that

$$\begin{aligned}\left\|h^{(1)}\right\|_{H(C_1,T)}^2 &= \left\|J_{2,1}^{\star}\left[h^{(1)}\right]\right\|_{H(C_2,T)}^2 + \left\|P_{\mathcal{N}[B_2]^{\perp}}B_2\left[h^{(1)}\right]\right\|_{H(C_1,T)}^2 \\ &= \left\|J_{2,1}^{\star}\left[h^{(1)}\right]\right\|_{H(C_2,T)}^2 + \left\|B_2\left[h^{(1)}\right]\right\|_{H(C_1,T)}^2 \\ &= \left\|J_{2,1}^{\star}\left[h^{(1)}\right]\right\|_{H(C_2,T)}^2 + \left\|\left\{I_{H(C_1,T)} - J_{2,1}J_{2,1}^{\star}\right\}^{1/2}\left[h^{(1)}\right]\right\|_{H(C_1,T)}^2.\end{aligned}$$

For the second equality, let

- $H_1 = H(C_1, T)$,
- $H_2 = H(C_2, T)$,
- $B = J_{2,1}^{\star}$.

For fixed, but arbitrary $h^{(2)} \in H(C_2, T)$, trivially,

$$h^{(2)} = J_{2,1}^{\star}J_{2,1}\left[h^{(2)}\right] + \left\{I_{H(C_2,T)} - J_{2,1}^{\star}J_{2,1}\right\}\left[h^{(2)}\right] = BB^{\star}\left[h^{(2)}\right] + B_2^2\left[h^{(2)}\right],$$

and then (Proposition) 3.2.24 yields the result. □

3.3 Intersections of Reproducing Kernel Hilbert Spaces

It is the "relative size" of the intersection of the RKHS's associated with the respective covariances of two processes that determines in part whether the laws of those processes are equivalent or orthogonal (Sect. 5.3.2). Hence this section!

Suppose C, C_1, and C_2 are covariances on T such that $C = C_1 + C_2$. Let the following somewhat shorter notation be used:

$$J_1 : H(C, T) \longrightarrow H(C_1, T) \text{ is defined using } J_1[C(\cdot, t)] = C_1(\cdot, t),$$

$$J_2 : H(C, T) \longrightarrow H(C_2, T) \text{ is defined using } J_2[C(\cdot, t)] = C_2(\cdot, t).$$

Define also

$$H_1 = J_1^\star \left[H\left(C_1, T\right)\right] \subseteq H\left(C, T\right),$$

$$H_2 = J_2^\star \left[H\left(C_2, T\right)\right] \subseteq H\left(C, T\right),$$

$$H_0 = H\left(C_1, T\right) \cap H\left(C_2, T\right).$$

Remark 3.3.1 H_1 and H_2 are thus, respectively, $H\left(C_1, T\right)$ and $H\left(C_2, T\right)$, regarded as submanifolds of $H\left(C, T\right)$, rather than Hilbert spaces in their own right. Similarly $H_1^\perp$ is a (closed) subspace of $H\left(C, T\right)$, but, as shall be established, it is also a subset of H_2, that is $H_1^\perp$ is a subset of $H\left(C_2, T\right)$. In an effort to maintain a distinction between subspaces and subsets, one shall thus write, whenever uncertainty of meaning is possible, and a distinction appears useful, for example, $H_1^\perp$ for the subspace, and $s\left[H_1^\perp\right]$ for the subset.

Proposition 3.3.2 *Let C, C_1, C_2 be covariances on T such that $C = C_1 + C_2$. Let $J_1, J_2, H_0, H_1, H_1^\perp, H_2$ be as described above.*

1. *The following equalities obtain:*

$$J_1^\star J_1 + J_2^\star J_2 = I_{H(C,T)},$$

and

$$\|h\|_{H(C,T)}^2 = \|J_1\left[h\right]\|_{H(C_1,T)}^2 + \|J_2\left[h\right]\|_{H(C_2,T)}^2.$$

2. *For $h \in H_1^\perp$, $J_2^\star J_2\left[h\right] = h$, and*

$$\|h\|_{H(C,T)} = \|J_2^\star J_2\left[h\right]\|_{H(C,T)} = \|J_2\left[h\right]\|_{H(C_2,T)}.$$

In particular, set-wise, $s\left[H_1^\perp\right] \subseteq J_2^\star\left[H\left(C_2, T\right)\right] = H_2$.

3. *For $h \in H_0$,*

$$\langle J_1 J_1^\star\left[h\right], h\rangle_{H(C_1,T)} = \|J_1^\star h\|_{H(C,T)}^2 = \|J_2^\star h\|_{H(C,T)}^2 = \langle J_2 J_2^\star\left[h\right], h\rangle_{H(C_2,T)}.$$

4. *For $h \in H\left(C, T\right)$, $\langle J_1^\star J_1\left[h\right], J_2 J_2^\star\left[h\right]\rangle_{H(C,T)} \geq 0$.*

Proof One must remember that $C_1 \ll C$, and that $C_2 \ll C$. Furthermore, for example,

$$C_1\left(t_1, t_2\right) = \langle J_1\left[C\left(\cdot, t_1\right)\right], J_1\left[C\left(\cdot, t_2\right)\right]\rangle_{H(C_1,T)}.$$

For checking item 1, one may thus rewrite the equality

$$C\left(t_1, t_2\right) = C_1\left(t_1, t_2\right) + C_2\left(t_1, t_2\right)$$

as

$$\begin{aligned}\langle C(\cdot,t_1),C(\cdot,t_2)\rangle_{H(C,T)} &= \langle J_1[C(\cdot,t_1)],J_1[C(\cdot,t_2)]\rangle_{H(C_1,T)}\\ &\quad+\langle J_2[C(\cdot,t_1)],J_2[C(\cdot,t_2)]\rangle_{H(C_2,T)}\\ &= \langle J_1^\star J_1[C(\cdot,t_1)],C(\cdot,t_2)\rangle_{H(C,T)}\\ &\quad+\langle J_2^\star J_2[C(\cdot,t_1)],C(\cdot,t_2)\rangle_{H(C,T)}\\ &= \langle\{J_1^\star J_1+J_2^\star J_2\}[C(\cdot,t_1)],C(\cdot,t_2)\rangle_{H(C,T)}.\end{aligned}$$

Linearity and continuity lead, for $(h_1,h_2)\in H(C,T)\times H(C,T)$, fixed, but arbitrary, to

$$\langle h_1,h_2\rangle_{H(C,T)} = \langle\{J_1^\star J_1+J_2^\star J_2\}[h_1],h_2\rangle_{H(C,T)},$$

which yields item 1.

For item 2, one may proceed as follows. For fixed, but arbitrary $h\in H_1^\perp$, and all $h_1\in H(C_1,T)$,

$$0=\langle h,J_1^\star[h_1]\rangle_{H(C,T)}=\langle J_1[h],h_1\rangle_{H(C_1,T)},$$

so that $J_1[h]=0$, and thus, from item 1, $h=J_2^\star J_2[h]$. Consequently

$$H_1^\perp\subseteq J_2^\star[H(C_2,T)]=H_2.$$

Furthermore, on one hand,

$$\|h\|_{H(C,T)}=\|J_2^\star J_2[h]\|_{H(C,T)},$$

and, on the other hand, from item 1,

$$\|h\|_{H(C,T)}=\|J_2[h]\|_{H(C_2,T)}.$$

Item 2 thus obtains.

Suppose now, for item 3, that $h\in H_0$. Then $J_1^\star[h]=J_2^\star[h]$, and

$$\|J_1^\star[h]\|_{H(C,T)}^2=\langle J_1^\star[h],J_1^\star[h]\rangle_{H(C,T)}=\langle J_1J_1^\star[h],h\rangle_{H(C_1,T)}.$$

Item 3 is thus true.

For item 4, one has that

$$\begin{aligned}\|J_1[h]\|_{H(C_1,T)}^2 &= \langle J_1^\star J_1[h],h\rangle_{H(C,T)}\\ &= \langle J_1^\star J_1[h],\{J_1^\star J_1+J_2^\star J_2\}[h]\rangle_{H(C,T)}\end{aligned}$$

$$= \|J_1^\star J_1[h]\|^2_{H(C,T)} + \langle J_1^\star J_1[h], J_2^\star J_2[h]\rangle_{H(C,T)}.$$

But, since $J_1^\star$ is a contraction, $\left\|J_1^\star J_1[h]\right\|^2_{H(C,T)} \leq \|J_1[h]\|^2_{H(C_1,T)}$. □

Fact 3.3.3 ([44, pp. 24–25]) *Suppose that H_1 and H_2 are closed subspaces of some Hilbert space H. Then*

1. $H_1 \cap H_2 = \left(H_1^\perp + H_2^\perp\right)^\perp$;
2. $(H_1 \cap H_2)^\perp \supseteq \overline{\left(H_1^\perp + H_2^\perp\right)}$;
3. $H_1^\perp \cap H_2^\perp = (H_1 + H_2)^\perp$;
4. $\left(H_1^\perp \cap H_2^\perp\right)^\perp = \overline{H_1 + H_2}$.

Proposition 3.3.4 *Let C, C_1, C_2 be covariances on T and suppose that $C = C_1 + C_2$. Let $J_1, J_2, H_0, H_1, H_1^\perp, H_2$ be as in* (Proposition) 3.3.2 *above. In particular $H_1^\perp$ is the orthogonal complement in $H(C,T)$.*

Let:

(a) $H_{1,2} = H_1^\perp \cap H_2$,
(b) $\overline{H}_0$ *be the closure of H_0 in $H(C_2,T)$,*
(c) $H_0^\perp = \overline{H}_0^\perp$ *be the orthogonal complement, in $H(C_2,T)$, of, respectively, H_0 and $\overline{H}_0$*

Then:

1. $s[H_{1,2}] = s\left[H_1^\perp\right]$.
2. *The manifold $s\left[H_1^\perp\right]$ is a* (closed) *subspace of $H(C_2,T)$: thus*

$$s[H_{1,2}] = s\left[H_1^\perp\right] \subseteq H(C_2,T)$$

as (closed) *subspaces of functions.*

3. $s\left[H_1^\perp\right] = H_0^\perp$.
4. *Define two covariances Γ_1 and Γ_2 as follows: for $(t,x) \in T \times T$, fixed, but arbitrary,*

$$\Gamma_1(x,t) = P_{\overline{H}_0}[C_2(\cdot,t)](x),$$

and

$$\Gamma_2(x,t) = P_{H_0^\perp}[C_2(\cdot,t)](x).$$

Then:

(i) *For $(t_1,t_2) \in T \times T$, fixed, but arbitrary,*

$$C_2(t_1,t_2) = \Gamma_1(t_1,t_2) + \Gamma_2(t_1,t_2),$$

and

$$H(\Gamma_1, T) \cap H(\Gamma_2, T) = \{0_{\mathbb{R}^T}\}.$$

(ii) *One has that*

$$H(\Gamma_2, T) \cap \quad H(C_1, T) \quad = \{0_{\mathbb{R}^T}\},$$

$$H(\Gamma_2, T) \cap H(C_1 + \Gamma_1, T) = \{0_{\mathbb{R}^T}\}.$$

(iii) $H_1^{\perp} = \{0_{\mathbb{R}^T}\}$ *if, and only if,* $\overline{H}_0 = H(C_2, T)$, *and then*

$$H(C_2, T) = H(\Gamma_1, T).$$

A sufficient condition is that $H(C_2, T) \subseteq H(C_1, T)$, *that is, that* C_2 *is dominated by some positive constant times* C_1 [(Proposition) 3.1.34].

(iv) $s\left[H_1^{\perp}\right] = H(C_2, T)$ *if, and only if,* $H_0 = \{0_{\mathbb{R}^T}\}$.

Then $H(C, T)$ *is isomorphic to* $H(C_1, T) \oplus H(C_2, T)$, *and the elements* $h \in H(C, T)$ *have the unique representation*

$$h = J_1^{\star} J_1 [h] + J_2^{\star} J_2 [h].$$

Furthermore $J_1^{\star} J_1$ *and* $J_2^{\star} J_2$ *are the projections onto* $H(C_1, T)$ *and* $H(C_2, T)$ *respectively.*

Proof One shall use tacitly the following properties of orthogonal complements of subsets of Hilbert spaces [266, p. 35]: for a set $S \subseteq H$, H a Hilbert space,

$$S^{\perp} = V[S]^{\perp} = \overline{V[S]}^{\perp}, \text{ and } S^{\perp\perp} = \overline{V[S]}.$$

As the orthogonal complement of $H_1 \subseteq H(C, T)$, $H_1^{\perp}$ is a (closed) subspace of $H(C, T)$. But item 2 of (Proposition) 3.3.2 asserts that $H_1^{\perp}$ is a subset of H_2, so that $H_1^{\perp} \cap H_2 = H_1^{\perp}$. Thus item 1 is true.

Since C dominates C_2, that $H_1^{\perp}$ is a convex and closed, and thus weakly closed [60, p. 126], subset of $H(C, T)$, it follows [(Corollary) 3.1.13] that $H_1^{\perp} \cap H_2$ is a closed subspace of $H(C_2, T)$, that is $s\left[H_1^{\perp}\right]$ is a closed subspace of $H(C_2, T)$. Assertion 2 is thus proved.

To prove item 3, one may proceed as follows. Any $h \in H(C, T)$ has a unique decomposition in the form $h = h_1 + h_1^{\perp}$, where h_1 belongs to $\overline{H}_1$, the closure of H_1 in $H(C, T)$, and $h_1^{\perp}$ belongs to $H_1^{\perp}$. If h is chosen in H_2, since [(Proposition) 3.3.2, item 2], as manifolds, $H_1^{\perp} \subseteq H_2$, $h_1 = h - h_1^{\perp} \in H_2$. Thus $h_1 \in \overline{H}_1 \cap H_2$. Consequently, every $h \in H_2$ has a unique decomposition in the following form:

$$h = h_1 + h_2, \; h_1 \in \overline{H}_1 \cap H_2, \; h_2 \in H_1^{\perp} = H_1^{\perp} \cap H_2.$$

Since $\overline{H}_1$ and $H_1^{\perp}$ are closed subspaces of $H(C,T)$, again because of (Corollary) 3.1.13, $\overline{H}_1 \cap H(C_2,T)$ and $H_1^{\perp} = H_1^{\perp} \cap H(C_2,T)$ are closed subspaces of $H(C_2,T)$. Thus the following equality

$$H_2 = \left[\overline{H}_1 \cap H_2\right] + H_1^{\perp}$$

can be read as

$$H(C_2,T) = E + F, \tag{$\star$}$$

where

$$E = \overline{H}_1 \cap H(C_2,T)\,,\ F = s\left[H_1^{\perp}\right],$$

with E and F closed in $H(C_2,T)$, $E \cap F = \{0_{\mathbb{R}^T}\}$. But then $H(C_2,T)$ is isomorphic to $E \oplus F$ [167, p. 184], which means that E and F are orthogonal subspaces in $H(C_2,T)$. One shall now check that $E = \overline{H}_0$, which means that F is its orthogonal complement, and thus that item 3 is true. Now, in $H(C_2,T)$, one has, using item 3 of (Fact) 3.3.3, that

$$\left(H_0^{\perp} + \left\{\overline{H}_1 \cap H(C_2,T)\right\}\right)^{\perp} = \left\{H_0^{\perp}\right\}^{\perp} \cap \left\{\overline{H}_1 \cap H(C_2,T)\right\}^{\perp}.$$

But $\left\{H_0^{\perp}\right\}^{\perp} = \overline{H}_0$, and, because of ($\star$), $\left\{\overline{H}_1 \cap H(C_2,T)\right\}^{\perp} = F = s\left[H_1^{\perp}\right]$. Thus

$$\left(H_0^{\perp} + \left\{\overline{H}_1 \cap H_2\right\}\right)^{\perp} = \overline{H}_0 \cap s\left[H_1^{\perp}\right]. \tag{$\star\star$}$$

Since $H_0 \subseteq \overline{H}_1 \cap H(C_2,T)$, a closed set in $H(C_2,T)$,

$$\overline{H}_0 \subseteq \overline{H}_1 \cap H(C_2,T)\,,$$

and, because of ($\star$), $s\left[H_1^{\perp}\right] \subseteq \overline{H}_0^{\perp}$. Thus

$$\overline{H}_0 \cap s\left[H_1^{\perp}\right] = \{0_{\mathbb{R}^T}\}\,,$$

and, because of ($\star\star$),

$$\left(H_0^{\perp} + \overline{H}_1 \cap H_2\right)^{\perp} = \left\{0_{\mathbb{R}^T}\right\},$$

or

$$H_0^{\perp} + \overline{H}_1 \cap H_2 = H(C_2,T)\,. \tag{$\star\star\star$}$$

Comparing ($\star$) and ($\star\star\star$) one has that

$$H_0^{\perp} = \overline{H}_0^{\perp} = s\left[H_1^{\perp}\right],$$

that is, item 3.

The first part of (i) in assertion 4 obtains by definition, and the second, because [(Proposition) 1.6.1]

$$H(\Gamma_1, T) = \overline{H}_0, \text{ and } H(\Gamma_2, T) = H_0^{\perp}.$$

The first assertion of (ii) follows from the fact that ($\overline{H}_1$ is the closure of H_1 in $H(C,T)$)

$$H(\Gamma_2, T) = H_0^{\perp} = s\left[H_1^{\perp}\right], \text{ and that } H(C_1, T) \subseteq \overline{H}_1.$$

For the second assertion, one has that, as a manifold of functions, the set $H(C_1 + \Gamma_1, T)$ is made of functions of the form $h = h_1 + h_2$, with $h_1 \in H(C_1, T)$, and $h_2 \in H(\Gamma_1, T)$. But, from (i) and the first part of (ii), $H(\Gamma_2, T)$ has only the zero function in common with $H(C_1, T)$ and $H(\Gamma_1, T)$.

One has seen that $s\left[H_1^{\perp}\right] = H_0^{\perp}$. Thus the relation $H_1^{\perp} = \{0_{\mathbb{R}^T}\}$ is equivalent to the relation $\overline{H}_0 = H(C_2, T)$, and thus to $H(C_2, T) = H(\Gamma_1, T)$. That is item (iii) of assertion 4.

Item (iv) follows from similar considerations, since $s\left[H_1^{\perp}\right] = H(C_2, T)$ if, and only if, $\overline{H}_0 = \{0_{\mathbb{R}^T}\}$. □

Remark 3.3.5 The "quirky" nature of (Corollary) 3.3.4 is due to the fact that, when computing the RKHS of a sum of covariances, the intersection of the RKHS's of both components gets "factored out."

Example 3.3.6 Consider again the example of (Example) 1.3.14, where $C_1 = f \otimes f$, and $C_2 = C$, so that C of (Proposition) 3.3.2 is C_f. Then, as seen, when f does not belong to $H(C,T)$,

$$H\left(C_f, T\right) \text{ is isomorphic to } H(f \otimes f, T) \oplus H(C, T),$$

and one is in case (iv), item 4 of (Corollary) 3.3.4. When f belongs to $H(C,T)$, since $H(f \otimes f, T)$ has dimension one, it is a (closed) subspace of $H\left(C_f, T\right)$, and, since $H_0 = H(f \otimes f, T) \cap H(C, T) = H(f \otimes f, T)$, $\overline{H_0} = H(f \otimes f, T)$. Consequently one has the decomposition of $H\left(C_f, T\right)$ into the following form:

$$H\left(C_f, T\right) = H_0 + H_0^{\perp}.$$

As $H_1 = H_0$, $H_1^{\perp} = H_0^{\perp}$, and (Corollary) 3.3.4 is "obvious," since, as sets, $H\left(C_f, T\right) = H(C, T)$. Proposition 3.3.4 says that an analogous situation always prevails.

3.4 Dominated Families of Covariances

The family of covariances which are dominated by a given, fixed, but arbitrary covariance, is the family within which one may find simultaneous reductions of covariances, a device useful for the discrimination of Gaussian probability laws [9, 147].

3.4.1 *Spectral Representation of Dominated Covariances*

Let C_0 be a fixed, but arbitrary covariance on T, and let $\mathcal{D}[C_0]$ be the family of covariances on T which are dominated by C_0, that is, $C \in \mathcal{D}[C_0]$ means that $C \ll C_0$. Let $\mathcal{L}_0[H(C_0,T)]$ be the family of linear and bounded operators B of $H(C_0,T)$, which are self-adjoint, and such that (they are then contractions)

$$0 \leq m_B \|h\|^2_{H(C_0,T)} \leq \langle B[h], h\rangle_{H(C_0,T)} \leq M_B \|h\|^2_{H(C_0,T)} \leq \|h\|^2_{H(C_0,T)},$$

where

$$m_B = \inf\left\{\langle B[h], h\rangle_{H(C_0,T)},\ h \in H(C_0,T) : \|h\|_{H(C_0,T)} = 1\right\},$$

and

$$M_B = \sup\left\{\langle B[h], h\rangle_{H(C_0,T)},\ h \in H(C_0,T) : \|h\|_{H(C_0,T)} = 1\right\}.$$

Proposition 3.4.1 *There is a bijection*

$$\beta : \mathcal{D}[C_0] \longrightarrow \mathcal{L}_0[H(C_0,T)]$$

which preserves order, that is such that, for fixed, but arbitrary C_1 and C_2 in $\mathcal{D}[C_0]$, $C_1 \ll C_2$ implies, for $h \in H(C_0,T)$, fixed, but arbitrary,

$$\langle \beta[C_1][h], h\rangle_{H(C_0,T)} \leq \langle \beta[C_2][h], h\rangle_{H(C_0,T)}.$$

Furthermore, for $C \in \mathcal{D}[C_0]$,

$$\beta[C] = B_C \in \mathcal{L}_0[H(C_0,T)],$$

where, for fixed, but arbitrary $(t_1,t_2) \in T \times T$,

$$C(t_1,t_2) = \langle B_C[C_0(\cdot,t_1)], C_0(\cdot,t_2)\rangle_{H(C_0,T)}.$$

Proof β *is well defined.*

Fix arbitrarily $C \in \mathcal{D}[C_0]$. Since $C \ll C_0$, one has available (Proposition 3.1.5) the map $J_{C_0,C} : H(C_0,T) \longrightarrow H(C,T)$ which is uniquely defined using the following relation: for $t \in T$, fixed, but arbitrary,

$$J_{C_0,C}[C_0(\cdot,t)] = C(\cdot,t).$$

Then $B_C = J^\star_{C_0,C} J_{C_0,C}$ belongs to $\mathcal{L}_0[H(C_0,T)]$, and thus $\beta[C] = B_C$ is well defined.

Proof *β is an injection.*
Indeed, $\beta[C_1] = \beta[C_2]$ implies $J^\star_{C_0,C_1} J_{C_0,C_1} = J^\star_{C_0,C_2} J_{C_0,C_2}$, so that

$$\begin{aligned} C_1(t_1,t_2) &= \langle J_{C_0,C_1}[C(\cdot,t_1)], J_{C_0,C_1}[C(\cdot,t_2)]\rangle_{H(C_1,T)} \\ &= \langle J^\star_{C_0,C_1} J_{C_0,C_1}[C(\cdot,t_1)], C(\cdot,t_2)\rangle_{H(C_0,T)} \\ &= \langle J^\star_{C_0,C_2} J_{C_0,C_2}[C(\cdot,t_1)], C(\cdot,t_2)\rangle_{H(C_0,T)} \\ &= \langle J_{C_0,C_2}[C(\cdot,t_1)], J_{C_0,C_2}[C(\cdot,t_2)]\rangle_{H(C_2,T)} \\ &= C_2(t_1,t_2). \end{aligned}$$

Proof *β is also a surjection.*
Suppose indeed that $B \in \mathcal{L}_0[H(C_0,T)]$. Define then $F : T \longrightarrow H(C_0,T)$ using

$$F(t) = B^{1/2}[C_0(\cdot,t)].$$

The range of $L_F : H(C_0,T) \longrightarrow \mathbb{R}^T$ is obtained [(Proposition) 1.1.15] as

$$L_F[h](t) = \langle h, F(t)\rangle_{H(C_0,T)} = \langle B^{1/2}[h], C_0(\cdot,t)\rangle_{H(C_0,T)} = B^{1/2}[h](t),$$

so that $B^{1/2}$ is a contraction, and its range, an RKHS, with kernel

$$C_B(t_1,t_2) = \langle F(t_1), F(t_2)\rangle_{H(C_0,T)} = \langle B[C_0(\cdot,t_1)], [C_0(\cdot,t_2)]\rangle_{H(C_0,T)}.$$

But then

$$\begin{aligned} \sum_{i=1}^n \sum_{j=1}^n \alpha_i \alpha_j C_B(t_i,t_j) &= \left\| B^{1/2}\left[\sum_{i=1}^n \alpha_i C_0(\cdot,t_i)\right]\right\|^2_{H(C_0,T)} \\ &\le \left\| \sum_{i=1}^n \alpha_i C_0(\cdot,t_i)\right\|^2_{H(C_0,T)} \\ &= \sum_{i=1}^n \sum_{j=1}^n \alpha_i \alpha_j C_0(t_i,t_j). \end{aligned}$$

Consequently $C_B \in \mathcal{D}[C_0]$. Furthermore

$$\begin{aligned}\langle J^{\star}_{C_0,C_B} J_{C_0,C_B}[C_0(\cdot,t)], C_0(\cdot,x)\rangle_{H(C_0,T)} &= \\ &= C_B(x,t) \\ &= \langle B[C_0(\cdot,t)], C_0(\cdot,x)\rangle_{H(C_0,T)},\end{aligned}$$

so that $J^{\star}_{C_0,C_B} J_{C_0,C_B} = B$, or $\beta[C_B] = B$. The map $C_B \mapsto B_C$ is thus a bijection.

Proof *β, as defined, preserves order.*
Suppose thus that

$$C_1 \ll C_2 \ll C_0.$$

Then, for $i = 1, 2$,

$$\begin{aligned}\langle B_{C_i}[C_0(\cdot,t)], C_0(\cdot,t)\rangle_{H(C_0,T)} &= \\ &= \langle J^{\star}_{C_0,C_i} J_{C_0,C_i}[C_0(\cdot,t)], C_0(\cdot,t)\rangle_{H(C_0,T)} \\ &= C_i(t,t).\end{aligned}$$

Thus

$$\langle B_{C_1}[C_0(\cdot,t)], C_0(\cdot,t)\rangle_{H(C_0,T)} \le \langle B_{C_2}[C_0(\cdot,t)], C_0(\cdot,t)\rangle_{H(C_0,T)},$$

which proves preservation of order. □

Remark 3.4.2 Let $H_0 \subseteq H(C_0,T)$ be a closed subspace with projection P_0. Let $\mathcal{H}_0$ be the kernel of H_0. Then [(Proposition) 1.6.1] $\beta[\mathcal{H}_0] = P_0$.

Definition 3.4.3 A C_0-covariance Γ is an element of $\mathcal{D}[C_0]$ such that $H(\Gamma,T)$ is a (closed) subspace of $H(C_0,T)$.

Proposition 3.4.4 *Let $C \in \mathcal{D}[C_0]$ be fixed, but arbitrary, and let m_C and M_C denote the bounds of $B_C = \beta[C]$, as defined above* [(Proposition) 3.4.1]. *There exists a unique family of C_0-covariances, say $\{\Gamma_\lambda, \lambda \in \mathbb{R}\}$, such that:*

1. *whenever $\lambda_1 \le \lambda_2$, $\Gamma_{\lambda_1} \ll \Gamma_{\lambda_2}$;*
2. *whenever $\lambda < m_C$, $\Gamma_\lambda = 0$;*
3. *whenever $\lambda \ge M_C$, $\Gamma_\lambda = C_0$;*
4. *for $t \in T$, fixed, but arbitrary, the map from $\mathbb{R}$ to $H(C_0,T)$ defined using the following relation:*

$$\lambda \mapsto \Gamma_\lambda(\cdot,t)$$

is continuous to the right;

5. *for* $(t_1, t_2) \in T \times T$, *the map* $\mu_C(\cdot \mid \{t_1, t_2\}) : \mathbb{R} \longrightarrow \mathbb{R}$ *defined using the following relation:*

$$\mu_C(\lambda \mid \{t_1, t_2\}) = \chi_{]0,1]}(\lambda)\, \Gamma_\lambda(t_1, t_2)$$

is of bounded variation on $[0, 1]$, *and, as a Riemann-Stieltjes integral,*

$$C(t_1, t_2) = \int_0^1 \lambda\, \mu_C(d\lambda \mid \{t_1, t_2\}).$$

Proof Let

$$B_C = \int_{m_{B_C}-}^{M_{B_C}} \lambda\, E_C(d\lambda)$$

be the spectral representation of B_C [266, p. 181], and Γ_λ be the reproducing kernel of the projection $E_C(\lambda)$, so that [(Remark) 3.4.2] $\beta[\Gamma_\lambda] = E_C(\lambda)$. By definition, $[m_{B_C}, M_{B_C}] \subseteq [0, 1]$. As the spectral representation is obtained as a limit of sums of Riemann-Stieltjes type, there may always be, in the sums approaching

$$f(B_C) = \int_{m_{B_C}-}^{M_{B_C}} f(\lambda)\, E_C(d\lambda),$$

an initial term of the following form, representing, in the integral, the minus sign following m_{B_C}: for $\lambda < m_{B_C}$, fixed, but arbitrary,

$$f(m_{B_C})\{E_C(m_{B_C}) - E_C(\lambda)\} = f(m_{B_C})E_C(m_{B_C}).$$

When the domain of integration is contained in $[0, 1]$, as is here the case,

$$f(B_C) = \int_0^1 f(\lambda)E_C(d\lambda),$$

as the intervals $[0, m_{B_C}[$ and $]M_{B_C}, 1]$ have measure zero. The notation used in the sequel is thus legitimate.

Because of the definition of Γ as the kernel of a projection, one has [(Proposition) 1.6.1] that, as an element of $H(C_0, T)$,

$$\Gamma_\lambda(\cdot, t) = E_C(\lambda)[C_0(\cdot, t)],$$

so that, for fixed, but arbitrary $(\lambda_1, \lambda_2, t) \in \mathbb{R} \times \mathbb{R} \times T$,

$$\|\Gamma_{\lambda_1}(\cdot, t) - \Gamma_{\lambda_2}(\cdot, t)\|_{H(C_0,T)} =$$
$$= \|E_C(\lambda_1)[C_0(\cdot, t)] - E_C(\lambda_2)[C_0(\cdot, t)]\|_{H(C_0,T)},$$

and, for fixed, but arbitrary $(\lambda, t_1, t_2) \in \mathbb{R} \times T \times T$, that

$$\Gamma_\lambda (t_1, t_2) = \langle E_C (\lambda) [C_0 (\cdot, t_1)], C_0 (\cdot, t_2)\rangle_{H(C_0,T)}.$$

Items 1 to 4 state thus properties of E_C in terms of Γ_λ.

One of the consequences of the spectral theorem [266, p. 175], reported in (Fact) 2.7.2, is that, for fixed, but arbitrary $(h_1, h_2) \in H (C_0, T) \times H (C_0, T)$,

$$\langle B_C [h_1], h_2\rangle_{H(C_0,T)} = \int_0^1 \lambda \langle E_C (d\lambda) [h_1], h_2\rangle_{H(C_0,T)}.$$

But $\langle E_C (d\lambda) [C_0 (\cdot, t_1)], C_0 (\cdot, t_2)\rangle_{H(C_0,T)} = \mu_C (d\lambda \mid \{t_1, t_2\})$. Thus item 5 is true because of (Proposition) 3.4.1. □

Example 3.4.5 Suppose P is a projection of $H (C_0, T)$. The spectral representation of P is [266, p. 184]

$$E_P (\lambda) = \chi_{[0,1[} (\lambda) \left(I_{H(C_0,T)} - P\right) + \chi_{[1,\infty[} (\lambda) I_{H(C_0,T)},$$

so that

$$E_P (\lambda) [C_0 (\cdot, t)] = \chi_{[0,1[} (\lambda) \{C_0 (\cdot, t) - P [C_0 (\cdot, t)]\} + \chi_{[1,\infty[} (\lambda) C_0 (\cdot, t).$$

Since $C_P (\theta, t) = P [C_0 (\cdot, t)] (\theta)$ is a kernel in $\mathcal{D} [C_0]$, one has, in $\mathcal{D} [C_0]$, that

$$\Gamma_\lambda (\theta, t) = E_P (\lambda) [C_0 (\cdot, t)] (\theta) = \begin{cases} \text{zero kernel} & \text{when } \lambda < 0, \\ C_0 (\theta, t) - C_P (\theta, t) & \text{when } 0 \le \lambda < 1, \\ C_0 (\theta, t) & \text{when } \lambda \ge 1. \end{cases}.$$

Example 3.4.6 One starts with the following formulae [120, p. 406]:

$$e^{-|\alpha|} = \int_0^\infty \cos [\alpha t] \frac{2}{\pi} \frac{dt}{1 + t^2};$$

$$\frac{\sin [\alpha]}{\pi \alpha} = \int_0^1 \cos [\alpha t] \frac{dt}{\pi}.$$

Let

$$q (x) = \frac{1 + x^2}{2},$$

$$\mu_{C_0} (dx) = \frac{1}{q (x)} \frac{dx}{\pi},$$

$$\mu_C (dx) = \chi_{[0,1]} (x) \frac{dx}{\pi}$$
$$= \chi_{[0,1]} (x) q (x) \mu_{C_0} (dx).$$

The measure μ_{C_0}, on the Borel sets of $\mathbb{R}_+$ is a probability (set $\alpha = 0$ in the integral expression for the exponential); by definition, μ_C has mass $1/\pi$, and is absolutely continuous with respect to μ_{C_0}, with $\chi_{[0,1]}\, q$ as Radon-Nikodým derivative. Let $L_2[\mu_{C_0}] = L_2(\mathbb{R}_+, \mathcal{B}(\mathbb{R}_+), \mu_{C_0})$. $L_2[\mu_C]$ is defined analogously. The map $J : L_2[\mu_{C_0}] \longrightarrow L_2[\mu_C]$, obtained using the following assignment:

$$J\left([f_0]_{L_2[\mu_{C_0}]}\right) = [f_0]_{L_2[\mu_C]}, \qquad (\star)$$

makes sense and produces a contraction as, by definition, $0 \leq q \leq 1$, and, furthermore,

$$\begin{aligned}\int_{\mathbb{R}_+} f_0^2(x)\,\mu_C(dx) &= \int_0^1 f_0^2(x) q(x)\,\mu_{C_0}(dx)\\ &\leq \int_0^1 f_0^2(x)\,\mu_{C_0}(dx)\\ &\leq \int_{\mathbb{R}_+} f_0^2(x)\,\mu_{C_0}(dx).\end{aligned}$$

With the notation just introduced, one has that

$$\int_0^\infty \cos[\alpha t]\,\frac{2}{\pi}\,\frac{dt}{1+t^2} = \int_0^\infty \cos[\alpha t]\,\mu_{C_0}(dt)\,;$$

$$\int_0^1 \cos[\alpha t]\,\frac{dt}{\pi} = \int_0^\infty \cos[\alpha t]\,\mu_C(dt)\,.$$

Let, for $t \in [0, \infty[$, $c_t(x) = \cos[tx]$ and $s_t(x) = \sin[tx]$, and denote, for example, $[c_t]_{C_0}$ and $[s_t]_{C_0}$ the equivalence classes of respectively c_t and s_t with respect to μ_{C_0}.

Result 1:

An L_2 representation for the RKHS of $C_0(t_1, t_2) = e^{-|t_1 - t_2|}$.

It is a consequence of the definitions and properties listed above that

$$C_0(t_1, t_2) = e^{-|t_1-t_2|} = \int_0^\infty \cos[(t_1 - t_2)x]\,\mu_{C_0}(dx)$$

has the following representation:

$$C_0(t_1, t_2) = \langle [c_{t_1}]_{C_0}, [c_{t_2}]_{C_0}\rangle_{L_2[\mu_{C_0}]} + \langle [s_{t_1}]_{C_0}, [s_{t_2}]_{C_0}\rangle_{L_2[\mu_{C_0}]}.$$

Let H_{C_0} be the direct sum of $L_2[\mu_{C_0}]$ with itself, and define

$$\underline{F}_0 : T \longrightarrow H_{C_0}, \quad \text{using } \underline{F}_0(t) = \left([c_t]_{C_0}, [s_t]_{C_0}\right).$$

Then

$$C_0(t_1, t_2) = \langle \underline{F}_0(t_1), \underline{F}_0(t_2) \rangle_{H_{C_0}}.$$

Let $L_{C_0} : H_{C_0} \longrightarrow \mathbb{R}^T$ be defined using $\underline{F}_0$, as usual:

$$\begin{aligned} L_{C_0}[(h_1, h_2)](t) &= \langle (h_1, h_2), \underline{F}_0(t) \rangle_{H_{C_0}} \\ &= \langle h_1, [c_t]_{C_0} \rangle_{L_2[\mu_{C_0}]} + \langle h_2, [s_t]_{C_0} \rangle_{L_2[\mu_{C_0}]}. \end{aligned}$$

Its range is the RKHS determined by C_0.

The following families of $L_2[\mu_{C_0}]$,

$$\left\{[c_t]_{C_0}, \ t \in T\right\} \text{ and } \left\{[s_t]_{C_0}, \ t \in T\right\},$$

are dense in $L_2[\mu_{C_0}]$. Indeed, these families are dense in L_1 with respect to Lebesgue measure, a consequence of Fourier's theorem for the sine and cosine transforms [108, pp. 389,404]. But then they are dense in $L_1[\mu_{C_0}]$, and, since $L_2[\mu_{C_0}] \subseteq L_1[\mu_{C_0}]$ (μ_{C_0} is a probability), they are dense in $L_2[\mu_{C_0}]$. Thus $\{\underline{F}_0(t), t \in T\}$ is total in H_{C_0}, and L_{C_0} is a unitary map between H_{C_0} and $H(C_0, T)$.

Result 2:

An L_2 representation for the RKHS of $C(t_1, t_2) = \frac{\sin(t_1 - t_2)}{\pi(t_1 - t_2)}$.

Mutatis mutandis, for

$$C(t_1, t_2) = \frac{\sin(t_1 - t_2)}{\pi(t_1 - t_2)} = \int_0^{\infty} \cos[(t_1 - t_2)x] \, \mu_C(dx),$$

one has that

$$C(t_1, t_2) = \langle \underline{F}(t_1), \underline{F}(t_2) \rangle_{H_C},$$

where $\underline{F}(t) = ([c_t]_C, [s_t]_C)$. The map $L_C : H_C \longrightarrow \mathbb{R}^T$ whose range is $H(C, T)$ follows. Since the trigonometric functions are also dense in $L_2[0, 1]$ [134, p. 123], $\{\underline{F}(t), t \in T\}$ is total in H_C, and L_C is also unitary.

Result 3:

C belongs to $\mathcal{D}[C_0]$, and thus $J_{C_0,C}$ of (Proposition) 3.1.5 *exists.*

Since for $x \in [0, 1]$, $\frac{1}{q(x)} \in [1, 2]$, $\frac{2}{\pi} \frac{1}{1+x^2} - \chi_{[0,1]}(x) \frac{1}{\pi} \geq 0$,

$$\begin{aligned} &C_0(t_1, t_2) - C(t_1, t_2) = \\ &= \int_0^{\infty} \{c_{t_1}(x) c_{t_2}(x) + s_{t_1}(x) s_{t_2}(x)\} \left\{ \frac{2}{\pi} \frac{1}{1+x^2} - \chi_{[0,1]}(x) \frac{1}{\pi} \right\} dx \end{aligned}$$

is symmetric and positive definite on T, so that $C \in \mathcal{D}[C_0]$.

Result 4:

A contraction $V_J : H_{C_0} \longrightarrow H_C$.

J being introduced at $(\star)$, define $V_J : H_{C_0} \longrightarrow H_C$ using the following equality: $V_J[(h_1, h_2)] = (J[h_1], J[h_2])$. Then

$$V_J\left[\underline{F}_0(t)\right] = \underline{F}(t).$$

Result 5:

Identification of B_C.

One has that

$$L_C V_J[\underline{F}_0(t)] = L_C[\underline{F}(t)] = C(\cdot, t).$$

Furthermore

$$J_{C_0,C} L_{C_0}[\underline{F}_0(t)] = J_{C_0,C}[C_0(\cdot, t)] = C(\cdot, t).$$

One has thus the following commuting diagram:

$$\begin{array}{ccc} H_{C_0} & \xrightarrow{L_{C_0}} & H(C_0, T) \\ {\scriptstyle V_J}\downarrow & & \downarrow{\scriptstyle J_{C_0,C}} \\ H_C & \xrightarrow{L_C} & H(C, T)\,, \end{array}$$

and $J_{C_0,C} = L_C V_J L_{C_0}^{\star}$. Then

$$\begin{aligned} C(t_1, t_2) &= \langle C(\cdot, t_1), C(\cdot, t_2)\rangle_{H(C,T)} \\ &= \langle J_{C_0,C}[C_0(\cdot, t_1)], J_{C_0,C}[C_0(\cdot, t_2)]\rangle_{H(C,T)} \\ &= \langle J_{C_0,C}^{\star} J_{C_0,C}[C_0(\cdot, t_1)], C_0(\cdot, t_2)\rangle_{H(C_0,T)} \\ &= \langle L_{C_0} V_J^{\star} L_C^{\star} L_C V_J L_{C_0}^{\star}[C_0(\cdot, t_1)], C(\cdot, t_2)\rangle_{H(C_0,T)}, \end{aligned}$$

so that, as L_C is unitary, $B_C = L_{C_0} V_J^{\star} V_J L_{C_0}^{\star}$.

Result 6:

Computation of $V_J^{\star}$.

One has that

$$\begin{aligned}\langle J[f_0], g\rangle_{L_2[\mu_C]} &= \int_0^1 \dot{f_0}(x)\dot{g}(x)\,\mu_C(dx)\\ &= \int_0^\infty \dot{f}_0(x)\dot{g}(x)\,\chi_{[0,1]}\,q(x)\,\mu_{C_0}(dx)\\ &= \left\langle f_0, [\chi_{[0,1]}\,q\dot{g}]_{C_0}\right\rangle_{L_2[\mu_{C_0}]}.\end{aligned}$$

Thus

$$J^\star[g] = [\chi_{[0,1]}\,q\dot{g}]_{C_0}.$$

Using the definition of the adjoint, one has that $V_J^\star = (J^\star, J^\star)$, so that

$$V_J^\star[(h_1, h_2)] = \left([\chi_{[0,1]}\,q\dot{h}_1]_{C_0}, [\chi_{[0,1]}\,q\dot{h}_2]_{C_0}\right).$$

Result 7:

Computation of m_{B_C} and M_{B_C}

To compute m_{B_C} and M_{B_C}, suppose that, for $(h_1, h_2) \in H_{C_0}$, $h \in H(C_0, T)$, fixed, but arbitrary,

$$L_{C_0}[(h_1, h_2)] = h.$$

Then, since L_{C_0} is unitary,

$$\begin{aligned}&\|B_C[h]\|^2_{H(C_0,T)} =\\ &\quad= \left\|L_{C_0} V_J^\star V_J L_{C_0}^\star [h]\right\|^2_{H(C_0,T)}\\ &\quad= \left\|L_{C_0} V_J^\star V_J [(h_1, h_2)]\right\|^2_{H(C_0,T)}\\ &\quad= \left\|V_J^\star V_J [(h_1, h_2)]\right\|^2_{H_{C_0}}\\ &\quad= \left\|V_J^\star \left([\dot{h}_1]_C, [\dot{h}_2]_C\right)\right\|^2_{H_{C_0}}\\ &\quad= \left\|\left([\chi_{[0,1]}q\dot{h}_1]_{C_0}, [\chi_{[0,1]}q\dot{h}_2]_{C_0}\right)\right\|^2_{H_C}\\ &\quad= \int_0^\infty \dot{h}_1^2(x)\,\chi_{[0,1]}(x)\,q^2(x)\,\mu_{C_0}(dx) + \int_0^\infty \dot{h}_2^2(x)\,\chi_{[0,1]}(x)\,q^2(x)\,\mu_{C_0}(dx)\\ &\quad= \int_0^\infty \dot{h}_1^2(x)\,q(x)\,\mu_C(dx) + \int_0^\infty \dot{h}_2^2(x)\,q(x)\,\mu_C(dx).\end{aligned}$$

For $i = 1, 2$, choose $h_i \in L_2[\mu_{C_0}]$ such that $\int_1^\infty \dot{h}_i^2(x)\,\mu_{C_0}(dx) > 0$, and let

$$\dot{k}_i(x) = \frac{\chi_{]1,\infty[}\dot{h}_i(x)}{\sqrt{\int_1^\infty \dot{h}_i^2(x)\,\mu_{C_0}(dx)}}.$$

Then $k_i = [\dot{k}_i]_{C_0} \in L_2[\mu_{C_0}]$ has norm equal to one. Furthermore, letting

$$L_{C_0}\left[\left(\frac{k_1}{\sqrt{2}}, \frac{k_2}{\sqrt{2}}\right)\right] = k \in H(C_0, T), \quad \left\|\left(\frac{k_1}{\sqrt{2}}, \frac{k_2}{\sqrt{2}}\right)\right\|_{H_{C_0}} = \|k\|_{H(C_0,T)} = 1,$$

and, using the expression for $\|B_C[h]\|^2_{H(C_0,T)}$ obtained above,

$$\|B_C[h]\|^2_{H(C_0,T)} = 0,$$

so that $m_{B_C} = 0$.

Let now $\epsilon \in]0, 1[$ be fixed, but arbitrary, and $\dot{h}_1 = \dot{h}_2 = \chi_{[1-\epsilon,1]}$. Let also

$$h_1 = [\dot{h}_1]_{C_0} \text{ and } h_2 = [\dot{h}_2]_{C_0}.$$

Then $L_{C_0}\left[\left(\frac{h_1}{\sqrt{2}}, \frac{h_2}{\sqrt{2}}\right)\right] = h \in H(C_0, T)$, and, using the mean value theorem,

$$\begin{aligned}
\|h\|^2_{H(C_0,T)} &= \|(h_1, h_2)\|^2_{H_{C_0}} \\
&= \int_{1-\epsilon}^{1} \mu_{C_0}(dx) \\
&= \int_{1-\epsilon}^{1} \frac{2}{\pi}\frac{dx}{1+x^2} \\
&= \epsilon\frac{2}{\pi}\frac{1}{1+u^2}, \quad 1-\epsilon < u < 1.
\end{aligned}$$

Similarly, using the expression for $\|B_C[h]\|^2_{H(C_0,T)}$ obtained above,

$$\begin{aligned}
\|B_C[h]\|^2_{H(C_0,T)} &= \int_{1-\epsilon}^{1} q(x)\,\mu_C(dx) \\
&= \int_{1-\epsilon}^{1} \frac{1+x^2}{2}\frac{dx}{\pi} \\
&= \epsilon\frac{1+v^2}{2\pi}, \quad 1-\epsilon < v < 1.
\end{aligned}$$

Consequently

$$\frac{\|B_C[h]\|^2_{H(C_0,T)}}{\|h\|^2_{H(C_0,T)}} = \frac{\frac{1}{2\pi}\left(1+v^2\right)}{\frac{2}{\pi}\frac{1}{1+u^2}},$$

which has limit, as $\epsilon \downarrow 0$, equal to one, and thus $M_{B_C} = 1$.

Result 9:

The spectral representation of B_C.

Using the representation $B_C = L_{C_0} V_J^\star V_J L_{C_0}^\star$, one has that

$$B_C^n = L_{C_0}\left(V_J^\star V_J\right)^n L_{C_0}^\star.$$

But, for example,

$$V_J^\star V_J\left[\left([\chi_{[0,1]}\, q\dot{h}_1]_{C_0}, [\chi_{[0,1]}\, q\dot{h}_2]_{C_0}\right)\right] = \left([\chi_{[0,1]}\, q^2\dot{h}_1]_{C_0}, [\chi_{[0,1]}\, q^2\dot{h}_2]_{C_0}\right),$$

so that, for fixed, but arbitrary $(h_1, h_2) \in H_{C_0}$, and integer $n \in \mathbb{N}$,

$$B_C^n L_{C_0}[(h_1, h_2)] = L_{C_0}\left[\left(\left[\chi_{[0,1]} q^n \dot{h}_1\right]_{C_0}, \left[\chi_{[0,1]} q^n \dot{h}_2\right]_{C_0}\right)\right].$$

Let $\tilde{q} = \chi_{[0,1]} q$. Then, for any polynomial p_n,

$$p_n(B_C)\, L_{C_0}[(h_1, h_2)] = L_{C_0}\left[\left(\left[p_n(\tilde{q})\,\dot{h}_1\right]_{C_0}, \left[p_n(\tilde{q})\,\dot{h}_2\right]_{C_0}\right)\right].$$

Thus

$$p_n(B_C)\left[C_0(\cdot, t)\right] = L_{C_0}\left[\left([p_n(\tilde{q})\, c_t]_{C_0}, [p_n(\tilde{q})\, s_t]_{C_0}\right)\right],$$

so that

$$\begin{aligned}
&\langle p_n(B_C)\left[C_0(\cdot, t_1)\right], C_0(\cdot, t_2)\rangle_{H(C_0,T)} = \\
&\quad = \left\langle L_{C_0}\left[\left([p_n(\tilde{q})\, c_{t_1}]_{C_0}, [p_n(\tilde{q})\, s_{t_1}]_{C_0}\right)\right], L_{C_0}\left[\left([c_{t_2}]_{C_0}, [s_{t_2}]_{C_0}\right)\right]\right\rangle_{H(C_0,T)} \\
&\quad = \int_0^\infty p_n(\tilde{q}(x))\, c_{t_1}(x)\, c_{t_2}(x)\, \mu_{C_0}(dx) + \int_0^\infty p_n(\tilde{q}(x))\, s_{t_1}(x)\, s_{t_2}(x)\, \mu_{C_0}(dx).
\end{aligned}$$

For fixed, but arbitrary $\lambda \in T$, let $\{p_n(\cdot \mid \lambda),\ n \in \mathbb{N}\}$ be a sequence of polynomials which decreases point-wise to the indicator $\chi_{[0,\lambda]}$ [109, p. 386]. Then, for (t_1, t_2) in $T \times T$, fixed, but arbitrary, using, successively, item 14 of (Fact) 2.7.2, [129, p. 232], and what precedes,

$$\begin{aligned}
&\langle E_{B_C}(\lambda)[C_0(\cdot,t_1)], C_0(\cdot,t_2)\rangle_{H(C,T)} = \\
&= \lim_n \langle p_n(B_C \mid \lambda)[C_0(\cdot,t_1)], C_0(\cdot,t_2)\rangle_{H(C_0,T)} \\
&= \int_0^\infty \chi_{[0,\lambda]}(\tilde{q}(x))\, c_{t_1}(x)\, c_{t_2}(x)\, \mu_{C_0}(dx) + \int_0^\infty \chi_{[0,\lambda]}(\tilde{q}(x))\, s_{t_1}(x)\, s_{t_2}(x)\, \mu_{C_0}(dx).
\end{aligned}$$

Since $\tilde{q}(x) = \frac{1+x^2}{2} \geq \frac{1}{2}$ when $x \in [0,1]$, and $\tilde{q}(x) = 0$ when $x > 1$,

- for $\lambda \in \left[0, \frac{1}{2}\right[$,

$$\chi_{[0,\lambda]}(\tilde{q}(x)) = 1 \text{ for } x > 1,$$

  and zero otherwise. Thus, as stated in the proof of (Proposition) 3.4.4,

$$\Gamma_\lambda(t_1,t_2) = \int_1^\infty c_{t_1}(x)\, c_{t_2}(x)\, \mu_{C_0}(dx) + \int_1^\infty s_{t_1}(x)\, s_{t_2}(x)\, \mu_{C_0}(dx);$$

- for $\lambda \in \left[\frac{1}{2}, 1\right]$,

$$\chi_{[0,\lambda]}(\tilde{q}(x)) = 1 \text{ for } x > 1 \text{ and for } x \in \left[0, (2\lambda-1)^{1/2}\right],$$

  and zero otherwise. Thus

$$\begin{aligned}
\Gamma_\lambda(t_1,t_2) &= \Gamma_0(t_1,t_2) \\
&+ \int_0^{(2\lambda-1)^{1/2}} c_{t_1}(x)\, c_{t_2}(x)\, \mu_{C_0}(dx) \\
&+ \int_0^{(2\lambda-1)^{1/2}} s_{t_1}(x)\, s_{t_2}(x)\, \mu_{C_0}(dx).
\end{aligned}$$

Example 3.4.7 Suppose B_C is a compact operator whose eigenvalues that are different from zero are denoted $\{\lambda_i,\ i \in I \subseteq \mathbb{N}\}$. Let $\lambda_0 = 0$, P_0 be the projection onto $\mathcal{N}[B_C]$, and P_i be the projection onto $\mathcal{N}[\lambda_i I_{H(C_0,T)} - B_C]$. The following formula [266, p. 184] yields then the spectral decomposition of B_C in $H(C_0,T)$: for $\lambda < 0$,

$$E_{B_C}(\lambda) = 0,$$

and, for $\lambda \geq 0$,

$$E_{B_C}(\lambda)[h] = P_0[h] + \sum_{i \in I:\lambda_i \leq \lambda} P_i[h].$$

Consequently,

$$
\begin{aligned}
&\Gamma_\lambda(t_1, t_2) = \\
&\quad = \langle E_{B_C}(\lambda)[C_0(\cdot, t_1)], C_0(\cdot, t_2)\rangle_{H(C_0,T)} \\
&\quad = \langle P_0[C_0(\cdot, t_1)], C_0(\cdot, t_2)\rangle_{H(C_0,T)} + \sum_{i\in I:\lambda_i\le\lambda} \langle P_i[C_0(\cdot, t_1)], C_0(\cdot, t_2)\rangle_{H(C_0,T)}.
\end{aligned}
$$

Proposition 3.4.8 *Let $\sigma(B)$ designate the spectrum of the operator B, and $\{E_B(\lambda), \lambda \in \mathbb{R}\}$, the spectral family of B. For fixed, but arbitrary covariances C_1 and C_2 on T, let*

(a) $C = C_1 + C_2$,
(b) $J_{C,C_1} : H(C,T) \longrightarrow H(C_1,T)$ *be defined using* $J_{C,C_1}[C(\cdot,t)] = C_1(\cdot,t)$,
(c) $B_1 : H(C,T) \longrightarrow H(C,T)$ *be defined using* $B_1 = J^\star_{C,C_1} J_{C,C_1}$.

The operators J_{C,C_2} and B_2 are defined analogously. Then C_2 dominates C_1 if, and only if, $\sigma(B_1) \subseteq \left[0, \frac{1}{2}\right]$.

Proof Given a bounded, selfadjoint linear operator B, and a continuous function f, defined on $\mathbb{R}$, the operator $f(B)$ has the following representation [129, p. 256]:

$$\int_{m_B-}^{M_B} f(\lambda)\, E_B(d\lambda),$$

and may be approximated, in operator norm, by expressions of the following form:

$$\sum_{i=1}^{n} f(\lambda_i)\{E_B[\mu_i] - E_B[\mu_{i-1}]\},$$

where

- $\mu_0 < m_B = \mu_1 < \mu_2 < \cdots < \mu_{n-1} < \mu_n = M_B$,
- $\lambda_i \in [\mu_{i-1}, \mu_i],\ 1 \le i \le n$,

provided the intervals $[\mu_{i-1}, \mu_i]$ are uniformly small. The projections

$$\{E_B[\mu_i] - E_B[\mu_{i-1}],\ 1 \le i \le n\}$$

are orthogonal.

Proof Suppose that $\sigma(B_1) \subseteq \left[0, \frac{1}{2}\right]$.

Because of (Proposition) 3.4.1, it suffices to check that $B_2 - B_1$ is a positive operator. To that end, let Θ_n represent the following set:

- $n \in \mathbb{N}$,
- $\mu_0 < \mu_1 = 0 < \cdots < \mu_n = \frac{1}{2} < \mu_{n+1} < \cdots < \mu_{2n} = 1$,
- $\lambda_i \in [\mu_{i-1}, \mu_i],\ 1 \le i \le 2n$.

Let

$$B_{\Theta_n} = \sum_{i=1}^{2n} \lambda_i \left\{E_{B_1}(\mu_i) - E_{B_1}(\mu_{i-1})\right\}.$$

It is an approximation to B_1. Then,

$$I_{H(C,T)} - 2B_{\Theta_n} = \sum_{i=1}^{2n} (1-2\lambda_i)\left\{E_{B_1}(\mu_i) - E_{B_1}(\mu_{i-1})\right\}$$

is an approximation to $I_{H(C,T)} - 2B_1$. By assumption, for $i \geq n$, one has that $E_{B_1}(\mu_i) = I_{H(C,T)}$, so that the corresponding terms vanish. Thus

$$I_{H(C,T)} - 2B_{\Theta_n} = \sum_{i=1}^{n} (1-2\lambda_i)\left\{E_{B_1}(\mu_i) - E_{B_1}(\mu_{i-1})\right\}.$$

But, for $i \leq n$, $1 - 2\lambda_i \geq 0$, so that $I_{H(C,T)} - 2B_{\Theta_n} \geq 0$, and thus, since $I_{H(C,T)} = B_1 + B_2$,

$$B_2 - B_1 = I_{H(C,T)} - 2B_1 = \lim_n \left\{I_{H(C,T)} - 2B_{\Theta_n}\right\} \geq 0.$$

Proof *Suppose that the interval* $\left[0, \frac{1}{2}\right]$ *does not contain the set* $\sigma(B_1)$.

Let $\lambda \in \left]\frac{1}{2}, 1\right] \cap \sigma(B_1)$ be fixed, but arbitrary. Let Θ_n be the following set:

- $n \in \mathbb{N}$,
- $\mu_0 < \mu_1 = 0 < \cdots < \mu_n = 1$ and $\sup_{1 \leq i \leq n} (\mu_i - \mu_{i-1}) = \frac{1}{n}$,
- there is, in $\{1, \ldots, n\}$, an index i_λ such that

$$\mu_{i_\lambda - 1} = \lambda - \frac{1}{2n} \text{ and } \mu_{i_\lambda} = \lambda + \frac{1}{2n},$$

- $\lambda_i \in \left]\mu_{i-1}, \mu_i\right]$, $1 \leq i \leq n$,
- $\lambda \in \{\lambda_1, \ldots, \lambda_n\}$.

B_{Θ_n} shall be defined as above, *mutatis mutandis*. Since $\lambda \in \left]\frac{1}{2}, 1\right] \cap \sigma(B_1)$,

$$E_{B_1}\left(\lambda + \frac{1}{2n}\right) - E_{B_1}\left(\lambda - \frac{1}{2n}\right)$$

is a projection of $H(C,T)$ whose range strictly contains the subspace made of the zero vector [266, p. 189]. One can thus find an element h_n in that range that has norm equal to one, for which

$$\left\{E_{B_1}\left(\lambda+\frac{1}{2n}\right)-E_{B_1}\left(\lambda-\frac{1}{2n}\right)\right\}[h_n]=h_n.$$

Then

$$\left\langle\left(I_{H(C,T)}-2B_{\Theta_n}\right)[h_n],h_n\right\rangle_{H(C,T)}=1-2\lambda.$$

Consequently, since $I_{H(C,T)}=B_1+B_2$,

$$\begin{aligned}
&\left|\langle(B_2-B_1)[h_n],h_n\rangle_{H(C,T)}-(1-2\lambda)\right|=\\
&=\left|\left\langle\left(I_{H(C,T)}-2B_1\right)[h_n],h_n\right\rangle_{H(C,T)}-(1-2\lambda)\right|\\
&=\left|\left\langle\left(I_{H(C,T)}-2B_1\right)[h_n],h_n\right\rangle_{H(C,T)}-\left\langle\left(I_{H(C,T)}-2B_{\Theta_n}\right)[h_n],h_n\right\rangle_{H(C,T)}\right|\\
&=2\left|\langle(B_{\Theta_n}-B_1)[h_n],h_n\rangle_{H(C,T)}\right|\\
&\le 2n^{-1},
\end{aligned}$$

so that, for n large enough, since $\lambda>\frac{1}{2}$,

$$(1-2\lambda)-\frac{2}{n}\le\langle(B_2-B_1)[h_n],h_n\rangle_{H(C,T)}\le(1-2\lambda))+\frac{2}{n}<0,$$

and B_2 cannot dominate B_1, nor C_2, C_1. □

3.4.2 Simultaneous Reduction of Covariances

As mentioned above, simultaneous reduction of covariances has, sometimes, applications to Gaussian discrimination problems [9, 147].

Definition 3.4.9 Suppose C_0 and C are covariances on T, and that C belongs to $\mathcal{D}(C_0)$, with associated operator B_C. When B_C is compact, with eigenvalues $\{\lambda_i, i\in I\}$, using the notation of (Example) 3.4.7, let $\left\{e_j,\ j\in J\right\}$ be a complete orthonormal set formed by a choice of basis in each of the subspaces

$$\mathcal{N}[B_C],\ \mathcal{N}[\lambda_i I_{H(C,T)}-B_C],\ i\in I.$$

Then

$$\begin{aligned} C_0(t_1,t_2) &= \langle C_0(\cdot,t_1), C_0(\cdot,t_2)\rangle_{H(C,T)} \\ &= \sum_j \left\langle e_j, C_0(\cdot,t_1)\right\rangle_{H(C,T)} \left\langle e_j, C_0(\cdot,t_2)\right\rangle_{H(C,T)} \\ &= \sum_j e_j(t_1)\, e_j(t_2) \end{aligned}$$

and

$$\begin{aligned} C(t_1,t_2) &= \langle B_C[C_0(\cdot,t_1)], C_0(\cdot,t_2)\rangle_{H(C,T)} \\ &= \sum_j \lambda_j \left\langle e_j, C_0(\cdot,t_1)\right\rangle_{H(C,T)} \left\langle e_j, C_0(\cdot,t_2)\right\rangle_{H(C,T)} \\ &= \sum_j \lambda_j e_j(t_1)\, e_j(t_2)\,. \end{aligned}$$

These series expansions of, respectively, C and C_0 are what is meant by the expression "simultaneous reduction of covariances."

It is shown below how one can obtain, more generally, an approximate reduction for any two covariances on T, C_1 and C_2.

Let $C = C_1 + C_2$, and $B_1 : H(C,T) \longrightarrow H(C,T)$ denote the operator

$$J^{\star}_{C,C_1} J_{C,C_1}\,.$$

For fixed, but arbitrary real numbers

$$\mu_0 < \mu_1 = 0 < \mu_2 < \cdots < \mu_{n-1} < \mu_n = 1,$$

and fixed, but arbitrary $\{i,j\} \subseteq [0:n]$ with $i < j$,

$$\begin{aligned} &(E_{B_1}(\mu_i) - E_{B_1}(\mu_{i-1}))\left(E_{B_1}(\mu_j) - E_{B_1}(\mu_{j-1})\right) = \\ &= E_{B_1}(\mu_i)E_{B_1}(\mu_j) - E_{B_1}(\mu_i)E_{B_1}(\mu_{j-1}) \\ &- E_{B_1}(\mu_{i-1})E_{B_1}(\mu_j) + E_{B_1}(\mu_{i-1})E_{B_1}(\mu_{j-1}) \\ &= E_{B_1}(\mu_i) - E_{B_1}(\mu_i) - E_{B_1}(\mu_{i-1}) + E_{B_1}(\mu_{i-1}) \\ &= 0, \end{aligned}$$

so that the family of projections

$$\{E_{B_1}(\mu_1) - E_{B_1}(\mu_0),\ E_{B_1}(\mu_2) - E_{B_1}(\mu_1), \ldots, E_{B_1}(\mu_n) - E_{B_1}(\mu_{n-1})\}$$

is made of orthogonal elements. Furthermore

$$E_{B_1}(\mu_n) - E_{B_1}(\mu_{n-1}) = I_{H(C,T)} - E_{B_1}(\mu_{n-1}).$$

Let $0 \leq \lambda_1 \leq \cdots \leq \lambda_n$ be fixed, but arbitrary. Then, for

$$B = \sum_{i=1}^{n} \lambda_i \{E_{B_1}(\mu_i) - E_{B_1}(\mu_{i-1})\},$$

one has that

$$\begin{aligned}
\langle B[h], h\rangle_{H(C,T)} &= \sum_{i=1}^{n} \lambda_i \|\{E_{B_1}(\mu_i) - E_{B_1}(\mu_{i-1})\}[h]\|^2_{H(C,T)} \\
&\leq \lambda_n \sum_{i=1}^{n} \|\{E_{B_1}(\mu_i) - E_{B_1}(\mu_{i-1})\}[h]\|^2_{H(C,T)} \\
&= \lambda_n \left\| \sum_{i=1}^{n} \{E_{B_1}(\mu_i) - E_{B_1}(\mu_{i-1})\}[h] \right\|^2_{H(C,T)} \\
&= \lambda_n \left\| \left\{ \sum_{i=1}^{n} E_{B_1}(\mu_i) - E_{B_1}(\mu_{i-1}) \right\} [h] \right\|^2_{H(C,T)} \\
&= \lambda_n \|E_{B_1}(\mu_n)[h]\|^2_{H(C,T)} \\
&= \lambda_n \left\|I_{H(C,T)}[h]\right\|^2_{H(C,T)} \\
&= \lambda_n \|h\|^2_{H(C,T)}.
\end{aligned}$$

When λ_1 is allowed to be zero, the same calculation yields that

$$\langle B[h], h\rangle_{H(C,T)} \leq \lambda_n \left\|\{I_{H(C,T)} - E_{B_1}(0)\}[h]\right\|^2_{H(C,T)} \leq \lambda_n \|h\|^2_{H(C,T)}.$$

One shall need the following family of operators of $H(C,T)$:

$$B_1^{l,n} = \sum_{i=1}^{2^n} \frac{i-1}{2^n} \left\{ E_{B_1}\left(\frac{i}{2^n}\right) - E_{B_1}\left(\frac{i-1}{2^n}\right) \right\},$$

and

$$B_1^{u,n} = \frac{1}{2^n} I_{H(C,T)} + B_1^{l,n}.$$

From the preceding remarks, and similar calculations, one has that:

- $0 \leq \left\langle B_1^{l,n}[h], h \right\rangle_{H(C,T)} \leq \left\langle B_1^{u,n}[h], h \right\rangle_{H(C,T)} \leq \|h\|_{H(C,T)}^2$,
- $0 \leq \left\langle \left\{ B_1^{u,n} - B_1^{l,n} \right\}[h], h \right\rangle_{H(C,T)} \leq \frac{1}{2^n} \|h\|_{H(C,T)}^2$,
- $B_1^{l,n} \leq B_1^{l,n+1} \leq B_1 \leq B_1^{u,n+1} \leq B_1^{u,n}$,
- $\left\| B_1 - B_1^{l,n} \right\| \leq \frac{1}{2^n}$, and $\left\| B_1 - B_1^{u,n} \right\| \leq \frac{1}{2^n}$.

As

$$B_2 = J_{C,C_2}^{\star} J_{C,C_2} = I_{H(C,T)} - B_1 = \int_{0-}^{1} (1-\lambda)\, E_{B_1}(d\lambda),$$

one has that:

- $0 \leq I_{H(C,T)} - B_1^{u,n} \leq I_{H(C,T)} - B_1^{u,n+1} \leq I_{H(C,T)} - B_1 = B_2$,
- $B_2 \leq I_{H(C,T)} - B_1^{l,n+1} \leq I_{H(C,T)} - B_1^{l,n}$,
- $\left\| B_2 - \left(I_{H(C,T)} - B_1^{l,n} \right) \right\| = \left\| B_1 - B_1^{l,n} \right\| \leq \frac{1}{2^n}$,
- $\left\| B_2 - \left(I_{H(C,T)} - B_1^{u,n} \right) \right\| = \left\| B_1 - B_1^{u,n} \right\| \leq \frac{1}{2^n}$.

Let H_1 be the range of $E_{B_1}\left(\frac{1}{2^n}\right)$, and H_i be that of

$$E_{B_1}\left(\frac{i}{2^n}\right) - E_{B_1}\left(\frac{i-1}{2^n}\right),\ 2 \leq i \leq 2^n.$$

An orthonormal basis $\{e_{n,i},\ i \in I\}$ for $H(C,T)$ is built by taking an orthonormal basis in each of the subspaces H_i, $1 \leq i \leq 2^n$, and the corresponding eigenvalue shall be denoted $\lambda_{n,i}$ so that

$$\{\lambda_{n,i},\ i \in I\} \subseteq \left\{0, \frac{1}{2^n}, \ldots, \frac{2^n-1}{2^n}\right\}.$$

Since the covariance C_B associated with the operator B of $H(C,T)$ is given by the following relation [(Proposition) 3.4.1]:

$$C_B(t_1,t_2) = \langle B[C(\cdot,t_1)], C(\cdot,t_2) \rangle_{H(C,T)},$$

one has that the covariances associated with, respectively,

$$B_1^{l,n},\ B_1^{u,n},\ I_{H(C,T)} - B_1^{l,n},\ I_{H(C,T)} - B_1^{u,n},$$

are, respectively,

$$C_{n,1}^{(l)}(t_1,t_2) = \sum_{i\in I} \lambda_{n,i} e_{n,i}(t_1)\, e_{n,i}(t_2),$$

$$C_{n,1}^{(u)}(t_1,t_2) = \sum_{i\in I} \left(\lambda_{n,i} + \frac{1}{2^n}\right) e_{n,i}(t_1)\, e_{n,i}(t_2)$$

$$= C_{n,1}^{(l)}(t_1,t_2) + \frac{1}{2^n} C(t_1,t_2),$$

$$C_{n,2}^{(l)}(t_1,t_2) = \sum_{i\in I} (1-\lambda_{n,i})\, e_{n,i}(t_1)\, e_{n,i}(t_2),$$

$$C_{n,2}^{(u)}(t_1,t_2) = \sum_{i\in I} \left(1 - \lambda_{n,i} - \frac{1}{2^n}\right) e_{n,i}(t_1)\, e_{n,i}(t_2)$$

$$= C_{n,2}^{(l)}(t_1,t_2) - \frac{1}{2^n} C(t_1,t_2).$$

One then has, for example, that

$$\left|C_1(t_1,t_2) - C_{n,1}^{(l)}(t_1,t_2)\right| =$$

$$= \left|\langle B_1[C_1(\cdot,t_1)], C_1(\cdot,t_2)\rangle_{H(C,T)} - \langle B_1^{l,n}[C_1(\cdot,t_1)], C_1(\cdot,t_2)\rangle_{H(C,T)}\right|$$

$$\leq \left\|B_1 - B_1^{l,n}\right\| C(t_1,t_1)\, C(t_2,t_2)$$

$$\leq \frac{1}{2^n} C(t_1,t_1)\, C(t_2,t_2).$$

One consequently has that:

- $\left\{C_{n,1}^{(l)}, C_{n,1}^{(u)}, C_{n,2}^{(l)}, C_{n,2}^{(u)}\right\} \subseteq \mathcal{D}(C)$;
- $\lim_n C_{n,1}^{(l)}(t_1,t_2) = \lim_n C_{n,1}^{(u)}(t_1,t_2) = C_1(t_1,t_2)$;
- $\lim_n C_{n,2}^{(l)}(t_1,t_2) = \lim_n C_{n,2}^{(u)}(t_1,t_2) = C_2(t_1,t_2)$.

Furthermore, since $\sigma\left(B_1^{n,l}\right) = \{\lambda_{n,i},\ i \in I\}$,

- when C_2 dominates C_1, $C_{n,2}^{(l)}$ dominates $C_{n,1}^{(l)}$, so that $\sigma\left(B_1^{n,l}\right) \subseteq \left[0, \frac{1}{2}\right]$,
- when $\sigma\left(B_1^{n,l}\right) \subseteq \left[0, \frac{1}{2}\right]$, $C_2 + \frac{1}{2^n} C$ dominates C_1.

As a summary, one may say that $C_{n,1}^{(l)}$ and $C_{n,2}^{(u)}$ have the following properties:

1. they have a simultaneous reduction;
2. $C_{n,1}^{(l)} \ll C_1 \ll C_{n,1}^{(l)} + \frac{1}{2^n} C$;
3. $C_{n,2}^{(u)} \ll C_2 \ll C_{n,2}^{(u)} + \frac{1}{2^n} C$;
4. $\lim_n C_{n,1}^{(l)}(t_1,t_2) = C_1(t_1,t_2)$, and $\lim_n C_{n,2}^{(u)}(t_1,t_2) = C_2(t_1,t_2)$;

5. the domination properties of C_1 and C_2 are close to the domination properties of $C_{n,1}^{(l)}$ and $C_{n,2}^{(u)}$.

Remark 3.4.10 For item 4 above, convergence occurs uniformly as soon as C_1 and C_2 are bounded on the diagonal of $T \times T$.

Chapter 4
Reproducing Kernel Hilbert Spaces and Paths of Stochastic Processes

The problem addressed in this chapter is that of giving conditions which insure that the paths of a stochastic process belong to a given RKHS, a requirement for likelihood detection problems not to be singular. All along $(\Omega, \mathcal{A}, P)$ shall denote a fixed, but arbitrary probability space, and $H(\mathcal{H}, T)$, a fixed, but arbitrary RKHS. The source of the following material is [176].

4.1 Random Elements with Values in a Reproducing Kernel Hilbert Space

Some of the definitions, and results, which follow, repeat,[1] and complete, to some extent, those of Sects. 1.7.4 and 2.1, where some of the proofs may be found.

Definition 4.1.1 On $H(\mathcal{H}, T)$ there are two σ-algebras of particular interest. The first, denoted $\mathcal{C}(\mathcal{H}, T)$, is generated by the continuous linear functionals of $H(\mathcal{H}, T)$, and the second, denoted $\mathcal{B}(\mathcal{H}, T)$, is generated by the open sets of $H(\mathcal{H}, T)$.

Fact 4.1.2 ([260, p.17]) *One has always that*

$$\mathcal{C}(\mathcal{H}, T) \subseteq \mathcal{B}(\mathcal{H}, T),$$

but, when $H(\mathcal{H}, T)$ is separable,

$$\mathcal{C}(\mathcal{H}, T) = \mathcal{B}(\mathcal{H}, T).$$

[1] *Repetitio est mater studiorum!*

A.F. Gualtierotti, *Detection of Random Signals in Dependent Gaussian Noise*,
DOI 10.1007/978-3-319-22315-5_4

Definition 4.1.3 A probability measure Π on $\mathcal{B}(\mathcal{H},T)$ is Radon when, for each $B \in \mathcal{B}(\mathcal{H},T)$,

$$\Pi(B) = \sup_{\{K \subseteq B : K \text{ compact}\}} \Pi(K).$$

Fact 4.1.4 ([260, p. 80]) *Every probability measure Π on $\mathcal{C}(\mathcal{H},T)$ has a unique Radon extension to $\mathcal{B}(\mathcal{H},T)$.*

Definition 4.1.5 A random element on $(\Omega,\mathcal{A},P)$, with values in $H(\mathcal{H},T)$, is a map $\xi : \Omega \longrightarrow H(\mathcal{H},T)$, which is adapted to $\mathcal{A}$ and $\mathcal{C}(\mathcal{H},T)$. When ξ is adapted to $\mathcal{A}$ and $\mathcal{B}(\mathcal{H},T)$, one says that ξ is a Borel random element.

Fact 4.1.6 ([260, p. 89]) *When a function $\xi : \Omega \longrightarrow H(\mathcal{H},T)$ has separable range, that is, when there exists a countable $H_c \subseteq H(\mathcal{H},T)$ whose closure, in $H(\mathcal{H},T)$, contains $\xi(\Omega)$, then the following statements are equivalent:*

1. *ξ is a random element;*
2. *ξ is a Borel random element ;*
3. *given any separating subset $H_s \subseteq H(\mathcal{H}.T)$, the set of functions*

$$\left\{\omega \mapsto \langle \xi(\omega), h\rangle_{H(\mathcal{H},T)}, h \in H_s\right\}$$

is made of random variables.

When the above obtain, the probability law of ξ, P_ξ, is Radon [260, p. 29].

Definition 4.1.7 Let ξ and η be random elements defined on $(\Omega,\mathcal{A},P)$, with values in $H(\mathcal{H},T)$. η is a version of ξ whenever, for each $h \in H(\mathcal{H},T)$,

$$P\left(\omega \in \Omega : \langle \xi(\omega), h\rangle_{H(\mathcal{H},T)} = \langle \eta(\omega), h\rangle_{H(\mathcal{H},T)}\right) = 1.$$

Fact 4.1.8 ([260, p. 219]) *Let ξ be a random element on $(\Omega,\mathcal{A},P)$, with values in $H(\mathcal{H},T)$. P_ξ extends to a Radon measure on $\mathcal{B}(\mathcal{H},T)$ if, and only if, ξ has a version η whose range is separable.*

Remark 4.1.9 Since probability measures on $\mathcal{C}(\mathcal{H},T)$ have Radon extensions to $\mathcal{B}(\mathcal{H},T)$ [(Fact) 4.1.4], random elements have versions whose range is separable, that is, Borel versions [(Definition) 4.1.5]. The induced probability is thus Radon [(Fact) 4.1.8].

Definition 4.1.10 Let p be fixed, but arbitrary in $]0,\infty[$. A probability Π on $\mathcal{C}(\mathcal{H},T)$ is weakly of order p when, for fixed, but arbitrary $h \in H(\mathcal{H},T)$,

$$\int_{H(\mathcal{H},T)} \left|\langle x, h\rangle_{H(\mathcal{H},T)}\right|^p \Pi(dx) < \infty.$$

A random element ξ, defined on $(\Omega, \mathcal{A}, P)$, with values in $H(\mathcal{H}, T)$, is weakly of order p when P_ξ is weakly of order p. A probability Π on $\mathcal{B}(\mathcal{H}, T)$ is strongly of order p when

$$\int_{H(\mathcal{H},T)} \|x\|^p_{H(\mathcal{H},T)} \, \Pi(dx) < \infty.$$

A random element ξ, defined on $(\Omega, \mathcal{A}, P)$, with values in $H(\mathcal{H}, T)$, is strongly of order p when P_ξ is strongly of order p.

Fact 4.1.11 ([260, p. 114]) *Let ξ be a random element defined on $(\Omega, \mathcal{A}, P)$, with values in $H(\mathcal{H}, T)$, and suppose it is weakly of order one. There exists then a unique $m_\xi \in H(\mathcal{H}, T)$ such that, for every $h \in H(\mathcal{H}, T)$,*

$$\int_\Omega \langle \xi(\omega), h \rangle_{H(\mathcal{H},T)} P(\omega) = \int_{H(\mathcal{H},T)} \langle x, h \rangle_{H(\mathcal{H},T)} P_\xi(dx) = \langle h, m_\xi \rangle_{H(\mathcal{H},T)}.$$

One then writes $m_\xi = E_P[\xi]$, and one says that m_ξ is the Pettis integral of ξ with respect to P.

Definition 4.1.12 Let Π be a probability of weak order two on $\mathcal{C}(\mathcal{H}, T)$. Given fixed, but arbitrary $(h_1, h_2) \in H(\mathcal{H}, T) \times H(\mathcal{H}, T)$, let

$$\Gamma_\Pi(h_1, h_2) = \int_{H(\mathcal{H},T)} \langle x, h_1 \rangle_{H(\mathcal{H},T)} \langle x, h_2 \rangle_{H(\mathcal{H},T)} \Pi(dx),$$

and

$$C_\Pi(h_1, h_2) = \Gamma_\Pi(h_1, h_2) - \langle h_1, m_\Pi \rangle_{H(\mathcal{H},T)} \langle h_2, m_\Pi \rangle_{H(\mathcal{H},T)}.$$

Fact 4.1.13 ([260, p. 169, 170, 171 and 172]) *Let Π be a probability of weak order two on $\mathcal{C}(\mathcal{H}, T)$. Then Γ_Π and C_Π are bilinear forms on $H(\mathcal{H}, T)$ which are positive, symmetric, and continuous. There exist then linear, positive, continuous, and self-adjoint operators of $H(\mathcal{H}, T)$, say R_Π and S_Π, such that*

$$\begin{aligned} \Gamma_\Pi(h_1, h_2) &= \langle R_\Pi[h_1], h_2 \rangle_{H(\mathcal{H},T)}, \\ C_\Pi(h_1, h_2) &= \langle S_\Pi[h_1], h_2 \rangle_{H(\mathcal{H},T)}. \end{aligned}$$

Definition 4.1.14 S_Π is the covariance operator of Π.

Fact 4.1.15 ([260, p. 173]) *Operators of $H(\mathcal{H}, T)$, which are linear, positive, continuous, self-adjoint, with separable range, are the covariance operators of Radon probability measures which are weakly of second order.*

Fact 4.1.16 ([260, p. 177]) *Operators of $H(\mathcal{H}, T)$ which are linear, positive, continuous, self-adjoint, with finite trace, are the covariance operators of Radon probability measures which are strongly of second order.*

4.2 Paths and Values, in Reproducing Kernel Hilbert Spaces, of Random Elements

Under conditions to be made precise, random elements with values in an RKHS produce processes with paths in them, and conversely. Elements and paths so related share properties to be listed.

Proposition 4.2.1 *Let $H(\mathcal{H}, T)$ be an RKHS, and X be a stochastic process on $(\Omega, \mathcal{A}, P)$, with index set T.*

Suppose that there exists $A \in \mathcal{A}$ such that

1. $P(A) = 1$,
2. *for* $\omega \in A$, $X(\omega, \cdot) \in H(\mathcal{H}, T)$.

The formula

$$\xi(\omega) = \chi_A(\omega) X(\omega, \cdot)$$

determines then a random element in $H(\mathcal{H}, T)$ such that

$$\langle \xi(\omega), \mathcal{H}(\cdot, t) \rangle_{H(\mathcal{H},T)} = \chi_A(\omega) X(\omega, t).$$

Proof One has that

$$\left\langle \xi(\omega), \sum_{i=1}^{n} \alpha_i \mathcal{H}(\cdot, t_i) \right\rangle_{H(\mathcal{H},T)} = \chi_A(\omega) \sum_{i=1}^{n} \alpha_i X(\omega, t_i),$$

and thus $\langle \xi(\cdot), h \rangle_{H(\mathcal{H},T)}$ is adapted for every $h \in V[\mathcal{H}]$. But $V[\mathcal{H}]$ is dense in $H(\mathcal{H}, T)$ [(Proposition) 1.1.5], and thus $\langle \xi(\cdot), h \rangle_{H(\mathcal{H},T)}$ is adapted for every $h \in H(\mathcal{H}, T)$, as the limit of a sequence of adapted functions. □

Proposition 4.2.2 *When ξ is a random element defined on $(\Omega, \mathcal{A}, P)$, with values in $H(\mathcal{H}, T)$, the formula*

$$X(\omega, t) = \langle \xi(\omega), \mathcal{H}(\cdot, t) \rangle_{H(\mathcal{H})}$$

determines a stochastic process on $(\Omega, \mathcal{A}, P)$, with index set T, and paths in $H(\mathcal{H}, T)$.

Proposition 4.2.3 *Let ξ be a random element defined on $(\Omega, \mathcal{A}, P)$, with values in $H(\mathcal{H}, T)$, and let X be the process it determines* [(Proposition) 4.2.2]. *Then:*

1. *when ξ is weakly of first order, X has a mean in $H(\mathcal{H}, T)$, equal to that of ξ:*

$$\mu_X(t) = E[X(\cdot, t)] = \langle m_\xi, \mathcal{H}(\cdot, t)\rangle_{H(\mathcal{H},T)} = m_\xi(t), \ t \in T;$$

2. *when ξ is weakly of second order, X is a second order process, and, for fixed, but arbitrary $(t_1, t_2) \in T \times T$,*

$$C_X(t_1, t_2) = \langle S_\xi[\mathcal{H}(\cdot, t_1)], \mathcal{H}(\cdot, t_2)\rangle_{H(\mathcal{H},T)},$$

where S_ξ is the covariance operator of P_ξ. Furthermore

$$C_X \ll \|S_\xi\| \mathcal{H}.$$

Proof One has, because of the definition of weak first order, that

$$E[|X(\cdot, t)|] = E\left[\left|\langle \xi, \mathcal{H}(\cdot, t)\rangle_{H(\mathcal{H},T)}\right|\right] < \infty.$$

For the same reason, using (Fact) 4.1.11,

$$E[X(\cdot, t)] = E\left[\langle \xi, \mathcal{H}(\cdot, t)\rangle_{H(\mathcal{H},T)}\right] = \langle m_\xi, \mathcal{H}(\cdot, t)\rangle_{H(\mathcal{H},T)}.$$

Similarly $E\left[X^2(\cdot, t)\right] = E\left[\langle \xi, \mathcal{H}(\cdot, t)\rangle^2_{H(\mathcal{H},T)}\right] < \infty$, and

$$\begin{aligned} C_X(t_1, t_2) &= E[X(\cdot, t_1) X(\cdot, t_2)] - E[X(\cdot, t_1)] E[X(\cdot, t_2)] \\ &= E\left[\langle \xi, \mathcal{H}(\cdot, t_1)\rangle_{H(\mathcal{H},T)} \langle \xi, \mathcal{H}(\cdot, t_2)\rangle_{H(\mathcal{H},T)}\right] \\ &- E\left[\langle \xi, \mathcal{H}(\cdot, t_1)\rangle_{H(\mathcal{H},T)}\right] E\left[\langle \xi, \mathcal{H}(\cdot, t_2)\rangle_{H(\mathcal{H},T)}\right] \\ &= \langle S_\xi[\mathcal{H}(\cdot, t_1)], \mathcal{H}(\cdot, t_2)\rangle_{H(\mathcal{H},T)}. \end{aligned}$$

Finally

$$\begin{aligned} \sum_{i,j=1}^{n} \alpha_i \alpha_j C_X(t_i, t_j) &= \left\langle S_\xi\left[\sum_{i=1}^{n} \alpha_i \mathcal{H}(\cdot, t_i)\right], \sum_{i=1}^{n} \alpha_i \mathcal{H}(\cdot, t_i)\right\rangle_{H(\mathcal{H},T)} \\ &\leq \|S_\xi\| \left\|\sum_{i=1}^{n} \alpha_i \mathcal{H}(\cdot, t_i)\right\|^2_{H(\mathcal{H},T)} \\ &\leq \|S_\xi\| \sum_{i,j=1}^{n} \alpha_i \alpha_j \mathcal{H}(t_i, t_j). \end{aligned}$$

□

Remark 4.2.4 A second order stochastic process may, or may not, determine a weak second order random element. Here is an example.

Example 4.2.5 It may be built from the following RKHS. Let $H \subseteq \mathbb{R}^\infty$ be the family of sequences $\underline{a} = \{a_1, a_2, a_3, \dots\}$ such that

$$\sum_n n^2 a_n^2 < \infty.$$

It is a real vector space as

$$\alpha \underline{a} + \beta \underline{b} = \{\alpha a_n + \beta b_n,\ n \in \mathbb{N}\},$$

and

$$\sum_n n^2 (\alpha a_n + \beta b_n)^2 \leq 2 \sum_n n^2 \left(\alpha^2 a_n^2 + \beta^2 b_n^2\right) < \infty.$$

The prescription

$$\langle \underline{a}, \underline{b} \rangle_H = \sum_n n^2 a_n b_n$$

obviously defines a bilinear, strictly positive form, and thus an inner product. For the norm determined by this inner product, H is complete. Let indeed

$$\lim_{m,n} \left\| \underline{a}_m - \underline{a}_n \right\|_H = 0.$$

As, for fixed, but arbitrary $i \in \mathbb{N}$,

$$\left(a_i^{(m)} - a_i^{(n)}\right)^2 \leq \sum_j j^2 \left(a_j^{(m)} - a_j^{(n)}\right)^2 = \left\| \underline{a}_m - \underline{a}_n \right\|_H^2,$$

there exists, for each $i \in \mathbb{N}$, $\alpha_i = \lim_n a_i^{(n)}$. One must show that $\sum_i i^2 \alpha_i^2 < \infty$. Now

$$\begin{aligned} \sum_i i^2 \alpha_i^2 &= \sum_i i^2 \left(\alpha_i - a_i^{(n)} + a_i^{(n)}\right)^2 \\ &\leq 2 \left\{ \sum_i i^2 \left(\alpha_i - a_i^{(n)}\right)^2 + \sum_i i^2 \left[a_i^{(n)}\right]^2 \right\}, \end{aligned}$$

so that, since a Cauchy sequence is bounded,

$$\sum_i i^2 \alpha_i^2 \leq 2 \left\{ \sum_i i^2 \left(\alpha_i - a_i^{(n)}\right)^2 + \left\| \underline{a}_n \right\|_H^2 \right\} \leq 2 \left\{ \sum_i i^2 \left(\alpha_i - a_i^{(n)}\right)^2 + \kappa^2 \right\}.$$

Choose then $n_1 \in \mathbb{N}$ such that, for $n \in \mathbb{N},\ n > n_1$,

$$\left(\alpha_1 - a_1^{(n)}\right)^2 \leq 2^{-1},$$

and, successively, for every $p \in \mathbb{N}$, $n_{p+1} > n_p$ such that, for $n > n_{p+1}$,

$$(p+1)^2 \left(\alpha_{p+1} - a_{p+1}^{(n)}\right)^2 \leq 2^{-(p+1)}.$$

Then

$$\sum_i i^2 \alpha_i^2 \leq 2\left\{1 + \kappa^2\right\} < \infty.$$

Thus $\{\alpha_n,\ n \in \mathbb{N}\} \in H$.

A reproducing kernel for H is obtained setting

$$\mathcal{H}(i,j) = i^{-2}\delta_{i,j}, \quad \text{with } \delta_{i,j} = \begin{cases} 1 \text{ when } i = j \\ 0 \text{ when } i \neq j \end{cases}.$$

Indeed

- $\sum_{i,j} \alpha_i \alpha_j \mathcal{H}(i,j) = \sum_i \alpha_i^2 i^{-2} \geq 0$;
- given that $\underline{e}_i$ is the element of $\mathbb{R}^\infty$ with all components equal to zero, except the ith, which is 1,

$$\mathcal{H}(\cdot, i) = i^{-2}\underline{e}_i \in H;$$

- $\langle \underline{a}, \mathcal{H}(\cdot, i)\rangle_H = \langle \underline{a}, i^{-2}\underline{e}_i\rangle_H = a_i$.

For the probability space, one chooses $\Omega = \mathbb{N}$, for the events, $\mathcal{A} = \mathcal{P}(\mathbb{N})$, the family of subsets of $\mathbb{N}$, and, for the probability P, any sequence

$$\{\pi_n > 0,\ n \in \mathbb{N}\}$$

such that $\sum_n \pi_n = 1$ and $\pi_n = P(\{n\})$. The index set of X is $T = \mathbb{N}$, and X is defined, for fixed, but arbitrary $n \in \mathbb{N}$, using the following relation:

$$X(\omega, n) = \begin{Bmatrix} 1 \text{ when } \omega = n \\ 0 \text{ when } \omega \neq n \end{Bmatrix} = \chi_{\{n\}}(\omega).$$

X is a second order process as

$$E\left[X^2(\cdot, n)\right] = E\left[\chi^2_{\{n\}}\right] = E\left[\chi_{\{n\}}\right] = P(\{n\}) = \pi_n, \qquad (\star)$$

and has paths in H as $X(\omega,\cdot) = \underline{e}_{\omega}$. The random element associated with X is thus $\xi(\omega) = X(\omega,\cdot) = \underline{e}_{\omega}$, and

$$\langle \xi(\omega), \underline{a} \rangle_H = \omega^2 a_{\omega}.$$

The order of ξ shall be determined by the behavior of the following expression:

$$\int_{\Omega} \left| \langle \xi(\omega), \underline{a} \rangle_H \right|^p P(d\omega) = \sum_{\omega\in\Omega} \left| \langle \underline{e}_{\omega}, \underline{a} \rangle_H \right|^p P(\omega) = \sum_{\omega} \omega^{2p} |a_{\omega}|^p \pi_{\omega}.$$

- *Case 1:* $\pi_n = \kappa n^{-3/2}$, $\kappa > 0$, and $\sum_n \pi_n = 1$.
  If ξ were weakly of first order, the following relation:

  $$E[X(\cdot,n)] = \langle m_{\xi}, \mathcal{H}(\cdot,n) \rangle_H = m_{\xi}(n)$$

  would yield (as for $\star$) $m_{\xi}(n) = \pi_n$. But $\sum_n n^2 \pi_n^2 = \kappa^2 \sum_n n^2 n^{-3} = \infty$, a contradiction. ξ can thus not be weakly of first order.
  One may notice that

  $$C_X(m,n) = E[X(\cdot,m)X(\cdot,n)] - E[X(\cdot,m)]E[X(\cdot,n)]$$
  $$= \frac{\kappa}{n^{3/2}} \delta_{m,n} - \frac{\kappa^2}{m^{3/2} n^{3/2}} = \frac{\kappa}{n^{3/2}} \left\{ \delta_{m,n} - \frac{\kappa}{m^{3/2}} \right\},$$

  so that

  $$\sum_m m^2 \left\{ \frac{\kappa}{n^{3/2}} \left\{ \delta_{m,n} - \frac{\kappa}{m^{3/2}} \right\} \right\}^2 = \frac{\kappa^2}{n^3} \sum_m m^2 \left\{ \delta_{m,n} - \frac{\kappa}{m^{3/2}} \right\}^2 = \infty.$$

  Thus $C_X(\cdot,n)$ does not belong to H, and $H(C_X,\mathbb{N})$ cannot be a subset of H.
- *Case 2:* $\pi_n = \kappa n^{-5/3}$, $\kappa > 0$, and $\sum_n \pi_n = 1$.
  ξ is weakly of first order as

  $$\int_{\Omega} \left| \langle \xi(\omega), \underline{a} \rangle_H \right| P(d\omega) = \sum_{\omega\in\Omega} \omega^2 |a_{\omega}| \pi_{\omega},$$

  and that $\underline{\pi} = \{\pi_n,\ n \in \mathbb{N}\} \in H$ as

  $$\sum_n n^2 \pi_n^2 = \kappa^2 \sum_n n^2 n^{-10/3} = \sum_n n^{-4/3} < \infty.$$

  Now since $\sum_n n^2 n^{-19/6} = \sum_n n^{-7/6} < \infty$, the sequence

  $$\underline{h} = h = \left\{ n^{-19/12},\ n \in I\!N \right\} \text{ belongs to } H,$$

but $\int_{\Omega} \left|\langle \xi(\omega), \underline{h}\rangle_H\right|^2 P(d\omega) =$

$$\sum_{\omega\in\Omega} \left\langle \underline{e}_\omega, \underline{h}\right\rangle^2 \pi_\omega = \sum_n \left\{n^2 h_n\right\}^2 \frac{\kappa}{n^{5/3}} = \kappa \sum_n \frac{1}{n^{5/6}} = \infty,$$

so that ξ is not weakly of second order.

Proposition 4.2.6 *Let X be a second order stochastic process on $(\Omega, \mathcal{A}, P)$, with index set T, and let $H(\mathcal{H}, T)$ be an RKHS. Let μ_X denote the mean of X, and C_X, its covariance. Suppose that*

(a) $\mu_X \in H(\mathcal{H}, T)$,
(b) *there exists $A \in \mathcal{A}$ such that $P(A) = 1$, and $X(\omega, \cdot) \in H(\mathcal{H}, T)$, $\omega \in A$,*
(c) *there exists $\kappa \geq 0$ such that $C_X \ll \kappa\mathcal{H} = \mathcal{H}_\kappa$.*

Let $\xi(\omega) = \chi_A(\omega) X(\omega, \cdot) \in H(\mathcal{H}, T)$, $\omega \in \Omega$. Then:

1. *ξ is of weak second order;*
2. *letting $J_{\mathcal{H}_\kappa, C_X} : H(\mathcal{H}_\kappa, T) \longrightarrow H(C_X, T)$ be defined, using Assumption* (c) *and* (Proposition) 3.1.5, *as*

$$J_{\mathcal{H}_\kappa, C_X}\left[\mathcal{H}_\kappa(\cdot, t)\right] = C_X(\cdot, t),$$

one has that

$$S_\xi = J^\star_{\mathcal{H}_\kappa, C_X} J_{\mathcal{H}_\kappa, C_X}.$$

Proof $H(\mathcal{H}, T)$ and $H(\mathcal{H}_\kappa, T)$ are equal as sets of functions, and their respective norms are equivalent [(Example) 1.3.12]. Thus $\mu_X \in H(\mathcal{H}_\kappa, T)$, and also $X(\omega, \cdot) \in H(\mathcal{H}_\kappa, T)$, $\omega \in A$. Assumption (a) allows one to subtract from X its mean, and thus suppose that the mean is the zero function.

Fixed, but arbitrary elements h_1 and h_2 in $V[\mathcal{H}_\kappa]$ may be represented as

$$h_1 = \sum_{i=1}^n \alpha_i^{(1)} \mathcal{H}_\kappa(\cdot, t_i), \quad \text{and} \quad h_2 = \sum_{i=1}^n \alpha_i^{(2)} \mathcal{H}_\kappa(\cdot, t_i)$$

(introducing when necessary coefficients equal to zero). Then

$$\langle \xi, h_1\rangle_{H(\mathcal{H}_\kappa, T)} = \chi_A \sum_{i=1}^n \alpha_i^{(1)} X(\cdot, t_i),$$

and

$$\langle \xi, h_2\rangle_{H(\mathcal{H}_\kappa, T)} = \chi_A \sum_{i=1}^n \alpha_i^{(2)} X(\cdot, t_i).$$

Consequently

$$E\left[\langle \xi, h_1\rangle_{H(\mathcal{H}_\kappa,T)} \langle \xi, h_2\rangle_{H(\mathcal{H}_\kappa,T)}\right] = \sum_{i=1}^{n}\sum_{j=1}^{n} \alpha_i^{(1)}\alpha_j^{(2)} C_X\left(t_i, t_j\right)$$

$$= \left\langle J^{\star}_{\mathcal{H}_\kappa,C_X} J_{\mathcal{H}_\kappa,C_X}\left[h_1\right], h_2\right\rangle_{H(\mathcal{H}_\kappa,T)}.$$

Let now, h_1 and h_2 be fixed, but arbitrary elements of $H(\mathcal{H}_\kappa, T)$, and let

$$\left\{h_n^{(1)},\ n \in \mathbb{N}\right\} \subseteq V\left[\mathcal{H}_\kappa\right], \text{ and } \left\{h_n^{(2)},\ n \in \mathbb{N}\right\} \subseteq V\left[\mathcal{H}_\kappa\right],$$

be such that, in $H(\mathcal{H}_\kappa, T)$, $\lim_n h_n^{(1)} = h_1$, and $\lim_n h_n^{(2)} = h_2$. A first consequence is that, for $\omega \in \Omega$,

$$\lim_n \left\langle \xi(\omega), h_n^{(1)}\right\rangle_{H(\mathcal{H}_\kappa,T)} = \langle \xi(\omega), h_1\rangle_{H(\mathcal{H}_\kappa,T)}.$$

But

$$E\left[\left\{\left\langle \xi(\cdot), h_n^{(1)}\right\rangle_{H(\mathcal{H}_\kappa,T)} - \left\langle \xi(\cdot), h_p^{(1)}\right\rangle_{H(\mathcal{H}_\kappa,T)}\right\}^2\right] =$$

$$= E\left[\left\langle \xi(\cdot), h_n^{(1)} - h_p^{(1)}\right\rangle^2_{H(\mathcal{H}_\kappa,T)}\right]$$

$$= \left\langle J^{\star}_{\mathcal{H}_\kappa,C_X} J_{\mathcal{H}_\kappa,C_X}\left[h_n^{(1)} - h_p^{(1)}\right], h_n^{(1)} - h_p^{(1)}\right\rangle_{H(\mathcal{H}_\kappa,T)},$$

which converges to zero. Thus, in $L_2(\Omega, \mathcal{A}, P)$, as convergence takes place for every $\omega \in \Omega$,

$$\lim_n \left[\left\langle \xi(\cdot), h_n^{(1)}\right\rangle_{H(\mathcal{H}_\kappa,T)}\right]_{L_2(\Omega,\mathcal{A},P)} = \left[\langle \xi(\cdot), h_1\rangle_{H(\mathcal{H}_\kappa,T)}\right]_{L_2(\Omega,\mathcal{A},P)}.$$

Consequently

$$E\left[\langle \xi(\cdot), h_1\rangle_{H(\mathcal{H}_\kappa,T)} \langle \xi(\cdot), h_2\rangle_{H(\mathcal{H}_\kappa,T)}\right] =$$

$$= \lim_{n,p} E\left[\left\langle \xi(\cdot), h_n^{(1)}\right\rangle_{H(\mathcal{H}_\kappa,T)} \left\langle \xi(\cdot), h_p^{(2)}\right\rangle_{H(\mathcal{H}_\kappa,T)}\right]$$

$$= \left\langle J^{\star}_{\mathcal{H}_\kappa,C_X} J_{\mathcal{H}_\kappa,C_X}\left[h_1\right], h_2\right\rangle_{H(\mathcal{H}_\kappa,T)}.$$

□

Corollary 4.2.7 *The assumptions are those of* (Proposition) 4.2.6. *Then:*

1. $S_\xi = J^{\star}_{\mathcal{H}_\kappa,C_X} J_{\mathcal{H}_\kappa,C_X}$ *has finite trace if, and only if,* ξ *has a Borel version* η *which is strongly of second order.*
2. *When* $H(\mathcal{H}_\kappa, T)$ *is separable,* $S_\xi = J^{\star}_{\mathcal{H}_\kappa,C_X} J_{\mathcal{H}_\kappa,C_X}$ *has finite trace if, and only if,* ξ *is strongly of second order.*

Proof [1] *Suppose S_ξ has finite trace.*

Since random elements have Borel versions [(Remark) 4.1.9], one may assume that ξ is Borel. So P_ξ is a Radon probability measure [(Fact) 4.1.6], with covariance S_ξ. It must thus [(Fact) 4.1.16] be of strong second order since its representation $J^\star_{\mathcal{H}_\kappa, C_X} J_{\mathcal{H}_\kappa, C_X}$ has the properties required for (Fact) 4.1.16 to apply.

Proof [1] *Suppose that η is a Borel version of ξ (whose range is thus separable), and which is of strong second order.*

P_η is then Radon [(Fact) 4.1.6], and the covariance operator of η has finite trace [(Fact) 4.1.16]. But it is equal to S_ξ as η is a version of ξ.

Item 2 follows directly from item 1. □

The trace-class property encountered above determines a form of "strong order" among covariances as encapsulated in the definition which follows.

Definition 4.2.8 Let $\mathcal{H}$ and $\mathcal{K}$ be reproducing kernels on T. When there exists $\kappa > 0$ such that

1. $\mathcal{H} \ll \kappa \mathcal{K} = \mathcal{K}_\kappa$,
2. $J_{\mathcal{K}_\kappa, \mathcal{H}} : H(\mathcal{K}_\kappa, T) \longrightarrow H(\mathcal{H}, T)$, defined using [(Proposition) 3.1.5]

$$J_{\mathcal{K}_\kappa, \mathcal{H}} [\mathcal{K}_\kappa (\cdot, t)] = \mathcal{H}(\cdot, t) \, , \; t \in T,$$

is such that $J^\star_{\mathcal{K}_\kappa, \mathcal{H}} J_{\mathcal{K}_\kappa, \mathcal{H}}$ has finite trace,

one shall use the notation $\mathcal{H} \ll_\tau \mathcal{K}$.

Remark 4.2.9 When $J^\star_{\mathcal{K}_\kappa, \mathcal{H}} J_{\mathcal{K}_\kappa, \mathcal{H}}$ has finite trace, there are [235] orthonormal e_i's, and strictly positive coefficients λ_i, for which $\sum_i \lambda_i < \infty$, and

$$J^\star_{\mathcal{K}_\kappa, \mathcal{H}} J_{\mathcal{K}_\kappa, \mathcal{H}} = \sum_i \lambda_i e_i \otimes e_i \, .$$

Furthermore

$$\left\{ J^\star_{\mathcal{K}_\kappa, \mathcal{H}} J_{\mathcal{K}_\kappa, \mathcal{H}} \right\}^{1/2} = \sum_i \lambda_i^{1/2} e_i \otimes e_i \, ,$$

and, for a partial isometry W,

$$J_{\mathcal{K}_\kappa, \mathcal{H}} = W \left\{ J^\star_{\mathcal{K}_\kappa, \mathcal{H}} J_{\mathcal{K}_\kappa, \mathcal{H}} \right\}^{1/2} .$$

Thus $J^\star_{\mathcal{K}_\kappa, \mathcal{H}} J_{\mathcal{K}_\kappa, \mathcal{H}}$ has finite trace if, and only if, $J_{\mathcal{K}_\kappa, \mathcal{H}}$ is Hilbert-Schmidt. Section 3.1 contains a discussion of the Hilbert-Schmidt properties of $J_{\mathcal{K}_\kappa, \mathcal{H}}$.

4.3 Paths and Values, in Reproducing Kernel Hilbert Spaces, of Gaussian Elements

Gaussian processes having paths in a RKHS must have covariances which are "strongly dominated" as explained below.

Definition 4.3.1 Let $(\Omega, \mathcal{A}, P)$ be a probability space, $H(\mathcal{H}, T)$ be an RKHS, and ξ be a random element defined on $(\Omega, \mathcal{A}, P)$, with values in $H(\mathcal{H}, T)$. ξ is Gaussian when $\omega \mapsto \langle \xi(\omega), h \rangle_{H(\mathcal{H},T)}$ is a Gaussian random variable for every $h \in H(\mathcal{H}, T)$.

Remark 4.3.2 Every Gaussian random element with values in an RKHS determines a Gaussian random process [(Propositions) 4.2.2, 4.2.3].

Remark 4.3.3 ξ is a Gaussian random element as soon as $\langle \xi, \mathcal{H}(\cdot, t) \rangle_{H(\mathcal{H},T)}$ is a Gaussian random variable for all $t \in T$. For then indeed $\langle \xi, h \rangle_{H(\mathcal{H},T)}$ is Gaussian for $h \in V[\mathcal{H}]$ as

$$\langle \xi, h \rangle_{H(\mathcal{H},T)} = \sum_{i=1}^{n} \alpha_i [h] \langle \xi, \mathcal{H}(\cdot, t_i [h]) \rangle_{H(\mathcal{H},T)} .$$

If now, in $H(\mathcal{H}, T)$, $h = \lim_n h_n,\ h_n \in V[\mathcal{H}],\ n \in \mathbb{N}$,

$$\langle \xi, h \rangle_{H(\mathcal{H},T)} = \lim_n \langle \xi, h_n \rangle_{H(\mathcal{H},T)} ,$$

which is Gaussian as the everywhere limit of Gaussian random variables [200, p. 16].

Fact 4.3.4 ([260, p. 213]) *The class of covariance operators of Gaussian Radon measures on $H(\mathcal{H}, T)$ coincides with that of linear, continuous, positive, self-adjoint operators of $H(\mathcal{H}, T)$ that have finite trace.*

Proposition 4.3.5 *Let X be a Gaussian random process for the probability space $(\Omega, \mathcal{A}, P)$, with index set T, mean μ_X, and covariance C_X, and let $H(\mathcal{H}, T)$ be an RKHS. Suppose that*

(a) $\mu_X \in H(\mathcal{H}, T)$,
(b) *there exists $A \in \mathcal{A}$ such that $P(A) = 1$ and $X(\omega, \cdot) \in H(\mathcal{H}, T)$, $\omega \in A$.*

Then:

1. *X determines, on $(\Omega, \mathcal{A}, P)$, a Gaussian random element ξ whose values are in $H(\mathcal{H}, T)$;*
2. $C_X \ll_\tau \mathcal{H}$.

Proof Because of Assumption (a), one may, as in (Proposition) 4.2.6, subtract from X its mean, and thus assume that $\mu_X = 0$.

ξ is defined as in (Proposition) 4.2.1:

$$\langle \xi(\omega), \mathcal{H}(\cdot, t)\rangle_{H(\mathcal{H},T)} = \chi_A(\omega) X(\omega, t).$$

ξ is thus a random element with values in $H(\mathcal{H}, T)$, and is Gaussian since $\langle \xi(\omega), \mathcal{H}(\cdot, t)\rangle_{H(\mathcal{H},T)}$ is Gaussian for $t \in T$ [(Remark) 4.3.3]. It is thus automatically of weak order two. Result (Proposition) 4.2.3 then yields that $C_X \ll \kappa \mathcal{H}$, some $\kappa \geq 0$, and that, for $(t_1, t_2) \in T \times T$, fixed, but arbitrary,

$$C_X(t_1, t_2) = \langle S_\xi[\mathcal{H}(\cdot, t_1)], \mathcal{H}(\cdot, t_1)\rangle_{H(\mathcal{H},T)}.$$

Result (Proposition) 4.2.6 yields then that $S_\xi = J^\star_{\mathcal{H}_\kappa, C_X} J_{\mathcal{H}_\kappa, C_X}$.

As ξ has a version η with a Radon law [(Fact) 4.1.11], which must be Gaussian, one has that the trace of the covariance operator of ξ is finite [(Fact) 4.3.4]. □

Fact 4.3.6 ([98, p. 8]) *Let X be a Gaussian process, defined on the probability space $(\Omega, \mathcal{A}, P)$, with index set T. It defines a Gaussian random element $\underline{X}$ with values in $\mathbb{R}^T$: $\omega \mapsto \underline{X}[\omega]$ is indeed adapted to $\mathcal{A}$, and the cylinder sets of $\mathbb{R}^T$. If V is any linear manifold of $\mathbb{R}^T$ which belongs to the σ-algebra of cylinder sets of $\mathbb{R}^T$, then either $P(\underline{X} \in V) = 0$ or $P(\underline{X} \in V) = 1$.*

4.4 Processes of Second Order with Paths in a Reproducing Kernel Hilbert Space

This section gives conditions for a second order process to have paths in an RKHS: "strong domination" is again the determining factor. Given a probability space $(\Omega, \mathcal{A}, P)$, its completion shall be denoted

$$(\Omega, \mathcal{A}^\circ, P^\circ).$$

Proposition 4.4.1 *Let $(\Omega, \mathcal{A}, P)$ be a probability space, and X be a second order stochastic process defined on $(\Omega, \mathcal{A}, P)$, with index set T. The mean of X shall be denoted μ_X, and its covariance, C_X. $\mathcal{H}$ shall be a reproducing kernel on T such that*

(a) $m_X \in H(\mathcal{H}, T)$,
(b) $C_X \ll_\tau \mathcal{H}$.

There exists then a stochastic process Y, defined on $(\Omega, \mathcal{A}, P)$, with T as index set, which has the following properties:

1. *Y is a version of X;*
2. *there is $\kappa \geq 0$ such that $Y(\omega, \cdot) \in H(\mathcal{H}_\kappa, T)$, $\omega \in \Omega$;*
3. *furthermore, when*

(a) $d_{\mathcal{H}_K}$ *is a metric on* T,
(b) $H(\mathcal{H}_K, T)$ *is separable,*
(c) $X(\omega, \cdot)$ *is continuous on* $(T, d_{\mathcal{H}_K})$, $\omega \in \Omega$,

there exists then $A^\circ \in \mathcal{A}^\circ$ *such that* $P^\circ(A^\circ) = 1$, *and, for* $\omega \in A^\circ$, *fixed, but arbitrary,*

$$X(\omega, \cdot) \in H(\mathcal{H}_K, T).$$

Proof Assumption (a), and the fact that $H(\mathcal{H}, T)$ and $H(\mathcal{H}_K, T)$ represent the same set of functions, and have equivalent norms [(Example) 1.3.12], allow one to subtract from X its mean, and thus assume that X has a mean equal to zero. One may furthermore assume that $H(\mathcal{H}_K, T)$ is separable [(Proposition) 3.1.16].

Let T_0 be a Hamel subset for $\mathcal{H}_K$ [(Definition) 1.1.36]. Denote $\mathcal{H}_{K,0}$ the restriction of $\mathcal{H}_K$ to $T_0 \times T_0$. Since $H(\mathcal{H}_K, T)$ is separable, $H(\mathcal{H}_{K,0}, T_0)$ is separable [(Proposition) 1.6.3]. Then $(T_0, d_{\mathcal{H}_{K,0}})$ is a separable metric space [(Corollary) 1.6.21], and there is thus $T_c \subseteq T_0$ which is countable and dense in $(T_0, d_{\mathcal{H}_{K,0}})$. Its n-th element shall be denoted $t_{c,n}$.

One shall use the following notation (already encountered, starting with (Fact) 1.6.11):

- $T_{c|n} = \{t_{c,1}, t_{c,2}, t_{c,3}, \ldots t_{c,n}\} \subseteq T_c \subseteq T_0$,
- $T_c = \cup_n T_{c|n}$,
- $\mathcal{H}_{K,n}$ for the restriction of $\mathcal{H}_K$ to $T_{c|n} \times T_{c|n}$,
- M_n for the matrix resulting from $\mathcal{H}_{K,n}$,
- $m_n(i,j)$ for the element in row i and column j of the inverse of M_n (which exists since T_0 is a Hamel subset),
- $\mathcal{H}_{K,c}$ for the restriction of $\mathcal{H}_K$ to $T_c \times T_c$,
- $X_{c|n}$ for the restriction of X to $T_{c|n}$,
- Σ_n for the covariance matrix of $X_{c|n}$,
- $\sigma_n(i,j)$ for the element in row i and column j of Σ_n,
- X_c for the restriction of X to T_c.

One shall first prove that there exists $A_c \in \mathcal{A}$ such that $P(A_c) = 1$, and, for $\omega \in A_c$,

$$X_c(\omega, \cdot) \in H(\mathcal{H}_{K,c}, T_c).$$

For fixed, but arbitrary $\omega \in \Omega$, $X_{c|n}$ yields a finite dimensional vector, so that one may display explicitly

$$Z_n(\omega) = \left\|X_{c|n}(\omega, \cdot)\right\|^2_{H(\mathcal{H}_{K,n}, T_{c|n})}$$

as [(Example) 1.1.20]

$$Z_n(\omega) = \sum_{i=1}^{n} \sum_{j=1}^{n} X_{c|n}(\omega, t_{c,i}) X_{c|n}(\omega, t_{c,j}) m_n(i,j).$$

Since $X_{c|n} = X_{c|n+1}^{|T_{c|n}}$, the following sequence of norms:

$$\left\{ \left\| X_{c|n}(\omega, \cdot) \right\|^2_{H(\mathcal{H}_{K,n}, T_{c|n})}, \ n \in \mathbb{N} \right\}$$

is, for fixed, but arbitrary $\omega \in \Omega$, increasing [(Proposition) 1.6.3, (Fact) 1.6.11, (Lemma) 1.6.13]. There is thus a random variable Z which is the limit of that sequence. Using successively the monotone convergence theorem, the explicit expression for Z_n given above, and result (Corollary) 3.1.20 with the domination assumption, one has that

$$\begin{aligned} E[Z] &= \lim_n E[Z_n] \\ &= \lim_n \sum_{i=1}^{n} \sum_{j=1}^{n} \sigma_n(i,j) m_n(i,j) \\ &= \lim_n \tau\left(\Sigma_n M_n^{-1}\right) \\ &= \tau\left(J^{\star}_{\mathcal{H}_K, C_X} J_{\mathcal{H}_K, C_X}\right) \\ &< \infty. \end{aligned}$$

Thus, if $A_c = \{\omega \in \Omega : Z(\omega) < \infty\}$, $P(A_c) = 1$. Consequently [(Corollary) 1.6.21], for $\omega \in A_c$,

$$X_c(\omega, \cdot) \in H(\mathcal{H}_{K,c}, T_c),$$

and there is thus [(Corollary) 1.6.21] a unique $h[\omega] \in H(\mathcal{H}_K, T)$ such that

$$h[\omega]^{|T_c} = X_c(\omega, \cdot).$$

A function $\xi : \Omega \longrightarrow H(\mathcal{H}_K, T)$ is defined letting

$$\xi(\omega) = \begin{cases} h[\omega] & \text{when } \omega \in A_c \\ 0_{\mathbb{R}^T} & \text{when } \omega \in A_c^c. \end{cases}.$$

Since $H(\mathcal{H}_K, T)$ is separable, ξ has separable range. Furthermore, for fixed, but arbitrary $t_{c,i} \in T_c$, and $\omega \in A_c$,

$$\langle \xi(\omega), \mathcal{H}_K(\cdot, t_{c,i}) \rangle_{H(\mathcal{H}_K, T)} = X_c(\omega, t_{c,i}) = X(\omega, t_{c,i}),$$

so that the following family:

$$\left\{ \langle \xi (\cdot) , \mathcal{H}_K (\cdot, t_{c,i}) \rangle_{H(\mathcal{H}_K,T)} = \chi_{A_c} X (\cdot, t_{c,i}) , \; t_{c,i} \in T_c \right\}$$

is a family of random variables. Since T_c is a determining set [(Corollary) 1.6.21], the family

$$\{ \mathcal{H}_K (\cdot, t_{c,i}) , \; t_{c,i} \in T_c \}$$

is separating (total [(Proposition) 1.6.10]), and consequently ξ is a Borel random element [(Fact) 4.1.6]. It follows that $\omega \mapsto \| \xi (\omega) \|_{H(\mathcal{H}_K,T)}$ is adapted. Now

$$\begin{aligned} \left\| \xi (\omega) \; |_{T_{c|n}} \right\|^2_{H\left(\mathcal{H}_{K,n},T_{c|n}\right)} &= \chi_{A_c} (\omega) \left\| X_{c|n} (\omega, \cdot) \right\|^2_{H\left(\mathcal{H}_{K,n},T_{c|n}\right)} \\ &= \chi_{A_c} (\omega) \, Z_n (\omega) , \end{aligned}$$

and, using successively (Proposition) 1.6.19, the monotone convergence theorem, and what has already been proved above,

$$\begin{aligned} E \left[\| \xi \|^2_{H(\mathcal{H}_K,T)} \right] &= \lim_n E \left[\left\| \xi \; |_{T_{c|n}} \right\|^2_{H\left(\mathcal{H}_{K,n},T_{c|n}\right)} \right] \\ &= \lim_n E \left[Z_n \right] \\ &= \tau \left(J^{\star}_{\mathcal{H}_K,C_X} J_{\mathcal{H}_K,C_X} \right) < \infty . \end{aligned}$$

ξ is thus of second order, strongly. It is then of weak first order, so that its mean $E[\xi]$ exists, and belongs to $H(\mathcal{H}_K, T)$. But then [(Proposition) 4.2.3]

$$\mu_\xi (t) = \langle E[\xi] , \mathcal{H}_K (\cdot, t) \rangle_{H(\mathcal{H}_K,T)} = E \left[\langle \xi, \mathcal{H}_K (\cdot, t) \rangle_{H(\mathcal{H}_K,T)} \right] , \; t \in T.$$

Choosing $t = t_{c,i} \in T_c$, one gets $\mu_\xi (t) = E[X(\cdot, t_{c,i})] = 0$, and, since T_c is a determining set, as already noticed, $\mu_\xi = 0$.

Let

$$Y (\omega, t) = \langle \xi (\omega) , \mathcal{H}_K (\cdot, t) \rangle_{H(\mathcal{H}_K,T)} .$$

One thus defines a second order process with paths in $H(\mathcal{H}_K, T)$ [(Proposition) 4.2.2]. One shall now prove that Y is a version of X.

As a first step, one shall establish that

$$\| X_t - Y_t \|_{L_2(\Omega,\mathcal{A},P)} = 0, \; t \in T_0,$$

where X_t and Y_t are the equivalence classes in $L_2(\Omega, \mathcal{A}, P)$ of, respectively, $X(\cdot, t)$ and $Y(\cdot, t)$. Since, for $t \in T_c$, by definition,

$$Y (\cdot, t) = \chi_{A_c} X (\cdot, t) ,$$

one need only consider $t \in T_0 \setminus T_c$. Suppose that such is the case, and let $t_c \in T_c$ be fixed, but arbitrary. Then

$$\begin{aligned}
&\|X_t - Y_t\|_{L_2(\Omega,\mathcal{A},P)} = \\
&\qquad \le \|X_t - X_{t_c}\|_{L_2(\Omega,\mathcal{A},P)} + \|X_{t_c} - Y_{t_c}\|_{L_2(\Omega,\mathcal{A},P)} + \|Y_{t_c} - Y_t\|_{L_2(\Omega,\mathcal{A},P)} \\
&\qquad = \|X_t - X_{t_c}\|_{L_2(\Omega,\mathcal{A},P)} + \|Y_{t_c} - Y_t\|_{L_2(\Omega,\mathcal{A},P)} \,.
\end{aligned}$$

Now, using the isometry which exists between the linear space of a process and the RKHS determined by its covariance [(Example) 1.1.26],

$$\begin{aligned}
\|X_t - X_{t_c}\|_{L_2(\Omega,\mathcal{A},P)} &= \|C_X(\cdot,t) - C_X(\cdot,t_c)\|_{H(C_X,T)} \\
&= \|J_{\mathcal{H}_\kappa,C_X}[\mathcal{H}_\kappa(\cdot,t) - \mathcal{H}_\kappa(\cdot,t_c)]\|_{H(C_X,T)} \\
&\le \|J_{\mathcal{H}_\kappa,C_X}\|\,\|\mathcal{H}_\kappa(\cdot,t) - \mathcal{H}_\kappa(\cdot,t_c)\|_{H(\mathcal{H}_\kappa,T)} \\
&= \|J_{\mathcal{H}_\kappa,C_X}\|\, d_{\mathcal{H}_\kappa}(t,t_c)\,.
\end{aligned}$$

Furthermore, using the definition of Y and [260, p. 175],

$$\begin{aligned}
\|Y_{t_c} - Y_t\|^2_{L_2(\Omega,\mathcal{A},P)} &= E\left[\langle \xi, \mathcal{H}_\kappa(\cdot,t_c) - \mathcal{H}_\kappa(\cdot,t)\rangle^2_{H(\mathcal{H}_\kappa,T)}\right] \\
&\le E\left[\|\xi\|^2_{H(\mathcal{H}_\kappa,T)}\,\|\mathcal{H}_\kappa(\cdot,t_c) - \mathcal{H}_\kappa(\cdot,t)\|^2_{H(\mathcal{H}_\kappa,T)}\right] \\
&= d^2_{\mathcal{H}_\kappa}(t,t_c)\, E\left[\|\xi\|^2_{H(\mathcal{H}_\kappa,T)}\right] \\
&= d^2_{\mathcal{H}_\kappa}(t,t_c)\, \tau\left(J^\star_{\mathcal{H}_\kappa,C_X} J_{\mathcal{H}_\kappa,C_X}\right).
\end{aligned}$$

Consequently, for fixed, but arbitrary $t \in T_0 \setminus T_c$ and $t_c \in T_c$,

$$\|X_t - Y_t\|_{L_2(\Omega,\mathcal{A},P)} \le \left\{\|J_{\mathcal{H}_\kappa,C_X}\| + \tau^{1/2}\left(J^\star_{\mathcal{H}_\kappa,C_X} J_{\mathcal{H}_\kappa,C_X}\right)\right\} d_{\mathcal{H}_\kappa}(t,t_c)\,.$$

Since T_c is dense in $(T_0, d_{\mathcal{H}_\kappa})$, for fixed, but arbitrary $t \in T_0$, $Y_t = X_t$.

Suppose finally that $t \in T \setminus T_0$ is fixed, but arbitrary. Since T_0 is a Hamel subset for $\mathcal{H}_\kappa$,

$$\mathcal{H}_\kappa(\cdot,t) = \sum_{i=1}^{n} \alpha_i \mathcal{H}_\kappa(\cdot,t_i)\,,\ t_i \in T_0,\ 1 \le i \le n.$$

Then

$$C_X(\cdot,t) = J_{\mathcal{H}_\kappa,C_X}[\mathcal{H}_\kappa(\cdot,t)] = \sum_{i=1}^{n} \alpha_i C_X(\cdot,t_i)\,.$$

Consequently, in $L_2(\Omega, \mathcal{A}, P)$, by isometry,

$$X_t = \sum_{i=1}^{n} \alpha_i X_{t_i}.$$

On the other hand,

$$\begin{aligned} Y(\cdot, t) &= \langle \xi, \mathcal{H}_K(\cdot, t)\rangle_{H(\mathcal{H}_K, T)} \\ &= \sum_{i=1}^{n} \alpha_i \langle \xi, \mathcal{H}_K(\cdot, t_i)\rangle_{H(\mathcal{H}_K, T)} \\ &= \sum_{i=1}^{n} \alpha_i Y(\cdot, t_i), \end{aligned}$$

so that

$$Y_t = \sum_{i=1}^{n} \alpha_i Y_{t_i}.$$

Since $Y_t = X_t$ for $t \in T_0$, the same is thus true for $t \in T$. Y is thus a version of X.

Suppose now that the assumptions in the second part of statement (Proposition) 4.4.1 obtain. The elements in $H(\mathcal{H}_K, T)$ are continuous on $(T, d_{\mathcal{H}_K})$ [(Proposition) 2.6.9], and, since $H(\mathcal{H}_K, T)$ is separable, and that $d_{\mathcal{H}_K}$ is a metric, $(T, d_{\mathcal{H}_K})$ is separable [(Corollary) 1.5.11]. Let T_c be a countable, dense subset of $(T, d_{\mathcal{H}_K})$. Since Y is a version of X, there exists $A_c \in \mathcal{A}$ such that $P(A_c) = 1$, and, for $\omega \in A_c$ and $t_c \in T_c$, fixed, but arbitrary,

$$X(\omega, t_c) = Y(\omega, t_c).$$

But $t \mapsto X(\omega, t)$, and $t \mapsto Y(\omega, t)$ are continuous for $(T, d_{\mathcal{H}_K})$ (X by assumption, and Y because it has paths in $H(\mathcal{H}_K, T)$), and, for $\omega \in A_c$, equal on a dense subset of $(T, d_{\mathcal{H}_K})$. Thus, for $\omega \in A_c$, $X(\omega, t) = Y(\omega, t)$, $t \in T$. Now the set

$$A_0 = \left\{\omega \in \Omega : X(\omega, \cdot) \in \mathbb{R}^T \setminus H(\mathcal{H}_K, T)\right\}$$

is a subset of A_c^c, so that same set A_0 belongs to the completion of $\mathcal{A}$ with respect to P, that is, $\mathcal{A}^\circ$. Consequently

$$\begin{aligned} P^\circ(\omega \in \Omega : X(\omega, \cdot) \in H(\mathcal{H}_K, T)) &\geq \\ &\geq P^\circ(\omega \in \Omega : X(\omega, t) = Y(\omega, t), \ t \in T) \\ &= P^\circ(A_c) \end{aligned}$$

$$= P(A_c)$$
$$= 1.$$

□

Corollary 4.4.2 *Let $\mathcal{H}$ and $\mathcal{K}$ be reproducing kernels on T, and k be a fixed, but arbitrary element of $H(\mathcal{K}, T)$.*

There exists a Gaussian process X, on some probability space $(\Omega, \mathcal{A}, P)$, such that

1. $m_X = k$,
2. $C_X = \mathcal{H}$,
3. *there exists $A \in \mathcal{A}$ such that $P(A) = 1$, and $X(\omega, \cdot) \in H(\mathcal{K}, T)$, $\omega \in A$,*

if, and only if, $\mathcal{H} \ll_\tau \mathcal{K}$.

Proof Suppose items 1 to 3 obtain. Result (Proposition) 4.3.5 then states that one must have $\mathcal{H} = C_X \ll_\tau \mathcal{K}$. Suppose conversely that $\mathcal{H} \ll_\tau \mathcal{K}$. One knows that there exists a Gaussian process X that has k as mean, and $\mathcal{H}$, as covariance [200, p. 39]. Since Gaussian processes are second order, given the assumption, there exists, because of (Proposition) 4.4.1, a version of X with paths in $H(\mathcal{K}, T)$. But a version of a Gaussian process is Gaussian. □

Proposition 4.4.3 *Let $\mathcal{H}$ and $\mathcal{K}$ be reproducing kernels on T, and suppose that, as sets, $H(\mathcal{H}, T) \subseteq H(\mathcal{K}, T)$, and that $H(\mathcal{H}, T)$ is a separable subset of $H(\mathcal{K}, T)$. There exists then a second order random process whose covariance is $\mathcal{H}$, and whose paths belong to $H(\mathcal{K}, T)$.*

Proof Since $H(\mathcal{H}, T) \subseteq H(\mathcal{K}, T)$, there exists [(Proposition) 3.1.34] a constant $\kappa \geq 0$ such that $\mathcal{H} \ll \kappa\, \mathcal{K} = \mathcal{K}_\kappa$. The map $J_{\mathcal{K}_\kappa, \mathcal{H}} : \mathcal{K}_\kappa(\cdot, t) \mapsto \mathcal{H}(\cdot, t)$ is thus well defined, and, by assumption, $J^\star_{\mathcal{K}_\kappa, \mathcal{H}} J_{\mathcal{K}_\kappa, \mathcal{H}}$ has separable range. It is thus [(Fact) 4.1.15] the covariance of a weak second order Radon probability μ on $\mathcal{B}(\mathcal{K}_\kappa, T)$, the Borel sets of $H(\mathcal{K}_\kappa, T)$. But then there exists a random element ξ with values in $H(\mathcal{K}_\kappa, T)$, whose law is μ: let indeed $\Omega = H(\mathcal{K}_\kappa, T)$, $\mathcal{A} = \mathcal{C}(\mathcal{K}_\kappa, T)$, $P = \mu$, and $\xi = I_{H(\mathcal{K}_\kappa, T)}[\xi]$, the identity. Let X be the process associated with ξ [(Proposition) 4.2.2]: it has paths in $H(\mathcal{K}_\kappa, T)$. The covariance C_X of X is then given by the following expression [(Proposition) 4.2.3]:

$$C_X(t_1, t_2) = \left\langle S_\xi\left[\mathcal{K}_\kappa(\cdot, t_1)\right], \mathcal{K}_\kappa(\cdot, t_2)\right\rangle_{H(\mathcal{K}_\kappa, T)}.$$

Since $S_\xi = J^\star_{\mathcal{K}_\kappa, \mathcal{H}} J_{\mathcal{K}_\kappa, \mathcal{H}}$, and $J_{\mathcal{K}_\kappa, \mathcal{H}}\left[\mathcal{K}_\kappa(\cdot, t)\right] = \mathcal{H}(\cdot, t)$,

$$C_X(t_1, t_2) = \langle \mathcal{H}(\cdot, t_1), \mathcal{H}(\cdot, t_2)\rangle_{H(\mathcal{H}, T)} = \mathcal{H}(t_1, t_2).$$

□

Corollary 4.4.4 *Let $\mathcal{H}$ and $\mathcal{K}$ be reproducing kernels on T such that $\mathcal{H} \ll \mathcal{K}$, and $H(\mathcal{H}, T)$ be separable. There exists then a second order random process whose covariance is $\mathcal{H}$, and whose paths belong to $H(\mathcal{K}, T)$.*

Proof It is sufficient to see that, when $H(\mathcal{H}, T)$ is a separable Hilbert space, it is a separable subset of $H(\mathcal{K}_\kappa, T)$ as the image in $H(\mathcal{K}_\kappa, T)$ of the continuous $J_{\mathcal{K}_\kappa, \mathcal{H}}$ [84, p. 175]. One then uses (Proposition) 4.4.3. □

Corollary 4.4.5 *Let $\mathcal{H}$ be a reproducing kernel on T such that $H(\mathcal{H}, T)$ is separable. There exists then a second order process with paths in $H(\mathcal{H}, T)$ whose covariance is $\mathcal{H}$.*

Proof One uses (Corollary) 4.4.4 with $\mathcal{K} = \mathcal{H}$. □

4.5 Dichotomies

For Gaussian laws, assertions tend to be either true, or false, that is, dichotomous. Another instance of that phenomenon is met below. $(\Omega, \mathcal{A}^\circ, P^\circ)$ still denotes the completion of $(\Omega, \mathcal{A}, P)$.

Proposition 4.5.1 *Let $(\Omega, \mathcal{A}, P)$ be a probability space, and, on it, let X be a Gaussian process, with T as index set. Let $H(\mathcal{H}, T)$ be an RKHS, and suppose that $\mu_X \in H(\mathcal{H}, T)$. When $C_X \ll_\tau \mathcal{H}$ does not obtain,*

1. $\{\omega \in \Omega : X(\omega, \cdot) \in H(\mathcal{H}, T)\} \in \mathcal{A}^\circ$;
2. $P^\circ(\{\omega \in \Omega : X(\omega, \cdot) \in H(\mathcal{H}, T)\}) = 0$.

Proof Let $T_c \subseteq T$ be a countably infinite subset of a Hamel set for T. X_c shall denote the restriction of X to T_c, and $\mathcal{H}_c$, that of $\mathcal{H}$ to $T_c \times T_c$. Then [(Proposition) 1.6.3]:

$$\{\omega \in \Omega : X(\omega, \cdot) \in H(\mathcal{H}, T)\} \subseteq \{\omega \in \Omega : X_c(\omega, \cdot) \in H(\mathcal{H}_c, T_c)\}.$$

It suffices thus to prove that the latter set has zero probability. Let

$$T_c = \{t_1, t_2, t_3, \dots\}, \quad \text{and} \quad T_{c|n} = \{t_1, t_2, t_3, \dots, t_n\}.$$

Let $\mathbb{Q}$ be the set of rationals, and let $n \in \mathbb{N}$, and $\{\alpha_1, \dots, \alpha_n\} \subseteq \mathbb{Q}$ be fixed, but arbitrary. Let

$$\mathcal{F}(n, p, \underline{\alpha}) = \left\{ f \in \mathbb{R}^T : \frac{\left\{\sum_{i=1}^n \alpha_i f(t_i)\right\}^2}{\left\|\sum_{i=1}^n \alpha_i \mathcal{H}(\cdot, t_i)\right\|^2_{H(\mathcal{H}, T)}} \le p,\ t_i \in T_{c|n},\ 1 \le i \le n \right\}.$$

It is a cylinder set in $\mathbb{R}^{\mathbb{N}}$. Then, because of (Proposition) 1.2.1, the following set,

$$H(\mathcal{H}_c, T_c) = \bigcup_{p \in \mathbb{N}} \bigcap_{n \in \mathbb{N}} \bigcap_{\underline{\alpha} \in \mathbb{Q}^n} \mathcal{F}(n, p, \underline{\alpha}),$$

belongs to the σ-algebra of cylinder sets of $\mathbb{R}^{\mathbb{N}}$. Since it is a linear manifold of $\mathbb{R}^{\mathbb{N}}$, its probability is zero or one [(Fact) 4.3.6]. Suppose it is one. Then, because of (Proposition) 4.3.5, $C_X \ll_\tau \mathcal{H}$. This is impossible by assumption, and thus the probability must be zero. Consequently

$$\{\omega \in \Omega : X(\omega, \cdot) \in H(\mathcal{H}, T)\}$$

is a subset of a set whose probability is zero. □

Corollary 4.5.2 *Let X be a Gaussian process on the probability space $(\Omega, \mathcal{A}, P)$, with index set T. Suppose that the dimension of $H(C_X, T)$ is infinite. Then*

$$P^\circ(\omega \in \Omega : X(\omega, \cdot) \in H(C_X, T)\} = 0.$$

Proof In (Proposition) 4.5.1, one takes $\mathcal{H} = C_X$. The map $J_{\mathcal{H}_\kappa, C_X}$ is then the identity, and $J^\star_{\mathcal{H}_\kappa, C_X} J_{\mathcal{H}_\kappa, C_X}$ cannot have finite trace. Thus (Proposition) 4.5.1 applies. □

Corollary 4.5.3 *Let X be a Gaussian process on the probability space $(\Omega, \mathcal{A}, P)$, with index set T. Let $\mathcal{H}$ be a reproducing kernel, and suppose that $\mu_X \in H(\mathcal{H}, T)$. Then:*

1. *when $C_X \ll_\tau \mathcal{H}$ obtains, there exists a version Y of X such that*

$$P^\circ(\omega \in \Omega : Y(\omega, \cdot) \in H(\mathcal{H}, T)) = 1;$$

2. *when $C_X \ll_\tau \mathcal{H}$ does not obtain,*

$$P^\circ(\omega \in \Omega : X(\omega, \cdot) \in H(\mathcal{H}, T)) = 0.$$

Proof Item 1 repeats (Proposition) 4.4.1, and item 2, (Proposition) 4.5.1. □

Corollary 4.5.4 *Let X be a Gaussian process on the probability space $(\Omega, \mathcal{A}, P)$, with index set T. Let $\mathcal{H}$ be a reproducing kernel. Suppose that $d_{\mathcal{H}}$ is a metric, that the paths of X are continuous on $(T, d_{\mathcal{H}})$, and that $H(\mathcal{H}, T)$ is separable. Then:*

1. *when $C_X \ll_\tau \mathcal{H}$ obtains,*

$$P^\circ(\omega \in \Omega : X(\omega, \cdot) \in H(\mathcal{H}, T)) = 1;$$

2. *when $C_X \ll_\tau \mathcal{H}$ does not obtain,*

$$P^\circ(\omega \in \Omega : X(\omega, \cdot) \in H(\mathcal{H}, T)) = 0.$$

Proof Item 1 repeats (Proposition) 4.4.1, and item 2, (Proposition) 4.5.1. □

Chapter 5
Reproducing Kernel Hilbert Spaces and Discrimination

In this chapter, it is examined to what extent RKHS's allow one to discriminate between probability laws, that is determine their equivalence or singularity.

5.1 Context of Discrimination

This section details the framework within which discrimination problems shall be considered.

Definition 5.1.1 Let S and T be sets, and

$$\mathcal{E} = \{\mathcal{E}_t : S \longrightarrow \mathbb{R},\ t \in T\}$$

be a family of maps. The following notation and definitions shall be used:

1. $\mathcal{S}$ for the σ-algebra of subsets of S, generated by $\mathcal{E}$;
2. $\mathcal{L}(\mathcal{E})$ for the linear manifold generated by $\mathcal{E}$: $\mathcal{L}(\mathcal{E}) = V[\mathcal{E}]$;
3. $\mathcal{M}(\mathcal{S})$ for the linear space of functions adapted to $\mathcal{S}$ and $\mathcal{B}(\mathbb{R})$;
4. $V(\mathcal{M})$ for a fixed, but arbitrary linear manifold of $\mathcal{M}(\mathcal{S})$.

Remark 5.1.2 Among the linear manifolds of $\mathcal{M}(\mathcal{S})$, there are two which shall be of special interest:

1. $\mathcal{L}(\mathcal{E})$ (it shall be called the "linear manifold of evaluations");
2. $\mathcal{Q}(\mathcal{E})$ which contains the functions representable in the following form:

$$\alpha + \sum_{i=1}^{m} \alpha_i \mathcal{E}_{t_i} + \sum_{j=1}^{n} \sum_{k=1}^{p} \alpha_{j,k} \mathcal{E}_{u_j} \mathcal{E}_{v_k},$$

A.F. Gualtierotti, *Detection of Random Signals in Dependent Gaussian Noise*,
DOI 10.1007/978-3-319-22315-5_5

where

$$\begin{aligned}
&\{m, n, p\} && \subseteq \mathbb{N}, \\
&\{\alpha, \alpha_1, \ldots, \alpha_m, \alpha_{1,1}, \ldots, \alpha_{n,p}\} && \subseteq \mathbb{R}, \\
&\{t_1, \ldots, t_m, u_1, \ldots, u_n, v_1, \ldots, v_p\} && \subseteq T
\end{aligned}$$

(it shall be called "the quadratic manifold of evaluations").

Given two functions of $\mathcal{L}(\mathcal{E})$, say f and g, one shall always be able to represent them in the following form:

$$f(s) = \sum_{i=1}^{n} \alpha_i^f \mathcal{E}_{t_i}(s), \quad g(s) = \sum_{i=1}^{n} \alpha_i^g \mathcal{E}_{t_i}(s),$$

with

$$n \in \mathbb{N}, \ \{\alpha_1^f, \ldots, \alpha_n^f, \alpha_1^g, \ldots, \alpha_n^g\} \subseteq \mathbb{R}, \ \{t_1, \ldots, t_n\} \subseteq T.$$

It suffices indeed to let some of the α's be zero.

The context just described covers several examples, which are those of usually greatest practical interest. They shall now be listed.

Example 5.1.3 T is a set, and $S = \mathbb{R}^T$. Then $\mathcal{E}_t(s) = s(t)$, and $\mathcal{S}$ is the σ-algebra generated by the cylinder sets.

Example 5.1.4 ([38, p. 19]) $T = [0, 1]$, and $S = C[0, 1]$, the Banach space of continuous functions over $[0, 1]$. Then $\mathcal{E}_t(s) = s(t)$, and $\mathcal{S}$ is the σ-algebra generated by the cylinder sets: it is thus the σ-algebra of Borel sets.

Example 5.1.5 ([38, p. 109]) $T = [0, 1]$, and $S = D[0, 1]$, the space of functions on $[0, 1]$, which are continuous to the right, and have limits to the left. The topology is that of Skorohod, and $D[0, 1]$ is then a complete metric space. Then $\mathcal{E}_t(s) = s(t)$, and $\mathcal{S}$ is the σ-algebra generated by the cylinder sets: it is thus the σ-algebra of Borel sets.

Example 5.1.6 H is a real Hilbert space, and $T = H$. $S = H$, and $\mathcal{E}$ is the Hilbert space of continuous linear functionals on H. $\mathcal{S}$ is thus the σ-algebra generated by the cylinder sets, and the σ-algebra of Borel sets when H is separable.

Remark 5.1.7 Any set X, and vector space $\mathcal{V}[\mathcal{F}(X)]$ of real valued functions, defined on X, will do, provided the σ-algebra used is $\sigma(\mathcal{V}[\mathcal{F}(X)])$, and P is Gaussian [220, 261, p. 376].

Assumptions 5.1.8 *The context is that of (Definition)* 5.1.1. *Let P be a probability on $\mathcal{S}$. It shall always be assumed that*

$$\int_S \mathcal{E}_t^2(s) P(ds) < \infty, \; t \in T,$$

so that one can define, for the "process" $\mathcal{E}$, a covariance, and thus an RKHS: for fixed, but arbitrary $(t_1, t_2) \in T \times T$,

$$C_{\mathcal{E},P}(t_1, t_2) = \int_S \mathcal{E}_{t_1}(s) \mathcal{E}_{t_2}(s) P(ds) - \int_S \mathcal{E}_{t_1}(s) P(ds) \int_S \mathcal{E}_{t_1}(s) P(ds).$$

Definition 5.1.9 $\left[\overline{V(\mathcal{M})}^P\right]$

The context is that of (Definition) 5.1.1. Given a linear manifold of functions whose square is integrable with respect to P, say $V(\mathcal{M})$, one shall write

$$\overline{V(\mathcal{M})}^P$$

for the closure, in $L_2(S, \mathcal{S}, P)$, of the linear manifold of equivalence classes of elements in $V(\mathcal{M})$.

Definition 5.1.10 $[V_1(\mathcal{M})]$

The context is that of (Definition) 5.1.1. $V_1(\mathcal{M})$ shall be the set of functions in $V(\mathcal{M})$ which have a square that is integrable with respect to P, and an L_2 norm equal to one.

Definition 5.1.11 $[\lambda_{P,V(\mathcal{M})}]$

The following function, defined for $\alpha \geq 0$, shall be of use:

$$\lambda_{P,V(\mathcal{M})}(\alpha) = \sup_{f \in V_1(\mathcal{M})} P(s \in S : |f|(s) \leq \alpha) = \sup_{f \in V_1(\mathcal{M})} P_f([-\alpha, \alpha]).$$

In the latter expression, $P_f = P \circ f^{-1}$, and the supremum is set to one when $V_1(\mathcal{M}) = \emptyset$.

Remark 5.1.12 The map $\alpha \mapsto \lambda_{P,V(\mathcal{M})}(\alpha)$ is obviously monotone increasing.

Definition 5.1.13 [Property Π_1, for $V(\mathcal{M})$ and P]

The context is that of (Definition) 5.1.1. P shall have property Π_1, for $V(\mathcal{M})$ and P, when all the elements in $V(\mathcal{M})$ have a square that is integrable with respect to P.

Definition 5.1.14 [Property Π_2, for $V(\mathcal{M})$ and P]

The context is that of (Definition) 5.1.1. P shall have property Π_2 for $V(\mathcal{M})$ when $\lim_{\alpha \downarrow 0} \lambda_{P,V(\mathcal{M})}(\alpha) = 0$.

Example 5.1.15 [Property Π_2 for $\mathcal{L}[\mathcal{E}]$]

Suppose (Fact) 5.1.8 obtains with P, a probability for which $\mathcal{E}$ is a Gaussian stochastic process, with mean $\mu_{\mathcal{E},P}$ and covariance $C_{\mathcal{E},P}$. Let $V(\mathcal{M}) = \mathcal{L}[\mathcal{E}]$. Then all elements of $\mathcal{L}[\mathcal{E}]$ are Gaussian, and, for

$$f = \sum_{i=1}^{n} \alpha_i \mathcal{E}_{t_i},$$

one has that

$$E_P[f] = \sum_{i=1}^{n} \alpha_i \mu_{\mathcal{E},P}(t_i),$$

$$V_P[f] = \sum_{i=1}^{n}\sum_{j=1}^{n} \alpha_i \alpha_j C_{\mathcal{E},P}(t_i, t_j).$$

If one supposes that $\mu_{\mathcal{E},P} \in H(C_{\mathcal{E},P}, T)$, then

$$E_P[f] = \langle \mu_{\mathcal{E},P}, \sum_{i=1}^{n} \alpha_i C_{\mathcal{E},P}(\cdot, t_i) \rangle_{H(C_{\mathcal{E},P},T)}.$$

Let

$$h_f = \sum_{i=1}^{n} \alpha_i C_{\mathcal{E},P}(\cdot, t_i).$$

Then, using the following relation: $E[X^2] = V[X] + E^2[X]$, one gets that

$$\int_S f^2(s) P(ds) = \|h_f\|^2_{H(C_{\mathcal{E},P},T)} + \langle \mu_{\mathcal{E},P}, h_f \rangle^2_{H(C_{\mathcal{E},P},T)}.$$

The requirement that $\int_S f^2(s) P(ds) = 1$ means thus that h_f is not the zero function, and that one has a family of Gaussian random variables whose mean μ and variance $\sigma^2 > 0$ are tied in the expression $\mu^2 + \sigma^2 = 1$. Let then, for $\alpha > 0$, but small,

$$f_\alpha(\mu) = \frac{\alpha - \mu}{\sqrt{1-\mu^2}}, \text{ and } g_\alpha(\mu) = \frac{-\alpha - \mu}{\sqrt{1-\mu^2}}.$$

One must study, as a function of $\mu \in]-1, 1[$, the expression

$$P_f([-\alpha, \alpha]) \stackrel{def}{=} \Psi_\alpha(\mu) \stackrel{def}{=} \Phi(f_\alpha(\mu)) - \Phi(g_\alpha(\mu)) > 0,$$

where Φ is the distribution function of the standard normal random variable. Choosing successively $\mu_n = -1 + \frac{1}{n}$, then $\mu_n = 1 - \frac{1}{n}$, one sees that $\lim_n \Psi_\alpha(\mu_n) = 0$. Furthermore $\Psi_\alpha(0) = \Phi(\alpha) - \Phi(-\alpha)$. Differentiating with respect to μ, one gets that

$$\left\{\frac{d}{d\mu}\Psi_\alpha\right\}(\mu) = \frac{\alpha\mu + 1}{\sqrt{1-\mu^2}}\frac{e^{-\frac{1}{2}g_\alpha^2(\mu)}}{\sqrt{2\pi}} + \frac{\alpha\mu - 1}{\sqrt{1-\mu^2}}\frac{e^{-\frac{1}{2}f_\alpha^2(\mu)}}{\sqrt{2\pi}}.$$

Setting to zero, one achieves $1 = \frac{1-\alpha\mu}{1+\alpha\mu}e^{2\alpha\mu}$, or $\mu = 0$. P has thus property Π_2 for $\mathcal{L}(\mathcal{E})$.

Remark 5.1.16 An adroiter calculation may be culled from (Lemma) 5.6.11 and (Corollary) 5.6.14.

Definition 5.1.17 [Property Π_3 for P]
The context is that of (Definition) 5.1.1. P shall have property Π_3 for P when there exists $\kappa_P \geq 0$ such that [(Definition) 5.1.9],

$$\int_S f^4(s)\,P(ds) \leq \kappa_P^2 \left\{\int_S f^2(s)\,P(ds)\right\}^2,$$

for $[f]_{L_2(S,\mathcal{S},P)} \in \overline{\mathcal{L}(\mathcal{E})}^P$, fixed, but arbitrary,

Remark 5.1.18 Let $\phi_X(p) = E_P[|X|^p]$. Then [229, p. 73] $\ln[\phi_X(p)]$ is convex in the interior of

$$\{p \in]0, \infty[: \phi_X(p) < \infty\}.$$

Consider the line that goes through $(2, \ln[\phi_X(2)])$ and $(4, \ln[\phi_X(4)])$. It cuts the vertical axis at the point $(0, 2\ln[\phi_X(2)] - \ln[\phi_X(4)])$. When

$$2\ln[\phi_X(2)] - \ln[\phi_X(4)] \geq -\ln\left[\kappa^2\right], \text{ class of } X \in \overline{\mathcal{L}(\mathcal{E})}^P, \text{ some } \kappa,$$

then (Definition) 5.1.17 obtains. One has thus a geometric rephrasing of that condition.

Remark 5.1.19 Definition 5.1.17 expresses a type of uniform hypercontractive domination [165, p. 72], a feature that allows domination of higher moments by lower ones. It is uniform, as it is required for a set of random variables, rather than for a single one. As hypercontractive domination is a feature of distributions (it holds, in particular, for Bernoulli and Gaussian series, [165, p. 73]) when the linear space of a process is made of (classes of) random variables of the same type, say Gaussian (see next remark), it suffices to examine hypercontractivity for one random variable. But the linear space of a process need not contain variables all of the same type [123].

Remark 5.1.20 All zero mean Gaussian processes have property Π_3. Indeed, for a normal random variable X, with mean zero, one has that [138, p. 291]

$$E\left[X^4\right] = 3E^2\left[X^2\right].$$

But, when $\mathcal{E}$ is Gaussian for P, with zero mean, expressions of the form $\sum_{i=1}^n \alpha_i \mathcal{E}_{t_i}$ are Gaussian, and L_2 limits of sequences of such linear combinations are also Gaussian [273, p. 246], and Π_3 thus obtains.

Remark 5.1.21 Every time a process exhibits a Gaussian feature, there is a good chance for (Definition) 5.1.17 to hold. A few examples follow.

Example 5.1.22 Let A_1, A_2, W_1, W_2 be independent elements, A_1 and A_2, random variables, and W_1 and W_2, standard Wiener processes. Let $X = A_1 W_1 + A_2 W_2$, and $P = P_X$, the measure induced by X on, say, $\mathbb{R}^{[0,1]}$. And let $\mathcal{E}$ denote the evaluation maps. Then, when $f = \sum_{i=1}^n \alpha_i \mathcal{E}_{t_i}$, using the inequality $(a+b)^n \leq 2^{n-1}(a^n + b^n)$, and then independence,

$$\begin{aligned}
E_P\left[f^4\right] &= \\
&= E\left[\left\{\sum_{i=1}^n \alpha_i X(\cdot, t_i)\right\}^4\right] \\
&= E\left[\left\{A_1 \sum_{i=1}^n \alpha_i W_1(\cdot, t_i) + A_2 \sum_{i=1}^n \alpha_i W_2(\cdot, t_i)\right\}^4\right] \\
&\leq 2^3 \left\{E\left[A_1^4\right] E\left[\left\{\sum_{i=1}^n \alpha_i W_1(\cdot, t_i)\right\}^4\right] + E\left[A_2^4\right] E\left[\left\{\sum_{i=1}^n \alpha_i W_2(\cdot, t_i)\right\}^4\right]\right\} \\
&= 3 \cdot 2^3 E^2\left[\left\{\sum_{i=1}^n \alpha_i W_1(\cdot, t_i)\right\}^2\right]\left(E\left[A_1^4\right] + E\left[A_2^4\right]\right).
\end{aligned}$$

But

$$E_P\left[f^2\right] = \left(E\left[A_1^2\right] + E\left[A_2\right]\right) E\left[\left\{\sum_{i=1}^n \alpha_i W_1(\cdot, t_i)\right\}^2\right],$$

so that (provided the required moments exist)

$$\kappa_P^2 = 3 \cdot 2^3 \frac{E\left[A_1^4\right] + E\left[A_2^4\right]}{\left(E\left[A_1^2\right] + E\left[A_2\right]\right)^2}.$$

In the same vein, let

$$X_t = S_t + W_t,$$
$$S_t = A_0 \cos(\omega_0 t + U),$$

with U, uniform on $[-\pi, \pi]$, and independent of W, standard Wiener. Then [215, p. 85]

$$E[S_t] = 0,$$
$$C_S(s,t) = \frac{A_0^2}{2} \cos(\omega_0 [t-s]).$$

The elements of $L_2[X]$, the linear space generated by X, are then of the following generic form:

$$X = \alpha \cos(U) + \beta \sin(U) + N = V + N,$$

where V and N are independent, and N is normal, with mean zero. Thus

$$E^2\left[(V+N)^2\right] = E^2\left[V^2\right] + 2E\left[V^2\right]E\left[N^2\right] + E^2\left[N^2\right],$$

and

$$E\left[(V+N)^4\right] = E\left[V^4\right] + 6E\left[V^2\right]E\left[N^2\right] + 3E^2\left[N^2\right].$$

Now, using the properties of trigonometric functions,

$$E^2\left[V^2\right] = $$
$$= \alpha^4 E^2\left[\cos^2(U)\right] + 2\alpha^2\beta^2 E\left[\cos^2(U)\right]E\left[\sin^2(U)\right] + \beta^4 E^2\left[\sin^2(U)\right],$$

and

$$E\left[V^4\right] = $$
$$= \alpha^4 E\left[\cos^4(U)\right] + 6\alpha^2\beta^2 E\left[\cos^2(U)\sin^2(U)\right] + \beta^4 E\left[\sin^4(U)\right].$$

Thus Π_3 still obtains.

Example 5.1.23 Strictly sub-Gaussian random elements provide examples which are not Gaussian, but have property Π_3.

A random variable X is sub-Gaussian when there exists $\kappa \geq 0$ such that, for $\lambda \in \mathbb{R}$, fixed, but arbitrary,

$$E_P\left[e^{\lambda X}\right] \leq e^{\frac{\kappa^2 \lambda^2}{2}}.$$

It is strictly sub-Gaussian when it is sub-Gaussian, and one can choose, for κ^2, the variance of X. Strictly sub-Gaussian random variables have the property that [49, p. 17]

$$E_P[X] = E_P\left[X^3\right] = 0, \; E_P\left[X^4\right] \leq 3E_P^2\left[X^2\right].$$

Furthermore [49, p. 19], when $\mathcal{X}$ is a family of independent, strictly sub-Gaussian random variables, the closure in mean square of its linear span,

$$\overline{V[\mathcal{X}]}^P,$$

contains only strictly sub-Gaussian random variables.

Let now $\{X_n, n \in \mathbb{N}\}$ be a family of independent, strictly sub-Gaussian random variables with variance one, and $\{f_n, n \in \mathbb{N}\}$, a sequence of functions such that, for $t \in T$, fixed, but arbitrary,

$$\sum_n f_n^2(t) < \infty.$$

The assignment

$$X_t = \sum_n f_n(t) X_n$$

provides a stochastic process whose linear space $L_2[X]$ is made of strictly sub-Gaussian random variables.

Analogously to strictly sub-Gaussian random variables, one defines [49, p. 189] strictly sub-Gaussian random vectors, that is, $\underline{X} \in \mathbb{R}^n$ is strictly sub-Gaussian when, given a fixed, but arbitrary $\underline{\lambda} \in \mathbb{R}^n$,

$$E_P\left[e^{\langle \underline{\lambda}, \underline{X} \rangle_{\mathbb{R}^n}}\right] \leq e^{\frac{1}{2}\langle \Sigma_{\underline{X}}[\underline{\lambda}], \underline{\lambda} \rangle_{\mathbb{R}^n}},$$

where $\Sigma_{\underline{X}}$ is the covariance matrix of $\underline{X}$. A stochastic process is strictly sub-Gaussian when all its finite dimensional distributions are strictly sub-Gaussian, and then its linear space is made of strictly sub-Gaussian random variables [49, p. 190]. Consequently, when $\{X_n, n \in \mathbb{N}\}$ above is strictly sub-Gaussian, the random series process it determines, provided it exists, is strictly sub-Gaussian (no independence required) [49, p. 190]. Those two properties extend to the following case [49, p. 191].

Let T be a connected, compact set of $\mathbb{R}^n$, X, a strictly sub-Gaussian process, indexed by T, continuous in quadratic mean, and $\{f_n, n \in \mathbb{N}\}$, a family of continuous functions. Set

$$X_n = \int_T X_t f_n(t)\, dt,$$

and, given the family of functions $\{g_n, n \in \mathbb{N}\}$,

$$Y_t^{(n)} = \sum_{i=1}^{n} g_i(t)\, X_i.$$

One thus obtains a family of strictly sub-Gaussian processes such that, when, for fixed, but arbitrary $t \in T$, in quadratic mean,

$$Z_t = \lim_n Y_t^{(n)}$$

exists, Z is also strictly sub-Gaussian.

Example 5.1.24 ([36, p. 268]) Let $X_t = \langle \underline{a}(t), \underline{X} \rangle_{\mathbb{R}^n}$, with $\underline{X}$ orthogonally invariant.
Orthogonally invariant vectors have the following properties [96, p. 26]:

(i) the characteristic function $\varphi_{\underline{X}}(\underline{\theta})$ of $\underline{X}$ has the following form:

$$\varphi_{\underline{X}}(\underline{\theta}) = \phi_{\underline{X}}\left(\|\underline{\theta}\|^2_{\mathbb{R}^n}\right), \; \phi_{\underline{X}}(\theta) = \int_0^{\infty} \Omega_n\left(\theta x^2\right) F_{\underline{X}}(dx),$$

where $F_{\underline{X}}$ is a distribution function, and, with S_n the area of the surface of the unit sphere of $\mathbb{R}^n$,

$$\Omega_n\left(\|\underline{\theta}\|^2_{\mathbb{R}^n}\right) = \int_{\left\{\underline{x}\in\mathbb{R}^n : \|\underline{x}\|^2_{\mathbb{R}^n}=1\right\}} e^{i\langle \underline{\theta}, \underline{x}\rangle_{\mathbb{R}^n}} \frac{dS}{S_n};$$

(ii) $\underline{X}$ has the same law as $R\underline{U}_n$, where R has distribution $F_{\underline{X}}$, and $\underline{U}_n$ has the uniform distribution on the surface of the unit sphere of $\mathbb{R}^n$; when the probability of the origin is zero, the distribution of R and that of the norm of $\underline{X}$ are the same, as well as those of $\underline{U}_n$ and the ratio of $\underline{X}$ to its norm;

(iii) the inner product $\langle \underline{a}, \underline{X} \rangle_{\mathbb{R}^n}$ and the random variable $\|\underline{a}\|_{\mathbb{R}^n} X_1$ have the same distribution;

(iv) $\underline{X}$ has zero mean, and, when it exists, covariance $\frac{E_P[R^2]}{n} I_n$.

The process X has thus zero mean, and its covariance, when it exists, has the following form:

$$C_X(t_1, t_2) = \frac{E_P\left[R^2\right]}{n} \langle \underline{a}(t_1), \underline{a}(t_2) \rangle_{\mathbb{R}^n}.$$

The elements of $H(C_X, T)$ are of the form $t \mapsto \sum_{i=1}^n x_i a_i(t)$, and the elements of $L_2[X]$, of the form $\langle \underline{a}, \underline{X}\rangle_{\mathbb{R}^n}$, whose law is that of $\|\underline{a}\|_{\mathbb{R}^n} X_1$. Consequently, provided the required moments exist,

$$\frac{E_P\left[\langle \underline{a}, \underline{X}\rangle_{\mathbb{R}^n}^4\right]}{E_P^2\left[\langle \underline{a}, \underline{X}\rangle_{\mathbb{R}^n}^2\right]} = \frac{E_P\left[X_1^4\right]}{E_P^2\left[X_1^2\right]},$$

and Π_3 obtains. Furthermore, the finite dimensional distributions of X are elliptically contoured [36, p. 270], and reference [96, Chapter 3] is rich in particular cases.

Remark 5.1.25 It is not too difficult to find cases that do not enjoy property Π_3. An example follows.

Example 5.1.26 Let $X_t = W_{t\wedge T}$, where T is an exponential random variable, independent of the Wiener process W. Let $\mathcal{A}_t = \sigma(T) \vee \sigma_t(W)$. Then T is a stopping time of that filtration, with respect to which W remains a Wiener process independent of T. Furthermore X is adapted to it [67, p. 209]. As

$$W_{t\wedge T} = \int_0^\infty \chi_{[0,t]}(\theta)\, \chi_{[0,T(\omega)]}(\theta)\, W(\omega, d\theta),$$

one has that X has mean zero, and covariance

$$\begin{aligned} C_X(s,t) &= E\left[\int_0^\infty \chi_{[0,s]}(\theta)\, \chi_{[0,t]}(\theta)\, \chi_{[0,T(\omega)]}(\theta)\, d\theta\right] \\ &= \int_0^\infty \chi_{[0,s]}(\theta)\, \chi_{[0,t]}(\theta)\, P(T \geq \theta)\, d\theta \\ &= \int_0^\infty \chi_{[0,s]}(\theta)\, \chi_{[0,t]}(\theta)\, e^{-\theta} d\theta. \end{aligned}$$

Consequently $L_2[X]$ and $H(C_X, \mathbb{R}_+)$ are isomorphic to

$$L_2[\mu] = L_2\left(\mathbb{R}_+, \mathcal{B}(\mathbb{R}_+), \mu_{\text{Exp}}\right),$$

where μ_{Exp} is the measure with exponential density. The elements of $L_2[X]$ have thus a representation of the following form:

$$\int_0^T f dW,\ f \in L_2\left[\mu_{\text{Exp}}\right].$$

Now

$$E\left[e^{i\alpha \int_0^T f\, dW}\right] = \int_0^\infty e^{-\frac{\alpha^2}{2}\int_0^\tau f^2(\theta)\, d\theta}\mu_{\text{Exp}}(d\tau).$$

Letting $X(f) = \int_0^T f\,dW$, differentiating with respect to α, one obtains that

$$E\left[X^2(f)\right] = \int_0^{\infty} \left\{ \int_0^{\tau} f^2(\theta)\, d\theta \right\} \mu_{\text{Exp}}(d\tau),$$

and that

$$E\left[X^4(f)\right] = 3 \int_0^{\infty} \left\{ \int_0^{\tau} f^2(\theta)\, d\theta \right\}^2 \mu_{\text{Exp}}(d\tau).$$

Choose now $f(\theta) = \theta^n$. Integrating, one obtains that

$$E\left[X^2(f)\right] = \int_0^{\infty} \frac{\tau^{2n+1}}{2n+1} \mu_{\text{Exp}}(d\tau) = \frac{(2n+1)!}{2n+1},$$

and that

$$E\left[X^4(f)\right] = 3 \int_0^{\infty} \left[\frac{\tau^{2n+1}}{2n+1} \right]^2 \mu_{\text{Exp}}(d\tau) = 3\frac{(4n+2)!}{(2n+1)^2}.$$

Consequently

$$\frac{E\left[X^4(f)\right]}{E^2\left[X^2(f)\right]} \geq 4n+2,$$

so that Π_3 does not obtain.

For integrals of the type $M = \int X\,dW$, there are generally valid inequalities for fourth moments [56, p. 128]:

$$E\left[M_t^4\right] \leq 6E\left[M_t^2 \langle M \rangle_t\right] \leq 36E\left[\langle M \rangle_t^2\right],$$

but they are not tight enough to imply the Π_3 property, as one would need, instead of $E\left[\langle M \rangle_t^2\right]$, $E^2\left[\langle M \rangle_t\right]$.

Remark 5.1.27 In some cases, when investigating absolute continuity, the methodology which follows may be inoperative, or dispensed with. Some examples follow.

Example 5.1.28 Let U and V be independent, with mean zero and variance one. Let f and g be two linearly independent functions, and set

$$X_t = f(t)\,U + g(t)\,V.$$

The elements of $L_2[X]$ are of the form $X = \alpha U + \beta V$. Consequently, provided moments exist,

$$E\left[X^4\right] = \alpha^4 E\left[U^4\right] + 6\alpha^2\beta^2 + \beta^4 E\left[V^4\right],$$

whereas

$$E^2\left[X^2\right] = \alpha^4 + 2\alpha^2\beta^2 + \beta^4,$$

so that condition Π_3 becomes: for real (α, β), fixed, but arbitrary,

$$\left(\kappa^2 - E\left[U^4\right]\right)\alpha^4 + \left(\kappa^2 - 6\right)\alpha^2\beta^2 + \left(\kappa^2 - E\left[V^4\right]\right)\beta^4 \geq 0,$$

which, when moments are finite, is always true for κ large enough. But one would usually attempt to study the equivalence problem using the following representation:

$$X_t = \Phi\left(\begin{bmatrix} U \\ V \end{bmatrix}\right)[t], \quad \Phi(\underline{x})[t] = f(t)x_1 + g(t)x_2.$$

In the same vein, let $\{X_n,\ n \in \mathbb{N}\}$ be a sequence of independent random variables, with mean zero and variance one. Let also $\{f_n : T \longrightarrow \mathbb{R},\ n \in \mathbb{N}\}$ be a sequence of functions such that $\{f_n(t),\ n \in \mathbb{N}\} \in l_2$ for all $t \in T$. The random variables

$$\left\{\tilde{X}_n = f_n(t)\, X_n,\ n \in \mathbb{N}\right\}$$

are then independent, with mean zero and respective variances $f_n^2(t)$, which form a summable sequence. Consequently [54, p. 113],

$$X_t = \sum_n f_n(t)\, X_n$$

represents a second order random process, with mean zero and covariance

$$C_X(t_1, t_2) = \sum_n f_n(t_1) f_n(t_2) = \langle \underline{f}(t_1), \underline{f}(t_2)\rangle_{l_2}.$$

The elements $h \in H(C_X, T)$ have then the following representation:

$$h(t) = \langle \underline{\alpha}, \underline{f}(t)\rangle_{l_2} = \sum_n \alpha_n f_n(t),$$

$\underline{\alpha} \in l_2$. Let, in l_2,

$$l_2[\underline{f}] = \overline{V\left[\left\{\underline{f}(t),\ t \in T\right\}\right]}.$$

Then

$$\langle h_1, h_2 \rangle_{H(C_X,T)} = \langle P_{l_2[\underline{f}]} [\underline{\alpha}_1], \underline{\alpha}_2 \rangle_{l_2}.$$

Let $\underline{c}$ be a sequence in l_2. Then $X = \sum_n c_n X_n$ has mean zero, and an integrable square. Furthermore

$$E\left[\left(X - \sum_{i=1}^{p} \alpha_i X_{t_i}\right)^2\right] = \left\| \underline{c} - \sum_{i=1}^{p} \alpha_i \underline{f}(t_i) \right\|_{l_2}^2.$$

Consequently, when $l_2[\underline{f}] = l_2$, $L_2[X]$ is generated by elements of type X. Condition Π_3 then becomes

$$E\left[\langle \underline{c}, \underline{X} \rangle_{l_2}^4\right] \leq \kappa^2 E^2 \left[\langle \underline{c}, \underline{X} \rangle_{l_2}^2\right], \text{ all } \underline{c} \in l_2.$$

Here is the assessment of Π_3 for a particular case, that of independent, identically distributed, Laplace random variables, that is, since one must have $E\left[X_n^2\right] = 1$,

$$f_{X_n}(x) = \frac{1}{\sqrt{2}} e^{-\sqrt{2}|x|},\ E\left[e^{\theta X_n}\right] = \frac{1}{1 - \frac{\theta^2}{2}}.$$

Let then

$$\varphi(\theta) = E\left[e^{\theta \langle \underline{c}, \underline{X} \rangle_{l_2}}\right] = \prod_n \frac{1}{1 - \frac{c_n^2 \theta^2}{2}}.$$

Taking logarithms, differentiating, and setting $\theta = 0$, one gets:

$$E\left[\langle \underline{c}, \underline{X} \rangle_{l_2}^4\right] = 3E^2 \left[\langle \underline{c}, \underline{X} \rangle_{l_2}^2\right] + \frac{3}{4} \sum_n c_n^4.$$

Consequently (Hölder's inequality)

$$\frac{E\left[\langle \underline{c}, \underline{X} \rangle_{l_2}^4\right]}{E^2\left[\langle \underline{c}, \underline{X} \rangle_{l_2}^2\right]} \leq 4.$$

Let $\Phi : \mathbb{R}^{\mathbb{N}} \longrightarrow \mathbb{R}^T$ be defined using the following relation:

$$\Phi(\underline{x})[t] = \limsup_n \sum_{i=1}^{n} x_i f_i(t).$$

Then $X_t = \Phi(\underline{X})$, and equivalence and singularity of such processes may be investigated using Kakutani's dichotomy theorem [163, p. 116], rather than the more cumbersome methods to follow.

Let $X_t = \int_0^t s(\cdot, \theta)\, d\theta + W_t$, s a random signal, and W a standard Wiener process. The elements of the linear space of X shall have the following form:

$$X[f] = \langle f, s\rangle_{L_2[T]} + \int_T f\, dW.$$

As the law of s is typically unknown, it will usually be very difficult to check condition Π_3. In that case, one has Girsanov's theorem [264, p. 250] to carry the day. Furthermore specific assumptions on the behavior of $\mathcal{E}$ with respect to P are necessary for the linear space determined by $\mathcal{E}$ and P to have an explicit form, generally a prerequisite to checking the requirements of the theory.

Remark 5.1.29 The assumption that the elements of $\mathcal{E}$ have a square that is integrable with respect to P means, for example (Example) 5.1.6, that a covariance operator exists. For it to have finite trace something more must be assumed of P, for example that the square of the norm is adapted and integrable. When Π_3 obtains [(Definition) 5.1.17], one assumes in particular that, with respect to P, the "process" $\mathcal{E}$ has fourth moments.

The properties that have been listed (Π_1 in (Definition) 5.1.13, Π_2 in (Definition) 5.1.14, and Π_3 in (Definition) 5.1.17) have the following consequences, stated as propositions.

Proposition 5.1.30 *The context is that of (Definition)* 5.1.1. *Suppose Π_1 and Π_2 obtain for P and $V(\mathcal{M})$, and that $A \in \mathcal{S}$ is such that $P(A) > 0$. There exists then $\kappa_A > 0$ such that, for $f \in \overline{V(\mathcal{M})}^P$, fixed, but arbitrary,*

$$\int_A \dot{f}^2(s)\, P(ds) \geq \kappa_A \int_S \dot{f}^2(s)\, P(ds).$$

Proof One may assume that $\int_S \dot{f}^2(s)\, P(ds) > 0$, for otherwise there is nothing to prove.

Assume that $f \in V(\mathcal{M})$. Then, letting $n_f = \left\{\int_S f^2(s)\, P(ds)\right\}^{1/2}$, one has simultaneously that

$$\frac{f}{n_f} \in V_1(\mathcal{M}), \text{ and that } \int_A f^2(s)\, P(ds) = n_f^2 \int_A \frac{f^2(s)}{n_f^2} P(ds).$$

For fixed, but arbitrary $\epsilon > 0$, set

$$S_{f,\epsilon} = \left\{ s \in S : \frac{f^2(s)}{n_f^2} > \epsilon \right\}.$$

Then

$$\int_A \frac{f^2(s)}{n_f^2} P(ds) \geq \epsilon\, P\left(A \cap S_{f,\epsilon}\right).$$

But one has that $0 < P(A) = P\left(A \cap S_{f,\epsilon}\right) + P\left(A \cap S_{f,\epsilon}^c\right)$, and also that

$$P\left(A \cap S_{f,\epsilon}^c\right) \leq P\left(S_{f,\epsilon}^c\right) = P\left(s \in S : \frac{|f(s)|}{n_f} \leq \epsilon^{1/2}\right).$$

Let $0 < \delta < \frac{P(A)}{2}$. Property Π_2 has thus the consequence that, for ϵ small enough,

$$P\left(s \in S : \frac{|f(s)|}{n_f} \leq \epsilon^{1/2}\right) < \delta,$$

independently of f, and thus that

$$P\left(A \cap S_{f,\epsilon}\right) > \frac{P(A)}{2},$$

independently of f. It thus suffices to choose $\kappa_A = \epsilon \frac{P(A)}{2}$ for the result to hold in case $f \in V(\mathcal{M})$.

Suppose now that $f \in \overline{V(\mathcal{M})}^P$, and that $\{f_n,\ n \in \mathbb{N}\} \subseteq V(\mathcal{M})$ is such that, in $L_2(S, \mathcal{S}, P)$, $\lim_n [f_n]_{L_2(S,\mathcal{S},P)} = f$. Then

$$\begin{aligned}\int_A \dot{f}^2(s)\, P(ds) &= \lim_n \int_A f_n^2(s)\, P(ds) \\ &\geq \lim_n \kappa_A \int_S f_n^2(s)\, P(ds) \\ &= \kappa_A \int_S \dot{f}^2(s)\, P(ds).\end{aligned}$$

□

Proposition 5.1.31 *The context is that of (Definition)* 5.1.1. *Suppose that P has property Π_3. It has then property Π_1 for $\mathcal{Q}(\mathcal{E})$, that is, "quadratic functionals" have a square that is integrable with respect to P.*

Proof For fixed, but arbitrary $(t_1, t_2) \in T \times T$,

$$\left\{\int_S \left(\mathcal{E}_{t_1}(s)\, \mathcal{E}_{t_2}(s)\right)^2 P(ds)\right\}^2 \leq$$

$$\leq \int_S \mathcal{E}_{t_1}^4(s) P(ds) \int_S \mathcal{E}_{t_2}^4(s) P(ds)$$

$$\leq \kappa_P^4 \left\{ \int_S \mathcal{E}_{t_1}^2(s) P(ds) \right\}^2 \left\{ \int_S \mathcal{E}_{t_2}^2(s) P(ds) \right\}^2.$$

□

Proposition 5.1.32 *The context is that of (Definition)* 5.1.1. *Suppose that P has property Π_3, and that $m \in \mathcal{M}(\mathcal{S})$ is such that*

$$\int_S m^4(s) P(ds) < \infty.$$

Let $M : \overline{\mathcal{L}(\mathcal{E})}^P \longrightarrow L_2(S, \mathcal{S}, P)$ be defined using the following assignment:

$$M[f] = [m\dot{f}]_{L_2(S,\mathcal{S},P)}.$$

Then:

1. *M is a multiplication operator which is linear and bounded.*
2. *Let $P_{\overline{\mathcal{L}(\mathcal{E})}^P}$ denote the projection onto $\overline{\mathcal{L}(\mathcal{E})}^P$ in $L_2(S, \mathcal{S}, P)$: the operator*

$$P_{\overline{\mathcal{L}(\mathcal{E})}^P} M$$

is linear and bounded, with domain equal to $\overline{\mathcal{L}(\mathcal{E})}^P$, and range in $\overline{\mathcal{L}(\mathcal{E})}^P$.

Proof The definition of M makes sense, and yields a bounded, linear operator, as

$$\int_S \left\{m(s)\dot{f}(s)\right\}^2 P(ds) \leq \left\{ \int_S m^4(s) P(ds) \right\}^{\frac{1}{2}} \left\{ \int_S \dot{f}^4(s) P(ds) \right\}^{\frac{1}{2}}$$

$$\leq \kappa_P \left\{ \int_S m^4(s) P(ds) \right\}^{\frac{1}{2}} \int_S \dot{f}^2(s) P(ds),$$

where $f \in \overline{\mathcal{L}(\mathcal{E})}^P$ and $\dot{f} \in f$ are fixed, but arbitrary. □

Remark 5.1.33 M may be looked at as the restriction, to the closure of $\mathcal{L}(\mathcal{E})$, with respect to P, of a standard multiplication operator with domain and range in $L_2(S, \mathcal{S}, P)$. That domain contains the measurable functions whose fourth power is integrable. The Π_3 assumption insures that the closure, with respect to P, of $\mathcal{L}(\mathcal{E})$ belongs to the domain of M.

5.2 Atoms and Reduced Measures

Atoms, "isolated" points, or sets, of positive measure, are a "nuisance," when one is interested in "continuous" measures. Reduced measures are measures for which such "perturbing" elements do not exist, or have been removed. This section contains a number of technical considerations on the subject. The "canonical" cases are those of Gaussian measures [(Proposition) 5.6.10].

Let $S_0 \in \mathcal{S}$ be fixed, but arbitrary, and let $\mathcal{S}_0 = \{A \cap S_0,\ A \in \mathcal{S}\}$. $\mathcal{M}(\mathcal{S}_0)$ shall be the family of functions $f : S_0 \longrightarrow \mathbb{R}$ which are adapted to $\mathcal{S}_0$ and $\mathcal{B}(\mathbb{R})$.

Lemma 5.2.1 *Let, as just described, $S_0 \in \mathcal{S}$, and $\mathcal{S}_0 = \{A \cap S_0,\ A \in \mathcal{S}\}$. Then*

$$\mathcal{M}(\mathcal{S}_0) = \left\{f^{|S_0},\ f \in \mathcal{M}(\mathcal{S})\right\},$$

where $f^{|S_0}$ is the restriction of f to S_0.

Proof Suppose that $f \in \mathcal{M}(\mathcal{S})$, and let f_0 be the restriction of f to S_0. Then

$$\begin{aligned}\{s \in S_0 : a < f_0(s) \leq b\} &= \{s \in S_0 : a < f(s) \leq b\} \\ &= S_0 \cap f^{-1}(]a, b]) \\ &\in \mathcal{S}_0.\end{aligned}$$

Suppose now that $f_0 \in \mathcal{M}(\mathcal{S}_0)$. Let

$$\Phi_f(s) = \begin{cases} f_0(s) & \text{when } s \in S_0 \\ 0 & \text{when } s \in S_0^c \end{cases}.$$

f_0 is the restriction of Φ_f to S_0, and

$$\Phi_f^{-1}(]a, b]) = \left\{\begin{array}{ll} f_0^{-1}(]a, b]) \in \mathcal{S}_0 \subseteq \mathcal{S}, & \text{when } 0 \in]a, b]^c \\ f_0^{-1}(]a, b]) \cup S_0^c \in \mathcal{S}, & \text{when } 0 \in]a, b] \end{array}\right\} \in \mathcal{S},$$

so that $\Phi_f \in \mathcal{M}(\mathcal{S})$. □

Here is a way to spot atoms. The restriction of $\mathcal{E}_t$ to S_0 shall be denoted ξ_t.

Proposition 5.2.2 *The context is that of Sect. 5.1. Let, as just described, $S_0 \in \mathcal{S}$, and $\mathcal{S}_0 = \{A \cap S_0,\ A \in \mathcal{S}\}$. Suppose that $(S_0, \mathcal{S}_0, \mu)$ is a finite measure space for which, given $t \in T$, fixed, but arbitrary,*

$$\int_{S_0} \xi_t^2(s)\, \mu(ds) < \infty.$$

If, for all $t \in T$, all $f \in \mathcal{M}(\mathcal{S}_0)$ such that $\int_{S_0} f^2(s)\,\mu(ds) < \infty$, one has that

$$\int_{S_0} \xi_t(s) f(s)\,\mu(ds) = 0,$$

then either $\mu(S_0) = 0$, or S_0 is an atom for μ (any measurable subset of S_0 has either measure zero or a measure equal to that of S_0).

Proof One shall first prove that, in case $\mu(S_0) > 0$, every function f, with values in $\mathbb{R}$, which is adapted to $\mathcal{S}_0$ and $\mathcal{B}(\mathbb{R})$, is constant on a measurable subset $O_f \in \mathcal{S}_0$ with the property that $\mu(O_f) = \mu(S_0)$.

Suppose thus that $\mu(S_0) > 0$. Let $f = \xi_t$, so that, by assumption,

$$0 = \int_{S_0} \xi_t(s) f(s)\,\mu(ds) = \int_{S_0} \xi_t^2(s)\,\mu(ds).$$

Then, since $\mu(S_0) > 0$, one has that $\xi_t(s) = 0$ for $s \in O_t \subseteq S_0$, some $O_t \in \mathcal{S}_0$, and $\mu(O_t) = \mu(S_0)$.

Let f be adapted to $\mathcal{S}_0$. From (Lemma) 5.2.1, there is a g adapted to $\mathcal{S}$ such that, for fixed, but arbitrary $s \in S_0$, $f(s) = g(s)$. But [41, p. 144] g has a representation in the following form:

$$g(s) = \Phi_g\left(\left\{\mathcal{E}_{t_i^g}(s),\ i \in \mathbb{N}\right\}\right),$$

$s \in S$, $\Phi_g : \mathbb{R}^\infty \longrightarrow \mathbb{R}$, Φ_g adapted. Consequently, for $s \in S_0$,

$$f(s) = g(s) = \Phi_g\left(\left\{\xi_{t_i^g}(s),\ i \in \mathbb{N}\right\}\right).$$

Let $O = \cap_{i \in \mathbb{N}} O_{t_i^g}$, where the $O_{t_i^g}$'s have the meaning given above to O_t. Then $O \subseteq S_0$, $O \in \mathcal{S}_0$, and $\mu(O) = \mu(S_0)$, as a countable intersection of sets of full measure has full measure. Furthermore, for $s \in O$, $f(s) = \Phi_g(\{0,\ i \in \mathbb{N}\})$.

What precedes is now used with functions which are indicators of sets. For $A \cap S_0 \in \mathcal{S}_0$, let $\Phi_{A \cap S_0} : \mathbb{R}^\infty \longrightarrow \mathbb{R}$ be the measurable function such that

$$\chi_{A \cap S_0}(s) = \Phi_{A \cap S_0}\left(\left\{\xi_{t_i^{A \cap S_0}}(s),\ i \in \mathbb{N}\right\}\right).$$

Let $O_{A \cap S_0} \in \mathcal{S}_0$ be such that $\mu(O_{A \cap S_0}) = \mu(S_0)$, and, for $s \in O_{A \cap S_0}$,

$$\chi_{A \cap S_0}(s) = \Phi_{A \cap S_0}(\{0,\ i \in \mathbb{N}\}),$$

a constant, denoted $c_{A\cap S_0}$, which takes its values in $\{0, 1\}$. Then

$$\begin{aligned}
0 &= \int_{O_{A\cap S_0}} \left|\left(\chi_{A\cap S_0}(s) - c_{A\cap S_0}\right)\right| \mu(ds) \\
&= \int_{S_0} \left|\left(\chi_{A\cap S_0}(s) - c_{A\cap S_0}\right)\right| \mu(ds) \\
&= \int_{A\cap S_0} \left|\left(\chi_{A\cap S_0}(s) - c_{A\cap S_0}\right)\right| \mu(ds) + \int_{A^c\cap S_0} \left|\left(\chi_{A\cap S_0}(s) - c_{A\cap S_0}\right)\right| \mu(ds) \\
&= \left(1 - c_{A\cap S_0}\right) \mu(A \cap S_0) + c_{A\cap S_0} \mu(A^c \cap S_0).
\end{aligned}$$

As $\mu(S_0) = \mu(A \cap S_0) + \mu(A^c \cap S_0)$,

- when $c_{A\cap S_0} = 0$,

$$\mu(A \cap S_0) = 0, \text{ and } \mu(A^c \cap S_0) = \mu(S_0) > 0;$$

- when $c_{A\cap S_0} = 1$,

$$\mu(A^c \cap S_0) = 0, \text{ and } \mu(A \cap S_0) = \mu(S_0) > 0.$$

□

Remark 5.2.3 When S is one of the Examples 5.1.3 to 5.1.6,

$$\begin{aligned}
\chi_{A\cap S_0}(0) &= \Phi_{A\cap S_0}\left(\left\{\mathcal{E}_{t_i}^{A\cap S_0}(0), \; i \in \mathbb{N}\right\}\right) \\
&= \Phi_{A\cap S_0}(\{0, \; i \in \mathbb{N}\}) \\
&= c_{A\cap S_0},
\end{aligned}$$

so that, choosing $A = S_0$, one gets, from the following expression (see proof of (Proposition) 5.2.2):

$$\left(1 - c_{A\cap S_0}\right) \mu(A \cap S_0) + c_{A\cap S_0} \mu(A^c \cap S_0) = 0,$$

that

$$\left(1 - c_{S_0}\right) \mu(S_0) = 0.$$

Thus, since $\mu(S_0) > 0$, $c_{S_0} = 1$, and $0 \in S_0$. Furthermore, when $0 \in A \cap S_0$, $c_{A\cap S_0} = 1$, and $\mu(A \cap S_0) = \mu(S_0) > 0$.

Remark 5.2.4 Let S be a separable metric space, with metric d; $\mathcal{S}$, the family of its Borel sets; and μ, a finite measure on $\mathcal{S}$ such that, whatever $A \in \mathcal{S}$, either $\mu(A) = 0$, or $\mu(A) = \mu(S)$. Let $\epsilon > 0$ be fixed, but arbitrary, and $\{s_n, n \in \mathbb{N}\}$ be a countable, ϵ-dense subset of S [84, p. 187], that is, the balls

$$B(s_n, \epsilon) = \{s \in S : d(s_n, s) < \epsilon\}, \; n \in \mathbb{N},$$

cover S. There is thus at least one whose measure equals $\mu(S)$. For each $p \in \mathbb{N}$, choose then $s_{n(p)}$ such that

$$\mu\left(B\left(s_{n(p)}, \frac{1}{p}\right)\right) = \mu(S).$$

Denote $\{B_p, p \in \mathbb{N}\}$ the sequence of those balls, and let

$$A_n = \cap_{p=1}^{n} B_p, \;\; A = \cap_{n \in \mathbb{N}} A_n.$$

Then $\mu(A) = \mu(S)$, and A is a singleton [240, p. 68].

The spaces of (Examples) 5.1.4 and 5.1.5 are separable metric spaces, and that of (Example) 5.1.6 is also separable when H is separable.

Here is a case when atoms are "irrelevant."

Fact 5.2.5 ([46, pp. 114,122]) *The context is that of Sect. 5.1. Suppose $s_0 \in S$ and δ_0 is the measure on $\mathcal{S}$ defined using the following assignment:*

$$\delta_0(A) = \begin{cases} 1 \text{ when } s_0 \in A \\ 0 \text{ when } s_0 \in A^c \end{cases}.$$

Then

$$\int_S f(s)\, \delta_0(ds) = f(s_0), \; f \text{ adapted to } \mathcal{S} \text{ and } \mathcal{B}(\mathbb{R}).$$

Proposition 5.2.6 *The context is that of Sect. 5.1. Let μ be a finite measure on $\mathcal{S}$ for which*

$$\int_S \mathcal{E}_t^2(s)\, \mu(ds) < \infty, \; t \in T.$$

Suppose $s_0 \in S$ is such that $\mathcal{E}_t(s_0) = 0, \; t \in T$. Let

(a) δ_0 *be the measure on $\mathcal{S}$ defined in (Fact) 5.2.5,*
(b) $p_0 = \inf\{\mu(A) : A \in \mathcal{S}, A \ni s_0\} \leq \mu(S) < \infty$,
(c) $\mu_0 = \mu - p_0\delta_0$.

Then, for fixed, but arbitrary $(t_1, t_2) \in T \times T$,

$$\mathcal{H}_{\mathcal{E},\mu}(t_1, t_2) = \int_S \mathcal{E}_{t_1}(s)\,\mathcal{E}_{t_2}(s)\,\mu(ds) = \int_S \mathcal{E}_{t_1}(s)\,\mathcal{E}_{t_2}(s)\,\mu_0(ds).$$

Proof Using (Fact) 5.2.5, for every $f : S \longrightarrow \mathbb{R}$, adapted to $\mathcal{S}$ and $\mathcal{B}(\mathbb{R})$,

$$\int_S f(s)\,\delta_0(ds) = f(s_0).$$

Thus

$$\int_S \mathcal{E}_{t_1}(s)\,\mathcal{E}_{t_2}(s)\,\delta_0(ds) = \mathcal{E}_{t_1}(s_0)\,\mathcal{E}_{t_2}(s_0) = 0.$$

□

Remark 5.2.7 The previous result applies in particular to the Examples 5.1.3 to 5.1.6 with s_0, the zero function (element).

Definition 5.2.8 The context is that of Sect. 5.1. Let μ be a finite measure on $\mathcal{S}$ for which

$$\int_S \mathcal{E}_t^2(s)\,\mu(ds) < \infty,\ t \in T.$$

Let

$$p_0 = \inf\{\mu(A) : A \in \mathcal{S},\ A \ni s_0\} \le \mu(S) < \infty.$$

When $p_0 = 0$, one then says that μ is reduced at s_0.

Proposition 5.2.9 *The context is that of Sect. 5.1. Let* μ *be a finite measure on* $\mathcal{S}$ *for which*

$$\int_S \mathcal{E}_t^2(s)\,\mu(ds) < \infty,\ t \in T.$$

If μ *is reduced at some* $s_0 \in S$, *and* ν *is a measure on* $\mathcal{S}$ *which is absolutely continuous with respect to* μ, *then* ν *is also reduced at* s_0.

Proof Since μ is reduced at s_0, there is a sequence $\{A_n,\ n \in \mathbb{N}\} \subseteq \mathcal{S}$ such that $s_0 \in A_n,\ n \in \mathbb{N}$, and $\lim_n \mu(A_n) = 0$. $\{A_n,\ n \in \mathbb{N}\}$ may be assumed to be decreasing. As

$$\mu\left(s \in S : \left|\chi_{A_n} - 0\right| > \epsilon\right) = \mu(A_n),$$

for the measure μ, the sequence $\{\chi_{A_n},\ n \in \mathbb{N}\}$ converges to zero in measure. Consequently, since

$$\nu(A_n) = \int_S \chi_{A_n}(s) \frac{d\nu}{d\mu}(s)\, \mu(ds),$$

$\lim_n \nu(A_n) = 0$. □

5.3 Dependence of the Lebesgue Decomposition on Some Related Reproducing Kernel Hilbert Spaces

Between any two measures there is a relation, that described by the Lebesgue decomposition. It is examined below to what extent knowledge of the RKHS's associated with those measures makes that relation more explicit.

5.3.1 Background

One finds here the form of the Lebesgue decomposition that shall be used in the sequel.

The context is that of Sect. 5.1. In particular, the maps $\mathcal{E}_t,\ t \in T$, have always a square that is integrable. Let P and Q be probabilities on $\mathcal{S}$. One may thus compute, for $(t_1, t_2) \in T \times T$ fixed, but arbitrary,

$$\mathcal{H}_{\mathcal{E},P}(t_1, t_2) = \int_S \mathcal{E}_{t_1}(s)\, \mathcal{E}_{t_2}(s)\, P(ds)$$

and

$$\mathcal{H}_{\mathcal{E},Q}(t_1, t_2) = \int_S \mathcal{E}_{t_1}(s)\, \mathcal{E}_{t_2}(s)\, Q(ds).$$

The RKHS $H(\mathcal{H}_{\mathcal{E},P}, T)$, associated with both P and $\mathcal{H}_{\mathcal{E},P}$, is the range of the map $L_P : L_2(S, \mathcal{S}, P) \longrightarrow \mathbb{R}^T$ defined using the following relation:

$$L_P[f](t) = \int_S \dot{f}(s)\, \mathcal{E}_t(s)\, P(ds).$$

L_P is a partial isometry whose initial set is $\overline{\mathcal{L}(\mathcal{E})}^P$, and whose range and final set is $H(\mathcal{H}_{\mathcal{E},P}, T)$. The null space of L_P is

$$\mathcal{N}[L_P] = \left\{\overline{\mathcal{L}(\mathcal{E})}^P\right\}^{\perp}.$$

The restriction of L_P to $\overline{\mathcal{L}(\mathcal{E})}^P$ is unitary and shall be denoted U_P. One has in particular that

$$U_P\left[[\mathcal{E}_t]_{L_2(S,\mathcal{S},P)}\right] = \mathcal{H}_{\mathcal{E},P}(\cdot, t),\ t \in T,$$

and, for $h_f = U_P[f]$,

$$\left\|h_f\right\|_{H(\mathcal{H}_{\mathcal{E},P},T)} = \|f\|_{L_2(S,\mathcal{S},P)}.$$

The Lebesgue decomposition of Q with respect to P says that, uniquely,

$$Q(A) = \int_A D_P(s)\,P(ds) + Q(A \cap N_{sP}),$$

with

$$\{A, N_{sP}\} \subseteq \mathcal{S},\ \text{and}\ P(N_{sP}) = 0,$$

where D_P is, up to equivalence, the largest (extended) real and measurable function D such that

$$\int_A D(s)\,P(ds) \leq Q(A),\ A \in \mathcal{S}.$$

The absolutely continuous part of Q with respect to P shall be denoted Q_{aP}, and the singular part, Q_{sP}. Then

$$D_P = \frac{dQ_{aP}}{dP}.$$

Remark 5.3.1 The Lebesgue decomposition can be obtained as follows [201, p. 47]. One first proves that there is a measurable D such that

$$Q(A) = \int_A D(s)\,[P+Q](ds),\ \mathcal{R}_D \subseteq [0, 1].$$

One then sets $N = \{s \in S : D(s) = 1\}$, and establishes that $P(N) = 0$. One consequently sets, arbitrarily on N,

$$\frac{dQ}{dP} = \begin{cases} \frac{D}{1-D} & \text{on } N^c \\ \infty & \text{on } N \end{cases}.$$

Finally

$$\begin{aligned} Q(A) &= Q(A \cap N) + Q(A \cap N^c) \\ &= Q(A \cap N) + \int_{A \cap N^c} \frac{dQ}{dP}(s)\, P(ds). \end{aligned}$$

But

$$\begin{aligned} \int_{A \cap N} \frac{dQ}{dP}(s)\, P(ds) &\leq \int_N \frac{dQ}{dP}(s)\, P(ds) \\ &= \infty \times P(N) \\ &= 0, \end{aligned}$$

so that

$$\int_A \frac{dQ}{dP}(s)\, P(ds) = \int_{A \cap N^c} \frac{dQ}{dP}(s)\, P(ds).$$

Also

$$\left\{ s \in S : \frac{dQ}{dP}(s) = 0 \right\} \subseteq N^c$$

so that

$$Q\left(\left\{ s \in S : \frac{dQ}{dP}(s) = 0 \right\} \right) = 0.$$

One shall write, as already stated, N_{sP} for N, and $\frac{dQ_{aP}}{dP}$ for $\frac{dQ}{dP}$. Then

$$\begin{aligned} Q_{aP}(A) &= \int_A \frac{dQ_{aP}}{dP}(s)\, P(ds), \\ Q_{sP}(A) &= Q(A \cap N_{sP}). \end{aligned}$$

Definition 5.3.2 $[S_0, S_\epsilon]$ The following sets shall be of frequent use ($\epsilon > 0$ is fixed, but arbitrary):

1. $S_0 = \left\{s \in S : \frac{dQ_{aP}}{dP}(s) = 0\right\} \subseteq N_{sP}^c$, so that

$$Q(S_0) = Q_{aP}(S_0) = 0;$$

then

$$Q(S_0 \cup N_{sP}) = Q(S_0) + Q(N_{sP}) = Q(N_{sP}).$$

2. $S_\epsilon = \left\{s \in S : 0 < \frac{dQ_{aP}}{dP}(s) \leq \frac{1}{\epsilon}\right\} \subseteq N_{sP}^c$, so that

$$Q(S_\epsilon) = Q_{aP}(S_\epsilon);$$

then

(i) for $0 < \epsilon_1 < \epsilon_2$, $S_{\epsilon_1}^c \subseteq S_{\epsilon_2}^c$;
(ii) $S_\epsilon^c = S_0 \cup \left\{s \in S : \frac{dQ_{aP}}{dP}(s) > \frac{1}{\epsilon}\right\}$;
(iii) $\cap_{\epsilon > 0} S_\epsilon^c = S_0 \cup N_{sP}$.

Fact 5.3.3 (Background Summary) *The context is that of Sect. 5.1.*
For two probabilities on $\mathcal{S}$, P and Q,

1. $Q(A) = Q_{aP}(A) + Q_{sP}(A)$ *where*

(i) $Q_{sP}(A) = Q(A \cap N_{sP})$,
(ii) $P(N_{sP}) = 0$,
(iii) $Q_{aP}(A) = \int_A \frac{dQ_{aP}}{dP}(s)\, P(ds)$,
(iv) $0 \leq \frac{dQ_{aP}}{dP}(s) < \infty$, $s \in N_{sP}^c$,
(v) $\frac{dQ_{aP}}{dP}(s) = \infty$, $s \in N_{sP}$.

2. $S_0 = \left\{s \in S : \frac{dQ_{aP}}{dP}(s) = 0\right\}$.
3. $S_\epsilon = \left\{s \in S : 0 < \frac{dQ_{aP}}{dP}(s) \leq \frac{1}{\epsilon}\right\}$.

5.3.2 Lebesgue Decomposition and "Sizes" of Reproducing Kernel Hilbert Spaces Intersections

It is shown, in this section, that, in the Lebesgue decomposition, the presence, or absence, of the term $Q(N_{sP})$ is linked to the "size" of the intersection of the RKHS's associated with, respectively, P and Q.

Lemma 5.3.4 *The background is that of Sect. 5.3.1. Let P and Q be probabilities on $\mathcal{S}$, and*

$$H_0 = H\left(\mathcal{H}_{\mathcal{E},P}, T\right) \cap H\left(\mathcal{H}_{\mathcal{E},Q}, T\right).$$

Let $h \in H\left(\mathcal{H}_{\mathcal{E},Q}, T\right)$ be fixed, but arbitrary, and $f_h = U_Q^{\star}[h] \in \overline{\mathcal{L}(\mathcal{E})}^Q$. Form, from f_h, the following functions:

(a) $\dot{f}_{h,P}(s) = \chi_{S_\epsilon}(s)\dot{f}_h(s)\frac{dQ_{aP}}{dP}(s)$,
(b) $\dot{f}_{h,Q}(s) = \chi_{S_\epsilon}(s)\dot{f}_h(s)$,
(c) $h_\epsilon(t) = \int_{S_\epsilon} \dot{f}_h(s)\,\mathcal{E}_t(s)\,Q(ds)$.

Then

1. $f_{h,P} \in L_2(S, \mathcal{S}, P)$,
2. $f_{h,Q} \in L_2(S, \mathcal{S}, Q)$,
3. $h_\epsilon(t) = \int_S \dot{f}_{h,P}(s)\,\mathcal{E}_t(s)\,P(ds) = \int_S \dot{f}_{h,Q}(s)\,\mathcal{E}_t(s)\,Q(ds)$,
4. $h_\epsilon \in H_0$.

Proof $f_{h,P} \in L_2(S, \mathcal{S}, P)$ since

$$\begin{aligned}
\int_S \dot{f}_{h,P}^2(s)\,P(ds) &= \int_{S_\epsilon} \dot{f}_h^2(s)\left[\frac{dQ_{aP}}{dP}\right]^2(s)\,P(ds) \\
&\le \frac{1}{\epsilon}\int_{S_\epsilon} \dot{f}_h^2(s)\,\frac{dQ_{aP}}{dP}(s)\,P(ds) \\
&= \frac{1}{\epsilon}\int_{S_\epsilon} \dot{f}_h^2(s)\,Q_{aP}(ds) \\
&= \frac{1}{\epsilon}\int_{S_\epsilon} \dot{f}_h^2(s)\,Q(ds) \\
&\le \frac{1}{\epsilon}\int_{S} \dot{f}_h^2(s)\,Q(ds) \\
&< \infty.
\end{aligned}$$

Similarly, $f_{h,Q} \in L_2(S, \mathcal{S}, Q)$ since

$$\int_S \dot{f}_{h,Q}^2(s)\,Q(ds) = \int_{S_\epsilon} \dot{f}_h^2(s)\,Q(ds) \le \int_S \dot{f}_h^2(s)\,Q(ds) < \infty.$$

Since

$$\int_{S_\epsilon} \dot{f}_h(s)\,\mathcal{E}_t(s)\,Q(ds) = \int_{S_\epsilon} \dot{f}_h(s)\,\mathcal{E}_t(s)\,Q_{aP}(ds)$$

$$= \int_{S_\epsilon} \dot{f}_h(s)\, \mathcal{E}_t(s) \frac{dQ_{aP}}{dP}(s)\, P(ds)$$

$$= \int_S \dot{f}_{h,P}(s)\, \mathcal{E}_t(s)\, P(ds),$$

$h_\epsilon(t) = \int_{S_\epsilon} \dot{f}_h(s)\, \mathcal{E}_t(s)\, Q(ds)$ has a first representation as

$$h_\epsilon(t) = \int_S \dot{f}_{h,P}(s)\, \mathcal{E}_t(s)\, P(ds), \text{ where } f_{h,P} \in L_2(S, \mathcal{S}, P),$$

and a second representation as

$$h_\epsilon(t) = \int_S \dot{f}_{h,Q}(s)\, \mathcal{E}_t(s)\, Q(ds), \text{ where } f_{h,Q} \in L_2(S, \mathcal{S}, Q).$$

Thus $h_\epsilon \in H_0$. □

Lemma 5.3.5 *The background is that of Sect. 5.3.1. Elements P, Q, h, f_h, S_ϵ, and h_ϵ are as in (Lemma)* 5.3.4. *Define*

$$h^{(\epsilon)}(t) = h(t) - h_\epsilon(t),\ t \in T.$$

Then:

1. $h^{(\epsilon)} \in H(\mathcal{H}_{\mathcal{E},Q}, T)$;
2. $h^{(\epsilon)}(t) = \int_{S_\epsilon^c} \dot{f}_h(s)\, \mathcal{E}_t(s)\, Q(ds)$;
3. $\|h^{(\epsilon)}\|^2_{H(\mathcal{H}_{\mathcal{E},Q},T)} = \|h\|^2_{H(\mathcal{H}_{\mathcal{E},Q},T)} - \|h_\epsilon\|^2_{H(\mathcal{H}_{\mathcal{E},Q},T)}$.

Proof h has been chosen in $H(\mathcal{H}_{\mathcal{E},Q}, T)$, and h_ϵ has been shown to belong to it [(Lemma) 5.3.4, item 4]. So item 1 is true. Now, still from (Lemma) 5.3.4,

$$h(t) - h_\epsilon(t) = \int_S \dot{f}_h(s)\, \mathcal{E}_t(s)\, Q(ds) - \int_S \dot{f}_{h,Q}(s)\, \mathcal{E}_t(s)\, Q(ds)$$

$$= \int_S \left(1 - \chi_{S_\epsilon}\right)(s) \dot{f}_h(s)\, \mathcal{E}_t(s)\, Q(ds),$$

so that

$$\|h^{(\epsilon)}\|^2_{H(\mathcal{H}_{\mathcal{E},Q},T)} = \int_S \left(1 - \chi_{S_\epsilon}\right)^2 (s) \dot{f}_h^2(s)\, Q(ds)$$

$$= \int_S \left(1 - \chi_{S_\epsilon}\right)(s) \dot{f}_h^2(s)\, Q(ds)$$

$$= \|h\|^2_{H(\mathcal{H}_{\mathcal{E},Q},T)} - \|h_\epsilon\|^2_{H(\mathcal{H}_{\mathcal{E},Q},T)}.$$

□

Proposition 5.3.6 *The background is that leading to Sect. 5.3.2* (in particular, Sect. 5.1). *Elements P, Q, h, f_h, S_ϵ, and h_ϵ are as in (Lemma)* 5.3.4, $h^{(\epsilon)}$ *as in (Lemma)* 5.3.5. *Let $H_0^\perp$ be the orthogonal complement, in $H\left(\mathcal{H}_{\mathcal{E},Q}, T\right)$, of H_0* (defined in (Lemma) 5.3.4). *When $H_0^\perp$ is not the trivial subspace $\{0_{\mathbb{R}^T}\}$, $Q(N_{sP}) > 0$.*

Proof Suppose that both $h \in H_0^\perp$ and $h \neq 0_{\mathbb{R}^T}$ obtain. Because of the inclusion $H_0^\perp \subseteq H\left(\mathcal{H}_{\mathcal{E},Q}, T\right)$, (Lemmas) 5.3.4 and 5.3.5 apply to h, and one may proceed with the element $h^{(\epsilon)} = h - h_\epsilon$. Since $h_\epsilon \in H_0$ and $h \in H_0^\perp$,

$$\langle h, h_\epsilon \rangle_{H(\mathcal{H}_{\mathcal{E},Q},T)} = 0.$$

Thus

$$\left\| h^{(\epsilon)} \right\|^2_{H(\mathcal{H}_{\mathcal{E},Q},T)} = \|h\|^2_{H(\mathcal{H}_{\mathcal{E},Q},T)} + \|h_\epsilon\|^2_{H(\mathcal{H}_{\mathcal{E},Q},T)} .$$

It then follows, from (Lemma) 5.3.5, item 3, that

$$\left\| h^{(\epsilon)} \right\|_{H(\mathcal{H}_{\mathcal{E},Q},T)} = \|h\|_{H(\mathcal{H}_{\mathcal{E},Q},T)} .$$

But, from (Lemma) 5.3.5, item 2, one has that

$$h^{(\epsilon)}(t) = \int_S \chi_{S_\epsilon^c}(s) \dot{f}_h(s)\, \mathcal{E}_t(s)\, Q(ds),$$

so that

$$\int_{S_\epsilon^c} \dot{f}_h^2(s)\, Q(ds) = \left\| h^{(\epsilon)} \right\|^2_{H(\mathcal{H}_{\mathcal{E},Q},T)} = \|h\|^2_{H(\mathcal{H}_{\mathcal{E},Q},T)} > 0.$$

As advertised in (Definition) 5.3.2, item 2, S_ϵ^c decreases with ϵ, and $\cap_{\epsilon>0} S_\epsilon^c$ is the disjoint union of S_0 and N_{sP}. Furthermore $Q(S_0) = 0$. Thus, from the above,

$$0 < \|h\|^2_{H(\mathcal{H}_{\mathcal{E},Q},T)} = \lim_{\epsilon \downarrow 0} \int_{S_\epsilon^c} \dot{f}_h^2(s)\, Q(df) = \int_{N_{sP}} \dot{f}_h^2(s)\, Q(ds) \leq \|h\|^2_{H(\mathcal{H}_{\mathcal{E},Q},T)} .$$

Consequently $\int_{N_{sP}^c} \dot{f}_h^2(s)\, Q(ds) = 0$, and

$$h(t) = \int_{N_{sP}} \dot{f}_h(s)\, \mathcal{E}_t(s)\, Q(ds).$$

Since $h \neq 0_{\mathbb{R}^T}$, $Q(N_{sP}) > 0$. □

Remark 5.3.7 Result (Proposition) 5.3.6 says in particular that, when H_0 is "small," P and Q cannot be equivalent. The next proposition says that when P and Q are equivalent, H_0 is "large."

Proposition 5.3.8 *The background is that leading to Sect. 5.3.2 (in particular, Sect. 5.1). Elements P, Q, h, f_h, S_ϵ, and h_ϵ are as in (Lemma)* 5.3.4. *Let $\overline{H}_0$ be the closure, in $H(\mathcal{H}_{\mathcal{E},Q}, T)$, of H_0. When $Q(N_{sP}) = 0$,*

$$\overline{H}_0 = H\left(\mathcal{H}_{\mathcal{E},Q}, T\right).$$

Proof Again, as $h \in H(\mathcal{H}_{\mathcal{E},Q}, T)$ by assumption, and that $h_\epsilon \in H_0$ by construction [(Lemma) 5.3.4], one has that [(Lemma) 5.3.5]

$$h(t) - h_\epsilon(t) = \int_{S_\epsilon^c} \dot{f}_h(s)\, \mathcal{E}_t(s)\, Q(ds).$$

Consequently [Sect. 5.3.1 and assumption],

$$\lim_{\epsilon \downarrow 0} \|h - h_\epsilon\|^2_{H(\mathcal{H}_{\mathcal{E},Q},T)} = \lim_{\epsilon \downarrow 0} \int_{S_\epsilon^c} \dot{f}_h^2(s)\, Q(ds) = \int_{N_{sP}} \dot{f}_h^2(s)\, Q(ds) = 0.$$

So every element in $H(\mathcal{H}_{\mathcal{E},Q}, T)$ is the strong limit of a sequence in H_0. □

Proposition 5.3.9 *The background is that leading to Sect. 5.3.2 (in particular, Sect. 5.1). S is any of the spaces of (Examples)* 5.1.3 *to* 5.1.6, *so that the family $\mathcal{E}$ is then either that of the evaluation maps, or that of the continuous linear functionals. Suppose also that P and Q are probabilities on $\mathcal{S}$, and that P is reduced at zero* [(Definition) 5.2.8]. *H_0 is as in (Lemma)* 5.3.4, *and $H_0^\perp$ is the orthogonal complement of H_0, in $H(\mathcal{H}_{\mathcal{E},Q}, T)$. Then,*

1. *when $Q_{aP}(S) > 0$, H_0 contains $\{0_{\mathbb{R}^T}\}$ strictly;*
2. *when $H_0^\perp = H(\mathcal{H}_{\mathcal{E},Q}, T)$, that is, when $H_0 = \{0_{\mathbb{R}^T}\}$, $Q \perp P$.*

Proof Let

$$\mathcal{S}_\epsilon = \{A \cap S_\epsilon,\ A \in \mathcal{S}\},$$

and $Q^{(\epsilon)}$ and $Q_{aP}^{(\epsilon)}$ be the restrictions of, respectively, Q and Q_{aP} to $\mathcal{S}_\epsilon$. Since $S_\epsilon \subseteq N_{sP}^c$,

$$Q^{(\epsilon)} = Q_{aP}^{(\epsilon)}.$$

Let ξ_t denote the restriction of $\mathcal{E}_t$ to S_ϵ. It is adapted to $\mathcal{S}_\epsilon$ [(Lemma) 5.2.1], and

$$\begin{aligned}\int_{S_\epsilon} \xi_t^2(s)\, Q_{aP}^{(\epsilon)}(ds) &= \int_{S_\epsilon} \mathcal{E}_t^2(s)\, Q_{aP}^{(\epsilon)}(ds)\\ &= \int_{S_\epsilon} \mathcal{E}_t^2(s)\, Q_{aP}(ds)\\ &\leq \int_S \mathcal{E}_t^2(s)\, Q_{aP}(ds)\end{aligned}$$

$$\leq \int_S \mathcal{E}_t^2(s)\, Q(ds)$$
$$< \infty.$$

Thus when $f \in L_2\left(S_\epsilon, \mathcal{S}_\epsilon, Q_{aP}^{(\epsilon)}\right)$ is fixed, but otherwise arbitrary, the following definition:

$$h_{f,\epsilon}(t) = \int_{S_\epsilon} \dot{f}(s)\, \xi_t(s)\, Q_{aP}^{(\epsilon)}(ds)$$

is legitimate, as it is based on the inner product of $L_2\left(S_\epsilon, \mathcal{S}_\epsilon, Q_{aP}^{(\epsilon)}\right)$.

One then proceeds below as in (Lemma) 5.3.4 to prove that $h_{f,\epsilon}$ belongs to H_0, and thus, to show that H_0 is not the trivial subspace, it suffices to prove that there is one $h_{f,\epsilon}$ that is not identically zero. That is achieved using (Proposition) 5.2.2. The fact that attention is restricted to (Examples) 5.1.3 to 5.1.6, and that P is reduced at zero, is essential: only in those cases can one identify the carrier of mass.

Let thus, for $f \in L_2\left(S_\epsilon, \mathcal{S}_\epsilon, Q_{aP}^{(\epsilon)}\right)$, fixed, but arbitrary,

$$\dot{f}_{Q,\epsilon}(s) = \chi_{S_\epsilon}(s)\dot{f}(s),$$
$$\dot{f}_{P,\epsilon}(s) = \dot{f}_{Q,\epsilon}(s)\, \frac{dQ_{aP}}{dP}(s).$$

Then:

- $f_{Q,\epsilon} \in L_2(S, \mathcal{S}, Q)$:
  indeed, $\dot{f}_{Q,\epsilon}$ is adapted to $\mathcal{S}$, and

$$\int_S \dot{f}_{Q,\epsilon}^2(s)\, Q(ds) = \int_{S_\epsilon} \dot{f}^2(s)\, Q^{(\epsilon)}(ds)$$
$$= \int_{S_\epsilon} \dot{f}^2(s)\, Q_{aP}^{(\epsilon)}(ds)$$
$$< \infty.$$

- $f_{P,\epsilon} \in L_2(S, \mathcal{S}, P)$:
  indeed, $\dot{f}_{P,\epsilon}$ is adapted to $\mathcal{S}$, and

$$\int_S \dot{f}_{P,\epsilon}^2(s)\, P(ds) = \int_{S_\epsilon} \dot{f}^2(s) \left[\frac{dQ_{aP}}{dP}\right]^2(s)\, P(ds)$$
$$\leq \frac{1}{\epsilon} \int_{S_\epsilon} \dot{f}^2(s)\, Q_{aP}^{(\epsilon)}(ds)$$
$$< \infty.$$

- $h_{f,\epsilon} \in H_0$:
  indeed $h_{f,\epsilon}$ has the following representations:

$$\begin{aligned} h_{f,\epsilon}(t) &= \int_S \dot{f}_{P,\epsilon}(s)\,\mathcal{E}_t(s)\,P(ds) \\ &= \int_S \dot{f}_{Q,\epsilon}(s)\,\mathcal{E}_t(s)\,Q(ds), \end{aligned}$$

as

$$\begin{aligned} h_{f,\epsilon}(t) &= \int_{S_\epsilon} \dot{f}(s)\,\xi_t(s)\,Q_{aP}^{(\epsilon)} \\ &= \int_S \chi_{S_\epsilon}(s)\dot{f}(s)\,\mathcal{E}_t(s)\,Q_{aP}(ds) \\ &= \int_S \chi_{S_\epsilon}(s)\dot{f}(s)\,\mathcal{E}_t(s)\,\frac{dQ_{aP}}{dP}(s)P(ds) \\ &= \int_S \dot{f}_{P,\epsilon}(s)\,\mathcal{E}_t(s)\,P(ds), \end{aligned}$$

and

$$\begin{aligned} h_{f,\epsilon}(t) &= \int_{S_\epsilon} \dot{f}(s)\,\xi_t(s)\,Q_{aP}^{(\epsilon)} \\ &= \int_S \chi_{S_\epsilon}(s)\dot{f}(s)\,\mathcal{E}_t(s)\,Q(ds) \\ &= \int_S \dot{f}_{Q,\epsilon}(s)\,\mathcal{E}_t(s)\,Q(ds). \end{aligned}$$

Suppose thus that $Q_{aP}(S) > 0$. One has then that

$$Q_{aP}(S_\epsilon) + Q_{aP}\left(S_\epsilon^c\right) = Q_{aP}(S) > 0.$$

As ϵ decreases to zero,

$$\epsilon \mapsto Q_{aP}(S_\epsilon) \quad \text{is a function that increases,}$$

$$\epsilon \mapsto Q_{aP}\left(S_\epsilon^c\right) \quad \text{is a function that decreases to } Q_{aP}(N_{sP}) = 0.$$

Consequently, there is an $\epsilon > 0$ such that $Q_{aP}^{(\epsilon)}(S_\epsilon) = Q_{aP}(S_\epsilon) > 0$.
Suppose then that, for all $t \in T$, and all $f \in L_2\left(S_\epsilon, \mathcal{S}_\epsilon, Q_{aP}^{(\epsilon)}\right)$,

$$h_{f,\epsilon}(t) = 0.$$

Since, by definition,

$$h_{f,\epsilon}(t) = \int_{S_\epsilon} \dot{f}(s)\,\xi_t(s)\,Q_{aP}^{(\epsilon)}(ds)\,,$$

and that

$$\int_{S_\epsilon} \xi_t^2(s)\,Q_{aP}^{(\epsilon)}(ds) = \int_{S_\epsilon} \xi_t^2(s)\,Q_{aP}(ds) \leq \int_S \mathcal{E}_t^2(s)\,Q(ds) < \infty,$$

one may apply (Proposition) 5.2.2 and (Remark) 5.2.4: S_ϵ contains the zero element of S, and Q_{aP} has a unique point mass at zero, with value $Q_{aP}(S_\epsilon)$. In other words,

$$Q_{aP}(A \cap S_\epsilon) = Q_{aP}(S_\epsilon)\,\delta_A(0_S)\,.$$

It follows then that

$$Q_{aP}(S_\epsilon)\,\delta_A(0_S) = \int_{A\cap S_\epsilon} \frac{dQ_{aP}}{dP}(s)\,P(ds)\,.$$

If A contains the zero element of S, the latter expression yields that

$$\epsilon\, Q_{aP}(S_\epsilon) \leq P(A)\,,$$

so that P cannot be reduced, a contradiction. Claim 1 is thus valid. Claim 2 follows from claim 1. □

5.4 Domination of Probabilities on Sub-manifolds

Domination is a way to compare the respective supports of probability measures (for instance, (Examples) 5.4.2 and 5.4.4 below), and that explains its presence in discrimination matters. It is also the seed that produces the Radon-Nikodým derivative [(Corollaries) 5.4.7 and 5.4.9].

5.4.1 *The General Case*

By "general case," one understands domination over an arbitrary vector space of measurable functions. The immediately useful "special cases" are $\mathcal{L}(\mathcal{E})$ and $\mathcal{Q}(\mathcal{E})$, which are presented separately.

Definition 5.4.1 The context is that of Sect. 5.3.1. Let $V(\mathcal{M})$ be a linear manifold in $\mathcal{M}(S)$, and suppose that P and Q are probabilities for which Π_1 obtains for

$V(\mathcal{M})$. If there exists $\kappa[V(\mathcal{M}), P, Q] \geq 0$ such that, for all $f \in V(\mathcal{M})$,

$$\left\|[f]_{L_2(S,\mathcal{S},Q)}\right\|_{L_2(S,\mathcal{S},Q)} \leq \kappa[V(\mathcal{M}), P, Q]\left\|[f]_{L_2(S,\mathcal{S},P)}\right\|_{L_2(S,\mathcal{S},P)},$$

one says that P dominates Q on $V(\mathcal{M})$, and writes, whenever useful,

$$Q \preceq P[V(\mathcal{M})].$$

Example 5.4.2 Let $T = \{0, 1\}$. A function with domain T is then given by an element in $\mathbb{R}^2$. Thus $S = \mathbb{R}^T = \mathbb{R}^2$, and $\mathcal{E}_0$ and $\mathcal{E}_1$, the evaluation maps, are the coordinate maps, and $\mathcal{M}$ is the family of all Borel measurable functions from $\mathbb{R}^2$ to $\mathbb{R}$. Assume that P and Q are measures on the Borel sets of $\mathbb{R}^2$, with finite support, and that $Q \preceq P[\mathcal{M}]$. Then, domination of Q by P, on $\mathcal{M}$, that is

$$\sum_{i=1}^{n_Q} q_i f^2(x_{Q,i}, y_{Q,i}) \leq \kappa[\mathcal{M}, P, Q] \sum_{j=1}^{n_P} p_i f^2(x_{P,i}, y_{P,i}), \ f \in \mathcal{M},$$

means in particular that the support of Q must be contained in that of P.

The result which follows says that domination of probabilities on the linear space of a process is equivalent to the domination of the associated covariances.

Proposition 5.4.3 *Suppose that P and Q are probabilities on $\mathcal{S}$. P dominates Q on $\mathcal{L}(\mathcal{E})$ if, and only if, there exists $\kappa[\mathcal{L}(\mathcal{E}), P, Q] \geq 0$ such that*

$$\mathcal{H}_{\mathcal{E},Q} \ll \kappa[\mathcal{L}(\mathcal{E}), P, Q]\mathcal{H}_{\mathcal{E},P}.$$

Proof Let $f = \sum_{i=1}^n \alpha_i \mathcal{E}_{t_i} \in \mathcal{L}(\mathcal{E})$. Then, for example,

$$\begin{aligned}\left\|[f]_{L_2(S,\mathcal{S},P)}\right\|^2_{L_2(S,\mathcal{S},P)} &= \int_S f^2(s)P(ds) \\ &= \sum_{i=1}^n \sum_{j=1}^n \alpha_i\alpha_j \int_S \mathcal{E}_{t_i}(s)\mathcal{E}_{t_j}(s)P(ds) \\ &= \sum_{i=1}^n \sum_{j=1}^n \alpha_i\alpha_j \mathcal{H}_{\mathcal{E},P}(t_1, t_2).\end{aligned}$$

□

Example 5.4.4 Let P and Q be probabilities on $\mathcal{S}$ which make of $\mathcal{E}$ a continuous, zero mean martingale, with index set $\mathbb{R}_+$, and let $V(\mathcal{M}) = \mathcal{L}(\mathcal{E})$ (letting, for example, P and Q be the measures induced, on the space of continuous functions, by continuous martingales). One then has, for example, that

$$\mathcal{H}_{\mathcal{E},P}(t_1,t_2) = E_P\left[\langle\mathcal{E}\rangle(t_1 \wedge t_2)\right]$$

where $\langle\mathcal{E}\rangle$ denotes the quadratic variation process. There is also [264, p. 101] a measure $\mu_{\mathcal{E}}$ on $\mathcal{S}\otimes\mathcal{B}(\mathbb{R}_+)$ such that

$$\int_{S\times\mathbb{R}_+} f(s,t)\,\mu_{\mathcal{E}}(ds,dt) = E_P\left[\int_{\mathbb{R}_+} f(\cdot,t)\,\langle\mathcal{E}\rangle(\cdot,dt)\right].$$

Let μ be defined as follows:

$$\mu([0,\alpha]) = \mu_{\mathcal{E}}(S\times[0,\alpha]) = E_P\left[\langle\mathcal{E}\rangle_\alpha\right].$$

Then, for example,

$$\begin{aligned}
\left\|\left[\sum_{i=1}^n \alpha_i \mathcal{E}_{t_i}\right]_{L_2(S,\mathcal{S},P)}\right\|^2_{L_2(S,\mathcal{S},P)} &= \sum_{i=1}^n\sum_{i=1}^n \alpha_i\alpha_j E_P\left[\mathcal{E}_{t_i}\mathcal{E}_{t_j}\right] \\
&= \sum_{i=1}^n\sum_{i=1}^n \alpha_i\alpha_j E_P\left[\mathcal{E}^2_{t_i\wedge t_j}\right] \\
&= \sum_{i=1}^n\sum_{i=1}^n \alpha_i\alpha_j E_P\left[\langle\mathcal{E}\rangle_{t_i\wedge t_j}\right] \\
&= \sum_{i=1}^n\sum_{i=1}^n \alpha_i\alpha_j \mu([0,t_i\wedge t_j]) \\
&= \sum_{i=1}^n\sum_{i=1}^n \alpha_i\alpha_j \int_0^\infty \chi_{[0,t_i]}(t)\chi_{[0,t_j]}(t)\,\mu(dt) \\
&= \int_0^\infty \left\{\sum_{i=1}^n \alpha_i \chi_{[0,t_i]}(t)\right\}^2 \mu(dt),
\end{aligned}$$

so that

$$\overline{\mathcal{L}(\mathcal{E})}^P = \left\{\int_{\mathbb{R}_+} f\,d\mathcal{E},\ f\in L_2(\mathbb{R}_+,\mathcal{B}(\mathbb{R}_+),\mu)\right\}.$$

The domination condition (Proposition) 5.4.3 becomes

$$E_Q\left[\langle\mathcal{E}\rangle^Q(\cdot,t_1\wedge t_2)\right] \ll \kappa\left[\mathcal{L}(\mathcal{E}),P,Q\right] E_P\left[\langle\mathcal{E}\rangle^P(\cdot,t_1\wedge t_2)\right],$$

which translates into

$$\int_{\mathbb{R}_+}\left\{\sum_{i=1}^{n}\alpha_i\chi_{[0,t_i]}(t)\right\}^2\mu_Q(dt)\leq\kappa\left[\mathcal{L}(\mathcal{E}),P,Q\right]\int_{\mathbb{R}_+}\left\{\sum_{i=1}^{n}\alpha_i\chi_{[0,t_i]}(t)\right\}^2\mu_P(dt).$$

Suppose that P is Wiener measure, and that Q is induced by a process X of the following form: $X(\omega,t)=\int_0^t f(\omega,x)\,W(\omega,dx)$, W a Wiener process. It can be shown [172, p. 33] that

$$\langle\mathcal{E}\rangle^Q(s,t)=\int_0^t\phi^2(s,x)\,dx,\quad\text{where}\quad\int_0^t\phi^2(X(\omega,\cdot),x)\,dx=\int_0^t f^2(\omega,x)\,dx.$$

In that particular case, the domination condition becomes

$$\int_{\mathbb{R}_+}\left\{\sum_{i=1}^{n}\alpha_i I_{[0,t_i]}(t)\right\}^2 E_Q\left[\phi^2(\cdot,t)\right]dt\leq$$
$$\leq\kappa\left[\mathcal{L}(\mathcal{E}),P,Q\right]\int_{\mathbb{R}_+}\left\{\sum_{i=1}^{n}\alpha_i I_{[0,t_i]}(t)\right\}^2 dt.$$

It will obtain, in particular, when $t\mapsto E_Q\left[\phi^2(\cdot,t)\right]$ is bounded.

Proposition 5.4.5 *Let P and Q be probabilities on $\mathcal{S}$, and let $V(\mathcal{M})$ be a fixed, but arbitrary manifold of $\mathcal{M}$. Suppose that, for $V(\mathcal{M})$, P has property Π_1, and Q has properties Π_1 and Π_2. When P does not dominate Q on $V(\mathcal{M})$, P and Q are orthogonal.*

Proof Suppose that $Q_{aP}(S)>0$. As can be seen in the proof of (Proposition) 5.3.9, it may be assumed that there is an $\epsilon>0$ such that $Q(S_\epsilon)=Q_{aP}(S_\epsilon)>0$. Then, since property Π_2 obtains for Q, there exists [(Proposition) 5.1.30] $\kappa_{Q(S_\epsilon)}>0$ such that,

$$\text{for all } f\in\overline{V(\mathcal{M})}^Q,\quad\int_{S_\epsilon}\dot{f}^2(s)\,Q(ds)\geq\kappa_{Q(S_\epsilon)}\,\|f\|^2_{L_2(S,\mathcal{S},Q)}.$$

But

$$\begin{aligned}\int_{S_\epsilon}\dot{f}^2(s)\,Q(ds)&=\int_{S_\epsilon}\dot{f}^2(s)\,Q_{aP}(ds)\\&=\int_{S_\epsilon}\dot{f}^2(s)\,\frac{dQ_{aP}}{dP}(s)\,P(ds)\\&\leq\frac{1}{\epsilon}\int_{S_\epsilon}\dot{f}^2(s)\,P(ds),\end{aligned}$$

so that

$$\kappa_{Q(S_\epsilon)} \|f\|^2_{L_2(S,\mathcal{S},Q)} \leq \frac{1}{\epsilon} \int_{S_\epsilon} f^2(s)\, P(ds)\,. \qquad (*)$$

Since P does not dominate Q on $V(\mathcal{M})$, given any $\lambda > 0$, there exists f_λ in $V(\mathcal{M})$ such that

$$\left\| [f_\lambda]_{L_2(S,\mathcal{S},Q)} \right\|^2_{L_2(S,\mathcal{S},Q)} > \lambda \left\| [f_\lambda]_{L_2(S,\mathcal{S},P)} \right\|^2_{L_2(S,\mathcal{S},P)} .$$

Consequently, using the latter in (*) above,

$$\lambda\, \kappa_{Q(S_\epsilon)} \left\| [f_\lambda]_{L_2(S,\mathcal{S},P)} \right\|^2_{L_2(S,\mathcal{S},P)} < \frac{1}{\epsilon} \left\| [f_\lambda]_{L_2(S,\mathcal{S},P)} \right\|^2_{L_2(S,\mathcal{S},P)} ,$$

or $\epsilon\, \kappa_{Q(S_\epsilon)} < \lambda^{-1}$. Since ϵ and $\kappa_{Q(S_\epsilon)}$ are fixed, but λ may be arbitrarily large, the latter inequality is impossible, and $Q_{aP}(S) > 0$ cannot be sustained. □

Proposition 5.4.6 *Let P and Q be probabilities on $\mathcal{S}$, and $V(\mathcal{M})$, a fixed, but arbitrary manifold of $\mathcal{M}$. Suppose that, for $V(\mathcal{M})$, P and Q have property Π_1, and that P dominates Q on $V(\mathcal{M})$.*

Let $B : \overline{V(\mathcal{M})}^P \longrightarrow \overline{V(\mathcal{M})}^Q$ be defined using the following relation: for $f \in V(\mathcal{M})$, fixed, but arbitrary,

$$B\left[[f]_{L_2(S,\mathcal{S},P)} \right] = [f]_{L_2(S,\mathcal{S},Q)} .$$

B is an operator which is linear and bounded, with

$$\mathcal{D}[B] = \overline{V(\mathcal{M})}^P, \text{ and } \mathcal{R}[B] \subseteq \overline{V(\mathcal{M})}^Q, \text{ densely.}$$

Proof The domination inequality of (Definition) 5.4.1 translates into

$$\left\| B\left[[f]_{L_2(S,\mathcal{S},P)} \right] \right\|_{L_2(S,\mathcal{S},Q)} \leq \kappa\, [V(\mathcal{M}), P, Q] \left\| [f]_{L_2(S,\mathcal{S},P)} \right\|_{L_2(S,\mathcal{S},P)} .$$

B is thus linear and bounded on a dense set, and, on that set, has values in $\overline{V(\mathcal{M})}^Q$.

Furthermore, when $g \in \overline{V(\mathcal{M})}^Q$ is orthogonal to $\mathcal{R}[B]$, for $f \in V(\mathcal{M})$, fixed, but arbitrary,

$$0 = \langle g, B\left[[f]_{L_2(S,\mathcal{S},P)} \right] \rangle_{L_2(S,\mathcal{S},Q)} = \langle g, [f]_{L_2(S,\mathcal{S},Q)} \rangle_{L_2(S,\mathcal{S},Q)},$$

and, as $V(\mathcal{M})$ is dense in $\overline{V(\mathcal{M})}^Q$, $g = 0_{L_2(S,\mathcal{S},Q)}$. □

Corollary 5.4.7 *Let $f_1 = 1_S$, and suppose that $f_1 \in V(\mathcal{M})$. Then*

$$B\left[[f_1]_{L_2(S,\mathcal{S},P)} \right] = [f_1]_{L_2(S,\mathcal{S},Q)} ,$$

and one may define $\Lambda : \overline{V(\mathcal{M})}^{P} \longrightarrow \mathbb{R}$ *using the following assignment: for* $f \in \overline{V(\mathcal{M})}^{P}$, *fixed, but arbitrary,*

$$\Lambda(f) = \langle f, B^{\star}\left[[f_1]_{L_2(S,\mathcal{S},Q)}\right]\rangle_{L_2(S,\mathcal{S},P)}.$$

The continuous linear functional Λ *has then the following properties:*

1. *Given* $f_Q \in B^{\star}\left[[f_1]_{L_2(S,\mathcal{S},Q)}\right]$, $\Lambda(f) = \int_S \dot{f}(s) f_Q(s) P(ds)$.
2. *For* $f \in V(\mathcal{M})$, *fixed, but arbitrary,*

$$\Lambda\left([f]_{L_2(S,\mathcal{S},P)}\right) = \int_S f(s) Q(ds),$$

and, in particular,

(i) $\Lambda\left([f_1]_{L_2(S,\mathcal{S},P)}\right) = 1$,
(ii) *when* f *is, with respect to* Q, *almost surely positive,*

$$\Lambda\left([f]_{L_2(S,\mathcal{S},P)}\right) \geq 0.$$

Proof Item 1 is a rewriting of the definition of Λ. Using the definition of B and Λ, one has that

$$\begin{aligned}
\Lambda\left([f]_{L_2(S,\mathcal{S},P)}\right) &= \\
&= \left\langle [f]_{L_2(S,\mathcal{S},P)}, B^{\star}\left[[f_1]_{L_2(S,\mathcal{S},Q)}\right]\right\rangle_{L_2(S,\mathcal{S},P)} \\
&= \left\langle B\left[[f]_{L_2(S,\mathcal{S},P)}\right], [f_1]_{L_2(S,\mathcal{S},Q)}\right\rangle_{L_2(S,\mathcal{S},Q)} \\
&= \left\langle [f]_{L_2(S,\mathcal{S},Q)}, [f_1]_{L_2(S,\mathcal{S},Q)}\right\rangle_{L_2(S,\mathcal{S},Q)} \\
&= \int_S f(s) f_1(s) Q(ds) \\
&= \int_S f(s) Q(ds).
\end{aligned}$$

□

The following proposition indicates that domination on a manifold implies domination on the respective closures.

Proposition 5.4.8 *Let* P *and* Q *be probabilities on* $\mathcal{S}$, *and* $V(\mathcal{M})$, *a fixed, but arbitrary manifold of* $\mathcal{M}$. *Suppose that, for* $V(\mathcal{M})$, P *and* Q *have property* Π_1, *and that* P *dominates* Q *on* $V(\mathcal{M})$. *Then, for every Cauchy sequence*

$$\left\{[f_n]_{L_2(S,\mathcal{S},P)}, f_n \in V(\mathcal{M}), n \in \mathbb{N}\right\} \subseteq L_2(S,\mathcal{S},P),$$

there exists $f \in \mathcal{M}(\mathcal{S})$ *such that*

1. $[f]_{L_2(S,\mathcal{S},P)} \in \overline{V(\mathcal{M})}^P$, *and, in* $L_2(S,\mathcal{S},P)$,

$$\lim_n [f_n]_{L_2(S,\mathcal{S},P)} = [f]_{L_2(S,\mathcal{S},P)},$$

2. $[f]_{L_2(S,\mathcal{S},Q)} \in \overline{V(\mathcal{M})}^Q$, *and, in* $L_2(S,\mathcal{S},Q)$,

$$\lim_n [f_n]_{L_2(S,\mathcal{S},Q)} = [f]_{L_2(S,\mathcal{S},Q)},$$

3. $B\left[[f]_{L_2(S,\mathcal{S},P)}\right] = [f]_{L_2(S,\mathcal{S},Q)}$, *and*

$$\left\|[f]_{L_2(S,\mathcal{S},Q)}\right\|_{L_2(S,\mathcal{S},Q)} \leq \kappa\left[V(\mathcal{M}),P,Q\right]\left\|[f]_{L_2(S,\mathcal{S},P)}\right\|_{L_2(S,\mathcal{S},P)}.$$

Proof In $L_2(S,\mathcal{S},P)$, let f_P be the limit of the sequence

$$\left\{[f_n]_{L_2(S,\mathcal{S},P)},\ f_n \in V(\mathcal{M}),\ n \in \mathbb{N}\right\}.$$

As, by assumption,

$$\begin{aligned}\left\|[f_m]_{L_2(S,\mathcal{S},Q)} - [f_n]_{L_2(S,\mathcal{S},Q)}\right\|_{L_2(S,\mathcal{S},Q)} &\leq \\ &\leq \kappa\left[V(\mathcal{M}),P,Q\right]\left\|[f_m]_{L_2(S,\mathcal{S},P)} - [f_n]_{L_2(S,\mathcal{S},P)}\right\|_{L_2(S,\mathcal{S},P)},\end{aligned}$$

the sequence

$$\left\{[f_n]_{L_2(S,\mathcal{S},Q)},\ f_n \in V(\mathcal{M}),\ n \in \mathbb{N}\right\}$$

is a Cauchy sequence in $L_2(S,\mathcal{S},Q)$. Let its limit be denoted $f_{P,Q}$.

Now, for fixed, but arbitrary $\epsilon > 0$, S_ϵ being as previously defined [(Definition) 5.3.2],

$$\lim_n \int_{S_\epsilon}\left[f_n(s) - \dot{f}_P(s)\right]^2 P(ds) \leq \lim_n \left\|[f_n]_{L_2(S,\mathcal{S},P)} - f_P\right\|^2_{L_2(S,\mathcal{S},P)} = 0.$$

But

$$\begin{aligned}\int_{S_0 \cup S_\epsilon}\left[f_n(s) - \dot{f}_P(s)\right]^2 Q(ds) &= \int_{S_\epsilon}\left[f_n(s) - \dot{f}_P(s)\right]^2 Q_{aP}(ds) \\ &= \int_{S_\epsilon}\left[f_n(s) - \dot{f}_P(s)\right]^2 \frac{dQ_{aP}}{dP}(s)\,P(ds) \\ &\leq \frac{1}{\epsilon}\int_{S_\epsilon}\left[f_n(s) - \dot{f}_P(s)\right]^2 P(ds).\end{aligned}$$

Consequently,

$$\lim_n \int_{S_0 \cup S_\epsilon} \left[f_n(s) - \dot{f}_P(s)\right]^2 Q(ds) = 0,$$

and, with respect to Q, on $S_0 \cup S_\epsilon$, almost surely, $\dot{f}_{P,Q}(s) = \dot{f}_P(s)$. Since $\epsilon > 0$ is arbitrary, the same is true on

$$\left\{s \in S : 0 \leq \frac{dQ_{aP}}{dP}(s) < \infty\right\} = N_{sP}^c.$$

Define thus

$$\dot{f}(s) = \begin{cases} \dot{f}_P(s) & \text{for } s \in N_{sP}^c \\ \dot{f}_{P,Q}(s) & \text{for } s \in N_{sP} \end{cases}.$$

Then, for $\delta > 0$, fixed, but arbitrary,

$$\begin{aligned}
P\left(s \in S : \left|\dot{f}(s) - \dot{f}_P(s)\right| > \delta\right) = \\
&= P\left(N_{sP} \cap \left\{s \in S : \left|\dot{f}(s) - \dot{f}_P(s)\right| > \delta\right\}\right) \\
&+ P\left(N_{sP}^c \cap \left\{s \in S : \left|\dot{f}(s) - \dot{f}_P(s)\right| > \delta\right\}\right) \\
&= P\left(N_{sP} \cap \left\{s \in S : \left|\dot{f}_{P,Q}(s) - \dot{f}_P(s)\right| > \delta\right\}\right) \\
&+ P\left(N_{sP}^c \cap \left\{s \in S : \left|\dot{f}_P(s) - \dot{f}_P(s)\right| > \delta\right\}\right) \\
&\leq P(N_{sP}) + P(\emptyset) = 0,
\end{aligned}$$

and

$$\begin{aligned}
Q\left(s \in S : \left|\dot{f}(s) - \dot{f}_{P,Q}(s)\right| > \delta\right) = \\
&= Q\left(N_{sP} \cap \left\{s \in S : \left|\dot{f}(s) - \dot{f}_{P,Q}(s)\right| > \delta\right\}\right) \\
&+ Q\left(N_{sP}^c \cap \left\{s \in S : \left|\dot{f}(s) - \dot{f}_{P,Q}(s)\right| > \delta\right\}\right) \\
&= Q\left(N_{sP} \cap \left\{s \in S : \left|\dot{f}_{P,Q}(s) - \dot{f}_{P,Q}(s)\right| > \delta\right\}\right) \\
&+ Q\left(N_{sP}^c \cap \left\{s \in S : \left|\dot{f}_P(s) - \dot{f}_{P,Q}(s)\right| > \delta\right\}\right) \\
&= Q(\emptyset) + 0 = 0.
\end{aligned}$$

□

Corollary 5.4.9 *Let P and Q be probabilities on $\mathcal{S}$, and $V(\mathcal{M})$, a fixed, but arbitrary manifold of $\mathcal{M}$. Suppose that, for $V(\mathcal{M})$, P and Q have property Π_1,* *and that P dominates Q on V* $(\mathcal{M})$. *Then, given*

$$f \in \overline{V(\mathcal{M})}^{P},$$

there is $\dot{f} \in f$ *such that*

$$B\left[\left[\dot{f}\right]_{L_2(S,\mathcal{S},P)}\right] = \left[\dot{f}\right]_{L_2(S,\mathcal{S},Q)},$$

and thus Λ, *defined in (Corollary)* 5.4.7, *has the representation*

$$\Lambda(f) = \int_S \dot{f}(s)\, Q(ds),\ f \in \overline{V(\mathcal{M})}^{P}.$$

5.4.2 Domination on the Linear Manifold of Evaluations

The manifold of evaluations is $V(\mathcal{M}) = \mathcal{L}(\mathcal{E})$.

In this section, P and Q are probabilities on $\mathcal{S}$ such that, on $\mathcal{L}(\mathcal{E})$, P dominates Q. One then knows [(Proposition) 5.4.3] that there is a constant $\kappa \geq 0$ for which

$$\mathcal{H}_{\mathcal{E},Q} \ll \kappa \mathcal{H}_{\mathcal{E},P}.$$

Let $\mathcal{H}^{\kappa}_{\mathcal{E},P}$ denote $\kappa \mathcal{H}_{\mathcal{E},P}$. One then has at disposal [(Proposition) 3.1.5] the contraction

$$J_{P,Q} : H\left(\mathcal{H}^{\kappa}_{\mathcal{E},P}, T\right) \longrightarrow H\left(\mathcal{H}_{\mathcal{E},Q}, T\right)$$

defined using the following assignment: for $t \in T$, fixed, but arbitrary,

$$J_{P,Q}\left[\mathcal{H}^{\kappa}_{\mathcal{E},P}(\cdot,t)\right] = \mathcal{H}_{\mathcal{E},Q}(\cdot,t).$$

Let also $M_\kappa : H(\mathcal{H}_{\mathcal{E},P}, T) \longrightarrow H\left(\mathcal{H}^{\kappa}_{\mathcal{E},P}, T\right)$ be defined using the following assignment:

$$M_{\kappa^{1/2}}[h] = \kappa^{1/2} h.$$

Proposition 5.4.10 *Let P and Q be probabilities on* $\mathcal{S}$ *such that P dominates Q on* $\mathcal{L}(\mathcal{E})$. *Let*

$$U_P : \overline{L(\mathcal{E})}^{P} \longrightarrow H(\mathcal{H}_{\mathcal{E},P}, T)$$

be the isometry defined using the following assignment: for $t \in T$, fixed, but arbitrary,

$$U_P\left[[\mathcal{E}_t]_{L_2(S,\mathcal{S},P)}\right] = \mathcal{H}_{\mathcal{E},P}(\cdot,t)\,.$$

U_Q is defined analogously. Then, for $f \in \mathcal{L}(\mathcal{E})$,

$$[f]_{L_2(S,\mathcal{S},Q)} = U_Q^\star J_{P,Q} M_{\kappa^{1/2}} U_P\left[[f]_{L_2(S,\mathcal{S},P)}\right],$$

and the operator $B_L : \overline{\mathcal{L}(\mathcal{E})}^P \longrightarrow \overline{\mathcal{L}(\mathcal{E})}^Q$, defined on $\mathcal{L}(\mathcal{E})$, using the following equality:

$$B_L = U_Q^\star J_{P,Q} M_{\kappa^{1/2}} U_P$$

has a bounded linear extension to $\overline{\mathcal{L}(\mathcal{E})}^P$.

Proof Successively,

$$\begin{aligned}
U_P\left[[\mathcal{E}_t]_{L_2(S,\mathcal{S},P)}\right] &= \mathcal{H}_{\mathcal{E},P}(\cdot,t)\,,\\
M_{\kappa^{1/2}} U_P\left[[\mathcal{E}_t]_{L_2(S,\mathcal{S},P)}\right] &= \kappa^{1/2}\mathcal{H}_{\mathcal{E},P}(\cdot,t) = \mathcal{H}^{\kappa}_{\mathcal{E},P}(\cdot,t)\,,\\
J_{P,Q} M_{\kappa^{1/2}} U_P\left[[\mathcal{E}_t]_{L_2(S,\mathcal{S},P)}\right] &= \mathcal{H}_{\mathcal{E},Q}(\cdot,t)\,,\\
U_Q^\star J_{P,Q} M_{\kappa^{1/2}} U_P\left[[\mathcal{E}_t]_{L_2(S,\mathcal{S},P)}\right] &= [\mathcal{E}_t]_{L_2(S,\mathcal{S},Q)}\,.
\end{aligned}$$

B is a product of operators which are linear and bounded, so that the extension proceeds by linearity and denseness. □

Proposition 5.4.11 *Let P and Q be probabilities on $\mathcal{S}$ such that*

(a) *P dominates Q on $\mathcal{L}(\mathcal{E})$,*
(b) *P and Q have property Π_2 for $\mathcal{L}(\mathcal{E})$.*

Then

1. *when one does not have, for some $0 < \gamma < \Gamma < \infty$,*

$$\gamma\,\mathcal{H}_{\mathcal{E},P} \ll \mathcal{H}_{\mathcal{E},Q} \ll \Gamma\,\mathcal{H}_{\mathcal{E},P},$$

P and Q are orthogonal;

2. *when one does have, for some $0 < \gamma < \Gamma < \infty$,*

$$\gamma\,\mathcal{H}_{\mathcal{E},P} \ll \mathcal{H}_{\mathcal{E},Q} \ll \Gamma\,\mathcal{H}_{\mathcal{E},P},$$

P and Q need neither be orthogonal, nor one be absolutely continuous with respect to the other.

Proof The inequalities of item 1 are equivalent to domination, on $\mathcal{L}(\mathcal{E})$, of P by Q, and Q by P [(Proposition) 5.4.3]. Thus if one of the inequalities of item 1 does not hold, either P does not dominate Q, or Q does not dominate P, which are sufficient conditions for P and Q to be orthogonal [(Proposition) 5.4.5].

Return to (Example) 5.4.2. Let P be a probability on the Borel sets of $\mathbb{R}^2$ that gives mean zero to the evaluation maps. Then, for example,

$$\mathcal{H}_{\mathcal{E},P}(0,0) = \int_{\mathbb{R}^2} \mathcal{E}_0(x,y)\,\mathcal{E}_0(x,y)\,P(d(x,y)) = E_P\left[\mathcal{E}_0^2\right],$$

$$\mathcal{H}_{\mathcal{E},P}(0,1) = \int_{\mathbb{R}^2} \mathcal{E}_0(x,y)\,\mathcal{E}_1(x,y)\,P(d(x,y)) = E_P\left[\mathcal{E}_0\mathcal{E}_1\right],$$

$$\mathcal{H}_{\mathcal{E},P}(1,1) = \int_{\mathbb{R}^2} \mathcal{E}_1(x,y)\,\mathcal{E}_1(x,y)\,P(d(x,y)) = E_P\left[\mathcal{E}_1^2\right]$$

Let now P have mass one-fifth at the points

$$(0,0),\ (1,0),\ (-1,0),\ (0,1),\ (0,-1),$$

and Q have mass one-fourth at the points

$$(1,0),\ (-1,0),\ (0,2),\ (0,-2).$$

Then P and Q are neither orthogonal, nor one is absolutely continuous with respect to the other. $\mathcal{E}_0$ and $\mathcal{E}_1$ have zero mean and correlation for P and Q. Finally they have finite variances, and their kernels dominate each other. □

Remark 5.4.12 For P and Q not to be orthogonal, the associated RKHS's must contain the same functions.

5.4.3 *Domination on the Quadratic Manifold of Evaluations*

The quadratic manifold of evaluations is $V(\mathcal{M}) = \mathcal{Q}(\mathcal{E})$.

In this part one shall consistently assume that:

1. P and Q are two probabilities on $\mathcal{S}$ which have properties Π_1 and Π_2 for $\mathcal{Q}(\mathcal{E})$ (Π_1 for $\mathcal{Q}(\mathcal{E})$ may be a consequence of Π_3: (Proposition) 5.1.31);
2. P dominates Q on $\mathcal{Q}(\mathcal{E})$.

Assumption 1 has the particular consequence that, for example,

$$[\mathcal{E}_{t_1}\mathcal{E}_{t_2}]_{L_2(S,\mathcal{S},P)} \in L_2(S,\mathcal{S},P),$$

and Assumption 2, that there is some $\kappa \geq 0$ for which

$$\mathcal{H}_{\mathcal{E},Q} \ll \kappa \mathcal{H}_{\mathcal{E},P}.$$

B_Q shall denote the operator of (Proposition) 5.4.6, with $V(\mathcal{M}) = \mathcal{Q}(\mathcal{E})$, and f_Q a function in the equivalence class of

$$B_Q^{\star}\left[[f_1]_{L_2(S,\mathcal{S},Q)}\right]$$

(the function $f_1 = 1_S$ belongs to $\mathcal{Q}(\mathcal{E})$).

Proposition 5.4.13 *$\mathcal{H}_{\mathcal{E},Q}$ has the representation*

$$\mathcal{H}_{\mathcal{E},Q}(t_1,t_2) = \int_S \mathcal{E}_{t_1}(s)\,\mathcal{E}_{t_2}(s) f_Q(s)\,P(ds).$$

Proof Since, because of Assumption 1, $[\mathcal{E}_{t_1}\mathcal{E}_{t_2}]_{L_2(S,\mathcal{S},P)} \in \overline{\mathcal{Q}(\mathcal{E})}^P$, then because of (Corollary) 5.4.7,

$$\Lambda\left([\mathcal{E}_{t_1}\mathcal{E}_{t_2}]_{L_2(S,\mathcal{S},P)}\right) = \int_S \mathcal{E}_{t_1}(s)\,\mathcal{E}_{t_2}(s) f_Q(s)\,P(ds),$$

and

$$\Lambda\left([\mathcal{E}_{t_1}\mathcal{E}_{t_2}]_{L_2(S,\mathcal{S},P)}\right) = \int_S \mathcal{E}_{t_1}(s)\,\mathcal{E}_{t_2}(s)\,Q(ds) = \mathcal{H}_{\mathcal{E},Q}(t_1,t_2).$$

□

Fact 5.4.14 *In what follows* [(Proposition) 5.4.15], *one shall need the following considerations. Let $M_{\kappa^{1/2}} : H(\mathcal{H}_{\mathcal{E},P},T) \longrightarrow H\left(\mathcal{H}^{\kappa}_{\mathcal{E},P},T\right)$ be defined using the following assignment:*

$$M_{\kappa^{1/2}}[h] = \kappa^{1/2} h.$$

$M_{\kappa^{1/2}}$ is a well-defined unitary operator [(Example) 1.3.12], *so that*

$$M^{\star}_{\kappa^{1/2}} M_{\kappa^{1/2}} = I_{H(\mathcal{H}_{\mathcal{E},P},T)},$$

$$M^{\star}_{\kappa^{1/2}}[h] = \kappa^{-1/2} h.$$

Since $H\left(\mathcal{H}_{\mathcal{E},Q},T\right) \ll \kappa H(\mathcal{H}_{\mathcal{E},P},T)$, one has available [(Proposition) 3.1.5] *the contraction*

$$J_{P,Q} : H\left(\mathcal{H}^{\kappa}_{\mathcal{E},P},T\right) \longrightarrow H\left(\mathcal{H}_{\mathcal{E},Q},T\right)$$

defined using

$$J_{P,Q}\left[\mathcal{H}^{\kappa}_{\mathcal{E},P}(\cdot,t)\right] = \mathcal{H}_{\mathcal{E},Q}(\cdot,t).$$

Thus, since $J^{\star}_{P,Q}$ is inclusion,

$$M^{\star}_{\kappa^{1/2}} J^{\star}_{P,Q} J_{P,Q} M_{\kappa^{1/2}}\left[\mathcal{H}_{\mathcal{E},P}(\cdot,t)\right] = \kappa^{-1}\mathcal{H}_{\mathcal{E},Q}(\cdot,t) \in H\left(\mathcal{H}_{\mathcal{E},P},T\right).$$

Proposition 5.4.15 *Suppose, in addition to the "standard" assumptions of this section, that P has property Π_3, and that*

$$\int_S f_Q^4(s)\,P(ds) < \infty,$$

so that multiplication by a constant multiple of the function f_Q on $\overline{\mathcal{L}(\mathcal{E})}^P$ makes sense [(Proposition) 5.1.32], *and yields an element of $L_2(S,\mathcal{S},P)$. Consider then the following objects:*

(a) $f_Q^0 = f_Q - f_1,\ f_1 = 1_S$;

(b) $M_{\kappa,Q} : \overline{\mathcal{L}(\mathcal{E})}^P \longrightarrow L_2(S,\mathcal{S},P)$, *the operator defined using the following relation:*

$$M_{\kappa,Q}[f] = \kappa^{-1}\left[\dot{f} f_Q\right]_{L_2(S,\mathcal{S},P)};$$

(c) $B_{\kappa,Q} = P_{\overline{\mathcal{L}(\mathcal{E})}^P} M_{\kappa,Q}$;

(d) $M^0_{\kappa,Q} : \overline{\mathcal{L}(\mathcal{E})}^P \longrightarrow L_2(S,\mathcal{S},P)$, *the operator defined using the following relation:*

$$M^0_{\kappa,Q}[f] = \kappa^{-1}\left[\dot{f} f^0_Q\right]_{L_2(S,\mathcal{S},P)};$$

(e) $B^0_{\kappa,Q} = P_{\overline{\mathcal{L}(\mathcal{E})}^P} M^0_{\kappa,Q}$.

Then

1. $\left[f^0_Q\right]_{L_2(S,\mathcal{S},P)} \in \overline{\mathcal{Q}(\mathcal{E})}^P$,
2. $\int_S f^0_Q(s)\,P(ds) = 0$,
3. *as operators of $H\left(\mathcal{H}_{\mathcal{E},P},T\right)$,*

$$M^{\star}_{\kappa^{1/2}} J^{\star}_{P,Q} J_{P,Q} M_{\kappa^{1/2}} = U_P B_{\kappa,Q} U^{\star}_P = \kappa^{-1/2} I_{H\left(\mathcal{H}_{\mathcal{E},P},T\right)} + U_P B^0_{\kappa,Q} U^{\star}_P.$$

Proof As already seen [(Proposition) 5.1.32], $M_{\kappa,Q}$ and, because [275, p. 378]

$$(a+b)^p \le 2^p(|a|^p + |b|^p), \text{ so that } \left\{f^0_Q\right\}^4 \le 2^4\left\{f_Q^4 + 1\right\},$$

$M^0_{\kappa,Q}$ also, are well defined. The function f_1 belongs to $\mathcal{Q}(\mathcal{E})$, so that its equivalence class, with respect to P, belongs to the closure, with respect to P, of $\mathcal{Q}(\mathcal{E})$. Consequently, as a sum of elements in that closure, the equivalence class (with respect to P) of f^0_Q belongs to

$$\overline{\mathcal{Q}(\mathcal{E})}^P,$$

and [(Corollary) 5.4.7]

$$\int_S f^0_Q(s)\,P(ds) = 0.$$

Let $h \in V[\mathcal{H}_{\mathcal{E},P}]$ be fixed, but arbitrary. Then, as

$$\mathcal{H}_{\mathcal{E},P}(\cdot,t) = U_P\left[[\mathcal{E}_t]_{L_2(S,\mathcal{S},P)}\right],$$

one has that

$$h = \sum_{i=1}^n \alpha_i^h \mathcal{H}_{\mathcal{E},P}(\cdot,t_i^h) = U_P\left[\left[\sum_{i=1}^n \alpha_i^h \mathcal{E}_{t_i^h}\right]_{L_2(S,\mathcal{S},P)}\right],$$

and that

$$\begin{aligned}
h(t) &= \left\langle \sum_{i=1}^n \alpha_i^h \mathcal{H}_{\mathcal{E},P}(\cdot,t_i^h), \mathcal{H}_{\mathcal{E},P}(\cdot,t)\right\rangle_{H(\mathcal{H}_{\mathcal{E},P},T)} \\
&= \left\langle U_P\left[\left[\sum_{i=1}^n \alpha_i^h \mathcal{E}_{t_i^h}\right]_{L_2(S,\mathcal{S},P)}\right], U_P\left[[\mathcal{E}_t]_{L_2(S,\mathcal{S},P)}\right]\right\rangle_{H(\mathcal{H}_{\mathcal{E},P},T)} \\
&= \left\langle \left[\sum_{i=1}^n \alpha_i^h \mathcal{E}_{t_i^h}\right]_{L_2(S,\mathcal{S},P)}, [\mathcal{E}_t]_{L_2(S,\mathcal{S},P)}\right\rangle_{L_2(S,\mathcal{S},P)} \\
&= \int_S \left\{\sum_{i=1}^n \alpha_i^h \mathcal{E}_{t_i^h}(s)\right\} \mathcal{E}_t(s)\,P(ds).
\end{aligned}$$

As already seen [(Fact) 5.4.14],

$$M^\star_{\kappa^{1/2}} J^\star_{P,Q} J_{P,Q} M_{\kappa^{1/2}}\left[\mathcal{H}_{\mathcal{E},P}(\cdot,t)\right] = \kappa^{-1}\mathcal{H}_{\mathcal{E},Q}(\cdot,t),$$

so that, by the same calculation as that which has just been performed, to obtain the integral representation of h with respect to P,

$$\begin{aligned} M^{\star}_{\kappa^{1/2}} J^{\star}_{P,Q} J_{P,Q} M_{\kappa^{1/2}} [h] (t) &= \kappa^{-1} \sum_{i=1}^{n} \alpha_i^h \mathcal{H}_{\mathcal{E},Q} \left(t, t_i^h\right) \\ &= \kappa^{-1} \int_S \left\{ \sum_{i=1}^{n} \alpha_i^h \mathcal{E}_{t_i^h} \right\} (s) \, \mathcal{E}_t (s) \, Q (ds) . \end{aligned}$$

Using (Proposition) 5.4.13 on $\left\{\sum_{i=1}^{n} \alpha_i^h \mathcal{E}_{t_i^h}\right\} \mathcal{E}_t \in \mathcal{Q}(\mathcal{E})$, one has that

$$\begin{aligned} &\kappa^{-1} \int_S \left\{ \sum_{i=1}^{n} \alpha_i^h \mathcal{E}_{t_i^h} \right\} (s) \, \mathcal{E}_t (s) \, Q (ds) = \\ &\quad = \int_S \kappa^{-1} f_Q (s) \left\{ \sum_{i=1}^{n} \alpha_i^h \mathcal{E}_{t_i^h} \right\} (s) \, \mathcal{E}_t (s) \, P (ds) . \end{aligned}$$

But $\kappa^{-1} f_Q (s) \left\{\sum_{i=1}^{n} \alpha_i^h \mathcal{E}_{t_i^h}\right\}$ is in the equivalence class of $M_{\kappa,Q} \left[U_P^{\star} [h]\right]$. Consequently

$$M^{\star}_{\kappa^{1/2}} J^{\star}_{P,Q} J_{P,Q} M_{\kappa^{1/2}} [h] (t) = \left\langle M_{\kappa,Q} U_P^{\star} [h] , [\mathcal{E}_t]_{L_2(S,\mathcal{S},P)} \right\rangle_{L_2(S,\mathcal{S},P)} .$$

Now

$$\begin{aligned} &\left\langle M_{\kappa,Q} U_P^{\star} [h] , [\mathcal{E}_t]_{L_2(S,\mathcal{S},P)} \right\rangle_{L_2(S,\mathcal{S},P)} = \\ &\quad = \left\langle P_{\overline{\mathcal{L}(\mathcal{E})}^P} M_{\kappa,Q} U_P^{\star} [h] , [\mathcal{E}_t]_{L_2(S,\mathcal{S},P)} \right\rangle_{L_2(S,\mathcal{S},P)} \\ &\quad = \left\langle B_{\kappa,Q} U_P^{\star} [h] , [\mathcal{E}_t]_{L_2(S,\mathcal{S},P)} \right\rangle_{L_2(S,\mathcal{S},P)} \\ &\quad = \left\langle U_P B_{\kappa,Q} U_P^{\star} [h] , \mathcal{H}_{\mathcal{E},P} (\cdot, t) \right\rangle_{H(\mathcal{H}_{\mathcal{E},P},T)} \\ &\quad = \left\{ U_P B_{\kappa,Q} U_P^{\star} [h] \right\} (t) . \end{aligned}$$

It should then be clear that

$$\begin{aligned} &M^{\star}_{\kappa^{1/2}} J^{\star}_{P,Q} J_{P,Q} M_{\kappa^{1/2}} [h] (t) - \kappa^{-1} h (t) = \\ &\quad = \int_S \kappa^{-1} \left(f_Q (s) - 1\right) \left\{ \sum_{i=1}^{n} \alpha_i^h \mathcal{E}_{t_i^h} \right\} (s) \, \mathcal{E}_t (s) \, P (ds) \\ &\quad = U_P P_{\overline{\mathcal{L}[\mathcal{E}]}^P} M^0_{\kappa,Q} U_P^{\star} [h] (t) \\ &\quad = \left\{ U_P B^0_{\kappa,Q} U_P^{\star} [h] \right\} (t) . \end{aligned}$$

Since $V[\mathcal{H}_{\mathcal{E},P}]$ is dense in $H(\mathcal{H}_{\mathcal{E},P},T)$,

$$M^{\star}_{\kappa^{1/2}}J^{\star}_{P,Q}J_{P,Q}M_{\kappa^{1/2}} = U_P B_{\kappa,Q} U^{\star}_P = \kappa^{-1} I_{H(\mathcal{H}_{\mathcal{E},P},T)} + U_P B^0_{\kappa,Q} U^{\star}_P.$$

□

Corollary 5.4.16 *One has that* $B_{\kappa,Q} = B^{\star}_L B_L$, *where* B_L *is the operator of* (Proposition) 5.4.10, *so that* $B_{\kappa,Q}$ *is positive.*

Corollary 5.4.17 *The function* f_Q *is, with respect to* P, *almost surely positive.*

Proof One has [(Corollary) 5.4.16] that

$$\begin{aligned} 0 &\le \left\langle B_{\kappa,Q}\left[\sum_{i=1}^{n}\alpha_i\mathcal{E}_{t_i}\right]_{L_2(S,\mathcal{S},P)}, \left[\sum_{i=1}^{n}\alpha_i\mathcal{E}_{t_i}\right]_{L_2(S,\mathcal{S},P)}\right\rangle_{L_2(S,\mathcal{S},P)} \\ &= \left\langle P_{\overline{\mathcal{L}[\mathcal{E}]}^P} M_{\kappa,Q}\left[\sum_{i=1}^{n}\alpha_i\mathcal{E}_{t_i}\right]_{L_2(S,\mathcal{S},P)}, \left[\sum_{i=1}^{n}\alpha_i\mathcal{E}_{t_i}\right]_{L_2(S,\mathcal{S},P)}\right\rangle_{L_2(S,\mathcal{S},P)} \\ &= \left\langle M_{\kappa,Q}\left[\sum_{i=1}^{n}\alpha_i\mathcal{E}_{t_i}\right]_{L_2(S,\mathcal{S},P)}, \left[\sum_{i=1}^{n}\alpha_i\mathcal{E}_{t_i}\right]_{L_2(S,\mathcal{S},P)}\right\rangle_{L_2(S,\mathcal{S},P)} \\ &= \int_S f_Q(s)\left\{\sum_{i=1}^{n}\alpha_i\mathcal{E}_{t_i}\right\}^2 P(ds). \end{aligned}$$

Thus, for fixed, but arbitrary $f \in \overline{\mathcal{L}[\mathcal{E}]}^P$, $\int_S f_Q(s)\dot{f}^2(s)P(ds) \ge 0$.

To have that f_Q is, with respect to P, almost surely positive, one must have that, whatever $S_0 \in \mathcal{S}$,

$$\int_{S_0} f_Q(s)P(ds) \ge 0.$$

So it is not immediately obvious that $B_{\kappa,Q}$ positive is sufficient.

But [201, p. 3], the closure, with respect to P, of $\mathcal{Q}(\mathcal{E})$ is of the form $L_2(S,\mathcal{S}_0,P)$, where $\mathcal{S}_0$ is a complete σ-algebra contained in $\mathcal{S}$. Since $\mathcal{E}_t$ belongs to $\mathcal{Q}(\mathcal{E})$, $\mathcal{S}_0 = \mathcal{S}$. Finally, because of Proposition 531, the elements $\sum_{i=1}^n (\alpha_i\mathcal{E}_{t_i})^2$ are dense in the set of indicator functions, and thus $\overline{\mathcal{Q}(\mathcal{E})}^P$. □

5.5 Discrimination of Translates

Translates represent deterministic signals. Discrimination for translates corresponds thus to detection of such signals.

5.5.1 Preliminaries

In this section, S shall be one of the spaces of (Examples) 5.1.3 to 5.1.6. $\mathcal{E}$ shall be the family of evaluation maps in cases (Examples) 5.1.3 to 5.1.5, and that of linear functionals in case (Example) 5.1.6. In the case of (Example) 5.1.6, one shall write, when useful,

$$\mathcal{E}_t(h) = h(t) = \langle h, t\rangle_H .$$

Suppose P is a probability on $\mathcal{S}$ (with property Π_1 for $\mathcal{L}(\mathcal{E})$), which has a mean equal to 0, that is, such that, for $t \in T$, fixed, but arbitrary,

$$m_P(t) = \int_S \mathcal{E}_t(s) P(ds) = 0.$$

Given a function $m : T \longrightarrow \mathbb{R}$, let, for $s \in S$, fixed, but arbitrary,

$$T_m(s) = s + m,$$

and, given an element $m \in H$, let, for $h \in H$, fixed, but arbitrary,

$$T_m(h) = h + m.$$

Since, for $t \in T$, fixed, but arbitrary, $\mathcal{E}_t \circ T_m$ is adapted to $\mathcal{S}$, T_m is adapted to $\mathcal{S}$. The following probability:

$$P_m = P \circ T_m^{-1}$$

is thus well defined, and

$$\begin{aligned}\int_S \mathcal{E}_t^2(s) P_m(ds) &= \int_S [s(t) + m(t)]^2 P(ds)\\ &\leq 2\left\{m^2(t) + \int_S \mathcal{E}_t^2(s) P(ds)\right\}.\end{aligned}$$

A similar calculation yields that, for t, t_1, and $t_2 \in T$, fixed, but arbitrary,

$$\int_S \mathcal{E}_t(s) P_m(ds) = m(t),$$

and that

$$\int_S \mathcal{E}_{t_1}(s)\mathcal{E}_{t_2}(s) P_m(ds) = m(t_1)m(t_2) + \int_S \mathcal{E}_{t_1}(s)\mathcal{E}_{t_2}(s) P(ds),$$

so that

$$\mathcal{H}_{\mathcal{E},P_m} = m \otimes m + \mathcal{H}_{\mathcal{E},P}.$$

5.5.2 *The Case of a Signal Lying Outside of the Reproducing Kernel Hilbert Space of Noise*

In such a case, the following facts obtain [(Examples) 1.1.21, 1.3.14]:

1. $H(\mathcal{H}_{\mathcal{E},P_m}, T) = H(m \otimes m, T) \oplus H(\mathcal{H}_{\mathcal{E},P}, T)$ (direct sum),
2. $m \in H(m \otimes m, T) \subseteq H(\mathcal{H}_{\mathcal{E},P_m}, T)$,
3. $\|m\|_{H(\mathcal{H}_{\mathcal{E},P_m},T)} = \|m\|_{H(m\otimes m,T)} = 1$.

Let $U_m : \overline{\mathcal{L}(\mathcal{E})}^{P_m} \longrightarrow H(\mathcal{H}_{\mathcal{E},P_m}, T)$ be the usual unitary map: for $t \in T$, fixed, but arbitrary,

$$U_m\left[[\mathcal{E}_t]_{L_2(S,\mathcal{S},P_m)}\right] = \mathcal{H}_{\mathcal{E},P_m}(\cdot, t).$$

Lemma 5.5.1 *Let* $f_m = U_m^\star[m]$. *Then, when* $f = \sum_{i=1}^n \alpha_i^f \mathcal{E}_{t_i^f}$,

$$\begin{aligned}\left\|[f]_{L_2(S,\mathcal{S},P_m)}\right\|^2_{L_2(S,\mathcal{S},P_m)} &= \\ &= \left\langle [f]_{L_2(S,\mathcal{S},P_m)}, f_m\right\rangle^2_{L_2(S,\mathcal{S},P_m)} + \left\|[f]_{L_2(S,\mathcal{S},P)}\right\|^2_{L_2(S,\mathcal{S},P)}.\end{aligned}$$

Proof One has (from Sect. 5.5.1, remembering that P has zero mean) the following equality:

$$\begin{aligned}\int_S f^2(s)\,P_m(ds) &= \int_S\left[\sum_{i=1}^n \alpha_i^f\left[m\left(t_i^f\right) + s\left(t_i^f\right)\right]\right]^2 P(ds) \\ &= \left[\sum_{i=1}^n \alpha_i^f m\left(t_i^f\right)\right]^2 + \int_S f^2(s)\,P(ds).\end{aligned}$$

But

$$\begin{aligned}\sum_{i=1}^n \alpha_i^f m\left(t_i^f\right) &= \sum_{i=1}^n \alpha_i^f \left\langle m, \mathcal{H}_{\mathcal{E},P_m}\left(\cdot, t_i^f\right)\right\rangle_{H(\mathcal{H}_{\mathcal{E},P_m},T)} \\ &= \left\langle m, \sum_{i=1}^n \alpha_i^f \mathcal{H}_{\mathcal{E},P_m}\left(\cdot, t_i^f\right)\right\rangle_{H(\mathcal{H}_{\mathcal{E},P_m},T)}\end{aligned}$$

$$= \left\langle U_m \left[f_m\right], U_m \left[[f]_{L_2(S,\mathcal{S},P_m)}\right]\right\rangle_{H(\mathcal{H}_{\mathcal{E},P_m},T)}$$

$$= \left\langle f_m, [f]_{L_2(S,\mathcal{S},P_m)}\right\rangle_{L_2(S,\mathcal{S},P_m)},$$

and thus the lemma obtains. □

Proposition 5.5.2 *Whenever* $f \in \overline{\mathcal{L}(\mathcal{E})}^{P_m}$, *there is* $\dot{f} \in f$ *such that:*

1. $\left[\dot{f}\right]_{L_2(S,\mathcal{S},P)} \in \overline{\mathcal{L}(\mathcal{E})}^{P}$;
2. $\|f\|^2_{L_2(S,\mathcal{S},P_m)} = \langle f, f_m\rangle^2_{L_2(S,\mathcal{S},P_m)} + \left\|\left[\dot{f}\right]_{L_2(S,\mathcal{S},P)}\right\|^2_{L_2(S,\mathcal{S},P)}$.

Proof Let $f \in \overline{\mathcal{L}(\mathcal{E})}^{P_m}$ be fixed, but arbitrary. There is then a sequence $\{f_n, n \in \mathbb{N}\} \subseteq \mathcal{L}(\mathcal{E})$ such that, in $L_2(S, \mathcal{S}, P_m)$, $\lim_n [f_n]_{L_2(S,\mathcal{S},P_m)} = f$, that is,

$$\lim_n \int_S \left[\dot{f}(T_m(s)) - f_n(T_m(s))\right]^2 P(ds) = 0. \qquad (\star)$$

Now, because of (Lemma) 5.5.1,

$$\left\{[f_n]_{L_2(S,\mathcal{S},P)}, n \in \mathbb{N}\right\}$$

is a Cauchy sequence in $L_2(S, \mathcal{S}, P)$. So there exists $f_P \in \overline{\mathcal{L}(\mathcal{E})}^{P}$ such that, in $L_2(S, \mathcal{S}, P)$,

$$\lim_n [f_n]_{L_2(S,\mathcal{S},P)} = f_P.$$

But there exists also a subsequence

$$\left\{f_{n_p}, p \in \mathbb{N}\right\} \subseteq \{f_n, n \in \mathbb{N}\}$$

such that, almost surely with respect to P, $\lim_p f_{n_p} = \dot{f}_P$. By Fatou's lemma, using (*),

$$\int_S \left\{\dot{f}(T_m(s)) - \dot{f}_P(T_m(s))\right\}^2 P(ds) =$$

$$= \int_S \liminf_p \left\{\dot{f}(T_m(s)) - f_{n_p}(T_m(s))\right\}^2 P(ds)$$

$$\leq \lim_n \int_S \left\{\dot{f}(T_m(s)) - f_{n_p}(T_m(s))\right\}^2 P(ds)$$

$$= 0.$$

Thus $\dot{f} = \dot{f}_P$, almost surely, with respect to P_m. Now (Lemma) 5.5.1 yields that

$$\left\|[f_n]_{L_2(S,\mathcal{S},P_m)}\right\|^2_{L_2(S,\mathcal{S},P_m)} = $$
$$= \left\langle [f_n]_{L_2(S,\mathcal{S},P_m)}, f_m\right\rangle^2_{L_2(S,\mathcal{S},P_m)} + \left\|[f_n]_{L_2(S,\mathcal{S},P)}\right\|^2_{L_2(S,\mathcal{S},P)}.$$

Taking limits one gets that

$$\|f\|^2_{L_2(S,\mathcal{S},P_m)} = \langle f, f_m\rangle^2_{L_2(S,\mathcal{S},P_m)} + \|\dot{f}_P\|^2_{L_2(S,\mathcal{S},P)},$$

with $\dot{f}_P \in f$. □

Corollary 5.5.3 *There is $\dot{f}_m \in f_m$ for which the following statements are valid:*

1. $\left[\dot{f}_m\right]_{L_2(S,\mathcal{S},P)} \in \overline{\mathcal{L}(\mathcal{E})}^P$,
2. $\left\|\left[\dot{f}_m\right]_{L_2(S,\mathcal{S},P)}\right\|_{L_2(S,\mathcal{S},P)} = 0.$

Proposition 5.5.4 *When m does not belong to $H(\mathcal{H}_{\mathcal{E},P}, T)$, P_m and P are orthogonal.*

Proof For the appropriate representative $\dot{f}_m \in f_m$ [(Corollary) 5.5.3], one has that

$$\|m\|^2_{H(\mathcal{H}_{\mathcal{E},P_m},T)} = \int_S \dot{f}^2_m(s)\, P_m(ds)$$
$$= \int_S \dot{f}^2_m(s)\, P_{m_{aP}}(ds) + \int_S \dot{f}^2_m(s)\, P_{m_{sP}}(ds),$$

with $\dot{f}_m$ almost surely zero with respect to P, and thus with respect to $P_{m_{aP}}$. Furthermore, the norm of m in $H(\mathcal{H}_{\mathcal{E},P_m}, T)$ is 1. Thus

$$1 = \int_S \dot{f}^2_m(s)\, P_{m_{sP}}(ds).$$

To have that $P_{m_{sP}}(S) = 1$, it suffices to show that $\dot{f}_m(s) = 1$, for almost every $s \in S$, with respect to P_m. Now, since $H(\mathcal{H}_{\mathcal{E},P_m}, T)$ is the range of the operator L_m, defined, for fixed, but arbitrary $t \in T$, and $f \in L_2(S, \mathcal{S}, P_m)$, using the following relation:

$$L_m[f](t) = \langle f, F_m(t)\rangle_{L_2(S,\mathcal{S},P_m)} = \left\langle f, [\mathcal{E}_t]_{L_2(S,\mathcal{S},P_m)}\right\rangle_{L_2(S,\mathcal{S},P_m)},$$

and, since the class of f_1 ($f_1 = 1_S$) belongs to $L_2(S, \mathcal{S}, P_m)$,

$$L_m\left[[f_1]_{L_2(S,\mathcal{S},P_m)}\right](t) = \left\langle [f_1]_{L_2(S,\mathcal{S},P_m)}, [\mathcal{E}_t]_{L_2(S,\mathcal{S},P_m)}\right\rangle_{L_2(S,\mathcal{S},P_m)}$$
$$= m(t)$$
$$= \left\langle f_m, [\mathcal{E}_t]_{L_2(S,\mathcal{S},P_m)}\right\rangle_{L_2(S,\mathcal{S},P_m)}.$$

But, since $P_{\overline{\mathcal{L}(\mathcal{E})}^{P_m}}\left[[\mathcal{E}_t]_{L_2(S,\mathcal{S},P_m)}\right] = [\mathcal{E}_t]_{L_2(S,\mathcal{S},P_m)}$,

$$\left\langle [f_1]_{L_2(S,\mathcal{S},P_m)}, [\mathcal{E}_t]_{L_2(S,\mathcal{S},P_m)}\right\rangle_{L_2(S,\mathcal{S},P_m)} =$$
$$= \left\langle P_{\overline{\mathcal{L}(\mathcal{E})}^{P_m}}\left[[f_1]_{L_2(S,\mathcal{S},P_m)}\right], [\mathcal{E}_t]_{L_2(S,\mathcal{S},P_m)}\right\rangle_{L_2(S,\mathcal{S},P_m)}.$$

Consequently $f_m = P_{\overline{\mathcal{L}(\mathcal{E})}^{P_m}}\left[[f_1]_{L_2(S,\mathcal{S},P_m)}\right]$, and, in $L_2(S, \mathcal{S}, P_m)$, orthogonally,

$$[f_1]_{L_2(S,\mathcal{S},P_m)} = f_m + f_m^{\perp},$$

so that

$$1 = \left\|[f_1]_{L_2(S,\mathcal{S},P_m)}\right\|^2_{L_2(S,\mathcal{S},P_m)} = \|f_m\|^2_{L_2(S,\mathcal{S},P_m)} + \left\|f_m^{\perp}\right\|^2_{L_2(S,\mathcal{S},P_m)}.$$

Since $\|f_m\|_{L_2(S,\mathcal{S},P_m)} = 1$, $\left\|f_m^{\perp}\right\|_{L_2(S,\mathcal{S},P_m)} = 0$, and $f_m = [f_1]_{L_2(S,\mathcal{S},P_m)}$. □

The proof of the most general discrimination result for translates requires the following lemmas.

Lemma 5.5.5 *Let $(\Omega_1, \mathcal{A}_1)$ and $(\Omega_2, \mathcal{A}_2)$ be two measurable spaces, and*

$$\Phi : \Omega_1 \longrightarrow \Omega_2$$

be an adapted map, that is, such that $\Phi^{-1}[\mathcal{A}_2] \subseteq \mathcal{A}_1$. Let

$$\mathcal{B}_1 = \Phi^{-1}[\mathcal{A}_2] \subseteq \mathcal{A}_1, \quad and \quad \mathcal{B}_2 = \left\{A_2 \subseteq \Omega_2 : \Phi^{-1}[A_2] \in \mathcal{A}_1\right\},$$

so that $\mathcal{A}_2 \subseteq \mathcal{B}_2$. Then:

1. *$\mathcal{B}_1$ and $\mathcal{B}_2$ are σ-fields,*
2. *Φ is adapted to $\mathcal{A}_1$ and $\mathcal{B}_2$,*
3. *for an injective Φ,*

   (i) *when $A_1 \in \mathcal{A}_1$, then $\Phi[A_1] \in \mathcal{B}_2$,*
   (ii) *when $\mathcal{B}_2 = \mathcal{A}_2$, then $\mathcal{B}_1 = \mathcal{A}_1$.*

Proof Item 1 is standard measurability fare [138, pp. 46,87]. Item 2 obtains by definition. For item 3, one may proceed as follows. When Φ is an injection, $\Phi^{-1}\Phi(\omega) = \omega$, so that $\Phi^{-1}[\Phi(A_1)] = A_1$, and thus, when $A_1 \in \mathcal{A}_1$, $\Phi(A_1)$ is in $\mathcal{B}_2$, that is, $\Phi(\mathcal{A}_1) \subseteq \mathcal{B}_2$. If now $\mathcal{B}_2 = \mathcal{A}_2$,

$$\mathcal{A}_1 = \Phi^{-1}[\Phi(\mathcal{A}_1)] \subseteq \Phi^{-1}(\mathcal{B}_2) = \Phi^{-1}(\mathcal{A}_2) = \mathcal{B}_1 \subseteq \mathcal{A}_1.$$

□

When P and Q are probabilities on the same σ-algebra, $P \ll Q$ means that P is absolutely continuous with respect to Q.

Lemma 5.5.6 *Let Φ be denoted Ψ when it is taken as a map adapted to $\mathcal{A}_1$ and $\mathcal{B}_2$ of (Lemma) 5.5.5. Let P and Q be probabilities on $\mathcal{A}_1$, and*

$$P_\Phi = P \circ \Phi^{-1},$$
$$P_\Psi = P \circ \Psi^{-1},$$
$$Q_\Phi = Q \circ \Phi^{-1},$$
$$Q_\Psi = Q \circ \Psi^{-1}.$$

Then:

1. *when P and Q are orthogonal, and Φ, injective, P_Ψ and Q_Ψ are orthogonal;*
2. *when $Q \ll P$, $Q_\Phi \ll P_\Phi$;*
3. *when $Q \ll P$ and $\mathcal{B}_1 = \mathcal{A}_1$,*

$$\frac{dQ_\Phi}{dP_\Phi} \circ \Phi = \frac{dQ}{dP}.$$

Proof When Φ is injective, and $A_1 \in \mathcal{A}_1$, $\Phi(A_1) \in \mathcal{B}_2$ and

$$A_1 = \Psi^{-1}[\Phi(A_1)],$$

so that

$$P(A_1) = P_\Psi[\Phi(A_1)].$$

Since Φ is injective, the choice of A_1 such that $P(A_1) = Q\left(A_1^c\right) = 1$ does the trick, for [229, p. 128] $\Phi(A_1) \cap \Phi\left(A_1^c\right) = \emptyset$. Item 2 is a consequence of the definition.

Since, by the change of variables formula,

$$Q\left(\Phi^{-1}(A)\right) = Q_\Phi(A) = \int_A \frac{dQ_\Phi}{dP_\Phi}\, dP_\Phi = \int_{\Phi^{-1}(A)} \left\{\frac{dQ_\Phi}{dP_\Phi} \circ \Phi\right\} dP,$$

only when $\Phi^{-1}(\mathcal{A}_2) = \mathcal{A}_1$ does

$$\frac{dQ_\Phi}{dP_\Phi} \circ \Phi = \frac{dQ}{dP}$$

obtain. □

Lemma 5.5.7 *Let $\Phi : S \longrightarrow S$ be a map adapted to $\mathcal{S}$, and P and Q be two probabilities on $\mathcal{S}$. If one assumes that*

(a) *Φ is an injection,*

(b) $Q \ll P$,
(c) Ψ *is the map defined in (Lemma)* 5.5.6 (here $\mathcal{A}_1 = \mathcal{A}_2 = \mathcal{S}$),

then:

1. $Q_\Psi \ll P_\Psi$,
2. $\frac{dQ_\Psi}{dP_\Psi}(s) = \begin{cases} \frac{dQ}{dP} \circ \Phi^{-1}(s) \text{ when } s \in \mathcal{R}[\Phi] \\ 0 \qquad\qquad \text{when } s \in \mathcal{R}[\Phi]^c \end{cases}$.

Proof Let $\Phi^- : S \longrightarrow S$ be defined using the following rule:

$$\Phi^-(s) = \begin{cases} \Phi^{-1}(s) & \text{when } \ s \in \mathcal{R}[\Phi] \\ 0_S & \text{when } \ s \in \mathcal{R}[\Phi]^c \end{cases}.$$

The map Φ^- is adapted to $\mathcal{B}_2$ and $\mathcal{S}$. Indeed, for fixed, but arbitrary $S_0 \in \mathcal{S}$,

$$\begin{aligned} (\Phi^-)^{-1}(S_0) &= \{s \in \mathcal{R}[\Phi] : \Phi^-(s) \in S_0\} \uplus \{s \in \mathcal{R}[\Phi]^c : \Phi^-(s) \in S_0\} \\ &= \left\{s \in \mathcal{R}[\Phi] : \Phi^{-1}(s) \in S_0\right\} \uplus \{s \in \mathcal{R}[\Phi]^c : 0_S \in S_0\} \\ &= \left\{s' \in S : \Phi^{-1}\left(\Phi\left(s'\right)\right) \in S_0\right\} \cup \{\emptyset \text{ or } \mathcal{R}[\Phi]^c\} \\ &= S_0 \\ &\text{or } \ S_0 \cup \mathcal{R}[\Phi]^c. \end{aligned}$$

Since $\Phi^{-1}(\mathcal{R}[\Phi]) = S$, and $\Phi^{-1}(\mathcal{R}[\Phi]^c) = \emptyset$, $\mathcal{R}[\Phi]$ and $\mathcal{R}[\Phi]^c$ belong to $\mathcal{B}_2$. Thus, since $\mathcal{S} \subseteq \mathcal{B}_2$, $(\Phi^-)^{-1}(S_0) \in \mathcal{B}_2$. Furthermore the following product:

$$\chi_{\mathcal{R}[\Phi]}(s)\left(\frac{dQ}{dP} \circ \Phi^-\right)(s) = \begin{cases} \frac{dQ}{dP} \circ \Phi^{-1}(s) \text{ when } s \in \mathcal{R}[\Phi] \\ 0 \qquad\qquad\qquad \text{when } s \in \mathcal{R}[\Phi]^c \end{cases}$$

is then adapted to $\mathcal{B}_2$, and, for $B_2 \in \mathcal{B}_2$ fixed, but arbitrary,

$$\begin{aligned} &\int_{B_2} \chi_{\mathcal{R}[\Phi]}(s)\left(\frac{dQ}{dP} \circ \Phi^-\right)(s)\, P_\Psi(ds) = \\ &\qquad = \int_{B_2 \cap \mathcal{R}[\Phi]} \left(\frac{dQ}{dP} \circ \Phi^{-1}\right) d\Psi \\ &\qquad = \int_{\Psi^{-1}(B_2 \cap \mathcal{R}[\Phi])} \left(\frac{dQ}{dP} \circ \Phi^{-1} \circ \Psi\right) dP \\ &\qquad = \int_{\Psi^{-1}(B_2) \cap \Psi^{-1}(\mathcal{R}[\Phi])} \left(\frac{dQ}{dP} \circ \Phi^{-1} \circ \Phi\right) dP \end{aligned}$$

$$= \int_{\Psi^{-1}(B_2)} \frac{dQ}{dP} dP$$
$$= Q\left(\Psi^{-1}(B_2)\right)$$
$$= Q_\Psi (B_2).$$

□

Example 5.5.8 Let $m \in S$ be fixed, but arbitrary, and let $T_m : S \longrightarrow S$ be defined using $T_m[s] = s + m$. As already stated [Sect. 5.5.1], T_m is adapted to $\mathcal{S}$. One has furthermore that:

1. $\mathcal{B}_1 = \mathcal{S}$.
   Indeed

$$\{\mathcal{E}_t \circ T_m\}^{-1}(]a,b]) = \{s \in S : s(t) + m(t) \in]a,b]\}$$
$$= \{s \in S : s(t) \in]a - m(t), b - m(t)]\}$$
$$= \mathcal{E}_t^{-1}(]a - m(t), b - m(t)]).$$

2. $\mathcal{B}_2 = \mathcal{S}$.
   By definition, $T_m^{-1}(B) = B - m$. Suppose $B - m \in \mathcal{S}$. Then, for some measurable $\Phi : \mathbb{R}^\infty \longrightarrow \mathbb{R}$,

$$\chi_{B-m}(s) = \Phi(\{\mathcal{E}_{t_i}[s], i \in N\}).$$

   Thus

$$\chi_B(s) = \chi_{B-m}(s - m)$$
$$= \Phi(\{\mathcal{E}_{t_i}[s - m], i \in N\})$$
$$= \Phi(\{\mathcal{E}_{t_i} \circ T_{-m}[s], i \in N\}),$$

   and $B \in \mathcal{S}$. Thus $\mathcal{B}_2 = \mathcal{S}$.

   T_m is an injection, $\mathcal{S} = \mathcal{B}_1 = \mathcal{B}_2$, and the results of (Lemmas) 5.5.5, 5.5.6, and 5.5.7 obtain for T_m.

Corollary 5.5.9 *Let P be a probability on $\mathcal{S}$* (for which Π_1 obtains for $\mathcal{L}(\mathcal{E})$). *Let m_1 and m_2 be two elements in S, and let*

$$P_1 = P \circ T_{m_1}^{-1}, \text{ and } P_2 = P \circ T_{m_2}^{-1}.$$

Then, when $m_1 - m_2$ does not belong to $H(\mathcal{H}_{\mathcal{E},P}, T)$, P_1 and P_2 are orthogonal.

Proof Because of (Proposition) 5.5.4, $P_{m_1-m_2}$ and P are orthogonal. Then $P_{m_1-m_2} \circ T_{m_2}^{-1}$ and $P \circ T_{m_2}^{-1} = P_2$ are orthogonal because of (Lemma) 5.5.6. But

$$\begin{aligned} P_{m_1-m_2} \circ T_{m_2}^{-1} &= P \circ T_{m_1-m_2}^{-1} \circ T_{m_2}^{-1} \\ &= P \circ \{T_{m_2} \circ T_{m_1-m_2}\}^{-1} \\ &= P \circ T_{m_1}^{-1} = P_1. \end{aligned}$$

□

5.6 Discrimination of Gaussian Laws

The context shall be that of Sect. 5.1. One is here concerned with detection of Gaussian random signals immersed in Gaussian noise.

5.6.1 Some Properties of Gaussian Laws

Reviewed here are those properties of Gaussian laws that shall be of use in the sequel.

Definition 5.6.1 A probability P on $\mathcal{S}$ is a Gaussian probability, or a Gaussian law, with a mean equal to zero, whenever, given fixed, but arbitrary $n \in \mathbb{N}$, and $\{t_1, \ldots, t_n\} \subseteq T$, the functions $\{\mathcal{E}_{t_1}, \ldots, \mathcal{E}_{t_n}\}$ have a normal law, with a mean equal to zero, and a covariance with entries $\mathcal{H}_{\mathcal{E},P}(t_i, t_j)$, $1 \leq i, j \leq n$, that is, given fixed, but arbitrary $\{\alpha_1, \ldots, \alpha_n\} \subseteq \mathbb{R}$, and $\iota = \sqrt{-1}$,

$$\int_S e^{\iota \sum_{j=1}^n \alpha_j \mathcal{E}_{t_j}(s)} P(ds) = e^{-\frac{1}{2}\sum_{i=1}^n \sum_{j=1}^n \alpha_i \alpha_j \mathcal{H}_{\mathcal{E},T}(t_i, t_j)}.$$

P has thus property Π_1 for $\mathcal{L}(\mathcal{E})$.

Remark 5.6.2 Suppose that $f = \sum_{i=1}^n \alpha_i \mathcal{E}_{t_i} \in \mathcal{L}(\mathcal{E})$. As

$$\begin{aligned} \sum_{i=1}^n \sum_{j=1}^n \alpha_i \alpha_j \mathcal{H}_{\mathcal{E},P}(t_i, t_j) &= \left\| \sum_{i=1}^n \alpha_i \mathcal{H}_{\mathcal{E},P}(\cdot, t_i) \right\|^2_{H(\mathcal{H}_{\mathcal{E},P},T)} \\ &= \left\| \left[\sum_{i=1}^n \alpha_i \mathcal{E}_{t_i} \right]_{L_2(S,\mathcal{S},P)} \right\|^2_{L_2(S,\mathcal{S},P)}, \end{aligned}$$

the definition rewrites: for $f \in \mathcal{L}(\mathcal{E})$,

$$\int_S e^{\iota f(s)} P(ds) = e^{-\frac{1}{2}\left\|[f]_{L_2(S,\mathcal{S},P)}\right\|^2_{L_2(S,\mathcal{S},P)}}$$

$$= e^{-\frac{1}{2}\left\|U_P\left[[f]_{L_2(S,\mathcal{S},P)}\right]\right\|^2_{H\left(\mathcal{H}_{\mathcal{E},P},T\right)}},$$

where $U_P : \overline{\mathcal{L}(\mathcal{E})}^P \longrightarrow H(\mathcal{H}_{\mathcal{E},P}, T)$ is the usual isometry.

Remark 5.6.3 Suppose $\{f_n,\ n \in \mathbb{N}\} \subseteq \mathcal{L}(\mathcal{E})$ is such that, in $L_2(S, \mathcal{S}, P)$,

$$\lim_n [f_n]_{L_2(S,\mathcal{S},P)} = f \in \overline{\mathcal{L}(\mathcal{E})}^P.$$

There is then a subsequence that converges almost surely to $\dot{f}$. One can thus take limits to obtain that the characterizing equation in (Remark) 5.6.2 is valid for the closure of $\mathcal{L}(\mathcal{E})$ in $L_2(S, \mathcal{S}, P)$.

Remark 5.6.4 All elements $f \in \overline{\mathcal{L}(\mathcal{E})}^P$ are classes of Gaussian random variables, with a mean equal to zero [(Remarks) 5.6.2 and 5.6.3]. Thus [66, p. 92]

$$E_P\left[\dot{f}^4\right] = 3E_P^2\left[\dot{f}^2\right].$$

Consequently P has property Π_3.

Remark 5.6.5 P has property Π_1 for $\mathcal{Q}(\mathcal{E})$. This is due, in particular, to the property of centered Gaussian variables [66, p. 92] that

$$E[G_1G_2G_3G_4] =$$
$$= E[G_1G_2]E[G_3G_4] + E[G_1G_3]E[G_2G_4] + E[G_1G_4]E[G_2G_3].$$

Remark 5.6.6 Let

$$\sigma\left(\overline{\mathcal{L}(\mathcal{E})}^P\right)$$

denote the σ-algebra generated by the random variables whose equivalence class belongs to the closure in $L_2(S, \mathcal{S}, P)$ of $\mathcal{L}(\mathcal{E})$. Then

$$\sigma\left(\overline{\mathcal{L}(\mathcal{E})}^P\right) = \overline{\sigma(\{\mathcal{E}_t,\ t \in T\})}^P = \overline{\mathcal{S}}^P,$$

where, for example, $\overline{\mathcal{S}}^P$ denotes the completion of $\mathcal{S}$ with respect to P.

Indeed, $\sigma\left(\overline{\mathcal{L}(\mathcal{E})}^P\right)$ contains all null sets, and $[\mathcal{E}_t]_{L_2(S,\mathcal{S},P)} \in \overline{\mathcal{L}(\mathcal{E})}^P$, so that

$$\overline{\sigma(\{\mathcal{E}_t,\ t \in T\})}^P \subseteq \sigma\left[\overline{\mathcal{L}(\mathcal{E})}^P\right].$$

Conversely, a random variable whose equivalence class belongs to $\overline{\mathcal{L}(\mathcal{E})}^P$ is almost surely equal to an element of $\mathcal{L}(\mathcal{E})$, or is the limit, in $L_2(S, \mathcal{S}, P)$, of a sequence of such random variables, and thus the reverse inclusion obtains also.

Remark 5.6.7 Since the conditional expectation of a square integrable random variable is a projection [200, p. 24], setting

$$e_2 = \left\| [f_2]_{L_2(S,\mathcal{S},P)} \right\|_{L_2(S,\mathcal{S},P)}^{-1} f_2,$$

$$E_P[f_1 \mid f_2] = \left\langle [f_1]_{L_2(S,\mathcal{S},P)}, [e_2]_{L_2(S,\mathcal{S},P)} \right\rangle_{L_2(S,\mathcal{S},P)} e_2.$$

5.6.2 Quadratic Forms of Normal Random Variables

One shall use, for the discrimination of Gaussian laws, domination based on the manifold $\mathcal{Q}(\mathcal{E})$. Its properties, when Gaussian laws obtain, are listed below. The symbol $IID[\mathcal{N}(0,1)]$ means independent, identically distributed, standard random variables, and $X \sim Y$ indicates random elements with the same law.

Lemma 5.6.8 *Every $f \in \mathcal{Q}(\mathcal{E})$ has a representation of the following form:*

$$f = \gamma + \sum_{i=1}^{n} \lambda_i Z_i + \sum_{i=1}^{n} \sigma_i Z_i^2,$$

where $\{Z_1, \ldots, Z_n\}$ is a set of $IID[\mathcal{N}(0,1)]$'s.

Proof By definition,

$$f = \alpha + \sum_{i=1}^{m} \alpha_i \mathcal{E}_{t_i} + \sum_{j=1}^{n} \sum_{k=1}^{p} \alpha_{j,k} \mathcal{E}_{u_j} \mathcal{E}_{v_k},$$

and one assumes, which is no restriction, that the evaluations involved produce a law that is not degenerate. Let $\tau_1, \ldots, \tau_\nu$ be the distinct values taken by the t's, the u's and the v's. Let $\underline{\mathcal{E}}_\nu$ be the vector with components $\mathcal{E}_{\tau_i}$, $1 \le i \le \nu$, and $\underline{v}_\nu$ be the vector that secures the following equality:

$$\sum_{i=1}^{m} \alpha_i \mathcal{E}_{t_i} = \langle \underline{v}_\nu, \underline{\mathcal{E}}_\nu \rangle_{\mathbb{R}^\nu}.$$

Let M_ν be the matrix that secures the following equality:

$$\sum_{j=1}^{n}\sum_{k=1}^{p}\alpha_{j,k}\mathcal{E}_{u_j}\mathcal{E}_{v_k} = \langle M\left[\underline{\mathcal{E}}_\nu\right], \underline{\mathcal{E}}_\nu\rangle_{\mathbb{R}^\nu}.$$

One has that

$$\langle M\left[\underline{\mathcal{E}}_\nu\right], \underline{\mathcal{E}}_\nu\rangle_{\mathbb{R}^\nu} = \left\langle \frac{M+M^\star}{2}\left[\underline{\mathcal{E}}_\nu\right], \underline{\mathcal{E}}_\nu\right\rangle_{\mathbb{R}^\nu},$$

so that M may be assumed to be symmetric. Consequently,

$$f = \alpha + \langle \underline{v}_\nu, \underline{\mathcal{E}}_\nu\rangle_{\mathbb{R}^\nu} + \langle M\left[\underline{\mathcal{E}}_\nu\right], \underline{\mathcal{E}}_\nu\rangle_{\mathbb{R}^\nu}.$$

Let Σ_ν be the matrix with entries $\mathcal{H}_{\mathcal{E},P}\left(\tau_i, \tau_j\right),\ 1 \le i,j \le \nu$. Set

$$\underline{Z}_\nu = \Sigma_\nu^{-1/2}\left[\underline{\mathcal{E}}_\nu\right].$$

One thus obtains a vector of independent, standard normal random variables, and

$$f = \alpha + \left\langle \Sigma_\nu^{1/2}\left[\underline{v}_\nu\right], \underline{Z}_\nu\right\rangle_{\mathbb{R}^\nu} + \left\langle \Sigma_\nu^{1/2} M \Sigma_\nu^{1/2}\left[\underline{Z}_\nu\right], \underline{Z}_\nu\right\rangle_{\mathbb{R}^\nu}.$$

The matrix $\Sigma_\nu^{1/2} M \Sigma_\nu^{1/2}$, being symmetric, has a representation of the following form: $P_\nu D_\nu P_\nu^\star$, P_ν being orthogonal, and D_ν diagonal. One has finally that

$$f = \alpha + \left\langle P_\nu^\star \Sigma_\nu^{1/2}\left[\underline{v}_\nu\right], P_\nu^\star\left[\underline{Z}_\nu\right]\right\rangle_{\mathbb{R}^\nu} + \langle D_\nu P_\nu^\star\left[\underline{Z}_\nu\right], P_\nu^\star\left[\underline{Z}_\nu\right]\rangle_{\mathbb{R}^\nu},$$

which is the required representation, since the vector $P_\nu^\star\left[\underline{Z}_\nu\right]$ has components that are independent, standard normal, random variables. □

Lemma 5.6.9 *When* $(Z_1, \ldots, Z_n) \sim IID\left[\mathcal{N}(0,1)\right]$, *the first two moments of*

$$f = \gamma + \sum_{i=1}^{n}\lambda_i Z_i + \sum_{i=1}^{n}\sigma_i Z_i^2$$

are, respectively,

$$E_P[f] = \gamma + \sum_{i=1}^{n}\sigma_i,$$

$$V_P[f] = \sum_{i=1}^{n}\lambda_i^2 + 2\sum_{i=1}^{n}\sigma_i^2.$$

Proof f has the form $C + L + Q$, a constant (C), plus a linear term (L), plus a quadratic term (Q). Then, since odd powers of standard normal random variables have moments with zero expectation [200, p. 12],

$$\begin{aligned} E_P\left[f^2\right] &= E_P\left[C^2 + L^2 + Q^2 + 2CL + 2CQ + 2LQ\right] \\ &= C^2 + E_P\left[L^2\right] + E_P\left[Q^2\right] + 2CE_P\left[Q\right]. \end{aligned}$$

But

$$E_P\left[L^2\right] = \sum_{i=1}^{n} \lambda_i^2, \quad E_P\left[Q\right] = \sum_{i=1}^{n} \sigma_i,$$

and

$$E_P\left[Q^2\right] = 3\sum_{i=1}^{n} \sigma_i^2 + \sum_{i \neq j} \sigma_i \sigma_j = 2\sum_{i=1}^{n} \sigma_i^2 + \left(\sum_{i=1}^{n} \sigma_i\right)^2.$$

Thus

$$E_P\left[f^2\right] = \left(\gamma + \sum_{i=1}^{n} \sigma_i\right)^2 + \sum_{i=1}^{n} \lambda_i^2 + 2\sum_{i=1}^{n} \sigma_i^2.$$

□

5.6.3 Gaussian Laws Have the Properties Required for Domination

The aim here is to check that Gaussian laws have property Π_2 for $\mathcal{Q}(\mathcal{E})$, which is required, for instance, by result (Proposition) 5.4.5. Gaussian measures may be degenerate, that is, concentrated at one point. When they are not, they are reduced, and what follows concerns reduced Gaussian measures.

Proposition 5.6.10 *Suppose S is one of the spaces (Examples)* 5.1.3 *to* 5.1.6. *A Gaussian law whose mean is the zero function is either reduced, or a point mass located at the origin.*

Proof Let $\{S_n,\ n \in \mathbb{N}\} \subseteq \mathcal{S}$ be a sequence such that $0_S \in S_n,\ n \in \mathbb{N}$, and

$$\lim_n P\left(S_n\right) = \inf\left\{P\left(A\right) : \ A \in \mathcal{S} \text{ and } 0_S \in A\right\}.$$

One may assume that the sequence is decreasing. Let $S_0 = \cap_n S_n$, and let $A \in \mathcal{S}$ contain the zero element. Then, since $A \cap S_0$ contains the zero element, by the choice of S_0,

$$P(S_0) \leq P(A \cap S_0) \leq P(A).$$

In particular, let $\alpha > 0$ be fixed, arbitrary, and

$$S_t^{(\alpha)} = \{s \in S : \mathcal{E}_t(s) \in]-\alpha, \alpha[\}.$$

Then $P(S_0) \leq P\left(S_t^{(\alpha)}\right)$. Letting $\alpha \downarrow 0$, one has that

$$P(S_0) \leq P(\mathcal{E}_t = 0).$$

Since $\mathcal{E}_t$ is a normal random variable, with mean equal to zero, the inequality $P(S_0) > 0$ means that $\mathcal{E}_t$ takes the value zero with probability 1.

Now, given $A \in \mathcal{S}$, fixed, but arbitrary [41, p. 144],

$$\chi_A = \Phi(\{\mathcal{E}_{t_i}, i \in \mathbb{N}\}), \ \Phi : \mathbb{R}^\infty \longrightarrow \{0, 1\}, \text{ measurable.}$$

Consequently

$$P(A) = E_P[\Phi(\{\mathcal{E}_{t_i}, i \in \mathbb{N}\})] = \Phi(\{0_i, i \in \mathbb{N}\}).$$

Thus $P(A)$ is either zero or one. □

Lemma 5.6.11 *Let Z be a standard normal random variable, and*

$$\varphi(\theta) = E\left[e^{\iota\theta\left[aZ + bZ^2\right]}\right].$$

Then

$$|\varphi(\theta)| \leq \frac{1}{\left\{1 + \left[\frac{a^2}{2} + 4b^2\right]\theta^2\right\}^{1/4}}.$$

Proof Writing

$$E\left[e^{\iota\theta\left[aZ + bZ^2\right]}\right]$$

as an integral with respect to a Gaussian density (with complex parameters), one gets that [29, 66, p. 88,75]

$$\varphi(\theta) = \frac{e^{-\frac{a^2\theta^2}{2(1-2\iota\theta b)}}}{(1-2\iota\theta b)^{1/2}} = \frac{e^{-\frac{a^2\theta^2}{2(1+4b^2\theta^2)}+\iota\left\{\frac{-a^2b\theta^3}{1+4b^2\theta^2}\right\}}}{(1-2\iota\theta b)^{1/2}}.$$

Consequently, using $e^x > 1 + x,\ x > 0$, in the form $(1 + x)^{-1} > e^{-x}$, and denoting α the expression $1 + 4b^2\theta^2$, one has that

$$|\varphi(\theta)| = \frac{e^{-\frac{a^2\theta^2}{2\alpha}}}{\alpha^{1/2}} \le \alpha^{-1/2}\left(1+\frac{a^2\theta^2}{2}\right)^{-1} = \alpha^{1/2}\left(\alpha+\frac{a^2\theta^2}{2}\right)^{-1}$$
$$\le \left(\alpha+\frac{a^2\theta^2}{2}\right)^{-1/2} \le \left(\alpha+\frac{a^2\theta^2}{2}\right)^{-1/4}$$

□

Lemma 5.6.12 *Let $f \in \mathcal{Q}(\mathcal{E})$, fixed, but arbitrary, have the representation [(Lemma)]* 5.6.8. *There is a constant κ such that*

$$\left|E_P\left[e^{\iota\theta f}\right]\right| \le \left(1+\kappa^2\theta^2\right)^{-1/4}:$$

in fact

$$\kappa^2 = \frac{1}{2}\sum_{i=1}^{n}\lambda_i^2 + 4\sum_{i=1}^{n}\sigma_i^2.$$

Proof f has the following representation [(Lemma) 5.6.8]:

$$f = \gamma + \sum_{i=1}^{n}\lambda_i Z_i + \sum_{i=1}^{n}\sigma_i Z_i^2,\ (Z_1, \ldots, Z_n) \sim IID\left[\mathcal{N}(0,1)\right].$$

Thus, using (Lemma) 5.6.11,

$$\left|E_P\left[e^{\iota\theta f}\right]\right| = \left|e^{i\gamma\theta}\prod_{i=1}^{n}E_P\left[e^{\iota\theta\left[\lambda_i Z_i+\sigma_i Z_i^2\right]}\right]\right| \le \prod_{i=1}^{n}\left(1+\left[\frac{\lambda_i^2}{2}+4\sigma_i^2\right]\theta^2\right)^{-1/4}.$$

Then the following obvious inequality:

$$\prod_{i=1}^{n}\left(1+c_i^2\right) = 1+\sum_i c_i^2+\sum_{i\neq j}c_i^2c_j^2+\cdots+c_1^2c_2^2\cdots c_n^2 \ge 1+\sum_{i=1}^{n}c_i^2,$$

applied to the latter product yields that

$$\left|E_P\left[e^{\iota\theta f}\right]\right| \leq \left(1+\left[\frac{1}{2}\sum_{i=1}^{n}\lambda_i^2+4\sum_{i=1}^{n}\sigma_i^2\right]\theta^2\right)^{-1/4}.$$

□

Lemma 5.6.13 *For $a > 0$,*

$$\int_{-\infty}^{\infty}\left|\frac{\sin u}{u}\right|\left(1+a^2u^2\right)^{-1/4}du < \infty.$$

Proof Let $f(u) = \frac{\sin u}{u}\left(1+a^2u^2\right)^{-1/4}$, and define

$$I_{-1} = \int_{-\infty}^{-1}|f(u)|\,du,\ I_0 = \int_{-1}^{1}|f(u)|\,du,\ I_1 = \int_{1}^{\infty}|f(u)|\,du.$$

One has that

$$I_{-1} \leq \int_{-\infty}^{-1}|u|^{-1}\left(1+a^2u^2\right)^{-1/4}du.$$

The change of variables $u = -v$ yields that

$$I_{-1} \leq \int_{1}^{\infty}\frac{1}{v}\left(1+a^2v^2\right)^{-1/4}dv \leq \frac{1}{\sqrt{a}}\int_{1}^{\infty}\frac{dv}{v^{3/2}} < \infty.$$

I_1 has the same bound. I_0 is bounded because [2, p. 135], for $0 < x < \frac{\pi}{2}$,

$$0 < \cos x < \frac{\sin x}{x} < \frac{1}{\cos x}.$$

□

Corollary 5.6.14 *For $\alpha > 0$, and f as in (Lemma)* 5.6.8,

$$P(|f| \leq \alpha) = \frac{1}{\pi}\int_{-\infty}^{\infty}\frac{\sin\theta}{\theta}E_P\left[e^{\iota\frac{\theta}{\alpha}f}\right]d\theta.$$

Proof One starts with the inversion formula for characteristic functions [55, p. 141]:

$$\begin{aligned}P(|f| \leq \alpha) &= P\left(\frac{|f|}{\alpha} \leq 1\right)\\ &= P\left(-1 < \frac{f}{\alpha} < 1\right) + \frac{1}{2}\left\{P\left(\frac{f}{\alpha} = -1\right) + P\left(\frac{f}{\alpha} = 1\right)\right\}\end{aligned}$$

$$= \lim_{C\uparrow\infty} \frac{1}{2\pi} \int_{-C}^{C} \frac{e^{\iota t} - e^{-\iota t}}{\iota t} \varphi_{\left\{\frac{f}{\alpha}\right\}}(t)\, dt$$

$$= \lim_{C\uparrow\infty} \frac{1}{\pi} \int_{-C}^{C} \frac{\sin t}{t} E_P\left[e^{\iota t \frac{f}{\alpha}}\right] dt.$$

Then (Lemmas) 5.6.12 and 5.6.13 are used. □

Lemma 5.6.15 *Let $\alpha \geq 0, f$, and μ be real numbers. Then $|f| \leq \alpha$ implies that*

$$|f - \mu| \geq |\mu| - \alpha.$$

Proof In the standard inequality $||a| - |b|| \leq |a - b|$ use $a = f - \mu$ and $b = -\mu$ to get

$$||f - \mu| - |\mu|| \leq |f| \leq \alpha$$

so that $-\alpha \leq |f - \mu| - |\mu|$. □

Proposition 5.6.16 *P has property Π_2 for $\mathcal{Q}(\mathcal{E})$.*

Proof Let $f \in \mathcal{Q}_1(\mathcal{E})$ be fixed, but arbitrary, so that $E_P\left[f^2\right] = 1$, and $\mu = E_P[f]$. Choose arbitrarily $\epsilon \in]0, 1[$.

Case: $|\mu| \geq 1 - \epsilon$.

From (Lemma) 5.6.15

$$P(s \in S : |f(s)| \leq \alpha) \leq P(s \in S : |f(s) - \mu| \geq |\mu| - \alpha).$$

Since $\mu = E_P[f]$, one gets that

$$P(s \in S : |f(s)| \leq \alpha) \leq \frac{V_P[f]}{(|\mu| - \alpha)^2}.$$

Choose $\alpha \in [0, 1 - \epsilon[$. Then, since

$$1 - \mu^2 \leq 1 - (1 - \epsilon)^2, \quad \text{and} \quad |\mu| - \alpha \geq 1 - \epsilon - \alpha,$$

one obtains that

$$P(s \in S : |f(s)| \leq \alpha) \leq \frac{1 - \mu^2}{(|\mu| - \alpha)^2} \leq \frac{1 - (1 - \epsilon)^2}{(1 - \epsilon - \alpha)^2}.$$

Case: $|\mu| < 1 - \epsilon$.
Let $\delta > 0$ be fixed, but arbitrary. Then [(Lemma) 5.6.13]

$$P(|f| \leq \alpha) \leq \frac{1}{\pi} \int_{|\theta| \leq \delta} \frac{\sin\theta}{\theta} E_P\left[e^{\iota \frac{\theta}{\alpha} f}\right] d\theta + \frac{1}{\pi} \int_{|\theta| > \delta} \frac{\sin\theta}{\theta} E_P\left[e^{\iota \frac{\theta}{\alpha} f}\right] d\theta.$$

Since $\left|\frac{\sin\theta}{\theta} E_P\left[e^{\iota \frac{\theta}{\alpha} f}\right]\right| \leq 1$,

$$\int_{|\theta| \leq \delta} \frac{\sin\theta}{\theta} E_P\left[e^{\iota \frac{\theta}{\alpha} f}\right] d\theta \leq 2\delta.$$

Since [(Lemma) 5.6.9] $1 = E_P\left[f^2\right] = \mu^2 + \sum_{i=1}^n \lambda_i^2 + 2\sum_{i=1}^n \sigma_i^2$,

$$\begin{aligned}
\kappa^2 &\overset{\text{Lemma 5.6.12}}{=} \frac{\sum_{i=1}^n \lambda_i^2}{2} + 4\sum_{i=1}^n \sigma_i^2 \\
&\geq \frac{\sum_{i=1}^n \lambda_i^2 + 2\sum_{i=1}^n \sigma_i^2}{2} \\
&= \frac{1 - \mu^2}{2} \\
&\geq \frac{1 - (1-\epsilon)^2}{2},
\end{aligned}$$

and [(Lemma) 5.6.12]

$$\begin{aligned}
\int_{|\theta| > \delta} \frac{\sin\theta}{\theta} E_P\left[e^{\iota \frac{\theta}{\alpha} f}\right] d\theta &\leq \int_{|\theta| > \delta} \left\{1 + \kappa^2 \frac{\theta^2}{\alpha^2}\right\}^{-1/4} \frac{d\theta}{|\theta|} \\
&\leq \int_{|\theta| > \delta} \left\{1 + \frac{1 - (1-\epsilon)^2}{2} \frac{\theta^2}{\alpha^2}\right\}^{-1/4} \frac{d\theta}{|\theta|} \\
&\leq 2\int_{\theta > \delta} \left\{\frac{1 - (1-\epsilon)^2}{2} \frac{\theta^2}{\alpha^2}\right\}^{-1/4} \frac{d\theta}{\theta} \\
&= 2\frac{2^{1/4}\alpha^{1/2}}{\left(1 - (1-\epsilon)^2\right)^{1/4}} \int_{\theta > \delta} \frac{d\theta}{\theta^{3/2}} \\
&= 4\frac{2^{1/4}\alpha^{1/2}}{\left(1 - (1-\epsilon)^2\right)^{1/4}} \frac{1}{\delta^{1/2}}.
\end{aligned}$$

Thus

$$\pi P(|f| \leq \alpha) \leq 2\delta + \frac{2^{9/4}}{\left(1-(1-\epsilon)^2\right)^{1/4}} \left\{\frac{\alpha}{\delta}\right\}^{1/2}.$$

Finally, in both cases considered, $\lim_{\alpha \downarrow 0} \sup_{f \in \mathcal{Q}_1(\mathcal{E})} P(|f| \leq \alpha)$ is arbitrarily small. □

5.6.4 *The Quadratic Manifold of Evaluations: A Source of Hilbert-Schmidt Operators*

In most Gaussian "signal-in-noise" discrimination problems, one finds operators which are the sum of the identity with a Hilbert-Schmidt operator. The reason for that fact, as shall presently be seen, is that domination based on $\mathcal{Q}(\mathcal{E})$ is used. The basic point is that, in that latter case, the Radon-Nikodým derivative acts as a multiplication operator, and that such operators have a representation in the form of the sum of the identity with a Hilbert-Schmidt operator.

In this section, unless stated otherwise, $f_0 \in \mathcal{Q}(\mathcal{E})$ shall mean that

$$f_0 = m_0 + L_0 + Q_0,$$

where, $Z_{0,1}, \ldots, Z_{0,n_0}$ being independent, standard normal random variables,

$$m_0 = E_P[f_0],$$

$$L_0 = \sum_{i=1}^{n_0} \lambda_{0,i} Z_{0,i},$$

$$Q_0 = \sum_{i=1}^{n_0} \sigma_{0,i}\left(Z_{0,i}^2 - 1\right).$$

When one needs to work with the equivalence class of $Z_{0,i}$, one shall use the somewhat less cumbersome notation $\zeta_{0,i}$.

Lemma 5.6.17 *Let $f_0 \in \mathcal{Q}(\mathcal{E})$ be fixed, but arbitrary. Then*

$$E_P\left[f_0^4\right] < \infty,$$

so that the following assignment: given $f \in \overline{\mathcal{L}(\mathcal{E})}^P$, fixed, but arbitrary,

$$M_0[f] = \left[f_0 \dot{f}\right]_{L_2(S,\mathcal{S},P)},$$

defines a linear and bounded operator, with

$$\mathcal{D}[M_0] = \overline{\mathcal{L}(\mathcal{E})}^P, \quad \text{and} \quad \mathcal{R}[M_0] \subseteq L_2(S, \mathcal{S}, P).$$

The following operator:

$$B_0 = P_{\overline{\mathcal{L}(\mathcal{E})}^P} M_0,$$

is then well defined, linear, and bounded. Furthermore

$$\mathcal{R}[B_0] \subseteq \overline{\mathcal{L}(\mathcal{E})}^P.$$

Finally, letting

$$\tilde{B}_0[f] = P_{\overline{\mathcal{L}(\mathcal{E})}^P} \left[Q_0 \dot{f}\right]_{L_2(S,\mathcal{S},P)},$$

B_0 *has the following representation:*

$$B_0[f] = \left\{m_0 I_{\overline{\mathcal{L}(\mathcal{E})}^P} + \tilde{B}_0\right\}[f].$$

Proof In f_0^4, the powers of $Z_{0,i}$ with highest order are of the form $Z_{0,i}^8$, and [200, p. 12]

$$E\left[Z_{0,i}^8\right] = 105.$$

The first part of (Lemma) 5.6.17 is then true because of (Proposition) 5.1.32. One must thus only check that the representation of B_0 that is advertised above obtains.

Now

$$\left[f_0 \dot{f}\right]_{L_2(S,\mathcal{S},P)} = m_0 \left[\dot{f}\right]_{L_2(S,\mathcal{S},P)} + \left[L_0 \dot{f}\right]_{L_2(S,\mathcal{S},P)} + \left[Q_0 \dot{f}\right]_{L_2(S,\mathcal{S},P)},$$

and, for $g \in \overline{\mathcal{L}(\mathcal{E})}^P$, fixed, but arbitrary, since one integrates normal random variables with a mean equal to zero,

$$\begin{aligned}
\left\langle P_{\overline{\mathcal{L}(\mathcal{E})}^P}\left[\left[L_0 \dot{f}\right]_{L_2(S,\mathcal{S},P)}\right], g\right\rangle_{L_2(S,\mathcal{S},P)} &= \left\langle \left[L_0 \dot{f}\right]_{L_2(S,\mathcal{S},P)}, g\right\rangle_{L_2(S,\mathcal{S},P)} \\
&= \int_S L_0(s) \dot{f}(s) \dot{g}(s) P(ds) \\
&= 0.
\end{aligned}$$

Thus

$$P_{\overline{\mathcal{L}(\mathcal{E})}^P}\left[\left[L_0\dot{f}\right]_{L_2(S,\mathcal{S},P)}\right] = [0]_{L_2(S,\mathcal{S},P)}.$$

□

Lemma 5.6.18 *Let $f_0 \in \mathcal{Q}(\mathcal{E})$ be fixed, but arbitrary. Let*

(a) *H_0 denote the subspace of $\overline{\mathcal{L}(\mathcal{E})}^P$ generated by $Z_{0,1}, \ldots, Z_{0,n_0}$* (the random variables from which f_0 is built: their equivalence classes are orthonormal random elements),
(b) *$H_0^{\perp}$ be the orthogonal complement of H_0 in $\overline{\mathcal{L}(\mathcal{E})}^P$.*

For any two f and g belonging to $\overline{\mathcal{L}(\mathcal{E})}^P$, one has, in $\overline{\mathcal{L}(\mathcal{E})}^P$ taken as Hilbert space, that (R stands for "remainder"):

$$\dot{f} = \sum_{i=1}^{n_0} f_i Z_{0,i} + \dot{R}_f,\ R_f \in H_0^{\perp},$$

$$\dot{g} = \sum_{i=1}^{n_0} g_i Z_{0,i} + \dot{R}_g,\ R_g \in H_0^{\perp},$$

and

$$E_P\left[Q_0\dot{f}\dot{g}\right] = 2\sum_{i=1}^{n_0}\sigma_{0,i} f_i g_i.$$

Proof The representations of f and g in the lemma's statement reflect the fact that independent, standard normal, random variables correspond to orthonormal elements in the Hilbert space of equivalence classes of random variables. Since one deals with orthogonal, and thus independent, Gaussian random variables that have a mean equal to zero, using [66, p. 92], one has that

$$\begin{aligned} E_P\left[\dot{f}\dot{g}Z_{0,i}^2\right] &= E_P\left[\dot{f}\dot{g}\right]E_P\left[Z_{0,i}^2\right] + 2E_P\left[\dot{f}Z_{0,i}\right]E_P\left[\dot{g}Z_{0,i}\right] \\ &= E_P\left[\dot{f}\dot{g}\right] + 2f_i g_i. \end{aligned}$$

Consequently, as $E_P\left[\dot{f}\dot{g}\left(Z_{0,i}^2 - 1\right)\right] = E_P\left[\dot{f}\dot{g}\right] + 2f_i g_i - E_P\left[\dot{f}\dot{g}\right]$,

$$E_P\left[Q_0\dot{f}\dot{g}\right] = \sum_{i=1}^{n_0}\sigma_{0,i}E_P\left[\dot{f}\dot{g}\left(Z_{0,i}^2 - 1\right)\right] = 2\sum_{i=1}^{n_0}\sigma_{0,i} f_i g_i.$$

□

Lemma 5.6.19 *Let* $f_0 \in \mathcal{Q}(\mathcal{E})$ *be fixed, but arbitrary, and* f *be as in (Lemma)* 5.6.18. *Then*

$$P_{\overline{\mathcal{L}(\mathcal{E})}^P}\left[\left[\mathcal{Q}_0\dot{f}\right]_{L_2(S,\mathcal{S},P)}\right] = 2\sum_{i=1}^{n_0}\sigma_{0,i}f_i\zeta_{0,i}.$$

Proof Let $g \in \overline{\mathcal{L}(\mathcal{E})}^P$ be fixed, but arbitrary. Then, using (Lemma) 5.6.18,

$$\begin{aligned}\left\langle P_{\overline{\mathcal{L}(\mathcal{E})}^P}\left[\left[\mathcal{Q}_0\dot{f}\right]_{L_2(S,\mathcal{S},P)}\right], g\right\rangle_{L_2(S,\mathcal{S},P)} &= \left\langle\left[\mathcal{Q}_0\dot{f}\right]_{L_2(S,\mathcal{S},P)}, g\right\rangle_{L_2(S,\mathcal{S},P)} \\ &= E_P\left[\mathcal{Q}_0\dot{f}\dot{g}\right] \\ &= 2\sum_{i=1}^{n_0}\sigma_{0,i}f_i g_i \\ &= 2\sum_{i=1}^{n_0}\sigma_{0,i}f_i\left\langle\zeta_{0,i}, g\right\rangle_{L_2(S,\mathcal{S},P)} \\ &= \left\langle\sum_{i=1}^{n_0}2\sigma_{0,i}f_i\zeta_{0,i}, g\right\rangle_{L_2(S,\mathcal{S},P)}.\end{aligned}$$

□

Proposition 5.6.20 *The following operator* (where the notation $[a \otimes b](x)$ stands for $\langle x, b\rangle a$, with the appropriate inner product):

$$\tilde{B}_{[0]} \overset{def}{=} \sqrt{2}\sum_{i=1}^{n_0}\sigma_{0,i}\zeta_{0,i}\otimes\zeta_{0,i},$$

is a Hilbert-Schmidt operator, with Hilbert-Schmidt operator norm equal to

$$2\sum_{i=1}^{n_0}\sigma_{0,i}^2 = E_P\left[\mathcal{Q}_0^2\right].$$

Proof By definition, $\sum_{i=1}^{n_0}\sqrt{2}\sigma_{0,i}\zeta_{0,i}\otimes\zeta_{0,i}$ is a Hilbert-Schmidt operator, and its Hilbert-Schmidt norm is $2\sum_{i=1}^{n_0}\sigma_{0,i}^2$ [235, p. 34]. On the other hand,

$$\begin{aligned}E_P\left[\mathcal{Q}_0^2\right] &= \sum_{i=1}^{n_0}\sum_{j=1}^{n_0}\sigma_{0,i}\sigma_{0,j}E_P\left[\left(Z_{0,i}^2-1\right)\left(Z_{0,j}^2-1\right)\right] \\ &= \sum_{i=1}^{n_0}\sigma_{0,i}^2E_P\left[\left(Z_{0,i}^2-1\right)^2\right],\end{aligned}$$

and

$$E_P\left[\left(Z_{0,i}^2-1\right)^2\right]=E_P\left[Z_{0,i}^4\right]-2E_P\left[Z_{0,i}^2\right]+1$$
$$=3-2+1$$
$$=2.$$

□

Proposition 5.6.21 *Let* $f_0 \in \overline{\mathcal{Q}(\mathcal{E})}^P$ *be fixed, but arbitrary, and let* (where $f_1(s)=1_S$)

$$\mu_0 = E_P\left[\dot{f_0}\right],$$
$$L_0 = P_{\overline{\mathcal{L}(\mathcal{E})}^P}\left[f_0\right],$$
$$Q_0 = f_0-\mu_0\left[f_1\right]_{L_2(S,\mathcal{S},P)}-L_0.$$

Define M_0 *to be the operator of multiplication by* f_0, *and* $B_0 = P_{\overline{\mathcal{L}(\mathcal{E})}^P}M_0$. *Then*

$$B_0=\mu_0 I_{\overline{\mathcal{L}(\mathcal{E})}^P}+\tilde{B}_0,$$

with $\tilde{B}_0$, *a Hilbert-Schmidt operator, whose Hilbert-Schmidt norm, indexed by HS, is given by the following formula:*

$$\left\|\tilde{B}_0\right\|_{HS}^2=E_P\left[\dot{Q}_0^2\right].$$

Proof Let $\{f_{0,n}, n\in\mathbb{N}\}\subseteq\mathcal{Q}(\mathcal{E})$ be such that, in $L_2(S,\mathcal{S},P)$,

$$\lim_n\left[f_{0,n}\right]_{L_2(S,\mathcal{S},P)}=f_0.$$

Let also

$$\mu_{0,n} = E_P\left[f_{0,n}\right],$$
$$L_{0,n} = P_{\overline{\mathcal{L}(\mathcal{E})}^P}\left[\left[f_{0,n}\right]_{L_2(S,\mathcal{S},P)}\right],$$
$$Q_{0,n} = \left[f_{0,n}\right]_{L_2(S,\mathcal{S},P)}-\mu_{0,n}\left[f_1\right]_{L_2(S,\mathcal{S},P)}-L_{0,n}.$$

Then:

- $\lim_n \mu_{0,n}=\mu_0$, a result which follows from the following inequalities:

$$\left|\mu_{0,n}-\mu_0\right|\leq E_P\left[\left|f_{0,n}-\dot{f_0}\right|\right]\leq\left\|\left[f_{0,n}\right]_{L_2(S,\mathcal{S},P)}-f_0\right\|_{L_2(S,\mathcal{S},P)};$$

- in $L_2(S, \mathcal{S}, P)$, $\lim_n L_{0,n} = L_0$, a result which follows from the following inequalities:

$$\|L_{0,n} - L_0\|_{L_2(S,\mathcal{S},P)} = \left\| P_{\overline{\mathcal{L}(\mathcal{E})}^P} \left[[f_{0,n}]_{L_2(S,\mathcal{S},P)} - f_0 \right] \right\|_{L_2(S,\mathcal{S},P)}$$
$$\leq \left\| [f_{0,n}]_{L_2(S,\mathcal{S},P)} - f_0 \right\|_{L_2(S,\mathcal{S},P)};$$

- in $L_2(S, \mathcal{S}, P)$, $\lim_n Q_{0,n} = Q_0$, a result which follows from the following inequalities:

$$\|Q_{0,n} - Q_0\|_{L_2(S,\mathcal{S},P)} \leq \left\| [f_{0,n}]_{L_2(S,\mathcal{S},P)} - f_0 \right\|_{L_2(S,\mathcal{S},P)}$$
$$+ |\mu_0 - \mu_{0,n}|$$
$$+ \|L_{0,n} - L_0\|_{L_2(S,\mathcal{S},P)}.$$

Let $M_{0,n}$ be the operator of multiplication by $f_{0,n}$ (which is well defined because of (Lemma) 5.6.17), and

$$B_{0,n} = P_{\overline{\mathcal{L}(\mathcal{E})}^P} M_{0,n}.$$

Then [(Lemma) 5.6.17], for $f \in \overline{\mathcal{L}(\mathcal{E})}^P$, fixed, but arbitrary, setting

$$\tilde{B}_{0,n}[f] = P_{\overline{\mathcal{L}(\mathcal{E})}^P} \left[[\dot{Q}_{0,n}\dot{f}]_{L_2(S,\mathcal{S},P)} \right],$$

one has that

$$B_{0,n} = \mu_{0,n} I_{\overline{\mathcal{L}(\mathcal{E})}^P} + \tilde{B}_{0,n}.$$

Now $\tilde{B}_{0,n} - \tilde{B}_{0,p}$ has the same representation as $\tilde{B}_{0,n}$. Indeed

$$\left(\tilde{B}_{0,n} - \tilde{B}_{0,p}\right)[f] = \tilde{B}_{0,n}[f] - \tilde{B}_{0,p}[f]$$
$$= P_{\overline{\mathcal{L}(\mathcal{E})}^P} \left[[\dot{Q}_{0,n}\dot{f}]_{L_2(S,\mathcal{S},P)} \right] - P_{\overline{\mathcal{L}(\mathcal{E})}^P} \left[[\dot{Q}_{0,p}\dot{f}]_{L_2(S,\mathcal{S},P)} \right]$$
$$= P_{\overline{\mathcal{L}(\mathcal{E})}^P} \left[[(\dot{Q}_{0,n} - \dot{Q}_{0,p})\dot{f}]_{L_2(S,\mathcal{S},P)} \right].$$

Furthermore,

$$Q_{0,n} - Q_{0,p} = [f_{0,n} - f_{0,p}]_{L_2(S,\mathcal{S},P)} - (\mu_{0,n} - \mu_{0,p}) I_{\overline{\mathcal{L}(\mathcal{E})}^P} - (L_{0,n} - L_{0,p}),$$

with $f_{0,n} - f_{0,p} \in \mathcal{Q}(\mathcal{E})$, and

$$\begin{aligned} L_{0,n} - L_{0,p} &= P_{\overline{\mathcal{L}(\mathcal{E})}^P}\left[[f_{0,n}]_{L_2(S,\mathcal{S},P)}\right] - P_{\overline{\mathcal{L}(\mathcal{E})}^P}\left[[f_{0,p}]_{L_2(S,\mathcal{S},P)}\right] \\ &= P_{\overline{\mathcal{L}(\mathcal{E})}^P}\left[[f_{0,n}]_{L_2(S,\mathcal{S},P)} - [f_{0,p}]_{L_2(S,\mathcal{S},P)}\right] \\ &= P_{\overline{\mathcal{L}(\mathcal{E})}^P}\left[[(f_{0,n} - f_{0,p})]_{L_2(S,\mathcal{S},P)}\right]. \end{aligned}$$

Consequently [(Proposition) 5.6.20]

$$\left\|\tilde{B}_{0,n} - \tilde{B}_{0,p}\right\|_{HS}^2 = E_P\left[\left(\dot{Q}_{0,n} - \dot{Q}_{0,p}\right)^2\right] = \left\|Q_{0,n} - Q_{0,p}\right\|_{L_2(S,\mathcal{S},P)}^2.$$

There is thus a Hilbert-Schmidt limit $\tilde{B}_0$ to the sequence $\{\tilde{B}_{0,n},\ n \in \mathbb{N}\}$, and

$$\tilde{B}_0[f] = P_{\overline{\mathcal{L}(\mathcal{E})}^P}\left[[\dot{Q}_0\dot{f}]_{L_2(S,\mathcal{S},P)}\right],$$

with, in $L_2(S, \mathcal{S}, P)$,

$$Q_0 = \lim_n [Q_{0,n}]_{L_2(S,\mathcal{S},P)}, \quad \text{and} \quad \left\|\tilde{B}_0\right\|_{HS}^2 = E_P\left[\dot{Q}_0^2\right].$$

□

5.6.5 Discrimination of Gaussian Translates

In the Gaussian case, discrimination, as exhibited in Sect. 5.4, can be successfully completed. A Gaussian translate is a Gaussian law, with mean zero, to which a mean has been added.

Lemma 5.6.22 *Let P be a Gaussian probability whose mean is zero. Suppose that*

$$m \in H(\mathcal{H}_{\mathcal{E},P}, T),$$

and let (Sect. 5.5.2) $f_m = U_P^\star[m]$. *Then, for $f \in \overline{\mathcal{L}(\mathcal{E})}^P$, fixed, but arbitrary,*

$$\int_S e^{\dot{f}_m(s) + \iota\dot{f}(s)} P(ds) = e^{-\frac{1}{2}\|f - \iota f_m\|_{L_2(S,\mathcal{S},P)}^2}.$$

Proof Let $\{f_n, n \in \mathbb{N}\} \subseteq \mathcal{L}(\mathcal{E})$, and $\{f_n^{(m)}, n \in \mathbb{N}\} \subseteq \mathcal{L}(\mathcal{E})$ be such that, in $L_2(S, \mathcal{S}, P)$,

$$f = \lim_n [f_n]_{L_2(S,\mathcal{S},P)}$$

$$f_m = \lim_n \left[f_n^{(m)}\right]_{L_2(S,\mathcal{S},P)}.$$

Some subsequence will converge almost surely, with respect to P. Suppose one has already chosen such a subsequence. One may assume, choosing some of the coefficients to be zero, that

$$f_n = \sum_{i=1}^{n} \alpha_i^{(n)} \mathcal{E}_{t_i^{(n)}}, \quad \text{and} \quad f_n^{(m)} = \sum_{i=1}^{n} \beta_i^{(n)} \mathcal{E}_{t_i^{(n)}}.$$

The sequence with terms $g_n = e^{f_n^{(m)} + \iota f_n}$ is uniformly integrable [192, p. 19]. Indeed, choosing $G(t) = t^{1+\alpha}$, $\alpha > 0$,

$$G(|g_n|) = e^{(1+\alpha) f_n^{(m)}},$$

and, $f_n^{(m)}$ being a Gaussian random variable (with a mean equal to zero),

$$E_P[G(|g_n|)] = e^{\frac{1}{2}(1+\alpha)^2 \left\| \left[f_n^{(m)}\right]_{L_2(S,\mathcal{S},P)} \right\|^2_{L_2(S,\mathcal{S},P)}},$$

which is uniformly bounded. Thus

$$\lim_n \int_S e^{f_n^{(m)}(s) + \iota f_n(s)} P(ds) = \int_S e^{\dot{f}_m(s) + \iota \dot{f}(s)} P(ds).$$

On the other hand, either by doing the computation, or by invoking a complex variables result on entire functions which are equal on the reals [66, 200, p. 75,11], one may justify the following formal trick:

$$\int_S e^{f_n^{(m)}(s) + \iota f_n(s)} P(ds) = \int_S e^{\iota\left(f_n(s) - \iota f_n^{(m)}(s)\right)} P(ds)$$

$$= e^{-\frac{1}{2} \left\| [f_n]_{L_2(S,\mathcal{S},P)} - \iota \left[f_n^{(m)}\right]_{L_2(S,\mathcal{S},P)} \right\|^2_{L_2(S,\mathcal{S},P)}}.$$

The last term has, as limit, the value given in the lemma's statement. □

Proposition 5.6.23 *Let S be one of the spaces* [(Example) 5.1.3] *to* [(Example) 5.1.6], *and let P be a Gaussian measure on $\mathcal{S}$, with mean equal to zero. Suppose that m belongs to $H(\mathcal{H}_{\mathcal{E},P}, T)$, and let $f_m = U_P^{\star}[m]$.*

Then P and $P_m = P \circ T_m^{-1}$, where $T_m(s) = s + m$, $s \in S$, are mutually absolutely continuous, and, almost surely, with respect to P,

$$\frac{dP_m}{dP}(s) = e^{\dot{f}_m(s) - \frac{1}{2}\|f_m\|^2_{L_2(S,\mathcal{S},P)}}.$$

Proof Let $f \in \overline{\mathcal{L}(\mathcal{E})}^P$ be fixed, but arbitrary. Then

$$\int_S e^{\iota\dot{f}(s)} \left\{ e^{\dot{f}_m(s) - \frac{1}{2}\|f_m\|^2_{L_2(S,\mathcal{S},P)}} \right\} P(ds) =$$

$$= e^{-\frac{1}{2}\|f_m\|^2_{L_2(S,\mathcal{S},P)}} \int_S e^{\dot{f}_m(s) + \iota\dot{f}(s)} P(ds)$$

$$= e^{-\frac{1}{2}\|f_m\|^2_{L_2(S,\mathcal{S},P)}} e^{-\frac{1}{2}\|f - \iota f_m\|^2_{L_2(S,\mathcal{S},P)}}$$

$$= e^{\iota\langle f, f_m\rangle_{L_2(S,\mathcal{S},P)} - \frac{1}{2}\|f\|^2_{L_2(S,\mathcal{S},P)}}.$$

Another limiting argument, similar to that used in the proof of (Lemma) 5.6.22, yields that

$$\int_S e^{\iota\dot{f}(s)} P_m(ds) = e^{\iota\langle f, f_m\rangle_{L_2(S,\mathcal{S},P)} - \frac{1}{2}\|f\|^2_{L_2(S,\mathcal{S},P)}}.$$

Thus

$$\int_S e^{\iota\dot{f}(s)} \left\{ e^{\dot{f}_m(s) - \frac{1}{2}\|f_m\|^2_{L_2(S,\mathcal{S},P)}} \right\} P(ds) = \int_S e^{\iota\dot{f}(s)} P_m(ds),\ f \in \overline{\mathcal{L}(\mathcal{E})}^P.$$

Since probabilities on $\mathcal{S}$ are determined by their finite dimensional distributions, the result is proved. □

Corollary 5.6.24 *Let S be one of the spaces* [(Example) 5.1.3] *to* [(Example) 5.1.6], *and let P be a Gaussian measure on $\mathcal{S}$, with mean equal to zero. Suppose that $\{m_1, m_2\} \subseteq S$, and let $P_{m_1} = P \circ T_{m_1}^{-1}$, and $P_{m_2} = P \circ T_{m_2}^{-1}$. Then:*

1. *when $m_1 - m_2 \in H(\mathcal{H}_{\mathcal{E},P}, T)^c$, P_{m_1} and P_{m_2} are orthogonal;*
2. *when $m_1 - m_2 \in H(\mathcal{H}_{\mathcal{E},P}, T)$, P_{m_1} and P_{m_2} are mutually absolutely continuous, and, almost surely with respect to P_{m_1},*

$$\frac{dP_{m_2}}{dP_{m_1}}(s) = \frac{dP_{m_2 - m_1}}{dP} \circ T_{m_1}^{-1}(s) = e^{\dot{f}_{m_2 - m_1}(s - m_1) - \frac{1}{2}\left\|f_{m_2 - m_1}\right\|^2_{L_2(S,\mathcal{S},P)}}.$$

Furthermore, when $m_1 \in H(\mathcal{H}_{\mathcal{E},P}, T)$,

$$\frac{dP_{m_2}}{dP_{m_1}}(s) = e^{(\dot{f}_{m_2} - \dot{f}_{m_1})(s) - \frac{1}{2}\left\{\|f_{m_2}\|^2_{L_2(S,\mathcal{S},P)} - \|f_{m_1}\|^2_{L_2(S,\mathcal{S},P)}\right\}}.$$

Proof Item 1 is an application of (Corollary) 5.5.9.

When $m_1 - m_2 \in H(\mathcal{H}_{\mathcal{E},P}, T)$, $m_2 - m_1$ is also in $H(\mathcal{H}_{\mathcal{E},P}, T)$, and one has that $P_{m_2 - m_1}$ and P are mutually absolutely continuous [(Proposition) 5.6.23]. Because of (Lemmas) 5.5.6 and 5.5.7, as $P_{m_2} = P_{m_2 - m_1} \circ T^{-1}_{m_1}$, and $P_{m_1} = P \circ T^{-1}_{m_1}$, P_{m_2} and P_{m_1} are mutually absolutely continuous, with the following Radon-Nikodým derivative:

$$\frac{dP_{m_2}}{dP_{m_1}} = \frac{dP_{m_2 - m_1}}{dP} \circ T^{-1}_{m_1}.$$

But, because of (Proposition) 5.6.23,

$$\frac{dP_{m_2 - m_1}}{dP} = e^{\dot{f}_{m_2 - m_1} - \frac{1}{2}\|f_{m_2 - m_1}\|^2_{L_2(S,\mathcal{S},P)}},$$

and the first form, given above, for the Radon-Nikodým derivative obtains.

Suppose now that m_1 belongs to $H(\mathcal{H}_{\mathcal{E},P}, T)$, and let $f = \sum_{i=1}^n \alpha_i \mathcal{E}_{t_i}$. Then $m_2 \in H(\mathcal{H}_{\mathcal{E},P}, T)$,

$$f(s - m_1) = f(s) - \sum_{i=1}^n \alpha_i m_1(t_i),$$

and, with $f_{m_1} = U_P^\star[m_1]$,

$$\begin{aligned}\sum_{i=1}^n \alpha_i m_1(t_i) &= \sum_{i=1}^n \alpha_i \langle m_1, \mathcal{H}_{\mathcal{E},P}(\cdot, t_i)\rangle_{H(\mathcal{H}_{\mathcal{E},P},T)} \\ &= \left\langle m_1, \sum_{i=1}^n \alpha_i \mathcal{H}_{\mathcal{E},P}(\cdot, t_i)\right\rangle_{H(\mathcal{H}_{\mathcal{E},P},T)} \\ &= \left\langle f_{m_1}, \sum_{i=1}^n \alpha_i \mathcal{E}_{t_i}\right\rangle_{L_2(S,\mathcal{S},P)} \\ &= \langle f_{m_1}, f\rangle_{L_2(S,\mathcal{S},P)}.\end{aligned}$$

Let now

$$\lim_n \|f - f_n\|_{L_2(S,\mathcal{S},P)} = 0, \quad \text{and} \quad \dot{f}_n = \sum_{i=1}^{p_n} \alpha_i^{(n)} \mathcal{E}_{t_i^{(n)}}.$$

By taking, if necessary, a subsequence, one may assume that convergence takes place almost surely, so that, almost surely with respect to P,

$$\begin{aligned}\dot{f} \circ T_{m_1}^{-1}(s) &= \lim_n \dot{f}_n \circ T_{m_1}^{-1}(s) \\ &= \lim_n \left\{\dot{f}_n(s) - \langle f_{m_1}, f_n\rangle_{L_2(S,\mathcal{S},P)}\right\} \\ &= \dot{f}(s) - \langle f_{m_1}, f\rangle_{L_2(S,\mathcal{S},P)} .\end{aligned}$$

But, because

$$U_P\left[f_{m_2-m_1}\right] = m_2 - m_1 = U_P\left[f_{m_2}\right] - U_P\left[f_{m_1}\right] = U_P\left[f_{m_2} - f_{m_1}\right],$$

one has that

$$\dot{f}_{m_2-m_1} - \frac{1}{2}\left\|f_{m_2-m_1}\right\|^2_{L_2(S,\mathcal{S},P)} = \left(\dot{f}_{m_2} - \dot{f}_{m_1}\right) - \frac{1}{2}\left\|f_{m_2} - f_{m_1}\right\|^2_{L_2(S,\mathcal{S},P)} .$$

Moreover

$$\dot{f}_{m_2} \circ T_{m_1}^{-1} = \dot{f}_{m_2} - \langle f_{m_1}, f_{m_2}\rangle_{L_2(S,\mathcal{S},P)},$$

and

$$\dot{f}_{m_1} \circ T_{m_1}^{-1} = \dot{f}_{m_1} - \langle f_{m_1}, f_{m_1}\rangle_{L_2(S,\mathcal{S},P)},$$

so that

$$\begin{aligned}&\left\{\left(\dot{f}_{m_2} - \dot{f}_{m_1}\right) \circ T_{m_1}^{-1} - \frac{1}{2}\left\|f_{m_2} - f_{m_1}\right\|^2_{L_2(S,\mathcal{S},P)}\right\} = \\ &\quad = \left(\dot{f}_{m_2} - \dot{f}_{m_1}\right) - \frac{1}{2}\left\{\left\|f_{m_2}\right\|^2_{L_2(S,\mathcal{S},P)} - \left\|f_{m_1}\right\|^2_{L_2(S,\mathcal{S},P)}\right\} .\end{aligned}$$

□

5.6.6 *Discrimination of Gaussian Laws*

In this section, S shall be one of the spaces (Examples) 5.1.3 to 5.1.6, and P and Q, fixed, but arbitrary Gaussian measures on $\mathcal{S}$, with a mean equal to zero. It shall also be assumed that, on $\mathcal{Q}(\mathcal{E})$, P dominates Q, and Q dominates P. These assumptions have, as seen, the following consequences:

1. Because of (Corollaries) 5.4.9 and 5.4.17, there exists $\left[f_Q\right]_{L_2(S,\mathcal{S},P)} \in \overline{\mathcal{Q}(\mathcal{E})}^P$ such that

(i) $f_Q \geq 0$, almost surely with respect to P,
(ii) $\int_S f_Q(s) P(ds) = 1$,
(iii) for all $f \in \overline{\mathcal{Q}(\mathcal{E})}^P$, there exists $\dot{f} \in f$ such that

$$\int_S \dot{f}(s) Q(ds) = \int_S \dot{f}(s) f_Q(s) P(ds).$$

2. Because of (Proposition) 5.6.21, with f_0 becoming f_Q, one may define three linear and bounded operators:

   (i) for $f \in \overline{\mathcal{L}(\mathcal{E})}^P$, $M_Q[f] = [\dot{f} f_Q]_{L_2(S,\mathcal{S},P)}$,
   (ii) $B_Q = P_{\overline{\mathcal{L}(\mathcal{E})}^P} M_Q$,
   (iii) $\tilde{B}_Q = B_Q - I_{\overline{\mathcal{L}(\mathcal{E})}^P}$.
   The constant μ_0, in (Proposition) 5.6.21, is one because of the fact that the function $f_0 = f_Q$ integrates to one.

3. B_Q is strictly positive and self-adjoint:
   Indeed, from (Proposition) 5.4.15, one has that B_Q has a multiplicative decomposition of the following form:

$$U^\star M^\star J_{P,Q}^\star J_{P,Q} M U,$$

   where U is unitary, and M, multiplication by a strictly positive constant. Thus

$$\langle B_Q[f], f\rangle_{L_2(S,\mathcal{S},P)} = \|J_{P,Q} M U[f]\|^2_{H(\mathcal{H}_{\mathcal{E},P})}.$$

   When $\langle B_Q[f], f\rangle_{L_2(S,\mathcal{S},P)} = 0$,

$$MU[f] \in \mathcal{N}[J_{P,Q}],$$

   which is the orthogonal complement of the range of $J_{P,Q}^\star$ in $H(\mathcal{H}_{\mathcal{E},P}, T)$, that is, the orthogonal complement of $H(\mathcal{H}_{\mathcal{E},Q}, T)$ in $H(\mathcal{H}_{\mathcal{E},P}, T)$. But mutual domination of P and Q on $\mathcal{Q}(\mathcal{E})$ has, among its consequences, that, as sets, $H(\mathcal{H}_{\mathcal{E},Q}, T) = H(\mathcal{H}_{\mathcal{E},P}, T)$. Consequently $J_{P,Q}$ is an injection and f is the class of the zero function.
   It is here that mutual domination comes in for the first time. The second is in (Remark) 5.6.32.
4. $\tilde{B}_Q$ is Hilbert-Schmidt.

   The following notation shall be used:

(a) for the orthonormal representation of $\tilde{B}_Q$:

$$\sum_{i=1}^{\infty} \beta_i b_i \otimes b_i$$

where $\beta_i > -1,\ i \in \mathbb{N}$ (from item 3 above, as B_Q is strictly positive);

(b) for the subspace of $\overline{\mathcal{L}(\mathcal{E})}^P$ generated by $\{b_i,\ i \in \mathbb{N}\}$: H_Q;

(c) for the orthogonal complement of H_Q in $\overline{\mathcal{L}(\mathcal{E})}^P$: $H_Q^{\perp}$.

The calculations leading to discrimination shall be "parceled" into successive steps.

Lemma 5.6.25 *For $x > -1$,*

$$\frac{e^{1/2\left(\frac{x}{1+x}\right)}}{\sqrt{1+x}} \leq e^{1/2}\left(1+x^2\right).$$

Proof For $y > 0$, $y < e^y \leq \left\{1 + \left(\frac{1}{y} - 1\right)^2\right\}^2 e^y$, so that, for $y > 0$,

$$ye^{-y} \leq \left\{1 + \left(\frac{1}{y} - 1\right)^2\right\}^2.$$

Letting $y = \frac{1}{1+x}$, $x > -1$, it follows that $\frac{1}{y} - 1 = x$, and that

$$\frac{e^{-\frac{1}{1+x}}}{1+x} \leq \left\{1 + x^2\right\}^2,$$

so that, taking the square root,

$$\frac{e^{-\frac{1}{2}\frac{1}{1+x}}}{\sqrt{1+x}} \leq 1 + x^2.$$

It then suffices to multiply with $e^{1/2}$. □

Lemma 5.6.26 *In a neighborhood of 0,*

$$\frac{e^{-x/2}}{\sqrt{1-x}} \leq 1 + x^2.$$

Proof Consider the function $f(x) = (1-x)\left(1+x^2\right)^2 e^x$: $f(0) = 1$. Now, D denoting the derivative with respect to x,

$$Df(x) = -x\left(1+x^2\right)\left(x^2+4x-3\right)e^x,$$
$$D^2f(x) = -e^x\left\{\left(1-x^2\right)\left(x^2+4x-3\right) + x\left(1+x^2\right)\left(x^2+6x+1\right)\right\},$$

so that $Df(0) = 0$, and $D^2 f(0) = 3$. Consequently, in a neighborhood of $0, f(x) \geq 1$, so that

$$\frac{e^{-x}}{1-x} \leq \left(1 + x^2\right)^2.$$

One then takes the square root of the latter expression. □

In the next lemma, one introduces the exponent value p. It is there to allow for the Radon-Nikodým derivative to be integrable to the power p.

Lemma 5.6.27 *Let* $p > 1$ *be such that* $\max_{\beta_i > 0} \beta_i < \frac{1}{p-1}$, *and*

(a) $\gamma_i = \frac{\beta_i}{2(1+\beta_i)}$,
(b) $\delta_i = \frac{e^{\gamma_i}}{\sqrt{1+\beta_i}}$,
(c) $D_i(s) = \delta_i e^{\gamma_i \{b_i^2(s) - 1\}}$,
(d) $D_n(s) = \prod_{i=1}^n D_i(s)$.

Then:

1. $\{\prod_{i=1}^\infty \delta_i\}^p = \prod_{i=1}^\infty \delta_i^p$ *is convergent;*
2. $E_P\left[D_n^p\right] \leq \kappa < \infty$, *so that the sequence* $\{D_n,\ n \in \mathbb{N}\}$ *is uniformly integrable.*

Proof Item 1 follows from the fact (which uses (Lemma) 5.6.25) that

$$\prod_{i=1}^\infty \delta_i^p = \left\{\prod_{i=1}^\infty \frac{e^{\frac{1}{2}\frac{\beta_i}{1+\beta_i}}}{(1+\beta_i)^{1/2}}\right\}^p \leq \left\{e^{1/2} \prod_{i=1}^\infty \left(1 + \beta_i^2\right)\right\}^p,$$

and that, from the Hilbert-Schmidt property, $\sum_{i=1}^\infty \beta_i^2 < \infty$.

Now, $X_1, \ldots, X_n$ being independent, standard normal random variables (they stand for the b_i's),

$$E_P\left[D_n^p\right] = \prod_{i=1}^n \delta_i^p \prod_{i=1}^n E_P\left[e^{p\gamma_i\left(X_i^2 - 1\right)}\right].$$

As $1 - 2p\gamma_i = \frac{1-(p-1)\beta_i}{1+\beta_i} > 0$, and that, for $\alpha < 1/2$,

$$\int_{-\infty}^{+\infty} e^{\alpha x^2} \left\{\frac{e^{-\frac{x^2}{2}}}{\sqrt{2\pi}}\right\} dx = \frac{1}{\sqrt{1-2\alpha}},$$

one has that

$$E\left[e^{p\gamma_i\left(X_i^2-1\right)}\right] = \frac{e^{-p\gamma_i}}{\sqrt{1-2p\gamma_i}}.$$

Let $\phi_i = \frac{p\beta_i}{1+\beta_i} = 2p\gamma_i$. Then, from (Lemma) 5.6.26,

$$\frac{e^{-p\gamma_i}}{\sqrt{1-2p\gamma_i}} = \frac{e^{-\frac{1}{2}\phi_i}}{\sqrt{1-\phi_i}} \leq 1+\phi_i^2,$$

and thus, as, for $x > 0$,

$$(1+x)^2 + p^2x^2 \leq (1+x)^2 + p^2(1+x)^2 = (1+p^2)(1+x)^2 \leq 2(1+p^2)(1+x^2),$$

so that

$$1 + \frac{p^2x^2}{(1+x)^2} \leq \frac{2(1+p^2)(1+x^2)}{(1+x)^2} \leq 2(1+p^2)(1+x^2),$$

then

$$1+\phi_i^2 \leq \kappa^2\left(1+\beta_i^2\right).$$

Consequently

$$\prod_{i=1}^{n} E_P\left[e^{p\gamma_i(X_i^2-1)}\right]$$

is dominated by a convergent product, and so $E_P\left[D_n^p\right]$ is dominated by a product of convergent products. □

Lemma 5.6.28 *The series* $\sum_{i=1}^{\infty}\gamma_i\left\{\dot{b}_i^2-1\right\}$ *is convergent in* $L_2(S, \mathcal{S}, P)$.

Proof Given $\epsilon > 0$, there is $n(\epsilon)$ such that $|\beta_n| < \epsilon$, $n > n(\epsilon)$. Consequently, for $n > n(\epsilon)$, $1+\beta_n > 1-\epsilon$, and $\gamma_n < \frac{\beta_n}{1-\epsilon}$. Thus

$$\lim_{p,q} E_P\left[\left\{\sum_{i=p}^{q}\gamma_i\left\{\dot{b}_i^2-1\right\}\right\}^2\right] = \lim_{p,q} 2\sum_{i=p}^{q}\gamma_i^2 = 0.$$

□

Proposition 5.6.29 *The notation being that of the previous lemmas, the sequence* $\{D_n,\ n\in\mathbb{N}\}$ *converges to* $D = \prod_{i=1}^{\infty} D_i$, *in* $L_p(S, \mathcal{S}, P)$, *and*

$$E_P\left[D^p\right] < \infty.$$

Proof Because of (Lemma) 5.6.28, $\left\{D_n^p,\ n\in\mathbb{N}\right\}$ converges in probability to D. That fact combined with uniform convergence does the trick. □

Lemma 5.6.30 *The notation being that of the previous lemmas, one has that:*

1. (i) $E_P[D_i] = 1$;
   (ii) $E_P\left[\dot{b}_i D_i\right] = 0$;
   (iii) $E_P\left[\dot{b}_i^2 D_i\right] = 1 + \beta_i$;
   (iv) $E_P[D] = 1$;
   (v) $E_P\left[\dot{b}_i D\right] = E_P\left[\dot{b}_i D_i\right]$.
2. *Let* $\{f, f_1, f_2\} \subseteq \overline{\mathcal{L}(\mathcal{E})}^P$ *be fixed, but arbitrary, and have the respective decompositions*

$$f = g + h,\ f_1 = g_1 + h_1,\ f_2 = g_2 + h_2,$$

where $\{g, g_1, g_2\} \subseteq H_Q$, $\{h, h_1, h_2\} \subseteq H_Q^{\perp}$: *then*

   (i) $E_P\left[\dot{f}^2 D\right] = E_P\left[\left(\dot{g}^2 + \dot{h}^2\right) D\right] < \infty$;
   (ii) $E_P\left[\dot{f} D\right] = E_P\left[\dot{g} D\right] = 0$;
   (iii) $\left\langle [f_1]_{L_2(S,\mathcal{S},Q)}, [f_2]_{L_2(S,\mathcal{S},Q)} \right\rangle_{L_2(S,\mathcal{S},Q)} = \int_S \dot{f}_1(s) \dot{f}_2(s) D(s) P(ds)$.

Proof For $1 - 2\alpha > 0$, the following function:

$$f_\alpha(x) = \sqrt{\frac{1 - 2\alpha}{2\pi}}\, e^{-\frac{1}{2}(1-2\alpha)x^2},$$

is a Gaussian density, whose mean is equal to zero, and variance, equal to $(1-2\alpha)^{-1}$. It may be written in the following form:

$$f_\alpha(x) = \sqrt{1 - 2\alpha}\, e^{\alpha}\, e^{\alpha\left(x^2-1\right)} \frac{e^{-\frac{1}{2}x^2}}{\sqrt{2\pi}},$$

so that, in particular, since $1 - 2\gamma_i = \frac{1}{1+\beta_i} > 0$,

$$f_{\gamma_i}(x) = \delta_i e^{\gamma_i\left(x^2-1\right)} \frac{e^{-\frac{1}{2}x^2}}{\sqrt{2\pi}}$$

is a Gaussian density, with a mean equal to zero and a variance equal to $1-2\gamma_i$. Thus, for any Borel measurable function F, with appropriate integrability properties, since the $\dot{b}_i$'s are, with respect to P, independent, standard normal random variables,

$$\int_S F\left(\dot{b}_i(s)\right) D_i(s) P(ds) = \int_{\mathbb{R}} F(x)\, \delta_i e^{\gamma_i\{x^2-1\}} \frac{e^{-\frac{1}{2}x^2}}{\sqrt{2\pi}} dx = \int_{\mathbb{R}} F(x) f_{\gamma_i}(x)\, dx.$$

Consequently,

(i) $F(x) = 1$ yields

$$E_P[D_i] = 1, \text{ and then, because of (Proposition) 5.6.29, } E_P[D] = 1;$$

(ii) $F(x) = x$ yields

$$E_P\left[\dot{b}_i D_i\right] = 0;$$

(iii) $F(x) = x^2$ yields

$$E_P\left[\dot{b}_i^2 D_i\right] = \frac{1}{1 - 2\gamma_i} = 1 + \beta_i.$$

Finally, point (v) of item 1 is true using independence. Item 1 thus obtains.

Let $f \in H_Q^{\perp}$ be fixed, but arbitrary. Then, because of independence,

$$E_P\left[\dot{f}^2 D\right] = E_P\left[\dot{f}^2\right] E_P[D] < \infty.$$

If now $f \in H_Q$, let $f_i = \langle f, b_i \rangle_{L_2(S,\mathcal{S},P)}$. Then, almost surely for P,

$$\dot{f} = \sum_{i=1}^{\infty} f_i \dot{b}_i,$$

and, since $\dot{b}_i^2 D$ is integrable with respect to P, for all i, as, using independence, $E_P[\dot{b}_i \dot{b}_j D] = E_P[\dot{b}_i^2 D]$, when $i = j$, and zero otherwise, then

$$\begin{aligned} E_P\left[\left\{\sum_{i=1}^{n} f_i \dot{b}_i\right\}^2 D\right] &= \sum_{i=1}^{n}\sum_{j=1}^{n} f_i f_j E_P\left[\dot{b}_i \dot{b}_j D\right] \\ &= \sum_{i=1}^{n} f_i^2 (1 + \beta_i) \\ &\le \left(1 + \max_i |\beta_i|\right) \|f\|_{L_2(S,\mathcal{S},P)}. \end{aligned}$$

Thus, by Fatou's lemma,

$$E_P\left[\dot{f}^2 D\right] \le \liminf_n E_P\left[\left\{\sum_{i=1}^{n} f_i \dot{b}_i\right\}^2 D\right] \le \left(1 + \max_i |\beta_i|\right) \|f\|_{L_2(S,\mathcal{S},P)}.$$

Since $(|a| - |b|)^2$ is positive, $2|ab| \leq a^2 + b^2$, so that

$$2E_P\left[|\dot{g}\dot{h}|D\right] \leq E_P[\dot{g}^2D] + E_P[\dot{h}^2D] < \infty.$$

As a consequence, using independence again, and the fact that P has zero mean,

$$E_P\left[\dot{g}\dot{h}D\right] = E_P\left[\dot{g}D\right]E_P\left[\dot{h}\right] = 0.$$

But, as $E_P[|\dot{g}|D] \leq E_P[\dot{g}^2D] + E_P[D] < \infty$, $\dot{g}D$ is integrable, and thus

$$\begin{aligned} E_P\left[\dot{g}D\right] &= \sum_{i=1}^{\infty} \langle g, b_i \rangle_{L_2(S,\mathcal{S},P)} E_P\left[\dot{b}_i D\right] \\ &= \sum_{i=1}^{\infty} \langle g, b_i \rangle_{L_2(S,\mathcal{S},P)} E_P\left[\dot{b}_i D_i\right] \\ &= 0. \end{aligned}$$

One has already been reminded that, for $f \in \overline{\mathcal{Q}(\mathcal{E})}^P$, there exists $\dot{f}$ such that

$$\int_S \dot{f}(s)\, Q(ds) = \int_S \dot{f}(s) f_Q(s)\, P(ds).$$

Consequently, as $\{f_1, f_2\} \subseteq \overline{\mathcal{L}(\mathcal{E})}^P$,

$$f_1 f_2 \in \overline{\mathcal{Q}(\mathcal{E})}^P,$$

and

$$\int_S \dot{f}_1(s)\dot{f}_2(s)\, Q(ds) = \int_S \dot{f}_1(s)\dot{f}_2(s) f_Q(s)\, P(ds).$$

The right-hand side of the latter equality "reads as"

$$\begin{aligned} \left\langle f_1, \left[f_2 f_Q\right]_{L_2(S,\mathcal{S},P)} \right\rangle_{L_2(S,\mathcal{S},P)} &= \langle f_1, M_Q[f_2] \rangle_{L_2(S,\mathcal{S},P)} \\ &= \left\langle f_1, P_{\overline{\mathcal{L}(\mathcal{E})}^P} M_Q[f_2] \right\rangle_{L_2(S,\mathcal{S},P)} \\ &= \left\langle f_1, \left\{ I_{\overline{\mathcal{L}(\mathcal{E})}^P} + \tilde{B}_0 \right\} [f_2] \right\rangle_{L_2(S,\mathcal{S},P)}. \end{aligned}$$

Now

$$I_{\overline{\mathcal{L}(\mathcal{E})}^P} + \tilde{B}_0 = \sum_{i=1}^{\infty} (1+\beta_i)\, b_i \otimes b_i + P_{H_Q^{\perp}},$$

and thus, replacing, in the left-hand side below, f_1 and f_2 by their respective decompositions $g_1 + h_1$ and $g_2 + h_2$,

$$\left\langle f_1, \left\{ I_{\overline{\mathcal{L}(\mathcal{E})}^P} + \tilde{B}_0 \right\} [f_2] \right\rangle_{L_2(S,\mathcal{S},P)} =$$

$$= \sum_{i=1}^{\infty} (1+\beta_i) \langle g_1, b_i \rangle_{L_2(S,\mathcal{S},P)} \langle g_2, b_i \rangle_{L_2(S,\mathcal{S},P)} + \langle h_1, h_2 \rangle_{L_2(S,\mathcal{S},P)} \cdot$$

It is this latter equality that yields part (iii) of item 2 in (Lemma) 5.6.30. Indeed

$$\begin{aligned}\int_S \dot{f}_1(s) \dot{f}_2(s) D(s) P(ds) &= \int_S \dot{g}_1(s) \dot{g}_2(s) D(s) P(ds) \\ &+ \int_S \dot{g}_1(s) \dot{h}_2(s) D(s) P(ds) \\ &+ \int_S \dot{h}_1(s) \dot{g}_2(s) D(s) P(ds) \\ &+ \int_S \dot{h}_1(s) \dot{h}_2(s) D(s) P(ds),\end{aligned}$$

and, using what has already been proved, and, in particular, independence,

$$\begin{aligned}\int_S \dot{h}_1(s) \dot{h}_2(s) D(s) P(ds) &= \int_S \dot{h}_1(s) \dot{h}_2(s) P(ds) \int_S D(s) P(ds) \\ &= \int_S \dot{h}_1(s) \dot{h}_2(s) P(ds),\end{aligned}$$

$$\begin{aligned}\int_S \dot{h}_1(s) \dot{g}_2(s) D(s) P(ds) &= \int_S \dot{h}_1(s) D(s) P(ds) \int_S \dot{g}_2(s) P(ds) \\ &= 0,\end{aligned}$$

and

$$\int_S \dot{g}_1(s) \dot{g}_2(s) D(s) P(ds) =$$

$$= \sum_{i=1}^{\infty} \sum_{j=1}^{\infty} \langle g_1, b_i \rangle_{L_2(S,\mathcal{S},P)} \left\langle g_2, b_j \right\rangle_{L_2(S,\mathcal{S},P)} \int_S \dot{b}_i(s) \dot{b}_j(s) D(s) P(ds)$$

$$= \sum_{i=1}^{\infty} \langle g_1, b_i \rangle_{L_2(S,\mathcal{S},P)} \langle g_2, b_i \rangle_{L_2(S,\mathcal{S},P)} (1 + \beta_i) .$$

Part (iii) of item 2 is thus established. □

Proposition 5.6.31 *The prescription $d\tilde{Q} = DdP$ defines a probability measure, and $\tilde{Q} = Q$, so that Q is absolutely continuous to P, with D as Radon-Nikodým derivative.*

Proof Since $D \geq 0$, and that $E_P[D] = 1$, $d\tilde{Q} = DdP$ defines a probability law. Furthermore

$$E_{\tilde{Q}}\left[e^{\iota\theta\dot{b}_i}\right] = E_P\left[e^{\iota\theta\dot{b}_i} D\right] = \int_{\mathbb{R}} e^{\iota\theta x} f_{\gamma_i}(x)\, dx = e^{-\frac{1}{2}(1+\beta_i)\theta^2},$$

so that, using independence,

$$E_{\tilde{Q}}\left[e^{\iota\theta\dot{f}}\right] = E_P\left[e^{\iota\theta\left(\dot{g}+\dot{h}\right)} D\right] = E_P\left[e^{\iota\theta\dot{g}} D\right] E_P\left[e^{\iota\theta\dot{h}}\right],$$

and

$$\begin{aligned} E_P\left[e^{\iota\theta\dot{g}} D\right] &= \prod_{i=1}^{\infty} E_P\left[e^{\iota\theta\langle g, b_i\rangle_{L_2(S,\mathcal{S},P)} \dot{b}_i} D_i(s)\right] \\ &= e^{-\frac{1}{2}\theta^2 \sum_{i=1}^{\infty}(1+\beta_i)\langle g, b_i\rangle^2_{L_2(S,\mathcal{S},P)}}. \end{aligned}$$

But

$$\sum_{i=1}^{\infty} (1 + \beta_i) \langle g, b_i \rangle^2_{L_2(S,\mathcal{S},P)} = \int_S \dot{g}^2(s)\, D(s)\, P(ds),$$

and, since, with respect to P, $\dot{h}$ is Gaussian with a mean that equals zero,

$$E_P\left[e^{\iota\theta\dot{h}}\right] = e^{-\frac{1}{2}E_P\left[\dot{h}^2\right]}.$$

Consequently (Corollary 5.4.9 for the last equality)

$$\begin{aligned} E_{\tilde{Q}}\left[e^{\iota\theta\dot{f}}\right] &= e^{-\frac{1}{2}\theta^2\left\{\int_S \dot{g}^2(s)D(s)P(ds) + E_P\left[\dot{h}^2\right]\right\}} \\ &= e^{-\frac{1}{2}\theta^2\|f\|^2_{L_2[Q]}} \\ &= E_Q\left[e^{\iota\theta\dot{f}}\right]. \end{aligned}$$

□

Remark 5.6.32 One knows, from (Proposition) 5.4.5, that, when P and Q do not dominate each other, they are orthogonal. Since, from (Proposition) 5.6.31, domination implies absolute continuity, one has a dichotomy: Gaussian laws are either equivalent or orthogonal.

Remark 5.6.33 In case $S = H$, a real and separable Hilbert space, and $T = H$, with $\mathcal{E}_t(h) = \langle h, t\rangle_H$, $\mathcal{S}$ is the family of Borel sets. Let P be a Gaussian measure on $\mathcal{S}$, with a mean equal to zero, and a covariance denoted R_P. Then

$$\begin{aligned}\mathcal{H}_{\mathcal{E}}(t_1, t_2) &= \int_S \mathcal{E}_{t_1}(s)\,\mathcal{E}_{t_2}(s)\,P(ds)\\ &= \int_S \langle s, t_1\rangle_H \langle s, t_2\rangle_H P(ds)\\ &= \langle R_P[t_1], t_2\rangle_H .\end{aligned}$$

Let

(i) $U_P : H \longrightarrow H^\star$ be defined using

$$U_P[x](x) = \langle x, [h]\rangle_H ,$$

(ii) $V_P : H(\mathcal{H}_{\mathcal{E},P}, T) \longrightarrow \overline{\mathcal{L}(\mathcal{E})}^P$ be defined using

$$V_P[\mathcal{H}_{\mathcal{E},P}(\cdot, t)] = [\mathcal{E}_t]_{L_2(S,\mathcal{S},P)} ,$$

(iii) $W_P : \overline{\mathcal{R}[R_P^{1/2}]} \longrightarrow H(\mathcal{H}_{\mathcal{E},P}, T)$ be defined using

$$W_P\left[R_P^{1/2}[t]\right] = \mathcal{H}_{\mathcal{E},P}(\cdot, t) .$$

U_P, V_P and W_P are unitary.

Suppose Q is another Gaussian measure on $\mathcal{S}$, with a mean equal to zero. The following relation (introduction to Sect. 5.6.6, item 1):

$$\int_S \dot{f}(s)\,Q(ds) = \int_S \dot{f}(s) f_Q(s)\,P(ds) , \; f \in \overline{\mathcal{Q}(\mathcal{E})}^P ,$$

applied to $\mathcal{E}_{t_1}\mathcal{E}_{t_2}$, yields, as seen in the proof of (Lemma) 5.6.30, that

$$\langle R_Q[t_2], t_2\rangle_H = \left\langle \left\{I_{\overline{\mathcal{L}(\mathcal{E})}^P} + \tilde{B}_Q\right\}\left[[\mathcal{E}_{t_1}]_{L_2(S,\mathcal{S},P)}\right], [\mathcal{E}_{t_1}]_{L_2(S,\mathcal{S},P)}\right\rangle_{L_2(S,\mathcal{S},P)} .$$

Since

$$[\mathcal{E}_t]_{L_2(S,\mathcal{S},P)} = V_P W_P R_P^{1/2}[t] ,$$

one obtains the usual condition, in terms of covariance operators, for equivalence on Hilbert spaces :

$$R_Q = R_P^{1/2}\,(I + B)\,R_P^{1/2}. \qquad (\star)$$

The same condition expressed in RKHS's terms comes out as

$$\mathcal{H}_{\mathcal{E},Q}\,(t_1, t_2) = \langle \{I + B\}\,[\mathcal{H}_{\mathcal{E},P}\,(\cdot, t_1)]\,, \mathcal{H}_{\mathcal{E},P}\,(\cdot, t_2)\rangle_{H(\mathcal{H}_{\mathcal{E},P},T)}\,.$$

It is thus the $I + B$ element that is significant, rather than the square roots of $(\star)$ which represent the RKHS ingredient of the problem. The form $I + B$ is ubiquitous in those matters [42, 104].

5.7 An Extension to Mixtures of Gaussian Laws

As, in practice, Gaussian laws are hard to come across, it is useful to be able to compute likelihoods when laws are not Gaussian. The simple "gimmick" of change of time allows one to extend results valid for Gaussian laws to laws that are not. As shall be seen in Chap. 16, in so doing, one gets, in some cases at least, the best one may expect. However, to check that the best is achieved may be hard, as that same chapter proves.

In what follows, one shall show that the expression

$$X_A\,(\omega, t) = X\,(\omega, A\,(\omega, t))$$

makes sense when X and A are independent stochastic processes, and that the law of X_A is a mixture of the laws of X and A. When X is the Wiener process, X_A is an Ocone martingale [205]. Those martingales are introduced and described in Sect. 16.7.

One must introduce the following ingredients:

1. $(T, \mathcal{T})$, a measurable space;
2. $\mathcal{E}$, S and $\mathcal{S}$, as in (Examples) 5.1.3 to 5.1.6;
3. $(\Omega_1, \mathcal{A}_1, P_1)$ and $(\Omega_2, \mathcal{A}_2, P_2)$, probability spaces;
4. $(\Omega, \mathcal{A}, P) = (\Omega_1 \times \Omega_2, \mathcal{A}_1 \otimes \mathcal{A}_2, P_1 \otimes P_2)$;
5. $\Pi_1 : \Omega \longrightarrow \Omega_1 : (\omega_1, \omega_2) \mapsto \omega_1$;
6. $\Pi_2 : \Omega \longrightarrow \Omega_2 : (\omega_1, \omega_2) \mapsto \omega_2$;
7. $X : \Omega_1 \times T \longrightarrow \mathbb{R}$, and $A : \Omega_2 \times T \longrightarrow T$, stochastic processes;
8. $\boldsymbol{X} : \Omega_1 \longrightarrow S$, defined using the following assignment:

$$\omega_1 \mapsto \boldsymbol{X}\,[\omega_1] = X\,(\omega_1, \cdot) = \{X(\omega_1, t), t \in T\}\,;$$

9. $\boldsymbol{A} : \Omega_2 \longrightarrow T^T$, defined, *mutatis mutandis,* as $\boldsymbol{X}$ is;

10. for $t \in T$, fixed, but arbitrary, $J_t : \Omega \longrightarrow \Omega_1 \times T$ defined using the following assignment:

$$J_t(\omega) = (\Pi_1(\omega), A(\Pi_2(\omega), t)) = (\omega_1, A(\omega_2, t));$$

11. $Y : \Omega \times T \longrightarrow \mathbb{R}$ defined using the following assignment:

$$Y(\omega, t) = X(\Pi_1(\omega), t) = X(\omega_1, t);$$

12. $\boldsymbol{Y} : \Omega \longrightarrow S$ defined, *mutatis mutandis,* as $\boldsymbol{X}$ is;
13. $Z : \Omega \times T \longrightarrow \mathbb{R}$ defined using the following assignment:

$$Z(\omega, t) = X \circ J_t(\omega) = X(\omega_1, A(\omega_2, t));$$

14. $\boldsymbol{Z} : \Omega \longrightarrow S$ defined, *mutatis mutandis,* as $\boldsymbol{X}$ is;
15. for $\omega_2^0 \in \Omega_2$, fixed, but arbitrary, $Z_{\omega_2^0}(\omega_1, t) = Z\left(\left(\omega_1, \omega_2^0\right), t\right)$;
16. $\boldsymbol{Z}_{\omega_2^0} : \Omega_1 \longrightarrow S$ defined, *mutatis mutandis,* as $\boldsymbol{X}$ is;
17. $T_n = \{t_1, \ldots, t_n) \subseteq T$, distinct points;
18. for $s \in S$,

$$\mathcal{E}_{T_n}[s] = \begin{bmatrix} s(t_1) \\ \vdots \\ s(t_n) \end{bmatrix}.$$

Lemma 5.7.1 *Let*

(a) $B \in \mathcal{B}(\mathbb{R}^n)$,
(b) $C = \mathcal{E}_{T_n}^{-1}(B) \subseteq S$,
(c) $C_\Omega = \boldsymbol{Z}^{-1}(C) \subseteq \Omega$.

Let also $C_\Omega\left[\omega_2^0\right]$ *be the section of* C_Ω *at* ω_2^0, *that is,*

$$C_\Omega\left[\omega_2^0\right] = \left\{\omega_1 \in \Omega_1 : \left(\omega_1, \omega_2^0\right) \in C_\Omega\right\} \subseteq \Omega_1.$$

Then

$$C_\Omega\left[\omega_2^0\right] = \boldsymbol{Z}_{\omega_2^0}^{-1}(C).$$

Proof One has that

$$\begin{aligned} \boldsymbol{Z}_{\omega_2^0}^{-1}(C) &= \left\{\omega_1 \in \Omega_1 : Z_{\omega_2^0}(\omega_1, \cdot) \in C\right\} \\ &= \left\{\omega_1 \in \Omega_1 : Z\left(\left(\omega_1, \omega_2^0\right), \cdot\right) \in C\right\} \end{aligned}$$

$$= \left\{\omega_1 \in \Omega_1 : \left(\omega_1, \omega_2^0\right) \in \mathbf{Z}^{-1}(C)\right\}$$
$$= C_\Omega\left[\omega_2^0\right].$$

□

Lemma 5.7.2 *Let $F : T \longrightarrow T$ be a map, and define $\Phi_F : S \longrightarrow S$ using the following rule:*

$$\Phi_F[s] = s \circ F, \ s \in S.$$

Φ_F is then adapted to $\mathcal{S}$.

Proof Let $t \in T$ be fixed, but arbitrary. It is sufficient to prove that $\mathcal{E}_t \circ \Phi_F$ is adapted. But

$$\mathcal{E}_t \circ \Phi_F(s) = \mathcal{E}_t(s \circ F) = s(F(t)) = \mathcal{E}_{F(t)}(s),$$

so that $\mathcal{E}_t \circ \Phi_F = \mathcal{E}_{F(t)}$, which is adapted. □

Lemma 5.7.3 *Let $F_{\omega_2^0} : T \longrightarrow T$ be defined using the following rule:*

$$F_{\omega_2^0}(t) = A\left(\omega_2^0, t\right).$$

Let also $P_X = P_1 \circ \mathbf{X}^{-1}$. Then $\mathbf{Z}_{\omega_2^0}$ is adapted, and

$$P_1 \circ \mathbf{Z}_{\omega_2^0}^{-1} = P_X \circ \Phi_{F_{\omega_2^0}}^{-1}.$$

Proof Let $t \in T$ be fixed, but arbitrary. It is sufficient to prove that the variable $\mathcal{E}_t \circ \mathbf{Z}_{\omega_2^0}$ is adapted. But

$$\left\{\mathcal{E}_t \circ \mathbf{Z}_{\omega_2^0}\right\}(\omega_1) = Z_{\omega_2^0}(\omega_1, t) = Z\left(\left(\omega_1, \omega_2^0\right), t\right) = X\left(\omega_1, A\left(\omega_2^0, t\right)\right),$$

so that

$$\mathcal{E}_t \circ \mathbf{Z}_{\omega_2^0} = X\left(\cdot, A\left(\omega_2^0, t\right)\right),$$

which is adapted. Furthermore

$$\left\{\Phi_{F_{\omega_2^0}}(X[\omega_1])\right\}(t) = \left\{X[\omega_1] \circ F_{\omega_2^0}\right\}(t)$$
$$= X\left(\omega_1, F_{\omega_2^0}(t)\right)$$
$$= X\left(\omega_1, A\left(\omega_2^0, t\right)\right)$$
$$= \left\{Z_{\omega_2^0}[\omega_1]\right\}(t),$$

so that $\Phi_{F_{\omega_2^0}} \circ X = Z_{\omega_2^0}$, and, consequently, that

$$P_1 \circ Z_{\omega_2^0}^{-1} = P_1 \circ \left\{\Phi_{F_{\omega_2^0}} \circ X\right\}^{-1} = P_X \circ \Phi_{F_{\omega_2^0}}^{-1}.$$

□

Lemma 5.7.4 *Given $B \in \mathcal{B}(\mathbb{R}^n)$, and $C = \mathcal{E}_{T_n}^{-1}(B)$, let*

$$Q_1^{\omega_2^0}(C) = P_1\left(C_\Omega\left[\omega_2^0\right]\right).$$

Then

$$P_1 \circ Z_{\omega_2^0}^{-1}(C) = Q_1^{\omega_2^0}(C).$$

Proof It has been shown (Lemma 5.7.1) that $Z_{\omega_2^0}^{-1}(C) = C_\Omega\left[\omega_2^0\right]$. □

Lemma 5.7.5 *Let $S_0 \in \mathcal{S}$ be fixed, but arbitrary. The map*

$$\omega_2 \mapsto P_1 \circ Z_{\omega_2}^{-1}(S_0)$$

is adapted to $\mathcal{A}_2$.

Proof Again, due to (Lemma) 5.7.1, for $B \in \mathcal{B}(\mathbb{R}^n)$, and $C = \mathcal{E}_{T_n}^{-1}(B)$, fixed, but arbitrary,

$$Z_{\omega_2}^{-1}(C) = C_\Omega[\omega_2]$$

so that

$$P_1 \circ Z_{\omega_2}^{-1}(C) = P_1(C_\Omega[\omega_2]).$$

But it is a property of product measures [113, p. 238] that $\omega_2 \mapsto P_1(C_\Omega[\omega_2])$ is adapted. That it is adapted for all $S_0 \in \mathcal{S}$ follows from [199, p. 31]. □

Proposition 5.7.6 *Let $A \in \mathcal{A}_2$, and $S_0 \in \mathcal{S}$, be fixed, but arbitrary. Then*

$$\begin{aligned} P(Z \in S_0, \Pi_2 \in A) &= \int_A P_1 \circ Z_{\omega_2}^{-1}(S_0)\, P_2(d\omega_2) \\ &= \int_A P_2(d\omega_2)\, P_X \circ \Phi_{F_{\omega_2}}^{-1}(S_0), \end{aligned}$$

so that the following assignment: $(S_0, \omega_2) \mapsto P_1 \circ Z_{\omega_2}^{-1}(S_0)$ yields a regular conditional probability of Z with respect to Π_2.

Proof Let $B \in \mathcal{B}(\mathbb{R}^n)$, and $S_0 = \mathcal{E}_{T_n}^{-1}(B)$, be fixed, but arbitrary. Let $\underline{Z}_n$ be the vector with components $Z(\omega, t_i)$, $1 \le i \le n$. Then, by Fubini's theorem,

$$\begin{aligned} P(\underline{Z} \in S_0, \Pi_2 \in A) &= P\left(\underline{Z}_n \in B, \Pi_2 \in A\right) \\ &= \int_{\Omega} P(d\omega)\, \chi_{B}\left(\underline{Z}_n(\omega)\right) \chi_{A}\left(\Pi_2(\omega)\right) \\ &= \int_{A} P_2(\omega_2) \int_{\Omega_1} P_1(d\omega_1)\, \chi_{B}\left(\underline{Z}_n(\omega_1, \omega_2)\right). \end{aligned}$$

Now, in that latter integral,

$$\begin{aligned} \chi_{B}\left(\underline{Z}_n(\omega_1, \omega_2)\right) &= \\ &= \chi_{B}\left(Z_{\omega_2}(\omega_1, t_1), \ \ldots \ , Z_{\omega_2}(\omega_1, t_n)\right) \\ &= \chi_{B}\left(\mathcal{E}_{T_n} \circ Z_{\omega_2}[\omega_1]\right). \end{aligned}$$

Consequently, as $C = \mathcal{E}_{T_n}^{-1}(B)$,

$$\int_{\Omega_1} P_1(d\omega_1)\, \chi_{B}\left(\underline{Z}_n(\omega_1, \omega_2)\right) = P_1\{\omega_1 \in \Omega_1 : Z_{\omega_2}[\omega_1] \in C\} = P_1 \circ \mathbf{Z}_{\omega_2}^{-1}(C).$$

The proposition is thus true for "cylinder sets," and, consequently, must be true [192, p. 11]. □

Example 5.7.7 Ocone martingales [(Fact) 10.3.45 and Sect. 16.7]

An Ocone martingale is a continuous martingale M whose Dambis-Dubins-Schwarz representation $W_M \diamond \langle M \rangle$ is such that the Brownian motion W_M and the quadratic variation $\langle M \rangle$ are independent. Item (Proposition) 5.7.6 leads to a formula for the law of M.

For item (Proposition) 5.7.6 to be useful, one must be able to compute likelihoods for those representations. The following results provide tools towards that goal.

Fact 5.7.8 (Product Measure Theorem [6, p. 97–104]) *Let S be one of the spaces of Examples* 5.1.3 to 5.1.6. *In the last case, assume that S is separable. Let $(X, \mathcal{X})$ be a measurable space. One shall suppose that, for the following maps:*

$$P_1 : S \times X \longrightarrow [0, 1], \quad \textit{and} \quad P_2 : S \times X \longrightarrow [0, 1],$$

these statements are true:

(a) *for $i \in \{1, 2\}$, and each $x \in X$, $P_i(\cdot, x)$ is a probability law on $\mathcal{S}$,*
(b) *for $i \in \{1, 2\}$, and each $S_0 \in \mathcal{S}$, $x \mapsto P_i(S_0, x)$ is adapted to $\mathcal{X}$.*

Given two probability laws on $\mathcal{X}$, say ξ_1 and ξ_2, one may define two probability laws on $\mathcal{S} \otimes \mathcal{X}$, say Q_1 and Q_2, setting, when $(S_0, X_0) \in \mathcal{S} \otimes \mathcal{X}$, $i \in \{1, 2\}$,

$$Q_i (S_0 \times X_0) = \int_{X_0} P_i (S_0, x) \, \xi_i (dx) .$$

Those probability laws have the following properties, for $i \in \{1, 2\}$:

1. *For $F \in \mathcal{S} \otimes \mathcal{X}$, fixed, but arbitrary, $Q_i (F) = \int_X P_i (F [x], x) \, \xi_i (dx)$.*
2. *For every $f \geq 0$, adapted to $\mathcal{S} \otimes \mathcal{X}$, the integral*

$$\int_S f (s, x) P_i (ds, x)$$

   exists, is adapted to $\mathcal{X}$, and

$$\int_{S \times X} f (s, x) Q_i (ds, dx) = \int_X \left\{ \int_S f (s, x) P_i (ds, x) \right\} \xi_i (dx) .$$

3. *When $\int_{S \times X} f (s, x) Q_i (ds, dx)$ exists (respectively, is finite), then:*
   (i) *there exists $X_0^{(i)} \in \mathcal{X}$ such that $\xi_i \left(X_0^{(i)} \right) = 1$, and, for $x \in X_0^{(i)}$,*

$$\int_S f (s, x) P_i (ds, x)$$

   exists (respectively, is finite);
   (ii) *setting*

$$\phi (x) = \begin{cases} \int_S f (s, x) P_i (ds, x) & \text{when } x \in X_0^{(i)} \\ 0 & \text{when } x \in \left\{ X_0^{(i)} \right\}^c \end{cases} ,$$

   one obtains a map adapted to $\mathcal{X}$;
   (iii) *one has that*

$$\int_{S \times X} f (s, x) Q_i (ds, dx) = \int_X \left\{ \int_S f (s, x) P_i (ds, x) \right\} \xi_i (dx) .$$

4. *When $\int_X \left\{ \int_S |f| (s, x) P_i (ds, x) \right\} \xi_i (dx) < \infty$,*

$$\int_{S \times X} f (s, x) Q_i (ds, dx)$$

   is finite.

Proposition 5.7.9 *Assume that context [(Fact)] 5.7.8 obtains. Then:*

1. *When* $Q_2 \ll Q_1$*, one has that*

   (i) $\xi_2 \ll \xi_1$,
   (ii) *there exists* $X_0 \in \mathcal{X}$ *such that*

   - $\xi_2 (X_0) = 1$,
   - *for* $x \in X_0$, $P_2(\cdot, x) \ll P_1(\cdot, x)$.

2. *When*

   (a) $\xi_2 \ll \xi_1$,
   (b) *there exists* $X_0 \in \mathcal{X}$ *such that*

   - $\xi_2 (X_0) = 1$
   - *for* $x \in X_0$, $P_2(\cdot, x) \ll P_1(\cdot, x)$,

   then $Q_2 \ll Q_1$.
3. *When items 1 and 2 obtain, there exits*

   (i) $D_P : S \times X \longrightarrow \mathbb{R}$, *adapted to* $\mathcal{S} \otimes \mathcal{X}$,
   (ii) $X_0 \in \mathcal{X}$ *with* $\xi_2 (X_0) = 1$,
   such that

   - *for* $x \in X_0$, $\frac{dP_2(\cdot,x)}{dP_1(\cdot,x)} (s) = D_P (s, x)$,
   - $\frac{dQ_2}{dQ_1} (s, x) = D_P (s, x) \frac{d\xi_2}{d\xi_1} (x)$.

Proof [1] Suppose that $Q_2 \ll Q_1$. Let

$$D = \frac{dQ_2}{dQ_1}, \quad \text{and } d(x) = \int_S D(s, x) P_1(ds, x).$$

Set

$$d^-(x) = \begin{cases} \frac{1}{d(x)} & \text{when } d(x) > 0 \\ \\ 0 & \text{when } d(x) \leq 0 \end{cases}.$$

Then, for $S_0 \in \mathcal{S}, X_0 \in \mathcal{X}$, fixed, but arbitrary,

$$\begin{aligned} Q_2 (S_0 \times X_0) &= \int_{X_0} \int_{S_0} D(s, x) P_1(ds, x) \xi_1(dx) \\ &= \int_{X_0} \left\{ d^-(x) \int_{S_0} D(s, x) P_1(ds, x) \right\} d(x) \xi_1(dx). \end{aligned}$$

Thus, setting S_0 to S,

$$\xi_2(X_0) = \int_{X_0} d(x)\,\xi_1(dx),$$

which means, in particular, that $\xi_2 \ll \xi_1$, and that

$$\frac{d\xi_2}{d\xi_1}(x) = d(x).$$

Consequently

$$\begin{aligned} Q_2(S_0 \times X_0) &= \int_{X_0} \left\{ d^-(x) \int_{S_0} D(s,x)\,P_1(ds,x) \right\} \xi_2(dx) \\ &= \int_{X_0} \left\{ \int_{S_0} D(s,x)\,d^-(x)\,P_1(ds,x) \right\} \xi_2(dx). \end{aligned}$$

In other "words," for fixed, but arbitrary $S_0 \in \mathcal{S}$, almost surely in X, with respect to ξ_2,

$$P_2(S_0,x) = \int_{S_0} D(s,x)\,d^-(x)\,P_1(ds,x).$$

As S is separable, there is a countable base $\{S_i,\ i \in I\}$ for $\mathcal{S}$. Let X_i be the set in $\mathcal{X}$ for which $\xi_2(X_i) = 1$, and

$$P_2(S_i,x) = \int_{S_i} D(s,x)\,d^-(x)\,P_1(ds,x),\ x \in X_i.$$

Let $X_0 = \cap_{i \in I} X_i$. Then $\xi_2(X_0) = 1$, and, for $x \in X_0$, $P_2(\cdot,x) \ll P_1(\cdot,x)$, and

$$D_P(s,x) = D(s,x)\,d^-(x).$$

Proof [2] Suppose that $\xi_2 \ll \xi_1$, that there exists a set $X_0 \in \mathcal{X}$ such that $\xi_2(X_0) = 1$, and that, for $x \in X_0$, $P_2(\cdot,x) \ll P_1(\cdot,x)$.

Let $F \in \mathcal{S} \otimes \mathcal{B}$ be fixed, but arbitrary. As, for $i \in 1,2$,

$$Q_i(F) = \int_X P_i(F[x],x)\,\xi_i(dx),\ F \in \mathcal{S} \otimes \mathcal{X},$$

when $Q_1(F) = 0$, there is a set $X_0 \in \mathcal{X}$ such that $\xi_1(X_0) = 1$, and, for $x \in X_0$, $P_1(F[x],x) = 0$, that is, $P_2(F[x],x) = 0$. But $\xi_2\left(X_0^c\right) = 0$, so that $\xi_2(X_0) = 1$. Consequently $Q_2(F) = 0$.

Proof [3] Item 3 has been obtained while proving item 1. □

Remark 5.7.10 Let $\mathcal{E}_S : S \times X \longrightarrow S$ be the projection map: $\mathcal{E}_S(s,x) = s$. It is adapted to $\mathcal{S} \otimes \mathcal{X}$ and $\mathcal{S}$. Thus $\mu_i = Q_i \circ \mathcal{E}_S^{-1}$ defines a law on $\mathcal{S}$ which is a mixture of laws. Result (Lemma) 5.5.6 then yields conditions for equivalence of mixtures. Since $\mathcal{E}_S$ is not an injection, result (Lemma) 5.5.7 does not produce the Radon-Nikodým derivative for μ_2 with respect to μ_1. One can however try to use, to obtain that derivative, the facts which follow.

Lemma 5.7.11 *Let P and Q be probability laws on $(X, \mathcal{X})$. Let $(Y, \mathcal{Y})$ be a measurable space, and $F : X \longrightarrow Y$ be a map adapted to $\mathcal{X}$ and $\mathcal{Y}$. Let finally P_F and Q_F be the laws induced on $\mathcal{Y}$ by F and, respectively, P and Q. Then, when $Q \ll P$, $Q_F \ll P_F$, and*

$$\frac{dQ_F}{dP_F} = E_P\left[\frac{dQ}{dP} \mid F\right].$$

Furthermore, when X is a separable metric space, and $\mathcal{X}$ its Borel σ-algebra, then

$$E_P\left[\frac{dQ}{dP} \mid F\right](y) = \int_X \frac{dQ}{dP}(x)\, P_{\mathcal{X}|F}(dx, y),$$

where $P_{\mathcal{X}|F}$ is the regular image conditional law of P with respect to F, that is [274, p. 484],

$$P\left(X_0 \cap F^{-1}(Y_0)\right) = \int_{Y_0} P_{\mathcal{X}|F}(X_0, y)\, P_F(dy),\ X_0 \in \mathcal{X},\ Y_0 \in \mathcal{Y}.$$

Proof Let $f : Y \longrightarrow \mathbb{R}$ be adapted to $\mathcal{Y}$, and bounded. By definition, for fixed, but arbitrary $Y_0 \in \mathcal{Y}$,

$$\int_{F^{-1}(Y_0)} \{f \circ F\}(x) \frac{dQ}{dP}(x)\, P(dx) = \int_{Y_0} E_P\left[\{f \circ F\} \frac{dQ}{dP} \mid F\right](y)\, P_F(dy).$$

But

$$E_P\left[\{f \circ F\} \frac{dQ}{dP} \mid F\right] = f E_P\left[\frac{dQ}{dP} \mid F\right],$$

and

$$\int_{F^{-1}(Y)} \{f \circ F\}(x) \frac{dQ}{dP}(x)\, P(dx) = \int_Y f(y)\, Q_F(dy).$$

Consequently

$$\int_Y f(y)\, Q_F(dy) = \int_Y f(y)\, E_P\left[\frac{dQ}{dP} \mid F\right](y)\, P_F(dy),$$

and

$$\frac{dQ_F}{dP_F} = E_P\left[\frac{dQ}{dP} \mid F\right].$$

The last part of the statement follows from the standard properties of conditional expectations and regular conditional probability laws [274, p. 485]. □

To use (Lemma) 5.7.11 with mixture laws, one must proceed to the following assignments ($\mathcal{E}_S$ is the projection map $(s, x) \mapsto s$, and $a \longleftarrow b$ means that a is replaced with b):

$$\begin{aligned}
X &\longleftarrow S \times X, \\
Y &\longleftarrow S, \\
F &\longleftarrow \mathcal{E}_S, \\
\mathcal{X} &\longleftarrow \mathcal{S} \otimes \mathcal{X}, \\
\mathcal{Y} &\longleftarrow \mathcal{S}, \\
P &\longleftarrow Q_1, \\
Q &\longleftarrow Q_2, \\
P_F &\longleftarrow Q_1 \circ \mathcal{E}_S^{-1} = \mu_1, \\
Q_F &\longleftarrow Q_2 \circ \mathcal{E}_S^{-1} = \mu_2.
\end{aligned}$$

Result (Lemma) 5.7.11 then yields that

$$\frac{d\mu_2}{d\mu_1}(s) = \int_{S \times X} \frac{dQ_2}{dQ_1}(\sigma, x)\, [Q_1]_{\mathcal{S}\otimes\mathcal{X}|\mathcal{E}_S}\left(d(\sigma, x), s\right),$$

where, given $A \in \mathcal{S} \otimes \mathcal{X}$, and $S_0 \in \mathcal{S}$, fixed, but arbitrary,

$$Q_1\left(A \cap \mathcal{E}_S^{-1}(S_0)\right) = \int_{S_0} [Q_1]_{\mathcal{S}\otimes\mathcal{X}|\mathcal{E}_S}(A, s)\, Q_1 \circ \mathcal{E}_S^{-1}(ds).$$

One may then state:

Corollary 5.7.12 *One has, with the assumptions of* (Proposition) 5.7.9, item 2, *and the notation of (Remark)* 5.7.10, *that*

$$\frac{d\mu_2}{d\mu_1}(s) = \int_{S \times X} \frac{dP_2(\cdot, x)}{dP_1(\cdot, x)}(\sigma)\, \frac{d\xi_2}{d\xi_1}(x)\, [Q_1]_{\mathcal{S}\otimes\mathcal{X}|\mathcal{E}_S}\left(d(\sigma, x), s\right).$$

The following remarks are meant to clarify some of the statements to follow. To that end, let $\{S_x,\ x \in X\} \subseteq \mathcal{S}$ be a partition of S, that is, $S_{x_1} \cap S_{x_2} = \emptyset$ when $x_1 \neq x_2$, and $\cup_{x \in X} S_x = S$.

Remark 5.7.13 Let $f_x : S_x \longrightarrow Y$ be a map for each $x \in X$. There is a unique $F : S \longrightarrow Y$ such that $F\,|_{S_x} = f_x$.

Remark 5.7.14 When for $x \in X$, f_x is adapted to $\mathcal{S}$, F need not be adapted as $F^{-1}(Y_0) = \cup_{x \in X} \left\{ f_x^{-1}(Y_0) \cap S_x \right\}$.

Remark 5.7.15 Let $\mathcal{E}_S : S \times X \longrightarrow S$ be the evaluation map: $\mathcal{E}_S(s, x) = s$. Define $f : S \longrightarrow S \times X$ setting $f(s) = (s, x)$, where x is the index of the set S_x containing s. It is a well-defined map as there is only one set S_x that contains s. Then $\mathcal{E}_S f(s) = s$, and f is injective.

Lemma 5.7.16 *Let the assumptions of* (Proposition) 5.7.9, item 2, *be valid, and notation be that of (Remark)* 5.7.10.

1. *When there exists a measure μ on $\mathcal{S}$ such that, for $x \in X$, $P_1(\cdot, x) \ll \mu$, then*

$$\frac{d\mu_2}{d\mu_1}(s) = \frac{\int_X D_P(s, x) \frac{dP_1(\cdot, x)}{d\mu}(s)\, \xi_2(dx)}{\int_X \frac{dP_1(\cdot, x)}{d\mu}(s)\, \xi_1(dx)}.$$

2. *When there exists $\{S_x,\ x \in X\} \subseteq \mathcal{S}$ such that,*
   (a) *for* $x_1 \neq x_2$, $S_{x_1} \cap S_{x_2} = \emptyset$,
   (b) *for* $x_1 \neq x_2$, $P_1(\cdot, x_1) \perp P_1(\cdot, x_2)$,
   (c) *for* $x_1 \neq x_2$, $P_1(S_{x_1}, x_1) = 1,\ P_1(S_{x_2}, x_1) = 0$,
   (d) $d(s) = \sum_{x \in X} \frac{dP_2(\cdot, x)}{dP_1(\cdot, x)}(s)\, \chi_{S_x}(s)$ *is adapted,*
   (e) $\xi_2 = \xi_1$,

   then

$$\frac{d\mu_2}{d\mu_1} = d.$$

Proof Let

$$D(s, x) = \frac{dP_1(\cdot, x)}{d\mu}(s).$$

One may assume that D is adapted to $\mathcal{S} \otimes \mathcal{X}$: indeed, letting $Q = \mu \otimes \xi_1$, one has that $Q_1 \ll Q$, and one may use (Proposition) 5.7.9, item 3, to get an adapted map. Then:

$$\begin{aligned} Q_1(S_0 \times X_0) &= \int_{X_0} P_1(S_0, x)\, \xi_1(dx) \\ &= \int_{X_0} \left\{ \int_{S_0} \frac{dP_1(\cdot, x)}{d\mu}(s)\, \mu(ds) \right\} \xi_1(dx) \\ &= \int_{X_0} \left\{ \int_{S_0} D(s, x)\, \mu(ds) \right\} \xi_1(dx) \\ &= \int_{S_0} \left\{ \int_{X_0} D(s, x)\, \xi_1(dx) \right\} \mu(ds). \end{aligned}$$

Setting X_0 to X, one gets that

$$\mu_1(S_0) = \int_{S_0} \left\{ \int_X D(s,x)\, \xi_1(dx) \right\} \mu(ds),$$

so that $\mu_1 \ll \mu$, and

$$\frac{d\mu_1}{d\mu}(s) = \int_X D(s,x)\, \xi_1(dx).$$

Since $P_2(\cdot,x) \ll P_1(\cdot,x) \ll \mu$,

$$\frac{dP_2(\cdot,x)}{d\mu}(s) = \frac{dP_2(\cdot,x)}{dP_1(\cdot,x)}(s) \frac{dP_1(\cdot,x)}{d\mu}(s) = D_P(s,x)\, D(s,x).$$

As above

$$\begin{aligned} Q_2(S_0 \times X_0) &= \int_{X_0} P_2(S_0,x)\, \xi_2(dx) \\ &= \int_{X_0} \left\{ \int_{S_0} \frac{dP_2(\cdot,x)}{d\mu}(s)\, \mu(ds) \right\} \xi_2(dx) \\ &= \int_{X_0} \left\{ \int_{S_0} D_P(s,x)\, D(s,x)\, \mu(ds) \right\} \xi_2(dx) \\ &= \int_{S_0} \left\{ \int_{X_0} D_P(s,x)\, D(s,x)\, \xi_2(dx) \right\} \mu(ds), \end{aligned}$$

so that, setting X_0 to X,

$$\frac{d\mu_2}{d\mu}(s) = \int_X D_P(s,x)\, D(s,x)\, \xi_2(dx).$$

Finally, as $\mu_2 \ll \mu_1 \ll \mu$,

$$\frac{d\mu_2}{d\mu_1} = \frac{\frac{d\mu_2}{d\mu}}{\frac{d\mu_1}{d\mu}} = \frac{\int_X D_P(s,x)\, D(s,x)\, \xi_2(dx)}{\int_X D(s,x)\, \xi_1(dx)}.$$

Item 1 is thus true. For item 2, given a function $f : S \longrightarrow \mathbb{R}$ adapted to $\mathcal{S}$, and bounded,

$$\int_S f(s)\, d(s)\, P_1(ds,x) = \int_S f(s)\, P_2(ds,x),$$

and thus

$$\int_S f(s)\,d(s)\,\mu_1(ds) = \int_X \int_S f(s)\,d(s)\,P_1(ds,x)\,\xi_1(dx)$$
$$= \int_X \int_S f(s)\,P_2(ds,x)\,\xi_1(dx)$$
$$= \int_S f(s)\,\mu_2(ds)\,.$$

□

Proposition 5.7.17 *Let $\{P_i(\cdot,x)\,,\ x \in X,\ i \in \{1,2\}\}$ be a family of probability laws on $\mathcal{S}$ such that:*

(A) *for fixed, but arbitrary $S_0 \in \mathcal{S}$, the map $x \mapsto P_i(S_0,x)$ is adapted to $\mathcal{X}$;*
(B) *there is a family $\{S_x,\ x \in X\} \subseteq \mathcal{S}$ such that:*

  (a) *when $x_1 \neq x_2$, $S_{x_1} \cap S_{x_2} = \emptyset$;*
  (b) *for every fixed, but arbitrary $X_0 \in \mathcal{X}$, $\cup_{x \in X_0} S_x \in \mathcal{S}$;*
  (c) *for $i \in \{1,2\}$,*

$$P_i(S_{x_1},x_2) = \begin{cases} 1 \text{ when } x_1 = x_2 \\ 0 \text{ when } x_1 \neq x_2 \end{cases}.$$

Let ξ_1 and ξ_2 be laws on $\mathcal{X}$ and set, for $S_0 \in \mathcal{S},\ i \in \{1,2\}$,

$$\mu_i(S_0) = \int_X P_i(S_0,x)\,\xi_i(dx)\,.$$

Then $\mu_2 \ll \mu_1$ if, and only if,

1. *$\xi_2 \ll \xi_1$,*
2. *there exists $X_0 \in \mathcal{X}$ such that*
   - *$\xi_2(X_0) = 1$,*
   - *for $x \in X_0$, $P_2(\cdot,x) \ll P_1(\cdot,x)$.*

Then, on S_x,

$$\frac{d\mu_2}{d\mu_1}(s) = \frac{dP_2(\cdot,x)}{dP_1(\cdot,x)}(s)\,\frac{d\xi_2}{d\xi_1}(x)\,.$$

Proof Let, when $s \in S_x, f(s) = (s,x). f$ is adapted as, for fixed, but arbitrary $S_0 \in \mathcal{S}$ and $X_0 \in \mathcal{X}$,

$$\{s \in S : f(s) \in S_0 \times X_0\} = \cup_{x \in X_0}\{S_x \cap S_0\} = \{\cup_{x \in X_0} S_x\} \cap S_0 \in \mathcal{S}.$$

Let $Q = \mu_1 \circ f^{-1}$. As

$$\begin{aligned} S_{x_0} \cap [\{\cup_{x \in X_0} S_x\} \cap S_0] &= S_{x_0} \cap [\cup_{x \in X_0} (S_x \cap S_0)] \\ &= \cup_{x \in X_0} (S_{x_0} \cap S_x \cap S_0) \\ &= \begin{cases} S_{x_0} \cap S_0 & \text{when } x_0 \in X_0 \\ \emptyset & \text{when } x_0 \in X_0^c \end{cases}, \end{aligned}$$

one has that

$$\begin{aligned} P_1 (\{\cup_{x \in X_0} S_x\} \cap S_0, x_0) &= P_1 (S_{x_0} \cap [\{\cup_{x \in X_0} S_x\} \cap S_0], x_0) \\ &= \chi_{X_0} (x_0) P_1 (S_0, x_0). \end{aligned}$$

Consequently,

$$\begin{aligned} Q (S_0 \times X_0) &= \mu_1 \left(f^{-1} (S_0 \times X_0)\right) \\ &= \mu_1 (\{\cup_{x \in X_0} S_x\} \cap S_0) \\ &= \int_X P_1 (\{\cup_{x \in X_0} S_x\} \cap S_0, \tilde{x}) \, \xi_1 (d\tilde{x}) \\ &= \int_X \chi_{X_0} (\tilde{x}) P_1 (S_0, \tilde{x}) \, \xi_1 (d\tilde{x}) \\ &= Q_1 (S_0 \times X_0). \end{aligned}$$

The conclusion is that $\mu_2 \ll \mu_1$ implies $Q_2 \ll Q_1$, and (Proposition) 5.7.9 obtains.

It remains to identify the Radon-Nikodým derivative. To that end, let

$$\mu (S_0) = \int_X P_1 (S_0, x) \, \xi_2 (dx), \quad S_0 \in \mathcal{S}.$$

From (Lemma) 5.7.16, one has that $\mu_2 \ll \mu$, and that, for $s \in S_x$,

$$\frac{d\mu_2}{d\mu} (s) = \frac{P_2 (\cdot, x)}{P_1 (\cdot, x)} (s).$$

Let $\phi : S \longrightarrow \mathbb{R}$ be adapted to $\mathcal{S}$, and bounded, and define $d : S \longrightarrow \overline{\mathbb{R}}_+$ using, when $s \in S_x$,

$$d (s) = \frac{d\xi_2}{d\xi_1} (x).$$

Then

$$\int_S \phi(s)\,\mu(ds) = \int_X \left\{ \int_S \phi(s)\, P_1(ds, x) \right\} \xi_2(dx)$$

$$= \int_X \left\{ \int_S \phi(s)\, P_1(ds, x) \right\} \frac{d\xi_2}{d\xi_1}(x)\,\xi_1(dx)$$

$$= \int_X \left\{ \int_S \phi(s)\, d(s)\, P_1(ds, x) \right\} \xi_1(dx)$$

$$= \int_S \phi(s)\, d(s)\, \mu_1(dx).$$

Thus $\mu \ll \mu_1$, and $\frac{d\mu}{d\mu_1} = d$. Consequently, $\mu_2 \ll \mu \ll \mu_1$, and

$$\frac{d\mu_2}{d\mu_1} = \frac{d\mu_2}{d\mu}\frac{d\mu}{d\mu_1},$$

which is the required result. □

Example 5.7.18 Let P_W be the standard Wiener measure on $C[0, 1]$, and f be adapted and bounded. Let, for $c \in C[0, 1]$, and $\mathcal{E}(c, t) = c(t)$,

$$X(c, t) = \int_0^t f(x)\,dx + \mathcal{E}(c, t).$$

Let A be a stochastic process with continuous, strictly increasing paths, denoted a, independent of $\mathcal{E}$ (it could have, for example, the form $A(\omega, t) = \int_0^t g^2(\omega, x)\,dx$ with g independent of the Brownian motion, and appropriate paths). One can form

$$X(c, A(\omega, t)) = \int_0^{A(\omega,t)} f(x)\,dx + \mathcal{E}(c, A(\omega, t)),$$

which may be written [(Fact) 10.3.36], using $\Pi_C[c, \omega] = c$, $\Pi_\Omega[c, \omega] = \omega$,

$$Y([c, \omega], t) = \int_0^t f(A(\Pi_\Omega[c, \omega], x))\,A(\Pi_\Omega[c, \omega], dx) + M(\Pi_C[c, \omega], t).$$

Let W_a be the Gaussian process with covariance $a(t_1 \wedge t_2)$: it is a Brownian motion with change of time a. The laws of W_a and W_b are orthogonal as soon as $a \neq b$. As $\langle M \rangle = A$, $C_a = \{c : \langle M \rangle = a\}$ provides a partition of $C[0, 1]$ of the type required in (Proposition) 5.7.17. One has thus that, on C_a, with respect to P_{W_a},

$$\frac{dP_Y}{dP_M}(c) = \int_0^1 f(a(x))\,\mathcal{E}(c, dx) - \frac{1}{2}\int_0^1 f^2(a(x))\,a(dx).$$

Remark 5.7.19 The latter example exhibits what shall be proved (in Part III, Chap. 16, with much effort) to be practically the most general case of absolute continuity with respect to a continuous martingale that is not Gaussian. It is in that respect, as already mentioned, that mixtures are useful: they produce the adequate results, though it may prove difficult to ascertain that they are the best one can achieve.

Part II
Cramér-Hida Representations

The sequence of words *"Cramér-Hida representation"* shall often be abbreviated in the sequel using the acronym *"CHR,"* and *"Cramér-Hida," "CH."* The Cramér-Hida representation is a representation of a second order stochastic process as a sum of outputs of causal filters whose inputs are orthogonal "white noises." Such representations are useful for two reasons: the first is that the elements in the sum have a "natural" interpretation, and the second, that they are mathematically simple and highly structured. All these representations are obtained as unitary maps from L_2 spaces, or Hilbert spaces which behave like L_2 spaces, to spaces of classes of random variables, or vectors. The elements in the ranges of these unitary maps are often interpreted as stochastic integrals as they mimic, in spaces of (classes of) random variables, L_2 spaces.

The Cramér-Hida representation is the result of the search of flexible structures, in an a priori "shapeless" space, the linear space generated by the values of a Hibert space valued function. Chapter 6 exhibits what is obtained when using the immediate structure provided by a basis, hence the "first principles" denomination. Chapter 7 contains the representation of the linear space as a direct integral, before a basis is used. Finally, Chap. 9 exhibits the finer structures of the same linear space in the form of martingales. Chapter 8 is about multiplicity one, which is helpful in applications. It also shows the limits of multiplicity in that respect. One also finds in there Goursat processes whose deployment in applications has not found justice … yet.

In the sequel, the main use of the Cramér-Hida representation shall be that it serves as the bridge which allows one to use stochastic calculus techniques with models that are not semimartingales.

One shall sometimes use the abbreviated "Cramér-Hida" for "Cramér-Hida representation." The latter will also be called the "Cramér-Hida decomposition."

Chapter 6
Cramér-Hida Representations from "First Principles"

The Cramér-Hida representation shall be, at first, the decomposition of a function, with values in a Hilbert space, into manageable parts, one continuous, and one discontinuous. The continuous part shall furthermore be decomposed into a possibly infinite sum of continuous and orthogonal functions, which have a representation as integrals, with respect to an orthogonal vector measure, itself determined by a function with orthogonal increments. The decomposition is obtained under some "natural" assumptions [Assumption 6.4.1].

6.1 Preliminaries

This section registers all that is needed to obtain the Cramér-Hida representation (CHR), explains and illustrates the underlying assumptions.

6.1.1 Context

The setup shall be as follows. The set $T \subseteq \mathbb{R}$ is an interval, and H is a real Hilbert space. The elements $t_l = \inf T$, and $t_r = \sup T$, are the boundaries of T, they may or may not belong to T, and they may or may not be finite. The object of study shall be $f : T \longrightarrow H$, a well-defined map. The notation $t \downarrow\downarrow t_n$ means that $\{t_n, n \in \mathbb{N}\} \subseteq T$ is a strictly decreasing sequence whose elements are strictly greater that t and has the latter as limit. Analogously one defines $t_n \uparrow\uparrow t$. When $\lim_{\theta \uparrow\uparrow t} f(\theta)$ makes sense, and exists, it shall be denoted $f^-(t)$. *Mutatis mutandis*, $f^+(t)$ denotes $\lim_{t \downarrow\downarrow \theta} f(\theta)$.

A.F. Gualtierotti, *Detection of Random Signals in Dependent Gaussian Noise*,
DOI 10.1007/978-3-319-22315-5_6

For $t \in T$, fixed, but arbitrary, $L_t[f] \subseteq H$ represents the (closed) linear subspace generated in H by the set $\{f(\theta), \theta \leq t\}$:

$$L_t[f] = \overline{V[\{f(\theta), \theta \in T, \theta \leq t\}]},$$

and P_t, the projection of H, with range $L_t[f]$. The following subspaces shall also be of interest:

$$L_t^-[f] = \bigvee_{\theta < t} L_\theta[f], \text{ and } L_t^+[f] = \bigcap_{\theta > t} L_\theta[f].$$

The corresponding projections shall be P_t^-, and P_t^+. These subspaces are "banded" below by

$$L_{\cap T}[f] = \bigcap_{t \in T} L_t[f], \text{ with projection } P_\cap,$$

and above, by

$$L_{\cup T}[f] = \bigvee_{t \in T} L_t[f], \text{ with projection } P_\cup.$$

When

$$L_{\cup T}[f] = L_{\cap T}[f],$$

f is deemed deterministic, and, when

$$L_{\cap T}[f] = \{0_H\},$$

purely nondeterministic. When f represents an evolution, if it is deterministic, then the evolution is completely known "at the start of time"; if it is purely nondeterministic, nothing is known of the evolution "at the start of time." The Cramér-Hida representation concentrates on functions of the latter type.

Remark 6.1.1 One shall make repeated use of the following property. For projections P and Q of H, with $P \leq Q$,

$$\mathcal{R}[Q-P] = \mathcal{R}[Q] \cap \mathcal{R}[P]^\perp = \{h \in H : P[h] = 0, Q[h] = h\}.$$

Indeed the first equality is a standard result about projections [8, p. 413], and the second rephrases the middle expression above.

Remark 6.1.2 The dimension of $L_t[f] \cap L_t^-[f]^\perp$ is at most one.

One has that

$$f(t) = P_t^-[f(t)] + h, \; h \in L_t[f], \; h \perp L_t^-[f]^\perp,$$

and thus

$$\overline{V[\{f(\theta), \theta \le t\}]} \subseteq L_t^-[f] \bigvee V[h].$$

Remark 6.1.3 When f^+ exists, $f^+(t) \in L_t^+[f]$.

Indeed, given $\epsilon > 0$, and $h \in L_{t+\epsilon}[f]^\perp$, fixed, but arbitrary, as soon as $\delta < \epsilon$,

$$\langle f(t+\delta), h\rangle_H = 0.$$

Consequently,

$$\langle f^+(t), h\rangle_H = \lim_{\delta \downarrow\downarrow 0} \langle f(t+\delta), h\rangle_H = 0.$$

Thus $f^+(t) \in L_{t+\epsilon}[f]$.

Remark 6.1.4 When $f^+(t)$ exists, let $L_t^{(+)}[f]$ be the subspace generated by $L_t[f]$ and $f^+(t)$. It may be strictly included in $L_t^+[f]$. Here is an example:

Example 6.1.5 Let $T = [0,1]$, and $\{e_1, e_2, e_3\} \subseteq H$ be orthonormal. Consider the following function:

$$f(t) = \begin{cases} 0_H & \text{when } t \in \left[0, \frac{1}{4}\right] \\ e_1 & \text{when } t \in \left]\frac{1}{4}, \frac{1}{2}\right] \\ e_2 + \left(t - \frac{1}{2}\right) e_3 & \text{when } t \in \left]\frac{1}{2}, 1\right] \end{cases}.$$

The function f is continuous, except for $t = \frac{1}{4}$ and $t = \frac{1}{2}$, where

$$f\left(\frac{1}{4}\right) = 0_H, \; f^+\left(\frac{1}{4}\right) = e_1, \; f\left(\frac{1}{2}\right) = e_1, \; f^+\left(\frac{1}{2}\right) = e_2.$$

Furthermore,

$$L_t[f] = \begin{cases} \{0_H\} & \text{when } t \in \left[0, \frac{1}{4}\right] \\ V[e_1] & \text{when } t \in \left]\frac{1}{4}, \frac{1}{2}\right] \\ V[e_1, e_2, e_3] & \text{when } t \in \left]\frac{1}{2}, 1\right] \end{cases}.$$

The subspace generated by $L_{\frac{1}{2}}[f]$ and $f^{+}\left(\frac{1}{2}\right)$ is thus $V[e_1, e_2]$, whereas

$$L^{+}_{\frac{1}{2}}[f] = V[e_1, e_2, e_3].$$

That example also shows that, for $\theta > t$,

$$P^{+}_{t}[f(\theta)] \text{ does not necessarily equal } f^{+}(t)$$

(it does when f has orthogonal increments—see below). Indeed, $f^{+}\left(\frac{1}{4}\right) = e_1$, while

$$P^{+}_{\frac{1}{4}}\left[f\left(\frac{3}{4}\right)\right] = P_{V[e_1]}\left[e_2 + \frac{1}{4}e_3\right] = 0_H.$$

6.1.2 Functions and Determinism

The result to follow shows that a map $f : T \longrightarrow H$ can always be decomposed into deterministic and purely nondeterministic parts, that is, a "static" contribution of the "past," and an "evolving" part. In practice only the latter is of real interest.

Proposition 6.1.6 *Let T be an interval, H a real Hilbert space, and*

$$f : T \longrightarrow H,$$

a well-defined map. One has the following unique representation: for $t \in T$, fixed, but arbitrary,

$$f(t) = g(t) + h(t),$$

with:

1. $g : T \longrightarrow H$ *purely deterministic, that is, for $t \in T$, fixed, but arbitrary,*

$$L_{\cup T}[g] = L_t[g] = L_{\cap T}[g];$$

2. $h : T \longrightarrow H$ *purely nondeterministic, that is,*

$$L_{\cap T}[h] = \{0_H\}.$$

Proof Let $P_{\cap}$ be the projection in $L_{\cup T}[f]$, whose range is $L_{\cap T}[f]$. Set

$$g(t) = P_{\cap}[f(t)], \text{ and } h(t) = \left\{I_{L_{\cup T}[f]} - P_{\cap}\right\}[f(t)].$$

By definition, for $\{t_1, t_2\} \subseteq T$, fixed, but arbitrary,

$$\langle g(t_1), h(t_2)\rangle_H = 0,$$

and, for $t \in T$, fixed, but arbitrary, since

$$P_t \{I_{L_{\cup T}[f]} - P_{\cap}\} [f(t)] = \{I_{L_{\cup T}[f]} - P_{\cap}\} [f(t)],$$

one has that

$$L_t[h] \subseteq L_t[f], \quad L_t[h] \perp \mathcal{R}[P_{\cap}], \text{ and } L_t[g] \subseteq \mathcal{R}[P_{\cap}].$$

Consequently,

$$\bigcap_{t\in T} L_t[h] \subseteq \bigcap_{t\in T} L_t[f] = \mathcal{R}[P_{\cap}], \text{ and } \bigcap_{t\in T} L_t[h] \perp \mathcal{R}[P_{\cap}],$$

so that

$$L_{\cap T}[h] = \bigcap_{t\in T} L_t[h] = \{0_H\},$$

and h is purely nondeterministic. Now, as for fixed, but arbitrary $t \in T$,

$$f(t) = g(t) + h(t), \; g(t) \perp h(t),$$

one has that

$$L_t[f] \subseteq L_t[g] \oplus L_t[h].$$

But, since $P_t P_{\cap}[f(t)] = P_{\cap}[f(t)]$, for fixed, but arbitrary $t \in T$,

$$L_t[g] \subseteq L_t[f],$$

and, as seen,

$$L_t[h] \subseteq L_t[f].$$

It follows that

$$L_t[f] = L_t[g] \oplus L_t[h].$$

In particular

$$\mathcal{R}[P_{\cap}] \subseteq L_t[g] \oplus L_t[h],$$

and, since $\mathcal{R}[P_\cap] \perp L_t[h]$,

$$\mathcal{R}[P_\cap] \subseteq L_t[g] \subseteq L_t[f],$$

so that

$$\bigcap_{t\in T} L_t[g] = \mathcal{R}[P_\cap].$$

As, by definition, for $t \in T$, fixed, but arbitrary, $L_t[g] \subseteq \mathcal{R}[P_\cap]$,

$$L_t[g] = \mathcal{R}[P_\cap].$$

g is thus purely deterministic.

Suppose that f has another decomposition of the form $f = g + h$, say

$$f = \tilde{g} + \tilde{h}.$$

Then, for $k \in \mathcal{R}[P_\cap]$, and $t \in T$, fixed, but arbitrary,

$$\langle g(t), k\rangle_H = \langle f(t), k\rangle_H = \langle \tilde{g}(t), k\rangle_H,$$

so that $g(t) = \tilde{g}(t)$, and thus $h(t) = \tilde{h}(t)$. □

6.1.3 *Functions with Orthogonal Increments*

Functions with orthogonal increments shall be the building blocs of the Cramér-Hida decomposition. Here are the salient features of those functions, for that context.

Definition 6.1.7 Let $f : T \longrightarrow H$ have the following property: for fixed, but arbitrary $\{t_1, t_2, t_3, t_4\} \subseteq T$, such that $t_1 < t_2 \leq t_3 < t_4$,

$$\langle f(t_2) - f(t_1), f(t_4) - f(t_3)\rangle_H = 0.$$

One then says that f has orthogonal increments.

Example 6.1.8 Let $f_h(t) = P_t[h], h \in H$. The function f_h has orthogonal increments.

Proposition 6.1.9 *Let $f : T \longrightarrow H$ have orthogonal increments, and $t_0 \in T$ be fixed, but arbitrary. Set, for $t \in T$, fixed, but arbitrary,*

$$F(t) = \begin{cases} \|f(t) - f(t_0)\|_H^2 & when\ t \geq t_0 \\ -\|f(t) - f(t_0)\|_H^2 & when\ t < t_0 \end{cases}.$$

Then, for fixed, but arbitrary $t_1 < t_2$, $\{t_1, t_2) \subseteq T$,

$$F(t_2) - F(t_1) = \|f(t_2) - f(t_1)\|_H^2,$$

so that F is a function with values in $\mathbb{R}$, *which is increasing.*

Proof Since the increments of f are orthogonal in H, for fixed, but arbitrary $t_1 < t_2$, $\{t_1, t_2) \subseteq T$,

- when $t_0 < t_1 < t_2$,

$$\begin{aligned} F(t_2) &= \|f(t_2) - f(t_0)\|_H^2 \\ &= \|f(t_2) - f(t_1) + f(t_1) - f(t_0)\|_H^2 \\ &= \|f(t_2) - f(t_1)\|_H^2 + F(t_1); \end{aligned}$$

- when $t_1 < t_0 < t_2$,

$$\begin{aligned} \|f(t_2) - f(t_1)\|_H^2 &= \|f(t_2) - f(t_0) + f(t_0) - f(t_1)\|_H^2 \\ &= \|f(t_2) - f(t_0)\|_H^2 + \|f(t_0) - f(t_1)\|_H^2 \\ &= F(t_2) - F(t_1); \end{aligned}$$

- when $t_1 < t_2 < t_0$,

$$\begin{aligned} -F(t_1) &= \|f(t_1) - f(t_0)\|_H^2 \\ &= \|f(t_0) - f(t_2) + f(t_2) - f(t_1)\|_H^2 \\ &= \|f(t_0) - f(t_2)\|_H^2 + \|f(t_2) - f(t_1)\|_H^2 \\ &= -F(t_2) + \|f(t_2) - f(t_1)\|_H^2. \end{aligned}$$

When $t_0 = t_1$, $F(t_1) = F(t_0) = 0$, and

$$F(t_2) = \|f(t_2) - f(t_0)\|_H^2 = \|f(t_2) - f(t_1)\|_H^2,$$

and, when $t_0 = t_2$, as above, $F(t_2) = 0$, and

$$-F(t_1) = \|f(t_1) - f(t_0)\|_H^2 = \|f(t_2) - f(t_1)\|_H^2.$$

Consequently, in all cases,

$$F(t_2) - F(t_1) = \|f(t_2) - f(t_1)\|_H^2.$$

□

Corollary 6.1.10 *Let $f : T \longrightarrow H$ have orthogonal increments. The limits $f^{-}(t)$ and $f^{+}(t)$ exist. Furthermore, F^{-} and F^{+} being defined analogously to f^{-} and f^{+} respectively,*

$$\begin{aligned} F(t) - F^{-}(t) &= \|f(t) - f^{-}(t)\|_H^2, \\ F^{+}(t) - F(t) &= \|f^{+}(t) - f(t)\|_H^2, \\ F^{+}(t) - F^{-}(t) &= \|f^{+}(t) - f^{-}(t)\|_H^2. \end{aligned}$$

Proof To prove, for example, the existence of $f^{-}(t)$, one may proceed as follows. Let $t_n \uparrow\uparrow t$ be fixed, but arbitrary. Then, as seen above [(Proposition) 6.1.9],

$$\left\|f(t_n) - f(t_p)\right\|_H^2 = F(t_n) - F(t_p).$$

Since $F(t_n) \leq F(t) < \infty$, and that F has limits to the left,

$$\lim_{n,p\to\infty} \left\|f(t_n) - f(t_p)\right\|_H^2 = 0.$$

Thus $\{f(t_n), n \in \mathbb{N}\}$ is a Cauchy sequence. Its limit is $f^{-}(t)$. □

Remark 6.1.11 When $f : T \longrightarrow H$ is purely nondeterministic, and has orthogonal increments, for $\{t_1, t_2\} \subseteq T, t_1 < t_2$, fixed, but arbitrary,

$$\|f(t_2) - f(t_1)\|_H^2 = \|f(t_2)\|_H^2 - \|f(t_1)\|_H^2.$$

For then indeed, when t_l does not belong to T, there exists $t_0 \in T$, with $t_l < t_0 < t_1$, and then

$$\begin{aligned} \langle f(t_2), f(t_1)\rangle_H &= \langle f(t_2) - f(t_1) + f(t_1), f(t_1) - f(t_0) + f(t_0)\rangle_H \\ &= \langle f(t_2) - f(t_1), f(t_0)\rangle_H + \|f(t_1)\|_H^2. \end{aligned}$$

Letting $t_0 \downarrow\downarrow t_l$, one has that

$$\langle f(t_2), f(t_1)\rangle_H = \|f(t_1)\|_H^2.$$

When $t_l \in T, f(t_l) = 0_H$, and

$$\langle f(t_2), f(t_1)\rangle_H = \langle f(t_2) - f(t_1) + f(t_1), f(t_1) - f(t_l)\rangle_H = \|f(t_1)\|_H^2.$$

Finally $\|f(t_2) - f(t_1)\|_H^2 = \|f(t_2)\|_H^2 - 2\langle f(t_2), f(t_1)\rangle_H + \|f(t_1)\|_H^2$.

Proposition 6.1.12 *Let $f : T \longrightarrow H$ have orthogonal increments. The function f^{+} of (Corollary)* 6.1.10 *has then orthogonal increments, and is continuous to the right.*

Proof Let $\{t_1, t_2, t_3, t_4\} \subseteq T, t_1 < t_2 \leq t_3 < t_4$, be fixed, but arbitrary, and let $\{\delta, \delta_1, \delta_4\} \subseteq \mathbb{R}_+ \setminus \{0\}$ be such that $t_1 + \delta_1 < t_2 + \frac{\delta}{2} \leq t_3 + \delta < t_4 + \delta_4$. Then

$$0 = \left\langle f\left(t_2 + \frac{\delta}{2}\right) - f(t_1 + \delta_1), f(t_4 + \delta_4) - f(t_3 + \delta)\right\rangle_H.$$

Letting $\delta, \delta_1, \delta_4 \downarrow\downarrow 0$, one obtains that

$$0 = \left\langle f^+(t_2) - f^+(t_1), f^+(t_4) - f^+(t_3)\right\rangle_H.$$

Also $\lim_{\delta \downarrow\downarrow 0} f^+(t+\delta) = \lim_{\delta \downarrow\downarrow 0} \lim_{\delta^\star \downarrow\downarrow 0} f(t+\delta+\delta^\star) = f^+(t)$. □

Lemma 6.1.13 *Let $f : T \longrightarrow H$ have orthogonal increments. Let $\delta > 0$ be fixed, but arbitrary, and*

$$L_t^\delta[f] = \overline{V[\{f(t+d) - f(t),\ 0 < d \leq \delta\}]}.$$

Then

$$L_{t+\delta}[f] = L_t[f] \oplus L_t^\delta[f].$$

Proof Let $t_1 < t_2$, $\{t_1, t_2\} \subseteq T$, be fixed, but arbitrary. Then

$$\begin{aligned} P_{t_2}[f(t_1)] &= f(t_1), \\ P_{t_1}[f(t_2)] &= P_{t_1}[(f(t_2) - f(t_1)) + f(t_1)] \\ &= P_{t_1}[f(t_1)] \\ &= f(t_1). \end{aligned}$$

Consequently, when $\delta > 0$ is fixed, but arbitrary, and $P_0 = P_{t+\delta} - P_t$, for fixed, but arbitrary $d \in]0, \delta]$,

$$\begin{aligned} P_0[f(t+d) - f(t)] &= P_{t+\delta}[f(t+d)] - P_{t+\delta}[f(t)] - P_t[f(t+d)] + P_t[f(t)] \\ &= f(t+d) - f(t) - f(t) + f(t) \\ &= f(t+d) - f(t). \end{aligned}$$

Thus $L_t^\delta[f] \subseteq \mathcal{R}[P_0]$.

If now $h \in \mathcal{R}[P_0]$, but $h \perp L_t^\delta[f]$, for fixed, but arbitrary $d \in]0, \delta]$,

$$\langle h, f(t+d) - f(t)\rangle_H = 0.$$

But P_0 has $L_{t+\delta}[f] \cap L_t[f]^{\perp}$ as range [(Remark) 6.1.1], so that $h \perp L_t[f]$ by its choice, and thus

$$\langle h, f(t+d)\rangle_H = 0, \ d \in]0, \delta].$$

In other words $h \perp L_{t+\delta}[f]$. As $h \in L_{t+\delta}[f]$, one must have $h = 0_H$, and thus $\mathcal{R}[P_0] = L_t^{\delta}[f]$, that is,

$$L_t^{\delta}[f] = L_{t+\delta}[f] \cap L_t[f]^{\perp}.$$

The assertion follows from the following equality: $P_{t+\delta} = P_t + (P_{t+\delta} - P_t)$. □

Lemma 6.1.14 *Let $f : T \longrightarrow H$ have orthogonal increments. For fixed, but arbitrary $\delta > 0$,*

$$L_{t+\delta}[f] \cap L_t^+[f]^{\perp} = \mathcal{R}[P_{t+\delta} - P_t^+] = \overline{V[\{f(t+d) - f^+(t), \ 0 < d \le \delta\}]}.$$

Proof One has, for $\theta > 0$, fixed, but arbitrary, that (see proof of (Lemma) 6.1.13)

$$P_t^+[f(t+\theta)] = \lim_{\epsilon \downarrow\downarrow 0} P_{t+\epsilon}[f(t+\theta)] = \lim_{\epsilon \downarrow\downarrow 0} f(t+\epsilon) = f^+(t),$$

and that

$$P_{t+\theta}\left[f^+(t)\right] = \lim_{\epsilon \downarrow\downarrow 0} P_{t+\theta}[f(t+\epsilon)] = \lim_{\epsilon \downarrow\downarrow 0} f(t+\epsilon) = f^+(t).$$

Consequently,

$$P_{t+\delta}\left[f(t+d) - f^+(t)\right] = f(t+d) - f^+(t),$$

and

$$P_t^+\left[f(t+d) - f^+(t)\right] = 0_H.$$

Using (Remark) 6.1.1, one obtains that

$$\overline{V[\{f(t+d) - f^+(t), \ 0 < d \le \delta\}]} \subseteq L_{t+\delta}[f] \cap L_t^+[f]^{\perp}.$$

Suppose $h \in L_{t+\delta}[f] \cap L_t^+[f]^{\perp}$, is fixed, but arbitrary, and that

$$h \perp \overline{V[\{f(t+d) - f^+(t), \ 0 < d \le \delta\}]}.$$

Then

$$\langle h, f(t+d) - f^+(t)\rangle_H = 0,$$

so that

$$\langle h, f(t+d)\rangle_H = \left\langle h, f^+(t)\right\rangle_H = 0.$$

Consequently, $h \perp f(t+d),\ 0 < d \leq \delta$. But $h \perp L_t^+[f]$ by its choice. Thus $h \perp L_{t+\delta}[f]$. But then $h = 0_H$. □

Corollary 6.1.15 *Let $f : T \longrightarrow H$ have orthogonal increments. One has that*

$$\bigcap_{\delta>0} \overline{V[\{f(t+d) - f^+(t),\ 0 < d \leq \delta\}]} = \{0_H\}.$$

Proof The left-hand side of the assertion is the range of the limit of the net of projections $\{P_{t+\delta} - P_t^+, \delta > 0\}$ [(Lemma) 6.1.14]. □

Proposition 6.1.16 *Let $f : T \longrightarrow H$ have orthogonal increments. One has that*

$$L_t^+[f] \cap L_t[f]^\perp = \mathcal{R}[P_t^+ - P_t] = V\left[f^+(t) - f(t)\right].$$

Proof Because of (Lemma) 6.1.13, $\mathcal{R}[P_t^+ - P_t] = \cap_{\delta>0} L_t^\delta[f]$. It suffices thus to check that

$$V\left[f^+(t) - f(t)\right] = \cap_{\delta>0} L_t^\delta[f].$$

But one has the following orthogonal decomposition:

$$f(t+d) - f(t) = \left(f(t+d) - f^+(t)\right) + \left(f^+(t) - f(t)\right),$$

which extends to linear combinations, and one has thus the orthogonal decomposition

$$\begin{aligned} V[\{f(t+d) - f(t),\ d \in]0,\delta]\}] &= V\left[\{f(t+d) - f^+(t),\ d \in]0,\delta]\}\right] \\ &+ V\left[f^+(t) - f(t)\right]. \end{aligned}$$

But then [266, p. 40],

$$L_t^\delta[f] = \overline{V[\{f(t+d) - f^+(t),\ d \in]0,\delta]\}]} \oplus V\left[f^+(t) - f(t)\right],$$

and, because, for decreasing nets of projections, $\lim_\delta (P_\delta + P) = (\lim_\delta P_\delta) + P$,

$$\cap_{\delta>0} L_t^\delta[f] = \cap_{\delta>0} \overline{V[\{f(t+d) - f^+(t),\ d \in]0,\delta]\}]} \oplus V\left[f^+(t) - f(t)\right].$$

However, because of (Corollary) 6.1.15, $\cap_{\delta>0} \overline{V[\{f(t+d) - f^+(t),\ d \in]0,\delta]\}]} = \{0_H\}$. □

Corollary 6.1.17 *Let $f : T \longrightarrow H$ have orthogonal increments. One has that $P_t^+ = P_t$ if, and only if, $f^+(t) = f(t)$.*

Remark 6.1.18 Let $t_0 \leq t$, $\{t, t_0\} \subseteq T$, and $\delta > 0$ be fixed, but arbitrary. Then

$$\begin{aligned}\|f(t+\delta) - f(t_0)\|_H^2 &= \|f(t+\delta) - f(t) + f(t) - f(t_0)\|_H^2 \\ &= \|f(t+\delta) - f(t)\|_H^2 + \|f(t) - f(t_0)\|_H^2 \\ &\geq \|f(t+\delta) - f(t)\|_H^2\end{aligned}$$

Letting δ go to zero, one gets that

$$\left\|f^+(t) - f(t_0)\right\|_H^2 \geq \left\|f^+(t) - f(t)\right\|_H^2.$$

Proposition 6.1.19 *Let $f : T \longrightarrow H$ have orthogonal increments, and h in $L_{\cup T}[f]$ be fixed, but arbitrary. Set $f_h(t) = P_t[h]$. Then:*

1. *f_h is a function with orthogonal increments;*
2. *when f is continuous to the left, so is f_h;*
3. *f_h^+ exists;*
4. *$\sup_{t \in T} \|f_h(t)\|_H < \infty$;*
5. *when $L_{\cap T}[f] = \{0_H\}$, $L_{\cap T}[f_h] = \{0_H\}$.*

Proof Since the range of P_t is $L_t[f] \subseteq H$, $f_h(t) \in H$, for $t \in T$. Furthermore, for $t_1 < t_2 \leq t_3 < t_4$, $\{t_1, t_2, t_3, t_4\} \subseteq T$, fixed, but arbitrary,

$$\begin{aligned}\langle f_h(t_2) - f_h(t_1), f_h(t_4) - f_h(t_3)\rangle_H &= \\ &= \langle \{P_{t_2} - P_{t_1}\}[h], \{P_{t_4} - P_{t_3}\}[h]\rangle_H \\ &= \langle \{P_{t_4} - P_{t_3}\}\{P_{t_2} - P_{t_1}\}[h], h\rangle_H.\end{aligned}$$

However

$$\begin{aligned}\{P_{t_4} - P_{t_3}\}\{P_{t_2} - P_{t_1}\} &= P_{t_4}P_{t_2} - P_{t_4}P_{t_1} - P_{t_3}P_{t_2} + P_{t_3}P_{t_1} \\ &= P_{t_2} - P_{t_1} - P_{t_2} + P_{t_1}.\end{aligned}$$

f_h has thus orthogonal increments. Items 2 and 3 follow from the properties of the family $\{P_t,\ t \in T\}$. Item 4 is immediate as

$$\|f_h(t)\|_H = \|P_t[h]\|_H \leq \|h\|_H < \infty.$$

Item 5 follows from the inclusion $L_t[f_h] \subseteq L_t[f]$. □

6.2 Hilbert Space Valued, Countably Additive, Orthogonally Scattered Measures: A Summary of Results

The acronym for "countably additive, orthogonally scattered measures" shall be CAOSM, and to emphasize that values are in the real Hilbert space H, H-CAOSM. Expressions of the form

$$\int f\,dm,$$

where f is a class of functions with scalar values, and m, a measure, with values in a real Hilbert space, say H, are the building blocs of the Cramér-Hida representation. A compendium of properties of such objects is presented below. The reference is [182].

6.2.1 *General Case*

It is the case for which the measure m is defined for a family of subsets of an arbitrary set.

Measures

Definition 6.2.1 Let S be a set which is not empty. A family $\mathcal{P}[S]$ of subsets of S is a pre-ring over S when, given S_1 and S_2 in $\mathcal{P}[S]$,

1. $S_1 \cap S_2 \in \mathcal{P}[S]$;
2. there are disjoint sets $\Sigma_1, \ldots, \Sigma_n$ in $\mathcal{P}[S]$ such that $S_1 - S_2 = \cup_{i=1}^{n} \Sigma_i$.

Definition 6.2.2 Let $\mathcal{P}[S]$ be a pre-ring over S; H, a real Hilbert space; and $m : \mathcal{P}[S] \longrightarrow H$, a map such that

1. given disjoint $\{S_n, n \in \mathbb{N}\}$ in $\mathcal{P}[S]$ with $\cup_{n=1}^{\infty} S_n \in \mathcal{P}[S]$,

$$\sum_{n=1}^{\infty} m(S_n)$$

   is unconditionally convergent towards $m(\cup_{n=1}^{\infty} S_n)$,
2. given disjoint S_1 and S_2 in $\mathcal{P}[S]$, $m(S_1) \perp m(S_2)$.

The map m is then called a CAOSM on $\mathcal{P}[S]$.

Example 6.2.3 Given $H = L_2(X, \mathcal{X}, \xi)$, and $X_0 \in \mathcal{X}$, let I_{X_0} be the equivalence class of χ_{X_0} in H. The assignment $X_0 \mapsto I_{X_0}$ defines a CAOSM on $\mathcal{X}$.

Fact 6.2.4 *Let $\mathcal{P}[S]$ be a pre-ring over S; H, a real Hilbert space; and the map $m : \mathcal{P}[S] \longrightarrow H$, a CAOSM on $\mathcal{P}[S]$. Let, for $S_0 \in \mathcal{P}[S]$, fixed, but arbitrary,*

$$M(S_0) = \|m(S_0)\|_H^2 .$$

M is a finite, positive, countably additive measure on $\mathcal{P}[S]$. M is called "the measure associated with m."

Fact 6.2.5 *Let $\mathcal{P}[S]$ be a pre-ring over S; H, a real Hilbert space; and the map $m : \mathcal{P}[S] \longrightarrow H$, a CAOSM on $\mathcal{P}[S]$. All the sets involved below, at which m is evaluated, are assumed to belong to $\mathcal{P}[S]$. The following relations obtain:*

1. $m(S_1 \cup S_2) = m(S_1) + m(S_2) - m(S_1 \cap S_2)$.
2. $m(S_1 - S_2) = m(S_1) - m(S_1 \cap S_2)$.
3. $m(S_1 \Delta S_2) = m(S_1) + m(S_2) - 2m(S_1 \cap S_2)$.
4. $M(S_1 \Delta S_2) = \|m(S_1) - m(S_2)\|_H^2$.
5. $\|m(S_1) - m(S_2)\|_H^2 = M(S_1) + M(S_2) - 2M(S_1 \cap S_2)$.
6. *When* $S_1 \subseteq S_2$,

   (i) $m(S_2 - S_1) = m(S_2) - m(S_1)$.
   (ii) $\|m(S_2) - m(S_1)\|_H^2 = M(S_2) - M(S_1)$;
   (iii) $\|m(S_1)\|_H \le \|m(S_2)\|_H$;
   (iv) *when* $m(S_2) = 0_H$, $m(S_1) = 0_H$.

Fact 6.2.6 *Let $\mathcal{P}[S]$ be a pre-ring over S; H, a real Hilbert space; M, a finite, positive, countably additive measure on $\mathcal{P}[S]$; and $m : \mathcal{P}[S] \longrightarrow H$, a map. The following statements are equivalent:*

1. *Given S_1 and S_2 in $\mathcal{P}[S]$, fixed, but arbitrary,*

$$\langle m(S_1), m(S_2)\rangle_H = M(S_1 \cap S_2).$$

2. *m is a CAOSM over $\mathcal{P}[S]$, with M as associated measure.*

Fact 6.2.7 *Let $\mathcal{P}[S]$ be a pre-ring over S; H, a real Hilbert space; the map $m : \mathcal{P}[S] \longrightarrow H$, a CAOSM on $\mathcal{P}[S]$; $M_{\sigma(\mathcal{P}[S])}$, the unique, positive, σ-finite, countably additive, extension of M to the σ-ring $\sigma(\mathcal{P}[S])$ generated by $\mathcal{P}[S]$; and*

$$\sigma_f^m(\mathcal{P}[S]) = \left\{S_0 \in \sigma(\mathcal{P}[S]) : M_{\sigma(\mathcal{P}[S])}(S_0) < \infty\right\}.$$

Then:

1. *m has a unique CAOSM extension* $m_{\sigma_f^m(\mathcal{P}[S])}$ *to* $\sigma_f^m(\mathcal{P}[S])$.
2. $M_{\sigma_f^m(\mathcal{P}[S])}$, *the measure associated with* $m_{\sigma_f^m(\mathcal{P}[S])}$, *is obtained as the restriction of* $M_{\sigma(\mathcal{P}[S])}$ *to* $\sigma_f^m(\mathcal{P}[S])$.

Definition 6.2.8 Let $\mathcal{R}[S]$ be a σ-ring over S; M, a positive, σ-finite, countably additive, measure on $\mathcal{R}[S]$;

$$\mathcal{R}_f[S] = \{R \in \mathcal{R}[S] : M(R) < \infty\};$$

the map $m : \mathcal{R}_f[S] \longrightarrow H$, a CAOSM; and $M_{\mathcal{R}_f[S]}$, the restriction of M to $\mathcal{R}_f[S]$. Then m is a H-CAOSM over $(S, \mathcal{R}[S], M)$.

The Linear Space of a CAOSM

Definition 6.2.9 Let $\mathcal{P}[S]$ be a pre-ring of subsets of S; H, a real Hilbert space; and the map $m : \mathcal{P}[S] \longrightarrow H$, a CAOSM on $\mathcal{P}[S]$. The (linear sub-) space generated by m is the following subspace of H :

$$\overline{V\left[(m(S_0), S_0 \in \mathcal{P}[S]\}\right]}.$$

It shall be denoted $L[m]$.

Fact 6.2.10 *Let m be a H-CAOSM over* $(S, \mathcal{R}[S], M)$, *and* $\mathcal{P}[S]$ *be a pre-ring such that* $\mathcal{P}[S] \subseteq \mathcal{R}_f[S] \subseteq \sigma(\mathcal{P}[S])$. *Then, letting* $m_{\mathcal{P}[S]}$ *be the restriction of m to* $\mathcal{P}[S]$,

1. $L[m] = L\left[m_{\mathcal{P}[S]}\right]$;
2. *when* $\mathcal{P}[S]$ *is countable,* $L[m]$ *is separable.*

Integration

Definition 6.2.11 Let m be a H-CAOSM over $(S, \mathcal{R}[S], M)$. Given a function f, let its equivalence class in $L_2(S, \mathcal{R}[S], M)$ be $[f]$. The class of a simple function $f = \sum_{i=1}^{n} \alpha_i \chi_{S_i}$, $\alpha_i \in \mathbb{R}$, $S_i \in \mathcal{R}_f[S]$, $i \in [1 : n]$, has the following integral:

$$\int [f] dm = \sum_{i=1}^{n} \alpha_i m(S_i).$$

Given $[f] \in L_2(S, \mathcal{R}[S], M)$, let $\{[f_n], n \in \mathbb{N}\}$ be a sequence of equivalence classes of simple functions in $L_2(S, \mathcal{R}[S], M)$, converging, in that same space, to $[f]$. Then

$$\int [f] dm = \lim \int [f_n] dm.$$

Fact 6.2.12 *The integral defined in (Definition)* 6.2.11 *has the following properties:*

1. $\int [f] dm \in H$.
2. $\left(\int [f] dm, \int [g] dm \right)_H = \int fg dM$.
3. *Given* $S_0, S_1, S_2 \in \mathcal{R}[S]$,

$$\int_{S_0} [f] dm = \int [\chi_{S_0} f] dm,$$

and

$$\left\langle \int_{S_1} [f] dm, \int_{S_2} [g] dm \right\rangle_H = \left\langle \int_{S_1 \cap S_2} [f] dm, \int_{S_1 \cap S_2} [g] dm \right\rangle_H .$$

4. $\int [\alpha f + \beta g] dm = \alpha \int [f] dm + \beta \int [g] dm$.
5. $[f_n] \to [f]$ *in* $L_2(S, \mathcal{R}[S], M)$ *if, and only if,* $\int [f_n] dm \to \int [f] dm$.
6. $L[m] = \left\{ \int [f] dm, [f] \in L_2(S, \mathcal{R}[S], M) \right\}$, *and*

$$U : L_2(S, \mathcal{R}[S], M) \longrightarrow H$$

defined using the following relation:

$$U [[f]] = \int [f] dm$$

is a unitary operator onto $L[m]$.

Projection onto the Linear Space of a CAOSM

One may be interested in $P_{L[m]}$, the projection in H, onto $L[m]$. It may be computed as follows. Let again m be a H-CAOSM on $(S, \mathcal{R}[S], M)$. Given $h \in H$, and $R \in \mathcal{R}_f[S]$, fixed, but arbitrary, let

$$\mu_h(R) = \langle h, m(R) \rangle_H .$$

It is a finite, countably additive measure on the ring $\mathcal{R}_f[S]$.

Now, given a pre-ring $\mathcal{P}[S]$, and a finite, countably additive measure μ, on $\mathcal{P}[S]$, with, on $\mathcal{P}[S]$, finite total variation $|\mu|$, the latter has [85, p. 136] a unique, positive, σ-finite, countably additive, extension on the σ-ring $\sigma(\mathcal{P}[S])$ generated by $\mathcal{P}[S]$, say

$$|\mu|_{\sigma(\mathcal{P}[S])} .$$

Furthermore, μ has a unique, finite, countably additive extension, say $\mu_{\mathcal{R}_f[S]}$, to the ring

$$\mathcal{R}_f[S] = \left\{S_0 \in \sigma(\mathcal{P}[S]) : |\mu|_{\sigma(\mathcal{P}[S])}(S_0) < \infty\right\}.$$

One may apply the latter considerations to the measure μ_h. $|\mu_h|$, which is defined on the ring $\mathcal{R}_f[S]$, has an extension to the σ-ring $\sigma(\mathcal{R}_f[S])$, which is $\mathcal{R}[S]$. Write $|\mu_h|_{\mathcal{R}[S]}$ for that extension. μ_h has then an extension to the following ring:

$$\left\{S_0 \in \mathcal{R}[S] : |\mu_h|_{\mathcal{R}[S]}(S_0) < \infty\right\}.$$

That latter ring shall be denoted $\mathcal{R}_h[S]$. One has that

$$\mathcal{R}_f[S] \subseteq \mathcal{R}_h[S].$$

Fact 6.2.13 *Let m be a H-CAOSM on $(S, \mathcal{R}[S], M)$. Then:*

1. *on $\mathcal{R}_h[S]$, $\mu_h \ll M$;*
2. $\left[\frac{d\mu_h}{dM}\right] \in L_2(S, \mathcal{R}[S], M)$;
3. $P_{L[m]}[h] = \int \left[\frac{d\mu_h}{dM}\right] dm$;
4. *given $[f] \in L_2(S, \mathcal{R}[S], M)$, fixed, but arbitrary,*

$$\left\langle h, \int [f] dm \right\rangle_H = \int f d\mu_h.$$

Remark 6.2.14 Let m be a H-CAOSM on $(S, \mathcal{R}[S], M)$. One can associate with m a RKHS as follows. Define $F : \mathcal{R}_f[S] \longrightarrow H$ using the following relation:

$$F(S_0) = m(S_0).$$

Then

$$\overline{V[\{F(S_0), S_0 \in \mathcal{R}_f[S]\}]} = \overline{V[\{m(S_0), S_0 \in \mathcal{R}_f[S]\}]} = L[m].$$

Let then $L_F : H \longrightarrow \mathbb{R}^{\mathcal{R}_f[S]}$, be defined using the following relation:

$$L_F[h](S_0) = \langle h, m(S_0)\rangle_H = \mu_h(S_0).$$

Thus the RKHS associated with m is a Hilbert space of measures whose inner product is obtained using the following relation:

$$\begin{aligned}\langle \mu_{h_1}, \mu_{h_2}\rangle_{RKHS} &= \langle L_F[h_1], L_F[h_2]\rangle_{RKHS} \\ &= \left\langle P_{L[m]}[h_1], P_{L[m]}[h_2]\right\rangle_H\end{aligned}$$

$$= \left\langle \int \left[\frac{d\mu_{h_1}}{dM} \right] dm, \int \left[\frac{d\mu_{h_2}}{dM} \right] dm \right\rangle_H$$

$$= \int \frac{d\mu_{h_1}}{dM} \frac{d\mu_{h_2}}{dM} dM.$$

The reproducing kernel is obtained using the following relation:

$$\begin{aligned} C_F(S_1, S_2) &= \langle F(S_1), F(S_2) \rangle_H \\ &= \langle m(S_1), m(S_2) \rangle_H \\ &= M(S_1 \cap S_2) \\ &= \int \chi_{S_1} \chi_{S_2} \, dM. \end{aligned}$$

Remark 6.2.15 Let m be a H-CAOSM on $(S, \mathcal{R}[S], M)$; P, a projection of $L[m]$; and Q a projection of $L_2(S, \mathcal{R}[S], M)$. Then $U^{\star}PU$, U as in (Fact) 6.2.12, is a projection of $L_2(S, \mathcal{R}[S], M)$, and $UQU^{\star}$, a projection of $L[m]$.

A case in point is as follows. Suppose one defines $U[\phi_t] = f(t)$. Let the inclusion $T_0 \subseteq T$ be strict. Let $L_{T_0}[f]$ be the subspace of $L_{UT}[f]$ generated linearly by the family $\{f(t), t \in T_0\}$. The associated projection shall be P_0. $L_{T_0}[\phi]$ and Q_0 are obtained similarly, *mutatis mutandis*. As, when $t \in T_0$,

$$P_0[f(t)] = f(t) = \int \phi_t dm = \int Q_0[\phi_t] dm,$$

one has, for $\tilde{f} \in L_{UT}[f]$, $\tilde{\phi} \in L_{UT}[\phi]$, $\tilde{f} = U[\tilde{\phi}]$, fixed, but arbitrary,

$$P_0 \left[\tilde{f} \right] = \int Q_0 \left[\tilde{\phi} \right] dm,$$

or $P_0 = UQ_0U^{\star}$.

Remark 6.2.16 Let m be a H-CAOSM on $(S, \mathcal{R}[S], M)$. Let $S_0 \in \mathcal{R}[S]$, and $M(S_0) < \infty$. Let $L_0[m]$ be the subspace generated linearly in $L[m]$ by the elements $m(S_{00})$, $S_{00} \in \mathcal{R}[S]$, $S_{00} \subseteq S_0$. Then, P_0 being the projection onto $L_0[m]$,

$$P_0 \left[\int [f] dm \right] = \int I_{S_0} [f] dm.$$

The integral on the right of the latter equality is indeed an element of $L[m]$, and

$$\left\langle \int [f] dm - \int I_{S_0} [f] dm, m(S_{00}) \right\rangle_H = \int \chi_{S \setminus S_0} \chi_{S_{00}} f \, dM = 0.$$

Isomorphisms of Linear Spaces

The next result introduces two objects of the same kind. Indices are used to distinguish the cases.

Fact 6.2.17 *Let m be a H-CAOSM on $(S, \mathcal{R}[S], M)$, K, a real Hilbert space, and $A : L[m] \longrightarrow K$, a bounded, linear operator. Set, for $R \in \mathcal{R}_f[S]$, fixed, but arbitrary,*

$$m_A(R) = A[m(R)].$$

Suppose that, for R_1 and R_2 in $\mathcal{R}_f[S]$, fixed, arbitrary, and disjoint,

$$m_A(R_1) \perp m_A(R_2).$$

Then:

1. *m_A is a K-CAOSM on $\mathcal{R}_f[S]$;*
2. *for $R \in \mathcal{R}_f[S]$, fixed, but arbitrary,*

$$M_A(R) \leq \|A\|^2 M(R);$$

3. *$L_2(S, \mathcal{R}[S], M) \subseteq L_2(S, \mathcal{R}[S], M_A)$;*
4. *for $[f] \in L_2(S, \mathcal{R}[S], M)$, fixed, but arbitrary,*

$$A\left[\int [f]dm\right] = \int [f]dm_A;$$

5. *when A is unitary (write U for A), m_U is a K-CAOSM over $(S, \mathcal{R}[S], M)$, and $L[m_A] = U[L[m]]$.*

Fact 6.2.18 *Let m be a H-CAOSM, and n be a K-CAOSM, both on $(S, \mathcal{R}[S], M)$. There exists then a unitary $U : L[m] \longrightarrow L[n]$ such that, for $R \in \mathcal{R}_f[S]$, fixed, but arbitrary, $n(R) = U[m(R)]$.*

Absolute Continuity

Fact 6.2.19 *Let m be a H-CAOSM on $(S, \mathcal{R}[S], M)$, and $\Phi : S \longrightarrow \mathbb{R}$, an adapted function. Set:*

(a) $\mathcal{R}_\Phi[S] = \left\{R \in \mathcal{R}[S] : [\Phi] \in L_2(R, \mathcal{R}[S] \cap R, M^{|\mathcal{R}[S] \cap R})\right\}$,

(b) *for* $R \in \mathcal{R}_\Phi[S]$, *fixed, but arbitrary,*

$$M_\Phi(R) = \int_R \Phi^2(s) M(ds),$$

and

$$m_\Phi(R) = \int_R [\Phi] dm.$$

Then, $M_\Phi^{\sigma(\mathcal{R}_\Phi[S])}$ *being the extension of* M_Φ *to* $\sigma(\mathcal{R}_\Phi[S])$,

1. m_Φ *is a H-CAOSM over* $(S, \sigma(\mathcal{R}_\Phi[S]), M_\Phi^{\sigma(\mathcal{R}_\Phi[S])})$;
2. *given* $[f] \in L_2(S, \sigma(\mathcal{R}_\Phi[S]), M_\Phi^{\sigma(\mathcal{R}_\Phi[S])})$, *fixed, but arbitrary,*

$$[f\Phi] \in L_2(S, \mathcal{R}[S], M),$$

and

$$\int [f] dm_\Phi = \int [f\Phi] dm.$$

Change of Measure

Fact 6.2.20 *Let* m *be a H-CAOSM over* $(S, \mathcal{R}[S], M)$, *and* $\Psi : S \longrightarrow T$, *a map. Let*

$$\mathcal{R}_\Psi(T) = \left\{T_0 \subseteq T : \Psi^{-1}(T_0) \in \mathcal{R}[S]\right\}.$$

Then $m_\Psi = m \circ \Psi^{-1}$ *is a H-CAOSM over* $(T, \mathcal{R}_\Psi[T], M_\Psi = M \circ \Psi^{-1})$. *When* f *is adapted to* $\mathcal{R}_\Psi(T)$, $f \circ \Psi$ *is adapted to* $\mathcal{R}[S]$, *and*

$$\int [f] dm_\Psi = \int [f \circ \Psi] dm,$$

in the sense that, if either integral exists, the other does too, and they are equal.

An Interchange of Integration Lemma

Fact 6.2.21 ([153, p. 49]) *Let* m *be a H-CAOSM over* $(S, \mathcal{R}[S], M)$; *the space* $L_2(S, \mathcal{R}[S], M)$ *be separable; and*

(a) $\Phi : S \times S \longrightarrow \mathbb{R}$ *be adapted to* $\mathcal{R}[S] \otimes \mathcal{R}[S]$, *and such that, for* $\sigma \in S$, *fixed, but arbitrary,*

$$[\Phi(\sigma, \cdot)] \in L_2(S, \mathcal{R}[S], M);$$

(b) $f(\sigma) = \int [\Phi(\sigma, \cdot)] \, dm$;

(c) *μ be a measure on $\mathcal{R}[S]$ such that, for $s \in S$, fixed, but arbitrary,*

$$|\Psi|(s) = \int_S |\Phi(\sigma, s)| \, \mu(d\sigma) < \infty;$$

(d) $\Psi(s) = \int_S \Phi(\sigma, s) \, \mu(d\sigma)$.

When $|\Psi| \in L_2(S, \mathcal{R}[S], M)$, f is weakly (Pettis) integrable, and

$$\int_S f(s) \, \mu(ds) = \int [\Psi] \, dm.$$

Proof Let indeed $g \in L[m]$, fixed, but arbitrary, correspond unitarily to $[\phi_g]$ in $L_2(S, \mathcal{R}[S], M)$. As

$$\langle g, f(\sigma) \rangle_H = \int_T M(ds) \, \phi_g(s) \, \Phi(\sigma, s),$$

$\sigma \mapsto \langle g, f(\sigma) \rangle_H$ is adapted [46, p. 164]. Furthermore

$$\begin{aligned} \int_S \mu(d\sigma) \, |\langle g, f(\sigma) \rangle_H| &= \int_S \mu(d\sigma) \left| \int_S M(ds) \, \phi_g(s) \, \Phi(\sigma, s) \right| \\ &\leq \int_S M(ds) \, |\phi_g(s)| \, |\Psi|(s) \\ &< \infty. \end{aligned}$$

Finally, $L[m]$ is separable, as it is unitarily isomorphic to $L_2(T, \mathcal{T}, M)$, which has been assumed separable. Since the range of f is in $L[m]$, f is thus weakly integrable [207, p. 111]. But then

$$\begin{aligned} \left\langle g, \int_S f(\sigma) \, \mu(d\sigma) \right\rangle_H &= \int_S \langle g, f(\sigma) \rangle_H \, \mu(d\sigma) \\ &= \int_S \mu(d\sigma) \int_S M(ds) \, \phi_g(s) \, \Phi(\sigma, s). \end{aligned}$$

The assumption on $|\Psi|$ allows one to proceed to an integration interchange [46, p. 166], so that

$$\left\langle g, \int_S f(\sigma) \, \mu(d\sigma) \right\rangle_H = \int_S M(ds) \, \phi_g(s) \, \Psi(s) = \left\langle g, \int [\Psi] \, dm \right\rangle_H.$$

Since g is arbitrary, one is done. ☐

Representation of Functions as Integrals with Respect to a CAOSM

When the covariance of a function $f : T \longrightarrow H$, H, a real Hilbert space, has an L_2 representation (Definition 2.3.1), an integral representation of f itself is obtained from its covariance representation, as asserted in the next result. In such a case, one is thus able to produce "explicitly" a measure of type m, that is, one which is a CAOSM.

Fact 6.2.22 ([122, p. 59]) *Let $f : T \longrightarrow H$ be fixed, but arbitrary, and, for $\{t_1, t_2\} \subseteq T$, fixed, but arbitrary, set*

$$C_f(t_1, t_2) = \langle f(t_1), f(t_2) \rangle_H .$$

Suppose that there exists $\{k_t \in L_2(X, \mathcal{X}, \xi), t \in T\}$ such that,

$$C_f(t_1, t_2) = \langle k_{t_1}, k_{t_2} \rangle_{L_2(X, \mathcal{X}, \xi)} .$$

There exists then $m : \mathcal{S} \longrightarrow H$, a CAOSM, such that

$$f(t) = \int k_t dm.$$

Furthermore:

1. $L_{UT}[f] \subseteq L[m]$;
2. $L_{UT}[f] = L[m]$ *if, and only if, the following family: $\{k_t, t \in T\}$ is total in $L_2(X, \mathcal{X}, \xi)$.*

Proof Let K be the (closed) subspace generated in $L_2(X, \mathcal{X}, \xi)$ by the family $\{k_t, t \in T\}$, and define $U : K \longrightarrow H$ using the following relation:

$$U[k_t] = f(t).$$

That definition makes sense. Indeed, by assumption,

$$\langle f(t_1), f(t_2) \rangle_H = C_f(t_1, t_2) = \langle k_{t_1}, k_{t_2} \rangle_{L_2(X, \mathcal{X}, \xi)},$$

from which one obtains that

$$\begin{aligned} \left\| \sum_{i=1}^{n} \alpha_i U[k_{t_i}] \right\|_H^2 &= \left\| \sum_{i=1}^{n} \alpha_i f(t_i) \right\|_H^2 \\ &= \sum_{i=1}^{n} \sum_{j=1}^{n} \alpha_i \alpha_j \langle f(t_i), f(t_j) \rangle_H \end{aligned}$$

$$= \sum_{i=1}^{n} \sum_{j=1}^{n} \alpha_i \alpha_j \left\langle k_{t_i}, k_{t_j} \right\rangle_{L_2(X, \mathcal{X}, \xi)}$$

$$= \left\| \sum_{i=1}^{n} \alpha_i k_{t_i} \right\|_{L_2(X, \mathcal{X}, \xi)}^2 .$$

U may thus be extended, using linearity and continuity, to a unitary map between K and $L_{\cup T}[f]$.

Proof ([2]) *Suppose that* $K = L_2(X, \mathcal{X}, \xi)$.

U is then a unitary map between $L_2(X, \mathcal{X}, \xi)$ and $L_{\cup T}[f]$. Let $X_0 \in \mathcal{X}$ be such that $I_{X_0} \in L_2(X, \mathcal{X}, \xi)$, and define $m : \mathcal{X} \longrightarrow H$ using the following relation:

$$m(X_0) = U[I_{X_0}] .$$

One thus obtains an orthogonal, additive, vector set function, and

$$M(M_0) = \|m(X_0)\|_H^2 = \xi(X_0) .$$

Consequently, for $g \in L_2(X, \mathcal{X}, \xi)$, fixed, but arbitrary,

$$\int g\, dm = U[g] ,$$

and in particular

$$\int k_t\, dm = f(t) .$$

$L[m]$ is thus $L_{\cup T}[f]$.

Proof ([1]) *Suppose that K is strictly contained in* $L_2(X, \mathcal{X}, \xi)$.

Let:

(a) $\{k_\theta^\perp, \theta \in \Theta\}$ be a total family in $K^\perp$;
(b) $C_\perp(\theta_1, \theta_2) = \left\langle k_{\theta_1}^\perp, k_{\theta_2}^\perp \right\rangle_{L_2(X, \mathcal{X}, \xi)}$;
(c) $\Lambda = T \uplus \Theta$ (disjoint union);
(d) $k_\lambda^\star = k_t$, when $\lambda = t \in T$, and $k_\lambda^\star = k_\theta^\perp$, when $\lambda = \theta \in \Theta$;
(e) $C_\star(\lambda_1, \lambda_2) = \left\langle k_{\lambda_1}^\star, k_{\lambda_2}^\star \right\rangle_{L_2(X, \mathcal{X}, \xi)}$.

$C_\perp$ is a covariance, and there is thus a Hilbert space $H_\perp$, and a map

$$f_\perp : \Theta \longrightarrow H_\perp ,$$

such that

$$\langle f_\perp(\theta_1), f_\perp(\theta_2)\rangle_{H_\perp} = C_\perp(\theta_1, \theta_2).$$

Let $H_\star = H \oplus H_\perp$, and

$$f^\star(\lambda) = \begin{cases} \left(f(t), 0_{H_\perp}\right) \text{ when } \lambda = t \in T \\ \left(0_H, f_\perp(\theta)\right) \text{ when } \lambda = \theta \in \Theta \end{cases}.$$

$f^\star$ has covariance $C_\star$, and the subspace generated by the family

$$\{k_\lambda^\star, \lambda \in \Lambda\}$$

is $L_2(X, \mathcal{X}, \xi)$. However $L_{UT}[f]$ is isomorphic to K. □

6.2.2 *Case of Intervals*

It is the case for which the map m is defined for subsets of intervals of $\mathbb{R}$, and its values are obtained from the increments of a function with values in H, and orthogonal increments.

Let thus $T \subseteq \mathbb{R}$ be an interval; H, a real Hilbert space; $g : T \longrightarrow H$, a map with orthogonal increments. $\mathcal{P}[S]$ shall be $\mathcal{I}[T]$, the set of intervals $]t_1, t_2] \subseteq T$. m_g is obtained setting

$$m_g(]t_1, t_2]) = g(t_2+) - g(t_1+),$$

and M_g, setting

$$M_g(]t_1, t_2]) = \|g(t_2+) - g(t_1+)\|_H^2.$$

m_g and M_g have unique extensions to the Borel sets of T, say $\mathcal{T}$. One shall not distinguish between respectively m_g and M_g, and their extensions. The extension of m_g can be computed for sets in $\mathcal{T}$ of finite, extended, M_g-measure. The result is a CAOSM on $(T, \mathcal{T}, M_g)$. The distinguishing feature of the present case is that the integral can be obtained as a Stieltjes one, with some extra formulae in the bargain, as follows.

Fact 6.2.23 *Suppose g has orthogonal increments on $\mathbb{R}$; m_g and M_g are the (extended) measures introduced above; ϕ is a real valued, continuous function, defined on $[t_1, t_2]$. Then*

$$\int_{t_1}^{t_2} [\phi]\, dm_g$$

may be obtained as the limit, in H, of expressions of the following form:

$$\sum_{i=1}^{n} \phi(\tau_i)\{g(\theta_i) - g(\theta_{i-1}\},$$

and one has that the integral is the sum of three terms, that is,

$$\int_{t_1}^{t_2} [\phi] dm_g = \phi(t_1)\{g(t_1+) - g(t_1)\} + \int_{]t_1,t_2[} [\phi] dm_g + \phi(t_2)\{g(t_2) - g(t_2-)\}.$$

Fact 6.2.24 *When ϕ in (Fact)* 6.2.23 *is absolutely continuous,*

$$\int_{t_1}^{t_2} [\phi] dm_g = \phi(t_2)g(t_2) - \phi(t_1)g(t_1) - \int_{[t_1,t_2]} \phi'(\theta)g(\theta)d\theta,$$

where the integral on the right-hand side of the equality is a Bochner integral.

What has been streamlined above shall be used within the framework of the Cramér-Hida representation, where the specific assumptions, which are adopted to that end, lead to a number of adjustments, presented below in the form of remarks. They also serve to fix notation. That representation involves additive vector set functions always obtained from "ordinary" functions.

Let thus $f : T \longrightarrow H$ be a well-defined function which is continuous to the left (an assumption that shall be made to obtain also the Cramér-Hida representation), and set

$$m_f([t_1, t_2[) = f(t_2) - f(t_1).$$

The extension of m_f to a measure may be based on [182, p. 69], or on the fact that m_f is sigma-additive, and has finite variation, if, and only if, f is continuous to the left, and has bounded variation [75, Paragraph 18]. It will have orthogonal increments if, and only if, f does, and then

$$M_f([t_1, t_2[) = \|f(t_2) - f(t_1)\|_H^2 = F(t_2) - F(t_1).$$

Remark 6.2.25 When $t_l \in T$, $m_f([t_l, t_l + \epsilon[) = f(t_l + \epsilon) - f(t_l)$, so that, when $f^+(t_l)$ exists

$$m_f(\{t_l\}) = f^+(t_l) - f(t_l).$$

Remark 6.2.26 When $t_2 = t$, and $t_1 \downarrow\downarrow t_l$ (so that $f^+(t_l)$ exists), one gets that

$$m_f(]t_l, t[) = f(t) - f^+(t_l).$$

Thus, when $f^+(t_l) = 0_H$, $L_{\cup T}[f]$ is isomorphic to

$$L_2\left(]t_l,\infty[\cap T,]t_l,\infty[\cap T, M_f^{|]t_l,\infty[\cap T}\right),$$

and, when $t_l \in T$, from (Remark) 6.2.25, $m_f(\{t_l\}) = -f(t_l)$. When $f(t_l) \neq 0_H$, one has an, in practice, awkward situation, as one would expect that $f(t_l) = 0_H$ when $f^+(t_l) = 0_H$. The requirement that f be purely nondeterministic insures that fact, and that will be a consequence of the assumptions that shall guide the Cramér-Hida representation. Then, in particular, $L_{\cup T}[f] = L[m_f]$, so that the former has the integral representation of the latter, which is often useful when computing with f.

Remark 6.2.27 Let f, continuous to the left, have orthogonal increments, with $f^+(t_l) = 0_H$ (take for instance $f(t) = 0_H, t \in T$), and $L_{\cup T}[f] \subset H$. Let $h \in H, h \neq 0_H$, be orthogonal to $L_{\cup T}[f]$, and set $g(t) = h + f(t)$. g is then continuous to the left, has orthogonal increments, $g^+(t_l) = h$, and $m_g = m_f$. However

$$L_{\cup T}[g] = V[h] \oplus L_{\cup T}[f], \text{ and } L[m_g] = L[m_f] = L_{\cup T}[f],$$

so that $L[m_g] \subset L_{\cup T}[g]$.

Remark 6.2.28 Let h_1 and h_2 be orthonormal in H, and set $k_1 = h_1 + h_2$, $k_2 = 2^{-1}\{h_1 + 3h_2\}$. Define: $T = [0,1]$, and

$$f(t) = \chi_{]0,\frac{1}{2}]}(t)\,k_1 + \chi_{]\frac{1}{2},1]}(t)\,k_2.$$

f is continuous to the left. It has orthogonal increments. Indeed, for fixed, but arbitrary $\{t_1,t_2,t_3,t_4\} \subseteq [0,1]$, $t_1 < t_2 \leq t_3 < t_4$, one has to consider the following "qualitative" cases:

0	$]0,\frac{1}{2}]$	$]\frac{1}{2},1]$	$f(t_2)-f(t_1)$	$f(t_4)-f(t_3)$
t_1	t_2,t_3,t_4		k_1	0_H
t_1	t_2,t_3	t_4	k_1	k_2-k_1
t_1	t_2	t_3,t_4	k_1	0_H
t_1		t_2,t_3,t_4	k_2	0_H
	t_1,t_2,t_3,t_4		0_H	0_H
	t_1,t_2,t_3	t_4	0_H	k_2-k_1
	t_1,t_2	t_3,t_4	0_H	0_H
	t_1	t_2,t_3,t_4	k_2-k_1	0_H
		t_1,t_2,t_3,t_4	0_H	0_H

As $\langle k_1, k_2 - k_1 \rangle_H = \left\langle h_1 + h_2, \frac{h_2 - h_1}{2} \right\rangle_H = 0,$

$$\langle f(t_2) - f(t_1), f(t_4) - f(t_3) \rangle_H = 0.$$

Furthermore $f^+(0) = k_1 \neq 0_H$. However $L\left[m_f\right] = L_{\cup T}[f] = V[h_1, h_2]$, so that the latter may obtain without $f^+(t_l) = 0_H$.

Remark 6.2.29 In the sequel one shall try to work as much as possible with the following generic notation: for a function, say $f : T \longrightarrow H$, the additive vector set function it determines is denoted m_f, or m^f when indices are used. And M_f and M^f denote the measures associated respectively with m_f and m^f, using their norm squared.

6.3 Boundedness, Limits, and Separability

Separability refers to that of $L_{\cup T}[f]$ and is basic to the Cramér-Hida representation. It is furthermore necessary in practice, for computing. It is thus useful to have "reasonable" conditions on f that insure separability of $L_{\cup T}[f]$, and explain what other restrictions one uses may mean. Schematically, and intuitively, $L_{\cup T}[f]$ should be separable when f is smooth enough. One meets below two ways to understand "smooth enough."

Lemma 6.3.1 *Let $t \in T$ be fixed, but arbitrary, and*

$$\Sigma[t] = \{\underline{t} = \{t_n, n \in \mathbb{N}\} \subseteq T : t \downdownarrows t_n\}.$$

Define, for $f : T \longrightarrow H$, fixed, but arbitrary,

$$\sigma(t) = \sup_{\{\underline{t}_1, \underline{t}_2\} \subseteq \Sigma[t]} \left\{ \limsup_n \left\| f\left(t_n^{(1)}\right) - f\left(t_n^{(2)}\right) \right\|_H \right\}.$$

Then $f^+(t)$ exists if, and only if, $\sigma(t) = 0$.

Proof Suppose that $f^+(t)$ exists.

Then, whatever $\underline{t} \in \Sigma[t]$, $f^+(t) = \lim_n f(t_n)$. Thus, given $\{\underline{t}_1, \underline{t}_2\} \subseteq \Sigma[t]$,

$$\limsup_n \left\| f\left(t_n^{(1)}\right) - f\left(t_n^{(2)}\right) \right\|_H = \lim_n \left\| f\left(t_n^{(1)}\right) - f\left(t_n^{(2)}\right) \right\|_H = 0,$$

so that $\sigma(t) = 0$.

Proof Suppose that $\sigma(t) = 0$.

Let $\underline{t} \in \Sigma[t]$ be fixed, but arbitrary, and choose $\underline{t_1}$ and $\underline{t_2}$ to be arbitrary, but fixed subsequences of $\underline{t}$. The assumption $\sigma(t) = 0$ then means that $\{f(t_n), n \in \mathbb{N}\}$ is a Cauchy sequence, so that it has a limit, which, by definition, is $f^+(t)$. □

For a set S, $|S|$ represents the (cardinal) number of its elements.

Lemma 6.3.2 *Suppose that, for fixed, but arbitrary $t \in T$, $f^-(t)$ exists. Then, for $t \in T$, fixed, but arbitrary, there is $\epsilon_t > 0$ such that*

$$|\{\theta \in [t - \epsilon_t, t] \cap T : \sigma(\theta) = \infty\}| < \aleph_0.$$

Proof Suppose not, that is, there is $t_0 \in T$ such that, for fixed, but arbitrary $\epsilon > 0$,

$$|\{t \in [t_0 - \epsilon, t_0] \cap T : \sigma(t) = \infty\}| \geq \aleph_0.$$

Let T_0 be the set of those t values, at, or to the left of t_0, at which $\sigma(t) = \infty$. Then t_0 must be an accumulation point of T_0. Because of the definition of σ,

$$\sigma(t) \leq 2 \sup_{\underline{t} \in \Sigma[t]} \limsup_n \|f(t_n)\|_H.$$

Thus, when $t \in T_0$, $\sigma(t) = \infty$, and, since $\limsup a_n > c$ implies that $a_n > c$ infinitely often, for any $\kappa > 0$, there is then a sequence $\{t_n^{(\kappa)}\} \subseteq T_0$, converging to t_0 from the left, such that for all $n \in \mathbb{N}$,

$$\left\|f\left(t_n^{(\kappa)}\right)\right\|_H > \kappa.$$

But then, choosing $\kappa > \|f^-(t_0)\|_H$,

$$\begin{aligned}
\left\|f\left(t_n^{(\kappa)}\right) - f^-(t_0)\right\|_H &\geq \left|\left\|f\left(t_n^{(\kappa)}\right)\right\|_H - \|f^-(t_0)\|_H\right| \\
&= \left|\left\{\left\|f\left(t_n^{(\kappa)}\right)\right\|_H - \kappa\right\} + \{\kappa - \|f^-(t_0)\|_H\}\right| \\
&= \left\{\left\|f\left(t_n^{(\kappa)}\right)\right\|_H - \kappa\right\} + \{\kappa - \|f^-(t_0)\|_H\} \\
&> \kappa - \|f^-(t_0)\|_H.
\end{aligned}$$

Consequently, $\left\|f\left(t_n^{(\kappa)}\right) - f^-(t_0)\right\|_H$ is arbitrarily large, uniformly in $n \in \mathbb{N}$, and that contradicts the existence of a limit to the left. □

Proposition 6.3.3 *Suppose that, for fixed, but arbitrary $t \in T$, $f^-(t)$ exists. Then the number of points $t \in T$ at which $f^+(t)$ does not exist is at most countable.*

Proof Because of (Lemma) 6.3.1,

$$T_d = \left\{t \in T : f^+(t) \text{ does not exist}\right\} = \{t \in T : \sigma(t) > 0\}.$$

Let

$$T_d^{(n)} = \left\{ t \in T : \sigma(t) > n^{-1} \right\}.$$

Then

$$\{t \in T : \sigma(t) > 0\} = \bigcup_n \left\{ t \in T : \sigma(t) > n^{-1} \right\} = \bigcup_n T_d^{(n)}.$$

Suppose that the assertion is false, that is, $|T_d| > \aleph_0$. There exists then n_0 such that $\left| T_d^{(n_0)} \right| > \aleph_0$. Let, for $p \in \mathbb{N}$, fixed, but arbitrary,

$$T_d^{(n,p)} = T_d^{(n)} \cap [p, p+1[.$$

As

$$T_d^{(n_0)} = \bigcup_p \left\{ T_d^{(n_0)} \cap [p, p+1[\right\} = \bigcup_p T_d^{(n_0,p)},$$

there exists p_0 such that $\left| T_d^{(n_0,p_0)} \right| > \aleph_0$: $T_d^{(n_0,p_0)}$ is an uncountable set of points $t \in [p_0, p_0 + 1[$ at which $\sigma(t) > n_0^{-1}$, and for which $f^+(t)$ does not exist.

As shall presently be seen, there exists then $t_0 \in]p_0, p_0 + 1]$ such that, for fixed, but arbitrary $\epsilon > 0$,

$$\left| [t_0 - \epsilon, t_0] \cap T_d^{(n_0)} \right| \geq \aleph_0, \tag{$\star$}$$

that is, each left neighborhood of t_0 contains infinitely many values of t for which $\sigma(t) > n_0^{-1}$. Suppose indeed, as a temporary assumption, that such is not the case: for $t \in]p_0, p_0 + 1]$, fixed, but arbitrary, there exists $\epsilon_t > 0$, such that

$$\left| [t - \epsilon_t, t] \cap T_d^{(n_0)} \right| < \aleph_0.$$

Let $\mathcal{I}_{p_0+1}$ be the family of intervals of the form $[p_0 + 1 - \epsilon, p_0 + 1]$ for which $0 < \epsilon < 1$, and

$$\left| [p_0 + 1 - \epsilon, p_0 + 1] \cap T_d^{(n_0)} \right| < \aleph_0.$$

That family is not void, because of the temporary assumption, when choosing $t = p_0 + 1$. So the following definition:

$$t_\star = \inf_{\epsilon \in]0,1[} (p_0 + 1 - \epsilon)$$

makes sense. But one must then have that $t_\star = p_0$. Indeed, if $t_\star > p_0$, since $t_\star$ is the smallest element for which the interval it determines with $p_0 + 1$ contains a finite number of values, and that $T_d^{(n_0,p_0)}$ is uncountable, all sets of the form

$$[t_\star - \delta, t_\star] \cap T_d^{(n_0)}, \ \delta > 0,$$

will contain, because of the choice of n_0 and p_0, infinitely many points, contrary to the temporary assumption. Consequently

$$\left| [p_0, p_0 + 1] \cap T_d^{(n_0)} \right|$$

is at most countable, but that is impossible given that $\left|T_d^{(n_0,p_0)}\right| > \aleph_0$.

Let t_0 be the value appearing in ($\star$). Since $f^-(t_0)$ exists, for $\epsilon > 0$, fixed, but arbitrary, there exists $\delta_\epsilon > 0$ small enough such that, for $t \in [t_0 - \delta_\epsilon, t_0]$,

$$\|f^-(t_0) - f(t)\|_H < \epsilon.$$

Using ($\star$) and (Lemma) 6.3.2, let then

$$\left\{t_p^{(0)}, p \in \mathbb{N}\right\}$$

be a sequence such that $t_p^{(0)} \uparrow\uparrow t_0$, and $n_0^{-1} < \sigma\left(t_p^{(0)}\right) < \infty$. By definition

$$\sigma\left(t_p^{(0)}\right) = \sup_{\{\underline{\theta}_1,\underline{\theta}_2\} \subseteq \Sigma\left[t_p^{(0)}\right]} \limsup_p \left\|f\left(\theta_p^{(1)}\right) - f\left(\theta_p^{(2)}\right)\right\|_H.$$

Since, for $i \in \{1,2\}$, eventually $t_p^{(0)} < \theta_p^{(i)} < t_0$,

$$\begin{aligned} \left\|f\left(\theta_p^{(1)}\right) - f\left(\theta_n^{(2)}\right)\right\|_H &\leq \left\|f\left(\theta_p^{(1)}\right) - f^-(t_0)\right\|_H + \left\|f^-(t_0) - f\left(\theta_n^{(2)}\right)\right\|_H \\ &\leq n_0^{-1}, \end{aligned}$$

and one ends up with a contradiction as the left-hand side of those inequalities yields $\sigma(t_p^{(0)}) > n_0^{-1}$. □

Proposition 6.3.4 *Suppose that, for fixed, but arbitrary $t \in T$, $f^-(t)$ exists. Then*

1. *the number of points $t \in T$ at which f is not continuous to the left is at most countable;*
2. *the number of points $t \in T$ at which f is not continuous to the right is at most countable.*

Proof Let $\delta(t) = \|f^-(t) - f(t)\|_H$, $T_d = \{t \in T : \delta(t) > 0\}$, and suppose that $|T_d| > \aleph_0$. Again, there are integers n_0 and p_0 such that

$$\left|T_d^{(n_0,p_0)}\right| = \left|]p_0, p_0 + 1] \cap \left\{t \in T : \delta(t) > n_0^{-1}\right\}\right| > \aleph_0.$$

Furthermore, $(\star)$ of (Proposition) 6.3.3 obtains: there exists $t_0 \in]p_0, p_0 + 1]$ such that each left neighborhood of t_0 contains infinitely many points t for which

$$\delta(t) > n_0^{-1}.$$

Let thus $t_p^{(0)} \uparrow\uparrow t_0$ be such that, for $p \in \mathbb{N}$, fixed, but arbitrary,

$$\delta\left(t_p^{(0)}\right) > n_0^{-1}.$$

Given a fixed, but arbitrary $\epsilon > 0$, choose, which is possible as left limits exist, $\theta_p > 0$ such that

$$t_p^{(0)} + \theta_p < t_{p+1}^{(0)}, \text{ and } \left\| f\left(t_p^{(0)} + \theta_p\right) - f^-\left(t_{p+1}^{(0)}\right)\right\|_H < \epsilon.$$

Then, for $\epsilon > 0$ small enough,

$$\begin{aligned}
&\left\| f\left(t_p^{(0)} + \theta_p\right) - f\left(t_{p+1}^{(0)}\right)\right\|_H = \\
&\quad = \left\| \left\{ f^-\left(t_{p+1}^{(0)}\right) - f\left(t_{p+1}^{(0)}\right)\right\} - \left\{ f^-\left(t_{p+1}^{(0)}\right) - f\left(t_p^{(0)} + \theta_p\right)\right\}\right\|_H \\
&\quad \geq \left| \left\| f^-\left(t_{p+1}^{(0)}\right) - f\left(t_{p+1}^{(0)}\right)\right\|_H - \left\| f^-\left(t_{p+1}^{(0)}\right) - f\left(t_p^{(0)} + \theta_p\right)\right\|_H \right| \\
&\quad > n_0^{-1} - \epsilon.
\end{aligned}$$

Since f has limits to the left, the first term of that sequence of inequalities goes to zero, a contradiction. The proof of the second statement mimics that of the first. □

Proposition 6.3.5 *Suppose that for $t \in T$, fixed, but arbitrary, $f^-(t)$ exists. The space $L_{\cup T}[f]$ is then separable.*

Proof Let $T_{\mathbb{Q}}$ be the set of rational numbers in T, and T_0, that of numbers in T at which f is not continuous to the left. T_0 is countable [(Proposition) 6.3.4]. The (closed) linear subspace generated by the countable family $\{f(t), t \in T_{\mathbb{Q}} \cup T_0\}$ contains the family $\{f(t), t \in T\}$, and thus $L_{\cup T}[f]$. □

Corollary 6.3.6 *Suppose that for $t \in T$, fixed, but arbitrary, $f^-(t)$ exists. The dimension of the subspace*

$$L_t^+[f] \cap L_t[f]^\perp$$

is then at most countable.

Proof As seen in (Proposition) 6.3.5, $L_{\cup T}[f]$, and thus $L_t^+[f] \cap L_t[f]^\perp$, is separable when f^- exists. □

Remark 6.3.7 When $f : T \longrightarrow H$ is bounded, the following considerations also yield that $L_{UT}[f]$ is separable.

Let $L_f : H \longrightarrow \mathbb{R}^T$ be defined using

$$L_f[h](t) = \langle h, f(t)\rangle_H, \qquad (\star)$$

and

$$C_f(t_1, t_2) = \langle f(t_1), f(t_2)\rangle_H.$$

Then [(Proposition) 1.1.15], the range of L_f is $H(C_f, T)$, and $H_f = L_{UT}[f]$. Furthermore, those two spaces are unitarily isomorphic. Let $B[T,\mathbb{R}]$ be the Banach space of bounded, real valued functions, with the supremum norm. Then, because of $(\star)$, and the assumption that f is bounded, $H(C_f, T) \subseteq B[T,\mathbb{R}]$. Let J be the inclusion of $H(C_f, T)$ into $B[T,\mathbb{R}]$. It is a linear map, and

$$\sup_{\theta\in T}\left|\sum_{i=1}^{n}\alpha_i C_f(\theta, t_i) - \sum_{i=1}^{p}\beta_j C_f(\theta, \tau_j)\right| =$$

$$= \sup_{\theta\in T}\left|\langle f(\theta), \sum_{i=1}^{n}\alpha_i f(t_i) - \sum_{j=1}^{p}\beta_j f(\tau_j)\rangle\right|$$

$$\le \sup_{\theta\in T}\|f(\theta)\|_H \left\|\sum_{i=1}^{n}\alpha_i f(t_i) - \sum_{j=1}^{p}\beta_j f(\tau_j)\right\|_H.$$

J is thus a continuous injection. $J^\star$ is then continuous [44, p. 31].

Let $B_f[T,\mathbb{R}]$ be the closure, in $B[T,\mathbb{R}]$, of $H(C_f, T)$. Since a subset of a separable metric space is separable, in the induced topology [84, pp. 187,176], when $B_f[T,\mathbb{R}]$ is separable, so is $\{C_f(\cdot, t), t \in T\}$. If the latter set is separable, the (closed) subspace generated by the resulting countable subfamily, which is $B_f[T,\mathbb{R}]$ [46, p. 240], is separable. Since continuous images of separable spaces are separable [84, p. 175], one has that:

$L_{UT}[f]$ is separable if, and only if, the metric subspace

$$\{C_f(\cdot, t), t \in T\} \subseteq B[T,\mathbb{R}]$$

is separable.

6.4 The Cramér-Hida Representation

Cramér-Hida representations are decompositions into orthogonal subspaces of the linear space generated by a "time" function. That decomposition is chosen to preserve the "time" dimension inherent to the problem (it would otherwise have little interest). Each subspace in the decomposition is furthermore isomorphic to an L_2 space, and its elements are obtained has integrals of (classes of) numerical functions with respect to vector measures, with orthogonal increments, a feature that provides a modicum of computational ease. The measures determining these L_2 spaces are ordered by absolute continuity, and the number of subspaces is the same, regardless of the decomposition.

In this section, the Cramér-Hida representation (acronym: CHR) for a function $f : T \longrightarrow H$, T an interval, H a real Hilbert space, shall be obtained from first principles following [15]. In its uses the CHR takes most often H to be an L_2 space.

The assumptions to which the previous Sects. 6.1–6.3 provide context are now listed (they shall be referred to using the following acronym: CHA). They basically assume that there is no knowledge at time zero, that at time t, all the past is known, and that the energy in the system is finite.

Assumptions 6.4.1 (CHA: The Cramér-Hida Assumptions)

CHA-1 : *$f : T \longrightarrow H$ is continuous to the left;*

CHA-2 : *for $t \in T$, fixed, but arbitrary, $f^+(t)$ exists whenever it makes sense;*

CHA-3 : $\sup_{t \in T} \|f(t)\|_H \le \kappa < \infty$;

CHA-4 : $L_{\cap T}[f] = \{0_H\}$.

Assumptions CHA have immediate consequences, expressed below as remarks.

Remark 6.4.2 $L_{\cup T}[f]$ and $H(C_f, T)$ are separable.

Remark 6.4.3 Let P_t denote the projection onto $L_t[f]$. From the properties of projections [266, p. 85], one has that:

1. when $t_n \uparrow\uparrow t$, $P_{t_n} \xrightarrow{s} P$, that is, for $h \in H$, fixed, but arbitrary,

$$P_{t_n}[h] \longrightarrow P[h],$$

where, when f is continuous to the left, P is the projection onto

$$\overline{\cup_n \mathcal{R}[P_{t_n}]} = \overline{\cup_n L_{t_n}[f]} = L_t[f],$$

that is, $P = P_t^- = P_t$;

2. when $t \downdownarrows t_n$, $P_{t_n} \xrightarrow{s} P$, where P is the projection onto

$$\overline{\cap_n \mathcal{R}[P_{t_n}]} = \overline{\cap_n L_{t_n}[f]} = L_t^+[f],$$

that is, $P = P_t^+$.

Remark 6.4.4 $P_\cap = O_{L_{\cup T}[f]}$ (null operator), and $P_\cup = I_{L_{\cup T}[f]}$ (identity operator).

Remark 6.4.5 There are cases for which $P_t^- = P_t$, $t \in T$, but $f^-(t) \neq f(t)$, some $t \in T$. An example follows.

Example 6.4.6 Let $T = [0, 1]$, W, the standard Wiener process, and W_t, the equivalence class of $W(\cdot, t)$ in the appropriate L_2 space. Let

$$f(t) = \begin{cases} -W_t & \text{when } t < 2^{-1} \\ +W_t & \text{when } t \geq 2^{-1} \end{cases}.$$

Then, when $t < 2^{-1}$,

$$\begin{aligned} \left\| f(t) - f\left(2^{-1}\right) \right\|_H^2 &= \| -W_t - W_{2^{-1}} \|_{L_2(\Omega, \mathcal{A}, P)}^2 \\ &= E_P\left[\left(\dot{W}_t + \dot{W}_{2^{-1}} \right)^2 \right] \\ &= 3t + 2^{-1}. \end{aligned}$$

Thus f is not continuous to the left, in quadratic mean, at $t = 2^{-1}$. However, for $t < 2^{-1}$,

$$\left\| f\left(2^{-1}\right) - \{-f(t)\} \right\|_H^2 = \| W_{2^{-1}} - W_t \|_{L_2(\Omega, \mathcal{A}, P)}^2 = 2^{-1} - t.$$

Thus

$$f\left(2^{-1}\right) \in \overline{V\left[\{f(t),\ t < 2^{-1}\}\right]}$$

and $L_{2^{-1}}^-[f] = L_{2^{-1}}[f]$.

Remark 6.4.7 Let $h \in L_{\cup T}[f]$ be fixed, but arbitrary, and $f_h(t) = P_t[h]$. Then, as a consequence of CHA and (Remark) 6.4.3,

1. f_h has orthogonal increments, is continuous to the left, and has limits to the right;
2. $L_t[f_h] \subseteq L_t[f]$, so that f_h is purely nondeterministic;
3. as $\| f_h(t) \|_H \leq \| h \|_H$, f_h is bounded.

Thus CHA obtains for f_h, and, setting

$$m_{f_h}([t_1, t_2[) = f_h(t_2) - f_h(t_1),$$

one has that [(Remark) 6.2.26] $L[m_{f_h}] = L_{\cup T}[f_h]$, so that every element $k \in L_{\cup T}[f_h]$ can be expressed in the following form:

$$k = \int \phi_k \, dm_{f_h}, \quad \text{some } \phi_k \in L_2(T, \mathcal{T}, M_{f_h}).$$

Furthermore, for $\{t_1, t_2\} \subseteq T, t_1 < t_2$, fixed, but arbitrary,

$$\begin{aligned} M_{f_h}([t_1, t_2[) &= \|f_h(t_2) - f_h(t_1)\|_H^2 \\ &= \|f_h(t_2)\|_H^2 - \|f_h(t_1)\|_H^2 \\ &= F_h(t_2) - F_h(t_1). \end{aligned}$$

6.4.1 Canonical Representations

A Cramér-Hida representation (CHR) is a particular decomposition of $L_t[f]$ into a direct sum of parts of a similar type. To achieve it, one needs the properties of infinite sums of subspaces, gathered in the following fact [8, p. 433]. Those will be used more or less tacitly.

Fact 6.4.8 *Let H be a real Hilbert space, and $\{H_\lambda, \lambda \in \Lambda\}$, a family of (closed) subspaces of H.*

1. *Let $\{h_\lambda, \lambda \in \Lambda\} \subseteq H$ be a collection of elements. It is summable, or commutatively convergent, to $h \in H$, when, given $\epsilon > 0$, fixed, but arbitrary, there is a finite $\Lambda_\epsilon \subseteq \Lambda$ such that, given any finite subset of Λ, say Λ_f, containing Λ_ϵ,*

$$\left\| \sum_{\lambda \in \Lambda_f} h_\lambda - h \right\|_H < \epsilon.$$

One then writes:

$$h = \sum_{\lambda \in \Lambda} h_\lambda.$$

$\sum_{\lambda \in \Lambda} H_\lambda$ *stands for the vector space of commutatively convergent sequences* $\{h_\lambda, \lambda \in \Lambda : h_\lambda \in H_\lambda, \lambda \in \Lambda\}$.

2. *One has that*

$$\overline{\cup_{\lambda\in\Lambda}H_\lambda}=\overline{\sum_{\lambda\in\Lambda}H_\lambda}.$$

3. *When the subspaces* $\{H_\lambda, \lambda \in \Lambda\}$ *are pair-wise orthogonal, one has that:*

   (i) $\overline{\cup_{\lambda\in\Lambda}H_\lambda}=\sum_{\lambda\in\Lambda}H_\lambda$. *One then writes*

$$\bigoplus_{\lambda\in\Lambda}H_\lambda \text{ for } \sum_{\lambda\in\Lambda}H_\lambda.$$

   (ii) *When* $H=\bigoplus_{\lambda\in\Lambda}H_\lambda$, *every* $h\in H$ *has the following representation, which is unique:*

$$h=\sum_{\lambda\in\Lambda}h_\lambda,\ h_\lambda\in H_\lambda.$$

   (iii) $\overline{\cup_{\lambda\in\Lambda}H_\lambda}=\sum_{\lambda\in\Lambda}H_\lambda$ *is equivalent to the following statement:*

$$h\perp\cup_{\lambda\in\Lambda}H_\lambda \text{ implies that } h=0_H.$$

4. *Let* B, *be a bounded, linear operator of* H. *Then, when* $h=\sum_{\lambda\in\Lambda}h_\lambda$, *one has that*

$$B[h]=\sum_{\lambda\in\Lambda}B[h_\lambda].$$

5. *Let* $\{B_\lambda, \lambda\in\Lambda\}$ *be a family of bounded, linear operators of* H. *It is summable to* B *when, for* $h\in H$, *fixed, but arbitrary,* $\sum_{\lambda\in\Lambda}B_\lambda[h]=B[h]$.
6. *Let* $\{P_\lambda, \lambda\in\Lambda\}$ *be a family of orthogonal projections of* H *which are pair-wise orthogonal. Then:*

   (i) $\sum_{\lambda\in\Lambda}P_\lambda=P_\oplus$ *is the orthogonal projection onto*

$$\overline{\cup_{\lambda\in\Lambda}H_\lambda}=\bigoplus_{\lambda\in\Lambda}H_\lambda,\ H_\lambda=\mathcal{R}[P_\lambda]$$

   (the converse is also true);

   (ii) *when* Q *commutes with each* P_λ, $QP_\oplus=\bigoplus_{\lambda\in\Lambda}QP_\lambda$ [126, p. 50].

Definition 6.4.9 Posit *CHA*, and suppose that there exists a sequence of functions

$$\mathcal{F}_f[T,H,I]=\{f_i:T\longrightarrow H, i\in I\subseteq\mathbb{N}\}$$

such that

(a) each $f_i, i \in I$, has orthogonal increments;

(b) for $\{i,j\} \subseteq I, i \neq j, \{t_1, t_2\} \subseteq T$, fixed, but arbitrary,

$$\langle f_i(t_1), f_j(t_2)\rangle_H = 0;$$

(c) for $t \in T$, fixed, but arbitrary, $L_t[f_i, i \in I]$ being the (closed) subspace of H generated linearly by the family $\{f_i(\theta), \theta \leq t, i \in I\}$, one has that (Fact 6.4.8)

$$L_t[f_i, i \in I] = \bigoplus_{i \in I} L_t[f_i].$$

When, for $t \in T$, fixed, but arbitrary,

1. $L_t[f] \subseteq L_t[f_i, i \in I]$, the family $\mathcal{F}_f[T, H, I]$ is called a sequence of innovations for f, and the number $|I| \leq \aleph_0$, the multiplicity of the sequence;
2. $L_t[f] = L_t[f_i, i \in I]$, the family $\mathcal{F}_f[T, H, I]$ is called a canonical representation of f.

Proposition 6.4.10 *Posit* CHA: *f has then a canonical representation.*

Proof Let $\{h_i, i \in I \subseteq \mathbb{N}\}$ be a complete orthonormal set for $L_{\cup T}[f]$. Define recursively the following objects, respectively, elements, subspaces, and projections:

$$
\begin{aligned}
1.\ k_1 &= h_1, \\
K_1 &= \overline{V[\{P_t[k_1], t \in T\}]}, \\
P_{K^1} &= P_{K_1}, \\
&\quad \text{(one has then that } h_1 \in K_1\text{);}
\end{aligned}
$$

$$
\begin{aligned}
2.\ k_2 &= h_2 - P_{K^1}[h_2], \\
K_2 &= \overline{V[\{P_t[k_2], t \in T\}]}, \\
P_{K^2} &= P_{K_1 \vee K_2}, \\
&\quad \text{(as } h_2 = k_2 + P_{K_1}[h_2], h_2 \in K_1 \vee K_2\text{);}
\end{aligned}
$$

$$
\begin{aligned}
3.\ k_3 &= h_3 - P_{K^2}[h_3], \\
K_3 &= \overline{V[\{P_t[k_3], t \in T\}]}, \\
P_{K^3} &= P_{K_1 \vee K_2 \vee K_3}, \\
&\quad \text{(as } h_3 = k_3 + P_{K^2}[h_3], h_3 \in K_1 \vee K_2 \vee K_3\text{);}
\end{aligned}
$$

4. ...

Such a procedure either yields, after a finite number of steps, $L_{\cup T}[f]$, or one has that $\{h_i, i \in I\} \subseteq \bigvee_{i\in I} K_i \subseteq L_{\cup T}[f]$, that is, in all cases,

$$L_{\cup T}[f] = \bigvee_{i\in I} K_i.$$

Let, for $t \in T$, fixed, but arbitrary,

$$K_t^i = \overline{V[\{P_\theta[k_i], \theta \leq t\}]}.$$

As presently seen, for fixed, but arbitrary $i_1 \neq i_2$, in I, and $t_1 \neq t_2$, in T, one has that

$$K_{t_1}^{i_1} \perp K_{t_2}^{i_2}.$$

Indeed, assuming that $i_1 < i_2$ and $t_1 < t_2$, which is no restriction, one has that:

(i) $k_{i_2} \perp K_1 \vee \cdots \vee K_{i_2-1} \supseteq K_{i_1} \supseteq K_t^{i_1}$;
(ii) $\left(I_{L_{\cup T}[f]} - P_t\right)[k_{i_2}] \perp L_t[f] \supseteq K_t^{i_1}$.

Thus, when $\theta \leq t$, in T, as $K_\theta^{i_1} \subseteq K_t^{i_1}$, using (ii),

$$P_t[k_{i_2}] = k_{i_2} - \left(I_{L_{\cup T}[f]} - P_t\right)[k_{i_2}] \perp K_\theta^{i_1}.$$

When $\theta > t$, in T, because of (i), and $P_t(P_\theta - P_t) = O_H$,

$$\begin{aligned}\langle P_t[k_{i_2}], P_\theta[k_{i_1}]\rangle_H &= \langle P_t[k_{i_2}], (P_\theta - P_t)[k_{i_1}]\rangle_H + \langle P_t[k_{i_2}], P_t[k_{i_1}]\rangle_H \\ &= 0.\end{aligned}$$

Thus, for $\theta \in T$, fixed, but arbitrary, $P_t[k_{i_2}] \perp P_\theta[k_{i_1}]$, and consequently,

$$\bigvee_{i\in I} K_t^i = \bigoplus_{i\in I} K_t^i.$$

Furthermore

$$\langle k_{i_1}, k_{i_2}\rangle_H = \lim_{\{t_1,t_2\}\subseteq T, t_1\neq t_2, t_1\uparrow, t_2\uparrow} \langle P_{t_1}[k_{i_1}], P_{t_2}[k_{i_2}]\rangle_H = 0,$$

so that $K_{i_1} \perp K_{i_2}$, and

$$L_{\cup T}[f] = \bigvee_{i\in I} K_i = \bigoplus_{i\in I} K_i.$$

Finally, $P_t[K_i] = K_t^i$, and, as P_t and P_{K_i} commute ($k_i \in L_{\cup T}[f]$, thus $P_t[k_i] \in L_t[f]$, so that $K_i \subseteq L_t[f]$), then [126, p. 50]

$$\bigoplus_{i \in I} K_t^i = \bigvee_{i \in I} K_t^i = \bigvee_{i \in I} P_t P_{K_i}[K_i] = P_t\left[\bigvee_{i \in I} K_i\right] = L_t[f].$$

It then suffices to define, for $i \in I$, fixed, but arbitrary,

$$f_i(t) = P_t[k_i],$$

so that $L_t[f_i] = K_t^i$, and $L_{\cup T}[f_i] = K_i$. □

Remark 6.4.11 Combining (Facts) 6.2.12, 6.2.19, and 6.4.8, one has that

$$f(t) = \sum_{i \in I} \int \phi_i(t)\, dm_i^f$$

where, for $i \in I$, and $\{t, t_1, t_2\} \subseteq T$, $t_1 < t_2$, fixed, but arbitrary,

- $m_i^f([t_1, t_2[) = f_i(t_2) - f_i(t_1)$,
- $M_i^f([t_1, t_2[) = \left\| m_i^f([t_1, t_2[)\right\|_H^2 = \|P_{t_2}[k_i]\|_H^2 - \|P_{t_1}[k_i]\|_H^2$,
- $\phi_i(t) \in L_2\left(T, \mathcal{T}, M_i^f\right)$,
- $\int \phi_i(t)\, dm_i^f \in L_t[f_i]$.

The latter property means [(Remark) 6.2.15] that, letting $T_t = T \cap]-\infty, t]$, the following obtains: $I_{T_t}\phi_i(t) = \phi_i(t)$.

Remark 6.4.12 The measures M_i^f shall sometimes be called the *basis measures* of the canonical representation. One shall also use, when convenient, the notation $L_t[k_i]$ for $L_t[f_i]$. Finally the spaces

$$L_2\left(T, \mathcal{T}, M_i^f\right) \text{ and } L_2\left(T_t, \mathcal{T}_t, M_i^{f|T_t}\right)$$

shall be called the *basis L_2-spaces.*

Remark 6.4.13 It is sometimes convenient to use the following shortened notation:

$$f(t) = \int \langle \boldsymbol{\phi}(t), \boldsymbol{dm} \rangle.$$

$\boldsymbol{\phi}$ is then interpreted as the function

$$\boldsymbol{\phi} : T \longrightarrow \bigoplus_{i \in I} L_2(T, \mathcal{T}, M_i),$$

whose value is determined by the expansion of $f(t)$ which is unique [8, p. 435].

Remark 6.4.14 Suppose one knows, instead of

$$L_t[f] = \bigoplus_{i \in I} L_t[f_i], \tag{$\star$}$$

that, for $t \in T$, fixed, but arbitrary,

$$f(t) = \sum_{i \in I} \int I_{T_t} \tilde{f}_i dm_{f_i}. \tag{$\star\star$}$$

Then ($\star\star$) implies ($\star$) when one of the following equivalent conditions obtains.

1. For $t \in T$, and $\theta \leq t$, fixed, but arbitrary,

$$\sum_{i \in I} \int_{T_\theta} \dot{\tilde{f}}_i(\theta)[x] g_i(x) M_{f_i}(dx) = 0$$

   implies that

$$\sum_{i \in I} \int_{T_t} g_i^2(x) M_{f_i}(dx) = 0.$$

2. Each family $\{\tilde{f}_i(\theta), \theta \leq t\}$, $i \in I$, is total in its respective

$$L_2(T_t, \mathcal{T} \cap T_t, M_{f_i}^{|\mathcal{T} \cap T_t}).$$

   space.

Relation ($\star\star$) yields indeed that $L_t[f] \subseteq \bigoplus_{i \in I} L_t[f_i]$. Suppose the inclusion is strict. There is then an element

$$G_t = \sum_{i \in I} \int I_{T_t}[g_i] dm_{f_i} \perp L_t[f], \ \|G_t\|_H^2 = \sum_{i \in I} \int_{T_t} g_i^2(x) M_{f_i}(dx) \neq 0.$$

But, for $\theta \leq t$, fixed, but arbitrary,

$$\langle G_t, f(\theta) \rangle_H = \int_{T_\theta} \dot{\tilde{f}}_i(\theta)[x] g_i(x) M_{f_i}(dx) = 0.$$

The above is, *mutatis mutandis,* a version of (Proposition) 6.2.22.

Remark 6.4.15 Let $\theta < t$ be fixed, but arbitrary. Then

$$P_\theta [f(t)] = \bigoplus_{i \in I} \int I_{T_\theta} \tilde{f}_i(t) dm_{f_i}.$$

Indeed, the latter's right-hand side belongs to $\bigoplus_{i\in I} L_\theta[f_i]$, and, for $u \leq \theta$, fixed, but arbitrary,

$$\left\langle f(t) - \bigoplus_{i\in I} \int I_{T_\theta} \tilde{f}_i(t) dm_{f_i}, \bigoplus_{i\in I} \int I_{T_u} \tilde{f}_i(u) dm_{f_i} \right\rangle_H =$$

$$= \bigoplus_{i\in I} \int_{[T_t \setminus T_\theta] \cap T_u} \dot{\tilde{f}}_i(t)[x] \dot{\tilde{f}}_i(u)[x] M_{f_i}(dx) = 0_H.$$

That result is thus a version of (Remark) 6.2.16.

6.4.2 *Proper Canonical Representations*

Proper canonical representations are refinements of canonical ones.

Definition 6.4.16 Proper canonical representations are canonical representations which distinguish themselves by

1. having the measures $\{M_i, i \in I\}$ ordered by absolute continuity,
2. the fact that the number of those measures is unique, and thus chosen to be the multiplicity of the representation.

Remark 6.4.17 Until item 2 of (Definition) 6.4.16 is established, a proper canonical representation shall be one for which item 1 of that same reference obtains. In its most detailed version, a proper canonical representation is obtained by splitting the function $f : T \longrightarrow H$ into continuous and discontinuous parts. The continuous part has a proper canonical representation whose components are continuous with basis measures ordered by absolute continuity. The discontinuous part is a superposition of step functions. The detailed definition of a proper canonical representation is that of statements (Propositions) 6.4.46 and 6.4.47.

Definition 6.4.18 (Notation) For $t \in T$, fixed, but arbitrary, one sets (O_H is the null operator)

$$\Delta_t [P] = P_t^+ - P_t,$$

$$T_d \quad = \{t \in T : \Delta_t [P] \neq O_H\}.$$

Lemma 6.4.19 *Let $t \in T$ be fixed, but arbitrary. The range of $\Delta_t [P]$ is equal to the two following subsets, and thus subspaces:*

H_t : *the set of those $h \in H$ for which $P_t [h] = 0_H$, and, for $\theta \in T, \theta > t$, fixed, arbitrary, $P_\theta[h] = h$;*

$H_t^\star$: *the set of those $h \in H$ for which $P_t [h] = 0_H$, and $P_t^+ [h] = h$.*

Proof When, for $\theta \in T, \theta > t$, fixed, but arbitrary, $P_\theta [h] = h$, then $P_t^+ [h] = h$, and, when the latter obtains,

$$P_\theta [h] = P_\theta \left[P_t^+ [h]\right] = P_t^+ [h] = h.$$

One then uses (Remark) 6.1.1. □

Proposition 6.4.20 *One has that $|T_d| \leq \aleph_0$, and, when $\{t_1, t_2\} \subseteq T_d, t_1 \neq t_2$, fixed, but arbitrary,*

$$\Delta_{t_1} [P] \perp \Delta_{t_2} [P]$$

so that $\Delta [P] = \sum_{t \in T_d} \Delta_t [P]$ is a projection (see also (Proposition) 6.3.4).

Proof Suppose that $t_1 < t_2$, which is no restriction, and that

$$0_H \neq h_1 \in \mathcal{R}[\Delta_{t_1} [P]], \ \ 0_H \neq h_2 \in \mathcal{R}[\Delta_{t_2} [P]].$$

Then $h_2 \perp L_{t_2} [f] \supseteq L_{t_1}^+ [f] \ni h_1$, that is $h_2 \perp h_1$. Thus $\Delta_{t_1} [P] \perp \Delta_{t_2} [P]$. Since $L_{\cup T} [f]$ is separable, it may contain at most a countable number of orthogonal elements. □

Lemma 6.4.21 *For $t \in T$, fixed, but arbitrary,*

1. *P_t and $\Delta [P]$ commute;*
2. *P_t and $I_{L_{\cup T}[f]} - \Delta [P]$ commute.*

Proof When $t \leq \theta$, in T,

$$\left(P_\theta^+ - P_\theta\right) P_t = O_H = P_t \left(P_\theta^+ - P_\theta\right),$$

and, when $t > \theta$, in T,

$$\left(P_\theta^+ - P_\theta\right) P_t = \left(P_\theta^+ - P_\theta\right) = P_t \left(P_\theta^+ - P_\theta\right),$$

so that, for $\{t, \theta\} \subseteq T$, fixed, but arbitrary, P_t and $P_\theta^+ - P_\theta$ commute. Then [126, p. 50]

$$\begin{aligned} P_t \Delta [P] &= P_t \sum_{t \in T_d} \Delta_t [P] \\ &= P_t \bigvee_{t \in T_d} \Delta_t [P] \\ &= \bigvee_{t \in T_d} \{P_t \Delta_t [P]\} \\ &= \bigvee_{t \in T_d} \{\Delta_t [P] P_t\} \end{aligned}$$

$$= \left\{ \bigvee_{t \in T_d} \Delta_t [P] \right\} P_t$$

$$= \left\{ \sum_{t \in T_d} \Delta_t [P] \right\} P_t$$

$$= \Delta [P] P_t.$$

□

Definition 6.4.22 (Notation) One sets:

1. $L_d [f]$ is the range of $\Delta [P]$;
2. $L_c [f]$ is the orthogonal complement of $L_d [f]$ in $L_{\cup T} [f]$;
3. given $h \in L_c [f]$, fixed, but arbitrary,

$$f_h (t) = P_t [h]$$

(then, automatically [(Remark) 6.4.7], f_h is purely nondeterministic, has orthogonal increments, and is continuous to the left, with limits to the right).

Lemma 6.4.23 *The map*

$$\|f_h\| : t \mapsto \|f_h (t)\|_H$$

is continuous if, and only if, for $t \in T$, fixed, but arbitrary,

$$\Delta_t [P] [h] = 0_H,$$

that is, if, and only if, f_h is continuous to the right.

Proof For $t_1 < t_2$, in T, fixed, but arbitrary, since f_h is purely nondeterministic, and has orthogonal increments,

$$\|f_h (t_2) - f_h (t_1)\|_H^2 = \|f_h (t_2)\|_H^2 - \|f_h (t_1)\|_H^2 .$$

Thus the continuity of $t \mapsto f_h (t)$ is equivalent to that of $t \mapsto \|f_h (t)\|_H$. Since, by definition, $t \mapsto f_h (t)$ is continuous to the left, it is continuous if, and only if, $\Delta_t [P] [h] = 0_H$, that is, if, and only if, it is continuous to the right. □

Proposition 6.4.24 *One has that $h \in L_c [f]$ if, and only if, the map*

$$\|f_h\| : t \mapsto \|f_h(t)\|_H$$

is continuous.

Proof *Suppose that $\|f_h\|$ is continuous.*

Because of (Lemma) 6.4.23, for $t \in T$, fixed, but arbitrary, $P_t^+ [h] = P_t [h]$.

Let then $t \in T_d$ be fixed, but arbitrary, and $\theta \in T$ be such that $\theta > t$. Let $k \in \mathcal{R}[\Delta_t [P]]$ be fixed, but arbitrary, so that $k = \Delta_t [P] [k] \neq 0_H$. Then

$$\langle h, k \rangle_H = \langle h, P_\theta [k] \rangle_H = \langle P_\theta [h], k \rangle_H ,$$

so that, since $P_t \Delta_t [P]$ is the zero operator,

$$\begin{aligned} \langle h, k \rangle_H &= \lim_{\theta \downarrow\downarrow t} \langle P_\theta [h], k \rangle_H \\ &= \langle P_t^+ [h], k \rangle_H \\ &= \langle P_t [h], k \rangle_H \\ &= \langle h, P_t [k] \rangle_H \\ &= \langle h, P_t \Delta_t [P][k] \rangle_H \\ &= 0. \end{aligned}$$

Consequently $h \perp k, k \in L_d [f]$, that is, $h \in L_c [f]$.

Proof Suppose that $h \in L_c [f]$.

Suppose that $t \mapsto \| f_h (t) \|_H$ is not continuous at $t \in T$. Then [(Lemma) 6.4.23] $\Delta_t [P] [h] \neq 0_H$, and

$$f_h^+ (t) = f_h (t) + \Delta_t [P] [h] .$$

Now

$$P_t^+ [\Delta_t [P] [h]] = \Delta_t [P] [h] ,$$

and

$$P_t [\Delta_t [P] [h]] = 0_H ,$$

so that [(Lemma) 6.4.19] $\Delta_t [P] [h] \in L_d [f]$. But

$$\begin{aligned} \langle h, \Delta_t [P] [h] \rangle_H &= \langle h, P_t^+ [\Delta_t [P] [h]] \rangle_H \\ &= \langle f_h^+ (t), \Delta_t [P] [h] \rangle_H \\ &= \langle f_h (t) + \Delta_t [P] [h], \Delta_t [P] [h] \rangle_H . \end{aligned}$$

Since, by definition, $\Delta_t [P] [h] \perp f_h (t)$,

$$\langle h, \Delta_t [P] [h] \rangle_H = \| \Delta_t [P] [h] \|_H^2 > 0.$$

h is thus not orthogonal to $L_d [f]$, and consequently not in $L_c [f]$, a contradiction.

□

Corollary 6.4.25 *Let $h \in L_c[f]$ be fixed, but arbitrary, and, for fixed, but arbitrary $t \in T$, let $f_h(t) = P_t[h]$. f_h has the properties of f, in particular CHA, as already seen* [(Remark) 6.4.7]. *But it has now the further property of being continuous.*

Proof Since f_h is purely nondeterministic, and has orthogonal increments,

$$\|f_h(t_2) - f_h(t_1)\|_H^2 = \|f_h(t_2)\|_H^2 - \|f_h(t_1)\|_H^2,$$

and then f_h is continuous because of (Proposition) 6.4.24. □

Corollary 6.4.26 *Let, for $t \in T$, fixed, but arbitrary, $f_c(t) = P_{L_c[f]}[f(t)]$. $L_{\cup T}[f_c]$ has then a canonical representation with continuous components, as defined in the proof to follow.*

Proof Since $P_{L_c[f]}$ is continuous and bounded, f_c has the continuity properties of f, and is bounded, since P_t and $P_{L_c[f]}$ are. Since P_t and $P_{L_c[f]}$ commute [(Lemma) 6.4.21],

$$f_c(t) = P_{L_c[f]}[f(t)] = P_{L_c[f]}P_t[f(t)] = P_tP_{L_c[f]}[f(t)] = P_t[f_c(t)] \in L_t[f],$$

so that f_c is purely nondeterministic. The CHA thus obtain, and one has that, for $t \in T$, fixed, but arbitrary,

$$L_t[f_c] = \bigoplus_{i \in I} L_t[f_i^c],$$

where, P_t^c being the projection with range $L_t[f_c]$,

$$f_i^c(t) = P_t^c[k_i^c],\ k_i^c \in L_{\cup T}[f_c] \subseteq L_c[f]. \qquad (\star)$$

The latter inclusion is due to the fact that, by definition, $f_c(t) \in L_c[f]$. Now, for $\theta \leq t$, in T, and $i \in I$, fixed, but arbitrary, again since the projections P_t and $P_{L_c[f]}$ commute [(Lemma) 6.4.21],

$$\begin{aligned}\langle k_i^c - P_t[k_i^c], f_c(\theta)\rangle_H &\overset{def}{=} \langle k_i^c - P_t[k_i^c], P_{L_c[f]}[f(\theta)]\rangle_H \\ &\overset{(\star)}{=} \langle k_i^c - P_t[k_i^c], f(\theta)\rangle_H \\ &\overset{(\star\star)}{=} 0.\end{aligned}$$

Equality $(\star\star)$ above expresses indeed the fact that P_t is the projection onto $L_t[f_c]$ [44, p. 80], so that $P_t^c[k_i^c] = P_t[k_i^c]$, and $f_i^c(t) = P_t[k_i^c]$. It follows then from (Corollary) 6.4.25 that f_i^c is continuous. □

Definition 6.4.27 (Notation) Let $t \in T_d$ be fixed, but arbitrary, and

$$M[t] = \dim\{\mathcal{R}[\Delta_t[P]]\}.$$

Because of the characterization of $\mathcal{R}[\Delta_t[P]]$ [(Remark) 6.1.1, (Lemma) 6.4.19], one has that

$$\begin{aligned} M[t] &= \dim\left\{L_t^+[f] \cap L_t[f]^\perp\right\} \\ &= \dim\left\{h \in L_{\cup T}[f] : P_t^+[h] = h,\ P_t[h] = 0_H\right\}. \end{aligned}$$

The result which follows provides an alternate version of the canonical representation (Proposition) 6.4.10, in the form of a sum of a continuous part, and a discontinuous one, and the ensuing corollaries describe the discontinuous part.

Proposition 6.4.28 *Assume that* CHA *obtains for* f. *Let*

$$f_c(t) = P_{L_c[f]}[f(t)], \quad and\ f_d(t) = P_{L_d[f]}[f(t)].$$

Then, for $t \in T$, *fixed, but arbitrary, in* $L_{\cup T}[f]$,

$$f(t) = f_c(t) + f_d(t).$$

$f_c(t)$ *has the following representation:*

$$f_c(t) = \int \left\langle \boldsymbol{\phi}^c(t), d\boldsymbol{m}_{f_c}\right\rangle.$$

$f_d(t)$ *has the following representation:*

$$f_d(t) = \sum_{\theta \in T_d, \theta < t} \sum_{j=1}^{M[\theta]} \left\langle f(t), h_j^{(\theta)}\right\rangle_H h_j^{(\theta)}.$$

The elements in those representations have the following properties:

1. *the vector measure components* $m_i^{f_c}$ *of* $\boldsymbol{m}_{f_c}$, $i \in I_{f_c}$, *are obtained from continuous, purely nondeterministic, orthogonal functions with orthogonal increments:*

$$f_i^c(\theta) = P_\theta\left[k_i^c\right], \theta \in T,$$

using the following formulae ($\{t_1, t_2\} \subseteq T, t_1 < t_2,$) :

$$m_i^{f_c}([t_1, t_2[) = f_i^c(t_2) - f_i^c(t_1),$$

$$M_i^{f_c}([t_1, t_2[) = \left\|m_i^{f_c}([t_1, t_2[)\right\|_H^2;$$

2. *the components* $\phi_i^c(t)$ *of* $\boldsymbol{\phi}^c(t)$, $i \in I_{f_c}$, *belong to* $L_2\left(T, \mathcal{T}, M_i^{f_c}\right)$;
3. *the set* $T_d \subseteq T$ *gathers the points of* T *at which* $t \mapsto P_t$ *is discontinuous;*

4. *for fixed, but arbitrary* $\theta \in T_d$, $M[\theta]$ *is the dimension of the range of* $P_\theta^+ - P_\theta$, *and*

$$\left\{h_j^{(\theta)}, j \in [1 : M[\theta]]\right\}$$

is a complete orthonormal set in that range, with the property that, for fixed, but arbitrary $j \in [1 : M[\theta]]$,

$$\text{when} \quad t > \theta, \quad P_t\left[h_j^{(\theta)}\right] = h_j^{(\theta)},$$

$$\text{and} \quad P_\theta\left[h_j^{(\theta)}\right] = 0_H.$$

Proof By definition, $L_c[f]$ and $L_d[f]$ form an orthogonal decomposition of $L_{\cup T}[f]$. The representation of f_c is due to (Corollary) 6.4.26, and that of $f_d(t)$, to (Proposition) 6.4.20, using the fact [(Lemma) 6.4.21] that, on the one hand, P_t and $P_{L_d[f]}$ commute, and, on the other hand, the following identities obtain:

$$\text{for } t \leq \theta, P_t\left(P_\theta^+ - P_\theta\right) = 0,$$

$$\text{for } t > \theta, P_t\left(P_\theta^+ - P_\theta\right) = P_\theta^+ - P_\theta.$$

□

Remark 6.4.29 Since [(Remark) 6.1.4] $L_t^{(+)}[f]$ may be a strict subspace of $L_t^+[f]$, it may at times be useful to build the basis of

$$L_t^+[f] \cap L_t[f]^\perp$$

using a basis of

$$L_t^{(+)}[f] \cap L_t[f]^\perp,$$

and then completing it. When doing that, one shall write:

$$h_j^{(+|\theta)}, j \in [1 : M[+ \mid \theta]],$$

for the basis of $L_t^{(+)}[f] \cap L_t[f]^\perp$, and

$$h_j^{(\theta|+)}, j \in [1 : M[\theta \mid +]],$$

for the elements of the completion.

Corollary 6.4.30 *Let $\theta \in T_d$ be fixed, but arbitrary, and $h_j^{(+|\theta)}$ be as defined in (Remark)* 6.4.29. *Then, for fixed, but arbitrary $j \in [1 : M[+ \mid \theta]]$,*

$$\left|\left\langle f^+(\theta), h_j^{(+|\theta)}\right\rangle_H\right| > 0.$$

Proof Let $j_0 \in [1 : M[+ \mid \theta]]$ be fixed, but arbitrary, and suppose that

$$\left\langle f^+(\theta), h_{j_0}^{(+|\theta)}\right\rangle_H = 0.$$

Since $h_{j_0}^{(+|\theta)}$ belongs to the range of $\Delta_\theta[P]$, and that $\Delta_\theta[P]$ and P_t are orthogonal, for $t \leq \theta$, fixed, but arbitrary,

$$\left\langle f(t), h_{j_0}^{(+|\theta)}\right\rangle_H = 0.$$

$h_{j_0}^{(+|\theta)}$ is thus orthogonal to $L_\theta^{(+)}[f]$, since the latter is generated by $L_\theta[f]$ and f_θ^+. But since $h_{j_0}^{(+|\theta)}$ belongs to $L_\theta^{(+)}[f]$, it must be 0_H, a contradiction. □

Corollary 6.4.31 *Let $\phi_d(t) = \Delta[P][f(t)]$. Then, for fixed, but arbitrary $\theta_0 \in T_d$, the map*

$$\|\phi_d\|^2 : t \mapsto \|\phi_d(t)\|_H^2$$

has a jump at θ_0 whose magnitude is equal to

$$\sum_{j=1}^{M[+|\theta_0]} \left\langle f^+(\theta_0), h_j^{(+|\theta_0)}\right\rangle_H^2.$$

Proof One has that [(Proposition) 6.4.20, (Definition) 6.4.22, (Proposition) 6.4.28]

$$\begin{aligned}\|\phi_d(t)\|_H^2 &= \|\Delta[P][f(t)]\|_H^2 \\ &= \sum_{\theta \in T_d, \theta < t} \|\Delta_\theta[P][f(t)]\|_H^2 \\ &= \sum_{\theta \in T_d, \theta < t} \sum_{i=1}^{M[\theta]} \left\langle f(t), h_j^{(\theta)}\right\rangle_H^2,\end{aligned}$$

so that, as soon as t runs past θ_0,

$$\|\phi_d(t)\|_H^2 - \|\phi_d(\theta_0)\|_H^2 = \sum_{i=1}^{M[\theta_0]} \left\langle f(t), h_j^{(\theta_0)}\right\rangle_H^2 = \left\|\left(P_{\theta_0}^+ - P_{\theta_0}\right)[f(t)]\right\|_H^2.$$

Thus

$$\lim_{t \downarrow\downarrow \theta_0} \|\phi_d(t)\|_H^2 - \|\phi_d(\theta_0)\|_H^2 = \left\| \left(P_{\theta_0}^{+} - P_{\theta_0}\right) [f^+(\theta_0)] \right\|_H^2$$
$$= \left\| \left(P_{\theta_0}^{(+)} - P_{\theta_0}\right) [f^+(\theta_0)] \right\|_H^2$$
$$= \sum_{j=1}^{M[+|\theta_0]} \left\langle f^+(\theta_0), h_j^{(+|\theta_0)} \right\rangle_H^2 .$$

□

Proposition 6.4.32 *Let f have orthogonal increments, and, as above,*

$$\phi_d(t) = \Delta[P][f(t)] .$$

Then, the assignment $t \mapsto \|\phi_d(t)\|_H$ produces a step function. It is the sum of "elementary" step functions $\{t \mapsto \|\phi_{d,\theta}(t)\|, \theta \in T_d\}$ obtained as follows:

$$\|\phi_{d,\theta}(t)\|^2 = \begin{cases} 0 & \text{when } t \le \theta \\ \sum_{j=1}^{M[+|\theta]} \left\langle f^+(\theta), h_j^{(+|\theta)} \right\rangle_H^2 & \text{when } t > \theta \end{cases} .$$

Proof Let $\phi_{d,\theta}(t) = \Delta_\theta[P][f(t)]$. One starts from the following expression:

$$\phi_d(t) = \Delta[P][f(t)] = \sum_{\theta \in T_d} \Delta_\theta[P][f(t)] = \sum_{\theta \in T_d} \phi_{d,\theta}(t) ,$$

and [(Proposition) 6.4.20, Definition 6.4.22, Proposition 6.4.28]

$$\|\phi_d(t)\|_H^2 = \sum_{\theta \in T_d} \|\phi_{d,\theta}(t)\|_H^2 = \sum_{\theta \in T_d} \sum_{i=1}^{M[\theta]} \left\langle f(t), h_j^{(\theta)} \right\rangle_H^2 .$$

When $t \le \theta$, as $h_j^{(\theta)} \in \mathcal{R}[\Delta_\theta[P]] \perp L_\theta[f]$,

$$\left\langle f(t), h_j^{(\theta)} \right\rangle_H = 0.$$

Let now $\{t_1, t_2\} \subseteq T$, and $\theta \in T_d$, be such that $t_2 > t_1 > \theta$. As (f has here orthogonal increments)

$$f(t_2) - f(t_1) \perp L_{t_1}[f] \supseteq L_\theta^+[f] ,$$

one has that

$$\left\langle f(t_2) - f(t_1), h_j^{(\theta)} \right\rangle_H = 0,\ j \in [1 : M[\theta]].$$

Consequently

$$\left\langle f(t_2), h_j^{(\theta)} \right\rangle_H = \left\langle f(t_1), h_j^{(\theta)} \right\rangle_H,\ j \in [1 : M[\theta]].$$

Letting $t_1 \downdownarrows \theta$, it follows that, for fixed, but arbitrary $t > \theta,\ t \in T$,

$$\left\langle f(t), h_j^{(\theta)} \right\rangle_H = \left\langle f^+(\theta), h_j^{(\theta)} \right\rangle_H,\ j \in [1 : M[\theta]].$$

Since $\left\langle f^+(t), h_j^{(\theta|+)} \right\rangle_H = 0$, the result follows. □

Getting the proper canonical representation requires that many details be checked. Most of these are presented as lemmas. The main idea is that weighted sums of orthogonal elements provide the key mechanism in bringing about absolute continuity of the basis measures.

Lemma 6.4.33 *Let $L_c[f]$ be the (closed) subspace defined in (Definition)* 6.4.22, *and $f_c(t) = P_{L_c[f]}[f(t)]$. Then*

1. $L_t[f_c] = P_{L_c[f]}[L_t[f]] \subseteq L_t[f]$;
2. $L_{\cup T}[f_c] = L_c[f]$.

Proof The definition of f_c yields that $P_{L_c[f]}[L_t[f]] \subseteq L_t[f_c]$. For $h \in L_t[f_c]$, orthogonal to $P_{L_c[f]}[L_t[f]]$, and, in T, $\theta \le t$, one has that

$$0 = \langle h, P_{L_c[f]}[f(\theta)] \rangle_H = \langle h, f_c(\theta) \rangle_H,$$

so that $h = 0_H$, and the inclusion of the proof's beginning is an equality. As P_t and $P_{L_c[f]}$ commute [(Lemma) 6.4.21],

$$f_c(t) = P_{L_c[f]}[f(t)] = P_{L_c[f]}P_t[f(t)] = P_tP_{L_c[f]}[f(t)] \in L_t[f],$$

and the inclusion stated in item 1 obtains also.

By definition $f_c(t) \in L_c[f]$, and thus $L_{\cup T}[f_c] \subseteq L_c[f]$. Suppose one has that $h \in L_c[f]$ is orthogonal to $L_{\cup T}[f_c]$. Then, for fixed, but arbitrary $t \in T$,

$$0 = \langle h, f_c(t) \rangle_H = \langle h, P_{L_c[f]}[f(t)] \rangle_H = \langle P_{L_c[f]}[h], f(t) \rangle_H = \langle h, f(t) \rangle_H,$$

so that, using item 1, $h = 0_H$. □

Lemma 6.4.34 *Let, as in* (Proposition) 6.4.10,

$$\{f_i : t \mapsto f_i(t) = P_t[k_i],\ t \in T,\ k_i \in H,\ i \in I \subseteq \mathbb{N}\}$$

be a family of orthogonal functions, with orthogonal increments, arising in a canonical decomposition, and let

$$\begin{aligned} M_i([t_1, t_2[) &= \|m_i([t_1, t_2[)\|_H^2 \\ &= \|f_i(t_2) - f_i(t_1)\|_H^2 \\ &= \|f_i(t_2)\|_H^2 - \|f_i(t_1)\|_H^2 . \end{aligned}$$

When $\{\alpha_i,\ i \in I\} \subseteq (\mathbb{R} \setminus \{0\})^I$ *is such that*

$$\sum_{i \in I} \alpha_i^2 \|k_i\|_H^2 < \infty,$$

then

$$f_\alpha(t) = \sum_{i \in I} \alpha_i f_i(t)$$

has orthogonal increments, and, letting,

$$\begin{aligned} M_\alpha([t_1, t_2[) &= \|m_\alpha([t_1, t_2[)\|_H^2 \\ &= \|f_\alpha(t_2) - f_\alpha(t_1)\|_H^2 \\ &= \|f_\alpha(t_2)\|_H^2 - \|f_\alpha(t_1)\|_H^2 , \end{aligned}$$

one has that

$$M_\alpha([t_1, t_2[) = \sum_{i \in I} \alpha_i^2 M_i([t_1, t_2[) .$$

Furthermore, one may always choose the family $\{\alpha_i,\ i \in I\}$ *so that, given a* $\kappa > 0$,

$$M_\alpha(T) \leq \kappa < \infty.$$

Proof The orthogonality properties of the functions involved allow one to write that

$$\begin{aligned} &\langle f_\alpha(t_2) - f_\alpha(t_1), f_\alpha(t_4) - f_\alpha(t_3)\rangle_H = \\ &\quad = \sum_{i \in I} \alpha_i^2 \langle f_i(t_2) - f_i(t_1), f_i(t_4) - f_i(t_3)\rangle_H . \end{aligned}$$

Furthermore, $M_\alpha(T)$ is the limit of $M_\alpha([t_1, t_2[)$, when t_1 and t_2 become large, as, for example, $f_i(t) = P_t[k_i]$, and, analogously, the same obtains for the components $M_i(T)$, so that

$$M_\alpha(T) = \sum_{i \in I} \alpha_i^2 M_i(T) = \sum_{i \in I} \alpha_i^2 \|k_i\|_H^2 .$$

One need then only "adjust" the α_i's ($\lambda \sum_{i \in I} \alpha_i^2 \|k_i\|_H^2 = \kappa$). □

Remark 6.4.35 The last assertion of (Lemma) 6.4.34 shall prove essential in the computation of the likelihood (part III).

Lemma 6.4.36 *Let*

$$L_t[f] = \bigoplus_{i \in I} L_t[f_i]$$

be a canonical representation of f, *that is* [(Proposition) 6.4.10, (Remark) 6.4.11], $f_i(t) = P_t[k_i]$, $\{k_i, i \in I\}$ *orthogonal, and*

$$\begin{aligned} M_i^f([t_1, t_2[) &= \left\| m_i^f([t_1, t_2[) \right\|_H^2 \\ &= \|f_i(t_2) - f_i(t_1)\|_H^2 \\ &= \|f_i(t_2)\|_H^2 - \|f_i(t_1)\|_H^2 . \end{aligned}$$

Consider the following items:

(A) α*-elements:*

(a) $\{\alpha_i, i \in I\} \in (\mathbb{R} \setminus \{0\})^I$ *such that* $\sum_{i \in I} \alpha_i^2 \|k_i\|_H^2 < \infty$;
(b) $k_\alpha = \sum_{i \in I} \alpha_i k_i$;
(c) $f_\alpha(t) = P_t[k_\alpha] = \sum_{i \in I} \alpha_i f_i(t)$;
(d) $M_\alpha([t_1, t_2[)$ *as in (Lemma)* 6.4.34;

(B) h*-elements:*

(a) $h \in L_{UT}[f]$ *with decomposition* $h = \sum_{i \in I} h_i$, $h_i \in L_{UT}[f_i]$, *that is,*

$$h_i = \int \phi_i^h dm_i^f, \ \phi_i^h \in L_2\left(T, \mathcal{T}, M_i^f\right);$$

(b) $g_i(t) = P_t[h_i]$;
(c) $M_i^g([t_1, t_2[)$ *as, mutatis mutandis,* $M_\alpha([t_1, t_2[)$ (the measure m_α becomes m_i^g, and the function f_α, g_i);
(d) $g_h(t) = \sum_{i \in I} g_i(t)$;
(e) $M_{g_h} = \sum_{i \in I} M_i^g$.

Then, for $i \in I$, *fixed, but arbitrary,* $M_i^g \ll M_i^f$, *and* $M_{g_h} \ll M_\alpha$.

Proof One should notice that the items of the statement are legitimate and make sense. Thus (A.b) follows from (A.a) and the fact that the k_i's are orthogonal. Also (A.d) is due to (A.c) and (Lemma) 6.4.34. In fact

$$M_\alpha\left([t_1, t_2[\right) = \sum_{i \in I} \alpha_i^2 M_i^f\left([t_1, t_2[\right).$$

The canonical representation [(Proposition) 6.4.10, (Fact) 6.2.12] justifies item (B.e), and the consequence that $M_i^g \ll M_i^f$. Item (B.h) follows from the orthogonality of the h_i's. The conclusion of the statement is then immediate. □

The following table may perhaps usefully display the ingredients of the previous lemma:

			P_t			
$L_t[f] = \bigoplus_{i \in I} L_t[f_i]$	f	k_i	f_i	m_i^f	M_i^f	
$k_\alpha = \sum_{i \in I} k_i$	f_α			m_α	M_α	
$h = \sum_{i \in I} h_i$		h_i	g_i	m_i^g	M_i^g	$\ll M_i^f$
	g_h			m_{g_h}	M_{g_h}	$\ll M_\alpha$

The result which follows provides the basic feature of the existence of a proper canonical representation. The main improvement with respect to the simpler canonical representations is the ordering by absolute continuity of the measures appearing in it. It is that feature which insures that proper canonical representations are essentially unique, in the sense that they have the same multiplicity.

Proposition 6.4.37 *Let* CHA *obtain for* f, *so that a canonical representation exists. There exists then a* (proper [(Remark) 6.4.17]) *canonical representation with orthogonal components* $\{f_i^p : T \longrightarrow H, i \in I\}$ *such that, for* $i \in I$, *fixed, but arbitrary,* $M_{i+1}^p \ll M_i^p$. *When the map* $g : T \longrightarrow H$ *is such that, for* $t \in T$, *fixed, but arbitrary,* $g(t) \in L_t[f]$, *then* $g(t) = \sum_{i \in I} g_i(t)$, *and*

$$g_i(t) = \int \phi_i^g(t)\, dm_i^{f^p},\ I_{T_t}\phi_i^g(t) = \phi_i^g(t).$$

Proof Since the representation of g is a consequence of any canonical representation, one need only produce a proper [(Remark) 6.4.17] canonical representation for f. Let thus

$$L_t[f] = \bigoplus_{i \in I} L_t[f_i]$$

be a canonical representation [(Proposition) 6.4.10]. One starts with a fixed, but arbitrary $k_1 \in L_{\cup T}[f]$ which is obtained as the k_α of (Lemma) 6.4.36. The latter's f_α is now denoted f_1^p, and M_α, M_1^p. One then knows, from (Lemma) 6.4.36, that, for fixed, but arbitrary $h \in L_{UT}[f]$, the measure M_{g_h} associated with it has the property that $M_{g_h} \ll M_\alpha = M_1^p$.

Let $L_{[1]}[f] = L_{UT}[f] \cap L_{UT}\left[f_1^p\right]^\perp$. Choose in $L_{[1]}[f]$ a complete orthonormal set $\{h_i^{(1)}, i \in I_1\}$, and build a sequence $\{k_i^{(1)}, i \in I_1\}$ as in (Proposition) 6.4.10, so that

$$L_{[1]}[f] = \bigoplus_{i \in I_1} L_{UT}\left[f_i^{(1)}\right], \; f_i^{(1)}(t) = P_t\left[k_i^{(1)}\right].$$

Choose then, in $(\mathbb{R} \setminus \{0\})^{I_1}$, a sequence $\{\alpha_i^{(1)}, i \in I_1\}$ such that

$$\sum_{i \in I_1} \left\{\alpha_i^{(1)}\right\}^2 \left\|k_i^{(1)}\right\|_H^2 < \infty.$$

Define then, as in (Lemma) 6.4.36,

(i) $k_2 = \sum_{i \in I_1} \alpha_i^{(1)} k_i^{(1)} \in L_{[1]}[f]$;

(ii) $f_2^p(t) = P_t[k_2] = \sum_{i \in I_1} \alpha_i^{(1)} f_i^{(1)}(t)$;

(iii) M_2^p, the measure associated with f_2^p.

One thus obtains that $M_2^p \ll M_1^p$, and that, for $h \in L_{[1]}[f]$, fixed, but arbitrary, and M_{g_h} built, *mutatis mutandis,* as in (Lemma) 6.4.36, $M_{g_h} \ll M_2^p$. Furthermore, as proven in (Proposition) 6.4.10, for $\{t_1, t_2\} \subseteq T$, fixed, but arbitrary,

$$L_{t_1}\left[f_1^p\right] \perp L_{t_2}\left[f_2^p\right].$$

Such a procedure may be continued "identically, and indefinitely," until $L_{UT}[f]$ is "emptied." Since the subspaces built in such a fashion are orthogonal, and that $L_{UT}[f]$ is separable, there shall be at most a countable number of them. Since subspaces are ordered by inclusion, that successive sums of orthogonal subspaces form an ordered chain with an upper bound, by Zorn's lemma, there shall be a maximal subspace, which can only be $L_{UT}[f]$, which has thus a representation, with components f_i^p, whose associated measures M_i^p behave as in the proof of the present proposition. □

Remark 6.4.38 When $g \in L_{UT}[f]$, $P_t[g] \in L_t[f]$, and (Proposition) 6.4.37 applies. Letting $g(t) = f(t)$, one gets a canonical decomposition for f, with accompanying measures that are ordered by absolute continuity. Letting $f = f_c$, and then $g(t) = f_c(t)$, one gets a canonical decomposition for f_c, with continuous components, and accompanying measures that are ordered by absolute continuity.

As stated, proper canonical representations are essentially unique. That is the content of the lemmas to follow. They require the introduction of a new concept, that of a maximal chain.

Definition 6.4.39 The family of measures $\{M_i,\ i \in I\}$ of (Proposition) 6.4.37 is said to be maximal, or to form a maximal chain. It is characterized by two facts:

1. for fixed, but arbitrary $i \in I$, $M_i \gg M_{i+1}$;
2. for $h \in L_{UT}[f]$, and g_h as in (Lemma) 6.4.36, $M_{g_h} \ll M_1$, and, for fixed, but arbitrary $i \in I$ and $h \perp \bigoplus_{l=1}^{i} L_{UT}[f_l]$, $M_{g_h} \ll M_{i+1}$.

Remark 6.4.40 Let $L_{UT}[f] = \bigoplus_{i \in I} L_{UT}[f_i]$ be such that, for $i \in I$, fixed, but arbitrary, $M_{i+1} \ll M_i$. Item 2 of (Definition) 6.4.39 is then automatically true.

Indeed, when $h \perp \bigoplus_{l=1}^{i} L_{UT}[f_l]$,

$$h \in \bigoplus_{i+1 \leq l \in I} L_{UT}[f_l],$$

so that

$$h = \sum_{i+1 \leq l \in I} \int \phi_l^h dm_l,\ \phi_l^h \in L_2(T, \mathcal{T}, M_i),\ i+1 \leq l \in I.$$

Consequently

$$dM_{g_h} = \sum_{n+1 \leq l \in I} \{\dot{\phi}_l^h\}^2 dM_l \ll M_{i+1}.$$

The result which follows serves as *deus ex machina* for proper canonical representations. It provides a useful consequence of the lack of absolute continuity.

Lemma 6.4.41 *Suppose that*

(A) *for $t \in T$, fixed, but arbitrary,*

$$L_t[f] = \bigoplus_{i \in I} L_t[f_i]$$

is a proper canonical representation, that is, a canonical representation for which the family $\{M_i^f, i \in I\}$ is a maximal chain;

(B) *it obtains that the following direct sum*

$$\bigoplus_{j \in J} L_{UT}[g_j] \subseteq L_{UT}[f]$$

has the properties of a proper [(Remark) 6.4.17] *canonical representation, that is, such that, for $j \in J$, fixed, but arbitrary, $M_j^g \gg M_{j+1}^g$;*

(C) *there exists $j_0 \in J$ such that, for $j \in [1 : j_0 - 1]$, fixed, but arbitrary,*

$$M_j^g \ll M_j^f,$$

but that $M^g_{j_0}$ *is not absolutely continuous with respect to* $M^f_{j_0}$ (however, since the M^f_i form a maximal chain, $M^g_j \ll M^f_1, j \geq 1$).

Since all measures that are used are finite, one may call on the Lebesgue decomposition [5, p. 68] *to write that*

$$M^g_{j_0} = M_0 + M_0^{\perp}, \ \textit{with}\ M_0 \ll M^f_{j_0}, \ \textit{and}\ M_0^{\perp} \perp M^f_{j_0}.$$

As, for $j \in [1 : j_0 - 1]$, *fixed, but arbitrary,*

$$M_0^{\perp} \ll M^g_{j_0} \ll M^g_j \ll M^f_j,$$

one may define the following items:

(a) *for* $j \in [1 : j_0]$, *fixed, but arbitrary,*

$$\begin{aligned} d^g_j &= \frac{dM_0^{\perp}}{dM^g_j}, \\ D^g_j &= \textit{equivalence class of}\ \sqrt{d^g_j}, \\ h^g_j &= \int D^g_j \, dm^g_j \in L_{UT}[g_j]; \end{aligned}$$

(b) *for* $j \in [1 : j_0 - 1]$, *fixed, but arbitrary,*

$$\begin{aligned} d^f_j &= \frac{dM_0^{\perp}}{dM^f_j}, \\ D^f_j &= \textit{equivalence class of}\ \sqrt{d^f_j}, \\ h^f_j &= \int D^f_j \, dm^f_j \in L_{UT}[f_j]. \end{aligned}$$

Let

$$\begin{aligned} \tilde{f}_j(t) &= P_t\left[h^f_j\right], \\ \tilde{g}_j(t) &= P_t\left[h^g_j\right]. \end{aligned}$$

The following assertions then obtain:

1. *for* $\{j, j_1, j_2\} \subseteq [1 : j_0]$, *fixed, but arbitrary,*

   (i) $M_j^{\tilde{g}} \ll M_0^{\perp}$ (in fact they are equal),

   (ii) *when* $j_1 \neq j_2$, *for fixed, but arbitrary* $\{t_1, t_2\} \subseteq T$,

$$L_{t_1}\left[\tilde{g}_{j_1}\right] \perp L_{t_2}\left[\tilde{g}_{j_2}\right]$$

   (in particular, $h_{j_1}^{g} \perp h_{j_2}^{g}$);

2. *for* $\{j, j_1, j_2\} \subseteq [1 : j_0 - 1]$, *fixed, but arbitrary,*

   (i) $M_j^{\tilde{f}} \ll M_0^{\perp}$ (in fact they are equal),

   (ii) *when* $j_1 \neq j_2$, *for fixed, but arbitrary* $\{t_1, t_2\} \subseteq T$,

$$L_{t_1}\left[\tilde{f}_{j_1}\right] \perp L_{t_2}\left[\tilde{f}_{j_2}\right]$$

   (in particular, $h_{j_1}^{f} \perp h_{j_2}^{f}$);

3. $\bigoplus_{j=1}^{j_0-1} L_{\cup T}\left[\tilde{f}_j\right] = \left\{h \in L_{\cup T}[f] : M_{g_h} \ll M_0^{\perp}\right\}$.

Proof Items 1 and 2 restate either immediate consequences of the definitions (absolute continuity), or properties of a canonical decomposition (orthogonality). Thus only item 3 requires a proof.

It is useful to notice the following two facts:

- since $M_0^{\perp} \perp M_{j_0}^{f}$, there exists $T_0 \in \mathcal{T}$ such that

$$M_{j_0}^{f}(T_0) = 0, \text{ and } M_0^{\perp}(T_0) = M_0^{\perp}(T);$$

- since $\left\{M_i^f, i \in I\right\}$ is a maximal chain, for $i > j_0$, $M_i^f \ll M_{j_0}^f$, so that, for $i \geq j_0$,

$$M_i^f(T_0) = 0, \quad \text{and} \quad M_i^f \perp M_0^{\perp}.$$

Proof Suppose that $h \in L_{\cup T}[f]$, *and that*

$$M_{g_h} \ll M_0^{\perp}. \qquad (\star)$$

Assumption (A: canonical representation, maximal chain) assures that [(Proposition) 6.4.37]:

- h has the decomposition $h = \sum_{i \in I} h_i$, with $h_i = P_{L_{\cup T}[f_i]}[h]$;
- h_i has the representation $h_i = \int \phi_i^h \, dm_i^f$, $\phi_i^h \in L_2\left(T, \mathcal{T}, M_i^f\right)$;
- $M_i^{g_h}$ is determined using the function $t \mapsto P_t[h_i]$ ($g_i(t)$ in (Lemma) 6.4.36), and M_{g_h}, the function $t \mapsto P_t[h]$;

- for $\tilde{T} \in \mathcal{T}$, fixed, but arbitrary, $M_i^{g_h}\left(\tilde{T}\right) = \int_{\tilde{T}} \left(\dot{\phi}_i^h\right)^2 dM_i^f$;
- for $\tilde{T} \in \mathcal{T}$, fixed, but arbitrary, $M_{g_h}\left(\tilde{T}\right) = \sum_{i \in I} M_i^{g_h}\left(\tilde{T}\right)$.

Since it is assumed ($\star$) that $M_{g_h} \ll M_0^{\perp}$, for $i \in I$, fixed, but arbitrary,

$$M_i^{g_h} \ll M_{g_h} \ll M_0^{\perp},$$

so that one may compute

$$d_i^h = \frac{dM_i^{g_h}}{dM_0^{\perp}}.$$

Then, for $\tilde{T} \in \mathcal{T}$, fixed, but arbitrary,

$$\int_{T_0 \cap \tilde{T}} d_i^h \, dM_0^{\perp} = M_i^{g_h}\left(T_0 \cap \tilde{T}\right) = \int_{T_0 \cap \tilde{T}} \left(\dot{\phi}_i^h\right)^2 dM_i^f. \qquad (\star\star)$$

As, when $i \geq j_0$, $M_i^f(T_0) = 0$, and $M_0^{\perp}(T_0) = M_0^{\perp}(T)$, ($\star\star$) rewrites as

$$\int_{\tilde{T}} d_i^h \, dM_0^{\perp} = 0,$$

so that d_i^h is almost surely zero, with respect to $M_0^{\perp}$. But then $M_i^{g_h}$ is a zero measure, that is, ϕ_i^h is the equivalence class of the zero function, which, in turn, means that $h_i = 0_H$. Consequently

$$h = \sum_{j=1}^{j_0 - 1} h_j.$$

Now, given (A), and the assumption ($\star$) that $M_{g_h} \ll M_0^{\perp}$, for $j \in [1 : j_0 - 1]$, fixed, but arbitrary,

$$M_j^{g_h} \ll M_{g_h} \ll M_0^{\perp} \ll M_0 + M_0^{\perp} = M_{j_0}^f \ll M_j^f,$$

so that, for $\tilde{T} \in \mathcal{T}$, fixed, but arbitrary,

$$M_j^{g_h}\left(\tilde{T}\right) = \begin{cases} \int_{\tilde{T}} \left(\dot{\phi}_i^h\right)^2 dM_j^f \\ \int_{\tilde{T}} \left\{\frac{dM_j^{g_h}}{dM_0^{\perp}}\right\} \left\{\frac{dM_0^{\perp}}{dM_j^f}\right\} dM_j^f = \int_{\tilde{T}} d_j^h d_j^f \, dM_j^f \end{cases}.$$

Consequently, almost surely with respect to M_j^f, $d_j^h d_j^f = \left(\dot{\phi}_i^h\right)^2$.

Given a function $\dot{f}$, with extended real values, let $s(f)$ be the equivalence class of the following function:

$$\dot{s}\left(\dot{f}\right)=\begin{cases}1 \text{ when } \dot{f}\geq 0\\ -1 \text{ when } \dot{f}<0\end{cases}.$$

Let also D_j^h be the equivalence class of $\sqrt{d_j^h}$. Then

$$\int s\left(\phi_j^h\right)D_j^h dm_j^{\tilde{f}}=\int s\left(\phi_j^h\right)D_j^h D_j^f dm_j^f=\int \phi_j^h dm_j^f=h_j.$$

Consequently $h_j\in L\left[m_j^{\tilde{f}}\right]=L_{\cup T}\left[\tilde{f}_j\right]$, so that

$$h\in\bigoplus_{j=1}^{j_0-1}L_{\cup T}\left[\tilde{f}_j\right],$$

and

$$\left\{h\in L_{\cup T}\left[f\right]:M_{f_h}\ll M_0^{\perp}\right\}\subseteq\bigoplus_{j=1}^{j_0-1}L_{\cup T}\left[\tilde{f}_j\right].$$

Proof *Suppose now that* $h\in\bigoplus_{j=1}^{j_0-1}L_{\cup T}\left[\tilde{f}_j\right]$.
Then

$$h=\sum_{j=1}^{j_0-1}\int\phi_j^h dm_j^{\tilde{f}},\ \phi_j^h\in L_2\left(T,\mathcal{T},M_j^{\tilde{f}}\right).$$

But

$$\int\phi_j^h dm_j^{\tilde{f}}=\int\phi_j^h D_j^f dm_j^f,$$

so that

$$dM_{g_h}=\sum_{j=1}^{j_0-1}\left(\dot{\phi}_j^h\right)^2 d_j^f dM_j^f=\sum_{j=1}^{j_0-1}\left(\dot{\phi}_j^h\right)^2 dM_0^{\perp},$$

that is, $M_{g_h}\ll M_0^{\perp}$. □

The next lemma says that, when the index is small enough, that is, in item (B) of (Lemma) 6.4.41, $k \in I \cap J$, absolute continuity prevails, that is $M_k^g \ll M_k^f$.

Lemma 6.4.42 *Suppose that items* (A) *and* (B) *of (Lemma)* 6.4.41 *obtain, and let* $\tilde{I} = |I|, \tilde{J} = |J|$. *Then, for* $k \leq \tilde{I} \wedge \tilde{J}$, *fixed, but arbitrary,*

$$M_k^g \ll M_k^f.$$

Proof Notation, and definitions, are those of (Lemma) 6.4.41. Let, for $j \in J$, fixed, but arbitrary, $g_j(t) = P_t\left[h_j\right], h_j \in L_{\cup T}[f]$. As

$$\left\{M_i^f, i \in I\right\}$$

is a maximal chain, and that, for $j \in J$, fixed, but arbitrary, $h_j \in L_{\cup T}[f]$,

$$h_j = \sum_{i \in I} P_{L_{\cup T}[f_i]}\left[h_j\right] = \sum_{i \in I} \int \phi_i^{h_j}\, dm_i^f,$$

so that

$$dM_j^g = \left(\dot{\phi}_i^{h_j}\right)^2 dM_i^f \ll dM_1^f,$$

and, in particular, $M_1^g \ll M_1^f$.

Suppose that $k_0 \leq \tilde{I} \wedge \tilde{J}$ is the first integer, if there is one, for which one does not have $M_{k_0}^g \ll M_{k_0}^f$. Then, because of (Lemma) 6.4.41,

$$\bigoplus_{k=1}^{k_0-1} L_{\cup T}\left[\tilde{f}_k\right] = \left\{h \in L_{\cup T}[f] : M_{g_h} \ll M_0^{\perp}\right\}.$$

Since [(Lemma) 6.4.41] $M_k^{\tilde{g}} \ll M_0^{\perp}$, $h_k^g \in \bigoplus_{l=1}^{k_0-1} L_{\cup T}\left[\tilde{f}_l\right]$, and thus

$$h_k^g = \sum_{l=1}^{k_0-1} \int \phi_l^{h_k^g}\, dm_l^{\tilde{f}},$$

so that

$$dM_k^{\tilde{g}} = \sum_{l=1}^{k_0-1} \left(\dot{\phi}_l^{h_k^g}\right)^2 dM_l^{\tilde{f}}.$$

Using the definitions of (Lemma) 6.4.41, one has, for example, that

$$dM_l^{\tilde{f}} = d_l^f\, dM_l^f = dM_0^{\perp}.$$

So, for fixed, but arbitrary

$$j \in [1 : k_0] : dM_j^{\tilde{g}} = dM_0^{\perp},$$
$$i \in [1 : k_0 - 1] : dM_i^{\tilde{f}} = dM_0^{\perp},$$

and thus, almost surely, with respect to $M_0^{\perp}$,

$$\sum_{l=1}^{k_0-1} \left(\dot{\phi}_l^{h_k^g}\right)^2 = 1.$$

On the other hand, when $l \neq \lambda$,

$$\begin{aligned} 0 &= \langle \tilde{g}_l(t), \tilde{g}_\lambda(t) \rangle_H \\ &= \left\langle \sum_{k=1}^{k_0-1} \int I_{T_t} \phi_k^{h_l^g} dm_k^{\tilde{f}}, \sum_{k=1}^{k_0-1} \int I_{T_t} \phi_k^{h_\lambda^g} dm_k^{\tilde{f}} \right\rangle_H \\ &= \sum_{k=1}^{k_0-1} \int_{T_t} \dot{\phi}_k^{h_l^g} \dot{\phi}_k^{h_\lambda^g} dM_k^{\tilde{f}} \\ &= \int_{T_t} \sum_{k=1}^{k_0-1} \dot{\phi}_k^{h_l^g} \dot{\phi}_k^{h_\lambda^g} dM_0^{\perp}. \end{aligned}$$

Consequently, almost surely, with respect to $M_0^{\perp}$,

$$\sum_{k=1}^{k_0-1} \dot{\phi}_k^{h_l^g} \dot{\phi}_k^{h_\lambda^g} = 0.$$

Let $\underline{v}_l(t) \in \mathbb{R}^{k_0-1}$ be the vector with components

$$\left\{\dot{\phi}_k^{h_l^g}(t), k \in [1 : k_0 - 1]\right\}.$$

From what precedes, almost surely for $t \in T$, with respect to $M_0^{\perp}$, the set

$$\{\underline{v}_1(t), \ldots, \underline{v}_{k_0}(t)\}$$

forms a family of k_0 orthonormal vectors in $\mathbb{R}^{k_0-1}$. This is possible only when $M_0^{\perp}(T) = 0$. □

The lemma which follows is the key to proving that multiplicities must be equal.

Lemma 6.4.43 *Suppose that items* (A) *and* (B) *of (Lemma)* 6.4.41 *obtain, and that*

$$|I| = \tilde{I} < \tilde{J} = |J|.$$

Then, because of (Lemma) 6.4.42,

$$M_i^g \ll M_i^f, i \in [1 : \tilde{I}].$$

Furthermore, because of Assumption (B), $M_{\tilde{I}+1}^g \ll M_{\tilde{I}}^g$, *so that*

$$M_{\tilde{I}+1}^g \ll M_{\tilde{I}}^f.$$

It makes thus sense to set:

(a) *for* $j \in \left[1 : \tilde{I} + 1\right]$, *fixed, but arbitrary,*

$$\begin{aligned} d_j^g &= \frac{dM_{\tilde{I}+1}^g}{dM_j^g}, \\ D_j^g &= \textit{equivalence class of } \sqrt{d_j^g}, \\ h_j^g &= \int D_j^g \, dm_j^g; \end{aligned}$$

(b) *for* $i \in \left[1 : \tilde{I}\right]$, *fixed, but arbitrary,*

$$\begin{aligned} d_i^f &= \frac{dM_{\tilde{I}+1}^g}{dM_i^f}, \\ D_i^f &= \textit{equivalence class of } \sqrt{d_i^f}, \\ h_i^f &= \int D_i^f \, dm_i^f. \end{aligned}$$

Then, with notation and definitions of (Lemma) 6.4.42,

1. $\tilde{I} < \infty$;
2. *for* $i \in \left[1 : \tilde{I}\right]$, *and* $j \in \left[1 : \tilde{I} + 1\right]$, *fixed, but arbitrary,*

$$M_j^{\tilde{g}} = M_{\tilde{I}+1}^g = M_i^{\tilde{f}};$$

3. $\bigoplus_{i \in I} L_{UT}\left[\tilde{f}_i\right] = \left\{h \in L_{UT}[f] : M_{g_h} \ll M_{\tilde{I}+1}^g\right\}$;
4. $\bigoplus_{j=1}^{\tilde{I}+1} L_{UT}\left[\tilde{g}_j\right] \subseteq \bigoplus_{i \in I} L_{UT}\left[\tilde{f}_i\right]$.

Proof Item 1 expresses the fact that J is at most countable, and that one has assumed $\tilde{I} < \tilde{J}$. Item 2 is a direct consequence of the definitions (in (a) and (b)) of the elements involved: for example $dM_j^{\tilde{g}} = d_j^g dM_j^g = dM_{\tilde{I}+1}^g$.

Proof ([3],⊆) Let $h \in \bigoplus_{i \in I} L_{\cup T}\left[\tilde{f}_i\right]$ be fixed, but arbitrary.
Then

$$h = \sum_{i \in I} h_i,\ h_i = P_{L_{\cup T}[\tilde{f}_i]}\left[h\right].$$

Thus

$$h_i = \int \phi_i^h dm_i^{\tilde{f}} = \int \phi_i^h D_i^f dm_i^f,$$

$$dM_i^{g_h} = \left(\phi_i^h\right)^2 dM_i^{\tilde{f}} = \left(\phi_i^h\right)^2 dM_{\tilde{I}+1}^g \text{ (item 2)},$$

and

$$\infty > \int \left(\dot{\phi}_i^h\right)^2 dM_i^{\tilde{f}} = \begin{cases} \int \left(\dot{\phi}_i^h\right)^2 d_i^f dM_i^f \\ \int \left(\dot{\phi}_i^h\right)^2 dM_{\tilde{I}+1}^g \end{cases}.$$

Consequently

$$h_i \in L\left[m_i^f\right] = L_{\cup T}\left[f_i\right] \subseteq L_{\cup T}\left[f\right].$$

Furthermore

$$M_i^{g_h} \ll M_{\tilde{I}+1}^g,$$

so that both $h \in L_{\cup T}[f]$ and $M_{g_h} \ll M_{\tilde{I}+1}^g$ obtain. The (⊆)-part of the equality of item 3 is thus true.

Proof ([3],⊇) Suppose that $h \in L_{\cup T}[f]$ is such that $M_{g_h} \ll M_{\tilde{I}+1}^g$.
Then

$$h = \sum_{i \in I} h_i,\ h_i = \int \phi_i^h dm_i^f \in L\left[m_i^f\right] = L_{\cup T}\left[f_i\right],$$

and thus

$$dM_i^{g_h} = \left(\dot{\phi}_i^h\right)^2 dM_i^f.$$

But, for $i \in I$, fixed, but arbitrary, first by definition, and then by assumption,

$$M_i^{g_h} \ll M_{g_h} \ll M_{\tilde{I}+1}^g.$$

Let then

$$\begin{aligned} d_i^h &= \frac{dM_i^{g_h}}{dM_{\tilde{I}+1}^g}, \\ D_i^h &= \text{equivalence class of } \sqrt{d_i^h}. \end{aligned}$$

One may thus write, using item 2, that

$$dM_i^{g_h} = d_i^h dM_{\tilde{I}+1}^g = d_i^h dM_i^{\tilde{f}} = d_i^h d_i^f dM_i^f,$$

with the consequence that, almost surely with respect to M_i^f,

$$\left(\dot{\phi}_i^h\right)^2 = d_i^h d_i^f.$$

One may then check, as in the proof of (Lemma) 6.4.42, that

$$h_i = \int s\left(\phi_i^h\right) D_i^h dm_i^{\tilde{f}} \in L\left[m_i^{\tilde{f}}\right] = L_{UT}\left[\tilde{f}_i\right].$$

The proof of item 4 proceeds then as follows. Let $k \in \bigoplus_{j=1}^{\tilde{I}+1} L_{UT}\left[\tilde{g}_j\right]$ be fixed, but arbitrary. Then

$$k = \sum_{j=1}^{\tilde{I}+1} k_j,\ k_j \in L_{UT}\left[\tilde{g}_i\right], \text{ or } k_j = \int \phi_j^k D_j^g dm_j^g \in L\left[m_j^g\right] = L_{UT}\left[g_j\right],$$

so that, using the Assumption (B), $k \in L_{UT}[f]$. Furthermore, using the definitions, and item 2,

$$M_{g_k} = \sum_{j=1}^{\tilde{I}+1} M_j^{\tilde{g}} \ll M_{\tilde{I}+1}^g.$$

But then, by item 3, $k \in \bigoplus_{\in I} L_{UT}\left[\tilde{f}_i\right]$. □

One shall now see that multiplicities of proper canonical representations dominate those of similar representations.

Lemma 6.4.44 *Suppose that items* (A) *and* (B) *of (Lemma)* 6.4.41 *obtain. Then:*

1. $\tilde{J} \leq \tilde{I}$;
2. *for* $j \in J$, *fixed, but arbitrary,* $M_j^g \ll M_j^f$.

Proof Suppose one establishes that

$$M^g_{\tilde{I}+1}(T) = 0.$$

As, by assumption, for $\{j_1, j_2\} \subseteq J, j_1 < j_2$, fixed, but arbitrary,

$$M^g_{j_2} \ll M^g_{j_1},$$

one shall then have that, for $j > \tilde{I}$, fixed, but arbitrary, $M_j^g(T) = 0$, that is

$$\tilde{J} \leq \tilde{I}.$$

And it follows then, from (Lemma) 6.4.42, that, for $j \in J$, fixed but arbitrary,

$$M_j^g \ll M_j^f.$$

Suppose thus that $\tilde{I} < \tilde{J}$. From (Lemma) 6.4.43, items 2 and 3, it follows respectively that:

(i) for $i \in [1 : \tilde{I}]$, and $j \in [1 : \tilde{I} + 1]$, fixed, but arbitrary,

$$M_j^{\tilde{g}} = M^g_{\tilde{I}+1} = M_i^{\tilde{f}};$$

(ii) $\bigoplus_{i \in I} L_{\cup T}[\tilde{f}_i] = \left\{ h \in L_{\cup T}[f] : M_{g_h} \ll M^g_{\tilde{I}+1} \right\}$.

Since, from (i), for $j \in [1 : \tilde{I} + 1]$, fixed, but arbitrary, $M_j^{\tilde{g}} \ll M^g_{\tilde{I}+1}$, then, from (ii),

$$h_j^g = \sum_{i=1}^{\tilde{I}} \int \phi_i^{h_j^g} \, dm_i^{\tilde{f}},$$

so that, with h_j^g, the element of H that determines the function $\tilde{g}_j$ in terms of the projections P_t,

$$dM_j^{\tilde{g}} = \sum_{i=1}^{\tilde{I}} \left(\dot{\phi}_i^{h_j^g} \right)^2 dM_i^{\tilde{f}}.$$

It follows from there, almost surely, with respect to $M^g_{\tilde{I}+1}$, that

$$\sum_{i=1}^{\tilde{I}} \left(\dot{\phi}_i^{h_j^g} \right)^2 = 1.$$

One has thus a situation analogous to that encountered in the proof of (Lemma) 6.4.42. *Mutatis mutandis* one must conclude that $M^g_{\tilde{I}+1}(T) = 0$. □

One sees next that multiplicity is unique.

Lemma 6.4.45 *Suppose that*

$$L_{\cup T}[f] = \bigoplus_{i \in I} L_{\cup T}[f_i] \bigoplus L_d[f],$$

with, for $i \in I$, fixed, but arbitrary,

- *the maps with orthogonal increments $f_i(t) = P_t[k_i]$ are continuous,*
- $M_i^f \gg M_{i+1}^f$.

If also

$$L_{\cup T}[f] = \bigoplus_{j \in J} L_{\cup T}[g_j] \bigoplus L_d[f]$$

with, for $j \in J$, fixed, but arbitrary, $M_j^g \gg M_{j+1}^g$, then

1. $\tilde{J} = \tilde{I}$;
2. *for $i \in I$, fixed, but arbitrary, $M_i^g \equiv M_i^f$.*

Proof The assumption amounts to the following equality:

$$\bigoplus_{i \in I} L_{\cup T}[f_i] = L_c[f] = \bigoplus_{j \in J} L_{\cup T}[g_j].$$

The conclusion follows from (Lemma) 6.4.44 provided one knows that the measures associated with the decompositions are maximal. But that follows from (Lemma) 6.4.40. □

It is now possible to state the two forms of the (proper, canonical) CHR, one due to Cramér, and the other, to Hida. The proofs have already been presented above, piecemeal.

Proposition 6.4.46 (Cramér's Representation) *Let CHA obtain for*

$$f : T \longrightarrow H.$$

There exists then $\{k_i, i \in I\} \subseteq H$ *such that, with* $f_i(t) = P_t[k_i]$,

1. *for* $t \in T$, *fixed, but arbitrary,* $L_t[f] = \bigoplus_{i \in I} L_t[f_i]$;
2. *for* $i \in I$, *fixed, but arbitrary,* $M_i^f \gg M_{i+1}^f$;
3. *when there exits* $\{\tilde{k}_i, \imath \in \tilde{I}\} \subseteq H$ *such that, with* $\tilde{f}_i(t) = P_t[\tilde{k}_i]$,
   (a) *for* $t \in T$, *fixed, but arbitrary,* $L_t[f] = \bigoplus_{i \in \tilde{I}} L_t[\tilde{f}_i]$;
   (b) *for* $i \in \tilde{I}$, *fixed, but arbitrary,* $M_i^{\tilde{f}} \gg M_{i+1}^{\tilde{f}}$,

   then

   (i) $\tilde{I} = I$;
   (ii) *for* $i \in I$, *fixed, but arbitrary,* $M_i^{\tilde{f}} \equiv M_i^f$;
4. *for* $h \in L_t[f]$, *fixed, but arbitrary,*

$$h = \sum_{i \in I} \int f_i^h \, dm_i^f, f_i^h \in L_2(T, \mathcal{T}, M_i^f), I_{T_t} f_i^h = f_i^h.$$

$\tilde{I}$ *is the multiplicity of* f.

Proposition 6.4.47 (Hida's Representation) *Let CHA obtain for*

$$f : T \longrightarrow H.$$

Then, for $t \in T$, *fixed, but arbitrary, in* H,

$$f(t) = \sum_{i \in I} \int \phi_i(t) \, dm_i^f + \sum_{t_i < t} \sum_{j=1}^{n[t_i]} \psi_i^j(t) \, h_i^j$$

where,

1. *for the sum of integrals,*
   (i) *for* $i \in I$, *fixed, but arbitrary, the additive vector set functions* m_i^f *are obtained from functions* $f_i : T \longrightarrow H$ *that are continuous, orthogonal, and have orthogonal increments;*
   (ii) *the numerical additive set functions* M_i^f *associated with* m_i^f *and* f_i *are ordered by absolute continuity: for* $i \in I$, *fixed, but arbitrary,*

$$M_i^f \gg M_{i+1}^f;$$

   (iii) *the functions*

$$\phi_i : T \longrightarrow L_2(T, \mathcal{T}, M_i^f), i \in I,$$

   have the following features: $I_{T_t} \phi_i(t) = \phi_i(t)$, $\sum_{i \in I} \int \dot{\phi}_i^2 \, dM_i^f < \infty$;

2. *for the sum of sums,*

   (i) *the points* t_i *are those elements* $t \in T$ *at which* $P_t^+ - P_t \neq O_H$ *(null projection);*
   (ii) $n[t_i]$ *is the dimension of the range of* $P_t^+ - P_t$ [(Remark) 6.1.4]*: it is either a strictly positive integer or infinity;*
   (iii) *the* h_i^j*'s are orthonormal elements of* H*; the family*

$$\{h_i^j, j \in [1 : n[t_i]]\}$$

   forms a basis for the range of $P_t^+ - P_t$*;*
   (iv) $\psi_i^j(t) = \langle f(t), h_i^j \rangle_H$*;*

3. *the sum of integrals and the sum of sums are orthogonal in* H*, and, for* $t \in T$*, fixed, but arbitrary,*

$$L_t[f] = \sum_{i \in I} L_t[f_i] \bigoplus \overline{V[\{h_i^j, j \in [1 : n[t_i]], t_i < t\}]};$$

4. *suppose that, for* $t \in T$*, fixed, but arbitrary, also*

$$f(t) = \sum_{j \in J} \int \tilde{\phi}_j(t)\, d\tilde{m}_j^f + \sum_{t_i < t} \sum_{k=1}^{n[t_i]} \tilde{\psi}_i^k(t)\, \tilde{h}_i^k,$$

   an expression to be interpreted as the original one above: then

   (i) $\tilde{I} = \tilde{J}$*;*
   (ii) *for* $i \in I$*, fixed, but arbitrary,* $\tilde{M}_i^f \equiv M_i^f$ *(mutual absolute continuity), where the relation of* $\tilde{M}_i^f$ *to* $\tilde{m}_i^f$ *and* $\tilde{f}_i$ *is that of* M_i^f *to* m_i^f *and* f_i*; furthermore, almost surely with respect to* M_i^f*,*

$$\tilde{\phi}_i = \phi_i \sqrt{\frac{dM_i^f}{d\tilde{M}_i^f}};$$

   (iii) *for* $i \in I$*, fixed, but arbitrary, there is a unitary map* U_i *on the range of* $P_{t_i}^+ - P_{t_i}$ *for which, for* $j \in [1 : n[t_i]]$*, fixed, but arbitrary,*

$$\tilde{h}_i^j = U_i[h_i^j].$$

The multiplicity of f *is the value* $\sup\{\tilde{I}, \dim(L_d[f])\}$.

Remark 6.4.48 Let μ and ν be equivalent (mutually absolutely continuous) measures on $(X, \mathcal{X})$, and define $U : L_2(X, \mathcal{X}, \mu) \longrightarrow L_2(X, \mathcal{X}, \nu)$, using the following assignment:

$$f \mapsto \left[\dot{f}\sqrt{\frac{d\mu}{d\nu}}\right]_{L_2(X,\mathcal{X},\nu)}.$$

As

$$\|U[f]\|^2_{L_2(X,\mathcal{X},\nu)} = \|f\|^2_{L_2(X,\mathcal{X},\mu)},$$

U is an isometry. But, if g is orthogonal to the range of U, one has, for fixed, but arbitrary $f \in L_2(X, \mathcal{X}, \mu)$, that

$$0 = \langle g, U[f]\rangle^2_{L_2(X,\mathcal{X},\nu)} = \int_X \dot{g}(x)\,\dot{f}(x)\,\sqrt{\frac{d\mu}{d\nu}}(x)\,\nu(dx).$$

Consequently, since, because of equivalence, the Radon-Nikodým derivative is almost surely different from 0, g is the zero element, and U is unitary. The CHR's are thus unique modulo unitary maps.

It remains to check that the multiplicities of the Cramér and the Hida representations are the same.

Lemma 6.4.49 *Definitions, and notation, are as in* (Proposition) 6.4.47. *Let*

$$\Theta = \{(i,j) \in \mathbb{N}\times\mathbb{N} : t_i \in T_d \text{ and } j \in [1 : n[t_i]]\},$$

and choose a set $\{\alpha_{1,\theta}, \theta \in \Theta\} \subseteq \mathbb{R}\setminus\{0\}$ *with* $\sum_{\theta\in\Theta}\alpha^2_{1,\theta} = 1$. *Define then*

$$k_1 = \sum_{\theta\in\Theta,\theta=(i,j)} \alpha_{1,\theta}\,h_i^j.$$

Continue producing orthonormal k_n*'s in the following fashion, given, for* $n \in \mathbb{N}$, $V_n[k] = L[k_1,\ldots,k_n]$, *the vector subspace with basis* $\{k_1,\ldots,k_n\}$:

(a) *when, given* $\theta = (i,j)$, $h_i^j \in V_n[k]$, $\alpha_{n+1,\theta} = 0$;
(b) *when, given* $\theta = (i,j)$, $h_i^j \notin V_n[k]$, $\alpha_{n+1,\theta} \in \mathbb{R}\setminus\{0\}$;
(c) $\sum_{\theta\in\Theta}\alpha^2_{n+1,\theta} = 1$;
(d) $k_{n+1} = \sum_{\theta\in\Theta,\theta=(i,j)}\alpha_{n+1,\theta}\,h_i^j$, *and* $k_{n+1} \perp V_n[k]$.

The procedure just described yields a basis for $L_d[f]$.

Proof It suffices to check that $\{h_i^j, (i,j) = \theta \in \Theta\} \subseteq \bigvee_n V_n[k]$. To that end, let

$$\Theta_0 = \left\{\theta \in \Theta, \theta = (i,j) : h_i^j \notin \bigvee_n V_n[k]\right\}.$$

Then, because of the way the sequence $\{k_n, n \in N_0 \subseteq \mathbb{N}\}$ was built, there is, in that sequence, an element k_{n_0} whose α-coefficients are different from zero when they weigh an element h_i^j such that $(i,j) = \theta \in \Theta_0$. Let Q be the projection onto $\bigvee_n V_n[k]$. Since

$$k_{n_0} = \sum_{\theta \in \Theta_0, \theta=(i,j)} \alpha_{n_0,(i,j)} h_i^j = \sum_{\theta \in \Theta_0, \theta=(i,j)} \alpha_{n_0,(i,j)} Q\left[h_i^j\right],$$

one has that

$$0_H = \sum_{\theta \in \Theta_0, \theta=(i,j)} \alpha_{n_0,(i,j)} \left\{h_i^j - Q\left[h_i^j\right]\right\}.$$

But

$$\left\langle h_i^j - Q\left[h_i^j\right], h_k^l - Q\left[h_k^l\right]\right\rangle_H = -\left\langle Q\left[h_i^j\right], Q\left[h_k^l\right]\right\rangle_H,$$

so that

$$\begin{aligned} 0 &= \left\| \sum_{\theta \in \Theta_0, \theta=(i,j)} \alpha_{n_0,(i,j)} \left\{h_i^j - Q\left[h_i^j\right]\right\} \right\|_H^2 \\ &= - \sum_{\theta \in \Theta_0, \theta=(i,j)} \sum_{\theta \in \Theta_0, \theta=(k,l)} \alpha_{n_0,(i,j)} \alpha_{n_0,(k,l)} \left\langle Q\left[h_i^j\right], Q\left[h_k^l\right]\right\rangle_H \\ &= - \left\| Q \left\{ \sum_{\theta \in \Theta_0, \theta=(i,j)} \alpha_{n_0,(i,j)} h_i^j \right\} \right\|_H^2 \\ &= - \left\| \sum_{\theta \in \Theta_0, \theta=(i,j)} \alpha_{n_0,(i,j)} h_i^j \right\|_H^2 \\ &= - \sum_{\theta \in \Theta_0, \theta=(i,j)} \alpha_{n_0,(i,j)}^2. \end{aligned}$$

Thus $k_{n_0} = 0$, which is impossible, as it belongs to an orthonormal set. □

Proposition 6.4.50 *The Cramér and Hida proper canonical representations have the same multiplicity.*

Proof Notation and definitions are as in (Propositions) 6.4.46, 6.4.47, and 6.4.49. Let

$$L_{\cup T}[f] = L_c[f] \oplus L_d[f]$$

and let

- $\{k_i^c, i \in I_c\} \subseteq H$ be a family such that, when $f_i^c(t) = P_t\left[k_i^c\right]$,

  $$L_c[f] = \bigoplus_{i \in I_c} L_{\cup T}\left[f_i^c\right]$$

  is a proper canonical representation;
- $\{k_i^d, i \in I_d\} \subseteq H$ be a family such that

  $$L_d[f] = \bigvee_{i \in I_d} V_i\left[k_i^d\right],$$

  as in (Lemma) 6.4.49.

  Set

$$k_i = \begin{cases} k_i^c + k_i^d \text{ when } i \leq \tilde{I}_c \wedge \tilde{I}_d \\ k_i^c \qquad\quad \text{when } \tilde{I}_d < i \leq \tilde{I}_c \\ k_i^d \qquad\quad \text{when } \tilde{I}_c < i \leq \tilde{I}_d \end{cases}.$$

  Since, for $(i,j) = \theta \in \Theta$, the measure associated with

  $$P_t\left[h_i^j\right],$$

has a one point support made of t_i, whose measure is one, the measures associated with the functions $f_i^d(t) = P_t\left[k_i^d\right], i \in I_d$, have decreasing supports contained in T_d. Consequently

$$L_{\cup T}[f] = \left\{\bigoplus_{i \in I_c} L_{\cup T}\left[f_i^c\right]\right\} \oplus \left\{\bigoplus_{i \in I_d} L_{\cup T}\left[f_i^d\right]\right\}$$

is a proper canonical representation in the sense of Cramér, and thus both multiplicities agree. □

Remark 6.4.51 "Current language" makes no distinction between the Cramér's and the Hida's representations as, most often, the processes concerned are at least second order continuous, and then the two types coincide. The common term is thus "Cramér-Hida" representation.

Remark 6.4.52 The CHR may be obtained "without work" using the Hellinger-Hahn theorem [249, Chapter 7], as done in [132] (for a complete exposition "under the same roof," see [143]). It is obviously a shorter way to the CHR, though, perhaps, a less informative one. It turns out [15] that within cosmetic rearrangements, the results obtained here for the CHR amount to a proof of the Hellinger-Hahn theorem. It all boils down to the fact the Cramér-Hida representation furnishes a left-continuous resolution of the identity.

Chapter 7
Cramér-Hida Representations via Direct Integrals

Direct integrals generalize direct sums. As the CHR is a direct sum decomposition (preserving the time structure), it is perhaps unsurprising that direct integrals have a part to play in the study of the CHR.

Direct integrals allow one to give a representation of the subspace generated linearly by the range of a function $f : T \longrightarrow H$ akin to that available for processes with orthogonal increments, for which the elements of $L_{\cup T}[f]$ are represented as integrals with respect to a H-CAOSM. In the general case the integrals are the so-called direct integrals, developed mostly for the analysis of operator classes [76, 203]. Such integrals preserve the time dimension inherent to f, and one has then at disposal a reasonably flexible computing instrument. One consequence of the representation of $L_{\cup T}[f]$ by a direct integral is the Cramér representation of f, with an emphasis on its global properties, rather than its local ones.

7.1 Direct Integrals

Direct integrals are Hilbert subspaces of uncountable products of Hilbert spaces which are not, "naturally," Hilbert spaces [269, p. 185]. One starts with measurable fields of Hilbert spaces, a particular class of submanifolds, and it is within those that one finds the direct integrals, using a measure on the space of indices of the uncountable product.

A.F. Gualtierotti, *Detection of Random Signals in Dependent Gaussian Noise*,
DOI 10.1007/978-3-319-22315-5_7

7.1.1 Measurable Fields of Hilbert Spaces

Definition 7.1.1 Let $(S, \mathcal{S})$ be a measurable space, and for each $s \in S$, let H_s be a real Hilbert space. One shall assume that $H_s \neq \{0_{H_s}\}$, $s \in S$. Let

$$H_S = \prod_{s \in S} H_s.$$

A measurable field of Hilbert spaces, defined on $(S, \mathcal{S})$, is a linear manifold $\mathcal{H}_S$ of H_S which has the following properties:

1. for fixed, but arbitrary $\{h_1, h_2\} \subseteq \mathcal{H}_S$, the map

$$s \mapsto \langle h_1(s), h_2(s) \rangle_{H_s}$$

   is adapted to $\mathcal{S}$;
2. whenever $f \in H_S$ is such that, for all $h \in \mathcal{H}_S$,

$$s \mapsto \langle f(s), h(s) \rangle_{H_s}$$

   is adapted to $\mathcal{S}$, then $f \in \mathcal{H}_S$;
3. there exists a family $\{h_n, n \in \mathbb{N}\} \subseteq \mathcal{H}_S$ such that, for fixed, but arbitrary $s \in S$, $\{h_n(s), n \in \mathbb{N}\}$ is total in H_s.

An element of $\mathcal{H}_S$ is called a measurable field of vectors.

Remark 7.1.2 Item 1 of (Definition) 7.1.1 has the consequence that, for $h \in \mathcal{H}_S$, fixed, but arbitrary, the map $s \mapsto \|h(s)\|_{H_s}$ is adapted.

Remark 7.1.3 Item 3 of (Definition) 7.1.1 has, as consequence, that, for fixed, but arbitrary $s \in S$, H_s is separable. Consequently its Borel sets are generated by the cylinder sets: $\mathcal{B}(H_s) = \mathcal{C}(H_s)$, $s \in S$.

Example 7.1.4 Let H be a real, separable, Hilbert space, and $H_s = H$, $s \in S$. Then $H_S = H^S$, and the linear manifold $\mathcal{M}$ of maps $s \mapsto f(s) \in H$ that are adapted to $\mathcal{S}$ and $\mathcal{B}(H)$ is a measurable field of Hilbert spaces, called the constant field associated with H and $(S, \mathcal{S})$.

Indeed, since $s \mapsto f(s)$ is adapted if, and only if, $s \mapsto \langle f(s), h \rangle_H$ is adapted for all $h \in H$ [(Remark) 7.1.3], the constant functions belong to $\mathcal{M}$.

Let $\{f_1, f_2\} \subseteq \mathcal{M}$ be fixed, but arbitrary. Then

$$s \mapsto \langle f_1(s), f_2(s) \rangle_{H_s} = \langle f_1(s), f_2(s) \rangle_H$$

is adapted to $\mathcal{S}$ as it is the composition of the following maps:

$$s \mapsto (f_1(s), f_2(s)) \in H \times H, \text{ and } (h_1, h_2) \mapsto \langle h_1, h_2 \rangle_H.$$

When $g \in H^S$ is such that

$$\left\{s \mapsto \langle g(s), f(s)\rangle_{H_s} = \langle g(s), f(s)\rangle_H, \, f \in \mathcal{M}\right\}$$

is a family of functions adapted to $\mathcal{S}$, then $g \in \mathcal{M}$: it suffices to restrict attention to the functions f that are constant.

Let $\{h_i, \ i \in I\}$ be a complete orthonormal set in H. Item 3 of (Definition) 7.1.1 is given by the family $\{s \mapsto h_i, \ i \in I\}$.

Example 7.1.5 Let $(X, \mathcal{X})$ be a measurable space, and let $\mathcal{X}$ be countably generated. Suppose that, for each $s \in S$, $\mu(s, \cdot)$ is a finite, positive measure on $(X, \mathcal{X})$, and that, for each $X_0 \in \mathcal{X}$, the map $s \mapsto \mu(s, X_0)$ is adapted to $\mathcal{S}$. Then, given f adapted to $\mathcal{S} \otimes \mathcal{X}$ and $\mathcal{B}(\mathbb{R})$,

$$s \mapsto \int_X f(s, x)\, \mu(s, dx)$$

is adapted to $\mathcal{S}$ as soon as the integral makes sense.

Let

$$H_S = \prod_{s \in S} L_2(X, \mathcal{X}, \mu(s, \cdot)),$$

and, in H_S, let $H_{\mathcal{S}}$ be the family of elements h for which there exists $\tilde{h}$, adapted to $\mathcal{S} \otimes \mathcal{X}$ and $\mathcal{B}(\mathbb{R})$, with the property that

$$\left[\tilde{h}(s, \cdot)\right]_{L_2(X, \mathcal{X}, \mu(s, \cdot))} = h(s), \ s \in S.$$

$H_{\mathcal{S}}$ is then a measurable field of Hilbert spaces.

Let indeed $\tilde{h}_1$ correspond to h_1, and $\tilde{h}_2$ to h_2. Then the following map:

$$s \mapsto \langle h_1(s), h_2(s)\rangle_{L_2(X, \mathcal{X}, \mu(s, \cdot))} = \int_X \tilde{h}_1(s, x)\, \tilde{h}_2(s, x)\, \mu(s, dx)$$

is adapted to $\mathcal{S}$.

Let $\mathcal{X}_c \subseteq \mathcal{X}$ be a countable family of sets of positive measure generating $\mathcal{X}$. For fixed, but arbitrary $X_c \in \mathcal{X}_c$, let

$$\tilde{h}_{X_c}(s, x) = \chi_{X_c}(x).$$

Then, by definition of $H_{\mathcal{S}}$, the following relation:

$$h_{X_c}(s) = \left[\tilde{h}_{X_c}(s, \cdot)\right]_{L_2(X, \mathcal{X}, \mu(s, \cdot))}$$

determines an element of $H_{\mathcal{S}}$. The family $\{h_{X_c}, \ X_c \in \mathcal{X}\}$ is the family required by item 3 of (Definition) 7.1.1.

Suppose finally that the following map: $f \in H_S$ is such that

$$s \mapsto \langle f(s), h(s) \rangle_{L_2(X,\mathcal{X},\mu(s,\cdot))}$$

is adapted to $\mathcal{S}$ for $h \in H_{\mathcal{S}}$. For fixed, but arbitrary $n \in \mathbb{N}$, let $\mathcal{X}_n \subseteq \mathcal{X}$ be a finite σ-algebra. Suppose that

$$\mathcal{X} = \sigma\{\mathcal{X}_n, \ n \in \mathbb{N}\}.$$

Then, for fixed, but arbitrary $s \in S$, by the appropriate martingale convergence theorem, almost surely with respect to $\mu(s,\cdot)$,

$$\lim_n E_{\mu(s,\cdot)}\left[\dot{f}(s) \mid \mathcal{X}_n\right] = \dot{f}(s).$$

Let

$$\mathcal{P}_n[X] = \left\{X_i^{(n)}, \ i \in I_n\right\}$$

be the finite partition of X generated by $\mathcal{X}_n$. The explicit formula for the conditional expectation of $\dot{f}(s)$ with respect to $\mathcal{X}_n$ is then [54, p. 215]:

$$\tilde{h}_n(s,x) = \sum_{X_i^{(n)} \in \mathcal{P}_n[X]} \left\langle f(s), h_{X_i^{(n)}}(s) \right\rangle_{L_2(X,\mathcal{X},\mu(s,\cdot))} \frac{\chi_{X_i^{(n)}}(x)}{\mu\left(s, X_i^{(n)}\right)}.$$

$\tilde{h}_n$ is adapted to $\mathcal{S} \otimes \mathcal{X}$ and, for $s \in S$, fixed, but arbitrary,

$$\left[\tilde{h}_n(s,\cdot)\right]_{L_2(X,\mathcal{X},\mu(s,\cdot))} = \left[E_{\mu(s,\cdot)}\left[\dot{f}(s) \mid \mathcal{X}_n\right]\right]_{L_2(X,\mathcal{X},\mu(s,\cdot))}.$$

Then the element

$$\tilde{h} = \limsup \tilde{h}_n$$

is adapted to $\mathcal{S} \otimes \mathcal{X}$ and, for $s \in S$, fixed, but arbitrary,

$$\left[\tilde{h}(s,\cdot)\right]_{L_2(X,\mathcal{X},\mu(s,\cdot))} = \dot{f}(s).$$

Item 2 of (Definition) 7.1.1 thus obtains.

Remark 7.1.6 Example 7.1.5 provides one possible construction for the Hilbert space of estimating functions in the sense of [174, 244]. Using a measure σ on $\mathcal{S}$ yields the Bayesian framework for estimation.

Lemma 7.1.7 *Let (Definition) 7.1.1 obtain. Let $I \subseteq \mathbb{N}$, and $\{h_i,\ i \in I\} \subseteq H_S$ be fixed, but arbitrary. Let $\{\phi_i,\ i \in I\}$ be a fixed, but arbitrary family of functions with values in $\mathbb{R}$ that are adapted to $\mathcal{S}$. When I is infinite, suppose that, for fixed, but arbitrary $s \in S$, the series*

$$\sum_{i \in I} \phi_i(s)\, h_i(s)$$

is convergent in H_s. The following assignment:

$$f(s) = \sum_{i \in I} \phi_i(s)\, h_i(s)$$

produces then an element of H_S.

Proof Let $h \in H_S$ be fixed, but arbitrary. Then

$$s \mapsto \langle f(s), h(s) \rangle_{H_s} = \sum_{i \in I} \phi_i(s) \langle h_i(s), h(s) \rangle_{H_s}$$

is an adapted function [(Definition) 7.1.1, item 1]. Consequently $f \in H_S$ [(Definition) 7.1.1, item 2]. □

Lemma 7.1.8 *Let H be a real Hilbert space, and let $\{h_n,\ n \in \mathbb{N}\} \subseteq H$ be any sequence that contains elements other than 0_H. For $n \in \mathbb{N}$, let $V_n[h]$ be the (closed) subspace generated linearly by $\{h_1, \ldots, h_n\}$. There exists then an orthonormal set, say $\{k_i,\ i \in I \subseteq \mathbb{N}\} \subseteq H$, with I either equal to $\mathbb{N}$, or to an "interval" of the form $[1:n]$, some $n \in \mathbb{N}$, such that, when $V_n[h] \neq \{0_H\}$,*

$$V_n[h] = V_p[k],$$

where $p = \dim V_n[h]$, and $V_p[k]$ is the (closed) subspace generated linearly by

$$\{k_1, \ldots, k_p\}.$$

Proof Below, the minimum of an empty set of integers shall be ∞. Since there is $n \in \mathbb{N}$ such that $\|h_n\|_H \neq 0$, the following number is finite:

$$n_1 = \min\{n \in \mathbb{N} : \|h_n\|_H \neq 0\}.$$

Set $k_1 = \|h_{n_1}\|_H^{-1} h_{n_1}$, and

$$h_n^{(1)} = h_n - \langle h_n, k_1 \rangle_H k_1,\ n \in \mathbb{N}.$$

Then

$$V[h_1, \ldots, h_{n_1}] = V[h_{n_1}] = V[k_1].$$

Let then $n_2 = \infty$ when

$$\left\{h_n^{(1)},\ n \in \mathbb{N} : \left\|h_n^{(1)}\right\|_H \neq 0\right\} = \emptyset,$$

and, otherwise,

$$n_2 = \min\left\{n \in \mathbb{N} : \left\|h_n^{(1)}\right\|_H \neq 0\right\}.$$

If $n_2 = \infty$, the process stops. Otherwise one sets

$$k_2 = \left\|\left\{h_{n_2}^{(1)}\right\}^{-1}\right\|_H h_{n_2}^{(1)},$$

and

$$h_n^{(2)} = h_n^{(1)} - \left\langle h_n^{(1)}, k_2\right\rangle_H k_2,\ n \in \mathbb{N}.$$

When $n_2 = \infty$, $h_n = \langle h_n, k_1\rangle_H k_1$, all n. Consequently

$$V[h_1, \ldots, h_n] = V[k_1],\ n \geq 1.$$

Otherwise

$$V\left[h_1, \ldots, h_p\right] = V[k_1],\ p < n_2, \text{ and } V[h_1, \ldots, h_{n_2}] = V[k_1, k_2].$$

One continues in similar fashion. If there is a $p \in \mathbb{N}$ such that

$$\left\{h_n^{(p)},\ n \in \mathbb{N} : \left\|h_n^{(p)}\right\|_H \neq 0\right\} = \emptyset,$$

the process stops with n_p. Otherwise the process will continue indefinitely. □

Remark 7.1.9 When there exists $p \in \mathbb{N}$ such that, for $i \in [1 : p]$, fixed, but arbitrary, $n_i < \infty$, but $n_p + 1 = \infty$, then

$$\dim \overline{V[\{h_n, n \in \mathbb{N}\}]} = p.$$

Remark 7.1.10 When, for $p \in \mathbb{N}$, fixed, but arbitrary, $n_p < \infty$, then

$$\dim \overline{V[\{h_n, n \in \mathbb{N}\}]} = \infty.$$

Proposition 7.1.11 *Let (Definition)* 7.1.1 *obtain. Then:*

1. *there exists* $\{e_i,\ i \in I \subseteq \mathbb{N}\} \subseteq \mathsf{H}_S$ *such that*
   (i) *when* $\dim\{H_s\} = \infty$, $\{e_i(s),\ i \in I\}$ *is a basis for* H_s,
   (ii) *when* $\dim\{H_s\} < \infty$,
      (a) $\{e_i(s),\ i \in [1 : \dim\{H_s\}]\}$ *is a basis for* H_s,
      (b) *for* $i > \dim\{H_s\}$, $e_i(s) = 0_{H_s}$;
2. *the map* $s \mapsto \dim\{H_s\}$ *is adapted to* $\mathcal{S}$;
3. $f \in \mathsf{H}_S$ *belongs to* H_S *if, and only if,* $s \mapsto \langle f(s), e_i(s)\rangle_{H_s}$ *is adapted for all* $i \in I$.

Proof Let $\{h_n,\ n \in \mathbb{N}\}$ be the family of item 3 in (Definition) 7.1.1. As seen in (Remark) 7.1.2, for $n \in \mathbb{N}$, fixed, but arbitrary, $s \mapsto \|h_n(s)\|_{H_s}$ is adapted. Let

$$n_1(s) = \min\left\{n \in \mathbb{N} : \|h_n(s)\|_{H_s} \neq 0\right\}.$$

$n_1(s) \in \mathbb{N}$ for every $s \in S$, as one has assumed that H_s has at least dimension one. Consequently

$$\begin{aligned} S_{1,n} &= \{s \in S : n_1(s) = n\} \\ &= \left\{s \in S : \|h_i(s)\|_{H_s} = 0, i < n, \|h_n(s)\|_{H_s} \neq 0\right\} \\ &\in \mathcal{S}. \end{aligned}$$

Because of (Lemma) 7.1.7, the following definition makes sense, yields a measurable field of vectors, each of norm one:

$$e_1(s) = \sum_{n \in \mathbb{N}} \chi_{S_{1,n}}(s)\, \|h_n(s)\|_{H_s}^{-1}\, h_n(s).$$

When $s \in S_{1,n}$, $h_n(s)$ is the first element in $\{h_n(s), n \in \mathbb{N}\}$ whose norm is not zero, so that $\|h_n(s)\|_{H_s}^{-1} h_n(s)$ corresponds, in H_s, to k_1 of (Lemma) 7.1.8, and thus $e_1(s)$ yields the equivalent of k_1 of (Lemma) 7.1.8 for arbitrary s. For $n \in \mathbb{N}$, let

$$h_n^{(1)}(s) = h_n(s) - \langle h_n(s), e_1(s)\rangle_{H_s} e_1(s).$$

One thus obtains, again because of (Lemma) 7.1.7, a sequence of measurable fields of vectors. Let

$$\mathcal{N}_1(s) = \left\{n \in \mathbb{N} : \left\|h_n^{(1)}(s)\right\|_{H_s} > 0\right\},$$

$$n_2(s) = \begin{cases} \infty & \text{when } \mathcal{N}_1(s) = \emptyset \\ \min \mathcal{N}_1(s) & \text{when } \mathcal{N}_1(s) \neq \emptyset \end{cases},$$

$$S_{2,n} = \{s \in S : n_2(s) = n\}.$$

Set again

$$e_2(s) = \sum_{n \in \mathbb{N}} \chi_{S_{2,n}}(s) \left\| h_n^{(1)}(s) \right\|_{H_s}^{-1} h_n^{(1)}(s).$$

The procedure yields adapted elements and may be continued indefinitely. The orthogonality properties of the $e_n(s)$-sequence follow, as seen, from (Lemma) 7.1.8.

Since $n_p(s)$ is finite when there are at least p orthonormal vectors in H_s,

$$\{s \in S : \dim \{H_s\} \geq p\} = \left\{s \in S : n_p(s) < \infty\right\}.$$

The map $s \mapsto \dim \{H_s\}$ is thus adapted.

To see that item 3 holds, one may proceed as follows. Let $h \in H_S$ be fixed but arbitrary. The map

$$s \mapsto \langle h(s), e_i(s) \rangle_{H_s}$$

is adapted to $\mathcal{S}$ as both h and e_i belong to H_S. If now $f \in H_s$ is fixed, but arbitrary, and that

$$s \mapsto \langle f(s), e_i(s) \rangle_{H_s}$$

is adapted to $\mathcal{S}$ for all $i \in I$, then, for fixed, but arbitrary $h \in H_S$,

$$s \mapsto \langle f(s), h(s) \rangle_{H_s}$$

is adapted to $\mathcal{S}$ as

$$\langle f(s), h(s) \rangle_{H_s} = \sum_{i \in I} \langle f(s), e_i(s) \rangle_{H_s} \langle e_i(s), h(s) \rangle_{H_s}.$$

Consequently, $f \in H_S$. □

Corollary 7.1.12 *Let (Definition)* 7.1.1 *obtain. Let* $\dim \{H_s\} = d$, *all* $s \in S$, *and let* H *be a real Hilbert space of dimension* d. *There exist then unitary maps*

$$U_s : H_s \longrightarrow H, \; s \in S,$$

such that $f \in H_S$ *if, and only if,* $s \mapsto U_s[f(s)]$ *is adapted to* $\mathcal{S}$ *and* $\mathcal{B}(H)$.

Proof Let $\{b_i,\ i \in I\}$ be a basis for H, $|I| = d$, and $\{e_i,\ i \in I\}$ be the family of bases of (Proposition) 7.1.11. The assignment $e_i(s) \mapsto b_i$ determines the isometry U_s. But then

$$\langle U_s[f(s)], b_i\rangle_H = \left\langle f(s), U_s^\star[b_i]\right\rangle_{H_s} = \langle f(s), e_i(s)\rangle_{H_s}.$$

Consequently the map f belongs to H_S if, and only if, $s \mapsto U_s[f(s)]$ is adapted to $\mathcal{B}(H)$ [(Remark) 7.1.3]. □

Remark 7.1.13 A similar result, with the same proof, but more complex notation, holds when the dimension is not constant: but then one has only an isometry between spaces of H_S type.

Lemma 7.1.14 *Let (Definition)* 7.1.1 *obtain. Let* $\{h_\lambda,\ \lambda \in \Lambda\} \subseteq H_S$ *have the following properties:*

(a) for fixed, but arbitrary $\{\lambda_1, \lambda_2\} \subseteq \Lambda$, *the map* $s \mapsto \langle h_{\lambda_1}(s), h_{\lambda_2}(s)\rangle_{H_s}$ *is adapted to* $\mathcal{S}$;

(b) there exists $\{h_{\lambda_n},\ n \in \mathbb{N}\} \subseteq \{h_\lambda,\ \lambda \in \Lambda\}$ *such that, for fixed, but arbitrary* $s \in S$, *the set* $\{h_{\lambda_n}(s),\ n \in \mathbb{N}\}$ *is total in* H_s.

Let $\{k_i,\ i \in I\}$ *be the sequence obtained from* $\{h_{\lambda_n},\ n \in \mathbb{N}\}$, *using (Lemma)* 7.1.8 *and the proof of* (Proposition) 7.1.11.

Then, for a fixed, but arbitrary $f \in H_S$, *one has that*

$$\mathcal{F} = \left\{s \mapsto \langle f(s), h_\lambda(s)\rangle_{H_s},\ \lambda \in \Lambda\right\}$$

is a family of functions adapted to $\mathcal{S}$ *if, and only if,*

$$\mathcal{G} = \left\{s \mapsto \langle f(s), k_i(s)\rangle_{H_s},\ i \in I\right\}$$

is a family of functions adapted to $\mathcal{S}$.

Proof Suppose that $\mathcal{F}$ contains adapted functions. The construction of the e_i's in (Proposition) 7.1.11 only requires items 1 and 3 of (Definition) 7.1.1, and those are items (a) and (b) of the present lemma. So the k_i's are well defined. They are obtained as series of elements h_{λ_n} multiplied by adapted, scalar functions. The maps $s \mapsto \langle f(s), k_i(s)\rangle_{H_s}$ are thus adapted.

Suppose conversely that $\mathcal{G}$ contains adapted functions. One has, for $\lambda \in \Lambda$, fixed, but arbitrary, that

$$\langle f(s), h_\lambda(s)\rangle_{H_s} = \sum_{i \in I} \langle f(s), k_i(s)\rangle_{H_s} \langle k_i(s), h_\lambda(s)\rangle_{H_s}.$$

The first terms in the products of that latter sum are adapted because they belong to $\mathcal{G}$. The map $s \mapsto \langle k_i(s), h_\lambda(s)\rangle_{H_s}$ is adapted as countable linear combinations of terms found in Assumption (a). The functions of $\mathcal{F}$ are thus adapted. □

Proposition 7.1.15 *Let the assumptions of (Lemma)* 7.1.14 *obtain, set*

$$H_S^\Lambda = \{h_\lambda, \lambda \in \Lambda\},$$

and let

$$\mathbb{H}_S^\Lambda = \left\{f \in \mathbb{H}_S : s \mapsto \langle f(s), h_\lambda(s)\rangle_{H_s} \text{ is adapted to } \mathcal{S},\ \lambda \in \Lambda\right\}.$$

Then:

1. $\mathbb{H}_S^\Lambda$ *is a measurable field of Hilbert spaces;*
2. $H_S^\Lambda \subseteq \mathbb{H}_S^\Lambda$;
3. *if* $K_S \subseteq \mathbb{H}_S$ *is any measurable field of Hilbert spaces that contains* H_S^Λ, *then* $K_S = \mathbb{H}_S^\Lambda$.

$\mathbb{H}_S^\Lambda$ *is called the (unique) measurable field of Hilbert spaces generated by the family* H_S^Λ.

Proof Let $\{k_i,\ i \in I\}$ be the sequence of (Lemma) 7.1.14.

$\mathbb{H}_S^\Lambda$ is a measurable field of Hilbert spaces for the following reasons:

- it is a vector space;
- let $\{f_1, f_2\} \subseteq \mathbb{H}_S^\Lambda$ be fixed, but arbitrary: then

$$s \mapsto \langle f_1(s), f_2(s)\rangle_{H_s} = \sum_{i\in I} \langle f_1(s), k_i(s)\rangle_{H_s} \langle f_2(s), k_i(s)\rangle_{H_s}$$

  is, because of (Lemma) 7.1.14, an at most countable sum of products of adapted terms, and is thus adapted;
- when $g \in \mathbb{H}_S$ is such that $s \mapsto \langle g(s), f(s)\rangle_{H_s}$ is adapted for all $f \in \mathbb{H}_S^\Lambda$, the same is true when choosing h_λ for f, and thus $g \in \mathbb{H}_S^\Lambda$ by definition;
- the existence of a family that produces total subsets in each H_s has been assumed.

Furthermore $\mathbb{H}_S^\Lambda$ contains $\{h_\lambda,\ \lambda \in \Lambda\}$ by its very definition, and the assumptions.

To see that $\mathbb{H}_S^\Lambda$ is unique, suppose that K_S is a measurable field of Hilbert spaces that contains the family $\{h_\lambda,\ \lambda \in \Lambda\}$. It then necessarily contains the k_i's, because of the assumption that those (assumptions) of (Lemma) 7.1.14 obtain.

Since K_S is a measurable field of Hilbert spaces that contains H_S^Λ, when $k \in K_S$ is fixed, but arbitrary, $s \mapsto \langle k(s), h_\lambda(s)\rangle_{H_s}$ is adapted to $\mathcal{S}$ for all $\lambda \in \Lambda$, that is $k \in \mathbb{H}_S^\Lambda$. Thus $K_S \subseteq \mathbb{H}_S^\Lambda$.

Suppose now that $f \in \mathbb{H}_S^\Lambda$ is fixed, but arbitrary. It follows that the maps $s \mapsto \langle f(s), k_i(s)\rangle_{H_s}$ are adapted to $\mathcal{S}$. But, for fixed, but arbitrary $k \in K_S$,

$$\langle f(s), k(s)\rangle_{H_s} = \sum_{i\in I} \langle f(s), k_i(s)\rangle_{H_s} \langle k_i(s), k(s)\rangle_{H_s}.$$

The first terms in the products of latter sum are adapted as just seen, and the second ones are because the k_i's belong to K_S. Consequently $f \in K_S$, and $H_S^{\Lambda} \subseteq H_S$. □

7.1.2 The Direct Integral: Existence

Direct integrals are subsets of $\prod_{s \in S} H_s$ which are given a Hilbert space structure. The term "integral" is possibly due to the fact that integrals are sums, and that direct integrals, as already mentioned, generalize the notion of direct sum. Furthermore "genuine" integrals are used to define direct integrals, and when, for fixed, but arbitrary $s \in S$, $H_s = H$, a Hilbert space, the direct integral can be identified with an actual integral (of functions with S as domain, and H as range).

Proposition 7.1.16 *Let (Definition) 7.1.1 obtain. Assume that σ is a σ-finite measure on $\mathcal{S}$. Let H_S^I be the subset of H_S of those elements h for which the map $s \mapsto \|h(s)\|_{H_s}^2$ is integrable for σ, and let h_1 and h_2 belong to H_S^I. When*

$$\int_S \|h_1(s) - h_2(s)\|_{H_s}^2 \, \sigma(ds) = 0,$$

h_1 and h_2 shall be in the same equivalence class. H_S^σ then denotes the resulting partition of H_S^I into equivalence classes. The equivalence class of $h \in H_S^I$ shall be denoted $[h]$ and, when $h \in H_S^\sigma$, $\dot{h}$ is a function in that class. One has that:

1. *H_S^σ is a real Hilbert space with inner product*

$$\langle h_1, h_2 \rangle_{H_S^\sigma} = \int_S \langle \dot{h}_1(s), \dot{h}_2(s) \rangle_{H_s} \sigma(ds), \ \{h_1, h_2\} \subseteq H_S^\sigma.$$

2. *Let $\{e_i, i \in I\}$ be the family of equivalence classes in H_S^σ of a sequence of elements $\dot{e}_i \in H_S$ with the properties listed in* (Proposition) 7.1.11. *For fixed, but arbitrary $i \in I$, let*

$$S_i = \{s \in S : \dim\{H_s\} \geq i\} = \{s \in S : \dot{e}_i(s) \neq 0_{H_s}\},$$

and $$\sigma_i(S_0) = \sigma(S_0 \cap S_i), \ S_0 \in \mathcal{S}.$$

Let $U_i : L_2(S, \mathcal{S}, \sigma_i) \longrightarrow H_S^\sigma$ be defined using the following assignment:

$$f^{(i)} \mapsto \left[\dot{f}^{(i)} \dot{e}_i\right] = f^{(i)} e_i,$$

where $f^{(i)}$ is a fixed, but arbitrary element in $L_2(S, \mathcal{S}, \sigma_i)$. Then

(i) *for fixed, but arbitrary $i \in I$, U_i is a partial isometry, whose initial set is $L_2(S, \mathcal{S}, \sigma_i)$, and final set, $\mathcal{R}_{U_i} \subseteq H_S^\sigma$;*

(ii) $H_S^\sigma = \bigoplus_{i \in I} U_i [L_2(S, \mathcal{S}, \sigma_i)]$.

Proof One must notice that, when $s \in S_i^c$, $\dot{e}_i(s) = 0_{H_s}$, and consequently that

$$\dot{f}(s)\dot{e}_i(s) = \chi_{S_i}(s)\dot{f}(s)\dot{e}_i(s).$$

The assignment $f^{(i)} \mapsto \left[\dot{f}^{(i)}\dot{e}_i\right]$ makes sense as (Lemma 7.1.7) $\dot{f}^{(i)}\dot{e}_i \in H_S$, and

$$\begin{aligned}\left\|U_i\left[f^{(i)}\right]\right\|_{H_S^\sigma}^2 &= \int_S \left\|\dot{f}^{(i)}(s)\dot{e}_i(s)\right\|_{H_s}^2 \sigma(ds)\\ &= \int_S \left\{\dot{f}^{(i)}\right\}^2(s)\left\|\dot{e}_i(s)\right\|_{H_s}^2 \chi_{S_i}(s)\,\sigma(ds)\\ &= \int_S \left\{\dot{f}^{(i)}\right\}^2(s)\,\sigma_i(ds)\\ &= \left\|f^{(i)}\right\|_{L_2(S,\mathcal{S},\sigma_i)}^2.\end{aligned}$$

That latter equality also proves that U_i is a partial isometry. Furthermore, as presently seen, when $\{i_1, i_2\} \subseteq I$ are such that $i_1 \neq i_2$, one has, for elements $f^{(i_1)} \in L(S, \mathcal{S}, \sigma_{i_1})$ and $f^{(i_2)} \in L(S, \mathcal{S}, \sigma_{i_2})$, fixed, but arbitrary, that

$$\begin{aligned}\left\langle U_{i_1}\left[f^{(i_1)}\right], U_{i_2}\left[f^{(i_2)}\right]\right\rangle_{H_S^\sigma} &= \int_S \dot{f}^{(i_1)}(s)\dot{f}^{(i_2)}(s)\left\langle \dot{e}_{i_1}(s), \dot{e}_{i_2}(s)\right\rangle_{H_s}\sigma(ds)\\ &= 0.\end{aligned}$$

Indeed $\langle \dot{e}_{i_1}(s), \dot{e}_{i_2}(s)\rangle_{H_s} = 0$ as, when $s \in S_{i_1} \cap S_{i_2}$, $\dot{e}_{i_1}(s) \perp \dot{e}_{i_2}(s)$, and, when $s \in S_{i_1}^c \cup S_{i_2}^c$, at least one of $\{\dot{e}_{i_1}(s), \dot{e}_{i_2}(s)\}$ is the zero element. Consequently $\mathcal{R}[U_{i_1}] \perp \mathcal{R}[U_{i_2}]$, and

$$\bigoplus_{i\in I} U_i\left[L_2(S, \mathcal{S}, \sigma_i)\right] \subseteq H_S^\sigma.$$

Let now $h \in H_S^\sigma$ be fixed, but arbitrary. Then, for $s \in S$, fixed, but arbitrary,

$$\dot{h}(s) = \sum_{i\in I}\left\langle \dot{h}(s), \dot{e}_i(s)\right\rangle_{H_s}\dot{e}_i(s).$$

As

$$\int_S \left\langle \dot{h}(s), \dot{e}_i(s)\right\rangle_{H_s}^2 \sigma_i(ds) \leq \int_S \left\|\dot{h}(s)\right\|_{H_s}^2 \sigma(ds) < \infty,$$

the map

$$s \mapsto \left\langle \dot{h}(s), \dot{e}_i(s)\right\rangle_{H_s}$$

belongs to $L_2(S, \mathcal{S}, \sigma_i)$, and, its equivalence class being denoted $f^{(i)}$,

$$\sum_{i\in I} U_i\left[f^{(i)}\right] = \sum_{i\in I} f^{(i)} e_i = \sum_{i\in I}\left[\left\langle \dot{h}(\cdot), \dot{e}_i(\cdot)\right\rangle_{H_s}\right] e_i = \left[\dot{h}\right] = h.$$

Thus

$$\bigoplus_{i\in I} U_i\left[L_2(S, \mathcal{S}, \sigma_i)\right] = H_S^\sigma.$$

But then H_S^σ is a Hilbert space. □

Remark 7.1.17 H_S^σ is called a direct, or Hilbertian, integral of the measurable field of Hilbert spaces H_S. One often sees the following notation for H_S^σ :

$$H_S^\sigma = \int_S^\oplus H_s \sigma(ds).$$

Remark 7.1.18 Since, by definition, $S_{i+1} \subseteq S_i$, one has that

$$\sigma_{i+1}(S_0) = \sigma(S_0 \cap S_{i+1}) \leq \sigma(S_0 \cap S_i) = \sigma_i(S_0).$$

Thus $\sigma_{i+1} \ll \sigma_i$.

Remark 7.1.19 Let $S_0 \in \mathcal{S}$ be fixed, but arbitrary. Define $P_{S_0} : H_S^\sigma \longrightarrow H_S^\sigma$ using the following assignment:

$$h \mapsto \left[\chi_{S_0}\dot{h}\right] = I_{S_0} h,\ \dot{h} \in h.$$

P_{S_0} is a projection. Its range shall be denoted $H_S^{\sigma|S_0}$. One has that

$$P_{S_0}^\perp = P_{S_0^c}.$$

In particular

$$\left\langle P_{S_0}[h_1], h_2\right\rangle_{H_S^\sigma} = \int_{S_0} \left\langle \dot{h}_1(s), \dot{h}_2(s)\right\rangle_{H_s} \sigma(ds).$$

Example 7.1.20 The Hilbertian integral of (Example) 7.1.4 is $L_2^H(S, \mathcal{S}, \sigma)$ which, when H has dimension one, is isomorphic to $L_2(S, \mathcal{S}, \sigma)$.

Example 7.1.21 The Hilbertian integral of (Example) 7.1.5 is isomorphic to

$$L_2(S \times X, \mathcal{S} \otimes \mathcal{X}, \tau), \text{ where } \tau(ds, dx) = \mu(s, dx)\sigma(ds).$$

The isomorphism is obtained using the formula

$$\int_{S\times X} f(s,x)\, g(s,x)\, \tau(ds,dx) = \int_S \langle [f(s,\cdot)], [g(s,\cdot)] \rangle_{L_2(X,\mathcal{X},\mu(s,\cdot))}\, \sigma(ds).$$

7.1.3 The Direct Integral: Properties

Definition 7.1.22 Let $H_S^\sigma \subseteq \prod_{s\in S} H_s$ and $K_S^\tau \subseteq \prod_{s\in S} K_s$ be two direct integrals, and, for fixed, but arbitrary $s \in S$, let $T_s : H_s \longrightarrow K_s$ be a linear, and bounded operator. The map $s \mapsto T_s$ is then called a field of linear, and bounded operators. For fixed, but arbitrary $h \in H_S$ and $s \in S$, let

$$k(s) = T_s[h(s)].$$

The field of linear and bounded operators $s \mapsto T_s$ is measurable when k is a measurable field of vectors for all $h \in H_S$.

Proposition 7.1.23 *Let H_S^σ and K_S^τ be two Hilbertian integrals, and suppose that*

$$U : H_S^\sigma \longrightarrow K_S^\tau$$

is a unitary operator, with the property that, for fixed, but arbitrary $S_0 \in \mathcal{S}$,

$$U\left[H_S^{\sigma|S_0}\right] = K_S^{\tau|S_0}.$$

Then $\sigma \equiv \tau$ (mutual absolute continuity), *and there is a measurable field of unitary operators*

$$U_s : H_s \longrightarrow K_s$$

such that, for $[h] \in H_S^\sigma$, fixed, but arbitrary,

$$\overset{\bullet}{\overbrace{U[h]}}(s) = \left\{\frac{d\sigma}{d\tau}(s)\right\}^{-1/2} U_s[h(s)].$$

Proof Let P_{S_0} be the projection of H_S^σ defined in (Remark) 7.1.19, and Q_{S_0} be the analogous projection in K_S^τ. The assumption "reads" as

$$UP_{S_0} = Q_{S_0} U.$$

Consequently, $\dot{e}_i$ being one member of the family appearing in (Proposition) 7.1.11,

$$UP_{S_0}[e_i] = \begin{cases} U\left[I_{S_0}e_i\right] \\ Q_{S_0}[U[e_i]] = I_{S_0}U[e_i] \end{cases},$$

so that, "taking norms" on both right-hand sides of the latter equality, one gets

$$\sigma(S_0) = \int_{S_0} \left\| \overbrace{U[e_i]}^{\cdot}(s) \right\|_{K_s}^2 \tau(ds).$$

Thus $\sigma \ll \tau$. This relation is symmetric in the sense that $P_{S_0}U^\star = U^\star Q_{S_0}$. Consequently $\tau \equiv \sigma$.

Let D denote the Radon-Nikodým derivative of σ with respect to τ. Then

$$\sigma(S_0) = \int_{S_0} \left\| \overbrace{U[e_i]}^{\cdot}(s) \right\|_{K_s}^2 D(s)\,\sigma(ds),$$

and thus, almost surely with respect to σ and τ,

$$\left\| \overbrace{U[e_i]}^{\cdot}(s) \right\|_{K_s}^2 D(s) = 1.$$

Define $U_s : H_s \longrightarrow K_s$ using the following assignment:

$$U_s[\dot{e}_i(s)] = D^{1/2}(s)\,\overbrace{U[e_i]}^{\cdot}(s).$$

As, for almost every $s \in S$,

$$\|U_s[\dot{e}_i(s)]\|_{K_s}^2 = D(s) \left\| \overbrace{U[e_i]}^{\cdot}(s) \right\|_{K_s}^2 = 1,$$

one may assume that U_s is an isometry for $s \in S$. Then

$$\left\{D^{-1/2}(s)\,U_s[\dot{e}_i(s)],\ s \in S\right\}$$

is a measurable field of vectors, the image by U of e_i. □

Lemma 7.1.24 ((Lemma) 2.1.11) *Suppose* $\{h_n,\ n \in \mathbb{N}\} \subseteq H_S^\sigma$ *is convergent to* h. *There is then a subsequence* $\{h_{n_p},\ p \in \mathbb{N}\} \subseteq \{h_n,\ n \in \mathbb{N}\}$ *such that*

$$\left\{\dot{h}_{n_p}(s),\ p \in \mathbb{N}\right\}$$

converges to $\dot{h}(s)$ *, for almost every* $s \in S$*, with respect to* σ.

Proof Since $\{h_n,\ n \in \mathbb{N}\}$ is convergent, it is Cauchy. One can thus find $\{h_{n_p},\ p \in \mathbb{N}\}$ such that

$$\sum_p \left\|h_{n_{p+1}} - h_{n_p}\right\|_{H_S^\sigma} < \infty.$$

Let $\{S_n,\ n \in N\} \subseteq \mathcal{S}$ be an at most countable decomposition of S into measurable sets of finite measure. Then, by Jensen's inequality,

$$\sum_p \int_{S_n} \left\|\dot{h}_{n_{p+1}}(s) - \dot{h}_{n_p}(s)\right\|_{H_s} \sigma(ds) \leq$$

$$\leq \{\sigma(S_n)\}^{1/2} \sum_p \left\{\int_{S_n} \left\|\dot{h}_{n_{p+1}}(s) - \dot{h}_{n_p}(s)\right\|_{H_s}^2 \sigma(ds)\right\}^{1/2}$$

$$\leq \{\sigma(S_n)\}^{1/2} \sum_p \left\|h_{n_{p+1}} - h_{n_p}\right\|_{H_S^\sigma} < \infty.$$

Consequently, almost surely with respect to σ,

$$\sum_p \left\|\dot{h}_{n_{p+1}}(s) - \dot{h}_{n_p}(s)\right\|_{H_s} < \infty,$$

so that, almost surely with respect to σ,

$$\dot{h}(s) = \dot{h}_{n_1}(s) + \sum_p \left\{\dot{h}_{n_{p+1}}(s) - \dot{h}_{n_p}(s)\right\}.$$

□

Proposition 7.1.25 *Let* $H = \{h_n, n \in \mathbb{N}\} \subseteq H_S$ *be the distinguished family of (Definition)* 7.1.1, *and let* $\{k_n,\ n \in \mathbb{N}\}$ *be a sequence of measurable fields of vectors. Let* K *be the set, in* H_S^σ, *formed by the classes of products of the following form: for* $n \in \mathbb{N}$, *and* $f : S \longrightarrow \mathbb{R}$, *adapted,*

$$s \mapsto \{f(s)\, k_n(s)\}.$$

When the classes of elements in H *belong to* H_S^σ, *and* K *is total in* H_S^σ, *almost surely with respect to* σ, $\{k_n(s),\ n \in \mathbb{N}\}$ *is total in* H_s.

Proof As, for fixed, but arbitrary $n \in \mathbb{N}$, $[h_n] \in H_S^{\sigma}$, there exists by assumption

$$\left\{k_p^{(n)}, \ p \in \mathbb{N}\right\} \subseteq V[K]$$

such that, in H_T^{σ}, $[h_n] = \lim_p k_p^{(n)}$. There is then a subsequence [(Lemma) 7.1.24], say

$$\left\{k_{p_q}^{(n)}, \ q \in \mathbb{N}\right\} \subseteq \left\{k_p^{(n)}, \ p \in \mathbb{N}\right\},$$

such that, almost surely with respect to σ, $h_n(s) = \lim_q \dot{k}_{p_q}^{(n)}(s)$. Now $\dot{k}_{p_q}^{(n)}(s)$ is a finite linear combination of elements in $\{k_n(s), \ n \in \mathbb{N}\}$. Thus, almost surely,

$$h_n(s) \in \overline{V[\{k_n(s), \ n \in \mathbb{N}\}]}.$$

Since there are only at most countable elements, almost surely,

$$\{h_n(s), \ n \in \mathbb{N}\} \subseteq \overline{V[\{k_n(s), \ n \in \mathbb{N}\}]}.$$

□

7.2 Representations of the Linear Closure of the Range of a Function with Values in a Hilbert Space

In this section, (Definition) 7.1.1 shall obtain. The first result is a representation of $L_{UT}[f]$ in the form of a direct integral, and the second, in the form of a direct sum of spaces, each of which is isomorphic to a L_2 space. Those representations preserve their structure in time. Notation, and definitions, are as in Sect. 7.1.

Proposition 7.2.1 *Let $f : T \longrightarrow H$ be a map for which CHA obtain. There is a direct integral H_T^{τ} with the property that $L_{UT}[f]$ and H_T^{τ} are unitarily isomorphic, and, for that same isomorphism, for $t \in T$, fixed, but arbitrary,*

$$H_T^{\tau|T_t} \ \text{and} \ L_t[f]$$

are unitarily isomorphic.

Proof For fixed, but arbitrary $h \in L_{UT}[f]$, let $f_h(t) = P_t[h]$. It is a function with orthogonal increments. Let M_h^f be the associated measure:

$$M_h^f([t_1, t_2[) = \left\|m_h^f([t_1, t_2[)\right\|_H^2 = \|f_h(t_2)\|_H^2 - \|f_h(t_1)\|_H^2.$$

Since, for fixed, but arbitrary $\{\alpha_1, \alpha_2\} \subseteq \mathbb{R}$ and $\{h_1, h_2\} \subseteq L_{\cup T}[f]$,

$$M^f_{\alpha_1 h_1 + \alpha_2 h_2}([t_1, t_2[) = \alpha_1^2 M^f_{h_1}([t_1, t_2[) + 2\alpha_1\alpha_2 \left\{ \langle f_{h_1}(t_2), f_{h_2}(t_2)\rangle_H - \langle f_{h_1}(t_1), f_{h_2}(t_1)\rangle_H \right\} + \alpha_2^2 M^f_{h_2}([t_1, t_2[),$$

one has that

$$M^f_{h_1+h_2}([t_1, t_2[) - M^f_{h_1-h_2}([t_1, t_2[) = = 4\left\{ \langle f_{h_1}(t_2), f_{h_2}(t_2)\rangle_H - \langle f_{h_1}(t_1), f_{h_2}(t_1)\rangle_H \right\}.$$

The latter's right-hand side is thus a measure. Its value divided by 4 shall be denoted $M^f_{h_1,h_2}$. One has, in particular, that $M^f_{h_1,h_1} = M^f_{h_1}$. As, for $t \in T$, fixed, but arbitrary, and $T_t = T \cap]-\infty, t[$,

$$\begin{aligned} M^f_{h_1}(T_t) + M^f_{h_2}(T_t) &= \|f_{h_1}(t)\|_H^2 + \|f_{h_2}(t)\|_H^2 \\ &\geq 2\|f_{h_1}(t)\|_H \|f_{h_2}(t)\|_H \\ &\geq 2\left|\langle f_{h_1}(t), f_{h_2}(t)\rangle_H\right| \\ &= 2\left|M^f_{h_1,h_2}(T_t)\right|, \end{aligned}$$

one has that $M^f_{h_1,h_2} \ll M^f_{h_1} + M^f_{h_2}$.

Let $\{h_i, i \in I\} \subseteq L_{\cup T}[f]$ be a complete orthonormal set, and let

$$M_f = \sum_{i \in I} 2^{-i} M^f_{h_i}.$$

Since, for $\{i, j\} \subseteq I$, fixed, but arbitrary, $M^f_{h_i,h_j} \ll M_f$, one can compute the following Radon-Nikodým derivative:

$$D_{i,j} = \frac{dM^f_{h_i,h_j}}{dM_f}.$$

Define then, for $t \in T$, fixed, but arbitrary, a tentative inner product, using the following relation:

$$[h_i, h_j]_t = D_{i,j}(t).$$

Since $\{h_i, i \in I\}$ is a basis, the conditions for the existence of a bilinear functional with assigned values on a given subset obtain [46, p. 30], and there is thus a bilinear

$\beta_t : L_{\cup T}[f] \times L_{\cup T}[f] \longrightarrow \mathbb{R}$ such that

$$\beta_t\left(h_i, h_j\right) = \left[h_i, h_j\right]_t = D_{i,j}(t).$$

If the map $h \mapsto \beta_t(h, h)$ is not strictly positive, let $\mathcal{N}[\beta_t]$ be its null space, and $V_t^0[\{h_i, i \in I\}]$ be the quotient of $V[\{h_i, i \in I\}]$ with respect to $\mathcal{N}[\beta_t]$. One can then define an inner product β_t^0 on $V_t^0[\{h_i, i \in I\}] \times V_t^0[\{h_i, i \in I\}]$ setting

$$\beta_t^0([h_1], [h_2]) = \beta_t(h_1, h_2),$$

where $[h]$ is the equivalence class of h. One may thus assume for simplicity's sake that β_t is already an inner product on $L_{\cup T}[f]$. H_t then designates the completion of $V[\{h_i, i \in I\}]$ for β_t. Let $[h]_t$ be the class of h in that completion. Then

$$\left\langle [h_i]_t, \left[h_j\right]_t \right\rangle_{H_t} = \beta_t\left(h_i, h_j\right) = D_{i,j}(t).$$

The family of spaces $\{H_t, t \in T\}$ has the following properties:

1. for $\{i, j\} \subseteq I$, fixed, but arbitrary, the map

$$t \mapsto \left\langle [h_i]_t, \left[h_j\right]_t \right\rangle_{H_t}$$

   is adapted, as $D_{i,j}$ is;
2. since $V[\{h_i, i \in I\}]$ is dense in its completion, $\{[h_i]_t, i \in I\}$ is total in H_t.

   One can thus build $H_{\mathcal{T}}$ [(Proposition) 7.1.15], and $H_{\mathcal{T}}^{M_f}$ [(Proposition) 7.1.16]. Let $U : L_{\cup T}[f] \longrightarrow H_{\mathcal{T}}^{M_f}$ be defined using the following assignment:

$$U[h_i] = [\{[h_i]_t, t \in T\}].$$

Then, for $t \in T$, fixed, but arbitrary,

$$\begin{aligned}
\left\langle P_t[h_i], P_t\left[h_j\right]\right\rangle_H &= \left\langle f_{h_i}(t), f_{h_j}(t)\right\rangle_H \\
&= M^f_{h_i,h_j}(T_t) \\
&= \int_{T_t} D_{i,j}(\theta)\, M_f(d\theta) \\
&= \int_{T_t} \left\langle [h_i]_\theta, \left[h_j\right]_\theta \right\rangle_{H_\theta} M_f(d\theta) \\
&= \left\langle P_{H_{\mathcal{T}}^{M_f|T_t}}[U[h_i]], P_{H_{\mathcal{T}}^{M_f|T_t}}[U[h_i]] \right\rangle_{H_{\mathcal{T}}^{M_f}},
\end{aligned}$$

so that

$$\langle P_t[h], P_t[k]\rangle_H = \left\langle P_{H_{\mathcal{T}}^{M_f|T_t}}[U[h]], P_{H_{\mathcal{T}}^{M_f|T_t}}[U[k]]\right\rangle_{H_{\mathcal{T}}^{M_f}}.$$

Thus $P_t = U^{\star} P_{H_{\mathcal{T}}^{M_f|T_t}} U$. □

Remark 7.2.2 For fixed, but arbitrary $h \in L_{\cup T}[f]$, $M_h^f \ll M_f$. Indeed

$$\begin{aligned}
M_h^f(T_t) &= \|f_h(t)\|_H^2 \\
&= \|P_t[h]\|_H^2 \\
&= \sum_{i\in I}\sum_{j\in I} \langle h, h_i\rangle_H \langle h, h_j\rangle_H \langle P_t[h_i], P_t[h_j]\rangle_H \\
&= \sum_{i\in I}\sum_{j\in I} \langle h, h_i\rangle_H \langle h, h_j\rangle_H \langle f_{h_i}(t), f_{h_j}(t)\rangle_H \\
&= \sum_{i\in I}\sum_{j\in I} \langle h, h_i\rangle_H \langle h, h_j\rangle_H M_{h_i,h_j}^f(T_t) \\
&\ll M_f(T_t),
\end{aligned}$$

and the same obtains for all sets in $\mathcal{T}$. Consequently, for fixed, but arbitrary $\{k_1, k_2\} \subseteq L_{\cup T}$, $M_{k_1,k_2}^f \ll M_f$.

Remark 7.2.3 For fixed, but arbitrary $\{k_1, k_2\} \subseteq L_{\cup T}[f]$, almost surely with respect to M_f,

$$\frac{dM_{k_1,k_2}^f}{dM_f}(t) = \langle [k_1]_t, [k_2]_t\rangle_{H_t}.$$

Indeed, the equality

$$\langle P_t[k_1], P_t[k_2]\rangle_H = \left\langle P_{H_{\mathcal{T}}^{M_f|T_t}}[U[k_1]], P_{H_{\mathcal{T}}^{M_f|T_t}}[U[k_2]]\right\rangle_{H_{\mathcal{T}}^{M_f}}$$

of (Proposition) 7.2.1 is the same as the equality

$$\int_{T_t} \frac{dM_{k_1,k_2}^f}{dM_f}(t)\, M_f(dt) = \int_{T_t} \langle [h_i]_t, [h_j]_t\rangle_{H_t} M_f(dt).$$

Remark 7.2.4 The range of $P_{H_{\mathcal{T}}^{M_f|T_0}}$ is the zero subspace if, and only if,

$$M_f(T_0) = 0.$$

Indeed, when $M_f(T_0) = 0$, since

$$\left\| P_{H_{\mathcal{T}}^{M_f | T_0}} [U[h]] \right\|^2_{H_{\mathcal{T}}^{M_f}} = \int_{T_0} \| [h]_t \|^2_{H_t} M_f(dt),$$

then

$$\left\| P_{H_{\mathcal{T}}^{M_f | T_0}} [U[h]] \right\|^2_{H_{\mathcal{T}}^{M_f}} = 0.$$

Suppose conversely that $M_f(T_0) > 0$, but that

$$\left\| P_{H_{\mathcal{T}}^{M_f | T_0}} [U[h]] \right\|^2_{H_{\mathcal{T}}^{M_f}} = 0, \; h \in L_{\cup T}[f].$$

Then, for $i \in I$, fixed, but arbitrary,

$$\begin{aligned} 0 &= \left\| P_{H_{\mathcal{T}}^{M_f | T_0}} [U[h_i]] \right\|^2_{H_{\mathcal{T}}^{M_f}} \\ &= \int_{T_0} \langle [h_i]_t, [h_i]_t \rangle^2_{H_t} M_f(dt) \\ &= \int_{T_0} D_{i,i}(t) M_f(dt) \\ &= M^f_{h_i,h_i}(T_0) \\ &= M^f_{h_i}(T_0). \end{aligned}$$

Thus $M^f_{h_i,h_i}(T_0) = 0$, all $i \in I$, that is $M_f(T_0) = 0$, a contradiction.

7.3 Cramér's Representation

The CHR is an immediate consequence of what precedes.

Proposition 7.3.1 *There exists an orthonormal family $\{h_i, \; i \in I\} \subseteq L_{\cup T}[f]$ such that, for $t \in T$, fixed, but arbitrary,*

$$L_t[f] = \bigoplus_{i \in I} L_t[f_{h_i}]$$

if, and only if,

$$\sup_{t \in T} (\dim H_t) \leq |I| = \tilde{I},$$

where the family of subspaces $\{H_t, t \in T\}$ is that of a Hilbertian integral representation of $L_{\cup T}[f]$. One has then a proper canonical representation.

Proof One may restrict attention to $L_{\cup T}[f] = \bigoplus_{i \in I} L_{\cup T}[f_{h_i}]$ as the equality $L_t[f] = \bigoplus_{i \in I} L_t[f_{h_i}]$ may be written in the form

$$L_{\cup T}[P_t[f]] = \bigoplus_{i \in I} L_{\cup T}[P_t[f_{h_i}]].$$

Proof Suppose that $L_{\cup T}[f] = \bigoplus_{i \in I} L_{\cup T}[f_{h_i}]$.

Let $H_T^{M_f}$ be the representation of $L_{\cup T}[f]$, given in (Proposition) 7.2.1, using the orthonormal family $\{h_i, i \in I\}$ of the proposition's statement, and let U be the associated isometry. Denote k_i the element $U[h_i]$. It is the equivalence class of a measurable field of vectors

$$\dot{k}_i = \left\{\dot{k}_i(t), t \in T\right\}.$$

Then

$$U[P_t[h_i]] = P_{H_T^{M_f|T_t}} U[h_i] = P_{H_T^{M_f|T_t}}[k_i],$$

which is [(Proposition) 7.3.1] the equivalence class of the measurable field of vectors

$$\chi_{T_t} \dot{k}_i.$$

Consequently $U[L_{\cup T}[h_i]]$ is generated by the equivalence classes of the elements in the family

$$\left\{\chi_{T_t} \dot{k}_i, t \in T\right\}.$$

But then $H_T^{M_f}$ is generated by the equivalence classes of the elements in the family

$$\left\{\chi_{T_t} \dot{k}_i, t \in T, i \in I\right\}.$$

From (Proposition) 7.1.25, one then knows that, for fixed, but arbitrary $t \in T$, almost surely with respect to M_f,

$$\left\{\dot{k}_i(t), i \in I\right\}$$

is total in H_t. On the exceptional set the dimension can be set arbitrarily to one.

Proof Suppose that $\sup_{t \in T} (\dim H_t) \leq |I| = \tilde{I}$.

Let

$$H_{\mathcal{T}}^{M_f} = \bigoplus_{j \in J} U_j \left[L_2 \left(T, \mathcal{T}, \tau_j \right) \right]$$

be the representation of (Proposition) 7.1.16. U_j is a partial isometry defined using the following relation:

$$U_j [f] = \left[\dot{f} e_j \right],$$

for which the e_j's are as in (Proposition) 7.1.11, and $\dim H_t \geq j$ if, and only if, $e_j(t) \neq 0_{H_t}$. Thus when $j > |I|$, $e_j(t) = 0_{H_t}$, $t \in T$, so that $J = |I|$.

Let $h_i = U^\star [e_i]$, $i \in I$. One has that

$$P_t [h_i] = P_t U^\star [e_i] = U^\star P_{H_{\mathcal{T}}^{M_f|T_t}} [e_i] = U^\star [I_{T_t} e_i] = U^\star U_i [I_{T_t}].$$

Consequently $L_{\cup T} [h_i]$ and $L_2 (T, \mathcal{T}, \tau_i)$ are unitarily equivalent. It follows that $\bigoplus_{i \in I} L_{\cup T} [h_i]$ and $\bigoplus_{i \in I} L_2 (T, \mathcal{T}, \tau_i)$ are unitarily equivalent. But the latter is unitarily equivalent to $L_{\cup T} [f]$.

As $P_t [h_i] = U^\star U_i [I_{T_t}]$,

$$M_{h_i}^f (T_t) = \| f_{h_i} (t) \|_H^2 = \| P_t [h_i] \|_H^2 = \tau_i (T_t).$$

Since the τ_i's are "ordered by inclusion" [(Remark) 7.1.18], so are the $M_{h_i}^f$'s, and the decomposition is proper canonical. □

Chapter 8
Some Facts About Multiplicity

Multiplicity is not easy to fathom. One way to acquire a feel for it, is to look at examples, and, in this chapter, multiplicity is looked at from different angles.

8.1 All Multiplicities May Occur

That all multiplicities may occur should not come as a surprise [(Example) 6.1.5]. Below is a typical example for that fact (for a further example, see (Example) 9.2.1). It is based on a construction involving Cantor-like sets.

Fact 8.1.1 ([275, p. 87]) *A Cantor-like set is a subset F of $[0,1]$ obtained as follows.*

Let $t \in]0,1[$ be fixed, but arbitrary. Let $I_{1,1}$ be the open interval of length $\frac{t}{2}$ at the center of $[0,1]$. When it is removed from $[0,1]$, one is left with two disjoint, closed intervals, $J_{1,1}$ and $J_{1,2}$.

Let $I_{2,1}$ be the open interval of length $\frac{1}{2} \times \frac{t}{2^2}$ at the center of the closed interval $J_{1,1}$, and $I_{2,2}$ be the open interval of length $\frac{1}{2} \times \frac{t}{2^2}$ at the center of the closed interval $J_{2,1}$. When both $I_{2,1}$ and $I_{2,2}$ are also removed from $[0,1]$, one is left with 2^2 closed, disjoint intervals $J_{2,1}, J_{2,2}, J_{2,3}, J_{2,4}$.

$J_{2,1}, J_{2,2}, J_{2,3}, J_{2,4}$ contain at their center 2^2 open intervals $I_{3,k}$, $1 \le k \le 2^2$, each of length $\frac{1}{4} \times \frac{t}{2^3}$. When these open intervals are also removed from $[0,1]$, one is left with 2^3 closed, disjoint intervals $J_{3,k}$, $1 \le k \le 2^3$.

Thus, at step 1, there is one I interval and two J intervals; at step 2, there are two I intervals and four J intervals; at step three, four I intervals and eight J intervals. Pursuing in the same way, at step $n \ge 2$, there are 2^{n-1} intervals of type I contained in the J intervals of step $n-1$, and 2^n intervals of type J. The length of the I intervals

A.F. Gualtierotti, *Detection of Random Signals in Dependent Gaussian Noise*,
DOI 10.1007/978-3-319-22315-5_8

at step n is

$$\frac{1}{2^{n-1}} \times \frac{t}{2^n}.$$

Let

$$G_n = \cup_{k=1}^{2^{n-1}} I_{n,k},$$
$$F_n = \cup_{k=1}^{2^n} J_{n,k}.$$

The G_n's are disjoint and open, and the F_n's are decreasing and closed. One sets

$$F = \cap_{n=1}^{\infty} F_n = [0, 1] \cap \left\{ \cup_{n=1}^{\infty} G_n \right\}^c.$$

Remark 8.1.2 Let *Leb* be Lebesgue measure, and $G = \cup_{n=1}^{\infty} G_n$. Then

$$Leb\,(G_n) = \frac{t}{2^n}, \quad \text{and} \quad Leb\left(\cup_{k=1}^{n} G_k\right) = \left(1 - \frac{1}{2^n}\right) t.$$

F_n is what is left of $[0, 1]$ when $\cup_{k=1}^{n} G_k$ is removed from it. It is made of 2^n intervals of equal length. Thus

$$Leb\,(J_{n,k}) = \frac{1}{2^n}\left\{1 - \left(1 - \frac{1}{2^n}\right) t\right\} \leq \frac{1}{2^n}.$$

Furthermore $Leb\,(G) = t$, and $Leb\,(F) = 1 - t$.

Remark 8.1.3 F is compact, as it is closed and bounded.

Remark 8.1.4 F contains no open interval, so that G is dense in $[0, 1]$.

Suppose indeed that I is an open interval of strictly positive length $Leb\,(I)$. There is an $n_0\,(I) \in \mathbb{N}$ such that, for $n \geq n_0\,(I)$, $\frac{1}{2^n} < Leb\,(I)$, and I cannot be contained in F_n, for $n \geq n_0\,(I)$. But then I cannot be contained in F.

Remark 8.1.5 When one starts with $]0, 1[$, instead of $[0, 1]$, the same procedure works, but one no longer may claim that F is closed.

Proposition 8.1.6 *There is, in* $[0, 1]$, *a subset of Lebesgue measure zero, whose complement, in* $[0, 1]$, *can be countably partitioned in such a way that the intersection of each set of that decomposition with any open, not void, subinterval of* $[0, 1]$, *has positive Lebesgue measure.*

Proof An interval of the form $I_{n,k}$ shall be called an interval of type I.

Let τ_1 be a fixed, but arbitrary value of $]0, 1[$. Let F_1 be a Cantor-like set in $]0, 1[$ such that $Leb\,(F_1) = \tau_1$.

The complement of F_1 in $[0, 1]$ is the union of the disjoint, open intervals $I_{n,k}$, $n \in \mathbb{N}$, $1 \leq k \leq 2^{n-1}$. In each set $I_{n,k}$, one may thus obtain a Cantor-like set $F_{n,k}$, whose

Lebesgue measure is $\tau_{n,k} > 0$. Let

$$F_2 = \cup_{n,k} F_{n,k}.$$

Sets, obtained in the way F_2 has been, shall be called sets of type F. F_1 and F_2 are disjoint by construction, and the complement of their union is again a union of disjoint intervals of type I. One can thus continue to build sets of type F and, adjusting the τ values, one will eventually have a disjoint sequence of sets of type F of strictly positive measure, whose combined measure is 1.

Each Cantor-like set F of that construction is contained in an interval of type I, say I_F. Let J be an open, not void, subinterval of I_F. Since F is nowhere dense in I_F, J must intersect one of the open intervals of $I_F \setminus F$.

Let $]a, b[\subseteq [0, 1]$ be fixed, but arbitrary, and let $\epsilon > 0$ be such that $a + \epsilon < b - \epsilon$. There is thus $I_{n,k}$ such that

$$J_1 =]a + \epsilon, b - \epsilon[\cap I_{n,k} \neq \emptyset.$$

Since $J_1 \subseteq I_{n,k}$, and is an open, not void subinterval, there is an interval of type I in $I_{n,k}$ such that its intersection with J_1 is a not void interval. Let that intersection be J_2. One thus builds a sequence $\{J_n, \ n \in \mathbb{N}\}$ of intervals whose sizes decrease to zero as they are contained in intervals of type I of strictly decreasing size. But then, when the size of these intervals of type I is strictly less than epsilon, one of them must be contained in $]a, b[$. Consequently as every open interval of $[0, 1]$ contains an interval of type I, it will contain an interval of type F of strictly positive measure.

Let $\mathbb{N} = \cup_p \mathbb{N}_p$, where $\mathbb{N}_p$ is a subsequence of $\mathbb{N}$, and $\mathbb{N}_p \cap \mathbb{N}_q = \emptyset$ when $p \neq q$. This can be achieved, for instance, representing the rationals as pairs of integers, and using the usual diagonal method to number the integers [230, p. 29]: the p-th row or the p-th column of the resulting table yield the sets $\mathbb{N}_p$. Let $\{F_n, \ n \in \mathbb{N}\}$ be the sequence of sets of type F that were sequentially built above. Let

$$T_p = \cup_{q \in \mathbb{N}_p} F_q.$$

These are disjoint by construction, and the measure of their sum is one. Furthermore, since any open interval of $[0, 1]$ shall contain one of the sets of type F that have strictly positive measure, the intersection of one of these intervals with an arbitrary T_p shall be positive. □

Remark 8.1.7 The proposition's validity is not restricted to the unit interval: cases of other intervals in $\mathbb{R}$ are somewhat more complicated to describe but have a very similar proof (see for example [227, p. 8], where one finds the same proof for any finite open interval; for infinite ones, one decomposes them into countable sums of disjoint, finite intervals).

Proposition 8.1.8 *Let $T = [0, 1]$, and $\{f_i : T \longrightarrow H, \ i \in I\}$ be a finite, or infinite, sequence of purely nondeterministic, orthogonal functions, with orthogonal increments. Suppose furthermore that those functions are continuous to the left. There is* *then a function $f : T \longrightarrow H$, whose canonical representation is obtained from the sequence $\{f_i : T \longrightarrow H,\ i \in I\}$.*

Proof The "distribution" associated with f_i shall be defined to be the following function:

$$F_i(t) = \|f_i(t)\|_H^2 .$$

Let [(Proposition) 8.1.6] $\{T_i,\ i \in I\}$ be a family of disjoint measurable subsets of $[0, 1]$ such that

- $Leb\left([0, 1] \setminus \biguplus_{i\in I} T_i\right) = 0$,
- given a fixed, but arbitrary interval of $T = [0, 1]$, say $]a, b[\,,\ a < b$, for fixed, but arbitrary $i \in I$, $Leb\left(]a, b[\cap T_i\right) > 0$.

For $i \in I$, fixed, but arbitrary, let $\sigma_i : T_i \longrightarrow \mathbb{R}$ be adapted, and strictly positive, and

$$g_i(t) = \chi_{T_i}(t)\,\sigma_i(t) f_i(t) .$$

Since f_i is continuous to the left, for $h \in H$, fixed, but arbitrary, the map $t \mapsto \langle f_i(t), h\rangle_H$ is continuous to the left, and thus adapted [128, p. 86]. Consequently, f_i is adapted, and thus so is g_i. As

$$\begin{aligned}\int \|g_i(t)\|_H\,dt &= \int_{T_i} \sigma_i(t)\,\|f_i(t)\|_H\,dt \\ &\le \left\{\int_{T_i} \sigma_i^2(t)\,dt\right\}^{1/2} \left\{\int_{T_i} F_i(t)\,dt\right\}^{1/2} \\ &\le F_i^{1/2}(1) \left\{\int_{T_i} \sigma_i^2(t)\,dt\right\}^{1/2},\end{aligned}$$

one sees that, adequately choosing the functions $\{\sigma_i,\ i \in I\}$, one obtains functions $f : T \longrightarrow H$, and $g : T \longrightarrow H$, when setting

$$g(t) = \sum_{i\in I} g_i(t)\,, \quad \text{and} f(t) = \int_0^t g(\theta)\,d\theta.$$

One must then prove that, for fixed, but arbitrary $t \in T$, $L_t[f] = \bigoplus_{i\in I} L_t[f_i]$.

By construction, for fixed, but arbitrary $\theta \in T$, $g(\theta) \in \bigoplus_{i\in I} L_t[f_i]$, and thus, its integral on $[0, t]$ belongs to $\bigoplus_{i\in I} L_t[f_i]$, as it is a limit of linear combinations of $g(\theta)$'s, $\theta \le t$. It then follows that $L_t[f] \subseteq \bigoplus_{i\in I} L_t[f_i]$.

Conversely, by definition, f satisfies the fundamental theorem of calculus [262, p. 238]. It is thus differentiable, and, almost surely, with respect to Lebesgue measure, the derivative of f, f' equals g [262, p. 241]. Since $f'(t)$ can be obtained as the limit of $-n^{-1}\{f(t - n^{-1}) - f(t)\}$, $f'(t) \in L_t[f]$. Consequently, for almost every $t \in T_i$, fixed, but arbitrary,

$$f'(t) = \sigma_i(t) f_i(t), \; \sigma_i(t) > 0.$$

Let T_i^0 be the exceptional set, and $t \in \{T_i^0\}^c$ be fixed, but arbitrary. Since [(Proposition) 8.1.6], for fixed, but arbitrary $\epsilon > 0$, $Leb\,(]t - \epsilon, t[\cap T_i) > 0$, for every $n \in \mathbb{N}$, there exits

$$t_n \in \left]t - \frac{1}{n}, t\right[\quad \text{such that} \quad f_i(t_n) \in L_t[f].$$

Since f_i is continuous to the left,

$$f_i(t) = \lim_n f_i(t_n) \in L_t[f].$$

Thus $f_i(t) \in L_t[f]$, $t \in T_i$, and, consequently, for $t \in \biguplus_{i \in I} T_i$. Since the latter is dense in T, the result obtains again because of the continuity properties of the f_i's.

□

8.2 Invariance of Multiplicity

Transformations of signals and, in particular, linear transformations are pervasive in communication theory. It is thus of interest to know how multiplicity behaves under transformations. In this section it is invariance with respect to linear transformations that is examined.

8.2.1 The Case of Projections

The fact that multiplicity is not preserved by projection shows that it is a feature whose embodiments are difficult to anticipate. The example given shows even more: the projection of a purely nondeterministic process of multiplicity one may yield a deterministic process of arbitrarily large multiplicity.

Example 8.2.1 (Projection Does Not Generally Preserve Multiplicity) Let $T = [0, 1]$, and $W : T \longrightarrow H = L_2(\Omega, \mathcal{A}, P)$ be a standard Wiener process (continuous in quadratic mean, with stationary, normal, independent increments, and variance one). It has multiplicity one since

$$W(t) = \int I_{[0,t]} \, dm_W,$$

and that the family $\{I_{[0,t]},\ t \in [0,1]\}$ is total in $L_2[0,1]$. It turns out that certain projections of W have arbitrarily large multiplicity. Indeed W has a representation in quadratic mean of the following form [271, p. 87]:

$$W(t) = \sum_{k=0}^{\infty} f_k(t)\, W_k,$$

where the W_k's are orthogonal, and the f_k's, which are orthonormal in $L_2[0,1]$, are given by relations of the following type:

$$f_k(t) = \sqrt{2}\sin\left(\left[k + \frac{1}{2}\right]\pi t\right).$$

One now needs the following facts:

[1] *Let $n \in \mathbb{N}$, and $0 < a_1 < a_2 < a_3 < \cdots < a_n$, be fixed, but arbitrary, and, with $k \in [0:n-1]$ and $l \in [1:n]$,*

$$M_n = \begin{bmatrix} a_1 & \cdots & a_1^{2k+1} & \cdots & a_1^{2n-1} \\ \vdots & & \vdots & & \vdots \\ a_l & \cdots & a_l^{2k+1} & \cdots & a_l^{2n-1} \\ \vdots & & \vdots & & \vdots \\ a_n & \cdots & a_n^{2k+1} & \cdots & a_n^{2n-1} \end{bmatrix}.$$

M_n is invertible, and its determinant is computed as indicated below.

The simplest way to see that the result obtains is to proceed inductively. M_1 has one entry, a_1, and is thus invertible with determinant equal to a_1. M_2 has a determinant equal to $a_1 a_2 \left(a_2^2 - a_1^2\right) > 0$. For M_3, consider the matrix

$$T_3 = \begin{bmatrix} 1 & -a_3^2 & 0 \\ 0 & 1 & -a_3^2 \\ 0 & 0 & 1 \end{bmatrix}.$$

Then

$$M_3 T_3 = \left[\begin{array}{c|c} \underline{a}_2 & D_2 M_2 \\ \hline a_3 & 0\ 0 \end{array}\right],$$

where $\underline{a}_2$ has entries a_1 and a_2, and

$$D_2 = \begin{bmatrix} a_1^2 - a_3^2 & 0 \\ 0 & a_2^2 - a_3^2 \end{bmatrix}.$$

Similarly, and iteratively,

$$M_n T_n = \left[\begin{array}{c|c} \underline{a}_{n-1} & D_{n-1} M_{n-1} \\ \hline a_n & \underline{0}^{\star}_{n1} \end{array} \right],$$

where

$$\underline{a}^{\star}_{n-1} = (a_1, \ldots, a_{n-1}),$$

$$T_n = \begin{bmatrix} 1 & -a_n^2 & 0 & 0 \ldots 0 & 0 & 0 \\ 0 & 1 & -a_n^2 & 0 \ldots 0 & 0 & 0 \\ 0 & 0 & 1 & -a_n^2 \ldots 0 & 0 & 0 \\ \vdots & \vdots & \vdots & \vdots & \vdots\ \vdots & \vdots \\ 0 & 0 & 0 & 0 \cdots 0 & 1 & -a_n^2 \\ 0 & 0 & 0 & 0 \cdots 0 & 0 & 1 \end{bmatrix},$$

and D_{n-1} is a diagonal matrix with entries $a_1^2 - a_n^2, \ldots, a_{n-1}^2 - a_n^2$. Then, denoting $|M_n|$ the determinant of M_n,

$$|M_n| = |M_n T_n| = (-1)^{n+1} a_n \prod_{l=1}^{n-1} \left(a_l^2 - a_n^2\right) |M_{n-1}| \neq 0.$$

[2] *For $n \in \mathbb{N}$, and $t \in]0, 1]$, fixed, but arbitrary, the following family of functions is linearly independent:*

$$s_k(\theta) = \sin\left([2k+1]\frac{\pi}{2}\theta\right), \ \textit{for}\ \theta \in [0, t],\ k \in [0 : n-1].$$

Suppose indeed that, for $\theta \in [0, t]$, fixed, but arbitrary,

$$f(\theta) = \sum_{k=0}^{n-1} \alpha_k s_k(\theta) = 0.$$

One has that, $s_k^{(l)}$ denoting the derivative of order l of s_k,

$$\begin{aligned} s_k^{(2l)}(0) &= 0, \\ s_k^{(2l+1)}(0) &= (-1)^{l+2} (2k+1)^{2l+1} \left(\frac{\pi}{2}\right)^{2l+1}. \end{aligned}$$

Consequently, for $l \in [0 : n-1]$, fixed, but arbitrary, the assumption that f is the zero function leads to the following identity:

$$f^{(2l+1)}(0) = 0$$

which translates to

$$\sum_{k=0}^{n-1} \alpha_k (2k+1)^{2l+1} = 0.$$

Fact [1] then yields that the α_k's are zero.

[3] *When the functions $f_1, \ldots, f_n$ are linearly independent on $]0, t[$, $t > 0$, that latter interval contains n distinct points $t_1, \ldots, t_n$ such that, $\underline{f}(t)$ being the vector in $\mathbb{R}^n$ with components $f_1(t), \ldots, f_n(t)$, the vectors $\underline{f}(t_1), \ldots \underline{f}(t_n)$ form a basis for $\mathbb{R}^n$.*

Linear independence means that the range of $\underline{f}$ is total in $\mathbb{R}^n$, that is, $\mathbb{R}^n$ is the set of linear combinations of the form

$$\sum_{i=1}^{n} \alpha_i \underline{f}(t_i).$$

The range of $\underline{f}$ is thus a basis of $\mathbb{R}^n$, and it then contains n linearly independent elements.

Now, given $t \in]0, 1]$, fixed, but arbitrary, one can thus find (facts [1] and [2] above), in $]0, t[$, points $\{t_1, \ldots, t_n\}$ such that $t_1 < \cdots < t_n$, and the matrix M_n, with entries $f_i(t_j)$, is invertible. Let then

$$W_n(t) = \sum_{i=0}^{n-1} f_i(t) W_i.$$

Denote $\underline{W}_n$ the vector with components $W_n(t_i)$, and $\underline{W}$ that with components W_i. Then

$$\underline{W}_n = M_n \underline{W}, \quad \text{so that} \quad \underline{W} = M_n^{-1} \underline{W}_n.$$

The process W_n is the projection of the process W onto the subspace generated by the classes of the random variables $W_1, \ldots, W_n$, and generates a subspace of the same dimension. Its multiplicity is thus n. And the vector with components $W_1, \ldots, W_n$ is present at the "origin of time."

8.2.2 *The Case of Unitary Transformations*

Unitary transformations ought to preserve multiplicity. Conditions that make that assertion true are given below.

Let H be a real Hilbert space, and $\{H_t,\ t \in T\}$ be a family of (closed) subspaces of H such that, for $t \in T$, fixed, but arbitrary,

$$H_t = \bigvee_{\{\theta \in T, \theta \leq t\}} H_\theta,$$

and

$$H = \bigvee_{t \in T} H_t.$$

The projection with range H_t is denoted P_t^H. $K,\ K_t,\ P_t^K$ are defined analogously.

Remark 8.2.2 Given H and the family $\{H_t, t \in T\}$, one obtains for it a canonical representation proceeding as in (Proposition) 6.4.10. One says that the orthonormal set used in so doing generates the canonical representation.

Definition 8.2.3 The families of (closed) subspaces

$$\{H_t,\ t \in T\} \text{ and } \{K_t,\ t \in T\}$$

are said to be isometric when there is a unitary operator $U : H \longrightarrow K$ such that, for $t \in T$, fixed, but arbitrary, $U[H_t] = K_t$.

Remark 8.2.4 When $\{H_t,\ t \in T\}$ and $\{K_t,\ t \in T\}$ are isometric, the restriction of U to H_t is unitary, and so is the restriction of $U^\star$ to K_t [266, p. 87].

Lemma 8.2.5 *Suppose that* $\{H_t,\ t \in T\}$ *and* $\{K_t,\ t \in T\}$ *are isometric. Then* $U\left[H_t^\perp\right] = K_t^\perp$, *and, for fixed, but arbitrary* $t \in T$, $P_t^K = UP_t^H U^\star$.

Proof $U\left[H_t^\perp\right] = K_t^\perp$:

Let $h_t^\perp \in H_t^\perp$, and $k_t \in K_t$, be fixed, but arbitrary. Then, by assumption, for some $h_t(k_t) \in H_t$, $k_t = U[h_t(k_t)]$, and

$$\langle U\left[h_t^\perp\right], k_t\rangle_K = \langle U\left[h_t^\perp\right], U\left[h_t(k_t)\right]\rangle_K = \langle h_t^\perp, h_t(k_t)\rangle_H = 0.$$

Consequently $U\left[H_t^\perp\right] \subseteq K_t^\perp$. Let thus $k_t^\perp \in K_t^\perp$ be such that, for fixed, but arbitrary $h_t^\perp \in H_t^\perp$,

$$\langle k_t^\perp, U\left[h_t^\perp\right]\rangle_K = 0.$$

Then $\langle U^\star [k_t^\perp], h_t^\perp\rangle_H = 0$, for all $h_t^\perp \in H_t^\perp$. Thus $U^\star [k_t^\perp] \in H_t$, and $k_t^\perp \in K_t$. So $k_t^\perp = 0_K$, and

$$U[H_t^\perp] = K_t^\perp.$$

Proof $P_t^K = UP_t^H U^\star$:

Suppose that $k_t \in K_t$ is fixed, but arbitrary. Then, because $U[H_t] = K_t$, $U^\star [k_t] \in H_t$, so that $P_t^H [U^\star [k_t]] = U^\star [k_t]$, or

$$UP_t^H U^\star [k_t] = k_t,\ k_t \in K_t.$$

Similarly, for fixed, but arbitrary $k_t^\perp \in K_t^\perp$, since $U[H_t^\perp] = K_t^\perp$, one has that $U^\star [k_t^\perp] \in H_t^\perp$, so that $P_t^H [U^\star [k_t^\perp]] = 0_H$, and thus

$$UP_t^H U^\star [k_t^\perp] = 0_K,\ k_t^\perp \in K_t^\perp.$$

Consequently $UP_t^H U^\star = P_t^K$. □

Proposition 8.2.6 *Isometric families of subspaces have the same multiplicity.*

Proof Suppose that $\{H_t,\ t \in T\}$ and $\{K_t,\ t \in T\}$ are isometric through the unitary U. Let $\{h_i,\ i \in I\}$ generate a canonical representation [(Remark) 8.2.2] for the family $\{H_t,\ t \in T\}$, and let, for $i \in I$, fixed, but arbitrary, $k_i = U[h_i]$. Then [(Lemma) 8.2.5]

$$P_t^K [k_i] = UP_t^H U^\star [k_i] = UP_t^H [h_i].$$

Since, for example,

$$L_t [k_i] = \overline{V[\{P_t^K [k_i],\ t \in T\}]},$$

one has that

$$L_t [k_i] = UL_t [h_i],\ t \in T.$$

Furthermore, for $i_1 \neq i_2$ in I, and t_1 and t_2 in T, fixed, but arbitrary, since one works with a canonical representation,

$$\begin{aligned} \langle P_{t_1}^K [k_{i_1}], P_{t_2}^K [k_{i_2}]\rangle_K &= \langle UP_{t_1}^H [h_{i_1}], UP_{t_2}^H [h_{i_2}]\rangle_K \\ &= \langle P_{t_1}^H [h_{i_1}], P_{t_2}^H [h_{i_2}]\rangle_H \\ &= 0. \end{aligned}$$

Thus

$$\bigoplus_{i\in I} L_t[k_i] = \bigoplus_{i\in I} UL_t[h_i] = U\left[\bigoplus_{i\in I} L_t[h_i]\right] = U[H_t] = K_t,$$

and $\{k_i,\ i \in I\}$ generates a canonical representation for the family $\{K_t, t \in T\}$. □

Remark 8.2.7 For the measures associated with the function $t \mapsto P_t^K[k_i]$, one has that

$$\begin{aligned} m_i^K([t_1, t_2[) &= P_{t_2}^K[k_i] - P_{t_1}^K[k_i] \\ &= P_{t_2}^K U[h_i] - P_{t_1}^K U[h_i] \\ &= U\left[P_{t_2}^H[h_i] - P_{t_1}^H[h_i]\right] \\ &= U\left[m_i^H([t_1, t_2[)\right], \end{aligned}$$

so that, for $i \in I$ and $[t_1, t_2[\subseteq T$, fixed, but arbitrary,

$$M_i^K([t_1, t_2[) = \left\|m_i^K([t_1, t_2[)\right\|_K^2 = \left\|m_i^H([t_1, t_2[)\right\|_H^2 = M_i^H([t_1, t_2[).$$

Proposition 8.2.8 *Canonical representations, with the same multiplicity, and mutually absolutely continuous basis measures lead to isometric families of subspaces.*

Proof Suppose that, for fixed, but arbitrary $t \in T$,

$$H_t = \bigoplus_{i\in I} L_t[h_i], \text{ and } K_t = \bigoplus_{i\in I} L_t[k_i],$$

and, furthermore, that, for $i \in I$, fixed, but arbitrary, $M_i^K \equiv M_i^H$. Let, for convenience, and for $i \in I$, fixed, but arbitrary,

$$f_i(t) = P_t^H[h_i],\ g_i(t) = P_t^K[k_i], \text{ and } dM_i^g = d_i dM_i^f.$$

As presently seen, it is no restriction to assume that the measures equivalence is an equality. Indeed

$$\int_T d_i^{-1} dM_i^g = M_i^f(T) < \infty,$$

so that one may set

$$D_i^{-1/2} = \left[d_i^{-1/2}\right]_{L_2\left(T, \mathcal{T}, M_i^g\right)}.$$

The following definitions then make sense:

$$dm_i^\star = D_i^{-1/2}\, dm_i^g,$$

and

$$g_i^\star(t) = m_i^\star(T_t) = \int \chi_{T_t}\, dm_i^\star = \int \chi_{T_t} D_i^{-1/2}\, dm_i^g.$$

Consequently, for $t \in T$, fixed, but arbitrary, $L_t\left[g_i^\star\right] \subseteq L_t[g_i]$. But if $k \in L_t[g_i]$ is orthogonal to $L_t\left[g_i^\star\right]$, then, for $\theta \in T$, $\theta \le t$, fixed, but arbitrary,

$$k = \int f^K\, dm_i^g,$$

and

$$\begin{aligned} 0 &= \langle k, g_i^\star(\theta)\rangle_K \\ &= \left\langle \int f^K\, dm_i^g, \int \chi_{T_\theta} D_i^{-1/2}\, dm_i^g \right\rangle_K \\ &= \int_{T_\theta} \dot{f}^K d_i^{-1/2}\, dM_i^g, \end{aligned}$$

so that f^K is the zero element in K, and the inclusion is an equality. Furthermore

$$m_i^{g^\star} = m_i^\star,$$

and

$$M_i^{g^\star} = M_i^\star = M_i^f.$$

But, since then the basis L_2 spaces are equal, the spaces L_t, which are unitary images of those, are in turn unitarily equivalent. □

Remark 8.2.9 The "trick" used in the proof of (Proposition) 8.2.8 has already been put to work many times!

8.2.3 *Other Linear Transformations Which Preserve Multiplicity*

Linear transformations which preserve multiplicity, but are not "obvious," require some work to devise. This section is devoted to the results of such a search. The end product says that multiplicity is preserved when covariances can be adequately factored, and that one can achieve such a factorization when the covariances are the sum of the identity and a Hilbert-Schmidt operator, a ubiquitous feature of the whole subject.

Definition 8.2.10 Let H and K be real Hilbert spaces, and $V[H]$ be a manifold, dense in H. Suppose $B : V[H] \longrightarrow K$ is a linear map such that, for $\{h_1, h_2\} \subseteq V[H]$, fixed, but arbitrary,

$$\langle B[h_1], B[h_2]\rangle_K = \langle C_B[h_1], h_2\rangle_H,$$

where $C_B : H \longrightarrow H$ is a weak covariance operator, that is [25, p. 267], a linear, bounded, positive, and self-adjoint operator. B is called a generalized process on H. One writes then $(B, V[H], K)$.

Remark 8.2.11 Since, on $V[H]$,

$$\|B[h]\|_K^2 = \left\|C_B^{1/2}[h]\right\|_H^2 \leq \left\|C_B^{1/2}\right\|^2 \|h\|_H^2,$$

B is necessarily bounded.

Lemma 8.2.12 (Proposition 3.2.5) *Let $A : H \longrightarrow L$, and $B : K \longrightarrow L$, be bounded, linear operators of Hilbert spaces. Then $\mathcal{R}[A] \subseteq \mathcal{R}[B]$ if, and only if, one of the following assertions obtains:*

1. *there exists a bounded, linear operator $C : H \longrightarrow K$ such that $A = BC$;*
2. *there exists $\kappa \in \mathbb{R}_+$ such that $\langle AA^\star[l], l\rangle_L \leq \kappa \langle BB^\star[l], l\rangle_L$.*

Proof The relation $A = BC$ obviously means range inclusion.

Suppose thus that $\mathcal{R}[A] \subseteq \mathcal{R}[B]$. Then $A[h] = B[k]$. Define $C : H \longrightarrow K$ using the following relation:

$$C[h] = P_{\overline{\mathcal{R}[B^\star]}}[k].$$

That makes sense. Suppose indeed that

$$C[h] = P_{\overline{\mathcal{R}[B^\star]}}[k_1], \text{ and that } C[h] = P_{\overline{\mathcal{R}[B^\star]}}[k_2].$$

Then [8, p. 363], $k_1 - k_2 \in \mathcal{N}[B]$. But

$$k_1 - k_2 = \left\{P_{\overline{\mathcal{R}[B^\star]}}[k_1] + P_{\mathcal{N}[B]}[k_1]\right\} - \left\{P_{\overline{\mathcal{R}[B^\star]}}[k_2] + P_{\mathcal{N}[B]}[k_2]\right\},$$

so that

$$k_1 - k_2 - P_{\mathcal{N}}[B][k_1] + P_{\mathcal{N}[B]}[k_2] = P_{\overline{\mathcal{R}[B^\star]}}[k_1] - P_{\overline{\mathcal{R}[B^\star]}}[k_2].$$

Since the left-hand side of the latter equality is in $\mathcal{N}[B]$, and the right-hand one, in $R_{\overline{\mathcal{R}[B^\star]}}$, its orthogonal complement [8, p. 363], one must have that

$$P_{\overline{\mathcal{R}[B^\star]}}[k_1] - P_{\overline{\mathcal{R}[B^\star]}}[k_2] = 0_K.$$

C is also linear. Suppose indeed that $A[h_1] = B[k_1]$, and that $A[h_2] = B[k_2]$, that is, $C[h_1] = P_{\overline{\mathcal{R}[B^\star]}}[k_1]$, and $C[h_2] = P_{\overline{\mathcal{R}[B^\star]}}[k_2]$. One has that

$$A[\alpha_1 h_1 + \alpha_2 h_2] = B[\alpha_1 k_1 + \alpha_2 k_2],$$

so that

$$C[\alpha_1 h_1 + \alpha_2 h_2] = \alpha_1 P_{\overline{\mathcal{R}[B^\star]}}[k_1] + \alpha_2 P_{\overline{\mathcal{R}[B^\star]}}[k_2] = \alpha_1 C[h_1] + \alpha_2 C[h_2].$$

C is closed. Suppose indeed that $h_n \to h$, and that

$$P_{\overline{\mathcal{R}[B^\star]}}[k_n] = C[h_n] \to k.$$

The latter equality means that $A[h_n] = B[k_n]$. Consequently, as [8, p. 363] $B[k_n] = B[P_{\overline{\mathcal{R}[B^\star]}}[k_n]]$, $A[h] = B[k]$. As $k \in \overline{\mathcal{R}[B^\star]}$, $C[h] = k$. It follows (closed graph theorem [8, p. 272]) that C is continuous.

When range inclusion obtains, from item 1,

$$\begin{aligned} \langle AA^\star[l], l\rangle_L &= \|A^\star[l]\|_H^2 \\ &= \|C^\star B^\star[l]\|_H^2 \\ &\le \|C^\star\|^2 \|B^\star[l]\|_K^2 \\ &= \|C^\star\|^2 \langle BB^\star[l], l\rangle_L . \end{aligned}$$

Suppose conversely that latter equality prevails (with κ instead of $\|C^\star\|^2$). Let $h \in H$ be fixed, but arbitrary. Let

$$\varphi_h (B^\star[l]) = \langle h, A^\star[l]\rangle_H .$$

One has that

$$|\varphi_h (B^\star[l])| \le \|h\|_H \|A^\star[l]\|_H^2 \le \kappa \|B^\star[l]\|_H^2 .$$

φ_h extends thus to a continuous, linear functional on the closure of the range of $B^\star$. There exists thus $k \in \overline{\mathcal{R}[B^\star]}$ such that $\langle h, A^\star[l]\rangle_H = \langle k, B^\star[l]\rangle_K$. Consequently,

$$\langle A[h], l\rangle_L = \langle B[k], l\rangle_L ,$$

or $A[h] = B[k]$, that is, $\mathcal{R}[A] \subseteq \mathcal{R}[B]$. □

Remark 8.2.13 One sees, using (Lemma) 8.2.12, that (Definition) 8.2.10 has the consequence that $\mathcal{R}[B^\star] = \mathcal{R}[C_B^{1/2}]$. (Lemma) 8.2.12 is another avatar of Douglas's range inclusion result [80].

Remark 8.2.14 The terminology used in (Definition) 8.2.10 comes from the following particular case.

Let T be an interval of reals, $f : T \longrightarrow H$, and $g : T \longrightarrow K$, maps. Suppose, with, for example, $C_f(t_1, t_2) = \langle f(t_1), f(t_2)\rangle_H$, that

$$C_g \ll \kappa^2 C_f.$$

There is then [(Proposition) 3.1.5] an operator $J_{f,g} : H\left(C_f, T\right) \longrightarrow H\left(C_g, T\right)$ for which, for $t \in T$, fixed, but arbitrary,

$$J_{f,g}\left[C_f(\cdot, t)\right] = C_g(\cdot, t).$$

Let $U_f : H\left(C_f, T\right) \longrightarrow L_{\cup T}[f]$ be the isometry that associates $f(t)$ with $C_f(\cdot, t)$. Then

$$g(t) = U_g J_{f,g} U_f^\star [f(t)] = B[f(t)],$$

and $(B, V[\{f(t), t \in T\}], K)$ is a generalized process for $L_\cup[f]$ and $L_\cup[g]$.

Typically H and K are L_2 spaces over a probability space, and f and g second order processes. One then deals with linear operations on second order processes of the following type: $Y_t = A[X_t]$, and invariance of multiplicity means then that certain structures of the input remain unaltered at the output.

Here are a couple of concrete cases illustrating (Definition) 8.2.10.

Example 8.2.15 Let $f : T \longrightarrow K$ be a map with orthogonal increments. For fixed, but arbitrary $h \in H = L_2\left(T, \mathcal{T}, M_f\right)$, let

$$B[h] = \int h\, dm_f.$$

Then

$$\langle B[h_1], B[h_2]\rangle_K = \langle h_1, h_2\rangle_{L_2(T,\mathcal{T},M_f)},$$

so that $V[H] = L_2\left(T, \mathcal{T}, M_f\right)$, and C_B is the identity operator of H.

Example 8.2.16 Let X be a second order, measurable stochastic process. Let its mean be zero, and its covariance be C_X. Suppose that C_X is a $(2,2)$-bounded kernel for $L_2(T, \mathcal{T}, Leb)$ [(Definition) 2.1.10]. The following assignment, valid for fixed, but arbitrary $h \in H = L_2(T, \mathcal{T}, Leb)$, produces then a generalized process with values in the space $K = L_2(\Omega, \mathcal{A}, P)$ (thus $B = X$):

$$\dot{X}[h](\omega) : (\omega, h) \mapsto \langle X[\omega], h\rangle_{L_2(T,\mathcal{T},Leb)},$$

for which

$$\langle X[h_1], X[h_2]\rangle_{L_2(\Omega,\mathcal{A},P)} = \langle \Gamma_X[h_1], h_2\rangle_{L_2(T,\mathcal{T},Leb)},$$

where Γ_X is the operator of H with kernel C_X.

Remark 8.2.17 Let $(B, V[H], K)$ be a generalized process on H, and let $\{V_t,\ t \in T\}$ be an increasing sequence of manifolds of H. Let

$$L_t[B] = \overline{V[\{B[h],\ h \in V_t\}]}.$$

There are operators $\Gamma_B : H \longrightarrow H$ such that

$$C_B = \Gamma_B^{\star}\Gamma_B$$

(for example $\Gamma_B = UC_B^{1/2}$, U a partial isometry). Let $H_t^{\Gamma_B}$ be the closure of the range of Γ_B restricted to V_t. Then

$$\langle B[h_1], B[h_2]\rangle_K = \langle \Gamma_B[h_1], \Gamma_B[h_2]\rangle_H,$$

and thus the families of subspaces

$$\{L_t[B],\ t \in T\} \text{ and } \{H_t^{\Gamma_B},\ t \in T\}$$

are isometric, that is, there is a unitary $U : L_{\cup T}[B] \longrightarrow H$ such that

$$U[L_t[B]] = H_t^{\Gamma_B}.$$

The search of operators B that make H_t, the closure of V_t, and $L_t[B]$ isometric reduces thus to the search for operators $\Gamma : H \longrightarrow H$ that make H_t, isometric to H_t^{Γ}, the closure of the range of Γ restricted to V_t, and have the property that $\Gamma^{\star}\Gamma = C_B$. It is the introduction of increasing families of subspaces, and their images, that, in such an instance, gives sense to the notion of generalized process. Otherwise it suffices to use Douglas's theorem [80] to have that B and Γ_B are unitarily related.

Example 8.2.18 For the generalized process of (Example) 8.2.16, the "obvious" choices for $V[H]$ and V_t are as follows. One has that $H = L_2(T, \mathcal{T}, Leb)$. For fixed, but arbitrary $\{h_1, \ldots, h_n\} \subseteq H$, and $\{\alpha_i,\ i \in [1:n]\}$, functions with compact support, and square that is integrable, $h(t) = \sum_{i=1}^{n} \alpha_i(t) h_i \in H$, and these functions form a manifold dense in H. $V[H]$ is that manifold, and V_t is made of the same functions restricted to T_t.

Definition 8.2.19 Let $(B, V[H], K)$ be a generalized process. Let

(a) $\{V_t,\ t \in T\}$ be a family of increasing manifolds of H,
(b) H_t be the closure of V_t,
(c) $H = \bigvee_{t\in T} H_t$,
(d) $L_t[B]$ be the closure of the manifold generated by the variables

$$\{B[h],\ h \in V_t\},$$

(e) $L_{\cup T}[B]$ be the closure of the manifold generated by $\{B[h],\ h \in V[H]\}$.

Then one says that

1. $(B, V[H], K)$ is regular when there is a unitary operator

$$U : H \longrightarrow L_{\cup T}[B]$$

such that, for $t \in T$, fixed, but arbitrary,

$$L_t[B] = UH_t;$$

2. the operator C_B of Definition 8.2.10 can be represented as a product of factors when it may be written in the following form:

$$C_B = \Gamma^{\star}\Gamma,$$

where $\Gamma : H \longrightarrow H$ is a bounded, linear operator such that, for $t \in T$, fixed, but arbitrary,

$$H_t^{\Gamma} = H_t,$$

where H_t^{Γ} is still the closure of the range of Γ, restricted to V_t.

Remark 8.2.20 The notion of regularity of a generalized process covers invariance of multiplicity [(Proposition) 8.2.6, (Remark) 8.2.17].

Remark 8.2.21 A regular generalized process has the property that the image of a specific sequence of increasing manifolds, by the operator of the process, essentially reproduces the structure of the starting sequence. One thus senses why regularity has something to do with the preservation of multiplicity.

Remark 8.2.22 The generalized process $(B, V[H], K)$ of (Definition) 8.2.19 is regular if, and only if, C_B can be represented as a product of factors. Invariance of multiplicity is thus a consequence of the ability to adequately factor covariances.

Indeed:

1. *when the process is regular:*

   One has then the following unitary relations:

$$L_t[B] = UH_t, \; t \in T.$$

   Let

$$\Gamma = UC_B^{1/2} : H \longrightarrow L_{UT}[B].$$

   By definition, the domain of Γ is H, and thus contains $V[H]$. As seen in (Remark) 8.2.17, there is a unitary operator $V : L_{UT}[B] \longrightarrow H$ such that $V[L_t[B]] = H_t^{\Gamma}$. Thus $UH_t = L_t[B] = V^{\star}H_t^{\Gamma}$, so that

$$H_t = U^{\star}V^{\star}H_t^{\Gamma} = \hat{H}_t^{\Gamma},$$

   where, as presently seen, $\hat{H}_t^{\Gamma}$ is the closure of the range of the operator $U^{\star}V^{\star}\Gamma$, restricted to V_t. Indeed, when $h = \lim_n U^{\star}V^{\star}\Gamma[h_n]$,

$$VU[h] = \lim_n \Gamma[h_n] \in H_t^{\Gamma},$$

   so that $h \in U^{\star}V^{\star}H_t^{\Gamma}$. Conversely, when

$$h = U^{\star}V^{\star}\left[\hat{h}\right], \text{ with } \hat{h} = \lim_n \Gamma\left[\hat{h}_n\right],$$

   $h = \lim_n U^{\star}V^{\star}\Gamma\left[\hat{h}_n\right] \in \hat{H}_t^{\Gamma}$.

   One then sets $\hat{\Gamma} = U^{\star}V^{\star}\Gamma$. Then, by the very definition of Γ,

   - $\hat{\Gamma}^{\star}\hat{\Gamma} = C_B$,
   - $\hat{\Gamma} : H \longrightarrow H$ is linear and bounded,
   - H_t is the closure of the range of $\hat{\Gamma}$, restricted to V_t.

2. *when C_B can be represented as a product of factors using the operator Γ:*

   As seen in (Remark) 8.2.17, there exists a unitary $V : L_{UT}[B] \longrightarrow H$ such that

$$V[L_t[B]] = H_t^{\Gamma}.$$

   But then, for $t \in T$, fixed, but arbitrary, $L_t[B] = UH_t^{\Gamma}$, with $U = V^{\star}$.

Remark 8.2.23 Any two representations of C_B as a product of factors in the sense of (Definition) 8.2.19 are equal within a unitary operator U such that $UH_t = H_t$, $t \in T$.

The polar decomposition yields indeed that

- $\Gamma = UC_B^{1/2}$, U a partial isometry with
  - initial set: $\overline{\mathcal{R}[C_B^{1/2}]}$,
  - final set: $\overline{\mathcal{R}[\Gamma]}$;
- $\hat{\Gamma} = \hat{U}C_B^{1/2}$, $\hat{U}$ a partial isometry with
  - initial set: $\overline{\mathcal{R}[C_B^{1/2}]}$,
  - final set: $\overline{\mathcal{R}[\hat{\Gamma}]}$.

Thus $U^\star\Gamma = \hat{U}^\star\hat{\Gamma}$ and, since $UU^\star$ is the projection onto the closure of the range of Γ [266, p. 86],

$$\Gamma = U\hat{U}^\star\hat{\Gamma}.$$

Consequently, given that H_t^Γ is the closure of the range of Γ, restricted to V_t, and $H_t^{\hat{\Gamma}}$, that of $\hat{\Gamma}$ restricted to V_t,

$$H_t^\Gamma = U\hat{U}^\star H_t^{\hat{\Gamma}},$$

which, given the assumptions, rewrites as

$$H_t = U\hat{U}^\star H_t,\ t \in T.$$

But then the partial isometry $U\hat{U}^\star$ must be unitary.

The result which follows provides generally sufficient conditions for a covariance to be represented as a product of factors.

Proposition 8.2.24 *Let H be a real Hilbert space, and $\{H_t,\ t \in T\}$ be a family of increasing subspaces of H such that $H = \bigvee_{t\in T} H_t$. Let $C : H \longrightarrow H$ be a bounded, linear operator that is positive and self-adjoint. Suppose that*

$$C = C_-C_+$$

with

(a) *C_- and C_+ linear, bounded, with bounded inverse,*
(b) *for $t \in T$, fixed, but arbitrary,*

$$C_+\left[H_t\right] = H_t,\ \ C_-\left[H_t^\perp\right] \subseteq H_t^\perp.$$

Then:

1. *$C_-^\star\left[H_t\right] \subseteq H_t$;*
2. *$C_+^{-1}\left[H_t\right] = H_t$;*

3. C can be represented as a product of factors.

Proof ([1]) Let $t \in T$, $h_1 \in H_t$, and $h_2 \in H_t^{\perp}$, be fixed, but arbitrary. Then, using Assumption (a)

$$\langle C_-^{\star}[h_1], h_2\rangle_H = \langle h_1, C_-[h_2]\rangle_H = 0.$$

Thus $C_-^{\star}[h_1] \in \{H_t^{\perp}\}^{\perp} = H_t$.

Proof ([2]) Let $h \in H_t$ be fixed, but arbitrary. Because of Assumption (b), there is $h_1 \in H_t$ such that $h = C_+[h_1]$. Then, because of Assumption (a), $C_+^{-1}[h] = h_1 \in H_t$, so that $C_+^{-1}[H_t] \subseteq H_t$. But, still because of Assumption (a), $h = C_+^{-1}C_+[h]$, so that, as, because of Assumption (b), $C_+[h] \in H_t$, C_+^{-1} is onto H_t.

Proof ([3]) Let $D = \{C_+^{-1}\}^{\star} CC_+^{-1}$. Then, since, by assumption, C is self-adjoint and $C = C_-C_+$,

$$\begin{aligned} D &= \{C_+^{\star}\}^{-1} C^{\star}C_+^{-1} \\ &= \{C_+^{\star}\}^{-1} C_+^{\star}C_-^{\star}C_+^{-1} \\ &= C_-^{\star}C_+^{-1}. \end{aligned}$$

Thus, using conclusion 1 only, $D[H_t] = C_-^{\star}C_+^{-1}[H_t] = C_-^{\star}[H_t] \subseteq H_t$. Suppose that $h \in H_t$, and also that $h \perp D[H_t]$. Then, for fixed, but arbitrary $h_1 \in H_t$, using the definition of D,

$$0 = \langle h, D[h_1]\rangle_H = \left\langle C\left[C_+^{-1}[h]\right], C_+^{-1}[h_1]\right\rangle_H.$$

Let $h_1 = h$. Since C is positive, self-adjoint, and has bounded inverse, as the product of two operators with bounded inverse, there is $\kappa > 0$ such that

$$\kappa \left\|C_+^{-1}[h]\right\|_H^2 \leq \left\langle C\left[C_+^{-1}[h]\right], C_+^{-1}[h_1]\right\rangle_H = 0.$$

Consequently $C_+^{-1}[h] = 0$. But C_+^{-1} has C_+ as inverse, and thus

$$h = C_+C_+^{-1}[h] = C_+[0_H] = 0_H,$$

or $D[H_t] = H_t$.

By definition, D is bounded, positive, and self-adjoint. It has thus a square root. Let $\Gamma = D^{1/2}C_+$. Then $\Gamma^{\star}\Gamma = C_+^{\star}DC_+ = C$. Furthermore, as presently seen, using Assumption (b),

$$\Gamma[H_t] = D^{1/2}C_+[H_t] = D^{1/2}[H_t] = H_t.$$

Indeed, the range of D is contained in that of its square root, so that $H_t = D[H_t] \subseteq D^{1/2}[H_t]$. But then, multiplying the latter relation by $D^{1/2}$, $D^{1/2}[H_t] \subseteq D[H_t] = H_t$, and thus $D^{1/2}[H_t] = H_t$. Consequently $C = \Gamma^\star \Gamma$ is a factorization of C in the sense of (Definition) 8.2.19. □

One shall prove below that operators of the following form: $C = I_H - R$, R Hilbert-Schmidt, may be represented as products of factors. The proof proceeds in two steps. The first one uses a decomposition of operators as finite sums which involve projections to secure that the assumptions of 8.2.25 below obtain. The second step amounts to a limiting procedure on those representations in terms of finite sums.

For what follows, one makes consistently the following assumptions:

Assumptions 8.2.25 *1. H is a real Hilbert space, and $\{H_t, t \in T\}$ is an increasing family of subspaces of H, with respective associated projections P_t, such that*

$$\bigvee_{t \in T} H_t = H.$$

2. $C : H \longrightarrow H$ is a linear, bounded, positive, and self-adjoint operator with bounded inverse, and a representation of the following form:

$$C = I_H - R.$$

The First Step: The Finite Sums Case

The first lemma serves to show that, for any projection Q, $I_H - QRQ$ is also positive, and has bounded inverse.

Lemma 8.2.26 *Assume that (Assumption) 8.2.25 obtains, and that Q is a projection of H. Then:*

1. $\|R\| = \sup_{\|h\|=1} \langle R[h], h\rangle_H = \rho < 1$;
2. $I_H - QRQ$ is positive, with bounded inverse, and

$$\|I_H - QRQ\|^{-1} \leq (1-\rho)^{-1}.$$

Proof As [(Fact) 1.3.17] $\langle R[h], h\rangle_H = \|R^{1/2}\|^2 = \left\{\|R\|^{1/2}\right\}^2$, the first part of item 1 follows from the definition. Since C has bounded inverse, there exists $\kappa > 0$ such that, for fixed, but arbitrary $h \in H$ such that $\|h\|_H = 1$,

$$0 < \kappa \leq \langle C[h], h\rangle_H = 1 - \langle R[h], h\rangle_H.$$

Thus

$$\sup_{\|h\|_H=1} \langle R[h], h\rangle_H = 1 - \inf_{\|h\|_H=1} \langle C[h], h\rangle_H \leq 1 - \kappa.$$

As $1-\kappa \geq 1$ implies $\kappa \leq 0$, one must have that

$$\sup_{\|h\|_H=1} \langle R[h], h\rangle_H = 1 - \inf_{\|h\|_H=1} \langle C[h], h\rangle_H = \rho < 1.$$

Now, as $\langle R[Q[h]], Q[h]\rangle_H \leq \|R\| \, \|Q[h]\|^2 \leq \|R\| \, \|h\|^2$,

$$\inf_{\|h\|_H=1} \langle (I_H - QRQ)[h], h\rangle_H = 1 - \sup_{\|h\|_H=1} \langle R[Qh], Q[h]\rangle_H \geq 1-\rho > 0.$$

Consequently $I_H - QRQ$ is a positive operator, with bounded inverse. Furthermore, when h has norm one, as

$$\langle (I_H - QRQ)[h], h\rangle_H \leq \|I_H - QRQ\|,$$

one gets $0 < 1-\rho \leq \|I_H - QRQ\|$. □

The following definition yields in fact the decomposition of operators that shall be used to reach the assigned goal.

Definition 8.2.27 Let $\{Q_i,\ i \in [0:n]\}$ be an increasing family of projections of H such that $Q_0 = O_H$, and $Q_n = I_H$. Then $\sum_{i=1}^n \{Q_i - Q_{i-1}\} = I_H$. Thus, given an operator B, one has that

$$\begin{aligned} B &= \left(\sum_{i=1}^n \{Q_i - Q_{i-1}\}\right) B \left(\sum_{i=1}^n \{Q_i - Q_{i-1}\}\right) \\ &= \sum_{i,j} \{Q_i - Q_{i-1}\} B \{Q_j - Q_{j-1}\} \\ &= \sum_{i<j} \{Q_i - Q_{i-1}\} B \{Q_j - Q_{j-1}\} + \sum_{i \geq j} \{Q_i - Q_{i-1}\} B \{Q_j - Q_{j-1}\}. \end{aligned}$$

One then sets

$$S_Q(B) = \sum_{i<j} (Q_i - Q_{i-1}) B (Q_j - Q_{j-1})$$

and

$$S_Q^c(B) = \sum_{i \geq j} (Q_i - Q_{i-1}) B (Q_j - Q_{j-1}).$$

Thus, by definition,

$$B = S_Q(B) + S_Q^c(B).$$

Remark 8.2.28 Let $B_{i,j} = \{Q_i - Q_{i-1}\} B \{Q_j - Q_{j-1}\}$. Then $S_Q(B)$ contains the components of the following table:

$S_Q(B)$	1	2	3	4	5	$\cdots$	$n-1$	n
1		$B_{1,2}$	$B_{1,3}$	$B_{1,4}$	$B_{1,5}$	$\cdots$	$B_{1,n-1}$	$B_{1,n}$
2			$B_{2,3}$	$B_{2,4}$	$B_{2,5}$	$\cdots$	$B_{2,n-1}$	$B_{2,n}$
3				$B_{3,4}$	$B_{3,5}$	$\cdots$	$B_{3,n-1}$	$B_{3,n}$
4					$B_{4,5}$	$\cdots$	$B_{4,n-1}$	$B_{4,n}$
5							$B_{5,n-1}$	$B_{5,n}$
$\vdots$							$\vdots$	$\vdots$
$n-1$								$B_{n-1,n}$

From that table, it is immediate that

$$S_Q(B) = \sum_{i=1}^{n-1} \sum_{j=i+1}^{n} B_{i,j} = \sum_{j=2}^{n} \sum_{i=1}^{j-1} B_{i,j},$$

$$S_Q^c(B) = \sum_{i=1}^{n-1} \sum_{j=1}^{i} B_{i,j}.$$

Let $B = \beta_1 B^{(1)} + \beta_2 B^{(2)}$. Then $B_{i,j} = \beta_1 B_{i,j}^{(1)} + \beta_2 B_{i,j}^{(2)}$, and, since the terms $B_{i,j}$ have the form $C_i B C_j$, $S_Q(B)$ and $S_Q^c(B)$ are linear in B.

Lemma 8.2.29 *In the context of (Definition)* 8.2.27 *one has that*

$$S_Q(B) = \sum_{i=1}^{n-1} Q_i B (Q_{i+1} - Q_i),$$

and that

$$S_Q^c(B) = \sum_{i=1}^{n-1} (Q_i - Q_{i-1}) B Q_i.$$

Proof Using the explicit expression for $B_{i,j}$,

$$\sum_{i=1}^{j-1} B_{i,j} = Q_{j-1} B (Q_j - Q_{j-1}).$$

The first formula then results from (Remark) 8.2.28. For the second one, one proceeds analogously. with

$$\sum_{j=1}^{i} B_{i,j} = \sum_{j=1}^{i} (Q_i - Q_{i-1}) B \left(Q_j - Q_{j-1}\right) = (Q_i - Q_{i-1}) BQ_i.$$

□

The following lemma shows that $S_P(B)$ and $S_P^c(B)$ provide operators of the form required in (Proposition) 8.2.24.

Lemma 8.2.30 *Assume that (Assumption)* 8.2.25 *obtains. Fix arbitrarily* $\{t_1, \ldots, t_n\}$
$\subseteq T$ *such that* $t_1 < \cdots < t_n$. *Set*

$$t_0 < t_1, \text{ and } P_{t_0} = O_H,$$

$$t_{n+1} > t_n, \text{ and } P_{t_{n+1}} = I_H.$$

Let B be an operator of H, and $S_P(B)$ *and* $S_P^c(B)$ *be as in (Lemma)* 8.2.29 (one chooses, as the Q projections, the projections P_t). *Then*

1. $S_P(B)[H_{t_1}] = \{0_H\}$ (zero subspace)*;*
2. *for* $k \in [2 : n]$,

$$S_P(B)[H_{t_k}] = \sum_{i=1}^{k-1} P_{t_i} B\left(P_{t_{i+1}} - P_{t_i}\right)[H_{t_k}] \subseteq H_{t_{k-1}};$$

3. *for* $k \in [1 : n]$,

$$S_P^c(B)\left[H_{t_k}^{\perp}\right] = \sum_{i=k}^{n} \left(P_{t_{i+1}} - P_{t_i}\right) BP_{t_{i+1}}\left[H_{t_k}^{\perp}\right] \subseteq H_{t_k}^{\perp}.$$

Proof One must have in mind the following fact [266, p. 84]: given two projections, P and Q, $Q - P$ is a projection if, and only if, the range of P is contained in that of Q, and, in that latter case, it obtains that

$$\mathcal{R}[Q - P] = \mathcal{R}[Q] \cap \mathcal{R}[P^{\perp}].$$

The range of $P_{t_{i+1}} - P_{t_i}$ is thus $H_{t_{i+1}} \cap H_{t_i}^{\perp}$, and then, for $i \geq k$, $P_{t_{i+1}} - P_{t_i}$, acting on H_{t_k}, yields the zero subspace. Thus, when $k = 1$, one has that $S_P(B)[H_{t_1}] = \{0_H\}$, and, when $k > 1$, as

$$S_P(B)[H_{t_k}] = \sum_{i=1}^{k-1} P_{t_i} B \left(P_{t_{i+1}} - P_{t_i}\right)[H_{t_k}],$$

the sum to the right in that latter expression shall have all its values in the range of $P_{t_{k-1}}$, that is, $H_{t_{k-1}}$. Items 1 and 2 thus obtain.

In a similar vein, P_{t_i} acting on $H_{t_k}^{\perp}$ yields the zero subspace when $i \leq k$. Thus (the sum goes to n as one has $P_{t_{n+1}} = I_H$)

$$S_P^c(B)\left[H_{t_k}^{\perp}\right] = \sum_{i=k+1}^{n} \left(P_{t_i} - P_{t_{i-1}}\right) B P_{t_i}\left[H_{t_k}^{\perp}\right].$$

Now $\left(P_{t_i} - P_{t_{i-1}}\right) B P_{t_i}\left[H_{t_k}^{\perp}\right]$ is a subset of $H_{t_{i-1}}^{\perp}$. But, since the subspaces $H_t^{\perp}$ are decreasing when t increases in T,

$$H_{t_{i-1}}^{\perp} \subseteq H_{t_k}^{\perp},\ i > k.$$

Consequently $S_P^c(B)\left[H_{t_k}^{\perp}\right] \subseteq H_{t_k}^{\perp}$. □

The following "appendix" to (Lemma) 8.2.30 is central to what follows.

Remark 8.2.31 Let $H_{t_{n+1}} = H$. For fixed, but arbitrary $k \in [1 : n+1]$, one has that

$$S_P(B)\,S_P(B)[H_{t_k}] \subseteq S_P(B)[H_{t_{k-1}}] \subseteq H_{t_{k-2}}.$$

Thus

$$S_P^{k-1}(B)[H_{t_k}] \subseteq H_{t_1},$$

and $S_P^k(B)[H_{t_k}]$ is the zero subspace. Consequently

$$S_P^{n+1}(B)\left[H_{t_{n+1}}\right] = S_P^{n+1}(B)[H] = \{0_H\}.$$

$S_P(B)$ is thus nilpotent [1, p. 244], and its spectrum is reduced to the zero value.

Lemma 8.2.32 *Assume that (Assumption)* 8.2.25 *obtains. Because of (Lemma)* 8.2.26, *the following definition makes sense:*

$$R_+ = \sum_{i<j} \left(I_H - P_{t_j} R P_{t_j}\right)^{-1} \left(P_{t_{i+1}} - P_{t_i}\right) R \left(P_{t_{j+1}} - P_{t_j}\right).$$

Then

$$S_P(R[I_H + R_+]) = R_+.$$

Proof Let

- $R_j = \left(I_H - P_{t_j} R P_{t_j}\right)^{-1}$,
- $R_{i,j} = B_{i,j}$ (R takes the place of B in the proof of (Lemma) 8.2.29),
- $\tilde{R}_{i,j} = R_j R_{i,j}$.

Then

$$R_+ = \sum_{i<j} R_j R_{i,j} = \sum_{i<j} \tilde{R}_{i,j}.$$

Consequently, as in (Remark) 8.2.28,

$$R_+ = \sum_{i=0}^{n-1} \sum_{j=i+1}^{n} \tilde{R}_{i,j} = \sum_{j=1}^{n} \sum_{i=0}^{j-1} \tilde{R}_{i,j}.$$

But

$$\sum_{i=0}^{j-1} \tilde{R}_{i,j} = \sum_{i=0}^{j-1} R_j R_{i,j} = R_j \sum_{i=0}^{j-1} R_{i,j} = R_j P_{t_j} R \left(P_{t_{j+1}} - P_{t_j}\right),$$

as in the proof of (Lemma) 8.2.29. Thus

$$R_+ = \sum_{i=1}^{n} R_i P_{t_i} R \left(P_{t_{i+1}} - P_{t_i}\right). \tag{$\star$}$$

Then, using (Lemma) 8.2.29, and the definition of R_+,

$$\begin{aligned} S_P\left(RR_+\right) &= \sum_{i=1}^{n} P_{t_i} \left(RR_+\right) \left(P_{t_{i+1}} - P_{t_i}\right) \\ &= \sum_{i=1}^{n} P_{t_i} R \left\{ \sum_{j=1}^{n} R_j P_{t_j} R \left(P_{t_{j+1}} - P_{t_j}\right) \right\} \left(P_{t_{i+1}} - P_{t_i}\right) \\ &= \sum_{i=1}^{n} P_{t_i} R \left\{ R_i P_{t_i} R \left(P_{t_{i+1}} - P_{t_i}\right) \right\}. \end{aligned}$$

But

$$\left(I_H - P_{t_i} R P_{t_i}\right) P_{t_i} = P_{t_i} - P_{t_i} R P_{t_i} = P_{t_i} \left(I_H - P_{t_i} R P_{t_i}\right),$$

and, from $AB = BA$, one has, successively, that, when A is invertible, that indeed $B = A^{-1}BA$, and then $BA^{-1} = A^{-1}B$. Thus R_i and P_{t_i} commute. Consequently, as a projection equals its square,

$$S_P(RR_+) = \sum_{i=1}^{n} \{P_{t_i}RP_{t_i}\} R_i P_{t_i} R \left(P_{t_{i+1}} - P_{t_i}\right).$$

Finally, since S_P is additive, and since $(I-A)(I-A)^{-1} = I$ yields

$$(I-A)^{-1} = I + A(I-A)^{-1}, \tag{$\star\star$}$$

letting $I = I_H$, and $A = P_{t_i}RP_{t_i}$,

$$(I-A)^{-1} = (I_H - P_{t_i}RP_{t_i})^{-1} = R_i, \tag{$\star\star\star$}$$

and

$$\begin{aligned} S_P(R[I_H + R_+]) &= S_P(R) + S_P(RR_+) \\ &= \sum_{i=1}^{n} \left[P_{t_i}R\left(P_{t_{i+1}} - P_{t_i}\right)\right] \\ &+ \sum_{i=1}^{n} \{P_{t_i}RP_{t_i}\} R_i \left[P_{t_i}R\left(P_{t_{i+1}} - P_{t_i}\right)\right] \\ &= \sum_{i=1}^{n} (I_H + \{P_{t_i}RP_{t_i}\} R_i) \left[P_{t_i}R\left(P_{t_{i+1}} - P_{t_i}\right)\right]. \end{aligned}$$

Now, using successively $(\star\star\star)$ and then $(\star\star)$,

$$I_H + \{P_{t_i}RP_{t_i}\} R_i = R_i,$$

so that

$$S_P(R[I_H + R_+]) = \sum_{i=1}^{n} R_i \left[P_{t_i}R\left(P_{t_{i+1}} - P_{t_i}\right)\right],$$

which, because of $(\star)$, is R_+. □

Corollary 8.2.33 *Assume that (Assumption)* 8.2.25 *obtains.* R_+ *shall be as in (Lemma)* 8.2.32. *Then*

$$\lambda \mapsto (R_+ - \lambda I_H)^{-1}$$

is, for $\lambda \neq 0$, *linear and bounded, and* $I_H + R_+$ *has bounded inverse.*

Proof Since R_+ is of the form $S_P(B)$, its spectrum reduces to zero, that is, $(R_+ - \lambda I_H)^{-1}$ is linear and bounded for $\lambda \neq 0$. Choosing $\lambda = -1$, one gets that $I_H + R_+$ has bounded inverse. □

Lemma 8.2.34 *Assume that (Assumption)* 8.2.25 *obtains.* R_+ *shall be as in (Lemma)* 8.2.32. *Let*

$$R_- = S_P^c\,(R\,[I_H + R_+])\,.$$

Then

$$C = I_H - R = (I_H - R_-)\,(I_H + R_+)^{-1}\,.$$

Proof The inverse of $I_H + R_+$ exists because of (Corollary) 8.2.33. From the following identity [(Definition) 8.2.27]:

$$B = S_P\,(B) + S_P^c\,(B)$$

one has that [(Corollary) 8.2.33]

$$\begin{aligned}(I_H - R)\,(I_H + R_+) &= (I_H + R_+) - R\,(I_H + R_+)\\ &= (I_H + R_+) - S_P\,(R\,[I_H + R_+]) - S_P^c\,(R\,[I_H + R_+])\\ &= (I_H + R_+) - R_+ - R_-\\ &= I_H - R_-.\end{aligned}$$

□

Lemma 8.2.35 *Let H be a Hilbert space, K, a closed subspace, and T, a bounded, linear operator of H, which leaves K invariant, that is, such that* $T[K] \subseteq K$. *Suppose that* $I_H + T$ *has an inverse that is linear and bounded. Then* $(I_H + T)^{-1}$ *leaves K invariant.*

Proof Since K is a subspace, $I_H + T$ leaves K invariant. As a necessary, and sufficient, condition for a subspace K to be invariant with respect to a bounded, linear operator B, is that $BP_K = P_K B P_K$ [236, p. 298],

$$P_K = (I_H + T)^{-1}\,[(I_H + T)\,P_K] = (I_H + T)^{-1}\,[P_K\,(I_H + T)\,P_K]\,.$$

Multiplying by P_K that latter expression, one also obtains that

$$P_K = P_K\,(I_H + T)^{-1}\,P_K\,(I_H + T)\,P_K.$$

Thus equating the right-hand sides of the latter two expressions, one obtains that

$$(I_H + T)^{-1}\,P_K\,[(I_H + T)\,P_K] = P_K\,(I_H + T)^{-1}\,P_K\,[(I_H + T)\,P_K]\,.$$

Consequently, on $\overline{\mathcal{R}[(I_H + T)\, P_K]}$,

$$(I_H + T)^{-1}\, P_K = P_K\, (I_H + T)^{-1}\, P_K.$$

But, since $I_H + T^\star$ must have bounded inverse [266, p. 71, 74], using [266, p. 71],

$$\overline{\mathcal{R}\,(I_H + T)\, P_K]} = \mathcal{N}[P_K\,(I_H + T^\star)]^\perp = K^{\perp\perp} = K.$$

The claim is thus true. □

Lemma 8.2.36 *Assume that (Assumption)* 8.2.25 *obtains.* R_+ *shall be as in (Lemma)* 8.2.32, *and* R_-, *as in (Lemma)* 8.2.34. *Let*

$$C_+ = (I_H + R_+)^{-1}\,, \text{ and } C_- = I_H - R_-.$$

Then [(Lemma) 8.2.34] $C = C_- C_+$ *can be represented as a product of factors for the family of subspaces* $\{H_{t_k},\ k \in [0, n+1]\}$ (H_{t_0} is the zero subspace).

Proof The definitions of C_+ and C_- insure that they are linear, bounded, with bounded inverse. To apply (Proposition) 8.2.24, one must check that

$$C_+\,[H_{t_k}] \subseteq H_{t_k}, \text{ and that } C_-\left[H_{t_k}^\perp\right] \subseteq H_{t_k}^\perp.$$

As R_+ is of the form $S_P\,(B)$ [(Lemma) 8.2.32], $R_+\,[H_{t_k}] \subseteq H_{t_{k-1}} \subseteq H_{t_k}$ [(Lemma) 8.2.30], and thus H_{t_k} is invariant for R_+, and, consequently, for $I_H + R_+$ [(Lemma) 8.2.35]. One then applies (Corollary) 8.2.33 and (Lemma) 8.2.35. Analogously, as R_- is of the form $S_P^c\,[B]$, $H_{t_k}^\perp$ is invariant for C_- [(Lemma) 8.2.30]. □

Lemma 8.2.36 shows that C's of the form $I_H - R$ can be represented as a product of factors when one restricts attention to a finite sample of subspaces H_t. Since the size of the sample is irrelevant, one may try to let it increase indefinitely. Let R_n^+ be R_+ for the sample considered below in (Fact) 8.2.40, and suppose that a sequence of such operators has a limit R_+, as the sample increases, and that $I_H + R_+$ is an operator with bounded inverse. Then R_n^- will also have a limit, and one will be able to represent C as a product of factors. What follows is geared to proving those facts.

The Second Step: The Limiting Procedure

What follows requires an added assumption:

Assumptions 8.2.37 *R is Hilbert-Schmidt.*

The properties of Hilbert-Schmidt operators may be found in [235], for example.

Lemma 8.2.38 *Assume that (Assumptions)* 8.2.25 *and* 8.2.37 *obtain. Let* $\{Q_{t_i},\ i \in [0, n]\}$ *be a family of projections as in (Definition)* 8.2.27, *and* $\{B_i,\ i \in [0:n]\}$ *be a family of bounded, linear operators of* H. *Then*

$$\left\| \sum_{i=1}^{n} B_i R \left(Q_{t_i} - Q_{t_{i-1}}\right) \right\|_{HS}^{2} = \sum_{i=1}^{n} \left\| B_i R \left(Q_{t_i} - Q_{t_{i-1}}\right) \right\|_{HS}^{2}.$$

Proof Let H_i be the range of $Q_{t_i} - Q_{t_{i-1}}$, and

$$\left\{ e_j^{(i)},\ j \in J_i \right\}$$

be a complete orthonormal set in H_i. Then, since the (orthogonal) sum of the H_i's is H,

$$\left\{ e_j^{(i)},\ j \in J_i,\ i \in [0:n] \right\}$$

is a complete orthonormal set in H. Since a finite sum of Hilbert-Schmidt operators is Hilbert-Schmidt, as well as the product of a Hilbert-Schmidt operator by a bounded one, $\sum_{i=1}^{n} B_i R \left(Q_{t_i} - Q_{t_{i-1}}\right)$ is a Hilbert-Schmidt operator. Thus, using the definition of the Hilbert-Schmidt norm, and considering the "effective domain" of the operators concerned, the H_i's,

$$\begin{aligned}
\left\| \sum_{i=1}^{n} B_i R \left(Q_{t_i} - Q_{t_{i-1}}\right) \right\|_{HS}^{2} &= \sum_{i \in [1:n], j \in J_i} \left\| B_i R \left(Q_{t_i} - Q_{t_{i-1}}\right) \left[e_j^{(i)} \right] \right\|_{H}^{2} \\
&= \sum_{i=1}^{n} \left\{ \sum_{j \in J_i} \left\| B_i R \left(Q_{t_i} - Q_{t_{i-1}}\right) \left[e_j^{(i)} \right] \right\|_{H}^{2} \right\} \\
&= \sum_{i=1}^{n} \left\| B_i R \left(Q_{t_i} - Q_{t_{i-1}}\right) \right\|_{HS}^{2}.
\end{aligned}$$

□

Lemma 8.2.39 *Assume that (Assumptions)* 8.2.25 *and* 8.2.37 *obtain. Let* B_1 *and* B_2 *be bounded, linear operators of* H. *Then*

$$\| B_1 R B_2 \|_{HS}^{2} = \| B_2^{\star} R B_1^{\star} \|_{HS}^{2},$$

and

$$\| B_1 R B_2 \|_{HS}^{2} \leq \| B_1 \|^{2} \, \| R \|_{HS}^{2} \, \| B_2 \|^{2}.$$

Proof R is self-adjoint and Hilbert-Schmidt. Thus B_1RB_2 is Hilbert-Schmidt, and so is its transpose $B_2^\star RB_1^\star$. As a Hilbert-Schmidt operator and its transpose have the same Hilbert-Schmidt norm, the first assertion obtains. The second is a consequence of the following inequality:

$$\|BR\|_{HS} \vee \|RB\|_{HS} \leq \|B\| \, \|R\|_{HS} ,$$

valid for Hilbert-Schmidt R and bounded B. □

Fact 8.2.40 *Let $T_n = \{t_i^{(n)} : t_i^{(n)} < t_{i+1}^{(n)}, \ i \in [0 : n]\}$ be a set of indices such that*

$$P_{t_0^{(n)}} = O_H, \ P_{t_{n+1}^{(n)}} = I_H, \ \text{and, for } i \in [1 : n-1] , \ t_i^{(n)} \in T.$$

In conformity with notation used in the first step [(Lemmas) 8.2.30, 8.2.32], *one shall write that*

$$R_{t_i^{(n)}} = \left(I_H - P_{t_i^{(n)}} R P_{t_i^{(n)}}\right)^{-1} ,$$

and that

$$R_n^+ = \sum_{i=1}^{n} R_{t_i^{(n)}} P_{t_i^{(n)}} R \left(P_{t_{i+1}^{(n)}} - P_{t_i^{(n)}}\right) .$$

Let $T_n \subseteq T_{n+p}$, and $t_{(i)}^{(n+p)} = t_i^{(n)}$. Then, J_i being the set of indices of the elements of T_{n+p} located between $t_i^{(n)}$ and $t_{i+1}^{(n)}$,

$$P_{t_{i+1}^{(n)}} - P_{t_i^{(n)}} = P_{t_{(i+1)}^{(n+p)}} - P_{t_{(i)}^{(n+p)}} = \sum_{j \in J_i} \left(P_{t_{j+1}^{(n+p)}} - P_{t_j^{(n+p)}}\right) ,$$

and

$$R_n^+ = \sum_{i=1}^{n} \sum_{j \in J_i} \left\{ R_{t_{(i)}^{(n+p)}} P_{t_{(i)}^{(n+p)}} R \left(P_{t_{j+1}^{(n+p)}} - P_{t_j^{(n+p)}}\right)\right\} .$$

Consequently

$$R_{n+p}^+ - R_n^+ =$$

$$= \sum_{i=1}^{n} \sum_{j \in J_i} \left\{ \left[R_{t_j^{(n+p)}} P_{t_j^{(n+p)}} - R_{t_{(i)}^{(n+p)}} P_{t_{(i)}^{(n+p)}}\right] R \left(P_{t_{j+1}^{(n+p)}} - P_{t_j^{(n+p)}}\right)\right\} .$$

Lemma 8.2.41 *Let Assumptions* 8.2.25, 8.2.37, *and Fact* 8.2.40 *obtain. The following equality prevails* (ρ is defined in (Lemma) 8.2.26):

$$\left\|R^+_{n+p} - R^+_n\right\|^2_{HS} \leq$$
$$\leq 2\sum_{i=1}^{n}\sum_{j\in J_i}\left\|R_{t_j^{(n+p)}} - R_{t_{(i)}^{(n+p)}}\right\|^2 \left\|R\left(P_{t_{j+1}^{(n+p)}} - P_{t_j^{(n+p)}}\right)\right\|^2_{HS}$$
$$+ \frac{2}{(1-\rho)^2}\sum_{i=1}^{n}\sum_{j\in J_i}\left\|\left(P_{t_{(i+1)}^{(n+p)}} - P_{t_{(i)}^{(n+p)}}\right) R \left(P_{t_{j+1}^{(n+p)}} - P_{t_j^{(n+p)}}\right)\right\|^2_{HS}.$$

Proof Adding and subtracting $R_{t_j^{(n+p)}} P_{t_{(i)}^{(n+p)}}$, one has that

$$R_{t_j^{(n+p)}} P_{t_j^{(n+p)}} - R_{t_{(i)}^{(n+p)}} P_{t_{(i)}^{(n+p)}} =$$
$$= R_{t_j^{(n+p)}}\left(P_{t_j^{(n+p)}} - P_{t_{(i)}^{(n+p)}}\right) + \left(R_{t_j^{(n+p)}} - R_{t_{(i)}^{(n+p)}}\right) P_{t_{(i)}^{(n+p)}}.$$

The $\{\cdots\}$ bracket, in the initial expression for $R^+_{n+p} - R^+_n$ [(Fact) 8.2.40], then becomes

$$R_{t_j^{(n+p)}}\left(P_{t_j^{(n+p)}} - P_{t_{(i)}^{(n+p)}}\right) R \left(P_{t_{j+1}^{(n+p)}} - P_{t_j^{(n+p)}}\right) +$$
$$+ \left(R_{t_j^{(n+p)}} - R_{t_{(i)}^{(n+p)}}\right) P_{t_{(i)}^{(n+p)}} R \left(P_{t_{j+1}^{(n+p)}} - P_{t_j^{(n+p)}}\right).$$

Using, with the Hilbert-Schmidt norm, the Hilbert space norm inequality

$$\|h+k\|^2 \leq 2\left\{\|h\|^2 + \|k\|^2\right\},$$

and then (Lemma) 8.2.39, one gets that the Hilbert-Schmidt norm of the $\{\cdots\}$ bracket [(Fact) 8.2.40] is dominated by 2 times

$$\left\|R_{t_j^{(n+p)}}\right\|^2 \left\|\left(P_{t_j^{(n+p)}} - P_{t_{(i)}^{(n+p)}}\right) R \left(P_{t_{j+1}^{(n+p)}} - P_{t_j^{(n+p)}}\right)\right\|^2_{HS} +$$
$$+ \left\|R_{t_j^{(n+p)}} - R_{t_{(i)}^{(n+p)}}\right\|^2 \left\|P_{t_{(i)}^{(n+p)}}\right\|^2 \left\|R\left(P_{t_{j+1}^{(n+p)}} - P_{t_j^{(n+p)}}\right)\right\|^2_{HS}.$$

The required inequality follows then from (Lemmas) 8.2.26 and 8.2.38. □

Lemma 8.2.42 *Let Assumptions* 8.2.25, 8.2.37, *and Fact* 8.2.40 *obtain. Let also, for fixed, but arbitrary* $t_1 < t_2$, $\{t_1, t_2\} \subseteq T$,

$$\mu_R([t_1, t_2[) = \|R(P_{t_2} - P_{t_1})\|_{HS}^2 .$$

Then μ_R *can be extended to a finite measure on the Borel sets of* T.

Proof Because of (Lemma) 8.2.38, μ_R is additive on the semiring formed by the intervals, and thus can be extended to an additive set function on the generated ring. Since the elements of the ring are finite unions of elements in the semiring, to have that μ_R is continuous from above at the empty set, and thus σ-additive, it suffices to check that $\mu_R([t_1, t_2[)$ goes to zero when the length of the interval does. Now, as R is Hilbert-Schmidt and self-adjoint, it has a representation of the following form:

$$R = \sum_{i \in I} \rho_i r_i \otimes r_i,$$

where $\{r_i,\ i \in I\}$ is an orthonormal family, and $\sum_{i \in I} \rho_i^2 < \infty$. Furthermore, for the (any) Hilbert-Schmidt operator $R(P_{t_2} - P_{t_1})$,

$$\begin{aligned}
\|R(P_{t_2} - P_{t_1})\|_{HS}^2 &= \langle R(P_{t_2} - P_{t_1}), R(P_{t_2} - P_{t_1})\rangle_{HS} \\
&= \left\langle \{R(P_{t_2} - P_{t_1})\}^\star, \{R(P_{t_2} - P_{t_1})\}^\star \right\rangle_{HS} \\
&= \|(P_{t_2} - P_{t_1}) R\|_{HS}^2 .
\end{aligned}$$

Thus, since the "effective domain" of R is spanned by its eigenvectors,

$$\begin{aligned}
\mu_R([t_1, t_2[) &= \|R(P_{t_2} - P_{t_1})\|_{HS}^2 \\
&= \|(P_{t_2} - P_{t_1}) R\|_{HS}^2 \\
&= \sum_{i \in I} \|(P_{t_2} - P_{t_1}) R[r_i]\|_H^2 \\
&= \sum_{i \in I} \rho_i^2 \|(P_{t_2} - P_{t_1})[r_i]\|_H^2 .
\end{aligned}$$

But $\lim_{t_1 \uparrow t_2} P_{t_1} = P_{t_2}$, and thus, by dominated convergence,

$$\lim_{t_1 \uparrow t_2} \mu_R([t_1, t_2[) = 0.$$

One then calls on the standard extension theorem for measures [83, p. 64]. □

Lemma 8.2.43 *Let Assumptions* 8.2.25, 8.2.37, *and Fact* 8.2.40 *obtain. Let, for fixed, but arbitrary* $t_1 < t_2,\ t_3 < t_4,\ \{t_1, t_2, t_3, t_4\} \subseteq T$,

$$\nu_R\left([t_1, t_2[\times [t_3, t_4[\right) = \|(P_{t_2} - P_{t_1}) R (P_{t_4} - P_{t_3})\|_{HS}^2 .$$

Then ν_R *can be extended to a finite measure on the Borel sets of* $T \times T$.

Proof The argument is that of (Lemma) 8.2.42. □

Lemma 8.2.44 *Let Assumptions* 8.2.25, 8.2.37, *and Fact* 8.2.40 *obtain. Let*

$$\{T_n,\ n \in \mathbb{N}\}$$

be a sequence of partitions as defined in (Fact) 8.2.40, *which have the property that* $T_n \subseteq T_{n+1},\ n \in \mathbb{N}$. *Then*

$$\lim_{n,p} \left\|R_n^+ - R_p^+\right\|_{HS}^2 = 0.$$

Proof One uses (Lemma) 8.2.41 as follows.

The estimate for the first term of Lemma 8.2.41:

Let $t_1 < t_2,\ \{t_1, t_2\} \subseteq T$, be fixed, but arbitrary, and

$$\phi(t) = \nu_R(T_t \times T_t) = \|P_t R P_t\|_{HS}^2 \le \|P_t\|^2 \|R\|_{HS}^2 \|P_t\|^2 \le \|R\|_{HS}^2 .$$

ϕ is thus a positive function, continuous to the left, increasing, and bounded by the square of Hilbert-Schmidt norm of R. As, using the properties of projections, and those of the Hilbert-Schmidt inner product,

$$\begin{aligned}
\langle P_{t_1} R P_{t_1}, P_{t_2} R P_{t_2} \rangle_{HS} &= \langle P_{t_1} R P_{t_1}, P_{t_1} R P_{t_2} \rangle_{HS} \\
&= \left\langle (P_{t_1} R P_{t_1})^\star , (P_{t_1} R P_{t_2})^\star \right\rangle_{HS} \\
&= \langle P_{t_1} R P_{t_1}, P_{t_2} R P_{t_1} \rangle_{HS} \\
&= \langle P_{t_1} R P_{t_1}, P_{t_1} R P_{t_1} \rangle_{HS} ,
\end{aligned}$$

one has that

$$\|P_{t_1} R P_{t_1} - P_{t_2} R P_{t_2}\|_{HS}^2 = \|P_{t_2} R P_{t_2}\|_{HS}^2 - \|P_{t_1} R P_{t_1}\|_{HS}^2 = \phi(t_2) - \phi(t_1) .$$

Now [(Lemma) 8.2.32, beginning of proof, for the notation]

$$R_{t_2}\left(R_{t_1}^{-1} - R_{t_2}^{-1}\right) R_{t_1} = R_{t_2} - R_{t_1},$$

and

$$R_{t_2}^{-1} - R_{t_1}^{-1} = P_{t_1} R P_{t_1} - P_{t_2} R P_{t_2}.$$

The latter difference is Hilbert-Schmidt, its Hilbert-Schmidt norm, as seen just above, has value $\phi(t_2) - \phi(t_1)$, and, using (Lemmas) 8.2.26 and 8.2.39,

$$\begin{aligned}\|R_{t_2} - R_{t_1}\|_{HS}^2 &= \left\|R_{t_2}\left(R_{t_1}^{-1} - R_{t_2}^{-1}\right)R_{t_1}\right\|_{HS}^2 \\ &\leq \frac{1}{(1-\rho)^2}\left\|R_{t_1}^{-1} - R_{t_2}^{-1}\right\|_{HS}^2 \\ &= \frac{1}{(1-\rho)^2}\{\phi(t_2) - \phi(t_1)\}.\end{aligned}$$

As the operator norm is dominated by the Hilbert-Schmidt norm, one finally has that

$$\|R_{t_2} - R_{t_1}\|^2 \leq \frac{1}{(1-\rho)^2}\{\phi(t_2) - \phi(t_1)\}.$$

Consequently

$$\sum_{i=1}^{n}\sum_{j\in J_i}\left\|R_{t_j^{(n+p)}} - R_{t_{(i)}^{(n+p)}}\right\|^2\left\|R\left(P_{t_{j+1}^{(n+p)}} - P_{t_j^{(n+p)}}\right)\right\|_{HS}^2 \leq$$

$$\leq \frac{1}{(1-\rho)^2}\sum_{i=1}^{n}\sum_{j\in J_i}\left\{\phi\left(t_j^{(n+p)}\right) - \phi\left(t_{(i)}^{(n+p)}\right)\right\}\mu_R\left(\left[t_j^{(n+p)}, t_{j+1}^{(n+p)}\right[\right).$$

This last sum is the product of a constant by the difference of two terms, say I_{n+p} and $\tilde{I}_{n+p}$. One has that

$$\begin{aligned}I_{n+p} &= \sum_{i=1}^{n}\sum_{j\in J_i}\phi\left(t_j^{(n+p)}\right)\mu_R\left(\left[t_j^{(n+p)}, t_{j+1}^{(n+p)}\right[\right) \\ &= \int_T\sum_{i=1}^{n}\sum_{j\in J_i}\phi\left(t_j^{(n+p)}\right)\chi_{\left[t_j^{(n+p)}, t_{j+1}^{(n+p)}\right[}(t)\,\mu_R(dt),\end{aligned}$$

with ϕ uniformly bounded by $\|R\|_{HS}^2$, and μ_R finite. Similarly

$$\tilde{I}_{n+p} = \int_T\sum_{i=1}^{n}\sum_{j\in J_i}\phi\left(t_{(i)}^{(n+p)}\right)\chi_{\left[t_j^{(n+p)}, t_{j+1}^{(n+p)}\right[}(t)\,\mu_R(dt) = I_n.$$

As n and p increase indefinitely, the elements I_{n+p} and $\tilde{I}_{n+p}$ form increasing sequences tending to the same finite limit, and thus the first term of (Lemma) 8.2.41 is dominated by a quantity whose limit is zero.

The estimate for the second term of Lemma 8.2.41:

Let

$$\Delta(\delta) = \{(t_1, t_2) \in T \times T : t_1 < t_2 < t_1 + \delta\} :$$

it is the "open strip" of width $\delta/\sqrt{2}$ above and adjacent to the diagonal in $T \times T$. Let

$$\delta = \max_{i \in [0:n]} \left(t_{i+1}^{(n)} - t_i^{(n)}\right).$$

Then, when

$$t_{(i)}^{(n+p)} \leq t_1 < t_j^{(n+p)}, \quad t_j^{(n+p)} \leq t_2 < t_{j+1}^{(n+p)}, \quad t_{j+1}^{(n+p)} - t_{(i)}^{(n+p)} < \delta,$$

one has that

$$t_1 < t_j^{(n+p)} \leq t_2 < t_{j+1}^{(n+p)} < \delta + t_{(i)}^{(n+p)} \leq \delta + t_1.$$

Consequently, when the "(n)-partition" in T has intervals of length at most δ, the intervals of the "$(n+p)$-partition" have the property that

$$\left[t_{(i)}^{(n+p)}, t_j^{(n+p)}\right[\times \left[t_j^{(n+p)}, t_{j+1}^{(n+p)}\right[\subseteq \Delta(\delta).$$

Now the second term of (Lemma) 8.2.41 is a constant times

$$\sum_{i=1}^{n} \sum_{j \in J_i} \left\| \left(P_{t_{(i+1)}^{(n+p)}} - P_{t_{(i)}^{(n+p)}}\right) R \left(P_{t_{j+1}^{(n+p)}} - P_{t_j^{(n+p)}}\right) \right\|_{HS}^2 =$$

$$= \sum_{i=1}^{n} \sum_{j \in J_i} \nu_R \left(\left[t_{(i)}^{(n+p)}, t_{(i+1)}^{(n+p)}\right[\times \left[t_j^{(n+p)}, t_{j+1}^{(n+p)}\right[\right)$$

$$= \nu_R \left(\bigcup_{i \in [1:n], j \in J_i} \left[t_{(i)}^{(n+p)}, t_{(i+1)}^{(n+p)}\right[\times \left[t_j^{(n+p)}, t_{j+1}^{(n+p)}\right[\right)$$

$$\leq \nu_R(\Delta(\delta)).$$

As n increases, δ decreases, and $\Delta(\delta)$ decreases to the empty set. □

Proposition 8.2.45 *Let Assumptions* 8.2.25, 8.2.37, *and Fact* 8.2.40 *obtain. In the Hilbert space of Hilbert-Schmidt operators,* $\lim_n R_n^+$ *exists, and is thus a Hilbert-Schmidt operator. It shall be denoted* R_+.

Proof The result is an immediate consequence of (Lemma) 8.2.44, and the fact that Hilbert-Schmidt operators form a Hilbert space for the Hilbert-Schmidt norm. □

Lemma 8.2.46 *Let Assumptions* 8.2.25, 8.2.37, *and Fact* 8.2.40 *obtain. Suppose that, for some* $\lambda \in \mathbb{C} \setminus \{0\}$, *but all* $n \in \mathbb{N}$,

$$R_\lambda^{(n)} = \left(R_n^+ - \lambda I_H\right)^{-1}, \text{ and } R_\lambda^+ = \left(R_+ - \lambda I_H\right)^{-1},$$

are bounded, linear operators. Then $\left\{R_\lambda^{(n)},\ n \in \mathbb{N}\right\}$ *converges uniformly* (operator norm) *to* R_λ^+.

Proof From $I_H = \left(R_+ - \lambda I_H\right)^{-1}\left(R_+ - \lambda I_H\right) = R_\lambda^+\left(R_+ - \lambda I_H\right)$, it follows that

$$I_H - R_\lambda^+ R_+ = -\lambda R_\lambda^+.$$

Adding $R_\lambda^+ R_n^+$ to the latter expression, one gets that

$$I_H - R_\lambda^+\left(R_+ - R_n^+\right) = R_\lambda^+\left(R_n^+ - \lambda I_H\right).$$

Inverting that latter equality, on obtains that

$$\left\{I_H - R_\lambda^+\left(R_+ - R_n^+\right)\right\}^{-1} = R_\lambda^{(n)}\left\{R_\lambda^+\right\}^{-1},$$

so that, finally,

$$\left\{I_H - R_\lambda^+\left(R_+ - R_n^+\right)\right\}^{-1} R_\lambda^+ = R_\lambda^{(n)}.$$

Consequently

$$R_\lambda^+ - R_\lambda^{(n)} = \left\{I_H - \left[I_H - R_\lambda^+\left(R_+ - R_n^+\right)\right]^{-1}\right\} R_\lambda^+.$$

Because of (Proposition) 8.2.45, for n large enough, $\left[I_H - R_\lambda^+\left(R_+ - R_n^+\right)\right]^{-1}$ has the Neumann series [129, p. 161] expansion

$$\left[I_H - R_\lambda^+\left(R_+ - R_n^+\right)\right]^{-1} = \sum_{k=0}^{\infty}\left\{R_\lambda^+\right\}^k\left(R_+ - R_n^+\right)^k.$$

Consequently

$$\begin{aligned} R_\lambda^+ - R_\lambda^{(n)} &= \left\{\sum_{k=1}^{\infty}\left\{R_\lambda^+\right\}^k\left(R_+ - R_n^+\right)^k\right\} R_\lambda^+ \\ &= R_\lambda^+\left(R_+ - R_n^+\right)\left\{\sum_{k=0}^{\infty}\left\{R_\lambda^+\right\}^k\left(R_+ - R_n^+\right)^k\right\} R_\lambda^+, \end{aligned}$$

so that, for $n \in \mathbb{N}$ large enough, still because of [129, p. 161],

$$\left\|R_\lambda^+ - R_n^+\right\| \leq \frac{\left\|R_\lambda^+\right\|^2 \left\|R_+ - R_n^+\right\|}{1 - \left\|R_\lambda^+ \left(R_+ - R_n^+\right)\right\|}.$$

The lemma's statement thus obtains because of (Proposition) 8.2.45, and the fact that the Hilbert-Schmidt norm dominates the operator one. □

Remark 8.2.47 The result which follows uses a property of analytic functions with values in a Banach space that shall now be stated [74, p. 241].

Definition 8.2.48 Let X_0 be a subset of the metric space X. A point $x_0 \in X_0$ is an isolated point of X_0 when there is a neighborhood of x_0 in X, say V_0, such that $V_0 \cap X_0 = \{x_0\}$ *(example: X_0 is the complement of a disc in the complex plane, centered at the origin, with positive radius, from which that origin has been removed).*

Fact 8.2.49 *Let*

(a) *X be a Banach space,*
(b) *C_0, an open subset of $\mathbb{C}$,*
(c) *z_0, an isolated point of $\mathbb{C} \setminus C_0$,*
(d) *$\rho > 0$, such that $\{z \in \mathbb{C} : |z - z_0| \leq \rho\} \setminus \{z_0\} \subseteq C_0$,*
(e) *$f : C_0 \longrightarrow X$, analytic.*

Then, when z is such that $0 < |z - z_0| < \rho$,

$$f(z) = \sum_{n=0}^{\infty} \alpha_n (z - z_0)^n + \sum_{n=1}^{\infty} \frac{\beta_n}{(z - z_0)^n}.$$

Both series converge, and, Γ_0 being the circle centered at z_0 whose radius is ρ,

$$\alpha_n = \frac{1}{2\pi i} \int_{\Gamma_0} \frac{f(z)}{(z - z_0)^{n+1}} dz,$$

$$\beta_n = \frac{1}{2\pi i} \int_{\Gamma_0} (z - z_0)^{n-1} f(z)\, dz.$$

Remark 8.2.50 When, in (Fact) 8.2.49, one can compute the integrals producing the β's, and obtain zero, it follows that f is analytic at z_0.

The result just stated shall be used with $f(\lambda) = R_\lambda^+$, $C_0 = \rho(R_+)$, the resolvent set of R_+, and $z_0 \in \sigma_p(R_+)$, the point spectrum of R_+.

Proposition 8.2.51 *Let Assumptions* 8.2.25, 8.2.37, *and Fact* 8.2.40 *obtain. Then*

$$\sigma(R_+) = \{0\}.$$

Proof Since R_+ is compact [(Proposition) 8.2.45],

$$\sigma(R_+) = \sigma_p(R_+) \cup \{0\}.$$

Let $\lambda_0 \in \sigma_p(R_+)$ be fixed, but arbitrary. It is an isolated point of the spectrum of R_+. There is thus a punctured disk contained in the resolvent set of R_+, centered at λ_0, with boundary Γ_0. Since R_n^+, defined in (Fact) 8.2.40, has the form of R_+ in (Lemma) 8.2.32, which, in turn, is of the form $S_P(B)$ [(Lemma) 8.2.29], and that operators of that form have a spectrum reduced to the zero point [(Remark) 8.2.31],

$$\lambda \mapsto R_\lambda^{(n)},$$

defined in (Lemma) 8.2.46, is analytic on $\mathbb{C} \setminus \{0\}$, and

$$\int_{\Gamma_0} \lambda^p R_\lambda^{(n)} d\lambda = 0.$$

Since $\{R_\lambda^{(n)},\ n \in \mathbb{N}\}$ converges uniformly to R_λ^+ [(Lemma) 8.2.46], and that $\lambda \mapsto R_\lambda^+$ is bounded on Γ_0 [266, p. 99],

$$\lim_n \int_{\Gamma_0} \lambda^p R_\lambda^{(n)} d\lambda = \int_{\Gamma_0} \lambda^p R_\lambda^+ d\lambda.$$

The latter integral must be zero, and thus the resolvent set of R_+ is $\mathbb{C} \setminus \{0\}$. □

Proposition 8.2.52 *Let Assumptions* 8.2.25, 8.2.37, *and Fact* 8.2.40 *obtain. The operator* $C = I_H - R$ *can be represented as the product* $C_- C_+$ *of the following factors:*

$$C_+ = (I_H + R_+)^{-1}, \text{ and } C_- = C(I_H + R_+).$$

Proof Since the spectrum of R_+ reduces to zero, $I_H + R_+$ has bounded inverse, and thus C_+ is well defined. Consequently

$$C_- C_+ = C(I_H + R_+)(I_H + R_+)^{-1} = C.$$

One must thus only check the invariance of the subspaces H_t and $H_t^\perp$.

Invariance of H_t*:*

Let $h \in H_t$ be fixed, but arbitrary. Since

$$H_t = \bigvee_{t_0 < t, t_0 \in T} H_{t_0},$$

there exists sequences

$$\{t_n,\ n \in \mathbb{N} : t_n \uparrow t\}\,,$$

and

$$\left\{h_n,\ n \in \mathbb{N} : h_n \in H_{t_n} \text{ and } \lim_n h_n = h\right\}.$$

Let

$$\left\{R_+^{p[n])},\ p[n] < p[n+1]\,,\ n \in \mathbb{N}\right\}$$

be a sequence such that the sample on which $R_+^{p[n])}$ is based contains t_n. Then

$$R_+^{p[n])}[h_n] \in H_{t_n} \subseteq H_t,$$

so that

$$R_+[h_n] = \lim_{p[n]} R_+^{p[n])}[h_n] \in H_t.$$

Consequently $R_+[h] = \lim_n R_+[h_n] \in H_t$. But then $(I_H + R_+)[H_t] \subseteq H_t$, and thus, using (Lemma) 8.2.35,

$$C_+[H_t] = (I_H + R_+)^{-1}[H_t] \subseteq H_t.$$

Invariance of $H_t^\perp$*:*

One proceeds analogously. Let indeed $C_n = C\left(I_H + R_n^+\right)$. Then

$$C_- = \lim_n C_n.$$

But [(Lemma) 8.2.34, using R_n^- for R_-],

$$C = \left(I_H - R_n^-\right)\left(I_H + R_n^+\right)^{-1},$$

so that

$$C_n = I_H - R_n^-.$$

Now R_n^- is of the form $S_P^c(B)$, for which the appropriate $H_t^\perp$ subspaces are invariant. □

Since, for Gaussian processes X and Y, the measures they determine, through the cylinder sets of $\mathbb{R}^T$, are equivalent if, and only if, there is

$$F : L_{\cup T}[X] \longrightarrow L_{\cup T}[Y],$$

with bounded inverse, such that $Y_t = FX_t$, and $I - F^{\star}F$ is Hilbert-Schmidt [97], one has that:

Proposition 8.2.53 *Equivalent Gaussian processes have the same multiplicity.*

Proof One is in the situation of (Remark) 8.2.14, and one applies (Remark) 8.2.22, since

$$\langle Y_{t_1}, Y_{t_2}\rangle_{L_2(\Omega,\mathcal{A},P)} = \langle FX_{t_1}, FX_{t_2}\rangle_{L_2(\Omega,\mathcal{A},P)} = \langle F^{\star}FX_{t_1}, X_{t_2}\rangle_{L_2(\Omega,\mathcal{A},P)}.$$

□

8.3 Smoothness and Multiplicity: Multiplicity One

The multiplicity one category of processes is mainly of use when choosing a model to adjust to data, as it is difficult, perhaps impossible, knowing the covariance, to obtain explicitly its factorization (see, for instance, example (Remark) 8.3.13), and then its analytical properties, such as continuity, or differentiability, of its components. The same is true for Goursat processes which broaden the class of models one may entertain. Thus, for practical purposes, multiplicity one is of primary importance. Hence the need to assess its scope and the constraints which bear on it.

The rule of thumb is that the smoother the process, the lower the multiplicity. That rule is illustrated below in three cases. It is shown, in the first, that integrals of very smooth integrands, with respect to processes with orthogonal increments, lead to multiplicity one. In the second, it is shown that processes with rather smooth predictions have multiplicity one. The third covers Goursat processes, which are linear combinations of orthogonal processes with orthogonal increments, and the coefficients of the linear combinations, functions. They have potentially "high" (> 1) multiplicity: one shall see that the smoothness of the coefficients reduces that multiplicity.

In expressions of the form

$$g(t) = \int \gamma(t)\, dm_f,$$

$\gamma(t)$ is an equivalence class in $L_2\left(T, \mathcal{T}, M_f\right)$, that of functions of the following form: $\gamma(t,\theta)$, $\theta \in T_t$. When working with examples, it is often easier to deal with integrals of Itô type, that is, with processes of the following form:

$$g(t,\omega) = \int_0^t \gamma(t,\theta)\, M(\omega, d\theta),$$

where M is a martingale in L_2, and γ is adequately measurable. That these two points of view are compatible follows from the following facts:

- the Itô stochastic integral may be defined as an isometric integral [150, p. 53] in the classes of which an appropriate version (stochastic process) is chosen;
- since

$$C_g(t_1, t_2) = \langle \gamma(t_1), \gamma(t_2) \rangle_{L_2(T, \mathcal{T}, M_f)},$$

when C_g is adapted, and the subspace linearly generated by $\{\gamma(t), t \in T\}$, separable, then, in the equivalence class $\gamma(t)$, an appropriate version may be chosen so that γ is adequately measurable as a function of two variables (Proposition 2.3.7: typically g will be continuous (in quadratic mean), and the L_2 spaces concerned, separable [41, p. 252]).

Thus, according to convenience, one or the other of these points of view will tacitly prevail.

When, below, f is a function of two variables, $D_1 f$ shall be the partial derivative of f with respect to the first variable, and $D_2 f$ that with respect to the second.

Also repeated use shall be made of the following formula [109, p. 222]:

Fact 8.3.1 *For*

$$\Gamma(t) = \int_a^t G(t,x)\, dx,$$

$$\frac{d\Gamma}{dt}(\theta) = \int_a^\theta D_1 G(\theta, x)\, dx + G(\theta, \theta).$$

8.3.1 Multiplicity One: Smoothness of Integrands

The following definition, tailored to its purpose, yields a family of multiplicity one processes.

Definition 8.3.2 Let T be an interval of the real line, $\mathcal{T}$ be its Borel sets, and H a real Hilbert space. Let $f : T \longrightarrow H$ be a purely nondeterministic function, with orthogonal increments, continuous to the left. Let M_f be the measure induced on $\mathcal{T}$ using f, and $F(t) = M_f(T_t)$. Assume that:

(A) There exists $\{t_1, t_2\} \subseteq T$, $t_1 < t_2$, such that $F(t_1) < F(t_2)$.

(B) There exists an integrable d_F, which has at most a finite number of discontinuities in any finite interval, such that $F(t) = \int_{T_t} d_F(x)\,dx$.

Let $\Delta = \{(t, x) \in T \times T : x \leq t\}$, and $\gamma : T \times T \longrightarrow \mathbb{R}$, a measurable map such that

(a) for fixed, but arbitrary $t \in T$,

$$\int_T \gamma^2(t, x)\, M_f(dx) < \infty;$$

(b) when $x > t$, $\gamma(t, x) = 0$;
(c) for fixed, but arbitrary $t \in T$,

$$\gamma(t, t) = 1;$$

(d) γ and $D_1\gamma$ are bounded, and continuous on Δ.

Let γ_t be the equivalence class of $x \mapsto \gamma(t, x)$, and

$$g(t) = \int I_{T_t} \gamma_t dm_f.$$

The function g is then said to be regular.

Example 8.3.3 Let $m = m_W$, W a standard Wiener process on $T = [0, 1]$, with $H = L_2(\Omega, \mathcal{A}, P)$, a probability space. Choose

$$\gamma(t, \theta) = \chi_{[0,t]}(\theta)(1 + t)(1 + \theta)^{-1},$$

and set

$$g(t) = \int \gamma_t dm_W.$$

g, as shall be seen, is a Goursat process. Referring to (Definition) 8.3.2, $F(t) = t$, $d_F = 1$, (a)–(d) are obviously true, and so g is regular. It has also multiplicity one, as

$$\int \gamma(t, \theta)\phi(\theta)d\theta = 0,\ t \in [0, 1],$$

yields that ϕ is in the class of the zero function.

Remark 8.3.4 As mentioned at the beginning of this section, (Definition) 8.3.2 has been concocted to facilitate the use of standard analysis results. In this particular case, multiplicity one means that, for $t \in T$, fixed, but arbitrary,

$$L_t[g] = L_t[m_f] = \left\{ \int I_{T_t}[\phi] dm_f, [\phi] \in L_2(T, \mathcal{T}, M_f) \right\}.$$

That, in turn, means that, for $h \in L_t[m_f]$, $h \perp L_t[g]$, and $\theta \leq t$, fixed, but arbitrary,

$$h = \int I_{T_t} \tilde{h} dm_f, \quad \tilde{h} \in L_2(T, \mathcal{T}, M_f),$$

and

$$0 = \langle g(\theta), h \rangle_H = \int \gamma(\theta, x) \dot{\tilde{h}}(x) M_f(dx) = \int \gamma(\theta, x) \dot{\tilde{h}}(x) d_F(x) dx.$$

Using formula (Fact) 8.3.1 above, one may differentiate, with respect to θ, to obtain that

$$\int_{T_\theta} D_1 \gamma(\theta, x) \dot{\tilde{h}}(x) d_F(x) dx + \dot{\tilde{h}}(\theta) d_F(\theta) = 0.$$

One thus gets a homogeneous Volterra equation which has only the trivial zero solution [69, p. 239]. g has indeed multiplicity one.

Remark 8.3.5 Definition 8.3.2 and its consequence remain valid for unbounded intervals [63], provided it is assumed that, for fixed, but arbitrary $t \in T$,

$$\int_{T_t} \left| \frac{\partial \gamma}{\partial t}(t, \theta) \right| d\theta < \infty.$$

One can indeed check that the method used to solve the Volterra equation for bounded intervals still applies.

Remark 8.3.6 Remark 8.3.4 is no longer true without the regularity conditions. Here is an example [63].

Example 8.3.7 Let $T =]a, b[$, $a < b$, and f be such that $M_f(T_t) = F(t)$, with $dF = d_F dt$, d_F bounded below (for example, $f = W$, a Wiener process with $a > 0$, or $f(t) = W_{A(t)}$, A an absolutely continuous function that is bounded below). Let

$$\gamma(t, \theta) = 2 \frac{\theta - a}{t - a} - 1.$$

Then

$$\frac{\partial \gamma}{\partial t}(t, \theta) = -2 \frac{\theta - a}{(t - a)^2},$$

which is not bounded when t is close to a. But, choosing ψ such that $\psi d_F = 1$, one gets a square integrable function that is not a null function, but for which

$$\int_{]a,t[} \gamma(t,\theta)\,\psi(\theta)\,d_F(\theta)\,d\theta = 0.$$

Thus $g(t) = \int I_{T_t}\gamma(t)\,dm_f$ is not a proper canonical representation.

Remark 8.3.8 Definition 8.3.2 is of limited scope. Indeed the process

$$g(t) = \int I_{[0,t]}\left[(t-\cdot)\right]_{L_2[0,1]} dm_f,$$

f a standard Wiener process, is such that $\gamma(t,t) = 0$, and (Definition) 8.3.2 does not apply. The representation is however proper canonical. Suppose indeed that

$$\int_0^t (t-u)\,\psi(u)\,du = 0.$$

Differentiating, or using (Fact) 8.3.1, one gets that

$$\int_0^t \psi(x)\,dx + t\psi(t) = t\psi(t),$$

so that $\psi = 0$.

Remark 8.3.9 The requirement in (Definition) 8.3.2 that $\gamma(t,t) = 1$, for regularity to obtain, may be replaced by $\gamma(t,t) > 0$.

Consider indeed the function $\theta \mapsto \gamma(\theta,\theta)$. It is, by assumption, continuous and bounded above. Its values are strictly positive, so that one may divide by $\gamma(t,t)$. The integral

$$\int \left[\gamma(\cdot,\cdot)\right]_{L_2(T,\mathcal{T},M_f)} dm_f$$

is thus well defined. Let m_γ be defined using the following relation:

$$m_\gamma(T_t) = \int I_{T_t}\left[\gamma(\cdot,\cdot)\right]_{L_2(T,\mathcal{T},M_f)} dm_f.$$

Since

$$\int_T \left[\frac{\gamma(t,\theta)}{\gamma(\theta,\theta)}\right]^2 M_\gamma(d\theta) < \infty,$$

the integral

$$\int \left[\frac{\gamma(t,\cdot)}{\gamma(\cdot,\cdot)}\right]_{L_2(T,\mathcal{T},M_f)} dm_\gamma$$

is well defined, and

$$g(t) = \int I_{T_t} \left[\frac{\gamma(t,\cdot)}{\gamma(\cdot,\cdot)}\right]_{L_2(T,\mathcal{T},M_f)} dm_\gamma.$$

Furthermore the ratio in the bracket past the integral sign of the latter expression is up to the requirements of (Definition) 8.3.2.

Remark 8.3.10 In (Remark) 8.3.9, $\gamma(t,t) > 0$ is meant for every t. It excludes thus

$$\gamma(t,x) = (n+1)x - nt,\ t \in [0,1],$$

for which $\gamma(t,t) = t$, and $\int_0^t x^{n-1}\gamma(t,x)\,dx = 0,\ t > 0$. In [115], one finds an example of a process

$$f(t) = \int I_{[0,t]}\gamma_1(t)\,dm_{f_1} + \int I_{[0,t]}\gamma_2(t)\,dm_{f_2}$$

of multiplicity 2 such that $\gamma_1(t,t) + \gamma_2(t,t) = 0$, all $t \in [0,1]$.

Proposition 8.3.11 *Let g be a regular function, as in (Definition)* 8.3.2. *Its covariance C_g is continuous, has continuous partial derivatives off the diagonal of $T \times T$ (except at a countable number of points), and a discontinuity on the diagonal, whose size is provided by the density d_F, that of the measure determined by the integrator in the representation of g as an isometric integral.*

Proof The representation of g as an integral yields that

$$C_g(t_1,t_2) = \int_{T_{t_1\wedge t_2}} \gamma(t_1,x)\,\gamma(t_2,x)\,d_F(x)\,dx.$$

Continuity of the covariance is that of the integral. Differentiability follows from formula (Fact) 8.3.1 which yields, for example, and when $\theta_1 < \theta_2$, using item (c) of (Definition) 8.3.2,

$$D_1C_g(\theta_1,\theta_2) = \int_{T_{\theta_1}} [D_1\gamma(\theta_1,x)]\,\gamma(\theta_2,x)\,d_F(x)\,dx + \gamma(\theta_2,\theta_1)\,d_F(\theta_1).$$

When, in that formula, one lets $\theta_1 \uparrow \theta_2$, one gets (except for an at most countable number of points)

$$D_1 C_g(\theta_2, \theta_2) = \int_{T_{\theta_1}} [D_1 \gamma(\theta_2, x)]\, \gamma(\theta_2, x)\, d_F(x)\, dx + d_F(\theta_2)\,.$$

But, when $\theta_1 > \theta_2$, the analogous formula yields only the first term in the derivative as $\gamma(\theta_1, \theta_2) = 0$. The difference is thus d_F. □

The main use of (Definition) 8.3.2 is that it provides a representation that is unique.

Proposition 8.3.12 *Let g be a regular function as in (Definition)* 8.3.2: *its representation is essentially unique.*

Proof Let, for $i = 1, 2$, $\gamma_t^{(i)}$ be the equivalence class of $\gamma_i(t, \cdot)$ in $L_2(T, \mathcal{T}, M_{f_i})$. Suppose that

$$g(t) = \int I_{T_t} \gamma_t^{(1)}\, dm_{f_1} = \int I_{T_t} \gamma_t^{(2)}\, dm_{f_2},$$

and that the assumptions of (Definition) 8.3.2 obtain for γ_1, γ_2, f_1, and f_2. Since the representations are proper canonical, for $\{t_0, t\} \subseteq T,\ t_0 < t$,

$$g(t) - P_{t_0}^g [g(t)] = \int I_{[t_0,t[} \gamma_t^{(1)} dm_{f_1} = \int I_{[t_0,t[} \gamma_t^{(2)} dm_{f_2}.$$

Letting t_0 be t, and t be $t + \theta,\ \theta > 0$, one has, for $i = 1, 2$, that

$$\int I_{[t_0,t[} \gamma_t^{(i)}\, dm_{f_i}$$

may be written in the following form:

$$\int I_{[t,t+\theta[} dm_{f_i} - \int I_{[t,t+\theta[} \left\{1 - \gamma_{t+\theta}^{(i)}\right\} dm_{f_i}.$$

Consequently, computing a squared norm as an inner product, one has, for $i = 1, 2$, that the following norm:

$$\left\| g(t+\theta) - P_t^g [g(t+\theta)] \right\|_H^2$$

equals

$$A - 2B + C,$$

where

$$A = \|f_i(t+\theta) - f_i(t)\|_H^2 ,$$

$$B = \left\langle \int I_{[t,t+\theta[} dm_{f_i}, \int I_{[t,t+\theta[} \left\{1 - \gamma_{t+\theta}^{(i)}\right\} dm_{f_i} \right\rangle_H$$

$$= \int_{[t,t+\theta[} \{1 - \gamma_i(t+\theta, x)\} M_{f_i}(dx) ,$$

$$C = \left\| \int I_{[t,t+\theta[} \left\{1 - \gamma_{t+\theta}^{(i)}\right\} dm_{f_i} \right\|_H^2$$

$$= \int_{[t,t+\theta[} \{1 - \gamma_i(t+\theta, x)\}^2 M_{f_i}(dx) .$$

Now, on one hand,

$$\|f_i(t+\theta) - f_i(t)\|_H^2 = M_{f_i}([t, t+\theta[) = F_i(t+\theta) - F_i(t) ,$$

and, on the other hand, since $\gamma_i(t,t) = 1$,

$$1 - \gamma_i(t+\theta, x) = \gamma_i(t,t) - \gamma_i(t+\theta, x) ,$$

so that, for θ small enough, because of the continuity assumptions on γ_i, for $i = 1, 2$, and the fact that (t,t) and $(t+\theta, x)$ belong to Δ,

$$|\gamma_i(t,t) - \gamma_i(t+\theta, x)| < \epsilon.$$

Consequently B and C are negligible, and, almost surely,

$$\lim_{\theta \downarrow 0} \frac{\left\|g(t+\theta) - P_t^g [g(t+\theta)]\right\|_H^2}{\theta} = \lim_{\theta \downarrow 0} \frac{F_i(t+\theta) - F_i(t)}{\theta} = d_{F_i}(t) ,$$

so that d_{F_1} and d_{F_2} belong to the same equivalence class with respect to Lebesgue measure.

Similarly, partitioning $[t_1, t_2[$ with intervals $\Delta_j = \left[t_{j-1}, t_j\right[$, and into increments $\Delta_j[f_i] = f_i\left(t_j\right) - f_i\left(t_{j-1}\right)$,

$$\|[f_1(t_2) - f_1(t_1)] - [f_2(t_2) - f_2(t_1)]\|_H^2 = \left\| \sum_j \{\Delta_j[f_1] - \Delta_j[f_2]\} \right\|_H^2 ,$$

and the right-hand side of the latter expression equals, adding and subtracting $\Delta_j[g]$,

$$\left\| \sum_j \left\{ \int \chi_{\Delta_j} \left\{ 1 - \gamma_{t_j}^{(1)} \right\} dm_{f_1} - \int \chi_{\Delta_j} \left\{ 1 - \gamma_{t_j}^{(2)} \right\} dm_{f_2} \right\} \right\|_H^2 =$$

$$= \left\| \int \left\{ \sum_j \chi_{\Delta_j} \left\{ 1 - \gamma_{t_j}^{(1)} \right\} \right\} dm_{f_1} - \int \left\{ \sum_j \chi_{\Delta_j} \left\{ 1 - \gamma_{t_j}^{(2)} \right\} \right\} dm_{f_2} \right\|_H^2$$

$$\leq \int \left\{ \sum_j \chi_{\Delta_j} \left\{ 1 - \gamma_1 \left(t_j, \cdot \right) \right\} \right\}^2 dM_{f_1} + \int \left\{ \sum_j \chi_{\Delta_j} \left\{ 1 - \gamma_2 \left(t_j, \cdot \right) \right\} \right\}^2 dM_{f_2}$$

$$= \int \sum_j \chi_{\Delta_j} \left\{ 1 - \gamma_1 \left(t_j, \cdot \right) \right\}^2 dM_{f_1} + \int \sum_j \chi_{\Delta_j} \left\{ 1 - \gamma_2 \left(t_j, \cdot \right) \right\}^2 dM_{f_2}.$$

Let M be the measure $M_{f_1} = M_{f_2}$ (it was checked above that d_{F_1} and d_{F_2} are in the same equivalence class). Since one works with finite intervals, one may assume that continuity is uniform so that, for small enough equal Δ_j's,

$$\| [f_1(t_2) - f_1(t_1)] - [f_2(t_2) - f_2(t_1)] \|_H^2 \leq 2\epsilon^2 M([t_1, t_2[) . \qquad (\star)$$

Let finally $\theta > 0$, and $u > t + \theta$. Then for $i = 1, 2$, considering temporarily $\gamma_i(u, t + \theta)$ as (the equivalence class of) a constant,

$$P_{t+\theta}^g [g(u)] - P_t^g [g(u)] =$$

$$= \int I_{[t,t+\theta[} \gamma_u^{(i)} dm_{f_i}$$

$$= \int I_{[t,t+\theta[} \left\{ \gamma_i(u, t+\theta) - \gamma_i(u, t+\theta) + \gamma_u^{(i)} \right\} dm_{f_i}$$

$$= \gamma_i(u, t+\theta) \int I_{[t,t+\theta[} dm_{f_i} - \int I_{[t,t+\theta[} \left\{ \gamma_i(u, t+\theta) - \gamma_u^{(i)} \right\} dm_{f_i},$$

so that

$$\| \gamma_1(u, t+\theta) \{ f_1(t+\theta) - f_1(t) \} - \gamma_2(u, t+\theta) \{ f_2(t+\theta) - f_2(t) \} \|_H^2 \qquad (\star\star)$$

equals $\|A - B\|_H^2$ where

$$A = \int I_{[t,t+\theta[} \left\{ \gamma_1(u, t+\theta) - \gamma_u^{(1)} \right\} dm_{f_1},$$

$$B = \int I_{[t,t+\theta[} \left\{ \gamma_2(u, t+\theta) - \gamma_u^{(2)} \right\} dm_{f_2}.$$

The expression $(\star\star)$ may be reformulated as the square of the norm of the difference of two terms,

$$\{\gamma_1(u,t+\theta)-\gamma_2(u,t+\theta)\}\{f_1(t+\theta)-f_1(t)\}, \qquad (\star\star\star)$$

and

$$\gamma_2(u,t+\theta)\{[f_1(t+\theta)-f_1(t)]-[f_2(t+\theta)-f_2(t)]\}.$$

By what precedes $(\star)$, the second of those terms is zero, so that $(\star\star)$ equals

$$\{\gamma_1(u,t+\theta)-\gamma_2(u,t+\theta)\}^2 \|f_1(t+\theta)-f_1(t)\|_H^2.$$

But $\|A-B\|_H^2$ is dominated by

$$2\int_{[t,t+\theta[}\left[\{\gamma_1(u,t+\theta)-\gamma_1(u,x)\}^2+\{\gamma_2(u,t+\theta)-\gamma_2(u,x)\}^2\right]M(dx),$$

and, because of the continuity of γ_1 and γ_2, that latter integral is dominated by

$$4\epsilon^2 M([t,t+\theta[) = 4\epsilon^2\|f_1(t+\theta)-f_1(t)\|_H^2.$$

Thus, in the end, using $(\star\star\star)$,

$$|\gamma_1(u,t)-\gamma_2(u,t)|\le 2\epsilon.$$

□

Remark 8.3.13 In practice, one is "given" C_g, and, as shall eventually be seen, needs to deduce γ from it (f is taken to be "white noise," denoted W below). In principle one should also be able to express f in terms of g.

Lévy [173, p. 146] provides a method of solution for the following equation:

$$C_g(s,t)=\int_0^{s\wedge t}\gamma(s,x)\,\gamma(t,x)\,dx,$$

when γ has some regularity properties, those used in the (informal) example below. That example illustrates rather well the difficulties encountered when trying to factor explicitly a covariance.

For a function $(v,u)\mapsto F(v,u)$, the equivalence class, for fixed v, of the following map: $u\mapsto F(v,u)$ shall be denoted F_v.

Example 8.3.14 (Part A) Let W be a wide sense Wiener process on $[0, 1]$, and

$$g(t) = \int I_{T_t} \gamma_t dW$$

be a regular process. The covariance of g is, for an L_2 probability space H,

$$C_g(t_1, t_2) = \langle g(t_1), g(t_2) \rangle_H.$$

Let $\tau_1 < \tau_2 \le t_1 < t_2$, and R be the rectangle with vertices $(\tau_1, t_1), (\tau_1, t_2), (\tau_2, t_1)$, and (τ_2, t_2). Then

$$\Delta_R C_g = C_g(\tau_2, t_2) - C_g(\tau_2, t_1) - C_g(\tau_1, t_2) + C_g(\tau_1, t_1),$$

so that

$$\Delta_R C_g = \langle g(\tau_2) - g(\tau_1), g(t_2) - g(t_1) \rangle_H. \qquad (1)$$

As g is absolutely continuous, so is G_g, which has thus a representation in the following form:

$$C_g(t_1, t_2) = \int_0^{t_1} \int_0^{t_2} \Gamma(x, y)\, dx dy.$$

So, for $\tau_2 - \tau_1$ and $t_2 - t_1$ small enough,

$$\Delta_R C_g \approx \Gamma(\tau_1, t_1)(\tau_2 - \tau_1)(t_2 - t_1). \qquad (2)$$

On the other hand, because regular functions have proper, canonical representations,

$$P_{L_{t_1}[g]}[g(t_2)] - g(t_1) = \int I_{T_{t_1}} \gamma_{t_2} dW - g(t_1) = \int I_{T_{t_1}} \{\gamma_{t_2} - \gamma_{t_1}\} dW, \qquad (3)$$

and

$$\begin{aligned} g(t_2) - g(t_1) &= \int I_{T_{t_1}} \gamma_{t_2} dW - g(t_1) + \int I_{]t_1, t_2]} \gamma_{t_2} dW \\ &= \{P_{L_{t_1}[g]}[g(t_2)] - g(t_1)\} + \int I_{]t_1, t_2]} \gamma_{t_2} dW. \end{aligned}$$

Let $\theta_j = \tau_1$, and $\theta_{j+1} = \tau_2$, in a partition of $[0, t_1]$, using θ_i's. As an element of $L_{t_1}[g]$,

$$P_{L_{t_1}[g]}[g(t_2)] - g(t_1) = \int I_{T_{t_1}} G_{t_1} dm_g, \qquad (4)$$

and

$$\int I_{T_{t_1}} G_{t_1}\, dm_g \approx \sum_{i=1}^{n} G(t_1, \theta_{i-1}) \{g(\theta_i) - g(\theta_{i-1}\}$$
$$= G(t_1, \theta_{j-1}) \{g(\theta_j - \theta_{j-1}\} + \sum_{i \neq j} G(t_1, \theta_{i-1}) \{g(\theta_i) - g(\theta_{i-1}\} .$$

Thus, since $g(\tau_2) - g(\tau_1) \perp L_{t_1}[g]^{\perp}$, using (3), (4), and the latter relation,

$$\langle g(\tau_2) - g(\tau_1), g(t_2) - g(t_1)\rangle_H =$$

$$= \langle g(\tau_2) - g(\tau_1), P_{L_{t_1}}[g(t_2)] - g(t_1)\rangle_H$$
$$= \left\langle g(\tau_2) - g(\tau_1), \int I_{T_{t_1}} G_{t_1}\, dm_g \right\rangle_H$$
$$\approx G(t_1, \tau_1)\, \|g(\tau_2) - g(\tau_1)\|_H^2 + \sum_{i \neq j} G(t_1, \theta_i)\, \Gamma(\tau_1, \theta_i)(\tau_2 - \tau_1)(\theta_{i+1} - \theta_i)$$
$$\approx G(t_1, \tau_1)\, \|g(\tau_2) - g(\tau_1)\|_H^2 + (\tau_2 - \tau_1) \int_0^{t_1} G(t_1, x)\, \Gamma(\tau_1, x)\, dx,$$

that is, using (1) and the latter expression,

$$\Delta_R C_g \approx G(t_1, \tau_1)\, \|g(\tau_2) - g(\tau_1)\|_H^2 + (\tau_2 - \tau_1) \int_0^{t_1} G(t_1, x)\, \Gamma(\tau_1, x)\, dx. \tag{5}$$

There are two more facts to notice. The first is that

$$\|g(\tau_2) - g(\tau_1)\|_H^2 = (\tau_2 - \tau_1)^2 \int_0^{\tau_1} \left\{ \frac{\gamma(t_2, x) - \gamma(t_1, x)}{\tau_2 - \tau_1} \right\}^2 dx + \int_{\tau_1}^{\tau_2} \gamma^2(t_2, x)\, dx,$$

and consequently that

$$\frac{\|g(\tau_2) - g(\tau_1)\|_H^2}{\tau_2 - \tau_1}$$

has, as limit, when τ_2 goes to τ_1, $\gamma^2(\tau_1, \tau_1)$. The second is that, using (3), and comparing with (4), one sees that

$$\gamma(t_2, x) - \gamma(t_1, x) = (t_2 - t_1) D_1 \gamma(t, x),$$

$G(t_1, x)$ has the form $(t_2 - t_1) G^{\star}(t, x)$. So, finally, using (2), and (5), and the last two remarks, the following equation emerges:

$$\begin{aligned}
&\Gamma(\tau_1, t_1)(\tau_2 - \tau_1)(t_2 - t_1) \approx \\
&\quad \approx G^{\star}(t_1, \tau_1) \frac{\|g(\tau_2) - g(\tau_1)\|_H^2}{\tau_2 - \tau_1} (\tau_2 - \tau_1)(t_2 - t_1) \\
&\quad + (\tau_2 - \tau_1)(t_2 - t_1) \int_0^{t_1} G^{\star}(t, x) \Gamma(\tau_1, x)\, dx.
\end{aligned}$$

Going to the limit yields that

$$\Gamma(t, \theta) = G(t, \theta) \gamma^2(\theta, \theta) + \int_0^t G(t, x) \Gamma(\theta, x)\, dx,$$

where the latter G stands for $G^{\star}$ above, for notational convenience. The unknown is G, as Γ and $\theta \longrightarrow \gamma^2(\theta, \theta)$ are obtained from the covariance C_g. For $t \in T$, fixed, but arbitrary, one has then a Volterra equation of the second kind, with kernel

$$\frac{\Gamma(\theta, \tau)}{\gamma^2(\theta, \theta)}.$$

It has a unique solution when the covariance is strictly positive definite.

Example 8.3.15 (Illustration of (Example) 8.3.14*)* Consider the following function:

$$C(s, t) = s \wedge t - st, \ 0 \le s, t \le 1.$$

It is a covariance: indeed

$$\begin{aligned}
&\sum_{i,j=1}^{n} \alpha_i \alpha_j \left(t_i \wedge t_j - t_i t_j\right) = \\
&\quad = \int_0^1 \left\{ \sum_{i=1}^{n} \alpha_i \chi_{[0, t_i]}(x) \right\}^2 dx - \left\{ \int_0^1 \sum_{i=1}^{n} \alpha_i \chi_{[0, t_i]}(x)\, dx \right\}^2
\end{aligned}$$

is a positive number as it is the variance of a random variable. Now, using (Fact) 8.3.1, one has, for $s \le t$, that

$$\frac{\partial C(s, t)}{\partial s} = \gamma(s, s) \gamma(t, s) + \int_0^s \frac{\partial \gamma(s, \theta)}{\partial s} \gamma(t, \theta)\, d\theta.$$

By differentiation under the integral sign one also has that

$$\frac{\partial C(s,t)}{\partial t} = \int_0^s \gamma(s,\theta) \frac{\partial \gamma(t,\theta)}{\partial t} d\theta.$$

Consequently

$$\lim_{s \uparrow\uparrow t} \left\{ \frac{\partial C(s,t)}{\partial s} - \frac{\partial C(s,t)}{\partial t} \right\} = \gamma^2(t,t).$$

Computing these values with $s \wedge t - st$ yields that $\gamma^2(t,t) = 1$. Finally $\Gamma(t,\theta) = -1$. The equation for G is thus

$$-1 = G(t,\theta) - \int_0^t G(t,\theta)\, d\theta.$$

For t fixed, but arbitrary, $G(t,\theta)$ is the constant $c_t = \frac{-1}{1-t}$.

Example 8.3.16 (Part B of (Example) 8.3.14*)* At this point G is "known." One then sets:

(a) $f(t) = \int I_{T_t} G_t dm_g$,
(b) $F(t) = \int_0^t f(\theta)\, d\theta$,
(c) $\Phi(t) = g(t) - F(t)$,
(d) $\phi(t) = \int I_{T_t} G_t dm_\Phi$.

From (c) above, Δ_t denoting an increment of size Δ at t,

$$\Delta_t \Phi = \Delta_t g - \Delta_t F = \Delta_t g - f(t)\Delta.$$

Now, using (4) in (3) of (Example) 8.3.14, one sees that

$$f(t) = \int I_{T_t} \{\gamma_{t_2} - \gamma_{t_1}\} dW.$$

As [47, p. 243]

$$\psi(s+\delta, t+\Delta) = \psi(s,t) + D_1\psi(s,t)\delta + D_2\psi(s,t)\Delta + R(\delta,\Delta),$$

one has that

$$\gamma(t+\Delta,\theta) = \gamma(t,t) + D_1\gamma(t,t)\Delta + D_2\gamma(t,t)(\theta - t) + R(\Delta, \theta - t),$$

and thus

$$\gamma(t+\Delta,\theta) - \gamma(t,\theta) = D_1(\gamma,\theta)\Delta.$$

That, in turn, yields that

$$\begin{aligned}\int_t^{t+\Delta} \gamma_{t+\Delta} dW &= \int_t^{t+\Delta} \{\gamma_{t+\Delta} - \gamma(t,t)\}\, dW + \int_t^{t+\Delta} \gamma(t,t)\, dW \\ &= \Delta \int_t^{t+\Delta} [D_1\gamma(t,\cdot)] dW + \gamma(t,t)\Delta_t W \\ &\approx \gamma(t,t)\Delta_t W.\end{aligned}$$

so that

$$\begin{aligned}\Delta_t g &= f(t)\Delta + \gamma(t,t)\Delta_t W, \\ \Delta_t \Phi &= \gamma(t,t)\Delta_t W.\end{aligned}$$

Now

$$\begin{aligned}\phi(t) &= \int_0^t G(t,\theta)\, m_\Phi(d\theta) \\ &= \int_0^t G(t,\theta) \left\{m_g(d\theta) - m_F(d\theta)\right\} \\ &= \int_0^t G(t,\theta) \left\{m_g(d\theta) - f(\theta)\, d\theta\right\} \\ &= f(t) - \int_0^t G(t,\theta) f(\theta)\, d\theta.\end{aligned}$$

One has thus a Volterra equation of the second kind with f as unknown. The solution of that equation [245, p. 26] has the following form:

$$f(t) = \phi(t) - \int_0^t R(t,\theta)\, \phi(\theta)\, d\theta,$$

where R is the resolvent kernel, that is, the solution of the following equation:

$$G(t,\theta) + R(t,\theta) = \int_\theta^t G(t,\tau) R(\tau,\theta)\, d\tau = \int_\theta^t R(t,\tau) G(\tau,\theta)\, d\tau.$$

But

$$\begin{aligned}\int_0^t R(t,\theta)\, \phi(\theta)\, d\theta &= \int_0^t R(t,\theta) \left\{\int_0^\theta G(\theta,\tau)\, dm_\Phi(d\tau)\right\} d\theta \\ &= \int_0^t m_\Phi(d\tau) \int_\tau^t R(t,\theta) G(\theta,\tau)\, d\theta\end{aligned}$$

$$= \int_0^t m_\Phi(d\tau)\left[R(t,\tau) + G(t,\tau)\right]$$

$$= \int_0^t m_\Phi(d\tau)\, R(t,\tau) + \phi(t),$$

that is,

$$\phi(t) - \int_0^t R(t,\theta)\phi(\theta)d\theta = -\int_0^t R(t,\tau) m_\Phi(d\tau),$$

and thus

$$f(t) = -\int_0^t R(t,\theta)\, m_\Phi(d\theta) = -\int_0^t R(t,\theta)\,\gamma(\theta,\theta)\, m_W(d\theta).$$

Then

$$\begin{aligned} g(t) &= F(t) + \Phi(t) \\ &= \int_0^t f(\theta)\,d\theta + \int_0^t \gamma(\theta,\theta)\, m_W(d\theta) \\ &= \int_0^t \left\{ -\int_0^\theta R(\theta,\tau)\,\gamma(\tau,\tau)\, m_W(d\tau) \right\} d\theta + \int_0^t \gamma(\theta,\theta)\, m_W(d\theta) \\ &= -\int_0^t m_W(d\theta)\,\gamma(\theta,\theta) \int_\theta^t R(\tau,\theta)\,d\tau + \int_0^t \gamma(\theta,\theta)\, m_W(d\theta) \\ &= \int_0^t m_W(d\theta)\,\gamma(\theta,\theta)\left\{ 1 - \int_\theta^t R(\tau,\theta)\,d\tau \right\}. \end{aligned}$$

The comparison with $g(t) = \int_0^t \gamma(t,\theta)\, m_W(d\theta)$ yields that

$$\gamma(t,\theta) = \gamma(\theta,\theta)\left\{ 1 - \int_\theta^t R(\tau,\theta)\,d\tau \right\} d\theta,$$

where $\gamma(\theta,\theta)$ and $R(\tau,\theta)$ are obtained from the covariance of g.

A partial justification for that procedure may be found in [150, p. 241]: there, there are no assumptions about existence and continuity of derivatives, but another setup (strict sense rather than wide sense) is required for the definition of the problem, in particular to make sense of the expression

$$\delta g_t = dt \int_0^t G(t,\theta)\, m_g(d\theta) + \gamma(t,t)\,\delta W_t.$$

A rather different approach, in the spirit of Lévy's paper however, is provided by Knight [157]. But the actual computations are then harder as one must have an explicit expression for the projection of $g(t)$ onto $L_\theta[g]$.

Example 8.3.17 (Illustration of (Example) 8.3.16*)* The resolvent equation is

$$\frac{-1}{1-t} + R(t,\theta) = -\int_\theta^t R(t,\tau)\frac{d\tau}{1-\tau}.$$

Keeping t fixed, one has thus to solve an equation of the form:

$$\kappa + \psi(\theta) = -\int_\theta^t \frac{\psi(\tau)}{1-\tau}d\tau$$

whose solution is $\psi(\theta) = \frac{1}{1-\theta}$. Thus, since, as seen, $\gamma(\theta,\theta) = 1$,

$$\gamma(t,\theta) = 1 - \int_\theta^t R(\tau,\theta)\,d\tau = 1 - \int_\theta^t \frac{d\tau}{1-\theta} = 1 - \frac{t-\theta}{1-\theta} = \frac{1-t}{1-\theta}.$$

Proposition 8.3.18 *Let $f(t) = \bigoplus_{i\in I} f_i(t)$ be a proper canonical representation of f for which all processes f_i are regular. Then $|I| = 1$.*

Proof Suppose that $|I| > 1$. Then at least $|I| = 2$. Let, for $i \in I$, fixed, but arbitrary,

$$f_i(t) = \int \phi_i(t)\,dm_i^f,\ \phi_i(t) \in L_2\left(T, \mathcal{T}, M_i^f\right),\ \phi_i(t) = I_{T_t}\phi_i(t).$$

It is no restriction to suppose that $M_1^f \neq 0$ and $M_2^f \neq 0$.

Let d_{F_1} and d_{F_2} be the "derivatives" of, respectively

$$F_1(t) = M_1^f(T_t) \text{ and } F_2(t) = M_2^f(T_t).$$

The assumptions on d_{F_1} and d_{F_2} allow one to assume that there is an interval $]a,b[\subseteq T$ such that $a < b$ and d_{F_1} and d_{F_2} are strictly positive on $]a,b[$. Let $\{\psi_i,\ i \in I\}$ be defined as follows:

$$\psi_1(t) = 0,\ t \in T \setminus]a,b[,\ \text{class of } \psi_1 \in L_2\left(T, \mathcal{T}, M_1^f\right),$$
$$\psi_2(t) = 0,\ t \in T \setminus]a,b[,\ \text{class of } \psi_2 \in L_2\left(T, \mathcal{T}, M_2^f\right),$$
$$\psi_i(t) = 0,\ t \in T,\ i \in I \setminus \{1,2\}.$$

Let $t_0 \in\ ]a,b[$ be fixed, but arbitrary, and let $\dot{\phi}_i(t,x)$ be a version of $\phi_i(t)$ that embodies the regularity conditions for f_i. Suppose that, for fixed, but arbitrary $t \in$

$]a,b[$, $t \leq t_0$,

$$\sum_{i \in I} \int_{T_t} \dot{\phi}_i(t,x)\, \psi_i(x)\, M_i^f(dx) = 0.$$

Then

$$\int_{]a,t[} \dot{\phi}_1(t,x)\, \psi_1(x)\, d_{F_1}(x)dx + \int_{]a,t[} \dot{\phi}_2(t,x)\, \psi_2(x)\, d_{F_2}(x)dx = 0.$$

The assumptions on the ϕ's allow one to differentiate that expression using the following formulae ($i \in [1:2]$) [108, p. 222]:

$$\begin{aligned} \xi_i(t,x) &= \dot{\phi}_i(t,x)\, \psi_i(x)\, d_{F_i}(x), \\ \Xi_i(t) &= \int_a^t \xi_i(t,x)\, dx, \\ \frac{d\Xi_i}{dt}(\theta) &= \int_a^\theta \frac{\partial \xi_i}{\partial t}(\theta,x)\, dx + \xi_i(\theta,\theta). \end{aligned}$$

Since $\phi_i(t,t) = 1$, one obtains that

$$\begin{aligned} 0 = {} & \psi_1(t)\, d_{F_1}(t) + \int_{]a,t[} \frac{\partial \dot{\phi}_1}{\partial t}(t,x)\, \psi_1(x)\, d_{F_1}(x)\, dx \\ & + \psi_2(t)\, d_{F_2}(t) + \int_{]a,t[} \frac{\partial \dot{\phi}_2}{\partial t}(t,x)\, \psi_2(x)\, d_{F_2}(x)\, dx. \end{aligned}$$

Let $\varphi_i = \psi_i d_{F_i}$, $i = 1,2$, and consider both equations

$$\varphi_1(t) = -1 - \int_{]a,t[} \frac{\partial \dot{\phi}_1}{\partial t}(t,x)\, \varphi_1(x)\, dx,$$

and

$$\varphi_2(t) = +1 - \int_{]a,t[} \frac{\partial \dot{\phi}_2}{\partial t}(t,x)\, \varphi_2(x)\, dx,$$

where the respective unknowns are φ_1 and φ_2. Those equations are Volterra equations of the second kind and have unique solutions [69, p. 235] that cannot be the zero function. But, since the densities d_{F_i}, $i = 1,2$, are strictly positive, the functions ψ_i, $i = 1,2$, must be different from zero. That means that the CHR representation is not proper canonical and thus contradicts the assumption. □

8.3.2 Multiplicity One from the Prediction Process' Behavior

The prediction process has been explicitly designated as such, it seems, by Knight (for example in [157]), and is defined, for H, a real Hilbert space, a map $f : T \longrightarrow H$, and $\{\theta, t\} \subseteq T$, $\theta \leq t$, fixed, but arbitrary, as

$$\pi_f(\theta \mid t) = P_\theta [f(t)],$$

the best prediction of $f(t)$ at time θ.

To Cramér, solutions based on Sect. 8.3.1 were clearly inadequate [64], as it is a very rare occurrence when γ can be obtained from C_g. Cramér also sensed, it seems, that the prediction process opens the way to a better solution. He was thus able to check that multiplicity one follows from smoothness conditions on the prediction process, conditions which shall presently be stated.

To that end some notation is required. Let thus T be finite, and T^δ be the subset of T made of points of the form $k\delta$, where k is an integer. Let then

$$T_t^\delta = T^\delta \cap T_t.$$

$L_t^\delta[f]$ shall be the linear space generated by $\{f(t), t \in T_t^\delta\}$, and the associated projection shall be denoted P_t^δ. Finally Δ_T is the "triangle" in $T \times T$ of elements $(t, \theta) \in T \times T$ such that $\theta \leq t$.

For a purely nondeterministic $f : T \longrightarrow H$, the conditions required for the developments to follow are now listed as a (long) list of assumptions. In many ways they are quite difficult to check, as one must be able to compute the prediction process, but represent some progress with respect to solving equations of the type illustrated in (Example) 8.3.14.

Assumptions 8.3.19 *1. There exists, for fixed, but arbitrary* $(t, \theta) \in \Delta_T$

$$(t, \theta) \mapsto \phi(t, \theta)$$

such that

(i) ϕ *is bounded;*
(ii) *for some* $\kappa > 0$, *and fixed, but arbitrary* $(t, \theta) \in \Delta_T$,

$$\phi(t, \theta) \geq \kappa;$$

(iii) *for fixed, but arbitrary* $(t, \theta) \in \overset{\circ}{\Delta}_T$, *the interior of* Δ_T, ϕ *has bounded, first order partial derivatives;*
(iv) *for fixed, but arbitrary* $\delta > 0$ *such that* $\{\theta - \delta, \theta, \theta + \delta, t\} \subseteq T$, *and* $\theta + \delta < t$,

$$\left\| \pi_f(\theta \mid t) - \pi_f(\theta - \delta \mid t) \right\|_H^2 = \delta \phi(t, \theta) + O\left(\delta^{3/2}\right)$$

and

$$\left\| \pi_f(\theta + \delta \mid t) - \pi_f(\theta \mid t) \right\|_H^2 = \delta \phi(t, \theta) + O\left(\delta^{3/2}\right).$$

2. *There exists, for fixed, but arbitrary* $(t, \theta) \in \Delta_T$,

$$(t, \theta) \mapsto \psi(t, \theta)$$

such that

(i) ψ *is bounded;*
(ii) *for fixed, but arbitrary* $(t, \theta) \in \overset{\circ}{\Delta}_T$, ψ *has bounded, first order partial derivatives;*
(iii) *for fixed, but arbitrary* $\delta > 0$ *such that* $\{\theta - \delta, \theta, t\} \subseteq T$, *and* $\theta < t$, *the inner product*

$$\left\langle \pi_f(\theta \mid t) - \pi_f(\theta - \delta \mid t), \pi_f(\theta \mid \theta) - \pi_f(\theta - \delta \mid \theta) \right\rangle_H$$

equals

$$\delta \psi(t, \theta) + O\left(\delta^{3/2}\right).$$

3. *For fixed, but arbitrary* $(k\delta, t) \in T^\delta \times T$, $k\delta \leq t$,

$$\left\| \pi_f(k\delta \mid t) - P_{k\delta}^\delta [f(t)] \right\|_H^2 = O\left(\delta^2\right).$$

The proof that Assumption 8.3.19 entails multiplicity one is facilitated by a number of preliminary remarks which depend on the following inequalities.

Lemma 8.3.20 *One shall make repeated use of the following inequalities (all elements considered below belong to the same real Hilbert space H):*

1. $\left\| \sum_{i=1}^n h_i \right\|_H^2 \leq n \sum_{i=1}^n \|h_i\|_H^2;$

2. $$\left\{ \|h_1\|_H^2 - \|h_2\|_H^2 \right\}^2 \leq 2\|h_1 - h_2\|_H^2 \left\{ \|h_1\|_H^2 + \|h_2\|_H^2 \right\}$$
$$\leq 6\|h_1 - h_2\|_H^2 \left\{ \|h_1\|_H^2 + \|h_1 - h_2\|_H^2 \right\};$$

3. $$\frac{1}{3} \left| \langle h_1, h_2 \rangle_H - \langle k_1, k_2 \rangle_H \right|^2 \leq \|h_1\|_H^2 \|h_2 - k_2\|_H^2$$
$$+ \|h_2\|_H^2 \|h_1 - k_1\|_H^2$$
$$+ \|h_1 - k_1\|_H^2 \|h_2 - k_2\|_H^2.$$

Proof Let X be a random variable which takes the values $\|h_1\|_H, \ldots, \|h_n\|_H$ with uniform probability. The inequality $E^2[X] \leq E[X^2]$ then rewrites as

$$\left\{\frac{\|h_1\|_H + \cdots + \|h_n\|_H}{n}\right\}^2 \leq \frac{\|h_1\|_H^2 + \cdots + \|h_n\|_H^2}{n},$$

which yields assertion 1:

$$\left\|\sum_{i=1}^n h_i\right\|_H^2 \leq (\|h_1\|_H + \cdots + \|h_n\|_H)^2 \leq n\left(\|h_1\|_H^2 + \cdots + \|h_n\|_H^2\right).$$

Similarly

$$\left\{\|h_1\|_H^2 - \|h_2\|_H^2\right\}^2 = \{\|h_1\|_H - \|h_2\|_H\}^2 \{\|h_1\|_H + \|h_2\|_H\}^2,$$

with

$$\{\|h_1\|_H + \|h_2\|_H\}^2 \leq 2\left\{\|h_1\|_H^2 + \|h_2\|_H^2\right\}. \qquad (\star)$$

But

$$\{\|h_1\|_H - \|h_2\|_H\}^2 = \{|\|h_1\|_H - \|h_2\|_H|\}^2 \leq \|h_1 - h_2\|_H^2,$$

and one has the first part of item 2. One then dominates, in $(\star)$, $\|h_2\|_H^2$ with

$$\|h_2\|_H^2 = \|(h_2 - h_1) + h_1\|_H^2 \leq 2\left\{\|h_2 - h_1\|_H^2 + \|h_1\|_H^2\right\},$$

and the second part of item 2 follows. For item 3, one has that

$$\begin{aligned} |\langle h_1, h_2\rangle_H - \langle k_1, k_2\rangle_H| &= |\langle h_1, h_2 - k_2\rangle_H + \langle h_1 - k_1, k_2\rangle_H| \\ &\leq \|h_1\|_H \|h_2 - k_2\|_H + \|h_1 - k_1\|_H \|k_2\|_H. \end{aligned}$$

One then dominates $\|k_2\|_H$ with

$$\|k_2\|_H = \|(k_2 - h_2) + h_2\|_H \leq \|k_2 - h_2\|_H + \|h_2\|_H.$$

□

The Assumptions of 8.3.19 have the following consequences.

Remark 8.3.21 From items (iv) of 1, and (iii) of 2 of Assumption 8.3.19, one has that

$$\delta\phi(t,t) + O(\delta^{3/2}) = \left\|\pi_f(t \mid t) - \pi_f(t-\delta \mid t)\right\|_H^2 = \delta\psi(t,t) + O(\delta^{3/2}).$$

Thus

$$\phi(t,t) - \psi(t,t) = \delta^{-1}\left(\delta\phi(t,t) - \delta\psi(t,t)\right) = \delta^{-1}O(\delta^{3/2}) = O(\delta^{1/2}).$$

Thus $\phi(t,t) = \Psi(t,t)$, and the difference between the left-hand and right-hand sides of the former equality is $O(\delta^{3/2})$.

Remark 8.3.22 ϕ and ψ are continuous [230, p. 239].

Remark 8.3.23 Let $t \in T$ be fixed, and $\theta \in T_t$ be arbitrary. $\pi_f(\cdot \mid t)$ has orthogonal increments: its associated "distribution" function is

$$F_t(\theta) = \left\|\pi_f(\theta \mid t)\right\|_H^2.$$

Then, because of Assumption 8.3.19, item 1,

$$F_t(\theta+\epsilon) - F_t(\theta) = \left\|\pi_f(\theta+\epsilon \mid t) - \pi_f(\theta \mid t)\right\|_H^2 = \epsilon\phi(t,\theta) + O\left(\epsilon^{3/2}\right),$$

and, consequently,

$$\frac{dF_t}{d\theta}(\theta) = \phi(t,\theta),$$

so that

$$M_{\pi_f(\cdot\mid t)}(d\theta) = \phi(t,\theta)\,d\theta.$$

Let ϕ_t be the equivalence class, with respect to $M_{\pi_f(\cdot\mid t)}$, of $x \mapsto \phi^{-1/2}(t,x)$, which is bounded below by assumption, and

$$\Phi_t(\theta) = \int I_{T_\theta}\phi_t\,dm_{\pi_f(\cdot\mid t)}.$$

Φ_t has orthogonal increments, and $M_{\Phi_t}(d\theta) = d\theta$. Furthermore, ϕ_t^{-1} being the equivalence class, with respect to M_{Φ_t}, of $x \mapsto \phi^{1/2}(t,x)$,

$$\int I_{T_\theta}\phi_t^{-1}dm_{\Phi_t} = \int I_{T_\theta}\phi_t^{-1}\phi_t\,dm_{\pi_f(\cdot\mid t)} = m_{\pi_f(\cdot\mid t)}(T_\theta) = \pi_f(\theta \mid t).$$

Remark 8.3.24 One has that

$$\left\|P^{\delta}_{k\delta}\left[f\left(t\right)\right]-P^{\delta}_{(k-1)\delta}\left[f\left(t\right)\right]\right\|^2_H=\delta\phi\left(t,k\delta\right)+O\left(\delta^{3/2}\right).$$

Let indeed $P^{\delta}_{k\delta}-P^{\delta}_{(k-1)\delta}$ be written as

$$\left\{P^{\delta}_{k\delta}-P_{k\delta}\right\}+\left\{P_{k\delta}-P_{(k-1)\delta}\right\}+\left\{P_{(k-1)\delta}-P^{\delta}_{(k-1)\delta}\right\}.$$

Then

$$\begin{aligned}
&\left\|P^{\delta}_{k\delta}\left[f\left(t\right)\right]-P^{\delta}_{(k-1)\delta}\left[f\left(t\right)\right]\right\|^2_H=\\
&=\left\|P^{\delta}_{k\delta}\left[f\left(t\right)\right]-P_{k\delta}\left[f\left(t\right)\right]\right\|^2_H\\
&+\left\|P_{k\delta}\left[f\left(t\right)\right]-P_{(k-1)\delta}\left[f\left(t\right)\right]\right\|^2_H\\
&+\left\|P_{(k-1)\delta}\left[f\left(t\right)\right]-P^{\delta}_{(k-1)\delta}\left[f\left(t\right)\right]\right\|^2_H\\
&+2\left\langle P^{\delta}_{k\delta}\left[f\left(t\right)\right]-P_{k\delta}\left[f\left(t\right)\right],P_{k\delta}\left[f\left(t\right)\right]-P_{(k-1)\delta}\left[f\left(t\right)\right]\right\rangle_H\\
&+2\left\langle P^{\delta}_{k\delta}\left[f\left(t\right)\right]-P_{k\delta}\left[f\left(t\right)\right],P_{(k-1)\delta}\left[f\left(t\right)\right]-P^{\delta}_{(k-1)\delta}\left[f\left(t\right)\right]\right\rangle_H\\
&+2\left\langle P_{k\delta}\left[f\left(t\right)\right]-P_{(k-1)\delta}\left[f\left(t\right)\right],P_{(k-1)\delta}\left[f\left(t\right)\right]-P^{\delta}_{(k-1)\delta}\left[f\left(t\right)\right]\right\rangle_H.
\end{aligned}$$

In the right-hand side of that latter expression, the first, third, and fifth terms are $O\left(\delta^2\right)$ because of Assumption 8.3.19, item 3. The second term is

$$\delta\phi\left(t,k\delta\right)+O\left(\delta^{3/2}\right)$$

because of Assumption 8.3.19, item 1. The two remaining terms are inner products of the form $2\langle a,b\rangle$ for which $|2\langle a,b\rangle|\leq 2\|a\|\|b\|$. Since $\|a\|$ is, because of Assumption 8.3.19, item 3, $O(\delta)$, and $\|b\|$, because of Assumption 8.3.19, item 1, $O(\delta^{1/2})$, they are thus $O(\delta^{3/2})$ terms.

Remark 8.3.25 An argument similar to that of (Remark) 8.3.24 which uses Assumption 8.3.19, item 2, also yields that

$$\left\langle\left\{P^{\delta}_{k\delta}-P^{\delta}_{(k-1)\delta}\right\}\left[f\left(t\right)\right],\left\{P^{\delta}_{k\delta}-P^{\delta}_{(k-1)\delta}\right\}\left[f\left(k\delta\right)\right]\right\rangle=\delta\psi\left(t,k\delta\right)+O\left(\delta^{3/2}\right).$$

Proposition 8.3.26 *Let T be a finite interval, H, a real Hilbert space, and $f:T\longrightarrow H$, a purely nondeterministic map for which (Assumption)* 8.3.19 *obtains. f has then multiplicity one.*

Proof Let $\{\theta, t\} \subseteq \overset{\circ}{\Delta}_T$, $\theta < t$, be fixed, but arbitrary, and $k[t]$ be the integer for which $k[t]\,\delta \le t < (k[t]+1)\,\delta$. Let also

$$g_\delta(k\delta) = f(k\delta) - P^\delta_{(k-1)\delta}[f(k\delta)] = \left\{P^\delta_{k\delta} - P^\delta_{(k-1)\delta}\right\}[f(k\delta)],$$

and

$$h_\delta(k\delta) = \delta^{1/2}\,\|g_\delta(k\delta)\|_H^{-1}\,g_\delta(k\delta).$$

$g_\delta(k\delta)$ is the "innovation" in $f(k\delta)$ with respect to the subspace spanned prior to that "time," and $h_\delta(k\delta)$ is its "standardization" with small "absolute value." The families

$$\{g_\delta(k\delta),\ k \le k[t]\} \text{ and } \{h_\delta(k\delta),\ k \le k[t]\}$$

are made, by definition, of orthogonal elements. Furthermore

$$f(k\delta) = g_\delta(k\delta) + P^\delta_{(k-1)\delta}[f(k\delta)],$$

so that each element $f(k\delta)$ is the sum of an "innovation" $g_\delta(k\delta)$ that is orthogonal to the subspace generated by the elements of the form $f(l\delta)$, $l < k$, and of its "prediction" in that subspace. Consequently

$$L^\delta_{k\delta}[f] = \overline{V[\{g_\delta(l\delta),\ l \le k\}]} = \overline{V[\{h_\delta(l\delta),\ l \le k\}]},$$

and, in particular,

$$f(k[t]\,\delta) = \sum_{k \le k[t]} \alpha(t,\delta,k)\,h_\delta(k\delta),$$

where

$$\alpha(t,\delta,k) = \langle f(k[t]\,\delta),\ h_\delta(k\delta)\rangle_H.$$

Consequently, for $k_0 \le k[t]$,

$$P^\delta_{k_0\delta}[f(k[t]\,\delta)] = \sum_{k \le k_0} \alpha(t,\delta,k)\,h_\delta(k\delta),$$

and thus, for $k_1 < k_2 \le k[t]$, fixed, but arbitrary,

$$\left\{P^\delta_{k_2\delta} - P^\delta_{k_1\delta}\right\}[f(k[t]\,\delta)] = \sum_{k_1 < k \le k_2} \alpha(t,\delta,k)\,h_\delta(k\delta).$$

From the following equality [(Remark) 8.3.23]:

$$P_\theta [f(t)] = \pi_f(\theta \mid t) = \int I_{T_\theta} \phi_t^{-1} dm_{\Phi_t},$$

one has that

$$P_{k_0\delta} [f(k[t]\delta)] = \int I_{T_{k_0\delta}} \phi_{k[t]\delta}^{-1} dm_{\Phi_{k[t]\delta}}.$$

It follows that

$$\{P_{k_2\delta} - P_{k_1\delta}\} [f(k[t]\delta)] = \int I_{T_{k_2\delta} \setminus T_{k_1\delta}} \phi_{k[t]\delta}^{-1} dm_{\Phi_{k[t]\delta}}.$$

Now, as Assumption 8.3.19 is assumed to obtain, one has the following approximation to the increments of the prediction process:

$$\left\| \left\{P^\delta_{k_2\delta} - P^\delta_{k_1\delta}\right\} [f(k[t]\delta)] - \{P_{k_2\delta} - P_{k_1\delta}\} [f(k[t]\delta)] \right\|_H^2 =$$

$$\left\| \left\{P^\delta_{k_2\delta} - P_{k_2\delta}\right\} [f(k[t]\delta)] - \left\{P^\delta_{k_1\delta} - P_{k_1\delta}\right\} [f(k[t]\delta)] \right\|_H^2 = O(\delta^2).$$

That latter equality rewrites then as

$$\left\| \sum_{k_1 < k \leq k_2} \alpha(t, \delta, k) h_\delta(k\delta) - \int I_{T_{k_2\delta} \setminus T_{k_1\delta}} \phi_{k[t]\delta}^{-1} dm_{\Phi_{k[t]\delta}} \right\|_H^2 = O\left(\delta^2\right), \qquad (1)$$

which evaluates the increments of the prediction process as a sum of orthogonal, "almost orthonormal" elements. The "unknowns" in that norm are the $\alpha(t, \delta, k)$'s. They are calculated in the next step.

As already acknowledged,

$$\left\{P^\delta_{k\delta} - P^\delta_{(k-1)\delta}\right\} [f(k[t]\delta)] = \alpha(t, \delta, k) h_\delta(k\delta),$$

so that

$$\left\langle \left\{P^\delta_{k\delta} - P^\delta_{(k-1)\delta}\right\} [f(k[t]\delta)], h_\delta(k\delta) \right\rangle_H = \alpha(t, \delta, k) \|h_\delta(k\delta)\|_H^2,$$

or

$$\alpha(t, \delta, k) = \left\langle \left\{P^\delta_{k\delta} - P^\delta_{(k-1)\delta}\right\} [f(k[t]\delta)], \|h_\delta(k\delta)\|_H^{-2} h_\delta(k\delta) \right\rangle_H.$$

But, by definition, $\|h_\delta(k\delta)\|_H = \delta^{1/2}$ so that

$$\|h_\delta(k\delta)\|_H^{-2} h_\delta(k\delta) = \frac{1}{\delta^{1/2}} \frac{g_\delta(k\delta)}{\|g_\delta(k\delta)\|_H},$$

and also

$$g_\delta(k\delta) = \left\{P^\delta_{k\delta} - P^\delta_{(k-1)\delta}\right\}[f(k\delta)].$$

So, finally,

$$\alpha(t,\delta,k) = \frac{\left\langle \left\{P^\delta_{k\delta} - P^\delta_{(k-1)\delta}\right\}[f(k[t]\delta)], \left\{P^\delta_{k\delta} - P^\delta_{(k-1)\delta}\right\}[f(k\delta)]\right\rangle_H}{\delta^{1/2}\left\|\left\{P^\delta_{k\delta} - P^\delta_{(k-1)\delta}\right\}[f(k\delta)]\right\|_H}.$$

Using (Remarks) 8.3.24 and 8.3.25, one may write that

$$\alpha(t,\delta,k) = \frac{\delta\psi(k[t]\delta, k\delta) + O(\delta^{3/2})}{\delta^{1/2}\{\delta\phi(k\delta,k\delta) + O(\delta^{3/2})\}^{1/2}} = \frac{\psi(k[t]\delta, k\delta) + O(\delta^{1/2})}{\{\phi(k\delta,k\delta) + O(\delta^{1/2})\}^{1/2}}.$$

Let

$$\eta(t,\theta) = \frac{\psi(t,\theta)}{\phi(\theta,\theta)^{1/2} + O(\delta^{1/2})}.$$

Omitting arguments, one has that

$$\alpha = \frac{\psi + O}{(\phi + O)^{1/2}}, \text{ and that } \eta = \frac{\psi}{\phi^{1/2} + O}.$$

ϕ being bounded and strictly positive by assumption, one has that

$$(\phi + O)^{1/2} = \phi^{1/2}(1+O)^{1/2} = \phi^{1/2}(1+O) = \phi^{1/2} + O.$$

Consequently

$$\alpha = \eta + \frac{O}{\phi^{1/2} + O} = \eta + O,$$

that is

$$\alpha(t,\delta,k) = \eta(k[t]\delta, k\delta) + O\left(\delta^{1/2}\right).$$

Thus, while, above, the α-term depends on k, the η-one does only in its argument, which shows that the approximation in (1) is by "white-noise-like" elements.

The aim is now to obtain, using (1), the relation marked below with (5). One proceeds inserting terms in (1) in such a way that the resulting differences except the one of interest may be neglected. Now when $\delta \downarrow 0$, t and $\theta < t$ shall remain fixed, and one shall constantly choose

- $k[t]$ such that $k[t] \leq t < (k[t]+1)\delta$,
- k_1 such that $k_1\delta \leq \theta < (k_1+1)\delta$,
- k_2 such that $k_2\delta \leq \theta + \sqrt{\delta} < (k_2+1)\delta$.

Such a choice entails that

$$k_2 - (k_1+1) < k_2 - \frac{\theta}{\delta} \leq \frac{1}{\delta^{1/2}} < (k_2+1) - \frac{\theta}{\delta} \leq (k_2+1) - k_1,$$

that is,

$$k_2 - k_1 - 1 < \frac{1}{\delta^{1/2}} < k_2 - k_1 + 1.$$

Then, as ϕ and ψ, and thus η, are continuous, for appropriate functions' arguments,

$$\phi(k[t], \theta_1)^{1/2} - \phi(k[t], \theta_2)^{1/2} = O\left(\delta^{1/2}\right),$$

and, as seen,

$$\alpha(t, \delta, k) - \eta(k[t], \theta) = O\left(\delta^{1/2}\right).$$

Consequently, as $k_2 - k_1$ behaves as $\frac{1}{\delta^{1/2}}$ (denoted $\approx$ below),

$$\left\| \sum_{k_1 < k \leq k_2} \alpha(t, \delta, k)\, h_\delta(k\delta) - \sum_{k_1 < k \leq k_2} \eta(k[t]\delta, \theta)\, h_\delta(k\delta) \right\|_H^2 \tag{2}$$

equals

$$\sum_{k_1 < k \leq k_2} \{\alpha(t, \delta, k) - \eta(k[t]\delta, \theta)\}^2 \|h_\delta(k\delta)\|_H^2 \approx (k_2 - k_1)\delta^2 = O\left(\delta^{3/2}\right).$$

Similarly, remembering the definitions in (Remark) 8.3.23,

$$\left\| \phi(k[t]\delta, \theta)^{1/2} \int I_{T_{k_2\delta} \backslash T_{k_1\delta}}\, dm_{\Phi_{k[t]\delta}} - \int I_{T_{k_2\delta} \backslash T_{k_1\delta}} \phi_{k[t]\delta}^{-1}\, dm_{\Phi_{k[t]\delta}} \right\|_H^2 \tag{3}$$

equals

$$\int_{k_1\delta}^{k_2\delta} \{\phi(k[t]\delta,\theta)^{1/2} - \phi(k[t]\delta,\tau)^{1/2}\}^2 M_{\Phi_{k[t]\delta}}(d\tau) =$$

$$= \int_{k_1\delta}^{k_2\delta} \{\phi(k[t]\delta,\theta)^{1/2} - \phi(k[t]\delta,\tau)^{1/2}\}^2 d\tau$$

$$\approx (k_2 - k_1)\delta^2 = O\left(\delta^{3/2}\right).$$

Writing (1) as the square of the norm of $A - B$, (2) as that of $A - C$, (3) as that of $D - B$, and (4) as that of $C - D$, one has that

$$A - B = (A - C) + (C - D) + (D - B),$$

and that the square of the norm of the term $C - D$, that is

$$\left\| \eta(k[t]\delta,\theta) \sum_{k_1<k\leq k_2} h_\delta(k\delta) - \phi(k[t]\delta,\theta)^{1/2} \int I_{T_{k_2\delta}\setminus T_{k_1\delta}} dm_{\Phi_{k[t]\delta}} \right\|_H^2 \tag{4}$$

is evaluated to be $O(\delta^{3/2})$. One has that

$$\|C\|_H^2 = \eta^2(k[t]\delta,\theta)(k_2 - k_1)\delta,$$

and that

$$\|D\|_H = \phi(k[t]\delta,\theta)(k_2 - k_1)\delta.$$

Thus

$$\delta^2(k_2 - k_1)^2\{\eta^2(k[t]\delta,\theta) - \phi(k[t]\delta,\theta)\}^2 = \left(\|C\|_H^2 - \|D\|_H^2\right)^2.$$

Using item 2 of (Lemma) 8.3.20, one has that the right-hand side of the latter equality is dominated by

$$2\left(\|C - D\|_H^2\right)\left\{\|C\|_H^2 + \|D\|_H^2\right\}$$

which equals

$$2\left(\|C - D\|_H^2\right)\{\eta^2(k[t]\delta,\theta) + \phi(k[t]\delta,\theta)\}\delta(k_2 - k_1).$$

Comparing with the left-hand side, using the fact that the ϕ and η functions are bounded, and that [(4)]

$$\|C - D\|_H^2 = O\left(\delta^{3/2}\right),$$

one obtains that

$$\eta^2 (k [t] \delta, \theta) - \phi (k [t] \delta, \theta) = O\left(\delta^{1/2}\right),$$

and then that

$$\|C\|_H^2 - \|D\|_H^2 = O(\delta).$$

From there, with sufficiently small δ,

$$\eta^2 (k [t] \delta, \theta) = \phi (k [t] \delta, \theta) + O\left(\delta^{1/2}\right) > \frac{\kappa}{2},$$

so that, from (4),

$$\left\| \sum_{k_1 < k \le k_2} h_\delta (k\delta) - \frac{\phi (k [t] \delta, \theta)^{1/2}}{|\eta (k [t] \delta, \theta)|} \int I_{T_{k_2\delta} \backslash T_{k_1\delta}} dm_{\Phi_{k[t]\delta}} \right\|_H^2 = O\left(\delta^{3/2}\right).$$

An argument similar to that used for (3) (it simplifies matters that here one does not distinguish function from equivalence class) yields that

$$\left\| \sum_{k_1 < k \le k_2} h_\delta (k\delta) - \int I_{T_{k_2\delta} \backslash T_{k_1\delta}} \frac{\phi (k [t] \delta, \cdot)^{1/2}}{|\eta (k [t] \delta, \cdot)|} dm_{\Phi_{k[t]\delta}} \right\|_H^2 = O\left(\delta^{3/2}\right). \tag{5}$$

As $k_2 - k_1 \approx \delta^{-1/2}$, in $T_{k[t]\delta}$, the number of equal intervals of length $(k_2 - k_1)\delta$ is of order

$$O\left(\delta^{-1/2}\right) \text{ as } \frac{t}{(k_2 - k_1)\delta} \approx t\delta^{-1/2}.$$

The expression

$$\left\| \sum_{k \le k[t]} h_\delta (k\delta) - \int I_{T_{k[t]\delta}} \frac{\phi (k [t] \delta, \cdot)^{1/2}}{|\eta (k [t] \delta, \cdot)|} dm_{\Phi_{k[t]\delta}} \right\|_H^2$$

is a sum of orthogonal terms of a type similar to that evaluated just above. Item 1 of (Lemma) 8.3.20 then says that it is dominated by the number of terms times the sum of the norms squared of those terms. It thus evaluates at

$$\left\{O\left(\delta^{-1/2}\right)\right\}^2 O\left(\delta^{3/2}\right) = O\left(\delta^{1/2}\right).$$

Since M_{Φ_t} is Lebesgue measure [(Remark) 8.3.23], that, above, η^2 is a strictly positive level, and that ϕ/η^2 is continuous, in H,

$$\lim_{\delta \downarrow 0} \int I_{T_{k[t]\delta}} \frac{\phi\left(k\left[t\right]\delta, \cdot\right)^{1/2}}{\left|\eta\left(k\left[t\right]\delta, \cdot\right)\right|} dm_{\Phi_{k[t]\delta}} = \int I_{T_t} \frac{\phi\left(t, \cdot\right)^{1/2}}{\left|\eta\left(t, \cdot\right)\right|} dm_{\Phi_t}.$$

As [(Remark) 8.3.23], $d\Phi_t = \phi_t \, dm_{\pi_f(\cdot|t)}$, it follows that, in H,

$$\lim_{\delta \downarrow 0} \sum_{k \leq k[t]} h_\delta\left(k\delta\right) = \int I_{T_t} \frac{1}{\left|\eta\left(t, \cdot\right)\right|} dm_{\pi_f(\cdot|t)}.$$

The latter integral belongs thus to $L_t\left[f\right]$, and, as the limit of orthogonal summands, defines a function with orthogonal increments. Let it be denoted W. One has then that

$$\int I_{T_t} \left|\eta\left(t, \cdot\right)\right| dW = \int I_{T_t} dm_{\pi_f(\cdot|t)} = f(t).$$

Since the family $\left\{\left|\eta\left(t, \cdot\right)\right|, \ t \in T\right\}$ is total (it is strictly positively bounded below), f has multiplicity one. □

Example 8.3.27 The following example illustrates the conditions of (Proposition) 8.3.26.

Let $T =]0, 1[$, and C be a triangular covariance:

$$C\left(t_1, t_2\right) = c_\wedge\left(t_1 \wedge t_2\right) c_\vee\left(t_1 \vee t_2\right).$$

It is, for example, the covariance of the function

$$f\left(t\right) = c_\vee\left(t\right) W\left(\frac{c_\wedge\left(t\right)}{c_\vee\left(t\right)}\right),$$

where W is a Wiener process. Let f have covariance C, that is

$$C_f\left(t_1, t_2\right) = \left\langle f\left(t_1\right), f\left(t_2\right)\right\rangle_H = C\left(t_1, t_2\right).$$

For fixed, but arbitrary $0 < \theta_1 \le \theta_2 \le t$ one has that

$$\left\langle f(t) - \frac{C(t,\theta_2)}{C(\theta_2,\theta_2)} f(\theta_2), f(\theta_1) \right\rangle_H = C(t,\theta_1) - \frac{C(t,\theta_2)}{C(\theta_2,\theta_2)} C(\theta_2,\theta_1)$$
$$= c_\wedge(\theta_1)\, c_\vee(t) - \frac{c_\wedge(\theta_2)\, c_\vee(t)}{c_\wedge(\theta_2)\, c_\vee(\theta_2)} c_\wedge(\theta_1)\, c_\vee(\theta_2)$$
$$= 0.$$

Thus

$$f(t) - \frac{C(t,\theta_2)}{C(\theta_2,\theta_2)} f(\theta_2) \perp f(\theta_1),\ \theta_1 \le \theta_2,$$

that is, for $\theta \le t$, fixed, but arbitrary,

$$\frac{C(t,\theta)}{C(\theta,\theta)} f(\theta) = P_\theta\,[f(t)]\,.$$

Furthermore, as

$$P_{k\delta}\,[f(t)] = \frac{C(t,k\delta)}{C(k\delta,k\delta)} f(k\delta),$$

$P^{\delta}_{k\delta}\,[f(t)] = P_{k\delta}\,[f(t)]$, so that Assumption 8.3.19, item 3, obtains. Let

$$\Gamma(t_1,t_2) = \frac{C(t_1,t_2)}{C^{1/2}(t_1,t_1)\, C^{1/2}(t_2,t_2)}.$$

As, for $\epsilon > 0$, $0 < \theta - \epsilon < \theta \le t$, fixed, but arbitrary, writing Γ in terms of C,

$$\Gamma(t,\theta)\,\Gamma(\theta,\theta-\epsilon) = \frac{C(t,\theta-\epsilon)}{\sqrt{C(t,t)\, C(\theta-\epsilon,\theta-\epsilon)}} \times \frac{C(t,\theta)\, C(\theta,\theta-\epsilon)}{C(\theta,\theta)\, C(t,\theta-\epsilon)},$$

and, writing the second term of the latter product in terms of $c_\wedge$ and $c_\vee$,

$$\frac{C(t,\theta)\, C(\theta,\theta-\epsilon)}{C(\theta,\theta)\, C(t,\theta-\epsilon)} = 1,$$

one has that (interchanging right-hand and left-hand sides)

$$\Gamma(t,\theta-\epsilon) = \Gamma(t,\theta)\,\Gamma(\theta,\theta-\epsilon)\,.$$

Let $\Gamma_t(\theta) = \Gamma(t,\theta)$, and $\gamma(\theta) = (D_2\Gamma)(\theta,\theta)$: the former equality yields, provided the appropriate differentiability assumptions obtain, that

$$\Gamma_t' = \gamma \Gamma_t,$$

an equation whose solution has the following form:

$$\Gamma_t(x) = \Gamma_t(x_0)\, e^{\int_{x_0}^{x} \gamma(\xi)\, d\xi}.$$

Letting $x_0 = \theta$ and $x = t$, one obtains that

$$\Gamma(t,\theta) = e^{-\int_\theta^t (D_2\Gamma)(x,x)\, dx}.$$

Now, letting $g(t) = f(t)/\|f(t)\|_H$, $\langle g(t), g(\theta)\rangle_H = \Gamma(t,\theta)$, and the following expression:

$$\|P_\theta[f(t)] - P_{\theta-\epsilon}[f(t)]\|_H^2 = \left\| \frac{C(t,\theta)}{C(\theta,\theta)} f(\theta) - \frac{C(t,\theta-\epsilon)}{C(\theta-\epsilon,\theta-\epsilon)} f(\theta-\epsilon) \right\|_H^2$$

can be rewritten as

$$\begin{aligned}
&\|\Gamma(t,\theta)\,\|f(t)\|_H\, g(\theta) - \Gamma(t,\theta-\epsilon)\,\|f(t)\|_H\, g(\theta-\epsilon)\|_H^2 = \\
&\quad = C(t,t)\,\Gamma^2(t,\theta) \left\| g(\theta) - \frac{\Gamma(t,\theta-\epsilon)}{\Gamma(t,\theta)} g(\theta-\epsilon) \right\|_H^2 \\
&\quad = C(t,t)\,\Gamma^2(t,\theta) \left\{ 1 - 2\frac{\Gamma(t,\theta-\epsilon)\,\Gamma(\theta,\theta-\epsilon)}{\Gamma(t,\theta)} + \frac{\Gamma^2(t,\theta-\epsilon)}{\Gamma^2(t,\theta)} \right\}.
\end{aligned}$$

Writing Γ in terms of $c_\wedge$ and $c_\vee$, and taking into account that $\theta \le t$, the last bracket becomes

$$1 - \frac{c_\wedge(\theta-\epsilon)\, c_\vee(\theta)}{c_\wedge(\theta)\, c_\vee(\theta-\epsilon)} = 1 - \Gamma^2(\theta,\theta-\epsilon).$$

But Γ is the exponential of the integral $-\int_{\theta-\epsilon}^{\theta} (D_2\Gamma)(x,x)\, dx$, which, provided $(D_2\Gamma)(x,x)$ is differentiable, by the mean value theorem, equals $-\epsilon$ times the derivative of $(D_2\Gamma)(x,x)$ at some point between $\theta - \epsilon$ and θ. Part of Assumption 8.3.19, item 1, is thus valid provided enough regularity assumptions are required from $c_\vee$ and $\frac{c_\wedge}{c_\vee}$. The other assumptions are checked analogously, and in particular those relating to the smoothness of

$$\langle \{P_\theta - P_{\theta-\epsilon}\}[f(t)], \{P_\theta - P_{\theta-\epsilon}\}[f(\theta)] \rangle_H = C(t,\theta)\left\{1 - \Gamma^2(\theta,\theta-\epsilon)\right\}.$$

The following result extends slightly the scope of (Proposition) 8.3.26.

Proposition 8.3.28 *Let T be an interval in $\mathbb{R}$; H, a real Hilbert space; and $f : T \longrightarrow H$, a purely nondeterministic map that is continuous to the left, and has limits to the right. When, for any fixed, but arbitrary finite interval $[t_1, t_2] \subseteq T$, (Assumption)* 8.3.19, *items 1 and* 2, *obtain, and (Assumption)* 8.3.19, *item* 3, *does also for*

$$g_{t_1}(t) = f(t) - P_{t_1}[f(t)],$$

f has then multiplicity one.

Proof Let $\epsilon > 0$ be arbitrarily small, but fixed, and $\{\theta, t\} \subseteq]t_1, t_2[$, $\theta < t$. Then

$$P_\theta[g_{t_1}(t)] - P_{\theta\pm\epsilon}[g_{t_1}(t)] = P_\theta[f(t)] - P_{\theta\pm\epsilon}[f(t)].$$

Thus Assumption 8.3.19 obtains for g_{t_1}. In particular, the functions ϕ, ψ (and thus η) appearing in Assumption 8.3.19 are independent of the interval over which they are considered, as they are "local" conditions that "transfer" to g_{t_1}. Furthermore the following representation obtains:

$$g_{t_1}(t) = \int I_{[t_1, t_2] \cap T_t}\, \eta(t, \cdot)\, dW_{t_1},$$

where W_{t_1} is a function with orthogonal increments obtained as the following limit in H:

$$W_{t_1}(\theta) = \lim_{\delta \downarrow 0} \sum_{t_1 < k\delta \leq \theta} h_\delta(k\delta).$$

When $t_1 < 0$, one has, for $\theta > 0$, that

$$W_{t_1}(\theta) = W_{t_1}(0) + \lim_{\delta \downarrow 0} \sum_{0 < k\delta \leq \theta} h_\delta(k\delta) = W_{t_1}(0) + W_0(\theta),$$

and, for $\theta \in]t_1, 0]$, that

$$W_{t_1}(\theta) = W_{t_1}(0) - \lim_{\delta \downarrow 0} \sum_{\theta < k\delta \leq 0} h_\delta(k\delta).$$

Thus $W_{t_1}(\theta)$ is always the sum of the element $W_{t_1}(0)$, independent of θ, and a process with orthogonal increments independent of t_1. Let it be denoted W. Then, for $\{\theta, t\} \subseteq]t_1, t_2[$, $\theta < t$, $\Delta W_{t_1} = \Delta W$, so that, using the integral representation of W_{t_1},

$$f(t) = P_{t_1}[f(t)] + \int I_{[t_1, t_2] \cap T_t}\, \eta(t, \cdot)\, dW.$$

The process being purely nondeterministic, letting $t_1 \downarrow$, one obtains that

$$f(t) = \int I_{T_t} \eta(t, \cdot)\, dW,$$

a canonical representation of multiplicity one. □

8.3.3 Approximation by Processes of Multiplicity One

For "practical" purposes, multiplicity one is very much easier to handle than higher multiplicity, as one is immediately confronted, in the latter case, as shall be seen when computing the likelihood, with a problem quite harder than that illustrated in (Remark) 8.3.13. So, asking for approximations having multiplicity one makes sense. A general result exists which provides such approximations. It is presented first.

Lemma 8.3.29 *Let H be a real Hilbert space, $T = [0, 1]$, and a fixed, but arbitrary map $f : T \longrightarrow H$ be given. Let $0 = \theta_0 < \theta_1 < \cdots < \theta_{n-1} < \theta_n = 1$, and suppose that $\{f(\theta_1), \ldots, f(\theta_n)\}$ are linearly independent. Let, for a noncausal map $T \times T \longrightarrow \mathbb{R}$: $[h(t, \theta) = 0, \theta > t)]h$ such that, for $t \in T$, fixed, but arbitrary, $h(t, t) \neq 0$,*

$$g(t) = \sum_{\theta_i \leq t} h(t, \theta_i) f(\theta_i).$$

g has multiplicity one.

Proof Let $\{f_1, \ldots, f_n\}$ be the orthonormal set obtained, using the Gram-Schmidt procedure, from the starting set $\{f(\theta_1), \ldots, f(\theta_n)\}$. One has that

(a) for $t < \theta_1$, $g(t) = 0_H$,
(b) for $t \in [\theta_1, \theta_2[$, $g(t) = h(t, \theta_1) f(\theta_1)$,
(c) for $t \in [\theta_2, \theta_3[$, $g(t) = h(t, \theta_1) f(\theta_1) + h(t, \theta, 2) f(\theta_2)$,
(d) etc.

As, by assumption, for $t \in T$, fixed, but arbitrary, $h(t, t) \neq 0$,

1. for $t < \theta_1$, $L_t[g] = \{0_H\}$,
2. for $t \in [\theta_1, \theta_2[$, as $h(\theta_1, \theta_1) \neq 0$,

$$L_t[g] = V[f(\theta_1)] = V[f_1],$$

3. for $t \in [\theta_2, \theta_3[$, again because $h(t, t) \neq 0$,

$$L_t[g] = V[f(\theta_1), f(\theta_2)] = V[f_1, f_2],$$

4. etc.

In particular, $L_{\cup T}[g] = V[f_1, \ldots, f_n]$.

The construction of a canonical representation of g proceeds, according to (Proposition) 6.4.10, as follows. One starts with $g_1 = f_1$, and

$$K_1 = \overline{V[P_t^g[g_1], t \in T]}.$$

But, for $t \geq \theta_1, f_1 \in L_t[g]$, so that $K_1 = V[f_1]$. One continues with the requirement that $g_2 = f_2 - P_{K_1}[f_2] = f_2$, and that

$$K_2 = \overline{V[P_t^g[g_2], t \in T]}.$$

But, when $t \geq \theta_2$, $P_t^g[g_2] = P_t^g[f_2] = f_2$. Consequently,

$$K_1 \vee K_2 = V[f_1] \oplus V[f_2].$$

g has thus a canonical representation of the following form

$$g(t) = \oplus_{i=1}^n g_i(t), \; g_i(t) = P_t^g[f_i].$$

To obtain a proper canonical representation, one follows (Proposition) 6.4.37, which begins with an element of the following form: $h_1 = \sum_{i=1}^n \alpha_i f_i, \; \alpha_i \neq 0, \; i \in [1:n]$. Now

$$P_t^g[h_1] = \sum_{i=1}^n \alpha_i P_t^g[f_i],$$

and, as seen above, $L_{\cup T}[h_1] = L_{\cup T}[g]$. So g has multiplicity one. □

Remark 8.3.30 Linear independence can be dispensed with because of (Lemma) 7.1.8. Lemma 8.3.29 illustrates rather well the difference between canonical and proper canonical representations. The multiplicity of the latter is always smaller than that of the former. That fact is mentioned in [242].

Proposition 8.3.31 ([91]) *Purely nondeterministic processes $f : T \longrightarrow H$ belonging to $L_2^H([0,1], \mathcal{B}([0,1], Leb)$ can be approximated, in that space, with processes of multiplicity one.*

Proof Let $h_n(t, \theta) = \chi_{]t-2^{-n}, t]}(\theta)$, and g_n be defined as in (Lemma) 8.3.29, with h_n in place of h. Then

$$g_n(t) = f(i2^{-n}),$$

where $i2^{-n} \in]t - 2^{-n}, t]$. When f is continuous, and thus uniformly continuous, f can be approximated by g_n. But continuous functions are dense in $L_2^H[T]$ [135, p. 86]. □

Result (Proposition) 8.3.31 is not particularly useful, as the approximating process has no structure (it is here perhaps that the shortcomings of multiplicity theory are easiest to see). What one needs in practice are approximations g typically of the following type (linear estimator) [40, p. 116]:

$$\int_0^t R(t,\theta)X(\omega, d\theta),$$

that is, of the form appearing in the (proper) canonical representation. When $L_{UT}[f]$ is separable, they always exist [(Remark) 1.5.13], but need not have multiplicity one [(Example) 9.2.1]. Multiplicity one is usually required when one must invert elements.

That question shall be illustrated by means of an (a class of) example(s) found in [137]. The conclusion is that, in general, and unsurprisingly, such approximations do not exist. Indeed the problem is akin to approximating two orthogonal vectors using a single one. In [137] the actual objective is assessment of multiplicity one for a simple signal-in-noise "model," rather than approximation, but it illustrates well the limits of multiplicity theories: it is difficult to find general results that could be of use in applications (with the exception perhaps of the calculation of likelihoods, as shall be seen in Part III).

The projection determined by the map $\phi \mapsto I_t\phi$ shall be denoted Q_t.

Let $T = [0, 1]$, and H be a real Hilbert space. Let $W : T \longrightarrow H$ be a wide sense Wiener process, and $f : T \longrightarrow H$ be a map such that (capital C's denote covariances)

$$C_\alpha = \alpha C_W \ll C_f \ll \beta C_W = C_\beta,\ \alpha > 0,\ \beta > 0.$$

Let:

- $L_\alpha : H(C_W, T) \longrightarrow H(C_\alpha, T)$ be the unitary map

$$h \mapsto \alpha^{1/2} h;$$

- $J_{f,\alpha} : H(C_f, T) \longrightarrow H(C_\alpha, T)$ be the map

$$C_f(\cdot, t) \mapsto C_\alpha(\cdot, t) = \alpha^{1/2} C_W(\cdot, t);$$

- $J_{\beta,f} : H(C_\beta, T) \longrightarrow H(C_f, T)$ be the map

$$C_\beta(\cdot, t) \mapsto C_f(\cdot, t).$$

As sets, one has that

$$H(C_W, T) = H(C_\alpha, T) \subseteq H(C_f, T) \subseteq H(C_\beta, T) = H(C_W, T).$$

One has furthermore the following diagram:

$$\begin{array}{ccc} H(C_\beta, T) & \xrightarrow{J_{\beta,f}} & H(C_f, T) \\ \uparrow L_\beta & & J_{f,\alpha} \downarrow \\ H(C_W, T) & \xrightarrow{L_\alpha} & H(C_\alpha, T) \end{array}$$

As, for $t \in T$, fixed, but arbitrary,

$$L_\alpha^\star J_{f,\alpha} J_{\beta,f} L_\beta [C_W(\cdot, t)] = C_W(\cdot, t), \text{ and } J_{\beta,f} L_\beta L_\alpha^\star J_{f,\alpha}[C_f(\cdot, t)] = C_f(\cdot, t),$$

$J_{W,f} = J_{\beta,f} L_\beta : H(C_W, T) \longrightarrow H(C_f, T)$ is thus a bounded linear operator with bounded inverse such that

$$J_{W,f}\left[C_W(\cdot, t)\right] = C_f(\cdot, t).$$

Finally, for $h \in H(C_f, T)$, fixed, but arbitrary, $J_{W,f}^\star[h] = \beta^{1/2} h \in H(C_W, T)$. Indeed, $J_{W,f}^\star = L_\beta^\star J_{\beta,f}^\star$. But $L_{\beta,f}^\star$ makes h an element of $H(C_\beta, T)$, and $L_\beta^\star$ makes of h the element $\beta^{1/2} h$ [(Proposition) 1.1.15, (Example) 1.3.12].

Let

$$U_W : H(C_W, T) \longrightarrow L_{\cup T}[W]$$

be the unitary map defined using the following relation:

$$U_W\left[C_W(\cdot, t)\right] = W(t).$$

The operator U_f is defined analogously: $U_f[C_f(\cdot, t)] = f(t)$. Let finally, $L_2[T]$ being the classes of functions whose square, over $T = [0, 1]$, is integrable,

$$V_W : L_2[T] \longrightarrow H(C_W, T)$$

be the unitary map defined using the following relation:

$$V_W[\phi](t) = \int_0^t \phi(\theta)\, d\theta.$$

One has that

$$W[\phi] = \int \phi \, dm_W = U_W V_W [\phi] .$$

In particular $W(t) = W(I_t) = U_W V_W [I_t]$. As $U_f J_{W,f} U_W^\star [W(t)] = f(t)$, one has, by analogy,

$$f[\phi] = U_f J_{W,f} U_W^\star [W[\phi]] .$$

Since $W[I_t] = W(t)$, then $f[I_t] = f(t)$, so that, for $t \in T$, fixed, but arbitrary,

$$B_{W,f} = U_f J_{W,f} U_W^\star$$

is an isomorphism between $L_t[W]$ and $L_t[f]$. One may notice that

$$f[\phi] = B_{W,f}[W[\phi]] = U_f J_{W,f} V_W [\phi] .$$

A second inner product, besides the standard one, shall be used on $L_2[T]$. It is defined as follows:

$$\langle \phi, \psi \rangle_f = \langle f[\phi], f[\psi] \rangle_H = \left\langle J_{W,f} V_W [\phi], J_{W,f} V_W [\psi] \right\rangle_{H(C_f, T)} .$$

Since U_f and V_W are unitary, and $J_{W,f}$ has bounded inverse, the resulting norm is equivalent to the standard norm of $L_2[T]$. One shall use the following map:

$$B_f = V_W^\star J_{W,f}^\star J_{W,f} V_W .$$

It is bounded, with bounded inverse. Furthermore, using the polar decomposition,

$$J_{W,f} V_W = U_{W,f} B_f^{1/2},$$

where $U_{W,f}$ is a partial isometry, whose initial set is the closure of the range of $B_f^{1/2}$, and final set, the closure of the range of $J_{W,f} V_W$. $U_{W,f}$ is thus unitary. As $J_{W,f} = U_f^\star B_{W,f} U_W$, one has that

$$B_f = V_W^\star U_W^\star B_{W,f}^\star U_f U_f^\star B_{W,f} U_W V_W = V_W^\star U_W^\star B_{W,f}^\star B_{W,f} U_W V_W,$$

and that

$$\langle f[\phi], f[\psi] \rangle_H = \langle \phi, \psi \rangle_f = \left\langle B_f[\phi], \psi \right\rangle_{L_2[T]} .$$

The following diagram summarizes the maps just introduced.

$$
\begin{array}{ccc}
L_2[T] & \xleftarrow[V_W^\star]{} & H(C_W,T) \\
\Big\downarrow {\scriptstyle V_W} & & {\scriptstyle J_{W,f}^\star} \Big\uparrow \\
H(C_W,T) & \xrightarrow{J_{W,f}} & H(C_f,T) \\
\Big\downarrow {\scriptstyle U_W} & & {\scriptstyle U_f} \Big\downarrow \\
L_{\cup T}[W] & \xrightarrow{B_{W,f}} & L_{\cup T}[f]
\end{array}
\qquad B_f = V_W^\star J_{W,f}^\star J_{W,f} V_W
$$

Remark 8.3.32 $P_t^f\left[f\left[\phi\right]\right] = f\left[Q_t\left[\phi\right]\right]$.
As, for $\theta \leq t$, fixed, but arbitrary, $f\left[Q_t\left[I_\theta\right]\right] = f\left(\theta\right)$, then

$$L_t\left[f\right] = \left\{f\left[Q_t\left[\phi\right]\right], \phi \in L_2\left[T\right]\right\}.$$

But, for $\theta \leq t$, fixed, but arbitrary in T,

$$\langle f(\phi), f(\theta)\rangle_H = \lim_{\tau \uparrow\uparrow 1} \langle f(Q_\tau[\phi]), f(\theta)\rangle_H.$$

However, for $\tau > t$, $\langle f(Q_\tau[\phi]), f(\theta)\rangle_H = \langle f(Q_t[\phi]), f(\theta)\rangle_H$.

Remark 8.3.33 One considers below a "signal-plus-noise" model, $f = s + W$. When C_W dominates C_s, but C_f does not dominate C_W, the above remains true, except for the existence of a bounded inverse, and the facts that require that inverse. In particular the range of $B_{W,f}$ may no longer equal $L_{\cup T}\left[f\right]$. But it is always dense in it.

Henceforth the map f shall have a specific form, namely,

$$f\left(t\right) = s\left(t\right) + W\left(t\right)$$

where s and W are orthogonal, and $C_s \ll \kappa C_W$, some $\kappa > 0$. Then, as

$$C_f = C_s + C_W,$$

C_f and C_W are equivalent, that is, each one dominates the other. Since s and W are orthogonal, the RKHS of C_f is the direct sum of those respectively of C_s and C_W.

The following subspaces shall be relevant:

- $L_t\left[s, W\right] = L_t\left[s\right] \oplus L_t\left[W\right]$;
- $L_{\cup T}\left[s, W\right] = L_{\cup T}\left[s\right] \oplus L_{\cup T}\left[W\right]$;
- $L_t^\perp\left[f\right] = L_{\cup T}\left[s, W\right] \ominus L_t\left[f\right]$.

Since C_s is dominated by κC_W, one is able to define, as above,

- $J_{W,s} = J_{\kappa,s} L_\kappa$,

- $B_{W,s} = U_s J_{W,s} U_W^{\star}$,
- $B_s = V_W^{\star} U_W^{\star} B_{W,s}^{\star} B_{W,s} U_W V_W = V_W^{\star} J_{W,s}^{\star} J_{W,s} V_W$,

with the same properties. Furthermore, since $f(t) = s(t) + W(t)$ may be expressed as

$$f[I_t] = s[I_t] + W[I_t],$$

one has that

$$f[\phi] = s[\phi] + W[\phi].$$

Since $C_f = C_s + C_W$, looking at $C_f(\cdot, t) = C_s(\cdot, t) + C_W(\cdot, s)$, one has that

$$J_{W,f} V_W = J_{W,s} V_W + V_W,$$

and, since the range of $J_{W,s}$ and that of V_W are orthogonal, $V_W^{\star} J_{W,s} V_W$ is the zero operator, and, using the definition of B_f,

$$B_f = I_{L_2[T]} + B_s,$$

an operator, linear and bounded, with bounded inverse. Furthermore [8, p. 359], since

$$B_s = (J_{W,s} V_W)^{\star} J_{W,s} V_W,$$

B_f^{-1} is positive definite and both the following obtain:

$$\left\| B_f^{-1} \right\| \leq 1, \qquad \left\| J_{W,s} V_W B_f^{-1} \right\| \leq 1.$$

Finally

$$\langle s[\phi], s[\psi] \rangle_H = \langle \phi, \psi \rangle_s = \langle B_s[\phi], \psi \rangle_{L_2[T]}.$$

The following abbreviations shall be used:

$$B_{s,t} = Q_t B_s Q_t, \quad B_{f,t} = Q_t + B_{s,t} = Q_t B_f Q_t.$$

Fact 8.3.34 *Here are two basic properties:*

1. Let $s[\phi] + W[\psi] \in L_{UT}[s, W]$ be fixed, but arbitrary. Then

$$s[\phi] + W[\psi] \in L_t^{\perp}[f] \text{ if, and only if, } Q_t[B_s[\phi] + \psi] = 0,$$

or, equivalently, for almost every $\theta \leq t$,

$$\psi(\theta) = -B_s[\phi](\theta).$$

Consequently, when $s[Q_t[\phi]] + W[Q_t[\psi]] \in L_t[s, W]$ *is fixed, but arbitrary, then*

$$s[Q_t[\phi]] + W[Q_t[\psi]] \in L_t^{\perp}[f] \text{ if, and only if, } Q_t[\psi] = -B_{s,t}[\phi].$$

2. *For* $t \in T$*, fixed, but arbitrary, the operator*

$$B_{f,t} = Q_t B_f Q_t = Q_t + B_{s,t}$$

is an isomorphism of $L_2[0,t]$*, and thus* $B_{f,t}^{-1}$ *is also an isomorphism of* $L_2[0,t]$*. In what follows one shall use tacitly the following equality:*

$$Q_t B_{f,t}^{-1} = B_{f,t}^{-1}.$$

The presence of Q_t *acts mainly as a reminder that one is "living" in* $L_2[T]$.

Proof The elements of $L_t[f]$ have the form $f[Q_t[\eta]]$, t fixed, η arbitrary. Given $s[\phi] + W[\psi]$, one must thus have, for all $\eta \in L_2[T]$,

$$\langle f[Q_t[\eta]], s[\phi] + W[\psi]\rangle_H = 0.$$

But

$$\begin{aligned}\langle f[Q_t[\eta]], s[\phi] + W[\psi]\rangle_H &= \langle s[Q_t[\eta]] + W[Q_t[\eta]], s[\phi] + W[\psi]\rangle_H \\ &= \langle B_s[Q_t[\eta]], \phi\rangle_{L_2[T]} + \langle Q_t[\eta], \psi\rangle_{L_2[T]} \\ &= \langle \eta, Q_t[B_s[\phi] + \psi]\rangle_{L_2[T]}.\end{aligned}$$

□

Fact 8.3.35 *Let* $h(t) = s[Q_t[\phi]] + W[Q_t[\psi]] \in L_t[s, W]$ *be fixed, but arbitrary. It may be expressed in the following form*

$$h(t) = h_f(t) + h_f^{\perp}(t),$$

with

$$h_f(t) \in L_t[f], \text{ and } h_f^{\perp}(t) \in L_t[s, W] \ominus L_t[f],$$

where

$$\begin{aligned}h_f(t) &= f\left[Q_t B_{f,t}^{-1}\{B_{s,t}[\phi] + Q_t[\psi]\}\right], \\ h_f^{\perp}(t) &= s\left[Q_t B_{f,t}^{-1} Q_t[\phi - \psi]\right] - W\left[B_{s,t} B_{f,t}^{-1} Q_t[\phi - \psi]\right].\end{aligned}$$

Proof A generic element of $L_t[f]$ has the following form: $f[Q_t[\eta]]$, and a generic term of $L_t[s, W] \ominus L_t[f]$ has, using (Fact) 8.3.34, the following one:

$$s[Q_t[\xi]] + W[Q_t[\eta]] = s[Q_t[\xi]] - W[B_{s,t}[\xi]].$$

Given the left-hand side, one must solve, for unknowns ξ, η, the following equation:

$$\begin{aligned} s[Q_t[\phi]] + W[Q_t[\psi]] &= f[Q_t[\eta]] + s[Q_t[\xi]] - W[B_{s,t}[\xi]] \\ &= s[Q_t[\eta]] + W[Q_t[\eta]] + s[Q_t[\xi]] - W[B_{s,t}[\xi]] \\ &= s[Q_t[\eta] + Q_t[\xi]] + W[Q_t[\eta] - B_{s,t}[\xi]], \end{aligned}$$

that is, the following system:

$$\begin{aligned} Q_t[\phi] &= Q_t[\eta] + Q_t[\xi], \\ Q_t[\psi] &= Q_t[\eta] - B_{s,t}[\xi]. \end{aligned}$$

Subtracting the second equation from the first, one obtains that

$$Q_t[\phi] - Q_t[\psi] = Q_t[\xi] + B_{s,t}[Q_t[\xi]] = B_{f,t}[\xi],$$

or

$$Q_t[\xi] = B_{f,t}^{-1} Q_t[\phi - \psi].$$

From the first equation one computes:

$$\begin{aligned} Q_t[\eta] &= Q_t[\phi] - Q_t[\xi] \\ &= Q_t[\phi] - B_{f,t}^{-1}[Q_t[\phi] - Q_t[\psi]] \\ &= B_{f,t}^{-1}[Q_t[\psi]] + Q_t[\phi] - B_{f,t}^{-1}[Q_t[\phi]] \\ &= B_{f,t}^{-1}\left[Q_t[\psi] + B_{f,t}[Q_t[\phi]] - Q_t[\phi]\right] \\ &= B_{f,t}^{-1}[Q_t[\psi] + B_{s,t}[Q_t[\phi]]]. \end{aligned}$$

□

Definition 8.3.36 The function $g : T \longrightarrow H$, $g(0) = 0_H$, is a wide sense martingale with respect to the function $f : T \longrightarrow H$, $f(0) = 0_H$, when, for $\theta \leq t$ in T, fixed, but arbitrary,

$$g(\theta) = P_\theta^f[g(t)].$$

One may always use $t = 1$. Then, given, arbitrarily in T, but fixed,

$$0 = \theta_0 < \theta_1 < \cdots < \theta_{n-1} < \theta_n = t,$$

one has that

$$\begin{aligned}
\sum_{i=1}^{n} \|g(\theta_i) - g(\theta_{i-1})\|_H^2 &= \sum_{i=1}^{n} \left\| \left\{ P^f_{\theta_i} - P^f_{\theta_{i-1}} \right\} [g(1)] \right\|_H^2 \\
&= \left\| \sum_{i=1}^{n} \left\{ P^f_{\theta_i} - P^f_{\theta_{i-1}} \right\} [g(1)] \right\|_H^2 \\
&= \|g(t)\|_H^2 \\
&\leq \|g(1)\|_H^2 .
\end{aligned}$$

The value $\|g(t)\|_H^2$ shall be called the wide-sense quadratic variation of g at t, for f, and denoted $\langle g \rangle_f(t)$. Quadratic variation is thus equivalent to the usual bounded variation [135, p. 59].

Remark 8.3.37 When g is a wide sense martingale with respect to f, and $t \geq \theta$, then $g(t) - g(\theta)$ is orthogonal to $L_\theta[f]$. Indeed,

$$\left\langle g(t) - g(\theta), P^f_\theta[h] \right\rangle_H = \left\langle \left\{ P^f_t - P^f_\theta \right\} [g(1)], P^f_\theta[h] \right\rangle_H = 0.$$

Example 8.3.38 Let $h \in H$ be fixed, but arbitrary, and $g(t) = P^f_t[h]$. g is a wide sense martingale with respect to f as $P^f_\theta[g(t)] = P^f_\theta[P^f_t[h]] = P^f_\theta[h] = g(\theta)$.

Fact 8.3.39 *Let $g(t) = P^f_t[W[Q_t[\gamma]]]$. g is a wide sense martingale with respect to f.*

Proof $g(t) = P^f_t[W[\gamma]]$ is a wide sense martingale with respect to f [(Example) 8.3.38]. But

$$P^f_t[W[\gamma]] = P^{s,W}_t[W[\gamma]],$$

and, for $\theta \leq t$, fixed, but arbitrary,

$$\begin{aligned}
\langle W[\gamma], s[Q_\theta[\phi]] + W[Q_\theta[\phi]] \rangle_H &= \langle W[\gamma], W[Q_\theta[\phi]] \rangle_H \\
&= \langle \gamma, Q_\theta[\phi] \rangle_{L_2[T]} \\
&= \langle Q_t[\gamma], Q_\theta[\phi] \rangle_{L_2[T]} \\
&= \langle W[Q_t[\gamma]], W[Q_\theta[\phi]] \rangle_H \\
&= \langle W[Q_t[\gamma]], s[Q_\theta[\phi]] + W[Q_\theta[\phi]] \rangle_H .
\end{aligned}$$

Now, as in (Remark) 8.3.32, $P_t^W[W[\gamma]] = W[Q_t[\gamma]]$. As $P_t^s[W[\gamma]] = 0_H$,

$$P_t^f[W[\gamma]] = P_t^{s,W}[W[\gamma]] = P_t^W[W[\gamma]] = W[Q_t[\gamma]],$$

so that

$$P_t^f[W[\gamma]] = W[Q_t[\gamma]] = P_t^f[W[Q_t[\gamma]]].$$

□

Fact 8.3.40 *The martingale g of (Fact)* 8.3.39 *has the following representation:*

$$g(t) = f\left[Q_t B_{f,t}^{-1} Q_t[\gamma]\right].$$

Proof Since $g(t) = P_t^f[W[Q_t[\gamma]]]$, (Fact) 8.3.35, with $\phi = 0_{L_2[T]}$ and $\psi = \gamma$, yields that

$$g(t) = h_f(t) = f\left[Q_t B_{f,t}^{-1} Q_t[\gamma]\right].$$

□

Fact 8.3.41 *The following formulae obtain.*

1. $P_t^f[f[\phi]] = f\left[Q_t B_{f,t}^{-1} Q_t B_f[\phi]\right]$, *and, in particular,*

$$P_t^f[f(t+\theta)] = f\left[Q_t B_{f,t}^{-1} Q_t B_f Q_{t+\theta}\left[I_{L_2[T]}\right]\right].$$

2. *Maps of the following form:*

$$g(t) = f\left[Q_t B_{f,t}^{-1} Q_t[\phi]\right]$$

are wide sense martingales with respect to f.

Proof Let $\theta \le t$ be fixed, but arbitrary. One has that

$$\begin{aligned}
&\langle f[\phi], f(\theta)\rangle_H = \\
&= \langle s[\phi], s[I_\theta]\rangle_H + \langle W[\phi], W[I_\theta]\rangle_H \\
&= \langle B_s[\phi], I_\theta\rangle_{L_2[T]} + \langle \phi, I_\theta\rangle_{L_2[T]} \\
&= \left\langle Q_t\left(B_s + I_{L_2[T]}\right)[\phi], I_\theta\right\rangle_{L_2[T]} \\
&= \left\langle B_{f,t}^{-1} Q_t B_f[\phi], B_{f,t}[I_\theta]\right\rangle_{L_2[T]} \\
&= \left\langle B_{f,t}^{-1} Q_t B_f[\phi], B_f[I_\theta]\right\rangle_{L_2[T]} \\
&= \left\langle B_s B_{f,t}^{-1} Q_t B_f[\phi], I_\theta\right\rangle_{L_2[T]} + \left\langle B_{f,t}^{-1} Q_t B_f[\phi], I_\theta\right\rangle_{L_2[T]} \\
&= \left\langle s\left[B_{f,t}^{-1} Q_t B_f[\phi]\right], s[I_\theta]\right\rangle_H + \left\langle W\left[B_{f,t}^{-1} Q_t B_f[\phi]\right], W[I_\theta]\right\rangle_H
\end{aligned}$$

$$= \left\langle f\left[B_{f,t}^{-1}Q_tB_f[\phi]\right], f(\theta)\right\rangle_H$$
$$= \left\langle f\left[Q_tB_{f,t}^{-1}Q_tB_f[\phi]\right], f(\theta)\right\rangle_H.$$

Now, since $B_{f,t+\theta} = Q_{t+\theta}B_fQ_{t+\theta}$.

$$\begin{aligned} P_t^f\left[g(t+\theta)\right] &= P_t^f\left[f\left[Q_{t+\theta}B_{f,t+\theta}Q_{t+\theta}[\phi]\right]\right] \\ &= f\left[Q_tB_{f,t}^{-1}Q_tB_f\left[Q_{t+\theta}B_{f,t+\theta}^{-1}Q_{t+\theta}[\phi]\right]\right] \\ &= f\left[Q_tB_{f,t}^{-1}Q_tQ_{t+\theta}B_f\left[Q_{t+\theta}B_{f,t+\theta}^{-1}Q_{t+\theta}[\phi]\right]\right] \\ &= f\left[Q_tB_{f,t}^{-1}Q_tQ_{t+\theta}[\phi]\right] \\ &= g(t). \end{aligned}$$

□

Fact 8.3.42 *Let $g : T \longrightarrow H$ be a wide sense martingale with respect to f such that $\mathcal{R}[g] \subseteq L_{UT}[f]$. There exists then $\gamma \in L_2[T]$ such that, for $t \in T$, fixed, but arbitrary,*

$$g(t) = f\left[Q_tB_{f,t}^{-1}Q_t[\gamma]\right].$$

Furthermore,

$$\langle g\rangle_f(t) = \left\langle B_{f,t}^{-1}Q_t[\gamma], Q_t[\gamma]\right\rangle_{L_2[T]}.$$

Proof As $g(1) \in L_{UT}[f]$, $g(1) = f[\phi]$. Choose $\gamma = B_f[\phi]$. Then, from (Fact) 8.3.41,

$$g(t) = P_t^f[g(1)] = P_t^f[f[\phi]] = f\left[Q_tB_{f,t}^{-1}Q_tB_f[\phi]\right] = f\left[Q_tB_{f,t}^{-1}Q_t[\gamma]\right].$$

Now, using $B_{f,t} = B_{s,t} + Q_t$, $Q_tB_{f,t}^{-1} = B_{f,t}^{-1}$, $B_{s,t} = Q_tB_sQ_t$,

$$\begin{aligned} &\left\langle B_{f,t}^{-1}Q_t[\gamma], Q_t[\gamma]\right\rangle_{L_2[T]} = \\ &= \left\langle B_{f,t}^{-1}Q_t[\gamma], B_{f,t}B_{f,t}^{-1}Q_t[\gamma]\right\rangle_{L_2[T]} \\ &= \left\langle B_{f,t}^{-1}Q_t[\gamma], B_{s,t}B_{f,t}^{-1}Q_t[\gamma]\right\rangle_{L_2[T]} + \left\langle B_{f,t}^{-1}Q_t[\gamma], B_{f,t}^{-1}Q_t[\gamma]\right\rangle_{L_2[T]} \\ &= \left\langle B_{f,t}^{-1}Q_t[\gamma], B_sB_{f,t}^{-1}Q_t[\gamma]\right\rangle_{L_2[T]} + \left\langle B_{f,t}^{-1}Q_t[\gamma], B_{f,t}^{-1}Q_t[\gamma]\right\rangle_{L_2[T]} \\ &= \left\langle s\left[B_{f,t}^{-1}Q_t[\gamma]\right], s\left[B_{f,t}^{-1}Q_t[\gamma]\right]\right\rangle_H + \left\langle W\left[B_{f,t}^{-1}Q_t[\gamma]\right], W\left[B_{f,t}^{-1}Q_t[\gamma]\right]\right\rangle_H \\ &= \left\|f\left[Q_tB_{f,t}^{-1}Q_t[\gamma]\right]\right\|_H^2 \\ &= \|g(t)\|_H^2. \end{aligned}$$

□

Fact 8.3.43 *One has that* $P_t^f\,[s\,[\eta]] = f\left[Q_t B_{f,t}^{-1} B_s\,[\eta]\right]$.

Proof The reason is the following calculation: for fixed, but arbitrary ϕ and ψ in $Q_t\,[L_2\,[T]]$,

$$\begin{aligned}
&\langle s\,[\eta]\,, s\,[\phi] + W\,[\phi]\rangle_H = \\
&= \langle s\,[\eta]\,, s\,[\phi]\rangle_H \\
&= \langle B_s\,[\eta]\,, \phi\rangle_{L_2[T]} \\
&= \left\langle B_{f,t}^{-1} B_s\,[\eta]\,, B_{f,t}\,[\phi]\right\rangle_{L_2[T]} \\
&= \left\langle B_{f,t}^{-1} B_s\,[\eta]\,, B_{s,t}\,[\phi]\right\rangle_{L_2[T]} + \left\langle B_{f,t}^{-1} B_s\,[\eta]\,, Q_t\,[\phi]\right\rangle_{L_2[T]} \\
&= \left\langle B_{f,t}^{-1} B_s\,[\eta]\,, B_s\,[\phi]\right\rangle_{L_2[T]} + \left\langle B_{f,t}^{-1} B_s\,[\eta]\,, [\phi]\right\rangle_{L_2[T]} \\
&= \left\langle s\left[B_{f,t}^{-1} B_s\,[\eta]\right], s\,[\phi]\right\rangle_H + \left\langle W\left[B_{f,t}^{-1} B_s\,[\eta]\right], W\,[\phi]\right\rangle_H \\
&= \left\langle f\left[B_{f,t}^{-1} B_s\,[\eta]\right], s\,[\phi] + W\,[\phi]\right\rangle_H .
\end{aligned}$$

□

Definition 8.3.44 Let W be a wide sense Wiener process with respect to f. f is said to have a single Brownian innovation when, for $t \in T$, fixed, but arbitrary, $L_t[f] = L_t[W]$ (f has thus multiplicity one, with a representation of the form $f(t) = W\,[\Phi_t]$).

Fact 8.3.45 *Given* $\phi \in L_2\,[T]$*, fixed, but arbitrary, let*

$$\begin{aligned}
F_\phi\,(t) &= \left\langle B_{f,t}^{-1} Q_t\,[\phi]\,, Q_t\,[\phi]\right\rangle_{L_2[T]} \\
&= \left\langle B_{f,t}\left[B_{f,t}^{-1} Q_t\,[\phi]\right], B_{f,t}^{-1}\,[Q_t\,[\phi]]\right\rangle_{L_2[T]} \\
&= \left\langle B_f\left[B_{f,t}^{-1} Q_t\,[\phi]\right], B_{f,t}^{-1}\,[Q_t\,[\phi]]\right\rangle_{L_2[T]} \\
&= \left\| f\left[Q_t B_{f,t}^{-1} Q_t\,[\phi]\right]\right\|_H^2 ,
\end{aligned}$$

which is [(Fact) 8.3.42] *a monotone increasing function.*

f has a single Brownian innovation if, and only if, there exists $\phi_f \in L_2\,[T]$ such that, for $t \in T$, fixed, but arbitrary,

1. *the measure induced by F_{ϕ_f} is equivalent to Lebesgue measure;*
2. *the following set of equalities*

$$\left\{\left\langle B_{f,\theta}^{-1} Q_\theta\left[\phi_f\right], Q_\theta\,[\phi]\right\rangle_{L_2[T]} = 0,\ \theta \le t\right\} \qquad (\star)$$

implies that, almost surely, $\phi\,(\theta) = 0,\ \theta \le t$.

Proof Suppose that f has the single Brownian innovation $\mathcal{W}$. Since $\mathcal{W}$ is a martingale in the wide sense with respect to f (as indeed one checks that $P^f_\theta[\mathcal{W}(t)] = P^{\mathcal{W}}_\theta[\mathcal{W}(t)] = \mathcal{W}(t)$), it has [(Fact) 8.3.42] a representation in the following form:

$$\mathcal{W}(t) = f\left[Q_t B^{-1}_{f,t} Q_t[\beta]\right],$$

with $\langle \mathcal{W} \rangle (t) = t$. Choose β as ϕ_f so that item 1 is automatically true.

Suppose that, for $\theta \leq t$, fixed, but arbitrary,

$$\left\langle B^{-1}_{f,\theta} Q_\theta[\beta], Q_\theta[\phi]\right\rangle_{L_2[T]} = 0.$$

One has then that

$$\begin{aligned}\left\langle B^{-1}_{f,\theta} Q_\theta[\beta], Q_\theta[\phi]\right\rangle_{L_2[T]} &= \left\langle B^{-1}_{f,\theta} Q_\theta[\beta], B_{f,\theta} B^{-1}_{f,\theta} Q_\theta[\phi]\right\rangle_{L_2[T]} \\ &= \left\langle B^{-1}_{f,\theta} Q_\theta[\beta], B^{-1}_{f,\theta} Q_\theta[\phi]\right\rangle_f \\ &= \left\langle f(Q_\theta B^{-1}_{f,\theta} Q_\theta[\beta]), f(Q_\theta B^{-1}_{f,\theta} Q_\theta[\phi])\right\rangle_H \\ &= \langle \mathcal{W}(\theta), g_\phi(\theta)\rangle_H,\end{aligned}$$

where g_ϕ is as in (Fact) 8.3.41, and in particular a martingale in the wide sense for f. Then when $\tau \leq \theta \leq t$, using condition ($\star$) in item 2, and the latter equality,

$$\langle \mathcal{W}(\tau), g_\phi(\theta)\rangle_H = \langle \mathcal{W}(\tau), g_\phi(\theta) - g_\phi(\tau)\rangle_H.$$

But $\mathcal{W}(\tau) \in L_\tau[f]$, and thus $\langle \mathcal{W}(\tau), g_\phi(\theta) - g_\phi(\tau)\rangle_H = 0$. The same relation holds when the order of θ and τ is reversed. Thus, for $\theta \leq t$, fixed, but arbitrary, $g_\phi(\theta)$ is orthogonal to $L_t[\mathcal{W}] = L_t[f]$, while belonging to it. It is thus 0_H, and then

$$0 = \langle g_\phi \rangle_f(\theta) = \left\| f\left[Q_\theta B^{-1}_{f,\theta} Q_\theta[\phi]\right]\right\|^2_H = \left\| B^{-1}_{f,\theta} Q_\theta[\phi]\right\|^2_{L_2[T]}.$$

But $B^{-1}_{f,t}$ is an isomorphism of $Q_t[L_2[T]]$. Consequently $Q_\theta[\phi] = 0$, almost surely for $\theta \leq t$.

Suppose conversely that items 1 and 2 above obtain. Let *Leb* denote Lebesgue measure, and, because of item 1,

$$\delta^2 = \frac{dLeb}{dF_{\phi_f}}.$$

Let g denote the martingale defined as [(Fact) 8.3.41]

$$g(t) = f\left[Q_t B^{-1}_{f,t} Q_t[\phi_f]\right].$$

Then $M_g = F_{\phi_f}$. Set $\mathcal{W}(t) = g[Q_t[\delta]]$, the "integral" of $Q_t[\delta]$ with respect to the martingale in the wide sense g. Since

$$\langle \mathcal{W}(t) \rangle = \|g(Q_t[\delta])\|_H^2 = \int \left\{ \overbrace{Q_t[\delta]}^{\cdot} \right\}^2 dM_g = t,$$

$\mathcal{W}$ is then a wide sense Wiener process, and, since F_{ϕ_f} is equivalent to Lebesgue measure, $L_t[\mathcal{W}] = L_t[g]$. Since item 2 says that $L_t[g] = L_t[f]$, f has a single Brownian motion. □

Remark 8.3.46 Since, g having orthogonal increments,

$$\left\|g\left[B_{f,t}^{-1} Q_t[\phi_f]\right]\right\|_H^2 = \left\|B_{f,t}^{-1} Q_t[\phi_f]\right\|_{L_2[\langle g \rangle_f]}^2,$$

one has, for the canonical representation of f,

$$f(t) = g\left[B_{f,t}^{-1} Q_t[\phi_f]\right] = g[\phi_t],$$

as item 2 in (Fact) 8.3.45 asserts that $\{\phi_t, t \in T\}$ is total in $L_2[T]$.

Remark 8.3.47 When B_s is Hilbert-Schmidt, B_f has an inverse of the following form:

$$\left(I_{L_2[T]} - L^\star\right)\left(I_{L_2[T]} - L\right),$$

where L is a Volterra, Hilbert-Schmidt operator [25, p. 130], and then the conditions of (Fact) 8.3.45 may be easier to check.

Example 8.3.48 Let $T = [0, 1]$, and H, a real Hilbert space. Let $\mathcal{W}, W : T \longrightarrow H$ be two wide sense, orthogonal Wiener processes, and $\sigma : T \longrightarrow \mathbb{R}$ be the square root map. Set

$$s(t) = \sigma(t)\mathcal{W}(t), \text{ and } f(t) = s(t) + W(t).$$

The map f has then multiplicity two [136].

One has that

$$C_s(t_1, t_2) = \sigma(t_1)\sigma(t_2)\{t_1 \wedge t_2\}.$$

For checking that $C_s \ll \kappa C_{\mathcal{W}}$, one may restrict attention to sets of strictly increasing "time" indices as, for example, when $t_1 = t_2 < t_3$ in T,

$$\sum_{i=1}^{3}\sum_{j=1}^{3} \alpha_i\alpha_j C_\cdot\left(t_i, t_j\right) =$$

$$= (\alpha_1 + \alpha_2)^2 C_\cdot(t_1, t_1) + 2(\alpha_1 + \alpha_2)\alpha_3 C_\cdot(t_1, t_3) + \alpha_3^2 C_\cdot(t_3, t_3).$$

Then, to be explicit, for

$$\Sigma_{\mathcal{W}} = \begin{bmatrix} t_1 & t_1 & t_1 & t_1 \\ t_1 & t_2 & t_2 & t_2 \\ t_1 & t_2 & t_3 & t_3 \\ t_1 & t_2 & t_3 & t_4 \end{bmatrix},$$

one has that $\Sigma_{\mathcal{W}} = MM^{\star}$, where

$$M = \begin{bmatrix} \sqrt{t_1} & 0 & 0 & 0 \\ \sqrt{t_1} & \sqrt{t_2 - t_1} & 0 & 0 \\ \sqrt{t_1} & \sqrt{t_2 - t_1} & \sqrt{t_3 - t_2} & 0 \\ \sqrt{t_1} & \sqrt{t_2 - t_1} & \sqrt{t_3 - t_2} & \sqrt{t_4 - t_3} \end{bmatrix}.$$

Let D_σ denote the diagonal matrix whose diagonal elements are respectively $\sigma(t_1)$, $\sigma(t_2)$, $\sigma(t_3)$, and $\sigma(t_4)$. Then

$$D_\sigma \Sigma_{\mathcal{W}} D_\sigma = D_\sigma M (D_\sigma M)^{\star} =$$

$$= \left[\begin{array}{c|c|c|c} \sigma(t_1)\sigma(t_1)t_1 & \sigma(t_1)\sigma(t_2)t_1 & \sigma(t_1)\sigma(t_3)t_1 & \sigma(t_1)\sigma(t_4)t_1 \\ \sigma(t_2)\sigma(t_1)t_1 & \sigma(t_2)\sigma(t_2)t_2 & \sigma(t_2)\sigma(t_3)t_2 & \sigma(t_2)\sigma(t_4)t_2 \\ \sigma(t_3)\sigma(t_1)t_1 & \sigma(t_3)\sigma(t_2)t_2 & \sigma(t_3)\sigma(t_3)t_3 & \sigma(t_3)\sigma(t_4)t_3 \\ \sigma(t_4)\sigma(t_1)t_1 & \sigma(t_4)\sigma(t_2)t_2 & \sigma(t_4)\sigma(t_3)t_3 & \sigma(t_4)\sigma(t_4)t_4 \end{array}\right].$$

Furthermore, *mutatis mutandis,*

$$\sum_{i=1}^{n}\sum_{j=1}^{n} \alpha_i \alpha_j C_{\mathcal{W}}(t_i, t_j) = \langle MM^{\star}\underline{\alpha}, \underline{\alpha}\rangle_{\mathbb{R}^n},$$

$$\sum_{i=1}^{n}\sum_{j=1}^{n} \alpha_i \alpha_j C_s(t_i, t_j) = \left\langle D_\sigma M (D_\sigma M)^{\star}\underline{\alpha}, \underline{\alpha}\right\rangle_{\mathbb{R}^n},$$

and, using Douglas's range inclusion result [80],

$$\left\langle D_\sigma M (D_\sigma M)^{\star}\underline{\alpha}, \underline{\alpha}\right\rangle_{\mathbb{R}^n} \le \kappa \langle MM^{\star}\underline{\alpha}, \underline{\alpha}\rangle_{\mathbb{R}^n}$$

if, and only if, $\mathcal{R}[D_\sigma M] \subseteq \mathcal{R}[M]$. But $\mathcal{R}[M] = \mathbb{R}^n$. Thus $C_s \ll \kappa C_{\mathcal{W}}$.

Below $M_{\sqrt{\cdot}}$ shall be multiplication by square root, and L, the Volterra operator with a constant kernel equal to one.

The computation of $H(C_s, T)$ proceeds as usual. Let $F : T \longrightarrow H(C_{\mathcal{W}}, T)$ be computed as

$$F(t) = t^{1/2} C_{\mathcal{W}}(\cdot, t),$$

and $L_F : H(C_{\mathcal{W}}, T) \longrightarrow \mathbb{R}^T$, as

$$L_F[h](\theta) = \langle h, F(\theta)\rangle_{H(C_{\mathcal{W}},T)} = \theta^{1/2} h(\theta).$$

L_F is unitary, and its range is $H(C_s, T)$. Furthermore

$$L_F\left[t^{1/2} C_{\mathcal{W}}(\cdot, t)\right] = t^{1/2} L_F\left[C_{\mathcal{W}}(\cdot, t)\right](\theta) = C_s(\theta, t) = J_{\mathcal{W},s}\left[C_{\mathcal{W}}(\cdot, t)\right].$$

Define $D : L_2[T] \longrightarrow L_2[T]$ requiring that

$$D[I_t] = t^{1/2} I_t.$$

As

$$\left\|\sum_i \alpha_i D[I_{t_i}]\right\|^2_{L_2[T]} = \left\|\sum_i \alpha_i C_s(\cdot, t)\right\|^2_{H(C_s,T)} = \left\|J_{\mathcal{W},s} V_{\mathcal{W}}\left[\sum_i \alpha_i I_{t_i}\right]\right\|^2_{H(C_s,T)},$$

D is a well-defined operator, linear and bounded. Furthermore, forgetting temporarily the distinction between functions and their classes, with $\Phi(t) = L[\phi](t)$,

$$\begin{aligned}
\langle D[I_t], \phi\rangle_{L_2[T]} &= \int_0^1 t^{1/2} I_t(\theta)\,\phi(\theta)\,d\theta \\
&= t^{1/2}\Phi(t) \\
&= \int_0^t \frac{d\theta}{2\theta^{1/2}}\Phi(\theta) + \int_0^t \theta^{1/2}\phi(\theta)\,d\theta \\
&= \frac{1}{2}\left\langle I_t, M_{\sqrt{\cdot}}^{-1} L[\phi]\right\rangle_{L_2[T]} + \left\langle I_t, M_{\sqrt{\cdot}}[\phi]\right\rangle_{L_2[T]} \\
&= \left\langle I_t, \left\{M_{\sqrt{\cdot}} + \frac{1}{2} M_{\sqrt{\cdot}}^{-1} L\right\}[\phi]\right\rangle_{L_2[T]}.
\end{aligned}$$

The above holds provided $M_{\sqrt{\cdot}}^{-1} L$ is bounded. But [25, p. 70]

$$\int_0^1 dx \left\{\frac{1}{x^{1/2}}\int_0^x |\phi(y)|\,dy\right\}^2 \le \int_0^1 dx\left\{\frac{1}{x}\int_0^x |\phi(y)|\,dy\right\}^2 \le 4\,\|\phi\|^2_{L_2[T]}.$$

Consequently

$$D^\star = M_{\sqrt{\cdot}} + \frac{1}{2} M_{\sqrt{\cdot}}^{-1} L,$$

and thus

$$D = M_{\sqrt{\cdot}} + \frac{1}{2} L^\star M_{\sqrt{\cdot}}^{-1}.$$

One has, using $L_F V_{\mathcal{W}} D = J_{\mathcal{W},s} V_{\mathcal{W}}$, that $B_s = D^\star D$.

Finally one has that [(Fact) 6.2.24]

$$s(t) = \sigma(t)\mathcal{W}(t) = \int_0^t \frac{\mathcal{W}(\theta)}{2\theta^{1/2}} d\theta + \int_0^t \theta^{1/2} m_{\mathcal{W}}(d\theta),$$

so that

$$f(t) = \int_0^t \frac{\mathcal{W}(\theta)}{2\theta^{1/2}} d\theta + \int_0^t \theta^{1/2} m_{\mathcal{W}}(d\theta) + W(t),$$

and, consequently,

$$f[Q_t[\phi]] = \int_0^t \frac{\mathcal{W}(\theta)}{2\theta^{1/2}} \phi(\theta)\, d\theta + \int_0^t \theta^{1/2} \phi(\theta)\, m_{\mathcal{W}}(d\theta) + \int_0^t \phi(\theta)\, m_W(d\theta).$$

Let

$$X(t) = \int_0^t \frac{\mathcal{W}(\theta)}{2\theta^{1/2}} \phi(\theta)\, d\theta,$$

$$Y(t) = \int_0^t \theta^{1/2} \phi(\theta)\, m_{\mathcal{W}}(d\theta),$$

$$Z(t) = \int_0^t \phi(\theta)\, m_W(d\theta).$$

One shall need the following evaluation:

$$\begin{aligned}
&\|X(1) + Y(1) + Z(1)\|_H^2 = \\
&= \|X(1)\|_H^2 + \|Y(1)\|_H^2 + \|Z(1)\|_H^2 + 2\langle X(1), Y(1)\rangle_H \\
&= \int_0^1 \int_0^1 dx dy \phi(x)\, \phi(y)\, \frac{x \wedge y}{4(xy)^{1/2}} \\
&+ 2 \int_0^1 dx \frac{\phi(x)}{2x^{1/2}} \int_0^x dy \phi(y)\, y^{1/2} \\
&+ \int_0^1 dx x \phi^2(x) + \int_0^1 dx \phi^2(x).
\end{aligned}$$

Now

$$\int_0^1 \int_0^1 dxdy\phi(x)\,\phi(y)\,\frac{x \wedge y}{4(xy)^{1/2}} =$$
$$= \frac{1}{4}\int_0^1 dx\phi(x)\left\{\frac{1}{x^{1/2}}\int_0^x dyy^{1/2}\phi(y) + x^{1/2}\int_x^1 dy\frac{\phi(y)}{y^{1/2}}\right\}$$
$$= \frac{1}{4}\left\langle M_{\sqrt{\cdot}}^{-1} L M_{\sqrt{\cdot}}[\phi] + M_{\sqrt{\cdot}} L^{\star} M_{\sqrt{\cdot}}^{-1}[\phi], \phi\right\rangle_{L_2[T]}$$
$$= \frac{1}{2}\left\langle M_{\sqrt{\cdot}}^{-1} L M_{\sqrt{\cdot}}[\phi], \phi\right\rangle_{L_2[T]}.$$

Identically,

$$\int_0^1 dx\frac{\phi(x)}{2x^{1/2}}\int_0^x dy\phi(y)\,y^{1/2} = \frac{1}{2}\left\langle M_{\sqrt{\cdot}}^{-1} L M_{\sqrt{\cdot}}[\phi], \phi\right\rangle_{L_2[T]}.$$

The above holds if $M_{\sqrt{\cdot}}^{-1} L M_{\sqrt{\cdot}}$ is bounded. But

$$\int_0^1 dx\left\{\frac{1}{x^{1/2}}\int_0^x dyy^{1/2}\phi(y)\right\}^2 \le \int_0^1 dx\left\{\int_0^x dy\,|\phi(y)|\right\}^2 \le \|\phi\|_{L_2[T]}^2.$$

Thus

$$\|X(1) + Y(1) + Z(1)\|_H^2 = \left\langle\left\{I_{L_2[T]} + M_{\sqrt{\cdot}}^2 + \frac{3}{2}M_{\sqrt{\cdot}}^{-1} L M_{\sqrt{\cdot}}\right\}[\phi], \phi\right\rangle_{L_2[T]}.$$

Let

$$\psi = \sum_{i=1}^{n} \alpha_i \chi_{]t_{i-1}, t_i]}.$$

A wide sense martingale with respect to f has the following form:

$$m_\phi(t) = f[Q_t[\phi]] = X(t) + Y(t) + Z(t),$$

so that

$$\int \psi\, dm_\phi = \sum_{i=1}^{n} \alpha_i \left\{m_\phi(t_i) - m_\phi(t_{i-1})\right\} = X(\psi) + Y(\psi) + Z(\psi),$$

and thus

$$\left\| \int \psi \, dm_\phi \right\|_H^2 = \left\langle \left\{ I_{L_2[T]} + M_{\sqrt{\cdot}}^2 + \frac{3}{2} M_{\sqrt{\cdot}}^{-1} L M_{\sqrt{\cdot}} \right\} [\psi\phi], \psi\phi \right\rangle_{L_2[T]}.$$

The functions that are integrable with respect to m_ϕ are thus those ψ for which $\psi\phi \in L_2[T]$:

$$L_2[M_\phi] = \{\psi : \psi\phi \in L_2[T]\}.$$

Let $\{F_t, t \in T\}$ be total in $L_2[M_\phi]$. The projection of an element of $L_{\cup T}[f]$ onto $L_t[f]$ has the form $m_\psi(t)$, some $\psi \in L_2[T]$, and thus

$$\left\| m_\psi(t) - \int F_t \, dm_\phi \right\|_H^2 =$$

$$\left\langle \left\{ I_{L_2[T]} + M_{\sqrt{\cdot}}^2 + \frac{3}{2} M_{\sqrt{\cdot}}^{-1} L M_{\sqrt{\cdot}} \right\} [Q_t[\psi] - F_t\phi], Q_t[\psi] - F_t\phi \right\rangle_{L_2[T]}.$$

As

$$\|Q_t[\psi_1] - F_t\phi\|_{L_2[T]} \geq \left| \|Q_t[\psi_1] - Q_t[\psi_2]\|_{L_2[T]} - \|Q_t[\psi_2] - F_t\phi\|_{L_2[T]} \right|,$$

a process of multiplicity one cannot approximate uniformly closely both processes

$$m_{\psi_1} \text{ and } m_{\psi_2}.$$

Any assumption of multiplicity one must thus be carefully validated.

8.4 Smoothness and Multiplicity: Goursat Maps

A Goursat map $f : T \longrightarrow H$, T an interval of $\mathbb{R}$, H a real Hilbert space, is a map of the following form:

$$f(t) = \sum_{i=1}^{n} a_i(t) h_i(t),$$

where

$$\{a_i : T \longrightarrow \mathbb{R}, i \in [1:n]\}$$

is a family of ordinary functions, and

$$\{h_i : T \longrightarrow H, i \in [1:n]\}$$

a family of functions with, in particular, orthogonal increments (more is in fact required: the h_i's are the components of a wide sense martingale, as in (Definition) 8.4.8 below). When H is the space $L_2(\Omega, \mathcal{A}, P)$, and $h_i(t)$ is the equivalence class of a random variable with a mean equal to zero, the usual term is "process." Goursat maps seem to be, at first sight, fairly simple objects. However their properties, and, in particular, multiplicity ones, are manifold. Their practical, and illustrative, usefulness stems from two facts:

(a) they are computationally tractable, and in particular, their prediction maps, items entering the study of multiplicity (as in Chap. 9), have relatively simple forms;
(b) they may be used as approximations to general CHR's: when

$$h_i(t) = \int I_{T_t}\phi_i dm_i,$$

then

$$\sum_{i=1}^{n} a_i(t)h_i(t) = \sum_{i=1}^{n} \int F_i(t,\cdot)\, dm_i, \quad F_i(t,\cdot) = a_i(t)I_{T_t}\phi_i,$$

and functions of form $a \times b$, where $a \in L_2(A, \mathcal{A}, \alpha)$, and $b \in L_2(B, \mathcal{B}, \beta)$, generate [200, p. 115] the space $L_2(A \times B, \mathcal{A} \otimes \mathcal{B}, \alpha \otimes \beta)$.

8.4.1 Hilbert Spaces from Matrix Measures

The work with Goursat maps is, unsurprisingly, given the results presented so far, based on L_2-type Hilbert spaces, which are obtained using a form of integration of maps with values in $\mathbb{R}^n$, with respect to measures which take matrices as their values. All measures considered in this section are assumed to be Borel measures, that is regular ones (as one uses Lusin's theorem), and statements involving measures of sets presuppose that the measures of the concerned sets are finite. The relevant facts are now listed. Proofs can be found in [86, p. 1337]. These integrals can also be seen as direct integrals (as in Sect. 7.1).

Assumptions 8.4.1 *1. T is an interval of $\mathbb{R}$.*
2. $\mathcal{T} = \mathcal{B}(\mathbb{R}) \cap T$.
3. For fixed, but arbitrary $\{i,j\} \subseteq [1:n]$, $\tau_{i,j}$ is a σ-additive, signed measure on $\mathcal{T}$.

4. *For fixed, but arbitrary* $T_0 \in \mathcal{T}$, $\boldsymbol{\tau}(T_0)$ *is the matrix with entries*

$$\tau_{i,j}(T_0), \ \{i,j\} \subseteq [1:n].$$

Definition 8.4.2 Let Assumption 8.4.1 obtain. The family $\{\tau_{i,j}, \{i,j\} \subseteq [1:n]\}$ forms a matrix measure when $\boldsymbol{\tau}(T_0)$ is symmetric, and positive definite, for all $T_0 \in \mathcal{T}$.

When the signed measures $\tau_{i,j}$ are all dominated by one, and the same, σ-finite measure, the resulting matrix of Radon-Nikodým derivatives has some indispensable properties that are now stated.

Fact 8.4.3 *Let (Assumption)* 8.4.1 *obtain. Suppose that, for some* τ, *a* σ*-finite measure,*

$$\tau_{i,j} \ll \tau, \ \{i,j\} \subseteq [1:n],$$

and let $D_\tau(t)$ *be a matrix of Radon-Nikodým derivatives of* $\tau_{i,j}$ *with respect to* τ, $\{i,j\} \subseteq [1:n]$.

1. *There is a choice of Radon-Nikodým derivatives for which, almost surely with respect to* τ, D_τ *is symmetric, and positive definite.*
2. *There is a decomposition of* D_τ *into adapted and integrable orthonormal eigenvector functions, and positive decreasing eigenvalue functions, that is valid almost surely, with respect to* τ:

$$D_\tau(t) = \sum_{i=1}^{n} \delta_i(t) \left\{\underline{d}_i(t) \otimes \underline{d}_i(t)\right\}.$$

Definition 8.4.4 Let (Definition) 8.4.2 and (Fact) 8.4.3 obtain. $\mathcal{L}_2(T, \mathcal{T}, \boldsymbol{\tau})$ shall denote the family of maps $t \mapsto \underline{h}(t) \in \mathbb{R}^n$ whose components are adapted to $\mathcal{T}$, and such that

$$\int_T \langle D_\tau(t)\, \underline{h}(t), \underline{h}(t)\rangle_{\mathbb{R}^n} \ \tau(dt) < \infty.$$

Fact 8.4.5 (A Hilbert Space of Vector Valued Functions)

1. *The choice made in (Fact)* 8.4.3 *for* D_τ *allows one to use Schwarz's inequality to obtain that*

$$\begin{aligned}&\left\langle D_\tau(t)\, \underline{h}_1(t), \underline{h}_2(t)\right\rangle_{\mathbb{R}^n}^2 \le \\ &\quad \le \left\langle D_\tau(t)\, \underline{h}_1(t), \underline{h}_1(t)\right\rangle_{\mathbb{R}^n} \left\langle D_\tau(t)\, \underline{h}_2(t), \underline{h}_2(t)\right\rangle_{\mathbb{R}^n}.\end{aligned}$$

2. *The integral*

$$\int_T \left\langle D_\tau(t)\,\underline{h}_1(t), \underline{h}_2(t)\right\rangle_{\mathbb{R}^n} \tau(dt)$$

thus exists, for fixed, but arbitrary elements $\{\underline{h}_1, \underline{h}_2\} \subseteq \mathcal{L}_2(T, \mathcal{T}, \boldsymbol{\tau})$*, and*

$$\left|\int_T \left\langle D_\tau(t)\,\underline{h}_1(t), \underline{h}_2(t)\right\rangle_{\mathbb{R}^n} \tau(dt)\right|^2 \leq$$

$$\leq \int_T \left\langle D_\tau(t)\,\underline{h}_1(t), \underline{h}_1(t)\right\rangle_{\mathbb{R}^n} \tau(dt) \int_T \left\langle D_\tau(t)\,\underline{h}_2(t), \underline{h}_2(t)\right\rangle_{\mathbb{R}^n} \tau(dt).$$

It follows that $\mathcal{L}_2(T, \mathcal{T}, \boldsymbol{\tau})$ *is a linear manifold.*

3. *Maps* $\underline{h} \in \mathcal{L}_2(T, \mathcal{T}, \boldsymbol{\tau})$ *such that*

$$\int_T \left\langle D_\tau(t)\,\underline{h}(t), \underline{h}(t)\right\rangle_{\mathbb{R}^n} \tau(dt) = 0$$

are called null functions, and, for the same reason that $\mathcal{L}_2(T, \mathcal{T}, \boldsymbol{\tau})$ *is a linear manifold, the set of null functions is a linear manifold in* $\mathcal{L}_2(T, \mathcal{T}, \boldsymbol{\tau})$*.*

4. *The quotient of* $\mathcal{L}_2(T, \mathcal{T}, \boldsymbol{\tau})$ *by the manifold of null functions shall be denoted* $L_2(T, \mathcal{T}, \boldsymbol{\tau})$*, and one shall write, for the equivalence class* $[\underline{h}]$ *of* $\underline{h}$*,*

$$\|[\underline{h}]\|^2_{L_2(T,\mathcal{T},\tilde{\tau})} = \int_T \left\langle D_\tau(t)\,\underline{h}(t), \underline{h}(t)\right\rangle_{\mathbb{R}^n} \tau(dt).$$

Similarly

$$\left\langle [\underline{h}_1], [\underline{h}_2]\right\rangle_{L_2(T,\mathcal{T},\tilde{\tau})} = \int_T \left\langle D_\tau(t)\,\underline{h}_1(t), \underline{h}_2(t)\right\rangle_{\mathbb{R}^n} \tau(dt).$$

One has that the (right-hand side) *of the expression defining the inner product*

$$\left\langle [\underline{h}_1], [\underline{h}_2]\right\rangle_{L_2(T,\mathcal{T},\tilde{\tau})}$$

is independent of τ*, and that the latter inner product is an inner product for which* $L_2(T, \mathcal{T}, \boldsymbol{\tau})$ *is a Hilbert space.*

Fact 8.4.6 *Let (Definition)* 8.4.2 *and (Fact)* 8.4.3 *obtain. Let* $t \mapsto \underline{c} \in \mathbb{R}^n$*, and* $t \mapsto \underline{d} \in \mathbb{R}^n$ *be constant functions. Then, for fixed, but arbitrary* $\{T_1, T_2\} \subseteq \mathcal{T}$*,*

$$\int_T \left\langle D_\tau(t)\, I_{T_1}(t)\,\underline{c}, I_{T_2}(t)\,\underline{d}\right\rangle_{\mathbb{R}^n} \tau(dt) =$$

$$= \sum_{i=1}^n \sum_{j=1}^n c_i d_j \tau_{i,j}(T_1 \cap T_2)$$

$$= \left\langle \boldsymbol{\tau}(T_1 \cap T_2)\,\underline{c}, \underline{d}\right\rangle_{\mathbb{R}^n},$$

so that constant functions are "integrable" on sets of finite measure (as they should be).

Example 8.4.7 The following example shows that some care is required when handling matrix measures. Let indeed μ be a measure over the Borel sets of T, and (since τ can be taken as μ)

$$\boldsymbol{\tau} = \begin{bmatrix} \mu & -\mu \\ -\mu & \mu \end{bmatrix}, \text{ so that } D_\tau = \begin{bmatrix} 1 & -1 \\ -1 & 1 \end{bmatrix}.$$

Then

$$\int_T \langle D_\tau(t)\underline{h}(t), \underline{h}(t)\rangle_{\mathbb{R}^2}\, \tau(dt) = \int_T (h_1(t) - h_2(t))^2\, \tau(dt),$$

so that $\underline{h}$ may belong to $L_2(T, \mathcal{T}, \boldsymbol{\tau})$ without its components belonging to $L_2(T, \mathcal{T}, \tau)$ $(\tau = \mu)$.

8.4.2 Martingales in the Wide Sense

The study of multiplicity for Goursat maps requires integral representations with respect to martingales in the wide sense analogous to those obtained for processes with orthogonal increments in Chap. 1. That is what one will find below.

Definition 8.4.8 Let T be an interval of the real line, and H be a real Hilbert space. For $t \in T$, let $\underline{h}(t)$ have components

$$h_i(t) \in H,\ i \in [1:n]$$

with the following property: for fixed, but arbitrary

$$\{t_1, t_2, t_3\} \subseteq T,\ t_1 \leq t_2 \leq t_3,\ \{i, j\} \subseteq [1:n],$$

one has that

$$\langle h_i(t_1), h_j(t_3) - h_j(t_2)\rangle_H = 0.$$

One shall say that $\underline{h} = \{\underline{h}(t),\ t \in T\}$ is a martingale in the wide sense of dimension n (the dimension shall usually be omitted).

Example 8.4.9 Let $\underline{h}(t)$ have two components made of the same wide-sense, standard Wiener process. It is a martingale in the wide sense of order two, but its components are not orthogonal.

Example 8.4.10 Let $T = [0, 1]$, $t \mapsto \Sigma(t)$ be a matrix valued function, with measurable components $t \mapsto \sigma_{i,j}(t)$, and bounded trace. Each $\Sigma(t)$ is a covariance matrix of dimension n. Let $\underline{W}(\cdot, t)$ be a Wiener process as defined in [33, p. 168]. It is a Gaussian process with a mean equal to zero, whose covariance properties are derived from the following expression: for $\{t_1, t_2\} \subseteq T$ and $\{\underline{\alpha}_1, \underline{\alpha}_2\} \subseteq \mathbb{R}^n$, fixed, but arbitrary,

$$E\left[\langle \underline{W}(\cdot, t_1), \underline{\alpha}_1\rangle_{\mathbb{R}^n} \langle \underline{W}(\cdot, t_2), \underline{\alpha}_2\rangle_{\mathbb{R}^n}\right] = \int_0^{t_1 \wedge t_2} \langle \Sigma(\theta)\underline{\alpha}_1, \underline{\alpha}_2\rangle_{\mathbb{R}^n}\, d\theta.$$

For matrix functions M_1 and M_2 of the appropriate dimensions, one has, as a consequence [33, p. 184]

$$E\left[\left\langle \int_0^{t_1} M_1(\theta)\underline{W}(\cdot, d\theta), \underline{\alpha}_1\right\rangle_{\mathbb{R}^n} \left\langle \int_0^{t_2} M_2(\theta)\underline{W}(\cdot, d\theta), \underline{\alpha}_2\right\rangle_{\mathbb{R}^n}\right] =$$

$$= \int_0^{t_1 \wedge t_2} \langle M_2(\theta)\Sigma(\theta)M_1^\star(\theta)\underline{\alpha}_1, \underline{\alpha}_2\rangle_{\mathbb{R}^n}\, d\theta.$$

Let $h_i(t)$ be the equivalence class of

$$\left\langle \int_0^t M(\theta)\, \underline{W}(\cdot, d\theta), \underline{e}_i\right\rangle_{\mathbb{R}^n},$$

where $\underline{e}_i$ is the i-th member of the standard basis of $\mathbb{R}^n$. Then indeed

$$\begin{aligned}
&\langle h_i(t_1), h_j(t_3) - h_j(t_2)\rangle_H = \\
&= E\left[\left\langle \int_0^{t_1} M(\theta)\, \underline{W}(\cdot, d\theta), \underline{e}_i\right\rangle_{\mathbb{R}^n}\right. \\
&\qquad \left.\left\{\left\langle \int_0^{t_3} M(\theta)\, \underline{W}(\cdot, d\theta), \underline{e}_j\right\rangle_{\mathbb{R}^n} - \left\langle \int_0^{t_2} M(\theta)\, \underline{W}(\cdot, d\theta), \underline{e}_j\right\rangle_{\mathbb{R}^n}\right\}\right] \\
&= 0.
\end{aligned}$$

As a particular case, let $n = 3$, and, with $\underline{v} \otimes \underline{v} = \underline{v}\,\underline{v}^\star$, κ, a constant, I_3, the identity matrix,

$$\Sigma(t) = \sigma(t) I_3 + \kappa \underline{\sigma}(t) \otimes \underline{\sigma}(t),$$

where σ and the components of $\underline{\sigma}$ are, say, continuous functions. Then

$$\langle \Sigma(t)\underline{\alpha}, \underline{\alpha}\rangle_{\mathbb{R}^3} = \sigma(t)\, \|\underline{\alpha}\|^2_{\mathbb{R}^3} + \kappa\, \langle \underline{\sigma}(t), \underline{\alpha}\rangle^2_{\mathbb{R}^3}.$$

For $\|\underline{\alpha}\|_{\mathbb{R}^3} > 0$, and $\underline{u}\left[\underline{\alpha}\right] = \underline{\alpha} / \|\underline{\alpha}\|_{\mathbb{R}^3}$,

$$\langle \Sigma(t)\underline{\alpha}, \underline{\alpha} \rangle_{\mathbb{R}^3} = \|\underline{\alpha}\|^2_{\mathbb{R}^3} \left\{ \sigma(t) + \kappa \langle \underline{\sigma}(t), \underline{u}\left[\underline{\alpha}\right] \rangle^2_{\mathbb{R}^3} \right\},$$

and thus $\Sigma(t)$ is a covariance matrix when $\sigma(t) + \kappa \langle \underline{\sigma}(t), \underline{u}\left[\underline{\alpha}\right] \rangle^2_{\mathbb{R}^3} \geq 0$. The spectral properties of $\Sigma(t)$ are as follows [121, p. 170 and 187]:

1. $\Sigma(t)$ has an inverse when $\sigma(t) \neq 0$ and $\kappa \neq -\sigma(t)/ \|\underline{\sigma}(t)\|^2_{\mathbb{R}^3}$: then

$$\{\Sigma(t)\}^{-1} = \frac{I_3}{\sigma(t)} - \frac{\kappa\sigma(t)}{\sigma(t) + \kappa \|\underline{\sigma}(t)\|^2_{\mathbb{R}^3}} \left\{ \frac{\underline{\sigma}(t)}{\sigma(t)} \otimes \frac{\underline{\sigma}(t)}{\sigma(t)} \right\};$$

2. $\Sigma(t)$ has two eigenvalues equal to $\sigma(t)$, and one equal to $\sigma(t) + \kappa \|\underline{\sigma}(t)\|^2_{\mathbb{R}^3}$; furthermore, a calculation shows that
   (i) $\Sigma(t)\underline{x} = \sigma(t)\underline{x}$ yields that $\underline{x} \perp \underline{\sigma}(t)$;
   (ii) $\Sigma(t)\underline{x} = \left\{\sigma(t) + \kappa \|\underline{\sigma}(t)\|^2_{\mathbb{R}^3}\right\} \underline{x}$ yields that $\underline{x} = \underline{\sigma}(t)$.

Choosing for M a diagonal matrix of functions, one already ends up with matrices whose eigenvalues are in general difficult to evaluate though the characteristic equation is known [121, p. 187].

Fact 8.4.11 *A martingale in the wide sense, of dimension one, is a process with orthogonal increments. A process h with orthogonal increments, and index set* $[0, T]$, *such that* $h(0) = 0_H$, *is a martingale in the wide sense of dimension one.*

Proof Indeed, from the definition, for fixed, but arbitrary $\{t_1, t_2, t_3, t_4\}$ in T, with $t_1 < t_2 \leq t_3 < t_4$, the following equality:

$$\langle h(t_1), h(t_4) - h(t_3) \rangle_H = \langle h(t_2), h(t_4) - h(t_3) \rangle_H = 0,$$

yields that

$$\langle h(t_2) - h(t_1), h(t_4) - h(t_3) \rangle_H = 0.$$

When $h(0) = 0_H$,

$$\langle h(t_1), h(t_4) - h(t_3) \rangle_H = \langle h(t_1) - h(0), h(t_4) - h(t_3) \rangle_H = 0.$$

□

A process with orthogonal increments need not be a martingale in the wide sense of dimension one.

Example 8.4.12 When $T = \{1, 2, 3, 4\}$, and h with the following values:

$$h(1) = h,\ h(2) = h,\ h(3) = k,\ h(4) = l,$$

one has a process with orthogonal increments. For it to be a martingale in the wide sense, one must have, for example that $\langle h, l-k\rangle_H = 0$, a condition that is easy to violate, for example, choosing $h = e_2, k = e_2, l = e_3$, orthonormal.

Fact 8.4.13 *When a process with orthogonal increments is purely nondeterministic, then it is a martingale in the wide sense of dimension one.*

Proof One lets, in $\langle h(t_2)-h(t_1), h(t_4)-h(t_3)\rangle_H = 0$, $h(t_1)$ decrease indefinitely, and applies the appropriate reverse martingale convergence theorem [78, p. 166]. □

Fact 8.4.14 *Let $\underline{h} = \{\underline{h}(t), t \in T\}$ be a martingale in the wide sense. For $t \in T$, fixed, but arbitrary, $\{h_1(t), \ldots, h_n(t)\}$ are linearly independent when none of $h_1(t), h_2(t) - h_1(t), \ldots, h_n(t) - h_{n-1}(t)$ is 0_H.*

Proof Indeed, letting $h_0(t) = 0_H$,

$$\sum_{i=1}^{n} \alpha_i h_i(t) = \sum_{i=1}^{n} \alpha_i \sum_{j=1}^{i} \{h_i(t) - h_{i-1}(t)\}$$

$$= \sum_{i=1}^{n} \left\{ \sum_{j=i}^{n} \alpha_j \right\} \{h_i(t) - h_{i-1}(t)\},$$

and $\{h_1(t), h_2(t) - h_1(t), \ldots, h_n(t) - h_{n-1}(t)\}$ are orthogonal. □

Fact 8.4.15 *Let $\underline{h} = \{\underline{h}(t), t \in T\}$ be a martingale in the wide sense. Let $\{t_1, t_2, t_3, t_4\} \subseteq T$, $t_1 < t_2$, $t_3 < t_4$, and $\{i, j\} \subseteq [1 : n]$ be fixed but arbitrary. Then, for*

$$[t_1, t_2[\cap [t_3, t_4[= [\alpha, \beta[\neq \emptyset,$$

one has that

$$\langle h_i(t_2) - h_i(t_1), h_j(t_4) - h_j(t_3)\rangle_H = \langle h_i(\beta) - h_i(\alpha), h_j(\beta) - h_j(\alpha)\rangle_H$$

(it is zero otherwise).

Proof For example, the inner product

$$\langle h_i(t_2) - h_i(t_1), h_j(t_4) - h_j(t_3)\rangle_H$$

equals

- 0 when $t_4 \leq t_1$, and when $t_2 \leq t_3$, and then

$$[t_1, t_2[\cap [t_3, t_4[= \emptyset,$$

- $\langle h_i(t_4) - h_i(t_1), h_j(t_4) - h_j(t_1)\rangle_H$ when $t_3 \leq t_1 < t_4 \leq t_2$, and then

$$[t_1, t_2[\cap [t_3, t_4[= [t_1, t_4[,$$

the other cases being treated similarly. □

Fact 8.4.16 *Let* $\underline{h} = \{\underline{h}(t), t \in T\}$ *be a martingale in the wide sense. Let*

$$L_t\,[\underline{h}] = \overline{V\,[\{h_i\,(\theta)\,,\ \theta \in T,\ \theta \leq t,\ i \in [1:n]\}]},$$

and P_t *be the projection with that range. Then the fact that* $\underline{h}$ *is a martingale in the wide sense can be expressed in the following succinct, and intuitively more expressive form:*

$$P_{t_1}\,[\underline{h}(t_2)] = \underline{h}(t_1),\ \{t_1, t_2\} \subseteq T,\ t_1 < t_2.$$

Proof The following relation, valid, when i is fixed, for all j and $\theta \leq t$:

$$\langle h_i(t_2) - h_i(t_1), h_j(\theta)\rangle_H = 0$$

means that $h_i(t_1)$ is the projection of $h_i(t_2)$ onto $L_{t_1}\,[\underline{h}]$. □

Lemma 8.4.17 *let* H *be a Hilbert space. All (closed) subspaces considered below, denoted* H *with an index are subsets of* H. $H_1 \bigvee H_2$ *denotes the intersection of all the (closed) subspaces which contain* $H_1 \cup H_2$. *The associated projection is denoted* $P_1 \bigvee P_2$.

1. *Let* S *be a subset of* H *generating* H_1: $\overline{V[S]} = H_1$. *Then:*

$$H_0 \cap H_1 = \overline{V[H_0 \cap S]}.$$

2. $\{\bigvee_{i=1}^n H_i\}^\perp = \cap_{i=1}^n H_i^\perp$.
3. $\left(\bigvee_{i=1}^n H_i\right) \cap H_0 = \bigvee_{i=1}^n (H_i \cap H_0)$.
4. *Let* $H_\lambda^{(1)}$ *be a member of a decreasing family of subsets, with intersection* H_1. $H_\lambda^{(2)}$ *and* H_2 *are defined analogously. Then:*

$$H_1 \bigvee H_2 = \bigcap_\lambda \left(H_\lambda^{(1)} \bigvee H_\lambda^{(2)}\right).$$

That latter relation extends to a finite number of terms.

5. *Let* $H_{1,i} \subseteq H_{2,i}$, $i \in [1:n]$, $H_1 = \bigvee_{i=1}^n H_{1,i}$, $H_2 = \bigvee_{i=1}^n H_{2,i}$. *Then:*

$$H_2 \cap H_1^\perp \subseteq \bigvee_{1=1}^n \{H_{2,i} \cap H_{1,i}^\perp\}.$$

Proof ([1]) Since $H_0 \cap H_1$ is a (closed) subspace which contains $H_0 \cap S$,

$$\overline{V[H_0 \cap S]} \subseteq H_0 \cap H_1 .$$

Also $H_1 = \overline{V[S]}$, so that $\overline{V[H_0 \cap S]}$ is a (closed) subspace containing $H_0 \cap H_1$.

Proof ([2]) When h is orthogonal to $\bigvee_{i=1}^{n} H_i$, it is orthogonal to each H_i, and thus belongs to $\cap_{i=1}^{n} H_i^{\perp}$. Conversely, when h belongs to $\cap_{i=1}^{n} H_i^{\perp}$, it is orthogonal to each H_i, and thus to $\bigvee_{i=1}^{n} H_i$.

Proof ([3]) Let $H_{0,1,2}$ be the subspace generated by

$$(H_1 \cup H_2) \cap H_0 = (H_1 \cap H_0) \cup (H_2 \cup H_0) .$$

Since $(H_1 \bigvee H_2) \cap H_0$ is a subspace which contains $(H_1 \cup H_2) \cap H_0$,

$$H_{0,1,2} \subseteq \left(H_1 \bigvee H_2\right) \cap H_0 .$$

But, by definition, $H_{0,1,2} = (H_1 \cap H_0) \bigvee (H_2 \cap H_0)$. Thus

$$(H_1 \cap H_0) \bigvee (H_2 \cap H_0) \subseteq \left(H_1 \bigvee H_2\right) \cap H_0 .$$

Now $(H_1 \cap H_0) \bigvee (H_2 \cap H_0)$ is a subspace which contains

$$(H_1 \cap H_0) \cup (H_2 \cap H_0) = (H_1 \cup H_2) \cap H_0 ,$$

and thus, because of item 1, $(H_1 \bigvee H_2) \cap H_0$. Finally

$$\begin{aligned}(H_1 \bigvee H_2 \bigvee H_3) \cap H_0 &= \left\{(H_1 \bigvee H_2) \cap H_0\right\} \bigvee (H_3 \cap H_0) \\ &= (H_1 \cap H_0) \bigvee (H_2 \cap H_0) \bigvee (H_3 \cap H_0) .\end{aligned}$$

Proof ([4]) One has that

$$H_1 \cup H_2 = \left(\cap_\lambda H_\lambda^{(1)}\right) \cup \left(\cap_\lambda H_\lambda^{(2)}\right) = \cap_{\lambda,\mu} \left(H_\lambda^{(1)} \cup H_\mu^{(2)}\right) .$$

But

$$\cap_{\lambda,\mu} \left(H_\lambda^{(1)} \cup H_\mu^{(2)}\right) = \cap_\lambda \left(H_\lambda^{(1)} \cup H_\lambda^{(2)}\right) . \qquad (\star)$$

Indeed, $H_\lambda^{(1)} \cup H_\mu^{(2)}$ is either $H_{\lambda\wedge\mu}^{(1)} \cup H_{\lambda\vee\mu}^{(2)}$, or $H_{\lambda\vee\mu}^{(1)} \cup H_{\lambda\wedge\mu}^{(2)}$. It thus contains $H_{\lambda\wedge\mu}^{(1)} \cup H_{\lambda\wedge\mu}^{(2)}$, so that

$$\cap_{\lambda,\mu}\left(H_\lambda^{(1)} \cup H_\mu^{(2)}\right) \supseteq \cap_\lambda \left(H_\lambda^{(1)} \cup H_\lambda^{(2)}\right).$$

The inclusion in the other direction is due to the fact that the right-hand side of $(\star)$ is the intersection of only part of the sets in the intersection of the left-hand side. Thus

$$H_1 \cup H_2 = \cap_\lambda \left(H_\lambda^{(1)} \cup H_\lambda^{(2)}\right).$$

The smallest (closed) subspace containing $\cap_\lambda \left(H_\lambda^{(1)} \cup H_\lambda^{(2)}\right)$ must be the following subspace: $\cap_\lambda \left(H_\lambda^{(1)} \bigvee H_\lambda^{(2)}\right)$, for, if L is the smallest, then the inclusion $\cap_\lambda \left(H_\lambda^{(1)} \cup H_\lambda^{(2)}\right) \subseteq L$ leads to

$$\cap_\lambda \left(H_\lambda^{(1)} \cup H_\lambda^{(2)}\right) \subseteq L \cap \left\{\cap_\lambda \left(H_\lambda^{(1)} \bigvee H_\lambda^{(2)}\right)\right\},$$

and, consequently,

$$H_1 \bigvee H_2 = \cap_\lambda \left(H_\lambda^{(1)} \bigvee H_\lambda^{(2)}\right).$$

Suppose that $H_1 = \cap_\lambda H_\lambda^{(1)}, H_2 = \cap_\lambda H_\lambda^{(2)}, H_3 = \cap_\lambda H_\lambda^{(3)}$. Let $H_0 = H_1 \bigvee H_2$, $H_\lambda^0 = H_\lambda^{(1)} \bigvee H_\lambda^{(2)}$. Then

$$H_1 \bigvee H_2 \bigvee H_3 = H_0 \bigvee H_3 = \cap_\lambda \left(H_\lambda^{(0)} \bigvee H_\lambda^{(3)}\right) = \cap_\lambda \left(H_\lambda^{(1)} \bigvee H_\lambda^{(2)} \bigvee H_\lambda^{(3)}\right).$$

Proof ([5]) Because of item 2, and then item 3,

$$\begin{aligned} H_2 \cap H_1^\perp &= \left\{\bigvee_{i=1}^n H_{2,i}\right\} \cap \left\{\cap_{i=1}^n H_{1,i}^\perp\right\} \\ &= \bigvee_{i=1}^n \left\{H_{2,i} \cap \left\{\cap_{i=1}^n H_{1,i}^\perp\right\}\right\} \\ &\subseteq \bigvee_{i=1}^n \left\{H_{2,i} \cap H_{1,i}^\perp\right\}. \end{aligned}$$

□

Fact 8.4.18 *Let* $\underline{h} = \{\underline{h}(t), t \in T\}$ *be a martingale in the wide sense.* $L_t[h_i]$ *is the subspace generated by* $\{h_i(\theta), \theta \leq t\}$, *and* $L_t[\underline{h}]$, *that generated by* $\{h_i(\theta), \theta \leq t, i \in [1:n]\}$. *One has that*

1. $L_t[\underline{h}] = \bigvee_{i=1}^n L_t[h_i]$,
2. $L_t^+[\underline{h}] = \bigvee_{i=1}^n L_t^+[h_i]$,
3. $L_t^+[\underline{h}] \cap L_t[\underline{h}]^\perp = \bigvee_{i=1}^n \{L_t^+[h_i] \cap L_t[h_i]^\perp\}$.

Proof ([1]) Let

$$\mathcal{H}_{i,t} = \{h_i(\theta), \theta \leq t\}, \qquad L_t[h_i] = \overline{V[\mathcal{H}_{i,t}]},$$

$$\mathcal{H}_t = \{h_i(\theta), \theta \leq t, i \in [1:n]\}, \quad L_t[\underline{h}] = \overline{V[\mathcal{H}_t]}.$$

One has that:

$$\mathcal{H}_{i,t} \subseteq \mathcal{H}_t \Longrightarrow L_t[h_i] \subseteq L_t[\underline{h}] \Longrightarrow \bigvee_{i=1}^n L_t[h_i] \subseteq L_t[\underline{h}],$$

and that

$$\mathcal{H}_t = \cup_{i=1}^n \mathcal{H}_{i,t} \subseteq \cup_{i=1}^n L_t[h_i] \subseteq \bigvee_{i=1}^n L_t[h_i].$$

Thus

$$L_t[\underline{h}] = \bigvee_{i=1}^n L_t[h_i]. \tag{$\star$}$$

Proof ([2]) Relation ($\star$) above is valid for $t \leftarrow t + \delta$. One then applies item 4 of (Lemma) 8.4.17.

Proof ([3]) The left-hand side of item 3 is contained in the right-hand side because of item 5 of (Lemma) 8.4.17. Now $h_i^+(t) - h_i(t)$ belongs to $L_t^+[\underline{h}]$, and is orthogonal to $L_t[\underline{h}]$ because of the martingale property of $\underline{h}$. The reverse inclusion thus obtains. □

Fact 8.4.19 *Let* $\underline{h} = \{\underline{h}(t), t \in T\}$ *be a martingale in the wide sense. Let* $P_{t_l}^+$ *denote the projection whose range is* $L_{t_l}^+[\underline{h}]$. *Let, for* $i \in [1:n]$, *fixed, but arbitrary,*

$$k_i(t) = h_i(t) - P_{t_l}^+[h_i(t)] = \{I_H - P_{t_l}^+\}[h_i(t)].$$

$\underline{k}$ *is then a martingale in the wide sense such that*

$$\underline{k}^+(t_l) = \underline{0}_{H^n}, \text{ and, when } t_l \in T,\ \underline{k}(t_l) = \underline{0}_{H^n}.$$

Proof Let indeed $\{i,j\} \subseteq [1:n]$, and $\{t_1,t_2,t_3\} \subseteq T,\ t_1 \leq t_2 \leq t_3$ be fixed, but arbitrary. Then

$$\begin{aligned}\langle k_i(t_1), k_j(t_3) - k_j(t_2)\rangle_H &= \\ &= \left\langle \left\{I_H - P_{t_l}^+\right\}[h_i(t_1)], h_j(t_3) - h_j(t_2)\right\rangle_H \\ &- \left\langle \left\{I_H - P_{t_l}^+\right\}[h_i(t_1)], P_{t_l}^+\left[h_j(t_3) - h_j(t_2)\right]\right\rangle_H.\end{aligned}$$

The second inner product on the right-hand side of the latter equality is zero, as its two components are orthogonal. The first is equal to the limit, as $\delta \downarrow 0,\ \delta > 0$, of

$$\left\langle \{I_H - P_{t_l+\delta}\}[h_i(t_1)], \left(h_j(t_3) - h_j(t_l+\delta)\right) - \left(h_j(t_2) - h_j(t_l+\delta)\right)\right\rangle_H.$$

Since

$$\left(h_j(t_2) - h_j(t_l+\delta)\right) \perp L_{t_l+\delta}[\underline{h}],\ \left(h_j(t_3) - h_j(t_l+\delta)\right) \perp L_{t_l+\delta}[\underline{h}],$$

the latter inner product equals

$$\langle h_i(t_1), h_j(t_3) - h_j(t_2)\rangle_H,$$

which is zero by assumption. □

To be in line with the assumptions made to obtain the CHR representation, one shall often assume that a martingale in the wide sense is purely nondeterministic, and continuous to the left. That will in particular allow one to write that, for a martingale in the wide sense $\underline{h}$, and fixed, but arbitrary $\{t_1,t_2\} \subseteq T,\ t_1 < t_2$,

$$\underline{h}(t_2) - \underline{h}(t_1) = \int I_{[t_1,t_2[}\, d\underline{m}_h.$$

Hence the definition which follows.

Definition 8.4.20 A CH-martingale is a martingale in the wide sense, that is purely nondeterministic, and continuous to the left.

Lemma 8.4.21 *Let $\underline{h} = \{\underline{h}(t),\ t \in T\}$ be a martingale in the wide sense. For $\{t_1,t_2\} \subseteq T,\ t_1 < t_2$, fixed, but arbitrary,*

$$\langle h_i(t_2) - h_i(t_1), h_j(t_2) - h_j(t_1)\rangle_H = \langle h_i(t_2), h_j(t_2)\rangle_H - \langle h_i(t_1), h_j(t_1)\rangle_H.$$

Proof Definition 8.4.8 yields that

$$\begin{aligned}\langle h_i(t_2) - h_i(t_1), h_j(t_2) - h_j(t_1)\rangle_H &= \langle h_i(t_2), h_j(t_2) - h_j(t_1)\rangle_H \\ &= \langle h_i(t_2), h_j(t_2)\rangle_H - \langle h_i(t_2), h_j(t_1)\rangle_H.\end{aligned}$$

Now

$$\langle h_i(t_2), h_j(t_1)\rangle_H = \langle (h_i(t_2) - h_i(t_1)) + h_i(t_1), h_j(t_1)\rangle_H ,$$

and one uses again (Definition) 8.4.8. □

Lemma 8.4.22 *Let* $\underline{h} = \{\underline{h}(t),\ t \in T\}$ *be a martingale in the wide sense. For* $\{i,j\} \subseteq [1:n]$*, and* $t \in T$*, fixed, but arbitrary, let*

$$F^{\underline{h}}_{i,j}(t) = \langle h_i(t), h_j(t)\rangle_H .$$

$F^{\underline{h}}_{i,j}$ *is locally a function of bounded variation, and, for fixed, but arbitrary* $\{i,j\} \subseteq [1:n]$*, and* $\{t_1, t_2\} \subseteq T,\ t_1 < t_2$*,*

$$F^{\underline{h}}_{i,j}(t_2) - F^{\underline{h}}_{i,j}(t_1) = \langle h_i(t_2) - h_i(t_1), h_j(t_2) - h_j(t_1)\rangle_H . \qquad (\star)$$

Proof The second assertion $(\star)$ follows from (Lemma) 8.4.21. Now, given fixed, but arbitrary $\{t_0, t_1, \ldots, t_p\} \subseteq T$ such that $t_0 < t_1 < \cdots < t_p$, using $(\star)$,

$$\left|F^{\underline{h}}_{i,j}(t_k) - F^{\underline{h}}_{i,j}(t_{k-1})\right| \le \|h_i(t_k) - h_i(t_{k-1})\|_H \left\|h_j(t_k) - h_j(t_{k-1})\right\|_H ,$$

and, invoking the inequality of Cauchy-Schwarz,

$$\sum_{k=1}^{p}\left|F^{\underline{h}}_{i,j}(t_k) - F^{\underline{h}}_{i,j}(t_{k-1})\right| \le \left\{\sum_{k=1}^{p}\|h_i(t_k) - h_i(t_{k-1})\|_H^2\right\}^{1/2} \times \left\{\sum_{k=1}^{p}\left\|h_j(t_k) - h_j(t_{k-1})\right\|_H^2\right\}^{1/2} .$$

Since the map h_i has orthogonal increments,

$$\|h_i(t_k) - h_i(t_{k-1})\|_H^2 = F^{\underline{h}}_{i,i}(t_k) - F^{\underline{h}}_{i,i}(t_{k-1}) ,$$

so that

$$\sum_{k=1}^{p}\|h_i(t_k) - h_i(t_{k-1})\|_H^2 = F^{\underline{h}}_{i,i}(t_p) - F^{\underline{h}}_{i,i}(t_1) = \left\|h_i(t_p) - h_i(t_1)\right\|_H^2 .$$

□

Example 8.4.23 Let $\underline{h}$ be the Wiener process of (Example) 8.4.10 with covariance

$$\Sigma(t) = \sigma(t) I_3 + \kappa\, \underline{\sigma}(t) \otimes \underline{\sigma}(t).$$

Then

$$F_{i,j}^{\underline{h}}(t) = \int_0^t \left\langle \Sigma(\theta)\underline{e}_i, \underline{e}_j \right\rangle_{\mathbb{R}^3} d\theta$$

$$= \begin{cases} \int_0^t \left\{\sigma(\theta) + \kappa\sigma_i^2(\theta)\right\} d\theta & \text{when } i = j \\ \kappa \int_0^t \sigma_i(\theta)\sigma_j(\theta) d\theta & \text{when } i \neq j \end{cases},$$

and, for the matrix $F_{\underline{h}}$, with entries $F_{i,j}^{\underline{h}}(t)$,

$$\left\langle F_{\underline{h}}(t)\underline{\alpha}, \underline{\alpha} \right\rangle_{\mathbb{R}^3} = \int_0^t \langle \Sigma(\theta)\underline{\alpha}, \underline{\alpha}\rangle_{\mathbb{R}^3} d\theta$$

$$= \|\underline{\alpha}\|_{\mathbb{R}^3}^2 \int_0^t \sigma(\theta) d\theta + \kappa \int_0^t \langle \underline{\sigma}(\theta), \underline{\alpha}\rangle_{\mathbb{R}^3}^2 d\theta.$$

Proposition 8.4.24 *Let* $\underline{h} = \{\underline{h}(t),\ t \in T\}$ *be a CH-martingale, and* $\{i, j\}$, *fixed, but arbitrary, belong to* $[1:n]$. *The function* $F_{i,j}^{\underline{h}}$ *determines a measure* $M_{i,j}^{\underline{h}}$ *on the Borel sets of* T, *that is,*

$$M_{i,j}^{\underline{h}}([t_1, t_2[) = F_{i,j}^{\underline{h}}(t_2) - F_{i,j}^{\underline{h}}(t_1),$$

and then:

$$M_{i,j}^{\underline{h}} \ll \mu_{\underline{h}} = \sum_{i=1}^{n} M_{i,i}^{\underline{h}}.$$

Proof For fixed, but arbitrary $\{t_1, t_2\} \subseteq T,\ t_1 < t_2$, using (Lemma) 8.4.22,

$$\left|F_{i,j}^{\underline{h}}(t_2) - F_{i,j}^{\underline{h}}(t_1)\right| \le \|h_i(t_2) - h_i(t_1)\|_H \left\|h_j(t_2) - h_j(t_1)\right\|_H$$

$$\le \frac{1}{2}\left\{\|h_i(t_2) - h_i(t_1)\|_H^2 + \left\|h_j(t_2) - h_j(t_1)\right\|_H^2\right\}$$

$$= \frac{1}{2}\left\{\left[F_{i,i}^{\underline{h}}(t_2) - F_{i,i}^{\underline{h}}(t_1)\right] + \left[F_{j,j}^{\underline{h}}(t_2) - F_{j,j}^{\underline{h}}(t_1)\right]\right\}.$$

□

Remark 8.4.25 With no regularity assumption on $F_{i,j}^{\underline{h}}$, denoted temporarily F, one has that [275, p. 508]

$$F([t_1, t_2[) = F^-(t_2) - F^-(t_1).$$

Hence the requirement that the martingale be a CH-martingale.

Remark 8.4.26 The measures of (Proposition) 8.4.24 are Borel measures as they are obtained from functions of bounded variation [263, p. 136].

Example 8.4.27 For (Example) 8.4.10 one has that:

$$M_{i,j}^{\underline{h}}\left([t_1, t_2[\right) = F_{i,j}^{\underline{h}}(t_2) - F_{i,j}^{\underline{h}}(t_1)$$

$$= \begin{cases} \int_{t_1}^{t_2} \{\sigma(\theta) + \kappa\sigma_i^2(\theta)\}\, d\theta & \text{when } i = j \\ \kappa \int_{t_1}^{t_2} \sigma_i(\theta)\sigma_j(\theta)\, d\theta & \text{when } i \neq j \end{cases}.$$

Thus

$$\mu_{\underline{h}}\left([t_1, t_2[\right) = \int_{t_1}^{t_2} \left\{3\sigma(\theta) + \kappa \sum_{i=1}^{3} \sigma_i^2(\theta)\right\} d\theta,$$

and

$$\frac{dM_{i,j}^{\underline{h}}}{d\mu_{\underline{h}}}(\theta) = \begin{cases} \frac{\sigma(\theta)+\kappa\sigma_i^2(\theta)}{3\sigma(\theta)+\kappa\sum_{i=1}^{3}\sigma_i^2(\theta)} & \text{when } i = j \\ \frac{\kappa\sigma_i(\theta)\sigma_j(\theta)}{3\sigma(\theta)+\kappa\sum_{i=1}^{3}\sigma_i^2(\theta)} & \text{when } i \neq j \end{cases}.$$

Definition 8.4.28

1. Let $\underline{h} = \{\underline{h}(t), t \in T\}$ be a CH-martingale, and $F_{\underline{h}}$ be the matrix with entries $F_{i,j}^{\underline{h}}$, $\{i, j\} \subseteq [1 : n]$. $F_{\underline{h}}$ is called the structure matrix of $\underline{h}$.
2. The matrix $\boldsymbol{M}_{\underline{h}}$, with entries $M_{i,j}^{\underline{h}}$, $\{i, j\} \subseteq [1 : n]$, is called the associated matrix valued measure.
3. $D_{\mu_{\underline{h}}}$ denotes the matrix of Radon-Nikodým derivatives

$$\frac{dM_{i,j}^{\underline{h}}}{d\mu_{\underline{h}}}, \ \{i, j\} \subseteq [1 : n].$$

4. $\underline{h}$ is said to be non-singular when $F_{\underline{h}}(t)$ is non-singular for $t \in T$.

Remark 8.4.29 For a CH-martingale $\underline{h}$, the structure matrix is a function which is zero at the origin and continuous to the left. Suppose that the map $t \mapsto F(t)$ has matrices as values, is continuous to the left, and has increments which are symmetric, and positive definite. There is then a multivariate Gaussian martingale whose structure matrix is F. Indeed, when, for $t_1 < t_2$ in T, fixed, but arbitrary, G_{t_1,t_2} denotes the Gaussian measure on $\mathbb{R}^n$, with a mean equal to zero, and covariance $F(t_2) - F(t_1)$, one has, for $t_1 < t_2 < t_3$ in T, that $G_{t_1,t_2} \star G_{t_2,t_3} = G_{t_1,t_3}$ ($\star$ denotes the convolution product). There is then a general result [67, p. 5] which asserts the existence of a multivariate Gaussian process, with independent increments, and F as covariance function.

Proposition 8.4.30 *Let* $\{\underline{h}(t), t \in T\}$ *be a CH-martingale. For fixed, but arbitrary* $\{t_1, t_2\} \subseteq T,\ t_1 < t_2$, *the matrix* $F_{\underline{h}}(t_2) - F_{\underline{h}}(t_1)$ *is positive definite.*

Proof One has, from (Lemma) 8.4.22, for fixed, but arbitrary $\{i, j\} \subseteq [1 : n]$, that

$$F^{\underline{h}}_{i,j}(t_2) - F^{\underline{h}}_{i,j}(t_1) = \langle h_i(t_2) - h_i(t_1), h_j(t_2) - h_j(t_1)\rangle_H .$$

Thus, for fixed, but arbitrary $\underline{\alpha} \in \mathbb{R}^n$,

$$\begin{aligned}
\langle [F_{\underline{h}}(t_2) - F_{\underline{h}}(t_1)]\,\underline{\alpha}, \underline{\alpha}\rangle_{\mathbb{R}^n} &= \sum_{i=1}^{n}\sum_{j=1}^{n} \alpha_i\alpha_j \left[F^{\underline{h}}_{i,j}(t_2) - F^{\underline{h}}_{i,j}(t_1)\right] \\
&= \sum_{i=1}^{n}\sum_{j=1}^{n} \alpha_i\alpha_j \langle h_i(t_2) - h_i(t_1), h_j(t_2) - h_j(t_1)\rangle_H \\
&= \left\| \sum_{i=1}^{n} \alpha_i \left[h_i(t_2) - h_i(t_1)\right] \right\|_H^2 .
\end{aligned}$$

□

Fact 8.4.31 *Let* $\{\underline{h}(t), t \in T\}$ *be a CH-martingale.* $D_{\mu_{\underline{h}}}$ *of (Definition)* 8.4.28 *may be chosen* [(Fact) 8.4.3], *and will be chosen, to be symmetric, and positive definite* (almost surely, with respect to $\mu_{\underline{h}}$) *so that the Hilbert space* $L_2(T, \mathcal{T}, \boldsymbol{M}_{\underline{h}})$ *is well defined.*

Remark 8.4.32 Let $\underline{h} = \{\underline{h}(t), t \in T\}$ be a martingale in the wide sense. For fixed, but arbitrary $\underline{\alpha} \in \mathbb{R}^n$, expressions of the form $[\underline{\alpha}, \underline{h}(t)]$ shall be understood as shorthand for the following linear combination in H:

$$\sum_{i=1}^{n} \alpha_i h_i(t).$$

They usually may be manipulated as inner products. For example, when A is a square matrix of dimension n,

$$[\underline{\alpha}, A[\underline{h}(t)]] = [A^\star[\underline{\alpha}], \underline{h}(t)] .$$

Remark 8.4.33 For fixed, but arbitrary $\underline{\alpha} \in \mathbb{R}^n$, one has that

$$\langle F_{\underline{h}}(t)\,\underline{\alpha}, \underline{\alpha}\rangle_{\mathbb{R}^n} = \sum_{i=1}^{n}\sum_{j=1}^{n} \alpha_i\alpha_j \langle h_i(t), h_j(t)\rangle_H = \|[\underline{\alpha}, \underline{h}(t)]\|_H^2 .$$

Remark 8.4.34 Choosing for $\underline{\alpha}$, in (Remark) 8.4.32, the i-th vector of the standard basis of $\mathbb{R}^n$, one obtains that

$$\left\langle F_{\underline{h}}(t)\,\underline{e}_i, \underline{e}_i\right\rangle_{\mathbb{R}^n} = \|h_i(t)\|_H^2 .$$

Consequently, when $F_{\underline{h}}(t)$ is non-singular, $h_i(t) \neq 0_H,\ i \in [1:n]$.

Remark 8.4.35 Let $\underline{h}$ be a CH-martingale. For $i \in [1:n]$, fixed, but arbitrary, let $\mu_i^{\underline{h}}$ be the measure defined using the following relation:

$$d\mu_i^{\underline{h}} = \delta_i^{\underline{h}}\, d\mu_{\underline{h}},$$

where $\delta_i^{\underline{h}}$ is the i-th eigenvalue function as in (Fact) 8.4.3. Then, as $\underline{h}$ is purely nondeterministic,

$$\begin{aligned}\left\langle F_{\underline{h}}(t)\,\underline{\alpha}, \underline{\alpha}\right\rangle_{\mathbb{R}^n} &= \sum_{i=1}^{n}\sum_{j=1}^{n} \alpha_i\alpha_j F_{i,j}^{\underline{h}}(t) \\ &= \int_{T_t} \left\langle D_{\mu_{\underline{h}}}(\theta)\,\underline{\alpha}, \underline{\alpha}\right\rangle_{\mathbb{R}^n} \mu_{\underline{h}}(d\theta) \\ &= \int_{T_t} \sum_{i=1}^{n} \delta_i^{\underline{h}}(\theta) \left\langle \underline{d}_i^{\underline{h}}(\theta), \underline{\alpha}\right\rangle_{\mathbb{R}^n}^2 \mu_{\underline{h}}(d\theta) \\ &\geq \mu_n^{\underline{h}}(T_t)\, \|\underline{\alpha}\|_{\mathbb{R}^n}^2 .\end{aligned}$$

Remark 8.4.36 Let $\underline{h}$ be a CH-martingale. For fixed, but arbitrary $\{t_1, t_2\}$ in T, $t_1 < t_2$, as $\underline{h}$ is purely nondeterministic, one has that

$$\begin{aligned}&\left\langle \left\{F_{\underline{h}}(t_2) - F_{\underline{h}}(t_1)\right\}\underline{\alpha}, \underline{\alpha}\right\rangle_{\mathbb{R}^n} = \\ &\quad = \sum_{i=1}^{n}\sum_{j=1}^{n} \alpha_i\alpha_j \left\{F_{i,j}^{\underline{h}}(t_2) - F_{i,j}^{\underline{h}}(t_1)\right\} \\ &\quad = \sum_{i=1}^{n}\sum_{j=1}^{n} \alpha_i\alpha_j \int_{[t_1,t_2[} \frac{dM_{i,j}^{\underline{h}}}{d\mu_{\underline{h}}}(\theta)\, \mu_{\underline{h}}(d\theta) \\ &\quad = \int_T \left\langle D_{\mu_{\underline{h}}}(\theta)\, \chi_{[t_1,t_2[}(\theta)\,\underline{\alpha}, \chi_{[t_1,t_2[}(\theta)\,\underline{\alpha}\right\rangle_{\mathbb{R}^n} \mu_{\underline{h}}(d\theta) .\end{aligned}$$

Thus, given $\{t_1, \dots, t_n\} \subseteq T,\ t_1 < \cdots < t_n, \{\underline{\alpha}_1, \dots, \underline{\alpha}_n\} \subseteq \mathbb{R}^n$, fixed, but arbitrary,

$$\begin{aligned}&\sum_{k=1}^{n} \left\langle \left\{F_{\underline{h}}(t_k) - F_{\underline{h}}(t_{k-1})\right\}\underline{\alpha}_k, \underline{\alpha}_k\right\rangle_{\mathbb{R}^n} = \\ &\quad = \int_T \left\langle D_{\mu_{\underline{h}}}(\theta) \sum_{k=1}^{n} \chi_{[t_{k-1},t_k[}(\theta)\,\underline{\alpha}_k, \sum_{k=1}^{n} \chi_{[t_{k-1},t_k[}(\theta)\,\underline{\alpha}_k\right\rangle_{\mathbb{R}^n} \mu_{\underline{h}}(d\theta) ,\end{aligned}$$

and the latter equality shows how to approximate the norm of elements in $L_2\left(T, \mathcal{T}, \boldsymbol{M}_{\underline{h}}\right)$.

Example 8.4.37 In (Example) 8.4.10, let M_1 and M_2 have zero entries except, respectively, in positions (i,i) and (j,j), where the values are, respectively, the functions ϕ_i and ϕ_j. One then obtains that

$$E\left[\int_0^{t_1} \phi_i(\theta)\, W_i\,(\cdot, d\theta) \int_0^{t_2} \phi_j(\theta)\, W_j\,(\cdot, d\theta)\right] = \int_0^{t_1\wedge t_2} \phi_i(\theta)\,\phi_j(\theta)\,\sigma_{i,j}(\theta)\,d\theta.$$

Let thus, for $i \in [1:n]$, fixed, but arbitrary, and appropriate ϕ_i,

$$h_i(t) = \int I_{T_t}\phi_i\, dm_i^W,$$

where m_i^W is the vector measure obtained from the i-th component of $\underline{W}$. One then has that

$$F_{i,j}^{\underline{h}}(t) = \int_0^t \phi_i(\theta)\,\phi_j(\theta)\,\sigma_{i,j}(\theta)\,d\theta,$$

and consequently that

$$M_{i,j}^{\underline{h}}(d\theta) = \phi_i(\theta)\,\phi_j(\theta)\,\sigma_{i,j}(\theta)\,d\theta = \frac{\phi_i(\theta)\,\phi_j(\theta)\,\sigma_{i,j}(\theta)}{\sum_{i=1}^n \phi_i^2(\theta)\,\sigma_{i,i}(\theta)}\,\mu_{\underline{h}}(d\theta).$$

Let Φ be the diagonal matrix with diagonal elements $\phi_i, i \in [1:n]$. Then the elements of $L_2\left(T, \mathcal{T}, \boldsymbol{M}_{\underline{h}}\right)$ are the equivalence classes of those functions $t \mapsto \underline{\alpha}(t)$ for which the integral

$$\int_T \langle \Phi(\theta)\,\Sigma(\theta)\,\Phi(\theta)[\underline{\alpha}(\theta)], \underline{\alpha}(\theta)\rangle_{\mathbb{R}^n}\, d\theta$$

is finite. When $\sigma_{i,j} = \sigma_i\sigma_j$, $\frac{dF_{\underline{h}}}{d\lambda}$ has the following form: $\underline{\Phi} \otimes \underline{\Phi}$, $\Phi_i = \phi_i\sigma_i$.

8.4.3 Integration with Respect to Cramér-Hida Martingales

Notation below is that of Sects. 8.4.1 and 8.4.2. Let $\underline{h} = \{\underline{h}(t), t \in T\}$ be a CH-martingale. One shall give meaning to objects of the following form (integrals of vector valued, deterministic functions, with respect to a vector measure, obtained from $\underline{h}$ as elements in the range of a unitary operator, whose domain is the

$L_2\left(T, \mathcal{T}, \boldsymbol{M}_{\underline{h}}\right)$ space):

$$\int \left[\underline{\alpha}, d\underline{m}_{\underline{h}}\right] = \sum_{i=1}^{n} \int \alpha_i dm_i^{\underline{h}}.$$

That meaning has its source in the following pair of relations, valid for fixed, but arbitrary $\{t_1, t_2, t_3, t_4\} \subseteq T$, $t_1 < t_2$, $t_3 < t_4$, and the standard basis of $\mathbb{R}^n$, $\{\underline{e}_1, \ldots, \underline{e}_n\}$:

$$\begin{aligned} M_{i,j}^{\underline{h}}\left([t_1, t_2[\cap [t_3, t_4[\right) &= \left\langle h_i(t_2) - h_i(t_1), h_j(t_4) - h_j(t_3)\right\rangle_H \\ &= \left\langle \int I_{[t_1,t_2[} dm_i^{\underline{h}}, \int I_{[t_3,t_4[} dm_j^{\underline{h}}\right\rangle_H, \end{aligned}$$

$$M_{i,j}^{\underline{h}}\left([t_1, t_2[\cap [t_3, t_4[\right) = \left\langle \chi_{[t_1,t_2[}\underline{e}_i, \chi_{[t_3,t_4[}\underline{e}_j\right\rangle_{L_2\left(T,\mathcal{T},\boldsymbol{M}_{\underline{h}}\right)}.$$

The first is valid because of (Fact) 8.4.15, and the fact that the h_i's have orthogonal increments. For the second, one has, by definition, that

$$\begin{aligned} &\left\langle \chi_{[t_1,t_2[}\underline{e}_i, \chi_{[t_3,t_4[}\underline{e}_j\right\rangle_{L_2\left(T,\mathcal{T},\boldsymbol{M}_{\underline{h}}\right)} = \\ &\quad = \int_T \left\langle D_{\mu_{\underline{h}}}(t)\, \chi_{[t_1,t_2[}(t)\, \underline{e}_i, \chi_{[t_3,t_4[}(t)\, \underline{e}_j\right\rangle_{\mathbb{R}^n} \mu_{\underline{h}}(dt). \end{aligned}$$

The latter equals

$$\int_{[t_1,t_2[\cap[t_3,t_4[} \left\langle D_{\mu_{\underline{h}}}(t)\, \underline{e}_i, \underline{e}_j\right\rangle_{\mathbb{R}^n} \mu_{\underline{h}}(dt),$$

which, because of (Fact) 8.4.6, equals

$$\left\langle \boldsymbol{M}_{\underline{h}}\left([t_1, t_2[\cap [t_3, t_4[\right) \underline{e}_i, \underline{e}_j\right\rangle_{\mathbb{R}^n}.$$

Consequently, for fixed, but arbitrary $\{\underline{\alpha}_1, \underline{\alpha}_2\} \subseteq \mathbb{R}^n$,

$$\left\langle \chi_{[t_1,t_2[}\underline{\alpha}_1, \chi_{[t_3,t_4[}\underline{\alpha}_2\right\rangle_{L_2\left(T,\mathcal{T},\boldsymbol{M}_{\underline{h}}\right)} = \left\langle \boldsymbol{M}_h\left([t_1, t_2[\cap [t_3, t_4[\right) \underline{\alpha}_1, \underline{\alpha}_2\right\rangle_{\mathbb{R}^n}.$$

The Integral with Respect to a Cramér-Hida Martingale

Define the assignment $U_{\underline{h}} : L_2\left(T, \mathcal{T}, \boldsymbol{M}_{\underline{h}}\right) \longrightarrow H$ using the following relation:

$$
\begin{aligned}
U_{\underline{h}}\left[I_{[t_1,t_2[}\underline{\alpha}\right] &= \sum_{i=1}^{n} \alpha_i \left(h_i(t_2) - h_i(t_1)\right) \\
&= \sum_{i=1}^{n} \alpha_i m_i^{\underline{h}} \left([t_1, t_2[\right) \\
&= \sum_{i=1}^{n} \alpha_i \int I_{[t_1,t_2[} dm_i^{\underline{h}} \\
&= \int \left[I_{[t_1,t_2[}\underline{\alpha}, d\underline{m}_{\underline{h}}\right].
\end{aligned}
$$

Then, as [(Fact) 8.4.15]

$$
\begin{aligned}
&\left\langle U_{\underline{h}}\left[I_{\left[t_1^{(i)},t_2^{(i)}\right[}\underline{\alpha}_i\right], U_{\underline{h}}\left[I_{\left[t_1^{(j)},t_2^{(j)}\right[}\underline{\alpha}_j\right]\right\rangle_H = \\
&= \sum_{\nu=1}^{n}\sum_{\eta=1}^{n} \alpha_\nu^{(i)}\alpha_\eta^{(j)} \left\langle h_\nu\left(t_2^{(i)}\right) - h_\nu\left(t_1^{(i)}\right), h_\eta\left(t_2^{(j)}\right) - h_\eta\left(t_1^{(j)}\right)\right\rangle_H \\
&= \sum_{\nu=1}^{n}\sum_{\eta=1}^{n} \alpha_\nu^{(i)}\alpha_\eta^{(j)} M_{\nu,\eta}^{\underline{h}} \left(\left[t_1^{(i)}, t_2^{(i)}\right[\cap \left[t_1^{(j)}, t_2^{(j)}\right[\right) \\
&= \left\langle \boldsymbol{M}_{\underline{h}}\left(\left[t_1^{(i)}, t_2^{(i)}\right[\cap \left[t_1^{(j)}, t_2^{(j)}\right[\right) \underline{\alpha}_i, \underline{\alpha}_j\right\rangle_{\mathbb{R}^n},
\end{aligned}
$$

one has that

$$
\begin{aligned}
&\left\| \sum_{j=1}^{p} x_j U_{\underline{h}}\left[I_{\left[t_1^{(j)},t_2^{(j)}\right[}\underline{\alpha}_j\right]\right\|_H^2 = \\
&= \sum_{j=1}^{p}\sum_{k=1}^{p} x_j x_k \left\langle \boldsymbol{M}_{\underline{h}}\left(\left[t_1^{(j)}, t_2^{(j)}\right[\cap \left[t_1^{(k)}, t_2^{(k)}\right[\right) \underline{\alpha}_j, \underline{\alpha}_k\right\rangle_{\mathbb{R}^n} \\
&= \sum_{j=1}^{p}\sum_{k=1}^{p} x_j x_k \int_T \left\langle D_{\mu_{\underline{h}}} \chi_{\left[t_1^{(j)},t_2^{(j)}\right[}(t)\,\underline{\alpha}_j, \chi_{\left[t_1^{(k)},t_2^{(k)}\right[}(t)\,\underline{\alpha}_k\right\rangle_{\mathbb{R}^n} d\mu_{\underline{h}} \\
&= \int_T \left\langle D_{\mu_{\underline{h}}}(t) \sum_{j=1}^{p} x_j \chi_{\left[t_1^{(j)},t_2^{(j)}\right[}(t)\,\underline{\alpha}_j, \sum_{j=1}^{p} x_j \chi_{\left[t_1^{(j)},t_2^{(j)}\right[}(t)\,\underline{\alpha}_j\right\rangle_{\mathbb{R}^n} d\mu_{\underline{h}}.
\end{aligned}
$$

Consequently

$$\sum_{j=1}^{p} x_j U_{\underline{h}} \left(I_{\left[t_1^{(j)}, t_2^{(j)}\right[} \underline{\alpha}_j \right) \text{ is zero}$$

if, and only if,

$$\sum_{j=1}^{p} x_j \chi_{\left[t_1^{(j)}, t_2^{(j)}\right[} (t) \, \underline{\alpha}_j \text{ is a zero function,}$$

and thus $U_{\underline{h}}$ has a unique (isometric) linear extension to the manifold generated by the classes of functions of the following type:

$$t \longrightarrow I_{[t_1, t_2[} (t) \, \underline{\alpha}.$$

The resulting isometry is what is called the (isometric) integral, and one writes

$$U_{\underline{h}} [\underline{\alpha}] = \int \left[\underline{\alpha}, d\underline{m}_{\underline{h}}\right].$$

Remark 8.4.38 One could define the integral in terms of orthogonally scattered measures with values in H^n, and then "return" to H, using inner products.

Properties of the Integral

The integral just defined has the properties to be expected:

1. Let $T_t =]-\infty, t[\cap T$, $\mathcal{T}_t =]-\infty, t[\cap \mathcal{T}$,

$$\boldsymbol{M}_{\underline{h}}^t = \boldsymbol{M}_{\underline{h}}^{|\mathcal{T}_t},$$

   and Q_t be the projection of $L_2 \left(T, \mathcal{T}, \boldsymbol{M}_{\underline{h}}\right)$, defined using the following relation:

$$Q_t [\underline{\alpha}] = I_{T_t} \underline{\alpha}.$$

   The range of Q_t and $L_2 \left(T_t, \mathcal{T}_t, \boldsymbol{M}_{\underline{h}}^t\right)$ are unitary images of each other.
2. Since $\underline{h}$ is a CH-martingale, and thus, in particular, purely nondeterministic, the relation defining $U_{\underline{h}}$ shows that $L_2 \left(T, \mathcal{T}, \boldsymbol{M}_{\underline{h}}\right)$ and $L_{\cup T} [\underline{h}]$ are unitary images of each other.
3. Similarly, for $t \in T$, fixed, but arbitrary,

$$L_2 \left(T_t, \mathcal{T}_t, \boldsymbol{M}_{\underline{h}}^t\right) \text{ and } L_t [\underline{h}]$$

   are unitary images of each other.

4. For $\{t, t_1, t_2\} \subseteq T, t_1 < t_2$, fixed, but arbitrary, the following relation:

$$U_{\underline{h}}\left[I_{[t_1,t_2[}\underline{\alpha}(t)\right] = \left[\underline{\alpha}(t), \underline{m}_{\underline{h}}([t_1, t_2[)\right]$$

leads to, h being defined by the first equality sign below,

$$h(t) = [\underline{\alpha}(t), \underline{h}(t)] = \left[\underline{\alpha}(t), \underline{m}_{\underline{h}}(T_t)\right] = \int \left[I_{T_t}\underline{\alpha}(t), d\underline{m}_{\underline{h}}\right],$$

so that

$$\begin{aligned}\left\langle h(t), \int \left[\underline{a}, d\underline{m}_{\underline{h}}\right]\right\rangle_H &= \langle I_{T_t}\underline{\alpha}(t), \underline{a}\rangle_{L_2(T,\mathcal{T},\boldsymbol{M}_{\underline{h}})} \\ &= \langle \underline{\alpha}(t), \underline{a}\,|_{T_t}\rangle_{L_2(T,\mathcal{T},\boldsymbol{M}_{\underline{h}})}.\end{aligned}$$

8.4.4 *Multiplicity of Martingales in the Wide Sense*

Basically the multiplicity of a CH-martingale is determined by the spectral properties of its covariance.

From what precedes one has that

$$\left\langle D_{\mu_{\underline{h}}}(t)\,\underline{\alpha}(t), \underline{\alpha}(t)\right\rangle_{\mathbb{R}^n} = \sum_{i=1}^{n} \delta_i^{\underline{h}}(t)\left\langle \underline{\alpha}(t), \underline{d}_i^{\underline{h}}(t)\right\rangle_{\mathbb{R}^n}^2,$$

so that its integral with respect to $\mu_{\underline{h}}$ may be written in the following form:

$$\begin{aligned}&\int_T \left\langle D_{\mu_{\underline{h}}}(t)\,\underline{\alpha}(t), \underline{\alpha}(t)\right\rangle_{\mathbb{R}^n} \mu_{\underline{h}}(dt) = \\ &= \sum_{i=1}^{n} \int_T \left\langle \underline{d}_i^{\underline{h}}(t) \otimes \underline{d}_i^{\underline{h}}(t)\left[\underline{\alpha}(t)\right], \underline{\alpha}(t)\right\rangle_{\mathbb{R}^n} \mu_i^{\underline{h}}(dt).\end{aligned}$$

It is that right-hand side which leads to the CHR of $\underline{h}$.

Let indeed

(a) $D_i^{\underline{h}}(t) = \underline{d}_i^{\underline{h}}(t) \otimes \underline{d}_i^{\underline{h}}(t)$;
(b) $H_i = L_2\left(T, \mathcal{T}, \mu_i^{\underline{h}}\right)$ be the Hilbert space built analogously to what is done in Sect. 8.4.1, using $D_i^{\underline{h}}$ and $\mu_i^{\underline{h}}$ in place of $D_{\mu_{\underline{h}}}$ and $\mu_{\underline{h}}$, and having inner product

$$\langle [\underline{\alpha}_1], [\underline{\alpha}_2]\rangle_{H_i} = \int_T \left\langle D_i^{\underline{h}}(t)\left[\underline{\alpha}_1(t)\right], \underline{\alpha}_2(t)\right\rangle_{\mathbb{R}^n} \mu_i^{\underline{h}}(dt);$$

(c) P_i be the projection of $L_2\left(T, \mathcal{T}, \boldsymbol{M}_{\underline{h}}\right)$ that sends the equivalence class of $\underline{\alpha}$ to the equivalence class of

$$\left\langle \underline{\alpha}\,(t)\,, \underline{d}_i^{\underline{h}}\,(t) \right\rangle_{\mathbb{R}^n} \underline{d}_i^{\underline{h}}\,(t)\,;$$

(d) $U_i : H_i \longrightarrow L_2\left(T, \mathcal{T}, \boldsymbol{M}_{\underline{h}}\right)$ be the map defined using the following assignment:

$$U_i\,[\underline{\alpha}] = P_i\,[\underline{\alpha}]\,.$$

It follows from those definitions that

$$\begin{aligned}
\langle [\underline{\alpha}_1]\,, [\underline{\alpha}_2] \rangle_{H_i} &= \int_T \left\langle D_i^{\underline{h}}[\underline{\alpha}_1(t)], \underline{\alpha}_2(t) \right\rangle_{\mathbb{R}^n} d\mu_i^{\underline{h}} \\
&= \int_T \left\langle \underline{\alpha}_1\,(t)\,, \underline{d}_i^{\underline{h}}\,(t) \right\rangle_{\mathbb{R}^n} \left\langle \underline{\alpha}_2\,(t)\,, \underline{d}_i^{\underline{h}}\,(t) \right\rangle_{\mathbb{R}^n} \mu_i^{\underline{h}}\,(dt) \\
&= \int_T \left\langle D_{\mu_{\underline{h}}}\,(t)\,[P_i\,[\underline{\alpha}_1]\,(t)]\,, P_i\,[\underline{\alpha}_2]\,(t) \right\rangle_{\mathbb{R}^n} \mu_{\underline{h}}\,(dt) \\
&= \langle P_i\,[\underline{\alpha}_1]\,, P_i\,[\underline{\alpha}_2] \rangle_{L_2(T, \mathcal{T}, \boldsymbol{M}_{\underline{h}})} \\
&= \langle U_i\,[\underline{\alpha}_1]\,, U_i\,[\underline{\alpha}_2] \rangle_{L_2(T, \mathcal{T} \boldsymbol{M}_{\underline{h}})}\,.
\end{aligned}$$

The map U_i is thus a partial isometry, and the map

$$U : \oplus_{i=1}^n H_i \longrightarrow L_2\left(T, \mathcal{T}, \boldsymbol{M}_{\underline{h}}\right),$$

defined using the following assignment:

$$\left([\underline{\alpha}_1]\,, \ldots, [\underline{\alpha}_n]\right) \mapsto U_1\,[\underline{\alpha}_1] + \cdots + U_n\,[\underline{\alpha}_n]\,,$$

is unitary. As the unitary relation is through "integrals," for fixed, but arbitrary $t \in T$,

$$L_2\left(T_t, \mathcal{T}_t, \boldsymbol{M}_{\underline{h}}^t\right) \text{ and } \oplus_{i=1}^n H_i^t$$

are the unitary image of each other (H_i^t is obtained as H_i by restricting functions to the domain T_t).

It remains to check that the spaces $L_t\,[\underline{h}]$ may be written as spaces generated by a vector map whose components are orthogonal. To that end let

$$U_{\underline{h}} : L_2(T, \mathcal{T}, \boldsymbol{M}_h) \longrightarrow H$$

be defined using the following relations:

$$k_i^{\underline{h}}(t) = U_{\underline{h}}\left[U_i\left[I_{T_t}\underline{d}_i^{\underline{h}}\right]\right] = \int \left[I_{T_t}\underline{d}_i^{\underline{h}}, d\underline{m}_{\underline{h}}\right].$$

One has then that

$$\left\langle k_i^{\underline{h}}(t_1), k_i^{\underline{h}}(t_2)\right\rangle_H = \mu_i^{\underline{h}}\left(T_{t_1 \wedge t_2}\right).$$

The map $t \mapsto k_i^{\underline{h}}(t)$ has orthogonal increments, and basis $\left(T, \mathcal{T}, \mu_i^{\underline{h}}\right)$. Also, for $i \neq j$, $\{i, j\} \subseteq [1 : n]$, fixed, but arbitrary,

$$k_i^{\underline{h}} \perp k_j^{\underline{h}}.$$

Consequently

$$L_{UT}\left[\underline{h}\right] = \oplus_{i=1}^{n} L_{UT}\left[k_i^{\underline{h}}\right],$$

and, for $t \in T$, fixed, but arbitrary,

$$L_t\left[\underline{h}\right] = \oplus_{i=1}^{n} L_t\left[k_i^{\underline{h}}\right].$$

One thus sees that the multiplicity of $\underline{h}$ is at most n and is given by the rank of $D_{\mu_{\underline{h}}}$.

$\underline{h}$ may then be given the following (CHR) representation. Let

$$D_{\mu_{\underline{h}}}(t) = D_{\underline{h}}^{\star}(t)\, \Delta_{\underline{h}}(t)\, D_{\underline{h}}(t),$$

where $D_{\underline{h}}$ is the matrix with the $\underline{d}_i^{\underline{h}}$'s as columns, and $\Delta_{\underline{h}}$ is the diagonal matrix whose diagonal elements are the $\delta_i^{\underline{h}}$'s. Then, with

$$m^{k_i^{\underline{h}}}$$

being the orthogonally scattered measure resulting from $k_i^{\underline{h}}$, and $M^{k_i^{\underline{h}}}$ the associated measure on $\mathcal{T}$, as

$$\frac{dM^{k_i^{\underline{h}}}}{d\mu_{\underline{h}}} = \Delta_{\underline{h}},$$

one has that

$$\left\langle \int \left[\underline{\alpha}_1, dm^{k_i^{\underline{h}}}\right], \int \left[\underline{\alpha}_2, dm^{k_i^{\underline{h}}}\right]\right\rangle_H =$$

$$= \int_T \langle \Delta_{\underline{h}}(t)\,\underline{\alpha}_1(t), \underline{\alpha}_2(t)\rangle_{\mathbb{R}^n} \mu_{\underline{h}}(dt)$$

$$= \int_T \left\langle D_{\mu_{\underline{h}}}(t)\, D_{\underline{h}}^{\star}(t)\,\underline{\alpha}_1(t), D_{\underline{h}}^{\star}(t)\,\underline{\alpha}_2(t)\right\rangle_{\mathbb{R}^n} \mu_{\underline{h}}(dt)$$

$$= \left\langle \int \left[D_{\underline{h}}^{\star}\underline{\alpha}_1, d\underline{m}_{\underline{h}}\right], \int \left[D_{\underline{h}}^{\star}\underline{\alpha}_2, d\underline{m}_{\underline{h}}\right]\right\rangle_H .$$

Consequently, when $\underline{\alpha}_1 = \underline{d}_i^{h}$,

$$\int \left[\underline{d}_i^{h}, dm^{k_i^{\underline{h}}}\right] = h_i,$$

and $\underline{h}$ is thus given a representation, with respect to a martingale in the wide sense, whose components are orthogonal.

8.4.5 Goursat Maps: Definition and Properties

Let $f : T \longrightarrow H$ be a map whose index set is an interval of $\mathbb{R}$, and range, a subset of H, a real Hilbert space.

Definition 8.4.39 f is *(or has a representation as)* a Goursat map *(or process, when $H = L_2(\Omega, \mathcal{A}, P)$, and the mean is zero)* of rank $n \in \mathbb{N}$ when it has the following form:

$$f(t) = [\underline{a}(t), \underline{h}(t)] \qquad (\star)$$

where

1. $t \mapsto \underline{a}(t)$ is a map with values in $\mathbb{R}^n$,
2. $\underline{h}$ is a martingale in the wide sense, with values in H^n,
3. the "inner product" of $(\star)$ means that, for fixed, but arbitrary $t \in T$,

$$f(t) = \sum_{i=1}^{n} a_i(t)\, h_i(t).$$

$\underline{h}$ is called the martingale associated with f.

Definition 8.4.40 When the martingale associated with the Goursat map f is a CH-martingale, f shall be called a CH-Goursat map.

Fact 8.4.41 *As, by definition, $L_t[f] \subseteq L_t[\underline{h}]$, a CH-Goursat map is purely nondeterministic.*

Remark 8.4.42 CH-Goursat maps are introduced for two reasons. The first is that the Cramér-Hida representation has been derived assuming non-determinism and continuity to the left. The second is that isometric integration has a simpler notation with those assumptions. Thus every time an isometric integral is used, implicitly the map involved will be of the CH kind.

Definition 8.4.43 When, for fixed, but arbitrary $t \in T$, $L_t[f] = L_t[\underline{h}]$, the Goursat map f is said to be proper.

As explained in the next statement, a Goursat map may always be given a proper form.

Proposition 8.4.44 *$f(t) = [\underline{a}(t), \underline{h}(t)]$ be a Goursat map of rank n, and P_t^f be the projection with range $L_t[f]$. Then:*

1. *$\underline{h}_f(t) = P_t^f[\underline{h}(t)]$ is a martingale in the wide sense, with respect to f, as well as with respect to itself;*
2. *$f(t) = \left[\underline{a}(t), \underline{h}_f(t)\right]$, and $\left[\underline{a}(t), \underline{h}_f(t)\right]$ is a proper Goursat map of rank n;*
3. *$f(t) = [\underline{a}(t), \underline{h}(t)]$ is proper if, and only if, for fixed, but arbitrary $t \in T$, $P_t^f[\underline{h}(t)] = \underline{h}(t)$.*

Proof Let $\{t_1, t_2\} \subseteq T$, $t_1 < t_2$, be fixed, but arbitrary. Then

$$P_{t_1}^f\left[\underline{h}_f(t_2)\right] = P_{t_1}^f P_{t_2}^f[\underline{h}(t_2)] = P_{t_1}^f[\underline{h}(t_2)].$$

As, for fixed, but arbitrary $t \in T$, $L_t[f] \subseteq L_t[\underline{h}]$, $P_t^{\underline{h}}$ denoting the projection whose range is $L_t[\underline{h}]$,

$$P_{t_1}^f[\underline{h}(t_2)] = P_{t_1}^f P_{t_1}^{\underline{h}}[\underline{h}(t_2)] = P_{t_1}^f[\underline{h}(t_1)] = \underline{h}_f(t_1).$$

Thus

$$P_{t_1}^f\left[\underline{h}_f(t_2)\right] = \underline{h}_f(t_1),$$

and, since $L_t\left[\underline{h}_f\right] \subseteq L_t[f]$, denoting $P_t^{\underline{h}_f}$ the projection whose range is $L_t\left[\underline{h}_f\right]$,

$$P_{t_1}^{\underline{h}_f}\left[\underline{h}_f(t_2)\right] = P_{t_1}^{\underline{h}_f} P_{t_1}^f\left[\underline{h}_f(t_2)\right] = P_{t_1}^{\underline{h}_f}\left[\underline{h}_f(t_1)\right] = \underline{h}_f(t_1).$$

Item 1 thus obtains.

From the definitions, one has furthermore that

$$f(t) = P_t^f[f(t)] = \sum_{i=1}^{n} a_i(t)\, P_t^f[h_i(t)] = \left[\underline{a}, \underline{h}_f(t)\right].$$

Then

$$L_t[f] \subseteq L_t\left[\underline{h}_f\right] \subseteq L_t[f],$$

and item 2 obtains.

Suppose f is proper. Then $P_t^f = P_t^{\underline{h}}$, and thus $P_t^f[\underline{h}(t)] = P_t^{\underline{h}}[\underline{h}(t)] = \underline{h}(t)$. Conversely, when $P_t^f[\underline{h}(t)] = \underline{h}(t)$, then

$$f(t) = [\underline{a}(t), \underline{h}(t)] = \left[\underline{a}(t), P_t^f[\underline{h}(t)]\right],$$

so that f is proper by item 2. □

Goursat maps have the following properties.

Fact 8.4.45 *A Goursat map f has a covariance of the following form:*

$$C_f(t_1, t_2) = \langle f(t_1), f(t_2)\rangle_H = \left\langle F_{\underline{h}}(t_1 \wedge t_2)\, \underline{a}(t_1), \underline{a}(t_2)\right\rangle_{\mathbb{R}^n}.$$

Fact 8.4.46 *For a Goursat map f, and $\{t_1, t_2\} \subseteq T$, $t_1 \leq t_2$, fixed, but arbitrary,*

$$P_{t_1}^f[f(t_2)] = [\underline{a}(t_2), \underline{h}(t_1)].$$

Proof Fix arbitrarily $\theta \in T$, $\theta \leq t_1$. Then indeed, using (Fact) 8.4.45,

$$\begin{aligned}\langle [\underline{a}(t_2), \underline{h}(t_1)], f(\theta)\rangle_H &= \langle [\underline{a}(t_2), \underline{h}(t_1)], [\underline{a}(\theta), \underline{h}(\theta)]\rangle_H \\ &= \left\langle F_{\underline{h}}(\theta)\, \underline{a}(\theta), \underline{a}(t_2)\right\rangle_{\mathbb{R}^n} \\ &= C_f(t_2, \theta) \\ &= \langle f(t_2), f(\theta)\rangle_H.\end{aligned}$$

□

The prediction from the past of a Goursat map has thus a rather simple expression provided one knows its ingredients. The RKHS properties of f are closely related to the theory of Goursat maps, as shown by the following result.

Fact 8.4.47 *Let $f : T \longrightarrow H$ be a map.*

1. *One may define a map $L_f : H \longrightarrow \mathbb{R}^T$ using the following assignment:*

$$t \mapsto L_f[h](t) = \langle h, f(t)\rangle_H.$$

The range of L_f is the RKHS of C_f, and L_f, restricted to $L_{\cup T}[f]$, is unitary, with

$$L_f[f(t)] = C_f(\cdot, t).$$

It follows that, when $g : T \longrightarrow H$ is another map such that

$$C_g(t_1, t_2) = C_f(t_1, t_2),$$

then $g(t)$ and $f(t)$ are unitarily related.

2. *Let f be the Goursat map $t \mapsto [\underline{a}(t), \underline{h}(t)]$. Let then*

$$F : T \longrightarrow L_2\left(T, \mathcal{T}, \boldsymbol{M}_{\underline{h}}\right)$$

be defined using the following relation:

$$F(t) = \underline{a}(t) I_{T_t},$$

and $L_F : L_2\left(T, \mathcal{T}, \boldsymbol{M}_{\underline{h}}\right) \longrightarrow \mathbb{R}^T$, using the following one:

$$t \mapsto L_F[\underline{\alpha}](t) = \langle \underline{\alpha}, F(t) \rangle_{L_2(T, \mathcal{T}, \boldsymbol{M}_{\underline{h}})} = \int_{T_t} \left\langle D_{\underline{h}}(\theta)[\underline{a}(t)], \underline{\alpha}(\theta) \right\rangle_{\mathbb{R}^n} \mu_{\underline{h}}(d\theta).$$

The range of L_F is $H(C_f, T)$, as [(Definition) 8.4.4, (Fact) 8.4.6]

$$\begin{aligned}
\langle F(t_1), F(t_2) \rangle_{L_2(T, \mathcal{T}, \boldsymbol{M}_{\underline{h}})} &= \\
&= \int_{T_{t_1} \cap T_{t_2}} \left\langle D_{\mu_{\underline{h}}}(\theta)\, \underline{a}(t_1), \underline{a}(t_2) \right\rangle_{\mathbb{R}^n} \mu_{\underline{h}}(d\theta) \\
&= \left\langle F_{\underline{h}}(t_1 \wedge t_2)\, \underline{a}(t_1), \underline{a}(t_2) \right\rangle_{\mathbb{R}^n} \\
&= C_f(t_1, t_2).
\end{aligned}$$

Let H_F be the (closed) *subspace generated linearly in $L_2\left(T, \mathcal{T}, \boldsymbol{M}_{\underline{h}}\right)$ by the following family: $\{\underline{a}(t) I_{T_t}, t \in T\}$. $H(C_f, T)$ is the unitary image of L_F restricted to H_F, and $H_F^{\perp}$ is the kernel of L_F.*

The proposition which follows characterizes proper CH-Goursat maps in terms of their ingredients.

Proposition 8.4.48 *Let the class of* $\underline{\alpha}$ *be in* $L_2\left(T, \mathcal{T}, \boldsymbol{M}_{\underline{h}}\right)$. *The CH-Goursat map* $f(t) = [\underline{a}(t), \underline{h}(t)]$ *is proper if, and only if, whenever, for fixed, but arbitrary* $t \in T$,

$$\int_{T_\theta} \left\langle D_{\mu_{\underline{h}}}(u)\, \underline{\alpha}(u), \underline{a}(\theta)\right\rangle_{\mathbb{R}^n} \mu_{\underline{h}}(du) = 0,\ \theta \leq t,$$

$\underline{\alpha}$ *is a null function.*

Proof The class of the "constant" $\underline{a}(\theta)$ belongs to $L_2\left(T, \mathcal{T}, \boldsymbol{M}_{\underline{h}}\right)$ [(Fact) 8.4.6] as

$$\int_{T_\theta} \left\langle D_{\mu_{\underline{h}}}(u)\, \underline{a}(\theta), \underline{a}(\theta)\right\rangle_{\mathbb{R}} \mu_{\underline{h}}(du) = \left\langle F_{\underline{h}}(\theta)\, \underline{a}(\theta), \underline{a}(\theta)\right\rangle_{\mathbb{R}} = C_f(\theta, \theta),$$

and thus the integral of the statement's condition is well defined. One has that $L_t[f] \subseteq L_t[\underline{h}]$. Let $k \in L_t[\underline{h}] \ominus L_t[f]$ be fixed, but arbitrary. Then [Sect. 8.4.3]:

$$k = \int \left[I_{T_t} \underline{\alpha}^k, d\underline{m}_{\underline{h}}\right],\ \underline{\alpha}^k \in L_2\left(T, \mathcal{T}, \boldsymbol{M}_{\underline{h}}\right),$$

$$f(\theta) = \int \left[I_{T_\theta} \underline{a}(\theta), d\underline{m}_{\underline{h}}\right],\ \theta \in T,\ \theta \leq t,$$

$$\langle k, f(\theta)\rangle_H = \int_{T_\theta} \left\langle D_{\mu_{\underline{h}}}(u)\, \underline{\alpha}^k(u), \underline{a}(\theta)\right\rangle_{\mathbb{R}^n} \mu_{\underline{h}}(du).$$

Thus the statement's condition is equivalent to $k = 0_H$, which is equivalent to $L_t[f] = L_t[\underline{h}]$. □

In the language of (Fact) 8.4.47, the Goursat map is proper when the map L_F is an injection.

Fact 8.4.49 *A proper Goursat map has the multiplicity properties of its associated martingale, and these are determined by the spectral properties of the latter's structure matrix. In particular, when* f *is a proper CH-Goursat map, the proper canonical CHR of* f*, that is, an expression of the form*

$$f(t) = \sum_{i=1}^{n} \int F_t^{(i)} dm_i,$$

is obtained setting

1. $m_i = m^{k_i^{\underline{h}}},\ k_i^{\underline{h}} = \int \left[\underline{d}_i^{\underline{h}}, d\underline{m}_{\underline{h}}\right]$,
2. $F_t^{(i)} = \sum_{j=1}^{n} a_j(t) \left[\chi_{T_t} d_{i,j}^{\underline{h}}\right]_{L_2\left(T, \mathcal{T}, \mu_i^{\underline{h}}\right)}$,

where the notation and meaning of the diverse ingredients are those of the present, and previous sections.

Remark 8.4.50 A CH-Goursat map f may always be expressed in the following form:

$$f(t) = \sum_{i=1}^{n} \int F_t^{(i)} dm_i^{\underline{h}}, \ F_t^{(i)} = a_i(t) I_{T_t}.$$

The latter is however not necessarily the CHR representation of f. It will be when f is proper, and the associated martingale has nonzero eigenvalue functions.

Fact 8.4.51 *For fixed, but arbitrary $\{t_1, t_2\} \subseteq T$, $t_1 < t_2$, adding and subtracting $[\underline{a}(t_2), \underline{h}(t_1)]$, one obtains the orthogonal decomposition*

$$f(t_2) - f(t_1) = [\underline{a}(t_2), \underline{h}(t_2) - \underline{h}(t_1)] + [\underline{a}(t_2) - \underline{a}(t_1), \underline{h}(t_1)],$$

from which it follows that

$$\begin{aligned} \|f(t_2) - f(t_1)\|_H^2 &= \|[\underline{a}(t_2), \underline{h}(t_2) - \underline{h}(t_1)]\|_H^2 \\ &+ \|[\underline{a}(t_2) - \underline{a}(t_1), \underline{h}(t_1)]\|_H^2 \\ &= \left\langle \left\{F_{\underline{h}}(t_2) - F_{\underline{h}}(t_1)\right\} \underline{a}(t_2), \underline{a}(t_2)\right\rangle_{\mathbb{R}^n} \\ &+ \left\langle F_{\underline{h}}(t_1)\{\underline{a}(t_2) - \underline{a}(t_1)\}, \{\underline{a}(t_2) - \underline{a}(t_1)\}\right\rangle_{\mathbb{R}^n}. \end{aligned}$$

Consequently:

1. $\|f(t_2) - f(t_1)\|_H^2 \geq \left\langle \left\{F_{\underline{h}}(t_2) - F_{\underline{h}}(t_1)\right\} \underline{a}(t_2), \underline{a}(t_2)\right\rangle_{\mathbb{R}^n}$.
2. *Let*
   (a) $F_{\underline{h}}(t) = \sum_{i=1}^{n} \phi_i^{\underline{h}}(t) \underline{f}_i^{\underline{h}}(t) \otimes \underline{f}_i^{\underline{h}}(t)$,
   (b) $\phi_m^{\underline{h}}(t)$ *denote the smallest, nonzero eigenvalue,*
   (c) Q_m^t *be the projection whose range is spanned by the set*

$$\left\{\underline{f}_1^{\underline{h}}(t), \ldots, \underline{f}_m^{\underline{h}}(t)\right\}:$$

   then $\|f(t_2) - f(t_1)\|_H^2 \geq \phi_m^{\underline{h}}(t_1) \left\|Q_m^{t_1}[\underline{a}(t_2) - \underline{a}(t_1)]\right\|_{\mathbb{R}^n}^2$.
3. *When f is a CH-Goursat map, and $\underline{a}$ is continuous to the left, f is continuous to the left.*

Fact 8.4.52 *The map $t \mapsto \phi_m^{\underline{h}}(t)$ is increasing.*

Proof Indeed, as [(Proposition) 8.4.30], for $\{t_1, t_2\} \subseteq T$, $t_1 < t_2$, fixed, but arbitrary, $F_{\underline{h}}(t_2) - F_{\underline{h}}(t_1)$ is positive definite, then, for fixed, but arbitrary $\underline{x} \in \mathbb{R}^n$,

$$\left\langle F_{\underline{h}}(t_2)\underline{x}, \underline{x}\right\rangle_{\mathbb{R}^n} \geq \left\langle F_{\underline{h}}(t_1)\underline{x}, \underline{x}\right\rangle_{\mathbb{R}^n},$$

so that, by the Rayleigh-Ritz theorem [170, p. 319], $\phi_m^{\underline{h}}(t_2) \geq \phi_m^{\underline{h}}(t_1)$. □

Fact 8.4.53 *When the martingale in the wide sense* $\underline{h}$*, associated with* f*, is non-singular, the following property obtains: when* f *is continuous (continuous to the left, satisfies a local Hölder condition),* $\underline{a}$ *is continuous (continuous to the left, satisfies a local Hölder condition), for then* $\phi_m^{\underline{h}}(t) > 0$*, and the validity of the claim follows from (Fact)* 8.4.51.

Let f be any function which has a limit to the left at every point of its domain. Then the function f^- is the function of limits to the left of f.

Fact 8.4.54 *When the martingale in the wide sense* $\underline{h}$*, associated with* f*, is non-singular, the following properties obtain:*

when, for $t \in T$*, the following limit exits in* H*:*

$$D^-[f](t) = \lim_{T \ni \theta \uparrow\uparrow t} \frac{f(t) - f(\theta)}{t - \theta},$$

then the following limit exists:

$$D^-[\underline{a}](t) = \lim_{T \ni \theta \uparrow\uparrow t} \frac{\underline{a}(t) - \underline{a}(\theta)}{t - \theta},$$

and, furthermore, in H*,*

$$D^-[f](t) = [D^-[\underline{a}](t), \underline{h}^-(t)].$$

Proof Since $\underline{h}$ is non-singular, one has, from (Fact) 8.4.51, and the differentiability assumption on f, that the family of ratios (t fixed, θ variable, $\theta < t$) of the form

$$\frac{\|\underline{a}(t) - \underline{a}(\theta)\|_{\mathbb{R}^n}}{t - \theta}$$

is bounded. There is thus an increasing sequence $\{\theta_n,\ n \in \mathbb{N}\} \subseteq T$ such that

$$\lim_{T \ni \theta_n \uparrow\uparrow t} \frac{\underline{a}(t) - \underline{a}(\theta_n)}{t - \theta_n}$$

exists. As $F_{\underline{h}}$ is monotone increasing, the following limit exists:

$$\lim_{T \ni \theta_n \uparrow\uparrow t} \underline{h}(\theta_n).$$

Thus

$$\lim_{T \ni \theta_n \uparrow\uparrow t} \frac{[\underline{a}(t) - \underline{a}(\theta_n), \underline{h}(\theta_n)]}{t - \theta_n}$$

exists. Subtracting, and then adding $[\underline{a}(\theta_n), \underline{h}(\theta_n)]$, defining k_n using the first equal sign below,

$$k_n = \frac{[\underline{a}(t), \underline{h}(t) - \underline{h}(\theta_n)]}{t - \theta_n} = \frac{f(t) - f(\theta_n)}{t - \theta_n} - \frac{[\underline{a}(t) - \underline{a}(\theta_n), \underline{h}(\theta_n)]}{t - \theta_n}. \qquad (\star)$$

Since [(Fact) 8.4.46] $[\underline{a}(t), \underline{h}(\theta_n)] = P^f_{\theta_n}[f(t)]$, the ratios in $(\star)$ belong to $L_t[f]$. From what precedes, they have a limit, say $k \in L_t[f]$. Now, for fixed, but arbitrary $t_0 \in T,\ t_0 < t$,

$$\begin{aligned}\langle f(t_0), k\rangle_H &= \lim_n \langle f(t_0), k_n\rangle_H \\ &= \lim_n \frac{\langle [\underline{a}(t_0), \underline{h}(t_0)], [\underline{a}(t), \underline{h}(t) - \underline{h}(\theta_n)]\rangle_H}{t - \theta_n}.\end{aligned}$$

As, for $n \in \mathbb{N}$ large enough, $\underline{h}(t) - \underline{h}(\theta_n) \perp \underline{h}(t_0)$,

$$k \perp f(t_0),\ t_0 < t.$$

Consequently $k \perp L_t^-[f]$, and, since, by the assumption that the derivative to the left exists, f is continuous to the left, $L_t^-[f] = L_t[f]$, and $k \perp L_t[f]$. But, as seen above, $k \in L_t[f]$. Thus $k = 0_H$, and

$$D^-[f](t) = \lim_n \frac{[\underline{a}(t) - \underline{a}(\theta_n), \underline{h}(\theta_n)]}{t - \theta_n}. \qquad (\star\star)$$

But that equality obtains irrespectively of the sequence chosen, and thus, letting

$$\begin{aligned}\Delta_{t,\theta}[\underline{a}] &= \frac{\underline{a}(t) - \underline{a}(\theta)}{t - \theta}, \\ D^-[f](t) &= \lim_{T \ni \theta \uparrow\uparrow t} [\Delta_{t,\theta}[\underline{a}], \underline{h}(\theta)], \\ \|D^-[f](t)\|_H^2 &= \lim_{T \ni \theta \uparrow\uparrow t} \langle F_{\underline{h}}(\theta)\, \Delta_{t,\theta}[\underline{a}], \Delta_{t,\theta}[\underline{a}]\rangle_{\mathbb{R}^n},\end{aligned}$$

- as

$$\begin{aligned}&\langle F_{\underline{h}}(\theta)\, \Delta_{t,\theta}[\underline{a}], \Delta_{t,\theta}[\underline{a}]\rangle_{\mathbb{R}^n} = \\ &\quad = \left\langle F_{\underline{h}}^-(t)\, \Delta_{t,\theta}[\underline{a}], \Delta_{t,\theta}[\underline{a}]\right\rangle_{\mathbb{R}^n} \\ &\quad - \left\langle \left\{F_{\underline{h}}^-(t) - F_{\underline{h}}(\theta)\right\} \Delta_{t,\theta}[\underline{a}], \Delta_{t,\theta}[\underline{a}]\right\rangle_{\mathbb{R}^n},\end{aligned}$$

- that the norm of $\Delta_{t,\theta}[\underline{a}]$ is bounded,

- that $\underline{h}$ has limits to the left,

$$\left\|D^{-}[f](t)\right\|_H^2 = \lim_{T \ni \theta \uparrow\uparrow t} \left\langle F_{\underline{h}}^{-}(t)\, \Delta_{t,\theta}[\underline{a}], \Delta_{t,\theta}[\underline{a}] \right\rangle_{\mathbb{R}^n}.$$

As $F_{\underline{h}}^{-}(t)$ is strictly positive definite, $\lim_{T \ni \theta \uparrow\uparrow t} \Delta_{t,\theta}[\underline{a}]$ exists. One finally uses ($\star\star$). □

Fact 8.4.55 *When the martingale in the wide sense $\underline{h}$, associated with f, is non-singular, the following property obtains:*

when f is differentiable n times, its n-th derivative $f^{(n)}$ has the following form:

$$f^{(n)}(t) = \left[\underline{a}^{(n)}(t), \underline{h}(t)\right],$$

$\underline{a}^{(n)}(t)$ denoting the n-th derivative of $\underline{a}$ at $t \in T$.

Proof Let $\{\theta, t, \tau\} \subseteq T$, $\theta < t, \tau$. Then

$$\begin{aligned} P_\theta^f[f(\tau) - f(t)] &= P_\theta^f[[\underline{a}(\tau), \underline{h}(\tau)]] - P_\theta^f[[\underline{a}(t), \underline{h}(t)]] \\ &= [\underline{a}(\tau), \underline{h}(\theta)] - [\underline{a}(t), \underline{h}(\theta)] \\ &= [\underline{a}(\tau) - \underline{a}(t), \underline{h}(\theta)]. \end{aligned}$$

Consequently

$$\left\|P_\theta^f\left[\frac{f(\tau) - f(t)}{\tau - t}\right]\right\|_H^2 = \left\langle F_{\underline{h}}(\theta) \frac{\underline{a}(\tau) - \underline{a}(t)}{\tau - t}, \frac{\underline{a}(\tau) - \underline{a}(t)}{\tau - t} \right\rangle_{\mathbb{R}^n},$$

so that (with $\underline{a}' = \underline{a}^{(1)}$) $\underline{a}'(t)$ exists, and

$$P_\theta^f[D[f](t)] = \left[\underline{a}'(t), \underline{h}(\theta)\right].$$

But

$$\begin{aligned} \left\langle \left[\underline{a}'(t), \underline{h}(\theta)\right], [\underline{a}(\theta), \underline{h}(\theta)] \right\rangle_H &= \left\langle F_{\underline{h}}(\theta)\, \underline{a}(\theta), \underline{a}'(t) \right\rangle_{\mathbb{R}^n} \\ &= \left\langle \left[\underline{a}'(t), \underline{h}(t)\right], [\underline{a}(\theta), \underline{h}(\theta)] \right\rangle_H, \end{aligned}$$

so that, using the operational definition of projection, for $\theta < t$, fixed, but arbitrary,

$$P_\theta^f[D[f](t)] = P_\theta^f\left[\left[\underline{a}'(t), \underline{h}(t)\right]\right].$$

Consequently, the projections of respectively $D[f](t)$ and $[\underline{a}'(t), \underline{h}(t)]$ onto $L_t^{-}[f]$ are equal. Since f is continuous, the projections onto $L_t[f]$ are equal. Since $D[f](t) = D^{-}[f](t)$, $D[f](t) \in L_t[f]$, and, since one may choose an $\underline{h}$ that yields a proper representation of f [(Proposition) 8.4.44], the "inner product" also belongs to $L_t[f]$, and thus the required equality in H obtains:

$$D[f](t) = \left[\underline{a}'(t), \underline{h}(t)\right].$$

□

Definition 8.4.56 When the Goursat map f has an associated non-singular martingale in the wide sense, and that, for fixed, but arbitrary $t < t_r$, there exists $\{t_1, \ldots, t_n\} \subseteq T$ such that

(a) $t \leq t_1 < t_2 < \cdots < t_{n-1} < t_n$,
(b) the matrix $A[t \mid t_1, \ldots, t_n] =$

$$\begin{bmatrix} a_1(t_1) & a_1(t_2) \cdots & a_1(t_{n-1}) & a_1(t_n) \\ a_2(t_1) & a_2(t_2) \cdots & a_2(t_{n-1}) & a_2(t_n) \\ \vdots & \vdots & \vdots & \vdots \\ a_{n-1}(t_1) & a_{n-1}(t_2) \cdots & a_{n-1}(t_{n-1}) & a_{n-1}(t_n) \\ a_n(t_1) & a_n(t_2) \cdots & a_n(t_{n-1}) & a_n(t_n) \end{bmatrix}$$

is non-singular,

then the Goursat map f is said to be non-singular.

Fact 8.4.57 *Suppose f has $i-1$ derivatives, and that, for $t \in T$, fixed, but arbitrary, the set*

$$\{f(t), f^{(1)}(t), \ldots, f^{(i-1)}(t)\}$$

is linearly independent. Then, when the associated martingale is non-singular:

1. *When $f^{(i-1)}$ is continuous, the dimension of the subspace $L_t^{+}[\underline{h}] \cap L_t^{\perp}[\underline{h}]$ is at most $n-i$.*
2. *When there exists $\alpha > \frac{1}{2}$ for which $f^{(i-1)}$ satisfies a Lipschitz condition of order α, the multiplicity of $\underline{h}$ is at most $n-i$.*

Proof Letting $f^{(0)}(t) = f(t)$, and $\underline{a}^{(0)}(t) = \underline{a}(t)$, one has that [(Fact) 8.4.55]

$$\sum_{j=0}^{i-1} \alpha_j f^{(j)}(t) = \left[\sum_{j=1}^{i-1} \alpha_j \underline{a}^{(j)}(t), \underline{h}(t)\right],$$

and thus that

$$\left\| \sum_{j=1}^{i-1} \alpha_j f^{(j)}(t) \right\|_H^2 = \left\langle F_{\underline{h}}(t) \sum_{j=1}^{i-1} \alpha_j \underline{a}^{(j)}(t), \sum_{j=1}^{i-1} \alpha_j \underline{a}^{(j)}(t) \right\rangle_{\mathbb{R}^n}.$$

Since $F_{\underline{h}}(t)$ is assumed strictly positive definite, linear independence of

$$\{f(t),f^{(1)}(t),\ldots,f^{(i-1)}(t)\}$$

is equivalent to linear independence of

$$\{\underline{a}(t),\underline{a}^{(1)}(t),\ldots,\underline{a}^{(i-1)}(t)\}.$$

Now, for $\delta > 0$, and $j \in [0 : i-1]$, fixed, but arbitrary, using (Facts) 8.4.51 and 8.4.55,

$$\begin{aligned}\left\|f^{(j)}(t+\delta)-f^{(j)}(t)\right\|_H^2 &= \\ &= \left\langle\{F_{\underline{h}}(t+\delta)-F_{\underline{h}}(t)\}\underline{a}^{(j)}(t+\delta),\underline{a}^{(j)}(t+\delta)\right\rangle_{\mathbb{R}^n} \\ &+ \left\langle F_{\underline{h}}(t)\{\underline{a}^{(j)}(t+\delta)-\underline{a}^{(j)}(t)\},\underline{a}^{(j)}(t+\delta)-\underline{a}^{(j)}(t)\right\rangle_{\mathbb{R}^n}.\end{aligned}$$

Since f and its derivatives are assumed to be continuous, the derivatives of $\underline{a}$ are continuous [(Fact) 8.4.53], and thus, letting δ go to zero, one has that

$$0 = \left\langle\left\{F_{\underline{h}}^{+}(t)-F_{\underline{h}}(t)\right\}\underline{a}^{(j)}(t),\underline{a}^{(j)}(t)\right\rangle_{\mathbb{R}^n}.$$

But then the kernel of

$$F_{\underline{h}}^{+}(t)-F_{\underline{h}}(t)$$

contains at least i independent vectors, and thus its rank is at most $n-i$. Let $\{\underline{x}_1,\ldots,\underline{x}_i\} \subseteq \mathbb{R}^n$ be orthonormal elements in the kernel of $F_{\underline{h}}^{+}(t)-F_{\underline{h}}(t)$, and

$$\{k_1,\ldots,k_p\} \subseteq L_t^{+}[\underline{h}] \cap L_t^{\perp}[\underline{h}]$$

be a basis [(Fact) 8.4.18].

Let $\{\underline{x},\underline{y}\} \subseteq \mathbb{R}^n$ be fixed, but arbitrary. Then

$$\left\langle\sum_{i=1}^{n} x_i\left(h_i^{+}(t)-h_i(t)\right),\sum_{i=1}^{n} y_i\left(h_i^{+}(t)-h_i(t)\right)\right\rangle_H = \left\langle\left[F_{\underline{h}}^{+}(t)-F_{\underline{h}}(t)\right]\underline{x},\underline{y}\right\rangle_{\mathbb{R}^n},$$

and, as

$$\begin{aligned}\sum_{i=1}^{n} x_i\left(h_i^{+}(t)-h_i(t)\right) &= \sum_{i=1}^{n} x_i\sum_{j=1}^{p}\alpha_j^{(i)}k_i \\ &= \sum_{j=1}^{p} k_j\sum_{i=1}^{n} x_i\alpha_j^{(i)}\end{aligned}$$

$$= \sum_{j=1}^{p} \left\langle \underline{x}, \underline{\alpha}_j \right\rangle_{\mathbb{R}^n} k_j,$$

$$\left\langle \sum_{i=1}^{n} x_i \left(h_i^+(t) - h_i(t)\right), \sum_{i=1}^{n} y_i \left(h_i^+(t) - h_i(t)\right) \right\rangle_H = \sum_{j=1}^{p} \left\langle \underline{x}, \underline{\alpha}_j \right\rangle_{\mathbb{R}^n} \left\langle \underline{y}, \underline{\alpha}_j \right\rangle_{\mathbb{R}^n}.$$

Thus

$$\left\langle \left[F_{\underline{h}}^+(t) - F_{\underline{h}} \right] \underline{x}, \underline{x} \right\rangle_{\mathbb{R}^n} = \sum_{j=1}^{p} \left\langle \underline{x}, \underline{\alpha}_j \right\rangle_{\mathbb{R}^n}^2 .$$

If one chooses the $\underline{x}_j$'s for $\underline{x}$, one has that $\{\underline{\alpha}_1, \ldots, \underline{\alpha}_p\}$ belongs to the subspace that is orthogonal to that spanned by $\{\underline{x}_1, \ldots, \underline{x}_i\}$, and thus $p \leq n - i$.

Suppose now that $f^{(i-1)}$ satisfies a Lipschitz condition of order $\alpha > \frac{1}{2}$. Generally, when, on $[A, B]$, φ has derivative φ' which satisfies a Lipschitz condition of order α,

- when $\alpha > 1$, φ is constant, as its derivative is zero, and thus φ satisfies a Lipschitz condition of order α;
- when $\alpha \leq 1$,

$$\begin{aligned} |\varphi(b) - \varphi(a)| &= \left| \int_a^b \varphi'(x) dx \right| \\ &\leq \int_a^b \left\{ \left|\varphi'(x) - \varphi'(a)\right| dx + \left|\varphi'(a)\right| \right\} dx \\ &\leq \kappa \int_a^b (x-a)^\alpha \, dx + \varphi'(a)(b-a) \\ &\leq \kappa \frac{(b-a)^{1+\alpha}}{1+\alpha} + \left|\varphi'(a)\right| (b-a) \\ &\leq (b-a)^\alpha \left\{ \kappa \frac{B-A}{1+\alpha} + \sup_{x \in [A,B]} \left|\varphi'(x)\right| (B-A)^{1-\alpha} \right\}, \end{aligned}$$

  and thus φ satisfies a Lipschitz condition of order α.

Thus, for $[t_l, t_r] \subseteq T$, and $j \in [0 : i-1]$, fixed, but arbitrary, there is $\kappa \in \mathbb{R}_+$ such that, for fixed, but arbitrary $\{t_1, t_2\} \subseteq [t_l, t_r]$,

$$\left\| f^{(j)}(t_1) - f^{(j)}(t_2) \right\|_H \leq \kappa \, |t_1 - t_2|^\alpha .$$

Then, using (Fact) 8.4.51,

$$\sum_{k:t_l\le t_k\le t_r} \left\langle \left\{F_{\underline{h}}(t_k) - F_{\underline{h}}(t_{k-1})\right\} \underline{a}^{(j)}(t_k), \underline{a}^{(j)}(t_k)\right\rangle_{\mathbb{R}^n} \le$$

$$\le \sum_{k:t_l\le t_k\le t_r} \left\|f^{(j)}(t_k) - f^{(j)}(t_{k-1})\right\|_H^2$$

$$\le \kappa^2 \sum_{k:t_l\le t_k\le t_r} |t_k - t_{k-1}|^{2\alpha}.$$

Since $2\alpha > 1$ by assumption, when the segments of the partition of $[t_l, t_r]$ by the t_k's have length which tends to zero, the right-hand side member of the latter inequality vanishes, and [(Remark) 8.4.36] the left term tends to the integral

$$\int_{[t_l,t_r]} \left\langle \boldsymbol{M}_{\underline{h}}(d\theta)\, \underline{a}^{(j)}(\theta), \underline{a}^{(j)}(\theta)\right\rangle_{\mathbb{R}^n} =$$

$$= \int_{[t_l,t_r]} \left\langle D_{\mu_{\underline{h}}}(\theta)\, \underline{a}^{(j)}(\theta), \underline{a}^{(j)}(\theta)\right\rangle_{\mathbb{R}^n} \mu_{\underline{h}}(d\theta).$$

Since $[t_l, t_r]$ is arbitrary, it is now the kernel of $D_{\mu_{\underline{h}}}(t)$ that contains at least i independent vectors (almost surely with respect to $\mu_{\underline{h}}$). □

8.4.6 *Goursat and Markov Maps of Order n in the Wide Sense*

The Markov property (of order n, in the wide sense) characterizes a particular feature of the prediction map: the projection of the "future" onto the "past" yields a subspace whose dimension remains constant through time. As that property incarnates in Goursat maps, one is therefore given, with those, a useful representation of Markov maps, of order $n \in \mathbb{N}$, in the wide sense, and of the scope of Goursat processes.

Definition 8.4.58 Let $T \subseteq \mathbb{R}$ be an interval, and H, a real Hilbert space. The map $f : T \longrightarrow H$ is Markov of order $n \in \mathbb{N}$ in the wide sense whenever, for fixed, but arbitrary $\{t_l, t_r\} \subseteq T$, $t_l \le t_r$, the set

$$\left\{P^f_{t_l}[f(t)],\ t \ge t_r\right\}$$

contains exactly n linearly independent elements.

Proposition 8.4.59 *$f : T \longrightarrow H$ is Markov of order n in the wide sense if, and only if, it is a proper, non-singular, Goursat map of rank n. Let that representation be*

$$f(t) = [\underline{a}(t), \underline{h}(t)].$$

When

$$f(t) = [\underline{b}(t), \underline{k}(t)]$$

is another proper Goursat representation of f, there exists an invertible matrix M such that, for $t \in T$, fixed, but arbitrary,

$$\underline{k}(t) = M[\underline{h}(t)],$$

and

$$\underline{b}(t) = \{M^{\star}\}^{-1}[\underline{a}(t)].$$

Proof *Suppose that $f(t) = [\underline{a}(t), \underline{h}(t)]$ is proper and non-singular, of rank n.*

As seen [(Fact) 8.4.46], for $\{\theta, t\} \subseteq T,\ \theta \leq t$, fixed, but arbitrary, one has that

$$P_{\theta}^{f}[f(t)] = [\underline{a}(t), \underline{h}(\theta)].$$

Then, given fixed, but arbitrary $\{\eta_1, \ldots \eta_p\} \subseteq T,\ \underline{\alpha} \in \mathbb{R}^p$,

$$\left\| \sum_{i=1}^{p} \alpha_i [\underline{a}(\eta_i), \underline{h}(\theta)] \right\|_H^2 = \left\langle F_{\underline{h}}(\theta) \sum_{i=1}^{p} \alpha_i \underline{a}(\eta_i), \sum_{i=1}^{p} \alpha_i \underline{a}(\eta_i) \right\rangle_{\mathbb{R}^p}.$$

Thus, since there are at most n linearly independent vectors of type $\underline{a}(\eta)$, then, for $\{t_l, t_r\} \subseteq T,\ t_l \leq t_r \leq t$, fixed, but arbitrary, the family

$$\left\{P_{t_l}^{f}[f(t)],\ t \geq t_r\right\} = \{[\underline{a}(t), \underline{h}(t_l)],\ t \geq t_r\}$$

contains at most n independent elements. But, as f is non-singular, $\underline{h}$ is non-singular, and has thus linearly independent components, and there exists a non-singular matrix $A[t_l \mid t_1, \ldots, t_n]$ [(Definition) 8.4.56]. As

$$A[t_l \mid t_1, \ldots, t_n][\underline{h}(t_l)] = \begin{bmatrix} [\underline{a}(t_1), \underline{h}(t_l)] \\ \vdots \\ [\underline{a}(t_n), \underline{h}(t_l)] \end{bmatrix} = \begin{bmatrix} P_{t_l}^{f}[f(t_1)] \\ \vdots \\ P_{t_l}^{f}[f(t_n)] \end{bmatrix},$$

the right-hand side of the latter equality has linearly independent components. f is thus Markov of order n in the wide sense.

Proof Suppose that f is Markov of order n in the wide sense.

Let $\{t_1, t_2\} \subseteq T,\ t_1 \leq t_2$, be fixed, but arbitrary. Define

$$\mathcal{L}[t_1, t_2] = \left\{P^f_{t_1}[f(t)],\ t \geq t_2\right\},$$

$$L[t_1, t_2] = \overline{V[\mathcal{L}[t_1, t_2]]}.$$

In particular, the set

$$\mathcal{L}[t, t] = \left\{P^f_t[f(\theta)],\ \theta \geq t\right\}$$

contains $f(t)$, and all the projections onto $L_t[f]$ of the values of $f(\theta)$, for θ following t. Thus

$$L[t, t] \subseteq L_t[f]. \qquad (\star)$$

Since, by assumption, $\mathcal{L}[t_1, t_2]$ contains exactly n linearly independent elements, $L[t_1, t_2]$ has dimension n.

Still with $t_1 \leq t_2$ in T, fixed, but arbitrary, $L[t_1, t_1]$ and $L[t_2, t_2]$ are subspaces of H of dimension n, and one may define $B_{t_2,t_1} : L[t_2, t_2] \longrightarrow L[t_1, t_1]$ using the following relation:

$$B_{t_2,t_1}\left[P^f_{t_2}[f(t)]\right] = P^f_{t_1}[f(t)],\ t \geq t_2.$$

Indeed

$$\left\|B_{t_2,t_1}\left[P^f_{t_2}[f(t)]\right]\right\|_H = \left\|P^f_{t_1}[f(t)]\right\|_H = \left\|P^f_{t_1}P^f_{t_2}[f(t)]\right\|_H \leq \left\|P^f_{t_2}[f(t)]\right\|_H,$$

so that B_{t_2,t_1} is well defined on $\mathcal{L}[t_2, t_2]$, linear and bounded. It has thus a linear and bounded extension to $L[t_2, t_2]$, and a bounded adjoint. But, for $\theta \geq t_1$ in T, fixed, but arbitrary,

$$\begin{aligned}\left\langle B_{t_2,t_1}\left[P^f_{t_2}[f(t)]\right], P^f_{t_1}[f(\theta)]\right\rangle_H &= \left\langle P^f_{t_1}[f(t)], P^f_{t_1}[f(\theta)]\right\rangle_H \\ &= \left\langle P^f_{t_2}[f(t)], P^f_{t_1}[f(\theta)]\right\rangle_H,\end{aligned}$$

so that the adjoint is the identity, and consequently B_{t_2,t_1} is onto [250, p. 81]. As

$$\dim L[t_1, t_1] = \dim L[t_2, t_2] = \dim L[t_1, t_2] = n,$$

B_{t_2,t_1} is a bijection [250, p. 81]. Furthermore, for $k \in L[t_2, t_2]$,

$$B_{t_2,t_1}[k] = P^f_{t_1}[k].$$

Let $t_0 \in T$ be fixed, but arbitrary, and let

$$\{k_{0,1}, \ldots, k_{0,n}\}$$

be a basis of $L[t_0, t_0]$. For fixed, but arbitrary $t \in T$, and $i \in [1:n]$, set

$$h_i(t) = P_t^f[k_{0,i}].$$

Since $(\star)$ $L[t_0, t_0] \subseteq L_{t_0}[f]$, for fixed, but arbitrary $t \geq t_0$,

$$h_i(t) = P_t^f[k_{0,i}] = k_{0,i}.$$

$\underline{h}$ shall be the map with components h_i. Those components being, at time t, in $L_t[f]$, one has that $L_t[\underline{h}] \subseteq L_t[f]$.

In T, for $t \geq t_0$, fixed, but arbitrary, $h_i(t) = k_{0,i}$, so that $\underline{h}(t)$ has linearly independent components. When $t < t_0$, $L[t, t]$ is the image by $B_{t_0,t}$ of $L[t_0, t_0]$, and

$$h_i(t) = P_t^f[k_{0,i}] = B_{t_0,t}[k_{0,i}],$$

so that again $\underline{h}(t)$ has linearly independent components. Thus the components of $\underline{h}(t)$ form a basis of $L[t, t]$, which in turn contains $f(t)$, so that $\underline{h}$ is non-singular, and, for some $\{a_1(t), \ldots, a_n(t)\}$,

$$f(t) = [\underline{a}(t), \underline{h}(t)].$$

In particular $f(t) \in L_t[\underline{h}]$, and f is a proper Goursat process, provided $\underline{h}$ is a martingale in the wide sense.

Let $t \in T$ be fixed, but arbitrary. Since f is assumed to be Markov of order n, there exits $\{t_1, \ldots, t_n\} \subseteq T$ such that

$$t_1 < \cdots < t_n,\ t_1 \geq t,\ \left\{P_t^f[f(t_1)], \ldots, P_t^f[f(t_n)]\right\} \text{ linearly independent.}$$

Furthermore, as $P_t^f\left[h_j(t_i)\right] = P_t^f P_{t_i}^f\left[k_{0,j}\right] = P_t^f\left[k_{0,j}\right] = h_j(t)$,

$$P_t^f[f(t_i)] = \sum_{j=1}^n a_j(t_i)\, P_t^f\left[h_j(t_i)\right] = \sum_{j=1}^n a_j(t_i) h_j(t),$$

and thus

$$\sum_{i=1}^n \alpha_i P_t^f[f(t_i)] = \sum_{i=1}^n \alpha_i \left\{\sum_{j=1}^n a_j(t_i) h_j(t)\right\} = \sum_{j=1}^n \left\{\sum_{i=1}^n \alpha_i a_j(t_i)\right\} h_j(t).$$

Since the components of $\underline{h}(t)$ are linearly independent, the independence of

$$\left\{P_t^f\left[f(t_1)\right], \ldots, P_t^f\left[f(t_n)\right]\right\}$$

is equivalent to that of $\{\underline{a}(t_1), \ldots, \underline{a}(t_n)\}$, so that, again, f is a non-singular Goursat process, provided $\underline{h}$ is a martingale in the wide sense.

It remains thus to check that $\underline{h}$ is a martingale in the wide sense. Let $\{t_1, t_2\}$ in T, $t_1 < t_2$, be fixed, but arbitrary. There are three cases to consider (t_0 is the value used for the definition of $\underline{h} : h_i(t_0) = P_{t_0}^f\left[k_{0,i}\right] = k_{0,i}$):

- $t_2 \leq t_0$
  One has, by definition, that

$$P_{t_1}^f\left[\underline{h}(t_2)\right] = P_{t_1}^f\left[P_{t_2}^f\left[\underline{h}(t_0)\right]\right] = P_{t_1}^f\left[\underline{h}(t_0)\right] = \underline{h}(t_1).$$

- $t_1 \leq t_0 < t_2$
  One has, by definition, that

$$P_{t_1}^f\left[\underline{h}(t_2)\right] = P_{t_1}^f\left[\underline{h}(t_0)\right] = \underline{h}(t_1).$$

- $t_0 < t_1$
  One has, by definition, that

$$P_{t_1}^f\left[\underline{h}(t_2)\right] = P_{t_1}^f\left[\underline{h}(t_0)\right] = \underline{h}(t_0) = \underline{h}(t_1).$$

Proof *Suppose that* $f(t) = \left[\underline{b}(t), \underline{k}(t)\right]$ *is also a proper representation.*

Let $\{\theta, t\} \subseteq T$, $\theta \leq t$, be fixed, but arbitrary. Then, since the representation is proper, and $\underline{h}$ is a martingale in the wide sense,

$$P_\theta^f\left[f(t)\right] = \left[\underline{a}(t), P_\theta^f\left[\underline{h}(t)\right]\right] = \left[\underline{a}(t), P_\theta^{\underline{h}}\left[\underline{h}(t)\right]\right] = \left[\underline{a}(t), \underline{h}(\theta)\right].$$

Choosing $\{\theta, t_1, \ldots, t_n\} \subseteq T$, $\theta < t_1 < \cdots < t_n$, fixed, but arbitrary, $\underline{t}$, the vector with components $t_1, \ldots, t_n$, $A\left[\underline{t}\right]^\star$ the matrix whose columns are $\underline{a}(t_1), \ldots, \underline{a}(t_n)$, one gets that

$$\begin{bmatrix} P_\theta^f\left[f(t_1)\right] \\ \vdots \\ P_\theta^f\left[f(t_n)\right] \end{bmatrix} = \begin{bmatrix} \left[\underline{a}(t_1), \underline{h}(\theta)\right] \\ \vdots \\ \left[\underline{a}(t_n), \underline{h}(\theta)\right] \end{bmatrix} = A\left[\underline{t}\right]\left[\underline{h}(\theta)\right].$$

Then, for fixed, but arbitrary $\underline{\alpha} \in \mathbb{R}^n$,

$$\sum_{i=1}^n \alpha_i P_\theta^f\left[f(t_i)\right] = \left[\sum_{i=1}^n \alpha_i \underline{a}(t_i), \underline{h}(\theta)\right].$$

Consequently, since f is Markov of order n in the wide sense, the matrix $A\,[\underline{t}]$ may be assumed to be invertible. A similar expression may be obtained for the representation $f(t) = [\underline{b}\,(t)\,, \underline{k}(t)]$, so that

$$B\,[\underline{t}]\,[\underline{k}(\theta)] = A\,[\underline{t}]\,[\underline{h}(\theta)]\,,\ B\,[\underline{t}]\ \text{non-singular},$$

or

$$\underline{k}(\theta) = \{B\,[\underline{t}]\}^{-1}\,A\,[\underline{t}]\,[\underline{h}(\theta)]\,.$$

Let now $\{\theta_0, \theta_1, \ldots, \theta_n, t_1, \ldots, t_n\} \subseteq T,\ \theta_0 < \theta_1 < \cdots < \theta_n < t_1 < \cdots < t_n$, be fixed, but arbitrary. Then, from the above,

$$\underline{k}(\theta_0) = \{B\,[\underline{\theta}]\}^{-1}\,A\,[\underline{\theta}]\,[\underline{h}(\theta_0)]\,,$$

and

$$\underline{k}(\theta_0) = \{B\,[\underline{t}]\}^{-1}\,A\,[\underline{t}]\,[\underline{h}(\theta_0)]\,,$$

so that

$$\left[\{B\,[\underline{\theta}]\}^{-1}\,A\,[\underline{\theta}] - \{B\,[\underline{t}]\}^{-1}\,A\,[\underline{t}]\right\}\,[\underline{h}(\theta_0)] = \underline{0}_{H^n}.$$

Since $\underline{h}$ is non-singular, the two matrix products in the latter relation are equal, and define a non-singular matrix M, independent of the argument. Then

$$\begin{aligned}[\underline{a}\,(t)\,, \underline{h}(t)] &= f(t)\\ &= [\underline{b}\,(t)\,, \underline{k}(t)]\\ &= [\underline{b}\,(t)\,, M\underline{h}(t)]\\ &= [M^{\star}\underline{b}\,(t)\,, \underline{h}(t)]\,.\end{aligned}$$

Since $\underline{h}$ is non-singular, $\underline{a}\,(t) = M^{\star}\underline{b}\,(t)$. □

Remark 8.4.60 Let f be a Markov map of order n in the wide sense with proper representation

$$f(t) = [\underline{a}\,(t)\,, \underline{h}(t)]\,.$$

Let now

$$f(t) = [\underline{b}\,(t)\,, \underline{k}(t)]$$

be any other representation of f, and

$$f(t) = [\underline{b}(t), \underline{l}(t)]$$

be its proper version, that is,

$$\underline{l}(t) = P_t^f [\underline{k}(t)] .$$

As there is an invertible matrix M such that

$$\underline{l}(t) = M [\underline{h}(t)] ,$$

and that

$$\underline{b}(t) = \{M^{\star}\}^{-1} [\underline{a}(t)] ,$$

it follows that

$$f(t) = \left[\underline{a}(t), M^{-1}\underline{l}(t)\right] = \left[\underline{a}(t), M^{-1}P_t^f [\underline{k}(t)]\right] .$$

Since M is invertible, $f(t) = \left[\underline{a}(t), M^{-1}P_t^f [\underline{k}(t)]\right]$ is a proper representation.

Furthermore, since, for fixed, but arbitrary $\underline{\alpha} \in \mathbb{R}^n$, and $\underline{\beta} = \{M^{\star}\}^{-1} \underline{\alpha}$,

$$\begin{aligned}
\left\langle F_{\left[M^{-1}P_t^f [\underline{k}(t)]\right]} \underline{\alpha}, \underline{\alpha} \right\rangle_{\mathbb{R}^n} &= \left\| \left[\underline{\alpha}, M^{-1}P_t^f [\underline{k}(t)]\right] \right\|_H^2 \\
&= \left\| \left[\underline{\beta}, P_t^f [\underline{k}(t)]\right] \right\|_H^2 \\
&= \sum_{i=1}^{n} \sum_{j=1}^{n} \beta_i \beta_j \left\langle P_t^f [k_i(t)], P_t^f \left[k_j(t)\right] \right\rangle_H \\
&= \left\| P_t^f \left[\sum_{i=1}^{n} \beta_i k_i(t) \right] \right\|_H^2 \\
&\leq \left\| \sum_{i=1}^{n} \beta_i k_i(t) \right\|_H^2 \\
&= \left\| \left[\underline{\beta}, \underline{k}(t)\right] \right\|_H^2 \\
&= \left\| \left[\underline{\alpha}, M^{-1}\underline{k}(t)\right] \right\|_H^2
\end{aligned}$$

$$= \left\langle F_{\left[M^{-1}\underline{k}(t)\right]}\underline{\alpha}, \underline{\alpha}\right\rangle_{\mathbb{R}^n},$$

it follows that

$$F_{\left[M^{-1}P^f_t\underline{k}(t)\right]} \ll F_{\left[M^{-1}\underline{k}(t)\right]}.$$

Thus, since $M^{-1}P^f_t\underline{k}(t) = P^f_tM^{-1}\underline{k}(t)$, the proper representations of f are obtained while choosing, for the solution F of the following equation:

$$F_{\underline{h}}\left[\underline{a}(t)\right] = F(t)\left[\underline{a}(t)\right],$$

one that is minimal for the order given by domination for matrices (denoted $\ll$).

8.4.7 *Covariance Kernels of Markov Maps of Order n in the Wide Sense*

This section's content is motivated by the following remark. Let $f : T \longrightarrow H$ be a map with a covariance kernel of the following form: for fixed, but arbitrary $\{t_1, t_2\} \subseteq T$,

$$C_f(t_1, t_2) = \langle \underline{v}(t_1 \wedge t_2), \underline{u}(t_1 \vee t_2)\rangle_{\mathbb{R}^n}.$$

Let $V[\underline{v} \mid \theta]$ be the subspace of dimension at most n generated by the real valued functions $\{v_1, \ldots, v_n\}$, when their domain is restricted to the interval $[0, \theta]$.

Let $F : T \longrightarrow H$ be given as $F(t) = f(t)$, and $L_F : H \longrightarrow \mathbb{R}^T$ as

$$L_F[h](\tau) = \langle h, F(\tau)\rangle_H.$$

Let, for $\{\theta, t\} \subseteq T$, $\theta \leq t$, fixed, but arbitrary,

$$\phi_{\theta,t}(\cdot) = L_F\left[P^f_\theta\left[f(t)\right]\right] = \left\langle P^f_\theta\left[f(t)\right], f(\cdot)\right\rangle_H \in H(C_f, T).$$

When $\tau \leq \theta$,

$$\begin{aligned}
\mathcal{E}_\tau\left[\phi_{\theta,t}\right] &= \phi_{\theta,t}(\tau) \\
&= \left\langle P^f_\theta\left[f(t)\right], f(\tau)\right\rangle_H \\
&= \langle f(t), f(\tau)\rangle_H \\
&= \sum_{i=1}^n u_i(t)\, v_i(\tau) \\
&= \mathcal{E}_\tau\left[\sum_{i=1}^n u_i(t)\, v_i\right].
\end{aligned}$$

When $\tau > \theta$,

$$P^f_\theta [f(\tau)] = \lim_p \sum \alpha_j^{(p)} [\tau] f\left(\theta_j^{(p)} [\tau]\right), \ \theta_j^{(p)} [\tau] \leq \theta,$$

so that

$$\begin{aligned}
\mathcal{E}_\tau [\phi_{\theta,t}] &= \phi_{\theta,t}(\tau) \\
&= \left\langle P^f_\theta [f(t)], f(\tau)\right\rangle_H \\
&= \left\langle f(t), P^f_\theta [f(\tau)]\right\rangle_H \\
&= \lim_p \sum \alpha_j^{(p)} [\tau] \left\langle f(t), f\left(\theta_j^{(p)} [\tau]\right)\right\rangle_H \\
&= \lim_p \sum \alpha_j^{(p)} [\tau] \sum_{i=1}^n u_i(t)\, v_i \left(\theta_j^{(p)} [\tau]\right) \\
&= \lim_p \mathcal{E}_{\theta_j^{(p)}[\tau]} \left[\sum \alpha_j^{(p)} [\tau] \sum_{i=1}^n u_i(t)\, v_i\right].
\end{aligned}$$

Thus $\phi_{\theta,t} \in V[\underline{v} \mid \theta]$. As a consequence, one has that the subspace generated by the following family:

$$\left\{P^f_\theta [f(t)], \ t \geq \theta\right\}$$

has dimension at most n. Indeed, since on elements of the form $P^f_\theta [f(t)]$, L_F is unitary, that subspace is isometric to a subspace of $V[\underline{v} \mid \theta]$.

Definition 8.4.61 Let $\{t_1, t_2\} \subseteq T$, $t \mapsto \underline{u}(t) \in \mathbb{R}^n$, and $t \mapsto \underline{v}(t) \in \mathbb{R}^n$ be fixed, but arbitrary. Let

$$C(t_1, t_2) = \langle \underline{v}(t_1 \wedge t_2), \underline{u}(t_1 \vee t_2)\rangle_{\mathbb{R}^n}.$$

Then:

1. $\underline{u}$ has property $\boldsymbol{\Pi}_1$
   when, given fixed, but arbitrary $t < t_r$, there exist $\{t_1, \ldots, t_n\} \subseteq T$ such that
   a. $t \leq t_1 < \cdots < t_n$,
   b. $U[t \mid t_1, \ldots, t_n]$, the matrix with entries $u_i(t_j)$, $\{i, j\} \subseteq [1 : n]$, is invertible;
2. $\underline{v}$ has property $\boldsymbol{\Pi}_2$
   when, given fixed, but arbitrary $t > t_l$, there exist $\{t_1, \ldots, t_n\} \subseteq T$ such that
   a. $t_1 < \cdots < t_n \leq t$,
   b. $V[t_1, \ldots, t_n \mid t]$, the matrix with entries $v_i(t_j)$, $\{i, j\} \subseteq [1 : n]$, is invertible;
3. C has property $\boldsymbol{\Pi}$
   when $\underline{u}$ has property $\boldsymbol{\Pi}_1$ and $\underline{v}$ has property $\boldsymbol{\Pi}_2$.

Proposition 8.4.62 *Let the covariance C have the following representation: for $\{t_1, t_2\} \subseteq T$ fixed, but arbitrary,*

$$C(t_1, t_2) = \langle \underline{v}(t_1 \wedge t_2), \underline{u}(t_1 \vee t_2) \rangle_{\mathbb{R}^n}.$$

C is the covariance of a Markov map of order n in the wide sense if, and only if, it has property $\boldsymbol{\Pi}$.

Proof *Suppose that C has property $\boldsymbol{\Pi}$.*

Let

$$s \leq \theta \leq t,$$
$$\{\sigma_1, \ldots, \sigma_n\} \subseteq T,\ \sigma_1 < \cdots < \sigma_n \leq s,$$
$$\{\tau_1, \ldots, \tau_n\} \subseteq T,\ t \leq \tau_1 < \cdots < \tau_n,$$

be fixed, but arbitrary. Let V_σ be the matrix whose columns are

$$\underline{v}(\sigma_1), \ldots, \underline{v}(\sigma_n),$$

and U_τ, that with columns $\underline{u}(\tau_1), \ldots, \underline{u}(\tau_n)$. The matrix with entries

$$C(\sigma_i, \tau_j),\ \{i,j\} \subseteq [1:n]$$

is then the matrix $V_\sigma^\star U_\tau$. Assumption ($\boldsymbol{\Pi}$) allows one to assume that V_σ, U_τ, and thus $V_\sigma^\star U_\tau$ have rank n.

Let X be a zero mean, Gaussian process with covariance C [273, p. 238]. Choose for $f(t)$, X_t, the equivalence class of the variable $X(\cdot, t)$. Let $\underline{X}_\sigma$ be the vector with components $X_{\sigma_1}, \ldots, X_{\sigma_n}$. $\underline{X}_\tau$ is defined similarly. The projection of $\underline{X}_\tau$ onto the subspace generated by $\underline{X}_\sigma$ is given by the following expression [29, p. 92]:

$$E[\underline{X}_\tau \mid \underline{X}_\sigma] = \Sigma_{\tau\sigma}\Sigma_{\sigma\sigma}^{-1}[\underline{X}_\sigma],$$

where $\Sigma_{\tau\sigma} = E[\underline{X}_\tau \underline{X}_\sigma^\star] = V_\sigma^\star U_\tau$, and $\Sigma_{\sigma\sigma} = E[\underline{X}_\sigma \underline{X}_\sigma^\star] = V_\sigma^\star V_\sigma$. Since $\Sigma_{\sigma\sigma}$ has full rank, the elements of $\underline{X}_\sigma$ are linearly independent, and the form of the conditional expectation exhibited above means that $\dim \mathcal{L}[\theta, t] \geq n$, where the latter symbol has the meaning given to it in (Proposition) 8.4.59. But, as noticed in the prelude to (Definition) 8.4.61, given the form of the covariance, $\dim \mathcal{L}[\theta, t] \leq n$. C is thus the covariance of a Markov map of order n in the wide sense.

Proof *Suppose that $C = C_f$, f a Markov map of order n in the wide sense.*

The following set, where $t \in T$ is fixed, but arbitrary:

$$\{P_\theta^f[f(t)],\ t \geq \theta\}$$

has thus dimension n, so that the functions $v_1, \ldots, v_n$, restricted to arguments $\tau \leq \theta$, are linearly independent [prelude to (Definition) 8.4.61], and so $\boldsymbol{\Pi}_2$ obtains.

Let $\{t_1, \ldots, t_n\} \subseteq T,\ \theta \le t_1 < \cdots < t_n$, be such that $P_\theta^f[f(t_1)], \ldots, P_\theta^f[f(t_n)]$ are linearly independent. From $(\tau \le \theta)$

$$\left\langle f(\tau), P_\theta^f[f(t_i)]\right\rangle_H = \langle \underline{v}(\tau), \underline{u}(t_i)\rangle_{\mathbb{R}^n},$$

one gets that

$$\left\langle f(\tau), \sum_{i=1}^n \alpha_i P_\theta^f[f(t_i)]\right\rangle_H = \left\langle \underline{v}(\tau), \sum_{i=1}^n \alpha_i \underline{u}(t_i)\right\rangle_{\mathbb{R}^n}.$$

Thus, when one assumes that $\sum_{i=1}^n \alpha_i \underline{u}(t_i) = \underline{0}_{\mathbb{R}^n}$,

$$\left\langle f(\tau), \sum_{i=1}^n \alpha_i P_\theta^f[f(t_i)]\right\rangle_H = 0.$$

But, since that is true for every $\tau \le \theta$, one must have that

$$\sum_{i=1}^n \alpha_i P_\theta^f[f(t_i)] = 0_H,$$

that is $\underline{\alpha} = \underline{0}_{\mathbb{R}^n}$, which means that $\boldsymbol{\Pi}_1$ obtains. □

Lemma 8.4.63 *Let $\{t_1, t_2\} \subseteq T$, $t \mapsto \underline{u}(t) \in \mathbb{R}^n$, and $t \mapsto \underline{v}(t) \in \mathbb{R}^n$, be fixed, but arbitrary. Let*

$$C(t_1, t_2) = \langle \underline{v}(t_1 \wedge t_2), \underline{u}(t_1 \vee t_2)\rangle_{\mathbb{R}^n}$$

have property $\boldsymbol{\Pi}$.

Suppose that there are $t \mapsto \underline{U}(t) \in \mathbb{R}^n$, and $t \mapsto \underline{V}(t) \in \mathbb{R}^n$, such that, for fixed, but arbitrary $\{t_1, t_2\} \subseteq T,\ t_1 \le t_2$,

$$C(t_1, t_2) = \langle \underline{V}(t_1), \underline{U}(t_2)\rangle_{\mathbb{R}^n}.$$

There exists then a unique, invertible matrix M, independent of $t \in T$, such that

1. *for $t \in T$, fixed, but arbitrary,*
   (i) $\underline{U}(t) = M[\underline{u}(t)]$,
   (ii) $\underline{v}(t) = M^\star[\underline{V}(t)]$;
2. (i) *$\underline{U}$ has property $\boldsymbol{\Pi}_1$,*
   (ii) *$\underline{V}$ has property $\boldsymbol{\Pi}_2$,*
3. *when $\underline{U} = \underline{u}$, $\underline{V} = \underline{v}$.*

Proof Let $t < t_r$ be fixed but arbitrary, and $t_1, \ldots, t_n$ be the points of assumption $(\boldsymbol{\Pi}_1)$. The vectors $\{\underline{u}(t_1), \ldots, \underline{u}(t_n)\}$ are thus linearly independent. Define the following map: for $i \in [1:n]$, fixed, but arbitrary,

$$M[\underline{u}(t_i)] = \underline{U}(t_i).$$

Trivially then, the following relation

$$\sum_{i=1}^{n} \alpha_i \underline{u}(t_i) = \underline{0}_{\mathbb{R}^n}$$

implies the following one (the α_i's are zero because of linear independence of $\underline{u}_i$'s):

$$\sum_{i=1}^{n} \alpha_i M[\underline{u}(t_i)] = \underline{0}_{\mathbb{R}^n},$$

and M is thus uniquely defined as a linear transformation [46, p. 26]. Then, because of the assumption on C, for $\theta \in T$, $\theta \leq t$, and $i \in [1:n]$, fixed, but arbitrary,

$$\begin{aligned}
\langle \underline{v}(\theta), \underline{u}(t_i)\rangle_{\mathbb{R}^n} &= \langle \underline{V}(\theta), \underline{U}(t_i)\rangle_{\mathbb{R}^n} \\
&= \langle \underline{V}(\theta), M[\underline{u}(t_i)]\rangle_{\mathbb{R}^n} \\
&= \langle M^\star[\underline{V}(\theta)], \underline{u}(t_i)\rangle_{\mathbb{R}^n},
\end{aligned}$$

so that $\underline{v}(\theta) = M^\star[\underline{V}(\theta)]$, $\theta \leq t$. Since the range of $M^\star$, by assumption $(\boldsymbol{\Pi}_2)$, contains n independent elements, that matrix is invertible. Consequently, since $\theta \leq t$ is arbitrary, condition $\boldsymbol{\Pi}_2$ obtains for $\underline{V}$. Furthermore, for fixed, but arbitrary $\{\theta, \tau\} \subseteq T$, $\theta \leq t \leq \tau$, using the assumption on C, and the equality $\underline{v}(\theta) = M^\star[\underline{V}(\theta)]$,

$$\begin{aligned}
\langle \underline{V}(\theta), \underline{U}(\tau)\rangle_{\mathbb{R}^n} &= \langle \underline{v}(\theta), \underline{u}(\tau)\rangle_{\mathbb{R}^n} \\
&= \langle M^\star[\underline{V}(\theta)], \underline{u}(\tau)\rangle_{\mathbb{R}^n} \\
&= \langle \underline{V}(\theta), M[\underline{u}(\tau)]\rangle_{\mathbb{R}^n}.
\end{aligned}$$

It follows, since $\boldsymbol{\Pi}_2$ obtains for $\underline{V}$, that, for fixed, but arbitrary $\tau \geq t$,

$$\underline{U}(\tau) = M[\underline{u}(\tau)].$$

Letting $t \downarrow t_l$, while keeping $\{t_1, \ldots, t_n\}$ fixed, shows that $\underline{U}(t) = M[\underline{u}(t)]$. But as M is invertible, and $\underline{u}$ "generates" n independent vectors, condition $\boldsymbol{\Pi}_1$ obtains for $\underline{U}$ also.

Finally, as above, but for $\tau > t$,

$$\begin{aligned}\langle \underline{v}(t), \underline{u}(\tau)\rangle_{\mathbb{R}^n} &= \langle \underline{V}(t), \underline{U}(\tau)\rangle_{\mathbb{R}^n} \\ &= \langle \underline{V}(t), M[\underline{u}(\tau)]\rangle_{\mathbb{R}^n} \\ &= \langle M^\star[\underline{V}(t)], \underline{u}(\tau)\rangle_{\mathbb{R}^n}.\end{aligned}$$

Thus, given $\boldsymbol{\Pi}_1$, $\underline{v}(t) = M^\star[\underline{V}(t)]$, and, since $t \in T$ is arbitrary, for $t \in T$,

$$\underline{v}(t) = M^\star[\underline{V}(t)].$$

□

Proposition 8.4.64 *For fixed, but arbitrary* $\{t_1, t_2\} \subseteq T$, *let*

$$C(t_1, t_2) = \langle \underline{v}(t_1 \wedge t_2), \underline{u}(t_1 \vee t_2)\rangle_{\mathbb{R}^n},$$

and suppose that C *has property* $\boldsymbol{\Pi}$. *Then* C *is positive definite if, and only if, there is a map* $t \mapsto \Gamma(t)$ *such that*

1. *for* $t \in T$, *fixed, but arbitrary,* $\Gamma(t)$ *is a positive definite matrix,*
2. *for* $\{t_1, t_2\} \subseteq T$, $t_1 < t_2$, *fixed, but arbitrary,* $\Gamma(t_1) \ll \Gamma(t_2)$,
3. *for* $t \in T$, *fixed, but arbitrary,* $\underline{v}(t) = \Gamma(t)[\underline{u}(t)]$.

Proof *Suppose that* C *is positive definite.*

Because of (Proposition) 8.4.62, C is then the covariance of a Markov map of order n in the wide sense, say f. Because of (Proposition) 8.4.59, one has that f is a proper, non-singular, Goursat map of rank n:

$$f(t) = [\underline{a}(t), \underline{h}(t)],$$

and thus, when $t_1 \leq t_2$,

$$C(t_1, t_2) = \left\langle F_{\underline{h}}(t_1)\,\underline{a}(t_1), \underline{a}(t_2)\right\rangle_{\mathbb{R}^n}.$$

Because of (Lemma) 8.4.63, there is an invertible matrix M such that, for $t \in T$, fixed, but arbitrary, $\underline{a}(t) = M[\underline{u}(t)]$, and $M^\star\left[F_{\underline{h}}(t)\,\underline{a}(t)\right] = \underline{v}(t)$. Thus

$$\underline{v}(t) = M^\star F_{\underline{h}}(t)\, M[\underline{u}(t)],$$

and $t \mapsto M^\star F_{\underline{h}}(t)\, M$ has the properties listed as items 1, 2, and 3 of the proposition's statement.

Proof *Suppose that items 1, 2, and 3 of the proposition's statement obtain.*

The map $(t_1, t_2) \mapsto \Gamma(t_1 \wedge t_2)$ is positive definite. Indeed, let, to be specific, $t_1 \leq t_2 \leq t_3$, and $\underline{\alpha}_1, \underline{\alpha}_2, \underline{\alpha}_3$ be fixed, but arbitrary elements in T and $\mathbb{R}^n$, respectively.

One must prove [106, p. 354] that

$$0 \leq \left\langle \begin{bmatrix} \Gamma(t_1) & \Gamma(t_1) & \Gamma(t_1) \\ \Gamma(t_1) & \Gamma(t_2) & \Gamma(t_2) \\ \Gamma(t_1) & \Gamma(t_2) & \Gamma(t_3) \end{bmatrix} \begin{bmatrix} \underline{\alpha}_1 \\ \underline{\alpha}_2 \\ \underline{\alpha}_3 \end{bmatrix}, \begin{bmatrix} \underline{\alpha}_1 \\ \underline{\alpha}_2 \\ \underline{\alpha}_3 \end{bmatrix} \right\rangle_{\mathbb{R}^{3n}}$$

$$= \langle \Gamma(t_1)[\underline{\alpha}_1], \underline{\alpha}_1 \rangle_{\mathbb{R}^n} + \langle \Gamma(t_2)[\underline{\alpha}_2], \underline{\alpha}_2 \rangle_{\mathbb{R}^n} + \langle \Gamma(t_3)[\underline{\alpha}_3], \underline{\alpha}_3 \rangle_{\mathbb{R}^n}$$

$$+ 2 \langle \Gamma(t_1)[\underline{\alpha}_1], \underline{\alpha}_2 \rangle_{\mathbb{R}^n} + 2 \langle \Gamma(t_1)[\underline{\alpha}_1], \underline{\alpha}_3 \rangle_{\mathbb{R}^n} + 2 \langle \Gamma(t_2)[\underline{\alpha}_2], \underline{\alpha}_3 \rangle_{\mathbb{R}^n}.$$

The latter sum of inner products may be written in the following form:

$$\begin{aligned} &\langle \Gamma(t_1)[\underline{\alpha}_1 + \underline{\alpha}_2 + \underline{\alpha}_3], \underline{\alpha}_1 + \underline{\alpha}_2 + \underline{\alpha}_3 \rangle_{\mathbb{R}^n} \\ &+ \langle \{\Gamma(t_2) - \Gamma(t_1)\}[\underline{\alpha}_2 + \underline{\alpha}_3], \underline{\alpha}_2 + \underline{\alpha}_3 \rangle_{\mathbb{R}^n} \\ &+ \langle \{\Gamma(t_3) - \Gamma(t_2)\}[\underline{\alpha}_3], \underline{\alpha}_3 \rangle_{\mathbb{R}^n}, \end{aligned}$$

which is positive because of items 1 and 2. There is thus a Gaussian vector map $t \mapsto \underline{X}_t$ with $\Gamma(\cdot \wedge \cdot)$ as covariance [106, p. 358]. Since, for $t_1 \leq t_2 \leq t_3$, and $\underline{e}_i$ and $\underline{e}_j$ fixed, but arbitrary elements in, respectively, T and the standard basis of $\mathbb{R}^n$,

$$E\left[\langle \underline{X}_{t_1}, \underline{e}_i \rangle_{\mathbb{R}^n} \left\langle \underline{X}_{t_3} - \underline{X}_{t_2}, \underline{e}_j \right\rangle_{\mathbb{R}^n}\right] = \left\langle \Gamma(t_1)[\underline{e}_i], \underline{e}_j \right\rangle_{\mathbb{R}^n} - \left\langle \Gamma(t_1)[\underline{e}_i], \underline{e}_j \right\rangle_{\mathbb{R}^n} = 0,$$

the map $t \mapsto \underline{X}_t$ is a martingale in the wide sense.

Let thus $\underline{h}$ be a martingale in the wide sense whose covariance is obtained from Γ, and set:

$$f(t) = [\underline{u}(t), \underline{h}(t)].$$

As, *in fine*, for $t_1 \leq t_2$, using item 3,

$$\begin{aligned} C_f(t_1, t_2) &= \langle [\underline{u}(t_1), \underline{h}(t_1)], [\underline{u}(t_2), \underline{h}(t_2)] \rangle_H \\ &= \sum_{i=1}^{n} \sum_{j=1}^{n} u_i(t_1) u_j(t_2) \langle h_i(t_1), h_j(t_2) \rangle_H \\ &= \sum_{i=1}^{n} \sum_{j=1}^{n} u_i(t_1) u_j(t_2) \Gamma_{i,j}(t_1) \\ &= \sum_{i=1}^{n} \sum_{j=1}^{n} u_i(t_1 \wedge t_2) u_j(t_1 \vee t_2) \Gamma(t_1 \wedge t_2) \end{aligned}$$

$$= \langle \Gamma(t_1 \wedge t_2)[\underline{u}(t_1 \wedge t_2)], \underline{u}(t_1 \vee t_2)\rangle_{\mathbb{R}^n}$$
$$= \langle \underline{v}(t_1 \wedge t_2)], \underline{u}(t_1 \vee t_2)\rangle_{\mathbb{R}^n},$$

and that, given the symmetry of $(t_1, t_2) \mapsto \Gamma(t_1 \wedge t_2)$, the same is true for $t_1 \geq t_2$, the covariance of f is C, which is thus positive definite. □

8.4.8 Multiplicity One for Goursat Maps

When the martingale associated with a Goursat map has multiplicity one, it becomes easier to check that the process itself is proper and has thus multiplicity one. This section thus explains first how multiplicity one follows from smoothness properties of the covariance of a Goursat map whose associated martingale has multiplicity one. Since establishing that the martingale associated with a Goursat map has multiplicity one may in turn prove difficult, it is shown, in a second step, that the determining factor is the behavior, with respect to Lebesgue measure, of the trace of the matrix measure associated with the structure matrix of the martingale. Multiplicity one occurs when there is equivalence, that is, when the involved martingale behaves much like a Wiener process. The procedure is made operationally effective transforming the problem into one of solving differential equations.

Proposition 8.4.65 *Let $f(t) = [\underline{a}(t), \underline{h}(t)]$ be a CH-Goursat process such that, almost surely for $t \in T$, with respect to $\mu_{\underline{h}}$,*

1. $D_{\mu_{\underline{h}}}(t) = \underline{d}(t) \otimes \underline{d}(t)$,
2. $\|\underline{d}(t)\|_{\mathbb{R}^n} = 1$.

Let ϕ belong to $\mathcal{L}_2(T, \mathcal{T}, \mu_{\underline{h}})$. Then f is proper if, and only if, for each fixed, but arbitrary $t \in T$, the following set of equalities:

$$\int_{T_\theta} \langle \underline{a}(\theta), \underline{d}(u)\rangle_{\mathbb{R}^n} \phi(u)\, \mu_{\underline{h}}(du) = 0,\ \theta \leq t,$$

means that ϕ is almost surely zero with respect to $\mu_{\underline{h}}$.

Proof Let $\underline{\alpha}(\theta) = \phi(\theta)\underline{d}(\theta) + \underline{d}(\theta)^{\perp}$. It is a zero function of

$$\mathcal{L}_2(T, \mathcal{T}, \boldsymbol{M}_{\underline{h}})$$

when ϕ is almost surely zero with respect to $\mu_{\underline{h}}$ as, in the context of the present proposition,

$$\int_{T_t} \langle D_{\mu_{\underline{h}}}(\theta)\,[\underline{\alpha}(\theta)], \underline{\alpha}(\theta)\rangle_{\mathbb{R}^n} \mu_{\underline{h}}(d\theta) = \int_{T_t} \phi^2(\theta)\,\mu_{\underline{h}}(d\theta).$$

But the condition of statement (Proposition) 8.4.65 is the translation of that of (Proposition) 8.4.48. □

Example 8.4.66 Let W be a standard, wide sense, Wiener process over $[0, 1]$, f and g, functions whose square is integrable over $[0, 1]$, with respect to Lebesgue measure. Let

$$h_1(t) = \int_0^t f(\theta) W(d\theta), \quad h_2(t) = \int_0^t g(\theta) W(d\theta).$$

Then:

$$F_{1,1}^{\underline{h}}(t) = \|h_1(t)\|_H^2 = \int_0^t f^2(\theta) d\theta,$$

$$F_{1,2}^{\underline{h}}(t) = \langle h_1(t), h_2(t)\rangle_H = \int_0^t f(\theta) g(\theta) d\theta,$$

$$F_{2,2}^{\underline{h}}(t) = \|h_2(t)\|_H^2 = \int_0^t g^2(\theta) d\theta,$$

so that

$$dM_{1,1}^{\underline{h}} = f^2 dLeb, \; dM_{1,2}^{\underline{h}} = fg dLeb, \; dM_{2,2}^{\underline{h}} = g^2 dLeb,$$

and, consequently that

$$d\mu_{\underline{h}} = dM_{1,1}^{\underline{h}} + dM_{2,2}^{\underline{h}} = \left\{f^2 + g^2\right\} dLeb.$$

Thus, as, for example, provided $f^2 + g^2 > 0$, almost surely, with respect to $d\mu_{\underline{h}}$,

$$dM_{1,1}^{\underline{h}} = f^2 dLeb = \frac{f^2}{f^2 + g^2} d\mu_{\underline{h}},$$

one has that

$$D_{\mu_{\underline{h}}} = \begin{bmatrix} \frac{f^2}{f^2+g^2} & \frac{fg}{f^2+g^2} \\ \\ \frac{fg}{f^2+g^2} & \frac{g^2}{f^2+g^2} \end{bmatrix}.$$

$D_{\mu_{\underline{h}}}$ is thus a matrix of the following form:

$$\begin{bmatrix} a & b \\ b & c \end{bmatrix}, \quad \text{with } a + c = 1, \text{ and } ac = b^2.$$

Such a matrix has eigenvalues 0 and 1. An eigenvector associated

$$\text{with } 0 \text{ is } \begin{bmatrix} 1 \\ -a/b \end{bmatrix}, \quad \text{with } 1 \text{ is } \begin{bmatrix} b/c \\ 1 \end{bmatrix}.$$

Letting $\underline{d}$ be the vector

$$c^{1/2} \begin{bmatrix} b/c \\ 1 \end{bmatrix},$$

one has that

$$\begin{bmatrix} a & b \\ b & c \end{bmatrix} = \underline{d} \otimes \underline{d},$$

that is,

$$D_{\mu_{\underline{h}}}(t) = \underline{d}(t) \otimes \underline{d}(t), \quad \text{with } \underline{d}(t) = \left\{ \frac{g^2(t)}{f^2(t) + g^2(t)} \right\}^{1/2} \begin{bmatrix} \frac{f(t)}{g(t)} \\ 1 \end{bmatrix}.$$

One must thus assume that g, with respect to $\mu_{\underline{h}}$, is endowed with adequate properties. The condition of (Proposition) 8.4.65 rewrites as ($s(g)$ is the function which delivers the sign of g)

$$0 = a_1(\theta) \int_0^\theta \phi(u) s(g(u)) f(u) \left\{f^2(u) + g^2(u)\right\}^{1/2} Leb(du)$$
$$+ a_2(\theta) \int_0^\theta \phi(u) \, |g(u)| \left\{f^2(u) + g^2(u)\right\}^{1/2} Leb(du).$$

Remark 8.4.67 In (Proposition) 8.4.65, when f is proper, it has multiplicity one. The latter example shows that, to be able to assert multiplicity one, one must have better tools at disposal. Some are explained in the material which follows.

Remark 8.4.68 Suppose that $D_{\mu_{\underline{h}}} = \underline{d} \otimes \underline{d}$, with $\underline{d}(\theta)$ of norm one, almost surely with respect to $\mu_{\underline{h}}$, and that $\underline{a}$ is adapted. Then, provided the integral exists, which will be the case, for instance, when $\underline{a}$ is continuous, over an appropriate domain,

$$\|\underline{a}\|^2_{L_2(T,\mathcal{T},\boldsymbol{M}_{\underline{h}})} = \int_T \langle \underline{d}(\theta), \underline{a}(\theta) \rangle^2_{\mathbb{R}^n} \, \mu_{\underline{h}}(d\theta).$$

The integrand in that latter integral is the norm of the projection of $\underline{a}(\theta)$ onto the subspace generated by $\underline{d}(\theta)$. When it is required below that the integrand be almost surely strictly positive with respect to $\mu_{\underline{h}}$, the restriction is mild as one could restrict attention to its support, though all statements would then be quite messier.

Remark 8.4.69 The function f is absolutely continuous with respect to the measure μ when one may write for ϕ integrable with respect to μ:

$$f(t) = \kappa + \int_{T_t} \phi(\theta)\mu(d\theta).$$

Then, for $t_1 < t_2$, fixed, but arbitrary in T,

$$f(t_2) - f(t_1) = \int_{[t_1, t_2[} \phi(\theta)\mu(d\theta).$$

One may thus define a measure μ_f setting $\mu_f([t_1, t_2[) = f(t_1) - f(t_1)$, and then $\mu_f \ll \mu$. Furthermore $\phi(\theta)$ may be obtained as the limit of ratios of the following type:

$$\frac{\mu_f(B(\theta, \epsilon))}{\mu(B(\theta, \epsilon))},$$

with shrinking sets $B(\theta, \epsilon)$, which are typically intervals [32, p. 378]. That property is used in many calculations which follow.

Lemma 8.4.70 *Let* $\mu_{\underline{h}}$ *be as in* (Proposition) 8.4.24, *and have distribution function* $F_{\mu_{\underline{h}}}$. *Define successively*

$$a(t) = \kappa_a + \int_{T_t} \alpha(\theta)\, \mu_{\underline{h}}(d\theta),$$

$$b(t) = \kappa_b + \int_{T_t} \beta(\theta)\, \mu_{\underline{h}}(d\theta),$$

$$V(t) = a(t)\, b(t).$$

V is then a function of bounded variation. Let μ_V be the measure that it determines. Then, almost surely with respect to $\mu_{\underline{h}}$,

$$\frac{d\mu_V}{d\mu_{\underline{h}}}(t) = \lim_{\epsilon \downarrow\downarrow 0} \frac{V(t+\epsilon) - V(t-\epsilon)}{F^+_{\mu_{\underline{h}}}(t+\epsilon) - F_{\mu_{\underline{h}}}(t-\epsilon)} = a(t)\,\beta(t) + \alpha(t)\, b(t).$$

Proof V is of bounded variation as a product of functions of bounded variation. As, for fixed, but arbitrary $t \in T$, and $\epsilon > 0$,

$$V(t+\epsilon) - V(t-\epsilon) =$$
$$= b(t+\epsilon)\{a(t+\epsilon) - a(t-\epsilon)\} + a(t-\epsilon)\{b(t+\epsilon) - b(t-\epsilon)\},$$

and that, almost surely with respect to $\mu_{\underline{h}}$,

$$\lim_{\epsilon \downarrow\downarrow 0} \frac{a(t+\epsilon) - a(t-\epsilon)}{F^{+}_{\mu_{\underline{h}}}(t+\epsilon) - F_{\mu_{\underline{h}}}(t-\epsilon)} = \alpha(t),$$

and

$$\lim_{\epsilon \downarrow\downarrow 0} \frac{b(t+\epsilon) - b(t-\epsilon)}{F^{+}_{\mu_{\underline{h}}}(t+\epsilon) - F_{\mu_{\underline{h}}}(t-\epsilon)} = \beta(t),$$

it follows that

$$\lim_{\epsilon \downarrow\downarrow 0} \frac{V(t+\epsilon) - V(t-\epsilon)}{F^{+}_{\mu_{\underline{h}}}(t+\epsilon) - F_{\mu_{\underline{h}}}(t-\epsilon)} = a(t)\,\beta(t) + \alpha(t)\,b(t).$$

□

Proposition 8.4.71 *Let* $f(t) = [\underline{a}(t), \underline{h}(t)]$ *be a CH-Goursat map such that*

(A) *almost surely with respect to* $\mu_{\underline{h}}$,

  (a) $D_{\mu_{\underline{h}}}(t) = \underline{d}(t) \otimes \underline{d}(t)$,
  (b) $\|\underline{d}(t)\|_{\mathbb{R}^n} = 1$;

(B) $\underline{a}(t) = \underline{\kappa}_a + \int_{T_t} \underline{\alpha}(\theta)\,\mu_{\underline{h}}(d\theta)$, *with, for* $i \in [1:n]$, *fixed, but arbitrary,*

$$\int_T |\alpha_i(\theta)|\,\mu_{\underline{h}}(d\theta) < \infty;$$

(C) *almost surely, with respect to* $\mu_{\underline{h}}$, $\langle \underline{d}(\theta), \underline{a}(\theta)\rangle^2_{\mathbb{R}^n} > 0$, *and, for* $t \in T$, *fixed, but arbitrary,*

$$\int_{T_t} \frac{\|\underline{\alpha}(\theta)\|^2_{\mathbb{R}^n}}{\langle \underline{d}(\theta), \underline{a}(\theta)\rangle^2_{\mathbb{R}^n}} d\mu_{\underline{h}}(d\theta) < \infty.$$

Then f is proper.

Proof One shall need the fact that $\underline{\alpha}$ has a norm whose square is integrable. That is seen as follows. Let

$$\kappa^2(t) = 2\left\{ \|\underline{\kappa}_a\|^2_{\mathbb{R}^n} + \sum_{i=1}^{n} \left\{ \int_{T_t} |\alpha_i(\eta)|\,\mu_{\underline{h}}(d\eta) \right\}^2 \right\}.$$

One has then that, with respect to $\mu_{\underline{h}}$, for almost every $\theta \in T,\ \theta \le t$,

$$\langle \underline{d}(\theta), \underline{a}(\theta)\rangle^2_{\mathbb{R}^n} \le \|\underline{a}(\theta)\|^2_{\mathbb{R}^n} \le \kappa^2(t),$$

which is a finite quantity independent of θ. Then, because of Assumption (C),

$$\int_{T_t} \frac{\|\underline{\alpha}(\theta)\|^2_{\mathbb{R}^n}}{\kappa^2(t)} \mu_{\underline{h}}(d\theta) \leq \int_{T_t} \frac{\|\underline{\alpha}(\theta)\|^2_{\mathbb{R}^n}}{\langle \underline{d}(\theta), \underline{a}(\theta)\rangle^2_{\mathbb{R}^n}} d\mu_{\underline{h}}(d\theta) < \infty,$$

and it follows that

$$\int_{T_t} \|\underline{\alpha}(\theta)\|^2_{\mathbb{R}^n} \mu_{\underline{h}}(d\theta) < \infty.$$

Because of (Proposition) 8.4.65, to check that f is proper, it suffices to establish that functions $\phi \in L_2(T, \mathcal{T}, \mu_{\underline{h}})$ which are the solutions of the following set of equations:

$$\int_{T_\theta} \langle \underline{a}(\theta), \underline{d}(u)\rangle_{\mathbb{R}^n} \phi(u)\, \mu_{\underline{h}}(du) = 0,\ \theta \leq t,$$

can only be zero. But, letting

$$\begin{aligned}\Delta(\epsilon) &= \int_{T_{\theta+\epsilon}} \langle \underline{a}(\theta+\epsilon), \underline{d}(u)\rangle_{\mathbb{R}^n} \phi(u)\, \mu_{\underline{h}}(du) \\ &\quad - \int_{T_{\theta-\epsilon}} \langle \underline{a}(\theta-\epsilon), \underline{d}(u)\rangle_{\mathbb{R}^n} \phi(u)\, \mu_{\underline{h}}(du),\end{aligned}$$

$$\begin{aligned}\Delta_1(\epsilon) &= \int_{T_{\theta+\epsilon}} \langle \underline{a}(\theta), \underline{d}(u)\rangle_{\mathbb{R}^n} \phi(u)\, \mu_{\underline{h}}(du) \\ &\quad - \int_{T_{\theta-\epsilon}} \langle \underline{a}(\theta), \underline{d}(u)\rangle_{\mathbb{R}^n} \phi(u)\, \mu_{\underline{h}}(du),\end{aligned}$$

$$\begin{aligned}\Delta_2(\epsilon) &= \int_{T_{\theta+\epsilon}} \langle \underline{a}(\theta+\epsilon) - \underline{a}(\theta), \underline{d}(u)\rangle_{\mathbb{R}^n} \phi(u)\, \mu_{\underline{h}}(du) \\ &\quad - \int_{T_{\theta-\epsilon}} \langle \underline{a}(\theta+\epsilon) - \underline{a}(\theta), \underline{d}(u)\rangle_{\mathbb{R}^n} \phi(u)\, \mu_{\underline{h}}(du)\end{aligned}$$

$$\Delta_3(\epsilon) = \int_{T_{\theta-\epsilon}} \langle \underline{a}(\theta+\epsilon) - \underline{a}(\theta-\epsilon), \underline{d}(u)\rangle_{\mathbb{R}^n} \phi(u)\, \mu_{\underline{h}}(du),$$

then

$$\Delta(\epsilon) = \Delta_1(\epsilon) + \Delta_2(\epsilon) + \Delta_3(\epsilon),$$

and

$$\Delta_2(\epsilon) = \sum_{i=1}^{n} \{a_i(\theta+\epsilon) - a_i(\theta)\} \times$$
$$\times \left\{ \int_{T_{\theta+\epsilon}} d_i(u)\,\phi(u)\,\mu_{\underline{h}}(du) - \int_{T_{\theta-\epsilon}} d_i(u)\,\phi(u)\,\mu_{\underline{h}}(du) \right\}.$$

One has thus that

$$\frac{\Delta(\epsilon)}{F^{+}_{\mu_{\underline{h}}}(\theta+\epsilon) - F_{\mu_{\underline{h}}}(\theta-\epsilon)}$$

is the sum of three terms such that, when $\epsilon \downarrow\downarrow 0$, then, almost surely with respect to $\mu_{\underline{h}}$,

- the first, whose numerator is $\Delta_1(\epsilon)$, has limit $\phi(\theta)\,\langle \underline{a}(\theta), \underline{d}(\theta)\rangle_{\mathbb{R}^n}$,
- the second, whose numerator is $\Delta_2(\epsilon)$, has, in the form exhibited above, a limit equal to zero (since $\underline{a}$ is continuous and the "remainder" has a limit),
- the third, whose numerator is $\Delta_3(\epsilon)$, has limit

$$\int_{T_\theta} \langle \underline{\alpha}(\theta), \underline{d}(u)\rangle_{\mathbb{R}^n}\,\phi(u)\,\mu_{\underline{h}}(du)$$

(as the square of the norm of $\underline{\alpha}$ is integrable).

The end result is the following equation, valid, with respect to $\mu_{\underline{h}}$, almost surely for $\theta \in T,\ \theta \le t$:

$$0 = \langle \underline{a}(\theta), \underline{d}(\theta)\rangle_{\mathbb{R}^n}\,\phi(\theta) + \int_{T_\theta} \langle \underline{\alpha}(\theta), \underline{d}(u)\rangle_{\mathbb{R}^n}\,\phi(u)\,\mu_{\underline{h}}(du),$$

which may be rewritten as:

$$-\int_{T_\theta} \frac{\langle \underline{\alpha}(\theta), \underline{d}(u)\rangle_{\mathbb{R}^n}}{\langle \underline{a}(\theta), \underline{d}(\theta)\rangle_{\mathbb{R}^n}}\,\phi(u)\,\mu_{\underline{h}}(du) = \phi(\theta).$$

But, as $\underline{d}$ has an Euclidean norm almost surely equal to one, with respect to $\mu_{\underline{h}}$,

$$\int_{T_t} \mu_{\underline{h}}(d\theta) \int_{T_t} \mu_{\underline{h}}(du) \left\{ \frac{\langle \underline{\alpha}(\theta), \underline{d}(u)\rangle_{\mathbb{R}^n}}{\langle \underline{a}(\theta), \underline{d}(\theta)\rangle_{\mathbb{R}^n}} \right\}^2 \le$$
$$\le \mu_{\underline{h}}(T_t) \int_{T_t} \mu_{\underline{h}}(d\theta)\,\frac{\|\underline{\alpha}(\theta)\|^2_{\mathbb{R}^n}}{\langle \underline{a}(\theta), \underline{d}(\theta)\rangle^2_{\mathbb{R}^n}}.$$

Since the right-hand side of the latter expression is finite by assumption, one has that ϕ is an eigenvector of a Hilbert-Schmidt Volterra equation. As such it must be zero [119, p. 70]. □

Remark 8.4.72 The integrability condition of (Proposition) 8.4.71 (item (C)) has a drawback: one must be able to compute $\underline{d}$, a task that is not necessarily easy as seen, for instance, in (Example) 8.4.10. Under further smoothness conditions, one is able to express that condition in terms of C_f, which is supposedly known. That is the aim of the calculations which follow. The end result is (Proposition) 8.4.79.

Lemma 8.4.73 *Let $\mu_{\underline{h}}$ be as in* (Proposition) 8.4.24, *and have distribution function $F_{\mu_{\underline{h}}}$. Suppose that, for $i \in [1:3]$,*

$$a_i(t) = \kappa_i + \int_{T_t} \alpha_i(\theta)\, \mu_{\underline{h}}(d\theta),$$

and that

$$a(t) = a_1(t)\, a_2(t)\, a_3(t).$$

a is then a function of bounded variation. Let μ_a be its associated measure. Then, almost surely with respect to $\mu_{\underline{h}}$,

$$\frac{d\mu_a}{d\mu_{\underline{h}}}(t) = \lim_{\epsilon \Downarrow 0} \frac{a(t+\epsilon) - a(t-\epsilon)}{F^{+}_{\mu_{\underline{h}}}(t+\epsilon) - F_{\mu_{\underline{h}}}(t-\epsilon)}$$

$$= \alpha_1(t)\, a_2(t)\, a_3(t) + a_1(t)\, \alpha_2(t)\, a_3(t) + a_1(t)\, a_2(t)\, \alpha_3(t).$$

Proof For $u < v$, fixed, but arbitrary,

$$\begin{aligned} a(v) - a(u) &= a_2(v)\, a_3(v) \{a_1(v) - a_1(u)\} \\ &\quad + a_1(u)\, a_3(v) \{a_2(v) - a_2(u)\} \\ &\quad + a_1(u)\, a_2(u) \{a_3(v) - a_3(u)\}. \end{aligned}$$

One finishes as in (Proposition) 8.4.70. □

Lemma 8.4.74 *Let $f(t) = [\underline{a}(t), \underline{h}(t)]$ be a CH-Goursat map with covariance*

$$C_f(t_1, t_2) = \left\langle F_{\underline{h}}(t_1 \wedge t_2)\, \underline{a}(t_1), \underline{a}(t_2) \right\rangle_{\mathbb{R}^n},$$

and suppose that

$$\underline{a}(t) = \underline{\kappa}_a + \int_{T_t} \underline{\alpha}(\theta)\, \mu_{\underline{h}}(d\theta).$$

Then

$$\frac{dC_f}{d\mu_{\underline{h}}}(t) = \lim_{\epsilon\downarrow\downarrow 0} \frac{C_f(t+\epsilon, t+\epsilon) - C_f(t-\epsilon, t-\epsilon)}{F^+_{\mu_{\underline{h}}}(t+\epsilon) - F_{\mu_{\underline{h}}}(t-\epsilon)}$$

$$= \left\langle D_{\mu_{\underline{h}}}(t)\,\underline{a}(t), \underline{a}(t)\right\rangle_{\mathbb{R}^n} + 2\left\langle F_{\underline{h}}(t)\,\underline{a}(t), \underline{\alpha}(t)\right\rangle_{\mathbb{R}^n}.$$

Proof Choose, in (Lemma) 8.4.73, $a_1 = F^{\underline{h}}_{i,j}$, $a_2 = a_i$, $a_3 = a_j$. Then

$$\lim_{\epsilon\downarrow\downarrow 0} \frac{a(t+\epsilon) - a(t-\epsilon)}{F^+_{\mu_{\underline{h}}}(t+\epsilon) - F_{\mu_{\underline{h}}}(t-\epsilon)} =$$

$$= \frac{dM^{\underline{h}}_{i,j}}{d\mu_{\underline{h}}}(t)\,a_i(t)\,a_j(t) + F^{\underline{h}}_{i,j}(t)\,\alpha_i(t)\,a_j(t) + F^{\underline{h}}_{i,j}(t)\,a_i(t)\,\alpha_j(t).$$

□

The following lemma entails prescriptions that reappear identically in some subsequent lemmas. They are thus headlined under an "assumptions" banner.

Lemma 8.4.75 *Let the following prescriptions prevail:*

`Assumptions:`

$f(t) = [\underline{a}(t), \underline{h}(t)]$ be a CH-Goursat map with covariance

$$C_f(t_1, t_2) = \left\langle F_{\underline{h}}(t_1 \wedge t_2)\,\underline{a}(t_1), \underline{a}(t_2)\right\rangle_{\mathbb{R}^n},$$

and suppose that

$$\underline{a}(t) = \underline{\kappa}_a + \int_{T_t} \underline{\alpha}(\theta)\,\mu_{\underline{h}}(d\theta).$$

Let

$$\underline{A}(t) = F_{\underline{h}}(t)\,\underline{a}(t).$$

The components of $\underline{A}(t)$ *have then bounded variation. Let* μ_{A_i} *denote the measure corresponding to* A_i, *and* $\underline{\mu}_A$, *the vector with components* μ_{A_i}. *Then, componentwise, almost surely with respect to* $\mu_{\underline{h}}$,

$$\frac{d\underline{\mu}_A}{d\mu_{\underline{h}}}(t) = \lim_{\epsilon\downarrow\downarrow 0} \frac{\underline{A}(t+\epsilon) - \underline{A}(t-\epsilon)}{F^+_{\mu_{\underline{h}}}(t+\epsilon) - F_{\mu_{\underline{h}}}(t-\epsilon)} = D_{\mu_{\underline{h}}}(t)\,\underline{a}(t) + F_{\underline{h}}(t)\,\underline{\alpha}(t).$$

Furthermore

$$\frac{dC_f}{d\mu_{\underline{h}}}(t) = \left\langle \frac{d\mu_{\underline{A}}}{d\mu_{\underline{h}}}(t), \underline{a}(t) \right\rangle_{\mathbb{R}^n} + \langle \underline{A}(t), \underline{\alpha}(t) \rangle_{\mathbb{R}^n}.$$

Proof By definition, for $i \in [1:n]$, fixed, but arbitrary,

$$\begin{aligned} A_i(t+\epsilon) - A_i(t-\epsilon) &= \sum_{j=1}^{n} \left\{ F^{\underline{h}}_{i,j}(t+\epsilon) a_j(t+\epsilon) - F^{\underline{h}}_{i,j}(t-\epsilon) a_j(t-\epsilon) \right\} \\ &= \sum_{j=1}^{n} \left\{ F^{\underline{h}}_{i,j}(t+\epsilon) - F^{\underline{h}}_{i,j}(t-\epsilon) \right\} a_j(t+\epsilon) \\ &+ \sum_{j=1}^{n} F^{\underline{h}}_{i,j}(t-\epsilon) \left\{ a_j(t+\epsilon) - a_j(t-\epsilon) \right\}. \end{aligned}$$

The first "derivation" formula of the statement follows. For the second, one has that

$$C_f(t,t) = \langle \underline{A}(t), \underline{a}(t) \rangle_{\mathbb{R}^n},$$

and also that

$$\frac{d \langle \underline{A}(\cdot), \underline{a}(\cdot) \rangle_{\mathbb{R}^n}}{d\mu_{\underline{h}}}(t) = \left\langle \frac{d\mu_{\underline{A}}}{d\mu_{\underline{h}}}(t), \underline{a}(t) \right\rangle_{\mathbb{R}^n} + \langle \underline{A}(t), \underline{\alpha}(t) \rangle_{\mathbb{R}^n}.$$

Replacing $\frac{d\mu_{\underline{A}}}{d\mu_{\underline{h}}}$ with the value just obtained yields that

$$\begin{aligned} &\left\langle \frac{d\mu_{\underline{A}}}{d\mu_{\underline{h}}}(t), \underline{a}(t) \right\rangle_{\mathbb{R}^n} + \langle \underline{A}(t), \underline{\alpha}(t) \rangle_{\mathbb{R}^n} = \\ &\quad = \left\langle D_{\mu_{\underline{h}}}(t)\, \underline{a}(t), \underline{a}(t) \right\rangle_{\mathbb{R}^n} + 2 \left\langle F_{\underline{h}}(t)\, \underline{\alpha}(t), \underline{a}(t) \right\rangle_{\mathbb{R}^n}, \end{aligned}$$

which is [(Lemma) 8.4.74] $\frac{dC_f}{d\mu_{\underline{h}}}(t)$. □

Lemma 8.4.76 *Let the* `Assumptions` *of (Lemma)* 8.4.75 *obtain. One has that*

$$\left\langle D_{\mu_{\underline{h}}}(t)\, \underline{a}(t), \underline{a}(t) \right\rangle_{\mathbb{R}^n} = \left\langle \frac{d\mu_{\underline{A}}}{d\mu_{\underline{h}}}(t), \underline{a}(t) \right\rangle_{\mathbb{R}^n} - \langle \underline{A}(t), \underline{\alpha}(t) \rangle_{\mathbb{R}^n}.$$

Proof One compares (Lemma) 8.4.74 with (Lemma) 8.4.75, using the definition of $\underline{A}$. □

Lemma 8.4.77 *Let the* `Assumptions` *of (Lemma)* 8.4.75 *obtain. Let also*

$$\Delta_{\epsilon,t}C_f = C_f(t+\epsilon, t+\epsilon) - 2C_f(t+\epsilon, t-\epsilon) + C_f(t-\epsilon, t-\epsilon).$$

When $F_{\mu_{\underline{h}}}$ *is continuous,*

$$\lim_{\epsilon \downarrow\downarrow 0} \frac{\Delta_{\epsilon,t}C_f}{F_{\mu_{\underline{h}}}(t+\epsilon) - F_{\mu_{\underline{h}}}(t-\epsilon)} = \left\langle D_{\mu_{\underline{h}}}(t)\,\underline{a}(t), \underline{a}(t)\right\rangle_{\mathbb{R}^n}.$$

Proof One has that

$$\begin{aligned}
\Delta_{\epsilon,t}C_f &= C_f(t+\epsilon, t+\epsilon) - 2C_f(t+\epsilon, t-\epsilon) + C_f(t-\epsilon, t-\epsilon) \\
&= \left\langle \left\{F_{\underline{h}}(t+\epsilon) - F_{\underline{h}}(t-\epsilon)\right\}\underline{a}(t+\epsilon), \underline{a}(t+\epsilon)\right\rangle_{\mathbb{R}^n} \\
&+ \left\langle F_{\underline{h}}(t-\epsilon)\left\{\underline{a}(t+\epsilon) - \underline{a}(t-\epsilon)\right\}, \left\{\underline{a}(t+\epsilon) - \underline{a}(t-\epsilon)\right\}\right\rangle_{\mathbb{R}^n}.
\end{aligned}$$

Thus

$$\begin{aligned}
&\frac{\Delta_{\epsilon,t}C_f}{F_{\mu_{\underline{h}}}(t+\epsilon) - F_{\mu_{\underline{h}}}(t-\epsilon)} = \\
&\quad= \frac{\left\langle \left\{F_{\underline{h}}(t+\epsilon) - F_{\underline{h}}(t-\epsilon)\right\}\underline{a}(t+\epsilon), \underline{a}(t+\epsilon)\right\rangle_{\mathbb{R}^n}}{F_{\mu_{\underline{h}}}(t+\epsilon) - F_{\mu_{\underline{h}}}(t-\epsilon)} \\
&\quad+ \frac{\left\langle F_{\underline{h}}(t-\epsilon)\left\{\underline{a}(t+\epsilon) - \underline{a}(t-\epsilon)\right\}, \left\{\underline{a}(t+\epsilon) - \underline{a}(t-\epsilon)\right\}\right\rangle_{\mathbb{R}^n}}{\left\{F_{\mu_{\underline{h}}}(t+\epsilon) - F_{\mu_{\underline{h}}}(t-\epsilon)\right\}^2} \times \\
&\quad\times \left\{F_{\mu_{\underline{h}}}(t+\epsilon) - F_{\mu_{\underline{h}}}(t-\epsilon)\right\}.
\end{aligned}$$

□

Lemma 8.4.78 *Let* $f(t) = [\underline{a}(t), \underline{h}(t)]$ *be a CH-Goursat map with covariance*

$$C_f(t_1, t_2) = \left\langle F_{\underline{h}}(t_1 \wedge t_2)\,\underline{a}(t_1), \underline{a}(t_2)\right\rangle_{\mathbb{R}^n},$$

and suppose that

$$\underline{a}(t) = \underline{\kappa}_a + \int_{T_t} \underline{\alpha}(\theta)\,\mu_{\underline{h}}(d\theta).$$

Let $\Delta_{\epsilon,t}C_f$ *be as defined in (Lemma)* 8.4.77. *Then:*

1. *when* $\mu_{\underline{h}}$ *is equivalent to Lebesgue measure (Leb), the following limits exist almost surely with respect to Leb and* $\mu_{\underline{h}}$*:*

$$\sigma^2(t) = \lim_{\epsilon \downarrow\downarrow 0} \frac{\Delta_{\epsilon,t}C_f}{Leb(t+\epsilon) - Leb(t-\epsilon)},$$

$$\underline{a}'(t) = \frac{d\underline{a}}{dLeb}(t)$$
$$= \lim_{\epsilon \downarrow\downarrow 0} \frac{\underline{a}(t+\epsilon) - \underline{a}(t-\epsilon)}{Leb(t+\epsilon) - Leb(t-\epsilon)};$$

2. *and, when also, almost surely with respect to* $\mu_{\underline{h}}$,

   (a) $D_{\mu_{\underline{h}}}(t) = \underline{d}(t) \otimes \underline{d}(t)$,
   (b) $\|\underline{d}(t)\|_{\mathbb{R}^n} = 1$,

   then

$$\int_{T_t} \frac{\|\underline{\alpha}(\theta)\|^2_{\mathbb{R}^n}}{\langle \underline{d}(\theta), \underline{a}(\theta)\rangle^2_{\mathbb{R}^n}} d\mu_{\underline{h}}(d\theta) = \int_{T_t} \frac{\|\underline{a}'(\theta)\|^2_{\mathbb{R}^n}}{\sigma^2(\theta)} Leb(d\theta).$$

Proof One has that

$$\frac{\Delta_{\epsilon,t} C_f}{Leb(t+\epsilon) - Leb(t-\epsilon)} =$$
$$= \frac{\Delta_{\epsilon,t} C_f}{F_{\mu_{\underline{h}}}(t+\epsilon) - F_{\mu_{\underline{h}}}(t-\epsilon)} \times \frac{F_{\mu_{\underline{h}}}(t+\epsilon) - F_{\mu_{\underline{h}}}(t-\epsilon)}{Leb(t+\epsilon) - Leb(t-\epsilon)}.$$

Since $\mu_{\underline{h}}$ is equivalent to *Leb*, $F_{\mu_{\underline{h}}}$ is continuous, and thus, from the definition of σ, and (Lemma) 8.4.77,

$$\sigma^2(t) = \left\langle D_{\mu_{\underline{h}}}(t)\, \underline{a}(t), \underline{a}(t)\right\rangle_{\mathbb{R}^n} \frac{d\mu_{\underline{h}}}{dLeb}(t) = \langle \underline{d}(t), \underline{a}(t)\rangle^2_{\mathbb{R}^n} \frac{d\mu_{\underline{h}}}{dLeb}(t).$$

Thus

$$\frac{\|\underline{\alpha}(\theta)\|^2_{\mathbb{R}^n}}{\langle \underline{d}(\theta), \underline{a}(\theta)\rangle^2_{\mathbb{R}^n}} = \frac{\|\underline{\alpha}(\theta)\|^2_{\mathbb{R}^n}}{\sigma^2(\theta)} \frac{d\mu_{\underline{h}}}{dLeb}(\theta).$$

But, as

$$\underline{a}(t) = \underline{\kappa}_a + \int_{T_t} \underline{\alpha}(\theta)\, \mu_{\underline{h}}(d\theta) = \underline{\kappa}_a + \int_{T_t} \underline{\alpha}(\theta) \frac{d\mu_{\underline{h}}}{dLeb}(\theta)\, Leb(d\theta),$$
$$\underline{a}'(t) = \frac{d\underline{a}}{dLeb}(t) = \underline{\alpha}(t) \frac{d\mu_{\underline{h}}}{dLeb}(t),$$

and the result follows. □

Result (Proposition) 8.4.79 below lists conditions under which a Goursat map is in fact a (wide sense) stochastic integral with respect to a (wide sense) Wiener process. It is an explicit version of (Proposition) 8.4.71. Again, as the assumptions must be repeated identically in a second statement, the assumptions of (Proposition) 8.4.79 below are highlighted.

Proposition 8.4.79 *Suppose that* $f(t) = [\underline{a}(t), \underline{h}(t)]$ *is a CH-Goursat map such that*

`Assumptions:`

(A) $\mu_{\underline{h}}$ is equivalent to Lebesgue measure;
(B) for $t \in T$, fixed, but arbitrary, $\underline{a}(t) = \underline{\kappa}_a + \int_{T_t} \underline{\alpha}(\theta)\, \mu_{\underline{h}}(d\theta)$;
(C) almost surely, with respect to $\mu_{\underline{h}}$ and Lebesgue measure,

(a) $\sigma^2(t)$ [*(Lemma)* 8.4.78] exists, and is strictly positive,
(b) $\int_{T_t} \frac{\|\underline{a}'(\theta)\|^2_{\mathbb{R}^n}}{\sigma^2(\theta)}\, Leb\,(d\theta) < \infty$;

(D) given that $\underline{A}(t) = F_{\underline{h}}(t)\, \underline{a}(t)$, $t \in T$, there exists a map $t \mapsto F(t)$ which has the same properties as those of the structure function of a CH-martingale and is such that

(a) for $t \in T$, fixed, but arbitrary, $\underline{A}(t) = F(t)\, \underline{a}(t)$,
(b) $\frac{dF}{dLeb}(t)$ exists and has rank one.

f has then multiplicity one.

Proof Let $\underline{h}_F$ be a CH-martingale with structure function F [(Remark) 8.4.29], and set

$$f_F(t) = \left[\underline{a}(t), \underline{h}_F(t)\right].$$

As f and f_F have the same covariance, the elements $f(t)$ and $f_F(t)$ are unitarily related [(Fact) 8.4.47], and thus, for $t \in T$, fixed, but arbitrary,

$$L_t[f] \text{ and } L_t[f_F]$$

are unitarily isomorphic. Consequently f and f_F have the same multiplicity properties.

Since $\frac{dF}{dLeb}(\theta)$ has rank one,

$$\frac{dF}{dLeb}(\theta) = \delta(\theta)\, \{\underline{d}(\theta) \otimes \underline{d}(\theta)\},$$

and

$$\begin{aligned}\left\langle \boldsymbol{M}_{\underline{h}_F}([t_1, t_2[)\, \underline{[x]}, \underline{y}\right\rangle_{\mathbb{R}^n} &= \int_{[t_1, t_2[} \left\langle \frac{dF}{dLeb}(\theta)\, \underline{[x]}, \underline{y}\right\rangle_{\mathbb{R}^n} dLeb \\ &= \int_{[t_1, t_2[} \delta(\theta)\, \langle \underline{d}(\theta), \underline{x}\rangle_{\mathbb{R}^n} \left\langle \underline{d}(\theta), \underline{y}\right\rangle_{\mathbb{R}^n} dLeb,\end{aligned}$$

so that [(Fact) 8.4.6]

$$M_{i,i}^{\underline{h}_F}([t_1, t_2[) = \int_{[t_1,t_2[} \delta(\theta) \left\langle \underline{d}(\theta), \underline{e}_i \right\rangle_{\mathbb{R}^n}^2 dLeb \, ,$$

and thus

$$\mu_{\underline{h}_F}([t_1, t_2[) = \int_{[t_1,t_2[} \delta(\theta) dLeb \, . \qquad (\star)$$

Consequently

$$D_{\mu_{\underline{h}_F}} = \underline{d} \otimes \underline{d}.$$

The assumption of rank one says that $\mu_{\underline{h}_F}$ is equivalent to Lebesgue measure. Result (Lemma) 8.4.78 thus applies, and, consequently, also result (Proposition) 8.4.71, as, using Assumption (A) and result ($\star$),

$$\underline{a}(t) = \underline{\kappa}_a + \int_{T_t} \underline{\alpha}(\theta) \frac{d\mu_{\underline{h}}}{dLeb}(\theta) \frac{1}{\delta(\theta)} \mu_{\underline{h}_F}(d\theta).$$

f_F has thus multiplicity one. □

Result (Proposition) 8.4.80 which follows shows that, when (Proposition) 8.4.79 obtains, the function $t \mapsto F(t)$ that one finds there is the solution of a differential equation.

Proposition 8.4.80 *Suppose that* $f(t) = [\underline{a}(t), \underline{h}(t)]$ *is a CH-Goursat map for which the* `Assumptions` *of* (Proposition) 8.4.79 *obtain. F of Assumption* (D) *is then the solution of the following differential equation:*

$$\frac{dF}{dLeb}(t) = \frac{\left\{\frac{d\underline{A}}{dLeb}(t) - F(t)\frac{d\underline{a}}{dLeb}(t)\right\}\left\{\frac{d\underline{A}}{dLeb}(t) - F(t)\frac{d\underline{a}}{dLeb}(t)\right\}^\star}{\left\langle \underline{a}(t), \left\{\frac{d\underline{A}}{dLeb}(t) - F(t)\frac{d\underline{a}}{dLeb}(t)\right\}\right\rangle_{\mathbb{R}^n}} \, .$$

Proof As $\underline{a}$ and F are differentiable, $\underline{A} = F\underline{a}$ is, and, differentiating that product, one has that

$$\frac{d\underline{A}}{dLeb}(t) = \frac{dF}{dLeb}(t)\,\underline{a}(t) + F(t)\frac{d\underline{a}}{dLeb}(t) \, ;$$

thus, moving the appropriate term, that

$$\frac{dF}{dLeb}(t)\,\underline{a}(t) = \frac{d\underline{A}}{dLeb}(t) - F(t)\frac{d\underline{a}}{dLeb}(t) \, . \qquad (1)$$

The derivative $\frac{dF}{dLeb}(t)$, having rank one, has a representation as

$$\frac{dF}{dLeb}(t) = \underline{d}(t) \otimes \underline{d}(t), \tag{2}$$

and thus

$$\frac{dF}{dLeb}(t)\,\underline{a}(t) = \langle \underline{a}(t), \underline{d}(t) \rangle_{\mathbb{R}^n} \underline{d}(t). \tag{3}$$

Taking the inner product of that last expression with $\underline{a}(t)$ yields that

$$\langle \underline{a}(t), \underline{d}(t) \rangle_{\mathbb{R}^n}^2 = \left\langle \frac{dF}{dLeb}(t)\,\underline{a}(t), \underline{a}(t) \right\rangle_{\mathbb{R}^n}. \tag{4}$$

But, as in the proof of (Lemma) 8.4.78, one has that

$$\begin{aligned}
\sigma^2(t) &= \lim_{\epsilon \downarrow\downarrow 0} \frac{\Delta_{\epsilon,t} C_f}{Leb(t+\epsilon) - Leb(t-\epsilon)} \\
&= \lim_{\epsilon \downarrow 0, \epsilon > 0} \frac{\Delta_{\epsilon,t} C_f}{F_{\mu_{\underline{h}}}(t+\epsilon) - F_{\mu_{\underline{h}}}(t-\epsilon)} \frac{F_{\mu_{\underline{h}}}(t+\epsilon) - F_{\mu_{\underline{h}}}(t-\epsilon)}{Leb(t+\epsilon) - Leb(t-\epsilon)} \\
&= \left\langle D_{\mu_{\underline{h}}}(t)\left[\underline{a}(t)\right], \underline{a}(t) \right\rangle_{\mathbb{R}^n} \frac{d\mu_{\underline{h}}}{dLeb}(t).
\end{aligned}$$

Using a definition analogous to that found in (Lemma) 8.4.75,

$$\frac{d\mu_{\underline{A}}}{dLeb} = \frac{d\underline{A}}{dLeb}. \tag{5}$$

Also, since $\mu_{\underline{h}}$ is equivalent to Lebesgue measure,

$$\frac{d\mu_{\underline{A}}}{dLeb} = \frac{d\mu_{\underline{A}}}{d\mu_{\underline{h}}} \frac{d\mu_{\underline{h}}}{dLeb}. \tag{6}$$

Thus

$$\begin{aligned}
&\left\langle D_{\mu_{\underline{h}}}(t)\left[\underline{a}(t)\right], \underline{a}(t) \right\rangle_{\mathbb{R}^n} \frac{d\mu_{\underline{h}}}{dLeb}(t) = \\
&\overset{8.4.76}{=} \left\{ \left\langle \frac{d\mu_{\underline{A}}}{d\mu_{\underline{h}}}(t), \underline{a}(t) \right\rangle_{\mathbb{R}^n} - \langle \underline{A}(t), \underline{\alpha}(t) \rangle_{\mathbb{R}^n} \right\} \frac{d\mu_{\underline{h}}}{dLeb}(t) \\
&\overset{(5),(6),(B)}{=} \left\langle \frac{d\underline{A}}{dLeb}(t), \underline{a}(t) \right\rangle_{\mathbb{R}^n} - \left\langle \underline{A}(t), \frac{d\underline{a}}{dLeb}(t) \right\rangle_{\mathbb{R}^n}
\end{aligned}$$

$$\begin{aligned}
&\overset{def\,\underline{A}}{=} \left\langle \frac{d\underline{A}}{dLeb}(t), \underline{a}(t)\right\rangle_{\mathbb{R}^n} - \left\langle \underline{a}(t), F(t)\frac{d\underline{a}}{dLeb}(t)\right\rangle_{\mathbb{R}^n} \\
&= \left\langle \frac{d\underline{A}}{dLeb}(t) - F(t)\frac{d\underline{a}}{dLeb}(t), \underline{a}(t)\right\rangle_{\mathbb{R}^n} \\
&\overset{(1)}{=} \left\langle \frac{dF}{dLeb}(t)\,\underline{a}(t), \underline{a}(t)\right\rangle_{\mathbb{R}^n} .
\end{aligned}$$

Consequently, using the expression for $\sigma^2(t)$ obtained above,

$$\sigma^2(t) = \left\langle \frac{dF}{dLeb}(t)\,\underline{a}(t), \underline{a}(t)\right\rangle_{\mathbb{R}^n} .$$

Having made the assumption that, almost surely with respect to *Leb*, one has that $\sigma^2(t) > 0$, one concludes, from equality (4), that, almost surely with respect to *Leb*,

$$\left|\langle \underline{a}(t), \underline{d}(t)\rangle_{\mathbb{R}^n}\right| > 0 .$$

Thus finally, from equality (3),

$$\underline{d}(t) = \left\{\langle \underline{a}(t), \underline{d}(t)\rangle_{\mathbb{R}^n}\right\}^{-1} \frac{dF}{dLeb}(t)\,\underline{a}(t),$$

and, from equality (2),

$$\frac{dF}{dLeb}(t) = \frac{\left\{\frac{dF}{dLeb}(t)\,\underline{a}(t)\right\} \otimes \left\{\frac{dF}{dLeb}(t)\,\underline{a}(t)\right\}}{\langle \underline{a}(t), \underline{d}(t)\rangle^2_{\mathbb{R}^n}} .$$

Because of equalities (1) and (4), F must indeed be the solution of the equation in statement (Proposition) 8.4.80. □

Result (Proposition) 8.4.82 which follows states that functions $t \mapsto F(t)$ that are solutions to the differential equation found in (Proposition) 8.4.80 yield functions that are required in the assumptions of (Proposition) 8.4.79.

Lemma 8.4.81 *Let $\underline{a}$ and $\underline{A}$ be absolutely continuous with respect to Lebesgue measure (Leb), and suppose that there exists $[t_l, t_r[\subseteq T$, and F, such that*

(A) $F(t_l)\,\underline{a}(t_l) = \underline{A}(t_l)$;
(B) *for $t \in [t_l, t_r[$, fixed, but arbitrary,*

(a) $\frac{dF}{dLeb}(t)$ *exists,*
(b) $\frac{dF}{dLeb}(t) = \frac{\left\{\frac{d\underline{A}}{dLeb}(t) - F(t)\frac{d\underline{a}}{dLeb}(t)\right\}\left\{\frac{d\underline{A}}{dLeb}(t) - F(t)\frac{d\underline{a}}{dLeb}(t)\right\}^\star}{\left\langle \underline{a}(t), \left\{\frac{d\underline{A}}{dLeb}(t) - F(t)\frac{d\underline{a}}{dLeb}(t)\right\}\right\rangle_{\mathbb{R}^n}}$.

Then $F(t)\,\underline{a}(t) = \underline{A}(t)$, $t \in [t_l, t_r[$.

Furthermore, when $\frac{dF}{dLeb}$ *is a function whose values are symmetric, positive definite matrices, for* $\{t_1, t_2\} \subseteq [t_l, t_r[\,,\ t_1 < t_2$, *fixed, but arbitrary,*

$$\frac{dF}{dLeb}(t_1) \ll \frac{dF}{dLeb}(t_2)\,.$$

Proof It follows from item (b) of Assumption (B) that

$$\frac{dF}{dLeb}(t)\,\underline{a}(t) = \frac{d\underline{A}}{dLeb}(t) - F(t)\,\frac{d\underline{a}}{dLeb}(t)\,.$$

Consequently

$$\frac{d\left[F(\cdot)\underline{a}(\cdot)\right]}{dLeb}(t) = \frac{d\underline{A}}{dLeb}(t)\,,$$

and thus

$$F(t)\,\underline{a}(t) = \underline{A}(t)\,,\ t \in [t_l, t_r[\,.$$

Furthermore, by the mean value theorem,

$$\langle\{F(t_2) - F(t_1)\}\,\underline{x}, \underline{x}\rangle_{\mathbb{R}^n} = \left\langle\frac{dF}{dLeb}(t_1 + \theta)\,\underline{x}, \underline{x}\right\rangle_{\mathbb{R}^n}\,.$$

□

The following result may be considered to be the "operational version" of (Proposition) 8.4.79.

Proposition 8.4.82 *Let* $f(t) = [\underline{a}(t)\,,\underline{h}(t)]$ *be a CH-Goursat process, and let* $A(t) = F_{\underline{h}}(t)\,\underline{a}(t)$, *so that, for fixed, but arbitrary* $\{\theta, t\} \subseteq T,\ \theta \le t$

$$C_f(\theta, t) = \langle\underline{A}(\theta)\,,\underline{a}(t)\rangle_{\mathbb{R}^n}\,.$$

Suppose that

(A) $\underline{a}$ *and* $\underline{A}$ *are absolutely continuous with respect to Lebesgue measure (Leb),*
(B) *almost surely with respect to Lebesgue measure,* σ^2 [as defined in (Lemma) 8.4.78] *exists, and is strictly positive,*
(C) *for* $t \in T$, *fixed, but arbitrary,*

$$\int_{T_t} \frac{\|\underline{a}'(\theta)\|^2_{\mathbb{R}^n}}{\sigma^2(\theta)}\,Leb\,(d\theta) < \infty,$$

1. *for $t \in T$, and symmetric, positive definite, constant matrix Φ such that*

$$\Phi\left[\underline{a}(t)\right] = \underline{A}(t),$$

both t and Φ fixed, but arbitrary, the following equation:

$$\begin{cases} \dfrac{dF}{dLeb}(t) = \dfrac{\left\{\frac{d\underline{A}}{dLeb}(t) - F(t)\frac{d\underline{a}}{dLeb}(t)\right\}\left\{\frac{d\underline{A}}{dLeb}(t) - F(t)\frac{d\underline{a}}{dLeb}(t)\right\}^{*}}{\left\langle \underline{a}(t), \left\{\frac{d\underline{A}}{dLeb}(t) - F(t)\frac{d\underline{a}}{dLeb}(t)\right\}\right\rangle_{\mathbb{R}^n}} \\ F(t) = \Phi \end{cases}$$

has a solution $F(t)$ on $[t, t+\epsilon[$, some $\epsilon > 0$ (F is thus absolutely continuous with respect to Lebesgue measure).

f has then multiplicity one, and $\mu_{\underline{h}}$ is equivalent to Lebesgue measure (abbreviated into the expression: "f has simple Lebesgue spectrum").

Proof The first step in the proof amounts to checking that it suffices to prove the result "locally," and the second, that the local problem has the required solution.

Suppose one has a solution on $[t, t+\epsilon[$, yielding an f with single Lebesgue spectrum. If $t+\epsilon \in T$, one can find $\delta > 0$ such that f has simple Lebesgue spectrum on $[t+\epsilon, t+\epsilon+\delta[$. Repeating that process as necessary, one will find a largest t_m such that f has simple Lebesgue spectrum on $[t, t_m[$. But then one must have that $t_m = t_r$. Since the choice of the starting $t \in T$ is arbitrary, one must have that the purely nondeterministic f has simple Lebesgue spectrum on T (when $t_r \in T$, the value of the integral representation at t_r is by continuity).

Because of (Proposition) 8.4.44, and of (Remark) 8.4.60, one may assume that the representation of f is proper. Let F be the solution of the statement's differential equation when the initial value is $\Phi = F_{\underline{h}}(t)$ (Assumption (D)). Let

$$\underline{b}(t) = \frac{d\underline{A}}{dLeb}(t) - F(t)\frac{d\underline{a}}{dLeb}(t) = \frac{dF}{dLeb}(t)\underline{a}(t).$$

Then (Assumption (D) again)

$$\left\langle \frac{dF}{dLeb}(t)\left[\underline{x}\right], \underline{x}\right\rangle_{\mathbb{R}^n} = \frac{\langle \underline{b}(t), \underline{x}\rangle_{\mathbb{R}^n}^2}{\langle \underline{a}(t), \underline{b}(t)\rangle_{\mathbb{R}^n}}.$$

From the proof of (Proposition) 8.4.80, one has, because of Assumption (B), that

$$0 < \sigma^2(t) = \left\langle \frac{dF}{dLeb}(t)\left[\underline{a}(t)\right], \underline{a}(t)\right\rangle_{\mathbb{R}^n} = \langle \underline{a}(t), \underline{b}(t)\rangle_{\mathbb{R}^n}.$$

The derivative $\frac{dF}{dLeb}(t)$ is thus positive definite. Then (Lemma) 8.4.81 applies, and one has that, for $\{\theta, \theta_1, \theta_2\} \subseteq\,]t, t+\epsilon[\,$, $\theta_1 < \theta_2$, fixed, but arbitrary,

- $\underline{A}(\theta) = F(\theta)\,\underline{a}(\theta)$,

- $F(\theta_1) \ll F(\theta_2)$.

  Define

$$G(\theta) = \begin{cases} F_{\underline{h}}(\theta) & \theta \in T_t \cup \{t\} \\ F(t) & \theta \in]t, t+\epsilon[\end{cases} .$$

Then

$$\underline{A}(\theta) = G(\theta)\, \underline{a}(\theta)\,,\ \theta \in T,\ \theta < t+\epsilon\,,$$

and G is increasing in the positive definite sense. Let $\underline{h}_G$ be a wide sense martingale defined for $\theta \in T,\ \theta < t+\epsilon$, with structure matrix G [(Remark) 8.4.29], and let

$$f_G(t) = \left[\underline{a}(t)\,, \underline{h}_G(t)\right].$$

Then, for fixed, but arbitrary $\{t_1, t_2\} \subseteq T,\ t_1 < t+\epsilon,\ t_2 < t+\epsilon,\ t_1 \leq t_2$, since $\underline{A}$ is the result of multiplying $\underline{a}$ by both F and $F_{\underline{h}}$,

$$C_{f_G}(t_1, t_2) = \langle \underline{A}(t_1)\,, \underline{a}(t_2)\rangle = C_f(t_1, t_2)\,,$$

so that, for fixed, but arbitrary $\theta \in T,\ \theta < t+\epsilon$, there is [(Fact) 8.4.47] a unitary map

$$U_\theta : L_\theta[f] \longrightarrow L_\theta[f_G]\,.$$

Since one has taken f to be proper, for fixed, but arbitrary $t \in T$, one has that $L_t[f] = L_t[\underline{h}]$. Since, for fixed, but arbitrary $\theta \in T,\ \theta \leq t,\ G(\theta) = F_{\underline{h}}(\theta)$, for fixed, but arbitrary $\theta \in T,\ \theta \leq t$, there is a unitary map

$$V_\theta : L_\theta[\underline{h}] \longrightarrow L_0\left[\underline{h}_G\right].$$

Consequently, for fixed, but arbitrary $\theta \in T,\ \theta \leq t$,

$$L_\theta[f_G] = L_\theta\left[\underline{h}_G\right],\ \theta \leq t.$$

That the same is true for $\theta \in]t, t+\epsilon[$ is seen as follows. Consider (Proposition) 8.4.65. The condition there yields that ϕ must be zero up to time t, since one knows that the representation is proper up to that time. Consequently it suffices to use that result on $[t, t+\epsilon[$. Rewriting its proof with, in that interval, for $\underline{d}(\theta)$, the vector

$$\frac{\underline{b}(\theta)}{\|\underline{b}(\theta)\|_{\mathbb{R}^n}}\,,$$

and, for $d\mu_{\underline{h}}$, the measure $\|\underline{b}(\theta)\|^2_{\mathbb{R}^n}\, dLeb$, one gets indeed the expression

$$0 = \int_{T_\theta} \langle \underline{b}(u), \underline{a}(\theta)\rangle_{\mathbb{R}^n} \phi(u)\, Leb(d\theta).$$

Since the inner product cannot be zero (its square is σ^2), the norm cannot be zero either, so that ϕ must be zero on $[t, t+\epsilon[$.

Now

$$\begin{aligned}
\mu_{\underline{h}_G}([t_1, t_2[) &= \sum_{i=1}^{n} M_{i,i}^{\underline{h}G}([t_1, t_2[) \\
&= \sum_{i=1}^{n} \{G_{i,i}(t_2) - G_{i,i}(t_1)\} \\
&= \sum_{i=1}^{n} \left\langle \{G(t_2) - G(t_1)\}\, \underline{e}_i, \underline{e}_i \right\rangle_{\mathbb{R}^n} \\
&= \sum_{i=1}^{n} \int_{[t_1, t_2[} \left\langle \frac{dG}{dLeb}(\theta)\, \underline{e}_i, \underline{e}_i \right\rangle_{\mathbb{R}^n} Leb(d\theta),
\end{aligned}$$

and (again the proof of (Proposition) 8.4.80)

$$\left\langle \frac{dG}{dLeb}(\theta)\, \underline{e}_i, \underline{e}_i \right\rangle_{\mathbb{R}^n} = \frac{\langle \underline{b}(\theta), \underline{e}_i\rangle_{\mathbb{R}^n}}{\langle \underline{a}(\theta), \underline{b}(\theta)\rangle^2_{\mathbb{R}^n}} = \frac{\langle \underline{b}(\theta), \underline{e}_i\rangle^2_{\mathbb{R}^n}}{\sigma^2(\theta)},$$

so that

$$\mu_{\underline{h}_G}([t_1, t_2[) = \int_{[t_1, t_2[} \frac{\|\underline{b}(\theta)\|^2_{\mathbb{R}^n}}{\sigma^2(\theta)} Leb(d\theta).$$

Thus, since

$$0 < \sigma^2(\theta) = \langle \underline{a}(\theta), \underline{b}(\theta)\rangle_{\mathbb{R}^n},$$

$\mu_{\underline{h}_G}$ and Leb are equivalent on $[t, t+\epsilon[$. f_G has then multiplicity one [(Proposition) 8.4.79]. Since f and f_G are unitarily related, the same is true for f, so that f has simple Lebesgue spectrum on T. □

Riccati Matrix Differential Equations

The differential equation of Assumption (D) of Proposition 8.4.82 may be expressed in the following from:

$$F' = \frac{\{\underline{A}' - F[\underline{a}']\}\{\underline{A}' - F[\underline{a}']\}^\star}{\sigma^2},$$

where σ^2 is obtained directly form the covariance. It can thus be seen as a Riccati matrix differential equation of the following form:

$$\begin{aligned}\dot{M}(t) &= [\underline{u}(t) - M(t)[\underline{v}(t)]]\,[\underline{u}(t) - M(t)[\underline{v}(t)]]^\star \\ &= U(t) - M(t)W(t) - W(t)^\star M(t)^\star + M(t)V(t)M(t)^\star,\end{aligned}$$

where $U(t) = \underline{u}(t)\,\underline{u}(t)^\star$, $V(t) = \underline{v}(t)\,\underline{v}(t)^\star$, $W(t) = \underline{v}(t)\,\underline{u}(t)^\star$. One furthermore wants a solution M which is symmetric and positive definite. Such an equation is known as a matrix Riccati differential equation. Investigations of solutions of such equations are present in many sources. One [272] is cited and used here as it matches well the present context.

Let, for $t \in \mathbb{R}_+$, fixed, but arbitrary, $M(t)$ be a square, symmetric matrix of dimension $n \in \mathbb{N}$. Let $\mathcal{E}[M(t)]$ be the matrix defined as the following formal infinite sum:

$$\begin{aligned}\mathcal{E}[M(t)] &= I_n \\ &+ \int_0^t M(\theta)\,d\theta \\ &+ \int_0^t M(\theta_1)\left\{\int_0^{\theta_1} M(\theta)\,d\theta\right\} d\theta_1 \\ &+ \int_0^t M(\theta_1)\left\{\int_0^{\theta_1} M(\theta_2)\left\{\int_0^{\theta_2} M(\theta)\,d\theta\right\} d\theta_2\right\} d\theta_1 + \cdots\end{aligned}$$

The matrix $\mathcal{F}[M(t)]$ is defined analogously, changing the products of type $M(\int M)$ to products of type $(\int M)M$. When M and $\int M$ commute, one has that:

$$\mathcal{E}[M(t)] = \mathcal{F}[M(t)] = e^{\int_0^t M(\theta)\,d\theta}.$$

Information about the relation $M'M = MM'$ may be found in [93]. When $t \mapsto M(t)$ is bounded and integrable on the interval $[0, T]$,

1. $\mathcal{E}[M(t)]$ and $\mathcal{F}[M(t)]$ define actual matrices which are invertible with

$$\mathcal{F}[M(t)]\mathcal{E}[-M(t)] = \mathcal{E}[-M(t)]\mathcal{F}[M(t)] = I_n;$$

2. the Riccati equation:

$$\dot{M}(t) = Q(t) + A(t)M(t) - M(t)B(t) - M(t)PM(t)$$

has the following solution: given that

$$\begin{bmatrix} M_1(t) \\ M_2(t) \end{bmatrix} = \mathcal{E}\left[\begin{bmatrix} A(t) & Q(t) \\ P(t) & B(t) \end{bmatrix}\right]\begin{bmatrix} M(0) \\ I_n \end{bmatrix},$$

then

$$M(t) = M_1(t)M_2(t)^{-1}.$$

Furthermore there is a way to a unique solution.

For the application at hand, one must set:

$$\begin{aligned} A(t) &\leftarrow -\underline{u}(t)\,\underline{v}(t)^\star, \\ B(t) &\leftarrow \ \ \underline{v}(t)\,\underline{u}(t)^\star, \\ P(t) &= -\underline{v}(t)\,\underline{v}(t)^\star, \\ Q(t) &\leftarrow \ \ \underline{u}(t)\,\underline{u}(t)^\star, \end{aligned}$$

to obtain that

$$\begin{bmatrix} A(t) & Q(t) \\ P(t) & B(t) \end{bmatrix} = -\begin{bmatrix} \underline{u}(t) \\ \underline{v}(t) \end{bmatrix}\left[\,\underline{v}(t)^\star \ \ -\underline{u}(t)^\star\,\right].$$

The above applies directly to the following case, as continuous coefficients on a compact interval are bounded:

Fact 8.4.83 *When $\underline{a}'$ and $\underline{A}'$ are continuous, the differential equation of* (Proposition) 8.4.82, *item (D), has a solution.*

Fact 8.4.84 *When the conditions listed below obtain, the differential equation of* (Proposition) 8.4.82, *item (D), has a solution:*

(A) *for $t \in T$, fixed, but arbitrary, there exists $\epsilon > 0$, and continuously differentiable*

$$\left\{t \mapsto \underline{a}_n(t)\,,\ t \mapsto \underline{A}_n(t)\,,\ n \in \mathbb{N}\right\}$$

such that

(a) *for fixed, but arbitrary $\{t_1, t_2\} \subseteq [t, t+\epsilon[$,*

$$C_n(t_1, t_2) = \left\langle \underline{A}_n(t_1 \wedge t_2)\,,\underline{a}_n(t_1 \vee t_2)\right\rangle_{\mathbb{R}^n}$$

is positive definite,

(b) *for fixed, but arbitrary* $n \in \mathbb{N}$, *almost surely (with respect to Lebesgue measure) on* $[t, t+\epsilon[$, $\sigma_n^2(\theta) > 0$, *where* $\sigma_n = \sigma$ *for* $\underline{a}_n = \underline{a}$ *and* $\underline{A}_n = \underline{A}$ *in* (Proposition) 8.4.82,

(B) *the following function is locally square integrable (with respect to Lebesgue measure):*

$$t \mapsto \frac{\|\underline{a}'(t)\|_{\mathbb{R}^n} + \|\underline{A}'(t)\|_{\mathbb{R}^n}}{\sigma(t)};$$

(C)

$$\lim_n \int_t^{t+\epsilon} d\theta \left\{ \left\| \frac{\underline{a}_n'(\theta)}{\sigma_n(\theta)} - \frac{\underline{a}'(\theta)}{\sigma(\theta)} \right\|_{\mathbb{R}^n}^2 + \left\| \frac{\underline{A}_n'(\theta)}{\sigma_n(\theta)} - \frac{\underline{A}'(\theta)}{\sigma(\theta)} \right\|_{\mathbb{R}^n}^2 \right\} = 0\,.$$

Proof That latter result is a consequence of the fact that the solution, in the continuous case, is a series of integrals, and that the assumptions allow those integrals, and the series they form, to converge. □

Fact 8.4.84 has the two applications (corollaries) which follow.

Corollary 8.4.85 *Let* $f(t) = [\underline{a}(t), \underline{h}(t)]$ *be a Goursat process of order n, and let* $\underline{A}(t) = F_{\underline{h}}(t)[\underline{a}(t)]$. *Suppose that*

(a) $\underline{a}$ *and* $\underline{A}$ *are absolutely continuous (with respect to Lebesgue measure),*
(b) $\underline{a}'$ *and* $\underline{A}'$ *are locally square integrable,*
(c) $F_{\underline{h}}$ *has a continuous derivative,*
(d) σ^2 *exists and is strictly positive.*

Then f has simple Lebesgue spectrum.

Proof Let indeed $\{\underline{a}_n,\ n \in \mathbb{N}\}$ be a sequence of continuously differentiable functions such that

$$\lim_n \int_t^{t+\epsilon} \left\|\underline{a}_n'(\theta) - \underline{a}'(\theta)\right\|_{\mathbb{R}^n}^2 d\theta = 0,$$

and let $\underline{A}_n(t) = F_{\underline{h}}(t)\left[\underline{a}_n(t)\right]$. Then:

- $C_n(t_1, t_2) = \left\langle \underline{A}_n(t_1 \wedge t_2), \underline{a}_n(t_1 \vee t_2)\right\rangle_{\mathbb{R}^n}$, $\{t_1, t_2\} \subseteq T$, is positive definite;
- $\underline{A}_n'(t) = F_{\underline{h}}'(t)\left[\underline{a}_n(t)\right] + F_{\underline{h}}(t)\left[\underline{a}_n'(t)\right]$ is continuous, and converges in L_2 to $\underline{A}'$;
- $\sigma_n^2(t) = \left\langle F_{\underline{h}}\left[\underline{a}_n(t)\right], \underline{a}_n(t)\right\rangle_{\mathbb{R}^n}$ converges uniformly to $\sigma^2(t) > 0$ on $[t, t+\epsilon[$.

The conditions of case (Fact) 8.4.84 thus obtain. □

Corollary 8.4.86 *Let* $f(t) = [\underline{a}(t), \underline{h}(t)]$ *be a Goursat process of order n, and let* $\underline{A}(t) = F_{\underline{h}}(t)[\underline{a}(t)]$. *Suppose that*

(a) $\underline{a}$ *and* $\underline{A}$ *are absolutely continuous (with respect to Lebesgue measure) and, for* $t \in T$, *fixed, but arbitrary,* $\underline{a}(t) \neq \underline{0}_{\mathbb{R}^n}$,

(b) *$\underline{a}'$ and $\underline{A}'$ have a norm whose square is locally integrable,*
(c) *$F_{\underline{h}}$ is absolutely continuous (with respect to Lebesgue measure),*
(d) *$F'_{\underline{h}}$ has a norm whose square is locally integrable and its smallest eigenvalue is locally bounded below.*

Then f has simple Lebesgue spectrum.

Proof Let $\underline{a}_p$ be as in (Corollary) 8.4.85, and F_p be a continuously differentiable approximation of $F_{\underline{h}}$ such that

- $$\lim_p \int_t^{t+\epsilon} \left\| F'_p(\theta) - F'_{\underline{h}}(\theta) \right\|_B^2 d\theta = 0,$$
- there exists $\delta > 0$ so that, for $t \in [t, t+\epsilon[$, fixed, but arbitrary,

$$F'_p(t) \geq \delta I_n.$$

Setting $\underline{A}_p(t) = F_p(t)\,\underline{a}_p(t)$, one finishes as in (Corollary) 8.4.85. □

Corollary 8.4.87 *Let $f(t) = [\underline{a}(t), \underline{h}(t)]$ be a Goursat process such that*

(a) *$\underline{h}$ is a standard Wiener process in the wide sense,*
(b) *for $t \in T$, fixed, but arbitrary, $\underline{a}(t) \neq \underline{0}_{\mathbb{R}^n}$,*
(c) *$\underline{a}$ is absolutely continuous (with respect to Lebesgue measure),*
(d) *for fixed, but arbitrary $t \in T$,*

$$\int_{T_t} \left\| \underline{a}'(\theta) \right\|_{\mathbb{R}^n}^2 d\theta < \infty.$$

Then f has simple Lebesgue spectrum.

Proof In that case indeed $F_{\underline{h}} = tI_n$, and (Corollary) 8.4.87 is a particular case of (Corollary) 8.4.86. □

Example 8.4.88 The case $n = 2$ may be made explicit as follows. The differential equation of Assumption (D) of (Proposition) 8.4.82 is then a set of four equations with three unknowns:

$$F_{1,1},\ F_{1,2} = F_{2,1},\ F_{2,2}.$$

Let $F_{\underline{h}}(t)\,\underline{a}(t) = F(t)\,\underline{a}(t)$, and $M = F - F_{\underline{h}}$. Since, by definition,

$$M(t)\,\underline{a}(t) = \underline{0}_{\mathbb{R}^n},$$

M may be expressed, as presently seen, as follows:

$$M(t) = \mu(t) \left[\begin{array}{r|r} a_2^2(t) & -a_1(t)\,a_2(t) \\ -a_1(t)\,a_2(t) & a_1^2(t) \end{array} \right].$$

One has indeed a system of equations of the following form:

$$\begin{array}{l} \alpha a_1 + \beta a_2 = 0 \\ \beta a_1 + \gamma a_2 = 0 \end{array}.$$

Multiplying the first by a_1, the second by a_2, one obtains that

$$\alpha = \frac{a_2^2}{a_1^2}\gamma \text{ and } \beta = -\frac{a_2}{a_1}\gamma.$$

μ is the function γa_1^{-2}.

Since f has simple Lebesgue spectrum, F' must have rank one, and thus

$$\det F'(t) = \det\left\{F'_{\underline{h}}(t) + M'(t)\right\} = 0.$$

This latter equation yields a Riccati equation of the following form:

$$\xi(t)\,\mu'(t) + \eta(t)\,\mu(t) + \zeta(t)\,\mu^2(t) = -\upsilon(t),$$

where

$$\xi(t) = a_2^2(t)\,\frac{dF^{\underline{h}}_{2,2}}{dt}(t) + a_1^2(t)\,\frac{dF^{\underline{h}}_{1,1}}{dt}(t) + 2a_1(t)\,a_2(t)\,\frac{dF^{\underline{h}}_{1,1}}{dt}(t),$$

$$\upsilon(t) = \frac{F^{\underline{h}}_{1,1}}{dt}(t)\,\frac{dF^{\underline{h}}_{2,2}}{dt}(t) - \left\{\frac{dF^{\underline{h}}_{1,2}}{dt}(t)\right\}^2,$$

$$\eta(t) = 2\left\{a_2(t)\,a_2'(t)\,\frac{F^{\underline{h}}_{2,2}}{dt}(t) + a_1(t)\,a_1'(t)\,\frac{dF^{\underline{h}}_{1,1}}{dt}(t)\right\}$$

$$+ 2\left\{\left[a_1(t)\,a_2'(t) + a_1'(t)\,a_2(t)\right]\frac{dF^{\underline{h}}_{1,2}}{dt}(t)\right\},$$

$$\zeta(t) = -\left[a_1(t)\,a_2'(t) + a_1'(t)\,a_2(t)\right]^2.$$

Applying the procedure just described to

$$f(t) = \cos(t)\,W_t^{(1)} + \sin(t)\,W_t^{(2)},$$

where $t \mapsto W_t^{(1)}$, and $t \mapsto W_t^{(2)}$, are orthogonal Wiener processes in the wide sense, the Riccati equation becomes

$$\mu'(t) - \mu^2(t) = -1.$$

When the initial condition is that $F(0)$ is the zero matrix, $\mu(0) = 0$, and $\mu(t) = -\tanh(t)$, so that

$$F(t) = \begin{bmatrix} t & 0 \\ 0 & t \end{bmatrix} - \tanh(t) \left[\begin{array}{r|r} \cos^2(t) & -\sin(t)\cos(t) \\ -\sin(t)\cos(t) & \sin^2(t) \end{array}\right].$$

8.4.9 Goursat Representations with Smooth, Deterministic Part, Are Proper

In this section it is explained how, when $\underline{a}$ is smooth, proper Goursat representations $f(t) = [\underline{a}(t), \underline{h}(t)]$ may be obtained.

The following notation shall be used.

1. Let $[\theta_l, \theta_r] \subseteq T$ be an interval. $\underline{\theta} \in \mathbb{R}^{n+1}$ is a finite partition of $[\theta_l, \theta_r]$ when, for some $n \in \mathbb{N}$,
$$\theta_l = \theta_0 < \theta_1 < \cdots \theta_{n-1} < \theta_n = \theta_r.$$
$\mathcal{P}[\theta_l, \theta_r]$ is the set of finite partitions of $[\theta_l, \theta_r]$.
2. Given a function f, defined on $[\theta_l, \theta_r]$, and a finite partition $\underline{\theta} \in \mathcal{P}[\theta_l, \theta_r]$,
$$\Delta_i[f, \underline{\theta}] = f(\theta_{i+1}) - f(\theta_i).$$
3. Given two functions f and g defined on $[\theta_l, \theta_r]$,
$$I\left[\frac{f^2}{g} \mid \theta_l, \theta_r\right] = \sup_{\underline{\theta} \in \mathcal{P}[\theta_l, \theta_r]} \sum_i \frac{\Delta_i^2[f, \underline{\theta}]}{\Delta_i[g, \underline{\theta}]}.$$
4. Given a map $f : T \longrightarrow H$, and $\{\theta_l, \theta_r\} \subseteq T,\ \theta_l < \theta_r$,
$$L_{]\theta_l, \theta_r]}[f] = \overline{V[\{f(t),\ t \in]\theta_l, \theta_r]\}]}.$$

Proposition 8.4.89 *Let*

(a) $\underline{h}$ *be a CH-martingale, and* $F^{\underline{h}}(t)$ *be the trace, at* $t \in T$, *of its structure function;*
(b) $\underline{a} : T \longrightarrow \mathbb{R}^n$ *be continuously differentiable* $n - 1$ *times;*

(c) $c : T \longrightarrow T$ *be continuous, positive, and such that, for arbitrary* $\theta_l < \theta_r$, $[\theta_l, \theta_r] \subseteq]t_l, t_r[$,

$$I\left[\frac{c^2}{F^{\underline{h}}} \mid \theta_l, \theta_r\right] = \infty;$$

(d) $f(t) = [\underline{a}(c(t)), \underline{h}(t)]$.

Then, given fixed, but arbitrary $\{\theta_l, \theta_r\} \subseteq T,\ \theta_l < \theta_r$, *and* $i \in [1 : n-1]$,

$$\left[\underline{a}^{(i)}(c(\theta_r)), \underline{h}(\theta_r)\right] \in L_{]\theta_l, \theta_r]}[f].$$

Proof Let $t \in]\theta_l, \theta_r[$ be fixed, but arbitrary and $\underline{\theta} \in \mathcal{P}[t, \theta_r]$ be such that, for i fixed, but arbitrary, $|c(\theta_{i+1}) - c(\theta_i)| > 0$. Set

$$\pi(\underline{\theta}) = \sum_i \frac{\Delta_i^2[c, \underline{\theta}]}{\Delta_i[F^{\underline{h}}, \underline{\theta}]},$$

$$\pi_i(\underline{\theta}) = \frac{1}{\pi(\underline{\theta})}\left[\frac{\Delta_i^2[c, \underline{\theta}]}{\Delta_i[F^{\underline{h}}, \underline{\theta}]}\right],$$

$$f[\underline{\theta}] = \sum_i \pi_i(\underline{\theta}) \frac{\Delta_i[f, \underline{\theta}]}{\Delta_i[c, \underline{\theta}]}.$$

The expression $f[\underline{\theta}]$ is thus a weighted average, with strictly positive weights summing up to one. Then, inserting and subtracting, in $f[\underline{\theta}]$, terms of the type $[\underline{a}(c(\theta_{i+1})), \underline{h}(\theta_i)]$, one has that

$$\begin{aligned} f(\theta_{i+1}) - f(\theta_i) &= [\underline{a}(c(\theta_{i+1})), \underline{h}(\theta_{i+1})] - [\underline{a}(c(\theta_i)), \underline{h}(\theta_i)] \\ &= [\underline{a}(c(\theta_{i+1})), \underline{h}(\theta_{i+1}) - \underline{h}(\theta_i)] + [\underline{a}(c(\theta_{i+1})) - \underline{a}(c(\theta_i)), \underline{h}(\theta_i)], \end{aligned}$$

so that

$$f[\underline{\theta}] = \sum_i \pi_i(\theta)\left[\underline{a}(c(\theta_{i+1})), \frac{\Delta_i[\underline{h}, \underline{\theta}]}{\Delta_i[c, \underline{\theta}]}\right] + \sum_i \pi_i(\underline{\theta})\left[\frac{\Delta_i[\underline{a} \circ c, \underline{\theta}]}{\Delta_i[c, \underline{\theta}]}, \underline{h}(\theta_i)\right].$$

Consequently, as, by the mean value theorem, there exists $\eta_i \in]\theta_i, \theta_{i+1}[$ such that

$$\frac{\Delta_i[\underline{a} \circ c, \underline{\theta}]}{\Delta_i[c, \underline{\theta}]} = \underline{a}'(c(\eta_i)),$$

one has that

$$\left\|f\left[\underline{\theta}\right]-\left[\underline{a}'\left(c\left(\theta_r\right)\right),\underline{h}(\theta_r)\right]\right\|_H \leq$$

$$\leq \left\|\sum_i \pi_i\left(\theta\right)\left[\underline{a}\left(c\left(\theta_{i+1}\right)\right),\frac{\Delta_i\left[\underline{h},\underline{\theta}\right]}{\Delta_i\left[c,\underline{\theta}\right]}\right]\right\|_H$$

$$+\left\|\sum_i \pi_i\left(\theta\right)\left[\underline{a}'\left(c\left(\eta_i\right)\right),\underline{h}(\theta_i)\right]-\left[\underline{a}'\left(c\left(\theta_r\right)\right),\underline{h}(\theta_r)\right]\right\|_H .$$

Let $A\left(\underline{\theta}\right)$ and $B\left(\underline{\theta}\right)$ be the first, respectively, the second term of the right-hand side of the latter inequality. One must prove that both can be made negligible.

Case of $B\left(\underline{\theta}\right)$: One has that

$$0 \leq B\left(\underline{\theta}\right) \leq \sum_i \pi_i\left(\theta\right)\left\|\left[\underline{a}'\left(c\left(\eta_i\right)\right),\underline{h}(\theta_i)\right]-\left[\underline{a}'\left(c\left(\theta_r\right)\right),\underline{h}(\theta_r)\right]\right\|_H .$$

Let

$$\kappa\left(\theta_l\right)=\sup_{s\in[\theta_l,\theta_r]}\left\|\left[\underline{a}'\left(c\left(s\right)\right),\underline{h}(\theta_i)\right]-\left[\underline{a}'\left(c\left(\theta_r\right)\right),\underline{h}(\theta_r)\right]\right\|_H .$$

Then $0 \leq B\left(\underline{\theta}\right) \leq \kappa\left(\theta_l\right)$, and $\lim_{\theta_l\uparrow\theta_r}\kappa\left(\theta_l\right)=0$.

Case of $A\left(\underline{\theta}\right)$: Since $\underline{h}$ is a martingale in the wide sense,

$$A^2\left(\underline{\theta}\right)=\sum_i \frac{\pi_i^2\left(\theta\right)}{\Delta_i^2\left[c,\underline{\theta}\right]}\left\langle\left\{F_{\underline{h}}\left(\theta_{i+1}\right)-F_{\underline{h}}\left(\theta_i\right)\right\}\left[\underline{a}\left(c\left(\theta_{i+1}\right)\right)\right],\underline{a}\left(c\left(\theta_{i+1}\right)\right)\right\rangle_{\mathbb{R}^n} .$$

Let

$$\kappa\left[\theta_l,\theta_r\right]=\sup_{s\in[\theta_l,\theta_r]}\left\|\underline{a}\left(c\left(s\right)\right)\right\|_{\mathbb{R}^n} .$$

As, for a symmetric, positive definite matrix $M=\sum_{i=1}^n \mu_i \underline{m}_1\otimes\underline{m}_i$,

$$0 \leq \left\langle M\underline{x},\underline{x}\right\rangle_{\mathbb{R}^n}=\sum_{i=1}^n \mu_i\left\langle\underline{m}_i,\underline{x}\right\rangle_{\mathbb{R}^n}^2 \leq \left\|\underline{x}\right\|_{\mathbb{R}^n}^2\sum_{i=1}^n \mu_i=\operatorname{trace}\left(M\right)\left\|\underline{x}\right\|_{\mathbb{R}^n}^2 ,$$

and, for matrices M_1, M_2,

$$\operatorname{trace}\left(M_1-M_2\right)=\operatorname{trace}\left(M_1\right)-\operatorname{trace}\left(M_2\right),$$

one has that

$$0 \leq \left\langle \left\{ F_{\underline{h}}\left(\theta_{i+1}\right) - F_{\underline{h}}\left(\theta_i\right) \right\} \left[\underline{a}\left(c\left(\theta_{i+1}\right)\right) \right], \underline{a}\left(c\left(\theta_{i+1}\right)\right) \right\rangle_{\mathbb{R}^n}$$
$$\leq \kappa^2 \left[\theta_l, \theta_r\right] \left\{ F^{\underline{h}}\left(\theta_{i+1}\right) - F^{\underline{h}}\left(\theta_i\right) \right\}.$$

Consequently, as $\pi(\underline{\theta})\pi_i(\underline{\theta}) = \frac{\Delta_i^2[c,\underline{\theta}]}{\Delta_i[F^{\underline{h}},\underline{\theta}]}$, and $\sum_i \pi_i(\underline{\theta}) = 1$,

$$A^2\left(\underline{\theta}\right) \leq \kappa^2\left[\theta_l, \theta_r\right] \sum_i \frac{\pi_i^2\left(\theta\right)}{\Delta_i^2\left[c, \underline{\theta}\right]} \left\{ F^{\underline{h}}\left(\theta_{i+1}\right) - F^{\underline{h}}\left(\theta_i\right) \right\} \leq \frac{\kappa^2\left[\theta_l, \theta_r\right]}{\pi\left(\underline{\theta}\right)}.$$

The assumption on $\pi\left(\underline{\theta}\right)$ (it increases to infinity) leads to

$$A\left(\underline{\theta}\right) \leq \sqrt{\theta_r - \theta_l}.$$

The same argument obtains for all derivatives. □

Remark 8.4.90 Further meaning for condition (c) in (Proposition) 8.4.89 may be obtained as follows. Let $\mathcal{T}_n\left(\left[\theta_l, \theta_r\right]\right)$ be the σ-algebra generated on $\left[\theta_l, \theta_r\right]$ by the dyadic intervals of order n. Let μ and ν be two measures on Borel σ-algebra of T, and μ_n and ν_n their respective restrictions to $\mathcal{T}_n\left(\left[\theta_l, \theta_r\right]\right)$. Let

$$\left\{ \Theta_n^{(i)},\ i \in I_n,\ \left|I_n\right| < \aleph_0 \right\}$$

be a partition of $\left[\theta_l, \theta_r\right]$ by sets of positive measure for μ_n in $\mathcal{T}_n\left(\left[\theta_l, \theta_r\right]\right)$ and generating it (modulo sets of measure zero), and let

$$f_n\left(t\right) = \sum_i \chi_{\Theta_n^{(i)}}\left(t\right) \frac{\nu_n\left(\Theta_n^{(i)}\right)}{\mu_n\left(\Theta_n^{(i)}\right)}.$$

f_n is the Radon-Nikodým derivative of ν_n with respect to μ_n. It can be shown [201, p. 50] that $\{f_n,\ n \in \mathbb{N}\}$ is almost surely convergent with respect to μ to the Radon-Nikodým derivative of ν with respect to μ. One has that

$$\int f_n^2 d\mu = \sum_i \frac{\nu_n^2\left(\Theta_n^{(i)}\right)}{\mu_n\left(\Theta_n^{(i)}\right)}.$$

Having $\Delta_i\left[c, \underline{\theta}\right]$ taking the part of $\nu_n(\Theta_n^{(i)})$, and $\Delta_i\left[F^{\underline{h}}, \underline{\theta}\right]$, that of $\mu_n(\Theta_n^{(i)})$, one has, when it obtains, that the Radon-Nykodým derivative of the measure determined by c, with respect to that determined by $F^{\underline{h}}$, is in L_1, but not in L_2.

Corollary 8.4.91 *Given the assumptions of* (Proposition) 8.4.89, *when the Wronskian of* $\underline{a}(t)$ *is different from zero at* $t = c\left(t_r\right)$, *its derivatives are linearly* *independent and, consequently,* $\underline{h}(t_r) \in L_{t_r}[f]$, *and the representation of* (Proposition) 8.4.89 *is proper.*

Example 8.4.92 Let $T = [0, 1]$, and $\underline{h}(t) = \underline{W}_t$ be a standard Wiener process in the wide sense, of dimension two. Then $F^{\underline{h}}(t) = 2t$. Let

$$c(t) = \sqrt{t} = \int_0^t \frac{1}{2\sqrt{\theta}} d\theta.$$

As, for $a < b$, fixed, but arbitrary,

$$\frac{\left(\sqrt{b} - \sqrt{a}\right)^2}{b-a} = \frac{\left(\int_a^b \frac{1}{2\sqrt{\theta}} d\theta\right)^2}{b-a} \geq \frac{1}{4b}(b-a) = \frac{1}{4}\left(1 - \frac{a}{b}\right),$$

when $a = \frac{i}{n}$ and $b = \frac{i+1}{n}$,

$$\frac{1}{4}\left(1 - \frac{a}{b}\right) = \frac{1}{4(i+1)},$$

and condition (c) of (Proposition) 8.4.89 obtains for any interval contained in T. Let

$$\underline{\alpha}(t) = \begin{bmatrix} 1 \\ e^t \end{bmatrix}.$$

Its Wronskian is e^t. Consequently

$$f(t) = W_t^{(1)} + e^{\sqrt{t}} W_t^{(2)}$$

is a proper Goursat representation $[\underline{a}, \underline{h}(t)]$, with $\underline{a} = \underline{\alpha} \circ c$ and $\underline{h}(t) = \underline{W}_t$.

Remark 8.4.93 The example just seen shows how to obtain explicitly Markov processes of order n in the wide sense, given a non-singular CH-martingale in the wide sense: one chooses c strictly increasing and continuous, so that condition (c) of (Proposition) 8.4.89 obtains, and then a vector of exponentials, say

$$a_i(t) = e^{\kappa(i-1)t}, \ i \in [1:n].$$

$f(t) = [\underline{a}(c(t)), \underline{h}(t)]$ is then a proper representation with a smooth, deterministic part.

Remark 8.4.94 Processes within the fold of (Proposition) 8.4.89 have the following property: for fixed, but arbitrary $\{t_l, t_r\} \subseteq T, \ t_l < t_r$,

$$L_{]t_l, t_r]}[f] = L_{]t_l, t_r]}[\underline{h}].$$

But then, t_r being fixed, but arbitrary,

$$\cap_{\{t_l<t_r\}} L_{]t_l,t_r]}\left[f\right] = \cap_{\{t_l<t_r\}} L_{]t_l,t_r]}\left[\underline{h}\right].$$

As $\underline{h}$ is continuous to the left, when $\underline{h}$ is non-singular,

$$\cap_{\{t_l<t_r\}} L_{]t_l,t_r]}\left[f\right] = L\left[\underline{h}(t_r)\right],$$

whose dimension is n.

Remark 8.4.95 Let $\underline{h}$ be a non-singular CH-martingale, and

$$f(t) = \left[\underline{a}\left(t\right), \underline{h}(t)\right].$$

When, for $t \in T$, fixed, but arbitrary,

$$\dim \cap_{\{\theta \in T, \theta < t\}} L_{]\theta,t]}\left[f\right] = n,$$

the representation of f is proper. Indeed, since, given the assumption on $\underline{h}$, $\dim L\left[\underline{h}(t)\right] = n$, and

$$\cap_{\{\theta \in T, \theta < t\}} L_{]\theta,t]}\left[f\right] \subseteq \cap_{\{\theta \in T, \theta < t\}} L_{]\theta,t]}\left[\underline{h}\right] = L\left[\underline{h}(t)\right],$$

the latter inclusion is an equality, and $\underline{h}(t) \in L_t\left[f\right]$.

Chapter 9
Cramér-Hida Representations via the Prediction Process

9.1 Introductory Remarks

The prediction process of Knight [157, 158, 160] has "wide sense" and "strict sense" interpretations, with the meaning of those terms that of Doob [78]. One shall thus use, in what follows, the term "process" for a map $t \mapsto f(t)$ as well as for a *bona fide* process $(\omega, t) \mapsto X(\omega, t)$.

Prediction processes provide yet another way to the CHR representation, with the added advantage of a bond between a process and its representation that is tighter than that provided by other ways to such a representation. Both types of results (wide sense and strict sense) make the rather demanding assumption, given the context, that explicit expressions for the projection process be available. That such an assumption is not utterly unreasonable is due to at least two reasons.

The first reason is that, with Goursat processes, one has a fairly wide class of processes for which the prediction process has an "immediate" expression, when the representation is proper, that is,

$$P_t\,[f(t+\theta)] = P_t\,[[\underline{a}(t+\theta), \underline{h}(t+\theta)]] = [\underline{a}(t+\theta), \underline{h}(t)]\,,$$

and computations with the prediction process are thus to a large extent computations with $\underline{a}$ (the variable is θ).

Remark 9.1.1 As already seen in Sect. 8.4, to get a proper representation may not be easy. In simple cases already, finding the prediction process amounts to solving explicitly differential equations. Here is a simple example. U and V are independent standard Wiener processes, s is sine, c, cosine. Set

$$X_t = s(t)\,U_t + c(t)\,V_t.$$

A.F. Gualtierotti, *Detection of Random Signals in Dependent Gaussian Noise*,
DOI 10.1007/978-3-319-22315-5_9

One may write

$$X_t = [\underline{a}(t), \underline{h}(t)],\ \underline{a}(t) = \begin{bmatrix} s(t) \\ c(t) \end{bmatrix},\ \underline{h}(t) = \begin{bmatrix} U_t \\ V_t \end{bmatrix}.$$

The proper representation of X is [(Proposition) 8.4.44]

$$[\underline{a}(t), \underline{h}_X(t)], \quad \text{with } \underline{h}_X(t) = P_t^X[\underline{h}(t)].$$

The elements of $L_t[X]$ have the following generic form [(Fact) 6.2.23]:

$$\int_0^t \phi(y) X(dy) =$$

$$= \int_0^t \phi(x) \{c(x) U_x dx + s(x) U(dx) - s(x) V_x dx + c(x) V(dx)\}.$$

To have a proper representation, one must obtain the projection of U_t and V_t onto $L_t[X]$. One must thus compute, for $\theta \leq t$,

$$\langle U_t, X_\theta \rangle_{L_2(\Omega, \mathcal{A}, P)} = \theta s(\theta) \ \text{ and } \ \langle \int_0^t \phi(y) X(dy), X_\theta \rangle_{L_2(\Omega, \mathcal{A}, P)}.$$

A calculation yields that the latter inner product is the right-hand side of the following equality, which, in principle, determines ϕ:

$$\theta s(\theta) = \int_0^\theta \phi(x) \{x s(\theta - x) + c(\theta - x)\} dx + \theta \int_\theta^t \phi(x) s(\theta - x) dx. \qquad (\star)$$

To simplify that expression, one differentiates it twice using the following formula [262, p. 255]: when

$$g(t) = \int_{a(t)}^{b(t)} f(t, x) dx,$$

then

$$g'(t) = \int_{a(t)}^{b(t)} \frac{\partial f}{\partial t}(t, x) dx + f(t, b(t)) b'(t) - f(t, a(t)) a'(t).$$

The result is:

$$2c(\theta) - \theta s(\theta) =$$
$$= \phi'(\theta) + 2\int_\theta^t \phi(x)c(\theta - x)\,dx$$
$$- \left\{ \int_0^\theta \phi(x)\,\{xs(\theta - x) + c(\theta - x)\}\,dx + \theta\int_\theta^t \phi(x)s(\theta - x)\,dx \right\}.$$

Since the right-hand side parenthesis is $\theta s(\theta)$, the end result is

$$2c(\theta) = \phi'(\theta) + 2\int_\theta^t \phi(x)\,c(\theta - x)\,dx.$$

Integrating the whole expression, and interchanging order of integration on the right-hand side, yields that

$$2s(t) = \phi(t) + 2\int_0^t \phi(x)\,s(s)\,dx,$$

and then, differentiating again, one has the following first order differential equation:

$$2c(t) = \phi'(t) + 2s(t)\phi(t).$$

Its solution, using the formula of [2, p. 310, Vol. I], is:

$$\phi_U(t) = 2e^{2c(t)}\int_0^t c(x)\,e^{-2c(x)}dx.$$

For V_t, the only thing that changes in the calculation is the left-hand side of ($\star$), where $\theta s(\theta)$ is replaced by $\theta c(\theta)$, and the result is

$$\phi_V(t) = -2e^{2c(t)}\int_0^t s(x)\,e^{-2c(x)}dx.$$

Thus the proper representation is

$$X_t = s(t)\int_0^t \phi_U\,dX + c(t)\int_0^t \phi_V\,dX.$$

The second reason why there is hope to obtain, in certain cases, an explicit expression for the projection process is that the projection $P_t\,[f(t+\theta)]$ can be obtained [(Corollary) 1.6.23] as the limit of expressions of the form

$$\left[\Sigma^{-1}\underline{c}, \underline{f}\right],$$

where $\underline{c}$ and Σ have components whose values are given by the covariance of f, and $\underline{f}$ is made of "time observations" of f.

The approach to CHR using the prediction process enables one to construct, from the projections $P_t\left[f(t+\theta)\right]$, the elements ϕ and h that enter the "multiplicity representation" of f, in the simplest case:

$$f(t) = \int I_{T_t}\phi_t dm_h = \int \phi_t dm_h.$$

The difference with the CHR representations obtained earlier is as follows. A CHR representation is generally an existence result: one knows that a ϕ and a h exist. It does not say how one constructs them directly from f. This is what the method based on the prediction process achieves to a certain extent (it yields more for multiplicity one than for higher multiplicity, and always requires a "starting" integral representation). The prediction process approach, and there lies perhaps its main interest, produces a family of martingales (those of (Definition) 9.2.9) that embody the original process' properties.

Formally, the prediction method consists in introducing an object that is reminiscent of the resolvent in the theory of Markov processes, and makes substantial use of the Laplace transform. Smoothness properties are more stringent than is the case for the general CHR representation.

9.2 The Case of Karhunen Representations

In this section one assumes that the map f has the so-called Karhunen representation:

$$f(t) = \int_0^t \phi_t dm_h,$$

where h has orthogonal increments, and ϕ_t is an equivalence class in

$$L_2(T, \mathcal{T}, M_h),$$

with the property that $I_{[0,t]}\phi_t = \phi_t$, of which one is reminded when an integral symbol whose bounds are 0 and t is used. The apparent simplicity of such a representation is rather deceptive, as evidenced by the following example [144], which shows that representing functions as integrals is, in this context, not much of a limitation. The requirement of multiplicity one is thus always stringent.

Example 9.2.1 Let, for $k \in \mathbb{N}_0$, fixed, but arbitrary,

$$I_k = \left]\frac{1}{2^{k+1}}, \frac{1}{2^k}\right], \quad \text{and } J_k = \left]2^k, 2^{k+1}\right].$$

One has, for strictly positive integer l,

$$\frac{1}{2^l} I_k = I_{k+l}, \quad \text{and } \frac{1}{2^l} J_k = \begin{cases} J_{k-l} \text{ when } k \geq l \\ I_{l-k} \text{ when } k < l \end{cases}.$$

Those intervals have a natural order: $\ldots, I_3, I_2, I_1, I_0, J_0, J_1, J_2, J_3, \ldots$ For $p \in \mathbb{N}$, fixed, but arbitrary, let these intervals be assembled as follows, to form disjoint subsets $S_0, S_1, \ldots, S_{p-1}$ (one considers, to be explicit, the case $p = 4$, and retains, to form one set, every fourth interval):

$$\begin{array}{cccc} \vdots & \vdots & \vdots & \vdots \\ I_{11} & I_{10} & I_9 & I_8 \\ I_7 & I_6 & I_5 & I_4 \\ I_3 & I_2 & I_1 & I_0 \\ J_0 & J_1 & J_2 & J_3 \\ J_4 & J_5 & J_6 & J_7 \\ J_8 & J_9 & J_{10} & J_{11} \\ \vdots & \vdots & \vdots & \vdots \\ \hline S_0 & S_1 & S_2 & S_3 \end{array}$$

Thus, letting $i \in \mathbb{N}_0$ and $j \in \mathbb{N}$, S_0 is the union of the sets J_{4i} and I_{4j-1}; S_1, J_{4i+1} and I_{4j-2}; S_2, J_{4i+2} and I_{4j-3}; S_3, J_{4i+3} and I_{4j-4}. Those sets have particular transformation properties as follows:

Case of S_0: $\frac{1}{2^1} S_0 = S_3$, $\frac{1}{2^2} S_0 = S_2$, $\frac{1}{2^3} S_0 = S_1$, $\frac{1}{2^4} S_0 = S_0$

Indeed

$$\frac{1}{2^k} I_{4i-1} = \left]\frac{1}{2^{4i+k}}, \frac{1}{2^{4i+k-1}}\right] = \begin{cases} k = 1 : I_{4i} & = I_{4(i+1)-4} \in S_3 \\ k = 2 : I_{4i+1} & = I_{4(i+1)-3} \in S_2 \\ k = 3 : I_{4i+2} & = I_{4(i+1)-2} \in S_1 \\ k = 4 : I_{4(i+1)-1} & \in S_0 \end{cases},$$

and

$$\frac{1}{2^k} J_{4i} = \left]2^{4i-k}, 2^{4i-k+1}\right] = \begin{cases} k=1, i=0 : I_0 & \in S_3 \\ k=1, i\geq 1 : J_{4i-1} \quad = J_{4(i-1)+3} & \in S_3 \\ k=2, i=0 : I_1 & \in S_2 \\ k=2, i\geq 1 : J_{4i-2} \quad = J_{4(i-1)+2} & \in S_2 \\ k=3, i=0 : I_2 & \in S_1 \\ k=3, i\geq 1 : J_{4i-3} \quad = J_{4(i-1)+1} & \in S_1 \\ k=4, i=0 : I_3 & \in S_0 \\ k=4, i\geq 1 : J_{4(i-1)} & \in S_0 \end{cases}.$$

Case of S_1: $\frac{1}{2^1}S_1 = S_0$, $\frac{1}{2^2}S_1 = S_3$, $\frac{1}{2^3}S_1 = S_2$, $\frac{1}{2^4}S_1 = S_1$

Indeed

$$\frac{1}{2^k} I_{4i-2} = \left]\frac{1}{2^{4i+k-1}}, \frac{1}{2^{4i+k-2}}\right] = \begin{cases} k=1 : I_{4i-1} & \in S_0 \\ k=2 : I_{4i} \quad = I_{4(i+1)-4} & \in S_3 \\ k=3 : I_{4i+1} \quad = I_{4(i+1)-3} & \in S_2 \\ k=4 : I_{4(i+1)-2} & \in S_1 \end{cases},$$

and

$$\frac{1}{2^k} J_{4i+1} = \left]2^{4i+1-k}, 2^{4i+2-k}\right] = \begin{cases} k=1 \quad : J_{4i} & \in S_0 \\ k=2, i=0 : I_0 & \in S_3 \\ k=2, i\geq 1 : J_{4i-1} \quad = J_{4(i-1)+3} & \in S_3 \\ k=3, i=0 : I_1 & \in S_2 \\ k=3, i\geq 1 : J_{4i-2} \quad = I_{4(i-1)+2} & \in S_2 \\ k=4, i=0 : I_2 & \in S_1 \\ k=4, i\geq 1 : J_{4(i-1)+1} & \in S_1 \end{cases}.$$

Case of S_2: $\frac{1}{2^1}S_2 = S_1$, $\frac{1}{2^2}S_2 = S_0$, $\frac{1}{2^3}S_2 = S_3$, $\frac{1}{2^4}S_2 = S_2$

Indeed

$$\frac{1}{2^k} I_{4i-3} = \left]\frac{1}{2^{4i+k-2}}, \frac{1}{2^{4i+k-3}}\right] = \begin{cases} k=1 : I_{4i-2} & \in S_1 \\ k=2 : I_{4i-1} & \in S_0 \\ k=3 : I_{4i} \quad = I_{4(i+1)-4} & \in S_3 \\ k=4 : I_{4(i+1)-3} & \in S_2 \end{cases},$$

and

$$\frac{1}{2^k}J_{4i+2} =]2^{4i+2-k}, 2^{4i+3-k}] = \begin{cases} k=1 & : J_{4i+1} & & \in S_1 \\ k=2 & : J_{4i} & & \in S_0 \\ k=3, i=0 & : I_0 & & \in S_3 \\ k=3, i\geq 1 & : J_{4i-1} & = I_{4(i-1)+3} & \in S_3 \\ k=4, i=0 & : I_1 & & \in S_2 \\ k=4, i\geq 1 & : J_{4(i-1)+2} & & \in S_2 \end{cases}.$$

Case of S_3: $\frac{1}{2^1}S_3 = S_2$, $\frac{1}{2^2}S_3 = S_1$, $\frac{1}{2^3}S_3 = S_0$, $\frac{1}{2^4}S_3 = S_3$

Indeed

$$\frac{1}{2^k}I_{4i-4} = \left]\frac{1}{2^{4i+k-3}}, \frac{1}{2^{4i+k-4}}\right] = \begin{cases} k=1 : I_{4i-3} & \in S_2 \\ k=2 : I_{4i-2} & \in S_1 \\ k=3 : I_{4i-1} & \in S_0 \\ k=4 : I_{4(i+1)-4} & \in S_3 \end{cases},$$

and

$$\frac{1}{2^k}J_{4i+3} =]2^{4i+3-k}, 2^{4i+4-k}] = \begin{cases} k=1 & : J_{4i+2} & \in S_2 \\ k=2 & : J_{4i+1} & \in S_1 \\ k=3 & : J_{4i} & \in S_0 \\ k=4, i=0 & : I_0 & \in S_3 \\ k=4, i\geq 0 & : J_{4(i-1)+3} & \in S_3 \end{cases}.$$

When, for example, $t \in S_0$, $t/2 \in S_0/2 = S_3$, and thus

$$\chi_{]0,\frac{t}{2}]\cap S_3} = \begin{cases} 1 & \text{when } t \in S_0 \\ 0 & \text{when } t \in S_0^c \end{cases}.$$

Generally,

$$\chi_{\left]0,\frac{t}{2^{p-k}}\right]\cap S_k} = \begin{cases} 1 \text{ when } t \in S_0 \\ 0 \text{ when } t \in S_0^c \end{cases}. \qquad (\star)$$

To summarize:

$\frac{1}{2^{4-k}}S_k$	S_0	S_1	S_2	S_3	$\chi_{\left]0,\frac{t}{2^{4-k}}\right]\cap S_k}(t)$	$t \in S_0$	$t \in S_1$	$t \in S_2$	$t \in S_3$
$\frac{1}{2^{4-3}}$	S_3	S_0	S_1	S_2	$k=3$	1	0	0	0
$\frac{1}{2^{4-2}}$	S_2	S_3	S_0	S_1	$k=2$	1	0	0	0
$\frac{1}{2^{4-1}}$	S_1	S_2	S_3	S_0	$k=1$	1	0	0	0
$\frac{1}{2^{4-4}}$	S_0	S_1	S_2	S_3	$k=0$	1	0	0	0

.

Let W be a Wiener process in the wide sense, and let, for $k \in [1:3]$, fixed, but arbitrary,

$$h_k(t) = \int \left[\chi_{\left]0, \frac{t}{2p-k}\right] \cap S_k} \right] dm_W.$$

Then

$$\langle h_k(t_1), h_l(t_2) \rangle =$$

$$= \int \left[\chi_{\left\{\left]0, \frac{t_1}{2p-k}\right] \cap \left]0, \frac{t_2}{2p-l}\right] \cap S_k \cap S_l\right\}} \right] dLeb$$

$$= \begin{cases} 0 & \text{when } l \neq k \\ \int \left[\chi_{\left\{\left]0, \frac{t_1 \wedge t_2}{2p-k}\right] \cap S_k\right\}} \right] dLeb & \text{when } l = k \end{cases}.$$

Consequently, when $l \neq k$, h_k, and h_l are orthogonal, and when $l = k$ and $t_1 \leq t_2$,

$$\langle h_k(t_1), h_k(t_2) \rangle = \langle h_k(t_1), h_k(t_1) \rangle,$$

so that $h_k(t_2) - h_k(t_1) \perp h_k(t_1)$: h_k is a martingale in the wide sense. Because of $(\star)$, $\|h_k(t)\|^2$ increases as t on S_0, and is constant on S_0^c, independently of k.

Let φ be a function with the properties of the function c in (Proposition) 8.4.89. Then

$$f(t) = \int_0^t \sum_{k=0}^{p-1} \left[\chi_{\left]0, \frac{t}{2p-k}\right] \cap S_k} \varphi^k(t) \right] dm_W = \sum_{k=0}^{p-1} \varphi^k(t) h_k(t),$$

so that f has multiplicity p [(Corollary) 8.4.91].

9.2.1 Notation, Assumptions, and Some Consequences

The reasons for the assumptions stated below shall emerge within the developments to follow.

Assumptions 9.2.2 *The tacit assumptions for the sequel shall be that:*

1. $T = \mathbb{R}_+$,
2. $f : T \longrightarrow H$ *is a continuous map with (continuous) covariance* C_f,

3. *for fixed, but arbitrary* $\lambda > 0$,

$$\Gamma_f(\lambda) = \int_T e^{-\lambda\theta} C_f(\theta, \theta)\, d\theta < \infty,$$

(an assumption based on the continuity of C_f*).*

Definition 9.2.3 In the sequel, repeated use of the following facts shall be made. Let Π_α^a denote the probability on $\mathcal{B}([a, \infty[)$ with density

$$\Pi_\alpha^a(d\theta) = \chi_{[a,\infty[}(\theta)\, \alpha\, e^{-\alpha(\theta - a)} d\theta.$$

For appropriate f,

$$E_{\Pi_\alpha^a}[f] = \alpha\, e^{\alpha a} \int_a^\infty e^{-\alpha\theta} f(\theta)\, d\theta = \alpha \int_0^\infty e^{-\alpha\theta} f(a + \theta)\, d\theta.$$

Consequently, for $f \geq 0$,

$$E_{\Pi_\alpha^a}[f] \leq E_{\Pi_\alpha^a}^{1/2}\left[f^2\right],$$

or

$$\sqrt{\alpha} \int_0^\infty e^{-\alpha\theta} f(a + \theta)\, d\theta \leq \left\{ \int_0^\infty e^{-\alpha\theta} f^2(a + \theta)\, d\theta \right\}^{1/2}.$$

Lemma 9.2.4

1. Γ_f *is continuous;*
2. $\int_T e^{-\alpha\theta} f(t + \theta)\, d\theta$ *may be used as a direct integral;*
3. $\int_T e^{-\alpha\theta} f(t + \theta)\, d\theta$ *may be used as a Bochner integral.*

Proof Item 1 is the consequence of the theorem on the continuity of integrals depending on a parameter [113, p. 136]. The other claims depend on the fact that

$$C_f(t, t) = \|f(t)\|_H^2.$$

$$\begin{aligned}
\left\| \int_T e^{-\alpha\theta} f(t + \theta)\, d\theta \right\|_{L_2^H(T, \mathcal{T}, Leb)}^2 &= \int_T \left\| e^{-\alpha\theta} f(t + \theta) \right\|_H^2 d\theta \\
&= \int_T e^{-2\alpha\theta} C_f(t + \theta, t + \theta)\, d\theta \\
&= e^{2\alpha t} \int_t^\infty e^{-2\alpha u} C_f(u, u)\, du \\
&\leq e^{2\alpha t} \Gamma_f(2\alpha),
\end{aligned}$$

and

$$\begin{aligned}\int_T \left\|e^{-\alpha\theta}f(t+\theta)\right\|_H d\theta &= \int_T e^{-\alpha\theta}\left\|f(t+\theta)\right\|_H d\theta \\ &= \alpha^{-1}E_{\Pi_\alpha^0}\left[\|f(t+\cdot)\|_H\right] \\ &\leq \alpha^{-1}E_{\Pi_\alpha^0}^{1/2}\left[\|f(t+\cdot)\|_H^2\right] \\ &= \alpha^{-1}\left\{\alpha\int_T e^{-\alpha\theta}C_f\left(t+\theta,t+\theta\right)d\theta\right\}^{1/2} \\ &= \alpha^{-1/2}\left\{e^{\alpha t}\int_t^\infty e^{-\alpha\theta}C_X\left(\theta,\theta\right)d\theta\right\}^{1/2} \\ &\leq \left\{\frac{e^{\alpha t}}{\alpha}\right\}^{1/2}\Gamma_f^{1/2}\left(\alpha\right).\end{aligned}$$

□

Remark 9.2.5 One of the consequences of Bochner integrability is that operators and integral signs commute [135, p. 83]: for projections, one has that

$$P_t\int=\int P_t.$$

Remark 9.2.6 Since f is continuous, $f(t)\in L_t^-\left[f\right]=L_t\left[f\right]$.

Remark 9.2.7 The proper canonical representation of f, when the multiplicity is one, shall be denoted

$$f(t)=\int_0^t F_t dm_h,$$

and the projection onto $L_t\left[f\right]$, P_t.

Lemma 9.2.8 *For $\lambda>0$, and $t\in T$, fixed, but arbitrary, let*

$$\begin{aligned}j\left(t,\lambda\right)&=\int_T e^{-\lambda\theta}f(t+\theta)d\theta,\\ q\left(t,\lambda\right)&=\lambda\int_T e^{-\lambda\theta}P_t\left[f(t+\theta)\right]d\theta.\end{aligned}$$

Then:

1. *$\theta\mapsto P_t\left[f(t+\theta)\right]$ is continuous;*
2. *$q(t,\lambda)=\lambda P_t\left[j(t,\lambda)\right]$;*
3. *$t\mapsto j\left(t,\lambda\right)$, and $\lambda\mapsto j(t,\lambda)$, are continuous;*
4. *$t\mapsto q\left(t,\lambda\right)$ is continuous to the left, and $\lambda\mapsto q(t,\lambda)$ is continuous.*

Proof Item 1 expresses the continuity of f, and item 2, the interchange property [(Remark) 9.2.5] of the Bochner integral. For item 3, one has that (Π_α^a is defined in (Definition) 9.2.3):

$$\begin{aligned}\|j(t+\tau,\lambda)-j(t,\lambda)\|_H &\le \int_T e^{-\lambda\theta}\,\|f(t+\tau+\theta)-f(t+\theta)\|_H\,d\theta\\ &= e^{\lambda t}\int_t^\infty e^{-\lambda u}\,\|f(\tau+u)-f(u)\|_H\,du\\ &\le e^{\lambda t}\int_T e^{-\lambda u}\,\|f(\tau+u)-f(u)\|_H\,du\\ &= \lambda^{-1}e^{\lambda t}E_{\Pi_\lambda^0}\left[\|f(\tau+\cdot)-f(\cdot)\|_H\right]\\ &\le \lambda^{-1}e^{\lambda t}E^{1/2}_{\Pi_\lambda^0}\left[\|f(\tau+\cdot)-f(\cdot)\|_H^2\right]\\ &= \lambda^{-1}e^{\lambda t}\left\{\lambda\int_T e^{-\lambda\theta}\,\|f(\tau+\theta)-f(\theta)\|_H^2\,d\theta\right\}^{1/2}\\ &= \frac{e^{\lambda t}}{\sqrt{\lambda}}\left\{\int_T e^{-\lambda\theta}\,\|f(\tau+\theta)-f(\theta)\|_H^2\,d\theta\right\}^{1/2}.\end{aligned}$$

Let

$$\begin{aligned}\phi_\tau(\theta) &= e^{-\lambda\theta}\,\|f(\tau+\theta)-f(\theta)\|_H^2\,,\\ \psi_\tau(\theta) &= 2e^{-\lambda\theta}\left\{\|f(\tau+\theta)\|_H^2+\|f(\theta)\|_H^2\right\}.\end{aligned}$$

Then

$$\lim_{\tau\downarrow 0}\phi_\tau(\theta)=0,\ \lim_{\tau\downarrow 0}\psi_\tau(\theta)=4e^{-\lambda\theta}\,\|f(\theta)\|_H^2\,,\ \phi_\tau(\theta)\le\psi_\tau(\theta).$$

Furthermore, as seen [proof of (Lemma) 9.2.4]

$$\int_T \left\|e^{-\lambda\theta}f(t+\theta)\right\|_H^2 d\theta = e^{2\lambda t}\int_t^\infty e^{-2\lambda\theta}C_f(\theta,\theta)\,d\theta,$$

so that

$$\begin{aligned}\lim_{\tau\downarrow 0}\int_T \psi_\tau(\theta)\,d\theta &= 2\lim_{\tau\downarrow 0}e^{2\lambda\tau}\int_\tau^\infty e^{-2\lambda\theta}C_f(\theta,\theta)\,d\theta\\ &\quad+2\int_0^\infty e^{-2\lambda\theta}C_f(\theta,\theta)\,d\theta\\ &= 4\int_0^\infty e^{-2\lambda\theta}C_f(\theta,\theta)\,d\theta\\ &= \int_0^\infty \lim_{\tau\downarrow 0}\psi_\tau(\theta)\,d\theta.\end{aligned}$$

The dominated convergence theorem [226, p. 232] yields then that

$$\lim_{\tau \downarrow 0} \int_T e^{-\lambda\theta} \|f(\tau+\theta) - f(\theta)\|_H^2 \, d\theta = 0,$$

independently of $t \in T$ (thus on bounded intervals, continuity is uniform). The continuity of j with respect to λ follows from the fact that it is differentiable [262, p. 253]. Now, for $\tau > 0$, fixed, but arbitrary, inserting and subtracting $P_{t-\tau}\,[j(t,\lambda)]$ to obtain the inequality,

$$\begin{aligned} &\|q(t-\tau,\lambda) - q(t,\lambda)\|_H = \\ &\quad = \lambda \, \|P_{t-\tau}\,[j(t-\tau,\lambda)] - P_t\,[j(t,\lambda)]\|_H \\ &\quad \leq \lambda \, \|j(t-\tau,\lambda) - j(t,\lambda)\|_H + \lambda \, \|(P_{t-\tau} - P_t)\,[j(t,\lambda)]\|_H \, . \end{aligned}$$

Since j is continuous, and P_t continuous to the left, q is continuous to the left. Continuity in λ follows from items 2 and 3. □

Definition 9.2.9 The following definition makes sense because of (Lemmas) 9.2.4 and 9.2.8:

$$h_\lambda(t) = q(t,\lambda) - q(0,\lambda) + \lambda \int_0^t \{f(\theta) - q(\theta,\lambda)\}\, d\theta.$$

Maps of type $h_\lambda(t)$ shall be called *Knight's martingales,* as they are, as shall be seen below, martingales in the wide sense.

Example 9.2.10 Let, for W, a standard, wide sense, Wiener process,

$$f(t) = a(t) \int_0^t b(\theta)\, W(d\theta) = a(t)B(t).$$

f is continuous in mean square when a is continuous and b adequately integrable. Then $P_t[f(t+\theta)] = a(t+\theta)B(t)$, so that, with a change of variables,

$$\int_0^\infty e^{-\lambda\theta} P_t[f(t+\theta)]\, d\theta = B(t) \int_0^\infty e^{-\lambda\theta} a(t+\theta)\, d\theta = e^{\lambda t} c(t) B(t),$$

where $c(t) = \int_t^\infty e^{-\lambda\theta} a(\theta)\, d\theta$. Then, letting $C(t) = \lambda\, e^{\lambda t} c(t)$,

$$h_\lambda(t) = C(t)B(t) + \lambda \int_0^t \{a(\theta) - C(\theta)\}\, B(\theta)\, d\theta.$$

Using formula (Fact) 6.2.24, with $\phi'(t) = \lambda \{a(t) - C(t)\}$, omitting notation for equivalence classes, one has that

$$\lambda \int_0^t \{a(\theta) - C(\theta)\} B(\theta)\, d\theta = \phi(t)B(t) - \int_0^t \phi(\theta)B(d\theta).$$

Thus

$$h_\lambda(t) = \{C(t) + \phi(t)\} B(t) - \int_0^t \phi(\theta)B(d\theta).$$

h_λ resembles indeed a wide sense martingale, the obstacle being the factor $C + \phi$. The theory to follow [(Proposition) 9.2.21] shall establish that fact.

Remark 9.2.11 Definition 9.2.9 makes sense as, using the assumption on f, and (Lemma) 9.2.8, item 2,

$$\int_0^t \|f(\theta) - q(\theta, \lambda)\|_H\, d\theta \le \int_0^t \|f(\theta) - \lambda j(\theta, \lambda)\|_H\, d\theta.$$

Remark 9.2.12 $P_t[h_\lambda(t)] = h_\lambda(t)$.

Remark 9.2.13 It is h_λ that provides the process with orthogonal increments of the representation of f based on the prediction process.

Proposition 9.2.16 below requires the following result [192, p. 154] which is now stated for convenience.

Fact 9.2.14 *Suppose that $\mathcal{F}$ is a separable σ-field (countably generated), that $(T, \mathcal{T})$ is a measurable space, and that $\{P_t,\ t \in T\}$ and $\{Q_t,\ t \in T\}$ are two measurable families of bounded, positive measures ($t \mapsto P_t(A)$ is adapted to $\mathcal{T}$ for each $A \in \mathcal{F}$) on $(\Omega, \mathcal{F})$ such that, for $t \in T$, fixed, but arbitrary, $Q_t \ll P_t$. There exists then a positive function $D(\omega, t)$, adapted to $\mathcal{F} \otimes \mathcal{T}$, such that, for fixed, but arbitrary $t \in T$,*

$$\frac{dQ_t}{dP_t}(\omega) = D(\omega, t).$$

Let $\mathcal{F}_n$ be generated by the first n sets of the family generating $\mathcal{F}$. These sets generate a measurable partition

$$\{A_i^{(n)},\ i \in [1 : p_n]\}$$

of Ω. Then the limit of

$$D_n(\omega, t) = \sum_{i=1}^{p_n} \frac{Q_t\left(A_i^{(n)}\right)}{P_t\left(A_i^{(n)}\right)} \chi_{A_i^{(n)}}(\omega),$$

when it exists, yields $D(\omega, t)$. Elsewhere D is zero.

Remark 9.2.15 Fact 9.2.14 obtains also when $\{Q_t,\ t \in T\}$ is a family of bounded, signed measures. It suffices to consider separately the elements of the Jordan decomposition of Q_t.

Proposition 9.2.16 *Let $h : T \longrightarrow H$ have orthogonal increments, and be continuous to the left, and m_h and M_h be as in Sect.* 6.2. *In the representation*

$$f(t) = \int_T \Phi_t dm_h,\ I_{T_t}\Phi_t = \Phi_t,$$

one may assume that Φ_t is the equivalence class, with respect to M_h, of the function $x \mapsto \Phi(t,x)$ obtained, for fixed, but arbitrary $t \in T$, as the section at t of a measurable $(t,x) \mapsto \Phi(t,x)$.

Proof Let $\alpha_t : T \longrightarrow \mathbb{R}$ be defined using the following relation:

$$\alpha_t(\theta) = \langle f(t), h(\theta)\rangle_H.$$

It is of bounded variation as (with the convention $0 \times \infty = 0$)

$$\begin{aligned}
\sum_i |\alpha_t(\theta_{i+1}) - \alpha_t(\theta_i)| &= \sum_i |\langle f(t), h(\theta_{i+1}) - h(\theta_i)\rangle_H| \\
&= \sum_i \left| \int_0^t \chi_{[\theta_i,\theta_{i+1}[}(\theta)\, \dot{\Phi}_t(\theta)\, M_h(d\theta) \right| \\
&\le \sum_i \int_0^t \chi_{[\theta_i,\theta_{i+1}[}(\theta)\, \left|\dot{\Phi}_t(\theta)\right| M_h(d\theta) \\
&= \int_0^t \left|\dot{\Phi}_t(\theta)\right| M_h(d\theta).
\end{aligned}$$

Using the inequality of Cauchy-Schwarz, the last integral is less than, or equal to

$$M_h^{1/2}([0,t])\, \|\Phi_t\|_{L_2[T,\mathcal{T},M_h)},$$

which is finite, as M_h is a measure derived from a monotone increasing function into the reals.

Let μ_t^α be the signed measure associated with α_t: it is absolutely continuous with respect to M_h, and bounded, since, for $\{t_1, t_2\} \subseteq T,\ t_1 < t_2$, fixed, but arbitrary,

$$\mu_t^\alpha([t_1,t_2[) = \langle f(t), h(t_2) - h(t_1)\rangle_H = \int_0^t \chi_{[t_1,t_2[}(\theta)\, \dot{\Phi}_t(\theta)\, M_h(d\theta),$$

and $t \mapsto \mu_t^{\alpha}(A)$ is adapted, since it is, as presently seen, continuous. Indeed

$$\begin{aligned}
\left|\mu_{t+\eta}^{\alpha}(A) - \mu_t^{\alpha}(A)\right| &= \left|\int_A \left\{\dot{\Phi}_{t+\eta}(\theta) - \dot{\Phi}_t(\theta)\right\} M_h(d\theta)\right| \\
&\leq \int_A \left|\dot{\Phi}_{t+\eta}(\theta) - \dot{\Phi}_t(\theta)\right| M_h(d\theta) \\
&\leq M_h^{1/2}(A \cap [0, t+\eta]) \left\|\Phi_{t+\eta} - \Phi_t\right\|_{L_2(T,\mathcal{T},M_h)} \\
&= M_h^{1/2}(A \cap [0, t+\eta]) \left\|f(t+\eta) - f(t)\right\|_H .
\end{aligned}$$

Φ_t is furthermore the equivalence class of the Radon-Nikodým derivative of μ_t^{α} with respect to M_h. But, in the context of (Fact) 9.2.14, the quantity $D(\theta, t)$ is the limit of a sequence of the following type:

$$\sum_{i=1}^{n-1} \chi_{\left[t_i^{(n)}, t_{i+1}^{(n)}\right[}(\theta) \frac{\mu_t^{\alpha}\left(\left[t_i^{(n)}, t_{i+1}^{(n)}\right[\right)}{M_h\left(\left[t_i^{(n)}, t_{i+1}^{(n)}\right[\right)},$$

and one may choose D to be Φ. □

Remark 9.2.17 In (Proposition) 9.2.16, one has, almost surely, with respect to M_h, that $\dot{\Phi}(t, 0) = 0$.

Remark 9.2.18 (Result (Proposition)) 9.2.16 shows how, knowing f and h, one obtains Φ.

Remark 9.2.19 Let W be a Wiener process. (Result (Proposition)) 9.2.16 allows one to switch "seamlessly" from an interpretation of $\int_0^t F(t, x)\, W(dx)$ as an isometric integral to its interpretation as a stochastic integral, and conversely.

Remark 9.2.20 It shall always be tacitly assumed in what follows that ϕ_t is obtained as in (Proposition) 9.2.16.

9.2.2 Properties of Knight's Martingales

Given the part that the Knight's martingales h_λ are going to play, their properties are required, the most important being that they are processes with orthogonal increments. Their second order properties are equally relevant.

Proposition 9.2.21 *h_λ has the following properties:*

1. *$h_\lambda(t) \in L_t[f]$;*
2. *$t \mapsto h_\lambda(t)$ is continuous to the left;*
3. *$\lambda \mapsto h_\lambda(t)$ is continuous;*
4. *for fixed, but arbitrary $\lambda > 0$, h_λ is a martingale in the wide sense with respect to f: for $\{t_1, t_2, t_3\} \subseteq T$, $t_1 \le t_2 \le t_3$, fixed, but arbitrary,*

$$P_{t_1}\left[h_\lambda(t_3) - h_\lambda(t_2)\right] = 0_H;$$

5. *for fixed, but arbitrary $\{\lambda, \mu\} \subseteq]0, \infty[$ and $\{t_1, t_2, t_3\} \subseteq T$, $t_1 \le t_2 \le t_3$,*

$$\langle h_\lambda(t_1), h_\mu(t_3) - h_\mu(t_2)\rangle_H = 0;$$

6. *$P_{t_1}\left[q(t_2, \lambda) - q(t_1, \lambda)\right] = \lambda P_{t_1}\left[\int_{t_1}^{t_2}\{q(\theta, \lambda) - f(\theta)\}\, d\theta\right]$.*

Proof Items 1 and 2 follow directly from the definition of $h_\lambda(t)$ [(Definition) 9.2.9, (Lemma) 9.2.8, (Remark) 9.2.11]. Item 3 also follows from (Definition) 9.2.9, and the fact that

$$\begin{aligned}\left\|\int_0^t q(\theta, \lambda + \epsilon)d\theta - \int_0^t q(\theta, \lambda)d\theta\right\|_H &= \\ &= \left\|(\lambda + \epsilon)\int_0^t P_\theta\left[j(\theta, \lambda + \epsilon)\right]d\theta - \lambda\int_0^t P_\theta\left[j(\theta, \lambda)\right]d\theta\right\|_H \\ &\le |\epsilon|\int_0^t \|j(\theta, \lambda + \epsilon)\|_H\, d\theta + \lambda\int_0^t \|j(\theta, \lambda + \epsilon) - j(\theta, \lambda)\|_H\, d\theta.\end{aligned}$$

From that same definition [(Definition) 9.2.9], one has that

$$h_\lambda(t_2) - h_\lambda(t_1) = q(t_2, \lambda) - q(t_1, \lambda) + \lambda\int_{t_1}^{t_2}\{f(\theta) - q(\theta, \lambda)\}\, d\theta. \qquad (\star)$$

Item 6 follows directly from item 4, applying P_{t_1} to that last equality $(\star)$, and item 5 follows from item 4, and the fact that $h_\lambda(t_1) = P_{t_1}\left[h_\lambda(t_1)\right]$. So one must prove item 4.

As [(Lemma) 9.2.8]

$$
\begin{aligned}
q(t,\lambda) &= \lambda P_t\left[j(t,\lambda)\right] \\
&= \lambda P_t\left[\int_0^\infty e^{-\lambda\theta} f(t+\theta)\,d\theta\right] \\
&= \lambda P_t\left[\int_t^\infty e^{-\lambda(\theta-t)} f(\theta)\,d\theta\right],
\end{aligned}
$$

the right-hand side of that same equality ($\star$) has the following form:

$$
P_{t_2}\left[\lambda\int_{t_2}^\infty e^{-\lambda(\theta-t_2)} f(\theta)\,d\theta\right] - P_{t_1}\left[\lambda\int_{t_1}^\infty e^{-\lambda(\theta-t_1)} f(\theta)\,d\theta\right] +
$$

$$
+\lambda\int_{t_1}^{t_2}\{f(\theta)-q(\theta,\lambda)\}\,d\theta.
$$

Let its successive terms be denoted A, B, and C, so that it is expressed as $A-B+C$. Because $P_t\int=\int P_t$, one has that:

- $P_{t_1}[A]=\lambda\int_{t_2}^\infty e^{-\lambda(\theta-t_2)}P_{t_1}[f(\theta)]\,d\theta$,
- $P_{t_1}[B]=\lambda\int_{t_1}^\infty e^{-\lambda(\theta-t_1)}P_{t_1}[f(\theta)]\,d\theta$,
- for $\theta\geq t_1$, $P_{t_1}[q(\theta,\lambda)]=\lambda\int_\theta^\infty e^{-\lambda(u-\theta)}P_{t_1}[f(u)]\,du$, so that

$$
P_{t_1}[C]=\lambda\int_{t_1}^{t_2}P_{t_1}[f(\theta)]\,d\theta-\lambda^2\int_{t_1}^{t_2}d\theta\int_\theta^\infty du\,e^{-\lambda(u-\theta)}P_{t_1}[f(u)].
$$

Consequently

$$
\begin{aligned}
P_{t_1}\left[h_\lambda(t_2)-h_\lambda(t_1)\right] &= \lambda\int_{t_2}^\infty\left\{e^{-\lambda(\theta-t_2)}-e^{-\lambda(\theta-t_1)}\right\}P_{t_1}[f(\theta)]\,d\theta \\
&- \lambda\int_{t_1}^{t_2}\left\{e^{-\lambda(\theta-t_1)}-1\right\}P_{t_1}[f(\theta)]\,d\theta \\
&- \lambda^2\int_{t_1}^{t_2}d\theta\int_\theta^{t_2}du\,e^{-\lambda(u-\theta)}P_{t_1}[f(u)] \\
&- \lambda^2\int_{t_1}^{t_2}d\theta\int_{t_2}^\infty du\,e^{-\lambda(u-\theta)}P_{t_1}[f(u)].
\end{aligned}
$$

Following an interchange of integration [262, p. 80], the fourth term on the right-hand side of the latter equality becomes

$$-\lambda \int_{t_2}^{\infty} du P_{t_1}\left[f(u)\right]\left\{e^{-\lambda(u-t_2)} - e^{-\lambda(u-t_1)}\right\},$$

so that it cancels the first term. Similarly the third term cancels the second. □

Lemma 9.2.22 *Let $h : T \longrightarrow H$ have orthogonal increments, and be continuous to the left, and m_h and M_h be as in Sect.* 6.2. *Suppose that*

$$f(t) = \int_T \left[\Phi(t,\cdot)\right] dm_h,$$

where Φ is as in (Proposition) 9.2.16, *and $\chi_{T_t}(\theta)\Phi(t,\theta) = \Phi(t,\theta)$. Let*

$$\Psi(t,\theta) = \int_T e^{-\lambda\tau}\Phi(t+\tau,\theta)\,d\tau.$$

Then $\Psi(0,0) = 0$, and

$$\int_T e^{-\lambda\tau} P_t\left[f(t+\tau)\right] d\tau = \int_0^t \left[\Psi(t,\cdot)\right] dm_h.$$

Proof The first claim follows from the properties of Φ [(Remark) 9.2.17]. As

$$P_t\left[f(t+\tau)\right] = \int_T I_{T_t}\left[\Phi\left(t+\tau,\cdot\right)\right] dm_h,$$

one has that

$$\int_T e^{-\lambda\tau} P_t\left[f(t+\tau)\right] d\tau = \int_T d\tau \left\{\int_T I_{T_t}\left[e^{-\lambda\tau}\Phi\left(t+\tau,\cdot\right)\right] dm_h\right\}.$$

Set

$$G_t(\tau,\theta) = e^{-\lambda\tau}\Phi(t+\tau,\theta)\,\chi_{T_t}(\theta).$$

Then

$$\int_T G_t(\tau,\theta)\,d\tau = \chi_{T_t}(\theta)\,\Psi(t,\theta).$$

The conclusion thus obtains provided the integration interchange is legitimate. Now one has that

$$\begin{aligned}
\left\{\int_T |G_t(\tau,\theta)|\,d\tau\right\}^2 &= \chi_{T_t}(\theta)\left\{\int_0^\infty e^{-\lambda\tau}\,|\Phi\,(t+\tau,\theta)|\,d\tau\right\}^2 \\
&= \chi_{T_t}(\theta)\lambda^{-2}E^2_{\Pi^0_\lambda}\left[|\Phi\,(t+\cdot,\theta)|\right] \\
&\le \chi_{T_t}(\theta)\lambda^{-2}E_{\Pi^0_\lambda}\left[\Phi^2\,(t+\cdot,\theta)\right] \\
&= \chi_{T_t}(\theta)\lambda^{-1}\int_T e^{-\lambda\tau}\Phi^2\,(t+\tau,\theta)\,d\tau,
\end{aligned}$$

so that

$$\int_T M_h(d\theta)\left\{\int_T |G_t(\tau,\theta)|\,d\tau\right\}^2 \le \lambda^{-1}\int_0^t\left\{\int_T e^{-\lambda\tau}\Phi^2\,(t+\tau,\theta)\,d\tau\right\}M_h\,(d\theta),$$

and, recalling (Lemma) 9.2.4,

$$\begin{aligned}
\int_0^t\left\{\int_T e^{-\lambda\tau}\Phi^2\,(t+\tau,\theta)\,d\tau\right\}M_h\,(d\theta) &= \int_T d\tau\,e^{-\lambda\tau}\int_0^t M_h\,(d\theta)\,\Phi^2\,(t+\tau,\theta) \\
&= \int_T d\tau\,e^{-\lambda\tau}\,\|P_t\,[f(t+\tau)]\|_H^2 \\
&\le \int_T d\tau\,e^{-\lambda\tau}\,\|f(t+\tau)\|_H^2 \\
&< \infty.
\end{aligned}$$

One may thus apply the interchange of integration (Lemma) 6.2.21. □

Proposition 9.2.23 *Let $h : T \longrightarrow H$ have orthogonal increments, and be continuous to the left, and m_h and M_h be as in Sect. 6.2. Suppose that*

(a) $f(t) = \int_T \Phi_t\,dm_h$, *where* $I_{T_t}\Phi_t = \Phi_t$,
(b) $\Psi(t,\theta) = \int_T e^{-\lambda\tau}\Phi(t+\tau,\theta)\,d\tau$,
(c) h_λ *is as in (Definition)* 9.2.9.

Then, for $\lambda > 0$ and $t \ge 0$, fixed, but arbitrary,

1. *almost surely, with respect to M_h, $\Psi(\theta,\theta)$ exists, and $\theta \mapsto \Psi(\theta,\theta)$ has an equivalence class that belongs to $L_2\,(T,\mathcal{T},M_h)$;*
2. *one has the following representation for h_λ:*

$$h_\lambda(t) = \lambda\int_0^t [\Psi(\cdot,\cdot)]\,dm_h.$$

Proof A change of variables yields that

$$\lambda \Psi(\theta,\theta) = \lambda \int_T e^{-\lambda\tau} \Phi(\theta+\tau,\theta)\,d\tau = \lambda \int_\theta^\infty e^{-\lambda(\eta-\theta)} \Phi(\eta,\theta)\,d\eta.$$

But [(Definition) 9.2.3]

$$\{\lambda \Psi(\theta,\theta)\}^2 = E^2_{\Pi^\theta_\lambda}[\Phi(\cdot,\theta)] \le E_{\Pi^\theta_\lambda}\left[\Phi^2(\cdot,\theta)\right] = \lambda \int_\theta^\infty e^{-\lambda(\eta-\theta)} \Phi^2(\eta,\theta)\,d\eta,$$

and then, proceeding with an interchange of integration, and remembering that $I_{[0,t]}\Phi_t = \Phi_t$,

$$\begin{aligned}
\int_T \{\lambda \Psi(\theta,\theta)\}^2 M_h(d\theta) &= \int_T \left\{\lambda \int_T e^{-\lambda\tau} \Phi(\theta+\tau,\theta)\,d\tau\right\}^2 M_h(d\theta) \\
&\le \int_T M_h(d\theta) \int_\theta^\infty d\eta\,\lambda e^{-\lambda(\eta-\theta)} \Phi^2(\eta,\theta) \\
&= \lambda \int_T d\eta\, e^{-\lambda\eta} \int_0^{t\wedge\eta} M_h(d\theta) e^{\lambda\theta} \Phi^2(\eta,\theta) \\
&\le \lambda e^{\lambda t} \int_T d\eta\, e^{-\lambda\eta} \int_T M_h(d\theta)\, \Phi^2(\eta,\theta) \\
&= \lambda e^{\lambda t} \int_T d\eta\, e^{-\lambda\eta} \left\|\Phi_\eta\right\|^2_{L_2(T,\mathcal{T},M_h)} \\
&= \lambda e^{\lambda t} \int_T e^{-\lambda\eta} C_f(\eta,\eta)\,d\eta.
\end{aligned}$$

The first assertion thus obtains. For the second, using (Lemma) 9.2.22, one has that

$$q(t,\lambda) = \lambda \int_T e^{-\lambda\theta} P_t[f(t+\theta)]\,d\theta = \lambda \int_0^t [\Psi(t,\cdot)]\,dm_h.$$

Then, since $\Psi(0,0) = 0$, using the definition of h_λ [(Definition) 9.2.9],

$$h_\lambda(t) \overset{(\star)}{=} \lambda \int_0^t [\Psi(t,\cdot)]\,dm_h + \lambda \int_0^t \left\{ \int_0^\theta [\Phi(\theta,\cdot)]\,dm_h - \lambda \int_0^\theta [\Psi(\theta,\cdot)]\,dm_h \right\} d\theta.$$

One must now proceed to an interchange of integration. Consider for an instant the following functions ($\tau \le \theta \le t$):

$$G_t(\theta, \tau) = \chi_{[0,t]}(\theta)\chi_{[0,\theta]}(\tau)\Upsilon(\theta, \tau),$$

$$\Gamma(t, \tau) = \int_T G_t(\theta, \tau)d\theta = \begin{cases} \int_\tau^t \Upsilon(\theta, \tau)d\theta & \text{when } \tau \le t \\ 0 & \text{when } \tau > t \end{cases},$$

$$g_t(\theta) = \int_T [G_t(\theta, \cdot)]\, dm_h.$$

If allowed, the interchange of integration yields that

$$\int_T g_t(\theta)d\theta = \int_T [\Gamma(t, \cdot)]\, dm_h = \int_0^t [\Gamma(t, \cdot)]\, dm_h.$$

Now $g_t(\theta) = I_{[0,t]} \int_T I_{[0,\theta]} [\Upsilon(\theta, \cdot)]\, dm_h$. Thus

$$\int_0^t d\theta \int_0^\theta [\Upsilon(\theta, \cdot)]\, dm_h = \int_T [\Gamma(t, \cdot)]\, dm_h = \int_0^t [\Gamma(t, \cdot)]\, dm_h.$$

Applied to the second term of the last expression obtained for h_λ, marked ($\star$), the latter interchange of integration yields that

$$h_\lambda(t) \overset{(\star\star)}{=} \lambda \int_0^t [\Psi(t, \cdot)]\, dm_h + \lambda \int_0^t dm_h \left\{ [\tilde{\Phi}(t, \cdot)] - \lambda \left[\tilde{\Psi}(t, \cdot)\right] \right\},$$

where $\tilde{\Phi}(t, \tau) = \int_\tau^t \Phi(\theta, \tau)d\theta$, and $\tilde{\Psi}(t, \tau) = \int_\tau^t \Psi(\theta, \tau)d\theta$. One has already seen [(Lemma) 9.2.22] that the interchange involving Ψ is allowed. That concerning Φ proceeds as follows:

$$\begin{aligned} \int_0^t M_h(dx) \left\{ \int_0^t d\theta\, |\Phi(\theta, x)| \right\}^2 &= \\ &= \int_0^t d\theta \int_0^t d\tau \int_0^t M_h(dx)\, |\Phi(\theta, x)|\, |\Phi(\tau, x)| \\ &\le \int_0^t d\theta \int_0^t d\tau\, \|[\Phi(\theta, \cdot)]\|_{L_2(T, \mathcal{T}, M_h)}\, \|[\Phi(\tau, \cdot)]\|_{L_2(T, \mathcal{T}, M_h)} \\ &= \left\{ \int_0^t d\theta\, \|[\Phi(\theta, \cdot)]\|_{L_2(T, \mathcal{T}, M_h)} \right\}^2, \end{aligned}$$

which is finite, since $t \mapsto [\Phi(t, \cdot)]$ is continuous, as f is continuous, by assumption. Now, successively, using the definition of Ψ, interchanging the order of integration, continuing with a change of variables, and then integration by parts, one obtains that

$$-\lambda \int_{\tau}^{t} d\theta \, \Psi(\theta, \tau) =$$

$$= \int_{T} \frac{d}{dx} \left[e^{-\lambda x} \right] \int_{\tau+x}^{t+x} \Phi(u, \tau) du$$

$$= \int_{\tau}^{t} \Phi(u, \tau) du - \int_{T} e^{-\lambda x} \left[\frac{d}{dx} \int_{\tau+x}^{t+x} \Phi(u, \tau) du \right].$$

But [262, p. 255]

$$\frac{d}{dx} \left\{ \int_{\tau+x}^{t+x} \Phi \, (u, \tau) \, du \right\} = \Phi(t + x, \tau) - \Phi(\tau + x, \tau).$$

Thus

$$-\lambda \tilde{\Psi}(t, \tau) = \tilde{\Phi}(t, \tau) - \int_{T} e^{-\lambda x} \Phi(t + x, \tau) dx + \int_{T} e^{-\lambda x} \Phi(\tau + x, \tau) dx,$$

that is,

$$-\lambda \tilde{\Psi}(t, \tau) = \tilde{\Phi}(t, \tau) - \Psi(t, \tau) + \Psi(\tau, \tau).$$

Inserting that in the last expression above for h_λ, marked ($\star\star$), one finally obtains item 2 of the proposition's statement. □

The following remarks ease the computations which follow, and those yield, *in fine* [(Corollary) 9.2.33], the covariance of h_λ.

Remark 9.2.24 For $\tau > 0$, fixed, but arbitrary, one has, by change of variables ($\theta + \tau = u$), that

$$\int_{T} e^{-\lambda \theta} f(t + \theta + \tau) d\theta = e^{\lambda \tau} \int_{\tau}^{\infty} e^{-\lambda u} f(t + u) du,$$

so that, with notation explained in Lemma 9.2.8, one has that

$$j\,(t + \tau, \lambda) - j\,(t, \lambda) = \left(e^{\lambda \tau} - 1\right) \int_{\tau}^{\infty} e^{-\lambda \theta} f(t + \theta) d\theta - \int_{0}^{\tau} e^{-\lambda \theta} f(t + \theta) d\theta.$$

Remark 9.2.25 One has that [(Definition) 9.2.3]

$$E_{\Pi_\lambda^0} \left[\|f(t + \cdot)\|_H^2 \right] = \lambda \int_{0}^{\infty} e^{-\lambda u} C_f \,(t + u, t + u) \, du.$$

The change of variables $t + u = \theta$ yields that the right-hand side of the latter expression equals [Assumption 9.2.2]

$$\lambda e^{\lambda t} \int_t^\infty e^{-\lambda\theta} C_f(\theta, \theta)\, d\theta \leq \lambda e^{\lambda t} \Gamma_f(\lambda).$$

Thus

$$E_{\Pi_\lambda^0}\left[\|f(t+\cdot)\|_H^2\right] \leq \lambda e^{\lambda t} \Gamma_f(\lambda).$$

Remark 9.2.26 Let $0 < \epsilon, \kappa < \infty$ be fixed, but arbitrary. Since f is continuous, it is uniformly continuous on $[0, \kappa]$, and thus there exists $\delta_f(\epsilon, \kappa)$ such that both $\{t_1, t_2\} \subseteq [0, \kappa]$ and $|t_1 - t_2| < \delta_f(\epsilon, \kappa)$ imply that

$$\|f(t_1) - f(t_2)\|_H < \epsilon.$$

Consequently, when $\tau < \delta_f(\epsilon, \kappa)$ and $n \in \mathbb{N}$,

$$\int_0^\tau \|f(t) - f(t+\theta)\|_H^n\, d\theta < \tau\epsilon^n.$$

There is furthermore $\kappa_f \in \mathbb{R}_+$ such that

$$\|f(t)\| \leq \kappa_f,\ 0 \leq t \leq \kappa.$$

Analogous values shall obtain for $t \mapsto q(t, \lambda)$ [(Lemma) 9.2.8], and shall be denoted $\delta_q(\epsilon, \kappa)$ and κ_q.

Remark 9.2.27

1. As $e^{-x} \geq 1 - x$, $\lambda \int_0^\tau e^{-\lambda\theta} d\theta = 1 - e^{-\lambda\tau} \leq \lambda\tau$.
2. For $x > 0$, fixed, but arbitrary, using the series expansion of e^x,

$$e^x \leq 1 + x + x^2 e^x, \text{ so that } \frac{e^x - 1 - x}{x} \leq x e^x.$$

Lemma 9.2.28 *j and q are as in (Lemma) 9.2.8. Let $0 < \epsilon, \kappa < \infty$ be fixed, but arbitrary. There exists then $\delta(\epsilon, \kappa) > 0$, and $\alpha \in \mathbb{R}_+$, such that, given constraints $0 < \tau < \delta(\epsilon, \kappa)$ and $\{t, t+\tau\} \subseteq [0, \kappa]$,*

$$\begin{aligned}\left\|\frac{P_t[q(t+\tau, \lambda) - q(t, \lambda)]}{\tau} - \lambda\{q(t, \lambda) - f(t)\}\right\|_H &\leq \\ \leq \lambda\left\|\frac{j(t+\tau, \lambda) - j(t, \lambda)}{\tau} - \lambda j(t, \lambda) + f(t)\right\|_H & \\ \leq \lambda(\epsilon + \alpha\tau), &\end{aligned}$$

where α is a constant.

Proof As $q(t,\lambda) = \lambda P_t[j(t,\lambda)]$,

$$\frac{P_t[q(t+\tau,\lambda) - q(t,\lambda)]}{\tau} - \lambda\{q(t,\lambda) - f(t)\} =$$

$$= \lambda P_t\left[\frac{j(t+\tau,\lambda) - j(t,\lambda)}{\tau} - \lambda j(t,\lambda) + f(t)\right],$$

hence the first inequality in the statement. As [(Remark) 9.2.24]

$$\frac{j(t+\tau,\lambda) - j(t,\lambda)}{\tau} =$$

$$= \lambda\frac{e^{\lambda\tau}-1}{\lambda\tau}\int_\tau^\infty e^{-\lambda\theta}f(t+\theta)d\theta - \frac{1}{\tau}\int_0^\tau e^{-\lambda\theta}f(t+\theta)d\theta,$$

and, using the definition of j,

$$\lambda j(t,\lambda) = \lambda\int_\tau^\infty e^{-\lambda\theta}f(t+\theta)d\theta + \lambda\int_0^\tau e^{-\lambda\theta}f(t+\theta)d\theta,$$

one may write, setting

$$A = \lambda\left\{\frac{e^{\lambda\tau}-1-\lambda\tau}{\lambda\tau}\right\}\int_\tau^\infty e^{-\lambda\theta}f(t+\theta)d\theta,$$

$$B = -\lambda\int_0^\tau e^{-\lambda\theta}f(t+\theta)d\theta,$$

$$C = f(t) - \frac{1}{\tau}\int_0^\tau e^{-\lambda\theta}f(t+\theta)d\theta,$$

that

$$\frac{j(t+\tau,\lambda) - j(t,\lambda)}{\tau} - \lambda j(t,\lambda) + f(t) = A + B + C.$$

For the evaluation of the norm of A, one uses (Remark) 9.2.27, item 2, to obtain that

$$\frac{e^{\lambda\tau}-1-\lambda\tau}{\lambda\tau} \le \lambda\tau e^{\lambda\tau}.$$

One also has that [(Remark) 9.2.25]

$$\lambda\int_\tau^\infty e^{-\lambda\theta}\|f(t+\theta)\|_H\,d\theta \le E_{\Pi_\lambda^0}[\|f(t+\theta)\|_H] \le \left\{\lambda e^{\lambda t}\Gamma_f(\lambda)\right\}^{1/2}.$$

So the following inequality prevails:

$$\|A\|_H \le \left\{\lambda^{3/2}e^{\lambda\{\tau+(t/2)\}}\Gamma_f^{1/2}(\lambda)\right\}\tau.$$

The norm of B is evaluated noticing that [(Remark) 9.2.26], for $\theta \leq \tau < \delta_f(\epsilon, \kappa)$,

$$\|f(t+\theta)\|_H \leq \|f(t+\theta) - f(t)\|_H + \|f(t)\|_H < \epsilon + \kappa_f,$$

so that

$$\|B\|_H \leq \lambda \int_0^\tau e^{-\lambda\theta} \|f(t+\theta)\|_H \, d\theta \leq \lambda \left(\epsilon + \kappa_f\right) \tau.$$

Finally, as

$$C = \frac{1}{\tau} \int_0^\tau \left\{f(t) - e^{-\lambda\theta} f(t+\theta)\right\} d\theta,$$

and that

$$\left\|f(t) - e^{-\lambda\theta} f(t+\theta)\right\|_H \leq$$
$$\leq \|f(t)\|_H \left(1 - e^{-\lambda\theta}\right) + e^{-\lambda\theta} \|f(t+\theta) - f(t)\|_H,$$

one has, for $0 < \tau < \delta_f(\epsilon, \kappa)$, that

$$\|C\|_H \leq \kappa_f \left(1 - e^{-\lambda\tau}\right) + \epsilon \leq \lambda \kappa_f \tau + \epsilon.$$

In fine one must take into account the λ that multiplies $A + B + C$. □

Lemma 9.2.29 *Let $0 < \epsilon, \kappa < \infty$ be fixed, but arbitrary. There exists then $\delta(\epsilon, \kappa) > 0$ and $\alpha \in \mathbb{R}_+$ such that, for $0 < \tau < \delta(\epsilon, \kappa)$ and $\{t, t+\tau\} \subseteq [0, \kappa]$,*

$$\|P_t [q(t+\tau) - q(t, \lambda)]\|_H \leq \lambda \tau \left\{\epsilon + \alpha\tau + \kappa_q + \kappa_f\right\}.$$

Proof One has that $\|P_t [q(t+\tau, \lambda) - q(t, \lambda)]\|_H$ is dominated by

$$\tau \left\{ \left\| \frac{P_t [q(t+\tau, \lambda) - q(t, \lambda)]}{\tau} - \lambda \{q(t, \lambda) - f(t)\} \right\|_H + \lambda \|q(t, \lambda) - f(t)\|_H \right\},$$

but, in the latter expression, the first norm is dominated by $\lambda(\epsilon + \alpha\tau)$ [(Lemma) 9.2.28], and the second, by $\lambda(\kappa_q + \kappa_f)$ [(Remark) 9.2.26]. □

Lemma 9.2.30 *Let $0 < \epsilon, \kappa < \infty$ be fixed, but arbitrary. There exists then $\delta(\epsilon, \kappa) > 0$ such that, for $0 < \tau < \delta(\epsilon, \kappa)$ and $\{t, t+\tau\} \subseteq [0, \kappa]$,*

$$\left\| \{q(t, \lambda) - f(t)\} - \frac{1}{\tau} \int_t^{t+\tau} \{q(\theta, \lambda) - f(\theta)\} \, d\theta \right\|_H \leq 2\epsilon.$$

Proof One has that

$$\left\| \frac{1}{\tau} \int_t^{t+\tau} \{q(\theta,\lambda) - f(\theta)\}\, d\theta - \{q(t,\lambda) - f(t)\} \right\|_H =$$

$$= \frac{1}{\tau} \left\| \int_t^{t+\tau} [q(\theta,\lambda) - q(t,\lambda)]\, d\theta + \int_t^{t+\tau} [f(t) - f(\theta)]\, d\theta \right\|_H$$

$$\leq \frac{1}{\tau} \left\{ \int_0^{\tau} \| q(t+\eta,\lambda) - q(t,\lambda) \|_H\, d\eta + \int_0^{\tau} \| f(t+\eta) - f(t) \|_H\, d\eta \right\}.$$

One must then choose [(Remark) 9.2.26] $\delta(\epsilon,\kappa) \leq \delta_f(\epsilon,\kappa) \wedge \delta_q(\epsilon,\kappa)$. □

Proposition 9.2.31 *For $\lambda > 0$ and $t \geq 0$, fixed, but arbitrary,*

$$\|h_\lambda(t)\|_H^2 = \|q(t,\lambda)\|_H^2 - \|q(0,\lambda)\|_H^2 + 2\lambda \int_0^t d\theta\, \langle f(\theta) - q(\theta,\lambda), q(\theta,\lambda) \rangle_H.$$

Proof It suffices to prove the result for $t = 1$, as the actual value of t is "irrelevant" to the claim. That choice however helps keeping the notation simpler.

One has, for $n \in \mathbb{N}$, fixed, but arbitrary, that

$$\|h_\lambda(1)\|_H^2 = \left\| \sum_{i=1}^{n} \left\{ h_\lambda\left(\frac{i}{n}\right) - h_\lambda\left(\frac{i-1}{n}\right) \right\} \right\|_H^2,$$

and thus [(Proposition) 9.2.21]

$$\|h_\lambda(1)\|_H^2 = \lim_n \left\| \sum_{i=1}^{n} \left\{ h_\lambda\left(\frac{i}{n}\right) - h_\lambda\left(\frac{i-1}{n}\right) \right\} \right\|_H^2.$$

It follows from the definition of h_λ [(Definition) 9.2.9] that

$$h_\lambda(t+\tau) - h_\lambda(t) = \{q(t+\tau,\lambda) - q(t,\lambda)\} + \lambda \int_t^{t+\tau} \{f(\theta) - q(\theta,\lambda)\}\, d\theta.$$

Let

$$A_{[t,\tau]} = \{q(t+\tau,\lambda) - q(t,\lambda)\} - P_t[q(t+\tau,\lambda) - q(t,\lambda)],$$

$$B_{[t,\tau]} = P_t[q(t+\tau,\lambda) - q(t,\lambda)] - \lambda\tau\{q(t,\lambda) - f(t)\},$$

$$C_{[t,\tau]} = \lambda\tau\{q(t,\lambda) - f(t)\} + \lambda \int_t^{t+\tau} \{f(\theta) - q(\theta,\lambda)\}\, d\theta.$$

Then

$$h_\lambda(t+\tau) - h_\lambda(t) = A_{[t,\tau]} + B_{[t,\tau]} + C_{[t,\tau]}.$$

Consequently, letting $t = \frac{i-1}{n}$, and $\tau = \frac{1}{n}$,

$$\sum_{i=1}^{n} \left\{ h_\lambda\left(\frac{i}{n}\right) - h_\lambda\left(\frac{i-1}{n}\right) \right\} =$$

$$= \sum_{i=1}^{n} A_{\left[\frac{i-1}{n},\frac{1}{n}\right]} + \sum_{i=1}^{n} B_{\left[\frac{i-1}{n},\frac{1}{n}\right]} + \sum_{i=1}^{n} C_{\left[\frac{i-1}{n},\frac{1}{n}\right]}.$$

Let

$$A_n = \sum_{i=1}^{n} A_{\left[\frac{i-1}{n},\frac{1}{n}\right]},\ \ B_n = \sum_{i=1}^{n} B_{\left[\frac{i-1}{n},\frac{1}{n}\right]},\ \ C_n = \sum_{i=1}^{n} C_{\left[\frac{i-1}{n},\frac{1}{n}\right]}.$$

One has that, as $\tau = 1/n$,

$$B_{\left[\frac{i-1}{n},\frac{1}{n}\right]} =$$

$$= \frac{1}{n} \left\{ \frac{P_{\frac{i-1}{n}}\left[q\left(\frac{i}{n},\lambda\right) - q\left(\frac{i-1}{n},\lambda\right)\right]}{\frac{1}{n}} - \lambda \left\{ q\left(\frac{i-1}{n},\lambda\right) - f\left(\frac{i-1}{n}\right) \right\} \right\},$$

so that, using (Lemma) 9.2.28,

$$\|B_n\|_H \leq \sum_{i=1}^{n} \left\| B_{\left[\frac{i-1}{n},\frac{1}{n}\right]} \right\|_H \leq \lambda \sum_{i=1}^{n} \frac{1}{n} \left\{ \epsilon + \frac{\alpha}{n} \right\} = \lambda \left(\epsilon + \frac{\alpha}{n} \right).$$

Analogously, using (Lemma) 9.2.30,

$$\|C_n\|_H \leq 2\lambda\epsilon.$$

Consequently

$$\lim_n \left\| \sum_{i=1}^{n} \left\{ h_\lambda\left(\frac{i}{n}\right) - h_\lambda\left(\frac{i-1}{n}\right) \right\} \right\|_H^2 = \lim_n \|A_n\|_H^2.$$

For temporary convenience, let

$$q_{\lambda,n,i} = q\left(\frac{i}{n},\lambda\right) - q\left(\frac{i-1}{n},\lambda\right).$$

One has that $\|A_n\|_H^2$ is a sum of terms of the form

$$\langle A_{\left[\frac{i-1}{n},\frac{1}{n}\right]}, A_{\left[\frac{j-1}{n},\frac{1}{n}\right]}\rangle_H,$$

where

$$A_{\left[\frac{i-1}{n},\frac{1}{n}\right]} = q_{\lambda,n,i} - P_{\frac{i-1}{n}}\left[q_{\lambda,n,i}\right] = P^{\perp}_{\frac{i-1}{n}}\left[q_{\lambda,n,i}\right].$$

Suppose $i < j$. Then

$$\left\langle A_{\left[\frac{i-1}{n},\frac{1}{n}\right]}, A_{\left[\frac{j-1}{n},\frac{1}{n}\right]}\right\rangle_H = \left\langle P^{\perp}_{\frac{i-1}{n}}\left[q_{\lambda,n,i}\right], P^{\perp}_{\frac{j-1}{n}}\left[q_{\lambda,n,j}\right]\right\rangle_H.$$

But

$$P^{\perp}_{\frac{j-1}{n}}\left[q_{\lambda,n,i}\right] = P^{\perp}_{\frac{j-1}{n}}\left[\lambda P_{\frac{i}{n}}\left[j\left(\frac{i}{n},\lambda\right)\right] - \lambda P_{\frac{i-1}{n}}\left[j\left(\frac{i-1}{n},\lambda\right)\right]\right] = 0_H.$$

Thus only inner products with the same index in both entries are different from zero, and

$$\begin{aligned}\left\|A_{\left[\frac{i-1}{n},\frac{1}{n}\right]}\right\|_H^2 &= \left\|q_{\lambda,n,i} - P_{\frac{i-1}{n}}\left[q_{\lambda,n,i}\right]\right\|_H^2 \\ &= \|q_{\lambda,n,i}\|_H^2 - \left\|P_{\frac{i-1}{n}}\left[q_{\lambda,n,i}\right]\right\|_H^2.\end{aligned}$$

As, from (Lemma) 9.2.29,

$$\left\|P_{\frac{i-1}{n}}\left[q_{\lambda,n,i}\right]\right\|_H^2 \le \frac{\lambda^2}{n^2}\left\{\epsilon + \frac{\alpha}{n} + \kappa_q + \kappa_f\right\}^2, \tag{$\star$}$$

one obtains that

$$\|h_\lambda(1)\|_H^2 = \lim_n \left\|\sum_{i=1}^n \left\{h_\lambda\left(\frac{i}{n}\right) - h_\lambda\left(\frac{i-1}{n}\right)\right\}\right\|_H^2 = \lim_n \sum_{i=1}^n \|q_{\lambda,n,i}\|_H^2.$$

But

$$\begin{aligned} q_{\lambda,n,i} &= \\ &= q\left(\frac{i}{n},\lambda\right) - q\left(\frac{i-1}{n},\lambda\right) \\ &= \left\{q\left(\frac{i}{n},\lambda\right) - \lambda P_{\frac{i-1}{n}}\left[j\left(\frac{i}{n},\lambda\right)\right]\right\} + \left\{\lambda P_{\frac{i-1}{n}}\left[j\left(\frac{i}{n},\lambda\right)\right] - q\left(\frac{i-1}{n},\lambda\right)\right\} \\ &= \lambda\left(P_{\frac{i}{n}} - P_{\frac{i-1}{n}}\right)\left[j\left(\frac{i}{n},\lambda\right)\right] + P_{\frac{i-1}{n}}\left[q_{\lambda,n,i}\right].\end{aligned}$$

Since the sum of the squared norms of second terms in the latter expression is negligible, as already seen ($\star$),

$$\|h_\lambda(1)\|_H^2 = \lim_n \sum_{i=1}^{n} \|q_{\lambda,n,i}\|_H^2 = \lambda^2 \lim_n \sum_{i=1}^{n} \left\| \left(P_{\frac{i}{n}} - P_{\frac{i-1}{n}}\right)\left[j\left(\frac{i}{n},\lambda\right)\right]\right\|_H^2 .$$

Now

$$\left\| \left(P_{\frac{i}{n}} - P_{\frac{i-1}{n}}\right)\left[j\left(\frac{i}{n},\lambda\right)\right]\right\|_H^2 = \left\| P_{\frac{i}{n}}\left[j\left(\frac{i}{n},\lambda\right)\right]\right\|_H^2 - \left\| P_{\frac{i-1}{n}}\left[j\left(\frac{i}{n},\lambda\right)\right]\right\|_H^2 ,$$

and, by inspection,

$$\sum_{i=1}^{n} (a_i - b_i) = (a_n - b_1) - \sum_{i=1}^{n-1} (b_{i+1} - a_i) .$$

Thus, letting

$$a_i = \left\| P_{\frac{i}{n}}\left[j\left(\frac{i}{n},\lambda\right)\right]\right\|_H^2 , \quad \text{and } b_i = \left\| P_{\frac{i-1}{n}}\left[j\left(\frac{i}{n},\lambda\right)\right]\right\|_H^2 ,$$

one obtains that

$$\begin{aligned}\|h_\lambda(1)\|_H^2 &= \lambda^2 \|P_1[j(1,\lambda)]\|_H^2 - \lambda^2 \lim_n \left\| P_0\left[j\left(\frac{1}{n},\lambda\right)\right]\right\|_H^2 \\ &\quad - \lambda^2 \lim_n \sum_{i=1}^{n-1} \left\{ \left\| P_{\frac{i}{n}}\left[j\left(\frac{i+1}{n},\lambda\right)\right]\right\|_H^2 - \left\| P_{\frac{i}{n}}\left[j\left(\frac{i}{n},\lambda\right)\right]\right\|_H^2 \right\} .\end{aligned}$$

But, in any real Hilbert space, $\|x\|^2 - \|y\|^2 = \langle x-y, x+y\rangle$, and thus

$$\left\| P_{\frac{i}{n}}\left[j\left(\frac{i+1}{n},\lambda\right)\right]\right\|_H^2 - \left\| P_{\frac{i}{n}}\left[j\left(\frac{i}{n},\lambda\right)\right]\right\|_H^2$$

is the inner product of

$$P_{\frac{i}{n}}\left[j\left(\frac{i+1}{n},\lambda\right)\right] - P_{\frac{i}{n}}\left[j\left(\frac{i}{n},\lambda\right)\right]$$

and

$$P_{\frac{i}{n}}\left[j\left(\frac{i+1}{n},\lambda\right)\right] + P_{\frac{i}{n}}\left[j\left(\frac{i}{n},\lambda\right)\right].$$

So, finally, j being continuous [(Lemma) 9.2.8], $\|h_\lambda(1)\|_H^2$ is $\|q(1,\lambda)\|_H^2 - \|q(0,\lambda)\|_H^2$ from which the following quantity is subtracted:

$$\lim_n \sum_{i=1}^{n-1} \langle P_{\frac{i}{n}}\left[q\left(\frac{i+1}{n},\lambda\right)\right] - q\left(\frac{i}{n},\lambda\right), P_{\frac{i}{n}}\left[q\left(\frac{i+1}{n},\lambda\right)\right] + q\left(\frac{i}{n},\lambda\right)\rangle_H.$$

However, from (Lemma) 9.2.28,

$$\left\|P_{\frac{i}{n}}\left[q\left(\frac{i+1}{n},\lambda\right)\right] - q\left(\frac{i}{n},\lambda\right) - \frac{\lambda}{n}\left\{q\left(\frac{i}{n},\lambda\right) - f\left(\frac{i}{n}\right)\right\}\right\|_H \le \frac{\lambda}{n}\left(\epsilon + \frac{\alpha}{n}\right),$$

and, from (Lemma) 9.2.29,

$$\left\|P_{\frac{i}{n}}\left[q\left(\frac{i+1}{n},\lambda\right)\right] - q\left(\frac{i}{n},\lambda\right)\right\|_H \le \frac{\lambda}{n}\left(\epsilon + \frac{\alpha}{n} + \kappa_f + \kappa_q\right).$$

Consequently, omitting negligible terms,

$$\sum_{i=1}^{n-1}\left\langle P_{\frac{i}{n}}\left[q\left(\frac{i+1}{n},\lambda\right)\right] - q\left(\frac{i}{n},\lambda\right), P_{\frac{i}{n}}\left[q\left(\frac{i+1}{n},\lambda\right)\right] + q\left(\frac{i}{n},\lambda\right)\right\rangle_H =$$

$$= 2\lambda \sum_{i=1}^{n-1} \frac{1}{n}\left\langle q\left(\frac{i}{n},\lambda\right) - f\left(\frac{i}{n}\right), q\left(\frac{i}{n},\lambda\right)\right\rangle_H,$$

whose limit is

$$2\lambda \int_0^1 \langle q(\theta,\lambda) - f(\theta), q(\theta,\lambda)\rangle_H d\theta.$$

□

Remark 9.2.32 A similar calculation, sketched below, yields an expression for

$$\|h_{\lambda_1}(t) \pm h_{\lambda_2}(t)\|_H^2.$$

Let, for $i \in \{1,2\}$,

$$A_{[t,\tau]}^{(i)} = \{q(t+\tau,\lambda_i) - q(t,\lambda_i)\} - P_t[q(t+\tau,\lambda_i) - q(t,\lambda_i)],$$

$$B_{[t,\tau]}^{(i)} = P_t[q(t+\tau,\lambda_i) - q(t,\lambda_i)] - \lambda_i\tau\{q(t,\lambda_i) - f(t)\},$$

$$C_{[t,\tau]}^{(i)} = \lambda_i\tau\{q(t,\lambda_i) - f(t)\} + \lambda_i \int_t^{t+\tau} \{f(\theta) - q(\theta,\lambda_i)\}\, d\theta,$$

so that

$$h_{\lambda_i}(t+\tau) - h_{\lambda_i}(t) = A^{(i)}_{[t,\tau]} + B^{(i)}_{[t,\tau]} + C^{(i)}_{[t,\tau]},$$

and thus

$$\begin{aligned} h_{\lambda_1}(t+\tau) + h_{\lambda_2}(t+\tau) - \{h_{\lambda_1}(t) + h_{\lambda_2}(t)\} &= \\ = \left\{A^{(1)}_{[t,\tau]} + A^{(2)}_{[t,\tau]}\right\} + \left\{B^{(1)}_{[t,\tau]} + B^{(2)}_{[t,\tau]}\right\} + \left\{C^{(1)}_{[t,\tau]} + C^{(2)}_{[t,\tau]}\right\}. \end{aligned}$$

Consequently, letting, for $i \in \{1,2\}$,

$$A^{(i)}_n = \sum_{j=1}^{n} A^{(i)}_{\left[\frac{j-1}{n},\frac{1}{n}\right]},$$

the following expression:

$$\lim_n \left\| \sum_{i=1}^{n} \left[h_{\lambda_1}\left(\frac{i}{n}\right) + h_{\lambda_2}\left(\frac{i}{n}\right) - \left\{ h_{\lambda_1}\left(\frac{i-1}{n}\right) + h_{\lambda_2}\left(\frac{i-1}{n}\right)\right\}\right]\right\|_H^2$$

may be computed as

$$\lim_n \left\| A^{(1)}_n + A^{(2)}_n \right\|_H^2,$$

whose elements are sums of terms of the following form:

$$\left\langle A^{(i)}_{\left[\frac{k-1}{n},\frac{1}{n}\right]}, A^{(j)}_{\left[\frac{l-1}{n},\frac{1}{n}\right]}\right\rangle_H.$$

Of these, only inner products with the same lower index in both entries are different from zero, that is terms of the form

$$\left\langle A^{(i)}_{\left[\frac{k-1}{n},\frac{1}{n}\right]}, A^{(j)}_{\left[\frac{k-1}{n},\frac{1}{n}\right]}\right\rangle_H,$$

which, using the representation

$$A^{(i)}_{\left[\frac{k-1}{n},\frac{1}{n}\right]} = q_{\lambda_i,n,k} - P_{\frac{k-1}{n}}\left[q_{\lambda_i,n,k}\right],$$

are equal to

$$\left\langle q_{\lambda_i,n,k}, q_{\lambda_j,n,k}\right\rangle_H - \left\langle P_{\frac{k-1}{n}}\left[q_{\lambda_i,n,k}\right], P_{\frac{k-1}{n}}\left[q_{\lambda_j,n,k}\right]\right\rangle_H.$$

Since the terms in the second inner product of the latter expression go to zero,

$$\|h_{\lambda_1}(t) + h_{\lambda_2}(t)\|_H^2 = \lim_n \sum_{i=1}^{2} \sum_{j=1}^{2} \sum_{k=1}^{n} \langle q_{\lambda_i,n,k}, q_{\lambda_j,n,k} \rangle_H$$

$$= \lim_n \sum_{k=1}^{n} \|q_{\lambda_1,n,k} + q_{\lambda_2,n,k}\|_H^2 .$$

The term $q(\lambda_i, n, k)$ can again be "reduced" to a term of the following form:

$$q(t+\tau, \lambda_i) - \lambda_i P_t [j(t+\tau, \lambda_i)] = \lambda_i \{P_{t+\tau} [j(t+\tau, \lambda_i)] - P_t [j(t+\tau, \lambda_i)]\},$$

so that

$$\lim_n \sum_{k=1}^{n} \|q_{\lambda_1,n,k} + q_{\lambda_2,n,k}\|_H^2 =$$

$$= \lim_n \sum_{k=1}^{n} \left\| \lambda_1 \left\{ P_{\frac{k}{n}} \left[j\left(\frac{k}{n}, \lambda_1\right)\right] - P_{\frac{k-1}{n}} \left[j\left(\frac{k}{n}, \lambda_1\right)\right]\right\} \right.$$

$$\left. + \lambda_2 \left\{ P_{\frac{k}{n}} \left[j\left(\frac{k}{n}, \lambda_2\right)\right] - P_{\frac{k-1}{n}} \left[j\left(\frac{k}{n}, \lambda_2\right)\right]\right\} \right\|_H^2 .$$

That latter norm expression is of the following form:

$$\|\alpha (a-b) + \beta (A-B)\|^2 = \alpha^2 \|a-b\|^2 + 2\alpha\beta \langle a-b, A-B\rangle + \beta^2 \|A-B\|^2 .$$

The particular values of a, b, A, and B produce the following equalities:

$$\|a-b\|^2 = \|a\|^2 - \|b\|^2 ,$$

$$\|A-B\|^2 = \|A\|^2 - \|B\|^2 ,$$

$$\langle a-b, A-B\rangle = \langle a, A\rangle - \langle b, B\rangle ,$$

so that

$$\|\alpha (a-b) + \beta (A-B)\|^2 = \|\alpha a + \beta A\|^2 - \|\alpha b + \beta B\|^2 .$$

Inserting the actual values of a, b, A and B, one finally obtains that the limit to be computed is that of a sum of terms of the following form:

$$\left\| P_{\frac{k}{n}} \left[\lambda_1 j \left(\frac{k}{n}, \lambda_1 \right) + \lambda_2 j \left(\frac{k}{n}, \lambda_2 \right) \right] \right\|_H^2 - \\ - \left\| P_{\frac{k-1}{n}} \left[\lambda_1 j \left(\frac{k}{n}, \lambda_1 \right) + \lambda_2 j \left(\frac{k}{n}, \lambda_2 \right) \right] \right\|_H^2 .$$

Introducing the possibility of a minus sign, that last expression has the following form:

$$\|A_1 \pm A_2\|^2 - \|B_1 \pm B_2\|^2 = \langle (A_1 + B_1) \pm (A_2 + B_2), (A_1 - B_1) \pm (A_2 - B_2) \rangle,$$

where

$$\begin{array}{ll} A_1 + B_1 & \text{leads to } 2\, q_1, \\ A_2 + B_2 & \text{leads to } 2\, q_2, \\ n\,(A_1 - B_1) & \text{leads to } \lambda_1\,(q_1 - f), \\ n\,(A_2 - B_2) & \text{leads to } \lambda_2\,(q_2 - f), \end{array}$$

and that form leads to the integral

$$2 \int \langle q_1 \pm q_2, \lambda_1\,(q_1 - f) \pm \lambda_2\,(q_2 - f) \rangle .$$

Hence the following corollary:

Corollary 9.2.33 *Letting*

$$\Psi\,(\theta) = \lambda_1\,(f(\theta) - q\,(\theta, \lambda_1)) \pm \lambda_2\,(f(\theta) - q\,(\theta, \lambda_2)),$$

one has that

$$\|h_{\lambda_1}(t) \pm h_{\lambda_2}(t)\|_H^2 = \|q\,(t, \lambda_1) \pm q\,(t, \lambda_2)\|_H^2 - \|q\,(0, \lambda_1) \pm q\,(0, \lambda_2)\|_H^2 \\ + 2 \int_0^t d\theta\, \langle \Psi\,(\theta), q\,(t, \lambda_1) \pm q\,(t, \lambda_2) \rangle_H ,$$

and, using the expression $\langle h, k \rangle = \frac{1}{4}\left\{\|h + k\|^2 - \|h - k\|^2\right\}$, *that the following inner product:*

$$\langle h_{\lambda_1}(t), h_{\lambda_2}(t) \rangle_H$$

is the sum of

$$\langle q(t,\lambda_1), q(t,\lambda_2)\rangle_H - \langle q(0,\lambda_1), q(0,\lambda_2)\rangle_H$$

and

$$\int_0^t d\theta \{\langle f(\theta), \lambda_1 q(\theta,\lambda_1) + \lambda_2 q(\theta,\lambda_2)\rangle_H - (\lambda_1+\lambda_2)\langle q(\theta,\lambda_1), q(\theta,\lambda_2)\rangle_H\}.$$

9.2.3 The Cramér-Hida Representation from the Prediction Process

In this section one assumes that

$$f(t) = \int_T [\Phi(t,\cdot)]\, dm_h$$

is a proper canonical representation (needed, in particular, in (Lemma) 9.2.39 below) with Φ as in (Proposition) 9.2.16. One will see how Φ and h may be obtained from f and q.

The following notation shall at times be used: the map $\theta \mapsto \Phi(t+\theta,t)$ is denoted Φ_t, and its Laplace transform, $\mathcal{L}_t^{\Phi}$:

$$\mathcal{L}_t^{\Phi}(\lambda) = \int_0^\infty e^{-\lambda\theta}\,\Phi(t+\theta,t)d\theta = \int_0^\infty e^{-\lambda\theta}\,\Phi_t(\theta)d\theta.$$

As $\mathcal{L}_t^{\Phi}(\lambda)$ is $\Psi(t,t)$ of (Lemma) 9.2.22, one has that [(Proposition) 9.2.23]

$$\int_T M_h(dt)\left[\int_0^\infty e^{-\lambda\theta}\,\Phi(t+\theta,t)d\theta\right]^2 < \infty,$$

and, since M_h is locally finite, so that functions locally in L_2 are locally in L_1, one has that:

Lemma 9.2.34 *For $\lambda > 0$, fixed, but arbitrary, for almost every $t \in T$, with respect to M_h,*

$$\int_0^\infty e^{-\lambda\theta}\,|\Phi(t+\theta,t)|\,d\theta < \infty.$$

Remark 9.2.35 Lemma 9.2.39 below "says" that, for almost every $t \in T$, with respect to M_h, $\mathcal{L}_t^{\Phi}$ has domain $]0, \infty[$. As

$$t \mapsto \int_0^{\infty} e^{-\lambda\theta} \left|\Phi\left(t+\theta, t\right)\right| d\theta$$

is measurable, one may and shall assume that the Laplace transform exists for every $t \in T$.

Fact 9.2.36 *One shall use the following property of the Laplace transform, which is a consequence of it is analyticity* [46, 268, p. 57, respectively p. 73]*:*

1. *when $\mathcal{L}_t^{\Phi}$ is zero at the values of a sequence in $]0, \infty[$ which converges to a point of $]0, \infty[$, then Φ_t is, with respect to Lebesgue measure, almost surely zero;*
2. *when Φ_t is not, with respect to Lebesgue measure, almost surely equal to the zero function, $\mathcal{L}_t^{\Phi}$ can be zero in a finite and closed interval only at a finite number of points.*

One shall need, for some ensuing proofs, two specific, measurable sets, denoted and defined respectively using the following expressions:

$$\Delta_t = \{\theta \in]0, t[: \text{WRT} \, Leb, \Phi_\theta(\tau) = 0, a.e.\tau\},$$
$$\Lambda_t = \left\{(\lambda, \theta) \in]0, \infty[\times]0, t[: \mathcal{L}_\theta^{\Phi}(\lambda) = 0\right\}.$$

Fact 9.2.37 *One has that Λ_t and its sections,*

$$\Lambda_t[\lambda] = \left\{\theta \in]0, t[: \mathcal{L}_\theta^{\Phi}(\lambda) = 0\right\},$$
$$\Lambda_t[\theta] = \left\{\lambda \in]0, \infty[: \mathcal{L}_\theta^{\Phi}(\lambda) = 0\right\},$$

are measurable sets.

Proof Indeed, the map $(\lambda, \theta, x) \mapsto e^{-\lambda x}\Phi\left(\theta + x, \theta\right)$ is measurable, so that the following map:

$$(\lambda, \theta) \mapsto \int_0^{\infty} e^{-\lambda x}\Phi\left(\theta + x, \theta\right) dx = \mathcal{L}_\theta^{\Phi}(\lambda)$$

is measurable. Consequently Λ_t is measurable, and so are its sections. □

Fact 9.2.38 *One has that*

$$\Delta_t = \left\{\theta \in]0, t[: Leb\left(\Lambda_t[\theta]\right) = \int_{]0,\infty[} \chi_{\Lambda_t}(\lambda, \theta)\, d\lambda > 0\right\}.$$

Proof The right-hand side of the latter equality lists those θ's for which $\mathcal{L}_\theta^{\Phi}$ is zero for a set of positive Lebesgue measure. They form a measurable set. Let $\Delta_t^\star$ denote that set. When Φ_θ is almost surely equal to zero, its Laplace transform is zero, and the section at θ of Λ_t equals $]0, \infty[$, so that $\Delta_t \subseteq \Delta_t^\star$. Suppose conversely that $\theta \in \Delta_t^\star$. Then $\Lambda_t[\theta]$ is a Borel set of strictly positive Lebesgue measure. Since

$$\bigcup_n \left\{ \Lambda_t[\theta] \cap \left[\frac{1}{n}, n\right] \right\} = \Lambda_t[\theta],$$

there is an n such that

$$Leb\left(\Lambda_t[\theta] \cap \left[\frac{1}{n}, n\right] \right) > 0.$$

Thus the set of points $\lambda \in [\frac{1}{n}, n]$ at which $\mathcal{L}_\theta^\Phi(\lambda) = 0$ is at least countable. So there is a sequence of zeros of $\mathcal{L}_t^\Phi$ that converges to some element of $[\frac{1}{n}, n]$. But then Φ_θ must be zero almost surely, and θ belongs to Δ_t. □

Lemma 9.2.39 *One has that* $M_h(\Delta_t) = 0$.

Proof Suppose that $M_h(\Delta_t) > 0$. One has, by the definition of Δ_t, that

$$\int_0^\infty M_h(d\theta)\, \chi_{\Delta_t}(\theta) \int_0^\infty d\tau\, e^{-\lambda\tau} \Phi^2(\theta + \tau, \theta) = 0.$$

A change of variables, and an interchange of order of integration, yield successively that

$$\begin{aligned} 0 &= \int_0^\infty M_h(d\theta)\, \chi_{\Delta_t}(\theta)\, e^{\lambda\theta} \int_\theta^\infty du\, e^{-\lambda u} \Phi^2(u, \theta) \\ &= \int_0^\infty du\, e^{-\lambda u} \int_0^u M_h(d\theta)\, \chi_{\Delta_t}(\theta)\, e^{\lambda\theta} \Phi^2(u, \theta). \end{aligned}$$

It follows that, for $u \leq t$, $\Phi(u, \cdot)$ is, with respect to M_h, on Δ_t, zero almost surely, and thus that

$$f(u) = \int I_{\Delta_t^c}[\Phi(u, \cdot)]\, dm_h.$$

But then $\int I_{\Delta_t} dm_h$ is a non-null element of $L_t[m_h]$ that is orthogonal to $L_t[f]$, contradicting thus the assumption that one is in presence of a proper canonical representation of f. □

Lemma 9.2.40 *For fixed, but arbitrary* $t > 0$, *the map*

$$\lambda \mapsto \mu(\lambda) = M_h(\Lambda_t[\lambda])$$

is upper semi-continuous.

Proof It suffices [5, p. 388] to establish that $\limsup_n \mu(\lambda_n) \leq \mu(\lambda)$, whenever $\lim_n \lambda_n = \lambda$. Now

$$\mu(\lambda) = M_h(\Lambda_t[\lambda]) = \int \chi_{\Lambda_t[\lambda]}(\theta) M_h(d\theta).$$

Since attention is restricted to the interval $[0, t]$, one may assume that M_h is finite. Since the sequence of functions

$$\left\{\chi_{\Lambda_t[\lambda_n]}, \ n \in \mathbb{N}\right\}$$

is uniformly bounded, it is uniformly integrable [5, p. 296]. One may thus use an extension of Fatou's lemma [5, p. 295] to obtain that

$$\limsup_n \left\{\int \chi_{\Lambda_t[\lambda_n]}(\theta) M_h(d\theta)\right\} \leq \int \limsup_n \left\{\chi_{\Lambda_t[\lambda_n]}(\theta)\right\} M_h(d\theta).$$

But [3, p. 12]

$$\limsup_n \left\{\chi_{\Lambda_t[\lambda_n]}\right\} = \chi_{\limsup_n \Lambda_t[\lambda_n]},$$

and, by definition,

$$\limsup_n \Lambda_t[\lambda_n] = \bigcap_{n=1}^{\infty} \bigcup_{p=n}^{\infty} \Lambda_t[\lambda_p].$$

Since, for fixed, but arbitrary $\theta > 0$, $\lambda \mapsto \mathcal{L}_\theta^\Phi(\lambda)$ is continuous (it is analytic),

$$\bigcap_{n=1}^{\infty} \bigcup_{p=n}^{\infty} \Lambda_t[\lambda_p] \subseteq \Lambda_t[\lambda].$$

□

Definition 9.2.41 A measure μ on the Borel sets of a subset of $\mathbb{R}$ is continuous when, for all x in that subset, $\mu(\{x\}) = 0$.

Lemma 9.2.42 *Let μ be a fixed, but arbitrary continuous measure on the Borel sets of $]0, \infty[$. Let $\theta > 0$ be fixed, but arbitrary. When the map Φ_θ is not zero, almost surely, with respect to Lebesgue measure,*

$$\mu(\Lambda_t[\theta]) = 0.$$

Proof The statement expresses the fact that a Laplace transform is analytic, and that, in any finite and closed interval, the kernel of an analytic function is either void or finite [(Fact) 9.2.36]. □

Lemma 9.2.43 *Let $t > 0$ be fixed, but arbitrary, and μ be a fixed, but arbitrary continuous measure on the Borel sets of* $]0, \infty[$. *Then, with respect to μ, for almost every $\lambda \in]0, \infty[$,*

$$M_h\left(\Lambda_t[\lambda]\right) = 0.$$

Proof Using Fubini's theorem one has that

$$\begin{aligned}\int_{]0,\infty[} \mu(d\lambda)\, M_h\left(\Lambda_t[\lambda]\right) &= \int_{]0,\infty[} \mu(d\lambda) \int_{]0,t[} M_h(d\theta)\, I_{\Lambda_t}(\lambda,\theta) \\ &= \int_{]0,t[} M_h(d\theta) \int_{]0,\infty[} \mu(d\lambda)\, I_{\Lambda_t}(\lambda,\theta) \\ &= \int_{]0,t[} M_h(d\theta)\, \mu\left(\Lambda_t[\theta]\right).\end{aligned}$$

Now $\Lambda_t[\theta] = \left\{\lambda \in]0,\infty[: \mathcal{L}^{\Phi}_{\theta}(\lambda) = 0\right\}$. But the set of elements θ's for which $\Lambda_t[\theta] =]0,\infty[$, that is the set Δ_t, has, because of (Lemma) 9.2.39, measure zero for M_h, and, for the θ's not in Δ_t, $\mu\left(\Lambda_t[\theta]\right) = 0$, because of (Lemma) 9.2.42. Consequently,

$$\int_{]0,\infty[} \mu(d\lambda)\, M_h\left(\Lambda_t[\lambda]\right) = 0.$$

□

Lemma 9.2.44 *Let B be a Borel subset of* $]0, t[$, *and*

$$\begin{aligned}D(t,\lambda) &= \lambda\, \mathcal{L}^{\Phi}_t(\lambda), \\ \mu_\lambda(B) &= \int_B D^2(\theta,\lambda)\, M_h(d\theta).\end{aligned}$$

There exists an at most countable Λ_0 such that, for λ in Λ_0^c, fixed, but arbitrary, the measures μ_λ and M_h are mutually absolutely continuous.

Proof It suffices to prove that, except for an at most countable number of λ's, for fixed, but arbitrary $\lambda > 0$, $D^2(\cdot,\lambda)$ is, with respect to M_h, almost surely strictly positive. As $D(\theta,\lambda) = \lambda\, \mathcal{L}^{\Phi}_{\theta}(\lambda)$, the zeros of $\theta \mapsto D(\theta,\lambda)$, for fixed, but arbitrary $\lambda > 0$, are the elements of the set $\Lambda_t[\lambda]$. It thus suffices to prove that $M_h\left(\Lambda_t[\lambda]\right) = 0$ except for an at most countable set of λ's.

Let $\epsilon > 0$ be fixed, but arbitrary. Since $\lambda \mapsto M_h(\Lambda_t[\lambda])$ is upper semi-continuous [(Lemma) 9.2.40], the set

$$B_\epsilon = \{\lambda \in]0, \infty[: M_h(\Lambda_t[\lambda]) \geq \epsilon\}$$

is closed [5, p. 388], and thus a Borel set. Consequently [(Lemma) 9.2.43], for any fixed, but arbitrary continuous measure μ,

$$\mu(B_\epsilon) = \mu(\{\lambda \in]0, \infty[: M_h(\Lambda_t[\lambda]) \geq \epsilon\}) = 0. \qquad (\star)$$

Suppose that B_ϵ is uncountable. By the Alexandroff-Hausdorff theorem [83, p. 386], B_ϵ includes a perfect set P_{B_ϵ}, homeomorphic to C, the Cantor set. Let $h_{B_\epsilon,C} : C \longrightarrow P_{B_\epsilon}$ be that homeomorphism. Let $L : [0,1] \longrightarrow [0,2]$ be the sum of the Cantor–Lebesgue function and the identity map of $[0,1]$. Then [275, p. 89] L and L^{-1} are continuous, strictly increasing functions, and $Leb(L(C)) = 1$. The measure

$$\mu = Leb \circ \left\{h_{B_\epsilon,C} \circ L^{-1}\right\}^{-1}$$

is continuous by construction, and

$$\mu(P_{B_\epsilon}) = Leb(L(C)) = 1.$$

But the latter is impossible because of $(\star)$. Choosing successively, for $\epsilon > 0$, the elements of the sequence $\left\{\frac{1}{n}, n \in \mathbb{N}\right\}$, one sees that the claim obtains. □

Remark 9.2.45 Because of (Proposition) 9.2.23, since, by definition,

$$\lambda \Psi(t,t) = \lambda \mathcal{L}_t^{\Phi}(\lambda) = D(\theta, \lambda),$$

μ_λ of (Lemma) 9.2.44 is the measure determined by the process with orthogonal increments h_λ. It shall be an adequate choice for h provided one can prove that $\lambda \in \Lambda_0^c$ [(Lemma) 9.2.44]. The next result states that such a choice is possible.

Proposition 9.2.46 *One can actually find a $\lambda^\star$ such that one may choose for h, in the canonical representation of f, the process*

$$t \mapsto h_{\lambda^\star}(t).$$

Proof Let $0 < a < b < \infty$ be fixed, but arbitrary. As, for fixed, but arbitrary Borel B, $\lambda \mapsto \mu_\lambda(B)$ is, by its very definition [(Lemma) 9.2.44], measurable, the following definition makes sense:

$$\nu(B) = \int_a^b \mu_\lambda(B)\, d\lambda,$$

and, because of (Lemma) 9.2.44, ν is equivalent to M_h. But then, since

$$\nu(B) = \int_B M_h(d\theta) \int_a^b D^2(\theta, \lambda)\, d\lambda,$$

it follows that, with respect to M_h, almost surely,

$$0 < \frac{d\nu}{dM_h}(\theta) = \int_a^b D^2(\theta, \lambda)\, d\lambda.$$

Let $\delta(\theta) = \int_a^b D^2(\theta, \lambda) d\lambda$, so that $d\nu = \delta\, dM_h$. One may then write that

$$\mu_\lambda(B) = \int_B \frac{D^2(\theta, \lambda)}{\delta(\theta)} \delta(\theta) M_h(d\theta) = \int_B \frac{D^2(\theta, \lambda)}{\delta(\theta)} \nu(d\theta),$$

so that

$$\frac{d\mu_\lambda}{d\nu}(\theta) = \frac{D^2(\theta, \lambda)}{\int_a^b D^2(\theta, \lambda)\, d\lambda},$$

which is, from its definition, and for fixed, but arbitrary θ, analytic in λ. Let

$$C(\lambda, t) = \left\{\theta \in]0, t[: \frac{d\mu_\lambda}{d\nu}(\theta) = 0\right\}.$$

Were one able to find a $\lambda > 0$ such that $\nu(C(\lambda, t)) = 0$, one would have a $\lambda > 0$ such that ($\equiv$ denotes mutual absolute continuity) $\mu_\lambda \equiv \nu \equiv M_h$.

Let then $\{C_n(t), n \in \mathbb{N}\}$ be the increasing sequence whose terms are

$$C_n(t) = \left\{\lambda > 0 : \nu(C(\lambda, t)) \geq \frac{1}{n}\right\}.$$

One checks, as in the proof of (Lemma) 9.2.44, that $C_n(t)$ is closed, and at most countable. Let I be a finite, closed interval, and suppose that $C_n(t) \cap I$ contains an infinite number of points. There exists then

$$\{\lambda_p^{(n)}, p \in \mathbb{N}\} \subseteq C_n(t) \cap I, \text{ and } \lambda^{(n)} \in I,$$

such that $\lim_p \lambda_p^{(n)} = \lambda^{(n)}$. Fix arbitrarily

$$\theta \in \limsup_p C\left(\lambda_p^{(n)}, t\right), \text{ that is, } \theta \in C\left(\lambda_{p(\theta)}^{(n)}, t\right) \text{ for infinitely many } p(\theta)'s,$$

or, since $\frac{d\mu_\lambda}{d\nu}$, $D^2(\cdot,\lambda)$, and $\mathcal{L}_\cdot^\Phi(\lambda)$ have the same zeroes,

$$\mathcal{L}_\theta\left(\lambda_{p(\theta)}^{(n)}\right) = 0, \text{ and } \lim_{p(\theta)} \mathcal{L}_\theta\left(\lambda_{p(\theta)}^{(n)}\right) = \mathcal{L}\left(\lambda^{(n)}\right).$$

Then [(Fact) 9.2.36], almost surely, with respect to Lebesgue measure, it obtains that $\Phi_\theta = 0$. Since, as in (Lemma) 9.2.40,

$$\nu\left(\limsup_p C\left(\lambda_{p(\theta)}^{(n)}, t\right)\right) \geq \limsup_p \nu\left(C\left(\lambda_{p(\theta)}^{(n)}, t\right)\right) \geq \frac{1}{n},$$

and $\nu \equiv M_h$,

$$M_h\left(\theta \in]0,t[: \text{ WRT } Leb, \Phi_\theta = 0 \text{ a.s.}\right) = M_h\left(\Delta_t\right) > 0,$$

which is impossible because of (Lemma) 9.2.39. The sets $C_n^c \cap I \supseteq C_{n+1}^c \cap I$ thus contain finite, closed intervals of positive length. Proceeding inductively, one thus obtains a nested sequence of closed intervals which contain at least one point $\lambda^\star$ such that

$$\nu\left(C\left(\lambda^\star, t\right)\right) = 0.$$

One then chooses $h = h_{\lambda^\star}$. □

Remark 9.2.47 To find a point in C_n^c, it "suffices" to compute $\nu\left(C\left(\lambda, t\right)\right)$ for the rational λ's aligned in the "k-sequence" $p \in \mathbb{N}$, $0 < \lambda = \frac{k}{p} \leq p$.

Remark 9.2.48 To obtain the representation of f in terms of $h_{\lambda^\star}$, one must identify the substitute for Φ. But how that is "done" has been shown in (Proposition) 9.2.16. One introduces the following measure:

$$\mu_t^\star\left([t_1, t_2[\right) = \langle f(t), h_{\lambda^\star}(t_2) - h_{\lambda^\star}(t_1)\rangle_H,$$

and computes the Radon-Nikodým derivative $\frac{d\mu_t^\star}{dM_h}$. But here, from (Proposition) 9.2.23, one has that

$$\mu_t^\star\left([t_1, t_2[\right) = \lambda^\star \int_0^t \chi_{[t_1, t_2[}\, \Phi(t,\theta)\, \Psi(\theta,\theta)\, M_h(d\theta).$$

When one does not know Φ, and thus Ψ, as is generally the case, one may try to use the representation of $h_{\lambda^\star}$ in terms of the prediction process:

$$h_{\lambda^\star}(t_2) - h_{\lambda^\star}(t_1) = q(t_2, \lambda^\star) - q(t_1, \lambda^\star) + \lambda^\star \int_{t_1}^{t_2} \left\{f(\theta) - q\left(\theta, \lambda^\star\right)\right\} d\theta,$$

and the equality:

$$\langle f(t), q(\tau, \lambda)\rangle_H = \lambda \int_T e^{-\lambda\theta} \langle P_\tau [f(t)], P_\tau [f(\tau + \theta)]\rangle_H d\theta,$$

which should be computable as one "knows" the prediction process.

Example 9.2.49 Let

$$X_t = \int_0^t [2t - \theta]\, W(d\theta),$$

with W a Wiener process in the wide sense (one thus writes dW for dm_W). As, for $s < t$, fixed, but arbitrary,

$$\int_0^s [2s - \theta]\,[2t - \theta]\, d\theta = 3s^2 t - \frac{2}{3}s^3,$$

C_X, the covariance of X, has the form:

$$C_X(t_1, t_2) = 4\,(t_1 \wedge t_2)^2\,(t_1 \vee t_2) - \frac{2}{3}\,(t_1 \wedge t_2)^3.$$

Such a process has multiplicity one. Indeed, let f be square integrable, and $F(t)$ be $\int_0^t f(\theta)\, d\theta$. If, for fixed, but arbitrary $t \geq 0$,

$$\int_0^t [2t - \theta] f(\theta)\, d\theta = 0,$$

an integration by parts with $dF = f d\theta$ yields that

$$tF(t) + \int_0^t F(\theta)\, d\theta = 0,$$

a differential equation of the form $t\frac{dg}{dt} + g = 0$, whose solution is the logarithm. As one must have $g(0) = 0$, only $f = 0$ will do.

The representation of the prediction process $P_t[X_{t+s}]$ in terms of X may be obtained as follows. One has that

$$P_t[X_{t+s}] = \int_0^t [2(t+s) - \theta]\, W(d\theta) = X_t + 2sW_t.$$

But, as [(Fact) 6.2.24]

$$\int_a^b f(\theta)\, W(d\theta) = f(b)\, W_b - f(a)\, W_a - \int_a^b \frac{df}{d\theta}(\theta)\, W_\theta\, d\theta,$$

one has, applying that formula to the integral defining X, that

$$X_t = tW_t + \int_0^t W_\theta \, d\theta.$$

Consequently, using integration by parts, one obtains that

$$\begin{aligned}\int_0^t X_\theta \, d\theta &= \int_0^t \theta \, W_\theta \, d\theta + \int_0^t d\theta \int_0^\theta W_u \, du \\ &= \int_0^t \theta \, d\left[\int_0^\theta W_u \, du\right] + \int_0^t d\theta \int_0^\theta W_u \, du \\ &= t \int_0^t W_\theta \, d\theta.\end{aligned}$$

Thus

$$W_t = \frac{1}{t}\left\{X_t - \frac{1}{t}\int_0^t X_\theta \, d\theta\right\},$$

and

$$P_t\left[X_{t+s}\right] = X_t + 2sW_t = X_t + 2s\left\{\frac{1}{t}\left\{X_t - \frac{1}{t}\int_0^t X_\theta \, d\theta\right\}\right\},$$

so that

$$P_t\left[X_{t+s}\right] = \left\{1 + \frac{2s}{t}\right\} X_t - \frac{2s}{t^2}\int_0^t X_\theta \, d\theta.$$

One shall now compute successively $q(t, \lambda)$ and h_λ. By definition

$$q(t, \lambda) = \lambda \int_0^\infty e^{-\lambda\theta} P_t\left[X_{t+\theta}\right] d\theta. \qquad (\star)$$

Replacing the generic prediction process by its explicit expression, and using the following formulae:

$$\lambda \int_0^\infty e^{-\lambda\theta} d\theta = 1, \text{ and } \lambda \int_0^\infty \theta e^{-\lambda\theta} d\theta = \frac{1}{\lambda},$$

one obtains that

$$q(t, \lambda) = \left\{1 + \frac{2}{\lambda t}\right\} X_t - \frac{2}{\lambda t^2}\int_0^t X_\theta \, d\theta.$$

Consequently, as $q(0, \lambda)$ is the zero element ($\star$),

$$h_\lambda(t) = q(t,\lambda) + \lambda \int_0^t \{X_\theta - q(\theta,\lambda)\}\, d\theta$$

$$= q(t,\lambda) + 2\int_0^t d\theta \left\{ \frac{1}{\theta^2}\int_0^\theta X_u du - \frac{X_\theta}{\theta} \right\}$$

$$= q(t,\lambda) + 2\int_0^t \left\{ d\left[-\frac{1}{\theta}\right] \int_0^\theta X_u du - d\theta \frac{X_\theta}{\theta} \right\}$$

$$= q(t,\lambda) - \frac{2}{t}\int_0^t X_\theta\, d\theta$$

$$= \left\{1 + \frac{2}{\lambda t}\right\} X_t - 2\left\{\frac{1}{t} + \frac{1}{\lambda t^2}\right\} \int_0^t X_\theta\, d\theta.$$

From that latter expression one may notice that

$$h_1(t) - h_2(t) = \frac{X_t}{t} - \frac{1}{t^2}\int_0^t X_\theta\, d\theta,$$

which, as seen above, is W_t. In principle one does not know that latter fact, since the only available knowledge is that of X and of the prediction process. Thus, to establish that $h_1(t) - h_2(t)$ is a Wiener process, one would have to compute the square of its norm, either from the knowledge of the covariance of X, or from that of h_λ.

It remains to obtain F to have the representation $X_t = \int_T F_t dm_{h_\lambda}$. Let

$$N_t = h_1(t) - h_2(t) = \frac{X_t}{t} - \frac{1}{t^2}\int_0^t X_\theta\, d\theta.$$

Setting $f(t) = \frac{1}{t}$, $g(t) = \int_0^t X_\theta\, d\theta$, one has that

$$N_t = f(t)\frac{dg}{dt}(t) + \frac{df}{dt}(t)\, g(t),$$

so that

$$\int_0^t N_\theta\, d\theta = \frac{1}{t}\int_0^t X_\theta\, d\theta,$$

and then

$$X_t = tN_t + \int_0^t N_\theta\, d\theta.$$

One has thus here, to ease the calculations, two considerable advantages over the "general case:" one is able to know that N is a Wiener process, and X can be expressed in terms of N. So one continues the computation as follows: for $a < b \leq t$, fixed, but arbitrary,

$$\begin{aligned}
\langle X_t, N_b - N_a\rangle_{L_2(\Omega,\mathcal{F},P)} &= t\langle N_t, N_b - N_a\rangle_{L_2(\Omega,\mathcal{F},P)} \\
&+ \int_0^t d\theta\, \langle N_\theta, N_b - N_a\rangle_{L_2(\Omega,\mathcal{F},P)} \\
&= t\,(b-a) \\
&+ \int_0^a d\theta\, \langle N_\theta, N_b - N_a\rangle_{L_2(\Omega,\mathcal{F},P)} \\
&+ \int_a^b d\theta\, \langle N_\theta, N_b - N_a\rangle_{L_2(\Omega,\mathcal{F},P)} \\
&+ \int_b^t d\theta\, \langle N_\theta, N_b - N_a\rangle_{L_2(\Omega,\mathcal{F},P)} \\
&= t\,(b-a) + 0 + \int_a^b (\theta - a)\, d\theta + \int_b^t (b-a)\, d\theta \\
&= t\,(b-a) + \frac{(b-a)^2}{2} + (b-a)\,(t-b) \\
&= (b-a)\left(2t - \frac{a+b}{2}\right).
\end{aligned}$$

Thus the measure determined by the function

$$\theta \mapsto \langle X_t, N_\theta\rangle_{L_2(\Omega,\mathcal{F},P)}$$

is equivalent to Lebesgue measure, and the Radon-Nikodým derivative is

$$\frac{d\left[\langle X_t, N_\theta\rangle_{L_2(\Omega,\mathcal{F},P)}\right]}{d\theta} = \frac{d\left[2t\theta - \frac{\theta^2}{2}\right]}{d\theta} = 2t - \theta.$$

The procedure used returns thus the proper canonical representation.

The example which follows starts with a process Y whose representation is not canonical, but with a covariance equal to that of X in (Example) 9.2.49, and returns the canonical one.

Example 9.2.50 Let

$$Y_t = \int_0^t [4\theta - 3t]\, dW_\theta.$$

As, for $s < t$, fixed, but arbitrary,

$$\int_0^s [4\theta - 3s]\,[4\theta - 3t]\,d\theta = 3s^2 t - \frac{2}{3}s^2,$$

Y has the same covariance as X of (Example) 9.2.49. The representation of Y is thus not proper canonical. In fact,

$$\int_0^t [4\theta - 3t] f(\theta)\,d\theta = 0$$

yields that

$$3t \int_0^t f(\theta)\,d\theta = 4\int_0^t \theta f(\theta)\,d\theta,$$

which, differentiating, becomes

$$3\int_0^t f(\theta)\,d\theta + 3tf(t) = 4tf(t).$$

That, in turn, yields the differential equation

$$3F(t) = t\frac{dF}{dt}(t),$$

whose solution is proportional to t^3, so that $f(t)$ is proportional to t^2.

The prediction process for Y may be obtained as follows. One has, using integration by parts, that

$$\begin{aligned} Y_t &= 4\int_0^t \theta\,dW_\theta - 3tW_t \\ &= 4\int_0^t \theta\,dW_\theta - 3\left\{\int_0^t \theta\,dW_\theta + \int_0^t W_\theta\,d\theta\right\} \\ &= \int_0^t \theta\,dW_\theta - 3\int_0^t W_\theta\,d\theta. \end{aligned}$$

Y is thus a semi-martingale (in the wide sense), and the following expression is licit:

$$\int_0^t f(\theta)\,dY_\theta = \int_0^t \theta f(\theta)\,dW_\theta - 3\int_0^t f(\theta)\,W_\theta\,d\theta.$$

The integral $\int_0^t f(\theta)\, dY_\theta$ is the prediction process whenever it solves the following equation, for all $u \leq t$,

$$\langle Y_u, Y_{t+s} \rangle_{L_2(\Omega,\mathcal{F},P)} = \langle Y_u, \int_0^t f(\theta)\, dY_\theta \rangle_{L_2(\Omega,\mathcal{F},P)}.$$

The left-hand side of the latter expression is $3u^2(s+t) - \frac{2}{3}u^3$. The right-hand side is computed as follows, using the notation $f_t(\theta) = 4\theta - 3t$:

$$\begin{aligned}
&\langle Y_u, \int_0^t f(\theta)\, dY_\theta \rangle_{L_2(\Omega,\mathcal{F},P)} = \\
&= \langle \int_0^t f_u(\theta)\, dW_\theta, \int_0^t \theta f(\theta)\, dW_\theta - 3\int_0^t f(\theta)\, W_\theta\, d\theta \rangle_{L_2(\Omega,\mathcal{F},P)} \\
&= \int_0^u \theta f_u(\theta) f(\theta)\, d\theta - 3\int_0^t f(\theta)\, d\theta\, \langle W_\theta, \int_0^u f_u(x)\, dW_x \rangle_{L_2(\Omega,\mathcal{F},P)} \\
&= \int_0^u \theta f_u(\theta) f(\theta)\, d\theta - 3\int_0^t f(\theta)\, d\theta \int_0^{\theta\wedge u} f_u(x)\, dx \\
&= \int_0^u \theta f_u(\theta) f(\theta)\, d\theta - 3\int_0^u f(\theta)\, d\theta \int_0^\theta f_u(x)\, dx - 3\int_u^t f(\theta)\, d\theta \int_0^u f_u(x)\, dx \\
&= \int_0^u d\theta f(\theta) \left[\theta f_u(\theta) - 3\int_0^\theta f_u(x)\, dx\right] - 3\int_u^t f(\theta)\, d\theta \int_0^u f_u(x)\, dx.
\end{aligned}$$

Two calculations yield that

$$\theta f_u(\theta) - 3\int_0^\theta f_u(x)\, dx = 6u\theta - 2\theta^2,$$

and that

$$\int_0^u f_u(x)\, dx = -u^2.$$

One thus reaches the following equality:

$$2\int_0^u d\theta f(\theta)\left[3u\theta - \theta^2\right] + 3u^2\int_u^t d\theta f(\theta) = 3u^2(s+t) - \frac{2}{3}u^3.$$

In the first integral to the left of the latter expression, let $\theta = ux$ and, in the second, $\theta = u + x$. One thus obtains that

$$2u^3\int_0^1 dx f(ux)\left[3x - x^2\right] + 3u^2\int_0^{t-u} dx f(u+x) = 3u^2(s+t) - \frac{2}{3}u^3.$$

Dividing by u^2, one has that

$$2u \int_0^1 dxf(ux)\left[3x - x^2\right] + 3\int_0^{t-u} dxf(u+x) = 3(s+t) - \frac{2}{3}u.$$

Taking the limit as $u \downarrow 0$, one finally gets that

$$\int_0^t f(x)\, dx = s + t,$$

whose solution is

$$f(x) = 1 + \frac{2s}{t^2}x.$$

The prediction process is thus

$$\begin{aligned}
\int_0^t \theta\left[1 + \frac{2s}{t^2}\theta\right] dW_\theta - 3\int_0^t \left[1 + \frac{2s}{t^2}\theta\right] W_\theta\, d\theta &= \\
= \int_0^t \theta\, dW_\theta - 3\int_0^t W_\theta\, d\theta + \frac{2s}{t^2}&\left\{\int_0^t \theta^2 dW_\theta - 3\int_0^t \theta\, W_\theta\, d\theta\right] \\
= Y_t + \frac{2s}{t^2}\int_0^t \theta\left[\theta\, dW_\theta - 3W_\theta\, d\theta\right]& \\
= Y_t + \frac{2s}{t^2}\int_0^t \theta\, dY_\theta.&
\end{aligned}$$

An integration by parts yields that

$$\int_0^t \theta\, dY_\theta = tY_t - \int_0^t Y_\theta\, d\theta,$$

so that the prediction process is

$$\left(1 + \frac{2s}{t}\right)Y_t - \frac{2s}{t^2}\int_0^t Y_\theta\, d\theta.$$

Remark 9.2.51 The prediction process of Y is thus obtained by replacing X with Y in the expression for the prediction process of X! That is perfectly sensible. Indeed, as X and Y have the same RKHS, X_t and Y_t are unitarily related, and every valid linear expression formed from one of the processes yields a valid linear expression for the other. In practice however one does not know a priori what the proper canonical representation of the process is, and thus the prediction process is not "cullable" as in (Example) 9.2.49, and must usually be computed by procedures that can be, as those used for Y show, tortuous. But, once the prediction process is available, Knight's method provides the proper canonical representation! Indeed all the calculations in (Example) 9.2.50, starting with q and h_λ, require only the covariance of the original process as well as that of the prediction process, and not the original form of Y (or X), in terms of a Wiener process. Consequently one can replace X with Y in the calculations, and the final result is a proper canonical representation of Y!

9.3 Cramér-Hida and Knight's Representations

To count the number of stationary processes which are at the basis of a Gaussian process, Knight [158] has devised an index of stationarity. It is always larger than the multiplicity of that process, and yields some more precise representations than the CHR does. All of it is at the expense of some generality, as one must work with Gaussian strict sense processes, rather than wide sense ones, filtrations that satisfy the "usual" conditions of completeness and continuity to the right, and integrability conditions. Those are difficult to ascertain, even for Goursat processes, so that the interest here for the index of stationarity lies in that it offers an upper bound on multiplicity, and that it adds information on the process' structure. It is based on the dimension of the linear spaces spanned by the processes h_λ of Sect. 9.2.

Whereas CHR derives multiplicity directly from the process whose multiplicity is of interest, here, multiplicity shall be obtained from the prediction process, and the martingales that may be obtained from it.

9.3.1 Notation, Modifications to the Assumptions, and Consequences

The problem's framework must be completed as follows.

Assumptions 9.3.1 *To Assumption* 9.2.2, *on adds:*

1. $(\Omega, \mathcal{A}, P)$ *is a complete probability space;*
2. *X, which is f of (Assumption)* 9.2.2, *is Gaussian with a mean equal to zero, and H is* $L_2(\Omega, \mathcal{A}, P)$.

In (Assumption) 9.2.2, f was assumed continuous, so X shall be continuous in quadratic mean. X_t shall denote the equivalence class of $X(\cdot, t)$. The symbol $\sigma(\mathcal{F})$, where $\mathcal{F}$ is a family of (equivalence classes) of functions, denotes the σ-algebra generated by the elements of $\mathcal{F}$ (the elements in the equivalence classes of $\mathcal{F}$). One shall need the following families of σ-algebras: for fixed, but arbitrary $t \in T$,

1. $\sigma_t(X)$ is the σ algebra generated by $\{X(\cdot, \theta), \theta \leq t\}$:

$$\sigma_t(X) = \sigma(\mathcal{F}), \ \mathcal{F} = \{X(\cdot, \theta), \ \theta \leq t\};$$

2. $\sigma_t^o(X)$ is generated by $\sigma_t(X)$ and the sets of $\mathcal{A}$ that have measure zero for P,
3. $\sigma_t^+(X) = \cap_{\epsilon>0}\sigma_{t+\epsilon}^o(X)$.

The Gaussian assumption has a number of consequences, as follows [200, 202, p. 22 and p. 25, respectively, p. 43].

1. $\sigma_t^o(X) = \sigma(L_t[X])$.
2. For fixed, but arbitrary $Y \in L_{\cup T}[X]$, the equivalence class of the following random variable: $E_P[Y \mid \sigma_t^o(X)]$ is the projection in $L_{\cup T}[X]$ of Y onto $L_t[X]$. It then follows that, for fixed, but arbitrary $\epsilon > 0$, the equivalence class of $E_P[Y \mid \sigma_{t+\epsilon}^o(X)]$ is $P_{t+\epsilon}[Y]$, P_t the projection of $L_{\cup T}[X]$ with range $L_t[X]$. So the equivalence class of
$$E_P[Y \mid \sigma_t^+(X)]$$
is $P_t^+[Y]$, and thus
$$\sigma_t^+(X) = \sigma(L_t^+[X]).$$
In particular, in that context, there is no difference between strict and wide sense martingales, as long as one is not concerned with path properties, and takes, as a representative of a projection, the conditional expectation.
3. X has a measurable modification.

One may furthermore replace, in all expressions of Sect. 9.2, P_t with P_t^+ without altering the fundamental relations. However continuity to the left becomes continuity to the right. One consequence is that h_λ, defined in Sect. 9.2, is a Gaussian martingale in the wide sense which is continuous to the right in quadratic mean. It has then [223, p. 173] a modification whose paths have limits to the left, and are continuous to the right, for the σ-algebras $\sigma_t^+(X)$. It is such a modification that shall be henceforth used.

One then needs that the terms defining h_λ be given "path-wise" meaning. As X is continuous in quadratic mean, it has a measurable, and integrable, modification that allows one to define
$$\int_0^t X(\cdot,\theta)\,d\theta$$
as a Lebesgue integral, and thus as a continuous function. The only term that has then to be defined "path-wise" is
$$q(t,\lambda) = \lambda P_t^+[j(t,\lambda)].$$

Writing the definition of h_λ as

$$\lambda^{-1}h_\lambda(t) = \lambda^{-1}q(t,\lambda) - \lambda^{-1}q(0,\lambda) + \int_0^t X_\theta\, d\theta - \lambda \int_0^t \lambda^{-1}q(\theta,\lambda)\, d\theta,$$

and introducing

$$f_\lambda(t) = \lambda^{-1}q(t,\lambda) = P_t^+\left[j(t,\lambda)\right],$$

one obtains the following expression:

$$f_\lambda(t) - \lambda \int_0^t f_\lambda(\theta)\, d\theta = \lambda^{-1}h_\lambda(t) + P_0^+\left[j(0,\lambda)\right] - \int_0^t X_\theta\, d\theta,$$

in which the right-hand side, as it depends on the process and its prediction, is assumed to be known. As such it shall be denoted g_λ, a function continuous to the right, with limits to the left. Taking f_λ to be the unknown, one is in the presence of a Volterra equation of the second kind:

$$f_\lambda(t) - \lambda \int_0^t f_\lambda(\theta)\, d\theta = g_\lambda(t).$$

Its unique solution is [25, p. 84]

$$f_\lambda(t) = g_\lambda(t) + \lambda \int_0^t e^{\lambda(t-\theta)} g_\lambda(\theta)\, d\theta,$$

so that q may be taken to be continuous to the right, and have limits to the left, and one has the following "path-wise" identity, where the dot in $\dot{c}$ indicates an element in the equivalence class c:

$$\dot{h}_\lambda(\omega,t) = \dot{q}(\omega,t,\lambda) - \dot{q}(\omega,0,\lambda) + \lambda \int_0^t \left\{X(\omega,\theta) - \dot{q}(\omega,\theta,\lambda)\right\} d\theta.$$

Lemma 9.3.2 *One has that, uniformly on finite intervals of t-indices,*

$$\lim_{\lambda\uparrow\infty} \|q(t,\lambda) - X_t\|^2_{L_2(\Omega,\mathcal{A},P)} = 0.$$

Proof One has that

$$\begin{aligned}
\|q(t,\lambda) - X_t\|^2_{L_2(\Omega,\mathcal{A},P)} &= \left\| P_t\left[\lambda \int_T e^{-\lambda\theta} X_{t+\theta}\, d\theta\right] - X_t \right\|^2_{L_2(\Omega,\mathcal{A},P)} \\
&\le \left\| \lambda \int_T e^{-\lambda\theta} X_{t+\theta}\, d\theta - X_t \right\|^2_{L_2(\Omega,\mathcal{A},P)}
\end{aligned}$$

$$
\begin{aligned}
&= E_P\left[\lambda \int_0^{\infty} e^{-\lambda\theta}\,\{X(\cdot, t+\theta) - X(\cdot, t)\}\right]^2 \\
&= E_P\left[E^2_{\Pi^0_\lambda}\,[X(\cdot, t+\theta) - X(\cdot, t)]\right] \\
&\leq E_P\left[E_{\Pi^0_\lambda}\left[\{X(\cdot, t+\theta) - X(\cdot, t)\}^2\right]\right] \\
&= E_{\Pi^0_\lambda}\left[E_P\left[\{X(\cdot, t+\theta) - X(\cdot, t)\}^2\right]\right] \\
&= \int_0^{\infty} \lambda\, e^{-\lambda\theta}\, \|X_{t+\theta} - X_t\|^2_{L_2(\Omega,\mathcal{A},P)}\, d\theta \\
&\overset{\tau=\lambda\theta}{=} \int_0^{\infty} d\tau\, e^{-\tau} \left\|X_{t+\frac{\tau}{\lambda}} - X_t\right\|^2_{L_2(\Omega,\mathcal{A},P)}.
\end{aligned}
$$

Since $t \mapsto X_t$ is continuous, it is uniformly continuous on closed and bounded intervals, so that, when λ increases indefinitely, that latter integral will be uniformly small on closed and bounded intervals. □

Remark 9.3.3 For large λ, $q\,(t, \lambda)$ is thus an approximation to X_t.

Remark 9.3.4 Let $\sigma^o_t\,(q)$ be the σ-algebra generated by the family

$$
\{\dot{q}\,(\cdot, \theta, \lambda)\,,\ \theta \in [0, t]\,,\ \lambda \text{ a positive integer, or rational number}\}\,,
$$

and the sets of $\mathcal{A}$ that have measure zero for P. Then:

$$
\sigma^o_t\,(X) \subseteq \sigma^o_t\,(q) \subseteq \sigma^+_t\,(X)\,.
$$

Example 9.3.5 Let Z be a (the class of a) standard normal random variable, and W be an independent Wiener process. Let

$$
X_t = Z + \int_0^t W_\theta\, d\theta.
$$

As X is differentiable with W as derivative, for fixed, but arbitrary $t \in T$,

$$
L_t\,[X] = V\,[Z] \oplus L_t\,[W]\,.
$$

Furthermore

$$
X_{t+\theta} = X_t + \int_t^{t+\theta} W_\tau\, d\tau.
$$

But [(Fact) 6.2.24]

$$\int_t^{t+\theta} W_\tau \, d\tau = (t+\theta)\, W_{t+\theta} - tW_t - \int_t^{t+\theta} \tau\, W(d\tau)$$

$$= (t+\theta)\,(W_{t+\theta} - W_t) + \theta W_t - \int_t^{t+\theta} \tau\, W(d\tau)\,.$$

The first and third terms on the right of the latter equality are orthogonal to $V[Z]$, and to $L_t[W]$, and the second belongs to the latter. Consequently

$$P_t[X_{t+\theta}] = X_t + \theta W_t.$$

But then

$$q(t,\lambda) = \lambda \int_T e^{-\lambda\theta} \{X_t + \theta W_t\}\, d\theta = X_t + \lambda^{-1} W_t,$$

$$X_t - q(t,\lambda) = \lambda^{-1} W_t,$$

and

$$h_\lambda(t) = \lambda^{-1} W_t.$$

Thus, for $t \in T$, fixed, but arbitrary, the family of (classes of) random variables

$$\{q(0,\lambda)\,,\; h_\lambda(\theta),\; \theta \in [0,t]\}$$

generates the same completed σ-algebra as the family

$$\{q(\theta,\lambda)\,,\; \theta \in [0,t]\}\,,$$

despite the fact that

$$\lambda \int_0^t \{X_\theta - q(\theta,\lambda)\}\, d\theta = \int_0^t W_\theta\, d\theta$$

does not vanish as $\lambda \uparrow\uparrow \infty$. That equality of σ-algebras is true in general as explained in the following section.

9.3.2 *Equalities of σ-Algebras*

It is established below that some σ-algebras that are central to multiplicity considerations *à la Knight* are equal. As consequence, one obtains a proper representation.

Lemma 9.3.6 *Let X be a progressively measurable process with values in $\mathbb{R}$, for the filtration $\underline{\mathcal{A}}$ of the complete probability space $(\Omega, \mathcal{A}, P)$, satisfying the usual conditions. Assume that*

$$E_P\left[\int_0^\infty |X(\cdot,t)|\,dt\right] < \infty.$$

Let

$$U(\cdot,t) = E_P\left[\int_0^\infty X(\cdot,\theta)\,d\theta \mid \mathcal{A}_t\right],$$

$$V(\cdot,t) = E_P\left[\int_t^\infty X(\cdot,\theta)\,d\theta \mid \mathcal{A}_t\right],$$

$$W(\cdot,t) = \int_0^t X(\cdot,\theta)\,d\theta.$$

Then:

1. $U = V + W$,
2. *U is a uniformly integrable martingale, which one may assume to be continuous to the right and have limits to the left;*
3. *V is a quasimartingale* [128, p. 213], *which one may assume to be continuous to the right and have limits to the left;*
4. *the process*

$$Y_\lambda(\cdot,t) = U(\cdot,0) + \int_0^t e^{\lambda\theta}\,U(\cdot,d\theta)$$

is a martingale, which one may assume to be continuous to the right and have limits to the left, and

$$Y_\lambda(\cdot,t) = e^{\lambda t}V(\cdot,t) + \int_0^t e^{\lambda\theta}\left(X(\cdot,\theta) - \lambda V(\cdot,\theta)\right)d\theta.$$

Proof Item 1 follows from the fact that W is adapted. U is the conditional expectation of an integrable random variable and thus uniformly integrable [128, p. 6]. Martingales have unique modifications that are continuous to the right and have limits to the left [223, p. 173]. So item 2 obtains. For item 3, one makes use of the following definitions:

$$
\begin{aligned}
x^+ &= \quad x \vee 0, \\
x^- &= -(x \wedge 0), \\
V^{\pm}(\cdot, t) &= E_P\left[\int_t^{\infty} X^{\pm}(\cdot, \theta)\, d\theta \mid \mathcal{A}_t\right].
\end{aligned}
$$

As, for fixed, but arbitrary $0 \le t_1 < t_2$,

$$
\begin{aligned}
E_P\left[V^{\pm}(\cdot, t_2) \mid \mathcal{A}_{t_1}\right] &= E_P\left[E_P\left[\int_{t_2}^{\infty} X^{\pm}(\cdot, \theta)\, d\theta \mid \mathcal{A}_{t_2}\right] \mid \mathcal{A}_{t_1}\right] \\
&= E_P\left[\int_{t_2}^{\infty} X^{\pm}(\cdot, \theta)\, d\theta \mid \mathcal{A}_{t_1}\right] \\
&\le E_P\left[\int_{t_1}^{\infty} X^{\pm}(\cdot, \theta)\, d\theta \mid \mathcal{A}_{t_1}\right] \\
&= V^{\pm}(\cdot, t_1),
\end{aligned}
$$

$V^{\pm}$ is thus a positive supermartingale. Since

$$
t \mapsto E_P\left[V^{\pm}(\cdot, t)\right] = E_P\left[\int_t^{\infty} X^{\pm}(\cdot, \theta)\, d\theta\right]
$$

is a continuous map, $V^{\pm}$ may be assumed to be continuous to the right, and have limits to the left [223, p. 173]. V is thus the difference of two positive supermartingales, which are continuous to the right, and have limits to the left, that is, a quasimartingale [128, p. 214]. As such, it has a unique decomposition into the sum of a local martingale, and a predictable process, whose paths start at zero, and have locally integrable variation [216, p. 118]. That decomposition is obtained as follows. Using integration by parts [128, p. 244] on

$$
\int_0^t e^{\lambda\theta} U(\cdot, d\theta),
$$

one gets that

$$
\int_0^t e^{\lambda\theta} U(\cdot, d\theta) = e^{\lambda t} U(\cdot, t) - U(\cdot, 0) - \lambda \int_0^t e^{\lambda\theta} U(\cdot, \theta)\, d\theta. \tag{1}
$$

Replacing U with $V + W$, and using integration by parts on $e^{\lambda t}W(\cdot, t)$, one obtains that

$$e^{\lambda t}U(\cdot, t) = e^{\lambda t}V(\cdot, t) + \int_0^t e^{\lambda\theta}W(\cdot, d\theta) + \lambda \int_0^t e^{\lambda\theta}W(\cdot, \theta)d\theta. \tag{2}$$

But, again as $U = V + W$,

$$\int_0^t e^{\lambda\theta}U(\cdot, \theta)d\theta = \int_0^t e^{\lambda\theta}V(\cdot, \theta)d\theta + \int_0^t e^{\lambda\theta}W(\cdot, \theta)d\theta, \tag{3}$$

so that

$$\begin{aligned} Y_\lambda(\cdot, t) &= U(\cdot, 0) + \int_0^t e^{\lambda\theta}U(\cdot, d\theta) \\ &\overset{(1)}{=} e^{\lambda t}U(\cdot, t) - \lambda \int_0^t e^{\lambda\theta}U(\cdot, \theta)d\theta \\ &\overset{(2)}{=} e^{\lambda t}V(\cdot, t) + \int_0^t e^{\lambda\theta}W(\cdot, d\theta) + \lambda \int_0^t e^{\lambda\theta}W(\cdot, \theta)d\theta \\ &\overset{(3)}{-} \lambda \left\{ \int_0^t e^{\lambda\theta}V(\cdot, \theta)d\theta + \int_0^t e^{\lambda\theta}W(\cdot, \theta)d\theta \right\} \\ &= e^{\lambda t}V(\cdot, t) + \int_0^t e^{\lambda\theta}\left\{W(\cdot, d\theta) - \lambda V(\cdot, \theta)d\theta\right\}, \end{aligned}$$

which, given the definition of W, yields the formula of 4. As a stochastic integral, Y_λ is a local martingale. To prove that it is a martingale, it suffices to prove that [264, p. 64], for any stopping time T, the family

$$\left\{ \int_0^S e^{\lambda\theta}U(\cdot, d\theta),\ S \leq T,\ S \text{ a bounded stopping time} \right\} \tag{4}$$

is uniformly integrable. Using the representation of the integral in (4), obtained from (1), letting σ be a finite bound for S, and A, an arbitrary measurable set, one has that

$$\begin{aligned} E_P\left[\chi_A \left| \int_0^S e^{\lambda\theta}U(\cdot, d\theta) \right| \right] &\leq e^{\lambda\sigma}E_P\left[\chi_A |U(\cdot, S)|\right] + E_P\left[\chi_A |U(\cdot, 0)|\right] \\ &\quad + \lambda \int_0^\sigma e^{\lambda\theta}E_P\left[\chi_A |U(\cdot, \theta)|\right]d\theta. \end{aligned}$$

Let $G = E_P\left[\chi_A \mid \mathcal{A}_S\right]$, and $H = \int_0^\infty X(\cdot,\theta)\,d\theta$, which exists by assumption. Then, as

$$U(\cdot,t) = E_P\left[H \mid \mathcal{A}_t\right],$$

one has that

$$\begin{aligned}
E_P\left[\chi_A\,|U(\cdot,S)|\right] &= E_P\left[\left(E_P\left[\chi_A\,|U(\cdot,S)|\right]\right) \mid \mathcal{A}_S\right] \\
&= E_P\left[G\,|U_S|\right] \\
&= E_P\left[G\,|E_P\left[H \mid \mathcal{A}_S\right]|\right] \\
&= E_P\left[|E_P\left[GH \mid \mathcal{A}_S\right]|\right] \\
&\leq E_P\left[E_P\left[G\,|H| \mid \mathcal{A}_S\right]\right] \\
&= E_P\left[G\,|H|\right] \\
&\leq E_P\left[G \int_0^\infty |X(\cdot,\theta)|\,d\theta\right].
\end{aligned}$$

The conditions [5, p. 296] for uniform integrability thus obtain. □

Lemma 9.3.7 *Let $D = D(\mathbb{R}_+)$ be the Skorohod space of all functions that are continuous to the right and have limits to the left: it is a complete, separable metric space, or Polish space. Denote $L^1_{loc}(\mathbb{R}_+)$ the separable Fréchet space of locally integrable functions* (functions whose restriction to compact intervals is integrable). *The following sets shall be of use* (λ is fixed, but arbitrary):

$$D_\lambda = \left\{ f \in D : \lim_{t \uparrow\uparrow \infty} e^{-\lambda t}\,|f(t)| = 0 \right\},$$

$$A = \left\{ f \in L^1_{loc}(\mathbb{R}_+) : \textit{for all } \lambda > 0,\ \int_0^\infty e^{-\lambda t}\,|f(t)|\,dt < \infty \right\},$$

$$B = A \times \prod_{n \in \mathbb{N}} D_n.$$

For fixed, but arbitrary $f \in A$, $n \in \mathbb{N}$, $h_n \in D_n$, one defines the function $\Phi : B \longrightarrow D^{\mathbb{N}}$ using the following two relations:

$$k_n(t) = h_n(t) + \int_0^t \left\{ f(\theta) - n h_n(\theta) \right\} d\theta,$$

and ($\underline{h}$ has components h_n, and $\underline{k}$, k_n)

$$\Phi(f, \underline{h}) = \underline{k}.$$

There exists then a measurable $\Psi : D^{\mathbb{N}} \longrightarrow B$ *such that, for fixed, but arbitrary* $(f, \underline{h}) \in B$,

$$\Psi \circ \Phi\,(f, \underline{h}) = (f, \underline{h})\,.$$

Proof As functions in D are determined by dense sets,

$$D_\lambda = \bigcap_n \bigcup_p \bigcap_{q \in \mathbb{Q}, q>0} \left\{ f \in D : e^{-\lambda(p+q)} \left| f\,(p+q) \right| < \frac{1}{n} \right\}.$$

But, $\mathcal{E}_t$ denoting evaluation at t,

$$\begin{aligned} \left\{ f \in D : e^{-\lambda(p+q)} \left| f\,(p+q) \right| < \frac{1}{n} \right\} &= \left\{ f \in D : \left| \mathcal{E}_{p+q}\,(f) \right| < \frac{e^{\lambda(p+q)}}{n} \right\} \\ &= \mathcal{E}_{p+q}^{-1} \left(\left] -\frac{e^{\lambda(p+q)}}{n}, \frac{e^{\lambda(p+q)}}{n} \right[\right), \end{aligned}$$

so that D_λ is measurable, as the Borel sets of D are generated by the evaluation maps [145, p. 328]. A Polish space is Lusin [239, p. 94], and every Borel set of a Lusin space is Lusin [239, p. 95]. As a closed manifold in a Frechet space [113, p. 192], A is similarly measurable, and thus Lusin. B is thus Lusin as a countable product of Lusin spaces [239, p. 94].

Φ is measurable by its very definition, since the defining equation may be looked at as the result of computing the evaluation at t and at θ of the various functions entering the formula. It is also an injection. Suppose indeed that, for fixed, but arbitrary $f \in A$, $h_\lambda \in D_\lambda$, $\lambda > 0$, for all $t \geq 0$,

$$h_\lambda(t) + \int_0^t \{f(\theta) - \lambda h_\lambda(\theta)\}\, d\theta = 0.$$

Then, letting

$$F\,(t) = \int_0^t f\,(\theta)\,, \text{ and } H_\lambda\,(t) = \int_0^t h_\lambda\,(\theta)\, d\theta,$$

one has that

$$h_\lambda\,(t) + F\,(t) - \lambda H_\lambda\,(t) = 0.$$

The assumptions on the behavior of f and h_λ (locally integrable) allow one [238, p. 238] to proceed with Laplace transforms. Thus, using the following formulae [217]:

$$\mathcal{L}\left(f'\right)(s) = sf(s) - f(0),$$

$$\mathcal{L}\left(\int_0^{\cdot} f\right)(s) = \frac{\mathcal{L}(f)(s)}{s},$$

$$\mathcal{L}\left(e^{\lambda \cdot}\right)(s) = \frac{1}{s-\lambda},$$

one has that

$$0 = s\frac{\mathcal{L}(h_\lambda)(s)}{s} + \frac{\mathcal{L}(f)(s)}{s} - \lambda\frac{\mathcal{L}(h_\lambda)(s)}{s},$$

that is,

$$\mathcal{L}(h_\lambda)(s) = -\frac{\mathcal{L}(f)(s)}{s-\lambda} = -\mathcal{L}(f)(s)\,\mathcal{L}\left(e^{\lambda \cdot}\right)(s).$$

One can now use the inversion formula

$$\mathcal{L}^{-1}\left(\mathcal{L}(f)\,\mathcal{L}(g)\right) = \int_0^t g(\theta) f(t-\theta)\, d\theta$$

to obtain that

$$h_\lambda(t) = -e^{\lambda t}\int_0^t e^{-\lambda\theta} f(\theta)\, d\theta.$$

Having assumed that $\lim_{t\uparrow\uparrow\infty} e^{-\lambda t}\left|h_\lambda(t)\right| = 0$, one has that $\mathcal{L}(f)(\lambda) = 0$. When $\lambda \in \mathbb{N}$, the change of variables $y = e^{-x}$ yields that

$$\int_0^\infty e^{-nx} f(x)\, dx = \int_0^1 y^{n-1} f(-\ln y)\, dy.$$

Since the set of functions $x \mapsto x^n$ is total in $L_2[0,1]$ [134, p. 10], f must be zero almost surely with respect to Lebesgue measure. But then h_n must be zero for $n \in \mathbb{N}$, for, otherwise, as it is proportional to e^{nt}, it does not belong to D_n.

The conclusion follows from the Souslin-Lusin theorem [70, p. 77]: $\Phi(B)$ is Borel in $D^{\mathbb{N}}$, and Φ is a Borel isomorphism from B to $\Phi(B)$. Ψ may be chosen as follows:

$$\Psi\left(\underline{h}\right) = \begin{cases} \Phi^{-1}\left(h\right) \text{ when } \underline{h} \in \Phi\left(B\right) \\ \left(f_0, \underline{h}_0\right) \text{ when } \underline{h} \in \Phi\left(B\right)^c \end{cases},$$

where $\left(f_0, \underline{h}_0\right)$ is any constant function. □

Remark 9.3.8 In what follows and beyond, $\mathcal{A}_t$ shall always be $\left\{\sigma_t^o(X)\right\}^+$.

Remark 9.3.9 When, in (Lemma) 9.3.6, X is replaced with $e^{-\lambda\cdot}X$, V gets transformed as follows. Since now

$$V_\lambda(\cdot, t) = E_P\left[\int_t^\infty e^{-\lambda\theta} X(\cdot, \theta)d\theta \mid \mathcal{A}_t\right],$$

letting $\eta = \theta - t$,

$$V_\lambda(\cdot, t) = e^{-\lambda t} E_P\left[\int_0^\infty e^{-\lambda\eta} X(\cdot, t+\eta)d\eta \mid \mathcal{A}_t\right].$$

Consequently $e^{\lambda t}V_\lambda(\cdot, t)$ corresponds to $P_t\left[j\left(t, \lambda\right)\right]$, $\lambda\, e^{\lambda t}V_\lambda(\cdot, t)$ corresponds to $q\left(t, \lambda\right)$, and, using (Lemma) 9.3.6,

$$\begin{aligned} \lambda Y_\lambda(\cdot, t) &= \lambda\, e^{\lambda t} V_\lambda(\cdot, t) + \lambda \int_0^t e^{\lambda\theta}\left\{e^{-\lambda\theta} X(\cdot, \theta) - \lambda\, V_\lambda(\cdot, \theta)\right\} d\theta \\ &= \lambda\, e^{\lambda t} V_\lambda(\cdot, t) + \lambda \int_0^t \left\{X(\cdot, \theta) - \lambda\, e^{\lambda\theta} V_\lambda(\cdot, \theta)\right\} d\theta, \end{aligned}$$

so that $\lambda Y_\lambda(\cdot, t)$ corresponds to $h_\lambda(t)$. According to (Proposition) 9.3.10 which follows, one has then a "pathwise" solution to the equation

$$\dot{h}_\lambda(\cdot, t) = \dot{q}\left(\cdot, t, \lambda\right) + \lambda \int_0^t \left\{X(\cdot, \theta) - \dot{q}\left(\cdot, \theta, \lambda\right)\right\} d\theta,$$

where, since $\dot{h}_\lambda$ and $\dot{q}$ are taken as known, X is the unknown.

Proposition 9.3.10 *Let X be a progressively measurable process with values in $\mathbb{R}$, for the filtration $\underline{\mathcal{A}}$ of the complete probability space $(\Omega, \mathcal{A}, P)$, satisfying the usual conditions. Assume that*

$$E_P\left[\int_0^\infty e^{-\lambda t} |X(\cdot,t)|\, dt\right] < \infty.$$

There is then a deterministic procedure which, for almost every $\omega \in \Omega$, reconstructs the paths $t \mapsto X(\omega,t)$ and $t \mapsto E[j(\cdot,t,\lambda) \mid \mathcal{A}_t](\omega)$ from the paths $t \mapsto Y_\lambda(\omega,t)$, where

$$Y_\lambda(\omega,t) = e^{\lambda t} V_\lambda(\omega,t) + \int_0^t \left(X(\omega,\theta) - \lambda e^{\lambda\theta} V_\lambda(\omega,\theta)\right) d\theta,$$

$$V_\lambda(\omega,t) = E_P\left[\int_t^\infty e^{-\lambda\theta} X(\cdot,\theta)\, d\theta \mid \mathcal{A}_t\right](\omega).$$

Proof The assumption says that, for fixed, but arbitrary $\lambda > 0$, for almost every $\omega \in \Omega$, with respect to P,

$$\int_0^\infty e^{-\lambda t} |X(\omega,t)|\, dt < \infty.$$

Since

$$V_\lambda(\omega,t) = E_P\left[\int_t^\infty e^{-\lambda\theta} X(\cdot,\theta)\, d\theta \mid \mathcal{A}_t\right](\omega)$$

can be expressed (as in (Lemma) 9.3.6) as the difference of two supermartingales, continuous to the right, whose expectations both tend to zero, because of the integrability assumption, each of these supermartingales has then a limit that is zero, and thus, almost surely with respect to P,

$$\lim_{t\uparrow\uparrow\infty} e^{-\lambda t}\left\{e^{\lambda t} V_\lambda(\cdot,t)\right\} = 0.$$

One may thus apply (Lemma) 9.3.7 with

$$f(t) = X(\omega,t),$$

$$h_n(t) = e^{nt} V_n(\omega,t),$$

$$k_n(t) = Y_n(\omega,t),$$

so that, $\underline{Y}(\omega,\cdot)$ having components $Y_n(\omega,\cdot)$, and $\underline{V}(\omega,\cdot)$, $e^{n\cdot} V_n(\omega,\cdot)$, almost surely, with respect to P,

$$\underline{Y}(\omega,\cdot) = \Phi\left(\begin{bmatrix} X(\omega,\cdot) \\ \underline{V}(\omega,\cdot) \end{bmatrix}\right), \quad \text{and} \quad \begin{bmatrix} X(\omega,\cdot) \\ \underline{V}(\omega,\cdot) \end{bmatrix} = \Psi\left(\underline{Y}(\omega,\cdot)\right).$$

□

Corollary 9.3.11 *Mutatis, mutandis,* (Proposition) 9.3.10 *is true for finite intervals.*

Proof Let $\tau > 0$ be fixed, but arbitrary. Let $\mathcal{A}_t^\tau = \mathcal{A}_{t\wedge\tau}$, and

$$X_\tau(\cdot,t) = \begin{cases} X(\cdot,t) & \text{when } t \le \tau \\ E_P\left[X(\cdot,t) \mid \mathcal{A}_\tau\right] & \text{when } t \ge \tau \end{cases}.$$

The usual conditions obtain for the σ-algebras of the form $\mathcal{A}_t^\tau$, and X_τ is progressively measurable with respect to them. Furthermore, the integrability condition of (Proposition) 9.3.10 obtains for X_τ also. The process

$$V_\lambda^\tau(\cdot,t) = E_P\left[\int_t^\infty e^{-\lambda\theta} X_\tau(\cdot,\theta)\,d\theta \mid \mathcal{A}_t^\tau\right]$$

is thus well defined, and so is Y_λ^τ. Now, for $t \le \tau$, by definition, one has that $X_\tau(\cdot,t) = X(\cdot,t)$, and $V_\lambda^\tau(\cdot,t) = V_\lambda(\cdot,t)$, so that $Y_\lambda^\tau(\cdot,t) = Y_\lambda(\cdot,t)$. When $t > \tau$, then $\mathcal{A}_t^\tau = \mathcal{A}_\tau^\tau$, and, since Y_λ^τ is a martingale with respect to the $\mathcal{A}_t^\tau$'s, one has that

$$Y_\lambda^\tau(\cdot,t) = E_P\left[Y_\lambda^\tau(\cdot,t) \mid \mathcal{A}_t^\tau\right] = E_P\left[Y_\lambda^\tau(\cdot,t) \mid \mathcal{A}_\tau^\tau\right] = Y_\lambda^\tau(\cdot,\tau) = Y_\lambda(\cdot,\tau).$$

That means that knowing $Y_\lambda(\omega,\cdot)$ on $[0,\tau]$, one knows $Y_\lambda^\tau(\omega,\cdot)$ on $\mathbb{R}_+$, that is $X_\tau(\omega,\cdot)$ on $\mathbb{R}_+$, and thus, $X(\omega,\cdot)$ on $[0,\tau]$. □

Remark 9.3.12 When using (Corollary) 9.3.11 within the framework of Sect. 9.2.1, the basic integrability assumption of (Proposition) 9.3.10 is covered by that of Sect. 9.2.1. Indeed

$$\begin{aligned} E_P\left[\int_0^\infty e^{-\lambda\theta}\,|X(\cdot,\theta)|\,d\theta\right] &\le \int_0^\infty e^{-\lambda\theta} E_P^{1/2}\left[X^2(\cdot,\theta)\right] d\theta \\ &\le \int_0^\infty e^{-\lambda\theta} E_P\left[X^2(\cdot,\theta)\right] d\theta \int_0^\infty e^{-\lambda\theta}\,d\theta \\ &< \infty. \end{aligned}$$

Corollary 9.3.13 *Let $t \in T$ and $p \in \mathbb{N}$ be fixed, but arbitrary. The following σ-algebras have completions in $\mathcal{A}_t$ that are equal:*

$$\begin{aligned}\sigma_t(V) &= \sigma\left(V_\lambda(\cdot,\theta), \theta \in [0,t], \lambda > 0\right), \\ \sigma_t(Y_{n>p}) &= \sigma\left(V_n(\cdot,0), Y_n(\cdot,\theta), \theta \in]0,t], n > p\right), \\ \sigma_t(Y) &= \sigma\left(V_n(\cdot,0), Y_n(\cdot,\theta), \theta \in]0,t], n \in \mathbb{N}\right).\end{aligned}$$

Proof In the sequel, $\underline{V}$ has components $V_n, n \in \mathbb{N}$, $\underline{Y}$, Y_n. The terms "inclusion of Y in V" shall mean, in what follows, that $\sigma_t(Y) \subseteq \sigma_t(V)$. One has that, almost surely, with respect to P,

$$Y_\lambda(\cdot,0) = V_\lambda(\cdot,0).$$

By definition, the σ-algebras considered are in $\mathcal{A}_t$, and the completion of $\sigma_t(Y)$ is in that of $\sigma_t\,(X, \underline{V})$, since $\underline{Y}$ is a function of X and $\underline{V}$ [(Proposition) 9.3.10, (Corollary) 9.3.11]. But, because of (Lemma) 9.3.2, X may be recovered from $\underline{V}$ (the difference between q and V is multiplication by λ). Thus the "inclusion Y in $\underline{V}$" obtains.

For the "inclusion $\underline{V}$ in Y," one may proceed using the following remarks. From (Proposition) 9.3.10 and (Corollary) 9.3.11, $\underline{V}$ is a measurable function of $\underline{Y}$. It will suffice to prove that, for $\lambda > 0$, fixed, but arbitrary, $V_\lambda(\cdot,t)$ is the limit in $L_1(\Omega, \mathcal{A}, P)$ of linear combinations of the form $\sum_{i=1}^{p} \alpha_i V_{n_i}(\cdot,t)$. Now

$$\begin{aligned}&V_\lambda(\cdot,t) - \sum_{i=1}^{p} \alpha_i V_{n_i}(\cdot,t) \\ &\quad = \int_0^\infty \left\{ e^{-\lambda(\theta+t)} - \sum_{i=1}^{p} \alpha_i e^{-n_i(\theta+t)} \right\} E_P\left[X(\cdot,t+\theta)|\mathcal{A}_t\right] d\theta.\end{aligned}$$

The expectation of the absolute value of the left-hand side of the latter equality is smaller than (with the obvious notation)

$$\int_0^\infty \left|f(\theta) - f_p(\theta)\right| E_P\left[|X(\cdot,t+\theta)|\right] d\theta,$$

where the expectation is locally square integrable by assumption. Since [268, p. 62] a locally integrable g is almost surely zero when, for all $n \geq p$, p arbitrary, but fixed,

$$\int_0^\infty e^{-n\theta} g(\theta)\, d\theta = 0,$$

choosing $g \in L_2\left[\mathbb{R}_+\right]$, one obtains that the family of exponentials

$$\left\{\theta \mapsto e^{-n\theta}, n \geq p\right\}$$

is total in $L_2[\mathbb{R}_+]$. Choose thus a sequence $\{f_p\}$ that converges in $L_2[\mathbb{R}_+]$ to f. Since, with $g_p = |f(\theta) - f_p(\theta)|$,

$$\int_0^\infty g_p(\theta)E_P[|X(\cdot, t+\theta)|]\, d\theta \leq$$

$$\leq \sum_n \int_n^{n+1} g_p(\theta)E_P[|X(\cdot, t+\theta)|]\, d\theta,$$

for each fixed, but arbitrary n, one may choose p_n such that $p_n > p_{n-1}$ and, for $p > p_n$,

$$\int_n^{n+1} g_p(\theta)E_P[|X(\cdot, t+\theta)|]\, d\theta \leq$$

$$\leq \|f - f_p\|^2_{L_2[\mathbb{R}_+]} \int_n^{n+1} E_P\left[X^2(\cdot, t+\theta)\right] d\theta \leq \frac{\epsilon}{2^n}.$$

The required convergence to zero follows. □

Proposition 9.3.14 *The completion $\sigma_t^o(q)$ of the σ-algebra generated by the family*

$$\{\dot{q}(\cdot, \theta, \lambda),\ \theta \in [0, t],\ \lambda > 0\}$$

is equal to $\sigma_t^+(X)$.

Proof From (Remark) 9.3.9, one has that $\dot{q}(\cdot, t, \lambda)$ is $\lambda\, e^{\lambda t} V_\lambda(\cdot, t)$, and $\dot{h}_\lambda(\cdot, t)$, $\lambda Y_\lambda(\cdot, t)$. Let $\{t, t_1, t_2\} \subseteq T$, $t_1 < t_2$, be fixed, but arbitrary. One has already seen [(Remark) 9.3.4] that

$$\sigma_t^o(X) \subseteq \sigma_t^o(q) \subseteq \sigma_t^+(X).$$

Consequently

$$\sigma_{t_1}^+(X) \subseteq \sigma_{t_2}^o(X) \subseteq \sigma_{t_2}^o(q).$$

If one is able to prove that $\sigma_t^+(q) = \sigma_t^o(q)$, then

$$\sigma_{t_1}^+(X) \subseteq \sigma_{t_1}^+(q) = \sigma_{t_1}^o(q) \subseteq \sigma_{t_1}^+(X).$$

Now [(Corollary) 9.3.13] $\sigma_t^o(q) = \sigma_t^o(h)$, the σ-algebra generated from the processes of form h_n. Furthermore [Sect. 9.3.1 and result (Proposition) 9.2.21, item 4] h_n is a Gaussian martingale with respect to the σ-algebras generated by X, that is

$$h_n(t_2) - h_n(t_1) \perp L_{t_1}^+[X],$$

so that $\dot{h}_n(\cdot, t_2) - \dot{h}_n(\cdot, t_1)$ is independent of $\sigma_{t_1}^{+}(X)$, and thus of $\sigma_{t_1}^{o}(q)$ that it contains, and consequently of $\sigma_{t_1}^{o}(h)$, as the latter is equal to $\sigma_{t_1}^{o}(q)$.

Let $A \in \sigma_{t_1}^{+}(q)$ be fixed, but arbitrary. As, for fixed, but arbitrary $\epsilon > 0$,

$$\dot{h}_n(\cdot, t_2) - \dot{h}_n(\cdot, t_1 + \epsilon)$$

is independent of $\sigma_{t_1+\epsilon}^{o}(q)$, it is independent of

$$\sigma_{t_1}^{+}(q)$$

that it contains. Consequently, as $\dot{h}_n$ is continuous to the right, the difference $\dot{h}_n(\cdot, t_2) - \dot{h}_n(\cdot, t_1)$ is independent of A.

Now, as $A \in \sigma_{t_2}^{o}(q) = \sigma_{t_2}^{o}(h)$, and that a function adapted to a completion is almost surely equal to a function adapted to the underlying σ-algebra [275, p. 97], $\chi_{_A}$ may be taken to be a function adapted to the σ-algebra generated by the family of elements

$$\dot{h}_n,$$

and, as such, has a representation of the following form, for an appropriately measurable Φ_A[41, p. 144]:

$$\chi_{_A} = \Phi_A\left(\dot{h}_{n_i}(\cdot, t_i),\ t_i \in [0, t_2],\ i \in I,\ |I| \leq \aleph_0\right).$$

As, for $t_i > t_1$, fixed, but arbitrary,

$$\dot{h}_n(\cdot, t_i) = \left\{\dot{h}_n(\cdot, t_i) - \dot{h}_n(\cdot, t_1)\right\} + \dot{h}_n(\cdot, t_1),$$

one may rewrite the representation of $\chi_{_A}$ as follows:

$$I = I_1 \cup I_2,\ I_1 = \{i \in I : t_i \leq t_1\},\ I_2 = \{i \in I : t_1 < t_i \leq t_2\},$$

$$\chi_{_A} = \Psi_A\left(\dot{h}_{n_i}(\cdot, t_i),\ i \in I_1,\ \dot{h}_{n_j}(\cdot, t_1),\ \dot{h}_{n_j}(\cdot, t_j) - \dot{h}_{n_j}(\cdot, t_1),\ j \in I_2\right),$$

Ψ_A still being appropriately measurable.

Let

$$\sigma_1 = \sigma\left(\dot{q}(\cdot, 0, \lambda),\ \dot{h}_\lambda(\cdot, t_i, \lambda),\ i \in I_1,\ \dot{h}_{n_j}(\cdot, t_1),\ j \in I_2\right),$$

$$\sigma_2 = \sigma\left(\dot{h}_{n_i}(\cdot, t_i) - \dot{h}_{n_i}(\cdot, t_1),\ i \in I_2\right).$$

Let $\underline{X}$ be the vector whose components generate σ_1, and $\underline{Y}$ be the vector whose components generate σ_2. Then $\chi_{_A} = \Psi_A(\underline{X}, \underline{Y})$. Let B be a fixed, but arbitrary element of $\mathcal{B}(\mathbb{R}^{l_1})$. Then

$$\begin{aligned}
\int_{\underline{X}^{-1}(B)} \chi_A \, dP &= \int_{\underline{X}^{-1}(B)} \Psi_A(\underline{X}, \underline{Y}) \, dP \\
&= \int_{B\times\mathbb{R}^{l_2}} \Psi_A\left(\underline{x}, \underline{y}\right) P_{(\underline{X},\underline{Y})}\left(d\left(\underline{x}, \underline{y}\right)\right) \\
&= \int_{B\times\mathbb{R}^{l_2}} \Psi_A\left(\underline{x}, \underline{y}\right) P_{\underline{X}} \otimes P_{\underline{Y}}\left(d\left(\underline{x}, \underline{y}\right)\right) \\
&= \int_B P_{\underline{X}}(d\underline{x}) \int_{\mathbb{R}^{l_2}} \Psi_A\left(\underline{x}, \underline{y}\right) P_{\underline{Y}}\left(d\underline{y}\right) \\
&= \int_B E_P\left[\Psi_A(\underline{x}, \underline{Y})\right] P_{\underline{X}}(d\underline{x}) \\
&= \int_{\underline{X}^{-1}(B)} E_P\left[\chi_A \mid \sigma_1\right] dP.
\end{aligned}$$

Consequently, almost surely,

$$\chi_A = E_P\left[\chi_A \mid \sigma_1\right],$$

so that A is in the completion of σ_1, which is contained in $\sigma_{t_1}^o(h) = \sigma_{t_1}^o(q)$. □

Remark 9.3.15 When the processes h_λ hold the part of Z in a representation of the following type: $X_t = \int F_t \, dZ$, the conjunction of (Corollary) 9.3.13 and (Proposition) 9.3.14 says that the representation of X by that integral with respect to h_λ is proper.

9.3.3 The Index of Stationarity of a Gaussian Process

The index of stationarity does, to a certain extent, for the martingales obtained from X, what multiplicity does for X itself. But its computation is based on the λ parameter rather than time. One shall see that the value of the index is related to the stationary features of X, whence its name.

Definition 9.3.16 The index of stationarity at t, of the process X, from which the process h_λ may be derived, as in (Definition) 9.2.9, is the dimension $s_X[t]$ of the (closed) subspace $\Lambda_t[X]$ generated by the family $\{h_\lambda(t), \lambda > 0\}$, that is

$$s_X[t] = \dim\{\Lambda_t(X)\} = \dim\left\{\overline{V[\{h_\lambda(t), \lambda > 0\}]}\right\}.$$

Proposition 9.3.17 *For fixed, but arbitrary* $\{t_1, t_2\} \subseteq T,\ t_1 < t_2$, *one has that:*

1. $s_X[t_1] \le s_X[t_2]$;
2. *when* $t_1 > 0$, $s_X[t_1] < \infty$, *and* Y *is the following map:* $t \mapsto Y_t = X_{t_1+t}$, *then*

$$s_X[t_2] - s_X[t_1] \le s_Y[t_2 - t_1] \le s_X[t_2].$$

Proof (1) Let $L : \Lambda_{t_2}[X] \longrightarrow \Lambda_{t_1}[X]$ be defined using the following relation, based on the martingale property of h_λ [Sect. 9.3.1 and result (Proposition) 9.2.21, item 4]:

$$L[h_\lambda(t_2)] = P_{t_1}[h_\lambda(t_2)] = h_\lambda(t_1).$$

The closure of the range of L is $\Lambda_{t_1}[X]$. The adjoint of L, $L^\star$, is thus [266, p. 71] a linear injection of $\Lambda_{t_1}[X]$ into $\Lambda_{t_2}[X]$, so that the latter's dimension must exceed that of the former's.

Proof (2) Let $\{t, t_1\} \subseteq\ ]0, \infty[$ be fixed, but arbitrary, and $|a, b|$ denote a finite interval of any type, open, closed, half open. Then

$$\{Y_\theta, \theta \in |a, b|\} = \{X_{t_1+\theta}, \theta \in |a, b|\} = \{X_\theta, \theta \in |t_1 + a, t_1 + b|\}.$$

Let k_λ be for Y the counterpart of h_λ for X. One shall use the following fact, to be established below:

$$k_\lambda(t) = h_\lambda(t_1 + t) - h_\lambda(t_1). \qquad (\star)$$

With that fact taken temporarily for granted, by definition,

$$k_\lambda(t_2 - t_1) = h_\lambda(t_2) - h_\lambda(t_1),$$

and

$$\Lambda_{t_2-t_1}(Y) = \overline{V[\{k_\lambda(t_2 - t_1),\ \lambda > 0\}]} = \overline{V[\{h_\lambda(t_2) - h_\lambda(t_1),\ \lambda > 0\}]}.$$

Since one has the following orthogonal decomposition:

$$h_\lambda(t_2) = \{h_\lambda(t_2) - h_\lambda(t_1)\} + h_\lambda(t_1) = k_\lambda(t_2 - t_1) + h_\lambda(t_1),$$

it follows that

$$\begin{aligned}\overline{V[\{h_\lambda(t_2),\ \lambda > 0\}]} &\subseteq \overline{V[\{h_\lambda(t_1),\ \lambda > 0\}]} \oplus \overline{V[\{h_\lambda(t_2) - h_\lambda(t_1),\ \lambda > 0\}]}\\ &= \overline{V[\{h_\lambda(t_1),\ \lambda > 0\}]} \oplus \overline{V[\{k_\lambda(t_2 - t_1),\ \lambda > 0\}]},\end{aligned}$$

and that translates to

$$s_X[t_2] \leq s_X[t_1] + s_Y[t_2 - t_1].$$

Furthermore, as in item 1, the map $L : \Lambda_{t_2}[X] \longrightarrow \Lambda_{t_2-t_1}[Y]$ defined using the following relation:

$$L[h_\lambda(t_2)] = \left\{I_{L_2(\Omega,\mathcal{A},P)} - P_{t_1}\right\}[h_\lambda(t_2)] = h_\lambda(t_2) - h_\lambda(t_1) = k_\lambda(t_2 - t_1)$$

is linear and continuous, and the closure of its range is $\Lambda_{t_2-t_1}(Y)$. Consequently its adjoint is an injection, and the second inequality in item 2 obtains.

One must then prove the initial assertion ($\star$). Let $L_{t_1,t}[X]$ denote the (closed) subspace generated linearly by $\{X_\theta, \theta \in [t_1, t_1 + t]\}$. Then, as

$$\{X_\theta, \theta \in [t_1, t_1 + t]\} = \{Y_\theta, \theta \in [0, t]\},$$

it follows that $L_{t_1,t}[X] = L_t[Y]$. Consequently

$$L_t^+[Y] = \bigcap_{\epsilon>0} L_{t_1,t+\epsilon}[X] \subseteq L_{t_1+t}^+[X].$$

As

$$\begin{aligned}\{X_\theta,\ \theta \in [0, t_1 + t]\} &= \{X_\theta, \theta \in [0, t_1]\} \cup \{X_\theta \in [t_1, t_1 + t]\} \\ &= \{X_\theta, \theta \in [0, t_1]\} \cup \{Y_\theta, \theta \in [0, t]\},\end{aligned}$$

one has that

$$L_{t_1+t}[X] \subseteq L_{t_1}^+[X] \vee L_t^+[Y],$$

so that [(Lemma) 8.4.17]

$$\begin{aligned}L_{t_1+t}^+[X] &\subseteq \bigcap_{\epsilon>0} \{L_{t_1}^+[X] \vee L_{t+\epsilon}^+[Y]\} \\ &= L_{t_1}^+[X] \vee \bigcap_{\epsilon>0} L_{t+\epsilon}^+[Y] \\ &= L_{t_1}^+[X] \vee L_t^+[Y].\end{aligned}$$

Consequently $L_{t_1+t}^+[X] = L_{t_1}^+[X] \vee L_t^+[Y]$. □

Let

- $L_{t_1+t}^+[X] = L_t^+[Y] \oplus H_t^{X,Y}$,
- P_t^Y be the projection with range $L_t^+[Y]$,

- $\Pi_t^{X,Y}$ be the projection with range $H_t^{X,Y}$,
- q_Y be, for Y, the counterpart of q for X.

Since every element in an orthogonal decomposition has a unique sum representation, and that, for $\theta \in [0, t]$, fixed, but arbitrary, both

$$L_{t_1+\theta}^+ [X] \subseteq L_{t_1+t}^+ [X], \text{ and } L_\theta^+ [Y] \subseteq L_t^+ [Y],$$

obtain,

$$H_\theta^{X,Y} \subseteq H_t^{X,Y}.$$

Then, with $Y_1 \in H_t^{X,Y}$,

$$\begin{aligned} q_Y (t, \lambda) &= \lambda P_t^Y \left[\int_0^\infty e^{-\lambda\theta} Y_{t+\theta} \, d\theta \right] \\ &= \lambda \left\{ P_{t_1+t}^+ - \Pi_t^{X,Y} \right\} \left[\int_0^\infty e^{-\lambda\theta} X_{t_1+t+\theta} \, d\theta \right] \\ &= q (t_1 + t, \lambda) - Y_1. \end{aligned}$$

Similarly, with $Y_2 \in H_0^{X,Y}$,

$$q_Y (0, \lambda) = q (t_1, \lambda) - Y_2,$$

and, with $Z_\theta \in H_\theta^{X,Y}$, and $Y_3 \in H_t^{X,Y}$,

$$\begin{aligned} \int_0^t \{Y_\theta - q_Y (\theta, \lambda)\} \, d\theta &= \int_0^t \{X_{t_1+\theta} - q (t_1 + \theta, \lambda) + Z_\theta\} \, d\theta, \\ &= \int_{t_1}^{t_1+t} \{X_\theta - q (\theta, \lambda)\} \, d\theta + Y_3. \end{aligned}$$

It follows then, from the definition of h_λ [(Definition) 9.2.9], applied to $k_\lambda(t)$, and the preceding calculation, that

$$k_\lambda(t) = h_\lambda(t_1 + t) - h_\lambda(t_1) + Y_3 + Y_2 - Y_1,$$

where $Y_3 + Y_2 - Y_1 \in H_t^{X,Y}$. Let

$$K = L_{t_1+t}^+ [X], \; K_X = L_{t_1}^+ [X], \text{ and } K_Y = L_t^+ [Y].$$

As already seen,

$$K = K_X \vee K_Y.$$

Write

$$K_0 = K_X \cap K_Y,\; K_1 = K_X \cap K_0^{\perp},\; K_2 = K_Y \cap K_0^{\perp}.$$

One has that

$$K = (K_1 \vee K_0) \vee (K_0 \vee K_2) = K_1 \vee K_0 \vee K_2,$$

and that

$$K_1 \cap K_2 = \{0_K\},$$

so that

$$K = K_1 + K_0 + K_2,$$

and thus that [167, p. 184] K_1 is the orthogonal complement in K of K_Y, and K_2, that of K_X. Since the difference $h_\lambda(t_1 + t) - h_\lambda(t_1)$ is orthogonal to $L_{t_1}^+[X]$, $k_\lambda(t)$ is thus the sum of an element in $L_t^+[Y]$ and an element in its orthogonal complement. As $k_\lambda(t)$ belongs to

$$L_t^+[Y],$$

one must have that $Y_3 + Y_2 - Y_1 = 0_{L_2(\Omega,\mathcal{A},P)}$. □

Definition 9.3.18 Let X have the properties listed in (Assumption) 9.2.2 and Sect. 9.3.1, h_n be defined as the expression preceding the statement of (Lemma) 9.3.2, and let, for $t \in T$, fixed, but arbitrary,

$$L_t[h] = \overline{V[\{h_n(\theta),\; \theta \le t,\; n \in \mathbb{N}\}]}.$$

Suppose that, for an yet unspecified $m_X[t] \in \{0\} \cup \mathbb{N} \cup \{\infty\}$,

$$\{X^{(i)},\; i \in [1 : m_X[t]]\}$$

is a family of independent (orthogonal) Gaussian processes, each of which has a mean equal to zero, independent increments, and paths that are continuous to the right. When $m_X[t]$ is the smallest integer such that

$$\overline{V\left[\{X_\theta^{(i)},\; \theta \le u,\; i \in [1 : m_X[t]]\}\right]} = L_u[h],\; u \le t,$$

it is called the index of multiplicity of X_t. When $L_t[h]$ is reduced to the zero element, one sets $m_X[t] = 0$.

Lemma 9.3.19 *Given a sequence*

$$\mathcal{M} = \{M^{(i)},\ i \in I \subseteq \mathbb{N}\}$$

of martingales in the wide sense, there exits a sequence

$$\mathcal{N} = \{N^{(i)},\ i \in J \subseteq I\}$$

of martingales in the wide sense which has orthogonal terms, and generates, with respect to "time" the same linear subspaces as $\mathcal{M}$.

Proof Let $N^{(1)} = M^{(1)}$. It is, by definition and assumption, a martingale in the wide sense whose basis measure shall be denoted $M_{N^{(1)}}$. Let

$P_t^{N(1)}$ be the projection onto $L_t\left[N^{(1)}\right]$,

$M_{N^{(1)}}^t$ be the restriction of $M_{N^{(1)}}$ to $\mathcal{T}_t$.

The projection of $M^{(2)}$ onto $L_t\,[N^{(1)}]$ has then the following generic form:

$$P_t^{N(1)}\left[M_t^{(2)}\right] = \int_0^t F_t^{(2,1)} dm_{N^{(1)}},\ F_t^{(2,1)} \in L_2\left(T_t, \mathcal{T}_t, M_{N^{(1)}}^t\right).$$

Then, by definition, for fixed, but arbitrary $\theta \leq t$, using the characterization of projection [44, p. 80],

$$\begin{aligned} 0 &= \langle M_t^{(2)} - P_t^{N(1)}\left[M_t^{(2)}\right], N_\theta^{(1)}\rangle_{L_2(\Omega,\mathcal{A},P)} \\ &= \langle M_t^{(2)}, N_\theta^{(1)}\rangle_{L_2(\Omega,\mathcal{A},P)} - \int_0^\theta \dot{F}_t^{(2,1)}(\tau)\, M_{N^{(1)}}(d\tau), \end{aligned}$$

so that the following measure on $\mathcal{T}_t$ is well defined:

$$\mu_t^{N(2,1)}([0,\theta]) = \langle M_t^{(2)}, N_\theta^{(1)}\rangle_{L_2(\Omega,\mathcal{A},P)} = \int_0^\theta \dot{F}_t^{(2,1)}(\tau)\, M_{N^{(1)}}(d\tau),$$

and thus, in terms of equivalence classes, with respect to $M_{N^{(1)}}$,

$$F_t^{(2,1)} = \frac{d\mu_t^{N(2,1)}}{dM_{N^{(1)}}}.$$

One then sets

$$N_t^{(2)} = M_t^{(2)} - \int_0^t \frac{d\mu_t^{N(2,1)}}{dM_{N^{(1)}}}\, dm_{N^{(1)}}.$$

$N^{(2)}$ is a martingale in the wide sense as the sum of two such objects.

Proceeding recursively one will be able to obtain that

$$N_t^{(p+1)} = M_t^{(p+1)} - \sum_{i=1}^{p} \int_0^t \frac{d\mu_t^{N(p+1,i)}}{dM_{N^{(i)}}} dm_{N^{(i)}},$$

where

- $M_{N^{(i)}}$ is the basis measure of $N^{(i)}$, and, for $i \in [1:p]$ and $\theta \leq t$, fixed, but arbitrary,

$$\mu_t^{N(p+1,i)}([0,\theta]) = \langle M_t^{N(p+1)}, N_\theta^{(i)} \rangle_{L_2(\Omega,\mathcal{A},P)},$$

- $N_t^{(p)}$ is Gaussian (since it is obtained using linear operations on Gaussian processes), and is a wide sense martingale (by construction),
- for $p \neq q$, fixed, but arbitrary, $N^{(p)}$ and $N^{(q)}$ are orthogonal (by construction),
- $L_t[N] = L_t[M]$ (again by construction).

□

Remark 9.3.20 Let M_i denote the basis measure of $M^{(i)}$, and, generically,

$$m[\phi] = \int \phi \, dm.$$

The formulae resulting from (Lemma) 9.3.19 produce the following set of equalities, where $f_{p+1,i}$ stands for the corresponding Radon-Nikodým derivative in the formulae:

$$\begin{aligned} M_1 &= M_{N^{(1)}}, \\ M_2 &= M_{N^{(1)}}\left[f_{2,1}^2\right] + M_{N^{(2)}}, \\ M_3 &= M_{N^{(1)}}\left[f_{3,1}^2\right] + M_{N^{(2)}}\left[f_{3,2}^2\right] + M_{N^{(3)}}, \\ \cdots &= \cdots \end{aligned}$$

so that positive linear combinations in the measures M_i are mutually absolutely continuous with respect to the same linear combinations in the measures N_i.

Remark 9.3.21 The next lemma uses the Lebesgue decomposition as follows (the measures involved are on $\mathcal{T}_t$, the Borel sets of $[0,t]$).

Let $\nu_1 = M_{N^{(1)}}$, and, for $i \geq 1$, given that

$$N_i = \sum_{j=1}^{i} M_{N^{(j)}},$$

write

$$dM_{N^{(i+1)}} = f_{i+1}dN_i + d\nu_{i+1},$$

where

$$\nu_{i+1} \perp N_i, \quad \text{and has support } S_{i+1}.$$

The following facts shall be used.

1. Since $N_j(S_{i+1}) = 0$, for $j \in [1:i]$, fixed, but arbitrary $M_{N^{(j)}}(S_{i+1}) = 0$.
2. One may assume that the supports S_{i+1} are decreasing.

   Let indeed

$$\Sigma_i = \bigcup_{j \geq i} S_i.$$

   Since $\nu_{i+1}(\Sigma_{i+1}) \geq \nu_{i+1}(S_{i+1}) = \nu_{i+1}([0,t])$, and simultaneously $j \leq i$ and $k \geq i+1$, using item 1,

$$N_i(\Sigma_{i+1}) = \sum_{j=i}^{i} M_{N^{(j)}}(\Sigma_{i+1}) \leq \sum_{j=i}^{i} \sum_{k \geq i+1} M_{N^{(j)}}(S_k) = 0,$$

   so that the Σ_i's are decreasing supports.
3. Suppose that the S_i's are decreasing. They may then be taken as disjoint.

   Let indeed

$$\Sigma_i = S_i \cap S_{i+1}^c.$$

   Then, since, given the definition of ν_{i+1}, and then using item 1,

$$\nu_{i+1}(S_{i+2}) \leq M_{N^{(i+1)}}(S_{i+2}) = 0,$$

   one has that

$$\nu_{i+1}(\Sigma_{i+1}) = \nu_{i+1}(\Sigma_{i+1}) + \nu_{i+1}(S_{i+1} \cap S_{i+2}) = \nu_{i+1}(S_{i+1}) = \nu_{i+1}([0,t]),$$

   and, using item 1 again,

$$N_i(\Sigma_{i+1}) = \sum_{j=1}^{i} M_{N^{(j)}}\left(S_{i+1} \cap S_{i+2}^c\right) \leq \sum_{j=1}^{i} M_{N^{(j)}}(S_{i+1}) = 0.$$

4. Letting, in item 3 above, S_0 to be the complement of $\cup_{i \geq 1} S_i$, one has a sequence of disjoint supports for the measures ν_i, and, because of the definitions of the N_i's and ν_i's, then for i fixed, but arbitrary, and all $j > i$, $M_{N^{(i)}}(S_j) = 0$.

5. One has, using recursion, and the definitions and properties of the diverse ingredients, the following sets of equalities:

   (i) $M_{\mathcal{N}^{(1)}} = \nu_1$;
   (ii) $M_{\mathcal{N}^{(2)}} = \nu_1 [f_2] + \nu_2$;
   (iii) $M_{\mathcal{N}^{(3)}} = \nu_1 [f_3 + f_2 f_3] + \nu_2 [f_3] + \nu_3$;
   (iv) $\cdots$

   and, referring to (Remark) 9.3.20,

   (i) $M_1 = \nu_1$;
   (ii) $M_2 = \nu_1 \left[f_{2,1}^2 + f_2\right] + \nu_2$;
   (iii) $M_3 = \nu_1 \left[f_{3,1}^2 + f_{3,2}^2 f_2 + f_3\right] + \nu_2 \left[f_{3,2}^2 + f_3\right] + \nu_3$;
   (iv) $\cdots$

Lemma 9.3.22 *The sequence $\mathcal{N}$ of (Lemma)* 9.3.19 *may be chosen so that for $i \in I$ fixed, but arbitrary, $M_{\mathcal{N}^{(i+1)}} \ll M_{\mathcal{N}^{(i)}}$.*

Proof Since the formulae which appear in the proof are complicated, one shall restrict it to the details of the first steps of the recursion to which it is reduced. The processes of the statement's conclusion shall be denoted with boldface. Below, as usual, when S is a set, χ_S is the indicator of S, and I_S its equivalence class with respect to whatever measure is considered. The S_i's are the supports of (Remark) 9.3.21, item 4, so they are disjoint, and t there shall be 1 here. So the proof covers the interval $[0, 1]$. At the end of the proof, one shall see how that first result gets extended.

a) *Definition of $\mathbf{N}^{(1)}$*:

Let, for $t \in [0, 1]$, fixed, but arbitrary,

$$\mathbf{N}_t^{(1)} = \int_0^t I_{S_1} \, dm_{\mathcal{N}^{(1)}} + \sum_{i \geq 2} \frac{\int_0^t I_{S_i} \, dm_{\mathcal{N}^{(i)}}}{2^i \sqrt{M_{\mathcal{N}^{(i)}}([0, 1])}}.$$

One thus assumes that $M_{\mathcal{N}^{(i)}}([0, 1]) > 0$. When that is not the case, as the numerator is the zero element, the term does not enter the expression. Then

$$\begin{aligned} \left\|\mathbf{N}_t^{(1)}\right\|_{L_2(\Omega, \mathcal{A}, P)}^2 &= \int_0^t \chi_{S_1}(\theta) \, M_{\mathcal{N}^{(1)}}(d\theta) \\ &\quad + \sum_{i \geq 2} \frac{1}{2^{2i} M_{\mathcal{N}^{(i)}}([0, 1])} \int_0^t \chi_{S_i}(\theta) \, M_{\mathcal{N}^{(i)}}(d\theta) \\ &\leq M_{\mathcal{N}^{(1)}}([0, t]) + \sum_{i \geq 2} \frac{1}{2^{2i}} \\ &< \infty, \end{aligned}$$

and thus, since [(Remark) 9.3.21] $M_{N^{(1)}}(S_1) = \nu_1(S_1) = \nu_1([0,t])$, and

$$dM_{N^{(i+1)}} = f_{i+1}dN_i + d\nu_{i+1}, \text{ with } N_i(S_{i+1}) = 0,$$

one obtains that

$$\begin{aligned} dM_{N^{(1)}} &= dM_{N^{(1)}} + \sum_{i\geq 2} \frac{1}{2^{2i}M_{N^{(i)}}([0,1])} \chi_{S_i} dM_{N^{(i)}} \\ &= d\nu_1 + \sum_{i\geq 2} \frac{1}{2^{2i}M_{N^{(i)}}([0,1])} d\nu_i. \end{aligned}$$

Consequently, for fixed, but arbitrary $i \geq 1$, $\nu_i \ll M_{N^{(1)}}$. As [still (Remark) 9.3.21]

$$M_{N^{(1)}} = \nu_1, \text{ and } M_{N^{(2)}} = \nu_1[f_2] + \nu_2,$$

whose summands are both absolutely continuous with respect to $M_{N^{(1)}}$, then

$$M_{N^{(2)}} \ll M_{N^{(1)}}.$$

Since [still (Remark) 9.3.21] $M_{N^{(3)}} = \nu_1[f_3 + f_2 f_3] + \nu_2[f_3] + \nu_3$, then $M_{N^{(3)}} \ll M_{N^{(1)}}$, and it should be clear that the just initiated induction procedure may be continued indefinitely. Since, by construction,

$$M_t^{(p+1)} = N_t^{(p+1)} + \sum_{i=1}^{p} \int_0^t \frac{d\mu_t^{N(p+1,i)}}{dM_{N^{(i)}}} dm_{N^{(i)}},$$

one has also that $M_i \ll M_{N^{(1)}}$.

b) *Definition of* $N^{(2)}$:

Let $P_t^{N^{(1)}}$ denote the projection onto $L_t\left[N^{(1)}\right]$. Then,

$$P_t^{N^{(1)}}\left[N_t^{(j)}\right] = \int_0^t G_t^{(j)} dm_{N^{(1)}}, \text{ some } G_t^{(j)} \in L_2\left(T_t, \mathcal{T}_t, M_{N^{(1)}}^{|\mathcal{T}_t}\right),$$

where $j \geq 2$ is fixed, but arbitrary. One must have (characterization of projection: [44, p. 80]), for $\theta \leq t$, that

$$\langle N_t^{(j)}, N_\theta^{(1)} \rangle_{L_2(\Omega,\mathcal{A},P)} = \int_0^\theta \dot{G}_t^{(j)}(x) M_{N^{(1)}}(dx).$$

But, for fixed, but arbitrary $i \geq 1$, using the definition of $\boldsymbol{N}_t^{(1)}$, the S_i's being disjoint,

$$M_{N^{(1)}}\left([0,t] \cap S_i\right) = \frac{1}{2^{2i} M_{N^{(i)}}\left([0,1]\right)} M_{N^{(i)}}\left([0,t] \cap S_i\right), \tag{$\star$}$$

and thus, still using the definition of $\boldsymbol{N}_t^{(1)}$, plus $(\star)$,

$$\begin{aligned} \langle N_t^{(j)}, \boldsymbol{N}_\theta^{(1)} \rangle_{L_2(\Omega,\mathcal{A},P)} &= \frac{1}{2^j \sqrt{M_{N^{(j)}}\left([0,1]\right)}} \int_0^\theta \chi_{S_j} \, dM_{N^{(j)}} \\ &= \int_0^\theta 2^j \sqrt{M_{N^{(j)}}\left([0,1]\right)} \, \chi_{S_j} \, dM_{N^{(1)}}. \end{aligned}$$

Consequently

$$\dot{G}_t^{(j)} = 2^j \sqrt{M_{N^{(j)}}\left([0,1]\right)} \, \chi_{S_j},$$

and, since

$$P_t^{N^{(1)}}\left[N_t^{(j)}\right] = \int_0^t G_t^{(j)} \, dm_{N^{(1)}} = \int_0^t 2^j \sqrt{M_{N^{(j)}}\left([0,1]\right)} I_{S_j} \, dm_{N^{(1)}},$$

it follows, again using $(\star)$, that

$$P_t^{N^{(1)}}\left[N_t^{(j)}\right] = \int_0^t I_{S_j} \, dm_{N^{(j)}}. \tag{$\star\star$}$$

Let now, for $j \geq 2$,

$$N_t^{(2,j)} = N_t^{(j)} - P_t^{N^{(1)}}\left[N_t^{(j)}\right] = \int_0^t I_{S_j^c} \, dm_{N^{(j)}}.$$

Now S_j^c is the disjoint union of S_k's, $k < j$, and S_l's, $l > j$. But, since [(Remark) 9.3.21] $M_{N^{(j)}}\left(S_l\right) = 0$ for $l > j$,

$$N_t^{(2,j)} = \int_0^t I_{\left[\biguplus_{k=1}^{j-1} S_k\right]} \, dm_{N^{(j)}}.$$

Thus the basis measure of $N^{(2,j)}$ is absolutely continuous with respect to $M_{N^{(j)}}$, and, since $M_{N^{(j)}} \ll M_{N^{(1)}}$,

$$M_{N^{(2,j)}} \ll M_{N^{(1)}}.$$

Consider then the sequence

$$\left\{ N_t^{[i]} = N_t^{(2,i+1)} = \int_0^t I_{\left[\biguplus_{j=1}^{i} S_j\right]} dm_{N^{(i+1)}} \right\}.$$

It is orthogonal by construction, and made of martingales in the wide sense. One may thus define, as above,

$$\boldsymbol{N}_t^{(2)} = \int_0^t I_{S_1} dm_{N^{[1]}} + \sum_{i \geq 2} \frac{\int_0^t I_{S_i} dm_{N^{[i]}}}{2^i \sqrt{M_{N^{[i]}}([0,1])}},$$

which rewrites as

$$\boldsymbol{N}_t^{(2)} = \int_0^t I_{S_1} dm_{N^{(2)}} + \sum_{i \geq 2} \frac{\int_0^t I_{S_i} dm_{N^{(i+1)}}}{2^i \sqrt{M_{N^{(i+1)}}([0,1])}}.$$

In particular

$$M_{\boldsymbol{N}^{(2)}}([0,t] \cap B) = M_{N^{(2)}}([0,t] \cap S_1 \cap B) + \sum_{i \geq 2} \frac{M_{N^{(i+1)}}([0,t] \cap S_i \cap B)}{2^i \sqrt{M_{N^{(i+1)}}([0,1])}}.$$

c) *Definition of* $\boldsymbol{N}^{(3)}$:

One proceeds along the lines followed for the definition of $\boldsymbol{N}^{(2)}$. The novelty is the shift of indices separating the S_i's and the $N^{(i)}$'s.

Let $P_t^{\boldsymbol{N}^{(2)}}$ denote the projection onto $L_t\left[\boldsymbol{N}^{(2)}\right]$. Then,

$$P_t^{\boldsymbol{N}^{(2)}}\left[N_t^{(j)}\right] = \int_0^t G_t^{(j)} dm_{\boldsymbol{N}^{(2)}}, \text{ some } G_t^{(j)} \in L_2\left(T_t, \mathcal{T}_t, M_{\boldsymbol{N}^{(2)}}^{|\mathcal{T}_t}\right),$$

where $j \geq 3$ is fixed, but arbitrary. One must have, for $\theta \leq t$, that

$$\langle N_t^{(j)}, \boldsymbol{N}_\theta^{(2)} \rangle_{L_2(\Omega,\mathcal{A},P)} = \int_0^\theta \dot{G}_t^{(j)}(x)\, M_{\boldsymbol{N}^{(2)}}(dx).$$

But, for fixed, but arbitrary $i > 1$, letting $M_{N^{(i+1)}}([0,1])$ be denoted μ_i,

$$M_{\boldsymbol{N}^{(2)}}([0,t] \cap S_i) = \frac{1}{2^{2i}\mu_i} M_{N^{(i+1)}}([0,t] \cap S_i),$$

and thus

$$\langle N_t^{(j)}, N_\theta^{(2)} \rangle_{L_2(\Omega,\mathcal{A},P)} = \frac{1}{2^{j-1}\sqrt{\mu_{j-1}}} \int_0^\theta \chi_{S_{j-1}} \, dM_{N^{(j)}}$$
$$= \int_0^\theta 2^{j-1}\sqrt{\mu_{j-1}}\,\chi_{S_{j-1}} \, dM_{N^{(2)}}.$$

Consequently

$$\dot{G}_t^{(j)} = 2^{j-1}\sqrt{\mu_{j-1}}\,\chi_{S_{j-1}},$$

and

$$P_t^{N^{(2)}}\left[N_t^{(j)}\right] = \int_0^t G_t^{(j)}\, dm_{N^{(2)}}$$
$$= \int_0^t 2^{j-1}\sqrt{\mu_{j-1}}\, I_{S_{j-1}}\, dm_{N^{(2)}}$$
$$= \int_0^t I_{S_{j-1}}\, dm_{N^{(j)}}.$$

Let now, for $j \geq 3$,

$$N_t^{(3,j)} = N_t^{(j)} - P_t^{N^{(2)}}\left[N_t^{(j)}\right] = \int_0^t I_{S_{j-1}^c}\, dm_{N^{(j)}},$$

and for fixed, but arbitrary i,

$$\Sigma^i = \uplus_{j=1}^i S_j, \;\; \Sigma_i = \uplus_{j \geq i} S_j.$$

Then

$$S_{j-1}^c = \Sigma^{j-2} \uplus \Sigma_j,$$

and, since [(Remark) 9.3.21] $M_{N^{(j)}}(S_k) = 0$ for $k > j$,

$$N_t^{(3,j)} = \int_0^t I_{\left[\Sigma^{j-2} \uplus S_j\right]}\, dm_{N^{(j)}}.$$

Thus the basis measure of $N^{(3,j)}$ is absolutely continuous with respect to $M_{N^{(j)}}$, and, since $M_{N^{(j)}} \ll M_{N^{(2)}}$,

$$M_{N^{(3,j)}} \ll M_{N^{(2)}}.$$

Consider then the sequence

$$\left\{ N_t^{[i]} = N_t^{(3,i+2)} = \int_0^t I_{\left[\Sigma^i \uplus S_{i+2}\right]} \, dm_{N^{(i+2)}} \right\}.$$

It is orthogonal by construction, and made of martingales in the wide sense. One may thus define, as above,

$$\boldsymbol{N}_t^{(3)} = \int_0^t I_{S_1} dm_{N^{[1]}} + \sum_{i \geq 2} \frac{\int_0^t I_{S_i} dm_{N^{[i]}}}{2^i \sqrt{M_{N^{[i]}}([0,1])}},$$

which rewrites as

$$\boldsymbol{N}_t^{(3)} = \int_0^t I_{S_1} dm_{N^{(3)}} + \sum_{i \geq 2} \frac{\int_0^t I_{S_i} dm_{N^{(i+2)}}}{2^i \sqrt{M_{N^{(i+2)}}([0,1])}}.$$

d) Conclusion: *the processes* $\boldsymbol{N}^{(1)}$, $\boldsymbol{N}^{(2)}$ *and* $\boldsymbol{N}^{(3)}$ *have the following list of properties.*

(i) They are wide sense martingales.

(ii) They are orthogonal (as the sets S_i are associated with $N^{(i)}$ in $\boldsymbol{N}^{(1)}$, with $N^{(i+1)}$ in $\boldsymbol{N}^{(2)}$, and with $N^{(i+2)}$ in $\boldsymbol{N}^{(3)}$).

(iii) Absolute continuity prevails: $M_{\boldsymbol{N}^{(3)}} \ll M_{\boldsymbol{N}^{(2)}} \ll M_{\boldsymbol{N}^{(1)}}$.

(iv) The element $N_t^{(1)}$ belongs to $L_t\left[\boldsymbol{N}^{(1)}\right]$ (as, using $(\star\star)$,

$$N_t^{(1)} - P_t^{\boldsymbol{N}^{(1)}}\left[N_t^{(1)}\right] = \int_0^t I_{S_1^c} \, dm_{N^{(1)}} = 0_{L_2(\Omega,\mathcal{A},P)},$$

since [(Remark) 9.3.21] $M_{N^{(1)}}(S_{1+i}) = 0$).

(v) The element $N_t^{(2)}$ belongs to $L_t\left[\boldsymbol{N}^{(1)}\right] \oplus L_t\left[\boldsymbol{N}^{(2)}\right]$ (as, analogously to what has been done in (iv),

$$\begin{aligned} N_t^{(2)} - P_t^{\boldsymbol{N}^{(1)}}\left[N_t^{(2)}\right] &= \int_0^t I_{S_2^c} \, dm_{N^{(2)}} \\ &= \int_0^t I_{S_1} \, dm_{N^{(2)}} \\ &= \int_0^t I_{S_1} \, dm_{N^{(2)}}). \end{aligned}$$

(vi) The element $N_t^{(3)}$ belongs to $L_t\left[N^{(1)}\right] \oplus L_t\left[N^{(2)}\right] \oplus L_t\left[N^{(3)}\right]$ (as

$$N_t^{(3)} - P_t^{N(2)}\left[N_t^{(3)}\right] = \int_0^t I_{S_1 \uplus S_3} \, dm_{N^{(3)}}$$

$$= \int_0^t I_{S_1} \, dm_{N^{(3)}} + \int I_{S_3} \, dm_{N^{(3)}}$$

$$= \int_0^t I_{S_1} \, dm_{N^{(3)}} + \int I_{S_3} \, dm_{N^{(1)}}).$$

That shows how to ascertain that the statement obtains in $[0, 1]$. To extend its validity to $]1, 2]$, one repeats the procedure with the process

$$M_t^{(n)} - M_1^{(n)}.$$

□

Remark 9.3.23 The construction performed above is such that

1. the measure $M_{N^{(1)}}$ is a linear combination of all the measures ν_i;
2. for i, fixed, but arbitrary, $M_i \gg M_{N^{(i)}} \gg \nu_i$.

Consequently, any measure of type M_N such that, for all i, $M_N \gg M_i$, or $M_N \gg M_{N^{(i)}}$, or $M_N \gg \nu_i$, is such that $M_N \gg M_{N^{(1)}}$. In that respect $M_{N^{(1)}}$ is minimal. Since the construction repeats identically, that feature persists for the other terms in the sequence.

Remark 9.3.24 One has already seen [(Lemma) 6.4.45] that the sequence of (Lemma) 9.3.22 is unique within equivalence classes for mutual absolute continuity.

Remark 9.3.25 The conclusion is that it makes sense to define the multiplicity $m_X[t]$ of (Definition) 9.3.18 at t as the smallest integer n for which $M_{N^{(n+1)}}([0, t]) = 0$.

Remark 9.3.26 Integrating the $\mathbf{N}^{(i)}$'s with respect to proper subsets allows one to express the $N^{(i)}$'s and thus the $M^{(i)}$'s with respect to the $\mathbf{N}^{(i)}$'s. Thus, in the simplest case, one has that

$$M_t^{(1)} = N_t^{(1)} = \int_0^t I_{S_1} \, dm_{N^{(1)}}.$$

Proposition 9.3.27 *The CHR representation becomes, in the present context, and in quadratic mean:*

$$X_t = P_0^X[X_t] + \sum_{i=1}^{m_X[t]} \int_0^t F_t^{(i)} \, dm_{N^{(i)}},$$

with $F_t^{(i)}$ *a Radon-Nikodým derivative, as in* (Proposition) 9.2.16. *Furthermore*

$$m_X[t] \leq s_X(t).$$

Proof The first assertion follows from (Corollary) 9.3.13, (Proposition) 9.3.14, and the fact that, in a Gaussian environment, σ-algebras are generated by subspaces [200, p. 22, Chapter II].

Since $\lambda \mapsto M_t^{(\lambda)}$ is continuous in quadratic mean,

$$\Lambda_t[M] = \overline{V\left\{M_t^{(q)},\ q \in \mathbb{Q}\right\}}.$$

Let $\{N^{(t,i)},\ i \in I\}$ be an orthonormal basis for $\Lambda_t[M]$ obtained by applying Gram-Schmidt to

$$\overline{V\left\{M_t^{(q)},\ q \in \mathbb{Q}\right\}},$$

and let

$$N_\theta^{(t,i)} = P_\theta\left[N^{(t,i)}\right].$$

As $M_t^{(q)} = \sum_{i \in I} \alpha_i^{(t,q)} N^{(t,i)}$,

$$M_\theta^{(q)} = P_\theta\left[M_t^{(q)}\right] = \sum_{i \in I} \alpha_i^{(t,q)} P_\theta\left[N^{(t,i)}\right] = \sum_{i \in I} \alpha_i^{(t,q)} N_\theta^{(t,i)},$$

so that the σ-algebra generated by M is the same as the σ-algebra generated by N. So one may use N to obtain the multiplicity. But then one has the required inequality since the integers are a subset of the rationals. □

Remark 9.3.28 Knight [160, p. 124] provides formulae that express the prediction process, and its related martingales, in terms of the $N^{(i)}$'s. They are obtained as those in Sect. 9.2.3. The examples to follow illustrate, for simple cases, the attending computations.

Example 9.3.29 Let, ignoring the difference, for instance, between $e^{-\beta(t-\theta)}$ and its class,

$$X_t = \int_0^t e^{-\beta(t-\theta)} m_W(d\theta).$$

It is a proper canonical representation since $\int_0^t e^{\beta\theta} f(\theta) d\theta = 0$ for all t has the zero function as solution. The representation has thus multiplicity one. The usual calculation using the operational definition of projection [44, p. 80] (or the general representation of projection for a canonical representation) yields that

$$P_t\left[X_{t+\theta}\right] = \int_0^t e^{-\beta(t+\theta-x)} m_W(dx) = e^{-\beta(t+\theta)} \int_0^t e^{\beta x} m_W(dx).$$

Then

$$\begin{aligned} q(t,\lambda) &= \lambda \int_0^\infty e^{-\lambda\theta} P_t\left[X_{t+\theta}\right] d\theta \\ &= \lambda e^{-\beta t} \int_0^\infty d\theta\, e^{-(\beta+\lambda)\theta} \int_0^t e^{\beta x} m_W(dx) \\ &= \frac{\lambda e^{-\beta t}}{\beta+\lambda} \int_0^t e^{\beta x} m_W(dx) \\ &= \frac{\lambda}{\beta+\lambda} X_t. \end{aligned}$$

Consequently $X_\theta - q(\theta,\lambda) = \frac{\beta}{\beta+\lambda} X_\theta$, and thus

$$\begin{aligned} M_t^{(\lambda)} &= q(t,\lambda) - q(0,\lambda) + \lambda \int_0^t \{X_\theta - q(\theta,\lambda)\}\, d\theta \\ &= \frac{\lambda}{\beta+\lambda} X_t + \frac{\beta\lambda}{\beta+\lambda} \int_0^t X_\theta\, d\theta \\ &= \frac{\lambda}{\beta+\lambda} X_t + \frac{\beta\lambda}{\beta+\lambda} \int_0^t d\theta\, e^{-\beta\theta} \int_0^\theta e^{\beta x} W(dx). \end{aligned}$$

This latter integral is of the following form: $\int_0^t F(d\theta)\, G(\theta)$, and may be computed as

$$F(t)\,G(t) - F(0)\,G(0) - \int_0^t G(d\theta)\, F(\theta),$$

which translates into

$$\left\{ \frac{e^{-\beta\theta}}{-\beta} \int_0^\theta e^{\beta x} W(dx) \right\} \Bigg|_0^t - \int_0^t \frac{e^{-\beta\theta}}{-\beta} e^{\beta\theta} W(d\theta),$$

and that, in turn, yields $W_t - \frac{1}{\beta} X_t$. Finally

$$M_t^{(\lambda)} = \frac{\lambda}{\beta+\lambda} W_t, \quad \text{and } X_t = \frac{\beta+\lambda}{\lambda} \int_0^t e^{-\beta(t-\theta)} m_{M^{(\lambda)}}(d\theta).$$

Thus $m_X[t] = s_X[t] = 1$. When X is defined on the real line, it is an Ornstein-Uhlenbeck (stationary) process.

Example 9.3.30 Let

$$X_t = \int_0^t (2t - \theta)\, W(d\theta).$$

It has already been checked [(Example) 9.2.49] that the multiplicity is one, and computed that

$$X_t = tW_t + \int_0^t W_\theta \, d\theta,$$

and that

$$M_t^{(\lambda)} = \left(1 + \frac{2}{\lambda t}\right) X_t - 2\left(\frac{1}{t} + \frac{1}{\lambda t^2}\right) \int_0^t X_\theta \, d\theta.$$

Now

$$\begin{aligned}\int_0^t X_\theta \, d\theta &= \int_0^t \left\{ \theta W_\theta + \int_0^\theta W_x dx \right\} d\theta \\ &= \int_0^t \theta W_\theta d\theta + \left\{ \theta \int_0^\theta W_x dx \right\}\Bigg|_0^t - \int_0^t \theta W_\theta d\theta \\ &= t \int_0^t W_\theta \, d\theta.\end{aligned}$$

Consequently

$$M_t^\lambda = \left(t + \frac{2}{\lambda}\right) W_t - \int_0^t W_\theta d\theta.$$

But the two terms on the right are linearly independent, for indeed

$$\left\| \alpha \left(t + \frac{2}{\lambda}\right) W_t + \beta \int_0^t W_\theta d\theta \right\|^2_{L_2(\Omega, \mathcal{A}, P)} = \left\{ \alpha \left(t + \frac{2}{\lambda}\right) + \frac{\beta}{2} t \right\}^2 + \beta^2 \frac{t^2}{4}.$$

Thus $m_X[t] < s_X[t]$.

9.3.4 The Case of Finite Multiplicity

When a process X has finite multiplicity, one is in a position to characterize the cases for which the index of stationarity is finite, and that yields a finer representation of X. Only multiplicity one shall be considered. The case of higher multiplicity is dealt with similarly in [158, p. 560].

In what follows one shall need a specific form [110, p. 149] of the Riesz representation theorem for continuous linear functionals which shall now be explained.

Fact 9.3.31 *Let $C(\mathbb{R}_+)$ be the space of real valued, continuous functions, with domain $\mathbb{R}_+$. Let $\Phi \subseteq C(\mathbb{R}_+)$ be made of functions with the following properties:*

(a) *for $\phi \in \Phi$, and $t \in \mathbb{R}_+$, fixed, but arbitrary, $\phi(t) \geq 0$;*
(b) *for fixed, but arbitrary $n \in \mathbb{N}$, and $\{\phi_1, \ldots, \phi_n\} \subseteq \Phi$, there exists $\phi \in \Phi$, and $\kappa > 0$, such that, for $i \in [1:n]$, and $t \in \mathbb{R}_+$, fixed, but arbitrary, $\phi_i(t) \leq \kappa\phi(t)$;*
(c) *for fixed, but arbitrary $t \in \mathbb{R}_+$, there exists $\phi \in \Phi$ such that $\phi(t) > 0$.*

[One shall use below the family $\Phi = \{t \mapsto e^{\lambda t}, \lambda > 0\}$. The result applies equally well to any subset of $\mathbb{R}_+$.]

$C_\Phi(\mathbb{R}_+)$ shall be the family of continuous functions $t \mapsto f(t)$ such that, for fixed, but arbitrary $\phi \in \Phi$,

$$\sup_{\mathbb{R}_+} \{\phi(t)\,|f(t)|\} < \infty.$$

One then defines a family of seminorms as follows: for $\phi \in \Phi$, fixed, but arbitrary,

$$n_\phi(f) = \sup_{\mathbb{R}_+} \{\phi(t)\,|f(t)|\}.$$

$C_\Phi(\mathbb{R}_+)$ shall also denote the linear space which emerges when one endows it with the family of seminorms $\{n_\phi, \phi \in \Phi\}$.

$C_\Phi^0(\mathbb{R}_+)$ shall denote the subspace of $C_\Phi(\mathbb{R}_+)$ obtained when restricting attention to those functions f for which ϕf vanishes at infinity. That subspace is complete and separable.

The Riesz representation theorem then states that a linear functional

$$\mathcal{L} : C_\Phi^0(\mathbb{R}_+) \longrightarrow \mathbb{R}$$

has, for $f \in C_\Phi^0(\mathbb{R}_+)$, and $\phi \in \Phi$, fixed, but arbitrary, the property that

$$|\mathcal{L}[f]| \leq \kappa\, n_\phi(f)$$

if, and only if, for some measure $\mu_\mathcal{L}$,

$$\mathcal{L}(f) = \int_0^\infty f(t)\,\mu_\mathcal{L}(dt),$$

with

$$\int_0^\infty \frac{1}{\phi(t)} |\mu_{\mathcal{L}}| (dt) \leq \kappa.$$

Lemma 9.3.32 *Let F and N be as in* (Proposition) 9.3.27, *and*

$$\phi(t) = \sum_{i=1}^{n} \phi_i e^{-\lambda_i t},$$

$$\Phi_\lambda(\theta) = \int_0^\infty e^{-\lambda x} \phi(x) \dot{F}(\theta + x, \theta)\, dx.$$

Then, $C_0(\mathbb{R}_+)$ denoting the family of continuous functions vanishing at infinity,

1. $\varphi_\lambda(t) = \int_0^t \left\{ \int_0^\infty e^{-\lambda x} \dot{F}^2(\theta + x, \theta)\, dx \right\} M_N(d\theta) < \infty$;
2. $\Phi_\lambda^2(\theta) \leq \lambda^{-1} \|\phi\|^2_{C_0(\mathbb{R}_+)} \int_0^\infty e^{-\lambda x} \dot{F}^2(\theta + x, \theta)\, dx$;
3. $\left\| \int_0^t \Phi_\lambda(\theta)\, m_N(d\theta) \right\|^2_{L_2(\Omega, \mathcal{A}, P)} = \int_0^t \Phi_\lambda^2(\theta) M_N(d\theta) \leq \lambda^{-1} \varphi_\lambda(t) \|\phi\|^2_{C_0(\mathbb{R}_+)}$;
4. $\left\| M_t^{(\lambda)} \right\|^2_{L_2(\Omega, \mathcal{A}, P)} \leq \varphi_\lambda(t)$.

Proof Item 1 is already in (Lemma) 9.2.22. Item 2 obtains since [(Definition) 9.2.3]

$$\begin{aligned}
\left\{ \int_0^\infty e^{-\lambda x} \phi(x) \dot{F}(\theta + x, \theta)\, dx \right\}^2 &= \left\{ \lambda^{-1} E_{\Pi_\lambda^0} \left[\phi \dot{F}(\theta + \cdot, \theta) \right] \right\}^2 \\
&\leq \lambda^{-2} E_{\Pi_\lambda^0} \left[\phi^2 \right] E_{\Pi_\lambda^0} \left[\dot{F}^2(\theta + \cdot, \theta) \right] \\
&\leq \lambda^{-1} \|\phi\|^2_{C_0(\mathbb{R}_+)} \int_0^\infty e^{-\lambda x} \dot{F}^2(\theta + x, \theta)\, dx.
\end{aligned}$$

Item 3 is by integration, using items 1 and 2, and item 4 is obtained as item 2. □

Lemma 9.3.33 *When $s_X[t] = 1$, for functions $f \in C_\Phi^0(\mathbb{R}_+)$, one has, in $L_2(\Omega, \mathcal{A}, P)$, that*

$$\int_0^t \left\{ \int_0^\infty f(x) \dot{F}(\theta + x, \theta)\, dx \right\} m_N(d\theta) = \kappa(f) M_t^{(1)},$$

where the notation $\kappa(f)$ means that it is a unique constant that depends on f only.

Proof When $s_X[t] = 1$, one may assume that $\Lambda_t[X]$ is generated by $M_t^{(1)}$. Thus

$$M_t^{(\lambda)} = \kappa(\lambda) M_t^{(1)},\ \lambda > 0,\ \text{some}\ \kappa(\lambda).$$

When $f \in C_{\Phi}^{0}(\mathbb{R}_{+})$, f belongs to $C_0(\mathbb{R}_{+})$. Thus, for fixed, but arbitrary $\epsilon > 0$, there exists, by the Stone-Wierstrass theorem,

$$\phi_{\epsilon}(t) = \sum_{i=1}^{n} \phi_i^{(\epsilon)} e^{-\lambda_i^{(\epsilon)} t}$$

such that, with $f_{\lambda}(t) = e^{\lambda t} f(t)$, uniformly

$$|f_{\lambda}(t) - \phi_{\epsilon}(t)| < \epsilon.$$

As [(Proposition) 9.2.23]

$$M_t^{(\lambda)} = \lambda \int_0^t \left\{ \int_0^{\infty} e^{-\lambda x} \dot{F}(\theta + x, \theta)\, dx \right\} m_N(d\theta),$$

and [$\Phi_{\lambda}^{(\epsilon)}$ is as Φ_{λ} in (Lemma) 9.3.32]

$$\Phi_{\lambda}^{(\epsilon)}(\theta) = \sum_{i=1}^{n} \phi_i^{(\epsilon)} \int_0^{\infty} e^{-x\left(\lambda + \lambda_i^{(\epsilon)}\right)} \dot{F}(\theta + x, \theta)\, dx,$$

one has that

$$\begin{aligned}
\int_0^t \Phi_{\lambda}^{(\epsilon)}(\theta)\, m_N(d\theta) &= \sum_{i=1}^{n} \phi_i^{(\epsilon)} \int_0^t \left\{ \int_0^{\infty} e^{-x\left(\lambda + \lambda_i^{(\epsilon)}\right)} \dot{F}(\theta + x, \theta)\, dx \right\} m_N(d\theta) \\
&= \sum_{i=1}^{n} \phi_i^{(\epsilon)} \left(\lambda + \lambda_i^{(\epsilon)}\right)^{-1} M_t^{\left(\lambda + \lambda_i^{(\epsilon)}\right)} \\
&= M_t^{(1)} \sum_{i=1}^{n} \kappa\left(\lambda + \lambda_i^{(\epsilon)}\right) \phi_i^{(\epsilon)} \left(\lambda + \lambda_i^{(\epsilon)}\right)^{-1} \\
&= \kappa_{\epsilon,\lambda} M_t^{(1)}.
\end{aligned}$$

Let $\Phi(\theta) = \int_0^{\infty} f(x)\, \dot{F}(\theta + x, \theta)\, dx$. Then

$$\begin{aligned}
\left\{\Phi(\theta) - \Phi_{\lambda}^{(\epsilon)}(\theta)\right\}^2 &= \left\{ \int_0^{\infty} e^{-\lambda x} [f_{\lambda}(x) - \phi_{\epsilon}(x)]\, \dot{F}(\theta + x, \theta)\, dx \right\}^2 \\
&\leq \epsilon \left\{ \int_0^{\infty} e^{-\lambda x} \left|\dot{F}(\theta + x, \theta)\right| dx \right\}^2 \\
&= \epsilon \lambda^{-2} \left\{ E_{\Pi_{\lambda}^{0}} \left[\left|\dot{F}(\theta + \cdot, \theta)\right|\right] \right\}^2 \\
&\leq \epsilon \lambda^{-1} \int_0^{\infty} e^{-\lambda x} \dot{F}^2(\theta + x, \theta)\, dx.
\end{aligned}$$

Consequently

$$\int_0^t \Phi^2(\theta) M_N(d\theta) =$$

$$= \int_0^t \left\{\left[\Phi(\theta) - \Phi_\lambda^{(\epsilon)}(\theta)\right] + \Phi_\lambda^{(\epsilon)}(\theta)\right\}^2 M_N(d\theta)$$

$$\leq 2\left\{\int_0^t \left\{\Phi(\theta) - \Phi_\lambda^{(\epsilon)}(\theta)\right\}^2 M_N(d\theta) + \int_0^t \left\{\Phi_\lambda^{(\epsilon)}\right\}^2(\theta) M_N(d\theta)\right\},$$

so that [(Lemma) 9.3.32] $\int_0^t \Phi^2(\theta) M_N(d\theta) < \infty$, and

$$\left\|\int_0^t \Phi(\theta) m_N(d\theta) - \int_0^t \Phi_\lambda^{(\epsilon)}(\theta) m_N(d\theta)\right\|^2_{L_2(\Omega,\mathcal{A},P)} =$$

$$= \int_0^t \left\{\Phi(\theta) - \Phi_\lambda^{(\epsilon)}(\theta)\right\}^2 M_N(d\theta)$$

$$\leq \epsilon\lambda^{-1} \int_0^t \left\{\int_0^\infty e^{-\lambda x} \dot{F}^2(\theta + x, \theta)\, dx\right\}^2 M_N(d\theta).$$

The integral $\int_0^t \Phi(\theta) m_N(d\theta)$ thus exists in $L_2(\Omega, \mathcal{A}, P)$, and, in $L_2(\Omega, \mathcal{A}, P)$,

$$\int_0^t \Phi(\theta) m_N(d\theta) = \lim_{\epsilon \downarrow\downarrow 0} \int_0^t \Phi_\lambda^{(\epsilon)}(\theta) m_N(d\theta).$$

Since, as already assessed, $\int_0^t \Phi_\lambda^{(\epsilon)} dm_N = \kappa_{\epsilon,\lambda} M_t^{(1)}$, and that the corresponding net converges, the net $\{\kappa_{\epsilon,\lambda}\}$ is convergent to some unique κ, and then

$$\int_0^t \Phi\, dm_N = \kappa M_t^{(1)}.$$

But Φ and κ have the form of the lemma's statement [(Lemma) 9.3.33]. □

Lemma 9.3.34 *The assumptions are those of this section: in particular, both indexes, that of multiplicity, and that of stationarity, have value one. For* $f \in C_\Phi^0(\mathbb{R}_+)$, *fixed, but arbitrary, let*

$$\mathcal{L}[f] = \kappa(f),$$

where $\kappa(f)$ *is defined in (Lemma) 9.3.33.* $\mathcal{L}$ *is, on* $C_\Phi^0(\mathbb{R}_+)$, *a continuous, linear functional.*

Proof From (Lemma) 9.3.33 one has that

$$\kappa(f)\, M_t^{(1)} = \int_0^t \left\{ \int_0^\infty f(x)\, \dot{F}(\theta + x, \theta)\, dx \right\} m_N(d\theta). \qquad (\star)$$

$\mathcal{L}$ is thus well defined and linear. Write n_λ for n_ϕ when $\phi(t) = e^{\lambda t}$. Computing the square of the norm of that latter expression $(\star)$ yields that

$$\begin{aligned}\kappa^2(f) \left\| M_t^{(1)} \right\|^2_{L_2(\Omega,\mathcal{A},P)} &= \int_0^t \left\{ \int_0^\infty e^{\lambda x} f(x) e^{-\lambda x} \dot{F}(\theta + x, \theta) dx \right\}^2 M_N(d\theta) \\ &\le n_\lambda^2(f) \int_0^t \left\{ \int_0^\infty e^{-\lambda x} \dot{F}(\theta + x, \theta) dx \right\}^2 M_N(d\theta) \\ &\le \lambda^{-1} \varphi_\lambda(t) n_\lambda^2(f).\end{aligned}$$

Consequently, with $c_{\lambda,t} = \lambda \varphi_\lambda^{1/2}(t) / \left\| M_t^{(1)} \right\|_{L_2(\Omega,\mathcal{A},P)}$,

$$|\kappa(f)| \le c_{\lambda,t} n_\lambda(f).$$

□

Proposition 9.3.35 *Suppose that* $m_X[t] = 1$. *Then* $s_X[t] = 1$ *if, and only if, in the representation*

$$X_t = P_0^X[X_t] + \int_0^t F_t dm_N,$$

one may choose $\dot{F}_t(\theta) = G(t - \theta)$, $\theta \in [0, t]$, *some* G.

Proof Suppose that one may chose $\dot{F}_t(\theta) = G(t - \theta)$.

One shall write $\dot{F}(t, \theta)$ for $\dot{F}_t(\theta)$. As [(Proposition) 9.2.23]

$$M_t^{(\lambda)} = \lambda \int_0^t \left\{ \int_0^\infty e^{-\lambda x} \dot{F}(\theta + x, \theta)\, dx \right\} m_N(d\theta),$$

one obtains

$$M_t^{(\lambda)} = \lambda \int_0^t \left\{ \int_0^\infty e^{-\lambda x} G(x)\, dx \right\} N(d\theta) = \left\{ \lambda \int_0^\infty e^{-\lambda x} G(x)\, dx \right\} N_t.$$

Thus $\Lambda_t[X]$ is generated by N_t, and has thus dimension one.

Proof Suppose that $s_X[t] = 1$.

Let $\Psi(\theta) = \int_0^\infty e^{-x}\dot{F}(\theta + x, \theta)\, dx$. Then

$$M_t^{(1)} = \int_0^t \Psi dm_N,$$

and the relation ($\star$) at the beginning of the proof of (Lemma) 9.3.34, and (Lemma) 9.3.34 itself, yield, for all fixed, but arbitrary $t \in \mathbb{R}_+$ and $f \in C^0_\Phi(\mathbb{R}_+)$, μ_κ being the measure in the Riesz representation [(Fact) 9.3.31],

$$\left\{\int_0^\infty f(x)\mu_\kappa(dx)\right\}\left\{\int_0^t \Psi(\theta)m_N(d\theta)\right\} =$$
$$= \int_0^t \left\{\int_0^\infty f(x)\dot{F}(\theta + x, \theta)dx\right\} m_N(d\theta).$$

Thus, on $[0, t]$, almost surely in θ, with respect to M_N, but depending on f,

$$\int_0^\infty f(x)\dot{F}(\theta + x, \theta)\, dx = \Psi(\theta)\int_0^\infty f(x)\,\mu_\kappa(dx).$$

As $C^0_\Phi(\mathbb{R}_+)$ is separable, restricting attention to a countable, dense set of f's, one may assume that the set of θ's for which equality obtains is the same for all f's. Consequently

$$F(\theta + x, \theta)\, dx = \Psi(\theta)\,\mu_\kappa(dx),$$

and thus μ_κ is absolutely continuous with respect to Lebesgue measure (*Leb*). Let $\frac{d\mu_\kappa}{dLeb}$ denote the Radon-Nikodým derivative. Then, except for a set of (θ, x)'s in $[0, t] \times \mathbb{R}_+$ of measure zero for $\mu_\kappa \otimes Leb$,

$$\dot{F}(\theta + x, \theta) = \Psi(\theta)\frac{d\mu_\kappa}{dLeb}(x).$$

This latter expression, setting $y = \theta + x$, may be given the following form:

$$\dot{F}(y, \theta) = \Psi(\theta)\frac{d\mu_\mathcal{L}}{dLeb}(y - \theta).$$

□

Part III
Likelihoods

The word *"likelihood"* shall often be abbreviated in the sequel using the acronym *"LKD."* In this part, one sees how an analytical form for the detection likelihood may be obtained, when a random signal, whose law is unknown, is hidden in a mean square continuous, dependent, Gaussian noise. One starts with some tools needed for the material to follow [Chap. 10]. The Cramér-Hida representation requires a specific form of stochastic calculus, which is presented in Chap. 11, and specific sample spaces, which are explained in Chap. 12. The Girsanov theory of the likelihood is covered in Chap. 13 with moments and paths conditions, the latter being the adequate form for detection as presented here. Chapters 14–16 are concerned with the adequacy of the Girsanov likelihood for the detection problems of interest. In the last chapter [Chap. 17], the preceding material is assembled to produce the Gaussian likelihood, and some comments are made as to the practical usefulness of the theory that has been developed.

The Cramér-Hida representation of a Gaussian noise process N that is continuous as a map from $\mathbb{R}_+$, or some finite interval $[0, T]$, to $L_2(\Omega, \mathcal{A}, P)$, that is,

$$N_t = \sum_{i=1}^{M_N} \int_0^t F_t^{(i)} dm_{B_i},$$

may be looked at, formally, as a map sending a vector $\underline{B}$ of Gaussian martingales B_i to N_t, that is,

$$N_t = \Phi_t [\underline{B}],$$

Φ_t being determined by the sequence

$$\left\{F_t^{(i)}, i \in [1 : M_N]\right\},$$

and a subsequent integration procedure. It is that view that allows one to produce an expression for the likelihood.

To make such a view operational, one needs some stochastic calculus properties for integrators $\underline{B}$ with a possibly infinite, but countable number of components. Since in practice, M_N is unknown, one should proceed as if it were infinite. The particular features of the Cramér-Hida representation, and the generality required by practice, dictate that all those features, but no more, be conscripted when developing the calculus. That explains the need to tailor the "usual" Itô calculus to those properties, and justifies the developments that follow. The law of the received signal is, in practice again, usually unknown. In such a context the only known path to the likelihood is the appropriate version of Girsanov's formula [116]. That explains the space allocated to its derivation and its properties. One shall thus have to deal with processes $\underline{B}$ and $\underline{S}[\underline{a}] + \underline{B}$, where $\underline{B}$ emerges from the Cramér-Hida decomposition. These processes have values in l_2, and paths in $\prod_{n=1}^{\infty} C[0, 1]$. It is that product which will support the laws of $\underline{B}$ and $\underline{S}[\underline{a}] + \underline{B}$. Thus the general scheme of things is to make sense of Girsanov's original approach for the Wiener process in that wider framework.

The "Girsanov method" works, in its premier context [116], as follows. Suppose that P is the basic probability, and that the received signal is

$$Y_t = \int_0^t s_\theta \, d\theta + W_t,$$

W a standard Brownian motion. Let Q be defined using the following formula:

$$dQ = \left\{ e^{\int_0^1 (-s_\theta) dW_\theta - \frac{1}{2} \int_0^T s_\theta^2 d\theta} \right\} dP.$$

It turns out that, with the appropriate assumptions, P and Q are mutually absolutely continuous and, with respect to Q, Y is a standard Brownian motion. Now, when P_Y and Q_Y denote the measures induced on $C\,[0, 1]$ by, respectively, P and Y, and Q and Y, one has that P_Y and Q_Y are mutually absolutely continuous. Since $Q_Y = P_W$, it follows that P_W and P_Y are mutually absolutely continuous, and the likelihood is obtained from $\frac{dQ}{dP}$ with its explicit form. The map Φ then preserves those properties.

Chapter 10
Bench and Tools

This chapter covers two topics. As explained below, the "usual conditions" [70, p. 183] of stochastic calculus are not adequate in the present context. There is thus a section on sets of measure zero, enlargements of algebras using such sets, and restrictions to their complement. Exponentials of continuous martingales, continuous processes with independent increments, and the Wiener process are all closely related, and there is thus a second section in which those relations are examined, when the processes used take their values in l_2, the Hilbert space of sequences, whose components have squares that sum up to a finite number. The presence of l_2 is a consequence of the Cramér-Hida representation.

10.1 Some Terminology, Notation, and Attending Facts

Gathered here, for convenience, are the definitions, and attending facts, from [264] (which shall be the main reference used, in what follows, for basic stochastic calculus), that are not always "standard," but appear frequently in the sequel.

Let $(\Omega, \underline{\mathcal{A}}, P)$ represent a probability space $(\Omega, \mathcal{A}, P)$ with a family of σ-algebras $\underline{\mathcal{A}} = \{\mathcal{A}_t \subseteq \mathcal{A}, t \in [0, 1]\}$, such that, for $t_1 < t_2$ in $[0, 1]$, fixed, but arbitrary, $\mathcal{A}_{t_1} \subseteq \mathcal{A}_{t_2}$ (such a family shall be called a filtration for $\mathcal{A}$). $\mathcal{N}(\mathcal{A}, P)$ shall be the family of sets in $\mathcal{A}$ for which have zero measure P. The restriction to $[0, 1]$ stems from practical considerations, as observed signals get monitored over finite time. Since, in [264], the set of times is most often $\mathbb{R}_+$, to apply its results, it suffices to take, for $t > 1$, the value at one, of whatever object one considers. Thus, for example, for $t > 1$, fixed, but arbitrary, one has that $\mathcal{A}_t^+ = \mathcal{A}_1$.

A wide sense stopping time (for $\underline{\mathcal{A}}$) is usually [70, p. 184] understood as a map $S : \Omega \longrightarrow [0, 1]$ such that, for fixed, but arbitrary $t \in [0, 1]$, $\{\omega \in \Omega : S(\omega) \leq t\} \in \mathcal{A}_t^+$. That is equivalent [264, p. 32] to the requirement that $\{\omega \in \Omega : S(\omega) < t\} \in \mathcal{A}_t$. However, in [264], a wide sense stopping time is called a stopping time. What is

A.F. Gualtierotti, *Detection of Random Signals in Dependent Gaussian Noise*,
DOI 10.1007/978-3-319-22315-5_10

usually [70, p.184] called a stopping time ($\{\omega \in \Omega : S(\omega) \leq t\} \in \mathcal{A}_t$), becomes, in [264], a strict stopping time.

For a process X, $X^S(\omega, t)$ stands for $X(\omega, t \wedge S(\omega))$.

A localizing sequence is a sequence of wide sense stopping times $\{S_n, n \in \mathbb{N}\}$ such that, almost surely, with respect to P, $\lim_n S_n(\omega) = 1$. In that case 1 plays the part taken by ∞ when the time span is $\mathbb{R}_+$. That such sequences are required shall be seen, for example, when some uniform boundedness properties will prove necessary in the construction of the stochastic integral [(Proposition) 11.2.3].

A process X is adapted when, for $t \in [0, 1]$, fixed, but arbitrary, $X(\cdot, t)$ is adapted to $\mathcal{A}_t$ and the Borel sets of $\mathbb{R}$. All processes considered shall have all their paths continuous to the right. A martingale in L_2, say X, is thus an adapted process, whose paths are continuous to the right, with the added properties that for $t \in [0, 1]$, fixed, but arbitrary, $E_P\left[X^2(\cdot, t)\right] < \infty$, and that X is a martingale. X is a martingale bounded in L_2 when it is a martingale in L_2 for which

$$\sup_{t\in[0,1]} E_P\left[X^2(\cdot, t)\right] < \infty.$$

However, since X^2 is a submartingale [264, p. 44], for $t_1 < t_2$ in $[0, 1]$, fixed, but arbitrary, $E_P\left[X^2(\cdot, t_1)\right] \leq E_P\left[X^2(\cdot, t_2)\right]$, and X is automatically bounded in L_2 when it is in L_2. X is locally a martingale in L_2 when there is a localizing sequence, say $\{S_n, n \in \mathbb{N}\}$, such that, for $n \in \mathbb{N}$, fixed, but arbitrary, $X^{S_n}(\cdot, \cdot) - X(\cdot, 0)$ is a martingale in L_2, and thus bounded in L_2 as well. X is a local martingale when there is a localizing sequence, say $\{S_n, n \in \mathbb{N}\}$, such that, for $n \in \mathbb{N}$, fixed, but arbitrary, the process $X^{S_n}(\cdot, \cdot) - X(\cdot, 0)$ is a uniformly integrable martingale. Obviously a martingale that is locally in L_2 is a local martingale, but, more importantly for the present considerations, every (continuous to the right by assumption) almost surely continuous local martingale is a martingale locally in L_2 with, as localizing sequence [264, p. 63],

$$S_n(\omega) = \inf\{t \in [0, 1] : |X(\omega, t) - X(\omega, 0)| > n\}.$$

When X has all its paths continuous, one may replace strict inequality by inequality, and obtain a strict stopping time [264, p. 38]. X is almost surely continuous when

$$\{\omega \in \Omega : X[\omega] \text{ not continuous}\} \subseteq N \in \mathcal{N}(\mathcal{A}, P).$$

Fact 10.1.1 ([264, p. 71]) *Each adapted and almost surely continuous process has an adapted version whose paths which are not continuous are continuous to the right.*

The predictable sets $\mathcal{P}$ are usually a σ-algebra over $\Omega \times]0, \infty[$ [264, p.112], but, for what follows, the predictable sets of $\Omega \times]0, 1]$.

Definition 10.1.2 In case N is a Gaussian process, continuous in quadratic mean, the Gaussian martingales B_n, produced by the Cramér-Hida decomposition, shall be called, when looked at as vectors $\underline{B}$, whose components are the real stochastic processes B_n, Cramér-Hida processes.

Cramér-Hida processes shall be central to building the likelihood.

10.2 Sets of Measure Zero

Let X be an adapted process, and $\mathcal{N}(\mathcal{A}, P)$, the family of sets in $\mathcal{A}$ whose measure for P is zero. Let $X[\omega]$ denote the path of X at ω, and $\boldsymbol{X}$, the map

$$\boldsymbol{X} : \omega \mapsto X[\omega].$$

$\boldsymbol{X}$ is adapted to $\mathcal{A}$ and $\mathcal{C}(\mathbb{R}^{[0,1]})$, the σ-algebra generated by the cylinder sets. One would expect the law of an almost surely continuous process to sit on $C[0, 1]$, the space of continuous functions. But, in order to have an adapted map, one must deal with the paths that are not continuous. Let thus

$$X_c = \chi_{N^c} X.$$

X_c has continuous paths, $X_c(\cdot, t)$, and thus $\boldsymbol{X}_c$, are adapted to $\mathcal{A}$, but not necessarily adapted (to $\underline{\mathcal{A}}$). Thus, $\mathcal{E}_t$ being the evaluation at t, $\mathcal{E}_t(\boldsymbol{X}_c)$ may not be adapted to $\mathcal{A}_t$, and $\mathcal{E}$, not a process adapted the σ-algebras generated by the evaluation maps. One response to that situation is to assume, whatever t,

$$\mathcal{N}(\mathcal{A}, P) \subseteq \mathcal{A}_t,$$

or, even more, that $\mathcal{A}_t$ contains, with each $N \in \mathcal{N}(\mathcal{A}, P)$, all its subsets. But such a presupposition means that, at time zero, one knows something that can only be assessed at time 1. As stated in [251, p. 97] (see also [264, p. 31]), … *this solution* [completion] *is not entirely suitable for us, since the completion is a function of the underlying measure P, and in our applications the measure P changes, whereas the σ-algebras do not.* Then [264, p. 31]: *As a consequence we have to distinguish e.g. between continuous processes and a.s. continuous processes. But the reader should keep in mind that there is always the obvious "nullset elimination argument": If a probabilistic statement holds, say, for every continuous process with some additional properties then one can expect it to carry over to the almost surely continuous case. One just substitutes the original space Ω by the set Ω \ N with the induced filtration and measure (where N is the exceptional set). On Ω \ N the continuous version of the result in question can be applied and usually the conclusion stays valid if one shifts back to Ω.* Those considerations are particularly relevant when investigating absolute continuity of measures, and one shall meet, farther, examples of such "null sets arguments." That is why the stochastic calculus text one shall peruse is [264], and any concept, or fact, not explained here should be looked up there.

In the latter context, it is useful to be aware of the part sets of measure zero play. And one must distinguish between measurable sets of measure zero and subsets of such, not, a priori, measurable. The rule of thumb is that enlarging a sub-σ-algebra of $\mathcal{A}$ with sets in $\mathcal{N}(\mathcal{A}, P)$ causes little trouble, but that, otherwise, one must be rather careful. Below $\mathcal{B}$ shall be a σ-algebra of strictly contained in $\mathcal{A}$, $\mathcal{C}$, an enlargement of $\mathcal{B}$ with, typically, measurable sets, and $\mathcal{D}$, an enlargement with sets, typically, not taken, *a priori*, to be measurable.

10.2.1 Adjunction of Sets to σ-Algebras

The aim is to enlarge $\mathcal{B}$, adding to it the sets of a given family $\mathcal{N}$, and, possibly, subsets of the latter's elements. Given two families of sets, $\mathcal{S}$ and $\mathcal{T}$, $\mathcal{S} \vee \mathcal{T}$ denotes the σ-algebra they generate, that is, the smallest one which contains them both.

Enlargement with One Set and, Possibly, Its Subsets

Below, $\mathcal{A}$ and $\mathcal{B}$ are σ-algebras of subsets of Ω, with $\mathcal{B} \subseteq \mathcal{A}$.

Fact 10.2.1 *Let $\Omega_0 \subseteq \Omega$ be a fixed, but, arbitrary subset of Ω that does not belong to $\mathcal{B}$. Let also*

$$\mathcal{C} = \{(B_1 \cap \Omega_0) \cup (B_2 \cap \Omega_0^c), B_1 \in \mathcal{B}, B_2 \in \mathcal{B}\}.$$

$\mathcal{C}$ is a σ-algebra containing both $\mathcal{B}$ and Ω_0.

Proof Choosing $B_1 = B_2 = B$, one sees that $\mathcal{B} \subseteq \mathcal{C}$. Choosing $B_1 = \Omega$ and $B_2 = \emptyset$, one sees that $\Omega_0 \in \mathcal{C}$.

Let $C \in \mathcal{C}$ be fixed, but arbitrary, that is $C = (B_1 \cap \Omega_0) \uplus (B_2 \cap \Omega_0^c)$, $B_1 \in \mathcal{B}$, and $B_2 \in \mathcal{B}$. Then, as

$$\begin{aligned} \chi_{C^c} &= 1 - \chi_C \\ &= \chi_{\Omega_0} + \chi_{\Omega_0^c} - \left\{\chi_{B_1}\chi_{\Omega_0} + \chi_{B_2}\chi_{\Omega_0^c}\right\} \\ &= \chi_{\Omega_0}\left\{1 - \chi_{B_1}\right\} + \chi_{\Omega_0^c}\left\{1 - \chi_{B_2}\right\}, \end{aligned}$$

one has that $C^c = \left(B_1^c \cap \Omega_0\right) \cup \left(B_2^c \cap \Omega_0^c\right) \in \mathcal{C}$. Let $\{C_n, n \in \mathbb{N}\} \subseteq \mathcal{C}$ be fixed, but arbitrary, that is

$$C_n = (B_1^n \cap \Omega_0) \cup (B_2^n \cap \Omega_0^c), B_1^n \in \mathcal{B}, \text{ and } B_2^n \in \mathcal{B}.$$

Then:

$$\begin{aligned}\cup_n C_n &= \cup_{n\in\mathbb{N}}\{(B_1^n \cap \Omega_0) \cup (B_2^n \cap \Omega_0^c)\}\\ &= \{\cup_{n\in\mathbb{N}}(B_1^n \cap \Omega_0)\} \cup \{\cup_{n\in\mathbb{N}}(B_2^n \cap \Omega_0^c)\}\\ &= \{[\cup_{n\in\mathbb{N}}B_1^n] \cap \Omega_0\} \cup \{[\cup_{n\in\mathbb{N}}B_2^n] \cap \Omega_0^c\}\\ &\in \mathcal{C}.\end{aligned}$$

□

Fact 10.2.2 *In* (Fact) 10.2.1, $\mathcal{C} = \mathcal{B} \vee \{\Omega_0\}$.

Proof Result (Fact) 10.2.1 implies $\mathcal{B} \vee \{\Omega_0\} \subseteq \mathcal{C}$. Since, when $B \in \mathcal{B}$, $B \cap \Omega_0$ and $B \cap \Omega_0^c$ belong to $\mathcal{B} \vee \{\Omega_0\}$, $\mathcal{C} \subseteq \mathcal{B} \vee \{\Omega_0\}$. □

Fact 10.2.3 *The family*

$$\mathcal{D} = \{B\Delta\Omega_{00}, B \in \mathcal{B}, \Omega_{00} \subseteq \Omega_0\}$$

is a σ-algebra which contains $\mathcal{B}$, and the subsets of Ω_0.

Proof Since $\emptyset \in \mathcal{B}$ is a subset of Ω_0, both $\mathcal{B}$ and Ω_{00} belong to $\mathcal{D}$, and thus so do $\emptyset$ and Ω.

Let $D = B\Delta\Omega_{00}$ be fixed, but arbitrary. Since [79, p. 8] Δ is associative, and $D^c = D\Delta\Omega$,

$$D^c = (B\Delta\Omega_{00})^c = \Omega\Delta(B\Delta\Omega_{00}) = (\Omega\Delta B)\Delta\Omega_{00} = B^c\Delta\Omega_{00} \in \mathcal{D}.$$

Let $\{D_n, n \in \mathbb{N}\} \subseteq \mathcal{D}$ be fixed, but arbitrary. Then:

$$D_n = B_n\Delta\Omega_{00}^n,\ B_n \in \mathcal{B},\ \Omega_{00}^n \subseteq \Omega_0,$$

and, using 2.7 of [79, p. 8], and the fact that $A\Delta B \subseteq A \cup B$,

$$(\cup_n B_n)\Delta(\cup_n \Omega_{00}^n) \subseteq \cup_n(B_n\Delta\Omega_{00}^n) \subseteq (\cup_n B_n) \cup (\cup_n \Omega_{00}^n),$$

so that, since $(A \cup B) \setminus (A\Delta B) = A \cap B$,

$$\cup_n(B_n\Delta\Omega_{00}^n) = \{(\cup_n B_n)\Delta(\cup_n\Omega_{00}^n)\} \uplus \Omega_{00}^\star,\ \Omega_{00}^\star \subseteq (\cup_n B_n) \cap (\cup_n\Omega_{00}^n) \subseteq \Omega_0.$$

As, for disjoint A and B, $A \cup B = A\Delta B$, and that $\Omega_{00}^\star$ is disjoint from the set $(\cup_n B_n)\Delta(\cup_n\Omega_{00}^n)$,

$$\cup_n(B_n\Delta\Omega_{00}^n) = (\cup_n B_n)\Delta\{(\cup_n\Omega_{00}^n)\Delta\Omega_{00}^\star\} \in \mathcal{D}.$$

□

Fact 10.2.4 *Let $\mathcal{C}$ be as in* (Fact) 10.2.1, *and $\mathcal{D}$, as in* (Fact) 10.2.3. *Let*

$$\mathcal{E} = \sigma(\{B\Delta\Omega_0, B \in \mathcal{B}\}).$$

Then: $\mathcal{E} = \mathcal{C} \subseteq \mathcal{D}$.

Proof By definition, $B\Delta\Omega_0 \in \mathcal{C}$. Thus $\mathcal{E} \subseteq \mathcal{C}$. Since $\Omega_0 = \emptyset\Delta\Omega_0$, $\Omega_0 \in \mathcal{E}$. Since Δ is associative, $(B\Delta\Omega_0)\Delta\Omega_0 = B \in \mathcal{E}$, that is, $\mathcal{B} \subseteq \mathcal{E}$, and, consequently, $\mathcal{C} = \mathcal{B} \vee \{\Omega_0\} \subseteq \mathcal{E}$. □

The following examples illustrate what may happen with enlargement. The first one shows that $\mathcal{C}$ may be as large as $\mathcal{A}$, the second, larger.

Example 10.2.5 Let $\Omega = \{a,b,c,d\}$, and $\mathcal{A}$ be the σ-algebra of its subsets. Let also $B = \{a,b\}$, and $\mathcal{B}$ be the σ-algebra generated by B. Let finally $\Omega_0 = \{b,c\}$. The following table yields $\mathcal{C}$. $\mathcal{B}$ is in one diagonal of the table, and the generators of $\mathcal{E}$ are in the other. Here $\mathcal{E} = \mathcal{C} = \mathcal{D} = \mathcal{A}$.

	$\Omega_0^c = \{a,d\}$				
$\Omega_0 = \{b,c\}$	$\cup$	$\emptyset \cap \Omega_0^c = \emptyset$	$B \cap \Omega_0^c = \{a\}$	$B^c \cap \Omega_0^c = \{d\}$	$\Omega \cap \Omega_0^c = \{a,d\}$
	$\emptyset \cap \Omega_0 = \emptyset$	$\emptyset$	$\{a\}$	$\{d\}$	$\boldsymbol{\Omega\Delta\Omega_0} = \{a,d\}$
	$B \cap \Omega_0 = \{b\}$	$\{b\}$	B	$\boldsymbol{B^c\Delta\Omega_0} = \{b,d\}$	$\{a,b,d\}$
	$B^c \cap \Omega_0 = \{c\}$	$\{c\}$	$\boldsymbol{B\Delta\Omega_0} = \{a,c\}$	B^c	$\{a,c,d\}$
	$\Omega \cap \Omega_0 = \{b,c\}$	$\boldsymbol{\emptyset\Delta\Omega_0} = \{b,c\}$	$\{a,b,c\}$	$\{b,c,d\}$	Ω

Example 10.2.6 Let $\Omega = \{a,b,c,d,\alpha,\beta,\gamma,\delta\}$,

$$\begin{aligned} A_1 &= \{a,b\} \quad , B_1 = \{\alpha,\beta\}, \\ A_2 &= \{c,d\} \quad , B_2 = \{\gamma,\delta\}, \\ A &= A_1 \uplus A_2 \ , B = B_1 \uplus B_2, \end{aligned}$$

and $\mathcal{A}$ be generated by A_1, A_2, B_1, B_2. It consists thus of

$$\emptyset, \Omega, A_1, A_2, B_1, B_2, A, B,$$

$$A_1 \uplus B_1, A_1 \uplus B_2, A_2 \uplus B_1, A_2 \uplus B_2, A \uplus B_1, A \uplus B_2, A_1 \uplus B, A_2 \uplus B.$$

Let $\mathcal{B}$ be generated by B_1 and B_2. It consists thus of

$$\emptyset, \Omega, B_1, B_2, A, B, A \uplus B_1, A \uplus B_2.$$

Let $\Omega_0 = \{a, c\}$. Then $\Omega_0^c = \{b, d\} \uplus B$. $\mathcal{C}$ is then made of $\mathcal{B}$ and

$$\{a, c\}, \{a, c\} \uplus B_1, \{a, c\} \uplus B_2, \{a, c\} \uplus B, \{b, d\}, \{b, d\} \uplus B_1, \{b, d\} \uplus B_2, \{b, d\} \uplus B.$$

Here $\mathcal{C} \subset \mathcal{D}$, and $\mathcal{C}$ is not contained in $\mathcal{A}$.

Remark 10.2.7 For A and B disjoint, as noticed, $A \Delta B = A \cup B$. Thus

$$B_2 = (B_2 \cap \Omega_0) \cup (B_2 \cap \Omega_0^c) = (B_2 \cap \Omega_0) \Delta (B_2 \cap \Omega_0^c),$$

and

$$B_2 \Delta \{(B_1 \cap \Omega_0) \cup (B_2 \cap \Omega_0^c)\} = \\ = \{(B_2 \cap \Omega_0) \Delta (B_2 \cap \Omega_0^c)\} \Delta \{(B_1 \cap \Omega_0) \Delta (B_2 \cap \Omega_0^c)\}.$$

Since Δ is associative, and $A \Delta A = \emptyset$,

$$B_2 \Delta \{(B_1 \cap \Omega_0) \cup (B_2 \cap \Omega_0^c)\} = (B_1 \cap \Omega_0) \Delta (B_2 \cap \Omega_0).$$

Because of the formula [79, p. 8] $A \cap (B \Delta C) = (A \cap B) \Delta (A \cap C)$,

$$(B_1 \cap \Omega_0) \Delta (B_2 \cap \Omega_0) = (B_1 \Delta B_2) \cap \Omega_0 \subseteq \Omega_0.$$

Thus $B_2 \Delta \left\{(B_1 \cap \Omega_0) \cup (B_2 \cap \Omega_0^c)\right\} \subseteq \Omega_0$, and, when Ω_0 is a set of measure zero, the difference between B_2 and the sets in $\mathcal{C}$ it produces according to its building rule are subsets of a set of measure zero.

Enlargement with a Family of Sets and, Possibly, with the Subsets of Those Sets

Definition 10.2.8 A set $\Omega_0 \subseteq \Omega$ is internally P-negligible when, whenever $A \subseteq \Omega_0$ and $A \in \mathcal{A}$, $P(A) = 0$.

Remark 10.2.9 Every set of $\mathcal{N}(\mathcal{A}, P)$ is internally P-negligible. $\mathcal{N}(\mathcal{A}, P)$ contains, with every sequence of its sets, their union. The σ-algebra generated by $\mathcal{B}$ and $\mathcal{N}(\mathcal{A}, P)$ is contained in $\mathcal{A}$. Every subset of a measurable set of measure zero is internally negligible. When N is internally negligible, and $N \in \mathcal{A}$, $P(N) = 0$, as N is a subset of itself.

Fact 10.2.10 *Let $\mathcal{N}$ be a family of subsets of Ω that are internally negligible. Suppose $\mathcal{N}$ is closed for countable unions. Let $\mathcal{M}$ be the family of subsets of sets in $\mathcal{N}$, and $\mathcal{D} = \{B\Delta M, B \in \mathcal{B}, M \in \mathcal{M}\}$. $\mathcal{D}$ is a σ-algebra.*

Proof It is *mutatis mutandis* that of (Fact) 10.2.3. □

Proposition 10.2.11 *Let $P^{|\mathcal{B}}$ be the restriction of P to $\mathcal{B}$. Let $\mathcal{N}$ be a family of internally negligible sets for P, closed for countable unions. $P^{|\mathcal{B}}$ can be extended uniquely to a probability $P_{\mathcal{B}}$ on $\mathcal{B} \vee \mathcal{N}$ in such a way that the elements in $\mathcal{N}$ have zero $P_{\mathcal{B}}$-probability.*

Proof Since $\mathcal{M}$ of (Fact) 10.2.10 contains the empty set, $\mathcal{B} \vee \mathcal{N} \subseteq \mathcal{D}$. Given the set $B\Delta M \in \mathcal{D}$, fixed, but arbitrary, let

$$Q(B\Delta M) = P^{|\mathcal{B}}(B) = P(B).$$

One must check that Q is well defined, that is, independent of the representation of $B\Delta M$. Suppose thus that $B_1\Delta M_1 = B_2\Delta M_2$. Then

$$\emptyset = (B_1\Delta M_1)\,\Delta\,(B_2\Delta M_2) = (B_1\Delta B_2)\Delta(M_1\Delta M_2),$$

and thus

$$P^{|\mathcal{B}}\left(B_1\Delta B_2\right) = Q\left((B_1\Delta M_1)\,\Delta\,(B_2\Delta M_2)\right) = Q(\emptyset).$$

But, when $B\Delta M = \emptyset$, $B = B \cap M = M \subseteq N \in \mathcal{N}$, and, since $B \in \mathcal{A}$, $P(B) = 0$. Thus $Q(\emptyset) = 0$. Consequently $P^{|\mathcal{B}}(B_1\Delta B_2) = 0$. Since

$$P^{|\mathcal{B}}(B_1) = P^{|\mathcal{B}}(B_1 \setminus (B_1 \cap B_2)) + P^{|\mathcal{B}}(B_1 \cap B_2) = P^{|\mathcal{B}}(B_1 \cap B_2),$$

and

$$P^{|\mathcal{B}}(B_2) = P^{|\mathcal{B}}(B_2 \setminus (B_1 \cap B_2)) + P^{|\mathcal{B}}(B_1 \cap B_2) = P^{|\mathcal{B}}(B_1 \cap B_2),$$

$P^{|\mathcal{B}}(B_1) = P^{|\mathcal{B}}(B_2)$, and thus $Q(B_1\Delta M_1) = Q(B_2\Delta M_2)$.

One must next check that Q is a probability on $\mathcal{D}$. One has just seen that $Q(\emptyset) = 0$, and the proof of (Fact) 10.2.3 yields that

$$\cup_n(B_n\Delta M_n) = (\cup_n B_n)\Delta M,$$

so that

$$Q(\cup_n(B_n\Delta M_n)) = P^{|\mathcal{B}}(\cup_n B_n).$$

Suppose that $(B_1\Delta M_1)\cap(B_2\Delta M_2)=\emptyset$. Since [34, p. 29]

$$(B_1\Delta M_1)\cap(B_2\Delta M_2)=(B_1\cap B_2)\Delta M,\ M\subseteq M_1\cup M_2,$$

it follows that

$$0=Q(\emptyset)=Q((B_1\Delta M_1)\cap(B_2\Delta M_2))=Q((B_1\cap B_2)\Delta M)=P^{|\mathcal{B}}(B_1\cap B_2),$$

so that

$$P^{|\mathcal{B}}(B_1\cup B_2)=P^{|\mathcal{B}}(B_1)+P^{|\mathcal{B}}(B_2).$$

One may then continue inductively. Thus the following relation:

$$\begin{aligned}(B_1\Delta M_1)\cap(B_2\Delta M_2)\cap(B_3\Delta M_3)&=\{(B_1\cap B_2)\Delta M\}\cap(B_3\Delta M_3)\\&=(B_1\cap B_2\cap B_3)\Delta M'\end{aligned}$$

yields that $P^{|\mathcal{B}}(B_1\cap B_2\cap B_3)=0$, and thus that

$$P^{|\mathcal{B}}(B_1\cup B_2\cup B_3)=P^{|\mathcal{B}}(B_1)+P^{|\mathcal{B}}(B_2)+P^{|\mathcal{B}}(B_3),$$

so that, in the end,

$$Q(\cup_n(B_n\Delta M_n))=P^{|\mathcal{B}}(\cup_n B_n)=\sum_n P^{|\mathcal{B}}(B_n)=\sum_n Q(B_n\Delta M_n).$$

$P_{\mathcal{B}}$ is the restriction of Q to $\mathcal{B}\vee\mathcal{N}$.

Let $\Pi_{\mathcal{B},\mathcal{N}}$ be another probability on $\mathcal{B}\vee\mathcal{N}$ such that $\Pi_{\mathcal{B},\mathcal{N}}^{|\mathcal{B}}=P$ and, when $N\in\mathcal{N}$, $\Pi_{\mathcal{B},\mathcal{N}}(N)=0$. Every set in $\mathcal{M}$ is internally negligible for $\Pi_{\mathcal{B},\mathcal{N}}$. $\Pi_{\mathcal{B},\mathcal{N}}$ can thus be extended to a probability Q_Π on $\mathcal{D}$ whose restriction to $\mathcal{B}$ is P, and which gives probability zero to the sets of $\mathcal{M}$. But then $\Pi_{\mathcal{B},\mathcal{N}}$ must be $P_{\mathcal{B},\mathcal{N}}$. □

Remark 10.2.12 $\mathcal{D}=\mathcal{B}\vee\mathcal{M}$.

Indeed, $B\Delta M\in\mathcal{B}\vee\mathcal{M}$, so that $\mathcal{D}\subseteq\mathcal{B}\vee\mathcal{M}$. On the other hand, $\emptyset\in\mathcal{B}\cap\mathcal{M}$, so that $\mathcal{B}\subseteq\mathcal{D}$ and $\mathcal{M}\subseteq\mathcal{D}$, and thus $\mathcal{B}\vee\mathcal{M}\subseteq\mathcal{D}$.

Remark 10.2.13 Let $B\in\mathcal{B}$ and $M\in\mathcal{M}$, $M\subseteq N$, $N\in\mathcal{N}$, be fixed, but arbitrary. One has that

$$B\cup M=B\uplus(M\setminus B)=B\Delta(M\cap B^c),$$

and that

$$B\Delta M = (B \setminus N) \uplus ((B \cap N)\Delta M).$$

Thus $\mathcal{D}$ is also generated by sets of the form $B \uplus M$.

Remark 10.2.14 Let $\mathcal{C} = \sigma(\{B\Delta N, B \in \mathcal{B}, N \in \mathcal{N}\})$. Then

$$\mathcal{C} = \mathcal{B} \vee \mathcal{N} \subseteq \mathcal{D},$$

and, when $\mathcal{N} \subseteq \mathcal{A}$, $\mathcal{C} \subseteq \mathcal{A} \cap \mathcal{D}$.

Again, $B\Delta N \in \mathcal{B} \vee \mathcal{N}$, so that $\mathcal{C} \subseteq \mathcal{B} \vee \mathcal{N}$. Also, as $\emptyset \in \mathcal{B}$, $\mathcal{N} \subseteq \mathcal{C}$, and, since $B = (B\Delta N)\Delta N$, $\mathcal{B} \subseteq \mathcal{C}$, that is, $\mathcal{B} \vee \mathcal{N} \subseteq \mathcal{C}$.

Remark 10.2.15 When $\mathcal{N} \subseteq \mathcal{A}$, $Q^{|c} = P_{\mathcal{B},\mathcal{N}} = P$.

Remark 10.2.16 Every $C \in \mathcal{C}$, as an element of $\mathcal{D}$, has a representation as $C = B\Delta M$ with $M \in \mathcal{C}$.

As an element of $\mathcal{D}$, $C = B\Delta M$, $B \in \mathcal{B}$, $M \subseteq N$, with N internally negligible. One may thus "compute," for C and B, elements of $\mathcal{C}$, $B\Delta C$, and obtain an element of $\mathcal{C}$. But $B\Delta C = M$.

Example 10.2.17 Let $\Omega = \{a, b, c, d, e\}$ and $\mathcal{A}$ be generated by

$$\begin{aligned} A_1 &= \{a\}, & P(A_1) &= 1/3, \\ A_2 &= \{b\}, & P(A_2) &= 1/3, \\ A_3 &= \{c, d\}, & P(A_3) &= 1/3, \\ A_4 &= \{e\}. \end{aligned}$$

Let $\mathcal{B}$ be generated by $\{a, b\}$. Let $N = \{d, e\}$. It is internally negligible and

$$\mathcal{M} = \{\emptyset, \{d\}, \{e\}, \{d, e\}\}.$$

Then:

- $\emptyset\Delta M$ produces $\emptyset, \{d\}, \{e\}, \{d, e\}$;
- $\{a, b\}\,\Delta M$ produces $\{a, b\}, \{a, b, d\}, \{a, b, e\}, \{a, b, d, e\}$;
- $\{c, d, e\}\,\Delta M$ produces $\{c, d, e\}, \{c, e\}, \{c, d\}, \{e\}$;
- $\Omega\Delta M$ produces $\Omega, \{a, b, d, e\}, \{a, b, c, e\}, \{a, b, c\}$;
- $B\Delta N$ produces $\{c\}, \{d, e\}, \{a, b, c\}, \{a, b, d, e\}$, and generates

$$\{\emptyset, \{c\}, \{a, b\}, \{d, e\}, \{a, b, c\}, \{d, e, c\}, \{a, b, d, e\}, \Omega\}.$$

P is defined neither on $\mathcal{C}$, nor on $\mathcal{D}$.

Example 10.2.18 Let $\Omega = \{a, b, c, d, e\}$ and $\mathcal{A}$ be generated by

$$\begin{aligned} A_1 &= \{a\}, \; P(A_1) = 1/3, \\ A_2 &= \{b\}, \; P(A_2) = 1/3, \\ A_3 &= \{c\}, \; P(A_3) = 1/3, \\ A_4 &= \{d\}, \\ A_5 &= \{e\}. \end{aligned}$$

Let $\mathcal{B}$ be generated by $\{a, b\}$. Let $N = \{d, e\}$. It is internally negligible and

$$\mathcal{M} = \{\emptyset, \{d\}, \{e\}, \{d, e\}\} \subseteq \mathcal{A}.$$

P is defined on $\mathcal{C}$, as well as on $\mathcal{D}$.

Example 10.2.19 Let $\Omega = \{a, b, c, d, e\}$ and $\mathcal{A}$ be generated by

$$\begin{aligned} A_1 &= \{a\}, & P(A_1) &= 1/3, \\ A_2 &= \{b\}, & P(A_2) &= 1/3, \\ A_3 &= \{c\}, & P(A_3) &= 1/3, \\ A_4 &= \{d, e\}. \end{aligned}$$

Let $\mathcal{B}$ be generated by $\{a, b\}$. Let $N = \{d, e\} \in \mathcal{A}$. It is internally negligible and

$$\mathcal{M} = \{\emptyset, \{d\}, \{e\}, \{d, e\}\}.$$

Here $\mathcal{C} \subseteq \mathcal{A}$, but $\mathcal{D}$ is not contained in $\mathcal{A}$. P is defined on $\mathcal{C}$, but not on $\mathcal{D}$.

Example 10.2.20 Let $\Omega = \{1, 2, 3, 4\}$, and

$\mathcal{A}$	$\Leftrightarrow$	$\emptyset$	$\{1\}$	$\{2\}$	$\{3, 4\}$	$\{1, 2\}$	$\{1, 3, 4\}$	$\{2, 3, 4\}$	Ω
P	$\Leftrightarrow$	0	1/3	1/3	1/3	2/3	2/3	2/3	1
Q	$\Leftrightarrow$	0	1/2	1/2	0	1	1/2	1/2	1

Let also $\mathcal{B} = \{\emptyset, \{1, 2\}, \{3, 4\}, \Omega\}$, and $S = \{1, 2\}$. Then $Q(A) = (3/2)P(A \cap S)$. Q is absolutely continuous with respect to P, and $Q(S^c) = 0$. The completions of $\mathcal{A}$ and $\mathcal{B}$ with respect to Q are obtained, respectively, as $\mathcal{A} \vee \{\{3\}, \{4\}\} = \mathcal{P}(\Omega)$, and $\{\emptyset, \{1, 2\}, \{3\}, \{4\}, \{3, 4\}, \{1, 2, 3\}, \{1, 2, 4\}, \Omega\}$. P is not defined on those completions.

Remark 10.2.21 Let the identity of Ω be denoted id_Ω. It is, in the following scheme, adapted:

$$(\Omega, \mathcal{D}, Q) \xrightarrow{id_\Omega} (\Omega, \mathcal{C}, Q^{|\mathcal{C}}) \xrightarrow{id_\Omega} (\Omega, \mathcal{B}, Q^{|\mathcal{B}}),$$

and

$$Q^{|\mathcal{C}} = Q \circ id_{\Omega}^{-1},\ Q^{|\mathcal{B}} = Q^{|\mathcal{C}} \circ id_{\Omega}^{-1}.$$

It follows that, for $B \in \mathcal{B}$, $B \Delta M \in \mathcal{C}$, $M \in \mathcal{C}$ [(Remark) 10.2.16], $f_{\mathcal{B}}$ adapted to $\mathcal{B}$, and $f_{\mathcal{C}}$ adapted to $\mathcal{C}$,

$$\int_B f_{\mathcal{B}}\, dP = \int_B f_{\mathcal{B}}\, dQ^{|\mathcal{B}} = \int_B f_{\mathcal{B}}\, dQ^{|\mathcal{C}} = \int_B f_{\mathcal{B}}\, dQ,$$

and, when $\mathcal{N} \subseteq \mathcal{A}$,

$$\int_B f_{\mathcal{C}}\, dP = \int_{B\Delta M} f_{\mathcal{C}}\, dP = \int_{B\Delta M} f_{\mathcal{C}}\, dQ^{|\mathcal{C}} = \int_{B\Delta M} f_{\mathcal{C}}\, dQ.$$

Proposition 10.2.22 *Let $f_{\mathcal{D}}$ be adapted to $\mathcal{D}$. There exists $f_{\mathcal{B}}$, adapted to $\mathcal{B}$, such that $Q(|f_{\mathcal{D}} - f_{\mathcal{B}}| > 0) = 0$.*

Proof Let $f_{\mathcal{D}} = \chi_{B\Delta M}$, and $f_{\mathcal{B}} = \chi_B$. Then:

$$\begin{aligned} Q(|f_{\mathcal{D}} - f_{\mathcal{B}}| > 0) &= Q(|\chi_{B\Delta M} - \chi_B| > 0) \\ &= Q(\chi_{(B\Delta M)\Delta B} > 0) \\ &= Q(M) \\ &= 0. \end{aligned}$$

Let $\mathcal{F}_{\mathcal{D}}$ be the class of functions for which the statement is true. It contains the indicators of sets in $\mathcal{D}$ which, being a σ-algebra, is a π-class [128, p. 3]. $\mathcal{F}_{\mathcal{D}}$ contains the functions that are constant. Let now

$$\{f_{\mathcal{D}}^{(n)} \in \mathcal{F}_{\mathcal{D}}, n \in \mathbb{N}\}$$

be a sequence of positive functions that increases to the finite f. As a limit of functions adapted to $\mathcal{D}$, f is adapted to $\mathcal{D}$.

Let $f_{\mathcal{B}}^{(n)}$ be adapted to $\mathcal{B}$ and have the property that

$$Q(|f_{\mathcal{D}}^{(n)} - f_{\mathcal{B}}^{(n)}| > 0) = 0.$$

Let $D_n \subseteq \Omega$ be the set on which $f_{\mathcal{D}}^{(n)}$ and $f_{\mathcal{B}}^{(n)}$ have a different value, and $D_0 = \cup_{n\in\mathbb{N}} D_n$. Let also

$$f_{\mathcal{B}} = \sup_{n\in\mathbb{N}} f_{\mathcal{B}}^{(n)}.$$

Then:

$$Q(|f - f_{\mathcal{B}}| > 0) = Q(\{|f - f_{\mathcal{B}}| > 0\} \cap D_0) + Q(\{|f - f_{\mathcal{B}}| > 0\} \cap D_0^c) = 0.$$

The result thus follows from a monotone class theorem [128, p. 4]. □

Remark 10.2.23 *Mutatis mutandis,* the same result, with the same proof, apply to $\mathcal{B}$ and $\mathcal{C}$. When $\mathcal{N} \subseteq \mathcal{A}$, $Q^{|\mathcal{C}}$ may be replaced with P.

Remark 10.2.24 Let f be adapted to $\mathcal{D}$, and be integrable. Then:

$$\begin{aligned}\int_{B\Delta N} E_Q[f \mid \mathcal{C}]\, dQ &= \int_{B\Delta N} f\, dQ \\ &= \int_B f\, dQ \\ &= \int_B E_Q[f \mid \mathcal{B}]\, dQ \\ &= \int_{B\Delta N} E_Q[f \mid \mathcal{B}]\, dQ.\end{aligned}$$

Thus $E_Q[f \mid \mathcal{B}]$ is in the class of $E_Q[f \mid \mathcal{C}]$. When $\mathcal{N} \subseteq \mathcal{A}$, since $Q^{|\mathcal{B}} = P$, one may replace $E_Q[f \mid \mathcal{B}]$ with $E_P[f \mid \mathcal{B}]$.

Remark 10.2.25 Let $\mathcal{N} \subseteq \mathcal{A}$, and

$$C = \{\omega \in \Omega : E_P[f \mid \mathcal{B}](\omega) \neq E_Q[f \mid \mathcal{C}](\omega)\} \in \mathcal{C},$$

and $A \in \mathcal{A}$ be a set such that $P(A) = 0$, and, for $\omega \in A^c$,

$$E_P[f \mid \mathcal{B}](\omega) = E_P[f \mid \mathcal{C}](\omega).$$

Then, as $A^c \subseteq C^c$, $C \subseteq A$.

10.2.2 Enlargement with All the Sets of Measure Zero and, Possibly, Their Subsets

That enlargement, a special case of the one in the previous section, is the one of usual interest, as it amounts to completion. So one uses then a specific notation as follows. $\mathcal{D}$ becomes $\mathcal{B}^o$, and $\mathcal{C}$, ${}^o\mathcal{B}$. Q becomes $\{P^{|\mathcal{B}}\}^o$, and $Q^{|\mathcal{C}}$, ${}^o\{P^{|\mathcal{B}}\}$, or, when $\mathcal{N} \subseteq \mathcal{A}$, simply P. When $\mathcal{B} = \mathcal{A}$, (Proposition) 10.2.22 has an improved version as follows [70, p. 49]: f is adapted to $\mathcal{A}^o$ if and only if there are g and h, adapted to $\mathcal{A}$, such that $g \leq f \leq h$ and $P(|g - h| > 0) = 0$.

Let the σ-algebra $\mathcal{B} \subseteq \mathcal{A}$ contain $\mathcal{N}(\mathcal{A}, P)$. If f and g are adapted to $\mathcal{A}$, and are, with respect to P, almost surely equal, then they are either both adapted to $\mathcal{B}$, or none is. Thus an equivalence class of functions adapted to $\mathcal{A}$ either contains only functions adapted to $\mathcal{B}$ or none. When $\mathcal{B}$ does not contain $\mathcal{N}(\mathcal{A}, P)$, let $\mathcal{C} = \mathcal{B} \vee \mathcal{N}(\mathcal{A}, P)$. Then f, adapted to $\mathcal{A}$, is adapted to $\mathcal{C}$ if, and only if, there exits g adapted to $\mathcal{B}$ such that, almost surely, with respect to $P, f = g$.

When $\mathcal{B}_1 \subset \mathcal{B}_2 \subset \mathcal{A}$, the same order shall prevail for the different enlargements, and the different probabilities shall be consistent. So, when one deals with a filtration, and $\mathcal{N} \subseteq \mathcal{A}$, there is only a need to distinguish between the σ-algebras, but not between the probabilities.

10.2.3 Restriction to a Subset of the Base Set

When the process X has paths that are continuous to the right, and that almost all of them are continuous, it may prove convenient, in order to work with strict stopping times, rather than wide sense ones, to restrict the process to a subset of Ω of probability one for which the paths are continuous. One now lists some of the facts that may be pertinent in such a situation.

Let $N \in \mathcal{A}$ be a set of measure zero, $\Omega_N = \Omega \setminus N$, and $J_N : \Omega_N \longrightarrow \Omega$ be the inclusion map. When f is a map with domain Ω, its restriction to Ω_N is $f \circ J_N$. Since, for an $\Omega_0 \subseteq \Omega$, fixed, but arbitrary,

$$J_N^{-1}(\Omega_0) = \Omega_N \cap \Omega_0,$$

any measurable structure on Ω yields a measurable structure on Ω_N by intersection. Thus, for example,

$$J_N^{-1}(\mathcal{A}) = \Omega_N \cap \mathcal{A}.$$

Functions adapted to $\Omega_N \cap \mathcal{A}$ are of the form $f \circ J_N, f$ adapted to $\mathcal{A}$. Indeed, since N is assumed to belong to $\mathcal{A}$, when g is adapted to $\Omega_N \cap \mathcal{A}$, it is adapted to $\mathcal{A}$, and thus $h = \chi_{N^c} g$ is adapted to $\mathcal{A}$. But $h \circ J_N = g$. One may define a probability P_N on $\Omega_N \cap \mathcal{A}$ using, for $A \in \mathcal{A}$, fixed, but arbitrary, the following relation: $P_N\left(J_N^{-1}(A)\right) = P(A)$. Then

$$\int_A f dP = \int_A f d(P_N \circ J_N^{-1}) = \int_{J_N^{-1}(A)} (f \circ J_N) dP_N.$$

One shall use, when convenient, the following notation: $\mathcal{A}_t^N = J_N^{-1}(\mathcal{A}_t)$, and $\underline{\mathcal{A}}_N$ for the corresponding filtration. As a preliminary, one needs the following proposition.

Proposition 10.2.26 *Let $\underline{\mathcal{A}}$ be a filtration for $\mathcal{A}$ in Ω, and S be a wide sense stopping time for $\underline{\mathcal{A}}$.*

1. *Let $\theta \in [0,1]$, and $A_\theta \in \mathcal{A}_\theta$ be fixed, but arbitrary. Let, for $\omega \in \Omega$, fixed, but arbitrary,*

$$S_\theta(\omega) = \chi_{A_\theta}(\omega)\theta + \chi_{A_\theta^c}(\omega).$$

S_θ is a strict stopping time.
2. [128, p. 80] *S is adapted to*

$$\begin{aligned}\mathcal{A}_S^- &= \mathcal{A}_0 \vee \sigma\left(A \cap \{S > t\} : A \in \mathcal{A}_t, t \in [0,1]\right)\\ &= \mathcal{A}_0 \vee \sigma\left(A \cap \{S > t\}, A \in \mathcal{A}_t^+, t \in [0,1]\right).\end{aligned}$$

3. *When S_0 is another wide sense stopping time for $\underline{\mathcal{A}}$, and*

$$\begin{aligned}A \in \mathcal{A}_S^+ &= \left\{A \in \mathcal{A} : t \in [0,1] \Rightarrow A \cap \{S \le t\} \in \mathcal{A}_t^+\right\}\\ &= \{A \in \mathcal{A} : t \in [0,1] \Rightarrow A \cap \{S < t\} \in \mathcal{A}_t\},\end{aligned}$$

then $A \cap \{S < S_0\} \in \mathcal{A}_{S_0}^-$.
4. *When S_0 is adapted to $\mathcal{A}_S^+$, $S_0 \ge S$, and $S_0 > S$ on $\{S < 1\}$, then S_0 is a strict stopping time.*
5. *One has that*

$$S_n(\omega) = \sum_{k=1}^{2^n} \chi_{\left[\frac{k-1}{2^n}, \frac{k}{2^n}\right[}(S(\omega))\frac{k}{2^n} + \chi_{\{1\}}(S(\omega))$$

produces a sequence of strict stopping times that decrease to S.
6. *S is the lower envelope of a countable family of stopping times of the type found in item 1.*

Proof (1) Let $t \in [0,1]$ be fixed, but arbitrary. Then

$$\begin{aligned}\{\omega \in \Omega : S_\theta \le t\} &= [\{\omega \in \Omega : S_\theta(\omega) \le t\} \cap A_\theta] \cup \left[\{\omega \in \Omega : S_\theta(\omega) \le t\} \cap A_\theta^c\right]\\ &= [\{\omega \in \Omega : \theta \le t\} \cap A_\theta] \cup \left[\{\omega \in \Omega : 1 \le t\} \cap A_\theta^c\right]\\ &= \begin{cases} \emptyset \in \mathcal{A}_t \text{ when } t < \theta \\ A_\theta \in \mathcal{A}_t \text{ when } \theta \le t < 1 \\ \Omega \in \mathcal{A}_t \text{ when } t = 1 \end{cases}.\end{aligned}$$

Proof (2) By definition $\{S > t\} \in \mathcal{A}_S^-$, and that is enough.

Proof (3) Since, by assumption, $A \in \mathcal{A}_S^+$, then, by definition of $\mathcal{A}_S^+$,

$$A \cap \{S \leq t\} \in \mathcal{A}_t^+.$$

By definition then, since S_0 is a wide sense stopping time,

$$[A \cap \{S \leq t\}] \cap \{S_0 > t\} \in \mathcal{A}_{S_0}^-.$$

But, r denoting a rational in $[0, 1]$,

$$A \cap \{S < S_0\} = \cup_r \{[A \cap \{S \leq r\}] \cap \{S_0 > r\}\},$$

a set that belongs to $\mathcal{A}_{S_0}^-$.

Proof (4) Since $\{S_0 \leq 1\} \in \mathcal{A}_1^+ = \mathcal{A}_1$, one may assume $t < 1$. By assumption, $\{S_0 \leq t\} \in \mathcal{A}_S^+$. Now, since, on $\{S < 1\}$, $S < S_0$, there $(S_0 \leq t) \Rightarrow (S < t)$, so that $\{S_0 \leq t\} \cap \{S \geq t\} = \emptyset$ (given that $S \leq S_0 \leq t < 1$). Thus

$$\{S_0 \leq t\} = [\{S_0 \leq t\} \cap \{S < t\}] \uplus [\{S_0 \leq t\} \cup \{S \geq t\}] = \{S_0 \leq t\} \cap \{S < t\}.$$

But then one may use item 3. Indeed, since, by assumption, S_0 is adapted to $\mathcal{A}_S^+$, $\{S_0 \leq t\} \in \mathcal{A}_S^+$. Letting in item 3, $A = \{S_0 \leq t\}$, and choosing S_0 there to be identically t, it follows that $A \cap \{S < t\} \in \mathcal{A}_t^- \subseteq \mathcal{A}_t$, and thus that $\{S_0 \leq t\} \in \mathcal{A}_t$.

Proof (5) By item 2, S is adapted to $\mathcal{A}_S^-$, and thus to $\mathcal{A}_S^+$. It is thus also the case for S_n. By definition $S_n \geq S$, and, on $\{S < 1\}$, $S_n > S$. Consequently, because of item 4, S_n is a strict stopping time. That S_n decreases to S follows also from the definition.

Proof (6) Let $I_{n,k} = \left[\frac{k-1}{2^n}, \frac{k}{2^n}\right[$ and

$$\Sigma_{n,k}(\omega) = \chi_{I_{n,k}}(S(\omega))\frac{k}{2^n} + \chi_{I_{n,k}^c}(S(\omega)).$$

From item 1, $\Sigma_{n,k}$ is a strict stopping time. Suppose that $S(\omega) \in I_{n,k_0}$. Then $S_n(\omega) = \frac{k_0}{2^n}$, and

$$\Sigma_{n,k_0} = \frac{k_0}{2^n}, \quad \text{while } \Sigma_{n,k} = 1 \text{ for } k \neq k_0.$$

Consequently $S_n = \bigwedge_{k=1}^{2^n} \Sigma_{n,k}$. Since $S = \lim S_n = \inf S_n$, S is the lower envelope of the stopping times $\Sigma_{n,k}$. □

"Navigation" between Ω and Ω_N is ruled by the following facts.

Proposition 10.2.27 *Let $S : \Omega \longrightarrow [0, 1]$ be a map, $N \in \mathcal{A}$ be a fixed, chosen set of zero probability, and $\mathcal{N}$, the family of subsets of N. Let $\mathcal{A}_t^{\circ N}$ be the* *completion of* $\mathcal{A}_t$ *with respect to* N (as in (Facts) 10.2.10 and 10.2.11), *that is,* $\mathcal{A}_t^{\circ N} = \{A \Delta N_0, A \in \mathcal{A}_t, N_0 \in \mathcal{N}\}$ *. Then:*

1. $\mathcal{A}_t^N \subseteq \mathcal{A}_t^{\circ N}$;
2. *when* S *is a wide (strict) sense stopping time of* $\underline{\mathcal{A}}$, *it is a wide (strict) sense stopping time of* $\underline{\mathcal{A}}_N$;
3. *when* S_N, *the restriction of* S *to* Ω_N, *is a wide (strict) sense stopping time of* $\underline{\mathcal{A}}_N$, *there is a wide sense stopping time* $\dot{S}$ *of* $\underline{\mathcal{A}}$ *such that, on* Ω_N, $\dot{S}_N = S_N$.

Proof (1) The assertion obtains since [(Fact) 10.2.2]

$$A \cap \Omega_N = (\emptyset \cap N) \cup (A \cap \Omega_N) \in \mathcal{A}_t \vee \{N\} \subseteq \mathcal{A}_t^{\circ N}.$$

Proof (2) The assertion is true, as, for example, $\{S_N < t\} = \{S < t\} \cap \Omega_N$.

Proof (3) One has that

$$\{S < t\} = [\{S < t\} \cap \Omega_N] \Delta [\{S < t\} \cap N],$$
$$\{S \leq t\} = [\{S \leq t\} \cap \Omega_N] \Delta [\{S \leq t\} \cap N].$$

But, by assumption, using item 1,

$$\{S < t\} \cap \Omega_N = \{S_N < t\} \in \mathcal{A}_t^N \subseteq \mathcal{A}_t^{\circ N},$$
$$\{S \leq t\} \cap \Omega_N = \{S_N \leq t\} \in \mathcal{A}_t^N \subseteq \mathcal{A}_t^{\circ N}.$$

Furthermore $\{S < t\} \cap N \in \mathcal{N} \subseteq \mathcal{A}_t^{\circ N}$, and $\{S \leq t\} \cap N \in \mathcal{N} \subseteq \mathcal{A}_t^{\circ N}$. S is thus a wide (strict) sense stopping time of $\underline{\mathcal{A}}^{\circ N}$.

Because of (Proposition) 10.2.26, item 6, S is the lower envelope, for $\underline{\mathcal{A}}^{\circ N}$, of a sequence of stopping times of type met in (Proposition) 10.2.26, item 1. Consider thus a stopping time for $\underline{\mathcal{A}}^{\circ N}$ of the type met in (Proposition) 10.2.26, item 1, that is,

$$S_\theta = \chi_{A_\theta} \theta + \chi_{A_\theta^c}, \; A_\theta \in \mathcal{A}_\theta^{\circ N}.$$

One has that

- $A_\theta = \dot{A}_\theta \Delta N_\theta, \dot{A}_\theta \in \mathcal{A}_\theta, N_\theta \in \mathcal{N}$,
- $A_\theta^c = \Omega \Delta A_\theta = (\Omega \Delta \dot{A}_\theta) \Delta N_\theta = \dot{A}_\theta^c \Delta N_\theta$,
- $\dot{A}_\theta \Delta A_\theta = N_\theta$,
- $\dot{A}_\theta^c \Delta A_\theta^c = N_\theta$,

Thus $\dot{S}_\theta = \chi_{\dot{A}_\theta}\theta + \chi_{\dot{A}_\theta^c}$ is a strict stopping time for $\underline{\mathcal{A}}$ [(Proposition) 10.2.26, item 1], and

$$\begin{aligned}\left|S_\theta(\omega) - \dot{S}_\theta(\omega)\right| &\le \theta\left|\chi_{A_\theta} - \chi_{\dot{A}_\theta}\right| + \left|\chi_{A_\theta^c} - \chi_{\dot{A}_\theta^c}\right| \\ &= \theta\chi_{A_\theta \Delta \dot{A}_\theta} + \chi_{A_\theta^c \Delta \dot{A}_\theta^c} \\ &= (1+\theta)\chi_{N_\theta} \\ &\le (1+\theta)\chi_N .\end{aligned}$$

Since the minimum of two wide (strict) sense stopping times is a wide (strict) sense stopping time [128, p. 80], the minimum of a finite number of stopping times of the type $\dot{S}_\theta$ is a strict stopping time for $\underline{\mathcal{A}}$. The countable envelope $\dot{S}$ of stopping times of type $\dot{S}_\theta$ is the limit of the minimum of a finite number of such stopping times, and is thus a wide sense stopping time for $\underline{\mathcal{A}}$ [70, p. 189]. Since S_θ and $\dot{S}_\theta$ only differ in N, S and $\dot{S}$ only differ in N. □

Proposition 10.2.28 *Let H be a real and separable Hilbert space, and the adapted process $X : \Omega \times [0,1] \longrightarrow H$ have paths continuous to the right* (so that it may be stopped: [264, p. 55]). *Let $X_N : \Omega_N \times [0,1] \longrightarrow H$ denote its restriction to $\Omega_N \times [0,1] : X_N(\omega, t) = X(J_N(\omega), t)$. The base probability space of X_N is $(\Omega_N, \underline{\mathcal{A}_N}, P_N)$. $S : \Omega \longrightarrow [0,1]$ is a map, and $S_N = S \circ J_N$. One has that X_N is adapted to $\underline{\mathcal{A}_N}$ and that*

1. *when X is a martingale for $(\underline{\mathcal{A}}, P)$, X_N is one for $(\underline{\mathcal{A}_N}, P_N)$;*
2. *when X is a local martingale for $(\underline{\mathcal{A}}, P)$, and $\{S_n, n \in \mathbb{N}\}$ a localizing sequence for it, X_N is a local martingale for $(\underline{\mathcal{A}_N}, P_N)$ with localizing sequence $\{(S_n)_N, n \in \mathbb{N}\}$;*
3. *when X is a martingale locally in L_2 for $(\underline{\mathcal{A}}, P)$, and $\{S_n, n \in \mathbb{N}\}$ a localizing sequence for it, X_N is a martingale locally in L_2 for $(\underline{\mathcal{A}_N}, P_N)$ with localizing sequence $\{(S_n)_N, n \in \mathbb{N}\}$;*
4. *when X_N is a martingale for $\underline{\mathcal{A}_N}$, X is a martingale for $\underline{\mathcal{A}}$;*
5. *when $\underline{X_N}$ is a local martingale for $\underline{\mathcal{A}_N}$, $\underline{X}$ is a local martingale for $\underline{\mathcal{A}}$;*
6. *when $\underline{X_N}$ is, locally, a martingale in L_2 for $\underline{\mathcal{A}_N}$, $\underline{X}$ is, locally, a martingale in L_2 for $\underline{\mathcal{A}}$;*

Proof (1) One has that

$$\begin{aligned}\int_{\Omega_N} \|X_N(\omega,t)\|_H \, P_N(d\omega) &= \int_{J_N^{-1}(\Omega)} \|X(J_N(\omega),t)\|_H \, P_N(d\omega) \\ &= \int_{\Omega} \|X(\omega,t)\|_H \, P_N \circ J_N^{-1}(d\omega)\end{aligned}$$

$$= \int_{\Omega} \|X((\omega,t)\|_H \, P(d\omega)$$
$$< \infty,$$

and, for $t_1 < t_2$, and $A \in \mathcal{A}_{t_1}^N$, fixed, but arbitrary, as $A = \dot{A} \cap \Omega_N$, $\dot{A}$ in $\mathcal{A}_{t_1}$,

$$\begin{aligned}\int_A X_N(\omega,t_2)P_N(d\omega) &= \int_{J_N^{-1}(\dot{A})} X(J_N(\omega),t_2)P_N(d\omega)\\ &= \int_{\dot{A}} X(\omega,t_2)P_N \circ J_N^{-1}(d\omega)\\ &= \int_{\dot{A}} X(\omega,t_2)P(d\omega)\\ &= \int_{\dot{A}} X(\omega,t_1)P(d\omega)\\ &= \int_A X_N(\omega,t_1)P_N(d\omega).\end{aligned}$$

Items 2 and 3 are checked analogously. Item 4 is proved as item 1. The proof of item 5 is along the following lines. Let $\{S_n^N, n \in \mathbb{N}\}$ be a localizing sequence for X_N, and obtain

$$\{\dot{S}_n, n \in \mathbb{N}\}$$

according to (Proposition) 10.2.27. Let also

$$\begin{aligned}A(\alpha,n,t) &= \left\{\omega \in \Omega : \left\|X^{\dot{S}_n}(\omega,t) - X(\omega,0)\right\|_H > \alpha\right\}\\ &= \left\{\omega \in \Omega : \left\|X(\omega,\dot{S}_n(\omega) \wedge t) - X(\omega,0)\right\|_H > \alpha\right\}.\end{aligned}$$

Since $J_N^{-1}\{A(\alpha,n,t)\}$ is equal to

$$\begin{aligned}&\left\{\omega \in \Omega_N : \left\|X(J_N(\omega),\dot{S}_n(J_N(\omega)) \wedge t) - X(J_N(\omega),0)\right\|_H > \alpha\right\} =\\ &\qquad = \left\{\omega \in \Omega : \left\|X_N(\omega,S_n^N(\omega) \wedge t) - X_N(\omega,0)\right\|_H > \alpha\right\},\end{aligned}$$

one has that

$$\int_{\{\omega\in\Omega:\|X^{\dot{S}_n}(\omega,t)-X(\omega,0)\|_H>\alpha\}} \|X^{\dot{S}_n}(\omega,t)-X(\omega,0)\|_H P(d\omega) =$$

$$= \int_{\left\{\omega\in\Omega:\left\|X_N^{S_n^N}(\omega,t)-X_N(\omega,0)\right\|_H>\alpha\right\}} \left\|X_N^{S_n^N}(\omega,t)-X_N(\omega,0)\right\|_H P_N(d\omega).$$

□

10.3 Stopping and Change of Times

At different stages in the sequel, one shall need some facts about the inversion of monotone increasing functions that are either continuous or only continuous to the right. Change of (stopping) time covers inversion for paths of stochastic processes and shall be important in evaluating the scope of the models used for deriving the likelihood.

10.3.1 *Inverting Monotone Increasing Functions*

When a function α is increasing, but not strictly, the usual notion of inverse does not apply, and one replaces it with a surrogate, either $\underline{\alpha}$ defined using $\underline{\alpha}(t) = \inf\{x : \alpha(x) \geq t\}$, or $\overline{\alpha}(t) = \inf\{x : \alpha(x) > t\}$. The former seems better suited to inverting distribution functions, the latter, increasing processes. Since one shall need both, both cases are presented.

The Case of Distribution Functions

Definition 10.3.1 The map $\alpha : \mathbb{R} \longrightarrow \mathbb{R}$ is increasing when, given $x < y$ in $\mathbb{R}$, fixed, but arbitrary, then $\alpha(x) \leq \alpha(y)$. It follows that $\lim_{x\downarrow\downarrow-\infty} \alpha(x)$ exists, and shall be denoted α_l, and $\lim_{x\uparrow\uparrow\infty} \alpha(x)$ exists as well, and shall be denoted α_r. One sets

$$\alpha(-\infty) = \alpha_l, \text{ and } \alpha(\infty) = \alpha_r.$$

Definition 10.3.2 For $t \in \mathbb{R}$, fixed, but arbitrary, let

$$\underline{I}_\alpha(t) = \{x \in \mathbb{R} : \alpha(x) \geq t\}, \text{ and } \overline{I}_\alpha(t) = \{x \in \mathbb{R} : \alpha(x) > t\}.$$

The functions $\underline{\alpha}, \overline{\alpha} : \mathbb{R} \longrightarrow \overline{\mathbb{R}}$ are defined using the following assignments:

$$\underline{\alpha}(t) = \begin{cases} \infty & \text{when } \underline{I}_\alpha(t) = \emptyset \\ \inf \underline{I}_\alpha(t) & \text{when } \underline{I}_\alpha(t) \neq \emptyset \end{cases},$$

$$\overline{\alpha}(t) = \begin{cases} \infty & \text{when } \overline{I}_\alpha(t) = \emptyset \\ \inf \overline{I}_\alpha(t) & \text{when } \overline{I}_\alpha(t) \neq \emptyset \end{cases}.$$

Remark 10.3.3 Since $\overline{I}_\alpha(t) \subseteq \underline{I}_\alpha(t)$, $\underline{\alpha}(t) \leq \overline{\alpha}(t)$.

Remark 10.3.4 When $\alpha(x) \geq t$ for all $x \in \mathbb{R}$, $\underline{I}_\alpha(t) = \mathbb{R}$, so that $\underline{\alpha}(t) = -\infty$. When $\underline{\alpha}(t) = -\infty$, but there exists $x_0 \in \mathbb{R}$ such that $\alpha(x_0) < t$, one has then that $\underline{\alpha}(t) \geq x_0 > -\infty$. Thus $\underline{\alpha}(t) = -\infty$ if, and only if, $\alpha(x) \geq t$ for all $x \in \mathbb{R}$.

Remark 10.3.5 $\underline{\alpha}(t) = \infty$ if, and only if, $\underline{I}_\alpha(t) = \emptyset$, so that $\underline{\alpha}(t) = \infty$ if, and only if, $\alpha(x) < t$ for all $x \in \mathbb{R}$.

Remark 10.3.6 When $\alpha(x) > t$ for all $x \in \mathbb{R}$, $\overline{I}_\alpha(t) = \mathbb{R}$, so that $\overline{\alpha}(t) = -\infty$. When $\overline{\alpha}(t) = -\infty$, but there exists $x_0 \in \mathbb{R}$ such that $\alpha(x_0) \leq t$, one has then that $\overline{\alpha}(t) \geq x_0 > -\infty$, so that $\overline{\alpha}(t) = -\infty$ if, and only if, $\alpha(x) > t$ for all $x \in \mathbb{R}$.

Remark 10.3.7 $\overline{\alpha}(t) = \infty$ if, and only if, $\overline{I}_\alpha(t) = \emptyset$, so that $\overline{\alpha}(t) = \infty$ if, and only if, $\alpha(x) \leq t$ for all $x \in \mathbb{R}$.

Remark 10.3.8 Since, for fixed, but arbitrary $t_1 < t_2$ in $\mathbb{R}$,

$$\underline{I}_\alpha(t_2) \subseteq \underline{I}_\alpha(t_1), \text{ and } \overline{I}_\alpha(t_2) \subseteq \overline{I}_\alpha(t_1),$$

$\underline{\alpha}$ and $\overline{\alpha}$ are increasing functions.

Fact 10.3.9

1. *When $\underline{\alpha}(t) \in \mathbb{R}$, then, at t, $\underline{\alpha}$ is continuous to the left, and has a limit to the right.*
2. *When $\overline{\alpha}(t) \in \mathbb{R}$, then at t, $\overline{\alpha}$ is continuous to the right, and has a limit to the left.*

Proof Let $\{t_n, n \in \mathbb{N}\}$ be a sequence increasing to t, and $\epsilon > 0$, fixed, but arbitrary, be given. When, in $\mathbb{R}$, $\inf\{z \in \mathbb{R} : \alpha(z) \geq x\} = \underline{\alpha}(x) = y$,

$$\alpha(y - \epsilon) < x \leq \alpha(y + \epsilon).$$

Let thus $\underline{\alpha}(t) = \theta$, and $\underline{\alpha}(t_n) = \theta_n$. Then, as just seen,

$$\alpha(\theta_n - \epsilon) < t_n \leq \alpha(\theta_n + \epsilon).$$

Since $\underline{\alpha}$ is increasing, $\theta_n \leq \theta < \infty$. Thus $\lim_n \theta_n = \theta_0 \leq \theta$ exists. Suppose that $\theta_0 < \theta$. Since $\theta_n \leq \theta_0$,

$$\theta_n - \frac{\theta_0}{2} \leq \frac{\theta_0}{2}, \text{ and } \theta_n + \frac{\theta - \theta_0}{2} \leq \frac{\theta + \theta_0}{2} = \theta - \frac{\theta - \theta_0}{2}.$$

Consequently, choosing $\epsilon = \frac{\theta - \theta_0}{2}$, one has, since α is increasing, that

$$t_n \leq \alpha(\theta_n + \epsilon) \leq \alpha(\theta - \epsilon) < t,$$

a contradiction. Suppose now that $\{t_n, n \in \mathbb{N}\}$ decreases to t. Since $\underline{\alpha}$ is increasing, $\underline{\alpha}(t) \leq \underline{\alpha}(t_n)$. Since the left-hand side of that latter inequality is finite, the sequence $\{\underline{\alpha}(t_n), n \in \mathbb{N}\}$ has a limit.

Suppose similarly that $\{t_n, n \in \mathbb{N}\}$ is a sequence decreasing to t, and $\epsilon > 0$, fixed, but arbitrary, is given. When, in $\mathbb{R}$, $\inf\{z \in \mathbb{R} : \alpha(z) > x\} = \overline{\alpha}(x) = y$,

$$\alpha(y - \epsilon) \leq x < \alpha(y + \epsilon).$$

Let $\overline{\alpha}(t) = \theta$, and $\overline{\alpha}(t_n) = \theta_n$. Then $\theta_n \geq \theta$. Let $\theta_0 = \lim_n \theta_n$. Suppose that $\theta_0 > \theta$. Set $\epsilon = \frac{\theta_0 - \theta}{2}$. Then, as $\theta_0 \leq \theta_n$,

$$\frac{\theta_0}{2} \leq \theta_n - \frac{\theta_0}{2}, \text{ and } \theta + \epsilon = \theta + \frac{\theta_0 - \theta}{2} = \frac{\theta + \theta_0}{2} \leq \theta_n - \frac{\theta_0 - \theta}{2} = \theta_n - \epsilon,$$

so that, as α is increasing, $t < \alpha(\theta + \epsilon) \leq \alpha(\theta_n - \epsilon) \leq t_n$, a contradiction. That $\overline{\alpha}$ has a limit to the left is due to the fact that it is increasing, and $\overline{\alpha}(t)$, finite. □

Fact 10.3.10 *One has that:*

1. $\underline{\alpha}(\alpha(t)) \leq t$;
2. *when* α *is strictly increasing,* $\underline{\alpha}(\alpha(t)) = t$;
3. $\overline{\alpha}(\alpha(t)) \geq t$;
4. *when* α *is strictly increasing,* $\overline{\alpha}(\alpha(t)) = t$.

Proof As $\underline{\alpha}(\alpha(t)) = \inf\{x \in \mathbb{R} : \alpha(x) \geq \alpha(t)\}$, and that $\alpha(t) \geq \alpha(t)$, obviously $\underline{\alpha}(\alpha(t)) \leq t$. When α is strictly increasing, $\alpha(x) \geq \alpha(t)$ implies $x \geq t$, and thus $\underline{\alpha}(\alpha(t)) \geq t$. Since $x < t$ implies $\alpha(x) \leq \alpha(t)$, $\alpha(x) > \alpha(t)$ implies $x \geq t$, and thus the third assertion. Finally, when $t < \overline{\alpha}(\alpha(t))$, let $t < \theta < \overline{\alpha}(\alpha(t))$. If α is strictly increasing, $\alpha(t) < \alpha(\theta) < \alpha(\overline{\alpha}(\alpha(t))$, so that $\overline{\alpha}(\alpha(t))$ is not the infimum of those x's for which $\alpha(x) > \alpha(t)$. □

Fact 10.3.11 *When* $t \in \mathcal{R}[\alpha]$, $\underline{\alpha}(t) = \inf \alpha^{-1}(\{t\})$.

Proof Since $\underline{I}_\alpha(t) = \alpha^{-1}(\{t\}) \cup \overline{I}_\alpha(t)$, and $\inf(A \cup B) \leq \inf(A) \wedge \inf(B)$,

$$\underline{\alpha}(t) = \inf\left(\alpha^{-1}(\{t\})\right) \wedge \overline{\alpha}(t).$$

But both $\alpha(x) = t$ and $\alpha(y) > t$ imply $x \leq y$ as the inequality $x > y$ yields $t = \alpha(x) \geq \alpha(y) > t$. □

Fact 10.3.12 *When* α *is continuous to the right, one has that:*

1. $\underline{\alpha}(t) < \infty$ *implies* $\alpha(\underline{\alpha}(t)) \geq t$, *and thus that* $\underline{I}_\alpha(t)$ *is an interval closed to the left by* $\underline{\alpha}(t)$;
2. $t \in \inf(\mathcal{R}[\alpha]) \cup \mathcal{R}[\alpha] \cup \sup(\mathcal{R}[\alpha])$ *implies* $\alpha(\underline{\alpha}(t)) = t$;

3. $t < \inf(\mathcal{R}[\alpha])$ *implies* $\alpha(\underline{\alpha}(t)) > t$;
4. $t > \sup(\mathcal{R}[\alpha])$ *implies* $\alpha(\underline{\alpha}(t)) < t$.

Proof When $\underline{\alpha}(t) < \infty$, $\underline{I}_\alpha(t) \neq \emptyset$ [(Remark) 10.3.5]. There exists thus a sequence $\{\theta_n, n \in \mathbb{N}\} \subseteq \underline{I}_\alpha(t)$, decreasing to $\underline{\alpha}(t)$. But α is continuous to the right, and $\alpha(\theta_n) \geq t$, so that $t \leq \lim_n \alpha(\theta_n) = \alpha(\underline{\alpha}(t))$. In particular $\underline{\alpha}(t) \in \underline{I}_\alpha(t)$.

Suppose that $t \in \mathcal{R}[\alpha]$. Then, from (Fact) 10.3.11, $\underline{\alpha}(t) = \inf \alpha^{-1}(\{t\})$. Let, in $\alpha^{-1}(\{t\})$, $\{\theta_n, n \in \mathbb{N}\}$ be a sequence which decreases to $\underline{\alpha}(t)$. Then

$$t = \lim_n \alpha(\theta_n) = \alpha(\underline{\alpha}(t)).$$

Suppose that $t \in \inf(\mathcal{R}[\alpha])$. If $t \in \mathcal{R}[\alpha]$, the result has been proved, so one may assume that t is not in $\mathcal{R}[\alpha]$. But then $\underline{\alpha}(t) = -\infty$. As α is increasing,

$$\alpha(\underline{\alpha}(t)) = \alpha(-\infty) = \inf\{\alpha(x), x \in \mathbb{R}\} = \inf(\mathcal{R}[\alpha]) = t.$$

Suppose similarly that $t = \sup(\mathcal{R}[\alpha])$, but t does not belong to $\mathcal{R}[\alpha]$. Then $\underline{\alpha}(t) = \infty$. Consequently,

$$\alpha(\underline{\alpha}(t)) = \alpha(\infty) = \sup\{\alpha(x), x \in \mathbb{R}\} = \sup(\mathcal{R}[\alpha]) = t.$$

When $t < \inf(\mathcal{R}[\alpha])$, $\underline{\alpha}(t) = -\infty$. Thus

$$\alpha(\underline{\alpha}(t)) = \alpha(-\infty) = \inf(\mathcal{R}[\alpha]) > t.$$

When $t > \sup(\mathcal{R}[\alpha])$, $\underline{\alpha}(t) = \infty$. Thus

$$\alpha(\underline{\alpha}(t)) = \alpha(\infty) = \sup(\mathcal{R}[\alpha]) < t.$$

□

Fact 10.3.13 *One has that:*

1. $\alpha(x) \geq t$ *implies* $x \geq \underline{\alpha}(t)$, *and* $\alpha(x) > t$ *implies* $x \geq \overline{\alpha}(x)$;
2. *when* α *is continuous to the right,* $x \geq \underline{\alpha}(t)$ *(and thus* $x \geq \overline{\alpha}(t)$*) implies* $\alpha(x) \geq t$;
3. $\alpha(x) < t$ *implies* $x \leq \underline{\alpha}(t)$ *(and thus* $x \leq \overline{\alpha}(t)$*).*

Proof Item 1 follows from the definitions respectively of $\underline{\alpha}$ and $\overline{\alpha}$. The inequality $\underline{\alpha}(t) \leq x$ implies $\alpha(\underline{\alpha}(t)) \leq \alpha(x)$. But as α is continuous to the right, $\alpha(\underline{\alpha}(t)) \geq t$, because of (Fact) 10.3.12, item 1. Suppose finally that $\alpha(x) < t$. If there is y such that $\alpha(y) \geq t$, then $y \geq x$, so that $\underline{\alpha}(t) \geq x$. If there is no such y, $\underline{\alpha}(t) = \infty \geq x$. □

Fact 10.3.14 *One has that:*

$$]\underline{\alpha}(t-), \underline{\alpha}(t+)[\subseteq \alpha^{-1}(\{t\}) \subseteq [\underline{\alpha}(t-), \underline{\alpha}(t+)] .$$

Proof Since $\underline{\alpha}$ is increasing [(Remark) 10.3.8], $\underline{\alpha}(t-)$ and $\underline{\alpha}(t+)$ exist, and, when, in $\mathbb{R}$, $t_1 < t < t_2$,

$$\underline{\alpha}(t_1) \leq \underline{\alpha}(t-) \leq \underline{\alpha}(t) \leq \underline{\alpha}(t+) \leq \underline{\alpha}(t_2). \qquad (\star)$$

The proof uses the following remark: given two sets, A and B, with $A \setminus B = \emptyset$, $A = (A \setminus B) \uplus (A \cap B) \subseteq B$. Suppose thus that

$$\theta \in \{]\underline{\alpha}(t-), \underline{\alpha}(t+)[\} \setminus \left\{\alpha^{-1}(\{t\})\right\},$$

and let

$$\delta_- = \theta - \underline{\alpha}(t-) > 0,\ \delta_+ = \underline{\alpha}(t+) - \theta > 0.$$

Since $\alpha(\theta) \neq t$, either $\alpha(\theta) < t$, or $\alpha(\theta) > t$. If $\alpha(\theta) < t$, let

$$t_1 = \frac{\alpha(\theta) + t}{2} \in]\alpha(\theta), t[.$$

Since $\alpha(\theta) < t_1$, (Fact) 10.3.13 implies that $\theta \leq \underline{\alpha}(t_1)$, and thus $(\star)$ that

$$\theta \leq \underline{\alpha}(t_1) \leq \alpha(t-) = \theta - \delta_-,$$

a contradiction. Suppose thus that $\alpha(\theta) > t$, and let $t_2 = \alpha(\theta)$. Then, as $\alpha(\theta) \geq \alpha(\theta)$ and

$$\underline{\alpha}(t_2) = \inf\{x \in \mathbb{R} : \alpha(x) \geq t_2\} = \inf\{x \in \mathbb{R} : \alpha(x) \geq \alpha(\theta)\},$$

$\underline{\alpha}(t_2) \leq \theta$, and thus $(\star)$

$$\theta + \delta_+ = \underline{\alpha}(t+) \leq \underline{\alpha}(t_2) = \theta,$$

a contradiction. Thus

$$\{]\underline{\alpha}(t-), \underline{\alpha}(t+)[\} \setminus \left\{\alpha^{-1}(\{t\})\right\}$$

is empty, and the statement's first inclusion obtains.

As to the second, when $\alpha^{-1}(\{t\})$ is empty, there is nothing to prove. Suppose thus that $\alpha(\theta) = t$. By definition, $\alpha^{-1}(\{t\}) \subseteq \underline{I}_\alpha(t)$, so that

$$\underline{\alpha}(t-) \leq \underline{\alpha}(t) = \inf(\underline{I}_\alpha(t)) \leq \inf\left(\alpha^{-1}(\{t\})\right). \qquad (\star\star)$$

Fix now $\theta > t$, and $x \in \alpha^{-1}(\{t\})$. Then $\alpha(x) = t < \theta$, so that [(Fact) 10.3.13 again], $\underline{\alpha}(\theta) \geq x$. Thus, for $\theta > t$, and $x \in \alpha^{-1}(\{t\})$, $\underline{\alpha}(\theta) \geq x$, that is, $\underline{\alpha}(\theta) \geq$

$\sup\left(\alpha^{-1}(\{t\})\right)$, so that

$$\underline{\alpha}(t+) \geq \sup\left(\alpha^{-1}(\{t\})\right). \qquad (\star\star\star)$$

The statement's second inclusion thus also obtains ($\star\star$ and $\star\star\star$). □

Fact 10.3.15 *One has that:*

1. *α is continuous if, and only if, $\underline{\alpha}$ is strictly increasing;*
2. *α is strictly increasing if, and only if, $\underline{\alpha}$ is continuous on $\mathcal{R}[\alpha]$.*

Proof It suffices to prove that α is not continuous if, and only if, $\underline{\alpha}$ is not strictly increasing. Suppose thus that α is discontinuous at x, that is, letting $t_1 = \alpha(x-), t = \alpha(x), t_2 = \alpha(x+)$, that $t_1 < t_2$, and one of $t_1 \leq t < t_2$, $t_1 < t \leq t_2$ obtains. Let θ_1 and θ_2 be such that $t < \theta_1 < \theta_2 < t_2$ in the former case, or $t_1 < \theta_1 < \theta_2 < t$, in the latter, and suppose the latter obtains. By its very definition, $\underline{\alpha}$ is constant over $[\theta_1, \theta_2]$, and is thus not strictly increasing. If, conversely, $\underline{\alpha}$ is not strictly increasing, there are $t_1 < t_2$ such that $\underline{\alpha}(\theta)$ is constant over $[t_1, t_2]$. Let $\kappa = \underline{\alpha}(\theta)$ for $\theta \in [t_1, t_2]$. Then, given $\epsilon > 0$, fixed, but arbitrary, because of the definition of $\underline{\alpha}$, one has that

$$\alpha(\kappa - \epsilon) < t_1 < t_2 < \alpha(\kappa + \epsilon).$$

Consequently $\alpha(\kappa-) \leq t_1 < t_2 \leq \alpha(\kappa+)$, and α is discontinuous at κ.

For the second statement, it suffices to prove that α is not strictly increasing if, and only if, $\underline{\alpha}$ is discontinuous on $\mathcal{R}[\alpha]$. Suppose first that α is not strictly increasing, that is, there exist $x_1 < x_2$ such that α is constant on $[x_1, x_2]$. Let, for $x \in [x_1, x_2]$, $\alpha(x) = \kappa$. Then $\alpha^{-1}(\{\kappa\}) \supseteq [x_1, x_2]$. It follows, from (Fact) 10.3.14, that $\underline{\alpha}(\kappa-) < \underline{\alpha}(\kappa+)$, that is, $\underline{\alpha}$ is not continuous at $\kappa \in \mathcal{R}[\alpha]$. Suppose conversely that the latter obtains, that is, there exists $t \in \mathcal{R}[\alpha]$ at which $\underline{\alpha}(t-) < \underline{\alpha}(t+)$. It follows, from (Fact) 10.3.14, that $\alpha^{-1}(\{t\})$ contains an open interval, that is, α is not strictly increasing. □

Fact 10.3.16 *Suppose that $\alpha, \beta : \mathbb{R} \longrightarrow \mathbb{R}$ are increasing and continuous to the right. Then*

$$\underline{\alpha \circ \beta} = \underline{\beta} \circ \underline{\alpha}.$$

Proof From (Fact) 10.3.13, in the presence of continuity to the right, $\alpha(x) \geq t$ is equivalent to $\underline{\alpha}(t) \leq x$. Thus:

$$\begin{aligned}
\left\{\underline{\alpha \circ \beta}\right\}(t) &= \inf\{x \in \mathbb{R} : \alpha(\beta(x)) \geq t\} \\
&= \inf\{x \in \mathbb{R} : \underline{\alpha}(t) \leq \beta(x)\} \\
&= \inf\left\{x \in \mathbb{R} : \underline{\beta}\left(\underline{\alpha}(t)\right) \leq x\right\} \\
&= \underline{\beta}\left(\underline{\alpha}(t)\right). \square
\end{aligned}$$

Let X be a random variable, F_X, its (continuous to the right) distribution function, and m_X, the measure generated by F_X. The support of m_X is the closed set S_X of points $x \in \mathbb{R}$ for which, given $\epsilon > 0$, fixed, but arbitrary, $m_X(]x-\epsilon, x+\epsilon[) > 0$. It is also the smallest closed set of mass one, and its complement, the largest open set of mass zero. Now when $F_X(t_2) - F_X(t_1) = 0$, $m_X(]t_1, t_2]) = 0$, so that $]t_1, t_2[\subseteq S_X^c$, and thus $\{t_1, t_2\} \subseteq S_X^c$. One may thus define the (essential) range of X as S_X, and then, on the range of X thus defined, F_X is strictly increasing. That amounts to ignoring the values of X "which have zero probability."

Fact 10.3.17 *Let F_X be the (continuous to the right) distribution function of the random variable X. Then:*

1. *when F_X is continuous, $F_X \circ X$ has a uniform distribution on* $[0, 1]$;
2. *when U has a uniform distribution on* $[0, 1]$, *the law of $\underline{F}_X \circ U$ is that of X, that is, F_X.*

Proof F_X being continuous, $\underline{F}_X$ is strictly increasing [(Fact) 10.3.15]. Thus

$$P\left(F_X \circ X \le t\right) = P\left(\underline{F}_X\left(F_X \circ X\right) \le \underline{F}_X(t)\right).$$

But, since one looks at the inequality $\underline{F}_X\left(F_X \circ X\right) \le \underline{F}_X(t)$ probabilistically, one may assume that F_X is strictly increasing so that, because of (Fact) 10.3.10, $\underline{F}_X(F_X(X)) = X$. Thus

$$P\left(F_X \circ X \le t\right) = P(X \le \underline{F}_X(t)) = F_X(\underline{F}_X(t)).$$

Finally, because of (Fact) 10.3.12, $F_X(\underline{F}_X(t)) = t$ when $t \in \overline{\mathcal{R}[F_X]}$. But there is no need to consider $t > 1$, and thus indeed $P\left(F_X(X) \le t\right) = t$, which characterizes the uniform distribution on the unit interval.

Also, because of (Fact) 10.3.13, $\underline{F}_X(U) \le t$ if, and only if, $U \le F_X(t)$, so that

$$P\left(\underline{F}_X(U) \le t\right) = P(U \le F_X(t)) = F_X(t).$$

□

Example 10.3.18 $\alpha(x) = \chi_{]0,\infty[}(x)$: then

$$\underline{\alpha}(t) = \begin{cases} -\infty & \text{for } t \in]-\infty, 0] \\ 0 & \text{for } t \in]0, 1] \\ \infty & \text{for } t \in]1, \infty[\end{cases}, \quad \overline{\alpha}(t) = \begin{cases} -\infty & \text{for } t \in]-\infty, 0[\\ 0 & \text{for } t \in [0, 1[, \\ \infty & \text{for } t \in [1, \infty[\end{cases}.$$

Example 10.3.19 $\alpha(x) = \chi_{[0,\infty[}(x)$: then

$$\underline{\alpha}(t) = \begin{cases} -\infty & \text{for } t \in]-\infty, 0] \\ 0 & \text{for } t \in]0, 1] \\ \infty & \text{for } t \in]1, \infty[\end{cases}, \quad \overline{\alpha}(t) = \begin{cases} -\infty & \text{for } t \in]-\infty, 0[\\ 0 & \text{for } t \in [0, 1[, \\ \infty & \text{for } t \in [1, \infty[\end{cases}.$$

Example 10.3.20 $\alpha(x) = \frac{e^x}{1+e^x}$:

$$\underline{\alpha}(t) = \overline{\alpha}(t) = \begin{cases} -\infty & \text{for } t \in]-\infty, 0] \\ \ln\left\{\frac{1}{1-t}\right\} & \text{for } t \in]0, 1[\\ \infty & \text{for } t \in [1, \infty[\end{cases} .$$

The Case of Increasing Paths

Let $\alpha : \mathbb{R}_+ \longrightarrow \mathbb{R}$ be monotone increasing. Setting, for $t < 0$, $\tilde{\alpha} = \alpha(0)$, and, for $t \geq 0$, $\tilde{\alpha}(t) = \alpha(t)$, a number of results obtained for distribution functions translate without further ado to α, though one may have to check each time what happens at zero. That shall not be done here, and the presentation below shall be specific to paths of increasing processes, at the cost of some redundancy, but perhaps with the benefit of clarity and simplicity.

Let $\alpha : \mathbb{R}_+ \longrightarrow \mathbb{R}_+$ be monotone increasing and continuous to the right, and $\alpha(\infty) = \lim_{t\to\infty} \alpha(t)$. The range of α, $\mathcal{R}[\alpha]$, may, or may not, contain $\alpha(\infty)$. α^- is the following function:

$$\alpha^-(0) = \alpha(0), \quad \text{and, for } t > 0,\ \alpha^-(t) = \lim_{\theta \uparrow\uparrow t} \alpha(\theta).$$

Let

$$I_\alpha(t) = \{\alpha < t\}, \ J_\alpha(t) = \{\alpha \leq t\}, \ K_\alpha(t) = \{\alpha = t\}.$$

Those are intervals (with every two points in the set, the closed interval determined by those two points is in the set), and, since α is continuous to the right, $I^c_\alpha(t)$ contains its infimum. Also $I_\alpha(t) \uplus K_\alpha(t) = J_\alpha(t)$, and

$$\begin{aligned} I_\alpha(t) &= \emptyset \text{ when } t \leq \alpha(0)\ , \ I_\alpha(t) = \mathbb{R}_+ \text{ when } t > \alpha(\infty), \\ J_\alpha(t) &= \emptyset \text{ when } t < \alpha(0)\ , \ J_\alpha(t) = \mathbb{R}_+ \text{ when } t \geq \alpha(\infty), \\ K_\alpha(t) &= \emptyset \text{ when } t \in \mathcal{R}[\alpha]^c. \end{aligned}$$

Let *Leb* denote Lebesgue measure, and, for $t \in \mathbb{R}_+$, fixed, but arbitrary,

$$\underline{\alpha}(t) = Leb\,(I_\alpha(t))\,,$$

$$\overline{\alpha}(t) = Leb\,(J_\alpha(t))\,.$$

In particular, $\underline{\alpha}(t) \leq \overline{\alpha}(t)$, and

$$\begin{aligned} &\text{when } t \leq \alpha(0),\ \underline{\alpha}(t) = 0 \quad, \text{ when } t > \alpha(\infty),\ \underline{\alpha}(t) = \infty, \\ &\text{when } t < \alpha(0),\ \overline{\alpha}(t) = 0, \quad, \text{ when } t \geq \alpha(\infty),\ \overline{\alpha}(t) = \infty. \end{aligned}$$

Fact 10.3.21 *$\underline{\alpha}$ is continuous to the left, and $\overline{\alpha}$, continuous to the right. Both are increasing.*

Proof The sets $I_\alpha(t)$ and $J_\alpha(t)$ are increasing with t; $[0,t[=\cup_n[0,t_n[$, $t_n < t$, t_n increases to t; $[0,t]=\cap_n[0,t_n]$, $t<t_n$, t_n decreases to t. The result follows thus from the properties of measure. □

Fact 10.3.22 *One has that:*

1. $I_\alpha^c(t) = [\underline{\alpha}(t), \infty[$;
2. $\underline{\alpha}(t) = \sup I_\alpha(t) = \inf I_\alpha^c(t)$;
3. $\overline{\alpha}(t) = \sup J_\alpha(t) = \inf J_\alpha^c(t)$.

Proof When $I_\alpha^c(t)$ is the empty set,

$$I_\alpha(t) = \mathbb{R}_+,\ Leb(I_\alpha(t)) = \infty,\ \sup I_\alpha(t) = \infty,\ \inf I_\alpha^c(t) = \infty.$$

When $I_\alpha(t) \neq \emptyset$, let $\theta_0 = \inf I_\alpha^c(t)$. Then $I_\alpha(t) = [0, \theta_0[$, $I_\alpha^c(t) = [\theta_0, \infty[$, so that

$$Leb(I_\alpha(t)) = \theta_0,\ \sup I_\alpha(t) = \theta_0,\ \inf I_\alpha^c(t) = \theta_0.$$

The same reasoning applies to $J_\alpha(t)$ and its complement. Indeed, as $J_\alpha(t)$ and $J_\alpha^c(t)$ are intervals with the property that the complement of one is the other, $\sup J_\alpha(t) = \inf J_\alpha^c(t)$. Letting, in the nondegenerate case,

$$\theta_0 = \sup J_\alpha(t) = \inf J_\alpha^c(t),$$

$Leb(J_\alpha(t)) = \theta_0$. □

Remark 10.3.23 Because $\underline{\alpha}$ is an infimum, when $\alpha(\theta) \geq t$, $\underline{\alpha}(t) \leq \theta$, and, for the same reason, when $\theta \geq \underline{\alpha}(t)$, $\alpha(\theta) \geq t$: thus

$$\underline{\alpha}(t) \leq \theta \Leftrightarrow \alpha(\theta) \geq t, \text{ so that } \underline{\alpha}(t) > \theta \Leftrightarrow \alpha(\theta) < t.$$

Fact 10.3.24 *One has that $(\overline{\alpha})^-(t) = \underline{\alpha}(t)$.*

Proof Let t_n increase to t. Since $\overline{\alpha} \geq \underline{\alpha}$, $\overline{\alpha}(t_n) \geq \underline{\alpha}(t_n)$. But

$$\underline{\alpha}(t) = \inf I_\alpha^c(t) \geq \inf J_\alpha^c(t_n) = \overline{\alpha}(t_n).$$

Since $\underline{\alpha}$ is continuous to the left, one is done. □

Fact 10.3.25 *One has that:*

1. *when $\underline{\alpha}(t) < \infty$, $t \leq \alpha(\underline{\alpha}(t)) \leq \alpha(\overline{\alpha}(t))$;*
2. $\alpha^-(\underline{\alpha}(t)) \leq \alpha^-(\overline{\alpha}(t)) \leq t$.

Proof One has that $\overline{\alpha} \geq \underline{\alpha}$. When $\underline{\alpha}(t) < \infty$, $\underline{\alpha}(t) \in I_\alpha^c(t)$, so that $\alpha(\underline{\alpha}(t)) \geq t$. Let $\theta_n < \overline{\alpha}(t)$ increase to $\overline{\alpha}(t) = \sup J_\alpha(t)$. Then $\alpha(\theta_n) \leq t$, so that $\alpha^-(\overline{\alpha}(t)) \leq t$. □

Remark 10.3.26 Suppose that $\underline{\alpha}(t) = \infty$. Then $\alpha(\theta) < t$, all θ's, so that $\alpha(\underline{\alpha}(t)) = \alpha(\infty) \leq t$. When $\underline{\alpha}(t) < \infty$,

1. $\alpha^-(\underline{\alpha}(t)) \leq t \leq \alpha(\underline{\alpha}(t))$,
2. $\alpha^-(\overline{\alpha}(t)) \leq t \leq \alpha(\overline{\alpha}(t))$.

Fact 10.3.27 *One has that:*

1. $\overline{\alpha}(t) < \theta$ *implies* $\alpha(\theta) > t$;
2. $\overline{\alpha}(t) \leq \theta$ *implies* $\alpha(\theta) \geq t$;
3. *when* α *is continuous, and* $\alpha(\theta) > t, \overline{\alpha}(t) < \theta$.

Proof Suppose that $\overline{\alpha}(t) < \theta$. Then $\underline{\alpha}(t) < \theta$, and, because of (Fact) 10.3.25, $t \leq \alpha(\overline{\alpha}(t))$. But, since α is increasing, $\alpha(\overline{\alpha}(t)) \leq \alpha(\theta)$. Suppose that $\alpha(\theta) = t$. Then $\overline{\alpha}(t) = \overline{\alpha}(\alpha(\theta))$, and, since $\overline{\alpha}(\alpha(\theta)) = \inf\{\tau \in \mathbb{R}_+ : \alpha(\tau) > \alpha(\theta)\} \geq \theta, \overline{\alpha}(t) = \overline{\alpha}(\alpha(\theta)) \geq \theta$, a contradiction.

When $\overline{\alpha}(t) \leq \theta$, since α is increasing, $\alpha(\overline{\alpha}(t)) \leq \alpha(\theta)$. But, because of (Fact) 10.3.25, since $\underline{\alpha}(t) \leq \theta < \infty, t \leq \alpha(\overline{\alpha}(t))$.

Suppose that α is continuous, and that $t < \alpha(\theta)$. Then, since

$$\overline{\alpha}(t) = \inf\{\tau \in \mathbb{R}_+ : \alpha(\tau) > t\},$$

$\theta \geq \overline{\alpha}(t)$. Suppose that $\theta = \overline{\alpha}(t)$. Then $\alpha(\overline{\alpha}(t)) = \alpha(\theta) > t$. But, because of (Fact) 10.3.25, when α is continuous, $\alpha(\overline{\alpha}(t)) \leq t$, and that leads to a contradiction. □

Fact 10.3.28 *One has that* $\alpha(t) = \overline{(\overline{\alpha})}(t)$.

Proof Indeed, $t < \overline{\alpha}(\theta) = \sup\{x \in \mathbb{R}_+ : \alpha(x) \leq \theta\}$ implies [(Fact) 10.3.22] that $\alpha(t) \leq \theta$. Thus

$$\overline{(\overline{\alpha})}(t) = \inf\{\theta \in \mathbb{R}_+ : \overline{\alpha}(\theta) > t\} \geq \inf\{\theta \in \mathbb{R}_+ : \alpha(t) \leq \theta\} = \alpha(t).$$

On the other hand, $\overline{\alpha}(\alpha(t)) = \inf\{\theta \in \mathbb{R}_+ : \alpha(\theta) > \alpha(t)\} \geq t$, so that

$$\overline{\alpha}(\alpha(t + \epsilon)) \geq t + \epsilon > t,$$

and thus $\alpha(t + \epsilon) \in \{\theta \in \mathbb{R}_+ : \overline{\alpha}(\theta) > t\}$, so that $\alpha(t + \epsilon) \geq \overline{(\overline{\alpha})}(t)$. But α is continuous to the right. □

Remark 10.3.29 In the following cases, the point t does not belong to $\mathcal{R}[\alpha]$ ($K_\alpha(t) = \emptyset$):

(a) $t < \alpha(0)$;
(b) θ is a discontinuity point of α, and $t \in [\alpha^-(\theta), \alpha(\theta)[$;
(c) $\alpha_{(\infty)} \in \mathcal{R}[\alpha], t > \alpha_{(\infty)}$;
(d) $\alpha_{(\infty)}$ not in $\mathcal{R}[\alpha]$, and $t \geq \alpha_{(\infty)}$.

Then:

1. In case a), $\underline{\alpha}(t) = \overline{\alpha}(t) = 0$,

$$[\underline{\alpha}(t), \overline{\alpha}(t)[= \emptyset = \{\alpha = t\} \subseteq [\underline{\alpha}(t), \overline{\alpha}(t)] = \{0\}\,.$$

2. In case b), $\underline{\alpha}(t) = \overline{\alpha}(t) = \theta$, and

$$[\underline{\alpha}(t), \overline{\alpha}(t)[= \emptyset = \{\alpha = t\} \subseteq [\underline{\alpha}(t), \overline{\alpha}(t)] = \{\theta\}\,.$$

3. In cases c) and d), $\underline{\alpha}(t) = \overline{\alpha}(t) = \infty$, and

$$[\underline{\alpha}(t), \overline{\alpha}(t)[= \emptyset = \{\alpha = t\} \subseteq [\underline{\alpha}(t), \overline{\alpha}(t)] = \{\infty\}\,.$$

When $t \in \mathcal{R}[\alpha]$, $\{\alpha = t\}$ is an interval closed to the left. It can be finite, or infinite. When it is finite, it can be open to the right (there is a jump at the end of the interval), or closed (α is continuous at the end of the interval). When it is infinite, it is open to the right, as, for example, for the indicator of the interval $[a, \infty[$, a closed set.

Fact 10.3.30 *One has that* $[\underline{\alpha}(t), \overline{\alpha}(t)[\subseteq \{\alpha = t\} \subseteq [\underline{\alpha}(t), \overline{\alpha}(t)]\,.$

Proof Because of (Remark) 10.3.29, one may assume that $t \in \mathcal{R}[\alpha]$. Since α is continuous to the right,

$$\{\tau \in \mathbb{R}_+ : \alpha(\tau) \geq t\} = [\underline{\alpha}(t), \infty[\,.$$

Thus, when $\alpha(\theta) = t$, $\theta \geq \underline{\alpha}(t)$, and thus

$$\{\theta \in \mathbb{R}_+ : \alpha(\theta) = t\} \subseteq [\underline{\alpha}(t), \infty[\,.$$

As $\overline{\alpha}(t) = \sup\{\tau \in \mathbb{R}_+ : \alpha(\tau) \leq t\}$, when $\alpha(\theta) = t$, $\theta \leq \overline{\alpha}(t)$, and thus

$$\{\theta \in \mathbb{R}_+ : \alpha(\theta) = t\} \subseteq [\underline{\alpha}(t), \overline{\alpha}(t)]\,.$$

Finally, when $\underline{\alpha}(t) \leq \theta$, by definition, $\alpha(\theta) \geq t$, and, when $\theta < \overline{\alpha}(t)$, by definition, $\alpha(\theta) \leq t$, that is, $\alpha(\theta) = t$. Thus, when $\theta \in [\underline{\alpha}(t), \overline{\alpha}(t)[$, $\alpha(\theta) = t$. □

Remark 10.3.31 Fact 10.3.30 designates $\underline{\alpha}$ as the smallest inverse of α, and $\overline{\alpha}$, the largest.

Fact 10.3.32 *When, for $t \in \mathcal{R}[\alpha]$, fixed, but arbitrary, $\{\alpha = t\}$ is closed*

$$\{\alpha = t\} = [\underline{\alpha}(t), \overline{\alpha}(t)] \cap \mathbb{R}_+.$$

The result thus applies in particular when α is strictly increasing, or continuous.

Proof By assumption, $\{\alpha(\theta) = t\} \neq \emptyset$. It is thus an interval closed to the left. It can have two forms: $[\theta_1, \theta_2]$, or $[\theta, \infty[$. In the first case, $\underline{\alpha}(t) = \theta_1$, $\overline{\alpha}(t) = \theta_2$, and the result follows from (Fact) 10.3.30. In the second case, $\underline{\alpha}(t) = \theta$, and $\overline{\alpha}(t) = \infty$. The result is again a consequence of (Fact) 10.3.30. □

Fact 10.3.33 *When α is continuous, for $t \in \mathbb{R}_+$, fixed, but arbitrary,*

$$\{\alpha \circ \overline{\alpha}\}(t) = \alpha(0) \vee (t \wedge \alpha(\infty)).$$

Proof When α is continuous, and $\underline{\alpha}(t) < \infty$, $\alpha(\overline{\alpha}(t)) = t$ [(Remark) 10.3.26]. Since $0 \leq \overline{\alpha}(t) \leq \infty$, and α is increasing, $\alpha(0) \leq \alpha(\overline{\alpha}(t)) \leq \alpha(\infty)$, the result is true when $\underline{\alpha}(t) < \infty$.

When $\underline{\alpha}(t) = \infty$, since $\underline{\alpha} \leq \overline{\alpha}$, $\overline{\alpha}(t) = \infty$, and $\alpha(\overline{\alpha}(t)) = \alpha(\infty)$. When α is continuous, (Remark) 10.3.26 yields that $\alpha(0) \leq \alpha(\overline{\alpha}(t)) \leq t$, so that the result is also true when $\underline{\alpha}(t) = \infty$. □

Definition 10.3.34 The function $f : \mathbb{R}_+ \longrightarrow \mathbb{R}$ is α-continuous when it is constant on $[\alpha^-(t), \alpha(t)]$ whenever $\alpha^-(t) < \alpha(t)$.

Fact 10.3.35 *When f is α-continuous, for $t \in \mathbb{R}_+$, fixed, but arbitrary,*

$$f(\alpha(\underline{\alpha}(\theta) \wedge t)) = \begin{cases} f(\alpha(0)) & \text{when } \theta \in [0, \alpha(0)[\\ \\ f(\theta \wedge \alpha(t)) & \text{when } \theta \geq \alpha(0) \end{cases}.$$

Proof When:

(i) $0 \leq \theta < \alpha(0) : \underline{\alpha}(\theta) = \inf\{x \in \mathbb{R}_+ : \alpha(x) \geq \theta\} = 0$, and, consequently,

$$f(\alpha(\underline{\alpha}(\theta) \wedge t)) = f(\alpha(0));$$

(ii) $\alpha(0) \leq \theta \leq \alpha(t)$: because of (Remark) 10.3.23, $\underline{\alpha}(\theta) \leq t$, so that $\underline{\alpha}(\theta) < \infty$, and

$$f(\alpha(\underline{\alpha}(\theta) \wedge t)) = f(\alpha(\underline{\alpha}(\theta)));$$

since, on $[\alpha^-(\underline{\alpha}(\theta)), \alpha(\underline{\alpha}(\theta))]$, f is constant, then, because of (Remark) 10.3.26,

$$f(\alpha(\underline{\alpha}(\theta)) = f(\theta);$$

(iii) $\theta > \alpha(t)$: as $\underline{\alpha}(\theta) = \inf\{x \in \mathbb{R}_+ : \alpha(x) \geq \theta\}$, $t < \underline{\alpha}(\theta)$, and

$$f(\alpha(\underline{\alpha}(\theta) \wedge t)) = f(\alpha(t)).$$

□

Fact 10.3.36 *Let*

(a) $\alpha : \mathbb{R}_+ \longrightarrow \mathbb{R}_+$ *be continuous to the right, and monotone increasing;*
(b) $\beta : \mathbb{R}_+ \longrightarrow \mathbb{R}_+$ *be continuous to the right, monotone increasing, and* α*-continuous;*
(c) $\gamma = \beta \circ \alpha$.

m_α *shall denote the Borel measure generated by* α*. Then:*

1. *for fixed, but arbitrary Borel set* B,

$$m_\gamma \circ \alpha^{-1}(B) = m_\beta\left([\alpha(0), \alpha(\infty)] \cap B\right);$$

2. *for fixed, but arbitrary* $\phi : \mathbb{R}_+ \longrightarrow \mathbb{R}$*, integrable for* m_β*, the integrals in the following equality exist, and equality obtains:*

$$\int_0^t \{\phi \circ \alpha\}\, dm_\gamma = \int_{\alpha(0)}^{\alpha(t)} \phi\, dm_\beta. \tag{$\star$}$$

Proof One must first notice that $\underline{\alpha}(t) < \overline{\alpha}(t)$ has, as consequence that

$$\gamma(\overline{\alpha}(t)+) - \gamma(\underline{\alpha}(t)-) = 0. \tag{$\star\star$}$$

Indeed, as $\gamma(\overline{\alpha}(t)+) = \beta(\alpha(\overline{\alpha}(t)+))$, and that α is continuous to the right, one has that $\gamma(\overline{\alpha}(t)+) = \beta(\alpha(\overline{\alpha}(t)))$. Furthermore

$$\gamma(\underline{\alpha}(t)-) = \beta(\alpha(\underline{\alpha}(t)-)) = \beta(\alpha^-(\underline{\alpha}(t))).$$

But [(Fact) 10.3.24] $\underline{\alpha}(t) = (\overline{\alpha}(t))^-$, so that $\alpha^-(\underline{\alpha}(t)) = \alpha^-((\overline{\alpha}(t))^-)$. As α^- is continuous to the left, $\alpha^-((\overline{\alpha}(t))^-) = \alpha^-(\overline{\alpha}(t))$. Thus $\gamma(\underline{\alpha}(t)-) = \beta(\alpha^-(\overline{\alpha}(t)))$. When $\alpha^-(\overline{\alpha}(t)) = \alpha(\overline{\alpha}(t))$, there is nothing left to prove, and, when, otherwise, $\alpha^-(\overline{\alpha}(t)) < \alpha(\overline{\alpha}(t))$, then it is the α-continuity of β that does the trick.

The value of $m_\gamma(\{\alpha = t\})$ is obtained as follows:
when $\{\alpha = t\}$ is

(i) *the empty set,* the measure is zero;
(ii) *a genuine interval,* that is, $[\underline{\alpha}(t), \overline{\alpha}(t)[\subseteq \{\alpha = t\}$, $\underline{\alpha}(t) < \overline{\alpha}(t)$ (otherwise $\{\alpha = t\}$ is at most a point), the measure is also zero: indeed, as seen above ($\star\star$),

$$m_\gamma(\{\alpha = t\}) \leq \gamma(\overline{\alpha}(t)+) - \gamma(\underline{\alpha}(t)-) = 0;$$

(iii) *a point,* the measure is equally zero as seen presently: for then [(Fact) 10.3.32] $\{\alpha = t\} = \{\underline{\alpha}(t)\}$, so that [275, p. 508]

$$m_\gamma(\{\alpha = t\}) = m_\gamma(\{\underline{\alpha}(t)\}) = \gamma(\underline{\alpha}(t)+) - \gamma(\underline{\alpha}(t)-),$$

where, given that γ is continuous to the right, $\gamma(\underline{\alpha}(t)+) = \gamma(\underline{\alpha}(t))$; furthermore, $\{\alpha = t\} = \{\underline{\alpha}(t)\}$ means that $\alpha(\underline{\alpha}(t)) = t$, so that

$$m_\gamma\left(\{\underline{\alpha}(t)\}\right) = \beta(t) - \gamma(\underline{\alpha}(t)-);$$

as $\underline{\alpha}(t) < \infty$, $\alpha^-(\underline{\alpha}(t)) \leq t \leq \alpha(\underline{\alpha}(t))$ [(Remark) 10.3.26]: one has either an equality, and there is nothing to prove, or there is strict inequality, and, again, the α-continuity of β does the trick.

Consequently

$$\begin{aligned} m_\gamma\left(\{\alpha \leq t\}\right) &= m_\gamma\left(\{\alpha < t\}\right) \\ &= m_\gamma\left([0, \underline{\alpha}(t)[\right) \\ &= \gamma\left(\underline{\alpha}(t)-\right) - \gamma(0-) \\ &= \beta(\alpha^-(\underline{\alpha}(t))) - \beta(\alpha(0-)). \end{aligned}$$

When $\underline{\alpha}(t) < \infty$, because of (Remark) 10.3.26 and α-continuity of β,

$$\beta(\alpha^-(\underline{\alpha}(t))) = \beta(t) = \beta(t+) = \beta(]0, t]) + \beta(0+).$$

Furthermore

$$\beta(\alpha(0)) = \beta(\alpha(0)+) = \beta(]0, \alpha(0)]) + \beta(0+).$$

Thus, since $\alpha(0) \leq \alpha(\theta) < t$,

$$m_\gamma\left(\{\alpha \leq t\}\right) = \beta(]0, t]) - \beta(]0, \alpha(0)]) = \beta([\alpha(0), t]).$$

When $\underline{\alpha}(t) = \infty$, $\beta(\alpha^-(\underline{\alpha}(t))) = \beta(\alpha(\infty)) = \beta(]0, \alpha(\infty)]) + \beta(0+)$. Thus

$$m_\gamma\left(\{\alpha \leq t\}\right) = \beta([\alpha(0), \alpha(\infty)]).$$

Consequently, as $t > \alpha(\infty)$ implies $\underline{\alpha}(t) = \infty$,

$$m_\gamma\left(\{\alpha \leq t\}\right) = \beta([0, t] \cap [\alpha(0), \alpha(\infty)]),$$

and that proves item 1.

Let $a = \alpha(0)$, and $b = \alpha(t)$. Then

$$\int_0^\infty \{\chi_{[a,b]}(\alpha)\phi(\alpha)\}\, dm_\gamma = \int_0^\infty \left(\chi_{[a,b]}\phi\right) d\left\{m_\gamma \circ \alpha^{-1}\right\}.$$

Since $[a, b] \subseteq [\alpha(0), \alpha(\infty)]$, in the right-hand side of the latter equality, the measure $m_\gamma \circ \alpha^{-1}$ may be replaced, using item 1, with m_β. That takes care of the right-hand side of ($\star$) in item 2. As for the left-hand side, one proceeds as follows. One has that $[0, t] \subseteq \{a \leq \alpha \leq b\}$, and thus

$$\int_0^\infty \{\chi_{[a,b]}(\alpha)\phi(\alpha)\} \, dm_\gamma = \int_{[0,t]} \phi(\alpha) \, dm_\gamma + \int_{\{a\leq\alpha\leq b\}\setminus[0,t]} \phi(\alpha) \, dm_\gamma .$$

But $\theta \in \{a \leq \alpha \leq b\} \setminus [0, t]$ is such that $\alpha(\theta) = \alpha(t)$, and one has already seen that the set $\{\alpha = \alpha(t)\}$ has m_γ-measure zero. □

Here are a few examples, mostly for $t \in [0, 1]$, the time set that is here of main interest.

Example 10.3.37 $\underline{\alpha}(t) < \overline{\alpha}(t)$ occurs typically when α is constant over an interval. Let indeed $t = 2^{-1}$, and

$$\alpha(\theta) = \begin{cases} 2\theta & \text{when } 0 \leq \theta \leq \frac{1}{4} \\ \frac{1}{2} & \text{when } \frac{1}{4} < \theta \leq \frac{3}{4} \\ 2\theta - 1 & \text{when } \frac{3}{4} < \theta \leq 1 \end{cases} .$$

Then $I_\alpha^c(t) = \left[4^{-1}, 1\right]$, and $\underline{\alpha}(t) = 4^{-1}$, while $J_\alpha^c(t) = \left]3 \cdot 4^{-1}, 1\right]$, and, consequently, $\overline{\alpha}(t) = 3 \cdot 4^{-1}$.

Example 10.3.38 Let $\alpha_t(\theta) = \alpha(\theta \wedge \underline{\alpha}(t))$, and $\dot{\alpha}_t(\theta) = \alpha(\theta \wedge \overline{\alpha}(t))$. α_t and $\dot{\alpha}_t$ are α stopped at $\underline{\alpha}(t)$ and $\overline{\alpha}(t)$ respectively: they are constant once they have reached the threshold for the first time. Usually one expects that α_t and $\dot{\alpha}(t)$ be bounded by t. But some restrictions are needed for the bounds to hold. Suppose thus that $t = 2^{-1}$, and that

$$\alpha(\theta) = \begin{cases} \frac{\theta}{2} \text{ when } 0 \leq \theta < \frac{1}{2} \\ \frac{1}{2} \text{ when } \frac{1}{2} \leq \theta < 1 \\ 1 \text{ when } \theta = 1 \end{cases} .$$

Then $I_\alpha^c(t) = \left[2^{-1}, 1\right]$, and $\underline{\alpha}(t) = 2^{-1}$, while $J_\alpha^c(t) = \{1\}$, and $\overline{\alpha}(t) = 1$, so that

$$\alpha_t(\theta) = \begin{cases} \frac{\theta}{2} \text{ when } 0 \leq \theta < \frac{1}{2} \\ \frac{1}{2} \text{ when } \frac{1}{2} \leq \theta \leq 1 \end{cases} ,$$

and thus that $\alpha_t \leq \frac{1}{2}$, but $\dot{\alpha}_t = \alpha$, which $\frac{1}{2}$ does not bound.

Increasing functions are required. Thus, when $\alpha(\theta) = 1 - \theta$, and $t = 2^{-1}$,

$$I_{\alpha}^{c}(t) = \left[0, 2^{-1}\right], \ J_{\alpha}^{c}(t) = \left[0, 2^{-1}\right[, \ \underline{\alpha}(t) = \overline{\alpha}(t) = 0,$$

and $\alpha_t(\theta) = \dot{\alpha}_t(\theta) = 1 > t$.

Example 10.3.39 Let $\alpha = \chi_{]1,\infty[}$. Then

$$\overline{\alpha}(1/2) = 1, \ \text{ and } \ \alpha(\overline{\alpha}(1/2)) = 0 < 1/2.$$

When $\alpha = \chi_{[1,\infty[}$,

$$\overline{\alpha}(1/2) = 1, \ \text{ and } \alpha(\overline{\alpha}(1/2)) = 1 > 1/2.$$

10.3.2 Change of Time for a Continuous Local Martingale

One shall need a statement of the Dambis-Dubins-Schwarz theorem [264, p. 213]. Here it is preceded by some definitions.

Definition 10.3.40 A time-change is an increasing net of wide sense stopping times, say $\mathcal{S} = \{S_t, t \in \mathbb{R}_+\}$. Given a process X, its time-changed transformation by $\mathcal{S}$ is the process $(\omega, t) \mapsto X(\omega, S_t(\omega))$.

Definition 10.3.41 Let S be a positive random variable, and M, an almost surely continuous local martingale for $\underline{\mathcal{A}}$, starting at 0. M is a Brownian motion stopped by S when $\langle M \rangle = t \wedge S$.

Remark 10.3.42 Definition 10.3.41 does not mean that M is a Brownian motion. An example follows [221, p. 624].

Example 10.3.43 W is a standard Brownian motion, and one considers the following stopping times:

$$\begin{aligned} S_1 &= \inf\{t : W(\cdot, t) = 1\}, \\ S_{(1/2)} &= \inf\{t > S_1 : W(\cdot, t) = 1/2\}, \\ S_{(3/2)} &= \inf\{t > S_1 : W(\cdot, t) = 3/2\}, \\ S &= S_{(1/2)} \wedge S_{(3/2)}. \end{aligned}$$

Let

$$
\begin{aligned}
a(\omega,t) &= \chi_{]0,S]}(\omega,t)\\
&\quad + \chi_{\{S=S_{(1/2)}\}}(\omega)\,\chi_{]S,\infty[}(\omega,t)\\
&\quad + 3\,\chi_{\{S<S_{(1/2)}\}}(\omega)\,\chi_{]S_{(1/2)},\infty[}(\omega,t).
\end{aligned}
$$

It is a predictable process. Then

$$M(\omega,t) = \int_0^t a(\omega,\theta)\,W(\omega,d\theta)$$

is a continuous martingale. When

- $S_{(1/2)} \le S_{(3/2)}$, that is, $S = S_{(1/2)}$,

$$a(\omega,\cdot) \equiv 1, \text{ and } M(\omega,\cdot) = W(\omega,\cdot);$$

- when $S_{(1/2)} > S_{(3/2)}$, that is $S < S_{(1/2)}$,
  - for $0 < t \le S_{(3/2)}(\omega)$,

$$a(\omega,t) = 1, \text{ and } M(\omega,t) = W(\omega,t),$$

  - for $S_{(3/2)}(\omega) < t \le S_{(1/2)}$,

$$a(\omega,t) = \chi_{]0,S_{(3/2)}]}(\omega,t), \text{ and } M(\omega,t) = 3/2,$$

  - for $t > S_{(1/2)}(\omega)$,

$$a(\omega,t) = \chi_{]0,S_{(3/2)}]}(\omega,t) + 3\,\chi_{]S_{(1/2)},\infty]}(\omega,t),$$

    and

$$M(\omega,t) = 3/2 + 3(W(\omega,t) - (1/2)) = 3W(\omega,t).$$

Thus, for stopping times $T \le S$, $M^T = W^T$, and M is a stopped Brownian motion.

Remark 10.3.44 At the cost of expanding the basic probability space, all stopped Brownian motions may be obtained by stopping an actual Brownian motion [264, p. 207].

Fact 10.3.45 (Dambis, Dubins, Schwarz (DDS) Factorization) *Let M be an almost surely continuous local martingale for $\underline{\mathcal{A}}$, and*

$$\begin{aligned} S_M(\omega,t) &= \inf\{\theta \in \mathbb{R}_+ : \langle M\rangle(\omega,\theta) > t\}, \\ W_M(\omega,t) &= M(\omega, S_M(\omega,t)) - M(\omega,0), \\ \mathcal{B}_t &= \mathcal{A}^+_{S_M(\cdot,t)}. \end{aligned}$$

Then:

1. *when $P(\langle M\rangle(\cdot,\infty) = \infty) = 1$,*

$$W_M \text{ is a Brownian motion for } \underline{\mathcal{B}};$$

2. *when $P(\langle M\rangle(\cdot,\infty) < \infty) > 0$,*

$$W_M \text{ is, for } \underline{\mathcal{B}}, \text{ a Brownian motion stopped by } \langle M\rangle(\cdot,\infty);$$

3. *for $t \geq 0$, fixed, but arbitrary, almost surely with respect to P,*

$$M(\cdot,t) - M(\cdot,0) = W_M(\cdot, \langle M\rangle(\cdot,t)),$$

and, for progressively measurable f,

$$\int_0^t f\,dM = \int_0^{\langle M\rangle(\cdot,t)} f(\cdot, S_M(\cdot,\theta))\, W_M(\cdot, d\theta),$$

where one side of the latter equality is defined when the other is, and $f = 0$ when $S_M(\cdot,t) = \infty$.

W_M is called the Dambis, Dubins, Schwarz (DDS) Brownian motion of M. One shall write $W \diamond V(\omega,t)$ for $W(\omega, V(\omega,t))$, and designate it as the DDS factorization of M.

Remark 10.3.46 $W[\omega]$ denotes the path $t \mapsto W(\omega,t)$, at the value ω. One way to look at $W \diamond V$ is then as follows: $W \diamond V[\omega] = W[\omega] \circ V[\omega]$, and one has thus the following diagram:

$$\omega \mapsto (W[\omega], V[\omega]) \mapsto W[\omega] \circ V[\omega].$$

Define

$$\mathbf{W} : \Omega \longrightarrow C(\mathbb{R}_+), \quad \text{and } \mathbf{V} : \Omega \longrightarrow C(\mathbb{R}_+),$$

using, respectively, the following assignments:

$$\boldsymbol{W}(\omega) = W[\omega], \text{ and } \boldsymbol{V}(\omega) = V[\omega].$$

Letting $T : \Omega \longrightarrow C(\mathbb{R}+) \times C(\mathbb{R}_+)$ be defined using the following assignment:

$$T(\omega) = (\boldsymbol{W}, \boldsymbol{V})(\omega) = (\boldsymbol{W}(\omega), \boldsymbol{V}(\omega)) = (W[\omega], V[\omega]),$$

and $\Gamma : C(\mathbb{R}_+) \times C(R_+) \longrightarrow C(\mathbb{R}_+)$, the following one:

$$\Gamma(f, g) = f \circ g,$$

one has that $W \diamond V[\omega] = \Gamma \circ T[\omega]$. If $\Gamma \circ T$ is adapted,

$$P_{W \diamond V} = P_{W,V} \circ \Gamma^{-1}.$$

$C(R_+)$ is a separable Fréchet space for the following distance [267]:

$$d(f, g) = \sum_{n=1}^{\infty} \frac{1}{2^n} \frac{\|f - g\|_n}{1 + \|f - g\|_n}, \quad \|f\|_n = \sup_{t \in [0,n]} |f(t)|,$$

and, for that distance, $f_p \to f$ if, and only if, for $n \in \mathbb{N}$, fixed, but arbitrary,

$$\lim_p \left\| f_p - f \right\|_n = 0.$$

$C(\mathbb{R}_+) \times C(\mathbb{R}_+)$ is given the product topology, so that $\{(f_n, g_n), n \in \mathbb{N}\}$ converges to (f, g) if, and only if, $\{f_n, n \in \mathbb{N}\}$ converges to f, and $\{g_n, n \in \mathbb{N}\}$ converges to g [74, p. 73]. The Borel space of the product is the product of the Borel spaces of the components [208, p. 6].

T is adapted since $\boldsymbol{W}$ and $\boldsymbol{V}$ are. Γ is continuous. Let indeed $n \in \mathbb{N}$, be fixed, but arbitrary, and let $\{f_p\}$ converge to f, and $\{g_p\}$, to g. Then

$$\left\| f \circ g - f_p \circ g_p \right\|_n \leq \left\| f \circ g - f \circ g_p \right\|_n + \left\| f \circ g_p - f_p \circ g_p \right\|_n.$$

The first term on the right-hand side of the latter expression goes to zero because of [109, p. 360], and the second, because of [109, p. 370]. $\Gamma \circ T$ is thus a measurable map.

It may be useful to look at an example.

Example 10.3.47 Let f be a strictly positive, progressively measurable process such that

$$\int_0^\infty f^2(\cdot, \theta) Leb(d\theta) < \infty,$$

and W, a standard Brownian motion. Suppose that $M = \int f\,dW$. Then

$$\langle M\rangle = \int f^2\,dLeb.$$

It is continuous, and strictly increasing, and thus there is a unique $S_M(\cdot,t)$ such that

$$\int_0^{S_M(\cdot,t)} f^2\,dLeb = t.$$

But, using (Fact) 10.3.36, with m_β, Lebesgue measure, one gets

$$\int_0^{S_M(\cdot,t)} f^2(\cdot,\theta)Leb(d\theta) = \int_0^t f^2(\cdot,S_M(\cdot,\theta))m_{S_M(\cdot,\cdot)}(d\theta).$$

It follows that

$$f^2(\cdot,S_M(\cdot,\theta))m_{S_M(\cdot,\cdot)}(d\theta) = d\theta,$$

or

$$S_M(\cdot,t) = \int_0^t \frac{d\theta}{f^2(\cdot,S_M(\cdot,\theta))}.$$

Choosing $f(\cdot,\theta) = \{2\theta\}^{1/2}$, one obtains that $S_M(\cdot,t) = t^{1/2}$, so that

$$W_M(\cdot,t) = M \diamond S_M(\cdot,t) = 2^{1/2}\int_0^{t^{1/2}} \theta^{1/2}W(\cdot,d\theta).$$

10.4 Exponential Martingales

Exponential martingales are very handy tools in the study of processes of the type encountered below. This section gathers the facts useful for the sequel.

10.4.1 Positive Local Martingales

Let M be a process with values in $\mathbb{R}$, and paths that are continuous to the right, and, with respect to P, almost surely continuous. When M is a local martingale, M has the representation $M(\cdot,t) = M(\cdot,0) + M_0(\cdot,t)$, M_0 a local martingale that is zero at the origin. The fact that M is a local martingale puts no restriction on $M(\cdot,0)$.

Example 10.4.1 ([184, p. 95]) Let $\Omega = [0, 1]$, $\mathcal{A}_t = \mathcal{B}([0, 1])$, $t \in [0, 1]$, and also $P =$ Lebesgue measure. Let $M(\omega, t) = X$, a random variable that is not integrable. M is a continuous local martingale as it is a continuous process, and that $M_0 = 0$. M is not a martingale, and $M(\cdot, 0)$ is not integrable.

The results which follow, on continuous, local martingales, are stated for the index set $\mathbb{R}_+$. When one restricts attention to $[0, 1]$, since processes are all assumed continuous to the right, one sets, for $t > 1$, $M(\omega, t) = M(\omega, 1)$, $\mathcal{A}_t = \mathcal{A}_1$, and then applies the results valid for $\mathbb{R}_+$. The continuity assumption is not necessary for some of the statements, but since one is only concerned with continuous processes, let that be it.

Fact 10.4.2 ([264, p. 103]) *Let M be a local martingale, continuous to the right, which is, almost surely, with respect to P, continuous, and let S be a wide sense stopping time such that $E_P[\langle M\rangle(\cdot, S)] < \infty$. Then $M^S(\cdot, \cdot) - M(\cdot, 0)$ is a martingale bounded in L_2, and, when $M(\cdot, 0)$ has an integrable square, M^S is a martingale bounded in L_2. As a consequence, when, for $t \in \mathbb{R}_+$, fixed, but arbitrary, $E_P[\langle M\rangle(\cdot, t)] < \infty$, then M is a martingale in L_2.*

Proposition 10.4.3 ([184, p. 101]) *Let M be a positive, local martingale, continuous to the right, and, with respect to P, almost surely continuous. As such, it is also a supermartingale.*

Proof Let $\{S_n, n \in \mathbb{N}\}$ be the sequence that makes of $M_0^{S_n}$, for $n \in \mathbb{N}$, fixed, but arbitrary, a uniformly integrable martingale [264, p. 63]. For $t \in \mathbb{R}$, fixed, but arbitrary, the sequence

$$\left\{M_0^{S_n}(\cdot, t) = M_0(\cdot, t \wedge S_n), n \in \mathbb{N}\right\}$$

converges almost surely, with respect to P, to $M_0(\cdot, t)$. Since conditional expectations for positive random variables are well defined, one has, using Fatou's lemma, that, for $t_1 < t_2$ in $[0, 1]$, fixed, but arbitrary,

$$\begin{aligned}
E_P[M(\cdot, t_2) \mid \mathcal{A}_{t_1}] &= E_P[\liminf_n M^{S_n}(\cdot, t_2) \mid \mathcal{A}_{t_1}] \\
&\leq \liminf_n E_P[M^{S_n}(\cdot, t_2) \mid \mathcal{A}_{t_1}] \\
&= M(\cdot, 0) + \liminf_n E[M_0^{S_n}(\cdot, t_2) \mid \mathcal{A}_{t_1}] \\
&= M(\cdot, 0) + \liminf_n M_0^{S_n}(\cdot, t_1) \\
&= M(\cdot, 0) + M_0(\cdot, t_1) \\
&= M(\cdot, t_1).
\end{aligned}$$

□

The consequence of (Proposition) 10.4.3 is the following corollary:

Corollary 10.4.4 $E_P[M(\cdot, t)] \leq E_P[M(\cdot, 0)]$.

Proposition 10.4.5 *Let M be a positive, local martingale, continuous to the right, and, with respect to P, almost surely continuous. Suppose that, for some* $\tau > 0$, $E_P[M(\cdot, \tau)] = E_P[M(\cdot, 0)] < \infty$. *Then M is a martingale on* $[0, \tau]$.

Proof Because of (Proposition) 10.4.3, for $t \in [0, \tau]$, fixed, but arbitrary,

$$E_P[M(\cdot, t)] = E_P[M(\cdot, 0)].$$

Let $t_1 < t_2$ in $[0, \tau]$, and $A \in \mathcal{A}_{t_1}$, be fixed, but arbitrary. The supermartingale property of M [(Proposition) 10.4.3] yields that

$$\int_A M(\cdot, t_2)\, dP \leq \int_A M(\cdot, t_1)\, dP,$$

$$\int_{A^c} M(\cdot, t_2)\, dP \leq \int_{A^c} M(\cdot, t_1)\, dP.$$

Since $E_P[M(\cdot, t_1)] = E_P[M(\cdot, t_2)]$, from the second of the latter inequalities, one gets that

$$\int_A M(\cdot, t_2)\, dP \geq \int_A M(\cdot, t_1)\, dP,$$

and inequality may be replaced with equality. □

10.4.2 Relations Between Martingales and Their Exponentials

It is often easier to deal with the exponentials of martingales than with the martingales themselves. Their relations are described below.

Proposition 10.4.6 ([189, 4–54]) *Below,* $\alpha \in \mathbb{R}$ *is fixed, but arbitrary. M and V are real, adapted processes that are continuous to the right and, with respect to P, almost surely continuous. V has, almost surely, with respect to P, increasing paths that start at 0. Set*

$$X_\alpha(\omega, t) = \alpha M(\omega, t) - \frac{1}{2}\alpha^2 V(\omega, t),$$

$$E_\alpha(\omega, t) = e^{X_\alpha(\omega, t)}.$$

The following obtain:

1. *E_α is a local martingale for $\underline{A}$ if, and only if, M is, locally, a martingale in L_2, starting at zero, and $\langle M\rangle = V$.*
2. *When M is, locally, a martingale in L_2, that $\langle M\rangle = V$, and that, for $t \in \mathbb{R}_+$, fixed, but arbitrary,*

$$E_P\left[\int_0^t e^{2\alpha M}dV\right] < \infty,$$

   then E_α is a martingale in L_2, continuous to the right, and, with respect to P, almost surely continuous.
3. *If E_α is a martingale, and there exists $\alpha_0 > 0$ such that, for $t \in \mathbb{R}_+$, fixed, but arbitrary,*

$$E_P\left[e^{\alpha_0|M(\cdot,t)|}\right] < \infty,$$

   M is a martingale.

Proof When M is, locally, a martingale in L_2, such that $V = \langle M\rangle$, that E_α is, locally, a martingale is established in [264, p. 183] as a consequence of Itô's formula. It is also shown there that E_α solves $dE_\alpha = \alpha E_\alpha dM$, which explains that

$$\langle E_\alpha\rangle = \alpha^2\int_0^{\cdot} e^{2X_\alpha}dV.$$

Half of item 1 is thus proved.

When the assumptions of item 2 are valid, one has that

$$E_P\left[\int_0^t e^{2X_\alpha}dV\right] \leq E_P\left[\int_0^t e^{2\alpha M}dV\right] < \infty,$$

that is, $\langle E_\alpha\rangle$ has finite expectation. Fact 10.4.2 then yields that E_α is a martingale bounded in L_2.

Suppose now that E_α is a martingale, and that $E_P\left[e^{\alpha_0|M(\cdot,1)|}\right] < \infty$ (that is, item 3). Then, for $t_1 < t_2$ in $[0,1]$ and $A \in \mathcal{A}_{t_1}$, fixed, but arbitrary,

$$\int_A E_\alpha(\cdot,t_1)dP = \int_A E_\alpha(\cdot,t_2)dP.$$

Differentiating twice with respect to α (supposing for the time being that it is allowed), one obtains that

$$\int_A \{M(\cdot,t_1) - \alpha V(\cdot,t_1)\} E_\alpha(\cdot,t_1)dP = \int_A \{M(\cdot,t_2) - \alpha V(\cdot,t_2)\} E_\alpha(\cdot,t_2)dP,$$

and that

$$\int_A \left\{[M(\cdot,t_1) - \alpha V(\cdot,t_1)]^2 - V(\cdot,t_1)\right\} E_\alpha(\cdot,t_1)dP =$$
$$= \int_A \left\{[M(\cdot,t_2) - \alpha V(\cdot,t_2)]^2 - V(\cdot,t_2)\right\} E_\alpha(\cdot,t_2)dP.$$

Setting α to zero, the first equation says that M is a martingale, and the second, that $M^2 - V$ is, thus identifying V as the quadratic variation of M.

To check that differentiation is allowed, let $\underline{U} : \Omega \longrightarrow \mathbb{R}^2$ have, at $\omega \in \Omega$, components $M(\omega, t)$ and $V(\omega, t)$, B, be a Borel set in $\mathbb{R}^2$, and

$$f_\alpha(x,y) = e^{\alpha x - \frac{1}{2}\alpha^2 y},$$
$$g_\alpha(x,y) = (x - \alpha y) f_\alpha(x,y).$$

Then

$$\int_B f_\alpha(x,y) P \circ \underline{U}^{-1}(dx,dy) = \int_{\underline{U}^{-1}(B)} E_\alpha(\omega,t) P(d\omega),$$

and

$$\int_B g_\alpha(x,y) P \circ \underline{U}^{-1}(dx,dy) =$$
$$= \int_{\underline{U}^{-1}(B)} \{M(\omega,t_1) - \alpha V(\omega,t_1)\} E_\alpha(\omega,t_1) P(d\omega),$$

where $\underline{U}^{-1}(B) \in \mathcal{A}_t$. One then applies the standard differentiation theorem [113, p. 137] to f_α and g_α. The main step (uniform domination) may be checked as follows.

When $e^{\alpha_0 |M(\cdot,1)|}$ has finite expectation, since $\psi(x) = e^{\alpha|x|}$ is convex, $\psi(M(\cdot,t))$ is a submartingale, and, in particular,

$$E_P[\psi(M(\cdot,t))] \leq E_P[\psi(M(\cdot,1))] < \infty.$$

The case of f_α is thus a direct consequence of the assumption. Also, the powers of $|M(\cdot,t)|$ are integrable, and, in particular, when $\alpha \leq \frac{\alpha_0}{2}$,

$$E_P^2\left[|M(\cdot,t)|\, e^{\alpha|M(\cdot,t)|}\right] \leq E_P\left[M^2(\cdot,t)\right] E_P\left[e^{2\alpha|M(\cdot,t)|}\right] < \infty.$$

Thus

$$x \mapsto |x| \, e^{\frac{\alpha_0}{2}|x|}$$

is integrable with respect to $P \circ \underline{U}^{-1}$. The map

$$y \mapsto |y| \, e^{-\frac{\alpha_0^2}{2}|y|}$$

is bounded, and thus

$$|g_\alpha(x, y)| \le \left\{ |x| + \frac{\alpha_0}{2} \, |y| \right\} e^{\frac{\alpha_0}{2}|x| - \frac{\alpha_0^2}{2}|y|}.$$

Let

$$\{S_n, n \in \mathbb{N}\}$$

be a sequence of wide sense stopping times that stops the largest of V and $|M|$ at n [264, p. 63], and

$$\left\{\dot{S}_n, n \in \mathbb{N}\right\},$$

a sequence that localizes E_α. Then

$$\left\{\Sigma_n = S_n \wedge \dot{S}_n, n \in \mathbb{N}\right\}$$

is also a localizing sequence for E_α [264, p. 65]. But then $E_\alpha^{\Sigma_n}$ is a martingale by item 2, and the choice of stopping times. Consequently $M_0^{\Sigma_n}$ is a martingale with $V_0^{\Sigma_n}$ as variation, and the proof of item 1 is complete. □

In the remarks which follow, let $M(\cdot, 0) = 0$, M and V be continuous to the right, and, with respect to P, almost surely continuous.

Remark 10.4.7 Let M be locally in L_2, and such that $\langle M \rangle = V$. Then E_α, being a local martingale, is a supermartingale [(Proposition) 10.4.3] such that

$$E_P[E_\alpha(\cdot, t)] \le 1.$$

Remark 10.4.8 When there exists $\theta \in]0, 1]$ such that $E_P[E_\alpha(\cdot, \theta)] = 1$, E_α is a martingale for $t \in [0, \theta]$ [(Proposition) 10.4.5].

Remark 10.4.9 Let E_α be, for $\alpha_0 > 0$ and $\alpha \in]-\alpha_0, \alpha_0[$, fixed, but arbitrary, locally, a martingale in L_2, continuous to the right and, with respect to P, almost surely continuous, such that $E_P[V(\cdot, 1)] < \infty$. Then M is bounded in L_2. The proposition says indeed that M is, locally, a martingale in L_2 with $\langle M \rangle = V$. One then applies (Fact) 10.4.2.

Proposition 10.4.10 ([186]) *Let M and V be adapted, continuous to the right and, with respect to P, almost surely continuous and zero at the origin, as in* (Proposition) 10.4.6. *Suppose that M is, locally, a martingale in L_2, that $\langle M\rangle = V$, and that there exists $\alpha_0 > 0$ such that*

$$E_P[e^{\alpha_0^2 V(\cdot,1)}] < \infty.$$

Then:

1. *for $\alpha^2 \in [0, 2\alpha_0^2]$, E_α is a martingale;*
2. *for $\alpha^2 \in \left[0, \frac{\alpha_0^2}{3+2\sqrt{2}}\right]$, E_α is a martingale bounded in L_2.*

Proof (1) Because of (Proposition) 10.4.6, item 1, E_α is a local martingale. Let thus $\{S_n, n \in \mathbb{N}\}$ be a sequence that localizes it. Suppose one is able to prove that there exists $\epsilon > 0$ such that, for α as in item 1,

$$\sup_n E_P\left[\left\{E_\alpha^{S_n}(\cdot,t)\right\}^{1+\epsilon}\right] < \infty,$$

that is, the uniform integrability of the sequence $\left\{E_\alpha^{S_n}(\cdot,t), n \in \mathbb{N}\right\}$ [192, p. 19]. Since that sequence converges also almost surely, with respect to P, to $E_\alpha(\cdot,t)$, the convergence takes place in L_1, and the limit is in L_1 [192, p. 18]. Since $E_P\left[E_\alpha^{S_n}(\cdot,0)\right] = E_P\left[E_\alpha^{S_n}(\cdot,1)\right]$, E_α is then a martingale because of (Proposition) 10.4.5.

Let thus $X_\alpha^{(n)} = \alpha M^{S_n} - \frac{1}{2}\alpha^2 V^{S_n}$. To isolate V, in order to use the integrability assumption, Hölder's inequality is applied as follows. One starts with the insertion of $(1+\epsilon)\frac{\alpha^2}{2}V^{S_n}$ in

$$\begin{aligned}\frac{1}{p}X_{\alpha p(1+\epsilon)}^{(n)} &= \alpha(1+\epsilon)M^{S_n} - \frac{\alpha^2 p(1+\epsilon)^2}{2}V^{S_n}\\ &= (1+\epsilon)X_\alpha^{(n)}\\ &\quad + \{1-p(1+\epsilon)\}(1+\epsilon)\frac{\alpha^2}{2}V^{S_n}.\end{aligned}$$

Thus

$$(1+\epsilon)X_\alpha^{(n)} = \frac{1}{p}X_{\alpha p(1+\epsilon)}^{(n)} + \frac{\alpha^2}{2}(1+\epsilon)\{p(1+\epsilon)-1\}V^{S_n},$$

and, consequently, since V is increasing, using Hölder's inequality,

$$\begin{aligned}E_P\left[\left\{E_\alpha^{S_n}(\cdot,t)\right\}^{1+\epsilon}\right] &= E_P\left[e^{(1+\epsilon)X_\alpha^{(n)}(\cdot,t)}\right]\\ &= E_P\left[e^{\frac{1}{p}X_{\alpha p(1+\epsilon)}^{(n)}(\cdot,t)} \times e^{\frac{\alpha^2}{2}(1+\epsilon)\{p(1+\epsilon)-1\}V^{S_n}(\cdot,t)}\right]\end{aligned}$$

$$\le E_P^{\frac{1}{p}}\left[e^{X^{(n)}_{\alpha p(1+\epsilon)}(\cdot,t)}\right]E_P^{\frac{1}{q}}\left[e^{q\frac{\alpha^2}{2}(1+\epsilon)\{p(1+\epsilon)-1\}V^{S_n}(\cdot,t)}\right]$$
$$= E_P^{\frac{1}{p}}\left[E^{S_n}_{\alpha p(1+\epsilon)}(\cdot,t)\right]E_P^{\frac{1}{q}}\left[e^{q\frac{\alpha^2}{2}(1+\epsilon)\{p(1+\epsilon)-1\}V(\cdot,t)}\right].$$

Now, since $E_{\alpha p(1+\epsilon)}$ is a supermartingale [(Remark) 10.4.7], and S_n is a wide sense stopping time,

$$E^{S_n}_{\alpha p(1+\epsilon)} \text{ is a supermartingale [264, p. 57], and thus } E_P^{\frac{1}{p}}\left[E^{S_n}_{\alpha p(1+\epsilon)}(\cdot,t)\right] \le 1.$$

To use the assumption, given that $\alpha^2 \le 2\alpha_0^2$, one must choose $\epsilon > 0, p$ and q so that

$$q\frac{\alpha^2}{2}(1+\epsilon)\{p(1+\epsilon)-1\} \le \alpha_0^2 \, .$$

Let

$$\alpha^2 = \frac{2\alpha_0^2}{1+\kappa}, \ \kappa \ge 0, \text{ so that } \alpha_0^2 = (1+\kappa)\frac{\alpha^2}{2}.$$

Then, since $\frac{1}{p}+\frac{1}{q}=1$, that is, $p=\frac{q}{q-1}$,

$$\begin{aligned}1+\kappa &\ge q(1+\epsilon)\{p(1+\epsilon)-1\}\\ &= q(1+\epsilon)\left\{\frac{q}{q-1}(1+\epsilon)-1\right\}\\ &= \frac{q(1+\epsilon)(1+\epsilon q)}{q-1}\\ &= \frac{q+\epsilon q+\epsilon q^2+\epsilon^2q^2}{q-1}\\ &= \frac{(q-1)+(1+\epsilon q)+(1+\epsilon)\epsilon q^2}{q-1}.\end{aligned}$$

One must thus have

$$\frac{1+\epsilon q+(1+\epsilon)\epsilon q^2}{q-1} \le \kappa.$$

Let $q-1=\frac{2}{\kappa}$ to obtain that $1+\epsilon\left(1+\frac{2}{\kappa}\right)+(1+\epsilon)\epsilon\left(1+\frac{1}{\kappa}\right)^2\leq 2$, so that

$$\epsilon\leq\left\{\left(1+\frac{2}{\kappa}\right)+(1+\epsilon)\left(1+\frac{2}{\kappa}\right)^2\right\}^{-1}\leq\left\{\left(1+\frac{2}{\kappa}\right)+\left(1+\frac{2}{\kappa}\right)^2\right\}^{-1}.$$

Proof (2) Suppose that one proves that

$$\sup_n E_P\left[\left\{E_\alpha^{S_n}(\cdot,t)\right\}^2\right]\leq E_P\left[e^{\alpha_0^2 V(\cdot,t)}\right].$$

One has then again, for the same reason as that given at the beginning of the proof, uniform integrability. Paired with almost sure convergence, that yields, as for the proof of item 1, that $E_\alpha^2(\cdot,t)$ is integrable. But the same condition as that of item 1, with $\epsilon=1$, provides the required bound. Indeed condition

$$q\frac{\alpha^2}{2}(1+\epsilon)\{p(1+\epsilon)-1\}\leq\alpha_0^2$$

becomes

$$\alpha^2(2p-1)q\leq\alpha_0^2.$$

Letting $\kappa=\frac{\alpha_0^2}{\alpha^2}$ and $p=\frac{q}{q-1}$, one obtains that

$$\left(\frac{2q}{q-1}-1\right)q\leq\kappa,$$

or, equivalently that

$$q^2+(1-\kappa)q+\kappa\leq 0.$$

As one must have $q>1$, let $q=1+\delta$. The last equation then becomes

$$\delta^2+\delta\,(3-\kappa)+2\leq 0$$

which requires $(3-\kappa)^2\geq 8$, that is $3-\kappa\geq 2\sqrt{2}$, or $\kappa-3\geq 2\sqrt{2}$. Since one needs a lower bound to κ, one must choose $\kappa\geq 3+2\sqrt{2}$. □

10.5 Random Elements with Values in the Hilbert Space of Sequences

As one shall deal with vectors $\underline{X}$ whose components are random variables X_n such that, almost surely with respect to some probability P, $\sum_n X_n^2$ is finite, the basic facts about such elements are listed below. One short reference is [201]. The standard basis of l_2 shall be denoted $\{\underline{e}_n, n \in \mathbb{N}\}$.

Let $\underline{X} : \Omega \longrightarrow l_2$ be a map. It is a random element when it is adapted to $\mathcal{A}$ and the Borel sets of l_2. That is equivalent to the maps $X_n = \langle \underline{X}, \underline{e}_n \rangle_{l_2}$ being adapted for all $n \in \mathbb{N}$. The norm of a random element is a random variable as the composition of a measurable map with a continuous one. A random element is (Bochner or strongly) integrable when its norm has finite expectation, that is

$$E_P\left[\|\underline{X}\|_{l_2}\right] = E_P\left[\left(\sum_n X_n^2\right)^{1/2}\right] < \infty.$$

Then $E_P[\underline{X}] \in l_2$, and $\|E_P[\underline{X}]\|_{l_2} \leq E_P\left[\|\underline{X}\|_{l_2}\right]$. Also, for every continuous linear functional $\varphi : l_2 \longrightarrow \mathbb{R}$, one has that

$$E_P[\varphi(\underline{X})] = \varphi(E_P[\underline{X}]).$$

Choosing $\varphi(\cdot) = \langle \cdot, \underline{e}_n \rangle_{l_2}$, one sees that the components of $E_P[\underline{X}]$ are the expectations of the components of $\underline{X}$. Furthermore, for $\underline{\alpha} \in l_2$, fixed, but arbitrary, one gets that

$$E_P\left[\langle \underline{\alpha}, \underline{X} \rangle_{l_2}\right] = E_P\left[\sum_{n=1}^{\infty} \alpha_n X_n\right] = \sum_{n=1}^{\infty} \alpha_n E_P[X_n] = \langle \underline{\alpha}, E_P[\underline{X}] \rangle_{l_2}.$$

Let $L_1^{l_2}(\Omega, \mathcal{A}, P)$ be the Banach space of equivalence classes of integrable random elements with domain Ω, and range in l_2, and $\mathcal{B}$ be a sub-σ-algebra of $\mathcal{A}$.

Fact 10.5.1 *The conditional expectation of $\underline{X}$ with respect to $\mathcal{B}$ is a linear operator, denoted $E_P[\cdot \mid \mathcal{B}]$, idempotent and contracting, from $L_1^{l_2}(\Omega, \mathcal{A}, P)$ into itself, with the following properties:*

1. *When $\underline{X}(\omega) = \sum_{i=1}^{n} \underline{x}_n \chi_{\{\underline{X}=\underline{x}_n\}}(\omega)$,*

$$E_P[\underline{X} \mid \mathcal{B}] = \sum_{\underline{x} \in \underline{X}[\Omega]} \underline{x} P(\underline{X} = \underline{x} \mid \mathcal{B}).$$

2. *The range of the conditional expectation operator is the (closed) subspace of $L_1^{l_2}(\Omega, \mathcal{A}, P)$ of the classes of elements adapted to $\mathcal{B}$.*
3. *Almost surely with respect to P, $\|E_P[\underline{X} \mid \mathcal{B}]\|_{l_2} \leq E_P\left[\|\underline{X}\|_{l_2} \mid \mathcal{B}\right]$.*

4. *For every continuous linear functional* φ *of* l_2,

$$E_P[\varphi(\underline{X}) \mid \mathcal{B}] = \varphi(E_P[\underline{X} \mid \mathcal{B}])$$

[so that, in particular, the components of $E_P[\underline{X} \mid \mathcal{B}]$ are the conditional expectations of the components of $\underline{X}$].

A random element in l_2, that is, whose "square" is integrable, is one for which the expectation of $\|\underline{X}\|_{l_2}^2$ exists. For random elements whose values are almost surely in l_2 (rather than all in l_2), the same properties obtain *mutatis mutandis* [264, p. 31 and Sect. 10.2.3].

10.5.1 Sequence Valued Martingales and Associated Exponentials

Cramér-Hida processes (Brownian motions) are, in particular, martingales with values in l_2. A few features of such processes are now explained.

Definition 10.5.2 Below, $\mathcal{B}_1(l_2)$ denotes the Banach space of operators of l_2 that have finite trace [the norm is the trace]. A partial order is given by the request that, for C_1 and C_2 in $\mathcal{B}_1(l_2)$, $C_1 \leq C_2$ if, and only if, for $\underline{\alpha} \in l_2$, fixed, but arbitrary, $\langle C_1[\underline{\alpha}], \underline{\alpha}\rangle_{l_2} \leq \langle C_2[\underline{\alpha}], \underline{\alpha}\rangle_{l_2}$.

One should perhaps remember [264, p. 71] that each adapted, and almost surely continuous process, has an adapted version, whose paths that are not continuous are continuous to the right.

Definition 10.5.3 Let $C : \Omega \times [0, 1] \longrightarrow \mathcal{B}_1(l_2)$ be an adapted process with (operator) paths continuous to the right, and, almost surely, with respect to P, the zero operator at the origin, and paths continuous, and monotone increasing. It is the quadratic variation of the martingale $\underline{X}$, with values in l_2, if, and only if, for $\{\underline{\alpha}_1, \underline{\alpha}_2\} \subseteq l_2$, fixed, but arbitrary, the following map:

$$(\omega, t) \mapsto \langle \underline{\alpha}_1, \underline{X}(\omega, t)\rangle_{l_2} \langle \underline{\alpha}_2, \underline{X}(\omega, t)\rangle_{l_2} - \langle C(\omega, t)[\underline{\alpha}_1], \underline{\alpha}_2\rangle_{l_2}$$

is a martingale. In operator notation (tensor product), one has that $\underline{X} \otimes \underline{X} - C$ is a martingale. One often writes, for C, $\langle\langle \underline{X} \rangle\rangle$. One shows that C exists and is unique [65, 81].

One shall mostly need deterministic quadratic variations, and use to that end the following terminology.

Definition 10.5.4 Let $X : \Omega \times [0, 1] \longrightarrow \mathbb{R}$ be a map, and $b : [0, 1] \longrightarrow \mathbb{R}_+$, a continuous, monotone increasing function that is zero at the origin. X is an

$(\underline{\mathcal{A}}, b)$-martingale when

1. the paths of X are continuous to the right, and almost surely continuous, with respect to P;
2. X is a local martingale for $\underline{\mathcal{A}}$;
3. $\langle X \rangle = b$.

X is a b-martingale when, in item 2, just above, $\underline{\mathcal{A}}$ is $\underline{\sigma}(X)$.

Definition 10.5.5 Let $t \mapsto C(t)$ be a continuous, monotone increasing map into $\mathcal{B}_1(l_2)$, with $C(0)$, the zero operator, and $C(t)$, positive and self-adjoint. Let also $\underline{X} : \Omega \times [0, 1] \longrightarrow l_2$ be a map. $\underline{X}$ is an $(\underline{\mathcal{A}}, C)$-martingale when

1. the paths of $\underline{X}$ are continuous to the right, and almost surely continuous, with respect to P;
2. $\underline{X}$ is a local martingale for $\underline{\mathcal{A}}$;
3. $\langle\langle \underline{X} \rangle\rangle = C$.

$\underline{X}$ is a C-martingale when, in item 2, just above, $\underline{\mathcal{A}}$ is $\underline{\sigma}(\underline{X})$.

Let $\underline{X} : \Omega \times [0, 1] \longrightarrow l_2$ be a map, and $\underline{\alpha} \in l_2$ be a fixed, but arbitrary element. X_α shall denote the process $(\omega, t) \mapsto \langle \underline{\alpha}, \underline{X}(\omega, t) \rangle_{l_2}$. When S is a strict, or wide sense, stopping time, and the expressions used make sense,

$$X_\alpha^S = \langle \underline{\alpha}, \underline{X}^S \rangle_{l_2}.$$

That such a formula is reasonable shall be seen below.

The exponential (formula (Proposition)) 10.4.6 is also useful, *mutatis mutandis,* for processes with values in l_2. It shall only be needed for deterministic quadratic variation, though its scope is wider [206].

Proposition 10.5.6 *Let $\underline{X} : \Omega \times [0, 1] \longrightarrow l_2$ be a process with the following properties: it is adapted to $\underline{\mathcal{A}}$, has paths continuous to the right, and, almost surely, with respect to P, starts at $\underline{0}_{l_2}$, and has continuous paths. Let also $C : \Omega \times [0, 1] \longrightarrow \mathcal{B}_1(l_2)$ be a process with the following properties: it is adapted to $\underline{\mathcal{A}}$, has paths continuous to the right, and, almost surely, with respect to P, $C(\cdot, 0)$ is the zero operator, and C has continuous, monotone increasing paths. The following statements are then equivalent:*

1. *$\underline{X}$ is a local martingale such that $\langle\langle \underline{X} \rangle\rangle = C$;*
2. *for $\underline{\alpha} \in l_2$, fixed, but arbitrary, letting*

$$X_\alpha = \langle \underline{\alpha}, \underline{X} \rangle_{l_2}, \quad \textit{and} \quad C_\alpha = \langle C(\cdot)[\underline{\alpha}], \underline{\alpha} \rangle_{l_2},$$

the following process:

$$E_\alpha = e^{X_\alpha - \frac{1}{2} C_\alpha}$$

is local martingale.

Proof Suppose that item 1 obtains. Let $\{S_n, n \in \mathbb{N}\}$ be a sequence of wide sense stopping times that reduces $\underline{X}$ to a sequence of uniformly integrable martingales. Then

$$E_P\left[\left|X_\alpha^{S_n}(\cdot, t)\right|\right] \leq \|\underline{\alpha}\|_{l_2} E_P\left[\left\|\underline{X}^{S_n}(\cdot, t)\right\|_{l_2}\right] < \infty,$$

and $E_P[\underline{X}(\cdot, S_n) \mid \mathcal{A}_t] = \underline{X}^{S_n}(\cdot, t)$, so that

$$\begin{aligned} X_\alpha^{S_n}(\cdot, t) &= \langle \underline{\alpha}, \underline{X}^{S_n}(\cdot, t)\rangle_{l_2} \\ &= \langle \underline{\alpha}, E_P\left[\underline{X}(\cdot, S_n) \mid \mathcal{A}_t\right]\rangle_{l_2} \\ &= E_P\left[\langle \underline{\alpha}, \underline{X}(\cdot, S_n)\rangle_{l_2} \mid \mathcal{A}_t\right] \\ &= E_P\left[X_\alpha(\cdot, S_n) \mid \mathcal{A}_t\right]. \end{aligned}$$

Thus the properties of $\underline{X}$ carry over to X_α. Furthermore, the quadratic variation is such that, for fixed, but arbitrary $\underline{\alpha}_1$ and $\underline{\alpha}_2$ in l_2,

$$\langle \underline{\alpha}_1, \underline{X}(\cdot, t)\rangle_2 \langle \underline{\alpha}_2, \underline{X}(\cdot, t)\rangle_{l_2} - \langle \langle\!\langle \underline{X} \rangle\!\rangle [\underline{\alpha}_1], \underline{\alpha}_2\rangle_{l_2}$$

is a local martingale. But then the assumption on $\langle\!\langle \underline{X} \rangle\!\rangle$ implies that

$$X_\alpha^2 - C_\alpha$$

is a local martingale. Consequently [(Proposition) 10.4.6], item 2 obtains.

Suppose conversely that item 2 obtains. Then [(Proposition) 10.4.6], X_α, is a continuous local martingale with C_α as quadratic variation. Let $\{S_n, n \in \mathbb{N}\}$ be the sequence that stops $\underline{X}$ when its norm crosses n, and $\{S_n^\alpha, n \in \mathbb{N}\}$ be a sequence that reduces X_α in L_2 [264, p. 63]. Then [264, p. 57], for $p \in \mathbb{N}$, fixed, but arbitrary,

$$X_\alpha^{S_n^\alpha \wedge S_p}$$

is a martingale. Since the process $\underline{X}^{S_n^\alpha \wedge S_p}$ is adapted [264, p. 41], and its expectation exists, from

$$E_P\left[X_\alpha^{S_n^\alpha \wedge S_p}(\cdot, t_2) \mid \mathcal{A}_{t_1}\right] = X_\alpha^{S_n^\alpha \wedge S_p}(\cdot, t_1),$$

one obtains, writing that equality explicitly, that

$$\begin{aligned} \langle \underline{\alpha}, \underline{X}^{S_p \wedge S_n^\alpha}(\cdot, t_1)\rangle_{l_2} &= E_P\left[\langle \underline{\alpha}, \underline{X}^{S_p \wedge S_n^\alpha}(\cdot, t_2)\rangle_{l_2} \mid \mathcal{A}_{t_1}\right] \\ &= \langle \underline{\alpha}, E_P\left[\underline{X}^{S_p \wedge S_n^\alpha}(\cdot, t_2) \mid \mathcal{A}_{t_1}\right]\rangle_{l_2}. \end{aligned}$$

Because of the continuity properties of paths, and the boundedness introduced by S_p, one may let n increase indefinitely, to obtain that

$$\langle \underline{\alpha}, \underline{X}^{S_p}(\cdot, t_1)\rangle_{l_2} = \langle \underline{\alpha}, E_P\left[\underline{X}^{S_p}(\cdot, t_2) \mid \mathcal{A}_{t_1}\right]\rangle_{l_2}.$$

But $\underline{\alpha}$ is arbitrary, and thus

$$\underline{X}^{S_p}(\cdot, t_1) = E_P\left[\underline{X}^{S_p}(\cdot, t_2) \mid \mathcal{A}_{t_1}\right].$$

Since

$$X_{\alpha_1} X_{\alpha_2} - \frac{1}{4}\{\langle X_{\alpha_1} + X_{\alpha_2}\rangle - \langle X_{\alpha_1} - X_{\alpha_2}\rangle\}$$

is a local martingale, and that, for example,

$$\langle X_{\alpha_1} + X_{\alpha_2}\rangle = \langle X_{\alpha_1+\alpha_2}\rangle = \langle C(\cdot)[\underline{\alpha}_1 + \underline{\alpha}_2], \underline{\alpha}_1 + \underline{\alpha}_2\rangle_{l_2},$$

the tensor quadratic variation of $\underline{X}$ is indeed provided by C. □

10.5.2 Sequence Valued Processes with Independent Increments

Cramér-Hida processes (Brownian motions) are, in particular, processes with values in l_2 and independent increments. A few features of such processes are now explained.

Definition 10.5.7 Let $(\Omega, \underline{\mathcal{A}}, P)$ be a probability space with a filtration, and $\underline{X} : \Omega \times [0, 1] \longrightarrow l_2$ be a process adapted to $\underline{\mathcal{A}}$. It has independent increments when, for fixed, but arbitrary $t_1 < t_2$ in $[0, 1]$, the increment $\underline{X}(\cdot, t_2) - \underline{X}(\cdot, t_1)$ is independent of $\mathcal{A}_{t_1}$.

Fact 10.5.8 *Let $\underline{X}$ have independent increments. Let $n \in \mathbb{N}$, and times $t_0 < t_1 < \cdots < t_n$ in $[0, 1]$ be fixed, but arbitrary. Then the following family of increments forms an independent family:*

$$\underline{X}(\cdot, t_1) - \underline{X}(\cdot, t_0), \ldots, \underline{X}(\cdot, t_n) - \underline{X}(\cdot, t_{n-1}).$$

Proof That is seen by successive conditioning. □

Fact 10.5.9 *Let $\underline{X}$ have independent increments. $X_\alpha = \langle \underline{\alpha}, \underline{X}\rangle_{l_2}$ has independent increments.*

When $\underline{\alpha}$ is one of the elements of an orthonormal basis, one shall write ξ_α for X_α, and, when α has an index n, ξ_n (so that, for example, $X_{\underline{e}_n} = \xi_n$).

Fact 10.5.10 *Let $\underline{X}$ have independent increments. Since the Borel sets of l_2 are generated by the continuous, linear functionals, and that those are generated by the continuous, linear functionals corresponding to any orthonormal basis,*

$$\sigma(\underline{X}(\cdot, t_2) - \underline{X}(\cdot, t_1)) = \sigma(\{\xi_n(\cdot, t_2) - \xi_n(\cdot, t_1), n \in \mathbb{N}\}).$$

Thus, given that the family $\{\xi_n, n \in \mathbb{N}\}$ is made of processes with independent increments, $\underline{X}$ will have independent increments.

Fact 10.5.11 *Let $\underline{X}$ have independent increments. When* (Fact) 10.5.8 *obtains, $\underline{X}$ has independent increments for the filtration it generates. The requirement for independent increments is then equivalent to the following (set of) condition(s):* [260, p. 259] *for*

$$n \in \mathbb{N},\ t_0 < t_1 < t_2 < t_3 < \cdots < t_n \text{ in } [0, 1],\ \{\underline{\theta}_1, \underline{\theta}_2, \underline{\theta}_3, \ldots \underline{\theta}_n\} \text{ in } l_2,$$

fixed, but arbitrary,

$$E_P\left[e^{i\sum_{k=1}^{n}\langle\underline{\theta}_k, \underline{X}(\cdot, t_k) - \underline{X}(\cdot, t_{k-1})\rangle_{l_2}}\right] = \prod_{k=1}^{n} E_P\left[e^{i\langle\underline{\theta}_k, \underline{X}(\cdot, t_k) - \underline{X}(\cdot, t_{k-1})\rangle_{l_2}}\right].$$

The following example shows the need for different $\underline{\theta}$-parameters in (Fact) 10.5.11.

Example 10.5.12 Let X and Y be independent, standard normal random variables and $\underline{e}_1, \underline{e}_2$ be the first two elements of the standard basis of l_2. Let

$$\underline{U} = X\underline{e}_1 + Y\underline{e}_2, \text{ and } \underline{V} = X\underline{e}_2 - Y\underline{e}_1.$$

When $\underline{a}$, $\underline{b}$, and $\underline{\theta}$, are arbitrary elements of l_2,

$$\begin{aligned}
\langle\underline{\theta}, \underline{U}\rangle_{l_2} &= \theta_1 X + \theta_2 Y \sim \mathcal{N}(0, \theta_1^2 + \theta_2^2), \\
\langle\underline{\theta}, \underline{V}\rangle_{l_2} &= \theta_2 X - \theta_1 Y \sim \mathcal{N}(0, \theta_1^2 + \theta_2^2), \\
\langle\underline{\theta}, \underline{U}\rangle_{l_2} + \langle\underline{\theta}, \underline{V}\rangle_{l_2} &= (\theta_1 + \theta_2)X + (\theta_2 - \theta_1)Y \sim \mathcal{N}(0, 2(\theta_1^2 + \theta_2^2)), \\
\langle\underline{a}, \underline{U}\rangle_{l_2} + \langle\underline{b}, \underline{V}\rangle_{l_2} &= (a_1 + b_2)X + (a_2 - b_1)Y \sim \mathcal{N}(0, (a_1 + b_2)^2 + (a_2 - b_1)^2).
\end{aligned}$$

Thus $\langle\underline{\theta}, \underline{U}\rangle_{l_2}$ and $\langle\underline{\theta}, \underline{V}\rangle_{l_2}$ are independent, but $\langle\underline{a}, \underline{U}\rangle_{l_2}$ and $\langle\underline{b}, \underline{V}\rangle_{l_2}$ are not.

Fact 10.5.13 *Let $\underline{X}$ have independent increments. Let ${}^{o}\!\mathcal{A}_t$ be the σ-algebra generated by $\mathcal{A}_t$ and the sets of $\mathcal{A}$ which have zero measure for P. Then, when $\underline{X}$ has independent increments with respect to $\underline{\mathcal{A}}$, it has also independent increments with respect to ${}^{o}\underline{\mathcal{A}}$.*

Proof Indeed, generally, a random element X is independent from the σ-algebra $\mathcal{X}$, if, and only if, for fixed, but arbitrary bounded and measurable f, almost surely, $E[f(X) \mid \mathcal{X}] = E[f(X)]$. But, using (Proposition) 10.2.24, one has, for fixed, but arbitrary $t_1 < t_2$ in $[0, 1]$, and f, bounded and measurable, that, almost surely:

$$\begin{aligned} E_P [f(\underline{X}(\cdot, t_2) - \underline{X}(\cdot, t_1)) \mid {}^o\!\mathcal{A}_t] &= E_P [f(\underline{X}(\cdot, t_2) - \underline{X}(\cdot, t_1)) \mid \mathcal{A}_t] \\ &= E_P [f(\underline{X}(\cdot, t_2) - \underline{X}(\cdot, t_1))] . \end{aligned}$$

The following standard formula, valid for σ-algebras $\mathcal{A}_1$ and $\mathcal{A}_2$, $\mathcal{A}_2 \subseteq \mathcal{A}_1$,

$$E[E[X \mid \mathcal{A}_2] \mid \mathcal{A}_1] = E[X \mid \mathcal{A}_2]$$

says that the converse is also true. □

Fact 10.5.14 *Let $\underline{X}$ have independent increments, and paths continuous to the right, and almost surely continuous. Let $N \in \mathcal{A}$ be a set such that $P(N) = 0$, and, for $\omega \in N^c$, $t \mapsto \underline{X}(\omega, t)$ is continuous. Notation shall be as in Sect.* 10.2.3. *Then $\underline{X}_N$ has independent increments with respect to $\underline{\mathcal{A}}_N$.*

Proof Indeed, for $t_1 < t_2$ in $[0, 1]$, $A_N \in \mathcal{A}^N_{t_1}$, and f bounded, and adapted to the Borel sets of l_2, all fixed, but arbitrary,

$$\begin{aligned} &\int_{A_N} E_{P_N} \left[f\left(\underline{X}_N(\cdot, t_2) - \underline{X}_N(\cdot, t_1)\right) \mid \mathcal{A}^N_{t_1}\right] dP_N = \\ &= \int_{A_N} f\left(\underline{X}_N(\cdot, t_2) - \underline{X}_N(\cdot, t_1)\right) dP_N \\ &= \int_A f\left(\underline{X}(\cdot, t_2) - \underline{X}(\cdot, t_1)\right) dP \\ &= \int_A E_P \left[f\left(\underline{X}(\cdot, t_2) - \underline{X}(\cdot, t_1)\right) \mid \mathcal{A}_{t_1}\right] dP \\ &= \int_A E_P \left[f\left(\underline{X}(\cdot, t_2) - \underline{X}(\cdot, t_1)\right)\right] dP \\ &= \int_{A_N} E_{P_N} \left[f\left(\underline{X}_N(\cdot, t_2) - \underline{X}_N(\cdot, t_1)\right)\right] dP_N. \end{aligned}$$

□

Corollary 10.5.15 *The process $\underline{X}_N$ of the latter fact is then Gaussian, as it has continuous paths, and independent increments* [114, 247].

Corollary 10.5.16 *An analogous calculation (assuming zero mean), using the fact that $\underline{X}_N$ is Gaussian (with, say, C as covariance), yields that $\underline{X}$ is Gaussian, and that*

$$E_P\left[e^{i\langle \underline{\theta}, \underline{X}(\cdot,t)\rangle_{l_2}}\right] = e^{-\frac{1}{2}\langle C(t)[\underline{\theta}], \underline{\theta}\rangle_{l_2}},$$

where $t \in [0,1]$, and $\underline{\theta} \in l_2$, are fixed, but arbitrary, and $C(t)$ is a bounded, linear operator of l_2 that is positive, self-adjoint, and has finite trace. $C(t)$ has a representation [259, p. 32] *as an infinite matrix of elements $c_{n,p}(t)$ such that the sum of positive elements, $\sum_{n=1}^{\infty} c_{n,n}(t)$, is finite. Thus, in particular, $\underline{X}$ has finite second moments.*

Fact 10.5.17 *Suppose that $\underline{X}$ has independent increments with respect to $\underline{\mathcal{A}}$, has integrable norm, and zero mean. Then it is a martingale. Thus, when $\underline{X}$ has independent increments, and paths continuous to the right, and almost surely continuous, with zero mean, it is a Gaussian martingale.*

Proposition 10.5.18 *Let $\underline{X}$ be a $(\underline{\mathcal{A}}, C)$ martingale, and $\underline{\alpha} \in l_2$ be fixed, but arbitrary. Then, with $X_\alpha = \langle \underline{\alpha}, \underline{X}\rangle_{l_2}$, and $C_\alpha(t) = \langle C(t)[\underline{\alpha}], \underline{\alpha}\rangle_{l_2}$,*

1. *$E_\alpha(\omega,t) = e^{X_\alpha(\omega,t) - \frac{1}{2}C_\alpha(t)}$ is a martingale whose square is integrable;*
2. *for $t_1 < t_2$ in $[0,1]$, fixed, but arbitrary, $\underline{X}(\cdot,t_2) - \underline{X}(\cdot,t_1)$ is independent of $\mathcal{A}_{t_1}$;*
3. *$\underline{X}$ is a Gaussian random process whose covariance operator stems from $(t_1,t_2) \mapsto C(t_1 \wedge t_2)$;*
4. *for $t_1 < t_2$ in $[0,1]$, fixed, but arbitrary, $\underline{X}(\cdot,t_2) - \underline{X}(\cdot,t_1)$ has covariance operator $C(t_2) - C(t_1)$.*

Proof As [(Fact) 10.4.2] X_α is a $(\underline{\mathcal{A}}, C_\alpha)$ martingale, then [(Proposition) 10.4.6] E_α is a local martingale, and

$$\langle E_\alpha\rangle(\omega,t) = \int_0^t E_\alpha^2(\omega,\theta)\langle X_\alpha\rangle(\omega,d\theta) = \int_0^t e^{2X_\alpha(\omega,\theta) - C_\alpha(\theta)} C_\alpha(d\theta).$$

But the exponential of $2X_\alpha - \frac{1}{2}4C_\alpha$ is a positive, local martingale whose value at zero is one. Thus

$$E_P\left[E_\alpha^2(\cdot,t)\right] = e^{C_\alpha(t)} E_P\left[e^{2X_\alpha(\cdot,t) - \frac{1}{2}4C_\alpha(t)}\right] \le e^{C_\alpha(t)},$$

and

$$E_P\left[\langle E_\alpha\rangle(\cdot,t)\right] = \int_0^t E_P\left[E_\alpha^2(\cdot,\theta)\right] C_\alpha(d\theta) \le \int_0^t e^{C_\alpha(\theta)} C_\alpha(d\theta). \quad (\star)$$

Since

$$C_\alpha(t) \le C_\alpha(1) = \langle C(1)[\underline{\alpha}], \underline{\alpha}\rangle_{l_2} \le \|C(1)\| \, \|\underline{\alpha}\|_{l_2}^2,$$

the second integral on the right-hand side of $E_P\left[\langle E_\alpha\rangle(\cdot,t)\right]$, that is ($\star$), is finite, and E_α is a martingale whose square is integrable [(Fact) 10.4.2].

One has, from item 1, that $E_P\left[E_\alpha(\cdot,t)\right] = 1$, so that

$$E_P\left[e^{X_\alpha(\cdot,t)}\right] = e^{\frac{1}{2}C_\alpha(t)},$$

and, consequently, that $X_\alpha(\cdot,t)$ is a Gaussian random variable with mean zero and variance $C_\alpha(t)$. The martingale property of E_α yields that, for $t_1 < t_2$ in $[0,1]$, fixed, but arbitrary,

$$E_P\left[e^{X_\alpha(\cdot,t_2)-\frac{1}{2}C_\alpha(t_2)} \mid \mathcal{A}_{t_1}\right] = e^{X_\alpha(\cdot,t_1)-\frac{1}{2}C_\alpha(t_1)},$$

so that

$$E_P\left[e^{\langle \underline{\alpha}, \underline{X}(\cdot,t_2)-\underline{X}(\cdot,t_1)\rangle_{l_2}} \mid \mathcal{A}_{t_1}\right] = e^{\frac{1}{2}\langle\{C(t_2)-C(t_1)\}[\underline{\alpha}],\underline{\alpha}\rangle_{l_2}}.$$

That proves items 2, 3, and 4. □

Corollary 10.5.19 *Let $\underline{X} : \Omega \times [0,1] \longrightarrow l_2$ be an adapted (to $\underline{\mathcal{A}}$) process that is zero at the origin, continuous to the right, and almost surely continuous. Then:*

1. *$\underline{X}$ has independent increments if, and only if, it is Gaussian with covariance $(t_1,t_2) \mapsto C(t_1 \wedge t_2)$, where $t \mapsto C(t)$ is a map into the family of operators of l_2 that have finite trace;*
2. *$\underline{X}$ is an $(\underline{\mathcal{A}}, C)$-martingale if, and only if, it is Gaussian with covariance $(t_1,t_2) \mapsto C(t_1 \wedge t_2)$, where $t \mapsto C(t)$ is a map into the family of operators of l_2 that have finite trace.*

Chapter 11
Calculus for Cramér-Hida Processes

One shall find below the stochastic calculus results required to state, and prove, a Girsanov's formula tailored to the Cramér-Hida representation. One attaches meaning to expressions on the following form:

$$I_{\underline{B}}\{\underline{a}\}(\omega, t),$$

which shall be interpreted as (stochastic) integrals of $\underline{a}$ with respect to $\underline{B}$. The process $\underline{a}$ (the "derivative" of the signal) shall have paths in a Hilbert space isomorphic to the RKHS of the process N that gives rise to $\underline{B}$, its associated Cramér-Hida process, that is, a vector whose countable components enjoy the properties of the Cramér-Hida representation. They are, in particular, independent, mean zero, almost surely continuous, Gaussian martingales, and the measures attached to their quadratic variations have absolute continuity properties. $I_{\underline{B}}\{\underline{a}\}$ shall in fact be the countable sum of usual stochastic integrals. However, since the components of integrators and integrands are to a certain extent "organically" tied, and that one may interpret $\underline{a}$ as a (random) functional on l_2, linear and continuous, acting on $\underline{B}$, one may have there a justification for notation and vocabulary as well.

11.1 Integrators: Cramér-Hida Processes

Cramér-Hida processes (Cramér-Hida Brownian motions) are Gaussian martingales with values in l_2, with added properties: their covariance operators are diagonal, so that their components are independent, and the measures corresponding to the quadratic variation of each component are ordered by absolute continuity. $(\Omega, \underline{A}, P)$ shall be a fixed, but arbitrary probability space with filtration.

A.F. Gualtierotti, *Detection of Random Signals in Dependent Gaussian Noise*,
DOI 10.1007/978-3-319-22315-5_11

Definition 11.1.1 A Cramér-Hida process is a map

$$\underline{B} : \Omega \times [0,1] \longrightarrow \mathbb{R}^{\infty},$$

with components $\{B_n, n \in \mathbb{N}\}$, and the following properties:

1. for $\{n, p\} \subseteq \mathbb{N}$, and $t \leq t_1 < t_2$ in $[0, 1]$, fixed, but arbitrary,
   (i) $B_n(\cdot, t)$ is adapted to $\underline{\mathcal{A}}$;
   (ii) $E_P[B_n(\cdot, t)] = 0$;
   (iii) $\mathcal{A}_t$ and $B_n(\cdot, t_2) - B_n(\cdot, t_1)$ are independent;
   (iv) $E_P[B_n^2(\cdot, t)] = b_n(t)$, a continuous, monotone increasing function such that $b_n(1) = 1/2^n$;
   (v) the processes $\{B_n, n \in \mathbb{N}\}$ are independent; they have paths continuous to the right, and almost all of their paths are continuous;
2. for $n \in \mathbb{N}$, and $t_1 < t_2$ in $[0, 1]$, fixed, but arbitrary, let M_n be the measure on $\mathcal{B}([0, 1])$ determined by b_n, that is, $M_n([t_1, t_2]) = b_n(t_2) - b_n(t_1)$: then M_{n+1} is absolutely continuous with respect to M_n [denoted $M_{n+1} \ll M_n$].

Remark 11.1.2 Feature (iv) of item 1 in (Definition) 11.1.1 is no restriction as seen in (Lemma) 6.4.34.

Fact 11.1.3 *Let* $b(t) = \sum_{n=1}^{\infty} b_n(t)$. *Then* $b(1) = 1$, *and*

$$E_P\left[\|\underline{B}(\cdot, t)\|_{l_2}^2\right] = b(t).$$

$\underline{B}(\cdot, t)$ *has thus almost all its values in* l_2, *and is a martingale in* L_2, *since all its components are martingales in* L_2 [(Fact) 10.5.1].

Fact 11.1.4 *Let* $\underline{B}_n$ *be the vector whose components are the first n components of* $\underline{B}$, *as well as the projection of* $\underline{B}$ *onto the subspace of* l_2 *generated by* $\{\underline{e}_1, \ldots, \underline{e}_n\}$ *(then the components following the n-th one are all zero). The path properties of* $\underline{B}$ *are those of* $\underline{B}_n$.

Proof With $\epsilon > 0$, fixed, but arbitrary, one has that

$$P\left(\omega \in \Omega : \sup_{t\in[0,1]} \|\underline{B}(\omega, t) - \underline{B}_n(\omega, t)\|_{l_2}^2 > \epsilon\right) =$$

$$= P\left(\omega \in \Omega : \sup_{t\in[0,1]} \sum_{i=n+1}^{\infty} B_i^2(\omega, t) > \epsilon\right).$$

Let

$$A_{n,p} = \left\{\omega \in \Omega : \sup_{t\in[0,1]} \sum_{i=n+1}^{n+p} B_i^2(\omega, t) > \epsilon\right\}.$$

Then, for $p \in \mathbb{N}$, fixed, but arbitrary, $A_{n,p} \subseteq A_{n,p+1}$. Let

$$A_n = \bigcup_{p \in \mathbb{N}} A_{n,p}.$$

Then

$$A_n = \left\{ \omega \in \Omega : \sup_{t \in [0,1]} \sum_{i=n+1}^{\infty} B_i^2(\omega, t) > \epsilon \right\},$$

and $P(A_n) = \lim_p P(A_{n,p})$. By Doob's maximal inequality [221, p. 54],

$$P(A_{n,p}) = P\left(\omega \in \Omega : \sup_{t \in [0,1]} \sum_{i=n+1}^{n+p} B_i^2(\omega, t) \geq \epsilon \right) \leq \frac{1}{\epsilon} E_P \left[\sum_{i=n+1}^{n+p} B_i^2(\cdot, 1) \right].$$

But

$$E_P \left[\sum_{i=n+1}^{n+p} B_i^2(\cdot, 1) \right] = \sum_{i=n+1}^{n+p} b_i(1) = \sum_{i=n+1}^{n+p} \frac{1}{2^i} = \frac{1}{2^n} \left(1 - \frac{1}{2^p} \right),$$

so that

$$P\left(\omega \in \Omega : \sup_{t \in [0,1]} \| \underline{B}(\omega, t) - \underline{B}_n(\omega, t) \|_{l_2}^2 > \epsilon \right) \leq \frac{1}{2^n \epsilon}.$$

□

Fact 11.1.5 *Let, for $t \in [0, 1]$, fixed, but arbitrary, $C(t) : l_2 \longrightarrow l_2$ be defined using the following relation:*

$$C(t) \left[\underline{e}_n \right] = b_n(t) \underline{e}_n.$$

Since $\left\langle C(t)[\underline{e}_n], \underline{e}_n \right\rangle_{l_2} = b_n(t)$, $C(t)$ has finite trace, equal to $b(t)$. Furthermore, since

$$E_P \left[\langle \underline{\alpha}_1, \underline{B}(\cdot, t_1) \rangle_{l_2} \langle \underline{\alpha}_2, \underline{B}(\cdot, t_2) \rangle_{l_2} \right] = \langle C(t_1 \wedge t_2) \underline{\alpha}_1, \underline{\alpha}_2 \rangle_{l_2},$$

$C(t)$ is the covariance operator of $\underline{B}(\cdot, t)$. Since b is continuous, $t \mapsto C(t)$ is continuous for the trace norm.

Fact 11.1.6 *Let $B(\omega, t) = \| \underline{B}(\omega, t) \|_{l_2}^2 = \sum_{n=1}^{\infty} B_n^2(\omega, t)$. B is a uniformly integrable submartingale, and, in the Doob-Meyer decomposition $B = M + V$, M a martingale, and V a monotone increasing process* [264, p. 145], *V is b.*

Proof Let indeed $t_1 < t_2$ in $[0, 1]$, and $A \in \mathcal{A}_{t_1}$ be fixed, but arbitrary. Then

$$\begin{aligned}\int_A B(\omega, t_2)P(d\omega) &= \sum_{n=1}^{\infty} \int_A \{[B_n(\omega, t_2) - B_n(\omega, t_1)] + B_n(\omega, t_1)\}^2 P(d\omega) \\ &\geq \sum_{n=1}^{\infty} \int_A B_n^2(\omega, t_1)P(d\omega) \\ &= \int_A B(\omega, t_1)P(d\omega).\end{aligned}$$

Furthermore, since $B_n^2(\cdot, t) - b_n(t) = B_n^2(\cdot, t) - \langle B_n \rangle (\cdot, t)$ is a martingale,

$$\begin{aligned}\int_A \{B(\omega, t_2) - b(t_2)\} P(d\omega) &= \sum_{n=1}^{\infty} \int_A \{B_n^2(\omega, t_2) - b_n(t_2)\} P(d\omega) \\ &= \sum_{n=1}^{\infty} \int_A \{B_n^2(\omega, t_1) - b_n(t_1)\} P(d\omega) \\ &= \int_A \{B(\omega, t_1) - b(t_1)\} P(d\omega).\end{aligned}$$

b is thus $\langle \underline{B} \rangle$, the first increasing process of $\underline{B}$ [190, p. 115]. □

Fact 11.1.7 *The quadratic variation process of* $\underline{B}$ *is* $\langle\langle \underline{B} \rangle\rangle (\omega, t) = C(t)$*, as, for* $\{\underline{\alpha}_1, \underline{\alpha}_2\} \subseteq l_2$*, fixed, but arbitrary,*

$$\langle \underline{B}(\cdot, t), \underline{\alpha}_1 \rangle_{l_2} \langle \underline{B}(\cdot, t), \underline{\alpha}_2 \rangle_{l_2} - \langle C(t)[\underline{\alpha}_1], \underline{\alpha}_2 \rangle_{l_2}$$

is a martingale [65, p. 81].

11.1.1 The σ-Algebras Generated by a Cramér-Hida Process

One shall use the following notation. For $t \in [0, 1]$, fixed, but arbitrary, $\mathcal{B}_t$ stands for $\sigma_t(\underline{B})$. For $t \in [0, 1]$, and $\underline{\theta} \in l_2$, fixed, but arbitrary, let

1. $\phi(\omega, \underline{\theta}, t) = e^{\iota \langle \underline{\theta}, \underline{B}(\omega, t) \rangle_{l_2}}$,
2. $\Phi(\underline{\theta}, t) = E_P[\phi(\cdot, \underline{\theta}, t)] = E_P\left[e^{\iota \langle \underline{\theta}, \underline{B}(\cdot, t) \rangle_{l_2}}\right] = e^{-\frac{1}{2}\langle C(t)[\underline{\theta}], \underline{\theta} \rangle_{l_2}} > 0$,
3. $Z(\omega, \underline{\theta}, t) = \frac{\phi(\omega, \underline{\theta}, t)}{\Phi(\underline{\theta}, t)}$.

Lemma 11.1.8 *Let* $t_1 < t_2$ *in* $[0,1]$ *be fixed, but arbitrary. Then* $Z(\cdot, \underline{\theta}, t_1)$ *is a version of* $E_P\left[Z(\cdot, \underline{\theta}, t_2) \mid \mathcal{B}_{t_1}\right]$.

Proof Indeed

$$\begin{aligned} E_P\left[e^{\iota\langle\underline{\theta}, \underline{B}(\cdot,t_2)\rangle_{l_2}} \mid \mathcal{B}_{t_1}\right] &= e^{\iota\langle\underline{\theta}, \underline{B}(\cdot,t_1)\rangle_{l_2}} E_P\left[e^{\iota\langle\underline{\theta}, \underline{B}(\cdot,t_2)-\underline{B}(\cdot,t_1)\rangle_{l_2}}\right] \\ &= e^{\iota\langle\underline{\theta}, \underline{B}(\cdot,t_1)\rangle_{l_2}} \frac{\Phi(\underline{\theta}, t_2)}{\Phi(\underline{\theta}, t_1)}. \end{aligned}$$

□

Lemma 11.1.9 *Let* $t_1 < t_2$ *in* $[0,1]$ *be fixed, but arbitrary. Then,*

$$E_P\left[e^{\iota\langle\underline{\theta}, \underline{B}(\cdot,t_2)\rangle_{l_2}} \mid \mathcal{B}_{t_1}\right] \text{ is a version of } E_P\left[e^{\iota\langle\underline{\theta}, \underline{B}(\cdot,t_2)\rangle_{l_2}} \mid \mathcal{B}_{t_1}^{+}\right].$$

Proof Let $t \in]t_1, t_2[$ be fixed, but arbitrary. From [(Lemma) 11.1.8]

$$E_P\left[Z(\cdot, \underline{\theta}, t_2) \mid \mathcal{B}_t\right] = Z(\cdot, \underline{\theta}, t),$$

one obtains that

$$E_P\left[e^{\iota\langle\underline{\theta}, \underline{B}(\cdot,t_2)\rangle_{l_2}} \mid \mathcal{B}_t\right] = \frac{\Phi(\underline{\theta}, t_2)}{\Phi(\underline{\theta}, t)} e^{\iota\langle\underline{\theta}, \underline{B}(\cdot,t)\rangle_{l_2}}. \qquad (\star)$$

Since $\underline{B}$ has paths that are continuous to the right, and Φ is continuous in t, letting $t \downarrow\downarrow t_1$, and using the appropriate martingale convergence theorem [201, p. 118], almost surely, with respect to P,

$$E_P\left[e^{\iota\langle\underline{\theta}, \underline{B}(\cdot,t_2)\rangle_{l_2}} \mid \mathcal{B}_{t_1}^{+}\right] = \frac{\Phi(\underline{\theta}, t_2)}{\Phi(\underline{\theta}, t_1)} e^{\iota\langle\underline{\theta}, \underline{B}(\cdot,t_1)\rangle_{l_2}}.$$

Using $(\star)$ above with t_1 replacing t, one gets the equality in the lemma's statement. □

Lemma 11.1.10 *Let* $t_1 < t_2 \leq t_3$ *in* $[0,1]$, *and* $\{\underline{\theta}_2, \underline{\theta}_3\} \subseteq l_2$ *be fixed, but arbitrary. Let also*

$$X(\omega) = \langle\underline{\theta}_2, \underline{B}(\omega, t_2)\rangle_{l_2} + \langle\underline{\theta}_3, \underline{B}(\omega, t_3)\rangle_{l_2}.$$

Then $E_P\left[e^{\iota X} \mid \mathcal{B}_{t_1}\right]$ *is a version of* $E_P\left[e^{\iota X} \mid \mathcal{B}_{t_1}^{+}\right]$.

Proof Let $t \in]t_1, t_2[$ be fixed, but arbitrary. Then, since $e^{\iota X}$ is the product $\phi(\cdot, \underline{\theta}_2, t_2)\phi(\cdot, \underline{\theta}_3, t_3)$, adding and subtracting $\langle \underline{\theta}_3, \underline{B}(\omega, t_2)\rangle_{l_2}$,

$$\begin{aligned} E_P\left[e^{\iota X} \mid \mathcal{B}_t\right] &= E_P\left[\phi(\cdot, \underline{\theta}_2, t_2)\phi(\cdot, \underline{\theta}_3, t_3) \mid \mathcal{B}_t\right] \\ &= E_P\left[E_P\left[\phi(\cdot, \underline{\theta}_2 + \underline{\theta}_3, t_2)\, e^{\iota\left\langle \underline{\theta}_3, \underline{B}(\cdot, t_3) - \underline{B}(\cdot, t_2)\right\rangle_{l_2}} \mid \mathcal{B}_{t_2}\right] \mid \mathcal{B}_t\right] \\ &= E_P\left[\phi(\cdot, \underline{\theta}_2 + \underline{\theta}_3, t_2)\, E_P\left[e^{\iota\left\langle \underline{\theta}_3, \underline{B}(\cdot, t_3) - \underline{B}(\cdot, t_2)\right\rangle_{l_2}}\right] \mid \mathcal{B}_t\right] \\ &= \frac{\Phi(\underline{\theta}_3, t_3)}{\Phi(\underline{\theta}_3, t_2)}\, E_P\left[\phi(\cdot, \underline{\theta}_2 + \underline{\theta}_3, t_2) \mid \mathcal{B}_t\right]. \end{aligned}$$

But

$$\phi(\cdot, \underline{\theta}_2 + \underline{\theta}_3, t_2) = \phi(\cdot, \underline{\theta}_2 + \underline{\theta}_3, t)\, e^{\iota\left\langle \underline{\theta}_2 + \underline{\theta}_3, \underline{B}(\cdot, t_2) - \underline{B}(\cdot, t)\right\rangle_{l_2}},$$

so that

$$\begin{aligned} E_P\left[\phi(\cdot, \underline{\theta}_2 + \underline{\theta}_3, t_2) \mid \mathcal{B}_t\right] &= \phi(\cdot, \underline{\theta}_2 + \underline{\theta}_3, t)\, \frac{\Phi(\underline{\theta}_2 + \underline{\theta}_3, t_2)}{\Phi(\underline{\theta}_2 + \underline{\theta}_3, t)} \\ &= \Phi(\underline{\theta}_2 + \underline{\theta}_3, t_2)\, Z(\cdot, \underline{\theta}_2 + \underline{\theta}_3, t). \end{aligned}$$

Consequently

$$E_P\left[e^{\iota X} \mid \mathcal{B}_t\right] = \frac{\Phi(\underline{\theta}_3, t_3)}{\Phi(\underline{\theta}_3, t_2)}\, \Phi(\underline{\theta}_2 + \underline{\theta}_3, t_2)\, Z(\cdot, \underline{\theta}_2 + \underline{\theta}_3, t).$$

The same calculation yields that

$$E_P\left[e^{\iota X} \mid \mathcal{B}_{t_1}\right] = \frac{\Phi(\underline{\theta}_3, t_3)}{\Phi(\underline{\theta}_3, t_2)}\, \Phi(\underline{\theta}_2 + \underline{\theta}_3, t_2)\, Z(\cdot, \underline{\theta}_2 + \underline{\theta}_3, t_1).$$

Since, when $t \downarrow\downarrow t_1$, as already seen, $E_P\left[e^{\iota X} \mid \mathcal{B}_t\right]$ goes to $E_P\left[e^{\iota X} \mid \mathcal{B}_{t_1}^+\right]$, and $Z(\cdot, \underline{\theta}_2 + \underline{\theta}_3, t)$, to $Z(\cdot, \underline{\theta}_2 + \underline{\theta}_3, t_1)$, the lemma proves true. □

Proposition 11.1.11 *One has, for* $t \in [0, 1]$, *fixed, but arbitrary, that*

$$\mathcal{B}_t^+ \subseteq {}^o\mathcal{B}_t,$$

so that the filtration $\{{}^o\mathcal{B}_t, t \in [0, 1]\}$ *is continuous to the right.*

Proof With a proof that is essentially that of (Lemma) 11.1.10, one may state that, for

$$n \in \mathbb{N},\ t, t_1 < t_2 < t_3 < \cdots < t_n \text{ in } [0,1], \text{ and } \{\underline{\theta}_1, \underline{\theta}_2, \underline{\theta}_3, \ldots, \underline{\theta}_n\} \subseteq l_2,$$

fixed, but arbitrary, $E_P\left[e^{\iota \sum_{j=1}^{n} \langle \underline{\theta}_j, \underline{B}(\cdot,t_j)\rangle_{l_2}} \mid \mathcal{B}_t\right]$ is a version of

$$E_P\left[e^{\iota \sum_{j=1}^{n} \langle \underline{\theta}_j, \underline{B}(\cdot,t_j)\rangle_{l_2}} \mid \mathcal{B}_t^+\right].$$

Let $\mathcal{F}$ be the set of bounded functions f for which

$$E_P\left[f(\langle \underline{\theta}_1, \underline{B}(\cdot,t_1)\rangle_{l_2}, \ldots, \langle \underline{\theta}_n, \underline{B}(\cdot,t_n)\rangle_{l_2}) \mid \mathcal{B}_t\right]$$

is a version of $E_P\left[f(\langle \underline{\theta}_1, \underline{B}(\cdot,t_1)\rangle_{l_2}, \ldots, \langle \underline{\theta}_n, \underline{B}(\cdot,t_n)\rangle_{l_2}) \mid \mathcal{B}_t^+\right]$. It is a λ-system [184, p. 547] which contains the π-system [184, p. 547] of trigonometric polynomials, and thus the bounded, real valued functions adapted to the σ-algebra generated by the trigonometric polynomials [184, p. 550]. Since trigonometric polynomials generate L_2 spaces over finite intervals [134, p. 123], indicator functions of Borel sets in $I\!R^n$ are included in the latter family.

The family of indicator functions of sets in

$$\bigcup_{n\in\mathbb{N}} \sigma\left(\langle \underline{\theta}_1, \underline{B}(\cdot,t_1)\rangle_{l_2}, \ldots, \langle \underline{\theta}_n, \underline{B}(\cdot,t_n)\rangle_{l_2}\right)$$

forms a π-system, as the product of two indicators is an indicator. From what has been seen above, it is contained in the family of indicator functions of measurable sets χ_C for which $E_P[\chi_C \mid \mathcal{B}_t]$ is a version of $E_P[\chi_C \mid \mathcal{B}_t^+]$. But that latter family forms a λ-system by its very definition, and, again because of the monotone class theorem, the property extends to bounded functions adapted to $\mathcal{B}_1$. One then applies Theorem 9.7.1 of [264, p. 238]. □

11.1.2 The Reproducing Kernel Hilbert Space of a Cramér-Hida Process

The aim here is to describe the RKHS of $\underline{B}$, and that of structurally similar processes. The latter reveal the increase of complexity arising when going from Gaussian martingales to Cramér-Hida processes.

What follows [51] allows one to define the RKHS of a process with values in l_2, whose covariance is a positive definite function with values in a space of operators of l_2. Let thus S be a set, and K be a real Hilbert space. A K-reproducing kernel Hilbert space (K-RKHS) is a Hilbert space of functions $H(S, K) \subseteq K^S$, with domain S, and range in K, for which, given $s \in S$, fixed, but arbitrary, there exists $\kappa(s)$ such that, for $h \in H$, fixed, but arbitrary,

$$\|\mathcal{E}_s[h]\|_K = \|h(s)\|_K \leq \kappa(s) \|h\|_{H(S,K)} \cdot$$

$\mathcal{E}_s : H(S, K) \longrightarrow K$, the evaluation map at $s \in S$, is thus an operator which is linear and bounded.

A K-covariance is a symmetric map $\mathcal{H} : S \times S \longrightarrow \mathcal{L}(K, K)$, the set of bounded, linear operators of K, such that, for $n \in \mathbb{N}$, $\{s_1, \ldots, s_n\} \subseteq S$, $\{\alpha_1, \ldots, \alpha_n\} \subseteq \mathbb{R}$, fixed, but arbitrary, for $k \in K$, fixed, but arbitrary,

$$\sum_{i=1}^{n} \sum_{j=1}^{n} \alpha_i \alpha_j \langle \mathcal{H}(s_j, s_i)[k], k \rangle_K = \left\langle \left\{ \sum_{i=1}^{n} \sum_{j=1}^{n} \alpha_i \alpha_j \mathcal{H}(s_j, s_i) \right\} [k], k \right\rangle_K \geq 0.$$

Given a K-RKHS $H(S, K)$, there is a K-covariance associated with it as follows. Let

$$\mathcal{H}(s_1, s_2) = \mathcal{E}_{s_1} \mathcal{E}_{s_2}^{\star} \in \mathcal{L}(K, K).$$

Then indeed

$$\sum_{i=1}^{n} \sum_{j=1}^{n} \alpha_i \alpha_j \langle \mathcal{H}(s_j, s_i)[k], k \rangle_K = \left\| \sum_{i=1}^{n} \alpha_i \mathcal{E}_{s_i}^{\star}[k] \right\|_H^2 .$$

Let $\Gamma : S \longrightarrow \mathcal{L}(K, H(S, K))$ be defined using the following relation:

$$\Gamma(s) = \mathcal{E}_s^{\star} .$$

Then $\mathcal{H}(s_1, s_2) = \Gamma(s_1)^{\star} \Gamma(s_2)$.

The following properties obtain.

1. $\mathcal{H}$ just defined reproduces the functions $h : S \longrightarrow K$ of $H(S, K)$ in the following sense. $\Gamma(s)$ sends $k \in K$ to a function of $H(S, K)$, so that one may compute

$$\Gamma(s)[k](x) = \mathcal{E}_x [\Gamma(s)[k]] = \mathcal{E}_x \left[\mathcal{E}_s^{\star}[k]\right] = \mathcal{H}(x, s)[k] = \mathcal{E}_x [\mathcal{H}(\cdot, s)[k]] .$$

Thus $\Gamma(s)[k]$ is the function $x \mapsto \mathcal{H}(x, s)[k]$. Furthermore, for h in $H(S, K)$, fixed, but arbitrary,

$$\langle h(s), k \rangle_K = \left\langle h, \mathcal{E}_s^{\star}[k] \right\rangle_{H(S,K)} = \langle h, \Gamma(s)[k] \rangle_{H(S,K)} = \langle h, \mathcal{H}(\cdot, s)[k] \rangle_{H(S,K)} ,$$

so that, given the orthonormal basis $\{e_i^K, i \in I\}$ for K,

$$h(s) = \sum_{i \in I} \langle h(s), e_i^K \rangle_K e_i^K = \sum_{i \in I} \langle h, \mathcal{H}(\cdot, s)[e_i^K] \rangle_{H(S,K)} e_i^K.$$

2. The following family of functions of $H(S, K)$,

$$\mathcal{F} = \{\mathcal{H}(\cdot, s)[k] = \mathcal{E}_s^\star[k] = \Gamma(s)[k], s \in S, k \in K\},$$

is total in $H(S, K)$, that is,

$$\left\{ \bigcup_{s \in S} \mathcal{R}[\mathcal{E}_s^\star] \right\}^\perp = 0_{H(S,K)}.$$

Indeed [266, p. 35], for a set A in a Hilbert space, and the linear manifold $V(A)$ it generates,

$$A^\perp = \{V(A)\}^\perp = \left\{\overline{V(A)}\right\}^\perp,$$

so that [126, 266, p. 24, respectively 71],

$$\left\{ \bigcup_{s \in S} \mathcal{R}[\mathcal{E}_s^\star] \right\}^\perp = \left\{ \bigvee_{s \in S} \mathcal{R}[\mathcal{E}_s^\star] \right\}^\perp = \bigcap_{s \in S} \mathcal{N}[\mathcal{E}_s]$$

only contains the zero function.

3. Convergence in $H(S, K)$ implies "point-wise" convergence, and uniform convergence over sets over which the diagonal of $\mathcal{H}$ is bounded. One has indeed that

$$\|h(s)\|_K = \|\mathcal{E}_s[h]\|_K \leq \|\mathcal{E}_s\| \, \|h\|_{H(S,K)},$$

and that

$$\|\mathcal{H}(s, s)\| = \left\|\mathcal{E}_s \mathcal{E}_s^\star\right\| = \left\|\mathcal{E}_s^\star\right\|^2 = \|\mathcal{E}_s\|^2.$$

4. Given a K-covariance $\mathcal{H}$, there is a unique K-RKHS whose kernel is $\mathcal{H}$.

   Let $H_0 \subseteq K^S$ be the manifold generated by the family

$$\{\mathcal{H}(\cdot, s)[k], s \in S, k \in K\}$$

(it is the family $\mathcal{F}$ of item 2). A bilinear form on H_0 may be defined as follows:

$$h_1 = \sum_{i=1}^{m} \alpha_i^{(1)} \mathcal{H}(\cdot, s_i^{(1)})[k_i^{(1)}],$$

$$h_2 = \sum_{i=1}^{m} \alpha_i^{(2)} \mathcal{H}(\cdot, s_i^{(2)})[k_i^{(2)}],$$

$$\begin{aligned}\langle h_1, h_2\rangle_{H_0} &= \left\langle \sum_{i=1}^{m} \alpha_i^{(1)} \mathcal{H}(\cdot, s_i^{(1)})[k_i^{(1)}], \sum_{j=1}^{n} \alpha_j^{(2)} \mathcal{H}(\cdot, s_j^{(2)})[k_j^{(2)}] \right\rangle_{H_0} \\ &= \sum_{i=1}^{m}\sum_{j=1}^{n} \alpha_i^{(1)}\alpha_j^{(2)} \left\langle \mathcal{H}(s_j^{(2)}, s_i^{(1)})[k_i^{(1)}], [k_j^{(2)}] \right\rangle_K .\end{aligned}$$

That indeed makes sense. Let to that end $\Phi_0 : \mathcal{F} \times \mathcal{F} \longrightarrow \mathbb{R}$ be defined using the following relation:

$$\Phi_0(\mathcal{H}(\cdot, s_1)[k_1], \mathcal{H}(\cdot, s_2)[k_2]) = \langle \mathcal{H}(s_2, s_1)[k_1], k_2\rangle_K .$$

Suppose that $\sum_{i=1}^{n} \alpha_i \mathcal{H}(\cdot, s_i)[k_i]$ is the zero function from S to K. Then

$$\begin{aligned}\sum_{i=1}^{n} \alpha_i \Phi_0(\mathcal{H}(\cdot, s_i)[k_i], \mathcal{H}(\cdot, s)[k]) &= \sum_{i=1}^{n} \alpha_i \langle \mathcal{H}(s, s_i)[k_i], k\rangle_K \\ &= \left\langle \sum_{i=1}^{n} \alpha_i \mathcal{H}(s, s_i)[k_1], k \right\rangle_K \\ &= 0.\end{aligned}$$

The bilinear form just defined also reproduces the functions of H_0. Indeed, letting $h(s) = \sum_{i=1}^{n} \alpha_i \mathcal{H}(s, s_i)[k_i]$,

$$\begin{aligned}\langle h, \mathcal{H}(\cdot, s)[k]\rangle_{H_0} &= \left\langle \sum_{i=1}^{n} \alpha_i \mathcal{H}(\cdot, s_i)[k_i], \mathcal{H}(\cdot, s)[k] \right\rangle_{H_0} \\ &= \left\langle \sum_{i=1}^{n} \alpha_i \mathcal{H}(s, s_i)[k_i], k \right\rangle_K \\ &= \left\langle \mathcal{E}_s \left[\sum_{i=1}^{n} \alpha_i \mathcal{H}(\cdot, s_i)[k_i] \right], k \right\rangle_K \\ &= \langle \mathcal{E}_s[h], k\rangle_K .\end{aligned}$$

It remains to check that the bilinear form is an inner product. Let $h \in H_0$ be fixed, but arbitrary. Suppose that $k \in K$ has norm one. Then

$$|\langle h(s), k\rangle_K| = \left|\langle h, \mathcal{E}_s^\star[k]\rangle_{H_0}\right| \leq \|h\|_{H_0} \left\|\mathcal{E}_s^\star\right\|,$$

so that $\|h(s)\|_K \leq \|h\|_{H_0} \left\|\mathcal{E}_s^\star\right\|$, and thus h is the zero function when its H_0-norm is zero.

The following inequality shall be used tacitly below:

$$\begin{aligned}\|\mathcal{H}(\cdot, s)[k]\|_{H_0}^2 &= \langle \mathcal{H}(\cdot, s)[k], \mathcal{H}(\cdot, s)[k]\rangle_{H_0}\\ &= \langle \mathcal{H}(s, s)[k], k\rangle_K\\ &\leq \|\mathcal{H}(s, s)\| \, \|k\|_K^2 .\end{aligned}$$

Let H be the completion of H_0 for the inner product just defined. Let $\Gamma(s) : K \longrightarrow H$ be defined using the following relation:

$$\Gamma(s)[k] = [\mathcal{H}(\cdot, s)[k]]_H,$$

where $[\cdot]_H$ denotes equivalence class. Since

$$\|\Gamma(s)[k]\|_H = \|\mathcal{H}(\cdot, s)[k]\|_{H_0} \leq \|\mathcal{H}(s, s)\|^{\frac{1}{2}} \, \|k\|_K ,$$

$\Gamma(s)$ is an operator that is linear and bounded. Define then $J : H \longrightarrow K^S$ using the following relation:

$$J[h](s) = \Gamma(s)^\star[h].$$

J is an injection. Suppose indeed that $J[h]$ is the zero function. Then, for all $s \in S$,

$$h \in \mathcal{N}[\Gamma(s)^\star] = \mathcal{R}[\Gamma(s)]^\perp,$$

and, since $\bigcup_{s\in S} \mathcal{R}[\Gamma(s)]$ generates the inclusion of H_0 in H, $h = 0_H$. Thus H may be identified with $H(S, K) = J[H] \subseteq K^S$. Let

$$\langle J[h_1], J[h_2]\rangle_{H(S,K)} = \langle h_1, h_2\rangle_H .$$

With that inner product, $H(S, K)$ is a Hilbert space, and J is unitary. Furthermore

$$\begin{aligned}\|\mathcal{E}_s\,[J[h]]\|_K &= \|J[h](s)\|_K\\ &= \|\Gamma(s)^\star[h]\|_K\\ &\leq \|\Gamma(s)^\star\| \, \|h\|_H\\ &= \|\Gamma(s)^\star\| \, \|J[h]\|_{H(S,K)} ,\end{aligned}$$

so that $H(S, K)$ is a K-RKHS. Furthermore, since

$$\mathcal{E}_s\left[J[h]\right] = J[h](s) = \Gamma(s)^\star[h] = \Gamma(s)^\star J^\star J[h],$$

$\mathcal{E}_s = \Gamma(s)^\star J^\star$. Consequently,

$$\mathcal{E}_{s_1}\mathcal{E}_{s_2}^\star = \Gamma(s_1)^\star \Gamma(s_2),$$

and then

$$\begin{aligned}\left\langle \mathcal{E}_{s_1}\mathcal{E}_{s_2}^\star[k_1], k_2\right\rangle_K &= \langle \Gamma(s_2)[k_1], \Gamma(s_1)[k_2]\rangle_H \\ &= \langle [\mathcal{H}(\cdot, s_2)[k_1]]_H, [\mathcal{H}(\cdot, s_1)[k_2]]_H\rangle_H \\ &= \langle \mathcal{H}(\cdot, s_2)[k_1], \mathcal{H}(\cdot, s_1)[k_2]\rangle_{H_0} \\ &= \langle \mathcal{H}(s_1, s_2)[k_1], k_2\rangle_K,\end{aligned}$$

so that $\mathcal{H}$ is the reproducing kernel of $H(S, K)$. The RKHS so obtained is unique as there is only one completion of H_0 [129, p. 21].

5. Let S be a set, H and K, real Hilbert spaces, and $L : H \longrightarrow K^S$, a map. The following statements are equivalent (**equivalences**, denoted **e**-):

   (i) For $s \in S$, fixed, but arbitrary, there is $\kappa(s)$ such that, for $h \in H$, fixed, but arbitrary,

   $$\|L[h](s)\|_K \leq \kappa(s)\,\|h\|_H.$$

   (ii) There is a map $\Gamma : S \longrightarrow \mathcal{L}(K, H)$ such that, for $s \in S$ and $h \in H$, fixed, but arbitrary,

   $$L[h](s) = \Gamma(s)^\star[h].$$

   (iii) L is a partial isometry from H onto a K-RKHS $H(S, K) \subseteq K^S$.

   When these statements are true, then (**consequences**, denoted **c**-):

   (i) $\mathcal{N}_L = \left\{\bigcup_{s\in S}\mathcal{R}[\Gamma(s)]\right\}^\perp$;
   (ii) the reproducing kernel of $H(S, K)$ is $\mathcal{H}(s_1, s_2) = \Gamma(s_1)^\star\Gamma(s_2)$;
   (iii) $\mathcal{E}_s = \Gamma(s)^\star L^\star$, where L is the partial isometry of **e**-(iii).

   Indeed:

   [e-(i)⇒e-(ii)] By definition, for $h \in H$, and $s \in S$, fixed, but arbitrary, $L[h](s) \in K$. Thus the assignment $\Gamma^\star(s) : H \longrightarrow K$ obtained when setting $\Gamma^\star(s)[h] = L[h](s)$ is well defined. Furthermore, by assumption,

   $$\|\Gamma^\star(s)[h]\|_K = \|L[h](s)\|_K \leq \kappa(s)\,\|h\|_H,$$

so that $\Gamma^\star(s) \in \mathcal{L}(H,K)$, and its transpose, denoted $\Gamma(s)$, is well defined, linear, and bounded. Since $\Gamma(s)^\star = \Gamma^\star(s)$, **e**-(ii) is true.

[e-(ii)⇒e-(i)] That is immediate from the assumption, and the fact that $\Gamma(s)^\star$ is bounded.

[e-(ii)⇒c-(i)] Suppose that $L[h]$ is the zero function. Then, as

$$L[h](s) = \Gamma(s)^\star[h],$$

using in succession [126, 266, 266, p. 71, respectively, 24, and 35],

$$h \in \bigcap_{s\in S} \mathcal{N}[\Gamma(s)^\star] = \bigcap_{s\in S} \overline{\mathcal{R}[\Gamma(s)]}^{\perp} = \left\{ \bigvee_{s\in S} \overline{\mathcal{R}[\Gamma(s)]} \right\}^{\perp} \subseteq \left\{ \bigcup_{s\in S} \mathcal{R}[\Gamma(s)] \right\}^{\perp},$$

so that

$$\mathcal{N}[L] \subseteq \left\{ \bigcup_{s\in S} \mathcal{R}[\Gamma(s)] \right\}^{\perp}.$$

Suppose conversely that $h \in \left\{\bigcup_{s\in S} \mathcal{R}[\Gamma(s)]\right\}^{\perp}$. Then, for $s \in S$, and $k \in K$, fixed, but arbitrary, $0 = \langle h, \Gamma(s)[k]\rangle_H = \langle \Gamma(s)^\star[h], k\rangle_K$, so that, for $s \in S$, fixed but arbitrary, $L[h](s) = \Gamma(s)^\star[h] = 0$. $L[h]$ is thus the zero function, and

$$\left\{ \bigcup_{s\in S} \mathcal{R}[\Gamma(s)] \right\}^{\perp} \subseteq \mathcal{N}[L].$$

In particular L is injective on $\mathcal{N}[L]^\perp$.

[e-(ii)⇒e-(iii)] Let $H(S,K) = \mathcal{R}[L]$ be given the following Hilbert space structure:

$$\langle L[h_1], L[h_2]\rangle_{H(S,K)} = \langle h_1, h_2\rangle_H .$$

Let Λ be L as a map from H onto $H(S,K)$. It is a partial isometry, with $\mathcal{N}[L]^\perp$ as initial set. Thus $\Lambda^\star\Lambda$ is the projection onto $\mathcal{N}[L]^\perp$. One must prove that $H(S,K)$, with its inner product, is a K-RKHS. To that end, let $h_\Lambda = \Lambda[h]$, $h = h_1 + h_2, h_1 \in \mathcal{N}[L], h_2 \in \mathcal{N}[L]^\perp$. Then

$$h_\Lambda(s) = \Lambda[h](s) = L[h_2](s) = \Gamma(s)^\star[h_2] = \Gamma(s)^\star\Lambda^\star\Lambda[h] = \Gamma(s)^\star\Lambda^\star[h_\Lambda].$$

It follows that $\mathcal{E}_s = \Gamma(s)^\star\Lambda^\star$ is continuous, and thus that $H(S,K)$ is a K-RKHS, so that **c**-(iii) obtains.

[e-(ii)⇒c-(ii)] Since **e**-(ii) implies **c**-(iii), the reproducing kernel of $H(S,K)$ is obtained as

$$\mathcal{E}_{s_1}\mathcal{E}_{s_2}^\star = \Gamma(s_1)^\star \Lambda^\star \Lambda \Gamma(s_2) = \Gamma(s_1)^\star P_{\mathcal{N}[L]^\perp} \Gamma(s_2) = \Gamma(s_1)^\star \Gamma(s_2).$$

[e-(iii)⇒e-(ii)] It suffices to prove that **e**-(i) follows from **e**-(iii). But

$$\|L[h](s)\|_K = \|\mathcal{E}_s\,[L[h]]\|_K \leq \|\mathcal{E}_s\|\,\|L[h]\|_{H(S,K)} \leq \|\mathcal{E}_s\|\,\|h\|_H\,.$$

The treatment of the case of $\underline{B}$ takes into account the properties specific to that process, and one then needs some results from calculus for operator valued functions. The reference is [106, 7.4, 7.8 and 12.5].

Some Calculus for Operator Valued Functions

One shall peruse below the "standard" classes of compact operators for Hilbert space (Hilbert-Schmidt and trace-class). The only facts needed here [235] are that the Hilbert-Schmidt operators form a Hilbert space (denoted $\mathcal{B}_2(H)$), and the trace-class ones, a Banach space (denoted $\mathcal{B}_1(H)$), with respective inner product, and norm, as used. The continuous linear functionals of $\mathcal{B}_1(H)$ are of the following form [235, p. 47]: $\phi_L(T) = \text{trace}(TL)$, where T has finite trace, and L is a bounded, linear operator.

Let $\mathcal{S}$ be a σ-algebra of subsets of a set S, and H, a real and separable Hilbert space. Suppose that $M : \mathcal{S} \longrightarrow \mathcal{B}_1(H)$ is additive, and of bounded variation (for the trace-class norm). Let $|M|_{TC}$ denote the variation of M (for the trace-class norm). There exists then $D_M : S \longrightarrow \mathcal{B}_1(H)$, adapted to $\mathcal{S}$, and the Borel sets of $\mathcal{B}_1(H)$, such that:

1. when the values of M are positive (self-adjoint) operators, one may choose the values of D_M to be positive (self-adjoint) operators;
2. almost surely, with respect to $|M|_{TC}$, $\|D_M(s)\|_{TC} = 1$, and , when M is positive and self-adjoint, $D_M^{1/2}(s)$ is Hilbert-Schmidt;
3. for $S_0 \in \mathcal{S}$, fixed, but arbitrary,

$$M(S_0) = \int_{S_0} D_M(s)\,|M|_{TC}\,(ds),$$

   where the integral is a weak integral, that is, for a fixed, but arbitrary (operator) functional $T \mapsto \phi_L(T)$,

$$\phi_L[M(S_0)] = \int_{S_0} \phi_L[D_M(s)]\,|M|_{TC}\,(ds);$$

4. $\|M(S_0)\|_{TC} \leq \int_{S_0} \|D_M(s)\|_{TC}\,|M|_{TC}\,(ds) = |M|_{TC}\,(S_0)$.

The following considerations depend on the assumption that the diagonal of $S \times S$ belongs to $\mathcal{S} \otimes \mathcal{S}$, and that is the case when S is a complete, separable metric space [106, p. 613]. Let $m : \mathcal{S} \longrightarrow \mathbb{R}$ be additive, with bounded variation $|m|$, and L, a map, defined on S, with values in the space of bounded, linear operators of H. It is assumed that, for $\{h_1, h_2\} \subseteq H$, fixed, but arbitrary, $s \mapsto \langle L(s)[h_1], h_2 \rangle_H$ is adapted, and that $s \mapsto \|L(t)\|$ is integrable with respect to $|m|$. Then there is a unique bounded, linear operator $I_m\{L\}$, the weak integral of L with respect to m, such that

1. for $\{h_1, h_2\} \subseteq H$, fixed, but arbitrary,
$$\langle I_m\{L\}[h_1], h_2 \rangle_H = \int_S \langle L(s)[h_1], h_2 \rangle_H \, m(ds);$$
2. $\|I_m\{L\}\| \leq \int_S \|L(s)\| \, |m|(ds)$;
3. when, for $s \in S$, fixed, but arbitrary, $L(s)$ is Hilbert-Schmidt, and the Hilbert-Schmidt norm of $L(s)$ is integrable with respect to $|m|$, the weak integral is a Hilbert-Schmidt operator, and one may replace operator norm with Hilbert-Schmidt norm; *mutatis mutandis,* one may replace Hilbert-Schmidt by trace-class, and then one has furthermore that $s \mapsto trace L(s)$ is adapted, and that
$$trace(I_m\{L\}) = \int_S trace(L(s)) m(ds).$$

Let again $s \mapsto L(s)$ be a map whose values are bounded, linear operators of H, and $S_0 \mapsto M(S_0)$ be a map whose values are positive operators with finite trace. Then, provided the required assumptions are made, one may define an integral $I_M\{L\}$ using the following relation:

$$I_M\{L\} = I_{|M|_{TC}}\{LD_M\}.$$

Manifolds and Subspaces Generated by Functions Whose Values Are Hilbert-Schmidt Operators

Let $(\Omega, \mathcal{A}, P)$ be a probability space, and H, a real and separable Hilbert space. Let $L_2^H(\Omega, \mathcal{A}, P)$ be the Hilbert space of equivalence classes of random elements $X : \Omega \longrightarrow H$ such that

$$E_P\left[\|X\|_H^2\right] < \infty.$$

The inner product of $L_2^H(\Omega, \mathcal{A}, P)$ is computed according to the following formula:

$$\langle X, Y \rangle_{L_2^H(\Omega, \mathcal{A}, P)} = E_P\left[\langle X, Y \rangle_H\right] = trace(C_{X,Y}),$$

where $C_{X,Y}$ is the (cross-)covariance operator of X and Y [12].

A subspace K of $L_2^H(\Omega, \mathcal{A}, P)$ is invariant when, for a fixed, but arbitrary, bounded, linear operator L of H, $L(K) \subseteq K$ ($L(K)$) is the set of the elements obtained as compositions of elements of K with L. Given a fixed, but arbitrary subset K_0 of $L_2^H(\Omega, \mathcal{A}, P)$, there is a smallest invariant subspace that contains it. It is obtained as the closure in $L_2^H(\Omega, \mathcal{A}, P)$ of the family of elements of the following form:

$$\sum_{i=1}^{n} L_i[X_i],$$

where:

1. $\{L_1, \ldots, L_n\}$ is a family of bounded, linear operators of H;
2. $\{X_1, \ldots, X_n\} \subseteq K_0$.

Suppose now that $H = \mathcal{B}_2(H)$, and that $P = |M|_{TC}$ (one assumes thus that M has total trace-class variation equal to one, which is the case for Cramér-Hida processes). The elements X are then (equivalence classes of) functions$s \mapsto L(s) \in \mathcal{B}_2(H)$, and

$$\begin{aligned}
\langle L_1, L_2 \rangle_{L_2^{\mathcal{B}_2(H)}(S, \mathcal{S}, |M|_{TC})} &= \int_S \langle L_1(s), L_2(s) \rangle_{HS} \, |M|_{TC}\,(ds) \\
&= \int_S trace\left[L_1(s) L_2^\star(s)\right] \, |M|_{TC}\,(ds).
\end{aligned}$$

Let K denote $L_2^{\mathcal{B}_2(H)}(S, \mathcal{S}, |M|_{TC})$, and $L[M] \subseteq K$ be the subset of those classes of functions $s \mapsto L(s)$ for which, for fixed, but arbitrary $h \in H$, when $D_M(s)[h] = 0_H$, almost surely, with respect to $|M|_{TC}$, then $L(s)[h] = 0_H$, that is, $\mathcal{N}[D_M(s)] \subseteq \mathcal{N}[L(s)]$. $L[M]$ is a subspace of K.

Let $L_0[M]$ be the subset of $L[M]$ made of the classes of elements of the following form:

$$L(s) = \left\{ \sum_{i=1}^{n} \chi_{S_i}(s) L_i \right\} \circ D_M^{1/2}(s),$$

where $\{S_1, \ldots, S_n\} \subseteq \mathcal{S}$ is a partition of S in $\mathcal{S}$, and $\{L_1, \ldots, L_n\}$ are bounded, linear operators of H. $L_0[M]$ is a dense manifold of $L[M]$.

As, for L_1 and L_2 Hilbert-Schmidt, $L_1 L_2^\star$ has finite trace, on K, one may also define the following bilinear map which is, in fact, an inner product:

$$[L_1, L_2]_M = \int_S L_1(s) L_2^\star(s) \, |M|_{TC}\,(ds),$$

and obtain an operator of H of finite trace, provided the associated norms are integrable.

The Case of $\underline{B}$

Let C be the covariance function of $\underline{B}$. It is a continuous function, with, for the trace norm, bounded variation. Let M be defined using the following relation: $M(]t_1, t_2]) = C(t_2) - C(t_1)$. M can be extended to a measure on the Borel sets of $[0, 1]$, with values in the Banach space of operators with finite trace [75, p. 208]. $|M|$ shall be the trace norm variation. The properties of absolute continuity attached to $\underline{B}$ allow one to identify D_M as a diagonal matrix, whose diagonal elements are the Radon-Nikodým derivatives $\frac{dM_n}{dM_b}$, where M_n is the measure attached to b_n, and M_b, that attached to b. That fact however is here of marginal relevance.

Let $\underline{X} : \Omega \times [0, 1] \longrightarrow l_2$ be a process with mean zero and covariance function $(t_1, t_2) \mapsto C(t_1 \wedge t_2)$. Let

$$M_{\underline{X}}(]t_1, t_2]) = \underline{X}(\cdot, t_2) - \underline{X}(\cdot, t_1).$$

$M_{\underline{X}}$ is thus defined on the semiring of intervals $]t_1, t_2]$. Furthermore

$$E_P^{1/2}\left[\left\|M_{\underline{X}}(]t_1, t_2])\right\|_{l_2}^2\right] \leq M_b(]t_1, t_2]).$$

Consequently $\underline{X}$, as a map with domain $[0, 1]$, and values in the Hilbert space $K = L_2^{l_2}(\Omega, \mathcal{A}, P)$, is continuous, and a map of bounded variation. $M_{\underline{X}}$ thus extends to a measure on the Borel sets of $[0, 1]$, with values in K. It has orthogonal values on disjoint sets because of the latter inequality.

Let $\{L_1, L_2\} \subseteq L_0[M]$ be fixed, but arbitrary. One may assume that (the sets S_i's are the same in both sums)

$$L_1(t) = \left\{\sum_{i=1}^n \chi_{S_i} L_i^{(1)}\right\} \circ D_M^{1/2}(t), \text{ and } L_2(t) = \left\{\sum_{i=1}^n \chi_{S_i} L_2^{(1)}\right\} \circ D_M^{1/2}(t).$$

Then, by definition,

$$[L_1, L_2]_M = \sum_{i=1}^n \int_{S_i} \left\{L_i^{(1)} D_M(t) \left(L_i^{(2)}\right)^\star\right\} |M| (dt).$$

Let $\Lambda_1(t) = \sum_{i=1}^n \chi_{S_i} L_i^{(1)}$, and define a simple integral using the following relation:

$$\underline{X}_1 = \int_0^1 \Lambda_1(t) M_{\underline{X}}(dt) = \sum_{i=1}^n L_i \left[M_{\underline{X}}(S_i)\right].$$

$\underline{X}_2$, and Λ_2, its integral, are defined analogously. Then:

$$
\begin{aligned}
E_P\left[\langle \underline{X}_1, \underline{\alpha}_1\rangle_{l_2} \langle \underline{X}_2, \underline{\alpha}_2\rangle_{l_2}\right] =&\\
&= E_P\left[\left\langle \sum_{i=1}^n L_i^{(1)}[M_{\underline{X}}(S_i)], \underline{\alpha}_1\right\rangle_{l_2} \left\langle \sum_{i=1}^n L_i^{(2)}[M_{\underline{X}}(S_i)], \underline{\alpha}_2\right\rangle_{l_2}\right]\\
&= \sum_{i=1}^n \sum_{j=1}^n E_P\left[\left\langle M_{\underline{X}}(S_i), \left(L_i^{(1)}\right)^\star [\underline{\alpha}_1]\right\rangle_{l_2} \left\langle M_{\underline{X}}(S_j), \left(L_j^{(2)}\right)^\star [\underline{\alpha}_2]\right\rangle_{l_2}\right]\\
&= \sum_{i=1}^n \left\langle M(S_i)\left[\left(L_i^{(1)}\right)^\star [\underline{\alpha}_2]\right], \left(L_i^{(2)}\right)^\star [\underline{\alpha}_1]\right\rangle_{l_2}\\
&= \left\langle \sum_{i=1}^n L_i^{(2)} M(S_i) \left(L_i^{(1)}\right)^\star [\underline{\alpha}_2], \underline{\alpha}_1\right\rangle_{l_2}\\
&= \left\langle \sum_{i=1}^n L_i^{(1)} M(S_i) \left(L_i^{(2)}\right)^\star [\underline{\alpha}_1], \underline{\alpha}_2\right\rangle_{l_2},
\end{aligned}
$$

so that

$$
\begin{aligned}
C_{\underline{X}_1, \underline{X}_2} &= \sum_{i=1}^n L_i^{(1)} M(S_i) \left(L_i^{(2)}\right)^\star\\
&= \sum_{i=1}^n \int_{S_i} L_i^{(1)} D_M(t) \left(L_i^{(2)}\right)^\star |M|(dt)\\
&= [L_1, L_2]_M.
\end{aligned}
$$

Let $K(\underline{X})$ be the invariant subspace generated by the values of $M_{\underline{X}}$, and

$$
U: \left\{\sum_{i=1}^n \chi_{S_i} L_i^{(1)}\right\} \circ D_M^{1/2} \mapsto \int_0^1 \sum_{i=1}^n \chi_{S_i} L_i^{(1)} dM_{\underline{X}}.
$$

From what precedes U is an unitary operator from $L_0(M)$, with the $[\cdot, \cdot]_M$ inner product, to a family generating the invariant subspace $K(\underline{X})$ with the covariance inner product. The extension of U yields the integral with respect to $M_{\underline{X}}$, that is, when L is a map with $[0, 1]$ as domain, and bounded, linear operators as values, such that

$$
LD_M^{1/2} \in L[M],
$$

then

$$\int_0^1 L(t)M_{\underline{X}}(dt) = U\left[LD_M^{1/2}\right].$$

To obtain the reproducing kernel one proceeds as follows. Let $F(t)$ be the class of

$$\chi_{[0,t]}D_M^{1/2} \in L[M].$$

It is a map from $[0, 1]$ into the space of Hilbert-Schmidt operators. Then, in particular,

$$[F(t_1), F(t_2)]_M = C(t_1 \wedge t_2).$$

The functions in the RKHS of the process $\underline{X}$, or the covariance $C(t_1 \wedge t_2)$, are obtained as follows. Let

$$L_F : L[M] \longrightarrow \mathcal{B}_2(l_2)^{[0,1]}$$

be defined by the following relation:

$$L_F\left[LD_M^{1/2}\right](t) = [LD_M^{1/2}, F(t)]_M = \int_0^1 \chi_{[0,t]}(\theta)L(\theta)D_M(\theta)\,|M|\,(d\theta).$$

Since the following identity: $[LD_M^{1/2}, F(t)]_M = 0_{\mathcal{B}_1(l_2)}$, for $t \in [0, 1]$, fixed, but arbitrary, implies that L is the zero operator, L_F is a unitary map, so that the RKHS is isomorphic to $L[M]$ which is, in turn, isomorphic to $K(\underline{X})$.

$H(S, K)$ is the range of L_F. Since $L_F\,[F(t)]\,(\theta) = [F(t), F(\theta)]_M = C(t \wedge \theta)$, $L_F[F(t)]$ is the map $\theta \mapsto C(t \wedge \theta)$, and

$$\begin{aligned}\left\langle L_F[LD_M^{1/2}], C(t \wedge \cdot)\right\rangle_{H(S,K)} &= \left\langle L_F[LD_M^{1/2}], L_F[F(t)]\right\rangle_{H(S,K)} \\ &= [LD_M^{1/2}, F(t)] \\ &= L_F[LD_M^{1/2}](t).\end{aligned}$$

11.2 Families of Integrands

In the sequel one shall integrate some families of processes (the $\underline{a}$'s in $I_{\mathcal{B}}\{\underline{a}\}$) which are listed once their path spaces are described. Those are:

1. $\mathcal{L}_2\,[b_n]$ and $L_2\,[b_n]$:
   Let $a : [0, 1] \longrightarrow \overline{\mathbb{R}}$ be adapted to $\mathcal{B}([0, 1])$ and $\mathcal{B}(\overline{\mathbb{R}})$, and have the property that

$$\int_0^1 a^2(t) M_n(dt) < \infty.$$

   a is an element of $\mathcal{L}_2\,[b_n]$. The equivalence classes of elements of $\mathcal{L}_2\,[b_n]$ form $L_2\,[b_n]$. The equivalence class of a shall be denoted $[a]_n$ and the norm of the latter,

$$\|[a]_n\|_n\,.$$

2. $\mathcal{L}_2\,[\underline{b}]$ and $L_2\,[\underline{b}]$:
   Let $\underline{a}$ denote the vector with components $a_n \in \mathcal{L}_n[b_n], n \in \mathbb{N}$, and $\underline{b}$ that with components b_n. $\mathcal{L}_2\,[\underline{b}]$ is the family of those $\underline{a}$'s for which

$$\sum_{n=1}^{\infty} \int_0^1 a_n^2(t) M_n(dt) < \infty.$$

   Let $[\underline{a}]$ be the vector with components $[a_n]_n$. When $\underline{a}$ "runs through" $\mathcal{L}_2[\underline{b}]$, $[\underline{a}]$ generates $L_2[\underline{b}]$. One has that ($\oplus$ denoting the Hilbert direct sum)

$$L_2\,[\underline{b}] = \bigoplus_{n=1}^{\infty} L_2\,[b_n]\,, \quad \text{and} \quad \|[\underline{a}]\|^2_{L_2[\underline{b}]} = \sum_{n=1}^{\infty} \|[a_n]_n\|_n^2\,.$$

Remark 11.2.1 Notation $L_2\,[\underline{b}]$ has the following rationale [46, p. 143].

- Let $\{(\Omega_\lambda, \mathcal{A}_\lambda), \lambda \in \Lambda\}$ be an indexed family of measurable spaces with the property that the sets Ω_λ are pairwise disjoint. Then by the direct sum

$$\bigoplus_{\lambda \in \Lambda} (\Omega_\lambda, \mathcal{A}_\lambda)$$

  is meant the measurable space $(\Omega, \mathcal{A})$, where $\Omega = \bigcup_{\lambda \in \Lambda} \Omega_\lambda$, and $\mathcal{A}$ is the σ-algebra consisting of all those sets A of Ω with the property that $A \cap \Omega_\lambda \in \mathcal{A}_\lambda$, for every index $\lambda \in \Lambda$.
- When there is given a measure μ_λ on $(\Omega_\lambda, \mathcal{A}_\lambda)$, for each index λ, the set function μ defined using the following relation:

$$\mu(A) = \sum_{\lambda \in \Lambda} \mu_\lambda(A \cap \Omega_\lambda),\ A \in \mathcal{A},$$

  is a measure on $(\Omega, \mathcal{A})$.

- The measure space $(\Omega, \mathcal{A}, \mu)$ is called the direct sum of the family

$$\{(\Omega_\lambda, \mathcal{A}_\lambda, \mu_\lambda), \lambda \in \Lambda\}.$$

- $(\Omega, \mathcal{A})$ contains each of the measure spaces $(\Omega_\lambda, \mathcal{A}_\lambda)$ as a subspace.
- A function f on Ω is adapted to $\mathcal{A}$ if, and only if, $f^{|\Omega_\lambda}$ is adapted to $\mathcal{A}_\lambda$, for every index λ.
- When f is adapted to $\mathcal{A}$, it is integrable for μ if, and only if, the indexed family

$$\left\{\int_{\Omega_\lambda} |f|\, d\mu_\lambda, \lambda \in \Lambda\right\}$$

is summable, in which case

$$\int_\Omega f d\mu = \sum_{\lambda \in \Lambda} \left\{\int_{\Omega_\lambda} f d\mu_\lambda\right\}.$$

When one lets Λ be $\mathbb{N}$, $\Omega_n = (n, [0,1])$, $\mathcal{A}_n = (n, \mathcal{B}([0,1]))$, and $\mu_n = M_n$, $L_2[\underline{b}]$ becomes a *bona fide* L_2-space.

One shall use, in the sequel, the following notation:

1. For $S : \Omega \longrightarrow [0,1]$ (random time),

$$[\![0, S]\!] = \{(\omega, t) \in \Omega \times [0,1] : 0 \le t \le S(\omega)\}.$$

2. $\underline{a}_{|t} = \chi_{[0,t]}\underline{a}$, and $\underline{a}_{|S} = \chi_{[\![0,S]\!]}\underline{a}$.

The following families of integrands shall be used:

1. $\mathcal{I}_0[\underline{b}]$:

   It shall be the family of stochastic processes $\underline{a}$, with components a_n, with the following properties:

   (i) $a_n : \Omega \times [0,1] \longrightarrow \overline{\mathbb{R}}$ is adapted to $(\Omega, \underline{A}, P)$, and progressively measurable;
   (ii) $P(\omega \in \Omega : [\underline{a}(\omega, \cdot)] \in L_2[\underline{b}]) = 1$, that is (it shall be the preferred notation),

$$P\left(\omega \in \Omega : \|[\underline{a}(\omega, \cdot)]\|^2_{L_2[\underline{b}]} < \infty\right) =$$
$$= P\left(\omega \in \Omega : \sum_{n=1}^{\infty} \int_0^1 a_n^2(\omega, t) M_n(dt) < \infty\right) = 1.$$

2. $\mathcal{I}_2\left[\underline{b}\right]$:
   It shall be the subset of $\mathcal{I}_0\left[\underline{b}\right]$ made of those elements $\underline{a}$ for which

$$E_P\left[\|[\underline{a}(\cdot,\cdot)]\|^2_{L_2[\underline{b}]}\right] = \sum_{n=1}^{\infty} \int_0^1 E_P\left[a_n^2(\cdot,t)\right] M_n(dt) < \infty.$$

3. $\mathcal{I}_2^{loc}\left[\underline{b}\right]$:
   It shall be the family of stochastic processes $\underline{a}$, with components a_n, with the following properties:

   (i) $a_n : \Omega\times[0,1] \longrightarrow \overline{\mathbb{R}}$ is adapted to $(\Omega, \underline{\mathcal{A}}, P)$, and progressively measurable;
   (ii) there is a localizing sequence $\{S_n, n\in\mathbb{N}\}$ for which $\underline{a}_{|S_n} \in \mathcal{I}_2\left[\underline{b}\right]$.

The assumption that the integrands are progressively measurable is essentially minimal, as an adapted, measurable process has a version that is progressively measurable [192, p. 68]. One shall see below that

$$\left(\mathcal{I}_2\left[\underline{b}\right] \subseteq\right) \mathcal{I}_0\left[\underline{b}\right] \subseteq \mathcal{I}_2^{loc}\left[\underline{b}\right].$$

Those inclusions are strict. $\mathcal{I}_0[\underline{b}]$ is the class of processes that enter "naturally" the developments related to the Cramér-Hida decomposition as it represents the class of signals that are almost surely in the RKHS of the noise, a central characterization of nonsingular detection. $\mathcal{I}_2[\underline{b}]$ and $\mathcal{I}_2^{loc}[\underline{b}]$ are required for the definition of a stochastic integral that adequately accommodates the present context of $\mathcal{I}_0[\underline{b}]$.

To prove the latter inclusion one needs the following lemma that allows one to "ignore" the fact that the infinite sum in (ii) of item 1 above is finite only almost surely.

Lemma 11.2.2 *Let $\underline{a} \in \mathcal{I}_0\left[\underline{b}\right]$ be fixed, but arbitrary, and*

$$A(\omega,t) = \sum_{n=1}^{\infty} \int_0^t a_n^2(\omega,\theta)M_n(d\theta).$$

1. *A is adapted, continuous to the left, and separable;*
2. *there exists a progressively measurable $\hat{\underline{a}}$ that equals $\underline{a}$, path by path, almost surely with respect to P, such that the "corresponding" A, denoted $\hat{A}$, has continuous paths.*

Whenever an A process shall be encountered, one shall assume that it is $\hat{A}$.

Proof Let $A_n(\omega,t) = \sum_{i=1}^n \int_0^t a_i^2(\omega,\theta)M_i(d\theta)$. Since A_n is adapted, so is A. The latter is continuous to the left, by definition and monotone convergence (but, for some ω's, could have an infinite left limit), and, consequently, separable [264, p. 36].

The map $t \mapsto A(\omega, t)$ is not continuous to the right at $t < 1$ when $A(\omega, t) < \infty$, but $A(\omega, t + \frac{1}{n}) = \infty$ for all sufficiently large $n \in \mathbb{N}$. Let

$$I(\omega, t) = \chi_{\mathbb{R}_+}(A(\omega, t)) = \begin{cases} 1 \text{ when } A(\omega, t) < \infty \\ 0 \text{ when } A(\omega, t) = \infty \end{cases},$$

and

$$\underline{\hat{a}}(\omega, t) = I(\omega, t)\underline{a}(\omega, t).$$

Since A is adapted and continuous to the left, $I(\omega, t)$ is adapted and continuous to the left, and $\underline{\hat{a}}$ is thus progressively measurable [264, p. 40]. One has that $\underline{a}(\omega, \cdot) \neq \underline{\hat{a}}(\omega, \cdot)$ when there is a $t \in [0, 1[$ such that $A(\omega, t) = \infty$, and, since $A(\omega, t) = \infty$ implies $A(\omega, 1) = \infty$,

$$\{\omega \in \Omega : \underline{\hat{a}}(\omega, \cdot) \neq \underline{a}(\omega, \cdot)\} \subseteq \{\omega \in \Omega : A(\omega, 1) = \infty\},$$

a set of probability zero, so that $\underline{a}$ and $\underline{\hat{a}}$ are almost surely equal. One then sets

$$\hat{A}(\omega, t) = \sum_{n=1}^{\infty} \int_0^t \hat{a}_n^2(\omega, \theta) M_n(d\theta).$$

For every ω, $t \mapsto \hat{A}(\omega, t)$ is continuous to the left, again by monotone convergence.

Suppose that $A(\omega, t) < \infty$, but $A(\omega, t + \frac{1}{n}) = \infty$ for all sufficiently large $n \in \mathbb{N}$. Then, given $u > t$, fixed, but arbitrary, there is $n \in \mathbb{N}$ such that $t + \frac{1}{n} < u$, and then, for n large enough, $A(\omega, t + \frac{1}{n}) \leq A(\omega, u) = \infty$, and $\underline{\hat{a}}(\omega, u) = 0$. Consequently,

$$\sum_{n=1}^{\infty} \int_{]t,1]} \hat{a}_n^2(\omega, \theta) M_n(d\theta) = 0,$$

and thus, for $u > t$,

$$\sum_{n=1}^{\infty} \int_0^u \hat{a}_n^2(\omega, \theta) M_n(d\theta) = \sum_{n=1}^{\infty} \int_0^t \hat{a}_n^2(\omega, \theta) M_n(d\theta).$$

□

Proposition 11.2.3 $\mathcal{I}_0[\underline{b}] \subseteq \mathcal{I}_2^{loc}[\underline{b}]$.

Proof Let $\underline{a} \in \mathcal{I}_0[\underline{b}]$ be fixed, but arbitrary. Let $\hat{A}$ be the continuous version of A, obtained in (Lemma) 11.2.2, and, for $n \in \mathbb{N}$, fixed, but arbitrary,

$$S_n(\omega) = \begin{cases} 1 & \text{when } \left\{t \in [0,1] : \hat{A}(\omega, t) \geq n\right\} = \emptyset \\ \inf\left\{t \in [0,1] : \hat{A}(\omega, t) > n\right\} & \text{when } \left\{t \in [0,1] : \hat{A}(\omega, t) \geq n\right\} \neq \emptyset \end{cases}.$$

Since $\hat{A}$ is continuous and adapted, S_n is a strict stopping time of $\underline{\mathcal{A}}$ [264, p. 38], and the sequence $\{S_n, n \in \mathbb{N}\}$ is increasing. Since, almost surely with respect to P, $A(\omega, 1)$ is finite, almost surely with respect to P, $\lim_n S_n(\omega) = 1$, and, since, almost surely with respect to P, $t \mapsto A(\omega, t)$ is continuous, almost surely with respect to P, $A(\omega, S_n(\omega)) \leq n$. Thus

$$E_P[A(\cdot, S_n(\cdot))] \leq n,$$

and that means $\underline{a}_{|S_n} \in \mathcal{I}_2[\underline{b}]$, or $\underline{a} \in \mathcal{I}_2^{loc}[\underline{b}]$. □

Remark 11.2.4 In the sequel, repeated use shall be made of the sequence $\{S_n, n \in \mathbb{N}\}$ defined in (Proposition) 11.2.3. The reference shall be: "the localizing sequence of (Proposition) 11.2.3."

Remark 11.2.5 The inclusion in (Proposition) 11.2.3 is strict. Here is an example. Let

$$b_n(t) = 2^{-n}t, \text{ and } a_n(\omega, t) = (1-t)^{-1}.$$

Since a_n is adapted and continuous (to the right at 0, to the left at 1), it is progressively measurable [264, p. 40]. $A(\omega, t)$ is $(1-t)^{-1}$, and it then follows that $S_n(\omega) = 1 - n^{-1}$. One thus obtains a localizing sequence. A computation yields that

$$E_P\left[\left\|\underline{a}_{|S_n}\right\|^2_{L_2[\underline{b}]}\right] = n, \text{ but } \|\underline{a}\|^2_{L_2[\underline{b}]} = \infty.$$

11.3 Some Stochastic Integrals and Their Properties

One shall call "CH-stochastic integral" the limit, in an appropriate sense, of sums of the form

$$\sum_{i=1}^{n} \int_0^t a_i(\omega, \theta) B_n(\omega, d\theta).$$

As indicated in the introduction to the chapter, that limit shall be denoted $I_{\underline{B}}\{\underline{a}\}$.

11.3.1 Definition of the Integral

The CH-stochastic integral shall be defined for integrands in $\mathcal{I}_2^{loc}[\underline{b}]$, which is enough for present purposes. It will be, locally, a martingale in L_2.

Let $\underline{a} \in \mathcal{I}_2^{loc}[\underline{b}]$, and $\{S_n, n \in \mathbb{N}\}$ be a localizing sequence. Then, since, for $p \in \mathbb{N}$, fixed, but arbitrary,

$$E_P\left[\left\| a_{p|S_n}\right\|^2_{L_2[b_p]}\right] \leq E_P\left[\left\| \underline{a}_{|S_n}\right\|^2_{L_2[\underline{b}]}\right] < \infty,$$

$\{S_n, n \in I\!N\}$ is a localizing sequence for a_p. The (one-dimensional) stochastic integral

$$I_{B_p}\{a_p\} = \int_0^{\cdot} a_p dB_p$$

is thus well defined [264, p. 116]. So is then the sum

$$I_{\underline{B}_p}\left\{\underline{a}_p\right\} = \sum_{i=1}^{p} \int_0^{\cdot} a_i dB_i.$$

Now, for $\{n, p, q\} \subseteq \mathbb{N}$, and $\epsilon > 0$, fixed, but arbitrary,[1]

$$\begin{aligned} P&\left(\omega \in \Omega : \sup_{t\in[0,S_n(\omega)]} \left| I_{\underline{B}_{p+q}}\left\{\underline{a}_{p+q}\right\}(\omega,t) - I_{\underline{B}_p}\left\{\underline{a}_p\right\}(\omega,t)\right| > \epsilon\right) \leq \\ &\leq \epsilon^{-2} E_P\left[\left\{\sup_{t\in[0,S_n]} \left| I_{\underline{B}_{p+q}}\left\{\underline{a}_{p+q}\right\}(\cdot,t) - I_{\underline{B}_p}\left\{\underline{a}_p\right\}(\cdot,t)\right|\right\}^2\right] \\ &\leq 4\epsilon^{-2} E_P\left[\sum_{i=1}^{q} \int_0^{S_n} a_{p+i}^2(\cdot,t) M_{p+i}(dt)\right]. \end{aligned}$$

The justification for those inequalities is as follows. The first one is Markov inequality [138, p. 164], and the second, Doob's [264, p. 104], taking into account that the quadratic variation of

$$\left(I_{\underline{B}_{p+q}}\left\{\underline{a}_{p+q}\right\}(\cdot,t) - I_{\underline{B}_p}\left\{\underline{a}_p\right\}(\cdot,t)\right)^{S_n} = \sum_{i=1}^{q} \int_0^{S_n} a_{p+i} dB_{p+i}$$

[1] One could invoke the maximal inequality relating probability and expectation, but, in [264], matters are so presented that the longer expression yields a shorter reference.

is

$$\sum_{i=1}^{q}\sum_{j=1}^{q}\int_0^{S_n} a_{p+i}a_{p+j}d\langle B_{p+i},B_{p+j}\rangle=\sum_{i=1}^{q}\int_0^{S_n} a_{p+i}^2 dM_{p+i}.$$

In that latter calculation, one uses the following facts about quadratic variation:

1. formulae 7.3.2 in [264, p. 162], and 7.4.3 in [264, p. 174];
2. for continuous martingales the brackets [ ] and $\langle\ \rangle$ are equal [264, p. 148];
3. $\langle B_{p+i},B_{p+j}\rangle=0$ when $i\neq j$.

Since

$$E_P\left[\left\|\underline{a}_{|S_n}\right\|^2_{L_2[\underline{b}]}\right]=E_P\left[\sum_{p=1}^{\infty}\int_0^{S_n} a_p^2 dM_p\right]$$

is finite,

$$\lim_{p,q} P\left(\omega\in\Omega:\sup_{t\in[0,S_n(\omega)]}\left|I_{\underline{B}_{p+q}}\left\{\underline{a}_{p+q}\right\}(\omega,t)-I_{\underline{B}_p}\left\{\underline{a}_p\right\}(\omega,t)\right|>\epsilon\right)=0.$$

Consequently [264, p. 69], the sequence

$$\left\{I_{\underline{B}_p}\left\{\underline{a}_p\right\},p\in\mathbb{N}\right\}$$

converges locally, uniformly in probability, to a process which shall be denoted $I_{\underline{B}}\{\underline{a}\}$. Since the elements in the sequence may be taken to be continuous to the right [264, p. 71], and, furthermore, almost surely continuous with respect to P [264, p. 116], the limit may be, consequently, taken to have the same property [264, p. 69].

Definition 11.3.1 The process $I_{\underline{B}}\{\underline{a}\}$ shall be called the CH-stochastic integral of $\underline{a}$ with respect to $\underline{B}$.

11.3.2 Properties of the Integral

Local, uniform convergence in probability insures that the "usual" properties of the stochastic one-dimensional integral obtain for integrals of type $I_{\underline{B}}\{\underline{a}\}$. Only those properties of use in the sequel shall be listed.

Lemma 11.3.2 *Let U and V be two positive random variables, and $\epsilon > 0$ be fixed, but arbitrary. Then*

$$P(U + V > \epsilon) \leq P(U > 0) + P(V > \epsilon).$$

Proof One has that

$$\begin{aligned} \{U + V > \epsilon\} &= \left[\{U + V > \epsilon\} \bigcap \{U > 0\}\right] \\ &\bigcup \left[\{U + V > \epsilon\} \bigcap \{U = 0\}\right] \\ &\subseteq \{U > 0\} \bigcup \{V > \epsilon\}. \end{aligned}$$

□

The CH-stochastic integral of (Definition) 11.3.1 has the properties listed below.

Fact 11.3.3 *It is locally a martingale in L_2* [264, p. 93].

Fact 11.3.4 *Let $\langle I_{\underline{B}}\{\underline{a}\}\rangle$ denote the quadratic variation of $I_{\underline{B}}\{\underline{a}\}$. Then*

$$\langle I_{\underline{B}}\{\underline{a}\}\rangle(\omega, t) = \left\|\underline{a}_{|t}(\omega, \cdot)\right\|^2_{L_2[\underline{b}]}.$$

Proof Let, the notation not defined presently being that of the beginning of this section,

$$X_p = I_{\underline{B}_p}\left\{\underline{a}_p\right\}, \quad A_p = \textstyle\sum_{i=1}^p \int_0^{\cdot} a_i^2 \, dM_i,$$

$$X = I_{\underline{B}}\{\underline{a}\}, \quad A = \textstyle\sum_{i=1}^\infty \int_0^{\cdot} a_i^2 \, dM_i.$$

As seen $\langle X_p\rangle = A_p$. By Markov inequality [138, p. 164], the probability

$$P\left(\omega \in \Omega : \sup_{t \in [0, S_n(\omega)]} \left(A_{p+q}(\omega, t) - A_p(\omega, t)\right) > \epsilon\right)$$

is dominated by

$$\epsilon^{-1} E_P \left[\sum_{i=1}^q \int_0^{S_n} a_{p+i}^2 \, dM_{p+i}\right],$$

which has limit zero, so that the sequence $\{A_p, p \in \mathbb{N}\}$ converges locally, uniformly in probability, to a process that can be taken to be continuous to the right and almost surely continuous, with respect to P, say $A^\star$ [264, p. 69]. But an analogous inequality establishes that the same sequence converges locally, uniformly in probability, to A.

One may thus assume that A has the path properties of $A^\star$. It now suffices to check that $X^2 - A$ is, locally, a martingale in L_2 [264, p. 101]. But, for $n \in \mathbb{N}$, and $t \in [0, 1]$, fixed, but arbitrary, the sequence

$$\{A_p^{S_n}(\cdot, t), p \in \mathbb{N}\}$$

converges in L_1 to $A^{S_n}(\cdot, t)$ [192, p. 18]. Furthermore [264, p. 93] the sequence

$$\{X_p^{S_n}(\cdot, t), p \in \mathbb{N}\}$$

converges in L_2 to $X^{S_n}(\cdot, t)$. Since, for $t_1 < t_2$ in $[0, 1]$, fixed, but arbitrary,

$$E_P\left[\left(X_p^{S_n}\right)^2(\cdot, t_2) - A_p^{S_n}(\cdot, t_2) \mid \mathcal{A}_{t_1}\right] = \left(X_p^{S_n}\right)^2(\cdot, t_1) - A_p^{S_n}(\cdot, t_1),$$

that relation shall be preserved when taking limits with respect to the index p. □

Fact 11.3.5 *Given* $\{\alpha_1, \alpha_2\} \subseteq \mathbb{R}$, *and* $\{\underline{a}_1, \underline{a}_2\} \subseteq \mathcal{I}_2^{loc}[\underline{b}]$, *fixed, but arbitrary,*

1. $\alpha_1 I_{\underline{B}}\{\underline{a}_1\} + \alpha_2 I_{\underline{B}}\{\underline{a}_1\} = I_{\underline{B}}\{\alpha_1 \underline{a}_1 + \alpha_2 \underline{a}_2\}$;
2. $\langle I_{\underline{B}}\{\underline{a}_1\}, I_{\underline{B}}\{\underline{a}_2\}\rangle(\cdot, t) = \langle \underline{a}_{1|t}, \underline{a}_{2|t}\rangle_{L_2[\underline{b}]}$;
3. *for wide sense stopping times S,*

$$I_{\underline{B}}^S\{\underline{a}\} = I_{\underline{B}}\{\underline{a}_{|S_n}\} = I_{\underline{B}^S}\{\underline{a}\}.$$

Proof For $i \in [1 : 2]$, fixed, but arbitrary, let

$$X_i = I_{\underline{B}}\{\underline{a}_i\}, \qquad X_{i,p} = I_{\underline{B}_p}\{\underline{a}_{i,p}\},$$

$$Y = I_{\underline{B}}\{\alpha_1 \underline{a}_1 + \alpha_2 \underline{a}_2\}, \quad Y_p = I_{\underline{B}_p}\{\alpha_1 \underline{a}_{1,p} + \alpha_2 \underline{a}_{2,p}\},$$

$$Z_{i,p} = X_i - X_{i,p}, \qquad Z_p = Y - Y_p.$$

For $p \in \mathbb{N}$, fixed, but arbitrary, one has that [264, p. 162]

$$\alpha_1 X_{1,p} + \alpha_2 X_{2,p} = Y_p.$$

But, inserting that latter equality,

$$\left\{\omega \in \Omega : \sup_{t \in [0, S_n(\omega)]} |\alpha_1 X_1(\omega, t) + \alpha_2 X_2(\omega, t) - Y(\omega, t)| > \epsilon\right\} =$$

$$= \left\{\omega \in \Omega : \sup_{t \in [0, S_n(\omega)]} \left|\alpha_1 Z_{1,p}(\omega, t) + \alpha_2 Z_{2,p}(\omega, t) + Z_p(\omega, t)\right| > \epsilon\right\}$$

$$\subseteq \left\{\omega \in \Omega : |\alpha_1| \sup_{t\in[0,S_n(\omega)]} \left|Z_{1,p}(\omega,t)\right| > \frac{\epsilon}{3}\right\}$$

$$\bigcup \left\{\omega \in \Omega : |\alpha_2| \sup_{t\in[0,S_n(\omega)]} \left|Z_{2,p}(\omega,t)\right| > \frac{\epsilon}{3}\right\}$$

$$\bigcup \left\{\omega \in \Omega : \sup_{t\in[0,S_n(\omega)]} \left|Z_p(\omega,t)\right| > \frac{\epsilon}{3}\right\},$$

so that, taking probabilities, and using locally, uniform convergence in probability, the first relation obtains. The second follows from [264, p. 165]:

$$\langle I_{\underline{B}}\{\underline{a}_1\}, I_{\underline{B}}\{\underline{a}_2\}\rangle = \frac{1}{4}\left\{\langle I_{\underline{B}}\{\underline{a}_1\} + I_{\underline{B}}\{\underline{a}_2\}\rangle - \langle I_{\underline{B}}\{\underline{a}_1\} - I_{\underline{B}}\{\underline{a}_2\}\rangle\right\},$$

and both (Facts) 11.3.4 and 11.3.5, item 1, above.

Since $I_{\underline{B}}\{\underline{a}\}$ is, locally, the uniform limit in probability of the sequence

$$\left\{I_{\underline{B}_p}\left\{\underline{a}_p\right\}, p \in \mathbb{N}\right\},$$

the same will be true for the processes stopped at S: $I^S_{\underline{B}}\{\underline{a}\}$ is, locally, the uniform limit in probability of the sequence

$$\left\{I^S_{\underline{B}_p}\left\{\underline{a}_p\right\}, p \in \mathbb{N}\right\}.$$

But the result is true [264, p. 96] for the elements of

$$\left\{I_{\underline{B}_p}\left\{\underline{a}_p\right\}, p \in \mathbb{N}\right\}.$$

Then $I_{\underline{B}}\left\{\underline{a}_{|S_n}\right\}$ is the limit of

$$\left\{I_{\underline{B}_p}\left\{\underline{a}_{p|S_n}\right\}, p \in \mathbb{N}\right\},$$

and $I_{\underline{B}^S}\{\underline{a}\}$ that of

$$\left\{I_{\underline{B}^S_p}\left\{\underline{a}_p\right\}, p \in \mathbb{N}\right\}.$$

Since the limit of each of these sequences is unique, the stated equalities obtain. □

Remark 11.3.6 (Property (Fact)) 11.3.5, item 2, has the following consequence: every Gaussian process N that is continuous in quadratic mean is an isonormal [204,

p. 4] process with $L_2[\underline{b}]$ as indexing Hilbert space. That allows time to be taken into account.

Fact 11.3.7 *Let α be a progressively measurable process such that*

$$(\alpha \underline{a}) \in \mathcal{I}_2^{loc}[\underline{b}].$$

Then:

$$\int \alpha \, dI_{\underline{B}} \{\underline{a}\} = I_{\underline{B}} \{\alpha \underline{a}\} .$$

Proof The result is true in the real case [264, p. 171], so that

$$I_{\underline{B}} \{\alpha \underline{a}\} = \lim_n I_{\underline{B}_n} \{\alpha \underline{a}_n\} = \lim_n \int \alpha \, dI_{\underline{B}_n} \{\underline{a}_n\} .$$

But [264, p. 171]

$$\int \alpha \, dI_{\underline{B}} \{\underline{a}\} - \int \alpha \, dI_{\underline{B}_n} \{\underline{a}_n\} = \int \alpha \, d \left\{ I_{\underline{B}} \{\underline{a}\} - I_{\underline{B}_n} \{\underline{a}_n\} \right\} ,$$

and, for the appropriate localizing sequence $\{S_n, n \in \mathbb{N}\}$,

$$\begin{aligned} P\left(\left| \int^{S_n} \alpha \, d \left\{ I_{\underline{B}} \{\underline{a}\} - I_{\underline{B}_n} \{\underline{a}_n\} \right\} \right| \geq \epsilon \right) &= \\ &\leq \epsilon^{-2} E_P \left[\int^{S_n} \alpha^2 d \langle I_{\underline{B}} \{\underline{a}\} - I_{\underline{B}_n} \{\underline{a}_n\} \rangle \right] \\ &\leq \epsilon^{-2} E_P \left[\int \alpha^2 \sum_{i=n+1}^{\infty} a_i^2 dM_i \right] . \end{aligned}$$

□

Fact 11.3.8 *Let $\underline{a} \in \mathcal{I}_0 [\underline{b}]$ be fixed, but arbitrary, and S be a wide sense stopping time. Then, for $\epsilon > 0$ and $\delta > 0$, fixed, but arbitrary,*

$$\begin{aligned} P\left(\omega \in \Omega : \sup_{t \in [0, S(\omega)]} \left| I_{\underline{B}} \{\underline{a}\} (\omega, t) \right| > \epsilon \right) &\leq \\ &\leq P \left(\omega \in \Omega : A(\omega, S(\omega)) \geq \delta \right) + 4 \frac{\delta}{\epsilon^2} . \end{aligned}$$

Proof Let $\{S_n, n \in \mathbb{N}\}$ be the sequence of strict stopping times of (Proposition) 11.2.3. The following integrals are well defined:

$$I_{\underline{B}}\left\{\underline{a}_{|S_n}\right\}(\cdot, t), \quad \text{and } I_{\underline{B}}\left\{\underline{a} - \underline{a}_{|S_n}\right\}(\cdot, t).$$

Then

$$\begin{aligned} \sup_{t\in[0,1]} \left| I_{\underline{B}}\{\underline{a}\}(\omega, t) \right| &\leq \\ &\leq \sup_{t\in[0,1]} \left\{ \left| I_{\underline{B}}\left\{\underline{a} - \underline{a}_{|S_n}\right\}(\omega, t)\right| + \left| I_{\underline{B}}\left\{\underline{a}_{|S_n}\right\}(\omega, t)\right| \right\} \\ &\leq \sup_{t\in[0,1]} \left| I_{\underline{B}}\left\{\underline{a} - \underline{a}_{|S_n}\right\}(\omega, t)\right| + \sup_{t\in[0,1]} \left| I_{\underline{B}}\left\{\underline{a}_{|S_n}\right\}(\omega, t)\right|. \end{aligned}$$

Since the integral processes involved are separable, the *suprema* are random variables, and

$$P\left(\omega \in \Omega : \sup_{t\in[0,1]} \left| I_{\underline{B}}\{\underline{a}\}(\omega, t)\right| > \epsilon\right) \leq P\left(\omega \in \Omega : U(\omega) + V(\omega) > \epsilon\right)$$

when

$$\begin{aligned} U(\omega) &= \sup_{t\in[0,1]} \left| I_{\underline{B}}\left\{\underline{a} - \underline{a}_{|S_n}\right\}(\omega, t)\right|, \\ V(\omega) &= \sup_{t\in[0,1]} \left| I_{\underline{B}}\left\{\underline{a}_{|S_n}\right\}(\omega, t)\right|. \end{aligned}$$

It follows from (Lemma) 11.3.2 that

$$\begin{aligned} P\left(\omega \in \Omega : \sup_{t\in[0,1]} \left| I_{\underline{B}}\{\underline{a}\}(\omega, t)\right| > \epsilon\right) &\leq \\ &\leq P\left(\omega \in \Omega : U(\omega) > 0\right) + P\left(\omega \in \Omega : V(\omega) > \epsilon\right). \end{aligned}$$

For $n \in \mathbb{N}$, fixed, but arbitrary, set

$$\Omega_n^{(1)} = \{\omega \in \Omega : A(\omega, 1) < n\}, \quad \text{and } \Omega_n^{(2)} = \{\omega \in \Omega : S_n(\omega) = 1\}.$$

The definition of S_n then implies that

$$\Omega_n^{(1)} \subseteq \Omega_n^{(2)}.$$

Furthermore, when $\omega \in \Omega_n^{(2)}$, $\underline{a}_{|S_n}(\omega,\cdot) = \underline{a}(\omega,\cdot)$. Consequently,

$$\Omega_n^{(2)} \subseteq \{\omega \in \Omega : U(\omega) = 0\},$$

and thus

$$\{\omega \in \Omega : U(\omega) > 0) = \{\omega \in \Omega : U(\omega) = 0\}^c \subseteq \left[\Omega_n^{(2)}\right]^c \subseteq \left[\Omega_n^{(1)}\right]^c,$$

so that

$$P(\omega \in \Omega : U(\omega) > 0) \le P\left(\left[\Omega_n^{(1)}\right]^c\right) = P(\omega \in \Omega : A(\omega,1) \ge n).$$

Using again Markov's [138, p. 164], and Doob's inequalities [264, p. 58], one obtains that

$$\begin{aligned} P(\omega \in \Omega : V(\omega) > \epsilon) &\le 4\epsilon^{-2} E_P\left[I_{\underline{B}}^2\left\{\underline{a}_{|S_n}\right\}(\cdot,1)\right] \\ &= 4\epsilon^{-2} E_P\left[I_{\underline{B}}^2\{\underline{a}\}(\cdot,S_n)\right] \\ &= 4\epsilon^{-2} E_P\left[A(\cdot,S_n)\right] \\ &\le 4\epsilon^{-2} n. \end{aligned}$$

Finally

$$P\left(\omega \in \Omega : \sup_{t\in[0,1]} \left|I_{\underline{B}}\{\underline{a}\}(\omega,t)\right| > \epsilon\right) \le P(\omega \in \Omega : A(\omega,1) \ge n) + 4\epsilon^{-2} n.$$

As

$$\begin{aligned} &P\left(\omega \in \Omega : \sup_{t\in[0,1]} \left|I_{\underline{B}}\{\underline{a}\}(\omega,t)\right| > \epsilon\right) = \\ &\qquad = P\left(\omega \in \Omega : \sup_{t\in[0,1]} \left|I_{\underline{B}}\left\{(n/\delta)^{1/2}\underline{a}\right\}(\omega,t)\right| > (n/\delta)^{1/2}\epsilon\right), \end{aligned}$$

it follows from the last inequality between probabilities that

$$\begin{aligned} &P\left(\omega \in \Omega : \sup_{t\in[0,1]} \left|I_{\underline{B}}\{\underline{a}\}(\omega,t)\right| > \epsilon\right) \le \\ &\qquad \le P(\omega \in \Omega : (n/\delta)A(\omega,1) \ge n) + 4\epsilon^{-2}\delta. \end{aligned}$$

Let $f : [0,1] \longrightarrow \mathbb{R}$ denote a fixed, but arbitrary function, and $t \in [0,1]$ be fixed, but arbitrary. Then

$$\sup_{\theta\in[0,t]} f(\theta) = \sup_{\theta\in[0,1]} f(\theta \wedge t),$$

so that

$$P\left(\omega \in \Omega : \sup_{t\in[0,S(\omega)]} \left|I_{\underline{B}}\left\{\underline{a}\right\}(\omega,t)\right| > \epsilon\right) =$$

$$= P\left(\omega \in \Omega : \sup_{t\in[0,1]} \left|I_{\underline{B}}\left\{\underline{a}_{|S}\right\}(\omega,t)\right| > \epsilon\right).$$

□

Fact 11.3.9 *Suppose that* $\left\{\underline{a}_1, \underline{a}_2\right\} \subseteq \mathcal{I}_2[\underline{b}]$ *are fixed, but arbitrary, and that, almost surely with respect to P,*

$$I_{\underline{B}}\left\{\underline{a}_1\right\}(\cdot,1) = I_{\underline{B}}\left\{\underline{a}_2\right\}(\cdot,1).$$

Then $\underline{a}_1 = \underline{a}_2$ *in* $\mathcal{I}_2[\underline{b}]$.

Proof Indeed,

$$E_P\left[\left\{I_{\underline{B}}\left\{\underline{a}_1\right\}(\cdot,1) - I_{\underline{B}}\left\{\underline{a}_2\right\}(\cdot,1)\right\}^2\right] = E_P\left[\left\|\underline{a}_1 - \underline{a}_2\right\|^2_{L_2[\underline{b}]}\right].$$

□

Fact 11.3.10 *Let* $\underline{\alpha} \in l_2$ *be fixed, but arbitrary, and let, for* $n \in \mathbb{N}$ *and* $t \in [0,1]$, *fixed, but arbitrary,* $a_n^{\underline{\alpha}}(t) = \alpha_n$, *the n-th component of* $\underline{\alpha}$. *Since*

$$\int_0^1 \left\{a_n^{\underline{\alpha}}\right\}^2 dM_n = \alpha_n^2 b_n(1) = 2^{-n}\alpha_n^2 \le \alpha_n^2,$$

$\underline{a}^{\underline{\alpha}}$ *with components* $\left\{a_n^{\underline{\alpha}}, n \in \mathbb{N}\right\}$ *belongs to* $\mathcal{I}_2[\underline{b}]$ *and*

$$I_{\underline{B}}\left\{\underline{a}^{\underline{\alpha}}\right\} = \langle\underline{\alpha}, \underline{B}\rangle_{l_2}.$$

11.3.3 Stochastic Integrals and Change of Space

One shall also need, for stochastic integrals, transformation formulae analogous to

$$\int_A f d(P \circ T^{-1}) = \int_{T^{-1}(A)} (f \circ T) dP.$$

To that end, let $(\Omega, \underline{\mathcal{A}}, P)$ and $(\Theta, \underline{\mathcal{B}}, Q)$ be two filtered spaces whose index set is $[0, 1]$. Let $f : \Omega \longrightarrow \Theta$ be a map such that:

- for $t \in [0, 1]$, fixed, but arbitrary, $f^{-1}(\mathcal{B}_t) \subseteq \mathcal{A}_t$;
- $Q = P \circ f^{-1}$.

Let X and Y be local martingales for, respectively, $(\underline{\mathcal{A}}, P)$ and $(\underline{\mathcal{B}}, Q)$, continuous to the right, almost surely continuous, and such that

$$X(\omega, t) = Y(f(\omega), t).$$

Remark 11.3.11 Let $F : \Omega \times [0, 1] \longrightarrow \Theta \times [0, 1]$ be the map

$$(\omega, t) \mapsto (f(\omega), t),$$

that is,

$$F = (f \circ \Pi_\Omega, id \circ \Pi_{[0,1]}),$$

where the Π maps are the projections onto the factors. For S_Θ and $S_{[0,1]}$, sets respectively in Θ and $[0, 1]$,

$$F^{-1}\left(S_\Theta \times S_{[0,1]}\right) = f^{-1}(S_\Theta) \times S_{[0,1]}.$$

When $S_\Theta \in \mathcal{B}_t$ and $S_{[0,1]} \in \mathcal{B}([0, t])$,

$$F^{-1}\left(S_\Theta \times S_{[0,1]}\right) \in \mathcal{A}_t \times \mathcal{B}([0, t]).$$

As [138, p. 46] $\sigma(\Phi^{-1}(\mathcal{S})) = \Phi^{-1}(\sigma(\mathcal{S}))$,

$$f^{-1}(\mathcal{B}_t) \otimes \mathcal{B}([0, t]) = F^{-1}(\mathcal{B}_t \otimes \mathcal{B}([0, t])) \subseteq \mathcal{A}_t \otimes \mathcal{B}([0, t]).$$

Finally, as $F^{-1}(S \cap \{\Theta \times [0, t]\}) = F^{-1}(S) \cap \{\Omega \times [0, t]\}$, F is adapted to the progressively measurable sets.

Looking at sets of the form $B \times]t_1, t_2]$, $B \in \mathcal{B}_{t_1}$, one sees analogously that F is adapted to the predictable sets.

Consequently when ϕ is progressively measurable, or predictable, for $\underline{\mathcal{B}}$, $\phi \circ F$ is then progressively measurable, respectively, predictable for $\underline{\mathcal{A}}$.

Remark 11.3.12 Suppose that $\langle X \rangle = \langle Y \rangle \circ F$, F as in (Remark) 11.3.11, and let

$$\mu(d\omega, d\xi) = P(d\omega)\langle X\rangle(\omega, d\xi),$$
$$\nu(d\theta, d\eta) = Q(d\theta)\langle Y\rangle(\theta, d\eta).$$

Then

$$\begin{aligned}
&\mu \circ F^{-1}\,(B\times]t_1, t_2]) = \\
&= \int_{\Omega} P(d\omega)\,\chi_{f^{-1}(B)}(\omega)\,\{\langle X\rangle(\omega, t_2) - \langle X\rangle(\omega, t_1)\} \\
&= \int_{\Omega} P(d\omega)\,\chi_B(f(\omega))\,\{\langle Y\rangle(f(\omega), t_2) - \langle Y\rangle(f(\omega), t_1)\} \\
&= \int_{\Theta} Q(d\theta)\,\chi_B(\theta)\,\{\langle Y\rangle(\theta, t_2) - \langle Y\rangle(\theta, t_1)\} \\
&= \nu\,(B\times]t_1, t_2]),
\end{aligned}$$

so that, from $\int_C \phi\, d\nu = \int_{F^{-1}(C)} \phi \circ F\, d\mu$, one gets that

$$\int_C \phi(\theta, \eta)\, Q(d\theta)\, \langle Y\rangle(\theta, d\eta) = \int_{F^{-1}(C)} \phi \circ F(\omega, \xi)\, P(d\omega)\, \langle X\rangle(\omega, d\xi),$$

and, when $C = \Theta \times [0, t]$,

$$E_Q\left[\int_0^t \phi(\cdot, \eta)\, \langle Y\rangle(\cdot, d\eta)\right] = E_P\left[\int_0^t \phi(f(\cdot), \xi)\, \langle X\rangle(\cdot, d\xi)\right].$$

Remark 11.3.13 Since

$$\begin{aligned}
&P\,(\omega \in \Omega : \chi_B(f(\omega))\,\{\langle Y\rangle(f(\omega), t_2) - \langle Y\rangle(f(\omega), t_1)\} \in C) = \\
&= Q\,(\theta \in \Theta : \chi_B(\theta)\,\{\langle Y\rangle(\theta, t_2) - \langle Y\rangle(\theta, t_1)\} \in C),
\end{aligned}$$

and that, for example,

$$\chi_B(\theta)\,\{\langle Y\rangle(\theta, t_2) - \langle Y\rangle(\theta, t_1)\} = \int_0^1 \chi_B(\theta)\chi_{]t_1, t_2]}(\eta)\langle Y\rangle(\theta, d\eta),$$

for appropriate ϕ,

$$P\left(\omega \in \Omega : \int_0^1 \phi(f(\omega), \xi)\langle X\rangle(\omega, d\xi) \in C\right) =$$
$$= Q\left(\theta \in \Theta : \int_0^1 \phi(\theta, \eta)\langle Y\rangle(\omega, d\eta) \in C\right).$$

Proposition 11.3.14 *One has that*

1. $\langle X\rangle = \langle Y\rangle \circ F$;
2. *for predictable ϕ such that, for $t \in [0, 1]$, fixed, but arbitrary,*

$$Q\left(\theta \in \Theta : \int_0^t \phi^2(\theta, x)\,\langle Y\rangle(\theta, dx) < \infty\right) = 1,$$

for $t \in [0, 1]$, fixed, but arbitrary, almost surely, with respect to P,

$$\int_0^t \phi(f(\omega), x) X(\omega, dx) = \left\{\int_0^t \phi(\cdot, x)\, Y(\cdot, dx)\right\} \circ f(\omega).$$

Proof One has [264, p. 101], for $t \in [0, 1]$, fixed, but arbitrary, locally, uniformly in probability for Q, for every sequence of partitions of $[0, t]$, whose mesh goes to zero,

$$\langle Y\rangle(\cdot, t) = \lim_n \sum_{i=1}^{p_n} \left\{Y\left(\cdot, t_{i+1}^{(n)}\right) - Y\left(\cdot, t_i^{(n)}\right)\right\}^2.$$

Using, when necessary, a subsequence, one may assume that convergence is almost sure. But then, almost surely, with respect to P,

$$\begin{aligned}\langle Y\rangle(f(\omega), t) &= \lim_n \sum_{i=1}^{p_n} \left\{Y\left(f(\omega), t_{i+1}^{(n)}\right) - Y\left(f(\omega), t_i^{(n)}\right)\right\}^2 \\ &= \lim_n \sum_{i=1}^{p_n} \left\{X\left(\omega, t_{i+1}^{(n)}\right) - X\left(\omega, t_i^{(n)}\right)\right\}^2 \\ &= \langle X\rangle(\omega, t).\end{aligned}$$

The continuity of paths assumption says then that the paths of $\langle X\rangle$ and $\langle Y\rangle \circ F$ cannot be distinguished.

Suppose first that $Y(\theta, 1) \in L_2[Q]$, and let

- $U : L_2[\nu] \longrightarrow L_2[\mu]$ be given by $\phi \mapsto \phi \circ F$, and
- $V : L_2[Q] \longrightarrow L_2[P]$, by $\zeta \mapsto \zeta \circ f$.

U is an isometry because of (Remark) 11.3.12. V is an isometry because of the change of variables formula. Let I_P be the stochastic integral isometry between $L_2[\mu]$ and $L_2[P]$, and I_Q, that between $L_2[\nu]$ and $L_2[Q]$. Item 2 amounts to proving that

$$V \circ I_Q = I_P \circ U.$$

Now

$$\begin{aligned}\int_0^t \chi_B(f(\omega))\chi_{]t_1,t_2]}(\xi)X(\omega,d\xi) &= \chi_B(f(\omega))\,[X(\omega,t_2\wedge t) - X(\omega,t_1)]\\ &= \chi_B(f(\omega))\,[Y(f(\omega),t_2\wedge t) - Y(f(\omega),t_1)]\\ &= \{\chi_B(\cdot)\,[Y(\cdot,t_2\wedge t) - Y(\cdot,t_1)]\}\circ f(\omega)\\ &= \left\{\int_0^t \chi_B(\cdot)\chi_{]t_1,t_2]}(\eta)Y(\cdot,d\eta)\right\}\circ f(\omega),\end{aligned}$$

and the processes of the form $\chi_B\chi_{]t_1,t_2]}$ are total in $L_2[\nu]$.

In the general case, one uses a localizing sequence $\{S_n, n \in \mathbb{N}\}$ that localizes Y as well as

$$\int_0^t \phi dY.$$

Now $\{S_n \circ f, n \in \mathbb{N}\}$ localizes X, and $X^{S_n\circ f}(\omega,t) = Y^{S_n}(f(\omega),t)$, so that, from the first part,

$$\int_0^t (\phi\circ f)dX^{S_n\circ f} = \left\{\int_0^t \phi dY^{S_n}\right\}\circ f.$$

The properties of the stochastic integral yield then that

$$\int_0^{t\wedge(S_n\circ f)} (\phi\circ f)dX = \left\{\int_0^{t\wedge S_n} \phi dY\right\}\circ f.$$

One then lets n increase indefinitely. □

Given a martingale M, continuous to the right, almost surely continuous, and zero at the origin, one may define [264, p. 117] the Föllmer-Doléans measure μ_M on $\mathcal{A}\otimes\mathcal{B}([0,1])$ using, for measurable ϕ, the following relation:

$$\int_{\Omega\times[0,1]} \phi d\mu_M = E_P\left[\int_0^1 \phi d\langle M\rangle\right].$$

Then [264, p. 126], given a progressively measurable ψ for which, for $t \in [0, 1]$, fixed, but arbitrary,

$$P\left(\int_0^t \psi^2 d\langle M\rangle < \infty\right) = 1,$$

there is a predictable ϕ such that $\mu_M(\phi \neq \psi) = 0$. One then defines [264, p. 127] the stochastic integral of ψ with respect to M using the following relation:

$$\int_0^t \psi \, dM = \int_0^t \phi \, dM.$$

One has thus the following corollary:

Corollary 11.3.15 *(Result* (Proposition)) 11.3.14 *is true for progressively measurable ϕ.*

11.4 Local Martingales of a Cramér-Hida Process

Farther, one shall have to use martingales obtained from conditioning Radon-Nikodým derivatives. To establish that these martingales have the adequate regularity properties, one relies on the fact that those properties hold when the conditioning fields are obtained from Gaussian martingales. As seen below Hermite polynomials may serve as the basic tool of the matter.

Definition 11.4.1 Let $\underline{B}$ be a Cramér-Hida process for $\underline{\mathcal{A}}$.

1. Let $\underline{\mathcal{B}} = \{\mathcal{B}_t, t \in [0, 1]\}$ be a filtration of $\mathcal{A}$. It is called a $\underline{B}$-Gaussian martingale filtration when the following holds: for $t \in [0, 1]$, fixed, but arbitrary,

$$\sigma_t(\underline{B}) \subseteq \mathcal{B}_t \subseteq {}^o\sigma_t(\underline{B})$$

where ${}^o\sigma_t(\underline{B})$ is the σ-algebra generated by $\sigma_t(\underline{B})$ and the sets of $\mathcal{A}$ that have measure zero for P, that is, $\mathcal{N}(\mathcal{A}, P)$.
2. A polynomial in $(t, x) \in [0, \infty[\times\mathbb{R}$, of the following form, where $n \in \{0\} \cup \mathbb{N}$ is fixed, but arbitrary:

$$H_n(t, x) = \frac{(-t)^n}{n!} e^{\frac{x^2}{2t}} D_x^n \left[e^{-\frac{x^2}{2t}}\right]$$

is called a (generalized) Hermite polynomial of degree n [D_x^n *represents the n-th derivative with respect to the variable x*].

Remark 11.4.2 Let B be a Gaussian martingale for $\underline{\mathcal{A}}$, with variance function b, and paths continuous to the right, and almost surely, with respect to P, zero at the origin, and continuous. Let $\underline{\mathcal{B}}$ be a B-Gaussian filtration such that, for $t \in [0, 1]$, fixed, but arbitrary,

$$\sigma_t(B) \subseteq \mathcal{B}_t \subseteq {}^o\sigma_t(B),$$

where ${}^o\sigma_t(B)$ is the σ-algebra generated by $\sigma_t(B)$, and the sets of $\mathcal{A}$ that have measure zero for P. B is then a mean zero, almost surely continuous, Gaussian martingale for $\underline{\mathcal{B}}$ with variance function b.

Let indeed X be a martingale for $\underline{\sigma}(B)$. As seen (Remark 10.2.14, and Sect. 10.2.2),

$${}^o\sigma_t(B) = \sigma(\mathcal{C}_t),\ \mathcal{C}_t = \{B\Delta N, B \in \mathcal{B}_t, N \in \mathcal{N}(\mathcal{A}, P)\}.$$

One has, for $t_2 > t_1$ in $[0, 1]$ and $B\Delta N$ in $\mathcal{C}_{t_1}$, fixed, but arbitrary, that

$$\begin{aligned}\int_{B\Delta N} X(\omega, t_2)P(d\omega) &= \int_B X(\omega, t_2)P(d\omega)\\ &= \int_B X(\omega, t_1)P(d\omega)\\ &= \int_{B\Delta N} X(\omega, t_1)P(d\omega).\end{aligned}$$

By the monotone class theorem one has that equality obtains for all sets in $\sigma(\mathcal{C}_{t_1})$. So X is a martingale for ${}^o\underline{\sigma}(B)$. But then it is a martingale for the "smaller" filtration $\underline{\mathcal{B}}$.

Fact 11.4.3 *Hermite polynomials have the following properties* [264, p. 231]*:*

1. $H_0(t, x) = 1$, *and*

$$\begin{aligned}H_1(t, x) &= x,\\ H_2(t, x) &= \frac{1}{2}(x^2 - t),\\ H_3(t, x) &= \frac{1}{6}x^3 - \frac{1}{2}tx,\\ H_4(t, x) &= \frac{1}{24}x^4 - \frac{1}{4}tx^2 + \frac{1}{8}t^2,\\ &\dots\quad ;\end{aligned}$$

2. *the leading term of* $H_n(t, x)$ *is* $\frac{x^n}{n!}$*;*

3. $D_t H_n(t,x) + \frac{1}{2} D_x^2 H_n(t,x) = 0$, *so that* [264, p. 185], *for almost surely continuous local martingales* M,

$$dH_n(\langle M\rangle, M) = H_{n-1}(\langle M\rangle, M)dM;$$

Remark 11.4.4 From (Fact) 11.4.3, item 2, one has that, for $t \in \mathbb{R}$, fixed, but arbitrary, every polynomial may be expressed as a linear combination of Hermite polynomials. The coefficients will depend on t. For example,

$$\alpha_0 + \alpha_1 x + \alpha_2 x^2 = (\alpha_0 + \alpha_2 t) + \alpha_1 H_1(t,x) + 2\alpha_2 H_2(t,x).$$

Lemma 11.4.5 *Let* B *be a Gaussian martingale for* $\underline{\mathcal{A}}$*, with variance function* b*, and paths continuous to the right, and almost surely, with respect to* P*, zero at the origin, and continuous. Let* $\mathcal{N}(0, b(t))$ *denote the law of a real, normal random variable with mean zero and variance* $b(t)$*. Then:*

1. *the process* $(\omega, t) \mapsto H_n(b(t), B(\omega, t))$ *is a martingale in* L_2 *for* $\underline{\mathcal{A}}$ *(and thus for* $\underline{\sigma}(B)$*) such that*

$$\|H_n(b(t), B(\cdot, t))\|^2_{L_2(\Omega,\mathcal{A},P)} = \frac{b^n(t)}{n!};$$

2. *for* $b(t) > 0$*, fixed, but arbitrary, the following family of polynomial functions:*

$$\left\{\sqrt{\frac{n!}{b^n(t)}} H_n(b(t), \cdot), n \in \{0\} \cup \mathbb{N}\right\}$$

is orthonormal, and complete in the space $L_2(\mathbb{R}, \mathcal{B}(\mathbb{R}), \mathcal{N}(0, b(t)))$*;*

3. *every polynomial function* $p(X_1, \ldots, X_n)$ *of independent, Gaussian, centered random variables* $X_1, \ldots, X_n$*, with respective variances* $\sigma_1^2, \ldots, \sigma_n^2$*, has a representation of the following form:*

$$\sum_{k_1,\ldots,k_m=0}^{n} a_{k_1,\ldots,k_m} \prod_{i=1}^{m} H_{k_i}(\sigma_i^2, X_i),$$

and the terms of the sum are orthogonal [264, p. 234]*;*

4. *for* $a \in [0, 1[$*, and* $t \in [0, 1-a]$*, fixed, but arbitrary,*

$$H_n(b(t) - b(a), B(\cdot, t) - B(\cdot, a)) =$$
$$= \int_a^{a+t} H_{n-1}(b(\theta) - b(a), B(\cdot, \theta) - B(\cdot, a)) B(\cdot, d\theta).$$

Proof (1) Since $H_0(t,x) = 1$, and $H_1(t,x) = x$, item 1 is true for those two cases. Suppose the formula is true for n_0. Then, from (Fact) 11.4.3, item 3, and [43, p. 337],

$$\begin{aligned} E_P\left[H_{n_0+1}^2(b(t), B(\cdot,t))\right] &= \int_0^t E_P\left[H_{n_0}^2(b(\theta), B(\cdot,\theta))\right] M_b(d\theta) \\ &= \int_0^t \frac{b^{n_0}(\theta)}{n_0!} M_b(d\theta) \\ &= \frac{b^{n_0+1}(\theta)}{n_0+1!}. \end{aligned}$$

Proof (2,3) It proceeds, *mutatis mutandis,* as in [264, pp. 233–235].

Proof (4) Let $W_a(\cdot,t) = B(\cdot,a+t) - B(\cdot,a)$. Using the definitions and (Corollary) 10.5.19, one obtains, for the filtration

$$\underline{\mathcal{W}}^a = \left\{\mathcal{W}_t^a, t \in [0, 1-a]\right\} = \left\{\mathcal{B}_{a+t}, t \in [0, 1-a]\right\},$$

a Gaussian martingale whose variance function is $t \mapsto b(a+t) - b(a)$. Furthermore, as presently seen,

$$\int_0^t f(\cdot,\theta) W_a(\cdot,d\theta) = \int_a^{a+t} f(\cdot,\theta-a) B(\cdot,d\theta). \qquad (\star)$$

Let indeed $S_1 \leq S_2$ be wide sense stopping times for $\underline{\mathcal{W}}^a$ with values in $[0, 1-a]$. Then, for $t \in [0, 1-a]$, fixed, but arbitrary,

$$\begin{aligned} \int_0^t \chi_{]\!]S_1,S_2]\!]}(\cdot,\theta) W_a(\cdot,d\theta) &= W_a(\cdot,t\wedge S_2) - W_a(\cdot,t\wedge S_1) \\ &= B(\cdot,a+t\wedge S_2) - B(\cdot,a+t\wedge S_1) \\ &= \int_0^1 \chi_{]\!]a+t\wedge S_1,a+t\wedge S_2]\!]}(\cdot,\theta) B(\cdot,d\theta) \\ &= \int_0^1 \chi_{]\!]t\wedge S_1,t\wedge S_2]\!]}(\cdot,\theta-a) B(\cdot,d\theta) \\ &= \int_0^1 \chi_{[0,t]}(\cdot,\theta-a)\chi_{]\!]S_1,S_2]\!]}(\theta-a) B(\cdot,d\theta) \\ &= \int_a^{a+t} \chi_{]\!]S_1,S_2]\!]}(\theta-a) B(\cdot,d\theta). \end{aligned}$$

A monotone class argument then completes the proof of $(\star)$ above. Since one has that [(Fact) 11.4.3, item 3]

$$H_n(\langle W_a\rangle(\cdot,t),W_a(\cdot,t))=\int_0^t H_{n-1}(\langle W_a\rangle(\cdot,\theta),W_a(\cdot,\theta))W_a(\cdot,d\theta),$$

it follows from $(\star)$ above that

$$\begin{aligned}&H_n(b(a+t)-b(a),B(\cdot,a+t)-B(\cdot,a))=\\&\quad=\int_a^{a+t} H_{n-1}(\langle W_a\rangle(\cdot,\theta-a),W_a(\cdot,\theta-a))B(\cdot,d\theta).\end{aligned}$$

But $W_a(\cdot,\theta-a)=B(\cdot,[\theta-a]+a)-B(\cdot,a)$ and

$$\langle W_a\rangle(\cdot,\theta-a)=b([\theta-a]+a)-b(a).$$

Item 4 thus obtains. □

Fact 11.4.6 *Let B be a Gaussian martingale for $\underline{\mathcal{A}}$, with variance function b, and paths continuous to the right, and almost surely, with respect to P, zero at the origin, and continuous. Then:*

1. [264, p. 234] *the polynomial functions form a family that is dense in the space*

$$L_2\left(\mathbb{R},\mathcal{B}(\mathbb{R}),\mathcal{N}\left(0,b(t)\right)\right);$$

2. [264, p. 235] *let $\mathcal{P}(B)$ be the linear span of the random variables of the following form:*

$$\prod_{i=1}^{n} p_i\left(B(\cdot,t_{i+1})-B(\cdot,t_i)\right),$$

with fixed, but arbitrary $0\le t_1<t_2<t_3<\cdots<t_{n+1}$ in $[0,1]$, and polynomials $p_1,\ldots,p_n$: one has that $\mathcal{P}(B)$ is dense in $L_2(\Omega,\sigma(B),P)$.

With the adjustments listed in (Lemma) 11.4.5, the following three results are proved as in [264, p. 239].

Lemma 11.4.7 *Let B be a Gaussian martingale for $\underline{\mathcal{A}}$, with variance function b, and paths continuous to the right, and almost surely, with respect to P, zero at the origin, and continuous. Let $\underline{\mathcal{B}}$ be a Gaussian martingale filtration for B, that is, for $t\in[0,1]$, fixed, but arbitrary, $\sigma_t(B)\subseteq\mathcal{B}_t\subseteq{}^o\sigma_t(B)$. Let the following times, and polynomials be fixed, but arbitrary:*

$$t_1<t_2<t_3<\cdots<t_n<t_{n+1}\in[0,1],\ \textit{and}\ p_1,\ldots,p_n.$$

Let

$$X = \prod_{i=1}^{n} p_i(B(\cdot, t_{i+1}) - B(\cdot, t_i)).$$

Then, almost surely, with respect to P,

$$E_P[X \mid \mathcal{B}_t] = \int_0^t \xi(\cdot, \theta) B(\cdot, d\theta),$$

where ξ is a predictable process whose square is integrable for the product measure $P \otimes M_b$.

Proof Let $X = Y \times Z$, where Y is the product of the first $n-1$ polynomial terms of X, and Z is the last one. Since p_n may be written as a combination of Hermite polynomials, X is the sum of terms of the following form:

$$Y_c H_q\left(b(t_{n+1}) - b(t_n), B(\cdot, t_{n+1}) - B(\cdot, t_n)\right),$$

where Y_c is Y times the ("constant") coefficient of the Hermite polynomial which follows. Y_c is adapted to $\sigma_{t_n}(B) \subseteq \mathcal{B}_{t_n}$. Also one may assume that the q in H_q is strictly positive (otherwise the Hermite polynomial would be a constant). For $t \geq t_n$, one has that (Lemma 11.4.5, item 4):

$$H_q(b(t) - b(t_n), B(\cdot, t) - B(\cdot, t_n)) =$$
$$= \int_{t_n}^t H_{q-1}(b(\theta) - b(t_n), B(\cdot, \theta) - B(\cdot, t_n)) B(\cdot, d\theta).$$

To make notation shorter, one shall temporarily write, for the latter equality,

$$V(\cdot, t) = \int_{t_n}^t U(\cdot, \theta) B(\cdot, d\theta).$$

Since Y_c is adapted to $\sigma_t(B) \subseteq \mathcal{B}_t$,

$$E_P[X \mid \mathcal{B}_t] = Y_c E_P[V(\cdot, t_{n+1}) \mid \sigma_t(B)].$$

Since V is a martingale for $\underline{\sigma}(B)$ [(Lemma) 11.4.5, item 1], it is one for $\underline{\mathcal{B}}$ [(Remark) 11.4.2], and

$$E_P[V(\cdot, t_{n+1}) \mid \mathcal{B}_t] = V(\cdot, t \wedge t_{n+1}).$$

But V has an integral representation:

$$V(\cdot, t \wedge t_{n+1}) = \int_{t_n}^{t\wedge t_{n+1}} U(\cdot,\theta)B(\cdot,d\theta).$$

Using [264, p. 155],

$$Y_c \int_{t_n}^{t\wedge t_{n+1}} U(\cdot,\theta)B(\cdot,d\theta) = \int_{t_n}^{t\wedge t_{n+1}} Y_c U(\cdot,\theta)B(\cdot,d\theta).$$

But

$$\int_{t_n}^{t\wedge t_{n+1}} Y_c U(\cdot,\theta)B(\cdot,d\theta) = \int_0^t \chi_{]t_n,t_{n+1}]}(\theta) Y_c U(\cdot,\theta)B(\cdot,d\theta),$$

and

$$\chi_{]t_n,t_{n+1}]}(\theta) Y_c U(\cdot,\theta)$$

is predictable. That its square is integrable follows from (Lemma) 11.4.5, item 1. □

Lemma 11.4.8 *Let, in (Lemma)* 11.4.7, *X be a random variable whose square is integrable. The conditional expectation of X with respect to* $\underline{\mathcal{B}}$ *maintains its representation as a stochastic integral.*

Proof Let $\{X_n, n \in \mathbb{N}\}$ be a sequence of random variables of the form given in (Lemma) 11.4.7, such that [(Fact) 11.4.6, item 2],

$$\lim_n E_P\left[(E_P[X \mid \mathcal{B}_1] - X_n)^2\right] = 0.$$

Then [(Lemma) 11.4.7]

$$E_P[X_n \mid \mathcal{B}_t] = \int_0^t \xi_n dB.$$

X_n closes that martingale, and thus

$$E_P\left[(X_p - X_n)^2\right] = E_P\left[\left(\int_0^1 \{\xi_p - \xi_n\}\, dB\right)^2\right] = E_P\left[\int_0^1 (\xi_p - \xi_n)^2\, dM_b\right].$$

$\{\xi_n, n \in \mathbb{N}\}$ is thus a Cauchy sequence of predictable processes in the L_2 space of the measure $P \otimes M_b$. It has thus a predictable limit in that space, say ξ. Then:

$$E_P[X \mid \mathcal{B}_t] = \lim_n E_P[X_n \mid \mathcal{B}_t] = \lim_n \int_0^t \xi_n dB = \int_0^t \xi\, dB.$$

□

Fact 11.4.9 [237, p. 421] *Let $\mathcal{B}$ and $\mathcal{C}$ be σ-algebras contained in $\mathcal{A}$. Suppose that X is a positive random variable, or that its conditional expectation with respect to $\mathcal{C}$ is finite. When $\mathcal{B}$ and $\mathcal{C} \vee \sigma(X)$ are independent,*

$$E_P[X \mid \mathcal{B} \vee \mathcal{C}] = E_P[X \mid \mathcal{C}].$$

Fact 11.4.10 [264, p. 239] *Let $\underline{B}_n$ be the projection onto $\mathbb{R}^n$ of a Cramér-Hida process, and $\underline{\mathcal{B}}^{(n)}$, a Gaussian martingale filtration for $\underline{B}_n$. Then, when $X \in L_2(\Omega, \mathcal{A}, P)$, $t \in [0, 1]$, fixed, but arbitrary, and*

$$\sigma_t(\underline{B}_n) \subseteq \mathcal{B}_t^{(n)} \subseteq {}^o\sigma_t(\underline{B}_n),$$

one has that

$$E_P\left[X \mid \mathcal{B}_t^{(n)}\right] = \sum_{i=1}^{n} \int_0^t a_i(\cdot, \theta) B_i(\cdot, d\theta) = I_{\underline{B}_n}\left\{\underline{a}_n\right\}(\cdot, t),$$

where, for $i \in [1 : n]$, fixed, but arbitrary, the process a_i is predictable, and a_i^2 is integrable with respect to $P \otimes M_i$, the latter product measure component being the measure determined by the variance function of B_i.

Proof Let $X = \prod_{i=1}^n X_i$, where X_i is adapted to $\sigma(B_i)$, the σ-algebra generated by the i-th component of $\underline{B}_n$. Suppose that X has a square that is integrable, and a strictly positive variance, which, given the independence of its components, means that each component has a square that is integrable. Let

$$Y_i(\cdot, t) = E_P\left[X_i \mid \sigma_t(B_i)\right], \text{ and } Y = \prod_{i=1}^{n} Y_i.$$

Since Y_i is, by definition, a martingale with respect to $\underline{\sigma}(B_i)$, it has the representation $\int_0^{\cdot} \eta_i \, dB_i$. As such, it has, in particular, almost surely continuous paths. Since the processes $\{Y_1, \cdots, Y_n\}$ are furthermore independent, their product is a local martingale, and their mutual quadratic variation is zero [264, p. 168]. It follows by Itô's formula that

$$dY = \sum_{i=1}^{n} \left\{ \prod_{j=1, j \neq i}^{n} Y_j \right\} dY_i,$$

and, consequently, that

$$dY = \sum_{i=1}^{n} \left\{ \prod_{j=1, j \neq i}^{n} Y_j \eta_i \right\} dB_i.$$

For $i \in [1:n]$, fixed, but arbitrary, let $\dot{B}_i \in \sigma_t(B_i)$ be fixed, but arbitrary. Then

$$\begin{aligned}\int_{\cap_{i=1}^n \dot{B}_i} Y(\cdot,t)\,dP &= \int_{\cap_{i=1}^n \dot{B}_i} \prod_{i=1}^n E_P[X_i \mid \sigma_t(B_i)]\,dP \\ &= \prod_{i=1}^n \int_{\dot{B}_i} X_i\,dP \\ &= \int_{\cap_{i=1}^n \dot{B}_i} X\,dP \\ &= \int_{\cap_{i=1}^n \dot{B}_i} E_P[X \mid \sigma_t(\underline{B}_n)]\,dP,\end{aligned}$$

so that, by the monotone class theorem,

$$Y = E_P[X \mid \sigma_t(\underline{B}_n)].$$

Since the elements X introduced at the beginning of the proof are total in the L_2 space generated by the process $\underline{B}_n$, the proof is complete in view of (Lemma) 11.4.8 and (Remark) 11.4.2. □

Fact 11.4.11 (264, p. 239) *Let $\underline{B}_n$ be the projection onto $\mathbb{R}^n$ of a Cramér-Hida process, and $\underline{\mathcal{B}}^{(n)}$, a Gaussian martingale filtration for $\underline{B}_n$. Then, when M is a local martingale for $\underline{\mathcal{B}}^{(n)}$ (as such, it is continuous to the right), zero at the origin, one has that*

$$M(\cdot,t) = \sum_{i=1}^n \int_0^t a_i(\cdot,\theta)B_i(\cdot,d\theta) = I_{\underline{B}_n}\{\underline{a}_n\}(\cdot,t),$$

where, for $i \in [1:n]$, fixed, but arbitrary, the process a_i is predictable, and a_i^2 is integrable with respect to $P \otimes M_i$, the latter product measure component being the measure determined by the variance function of B_i.

Proof Let $\{S_p, p \in \mathbb{N}\}$ be a localizing sequence for M. Let X_p be a random variable such that its square is integrable and

$$\lim_p E_P\left[\left|M(\cdot,S_p) - X_p\right|\right] = 0.$$

Doob's inequality [192, p. 93] then yields that

$$\begin{aligned}\lambda P\left(\sup_{t\in[0,1]} \left|M^{S_p}(\cdot,t) - E_P[X_p \mid \mathcal{B}_t]\right| \geq \lambda\right) &\leq \\ &\leq \sup_{t\in[0,1]} E_P\left[\left|M^{S_p}(\cdot,t) - E_P[X_p \mid \mathcal{B}_t]\right|\right] \\ &= E_P\left[\left|M(\cdot,S_p) - X_p\right|\right].\end{aligned}$$

Since $E_P[X_p \mid \mathcal{B}_t]$ is a process whose paths are almost surely continuous, so is M, which is thus locally in L_2. One may then assume that M^{S_p} is a martingale in L_2. As such it has a representation as a stochastic integral:

$$M^{S_p}(\cdot,t) = I_{\underline{B}_n}\left\{\chi_{]\!]0,S_p]\!]}\underline{a}^{(p)}\right\}(\cdot,t).$$

Then

$$M = I_{\underline{B}_n}\{\underline{a}\}, \quad \text{with } \underline{a} = \sum_{p=1}^{\infty} \chi_{]\!]S_{p-1},S_p]\!]}\underline{a}^{(p)}.$$

□

Lemma 11.4.12 *Let $\underline{B}$ be a Cramér-Hida process, and $\underline{\mathcal{B}}$, a Gaussian martingale filtration for $\underline{B}$. Then, when $X \in L_2(\Omega, \mathcal{A}, P)$, there exists $\underline{a} \in \mathcal{I}_2[\underline{b}]$, with predictable components, such that*

$$E_P[X \mid \mathcal{B}_1] = I_{\underline{B}}\{\underline{a}\}(\cdot,1).$$

Proof Let $\mathcal{B}_1^{(n)} = \mathcal{B}_1 \cap {}^o\sigma_1(B_1,\ldots,B_n)$, and $\underline{B}_n$ be the usual restriction-projection of $\underline{B}$. It follows then from (Lemma) 11.4.8 that, for some $\underline{a}_n$,

$$E_P\left[X \mid \mathcal{B}_1^{(n)}\right] = I_{\underline{B}_n}\{\underline{a}_n\}(\cdot,1).$$

Letting $a_p^{(n)} = 0$ for $p > n$, keeping the same notation, one may replace

$$I_{\underline{B}_n}\{\underline{a}_n\}(\cdot,1) \text{ with } I_{\underline{B}}\{\underline{a}_n\}(\cdot,1).$$

Now [201, p. 29], almost surely with respect to P, as well as in quadratic mean,

$$\lim_n E_P\left[X \mid \mathcal{B}_1^{(n)}\right] = E_P[X \mid \mathcal{B}_1].$$

In particular,

$$\left\{E_P\left[X \mid \mathcal{B}_1^{(n)}\right], n \in \mathbb{N}\right\}$$

is a Cauchy sequence in quadratic mean so that

$$E_P\left[\left\{E_P\left[X \mid \mathcal{B}_1^{(n+p)}\right] - E_P\left[X \mid \mathcal{B}_1^{(n)}\right]\right\}^2\right] = E_P\left[\left\|\underline{a}_{n+p} - \underline{a}_n\right\|_{L_2[\underline{b}]}^2\right]$$

has limit zero as n and p increase indefinitely. Consequently $\{\underline{a}_n, n \in \mathbb{N}\}$ is a Cauchy sequence in the Hilbert space $\mathcal{I}_2[\underline{b}]$, that is,

$$\bigoplus_{n=1}^{\infty} L_2(\Omega \times [0,1], \mathcal{A} \otimes \mathcal{B}([0,1]), P \otimes M_n),$$

and has thus a limit, say $\underline{a}$. But, because of [264, p. 124] (with Lebesgue measure replaced by M_i), each a_i, $i \in \mathbb{N}$, may be taken to be predictable. Finally

$$\begin{aligned} E_P\left[\{E_P[X \mid \mathcal{B}_1] - I_{\underline{B}}\{\underline{a}\}(\cdot,1)\}^2\right] &= \\ = E_P\Big[\{(E_P[X \mid \mathcal{B}_1] - E_P[X \mid \mathcal{B}_1^{(n)}]) &+ (I_{\underline{B}}\{\underline{a}_n\}(\cdot,1) - I_{\underline{B}}\{\underline{a}\}(\cdot,1))\}^2\Big] \\ \leq 2\left\{E_P\left[\{E_P[X \mid \mathcal{B}_1] - E_P[X \mid \mathcal{B}_1^{(n)}]\}^2\right] \right. &+ \left. E_P\left[I_{\underline{B}}^2\{\underline{a} - \underline{a}_n\}(\cdot,1)\right]\right\}, \end{aligned}$$

and, as seen, the right-hand side of the latter inequality goes to zero. □

Remark 11.4.13 Predictability in (Lemma) 11.4.12 is for the filtration $\underline{\mathcal{B}}$.

Proposition 11.4.14 *Let M be a martingale in L_2 for $\underline{\mathcal{B}}$ that is zero at the origin. There exists then $\underline{a}_M \in \mathcal{I}_2[\underline{b}]$, with predictable components (for $\underline{\mathcal{B}}$), such that, for $t \in [0,1]$, fixed, but arbitrary, almost surely with respect to P,*

$$M(\cdot,t) = I_{\underline{B}}\{\underline{a}_M\}(\cdot,t).$$

M may thus be assumed to be continuous to the right and, with respect to P, almost surely continuous.

Proof From (Lemma) 11.4.12, since $M(\cdot,1)$ is adapted to $\mathcal{B}_1$, and has a square that is integrable,

$$M(\cdot,1) = I_{\underline{B}}\{\underline{a}_M\}(\cdot,1).$$

Since both sides of the latter equality are martingales, the result follows by conditioning. □

Proposition 11.4.15 *Let M be, for $\underline{\mathcal{B}}$, a martingale locally in L_2, that is zero at the origin. There exists then $\underline{a}_M \in \mathcal{I}_2^{loc}[\underline{b}]$, with predictable components (for $\underline{\mathcal{B}}$), such that, for $t \in [0,1]$, fixed, but arbitrary, almost surely with respect to P,*

$$M(\cdot,t) = I_{\underline{B}}\{\underline{a}_M\}(\cdot,t).$$

M may thus be assumed to be, with respect to P, almost surely continuous.

Proof Let $\{S_n, n \in \mathbb{N}\}$ be a localizing sequence for M. (Result (Proposition)) 11.4.14 applied to M^{S_n} yields a $\underline{a}_n \in \mathcal{I}_2[\underline{b}]$ such that

$$M^{S_n}(\cdot, t) = I_{\underline{B}}\left\{\underline{a}_n\right\}(\cdot, t).$$

Now, for $p \in \mathbb{N}$, fixed, but arbitrary, one has that

$$I_{\underline{B}}\left\{\underline{a}_n\right\}(\cdot, t) = M^{S_n}(\cdot, t) = M^{S_n \wedge S_{n+p}}(\cdot, t) = I_{\underline{B}}\left\{\chi_{]0,S_n]}\underline{a}_{n+p}\right\}(\cdot, t),$$

and thus, because of (Fact) 11.3.9,

$$\chi_{]0,S_n]}\underline{a}_{n+p} = \underline{a}_n.$$

One may consequently define a_i setting $\chi_{]0,S_n]}\underline{a} = \underline{a}_n$ to obtain that

$$\underline{a}_M = \chi_{]]0,S_1]]}\underline{a}_1 + \sum_{n=2}^{\infty} \chi_{]]S_{n-1},S_n]]}\underline{a}_n.$$

Stochastic intervals of the type used in that last expression are predictable [264, p. 112], so that one is summing products of predictable processes, and the result is predictable. The process $\underline{a}$ belongs to $\mathcal{I}_2^{loc}[\underline{b}]$ by construction. □

Proposition 11.4.16 *Let M be, for $\underline{\mathcal{B}}$, a local martingale that is zero at the origin. There exists then $\underline{a}_M \in \mathcal{I}_2^{loc}[\underline{b}]$, with predictable components (for $\underline{\mathcal{B}}$), such that, for $t \in [0, 1]$, fixed, but arbitrary, almost surely with respect to P,*

$$M(\cdot, t) = I_{\underline{B}}\left\{\underline{a}_M\right\}(\cdot, t).$$

M may thus be assumed to be, with respect to P, almost surely continuous.

Proof Let $\{S_n, n \in \mathbb{N}\}$ be a localizing sequence for M, so that, for $n \in \mathbb{N}$, fixed, but arbitrary, M^{S_n} is a uniformly integrable martingale. The random variable $M(\cdot, S_n(\cdot))$ is, in particular, adapted to $\mathcal{B}_1$, and integrable. There is thus a sequence $\left\{M_{n,p}, p \in \mathbb{N}\right\}$ of simple functions, adapted to $\mathcal{B}_1$, and integrable, with the property that, in L_1,

$$\lim_p M_{n,p} = M(\cdot, S_n(\cdot)).$$

Let $\left\{\epsilon_q > 0, q \in \mathbb{N}\right\}$ be such that $\lim_q \epsilon_q = 0$,

$$E_P\left[\left|M(\cdot, S_n(\cdot)) - M_{n,q}\right|\right] < \epsilon_q, \text{ and } U_{n,q}(\cdot, t) = E_P\left[M_{n,q} - E_P\left[M_{n,q}\right] \mid \mathcal{B}_t\right].$$

$U_{n,q}$ is a martingale in L_2 for $\underline{\mathcal{B}}$, zero at the origin, and thus (Proposition 11.4.14) continuous to the right, and almost surely continuous. But then so is

$$V_{n,q}(\cdot, t) = E_P\left[M_{n,q} \mid \mathcal{B}_t\right].$$

Let

$$W_{n,q}(\cdot, t) = M^{S_n}(\cdot, t) - V_{n,q}(\cdot, t).$$

Doob's maximal inequality [192, p. 93] says that

$$\lambda P\left(\omega \in \Omega : \sup_{t \in [0,1]} \left|W_{n,q}(\omega, t)\right| > \lambda\right) \leq E_P\left[\left|W_{n,q}(\cdot, 1)\right|\right] < \epsilon_q.$$

Consequently, $\{V_{n,q}, q \in \mathbb{N}\}$ is a sequence of martingales that are continuous to the right, and almost surely continuous, with respect to P, that converges uniformly in probability towards the martingale M^{S_n} that inherits thus the same path properties. But then, M is locally in L_2 [264, p. 63]. The result then follows from (Proposition) 11.4.15. □

Remark 11.4.17 In the proposition to follow the following fact is used. Let the map $f : [0, 1] \longrightarrow \mathbb{R}_+$ be continuous with $f(0) = 0$. Let

$$f^\star(t) = \sup_{\theta \in [0,t]} f(\theta),$$

and t_f be the smallest abscissa of those t's at which f is maximum. $f^\star$ increases from 0 to t_f and then is constant with value $f(t_f)$. Let n be a fixed, but arbitrary integer. If $n > f(t_f)$, let $t_n = 1$. Otherwise, let t_n be the smallest t such that $f(t) \geq n$. The interval $[0, t_n]$ is obtained as the set of t's for which $f^\star(t) \leq n$.

Proposition 11.4.18 *Let M be, for $\underline{\mathcal{B}}$, a martingale that is zero at the origin. There exists then $\underline{a}_M \in \mathcal{I}_0[\underline{b}]$, with predictable components (for $\underline{\mathcal{B}}$), such that, for $t \in [0, 1]$, fixed, but arbitrary, almost surely with respect to P,*

$$M(\cdot, t) = I_{\underline{B}}\left\{\underline{a}_M\right\}(\cdot, t).$$

M may thus be assumed to be continuous to the right, and, with respect to P, almost surely continuous.

Proof Since $\underline{\mathcal{B}}$ is continuous to the right, its martingales have modifications continuous to the right [264, p. 53]. Thus martingales for that filtration are local martingales [264, p. 63]. One may assume [(Proposition) 11.4.16] that M is continuous to the right, and almost surely continuous. It is no restriction to suppose that it is continuous [(Proposition) 10.2.28]. Let thus $I_{M,n}(\omega) = \{t \in [0,1] : |M(\omega,t)| \geq n\}$, and

$$S_n(\omega) = \begin{cases} 1 & \text{when } I_{M,n}(\omega) = \emptyset \\ \inf I_{M,n}(\omega) & \text{when } I_{M,n}(\omega) \neq \emptyset \end{cases}.$$

S_n is a strict stopping time. M^{S_n} is then a martingale for $\underline{\mathcal{B}}$, uniformly bounded by n, and, as such, has a representation of the following form:

$$M^{S_n}(\omega,t) = I_{\underline{B}}\left\{\underline{a}_n\right\}(\omega,t) \text{ with } \underline{a}_n \in \mathcal{I}_2[\underline{b}].$$

Let

$$M^\star(\omega,t) = \sup_{\theta\in[0,t]} |M(\omega,\theta)|\,, \text{ and } \Theta_n = \{(\omega,t) \in \Omega \times [0,1] : M^\star(\omega,t) \leq n\}\,.$$

One has, for $n \in \mathbb{N}$, fixed, but arbitrary, that $\Theta_n \subseteq \Theta_{n+1}$. Furthermore, for $p > n$, in $\mathbb{N}$, fixed, but arbitrary, $M^{S_p \wedge S_n} = M^{S_p}$ yields that

$$I_{\underline{B}}\left\{\underline{a}_{p|S_n}\right\} = I_{\underline{B}}\left\{\underline{a}_n\right\},$$

and [(Remark) 11.4.17] $\underline{a}_{p|S_n} = \chi_{\Theta_n}\underline{a}_p$. Thus [(Fact) 11.3.9], for $p > n$, in $\mathbb{N}$, fixed, but arbitrary,

$$\chi_{\Theta_n}\underline{a}_p = \underline{a}_n.$$

It thus makes sense to define $\underline{a}_M$ setting $\chi_{\Theta_n}\underline{a}_M = \underline{a}_n$. But then

$$\begin{aligned} P\left(\omega \in \Omega : \left\|\underline{a}_M(\omega,\cdot)\right\|^2_{L_2[\underline{b}]} = \infty\right) &\leq \\ &\leq P\left(\omega \in \Omega : \left\|\underline{a}_M(\omega,\cdot) - \underline{a}_n(\omega,\cdot)\right\|^2_{L_2[\underline{b}]} > 0\right) \\ &\leq P\left(\omega \in \Omega : M^\star(\omega,1) \geq n\right). \end{aligned}$$

Since M has continuous paths, the latter probability goes to zero when n increases indefinitely. Let $Y = I_{\underline{B}}\left\{\underline{a}_M\right\}$. Since [(Fact) 11.3.8]

$$\begin{aligned} P\left(\omega \in \Omega : \sup_{t\in[0,1]} \left|I_{\underline{B}}\left\{\underline{a}_M - \underline{a}_n\right\}(\omega,t)\right| > \epsilon\right) &\leq \\ &\leq P\left(\omega \in \Omega : \left\|\underline{a}_M(\omega,\cdot) - \underline{a}_n(\omega,\cdot)\right\|^2_{L_2[\underline{b}]} \geq \delta\right) + 4\frac{\delta}{\epsilon^2}, \end{aligned}$$

Y is the uniform limit, in probability, of M^{S_n} which converges, path by path, to M.

□

Chapter 12
Sample Spaces

The paths of $\underline{B}$, a Cramér-Hida process, belong, almost surely, to the set

$$C = \prod_{n=1}^{\infty} C_n[0,1], \; C_n[0,1] = C[0,1],$$

so that $P_{\underline{B}}$, the law determined by $\underline{B}$, will be defined for subsets of C. This chapter contains material on that subject. C is taken to be either a Banach space or a Fréchet space, and the properties of the "natural" sets of these spaces, that is, the Borel sets, are presented. There are also results on the existence of measures, a topic required when searching for the likelihood in the presence of Gaussian noise.

12.1 Topologies for Sample Spaces

Let Λ be a compact, metric space, typically $[0,1]$. As such, it is complete [111, p. 246], and separable [226, p. 163]. $C(\Lambda, l_2)$, the family of those functions $\underline{c} : \Lambda \longrightarrow l_2$ that are continuous, is a Banach space [111, p. 352]. It is furthermore separable [248, p. 63]. The countable product of copies of the Banach space $C(\Lambda, \mathbb{R})$ is a separable Fréchet space [185, p. 40]. In $C(\Lambda, l_2)$, convergence is determined by a norm (the supremum), in the Fréchet space case, by a quasinorm [46, p. 223]. The contrast between the two cases is analogous to that between $C[0,1]$ and $\mathbb{R}^{[0,1]}$. It is often useful to see $C(\Lambda, l_2)$ as the injective tensor product $C(\Lambda)\hat{\otimes} l_2$, where $C(\Lambda)$ is the space of real valued functions on Λ [231, p. 49]. It allows in particular the representation of continuous linear functionals on $C(\Lambda, l_2)$ [73, 231, p. 58, respectively, p. 182].

A.F. Gualtierotti, *Detection of Random Signals in Dependent Gaussian Noise*,
DOI 10.1007/978-3-319-22315-5_12

12.1.1 Fréchet Spaces

Some of the facts about Fréchet spaces needed in the sequel are gathered below. Some references are [100, p. 3] and [185, p. 40 and p. 294].

The Fréchet spaces one shall meet are obtained as follows. Let $\{E_n, n \in \mathbb{N}\}$ be a sequence of normed spaces with respective norms denoted $\|\cdot\|_n$. Let

$$E = \prod_{n \in \mathbb{N}} E_n,$$

with elements $\underline{x}$ with p-th component x_p in E_p. On E let

$$d(\underline{x}_1, \underline{x}_2) = \sum_{n \in \mathbb{N}} \frac{1}{2^n} \left\{ \frac{\left\|x_n^{(1)} - x_n^{(2)}\right\|_n}{1 + \left\|x_n^{(1)} - x_n^{(2)}\right\|_n} \right\}.$$

d is a metric and

1. a sequence $\{\underline{x}_n, n \in \mathbb{N}\} \subseteq E$ is Cauchy (convergent) if, and only if, for $p \in \mathbb{N}$, fixed, but arbitrary,

$$\{x_p^{(n)}, n \in \mathbb{N}\} \subseteq E_p$$

   is Cauchy (convergent);
2. (E, d) is a locally convex, metric, linear space;
3. when, for $n \in \mathbb{N}$, fixed, but arbitrary, $(E_n, \|\cdot\|_n)$ is complete, that is, a Banach space, (E, d) is complete, so that it is a Fréchet space, but, except when only a finite number of the Banach spaces entering the product are different from the zero space, (E, d) is not a Banach space.
4. $\|\underline{x}\|_E = d(\underline{x}, \underline{0})$ is a quasinorm [46, p. 225] that induces the product topology on E, that is, the coarsest topology on E for which all the canonical maps $\mathcal{E}_n : E \longrightarrow E_n$ are continuous ($\mathcal{E}_n(\underline{x}) = x_n$);
5. E is separable if, and only if, E_n is, for all $n \in \mathbb{N}$;
6. the dual of E is the direct sum $\bigoplus_{n \in \mathbb{N}} E_n^\star$.

Separable Fréchet spaces have the following useful properties [100, p. 3].

1. E is Polish.
2. The Borel sets of E, $\mathcal{B}(E)$ are generated by the continuous, linear functionals of E.
3. $\mathcal{B}(E)$ has a countable number of generators.
4. $\mathcal{B}(E)$ is generated by any sequence of measurable sets that separate points of E.
5. Any probability on $\mathcal{B}(E)$ is inner regular with respect to compact sets.
6. From any open covering of E, one may extract a countable one.
7. Any continuous map into a Polish space, with E as domain, has a universally measurable image.

8. The properties of E and $\mathcal{B}(E)$ remain true for any measurable subset of E endowed with the relative topology and σ-algebra.
9. The closed, convex hull of any compact subset of E is compact.

The Fréchet Space of Sequences

$\mathbb{R}^\infty$ shall be taken to be [185, p. 36] the Fréchet space obtained as $\prod_{n=1}^{\infty} \mathbb{R}$ with the distance

$$d_{\mathbb{R}^\infty}(\underline{u}, \underline{v}) = \sum_{n=1}^{\infty} \frac{1}{2^n} \left\{ \frac{|u_n - v_n|}{1 + |u_n - v_n|} \right\}, \; \{\underline{u}, \underline{v}\} \subseteq \mathbb{R}^\infty .$$

The Borel sets of $\mathbb{R}^\infty$, denoted $\mathcal{B}(\mathbb{R}^\infty)$, are obtained as the product of the Borel sets of $\mathbb{R}$ [208, p. 6], or as the sets in the σ-algebra generated by the evaluation maps $\{\mathcal{E}_n, n \in \mathbb{N}\}$, which are continuous in the product topology. A detailed description of the properties of $\mathbb{R}^\infty$ may be found in [259, p. 2].

The Borel sets of l_p shall be denoted $\mathcal{B}(l_p)$. They are generated by the family of continuous, linear functionals $\{\mathcal{E}_n, n \in \mathbb{N}\}$. Since, for $\underline{x} \in l_2$, fixed, but arbitrary, the map $t \mapsto \frac{t}{1+t} \leq t$ being monotone increasing,

$$\sum_{n=1}^{\infty} \frac{2^{-n} |x_n|}{1 + |x_n|} \leq \frac{\|\underline{x}\|_{l_2}}{1 + \|\underline{x}\|_{l_2}} \leq \|\underline{x}\|_{l_2} ,$$

the inclusion $J_{l_2, \mathbb{R}^\infty}$ of l_2 into $\mathbb{R}^\infty$ is well defined, injective, and a contraction. Furthermore, when $1 \leq p < q \leq \infty$, $l_p \subseteq l_q$ [275, p. 424], and, on l_p, it holds that $\|\underline{x}\|_q \leq \|\underline{x}\|_p$. J_{l_p, l_q} is thus equally well defined, and a contraction.

12.1.2 Norms, Quasinorms, and Distances on Sample Spaces

There are several "natural" ways to look at the path space of processes with an infinite, but countable number of continuous components of the type encountered so far, each corresponding to a specific norm, quasinorm, or distance. Let thus $\underline{c}$ be a vector with a countably infinite number of components denoted c_n, each of which is a continuous function on $[0, 1]$. The norm of a continuous function c on $[0, 1]$ shall be denoted

$$\|c\| = \sup_{t \in [0,1]} |c(t)| .$$

Let then

$$\begin{aligned}
\|\underline{c}\|_p &= \left\{\sum_{n=1}^{\infty} \|c_n\|^p\right\}^{1/p} = \left\{\sum_{n=1}^{\infty} \left\{\sup_{t\in[0,1]} |c_n(t)|\right\}^p\right\}^{1/p}, \\
\|\underline{c}\|_{C_{l_2}} &= \sup_{t\in[0,1]} \|\underline{c}(t)\|_{l_2} = \sup_{t\in[0,1]} \left\{\sum_{n=1}^{\infty} [c_n(t)]^2\right\}^{1/2}, \\
\|\underline{c}\|_{C_F} &= \sum_{n=1}^{\infty} \frac{2^{-n}\|c_n\|}{1+\|c_n\|} = \sum_{n=1}^{\infty} \frac{2^{-n}\sup_{t\in[0,1]}|c_n(t)|}{1+\sup_{t\in[0,1]}|c_n(t)|}.
\end{aligned}$$

For $n \in \mathbb{N}$, fixed, but arbitrary, let

$$C_n = C[0,1], \text{ and } C = \prod_{n\in\mathbb{N}} C_n.$$

C_S^p, C_{l_2}, and C_F shall denote the subset of C of elements $\underline{c}$ for which, respectively,

$$\|\underline{c}\|_p < \infty, \; \|\underline{c}\|_{C_{l_2}} < \infty, \; \|\underline{c}\|_{C_F} < \infty.$$

In fact, $C_F = C$ [185, p. 11], and, by definition, $\|\underline{c}\|_{C_F} = [\![\underline{c}]\!]_C$, where the notation on the right-hand side of the latter equality denotes quasinorm [46, p. 223]. It is a Fréchet space that is not a Banach space [185, p. 40]. C_S^p is a Banach space [60, p. 72]. C_{l_2} is the Banach space of continuous functions $\underline{c} : [0,1] \longrightarrow l_2$.

Those norms and quasinorm are related as follows. Since

$$\|c_n\| = \sup_{t\in[0,1]} |c_n(t)| \leq \sup_{t\in[0,1]} \|\underline{c}(t)\|_{l_2},$$

and that $t \mapsto \frac{t}{1+t} \leq t$ is increasing,

$$\|\underline{c}\|_{C_F} \leq \|\underline{c}\|_{C_{l_2}}.$$

Since, for $p \in [1,2]$, on l_p, $\|\underline{c}(t)\|_{l_2} \leq \|\underline{c}(t)\|_{l_p} \leq \|\underline{c}\|_p$, one has that

$$\|\underline{c}\|_{C_{l_2}} \leq \|\underline{c}\|_p.$$

One has thus that, for $p \in [1,2]$, fixed, but arbitrary,

$$C_S^p \subseteq C_{l_2} \subseteq C_F \text{ with } \|\underline{c}\|_{C_F} \leq \|\underline{c}\|_{C_{l_2}} \leq \|\underline{c}\|_p.$$

The inclusions are strict: Suppose indeed that $\underline{c}$ has each component equal to the same constant, say γ. Then $\|c_n\| = \gamma$, so that

$$\|\underline{c}\|_{C_F} = \frac{\gamma}{1+\gamma}, \text{ but } \|\underline{c}\|_{C_{l_2}} = \infty.$$

Let now

$$E = \left[\frac{1}{3}, \frac{2}{3}\right], \ \epsilon(i) \in \{0, 1\}, i \in [1:n],$$

and $E_{\epsilon(1),\ldots,\epsilon(n)}$ be the following interval:

$$\left[\sum_{i=1}^{n} \epsilon(i)\frac{2}{3^i} + \frac{1}{3^{n+1}}, \sum_{i=1}^{n} \epsilon(i)\frac{2}{3^i} + \frac{2}{3^{n+1}}\right].$$

Thus, for $n = 1$,

$$E_{\epsilon(1)} = \left[\epsilon(1)\frac{2}{3} + \frac{1}{9}, \epsilon(1)\frac{2}{3} + \frac{2}{9}\right]$$

yields the intervals

$$\left[\frac{1}{9}, \frac{2}{9}\right] \text{ when } \epsilon(1) = 0, \text{ and } \left[\frac{7}{9}, \frac{8}{9}\right] \text{ when } \epsilon(1) = 0.$$

Those are intervals that enter the construction of the Cantor set. Let c_1 be the function whose nonzero part has, as graph, the isosceles triangle with height one, and base $[\frac{1}{3}, \frac{2}{3}]$. The other elements in the sequence are built analogously: c_2 has base $[\frac{1}{9}, \frac{2}{9}]$, c_3, $[\frac{7}{9}, \frac{8}{9}]$, and so forth, using the formula for $E_{\epsilon(1),\ldots,\epsilon(n)}$ from "left to right." $\underline{c}$ has those components. Since $\|c_n\| = 1$, independently of n,

$$\|\underline{c}\|_{C_{l_p}} = \infty,$$

but, since, for $t \in [0, 1]$, fixed, but arbitrary, only one component of $\underline{c}(t)$ is not zero, and $\|\underline{c}(t)\|_{l_2} \leq 1$,

$$\|\underline{c}\|_{C_{l_2}} < \infty.$$

The different spaces are not closed in the larger spaces: Let $\underline{c}$ be the vector of "isosceles triangle functions" described above. Define

$$c_k^{(n)}(t) = \begin{cases} \frac{c_k(t)}{\sqrt[p]{n}} & \text{for } k \leq n \\ 0 & \text{for } k > n \end{cases}.$$

Then $\left\|c_k^{(n)}\right\|^p = n^{-1}$, and $\left\|\underline{c}_n\right\|_p = 1$. But $\left\|\underline{c}_n(t)\right\|_{l_2}^2 \leq n^{-\frac{2}{p}}$. Let finally

$$c_k^{(n)}(t) = \begin{cases} t^n \text{ for } k = n \\ 0 \text{ for } k \neq n \end{cases}.$$

Then $\left\|\underline{c}_n(t)\right\|_{l_2}^2 = t^{2n}$, and $\left\|\underline{c}_n\right\|_{C_{l_2}} = 1$. However $\left\|\underline{c}_n\right\|_{C_F} = 2^{-(n+1)}$.

The injections

$$J_{p,2} : C_S^p \longrightarrow C_{l_2} \text{ and } J_{2,F} : C_{l_2} \longrightarrow C_F$$

are continuous. C_S^p and C_{l_2} are FH spaces for $H = C_F$ [269, p. 202]. It follows [269, p. 203] that the topology of C_S^p is strictly stronger than the topology of C_{l_2}, which is, in turn, strictly stronger than that of C_F, and that [269, p. 204] C_S^p is of first category in C_{l_2}, which, in turn, is of first category in C_F.

12.2 Measurable Subsets of Sample Spaces

C_{l_2} and C_F are the "natural" spaces in the present context. In the sequel, no distinction shall be required, and one shall deal with a "generic" space that covers both cases. It shall be denoted K. When necessary, K shall be substituted with one of those C-spaces, or their respective index. One also uses the terms "Fréchet case" and "Banach case." The range space of the function $t \mapsto \underline{c}(t)$, $\underline{c} \in K$, shall be $\mathbb{R}^\infty$, or some l_p. Again no distinction is necessary as long as each is properly associated with its relevant space, that is, $\mathbb{R}^\infty$ "goes" with C_F, l_p with C_S^p, and l_2 with C_{l_2}. The "generic" form for those spaces of sequences shall be s.

A family $\mathcal{F}$ of functions $f : X \longrightarrow \mathbb{R}$ separates points of X when, given $\{x_1, x_2\} \subseteq X, x_1 \neq x_2$, fixed, but arbitrary, there exists $f \in \mathcal{F}$ such that $f(x_1) \neq f(x_2)$. Let $\sigma(\mathcal{F})$ be the σ-algebra generated on X by $\mathcal{F}$. Suppose that X is Polish (which includes complete metric spaces), and that $\mathcal{F}$, a family of functions, measurable for the Borel sets of X, separates points of X. Then [260, p. 6]:

1. when $\mathcal{F}$ is countable, $\sigma(\mathcal{F}) = \mathcal{B}(X)$ (the Borel sets of X);
2. when $\mathcal{F}$ is made of continuous functions, $\sigma(\mathcal{F}) = \mathcal{B}(X)$.

Those two facts are useful when $\mathcal{F}$ is made of evaluation maps on a space of functions. One shall use the following evaluations:

1. for $T \subseteq \mathbb{R}$, $t \in T$, X some set, and a function $f : T \longrightarrow X$, $\mathcal{E}_t(f) = f(t)$;
2. for $\underline{c} \in C$, $\mathcal{E}_n^C(\underline{c}) = c_n$, the n-th component function of $\underline{c}$;
3. for $t \in T$ and $\underline{c} \in C$, $\underline{\mathcal{E}}_t^C(\underline{c}) = \underline{c}(t)$.

They all separate points.

12.2.1 *Evaluation Maps and Borel Sets in the Fréchet Case*

The topology of C_F is the coarsest on C that makes the evaluations $\mathcal{E}_n^C$ continuous. Thus

$$\sigma\left(\{\mathcal{E}_n^C, n \in \mathbb{N}\}\right) = \mathcal{B}(C_F).$$

Since $(C_i[0,1] = C[0,1])$

$$\left\{\mathcal{E}_n^C\right\}^{-1}(\mathcal{B}(C[0,1])) = \prod_{i=1}^{n-1} C_i[0,1] \times \mathcal{B}(C[0,1]) \times \prod_{i=n+1}^{\infty} C_i[0,1],$$

$\mathcal{B}(C_F)$ is the product σ-algebra of a countably infinite number of copies of $\mathcal{B}(C[0,1])$.

Since $\mathcal{E}_t$ is continuous on $C[0,1]$, $\mathcal{E}_t \circ \mathcal{E}_n^C$ is continuous on C_F, and thus these evaluation maps also generate the Borel sets of C_F.

Finally, since

$$\|\underline{c}(t)\| = \sum_{n=1}^{\infty} \frac{1}{2^n} \frac{|c_n(t)|}{1+|c_n(t)|} \leq \sum_{n=1}^{\infty} \frac{1}{2^n} \frac{\|c_n\|}{1+\|c_n\|} = \|\underline{c}\|_C,$$

$\underline{\mathcal{E}}_t^C$ is also continuous, and $\mathcal{B}(C_F)$ is also generated by those maps.

Remark 12.2.1 Let $\sigma_t(C_F)$ be the σ algebra generated by one of the following family of maps:

$$\left\{\mathcal{E}_\theta \mathcal{E}_n^C, \theta \leq t, n \in \mathbb{N}\right\}, \text{ or } \left\{\underline{\mathcal{E}}_\theta^C, \theta \leq t\right\}.$$

Let also, for $n \in \mathbb{N}$, fixed, but arbitrary, $\mathcal{C}_t^{(n)}$ be the σ-algebra of the set of functions $C_n[0,1] = C[0,1]$ generated by the following family of evaluations: $\{\mathcal{E}_\theta, \theta \leq t\}$. Then $\sigma_t(C_F)$ is a product σ-algebra, that is,

$$\sigma_t(C_F) = \bigotimes_{n=1}^{\infty} \mathcal{C}_t^{(n)}.$$

Remark 12.2.2 Let, for $n \in \mathbb{N}$, fixed, but arbitrary,

$$P_{B_n}^{C_n} = P \circ B_n^{-1}, \text{ and } P_B^{C_F} = P \circ \underline{B}^{-1},$$

$\underline{B}$ being a Cramér-Hida process. One has that

$$P_B^{C_F} = \bigotimes_{n=1}^{\infty} P_{B_n}^{C_n}.$$

12.2.2 Evaluation Maps and Borel Sets in the Banach Case

Let $\underline{e}_n$ be the n-th element of the standard basis of l_2. Then

$$\mathcal{E}_n^C(\underline{c}) = \langle \underline{c}(\cdot), \underline{e}_n \rangle_{l_2} = c_n.$$

Since

$$|c_n(t)| \leq \|\underline{c}(t)\|_{l_2},$$

$\mathcal{E}_n^C$ is continuous on C_{l_2}. $\mathcal{E}_t \circ \mathcal{E}_n^C$ is again continuous, and so is $\underline{\mathcal{E}}_t^C$. Thus the Borel sets of C_{l_2} are also generated by the evaluation maps.

12.3 Measures for Sample Spaces

Farther, when dealing with the likelihood for general Gaussian noise, one shall need some results which secure the existence of a measure on the Borel sets of K. One shall find here the transcription for K (norm and quasinorm replace absolute value) of those weak convergence results for $C[0, 1]$ that are used there. The source is [38].

The basic tool is a form of the Arzelà-Ascoli theorem, which, in turn, uses the modulus of continuity. Let thus $\delta \in]0, 1[$ and $\underline{k} \in K$ be fixed, but arbitrary. Then, the modulus of continuity of $\underline{k}$ is measured with the help of the following index of continuity:

$$w(\underline{k}, \delta) = \sup_{\substack{\{u,v\} \subseteq [0,1] \\ |u-v| < \delta}} \|\underline{k}(u) - \underline{k}(v)\|_s .$$

Since $|\sup_S f - \sup_S g| \leq \sup_S |f - g|$ and that

$$\left\|\underline{k}_1(u) - \underline{k}_1(v)\right\|_s \leq \left\|\underline{k}_1(u) - \underline{k}_2(u)\right\|_s + \left\|\underline{k}_2(u) - \underline{k}_2(v)\right\|_s + \left\|\underline{k}_2(v) - \underline{k}_1(v)\right\|_s,$$

one has, for fixed, but arbitrary $\delta \in]0, 1[$, that

$$\left|w(\underline{k}_1, \delta) - w(\underline{k}_2, \delta)\right| \leq 2 \left\|\underline{k}_1 - \underline{k}_2\right\|_K,$$

so that $\underline{k} \mapsto w(\underline{k}, \delta)$ is continuous, and thus measurable. Furthermore, since the elements in K are uniformly continuous,

$$\lim_{\delta \to 0} w(\underline{k}, \delta) = 0,$$

and the latter establishes the uniform continuity of $\underline{k}$.

Proposition 12.3.1 *A subset $K_0 \subseteq K$ has compact closure if, and only if,*

1. $\sup_{\underline{k} \in K_0} \|\underline{k}(0)\|_s < \infty$;
2. $\lim_{\delta \downarrow\downarrow 0} \sup_{\underline{k} \in K_0} w(\underline{k}, \delta) = 0$.

Proof Let $\overline{K}_0$ be the closure of K_0. Since $\mathcal{E}_0^C$ is continuous, the image by it, in s, of $\overline{K}_0$, is compact, and, in a metric space, compact sets are closed and bounded

[84, p. 233]. Now $w(\underline{k}\,, \frac{1}{n})$ is continuous in $\underline{k}$, and decreasing to zero when n increases, so that convergence is uniform on compact sets [38, p. 218], and, in particular, on $\overline{K}_0$.

Suppose now that items 1 and 2 obtain. Choose n large enough so that

$$\sup_{\underline{k}\in K_0} w\left(\underline{k}, 1/n\right)$$

is finite. Since

$$\left\|\underline{k}(t)\right\|_s \leq \left\|\underline{k}(0)\right\|_s + \sum_{i=1}^{n} \left\| \underline{k}\left(\left[\frac{i}{n}\right] t\right) - \underline{k}\left(\left[\frac{i-1}{n}\right] t\right) \right\|_s ,$$

one has that

$$\sup_{t\in[0,1]} \sup_{\underline{k}\in K_0} \left\|\underline{k}(t)\right\|_s < \infty.$$

Let d_K denote the distance on K determined by its norm or quasinorm, and $\underline{\kappa} = \left\{\underline{k}_n, n \in \mathbb{N}\right\}$ be a sequence of points dense in K. The map $J_{\underline{\kappa}} : K \longrightarrow \mathbb{R}^\infty$ is defined using the following relation:

$$J_{\underline{\kappa}}(\underline{k}) = \left(d_K(\underline{k}, \underline{k}_1), d_K(\underline{k}, \underline{k}_1), d_K(\underline{k}, \underline{k}_1), \ldots\right).$$

$J_{\underline{\kappa}}(K)$ is homeomorphic to a subset of $\mathbb{R}^\infty$ [38, p. 219]. Since $J_{\underline{\kappa}}(K_0)$ has components

$$d_K(\underline{k}, \underline{k}_n) \leq \left\|\underline{k}\right\|_K + \left\|\underline{k}_n\right\|_K \leq 2 \sup_{\underline{k}\in K_0} \sup_{t\in[0,1]} \left\|\underline{k}(t)\right\|_s < \infty,$$

$J_{\underline{\kappa}}(K_0)$ has compact closure [38, p. 219]. Since $J_{\underline{\kappa}}$ is continuous, using [84, p. 80],

$$\overline{J_{\underline{\kappa}}^{-1}(J_{\underline{\kappa}}(K_0))} \subseteq J_{\underline{\kappa}}^{-1}\left(\overline{J_{\underline{\kappa}}(K_0)}\right),$$

and it follows that K_0 has compact closure. □

Proposition 12.3.2 *Let $\left\{P^{\mathcal{K}}, P_n^{\mathcal{K}}, n \in \mathbb{N}\right\}$ be probability measures on $\mathcal{K}$. When the finite dimensional distributions of $P_n^{\mathcal{K}}$ converge weakly to those of $P^{\mathcal{K}}$, and the family $\left\{P_n^{\mathcal{K}}, n \in \mathbb{N}\right\}$ is tight, then the latter converges weakly to $P^{\mathcal{K}}$.*

Proof According to Prohorov's theorem [38, p. 37], since K is separable and complete, relative compactness and tightness of measures are equivalent properties. The family $\left\{P_n^{\mathcal{K}}, n \in \mathbb{N}\right\}$ is thus relatively compact, and then [38, p. 35] each subsequence

$$\left\{P_{n_p}^{\mathcal{K}}, p \in \mathbb{N}\right\}$$

contains a weakly convergent subsequence

$$\left\{P^K_{n_{pq}}, q \in \mathbb{N}\right\},$$

with limit say Q^K. The finite-dimensional distributions of Q^K must be the weak limit of the corresponding finite-dimensional distributions of the elements in the sequence converging to Q^K [38, p. 29 and 35]. But then they must coincide with those of P^K. Since the cylinders form a determining class, one must have that $Q^K = P^K$. Thus each subsequence of the original sequence contains a subsequence converging to P^K. That means [38, p. 16] that the original sequence converges weakly to P^K. □

Proposition 12.3.3 *Let $\{P_n, n \in \mathbb{N}\}$ be a sequence of probability measures on $\mathcal{K}$. It is a tight sequence if, and only if,*

1. *for $\epsilon > 0$, fixed, but arbitrary, there exists λ such that, for all $n \in \mathbb{N}$,*

$$P_n\left(\underline{k} \in K : \left\|\mathcal{E}_0^C(\underline{k})\right\|_s > \lambda\right) \leq \epsilon;$$

2. *for $\epsilon > 0$ and $\delta > 0$, fixed, but arbitrary, there exists $\eta \in]0, 1[$, and $p \in \mathbb{N}$ such that, for all $n \geq p$,*

$$P_n\left(\underline{k} \in K : w(\underline{k}, \eta) \geq \delta\right) \leq \epsilon.$$

Proof Suppose first that the given sequence is tight, that is, for $\epsilon > 0$, fixed, but arbitrary, there is a compact $K_\epsilon \subseteq K$ such that, for all $n \in \mathbb{N}$, $P_n(K_\epsilon) > 1 - \epsilon$.

Since $\mathcal{E}_0^C$ is continuous and K_ϵ, compact,

$$\kappa_\epsilon = \sup_{\underline{k} \in K_\epsilon} \left\|\mathcal{E}_0^C(\underline{k})\right\|_s$$

is finite (it is a maximum). Thus

$$1 - \epsilon < P_n(K_\epsilon) \leq P_n\left(\underline{k} \in K : \left\|\mathcal{E}_0^C(\underline{k})\right\|_s \leq \kappa_\epsilon\right).$$

Since K_ϵ is uniformly equicontinuous [154, p. 233], given $\delta > 0$, there is $\eta > 0$ such that $w(\underline{k}, \eta) < \delta$ whatever $\underline{k} \in K_\epsilon$. Consequently, $K_\epsilon \subseteq \{\underline{k} \in K : w(\underline{k}, \eta) < \delta\}$, and

$$1 - \epsilon < P_n(K_\epsilon) \leq P_n\left(\underline{k} \in K : w(\underline{k}, \eta) < \delta\right).$$

Suppose now that items 1 and 2 of the statement obtain. One may assume that $p = 1$. Indeed, a single probability is tight [38, p. 10]. Thus, for $i \in [1 : p - 1]$, let K_i be a compact set such that $P_i(K_i) > 1 - \epsilon$. Since K_i is uniformly equicontinuous, there is $\eta_i > 0$ such that $K_i \subseteq \{\underline{k} \in K : w(\underline{k}, \eta_i) < \delta\}$, and $P_i(\underline{k} \in K : w(\underline{k}, \eta_i) \geq \delta) \leq \epsilon$. Let η_0 be the smallest of $\eta_1, \ldots, \eta_{p-1}, \eta$. Then

$$\{\underline{k} \in K : w(\underline{k}, \eta_0) \geq \delta\} \subseteq$$

$$\subseteq \{\underline{k} \in K : w(\underline{k}, \eta) \geq \delta\} \cap \left[\bigcap_{i=1}^{p-1} \{\underline{k} \in K : w(\underline{k}, \eta_i) \geq \delta\}\right],$$

and, for $n \in \mathbb{N}$, fixed, but arbitrary, $P_n\left(\underline{k} \in K : w(\underline{k}, \eta_0) \geq \delta\right) \leq \epsilon$.

Because of item 1, given $\epsilon > 0$, one may choose λ so that, setting

$$K_0 = \left\{\underline{k} \in K : \left\|\mathcal{E}_0^c(\underline{k})\right\|_s \leq \lambda\right\},$$

then, for all $n \in \mathbb{N}$,

$$P_n\left(K_0\right) \geq 1 - \frac{\epsilon}{2},$$

and, because of item 2, η_q so that, setting

$$K_q = \left\{\underline{k} \in K : w(\underline{k}, \eta_q) < \frac{1}{q}\right\},$$

then, for all $n \in \mathbb{N}$, one has that

$$P_n(K_q) \geq 1 - \frac{\epsilon}{2^{q+1}}.$$

Let $K_0 \cap \left(\cap_{q=1}^{\infty} K_q\right)$ have closure K_ϵ. Then, for all $n \in \mathbb{N}$, looking at the complement of K_ϵ,

$$P_n(K_\epsilon) \geq 1 - \epsilon.$$

But, because of (Proposition) 12.3.1, K_ϵ is compact. □

Proposition 12.3.4 *Let $\{P_n, n \in \mathbb{N}\}$ be probability measures on $\mathcal{K}$, and, for $t \in [0, 1]$ and $\eta \geq 0$, fixed, but arbitrary,*

$$I(t, \eta) = \{\theta \in [0, 1] : t \leq \theta \leq (t + \eta) \wedge 1\}.$$

Suppose that

1. *for $\epsilon > 0$, fixed, but arbitrary, there exists λ such that, for all $n \in \mathbb{N}$,*

$$P_n\left(\underline{k} \in K : \left\|\mathcal{E}_0^c(\underline{k})\right\|_s > \lambda\right) \leq \epsilon;$$

2. *for $\epsilon > 0$ and $\delta > 0$, fixed, but arbitrary, there exists $\eta \in]0,1[$, and $p \in \mathbb{N}$ such that, for $t \in [0,1]$, fixed, but arbitrary, for all $n \geq p$,*

$$\frac{1}{\eta} P_n \left(\underline{k} \in K : \sup_{I(t,\eta)} \|\underline{k}(\theta) - \underline{k}(t)\|_s \geq \epsilon \right) \leq \delta.$$

Then $\{P_n, n \in \mathbb{N}\}$ is tight.

Proof Fix $\eta \in]0,1[$ and let

$$K_{t,\eta} = \left\{ \underline{k} \in K : \sup_{I(t,\eta)} \|\underline{k}(\theta) - \underline{k}(t)\|_s \geq \epsilon \right\}.$$

For v fixed, the u's for which $|u - v| < \eta$ are the union of $]v - \eta, v]$ and $[v, v + \eta[$. For some integer i_v,

$$v \in [i_v \eta, (i_v + 1)\eta], \quad \text{denoted } I_v,$$

and, for some integer i_u,

$$u \in [i_u \eta, (i_u + 1)\eta], \quad \text{denoted } I_u.$$

When $u \in [v, v + \eta]$, one must have that u and v belong to $I_u \cup I_v$, I_u and I_v abutting. The situation for $]v - \eta, v]$ is symmetric, so that, when v is fixed and u belongs to $]v - \eta, v + \eta[$, u and v belong to three successive and abutting intervals of the form $[i\eta, (i+1)\eta]$. Consequently, for v fixed, $w(\underline{k}, \eta) \geq 3\epsilon$ implies that $\underline{k}$ is in the union of three successive sets of the form $K_{i,\eta}$. For example, when $i\eta < v < (i+1)\eta < u < v + \eta < (i+2)\eta$,

$$\begin{aligned}
\alpha &\leq \|\underline{k}(v) - \underline{k}(u)\|_s \\
&= \|\underline{k}(v) - \underline{k}(i\eta) + \underline{k}(i\eta) - \underline{k}((i+1)\eta) + \underline{k}((i+1)\eta) - \underline{k}(u)\|_s \\
&\leq \|\underline{k}(v) - \underline{k}(i\eta)\|_s + \|\underline{k}(i\eta) - \underline{k}((i+1)\eta)\|_s + \|\underline{k}((i+1)\eta) - \underline{k}(u)\|_s \\
&= a + b + c
\end{aligned}$$

implies that at least one of a, b, and c must exceed $\alpha/3$. But v is arbitrary so that, for $i\eta < 1$,

$$\{\underline{k} \in K : w(\underline{k}, \eta) \geq 3\epsilon\} \subseteq \cup_{i\eta} K_{i,\eta},$$

and thus

$$P_n(\underline{k} \in K : w(\underline{k}, \eta) \geq 3\epsilon) \leq P\left(\cup_{i\eta} K_{i,\eta}\right) \leq \sum_{i\eta} P\left(K_{i,\eta}\right).$$

Now, by assumption, $P(K_{i,\eta}) \leq \delta\eta$. Consequently, the sum of the latter probabilities will be smaller that $(1 + [\frac{1}{\eta}])\delta\eta$, and, since $\eta < 1$, $(1 + [\frac{1}{\eta}])\eta \leq 2$, one has that

$$P_n(\underline{k} \in K : w(\underline{k}, \eta) \geq 3\epsilon) \leq 2\delta.$$

□

Corollary 12.3.5 *Let, in* $[0, 1]$, $0 = t_0 < t_1 < \cdots < t_{n-1} < t_n = 1$ *be a partition such that, for* $i \in [2 : n-1]$, $t_i - t_{i-1} \geq \eta$. *Then*

$$P\left(\underline{k} \in K : w(\underline{k}, \eta) \geq 3\epsilon\right) \leq \sum_{i=1}^{n} P\left(\underline{k} \in K : \sup_{\theta \in [t_{i-1}, t_i]} \|\underline{k}(\theta) - \underline{k}(t_{i-1})\|_s \geq \epsilon\right).$$

Proof Since the intervals $[t_i, t_{i+1}]$ have length at least η, an interval of length 2η will be contained in at most two abutting intervals of the latter type. Thus, when $|u - v| < \eta$, for some i, one will have, for example,

$$t_{i-1} \leq u \leq t_i \leq v \leq t_{i+1},$$

so that

$$\|\underline{k}(u) - \underline{k}(v)\|_s = \|\underline{k}(u) - \underline{k}(t_{i-1}) + \underline{k}(t_{i-1}) - \underline{k}(t_i) + \underline{k}(t_i) - \underline{k}(v)\|_s.$$

□

Application of the previous results to the existence of measures requires the evaluation of partial sums which follows. Let thus $\underline{X}_1, \ldots, \underline{X}_n$ be random elements with values in s,

$$\underline{S}_0 = \underline{0}_s, \text{ and } \underline{S}_n = \underline{X}_1 + \cdots + \underline{X}_n.$$

Let also

$$\mu_n = \max_{i \in [0,n]} \|\underline{S}_i\|_s, \quad \nu_n = \max_{i \in [0:n]} \left\{\|\underline{S}_i\|_s \wedge \|\underline{S}_n - \underline{S}_i\|_s\right\}.$$

From the definition,

$$\nu_n \leq \mu_n.$$

Furthermore, since

$$\|\underline{S}_i\|_s = \|\underline{S}_i - \underline{S}_n + \underline{S}_n\|_s \leq \|\underline{S}_i - \underline{S}_n\|_s + \|\underline{S}_n\|_s,$$

and $(a+b)\wedge(a+c) = a + (b\wedge c)$, one has that

$$\begin{aligned}\|\underline{S}_i\|_s &\leq \{\|\underline{S}_n\|_s + \|\underline{S}_i\|_s\} \wedge \{\|\underline{S}_n\|_s + \|\underline{S}_n - \underline{S}_i\|_s\} \\ &= \|\underline{S}_n\|_s + \{\|\underline{S}_i\|_s \wedge \|\underline{S}_n - \underline{S}_i\|_s\},\end{aligned}$$

and thus

$$\mu_n \leq \nu_n + \|\underline{S}_n\|_s .$$

Lemma 12.3.6 *Let, for strictly positive a, b, x, y, and positive α,*

$$f(x,y) = \frac{a}{x^{2\alpha}} + \frac{b}{y^{2\alpha}} .$$

Let also $\gamma = (2\alpha+1)^{-1}$. Then

$$\min_{x+y=z} f(x,y) = (a^\gamma + b^\gamma)^{\frac{1}{\gamma}} \frac{1}{z^{2\alpha}} .$$

Proof Let $g(x,y) = f(x,y) - c(x+y)$. As, for example,

$$\frac{\partial g}{\partial x} = -\left(c + \frac{2\alpha a}{x^{2\alpha+1}}\right),$$

equating the partial derivatives to zero yields that

$$\frac{a}{x^{2\alpha+1}} = \frac{b}{y^{2\alpha+1}}, \quad \text{or} \quad \frac{x}{y} = \left(\frac{a}{b}\right)^\gamma .$$

As $x+y=z$, $x\left(1+\frac{y}{x}\right) = z$, so that

$$x = z\left(1+\frac{y}{x}\right)^{-1} = z\left(1+\left(\frac{b}{a}\right)^\gamma\right)^{-1} = z\left(\frac{a^\gamma+b^\gamma}{a^\gamma}\right)^{-1} = \frac{a^\gamma}{a^\gamma+b^\gamma} z,$$

and

$$y = z - x = \frac{b^\gamma}{a^\gamma+b^\gamma} z .$$

The minimum of f is then

$$\begin{aligned}\frac{a}{\left\{\frac{za^\gamma}{a^\gamma+b^\gamma}\right\}^{2\alpha}} + \frac{b}{\left\{\frac{za^\gamma}{a^\gamma+b^\gamma}\right\}^{2\alpha}} &= \frac{(a^\gamma+b^\gamma)^{2\alpha}}{z^{2\alpha}}\left\{\frac{a}{a^{2\alpha\gamma}} + \frac{b}{b^{2\alpha\gamma}}\right\} \\ &= \frac{(a^\gamma+b^\gamma)^{2\alpha+1}}{z^{2\alpha}} .\end{aligned}$$

□

Proposition 12.3.7 *Let $\alpha \geq 0$ and $\beta > \frac{1}{2}$ be fixed, but arbitrary. Suppose that for all $\lambda > 0$, and $0 \leq i \leq j \leq k \leq n$, there exists $\{\kappa_1, \dots, \kappa_n\} \subseteq \mathbb{R}_+$ such that*

$$P\left(\left\|\underline{S}_j - \underline{S}_i\right\|_s \geq \lambda, \left\|\underline{S}_k - \underline{S}_j\right\|_s \geq \lambda\right) \leq \frac{1}{\lambda^{2\alpha}} \left\{\sum_{l \in]i:k]} \kappa_l\right\}^{2\beta}.$$

There exits then $\kappa(\alpha, \beta)$ such that, for all $\lambda > 0$,

$$P\left(\nu_n \geq \lambda\right) \leq \kappa(\alpha, \beta) \frac{\{\kappa_1 + \dots + \kappa_n\}^{2\beta}}{\lambda^{2\alpha}}.$$

Proof The proof shall be by induction. Let $\gamma = (2\alpha + 1)^{-1}$: $\gamma \in]0, 1]$. Let also, for $\kappa > 0$,

$$f(\kappa) = 2^\gamma \left\{\frac{1}{2^{2\beta\gamma}} + \frac{1}{\kappa^\gamma}\right\}.$$

f is decreasing, and its limit is $2^{-\gamma(2\beta-1)} < 1$. Choose κ such that $\kappa \geq 1$, and $f(\kappa) \leq 1$. Let $c = \kappa_1 + \dots + \kappa_n$, $c_0 = 0$, and $c_i = c^{-1}(\kappa_1 + \dots + \kappa_i)$.

The case $n = 1$: As, by definition, $\nu_1 = 0$, $P(\nu_1 \geq \lambda) = 0$, which is smaller than any positive quantity, so that the result is true.
The case $n = 2$: The only term entering the definition of ν_2 which is possibly different from zero is $\left\|\underline{S}_1\right\|_s \wedge \left\|\underline{S}_2 - \underline{S}_1\right\|_s$, so that ν_2 equals the latter, and the assumption says, since $\underline{S}_0 = \underline{0}_s$, that, with $\kappa \geq 1$,

$$\begin{aligned}
P(\nu_2 \geq \lambda) &= P(\left\|\underline{S}_1\right\|_s \wedge \left\|\underline{S}_2 - \underline{S}_1\right\|_s \geq \lambda) \\
&\leq P(\left\|\underline{S}_1 - \underline{S}_0\right\|_s \geq \lambda, \left\|\underline{S}_2 - \underline{S}_1\right\|_s \geq \lambda) \\
&\leq \frac{(\kappa_1 + \kappa_2)^{2\beta}}{\lambda^{2\alpha}} \\
&\leq \kappa \frac{(\kappa_1 + \kappa_2)^{2\beta}}{\lambda^{2\alpha}}.
\end{aligned}$$

Suppose that the result obtains up to $n-1$. The numbers $c_i = \frac{\kappa_1 + \dots + \kappa_i}{c}$ are increasing to one. So there is i_0 such that $c_{i_0-1} \leq \frac{1}{2} \leq c_{i_0}$.

One shall dominate ν_n with the help of the following variables:

$$U_1 = \max_{j \in [0:i_0-1]} \left\{\left\|\underline{S}_j\right\|_s \wedge \left\|\underline{S}_{i_0-1} - \underline{S}_j\right\|_s\right\},$$

$$U_2 = \max_{j \in [i_0,n]} \left\{\left\|\underline{S}_j - \underline{S}_{i_0}\right\|_s \wedge \left\|\left(\underline{S}_n - \underline{S}_{i_0}\right) - \left(\underline{S}_j - \underline{S}_{i_0}\right)\right\|_s\right\}$$

$$= \max_{j\in[i_0,n]} \left\{ \left\|\underline{S}_j - \underline{S}_{i_0}\right\|_s \wedge \left\|\underline{S}_n - \underline{S}_j\right\|_s \right\},$$

$$V_1 = \left\|\underline{S}_{i_0-1}\right\|_s \wedge \left\|\underline{S}_n - \underline{S}_{i_0-1}\right\|_s,$$

$$V_2 = \left\|\underline{S}_{i_0}\right\|_s \wedge \left\|\underline{S}_n - \underline{S}_{i_0}\right\|_s.$$

Domination of U_1: When $i_0 = 1$, U_1 is zero, and one need not be concerned. Suppose thus that $i_0 > 1$. Since $i_0 - 1 < n$, the result is true for $\underline{X}_1, \ldots, \underline{X}_{i_0-1}$, so that, since $c_{i_0-1} \leq \frac{1}{2}$,

$$P(U_1 \geq \lambda) \leq \kappa \frac{(c_{i_0-1}c)^{2\beta}}{\lambda^{2\alpha}} \leq \frac{\kappa}{\lambda^{2\alpha}} \left(\frac{c}{2}\right)^{2\beta}.$$

Domination of U_2: When $i_0 = n$, $U_2 = 0$, and again there is no need to be concerned. Otherwise, since $n - i_0 < n$, the result is also true for

$$\underline{\tilde{X}}_1 = \underline{X}_{i_0+1}, \ldots, \underline{\tilde{X}}_{n-i_0} = \underline{X}_n,$$

using $\tilde{\kappa}_1 = \kappa_{i_0+1}, \ldots, \tilde{\kappa}_{n-i_0} = \kappa_n$, so that

$$P(U_2 \geq \lambda) \leq \frac{\kappa}{\lambda^{2\alpha}} \left(\frac{c}{2}\right)^{2\beta}.$$

Domination of V_1 *and* V_2: When $i_0 = 1$, $V_1 = 0$, and, when $i_0 = n$, $V_2 = 0$. One must then only consider the case $1 < i_0 < n$. But then the assumption yields directly, using again the fact that $\underline{S}_0 = \underline{0}_s$, that

$$P(V_1 \geq \lambda) \leq \frac{c^{2\beta}}{\lambda^{2\alpha}}, \quad \text{and } P(V_2 \geq \lambda) \leq \frac{c^{2\beta}}{\lambda^{2\alpha}}.$$

Domination of v_n *using* U_1, V_1, U_2, V_2:

- When $j \in [0 : i_0 - 1]$,

$$\xi_j = \left\|\underline{S}_j\right\|_s \wedge \left\|\underline{S}_n - \underline{S}_j\right\|_s \leq U_1 + V_1.$$

  Indeed:

  - when $\left\|\underline{S}_j\right\|_s \leq U_1$,

$$\xi_j \leq \left\|\underline{S}_j\right\|_s \leq U_1 \leq U_1 + V_1;$$

  - when $\left\|\underline{S}_{i_0-1} - \underline{S}_j\right\|_s \leq U_1$ and

- $\left\|\underline{S}_{i_0-1}\right\|_s = V_1$, then

$$\xi_j \le \left\|\underline{S}_j\right\|_s \le \left\|\underline{S}_{i_0-1} - \underline{S}_j\right\|_s + \left\|\underline{S}_{i_0-1}\right\|_s \le U_1 + V_1;$$

- $\left\|\underline{S}_n - \underline{S}_{i_0-1}\right\|_s = V_1$, then

$$\xi_j \le \left\|\underline{S}_n - \underline{S}_j\right\|_s \le \left\|\underline{S}_{i_0-1} - \underline{S}_j\right\|_s + \left\|\underline{S}_n - \underline{S}_{i_0-1}\right\|_s \le U_1 + V_1.$$

- When $j \in [i_0 : n]$,

$$\xi_j = \left\|\underline{S}_j\right\|_s \wedge \left\|\underline{S}_n - \underline{S}_j\right\|_s \le U_2 + V_2.$$

Indeed:

– when $\left\|\underline{S}_j\right\|_s \le U_2$,

$$\xi_j \le \left\|\underline{S}_j\right\|_s \le U_2 \le U_2 + V_2;$$

– when $\left\|\underline{S}_{i_0} - \underline{S}_j\right\|_s \le U_2$ and

- $\left\|\underline{S}_{i_0}\right\|_s = V_2$, then

$$\xi_j \le \left\|\underline{S}_j\right\|_s \le \left\|\underline{S}_{i_0} - \underline{S}_j\right\|_s + \left\|\underline{S}_{i_0}\right\|_s \le U_2 + V_2;$$

- $\left\|\underline{S}_n - \underline{S}_{i_0}\right\|_s = V_1$, then

$$\xi_j \le \left\|\underline{S}_n - \underline{S}_j\right\|_s \le \left\|\underline{S}_{i_0} - \underline{S}_j\right\|_s + \left\|\underline{S}_n - \underline{S}_{i_0}\right\|_s \le U_2 + V_2.$$

Consequently,

$$\nu_n \le \{U_1 + V_1\} \vee \{U_2 + V_2\},$$

so that, since $a \vee b \ge \lambda$ implies either $a \ge \lambda$, or $b \ge \lambda$, or both,

$$P(\nu_n \ge \lambda) \le P(U_1 + V_1 \ge \lambda) + P(U_2 + V_2 \ge \lambda).$$

Domination of $P(U_1 + V_1 \ge \lambda)$: Let $\lambda > 0$ be decomposed arbitrarily into $\lambda_u + \lambda_v$, with $\lambda_u > 0$ and $\lambda_v > 0$. Then the above established probability bounds yield that

$$P(U_1 + V_1 \ge \lambda) \ge P(U_1 \ge \lambda_u) + P(V_1 \ge \lambda_v) \le \frac{\kappa}{\lambda_u^{2\alpha}}\left(\frac{c}{2}\right)^{2\beta} + \frac{c^{2\beta}}{\lambda_v^{2\alpha}}.$$

One may thus apply the lemma, with $a = \kappa\left(\frac{c}{2}\right)^{2\beta}$ and $b = c^{2\beta}$, to obtain that

$$P(U_1 + V_1 \geq \lambda) \leq \frac{c^{2\beta}}{\lambda^{2\alpha}}\left\{1 + \left(\frac{\kappa}{2^{2\beta}}\right)^{\gamma}\right\}^{\frac{1}{\gamma}}.$$

The induction proof may now be completed. The same inequality obtains for $P(U_2 + V_2 \geq \lambda)$ as for $P(U_1 + V_1 \geq \lambda)$, so that

$$P(\nu_n \geq \lambda) \leq 2\frac{c^{2\beta}}{\lambda^{2\alpha}}\left\{1 + \left(\frac{\kappa}{2^{2\beta}}\right)^{\gamma}\right\}^{\frac{1}{\gamma}}.$$

Since, by choice,

$$\frac{2^{\gamma}}{\kappa^{\gamma}}\left\{1 + \left(\frac{\kappa}{2^{2\beta}}\right)^{\gamma}\right\} = 2^{\gamma}\left(\frac{1}{2^{2\beta\gamma}} + \frac{1}{\kappa^{\gamma}}\right) \leq 1,$$

and that $\gamma^{-1} = 2\alpha + 1 \geq 1$, it follows that

$$\frac{2}{\kappa}\left\{1 + \left(\frac{\kappa}{2^{2\beta}}\right)^{\gamma}\right\}^{\frac{1}{\gamma}} \leq 1,$$

so that

$$P(\nu_n \geq \lambda) \leq \kappa\frac{c^{2\beta}}{\lambda^{2\alpha}},$$

and the induction is complete. □

Proposition 12.3.8 *Suppose that, for* $\alpha \geq 0$, $\beta > 1$, $\{i, j\} \subseteq [0:n]$ *such that* $i \leq j$, *fixed, but arbitrary, for all* $\lambda \geq 0$,

$$P\left(\left\|\underline{S}_j - \underline{S}_i\right\|_s \geq \lambda\right) \leq \frac{1}{\lambda^{\alpha}}\left(\sum_{k\in]i:j]} c_k\right)^{\beta},$$

then, for all $\lambda \geq 0$,

$$P(\mu_n \geq \lambda) \leq \tilde{\kappa}(\alpha, \beta)\frac{c^{\beta}}{\lambda^{\alpha}}.$$

In fact, one may choose $\tilde{\kappa}(\alpha, \beta)$ *equal to* $2^{\alpha}\left\{1 + \kappa\left(\frac{\alpha}{2}, \frac{\beta}{2}\right)\right\}$.

Proof By Schwarz's inequality, $P(A_1 \cap A_2) \le P^{\frac{1}{2}}(A_1)P^{\frac{1}{2}}(A_2)$. The assumption, and the fact that $ab \le (a+b)^2$, yield that

$$P\left(\left\|\underline{S}_j - \underline{S}_i\right\|_s \ge \lambda, \left\|\underline{S}_k - \underline{S}_j\right\|_s \ge \lambda\right) \le$$

$$\le \frac{1}{\lambda^{\frac{\alpha}{2}}}\left(\sum_{l\in]i:j]} c_l\right)^{\frac{\beta}{2}} \frac{1}{\lambda^{\frac{\alpha}{2}}}\left(\sum_{l\in]j:k]} c_l\right)^{\frac{\beta}{2}} \le \frac{1}{\lambda^{\alpha}}\left(\sum_{k\in]i:j]} c_k\right)^{\beta}.$$

Thus (Proposition) 12.3.7 obtains when α becomes $\frac{\alpha}{2}$, and β, $\frac{\beta}{2}$, that is,

$$P(\nu_n \ge \lambda) \le \kappa\left(\frac{\alpha}{2}, \frac{\beta}{2}\right)\frac{c^\beta}{\lambda^\alpha}.$$

The assumption says again that

$$P\left(\left\|\underline{S}_n\right\|_s \ge \lambda\right) \le \frac{c^\beta}{\lambda^\alpha}.$$

As

$$P(\mu_n \ge \lambda) \le P\left(\nu_n \ge \frac{\lambda}{2}\right) + P\left(\left\|\underline{S}_n\right\|_s \ge \frac{\lambda}{2}\right),$$

the proof is finished. □

Proposition 12.3.9 *Let $\{\underline{R}_n, n \in \mathbb{N}\}$ be a sequence of random elements in K. It is tight when the following obtain:*

1. *the sequence $\{\underline{R}_n(\cdot, 0), n \in \mathbb{N}\}$ is tight;*
2. *there exist $\alpha \ge 0$, $\beta > 1$, and a real, monotone increasing function ρ, with domain $[0, 1]$, such that, for all t_1, t_2, n, and positive λ's,*

$$P\left(\omega \in \Omega : \|\underline{R}_n(\omega, t_1) - \underline{R}_n(\omega, t_2)\|_s \ge \lambda\right) \le \frac{1}{\lambda^\alpha}|\rho(t_2) - \rho(t_1)|^\beta.$$

The following inequality implies item 2:

$$E_P\left[\|\underline{R}_n(\omega, t_1) - \underline{R}_n(\omega, t_2)\|_s^{\kappa_1}\right] \le |\rho(t_2) - \rho(t_1)|^{\kappa_2}.$$

Proof Because of (Proposition) 12.3.3, it suffices, given $\epsilon > 0$ and $\delta > 0$, fixed, but arbitrary, to produce an $\eta \in]0, 1[$ such that, for all n,

$$P\left(w(\underline{R}_n, \eta) \ge 3\epsilon\right) \le \delta.$$

Because of (Corollary) 12.3.5, that will obtain provided η^{-1} is an integer, and

$$\sum_{i<\frac{1}{\eta}} P\left(\sup_{t\in[i\eta:(i+1)\eta]} \|\underline{R}_n(\cdot,t) - \underline{R}_n(\cdot,i\eta)\|_s \geq \epsilon\right) \leq \delta.$$

So the latter shall be checked. To that end, let n, η and i be, for the moment, fixed, and p be an arbitrary integer in $\mathbb{N}$. Set, for $k \in [1 : p]$,

$$\underline{X}_k = \underline{R}_n\left(\cdot, i\eta + \frac{k}{p}\eta\right) - \underline{R}_n\left(\cdot, i\eta + \frac{k-1}{p}\eta\right),$$

$$c_k = \rho\left(i\eta + \frac{k}{p}\eta\right) - \rho\left(i\eta + \frac{k-1}{p}\eta\right).$$

From the definitions, one has that

$$\underline{S}_k = \underline{R}_n\left(\cdot, i\eta + \frac{k}{p}\eta\right) - \underline{R}_n\left(\cdot, i\eta\right),$$

and

$$c_1 + \cdots + c_k = \rho\left(i\eta + \frac{k}{p}\eta\right) - \rho(i\eta),$$

so that

$$\underline{S}_l - \underline{S}_k = \underline{R}_n\left(\cdot, i\eta + \frac{l}{p}\eta\right) - \underline{R}_n\left(\cdot, i\eta + \frac{k}{p}\eta\right),$$

and

$$c_l - c_k = \rho\left(i\eta + \frac{l}{p}\eta\right) - \rho\left(i\eta + \frac{k}{p}\eta\right).$$

Thus, using the assumption,

$$\begin{aligned}
P\left(\|\underline{S}_l - \underline{S}_k\|_s \geq \epsilon\right) &= P\left(\left\|\underline{R}_n\left(\cdot, i\eta + \frac{l}{p}\eta\right) - \underline{R}_n\left(\cdot, i\eta + \frac{k}{p}\eta\right)\right\|_s \geq \epsilon\right) \\
&\leq \frac{1}{\epsilon^\alpha}\left|\rho\left(i\eta + \frac{l}{p}\eta\right) - \rho\left(i\eta + \frac{k}{p}\eta\right)\right|^\beta \\
&= \frac{1}{\epsilon^\alpha}\left(\sum_{m\in]k:l]} c_m\right)^\beta.
\end{aligned}$$

One may thus use (Proposition) 12.3.8 to assert that

$$P\left(\max_{k\in[0,p]}\left\|\underline{R}_n\left(\cdot, i\eta+\frac{k}{p}\eta\right)-\underline{R}_n(\cdot, i\eta)\right\|_s \geq \epsilon\right) \leq$$
$$\leq \kappa(\alpha,\beta)\frac{(\rho((i+1)\eta)-\rho(i\eta))^\beta}{\epsilon^\alpha}.$$

Since $\underline{R}_n$ has continuous paths,

$$\lim_p \max_{k\in[0,p]}\left\|\underline{R}_n\left(\cdot, i\eta+\frac{k}{p}\eta\right)-\underline{R}_n(\cdot, i\eta)\right\|_s = \sup_{t\in[i\eta:(i+1)\eta]}\|\underline{R}_n(\cdot,t)-\underline{R}_n(\cdot,i\eta)\|_s .$$

Consequently

$$\sum_{i<\frac{1}{\eta}} P\left(\sup_{t\in[i\eta:(i+1)\eta]}\|\underline{R}_n(\cdot,t)-\underline{R}_n(\cdot,i\eta)\|_s \geq \epsilon\right)$$

is dominated by

$$\frac{\kappa(\alpha,\beta)\,(\rho(1)-\rho(0))}{\epsilon^\alpha}\max_{i<\frac{1}{\eta}}(\rho((i+1)\eta)-\rho(i\eta))^{\beta-1}.$$

Since ρ is continuous, and $\beta > 1$, the statement's assertion is true taking $\delta = p^{-1}$, p a large integer. □

Proposition 12.3.10 *Let, for all $n \in \mathbb{N}$, $\underline{t}_n = \{t_1,\ldots,t_n\} \subseteq [0,1]$, $P_{\underline{t}_n}$ be a probability on the Borel sets of s^n. Suppose that the resulting family is consistent, and that there exists $\alpha \geq 0$, $\beta > 1$, and $\rho : [0,1] \longrightarrow \mathbb{R}_+$, monotone and continuous, such that, for all t_1, t_2 and $\lambda \geq 0$,*

$$P_{\underline{t}_2}\left((\underline{k}_1,\underline{k}_2)\in s\times s : \left\|\underline{k}_1-\underline{k}_2\right\|_s \geq \lambda\right) \leq \frac{1}{\lambda^\alpha}\left|\rho(t_1)-\rho(t_2)\right|^\beta .$$

A sufficient condition is

$$\int_{s\times s}\left\|\underline{k}_1-\underline{k}_2\right\|_s^\lambda P_{\underline{t}_2}(d\underline{k}_1,d\underline{k}_2) \leq \left|\rho(t_1)-\rho(t_2)\right|^\beta .$$

There exists then in K a random element whose finite dimensional projections have the $P_{\underline{t}_n}$'s as distributions.

Proof Let $s^{[0,1]}$ be the set of functions with domain $[0,1]$, and range in s. If

$$\mathcal{C}(s^{[0,1]})$$

is the σ-algebra of $s^{[0,1]}$ generated by the evaluation maps, that is, the product σ-algebra of the Borel sets of s, there exists a probability measure P on it whose marginal distributions are the $P_{\underline{t}_n}$'s [70, p. 111].

Let $n \in \mathbb{N}$ be fixed, but arbitrary, and, for $i \in [0 : 2^n]$, $t_i^{(n)} = \frac{i}{2^n}$. The evaluation maps

$$\mathcal{E}_i^{s,n} : s^{[0,1]} \longrightarrow s$$

that assign to $t \mapsto \underline{k}(t)$ the value $\underline{k}(t_i^{(n)})$, define random elements whose joint law is $P_{\underline{t}_n}$. Let thus the random process $\underline{R}_n$ be defined as follows on $s^{[0,1]}$:

- $\underline{R}_n(\underline{k}, t_i^{(n)}) = \mathcal{E}_i^{s,n}(\underline{k})$,
- when $\theta \in]t_{i-1}^{(n)}, t_i^{(n)}[$, (linear interpolation)

$$\underline{R}_n(\underline{k}, \theta) = \frac{t_i^{(n)} - \theta}{t_i^{(n)} - t_{i-1}^{(n)}} \underline{R}_n(\underline{k}, t_{i-1}^{(n)}) + \frac{\theta - t_{i-1}^{(n)}}{t_i^{(n)} - t_{i-1}^{(n)}} \underline{R}_n(\underline{k}, t_i^{(n)}).$$

By construction, $\underline{R}_n$ belongs to K, and the law of

$$\left\{\underline{R}_n(\cdot, 0), \underline{R}_n\left(\cdot, \frac{1}{2^n}\right), \ldots, \underline{R}_n\left(\cdot, \frac{2^n - 1}{2^n}\right), \underline{R}_n(\cdot, 1)\right\}$$

is $P_{\underline{t}_n}$. When $t_1 = \frac{i}{2^n}$ and $t_2 = \frac{j}{2^n}$, then, by assumption,

$$P\left(\|\underline{R}_n(\cdot, t_1) - \underline{R}_n(\cdot, t_2)\|_s \geq \lambda\right) \leq \frac{1}{\lambda^\alpha} |\rho(t_1) - \rho(t_2)|^\beta .$$

As in the proof of the previous proposition, let n, η, and i be fixed, with η^{-1} an integer. If the points of the form $i\eta + \frac{k}{p}\eta$ are also of the form $\frac{j}{2^n}$, then the expression of the previous proposition involving the maximum obtains equally. Suppose furthermore that $\eta 2^n$ is also an integer, and let $p = \eta 2^n$. Then the points of the form $i\eta + \frac{k}{p}\eta$ are indeed of the form $\frac{j}{2^n}$, and they partition the intervals $[i\eta, (i+1)\eta]$. It follows that the conclusions of the previous proof involving suprema obtain also. But, if $\eta = 2^q$, q an integer large enough for the right-hand side of the last relation of the previous proof be less than δ, the $\underline{R}_n$'s for n beyond a certain fixed value form a tight sequence. There is thus, by Prohorov's theorem a random element $\underline{R}$ in K, and a subsequence with elements of the form $\underline{R}_{n_p}$ that converges in distribution (weakly) to it.

$\underline{R}$ is the required random element. Indeed, when $\{t_1, \ldots, t_m\}$ are dyadic rationals, because of the consistency assumption, the law of

$$\{\underline{R}_n(\cdot, t_1), \ldots, \underline{R}_n(\cdot, t_m)\}$$

is $P_{\underline{t}_m}$ for large enough n, and it follows that the latter is also the law of $\{\underline{R}(\cdot, t_1), \ldots, \underline{R}(\cdot, t_m)\}$. When $\{t_1, \ldots, t_m\}$ are not dyadic rationals, one chooses dyadic rationals $\{d_{1,q}, \ldots, d_{m,q}\}$ such that $\lim_q d_{q,i} = t_i$, $i \in [1, m]$. Then $\{\underline{R}(\cdot, d_{q,1}), \ldots, \underline{R}(\cdot, d_{q,m})\}$ converges in distribution to

$$\{\underline{R}(\cdot, t_1), \ldots, \underline{R}(\cdot, t_m)\} .$$

But, because of the assumption and the continuity of ρ, $P_{\underline{d}_{q,m}}$ converges weakly to $P_{\underline{t}_m}$, so that

$$\{\underline{R}(\cdot, t_1), \ldots, \underline{R}(\cdot, t_m)\}$$

has law $P_{\underline{t}_m}$. □

Chapter 13
Likelihoods for Signal Plus "White Noise" Versus "White Noise"

In this chapter, one obtains the likelihood for a "signal plus noise" model for which the noise is a Cramér-Hida process. The "white noise" of the chapter's title is a convenience: for a short glimpse at "real white noise," one may, for example, look at [164, p. 260]. As for the finite dimensional case, the road to the likelihood is based on a version of Girsanov's theorem, and the likelihood itself follows when the "signal plus noise," that is, the observation process, has a representation as the solution of a stochastic differential equation.

One shall use the following acronyms: SPWN shall mean "signal plus white noise," SPGN, "signal plus Gaussian noise" (that is, Gaussian, but not "white").

13.1 A Version of Girsanov's Theorem

The version to follow shall allow one to implement, for Cramér-Hida processes (noises), the procedure delineated at the beginning of this part (Part III) of the book.

13.1.1 Framework

The framework for the sequel is now described. When using simultaneously probabilities P and Q, one shall write respectively

$$\mathcal{I}_0^P[\underline{b}] \text{ and } \mathcal{I}_0^Q[\underline{b}]$$

for the corresponding $\mathcal{I}_0[\underline{b}]$. When $Q \ll P$, $\mathcal{I}_0^P[\underline{b}] \subseteq \mathcal{I}_0^Q[\underline{b}]$.

Let $(\Omega, \mathcal{A}, P)$ be the basic probability space, and $\underline{B} : \Omega \times [0,1] \longrightarrow s$ be a Cramér-Hida process whose covariance function is C [(Fact) 11.1.5]. Let $\underline{a} \in \mathcal{I}_0[\underline{b}]$

A.F. Gualtierotti, *Detection of Random Signals in Dependent Gaussian Noise*,
DOI 10.1007/978-3-319-22315-5_13

be fixed, but arbitrary. $\underline{X}$ is the process, with values in s, whose components X_n have the following form: for $n \in \mathbb{N}$, and $t \in [0, 1]$, fixed, but arbitrary,

$$X_n(\omega, t) = \int_0^t a_n(\omega, \theta) M_n(d\theta) + B_n(\omega, t) = S_n[\underline{a}](\omega, t) + B_n(\omega, t).$$

The integral in that latter expression yields an adapted process (Fubini). One shall most often use the abridged notation $\underline{X} = \underline{S}[\underline{a}] + \underline{B}$. Now, since, for $t_1 < t_2$ in $[0, 1]$, fixed, but arbitrary, by Schwarz's inequality,

$$\begin{aligned} \left\{ \int_{t_1}^{t_2} a_n(\omega, \theta) M_n(d\theta) \right\}^2 &\leq \{b_n(t_2) - b_n(t_1)\} \, \|a_n(\omega, \cdot)\|^2_{L_2[b_n]} \\ &\leq \{b(t_2) - b(t_1)\} \, \|a_n(\omega, \cdot)\|^2_{L_2[b_n]}, \end{aligned}$$

one has that

$$\sum_{n=1}^{\infty} \left\{ \int_{t_1}^{t_2} a_n(\omega, \theta) M_n(d\theta) \right\}^2 \leq \{b(t_2) - b(t_1)\} \, \|\underline{a}(\omega, \cdot)\|^2_{L_2[\underline{b}]} .$$

Since $\underline{a}$ can be chosen to have finite norm for every $\omega \in \Omega$ [(Lemma) 11.2.2], $\underline{S}[\underline{a}]$ has paths in K.

Given a fixed, but arbitrary $\underline{\alpha} \in \mathcal{I}_0[\underline{b}]$, the meaning of $I_{\underline{X}}\{\underline{\alpha}\}(\omega, t)$ shall be the following expression:

$$I_{\underline{X}}\{\underline{\alpha}\}(\omega, t) = \langle \underline{\alpha}_{|t}(\omega, \cdot), \underline{a}_{|t}(\omega, \cdot) \rangle_{L_2[\underline{b}]} + I_{\underline{B}}\{\underline{\alpha}\}(\omega, t).$$

When probabilities P and Q must be entertained, one shall write, for such an integral, when distinction is required,

$$I^P_{\underline{X}}\{\underline{\alpha}\}(\omega, t), \quad \text{and} \quad I^Q_{\underline{X}}\{\underline{\alpha}\}(\omega, t).$$

The probability determined, on $\mathcal{K}$, by a process $\underline{X}$, and a probability P, shall be denoted $P^{\mathcal{K}}_{\underline{X}}$. When Q is a probability (measure) absolutely continuous with respect to the probability (measure) P, one writes sometimes $Q \ll P$; when P and Q are mutually absolutely continuous, one sometimes writes $Q \equiv P$.

13.1.2 Tools for Absolutely Continuous Change of Measure

A few required facts from martingale theory are now listed.

Fact 13.1.1 ([264, pp. 246–247]) *Let Q be a probability on $\mathcal{A}$ such that $Q \ll P$ (absolute continuity), and let D denote the Radon-Nikodým derivative. When D has paths that are continuous to the right,*

1. *given a wide sense stopping time S, denoting P_S and Q_S the restrictions to $\mathcal{A}_S$ of, respectively, P and Q, one has that*

$$\frac{dQ_S}{dP_S}(\omega) = D(\omega, S(\omega));$$

2. *given a process X that is adapted, and continuous to the right, it is a local martingale with respect to Q if, and only if, the process $X \times D$ is a local martingale with respect to P, and, when D is almost surely continuous, one may replace "local martingale" with "a martingale locally in L_2."*

One also needs an approximation result based on properties of convolution. The convolution of two functions, f and g, with $\mathbb{R}$ as domain, is the function obtained, when it makes sense, using the following formula:

$$\{f \star g\}(t) = \int_{\mathbb{R}} f(\theta) g(t-\theta) d\theta = \int_{\mathbb{R}} f(t-\theta) g(\theta) d\theta.$$

When $f(t)$ and $g(t)$ are zero for $t < 0$, $\{f \star g\}(t) = 0$ for $t \leq 0$, and, for $t > 0$,

$$\{f \star g\}(t) = \int_0^t f(\theta) g(t-\theta) d\theta.$$

When $f \in L_1[\mathbb{R}]$, and $g \in L_\infty[\mathbb{R}]$, one has that [108, p. 345]:

1. $f \star g(t)$ makes sense for $t \in \mathbb{R}$;
2. the class of $f \star g$ is in $L_\infty[\mathbb{R}]$;
3. $\|f \star g\| \leq \|f\|_{L_1[\mathbb{R}]} \|g\|_{L_\infty[\mathbb{R}]}$;
4. $f \star g \in C[\mathbb{R}]$, the space of continuous functions.

A function is locally in a space of integrable functions when its restriction to every compact set is in the corresponding space of integrable functions. An approximate unit for convolution is a sequence $\{\delta_n, n \in \mathbb{N}\}$ of (classes of) functions of $L_1[\mathbb{R}]$ such that, given $f \in L_1^{loc}[\mathbb{R}]$, or $f \in L_2^{loc}[\mathbb{R}]$, fixed, but arbitrary, $\{\delta_n \star f, n \in \mathbb{N}\}$ converges in $L_1^{loc}[\mathbb{R}]$, respectively $\in L_2^{loc}[\mathbb{R}]$. To obtain sequences of type $\{\delta_n, n \in \mathbb{N}\}$, it suffices [108, p. 372] to choose probability densities $\delta \in L_1[\mathbb{R}]$, with compact support, and then to set $\delta_n(t) = n\delta(nt)$. One then has, when choosing $f_n = \delta_n \star f$, the following lemma:

Lemma 13.1.2 *Let $f : \mathbb{R}_+ \longrightarrow \mathbb{R}$ be Borel measurable, and bounded by κ. There is a sequence $\{f_n, n \in \mathbb{N}\}$ of continuous functions, bounded by κ, such that, on every finite interval, say $[a, b]$,*

$$\lim_n \int_a^b \{f - f_n\}^2 \, dLeb = 0.$$

Lemma 13.1.3 *Let X be a progressively measurable process on $(\Omega, \underline{A}, P)$, with index set $[0, 1]$, whose square is integrable with respect to $P \otimes Leb$. There exists then a sequence $\{X_n, n \in \mathbb{N}\}$ of (uniformly) bounded simple, progressively measurable processes such that*

$$\lim_n \int_\Omega \int_0^t \{X(\omega, t) - X_n(\omega, t)\}^2 \, P \otimes Leb(d\omega, dt) = 0.$$

Proof It is no restriction, when necessary, to assume that the index set is $\mathbb{R}$, or $\mathbb{R}_+$, as, with respect to $P \otimes Leb$, $X = \chi_{]0,1]} X$. The proof consists in recognizing that any process may be approximated by bounded ones, that bounded ones may be approximated by continuous ones, and continuous ones, by simple functions.

Suppose that X is a (uniformly) bounded, adapted process with continuous paths (it is thus progressively measurable [264, p. 71]). Let

$$X_n(\omega, t) = \sum_{n=0}^{n-1} X(\omega, t_i) \, \chi_{]t_i, t_{i+1}]}(t),$$

X_n is by definition progressively measurable, and (uniformly) bounded. Thus, by dominated convergence, for each $\omega \in \Omega$, since X has continuous paths,

$$\lim_n \int_0^1 |X(\omega, t) - X_n(\omega, t)|^2 \, Leb(dt) = 0.$$

The same is true for the expectation.

Suppose now that X is progressively measurable and (uniformly) bounded. When applying (Lemma) 13.1.3 to $X(\omega, \cdot)$, one may assume that the latter is f, and thus X bounded yields f integrable. Furthermore f_n involves the integral from 0 to t, and it is bounded. Let thus

$$X_n(\omega, t) = \int_0^t \delta_n(t - \theta) X(\omega, \theta) Leb(d\theta).$$

One has that X_n has continuous paths, and is bounded by the same bound as that which obtains for X. Also, for each $\omega \in \Omega$,

$$\lim \int_0^1 \{X(\omega, t) - X_n(\omega, t)\}^2 \, Leb(dt).$$

The same is true for expectations.

Finally $X_n = \chi_{[-n,n[}(X)$ provides a bounded approximation to X. □

Lemma 13.1.4 *Let M_b be a measure on the Borel sets of $[0, 1]$, generated by the continuous, increasing function b, which is zero at the origin. Let a be a real valued,* *progressively measurable process on* $(\Omega\ \underline{\mathcal{A}}, P)$, $\underline{\mathcal{A}}$ *indexed by* $[0, 1]$, *such that*

$$\int_{\Omega}\int_0^1 |a(\omega,t)|^2\, P\otimes M_b(d\omega,dt) < \infty.$$

There exists a sequence of simple, progressively measurable processes a_n *on* $(\Omega\ \underline{\mathcal{A}}, P)$ *such that*

$$\lim_n \int_{\Omega}\int_0^1 |a(\omega,t) - a_n(\omega,t)|^2\, P\otimes M_b(d\omega,dt) = 0.$$

Proof Let b_e be the extension of b to $\mathbb{R}_+$ obtained using, when $t > 1$, the following assignment: $b_e(t) = b(1)$. The σ-algebras in $\underline{\mathcal{A}}$ and any function f are extended similarly to obtain, respectively, $\underline{\mathcal{A}}_e$ and f_e ($\mathcal{A}_\infty = \mathcal{A}_1$ and $f_e(\infty) = f(1)$). Let $\overline{b}_e(t) = \inf\{\theta \in \mathbb{R}_+ : b_e(\theta) > t\}$. It is a strict stopping time [(Fact) 10.3.22], and an increasing function [(Fact) 10.3.21], continuous to the right. The following process: $(\omega,t) \mapsto a(\omega,\overline{b}_e(t))$ is a progressively measurable process for the filtration with elements $\mathcal{A}_{\overline{b}_e(t)}$ [192, p. 73]. Since b_e is continuous, because of (Fact) 10.3.33,

$$b_e\left(\overline{b}_e(t)\right) = t.$$

As

$$b_e\left(\overline{b}_e(t)\right) - b_e\left(\overline{b}_e(t-\Delta)\right) = \Delta,$$

b_e is $\overline{b}_e$-continuous [(Definition) 10.3.34]. One may thus apply, path by path, the change of variables formula [(Fact) 10.3.36], with $\phi = a(\omega,\cdot)$, $\alpha = \overline{b}_e$, $\beta = b_e$, and $\gamma = Leb$, to obtain that

$$\int_0^t \left\{\phi\circ\overline{b}_e\right\}^2 dLeb = \int_0^{\overline{b}_e(t)} \phi^2\, dm_{b_e}.$$

Let now a have an integrable square with respect to $P\otimes M$. Then $\phi\circ\overline{b}_e$ has a square that is integrable with respect to Lebesgue measure. There is thus [150, p. 60] a sequence of processes of the following form:

$$\Phi_n(\omega,t) = \sum_{i=1}^{n-1} a(\omega,\overline{b}_e(t_i))\,\chi_{[t_i,t_{i+1}[}(t)$$

such that

$$\lim_n E_P\left[\int \left\{a\diamond\overline{b}_e - \Phi_n\right\}^2 dLeb\right] = 0.$$

Since $b_e(\overline{b}_e(t)) = t$, $\Phi_n(\omega, t) = \Phi_n(\omega, b_e(\overline{b}_e(t)))$. One may thus apply the change of variable formula [(Fact) 10.3.36] in reverse to obtain that

$$\lim_n E_P \left[\int \{a - \Phi_n \diamond b_e\}^2 \, dM_{b_e} \right] = 0.$$

Since [(Fact) 10.3.28] $b_e = \overline{\overline{b}}_e$, b_e is a change of time for the σ-algebras of type

$$\mathcal{A}_{\overline{b}_e(t)},$$

that is, $\mathcal{A}_t$. Consequently $\Phi_n \diamond b_e$ is progressively measurable for $\underline{\mathcal{A}}$ [192, p. 73]. It remains to see that the M_{b_e}-measure of $[1, \infty]$ is zero. □

Corollary 13.1.5 *Add to the assumptions of (Lemma)* 13.1.4 *that Q is absolutely continuous with respect to P, and that*

$$\int_\Omega \int_0^1 |a(\omega, t)|^2 \, Q \otimes M_b(d\omega, dt) < \infty.$$

There is then a sequence of simple, progressively measurable processes a_n on $(\Omega\ \underline{\mathcal{A}}, P)$ such that

$$\lim_n \int_\Omega \int_0^1 |a(\omega, t) - a_n(\omega, t)|^2 \, P \otimes M_b(d\omega, dt) = 0,$$

and

$$\lim_n \int_\Omega \int_0^1 |a(\omega, t) - a_n(\omega, t)|^2 \, Q \otimes M_b(d\omega, dt) = 0$$

Proof Let $\Pi = (P + Q)/2$. When a is progressively measurable and has a square that is integrable with respect to both $P \otimes M_b$ and $Q \otimes M_b$, the same is true with respect to Π. Applying (Lemma) 13.1.4 with Π in place of P, one gets the required sequence as, for example,

$$\int a^2 d(P \otimes M_b) \leq 2 \int a^2 d\Pi.$$

□

Definition 13.1.6 Let B be a zero mean, almost surely continuous, Gaussian martingale with variance function b. Let M_b be the measure corresponding to b, and a, a process whose paths are almost surely in $L_2([0, 1], \mathcal{B}, M_b)$. Given

$$X(\cdot, t) = \int_0^t a(\cdot, \theta) M(d\theta) + B(\cdot, t),$$

one sets [264, p. 151], for α, progressively measurable, with paths almost surely in $L_2([0,1], \mathcal{B}, M_b)$ [264, p. 152],

$$\int_0^t \alpha(\cdot,\theta)X(\cdot,d\theta) = \int_0^t \alpha(\cdot,\theta)a(\cdot,\theta)M(d\theta) + \int_0^t \alpha(\cdot,\theta)B(\cdot,d\theta).$$

Proposition 13.1.7 *Let a_P, a_Q, B_P, B_Q have the meaning of a and B in (Definition) 13.1.6. Suppose that Q is a probability which is absolutely continuous with respect to P, and that, with respect to P, X has the following representation:*

$$X_P(\omega,t) = \int_0^t a_P(\cdot,\theta)M(d\theta) + B_P(\cdot,t),$$

while, with respect to Q, it has the following representation:

$$X_Q(\omega,t) = \int_0^t a_Q(\cdot,\theta)M(d\theta) + B_Q(\cdot,t),$$

where B_P and B_Q have the law of B in (Definition) 13.1.6. Let α be such that its integral with respect to X makes sense for P as well as for Q. Then, almost surely with respect to Q,

$$\int_0^t \alpha\, dX_P = \int_0^t \alpha\, dX_Q.$$

Proof Let $\{\alpha_n, n \in \mathbb{N}\}$ be a sequence as in (Corollary) 13.1.5. Then, since $\int \alpha_n\, dX$ does depend neither on P, nor on Q, the claim is valid for each α_n. Now, for example, for $\alpha \in L_2[P \otimes M_b]$,

$$E_P\left[\left\{\int_0^t \alpha\, dX_P - \int_0^t \alpha_n\, dX_P\right\}^2\right] \leq$$
$$\leq 2\left\{1 + E_P\left[\int_0^t a^2 dM_b\right]\right\} E_P\left[\int_0^t \{\alpha - \alpha_n\}^2\, dM_b\right].$$

One can thus find a subsequence of $\{\alpha_n, n \in \mathbb{N}\}$, say $\{\alpha_{n_p}, p \in \mathbb{N}\}$, such that, almost surely, with respect to P, and thus with respect to Q,

$$\lim_n \int_0^t \alpha_{n_p}\, dX = \lim_n \int_0^t \alpha_{n_p}\, dX_P = \int_0^t \alpha\, dX_P,$$

and, almost surely, with respect to Q,

$$\lim_n \int_0^t \alpha_{n_p}\, dX = \lim_n \int_0^t \alpha_{n_p}\, dX_Q = \int_0^t \alpha\, dX_Q.$$

Consequently, the claim obtains for α's such that

$$E_P\left[\int_0^1 \alpha^2 dM_P\right] \vee E_Q\left[\int_0^1 \alpha^2 dM_Q\right] < \infty.$$

For an α that is almost surely square integrable with respect to both P and Q, one can find a family of stopping times $\{S_n, n \in \mathbb{N}\}$ such that

$$E_P\left[\int_0^1 \chi_{[\![0,S_n]\!]}\alpha^2 dM_P\right] \vee E_Q\left[\int_0^1 \chi_{[\![0,S_n]\!]}\alpha^2 dM_Q\right] < \infty.$$

A modification of α as in (Lemma) 11.2.2 allows one to even choose strict stopping times. From the L_2 case, one has that, almost surely with respect to Q,

$$\int_0^{t\wedge S_n} \alpha\, dX_P = \int_0^{t\wedge S_n} \alpha\, dX_Q.$$

□

13.1.3 A Girsanov Type Theorem

One proves below a Girsanov's type theorem for the “model”

$$\underline{X} = \underline{S}[\underline{a}] + \underline{B}.$$

The proof requires the following result:

Proposition 13.1.8 *Suppose that Q is a probability, absolutely continuous with respect to P, and that $\underline{X}$ is a process, adapted to $\underline{\mathcal{A}}$, which has, for P and Q, the following respective representations:*

$$\underline{X}_P = \underline{S}\left[\underline{a}_P\right] + \underline{B}_P,$$
$$\underline{X}_Q = \underline{S}\left[\underline{a}_Q\right] + \underline{B}_Q,$$

for which, $\mathcal{L}(X)$ denoting the law of the process X,

$$\mathcal{L}(\underline{B}_P) = \mathcal{L}(\underline{B}_Q),$$

and

$$P\left(\omega \in \Omega : \left\|\underline{a}_P(\omega,\cdot)\right\|_{L_2[\underline{b}]} < \infty\right) = Q\left(\omega \in \Omega : \left\|\underline{a}_Q(\omega,\cdot)\right\|_{L_2[\underline{b}]} < \infty\right) = 1.$$

Suppose that $\underline{\alpha}$ is a process adapted to $\underline{A}$, which belongs to $\mathcal{I}_0^P[\underline{b}]$, and thus also to $\mathcal{I}_0^Q[\underline{b}]$. Then, almost surely, with respect to Q,

$$I^Q_{\underline{X}_Q}\{\underline{\alpha}\} = I^P_{\underline{X}_P}\{\underline{\alpha}\}.$$

Proof Since the integrals in the proposition are limits, locally, uniformly in probability, of finite sums, and that the result is true for finite sums [(Proposition) 13.1.7], it remains true for the integrals themselves. □

Proposition 13.1.9 *Let $\underline{a} \in \mathcal{I}_0^P[\underline{b}]$ be fixed, but arbitrary, and $\underline{X} = \underline{S}[\underline{a}] + \underline{B}$, as in Sect.* 13.1.1. *Let the processes D and U, and the probability Q, be defined as follows:*

$$U(\omega, t) = -\left\{I_{\underline{B}}\{\underline{a}\}(\omega, t) + (1/2)\langle I_{\underline{B}}\{\underline{a}\}\rangle(\omega, t)\right\},$$

$$\ln D = U, \text{ and } dQ = D(\cdot, 1)dP.$$

Suppose that $E_P[D(\cdot, 1)] = 1$. Then:

1. $Q^K_X = P^K_B$;
2. $P^K_B \ll P^K_X$, *and, almost surely, with respect to* P^K_X,

$$\frac{dP^K_B}{dP^K_X}(\underline{k}) = E_P[D(\cdot, 1) \mid \underline{X} = \underline{k}];$$

3. $P^K_X \ll P^K_B$, *and, almost surely, with respect to* P^K_B,

$$\frac{dP^K_X}{dP^K_B}(\underline{k}) = E_Q\left[\{D(\cdot, 1)\}^{-1} \mid \underline{X} = \underline{k}\right].$$

Proof (1) One must check that, with respect to Q, $\underline{X}$ is a Cramér-Hida process whose covariance function is C. According to (Corollary) 10.5.19, it suffices to prove that, with respect to Q, $\underline{X}$ is an $(\underline{A}, C)$-martingale. Because of (Proposition) 10.5.6, it is enough to check that, for $\underline{\alpha} \in l_2$, fixed, but arbitrary,

$$E_\alpha(\omega, t) = e^{X_\alpha(\omega,t) - \frac{1}{2}C_\alpha(t)} = e^{\langle \underline{\alpha}, \underline{X}(\omega,t)\rangle_{l_2} - \frac{1}{2}\langle C(t)[\underline{\alpha}], \underline{\alpha}\rangle_{l_2}}$$

is a local martingale with respect to Q. But that will be the case [(Fact) 13.1.1, item 2] if the process $E_\alpha \times D$ is a local martingale with respect to P, and that will obtain [(Proposition) 10.4.6] if it is possible to express $E_\alpha \times D$ as the exponential of a local martingale minus half its quadratic variation.

Now

- D is the exponential of $-I_{\underline{B}}\{\underline{a}\} - \frac{1}{2}\langle I_{\underline{B}}\{\underline{a}\}\rangle$ with

$$\langle I_{\underline{B}}\{\underline{a}\}\rangle(\omega, t) = \left\|\underline{a}_{|t}(\omega, \cdot)\right\|^2_{L_2[\underline{b}]};$$

- E_α is the exponential of $\langle \underline{\alpha}, \underline{X} \rangle_{l_2} - \frac{1}{2} \langle C(\cdot)[\underline{\alpha}], \underline{\alpha} \rangle_{l_2}$ with [(Definition) 13.1.6]

$$\langle \underline{\alpha}, \underline{X} \rangle_{l_2} = \langle \underline{\alpha}, \underline{S}[\underline{a}] \rangle_{l_2} + \langle \underline{\alpha}, \underline{B} \rangle_{l_2}.$$

With notation and (property (Fact)) 11.3.10, $\langle \underline{\alpha}, \underline{B} \rangle_{l_2} = I_{\underline{B}} \{\underline{a}^{\underline{\alpha}}\}$, and

$$\langle I_{\underline{B}} \{\underline{a}^{\underline{\alpha}}\} \rangle = \| \underline{a}^{\underline{\alpha}} \|^2_{L_2[\underline{b}]} = \langle C(\cdot)[\underline{\alpha}], \underline{\alpha} \rangle_{l_2}.$$

Furthermore

$$\langle \underline{\alpha}, \underline{S}[\underline{a}] \rangle_{l_2} = \langle \underline{a}^{\underline{\alpha}}, \underline{a} \rangle_{L_2[\underline{b}]}.$$

Putting the pieces together, one obtains that the process $E_\alpha \times D$ is the exponential of

$$I_{\underline{B}} \{\underline{a}^{\underline{\alpha}} - \underline{a}\} - (1/2) \langle I_{\underline{B}} \{\underline{a}^{\underline{\alpha}} - \underline{a}\} \rangle,$$

and is thus a local martingale with respect to P.

Proof (2) Since $Q \ll P$, $Q_{\underline{X}}^K \ll P_{\underline{X}}^K$. But, as seen, $Q_{\underline{X}}^K = P_{\underline{B}}^K$, so that

$$P_{\underline{B}}^K \ll P_{\underline{X}}^K.$$

Furthermore, for measurable $K_0 \subseteq K$,

$$\begin{aligned} P_{\underline{B}}^K(K_0) &= Q_{\underline{X}}^K(K_0) \\ &= Q\left(\underline{X}^{-1}(K_0)\right) \\ &= \int_{\underline{X}^{-1}(K_0)} D(\cdot, 1)\, dP \\ &= \int_{K_0} E_P\left[D(\cdot, 1) \mid \underline{X} = \underline{k}\right] P_{\underline{X}}^K(d\underline{k}). \end{aligned}$$

Proof (3) One must first check that, with respect to Q, almost surely,

$$D(\cdot, 1) > 0.$$

But, by assumption, $\underline{a} \in \mathcal{I}_0^P[\underline{b}]$ so that

$$P\left(\omega \in \Omega : \langle I_{\underline{B}} \{\underline{a}\} \rangle(\omega, 1) < \infty\right) = 1.$$

Since $Q \ll P$, one has, as well, that

$$Q\left(\omega \in \Omega : \langle I_{\underline{B}}\{\underline{a}\}\rangle(\omega, 1) < \infty\right) = 1.$$

One must thus secure that $I_{\underline{B}}\{\underline{a}\}$ is, with respect to Q, almost surely finite. But that is a consequence of (Proposition) 13.1.8. One may thus write that

$$dP = D^{-1}(\cdot, 1)D(\cdot, 1)dP = D^{-1}(\cdot, 1)dQ.$$

Consequently $P \ll Q$, and thus $P_{\underline{X}}^{K} \ll Q_{\underline{X}}^{K} = P_{\underline{B}}^{K}$. Finally, for measurable $K_0 \subseteq K$,

$$\begin{aligned}
P_{\underline{X}}^{K}(K_0) &= P\left(\underline{X}^{-1}(K_0)\right) \\
&= \int_{\underline{X}^{-1}(K_0)} D^{-1}(\cdot, 1)dQ \\
&= \int_{K_0} E_Q\left[D^{-1}(\cdot, 1) \mid \underline{X} = \underline{k}\right] Q_{\underline{X}}^{K}(d\underline{k}) \\
&= \int_{K_0} E_Q\left[D^{-1}(\cdot, 1) \mid \underline{X} = \underline{k}\right] P_{\underline{B}}^{K}(d\underline{k}).
\end{aligned}$$

□

Remark 13.1.10 The condition $E_P[D(\cdot, 1)] = 1$ of (Proposition) 13.1.9 obtains in particular when the map $\omega \mapsto \|\underline{a}(\omega, \cdot)\|_{L_2[\underline{b}]}$ is, with respect to P, almost surely bounded [(Proposition) 10.4.10].

Proposition 13.1.11 *When one omits, in* (Proposition) 13.1.9, *the requirement that*

$$E_P[D(\cdot, 1)] = 1,$$

one still has that $P_{\underline{X}}^{K} \ll P_{\underline{B}}^{K}$.

Proof Let $\{S_n, n \in \mathbb{N}\}$ be the localizing sequence of (Proposition) 11.2.3, and $\hat{\underline{a}}$, the modification of $\underline{a}$ defined there. Set

$$\hat{\underline{X}}^{(n)}(\omega, t) = \underline{S}\left[\hat{\underline{a}}_{|S_n}\right] + \underline{B}.$$

Since the norm of $\hat{\underline{a}}_{|S_n}$ is, by definition, bounded by n, $P_{\hat{\underline{X}}^{(n)}}^{K} \equiv P_{\underline{B}}^{K}$, a consequence of (Proposition) 13.1.9 and (Remark) 13.1.10. Let then

$$\Omega_n = \{\omega \in \Omega : S_n(\omega) = 1\}, \text{ and } \Omega_{\cup} = \cup_{n \in \mathbb{N}} \Omega_n.$$

By definition $P(\Omega_{\cup}) = 1$. Let K_0 be a measurable set. Then

$$\begin{aligned} P_X^K(K_0) &= P\left(\{\omega \in \Omega : \underline{X}(\omega,\cdot) \in K_0\} \cap \Omega_{\cup}\right) \\ &= P\left(\cup_{n\in\mathbb{N}} \left(\{\omega \in \Omega : \underline{X}(\omega,\cdot) \in K_0\} \cap \Omega_n\right)\right). \end{aligned}$$

However, on Ω_n, almost surely with respect to P, $\underline{X} = \underline{\hat{X}}^{(n)}$, so that

$$P_X^K(K_0) = P\left(\cup_{n\in\mathbb{N}} \left(\left\{\omega \in \Omega : \underline{\hat{X}}^{(n)}(\omega,\cdot) \in K_0\right\} \cap \Omega_n\right)\right).$$

Suppose now that $P_B^K(K_0) = 0$. Then, because of mutual absolute continuity, $P_{\hat{X}^{(n)}}^K(K_0) = 0$, and consequently $P_X^K(K_0) = 0$. □

Remark 13.1.12 A form of the likelihood has been obtained in (Proposition) 13.1.9. However, it depends functionally on the paths of the SPWN process, and, when detecting, that is what one tries to discover, rather than assume. One must thus decouple the likelihood from the underlying processes, and that is achieved through a preliminary series of decompositions.

13.2 Decomposition of Processes

Let $(\Omega, \mathcal{A}, P)$ be the base probability space, and $\mathcal{N}(\mathcal{A}, P)$, the sets of $\mathcal{A}$ of P-probability zero. Let $\mathcal{B} \subseteq \mathcal{A}$ be a σ-algebra. Then

$${}^o\mathcal{B} = \sigma\left(\mathcal{B}, \mathcal{N}(\mathcal{A}, P)\right).$$

When $\mathcal{A}$ is complete, one shall write $\mathcal{B}^o$ for ${}^o\mathcal{B}$.

The evaluations $\left\{\mathcal{E}_t^K, t \in [0,1]\right\}$ shall often be looked at as a process on the space $(K, \mathcal{K})$, and, to indicate that they are considered as a process for a given measure, say μ, one shall write $\underline{\mathcal{E}}_\mu$ to mean that the operating measure is μ, and that

$$\underline{\mathcal{E}}_\mu(\underline{k}, t) = \mathcal{E}_t^K(\underline{k}) = \underline{k}(t) \in s.$$

Those evaluation maps generate a filtration on K, denoted $\underline{\mathcal{K}}$:

$$\mathcal{K}_t = \sigma_t\left(\mathcal{E}^K\right).$$

Given a process $\underline{X}$, $\underline{\boldsymbol{X}}$ shall denote the random element whose value at ω is the path $\underline{X}[\omega] \in K$:

$$\underline{X}[\omega] = \{\underline{X}(\omega, t), t \in [0,1]\} = [t \mapsto \underline{X}(\omega, t)].$$

One then defines $\Phi_X : \Omega \times [0,1] \longrightarrow K \times [0,1]$ using the following assignment:

$$\Phi_X(\omega, t) = (\underline{X}[\omega], t)\,.$$

d_s shall denote the distance of s, obtained with the l_2 norm when $s = l_2$, and the Fréchet distance when $s = \mathbb{R}^\infty$.

When $\underline{a}^K : K \times [0,1] \longrightarrow s$, the expression $\left\| \underline{a}^K_{|t}(\underline{k}, \cdot) \right\|^2_{L_2[\underline{b}]}$ shall mean

$$\sum_{n=1}^{\infty} \int_0^t \left\{a_n^K(\underline{k}, \theta)\right\}^2 M_n(d\theta).$$

Furthermore

$$\underline{a}^K \Box \Phi_X(\omega, t) = \underline{a}^K(\underline{X}[\omega], t)\,.$$

Fact 13.2.1 ([138, p. 443]) *Let $f : E \longrightarrow (F, \mathcal{F})$, and $g : E \longrightarrow (G, \mathcal{G})$, be two maps into measurable spaces. Suppose that f is adapted to $\sigma(g)$. There exists $\phi : G \longrightarrow F$, adapted to $\mathcal{G}$ and $\mathcal{F}$, such that $f = \phi \circ g$ in each of the following two cases:*

1. *F is a complete, and separable, metric space, and $\mathcal{F} = \mathcal{B}(F)$;*
2. *$\mathcal{F}$ separates points in F, and $g(E) \in \mathcal{G}$.*

Proposition 13.2.2 (Decomposing "White Noise") *Suppose that P^o denotes the probability obtained when completing $\mathcal{A}$ with respect to P and that $\underline{B}$ is adapted to the filtration $\underline{\sigma}_t^o(\underline{X})$. There exists then, on $\left(K, \underline{\mathcal{K}}, P_X^K\right)$, an adapted process $\underline{B}_X$ such that*

1. *for $t \in [0,1]$, fixed, but arbitrary, almost surely, with respect to P^o,*

$$\underline{B}_X \Box \Phi_X(\omega, t) = \underline{B}(\omega, t);$$

2. $P^K_{\underline{B}_X} = P^K_{\underline{B}}$.

Proof Let $t \in [0,1]$ be fixed, but arbitrary, and let $\underline{B}^\star(\omega, t)$ be adapted to $\sigma_t(\underline{X})$, and almost surely equal to $\underline{B}(\cdot, t)$ [(Proposition) 10.2.22]. $\underline{\boldsymbol{X}}$ is adapted to $\sigma_t(\underline{X})$ and $\mathcal{K}_t$, as, for $\theta \leq t$, fixed, but arbitrary, $\mathcal{E}_\theta^K \circ \underline{\boldsymbol{X}} = \underline{X}(\cdot, \theta)$ [192, p. 11]. Furthermore, $\underline{\boldsymbol{X}}^{-1}(\mathcal{K}_t) = \sigma_t(\underline{X})$. One has thus the following pattern:

$$\begin{array}{lll} \underline{B}^\star(\cdot, t) & : \quad \Omega \longrightarrow (s, \mathcal{B}(s)) & \text{(it is } f \text{ in (Fact) 13.2.1)}, \\ \underline{\boldsymbol{X}} & : \quad \Omega \longrightarrow (K, \mathcal{K}_t) & \text{(it is } g \text{ in (Fact) 13.2.1)}, \\ \underline{B}^\star(\cdot, t) & \text{is adapted to} & \sigma_t(\underline{X}) = \underline{\boldsymbol{X}}^{-1}(\mathcal{K}_t). \end{array}$$

There is thus [(Fact) 13.2.1] an adapted $\underline{B}_X(\cdot,t) : (K,\mathcal{K}_t) \longrightarrow (s,\mathcal{B}(s))$ such that

$$\underline{B}^{\star}(\omega,t) = \underline{B}_X(\underline{X}[\omega],t) = \underline{B}_X \Box \Phi_X(\omega,t).$$

Let $n \in \mathbb{N}$, and $\{B_1^s,\ldots,B_n^s\} \subseteq \mathcal{B}(s)$ be fixed, but arbitrary Borel sets. Then

$$\begin{aligned} P_X^K\left(\underline{k} \in K : \underline{B}_X(\underline{k},t_1) \in B_1^s,\ldots,\underline{B}_X(\underline{k},t_n) \in B_n^s\right) &= \\ &= P\left(\omega \in \Omega : \underline{B}^{\star}(\omega,t_1) \in B_1^s,\ldots,\underline{B}^{\star}(\omega,t_n) \in B_n^s\right) \\ &= P^o\left(\omega \in \Omega : \underline{B}(\omega,t_1) \in B_1^s,\ldots,\underline{B}(\omega,t_n) \in B_n^s\right). \end{aligned}$$

One must now insure that $\underline{B}_X$ has the proper path continuity properties. To that end, let

- $\mathcal{D}$ denote the dyadic rationals of $[0,1]$,
- $I_i^{(n)} = \left[\frac{i}{2^n},\frac{i+1}{2^n}\right]$, $i \in [0:2^n-1]$, $n \in \mathbb{N}$,
- $\underline{B}_X^{\mathcal{D}}$ denote the restriction of $\underline{B}_X$ to $\mathcal{D}$.

Set

- $\mathcal{D}_n = \left\{\frac{k}{2^n}, k \in [0:2^n]\right\}$: then $\mathcal{D}_n \subseteq \mathcal{D}_{n+1}$, $\mathcal{D} = \cup_n \mathcal{D}_n$;
- $D_n = \left\{(d_i,d_j) \in \mathcal{D}\times\mathcal{D} : \left|d_i - d_j\right| < 2^{-n}\right\}$: then $D_{n+1} \subseteq D_n$;
- $D_{n,i} = \mathcal{D} \cap I_i^{(n)}$;
- $U_n = \sup_{D_n} d_s\left(\underline{B}_X^{\mathcal{D}}(\cdot,d_i),\underline{B}_X^{\mathcal{D}}(\cdot,d_j)\right)$: then $U_{n+1} \le U_n$;
- $V_i^{(n)} = \sup_{D_{n,i}} d_s\left(\underline{B}_X^{\mathcal{D}}(\cdot,d),\underline{B}_X^{\mathcal{D}}(\cdot,\frac{i}{2^n})\right)$,
- $V_n = \max_i V_i^{(n)}$.

Let $(d_i,d_j) \in D_n$ be fixed, but arbitrary. Since

$$\mathcal{D} = \cup_k(\mathcal{D} \cap I_k^{(n)}),$$

$d_i \in I_{k_i}^{(n)}$ and $d_j \in I_{k_j}^{(n)}$. $I_{k_i}^{(n)}$ and $I_{k_j}^{(n)}$ must be adjacent since otherwise one would have $\left|d_i - d_j\right| \ge 2^{-n}$. Thus

$$\left|\frac{k_i}{2^n} - \frac{k_j}{2^n}\right| < \frac{1}{2^n}.$$

Then

$$\begin{aligned} d_s\left(\underline{B}_X^{\mathcal{D}}(\cdot,d_i),\underline{B}_X^{\mathcal{D}}(\cdot,d_j)\right) &\le d_s\left(\underline{B}_X^{\mathcal{D}}(\cdot,d_i),\underline{B}_X^{\mathcal{D}}(\cdot,\frac{k_i}{2^n})\right) \\ &+ d_s\left(\underline{B}_X^{\mathcal{D}}(\cdot,\frac{k_i}{2^n}),\underline{B}_X^{\mathcal{D}}(\cdot,\frac{k_j}{2^n})\right) \\ &+ d_s\left(\underline{B}_X^{\mathcal{D}}(\cdot,\frac{k_j}{2^n}),\underline{B}_X^{\mathcal{D}}(\cdot,d_j)\right) \end{aligned}$$

$$\leq \sup_{D_{n,k_i}} d_s\left(\underline{B}_X^{\mathcal{D}}(\cdot, d_i), \underline{B}_X^{\mathcal{D}}(\cdot, \frac{k_i}{2^n})\right)$$

$$+ \sup_{D_{n,k_i}} d_s\left(\underline{B}_X^{\mathcal{D}}(\cdot, \frac{k_i}{2^n}), \underline{B}_X^{\mathcal{D}}(\cdot, \frac{k_j}{2^n})\right)$$

$$+ \sup_{D_{n,k_j}} d_s\left(\underline{B}_X^{\mathcal{D}}(\cdot, \frac{k_j}{2^n}), \underline{B}_X^{\mathcal{D}}(\cdot, d_j)\right)$$

$$= 2V_{k_i}^{(n)} + V_{k_j}^{(n)}$$

$$\leq 3V_n.$$

Consequently, $U_n \leq 3V_n$, and $\{\underline{k} \in K : U_n(\underline{k}) > \epsilon\} \subseteq \left\{\underline{k} \in K : V_n(\underline{k}) > \frac{\epsilon}{3}\right\}$. It follows from the equalities relating finite dimensional distributions, seen above, that

$$P_X^K\left(\left\{\underline{k} \in K : V_n(\underline{k}) > \frac{\epsilon}{3}\right\}\right) =$$

$$= P^o\left(\left\{\max_i \sup_{D_{n,i}} d_s\left(\underline{B}(\cdot, d), \underline{B}\left(\cdot, \frac{i}{2^n}\right)\right) > \frac{\epsilon}{3}\right\}\right),$$

and, since $\underline{B}$ has, with respect to P^o, paths that are almost surely uniformly continuous, the right-hand side of the latter equality does have a zero limit, so that the left-hand side does too, which means that $P_X^K(\{U_n > \epsilon\})$ has a zero limit also. That latter limit says that $\underline{B}_X^{\mathcal{D}}$ has, almost surely, uniformly continuous paths.

But then, path by path, $\underline{B}_X^{\mathcal{D}}$ can be extended uniquely to a continuous function on $[0, 1]$ [84, p. 302]. The resulting process is adapted as the procedure used above may be restricted to $[0, t]$ without change other than notational. That extension shall be denoted $\underline{B}_X^{ext}$. Then, when $t \in [0, 1]$ is not in $\mathcal{D}$, given a fixed, but arbitrary $\epsilon > 0$,

$$P_X^K\left(\underline{k} \in K : d_s\left(\underline{B}_X^{ext}(\underline{k}, t), \underline{B}_X(\underline{k}, t)\right) > \epsilon\right) =$$

$$= P_X^K\left(\underline{k} \in K : d_s\left(\lim_{[d \in \mathcal{D}, d \uparrow\uparrow t]} \underline{B}_X^{\mathcal{D}}(\underline{k}, d), \underline{B}_X(\underline{k}, t)\right) > \epsilon\right)$$

$$= P\left(\omega \in \Omega : d_s\left(\lim_{[d \in \mathcal{D}, d \uparrow\uparrow t]} \underline{B}_X^{\mathcal{D}}(\underline{X}[\omega], d), \underline{B}_X(\underline{X}[\omega], t)\right) > \epsilon\right)$$

$$= P\left(\omega \in \Omega : d_s\left(\lim_{[d \in \mathcal{D}, d \uparrow\uparrow t]} \underline{B}^\star(\omega, d), \underline{B}^\star(\omega, t)\right) > \epsilon\right)$$

$$= P^o\left(\omega \in \Omega : d_s\left(\lim_{[d \in \mathcal{D}, d \uparrow\uparrow t]} \underline{B}(\omega, d), \underline{B}(\omega, t)\right) > \epsilon\right)$$

$$= 0.$$

Thus

$$\begin{aligned}P^o\left(\omega \in \Omega : d_s\left(\underline{B}_X^{ext}\left(\underline{X}[\omega], t\right), \underline{B}(\omega, t)\right) > \epsilon\right) = \\ &= P\left(\omega \in \Omega : d_s\left(\underline{B}_X^{ext}\left(\underline{X}[\omega], t\right), \underline{B}^{\star}(\omega, t)\right) > \epsilon\right) \\ &= P\left(\omega \in \Omega : d_s\left(\underline{B}_X^{ext}\left(\underline{X}[\omega], t\right), \underline{B}_X\left(\underline{X}[\omega], t\right)\right) > \epsilon\right) \\ &= P_X^K\left(\underline{k} \in K : d_s\left(\underline{B}_X^{ext}\left(\underline{k}, t\right), \underline{B}_X\left(\underline{k}, t\right)\right) > \epsilon\right) \\ &= 0.\end{aligned}$$

□

The $\underline{B}_X^{ext}$ process shall be denoted $\underline{B}_X$.

Remark 13.2.3 One may adapt the proof of [264, p. 115] to obtain that measurable processes on $K \times [0, 1]$, adapted to the filtrations generated by the evaluation maps, are predictable.

Proposition 13.2.4 (Decomposing Stochastic Differential Equations) *Suppose that*

(a) $\underline{a}^K : K \times [0, 1] \longrightarrow \mathbb{R}^\infty$ *has progressively measurable components;*
(b) *there exists* $\underline{X} : (\Omega, \underline{A}, P) \times [0, 1] \longrightarrow \mathbb{R}^\infty$ *such that, for* $t \in [0, 1]$*, fixed, but arbitrary, almost surely, with respect to* P,

$$\underline{X}(\omega, t) = \underline{S}\left[\underline{a}^K \square \Phi_X\right](\omega, t) + \underline{B}(\omega, t);$$

(c) $P_X^K\left(\underline{k} \in K : \|\underline{a}^K\left(\underline{k}, \cdot\right)\|_{L_2[\underline{b}]}^2 < \infty\right) = 1.$

There exists then a Cramér-Hida process $\underline{B}_X$*, with base* $\left(K, \underline{K}, P_X^K\right)$*, such that*

1. $P_{\underline{B}_X}^K = P_{\underline{B}}^K$*;*
2. *for* $t \in [0, 1]$*, fixed, but arbitrary, almost surely, with respect to* P_X^K*,*

$$\underline{\mathcal{E}}_{P_X^K}(\underline{k}, t) = \underline{S}\left[\underline{a}^K\right](\underline{k}, t) + \underline{B}_X(\underline{k}, t).$$

Proof Let

$$\underline{B}_X(\underline{k}, t) = \underline{\mathcal{E}}_{P_X^K}(\underline{k}, t) - \underline{S}\left[\underline{a}^K\right](\underline{k}, t).$$

That yields an adapted process whose paths may be assumed [(Lemma) 11.2.2] to be continuous. It follows that, almost surely with respect to P,

$$\begin{aligned}\underline{B}_X(\underline{X}[\omega], t) &= \underline{\mathcal{E}}_{P_X^K}(\underline{X}[\omega], t) - \underline{S}\left[\underline{a}^K\right](\underline{X}[\omega], t) \\ &= \underline{X}(\omega, t) - \underline{S}\left[\underline{a}^K \square \Phi_X\right](\omega, t) \\ &= \underline{B}(\omega, t),\end{aligned}$$

so that, with respect to P_X^K, $\underline{B}_X$ is a Cramér-Hida process with law P_B^K. □

Proposition 13.2.5 (Decomposing Stochastic Integrals) *Let the framework be that of* (Proposition) 13.2.4. *The following integrals make then sense:*

1. $I_{\underline{B}}\{\underline{a}^K \square \Phi_X\}$, *for* $\underline{\sigma}(\underline{X})$ *and* P,
2. $I_{\underline{B}_X}\{\underline{a}^K\}$, *for* $\underline{\mathcal{K}}$ *and* P_X^K,

and, for $t \in [0,1]$, *fixed, but arbitrary, almost surely with respect to* P,

$$I_{\underline{B}_X}\left\{\underline{a}^K\right\} \square \Phi_X(\omega, t) = I_{\underline{B}}\left\{\underline{a}^K \square \Phi_X\right\}(\omega, t).$$

Proof Within the framework of Sect. 11.3.3, let $(K, \underline{\mathcal{K}}, P_X^K)$ take the part of $(\Theta, \underline{\mathcal{B}}, Q)$, and $\underline{\boldsymbol{X}}$, that of f. Since $\mathcal{E}_t^K \circ \underline{\boldsymbol{X}} = \underline{X}(\cdot, t)$, the requirements for the application of (Proposition) 11.3.14 are met. Let $\underline{\alpha}$ be a fixed, but arbitrary element in l_2, and set:

$$B_\alpha = \langle \underline{\alpha}, \underline{B} \rangle_{l_2}, \quad \text{and} \quad B_\alpha^X = \langle \underline{\alpha}, \underline{B}_X \rangle_{l_2}.$$

One has then [(Proposition) 11.3.14] that, for appropriate ψ, with domain $K \times [0,1]$,

$$\int_0^t \psi(\underline{X}[\omega], \theta) B_\alpha(\omega, d\theta) = \left\{\int_0^t \psi(\underline{k}, \theta) B_\alpha^X(\underline{k}, d\theta)\right\} \circ \underline{X}[\omega].$$

But [(Fact) 11.3.10] $B_\alpha = I_{\underline{B}}\left\{\underline{a}_K^\alpha\right\}$, so that $dB_\alpha = \langle \underline{\alpha}, d\underline{B} \rangle_{l_2}$, and, consequently, choosing, for $\underline{\alpha}$, the element $\underline{e}_n$, and, for ψ, the element a_n^K, one obtains that

$$\int_0^t a_n^K(\underline{X}[\omega], \theta) B_n(\omega, d\theta) = \left\{\int_0^t a_n^K(\underline{k}, \theta) B_n^X(\underline{k}, d\theta)\right\} \circ \underline{X}[\omega],$$

that is,

$$I_{\underline{B}}\left\{\underline{a}^K \square \Phi_X\right\} = I_{\underline{B}_X}\left\{\underline{a}^K\right\} \square \Phi_X.$$

□

13.3 Likelihoods with Moment Conditions

The likelihood shall now be produced for a SPWN model, whose noise is, of course, a Cramér-Hida process, and whose observations (signal plus noise) are the solution of a stochastic differential equation. The moment conditions are those of item (c) of (Proposition) 13.3.1 just below.

Proposition 13.3.1 *Suppose that, for* $t \in [0,1]$, *fixed, but arbitrary, almost surely with respect to* P,

$$\underline{X}(\omega, t) = \underline{S}\left[\underline{a}^K \square \Phi_X\right](\omega, t) + \underline{B}(\omega, t)$$

where

(a) $\underline{B}$ *is a Cramér-Hida process;*
(b) $P_X^K\left(\underline{k} \in K : \|\underline{a}^K(\underline{k}, \cdot)\|_{L_2[\underline{b}]}^2 < \infty\right) = 1;$
(c) $E_P[D] = 1$, *with*

$$ln\,[D(\omega)] = -I_{\underline{B}}\left\{\underline{a}^K \Box \Phi_X\right\}(\omega, 1) - \frac{1}{2}\left\|\underline{a}^K \Box \Phi_X(\omega, \cdot)\right\|_{L_2[\underline{b}]}^2.$$

Then

1. *almost surely, with respect to* P_X^K,

$$\ln\left[\frac{dP_B^K}{dP_X^K}(\underline{k})\right] = -I_{\underline{B}_X}\left\{\underline{a}^K\right\}(\underline{k}, 1) - \frac{1}{2}\left\|\underline{a}^K(\underline{k}, \cdot)\right\|_{L_2[\underline{b}]}^2;$$

2. *almost surely, with respect to* P_X^K,

$$\ln\left[\frac{dP_B^K}{dP_X^K}(\underline{k})\right] = -I_{\underline{\varepsilon}_{P_X^K}}\left\{\underline{a}^K\right\}(\underline{k}, 1) + \frac{1}{2}\left\|\underline{a}^K(\underline{k}, \cdot)\right\|_{L_2[\underline{b}]}^2;$$

3. *almost surely, with respect to* P_B^K,

$$\ln\left[\frac{dP_B^K}{dP_X^K}(\underline{k})\right] = -I_{\underline{\varepsilon}_{P_B^K}}\left\{\underline{a}^K\right\}(\underline{k}, 1) + \frac{1}{2}\left\|\underline{a}^K(\underline{k}, \cdot)\right\|_{L_2[\underline{b}]}^2;$$

4. *almost surely, with respect to* P_B^K,

$$\ln\left[\frac{dP_X^K}{dP_B^K}(\underline{k})\right] = I_{\underline{\varepsilon}_{P_B^K}}\left\{\underline{a}^K\right\}(\underline{k}, 1) - \frac{1}{2}\left\|\underline{a}^K(\underline{k}, \cdot)\right\|_{L_2[\underline{b}]}^2;$$

5. *almost surely, with respect to* P_X^K,

$$\ln\left[\frac{dP_X^K}{dP_B^K}(\underline{k})\right] = I_{\underline{\varepsilon}_{P_X^K}}\left\{\underline{a}^K\right\}(\underline{k}, 1) - \frac{1}{2}\left\|\underline{a}^K(\underline{k}, \cdot)\right\|_{L_2[\underline{b}]}^2.$$

Proof (1) Because of Assumption (b), $I_{\underline{B}_X}\left\{\underline{a}^K\right\}$ is well defined. Let thus

$$\ln\left[D_K(\underline{k})\right] = -I_{\underline{B}_X}\left\{\underline{a}^K\right\}(\underline{k}, 1) - \frac{1}{2}\left\|\underline{a}^K(\underline{k}, \cdot)\right\|_{L_2[\underline{b}]}^2.$$

Then

$$\ln\left[D_K \circ \underline{X}[\omega]\right] = -I_{\underline{B}_X}\left\{\underline{a}^K\right\} \Box \Phi_X(\omega, 1) - \frac{1}{2}\left\|\underline{a}^K \Box \Phi_X(\omega, \cdot)\right\|_{L_2[\underline{b}]}^2.$$

However [(Proposition) 13.2.5] $I_{\underline{B}_X}\{\underline{a}^K\}\,\square\Phi_X = I_{\underline{B}}\{\underline{a}^K\square\Phi_X\}$, so that

$$D_K \circ \underline{X} = D.$$

Let now $K_0 \in \mathcal{K}$ be fixed, but arbitrary, and $dQ = DdP$. From Girsanov's theorem, one has that $P_B^K = Q_X^K$, so that

$$\begin{aligned} P_B^K(K_0) &= Q\left(\underline{X}^{-1}(K_0)\right) \\ &= \int_{\underline{X}^{-1}(K_0)} D(\omega)P(d\omega) \\ &= \int_{\underline{X}^{-1}(K_0)} D_K\left(\underline{X}[\omega]\right) P(d\omega) \\ &= \int_{K_0} D_K(\underline{k})P_X^K(d\underline{k}). \end{aligned}$$

Proof (2) Because of the assumptions, and (Proposition) 13.2.4, on $K \times [0, 1]$,

$$\underline{\mathcal{E}}_{P_X^K} = \underline{S}\left[\underline{a}^K\right] + \underline{B}_X,$$

so that [(Proposition) 13.1.7] $\underline{a}^K$ is integrable with respect to $\underline{\mathcal{E}}_{P_X^K}$, and

$$I_{\underline{\mathcal{E}}_{P_X^K}}\left\{\underline{a}^K\right\} = \left\|\underline{a}^K\left(\underline{k}, \cdot\right)\right\|_{L_2[\underline{b}]}^2 + I_{\underline{B}_X}\left\{\underline{a}^K\right\}.$$

It then suffices to use that latter formula in the expression of item 1 to obtain that of item 2.

Proof (3) Since $Q_X^K \ll P_X^K$, and because of (Proposition) 13.1.8, almost surely, with respect to Q_X^K, that is, P_B^K,

$$I_{\underline{\mathcal{E}}_{P_X^K}}\left\{\underline{a}^K\right\} = I_{\underline{\mathcal{E}}_{P_B^K}}\left\{\underline{a}^K\right\}.$$

Exchanging integrals in the expression of item 2 yields that of item 3.

Proof (4) Let now

$$\ln\left[D^\star(\omega)\right] = I_{\underline{X}}\left\{\underline{a}^K\square\Phi_X\right\}(\omega, 1) - \frac{1}{2}\left\|\underline{a}^K\square\Phi_X(\omega, \cdot)\right\|_{L_2[\underline{b}]}^2.$$

Since

$$I_{\underline{X}}\left\{\underline{a}^K\square\Phi_X\right\} = \left\|\underline{a}^K\square\Phi_X\right\|_{L_2[\underline{b}]}^2 + I_{\underline{B}}\left\{\underline{a}^K\square\Phi_X\right\},$$

one has that $D^\star dQ = D^\star D dP = dP$. Let

$$\ln\left[D_K^\star(\underline{k})\right] = I_{\underline{\mathcal{E}}_{P_B^K}}\left\{\underline{a}^K\right\}(\underline{k},1) - \frac{1}{2}\left\|\underline{a}^K(\underline{k},\cdot)\right\|_{L_2[\underline{b}]}^2 .$$

Since, an evaluation process being an evaluation, almost surely, with respect to P, $\underline{\mathcal{E}}_{P_B^K} \circ \underline{X} = \underline{X}$, so that $D_K^\star \circ \underline{X} = D^\star$, one has, for $K_0 \in \mathcal{K}$, fixed, but arbitrary, that

$$\begin{aligned}
P_X^K(K_0) &= \int_{\underline{X}^{-1}(K_0)} D^\star(\omega)\, Q(d\omega) \\
&= \int_{\underline{X}^{-1}(K_0)} D_K^\star(\underline{X}[\omega])\, Q(d\omega) \\
&= \int_{K_0} D_K^\star(\underline{k})\, Q_X^K(d\underline{k}) \\
&= \int_{K_0} D_K^\star(\underline{k})\, P_B^K(d\underline{k}),
\end{aligned}$$

which yields the expression of item 4.

Proof (5) The expression of item 4, with the equality of integrals of item 3, yields the expression of item 5. □

13.4 Likelihoods with Paths Conditions

Assumption (c) in (Proposition) 13.3.1 (the moment conditions) is rather inconvenient, and one would like to stay within the assumption that the paths of the signal have finite energy. To what extent that is possible gets investigated in the present section. The following lemmas are required in the sequel.

Lemma 13.4.1 *Let M be a real valued, continuous to the right, almost surely continuous, local martingale, with a quadratic variation $\langle M\rangle$ that is zero at the origin. Let $\alpha > 0$, and $\beta > 0$, be fixed, but arbitrary. Then, for $\Omega_0 \in \mathcal{A}$, fixed, but arbitrary,*

$$\begin{aligned}
&P(\omega \in \Omega : \chi_{\Omega_0}(\omega)\,|M(\omega,t)| > \alpha) \le \\
&\qquad \le P(\omega \in \Omega : \chi_{\Omega_0}(\omega)\langle M\rangle(\omega,t) > \beta) + 2e^{-(1/2)(\alpha^2/\beta)}.
\end{aligned}$$

Proof Let at first $\Omega_0 = \Omega$. Since $|a| > b$ means $a > b$ or $-a > b$,

$$\begin{aligned}
P(\omega \in \Omega : |M(\omega,t)| > \alpha) &\le P(\omega \in \Omega : M(\omega,t) > \alpha \text{ and } \langle M\rangle(\omega,t) \le \beta) \\
&\quad + P(\omega \in \Omega : -M(\omega,t) > \alpha \text{ and } \langle M\rangle(\omega,t) \le \beta) \\
&\quad + P(\omega \in \Omega : \langle M\rangle(\omega,t) > \beta).
\end{aligned}$$

Let $\gamma > 0$ be fixed, but arbitrary. Since $M(\omega, t) > \alpha$, and $\langle M \rangle(\omega, t) \leq \beta$, imply

$$\gamma M(\omega, t) - (\gamma^2/2)\langle M \rangle(\omega, t) > \gamma\alpha - (\gamma^2/2)\beta,$$

one has that

$$P(\omega \in \Omega : M(\omega, t) > \alpha \text{ and } \langle M \rangle(\omega, t) \leq \beta)$$

is less or equal to

$$P\left(\omega \in \Omega : e^{\gamma M(\omega,t)-(\gamma^2/2)\langle M \rangle(\omega,t)} > e^{\gamma\alpha-(\gamma^2/2)\beta}\right).$$

But the process $e^{\gamma M-(\gamma^2/2)\langle M \rangle}$ is a supermartingale [(Propositions) 10.4.6, 10.4.3] and thus, by Doob's inequality, that latter probability is less than

$$e^{-\gamma\alpha+(\gamma^2/2)\beta} E_P\left[e^{\gamma M(\cdot,0)-(\gamma^2/2)\langle M \rangle(\cdot,0)}\right].$$

So

$$P(\omega \in \Omega : M(\omega, t) > \alpha \text{ and } \langle M \rangle(\omega, t) \leq \beta) \leq e^{-\gamma\alpha+(\gamma^2/2)\beta}.$$

The minimum value of the right-hand side of that latter inequality is obtained for $\gamma = \alpha/\beta$, and is the exponential of $-\alpha^2/(2\beta)$.

The same calculation works for the second term of the initial inequality of the current proof. The general case is obtained noticing that

$$P(\omega \in \Omega : \chi_{\Omega_0}(\omega) |M(\omega, t)| > \alpha) = P(\{\omega \in \Omega : |M(\omega, t)| > \alpha\} \cap \Omega_0).$$

□

Lemma 13.4.2 *Let P and Q be two probabilities on $(\Omega, \mathcal{A})$, and $\Omega_0 \in \mathcal{A}$ be a set such that $P(\Omega_0) = Q(\Omega_0) = 1$. Let then*

1. *$\mathcal{A}_0 = \mathcal{A} \cap \Omega_0$, and, for $A \in \mathcal{A}$, fixed, but arbitrary, $A_0 = A \cap \Omega_0$;*
2. *$P_0(A_0) = P(A \cap \Omega_0)$, and $Q_0(A_0) = Q(A \cap \Omega_0)$.*

When P_0 and Q_0 are mutually absolutely continuous, so are P and Q, and, almost surely, with respect to P and Q,

$$\frac{dQ}{dP} = \frac{dQ_0}{dP_0}.$$

Proof Let $J_0 : \Omega_0 \longrightarrow \Omega$ be the injection map. Since, for $A \in \mathcal{A}$, fixed, but arbitrary,

$$J_0^{-1}(A) = A \cap \Omega_0,$$

J_0 is adapted. Furthermore

$$P_0 \circ J_0^{-1}(A) = P_0(A \cap \Omega_0) = P(A \cap \Omega_0) = P(A).$$

Let, for $\omega \in \Omega$, fixed, but arbitrary,

$$D(\omega) = \chi_{\Omega_0}(\omega)\frac{dQ_0}{dP_0}(\omega).$$

D is adapted to $\mathcal{A}$, and, for $A \in \mathcal{A}$, fixed, but arbitrary,

$$\begin{aligned}\int_A D dP &= \int_A D d(P_0 \circ J_0^{-1}) \\ &= \int_{A_0} (D \circ J_0) dP_0 \\ &= \int_{A_0} \frac{dQ_0}{dP_0} dP_0 \\ &= Q_0(A_0) \\ &= Q(A).\end{aligned}$$

Finally, since $\frac{dQ_0}{dP_0}$ is almost surely strictly positive, so is D. □

Lemma 13.4.3 ([70, p. 230 et 234]) *Let E be a Polish space, and $\Omega \subseteq E^{[0,1]}$* [the latter with the uniform topology]. *Set*

- $X(\omega, t) = \mathcal{E}_t(\omega) = \omega(t)$;
- $\mathcal{A}_t = \sigma(X(\cdot, \theta), \theta \leq t) = \sigma(\mathcal{E}_\theta, \theta \leq t)$;
- *for $t \in [0, 1]$, fixed, but arbitrary, let $T_t : \Omega \longrightarrow \Omega$ be defined using the following relation:*

$$T_t[\omega](\theta) = \omega(\theta \wedge t).$$

Then:

1. $T_t^{-1}(\mathcal{A}_1) = \mathcal{A}_t$;
2. *a random variable V* [that is, a real valued function adapted to $\mathcal{A}_1$] *is adapted to $\mathcal{A}_t$ if, and only if, $V = V \circ T_t$;*
3. $S : \Omega \longrightarrow [0, 1]$, *adapted to $\mathcal{A}_1$, is a strict sense stopping time if, and only if, for $t \in [0, 1]$, fixed, but arbitrary, the following conditions:*

   - $S(\omega_1) \leq t$,
   - $X(\omega_1, \theta) = X(\omega_2, \theta)$ *for $\theta \in [0, t]$,*

   imply that $S(\omega_1) = S(\omega_2)$.

Proof (1) From the definitions, for t and θ in $[0, 1]$, and Borel E_0, fixed, but arbitrary, one has that

$$T_t^{-1}\left(\{\omega \in \Omega : \mathcal{E}_\theta(\omega) \in E_0\}\right) = \{\omega \in \Omega : \mathcal{E}_{\theta\wedge t}(\omega) \in E_0\}.$$

But, generally [275, p. 8], $f^{-1}(\sigma(\mathcal{S})) = \sigma(f^{-1}(\mathcal{S}))$.

Proof (2) V is adapted to $\mathcal{A}_t$ if, and only if [41, p. 144],

$$V = \phi(\mathcal{E}_{\theta_i}, \theta_i \le t, i \in I, |I| \le \aleph_0).$$

Then

$$\begin{aligned} V(\omega) &= \phi(\omega(\theta_i), \theta_i \le t, i \in I, |I| \le \aleph_0) \\ &= \phi(\omega(\theta_i \wedge t), \theta_i \le t, i \in I, |I| \le \aleph_0) \\ &= \phi(\mathcal{E}_{\theta_i}(T_t[\omega]), \theta_i \le t, i \in I, |I| \le \aleph_0) \\ &= V(T_t[\omega]). \end{aligned}$$

Conversely, when $V = V \circ T_t$,

$$V^{-1}(B) = T_t^{-1}(V^{-1}(B)) \in \mathcal{A}_t,$$

because of item 1.

Proof (3) The implication of the statement is equivalent to the following (weaker) one: the sufficient conditions of item 3 imply that $S(\omega_2) \le t$. That weaker implication obviously follows from the initial one.

Suppose now that second implication obtains. Let $\tau = S(\omega_1) \le t$. Then, given that both

$$S(\omega_1) \le \tau \text{ and } X(\omega_1, \theta) = X(\omega_2, \theta), \ \theta \in [0, \tau]$$

imply $S(\omega_2) \le \tau = S(\omega_1) \le t$, $S(\omega_2) \le S(\omega_1)$. But then one can interchange the roles of ω_1 and ω_2 to obtain $S(\omega_1) \le S(\omega_2)$. Consequently $S(\omega_1) = S(\omega_2)$.

Assume now the weaker implication $S(\omega_2) \le t$. One shall see that S is a strict sense stopping time. As, for $\theta \le t$ in $[0, 1]$,

$$X(\omega, \theta) = \omega(\theta) = \omega(\theta \wedge t) = T_t[\omega](\theta),$$

the following relation:

$$X(\omega_1, \theta) = X(\omega_2, \theta), \ \theta \in [0, t],$$

means that $T_t[\omega_1] = T_t[\omega_2]$. Let the latter equality mean that ω_1 and ω_2 are equivalent ($\omega_1 \sim \omega_2$ with equivalence class $[\omega_1] = [\omega_2]$). The weaker implication then means that

$$S(\omega_1) \leq t \text{ and } \omega_1 \sim \omega_2 \text{ imply } S(\omega_2) \leq t,$$

that is, if $\omega_1 \in \{\omega \in \Omega : S(\omega) \leq t\}$, then

$$[\omega_1] \subseteq \{\omega \in \Omega : S(\omega) \leq t\},$$

so that $\{\omega \in \Omega : S(\omega) \leq t\}$ is the union of the equivalence classes of its elements. But, since

$$T_t[\{T_t[\omega]\}](\theta) = \{T_t[\omega]\}(\theta \wedge t) = \omega(\{\theta \wedge t\} \wedge t) = T_t[\omega](\theta),$$

$T_t[\omega] \sim \omega$, and thus, since $\{\omega \in \Omega : S(T_t[\omega]) \leq t\} = T_t^{-1}(\{\omega \in \Omega : S(\omega) \leq t\})$,

$$\{S \leq t\} = T_t^{-1}(\{S \leq t\}).$$

Since $\{S \leq t\} \in \mathcal{A}_1$, it belongs, because of item 1, to $\mathcal{A}_t$, and S is a strict stopping time.

Suppose now that S is a strict stopping time. Then, because of item 2, one has the following equality: $S = S \circ T_t$, and thus, when $T_t[\omega_1] = T_t[\omega_2]$,

$$S(\omega_1) = S(T_t[\omega_1]) = S(T_t[\omega_2]) = S(\omega_2).$$

□

Proposition 13.4.4 *Suppose that, for $t \in [0, 1]$, fixed, but arbitrary, almost surely with respect to P,*

$$\underline{X}(\omega, t) = \underline{S}\left[\underline{a}^K \square \Phi_X\right](\omega, t) + \underline{B}(\omega, t),$$

where

(a) *$\underline{B}$ is a Cramér-Hida process,*
(b) $P_B^K\left(\underline{k} \in K : \|\underline{a}^K(\underline{k}, \cdot)\|_{L_2[\underline{b}]}^2 < \infty\right) = 1,$
(c) $P_X^K\left(\underline{k} \in K : \|\underline{a}^K(\underline{k}, \cdot)\|_{L_2[\underline{b}]}^2 < \infty\right) = 1,$
(d) $\ln[\Psi(\underline{k})] = -I_{\underline{\mathcal{E}}_{P_X^K}}\{\underline{a}^K\}(\underline{k}, 1) + \frac{1}{2}\|\underline{a}^K(\underline{k}, \cdot)\|_{L_2[\underline{b}]}^2.$

Then:

1. $E_{P_X^K}[\Psi] = 1$;
2. *P_X^K and P_B^K are mutually absolutely continuous.*

Proof The assumptions, and (Proposition) 13.2.4, yield that

$$\underline{\mathcal{E}}_{P_X^K}(\underline{k}, t) = \underline{S}\left[\underline{a}^K\right](\underline{k}, t) + \underline{B}_X(\underline{k}, t).$$

Step 1: One may assume that the paths of $\underline{S}\left[\underline{a}^K\right]$ *are continuous.*

Proof As in (Lemma) 11.2.2, let

$$A(\underline{k}, t) = \sum_{n=1}^{\infty} \int_0^t \left\{a_n^K(\underline{k}, \theta)\right\}^2 M_n(d\theta)$$

and

$$\hat{\underline{a}}^K(\underline{k}, t) = \chi_{\mathbb{R}_+}(A(\underline{k}, t))\underline{a}^K(\underline{k}, t).$$

The latter is progressively measurable,

$$\underline{S}\left[\hat{\underline{a}}^K\right]$$

has continuous paths, and the following probabilities are equal, and both are equal to one:

$$P_B^K\left(\underline{k} \in K : \left\|\hat{\underline{a}}^K(\underline{k}, \cdot)\right\|_{L_2[\underline{b}]}^2 < \infty\right), \; P_X^K\left(\underline{k} \in K : \left\|\hat{\underline{a}}^K(\underline{k}, \cdot)\right\|_{L_2[\underline{b}]}^2 < \infty\right).$$

Furthermore, for $t \in [0, 1]$, fixed, but arbitrary, almost surely, with respect to P_X^K,

$$\underline{\mathcal{E}}_{P_X^K}(\underline{k}, t) = \underline{S}\left[\hat{\underline{a}}^K\right](\underline{k}, t) + \underline{B}_X(\underline{k}, t).$$

Step 2: An restriction scheme

One reduces the continuous signal to a uniformly bounded one, using a localizing sequence. The aim is to apply Girsanov's theorem, the moment condition being a consequence of boundedness.

Define, as in (Proposition) 11.2.3, for $\underline{k} \in K$, fixed, but arbitrary,

$$T[\underline{k}] = \left\{t \in [0, 1] : \left\|\underline{a}_{|t}^K(\underline{k}, \cdot)\right\|_{L_2[\underline{b}]}^2 \geq n\right\},$$

and let

$$S_n^K(\underline{k}) = \begin{cases} 1 & \text{when } T[\underline{k}] = \emptyset \\ \inf T[\underline{k}] & \text{when } T[\underline{k}] \neq \emptyset \end{cases}.$$

One thus obtains a strict stopping time for $\underline{\mathcal{K}}$ [264, p. 38].

Let (the signal should be continuous *and* finite)

- $K_f = \left\{ \underline{k} \in K : \|\underline{a}^K(\underline{k}, \cdot)\|^2_{L_2[\underline{b}]} < \infty \right\};$
- $J_f : K_f \longrightarrow K$ be the inclusion map;
- $\mathcal{K}^f_t = J_f^{-1}(\mathcal{K}_t) = \mathcal{K}_t \cap K_f;$
- for fixed, but arbitrary $K_0 \in \mathcal{K}$, $K^f_0 = J_f^{-1}(K_0) = K_0 \cap K_f$;
- $P^{K_f}_X(K^f_0) = P^{K_f}_X(J_f^{-1}(K_0)) = P^K_X(K_0)$, so that $P^{K_f}_X \circ J_f^{-1} = P^K_X$;
- on the base $\left(K_f, \underline{\mathcal{K}}_f, P^{K_f}_X\right)$, the process $\underline{Y}^K_n$ obtained as follows:
  - when $(\underline{k}^f, t) \in [\![0, S^K_n]\!] \cap \left(K_f \times [0, 1]\right)$,
    $\underline{Y}^K_n(\underline{k}^f, t) = \underline{\mathcal{E}}_{P^K_X}(\underline{k}^f, t);$
  - when $(\underline{k}^f, t) \in [\![0, S^K_n]\!]^c \cap \left(K_f \times [0, 1]\right)$,
    $\underline{Y}^K_n(\underline{k}^f, t) = \underline{\mathcal{E}}_{P^K_X}(\underline{k}^f, t) - \left\{\underline{S}\left[\underline{a}^K\right](\underline{k}^f, t) - \underline{S}\left[\underline{a}^K\right](\underline{k}^f, S^K_n(\underline{k}^f))\right\}$

  (so that, when the signal becomes too large, it is replaced by a "constant");
- finally
  - $\underline{a}^K_n(\underline{k}, t) = \chi_{[\![0, S^K_n]\!]}(\underline{k}, t)\, \underline{a}^K(\underline{k}, t);$
  - $\Phi_{Y^K_n} : K_f \times [0, 1] \longrightarrow K \times [0, 1]$ be "computed" as

$$\Phi_{Y^K_n}(\underline{k}^f, t) = \left(\underline{Y}^K_n[\underline{k}^f], t\right).$$

Then the following assertions obtain:

- *for* $\underline{k}^f \in K_f$, $S^K_n\left(\Phi_{Y^K_n}[\underline{k}^f]\right) = S^K_n(\underline{k}^f)$;
- *for* $t \in [0, 1]$, *fixed, but arbitrary, almost surely, with respect to* $P^{K_f}_X$,

$$\underline{Y}^K_n(\underline{k}^f, t) = \underline{S}\left[\underline{a}^K_n \square \Phi_{Y^K_n}\right](\underline{k}^f, t) + \underline{B}^f_X(\underline{k}^f, t),$$

where $\underline{B}^f_X$ *is the restriction of* $\underline{B}_X$ *to* K_f.

Proof As a separable Fréchet space is Polish, s is Polish, and (Lemma) 13.4.3 applies, with $E = s$ (Ω is K). For $\underline{k}^f \in K_f$,

$$\left\|\underline{a}^K(\underline{k}^f, \cdot)\right\|^2_{L_2[\underline{b}]} < \infty,$$

so that, from its definition,

$$t \mapsto \underline{Y}^K_n(\underline{k}^f, t)$$

is a continuous function, and thus $\underline{Y}^K_n[\underline{k}^f] \in K$. Since S^K_n is a stopping time, and that, when $S^K_n(\underline{k}^f) = t$, on $[0, t]$, $\underline{Y}^K_n[\underline{k}^f] = \underline{k}^f$,

$$S^K_n(\underline{k}^f) = S^K_n(\underline{Y}^K_n[\underline{k}^f]).$$

Using that latter equality, and the definition of $\underline{a}_n^K$,

$$\underline{a}_n^K = \underline{a}_n^K \Box \Phi_{Y_n^K}.$$

Finally, using the following facts:

- $\underline{\mathcal{E}}_{P_X^K} = \underline{S}[\underline{a}^K] + \underline{B}_X$,
- when $\underline{k}^f \in K_f$, fixed, but arbitrary, and $S_n^K(\underline{k}^f) \geq t$, then $\underline{a}_n^K = \underline{a}^K$, and, having that

$$\underline{a}_n^K = \underline{a}_n^K \Box \Phi_{Y_n^K},$$

it follows that

- when $t \leq S_n^K(\underline{k}^f)$,

$$\begin{aligned}
\underline{Y}_n^K(\underline{k}^f, t) &= \mathcal{E}_{P_X^K}(\underline{k}^f, t) \\
&= \underline{S}\left[\underline{a}^K\right](\underline{k}^f, t) + \underline{B}_X^f(\underline{k}^f, t) \\
&= \underline{S}\left[\underline{a}_n^K\right](\underline{k}^f, t) + \underline{B}_X^f(\underline{k}^f, t) \\
&= \underline{S}\left[\underline{a}_n^K \Box \Phi_{Y_n^K}\right](\underline{k}^f, t) + \underline{B}_X^f(\underline{k}^f, t),
\end{aligned}$$

- when $t > S_n^K(\underline{k}^f)$,

$$\begin{aligned}
\underline{Y}_n^K(\underline{k}^f, t) &= \mathcal{E}_{P_X^K}(\underline{k}^f, t) - \left\{\underline{S}\left[\underline{a}^K\right](\underline{k}^f, t) - \underline{S}\left[\underline{a}^K\right](\underline{k}^f, S_n^K(\underline{k}^f))\right\} \\
&= \underline{S}\left[\underline{a}^K\right](\underline{k}^f, S_n^K(\underline{k}^f)) + \underline{B}_X^f(\underline{k}^f, t) \\
&= \underline{S}\left[\underline{a}_n^K\right](\underline{k}^f, t) + \underline{B}_X^f(\underline{k}^f, t) \\
&= \underline{S}\left[\underline{a}_n^K \Box \Phi_{Y_n^K}\right](\underline{k}^f, t) + \underline{B}_X^f(\underline{k}^f, t).
\end{aligned}$$

Step 3: When

$$\ln\left[\Psi_n^{K_f}(\underline{k}^f)\right] = -I_{\underline{B}_X^f}\left\{\underline{a}_n^K \Box \Phi_{Y_n^K}\right\}(\underline{k}^f, 1) - \frac{1}{2}\left\|\underline{a}_n^K \Box \Phi_{Y_n^K}(\underline{k}^f, \cdot)\right\|_{L_2[\underline{b}]}^2,$$

then

$$E_{P_X^{K_f}}\left[\Psi_n^{K_f}\right] = 1.$$

Proof Since $\underline{B}_X$ is a Cramér-Hida process for P_X^K, so is

$$\underline{B}_X^f = \underline{B}_X(J_f(\cdot), \cdot),$$

for $P_X^{K_f}$. Since the "signal" is bounded, the assertion is a consequence of the properties of exponential martingales.

Step 4: $\left\{\Psi_n^{K_f}, n \in \mathbb{N}\right\}$ *converges almost surely, with respect to* $P_X^{K_f}$, *to* $\Psi^{|K_f}$.

The exponent of $\Psi_n^{K_f}$ has been defined as

$$-I_{\underline{B}_X^f}\left\{\underline{a}_n^K \square \Phi_{Y_n^K}\right\}(\underline{k}^f, 1) - \frac{1}{2}\left\|\underline{a}_n^K \square \Phi_{Y_n^K}(\underline{k}^f, \cdot)\right\|_{L_2[\underline{b}]}^2.$$

It follows from the definitions that

$$\underline{a}_n^K \square \Phi_{Y_n^K}(\underline{k}^f, t) = \chi_{[0,S_n]}(\underline{k}^f, t)\underline{a}^K(\underline{k}^f, t).$$

Then, using (Proposition) 13.2.5, one sees that the first term of the exponent of $\Psi_n^{K_f}$ is one entry in the sequence whose limit is the integral of $\underline{a}^K$ with respect to $\underline{B}_X$. The second ("norm") term is

$$\left\|\underline{a}_n^K \square \Phi_{Y_n^K}(\underline{k}^f, \cdot)\right\|_{L_2[\underline{b}]}^2 = \left\|\chi_{[0,S_n]}(\underline{k}^f, \cdot)\underline{a}^K(\underline{k}^f, \cdot)\right\|_{L_2[\underline{b}]}^2,$$

so that monotone convergence applies to produce a limit which is finite by definition.

The limit of the exponent of $\Psi_n^{K_f}$ is thus

$$-I_{\underline{B}_X}\left\{\underline{a}^K\right\}(\underline{k}^f, 1) - \frac{1}{2}\left\|\underline{a}^K(\underline{k}^f, \cdot)\right\|_{L_2[\underline{b}]}^2.$$

Since

$$\underline{\mathcal{E}}_{P_X^K}(\underline{k}, t) = \underline{S}\left[\underline{a}^K\right](\underline{k}, t) + \underline{B}_X(\underline{k}, t),$$

one has that

$$I_{\underline{\mathcal{E}}_{P_X^K}}\left\{\underline{a}^K\right\} = \left\|\underline{a}^K\right\|_{L_2[\underline{b}]}^2 + I_{\underline{B}_X}\left\{\underline{a}^K\right\},$$

and thus that the limit of the exponent of $\Psi_n^{K_f}$ is

$$-I_{\underline{\mathcal{E}}_{P_X^K}}\left\{\underline{a}^K\right\} + \frac{1}{2}\left\|\underline{a}^K\right\|_{L_2[\underline{b}]}^2,$$

that is, the exponent of Ψ in Assumption (d) of the statement, restricted to K_f.

Step 5: The sequence $\left\{\Psi_n^{K_f}, n \in \mathbb{N}\right\}$ *is uniformly integrable for* $P_X^{K_f}$.

Proof Let $\sigma_t^o\left(\underline{Y}_n^K\right)$ be the completion, with respect to $P_X^{K_f}$, of $\sigma_t\left(\underline{Y}_n^K\right)$.

The process $\underline{B}_X$ is, for P_X^K, a Cramér-Hida process. Thus, for $P_X^{K_f}$, $\underline{B}_X^f$ is a Cramér-Hida process with law P_B^K. From step 2,

$$\underline{B}_X^f(\underline{k}^f, t) = \underline{Y}_n^K(\underline{k}^f, t) - \underline{S}\left[\underline{a}_n^K \square \Phi_{Y_n^K}\right](\underline{k}^f, t).$$

It follows that $\underline{B}_X^f$ is adapted to $\sigma_t^o\left(\underline{Y}_n^K\right)$. One may thus use (Proposition) 13.2.2 to decompose $\underline{B}_X^f$ as follows:

$$\underline{B}_X^f(\underline{k}^f, t) = \underline{B}_X^{K,n} \circ \underline{Y}_n^K(\underline{k}^f, t), \text{ almost surely with respect to } \left\{P_X^{K_f}\right\}^o,$$

where, for

$$Q_n^K = P_X^{K_f} \circ \left\{\underline{Y}_n^K\right\}^{-1},$$

$\underline{B}_X^{K,n}$, on the base $(K, \mathcal{K}, Q_n^K)$, has law P_B^K.

On $\left(K, \underline{\mathcal{K}}, Q_n^K\right)$, let

$$\ln\left[\Upsilon_n(\underline{k})\right] = -I_{\underline{B}_X^{K,n}}\left\{\underline{a}_n^K\right\}(\underline{k}, 1) - \frac{1}{2}\left\|\underline{a}_n^K(\underline{k}, \cdot)\right\|_{L_2[\underline{b}]}^2.$$

Then, because of (Proposition) 13.2.5,

$$\begin{aligned}
\ln\left[\Upsilon_n\left(\underline{Y}_n^K[\underline{k}^f]\right)\right] &= -I_{\underline{B}_X^{K,n}}\left\{\underline{a}_n^K\right\}\left(\underline{Y}_n^K[\underline{k}^f], 1\right) - \frac{1}{2}\left\|\underline{a}_n^K\left(\underline{Y}_n^K[\underline{k}^f], \cdot\right)\right\|_{L_2[\underline{b}]}^2 \\
&= -I_{\underline{B}_X^{K,n}}\left\{\underline{a}_n^K\right\}\Box\Phi_{Y_n^K}(\underline{k}^f, 1) - \frac{1}{2}\left\|\underline{a}_n^K\Box\Phi_{Y_n^K}(\underline{k}^f, \cdot)\right\|_{L_2[\underline{b}]}^2 \\
&= -I_{\underline{B}_X^f}\left\{\underline{a}_n^K\Box\Phi_{Y_n^K}\right\}(\underline{k}^f, 1) - \frac{1}{2}\left\|\underline{a}_n^K\Box\Phi_{Y_n^K}(\underline{k}^f, \cdot)\right\|_{L_2[\underline{b}]}^2 \\
&= \ln\left[\Psi_n^{K_f}(\underline{k}^f)\right],
\end{aligned}$$

and thus (step 3)

$$E_{Q_n^K}\left[\Upsilon_n\right] = E_{P_X^{K_f}}\left[\Upsilon_n \circ \underline{Y}_n^K\right] = E_{P_X^{K_f}}\left[\Psi_n^{K_f}\right] = 1.$$

Furthermore, because of (Proposition) 13.2.4, the following equation from step 2:

$$\underline{Y}_n^K = \underline{S}[\underline{a}_n^K\Box\Phi_{Y_n^K}] + \underline{B}_X$$

decomposes as

$$\underline{\mathcal{E}}_{Q_n^K} = \underline{S}[\underline{a}_n^K] + \underline{B}_X^{K,n},$$

so that one may apply Girsanov's theorem, and its consequences, to obtain that Q_n^K and P_B^K are mutually absolutely continuous, and that, almost surely, with respect to Q_n^K,

$$\frac{dP_B^K}{dQ_n^K} = \Upsilon_n.$$

But, with respect to P_B^K, Υ_n has the following representation [(Proposition) 13.3.1, item 3]:

$$\ln\left[\Upsilon_n\left(\underline{k}\right)\right] = -I_{\mathcal{E}_{P_B^K}}\left\{\underline{a}_n^K\right\}\left(\underline{k},1\right) + \frac{1}{2}\left\|\underline{a}_n^K(\underline{k},\cdot)\right\|_{L_2[\underline{b}]}^2.$$

As $-a+\frac{b}{2} > \alpha, b \geq 0$, implies either $|a| > \frac{\alpha}{2}$, or $b > \alpha$, one then has, given $\gamma > 0$, fixed, but arbitrary, that

$$\begin{aligned}
&\int_{\left\{\Psi_n^{K_f} > \gamma\right\}} \Psi_n^{K_f}\, dP_X^{K_f} = \\
&= \int_{\{\Upsilon_n > \gamma\}} \Upsilon_n dQ_n^K \\
&= P_B^K\left(\underline{k} \in K : \Upsilon_n(\underline{k}) > \gamma\right) \\
&= P_B^K\left(\underline{k} \in K : -I_{\mathcal{E}_{P_B^K}}\left\{\underline{a}_n^K\right\}\left(\underline{k},1\right) + \frac{1}{2}\left\|\underline{a}_n^K(\underline{k},\cdot)\right\|_{L_2[\underline{b}]}^2 > \ln\gamma\right) \\
&= P\left(\omega \in \Omega : -I_{\mathcal{E}_{P_B^K}}\left\{\underline{a}_n^K\right\}\left(\underline{B}[\omega],1\right) + \frac{1}{2}\left\|\underline{a}_n^K(\underline{B}[\omega],\cdot)\right\|_{L_2[\underline{b}]}^2 > \ln\gamma\right) \\
&\leq P\left(\omega \in \Omega : \left|I_{\mathcal{E}_{P_B^K}}\left\{\underline{a}_n^K\right\}\left(\underline{B}[\omega],1\right)\right| > \frac{\ln\gamma}{2}\right) \\
&+ P\left(\omega \in \Omega : \left\|\underline{a}_n^K(\underline{B}[\omega],\cdot)\right\|_{L_2[\underline{b}]}^2 > \ln\gamma\right).
\end{aligned}$$

But, according to (Lemma) 13.4.1, with $\alpha = \frac{\ln\gamma}{2}$, and $\beta = \ln\gamma$,

$$P\left(\omega \in \Omega : \left|I_{\mathcal{E}_{P_B^K}}\left\{\underline{a}_n^K\right\}\left(\underline{B}[\omega],1\right)\right| > \frac{\ln\gamma}{2}\right)$$

is smaller than

$$P\left(\omega \in \Omega : \left\|\underline{a}_n^K(\underline{B}[\omega],\cdot)\right\|_{L_2[\underline{b}]}^2 > \ln\gamma\right) + 2\gamma^{-\frac{1}{8}}.$$

Consequently

$$\begin{aligned}
\lim_{\gamma\uparrow\uparrow\infty}\int_{\left\{\Psi_n^{K_f} > \gamma\right\}} \Psi_n^{K_f}\, dP_X^{K_f} &\leq 2\lim_{\gamma\uparrow\uparrow\infty} P_B^K\left(\underline{k} \in K : \left\|\underline{a}^K(\underline{k},\cdot)\right\|_{L_2[\underline{b}]}^2 > \ln\gamma\right) \\
&+ 2\lim_{\gamma\uparrow\uparrow\infty}\gamma^{-\frac{1}{8}} \\
&= 0.
\end{aligned}$$

Step 6: Convergence in L_1

Proof Steps 4 and 5 insure convergence in L_1 [192, p. 18], so that

$$E_{P_X^{K_f}}\left[\Psi^{|K_f}\right] = 1.$$

Step 7: $E_{P_X^K}[\Psi] = 1.$

Proof That is the consequence of step 6 and of (Lemma) 13.4.2. □

Proposition 13.4.5 *In* (Proposition) 13.4.4,

1. *Assumption (b) is necessary and sufficient for P_X^K and P_B^K to be mutually absolutely continuous;*
2. *when Assumption (b) does not obtain, one has that*

   (i) *P_X^K is absolutely continuous with respect to P_B^K, and*
   (ii) *almost surely, with respect to P_X^K,*

$$\ln\left[\frac{dP_X^K}{dP_B^K}(\underline{k})\right] = I_{\mathcal{E}_{P_X^K}}\left\{\underline{a}^K\right\}(\underline{k}, 1) - \frac{1}{2}\left\|\underline{a}^K(\underline{k}, \cdot)\right\|^2_{L_2[\underline{b}]}.$$

Proof (Result (Proposition)) 13.1.11 says that, given a Cramér-Hida process $\underline{B}$, a SPWN model $\underline{X} = \underline{S}[\underline{a}] + \underline{B}$, with $\underline{a} \in L_2[\underline{b}]$, almost surely, with respect to P, has the property that $P_X^K \ll P_B^K$. Within (Proposition) 13.4.4, those are Assumptions (a) and (c), and they yield claim 2. Since Assumption (b) insures that Girsanov's theorem obtains, Assumption (b) implies that $P_X^K \equiv P_B^K$. The reverse is obvious. So it is item 3 that must be established. One proves that the Radon-Nikodým derivative is the limit, in $L_1[P_B^K]$, of a sequence whose limit in probability, with respect of P_X^K, is the density of item 3.

Retain, from (Proposition) 13.4.4, the following elements:

- $\left\{S_n^K, n \in \mathbb{N}\right\}$, the sequence of strict stopping times;
- $K_n = \left\{\underline{k} \in K : S_n^K(\underline{k}) = 1\right\}$;
- for $n \in \mathbb{N}$, fixed, but arbitrary,

$$\underline{a}_n^K = \chi_{[\![0, S_n^K]\!]}\,\underline{a}^K;$$

- $\underline{Y}_n^K$, the process on K_f, with law Q_n^K, with respect to $P_X^{K_f}$.

Let, with respect to P_B^K,

$$\ln\left[D_n(\underline{k})\right] = I_{\mathcal{E}_{P_B^K}}\left\{\underline{a}_n^K\right\}(\underline{k}, 1) - \frac{1}{2}\left\|\underline{a}_n^K(\underline{k}, \cdot)\right\|^2_{L_2[\underline{b}]}.$$

For $K_0 \in \mathcal{K}$, fixed, but arbitrary,

$$Q_n^K(K_0 \cap K_f \cap K_n) = P_X^{K_f}\left(\left\{\underline{Y}_n^K\right\}^{-1}\left\{K_0 \cap K_f \cap K_n\right\}\right).$$

But, as seen in (Proposition) 13.4.4, on $[\![0, S_n(\underline{k})]\!]$, $\underline{Y}_n^K(\underline{k}, t) = \underline{k}(t)$, so that

$$Q_n^K(K_0 \cap K_f \cap K_n) = P_X^{K_f}\left(K_0 \cap K_f \cap K_n\right),$$

and, furthermore, Q_n^K and P_B^K are mutually absolutely continuous, with Radon-Nikodým derivative

$$\frac{dQ_n^K}{dP_B^K} = D_n(\underline{k}).$$

Consequently

$$\begin{aligned}
P_X^K(K_0) &= \lim_n P_X^K(K_0 \cap K_n) \\
&= \lim_n P_X^K(K_0 \cap K_n \cap K_f) \\
&= \lim_n Q_n^K(K_0 \cap K_n \cap K_f) \\
&= \lim_n \int_{K_0 \cap K_n} \frac{dQ_n^K}{dP_B^K}\, dP_B^K \\
&= \lim_n \int_{K_0 \cap K_n} D_n\, dP_B^K.
\end{aligned}$$

Let

$$\ln\left[D(\underline{k})\right] = I_{\mathcal{E}_{P_X^K}}\left\{\underline{a}^K\right\}(\underline{k}, 1) - \frac{1}{2}\left\|\underline{a}^K(\underline{k}, \cdot)\right\|_{L_2[\underline{b}]}^2.$$

The proof now proceeds in steps. Let $\mathcal{D} = \left\{\chi_{K_n} D_n, n \in \mathbb{N}\right\}$.

Step 1: With respect to P_B^K, the sequence $\mathcal{D}$ converges in probability.

When $\underline{k} \in K_n$, as $S_n^K(\underline{k}) = 1$,

$$\left\|\underline{a}_n^K(\underline{k}, \cdot)\right\|_{L_2[\underline{b}]}^2 = \left\|\underline{a}^K(\underline{k}, \cdot)\right\|_{L_2[\underline{b}]}^2 < \infty,$$

that is, $\underline{k} \in K_f$. Thus

$$\chi_{K_n}(\underline{k}) D_n(\underline{k}) = \chi_{K_n}(\underline{k})\, e^{\chi_{K_f}(\underline{k})\, I_{\mathcal{E}_{P_B^K}}\left\{\underline{a}_n^K\right\}(\underline{k}, 1) - \frac{1}{2}\chi_{K_f}(\underline{k})\left\|\underline{a}_n^K(\underline{k}, \cdot)\right\|_{L_2[\underline{b}]}^2}.$$

Because of (Lemma) 13.4.1, with K_f for Ω_0, and $I_{\underline{\mathcal{E}}_{P_B^K}}$ for M, one has that

$$P_B^K\left(\underline{k} \in K : \chi_{K_f}(\underline{k})\left|I_{\underline{\mathcal{E}}_{P_B^K}}\left\{\underline{a}_n^K\right\}(\underline{k},1) - I_{\underline{\mathcal{E}}_{P_B^K}}\left\{\underline{a}_{n+p}^K\right\}(\underline{k},1)\right| > \alpha\right) =$$

$$= P_B^K\left(\underline{k} \in K : \chi_{K_f}(\underline{k})\left|I_{\underline{\mathcal{E}}_{P_B^K}}\left\{\underline{a}_n^K - \underline{a}_{n+p}^K\right\}(\underline{k},1)\right| > \alpha\right)$$

$$\leq P_B^K\left(\underline{k} \in K : \chi_{K_f}(\underline{k})\left\|\left\{\underline{a}_n^K - \underline{a}_{n+p}^K\right\}(\underline{k},\cdot)\right\|^2_{L_2[\underline{b}]} > \beta\right) + 2e^{-\frac{\alpha^2}{2\beta}},$$

and thus that the following sequence:

$$\left\{I_{\underline{\mathcal{E}}_{P_B^K}}\left\{\underline{a}_n^K\right\}(\cdot,1), n \in \mathbb{N}\right\}$$

is, with respect to P_B^K, convergent in probability. Let its limit be denoted J_B. Since, furthermore, almost surely, with respect to P_B^K,

$$\lim_n \chi_{K_f} \left\|\underline{a}_n^K\right\|^2_{L_2[\underline{b}]} = \chi_{K_f} \left\|\underline{a}^K\right\|^2_{L_2[\underline{b}]},$$

it follows that, in probability, with respect to P_B^K,

$$\lim_n \chi_{K_n} D_n = e^{J_B - \frac{1}{2}\chi_{K_f}\left\|\underline{a}_K\right\|^2_{L_2[\underline{b}]}}.$$

Step 2: Identification of the limit in probability, with respect to P_X^K, of the sequence $\mathcal{D}$.

From the following relation [(Proposition) 13.2.4]:

$$\underline{\mathcal{E}}_{P_X^K} = \underline{S}[\underline{a}^K] + \underline{B}_X,$$

valid with respect to P_X^K, and, since P_X^K is absolutely continuous with respect to P_B^K, because of (Proposition) 13.1.8, almost surely, with respect to P_X^K,

$$I_{\underline{\mathcal{E}}_{P_X^K}}\left\{\underline{a}_n^K\right\}(\cdot,1) = I_{\underline{\mathcal{E}}_{P_B^K}}\left\{\underline{a}_n^K\right\}(\cdot,1),$$

one has the following alternative representation of D_n:

$$D_n(\underline{k}) = e^{I_{\underline{\mathcal{E}}_{P_X^K}}\left\{\underline{a}_n^K\right\}(\underline{k},1) - \frac{1}{2}\left\|\underline{a}_n^K(\underline{k},\cdot)\right\|_{L_2[\underline{b}]}^2}.$$

Since, with respect to P_X^K, $\left\{\underline{a}_n^K, n \in \mathbb{N}\right\}$ also converges, in $L_2[\underline{b}]$, to $\underline{a}^K$, the sequence $\{D_n, n \in \mathbb{N}\}$ converges in probability, with respect to P_X^K, to D, which has the following expression:

$$D(\underline{k}) = e^{I_{\mathcal{E}_{P_X^K}}\{\underline{a}^K\}(\underline{k},1) - \frac{1}{2}\left\|\underline{a}^K(\underline{k},\cdot)\right\|^2_{L_2[\underline{b}]}}.$$

Step 3: The sequence $\mathcal{D}$ is, with respect to P_B^K, uniformly integrable.

Since Q_n^K and P_B^K are mutually absolutely continuous, with derivative D_n, given that $\{\chi_{K_n} D_n > \gamma\} = \{D_n > \gamma\} \cap K_n$, one has that

$$\begin{aligned}
\int_{\{\chi_{K_n} D_n > \gamma\}} \chi_{K_n} D_n dP_B^K &\leq \int_{\{D_n > \gamma\}} D_n dP_B^K \\
&= Q_n^K\left(\underline{k} \in K : D_n(\underline{k}) > \gamma\right) \\
&= P_X^{K_f}\left(\underline{k} \in K_f : [D_n \circ \underline{\mathbf{Y}}_n^K][\underline{k}] > \gamma\right).
\end{aligned}$$

Now, because of relation (Proposition) 11.3.14, symbolically

$$\left\{\int \phi dY\right\} \circ F = \int (\phi \circ F) d(Y \circ F),$$

one has (using again, locally, uniform convergence in probability) that

$$I_{\mathcal{E}_{P_X^K}}\left\{\underline{a}_n^K\right\} \circ \underline{Y}_n^K = I_{\mathcal{E}_{P_X^K} \circ \underline{Y}_n^K}\left\{\underline{a}_n^K \Box \Phi_{Y_n^K}\right\} = I_{\underline{Y}_n^K}\left\{\underline{a}_n^K \Box \Phi_{Y_n^K}\right\}.$$

But [(Proposition) 13.4.4, step 2], with respect to $P_X^{K_f}$,

$$\underline{Y}_n^K = \underline{S}[\underline{a}_n^K \Box \Phi_{Y_n^K}] + \underline{B}_X,$$

so that

$$\begin{aligned}
&P_X^{K_f}\left(\underline{k} \in K_f : [D_n \circ \underline{\mathbf{Y}}_n^K][\underline{k}] > \gamma\right) = \\
&= P_X^{K_f}\left(\underline{k} \in K_f : I_{\underline{B}_X}\left\{\underline{a}_n^K \Box \Phi_{Y_n^K}\right\}(\underline{k}, 1) + \frac{1}{2}\left\|\underline{a}_n^K \Box \Phi_{Y_n^K}(\underline{k}, \cdot)\right\|^2_{L_2[\underline{b}]} > \ln \gamma\right).
\end{aligned}$$

But, as in the final part of (Proposition) 13.4.4 (step 4), that latter probability is dominated by

$$2P_B^K\left(\underline{k} \in K : \left\|\underline{a}^K(\underline{k}, \cdot)\right\|^2_{L_2[\underline{b}]} > \ln \gamma\right) + 2\gamma^{-\frac{1}{8}}.$$

□

Chapter 14
Scope of the Signal Plus "White Noise" Model (I)

The model $\underline{X} = \underline{S}[\underline{a}] + \underline{B}$ (the SPWN model of the title) used so far has several potential limitations, summarized as the following items: a priori

(a) the interference of signal and noise could be other that additive;
(b) the signal could have paths outside the RKHS of the noise;
(c) the observation $\underline{X}$ need not solve a stochastic differential equation;
(d) the noise could be a martingale other than a Cramér-Hida process, a continuous martingale, for example, as continuous martingales are time changed Brownian motions.

The severity of such limitations will be, to some extent, examined in this, and the two chapters which follow. It is proved below that when the noise is a Cramér-Hida process, the additive form of interference between signal and noise, with signal in the RKHS of the noise, is not a limitation. That covers points (a) and (b) above. Point (c) is discussed in the next chapter, and (d), in the one which follows the latter.

As a preliminary step one must know something about weak solutions of stochastic differential equations.

14.1 Weak Solutions of Stochastic Differential Equations

Definition 14.1.1 Let $\underline{a}_K \in \mathcal{I}_0[\underline{b}]$ be given, and $\underline{B}$ be a Cramér-Hida process. A weak solution of the following formal equation

$$\underline{X} = \underline{S}[\underline{a}_K \Box \Phi_X] + \underline{B}\,,$$

where $\underline{X}$ represents the unknown, is a couple $(Q_K, \underline{N})$ where Q_K is a probability on $\mathcal{K}$, with respect to which $\underline{N}$, defined on $K \times [0,1]$, is a Cramér-Hida process such that

A.F. Gualtierotti, *Detection of Random Signals in Dependent Gaussian Noise*,
DOI 10.1007/978-3-319-22315-5_14

$Q_K \circ \underline{N}^{-1} = P_B^K$, and, for $t \in [0, 1]$, fixed, but arbitrary, almost surely with respect to Q_K,

$$\underline{\mathcal{E}}_{Q_K}(\underline{k}, t) = \underline{S}[\underline{a}_K](\underline{k}, t) + \underline{N}(\underline{k}, t).$$

When Q_K is the unique probability with respect to which what precedes obtains, the solution is deemed unique.

Remark 14.1.2 One should perhaps notice that, in relation to (Definition) 14.1.1, the value of the process $\underline{\mathcal{E}}_{Q_K}$ at $(\underline{k}, t)$ is $\underline{k}(t)$, so that one has indeed the "standard" stochastic differential equation form.

Proposition 14.1.3 *Let*

(a) $\underline{a}_K : K \times [0, 1] \longrightarrow$ *s have progressively measurable components for* $\underline{K}$*, and the property that,* for every $\underline{k} \in K$,

$$\left\|\underline{a}_K(\underline{k}, \cdot)\right\|_{L_2[\underline{b}]} < \infty;$$

(b) $\ln [D(\underline{k})] = I_{\underline{\mathcal{E}}_{P_B^K}} \{\underline{a}_K\} (\underline{k}, 1) - \frac{1}{2} \left\|\underline{a}_K(\underline{k}, \cdot)\right\|^2_{L_2[\underline{b}]}.$

The integral equation of (Definition) 14.1.1 *has a weak solution if, and only if,*

$$E_{P_B^K}[D] = 1.$$

When a solution exists, it is unique.

Proof *Suppose that* $E_{P_B^K}[D] = 1$.

Define then $\underline{N}$, with respect to P_B^K, as follows:

$$\underline{N}(\underline{k}, t) = \underline{S}[-\underline{a}_K](\underline{k}, t) + \underline{\mathcal{E}}_{P_X^K}(\underline{k}, t).$$

Let $dQ_K = DdP_B^K$. Since D may be seen as follows:

$$\ln [D(\underline{k})] = -I_{\underline{\mathcal{E}}_{P_B^K}} \{-\underline{a}_K\} (\underline{k}, 1) - \frac{1}{2} \left\|-\underline{a}_K(\underline{k}, \cdot)\right\|^2_{L_2[\underline{b}]},$$

given the assumption, one may invoke Girsanov's theorem to establish that, with respect to Q_K, $\underline{N}$ is a Cramér-Hida process whose covariance function is that of $\underline{B}$. But then $\underline{\mathcal{E}}_{P_B^K} = \underline{\mathcal{E}}_{Q_K}$, so that

$$\underline{\mathcal{E}}_{Q_K}(\underline{k}, t) = \underline{S}[\underline{a}_K](\underline{k}, t) + \underline{N}(\underline{k}, t),$$

and one thus indeed has a weak solution of the initial stochastic differential equation.

Proof Suppose that a weak solution exists.

By definition [(Remark) 14.1.2], it is then $\underline{\mathcal{E}}_{Q_K}$. Conditions

$$P_B^K\left(\underline{k} \in K : \|\underline{a}_K(\underline{k}, \cdot)\|^2_{L_2[\underline{b}]} < \infty\right) = 1,$$

and

$$Q_K\left(\underline{k} \in K : \|\underline{a}_K(\underline{k}, \cdot)\|^2_{L_2[\underline{b}]} < \infty\right) = 1,$$

are equally free of charge. It follows then [(Proposition) 13.4.4] that P_B^K and Q_K are mutually absolutely continuous, and that the Radon-Nikodým derivative

$$\frac{dQ_K}{dP_B^K}$$

is D, so that $E_{P_B^K}[D] = 1$.

Proof When a solution exists, it is unique.

Let $(Q_K^\star, \underline{N}^\star)$ be another solution. One will similarly have that

$$\frac{dQ_K^\star}{dP_B^K} = D,$$

from which it follows that $Q_K^\star = Q_K$. □

Remark 14.1.4 Suppose that item (a) of (Proposition) 14.1.3 is modified to

$$P_B^K\left(\underline{k} \in K : \|\underline{a}_K(\underline{k}, \cdot)\|^2_{L_2[\underline{b}]} < \infty\right) = 1.$$

When $E_{P_B^K}[D] = 1$, the same proof yields the existence of a solution, but one does not know whether

$$Q_K\left(\underline{k} \in K : \|\underline{a}_K(\underline{k}, \cdot)\|^2_{L_2[\underline{b}]} < \infty\right) = 1$$

or not, so that uniqueness does not follow.

Fact 14.1.5 ([128, p. 61]) *Let X be a positive supermartingale, with index set $\mathbb{R}_+$, which is continuous to the right. Let*

$$S_n = \inf\left\{t : X(\cdot, t) < \frac{1}{n}\right\}, \quad S = \sup_n S_n.$$

Then:

1. *$\{S_n, n \in \mathbb{N}\}$ and S are wide sense stopping times;*
2. *for almost all $\omega \in \{S < \infty\}$, $t \geq S(\omega)$ implies that $X(\omega, t) = 0$;*

3. *for all* $\omega \in \{S > 0\}$*, and* $t < S(\omega)$*,*

$$X(\omega, t) > 0, \quad \text{and} \quad \liminf_{\theta \uparrow\uparrow t} X(\omega, \theta) > 0.$$

S is called the time X attains zero.

Remark 14.1.6 Let X be a positive supermartingale, with index set $[0, 1]$, which is continuous to the right. Since, for $t_1 < t_2 \in [0, 1]$, fixed, but arbitrary, almost surely, with respect to P, $E_P[X(\cdot, t_2) \mid \mathcal{A}_{t_1}](\omega) \leq X(\omega, t_1)$, the inequality $X(\cdot, t_1) < \alpha$ implies, almost surely, with respect to P, $X(\cdot, t_2) < \alpha$.

Let, on $[0, 1]$, $\mathcal{A}_t^e$ be $\mathcal{A}_t$, and X_e be X; and, on $]1, \infty[$, $\mathcal{A}_t^e$ be $\mathcal{A}_1$, and X_e be $X(\cdot, 1)$. X_e is then a positive supermartingale with index set $\mathbb{R}_+$ which is continuous to the right. Let S and S_n be the stopping times of (Fact) 14.1.5 for X_e, and $\dot{S}$ and $\dot{S}_n$, the analogous stopping times for X. Let

$$B_n(\omega) = \left\{t \in [0, 1[: X(\omega, t) < \frac{1}{n}\right\},$$

$$\dot{A}_n(\omega) = \left\{t \in [0, 1] : X(\omega, t) < \frac{1}{n}\right\},$$

$$A_n(\omega) = \left\{t \in R_+ : X_e(\omega, t) < \frac{1}{n}\right\}.$$

When $X(\omega, 1) \geq \frac{1}{n}$,

$$B_n(\omega) = \dot{A}_n(\omega) = A_n(\omega),$$

but, when $X(\omega, 1) < \frac{1}{n}$, then

$$\dot{A}_n(\omega) = B_n(\omega) \cup \{1\}, \quad \text{and} \quad A_n(\omega) = B_n(\omega) \cup [1, \infty[.$$

Thus

$$B_n(\omega) \subseteq \dot{A}_n(\omega) \subseteq A_n(\omega).$$

One has the following cases:

- when $A_n(\omega) = \emptyset$, $\dot{A}_n(\omega) = \emptyset$, thus $\dot{S}_n(\omega) = 1$, but $S_n(\omega) = \infty$;
- when $A_n(\omega) \neq \emptyset$, and $X(\omega, 1) < \frac{1}{n}$, then
  - in case $B_n(\omega) = \emptyset$, $\dot{S}_n(\omega) = S_n(\omega) = 1$,
  - in case $B_n(\omega) \neq \emptyset$, $\dot{S}_n(\omega) = S_n(\omega) < 1$;
- when $A_n(\omega) \neq \emptyset$, and $X(\omega, 1) \geq \frac{1}{n}$, then $B_n(\omega) \neq \emptyset$, and, again,

$$\dot{S}_n(\omega) = S_n(\omega) < 1.$$

Thus $\dot{S}_n = S_n \wedge 1$, and $\dot{S}_n(\omega) \neq S_n(\omega)$ only when $A_n(\omega) = \emptyset$, that is, when $S_n(\omega) = \infty$.

Since the sets $A_n(\omega)$ and $\dot{A}_n(\omega)$ are decreasing with n, the times $S_n(\omega)$ and $\dot{S}_n(\omega)$ are increasing with n. Thus $S(\omega)$ and $\dot{S}(\omega)$ are the limits of increasing sequences. Furthermore

$$\dot{S}(\omega) = \sup_n \dot{S}_n(\omega) = \sup_n \{S_n(\omega) \wedge 1\} \leq S(\omega) \wedge 1.$$

Suppose that $\dot{S}(\omega) < S(\omega) \wedge 1$. Then, for all n, $\dot{S}_n(\omega) \leq \dot{S}(\omega) < 1$, that is, $B_n(\omega) \neq \emptyset$. But then $S_n(\omega) < 1$ for all n, that is $S(\omega) \leq 1$, and thus $\dot{S}(\omega) \leq S(\omega) \leq 1$. But then, for all n, $\dot{S}_n(\omega) = S_n(\omega)$, that is $\dot{S}(\omega) = S(\omega)$, a contradiction. So $\dot{S}(\omega) = S(\omega) \wedge 1$.

$S(\omega) = \infty$ if, and only if, there exists n with $S_n(\omega) = \infty$. Indeed, since $S_n \leq S$, when S_n is infinite, so is S. Suppose now that $S(\omega) = \infty$, but that $S_n(\omega) < \infty$, all n. Since $S_n(\omega) < \infty$ means $S_n(\omega) \leq 1$, $S(\omega) \leq 1$, a contradiction.

When $S(\omega) < \infty$, $S(\omega) = \dot{S}(\omega)$, and, according to (Fact) 14.1.5,

$$t \geq S(\omega) \text{ implies } X(\omega, t) = 0.$$

Thus, when $X(\omega, 1) > 0$, either $S(\omega) < \infty$, and then $S(\omega) > 1$ (which means $S(\omega) = \infty$, leading to a contradiction), or $S(\omega) = \infty$. Thus $S(\omega) = \infty$. Since $S(\omega)$ is the increasing limit of the $S_n(\omega)$'s, and that those are either smaller than, or equal to, one, or ∞, there is an n such that $S_n(\omega) = \infty$. But then $A_n(\omega) = \emptyset$, and $X(\omega, t) \geq \frac{1}{n}$ for all t.

14.2 A Signal Plus "White Noise" Model Is No Restriction

One should remark that "no restriction" is claimed only in case one decrees that a likelihood is required, and that the noise is "white," Gaussian.

Proposition 14.2.1 *Let $\underline{X} : \Omega \times [0, 1] \longrightarrow s$ be a process with paths in K, and assume that P_X^K and P_B^K are mutually absolutely continuous. There exists then a process $\underline{a}^K : K \times [0, 1] \longrightarrow s$, with base $(K, \underline{K}, P_X^K)$, and progressively measurable components, and a Cramér-Hida process $\underline{N}_X$, with base $(\Omega, \underline{\sigma}(\underline{X}), P)$, and law $P_{N_X}^K = P_B^K$ such that:*

1. *for $t \in [0, 1]$, fixed, but arbitrary, almost surely, with respect to P,*

$$\underline{X}(\omega, t) = \underline{S}\left[\underline{a}^K \square \Phi_X\right](\omega, t) + \underline{N}_X(\omega, t);$$

2. $P_B^K\left(\underline{k} \in K : \|\underline{a}^K(\underline{k}, \cdot)\|_{L_2[\underline{b}]}^2 < \infty\right) = 1;$
3. $P_X^K\left(\underline{k} \in K : \|\underline{a}^K(\underline{k}, \cdot)\|_{L_2[\underline{b}]}^2 < \infty\right) = 1.$

Proof Let

$$M(\underline{k},t) = E_{P_B^K}\left[\frac{dP_X^K}{dP_B^K} \mid \mathcal{K}_t\right].$$

M is a martingale for $\underline{\mathcal{K}}$, which is a $\underline{B}$-Gaussian martingale filtration [(Definition) 11.4.1]. As such it has a representation as a stochastic integral [(Proposition) 11.4.18]:

$$M(\underline{k},t) = 1 + I_{\mathcal{E}_{P_B^K}}\left\{\underline{a}^K\right\}(\underline{k},t),$$

where

- $\underline{a}^K$ is predictable for $\underline{\mathcal{K}}$;
- $P_B^K\left(\underline{k} \in K : \|\underline{a}^K(\underline{k},\cdot)\|_{L_2[\underline{b}]}^2 < \infty\right) = 1.$

It follows, in particular, that M is continuous to the right, and almost surely continuous.

Given that, by assumption, P_X^K and P_B^K are mutually absolutely continuous, $M(\cdot, 1) > 0$, almost surely, with respect to P_B^K, so that [(Remark) 14.1.6]

$$P_B^K\left(\underline{k} \in K : \inf_{t\in[0,1]} M(\underline{k},t) > 0\right) = 1.$$

Consequently, with respect to P_B^K, $\ln M$ is well defined (almost surely). Then Itô's formula yields that

$$\ln M(\underline{k},t) = I_{\mathcal{E}_{P_B^K}}\left\{\frac{\underline{a}^K}{M}\right\}(\underline{k},t) - \frac{1}{2}\left\|\left[\frac{\underline{a}^K(\underline{k},\cdot)}{M(\underline{k},\cdot)}\right]_{|t}\right\|_{L_2[\underline{b}]}^2.$$

Let

$$\underline{a}_M^K = \frac{\underline{a}^K}{M}.$$

As

$$\sum_{n=1}^{\infty}\int_0^1\left\{\frac{a_n^K(\underline{k},\theta)}{M(\underline{k},\theta)}\right\}^2 M_n(d\theta) \leq \frac{\|\underline{a}^K(\underline{k},\cdot)\|_{L_2[\underline{b}]}^2}{\inf_{t\in[0,1]} M^2(\underline{k},t)},$$

it follows that

$$P_B^K\left(\underline{k} \in K : \|\underline{a}_M^K(\underline{k},\cdot)\|_{L_2[\underline{b}]}^2 < \infty\right) = 1,$$

and, because of equivalence, also that

$$P_X^K\left(\underline{k} \in K : \left\|\underline{a}_M^K(\underline{k}, \cdot)\right\|_{L_2[\underline{b}]}^2 < \infty\right) = 1.$$

Since, by definition, $E_{P_B^K}[M(\cdot, 1)] = 1$, one knows from (Proposition) 14.1.3 that the formal equation of (Definition) 14.1.1 has a weak solution. One then sets, as in the proof of (Proposition) 14.1.3, for $(K, \underline{\mathcal{K}}, P_B^K)$,

$$dQ_K = M(\cdot, 1)\,dP_B^K, \quad \text{and} \quad \underline{N} = \underline{S}\left[-\underline{a}_M^K\right] + \underline{\mathcal{E}}_{P_X^K}.$$

Girsanov's theorem says that $\underline{N}$ is, with respect to Q_K, a Cramér-Hida process, with law P_B^K. Since $M(\cdot, 1)$ is a version of the Radon-Nikodým derivative of P_X^K with respect to P_B^K, $Q_K = P_X^K$. One finally composes the equation on K with $\underline{X} : \underline{N}_X = \underline{N} \Box \Phi_X$. □

As shall be seen below, (Proposition) 14.2.1 has a stronger version when, instead of the equivalence of P_X^K and P_B^K, one assumes only that P_X^K is absolutely continuous with respect to P_B^K. A number of features of martingales are required for the proof of that result, and they are listed below.

Remark 14.2.2 For $t \in [0, 1]$, fixed, but arbitrary, $S_t : K \longrightarrow [0, 1]$ with $S_t(\underline{k}) = t$, all $\underline{k} \in K$, is a strict, and thus also wide sense, stopping time for $\underline{\mathcal{K}}$ [264, p. 32]. Then [264, p. 33] $\mathcal{K}_{S_t} = \mathcal{K}_t$, and $\mathcal{K}_{S_t}^+ = \mathcal{K}_t^+$.

Remark 14.2.3 Let M be the uniformly integrable [264, p. 51], locally in L_2 [264, p. 63] martingale of (Proposition) 14.2.1, and $S_n(\underline{k}) = \inf\{t \in [0, 1] : M(\underline{k}, t) < n^{-1}\}$. M^{S_n} is a uniformly integrable martingale [264, p. 57], and it is also, locally, in L_2 [264, p. 63].

Remark 14.2.4 Since the minimum of two wide sense stopping times is a wide sense stopping time [264, p. 34], $S_{n,t} = S_n \wedge t$ is a wide sense stopping time. Then [264, p. 34]

$$\mathcal{K}_{S_{n,t}}^+$$

is a σ-algebra, and the random variable V is adapted to $\mathcal{K}_{S_{n,t}}^+$ if, and only if, for $\theta \in [0, 1]$, fixed, but arbitrary,

$$\chi_{\{S_{n,t} < \theta\}} V$$

is adapted to $\mathcal{K}_\theta$. Finally [264, p. 34],

$$\mathcal{K}_{S_{n,t}}^+ \subseteq \mathcal{K}_{S_t}^+ = \mathcal{K}_t^+ \quad \text{and} \quad \mathcal{K}_{S_{n,t}}^+ = \mathcal{K}_{S_n}^+ \cap \mathcal{K}_{S_t}^+ = \mathcal{K}_{S_n}^+ \cap \mathcal{K}_t^+.$$

Remark 14.2.5 When one has, on $\mathcal{K}$, the measure P_B^K, $\underline{\mathcal{K}}$ is the filtration generated by a Cramér-Hida process, and ${}^o\underline{\mathcal{K}}$ is thus continuous to the right

[(Proposition) 11.1.11]. Consequently, using the definitions [264, p. 33], the items of (Remark) 14.2.4 for ${}^{o}\underline{\mathcal{K}}$ are true without the "+" exponent. For example,

$$ {}^{o}\mathcal{K}_{S_n \wedge t} = {}^{o}\mathcal{K}_{S_n} \cap {}^{o}\mathcal{K}_t. $$

Remark 14.2.6 As M may be assumed to be continuous to the right, $M(\cdot, S_{n,t})$ is adapted to $\mathcal{K}_{S_{n,t}}$ [264, p. 41].

Proposition 14.2.7 *In* (Proposition) 14.2.1, *suppose that* P_X^K *is absolutely continuous with respect to* P_B^K, *rather than equivalent. Then items 1 and 3 of its conclusion obtain.*

Proof The beginning of it is the same as that of (Proposition) 14.2.1, up to the definition of the wide sense stopping time S_n. In particular $\underline{a}^K$ is almost surely in $L_2[\underline{b}]$, with respect to P_B^K, and thus also with respect to P_X^K.

One must deal with the fact that M of (Proposition) 14.2.1 can take the zero value with strictly positive P_B^K probability. Now, for $K_0 \in \mathcal{K}_{S_{n,t}}$, fixed, but arbitrary,

$$ \begin{aligned} P_X^K(K_0) &= \int_{K_0} \frac{dP_X^K}{dP_B^K}\, dP_B^K \\ &= \int_{K_0} E_{P_B^K}\left[\frac{dP_X^K}{dP_B^K} \mid \mathcal{K}_{S_{n,t}}\right] dP_B^K \\ &= \int_{K_0} M(\cdot, S_{n,t})\, dP_B^K, \end{aligned} $$

so that, on $\mathcal{K}_{S_{n,t}}$,

$$ dP_X^K = M^{S_n}(\cdot, t)\, dP_B^K. $$

Since $M^{S_n}(\cdot, t) \geq \frac{1}{n}$, one has also, on $\mathcal{K}_{S_{n,t}}$,

$$ \left\{M^{S_n}(\cdot, t)\right\}^{-1} dP_X^K = dP_B^K. $$

Consequently, as $K \in \mathcal{K}_{S_{n,t}}$,

$$ E_{P_X^K}\left[\left\{M^{S_n}(\cdot, t)\right\}^{-1}\right] = P_B^K(K) = 1. $$

The continuity properties of M, with respect to P_B^K, and thus to P_X^K, and Fatou's lemma, yield that [(Fact) 14.1.5]

$$ E_{P_X^K}\left[\left\{M^{S}(\cdot, t)\right\}^{-1}\right] = E_{P_X^K}\left[\liminf_{n\in\mathbb{N}} \left\{M^{S_n}(\cdot, t)\right\}^{-1}\right] $$

$$\leq \liminf_{n \in \mathbb{N}} E_{P_X^K}\left[\left\{M^{S_n}(\cdot, t)\right\}^{-1}\right]$$
$$= 1.$$

As $M(\cdot, S) = 0$ [(Fact) 14.1.5], one must have $S = 1$, almost surely with respect to P_X^K.

Now

$$\sum_{p=1}^{\infty} \int_0^{S_n} \left\{\frac{a_p^K(\underline{k}, \theta)}{M(\underline{k}, \theta)}\right\}^2 M_p(d\theta) \leq n^2 \left\|\underline{a}^K(\underline{k}, \cdot)\right\|_{L_2[\underline{b}]}^2,$$

so that, since, with respect to P_X^K, $\|\underline{a}^K(\underline{k}, \cdot)\|_{L_2[\underline{b}]}^2$ is finite almost surely,

$$P_X^K\left(\underline{k} \in K : \sum_{p=1}^{\infty} \int_0^1 \left\{\frac{a_p^K(\underline{k}, \theta)}{M(\underline{k}, \theta)}\right\}^2 M_p(d\theta) < \infty\right) = 1.$$

That is item 3 of the proposition's statement. One shall write again $\underline{a}_M^K$ for the ratio $\underline{a}^K/M$.

Let, on $(K, \underline{\mathcal{K}})$, P_n^K be defined setting

$$dP_n^K = M(\cdot, S_n)\, dP_B^K.$$

Then, for $K_0 \in \mathcal{K}_t$, fixed, but arbitrary,

$$P_n^K(K_0) = \int_{K_0} M(\cdot, S_n)\, dP_B^K$$
$$= \int_{K_0} E_{P_B^K}[M(\cdot, S_n) \mid \mathcal{K}_t]\, dP_B^K$$
$$= \int_{K_0} M(\cdot, S_{n,t})\, dP_B^K,$$

so that,

$$dP_n^{K|\mathcal{K}_t} = M(\cdot, S_{n,t})\, dP_B^K = M^{S_n}(\cdot, t)\, dP_B^K.$$

Furthermore, since M^{S_n} is a uniformly integrable martingale for P_B^K,

$$P_n^K(K) = E_{P_B^K}[M^{S_n}(\cdot, 1)] = 1.$$

Let, on the base space $(K, \underline{\mathcal{K}}, P_B^K)$,

$$\underline{N}_n(\underline{k}, t) = \underline{S}\left[-\underline{a}^K_{M|S_n}\right](\underline{k}, t) + \underline{\mathcal{E}}_{P_B^K}(\underline{k}, t).$$

As presently seen, the process $\underline{N}_n$ is, for the base $(K, \underline{\mathcal{K}}, P_n^K)$, a Cramér-Hida process with law P_B^K. Indeed, since, almost surely with respect to P_B^K, $M^{S_n}(\cdot, t) \geq \frac{1}{n}$, one is allowed to compute its logarithm, and use Itô's formula to obtain that, almost surely, with respect to P_B^K,

$$\ln\left[M^{S_n}(\cdot, t)\right] = I_{\underline{\mathcal{E}}_{P_B^K}}\left\{\underline{a}^K_{M|S_n}\right\}(\cdot, t) - \frac{1}{2}\left\|\underline{a}_{M|S_n\wedge t}\right\|^2_{L_2[\underline{b}]}.$$

Since

$$E_{P_B^K}\left[M(\cdot, S_n)\right] = E_{P_B^K}\left[M^{S_n}(\cdot, 1)\right] = 1,$$

one may use Girsanov's theorem to assert that, with respect to P_n^K, $\underline{N}_n$ is a Cramér-Hida process with law P_B^K. As $\underline{N}^{S_n}_{n+1} = \underline{N}_n$, one may define, with respect to P_B^K,

$$\underline{N}(\underline{k}, t) = \underline{S}\left[-\underline{a}^K_{M|S}\right](\underline{k}, t) + \underline{\mathcal{E}}_{P_B^K}(\underline{k}, t).$$

Since, almost surely, with respect to P_X^K, $S = 1$, and that $P_X^K \ll P_B^K$, then, almost surely, with respect to P_X^K,

$$\underline{N}(\underline{k}, t) = \underline{S}\left[-\underline{a}^K_M\right](\underline{k}, t) + \underline{\mathcal{E}}_{P_X^K}(\underline{k}, t).$$

In the final step one must establish that $\underline{N}$ is a Cramér-Hida process with law P_B^K. Now $\underline{N}^{S_n}M^{S_n} = \underline{N}_n M^{S_n}$. Since, with respect to P_n^K, $\underline{N}_n$ is a Cramér-Hida process, for $t_1 < t_2$ in $[0, 1]$, and $K_0 \in \mathcal{K}_{t_1}$, fixed, but arbitrary,

$$\int_{K_0} \underline{N}_n(\cdot, t_2)\, dP_n^K = \int_{K_0} \underline{N}_n(\cdot, t_1)\, dP_n^K.$$

As, on $\mathcal{K}_t$, $dP_n^K = M^{S_n}(\cdot, t)\, dP_B^K$, and that $K_0 \in \mathcal{K}_{t_1} \subseteq \mathcal{K}_{t_2}$,

$$\int_{K_0} \underline{N}_n(\cdot, t_2) M^{S_n}(\cdot, t_2)\, dP_B^K = \int_{K_0} \underline{N}_n(\cdot, t_1) M^{S_n}(\cdot, t_1)\, dP_B^K, \qquad (\star)$$

that is, $\underline{N}^{S_n}M^{S_n}$ is a martingale for $\underline{\mathcal{K}}$ and P_B^K. It is thus, since $\mathcal{K}_{S_{n,t}} \subseteq \mathcal{K}_t$, a martingale for the filtration of the σ-algebras $\mathcal{K}_{S_{n,t}}$. But, as seen earlier in the proof, on $\mathcal{K}_{S_{n,t}}$

$$M^{S_n}(\cdot, t)\, dP_B^K = dP_X^K.$$

Thus, choosing $K_0 \in \mathcal{K}_{S_{n,t_1}}$ in the last equality ($\star$) above, one has that $\underline{N}_n$ is a martingale for the filtration of the σ-algebras $\mathcal{K}_{S_{n,t}}$ and the probability P_X^K. Let $K_t \in \mathcal{K}$ be fixed, but arbitrary. One has that

$$K_t = [K_t \cap \{S_n < t\}] \cup [K_t \cap \{S_n \geq t\}] \in \mathcal{K}_{S_{n,t}}.$$

Consequently $\underline{N}^{S_n}$ is a martingale for $\underline{\mathcal{K}}$ and P_X^K.

Let $N_\alpha = \langle \underline{\alpha}, \underline{N} \rangle_{l_2}$, so that $N_\alpha^{S_n} = \langle \underline{\alpha}, \underline{N}_n \rangle_{l_2}$. Since, with respect to P_n^K, $\underline{N}_n$ is a Cramér-Hida process with law P_B^K, then, with respect to the law P_n^K, $\langle N_\alpha^{S_n} \rangle = \langle C(\cdot)\underline{\alpha}, \underline{\alpha} \rangle_{l_2}$. Since, on $\mathcal{K}_{S_{n,t}}$, $P_X^K \ll P_n^K$, the same is true with respect to P_X^K [264, p. 245]. Since $\langle C(\cdot)[\underline{\alpha}], \underline{\alpha} \rangle_{l_2}$ is bounded, $N_\alpha^{S_n}$ is uniformly integrable, and N_α is a local martingale with $\langle C(\cdot)[\underline{\alpha}], \underline{\alpha} \rangle_{l_2}$ as quadratic variation. The result then follows from (Proposition) 10.5.6 and (Corollary) 10.5.19. □

Chapter 15
Scope of the Signal Plus "White Noise" Model (II)

When one does not know that the SPWN model has a representation in the form of a stochastic differential equation, which is generally the case within the Cramér-Hida framework, it becomes important to know that such a representation exists, as it is that representation which allows an explicit form for the likelihood. In the previous chapter, it was seen that the existence of the likelihood has the consequence that the observations are represented in the form of a stochastic differential equation. The result is however an existence result which gives no hint as to the form of the resulting signal. When one is willing, or able, to assume some integrability conditions on the signal (that is typically the case when adjusting a model to data: one makes do with what is available, as long as the procedure is reasonable, and can be seriously evaluated), one then obtains a stochastic differential equation form in which the signal is a conditional expectation with respect to the observations. That latter stochastic differential equation representation is known under the appellation of "innovation representation," and makes up the next topic. In the framework of the Cramér-Hida representation, assumptions on the integrability of the signal in the derived "white noise model" may be difficult, nay, impossible to justify. It is nevertheless useful to know that the stochastic differential equation representation is related to conditional expectations with respect to the observations process.

The following "generic model" is assumed thereafter, and assumptions about it shall be explained as needed, as an innovation representation may be obtained under diverse assumptions on the signal and the noise:

$$\underline{X} = \underline{S}[\underline{a}] + \underline{N}.$$

In fine, $\underline{N}$ shall be $\underline{B}$.

A.F. Gualtierotti, *Detection of Random Signals in Dependent Gaussian Noise*,
DOI 10.1007/978-3-319-22315-5_15

15.1 The Case of Real Processes

The model is thus $X = S[a] + N$, with processes having values in $\mathbb{R}$. The idea is to condition $S[a] + N$ on X in the hope to obtain a relation

$$X = S[a \square \Phi_X] + N_X,$$

where that latter derived equation has, *mutatis mutandis,* the properties of the initial relation $X = S[a] + N$.

15.1.1 Measurable Versions of Conditional Expectations

Conditioning on X, for $t \in [0, 1]$, fixed, though arbitrary, produces a well-defined object. Needed are regularity properties, with respect to time, of the resulting objects. The following facts shall be assumed to obtain:

Assumptions 15.1.1 *The base assumptions are:*

1. *M, is a measure on the Borel sets of $\mathcal{B}([0, 1])$, obtained from a deterministic, monotone increasing, continuous function;*
2. *a is a progressively measurable process for $\underline{\sigma}(\underline{X})$;*
3. $E_P\left[\int_0^1 |a(\omega, t)| M(dt)\right] < \infty$.

The following lemmas are needed in the sequel.

Lemma 15.1.2 *Let M be a measure on $\mathcal{B}([0, 1])$, and*

(a) *$\mathcal{B} \subseteq \mathcal{A} \otimes \mathcal{B}([0, 1])$ be a σ-algebra;*
(b) *for $t \in [0, 1]$, fixed, but arbitrary, $a_t : \Omega \longrightarrow \mathbb{R}$ is adapted to $\mathcal{A}$ and $\mathcal{B}(\mathbb{R})$;*
(c) *for $n \in \mathbb{N}$, fixed, but arbitrary, $\alpha_n : \Omega \times [0, 1] \longrightarrow \mathbb{R}$ is adapted to $\mathcal{B}$ and $\mathcal{B}(\mathbb{R})$;*
(d) *for $t \in [0, 1]$, fixed, but arbitrary, almost surely, with respect to P,*

$$\lim_n \alpha_n(\omega, t) = a_t(\omega);$$

There exists then $\alpha : \Omega \times [0, 1] \longrightarrow \mathbb{R}$, adapted to $\mathcal{B}$ and $\mathcal{B}(\mathbb{R})$, such that

1. *almost surely with respect to $P \otimes M$, $\lim_n \alpha_n(\omega, t) = \alpha(\omega, t)$;*
2. *for $t \in [0, 1]$, fixed, but arbitrary, almost surely, with respect to P,*

$$\alpha(\omega, t) = a_t(\omega).$$

Proof Let

$$B_0 = \left\{(\omega, t) \in \Omega \times [0, 1] : \lim_n \alpha_n(\omega, t) \text{ does not exist}\right\}.$$

Then $B_0 \in \mathcal{B}$ [113, p. 93]. Let $B_0[t]$ be the section at t of B_0: it is also a measurable set [113, p. 238]. One has that

$$B_0[t] \subseteq \{\omega \in \Omega : \{\alpha_n(\omega, t), n \in \mathbb{N}\} \text{ does not converge to } a_t(\omega)\}.$$

Thus $P(B_0[t]) = 0$, and

$$[P \otimes M](B_0) = \int_0^1 P(B_0[t]) \, M(dt) = 0.$$

That is claim 1. For claim 2, one observes that, for $t \in [0, 1]$, and $\omega \in B_0[t]^c$, fixed, but arbitrary, $\lim_n \alpha(\omega, t)$ exists. Let it be denoted $\alpha(\omega, t)$. Then, for $t \in [0, 1]$, fixed, but arbitrary, almost surely with respect to P, $\alpha(\omega, t) = a_t(\omega)$. □

Lemma 15.1.3 *Let $t_{n,i} = i2^{-n}, i \in [0 : 2^n]$, and $Y : \Omega \times [0, 1] \longrightarrow \mathbb{R}$, a process such that*

(a) *$Y(\omega, t) = 0$ when $(\omega, t) \in (\Omega \times]t_{n,i}, t_{n,i+1}])^c$;*
(b) *for fixed, but arbitrary (n, i), Y is adapted to $\mathcal{A} \otimes \mathcal{B}(]t_{n,i}, t_{n,i+1}])$;*
(c) *for $t \in [0, 1]$, fixed, but arbitrary, $E_P[|Y(\cdot, t)|] < \infty$.*

Then

$$(\omega, t) \mapsto E_P\left[Y(\cdot, t) \mid \sigma_{t_{n,i}}(\underline{X})\right]$$

has a version $Y_{n,i}$ that is adapted to $\sigma_{t_{n,i}}(\underline{X}) \otimes \mathcal{B}(]t_{n,i}, t_{n,i+1}])$.

Proof It is no restriction to suppose the process Y positive.

Let Y have the following form, where $A \in \mathcal{A}_t$, and $t_{n,i} \leq t_1 < t_2 \leq t_{n,i+1}$:

$$Y(\omega, t) = \chi_A(\omega) \chi_{]t_1, t_2]}(t).$$

Then, as

$$E_P\left[Y(\cdot, t) \mid \sigma_{t_{n,i}}(\underline{X})\right] = \chi_{]t_1, t_2]}(t) E_P\left[\chi_A \mid \sigma_{t_{n,i}}(\underline{X})\right],$$

the claim is true for that type of Y, and thus for any linear combination of Y's of such type. Let $\mathcal{V}$ denote the vector space thus defined.

Let $\{Y_p, p \in \mathbb{N}\} \subseteq \mathcal{V}$ be an increasing sequence converging to Y, and define the following maps: for $p \in \mathbb{N}$, fixed, but arbitrary,

$$\alpha_p : \Omega \times [0, 1] \longrightarrow \mathbb{R}$$

is adapted to $\sigma_{t_{n,i}}(\underline{X}) \otimes \mathcal{B}(]t_{n,i}, t_{n,i+1}])$, and such that, for $t \in [0, 1]$, fixed, but arbitrary, almost surely, with respect to P,

$$\alpha_p(\omega, t) = E_P\left[Y_p(\cdot, t) \mid \sigma_{t_{n,i}}(\underline{X})\right](\omega).$$

Since $\{Y_p, p \in \mathbb{N}\}$ is increasing, and, for $t \in [0,1]$, fixed, but arbitrary, $Y(\cdot,t)$ is integrable, almost surely, with respect to P,

$$\lim_p E_P\left[Y_p(\cdot,t) \mid \sigma_{t_{n,i}}(\underline{X})\right](\omega) = E_P\left[Y(\cdot,t) \mid \sigma_{t_{n,i}}(\underline{X})\right](\omega).$$

It then suffices to choose, in (Lemma) 15.1.2,

$$a_t(\omega) = E_P\left[Y(\cdot,t) \mid \sigma_{t_{n,i}}(\underline{X})\right](\omega), \text{ and } \mathcal{B} = \sigma_{t_{n,i}}(\underline{X}) \otimes \mathcal{B}\left(]t_{n,i}, t_{n,i+1}]\right).$$

□

Proposition 15.1.4 *When (Assumption)* 15.1.1 *obtains, there exists* α*, predictable for* $\underline{\sigma}(\underline{X})$*, such that, almost surely, with respect to* M*,* $\alpha(\cdot,t)$ *is a version of* $E_P[a(\cdot,t) \mid \sigma_t(\underline{X})]$.

Proof Suppose first that $\underline{X}$ is continuous. Since, by assumption,

$$\int_0^1 M(dt)E_P\left[|a(\cdot,t)|\right] < \infty,$$

almost surely, with respect to M, $E_P[|a(\cdot,t)|] < \infty$, and $E_P[a(\cdot,t) \mid \sigma_t(\underline{X})]$ exists. Let

$$I = \{t \in [0,1] : E_P[|a(\cdot,t)|] < \infty\}.$$

I is measurable, has M-measure equal to one, and $a_I = \chi_I a$ is measurable, and adapted. It thus has a progressively measurable modification, say a_{pm} [192, p. 68]. a_{pm} has the same integrability properties as a, and thus the conditional expectation of a_{pm} exists for $t \in [0,1]$. a_{pm} may thus replace a.

Let $T_n = \{t_{n,i} \in [0,1] : t_{n,i} = i2^{-n}, i \in [0:2^n]\}$, and, for $t \in [0,1]$, and $n \in \mathbb{N}$, fixed, but arbitrary,

$$\text{when } t \in \left]t_{n,i}, t_{n,i+1}\right],\ t_{n,i(t)} = t_{n,i}.$$

The sequence $\{t_{n,i(t)}, n \in \mathbb{N}\}$ increases to t. As $\underline{X}$ has continuous paths, $\underline{\sigma}(\underline{X})$ is continuous to the left, and thus

$$\sigma_t(\underline{X}) = \bigvee_{n \in \mathbb{N}} \sigma_{t_{n,i(t)}}(\underline{X}).$$

Let

$$a_{n,i}(\omega,t) = \chi_{]t_{n,i},t_{n,i+1}]}(t)\,a_{pm}(\omega,t).$$

Result (Lemma) 15.1.3 applies to $a_{n,i}$, so that there exits $\alpha_{n,i}$, adapted to the algebra $\sigma_{t_{n,i}}(\underline{X}) \otimes \mathcal{B}\left(]t_{n,i}, t_{n,i+1}]\right)$, and thus predictable, such that, almost surely, with respect to P,

$$\alpha_{n,i}(\omega, t) = E_P\left[a_{n,i}(\cdot, t) \mid \sigma_{t_{n,i}}(\underline{X})\right](\omega).$$

Let then

$$\alpha_n(\omega, t) = \sum_{i=0}^{2^n-1} \alpha_{n,i}(\omega, t).$$

α_n is adapted to $\bigvee_{i=0}^{2^n-1}\left\{\sigma_{t_{n,i}}(\underline{X}) \otimes \mathcal{B}\left(]t_{n,i}, t_{n,i+1}]\right)\right\}$, and thus predictable. Furthermore, almost surely, with respect to P,

$$\alpha_n(\omega, t) = E_P\left[a_{pm}(\cdot, t) \mid \sigma_{n,i(t)}(\underline{X})\right].$$

A martingale convergence theorem [201, p. 29] yields that, almost surely, with respect to P, as n increases indefinitely,

$$E_P\left[a_{pm}(\cdot, t) \mid \sigma_{n,i(t)}(\underline{X})\right]$$

converges to

$$E_P\left[a_{pm}(\cdot, t) \mid \bigvee_{n=1}^{\infty} \sigma_{n,i(t)}(\underline{X})\right] = E_P\left[a_{pm}(\cdot, t) \mid \sigma_t(\underline{X})\right].$$

Consequently, for $t \in [0, 1]$, fixed, but arbitrary, almost surely, with respect to P,

$$\lim_n \alpha_n(\omega, t) = E_P\left[a_{pm}(\cdot, t) \mid \sigma_t(\underline{X})\right].$$

Letting, in (Lemma) 15.1.2, $\mathcal{B}$ be the predictable sets, and $a_t = E_P\left[a_{pm}(\cdot, t) \mid \sigma_t(\underline{X})\right]$, one obtains an α such that, for $t \in [0, 1]$, fixed, but arbitrary, almost surely with respect to P,

$$\alpha(\omega, t) = E_P\left[a_{pm}(\cdot, t) \mid \sigma_t(\underline{X})\right].$$

Furthermore, for $t \in I$, fixed, but arbitrary,

$$\alpha(\omega, t) = E_P\left[a_{pm}(\cdot, t) \mid \sigma_t(\underline{X})\right] = E_P\left[a(\cdot, t) \mid \sigma_t(\underline{X})\right].$$

Suppose now that $\underline{X}$ is only almost surely continuous. Let $\underline{X}_N$ be defined as in Sect. 10.2.3. There exists then α, predictable for

$$\sigma_t(\underline{X}_N) = \sigma_t(\underline{X}) \cap \Omega_N \subseteq \sigma_t(\underline{X}) \vee \{\Omega_N\},$$

and thus for the latter family of σ-algebras, such that, almost surely with respect to M, for t fixed, but arbitrary, $\alpha(\cdot, t)$ is a version of

$$E_P\left[a(\cdot, t) \mid \sigma_t(\underline{X}) \cap \Omega_N\right].$$

Now $E_P\left[a(\cdot, t) \mid \sigma_t(\underline{X})\right]$ is a version of $E_P\left[a(\cdot, t) \mid \sigma_t(\underline{X}) \vee \{\Omega_N\}\right]$. But, for arbitrary measurable sets A and B,

$$\begin{aligned}
\chi_{A\Delta\Omega_N} &= \left|\chi_A - \chi_{\Omega_N}\right| \\
&= \left|\chi_{A\cap N} + \chi_{A\cap\Omega_N} - \chi_{\Omega_N}\right| \\
&= \chi_{A\cap N} + \chi_{\Omega_N}\chi_{A^c}, \\
\chi_{A\Delta\Omega_N}\chi_{B\Delta\Omega_N} &= \chi_{A\cap B\cap N} + \chi_{\Omega_N}\chi_{A^c\cap B^c},
\end{aligned}$$

and also $\sigma_t(\underline{X}) \vee \{\Omega_N\} = \sigma(X_{t,0}\Delta\Omega_N, X_{t,0} \in \sigma_t(\underline{X}))$, so that

$$\begin{aligned}
\int_{X_{t,0}\Delta\Omega_N} E_P\left[a(\cdot, t) \mid \sigma_t(\underline{X}) \vee \{\Omega_N\}\right] &= \int_{X_{t,0}^c\cap\Omega_N} a(\cdot, t)\, dP \\
&= \int_{X_{t,0}^c\cap\Omega_N} \alpha(\cdot, t)\, dP \\
&= \int_{X_{t,0}\Delta\Omega_N} \alpha(\cdot, t)\, dP.
\end{aligned}$$

The variable $\alpha(\cdot, t)$ is thus a version of $E_P\left[a(\cdot, t) \mid \sigma_t(\underline{X}) \cap \{\Omega_N\}\right]$, and thus of $E_P\left[a(\cdot, t) \mid \sigma_t(\underline{X})\right]$. □

Remark 15.1.5 Since, for $t \in I$, $\alpha(\omega, t) = E_P\left[a(\cdot, t) \mid \sigma_t(\underline{X})\right]$, for $t \in I$, a set of M-measure one,

$$\int_0^t \alpha(\omega, \theta) M(d\theta)$$

exists, and is finite.

Remark 15.1.6 $E_P\left[\int_0^1 |\alpha(\cdot, \theta)|\, M(d\theta)\right] < \infty.$

Remark 15.1.7 When a is square integrable for $P \otimes M$, an analogous result obtains, and in particular,

$$E_P\left[\int_0^1 \alpha^2(\cdot,\theta)M(d\theta)\right] < \infty.$$

15.1.2 Innovations for Product Square Integrable Signals

The following facts shall be assumed to obtain:

Assumptions 15.1.8 *The assumptions are:*

1. *N is a continuous, square integrable martingale for $\underline{\mathcal{A}}$;*
2. *X is a continuous process adapted to $\underline{\mathcal{A}}$;*
3. *a is progressively measurable for $\underline{\mathcal{A}}$;*
4. *for $t \in [0,1]$, fixed, but arbitrary, almost surely with respect to P,*

$$X(\omega,t) = \int_0^t a(\omega,\theta)M(d\theta) + N(\omega,t),$$

5. $E_P\left[\int_0^1 a^2(\cdot,t)M(dt)\right] < \infty.$

Proposition 15.1.9 *Let $\underline{\mathcal{B}}$ be a filtration such that, for $t \in [0,1]$, fixed, but arbitrary, $\sigma_t(X) \subseteq \mathcal{B}_t \subseteq \mathcal{A}_t$ (one writes $\underline{\mathcal{B}} \subseteq \underline{\mathcal{A}}$), and*

$$I(\omega,t) = X(\omega,t) - \int_0^t \alpha(\omega,\theta)M(d\theta),$$

where α is a progressively measurable version of the conditional expectation of a with respect to $\underline{\mathcal{B}}$ (one thus supposes that α exists). I is then a continuous martingale for $\underline{\mathcal{B}}$, whose square is integrable, and whose square bracket is that of N. That statement applies in particular when $\mathcal{B}_t = \sigma_t(\underline{X})$, and X is one of the components of $\underline{X}$.

Proof By definition,

$$I(\omega,t) = \int_0^t \{a(\omega,\theta) - \alpha(\omega,\theta)\}M(d\theta) + N(\omega,t).$$

Hence, for $t_1 < t_2$, in $[0,1]$, fixed, but arbitrary,

$$E_P[I(\cdot,t_2) - I(\cdot,t_1) \mid \mathcal{B}_{t_1}] =$$
$$= E_P[N(\cdot,t_2) - N(\cdot,t_1) \mid \mathcal{B}_{t_1}] + E_P\left[\int_{t_1}^{t_2} \{a(\cdot,\theta) - \alpha(\cdot,\theta)\}M(d\theta) \mid \mathcal{B}_{t_1}\right].$$

The term involving N is zero, since it is a martingale for $\underline{\mathcal{A}}$, and that $\underline{\mathcal{B}} \subseteq \underline{\mathcal{A}}$. Let B be a fixed, but arbitrary element of $\mathcal{B}_{t_1}$. Then, since, in $[0, 1]$, for $\theta \geq t_1$, fixed, but arbitrary, $B \in \mathcal{B}_{t_1} \subseteq \mathcal{B}_\theta$,

$$\int_B E_P\left[\int_{t_1}^{t_2} \{a(\cdot,\theta) - \alpha(\cdot,\theta)\} M(d\theta) \mid \mathcal{B}_{t_1}\right] dP =$$

$$= \int_B dP \int_{t_1}^{t_2} \{a(\cdot,\theta) - \alpha(\cdot,\theta)\} M(d\theta)$$

$$= \int_{t_1}^{t_2} M(d\theta) \int_B dP \{a(\cdot,\theta) - \alpha(\cdot,\theta)\}$$

$$= \int_{t_1}^{t_2} M(d\theta) \int_B dP E_P\left[\{a(\cdot,\theta) - \alpha(\cdot,\theta)\} \mid \mathcal{B}_\theta\right].$$

But the inside integral of that latter double integral is zero by assumption, and I is a square integrable martingale for $\underline{\mathcal{B}}$ (definition of I, and, for α, (Remark) 15.1.7). Let π be a partition of $[0, t]$, using $t_0, t_1, \ldots, t_n$, that gets finer as n increases indefinitely. Then

$$\sum_{i=1}^{n} \{I(\cdot,t_i) - I(\cdot,t_{i-1})\}^2 =$$

$$= \sum_{i=1}^{n} \{N(\cdot,t_i) - N(\cdot,t_{i-1})\}^2$$

$$+ 2\sum_{i=1}^{n} \{N(\cdot,t_1) - N(\cdot,t_{i-1})\} \left\{\int_{t_{i-1}}^{t_i} \{a(\cdot,\theta) - \alpha(\cdot,\theta)\} M(d\theta)\right\}$$

$$+ \sum_{i=1}^{n} \left\{\int_{t_{i-1}}^{t_i} \{a(\cdot,\theta) - \alpha(\cdot,\theta)\} M(d\theta)\right\}^2.$$

The left-hand side of the latter equality, when the filtration is $\underline{\mathcal{B}}$, converges in probability [264, p. 164] to $\langle I \rangle$. For the same reason, the first term on the right-hand side, when the filtration is $\underline{\mathcal{A}}$, converges in probability to $\langle N \rangle$. Let

$$\xi(\omega,\pi) = \max_{\pi} \int_{t_{i-1}}^{t_i} |a(\omega,\theta) - \alpha(\omega,\theta)| M(d\theta).$$

The last right-hand term in the approximation to the quadratic variation of I is dominated by $\xi(\omega,\pi) \int_0^t |a(\omega,\theta) - \alpha(\omega,\theta)| M(d\theta)$, which converges to zero almost surely, with respect to P. Let

$$\eta(\omega,t) = \max_{\pi} |N(\omega,t_i) - N(\omega,t_{i-1})|.$$

The middle term on the right-hand side in the approximation to the quadratic variation of I is dominated by

$$2\eta(\omega, t) \int_0^t |a(\omega, \theta) - \alpha(\omega, \theta)| \, M(d\theta).$$

But, since N is continuous, $\eta(\omega, t)$ converges to zero almost surely, with respect to P. One consequence is that $\langle N \rangle$ is adapted to $\underline{\mathcal{B}}$, hence the stated equality of brackets. □

Remark 15.1.10 The same proof applies when the processes take their values in $\mathbb{R}^n$. When $\underline{N}$ is the vector whose components are the first n components of $\underline{B}$, a Cramér-Hida process, one shall have that the components of the innovation are independent, and that matters in the Cramér-Hida context.

15.1.3 Innovations for Product Integrable Signals

The following facts shall be assumed to obtain:

Assumptions 15.1.11 *The assumptions are those of (Assumption)* 15.1.8, *except that item 5 becomes*

$$E_P\left[\int_0^1 |a(\cdot, t)| \, M(dt)\right] < \infty. \tag{$\star$}$$

Proposition 15.1.12 *(Result* (Proposition)*)* 15.1.9 *obtains given that latter ($\star$) integrability assumption.*

Proof The only part of the proof of (Proposition) 15.1.9 that must be validated is that I has a square that is integrable. Let $B_n(\omega) = \{t \in [0, 1] : |I(\omega, t)| \geq n\}$, and

$$S_n = \begin{cases} 1 & \text{when } B_n = \emptyset \\ \inf B_n & \text{when } B_n \neq \emptyset \end{cases}.$$

S_n is a strict stopping time when restricted to the subset Ω_N of Ω for which the paths of I are continuous, and there is [(Proposition) 10.2.27] a wide sense stopping time for $\underline{\mathcal{B}}$ that is equal to S_n on Ω_N. Let the latter be denoted S_n also. I^{S_n} is a martingale, adapted to $\underline{\mathcal{B}}$, whose square is integrable. Using (Proposition) 15.1.9 and [264, p. 166], one has that

$$\langle I^{S_n} \rangle = \langle N^{S_n} \rangle = \langle N \rangle^{S_n} \leq \langle N \rangle.$$

Now, for $t \in [0,1]$, fixed, but arbitrary, $\lim_n I^{S_n}(\cdot,t) = I(\cdot,t)$, almost surely with respect to P, and, by Fatou's lemma,

$$\begin{aligned} E_P\left[N^2(\cdot,t)\right] &= E_P\left[\langle N\rangle(\cdot,t)\right] \\ &\geq \liminf_n E_P\left[(I^{S_n}(\cdot,t))^2\right] \\ &\geq E_P\left[\liminf_n (I^{S_n}(\cdot,t))^2\right] \\ &= E_P\left[I^2(\cdot,t)\right]. \end{aligned}$$

□

15.1.4 *Innovations for Signals That Are in the Reproducing Kernel Hilbert Space of the Noise*

That is the "natural" case in the Cramér-Hida context, but, unfortunately a case for which the "natural" assumptions seem insufficient. Indeed, when the signal is almost surely in the RKHS of a noise that is a Brownian motion, as shall be seen, a candidate for the conditional expectation exists. But, for that candidate to actually be the conditional expectation, one must know "independently" that that conditional expectation exists. That existence can be established under added assumptions, which, in practice, are hard to validate.

15.2 The Case of Vector Processes

One shall assume that the following facts obtain:

Assumptions 15.2.1 *They are as follows:*

1. *the process $\underline{a}$ shall belong to $\mathcal{I}_2[\underline{b}]$* (rather than to $\mathcal{I}_0[\underline{b}]$: that restriction has its origin in Sect. 15.1.4 above);
2. $\underline{X} = \underline{S}[\underline{a}] + \underline{B}$, *where $\underline{B}$, is a Cramér-Hida process.*

Proposition 15.2.2 *There exists a process $\underline{I}$, adapted to $\underline{\sigma}(\underline{X})$, whose law is P_B^K, and a predictable process $\underline{a}^K$, with base $(K, \underline{\mathcal{K}}, P_X^K)$, such that $\underline{a}^K \square \Phi_X$ belongs to $\mathcal{I}_2[\underline{b}]$, and, for $t \in [0,1]$, fixed, but arbitrary, almost surely, with respect to P,*

$$\underline{X}(\omega,t) = \underline{S}\left[\underline{a}_K \square \Phi_X\right](\omega,t) + \underline{I}(\omega,t).$$

That representation is unique.

Proof Since $X_n(\omega,t) = \int_0^t a_n(\omega,\theta)M_n(d\theta) + B_n(\omega,t)$, one may use (Proposition) 15.1.9 to obtain that

$$X_n(\omega,t) = \int_0^t \alpha_n(\omega,\theta)M_n(d\theta) + I_n(\omega,t),$$

where α_n is a version of the conditional expectation of a_n with respect to $\underline{\sigma}\,(\underline{X})$, and I_n has the same law as B_n. Furthermore $\underline{I}$ shall have independent components [(Remark) 15.1.10].

One must now decompose α_n to obtain a_n^K. One proceeds as in (Proposition) 15.1.9, using the partition of [0, 1] with points $t_{p,k}$. One has that

$$\alpha_n = \lim_p \alpha_p^{(n)},$$

$$\alpha_p^{(n)} = \sum_i \alpha_{p,i}^{(n)},$$

$\alpha_{p,i}^{(n)}$ adapted to $\sigma_{t_{p,i}}(\underline{X}) \otimes \mathcal{B}\left(\left]t_{p,i}, t_{p,i+1}\right]\right)$, and

$$\Phi_X^{-1}\left\{\mathcal{K}_{t_{p,i}} \otimes \mathcal{B}\left(\left]t_{p,i}, t_{p,i+1}\right]\right)\right\} = \sigma_{t_{p,i}}(\underline{X}) \otimes \mathcal{B}\left(\left]t_{p,i}, t_{p,i+1}\right]\right).$$

One has thus the following situation: on

$$\left(\Omega \times [0,1], \sigma_{t_{p,i}}(\underline{X}) \otimes \mathcal{B}\left(\left]t_{p,i}, t_{p,i+1}\right]\right)\right),$$

there are two maps,

- the first, Φ_X, adapted to the range $\left(K \times [0,1], \mathcal{K}_{t_{p,i}} \otimes \mathcal{B}\left(\left]t_{p,i}, t_{p,i+1}\right]\right)\right)$,
- and the second, $\alpha_{p,i}^{(n)}$, adapted to the range $(\mathbb{R} \times [0,1], \mathcal{B}(\mathbb{R}) \otimes \mathcal{B}([0,1]))$.

The factorization theorem [138, p. 443] then says that $\alpha_{p,i}^{(n)} = a_{n,p,i}^K \Box \Phi_X$ for some adapted

$$a_{n,p,i}^{(n)} : \left(K \times [0,1], \mathcal{K}_{t_{p,i}} \otimes \mathcal{B}\left(\left]t_{p,i}, t_{p,i+1}\right]\right)\right) \longrightarrow (\mathbb{R} \times [0,1], \mathcal{B}(\mathbb{R}) \otimes \mathcal{B}([0,1])).$$

One then sets

$$a_{n,p}^K = \sum_i a_{n,p,i}^K,$$

with the property that

$$a_{n,p}^K \Box \Phi_X = \alpha_p^{(n)}.$$

$a^K_{n,p}$ has base $(K, \underline{K}, P^K_X)$, and is adapted to $\bigvee_i \mathcal{K}_{t_{p,i}} \otimes \mathcal{B}\left(]t_{p,i}, t_{p,i+1}]\right)$. It is thus predictable.

Let

$$L_n = \left\{(\omega, t) \in \Omega \times [0, 1] : \lim_p a^K_{n,p} \square \Phi_X(\omega, t) \text{ does not exist}\right\}$$
$$= \left\{(\omega, t) \in \Omega \times [0, 1] : \lim_p \alpha^{(n)}_p(\omega, t) \text{ does not exist}\right\} .$$

By construction [(Lemma) 15.1.2], $\{P \otimes M_n\}(L_n) = 0$. Let

$$L^K_n = \left\{(\underline{k}, t) \in K \times [0, 1] : \lim_p a^K_{n,p}(\underline{k}, t) \text{ does not exist}\right\} .$$

One has that $\left\{P^K_X \otimes M_n\right\}\left(L^K_n\right) = \{P \otimes M_n\}(L_n) = 0$. One may thus define a^K_n setting

$$a^K_n(\underline{k}, t) = \lim_p a^K_{n,p}(\underline{k}, t).$$

a^K_n is thus predictable, and, for $t \in [0, 1]$, fixed, but arbitrary, almost surely, with respect to P,

$$a^K_n \square \Phi_X(\omega, t) = \alpha_n(\omega, t) = E_P\left[a_n(\cdot, t) \mid \sigma_t(\underline{X})\right](\omega),$$

and thus

$$E_P\left[\left\|\underline{a}^K \square \Phi_X\right\|^2_{L_2[\underline{b}]}\right] < \infty.$$

That the representation is unique is seen as follows. Let

$$\underline{X}(\omega, t) = \underline{S}[\underline{\tilde{a}}^K \square \Phi_X](\omega, t) + \underline{\tilde{I}}(\omega, t)$$

be another representation. Then, for $n \in \mathbb{N}$, and $t \in [0, 1]$, fixed, but arbitrary, almost surely with respect to P,

$$I_n(\omega, t) - \tilde{I}_n(\omega, t) = \int_0^t \left\{\tilde{a}^K_n(\omega, \theta) - a^K_n(\omega, \theta)\right\} M_n(d\theta).$$

On the left-hand side of that latter equality, one has a continuous martingale, and, on the right-hand side, a process of bounded variation. Both sides must then be zero.

□

Remark 15.2.3 Representation (Proposition) 15.2.2 allows one to obtain, *mutatis mutandis,* the likelihoods of Chap. 13.

15.3 Strong Solutions of Stochastic Equations

From the point of view of modeling, it is important to know that, when writing down a model, that model makes sense, here, that the given probability space, the given signal, and the given noise lead to a stochastic differential equation with a solution on the space on which they are defined: *In signal processing applications, one must deal with the noisy signal that one is given, so strong solutions are required* [87, p. 197]. When the SPWN model results from the Cramér-Hida representation, requiring Lipschitz coefficients, as is often done to obtain strong solutions, is not a realistic option. The premises of an approach that is better suited to the Cramér-Hida framework (signal in the RKHS of the noise) is presented below.

Remark 15.3.1 The notation being that of the previous sections, let

- $\mathcal{N}_X = \underline{\boldsymbol{X}}^{-1}\left(\mathcal{N}(\mathcal{K}, P_X^{\mathcal{K}})\right)$,
- $\mathcal{K}_X = \left\{K_0 \Delta M_X, K_0 \in \mathcal{K}, M_X \subset N_X \in \mathcal{N}(\mathcal{K}, P_X^{\mathcal{K}})\right\}$,
- for $\underline{k} \in K$, $\Omega[\underline{k}] = \underline{\boldsymbol{X}}^{-1}(\{\underline{k}\})$.

As, for $K_0 \subseteq K$, $\underline{\boldsymbol{X}}^{-1}(K_0) = \cup_{\underline{k} \in K_0} \Omega[\underline{k}]$, any subset of Ω that is not in the latter form cannot be the inverse image of a set in K. Since inverse images preserve set operations [84, p. 11],

$$\underline{\boldsymbol{X}}^{-1}(K_0 \Delta M_X) = \underline{\boldsymbol{X}}^{-1}(K_0) \Delta \underline{\boldsymbol{X}}^{-1}(M_X).$$

Furthermore, assuming P to be complete,

$$P(\underline{\boldsymbol{X}}^{-1}(M_X)) \leq P(\underline{\boldsymbol{X}}^{-1}(N_X)) = P_X^{\mathcal{K}}(N_X) = 0.$$

But a subset of a set in $\mathcal{N}_X$ need not be of the form $\underline{\boldsymbol{X}}^{-1}(M_X)$ (one has only that $A \subseteq f^{-1}(f(A))$ [84, p. 12], so that

$$\underline{\boldsymbol{X}}^{-1}(\mathcal{K}_X) \subseteq \{A \Delta M, A \in \sigma(\underline{X}), M \subseteq N \in \mathcal{N}_X\},$$

where strict inclusion is possible. However, because of the relation [138, p. 46] $\sigma(f^{-1}(\mathcal{H})) = f^{-1}(\sigma(\mathcal{H}))$,

$$\underline{\boldsymbol{X}}^{-1}(\mathcal{K} \vee \mathcal{N}(\mathcal{K}, P_X^{\mathcal{K}})) = \sigma(\underline{X}) \vee \mathcal{N}_X.$$

Suppose now that $(\Omega, \underline{\mathcal{A}}, P) = (K, \underline{\mathcal{K}}, Q^{\mathcal{K}})$, and that $\underline{X} = \mathcal{E}_{Q^{\mathcal{K}}}$. One has then that $P_X^{\mathcal{K}} = Q^{\mathcal{K}}$, and that $\underline{\boldsymbol{X}}$ is the identity. The particularities described above are no longer needed.

Lemma 15.3.2 *Let $\underline{W}$ be a Cramér-Hida process on $(K, \underline{\mathcal{K}}, Q)$, and $t \in]0, 1[$ be fixed but arbitrary.*

1. *Let $\sigma_{[t,1]}(\underline{W})$ be the σ-algebra generated by the following vectors:*

$$\{\underline{W}(\cdot, t_2) - \underline{W}(\cdot, t_1), \{t_1, t_2\} \subseteq [0, 1] : t \leq t_1 < t_2\}.$$

Then

$$\mathcal{K}_t \text{ and } \sigma_{[t,1]}(\underline{W})$$

are independent.

2. *Let V be an integrable variable adapted to $\mathcal{K}_t$. Then, almost surely, with respect to Q,*

$$E_Q[V \mid \sigma_t(\underline{W})] = E_Q[V \mid \sigma(\underline{W})].$$

Proof (1) Let $K_{[t,1]} \in \sigma_{[t,1]}(\underline{W})$ be fixed, but arbitrary. As seen in the proof of item 4 of (Lemma) 11.4.5,

$$\underline{W}(\cdot, t + \theta) - \underline{W}(\cdot, t)$$

is a Cramér-Hida process on $[t, 1]$, and the conditional expectation

$$E_Q\left[\chi_{K_{[t,1]}} - Q(K_{[t,1]}) \mid \sigma_{[t,1]}(\underline{W})\right]$$

has a representation as a stochastic integral. Using the fact that

$$\chi_{K_{[t,1]}} - Q(K_{[t,1]})$$

is adapted, one obtains that

$$\chi_{K_{[t,1]}} = Q(K_{[t,1]}) + I_{\underline{W}}\left\{\underline{a}^K_{|[t,1]}\right\}.$$

Let $K_t \in \mathcal{K}_t$ be fixed, but arbitrary. Then

$$E_Q\left[\chi_{K_t}\chi_{K_{[t,1]}}\right] = Q(K_t)Q(K_{[t,1]}) + E_Q\left[\chi_{K_t} I_{\underline{W}}\left\{\underline{a}^K_{|[t,1]}\right\}\right].$$

Because of [264, p. 155], and the definition of $I_{\underline{W}}$,

$$\chi_{K_t} I_{\underline{W}}\left\{\underline{a}^K_{|[t,1]}\right\} = I_{\underline{W}}\left\{\chi_{K_t}\underline{a}^K_{|[t,1]}\right\},$$

and the right-hand side is zero, as the integrand is zero [264, p. 35].

Proof (2) As presently seen, one has that $\sigma(\underline{W}) = \sigma_t(\underline{W}) \vee \sigma_{[t,1]}(\underline{W})$. Let indeed $\mathcal{R}$ be the ring generated by sets of the form $R_t \cap R_{[t,1]}$, with R_t in $\sigma_t(\underline{W})$, and $R_{[t,1]}$ in $\sigma_{[t,1]}(\underline{W})$. It suffices to prove that the following measures:

$$d\mu = E_Q[V \mid \sigma_t(\underline{W})]\, dQ, \text{ and } d\nu = E_Q[V \mid \sigma(\underline{W})]\, dQ$$

are equal on $\mathcal{R}$. Now:

$$\begin{aligned}
\int_{R_t \cap R_{[t,1]}} E_Q[V \mid \sigma(\underline{W})]\, dQ &= E_Q\left[\chi_{R_t}\chi_{R_{[t,1]}} E_Q[V \mid \sigma(\underline{W})]\right] \\
&= E_Q\left[E_Q\left[\chi_{R_t}\chi_{R_{[t,1]}} V \mid \sigma(\underline{W})\right]\right] \\
&= E_Q\left[\chi_{R_t}\chi_{R_{[t,1]}} V\right] \\
&\overset{\text{item 1}}{=} E_Q\left[\chi_{R_{[t,1]}}\right] E_Q\left[\chi_{R_t} V\right] \\
&= E_Q\left[\chi_{R_{[t,1]}}\right] E_Q\left[E_Q\left[\chi_{R_t} V \mid \sigma_t(\underline{W})\right]\right] \\
&= E_Q\left[\chi_{R_{[t,1]}}\right] E_Q\left[\chi_{R_t} E_Q[V \mid \sigma_t(\underline{W})]\right] \\
&= E_Q\left[\chi_{R_{[t,1]}}\chi_{R_t} E_Q[V \mid \sigma_t(\underline{W})]\right] \\
&= \int_{R_t \cap R_{[t,1]}} E_Q[V \mid \sigma_t(\underline{W})]\, dQ.
\end{aligned}$$

□

Proposition 15.3.3 *Let* $(\Omega, \underline{\mathcal{A}}, P)$ *be a complete probability space such that* $\mathcal{N}(\mathcal{A}, P) \subseteq \mathcal{A}_t$ *for* $t \in [0, 1]$*, fixed, but arbitrary. Let* $\underline{B}$ *and* $\underline{X}$ *be adapted processes related in the following way:*

(a) *for* $t \in [0, 1]$*, fixed, but arbitrary, almost surely, with respect to P,*

$$\underline{X}(\omega, t) = \underline{S}[\underline{a}^K \square \Phi_X](\omega, t) + \underline{B}(\omega, t),$$

where $\underline{a}^K$ *is progressively measurable for* $(K, \underline{\mathcal{K}})$*, and, for* $\underline{k} \in K$*, fixed, but arbitrary,*

$$\left\|\underline{a}^K(\underline{k}, \cdot)\right\|^2_{L_2[\underline{b}]} < \infty;$$

(b) $\underline{B}$ *is a Cramér-Hida process, adapted to* $\underline{\sigma}^o(\underline{X})$*, with decomposition*

$$\underline{B} = \underline{B}_X \square \Phi_X,$$

where [(Proposition) 13.2.2] $\underline{B}_X$ *is a Cramér-Hida process, with law* P^K_B *for* $(K, \underline{\mathcal{K}}, P^K_X)$.

Suppose that, when $\sigma(\underline{B}_X)$ does not equal $\mathcal{K}$, there exists a probability Q^K on $\mathcal{K}$ such that

(A) $Q^K \neq P_X^K$,
(B) $\{Q^K\}_{|\sigma(\underline{B}_X)} = \{P_X^K\}_{|\sigma(\underline{B}_X)}$,
(C) $\underline{B}_X$ *is a martingale for* $(K, \underline{\mathcal{K}}, Q^K)$.

Let $\mathcal{N}_t^X = \underline{\boldsymbol{X}}^{-1}(\mathcal{N}(\mathcal{K}_t, P_X^K)) = \mathcal{N}(\sigma_t(\underline{X}), P)$,

$$\mathcal{C}_t = \sigma_t(\underline{B}) \vee \mathcal{N}_t^X, \quad \textit{and} \quad \mathcal{C}_t^X = \sigma_t(\underline{B}_X) \vee \mathcal{N}(\mathcal{K}_t, P_X^K),$$

$$\mathcal{D}_t = \sigma_t(\underline{X}) \vee \mathcal{N}_t^X, \quad \textit{and} \quad \mathcal{D}_t^X = \quad \mathcal{K}_t \quad \vee \mathcal{N}(\mathcal{K}_t, P_X^K).$$

Then, for $t \in [0, 1]$, *fixed, but arbitrary,*

$$\mathcal{C}_t = \mathcal{D}_t.$$

Proof It shall be done in two steps.

Step 1: $\mathcal{C}_1 = \mathcal{D}_1$.

As seen in (Remark) 15.3.1,

$$\mathcal{C}_t = \underline{\boldsymbol{X}}^{-1}(\mathcal{C}_t^X), \text{ and } \mathcal{D}_t = \underline{\boldsymbol{X}}^{-1}(\mathcal{D}_t^X).$$

It is sufficient then to prove that $\mathcal{C}_1^X = \mathcal{D}_1^X$. The assumption, and (Proposition) 13.2.4, allow one to write that

$$\underline{\mathcal{E}}_{P_X^K} - \underline{S}[\underline{a}^K] = \underline{B}_X,$$

which indicates that $(\underline{B}_X, P_X^K)$ is the unique weak solution of the following formal equation, with unknown $\underline{Y}$: $\underline{Y} = \underline{S}[\underline{a}^K \square \Phi_Y] + \underline{B}$.

Suppose that $\mathcal{C}_1^X = \mathcal{D}_1^X$ does not hold. Then, according to the assumption, there is Q^K such that $\underline{B}_X$ is a martingale on $(K, \underline{\mathcal{K}}, Q^K)$. Since the restrictions to $\sigma(\underline{B}_X)$ of Q^K and P_X^K are, by assumption, equal, $\underline{B}_X$ is a also a Cramér-Hida process with respect to Q^K. But then $\underline{Y} = \underline{S}[\underline{a}^K \square \Phi_Y] + \underline{B}$ has two solutions, a contradiction.

Proof Step 2: For $t \in [0, 1[$, fixed, but arbitrary, $\mathcal{C}_t = \mathcal{D}_t$.

Let V be a random variable adapted to $\mathcal{K}_t$. Because of (Lemma) 15.3.2,

$$E_{P_X^K}\left[V \mid \mathcal{C}_t^X\right] = E_{P_X^K}\left[V \mid \mathcal{C}_1^X\right].$$

Since, by step 1, $\mathcal{C}_1 = \mathcal{D}_1$,

$$E_{P_X^K}\left[V \mid \mathcal{C}_t^X\right] = V,$$

and thus, since $\mathcal{C}_t^X \subseteq \mathcal{D}_t^X$, $\mathcal{C}_t^X = \mathcal{D}_t^X$. The result follows by taking the inverse image by $\underline{X}$. □

Definition 15.3.4 Let $(\Omega, \underline{\mathcal{A}}, P)$ be a probability space with filtration, $\underline{B}$ an adapted Cramér-Hida process, and $\underline{a}^K$ a progressively measurable process for $\underline{\mathcal{K}}$. A strong solution to the formal equation $\underline{Y} = \underline{S}[\,\underline{a}_K \square \Phi_Y] + \underline{B}$ is a process $\underline{X}$, adapted to $(\Omega, \underline{\mathcal{A}}, P)$, such that, for $t \in [0, 1]$, fixed, but arbitrary, almost surely, with respect to P,

$$\underline{X}(\omega, t) = \underline{S}[\underline{a}^K \square \Phi_X](\omega, t) + \underline{B}(\omega, t).$$

A strong solution is unique when, given two solutions, $\underline{X}_1$ and $\underline{X}_2$, almost surely, with respect to P, $\underline{X}_1[\omega] = \underline{X}_2[\omega]$.

Lemma 15.3.5 *Let $f : (\Omega, \mathcal{A}) \longrightarrow (X, \mathcal{X})$ be an adapted map between measurable spaces, and let $\mathcal{B} = f^{-1}(\mathcal{X})$. Suppose that Λ is a compact, metrizable space, and that, for $\lambda \in \Lambda$, fixed, but arbitrary, $g_\lambda : (\Omega, \mathcal{B}) \longrightarrow (l_2, \mathcal{B}(l_2))$ is adapted. Suppose finally that, for $\omega \in \Omega$, fixed, but arbitrary, $\lambda \mapsto g_\lambda(\omega)$ is continuous. There exists then adapted maps $\gamma_\lambda : (X, \mathcal{X}) \longrightarrow (l_2, \mathcal{B}(l_2))$ such that*

1. *$g_\lambda = \gamma_\lambda \circ f$;*
2. *for $x \in X$, fixed, but arbitrary, $\lambda \mapsto \gamma_\lambda(x)$ is continuous.*

Proof Let, as seen in Sects. 12.1 and 12.2, $C(\Lambda, l_2)$ be the Banach space of continuous functions, with domain Λ, and range in l_2, with the supremum norm. It is a separable space. The Borel σ-algebra of that Banach space shall be denoted $\mathcal{C}$. It is generated by the evaluation maps $\underline{\mathcal{E}}_\lambda(\underline{c}) = \underline{c}(\lambda) \in l_2$.

An elementary function ϕ from a measurable space $(M, \mathcal{M})$ to a metric space S is a function whose range is at most countable [260, p. 11]. When S is separable, every measurable function with range in S is the uniform limit of a sequence of elementary functions [260, p. 12]. These elementary functions are obtained as follows. Suppose that $\phi : M \longrightarrow S$. Let $\{s_n, n \in \mathbb{N}\}$ be a set dense in S. Set

$$\begin{aligned}
S_{n,1} &= \left\{s \in S : d_S(s, s_1) < n^{-1}\right\}, \\
S_{n,2} &= \left\{s \in S : d_S(s, s_1) \geq n^{-1}, d_S(s, s_2) < n^{-1}\right\}, \\
&\dots \\
S_{n,p} &= \left\{s \in S : d_S(s, s_1) \geq n^{-1}, \dots, d_S(s, s_{p-1}) \geq n^{-1}, d_S(s, s_p) < n^{-1}\right\}, \\
&\dots
\end{aligned}$$

Let, when $s \in S_{n,p}$, $\sigma_n^S(s) = s_p$. The uniform limit of the sequence σ_n^S is the identity. One then chooses $\phi_n^S = \sigma_n^S \circ \phi$. Thus

$$(\phi_n^S)^{-1}(\{s_p\}) = \phi^{-1}(S_{n,p}).$$

Let $\boldsymbol{g} : \Omega \longrightarrow C(\Lambda, l_2)$ be defined using the following assignment:

$$\boldsymbol{g}(\omega) = \{g_\lambda(\omega), \lambda \in \Lambda\}.$$

Since $\underline{\mathcal{E}}_\lambda(\boldsymbol{g}(\omega)) = g_\lambda(\omega)$, $\boldsymbol{g}$ is adapted to $\mathcal{B}$ and $\mathcal{C}$. Let thus $\Gamma = \{\underline{c}_n, n \in \mathbb{N}\}$ be a set dense in $C(\Lambda, l_2)$, $C^\Gamma_{n,p}$, the set corresponding to $S_{n,p}$, introduced above on S, and ξ^Γ_n, one of the maps, on $C(\Lambda, l_2)$, corresponding to the σ^S_n's above. Let also

$$\boldsymbol{g}_n = \xi^\Gamma_n \circ \boldsymbol{g}.$$

As above, $\boldsymbol{g}_n(\omega) = \underline{c}_p$ on $\boldsymbol{g}^{-1}(C^\Gamma_{n,p})$. But, since $\mathcal{B} = f^{-1}(\mathcal{X})$, for some set $X_{n,p} \in \mathcal{X}$,

$$\boldsymbol{g}^{-1}(C^\Gamma_{n,p}) = f^{-1}(X_{n,p}).$$

Let, for $x \in X$, fixed, but arbitrary, $\boldsymbol{\gamma}_n(x) = \underline{c}_p$ when $x \in X_{n,p}$, and, outside of $\cup_p X_{n,p}$, when necessary, the zero function of $C(\Lambda, l_2)$. Let $\omega[x]$ be an element of Ω such that $f(\omega[x]) = x$. When $x \in X_{n,p}$, $\boldsymbol{\gamma}_n(x) = \underline{c}_p$, and

$$\omega[x] \in f^{-1}(X_{n,p}) = \boldsymbol{g}^{-1}(C^\Gamma_{n,p}),$$

so that $\boldsymbol{g}_n(\omega[x]) = \underline{c}_p$, and $\boldsymbol{\gamma}_n(x) = \boldsymbol{g}_n(\omega[x])$. Since

$$f^{-1}\left(\cup_p X_{n,p}\right) = \boldsymbol{g}^{-1}\left(C(\Lambda, l_2)\right) = \Omega,$$

$\boldsymbol{\gamma}_n(x) = \boldsymbol{g}_n(\omega[x])$ whenever $f(\omega[x]) = x$, that is, $\boldsymbol{\gamma}_n(f(\omega[x])) = \boldsymbol{g}_n(\omega[x])$.

Let X^Γ_u be the set of x's at which $\{\boldsymbol{\gamma_n}, n \in \mathbb{N}\}$ converges uniformly on Λ. Since

$$X^\Gamma_u = \bigcap_m \bigcup_n \bigcap_{p \geq n} \left\{ x \in X : \sup_\lambda \left\| \underline{\mathcal{E}}_\lambda(\boldsymbol{\gamma_p}(x)) - \underline{\mathcal{E}}_\lambda(\boldsymbol{\gamma_n}(x)) \right\|_{l_2} \leq m^{-1} \right\},$$

$X^\Gamma_u \in \mathcal{X}$. One has furthermore, as presently seen, that $\mathcal{R}_f \subseteq X^\Gamma_u$. Indeed, given a fixed, but arbitrary $x \in \mathcal{R}_f$, and $\omega[x] \in f^{-1}(x)$, since $\{\boldsymbol{g_n}(\omega[x]), n \in \mathbb{N}\}$ converges uniformly to $\boldsymbol{g}(\omega[x])$, and that $\boldsymbol{\gamma_n}(x) = \boldsymbol{g_n}(\omega[x])$, then $x \in X^\Gamma_c$.

Setting

$$\underline{\mathcal{E}}_\lambda(\boldsymbol{\gamma}(x)) = \begin{cases} \lim_n \underline{\mathcal{E}}_\lambda(\boldsymbol{\gamma_n}(x)) & \text{when } x \in X_c \\ \underline{0}_{l_2} & \text{when } x \in X^c_c \end{cases},$$

one has the required decomposition. □

Proposition 15.3.6 *Let $(\Omega, \underline{\mathcal{A}}, P)$ be a complete probability space such that, for $t \in [0,1]$, $\mathcal{N}(\mathcal{A}, P) \subseteq \mathcal{A}_t$. Let $\underline{B} : \Omega \times [0,1]$ be a Cramér-Hida process adapted to $\underline{\mathcal{A}}$. Suppose that:*

(a) *$\underline{a}^K$ is progressively measurable for $\underline{\mathcal{K}}$, and such that, for $\underline{k} \in K$, fixed, but arbitrary,*

$$\left\|\underline{a}^K(\underline{k}, \cdot)\right\|^2_{L_2[\underline{b}]} < \infty,$$

(b) $E_{P^K_B}\left[e^{I_{\mathcal{E}_{P^K_B}}\{\underline{a}^K(\cdot,1)\} - \frac{1}{2}\left\|\underline{a}^K(\cdot,\cdot)\right\|^2_{L_2[\underline{b}]}}\right] = 1,$

so that a weak solution $(\underline{N}, Q^K)$ to the formal equation

$$\underline{Y} = \underline{S}[\underline{a}_K \square \Phi_Y] + \underline{N}$$

exists and is unique [(Proposition) 14.1.3], *that is,*

(i) *$\underline{N}$ is a Cramér-Hida process for $(K, \underline{\mathcal{K}}, Q^K)$;*
(ii) *for $t \in [0,1]$, fixed, but arbitrary, almost surely with respect to Q^K,*

$$\underline{\mathcal{E}}_{Q^K}(\underline{k}, t) = \underline{S}[\underline{a}^K](\underline{k}, t) + \underline{N}(\underline{k}, t).$$

Suppose furthermore that when $\sigma(\underline{N})$ does not equal $\mathcal{K}$, there exists a probability Π^K on $\mathcal{K}$ such that

(A) $\Pi^K \neq Q^K$,
(B) $\{\Pi^K\}_{|\sigma(\underline{N})} = \{Q^K\}_{|\sigma(\underline{N})}$,
(C) *$\underline{N}$ is a martingale for $(K, \underline{\mathcal{K}}, \Pi^K)$.*

There is then a strong solution $\underline{X}$ for $(\Omega, \underline{\mathcal{A}}, P)$ and $\underline{B}$.

Proof Result (Proposition) 15.3.3 yields that $\underline{\sigma}^o(\underline{N}) = \underline{\mathcal{K}}^o$, the completion being with respect to Q^K, and (Remark) 15.3.1 that $\underline{\sigma}^o(\underline{N}) = \underline{\boldsymbol{N}}^{-1}(\underline{\mathcal{K}}^o)$. There is then, because of (Lemma) 15.3.5, a decomposition of $\underline{\mathcal{E}}_{Q^K}$ in the following form:

$$\underline{\mathcal{E}}_{Q^K}(\underline{k}, t) = \underline{N}_{\mathcal{E}}\left(\underline{N}[\underline{k}], t\right),$$

with the property that, almost surely, with respect to Q^K,

$$\underline{\boldsymbol{\mathcal{E}}}_{Q^K} = \underline{\boldsymbol{N}}_{\mathcal{E}} \circ \underline{\boldsymbol{N}}.$$

The equation

$$\underline{\mathcal{E}}_{Q^K}(\underline{k}, t) = \underline{S}[\underline{a}^K](\underline{k}, t) + \underline{N}(\underline{k}, t)$$

yields then that

$$\underline{N}(\underline{k}, t) = -\underline{S}[\underline{a}^{K}](\underline{k}, t) + \underline{\mathcal{E}}_{Q^K}(\underline{k}, t)$$

has the form

$$\underline{N}(\underline{k}, t) = -\underline{S}[\underline{a}^{K}](\underline{\boldsymbol{N}}_{\mathcal{E}} \circ \underline{N}[\underline{k}], t) + \underline{N}_{\mathcal{E}}\,(\underline{N}[\underline{k}], t)\,,$$

which means that, almost surely, with respect to P_B^K,

$$\underline{\mathcal{E}}_{P_B^K}(\underline{k}, t) = -\underline{S}[\underline{a}^{K}](\underline{N}_{\mathcal{E}}[\underline{k}], t) + \underline{N}_{\mathcal{E}}(\underline{k}, t).$$

Let now $\underline{B}$ be a Cramér-Hida process for $(\Omega, \underline{\mathcal{A}}, P)$. It yields, when combined with the last equality, that, almost surely, with respect to P,

$$\underline{B}(\omega, t) = -\underline{S}[\underline{a}^{K}](\underline{\boldsymbol{N}}_{\mathcal{E}} \circ \underline{B}[\omega], t) + \underline{N}_{\mathcal{E}}(\underline{B}[\omega], t).$$

The process $\underline{X}(\omega, t) = \underline{N}_{\mathcal{E}}(\underline{B}[\omega], t)$ provides then the strong solution. □

Corollary 15.3.7 *Given the first two conditions of* (Proposition) 15.3.6 *insuring the existence of a weak solution, and the formal equation* $\underline{Y} = \underline{S}[\underline{a}^K \square \Phi_Y] + \underline{N}$, *one has that the following statements are equivalent* ($\underline{B}_X$ is the $\underline{N}$ of the weak solution $\underline{X}$):

1. *given a fixed, but arbitrary Cramér-Hida process* $\underline{B}$ *with law* P_B^K *for the probability space* $(\Omega, \underline{\mathcal{A}}, P)$, *the equation* $\underline{Y} = \underline{S}[\,\underline{a}^K \square\, \Phi_Y] + \underline{N}$ *admits* $\underline{X}$ *as strong solution;*
2. *for* $t \in [0, 1]$, *fixed, but arbitrary,* $\mathcal{K}_t$ *and* $\sigma_t(\underline{B}_X) \vee \mathcal{N}(\mathcal{K}_t, P_X^K)$ *have equal completions with respect to* P_B^K.

Proof Item 2 implies item 1 since the assumption of item 2 is the consequence of the third assumption of (Proposition) 15.3.6 that leads to the existence of a strong solution.

Suppose that item 1 obtains. Given $(\Omega, \underline{\mathcal{A}}, P)$ and $\underline{B}$, let $\underline{X}$ be a strong solution. Consider the filtration built as follows:

$$\sigma_t(\underline{B}) \vee \mathcal{N}(\sigma_t(\underline{X}), P).$$

$\underline{X}$ is also a strong solution with respect to the latter filtration. Consequently, since a solution is adapted, $\sigma_t(\underline{X}) \subseteq \sigma_t(\underline{B}) \vee \mathcal{N}(\sigma_t(\underline{X}), P)$. The reverse is true since a strong solution is also a weak solution. One then uses the fact that those filtrations are inverse images by $\underline{\boldsymbol{X}}$ of the filtrations on K given in the statement of item 2. □

One has also the following result [187, p. 41]:

Proposition 15.3.8 *In the statement of* (Proposition) 15.3.6 *omit the second condition, that which requires the expectation of the exponential to be one. Suppose that* $(\Omega, \underline{\mathcal{A}}, P)$ *is complete, and that, for* $t \in [0, 1]$, *fixed, but arbitrary,* $\mathcal{N}(\mathcal{A}, P)$ *is included in* $\mathcal{A}_t$. *Then, when a strong solution to* $\underline{Y} = \underline{S}[\underline{a}^K \square \Phi_Y] + \underline{B}$ *of (Corollary)* 15.3.7 *exists, it is unique.*

Chapter 16
Scope of Signal Plus "White Noise" Model (III)

Cramér-Hida processes have components that are continuous, Gaussian martingales. One may wonder what happens when one drops the Gaussian assumption, since, as seen, continuous martingales are time changed Wiener processes [264, p. 213]. One shall see that Girsanov's theorem loses then its strength to the point that the family of signals admissible for the computation of likelihoods becomes practically useless (signals must depend on the quadratic variation of the noise). But the road to establishing that fact is long and tortuous, though quite interesting. It also often requires the assumption that probability spaces are complete. *The standard assumption for the section shall thus be that basic probability spaces are complete, and that σ-algebras contained in the "mother" σ-algebras contain the sets of measure zero of the latter.*

16.1 Separable Families of Sets

To a large extent, what follows requires separable σ-algebras, that is, σ-algebras generated by countable families of elements. There are two notions of separability, one without probability, and one with probability. The difference is that zero measure sets are admitted in the second case, and one then speaks of σ-algebras that are essentially separable.

Let Ω be a set. A paving of Ω is a family of subsets of Ω that contains at least one set. A σ-algebra $\mathcal{A}$ of subsets of Ω is separable when there is a countable paving $\mathcal{B}$ of Ω such that $\sigma(\mathcal{B}) = \mathcal{A}$ [138, p. 66]. One has that:

1. [138, p. 66] When $\mathcal{A}$ is separable, it is generated by a countable algebra.
2. [138, p. 91] $\mathcal{A}$ is separable if, and only if, there exits a measurable function $f : \Omega \longrightarrow \mathbb{R}$ such that $\mathcal{A} = f^{-1}(\mathcal{B}(\mathbb{R}))$. A function f that is frequently used in such a context is the Marczewski function, an infinite linear combination of indicator functions of sets, basically those which generate the σ-algebra.

A.F. Gualtierotti, *Detection of Random Signals in Dependent Gaussian Noise*,
DOI 10.1007/978-3-319-22315-5_16

3. [138, p. 65] Given the paving $\mathcal{B}$, let

   (a) $\mathcal{E}(\mathcal{B}) = \{B_1 \cup B_2^c, B_1 \text{ and } B_2 \in \mathcal{B}\}$,
   (b) $\mathcal{B}_0 = \mathcal{B}$,
   (c) $\mathcal{B}_1 = \mathcal{B}_0 \cup \{\emptyset, \Omega\}$,
   (d) for integer $n \geq 1$, $\mathcal{B}_{n+1} = \mathcal{E}(\mathcal{B}_n)$.

   Then:

   (i) for $n \in \{0\} \cup \mathbb{N}$, $\mathcal{B}_n \subseteq \mathcal{B}_{n+1}$,
   (ii) the algebra generated by $\mathcal{B}$ is $\cup_{n=1}^{\infty} \mathcal{B}_n$;
   (iii) when $\mathcal{B}$ is countable, so is the algebra it generates.

4. [138, p. 66] Let $\mathcal{B}$ be the paving of Ω whose members are its countable subsets, and subsets whose complement is countable. Then:

   (i) $\mathcal{B}$ is a σ-algebra;
   (ii) when Ω is uncountable, $\mathcal{B}$ is not separable;
   (iii) [224, p. 15] let $\Omega = \mathbb{R}$, $\mathcal{A}$ be the Borel sets of $\mathbb{R}$, and $\mathcal{B}$, the paving just considered: one has thus a separable σ-algebra containing a σ-algebra that is not separable.

Suppose that $\mathcal{B}$ is a σ-algebra contained in the separable σ-algebra $\mathcal{A}$. To be able to conclude that $\mathcal{B}$ is separable also, one must use another definition of separability, that given in the next section, which is obtained integrating sets of measure zero into the definition. Thus, when one does not want to integrate those sets, one must assume separately that $\mathcal{B}$ is separable. It shall be assumed that probability spaces are complete, as farther use shall require, though that assumption is not required, immediately, and everywhere.

Let $(\Omega, \mathcal{A}, P)$ be a complete probability space; $\mathcal{L}_0(\Omega, \mathcal{A}, P)$ be the family of adapted maps $f : \Omega \longrightarrow \overline{\mathbb{R}}$ such that $\{\omega \in \Omega : |f(\omega)| = \infty\}$ has probability zero; $\mathcal{N}_0(\Omega, \mathcal{A}, P)$, that of maps such that $\{\omega \in \Omega : f(\omega) \neq 0\}$ has probability zero; and $L_0(\Omega, \mathcal{A}, P)$, the quotient of $\mathcal{L}_0(\Omega, \mathcal{A}, P)$ by $\mathcal{N}_0(\Omega, \mathcal{A}, P)$. The equivalence class of f is denoted $[f]_{0,P}$, and, when ϕ denotes an equivalence class,

$$\dot{\phi} \in \phi$$

denotes one of its elements. The topology of $L_0(\Omega, \mathcal{A}, P)$ is that of convergence in probability and is given by a distance. One convenient choice is the following one [41, p. 406]:

$$d_{0,P}([f]_{0,P}, [g]_{0,P}) = \int_{\Omega} \{|f(\omega) - g(\omega)| \wedge 1\} \, P(d\omega).$$

$L_0(\Omega, \mathcal{A}, P)$ is thus a complete, metric linear space (it is also a complete F-lattice, [276, p. 369]) and that distance corresponds to the obvious quasi-norm (let g be zero) [46, p. 226].

Another way to introduce a metric space related to $(\Omega, \mathcal{A}, P)$ is as follows [46, p. 173]. Two sets A_1 and A_2 of $\mathcal{A}$ are (P-)equivalent when $P(A_1 \Delta A_2) = 0$. Let $[A]_P$ denote the equivalence class of $A \in \mathcal{A}$. The collection of equivalence classes shall be denoted $\mathcal{A}^{eP}$, and P^{eP} is defined on $\mathcal{A}^{eP}$ setting

$$P^{eP}([A]_P) = P(A).$$

One thus gets the probability algebra associated with $(\Omega, \mathcal{A}, P)$. All Boolean operations make then sense. A distance may be defined on $\mathcal{A}^{eP}$ setting

$$\delta_P([A_1]_P, [A_2]_P) = P^{eP}([A_1 \Delta A_2]_P).$$

The resulting metric space is complete [46, p. 190].

Let $J_P : (\mathcal{A}^{eP}, P^{eP}) \longrightarrow L_0(\Omega, \mathcal{A}, P)$ be defined using the following relation:

$$J_P([A]_P) = [\chi_A]_{0,P}.$$

J_P is an isometry as

$$\begin{aligned}
\delta_P([A_1]_P, [A_2]_P) &= P(A_1 \Delta A_2) \\
&= E_P\left[\left|\chi_{A_1} - \chi_{A_2}\right|\right] \\
&= E_P\left[\left|\chi_{A_1} - \chi_{A_2}\right| \wedge 1\right] \\
&= d_{0,P}\left(\left[\chi_{A_1}\right]_{0,P}, \left[\chi_{A_2}\right]_{0,P}\right) \\
&= d_{0,P}\left(J_P\left([A_1]_{0,P}\right), J_P\left([A_2]_{0,P}\right)\right).
\end{aligned}$$

As measurable functions are limits of simple ones, $\mathcal{R}[J_P]$ is total in $L_0(\Omega, \mathcal{A}, P)$. As a consequence both

$$\left(\mathcal{A}^{eP}, P^{eP}\right), \quad \text{and} \quad L_0(\Omega, \mathcal{A}, P)$$

are simultaneously separable or not [46, p. 376].

Definition 16.1.1 $(\Omega, \mathcal{A}, P)$ is essentially separable when both

$$\left(\mathcal{A}^{eP}, P^{eP}\right) \text{ and } L_0(\Omega, \mathcal{A}, P)$$

are separable.

Remark 16.1.2 $(\Omega, \mathcal{A}, P)$ is essentially separable if, and only, if $\mathcal{A}$ is generated by a random variable, for example the Marczewski function (and zero measure sets).

16.2 Morphisms and Embeddings for Probability Spaces

Isomorphisms provide the necessary flexibility, in matters to be considered.

Definition 16.2.1 Let $(\Omega, \mathcal{A}, P)$ and $(\Theta, \mathcal{B}, Q)$ be two complete probability spaces. Let $\Psi : L_0(\Omega, \mathcal{A}, P) \longrightarrow L_0(\Theta, \mathcal{B}, Q)$ be a map such that, for fixed, but arbitrary,

$$\left[\begin{array}{l} n \in \mathbb{N}, \\ \phi : \mathbb{R}^n \longrightarrow \mathbb{R}, \text{ adapted to the Borel sets,} \\ \{f_1, \ldots, f_n\} \subseteq L_0(\Omega, \mathcal{A}, P), \end{array}\right.$$

one has that, $\dot{\Psi}(f)$ denoting an element in the class $\Psi(f)$,

$$\Psi\left(\left[\phi(\dot{f}_1, \ldots, \dot{f}_n)\right]_{0,P}\right) = \left[\phi\left(\dot{\Psi}(f_1), \ldots, \dot{\Psi}(f_n)\right)\right]_{0,Q}.$$

Ψ is called an almost sure morphism of $(\Omega, \mathcal{A}, P)$ towards $(\Theta, \mathcal{B}, Q)$.

Remark 16.2.2 Definition 16.2.1 asserts a form of commutative diagram. Let indeed Ψ_n send $(f_1, \ldots, f_n)$ to $(\Psi(f_1), \ldots, \Psi(f_n))$, dot_n be the operation that chooses an equivalence class of n functions, and *class* be the operation that sends an element to its equivalence class. Then

$$\Psi \circ (class \circ \phi \circ dot_n) = (class \circ \phi \circ dot_n) \circ \Psi_n.$$

Example 16.2.3 Let $(\Theta, \mathcal{B}, Q) = (\Omega, \mathcal{A}, Q)$, with $Q \ll P$. One has then that $\mathcal{N}_{0,P}(\Omega, \mathcal{A}, P) \subseteq \mathcal{N}_{0,Q}(\Omega, \mathcal{A}, Q)$, so that $[f]_{0,P} \subseteq [f]_{0,Q}$. The following map:

$$\Psi\left([f]_{0,P}\right) = [f]_{0,Q}$$

is thus well defined, and

$$\Psi\left(\left[\phi\left(\dot{f}_1, \ldots, \dot{f}_n\right)\right]_{0,P}\right) = \left[\phi\left(\dot{f}_1, \ldots, \dot{f}_n\right)\right]_{0,Q} = \left[\phi\left(\dot{\Psi}(f_1), \ldots, \dot{\Psi}(f_n)\right)\right]_{0,Q}.$$

Example 16.2.4 Let $(\Theta, \mathcal{B}, Q) = (\Omega, \mathcal{B}, P)$, with $\mathcal{A} \subseteq \mathcal{B}$. Let f be adapted to $\mathcal{A}$, and g be almost surely equal to f, with respect to P. g is then adapted to $\mathcal{A}$. Consequently $[f]_{0,P}$ is also an element of $L_0(\Omega, \mathcal{B}, P)$, and Ψ is the inclusion map.

Example 16.2.5 Let $f : \Omega \longrightarrow X$ be a map into some space X, and g, also with values in X, be similarly defined on Θ. Let $\mathcal{A}$ and $\mathcal{B}$ be generated by f, respectively g, and some σ-algebra $\mathcal{X}$ on X. Suppose that $Q_g = Q \circ g^{-1} \ll P_f = P \circ f^{-1}$. Let, for a random variable ϕ, adapted to $\mathcal{X}$,

$$\Psi\left([\phi(f)]_{0,P}\right) = [\phi(g)]_{0,Q}.$$

Ψ is well defined. Indeed, when, almost surely, with respect to P, $\phi(f) = \psi(f)$,

$$P_f(\phi \neq \psi) = P(\phi(f) \neq \psi(f)) = 0,$$

so that $0 = Q_g(\phi \neq \psi) = Q(\phi(g) \neq \psi(g))$, and thus, almost surely, with respect to Q, $\phi(g) = \psi(g)$.

Proposition 16.2.6 *The map* Ψ *of (Definition)* 16.2.1 *has the following properties:*

1. Ψ *is linear, increasing, and sends constants to constants;*
2. Ψ *is continuous for convergence in probability;*
3. *when* f *is the class of a random variable, the law of* $\Psi(f)$ *is absolutely continuous with respect to that of* f*;*
4. *when* f *stems from the indicator of a set,* $\Psi(f)$ *stems also from the indicator of a set;*
5. *the composition of two such morphisms is a morphism.*

Proof Let, for any set S, χ_S denote the indicator of S. Given any adapted function $f : \Omega \longrightarrow \overline{\mathbb{R}}$ which is almost surely finite, the set

$$\{\omega \in \Omega : \chi_\Omega(\omega) - \chi_{\mathbb{R}}(f(\omega)) \neq 0\}$$

has P-probability zero. Thus

$$[\chi_\Omega]_{0,P} = [\chi_{\mathbb{R}}(f)]_{0,P},$$

so that

$$\Psi\left([\chi_\Omega]_{0,P}\right) = \Psi\left([\chi_{\mathbb{R}}(f)]_{0,P}\right) = \left[\chi_{\mathbb{R}}\left(\dot{\Psi}\left([f]_{0,P}\right)\right)\right]_{0,Q} = [\chi_\Theta]_{0,Q}.$$

Thus, if Ψ is proved linear, it will send constants to constants, and in particular zero to zero. Let now, in (Definition) 16.2.1, $\phi(x, y) = ax + by$. Then

$$\begin{aligned}
\Psi(\alpha f_1 + \beta f_2) &= \Psi\left(\left[\alpha \dot{f_1} + \beta \dot{f_2}\right]_{0,P}\right) \\
&= \Psi\left(\left[\phi(\dot{f_1}, \dot{f_2})\right]_{0,P}\right) \\
&= \left[\phi\left(\dot{\Psi}(f_1), \dot{\Psi}(f_2)\right)\right]_{0,Q} \\
&= \left[\alpha \dot{\Psi}(f_1) + \beta \dot{\Psi}(f_2)\right]_{0,Q} \\
&= \alpha\left[\dot{\Psi}(f_1)\right]_{0,Q} + \beta\left[\dot{\Psi}(f_2)\right]_{0,Q} \\
&= \alpha \Psi(f_1) + \beta \Psi(f_2).
\end{aligned}$$

Thus Ψ is linear.

The relation $f \leq g$ extends to equivalence classes: $[f]_{0,P} \leq [g]_{0,P}$ if, and only if, $f \leq g$, except possibly on a set of P-probability zero. The relation $\leq$ is represented by the half space H of $\mathbb{R}^2$, situated above, and to the left, of the line $x = y$: $x \leq y$ if, and only if, $\chi_H(x, y) = 1$, a Borel measurable function. Thus $\chi_H(f(x), g(x)) = 1$ if, and only if, $f(x) \leq f(y)$. So

$$\Psi\left([\chi_H(f, g)]_{0,P}\right) = \left[\chi_H(\dot{\Psi}\left([f]_{0,P}\right), \dot{\Psi}\left([f]_{0,P}\right)\right]_{0,Q},$$

and both, argument and value, are simultaneously one. Ψ is thus increasing.

As one works in a linear, metric space, and that Ψ is linear, to prove its continuity, it suffices [228, p. 23] to consider a sequence $\{f_n, n \in \mathbb{N}\}$ that converges to zero in probability, and to prove that the following sequence:

$$\left\{\dot{\Psi}\left([f_n]_{0,P}\right), n \in \mathbb{N}\right\}$$

converges to zero in probability. In fact, it suffices to prove that each subsequence of

$$\left\{\dot{\Psi}\left([f_n]_{0,P}\right), n \in \mathbb{N}\right\}$$

contains a subsequence that converges almost surely [39, p. 231]. Suppose, which is no restriction, that

$$\left\{\dot{\Psi}\left([f_n]_{0,P}\right), n \in \mathbb{N}\right\}$$

is such a subsequence. Since $\lim_n E_P\left[|f_n| \wedge 1\right] = 0$, there is a subsequence such that

$$\sum_{p=1}^{\infty} E_P\left[\left|f_{n_p}\right| \wedge 1\right] < \infty.$$

But then, almost surely, with respect to P,

$$\sum_{p=1}^{\infty} \left\{\left|f_{n_p}\right| \wedge 1\right\} < \infty.$$

Since the general term of a convergent series of positive terms must go to zero, eventually $\left|f_{n_p}\right| \wedge 1 = \left|f_{n_p}\right|$, so that, almost surely,

$$\sum_{p=1}^{\infty} \left|f_{n_p}\right| < \infty,$$

and, letting S denote that latter sum, for $m \in \mathbb{N}$, fixed, but arbitrary,

$$\sum_{i=1}^{m} |f_{n_i}| \leq S.$$

Since Ψ preserves order, with $\phi(x_1, \ldots, x_m) = \sum_{i=1}^{m} |x_i|$,

$$\sum_{i=1}^{m} \left|\dot{\Psi}\left([f_{n_i}]_{0,P}\right)\right| \leq \dot{\Psi}\left([S]_{0,P}\right),$$

so that, almost surely, with respect to Q,

$$\sum_{i=1}^{\infty} \left|\dot{\Psi}\left([f_{n_i}]_{0,P}\right)\right| < \infty,$$

that is,

$$\lim_{p} \dot{\Psi}\left([f_{n_i}]_{0,P}\right) = 0.$$

Suppose that B is a Borel set for which $P(\dot{f} \in B) = 0$, that is, almost surely, with respect to P, $\chi_B(\dot{f}) = 0$. Then, since

$$\Psi\left(\left[\chi_B(\dot{f})\right]_{0,P}\right) = \left[\chi_B\left(\dot{\Psi}(f)\right)\right]_{0,Q},$$

and that zero is preserved by the morphism, almost surely, with respect to Q,

$$\chi_B\left(\dot{\Psi}(f)\right) = 0.$$

Thus the law of $\dot{\Psi}(f)$ is absolutely continuous with respect to the law of $\dot{f}$.

The law of an indicator function χ_A has a probability distribution that is a step function, with a jump at 0 of value $P(A^c)$, and a jump at 1, with value $P(A)$. Any law that is absolutely continuous with respect to that of an indicator must have a distribution that is a step function with jumps at the points 0 and 1 only, and is thus the law of an indicator function. □

Remark 16.2.7 Let Ψ be an almost sure morphism, and $\phi(x, y) = xy$. The following expression:

$$\Psi\left(\left[\phi\left(\chi_{A_1}, \chi_{A_2}\right)\right]_{0,P}\right) = \left[\phi\left(\dot{\Psi}\left(\left[\chi_{A_1}\right]_{0,P}\right), \dot{\Psi}\left(\left[\chi_{A_2}\right]_{0,P}\right)\right)\right]_{0,Q}$$

then translates to

$$\Psi\left(\left[\chi_{A_1\cap A_2}\right]_{0,P}\right)=\left[\dot{\Psi}\left(\left[\chi_{A_1}\right]_{0,P}\right)\dot{\Psi}\left(\left[\chi_{A_1}\right]_{0,P}\right)\right]_{0,Q}.$$

Since images of indicators by morphisms are indicators, letting

$$\Psi\left(\left[\chi_{A_1}\right]_{0,P}\right)=\left[\chi_{B_1}\right]_{0,Q},\quad\text{and}\quad\Psi\left(\left[\chi_{A_2}\right]_{0,P}\right)=\left[\chi_{B_2}\right]_{0,Q},$$

one obtains that

$$\Psi\left(\left[\chi_{A_1\cap A_2}\right]_{0,P}\right)=\left[\chi_{B_1\cap B_2}\right]_{0,Q}.$$

Define

$$\Phi_\Psi:\left(\mathcal{A}^{eP},P^{eP}\right)\longrightarrow\left(\mathcal{B}^{eQ},Q^{eQ}\right),$$

using the following assignment: given that $\Psi\left([\chi_A]_{0,P}\right)=[\chi_B]_{0,Q}$,

$$\Phi_\Psi\left([A]_{0,P}\right)=[B]_{0,Q}.$$

Φ_Ψ is well defined as Ψ is linear, and sends constants to constants.

Since

$$\begin{aligned}\Psi\left(\left[\chi_{A_1\setminus A_2}\right]_{0,P}\right)&=\Psi\left(\left[\chi_{A_1}\right]_{0,P}-\left[\chi_{A_1\cap A_2}\right]_{0,P}\right)\\&=\left[\chi_{B_1}\right]_{0,Q}-\left[\chi_{B_1\cap B_2}\right]_{0,Q}\\&=\left[\chi_{B_1\setminus B_2}\right]_{0,Q},\end{aligned}$$

it follows that

$$\begin{aligned}\Phi_\Psi\left([A_1]_{0,P}\setminus[A_2]_{0,P}\right)&=\Phi_\Psi\left([A_1\setminus A_2]_{0,P}\right)\\&=[B_1\setminus B_2]_{0,Q}\\&=[B_1]_{0,Q}\setminus[B_2]_{0,Q}\\&=\Phi_\Psi\left([A_1]_{0,P}\right)\setminus\Phi_\Psi\left([A_1]_{0,P}\right).\end{aligned}$$

Similarly,

$$\Phi_\Psi\left([A_1]_{0,P}\vee[A_2]_{0,P}\right)=\Phi_\Psi\left([A_1]_{0,P}\right)\vee\Phi_\Psi\left([A_2]_{0,P}\right),$$

and, when $[A_1]_P \leq [A_2]_P$,

$$\Phi_\Psi([A_1]_P) \leq \Phi_\Psi([A_2]_P).$$

As a consequence,

$$\Phi_\Psi\left(\vee_{n=1}^{\infty}[A_n]_P\right) = \vee_{n=1}^{\infty}\Phi_\Psi([A_n]_P),$$

and

$$\Phi_\Psi\left(\wedge_{n=1}^{\infty}[A_n]_P\right) = \wedge_{n=1}^{\infty}\Phi_\Psi([A_n]_P).$$

In particular Φ_Ψ is a σ-lattice homomorphism [226, p. 318], and the image of $\mathcal{A}$ by Ψ is a σ-algebra.

Remark 16.2.8 It follows from the definitions that $\Psi \circ J_P = J_Q \circ \Phi_\Psi$, where J_P and J_Q are as in Sect. 16.1, and Φ_Ψ, as in (Remark) 16.2.7.

Remark 16.2.9 ([46, p. 179]) Suppose $\Phi : (\mathcal{A}^{eP}, P^{eP}) \longrightarrow (\mathcal{B}^{eQ}, Q^{eQ})$ is such that

$$\Phi([A_1]_P \setminus [A_2]_P) = \Phi([A_1]_P) \setminus \Phi([A_1]_P),$$
$$\Phi(\vee_n [A_n]_P) = \vee_n \Phi([A_n]_P).$$

Let $\mathcal{A}_0^{eP} = \{[A_1]_P, \ldots, [A_n]_P\} \subseteq \mathcal{A}^{eP}$, be fixed, but arbitrary, and the image of that family by Φ, be

$$\mathcal{B}_0^{eQ} = \left\{[B_1]_Q, \ldots, [B_n]_Q\right\} \subseteq \mathcal{B}^{eQ}.$$

Then Φ carries the equivalence classes of the sets belonging to the partition of Ω determined by $\{A_1, \ldots, A_n\}$ of the following form:

$$[S_1 \cap \cdots \cap S_n]_P, S_i = A_i \text{ or } S_i = A_i^c,$$

onto the equivalence classes of the sets belonging to the partition of Θ determined by $\{B_1, \ldots, B_n\}$ in such a manner that, when $[A]_P$ is in the former partition, and that $\Phi([A]_P) = [B]_Q$, then

$$[A]_P \leq [A_i]_P, \text{ some } i \in [1:n] \text{ if, and only if, } [B]_Q \leq [B_i]_Q.$$

Let $S_0(\Omega, \mathcal{A}, P)$ denote the linear submanifold of $L_0(\Omega, \mathcal{A}, P)$ consisting of all equivalence classes of the form $[s]_{0,P}$, where s is a simple function. Analogous definitions for $(\Theta, \mathcal{B}, Q)$ are labeled accordingly. Let the map (J_P is as in Sect. 16.1)

$$\Psi_\Phi : \mathcal{R}[J_P] \longrightarrow \mathcal{R}[J_Q]$$

be defined using the following relation: when $\Phi\left([A]_P\right) = [B]_Q$,

$$\Psi_\Phi\left([\chi_A]_{0,P}\right) = [\chi_B]_{0,Q}\,.$$

One has that

$$\sum_{i=1}^{n} a_i \left[\chi_{A_i}\right]_{0,P} = \left[\sum_{i=1}^{n} a_i \chi_{A_i}\right]_{0,P}.$$

Let $s = \sum_{i=1}^{n} a_i \chi_{A_i}$, $\left\{\alpha_1, \ldots, \alpha_p\right\}$ be the distinct values taken by s, and

$$A_j^\alpha = \left\{\omega \in \Omega : s(\omega) = \alpha_j\right\}.$$

Then

$$s = \sum_{j=1}^{p} \alpha_j \chi_{A_j^\alpha}\,.$$

One may thus assume that the a_i's are distinct, and the A_i's disjoint. When s is zero almost surely, one may assume that the A_i's have positive probability, so that the a_i's must be zero. But then

$$\sum_{i=1}^{n} a_i \left[\chi_{B_i}\right]_{0,Q} = [0]_{0,Q}\,.$$

Ψ_Φ has thus a unique linear extension to $S_0\left(\Omega, \mathcal{A}, P\right)$ [46, p. 26], so that

$$\begin{aligned}\Psi_\Phi\left(\left[\sum_{i=1}^{n} a_i \chi_{A_i}\right]_{0,P}\right) &= \Psi_\Phi\left(\sum_{i=1}^{n} a_i \left[\chi_{A_i}\right]_{0,P}\right)\\ &= \sum_{i=1}^{n} a_i \left[\chi_{B_i}\right]_{0,Q}\\ &= \left[\sum_{i=1}^{n} a_i \chi_{B_i}\right]_{0,Q}.\end{aligned}$$

The lattice properties of Φ are recovered by Ψ_Φ, and, when, s_Ω being a simple function on Ω, and s_Θ, one on Θ,

$$\Psi_\Phi\left([s_\Omega]_{0,P}\right) = [s_\Theta]_{0,Q}\,,$$

one has that

$$\Psi_{\Phi}\left([|s_{\Omega}|]_{0,P}\right) = [|s_{\Theta}|]_{0,P}\,.$$

When Φ is furthermore a bijection, that is, when Φ is a weak measure ring isomorphism, so is Ψ_{Φ} a bijection onto $S_0(\Theta, \mathcal{B}, Q)$. Finally, when, in addition,

$$Q^{eQ}\left(\Phi\left([A]_P\right)\right) = P^{eP}\left([A]_P\right),$$

that is, when Φ is a measure ring isomorphism, one has, since the difference of two simple functions is a simple function, that

$$\begin{aligned}
E_Q\left[\left|\dot{\Psi}_{\Phi}\left(\left[\sum_{i=1}^{n} a_i\chi_{A_i}\right]_{0,P}\right)\right|\wedge 1\right] &= E_Q\left[\left|\sum_{i=1}^{n} a_i\dot{\Psi}_{\Phi}\left(\left[\chi_{A_i}\right]_{0,P}\right)\right|\wedge 1\right] \\
&= E_Q\left[\left|\sum_{i=1}^{n} a_i\chi_{B_i}\right|\wedge 1\right] \\
&= \sum_{i=1}^{n} (|a_i|\wedge 1)\, E_Q\left[\chi_{B_i}\right] \\
&= \sum_{i=1}^{n} (|a_i|\wedge 1)\, Q(B_i) \\
&= \sum_{i=1}^{n} (|a_i|\wedge 1)\, P(A_i) \\
&= E_P\left[\left|\sum_{i=1}^{n} a_i\chi_{A_i}\right|\wedge 1\right],
\end{aligned}$$

so that Ψ_{Φ} is a linear isometry, and, since the S_0-manifolds are dense in their respective L_0-spaces, Ψ_{Φ} extends to an isometry between those L_0-spaces.

Definition 16.2.10 An embedding of the probability space $(\Omega, \mathcal{A}, P)$ into the probability space $(\Theta, \mathcal{B}, Q)$ is an almost sure morphism Ψ such that, for fixed, but arbitrary $f \in L_0(\Omega, \mathcal{A}, P)$, the law of $\dot{\Psi}(f)$ is that of f (a function and its image have the same law).

Remark 16.2.11 When the law of $\dot{\Psi}(f)$ is that of f, and that

$$\Psi\left([\chi_A]_{0,P}\right) = [\chi_B]_{0,Q}\,,$$

then

$$P(A) = Q(B).$$

Thus

$$E_P[\chi_A] = E_Q\left[\dot{\Psi}\left([\chi_A]_{0,P}\right)\right],$$

and, since Ψ is linear, Ψ will be an isometry between the L_p-spaces based, respectively, on $(\Omega, \mathcal{A}, P)$ and $(\Theta, \mathcal{C}, Q)$, where $\mathcal{C}$ is the σ-algebra resulting from the range of Ψ. Furthermore, since, as seen [(Remark) 16.2.7],

$$E_P\left[\chi_A \chi_{A_0}\right] = E_Q\left[\dot{\Psi}\left([\chi_A]_{0,P}\right) \dot{\Psi}\left(\left[\chi_{A_0}\right]_{0,P}\right)\right],$$

conditional expectations will be carried by Ψ to conditional expectations.

Example 16.2.12 Let $\mathcal{A} = \sigma(f)$. Let g be an element of $L_0(\Theta, \mathcal{B}, Q)$ such that the law of $\dot{g}$ is that of $\dot{f}$. The following assignment:

$$\Psi\left(\left[\phi(\dot{f})\right]_{0,P}\right) = [\phi(\dot{g})]_{0,Q}$$

yields an embedding.

Remark 16.2.13 The composition of two embeddings is an embedding. An embedding is linear, continuous for convergence in probability, and carries set indicators to set indicators.

Proposition 16.2.14 *An embedding is injective.*

Proof Suppose that $\Psi(f)$ is the class of the zero function on Θ. Its elements have then a law that is a point mass at the origin. But then the elements of f have a law that is a point mass at the origin. f is thus the class of the zero function on Ω. □

Remark 16.2.15 When Ψ is an embedding, Φ_Ψ of (Remark) 16.2.7 is an injection.

Proposition 16.2.16 *When Ψ is an embedding, and $\mathcal{A}_0 \subseteq \mathcal{A}$, a σ-algebra, there is a unique (within isomorphisms) σ-algebra $\mathcal{B}_0 \subset \mathcal{B}$, denoted $\Psi(\mathcal{A}_0)$, such that Ψ is a bijection between $L_0(\Omega, \mathcal{A}_0, P)$ and $L_0(\Theta, \mathcal{B}_0, Q)$. The following relations are furthermore equivalent:*

$$[X_0]_{0,P} = [E_P[X \mid \mathcal{A}_0]]_{0,P},$$

$$\Psi\left([X_0]_{0,P}\right) = \left[E_Q\left[\dot{\Psi}\left([X_0]_{0,P}\right) \mid \mathcal{B}_0\right]\right]_{0,Q}.$$

Proof That is a consequence of (Remark) 16.2.11 and (Proposition) 16.2.14. □

Remark 16.2.17 The equivalent relations of (Proposition) 16.2.16 are easier to read when expressed as follows, as in [28]:

$$Y = E_P[Z|\mathcal{B}] \text{ if, and only if, } \Psi(Y) = E_Q[\Psi(Z)|\Psi(\mathcal{B}].$$

As a consequence of (Proposition) 16.2.16, and the isometric properties of Ψ [(Remark) 16.2.11], martingales and stochastic calculus on $(\Omega, \mathcal{A}, P)$ carry over to martingales and stochastic calculus on $(\Theta, \mathcal{B}, Q)$.

Proposition 16.2.18 *Let* $\Psi : L_0(\Omega, \mathcal{A}, P) \longrightarrow L_0(\Theta, \mathcal{B}, Q)$ *be an almost sure morphism. There is a probability* Π *on* $\mathcal{A}$*, absolutely continuous with respect to* P*, such that* Ψ *is an embedding of* $(\Omega, \mathcal{A}, \Pi)$ *into* $(\Theta, \mathcal{B}, Q)$.

Proof Let

$$\Pi(A) = E_Q\left[\dot{\Psi}\left([\chi_A]_{0,P}\right)\right].$$

Π is well defined, and, when $P(A) = 0$, because of (Proposition) 16.2.6,

$$E_Q\left[\dot{\Psi}\left([\chi_A]_{0,P}\right)\right] = 0.$$

□

The following proposition is a verification tool very similar in content to that of (Remark) 16.2.9.

Proposition 16.2.19 *Let* $(\Omega, \mathcal{A}, P)$ *and* $(\Theta, \mathcal{B}, Q)$ *be probability spaces,* $\mathcal{A}_0$ *be a Boolean algebra generating* $\mathcal{A}$*, and* Φ*, a map from* $\mathcal{A}_0$ *into* $\mathcal{B}$*, such that*

(a) Φ *commutes with the Boolean operations,*
(b) *given the fixed, but arbitrary* $n \in \mathbb{N}$, $\{A_1^0, \ldots, A_n^0\} \subseteq \mathcal{A}_0$*, and a Boolean operation* β *involving* n *elements, then, almost surely, with respect to* P*,*

$$\Phi\left(\beta\left(A_1^0, \ldots, A_n^0\right)\right) = \beta\left(\Phi\left(A_1^0\right), \ldots, \Phi\left(A_n^0\right)\right),$$

(c) *for* $A_0 \in \mathcal{A}_0$*, fixed, but arbitrary,* $P(A_0) = Q(\Phi(A_0))$.

There exist then a unique embedding Ψ_Φ *of* $(\Omega, \mathcal{A}, P)$ *into* $(\Theta, \mathcal{B}, Q)$ *such that, for* $A_0 \in \mathcal{A}_0$*, fixed, but arbitrary,*

$$\Psi_\Phi\left(\left[\chi_{A_0}\right]_{0,P}\right) = \left[\chi_{\Phi(A_0)}\right]_{0,P}.$$

Proof Let $\{\alpha_1, \ldots, \alpha_n\} \subseteq \mathbb{R}$ be fixed, but arbitrary, and suppose that, almost surely, with respect to P,

$$s = \sum_{i=1}^{n} \alpha_i \chi_{A_i^0} = 0. \qquad (\star)$$

Suppose that $I \subseteq [1:n]$ is such that $\sum_{i\in I}\alpha_i \neq 0$. Let $J = [1:n] \setminus I$, and

$$A_0^I = \cap_{i\in I} A_i^0,\ A_0^J = \cap_{j\in J}\left\{A_j^0\right\}^c,\ A_0^0 = A_0^I \cap A_0^J.$$

Since $A_0^0 \subseteq A_0^I$, on A_0^0, s is different from zero, so that $P(A_0^0) = 0$. Consequently, since Φ preserves probabilities, using Assumption (b),

$$\left\{\cap_{i\in I}\Phi\left(A_i^0\right)\right\} \cap \left\{\cap_{j\in J}\Phi\left(\left\{A_j^0\right\}^c\right)\right\}$$

has Q-probability zero. The following assignment thus makes sense:

$$\Psi_\Phi\left(\left[\sum_{i=1}^n \alpha_i \chi_{A_i^0}\right]_{0,P}\right) = \left[\sum_{i=1^n} \alpha_i \chi_{\Phi(A_i^0)}\right]_{0,Q}.$$

Letting $S_0(\Omega, \mathcal{A}_0, P)$ denote the equivalence classes of functions of type s, as in $(\star)$, one thus obtains a linear map $\Psi_\Phi : S_0(\Omega, \mathcal{A}_0, P) \longrightarrow S_0(\Theta, \mathcal{B}, Q)$ which maintains probability laws, and for which, given $\{f_1, \ldots, f_n\} \subseteq S_0(\Omega, \mathcal{A}_0, P)$, fixed, but arbitrary,

$$\Psi_\Phi\left(\left[\phi\left(\dot{f}_1, \ldots, \dot{f}_n\right)\right]_{0,P}\right) = \left[\phi\left(\dot{\Psi}_\Phi\left(f_1\right), \ldots, \dot{\Psi}_\Phi\left(f_n\right)\right]_{0,Q}\right).$$

Since probabilities are maintained,

$$E_P\left[\{|s| \wedge 1\}\right] = E_Q\left[\left\{\dot{\Psi}_\Phi\left([|s|]_{0,P}\right) \wedge 1\right\}\right],$$

so that Ψ_Φ is continuous for convergence in probability. Since simple functions are dense, and the L_0-spaces involved complete, Ψ_Φ has a unique, injective, isometric extension to $\mathcal{A}$. To check the validity of the formula characterizing morphisms, one may start with a continuous ϕ, and notice that such continuity makes the formula, valid for simple functions based on $\mathcal{A}_0$, valid also for $\{f_1, \ldots, f_n\} \subseteq L_0(\Omega, \mathcal{A}, P)$. To obtain that the formula is valid for Borel measurable ϕ, one argues that the family of ϕ's for which the formula is valid is closed for simple limits, and, as it contains the continuous functions, for all Borel measurable functions. □

Definition 16.2.20 An isomorphism of the probability spaces $(\Omega, \mathcal{A}, P)$ and $(\Theta, \mathcal{B}, Q)$ is a surjective embedding $\Psi : L_0(\Omega, \mathcal{A}, P) \longrightarrow L_0(\Theta, \mathcal{B}, Q)$.

One shall use below the following `convention`: $(\Omega \times \Theta, \mathcal{A}\overline{\otimes}\mathcal{B}, P\overline{\otimes}Q)$ stands for $(\Omega \times \Theta, \mathcal{A} \otimes \mathcal{B}, P \otimes Q)$ completed for $P \otimes Q$.

Lemma 16.2.21 ([168, p. 131]) *Let $\mathcal{C}$ and $\mathcal{D}$ be two σ-algebras, and f be a random variable adapted to $\mathcal{C} \vee \mathcal{D}$. There exists then variables Γ, adapted to $\mathcal{C}$, and Δ, adapted to $\mathcal{D}$, such that $\sigma(f) \subseteq \sigma(\Gamma, \Delta)$. There is in particular a Borel ϕ such that $f = \phi(\Gamma, \Delta)$. Γ and Δ may be chosen to be bounded.*

Proof Let $\Sigma = \cup_{\{\Gamma \in \mathcal{C}, \Delta \in \mathcal{D}\}} \sigma(\Gamma, \Delta)$. It is a σ-algebra. Indeed, $\emptyset$ and Ω are in each $\sigma(\Gamma, \Delta)$. Furthermore, when $A \in \Sigma$, A belongs to some $\sigma(\Gamma_A, \Delta_A)$, so that $A^c \in \sigma(\Gamma_A, \Delta_A)$ as well. When $\{A_n, n \in \mathbb{N}\} \subseteq \Sigma$, A_n belongs to some $\sigma(\Gamma_n, \Delta_n)$. Since the latter is separable, there is [Sect. 16.1] a family

$$\{C_p^n, D_p^n, p \in \mathbb{N}\}$$

such that the family $\{C_p^n \cap D_p^n, p \in \mathbb{N}\}$ generates $\sigma(\Gamma_n, \Delta_n)$. Let thus

$$\mathcal{F} = \{C_p^n, \{n, p\} \in \mathbb{N} \times \mathbb{N}\}, \ \mathcal{G} = \{D_p^n, \{n, p\} \in \mathbb{N} \times \mathbb{N}\}.$$

Define a random variable U generating the same σ-algebra as does $\mathcal{F}$, and a random variable V generating the same σ-algebra as does $\mathcal{G}$ (the Marczewski function for example). Then

$$\{A_n, n \in \mathbb{N}\} \subseteq \sigma(U, V) \subseteq \Sigma.$$

Since $\Sigma \subseteq \cup_{\{\Gamma \in \mathcal{C}, \Delta \in \mathcal{D}\}} \sigma(\Gamma) \vee \sigma(\Delta) \subseteq \mathcal{C} \vee \mathcal{D}$, and that the $\sigma(\chi_C, \chi_D)$'s generate $\mathcal{C} \vee \mathcal{D}$, the lemma is true. □

Proposition 16.2.22 *In $\mathcal{A}$, let $\mathcal{A}_1$ and $\mathcal{A}_2$ be two independent σ-algebras. There is a unique, almost sure morphism*

$$\Psi : L_0(\Omega, \mathcal{A}_1 \vee \mathcal{A}_2, P) \longrightarrow L_0(\Omega \times \Omega, \mathcal{A}_1 \overline{\otimes} \mathcal{A}_2, P \overline{\otimes} P)$$

such that, when f_1 is adapted to $\mathcal{A}_1$, and f_2, to $\mathcal{A}_2$, almost surely, with respect to $P \overline{\otimes} P$,

$$\dot{\Psi}\left([f_1 f_2]_{0,P}\right)(\omega_1, \omega_2) = f_1(\omega_1) f_2(\omega_2). \qquad (\star)$$

That morphism is in fact an isomorphism.

Proof A random variable adapted to $\mathcal{A}_1 \vee \mathcal{A}_2$ is, as seen [(Lemma) 16.2.21], of the following form: $\phi(A_1, A_2)$, A_1 adapted to $\mathcal{A}_1$, A_2, to $\mathcal{A}_2$. Let

$$A_1^{\otimes}(\omega_1, \omega_2) = A_1(\omega_1), \ A_2^{\otimes}(\omega_1, \omega_2) = A_2(\omega_2).$$

$A_1^{\otimes}$ and $A_2^{\otimes}$ are adapted to $\mathcal{A}_1 \overline{\otimes} \mathcal{A}_2$. When $\mathcal{A}_1$ and $\mathcal{A}_2$ are independent, one has, because of $(\star)$, using a monotone class theorem, that

$$E_P\left[|\phi(A_1, A_2) \wedge 1|\right] = E_{P \overline{\otimes} P}\left[\left|\phi(A_1^{\otimes}, A_2^{\otimes}) \wedge 1\right|\right].$$

The same equality obtains for expressions of the following form:

$$\phi\left(\phi_1(A_1^{(1)}, A_2^{(1)}), \ldots, \phi_n(A_1^{(n)}, A_2^{(n)})\right).$$

One thus defines a linear isometry between the respective L_0 spaces. That it is onto follows from the fact that measurable functions are approximated by simple ones, and that the indicators of $A_1 \cap A_2$ and $A_1 \times A_2$ are isometrically related. □

Lemma 16.2.23 *Let f_1 and f_2 be adapted to $\mathcal{A}$, and have the same law. The probability spaces $(\Omega, \sigma(f_1), P)$ and $(\Omega, \sigma(f_2), P)$ are isomorphic.*

Proof The isomorphism is defined using the following relation:

$$\Psi\left([\phi(f_1)]_{0,P}\right) = [\phi(f_2)]_{0,P}\,.$$

□

Lemma 16.2.24 *For $i \in \{1, 2\}$, let $(\Omega_i, \mathcal{A}_i, P_i)$ and $(\Theta_i, \mathcal{B}_i, Q_i)$ be isomorphic. Then $(\Omega_1 \times \Omega_2, \mathcal{A}_1 \overline{\otimes} \mathcal{A}_2, P_1 \overline{\otimes} P_2)$ and $(\Theta_1 \times \Theta_2, \mathcal{B}_1 \overline{\otimes} \mathcal{B}_2, Q_1 \overline{\otimes} Q_2)$ are isomorphic.*

Proof The isomorphism is defined using the following relation:

$$\Psi\left([\phi_1(f_1)\phi_2(f_2)]_{0,P_1\overline{\otimes}P_2}\right) = \left[\phi_1\left(\dot{\Psi}_1\left([f_1]_{0,P_1}\right)\right)\phi_2\left(\dot{\Psi}_2\left([f_2]_{0,P_2}\right)\right)\right]_{0,P_1\overline{\otimes}P_2}\,.$$

□

Corollary 16.2.25 *Let the random variables f_1 and f_2 be adapted to $\mathcal{A}$, and have the same law, and $\mathcal{A}_0 \subseteq \mathcal{A}$ be a σ-algebra independent of f_1 and f_2. There exists a unique isomorphism between $(\Omega, \mathcal{A}_0 \vee \sigma(f_1), P)$ and $(\Omega, \mathcal{A}_0 \vee \sigma(f_2), P)$ which acts as the identity on $(\Omega, \mathcal{A}, P)$, and sends the class of f_1 to that of f_2.*

Proof From (Proposition) 16.2.22,

$$(\Omega, \mathcal{A}_0 \vee \sigma(f_1), P) \text{ is isomorphic to } \left(\Omega \times \Omega, \mathcal{A}_0 \overline{\otimes} \sigma(f_1), P \overline{\otimes} P\right),$$

and

$$(\Omega, \mathcal{A}_0 \vee \sigma(f_2), P) \text{ is isomorphic to } \left(\Omega \times \Omega, \mathcal{A}_0 \overline{\otimes} \sigma(f_2), P \overline{\otimes} P\right).$$

Since, because of (Lemmas) 16.2.23 and 16.2.24,

$$\left(\Omega \times \Omega, \mathcal{A}_0 \overline{\otimes} \sigma(f_1), P \overline{\otimes} P\right) \text{ and } \left(\Omega \times \Omega, \mathcal{A}_0 \overline{\otimes} \sigma(f_2), P \overline{\otimes} P\right)$$

are isomorphic, so are $(\Omega, \mathcal{A}_0 \vee \sigma(f_1), P)$ and $(\Omega, \mathcal{A}_0 \vee \sigma(f_2), P)$. □

Proposition 16.2.26 *Let, for $(\Omega, \mathcal{A}, P)$, $\mathcal{C}$, $\mathcal{D}$, and $\mathcal{E}$ be σ-algebras contained in $\mathcal{A}$, having the following properties:*

(a) *$\mathcal{E}$ is essentially separable,*
(b) *$\mathcal{E} \subseteq \mathcal{C} \vee \mathcal{D}$,*
(c) *$\mathcal{C}$ is independent of $\mathcal{D}$ and $\mathcal{E}$.*

There exists then an embedding of $(\Omega, \mathcal{E}, P)$ into $(\Omega, \mathcal{D}, P)$.

Proof One may assume, without restricting validity, that $\mathcal{C} \vee \mathcal{D} = \mathcal{A}$. Let $\Omega_1 = \Omega_2 = \Omega$, and $P_1 = P^{|\mathcal{C}}$, $P_2 = P^{|\mathcal{D}}$. Let c be an equivalence class in $L_0(\Omega_1, \mathcal{C}, P_1)$, d, one in $L_0(\Omega_2, \mathcal{D}, P_2)$; Π_1 be the projection of $\Omega_1 \times \Omega_2$ onto its first component, Π_2, the projection onto its second one; $c \circ \Pi_1$ is the class of $\dot{c}$ composed with Π_1, $d \circ \Pi_2$, that of $\dot{d} \circ \Pi_2$. Since $\mathcal{C}$ is independent of $\mathcal{D}$, there is [(Proposition) 16.2.22] an almost sure morphism (an isomorphism in fact),

$$\Psi : L_0(\Omega, \mathcal{C} \vee \mathcal{D}, P) \longrightarrow L_0\left(\Omega_1 \times \Omega_2, \mathcal{C}\overline{\otimes}\mathcal{D}, P_1\overline{\otimes}P_2\right) (= L_0(\Theta, \mathcal{B}, Q)),$$

such that

$$\Psi(cd) = (c \circ \Pi_1)(d \circ \Pi_2).$$

Let $\mathcal{C}_\Psi = \Psi(\mathcal{C})$, and $\mathcal{D}_\Psi = \Psi(\mathcal{D})$. Since Ψ sends products to products, letting $\mathcal{T}$ denote the trivial σ-algebra,

$$\mathcal{C}_\Psi = \mathcal{C}\overline{\otimes}\mathcal{T}, \text{ and } \mathcal{D}_\Psi = \mathcal{T}\overline{\otimes}\mathcal{D}.$$

$\mathcal{C}_\Psi$ and $\mathcal{D}_\Psi$ are independent. Finally, when g is an essentially bounded element of $L_0(\Theta, \mathcal{B}, Q)$, let $\dot{g} \in g$ be adapted to $\mathcal{C} \otimes \mathcal{D}$. The characteristic equation for $E_Q[\dot{g} \mid \mathcal{C}_\Psi]$ is, with $C \in \mathcal{C}$, fixed, but arbitrary,

$$\int_{C\times\Omega} \dot{g}\, dQ = \int_{C\times\Omega} E_Q[\dot{g} \mid \mathcal{C}_\Psi]\, dQ.$$

But, using Fubini's theorem,

$$\begin{aligned}\int_{C\times\Omega} \dot{g}(\omega_1, \omega_2) Q(d\omega_1, d\omega_2) &= \int_C P_1(d\omega_1) \int_\Omega P_2(d\omega_2)\dot{g}(\omega_1, \omega_2)\\ &= \int_C P_1(d\omega_1) E_{P_2}[\dot{g}(\omega_1, \cdot)].\end{aligned}$$

Thus, almost surely, with respect to Q,

$$E_Q[\dot{g} \mid \mathcal{C}_\Psi] = E_{P_2}[\dot{g}(\Pi_1, \cdot)]. \qquad (\star)$$

Let $\mathcal{E}_0 \subseteq \mathcal{E}$ be a countable, dense algebra. Given $E_0 \in \mathcal{E}_0$, let

$$E_0^{\Psi} = \Psi([E_0]_P), \quad \text{and} \quad \dot{E}_0^{\Psi} \in \mathcal{C} \otimes \mathcal{D}.$$

The conditional expectation formula referred to above by ($\star$), with

$$\dot{g} = \chi_{\dot{E}_0^{\Psi}},$$

yields, $S[\omega]$ denoting the section of S at ω, that, almost surely, with respect to Q,

$$Q\left(\dot{E}_0^{\Psi} \mid \mathcal{C}_{\Psi}\right) = P_2\left(\dot{E}_0^{\Psi}[\Pi_1]\right).$$

But $\mathcal{C}$ and $\mathcal{E}$ being independent, so are $\mathcal{C}_{\Psi}$ and $\mathcal{E}_{\Psi}$, and thus,

$$Q\left(\dot{E}_0^{\Psi} \mid \mathcal{C}_{\Psi}\right) = Q\left(\dot{E}_0^{\Psi}\right).$$

Now, Ψ being an isomorphism,

$$Q\left(\dot{E}_0^{\Psi}\right) = P(E_0).$$

Thus, almost surely, with respect to Q,

$$P_2\left(\dot{E}_0^{\Psi}[\Pi_1]\right) = P(E_0).$$

Since $\mathcal{E}_0$ is countable, there exists $C_0 \in \mathcal{C}$ such that $P(C_0) = 0$, and, for $\omega_1 \in C_0^c$, fixed, but arbitrary, one has that:

(i) for $E_0 \in \mathcal{E}_0$, fixed, but arbitrary, $P_2(\dot{E}_0^{\Psi}[\omega_1]) = P(E_0)$;
(ii) for $n \in \mathbb{N}$, $\{E_{0,1}, \ldots, E_{0,n}\} \subseteq \mathcal{E}_0$, and ρ_n, a Boolean relation of n arguments, for which, almost surely, with respect to P_2,

$$\rho_n\left(\chi_{E_{0,1}}(\omega), \ldots, \chi_{E_{0,n}}(\omega)\right)$$

obtains, then, for almost all ω_2, with respect to P,

$$\rho_n\left(\chi_{\dot{E}_{0,1}^{\Psi}}(\omega_1, \omega_2), \ldots, \chi_{\dot{E}_{0,n}^{\Psi}}(\omega_1, \omega_2)\right)$$

obtains.

The embedding one looks for is obtained as follows. Let $\omega_1 \in C_0^c$, and $E_0 \in \mathcal{E}_0$ be fixed, but arbitrary. Since $\dot{E}_0^{\Psi} \in \mathcal{C} \otimes \mathcal{D}$, $\dot{E}_0^{\Psi}[\omega_1] \in \mathcal{D}$, and one thus sets:

$$\Psi_0\left([E_0]_P\right) = \left[\dot{E}_0^{\Psi}[\omega_1]\right]_{P_2}.$$

Because of the first property (i), itemized above, Ψ_0 preserves the law. For fixed, but arbitrary $\{E_{0,1}, \dots, E_{0,n}\} \subseteq \mathcal{E}_0$, and relation ρ_n, let

$$R = \rho_n (E_{0,1}, \dots, E_{0,n}) .$$

Let furthermore

$$\rho_{n+1} \equiv [r = \rho_n(e_1, \dots, e_n)] .$$

Because of the second property (ii), itemized above, used with ρ_{n+1}, one has that

$$\Psi_0 ([R]_P) = \left[\rho_n \left(\dot{\Psi}_0 \left([E_{0,1}]_P\right), \dots, \dot{\Psi}_0 \left([E_{0,1}]_P\right)\right)\right]_{P_2} .$$

(Result (Proposition)) 16.2.19 allows one to conclude, that is, obtain an embedding of $(\Omega, \mathcal{E}, P)$ into $(\Omega_2, \mathcal{D}, P_2) = (\Omega, \mathcal{D}, P)$. □

16.3 Morphisms and Inclusions for Filtrations

A filtered probability space shall have the following form: $(\Omega, \mathcal{A}, P, \underline{\mathcal{A}})$, where $\underline{\mathcal{A}}$ is an increasing family of σ-algebras, contained in $\mathcal{A}$, continuous to the right, whose elements each contain the subsets of $\mathcal{N}(\mathcal{A}, P)$. $\mathcal{A}_\infty$ is the σ-algebra generated by the filtration, and it is often no restriction to assume that $\mathcal{A}_\infty = \mathcal{A}$, but it is perhaps clearer to keep those σ-algebras separated in what follows.

Definition 16.3.1 Let $(\Omega, \mathcal{A}, P, \underline{\mathcal{A}})$ and $(\Theta, \mathcal{B}, Q, \underline{\mathcal{B}})$ be two filtered probability spaces. $\underline{\mathcal{A}}$ and $\underline{\mathcal{B}}$ are isomorphic, and one writes $\underline{\mathcal{A}} \approx \underline{\mathcal{B}}$ when there exists an isomorphism Ψ between $(\Omega, \mathcal{A}_\infty, P)$ and $(\Theta, \mathcal{B}_\infty, Q)$ such that, for $t \in \mathbb{R}_+$, fixed, but arbitrary, $\Psi(\mathcal{A}_t) = \mathcal{B}_t$.

Definition 16.3.2 $\underline{\mathcal{B}}$ and $\underline{\mathcal{C}}$ being two filtrations for $(\Omega, \mathcal{A}, P)$, the filtration $\underline{\mathcal{B}}$ is included in the filtration $\underline{\mathcal{C}}$ when, for $t \in \mathbb{R}_+$, fixed, but arbitrary, $\mathcal{B}_t \subseteq \mathcal{C}_t$. The notation shall be $\underline{\mathcal{B}} \sqsubseteq \underline{\mathcal{C}}$.

Definition 16.3.3 Let $(\Omega, \mathcal{A}, P, \underline{\mathcal{A}})$ and $(\Theta, \mathcal{B}, Q, \underline{\mathcal{B}})$ be two filtered probability spaces. $\underline{\mathcal{A}}$ is includable in $\underline{\mathcal{B}}$ when there exists a probability space $(E, \mathcal{E}, M)$ and two filtrations in $\mathcal{E}$, $\underline{\mathcal{F}}$ and $\underline{\mathcal{G}}$, such that

1. $\underline{\mathcal{F}} \sqsubseteq \underline{\mathcal{G}}$;
2. $(\Omega, \mathcal{A}, P, \underline{\mathcal{A}})$ and $(E, \mathcal{E}, M, \underline{\mathcal{F}})$ are isomorphic;
3. $(\Theta, \mathcal{B}, Q, \underline{\mathcal{B}})$ and $(E, \mathcal{E}, M, \underline{\mathcal{G}})$ are isomorphic (one may always use the former for the latter).

The notation shall be: $\underline{\mathcal{A}} \succ \underline{\mathcal{B}}$.

16.4 Multiplicity for Algebras of Sets

Multiplicity shall be of use when determining the uniqueness class of continuous local martingales [Sect. 16.8], but it has nobler purposes [28]![1]

In the sequel, $(\Omega, \mathcal{A}, P)$ shall be fixed, but arbitrary. Let $A \in \mathcal{A}$ be fixed, but arbitrary. $\mathcal{A}[A]$ shall be the σ-algebra

$$\{A_0 \in \mathcal{A} : A_0 \subseteq A\}.$$

Let $\mathcal{P}[\mathcal{A}]$ be the family of finite partitions of Ω with sets in $\mathcal{A}$. When $\wp \in \mathcal{P}[\mathcal{A}]$, $|\wp|$ denotes the number of sets in the partition $\wp$.

Definition 16.4.1 Let $\mathcal{A}_0 \subseteq \mathcal{A}$ be a σ-algebra, and $\wp \in \mathcal{P}[\mathcal{A}]$ be fixed, but arbitrary. The support of $\wp$ with respect to $\mathcal{A}_0$ is the set:

$$\begin{aligned} S(\wp \mid \mathcal{A}_0) &= \{\omega \in \Omega : \forall A \in \wp, P(A \mid \mathcal{A}_0)(\omega) > 0\} \\ &= \left\{\omega \in \Omega : \prod_{A \in \wp} P(A \mid \mathcal{A}_0) > 0\right\}. \end{aligned}$$

Remark 16.4.2 Since conditional expectations of the same element, with respect to the same σ-algebra, are in the same equivalence class, the support is defined within a set of zero probability. From the definition, $S(\wp \mid \mathcal{A}_0) \in \mathcal{A}_0$.

Lemma 16.4.3 *Let $\{x_1, \ldots, x_n\}$ and $\{y_1, \ldots, y_n\}$ be fixed, but arbitrary subsets of $[0, 1]$. Then*

$$\prod_{i=1}^{n} x_i \leq \sum_{i=1}^{n} x_i,$$

and, when $x_i \leq y_i, i \in [1 : n]$,

$$\prod_{i=1}^{n} x_i \leq \prod_{i=1}^{n} y_i.$$

---

[1]Si $(\Omega, \mathcal{A}, P)$ est un espace probabilisé, la dimension de l'espace vectoriel $L_0(\Omega, \mathcal{A}, P)$ est aussi le nombre maximal de valeurs différentes que peut prendre une variable aléatoire, ou encore le nombre maximal d'événements non négligeables et deux-à-deux disjoints. Si maintenant $\mathcal{B}$ est une sous-tribu de $\mathcal{A}$, ces équivalences subsistent conditionnellement à $\mathcal{B}$, c'est-à-dire en considérant comme constantes les variables aléatoires mesurables pour $\mathcal{B}$; la dimension devient alors elle-même une variable aléatoire mesurable pour $\mathcal{B}$; elle est introduite ci-dessous sous le nom de multiplicité conditionnelle de $\mathcal{A}$ par rapport à $\mathcal{B}$.

Proof For example

$$\begin{aligned} y_1y_2y_3y_4 - x_1x_2x_3x_4 &= (y_1y_2y_3y_4 - y_1y_2y_3x_4) + (y_1y_2y_3x_4 - y_1y_2x_3x_4) \\ &+ (y_1y_2x_3x_4 - y_1x_2x_3x_4) + (y_1x_2x_3x_4 - x_1x_2x_3x_4) \\ &= y_1y_2y_3(y_4 - x_4) + y_1y_2x_4(y_3 - x_3) \\ &+ y_1x_3x_4(y_2 - x_2) + x_2x_3x_4(y_1 - x_1) \\ &\geq 0. \end{aligned}$$

Then $x_1(1 - x_2x_3x_4) + x_2 + x_3 + x_4 \geq 0$, that is,

$$x_1 + x_2 + x_3 + x_4 \geq x_1x_2x_3x_4.$$

□

Remark 16.4.4 When $\wp_f$ is a partition that is finer than $\wp$, since

$$\prod_{A_f \in \wp_f} P(A_f \mid \mathcal{A}_0) \leq \prod_{A \in \wp} P(A \mid \mathcal{A}_0),$$

the support of $\wp_f$ is contained in that of $\wp$. Indeed, any set A of $\wp$ is of the following form (disjoint union): $A = A_1 \cup \cdots \cup A_n$, where $A_i \in \wp_f$, and $i \in [1 : n]$. Then (almost surely) $P(A \mid \mathcal{A}_0) = \sum_{i=1}^{n} P(A_i \mid \mathcal{A}_0)$. One then uses (Lemma) 16.4.3.

Remark 16.4.5 When $\wp$ contains a set $A_0 \in \mathcal{A}_0$, $S(\wp \mid \mathcal{A}_0) \subseteq A_0$. Indeed, $P(A_0 \mid \mathcal{A}_0) = \chi_{A_0}$.

Example 16.4.6 Let Ω be the disjoint union of A_1, A_2, and A_3, all of positive probability. Let $\mathcal{A}_0$ be generated by A_3 and $\wp$ be the partition made of A_1 and its complement. Then

$$P(A_1 \mid \mathcal{A}_0) = \frac{P(A_1)}{P(A_1) + P(A_2)} \chi_{A_3^c},$$

$$P(A_1^c \mid \mathcal{A}_0) = \frac{P(A_2)}{P(A_1) + P(A_2)} \chi_{A_3^c} + \chi_{A_3}.$$

The support of $\wp$ with respect to $\mathcal{A}_0$ is thus A_3^c.

Definition 16.4.7 ([128, p. 8]) Let $(\Omega, \mathcal{A}, P)$ be a probability space, and $\mathcal{F}$, a nonempty family of random variables. A random variable ϕ is called the essential supremum of $\mathcal{F}$ when both conditions which follow obtain:

(a) almost surely, with respect to $P, f \leq \phi$ whenever $f \in \mathcal{F}$;
(b) when ψ is another random variable for which *(a)* obtains, then, almost surely, with respect to P, $\phi \leq \psi$.

Fact 16.4.8 ([128, pp. 9–10]) *One has that:*

1. *the essential supremum exists and is unique;*
2. *there are at most a countable number of functions of $\mathcal{F}$, say*

$$\{f_i, i \in I \subseteq \mathbb{N}\}$$

such that

$$ess\ \sup \mathcal{F} = \bigvee_{i \in I} f_i;$$

3. *when $\mathcal{F}$ is closed for the $\bigvee$-operation, the f_i's may be chosen to be monotone increasing;*
4. *let $\mathcal{B} \subseteq \mathcal{A}$ be not-empty and $\mathcal{F} = \{\chi_B, B \in \mathcal{B}\}$: there is then*

$$\{B_i, i \in I \subseteq \mathbb{N}\}$$

such that

$$ess\ \sup \mathcal{F} = \bigvee_{i \in I} \chi_{B_i} = \chi_{\{\cup_{i \in I} B_i\}}.$$

Definition 16.4.9 The conditional multiplicity of $\mathcal{A}$ given $\mathcal{A}_0$ is defined as

$$mult[\mathcal{A} \mid \mathcal{A}_0] = \text{ess sup}_{\wp \in \mathcal{P}[\mathcal{A}]} \left\{ \chi_{S(\wp \mid \mathcal{A}_0)} |\wp| \right\} \in \mathbb{N} \cup \{\infty\}.$$

When it is necessary to take the probability into account, one shall write:

$$mult_P[\mathcal{A} \mid \mathcal{A}_0].$$

Remark 16.4.10 The essential supremum is a random variable [(Fact) 16.4.8], and it is thus adapted to $\mathcal{A}_0$. Since Ω is a partition by itself, and since

$$P(\Omega \mid \mathcal{A}_0) = 1,$$

it follows that $mult[\mathcal{A} \mid \mathcal{A}_0]$ is bounded below [(Remark) 16.4.4], almost surely, with respect to P, by the constant function whose value is one.

When $\mathcal{A}_0 = \mathcal{A}$, $P(A \mid \mathcal{A}_0) = \chi_A$, so that

$$\{\omega \in \Omega : \forall A \in \wp, P(A \mid \mathcal{A}_0)(\omega) > 0\} = \{\omega \in \Omega : \forall A \in \wp, \chi_A > 0\} = \emptyset,$$

for all partitions, except that made only of Ω, and then *mult* is the constant function with value one.

Example 16.4.11 In (Example) 16.4.6, $\chi_{A_3^c} |\wp|$ is a random variable that is zero on A_3, and 2 on A_3^c.

Proposition 16.4.12 *Let $n \in \mathbb{N}$ be fixed, but arbitrary. Then, almost surely, with respect to P, $mult[\mathcal{A} \mid \mathcal{A}_0] \leq n$ if, and only if, given disjoint*

$$\{A_1, \ldots, A_n, A_{n+1}\} \subseteq \mathcal{A},$$

one has that, almost surely, with respect to P,

$$\prod_{i=1}^{n+1} P(A_i \mid \mathcal{A}_0) = 0.$$

Proof Suppose that $mult[\mathcal{A} \mid \mathcal{A}_0] \leq n$, almost surely, with respect to P. Let $A_1, \ldots, A_n, A_{n+1}$ be given, pairwise disjoint, fixed, but arbitrary. Let $A_0 = (A_1 \cup \cdots \cup A_n)^c$. Then $A_0, A_1, \ldots, A_n$ form a partition, and, given the assumption,

$$\prod_{i=0}^{n} P(A_i \mid \mathcal{A}_0) = 0.$$

But $P(A_{n+1} \mid \mathcal{A}_0) \leq P(A_0 \mid \mathcal{A}_0)$, and thus

$$\prod_{i=1}^{n+1} P(A_i \mid \mathcal{A}_0) \leq \prod_{i=0}^{n} P(A_i \mid \mathcal{A}_0) = 0.$$

Suppose conversely that the statement's condition obtains. When none of $A_1, \cdots, A_n, A_{n+1}$ is void, there are two cases. Either it is a partition, and when not, $A_0, A_1, \ldots, A_n$ is a partition. By assumption, in both cases, one has a partition leading to a zero product, so that, by definition, $mult[\mathcal{A} \mid \mathcal{A}_0] \leq n$. □

Proposition 16.4.13 $mult[\mathcal{A} \mid \mathcal{A}_0] = 1$ *if, and only if,* $\mathcal{A} = \mathcal{A}_0$.

Proof Since always [(Remark) 16.4.4] $mult[\mathcal{A} \mid \mathcal{A}_0] \geq 1$, then $mult[\mathcal{A} \mid \mathcal{A}_0] = 1$ if, and only if, $mult[\mathcal{A} \mid \mathcal{A}_0] \leq 1$. Suppose the latter. Then, because of (Proposition) 16.4.12, for $A \in \mathcal{A}$, fixed, but arbitrary, almost surely, with respect to P,

$$P(A \mid \mathcal{A}_0) \times P(A^c \mid \mathcal{A}_0) = 0.$$

Let

$$S_1 = \{\omega \in \Omega : P(A \mid \mathcal{A}_0) > 0\}, \; S_2 = \{\omega \in \Omega : P(A^c \mid \mathcal{A}_0) = 0\}.$$

All the inclusions below shall be within a set of probability zero. Since, by the definition of conditional probability,

$$0 = \int_{S_1^c} P(A \mid \mathcal{A}_0) dP = \int_{S_1^c} \chi_A \, dP = P(A \setminus S_1),$$

then, almost surely with respect to P, $A \subseteq S_1$. But, given that the product of $P(A \mid \mathcal{A}_0)$ and $P(A^c \mid \mathcal{A}_0)$ is zero, $S_1 \subseteq S_2$. Finally, since

$$0 = \int_{S_2} P(A^c \mid \mathcal{A}_0) dP = \int_{S_2} \chi_{A^c} \, dP = P(S_2 \setminus A),$$

then $S_2 \subseteq A$. Thus $A \subseteq S_1 \subseteq S_2 \subseteq A$, and A is then almost surely equal to a set in $\mathcal{A}_0$. Consequently, $\mathcal{A} \subseteq \mathcal{A}_0$. But $\mathcal{A}_0 \subseteq \mathcal{A}$.

Suppose now that $\mathcal{A} = \mathcal{A}_0$. Then, for $A \in \mathcal{A}$,

$$P(A \mid \mathcal{A}_0) \times P(A^c \mid \mathcal{A}_0) = \chi_A \chi_{A^c} = 0,$$

and thus $mult[\mathcal{A} \mid \mathcal{A}_0] \leq 1$. □

Lemma 16.4.14 *The set* $S_1 = \{\omega \in \Omega : P(A \mid \mathcal{A}_0) > 0\}$ *is, within sets of measure zero for* P*, the smallest one in* $\mathcal{A}_0$ *which contains* A*.*

Proof By definition, $S_1 \in \mathcal{A}_0$. If now $A_0 \in \mathcal{A}_0$ contains A, $\chi_A \leq \chi_{A_0}$, so that

$$P(A \mid \mathcal{A}_0) = E_P[\chi_A \mid \mathcal{A}_0] \leq E_P\left[\chi_{A_0} \mid \mathcal{A}_0\right] = \chi_{A_0}.$$

Thus, $P(A \mid \mathcal{A}_0)$ is zero outside of A_0, and when it is positive, it is in A_0. □

Proposition 16.4.15 *When* Q *is a probability measure equivalent to* P*,*

$$mult_Q[\mathcal{A} \mid \mathcal{A}_0] = mult_P[\mathcal{A} \mid \mathcal{A}_0].$$

Proof The conclusion of (Lemma) 16.4.14 is independent of the actual choice of P. S_1 thus remains the smallest set in $\mathcal{A}_0$, containing A, with respect to probability measures equivalent to P. But then the same is true for supports, and hence for conditional multiplicity. □

Lemma 16.4.16 *Let* Q *be absolutely continuous with respect to* P*, but not equivalent. There exists* S *such that* $0 < P(S) < 1$*, and then*

$$Q_S(A) = \frac{P(A \cap S)}{P(S)}$$

determines a probability, absolutely continuous with respect to P*, mutually absolutely continuous with respect to* Q*.*

Proof Since $Q \ll P$, there exists an essentially unique, adapted f such that, for $A \in \mathcal{A}$, fixed, but arbitrary,

$$Q(A) = \int_A f dP.$$

Let $S = \{\omega \in \Omega : f(\omega) > 0\} \in \mathcal{A}$. Then

$$Q(A) = \int_A f dP = \int_{A \cap S} f dP. \tag{$\star$}$$

Suppose that $P(S) = 0$. Then, because of $(\star)$, for $A \in \mathcal{A}$, fixed, but arbitrary, $Q(A) = 0$, which is impossible as Q is a probability. Suppose that $P(S) = 1$. Again, because of $(\star)$, $Q(A) = 0$ implies that $P(A \cap S) = 0$, so that $P(A) = 0$, which is impossible, as P and Q are not mutually absolutely continuous.

Then the following definition:

$$A \in \mathcal{A},\ Q_S(A) = \frac{P(A \cap S)}{P(S)}$$

makes sense and defines a probability. Furthermore $dQ_S = P(S)^{-1}\chi_S dP$. By definition, $Q_S \ll P$. Suppose then that $Q(A) = 0$. Then, because of $(\star)$, $P(A \cap S) = 0$, that is, $Q_S(A) = 0$. Suppose that the latter is true. Then, by definition, $P(A \cap S) = 0$. Furthermore $P(S^c) = 0$. Thus, since Q is absolutely continuous with respect to P, $Q(A \cap S) = Q(A \cap S^c) = 0$, and $Q(A) = Q(A \cap S) + Q(A \cap S^c) = 0$. □

Lemma 16.4.17 $\mathcal{N}(\mathcal{A}, Q_S) = \{N \cup C, N \in \mathcal{N}(\mathcal{A}, P), C \in \mathcal{A} \cap S^c)\}$.

Proof The elements of the form $N \cup C$ of the lemma's statement are in $\mathcal{N}(\mathcal{A}, Q_S)$ as $Q_S(N \cup C) \leq Q_S(N) + Q_S(C) = 0$, as $Q_S \ll P$ and $C \subseteq S^c$. As $A = (A \cap S) \cup (A \cap S^c)$, $Q_S(A) = 0$ implies A is of the form $N \cup C$, where $N = A \cap S$, and $C = A \cap S^c$. □

Lemma 16.4.18 *Let $\mathcal{A}^{Q_S}$ be the completion of $\mathcal{A}$ with respect to Q_S, Q_S^o be the extension of Q_S to $\mathcal{A}^{Q_S}$, and $\mathcal{A}_0^{Q_S}$ be generated by $\mathcal{A}_0$ and the subsets of sets in $\mathcal{N}(\mathcal{A}, Q_S)$. Then*

1. $S \in \mathcal{A}_0^{Q_S}$,
2. $\mathcal{A}^{Q_S} = \{\Omega_0 \subseteq \Omega : \Omega_0 \cap S \in \mathcal{A}\}$,
3. $\mathcal{A}_0^{Q_S} = \{\Omega_0 \subseteq \Omega : \Omega_0 \cap S = A_0 \cap S$, *some* $A_0 \in \mathcal{A}_0\}$,
4. *for* $A^{Q_S} \in \mathcal{A}^{Q_S}$, *fixed, but arbitrary,* $Q_S^o(A^{Q_S}) = Q_S(A^{Q_S} \cap S)$.

Proof [1] As $Q_S(S^c) = 0$, $S^c \in \mathcal{A}_0^{Q_S}$, so that $S \in \mathcal{A}_0^{Q_S}$.

Proof [2] Let $\mathcal{A}_S = \{\Omega_0 \subseteq \Omega : \Omega_0 \cap S \in \mathcal{A}\}$. It is a σ-algebra. Indeed, one has that $\emptyset \cap S = \emptyset \in \mathcal{A}$, and that $\Omega \cap S = S \in \mathcal{A}$. Suppose now that $\Omega_0 \in \mathcal{A}_S$. Then

$$\Omega_0^c \cap S = (\Omega \Delta \Omega_0) \cap S = (\Omega \cap S) \Delta (\Omega_0 \cap S) \in \mathcal{A}.$$

Finally, when $\{\Omega_n, n \in \mathbb{N}\} \subseteq \mathcal{A}_S$,

$$(\cup_n \Omega_n) \cap S = \cup_n(\Omega_n \cap S) \in \mathcal{A}.$$

Since, for $A \in \mathcal{A}$, fixed, but arbitrary, $A \cap S \in \mathcal{A}$, $\mathcal{A} \subseteq \mathcal{A}_S$. If now M is a subset of a set N of measure zero for Q_S, $M \cap S \subseteq N \cap S \in \mathcal{A}$, and $Q_S(N) = 0$, so that $P(N \cap S) = 0$, and $M \cap S \in \mathcal{A}$, as $\mathcal{A}$ is complete. Thus $\mathcal{A}_S \supseteq \mathcal{A}^{Q_S}$.

But, when $\Omega_0 \in \mathcal{A}_S$, as $\Omega_0 = \Omega_0 \cap S + \Omega_0 \cap S^c$, Ω_0 is the set sum of a set in $\mathcal{A}$, and a subset of a set of measure zero for Q_S, S^c. So $\mathcal{A}_S \subseteq \mathcal{A}^{Q_S}$. That ends he proof of item 2.

Proof [3] Let $\mathcal{A}_S = \{\Omega_0 \subseteq \Omega : \Omega_0 \cap S = A_0 \cap S, \text{ some } A_0 \in \mathcal{A}_0\}$. It is a σ-algebra. Indeed, since $\emptyset \in \mathcal{A}_0$, $\emptyset \in \mathcal{A}_S$. The same is true of Ω. If now $\Omega_0 \in \mathcal{A}_S$, $\Omega_0 \cap S = A_0 \cap S$, $A_0 \in \mathcal{A}_0$, and

$$\begin{aligned} \Omega_0^c \cap S &= (\Omega \Delta \Omega_0) \cap S \\ &= (\Omega \cap S) \Delta (\Omega_0 \cap S) \\ &= (\Omega \cap S) \Delta (A_0 \cap S) \\ &= (\Omega \Delta A_0) \cap S \\ &= A_0^c \cap S. \end{aligned}$$

Finally, when $\{\Omega_n, n \in \mathbb{N}\} \subset \mathcal{A}_S$,

$$(\cup_n \Omega_n) \cap S = \cup_n (\Omega_n \cap S) = \cup_n (A_n \cap S) = (\cup_n A_n) \cap S.$$

Since each A_n is in $\mathcal{A}_0$, $\cup_n A_n$ is in $\mathcal{A}_0$, and $\cup_n \Omega_n$, in $\mathcal{A}_S$.

Obviously $\mathcal{A}_0 \subseteq \mathcal{A}_S$, and every subset M of a set N of measure zero for Q_S is in $\mathcal{A}_S$, since $M \cap S$ is a subset of a set of measure zero in $\mathcal{A}$, and that those are in $\mathcal{A}_0$. So, again, $\mathcal{A}_0^{Q_S} \subset \mathcal{A}_S$. Finally, since

$$\Omega_0 = \Omega_0 \cap S + \Omega_0 \cap S^c = A_0 \cap S + \Omega_0 \cap S^c,$$

Ω_0 is a set sum of a set in $\mathcal{A}_0$ and a subset of a set of measure zero for Q_S, and thus $\mathcal{A}_S \subseteq \mathcal{A}_0^{Q_S}$. Item 3 thus obtains.

Proof [4] Since $S \in \mathcal{A}_0^{Q_S}$, that S^c has measure zero for Q_S, and that, from item 2, $A^{Q_S} \cap S \in \mathcal{A}$,

$$Q_S^o(A^{Q_S}) = Q_S^o(A^{Q_S} \cap S) + Q_S^o(A^{Q_S} \cap S^c) = Q_S^o(A^{Q_S} \cap S) = Q_S(A^{Q_S} \cap S).$$

□

Lemma 16.4.19 *Given $A_0 \in \mathcal{A}_0$, fixed, but arbitrary, and $\wp = \{A_1, \ldots A_n\}$,*

$$S(\wp \mid \mathcal{A}_0) \cap A_0 = S(\wp_0 \mid \mathcal{A}_0),$$

where $\wp_0 = \{A_1 \cap A_0, \ldots, A_n \cap A_0\}$.

Proof One has that $P(A \mid \mathcal{A}_0)\,\chi_{A_0} = P(A \cap A_0 \mid \mathcal{A}_0)$. Thus

$$\{\omega \in \Omega : P(A \cap A_0 \mid \mathcal{A}_0) > 0\} = \{\omega \in \Omega : P(A \mid \mathcal{A}_0) > 0\} \cap A_0.$$

Let $C_i = \{\omega \in \Omega : P(A_i \mid \mathcal{A}_0) > 0\}$. Then $S(\wp \mid \mathcal{A}_0) = \cap_{i=1}^{n} C_i$. Thus

$$S(\wp \mid \mathcal{A}_0) \cap A_0 = \cap_{i=1}^{n} \{C_i \cap A_0\} = S(\wp_0 \mid \mathcal{A}_0).$$

□

Proposition 16.4.20 *Let $(\Omega, \mathcal{A}, P)$ be a complete probability space, Q, a probability on $\mathcal{A}$, absolutely continuous with respect to P, $\mathcal{A}_0 \subseteq \mathcal{A}$, a σ-algebra (containing the sets of measure zero for P in $\mathcal{A}$). Then, almost surely, with respect to Q,*

$$mult_Q[\mathcal{A}^Q \mid \mathcal{A}_0^Q] \leq mult_P[\mathcal{A} \mid \mathcal{A}_0],$$

where $\mathcal{A}^Q$ is the completion of $\mathcal{A}$ with respect to Q, and $\mathcal{A}_0^Q$ is the σ-algebra generated by $\mathcal{A}_0$ and the sets of measure zero of the extension of Q to $\mathcal{A}^Q$.

Proof Because of (Proposition) 16.4.15, one may assume that Q is Q_S of (Lemma) 16.4.16. Let thus

$$\wp^{Q_S} = \{A_1^{Q_S}, \ldots, A_n^{Q_S}\}$$

be a partition of Ω in $\mathcal{A}^{Q_S}$. Then

$$\wp_S^{Q_S} = \{A_1^{Q_S} \cap S, \ldots, A_n^{Q_S} \cap S\}$$

is a partition of S in $\mathcal{A}$ because of (Lemma) 16.4.18, item 2. Because of (Lemma) 16.4.18, item 4, and the definition of Q_S, when, for some $i \in [1:n]$, $P(A_i^{Q_S} \cap S) = 0$, $Q_S^o(A_i^{Q_S}) = 0$, and

$$\left|\wp^{Q_S}\right| \chi_{S\left(\wp^{Q_S} \mid \mathcal{A}_0^{Q_S}\right)} = 0.$$

One may thus assume that $P(A_i^{Q_S} \cap S) > 0, i \in [1:n]$. Since [(Lemma) 16.4.18, item 2]

$$A_n^{Q_S} \cup S^c = ([A_n^{Q_S} \cap S] \cup [A_n^{Q_S} \cap S^c]) \cup S^c = [A_n^{Q_S} \cap S] \cup S^c$$

belongs to $\mathcal{A}$, then

$$\wp^{Q_S}_{S,+} = \{A^{Q_S}_1 \cap S, \ldots, A^{Q_S}_{n-1} \cap S, A^{Q_S}_n \cup S^c\}$$

is a partition of Ω in $\mathcal{A}$. Since $A^{Q_S}_n \cup S^c = [A^{Q_S}_n \cap S] \cup S^c$, that, almost surely, with respect to P,

$$0 < P(S^c \mid \mathcal{A}_0) \leq P(A^{Q_S}_n \cup S^c \mid \mathcal{A}_0),$$

and finally that, using $A^{Q_S}_i = [A^{Q_S}_i \cap S] \cup [A^{Q_S}_i \cap S^c]$,

$$Q^o_S(A^{Q_S}_i \mid \mathcal{A}^{Q_S}_0) = Q^o_S(A^{Q_S}_i \cap S \mid \mathcal{A}^{Q_S}_0),$$

it will be sufficient to prove the following inclusion:

$$S(\wp^{Q_S} \mid \mathcal{A}^{Q_S}_0) \cap S \subset S(\wp^{Q_S}_{S,+} \mid \mathcal{A}_0),$$

or the following one: for $A \in \mathcal{A}$, fixed, but arbitrary,

$$\{\omega \in \Omega : Q^o_S(A \mid \mathcal{A}^{Q_S}_0) > 0\} \cap S \subset \{\omega \in \Omega : P(A \mid \mathcal{A}_0) > 0\}.$$

Let $X = \{\omega \in \Omega : Q^o_S(A \mid \mathcal{A}^{Q_S}_0) > 0\}$, and $Y = \{\omega \in \Omega : P(A \mid \mathcal{A}_0) > 0\}$. As seen [(Lemma) 16.4.14], Y is, within sets of probability zero for P, the smallest set in $\mathcal{A}_0$ which contains A, almost surely, with respect to P, and X is, within sets of probability zero for Q^o_S, the smallest set in $\mathcal{A}^{Q_S}_0$ which contains A, almost surely, with respect to Q^o_S. Now Y is also in $\mathcal{A}^{Q_S}_0$, and contains A, almost surely, with respect to Q_S, and thus with respect to Q^o_S. Consequently, since X is smallest, X is contained in Y, almost surely with respect to Q^o_S, and then $X \setminus Y$ is a set of measure zero for Q^o_S. So

$$0 = Q^o_S(X \setminus Y) = Q_S([X \setminus Y] \cap S) = P(S)^{-1} P([X \setminus Y] \cap S).$$

But, as $[X \cap S] \setminus [Y \cap S] = [X \cap S] \cap [Y^c \cup S^c] = [X \cap S] \cap Y^c = [X \setminus Y] \cap S$, then

$$\begin{aligned} X \cap S &= ([X \cap S] \setminus [Y \cap S]) \cup ([X \cap S] \cap [Y \cap S]) \\ &= ([X \setminus Y] \cap S) \cup ([X \cap S] \cap [Y \cap S]). \end{aligned}$$

Consequently, almost surely, with respect to P, thus to Q_S, and consequently to Q^o_S, $X \cap S \subseteq Y \cap S \subset Y$. □

Proposition 16.4.21 *When, for $n \in \mathbb{N} \cup \{\infty\}$, fixed, but arbitrary,*

$$P\left(\{mult[\mathcal{A} \mid \mathcal{A}_0] \geq n\}\right) > 0,$$

there exists $\wp$ such that $|\wp| = n$, and, almost surely, with respect to P,

$$S(\wp \mid \mathcal{A}_0) = \{mult[\mathcal{A} \mid \mathcal{A}_0] \geq n\}.$$

Proof Write M for $mult[\mathcal{A} \mid \mathcal{A}_0]$. When $M(\omega) = n$, there is at least a partition of Ω with n sets in $\mathcal{A}$, whose conditional probabilities with respect to $\mathcal{A}_0$ are strictly positive, and, furthermore, ω is in the support, with respect to $\mathcal{A}_0$, of that partition. There is thus a countable family of partitions, say $\{\wp_q, q \in \mathbb{N}\}$ such that

$$\{M \geq n\} \subseteq \cup_q S(\wp_q \mid \mathcal{A}_0).$$

Taking the intersection of the latter inclusion with $\{M \geq n\}$, a set in $\mathcal{A}_0$, one may assume [(Lemma) 16.4.19] that the inclusion is an equality; finally, since the sets $\{M = n\}$ are disjoint, one may assume that the union is a disjoint union. For each q, let $\{A_1^{(q)}, \ldots, A_n^{(q)}\}$ be a partition of Ω in $\mathcal{A}$, with n elements, which is not finer than $\wp_q$. Let

$$\tilde{A}_i^{(q)} = A_i^{(q)} \cap S(\wp_q \mid \mathcal{A}_0),$$
$$\hat{A}_i = \cup_q \tilde{A}_i^{(q)}.$$

One thus obtains, respectively, a partition of $S(\wp_q \mid \mathcal{A}_0)$, and of $\{M \geq n\}$. Now

$$\hat{A}_i \cap S(\wp_q \mid \mathcal{A}_0) = \tilde{A}_i^{(q)} = A_i^{(q)} \cap S(\wp_q \mid \mathcal{A}_0),$$

so that,

$$P(\hat{A}_i \mid \mathcal{A}_0) = \sum_q P(\tilde{A}_i^{(q)} \mid \mathcal{A}_0) > 0.$$

The following set of events:

$$\left\{\hat{A}_1 \cup \{M < n\}, \hat{A}_2, \ldots, \hat{A}_n\right\}$$

forms a partition $\wp$ of Ω in $\mathcal{A}$, of which each element has, with respect to $\mathcal{A}_0$, a strictly positive probability. Furthermore $\{M \geq n\} \subseteq S(\wp \mid \mathcal{A}_0)$, and, consequently, $\{M \geq n\} = S(\wp \mid \mathcal{A}_0)$. □

Corollary 16.4.22 *Using the notation of* (Proposition) 16.4.21, *when* $P(\{M \geq 2\}) > 0$*, there exists* $A \in \mathcal{A}$ *with*

$$\{M \geq 2\} \subseteq \{P(A \mid \mathcal{A}_0) > 0\} \cap \{P(A^c \mid \mathcal{A}_0) > 0\}.$$

Proof From (Proposition) 16.4.21, there exists $A \in \mathcal{A}$ such that, for example,

$$\{M \geq 2\} = \{P(A \mid \mathcal{A}_0) P(A^c \mid \mathcal{A}_0) > 0\} \subseteq \{P(A \mid \mathcal{A}_0) > 0\}.$$

□

16.5 Vershik's Lacunary Isomorphism Theorem

That theorem shall perform as a *Deus ex machina* in the matter which follows, as it finds, in, at first superficial sight, amorphous filtrations, the explicit structure that shall prove indispensable.

Definition 16.5.1 Let $(\Omega, \mathcal{A}, P)$ be a probability space, and let $\underline{\mathcal{A}}$ and $\underline{\mathcal{B}}$ be two filtrations in $\mathcal{A}$. $\underline{\mathcal{A}} \vee \underline{\mathcal{B}}$ shall be the filtration with entries $\mathcal{A}_t \vee \mathcal{B}_t$. One then says that:

1. $\underline{\mathcal{B}}$ is included in $\underline{\mathcal{A}}$ when, for $t \in \mathbb{R}_+$, fixed, but arbitrary, $\mathcal{B}_t \subseteq \mathcal{A}_t$.
2. $\underline{\mathcal{B}}$ is immersed in $\underline{\mathcal{A}}$ when every martingale for $\underline{\mathcal{B}}$ is one for $\underline{\mathcal{A}}$.
3. $\underline{\mathcal{B}}$ and $\underline{\mathcal{A}}$ are jointly immersed when both are immersed in $\underline{\mathcal{A}} \vee \underline{\mathcal{B}}$.
4. Let $\mathcal{B} \subseteq \mathcal{A}$ be a σ-algebra. It is saturated for $\underline{\mathcal{A}}$ when $\mathcal{B} \subseteq \mathcal{A}_\infty$, and, for fixed, but arbitrary $t \in \mathbb{R}_+$ and $B \in \mathcal{B}$, $P(B \mid \mathcal{A}_t)$ is adapted to $\mathcal{B}$.

Remark 16.5.2 (Immersion Implies Inclusion) Indeed, when $B \in \mathcal{B}_\theta$ is fixed, but arbitrary, $M(\cdot, t) = E_P[\chi_B \mid \mathcal{B}_t]$ is a martingale for which, when $t \geq \theta$, $M(\cdot, t) = \chi_B$. Since M is a martingale for $\underline{\mathcal{A}}$, $B \in \mathcal{A}_t$ for $t \geq \theta$.

Remark 16.5.3 Let $\underline{\mathcal{F}}$ and $\underline{\mathcal{G}}$ be independent filtrations, meaning that, when $t \in T$, is fixed, but arbitrary, $\mathcal{F}_t$ and $\mathcal{G}_t$ are independent. Then they are jointly immersed in $\{\mathcal{F}_t \vee \mathcal{G}_t, t \in T\}$. That follows from the fact that, for example, for a martingale M with respect $\underline{\mathcal{F}}$, $t_1 < t_2$ in T, $F \in \mathcal{F}_{t_1}$, $G \in \mathcal{G}_{t_1}$, fixed, but arbitrary, using the independence of χ_G and $\chi_F M$,

$$\int_{F \cap G} M(\cdot, t_2)\, dP = \int_{F \cap G} M(\cdot, t_1)\, dP.$$

Proposition 16.5.4 *Let* $(\Omega, \mathcal{A}, P)$ *be a probability space, and* $\underline{\mathcal{A}}$ *be a fixed, but arbitrary filtration of* $\mathcal{A}$*. Let* $\mathcal{F}_{\mathcal{I}}(\underline{\mathcal{A}})$ *denote the family of filtrations immersed in* $\underline{\mathcal{A}}$*, and* $\mathcal{S}(\underline{\mathcal{A}})$*, that of* σ*-algebras of* $\mathcal{A}$ *that are saturated for* $\underline{\mathcal{A}}$*.*

1. *When* $\underline{\mathcal{F}}$ *is immersed in* $\underline{\mathcal{A}}$*, and* $\mathcal{N}(\mathcal{A}_t, P) \subseteq \mathcal{F}_t$*, for* $t \in \mathbb{R}_+$*, fixed, but arbitrary,* $\mathcal{F}_t = \mathcal{F}_\infty \cap \mathcal{A}_t$*.*

2. *Let $\Phi : \mathcal{F}_{\mathcal{I}}(\underline{\mathcal{A}}) \longrightarrow \mathcal{S}(\underline{\mathcal{A}})$ be defined using the following rule:*

$$\underline{\mathcal{F}} \mapsto \mathcal{F}_\infty.$$

When $\mathcal{N}(\mathcal{A}, P) \subseteq \mathcal{F}_0$, Φ is a bijection, and its inverse is the following map:

$$\mathcal{S} \mapsto \{\mathcal{F}_t = \mathcal{S} \cap \mathcal{A}_t, t \in \mathbb{R}_+\}.$$

Proof [Step 1] *When $\underline{\mathcal{F}}$ be immersed in $\underline{\mathcal{A}}$, $\mathcal{F}_\infty$ is saturated for $\underline{\mathcal{A}}$, and Φ has indeed range in $\mathcal{S}(\underline{\mathcal{A}})$.*

Since immersion implies inclusion [(Remark) 16.5.2], $\mathcal{F}_\infty \subseteq \mathcal{A}_\infty$. Let $S \in \mathcal{F}_\infty$ be fixed, but arbitrary. Then

$$M(\cdot, t) = P(S \mid \mathcal{F}_t) = E_P[\chi_S \mid \mathcal{F}_t]$$

is a martingale for $\underline{\mathcal{F}}$. Thus [201, p. 96]

$$M_\infty = \lim_t M(\cdot, t) = E_P[\chi_S \mid \mathcal{F}_\infty] = \chi_S.$$

Let also

$$N(\cdot, t) = P(S \mid \mathcal{A}_t) = E_P[\chi_S \mid \mathcal{A}_t],$$

a martingale for $\underline{\mathcal{A}}$. Since $\mathcal{F}_\infty \subseteq \mathcal{A}_\infty$,

$$N_\infty = \lim_t N(\cdot, t) = E_P[\chi_S \mid \mathcal{A}_\infty] = \chi_S.$$

Since $\underline{\mathcal{F}}$ is immersed in $\underline{\mathcal{A}}$, M is a martingale for $\underline{\mathcal{A}}$. $M - N$ is then a martingale for $\underline{\mathcal{A}}$, so that $|M - N|$ is a submartingale for $\underline{\mathcal{A}}$. Since $M_\infty = N_\infty = \chi_S$, outside of a set $S_0 \in \mathcal{N}(\mathcal{A}_t, P)$, $M(\cdot, t) = N(\cdot, t)$. Consequently

$$P(S \mid \mathcal{A}_t) = \chi_{S_0^c} P(S \mid \mathcal{A}_t) + \chi_{S_0} P(S \mid \mathcal{A}_t) = \chi_{S_0^c} P(S \mid \mathcal{F}_t) + \chi_{S_0} P(S \mid \mathcal{A}_t).$$

Thus, when $\mathcal{N}(\mathcal{A}_t, P) \subseteq \mathcal{F}_t$, $P(S \mid \mathcal{A}_t)$ is adapted to $\mathcal{F}_t$, and a fortiori, to $\mathcal{F}_\infty$.

Proof [Step 2] *Given that, when $\underline{\mathcal{F}}$ is immersed in $\underline{\mathcal{A}}$, $\mathcal{F}_\infty$ is saturated for $\underline{\mathcal{A}}$, one has that $\mathcal{F}_t = \mathcal{F}_\infty \cap \mathcal{A}_t$.*

When $S \in \mathcal{F}_\infty \cap \mathcal{A}_t$, given the immersion and the general completion assumptions, from step 1,

$$E_P[\chi_S \mid \mathcal{F}_t] = E_P[\chi_S \mid \mathcal{A}_t] = \chi_S,$$

so that $S \in \mathcal{F}_t$, that is, $\mathcal{F}_\infty \cap \mathcal{A}_t \subseteq \mathcal{F}_t$. If now $F \in \mathcal{F}_t$ is fixed, but arbitrary, $F \in \mathcal{F}_\infty$, so that, as above, $E_P[\chi_F \mid \mathcal{A}_t] = E_P[\chi_F \mid \mathcal{F}_t] = \chi_F$, and $F \in \mathcal{A}_t$. Consequently $\mathcal{F}_t = \mathcal{F}_\infty \cap \mathcal{A}_t$.

It has thus been checked that, when $\underline{\mathcal{F}}$ immersed in $\underline{\mathcal{A}}$, $\mathcal{F}_\infty$ is saturated for $\underline{\mathcal{A}}$, and $\mathcal{F}_t = \mathcal{F}_\infty \cap \mathcal{A}_t$. So item 1 obtains.

Proof [Step 3] *When $\mathcal{S}$ is a σ-algebra saturated for $\underline{\mathcal{A}}$, there is a filtration $\underline{\mathcal{F}}$ such that $\mathcal{S} = \mathcal{F}_\infty$, and $\mathcal{F}_t = \mathcal{S} \cap \mathcal{A}_t$ (Φ is thus well defined and onto).*

By assumption, $\mathcal{S} \subseteq \mathcal{A}_\infty$, and, for $t \in \mathbb{R}_+$ and $S \in \mathcal{S}$, $P(S \mid \mathcal{A}_t)$ is adapted to $\mathcal{S}$. Then, in particular, $P(S \mid \mathcal{A}_t)$ is adapted to $\mathcal{S} \cap \mathcal{A}_t$. Let $\mathcal{F}_t = \mathcal{S} \cap \mathcal{A}_t \subseteq \mathcal{S}$, so that $\mathcal{F}_\infty \subseteq \mathcal{S}$.

Let X be bounded, and adapted to $\mathcal{S}$, and $M(\cdot, t) = E_P[X \mid \mathcal{A}_t]$. Because of the previous paragraph, by the monotone class theorem, M is adapted to $\underline{\mathcal{F}}$, and is thus a martingale with respect to that latter filtration: indeed, for $t_1 < t_2$, fixed, but arbitrary, when $M(\cdot, t_1)$ is adapted to $\mathcal{F}_{t_1}$,

$$\begin{aligned} E_P[M(\cdot, t_2) \mid \mathcal{F}_{t_1}] &= E_P[E_P[M(\cdot, t_2) \mid \mathcal{A}_{t_1}] \mid \mathcal{F}_{t_1}] \\ &= E_P[M(\cdot, t_1) \mid \mathcal{F}_{t_1}] \\ &= M(\cdot, t_1). \end{aligned}$$

Now, M being a martingale with respect to $\underline{\mathcal{F}}$, since $\mathcal{S} \subseteq \mathcal{A}_\infty$,

$$M(\cdot, \infty) = \lim_t E_P[X \mid \mathcal{A}_t] = E_P[X \mid \mathcal{A}_\infty] = X,$$

and, since $M(\cdot, \infty) \in \mathcal{F}_\infty$, $X \in \mathcal{F}_\infty$, that is, $\mathcal{S} \subseteq \mathcal{F}_\infty$. Consequently $\mathcal{S} = \mathcal{F}_\infty$.

Proof [Step 4] *The filtration $\underline{\mathcal{F}}$, defined in step 3, is immersed in $\underline{\mathcal{A}}$.*

According to the proof of step 3, given $\mathcal{S}$ saturated for $\underline{\mathcal{A}}$, one may define $\underline{\mathcal{F}}$ so that, for every X, bounded and adapted to $\mathcal{S}$, there is a process M that is simultaneously a martingale for $\underline{\mathcal{F}}$ and $\underline{\mathcal{A}}$. That result remains true for integrable X, using truncation at n, because of [56, p. 13], provided that $\mathcal{N}(\mathcal{A}, P) \subseteq \mathcal{F}_0$. When dealing with a general martingale, one applies the same procedure to $M^n(\cdot, t) = E_P[M(\cdot, n) \mid \mathcal{F}_t]$. □

Lemma 16.5.5 *Let $\mathcal{A}_n$ be, for $n \in \mathbb{N}$, a σ-algebra of subsets of Ω. Let $\mathcal{B}_n = \sigma(\mathcal{A}_k, k \le n)$, and $\mathcal{C} = \sigma(\mathcal{B}_n, n \in \mathbb{N})$. A real valued function adapted to $\mathcal{C}$ has the following form: $\Phi(\alpha_1, \ldots, \alpha_n, \ldots)$, where Φ is real valued and adapted to the Borel sets of $\mathbb{R}^\infty$, and, for $n \in \mathbb{N}$, α_n is real valued, and adapted to $\mathcal{A}_n$.*

Proof One has that $\vee_{n=1}^\infty \mathcal{A}_n = \vee_{n=1}^\infty \{\vee_{i=1}^n \mathcal{A}_i\}$. One uses then result [41, p. 144] to conclude. □

Proposition 16.5.6 *Let $(\Omega, \mathcal{A}, P)$ be a probability space, and $\underline{\mathcal{A}}$ be a filtration indexed by $\mathbb{Z}_0^-$, the set of negative integers and zero. Suppose that, for $n \in \mathbb{Z}_0^-$, fixed, but arbitrary,*

(a) *$\mathcal{B}_n$ is a σ-algebra,*
(b) *$\mathcal{B}_n \subseteq \mathcal{A}_n$,*
(c) *$\mathcal{B}_n$ is independent of $\mathcal{A}_{n-1}$.*

Let $\mathcal{C}_n = \sigma(\{\mathcal{B}_k, k \leq n\})$. Then:

1. *$\underline{\mathcal{C}}$ is immersed in $\underline{\mathcal{A}}$;*
2. *$\mathcal{C}_\infty = \sigma(\{\mathcal{C}_n, n \in \mathbb{Z}_0^-\})$ is saturated for $\underline{\mathcal{A}}$.*

Proof Let B_n denote a random variable adapted to $\mathcal{B}_n$. Every bounded random variable X, adapted to $\mathcal{C}_\infty$, has, for some measurable Φ, a representation of the following form [(Lemma) 16.5.5]:

$$X = \Phi(B_n, n \in \mathbb{Z}_0^-).$$

The family $\underline{B}_n = \{B_k, k \leq n\}$ is adapted to $\mathcal{A}_n$, and, furthermore, the family $\underline{B}^{(n)} = \{B_k, k \geq n+1\}$ is independent of $\mathcal{A}_n$. Thus [138, p. 452], setting $\Psi(\underline{b}_n) = E_P\left[\Phi\left(\underline{b}_n, \underline{B}^{(n)}\right)\right]$,

$$E_P[X \mid \mathcal{A}_n] = E_P\left[\Phi\left(\underline{B}_n, \underline{B}^{(n)}\right) \mid \mathcal{A}_n\right] = \Psi(\underline{B}_n).$$

So $E_P[X \mid \mathcal{A}_n]$ is adapted to $\mathcal{C}_n$. Furthermore, as $\mathcal{C}_n \subseteq \mathcal{A}_n$, for $C_n \in \mathcal{C}_n$, fixed, but arbitrary,

$$\int_{C_n} X dP = \int_{C_n} E_P[X \mid \mathcal{A}_n] dP,$$

and, by definition,

$$\int_{C_n} X dP = \int_{C_n} E_P[X \mid \mathcal{C}_n] dP.$$

Consequently $E_P[X \mid \mathcal{C}_n] = E_P[X \mid \mathcal{A}_n]$. So $\underline{\mathcal{C}}$ is immersed in $\underline{\mathcal{A}}$, and $\mathcal{C}_\infty$ is saturated for $\underline{\mathcal{A}}$ [(Proposition) 16.5.4, step 5]. □

Lemma 16.5.7 *A function $f : \mathbb{R}_+ \longrightarrow \mathbb{R}$ which is continuous to the right is Borel measurable.*

Proof For $\{k, n\} \subseteq \{0\} \cup \mathbb{N}$, let $t_{n,k} = \frac{k}{2^n}$, and $I_{n,k} = \left[\frac{k}{2^n}, \frac{k+1}{2^n}\right[$. Set

$$f_n(x) = \sum_k f\left((k+1)2^{-n}\right) \chi_{I_{n,k}}(x).$$

Given $x \in \mathbb{R}_+$, fixed, but arbitrary, there is a monotone decreasing sequence of intervals of type $I_{n,k}$, say $I^{(x)}_{n,k}$, which contain x. The intersection of those intervals is x. But

$$\lim_n \chi_{I^{(x)}_{n,k}}(x)\,|f(x) - f_n(x)| = \lim_n \chi_{I^{(x)}_{n,k}}\,|f(x) - f\,((k(x)+1)2^{-n})| = 0.$$

□

Lemma 16.5.8 *Let $(M, \mathcal{M})$ be a measurable space, and $f : M \times \mathbb{R} \longrightarrow \mathbb{R}$, a function such that, for $m \in M$, fixed, but arbitrary, $x \mapsto f(m,x)$ is continuous to the right, and, for fixed, but arbitrary $x \in \mathbb{R}$, $m \mapsto f(m,x)$ is adapted. Then f is adapted to $\mathcal{M} \otimes \mathcal{B}(\mathbb{R})$. The same is true when continuity is to the left.*

Proof Let $f_n(m,x) = \sum_k f\,(m,(k+1)2^{-n})\,\chi_{M\times I_{n,k}}(m,x)$. □

Lemma 16.5.9 *Let $(\Omega, \mathcal{A}, P)$ be a complete probability space, $\mathcal{C}_0 \subseteq \mathcal{A}$, a separable σ-algebra* [so that, [138, p. 91] for some f, $\mathcal{C}_0 = f^{-1}(\mathcal{B}(\mathbb{R}))$], *and also $\mathcal{C}$ be generated by $\mathcal{C}_0$ and the subsets of $\mathcal{N}(\mathcal{A}, P)$. Let $\hat{\mathcal{B}}(\mathbb{R})$ denote the universally measurable sets for $\mathcal{B}(\mathbb{R})$* [70, p. 50], *and $\phi : \mathbb{R} \longrightarrow D$ be adapted to $\hat{\mathcal{B}}(\mathbb{R})$ and $\mathcal{D}$. Then $\phi \circ f$ is adapted to $\mathcal{C}$ and $\mathcal{D}$.*

Proof Let

$$\mathcal{B}_f(\mathbb{R}) = \left\{S \subseteq \mathbb{R} : f^{-1}(S) \in \mathcal{C}_0\right\}.$$

It is a σ-algebra [30, p. 17], and f is adapted to $\mathcal{C}_0$ and $\mathcal{B}_f(\mathbb{R})$. The following relation: $P_f = P \circ f^{-1}$ then defines a probability on $\mathcal{B}_f(\mathbb{R})$. Let P^o_f and $\mathcal{B}^o_f(\mathbb{R})$ be the corresponding completions. One has that

$$\hat{\mathcal{B}}(\mathbb{R})) \subseteq \mathcal{B}^o_f(\mathbb{R}),$$

and ϕ is thus adapted to $\mathcal{B}^o_f(\mathbb{R})$. Let $D_0 \in \mathcal{D}$ be fixed, but arbitrary. Then

$$\phi^{-1}(D_0) = B_0 \Delta M_0,\ M_0 \subseteq N_0 \in \mathcal{N}(\mathcal{B}_f(\mathbb{R}), P_f).$$

Now

$$f^{-1}(B_0 \Delta M_0) = f^{-1}(B_0) \Delta f^{-1}(M_0),\ f^{-1}(M_0) \subseteq f^{-1}(N_0).$$

But $P(f^{-1}(N_0)) = P_f(N_0) = 0$. Thus $(\phi \circ f)^{-1}(D_0) \in \mathcal{C}$, and $\phi \circ f$ is adapted, as claimed. □

Lemma 16.5.10 *Let F be a continuous distribution function with associated measure μ_F. Then* (notation is from (Definition) 10.3.2):

1. $\mu_F(F^{-1}([0,t])) = t$;
2. $\overline{F}(F(t)) = t$.

Proof Since F is continuous, $\{x \in \mathbb{R} : F(x) \leq t\}$ is a closed interval. Its right-hand limit is $\overline{F}(t)$. The interval $[\underline{F}(t), \overline{F}(t)]$ has zero measure, and, consequently, $F(\overline{F}(t)) = F(\underline{F}(t)) = t$, because of (Fact) 10.3.12, item 2. Also, still because F is continuous, $\overline{F}(F(t)) = \sup\{x \in \mathbb{R} : F(x) \leq F(t)\} = t$. □

Proposition 16.5.11 *$\mathcal{L}(\cdot)$ denotes the law of element $(\cdot)$, and $\mathcal{U}[0, 1]$, the uniform distribution on $[0, 1]$. Let $\mathcal{B}$ and $\mathcal{C}$ be σ-algebras in $\mathcal{A}$, with $\mathcal{C} \subseteq \mathcal{B}$, and $\mathcal{B}$, essentially separable. Consider the following statements:*

1. *There exists $X : \Omega \longrightarrow \mathbb{R}$, adapted to $\mathcal{B}$, such that, for every $Y : \Omega \longrightarrow \mathbb{R}$, adapted to $\mathcal{C}$, $P(\omega \in \Omega : X(\omega) = Y(\omega)) = 0$.*
2. *There exists $X : \Omega \longrightarrow \mathbb{R}$, adapted to $\mathcal{B}$, independent of $\mathcal{C}$, such that $\mathcal{L}(X)$ is diffuse.*
3. *There exists $X : \Omega \longrightarrow \mathbb{R}$, adapted to $\mathcal{B}$, independent of $\mathcal{C}$, such that $\mathcal{L}(X) = \mathcal{U}[0, 1]$ and $\mathcal{C} \vee \sigma(X) = \mathcal{B}$.*
4. *Whenever $Z : \Omega \longrightarrow \mathbb{R}$ is such that $\mathcal{B} = \mathcal{C} \vee \sigma(Z)$, it has a law that is diffuse.*

Then, when the σ-algebras concerned contain the subsets of sets in $\mathcal{N}(\mathcal{A}, P)$, items 1–4 are equivalent.

Proof Obviously item 3 implies item 2.

Proof [2 ⇒ 1] Let $\Delta(\mathbb{R}^2)$ be the diagonal of $\mathbb{R}^2$. It is measurable [138, p. 92]. Let Y be an arbitrary random variable adapted to $\mathcal{C}$. Then, given X adapted to $\mathcal{B}$, independent of $\mathcal{C}$, with a diffuse law,

$$P(\omega \in \Omega : X(\omega) = Y(\omega)) = \int_{\Delta(\mathbb{R}^2)} P_X(dx)P_Y(dy) = \int_{\mathbb{R}} P_X(\{y\})P_Y(dy).$$

Since P_X is diffuse, one integrates the zero function.

Proof [1 ⇒ 3] Let the separable $\mathcal{B}$ be generated by the random variable B, and the separable $\mathcal{C}$, by C [138, p. 91]. Let also $P_{B|C}$ be a Markov kernel providing a regular version of the conditional law of B given C [30, p. 308]: one has that

$$P(B \in I, C \in J) = \int_J P_{B|C}(I, c)P_C(dc) = \int_{C^{-1}(J)} P_{B|C}(I, C(\omega))P(d\omega).$$

For $I =]-\infty, x]$, one shall write $F_c(x)$ for $P_{B|C}(I, c)$. Thus F_c is a distribution function, and, as such, it has limits from the left, and is continuous to the right. Furthermore, for x fixed, but arbitrary, $c \mapsto F_c(x)$ is adapted. Consequently, $F(x, c) = F_c(x)$ is jointly measurable [(Lemma) 16.5.8].

Since X is adapted to $\mathcal{B}$, one has that $X = \phi_X \circ B$, some measurable ϕ_X [138, p. 443]. Let Y be adapted to $\mathcal{C}$. Then, given the assumption,

$$\begin{aligned} P(\omega \in \Omega : B(\omega) = Y(\omega)) &\leq P(\omega \in \Omega : \phi_X \circ B(\omega) = \phi_X \circ Y(\omega)) \\ &= P(\omega \in \Omega : X(\omega) = \phi_X \circ Y(\omega)) \\ &= 0. \end{aligned}$$

Since $F_c^-(x)$ is continuous to the left, it is jointly measurable [(Lemma) 16.5.8]. For fixed, but arbitrary $\epsilon > 0$, the set $\{(c, x) \in \mathbb{R}^2 : F_c(x) - F_c^-(x) \geq \epsilon\}$ is jointly measurable, and thus, setting $J_\epsilon(c) = \{x \in \mathbb{R} : F_c(x) - F_c^-(x) \geq \epsilon\}$, one has that the following map:

$$a_\epsilon(c) = \begin{cases} \infty & \text{when } J_\epsilon(c) = \emptyset \\ \inf J_\epsilon(c) & \text{when } J_\epsilon(c) \neq \emptyset \end{cases}$$

is universally measurable [128, p. 110]. Since $J_\epsilon(c)$ is finite, $a_\epsilon(c)$ yields the smallest point at which the jump of F_c is greater than, or equal to, ϵ.

It follows from (Lemma) 16.5.9 that $a_\epsilon \circ C$ is adapted. Now, from what has been assessed above,

$$P(\omega \in \Omega : B(\omega) = a_\epsilon \circ C(\omega)) = 0,$$

that is,

$$P_{B,C}\left((x, y) \in \mathbb{R}^2 : x = a_\epsilon(y)\right) = 0,$$

so that, almost surely, there is no atom of size at least ϵ. Consequently F_C is diffuse almost surely.

Because of (Lemma) 16.5.10, $P_{B|C}\left(F_c^{-1}([0, t]), c\right) = t$. Let $Z = F(B, C) = F_C(B)$. Since $\mathcal{C} \subseteq \mathcal{B}$, Z is adapted to $\mathcal{B}$. Furthermore, for fixed, but arbitrary $t \in [0, 1]$, and measurable ϕ,

$$\begin{aligned} E_P\left[\{\phi \circ C\}\left\{\chi_{[0,t]}(Z)\right\}\right] &= \int_{\mathbb{R}^2} \phi(c)\, \chi_{[0,t]}(F(x, c))\, P_{B,C}(dx, dc) \\ &= \int_{\mathbb{R}} \phi(c) P_C(dc) \int_{\mathbb{R}} \chi_{[0,t]}(F(x, c))\, F_c(dx) \\ &= \int_{\mathbb{R}} \phi(c) P_C(dc) P_{B|C}(F_c^{-1}([0, t]), c) \\ &= t E_P[\phi(C)]. \end{aligned}$$

Thus Z and C are independent, and Z is uniformly distributed on the unit interval. Furthermore, using (Lemma) 16.5.10 again, one has, almost surely, that

$$B = \overline{F}_C(F_C(B)) = \overline{F}_C \circ Z$$

is adapted to $\mathcal{C} \vee \sigma(Z)$, and, consequently, $\mathcal{B} = \mathcal{C} \vee \sigma(Z)$.

Proof [1 $\Rightarrow$ 4] Suppose that Z is such that $\mathcal{B} = \mathcal{C} \vee \sigma(Z)$. For X of the assumption, adapted to $\mathcal{B}$, for some ϕ, $X = \phi(C, Z)$ [(Lemma) 16.2.21]. Thus, when $Z(\omega) = z$, $\phi(C(\omega), z) = X(\omega)$, and then

$$P(\omega \in \Omega : Z(\omega) = z) \leq P\left(\omega \in \Omega : X(\omega) = \phi(C(\omega), z)\right),$$

which, by assumption, is zero.

Proof [4 $\Rightarrow$ 1] One may assume that B generating $\mathcal{B}$ is strictly positive [138, p. 91]. Let Y be adapted to $\mathcal{C}$, and $Z = \chi_{\{Y \neq B\}} B$. Since $\mathcal{C} \subseteq \mathcal{B}$, Z is adapted to $\mathcal{B}$. Furthermore $\{Z = 0\} = \{Y = B\}$, so that

$$B = \chi_{\{Y \neq B\}} B + \chi_{\{Y = B\}} B = Z + \chi_{\{Z=0\}} Y,$$

and, consequently, $\mathcal{B} = \mathcal{C} \vee \sigma(Z)$. But then, by assumption, Z is diffuse. Thus $P(Y = B) = P(Z = 0) = 0$, and the required X of item 1 is B. □

Remark 16.5.12 (M. Émery) In the above, one may extend the results true for random variables to random elements, whenever the measure spaces involved are, as measure spaces, isomorphic to $\mathbb{R}$ and its Borel sets. One may also notice that the proof that item 1 implies item 4 applies to vectors as well.

Definition 16.5.13 When assertions 1–4 of (Proposition) 16.5.11 obtain, $\mathcal{B}$ is said to be conditionally nonatomic given $\mathcal{C}$, and any X for which item 3 obtains is said to be a complement to $\mathcal{C}$ in $\mathcal{B}$.

Proposition 16.5.14 *Let $\mathcal{B}$, $\mathcal{C}$, and $\mathcal{D}$ be σ-algebras in the essentially separable $\mathcal{A}$, and assume that $\mathcal{C} \subseteq \mathcal{B}$. Then,*

1. *when $\mathcal{B} \vee \mathcal{D}$ is conditionally nonatomic given $\mathcal{C} \vee \mathcal{D}$, then $\mathcal{B}$ is conditionally nonatomic given $\mathcal{C}$;*
2. *when $\mathcal{B}$ is conditionally nonatomic given $\mathcal{C}$, and $\mathcal{B}$ and $\mathcal{D}$ are independent, then $\mathcal{B} \vee \mathcal{D}$ is conditionally nonatomic given $\mathcal{C} \vee \mathcal{D}$, and every complement of $\mathcal{C}$ in $\mathcal{B}$ is a complement of $\mathcal{C} \vee \mathcal{D}$ is $\mathcal{B} \vee \mathcal{D}$.*

Proof Let Z be a random variable such that $\mathcal{C} \vee \sigma(Z) = \mathcal{B}$. One has then that $\mathcal{C} \vee \mathcal{D} \vee \sigma(Z) = \mathcal{B} \vee \mathcal{D}$, and the assumption, coupled with (Proposition) 16.5.11, item 4, yields that Z is diffuse. That same condition applied to $\mathcal{B}$, $\mathcal{C}$, and Z does the trick.

Let X be a complement of $\mathcal{C}$ in $\mathcal{B}$. It is independent of $\mathcal{C}$ [(Proposition) 16.5.11, item 3]. Furthermore, $\mathcal{D}$ being independent of $\mathcal{B}$, is independent of $\mathcal{C} \vee \sigma(X)$. So X is independent of $\mathcal{C} \vee \mathcal{D}$. As $\mathcal{C} \vee \sigma(X) \vee \mathcal{D} = \mathcal{B} \vee \mathcal{D}$, X is a complement of $\mathcal{C} \vee \mathcal{D}$ in $\mathcal{B} \vee \mathcal{D}$. □

Both examples which follow warn against possible pitfalls in the use of the previous concept.

Example 16.5.15 (Pitfall 1) Given that $\mathcal{B}$ is conditionally nonatomic given $\mathcal{C}$, that X is adapted to $\mathcal{B}$, is independent of $\mathcal{C}$, and has a diffuse law, there may exist no complement Y of $\mathcal{C}$ in $\mathcal{B}$ such that $\sigma(X) \subseteq \sigma(Y)$.

Let C, X and Z be independent, uniform on $[0, 1]$, and use the following notation: $\Gamma = \left\{C < \frac{1}{2}\right\}$. Set

$$\mathcal{C} = \sigma(C),\ \mathcal{B} = \sigma(C, X, \chi_\Gamma Z).$$

As X is adapted to $\mathcal{B}$, independent of $\mathcal{C}$, and diffuse, $\mathcal{B}$ is conditionally nonatomic given $\mathcal{C}$ because of (Proposition) 16.5.11, item 2, and (Definition) 16.5.13.

Let Y be adapted to $\mathcal{B}$, be independent of $\mathcal{C}$, and such that $\sigma(Y) \supseteq \sigma(X)$.

Suppose for a moment that $\sigma(Y) = \sigma(X)$. As

$$E_P[\chi_\Gamma Z \mid \sigma(C, X)] = \chi_\Gamma E_P[Z] = \chi_\Gamma \frac{1}{2},$$

$\chi_\Gamma Z$ is not adapted to $\sigma(C, X)$, so that $\sigma(C) \vee \sigma(Y) = \sigma(C) \vee \sigma(X) = \sigma(C, X) \subset \mathcal{B}$, and the example is confirmed.

One shall now check that, indeed, $\sigma(Y) = \sigma(X)$, for candidates Y to the role of complement. Since Y is adapted to $\mathcal{B} = \sigma(C, X, \chi_\Gamma Z)$, and that, by assumption, $\sigma(Y) \supseteq \sigma(X)$, there exists a measurable N, Borel Φ and Ψ, such that $P(N) = 0$, and, outside of N, both $X = \Phi(Y)$, and $Y = \Psi(C, X, \chi_\Gamma Z)$ obtain. Let $\xi(c, x) = \Psi(c, x, 0)$. Then, outside of N, since $C < \frac{1}{2}$,

$$\chi_{\left[\frac{1}{2}, 1\right]}(C) Y = \chi_{\left[\frac{1}{2}, 1\right]}(C) \xi(C, \Phi(Y)),$$

so that, as Y is adapted to $\mathcal{B}$, and independent of $\mathcal{C}$,

$$\begin{aligned} 0 &= E_P\left[\chi_{\left[\frac{1}{2}, 1\right]}(C) \left|Y - \xi(C, \Phi(Y))\right|\right] \\ &= \int_{\frac{1}{2}}^{1} \int_{\mathbb{R}} |y - \xi(c, \Phi(y)| P_{C,Y}(dc, dy) \\ &= \int_{\frac{1}{2}}^{1} P_C(dc) \int_{\mathbb{R}} |y - \xi(c, \Phi(y))| P_Y(dy), \end{aligned}$$

that is, for almost every $c \geq \frac{1}{2}$, with respect to P_C, almost surely, with respect to P_Y, $y = \xi(c, \Phi(y))$, which means that, for some $c \geq \frac{1}{2}$,

$$Y = \xi(c, \Phi(Y)) = \xi(c, X),$$

or that Y is adapted to $\sigma(X)$. Thus $\sigma(Y) = \sigma(X)$, and, from the first part, one has that $\sigma(C, X) \subset \mathcal{B}$.

Example 16.5.16 (Pitfall 2) When $\mathcal{B}$ is conditionally nonatomic given $\mathcal{C}$, and $\mathcal{D}$ such that $\mathcal{C} \vee \mathcal{D} = \mathcal{B}$, there may exist no $\mathcal{D}$-adapted complement to $\mathcal{C}$ in $\mathcal{B}$.

Let

$$\Omega_1 = \left[0, \frac{1}{2}\right] \times [0, 1], \; \Omega_2 = \left[\frac{1}{2}, 1\right] \times \left[0, \frac{1}{2}\right], \; \Omega_3 = \left[1, \frac{3}{2}\right] \times \left[\frac{1}{2}, 1\right],$$

and $\Omega = \Omega_1 \cup \Omega_\cup \Omega_3$. P shall be Lebesgue measure restricted to Ω. Let

$$C(x, y) = x, \; D(x, y) = y, \; X = \left|D - \frac{1}{2}\right|,$$

$$\mathcal{C} = \sigma(C), \; \mathcal{D} = \sigma(D), \; \mathcal{B} = \sigma(C, D).$$

C and D are not independent as, for example,

$$P\left(C \in \left[\frac{1}{2}, 1\right], D \in \left[\frac{1}{2}, 1\right]\right) = 0,$$

but

$$P\left(C \in \left[\frac{1}{2}, 1\right]\right) P\left(D \in \left[\frac{1}{2}, 1\right]\right) = \frac{1}{8}.$$

However, C and X are independent: indeed

$$P(C \leq \alpha, X \leq \beta) = \begin{cases} 2\beta \times \alpha & \text{when } \alpha \in \left[0, \frac{1}{2}\right] \\ 2\beta \times \left(\frac{1}{4} + \frac{\alpha}{2}\right) & \text{when } \alpha \in \left[\frac{1}{2}, \frac{3}{2}\right] \end{cases}$$

$$= P(C \leq \alpha) \, P(X \leq \beta).$$

Since X is adapted to $\mathcal{B}$, is independent of $\mathcal{C}$, has a uniform distribution, and is thus diffuse, $\mathcal{B}$ is conditionally nonatomic given $\mathcal{C}$ [(Proposition) 16.5.11, item 2]. Let Z be any random variable adapted to $\mathcal{D}$, such that $\mathcal{B} = \mathcal{C} \vee \sigma(Z)$. Suppose that $\sigma(Z) = \mathcal{D}$: then, since C and D are not independent, there is no complement to $\mathcal{C}$ in $\mathcal{B}$ adapted to $\mathcal{D}$.

Now, as in (Example) 16.5.15, almost surely, $Z = \Phi(D)$, and $D = \Psi(C, Z) = \Psi(C, \Phi(D))$, so that, because, choosing $C < \frac{1}{2}$, one works in a "rectangular" subset of Ω_1,

$$\begin{aligned}0 &= E_P\left[\chi_{\{C<\frac{1}{2}\}} |D - \Psi(C, \Phi(D))|\right] \\ &= \int_{[0,\frac{1}{2}[} \int_0^1 |y - \Psi(x, \Phi(y))| \, P_{C,D}(dx, dy) \\ &= \int_0^{\frac{1}{2}} dx \int_0^1 dy \, |y - \Psi(x, \Phi(y))| \, .\end{aligned}$$

There is thus an $x \in [0, \frac{1}{2}[$ such that, almost surely in $[0, 1]$, $y = \Psi(x, \Phi(y))$, that is $D = \Psi(x, Z)$, and thus $\sigma(Z) = \mathcal{D}$.

Example 16.5.15 involves a Y whose law is diffuse. When Y is discrete, a complement X always exists.

Proposition 16.5.17 *Suppose that $\mathcal{C} \subseteq \mathcal{B}$ are essentially separable σ-algebras in $\mathcal{A}$, and that $\mathcal{B}$ is conditionally nonatomic given $\mathcal{C}$. Then:*

1. *when X is adapted to $\mathcal{B}$, and takes at most countably many values, then $\mathcal{B}$ is conditionally nonatomic given $\mathcal{C} \vee \sigma(X)$;*
2. *when furthermore X is independent of $\mathcal{C}$, there exists a complement U to $\mathcal{C}$ in $\mathcal{B}$ such that $\sigma(X) \subseteq \sigma(U)$.*

Proof Suppose that Z is such that $\mathcal{B} = \mathcal{C} \vee \sigma(X) \vee \sigma(Z)$. Because of (Remark) 16.5.12, (Proposition) 16.5.11 applies to vectors, and, because of the assumption in the current statement, (Proposition) 16.5.11, item 4, is true. The law of the vector (X, Z) is thus diffuse. Since X takes at most a countable number of values,

$$P(\omega \in \Omega : Z(\omega) = z) = \sum_i P(\omega \in \Omega : Z(\omega) = z, X(\omega) = x_i) = 0.$$

Z is thus diffuse, and, consequently [(Proposition) 16.5.11, item 4], $\mathcal{B}$ is conditionally nonatomic given $\mathcal{C} \vee \sigma(X)$.

Suppose furthermore that X is independent of $\mathcal{C}$. Because of item 1 of the current statement, (Proposition) 16.5.11, item 3, obtains, and there exists Z, whose law is uniform, adapted to $\mathcal{B}$, independent of $\mathcal{C} \vee \sigma(X)$, such that

$$\{\mathcal{C} \vee \sigma(X)\} \vee \sigma(Z) = \mathcal{B}.$$

Let Y generate $\sigma(X, Z)$. Then $\sigma(Y) \supseteq \sigma(X)$ and $\sigma(Y) \supseteq \sigma(Z)$. Consequently $Z = \Phi(Y)$, and, as Z is diffuse,

$$P(Y = y) \leq P(\Phi(Y) = \Phi(y)) = P(Z = \Phi(y)) = 0.$$

So Y is diffuse. It follows that F_Y is continuous, and then that $U = F_Y(Y)$ has a uniform distribution. Thus U is adapted to $\mathcal{B}$, is independent of $\mathcal{C}$, has a uniform distribution, and $\mathcal{C} \vee \sigma(U) = \mathcal{B}$. U is thus a complement to $\mathcal{C}$ in $\mathcal{B}$. □

Remark 16.5.18 F_Y being continuous, $F_Y^{-1}(\{c\})$ is closed. But then it is a closed interval $[a, b], a \leq b$. When $a < b$, the subsets of $[a, b]$ are not in $F_Y^{-1}(\mathcal{B}(\mathbb{R}))$, hence the need to work with completed σ-algebras.

One shall need below the following considerations. Let K be a compact metric space with distance d_K, and $\mathcal{P}_K$ be the space of probability measures defined on the Borel sets of K. For weak convergence, $\mathcal{P}_K$ is a compact metric space [208, p. 45]. Denote $\mathcal{V}(\mathcal{A}, K)$ the family of random elements with values in K, adapted to $\mathcal{A}$. The distance between two elements X and Y of $\mathcal{V}(\mathcal{A}, K)$ is defined to be

$$d_{\mathcal{V}(\mathcal{A},K)}(X, Y) = E_P\left[d_K(X, Y)\right].$$

Let now $\mathcal{B}$ be a σ-algebra of $\mathcal{A}$, and $P_{X|\mathcal{B}}$ be the regular conditional law of X in $\mathcal{V}(\mathcal{A}, K)$, with respect to $\mathcal{B}$. It is an element of $\mathcal{V}(\mathcal{B}, \mathcal{P}_K)$. The Kantorovich-Rubinshtein (abbreviated KR) distance d_{KR} on $\mathcal{P}_K$ is defined as follows: let Π be a probability on (the Borel sets of) $K \times K$, and J_1 and J_2 be, respectively, the projections of $K \times K$ onto its first and second component. Then:

$$d_{KR}(P_1, P_2) = \inf_{\{\Pi : \Pi \circ J_1^{-1} = P_1, \Pi \circ J_2^{-1} = P_2\}} \int_{K \times K} d_K(k_1, k_2)\, \Pi(dk_1, dk_2).$$

Convergence for d_{KR} is equivalent to weak convergence [151].

Suppose now that $K = \{k_1, \ldots, k_n\}$, and that a minimum Π for $d_{KR}(P_1, P_2)$ exists. In the following table, the margins are obtained by summing the rows and the columns:

$$\begin{array}{ccc|c} \Pi(k_1, k_1) & \cdots & \Pi(k_1, k_n) & P_1(k_1) \\ \vdots & \vdots & \vdots & \vdots \\ \Pi(k_n, k_1) & \cdots & \Pi(k_n, k_n) & P_1(k_n) \\ \hline P_2(k_1) & \cdots & P_2(k_n) & 1 \end{array}.$$

The different values of Π may then be looked at as a solution of the system of linear equations $M\underline{\pi} = \underline{p}$, where $\pi_{i,j} = \Pi(k_i, k_j)$, and p_i represents a probability in the margins of the latter table. The matrix M "sums" its rows and columns. The solutions of the system have the following form [121, p. 143]:

$$\underline{\pi} = M^{-}\underline{p} + (I - M^{-}M)\underline{x},$$

where M^- is the generalized inverse of M, and $\underline{x}$ is any vector in $\mathbb{R}^{2n}$. $M^-\underline{p}$ is thus always a solution, and consequently Π may be chosen to depend "arithmetically" on P_1 and P_2.

The coming proposition expresses distance between random elements as distance between probability measures.

Proposition 16.5.19 *Let $(\Omega, \mathcal{A}, P)$ be a probability space. Suppose that:*

(a) *$\mathcal{C} \subseteq \mathcal{B}$ are essentially separable σ-algebras of $\mathcal{A}$,*
(b) *$\mathcal{B}$ is, given $\mathcal{C}$, conditionally nonatomic,*
(c) *$X \in \mathcal{V}(\mathcal{B}, K)$ and $Q \in \mathcal{V}(\mathcal{C}, \mathcal{P}_K)$ are simple random elements,*
(d) *the range of Q contains only probability measures with finite support.*

There exists then a random element Z in $\mathcal{V}(\mathcal{B}, K)$ such that

1. $P_{Z|\mathcal{C}} = Q$;
2. $d_{\mathcal{V}(\mathcal{B},K)}(X, Z) = E_P\left[d_{KR}(P_{X|\mathcal{C}}, Q)\right]$.

Proof Let

$$K_0 = \{k_1, \ldots, k_m\} \subseteq K,\ \{B_1, \ldots, B_m\} \subseteq \mathcal{B},\ X = \sum_{i=1}^{m} k_i \chi_{B_i},$$

$$\{C_1, \ldots, C_n\} \subseteq \mathcal{C}, \{P_1, \ldots, P_n\} \subseteq \mathcal{P}_K,\ Q = \sum_{j=1}^{n} P_j \chi_{C_j},$$

and P_j have support

$$K_j = \left\{k_1^{(j)}, \ldots, k_{n(j)}^{(j)}\right\} \subseteq K.$$

Set

$$K_f = K_0 \cup \left\{\cup_{j=1}^{n} K_j\right\}.$$

Since K_f is compact, and any distance is continuous (for the product topology, on the product with itself, of the space over which it is defined [84, p. 184]), $d_{KR}(P_{X|\mathcal{C}}, Q)$ has a minimum, say $\Pi_{X,Q}$, for every pair of fixed, but arbitrary random elements (X, Q). $\Pi_{X,Q}$ may be chosen, as seen, adapted to $\mathcal{C}$. The relations that follow thus obtain:

$$\begin{aligned}
\textstyle\sum_{k\in K_f} \Pi_{X,Q}(\omega, \cdot, k) &= P_{X|\mathcal{C}}(\omega, \cdot),\\
\textstyle\sum_{k\in K_f} \Pi_{X,Q}(\omega, k, \cdot) &= Q[\omega](\cdot),\\
\textstyle\sum_{(k,\tilde{k})\in K_f\times K_f} d_K(k, \tilde{k})\, \Pi_{X,Q}(\omega, k, \tilde{k}) &= d_{KR}\left(P_{X|\mathcal{C}}(\omega, \cdot), Q[\omega]\right).
\end{aligned}$$

Let

$$\begin{aligned} U(\omega) &= \sum_{k \in K_f} \Pi_{X,Q}(\omega, X(\omega), k) \\ &= \sum_{i=1}^{m} \chi_{B_i}(\omega) \Pi_{X,Q}(\omega, k_i, k) \\ &= \sum_{i=1}^{m} \chi_{B_i}(\omega) P_{X|\mathcal{C}}(\omega, k_i) \,. \end{aligned}$$

Thus, tentatively,

$$U(\omega)^{-1} = \sum_{i=1}^{m} \frac{\chi_{B_i}(\omega)}{P_{X|\mathcal{C}}(\omega, k_i)} \,.$$

But

$$\begin{aligned} P(U = 0, X = k_i \mid \mathcal{C}) &= P(\{U = 0\} \cap B_i \mid \mathcal{C}) \\ &= P\left(\left\{P_{X|\mathcal{C}}(\cdot, k_i) = 0\right\} \cap B_i \mid \mathcal{C}\right) \\ &= \chi_{\left\{P_{X|\mathcal{C}}(\cdot, k_i) = 0\right\}} P(B_i \mid \mathcal{C}) \\ &= \chi_{\left\{P_{X|\mathcal{C}}(\cdot, k_i) = 0\right\}} P_{X|\mathcal{C}}(\cdot, k_i) \\ &= 0. \end{aligned}$$

Consequently $P(U = 0) = 0$, and $U > 0$, almost surely, with respect to P.

Let $\{\kappa_1, \ldots, \kappa_p\}$ be an enumeration of the points of K_f. Let $V_0 \equiv 0$, and, for $l \in [1 : p]$, fixed, but arbitrary, set

$$V_l(\omega) = U(\omega)^{-1} \left\{ \sum_{i=1}^{l} \Pi_{X,Q}(\omega, X(\omega), \kappa_i) \right\} .$$

V_l is adapted to $\mathcal{C} \vee \sigma(X)$, and $\{V_l, l \in [0 : p]\}$ is an increasing sequence, starting at zero, and ending at one. Since X is simple, because of (Proposition) 16.5.17, $\mathcal{B}$ is, given $\mathcal{C} \vee \sigma(X)$, conditionally nonatomic, and, because of (Proposition) 16.5.11, there exists a complement W to $\mathcal{C} \vee \sigma(X)$ in $\mathcal{B}$, that is, W is adapted to $\mathcal{B}$, is independent of $\mathcal{C} \vee \sigma(X)$, has uniform law with support $[0, 1]$, and furthermore $\mathcal{C} \vee \sigma(X) \vee \sigma(W) = \mathcal{B}$. Let then, whenever $W(\omega) \in]V_{l-1}(\omega), V_l(\omega)]$,

$$Z(\omega) = \kappa_l.$$

As

$$P(Z = \kappa_l \mid \mathcal{C} \vee \sigma(X)) = E_P\left[\chi_{]V_{l-1},V_l]}(W) \mid \mathcal{C} \vee \sigma(X)\right],$$

that W is independent of the conditioning σ-algebra, letting

$$\phi(\underline{V}, W) = \chi_{]V_{l-1},V_l]}(W),$$

and

$$\psi(\underline{v}) = E_P\left[\phi(\underline{v}, W)\right] = \int_{\mathbb{R}} \phi(\underline{v}, w) P_W(dw) = \int_0^1 \phi(\underline{v}, w)\, dw,$$

one has that [138, p. 452]

$$\begin{aligned} E_P\left[\phi(\underline{V}, W) \mid \mathcal{C} \vee \sigma(X)\right] &= \psi(\underline{V}) \\ &= \int_0^1 \phi(\underline{V}, w)\, dw \\ &= \int_0^1 \chi_{]V_{l-1},V_l]}(w)\, dw \\ &= V_l - V_{l-1}. \end{aligned}$$

Thus, given the definition of U,

$$\begin{aligned} P\left(Z = \kappa_l \mid \mathcal{C} \vee \sigma(X)\right)(\omega) &= U(\omega)^{-1} \Pi_{X,Q}(\omega, X(\omega), \kappa_l), \\ P\left(\{X = k_i\} \cap \{Z = \kappa_l\} \mid \mathcal{C} \vee \sigma(X)\right)(\omega) &= \chi_{B_i}(\omega) \frac{\Pi_{X,Q}(\omega, k_i, \kappa_l)}{P_{X|\mathcal{C}}(\omega, k_i)}, \\ P\left(\{X = k_i\} \cap \{Z = \kappa_l\} \mid \mathcal{C}\right)(\omega) &= \frac{\Pi_{X,Q}(\omega, k_i, \kappa_l)}{P_{X|\mathcal{C}}(\omega, k_i)} E_P\left[\chi_{B_i} \mid \mathcal{C}\right] \\ &= \Pi_{X,Q}(\omega, k_i, \kappa_l). \end{aligned}$$

The law of (X, Z), given $\mathcal{C}$, is thus $\Pi_{X,Q}$. Hence, the law of Z, given $\mathcal{C}$, is Q. Furthermore:

$$E_P\left[d_K(X, Z) \mid \mathcal{C}\right] = \sum_{(k,\tilde{k}) \in K_f \times K_f} d_K(k, \tilde{k})\, \Pi_{X,Q}(\cdot, k, \tilde{k}) = d_{KR}\left(P_{X|\mathcal{C}}, Q\right),$$

hence $E_P\left[d_K(X, Z)\right] = E_P\left[d_{KR}(P_{X|\mathcal{C}}, Q)\right]$. □

The following remarks are useful for the next assertion.

Remark 16.5.20 Let $a^+ = \max(a, 0)$, and $a^- = -\min(a, 0)$. It then follows that $a^+ + a^- = |a|$. So

$$(a-b)^+ = a - [a \wedge b\}\,, \text{ and } (a-b)^- = b - \{a \wedge b\}\,.$$

Remark 16.5.21 Let $S = \{s_1, \ldots, s_n\}$: for $\{s_i, s_j\} \subseteq S$, the following relation:

$$d_S(s_i, s_j) = \begin{cases} 0 \text{ when } s_i = s_j \\ 1 \text{ when } s_i \neq s_j \end{cases}$$

defines a metric on S.

Remark 16.5.22 Let P_1 and P_2 be probabilities on the set S of (Remark) 16.5.21, and define, for $i \in [1:n]$, fixed, but arbitrary,

$$\Pi_1(s_i) = P_1(s_i) - \{P_1(s_i) \wedge P_2(s_i)\} = (P_1(s_i) - P_2(s_i))^+\,,$$
$$\Pi_2(s_i) = P_2(s_i) - \{P_1(s_i) \wedge P_2(s_i)\} = (P_1(s_i) - P_2(s_i))^-\,,$$

so that

$$\Pi_1(s_i) + \Pi_2(s_i) = |P_1(s_i) - P_2(s_i)|\,,$$

and

$$\sum_{i=1}^{n} \Pi_1(s_i) = 1 - \sum_{i=1}^{n} \{P_1(s_i) \wedge P_2(s_i)\} = \sum_{i=1}^{n} \Pi_2(s_i).$$

Let π denote the common value of the last set of equalities: then

$$2\pi = \sum_{i=1}^{n} |P_1(s_i) - P_2(s_i)|\,.$$

Remark 16.5.23 Let

$$\Pi_{1,2}(s_i, s_j) = \begin{cases} \pi^{-1} \Pi_1(s_i) \Pi_2(s_j) & \text{when } i \neq j \\ \pi^{-1} \Pi_1(s_i) \Pi_2(s_i) + \{P_1(s_i) \wedge P_2(s_i)\} & \text{when } i = j \end{cases}.$$

Then

$$\sum_{i,j=1}^{n} \Pi_{1,2}(s_i, s_j) = \pi^{-1} \sum_{i=1}^{n} \Pi_1(s_i) \sum_{j=1}^{n} \Pi_2(s_j) + \sum_{i=1}^{n} \{P_1(s_i) \wedge P_2(s_i)\} = 1,$$

so that $\Pi_{1,2}$ is a probability. Furthermore

$$\int_{S\times S} d_S(s,\tilde{s})\,\Pi_{1,2}(ds,d\tilde{s}) = \sum_{\substack{i,j=1\\ i\neq j}}^{n} \Pi_{1,2}(s_i,s_j) = \pi.$$

Remark 16.5.24 Let Π be any probability on $S\times S$. Then

$$\int_{S\times S} d_S(s,\tilde{s})\,\Pi(ds,d\tilde{s}) = \sum_{\substack{i,j=1\\ i\neq j}}^{n} \Pi(s_i,s_j) = 1 - \sum_{i=1}^{n} \Pi(s_i,s_i),$$

and the left-hand side of the latter relation is smallest when the second term on the right is largest. But, since one must have that

$$\Pi(s_i,s_j) \leq P_1(s_i)\wedge P_2(s_j),$$

$\Pi_{1,2}$ of (Remark) 16.5.23 yields a minimum.

Proposition 16.5.25 *$(\Omega,\mathcal{A},P)$ is the basic probability space. Let $\mathcal{C}\subseteq\mathcal{B}$ be essentially separable σ-algebras in $\mathcal{A}$, and $\mathcal{B}$ be, conditionally on $\mathcal{C}$, nonatomic. Let S be a finite set, and $X:\Omega\longrightarrow S$ be adapted to $\mathcal{B}$. There exists then $Z:\Omega\longrightarrow S$, adapted to $\mathcal{B}$, which is independent of X, and such that*

$$P(Z\neq X) = \frac{1}{2}\sum_{i=1}^{n} E_P\left[\left|P_{X|\mathcal{C}}(\cdot,s_i) - P(X=s_i)\right|\right].$$

Proof The context is that of (Proposition) 16.5.19. Choose for S, the set K_f; for d_K, d_S; and, for Q of (Proposition) 16.5.19, the constant function $Q[\omega]=P_X$. The Z of (Proposition) 16.5.19 is thus such that $P_{Z|\mathcal{C}}(\omega,\cdot)=Q[\omega]=P_X$, so that Z is independent of $\mathcal{C}$. Furthermore, using the preceding remarks, item 2 of the conclusion of (Proposition) 16.5.19 yields the second statement of the proposition. □

Proposition 16.5.26 *$(\Omega,\mathcal{A},P)$ is the basic probability space. Let $\mathbb{Z}_0^-$ be the set $\{\ldots,-2,-1,0\}$, and*

$$\{\mathcal{B}_n\subseteq\mathcal{A}, n\in\mathbb{Z}_0^-\}\subseteq\mathcal{A}$$

be a filtration such that:

(a) *$\mathcal{B}_0$ is essentially separable,*
(b) *each $\mathcal{B}_n$ is, conditionally on $\mathcal{B}_{n-1}$, nonatomic,*
(c) *$\mathcal{B}_{-\infty}=\bigcap_{[n\in\mathbb{Z}_0^-]}\mathcal{B}_n$ is essentially degenerate.*

Given $\epsilon > 0$, and a simple random variable X, adapted to $\mathcal{B}_0$, fixed, but arbitrary, there exists $n_{\epsilon,X} \in \mathbb{Z}_0^- \setminus \{0\}$, and a random variable $Y_{\epsilon,X}$, adapted to $\mathcal{B}_0$, independent of $\mathcal{B}_{n_{\epsilon,X}}$, such that

$$P(Y_{\epsilon,X} \neq X) < \epsilon.$$

Proof Let $\mathcal{R}[X]$ denote the range of X, and $x \in \mathcal{R}[X]$ be fixed, but arbitrary. Since $\mathcal{B}_{-\infty}$ is essentially degenerate, because of the reverse martingale convergence theorem [201, p. 118], in L_1,

$$\lim_n P_{X|\mathcal{B}_n}(\cdot, x) = \lim_n P(X = x \mid \mathcal{B}_n) = P(X = x).$$

There is thus $n_{\epsilon,X} < 0$ such that

$$\sum_{x \in \mathcal{R}[X]} E_P\left[\left|P_{X|\mathcal{B}_{n_{\epsilon,X}}}(\cdot, x) - P(X = x)\right|\right] < \epsilon.$$

The $Y_{\epsilon,X}$ one looks for is then provided by (Proposition) 16.5.25. □

Proposition 16.5.27 *$(\Omega, \mathcal{A}, P)$ is the basic probability space. Let*

$$\{\mathcal{B}_n \subseteq \mathcal{A}, n \in \mathbb{Z}_0^-\}$$

be a filtration as in (Proposition) 16.5.26. *Let $p \in \mathbb{Z}_0^- \setminus \{0\}$ be fixed, but arbitrary, and*

$$\underline{X}_p = \{X_{p+1}, \ldots, X_{-1}, X_0\}$$

be random variables such that, for $k \in [p+1 : 0]$, fixed, but arbitrary, X_k is a complement to $\mathcal{B}_{k-1}$ in $\mathcal{B}_k$. Let $\epsilon > 0$, and X, a random variable adapted to $\mathcal{B}_0$, be fixed, but arbitrary. There exists then, in $\mathbb{Z}_0^-$, $n_{\epsilon,p} < p$, a complement to $\mathcal{B}_{n_{\epsilon,p}}$ in $\mathcal{B}_p$, say $Y_{\epsilon,p}$, and a random variable Z, adapted to $\sigma(Y_{\epsilon,p}, \underline{X}_p)$, such that

$$P(Z \neq X) < \epsilon.$$

Proof That X_k is a complement to $\mathcal{B}_{k-1}$ in $\mathcal{B}_k$ means that [(Definition) 16.5.13] X_k is adapted to $\mathcal{B}_k$, that it is independent of $\mathcal{B}_{k-1}$, has a law that is uniform on the unit interval, and that $\mathcal{B}_{k-1} \vee \sigma(X_k) = \mathcal{B}_k$. One may thus write that

$$\mathcal{B}_0 = \mathcal{B}_p \vee \sigma(X_{p+1}, \ldots, X_0).$$

Since $\mathcal{B}_0$ is essentially separable, each $\mathcal{B}_n$, for $n \in \mathbb{Z}_0^-$, has the same property, and is thus the union of essentially finite σ-algebras. One may thus assume, since a "general" U is the almost sure limit of simple functions whose "base" is a finite

σ-algebra, that

$$X = \Phi\left(U, X_{p+1}, \ldots, X_0\right),$$

where Φ is a Borel measurable function, and U, a simple random variable, adapted to $\mathcal{B}_p$.

Let, for $q \in \mathbb{Z}_0^-$, fixed, but arbitrary, $\mathcal{C}_q = \mathcal{B}_{p+q}$. One may apply (Proposition) 16.5.26 to the filtration

$$\left\{\mathcal{C}_q, q \in \mathbb{Z}_0^-\right\},$$

and the simple random variable U, to obtain, given a fixed, but arbitrary $\epsilon > 0$, in $\mathbb{Z}_0^-$, a $q_\epsilon < p$, and a simple random variable V, adapted to $\mathcal{B}_p$, and independent of $\mathcal{B}_{q_\epsilon}$, such that $P(V \neq U) < \epsilon$. Consequently

$$P\left(\Phi(V, X_{p+1}, \ldots, X_0) \neq X\right) < \epsilon.$$

Now, because of (Proposition) 16.5.17, there is a complement W to $\mathcal{B}_{q_\epsilon}$ in $\mathcal{B}_p$ such that $\sigma(W) \supseteq \sigma(V)$. Thus $V = \psi(W)$, and $P\left(\Phi(\psi(W), X_{p+1}, \ldots, X_0) \neq X\right) < \epsilon$. □

Definition 16.5.28 Let $\left\{\mathcal{B}_n \subseteq \mathcal{A}, n \in \mathbb{Z}_0^-\right\}$ be a filtration. It is standard, nonatomic, when it is generated by a family $\left\{X_n, n \in \mathbb{Z}_0^-\right\}$ of independent, identically distributed random variables, whose law is uniform on the unit interval.

Definition 16.5.29 A filtration is standard when it may be immersed in a standard, nonatomic filtration.

Proposition 16.5.30 *Let* $\left\{\mathcal{B}_n \subseteq \mathcal{A}, n \in \mathbb{Z}_0^-\right\}$ *be a filtration as in* (Proposition) 16.5.26. *There exists a strictly increasing map* $f : \mathbb{Z}_0^- \longrightarrow \mathbb{Z}_0^-$ *such that the filtration whose elements are* $\mathcal{C}_n = \mathcal{B}_{f(n)}$ *is standard, nonatomic.*

Proof Let B be a random variable generating $\mathcal{B}_0$ [138, p. 91], and ϵ, a strictly positive sequence $\left\{\epsilon(n), n \in \mathbb{Z}_0^-\right\}$ such that $\epsilon(0) = 1$, and $\lim_{n\downarrow\downarrow-\infty} \epsilon(n) = 0$.

Since $\mathcal{B}_0$ is, conditionally on $\mathcal{B}_{-1}$, nonatomic, there exists [(Proposition) 16.5.26] a complement X_0 to $\mathcal{B}_{-1}$ in $\mathcal{B}_0$. It may be chosen to be adapted to $\mathcal{B}_0$, to be independent of $\mathcal{B}_{-1}$, to have a law which is uniform on the unit interval, and the property that $\mathcal{B}_{-1} \vee \sigma(X_0) = \mathcal{B}_0$. There exists then [(Proposition) 16.5.27] $p < -1$, a complement X_p to $\mathcal{B}_p$ in $\mathcal{B}_{-1}$ (with, *mutatis mutandis,* the properties of X_0), and a random variable Y_p, adapted to $\sigma(X_p, X_0)$, such that, for arbitrary a priori $\delta > 0$, $P(Y_p \neq B) < \delta$.

Choose thus

- $f(0) = 0 : \mathcal{C}_0 = \mathcal{B}_{f(0)} = \mathcal{B}_0$;
- $f(-1) = -1 : \mathcal{C}_{-1} = \mathcal{B}_{f(-1)} = \mathcal{B}_{-1}$, and $C_0 = X_0$, a complement of $\mathcal{C}_{-1}$ in $\mathcal{C}_0$;
- $f(-2) = p : \mathcal{C}_{-2} = \mathcal{B}_p$, $C_{-1} = X_p$, a complement of $\mathcal{C}_{-2}$ in $\mathcal{C}_{-1}$, and $D_{-1} = Y_p$ is adapted to $\sigma(C_{-1}, C_0)$, and such that $P(D_{-1} \neq B) < \epsilon(-1)$.

Continuing in the same vein, with the help of (Proposition) 16.5.27, one may construct, by induction, and for $n \in \mathbb{Z}_0^-$, fixed, but arbitrary,

- a strictly increasing map $f : \mathbb{Z}_0^- \longrightarrow \mathbb{Z}_0^-$, and a filtration with elements $\mathcal{C}_n = \mathcal{B}_{f(n)}$,
- a random variable C_n which is a complement to $\mathcal{C}_{n-1}$ in $\mathcal{C}_n$,
- a random variable D_n adapted to $\sigma(C_n, \ldots, C_0)$ and such that

$$P(D_n \neq B) < \epsilon(n).$$

Let $\mathcal{F}_n = \sigma(C_p, p \leq n)$. In (Proposition) 16.5.6, let $\mathcal{A}_n$ be $\mathcal{C}_n$ (as just above), and $\mathcal{B}_n$ be $\sigma(C_n)$, as just built. Then $\mathcal{C}_n$ of (Proposition) 16.5.6 becomes $\mathcal{F}_n$. One thus defines a filtration which is immersed in the filtration of the $\mathcal{C}_n$'s, for which $\mathcal{F}_0$ is saturated. By construction, B is the limit, in probability, of the sequence $\{D_n, n \in \mathbb{Z}_0^- \setminus \{0\}\}$. It is thus adapted to $\mathcal{F}_0$. But then, since B generates $\mathcal{B}_0$, $\mathcal{F}_0 = \mathcal{B}_0$. As $\mathcal{C}_0 = \mathcal{B}_0$, $\mathcal{F}_0 = \mathcal{C}_0$. By construction, the filtration of the $\mathcal{C}_n$'s is immersed in itself and saturated for it. Because of (Proposition) 16.5.4, the filtrations of the $\mathcal{F}_n$'s and $\mathcal{C}_n$'s are thus equal, and, since that of the $\mathcal{F}_n$'s is, by construction, standard, nonatomic, so is that of the $\mathcal{C}_n$'s. □

Proposition 16.5.31 (Vershik's Lacunary Isomorphism Theorem) *Let a filtration $\{\mathcal{A}_n, n \in \mathbb{Z}_0^-\}$ be given, for which $\mathcal{A}_0$ is essentially separable, and $\mathcal{A}_{-\infty}$, essentially degenerate. There exists then a strictly increasing $f : \mathbb{Z}_0^- \longrightarrow \mathbb{Z}_0^-$ such that the filtration $\{\mathcal{B}_n, n \in \mathbb{Z}_0^-\}$ obtained as $\mathcal{B}_n = \mathcal{A}_{f(n)}$ is standard.*

Proof Let $\{\mathcal{C}_n, n \in \mathbb{Z}_0^-\}$ be a standard, nonatomic filtration, independent of the filtration of the $\mathcal{A}_n$'s. It always exists for, when not, one enlarges the sample space. Let $\{\mathcal{D}_n, n \in \mathbb{Z}_0^-\}$ be the filtration for which $\mathcal{D}_n = \mathcal{A}_n \vee \mathcal{C}_n$.

The filtration of the $\mathcal{D}_n$'s has the same properties as that of the $\mathcal{A}_n$'s. Indeed, the σ-algebra generated by $\mathcal{A}_0$ and $\mathcal{C}_0$ is essentially separable, since both σ-algebras are. Using the formula [53, p. 29]

$$\mathcal{F} \vee \{\cap_n \mathcal{G}_n\} = \cap_n \{\mathcal{F} \vee \mathcal{G}_n\},$$

where $\{G_n, n \in \mathbb{N}\}$ is decreasing, and $\mathcal{F}$ and $\mathcal{G}_1$ are independent, one sees that

$$\mathcal{A}_n \vee \{\cap_p \mathcal{C}_p\} = \cap_p \{\mathcal{A}_n \vee \mathcal{C}_p\} \supseteq \cap_p \{\mathcal{A}_{n \wedge p} \vee \mathcal{C}_{n \wedge p}\}.$$

Since $\cap_p \mathcal{C}_p$ is essentially degenerate (Kolmogorov's zero-one law), the left-hand side of the latter expression is $\mathcal{A}_n$. Intersecting the $\mathcal{A}_n$'s, one gets the degenerate σ-algebra on the left, and, on the right, $\cap_m \{\mathcal{A}_m \vee \mathcal{C}_m\}$. Furthermore, because of (Proposition) 16.5.11, $\mathcal{D}_n$ is, conditionally on $\mathcal{D}_{n-1}$, nonatomic. One may thus apply (Proposition) 16.5.30 to the filtration of the $\mathcal{D}_n$'s to obtain a filtration of elements $\mathcal{E}_n = \mathcal{D}_{f(n)}$ that is standard, nonatomic. Let $\mathcal{B}_n = \mathcal{A}_{f(n)}$. Because of (Remark) 16.5.3, the filtration of the $\mathcal{B}_n$'s is immersed in that of the $\mathcal{D}_n$'s. It is thus, by definition, standard. □

Lacunary isomorphism supports in an essential way the next proposition, which, in turn, provides an indispensable link to the sequel. But that proposition requires acquaintance with the construction which follows, given as a remark.

Remark 16.5.32 Let U be a random variable that is uniformly distributed on $[0, 1]$. Using the nonterminating dyadic expansion of numbers in $[0, 1]$, one may write [39, p. 3] that

$$U = \sum_n \frac{1}{2^n} D_n,$$

where the D_n's are 0–1, independent random variables. Using the proof [218, p. 58] that the cardinality of $\mathbb{N}^2$ is that of $\mathbb{N}$, one may write $\mathbb{N}$ as a countable union of sequences

$$\left\{\left\{n_{i,j}, j \in \mathbb{N}\right\}, i \in \mathbb{N}\right\}.$$

Defining [200, p. 27]

$$U_i = \sum_j \frac{1}{2^j} D_{n_{i,j}},$$

one obtains independent, uniformly distributed random variables. Evaluating the inverse of the distribution function of a standard normal random variable at the U_i's, one obtains independent, standard normal random variables, say $\underline{N} = \left\{N_p, p \in \mathbb{N}\right\}$, generating the same σ-algebra as U. Given $\underline{N}$, one can manufacture [166, p. 104] a Brownian motion, which generates the same σ-algebra as U, respectively $\underline{N}$. Such a Brownian motion may be defined on any compact interval, adjusting the Haar system to be a basis of the L_2-space of that interval. Thus, given a family of independent random variables $\underline{U} = \{U_n, n \in \mathbb{N}\}$, which are uniformly distributed, one can manufacture a family of independent Brownian motions, each of which generates the same σ-algebra as that generated by the corresponding uniform random variable. Consequently, given a family of intervals, each associated with an independent, uniformly distributed random variable, one can juxtapose independent Brownian motions, according to the pattern of [166, p. 107], to obtain a single Brownian motion which will generate the same σ-algebra as the corresponding uniform variables do, and preserve the order resulting from the intervals entering the construction.

Proposition 16.5.33 ([89, p. 201]) *Let $\underline{\mathcal{F}}$ be some filtration, and $\underline{\mathcal{B}}$ be the filtration of a one-dimensional Brownian motion started at the origin (such a filtration is unique up to isomorphism). Then* [notation in Sect. 16.3] $\underline{\mathcal{F}} \succ \underline{\mathcal{B}}$ *if, and only if, $\mathcal{F}_0$ is essentially degenerate, and $\mathcal{F}_\infty$, essentially separable.*

Proof Let $\underline{\mathcal{H}}$ be the filtration made, for $n = 0, 1, 2, 3, \ldots$, of the elements $\mathcal{F}_{1/n}$. As $\cap_n \mathcal{H}_n = \mathcal{F}_0$ is essentially degenerate, and $\mathcal{H}_0 = \mathcal{F}_\infty$, essentially separable,

Vershik's theorem yields a standard subfiltration $\underline{\mathcal{K}}$ of $\underline{\mathcal{H}}$, with elements $\mathcal{F}_{1/n_p}, p \geq 0$, that is, one for which there is an isomorphism Ψ which sends $\underline{\mathcal{K}}$ to a filtration $\underline{\mathcal{L}}$, which is immersed into a filtration generated by a family of independent, uniformly distributed random variables. The elements of the latter filtration have the following form: $\mathcal{U}_n = \sigma(U_n, U_{n+1}, U_{n+2}, \ldots)$. Ψ sends $\underline{\mathcal{F}}$ to a filtration $\underline{\mathcal{F}}^\star$ whose element $\mathcal{F}^\star_{1/n_p} \subseteq \mathcal{U}_p$ is the corresponding element of $\underline{\mathcal{L}}$. Let $t_p = n_p^{-1}$, so that $t_0 = \infty$, and $\lim_p t_p \downarrow 0$. Proceeding as in (Remark) 16.5.32, let B be a Brownian motion, such that, for $p \geq 0$, fixed, but arbitrary,

$$\{B(\cdot, t) - B(\cdot, t_{p+2}), t \in [t_{p+2}, t_{p+1}]\}$$

generates the same σ-algebra as U_p. Thus, when $p = 0$,

$$\{B(\cdot, t) - B(\cdot, t_2), t \in [t_2, t_1]\}$$

generates the same σ-algebra as U_0 does, and then, when $t_{p+1} = t_1$, all the U_p's have entered the picture. One obtains a Brownian motion stopped at t_1, whose natural filtration contains $\underline{\mathcal{F}}^\star$. □

16.6 Girsanov's Theorem for Real, Continuous, Local Martingales

Here is a version of Girsanov's theorem valid for real, continuous, local martingales. What it misses, from the version valid for the Wiener process, is the part asserting that there is a property of law invariance.

Proposition 16.6.1 *Let $(\Omega, \mathcal{A}, P)$ be a probability space, with filtration $\underline{\mathcal{A}}$, indexed by $\mathbb{R}_+$; M, a real, continuous, local martingale for that filtered space; f, a progressively measurable process for $\underline{\mathcal{A}}$, such that, for $t \in \mathbb{R}_+$, fixed, but arbitrary, almost surely, with respect to P,*

$$\int_0^t f^2(\omega, \theta)\langle M\rangle(\omega, d\theta) < \infty.$$

Let $\tau \in \mathbb{R} \setminus \{0\}$ be fixed, but arbitrary. Let also, for $t \in [0, \tau]$, fixed, but arbitrary,

(a) $\ln[E_f(\omega, t)] = \int_0^t f(\omega, \theta) M(\omega, d\theta) - \frac{1}{2}\int_0^t f^2(\omega, \theta)\langle M\rangle(\omega, d\theta)$,
(b) $dQ_t = E_f(\cdot, t)dP$,

and suppose that $E_P\left[E_f(\cdot, t)\right] = 1$, so that Q_t is a probability measure on $\mathcal{A}_t$, absolutely continuous with respect to P, restricted to $\mathcal{A}_t$.

Consider then following probability space where $\underline{\mathcal{A}}_\tau = \{\mathcal{A}_t, t \in [0, \tau]\}$:

$$(\Omega, \underline{\mathcal{A}}_\tau, Q_\tau).$$

The process X, with index set $[0, \tau]$, for that probability space, defined using the following relation:

$$X(\omega,t) = M(\omega,t) - \int_0^t f(\omega,\theta)\langle M\rangle(\omega,d\theta),$$

is a real, continuous, local martingale such that $\langle X\rangle = \langle M\rangle$.

Proof $t \in [0,\tau]$ is fixed, but arbitrary. Since, because of the assumption of finite expectation, and (Remark) 10.4.8, E_f is a martingale for P, then, for $\theta < t$, fixed, but arbitrary,

$$E_P\left[\frac{dQ_t}{dP} \mid \mathcal{A}_\theta\right] = E_f(\cdot,\theta).$$

Since $E_f(\cdot,t)$ is an exponential, $P\left(E_f(\cdot,t) = 0\right) = 0$, so that P and Q_t are mutually absolutely continuous on $\mathcal{A}_t$.

Let, for $\alpha \in \mathbb{R}$, fixed, but arbitrary,

$$\ln[Y_\alpha(\omega,t)] = \alpha X(\omega,t) - \frac{\alpha^2}{2}\langle M\rangle(\omega,t).$$

Since

$$\alpha X - \frac{\alpha^2}{2}\langle M\rangle + \int f\,dM - \frac{1}{2}\int f^2\,d\langle M\rangle = \int \{\alpha+f\}\,dM - \frac{1}{2}\int \{\alpha+f\}^2\,d\langle M\rangle,$$

$Y_\alpha E_f$ may be expressed as $Y_\alpha E_f = Z_{\alpha,f}$, where

$$\ln[Z_{\alpha,f}] = N - \frac{1}{2}\langle N\rangle, \text{ and } N = \int \{\alpha+f\}\,dM.$$

The process $Z_{\alpha,f}$ is, with respect to P, a local martingale [(Proposition) 10.4.6]. Let

$$\{S_n^\circ, n \in I\!N\}$$

be the sequence of wide sense stopping times defined as follows:

$$S_n^\circ(\omega) = \inf\left\{t \in \mathbb{R}_+ : E_f(\omega,t) > n, Y_\alpha(\omega,t) > n, Z_{\alpha,f}(\omega,t) > n\right\}$$

($S_n^\circ(\omega) = 1$ when the defining set is empty). The sequence is increasing, almost surely, as the processes involved are almost surely finite. Let

$$\{S_n^\bullet, n \in N\}$$

be a sequence of wide sense stopping times that makes of $Z_{\alpha,f}$ a martingale bounded in L_2, and set

$$S_n = S_n^{\circ} \wedge S_n^{\bullet}.$$

One has then a sequence of wide sense stopping times, increasing indefinitely, and such that $Y_\alpha^{S_n}$ is a martingale in L_2 (in fact L_p) [264, p. 57]. Let $A \in \mathcal{A}_{t\wedge S_n}$ be fixed, but arbitrary. Then, using $Z_{\alpha,f} = Y_\alpha E_f$,

$$\int_A Y_\alpha(\omega, t \wedge S_n) E_f(\omega, t \wedge S_n) P(d\omega) = \int_A Z_{\alpha,f}(\omega, t \wedge S_n) P(d\omega).$$

Since $E_f^{S_n}$ is a martingale, it is a martingale for the filtrations of the following type: $\mathcal{A}_{t\wedge S_n} \subseteq \mathcal{A}_t$, and thus, for $S_p \geq S_n$,

$$\int_A Y_\alpha(\omega, t \wedge S_n) E_f(\omega, t \wedge S_p) P(d\omega) = \int_A Y_\alpha(\omega, t \wedge S_n) E_f(\omega, t \wedge S_n) P(d\omega).$$

Let $\{\theta, t\} \subseteq [0, \tau]$ be fixed, but arbitrary. The limit in L_1, with respect to n, of $E_f(\cdot, t \wedge S_n)$ is $E_f(\cdot, t)$, and $Y_\alpha^{S_n}$ is bounded. Thus, because of weak convergence in L_1,

$$\begin{aligned}
&\int_A Y_\alpha(\omega, t \wedge S_n) E_f(\omega, t \wedge S_n) P(d\omega) = \\
&\qquad = \lim_p \int_A Y_\alpha(\omega, t \wedge S_n) E_f(\omega, t \wedge S_p) P(d\omega) \\
&\qquad = \int_A Y_\alpha(\omega, t \wedge S_n) \lim_p E_f(\omega, t \wedge S_p) P(d\omega) \\
&\qquad = \int_A Y_\alpha(\omega, t \wedge S_n) E_f(\omega, t) P(d\omega),
\end{aligned}$$

and, consequently,

$$\begin{aligned}
&\int_A Z_{\alpha,f}(\omega, t \wedge S_n) P(d\omega) = \\
&\qquad = \int_A Y_\alpha(\omega, t \wedge S_n) E_f(\omega, t) P(d\omega)) \\
&\qquad = \int_A Y_\alpha(\omega, t \wedge S_n) Q_\tau(d\omega).
\end{aligned}$$

Finally, for $\theta > t$,

$$\begin{aligned}\int_A Y_\alpha(\omega,\theta\wedge S_n)Q_\tau(d\omega) &= \int_A Z_{\alpha,f}(\omega,\theta\wedge S_n)P(d\omega)\\ &= \int_A Z_{\alpha,f}(\omega,t\wedge S_n)P(d\omega)\\ &= \int_A Y_{\alpha,f}(\omega,t\wedge S_n)Q_\tau(d\omega).\end{aligned}$$

Thus $Y_\alpha^{S_n}$ is a martingale for the σ-algebras of type $\mathcal{A}_{t\wedge S_n}$, and the probability Q_τ. It is also, as presently seen, a martingale for the initial σ-algebras, and the same probability. Indeed, given $t_1 < t_2$ in $\mathbb{R}_+$ and A in $\mathcal{A}_{t_1}$, fixed, but arbitrary,

$$\begin{aligned}\int_A Y_\alpha^{S_n}(\omega,t_1)P(d\omega) &=\\ &= \int_{A\cap\{S_n<t_1\}} Y_\alpha^{S_n}(\omega,t_1)P(d\omega)\\ &+ \int_{A\cap\{S_n\geq t_1\}} Y_\alpha^{S_n}(\omega,t_1)P(d\omega).\end{aligned}$$

On $\{S_n < t_1\}$,

$$Y_\alpha^{S_n}(\cdot,t_1) = Y_\alpha(\cdot,S_n) = Y_\alpha(\cdot,t_2\wedge S_n), \quad \text{and } A\cap\{S_n<t_1\}\in\mathcal{A}_{t_1},$$

so that

$$\int_{A\cap\{S_n<t_1\}} Y_\alpha^{S_n}(\omega,t_1)P(d\omega) = \int_{A\cap\{S_n<t_1\}} Y_\alpha^{S_n}(\omega,t_2)P(d\omega).$$

Finally, since $A\cap\{S_n\geq t_1\}\in\mathcal{A}_{t_1\wedge S_n}$, using the martingale property of $Y_\alpha^{S_n}$,

$$\int_{A\cap\{S_n\geq t_1\}} Y_\alpha^{S_n}(\omega,t_1)P(d\omega) = \int_{A\cap\{S_n\geq t_1\}} Y_\alpha^{S_n}(\omega,t_2)P(d\omega).$$

But then [(Proposition) 10.4.6] X is local martingale with respect to Q_τ, such that

$$\langle X\rangle = \langle M\rangle.$$

□

Remark 16.6.2 Girsanov's theorem for continuous martingales that are not Gaussian states thus that it is the quadratic variation, rather than the law, that is invariant under a random translation. The natural question is then: when does invariance of quadratic variation mean invariance of law? As one shall see, one must deal with a special class of martingales, called Ocone martingales, whose main characteristic is that, in the "universal" equality $W_M = M \diamond S_M$ [(Fact) 10.3.45], W_M and S_M are independent (in fact, as shall be seen, even more is required). Hence the next topic.

16.7 Ocone Martingales

In the sequel, M shall always be a continuous to the right, almost surely continuous, local martingale, zero at the origin, and $W_M = M \diamond S_M$ [(Fact) 10.3.45]. $C[0, \infty[$ is the family of real, continuous functions, which start at zero, and its elements are denoted c. $\mathcal{E}_t$ is the evaluation map at t, and $\mathcal{C}_t = \sigma_t(\mathcal{E})$, $\mathcal{C} = \vee_t \mathcal{C}_t$. $M[\omega]$ denotes the path of M at ω; $\boldsymbol{M}$, the map $\omega \mapsto M[\omega]$. A similar notation is adopted for $\langle M \rangle$. The notation $\mathcal{L}(\cdot)$ designates the law of the item represented by the dot.

16.7.1 Definitions, Characterization, and Properties

Definition 16.7.1 M is an Ocone martingale when, in the representation $M = W_M \diamond \langle M \rangle$, W_M is independent of the quadratic variation $\langle M \rangle$ (and thus S_M).

One shall need the following (classes of) functions, and their integrals:

Definition 16.7.2

1. $\mathcal{F}$ is the class of functions of the following form:

$$f_\tau(t) = \chi_{[0,\tau]}(t) - \chi_{]\tau,\infty[}(t).$$

2. Φ is the class of processes X that are adapted to $\mathcal{C} \otimes \mathcal{B}(\mathbb{R}_+)$ and $\underline{\mathcal{C}}$ *(they are then predictable* [264, p. 115]), and have range in the set $\{-1, 1\}$.

One shall write $I_M\{f_\tau\}$ for the integral of $f_\tau \in \mathcal{F}$ with respect to M. It has the following properties:

Fact 16.7.3

1. $I_M\{f_\tau\}(\cdot, t) = M(\cdot, t \wedge \tau) - \{M(\cdot, t) - M(\cdot, t \wedge \tau)\}$*;*
2. using item 1,

$$I_M^\tau\{f_\tau\}(\cdot, t) = M^\tau(\cdot, t);$$

3. again using item 1, for $t_1 < t_2$*, fixed, but arbitrary,*

$$I_M\{f_{t_1}\}(\cdot, t_2) - I_M\{f_{t_1}\}(\cdot, t_1) = -\{M(\cdot, t_2) - M(\cdot, t_1)\}.$$

Remark 16.7.4 When $\theta = \tau$, $I_{I_M\{f_\tau\}}\{f_\theta\} = M(\cdot, t)$. Let

$$\begin{aligned} N(\omega, t) &= I_M\{f_\tau\}(\omega, t) \\ &= \int_0^t f_\tau(x) M(\omega, dx) \\ &= 2M(\omega, t \wedge \tau) - M(\omega, t). \end{aligned}$$

Then

$$\begin{aligned} I_N\{f_\theta\} &= 2N(\omega, t \wedge \theta) - N(\omega, t) \\ &= 2\{2M(\omega, t \wedge \theta \wedge \tau) - M(\omega, t \wedge \theta)\} - \{2M(\omega, t \wedge \tau) - M(\omega, t)\} \\ &\overset{\theta=\tau}{=} 4M(\omega, t \wedge \theta) - 2M(\omega, t \wedge \theta) - 2M(\omega, t \wedge \theta) + M(\omega, t). \end{aligned}$$

Definition 16.7.5 M is $\mathcal{F}$-invariant when, for $\tau \in \mathbb{R}_+$, fixed, but arbitrary,

$$\mathcal{L}(I_M\{f_\tau\}) = \mathcal{L}(M).$$

Remark 16.7.6 Because of (Fact) 16.7.3, item 3, $\mathcal{F}$-invariance is a symmetry requirement.

Definition 16.7.7 $J_M\{\phi\}(\omega, t)$ denotes

$$\int_0^t \phi(M[\omega], \theta) M(\omega, d\theta), \ \phi \in \Phi.$$

M is Φ-invariant when, for $\phi \in \Phi$, fixed, but arbitrary,

$$\mathcal{L}(J_M\{\phi\}) = \mathcal{L}(M).$$

Remark 16.7.8 Φ-invariance implies $\mathcal{F}$ invariance.

Definition 16.7.9 $P_{M|\langle M\rangle}$ shall be a regular conditional probability of $\boldsymbol{M}$ given $\langle \boldsymbol{M}\rangle$, that is,

1. for $\omega \in \Omega$, fixed, but arbitrary, $C \mapsto P_{M|\langle M\rangle}(\omega, C)$ is a probability, denoted $P^{\omega}_{M|\langle M\rangle}$, and,
2. for $C \in \mathcal{C}$, fixed, but arbitrary, almost surely, with respect to P,

$$P_{M|\langle M\rangle}(\cdot, C) = P(\boldsymbol{M} \in C \mid \langle \boldsymbol{M}\rangle).$$

Definition 16.7.10 M is, given $\langle M\rangle$, a conditionally Gaussian martingale when, for almost every $\omega \in \Omega$, with respect to P, the process

$$\mathcal{E}_{P^{\omega}_{M|\langle M\rangle}}$$

is a Gaussian martingale for $\underline{\mathcal{C}}$, and that, for $t \in \mathbb{R}_+$, fixed, but arbitrary,

$$E_{P^{\omega}_{M|\langle M\rangle}}\left[\mathcal{E}^2_{P^{\omega}_{M|\langle M\rangle}}(\cdot, t)\right] = \langle M\rangle(\omega, t).$$

Lemma 16.7.11 *When M_1 and M_2 are continuous, local martingales, with the same law, the couples $(M_1, \langle M_1\rangle)$ and $(M_2, \langle M_2\rangle)$ have the same law.*

Proof Taking continuity, and zero initial value into account, one has that $\langle M\rangle$ is [264, p. 101], locally uniformly, the limit in probability of sums of squared differences in M. □

Proposition 16.7.12 *When the continuous, local martingale M is Gaussian, conditionally on $\langle M\rangle$, it is Φ-invariant.*

Proof Let $\phi \in \Phi$, and $\omega \in \Omega$ be fixed, but arbitrary, and set:

$$X_{\phi,\omega}(c, t) = \int_0^t \phi(c, \theta)\mathcal{E}_{P^{\omega}_{M|\langle M\rangle}}(c, d\theta).$$

Since, by assumption, the process

$$\mathcal{E}_{P^{\omega}_{M|\langle M\rangle}}$$

is a Gaussian martingale with (deterministic) quadratic variation

$$t \mapsto \langle M\rangle(\omega, t),$$

and that, identically, $\phi^2 = 1$, it follows that

$$\langle X_{\phi,\omega}\rangle(c, t) = \int_0^t \phi^2(c, \theta)\langle\mathcal{E}_{P^{\omega}_{M|\langle M\rangle}}\rangle(c, d\theta) = \int_0^t \langle\mathcal{E}_{P^{\omega}_{M|\langle M\rangle}}\rangle(c, d\theta) = \langle M\rangle(\omega, t).$$

Thus $X_{\phi,\omega}$ is a continuous martingale, with (deterministic) quadratic variation $t \mapsto \langle M\rangle(\omega, t)$. It is thus Gaussian and has the same law as

$$\mathcal{E}_{P^{\omega}_{M|\langle M\rangle}}$$

(with respect to $P^{\omega}_{M|\langle M\rangle}$).

Suppose that one chooses a ϕ of the following (predictable) form:

$$\phi(c, t) = \sum_{i=0}^{n} \phi_i(c)\,\chi_{]\theta_i, \theta_{i+1}]}(t),$$

and sets

$$X(c,t) = \sum_{i=0}^{n} \phi_i(c) \{\mathcal{E}(c, t \wedge \theta_{i+1}) - \mathcal{E}(c, t \wedge \theta_i)\}.$$

Then:

$$X(M[\omega], t) = J_M \{\phi\} (\omega, t).$$

Let $C \in \mathcal{C}$ be fixed, but arbitrary. Then, using the following formula [139, p. 120]:

$$E[f(S) \mid T] = \int f(s) P_{S|T}(ds, T),$$

and then the identity in law, for $P^{\cdot}_{M|\langle M\rangle}$, of $X_{\phi,\cdot}$ and $\mathcal{E}_{P^{\cdot}_{M|\langle M\rangle}}$, established just above,

$$\begin{aligned}
P\left(J_M \{\phi\}[\cdot] \in C\right) &= E_P\left[\chi_C\left(J_M \{\phi\}[\cdot]\right)\right] \\
&= E_P\left[\chi_C\left(X\left[M[\cdot]\right]\right)\right] \\
&= E_P\left[E_P\left[\chi_C\left(X\left[M[\cdot]\right]\right) \mid \langle \boldsymbol{M}\rangle\right]\right] \\
&= E_P\left[\int_{C[0,\infty[} \chi_C\left(X[c]\right) P^{\cdot}_{M|\langle M\rangle}(dc)\right] \\
&= E_P\left[\int_{C[0,\infty[} \chi_C\left(\mathcal{E}_{P^{\cdot}_{M|\langle M\rangle}}[c]\right) P^{\cdot}_{M|\langle M\rangle}(dc)\right] \\
&= E_P\left[E_P\left[\chi_C\left(\mathcal{E}_{P^{\cdot}_{M|\langle M\rangle}}\left[M[\cdot]\right]\right) \mid \langle \boldsymbol{M}\rangle\right]\right] \\
&= E_P\left[\chi_C\left(M[\cdot]\right)\right] \\
&= P(M[\cdot] \in C).
\end{aligned}$$

Given the uniform in probability approximation property of stochastic integrals using elementary processes [264, p. 152], one has, for every elementary process ϕ_n approximating ϕ,

$$\mathcal{L}(J_M \{\phi_n\}) = \mathcal{L}(M),$$

and

$$\lim_n P(\boldsymbol{J}_M(\{\phi_n\}) \in C) = P(\boldsymbol{J}_M(\{\phi\}) \in C).$$

□

Proposition 16.7.13 *When the continuous, local martingale M is $\mathcal{F}$-invariant, it is, given $\langle M\rangle$, a conditionally Gaussian martingale.*

Proof To shorten notation, M_τ shall stand for $I_M\{f_\tau\}$.

Proof [Step 1] *Given* f_τ*, fixed, but arbitrary,* $\mathcal{L}(M_\tau, \langle M\rangle) = \mathcal{L}(M, \langle M\rangle)$.

Since, by assumption, $\mathcal{L}(M_\tau) = \mathcal{L}(M)$, then, because of (Lemma) 16.7.11,

$$\mathcal{L}(M_\tau, \langle M_\tau\rangle) = \mathcal{L}(M, \langle M\rangle).$$

But $\langle M_\tau\rangle = \langle M\rangle$ since the square of f_τ is the constant function with value one.

Proof [Step 2] *For almost every* $\omega \in \Omega$*, with respect to P, the process*

$$\mathcal{E}_{P^{\omega}_{M|\langle M\rangle}}$$

is a local martingale for $\underline{\mathcal{C}}$.

Let $\{S_n, n \in \mathbb{N}\}$ be the sequence of wide sense stopping times that stops $\langle M\rangle$ at n [264, p. 37]: $S_n = \inf\{t \in \mathbb{R}_+ : \langle M\rangle(\cdot, t) > n\}$. S_n^τ does the same for $\langle M_\tau\rangle$, and equals S_n, since the quadratic variation of the latter integral is that of M [step 1]. Then, in particular, M^{S_n} is an L_2-bounded martingale [264, p. 103].

Let $0 < t_1 < t_2$ in $\mathbb{R}_+$, $V_0 \in \sigma(\langle \boldsymbol{M}\rangle)$, and $C \in \mathcal{C}_{t_1}$, be fixed, but arbitrary. Because of the assumption of equality in law between M and M_τ, one has, *in law,* that:

(i) $\chi_C(\boldsymbol{M}) = \chi_C(\boldsymbol{M}_\tau)$;
(ii) $M^{S_n}(\cdot, t) = M_\tau^{S_n}(\cdot, t) = I_{M^{S_n}}\{f_\tau\}(\cdot, t)$, as, for example,

$$P\left(M^{S_n}(\cdot, t) \in B\right) =$$
$$= P\left(\{M(\cdot, t) \in B\} \cap \{t \le S_n\}\right) + P\left(\{M(\cdot, t) \in B\} \cap \{t > S_n\}\right),$$

and that

$$P\left(\{M(\cdot, t) \in B\} \cap \{t \le S_n\}\right) = P\left(\{M(\cdot, t) \in B\} \cap \{t \le S_n^\tau\}\right);$$

(iii) $\chi_{V_0} = \psi(\langle \boldsymbol{M}\rangle) = \psi(\langle \boldsymbol{M}_\tau\rangle)$ for some functional ψ adapted to $\mathcal{C}$.

One may thus replace, in expressions valid in law, containing elements (i)–(iii), M with M_τ. The following expressions are thus equal in law:

$$\psi(\langle \boldsymbol{M}\rangle)\, \chi_C(\boldsymbol{M}) \left(M^{S_n}(\cdot, t_2) - M^{S_n}(\cdot, t_1)\right), \qquad (\star)$$

and

$$\psi(\langle \boldsymbol{M}_\tau\rangle)\, \chi_C(\boldsymbol{M}_\tau) \left(M_\tau^{S_n^\tau}(\cdot, t_2) - M_\tau^{S_n^\tau}(\cdot, t_1)\right). \qquad (\star\star)$$

But one has also that [(Definition) 16.7.2, item 1]

$$M_\tau(\cdot, t) = M(\cdot, t \wedge \tau) - (M(\cdot, t) - M(\cdot, t \wedge \tau)),$$

and, consequently, using that latter expression with $\tau = t_1$, and $S_n^\tau = S_n$, that

$$M_{t_1}(\cdot, t_2 \wedge S_n) - M_{t_1}(\cdot, t_1 \wedge S_n) = -\{M(\cdot, t_2 \wedge S_n) - M(\cdot, t_1 \wedge S_n)\}.$$

The equality in law established above ($(\star) \overset{\mathcal{L}}{=} (\star\star)$) becomes that the following variables:

$$\chi_{V_0}\chi_C(\boldsymbol{M})\left(M^{S_n}(\cdot, t_2) - M^{S_n}(\cdot, t_1)\right),$$

and

$$-\chi_{V_0}\chi_C(\boldsymbol{M})\left(M^{S_n}(\cdot, t_2) - M^{S_n}(\cdot, t_1)\right),$$

are equal in law, so that

$$E_P\left[\chi_{V_0}\chi_C(\boldsymbol{M})\left(M^{S_n}(\cdot, t_2) - M^{S_n}(\cdot, t_1)\right)\right] = 0 \qquad (\star\star\star)$$

(the expectation exits because of the stopping at S_n). But then, the expectation, with respect to P, of the following expression, is zero:

$$\chi_{V_0}(\cdot)\int_{C[0,\infty[}\chi_C(c)\left\{\mathcal{E}_{P^{\cdot}_{M|\langle M\rangle}}(c, t_2 \wedge S_n) - \mathcal{E}_{P^{\cdot}_{M|\langle M\rangle}}(c, t_1 \wedge S_n)\right\} P^{\cdot}_{M|\langle M\rangle}(\cdot, dc).$$

Thus, given that one deals with a separable σ-algebra $\mathcal{C}_t$, and a continuous process (that of evaluation maps), almost surely, with respect to P, for all $t_1 < t_2$, c, and n,

$$E_{P^{\omega}_{M|\langle M\rangle}}\left[\mathcal{E}_{P^{\omega}_{M|\langle M\rangle}}(c, t_2 \wedge S_n(\omega)) - \mathcal{E}_{P^{\omega}_{M|\langle M\rangle}}(c, t_1 \wedge S_n(\omega) \mid \mathcal{C}_t\right] = 0, \qquad (\star\star\star\star)$$

that is, $\mathcal{E}_{P^{\omega}_{M|\langle M\rangle}}$ is a local martingale.

Proof [Step 3] $\mathcal{E}_{P^{\omega}_{M|\langle M\rangle}}$ *has* $\langle M\rangle(\omega, \cdot)$ *as quadratic variation.*

Relation $(\star\star\star)$ means that M is a continuous, local martingale with respect to the σ-algebras of the following type: $\sigma_t(M) \vee \sigma_t(\langle M\rangle)$. Thus the following process: $M^2 - \langle M\rangle = \frac{1}{2}\int M\, dM$ is a continuous, local martingale, with respect to the same filtrations. A derivation analogous to that leading to relation $(\star\star\star\star)$ says that the conditional expectation, with respect to $\mathcal{C}_t$, and for $P^{\omega}_{M|\langle M\rangle}$, of

$$\mathcal{E}^2_{P^{\omega}_{M|\langle M\rangle}}(\cdot, t_2 \wedge S_n(\omega)) - \mathcal{E}^2_{P^{\omega}_{M|\langle M\rangle}}(\cdot, t_1 \wedge S_n(\omega))$$

minus

$$\langle M\rangle(\omega, t_2 \wedge S_n(\omega)) - \langle M\rangle(\omega, t_1 \wedge S_n(\omega))$$

is zero, almost surely with respect to P, which shows that the evaluation map has, with respect to $P^{\omega}_{M|\langle M\rangle}$, quadratic variation $\langle M\rangle(\omega,\cdot)$. It is thus a Gaussian martingale. □

Remark 16.7.14 Since [(Proposition) 16.7.12] "conditionally Gaussian" implies Φ-invariant, that Φ-invariant implies $\mathcal{F}$-invariant [(Remark) 16.7.8], and that $\mathcal{F}$-invariant implies "conditionally Gaussian," [(Proposition) 16.7.13] one has that "conditionally Gaussian," Φ-invariant, and $\mathcal{F}$-invariant are equivalent properties.

Remark 16.7.15 The property of being "conditionally Gaussian" is equivalent to the following equality: for α measurable, and adapted to $\underline{\mathcal{C}}$ (and thus predictable [264, p. 115]), fixed, but arbitrary,

$$E_P\left[e^{\iota\int_0^\infty \alpha(\langle M\rangle[\cdot],\theta)M(\cdot,d\theta)} \mid \sigma(\langle \boldsymbol{M}\rangle)\right] = e^{-\frac{1}{2}\int_0^\infty \alpha^2(\langle M\rangle[\cdot],\theta)\langle M\rangle(\cdot,d\theta)}. \qquad (\star)$$

Indeed, because of (Proposition) 11.3.14, one has that

$$\langle M\rangle(\omega,t) = \langle \mathcal{E}_{P_M}\rangle(M[\omega],t),$$

and that

$$\int_0^t \alpha(\langle M\rangle[\cdot],\theta)M(\cdot,d\theta) = \left\{\int_0^t \alpha\left(\langle \mathcal{E}_{P_M}\rangle[\cdot],\theta\right)\mathcal{E}_{P_M}(\cdot,d\theta)\right\} \circ \boldsymbol{M}.$$

The left-hand side of $(\star)$ has thus the form $E_P\left[f(\boldsymbol{M}) \mid \sigma(\langle \boldsymbol{M}\rangle)\right]$. The following formula [139, p. 120], valid for regular conditional probabilities, yields that

$$E_P\left[f(\boldsymbol{M}) \mid \langle \boldsymbol{M}\rangle = m\right] = E_{P^m_{M|\langle M\rangle}}[f].$$

When M is conditionally Gaussian on $\langle M\rangle$,

$$\begin{aligned} E_{P^m_{M|\langle M\rangle}}[f] &= E_{P^m_{M|\langle M\rangle}}\left[e^{\iota\int_0^\infty \alpha\left(\left\langle \mathcal{E}_{P^m_{M|\langle M\rangle}}\right\rangle[\cdot],\theta\right)\left\langle \mathcal{E}_{P^m_{M|\langle M\rangle}}\right\rangle(\cdot,d\theta)}\right] \\ &= e^{-\frac{1}{2}\int_0^\infty \alpha^2\left(\left\langle \mathcal{E}_{P^m_{M|\langle M\rangle}}\right\rangle[\cdot],\theta\right)\left\langle \mathcal{E}_{P^m_{M|\langle M\rangle}}\right\rangle(\cdot,d\theta)}. \end{aligned}$$

But then [138, p. 449]

$$E_P\left[f(\boldsymbol{M}) \mid \langle \boldsymbol{M}\rangle\right] = e^{-\frac{1}{2}\int_0^\infty \alpha^2\left(\left\langle \mathcal{E}_{P^m_{M|\langle M\rangle}}\right\rangle[\cdot],\theta\right)\left\langle \mathcal{E}_{P^m_{M|\langle M\rangle}}\right\rangle(\cdot,d\theta)} \circ \langle \boldsymbol{M}\rangle,$$

that is, the right-hand side of $(\star)$. Finally, when the stated formula $(\star)$ obtains, M is indeed, conditionally on $\langle M\rangle$, Gaussian, as, α being arbitrary, the left-hand side of $(\star)$ yields the characteristic function of M, given $\langle M\rangle$, and the right one, the normal characteristic function.

Proposition 16.7.16 *Let M start at zero, and be divergent, that is,*

$$\lim_t \langle M \rangle(\cdot, t) = \infty,$$

almost surely. Formula ($\star$) of (Remark) 16.7.15 *is necessary, and sufficient, for M to be an Ocone martingale.*

Proof Suppose that, in $W_M = M \diamond S_M$, W_M and S_M are independent.

One shall use the following formula [138, p. 452]: on $(\Omega, \mathcal{A}, P)$, let X, and $\mathcal{B} \subseteq \mathcal{A}$, be independent, and Y be adapted to $\mathcal{B}$. Define, for appropriate Ψ,

$$\psi(y) = E_P\left[\Psi(X, y)\right].$$

Then, almost surely, with respect to P,

$$E_P\left[\Psi(X, Y) \mid \mathcal{B}\right] = \psi(Y).$$

One shall also use the following property [264, p. 213]:

$$\int_0^\infty f(\langle M \rangle[\cdot], t) M(\cdot, dt) = \int_0^\infty f(\langle M \rangle[\cdot], \mathcal{E}_t(S_M[\cdot])\, W_M(\cdot, t).$$

Let then $\Psi(\boldsymbol{W}_M, \langle \boldsymbol{M} \rangle)$ be the exponential of

$$\iota \int_0^\infty f(\langle M \rangle[\cdot], \mathcal{E}_t(S_M[\cdot]) W_M(\cdot, dt).$$

One obtains, with $\nu(t) = \inf\{\theta \in \mathbb{R}_+ : v(\theta) > t\}$, that

$$\psi(v) = E_P\left[\Psi(\boldsymbol{W}_M, v)\right] = e^{\iota \int_0^\infty f(v, \mathcal{E}_t(\nu)) W_M(\cdot, dt)} = e^{-\frac{1}{2}\int_0^\infty f^2(v, \mathcal{E}_t(\nu)) dt}.$$

Consequently, using the change of variables formula [(Fact) 10.3.36],

$$\psi(\langle \boldsymbol{M} \rangle) = e^{-\frac{1}{2}\int_0^\infty f^2(\langle M \rangle[\cdot], S_M(\cdot, t)) dt} = e^{-\frac{1}{2}\int_0^\infty f^2(\langle M \rangle[\cdot], t) \langle M \rangle(\cdot, dt)}.$$

Proof Suppose that formula ($\star$) of (Remark) 16.7.15 *obtains.*

Let W be a Brownian motion independent of $\langle M \rangle$, and let $N = W \diamond \langle M \rangle$. As in the first part of the proof, formula ($\star$) of (Remark) 16.7.15 obtains for N. Since $\langle N \rangle = \langle M \rangle$, M and N have the same law. Consequently $(W_M, \langle M \rangle)$ and $(W_N, \langle N \rangle) = (W, \langle M \rangle)$ have the same law because of that formula. But then W_M and $\langle M \rangle$ are independent. □

Proposition 16.7.17 *Let M be a divergent, continuous local martingale starting at zero. It is an Ocone martingale if, and only if, for fixed, but arbitrary $\sum_{k=1}^{n} \alpha_k \chi_{[0,t_k]}$,*

$$E_P\left[e^{\iota \int_0^\infty \left\{\sum_{k=1}^{n} \alpha_k \chi_{[0,t_k]}\right\} dM}\right] = E_P\left[e^{-\frac{1}{2}\int_0^\infty \left\{\sum_{k=1}^{n} \alpha_k \chi_{[0,t_k]}\right\}^2 d\langle M\rangle}\right].$$

Proof If M is an Ocone martingale, one uses (Proposition) 16.7.16, and then takes expectation. Suppose conversely that the given equality obtains. Let W be a standard Brownian motion independent of $\langle M\rangle$, and $N = W \diamond \langle M\rangle$. Since N is an Ocone martingale, the given equality applies to N. Since $\langle N\rangle = \langle M\rangle$, the right-hand side of the given equality is the same for M and N, which means that M and N have the same law. Since $W_M = M \diamond S_M$ and $W_N = N \diamond S_N = W \diamond S_M$, $(W_M, \langle M\rangle)$ and $(W_N, \langle N\rangle)$ have the same law, and that proves that W_M and $\langle M\rangle$ are independent, that is, M is an Ocone martingale. □

16.7.2 Ocone Martingales and Exponentials

Given a divergent, continuous, local martingale M, starting at zero, and a progressively measurable process α, one shall use the following notation:

- $\ln[E_\alpha(\omega, t)] = \int_0^t \alpha(\omega,\theta) M(\omega, d\theta) - \frac{1}{2}\int_0^t \alpha^2(\omega,\theta)\langle M\rangle(\omega, d\theta)$,
- $M_\alpha(\omega, t) = M(\omega,\theta) - \int_0^t \alpha(\omega,\theta)\langle M\rangle(\omega, d\theta)$,
- $dP_\alpha^{|A_t} = E_\alpha(\cdot, t)\, dP^{|A_t}$.

Proposition 16.7.18 *Let α denote a bounded function, which does not depend on ω, and is adapted to the Borel sets. When, for fixed, but arbitrary such α, E_α is a martingale such that, with respect to P_α, M_α is a martingale with the same law as M, with respect to P, then M is an Ocone martingale.*

Proof Let $f : C(\mathbb{R}_+) \longrightarrow \mathbb{R}$ be positive, and adapted to C_t. By definition, for α of the assumption,

$$E_{P_\alpha}[f(\langle M\rangle)] = E_P[f(\langle M\rangle)E_\alpha(\cdot, t)].$$

The identity in law of the assumption implies that

$$E_{P_\alpha}[f(\langle M\rangle)] = E_P[f(\langle M\rangle)].$$

Consequently

$$E_P[f(\langle M\rangle)E_\alpha(\cdot, t)] = E_P[f(\langle M\rangle)].$$

The latter equality is equivalent to the following one: almost surely, with respect to P,

$$E_P\left[E_\alpha(\cdot, t) \mid \sigma_t(\langle M \rangle)\right] = 1,$$

which means that

$$E_P\left[e^{\int_0^t \alpha(\theta)\, M(\cdot, d\theta)} \mid \sigma_t(\langle M \rangle)\right] = e^{\frac{1}{2}\int_0^t \alpha^2(\theta)\, \langle M \rangle(d\theta)}.$$

Let W be a Brownian motion independent of $\langle M \rangle$, and set $N = W \diamond \langle M \rangle$. Then

$$E_P\left[e^{\int_0^t \alpha(\theta)\, N(\cdot, d\theta)} \mid \sigma_t(\langle M \rangle)\right] = e^{\frac{1}{2}\int_0^t \alpha^2(\theta)\, \langle M \rangle(d\theta)},$$

so that,

$$E_P\left[e^{\int_0^t \alpha(\theta)\, M(\cdot, d\theta)} \mid \sigma_t(\langle M \rangle)\right] = E_P\left[e^{\int_0^t \alpha(\theta)\, N(\cdot, d\theta)} \mid \sigma_t(\langle M \rangle)\right],$$

that is, M and N have the same law. But then [(Lemma) 16.7.11], with respect to P, $\mathcal{L}(M, \langle M \rangle) = \mathcal{L}(N, \langle M \rangle)$. Now $M = W_M \diamond \langle M \rangle$ so that, using the time change S_M, one obtains that, with respect to P,

$$\mathcal{L}(W_M, \langle M \rangle) = \mathcal{L}(W, \langle M \rangle),$$

that is, W_M and $\langle M \rangle$ are independent. □

Remark 16.7.19 In (Proposition) 16.7.18, let $N_\alpha = \int \alpha\, dM$, so that $\langle N_\alpha \rangle = \int \alpha^2\, d\langle M \rangle$. One conclusion of the proof's first paragraph, when multiplying α with θ, a constant, is that

$$E_P\left[e^{\theta N_\alpha(\cdot, t)}\right] = E_P\left[e^{\frac{\theta^2}{2}\langle N_\alpha \rangle(\cdot, t)}\right]. \qquad (\star)$$

Stopping N_α when $|N_\alpha|$ crosses n, one may assume a bounded process. Then, for t fixed, but arbitrary, and z complex, one has that $z \mapsto E_P\left[e^{zN_\alpha(\cdot, t)}\right]$ is entire. Thus $(\star)$ extends to complex values, and one may use (Proposition) 16.7.17 to conclude.

Proposition 16.7.20 *When*

(a) *M is an Ocone martingale,*
(b) *$f : C(\mathbb{R}_+) \longrightarrow \mathbb{R}$, is a fixed, but arbitrary, bounded, progressively measurable process for $\underline{\mathcal{C}}$,*
(c) *$\alpha = f(\langle M \rangle[\cdot], \cdot)$,*

then E_α is a martingale, and the law of M_α with respect to P_α is that of M with respect to P.

Proof One shall use the transformation formula [(Fact) 10.3.36]

$$\int_0^t \{\phi \circ \alpha\}\, dm_\gamma = \int_{\alpha(0)}^{\alpha(t)} \phi\, dm_\beta,\ \gamma = \alpha \circ \beta,$$

with m_β, Lebesgue measure, and $\alpha = \langle M\rangle$, on an expression of the form $\int \psi\, d\langle M\rangle$. ψ must thus be written as $\psi(S_M \diamond \langle M\rangle)$. But that is possible only when $\langle M\rangle$ is strictly increasing. However, when it is constant, the interval of constancy has a measure determined by $\langle M\rangle$ that is zero. So it is not a restriction to assume $\langle M\rangle$ strictly increasing.

Let $\psi : C(\mathbb{R}_+) \longrightarrow \mathbb{R}$ be a positive functional adapted to $\mathcal{C}_t$. Then, by definition,

$$E_{P_\alpha}\left[\psi(M_\alpha[\cdot])\right] = E_P\left[\psi(M_\alpha[\cdot])\, e^{\int_0^t \alpha(\cdot,\theta)\, M(\cdot,d\theta) - \frac{1}{2}\int_0^t \alpha^2(\cdot,\theta)\, \langle M\rangle(\cdot,d\theta)}\right].$$

But, using the definitions of α and M_α, and the transformation formula repeated just above,

$$M_\alpha(\cdot,t) = W_M \diamond \langle M\rangle(\cdot,t) - \int_0^{\langle M\rangle(\cdot,t)} f(\langle M\rangle[\cdot], S_M(\cdot,\theta))\, d\theta,$$

and

$$\int_0^t \alpha^2(\cdot,\theta)\, \langle M\rangle(\cdot,d\theta) = \int_0^{\langle M\rangle(\cdot,t)} f^2(\langle M\rangle[\cdot], S_M(\cdot,\theta))\, d\theta.$$

Furthermore, because of the choice for α and (Fact) 10.3.45,

$$\int_0^t f(\langle M\rangle[\cdot],\theta) M(\cdot,d\theta) = \int_0^{\langle M\rangle(\cdot,t)} f(\langle M\rangle[\cdot], S_M(\cdot,\theta)) W_M(\cdot,d\theta).$$

Thus

$$\psi(M_\alpha[\cdot])\, e^{\int_0^t \alpha(\cdot,\theta)\, M(\cdot,d\theta) - \frac{1}{2}\int_0^t \alpha^2(\cdot,\theta)\, \langle M\rangle(\cdot,d\theta)}$$

has the form

$$\Psi(\mathbf{W}_M, \langle \mathbf{M}\rangle).$$

Now, letting

$$W_M^v(\omega,t) = W_M(\omega, v(t)),$$

and

$$W^v_\alpha(\omega, t) = W^v_M(\omega, t) - \int_0^t \alpha(v, \theta)\, v(d\theta),$$

one has, using Girsanov's theorem for Brownian motion, that

$$\begin{aligned} E_P[\Psi(\mathbf{W}_M, v)] &= E_P\left[\psi(\mathbf{W}^v_\alpha)\, e^{\int_0^t \alpha(v,\theta) W^v_M(\cdot, d\theta) - \frac{1}{2}\int_0^t \alpha^2(v,\theta) v(d\theta)}\right] \\ &= E_P\left[\psi(\mathbf{W}^v_M)\right]. \end{aligned}$$

Consequently, $E_P\left[\Psi(\mathbf{W}_M, \langle \boldsymbol{M}\rangle)\right] = E_P\left[\psi(\boldsymbol{M})\right]$, that is

$$E_{P_\alpha}\left[\psi\left(M_\alpha[\cdot]\right)\right] = E_P\left[\psi\left(M[\cdot]\right)\right],$$

so that M_α, with respect to P_α, has the same law as M with respect to P. □

Corollary 16.7.21 *The assumption of* (Proposition) 16.7.18, *and the conclusion of* (Proposition) 16.7.20, *provide equivalent conditions for M to be an Ocone martingale.*

Proof Indeed, the assumption of (Proposition) 16.7.18 is a particular case of the conclusion of (Proposition) 16.7.20. □

Proposition 16.7.22 *Another property equivalent to those of (Corollary)* 16.7.21 *for M to be an Ocone martingale is that, for fixed, but arbitrary α that does not depend on ω, bounded and adapted to the Borel sets, $\langle M\rangle$ has the same law with respect to P_α and P.*

Proof The proof of (Proposition) 16.7.18 uses the equality in law with respect to P_α and P of $\langle M\rangle$. The reverse claim is a consequence of (Proposition) 16.7.20. □

16.7.3 Ocone Martingales, More Properties, and Some Examples

One should keep in mind that only continuous (local) martingales are dealt with.

Proposition 16.7.23 *Let M be an Ocone martingale, and N be a local martingale for $\underline{\sigma}(\langle M\rangle)$. Then:*

1. *N is a local martingale for $\underline{\sigma}(M)$.*
2. *N is orthogonal to M, that is MN is a local martingale.*
3. *M is extremal if, and only if, $\langle M\rangle$ is deterministic.*

Proof [1] Since there is an underlying assumption of continuity, it is sufficient to prove the result for uniformly integrable martingales N, and, in that case, that

$E_P[N(\cdot,\infty) \mid \sigma_t(M)] = N(\cdot,t)$. Let ψ be a functional adapted to $\mathcal{C}_t$. Then, P_W being Wiener measure for continuous functions, and $\gamma(\omega,t) = c(\langle M\rangle(\omega,t))$, using first Ocone independence, and second, the martingale property of N with respect to $\underline{\sigma}(\langle M\rangle)$,

$$\begin{aligned}E_P[N(\cdot,\infty)\psi(M[\cdot])] &= E_P[N(\cdot,\infty)\psi(W_M \diamond \langle M\rangle[\cdot])] \\ &= \int_{C(\mathbb{R}_+)} P_W(dc)E_P[N(\cdot,\infty)\psi(\gamma[\cdot])] \\ &= \int_{C(\mathbb{R}_+)} P_W(dc)E_P[N(\cdot,t)\psi(\gamma[\cdot])] \\ &= E_P[N(\cdot,t)\psi(M[\cdot])].\end{aligned}$$

Proof [2] Let $t_1 < t_2$ be fixed, but arbitrary, and ψ be adapted to $\mathcal{C}_{t_1}$. Then

$$\begin{aligned}&E_P[M(\cdot,t_2)N(\cdot,t_2)\psi(M[\cdot])] = \\ &\qquad = E_P[W_M(\cdot,\langle M\rangle(\cdot,t_2))N(\cdot,t_2)\psi(W_M \diamond \langle M\rangle[\cdot])].\end{aligned}$$

The right-hand side of the latter equality is an expression of the following form:

$$E_P[f(\boldsymbol{W_M},\langle \boldsymbol{M}\rangle)] = E_P[E_P[f(\boldsymbol{W_M},\langle \boldsymbol{M}\rangle) \mid \sigma(\langle \boldsymbol{M}\rangle)]],$$

where the arguments of f are independent. Let

$$F(m) = E_P[f(\boldsymbol{W_M},\langle \boldsymbol{M}\rangle) \mid \langle \boldsymbol{M}\rangle = m] = E_P[f(\boldsymbol{W_M},m)].$$

Since N is a function of $\langle M\rangle$, one uses the martingale property of W_M to obtain that one may replace, in $W_M(\cdot,\langle M\rangle(\cdot,t_2))$, t_2 with t_1, so that

$$E_P[f(\boldsymbol{W_M},m)] = E_P[W_M(\cdot,m(t_1))N(\cdot,t_2)\psi(W_M \diamond m[\cdot])].$$

Then

$$\begin{aligned}&E_P[f(\boldsymbol{W_M},\langle \boldsymbol{M}\rangle) \mid \sigma(\langle \boldsymbol{M}\rangle)] = \\ &\qquad = F(\langle \boldsymbol{M}\rangle) \\ &\qquad = E_P[W_M \diamond \langle M\rangle(\cdot,t_1)N(\cdot,t_2)\psi(W_M \diamond \langle M\rangle[\cdot]) \mid \sigma(\langle \boldsymbol{M}\rangle)],\end{aligned}$$

and thus

$$E_P[M(\cdot,t_2)N(\cdot,t_2)\psi(M[\cdot])] = E_P[M(\cdot,t_1)N(\cdot,t_2)\psi(M[\cdot])].$$

But

$$
\begin{aligned}
&E_P\left[M(\cdot,t_1)N(\cdot,t_2)\,\psi(M[\cdot])\right] = \\
&\quad = E_P\left[W_M(\cdot,\langle M\rangle(\cdot,t_1))N(\cdot,t_2)\,\psi(W_M\diamond\langle M\rangle[\cdot])\right] \\
&\quad = \int_{C(\mathbb{R}_+)} P_W(dc)E_P\left[c(\langle M\rangle(\cdot,t_1))N(\cdot,t_2)\,\psi(c(\langle M\rangle[\cdot]))\right].
\end{aligned}
$$

Since N is a martingale with respect to the σ-algebras generated by $\langle M\rangle$, one may replace, in the latter integral, $N(\cdot,t_2)$ with $N(\cdot,t_1)$, and, consequently,

$$E_P\left[M(\cdot,t_2)N(\cdot,t_2)\,\psi(M[\cdot])\right] = E_P\left[M(\cdot,t_1)N(\cdot,t_1)\,\psi(M[\cdot])\right].$$

Proof [3] For an Ocone martingale, and Γ, a fixed, but arbitrary measurable set of continuous functions,

$$
\begin{aligned}
P_M(\Gamma) &= \int_{C(\mathbb{R}_+)} P_{\langle M\rangle}(dc)P(\mathbf{W}_M\diamond\langle\mathbf{M}\rangle\in\Gamma \mid \langle\mathbf{M}\rangle = c) \\
&= \int_{C(\mathbb{R}_+)} P_{\langle M\rangle}(dc)P(\mathbf{W}_M\diamond c\in\Gamma).
\end{aligned}
$$

When $P_{\langle M\rangle}$ is a point mass, P_M is the law of a deterministic time changed Brownian motion, and those processes are extremal martingales [221, pp. 213–214]. When P_M is extremal, and C is a measurable set, $P_{\langle M\rangle}(C)$ is either zero or one, as otherwise P_M could be represented as a mixture. But then $P_{\langle M\rangle}$ is Dirac measure, and again P_M is the law of a deterministic time changed Brownian motion. □

Example 16.7.24 Let W be a Brownian motion for $\underline{\mathcal{A}}$, and α, a continuous to the right process, which is almost surely continuous, and adapted to $\underline{\mathcal{A}}$, such that

- $|\alpha| > 0$, almost surely with respect to the product of P and Lebesgue measure,
- almost surely, with respect to P, $\int_0^\infty \alpha^2(\cdot,\theta)d\theta = \infty$.

Let s be the sign function, and

$$
\begin{aligned}
M(\cdot,t) &= \int_0^t \alpha(\cdot,\theta)\,W(\cdot,d\theta), \\
B(\cdot,t) &= \int_0^t \{s\circ\alpha(\cdot,\theta)\}\,W(\cdot,d\theta).
\end{aligned}
$$

Then M is an Ocone martingale if, and only if, the Brownian motion B and the following σ-algebra:

$$\sigma(|\boldsymbol{\alpha}|) = \sigma(\langle\boldsymbol{M}\rangle)$$

are independent.

The claim may be checked as follows. Let ψ be a progressively measurable process for $\underline{\mathcal{C}}$. Then, as $dM = \alpha dW = \{s \circ \alpha\} \alpha dB$,

$$\int_0^\infty \{\psi \diamond \Phi_{\langle M \rangle}\} dM = \int_0^\infty \{\psi \diamond \Phi_{\langle M \rangle}\} \{s \circ \alpha\} \alpha dB,$$

and, provided

$$\int_0^\infty \{\psi \diamond \Phi_{\langle M \rangle}\}^2 (\cdot, \theta) \alpha^2(\cdot, \theta) d\theta < \infty,$$

when B and $\sigma(|\boldsymbol{\alpha}|) = \sigma(\langle \boldsymbol{M} \rangle)$ are independent,

$$E_P\left[e^{\iota \int_0^\infty \{\psi \diamond \Phi_{\langle M \rangle}\}\{s \circ \alpha\} \alpha \, dB} \mid \sigma(\langle \boldsymbol{M} \rangle)\right] = e^{-\frac{1}{2}\int_0^\infty \{\psi \diamond \Phi_{\langle M \rangle}\}^2 (\cdot, \theta) \alpha^2(\cdot, \theta) d\theta},$$

which rewrites as

$$E_P\left[e^{\iota \int_0^\infty \{\psi \diamond \Phi_{\langle M \rangle}\} dM} \mid \sigma(\langle \boldsymbol{M} \rangle)\right] = e^{-\frac{1}{2}\int_0^\infty \{\psi \diamond \Phi_{\langle M \rangle}\}^2 d\langle M \rangle}.$$

But the latter equality means that M is an Ocone martingale [(Example) 16.7.28].

Conversely, when M is an Ocone martingale, the latter equality prevails, and, choosing f to be a generic, simple, deterministic function with compact support, setting

$$\psi \diamond \Phi_{\langle M \rangle} = \frac{f}{|\alpha|},$$

one obtains that

$$E_P\left[e^{\iota \int_0^\infty f \, dB} \mid \sigma(\langle \boldsymbol{M} \rangle)\right] = e^{-\frac{1}{2}\int_0^\infty f^2(\theta) d\theta},$$

which proves that B and $\sigma(\langle \boldsymbol{M} \rangle)$ are independent [138, p. 452].

Definition 16.7.25 Let $N = W_N \diamond \langle N \rangle$ be a local martingale for $\underline{\mathcal{A}}$, starting at zero, and divergent. When $\sigma_\infty(N) = \sigma_\infty(W_N)$, N is said to be a pure martingale.

Example 16.7.26 $M = \int B dB$ is a pure martingale when B is a standard Brownian motion. That claim may be checked as follows [252, p. 597]. One has that $\langle M \rangle = \int B^2$. Thus, by the change of time formula,

$$t = \int_0^t \frac{\langle M \rangle(\cdot, d\theta)}{B^2(\cdot, \theta)} = \int_0^{\langle M \rangle(\cdot, t)} \frac{d\theta}{B^2 \diamond S_M(\cdot, \theta)}.$$

Furthermore (Itô's formula) $B^2 = 2M + t$. Let $N = B^2 \diamond S_M$. Then

$$S_M = \int_0^{\cdot} \frac{d\theta}{N(\cdot,\theta)}, \quad \text{and } N = 2W_M + S_M.$$

Let $f(x,y) = (2x+y)^2$. Then $f(W_M, S_M) = N^2$, so that (Itô's formula again)

$$\begin{aligned} N^2(\cdot,t) &= \int_0^t \frac{\partial f}{\partial x}(W_M,S_M)dW_M + \int_0^t \frac{\partial f}{\partial y}(W_M,S_M)dS_M \\ &+ \frac{1}{2}\int_0^t \frac{\partial^2 f}{\partial x^2}(W_M,S_M)dt \\ &= 4\int_0^t N dW_M + 6t. \end{aligned}$$

By definition, $N \geq 0$, so that $N^2(\cdot,t) = 4\int_0^t \sqrt{N^2}dW_M + 6t$, and N^2 is the solution of a stochastic differential equation which has the property that [223, Vol. 2, p. 69] N^2 is adapted to the completion of the σ-algebras generated by W_M, say $\underline{\mathcal{B}}^M$. But then so is N, and, since $N = 2W_M + S_M$, so is S_M also. Since $S_M(\omega,\theta) \leq t$ is equivalent to $\langle M\rangle(\omega,t) \geq \theta$, and since $S_M(\cdot,\theta)$ is a wide sense stopping time for θ, fixed, but arbitrary, $\langle M\rangle(\cdot,t)$ is adapted to $\mathcal{B}_t^M$. Consequently so is $M = W_M \diamond S_M$.

Example 16.7.27 The same calculation as that of (Example) 16.7.26 applies to

$$M = M_1 + M_2,\ M_i = \int B_i dB_i,\ i \in \{1,2\},$$

where B_1 and B_2 are independent, standard Brownian motions. One has indeed, with $B^{\{2\}} = B_1^2 + B_2^2$, that $\langle M\rangle = \int_0^{\cdot} B^{\{2\}}$, and that

$$B^{\{2\}} = B_1^2 + B_2^2 = \left\{2\int B_1 dB_1 + t\right\} + \left\{\int B_2 dB_2 + t\right\} = 2M + 2t.$$

One may thus proceed replacing B^2 with $B^{\{2\}}$.

Example 16.7.28 Suppose that N is a pure, continuous local martingale. When the martingale M is orthogonal to N, and $\langle M\rangle$ is adapted to $\sigma(N)$, it is then an Ocone martingale. That claim may be checked as follows. Since M and N are orthogonal, it follows from [264, p. 216] that W_M and W_N are independent. Since furthermore $\sigma_\infty(N) = \sigma_\infty(W_N)$, W_M and N are independent, and thus W_M and $\langle M\rangle$ are independent.

Example 16.7.29 $M(\cdot,t) = \int_0^t B_1(\cdot,\theta)B_2(\cdot,d\theta)$ is an Ocone martingale when B_1 and B_2 are independent Brownian motions, but not when $B_1 = B_2$. Indeed, B_1 is pure, and $\langle M\rangle(\cdot,t) = \int_0^t B_1^2(\cdot,\theta)\,d\theta$ is adapted to the σ-algebras generated by B_1. Also B_1 and M are orthogonal [(Example) 16.7.28]. On the other hand, when

$B_1 = B_2 = B$, $M = B^2 - t$ and $d\langle M\rangle = B^2 dt$, so that $dS_M = B^{-2}dt$ and $W_M = B^2 \diamond S_M - S_M$. Thus W_M is adapted to B^2, and thus to M. So M is pure. But W_M and S_M are not independent.

Example 16.7.30 ([255]) Let $\underline{F} : R^n \longrightarrow \mathbb{R}^n$ be a map, $D[\underline{F}]$, be its derivative, and $\Delta[\underline{F}] = D[\underline{F}]^\star$. Suppose that, for $\underline{x} \in \mathbb{R}^n$, fixed, but arbitrary,

$$\Delta[\underline{F}](\underline{x})\underline{F}(\underline{x}) = \underline{x}, \text{ and } \langle \underline{F}(\underline{x}), \underline{x}\rangle_{\mathbb{R}^n} = 0.$$

Let $\underline{B}$ be a standard Brownian motion with values in $\mathbb{R}^n$, and set

$$M(\omega, t) = \sum_{i=1}^{n} \int_0^t F_i\left(\underline{B}(\omega, \theta)\right) B_i(\omega, d\theta).$$

M is an Ocone martingale. A particular case is $\underline{F}(\underline{x}) = A(\underline{x})$, where A is orthogonal and antisymmetric, for example

$$A = \begin{bmatrix} 0 & -1 \\ 1 & 0 \end{bmatrix}.$$

That claim may be checked as follows. One has that

$$\langle M\rangle(\omega, t) = \int_0^t \sum_{i=1}^{n} F_i^2\left(\underline{B}(\omega, \theta)\right) d\theta.$$

Let

$$\Phi(\underline{x}) = \sum_{i=1}^{n} F_i^2(\underline{x}).$$

Then

$$\frac{\partial \Phi}{\partial x_k}(\underline{\xi}) = 2\sum_{i=1}^{n} F_i(\underline{\xi})\frac{\partial F_i}{\partial x_k}(\underline{\xi}),$$

so that, given the assumptions, $D[\Phi](\underline{\xi}) = 2\underline{F}^\star(\underline{\xi})D[\underline{F}](\underline{\xi}) = 2\underline{\xi}^\star$. Consequently, $D^2[\Phi] = 2I_n$, and Itô's formula [56, p. 111] yields that

$$\Phi(\underline{B})(\omega, t) = 2\sum_{i=1}^{n} \int_0^t B_i(\omega, \theta) B_i(\omega, d\theta) + nt.$$

Finally

$$\langle M\rangle(\omega, t) = \int_0^t \Phi\left(\underline{B}(\omega, \theta)\right) d\theta.$$

Let

$$N(\omega, t) = \Phi(\underline{B})(\omega, t) - nt = 2\sum_{i=1}^{n} \int_0^t B_i(\omega, \theta) B_i(\omega, d\theta).$$

As seen in (Example) 16.7.27, N is a pure martingale, and $\langle M\rangle$ is adapted to the filtration generated by N. Thus, if M and N are orthogonal, M is an Ocone martingale because of (Example) 16.7.28. But

$$\langle M, N\rangle(\cdot, t) = \int_0^t \langle \underline{F}(\underline{B}(\cdot, \theta)), \underline{B}(\cdot, \theta)\rangle_{\mathbb{R}^n} d\theta = 0.$$

Example 16.7.31 Example 16.7.30 applies to the process $\frac{1}{2}\int_0^t \{B_1 dB_2 - B_2 dB_1\}$, an Ocone martingale.

16.8 The Uniqueness Class of Continuous Local Martingales

One finds below an answer to the following question, related to the second fundamental property of the original Girsanov's theorem: when does a continuous local martingale have a law which is determined by its quadratic variation?

Definition 16.8.1 (Class $\mathcal{M}$) All the martingales considered entering this section shall be zero at the origin, have (almost surely) continuous paths, and be divergent. They form the class $\mathcal{M}$.

Definition 16.8.2 (Class $\mathcal{P}$) $\mathcal{P}$ shall be the class of laws $\mathcal{L}(M)$, $M \in \mathcal{M}$, with the following property:

$$[N \in \mathcal{M} \text{ and } \mathcal{L}(\langle N\rangle) = \mathcal{L}(\langle M\rangle)] \Longrightarrow [\mathcal{L}(N) = \mathcal{L}(M)].$$

Remark 16.8.3 The probability space for N in (Definition) 16.8.2 may be different from that for M.

Proposition 16.8.4 *The following obtain:*

1. *when $\mathcal{L}(M) \in \mathcal{P}$, M is an Ocone martingale;*
2. *$\mathcal{L}(M) \in \mathcal{P}$ if, and only if,*

$$[N \in \mathcal{M} \textit{ and } \mathcal{L}(\langle N\rangle) = \mathcal{L}(\langle M\rangle)] \Longrightarrow N \textit{ is an Ocone martingale.}$$

Proof Suppose that $\mathcal{L}(M) \in \mathcal{P}$, and let $N = J_M\{\phi\}$, as defined in (Definition) 16.7.7. Since then, as $\phi^2 = 1$, $\langle N \rangle = \langle M \rangle$, and that $\mathcal{L}(M) \in \mathcal{P}$, it follows that $\mathcal{L}(N) = \mathcal{L}(M)$, so that M is Φ invariant. But then, because of (Example) 16.7.28, (Remark) 16.7.15, and (Proposition) 16.7.16, M is an Ocone martingale.

Suppose now that $\mathcal{L}(M) \in \mathcal{P}$, and that $N \in \mathcal{M}$ is such that $\mathcal{L}(\langle N \rangle) = \mathcal{L}(\langle M \rangle)$. By assumption and definition $\mathcal{L}(N) = \mathcal{L}(M)$. Using (Proposition) 16.7.17, one thus sees that N is an Ocone martingale.

Suppose finally that both $N \in \mathcal{M}$ and $\mathcal{L}(\langle N \rangle) = \mathcal{L}(\langle M \rangle)$ imply that N is an Ocone martingale. Choosing for N, $I_M\{f_\tau\}$, as defined in (Definition) 16.7.1, one has that the latter is an Ocone martingale, and, as such, $\mathcal{F}$-invariant. Thus [(Proposition) 16.7.16], for appropriate θ, $I_{I_M\{f_\tau\}}\{f_\theta\}$ has the same law as $I_M\{f_\tau\}$. But [(Remark) 16.7.4], when $\theta = \tau$, $I_{I_M\{f_\tau\}}\{f_\theta\} = M(\cdot, t)$. M is thus an Ocone martingale [(Proposition) 16.7.16]. However, two Ocone martingales, whose associated increasing processes have the same law, have themselves the same law. Thus $\mathcal{L}(M) \in \mathcal{P}$. □

Definition 16.8.5 One shall use below the following maps:

1. T_a: $a \in \mathbb{R}$, fixed, but arbitrary,

$$T_a(x) = x + a, \; x \in \mathbb{R};$$

2. F_ϵ: $\epsilon > 0, f : \mathbb{R}_+ \longrightarrow \mathbb{R}$, fixed, but arbitrary,

$$F_\epsilon[f](t) = \chi_{[\epsilon,\infty[}(t) f(T_{-\epsilon}(t));$$

3. G_ϵ: $G_\epsilon[f](t) = f(T_\epsilon(t))$.

Remark 16.8.6 In (Definition) 16.8.5, for $F_\epsilon[f]$ to yield a continuous function, one must have $f(0) = 0$. Furthermore $G_\epsilon \circ F_\epsilon$ is the identity.

Proposition 16.8.7 *Let $\epsilon > 0$, $M \in \mathcal{M}$ be fixed, but arbitrary, and*

$$\langle M \rangle_\epsilon(\omega, t) = F_\epsilon[\langle M \rangle(\omega, \cdot)](t).$$

Suppose that $N \in \mathcal{M}$ is such that $\mathcal{L}(\langle N \rangle) = \mathcal{L}(\langle M \rangle_\epsilon)$. Then $\mathcal{L}(M) \in \mathcal{P}$ implies $\mathcal{L}(N) \in \mathcal{P}$.

Proof The definition of $\langle M \rangle_\epsilon$ makes sense as $\langle M \rangle(\cdot, 0) = 0$. Let, for the filtration $\underline{\sigma}(N)$, $N = W_N \diamond \langle N \rangle$. Set

$$\begin{aligned} \langle N \rangle^\epsilon(\omega, t) &= G_\epsilon[\langle N \rangle(\omega, \cdot)](t), \\ \mathcal{B}_t^N &= \sigma^+_{S_N(\cdot,t)}(N). \end{aligned}$$

For $t \in \mathbb{R}_+$, fixed, but arbitrary, $\langle N \rangle^\epsilon(\cdot, t)$ is a strict stopping time for the filtration $\underline{\mathcal{B}}^N$. Indeed, one must have, for $\alpha \in \mathbb{R}_+$, fixed, but arbitrary,

$$\{\omega \in \Omega : \langle N \rangle^\epsilon(\omega, t) \leq \alpha\} \in \mathcal{B}^N_\alpha.$$

That is equivalent to [(Proposition) 10.2.26] for fixed, but arbitrary $\theta \in \mathbb{R}_+$,

$$\{\omega \in \Omega : \langle N \rangle^\epsilon(\omega, t) \leq \alpha\} \cap \{\omega \in \Omega : S_N(\omega, \alpha) < \theta\} \in \sigma_\theta(N).$$

Since $S_N(\cdot, \alpha)$ is a wide sense stopping time, the second set in the latter intersection is indeed in $\sigma_\theta(N)$. Since $\langle N \rangle$ is continuous [(Fact) 10.3.27],

$$\{\omega \in \Omega : S_N(\omega, \alpha) < \theta\} = \{\omega \in \Omega : \langle N \rangle(\omega, \theta) > \alpha\}.$$

It follows that $t + \epsilon \leq \theta$, and thus the first set in the latter intersection is in $\sigma_\theta(N)$.

The process $N_\epsilon = W_N \diamond \langle N \rangle^\epsilon$ is thus well defined, belongs to $\mathcal{M}$, and has quadratic variation $\langle N \rangle^\epsilon$. Furthermore

$$P_{\langle M \rangle_\epsilon} = P_{F_\epsilon \circ \langle M \rangle} = P_{\langle M \rangle} \circ F_\epsilon^{-1},$$

and

$$Q_{\langle N_\epsilon \rangle} = Q_{G_\epsilon \circ \langle N \rangle} = Q_{\langle N \rangle} \circ G_\epsilon^{-1}.$$

But, by assumption, $Q_{\langle N \rangle} = P_{\langle M \rangle_\epsilon}$ so that

$$Q_{\langle N_\epsilon \rangle} = P_{\langle M \rangle} \circ F_\epsilon^{-1} \circ G_\epsilon^{-1} = P_{\langle M \rangle} \circ \{G_\epsilon \circ F_\epsilon\}^{-1} = P_{\langle M \rangle}.$$

Thus, for $\epsilon > 0$, fixed, but arbitrary, N_ϵ is an Ocone martingale [(Proposition) 16.8.4]. It follows, taking the limit, as ϵ goes to zero, using (Proposition) 16.7.17, that N is an Ocone martingale, and thus that $\mathcal{L}(N) \in \mathcal{P}$ [(Proposition) 16.8.4 again]. □

Lemma 16.8.8 *Let M and N belong to $\mathcal{M}$. When $\mathcal{L}(\langle M \rangle) = \mathcal{L}(\langle N \rangle)$,*

1. *$\mathcal{L}(S_M) = \mathcal{L}(S_N)$;*
2. *when $\sigma_\infty(\langle M \rangle) = \sigma_0(S_M)$, then, provided the σ-algebras involved are complete, $\sigma_\infty(\langle N \rangle) = \sigma_0(S_N)$.*

Proof The first item follows from the fact that $\langle M \rangle(\cdot, t) > \theta$ if, and only if, $S_M(\cdot, \theta) < t$ [(Fact) 10.3.27], the second from equalities of the following type (equality in law may occur for different probability spaces):

$$\begin{aligned} 0 &= E_P\left[\left|\chi_{B_1}(\langle M \rangle(\cdot, t)) - \chi_{B_2}(S_M(\cdot, 0))\right|\right] \\ &= E_Q\left[\left|\chi_{B_1}(\langle N \rangle(\cdot, t)) - \chi_{B_2}(S_N(\cdot, 0))\right|\right], \end{aligned}$$

and the fact that $S_M(\cdot, 0) = \inf\{\theta \geq 0 : \langle M \rangle(\cdot, \theta) > 0\}$. □

Proposition 16.8.9 *Let $M \in \mathcal{M}$ be fixed, but arbitrary. When*

$$\sigma_\infty(\langle M\rangle) = \sigma_0(S_M),$$

then $\mathcal{L}(M) \in \mathcal{P}$.

Proof Since W_M, the DDS Brownian motion of M, is a standard Brownian motion with respect to the family

$$\left\{\mathcal{B}_t^M = \mathcal{A}^+_{S_M(\cdot,t)}, t \geq 0\right\},$$

it is independent of $\mathcal{B}_0^M$. But then, by assumption, since $\sigma_0(S_M) \subseteq \mathcal{B}_0^M$, it is independent of $\sigma_\infty(\langle M\rangle)$. M is thus an Ocone martingale. Let $N \in \mathcal{M}$ be such that $\mathcal{L}(\langle N\rangle) = \mathcal{L}(\langle M\rangle)$. Because of (Lemma) 16.8.8, it is an Ocone martingale for the same reason M is one. But then, because of (Proposition) 16.8.4, $\mathcal{L}(M) \in \mathcal{P}$. □

Example 16.8.10 Let $f, g : \mathbb{R}_+ \longrightarrow \mathbb{R}_+$ be continuous, increasing functions, starting at zero, but not identically equal to it, nor equal, and let

$$n_f = \inf\{\theta \geq 0 : f(\theta) > 0\},\ n_g = \inf\{\theta \geq 0 : g(\theta) > 0\}.$$

Let X be a random variable such that $P(X = 0) = P(X = 1) = \frac{1}{2}$. Define the following sets: $\Omega_0 = \{\omega \in \Omega : X(\omega) = 0\}$, $\Omega_1 = \{\omega \in \Omega : X(\omega) = 1\}$. Set

$$V(\omega, t) = \chi_{\Omega_0}(\omega) f(t) + \chi_{\Omega_1}(\omega) g(t),$$

and choose a standard Brownian motion W, independent of X, to define the following Ocone martingale:

$$M = W \diamond V.$$

Then $\langle M\rangle = V$. Denote ϕ the inverse of f ($\phi(t) = \inf\{\theta : f(\theta) > t\}$), and γ, the inverse of g ($\gamma(t) = \inf\{\theta : g(\theta) > t\}$). Then:

$$S_M(\omega, t) = \inf\{\theta \geq 0 : \langle M\rangle(\omega, \theta) > t\} = \chi_{\Omega_0}(\omega)\phi(t) + \chi_{\Omega_1}(\omega)\gamma(t).$$

In particular $S_M(\omega, 0) = \chi_{\Omega_0}(\omega) n_f + \chi_{\Omega_1}(\omega) n_g$.

Since, for some $t > 0$, $\langle M\rangle(\omega, t) > 0$, $\sigma_t(\langle M\rangle) = \sigma(X)$. When either one of n_f and n_g is different from zero, $\sigma_0(S_M) = \sigma(X)$ (otherwise $\sigma_0(S_M)$ is the trivial σ-field).

Finally, when P_f is the law of the following map: $(\omega, t) \mapsto W(\omega, f(t))$, and P_g, that of the following one: $(\omega, t) \mapsto W(\omega, g(t))$, $P_M = \frac{1}{2}\left\{P_f + P_g\right\}$, a law that is not

Gaussian as, for example,

$$E_P\left[e^{\iota\theta M(\cdot,t)}\right] = \frac{1}{2}\left\{e^{-\frac{\theta^2}{2}f(t)} + e^{-\frac{\theta^2}{2}g(t)}\right\}.$$

Also $\mathcal{L}(\langle M\rangle) = \frac{1}{2}\left\{\delta_f + \delta_g\right\}$.

Suppose now that $N \in \mathcal{M}$, on $(\Theta, \underline{\mathcal{B}}, Q)$, is such that $\mathcal{L}(\langle N\rangle) = \frac{1}{2}\left\{\delta_f + \delta_g\right\}$. Let $\Theta_f = \{\theta \in \Theta : \langle N\rangle(\theta,\cdot) = f\}$, and define Θ_g analogously. Let also $Y = 0$ on Θ_f, and $Y = 1$ on Θ_g. Y is not obviously independent of W_N. But, as seen, one has that $\sigma_\infty(\langle M\rangle) = \sigma_0(S_M)$, so that [(Proposition) 16.8.9] $\mathcal{L}(M) \in \mathcal{P}$, and thus $\mathcal{L}(N) = \mathcal{L}(M)$. Consequently, W_N and Y are indeed independent.

Let now ϵ be strictly positive, and N be an Ocone martingale for which $\mathcal{L}(\langle N\rangle)$ is $\mathcal{L}(\langle M\rangle_\epsilon)$. Then [(Proposition) 16.8.7] $\mathcal{L}(N) \in \mathcal{P}$. Thus, since $\mathcal{L}(\langle N\rangle) = \mathcal{L}(\langle M\rangle_\epsilon)$, $S_N(\cdot,0)$ equals either $\epsilon + n_f$ or $\epsilon + n_g$. Now $\sigma_0(S_N) = \sigma(X)$. And $\sigma_0(\langle N\rangle)$ is the trivial σ-algebra. Thus $\sigma_0(\langle N\rangle) \neq \sigma_0(S_N)$, but $\sigma_\infty(\langle N\rangle) = \sigma_0(S_N)$, and thus $\sigma_\infty(\langle N\rangle) \neq \sigma_0(\langle N\rangle)$.

Proposition 16.8.11 *Let $M \in \mathcal{M}$, for the filtration $\underline{\mathcal{B}}$, be such that*

$$\langle M\rangle(\cdot,t) = \int_0^t f^2(\cdot,\theta)\,d\theta,$$

where f is progressively measurable for $\underline{\mathcal{B}}$, and, almost surely, with respect to the product of P with Lebesgue measure, $f(\omega,t) > 0$. Then, when $\mathcal{L}(M) \in \mathcal{P}$, and B is any Brownian motion for $\underline{\mathcal{B}}$, $\langle M\rangle$ and B are independent.

Proof Let W be a Brownian motion independent of $\mathcal{B}_\infty$. One may make such an assumption as, when necessary, one can embed the basic space Ω into $\Omega \times C(\mathbb{R}_+)$, and use, on the latter, the process of evaluation maps with respect to P_W, the Wiener measure. Let

$$\mathcal{F}_t = \mathcal{B}_t \vee \sigma_t(W),$$

and

$$N_B(\cdot,t) = \int_0^t f(\cdot,\theta)B(\cdot,d\theta),\ \ N_W(\cdot,t) = \int_0^t f(\cdot,\theta)W(\cdot,d\theta),$$

defined as integrals with respect to $\underline{\mathcal{F}}$, which makes sense, as $\mathcal{B}_\infty$ and W are independent. One has, [237, p. 421], for example, that, for $t_1 < t_2$, fixed, but arbitrary,

$$E_P\left[N_B(\cdot,t_2) \mid \mathcal{F}_{t_1}\right] = E_P\left[N_B(\cdot,t_2) \mid \mathcal{B}_{t_1}\right] = N_B(\cdot,t_1).$$

Thus N_B and N_W are martingales with respect to $\underline{\mathcal{F}}$. Furthermore

$$\langle N_B\rangle(\cdot,t) = \int_0^t f^2(\cdot,\theta)\,d\theta,\ \ \langle N_W\rangle(\cdot,t) = \int_0^t f^2(\cdot,\theta)\,d\theta, \qquad (\star)$$

and

$$B(\cdot,t) = \int_0^t \frac{N_B(\cdot,d\theta)}{f(\cdot,\theta)},\ \ W(\cdot,t) = \int_0^t \frac{N_W(\cdot,d\theta)}{f(\cdot,\theta)}\,.$$

Because of $(\star)$, f is adapted to $\underline{\sigma}(N_B)$ and to $\underline{\sigma}(N_W)$. Since $f > 0$, $1/f$ is locally bounded, and the integrals yielding B and W may be taken as integrals with respect to the σ-fields generated by N_B and N_W [216, p. 175], so that B is adapted to $\underline{\sigma}(N_B)$, and W, to $\underline{\sigma}(N_W)$. Given the assumption $\mathcal{L}(M) \in \mathcal{P}$, and that $(\star)$ $\langle N_B\rangle = \langle N_W\rangle = \langle M\rangle$,

$$\mathcal{L}(N_B) = \mathcal{L}(M) = \mathcal{L}(N_W).$$

Consequently [(Lemma) 16.7.11]

$$\mathcal{L}(N_B, \langle N_B\rangle) = \mathcal{L}(M, \langle M\rangle) = \mathcal{L}(N_W, \langle N_W\rangle).$$

Since B is adapted to $\underline{\sigma}(N_B)$, and W, to $\underline{\sigma}(N_W)$, B is an "explicit" functional of N_B, and W, one of N_W, so that

$$\mathcal{L}(B, \langle M\rangle) = \mathcal{L}(W, \langle M\rangle).$$

Indeed [141, p. 345], f^2 is the limit of expressions of the following type:

$$n\left\{\langle M\rangle(\omega, t + n^{-1}) - \langle M\rangle(\omega, t)\right\}.$$

But then B and $\langle M\rangle$ are independent as W and $\langle M\rangle$ are. □

Lemma 16.8.12 *Let X and Y be random elements, and suppose that $E[X]$ makes sense. Then, both*

(a) *X and Y independent, and*
(b) *X adapted to $\sigma(Y)$,*

mean X constant.

Proof Independence means that $E[X \mid \sigma(Y)] = E[X]$. Adaptation means that $E[X \mid \sigma(Y)] = X$. Thus $X = E[X]$. □

Proposition 16.8.13 *Let $M \in \mathcal{M}$ have the following properties:*

1. $\mathcal{L}(M) \in \mathcal{P}$;
2. $\langle M\rangle(\cdot, t) = \int_0^t f^2(\cdot, \theta)\, d\theta$, *with f progressively measurable, and, with respect to the product of P and Lebesgue measure, almost surely strictly positive;*
3. $\langle M\rangle$ *is adapted to the filtration generated by a finite, or countable, number of independent standard Brownian motions (denoted B_i).*

Then $\langle M\rangle$ is a function of $t \in \mathbb{R}_+$ only.

Proof Let n be fixed, finite, and smaller than the number of Brownian motions of the assumption, and $\underline{f} : \mathbb{R}_+ \longrightarrow \mathbb{R}^n$ be a vector of bounded measurable functions f_i, $i \in [1 : n]$, with the property that, for $t \in \mathbb{R}_+$, fixed, but arbitrary,

$$\left\|\underline{f}(t)\right\|_{\mathbb{R}^n} > 0\,.$$

Let $\phi_i(t) = f_i(t) / \left\|\underline{f}(t)\right\|_{\mathbb{R}^n}$, and $W(\cdot, t) = \sum_{i=1}^n \int_0^t \phi_i(\theta)\, B_i(\cdot, \theta)$. W is a standard Brownian motion. Because of (Proposition) 16.8.11, W is independent of $\langle M\rangle$, and so then is

$$\left\|\underline{f}(t)\right\|_{\mathbb{R}^n} W(\cdot, t) = \sum_{i=1}^n \int_0^t f_i(\theta) B_i(\cdot, \theta).$$

For $t \in \mathbb{R}_+$, fixed, but arbitrary, let $F_t : C(\mathbb{R}_+) \longrightarrow \mathbb{R}$ be a positive functional adapted to $\mathcal{C}_t$. Then, because of independence,

$$E_P\left[F_t(\langle M\rangle)\, e^{\sum_{i=1}^n \int_0^t f_i(\theta)\, B_i(d\theta)}\right] = E_P\left[F_t(\langle M\rangle)\right] E_P\left[e^{\sum_{i=1}^n \int_0^t f_i(\theta)\, B_i(d\theta)}\right].$$

$\langle M\rangle$ and $\underline{B}$ are thus independent. Since $\langle M\rangle$ has been assumed to be adapted to the σ-algebras generated by the B_i's, $\langle M\rangle(\cdot, t) = \langle M\rangle(t)$ [(Lemma) 16.8.12]. □

Proposition 16.8.14 *Let M be a continuous local martingale such that*

(a) $\sigma_0(S_M)$ *is essentially degenerate;*
(b) $\mathcal{L}(M) \in \mathcal{P}$.

Then $\langle M\rangle$ is deterministic.

Proof Since M is continuous, $\sigma_\infty(S_M)$ is essentially separable. Then, because of (Proposition) 16.5.33, there is W, a Brownian motion such that $\sigma_t(S_M) \subseteq \sigma_t(W)$. Since $\langle M\rangle$ is a change of time for $\underline{\sigma}(S_M)$ [128, p. 103], it is one for $\underline{\sigma}(W)$. The local martingale $\tilde{M} = W \diamond \langle M\rangle$ has $\langle M\rangle$ as increasing process. But then (Proposition 16.8.4 and Assumption (b)), $\tilde{M}$ is an Ocone martingale, that is, $\langle M\rangle$ is independent of W. Since $\langle M\rangle$ is adapted to $\sigma_\infty(W)$, it must be deterministic [(Lemma) 16.8.12]. □

Proposition 16.8.15 *Let M be a continuous, local martingale, for the space $(\Omega, \underline{\sigma}(M), P)$, and $Q \ll P$ be such that $\Delta = \frac{dQ}{dP}$ is adapted to $\sigma_\infty(\langle M \rangle)$. When M is an Ocone martingale for P, it is also one for Q.*

Proof Let $D_M(\cdot, t) = E_P[\Delta \mid \sigma_t(M)]$, and $D_{\langle M \rangle}(\cdot, t) = E_P[\Delta \mid \sigma_t(\langle M \rangle)]$. Since M is an Ocone local martingale, and $D_{\langle M \rangle}$ is adapted to $\underline{\sigma}(\langle M \rangle)$, then because of (Proposition) 16.7.23, $D_{\langle M \rangle}$ is a martingale for $\underline{\sigma}(M)$, and $\langle D_{\langle M \rangle}, M \rangle = 0$. In particular, for $t_1 < t_2$, fixed, but arbitrary,

$$E_P\left[D_{\langle M \rangle}(\cdot, t_2) \mid \sigma_{t_1}(M)\right] = D_{\langle M \rangle}(\cdot, t_1).$$

Letting t_2 increase indefinitely, since Δ is adapted to $\sigma_\infty(\langle M \rangle)$, $D_{\langle M \rangle}(\cdot, t_2)$ converges to Δ in L_1, and thus $D_M(\cdot, t) = D_{\langle M \rangle}(\cdot, t)$. Then, as a consequence, as, as seen above, $\langle D_{\langle M \rangle}, M \rangle = 0$, $\langle D_M, M \rangle = 0$, and $D_M M$ is a local martingale. But then M is a local martingale for Q [128, p. 339].

Let $U \in \sigma_\infty(W_M)$, and $V \in \sigma_\infty(\langle M \rangle)$, be bounded, fixed, but arbitrary. Then

$$\begin{aligned} E_Q[UV] &\overset{(1)}{=} E_P[UV\Delta] \\ &\overset{(2)}{=} E_P[U]\,E_P[V\Delta] \\ &\overset{(3)}{=} E_P[U]\,E_P[V\Delta]\,E_P[\Delta] \\ &\overset{(4)}{=} E_P[U\Delta]\,E_P[V\Delta] \\ &\overset{(5)}{=} E_Q[U]\,E_Q[V], \end{aligned}$$

where equalities (1) and (5) follow from the definition of Q, (2) and (4), from independence (of U from $V\Delta$, as Δ is adapted to $\sigma_\infty(\langle M \rangle)$), and (3), from the fact that the expectation of Δ is one. Thus W_M and $\langle M \rangle$ are independent with respect to Q, and M is an Ocone martingale for Q. □

Proposition 16.8.16 *Let M be a continuous, local martingale for the space $(\Omega, \underline{\sigma}(M), P)$, and Q, a probability measure equivalent to P (mutually absolutely continuous), with Radon-Nikodým derivative $\Delta = \frac{dQ}{dP}$, adapted to $\sigma_\infty(\langle M \rangle)$. Then $\mathcal{L}_P(M) \in \mathcal{P}$ if, and only if, $\mathcal{L}_Q(M) \in \mathcal{P}$.*

Proof By assumption, for some adapted Φ,

$$\Delta = \Phi(\langle M \rangle(\cdot, t_i), i \in I \subseteq \mathbb{N}).$$

Let $\tilde{M}$ be a continuous, local martingale for the space $(\Theta, \underline{\sigma}(\tilde{M}), \tilde{Q})$, such that $\mathcal{L}_{\tilde{Q}}(\langle \tilde{M} \rangle) = \mathcal{L}_Q(\langle M \rangle)$. One must prove that $\mathcal{L}_{\tilde{Q}}(\tilde{M}) \in \mathcal{P}$.

To that end, let

$$\begin{aligned}\tilde{\Delta} &= \Phi(\langle\tilde{M}\rangle(\cdot,t_i), i \in I \subseteq \mathbb{N}),\\ d\tilde{P} &= \tilde{\Delta}^{-1} d\tilde{Q},\\ \tilde{D}(\cdot,t) &= E_{\tilde{Q}}\left[\tilde{\Delta}^{-1} \mid \sigma_t(\tilde{M})\right].\end{aligned}$$

The definition of $\tilde{P}$ makes sense as

$$\begin{aligned}\tilde{Q}\left(\{\Phi(\langle\tilde{M}\rangle(\cdot,t_i), i \in I \subseteq \mathbb{N}) > 0\}\right) &= Q\left(\{\Phi(\langle M\rangle(\cdot,t_i), i \in I \subseteq \mathbb{N}) > 0\}\right)\\ &= P\left(\{\Phi(\langle M\rangle(\cdot,t_i), i \in I \subseteq \mathbb{N}) > 0\}\right)\\ &= 1.\end{aligned}$$

For bounded, fixed, but arbitrary $\Psi(\langle\tilde{M}\rangle(\cdot,\theta_j), j \in J \subseteq \mathbb{N})$, one has that

$$\begin{aligned}E_{\tilde{P}}\left[\Psi(\langle\tilde{M}\rangle(\cdot,\theta_j), j \in J \subseteq \mathbb{N})\right] &= E_{\tilde{Q}}\left[\Psi(\langle\tilde{M}\rangle(\cdot,\theta_j), j \in J \subseteq \mathbb{N})\tilde{\Delta}^{-1}\right]\\ &= E_Q\left[\Psi(\langle M\rangle(\cdot,\theta_j), j \in J \subseteq \mathbb{N})\Delta^{-1}\right]\\ &= E_P\left[\Psi(\langle M\rangle(\cdot,\theta_j), j \in J \subseteq \mathbb{N})\right],\end{aligned}$$

that is,

$$\mathcal{L}_{\tilde{P}}(\langle\tilde{M}\rangle) = \mathcal{L}_P(\langle M\rangle). \qquad (\star)$$

Since $\tilde{P}$ is equivalent to $\tilde{Q}$, with derivative $\tilde{\Delta}^{-1}$, the process

$$\tilde{N}(\cdot,t) = \tilde{M}(\cdot,t) - \int_0^t \frac{1}{\tilde{D}(\cdot,\theta)} d\langle\tilde{M},\tilde{D}\rangle$$

is, with respect to $\tilde{P}$, by Girsanov's theorem, a local martingale such that $\langle\tilde{N}\rangle = \langle\tilde{M}\rangle$. But then, from ($\star$), $\mathcal{L}_{\tilde{P}}(\langle\tilde{N}\rangle) = \mathcal{L}_P(\langle M\rangle)$. Since $\mathcal{L}_P(M) \in \mathcal{P}$, $\mathcal{L}_{\tilde{P}}(\tilde{N}) = \mathcal{L}_P(M)$, and, because of (Proposition) 16.8.4, $\tilde{N}$ is an Ocone martingale for $\tilde{P}$. Consequently [(Proposition) 16.8.15], $\tilde{N}$ is an Ocone martingale for $\tilde{Q}$. Furthermore, since, with respect to $\tilde{P}$, $\langle\tilde{N}\rangle = \langle\tilde{M}\rangle$, and that $\tilde{P}$ and $\tilde{Q}$ are equivalent, equality is also valid with respect to $\tilde{Q}$, and thus, from (Proposition) 16.8.15 again, $\mathcal{L}_{\tilde{Q}}(\tilde{M}) \in \mathcal{P}$. □

Remark 16.8.17 Let $\mathcal{C}_t$ denote the σ-algebra of $C(\mathbb{R}_+)$ generated by the evaluation maps up to time t. $\mathcal{C}_\infty$ denotes $\sigma(\cup_t \mathcal{C}_t)$. Given a probability P, $\mathcal{N}_s(\mathcal{C}_\infty, P)$ is the family of subsets of sets in $\mathcal{C}_\infty$ which have P-measure zero. Given a filtration of elements $\mathcal{F}_t$, in $\mathcal{C}_\infty$, the resulting "usual conditions" filtration has elements [70, p. 183] $\mathcal{F}_t^P = \mathcal{F}_t^+ \vee \mathcal{N}_s(\mathcal{C}_\infty, P)$.

Let

$$q(c,t) = \limsup_{n} \sum_{i} \left\{c\left(t_{i+1}^{(n)} \wedge t\right) - c\left(t_i^{(n)} \wedge t\right)\right\}^2$$
$$= \limsup_{n} \sum_{i} \left\{\mathcal{E}_{t_{i+1}^{(n)} \wedge t}(c) - \mathcal{E}_{t_i^{(n)} \wedge t}(c)\right\}^2,$$

where the $t_i^{(n)}$'s form a partition of $\mathbb{R}_+$ whose largest step goes to zero with n. With respect to any probability Π for which the evaluation maps are a local martingale [264, p. 101],

$$q(\mathcal{E}_\Pi(c,\cdot),t) = \langle \mathcal{E}_\Pi \rangle (c,t).$$

Set

$$S_{\mathcal{E}_\Pi}(c,t) = \inf\{\theta \geq 0 : \langle \mathcal{E}_\Pi \rangle (c,\theta) > t\},\ \mathcal{S}_t = \sigma_t(S_{\mathcal{E}_\Pi}).$$

$\mathcal{S}_t^\Pi$ has a part, $\mathcal{S}_t^+$, which is independent of Π (but it is the presence of Π which gives q its meaning), and a part, $\mathcal{N}_s(\mathcal{C}_\infty, \Pi)$ which is strictly dependent on Π. Thus, when given two probabilities, P and Q, on $\mathcal{C}_\infty$, with $Q \ll P$, one shall have $\mathcal{S}_t^P \subseteq \mathcal{S}_t^Q$. The σ-algebra $\sigma_0(S_{\mathcal{E}_P})$ is, in fact, $\mathcal{S}_0^P$. The latter notation is hopefully useful for the proof which follows.

Proposition 16.8.18 *Let $M[C(\mathbb{R}_+)]$ be the family (convex set) of probability measures for which the evaluation map process $\mathcal{E}$ is a local martingale. When the measure is P, $\mathcal{E}$ is then denoted $\mathcal{E}_P$, so that $\mathcal{L}(\mathcal{E}_P) = P$.*

Let $P \in M[C(\mathbb{R}_+)]$ be fixed, but arbitrary, and Q be a probability on $\sigma_\infty(\mathcal{E}_P)$ that is absolutely continuous with respect to P. Suppose that $\Delta = \frac{dQ}{dP}$ is adapted to $\sigma_0(S_{\mathcal{E}_P})$. Then:

1. *$Q \in M[C(\mathbb{R}_+)]$;*
2. *$P \in \mathcal{P}$ implies $Q \in \mathcal{P}$.*

Proof Let [(Fact) 10.3.45] $W_{\mathcal{E}_P} = \mathcal{E}_P \diamond S_{\mathcal{E}_P}$ be the DDS Brownian motion for $\mathcal{E}_P$; $t_1 < t_2$ be fixed, but arbitrary in $\mathbb{R}_+$; and Φ be a bounded functional adapted to $\sigma_{S_{\mathcal{E}_P}(\cdot,t_1)}(\mathcal{E}_P)$. Then, since Δ is adapted to $\sigma_0(S_{\mathcal{E}_P})$,

$$E_Q\left[\{W_{\mathcal{E}_P}(\cdot,t_2) - W_{\mathcal{E}_P}(\cdot,t_1)\}\,\Phi\right] = E_P\left[\{W_{\mathcal{E}_P}(\cdot,t_2) - W_{\mathcal{E}_P}(\cdot,t_1)\}\,\Phi\Delta\right] = 0.$$

Thus, with respect to Q, and the filtration with entries $\sigma_t(S_{\mathcal{E}_P})$, $W_{\mathcal{E}_P}$ is a continuous martingale. For the same reason, $W_{\mathcal{E}_P}^2(\cdot,t) - t$ is a martingale for Q. But then the evaluation process is a continuous, local martingale, as a time changed Wiener process.

Item 2 has already been proved [(Proposition) 16.8.16] for the case of mutual absolute continuity. Suppose thus that $P(\Delta = 0) > 0$.

Let $\tilde{Q} \in M[C(\mathbb{R}_+)]$ have the property that

$$\mathcal{L}_{\tilde{Q}}(\langle \mathcal{E}_{\tilde{Q}} \rangle) = \mathcal{L}_Q(\langle \mathcal{E}_Q \rangle). \tag{$\star$}$$

Since Δ is adapted to $\mathcal{S}_0^P$, it is, as seen [(Remark) 16.8.17], adapted to $\mathcal{S}_0^Q$, and there are [70, p. 49] Δ_l and Δ_u, adapted to $\mathcal{S}_0^+$, for which $\Delta_l \leq \Delta \leq \Delta_u$ and

$$Q(\Delta_u - \Delta_l > 0) = 0.$$

But, because of assumption ($\star$), $\tilde{Q}(\Delta_u - \Delta_l > 0) = 0$, so that Δ is adapted to $\mathcal{S}_0^{\tilde{Q}}$, and the following assignments make sense:

$$\begin{aligned} d\hat{Q} &= \tfrac{\chi_{\{\Delta>0\}}}{\Delta} d\tilde{Q} + \chi_{\{\Delta=0\}} dP, \\ d\hat{Q}_1 &= \tfrac{\chi_{\{\Delta>0\}}}{P(\Delta>0)\Delta} d\tilde{Q}, \\ d\hat{Q}_2 &= \tfrac{\chi_{\{\Delta=0\}}}{P(\Delta=0)} dP, \end{aligned}$$

so that

$$d\hat{Q} = P(\Delta > 0) d\hat{Q}_1 + P(\Delta = 0) d\hat{Q}_2.$$

Since, by assumption, $\tilde{Q}$ belongs to $M[C(\mathbb{R}_+)]$, that $\hat{Q}_1$ is absolutely continuous with respect to $\tilde{Q}$, and that, by construction, the corresponding Radon-Nikodým derivative is adapted to

$$\sigma_0(S_{\mathcal{E}_{\tilde{Q}}}),$$

because of item 1, $\hat{Q}_1$ belongs to $M[C(\mathbb{R}_+)]$. For the same reasons, $\hat{Q}_2$ belongs to $M[C(\mathbb{R}_+)]$, so that, since the latter is convex, $\hat{Q}$ belongs to $M[C(\mathbb{R}_+)]$.

Let $\Gamma = \{q(c, \theta_1) \in B_1, \ldots, q(c, \theta_p) \in B_p\}$, the B_i's being Borel. Then, since it is assumed that $\mathcal{L}_{\tilde{Q}}(\langle \mathcal{E} \rangle_{\tilde{Q}}) = \mathcal{L}_Q(\langle \mathcal{E} \rangle_Q)$,

$$\begin{aligned} \hat{Q}(\Gamma) &= \int_{\Gamma \cap \{\Delta>0\}} \frac{d\tilde{Q}}{\Delta} + P(\Gamma \cap \{\Delta = 0\}) \\ &= \int_{\Gamma \cap \{\Delta>0\}} \frac{dQ}{\Delta} + P(\Gamma \cap \{\Delta = 0\}) \\ &= \int_{\Gamma \cap \{\Delta>0\}} \frac{\Delta dP}{\Delta} + P(\Gamma \cap \{\Delta = 0\}) \\ &= P(\Gamma). \end{aligned}$$

Consequently, $\mathcal{L}_{\hat{Q}}(\langle\mathcal{E}_{\hat{Q}}\rangle) = \mathcal{L}_P(\langle\mathcal{E}_P\rangle)$. But then, when $P \in \mathcal{P}$, $\hat{Q} \in \mathcal{P}$, and, since there is only one probability on $C(\mathbb{R}_+)$ which makes the evaluation process a Wiener process, $\hat{Q} = P$. Thus

$$dP = \frac{\chi_{\{\Delta>0\}}}{\Delta} d\tilde{Q} + \chi_{\{\Delta=0\}} dP,$$

so that

$$\chi_{\{\Delta>0\}} \Delta dP = \chi_{\{\Delta>0\}} d\tilde{Q},$$

or

$$\chi_{\{\Delta>0\}} dQ = \chi_{\{\Delta>0\}} d\tilde{Q}.$$

But $\{\Delta > 0\}$ is the support of Q, so that $1 = Q(\{\Delta > 0\}) = \tilde{Q}(\{\Delta > 0\})$, and $\{\Delta > 0\}$ is also the support of $\tilde{Q}$, hence $\tilde{Q} = Q$, or $Q \in \mathcal{P}$. □

Definition 16.8.19 The sign function $s[\cdot]$ is the function

$$s[x] = \chi_{]0,\infty[}(x) - \chi_{]-\infty,0]}(x).$$

Given a Brownian motion W, and a decreasing sequence of indices, say $\{t_n, n \in \mathbb{Z}_0^-\} \subseteq]0,\infty[$, one shall write $V_0 = 1$ and

$$V_n = s[W(\cdot, t_{n+1}) - W(\cdot, t_n)]. \qquad (\star)$$

Lemma 16.8.20 *Let $t_1 < t_2$ in $]0,\infty[$, be fixed, but arbitrary; U be a random variable for which $P(U = 1) = P(U = -1) = \frac{1}{2}$; and W be a Brownian motion independent of U. There exists then a Brownian motion B such that*

1. $s[B(\cdot, t_2) - B(\cdot, t_1)] = U$,
2. $\sigma_\infty(B) = \sigma_\infty(W) \vee \sigma(U)$,
3. *W is a Brownian motion for $\underline{\sigma}(B)$.*

Proof Let $\{t_n, n \in \mathbb{Z}_0^-\} \subseteq]0,\infty[$ be a decreasing sequence with $t_0 = t_2$ and $t_{-1} = t_1$. Let also, V_n being defined at $(\star)$ of (Definition) 16.8.19, for $n \in \mathbb{Z}_0^-$,

$$\tilde{V}_0 = 1,$$
$$\tilde{V}_n = V_{-1} \cdots V_n,$$
$$U_0 = U,$$
$$U_n = U\tilde{V}_n.$$

U_n is thus a function of U and

$$W(\cdot, t_0) - W(\cdot, t_{-1}), W(\cdot, t_{-1}) - W(\cdot, t_{-2}), \dots, W(\cdot, t_{n+1}) - W(\cdot, t_n).$$

As a first step, one proves that U_n and W are independent. For bounded, adapted functionals, denoted Φ and Ψ, one has that

$$\begin{aligned} E_P[\Phi(W)\Psi(U_n)] &= \\ &= E_P\left[\Phi(W)\Psi(1)\left\{\chi_{\{1\}}(U)\chi_{\{1\}}(\tilde{V}_n) + \chi_{\{-1\}}(U)\chi_{\{-1\}}(\tilde{V}_n)\right\}\right] \\ &+ E_P\left[\Phi(W)\Psi(-1)\left\{\chi_{\{1\}}(U)\chi_{\{-1\}}(\tilde{V}_n) + \chi_{\{-1\}}(U)\chi_{\{1\}}(\tilde{V}_n)\right\}\right] \\ &= \Psi(1)E_P\left[\Phi(W)\left\{\chi_{\{1\}}(U)\chi_{\{1\}}(\tilde{V}_n) + \chi_{\{-1\}}(U)\chi_{\{-1\}}(\tilde{V}_n)\right\}\right] \\ &+ \Psi(-1)E_P\left[\Phi(W)\left\{\chi_{\{1\}}(U)\chi_{\{-1\}}(\tilde{V}_n) + \chi_{\{-1\}}(U)\chi_{\{1\}}(\tilde{V}_n)\right\}\right]. \end{aligned}$$

Now, for example,

$$\begin{aligned} &E_P\left[\Phi(W)\left\{\chi_{\{1\}}(U)\chi_{\{1\}}(\tilde{V}_n) + \chi_{\{-1\}}(U)\chi_{\{-1\}}(\tilde{V}_n)\right\}\right] \\ &\quad = E_P\left[\chi_{\{1\}}(U)\right]E_P\left[\Phi(W)\chi_{\{1\}}(\tilde{V}_n)\right] \\ &\quad + E_P\left[\chi_{\{-1\}}(U)\right]E_P\left[\Phi(W)\chi_{\{-1\}}(\tilde{V}_n)\right] \\ &\quad = \frac{1}{2}E_P\left[\Phi(W)\left\{\chi_{\{1\}}(\tilde{V}_n) + \chi_{\{-1\}}(\tilde{V}_n)\right\}\right] \\ &\quad = \frac{1}{2}E_P[\Phi(W)]. \end{aligned}$$

The same calculation yields that $E_P[\Psi(U_n)] = E_P[\Psi(U)]$, so that

$$E_P[\Phi(W)\Psi(U_n)] = E_P[\Phi(W)]\,E_P[\Psi(U_n)].$$

Let

$$\Upsilon(\omega, t) = \sum_{n\in\mathbb{Z}_0^-} \chi_{]t_n, t_{n+1}]}(t)\,U_n + \chi_{]t_0,\infty[}(t)\,U_0,$$

$$B(\cdot, t) = \int_0^t \Upsilon(\cdot, \theta)\,W(\cdot, d\theta).$$

Let $p(t)$ be the integer for which $t \in]t_{p(t)}, t_{p(t)+1}]$ (when $t \in]t_0, \infty[$, one sets $t_{p(t)} = t_0$, and $t_{p(t)+1} = \infty$). By definition of the integral,

$$B(\cdot, t) = U_{p(t)}\left\{W(\cdot, t) - W(\cdot, t_{p(t)})\right\} + \sum_{i<p(t)} U_i\left\{W(\cdot, t_{i+1}) - W(\cdot, t_i)\right\}.$$

Thus, when $\theta_1 < \theta_2$,

$$\begin{aligned}B(\cdot,\theta_2) - B(\cdot,\theta_1) &= U_{p(\theta_2)}\left\{W(\cdot,\theta_2) - W(\cdot,t_{p(\theta_2)}\right\} \\ &\quad + \sum_{i=p(\theta_1)+1}^{p(\theta_2)-1} U_i\left\{W(\cdot,t_{i+1} - W(\cdot,t_i)\right\} \\ &\quad + U_{p(\theta_1)}\left\{W(\cdot,t_{p(\theta_1)+1} - W(\cdot,\theta_1)\right\}.\end{aligned}$$

In particular,

$$\begin{aligned}&B(\cdot,t_0) - B(\cdot,t_{-1}) = \\ &\quad = U_{-1}\left(W(\cdot,t_0) - W(\cdot,t_{-1})\right) \\ &\quad = U\left\{s\left[W(\cdot,t_0) - W(\cdot,t_{-1})\right]\right\}\left(W(\cdot,t_0) - W(\cdot,t_{-1})\right),\end{aligned}$$

so that $s[B(\cdot,t_0) - B(\cdot,t_{-1}] = s(U) = U_0$. Also

$$\begin{aligned}&B(\cdot,t_{-1}) - B(\cdot,t_{-2}) = \\ &\quad = U_{-2}\left(W(\cdot,t_{-1}) - W(\cdot,t_{-2})\right) \\ &\quad = U_{-1}V_{-2}\left(W(\cdot,t_{-1}) - W(\cdot,t_{-2})\right) \\ &\quad = U_{-1}\left\{s\left[W(\cdot,t_{-1}) - W(\cdot,t_{-2})\right]\right\}\left(W(\cdot,t_{-1}) - W(\cdot,t_{-2})\right),\end{aligned}$$

so that $s[B(\cdot,t_{-1}) - B(\cdot,t_{-2})] = s(U_{-1}) = U_{-1}$. And so forth ... Consequently

$$s\left[B(\cdot,t_n) - B(\cdot,t_{n-1})\right] = U_n,$$

and thus U_n is adapted to $\sigma_{t_n}(B)$.

Furthermore, when t belongs to $]t_n, t_{n+1}]$,

$$B(\cdot,t) - B(\cdot,t_n) = U_n\left(W(\cdot,t) - W(\cdot,t_n)\right).$$

Let, when $t \in [t_n, t_{n+1}[$,

$$\mathcal{B}_t = \sigma_t(W) \vee \sigma(U_n),$$

where $\sigma_t(W)$ and $\sigma(U_n)$ are complete. Then, since W and U_n are independent, one has that [53, p. 29]

$$\cap_{\epsilon>0}\left\{\sigma_{t+\epsilon}(W) \vee \sigma(U_n)\right\} = \left\{\cap_{\epsilon>0}\sigma_{t+\epsilon}(W)\right\} \vee \sigma(U_n) = \sigma_t(W) \vee \sigma(U_n),$$

and $\underline{\mathcal{B}}$ satisfies the usual conditions.

One must now check that B is a Brownian motion for $\underline{\mathcal{B}}$. To that end one applies the procedure used in the interval by interval construction of Brownian motion, as described in (Remark) 16.5.32. Let $t \in]t_{p(t)}, t_{p(t)+1}]$, and $u \in \{-1, 1\}$, be fixed, but arbitrary. Then, given a functional $\Phi_{t_{p(t)}}(W)$, adapted to $\sigma_{t_{p(t)}}(W)$, using the fact that U_p and W are independent,

$$
\begin{aligned}
E_P\left[\{B(\cdot,t) - B(\cdot,t_{p(t)})\}\, \chi_{\{U_{p(t)}=u\}}\, \Phi_{t_{p(t)}}(W)\right] &= \\
&= E_P\left[U_{p(t)}\, \chi_{\{U_{p(t)}=u\}}\, \{W(\cdot,t) - W(\cdot,t_{p(t)})\}\, \Phi_{t_{p(t)}}(W)\right] \\
&= E_P\left[U_{p(t)}\, \chi_{\{U_{p(t)}=u\}}\right] E_P\left[\{W(\cdot,t) - W(\cdot,t_{p(t)})\}\, \Phi_{t_{p(t)}}(W)\right] \\
&= 0.
\end{aligned}
$$

B is, by construction, continuous, and its quadratic variation is t. It is thus a Brownian motion. One consequence is that $\sigma_t(B) \subseteq \mathcal{B}_t$.

Since, when $t \in]t_{p(t)}, t_{p(t)+1}]$, $\Upsilon(\cdot,t) = U_{p(t)}$, and that $\mathcal{B}_t = \sigma_t(W) \vee \sigma(U_{p(t)})$, the process Υ is adapted to $\underline{\mathcal{B}}$. Then the following integral is well defined, and its value is, as $\Upsilon^2 = 1$,

$$\int \Upsilon\, dB = \int \Upsilon^2 dW = W.$$

But, when $t \in]t_{p(t)}, t_{p(t)+1}]$,

$$
\begin{aligned}
W(\cdot,t) &= \int_0^t \Upsilon(\cdot,\theta) B(\cdot,d\theta) \\
&= U_{p(t)}(\cdot)\{B(\cdot,t) - B(\cdot,t_n)\} + \sum_{i<p(t)} U_i(\cdot)\{B(\cdot,t_{i+1}) - B(\cdot,t_i)\},
\end{aligned}
$$

so that, given that $U_{p(t)}$ is adapted to $\sigma_{p(t)}(B)$, as seen above, $W(\cdot,t)$ is adapted to $\sigma_t(B)$. Consequently,

$$\sigma_t(B) \subseteq \mathcal{B}_t = \sigma_t(W) \vee \sigma(U_{p(t)}) \subseteq \sigma_t(B).$$

W is thus a Brownian motion for $\underline{\sigma}(B)$. Finally, since, for $t > t_0$,

$$\mathcal{B}_t = \sigma_t(W) \vee \sigma(U_{p(t)}) = \sigma_t(W) \vee \sigma(U) \subseteq \sigma_\infty(W) \vee \sigma(U),$$

$\mathcal{B}_\infty \subseteq \sigma_\infty(W) \vee \sigma(U)$. Furthermore $\sigma(U) \subseteq \mathcal{B}_t \subseteq \mathcal{B}_\infty$, and, since $\sigma_t(W) \subseteq \mathcal{B}_t$, $\sigma_\infty(W) \subseteq \mathcal{B}_\infty$, so that

$$\sigma_\infty(W) \vee \sigma(U) \subseteq \mathcal{B}_\infty.$$

□

Lemma 16.8.21 *Let $\epsilon > 0$ be fixed, but arbitrary, and M be a continuous, local martingale, with $\mathcal{L}(M) \in \mathcal{P}$. Let $M_\epsilon = W_{M_\epsilon} \diamond \langle M_\epsilon \rangle$ be a continuous, local martingale (which may be based on a distinct probability space) such that $\langle M_\epsilon \rangle$ has the same law as*

$$V(\cdot, t) = \chi_{[0,\epsilon]}(t)t + \chi_{]\epsilon,\infty]}(t)\{\epsilon + \langle M \rangle(\cdot, t - \epsilon)\}.$$

Then $\langle M_\epsilon \rangle$ is independent of $\sigma(W_{M_\epsilon}(\cdot, t + \epsilon) - W_{M_\epsilon}(\cdot, \epsilon), t \in \mathbb{R}_+)$.

Proof Result (Fact) 10.3.45, given the basic prevailing assumptions for the martingales one works with, and, in particular, the fact that filtrations obey the usual conditions, says that W_{M_ϵ} is a Brownian motion for the filtration with entries $\sigma_{S_{M_\epsilon}(\cdot,t)}(M_\epsilon)$ [221, p. 181], and, for the latter, $\langle M_\epsilon \rangle(\cdot, t)$ is a stopping time [216, p. 99]. Then [strong Markov property of Brownian motion] [216, p. 23]:

$$X(\cdot, t) = W_{M_\epsilon}(\cdot, t + \epsilon) - W_{M_\epsilon}(\cdot, \epsilon)$$

is a Brownian motion for the filtration with entries $\sigma_{S_{M_\epsilon}(\cdot,t+\epsilon)}(M_\epsilon)$, and, for the latter, $\langle M_\epsilon \rangle(\cdot, t + \epsilon)$ is a stopping time. Thus

$$\{\langle M_\epsilon \rangle(\cdot, t + \epsilon) - \epsilon \le \alpha\} = \{\langle M_\epsilon \rangle(\cdot, t + \epsilon) \le \alpha + \epsilon\} \in \sigma_{S_{M_\epsilon}(\cdot,\alpha+\epsilon)}(M_\epsilon),$$

and

$$Y(\cdot, t) = X(\cdot, \langle M_\epsilon \rangle(\cdot, t + \epsilon) - \epsilon)$$

is a change of time on a Brownian motion, and thus a continuous, local martingale, so that $\langle Y \rangle$ is well defined, with value $\langle M_\epsilon \rangle(\cdot, \cdot + \epsilon) - \epsilon$. Now, by assumption,

$$\langle M_\epsilon \rangle(\cdot, t + \epsilon) - \epsilon \text{ and } V(\cdot, t + \epsilon) - \epsilon = \langle M \rangle(\cdot, t)$$

have the same law. Thus $\mathcal{L}(\langle Y \rangle) = \mathcal{L}(\langle M \rangle)$, and, since, by assumption, the martingale M belongs to $\mathcal{P}$, Y is an Ocone martingale [(Proposition) 16.8.4]. Thus $\langle M_\epsilon \rangle(\cdot, \cdot + \epsilon) - \epsilon$ is independent of X. □

What follows is one characterization property of continuous, divergent, local martingales whose law is determined by their associated increasing process: they are Ocone martingales with a restriction on the related increasing process. A second one follows which involves the predictable representation property.

Proposition 16.8.22 *Let M be a continuous, divergent, local martingale. Then $\mathcal{L}_P(M) \in \mathcal{P}$ if, and only if, $\sigma_\infty(\langle M \rangle) = \sigma_0(S_M)$. Furthermore $\mathcal{L}_P(M)$ is Gaussian if, and only if, $\sigma_0(S_M)$ is essentially degenerate.*

Proof The Gaussian part of the statement follows from the first part, and the fact that a Gaussian martingale has a quadratic variation which is deterministic (in the sense of not random). Because of (Proposition) 16.8.9, one need only prove that $\mathcal{L}_P(M) \in$

$\mathcal{P}$ implies the equality $\sigma_\infty(\langle M\rangle) = \sigma_0(S_M)$. Since $\langle M\rangle$ is continuous, $S_M(\cdot, t) < \theta$ if, and only if, $\langle M\rangle(\cdot, \theta) > t$ [(Fact) 10.3.27], and thus $\sigma_\infty(\langle M\rangle) = \sigma_\infty(S_M)$. The conclusion then becomes $\sigma_\infty(S_M) = \sigma_0(S_M)$.

Proof Assume, to begin with, that $\sigma_0(S_M)$ is essentially degenerate.

As just seen, it suffices to prove that, for $t > 0$, fixed, but arbitrary, $\sigma_t(S_M)$ is essentially degenerate.

Suppose that there exists $t_0 > 0$, and $\Omega_{t_0} \in \sigma_{t_0}(S_M)$, such that

$$0 < P(\Omega_{t_0}) < 1.$$

Let

$$\Delta = \frac{1}{2}\left\{\frac{\chi_{\Omega_{t0}}}{P(\Omega_{t_0})} + \frac{\chi_{\Omega^c_{t0}}}{P(\Omega^c_{t_0})}\right\},$$

and

$$dQ = \Delta dP.$$

Since Δ is strictly positive, P and Q are mutually absolutely continuous. Since Δ is adapted to $\sigma_{t_0}(S_M)$, it is to $\sigma_\infty(S_M)$, which is, as seen above $\sigma_\infty(\langle M\rangle)$. But then, since it is assumed that $\mathcal{L}_P(M) \in \mathcal{P}$, because of (Proposition) 16.8.16, $\mathcal{L}_Q(M) \in \mathcal{P}$.

From the definition of Δ one has also that $Q(\Omega_{t_0}) = \frac{1}{2}$.

Let $0 < t_{-1} < t_0$ be fixed, but arbitrary, and

$$L(\cdot, t) = \chi_{[0,t_{-1}]}(t)t + \chi_{]t_{-1},\infty[}(t)\left\{\langle M\rangle(\cdot, t - t_{-1}) + t_{-1}\right\}.$$

The symbol Λ shall denote the inverse of L. By definition, $\Lambda(\cdot, t_{-1}) = S_M(\cdot, 0)$, and, since $\sigma_0(S_M)$ is, by assumption, essentially degenerate, $\Omega_{t_0} \in \sigma_{t_0}(S_M)$ is independent of $\sigma_0(S_M) = \sigma_{t_{-1}}(\Lambda)$.

Let

$$U = \chi_{\Omega_{t0}} - \chi_{\Omega^c_{t0}}.$$

One has that $Q(U = 1) = Q(U = -1) = \frac{1}{2}$. Also, with respect to Q, U and W_M are independent. Indeed, since $\mathcal{L}_Q(M) \in \mathcal{P}$, M is an Ocone martingale for Q [(Proposition) 16.8.4], and thus, for Q, W_M and S_M are independent. But $\Omega_{t_0} \in \sigma_{t_0}(S_M)$.

Let B be the Brownian motion obtained from U and W_M as in (Lemma) 16.8.20 (the notation which follows is as explained there).

Let $C_t = \sigma_t(B) \vee \sigma_t(\Lambda)$ (the latter completed, and made continuous to the right). Since Ω_{t_0} is a set independent of $\sigma_{t_{-1}}(\Lambda)$, one shall presently see that B is a Brownian motion for $\underline{C}$. Let indeed $t \in]t_n, t_{n+1}]$, Φ_{Λ,t_n} be bounded and adapted to $\sigma_{t_n}(\Lambda)$, Ψ_{W_M,t_n} be bounded and adapted to $\sigma_{t_n}(W_M)$, Υ be bounded and adapted to the Borel sets of the reals. For $n \in \mathbb{Z}_0^-$, let $U_n = U\tilde{V}_n$, where

$$\tilde{V}_n = \prod_{n=0}^{n+1} s\left[B(\cdot, t_n) - B(\cdot, t_{n-1})\right].$$

Then, using properties evidenced in the proof of (Lemma) 16.8.20,

$$\begin{aligned}
X_{n,t} &= E_Q\left[(B(\cdot,t) - B(\cdot,t_n))\, \Phi_{\Lambda,t_n}\, \Psi_{W_M,t_n}\, \Upsilon(U_n)\right] \\
&= E_Q\left[U_n\,(W_M(\cdot,t) - W_M(\cdot,t_n))\, \Phi_{\Lambda,t_n}\, \Psi_{W_M,t_n}\, \Upsilon(U_n)\right] \\
&= E_Q\left[\chi_{\Omega_{t_0}}\, \tilde{V}_n\,(W_M(\cdot,t) - W_M(\cdot,t_n))\, \Phi_{\Lambda,t_n}\, \Psi_{W_M,t_n}\, \Upsilon(\tilde{V}_n)\right] \\
&- E_Q\left[\chi_{\Omega_{t_0}^c}\, \tilde{V}_n\,(W_M(\cdot,t) - W_M(\cdot,t_n))\, \Phi_{\Lambda,t_n}\, \Psi_{W_M,t_n}\, \Upsilon(-\tilde{V}_n)\right] \\
&= E_Q\left[\chi_{\Omega_{t_0}}\, \Phi_{\Lambda,t_n}\right] E_Q\left[\tilde{V}_n\,(W_M(\cdot,t) - W_M(\cdot,t_n))\, \Psi_{W_M,t_n}\, \Upsilon(\tilde{V}_n)\right] \\
&- E_Q\left[\chi_{\Omega_{t_0}^c}\, \Phi_{\Lambda,t_n}\right] E_Q\left[\tilde{V}_n\,(W_M(\cdot,t) - W_M(\cdot,t_n))\, \Psi_{W_M,t_n}\, \Upsilon(-\tilde{V}_n)\right].
\end{aligned}$$

As $\tilde{V}_0 = 1, X_{0,t}$ has $E_Q\left[(W_M(\cdot,t) - W_M(\cdot,t_0))\, \Psi_{W_M,t_0}\right]$ as factor, which is zero. When $n \leq -1$, as Ω_{t_0} is independent of $\sigma_{t_{-1}}(\Lambda)$,

$$E_Q\left[\chi_{\Omega_{t_0}}\, \Phi_{\Lambda,t_n}\right] = \frac{1}{2} E_Q\left[\Phi_{\Lambda,t_n}\right],$$

so that

$$\begin{aligned}
X_{n,t} = \frac{1}{2} E_Q\left[\Phi_{\Lambda,t_n}\right] \big\{ & E_Q\left[\tilde{V}_n\,(W_M(\cdot,t) - W_M(\cdot,t_n))\, \Psi_{W_M,t_n}\, \Upsilon(\tilde{V}_n)\right] \\
& - E_Q\left[\tilde{V}_n\,(W_M(\cdot,t) - W_M(\cdot,t_n))\, \Psi_{W_M,t_n}\, \Upsilon(-\tilde{V}_n)\right] \big\}.
\end{aligned}$$

But U is independent of W_M, so that, because of (Lemma) 16.8.20, and as in its proof,

$$\begin{aligned}
X_{n,t} &= E_Q\left[\Phi_{\Lambda,t_n}\right] E_Q\left[U_n\,(W_M(\cdot,t) - W_M(\cdot,t_n))\, \Psi_{W_M,t_n}\, \Upsilon(U_n)\right] \\
&= E_Q\left[\Phi_{\Lambda,t_n}\right] E_Q\left[(B(\cdot,t) - B(\cdot,t_n))\, \Psi_{W_M,t_n}\, \Upsilon(U_n)\right] \\
&= E_Q\left[\Phi_{\Lambda,t_n}\right] \times 0 \\
&= 0.
\end{aligned}$$

Now, since L is a continuous change of time, $\{L(\cdot,\theta) \leq t\} = \{\Lambda(\cdot,t) \geq \theta\}$, so that L is a continuous change of time for $\underline{\mathcal{C}}$. Hence, since [149, p. 344], for time changes τ,

$$\left\{\int V dX\right\} \diamond \tau = \int \{V \diamond \tau\} d\{X \diamond \tau\},$$

and that $B = \int \Upsilon \, dW_M$, $M_{t_{-1}} = B \diamond L$ is a continuous, local martingale, and $\langle M_{t_{-1}} \rangle = L$. Thus [(Fact) 10.3.45], $B \diamond L$ is the Dambis-Dubins-Schwarz representation of $M_{t_{-1}}$, and, because of (Lemma) 16.8.21, $\langle M_{t_{-1}} \rangle$ is independent of the σ-algebra

$$\begin{aligned}\sigma(W_{M_{t_{-1}}}(\cdot,\theta+t_{-1}) - W_{M_{t_{-1}}}(\cdot,t_{-1}), \theta \in \mathbb{R}_+) = \\ = \sigma(B(\cdot,\theta+t_{-1}) - B(\cdot,t_{-1}), \theta \in \mathbb{R}_+),\end{aligned}$$

the latter, as $B = W_{M_{t_{-1}}}$, which was acknowledged above. But, since one has that $L(\cdot,\theta+t_{-1}) = \langle M\rangle(\cdot,\theta)$,

$$\langle M\rangle \text{ and } \sigma(B(\cdot,\theta+t_{-1}) - B(\cdot,t_{-1}), \theta \in \mathbb{R}_+)$$

are independent, and that contradicts the fact that, choosing

$$\Omega_{t_0} = \{B(\cdot,t_0) - B(\cdot,t_{-1}) > 0\},$$

one has that, using a remark made in the first paragraph of the proof,

$$\Omega_{t_0} \in \sigma_{t_0}(S_M) \subseteq \sigma_\infty(S_M) = \sigma_\infty(\langle M\rangle).$$

The claim is thus true when $\sigma_0(S_M)$ is essentially degenerate.

Proof The general case: no a priori assumption on $\sigma_0(S_M)$.

Since, as seen in the first paragraph of the proof, $\sigma_\infty(\langle M\rangle) = \sigma_\infty(S_M)$, it is sufficient [(Proposition) 16.4.13] to prove that, almost surely, with respect to P,

$$\{\omega \in \Omega : mult\,[\sigma_\infty(S_M) \mid \sigma_0(S_M)]\,(\omega) > 1\} = \emptyset.$$

One shall work within the framework of (Proposition) 16.8.18. Let, to that end, $M[\omega]$ be the path of M at ω. Then, for $P_M = P \circ \boldsymbol{M}^{-1}$, and the natural filtration, $\mathcal{E}$, the evaluations process on $C(\mathbb{R}_+)$, is a continuous, local martingale. The notation shall be $\mathcal{E}_{P_M}$. Furthermore $\langle M\rangle = \langle \mathcal{E}_{P_M}\rangle \circ \boldsymbol{M}$, and $S_M(\omega,t) = S_{\mathcal{E}_{P_M}}(M[\omega],t)$. Consequently

$$\sigma_0(S_M) = \sigma(S_{\mathcal{E}_{P_M}}(M[\cdot],0)) = \boldsymbol{M}^{-1}\left(\sigma_0(S_{\mathcal{E}_{P_M}})\right),$$

and

$$\sigma_\infty(S_M) = \sigma_\infty(S_{\mathcal{E}_{P_M}} \circ \boldsymbol{M}) = \boldsymbol{M}^{-1}(\sigma_\infty(S_{\mathcal{E}_{P_M}})).$$

It thus suffices to prove the claim on path space.

Suppose there exists $t_0 > 0$ such that with respect to P_M,

$$C_0 = \left\{c \in C(\mathbb{R}_+) : mult[\sigma_{t_0}(S_{\mathcal{E}_{P_M}}) \mid \sigma_0(S_{\mathcal{E}_{P_M}})](c) > 1\right\} \text{ has positive probability}$$

Let

$$dP_0 = \frac{\chi_{C_0}}{P_M(C_0)} dP_M.$$

By definition, the Radon-Nikodým derivative $\frac{dP_0}{dP_M}$ is adapted to $\sigma_0(S_{\mathcal{E}_{P_M}})$. Thus, because of (Proposition) 16.8.18, and the assumption that $M \in \mathcal{P}$, $P_0 \in \mathcal{P}$.

According to (Corollary) 16.4.22, there exists $\Omega_{t_0} \in \sigma_{t_0}(S_{\mathcal{E}_{P_M}})$ such that

$$C_0 \subseteq \left\{P_M(\Omega_{t_0} \mid \sigma_0(S_{\mathcal{E}_{P_M}}) > 0\right\} \cap \left\{P_M(\Omega_{t_0}^c \mid \sigma_0(S_{\mathcal{E}_{P_M}}) > 0\right\}.$$

Since $P_0 \ll P_M$, $\mathcal{E}_{P_0}$ and $\mathcal{E}_{P_M}$ have the same quadratic variation [(Proposition) 16.6.1], and thus $\sigma_0(S_{\mathcal{E}_{P_M}}) \subseteq \sigma_0(S_{\mathcal{E}_{P_0}})$, so that, given $C \in \sigma_0(S_{\mathcal{E}_{P_M}})$, fixed, but arbitrary,

$$P_0(\Omega_{t_0} \cap C) = \int_C P_0(\Omega_{t_0} \mid \sigma_0(S_{\mathcal{E}_{P_0}}))dP_0,$$

and

$$\begin{aligned} P_0(\Omega_{t_0} \cap C) &= P_M(C_0)^{-1} P_M(\Omega_{t_0} \cap C \cap C_0) \\ &= P_M(C_0)^{-1} \int_{C \cap C_0} P_M(\Omega_{t_0} \mid \sigma_0(S_{\mathcal{E}_{P_M}}))dP_M \\ &= \int_C P_M(\Omega_{t_0} \mid \sigma_0(S_{\mathcal{E}_{P_M}}))dP_0. \end{aligned}$$

But that equality of integrals over sets in $\sigma_0(S_{\mathcal{E}_{P_M}})$ extends to sets in $\sigma_0(S_{\mathcal{E}_{P_0}})$, so that

$$P_M(\Omega_{t_0} \mid \sigma_0(S_{\mathcal{E}_{P_M}})) \text{ is a version of } P_0(\Omega_{t_0} \mid \sigma_0(S_{\mathcal{E}_{P_0}})).$$

Consequently, almost surely with respect to P_0,

$$C_0 \subseteq \left\{P_0(\Omega_{t_0} \mid \sigma_0(S_{\mathcal{E}_{P_0}})) > 0\right\}.$$

But $P_0(C_0) = 1$. The following definitions make thus sense:

$$\Delta = \frac{1}{2}\left\{\frac{\chi_{\Omega_{t0}}}{P_0(\Omega_{t0} \mid \sigma_0(S_{\mathcal{E}_{P_0}}))} + \frac{\chi_{\Omega^c_{t0}}}{P_0(\Omega^c_{t0} \mid \sigma(S_{\mathcal{E}_{P_0}}))}\right\},$$

and

$$dQ = \Delta dP_0.$$

Since Δ is strictly positive, P_0 and Q are mutually absolutely continuous. Thus [(Proposition) 16.6.1] $\sigma_0(S_{\mathcal{E}_Q}) = \sigma_0(S_{\mathcal{E}_{P_0}})$. By construction,

$$\begin{aligned} Q(\Omega_{t0}) &= \frac{1}{2}E_{P_0}\left[\frac{\chi_{\Omega_{t0}}}{P_0(\Omega_{t0} \mid \sigma_0(S_{\mathcal{E}_{P_0}}))}\right] \\ &= \frac{1}{2}E_{P_0}\left[\frac{1}{P_0(\Omega_{t0} \mid \sigma_0(S_{\mathcal{E}_{P_0}}))}E_{P_0}\left[\chi_{\Omega_{t0}} \mid \sigma_0(S_{\mathcal{E}_{P_0}})\right]\right] \\ &= \frac{1}{2}. \end{aligned}$$

Similarly, when $C \in \sigma_0(S_{\mathcal{E}_Q})$ is fixed, but arbitrary, $Q(C) = P_0(C)$, so that

$$Q(\Omega_{t0} \cap C) = \frac{1}{2}E_{P_0}\left[\chi_C \frac{\chi_{\Omega_{t0}}}{P_0(\Omega_{t0} \mid \sigma_0(S_{\mathcal{E}_{P_0}}))}\right] = \frac{1}{2}P_0(C) = Q(\Omega_{t0})Q(C),$$

so that Ω_{t0} and C are independent for Q.

One may then proceed as in the first part of the proof, the probabilities P_0 and Q taking, respectively, the parts taken there by P and Q. □

Proposition 16.8.23 *$\mathcal{L}(M) \in \mathcal{P}$ if, and only if, M is an Ocone martingale, and W_M, the DDS Brownian motion of M, has the predictable representation property for the filtration whose elements are $\sigma_t(W_M) \vee \sigma_t(S_M)$.*

Proof Suppose to start with that $\mathcal{L}(M) \in \mathcal{P}$. Because of (Proposition) 16.8.22,

$$\sigma_0(S_M) = \sigma_\infty(\langle M \rangle).$$

Since M is an Ocone martingale, and $\sigma_\infty(\langle M \rangle) = \sigma_\infty(S_M)$,

$$\sigma_t(W_M) \vee \sigma_t(S_M) = \sigma_t(W_M) \vee \sigma_0(S_M),$$

with $\sigma_0(S_M) = \sigma_\infty(\langle M \rangle)$ independent of $\sigma_t(W_M)$. The predictable representation property is thus that of Brownian motion.

Suppose now that the predictable representation property obtains. Since M is Ocone, $\underline{\sigma}(W_M)$ and $\underline{\sigma}(S_M)$ are immersed [(Definition) 16.5.1] in $\underline{\sigma}(W_M) \vee \underline{\sigma}(S_M)$. Thus [145, p. 189], when N is a continuous, local martingale for $\underline{\sigma}(S_M)$,

$$N(\cdot, t) = N(\cdot, 0) + \int_0^t f_N(\cdot, \theta) W_M(\cdot, d\theta),$$

where f_N is predictable for $\underline{\sigma}(W_M) \vee \underline{\sigma}(S_M)$. Now, because of independence, $\langle N, W_M \rangle = 0$, that is, $\int_0^t f_N(\cdot, \theta)\, d\theta = 0$. Consequently $\int_0^t f_N^2(\cdot, \theta)\, d\theta = 0$. But then $N(\cdot, t) = N(\cdot, 0)$, and, consequently, $\sigma_0(S_M) = \sigma_\infty(S_M) = \sigma_\infty(\langle M \rangle)$, so that $\mathcal{L}(M) \in \mathcal{P}$. □

Chapter 17
Likelihoods for Signal Plus Gaussian Noise Versus Gaussian Noise

Here, "Gaussian noise" means noise which is Gaussian, but not "white."

The basic setup shall be as follows. The time index is $[0, 1]$, as it is required that signals have finite energy for finite time periods. N is a stochastic process, defined on the probability space $(\Omega, \mathcal{A}, P)$, with values in $\mathbb{R}$: it is Gaussian, with mean zero; it is continuous in L_2, and almost surely zero at the origin. The measure induced by N, on the σ-algebra generated by the cylinder sets of $\mathbb{R}^{[0,1]}$, denoted $\mathcal{C}(\mathbb{R}^{[0,1]})$, shall be written P_N, that induced on the Borel sets of $L_2[0, 1]$, P_N^2, and that induced on the Borel sets of $C[0, 1]$, in case it has continuous paths, P_N^c. P_N^2 and P_N^c are assumed to have a support that has infinite dimension. The Cramér-Hida decomposition of N is, for $t \in [0, 1]$, fixed, but arbitrary,

$$N_t \overset{L_2}{=} \sum_{k=1}^{M_N} \int_{[0,1]} \phi_k(t)\, dm_k^N, \qquad (\star)$$

where m_k^N is the vector measure with values in L_2, generated by a process with orthogonal increments, say B_k^N. $\underline{B}_N$ denotes the family of those processes, and the Cramér-Hida representation is such that $L_t[N] = L_t[\underline{B}_N]$, for $t \in [0, 1]$, fixed, but arbitrary. As a consequence $\sigma_t^\circ(N) = \sigma_t^\circ(\underline{B}_N)$. But then the usual conditions are satisfied [264, p. 238].

Now a process continuous in quadratic mean is not necessarily separable, as may be seen with the simplest example of a process which is not separable [258, p. 21]: $X(\omega, t) = \chi_{\{t\}}(\omega)$, $(t, \omega) \in [0, 1]$, the probability being Lebesgue measure. Then $E_P\left[(X(\cdot, t) - X(\cdot, \theta))^2\right] = 0$. Taking both sides of $(\star)$ separable, which is always possible when quadratic mean continuity prevails [199, p. 91], one may claim that the following representation obtains (almost surely):

$$N(\cdot, t) \overset{path-wise}{=} \sum_{k=1}^{M_N} \int_0^t F_k(t, \theta)\, B_k^N(\cdot, d\theta). \qquad (\star\star)$$

A.F. Gualtierotti, *Detection of Random Signals in Dependent Gaussian Noise*,
DOI 10.1007/978-3-319-22315-5_17

Given a stochastic process X, its path at ω shall be denoted $X[\omega]$. The Cramér-Hida map looks at the latter equality $(\star\star)$ as having the following form:

$$N[\omega] = \Phi\left[\underline{B}_N[\omega]\right],$$

and takes advantage of it to obtain the likelihood for a Gaussian noise that is no longer a martingale.

Since the value of M is usually unknown, one shall assume $M = \infty$. When M is finite, the extra components are set to be zero.

17.1 An Introductory but Instructive Example

One may see, on the basis of an appropriate example, that of a Goursat process with one component, what the solution to the **SPGN** case can be, once the **SPWN** case has been solved. That is the content of the present section.

Let W be a standard Brownian motion, and

$$N(\omega,t) = f(t)\int_0^t g(\theta)\,W(\omega,d\theta).$$

Thus, the Cramér-Hida representation of N uses

$$F(t,\theta) = f(t)\,\chi_{[0,t]}(\theta)\,g(\theta).$$

As

$$E_P\left[N^2(\cdot,t)\right] = f^2(t)\int_0^t g^2(\theta)\,d\theta,$$

one should expect g to have an integrable square. As, for $t_1 < t_2$,

$$\begin{aligned}&N(\cdot,t_2) - N(\cdot,t_1) =\\ &\quad= \{f(t_2) - f(t_1)\}\int_0^{t_1} g(\theta)W(\cdot,d\theta) + f(t_2)\int_{t_1}^{t_2} g(\theta)W(\cdot,d\theta),\end{aligned}$$

one has that

$$\begin{aligned}&E_P\left[\{N(\cdot,t_2) - N(\cdot,t_1)\}^2\right] =\\ &\quad= \{f(t_2) - f(t_1)\}^2\int_0^{t_1} g^2(\theta)W(\cdot,d\theta) + f^2(t_2)\int_{t_1}^{t_2} g^2(\theta)\,d\theta,\end{aligned}$$

and one should expect f to be continuous. Let, for $\phi \in L_2[0,1]$,

$$f(t)\int_0^t \phi(\theta)g(\theta)d\theta = \langle F(t,\cdot),\phi\rangle_{L_2[0,1]} = 0,\ t \in [0,1].$$

One should thus expect $f(t) \neq 0,\ t \in [0,1]$; g not to be zero, almost surely; and notice that one has, to the left of the zero, a function in the RKHS of N.

Let thus N be as above, with $f > 0$ and continuous, and g have an integrable square, and be different from zero, almost surely. Suppose furthermore that g is absolutely continuous with derivative γ. Then (Itô's formula, for example),

$$N(\omega,t) = f(t)\left\{g(t)W(\omega,t) - \int_0^t \gamma(\theta)W(\omega,\theta)d\theta\right\}.$$

Let $\Phi : C[0,1] \longrightarrow C[0,1]$ be defined using the following rule:

$$\Phi(c)(t) = f(t)\left\{g(t)c(t) - \int_0^t c(\theta)\gamma(\theta)d\theta\right\}. \qquad (\star\star\star)$$

One has that $N[\omega] = \Phi(W[\omega])$, that is, the Cramér-Hida representation of N is a functional transformation of a Cramér-Hida process. Furthermore, when s is progressively measurable, and has a square that is, almost surely, integrable with respect to Lebesgue measure,

$$\begin{aligned}
&\Phi\left(t \mapsto \int_0^t s(\omega,\theta)d\theta\right)(t) = \\
&= f(t)\left\{g(t)\int_0^t s(\omega,\theta)d\theta - \int_0^t\left\{\int_0^\theta s(\omega,x)dx\right\}\gamma(\theta)d\theta\right\} \\
&= f(t)\left\{g(t)\int_0^t s(\omega,\theta)d\theta - \left\{g(t)\int_0^t s(\omega,x)dx - \int_0^t s(\omega,x)g(x)dx\right\}\right\} \\
&= f(t)\int_0^t s(\omega,\theta)g(\theta)d\theta \\
&= \langle F(t,\cdot), s(\omega,\cdot)\rangle_{L_2[0,1]}.
\end{aligned}$$

Consequently Φ sends the paths

- of W to those of N,
- of $\int s + W$ to those of $\langle F, s\rangle_{L_2[0,1]} + N$.

Since one has, with appropriate assumptions, that the law of $\int s + W$ is absolutely continuous with respect to that of W, the same is true for

$$\Phi\left(\int s + W\right) = \langle F, s\rangle_{L_2[0,1]} + N = S + N, \quad \text{and} \quad \Phi(W) = N.$$

Then, formally,

$$\begin{aligned}
P_{S+N}(A) &= P \circ (S+N)^{-1}(A) \\
&= P \circ \left\{\Phi\left(\int s + W\right)\right\}^{-1}(A) \\
&= P \circ \left(\int s + W\right)^{-1}(\Phi^{-1}(A)) \\
&= \int_{\Phi^{-1}(A)} \left\{\frac{dP_{\int s+W}}{dP_W}\right\} dP_W \\
&= \int_{\Phi^{-1}(A)} \left\{\frac{dP_{\int s+W}}{dP_W} \circ (\Phi^{-1} \circ \Phi)\right\} dP_W \\
&= \int_A \left\{\frac{dP_{\int s+W}}{dP_W} \circ \Phi^{-1}\right\} dP_W \circ \Phi^{-1} \\
&= \int_A \left\{\frac{dP_{\int s+W}}{dP_W} \circ \Phi^{-1}\right\} dP_N.
\end{aligned}$$

So, knowing the likelihood for the SP**W**N model, and the inverse of Φ, one obtains the likelihood for the SP**G**N model. The objective of this chapter is the calculation of Φ and its inverse. There are several cases, depending on the assumptions. All the resulting Φ's shall be called "Cramér-Hida maps."

Supposing that the operations performed below are legitimate, one may try to invert Φ as follows. Let the left-hand side of $(\star\star\star)$ above, divided by f, be denoted b, a now known function. Let $a = \gamma^{-1}g$, and

$$F(t) = \int_0^t \gamma(x)c(x)\,dx.$$

Equation $(\star)$ above may now be given the following form:

$$b(t) = a(t)F'(t) - F(t),$$

and may be solved explicitly for F [211, p. 604]. Then $c = \gamma^{-1}F'$. For a Cramér-Hida process, such a procedure will not generally be possible, as the required assumptions shall not be available, but one may consider an analogous procedure to find the inverse of Φ, and that is the computation of W, conditional on N, in the probability sense. That such a procedure is feasible may be seen in the result of the following computations.

Let $0 < t_1 < \cdots < t_n \leq 1$ be fixed, but arbitrary, and

$$X_1 = \int_0^{t_1} g dW, \; X_2 = \int_{t_1}^{t_2} g dW, \; X_3 = \int_{t_2}^{t_3} g dW, \; \ldots$$

Then

$$\begin{aligned} N(\cdot, t_1) &= f(t_1) X_1, \\ N(\cdot, t_2) &= f(t_2) \{X_1 + X_2\}, \\ N(\cdot, t_3) &= f(t_3) \{X_1 + X_2 + X_3\}, \\ \cdots &= \cdots, \end{aligned}$$

and thus

$$\underline{N} = D_f L_n [\underline{X}],$$

where D_f is the diagonal matrix whose diagonal terms are the values of f at the time points, and L_n is a square matrix of dimension n, with the following form:

$$L_n = \begin{bmatrix} 1\,0\,0\,0\,0 \cdots 0 \\ 1\,1\,0\,0\,0 \cdots 0 \\ 1\,1\,1\,0\,0 \cdots 0 \\ \vdots\;\vdots\;\vdots\;\vdots\;\vdots \quad \vdots \\ 1\,1\,1\,1\,1 \cdots 1 \end{bmatrix}.$$

Σ_N, the covariance matrix of $\underline{N}$, is thus

$$\Sigma_N = D_f L_n V_X L_n^\star D_f,$$

where V_X is the covariance matrix of $\underline{X}$. It is diagonal, with diagonal elements of the following form ($\Pi_S[g]$ is the class of $\chi_S \dot{g}$):

$$v_i^2 = \left\| \Pi_{]t_{i-1}, t_i]}[g] \right\|_{L_2[0,1]}^2.$$

The generic form of Σ_N is thus $DL\Delta L^\star D$, with D and Δ diagonal. The inverse is then

$$D^{-1} \left(L^\star\right)^{-1} \Delta^{-1} L^{-1} D^{-1},$$

with

$$L^{-1} = \begin{bmatrix} 1 & 0\,0\,0\,0 \cdots & 0\,0 \\ -1 & 1\,0\,0\,0 \cdots & 0\,0 \\ 0 & -1\,1\,0\,0 \cdots & 0\,0 \\ \vdots & \vdots\,\vdots\,\vdots\,\vdots & \vdots\,\vdots \\ 0 & 0\,0\,0\,0 \cdots & -1\,1 \end{bmatrix}.$$

To be explicit, let $n = 9$, and set $N_i = N(\cdot, t_i)$, $W_i = W(\cdot, t_i)$. $\underline{N}$ has components $N_1, \ldots, N_9$, and $\underline{W}$, components W_3, W_4, W_5. The law of $\underline{W}$, given $\underline{N}$, is Gaussian, with mean

$$\Sigma_{W,N} \Sigma_N^{-1} \underline{N},$$

where $\Sigma_{W,N} = E_P\left[\underline{W}\,\underline{N}^\star\right]$, $\Sigma_N = E_P\left[\underline{N}\,\underline{N}^\star\right]$, and variance

$$\Sigma_W - \Sigma_{W,N} \Sigma_N^{-1} \Sigma_{N,W}.$$

One may write $\underline{W} = M[\underline{Y}]$, where $\underline{Y}$ contains the increments of W: for example $W_3 = Y_1 + Y_2 + Y_3$, and $Y_3 = W(\cdot, t_3) - W(\cdot, t_2)$. Now, with

$$M = \begin{bmatrix} 1\,1\,1\,0\,0 \\ 1\,1\,1\,1\,0 \\ 1\,1\,1\,1\,1 \end{bmatrix},$$

one has that

$$E_P\left[\underline{W}\,\underline{N}^\star\right] = E_P\left[M \underline{Y}\,\underline{X}^\star L_9^\star D_f\right] = M E_P\left[\underline{Y}\,\underline{X}^\star\right] L_9^\star D_f,$$

and the matrix $E_P\left[\underline{Y}\,\underline{X}^\star\right]$ has entries different from zero only in the positions (i, i), $i = 1, 2, 3, 4, 5$, and then

$$u_i = E_P[Y_i X_i] = \langle \Pi_{]t_{i-1}, t_i]} g, 1 \rangle_{L_2[0,1]}.$$

Consequently,

$$\begin{aligned} \Sigma_{W,N} \Sigma_N^{-1} &= \left\{ M E_P\left[\underline{Y}\,\underline{X}^\star\right] L_9^\star D_f \right\} \left\{ D_f^{-1} (L_9^\star)^{-1} V_X^{-1} L_9^{-1} D_f^{-1} \right\} \\ &= M E_P\left[\underline{Y}\,\underline{X}^\star\right] V_X^{-1} L_9^{-1} D_f^{-1}. \end{aligned}$$

But $E_P[\underline{Y}\,\underline{X}^\star] = [D_u\ O_{5,4}]$ ($O_{p,q}$ being a matrix of zeroes of dimensions (p,q)), so that $E_P[\underline{Y}\,\underline{X}^\star]V_X^{-1} = \left[D_u D_{v^2}^{-1}\ O_{5,4}\right]$. Since, for example,

$$\begin{bmatrix} d_1 & 0 & 0 & 0 \\ 0 & d_2 & 0 & 0 \\ 0 & 0 & d_3 & 0 \\ 0 & 0 & 0 & d_4 \end{bmatrix} \begin{bmatrix} 1 & 0 & 0 & 0 \\ -1 & 1 & 0 & 0 \\ 0 & -1 & 1 & 0 \\ 0 & 0 & -1 & 1 \end{bmatrix} = \begin{bmatrix} d_1 & 0 & 0 & 0 \\ -d_2 & d_2 & 0 & 0 \\ 0 & -d_3 & d_3 & 0 \\ 0 & 0 & -d_4 & d_4 \end{bmatrix},$$

one obtains that

$$\Sigma_{W,N}\Sigma_N^{-1} = M\left[\begin{array}{ccccc|c} \frac{u_1}{f(t_1)v_1^2} & 0 & 0 & 0 & 0 & \\ -\frac{u_2}{f(t_1)v_2^2} & \frac{u_2}{f(t_2)v_2^2} & 0 & 0 & 0 & \\ 0 & -\frac{u_3}{f(t_2)v_3^2} & \frac{u_3}{f(t_3)v_3^2} & 0 & 0 & O_{5,4} \\ 0 & 0 & -\frac{u_4}{f(t_3)v_4^2} & \frac{u_4}{f(t_4)v_4^2} & 0 & \\ 0 & 0 & 0 & -\frac{u_5}{f(t_4)v_5^2} & \frac{u_5}{f(t_5)v_5^2} & \end{array}\right],$$

and thus, with $h_i = u_i/v_i$, that $\Sigma_{W,N}\Sigma_N^{-1}[\underline{N}]$ has, for example, as first line, m_1 denoting the conditional mean of W_1 given $\underline{N}$, the following expression:

$$\begin{aligned} m_1 = &\frac{\langle h_1, 1\rangle_{L_2[0,1]}}{\left\|\Pi_{]0,t_1]}g\right\|_{L_2[0,1]}}\left\{\frac{N(\cdot,t_1)}{f(t_1)}\right\} \\ &+ \frac{\langle h_2, 1\rangle_{L_2[0,1]}}{\left\|\Pi_{]t_1,t_2]}g\right\|_{L_2[0,1]}}\left\{\frac{N(\cdot,t_2)}{f(t_2)} - \frac{N(\cdot,t_1)}{f(t_1)}\right\} \\ &+ \frac{\langle h_3, 1\rangle_{L_2[0,1]}}{\left\|\Pi_{]t_2,t_3]}g\right\|_{L_2[0,1]}}\left\{\frac{N(\cdot,t_3)}{f(t_3)} - \frac{N(\cdot,t_2)}{f(t_2)}\right\}. \end{aligned}$$

One then has, for example, that

$$\frac{1}{\left\|\Pi_{]t_1,t_2]}g\right\|_{L_2[0,1]}}\left\{\frac{N(\cdot,t_2)}{f(t_2)} - \frac{N(\cdot,t_1)}{f(t_1)}\right\}$$

may be looked at as the following product:

$$\frac{\int_{t_1}^{t_2} g(\theta)\,W(\cdot,d\theta)}{W(\cdot,t_2) - W(\cdot,t_1)} \times \frac{W(\cdot,t_2) - W(\cdot,t_1)}{(t_2 - t_1)^{1/2}} \times \left(\frac{\int_{t_1}^{t_2} g^2(\theta)\,d\theta}{t_2 - t_1}\right)^{-1/2}.$$

When $t_2 - t_1$ is small, the first term of the latter product is approximately g [221, p. 143]. The third is approximately $|g|^{-1}$, so that, conditioning on N, one recovers a large linear combination of independent normal random variables with small coefficients, that is something that looks like Brownian motion. Indeed,

$$W_t = \sum_i (t_i - t_{i-1})^{1/2} \frac{W_{t_i} - W_{t_{i-1}}}{(t_i - t_{i-1})^{1/2}}.$$

Suppose now that

$$\begin{aligned} N(\omega, t) &= \\ &= f_1(t) \int_0^t g_1(\theta)\, W_1(\omega, d\theta) + f_2(t) \int_0^t g_2(\theta)\, W_2(\omega, d\theta) \\ &= N_1(\omega, t) + N_2(\omega, t), \end{aligned}$$

where the f's and g's are as in the first example, and the Brownian motions are independent. Again, when g_1 and g_2 are absolutely continuous,

$$\begin{aligned} N(\omega, t) = f_1(t) &\left\{ g_1(t) W_1(\omega, t) - \int_0^t \gamma_1(\theta) W_1(\omega, \theta)\, d\theta \right\} \\ + f_2(t) &\left\{ g_2(t) W_2(\omega, t) - \int_0^t \gamma_2(\theta) W_2(\omega, \theta)\, d\theta \right\}, \end{aligned}$$

and $N = \Phi\left(\begin{bmatrix} W_1 \\ W_2 \end{bmatrix}\right)$, where

$$\Phi\left(\begin{bmatrix} c_1 \\ c_2 \end{bmatrix}\right) = f_1 \left\{ g_1 c_1 - \int_0^{\cdot} \gamma_1(\theta) c_1(\theta)\, d\theta \right\} + f_2 \left\{ g_2 c_2 - \int_0^{\cdot} \gamma_2(\theta) c_2(\theta)\, d\theta \right\}.$$

Computing explicitly the conditional law becomes more difficult, in practice rather than principle. Let again, *mutatis mutandis,*

$$\underline{N} = \underline{N}_1 + \underline{N}_2,\ \underline{N}_i = D_i L_n \left[\underline{X}_i\right],\ i = 1, 2,$$

so that $\Sigma_N = \Sigma_1 + \Sigma_2$. It follows from [278, p. 44] that

$$\Sigma_N^{-1} = \Sigma_1^{-1} \left(\Sigma_1^{-1} + \Sigma_2^{-1}\right)^{-1} \Sigma_2^{-1} = \Sigma_2^{-1} \left(\Sigma_1^{-1} + \Sigma_2^{-1}\right)^{-1} \Sigma_1^{-1}.$$

Since the inverse of Σ_i is a triangular matrix, the parenthesis is a triangular matrix whose inverse may be explicitly computed recursively. It has even a form $DL^\star \Delta LD$, with D and Δ diagonal [27]. Let now $\underline{W}_1$ be the vector with components $W_1(\cdot, t_3), W_1(\cdot, t_4), W_1(\cdot, t_5)$. $\underline{W}_2$ is built analogously. Then

$$E_P\left[\begin{bmatrix} \underline{W}_1 \\ \underline{W}_2 \end{bmatrix} \left[\underline{N}_1 + \underline{N}_2\right]^\star\right] = \begin{bmatrix} E_P\left[\underline{W}_1 \underline{N}_1^\star\right] \\ E_P\left[\underline{W}_2 \underline{N}_2^\star\right] \end{bmatrix},$$

so that

$$\Sigma_{W,N}\Sigma_N^{-1}[\underline{N}] = \begin{bmatrix} E_P\left[\underline{W}_1\underline{N}_1^\star\right]\Sigma_1^{-1}\left(\Sigma_1^{-1}+\Sigma_2^{-1}\right)^{-1}\Sigma_2^{-1}[\underline{N}] \\ E_P\left[\underline{W}_2\underline{N}_2^\star\right]\Sigma_2^{-1}\left(\Sigma_1^{-1}+\Sigma_2^{-1}\right)^{-1}\Sigma_1^{-1}[\underline{N}] \end{bmatrix}.$$

One thus sees that $E_P\left[\underline{W}_1\underline{N}_1^\star\right]\Sigma_1^{-1}$ is the same term as in the first part of the example, and that $\underline{N}$ gets replaced with

$$\underline{\tilde{N}} = \left(\Sigma_1^{-1}+\Sigma_2^{-1}\right)^{-1}\Sigma_2^{-1}[\underline{N}].$$

The procedure of the first part of the example thus yields, modulo computational complications, the same conclusion.

17.2 The Cramér-Hida Maps

From the Cramér-Hida representation, one has, for $t_1 < t_2$ in $[0,1]$, fixed, but arbitrary, that

$$E_P\left[\{N(\cdot,t_2)-N(\cdot,t_1)\}^2\right] = \|\underline{F}(t_2)-\underline{F}(t_1)\|^2_{L_2[\underline{b}]},$$

where $\underline{F} : [0,1] \longrightarrow L_2[\underline{b}]$ has, for components, the equivalence classes of the $F_n(t,\cdot)$'s. $\underline{F}$ is thus continuous as N is. Let $L_F : L_2[\underline{b}] \longrightarrow \mathbb{R}^{[0,1]}$ be defined using the following relation:

$$L_F[\underline{a}](t) = \langle \underline{a}, \underline{F}(t)\rangle_{L_2[\underline{b}]}.$$

As

$$C_N(t_1,t_2) = \langle \underline{F}(t_1), \underline{F}(t_2)\rangle_{L_2[\underline{b}]},$$

the range of L_F is the RKHS of C_N, and, as the representation is proper canonical,

$$\langle L_F[\underline{a}_1], L_F[\underline{a}_2]\rangle_{H(C_N,[0,1])} = \langle \underline{a}_1, \underline{a}_2\rangle_{L_2[\underline{b}]}.$$

L_F, as an operator into the RKHS of N, is then unitary. In the sequel, the following notation, which makes sense, shall be used:

$$\begin{aligned} \underline{X} &= S[\underline{a}] + \underline{B}_N, \\ I_{\underline{X}}\{\underline{F}\} &= \langle \underline{F}, \underline{a}\rangle_{L_2[\underline{b}]} + I_{\underline{B}_N}\{\underline{F}\}. \end{aligned}$$

There shall be three Cramér-Hida maps, corresponding to paths of N lying, respectively, in $\mathbb{R}^{[0,1]}$, $\mathcal{L}_2[0,1]$, and $C[0,1]$. Each has its advantages, ... and limitations. When avoidable, no distinction shall be made between a random variable and its class.

Proposition 17.2.1 (Real Functions as Paths)

1. *There exists a map $\Phi : K \longrightarrow \mathbb{R}^{[0,1]}$, adapted to $\mathcal{K}$ and $\mathcal{C}(\mathbb{R}^{[0,1]})$, such that, for $t \in [0,1]$, fixed, but arbitrary, almost surely, with respect to P,*

$$N(\omega, t) = \mathcal{E}_t\left(\Phi(\underline{B}_N[\omega])\right).$$

2. *Suppose that $P_X^{\mathcal{K}}$ is absolutely continuous with respect to $P_{B_N}^{\mathcal{K}}$. Then, for $t \in [0,1]$, fixed, but arbitrary, almost surely, with respect to P,*

$$I_{\underline{X}}\{\underline{F}\}(\omega, t) = \mathcal{E}_t\left(\Phi(\underline{X}[\omega])\right).$$

Proof Let i, j, m, n, and p belong to $\mathbb{N}$, with $m \in [1 : 2^n]$, fixed, but arbitrary. One shall use the following type of step functions, with domain $[0,1]$, where $\kappa_j \in \mathbb{R}$, $0 \leq \lambda_j < \mu_j \leq 1$:

$$F_{i,m,n}(t) = \sum_{j=1}^{p} \kappa_j \chi_{]\lambda_j, \mu_j]}(t),$$

chosen, because of (uniform) continuity of $\underline{F}$, so that

$$\left\| F_i\left(\frac{m}{2^n}, \cdot\right) - F_{i,m,n} \right\|_{L_2[M_i]}^2 \leq \frac{1}{2^{n+i}}. \qquad (\star)$$

One shall set, for arbitrary $f : [0,1] \longrightarrow \mathbb{R}$:

$$\int_0^1 F_{i,m,n}(t) f(dt) = \sum_{j=1}^{p} \kappa_j \left\{ f(\mu_j) - f(\lambda_j) \right\}. \qquad (\star\star)$$

Let then

$$\tilde{F}_{n,i}(t, \theta) = \textstyle\sum_{m=1}^{2^n} \chi_{]\frac{m-1}{2^n}, \frac{m}{2^n}]}(t) F_{i,m,n}(\theta),$$

$$\underline{\tilde{F}}_n(t, \theta) \qquad \text{have components } \tilde{F}_{n,i}(t, \theta),$$

$$\underline{F}^{(n)}(t, \theta) = \textstyle\sum_{m=1}^{2^n} \chi_{]\frac{m-1}{2^n}, \frac{m}{2^n}]}(t) \underline{F}\left(\frac{m}{2^n}, \theta\right).$$

Define also $\Phi_n : K \longrightarrow \mathbb{R}^{[0,1]}$ using the following relation ($\underline{k} \in K$):

$$\Phi_n \left[\underline{k}\right](t) = \sum_{i=1}^{n} \int_0^1 \tilde{F}_{n,i}(t,\theta) k_i(d\theta),$$

where k_i is the i-th component of $\underline{k}$. The definition makes sense because of ($\star\star$) and the relation which follows it.

Step 1: Φ_n *is adapted to* $\mathcal{K}$ *and* $\mathcal{C}(\mathbb{R}^{[0,1]})$.

Indeed, for fixed, but arbitrary i and θ, $\underline{k} \mapsto k_i(\theta) = \mathcal{E}_\theta \circ \mathcal{E}_i^K(\underline{k})$ is adapted to $\mathcal{K}$ and $\mathcal{C}(\mathbb{R}^{[0,1]})$. Thus, given the definition of the elements involved, for fixed, but arbitrary t, $\underline{k} \mapsto \Phi_n \left[\underline{k}\right](t)$ is adapted to $\mathcal{K}$ and $\mathcal{C}(\mathbb{R}^{[0,1]})$, as it is a linear combination of terms of type $\mathcal{E}_\theta \circ \mathcal{E}_i^K(\underline{k})$.

Step 2: $\{\Phi_n, n \in \mathbb{N}\}$ *is, for* $P_{\underline{B}_N}^K$*, a Cauchy sequence in probability.*

Given the definitions,

$$\Phi_n \left[\underline{B}_N\right](t) = \sum_{i=1}^{n} \int_0^1 \tilde{F}_{n,i}(t,\theta) B_i^N(\cdot, d\theta).$$

Thus, for $q > n$, fixed, but arbitrary, using Markov's inequality [138, p. 164],

$$\begin{aligned}
&P_{\underline{B}_N}^K \left(\underline{k} \in K : \left|\Phi_q \left[\underline{k}\right](t) - \Phi_n \left[\underline{k}\right](t)\right| > \epsilon\right) \leq \\
&\quad \leq P\left(\omega \in \Omega : \left|\sum_{i=1}^{n} \int_0^1 \left\{\tilde{F}_{q,i}(t,\theta) - \tilde{F}_{n,i}(t,\theta)\right\} B_i^N(\omega, d\theta)\right| > \frac{\epsilon}{2}\right) \\
&\quad + P\left(\omega \in \Omega : \left|\sum_{i=n+1}^{q} \int_0^1 \tilde{F}_{q,i}(t,\theta) B_i^N(\omega, d\theta)\right| > \frac{\epsilon}{2}\right) \\
&\quad \leq \frac{4}{\epsilon^2} \sum_{i=1}^{n} \int_0^1 \left\{\tilde{F}_{q,i}(t,\theta) - \tilde{F}_{n,i}(t,\theta)\right\}^2 M_i(d\theta) \\
&\quad + \frac{4}{\epsilon^2} \sum_{i=n+1}^{q} \int_0^1 \tilde{F}_{q,i}^2(t,\theta) M_i(d\theta).
\end{aligned}$$

Now, using $E^2[X] \leq E[X^2]$ applied to $X = 4^{-1}(a+b+c+d)$, one has that $(a+b+c+d)^2 \leq 4(a^2+b^2+c^2+d^2)$, so that, inserting and deleting the appropriate terms,

$$\begin{aligned}
&\int_0^1 \left\{\tilde{F}_{q,i}(t,\theta) - \tilde{F}_{n,i}(t,\theta)\right\}^2 M_i(d\theta) \leq \\
&\qquad \leq 4 \int_0^1 \left\{\tilde{F}_{q,i}(t,\theta) - F_i^{(q)}(t,\theta)\right\}^2 M_i(d\theta)
\end{aligned}$$

$$
\begin{aligned}
&+ 4\int_0^1 \left\{F_i^{(q)}(t,\theta) - F_i(t,\theta)\right\}^2 M_i(d\theta) \\
&+ 4\int_0^1 \left\{F_i(t,\theta) - F_i^{(n)}(t,\theta)\right\}^2 M_i(d\theta) \\
&+ 4\int_0^1 \left\{F_i^{(n)}(t,\theta) - \tilde{F}_{n,i}(t,\theta)\right\}^2 M_i(d\theta).
\end{aligned}
$$

Using the definitions, and in particular ($\star$), one has also that

$$
\begin{aligned}
&\int_0^1 \left\{\tilde{F}_{q,i}(t,\theta) - F_i^{(q)}(t,\theta)\right\}^2 M_i(d\theta) = \\
&= \sum_{m=1}^{2^q} \chi_{\left]\frac{m-1}{2^q},\frac{m}{2^q}\right]}(t) \int_0^1 \left\{F_{i,m,q}(\theta) - F_i\left(\frac{m}{2^q},\theta\right)\right\}^2 M_i(d\theta) \\
&\le \sum_{m=1}^{2^q} \chi_{\left]\frac{m-1}{2^q},\frac{m}{2^q}\right]}(t) \frac{1}{2^{q+i}} \\
&\le \frac{1}{2^{q+i}}.
\end{aligned}
$$

Furthermore, still because of the definitions and ($\star$),

$$
\begin{aligned}
&\sum_{i=n+1}^{q} \int_0^1 \tilde{F}_{q,i}^2(t,\theta) M_i(d\theta) = \\
&= \sum_{i=n+1}^{q} \left\|\tilde{F}_{q,i}(t,\cdot)\right\|_{L_2[M_i]}^2 \\
&= \sum_{i=n+1}^{q} \sum_{m=1}^{2^q} \chi_{\left]\frac{m-1}{2^q},\frac{m}{2^q}\right]}(t) \left\|F_{i,m,q}(t,\cdot)\right\|_{L_2[M_i]}^2 \\
&\le \sum_{i=n+1}^{q} \sum_{m=1}^{2^q} \chi_{\left]\frac{m-1}{2^q},\frac{m}{2^q}\right]}(t) \left\{\frac{1}{2^{\frac{q+i}{2}}} + \left\|F_i\left(\frac{m}{2^q},\cdot\right)\right\|_{L_2[M_i]}\right\}^2 \\
&\le 2\sum_{i=n+1}^{q} \sum_{m=1}^{2^q} \chi_{\left]\frac{m-1}{2^q},\frac{m}{2^q}\right]}(t) \left\{\frac{1}{2^{q+i}} + \left\|F_i\left(\frac{m}{2^q},\cdot\right)\right\|_{L_2[M_i]}^2\right\} \\
&= \frac{1}{2^{q-1}} \sum_{i=n+1}^{q} \frac{1}{2^i} + \sum_{i=n+1}^{q} \left\|F_i^{(q)}(t,\cdot)\right\|_{L_2[M_i]}^2.
\end{aligned}
$$

Consequently

$$
\begin{aligned}
P^{K}_{\underline{B}_N}\left(\underline{k} \in K : \left|\Phi_q\left[\underline{k}\right](t) - \Phi_n\left[\underline{k}\right](t)\right| > \epsilon\right) \leq \\
&\leq \frac{16}{\epsilon^2}\left\{\frac{1}{2^q} + \left\|\underline{F}^{(q)}(t) - \underline{F}(t)\right\|^2_{L_2[\underline{b}]}\right\} \\
&+ \frac{16}{\epsilon^2}\left\{\left\|\underline{F}^{(n)}(t) - \underline{F}(t)\right\|^2_{L_2[\underline{b}]} + \frac{1}{2^n}\right\} \\
&+ \frac{4}{\epsilon^2}\left\{\frac{1}{2^{q-1}}\sum_{i=n+1}^{q}\frac{1}{2^i} + \sum_{i=n+1}^{q}\left\|F_i^{(q)}(t,\cdot)\right\|^2_{L_2[M_i]}\right\},
\end{aligned}
$$

a quantity that, for fixed t, goes to zero as n and q increase indefinitely. Φ shall denote the limit in probability, with respect to $P^{K}_{\underline{B}_N}$, of the sequence $\{\Phi_n, n \in \mathbb{N}\}$.

Step 3: The first part of statement (Proposition) 17.2.1 *obtains.*

One has, for $t \in [0, 1]$, fixed, but arbitrary, that

$$
\begin{aligned}
P\left(\omega \in \Omega : \left|N(\omega,t) - \mathcal{E}_t\left(\Phi(\underline{B}_N(\omega))\right)\right| > \epsilon\right) \leq \\
&\leq P\left(\omega \in \Omega : \left|N(\omega,t) - \mathcal{E}_t\left(\Phi_n(\underline{B}_N(\omega))\right)\right| > \frac{\epsilon}{2}\right) \\
&+ P\left(\omega \in \Omega : \left|\mathcal{E}_t\left(\Phi_n(\underline{B}_N(\omega))\right) - \mathcal{E}_t\left(\Phi(\underline{B}_N(\omega))\right)\right| > \frac{\epsilon}{2}\right) \\
&= p_n + q_n.
\end{aligned}
$$

By step 2, $\lim_n q_n = 0$. Now, using on the second $[\{\cdots\}^2]$ expression below, $(a+b)^2 \leq 2(a^2+b^2)$, one has that

$$
\begin{aligned}
p_n &= P\left(\omega \in \Omega : \left|N(\omega,t) - \mathcal{E}_t\left(\Phi_n(\underline{B}_N(\omega))\right)\right| > \frac{\epsilon}{2}\right) \\
&\leq \frac{4}{\epsilon^2} E_P\left[\left\{N(\omega,t) - \mathcal{E}_t\left(\Phi_n(\underline{B}_N(\omega))\right)\right\}^2\right] \\
&= \frac{4}{\epsilon^2} E_P\left[\left\{\sum_{i=1}^{n}\int_0^1 \left(F_i(t,\theta) - \tilde{F}_{n,i}(t,\theta)\right) B_i^N(\cdot, d\theta)\right.\right. \\
&\qquad \left.\left. + \sum_{i=n+1}^{\infty}\int_0^1 F_i(t,\theta) B_i^N(\cdot, d\theta)\right\}^2\right] \\
&\leq \frac{8}{\epsilon^2}\left\{\sum_{i=1}^{n}\int_0^1 \left\{F_i(t,\theta) - \tilde{F}_{n,i}(t,\theta)\right)\right\}^2 M_i(d\theta) \\
&+ \frac{8}{\epsilon^2}\sum_{i=n+1}^{\infty}\int_0^1 F_i^2(t,\theta) M_i(d\theta).
\end{aligned}
$$

As

$$F_i - \tilde{F}_{n,i} = \left[F_i - F_i^{(n)}\right] + \left[F_i^{(n)} - \tilde{F}_{n,i}\right],$$

one may proceed as in step 2 to obtain that $\lim_n p_n = 0$.

Step 4: For $t \in [0, 1]$, fixed, but arbitrary, the sequence $\{\mathcal{E}_t(\Phi_n(\underline{X})), n \in \mathbb{N}\}$ converges in probability, with respect to P, to $I_{\underline{X}}\{\underline{F}\}(\cdot, t)$.

One has that

$$\begin{aligned} I_{\underline{X}}\{\underline{F}\}(\cdot,t) - \mathcal{E}_t(\Phi_n(\underline{X}(\cdot))) &= \sum_{i=1}^{n} \int_0^t \left\{F_i(t,\theta) - \tilde{F}_{n,i}(t,\theta)\right\} a_i(\cdot,\theta) M_i(d\theta) \\ &+ \sum_{i=n+1}^{\infty} \int_0^t F_i(t,\theta) a_i(\cdot,\theta) M_i(d\theta) \\ &+ N(\cdot,t) - \mathcal{E}_t\left(\Phi_n(\underline{B}_N[\cdot])\right). \end{aligned}$$

As seen in the proof of item 1, the third term on the right of the latter expression converges to zero in probability, with respect to P. The second term converges equally to zero since $\underline{F}$ and $\underline{a}$ belong to $L_2[\underline{b}]$, at least almost surely for $\underline{a}$, with respect to P. Thus only the first term requires attention. But

$$\begin{aligned} &\left| \sum_{i=1}^{n} \int_0^t \left\{F_i(t,\theta) - \tilde{F}_{n,i}(t,\theta)\right\} a_i(\omega,\theta) M_i(d\theta) \right| \leq \\ &\leq \sum_{i=1}^{n} \int_0^1 \left|F_i(t,\theta) - \tilde{F}_{n,i}(t,\theta)\right| |a_i(\omega,\theta)| M_i(d\theta) \\ &\leq \sum_{i=1}^{n} \left\{ \int_0^1 \left\{F_i(t,\theta) - \tilde{F}_{n,i}(t,\theta)\right\}^2 M_i(d\theta) \right\}^{1/2} \left\{ \int_0^1 a_i^2(\omega,\theta) M_i(d\theta) \right\}^{1/2} \\ &\leq \left\{ \sum_{i=1}^{n} \int_0^1 \left\{F_i(t,\theta) - \tilde{F}_{n,i}(t,\theta)\right\}^2 M_i(d\theta) \right\} \left\{ \sum_{i=1}^{n} \int_0^1 a_i^2(\omega,\theta) M_i(d\theta) \right\} \\ &\overset{(step\ 2)}{\leq} 2 \left\{ \frac{1}{2^n} + \left\|\underline{F}(t) - \underline{F}^{(n)}(t)\right\|_{L_2[\underline{b}]}^2 \right\} \|\underline{a}(\omega,\cdot)\|_{L_2[\underline{b}]}^2 . \end{aligned}$$

Convergence of the last term in the right-hand side of the latter expression is almost surely, with respect to P, uniform in t.

Step 5: For $t \in [0, 1]$, fixed, but arbitrary, almost surely, with respect to P,

$$I_{\underline{X}}\{\underline{F}\}(\cdot,t) = \mathcal{E}_t(\Phi(\underline{X}[\cdot])).$$

From step 2, one has that, for $t \in [0,1]$, fixed, but arbitrary, the sequence $\{\mathcal{E}_t(\Phi_n), n \in \mathbb{N}\}$ converges in probability to $\mathcal{E}_t(\Phi)$, with respect to $P^K_{B_N}$. By taking a subsequence if necessary, one may assume that convergence is almost sure. But since, by assumption,

$$P^K_X \ll P^K_{B_N},$$

convergence takes place almost surely with respect to P^K_X. Step 4 then insures that item 2 of the statement obtains. □

Proposition 17.2.2 (Square Integrable Functions as Paths) *There is an adapted map $\Phi_2 : K \longrightarrow L_2[0,1]$ such that*

1. *almost everywhere, with respect to P, $N[\omega] = \Phi_2(\underline{B}_N[\omega])$;*
2. *when $P^K_X \ll P^K_{B_N}$, almost everywhere, with respect to P,*

$$I_{\underline{X}}\{\underline{F}\}[\omega] = \Phi_2(\underline{X}[\omega]).$$

Proof As seen [(Remark) 11.2.1], $L_2[\underline{b}]$ may be looked at as an L_2-space, and thus the map $\underline{F} : [0,1] \longrightarrow L_2[\underline{b}]$ may be seen as a square integrable map into it. But the space $L_2^H([0,1])$, $H = L_2[\underline{b}]$, is isomorphic to an L_2-space built on a product of measure spaces [202, p. 115], and, in the present case, that corresponding to Lebesgue measure on $[0,1]$ and that corresponding to the direct sum of the L_2 spaces of the form $L_2([0,1], \mathcal{B}([0,1]), M_n)$.

Let $\{\mathcal{B}_n([0,1]), n \in \mathbb{N}\}$ be an increasing sequence of finite σ-algebras generated by intervals whose union generates $\mathcal{B}([0,1])$ (for instance, the dyadic intervals of length 2^{-n}). Let $\mathcal{A}_n$ be the σ-algebra generated in the direct sum using the family $\{\mathcal{B}_n([0,1]), n \in \mathbb{N}\}$, that is, $A \in \mathcal{A}_n$ if and only if $A \cap [0,1]$ belongs to $\mathcal{B}_n([0,1])$ for each $n \in \mathbb{N}$. Because of [177, p. 180], there is then a sequence of "simple" functions

$$\{\underline{F}^{(n)}, n \in \mathbb{N}\}$$

that converges in L_2 to $\underline{F}$. The generic form of an entry in an element of that sequence has the following aspect:

$$F_m^{(n)}(t,\theta) = \sum_{i=1}^{p_n} \kappa_{i,m}^{(n)} \chi_{\left]t_{i,l}^{(n)}, t_{i,r}^{(n)}\right]}(t)\, \chi_{\left]\theta_{i,l}^{(n)}, \theta_{i,r}^{(n)}\right]}(\theta).$$

It has the property that

$$\begin{aligned}\left\|F_m^{(n)}(t,\cdot)\right\|^2_{L_2[b_m]} &= \int_0^1 \left\{F_m^{(n)}(t,\theta)\right\}^2 M_m(d\theta)\\ &= \sum_{i=1}^{p_n} \chi_{\left]t_{i,l}^{(n)}, t_{i,r}^{(n)}\right]}(t) \left(\left\{\kappa_{i,m}^{(n)}\right\}^2 \left\{M_m\left(\theta_{i,r}^{(n)}\right) - M_m\left(\theta_{i,l}^{(n)}\right)\right\}\right).\end{aligned}$$

Consequently, taking into account the fact that $\left\|\underline{F}^{(n)}(t)\right\|_{L_2[\underline{b}]}$ must be finite,

$$\begin{aligned}\left\|\underline{F}^{(n)}(t)\right\|^2_{L_2[\underline{b}]} &= \sum_{m=1}^{\infty} \left\|F_m^{(n)}(t,\cdot)\right\|^2_{L_2[b_m]} \\ &= \sum_{i=1}^{p_n} \chi_{\left]t_{i,l}^{(n)},t_{i,r}^{(n)}\right]}(t) \left(\sum_{m=1}^{\infty} \left\{\kappa_{i,m}^{(n)}\right\}^2 \left\{M_m\left(\theta_{i,r}^{(n)}\right) - M_m\left(\theta_{i,l}^{(n)}\right)\right\}\right).\end{aligned}$$

Define then $\Phi_2^{(n)} : K \longrightarrow L_2[0,1]$ using the following formula:

$$\overbrace{\Phi_2^{(n)}[\underline{k}](t)}^{\cdot} = \sum_{i=1}^{p_n} \chi_{\left]t_{i,l}^{(n)},t_{i,r}^{(n)}\right]}(t) \sum_{m=1}^{\infty} \kappa_{i,m}^{(n)} \left\{k_m(\theta_{i,r}^{(n)}) - k_m(\theta_{i,l}^{(n)})\right\}.$$

So one has that

$$\begin{aligned}&E_{P_{B_N}^K}\left[\left\|\Phi_2^{(n)}\right\|^2_{L_2[0,1]}\right] = \\ &= E_P\left[\left\|\Phi_2^{(n)}[\underline{B}_N]\right\|^2_{L_2[0,1]}\right] \\ &= E_P\left[\int_0^1 \left\{\sum_{i=1}^{p_n} \chi_{\left]t_{i,l}^{(n)},t_{i,r}^{(n)}\right]}(t) \sum_{m=1}^{\infty} \kappa_{i,m}^{(n)} \left\{B_m^N(\cdot,\theta_{i,r}^{(n)}) - B_m^N(\cdot,\theta_{i,l}^{(n)})\right\}\right\}^2 dt\right] \\ &= E_P\left[\int_0^1 \sum_{i=1}^{p_n} \chi_{\left]t_{i,l}^{(n)},t_{i,r}^{(n)}\right]}(t) \left\{\sum_{m=1}^{\infty} \kappa_{i,m}^{(n)} \left\{B_m^N(\cdot,\theta_{i,r}^{(n)}) - B_m^N(\cdot,\theta_{i,l}^{(n)})\right\}\right\}^2 dt\right] \\ &= \int_0^1 \sum_{i=1}^{p_n} \chi_{\left]t_{i,l}^{(n)},t_{i,r}^{(n)}\right]}(t) \left\{\sum_{m=1}^{\infty} \left\{\kappa_{i,m}^{(n)}\right\}^2 \left\{M_m(\theta_{i,r}^{(n)}) - M_m(\theta_{i,l}^{(n)})\right\}\right\} dt \\ &= \int_0^1 \left\|\underline{F}^{(n)}(t)\right\|^2_{L_2[\underline{b}]} dt.\end{aligned}$$

Since $\Phi_2^{(n)} - \Phi_2^{(q)}$ has the same generic aspect as $\Phi_2^{(n)}$,

$$E_{P_{B_N}^K}\left[\left\|\Phi_2^{(n)} - \Phi_2^{(q)}\right\|^2_{L_2[0,1]}\right] = \int_0^1 \left\|\underline{F}^{(n)} - \underline{F}^{(q)}\right\|^2_{L_2[\underline{b}]} dt,$$

and $\left\{\Phi_2^{(n)}, n \in \mathbb{N}\right\}$ is a Cauchy sequence in

$$L_2^{L_2[0,1]}\left(K, \mathcal{K}, P_{B_N}^K\right).$$

Let Φ_2 be its limit. By, when necessary, taking a subsequence, one may assume that convergence takes also place almost surely, with respect to $P^{\kappa}_{B_N}$. Finally

$$\frac{1}{2}E_P\left[\|N-\Phi_2(\underline{B}_N)\|^2_{L_2[0,1]}\right] \le E_P\left[\left\|N-\Phi_2^{(n)}(\underline{B}_N)\right\|^2_{L_2[0,1]}\right]$$
$$+ E_P\left[\left\|\Phi_2^{(n)}(\underline{B}_N)-\Phi_2(\underline{B}_N)\right\|^2_{L_2[0,1]}\right]$$
$$= \int_0^1 \left\|\underline{F}(t)-\underline{F}^{(n)}(t)\right\|^2_{L_2[\underline{b}]}dt$$
$$+ E_{P^{\kappa}_{B_N}}\left[\left\|\Phi_2^{(n)}-\Phi_2\right\|^2_{L_2}\right].$$

Thus, almost surely, with respect to P, in $L_2[0,1]$, $N[\omega]=\Phi_2(\underline{B}_N[\omega])$.

Let $\kappa \ge 1$ be fixed, but arbitrary, and

$$\Omega_\kappa = \left\{\omega\in\Omega : \|\underline{a}(\omega,\cdot)\|_{L_2[\underline{b}]}\le\kappa\right\}.$$

One has that

$$\overbrace{\Phi_2^{(n)}(\underline{X}[\omega])}^{\cdot}(t) =$$
$$= \sum_{i=1}^{p_n}\chi_{\left]t^{(n)}_{i,l},t^{(n)}_{i,r}\right]}(t)\sum_{m=1}^{\infty}\kappa^{(n)}_{i,m}\left\{X_m(\omega,\theta^{(n)}_{i,r})-X_m(\omega,\theta^{(n)}_{i,l})\right\}$$
$$= \sum_{i=1}^{p_n}\chi_{\left]t^{(n)}_{i,l},t^{(n)}_{i,r}\right]}(t)\sum_{m=1}^{\infty}\kappa^{(n)}_{i,m}\int_{\theta^{(n)}_{i,l}}^{\theta^{(n)}_{i,r}}a_m(\omega,\theta)M_m(d\theta)+\overbrace{\Phi_n^{(n)}(\underline{B}_N[\omega])}^{\cdot}(t)$$
$$= \left\langle\underline{F}^{(n)}(t),\underline{a}(\omega,\cdot)\right\rangle_{L_2[\underline{b}]}+\overbrace{\Phi_n^{(n)}(\underline{B}_N[\omega])}^{\cdot}(t).$$

Consequently,

$$I_{\underline{X}}\{\underline{F}\}(\omega,t)-\overbrace{\Phi_2^{(n)}(\underline{X}[\omega])}^{\cdot}(t) =$$
$$= \left\langle\underline{F}(t)-\underline{F}^{(n)}(t),\underline{a}(\omega,\cdot)\right\rangle_{L_2[\underline{b}]}+\left\{N(\omega,t)-\overbrace{\Phi_2^{(n)}(\underline{B}_N[\omega])}^{\cdot}(t)\right\},$$

so that

$$\left\|I_{\underline{X}}\{\underline{F}\}(\omega,\cdot)-\Phi_2^{(n)}(\underline{X}[\omega])\right\|^2_{L_2[0,1]} =$$
$$= \left\|\left\langle\underline{F}(\cdot)-\underline{F}^{(n)}(\cdot),\underline{a}(\omega,\cdot)\right\rangle_{L_2[\underline{b}]}+\left\{N[\omega]-\Phi_2^{(n)}(\underline{B}_N[\omega])\right\}\right\|^2_{L_2[0,1]}$$

$$\leq 2\left\|\underline{a}(\omega,\cdot)\right\|_{L_2[\underline{b}]}^2 \int_0^1 \left\|\underline{F}(t)-\underline{F}^{(n)}(t)\right\|_{L_2[\underline{b}]}^2 dt$$
$$+ 2\left\|N[\omega]-\Phi_2^{(n)}(\underline{B}_N[\omega])\right\|_{L_2[0,1]}^2,$$

and thus

$$E_P\left[\chi_{\Omega_\kappa}\left\|I_{\underline{X}}\{\underline{F}\}-\Phi_2^{(n)}(\underline{X})\right\|_{L_2[0,1]}^2\right] \leq 2\left(1+\kappa^2\right)\int_0^1 \left\|\underline{F}(t)-\underline{F}^{(n)}(t)\right\|_{L_2[\underline{b}]}^2 dt,$$

that is, for fixed, but arbitrary $\kappa \geq 1$, almost surely on Ω_κ, with respect to P, in $L_2[0,1]$,

$$\lim_n \Phi_2^{(n)}(\underline{X}) = I_{\underline{X}}\{\underline{F}\}.$$

Since, with respect to $P_{\underline{B}_N}^K$, $\Phi_2^{(n)}$ converges almost surely to Φ_2, and that, furthermore, by assumption, $P_{\underline{X}}^K \ll P_{\underline{B}_N}^K$, that convergence obtains also with respect to $P_{\underline{X}}^K$. Hence, almost surely, with respect to P,

$$\lim_n \Phi_2^{(n)}(\underline{X}[\omega]) = \Phi_2[\underline{X}(\omega]).$$

It follows that, almost surely on Ω_κ, with respect to P,

$$I_{\underline{X}}\{F\}[\omega] = \Phi_2(\underline{X}[\omega]).$$

□

Proposition 17.2.3 (Continuous Functions as Paths) *When N has continuous paths, there is an adapted map $\Phi_c : K \longrightarrow C[0,1]$ such that*

1. *almost everywhere, with respect to P, $N[\omega] = \Phi_c(\underline{B}_N[\omega])$;*
2. *when $P_{\underline{X}}^K \ll P_{\underline{B}_N}^K$, almost everywhere, with respect to P,*

$$I_{\underline{X}}\{\underline{F}\}[\omega] = \Phi_c(\underline{X}[\omega]).$$

Proof Let $J : C[0,1] \longrightarrow L_2[0,1]$ be the inclusion map which associates with a continuous function its equivalence class. It is a continuous injection since

$$\int_0^1 \{c_1(t)-c_2(t)\}^2 dt \leq \sup_{t\in[0,1]} \{c_1(t)-c_2(t)\}^2.$$

Since $C[0,1]$ is a Borel set, because of [208, p. 22], $\mathcal{R}[J]$ is a Borel set, and J an isomorphism of $C[0,1]$ onto $\mathcal{R}[J]$. Let

$$\Phi_c = J^{-1} \circ \left\{\chi_{\mathcal{R}[J]}(\Phi_2)\,\Phi_2\right\}.$$

Φ_c is adapted as a composition of adapted functions. Furthermore, because of the continuity assumption, $\chi_{\mathcal{R}[J]}(\Phi_2)\,\Phi_2 = \Phi_2$, so that the result is true. □

Definition 17.2.4 The maps defined above, that is, Φ, Φ_2, and Φ_c shall be called the Cramér-Hida maps.

Remark 17.2.5 As the proofs above show, the Cramér-Hida maps are linear, and "differentiate," in the sense that they send vectors with components of type $\int_0^t s_n(\cdot,\theta)M_n(d\theta)$ to inner products with terms of type

$$\int_0^t F_n(t,\theta)s_n(\cdot,\theta)M_n(d\theta).$$

17.3 Inverse Cramér-Hida Maps

As seen [Sect. 17.1], expressions for the likelihood do involve the inverses of the Cramér-Hida maps. Those are obtained below.

17.3.1 The Inverse for Square Integrable Paths

Let $S : L_2[\underline{b}] \longrightarrow L_2[0,1]$ be the operator obtained according to the following rule: $S[\underline{a}]$ is the equivalence class of

$$t \mapsto \langle \underline{F}(t), \underline{a}\rangle_{L_2[\underline{b}]}\,.$$

S is thus the composition of the unitary identification of $L_2[\,\underline{b}\,]$ with the RKHS $H(C_N,[0,1])$, followed by the imbedding of the latter into $L_2[0,1]$. Since

$$\int_0^1 \langle \underline{F}(t), \underline{a}\rangle^2_{L_2[\underline{b}]}\,dt \leq \|\underline{a}\|^2_{L_2[\underline{b}]} \int_0^1 \|\underline{F}(t)\|^2_{L_2[\underline{b}]}\,dt,$$

S is well defined, linear, and bounded.

When $f \in \mathcal{L}_2[0,1]$, for almost every $t \in [0,1]$, with respect to Lebesgue measure, the element $f(t)\underline{F}(t)$ belongs to $L_2[\underline{b}]$ since

$$\|f(t)\underline{F}(t)\|^2_{L_2[\underline{b}]} = f^2(t)\,\|\underline{F}(t)\|^2_{L_2[\underline{b}]} \leq \max_{t\in[0,1]} \|\underline{F}(t)\|^2_{L_2[\underline{b}]} f^2(t).$$

The map $t \mapsto f(t)\underline{F}(t)$ is Bochner integrable. Indeed,

$$\int_0^1 \|f(t)\underline{F}(t)\|_{L_2[\underline{b}]}\,dt = \int_0^1 |f(t)|\,\|\underline{F}(t)\|_{L_2[\underline{b}]}\,dt \\ \leq \|f\|_{L_2[0,1]} \left\{\int_0^1 \|\underline{F}(t)\|^2_{L_2[\underline{b}]}\,dt\right\}^{1/2}.$$

Consequently

$$\left\langle \int_0^1 \underline{F}(t)f(t)\,dt, \underline{a} \right\rangle_{L_2[\underline{b}]} = \int_0^1 \langle \underline{F}(t), \underline{a}\rangle_{L_2[\underline{b}]} f(t)\,dt = \langle S[\underline{a}], f\rangle_{L_2[0,1]}\,.$$

$S^\star[f]$ is thus the equivalence class of

$$f \mapsto \int_0^1 \underline{F}(t)\,\dot{f}(t)\,dt.$$

Furthermore

$$\overbrace{SS^\star[f]}^{\cdot}(t) = \overbrace{S[\int_0^1 \underline{F}(\theta)f(\theta)\,d\theta](t)}^{\cdot} \\ = \left\langle \underline{F}(t), \int_0^1 \underline{F}(\theta)f(\theta)\,d\theta \right\rangle_{L_2[\underline{b}]} \\ = \int_0^1 \langle \underline{F}(t), \underline{F}(\theta)\rangle_{L_2[\underline{b}]}\,\dot{f}(\theta)\,d\theta \\ = \overbrace{R_N[f]}^{\cdot}(t),$$

so that $SS^\star = R_N : L_2[0,1] \longrightarrow L_2[0,1]$, the covariance operator associated with the covariance $C_N(t_1,t_2) = \langle \underline{F}(t_1), \underline{F}(t_2)\rangle_{L_2[\underline{b}]}$ of the noise N. The polar decomposition yields that

$$S^\star = U\left((S^\star)^\star S^\star\right)^{1/2} = UR_N^{1/2},$$

where U is a partial isometry whose

- initial set is $\overline{\mathcal{R}[R_N^{1/2}]}$, with closure in $L_2[0,1]$,
- and final set, $\overline{R[S^\star]}$, with closure in $L_2[\underline{b}]$.

Since, H_F being the subspace generated linearly in $L_2[\ \underline{b}\ ]$ by the family $\{\underline{F}(t), t \in [0,1]\}$, $\mathcal{N}[S] = H_F^\perp$, and, since one works with a proper canonical decomposition,

$$\mathcal{N}[S] = \left\{ \underline{0}_{L_2[\underline{b}]} \right\}.$$

Since the support of a Gaussian measure on a Hilbert space is the closure of the square root of its covariance operator [142], and that one may assume, without restricting generality, full support,

$$\overline{\mathcal{R}[R_N^{1/2}]} = L_2[0,1],$$

and U is unitary.

Let $\{e_n, n \in \mathbb{N}\}$ and $\{\lambda_n, n \in \mathbb{N}\}$ be, respectively, the eigenvectors and the eigenvalues of R_N, a compact operator with finite trace. Let

$$J_n : L_2[M_n] \longrightarrow L_2[\underline{b}]$$

be the imbedding of $L_2[M_n]$ into $L_2[\underline{b}]$, and $\Pi_n = J_n J_n^\star$. Then $J_n^\star$ is the coordinate map, and Π_n, the projection onto $\mathcal{R}[J_n]$. Let $\underline{1} \in L_2[\underline{b}]$ be the class of the element whose components are the equivalence classes of the constant function equal to one. Then

$$\begin{aligned}
\left\langle S[\Pi_n[\chi_{[0,t]}\underline{1}]], e_p \right\rangle_{L_2[0,1]} &= \left\langle \Pi_n[\chi_{[0,t]}\underline{1}], S^\star[e_p] \right\rangle_{L_2[\underline{b}]} \\
&= \left\langle J_n J_n^\star[\chi_{[0,t]}\underline{1}], U R_N^{1/2}[e_p] \right\rangle_{L_2[\underline{b}]} \\
&= \lambda_p^{1/2} \left\langle J_n^\star\left[\chi_{[0,t]}\underline{1}\right], J_n^\star[U[e_p]] \right\rangle_{L_2[M_n]} \\
&= \lambda_p^{1/2} \int_0^t \overbrace{J_n^\star[U[e_p]]}^{\cdot}(\theta) M_n(d\theta).
\end{aligned}$$

The map $t \mapsto \int_0^t \overbrace{J_n^\star[U[e_p]]}^{\cdot}(\theta) M_n(d\theta)$ shall be denoted $\phi_{n,p}$. It is a continuous function, and, for $\{n,p\} \subseteq \mathbb{N}$ and $t_1 < t_2$ in $[0,1]$, fixed, but arbitrary, as $J_n^\star U$ is a contraction,

$$\left|\phi_{n,p}(t_2) - \phi_{n,p}(t_1)\right| \leq b_n(t_2) - b_n(t_1).$$

With respect to P_N^2, for $f \in L_2[0,1]$, and $p \in \mathbb{N}$, fixed, but arbitrary, the map $f \mapsto \langle f, e_p \rangle_{L_2[0,1]}$ is a normal random variable, with mean zero, and variance λ_p, so that

$$X_p = \lambda_p^{-\frac{1}{2}} \langle \cdot, e_p \rangle_{L_2[0,1]}$$

is $N(0,1)$. The family $\{X_p, p \in \mathbb{N}\}$ is made of independent random variables. Let then, for $f \in (L_2[0,1], \mathcal{B}(L_2[0,1]), P_N^2)$, fixed, but arbitrary,

$$V_{n,p}(f,t) = \lambda_p^{-1} \left\langle S\Pi_n[\chi_{[0,t]}\underline{1}], e_p\right\rangle_{L_2[0,1]} \left\langle f, e_p\right\rangle_{L_2[0,1]}.$$

From what has been acknowledged above,

$$V_{n,p}(f,t) = \phi_{n,p}(t)X_p(f).$$

One has thus that, for $n \in \mathbb{N}$, and $t \in [0,1]$, fixed, but arbitrary, the family $\{V_{n,p}, p \in \mathbb{N}\}$ is made of independent, normal random variables, with mean zero, and variance $\phi_{n,p}^2(t)$. Furthermore

$$\sum_{p=1}^{\infty} \phi_{n,p}^2(t) = \sum_{p=1}^{\infty} \left\langle \Pi_n[\chi_{[0,t]}\underline{1}], U[e_p]\right\rangle_{L_2[\underline{b}]}^2 = \left\|\Pi_n[\chi_{[0,t]}\underline{1}]\right\|_{L_2[\underline{b}]}^2 = b_n(t).$$

It follows that [199, p. 146], for $n \in \mathbb{N}$ and $t \in [0,1]$, fixed, but arbitrary,

$$\begin{aligned} m_n(f,t) &= \sum_{p=1}^{\infty} V_{n,p}(f,t) \\ &= \sum_{p=1}^{\infty} \phi_{n,p}(t)X_p(f) \\ &= \sum_{p=1}^{\infty} \lambda_p^{-1} \left\langle S[\Pi_n[\chi_{[0,t]}\underline{1}]], e_p\right\rangle_{L_2[0,1]} \left\langle \cdot, e_p\right\rangle_{L_2[0,1]} \end{aligned}$$

is well defined. It is a Gaussian process, as a limit of Gaussian random variables [241, p. 304]. It has, almost surely, with respect to P_N^2, continuous paths. Let indeed

$$m_{n,p}(f,t) = \sum_{i=1}^{p} \phi_{n,i}(t)X_i(f).$$

Then

$$m_n(f,t) - m_{n,p}(f,t) = \sum_{i=p+1}^{\infty} \phi_{n,i}(t)X_i(f).$$

The latter sum is a Gaussian random variable, with mean zero, and variance

$$\sum_{i=p+1}^{\infty} \phi_{n,i}^2(t) = \sum_{i=p+1}^{\infty} \left\langle \Pi_n[\chi_{[0,t]}\underline{1}], U[e_i]\right\rangle_{L_2[\underline{b}]}^2.$$

Let $\tilde{\Pi}_n$ be the projection in $L_2[\underline{b}]$ generated by $\{U[e_1], \ldots, U[e_n]\}$. One has that

$$\sum_{i=p+1}^{\infty} \langle \Pi_n[\chi_{[0,t]}\underline{1}], U[e_i]\rangle^2_{L_2[\underline{b}]} = \left\|\tilde{\Pi}_p^{\perp}\Pi_n[\chi_{[0,t]}\underline{1}]\right\|^2_{L_2[\underline{b}]}.$$

Let, when the latter norm object is different from zero,

$$X(f) = \left\{\sum_{i=p+1}^{\infty} \phi_{n,i}(t)X_i(f)\right\} / \left\{\left\|\tilde{\Pi}_p^{\perp}\Pi_n[\chi_{[0,t]}\underline{1}]\right\|_{L_2[\underline{b}]}\right\}.$$

X is, with respect to P_N^2, a standard normal random variable. Then

$$\sum_{i=p+1}^{\infty} \phi_{n,i}(t)X_i(f) = \left\|\tilde{\Pi}_p^{\perp}\Pi_n[\chi_{[0,t]}\underline{1}]\right\|_{L_2[\underline{b}]} X(f),$$

and

$$P_N^2\left(f \in L_2[0,1] : \sup_{t\in[0,1]} \left|m_n(f,t) - m_{n,p}(f,t)\right| > \epsilon\right) =$$

$$= P_N^2\left(f \in L_2[0,1] : \sup_{t\in[0,1]} \left|\left\|\tilde{\Pi}_p^{\perp}\Pi_n[\chi_{[0,t]}\underline{1}]\right\|_{L_2[\underline{b}]} X(f)\right| > \epsilon\right)$$

$$\leq \frac{1}{\epsilon^2} \sup_{t\in[0,1]} \left\|\tilde{\Pi}_p^{\perp}\Pi_n[\chi_{[0,t]}\underline{1}]\right\|^2_{L_2[\underline{b}]}.$$

Let $\xi_p(t) = \left\|\tilde{\Pi}_p^{\perp}\Pi_n[\chi_{[0,t]}\underline{1}]\right\|_{L_2[\underline{b}]}$. One has that

- ξ_p is continuous;
- $\xi_p \geq \xi_{p+1}$;
- $\lim_p \xi_p(t) = 0$.

Consequently, ξ_p converges uniformly to zero [230, p. 150], and m_n is the uniform limit in probability of the following sequence of continuous functions: $\{m_{n,p}, p \in \mathbb{N}\}$. It is thus, with respect to P_N^2, almost surely continuous, and thus separable. In the sequel one shall consider the following process:

$$\underline{m}(f,t) : L_2[0,1] \times [0,1] \longrightarrow \mathbb{R}^{\infty},$$

with components $m_n(f,t)$. It has values in l_2 as

$$E_{P_N^2}\left[\|\underline{m}(\cdot,t)\|^2_{l_2}\right] = \left\|\chi_{[0,t]}\underline{1}\right\|^2_{L_2[\underline{b}]}.$$

The process $\underline{m}$ is, as shall be seen, the inverse of Φ_2.

Fact 17.3.1 *For $\{n, p\} \subseteq \mathbb{N}$, and $t \in [0, 1]$, fixed, but arbitrary,*

$$E_P\left[B_n(\cdot, t)\left\langle N[\cdot], e_p\right\rangle_{L_2[0,1]}\right] = \left\langle S[\Pi_n[\chi_{[0,t]}\underline{1}]], e_p\right\rangle_{L_2[0,1]} = \lambda_p^{1/2}\phi_{n,p}(t).$$

Proof One has indeed that

$$\begin{aligned}
&E_P\left[B_n(\cdot, t)\left\langle N[\cdot], e_p\right\rangle_{L_2[0,1]}\right] = \\
&= E_P\left[B_n(\cdot, t)\int_0^1 N(\cdot, \theta)\dot{e}_p(\theta)\,d\theta\right] \\
&= \int_0^1 d\theta\,\dot{e}_p(\theta)\,E_P\left[B_n(\cdot, t)N(\cdot, \theta)\right] \\
&= \int_0^1 d\theta\,\dot{e}_p(\theta)\,E_P\left[B_n(\cdot, t)\int_0^1 F_n(\theta, x)B_n(\cdot, dx)\right] \\
&= \int_0^1 d\theta\,\dot{e}_p(\theta)\int_0^1 \chi_{[0,t]}(x)F_n(\theta, x)M_n(dx) \\
&= \int_0^1 d\theta\,\dot{e}_p(\theta)\left\langle J_n^\star[\chi_{[0,t]}\underline{1}], J_n^\star[\underline{F}(\theta)]\right\rangle_{L_2[\underline{b}]} \\
&= \int_0^1 d\theta\,\dot{e}_p(\theta)\overbrace{S[\Pi_n[\chi_{[0,t]}\underline{1}]]}^{\cdot}(\theta) \\
&= \left\langle S[\Pi_n[\chi_{[0,t]}\underline{1}]], e_p\right\rangle_{L_2[0,1]}.
\end{aligned}$$

□

Fact 17.3.2 $E_P\left[\{m_n(N[\cdot], t) - B_n(\cdot, t)\}^2\right] = 0.$

Proof Expanding the square, the latter expectation is

$$E_{P_N^2}\left[m_n^2(\cdot, t)\right] - 2E_P\left[m_n(N[\cdot], t)B_n(\cdot, t)\right] + E_P\left[B_n^2(\cdot, t)\right].$$

One has already seen that $E_{P_N^2}\left[m_n^2(\cdot, t)\right] = b_n(t) = E_P\left[B_n^2(\cdot, t)\right]$. But, because of the L_2 convergence in the definition of m_n, and (Fact) 17.3.1,

$$\begin{aligned}
&E_P\left[m_n(N[\cdot], t)B_n(\cdot, t)\right] = \\
&= E_P\left[\sum_{i=1}^{\infty}\lambda_i^{-1}\left\langle S[\Pi_n[\chi_{[0,t]}\underline{1}]], e_i\right\rangle_{L_2[0,1]}\left\langle N[\cdot], e_i\right\rangle_{L_2[0,1]}B_n(\cdot, t)\right]
\end{aligned}$$

$$= \sum_{i=1}^{\infty} \lambda_i^{-1} \left\langle S[\Pi_n[\chi_{[0,t]}\underline{1}]], e_i \right\rangle_{L_2[0,1]} E_P \left[\langle N[\cdot], e_i \rangle_{L_2[0,1]} B_n(\cdot, t) \right]$$

$$= \sum_{i=1}^{\infty} \lambda_i^{-1} \left\langle S[\Pi_n[\chi_{[0,t]}\underline{1}]], e_i \right\rangle_{L_2[0,1]}^2$$

$$= \sum_{i=1}^{\infty} \left\langle \Pi_n[\chi_{[0,t]}\underline{1}], U[e_i] \right\rangle_{L_2[M_n]}^2$$

$$= \left\| \Pi_n[\chi_{[0,t]}\underline{1}] \right\|_{L_2[\underline{b}]}^2$$

$$= b_n(t).$$

□

Fact 17.3.3 *As a consequence of (Fact) 17.3.2, one has that:*

1. *whenever* $n \in \mathbb{N}$, $0 < t_1 < \cdots < t_n \leq 1$, *and* $\{\theta_1, \ldots, \theta_n\} \subseteq \mathbb{R}$,

$$E_{P_N^2} \left[e^{i \sum_{j=1}^n \theta_j m_n(\cdot, t_j)} \right] = E_P \left[e^{i \sum_{j=1}^n \theta_j B_n(\cdot, t_j)} \right],$$

so that, with respect to P_N^2, m_n *has the same law as* B_n *with respect to* P;

2. *with respect to* P_N^2, m_n *has independent increments;*
3. *whenever* $\{t_1, t_2\} \subseteq [0, 1]$,

$$E_{P_N^2} [m_n(\cdot, t_1) m_n(\cdot, t_2)] = b_n(t_1 \wedge t_2),$$

$$E_{P_N^2} \left[\{m_n(\cdot, t_1) - m_n(\cdot, t_2)\}^2 \right] = b_n(t_1 \vee t_2) - b_n(t_1 \wedge t_2);$$

4. *with respect to* P_N^2, *for the filtrations it generates,* m_n *is a square integrable martingale;*

One may thus state the following "summary":

Proposition 17.3.4 *Almost surely, with respect to* P_N^2, $\underline{m}$ *has the properties of* $\underline{B}$, *and, almost surely, with respect to* $P_{B_N}^K$, $\underline{m} \circ \Phi_2$ *is the identity of* K.

Proof Let $\underline{B}_N$ be the Cramér-Hida process obtained from N, using the Cramér-Hida representation: it has components B_n^N. One has then that

$$P_{B_N}^K \left(\underline{k} \in K : d_K(\underline{m}[\Phi_2[\underline{k}]], \underline{k}) > \epsilon \right) =$$

$$= P\left(\omega \in \Omega : d_K(\underline{m}[\Phi_2[\underline{B}_N[\omega]]], \underline{B}_N[\omega]) > \epsilon \right)$$

$$= P\left(\omega \in \Omega : d_K\left(\underline{m}[N[\omega]], \underline{B}_N[\omega] \right) > \epsilon \right).$$

When K is taken to be Fréchet (as opposed to Banach), it suffices to prove that, for $n \in \mathbb{N}$, fixed, but arbitrary,

$$P\left(\omega \in \Omega : \sup_{t\in[0,1]} \left|m_n(N[\omega], t) - B_n^N(\omega, t)\right| > \epsilon\right) = 0.$$

But, since m_n is a separable, continuous martingale, one may use Doob's L_2 inequality to obtain that the latter probability is dominated by

$$\frac{1}{\epsilon^2} \sup_{t\in[0,1]} E_P\left[\left\{m_n(N[\cdot], t) - B_n^N(\cdot, t)\right\}^2\right] = 0.$$

When K is taken to be $C^{l_2}([0, 1])$, since

$$d_K\left(\underline{m}(N[\omega], t), \underline{B}_N(\omega, t)\right) = \sum_{n=1}^{\infty} \left\{m_n(N[\omega], t) - B_n^N(\omega, t)\right\}^2,$$

the distance d_K will be zero for the reason already stated. □

Fact 17.3.5 *$\underline{B}_N$ has, with respect to N, a regular conditional distribution concentrated at the point $\underline{m} \circ N$.*

Proof Let $K_0 \in \mathcal{K}$, and $L_0 \in \mathcal{B}(L_2[0, 1])$, be fixed, but arbitrary. Since, when

$$P\left(\underline{B}_N \in K_0 \mid N = f\right) = \chi_{K_0}\left(\underline{m}[f]\right),$$

one has that

$$\begin{aligned} P\left(\underline{B}_N \in K_0, N \in L_0\right) &= \int_{L_0} P\left(\underline{B}_N \in K_0 \mid N = f\right) P_N^2(df) \\ &= \int_{L_0} \chi_{K_0}\left(\underline{m}[f]\right) P_N^2(df) \\ &= \int_{N^{-1}(L_0)} \chi_{K_0}\left(\underline{m}[N]\right) dP, \end{aligned}$$

one must check that

$$P\left(\underline{B}_N \in K_0, N \in L_0\right) = P\left(\underline{m}[N] \in K_0, N \in L_0\right).$$

That equality will be true as soon as it obtains for generating sets. Let thus

- $0 \le t_1 < \cdots < t_n \le 1$,
- $B_i \in \mathcal{B}(l_2),\ i \in [1 : n]$,
- $K_0 = \left\{\underline{k} \in K : \mathcal{E}_{t_1}^K(\underline{k}) \in B_1, \ldots, \mathcal{E}_{t_n}^K(\underline{k}) \in B_n\right\}$,

- $\{f_1, \dots, f_p\} \subseteq L_2[0,1]$,
- $C_j \in \mathcal{B}(\mathbb{R}),\ j \in [1:p]$,
- $L_0 = \left\{f \in L_2[0,1] : \langle f, f_1\rangle_{L_2[0,1]} \in C_1, \dots, \langle f, f_p\rangle_{L_2[0,1]} \in C_p\right\}$,

be fixed, but arbitrary. The required probability equality is then obviously true as, in L_2, for t fixed, but arbitrary, $\underline{B}_N(\cdot, t) = \underline{m}(N[\cdot], t)$. □

Remark 17.3.6 As seen above, for $n \in \mathbb{N}$, fixed, but arbitrary, the family $\{\phi_{n,p}, p \in \mathbb{N}\}$ is equicontinuous. The family

$$\{\lambda_p^{1/2}\phi_{n,p}, (n,p) \in \mathbb{N} \times \mathbb{N}\}$$

is also equicontinuous. Indeed

$$\left|\lambda_p^{1/2}\phi_{n,p}\right| = \left|\langle \Pi_n[\chi_{[0,t]}\underline{1}]], S^\star[e_i - e_j]\rangle_{L_2[\underline{b}]}\right| \leq b_n(t)\left\|S^\star[e_i - e_j]\right\|_{L_2[\underline{b}_n]}.$$

The sequence of eigenvectors of R_N converges weakly to zero, and S is compact. Thus $\{S^\star[e_i], i \in \mathbb{N}\}$ converges to zero.

17.3.2 The Inverses for Real and Continuous Paths

Since, in the Φ_2 case, $\underline{m}$ has been identified as the support of the conditional law of $\underline{B}_N$ given N, one shall, in the Φ and Φ_c cases, compute that conditional law, and then check that it is the inverse.

Since $N = \Phi(\underline{B}_N)$, the conditional law of $\underline{B}_N$ given N should be a point mass. Indeed, for $K_0 \in \mathcal{K}$, and $L_0 \in \mathcal{B}(L_2[0,1])$, fixed, but arbitrary, since $\underline{B}_N$ and N generate the same σ-algebras,

$$\{\underline{B}_N \in K_0\} = \{\underline{B}_N \in \Phi^{-1}(\tilde{K}_0)\},$$

some $\tilde{K}_0 \in \mathcal{B}(L_2[0,1])$. One then has that

$$\begin{aligned} P(\underline{B}_N \in K_0, N \in L_0) &= E_P\left[\chi_{K_0}(\underline{B}_N)\chi_{L_0}(\Phi(\underline{B}_N))\right] \\ &= \int_{K_0} \chi_{L_0}(\Phi(\underline{k}))P^K_{B_N}(d\underline{k}) \\ &= \int_{\Phi^{-1}(\tilde{K}_0)} \chi_{L_0}(\Phi(\underline{k}))P^K_{B_N}(d\underline{k}) \\ &= \int_{\tilde{K}_0} \chi_{L_0}(f)P_N(df) \\ &= \int_{L_0} \chi_{\tilde{K}_0}(f)P_N(df). \end{aligned}$$

However, what one needs, for practical purposes, is an explicit form of the conditional law. That is the object of the next proposition. Since no particular structure is assumed, one must trek through finite dimensional distributions. Since those require quite a bit of accounting, one shall start with a number of preliminaries about notation and implicit assumptions.

Since N is continuous in L_2, the conditional law, given N, is the same as the conditional law given $\underline{N}$, a vector of values of N obtained at a dense set of times, and the latter shall be the limit, as the subsets increase, of finite subsets of components of $\underline{N}$. For the finite dimensional distributions of $\underline{B}_N$, one will have to proceed with a finite number of components, and times of the latter. Hence the following notation.

Preliminary 17.3.7 (Indices)

1. For $\{p, n_1, \ldots, n_p\} \subseteq \mathbb{N}$ and $n_i < n_{i+1}, i \in [1 : p-1]$, fixed, but arbitrary,

$$\mathbb{N}_{(p)} = \{n_1, \ldots, n_p\};$$

2. for n and $p[n]$ in $\mathbb{N}$, fixed, but arbitrary, and $i \in [1 : p[n]-1]$,

$$\text{in }]0,1],\ t_i^{(n)} < t_{i+1}^{(n)},\ \text{ and } T_n = \{t_{n,1}, \ldots, t_{n,p[n]}\};$$

3. for $n \in \mathbb{N}$, fixed, but arbitrary, $T_n \subset T_{n+1}$ (strict inclusion);
4. $T = \bigcup_{n \in \mathbb{N}} T_n$, T dense in $[0, 1]$;
5. T shall be ordered as follows: the first $p[1]$ elements are those of T_1, the next those of $T_2 \setminus T_1$, in the order in which they appear in T_2, the next those of $T_3 \setminus T_2$, in the order in which they appear in T_3, and so forth, until T be exhausted, and the result is $T = \{t_1, t_2, t_3, \ldots\}$; thus, when $T_m \subset T_n$, T_n will contain, in the order given, the following elements:

$$t_{m,1}, \ldots, t_{m,p[m]}, \tau_1, \ldots, \tau_q,$$

where $\{\tau_1, \ldots, \tau_q\} = T_n \setminus T_m$, and $q = p[n] - p[m]$;
6. for $n \in \mathbb{N}$, fixed, but arbitrary,

$$\text{in }]0,1],\ \theta_i < \theta_{i+1}, i \in [1 : n-1],\ \text{and } \Theta_n = \{\theta_1, \ldots, \theta_n\}.$$

Preliminary 17.3.8 (Intervals)

1. For $n \in \mathbb{N}$ and $\{\underline{a}, \underline{b}\} \subseteq \mathbb{R}^n$, fixed, but arbitrary, $\underline{a} < \underline{b}$ ($\underline{a} \leq \underline{b}$) means that $a_i < b_i$ ($a_i \leq b_i$) for $i \in [1 : n]$;
2. $]\underline{a}, \underline{b}] = \{\underline{x} \in \mathbb{R}^n : \underline{a} < \underline{x} \leq \underline{b}\}$;
3. for $p \in \mathbb{N}$ and, in $\mathbb{R}^n$, $\underline{a}_j < \underline{b}_j, j \in [1 : p]$, fixed, but arbitrary,

$$]\underline{a}, \underline{b}]_n^p = \prod_{j=1}^{p}]\underline{a}_j, \underline{b}_j];$$

4. given that, for example, $a_i^{(j)}$ is the j-th component of $\underline{a}_i$,

$$]\underline{a}, \underline{b}]_{n,i}^p = \prod_{j=1}^{p}]a_i^{(j)}, b_i^{(j)}].$$

Preliminary 17.3.9 (Evaluation Maps)

One shall need a number of maps that are evaluations, as follows.

1. $\mathcal{E}_{N(q)}^K : K \longrightarrow \{C([0,1])\}^q$,
   which retains, from the vector of countable continuous functions $\underline{k}$, its components $k_{n_1}, \ldots, k_{n_q}$;
2. $\mathcal{E}_{\Theta_p}^q : \{C([0,1])\}^q \longrightarrow \mathbb{R}^{pq}$,
   which retains, from a vector $\underline{c}_q \in \{C([0,1])\}^q$, the following values:

$$\begin{matrix} c_1(\theta_1) & \cdots & c_1(\theta_p) \\ \vdots & & \vdots \\ c_q(\theta_1) & \cdots & c_q(\theta_p) \end{matrix};$$

3. $\mathcal{E}_{\Theta_p}^K : K \longrightarrow s^p$,
   which retains, from the vector of countable continuous functions $\underline{k}$, p elements of s, $\{\underline{k}(\theta_1), \ldots, \underline{k}(\theta_p)\}$;
4. $\mathcal{E}_{N(q)}^p : s^p \longrightarrow \mathbb{R}^{pq}$,
   which retains, from p vectors in s, $\underline{x}_1, \ldots, \underline{x}_p$, the following values:

$$\begin{matrix} x_{n_1}^{(1)} & \cdots & x_{n_1}^{(p)} \\ \vdots & & \vdots \\ x_{n_q}^{(1)} & \cdots & x_{n_q}^{(p)} \end{matrix};$$

5. for $n \in \mathbb{N}$, $p \in \mathbb{N} \cup \{\infty\}$, $n < p$, fixed, but arbitrary, $\mathcal{E}_{p,n} : \mathbb{R}^p \longrightarrow \mathbb{R}^n$,
   which retains the first n components of $\underline{x} \in \mathbb{R}^p$;
6. for a set S, a set of indices $T_0 \subseteq [0,1]$, and a function

$$f : [0,1] \longrightarrow S,$$

   fixed, but arbitrary, $\mathcal{E}_{T_0}(f) \in S^{T_0}$ is the restriction to T_0 of f:

$$\mathcal{E}_{T_0}(f) = f^{T_0}.$$

   One has in particular that

$$\mathcal{E}_{\Theta_p}^q \circ \mathcal{E}_{N(q)}^K = \mathcal{E}_{N(q)}^p \circ \mathcal{E}_{\Theta_p}^K.$$

Preliminary 17.3.10 (Projections in s)

Let $\{\underline{e}_n, n \in \mathbb{N}\} \subseteq s$ be the standard basis of l_2. It is also a Schauder basis for the Fréchet $\mathbb{R}^\infty$. Let, for $\underline{x} \in s$, $\langle \underline{x}, \underline{e}_n \rangle_s$ denote the usual inner product of l_2 when $\underline{x} \in l_2$, and the coordinate functional otherwise. In the latter case it is a continuous linear functional as well. Thus, letting $\underline{x}_n = \sum_{i=1}^n \langle \underline{x}, \underline{e}_i \rangle_s \underline{e}_i$,

$$\lim_n \left\| \underline{x} - \underline{x}_n \right\|_s = 0,$$

and $\underline{x}_n$ may be used as the projection of $\underline{x}$ on the subspace spanned by $\underline{e}_1, \ldots, \underline{e}_n$. One shall write $\Pi^s_{N(q)}$ for the projection onto the subspace of s spanned by $\underline{e}_{n_1}, \ldots, \underline{e}_{n_q}$. The projection

$$(\Pi^s_{N(q)}, \Pi^s_{N(q)})$$

shall be denoted $\Pi^{2s}_{N(q)}$. Finally, when $N(q_1) \subseteq N(q_2)$,

$$\left\| \Pi^s_{N(q_1)} \right\|_s \leq \left\| \Pi^s_{N(q_2)} \right\|_s .$$

For $\lambda > 0$, let

- $f_\lambda : s^2 \longrightarrow \mathbb{R}_+$ be defined as

$$f_\lambda(\underline{u}, \underline{v}) = \|\underline{u} - \underline{v}\|_s^\lambda ,$$

- $\phi_\lambda : \mathbb{R}^{2q} \longrightarrow \mathbb{R}_+$, as, *mutatis mutandis*, f_λ on s^2.

Letting $x_i = \langle \underline{u}, \underline{e}_{n_i} \rangle_s$ and $y_i = \langle \underline{v}, \underline{e}_{n_i} \rangle_s$, one has that

$$f_\lambda(\Pi^s_{N(q)}[\underline{u}], \Pi^s_{N(q)}[\underline{v}]) = \phi_\lambda(\underline{x}, \underline{y}) = \phi_\lambda(\mathcal{E}^s_{N(q)}(\underline{u}), \mathcal{E}^s_{N(q)}(\underline{v})).$$

Preliminary 17.3.11 (Sets)

One must manipulate N at times T_n, and the coordinates B^N_i of $\underline{B}_N$, at times Θ_p:

1. for $\underline{B}_N$, $K_c = \left\{\mathcal{E}^q_{\Theta_p} \circ \mathcal{E}^K_{\mathbb{N}(q)}\right\}^{-1} \left(]\underline{a}, \underline{b}]^q_p\right)$,
   so that $\underline{B}_N^{-1}(K_c)$ is the set of ω's in Ω such that, for

   $$\theta_i \in \Theta_p,\ i \in [1:p],\ n_j \in \mathbb{N}_{(q)},\ j \in [1:q],$$

   fixed, but arbitrary,

   $$B^N_{n_j}(\omega, \theta_i) \in \left]a_i^{(j)}, b_i^{(j)}\right] ;$$

2. for N, $D_c = \mathcal{E}_{T_n}^{-1}\left(]\underline{\lambda}, \underline{\mu}]_1^n\right)$, so that $\underline{N}^{-1}(D_c)$ is the set of ω's in Ω such that, for

$$t_i^{(n)} \in T_n,\ i \in [1 : p[n]],$$

fixed, but arbitrary,

$$N(\omega, t_i^{(n)}) \in]\lambda_i, \mu_i].$$

Preliminary 17.3.12 (Vectors of Normal Random Variables)

1. For $n_i \in \mathbb{N}_{(q)}$, fixed, but arbitrary, $\underline{X}_i^p$ is the vector with components $B_{n_i}^N(\cdot, \theta_j)$, $\theta_j \in \Theta_p$;
2. $\underline{X}_{p,q}$ is the vector with components $\underline{X}_1^p, \dots, \underline{X}_q^p$;
3. $\underline{Y}_n$ is the vector with components $N(\cdot, t_{n,k})$, $t_{n,k} \in T_n$, $k \in [1 : p[n]]$;
4. one has the following relation: $\mathcal{E}_{\infty,n} \circ \mathcal{E}_T(N) = \mathcal{E}_{\infty,n}(\underline{N}) = \underline{Y}_n$.

Preliminary 17.3.13 (Covariance Matrices)

1. for $p \in \mathbb{N}$ and $\{i,j\} \in [1 : q]$ fixed, but arbitrary,

$$\Sigma_X(p,i,j) = E_P\left[\underline{X}_i^p(\underline{X}_j^p)^\star\right];$$

that matrix has dimensions (p,p); it is either the zero matrix ($i \neq j$), or a matrix with the following form ($i = j$):

$$\begin{pmatrix} b_{n_i}(\theta_1) & b_{n_i}(\theta_1) & b_{n_i}(\theta_1) & \cdots \\ b_{n_i}(\theta_1) & b_{n_i}(\theta_2) & b_{n_i}(\theta_2) & \cdots \\ b_{n_i}(\theta_1) & b_{n_i}(\theta_2) & b_{n_i}(\theta_3) & \cdots \\ \vdots & \vdots & \vdots & \vdots \end{pmatrix},$$

and, since, for $\{\theta_k, \theta_l\} \subseteq \Theta_p$, fixed, but arbitrary,

$$\left\langle \Pi_{n_i}[\chi_{[0,\theta_k]}\underline{1}], \Pi_{n_j}[\chi_{[0,\theta_l]}\underline{1}]\right\rangle_{L_2[\underline{b}]} = \delta_{n_i,n_j} b_{n_i}(\theta_k \wedge \theta_l),$$

the matrix $\Sigma_X(p,i,j)$ is more usefully described as having entries given by the left-hand side of the latter set of equalities, that is, in terms of projections;
2. the following matrix has dimensions (pq,pq):

$$\Sigma_X^{p,q} = \begin{pmatrix} \Sigma_X(p,1,1) & \Sigma_X(p,1,2) & \Sigma_X(p,1,3) & \cdots \\ \Sigma_X(p,2,1) & \Sigma_X(p,2,2) & \Sigma_X(p,2,3) & \cdots \\ \Sigma_X(p,3,1) & \Sigma_X(p,3,2) & \Sigma_X(p,3,3) & \cdots \\ \vdots & \vdots & \vdots & \vdots \end{pmatrix},$$

and is thus a diagonal matrix with matrices in the diagonal as given in item 1; it is the covariance matrix of the vector $\underline{X}_{p,q}$;

3. the matrix

$$\Sigma_{X,Y}(n,p,i) = E_P\left[\underline{X}_i^p \underline{Y}_n^\star\right]$$

has dimensions $(p, p[n])$; its entry, in row k and column l, is

$$\begin{aligned} E_P\left[B_{n_i}^N(\cdot,\theta_k)N(\cdot,t_{n,l})\right] &= \\ &= \int_0^{\theta_k} F_{n_i}(t_{n,l},x)M_{n_i}(dx) \\ &= \left\langle \Pi_{n_i}[\chi_{[0,\theta_k]}\underline{1}], \underline{F}(t_{n,l})\right\rangle_{L_2[\underline{b}]}; \end{aligned}$$

4. the following matrix has dimensions $(pq, p[n])$:

$$\Sigma_{X,Y}^{n,p,q} = \begin{pmatrix} \Sigma_{X,Y}(n,p,1) \\ \Sigma_{X,Y}(n,p,2) \\ \Sigma_{X,Y}(n,p,3) \\ \vdots \end{pmatrix};$$

5. the matrix

$$\Sigma_Y^n = E_P\left[\underline{Y}_n \underline{Y}_n^\star\right]$$

has dimensions $(p[n], p[n])$, and entries of the following form, in row k and column l:

$$\begin{aligned} E_P\left[N(\cdot,t_{n,k})N(\cdot,t_{n,l})\right] &= C_N(t_{n,k},t_{n,l}) \\ &= \langle \underline{F}(t_{n,k}), \underline{F}(t_{n,l})\rangle_{L_2[\underline{b}]}; \end{aligned}$$

it shall be assumed that there is no interval of positive length over which N has paths that are almost surely constant, and thus that T may be chosen so that Σ_Y^n is always invertible: indeed the following relation:

$$E_P\left[\{N(\cdot,u) - N(\cdot,v)\}^2\right] = \|\underline{F}(u) - \underline{F}(v)\|_{L_2[\underline{b}]}^2$$

would imply, in case of constancy, that $\{\underline{F}(t), t \in [0,1]\}$ is not total;

6. the matrix

$$\Sigma_Z^{n,p,q} = \left(\begin{array}{c|c} \Sigma_X^{p,q} & \Sigma_{X,Y}^{n,p,q} \\ \hline \left(\Sigma_{X,Y}^{n,p,q}\right)^\star = \Sigma_{Y,X}^{n,p,q} & \Sigma_Y^n \end{array}\right)$$

has dimensions $(pq+p[n], pq+p[n])$, and is the covariance matrix of the Gaussian vector

$$\underline{Z}_{n,p,q} = \begin{bmatrix} \underline{X}_{p,q} \\ \underline{Y}_n \end{bmatrix}.$$

Preliminary 17.3.14 (Measures and Densities)

1. The Gaussian measure on the Borel sets of $\mathbb{R}^n$, with mean $\underline{m}$ and covariance C, shall be denoted, for Borel B, fixed, but arbitrary,

$$\Gamma_n(B \mid \underline{m}, C);$$

2. the vector $\underline{Z}_{n,p,q}$ has, as law, $\Gamma_{pq+p[n]}\left(\cdot \mid \underline{0}_{pq+p[n]}, \Sigma_Z^{n,p,q}\right)$;
3. the conditional law of $\underline{X}_{p,q}$ when $\underline{Y}_n = \underline{y}_{p[n]}$ is Gaussian with

$$\text{mean} \quad : m_{\left[\underline{X}_{p,q} \mid \underline{Y}_n = \underline{y}_{p[n]}\right]} = \Sigma_{X,Y}^{n,p,q} (\Sigma_Y^n)^{-1} \left(\underline{y}_{p[n]}\right),$$

$$\text{variance} : \Sigma_{\left[\underline{X}_{p,q} \mid \underline{Y}_n = \underline{y}_{p[n]}\right]} = \Sigma_X^{p,q} - \Sigma_{X,Y}^{n,p,q} (\Sigma_Y^n)^{-1} \Sigma_{Y,X}^{n,p,q}.$$

Proposition 17.3.15 *The conditional law $P_{\underline{B}_N|N}(f, \cdot)$ of $\underline{B}_N$, given that $N = f$, is provided by the following relation: almost surely, with respect to P_N, for $K_0 \in \mathcal{K}$, fixed, but arbitrary,*

$$P_{\underline{B}_N|N}(f, K_0) = \chi_{K_0}(\underline{m}(f, \cdot)),$$

where

(i) *$\underline{m}$ is a continuous, Gaussian process, defined on $(\mathbb{R}^{[0,1]}, \mathcal{C}(\mathbb{R}^{[0,1]}), P_N)$, whose law is the same as that of $\underline{B}_N$, with respect to P;*
(ii) *$\underline{m}$ is the weak limit of a sequence $\{\underline{m}_n, n \in \mathbb{N}\}$ of continuous, Gaussian processes, defined on $(\mathbb{R}^{[0,1]}, \mathcal{C}(\mathbb{R}^{[0,1]}, P_N)$, whose components have the following form:*

$$m_{n,p}(f, t) = \left\langle \mathcal{E}_{T_n}\left(L_F \Pi_p[\chi_{[0,t]}\underline{1}]\right), \Sigma_{N,n}^{-1} \mathcal{E}_{\infty,n} \mathcal{E}_T(f) \right\rangle_{\mathbb{R}^n},$$

where Π_n is as in Sect. 17.3.1, and $L_F : L_2[\underline{b}] \longrightarrow H(C_N, [0, 1])$, the map whose range is the RKHS associated with N.

Proof Let $\underline{N} = \mathcal{E}_T(N)$. Since N is continuous in L_2, N and $\underline{N}$ generate the same σ-algebra, so that, with

$$\underline{N}^{-1}(C_0) = N^{-1}\left(\mathcal{E}_T^{-1}(C_0)\right),$$

it is sufficient to consider the following expression:

$$
\begin{aligned}
P(\underline{B}_N \in K_0, N \in \mathcal{E}_T^{-1}(C_0)) &= P(\underline{B}_N \in K_0, \underline{N} \in C_0) \\
&= \int_{C_0} P_{\underline{B}_N|\underline{N}}(\underline{x}, K_0) P_{\underline{N}}(d\underline{x}) \\
&= \int_{\mathcal{E}_T^{-1}(C_0)} P_{\underline{B}_N|\underline{N}}(\mathcal{E}_T(f), K_0) P_N(df).
\end{aligned}
$$

One starts by computing the following probability:

$$
P\left(B_{n_i}(\cdot, \theta_j) \in]a_i^{(j)}, b_i^{(j)}], n_i \in \mathbb{N}_{(q)}, \theta_j \in \Theta_p, N(\cdot, t_{n,k}) \in]\lambda_k, \mu_k], t_{n,k} \in T_n\right),
$$

which is equal to the following expression [(Preliminary) 17.3.11]:

$$
P\left(\underline{B}_N \in K_c, \underline{N} \in D_c\right) = P\left(\underline{X}_{p,q} \in]\underline{a}, \underline{b}]_p^q, \underline{Y}_n \in]\underline{\lambda}, \underline{\mu}]_1^{p[n]}\right).
$$

Now, since $\underline{Y}_n = \mathcal{E}_{\infty,p[n]}(\underline{N})$ [(Preliminary) 17.3.12],

$$
\begin{aligned}
&P\left(\underline{X}_{p,q} \in]\underline{a}, \underline{b}]_p^q, \underline{Y}_n \in]\underline{\lambda}, \underline{\mu}]_1^{p[n]}\right) = \\
&= \int_{]\underline{\lambda}, \underline{\mu}]_1^{p[n]}} \Gamma_{p[n]}(d\underline{y}_{p[n]} \mid \underline{0}_{p[n]}, \Sigma_Y^n) \times \\
&\quad \times \Gamma_{pq}\left( ]\underline{a}, \underline{b}]_p^q \mid \underline{m}_{\left[\underline{X}_{p,q}|\underline{Y}_n = \underline{y}_{p[n]}\right]}, \Sigma_{\left[\underline{X}_{p,q}|\underline{Y}_n = \underline{y}_{p[n]}\right]}\right) \\
&= \int_{D_c} P_{\underline{N}}(d\underline{x}) \times \\
&\quad \times \Gamma_{pq}\left( ]\underline{a}, \underline{b}]_p^q \mid \underline{m}_{[\underline{X}_{p,q}|\underline{Y}_n = \mathcal{E}_{\infty,p[n]}(\underline{x})]}, \Sigma_{[\underline{X}_{p,q}|\underline{Y}_n = \mathcal{E}_{\infty,p[n]}(\underline{x})]}\right).
\end{aligned}
$$

The proof consists in studying the behavior of

$$
\Gamma_{pq}[\underline{x}](n) = \Gamma_{pq}\left( ]\underline{a}, \underline{b}]_p^q \mid \underline{m}_{[\underline{X}_{p,q}|\underline{Y}_n = \mathcal{E}_{\infty,p[n]}(\underline{x})]}, \Sigma_{[\underline{X}_{p,q}|\underline{Y}_n = \mathcal{E}_{\infty,p[n]}(\underline{x})]}\right),
$$

as the number $p[n]$ of evaluation points for N increases indefinitely.

Let H_F be the subspace of $L_2[\,\underline{b}\,]$ generated by $\{\underline{F}(t), t \in [0,1]\}$ (one thus does not yet take into account the assumption that the representation of N is proper canonical), and Π_F be the associated projection. The matrix $\Sigma_{F,q}(\theta_i, \theta_j)$, of dimensions (q, q), shall have entries, in position (k, l), given by the following expression:

$$
\left\langle \Pi_F^{\perp}[\Pi_{n_k}[\chi_{[0,\theta_i]}\underline{1}]], \Pi_{n_l}[\chi_{[0,\theta_j]}\underline{1}] \right\rangle_{L_2[\,\underline{b}\,]}.
$$

Thus, when the representation is proper canonical, $\Sigma_{F,q}(\theta_i, \theta_j)$ is a matrix with all entries equal to zero.

$\Sigma_F^{(pq)}$ shall be the $p \times p$ matrix of the $q \times q$ matrices $\Sigma_{F,q}(\theta_i, \theta_j)$: it has thus dimension $pq \times pq$. Thus, for example, when $p = 2$,

$$\Sigma_F^{(2q)} = \begin{bmatrix} \Sigma_{F,q}(\theta_1, \theta_1) & \Sigma_{F,q}(\theta_1, \theta_2) \\ \Sigma_{F,q}(\theta_2, \theta_1) & \Sigma_{F,q}(\theta_2, \theta_2) \end{bmatrix}.$$

If then, I_p being the (p, p) identity matrix, $M = \left[I_q \mid -I_q\right]$,

$$M\Sigma_F^{(2q)}M^\star = \Sigma_{F,q}(\theta_1, \theta_1) - \Sigma_{F,q}(\theta_1, \theta_2) - \Sigma_{F,q}(\theta_2, \theta_1) + \Sigma_{F,q}(\theta_2, \theta_2),$$

whose term, in position (k, l), is

$$\left\langle \Pi_F^\perp \Pi_{n_k}[\chi_{]\theta_1,\theta_2]}\underline{1}], \Pi_{n_l}[\chi_{]\theta_1,\theta_2]}\underline{1}]\right\rangle_{L_2[\underline{b}]}.$$

Let, for $\underline{x} \in \mathbb{R}^\infty$, and $t \in [0, 1]$, fixed, but arbitrary,

$$m_i^{(n)}(\underline{x}, t) = \left\langle \mathcal{E}_{T_n} \circ L_F \circ \Pi_i[\chi_{[0,t]}\underline{1}], (\Sigma_Y^n)^{-1}\mathcal{E}_{\infty,p[n]}(\underline{x})\right\rangle_{\mathbb{R}^{p[n]}}$$

be the components of $\underline{m}_n$. Suppose that one thus defines a process with paths in K. Then

$$P_{\underline{N}}\left(\mathcal{E}_{\theta_1} \circ \mathcal{E}_{n_1}^K\left(\underline{m}_n\right) \in B_1, \ldots, \mathcal{E}_{\theta_p} \circ \mathcal{E}_{n_p}^K\left(\underline{m}_n\right) \in B_p\right) =$$

$$= P_{\underline{N}}\left(m_{n_1}^{(n)}(\cdot, \theta_1) \in B_1, \ldots, m_{n_p}^{(n)}(\cdot, \theta_p) \in B_p\right)$$

is a finite dimensional distribution of $\underline{m}_n$, for the cylinder sets of $\mathbb{R}^\infty$ and $P_{\underline{N}}$. Were one to show that the sequence of $\underline{m}_n$'s is tight, and that the finite dimensional distributions converge weakly, one would then have established the existence of a process $\underline{m}$ which is the weak limit of the $\underline{m}_n$'s [38, p. 35].

Step 1: With respect to $P_{\underline{N}}$, for almost every $\underline{x} \in \mathbb{R}^\infty$, as $n \in \mathbb{N}$ increases indefinitely, $\Gamma_{pq}\boxed{x}(n)$ converges weakly to

$$\Gamma_{pq}(\cdot \mid \underline{m}_{p,q}\boxed{x}, \Sigma_F^{(pq)}),$$

where $\underline{m}_{p,q}\boxed{x}$ is as determined below.

Since one only deals with Gaussian measures on Euclidean spaces, it suffices to establish the convergence of means and covariance matrices. Now, the entries of

$$\Sigma_{X,Y}^{n,p,q}(\Sigma_Y^n)^{-1}\Sigma_{Y,X}^{n,p,q}$$

have the following form:

$$\underline{u}_{p[n]}^{\star}(\Sigma_Y^n)^{-1}\underline{v}_{p[n]} = \left\langle (\Sigma_Y^n)^{-1}\underline{v}_{p[n]}, \underline{u}_{p[n]} \right\rangle_{\mathbb{R}^{p[n]}},$$

where $\underline{u}_{p[n]}$ and $\underline{v}_{p[n]}$ are columns of $\Sigma_{Y,X}^{n,p,q}$. The latter are the columns of the successive

$$E_P[\underline{Y}_n(\underline{X}_i^p)^{\star}]\text{'s},$$

which, as seen [(Preliminary) 17.3.13], have the following form (for the k-th column):

$$\mathcal{E}_{T_n} \circ L_F \circ \Pi_{n_i}\left[\chi_{[0,\theta_k]}\underline{1}\right].$$

The entries of $\Sigma_{X,Y}^{n,p,q}(\Sigma_Y^n)^{-1}\Sigma_{Y,X}^{n,p,q}$ are thus of the following form:

$$\left\langle (\Sigma_Y^n)^{-1}\left\{\mathcal{E}_{T_n} \circ L_F \circ \Pi_{n_i}\left[\chi_{[0,\theta_k]}\underline{1}\right]\right\}, \mathcal{E}_{T_n} \circ L_F \circ \Pi_{n_j}\left[\chi_{[0,\theta_l]}\underline{1}\right]\right\rangle_{\mathbb{R}^{p[n]}}. \qquad (\star)$$

Let $H_n(C_N, [0,1])$ be the subspace of $H(C_N, [0,1])$ generated by the functions

$$\left\{t \mapsto C_N(\cdot, t_i^{(n)}), t_i^{(n)} \in T_n, i \in [1:p[n]]\right\},$$

and

$$\Pi_{H_n(C_N,[0,1])}$$

be the associated projection. Since the range of L_F is the RKHS of N, the latter inner product ($\star$) is [(Proposition) 1.6.22] the explicit expression for

$$\Big\langle \Pi_{H_n(C_N,[0,1])} \circ L_F \circ \Pi_{n_i}\left[\chi_{[0,\theta_k]}\underline{1}\right], \\ \Pi_{H_n(C_N,[0,1])} \circ L_F \circ \Pi_{n_j}\left[\chi_{[0,\theta_l]}\underline{1}\right]\Big\rangle_{H(C_N,[0,1])}.$$

Since N is continuous in L_2, and that T is dense in $[0,1]$, the strong limit in $H(C_N, [0,1])$ of the sequence

$$\left\{\Pi_{H_n(C_N,[0,1])}, n \in \mathbb{N}\right\}$$

is the identity of $H(C_N, [0,1])$, and thus

$$\lim_n \Sigma_{X,Y}^{n,p,q}(\Sigma_Y^n)^{-1}\Sigma_{Y,X}^{n,p,q}$$

has entries equal to

$$\left\langle L_F \circ \Pi_{n_i}\left[\chi_{[0,\theta_k]}\underline{1}\right], L_F \circ \Pi_{n_j}\left[\chi_{[0,\theta_l]}\underline{1}\right]\right\rangle_{H(C_N,[0,1])} =$$
$$= \left\langle \Pi_F \circ \Pi_{n_i}\left[\chi_{[0,\theta_k]}\underline{1}\right], \Pi_F \circ \Pi_{n_j}\left[\chi_{[0,\theta_l]}\underline{1}\right]\right\rangle_{L_2[\underline{b}]}.$$

Since $\Sigma_X^{p,q}$ has, as seen [(Preliminary) 17.3.13], entries of the following form:

$$\left\langle \Pi_{n_i}\left[\chi_{[0,\theta_k]}\underline{1}\right], \Pi_{n_j}\left[\chi_{[0,\theta_l]}\underline{1}\right]\right\rangle_{L_2[\underline{b}]},$$

it ensues that

$$\lim_n \left\{\Sigma_X^{p,q} - \Sigma_{X,Y}^{n,p,q}(\Sigma_Y^n)^{-1}\Sigma_{Y,X}^{n,p,q}\right\}$$

has entries of the following form:

$$\left\langle \Pi_F^{\perp}\left[\Pi_{n_i}\left[\chi_{[0,\theta_k]}\underline{1}\right]\right], \Pi_F^{\perp}\left[\Pi_{n_j}\left[\chi_{[0,\theta_l]}\underline{1}\right]\right]\right\rangle_{L_2[\underline{b}]},$$

which are those of $\Sigma_F^{(pq)}$.

Regarding the mean, one has that

$$\Sigma_{X,Y}(n,p,i)(\Sigma_Y^n)^{-1}\mathcal{E}_{\infty,p[n]}(\underline{x})$$

is a vector in $\mathbb{R}^p$ whose k-th entry has the following form:

$$\underline{v}^{\star}_{p[n]}(\Sigma_Y^n)^{-1}\mathcal{E}_{\infty,p[n]}(\underline{x}) = \left\langle (\Sigma_Y^n)^{-1}\mathcal{E}_{\infty,p[n]}(\underline{x}), \underline{v}_{p[n]}\right\rangle_{\mathbb{R}^{p[n]}},$$

where $\underline{v}_{p[n]}$ is the k-th column of $\Sigma^{\star}_{X,Y}(n,p,i) = \Sigma_{Y,X}(n,p,i)$, and is, as already seen [(Preliminary) 17.3.13],

$$\mathcal{E}_{T_n} \circ L_F \circ \Pi_{n_i}\left[\chi_{[0,\theta_k]}\underline{1}\right].$$

Consequently, the rows of the conditional mean have the following form:

$$m^{(n)}_{n_i,k}[\underline{x}] = \left\langle (\Sigma_Y^n)^{-1}\mathcal{E}_{\infty,p[n]}(\underline{x}), \mathcal{E}_{T_n} \circ L_F \circ \Pi_{n_i}\left[\chi_{[0,\theta_k]}\underline{1}\right]\right\rangle_{\mathbb{R}^{p[n]}}.$$

Now

$$m^{(n)}_{n_i,k}[\underline{N}] = \left\langle \underline{Y}_n, (\Sigma_Y^n)^{-1} \circ \mathcal{E}_{T_n} \circ L_F \circ \Pi_{n_i}\left[\chi_{[0,\theta_k]}\underline{1}\right]\right\rangle_{\mathbb{R}^{p[n]}},$$

so that [$(\star)$ again]

$$E_{P_{\underline{N}}}\left[\left\{m^{(n)}_{n_i,k}\right\}^2\right] = \left\|\Pi_{H_n(C_N,[0,1])} \circ L_F \circ \Pi_{n_i}[\chi_{[0,\theta_k]}\underline{1}]\right\|^2_{H(C_N,[0,1])}.$$

Furthermore, given $n < p$ in $\mathbb{N}$, and $\underline{v}_n \in \mathbb{R}^n$, fixed, but arbitrary, let $\underline{v}_{n,0}$ be the vector of $\mathbb{R}^p$ obtained by padding $\underline{v}_n$ with $p - n$ zeroes. Then, the transpose of $\mathcal{E}_{p,n}$ is obtained as

$$\mathcal{E}^\star_{p,n}(\underline{v}_n) = \underline{v}_{n,0}.$$

Thus, for $n < p$ in $\mathbb{N}$, using $(\star)$,

$$\begin{aligned}
&E_P\left[m^{(n)}_{n_i,k}(\underline{N})\, m^{(p)}_{n_i,k}(\underline{N})\right] = \\
&= E_P\left[\left\langle \underline{Y}_p, \mathcal{E}^\star_{p[p],p[n]}(\Sigma^n_Y)^{-1} \circ \mathcal{E}_{T_n} \circ L_F \circ \Pi_{n_i}\left[\chi_{[0,\theta_k]}\underline{1}\right]\right\rangle_{\mathbb{R}^{p[n]}} \times \right.\\
&\qquad \left. \times \left\langle \underline{Y}_p, (\Sigma^p_Y)^{-1} \circ \mathcal{E}_{T_p} \circ L_F \circ \Pi_{n_i}\left[\chi_{[0,\theta_k]}\underline{1}\right]\right\rangle_{\mathbb{R}^{p[p]}}\right] \\
&= \left\langle (\Sigma^n_Y)^{-1} \circ \mathcal{E}_{T_n} \circ L_F \circ \Pi_{n_i}\left[\chi_{[0,\theta_k]}\underline{1}\right], \right.\\
&\qquad \left. \mathcal{E}_{p[p],p[n]} \circ \mathcal{E}_{T_p} \circ L_F \circ \Pi_{n_i}\left[\chi_{[0,\theta_k]}\underline{1}\right]\right\rangle_{\mathbb{R}^{p[n]}} \\
&= \left\langle (\Sigma^n_Y)^{-1} \circ \mathcal{E}_{T_n} \circ L_F \circ \Pi_{n_i}\left[\chi_{[0,\theta_k]}\underline{1}\right], \mathcal{E}_{T_n} \circ L_F \circ \Pi_{n_i}\left[\chi_{[0,\theta_k]}\underline{1}\right]\right\rangle_{\mathbb{R}^{p[n]}} \\
&= \left\| \Pi_{H_n(C_N,[0,1])} \circ L_F \circ \Pi_{n_i}[\chi_{[0,\theta_k]}\underline{1}]\right\|^2_{H(C_N,[0,1])}.
\end{aligned}$$

Consequently

$$\begin{aligned}
&E_{P_{\underline{N}}}\left[\left\{m^{(p)}_{n_i,k} - m^{(n)}_{n_i,k}\right\}^2\right] = \\
&\qquad = \left\| \left\{\Pi_{H_p(C_N,[0,1])} - \Pi_{H_n(C_N,[0,1])}\right\}\left[L_F \circ \Pi_{n_i}[\chi_{[0,\theta_k]}\underline{1}]\right]\right\|^2_{H(C_N,[0,1])},
\end{aligned}$$

so that

$$\left\{m^{(n)}_{n_i,k}, n \in \mathbb{N}\right\}$$

is a Cauchy sequence in L_2 for $P_{\underline{N}}$. Choosing, when necessary, a subsequence, one may assume that convergence is almost sure. The limit shall be $m_{n_i,k}$, and $\underline{m}_{p,q}$ is the vector with those components. Since $\left\{\Gamma_{pq}[\underline{x}](n), n \in \mathbb{N}\right\}$ is, with respect to $P_{\underline{N}}$, for almost every $\underline{x}$, weakly convergent, the corresponding characteristic functions are almost surely convergent, and, by dominated convergence, the second requirement, reviewed in the lines preceding the statement of step 1, for $\left\{\underline{m}_n, n \in \mathbb{N}\right\}$ to be tight, is met.

Step 2: *A candidate for* $P_{\underline{B}_N|\underline{N}}$.

Let $m^{(n)}_i : \mathbb{R}^\infty \times [0,1] \longrightarrow \mathbb{R}$ be defined using the following relation:

$$m^{(n)}_i(\underline{x}, t) = \left\langle \mathcal{E}_{T_n} \circ L_F \circ \Pi_i[\chi_{[0,t]}\underline{1}], (\Sigma^n_Y)^{-1}\mathcal{E}_{\infty,p[n]}(\underline{x})\right\rangle_{\mathbb{R}^{p[n]}}. \qquad (\star\star)$$

Then

$$E_{P_{\underline{N}}}\left[\left\{m_i^{(n)}(\cdot,t)\right\}^2\right] =$$
$$= \left\langle (\Sigma_Y^n)^{-1}\mathcal{E}_{T_n} \circ L_F \circ \Pi_i[\chi_{[0,t]}\underline{1}], \mathcal{E}_{T_n} \circ L_F \circ \Pi_i[\chi_{[0,t]}\underline{1}]\right\rangle_{\mathbb{R}^{p[n]}}$$
$$= \left\| \Pi_{H_n(C_N,[0,1])}\left[L_F \circ \Pi_i[\chi_{[0,t]}\underline{1}]\right]\right\|^2_{H(C_N,[0,1]}$$
$$\leq \left\|\Pi_i[\chi_{[0,t]}\underline{1}]\right\|^2_{L_2[\underline{b}]}$$
$$= b_i(t).$$

$\underline{m}_n$ shall be the process, with values in l_2, whose components are the $m_i^{(n)}$'s.

To have that $\underline{m}_n$ is, for $P_{\underline{N}}$, a Gaussian process, one must check that, for $q \in \mathbb{N}$, $\{\underline{\alpha}_1, \ldots, \underline{\alpha}_q\} \subseteq l_2$, and $\{t_1, \ldots, t_q\} \subseteq \mathbb{R}$, fixed, but arbitrary,

$$V = \sum_{i=1}^{q} \langle \underline{\alpha}_i, \underline{m}_n(\underline{N}, t_i)\rangle_{l_2}$$

is Gaussian [65, p. 83]. But

$$\langle \underline{\alpha}, \underline{m}_n(\underline{N}, t)\rangle_{l_2} =$$
$$= \sum_{k=1}^{\infty} \alpha_k m_k^{(n)}(\underline{N}, t)$$
$$= \sum_{k=1}^{\infty} \alpha_k \left\langle \mathcal{E}_{T_n} \circ L_F \circ \Pi_k[\chi_{[0,t]}\underline{1}], (\Sigma_Y^n)^{-1}\mathcal{E}_{\infty,p[n]}(\underline{N})\right\rangle_{\mathbb{R}^{p[n]}}$$
$$= \left\langle \mathcal{E}_{T_n} \circ L_F\left[\left(\sum_{k=1}^{\infty} \alpha_k \Pi_k\right)[\chi_{[0,t]}\underline{1}]\right], (\Sigma_Y^n)^{-1}\mathcal{E}_{\infty,p[n]}(\underline{N})\right\rangle_{\mathbb{R}^{p[n]}}.$$

Let $\Pi_{\underline{\alpha}} = \sum_{k=1}^{\infty} \alpha_k \Pi_k$. Then

$$V = \left\langle \mathcal{E}_{T_n} \circ L_F\left[\sum_{i=1}^{q}\left[\Pi_{\underline{\alpha}_i}[\chi_{[0,t_i]}\underline{1}]\right]\right], (\Sigma_Y^n)^{-1}\mathcal{E}_{\infty,p[n]}(\underline{N})\right\rangle_{\mathbb{R}^{p[n]}},$$

a Gaussian random variable. The mean of $\underline{m}_n(\cdot, t)$ is, with respect to $P_{\underline{N}}$, zero, and

$$E_{P_{\underline{N}}}\left[\langle \underline{\alpha}, \underline{m}_n(\cdot, t_1)\rangle_{l_2} \langle \underline{\beta}, \underline{m}_n(\cdot, t_2)\rangle_{l_2}\right] =$$
$$= E_P[$$
$$\left\langle (\Sigma_Y^n)^{-1}\left[\mathcal{E}_{T_n} \circ L_F \circ \Pi_{\underline{\alpha}}[\chi_{[0,t_1]}\underline{1}]\right], \mathcal{E}_{T_n} \circ L_F \circ \Pi_{\underline{\alpha}}[\chi_{[0,t_1]}\underline{1}]\right\rangle_{\mathbb{R}^{p[n]}}$$

$$\left\langle (\Sigma_Y^n)^{-1}\left[\mathcal{E}_{T_n}\circ L_F\circ \Pi_{\underline{\beta}}[\chi_{[0,t_2]}\underline{1}]\right], \mathcal{E}_{T_n}\circ L_F\circ \Pi_{\underline{\beta}}[\chi_{[0,t_2]}\underline{1}]\right\rangle_{\mathbb{R}^{p[n]}}$$

]

$$= \left\langle \Pi_{H_n(C_N,[0,1])}\left[L_F\left[\Pi_{\underline{\alpha}}[\chi_{[0,t_1]}\underline{1}]\right]\right], L_F\left[\Pi_{\underline{\beta}}[\chi_{[0,t_2]}\underline{1}]\right]\right\rangle_{H(C_N,[0,1])}.$$

Now, given $\{u, v\} \subseteq [0, 1]$, $u \neq v$, fixed, but arbitrary, let $x = u \wedge v$ and $y = u \vee v$. Then, using the definition of $m_i^{(n)}$ given at $(\star\star)$,

$$\left\|\underline{m}_n(\underline{N}, u) - \underline{m}_n(\underline{N}, v)\right\|_{l_2}^2 =$$

$$= \sum_{i=1}^{\infty}\left\langle \mathcal{E}_{T_n}\circ L_F\left[\Pi_i[\chi_{]x,y]}\underline{1}]\right], (\Sigma_Y^n)^{-1}\mathcal{E}_{\infty,p[n]}(\underline{N})\right\rangle_{\mathbb{R}^{p[n]}}^2.$$

Let $f_{u,v} : \mathbb{N} \longrightarrow \mathbb{R}$ be the following function:

$$f_{u,v}(i) = \left\langle \mathcal{E}_{T_n}\circ L_F\left[\Pi_i[\chi_{]x,y]}\underline{1}]\right], (\Sigma_Y^n)^{-1}\mathcal{E}_{\infty,p[n]}(\underline{N})\right\rangle_{\mathbb{R}^{p[n]}}^2.$$

Since the evaluations over a RKHS are continuous functionals, and that L_F is a partial isometry, the following function is well defined:

$$g_{u,v}(i) = \|\mathcal{E}_{T_n}\|^2 \left\|(\Sigma_Y^n)^{-1}\underline{Y}_n\right\|_{\mathbb{R}^{p[n]}}^2 \left\|\Pi_i[\chi_{]x,y]}\underline{1}]\right\|_{L_2[\underline{b}]}^2.$$

But

$$\left\|\Pi_i[\chi_{]x,y]}\underline{1}]\right\|_{L_2[\underline{b}]}^2 = b_i(y) - b_i(x),$$

so that, for fixed, but arbitrary $i \in \mathbb{N}$, $f_{u,v}(i) \leq g_{u,v}(i)$, and

$$\lim_{|u-v|\downarrow 0} f_{u,v}(i) = \lim_{|u-v|\downarrow 0} g_{u,v}(i) = 0.$$

Finally

$$\sum_{i=1}^{\infty} g_{u,v}(i) = \|\mathcal{E}_{T_n}\|^2 \left\|(\Sigma_Y^n)^{-1}\underline{Y}_n\right\|_{\mathbb{R}^{p[n]}}^2 \sum_{i=1}^{\infty}\left\|\Pi_i[\chi_{]x,y]}\underline{1}]\right\|_{L_2[\underline{b}]}^2$$

$$= \|\mathcal{E}_{T_n}\|^2 \left\|(\Sigma_Y^n)^{-1}\underline{Y}_n\right\|_{\mathbb{R}^{p[n]}}^2 (b(y) - b(x)).$$

Consequently, using a general dominated convergence theorem [226, p. 232],

$$\lim_{|u-v|\downarrow 0}\sum_{i=1}^{\infty} f_{u,v}(i) = 0.$$

It follows that $\underline{m}_n$ is continuous.

To obtain that the sequence $\{\underline{m}_n, n \in \mathbb{N}\}$ is tight one may proceed as follows.

(a) *Case of $s = l_2$:* One has, for Gaussian elements with values in appropriate Banach spaces, among which l_2, that [171, p. 60], for fixed, but arbitrary $\{p, q\} \subseteq]0, \infty[$,

$$E_P^{1/p}\left[\|\underline{X}\|_{l_2}^p\right] \leq \kappa(p, q) E_P^{1/q}\left[\|\underline{X}\|_{l_2}^q\right].$$

Let now $q = 2$, and $\underline{X} = \underline{m}_n(\cdot, t_1) - \underline{m}_n(\cdot, t_2)$. One obtains that

$$\begin{aligned} E_P\left[\left\|\underline{m}_n(\cdot, t_1) - \underline{m}_n(\cdot, t_2)\right\|_{l_2}^p\right] &\leq \kappa^p(p, 2) E_P^{p/2}\left[\left\|\underline{m}_n(\cdot, t_1) - \underline{m}_n(\cdot, t_2)\right\|_{l_2}^2\right] \\ &\leq \kappa^p(p, 2)\left|b(t_1) - b(t_2)\right|^{p/2}. \end{aligned}$$

It suffices thus to take $p > 2$ and to apply the tightness (result (Proposition)) 12.3.10.

(b) *Case of $\mathbb{R}^\infty$:* Let $\underline{e}_n \in \mathbb{R}^\infty$ have all its components zero, except the n-th one that has value one. The set $\{\underline{e}_n, n \in \mathbb{N}\}$ is a Schauder basis for the Fréchet space $s = \mathbb{R}^\infty$. The coordinate functionals $\mathcal{E}_n^s(\underline{x}) = x_n$ are continuous, and the dual of s is obtained using the finite linear combinations of evaluation functionals. Furthermore, as seen [Sect. 12.1.2], when $s = \mathbb{R}^\infty$, and $\underline{x} \in l_2$, $\|\underline{x}\|_s \leq \kappa \|\underline{x}\|_{l_2}$. But the latter obtains almost surely [259, p. 15], and consequently the result for l_2 obtains also for s.

It may thus be taken as true that, with respect to $P_{\underline{N}}$, $\{\underline{m}_n, n \in \mathbb{N}\}$ converges weakly to some process $\underline{m}$ which is continuous and Gaussian, with, using step 1, a mean equal to zero, and a covariance with the following entries:

$$\left\langle C_{\underline{m}}(t_1, t_2)\left[\underline{\alpha}\right], \underline{\beta}\right\rangle_s = \left\langle L_F\left[\Pi_{\underline{\alpha}}\left[\chi_{[0,t_1]}\underline{1}\right]\right], L_F\left[\Pi_{\underline{\beta}}\left[\chi_{[0,t_2]}\underline{1}\right]\right]\right\rangle_{H(C_N,[0,1])},$$

where, for $s = l_2$, $C_{\underline{m}}(t_1, t_2)$ is an actual covariance matrix, whereas, when $s = \mathbb{R}^\infty$, the formal inner product denotes a bilinear functional, and, for example, $\underline{\alpha}$, the coefficients, finite in number, of an element in the dual of s. Furthermore, the components $m_{n_i,k}[\underline{x}]$ of $\underline{m}[\underline{x}]$ of step 1 are, almost surely, with respect to $P_{\underline{N}}$, equal to $m_{n_i}(\underline{x}, \theta_k)$, where m_{n_i} is the n_i-th component of $\underline{m}$.

Let Q be the following cylindrical probability on K [(Preliminary) 17.3.11, item 1, (Preliminary) 17.3.14, items 1 and 2]:

$$Q(K_c) = Q \circ \left\{\mathcal{E}_{\Theta_p}^q \circ \mathcal{E}_{N(q)}^K\right\}^{-1}\left(\left]\underline{a}, \underline{b}\right]_p^q\right) = \Gamma_{p,q}\left(\left]\underline{a}, \underline{b}\right]_p^q \,\middle|\, \underline{0}_{\mathbb{R}^{pq}}, \Sigma_F^{(pq)}\right).$$

One has, the notation being that of (Preliminary) 17.3.10, and (⋆⋆⋆) below referring to (Preliminary) 17.3.9, item 6, that

$$\int_{s\times s} f_\lambda(\underline{u},\underline{v})\,Q\circ\left\{\mathcal{E}^K_{\Theta_2}\right\}^{-1}(d\underline{u},d\underline{v}) =$$

$$\begin{aligned}
&= \lim_q \int_{s\times s} f_\lambda(\Pi^s_{N(q)}[\underline{u}],\Pi^s_{N(q)}[\underline{v}])\,Q\circ\left\{\mathcal{E}^K_{\Theta_2}\right\}^{-1}(d\underline{u},d\underline{v})\\
&= \lim_q \int_{s\times s} \phi_\lambda(\mathcal{E}^s_{N(q)}(\underline{u}),\mathcal{E}^s_{N(q)}(\underline{v}))\,Q\circ\left\{\mathcal{E}^K_{\Theta_2}\right\}^{-1}(d\underline{u},d\underline{v})\\
&= \lim_q \int_{\mathbb{R}^{2q}} \phi_\lambda(\underline{x},\underline{y})\,Q\circ\left\{\mathcal{E}^s_{N(q)}\circ\mathcal{E}^K_{\Theta_2}\right\}^{-1}(d\underline{x},d\underline{y})\\
&\overset{\star\star\star}{=} \lim_q \int_{\mathbb{R}^{2q}} \phi_\lambda(\underline{x},\underline{y})\,Q\circ\left\{\mathcal{E}^q_{\Theta_2}\circ\mathcal{E}^K_{N(q)}\right\}^{-1}(d\underline{x},d\underline{y})\\
&= \lim_q \int_{\mathbb{R}^{2q}} \phi_\lambda(\underline{x},\underline{y})\,\Gamma_{2q}\left(d\underline{x},d\underline{y}\,|\,\underline{0}_{\mathbb{R}^{2q}},\Sigma_F^{(2q)}\right)\\
&= \lim_q E_{\Gamma_{2q}\left(\cdot|\underline{0}_{\mathbb{R}^{2q}},\Sigma_F^{(2q)}\right)}[\phi_\lambda]\,.
\end{aligned}$$

Now, as above,

$$E_{\Gamma_{2q}\left(\cdot|\underline{0}_{\mathbb{R}^{2q}},\Sigma_F^{(2q)}\right)}[\phi_p] \le \kappa(p,2)^p E^{p/2}_{\Gamma_{2q}\left(\cdot|\underline{0}_{\mathbb{R}^{2q}},\Sigma_F^{(2q)}\right)}[\phi_2]\,.$$

But, with $\underline{z}$, obtained by stacking $\underline{x}$ and $\underline{y}$, and $M = [I_q \mid -I_q]$, one has that $\underline{x}-\underline{y} = M[\underline{z}]$, so that

$$\phi_2(\underline{x},\underline{y}) = \left\|\underline{x}-\underline{y}\right\|^2_{\mathbb{R}^p} = \langle M^\star M[\underline{z}],\underline{z}\rangle_{\mathbb{R}^{2q}}\,.$$

Using the formula for the expectation of a quadratic form [121, p. 243], one obtains that the expectation of ϕ_2 is

$$\operatorname{trace}\left(M^\star M\Sigma_F^{(2q)}\right) = \operatorname{trace}\left(M\Sigma_F^{(2q)}M^\star\right).$$

But, as seen above, $M\Sigma_F^{(2q)}M^\star$ is a $q\times q$ matrix with entries of the following form (as $\theta_1<\theta_2$):

$$\left\langle \Pi_F^\perp \Pi_{n_k}\left[\chi_{]\theta_1,\theta_2]}\underline{1}\right],\Pi_{n_l}\left[\chi_{]\theta_1,\theta_2]}\underline{1}\right]\right\rangle_{L_2[\underline{b}]}\,.$$

The trace is thus the sum of the diagonal terms ($k=l$), and that sum is dominated by $b(\theta_2)-b(\theta_1)$. Consequently (Proposition) 12.3.10 applies again, and Q extends to a unique probability measure on $\mathcal{K}$, also denoted Q. When the representation is proper canonical, Q is a point mass at $\underline{0}_K$.

For $\underline{\kappa} \in K$, fixed, but arbitrary, let $T_{\underline{\kappa}} : K \longrightarrow K$ be translation by $\underline{\kappa}$:

$$T_{\underline{\kappa}}(\underline{k}) = \underline{k} + \underline{\kappa}.$$

It is an adapted map, and thus, for appropriate $\underline{x} \in \mathbb{R}^{\infty}$, fixed, but arbitrary,

$$Q_{\underline{m}(\underline{x},\cdot)} = Q \circ T^{-1}_{\underline{m}(\underline{x},\cdot)}$$

is a well-defined probability, and a transition function. For appropriate $\underline{x} \in \mathbb{R}^{\infty}$ and $K_0 \in \mathcal{K}$, fixed, but arbitrary, let the candidate conditional probability be

$$Q_c(\underline{x}, K_0) = Q_{\underline{m}(\underline{x},\cdot)}(K_0).$$

When the representation is proper canonical, Q_c is a point mass at the proper value of $\underline{m}$.

Step 3: *For $K_0 \in \mathcal{K}$ and $C_0 \in \mathcal{B}(\mathbb{R}^{\infty})$, fixed, but arbitrary,*

$$P(\underline{B}_N \in K_0, \underline{N} \in C_0) = \int_{C_0} Q_c(\underline{x}, K_0) P_{\underline{N}}(d\underline{x}).$$

As seen above,

$$P\left(\underline{B}_N \in K_c, \underline{N} \in D_c\right) = \int_{D_c} P_{\underline{N}}(d\underline{x})\, \Gamma_{p,q}[\underline{x}](n),$$

and $\{\Gamma_{p,q}[\underline{x}](n), n \in \mathbb{N}\}$ converges weakly to $\Gamma_{p,q}(\cdot \mid \underline{m}[\underline{x}], \Sigma_F^{(p,q)})$. But, when the representation is proper canonical, the limit is a point mass at $\underline{m}[\underline{x}]$. When $I =]\underline{a}, \underline{b}]_p^q$ is a continuity set, by dominated convergence,

$$P(\underline{B}_N \in K_c, \underline{N} \in D_c) = \int_{D_c} P_{\underline{N}}(d\underline{x})\, \chi_{\left\{]\underline{a},\underline{b}]_p^q\right\}}(\underline{m}[\underline{x}]).$$

When $\underline{m}[\underline{x}]$ is on the boundary of I, it will not be on the boundary of $(\epsilon > 0)$ $](1-\epsilon)\underline{a}, (1+\epsilon)\underline{b}]_p^q$. The latter relation is true for the latter interval, and then one goes to zero with ϵ. So it is true for all intervals, and thus for all Borel sets. □

Proposition 17.3.16 *$\underline{m}$ is the inverse of Φ.*

Proof Since $\underline{m}$ is continuous, the following expression makes sense:

$$P_{B_N}^K\left(\underline{k} \in K : \sup_{t\in[0,1]} \left|\mathcal{E}_t \circ \mathcal{E}_n^K\left(\underline{m}[\Phi(\underline{k})]\right) - \mathcal{E}_t \circ \mathcal{E}_n^K\left(\underline{k}\right)\right| \geq \epsilon\right) =$$

$$= P\left(\omega \in \Omega : \sup_{t\in[0,1]} \left|m_n(N[\omega], t) - B_n^N(\omega, t)\right| \geq \epsilon\right).$$

Since $\underline{m}$ has, with respect to P_N, the same law as $\underline{B}_N$ with respect to P, one may use Doob's inequality to obtain that

$$P\left(\omega \in \Omega : \sup_{t\in[0,1]} \left|m_n(N[\omega],t) - B_n^N(\omega,t)\right| \geq \epsilon\right) \leq$$

$$\leq \epsilon^{-2} E_P\left[\left(m_n(N[\cdot],1) - B_n^N(\cdot,1)\right)^2\right].$$

Now $m_n(N[\cdot],1)$ is the limit in $L_2[P]$ of a sequence of terms of the following type:

$$m_n^{(m)}(N[\cdot],1) = \left\langle \mathcal{E}_{T_m} \circ L_F \circ \Pi_n\left[\chi_{[0,1]}\underline{1}\right], \left(\Sigma_Y^m\right)^{-1} \mathcal{E}_{\infty,p[m]} \circ \mathcal{E}_T(N[\cdot])\right\rangle_{\mathbb{R}^{p[m]}}.$$

Consequently

$$E_P\left[m_n(N[\cdot],1)B_n^N(\cdot,1)\right] = \lim_m E_P\left[m_n^{(m)}(N[\cdot],1)B_n^N(\cdot,1)\right].$$

But, since N is the sum of independent integral terms,

$$E_P\left[m_n^{(m)}(N[\cdot],1)B_n^N(\cdot,1)\right] =$$

$$= E_P\left[m_n^{(m)}\left(\int_0^{\cdot} F_n(\cdot,\theta)B_n^N(\cdot,d\theta),1\right)B_n^N(\cdot,1)\right].$$

Let $\underline{Z}_m$ have components of the following type:

$$Z_{m,k} = \int_0^{t_k^{(m)}} F(t_k^{(m)},\theta)B_n^N(\cdot,d\theta).$$

Then $m_n^{(m)}\left(\int_0^{\cdot} F_n(\cdot,\theta)B_n^N(\cdot,d\theta),1\right) B_n^N(\cdot,1)$ is

$$\left\langle \mathcal{E}_{T_m} \circ L_F \circ \Pi_n\left[\chi_{[0,1]}\underline{1}\right], \left(\Sigma_Y^m\right)^{-1}\left(B_n^N(\cdot,1)\underline{Z}_m\right)\right\rangle_{\mathbb{R}^{p[m]}}.$$

Taking the expectation of that latter expression has, as consequence, that the vector $B_n(\cdot,1)\underline{Z}_m$ is replaced by the vector with components of the following form (M_n is the measure determined by b_n):

$$\int_0^{t_k^{(m)}} F_n(t_k^{(m)},\theta)M_n(d\theta).$$

But those are obtained as

$$L_F\left[\Pi_n[\chi_{[0,1]}\underline{1}]\right](t_k^{(m)}) = \left\langle \Pi_n[\chi_{[0,1]}\underline{1}], \underline{F}(t_k^{(m)})\right\rangle_{L_2[\underline{b}]}.$$

Consequently $E_P\left[m_n^{(m)}(N[\cdot],1)B_n^N(\cdot,1)\right]$ equals

$$\left\langle \mathcal{E}_{T_m}\circ L_F\circ \Pi_n\left[\chi_{[0,1]}\underline{1}\right], \left(\Sigma_Y^n\right)^{-1}\mathcal{E}_{T_m}\circ L_F\circ \Pi_n\left[\chi_{[0,1]}\underline{1}\right]\right\rangle_{\mathbb{R}^{p[m]}},$$

which is

$$\left\|\Pi_{H_m(C_N,[0,1]}L_F\left[\Pi_n[\chi_{[0,1]}\underline{1}]\right]\right\|^2_{H(C_N,[0,1])}.$$

Thus

$$E_P\left[m_n(N[\cdot],1)B_n^N(\cdot,1)\right]=\left\|L_F\left[\Pi_n[\chi_{[0,1]}\underline{1}]\right]\right\|^2_{H(C_N,[0,1])}=b_n(1).$$

Expanding $E_P\left[\left(m_n(N[\cdot],1)-B_n^N(\cdot,1)\right)^2\right]$, one sees that it is thus zero. □

Remark 17.3.17 The proof just given works also for Φ_c.

Remark 17.3.18 At this point, one has proven that $\Psi\circ\Phi$ is, almost surely, with respect to $P_{B_N}^K$, the identity of K, provided Ψ is the appropriate inverse of Φ, which means that, almost surely, with respect to P, $\Psi\circ\Phi(\underline{B}_N[\omega])=\underline{B}_N[\omega]$. But then, $\Phi\circ\Psi\circ\Phi(\underline{B}_N[\omega])=\Phi(\underline{B}_N([\omega])$, that is $\Phi\circ\Psi(N[\omega])=N[\omega]$, so that $\Phi\circ\Psi$ is an almost sure identity for P_N.

Remark 17.3.19 The choice of $[0,1]$ to define Φ and Ψ is a convenience. The same proofs apply to any closed interval. The consequence is that Φ and Ψ are measurable when restricted to $\mathbb{R}^{[0,t]}$ and $K_{|t}$ (the functions of K restricted to $[0,t]$).

17.4 Absolute Continuity and Likelihoods for the Signal Plus Gaussian Noise Case

Absolute continuity and likelihoods for Gaussian noises follow from the respective results valid for Gaussian martingale noises and the Cramér-Hida maps. In the statements below there appears an independent process whose *raison d'être* is applications to information theory: it represents the message that is sent. It is independent of channel noise. The transmitted signal depends in a causal manner on the message and the channel output.

Proposition 17.4.1 *Let Ξ be independent of the Gaussian noise N, assumed to be continuous as a map $t\mapsto N_t\in L_2[P]$, and have a mean equal to zero. S is a signal, adapted to $\underline{\sigma}^\circ(N)\vee\underline{\sigma}(\Xi)$, with paths that are almost surely, with respect to P, in the RKHS of N. Let $\underline{F}_N$ and $\underline{B}_N$ be the ingredients of a proper, canonical Cramér-Hida representation of N, with $\underline{F}_N$ having components F_n^N. Then:*

1. *almost surely, with respect to* P, $S(\omega, t) = \langle \underline{F}_N(t), \underline{s}(\omega, \cdot)\rangle_{L_2[\underline{b}]}$, *where the components of* $\underline{s}$ *are predictable;*
2. *the process* $\underline{Y} = \underline{S}[\underline{s}] + \underline{B}_N$ *is such that* $P_{\underline{Y}}^K \ll P_{\underline{B}_N}^K$ (the corresponding Radon-Nikodým derivatives are to be found in Chap. 13);
3. *when, for* $t \in [0, 1]$, *fixed, but arbitrary, almost surely, with respect to* P, $X(\cdot, t) = S(\cdot, t) + N(\cdot, t)$, *then* $P_X \ll P_N$, *and*

$$\frac{dP_X}{dP_N} = \frac{dP_{\underline{Y}}^K}{dP_{\underline{B}_N}^K} \circ \underline{m},$$

where $\underline{m}$ *is to be found in* (Proposition) 17.3.15;

4. *when, almost surely with respect to* $P \otimes Leb$, $X = S + N$, *then* $P_X^2 \ll P_N^2$, *and*

$$\frac{dP_X^2}{dP_N^2} = \frac{dP_{\underline{Y}}^K}{dP_{\underline{B}_N}^K} \circ \underline{m},$$

where $\underline{m}$ *is to be found in Sect.* 17.3.1;

5. *when* N *has continuous paths, and, for* $t \in [0, 1]$, *fixed, but arbitrary, almost surely with respect to* P, $X(\cdot, t) = S(\cdot, t) + N(\cdot, t)$, *then* $P_X^c \ll P_N^c$, *and*

$$\frac{dP_X^c}{dP_N^c} = \frac{dP_{\underline{Y}}^K}{dP_{\underline{B}_N}^K} \circ \underline{m},$$

where $\underline{m}$ *is to be found in* (Proposition) 17.3.15.

Proof It consists essentially in proving that X and N have representations that entail absolute continuity for a related white Gaussian noise case, and then to use the appropriate Cramér-Hida map. One always works with a proper canonical representation of N, which yields the components $\underline{F}_N$ used in the proof to follow.

Step 1: *There exists a measurable* $\underline{s}$ *such that*

(i) *almost surely, with respect to* P, $\underline{s}[\omega] \in L_2[\underline{b}]$,
(ii) *and then,* $S(\omega, t) = \langle \underline{s}[\omega], \underline{F}_N(t)\rangle_{L_2[\underline{b}]}$.

Let $\underline{\Sigma} : \Omega \longrightarrow L_2[\underline{b}]$ be the following map:

$$\underline{\Sigma}(\omega) = \begin{cases} L_F^\star(S[\omega]) & \text{when } S[\omega] \in H(C_N, [0, 1]) \\ \underline{0}_{L_2[\underline{b}]} & \text{when } S[\omega] \in H(C_N, [0, 1])^c \end{cases}.$$

Let $\underline{a} \in L_2[\underline{b}]$ be fixed, but arbitrary. Since the representation is proper canonical, $\underline{a}$ is the limit of a sequence in $L_2[\underline{b}]$, of elements $\underline{a}_n$, such that

$$\underline{a}_n = \sum_{i=1}^{p[n]} \alpha_{n,i} \underline{F}_N(t_i^{(n)}).$$

Then, almost surely, with respect to P,

$$\begin{aligned}\langle \underline{a}, \underline{\Sigma}[\omega]\rangle_{L_2[\underline{b}]} &= \lim_n \langle \underline{a}_n, \underline{\Sigma}[\omega]\rangle_{L_2[\underline{b}]} \\ &= \lim_n \sum_{i=1}^{p[n]} \alpha_{n,i}\langle \underline{F}_N(t_i^{(n)}), \underline{\Sigma}[\omega]\rangle_{L_2[\underline{b}]} \\ &= \lim_n \sum_{i=1}^{p[n]} \alpha_{n,i}\langle S[\omega], L_F\left[\underline{F}_N(t_i^{(n)})\right]\rangle_{H(C_N,[0,1])} \\ &= \lim_n \sum_{i=1}^{p[n]} \alpha_{n,i}\langle S[\omega], C_N(\cdot, t_i^{(n)})\rangle_{H(C_N,[0,1])} \\ &= \lim_n \sum_{i=1}^{p[n]} \alpha_{n,i} S(\omega, t_i^{(n)}).\end{aligned}$$

Thus

(a) $\underline{\Sigma}$ is adapted,
(b) each component of $\underline{\Sigma}$ is adapted,
(c) $\langle \underline{a}, \underline{\Sigma}\rangle_{L_2[\underline{b}]}$ is adapted to $\sigma_t^\circ(N) \vee \sigma_t(\Xi)$ when $\underline{a}$ belongs to $L_t[\underline{F}_N]$, the subspace of $L_2[\underline{b}]$ spanned by $\{\underline{F}_N(\theta), \theta \leq t\}$.

For $n \in \mathbb{N}$, fixed, but arbitrary, let $\{e_{n,k}, k \in \mathbb{N}\}$ be a complete orthonormal set in $L_2[M_n]$, and let

$$s_{n,p}(\omega) = \sum_{i=1}^{p} \langle \Sigma_n(\omega), e_{n,i}\rangle_{L_2[M_n]} e_{n,i}.$$

Since

$$\lim_{p,q} E_P\left[\left\| s_{n,p} - s_{n,p+q}\right\|_{L_2[M_n]}^2\right] = \lim_{p,q} E_P\left[\sum_{i=p+1}^{q} \langle \Sigma_n, e_{n,i}\rangle_{L_2[M_n]}^2\right] = 0,$$

there exist a subsequence $\left\{\dot{s}_{n,p_k}, k \in \mathbb{N}\right\}$ which converges, almost surely with respect to $P \otimes M_n$, to some measurable $\dot{s}_n$. Then [229, p. 150], $s_n^c(\omega)$ denoting the class, in $L_2[M_n]$, of $\dot{s}_n(\omega, \cdot)$, $s_n^c = \Sigma_n$, almost surely, with respect to P. Thus, almost surely, with respect to P,

$$\underline{s}^c[\omega] = L_F^\star(S[\omega]).$$

Step 2: For $n \in \mathbb{N}$ and $t \in [0, 1]$, fixed, but arbitrary,

$$(\omega, t) \mapsto \int_0^t s_n^c(\omega, \theta) M_n(d\theta)$$

is adapted to $\sigma_t^\circ(N) \vee \sigma_t(\Xi)$.

Let $\Pi_{n,t}$ be the projection of $L_2[M_n]$ that sends the class of f to that of $\chi_{[0,t]}f$. Since $\{F_n^N(\theta, \cdot), \theta \leq t\}$ is dense in the range of $\Pi_{n,t}$, the class of $\chi_{[0,t]}$ may be approximated, in $L_2[M_n]$, by a sequence whose elements have the following form:

$$F_{n,p} = \sum_{i=1}^{q[p]} \alpha_{p,i} F_n^N(t_{p,i}, \cdot),$$

where $t_{p,i} \leq t$. Then, on Ω,

$$\begin{aligned}\int_0^t s_n^c(\omega, \theta) M_n(d\theta) &= \lim_p \int_0^1 \dot{F}_{n,p}(\theta) s_n^c(\omega, \theta) M_n(d\theta) \\ &= \lim_p \sum_{i=1}^{q[p]} \alpha_{p,i} \int_0^{t_{p,i}} F_n^N(t_{p,i}, \theta) s_n^c(\omega, \theta) M_n(d\theta),\end{aligned}$$

so that, using (c) of step 1, $\int_0^t s_n^c(\omega, \theta) M_n(d\theta)$ is adapted to $\sigma_t^\circ(N) \vee \sigma_t(\Xi)$.

Step 3: For $n \in \mathbb{N}$, fixed, but arbitrary, s_n may be taken to be predictable for $\underline{\sigma}^\circ(N) \vee \underline{\sigma}(\Xi)$.

For notational convenience, one shall henceforward write s_n for s_n^c. Given a function f, f^+ shall denote $f \vee 0$, and f^-, $(-f) \vee 0$. Let then

$$U_n^+(\omega, t) = \int_0^t s_n^+(\omega, \theta) M_n(d\theta).$$

It is, as seen, an adapted process. Since s_n^+ is almost surely integrable, U_n^+ has paths continuous to the left, and almost all of them are continuous. An argument similar to that used in (Lemma) 11.2.2 allows one to assume that U_n^+ has continuous paths. U_n^+ is, in particular, predictable.

Let $T_{n,p}^+(\omega) = \inf\{t \in [0, 1] : U_n^+(\omega, t) \geq p\}$. Since $T_{n,p}^+(\omega) \leq t$ if, and only if, $U_n^+(\omega, t) \geq p$, $T_{n,p}^+$ is a strict stopping time. Let

$$\Omega_{n,p} = \{\omega \in \Omega : T_{n,p}^+(\omega) = 1\}.$$

Since $T_{n,p}^+ \leq T_{n,p+1}^+$, $\Omega_{n,p} \subseteq \Omega_{n,p+1}$. Let $\Omega_n = \cup_p \Omega_{n,p}$. Since, almost surely, with respect to P, $U_n^+(\cdot, 1) < \infty$, $P(\Omega_n) = 1$.

Let, for product measurable G,

$$\mu_n^+(G) = E_P\left[\int_0^1 \chi_G(\cdot,t)s_n^+(\cdot,t)M_n(dt)\right].$$

Since $\mu_n^+([\![0,T_{n,p}^+]\!]) \le p$, μ_n^+ is a σ-finite measure on the measurable sets, as well as on the predictable ones, which is absolutely continuous with respect to $P \otimes M_n$. Let Δ_n^+ be the corresponding predictable, Radon-Nikodým derivative, and

$$V_n^+(\omega,t) = \int_0^t \Delta_n^+(\omega,\theta)M_n(d\theta).$$

Since $E_P\left[\int_0^1 \chi_{[\![0,T_{n,p}^+]\!]}\Delta_n^+ dM_n\right] = E_P\left[\int_0^1 \chi_{[\![0,T_{n,p}^+]\!]} s_n^+ dM_n\right] \le p$,

$$\int_0^{T_{n,p}^+} \Delta_n^+ dM_n$$

is almost surely finite, so that

$$\chi_{\Omega_{n,p}} \int_0^1 \Delta_n^+ dM_n$$

is almost surely finite, and thus so is $\int_0^1 \Delta_n^+ dM_n$. Consequently V_n^+ is a process with the same properties as U_n^+. Since U_n^+ and V_n^+ induce the same measure, for $t \in [0,1]$, fixed, but arbitrary, $V_n^+(\cdot,t)$ is almost surely equal to $U_n^+(\cdot,t)$. But then V_n^+ cannot be distinguished from U_n^+.

Step 4: Absolute continuity

Steps 1–3 allow one to write $S(\omega,t) = \langle \underline{F}_N(t), \underline{s}(\omega,\cdot)\rangle_{L_2[\underline{b}]}$. The Cramér-Hida representation allows one to write that $N = \Phi(\underline{B}_N)$. Girsanov's theorem yields that the law of $\underline{Y} = \underline{S}[\underline{s}] + \underline{B}_N$ is absolutely continuous with respect to that of $\underline{B}_N$, which in turn allows one to write that $X = \Phi(\underline{Y})$. The same is true, *mutatis mutandis,* when $N = \Phi_2(\underline{B}_N)$.

Step 5: The *Radon-Nikodým derivatives* are the consequence of the lemma which follows. □

Lemma 17.4.2 *Suppose that, on $(\Omega,\mathcal{A})$, one is given two probabilities, P and Q, with $Q \ll P$. Suppose also that $f : \Omega \longrightarrow X$ is adapted to $\mathcal{A}$ and $\mathcal{X}$, a σ-algebra of subsets of X. Suppose there exists a measurable $g : X \longrightarrow \Omega$ such that, almost surely with respect to P, $g \circ f(\omega) = \omega$. Then $Q_X = Q \circ f^{-1}$ is absolutely continuous with respect to $P_X = P \circ f^{-1}$, and*

$$\frac{dQ_X}{dP_X} = \frac{dQ}{dP} \circ g.$$

Proof Let $X_0 \in \mathcal{X}$ be fixed, but arbitrary. Then, by the standard image of measure property,

$$
\begin{aligned}
Q_X(X_0) &= Q\left(f^{-1}(X_0)\right) \\
&= \int_{f^{-1}(X_0)} \frac{dQ}{dP}(\omega) P(d\omega) \\
&= \int_{f^{-1}(X_0)} \frac{dQ}{dP}(g \circ f(\omega)) P(d\omega) \\
&= \int_{X_0} \frac{dQ}{dP} \circ g(x) P_X(dx).
\end{aligned}
$$

□

17.5 Scope of the Signal Plus Gaussian Noise Model

When the noise is Gaussian, not "white," the existence of the likelihood imposes again, as seen below, an additive model for the received signal.

Proposition 17.5.1 *$(\Omega, \underline{A}, P)$ is the basic probability space. Let N denote a process, continuous in mean of order two, with Cramér-Hida representation $\Phi\left(\underline{B}_N^K\right)$. Let X be a process, adapted to $\underline{\mathcal{F}} = \underline{\sigma}^\circ(N) \vee \underline{\sigma}(\Xi)$, such that, on $\mathcal{C}\left(\mathbb{R}^{[0,1]}\right)$, $P_X \ll P_N$. Then, for $t \in [0, 1]$, fixed, but arbitrary, almost surely, with respect to P,*

$$X(\omega, t) = S(\omega, t) + N_X(\omega, t),$$

where

(i) *S is, almost surely, with respect to P, path-wise, in the reproducing kernel of the noise, and is adapted to the σ-algebras generated by X;*
(ii) *N_X, with law P_N, is also adapted to the σ-algebras generated by X.*

Proof Let Ψ be the inverse of Φ, $\mathcal{E}_t^K : K \longrightarrow s$, the evaluation map at t, and $\underline{Y}(\omega, t) = \mathcal{E}_t^K\left(\Psi\left(X[\omega]\right)\right)$. Since $\boldsymbol{X}$ and Ψ are adapted to $\mathcal{C}_t(\mathbb{R}^{[0,t]})$ and $\mathcal{K}_t$, $\underline{Y}$ is adapted to the σ-algebras generated by X, and thus to $\underline{\mathcal{F}}$. Then $P_Y = P_X \circ \Psi^{-1}$. Since $P_X \ll P_N$, $P_Y \ll P_N \circ \Psi^{-1} = P_{B_N}^K$. One is thus within the framework of (Proposition) 14.2.7: for $t \in [0, 1]$, fixed, but arbitrary, almost surely, with respect to P,

$$\underline{Y}(\omega, t) = \underline{S}\left[\underline{a}^K \square \Phi_Y\right] + \underline{B}_Y(\omega, t). \qquad (\star)$$

In that latter equality, $\underline{a}^K$ is adapted to $(K, \underline{\mathcal{K}}, P_Y^K)$, and $\underline{B}_Y$, of law $P_{B_N}^K$, to $(\Omega, \underline{\sigma}(\underline{Y}), P)$. Furthermore, with respect to P_Y^K, almost surely, the paths of $\underline{a}^K$ are in $L_2[\,\underline{b}\,]$. Now $\Phi\underline{\boldsymbol{Y}} = \Phi \circ \Psi \circ \boldsymbol{X}$, and, since [(Remark) 17.3.18], $\Phi \circ \Psi$ is an almost sure identity with respect to P_N, it is also one with respect to P_X (the latter is assumed to be absolutely continuous with respect to the former). Thus $\Phi \circ \underline{Y} = \boldsymbol{X}$. Since $\underline{\boldsymbol{B}}_Y$ has law $P^K_{B_N}$, $\Phi(\underline{\boldsymbol{B}}_Y)$ has law P_N. Finally,

$$\underline{a}^K \Box \Phi_Y(\omega, t) = \underline{a}^K(\underline{Y}[\omega], t) = \underline{a}^K(\Psi(X[\omega])), t) = \underline{a}^K_\Psi \Box \Phi_X(\omega, t).$$

Applying Φ to ($\star$), as it "differentiates," one gets

$$X(\omega, t) = \langle \underline{F}, \underline{a}^K_\Psi \Box \Phi_X \rangle_{L_2[\underline{b}]}(\omega, t) + N_X(\omega, t).$$

□

Proposition 17.5.2 *Let X, in* (Proposition) 17.5.1, *be separable for closed sets, and adapted to* $\underline{F}$. *Then:*

1. *when* P_X *is absolutely continuous with respect to* P_N, *X has paths whose square is, almost surely, with respect to P, integrable with respect to Lebesgue measure, and* P^2_X *is absolutely continuous with respect to* P^2_N, *but the converse is false;*
2. *when N has continuous paths, and X is any process with continuous paths, the following equivalences obtain:*

   a) $P_X \ll P_N \Longleftrightarrow P^c_X \ll P^c_N \Longleftrightarrow P^2_X \ll P^2_N$;
   b) $P_N \ll P_X \Longleftrightarrow P^c_N \ll P^c_X \Longleftrightarrow P^2_N \ll P^2_X$;

Proof [1] $P_X \ll P_N$ means that [(Proposition) 17.5.1], for $t \in [0, 1]$, fixed, but arbitrary, almost surely, with respect to P, $X(\cdot, t) = S(\cdot, t) + N_X(\cdot, t)$, where N and N_X have the same law, S is adapted to $\underline{\sigma}(X)$, and belongs, almost surely, with respect to P, to the RKHS of N. But X is assumed separable, so that, almost surely, with respect to $P \otimes Leb$, $X(\omega, t) = S(\omega, t) + N_X(\omega, t)$. Since S has continuous paths, those of X have a square that is integrable, almost surely, with respect to P. But then [(Proposition) 17.4.1] $P^2_X \ll P^2_{N_X}$, that is, $P^2_X \ll P^2_N$, as N and N_X have the same law.

That the converse is false is seen on an example, as follows. Let h be an element in the RKHS of N. Define, for $t \in]0, 1], f(t) = h(t)$, but $f(0) \neq 0$. Let $X = f + N$. Since f does not belong to the RKHS of N, P_X is orthogonal to P_N. But the class of f belongs to the square root of the covariance operator of N, and thus P^2_X and P^2_N are equivalent.

Proof [2] Let $\mathcal{C}_f$ denote the cylinder sets of $\mathbb{R}^{[0,1]}$ and $\mathcal{C}_c$, those of $C_0([0, 1])$. Let also $t_1 < t_2 < \cdots < t_{n-1} < t_n$ in $[0, 1]$, and $B \in \mathcal{B}(\mathbb{R}^n)$, be fixed, but arbitrary. When

$$C_f = \left\{f \in \mathbb{R}^{[0,1]} : (f(t_1), \ldots, f(t_n)) \in B\right\} \in \mathcal{C}_f,$$
$$C_c = \{c \in C_0([0, 1]) : (c(t_1), \ldots, c(t_n)) \in B\} \in \mathcal{C}_c,$$

one has that $C_c = C_f \cap C_0([0, 1])$, and

$$P_X(C_f) = P(X \in C_f)$$
$$= P(X \in \left\{C_f \cap C_0([0, 1])\right\})$$

$$= P^c_X(C_c),$$
$$P_N(C_f) = P^c_N(C_c).$$

Consequently, given a fixed, but arbitrary $\epsilon > 0$, there exists a $\delta > 0$ such that $P_X(C_f) < \epsilon$ whenever $P_N(C_f) < \delta$ if, and only if, $P^c_X(C_c) < \epsilon$ whenever $P^c_N(C_c) < \delta$. Since the families $\mathcal{C}_f$ and $\mathcal{C}_c$ are generating families for the corresponding σ-algebras, the relations valid for the algebras carry over, and thus absolute continuity obtains.

Let $J : C[0, 1] \longrightarrow L_2[0, 1]$ be the injection that sends the continuous c to its equivalence class $[c] \in L_2[0, 1]$. It is a continuous map. The process X defines a map $\omega \mapsto X[\omega] \in C[0, 1]$, and a map $\omega \mapsto [X][\omega] \in L_2[0, 1]$, the latter being the equivalence class of $X[\omega]$, that is, $[X][\omega] = J \circ X[\omega]$. Consequently

$$P^2_X = P^c_X \circ J^{-1}.$$

By a theorem of Kuratowski [208, p. 21], J carries Borel sets isomorphically onto Borel sets, and thus, for Borel B in $C[0, 1]$,

$$\begin{aligned} P^c_X(B) &= P(\omega \in \Omega : X[\omega] \in B) \\ &= P(\omega \in \Omega : X[\omega] \in J^{-1}(J(B))) \\ &= P^c_X \circ J^{-1}(J(B)) \\ &= P^2_X(J(B)). \end{aligned}$$

Similar relations obtain for P^c_N and P^2_N, and the absolute continuity statements are immediate consequences. □

Remark 17.5.3 Item 2 of (Proposition) 17.5.2 does neither require that N be Gaussian, nor that it starts at zero almost surely.

Proposition 17.5.4 *Suppose that N is Gaussian, $t \mapsto N_t \in L_2(\Omega, \mathcal{A}, P)$ is continuous, and $t \mapsto E_P[N(\cdot, t)]$ is the zero function. X, in* (Proposition) 17.5.1, *is adapted to $\underline{\mathcal{F}}$. Then:*

1. *one may have $P_X \ll P_N$ (respectively $P^2_X \ll P^2_N$), but not, for $t \in [0, 1]$, fixed, but arbitrary, almost surely, with respect to P (respectively, almost surely with respect to $P \otimes Leb$), $X(\cdot, t) = S(\cdot, t) + N(\cdot, t)$ with S adapted to $\underline{\sigma}(X)$, and paths that are, almost surely, with respect to P, in the RKHS of N (respectively the range of the square root of the covariance operator of N);*
2. *when N and X have continuous paths, almost surely, with respect to P, one may have $P^c_X \ll P^c_N$, but no representation as $S + N$.*

Proof Let Ξ be a continuous paths, Gaussian process whose mean is zero. Those paths will be in the range of the square root of the covariance operator of N, C_N, or, equivalently, in the RKHS of N, if, and only if, its covariance operator, C_Ξ, has a representation of the following form:

$$C_{\Xi} = C_N^{1/2} T C_N^{1/2},$$

T having finite trace. Let, for $t \in [0, 1]$, fixed, but arbitrary,

$$X(\cdot, t) = \Xi(\cdot, t) + N(\cdot, t)$$

(X is thus adapted to $\underline{\mathcal{F}}$). Then [97], $P_X \ll P_N$ and $P_X^2 \ll P_N^2$, if, and only if, T is Hilbert-Schmidt.

Thus, when T is Hilbert-Schmidt, but not trace-class, one has absolute continuity (and $P_X^c \ll P_N^c$ when paths are continuous) without S having paths almost surely in the RKHS of the noise. □

17.6 From Theory to Practice: Some Comments

The results pertaining to the likelihood, when derived using the Cramér-Hida representation, presuppose knowledge that is not generally available in applications. Obtaining the eigenvectors and eigenvalues of integral operators, or Cramér-Hida decompositions, are elusive tasks, especially when unique, discrete, and finite data sets are at hand. Effective deployment, and employment, of those results in operational systems require that discrete time approximations be developed, and that such approximations be given in terms of quantities that can reasonably be expected to be available from either prior knowledge, or estimation from observed data.

One item that can usually be obtained, with reasonable accuracy, is a reliable covariance matrix of the noise, for the proper environment, and it shall be assumed below that such is the case.

There are desirable criteria that detection algorithms should meet. Among those are the following: the algorithms should

1. *be based on the likelihood-ratio* [the latter is at the core of all optimal procedures];
2. *preserve information* [the solution should begin with the analysis of the original continuous-time problem; the likelihood for the continuous-time problem should then be approximated as well as possible in forming a discrete-time algorithm; independent sampling, which typically destroys information, should not be used unless absolutely necessary];
3. *be implementable* [although the analytical expression of the likelihood should be part of the process leading to the algorithms, the latter must be developed into expressions that can reasonably be expected to be put to work, and thus must not require information that is not accessible in applications];
4. *be adaptive* [the algorithms should use data as soon as it becomes available].

Algorithms are consequently obtained through a drastic reduction of mathematical complexity. Qualitative justification for the diverse procedures presented below may be found (and should be read) in [14, 16]. Here only the mechanics are given, in order to illustrate what practice may mean in mathematical terms.

17.6.1 Framework

One sets:

1. B, W, standard Wiener processes;
2. $N(\omega,t) = \int_0^t F(t,\theta)\, W(\omega, d\theta)$, and multiplicity one;
3. $X(\omega,t) = \int_0^t F(t,\theta) Z(\omega, d\theta)$, with

$$Z(\omega,t) = \int_0^t \alpha(Z[\omega],\theta)\,d\theta + B(\omega,t),$$

and

$$P\left(\int_0^1 \alpha^2(Z[\omega],\theta)\,d\theta < \infty\right) = 1.$$

Let $T_n = \left\{t_0^{(n)}, \ldots, t_n^{(n)}\right\} \subseteq [0,1]$ be the observation times, where

$$t_0^{(n)} = 0,\; t_i^{(n)} < t_{i+1}^{(n)},\; t_n^{(n)} = 1.$$

Mimicking the Euler scheme (here the values of the process are "known" and α is unknown),

$$\begin{aligned} Z(\cdot,0) &= 0, \\ Z(\cdot,t_i^{(n)}) &= Z(\cdot,t_{i-1}^{(n)}) \\ &\quad + \alpha\left(Z(\cdot,t_{i-1}^{(n)})\right)\left(t_i^{(n)} - t_{i-1}^{(n)}\right) \\ &\quad + B(\cdot,t_i^{(n)}) - B(\cdot,t_{i-1}^{(n)}). \end{aligned}$$

Stochastic integrals shall be approximated using the following procedure:

$$\int_0^1 a(t)\,\psi(dt) \approx \sum_{i=1}^{n} a(t_{i-1}^{(n)})\left\{\psi(t_i^{(n)}) - \psi(t_{i-1}^{(n)})\right\} = \langle \underline{a}_n, \underline{\Delta}_n \psi\rangle_{\mathbb{R}^n},$$

where

$$\underline{a}_n^\star = [a(t_0^{(n)}), a(t_1^{(n)}), \ldots, a(t_{n-1}^{(n)})],$$

$$\underline{\Delta}_n^\star \psi = [\psi(t_1^{(n)}) - \psi(t_0^{(n)}), \psi(t_2^{(n)}) - \psi(t_1^{(n)}), \ldots, \psi(t_n^{(n)}) - \psi(t_{n-1}^{(n)})].$$

Then the logarithm of the likelihood may be approximated at the sample s using the following sum:

$$\sum_{i=1}^{n} \alpha(m(s, t_{i-1}^{(n)}), t_{i-1}^{(n)}) \left\{m(s, t_i^{(n)}) - m(s, t_{i-1}^{(n)})\right\} -$$

$$- \frac{1}{2} \sum_{i=1}^{n} \alpha^2(m(s, t_{i-1}^{(n)}), t_{i-1}^{(n)}) \left\{t_i^{(n)} - t_{i-1}^{(n)}\right\}.$$

The vector of evaluations of the function f at the points of T_n shall be denoted $\underline{\mathcal{E}}_{T_n}(f)$. One shall write

- $\alpha_{n,i}[f]$ for $\alpha(m(f, t_i^{(n)}), t_i^{(n)})$;
- $\underline{\alpha}_{T_n}[f]$ for the vectors whose entries are the $\alpha_{n,i}[f]$'s;
- $\Delta_{n,i}m[f]$ for $m(f, t_i^{(n)}) - m(f, t_{i-1}^{(n)})$;
- $\underline{\Delta}_{T_n}[f]$ for the vector whose components are the $\Delta_{n,i}m[f]$'s.

The approximation to the logarithm of the likelihood may thus be given the following familiar form, s denoting the received signal, and D_{T_n}, the diagonal matrix whose diagonal elements are the differences $t_{i+1}^{(n)} - t_i^{(n)}$:

$$\Lambda_n \left\{\underline{\mathcal{E}}_{T_n}(m[s])\right\} = \langle \underline{\alpha}_{T_n}[s], \underline{\Delta}_{T_n} m[s] \rangle_{\mathbb{R}^n} - \frac{1}{2} \langle D_{T_n} \left(\underline{\alpha}_{T_n}[s]\right), \underline{\alpha}_{T_n}[s] \rangle_{\mathbb{R}^n}.$$

Usually m and α are unknown.

17.6.2 The Approximation's Law for Noise Only

One shall do the calculations with uniform sampling: $t_i^{(n)} = i\Delta_n$. One has that

$$\begin{aligned} E_{P_N}\left[e^{\iota \langle \underline{\theta}_n, \underline{\mathcal{E}}_{T_n}(m[\cdot]) \rangle_{\mathbb{R}^n}}\right] &= E_P\left[e^{\iota \langle \underline{\theta}_n, \underline{\mathcal{E}}_{T_n}(m[N]) \rangle_{\mathbb{R}^n}}\right] \\ &= E_P\left[e^{\iota \langle \underline{\theta}_n, \underline{\mathcal{E}}_{T_n}(m[\Phi_2(W)]) \rangle_{\mathbb{R}^n}}\right] \\ &= E_{P_W}\left[e^{\iota \langle \underline{\theta}_n, \underline{\mathcal{E}}_{T_n}(m[\Phi_2]) \rangle_{\mathbb{R}^n}}\right] \end{aligned}$$

$$= E_{P_W}\left[e^{\iota\langle \underline{\theta}_n, \underline{\mathcal{E}}_{T_n}(\cdot)\rangle_{\mathbb{R}^n}}\right]$$

$$= E_P\left[e^{\iota\langle \underline{\theta}_n, \underline{\mathcal{E}}_{T_n}(W)\rangle_{\mathbb{R}^n}}\right]$$

Let $\underline{\Delta}_n W$ be the vector of differences of the following type:

$$W(\cdot, i\Delta_n) - W(\cdot, (i-1)\Delta_n),$$

and L_n be the summation operator:

$$L_n = \begin{bmatrix} 1\,0\,0 \cdots 0\,0 \\ 1\,1\,0 \cdots 0\,0 \\ 1\,1\,1 \cdots 0\,0 \\ \vdots\;\vdots\;\vdots\;\;\vdots\;\;\vdots\;\vdots \\ 1\,1\,1 \cdots 1\,1 \end{bmatrix}.$$

Then $L_n[\underline{\Delta}_n W]$ is the vector with components of type $W(\cdot, i\Delta_n)$, and its characteristic function is the exponential of $-\frac{\Delta_n}{2}\left\|L_n^\star[\underline{\theta}_n]\right\|^2_{\mathbb{R}^n}$. Thus

$$E_{P_N}\left[e^{\iota\langle \underline{\theta}_n, \underline{\mathcal{E}}_{T_n}(m[\cdot])\rangle_{\mathbb{R}^n}}\right] = e^{-\frac{\Delta_n}{2}\left\|L_n^\star[\underline{\theta}_n]\right\|^2_{\mathbb{R}^n}}.$$

The aim is now to replace $\underline{\mathcal{E}}_{T_n}(m[\cdot])$ with an expression that is computable, and has, with respect to P_N, the same law.

As seen, R_N, the covariance operator of N, has a decomposition $R_N = SS^\star$, where S is the integral operator with F as kernel. The eigenvectors and eigenvalues of R_N are denoted, respectively, e_i and λ_i. The process m has then, in the L_2 case, the following representation:

$$m(f,t) = \sum_{n=1}^{\infty} \lambda_n^{-1}\langle S\left[\chi_{[0,t]}\right], e_n\rangle_{L_2[0,1]}\langle f, e_n\rangle_{L_2[0,1]}.$$

Let L be the Volterra operator

$$L[f](t) = \int_0^t f(\theta)\,d\theta.$$

Then

$$\langle S\left[\chi_{[0,t]}\right], e_n\rangle_{L_2[0,1]} = \int_0^1 \left\{\int_0^t F(t,\theta)\,d\theta\right\} \dot{e}_n(t)\,dt$$

$$= \int_0^t \left\{ \int_0^1 F(t,\theta)\dot{e}_n(t)\,dt \right\} d\theta$$
$$= LS^\star[e_n](t),$$

and, Π_n denoting the projection whose range is spanned by $\{e_1, \ldots, e_n\}$,

$$\begin{aligned}
m(s,t) &= \lim_n \sum_{i=1}^n \lambda_i^{-1} LS^\star[e_i](t)\langle s, e_i\rangle_{L_2[0,1]} \\
&= \lim_n LS^\star\left[\sum_{i=1}^n \langle s, e_i\rangle_{L_2[0,1]}\lambda_i^{-1} e_i\right](t) \\
&= \lim_n LS^\star\left[\sum_{i=1}^n \langle s, e_i\rangle_{L_2[0,1]} R_N^{-1}[e_i]\right](t) \\
&= \lim_n LS^\star R_N^{-1}\left[\sum_{i=1}^n \langle s, e_i\rangle_{L_2[0,1]} e_i\right](t) \\
&= \lim_n LS^\star\,(SS^\star)^{-1}\,\Pi_n[s](t) \\
&= \lim_n LS^{-1}\Pi_n[s](t).
\end{aligned}$$

With probability one, s shall not be in the domain of S^{-1} [11], but one may proceed as follows. Let $R_{N,n}$ be the matrix with entries

$$R_{N,n}(i\Delta_n, j\Delta_n) = \int_0^{(i\wedge j)\Delta_n} F(i\Delta,\theta)F(j\Delta,\theta)\,d\theta.$$

$R_{N,n}$ may be written as $\Delta_n T_{N,n} T_{N,n}^\star$, with $T_{N,n}$ lower triangular [121, p. 189]. Then, with $\underline{N}_n$ having entries of type $N(\cdot, j\Delta_n)$,

$$\begin{aligned}
E_{P_N}\left[e^{\iota\langle \underline{\theta}_n, L_n T_{N,n}^{-1}[\underline{\mathcal{E}}_{T_n}(\cdot)]\rangle_{\mathbb{R}^n}}\right] &= E_P\left[e^{\iota\langle \underline{\theta}_n, L_n T_{N,n}^{-1}[\underline{N}_n]\rangle_{\mathbb{R}^n}}\right] \\
&= e^{-\frac{1}{2}\langle R_{N,n}(T_{N,n}^{-1})^\star L_n^\star[\underline{\theta}_n], (T_{N,n}^{-1})^\star L_n^\star[\underline{\theta}_n]\rangle_{\mathbb{R}^n}} \\
&= e^{-\frac{\Delta_n}{2}\left\|L_n^\star[\underline{\theta}_n]\right\|^2_{\mathbb{R}^n}}.
\end{aligned}$$

In the approximation to the likelihood, one may thus replace the vector $\mathcal{E}_{T_n}(m[\cdot])$ with the vector $L_n T_{N,n}^{-1}[\underline{s}_n]$.

In practice, one shall use $\Sigma_{N,n}$, the available covariance matrix of the noise. Factoring it as

$$\Sigma_{N,n} = \Delta_n \hat{T}_{N,n}\hat{T}_{N,n}^\star,$$

one may use $\hat{T}_{N,n}$ in place of $T_{N,n}$.

Example 17.6.1 For the case

$$N(\cdot, k\Delta_n) = f(k\Delta_n) \int_0^{k\Delta_n} g(\theta)\, W(\cdot, d\theta),$$

it has already been noticed that

$$R_{N,n} = D_n L_n \Gamma_n L_n^\star D_n,$$

where

- D_n is diagonal, with diagonal terms of the following form: $f(k\Delta_n)$;
- Γ_n is diagonal, with diagonal terms of the following form:

$$\gamma_k^2 = \left\| \Pi_{](k-1)\Delta_n, k\Delta_n]} g \right\|^2_{L_2[0,1]},$$

so that $T_{N,n} = \Delta_n^{-1/2} D_n L_n \Gamma_n^{1/2}$.

17.6.3 The Approximation's Law for Signal-Plus-Noise

Again, with X, the signal-plus-noise process, having the representation $\Phi_2[Z]$,

$$\begin{aligned} E_{P_{S+N}}\left[e^{\iota\langle \underline{\theta}_n, \underline{\mathcal{E}}_{T_n}(m[\cdot])\rangle_{\mathbb{R}^n}}\right] &= E_P\left[e^{\iota\langle \underline{\theta}_n, \underline{\mathcal{E}}_{T_n}(m[X])\rangle_{\mathbb{R}^n}}\right] \\ &= E_P\left[e^{\iota\langle \underline{\theta}_n, \underline{\mathcal{E}}_{T_n}(m[\Phi_2(Z)])\rangle_{\mathbb{R}^n}}\right] \\ &= E_{P_Z}\left[e^{\iota\langle \underline{\theta}_n, \underline{\mathcal{E}}_{T_n}(m[\Phi_2])\rangle_{\mathbb{R}^n}}\right]. \end{aligned}$$

Since $P_Z \ll P_W$, $m[\Phi_2]$ is also the identity with respect to P_Z, and the law of $\underline{\mathcal{E}}_{T_n}(m)$ with respect to P_{S+N} is that of Z.

Suppose now that, for $t \in [0,1]$, fixed, but arbitrary, $\theta \mapsto F(t,\theta)$ is smooth enough so that

$$\int_0^{(i\wedge j)\Delta_n} F(i\Delta_n, \theta) F(j\Delta_n, \theta)\, d\theta \approx \Delta_n \sum_{k=1}^{i\wedge j} F(i\Delta_n, k\Delta_n) F(j\Delta_n, k\Delta_n).$$

Example 17.6.2 With the noise chosen as in (Example) 17.6.1, one must thus have, as

$$F(i\Delta_n, k\Delta_n) = f(i\Delta_n)\, \chi_{[0,i\Delta_n]}(k\Delta_n)\, g(k\Delta_n),$$

that

$$f(i\Delta_n)f(j\Delta_n)\int_0^{(i\wedge j)\Delta_n} g^2(\theta)\,d\theta = f(i\Delta_n)f(j\Delta_n)\sum_{k=1}^{i\wedge j}\int_{(k-1)\Delta_n}^{k\Delta_n} g^2(\theta)d\theta$$
$$\approx \Delta_n f(i\Delta_n)f(j\Delta_n)\sum_{k=1}^{i\wedge j} g^2(k\Delta_n)$$
$$= \Delta_n \sum_{k=1}^{i\wedge j} F(i\Delta_n, k\Delta_n)F(j\Delta_n, k\Delta_n),$$

that is, systematically, $\int_{(k-1)\Delta_n}^{k\Delta_n} g^2(\theta)d\theta \approx \Delta_n g^2(k\Delta_n)$.

Let $\tilde{T}_{N,n}$ be the lower triangular matrix with entries $F(i\Delta_n, j\Delta_n)$. In happy circumstances, one shall have that

$$\tilde{T}_{N,n} \approx T_{N,n}.$$

Example 17.6.3 In the case of the noise chosen as example,

$$T_{N,n} = D_n L_n \left\{\Delta_n^{-1/2}\Gamma_n^{1/2}\right\}$$
$$\tilde{T}_{N,n} = D_n L_n \{G_n\},$$

where G_n has diagonal elements of the form $|g(k\Delta_n)|$.

Let $\underline{\Delta}_n Z$ be the vector with entries $\Delta_i Z = Z(\cdot, i\Delta_n) - Z(\cdot, (i-1)\Delta_n)$, and $\underline{X}_n$, the vector with entries

$$X(\cdot, i\Delta_n) = \int_0^{i\Delta_n} F(i\Delta_n, \theta)Z(\cdot, d\theta) \approx \sum_{k=1}^{i\Delta_n} F(i\Delta_n, (k-1)\Delta_n)\Delta_k Z.$$

When one may choose

$$\sum_{k=1}^{i\Delta_n} F(i\Delta_n, (k-1)\Delta_n)\Delta_k Z \approx \sum_{k=1}^{i\Delta_n} F(i\Delta_n, k\Delta_n)\Delta_k Z,$$

then

$$\underline{X}_n \approx \tilde{T}_{N,n}\underline{\Delta}_n Z \approx T_{N,n}\underline{\Delta}_n Z,$$

so that

$$L_n T_{N,n}^{-1}\underline{X}_n \approx L_n \underline{\Delta}_n Z = \underline{Z}_n,$$

the latter being the vector with entries $Z(\cdot, i\Delta_n)$. In the case of the noise chosen as example [(Example) 17.6.1], one makes the assumption that the variation of g is modest over each observation interval (it does not seem reasonable to expect that the variation of g be mitigated by that of Z as the latter is more of an unknown than g).

As a consequence, assuming α and F known and smooth, the probability of false alarm, calculated under the chosen approximation, will exactly be that which one would obtain with a "discretized" version of the exact, continuous-time likelihood (provided of course that the simplifying assumptions relate to the subjacent reality).

17.6.4 A Recursive Approximation to the Likelihood

To write the expression for the approximation to the likelihood, one must remark the following. The vector $\underline{e}_i$ being the i-th basis vector of the appropriate $\mathbb{R}^n$, $\underline{e}_i^\star M$ yields the i-th line of the matrix M. Consequently

$$\underline{e}_{i+1}^\star L_n T_{N,n}^{-1} - \underline{e}_i^\star L_n T_{N,n}^{-1} = \left\{\underline{e}_{i+1}^\star L_n - \underline{e}_i^\star L_n\right\} T_{N,n}^{-1} = \underline{e}_{i+1}^\star T_{N,n}^{-1},$$

and thus

$$\langle L_n T_{N,n}^{-1}[\underline{s}_n], \underline{e}_{i+1}\rangle_{\mathbb{R}^n} - \langle L_n T_{N,n}^{-1}[\underline{s}_n], \underline{e}_i\rangle_{\mathbb{R}^n},$$

the substitute for $m(s, t_{i+1}^{(n)}) - m(s, t_i^{(n)})$, has value

$$\langle T_{N,n}^{-1}[\underline{s}_n], \underline{e}_{i+1}\rangle_{\mathbb{R}^n},$$

so that the approximation to the logarithm of the likelihood shall have the following form:

$$\begin{aligned}\Lambda_n(\underline{s}_n) = &\sum_{i=1}^{n-1} \alpha\left(\langle L_n T_{N,n}^{-1}[\underline{s}_n], \underline{e}_i\rangle_{\mathbb{R}^n}, i\Delta_n\right) \langle T_{N,n}^{-1}[\underline{s}_n], \underline{e}_{i+1}\rangle_{\mathbb{R}^n} \\ &- \frac{\Delta_n}{2}\sum_{i=1}^{n-1} \alpha^2\left(\langle L_n T_{N,n}^{-1}[\underline{s}_n], \underline{e}_i\rangle_{\mathbb{R}^n}, i\Delta_n\right).\end{aligned}$$

Suppose that a new data point is observed, that is, the observation interval becomes $[0, 1 + \Delta_n]$. The approximation has then, as shall be seen, a recursive expression. One has that

$$R_{N,n+1} = \left[\begin{array}{c|c} R_{N,n} & \underline{r}_{N,n} \\ \hline \underline{r}_{N,n}^\star & r_{N,n+1}\end{array}\right] = \Delta_n T_{N,n+1} T_{N,n+1}^\star,$$

and

$$T_{N,n+1} = \left[\begin{array}{c|c} T^{(n)}_{N,n+1} & \underline{0}_n \\ \hline \underline{t}^{(n)\star}_{N,n+1} & t_{N,n+1} \end{array}\right].$$

Thus

$$\left[\begin{array}{c|c} \Delta_n T_{N,n} T^\star_{N,n} & \underline{r}_{N,n} \\ \hline \underline{r}^\star_{N,n} & r_{N,n+1} \end{array}\right] = \Delta_n \left[\begin{array}{c|c} T^{(n)}_{N,n+1} T^{(n)\star}_{N,n+1} & T^{(n)}_{N,n+1}\underline{t}^{(n)}_{N,n+1} \\ \hline \underline{t}^{(n)\star}_{N,n+1} T^{(n)\star}_{N,n+1} & t^2_{N,n+1} + \underline{t}^{(n)\star}_{N,n+1}\underline{t}^{(n)}_{N,n+1} \end{array}\right].$$

Since the triangular decomposition of a positive definite matrix is unique up to signs, one may use

$$T^{(n)}_{N,n+1} = T_{N,n}.$$

Furthermore

$$\left[\begin{array}{c|c} A & \underline{0}_n \\ \hline \underline{a}^\star_n & \alpha \end{array}\right]^{-1} = \left[\begin{array}{c|c} A^{-1} & \underline{0}_n \\ \hline -\alpha \underline{a}^\star_n A^{-1} & \alpha^{-1} \end{array}\right],$$

so that

$$T^{-1}_{N,n+1} = \left[\begin{array}{c|c} T^{-1}_{N,n} & \underline{0}_n \\ \hline -t^{-1}_{N,n+1}\underline{t}^{(n)\star}_{N,n+1} T^{-1}_{N,n} & t^{-1}_{N,n+1} \end{array}\right].$$

Consequently

$$T^{-1}_{N,n+1}[\underline{s}_{n+1}] = \left[\begin{array}{c} T^{-1}_{N,n}[\underline{s}_n] \\ \hline t^{-1}_{N,n+1}\left\{s_{n+1} - \langle T^{-1}_{N,n}[\underline{s}_n], \underline{t}^{(n)}_{N,n+1}\rangle_{\mathbb{R}^n}\right\} \end{array}\right],$$

and

$$L_{n+1} T^{-1}_{N,n+1}[\underline{s}_{n+1}] =$$

$$= \left[\begin{array}{c} L_n T^{-1}_{N,n}[\underline{s}_n] \\ \hline \langle \underline{1}_n, T^{-1}_{N,n}[\underline{s}_n]\rangle + t^{-1}_{N,n+1}\left\{s_{n+1} - \langle T^{-1}_{N,n}[\underline{s}_n], \underline{t}^{(n)}_{N,n+1}\rangle_{\mathbb{R}^n}\right\} \end{array}\right],$$

so that

$$\langle L_{n+1} T^{-1}_{N,n+1}[\underline{s}_{n+1}], \underline{e}_i\rangle_{\mathbb{R}^{n+1}} =$$

$$= \begin{cases} \langle L_n T^{-1}_{N,n}[\underline{s}_n], \underline{e}_i\rangle_{\mathbb{R}^n} & \text{when } i \le n \\ \langle \underline{1}_n, T^{-1}_{N,n}[\underline{s}_n]\rangle + t^{-1}_{N,n+1}\left\{s_{n+1} - \langle T^{-1}_{N,n}[\underline{s}_n], \underline{t}^{(n)}_{N,n+1}\rangle_{\mathbb{R}^n}\right\} & \text{when } i = n+1 \end{cases},$$

and

$$\langle T_{N,n+1}^{-1}[\underline{s}_{n+1}], \underline{e}_{i+1}\rangle_{\mathbb{R}^{n+1}} =$$

$$= \begin{cases} \langle T_{N,n}^{-1}[\underline{s}_n], \underline{e}_{i+1}\rangle_{\mathbb{R}^n} & \text{when } i \leq n-1 \\ t_{N,n+1}^{-1}\left\{s_{n+1} - \langle T_{N,n}^{-1}[\underline{s}_n], \underline{t}_{N,n+1}^{(n)}\rangle_{\mathbb{R}^n}\right\} & \text{when } i = n \end{cases}.$$

Finally, noting that $\Delta_{n+1} = \Delta_n$, since one has added Δ_n to the observation time,

$$\begin{aligned} \Lambda_n(\underline{s}_{n+1}) &= \Lambda_n(\underline{s}_n) \\ &+ \alpha\left(\langle L_n T_{N,n}^{-1}[\underline{s}_n], \underline{e}_n\rangle_{\mathbb{R}^n}, 1\right) \frac{s_{n+1} - \langle L_n T_{N,n}^{-1}[\underline{s}_n], \underline{t}_{N,n+1}^{(n)}\rangle_{\mathbb{R}^n}}{t_{N,n+1}} \\ &- \frac{\Delta_n}{2}\alpha^2\left(\langle L_n T_{N,n}^{-1}[\underline{s}_n], \underline{e}_n\rangle_{\mathbb{R}^n}, 1\right). \end{aligned}$$

Remark 17.6.4 The approximation to the likelihood that is presented above is not in general the likelihood one would obtain were one to know the law of the signal-plus-noise at the instants it is observed. But that is the case, as shall be seen, when $\alpha(t,c) = \alpha(t)\,c(t)$, that is, when the signal-plus-noise is Gaussian. That provides some evidence that the procedure developed above makes some sense.

Suppose that the model is $\alpha(t, Z[\cdot]) = \alpha(t)Z(\cdot, t)$. The following equation:

$$Z(\cdot, k\Delta_n) = \sum_{i=1}^{k} \int_{(i-1)\Delta_n}^{i\Delta_n} \alpha(Z[\cdot], \theta)d\theta + B(\cdot, k\Delta_n)$$

yields equalities of the following type:

$$Z_k = \Delta_n \sum_{j=0}^{k-1} \alpha_j Z_j + B_k. \qquad (\star)$$

One observes, since, using $(\star)$, $B_k = Z_k - \Delta_n \sum_{j=0}^{k-1} \alpha_j Z_j$, that

$$\begin{aligned} \Delta_n \alpha_1 Z_1 + B_1 &= (1 + \Delta_n \alpha_1) Z_1, \\ \Delta_n\{\alpha_1 Z_1 + \alpha_2 Z_2\} + B_2 &= (1 + \Delta_n \alpha_2) Z_2, \\ \dots \quad & \dots \end{aligned}$$

that is, a matrix expression of the following form (one writes I for the identity matrix, Δ for Δ_n, A for the diagonal matrix whose entries are the α_i's):

$$\Delta LA\underline{Z} + \underline{B} = (I + \Delta A)\underline{Z}.$$

Thus

$$\underline{Z} = \{I + \Delta A - \Delta LA\}^{-1}\,\underline{B},$$

and, consequently, Σ_Z being the covariance matrix of $\underline{Z}$,

$$\Sigma_Z^{-1} = (I + \Delta A - \Delta AL^\star)\Sigma_B^{-1}(I + \Delta A - \Delta LA).$$

Letting $M = \Delta A - \Delta LA$, one obtains that

$$\Sigma_Z^{-1} - \Sigma_B^{-1} = M^\star\,\Sigma_B^{-1} + \Sigma_B^{-1}M + M^\star\,\Sigma_B^{-1}M,$$

which, using $\Sigma_B^{-1} = \Delta^{-1}\,\{L^\star\}^{-1}\,L^{-1}$, yields that

$$\begin{aligned}
M^\star\,\Sigma_B^{-1} + \Sigma_B^{-1}M + M^\star\,\Sigma_B^{-1}M &= \\
&= (\Delta A - \Delta AL^\star)\Delta^{-1}\,\{L^\star\}^{-1}\,L^{-1} \\
&+ \Delta^{-1}\,\{L^\star\}^{-1}\,L^{-1}(\Delta A - \Delta LA) \\
&+ (\Delta A - \Delta AL^\star)\Delta^{-1}\,\{L^\star\}^{-1}\,L^{-1}(\Delta A - \Delta LA) \\
&= A\,\{L^\star\}^{-1}\,L^{-1} - AL^{-1} \\
&+ \{L^\star\}^{-1}\,L^{-1}A - \{L^\star\}^{-1}\,A \\
&+ \Delta A\,\{L^\star\}^{-1}\,L^{-1}A \\
&- \Delta AL^{-1}A - \Delta A\,\{L^\star\}^{-1}\,A + \Delta A^2.
\end{aligned}$$

Then

$$\begin{aligned}
&\langle(\Sigma_Z^{-1} - \Sigma_B^{-1})[\underline{x}], \underline{x}\rangle_{\mathbb{R}^n} = \\
&= \Delta\,\left\|L^{-1}A[\underline{x}]\right\|_{\mathbb{R}^n}^2 + 2\langle L^{-1}[\underline{x}], L^{-1}A[\underline{x}]\rangle_{\mathbb{R}^n} - 2\langle L^{-1}[\underline{x}], A[\underline{x}]\rangle_{\mathbb{R}^n} \\
&\quad -2\Delta\langle A[\underline{x}], L^{-1}A[\underline{x}]\rangle_{\mathbb{R}^n} + \Delta\,\|A[\underline{x}]\|_{\mathbb{R}^n}^2 \\
&= \Delta\,\left\|L^{-1}A[\underline{x}] - A[\underline{x}]\right\|_{\mathbb{R}^n}^2 + 2\langle L^{-1}[\underline{x}], L^{-1}A[\underline{x}] - A[\underline{x}]\rangle_{\mathbb{R}^n} \\
&= \Delta\sum_{i=1}^{n-1}\alpha_i^2 x_i^2 - 2\sum_{i=1}^{n-1}\alpha_i x_i(x_{i+1} - x_i),
\end{aligned}$$

so that

$$-\frac{1}{2}\langle(\Sigma_Z^{-1} - \Sigma_B^{-1})[\underline{x}], \underline{x}\rangle_{\mathbb{R}^n} = \sum_{i=1}^{n-1} \alpha_i x_i (x_{i+1} - x_i) - \frac{\Delta}{2}\sum_{i=1}^{n-1} \alpha_i^2 x_i^2.$$

The approximation to the likelihood is indeed a Gaussian likelihood.

17.6.5 Estimating the Drift Parameter Function

It cannot be expected that α be known, and thus it must be estimated on the basis of the available data. The latter may be in the form of ensembles of independent samples of representative signal-plus-noise data, of a long segment of data, longer than the observation time over which the detection algorithm is to perform, or of a single, observed, sample vector. The adequate procedure is thus data dependent, and must furthermore take into account the sampling rate. Implementations, for particular applications, must then be preceded by much numerical experimentation, and it is unlikely that one may obtain a "one size fits all" method.

In the available statistical literature, there is much which pertains to cases $\alpha(c,t) = \alpha(t(c))$, or $\alpha(c,t) = \alpha(c, t(t))$ little, or nothing, when α is a genuine functional of c, such as, for example,

$$\alpha(c,t) = \int_0^t f(c(\theta))\, d\theta.$$

Furthermore that literature deals mostly with homogeneous diffusions, is asymptotic in nature, and requires that the processes be sometimes recurrent, sometimes stationary [26, 214]. The inhomogeneous diffusion case seems to be given short shrift. Thus, it is commented in [95] that, *since only a trajectory of the process is observed ... there is not sufficient information to estimate the bivariate function,* [that is, α,] *without further restrictions,* and its author limits attention to the time dependent affine case, with the conclusion that *coefficient functions ... cannot be estimated reliably due to the collinearity effect in local estimation...*, whereas [213] simply adapts the procedure for the homogeneous case to the inhomogeneous one, adding a kernel for time, without further ado. All these cases use indeed kernel smoothing estimators. There is another all purpose method, that of sieves [122], in which one typically expands α into a finite number of elements of a basis, and uses maximum likelihood. Properties of sieve estimators are obtained by balancing out the increase in the amount of data available with that of the number of parameters, so that again, with fixed data, effective comparisons of the efficiencies of the different approaches are unknown.

As an illustration, remarking that [200, p. 115] the Hilbert tensor product $L_2(\Omega_1, \mathcal{A}_1, \mu_1) \otimes L_2(\Omega_2, \mathcal{A}_2, \mu_2)$ is isomorphic to $L_2(\Omega_1 \times \Omega_2, \mathcal{A}_1 \otimes \mathcal{A}_2, \mu_1 \otimes \mu_2)$, and that the latter is generated by linear combinations of maps of "product type," that is, $f_1(\omega_1)f_2(\omega_2)$, one shall consider an α expressed as

$$\alpha(c,t) = \sum_{k=0}^{n-1} \theta_k a_k(t)\,\beta_k(c)\,\chi_{[t_k,1]}(t),$$

$$\beta_k(c) = \prod_{j=1}^{k} b_{k,j}(c(j\Delta_n)),$$

where the a_k's and the $b_{k,j}$'s are function basis elements, chosen to validate the sufficient conditions required for the existence of solutions of stochastic differential equations with functional drift [191, 216, 264]. Suppose that

$$l\Delta_n \leq t < (l+1)\Delta_n.$$

Then

$$\alpha(c,t) = \sum_{k=0}^{l} \theta_k a_k(c)\,\beta_k(t),$$

and α is properly adapted. The logarithm of the likelihood is

$$\int_0^1 \alpha(c,t)\,c(dt) - \frac{1}{2}\int_0^1 \alpha^2(c,t)\,dt.$$

Let

$$A_k(c) = \int_{k\Delta_n}^{1} \alpha_k(t)\,c(dt),\ A_{k,l} = \int_{(k\vee l)\Delta_n}^{1} \alpha_k(t)\alpha_l(t)\,dt.$$

When the integral defining $A_k(c)$ is computed using a finite number of observations, the notation shall be $A_k(c \mid \delta)$. One has then that

$$\int_0^1 \alpha(c,t)\,c(dt) = \sum_{k=0}^{n-1} \theta_k \beta_k(c) \int_0^1 \alpha_k(t)\,\chi_{[t_k,1]}(t)\,c(dt) = \sum_{k=0}^{n-1} \theta_k \beta_k(c) A_k(c),$$

and that

$$\int_0^1 \alpha^2(c,t)\,dt = \sum_{k=0}^{n-1}\sum_{l=0}^{n-1} \theta_k\theta_l\,\beta_k(c)\,\beta_l(c)\,A_{k,l}.$$

Since, for appropriate matrices and vectors,

$$\langle A[\underline{x}], \underline{x}\rangle - 2\langle \underline{c}, \underline{x}\rangle = \left\| A^{\frac{1}{2}}[\underline{x}] - A^{-\frac{1}{2}}[\underline{c}] \right\|^2 - \left\| A^{-\frac{1}{2}}[\underline{c}] \right\|^2,$$

the logarithm of the likelihood shall be

$$\langle \underline{m}(c), \underline{\theta}\rangle - \frac{1}{2}\langle M(c)[\underline{\theta}], \underline{\theta}\rangle =$$

$$= -\frac{1}{2}\left\| M^{\frac{1}{2}}(c)[\underline{\theta}] - M^{-\frac{1}{2}}(c)[\underline{m}(c)] \right\|_{\mathbb{R}^n}^2 + \left\| M^{-1}(c)[\underline{m}(c)] \right\|_{\mathbb{R}^n}^2,$$

and thus

$$\hat{\underline{\theta}}(c) = M^{-1}(c)[\underline{m}(c)],$$

where the components of $\underline{m}(c)$ are the $\beta_k(c)\,A_k(c)$'s, and the entries of $M(c)$, the $\beta_k(c)\,\beta_l(c)A_{k,l}$'s. When dealing with a finite number of observation, one need only replace $A_k(c)$ with $A_k(c \mid \delta)$ to formally obtain the same estimate.

Remark 17.6.5 The reduced model used in previous sections has a form that seems tailored for the estimation of the signal as an inverse problem. The inverse solution however yields the values of α rather that its functional form. But, using an expansion similar to that presented above, one may expect to achieve similar results. Conversely, such a solution may also serve as a validating instrument, and one may see the likelihood approximation as the solution of that inverse problem. On the downside, to obtain mathematical results, one needs more assumptions on the signal, and there is no indication of how the prior should be picked.

What follows is from [62]. It describes *an abstract Bayesian framework for inverse problems in which the unknown is a function and the data are finite. These problems are hence underdetermined. They are also frequently ill-posed in a classical sense. We describe three key ideas:*

1. *we prove a version of the Bayes theorem relevant to this function space setting, showing that the posterior measure is absolutely continuous with respect to the prior measure, and identifying the Radon-Nikodým derivative as the data likelihood;*
2. *we demonstrate a form of well posedness by proving that the posterior measure is Lipschitz continuous in the data, when Hellinger metric is used as a metric on the posterior measure;*
3. *we show that the maximum a posteriori estimator for the posterior measure (the posterior probability maximizer) is well defined whenever the posterior measure is.*

One starts with the observations

$$X(\cdot, t_i) = \langle F(t_i, \cdot), s[\cdot]\rangle_{L_2[0,1]} + N(\cdot, t_i),$$

written in the following format:

$$\underline{X} = \Psi(s) + \underline{N},$$

where $\Psi : L_2[0, 1] \longrightarrow \mathbb{R}^n$ is the operator for which

$$\langle \Psi(f), \underline{e}_i\rangle_{\mathbb{R}^n} = \langle F(t_i, \cdot), f\rangle_{L_2[0,1]}.$$

Since

$$\|\Psi(f)\|^2_{\mathbb{R}^n} = \sum_{i=1}^{n} \langle F(t_i, \cdot), f\rangle^2_{L_2[0,1]} \leq \|f\|^2_{L_2[0,1]} \sum_{i=1}^{n} \|F(t_i, \cdot)\|^2_{L_2[0,1]},$$

Ψ is continuous, and thus measurable.

Let

- $G_N^{(n)}$ be the Gaussian law on the Borel sets of $\mathbb{R}^n$ determined by $\underline{N}$, and
- $G_{N,f}^{(n)}$ that of the translation of $\underline{N}$ by $\Psi(f)$.

Then, with Σ_N as the covariance of $\underline{N}$, and $\|\underline{x}\|_{\Sigma_N} = \left\| \Sigma_N^{-1/2}[\underline{x}] \right\|_{\mathbb{R}^n}$,

$$\ln \left\{ \frac{dG_{N,f}^{(n)}}{dG_N^{(n)}}(\underline{x}) \right\} = \frac{1}{2}\|\underline{x}\|^2_{\Sigma_N} - \frac{1}{2}\|\underline{x} - \Psi(f)\|^2_{\Sigma_N}.$$

Let P_s be the measure on $\mathcal{B}(L_2[0, 1])$ generated by s, and set

$$\Omega = L_2[0, 1] \times \mathbb{R}^n, \qquad \mathcal{A} = \mathcal{B}(L_2[0, 1]) \otimes \mathcal{B}(\mathbb{R}^n),$$

$$\Pi_1(f, \underline{x}) = f, \qquad \Pi_2(f, \underline{x}) = \underline{x},$$

$$P(df, d\underline{x}) = G_N^{(n)}(d\underline{x})P_s(df), \qquad Q(df, d\underline{x}) = G_{N,f}^{(n)}(d\underline{x})P_s(df).$$

Since Ψ is measurable, $G_{N,f}^{(n)}(\underline{x})$ is a Markov kernel, and thus Q is well defined. As

$$G_{N,f}^{(n)}(d\underline{x})P_s(df) = \frac{G_{N,f}^{(n)}}{G_N^{(n)}}(\underline{x})P_s(df)\,G_N^{(n)}(d\underline{x}),$$

Q is absolutely continuous with respect to P, with density

$$\Delta(f, \underline{x}) = \frac{G^{(n)}_{N,f}}{G^{(n)}_{N}}(\underline{x}).$$

Let, for $B_1 \in \mathcal{B}(L_2[0,1])$,

$$\Delta_{\Pi_1}(f) = \int_{\mathbb{R}^n} \Delta(f, \underline{x}) G^{(n)}_{N}(d\underline{x}),$$

$$\Delta_{\Pi_2}(\underline{x}) = \int_{L_2[0,1]} \Delta(f, \underline{x}) P_s(f),$$

$$\mathcal{D} = \left\{\underline{x} \in \mathbb{R}^n : 0 < \Delta_{\Pi_2}(\underline{x}) < \infty\right\},$$

$$\delta(f \mid \underline{x}) = \begin{cases} \frac{\Delta(f,\underline{x})}{\Delta_{\Pi_2}(\underline{x})} & \text{when } \underline{x} \in \mathcal{D} \\ \\ \Delta_{\Pi_1}(f) & \text{when } \underline{x} \in \mathbb{R}^n \setminus \mathcal{D} \end{cases},$$

$$P^{\Pi_2}_{\Pi_1}(B_1 \mid \underline{x}) = \int_{B_1} \delta(f \mid \underline{x}) P_s(df).$$

Then [139, p. 124], $P^{\Pi_2}_{\Pi_1}(B_1 \mid \underline{x})$ is a regular conditional distribution of Π_1 given Π_2, that is,

$$P^{\Pi_2}_{\Pi_1}(df \mid \underline{x}) = \delta(f \mid \underline{x}) P_s(df),$$

and the logarithm of the likelihood of the signal f, given the observation $\underline{x}$, is a constant minus $\frac{1}{2}\left\|\underline{x} - \Psi(f)\right\|^2_{\Sigma_N}$.

Let

$$\Upsilon(\underline{x}, f) = \frac{1}{2}\left\|\underline{x} - \Psi(f)\right\|^2_{\Sigma_N}.$$

The latter expression has the following properties:

1. when $\|\underline{x}\|_{\mathbb{R}^n} < \kappa$, letting $K_1 = \left\|\Sigma_N^{-1/2}\right\|^2 \left\{\kappa^2 + \sum_{i=1}^n \|F(t_i,\cdot)\|^2_{L_2[0,1]}\right\}$,

$$\begin{aligned} \Upsilon(\underline{x}, f) &\le \frac{1}{2}\left\|\Sigma_N^{-1/2}\right\|^2 \left\|\underline{x} - \Psi(f)\right\|^2_{\mathbb{R}^n} \\ &\le \left\|\Sigma_N^{-1/2}\right\|^2 \left\{\|\underline{x}\|^2_{\mathbb{R}^n} + \|f\|^2_{L_2[0,1]} \sum_{i=1}^n \|F(t_i,\cdot)\|^2_{L_2[0,1]}\right\} \\ &\le K_1 \left\{1 + \|f\|^2_{L_2[0,1]}\right\}; \end{aligned}$$

2. when $\max\left\{\|\underline{x}\|_{\mathbb{R}^n}, \|f_1\|_{L_2[0,1]}, \|f_2\|_{L_2[0,1]}\right\} < \kappa$, letting

$$K_2 = 2^{1/2} K_1^{1/2} \left\{1+\kappa^2\right\}^{1/2} \left\{\sum_{i=1}^{n} \|F(t_i,\cdot)\|^2_{L_2[0,1]}\right\}^{1/2},$$

one has, using property 1 above, that

$$\begin{aligned}
&|\Upsilon(\underline{x},f_1) - \Upsilon(\underline{x},f_2)| \leq \\
&\leq \frac{1}{2}\left\{\|\underline{x} - \Psi(f_1)\|_{\mathbb{R}^n} + \|\underline{x} - \Psi(f_2)\|_{\mathbb{R}^n}\right\} \|\Psi(f_2) - \Psi(f_1)\|_{\mathbb{R}^n} \\
&\leq K_2 \|f_1 - f_2\|_{L_2[0,1]};
\end{aligned}$$

3. when $\max\left\{\left\|\underline{x}_1\right\|_{\mathbb{R}^n}, \left\|\underline{x}_2\right\|_{\mathbb{R}^n}\right\} < \kappa$, letting

$$K_3 = 2^{1/2} K_1^{1/2},$$

one has, using property 1 above, that

$$\begin{aligned}
\left|\Upsilon(\underline{x}_1,f) - \Upsilon(\underline{x}_2,f)\right| &\leq \frac{1}{2}\left\{\left\|\underline{x}_1 - \Psi(f)\right\|_{\mathbb{R}^n} + \left\|\underline{x}_2 - \Psi(f)\right\|_{\mathbb{R}^n}\right\} \left\|\underline{x}_1 - \underline{x}_2\right\|_{\mathbb{R}^n} \\
&\leq K_3 \left\{1 + \|f\|^2_{L_2[0,1]}\right\}^{1/2} \left\|\underline{x}_1 - \underline{x}_2\right\|_{\mathbb{R}^n} \\
&\leq K_3 \left\{1 + \|f\|^2_{L_2[0,1]}\right\} \left\|\underline{x}_1 - \underline{x}_2\right\|_{\mathbb{R}^n}.
\end{aligned}$$

A continuity property of $\underline{x} \mapsto P_{\Pi_1}^{\Pi_2}(\cdot \mid \underline{x})$ shall be demonstrated next. To that end one needs the Hellinger distance for probability measures, defined as follows ($P \ll M, Q \ll M$):

$$d(P,Q) = \left\{\frac{1}{2}\int \left(\left[\frac{dP}{dM}\right]^{\frac{1}{2}} - \left[\frac{dQ}{dM}\right]^{\frac{1}{2}}\right)^2 dM\right\}^{\frac{1}{2}},$$

which will be used with $P = P_{\Pi_1}^{\Pi_2}(\cdot \mid \underline{x})$, $Q = P_{\Pi_1}^{\Pi_2}(\cdot \mid \underline{y})$, $M = P_s$. Let then

$$C(\underline{x}) = \int_{L_2[0,1]} e^{-\Upsilon(\underline{x},f)} P_s(df) \leq 1,$$

so that, when $C(\underline{x}) > 0$,

$$\delta(f \mid \underline{x}) = \frac{e^{-\Upsilon(\underline{x},f)}}{C(x)}.$$

Set then

$$B_{L_2[0,1]}(0,1) = \left\{ f \in L_2[0,1] : \|f\|_{L_2[0,1]} \leq 1 \right\},$$

and assume that $\|\underline{x}\|_{\mathbb{R}^n} \vee \left\|\underline{y}\right\|_{\mathbb{R}^n} < \kappa$. Using the first of the inequalities displayed above,

$$\begin{aligned} C(\underline{x}) &\geq \int_{B_{L_2[0,1]}(0,1)} e^{-\Upsilon(\underline{x},f)} P_s(df) \\ &\geq \int_{B_{L_2[0,1]}(0,1)} e^{-2K_1} P_s(df) \\ &= e^{-2K_1} P_s\left(B_{L_2[0,1]}(0,1)\right). \end{aligned}$$

When P_s is a probability with full support, then $C(\underline{x}) > 0$. Now, since [218, p. 383], when ϕ is continuous on $[a,b]$, and ϕ', its derivative, exists on $]a,b[$, there exists $c \in]a,b[$ such that

$$|\phi(b) - \phi(a)| \leq \left|\phi'(c)\right| (b-a),$$

one has, using for ϕ, the exponential in the integral defining C, and the third inequality above (it imposes that $\underline{x}$ and $\underline{y}$ belong to a fixed ball),

$$\left|C(\underline{x}) - C(\underline{y})\right| \leq K_3 \left\|\underline{x} - \underline{y}\right\|_{\mathbb{R}^n} \int_{L_2[0,1]} \left\{1 + \|f\|^2_{L_2[0,1]}\right\} P_s(df).$$

Thus, when P_s has strong order 2 (the norm is square integrable), which is compatible with the assumption that signals have finite energy, C is continuous (Lipschitz).

One has furthermore that

$$\begin{aligned} \left(\frac{e^{-\frac{x}{2}}}{a} - \frac{e^{-\frac{y}{2}}}{b}\right)^2 &= e^{-y}\left\{\left(\frac{e^{-\frac{x-y}{2}}}{a} - \frac{1}{a}\right) + \left(\frac{1}{a} - \frac{1}{b}\right)\right\}^2 \\ &\leq 2e^{-y}\left\{\frac{\left(e^{-\frac{x-y}{2}} - 1\right)^2}{a^2} + \left(\frac{1}{a} - \frac{1}{b}\right)^2\right\} \\ &= \frac{2}{a^2}\left(e^{-\frac{x}{2}} - e^{-\frac{y}{2}}\right)^2 + 2\left(\frac{1}{a} - \frac{1}{b}\right)^2 e^{-y}. \end{aligned}$$

Consequently

$$2d^2\left(P_{\Pi_1}^{\Pi_2}(\cdot \mid \underline{x}), P_{\Pi_1}^{\Pi_2}(\cdot \mid \underline{y})\right) \leq$$

$$\leq \frac{2}{C(\underline{x})}\int_{L_2[0,1]}\left\{e^{-\frac{1}{2}\Upsilon(\underline{x},f)} - e^{-\frac{1}{2}\Upsilon(\underline{y},f)}\right\}^2 P_s(df)$$

$$+ 2\left(\frac{1}{C^{\frac{1}{2}}(\underline{x})} - \frac{1}{C^{\frac{1}{2}}(\underline{y})}\right)^2 \int_{L_2[0,1]} e^{-\Upsilon(\underline{y},f)} P_s(df).$$

The inequality $\left|e^{-\frac{x}{2}} - e^{-\frac{y}{2}}\right| \leq \frac{1}{2}|x-y|$, and the third one given above, yield that the first integral on the right is dominated by

$$\frac{K_3}{4}\left\|\underline{x}-\underline{y}\right\|_{\mathbb{R}^n}^2 \int_{L_2[0,1]}\left\{1 + \|f\|_{L_2[0,1]}^2\right\}^2 P_s(df).$$

One must thus assume that the norm has integrable moments up to order four. That restricts the family of signal measures that are allowed. That family contains at least the Gaussian ones. Now

$$\left|\frac{1}{\sqrt{x}} - \frac{1}{\sqrt{y}}\right| = \left|\frac{\sqrt{y}-\sqrt{x}}{\sqrt{xy}}\right| = \frac{|y-x|}{y\sqrt{x}+x\sqrt{y}},$$

so that

$$\left(C^{-\frac{1}{2}}(\underline{x}) - C^{-\frac{1}{2}}(\underline{y})\right)^2 \leq \frac{\left(C(\underline{x}) - C(\underline{y})\right)^2}{\min\left\{C^3(\underline{x}), C^3(\underline{y})\right\}}.$$

Finally to estimate the signal, one could try to minimize Υ, given the observations. However, minimizing sequences in $L_2[0,1]$ for Υ may fail to converge, and thus a term is added to it, to force convergence. That can be done assuming that the a priori law of the signal is Gaussian, with a mean m_s in the range of the square root of its covariance operator C_s (denoted $G(m_s, C_s)$). Presumably, the choice of a Gaussian law, besides the fact that it is "easy" to compute with such measures, is based on the fact that the Gaussian distribution maximizes entropy over laws with the same covariance. The positioning of the mean, in the range of the square root, insures the existence of the likelihood of $G(m_s, C_s)$ with respect to $G(0_{L_2[0,1]}, C_s)$. One is thus led to consider the minimum of the following functional, over the range of the square root of the covariance operator C_s (the norm in $\mathcal{R}[C_s^{1/2}]$ shall be denoted $\|\cdot\|_{C_s}$):

$$\varphi(f \mid \underline{x}) = \frac{1}{2}\|f - m_s\|_{C_s}^2 + \frac{1}{2}\|\underline{x} - \Psi(f)\|_{\Sigma_N}^2, \ \{m_s, f\} \subseteq \mathcal{R}[C_s^{1/2}].$$

It is checked below that a minimum of φ, denoted $\tilde{\varphi}$, exists, and, if f_n belongs to a sequence for which $\{\varphi(f_n \mid \underline{x})\}$ tends to the minimum value of φ, then $\{f_n\}$ converges to the value at which the minimum is attained.

Let $\epsilon > 0$ be fixed, but arbitrary, and let n_ϵ be such that, for $n \geq n_\epsilon$,

$$\tilde{\varphi}(\underline{x}) = \inf_{\mathcal{R}[C_s^{1/2}]} \varphi(f \mid \underline{x}) \leq \varphi(f_n \mid \underline{x}) \leq \tilde{\varphi}(\underline{x}) + \epsilon,$$

so that

$$\frac{1}{2} \|f_n - m_s\|_{C_s}^2 \leq \tilde{\varphi}(\underline{x}) + \epsilon,$$

which proves, since $m_s \in \mathcal{R}[C_s^{1/2}]$, that the sequence $\left\{\|f_n\|_{C_s}, n \in \mathbb{N}\right\}$ is bounded. One may thus assume that there exists $f_l \in \mathcal{R}[C_s^{1/2}]$ which is the weak limit of the sequence $\{f_n, n \in \mathbb{N}\}$. But the embedding of $\mathcal{R}[C_s^{1/2}]$ into $L_2[0,1]$ is compact [21], so that the sequence is convergent in $L_2[0,1]$. But Υ is, as seen above (item 2), Lipshitz continuous in the f variable, so that it is weakly continuous on $\mathcal{R}[C_s^{1/2}]$. As the norm is weakly lower semicontinuous [71, p. 140], $\frac{1}{2}\|f - m_s\|_{C_s}^2$ is a lower semi-continuous functional on $\mathcal{R}[C_s^{1/2}]$, so that

$$f \mapsto \varphi(f \mid \underline{x})$$

is weakly lower semi-continuous on $\mathcal{R}[C_s^{1/2}]$. It is there furthermore convex, since, generically, expanding, then using Cauchy's inequality, and finally the fact that λ defines a probability,

$$\begin{aligned} \|\lambda h + (1-\lambda)k - c\|^2 &= \|\lambda(h-c) + (1-\lambda)(k-c)\|^2 \\ &\leq (\lambda\|h-c\| + (1-\lambda)\|k-c\|)^2 \\ &\leq \lambda\|h-c\|^2 + (1-\lambda)\|k-c\|^2. \end{aligned}$$

It is finally coercive. It has consequently a minimum [71, p. 226].

Generically again

$$\begin{aligned} \left\|\frac{h+k}{2} - c\right\|^2 &= \left\|\frac{h-c}{2} + \frac{k-c}{2}\right\|^2 \\ &= \frac{1}{4}\|h-c\|^2 + \frac{1}{2}\langle h-c, k-c\rangle + \frac{1}{4}\|h-c\|^2, \end{aligned}$$

so that

$$-2\langle h-c, k-c\rangle = \|h-c\|^2 + \|k-c\|^2 - 4\left\|\frac{h+k}{2} - c\right\|^2,$$

and thus that

$$\begin{aligned}\|h-k\|^2 &= \|(h-c)-(k-c)\|^2 \\ &= 2\|h-c\|^2 - 4\left\|\frac{h+k}{2}-c\right\|^2 + 2\|k-c\|^2 .\end{aligned}$$

Consequently, for n and p large enough,

$$\begin{aligned}&\frac{1}{4}\left\|f_n - f_p\right\|^2_{C_s} = \\ &= \frac{1}{2}\left\|f_n - m_s\right\|^2_{C_s} + \frac{1}{2}\left\|f_p - m_s\right\|^2_{C_s} - \left\|\frac{f_n+f_p}{2} - m_s\right\|^2_{C_s} \\ &= \varphi(f_n \mid \underline{x}) + \varphi(f_p \mid \underline{x}) - 2\varphi\left(\frac{f_n+f_p}{2} \mid \underline{x}\right) \\ &\quad -\Upsilon(\underline{x}, f_n) - \Upsilon(\underline{x}, f_p) + 2\Upsilon\left(\underline{x}, \frac{f_n+f_p}{2}\right) \\ &\leq 2\{\tilde{\varphi}+\epsilon)\} - 2\tilde{\varphi} - \Upsilon(\underline{x}, f_n) - \Upsilon(\underline{x}, f_p) + 2\Upsilon\left(\underline{x}, \frac{f_n+f_p}{2}\right) \\ &= 2\epsilon - \Upsilon(\underline{x}, f_n) - \Upsilon(\underline{x}, f_p) + 2\Upsilon\left(\underline{x}, \frac{f_n+f_p}{2}\right).\end{aligned}$$

Since the sequence $\{f_n, n \in \mathbb{N}\}$ converges in $L_2[0,1]$ to f_l, and that Υ is continuous in its second argument, that sequence is Cauchy in $\mathcal{R}[C_s^{1/2}]$, and thus has a limit.

17.6.6 Epilogue or ... Le mot de la fin

It shall be from [24]: *The key problem in developing effective signal detection algorithms based on* [the material in this book] *is the determination of good approximations to the functionals* [appearing in the martingale likelihood ratio]. *A "bullet-proof" general approach is still to be determined and seems unlikely to exist. However, in limited studies to date, algorithms based on such representations have generally outperformed classical methods using both active and passive sonar data. These representations have for the most part been determined by a combination of examining representative data and consideration of the physical properties of the noise, the target and the environment.*

As can be seen, the theory developed here does not provide, as do approaches such as matched filtering, a defined algorithm that can be implemented in a straightforward manner for a given system. Rather, it provides a likelihood ratio-based framework within which one can seek an effective implementation. As such, the further development of effective algorithms based on the theory is far more dependent on serious analysis of data properties and representation than is generally performed in development of contemporary signal processing methods. This lack of explicit solutions and the need for problem-by-problem approach to developing a solution are the negatives of the theoretical results so far as computational implementations are concerned. The positives are the existence of a likelihood ratio-based approach within which one can proceed with confidence and which applies to very general signal-plus-noise processes.

References

1. Abramovich, Y.A., Alpirantis, C.D.: An Invitation to Operator Theory. American Mathematical Society, Providence, RI (2002)
2. Apostol, T.M.: Calculus I. Wiley, New York (1967)
3. Arino, O., Delode, C., Genet, J.: Mesure et intégration. Vuibert, Paris (1976)
4. Aronszajn, N.: Theory of reproducing kernels. Trans. Am. Math. Soc. **68**, 307–404 (1950)
5. Ash, R.B.: Real Analysis and Probability. Academic, New York (1972)
6. Ash, R.B., Doléans-Dade, C.A.: Probability and Measure Theory. Harcourt, San Diego (2000)
7. Attéia, M., Gaches, J.: Approximation Hilbertienne. PUG, Grenoble (1999)
8. Bachman, G., Narici, L.: Functional Analysis. Academic, New York (1968)
9. Baker, C.R.: Complete simultaneous reduction of covariance operators. SIAM J. Appl. Math. **17**, 972–983 (1969)
10. Baker, C.R.: Detection and information theory. In: Third Southeastern Symposium on System Theory. Georgia Institute of Technology, Atlanta (1971)
11. Baker, C.R.: On equivalence of probability measures. Ann. Probab. **1**, 690–698 (1973)
12. Baker, C.R.: Joint measures and cross-covariance operators. Trans. Am. Math. Soc. **186**, 273–289 (1973)
13. Baker, C.R.: Capacity of the mismatched Gaussian channel. IEEE IT **33**, 802–812 (1987)
14. Baker, C.R.: Algorithms for optimum detection of signals in Gaussian noise. Technical Report (LISS 48). UNC, Department of Statistics, Chapel Hill, NC (1991)
15. Baker, C.R.: Spectral multiplicity of second-order stochastic processes. Technical Report (1991, DTIC). UNC, Department of Statistics (and Office of Naval Research), Chapel Hill, NC (1992)
16. Baker, C.R.: Array Modeling and Signal Detection for Non-Gaussian Data, 7 pages, OCEANS 94 OSATES, Brest (1994)
17. Baker, C.R., Gualtierotti, A.F.: Detection of non-Gaussian signals in Gaussian noise. In: Wegman, E.J., Smith, J.G. (eds.) Statistical Signal Processing, pp. 107–116. Dekker, New York (1984)
18. Baker, C.R., Gualtierotti, A.F.: Discrimination with respect to a Gaussian process. Probab. Theory Relat. Fields **71**, 159–182 (1986)
19. Baker, C.R., Gualtierotti, A.F.: Absolute continuity and mutual information for Gaussian mixtures. Stoch. Stoch. Rep. **39**, 139–157 (1992)
20. Baker, C.R., Gualtierotti, A.F.: Likelihood-ratio detection of stochastic signals. In: Poor, H.V., Thomas, J.B. (eds.) Advances in Statistical Signal Processing, vol. 2, pp. 1–34. JAI Press, Greenwich, CT (1993)

A.F. Gualtierotti, *Detection of Random Signals in Dependent Gaussian Noise*,
DOI 10.1007/978-3-319-22315-5

21. Baker, C.R., McKeague, I.W.: Compact covariance operators. Proc. Am. Math. Soc. **83**, 590–593 (1981)
22. Baker, C.R., Frey, M.R., Persons, C.E.: A computational evaluation of new detection algorithms. US Navy J. Underw. Acoust. **37**, 457–476 (1991)
23. Baker, C.R., Cherifi, H., Dankel, T.G., Frey, M.R.: Analysis, modeling and signal detection for a set of passive sonar data. Technical Report (LISS-50). UNC, Department of Statistics, Chapel Hill, NC (1992)
24. Baker, C.R., Climescu-Haulica, A., Gualtierotti, A.F.: Likelihood ratio methods for underwater acoustics signal detection. In: McWhirter, J.G., Proudler, I.K. (eds.) Mathematics in Signal Processing V. Institute of Mathematics and Its Applications Conference Series, vol. 71, pp. 339–348. Clarendon Press, Oxford (2002)
25. Balakrishnan, A.V.: Applied Functional Analysis. Springer, New York (1976)
26. Bandi, F.M., Phillips, P.C.P.: Fully nonparametric estimation of scalar diffusion models, vol. 1050. Yale University, Cowles Foundation Paper, New Haven, CT (2003)
27. Baranger, J., Duc-Jacquet, M.: Matrices tridiagonales symétriques et matrices factorisables. R.I.R.O. **5**ème année, no. **R-3**, 61–66 (1971)
28. Barlow, M.T., Émery, M., Knight, F.B., Song, S., Yor, M.: Autour d'un théorème de Tsirelson sur des filtrations browniennes et non browniennes. In: Séminaire des probabilités (Strasbourg), vol. 32, pp. 264–305. Springer, Berlin (1998)
29. Barra, J.-R.: Notions fondamentales de statistique mathématique. Dunod, Paris (1971)
30. Bauer, H.: Wahrscheinlichkeitstheorie und Grundzüge des Masstheorie, 2. Auflage. de Gruyter, Berlin (1974)
31. Beghdadi-Sakrani, S.: The uniqueness class of continuous local martingales. Bernoulli **8**, 207–217 (2002)
32. Benedetto, J.J., Czaja, W.: Integration and Modern Analysis. Birkhäuser, Boston (2009)
33. Bensoussan, A.: Filtrage optimal des systèmes linéaires. Dunod, Paris (1971)
34. Berberian, F.K.: Measure and integration. New York: MacMillan (1967)
35. Berlinet, A., Thomas-Agnan, C.: Reproducing Kernel Hilbert Spaces in Probability and Statistics. Kluwer, Boston (2004)
36. Berman, S.M.: Sojourns and Extremes of Stochastic Processes. Wadsworth & Brooks, Pacific Grove, CA (1992)
37. Bernstein, D.S.: Matrix Mathematics. Princeton University Press, Princeton (2005)
38. Billingsley, P.: Convergence of Probability Measures. Wiley, New York (1968)
39. Billingsley, P.: Probability and Measure. Wiley, New York (1979)
40. Blanc-Lapierre, A., Picinbono, B.: Fonctions aléatoires. Masson, Paris (1981)
41. Bogachev, V.I.: Measure Theory I. Springer, Berlin (2007)
42. Bogachev, V.I., Kolesnikov, A.V., Medvedev, K.V.: Triangular transformations of measures. Sb. Math. **196**(3), 3–30 (2005)
43. Brémaud, P.: Point processes and queues. New York (1981)
44. Brezis, H.: Analyse fonctionnelle. Masson, Paris (1983)
45. Brown, A.: A version of multiplicity theory. In: Pearcy, C. (ed.) Topics in Operator Theory. American Mathematical Society, Providence, RI (1974)
46. Brown, A., Pearcy, C.: Introduction to Operator Theory I. Springer, New York (1977)
47. Buck, C.R.: Advanced Calculus. McGraw-Hill, New York (1965)
48. Bulatović, J.: The separability of the Hilbert space generated by a stochastic process. J. Multivar. Anal. **7**, 215–219 (1977)
49. Buldygin, V.V., Kozachenko, Y.V.: Metric Characterization of Random Variables and Random Processes. Translation of Mathematical Monographs, vol. 188. American Mathematical Society, Providence, RI (2000)
50. Carmeli, C., De Vito, E., Toigo, A.: Reproducing kernel Hilbert spaces and Mercer theorem. arXiv:math.FA/0504071 v1, 5 Apr 2005
51. Carmeli, C., De Vito, E., Toigo, A.: Vector valued reproducing kernel Hilbert spaces of integrable functions and Mercer theorem. Anal. Appl. **4**, 377–408. World Scientific, Singapore (2006)

52. Chatterji, S.D., Mandrekar, V.S.: Equivalence and singularity of Gaussian measures and applications. In: Probabilistic Analysis and Related Topics, vol. 1. Academic, New York (1978)
53. Chaumont, L., Yor, M.: Exercises in Probability. Cambridge University Press, Cambridge (2003)
54. Chow, Y.S., Teicher, H.: Probability Theory. Springer, New York (1997)
55. Chung, K.L.: A Course in Probability Theory. Harcourt, Brace and World, New York (1968)
56. Chung, K.L., Williams, R.J.: Introduction to Stochastic Integration, 2nd edn. Birkhäuser, Boston (1990)
57. Clark, J.M.C.: On design of robust approximations to the stochastic differential equations of nonlinear filtering. And: Panel discussion: Are Itô calculus and martingale theory useful in parctice? In: Skwirzynski, J.K. (ed.) Proceedings NATO Advanced Study Institute on Communication Systems snd Random Process Theory, pp. 721–734/753–761. Sijthoff and Noordhoff, Alphen aan den Rijn (1978)
58. Climescu-Haulica, A.: Calcul stochastique appliqué aux problèmes de détection des signaux aléatoires. École polytechnique fédérale (EPFL), Thèse N° 2100, Lausanne (1999)
59. Climescu-Haulica, A., Gualtierotti, A.F.: Likelihood ratio detection of random signals: the case of causally filtered and weighted Wiener and Poisson processes. Int. J. Pure Appl. Math. **2**, 155–217 (2002)
60. Conway, J.B.: A Course in Functional Analysis, 2nd edn. Springer, New York (1990)
61. Conway, J.B.: A Course in Operator Theory. American Mathematical Society, Providence, RI (2000)
62. Cotter, S.L., Dashti, M., Robinson, J.C., Stuart, A.M.: Bayes inverse problems for functions and applications to fluid mechanics. Inverse Prob. **25**, 115009 (2009)
63. Cramér, H.: Structural and Statistical Problems for a Class of Stochastic Processes. Princeton University Press, Princeton (1971)
64. Cramér, H.: On some points of the theory of stochastic processes. Sankhyā A **40**, 91–115 (1978)
65. da Prato, G., Zabczyk, J.: Stochastic Equations in Infinite Dimensions. Cambridge University Press, Cambridge (1992)
66. Dacunha-Castelle, D., Duflo, M.: Exercices de probabilités et statistiques 1. Masson, Paris (1982)
67. Dacunha-Castelle, D., Duflo, M.: Probabilités et statistiques 2. Masson, Paris (1983)
68. de Brucq, D.: Processus Gaussiens multivariates et extraction d'un signal dans un bruit. Paris: Revue du CETHEDEC **49**, 67–110 (1976)
69. Debnath, L., Mikusiński, P.: Introduction to Hilbert spaces with applications. Academic, San Diego (1999)
70. Dellacherie, C., Meyer, P.-A.: Probabilités et potentiel I-IV. Hermann, Paris (1975)
71. Demengel, F., Demengel, G.: Convexité dans les espaces fonctionnels. Ellipses, Paris (2004)
72. de Sam Lazaro, J., Meyer, P.A.: Questions de théorie des flots. Séminaire des probabilités (Strasbourg). vol. 9, pp. 1–96. Springer, Berlin (1975)
73. Diestel, J., Uhl, J.J.: Vector Measures. American Mathematical Society, Providence, RI (1977)
74. Dieudonné, J.: Foundations of Modern Analysis. Academic, New York (1969)
75. Dinculeanu, N.: Vector Integration and Stochastic Integration in Banach Space. Wiley, New York (2000)
76. Dixmier, J.: Les algèbres d'opérateurs dans l'espace Hilbertien, 2nd edn. Gauthier-Villars, Paris (1969)
77. Donoghue, W.F., Jr.: Orthonormal sets in reproducing kernel Hilbert spaces and functional completion. Note di Matematica. **X**(Suppl. 1), 223–227 (1990)
78. Doob, J.L.: Stochastic Processes. Wiley, New York (1953)
79. Doob, J.L.: Measure Theory. Springer, New York (1994)
80. Douglas, R.G.: On majorization, factorization, and range inclusion of operators on Hilbert space. Proc. Am. Math. Soc. **17**, 413–415 (1966)
81. Douglas, R.G.: Banach Algebra Techniques in Operator Theory. Academic, New York (1972)

82. Duc-Jacquet, M.: Approximation des fonctionnelles linéaires sur les espaces hilbertiens autoreproduisants. Thèse, Université de Grenoble, Grenoble (1973)
83. Dudley, R.M.: Real Analysis and Probability. Wadsworth, Pacific Grove, CA (1989)
84. Dugundji, J.: Topology. Allyn and Bacon, Boston (1968)
85. Dunford, N., Schwartz, J.T.: Linear Operators I. Interscience, New York (1967)
86. Dunford, N., Schwartz, J.T.: Linear Operators II. Wiley, New York (1988)
87. Durret, R.: Stochastic Calculus. CRC Press, Boca Raton (1996)
88. Embrechts, P., Hofert, M.: A Note on Generalized Inverses. ETH, Zürich (2011)
89. Émery, M.: Sandwiched filtrations and Lévy processes. In: Séminaire des probabilités (Strasbourg), vol. 39, pp. 197–208. Springer, Berlin (2006)
90. Émery, M., Schachermayer, W.: On Verhik's standardness criterion and Tsirelson's notion of cosiness. In: Séminaire des probabilités (Strasbourg), vol. 35, pp. 265–305. Springer, Berlin (2001)
91. Ephremides, A.: A property of random processes with unit multiplicity. J. Multivar. Anal. **7**, 525–534 (1977)
92. Eubank, R.L., Hsing, T.: Canonical correlation for stochastic processes. Stoch. Process. Appl. **118**, 1634–1661 (2008)
93. Evard, J.C.: Commutation d'une matrice avec sa dérivée. Thèse, EPFL, Lausanne (1985)
94. Falk, H.: Prolog to detection of non-Gaussian signals: a paradigm for modern statistical processing. IEEE Proc. **82**, 1060 (1994)
95. Fan, J.: A selective overview of nonparametric methods in financial econometrics. Stat. Sci. **20**, 317–337 (2005)
96. Fang, K.-T., Kotz, S., Ng, K.-W.: Symmetric multivariate and related distributions. Chapman and Hall, London (1990)
97. Feldman, J.: Equivalence and perpendicularity of equivalent Gaussian processes. Pac. J. Math. **4**, 699–708 (1958)
98. Fernique, X.M.: Régularité des trajectoires des fonctions aléatoires gaussiennes. Lecture Notes in Mathematics, vol. 480, pp. 1–96. Springer, Berlin (1975)
99. Fernique, X.M.: Sur le théorème de Hajek-Feldman et de Cameron-Marin. C. R. Acad. Sci. Paris Sér. I 299(8), 355–358 (1984)
100. Fernique, X.M.: Fonctions aléatoires gaussiennes, vecteurs aléatoires gaussiens. CRM Publications, Montréal (1997)
101. Fernique, X.: Extension du théorème de Cameron-Martin aux translations aléatoires. Ann. Probab. **31**, 1296–1304 (2003)
102. Fernique, X.: Extension du théorème de Cameron-Martin aux translations aléatoires. II. Intégrabilité des densités. In: Progress in Probability, vol. 55, pp. 95–102. Birkäuser, Basel (2003)
103. Ferreira, J.C., Menegatto, V.A.: Eigenvalues of integral operators defined by smooth positive definite kernels. Integr. Equ. Oper. Theory **64**, 61–81 (2009)
104. Feyel, D., Üstunel, A.S., Zakai, M.: The realization of positive random variables via absolutely continuous transformations of measure on Wiener space. Probab. Surv. **3**, 170–205 (2006). Institute of Mathematical statistics and Bernoulli Society
105. Fillmore, P.A., Williams, J.P.: On operator ranges. Adv. Math. **7**, 254–281 (1971)
106. Fortet, R.M.: Vecteurs, fonctions et distributions aléatoires dans les espaces de Hilbert. Hermès, Paris (1995)
107. Fuhrmann, P.A.: Linear Systems and Operators in Hilbert Space. McGraw-Hill, New York (1981)
108. Garnir, H.G.: Fonctions de variables rélles II. Vander, Louvain (1965)
109. Garnir, H.G.: Fonctions de variables rélles I. Vander, Louvain (1970)
110. Garnir, H.G., de Wilde, M., Schmets, J.: Analyse fonctionnelle III, Espaces de fonctions usuels. Birkhäuser, Basel (1973)
111. Garsoux, J.: Analyse mathématique. Dunod, Paris (1968)
112. Garth, L.M., Poor, H.V.: Detection of non-Gaussian signals: a paradigm for modern statistical processing. IEEE Proc. **82**, 1061–1095 (1994)

113. Genet, J.: Mesure et intégration. Vuibert, Paris (1976)
114. Gikhman, I.I., Skorokhod, A.V.: The theory of stochastic processes II. Springer, Berlin (1983)
115. Gill, J., Salehi, H.: A note concerning spectral multiplicity one. Publications de l'Institut mathématique, Nouvelle série (Beograd). **44**(58), 121–126 (1988)
116. Girsanov, I.W.: On transforming a certain class of stochastic processes by absolutely continuous substitution of measures. Theory Probab. Appl. **5**, 285–301 (1960)
117. Goffman, C., Pedrick, G.: First course in functional analysis. Prentice-Hall, Englewood Cliffs (1965)
118. Gohberg, I., Goldberg, S., Kaashoek, M.A.: Classes of Linear Operators I. Birkhäuser, Basel (1990)
119. Gohberg, I., Goldberg, S., Krupnik, N.: Traces and Determinants of Linear Operators. Birkhäuser, Basel (2000)
120. Gradshteyn, I.S., Ryzhik, I.M.: Table of Integrals, Series and Products. Academic, New York (1980)
121. Graybill, F.A.: Introduction to Matrices with Applications in Statistics. Wadsworth, Belmont (1969)
122. Grenander, U.: Abstract Inference. Wiley, New York (1981)
123. Gualtierotti, A.F.: Skew-normal processes as models for random signals corrupted by Gaussian noise. Int. J. Pure Appl. Math. **20**, 109–142 (2005)
124. Guelfand, I.M., Vilenkin, N.Y.: Les distributions 4. Dunod, Paris (1967)
125. Hackenbroch, W., Thalmaier, A.: Stochastische Analysis. Teubner, Stuttgart (1994)
126. Halmos, P.R.: Introduction to Hilbert Space and the Theory of Spectral Multiplicity, 2nd edn. Chelsea Publishing, New York (1957)
127. Harville, D.A.: Matrix Algebra from a Statistician's Perspective. Springer, New York (1997)
128. He, S.-W., Wang, J.-G., Yan, J.-A.: Semimartingale Theory and Stochastic Calculus. CRC Press, Boca Raton (1992)
129. Helmberg, G.: Introduction to Spectral Theory in Hilbert Space. North-Holland, Amsterdam (1969)
130. Helstrom, C.W.: Statistical Theory of Signal Detection. Pergamon Press, London (1960)
131. Hewitt, E., Stromberg, K.: Real and Abstract Analysis. Springer, New York (1965)
132. Hida, T.: Canonical representations of Gaussian processes and their applications. Mem. Coll. Sci. Univ. Kyoto Ser. A Math. **XXXIII**(1), 109–155 (1960)
133. Hida, T., Hitsuda, M.: Gaussian Processes. American Mathematical Society, Providence, RI (1993)
134. Higgins, J.R.: Completeness and Basis Properties of Sets of Special Functions. Cambridge University Press, Cambridge (1977)
135. Hille, E., Phillips, R.S.: Functional Analysis and Semi-Groups. American Mathematical Society, Providence, RI (1965)
136. Hitsuda, M.: Multiplicity of some classes of Gaussian processes. Nagoya Math. J. **52**, 39–46 (1973)
137. Hitsuda, M.: Wiener-like integrals for Gaussian processes and the linear estimation problem. In: Pinsky, M.A. (ed.) Stochastic Analysis and Applications, pp. 167–177. Dekker, New York (1984)
138. Hoffmann-Jørgensen, J.: Probability with a View Towards Statistics I. Chapman & Hall, New York (1994)
139. Hoffmann-Jørgensen, J.: Probability with a View Towards Statistics II. Chapman & Hall, New York (1994)
140. Horton, C.W. Sr.: Signal Processing of Underwater Acoustic Waves. United States Government Printing Office, Washington (1969)
141. Hu, S.-T.: Elements of Real Analysis. Holden-Day, San Francisco (1967)
142. Itô, K.: The topological support of Gauss measure on Hilbert space. Nagoya Math. J. **38**, 181–183 (1970)
143. Ivković, Z.A., Bulatović, J., Vukmirović, J., Živanović, S.: Application of Spectral Multiplicity in Separable Hilbert Space to Stochastic Processes. Matematički Institut, Beograd (1974)

144. Ivković, Z.A.: Arbitrarines of multiplicity for a linear non-anticipative transformation of the Wiener process. Univ. u Novom Sadu, Zb. Rad. Prirod.-Mat. Fak., Ser. Mat. **20**(2), 91–94 (1990)
145. Jacod, J., Shiryaev, A.N.: Limit Theorems for Stochastic Processes. Springer, Berlin (2002)
146. Jørsboe, O.G.: Equivalence or singularity of Gaussian measures on function spaces. Matematisk Institut, Aarhus Universitet, Aarhus (1968)
147. Kadota, T.: Simultaneous diagonalization of two covariance kernels and application to second order stochastic processes. SIAM J. Appl. Math. **15**, 1470–1480 (1967)
148. Kailath, T., Poor, H.V.: Detection of stochastic processes. IEEE Trans. Inf. Theory **44**, 2230–2259 (1998)
149. Kallenberg, O.: Foundations of Modern Probability, 2nd edn. Springer, New York (2002)
150. Kallianpur, G.: Stochastic Filtering Theory. Springer, New York (1980)
151. Kantorovich, L., Rubinstein, G.: On a space of completely additive functions. Vestnik Leningrad. Univ. **13**, 52–59 (1958)
152. Kanwal, R.P.: Generalized Functions. Birkäuser, Boston (2004)
153. Karhunen, K.: Über lineare Methoden in der Wahrscheinlichkeitsrechnung. Annales Academiae Scientiarum Fennicae A. I. **37**, 1–79 (1947)
154. Kelley, J.L.: General Topology. Van Nostrand, New York (1969)
155. Kingman, J.F.C., Taylor, S.J.: Introduction to Measure and Probability. Cambridge University Press, Cambridge (1966)
156. Klebaner, F., Lipster, R.: When a stochastic exponential is a true martingale. Extension of a method of Benês. Theory Probab. Appl. **58**, 38–62 (2014)
157. Knight, F.B.: A transformation from prediction to past of an L^2-stochastic process. In: Séminaire des probabilités (Strasbourg), vol. 17, pp. 1–14. Springer, Berlin (1983)
158. Knight, F.B.: A post-predictive view of Gaussian processes. Annales scientifiques de l' École Normale Supérieure, 4e série, **16**, 541–566 (1983)
159. Knight, F.B.: On strict-sense forms of the Hida-Cramer representation. In: Seminar on Stochastic Processes, 1984. Birkhäuser, Boston (1986)
160. Knight, F.B.: Foundations of the Prediction Process. Clarendon Press, Oxford (1992)
161. Knight, F.B.: An impression of P.A. Meyer as deus ex machina. In: Séminaire des probabilités (Strasbourg), vol. 37, pp. IX–XI. Springer, Berlin (2003)
162. Kubrusly, C.S.: Hilbert Space Operators. Birkhäuser, Boston (2003)
163. Kuo, H.-H.: Gaussian Measures in Banach Spaces. Springer, Berlin (1975)
164. Kuo, H.-H.: Introduction to Stochastic Integration. Springer, New York (2006)
165. Kwapień, S., Woyczyński, W.A.: Random Series and Stochastic Integrals: Single and Multiple. Birkäuser, Boston (1992)
166. Lamperti, J.: Probability. Benjamin, New York (1966)
167. Lang, S.: Analysis II. Addison-Wesley, Reading, MA (1969)
168. Laurent, S.: On standardness and I-cosiness. In: Séminaire des probabilités (Strasbourg), vol. 43, pp. 127–186. Springer, Berlin (2011)
169. Laurent, S.: On Vershikian and I-cosy random variiables and filtrations. Theory Probab. Appl. **55**, 54–76 (2011)
170. Lax, P.D.: Functional Analysis. Wiley, New York (2002)
171. Ledoux, M., Talagrand, M.: Probability in Banach Spaces. Springer, Berlin (1991)
172. Lenglart, E.: Sur quelques points remarquables de la théorie des martingales locales et applications. Thèse, Faculté des Sciences de l'Univerité de Rouen, Rouen (1976)
173. Lévy, P.: Sur une classe de courbes de l'espace de Hilbert et sur une équation intégrale nonlinéaire. Annales Scientifiques de l'École Normale Supérieure **73**, 121–156 (1956)
174. Li, B., McCullagh, P.: Potential functions and conservative estimating functions. Ann. Stat. **22**, 340–356 (1994)
175. Luenberger, D.G.: Optimization by Vector Space Methods. Wiley, New York (1969)
176. Lukić, M.N., Beder, J.H.: Stochastic processes with sample paths in reproducing kernel Hilbert spaces. Am. Math. Soc. Trans. **353**, 3945–3969 (2001)

177. Malliavin, P.: Intégration et probabilités, analyse de Fourier et analyse spectrale. Masson, Paris (1982)
178. Malliavin, P.: Integration and Probability. Springer, New York (1995)
179. Mandrekar, V.S.: On multivariate wide-sense Markov processes. Nagoya J. Math. **33**, 7–20 (1968)
180. Mandrekar, V.S.: On the multiple Markov property of Levy-Hida for Gaussian processes. Nagoya Math. J. **54**, 69–78 (1974)
181. Marage, J.-P., Mori, Y.: Sonars et acoustique sous-marine. Lavoisier, Paris (2009)
182. Masani, P.: Orthogonally Scattered Measures. Adv. Math. **2**, 61–117 (1968)
183. McDonald, K.F., Blum, R.S.: A statistical and physical mechanisms-based interference and noise model for array observations. IEEE Trans. Signal Process. **48**, 2044–2056 (2000)
184. Medvegyev, P.: Stochastic Integration Theory. Oxford Univerity Press, Oxford (2007)
185. Meise, R., Vogt, D.: Introduction to Functional Analysis. Clarendon Press, Oxford (1997)
186. Mémin, J.: Le théorème de Girsanov. Rapport, Département de mathématiques et d'informatique, Université de Rennes, Rennes (1972)
187. Memin, J.: Sur quelques problèmes fondamentaux de la théorie du filtrage. Thèse, Université de Rennes, Rennes (1974)
188. Meschkowski, H.: Hilbertsche Räume mit Kernfunktion. Springer, Berlin (1962)
189. Métivier, M.: Processus de Markov. Département de mathématiques et d'informatique, Université de Rennes, Rennes (1972)
190. Métivier, M.: Semimartingales. de Gruyter, Berlin (1982)
191. Métivier, M., Pellaumail, J.: Stochastic Integration. Academic, New York (1980)
192. Meyer, P.A.: Probability and Potentials. Blaisdell, Waltham (1966)
193. Meyer, P.A.: A remark on F. Knight's paper. Annales Scientifiques de l'École Normale Supérieure, 4e série. **16**, 567–569 (1983)
194. Middleton, D.: A statistical theory of reverberation and similar first-order scattered fields. IEEE Trans. Inf. Theory **13**, 372–414 (1967)
195. Middleton, D.: Non-Gaussian noise models in signal processing for telecommunications. IEEE-IT **45**, 1129–1149 (1999)
196. Moché, R.: Une représentation des covariances à espace autoreproduisant séparable. Comptes rendus de l'Académie des Sciences A **274**, 346–349 (1972)
197. Moché, R.: Représentation canonique des covariances plus petites qu'une covariance donnée: applications. Comptes rendus de l'Académie des Sciences A **277**, 133–136 (1973)
198. Moché, R.: Études sur la stationnarité des fonctions aléatoires du second ordre. Thèse, Partie 2, Université des sciences et techniques de Lille, Lille (1977)
199. Neveu, J.: Mathematical Foundations of the Calculus of Probability. Holden-Day, San Francisco (1965)
200. Neveu, J.: Processus aléatoires gaussiens. Presses de l'Université de Montréal, Montréal (1968)
201. Neveu, J.: Martingales à temps discert. Masson, Paris (1972)
202. Neveu, J.: Processus aléatoires Gaussiens. Laboratoire de calcul des probabilités, Jussieu-Tour 56, Paris (1972)
203. Nielsen, O.A. Direct integral theory. Dekker, New York (1980)
204. Nualart, D.: The Malliavin Calculus and Related Topics, 2nd edn. Springer, Berlin (2006)
205. Ocone, D.L.: A symmetry characterization of conditionally independent increment martingales. In: Nualart, D., Sanz-Solé, M. (eds.) Barcelona Seminar on Stochastic Analysis. Progress in Probability, vol. 32, pp. 147–167. Birkhäuser, Basel (1993)
206. Ouvrard, J.-Y.: Martingales locales et théorème de Girsanov dans les espaces de Hilbert réels séparables. Ann. Inst Henri Poincaré **IX**, 351–368 (1973)
207. Pallu de la Barrière, R.: Intégration. Ellipses, Paris (1997)
208. Parthasarathy, K.R.: Probability Measures on Metric Spaces. Academic, New York (1967)

209. Parthasarathy, K.R., Schmidt, K.: Positive Definite kernels, Continuous Tensor Products, and Central Limit Theorems of Probability Theory. Springer, Berlin (1972)
210. Pietsch, A.: Nuclear Locally Convex Spaces. Springer, New York (1972)
211. Pisot, C., Zamansky, M.: Mathématiques générales. Dunod, Paris (1969)
212. Pitt, L.D.: Hida-Cramér multiplicity theory for multiple Markov processes and Goursat representations. Nagoya Math. J. **57**, 199–228 (1975)
213. Pons, O.: Functional Estimation for Density, Regression, Models and Processes. World Scientific, Hackensack (2011)
214. Prakasa Rao, B.L.S.: Statistical inference for diffusion type processes. Arnold, London (1999)
215. Primak, S., Kontorovich, V., Lyandres, V.: Stochastic Methods and Their Applications to Communications. Wiley, Chichester (2004)
216. Protter, P.E.: Stochastic Integration and Stochastic Differential Equations, 2nd edn. Springer, Berlin (2004)
217. Rainville, E.D.: The Laplace Transform. Macmillan, New York (1963)
218. Randolph, J.F.: Basic Real and Abstract Analysis. Academic, New York (1968)
219. Rao, C.R., Varadarajan, V.S.: Discrimination of Gaussian processes. Sankhyā **A25**, 303–330 (1963)
220. Rao, M.M. (Ed.): Real and Stochastic Analysis. CRC Press, Boca Raton (1997)
221. Revuz, D., Yor, M.: Continuous Martingales and Brownian Motion, 3rd edn. Springer, Berlin (1999)
222. Robertson, A.P., Robertson, W.J.: Topologische Vektorräume. Bibliographishes Institut, Mannheim (1967)
223. Rogers, L.C.G., Williams, D. : Diffusions, Markow Processes, and Martingales. Cambridge University Press, Cambridge (2000)
224. Romano, J.P., Siegel, A.F.: Counterexamples in Probability and Statistics. Chapman & Hall, New York (1986)
225. Root, W.L.: Singular gaussian measures in detection theory. In: Proceedings of the Symposium on Time Series Analysis, pp. 292–315. Wiley, New-York (1963)
226. Royden, H.L.: Real Analysis. Macmillan, London (1969)
227. Rozanov, Y.A.: Innovation Processes. Winston, Washington (1977)
228. Rudin, W.: Functional Analysis. McGraw-Hill, New York (1973)
229. Rudin, W.: Real and Complex Analysis. McGraw-Hill, New York (1974)
230. Rudin, W.: Principles of Mathematical Analysis. McGraw-Hill, New York (1976)
231. Ryan, R.A.: Introduction to Tensor Products of Banach Spaces. Springer, London (2002)
232. Saitoh, S.: Theory of Reproducing Kernels and Its Applications. Longman, Harlow (1988)
233. Sarason, D.: Sub-Hardy Hilbert Spaces in the Unit Disk. Wiley, New York (1994)
234. Schaefer, H.H.: Topological Vector Spaces. Macmillan, New York (1966)
235. Schatten, R.: Norm Ideals of Completely Continuous Operators. Springer, Berlin (1960)
236. Schechter, M.: Principles of Functional Analysis. American Mathematical Society, Providence, RI (2002)
237. Schmidt, K.D.: Mass und Wahrscheinlichkeit. Springer, Berlin (2009)
238. Schwartz, L.: Méthodes mathématiques pour les sciences physiques. Hermann, Paris (1965)
239. Schwartz, L.: Radon Measures on Arbitrary Topological Spaces and Cylindrical Measures. Oxford University Press, London (1973)
240. Searcóid, M.Ó.: Metric Spaces. Springer, London (2007)
241. Shiryaev, A.N.: Probability, 2nd edn. Springer, New York (1996)
242. Siraya, T.N.: On canonical representations for stochastic processes of multiplicities one and two. Theory Probab. Appl. **18**, 153–158 (1973)
243. Skorohod, A.V.: Integration in Hilbert Space. Springer, Berlin (1974)
244. Small, C.M., McLeish, D.L.: Hilbert Space Methods in Probability and Statistical Inference. Wiley, New York (1994)
245. Smithies, F.: Integral Equations. Cambridge University Press, Cambridge (1965)
246. Sobczyk, K.: Stochastic Wawe Propagation. Elsevier, Amsterdam (1985)

247. Spielman, J.L.: Stochastic processes with independent increments, taking values in a Hilbert space. J. Multivar. Anal. **6**, 426–446 (1976)
248. Srivastava, S.M.: A Course on Borel Sets. Springer, New York (1998)
249. Stone, M.H.: Linear Transformations in Hilbert Space and Applications in Analysis, vol. 15. American Mathematical Society Colloquium Publications, Providence, RI (1932)
250. Strang, G.: Linear Algebra and Applications. Academic, New York (1976)
251. Stroock, D.W., Varadhan, S.R.S.: Multidimensional Diffusion Processes. Springer, Berlin (1979)
252. Stroock, D.W., Yor, M.: Some remarkable martingales. In: Séminaire des probabilités (Strasbourg), vol. 15, pp. 590–603. Springer, Berlin (1981)
253. Stuart, A.M.: Inverse problems: a Bayesian prerspective. Acta Numer. **19**, 451–559 (2010)
254. Suquet, C.: Distance euclidiennes sur les mesures signées et application à des théorèmes de Berry-Esséen. Bull. Belg. Math. Soc. Simon Stevin **2**, 161–181 (1995)
255. Székely, B.: Strong approximation of stochastic processes using random walks. Thesis, Budapest University of Technology and Economics, Institue of Mathematics, Budapest (2004)
256. Tao, T.: What is good mathematics? Am. Math. Soc. Bull. **44**, 623–634 (2007)
257. Taylor, A.E.: Introduction to Functional Analysis. Wiley, New York (1958)
258. Todorovich, P.: An Introduction to Stochastic Processes and Their Applications. Springer, New York (1992)
259. Vakhania, N.N.: Probability Distributions on Linear Spaces. North Holland, New York (1981)
260. Vakhania, N.N., Tarieladze, V.I., Chobanyan, S.A.: Probability Distributions on Banach Spaces. Reidel, Dordrecht (1987)
261. Vakhania, N.N., Tarieladze, V.I.: On singularity and equivalence of Gaussian measures. In: Rao, M.M. (ed.) Real and Stochastic Analysis, pp. 367–389. CRC Press, Boca Raton (1997)
262. Väth, M.: Integration Theory. World Scientific, Singapore (2002)
263. Verley, J.-L.: Théorie élémentaire de l'intégration. Centre de documentation universitaire, Paris (1967)
264. von Weizsäcker, H., Winkler, G.: Stochastic Integrals. Vieweg, Braunschweig (1990)
265. Vostrikova, L., Yor, M.: Some invariance properties (of the laws) of Ocone's martingales. In: Séminaire des probabilités (Strasbourg), vol. 34, pp. 417–431. Springer, Berlin (2000)
266. Weidmann, J.: Lineare Opertoren in Hilberträumen. Teubner, Stuttgart (1976)
267. Whitt, W.: Weak convergence of probability measures on the function space $C[0, \infty[$. Ann. Math. Stat. **41**, 939–944 (1970)
268. Widder, D.V.: The Laplace Transform. Princeton University Press, Princeton (1966)
269. Wilansky, A.: Functional Analysis. Blaisdell, New York (1964)
270. Willard, S.: General Topology. Addison-Wesley, Reading, MA (1968)
271. Wong, E.: Stochastic Processes in Information and Dynamical Systems. McGraw-Hill, New York (1971)
272. Yan, Y.: The general solutions of linear ODE and Riccati equation by integral series. p. 9 (2011). arXiv:1006.4804v4 [math.CA]
273. Yeh, J.: Stochastic Processes and the Wiener Integral. Dekker, New York (1973)
274. Yeh, J.: Martingales and Stochastic Analysis. World Scientific, Singapore (1995)
275. Yeh, J.: Real Analysis, 2nd edn. World Scientific, Hackensack (2006)
276. Yosida, K.: Functional Analysis. Springer, Berlin (1966)
277. Zeidler, E.: Applied Functional Analysis II. Springer, New York (1995)
278. Zhang, F.: Matrix Theory. Springer, New York (1999)

Index

A.F. Gualtierotti, *Detection of Random Signals in Dependent Gaussian Noise*,
DOI 10.1007/978-3-319-22315-5

GPSR Compliance

The European Union's (EU) General Product Safety Regulation (GPSR) is a set of rules that requires consumer products to be safe and our obligations to ensure this.

If you have any concerns about our products, you can contact us on ProductSafety@springernature.com

In case Publisher is established outside the EU, the EU authorized representative is:

Springer Nature Customer Service Center GmbH
Europaplatz 3
69115 Heidelberg, Germany

Batch number: 10614898

Printed by Printforce, the Netherlands